Springer Handbook
of Computational Intelligence

Springer Handbooks provide a concise compilation of approved key information on methods of research, general principles, and functional relationships in physical and applied sciences. The world's leading experts in the fields of physics and engineering will be assigned by one or several renowned editors to write the chapters comprising each volume. The content is selected by these experts from Springer sources (books, journals, online content) and other systematic and approved recent publications of scientific and technical information.

The volumes are designed to be useful as readable desk reference book to give a fast and comprehensive overview and easy retrieval of essential reliable key information, including tables, graphs, and bibliographies. References to extensive sources are provided.

Springer Handbook
of Computational Intelligence

Janusz Kacprzyk, Witold Pedrycz (Eds.)

With 534 Figures and 115 Tables

 Springer

Editors
Janusz Kacprzyk
Polish Academy of Sciences
Systems Research Inst.
ul. Newelska 6
01-447 Warsaw, Poland
kacprzyk@ibspan.waw.pl

Witold Pedrycz
University of Alberta
Dep. Electrical and Computer Engineering
116 Street 9107
T6J 2V4, Edmonton, Alberta, Canada
wpedrycz@ualberta.ca

ISBN: 978-3-662-43504-5   e-ISBN: 978-3-662-43505-2
DOI 10.1007/978-3-662-43505-2
Springer Dordrecht Heidelberg London New York

Library of Congress Control Number: 2015936335

Production and typesetting: le-tex publishing services GmbH, Leipzig
Senior Manager Springer Handbook: Dr. W. Skolaut, Heidelberg
Typography and layout: schreiberVIS, Seeheim
Illustrations: Hippmann GbR, Schwarzenbruck
Cover design: eStudio Calamar Steinen, Barcelona
Cover production: WMXDesign GmbH, Heidelberg
Printing and binding: Printer Trento s.r.l., Trento

Printed on acid free paper

Springer is part of Springer Science+Business Media (www.springer.com)

Preface

We are honored and happy to be able to make available this *Springer Handbook of Computational Intelligence*, a large and comprehensive account of both the state-of-the-art of the research discipline, complemented with some historical remarks, main challenges, and perspectives of the future. To follow a predominant tradition, we have divided this Springer Handbook into parts that correspond to main fields that are meant to constitute the area of computational intelligence, that is, fuzzy sets theory and fuzzy logic, rough sets, evolutionary computation, neural networks, hybrid approaches and systems, all of them complemented with a thorough coverage of some foundational issues, methodologies, tools, and techniques.

We hope that the handbook will serve as an indispensable and useful source of information for all readers interested in both the theory and various applications of computational intelligence. The formula of the Springer Handbook as a convenient single-volume publication project should help the potential readers find a proper tool or technique for solving their problems just by simply browsing through a clearly composed and well-indexed contents. The authors of the particular chapters, who are the best known specialists in their respective fields worldwide, are the best assurance for the handbook to serve as an excellent and timely reference.

On behalf of the entire computational intelligence community, we wish to express sincere thanks, first of all, to the Part Editors responsible for the scope, authors, and composition of the particular parts for their great job to arrange the most appropriate topics, their coverage, and identify expert authors. Second, we wish to thank all the authors for their great contributions in the sense of clarity, comprehensiveness, novelty, vision, and – above all – understanding of the real needs of readers of diverse interests.

All that efforts would not end up with the success without a total and multifaceted publisher's dedication and support. We wish to thank very much Dr. Werner Skolaut, Ms. Constanze Ober, and their collaborators from Springer, Heidelberg, and le-tex publishing GmbH, Leipzig, respectively, for their extremely effective and efficient handling of this huge and difficult project.

September 2014
Janusz Kacprzyk
Witold Pedrycz

Warsaw
Edmonton

About the Editors

Janusz Kacprzyk graduated from the Department of Electronics, Warsaw University of Technology, Poland with an MSc in Automatic Control, a PhD in Systems Analysis and a DSc (*Habilitation*) in Computer Science from the Polish Academy of Sciences. He is Professor of Computer Science at the Systems Research Institute, Polish Academy of Sciences, Professor of Computerized Management Systems at WIT – Warsaw School of Information Technology, and Professor of Automatic Control at PIAP – Industrial Institute of Automation and Measurements, in Warsaw, Poland, and Department of Electrical and Computer Engineering, Cracow University of Technology, Poland. He is the author of 5 books, (co)editor of ca. 70 volumes, (co)author of ca. 500 papers. He is Editor-in-Chief of 6 book series and of 2 journals, and on the Editorial Boards of more than 40 journals.

Witold Pedrycz is a Professor and Canada Research Chair (CRC) in Computational Intelligence in the Department of Electrical and Computer Engineering, University of Alberta, Edmonton, Canada. He is also with the Systems Research Institute of the Polish Academy of Sciences, Warsaw. He also holds an appointment of special professorship in the School of Computer Science, University of Nottingham, UK. His main research directions involve computational intelligence, fuzzy modeling and granular computing, knowledge discovery and data mining, fuzzy control, pattern recognition, knowledge-based neural networks, relational computing, and software engineering. He has published numerous papers and is the author of 15 research monographs covering various aspects of computational intelligence, data mining, and software engineering. He currently serves as an Associate Editor of IEEE Transactions on Fuzzy Systems and is a member of a number of Editorial Boards of other international journals.

About the Part Editors

Cesare Alippi

Politecnico di Milano
Dip. Elettronica, Informazione e
Ingegneria
20133 Milano, Italy
alippi@elet.polimi.it

Part D

Cesare Alippi received his PhD in 1995 from Politecnico di Milano, Italy. Currently, he is Professor at the same institution. He has been a visiting researcher at UCL (UK), MIT (USA), ESPCI (F), CASIA (RC), USI (CH). Alippi is an IEEE Fellow, Vice-President Education of the IEEE Computational Intelligence Society, Associate Editor of the IEEE Computational Intelligence Magazine, Past Editor of the IEEE-TIM and IEEE-TNN(LS). In 2004 he received the IEEE Instrumentation and Measurement Society Young Engineer Award and in 2013 the IBM Faculty Award. His current research focuses on learning in non-stationary environments and intelligence for embedded systems. He holds 5 patents, has published 1 monograph book, 6 edited books and about 200 papers in international journals and conference proceedings.

Thomas Bartz-Beielstein

Cologne University of Applied Sciences
Faculty of Computer Science and
Engineering Science
51643 Gummersbach, Germany
thomas.bartz-beielstein@fh-koeln.de

Part E

Thomas Bartz-Beielstein is a Professor of Applied Mathematics at Cologne University of Applied Sciences (CUAS). His expertise lies in optimization, simulation, and statistical analysis of complex real-world problems. He has more than 100 publications on computational intelligence, optimization, simulation, and experimental research. He has been on the program committees of several international conferences and organizes the prestigious track *Evolutionary Computation in Practice* at GECCO. His books on experimental research are considered as milestones in this emerging field. He is speaker of the research center *Computational Intelligence plus* at CUAS and head of the SPOTSeven team.

Christian Blum

University of the Basque Country
Dep. Computer Science and Artificial
Intelligence
20018 San Sebastian, Spain
christian.blum@ehu.es

Part F

Christian Blum holds a Master's Degree in Mathematics (1998) from the University of Kaiserslautern, Germany, and a PhD degree in Applied Sciencies (2004) from the Free University of Brussels, Belgium. He currently occupies a permanent post as Ikerbasque Research Professor at the University of the Basque Country, San Sebastian, Spain. His research interests include the development of swarm intelligence techniques and the combination of metaheuristics with exact approaches for solving difficult optimization problems. So far he has co-authored about 150 research papers.

Oscar Castillo

Tijuana Institute of Technology
22379 Tijuana, Mexico
ocastillo@tectijuana.mx

Part G

Oscar Castillo holds the Doctor in Science degree in Computer Science from the Polish Academy of Sciences. He is a Professor of Computer Science in the Graduate Division, Tijuana Institute of Technology, Tijuana, Mexico. In addition, he serves as Research Director of Computer Science. Currently, he is Vice-President of HAFSA (Hispanic American Fuzzy Systems Association) and served as President of IFSA (International Fuzzy Systems Association). He belongs to the Mexican Research System with level II and is also a member of NAFIPS, IFSA, and IEEE. His research interests are in type-2 fuzzy logic, fuzzy control, and neuro-fuzzy and genetic-fuzzy hybrid approaches.

Carlos A. Coello Coello

CINVESTAV-IPN
Dep. Computación
D.F. 07300, México, Mexico
ccoello@cs.cinvestav.mx

Part E

Carlos A. Coello Coello received a PhD in Computer Science from Tulane University in 1996. He has made pioneering contributions to the research area currently known as evolutionary multi-objective optimization, mainly regarding the development of new algorithms. He is currently Professor at the Computer Science Department of CINVESTAV-IPN (Mexico City, México). He has co-authored more than 350 publications (his h-index is 62). He is Associate Editor of several journals, including *IEEE Transactions on Evolutionary Computation* and *Evolutionary Computation*. He has received Mexico's National Medal of Science in *Exact Sciences* and the IEEE Kiyo Tomiyasu Award. He is also an IEEE Fellow.

Bernard De Baets

Ghent University
Dep. Mathematical Modelling, Statistics and Bioinformatics
9000 Ghent, Belgium
bernard.debaets@ugent.be

Part A

Bernard De Baets (1966) holds an MSc degree in Mathematics, a postgraduate degree in Knowledge Technology, and a PhD degree in Mathematics. He is a full professor at UGent (Belgium), where he leads KERMIT, an interdisciplinary team in mathematical modeling, having delivered 50 PhD graduates to date. His bibliography comprises nearly 400 journal papers, 60 book chapters, and 300 conference contributions. He acts as Co-Editor-in-Chief (2007) of Fuzzy Sets and Systems. He is a recipient of a Government of Canada Award, Honorary Professor of Budapest Tech (Hungary), Fellow of the International Fuzzy Systems Association, and has been nominated for the Ghent University Prometheus Award for Research.

Roderich Groß

University of Sheffield
Dep. Automatic Control and Systems Engineering
Sheffield, S1 3JD, UK
r.gross@sheffield.ac.uk

Part F

Roderich Groß received a Diploma degree in Computer Science from TU Dortmund University in 2001 and a PhD degree in Engineering Sciences from the Université libre de Bruxelles in 2007. From 2005 to 2009 he was a fellow of the Japan Society for the Promotion of Science, a Research Associate at the University of Bristol, a Marie Curie Fellow at Unilever, and a Marie Curie Fellow at EPFL. Since 2010 he has been with the Department of Automatic Control and Systems Engineering at the University of Sheffield, where he is currently Senior Lecturer. His research interests include evolutionary and distributed robotics. He has authored over 60 publications on these topics. He is a Senior Member of the IEEE.

Enrique Herrera Viedma

University of Granada
Dep. Computer Science and Artificial Intelligence
18003 Granada, Spain
viedma@decsai.ugr.es

Part B

Enrique Herrera-Viedma received his PhD degree in Computer Science from Granada University in 1996. He is Professor at Granada University in the Depaartment of Computer Science and Artificial Intelligence and a member of the BoG in IEEE SMC. His interest topics are computing with words, fuzzy decision making, consensus, aggregation, social media, recommender systems, libraries, and bibliometrics. His h-index is 44 and presents over 7000 citations (WoS). In 2014 he was identified as Highly Cited Researcher by Thomson Reuters and Top Author in Computer Science according to Microsoft Academic Search.

Luis Magdalena

European Centre for Soft Computing
33600 Mieres, Spain
luis.magdalena@softcomputing.es

Part B

Luis Magdalena received the MS and PhD degrees in Telecommunication Engineering from the Technical University of Madrid, Spain, in 1988 and 1994. He has been Assistant (1990–1995) and Associate Professor (1995–2006) in Computer Science at the Technical University of Madrid. Since 2006 he has been Director General of the European Center for Soft Computing. His research interests include soft computing and its application. He has authored over 150 publications in the field. He has been President of the *European Society for Fuzzy Logic and Technologies* (2001–2005), Vice-President of the International Fuzzy Systems Association (2007–2011), and member of the IEEE Computational Intelligence Society AdCom (2011–2013).

Jörn Mehnen

Cranfield University
Manufacturing Dep.
Cranfield, MK43 0AL, UK
j.mehnen@cranfield.ac.uk

Part E

Dr Jörn Mehnen is Reader in Computational Manufacturing at Cranfield University, UK and Privatdozent at TU Dortmund, Germany. He is also Deputy Director of the EPSRC Centre in Through-life Engineering Services at Cranfield University. His research activities are in real-world applications of computer sciences in mechanical engineering with special focus on evolutionary optimization, cloud manufacturing, and additive manufacturing.

Patricia Melin

Tijuana Institute of Technology
Dep. Computer Science
Chula Vista, CA 91909, USA
pmelin@tectijuana.mx

Part G

Patricia Melin holds the Doctor in Science degree in Computer Science from the Polish Academy of Sciences. She has been a Professor of Computer Science in the Graduate Division, Tijuana Institute of Technology, Tijuana, Mexico since 1998. She serves as Director of Graduate Studies in Computer Science. Currently, she is Vice President of HAFSA (Hispanic American Fuzzy Systems Association). She is the founding Chair of the Mexican Chapter of the IEEE Computational Intelligence Society. She is member of NAFIPS, IFSA, and IEEE and belongs to the Mexican Research System with level III. Her research interests are in type-2 fuzzy logic, modular neural networks, pattern recognition, fuzzy control, and neuro-fuzzy and genetic-fuzzy hybrid approaches. She has published over 200 journal papers, 6 authored books, 20 edited books, and 200 papers in conference proceedings.

Peter Merz

University of Applied Sciences and Arts, Hannover
Dep. Business Administration and Computer Science
30459 Hannover, Germany
peter.merz@hs-hannover.de

Part E

Peter Merz received his PhD degree in Computer Science from the University of Siegen, Germany in 2000. Since 2009, he has been Professor at the University of Applied Sciences and Arts in Hannover. He is a well-known scientist in the field of evolutionary computation and meta-heuristics. His research interests center on fitness landscapes of combinatorial optimization problems and their analysis.

Radko Mesiar

STU in Bratislava
Dep. Mathematics and Descriptive Geometry
813 68 Bratislava, Slovakia
radko.mesiar@stuba.sk

Part A

Radko Mesiar received his PhD from Comenius University, Faculty of Mathematics and Physics, in 1979. He has been a member of the Department of Mathematics in the Faculty of Civil Engineering, STU Bratislava since 1978. He received his DSc in 1996 from the Czech Academy of Sciences. He has been a full professor since 1998. He is a fellow member of the Institute of Information and Automation at the Czech Academy of Sciences and of IRAFM, University of Ostrava (Czech Republic). He is co-author of two scientific monographs and five edited volumes. He is the author of more than 200 papers in WOS in leading journals. He is the co-founder of conferences AGOP, FSTA, ABLAT, and ISCAMI.

Frank Neumann

The University of Adelaide
School of Computer Science
Adelaide, SA 5005, Australia
frank.neumann@adelaide.edu.au

Part E

Frank Neumann received his diploma and PhD from the University of Kiel in 2002 and 2006, respectively. Currently, he is an Associate Professor and leader of the Optimisation and Logistics Group at the School of Computer Science, The University of Adelaide, Australia. He is the General Chair of ACM GECCO 2016. He is Vice-Chair of the IEEE Task Force on Theoretical Foundations of Bio-Inspired Computation, and Chair of the IEEE Task Force on Evolutionary Scheduling and Combinatorial Optimization. In his work, he considers algorithmic approaches and focuses on theoretical aspects of evolutionary computation as well as high impact applications in the areas of renewable energy, logistics, and sports.

Marios Polycarpou

University of Cyprus
Dep. Electrical and Computer
Engineering and KIOS Research Center
for Intelligent Systems and Networks
1678 Nicosia, Cyprus
mpolycar@ucy.ac.cy

Part D

Marios Polycarpou is Professor of Electrical and Computer Engineering and the Director of the KIOS Research Center for Intelligent Systems and Networks at the University of Cyprus. His research expertise is in the areas of intelligent systems and control, computational intelligence, fault diagnosis, cooperative and adaptive control, and distributed agents. He is a Fellow of the IEEE. He has participated in more than 60 research projects/grants, funded by several agencies and industries in Europe and the United States. In 2011, he was awarded the prestigious European Research Council (ERC) Advanced Grant.

Günther Raidl

Vienna University of Technology
Inst. Computer Graphics and Algorithms
1040 Vienna, Austria
raidl@ads.tuwien.ac.at

Part E

Günther Raidl is Professor at the Vienna University of Technology, Austria, and heads the Algorithms and Data Structures Group. He received his PhD in 1994 and completed his Habilitation in Practical Computer Science in 2003. In 2005 he received a professorship position for combinatorial optimization. His research interests include algorithms and data structures in general and combinatorial optimization in particular, with a specific focus on metaheuristics, mathematical programming, and hybrid optimization approaches.

Oliver Schütze

CINVESTAV-IPN
Dep. Computación
D.F. 07300, México, Mexico
schuetze@cs.cinvestav.mx

Part E

Oliver Schütze received a PhD in Mathematics from the University of Paderborn, Germany in 2004. He is currently Professor at Cinvestav-IPN in Mexico City (Mexico). His research interests focus on numerical and evolutionary optimization where he addresses scalar and multi-objective optimization problems. He has co-edited 5 books and is co-author of more than 90 papers. He is a co-founder of SON (Set Oriented Numerics) and founder of the NEO (Numerical and Evolutionary Optimization) workshop series.

Roman Słowiński

Poznań University of Technology
Inst. Computing Science
60-965 Poznań, Poland
Roman.Slowinski@cs.put.poznan.pl

Part C

Roman Słowiński is Professor and Founding Chair of the Laboratory of Intelligent Decision Support Systems at Poznań University of Technology. He is Academician and President of the Poznań Branch of the Polish Academy of Sciences and a member of Academia Europaea. In his research, he combines operations research and computational intelligence. He is renowned for his seminal research on using rough sets in decision analysis. He was laureate of the EURO Gold Medal (1991) and won the 2005 Prize of the Foundation for Polish Science. He is Doctor *Honoris Causa* of Polytech'Mons (2000), the University Paris Dauphine (2001), and the Technical University of Crete (2008).

Carsten Witt

Technical University of Denmark
DTU Compute, Algorithms, Logic and
Graphs
2800 Kgs., Lyngby, Denmark
cawi@imm.dtu.dk

Part E

Carsten Witt is Associate Professor at the Technical University of Denmark. He received his PhD in Computer Science from the Technical University of Dortmund in 2004. His main research interests are the theoretical aspects of nature-inspired algorithms, in particular evolutionary algorithms, ant colony optimization and particle swarm optimization. He is a member of the Editorial Boards of Evolutionary Computation and Theoretical Computer Science and has co-authored a textbook.

Yiyu Yao

University of Regina
Dep. Computer Science
Regina, Saskatchewan, S4S 0A2, Canada
yyao@cs.uregina.ca

Part C

Yiyu Yao is Professor of Computer Science in the Department of Computer Science, the University of Regina, Canada. His research interests include three-way decisions, rough sets, fuzzy sets, interval sets, granular computing, information retrieval, Web intelligence, and data mining. He is currently working on a triarchic theory of granular computing, a theory of three-way decisions and generalized rough sets.

List of Authors

Enrique Alba
Universidad de Malaga
E.T.S.I. Informática
Campus de Teatinos (3.2.12)
29071 Málaga, Spain
e-mail: *eat@lcc.uma.es*

Jose M. Alonso
European Centre for Soft Computing
Cognitive Computing
33600 Mieres, Spain
e-mail: *jose.alonso@softcomputing.es*

Jhon Edgar Amaya
Universidad Nacional Experimental del Táchira
Dep. Electronic Engineering
Av. Universidad. Paramillo
San Cristóbal, Venezuela
e-mail: *jedgar@unet.edu.ve*

Plamen P. Angelov
Lancaster University
School of Computing and Communications
Bailrigg, Lancaster, LA1 4YW, UK
e-mail: *p.angelov@lancaster.ac.uk*

Dirk V. Arnold
Dalhousie University
Faculty of Computer Science
6050 University Avenue
Halifax, Nova Scotia, B3H 4R2, Canada
e-mail: *dirk@cs.dal.ca*

Anne Auger
University Paris-Sud Orsay
CR Inria
LRI (UMR 8623)
91405 Orsay Cedex, France
e-mail: *anne.auger@inria.fr*

Davide Bacciu
Università di Pisa
Dip. Informatica
L.Go B. Pontecorvo, 3
56127 Pisa, Italy
e-mail: *bacciu@di.unipi.it*

Michał Baczynski
University of Silesia
Inst. Mathematics
Bankowa 14
40-007 Katowice, Poland
e-mail: *michal.baczynski@us.edu.pl*

Edurne Barrenechea
Universidad Pública de Navarra
Dep. Automática y Computación
31006 Pamplona (Navarra), Spain
e-mail: *edurne.barrenechea@unavarra.es*

Thomas Bartz-Beielstein
Cologne University of Applied Sciences
Faculty of Computer Science and Engineering
Science
Steinmüllerallee 1
51643 Gummersbach, Germany
e-mail: *thomas.bartz-beielstein@fh-koeln.de*

Lubica Benuskova
University of Otago
Dep. Computer Science
133 Union Street East
9016 Dunedin, New Zealand
e-mail: *lubica@cs.otago.ac.nz*

Dirk Biermann
TU Dortmund University
Dep. Mechanical Engineering
Baroper Str. 303
44227 Dortmund, Germany
e-mail: *biermann@isf.de*

Sašo Blažič
University of Ljubljana
Faculty of Electrical Engineering
Tržaška 25
1000 Ljubljana, Slovenia
e-mail: *saso.blazic@fe.uni-lj.si*

Christian Blum
University of the Basque Country
Dep. Computer Science and Artificial Intelligence
Paseo Manuel Lardizabal 1
20018 San Sebastian, Spain
e-mail: *christian.blum@ehu.es*

Andrea Bobbio
Università del Piemonte Orientale
DiSit – Computer Science Section
Viale Teresa Michel, 11
15121 Alessandria, Italy
e-mail: *andrea.bobbio@unipmn.it*

Josh Bongard
University of Vermont
Dep. Computer Science
33 Colchester Ave.
Burlington, VT 05405, USA
e-mail: *josh.bongard@uvm.edu*

Piero P. Bonissone
Piero P. Bonissone Analytics, LLC
3103 28th Street
San Diego, CA 92104, USA
e-mail: *bonissone@gmail.com*

Dario Bruneo
Universita' di Messina
Dip. Ingegneria Civile, Informatica
Contrada di Dio – S. Agata
98166 Messina, Italy
e-mail: *dbruneo@unime.it*

Alberto Bugarín Diz
University of Santiago de Compostela
Research Centre for Information Technologies
15782 Santiago de Compostela, Spain
e-mail: *alberto.bugarin.diz@usc.es*

Humberto Bustince
Universidad Pública de Navarra
Dep. Automática y Computación
31006 Pamplona (Navarra), Spain
e-mail: *bustince@unavarra.es*

Martin V. Butz
University of Tübingen
Computer Science, Cognitive Modeling
Sand 14
72076 Tübingen, Germany
e-mail: *martin.butz@uni-tuebingen.de*

Alexandre Campo
Université Libre de Bruxelles
Unit of Social Ecology
Boulevard du triomphe,
Campus de la Plaine
1050 Brussels, Belgium
e-mail: *alexandre.campo@ulb.ac.be*

Angelo Cangelosi
Plymouth University
Centre for Robotics and Neural Systems
Drake Circus
Plymouth, PL4 8AA, UK
e-mail: *A.Cangelosi@plymouth.ac.uk*

Robert Carrese
LEAP Australia Pty. Ltd.
Clayton North, Australia
e-mail: *robert.carrese@leapaust.com.au*

Ciro Castiello
University of Bari
Dep. Informatics
via E. Orabona, 4
70125 Bari, Italy
e-mail: *ciro.castiello@uniba.it*

Oscar Castillo
Tijuana Institute of Technology
Calzada Tecnolo-gico s/n
22379 Tijuana, Mexico
e-mail: *ocastillo@tectijuana.mx*

Davide Cerotti
Politecnico di Milano
Dip. Elettronica, Informazione e Bioingegneria
Via Ponzio 34/5
20133 Milano, Italy
e-mail: *davide.cerotti@polimi.it*

Badong Chen
Xi'an Jiaotong University
Inst. Artificial Intelligence and Robotics
710049 Xi'an, China
e-mail: *chenbd@mail.xjtu.edu.cn*

Ke Chen
The University of Manchester
School of Computer Science
G10 Kilburn Building, Oxford Road
Manchester, M13 9PL, UK
e-mail: *chen@cs.manchester.ac.uk*

Davide Ciucci
University of Milano-Bicocca
Dep. Informatics, Systems and Communications
viale Sarca 336/14
20126 Milano, Italy
e-mail: *ciucci@disco.unimib.it*

Carlos A. Coello Coello
CINVESTAV-IPN
Dep. Computación
Av. Instituto Politécnico Nacional No. 2508, Col.
San Pedro Zacatenco
D.F. 07300, México, Mexico
e-mail: *ccoello@cs.cinvestav.mx*

Chris Cornelis
Ghent University
Dep. Applied Mathematics and Computer Science
Krijgslaan 281 (S9)
9000 Ghent, Belgium
e-mail: *chriscornelis@ugr.es*

Nikolaus Correll
University of Colorado at Boulder
Dep. Computer Science
Boulder, CO 80309, USA
e-mail: *ncorrell@colorado.edu*

Carlos Cotta Porras
Universidad de Málaga
Dep. Lenguajes y Ciencias de la Computación
Avda Louis Pasteur, 35
29071 Málaga, Spain
e-mail: *ccottap@lcc.uma.es*

Damien Coyle
University of Ulster
Intelligent Systems Research Centre
Northland Rd
Derry, Northern Ireland, BT48 7JL, UK
e-mail: *dh.coyle@ulster.ac.uk*

Guy De Tré
Ghent University
Dep. Telecommunications and
Information Processing
Sint-Pietersnieuwstraat 41
9000 Ghent, Belgium
e-mail: *guy.detre@ugent.be*

Kalyanmoy Deb
Michigan State University
Dep. Electrical and Computer Engineering
428 S. Shaw Lane
East Lansing, MI 48824, USA
e-mail: *kdeb@egr.msu.edu*

Clarisse Dhaenens
University of Lille
CRIStAL laboratory
M3 building – Cité scientifique
59655 Villeneuve d'Ascq Cedex, France
e-mail: *clarisse.dhaenens@univ-lille1.fr*

Luca Di Gaspero
Università degli Studi di Udine
Dip. Ingegneria Elettrica,
Gestionale e Meccanica
via delle Scienze 208
33100 Udine, Italy
e-mail: *luca.digaspero@uniud.it*

Didier Dubois
Université Paul Sabatier
IRIT – Equipe ADRIA
118 route de Narbonne
31062 Toulouse Cedex 9, France
e-mail: *dubois@irit.fr*

Antonio J. Fernández Leiva
Universidad de Málaga
Dep. Lenguajes y Ciencias de la Computación
Avda Louis Pasteur, 35
29071 Málaga, Spain
e-mail: *afdez@lcc.uma.es*

Javier Fernández
Universidad Pública de Navarra
Dep. Automática y Computación
31006 Pamplona (Navarra), Spain
e-mail: *fcojavier.fernandez@unavarra.es*

Martin H. Fischer
University of Potsdam
Psychology Dep.
Karl-Liebknecht-Str. 24/25
14476 Potsdam OT Golm, Germany
e-mail: *martinf@uni-potsdam.de*

János C. Fodor
Óbuda University
Dep. Applied Mathematics
Bécsi út 96/b
1034 Budapest, Hungary
e-mail: *fodor@uni-obuda.hu*

Jairo Alonso Giraldo
Universidad de los Andes
Dep. Electrical and Electronics Engineering
Cra 1Este # 19A-40
111711 Bogotá, Colombia
e-mail: *ja.giraldo908@uniandes.edu.co*

Siegfried Gottwald
Leipzig University
Inst. Philosophy
Beethovenstr. 15
04107 Leipzig, Germany
e-mail: *gottwald@uni-leipzig.de*

Salvatore Greco
University of Catania
Dep. Economics and Business
Corso Italia 55
95129 Catania, Italy
e-mail: *salgreco@unict.it*

Marco Gribaudo
Politecnico di Milano
Dip. Elettronica, Informazione e Bioingegneria
Via Ponzio 34/5
20133 Milano, Italy
e-mail: *marco.gribaudo@polimi.it*

Roderich Groß
University of Sheffield
Dep. Automatic Control and Systems Engineering
Mappin Street
Sheffield, S1 3JD, UK
e-mail: *r.gross@sheffield.ac.uk*

Jerzy W. Grzymala-Busse
University of Kansas
Dep. Electrical Engineering and Computer Science
3014 Eaton Hall, 1520 W. 15th St.
Lawrence, KS 66045-7621, USA
e-mail: *jerzygb@ku.edu*

Hani Hagras
University of Essex
The Computational Intelligence Centre
Wivenhoe Park
Colchester, CO4 3SQ, UK
e-mail: *hani@essex.ac.uk*

Heiko Hamann
Universtity of Paderborn
Dep. Computer Science
Zukunftsmeile 1
33102 Paderborn, Germany
e-mail: *heiko.hamann@uni-paderborn.de*

Thomas Hammerl
Westbahnstraße 25/1/7
1070 Vienna, Austria
e-mail: *thomas.hammerl@gmail.com*

Julie Hamon
Ingenomix
Dep. Research and Development
Pole de Lanaud
87220 Boisseuil, France
e-mail: *julie.hamon@ingenomix.fr*

Nikolaus Hansen
Universitè Paris-Sud
Machine Learning and Optimization Group (TAO)
Rue Noetzlin
91405 Orsay Cedex, France
e-mail: *hansen@lri.fr*

Mark W. Hauschild
University of Missouri–St. Louis
Dep. Mathematics and Computer Science
1 University Blvd
St. Louis, MO 314-972-2419, USA
e-mail: *markhauschild@gmail.com*

Sebastien Hélie
Purdue University
Dep. Psychological Sciences
703 Third Street
West Lafayette, IN 47907-2081, USA
e-mail: *shelie@purdue.edu*

Jano I. van Hemert
Optos
Queensferry House, Carnegie Business Park
Dunfermline, KY11 8GR, UK
e-mail: *jano@vanhemert.co.uk*

Holger H. Hoos
University of British Columbia
Dep. Computer Science
2366 Main Mall
Vancouver, BC V6T 1Z4, Canada
e-mail: *hoos@cs.ubc.ca*

Tania Iglesias
University of Oviedo
Dep. Statistics and O.R.
3360 Oviedo, Spain
e-mail: *iglesiasctania@uniovi.es*

Giacomo Indiveri
University of Zurich and ETH Zurich
Inst. Neuroinformatics
Zurich, Switzerland
e-mail: *giacomo@ini.uzh.ch*

Masahiro Inuiguchi
Osaka University
Dep. Systems Innovation, Graduate School of
Engineering Science
1-3 Machikaneyama-cho
560-8531 Toyonaka, Osaka, Japan
e-mail: *inuiguti@sys.es.osaka-u.ac.jp*

Hisao Ishibuchi
Osaka Prefecture University
Dep. Computer Science and Intelligent Systems,
Graduate School of Engineering
1-1 Gakuen-Cho, Sakai
599-8531 Osaka, Japan
e-mail: *hisaoi@cs.osakafu-u.ac.jp*

Emiliano Iuliano
CIRA, Italian Aerospace Research Center
Fluid Dynamics Lab.
Via Maiorise
81043 Capua (CE), Italy
e-mail: *e.iuliano@cira.it*

Julie Jacques
Alicante LAB
50, rue Philippe de Girard
59113 Seclin, France
e-mail: *julie.jacques@alicante.fr*

Andrzej Jankowski
Knowledge Technology Foundation
Nowogrodzka 31
00-511 Warsaw, Poland
e-mail: *andrzej.adgam@gmail.com*

Balasubramaniam Jayaram
Indian Institute of Technology Hyderabad
Dep. Mathematics
ODF Estate, Yeddumailaram
502 205 Hyderabad, India
e-mail: *jbala@iith.ac.in*

Laetitia Jourdan
University of Lille 1
INRIA/UFR IEEA/laboratory CRIStAL/CNRS
59655 Lille, France
e-mail: *laetitia.jourdan@univ-lille1.fr*

Nikola Kasabov
Auckland University of Technology
KEDRI − Knowledge Engineering and
Discovery Research Inst.
120 Mayoral Drive
Auckland, New Zealand
e-mail: *nkasabov@aut.ac.nz*

Petra Kersting
TU Dortmund University
Dep. Mechanical Engineering
Baroper Str. 303
44227 Dortmund, Germany
e-mail: *pkersting@isf.de*

Erich P. Klement
Johannes Kepler University
Dep. Knowledge-Based Mathematical Systems
Altenberger Strasse 69
4040 Linz, Austria
e-mail: *ep.klement@jku.at*

Anna Kolesárová
Slovak University of Technology in Bratislava
Faculty of Chemical and Food Technology
Radlinského 9
812 37 Bratislava, Slovakia
e-mail: *anna.kolesarova@stuba.sk*

Magda Komorníková
Slovak University of Technology
Dep. Mathematics
Radlinského 11
813 68 Bratislava, Slovakia
e-mail: *magda@math.sk*

Mark Kotanchek
Evolved Analytics LLC
3411 Valley Drive
Midland, MI 48640, USA
e-mail: *mark@evolved-analytics.com*

Robert Kozma
University of Memphis
Dep. Mathematical Sciences
Memphis, TN 38152, USA
e-mail: *rkozma@memphis.edu*

Tomáš Kroupa
Institute of Information Theory and Automation
Dep. Decision-Making Theory
Pod Vodárenskou věží 4
182 08 Prague, Czech Republic
e-mail: *kroupa@utia.cas.cz*

Rudolf Kruse
University of Magdeburg
Faculty of Computer Science
Universitätsplatz 2
39114 Magdeburg, Germany
e-mail: *kruse@iws.cs.uni-magdeburg.de*

Tufan Kumbasar
Istanbul Technical University
Control Engineering Dep.
34469 Maslak, Istanbul, Turkey
e-mail: *kumbasart@itu.edu.tr*

James T. Kwok
Hong Kong University of Science and Technology
Dep. Computer Science and Engineering
Clear Water Bay
Hong Kong, Hong Kong
e-mail: *jamesk@cse.ust.edu.hk*

Rhyd Lewis
Cardiff University
School of Mathematics
Cardiff, CF10 4AG, UK
e-mail: *lewisR9@cf.ac.uk*

Xiaodong Li
RMIT University
School of Computer Science and
Information Technology
Melbourne, 3001, Australia
e-mail: *xiaodong.li@rmit.edu.au*

Paulo J.G. Lisboa
Liverpool John Moores University
Dep. Mathematics & Statistics
Byrom St
Liverpool, L3 3AF, UK
e-mail: *p.j.lisboa@ljmu.ac.uk*

Weifeng Liu
Jump Trading
600 W. Chicago Ave.
Chicago, IL 60654, USA
e-mail: *weifeng@ieee.org*

Fernando G. Lobo
Universidade do Algarve
Dep. Engenharia Electrónica e Informática
Campus de Gambelas
8005-139 Faro, Portugal
e-mail: *fernando.lobo@gmail.com*

Antonio López Jaimes
CINVESTAV-IPN
Dep. Computación
Av. Instituto Politécnico Nacional No. 2508, Col.
San Pedro Zacatenco
D.F. 07300, México, Mexico
e-mail: *tonio.jaimes@gmail.com*

Francisco Luna
Centro Universitario de Mérida
Santa Teresa de Jornet 38
06800 Mérida, Spain
e-mail: *fluna@unex.es*

Luis Magdalena
European Centre for Soft Computing
Gonzalo Gutiérrez Quirós s/n
33600 Mieres, Spain
e-mail: *luis.magdalena@softcomputing.es*

Sebastia Massanet
University of the Balearic Islands
Dep. Mathematics and Computer Science
Crta. Valldemossa km. 7,5
07122 Palma de Mallorca, Spain
e-mail: *s.massanet@uib.es*

Benedetto Matarazzo
University of Catania
Dep. Economics and Business
Corso Italia 55
95129 Catania, Italy
e-mail: *matarazz@unict.it*

Sergi Mateo Bellido
Polytechnic University of Catalonia
Dep. Computer Architecture
08034 Barcelona, Spain
e-mail: *sergim@ac.upc.edu*

James McDermott
University College Dublin
Lochlann Quinn School of Business
Belfield
Dublin 4, Ireland
e-mail: *jmmcd@jmmcd.net*

Patricia Melin
Tijuana Institute of Technology
Dep. Computer Science
Chula Vista, CA 91909, USA
e-mail: *pmelin@tectijuana.mx*

Corrado Mencar
University of Bari
Dep. Informatics
via E. Orabona, 4
70125 Bari, Italy
e-mail: *corrado.mencar@uniba.it*

Radko Mesiar
STU in Bratislava
Dep. Mathematics and Descriptive Geometry
Radlinskeho 11
813 68 Bratislava, Slovakia
e-mail: *radko.mesiar@stuba.sk*

Ralf Mikut
Karlsruhe Institute of Technology (KIT)
Inst. Applied Computer Science
Hermann-von-Helmholtz-Platz 1
76344 Eggenstein-Leopoldshafen, Germany
e-mail: *ralf.mikut@kit.edu*

Ali A. Minai
University of Cincinnati
School of Electronic & Computing Systems
2600 Clifton Ave.
Cincinnati, OH 45221-0030, USA
e-mail: *ali.minai@uc.edu*

Sadaaki Miyamoto
University of Tsukuba
Risk Engineering
1-1-1 Tennodai
305-8573 Tsukuba, Japan
e-mail: *miyamoto@risk.tsukuba.ac.jp*

Christian Moewes
University of Magdeburg
Faculty of Computer Science
Universitätsplatz 2
39114 Magdeburg, Germany
e-mail: *cmoewes@ovgu.de*

Javier Montero
Complutense University, Madrid
Dep. Statistics and Operational Research
Plaza de las Ciências, 3
28040 Madrid, Spain
e-mail: *monty@mat.ucm.es*

Ignacio Montes
University of Oviedo
Dep. Statistics and O.R.
3360 Oviedo, Spain
e-mail: *imontes@uniovi.es*

Susana Montes
University of Oviedo
Dep. Statistics and O.R.
3360 Oviedo, Spain
e-mail: *montes@uniovi.es*

Oscar H. Montiel Ross
Av. del Parque No. 131º
B.C. 22414, Mesa de Otay, Tijuana, Mexico
e-mail: *oross@citedi.mx*

Manuel Mucientes
University of Santiago de Compostela
Research Centre for Information Technologies
15782 Santiago de Compostela, Spain
e-mail: *manuel.mucientes@usc.es*

Nysret Musliu
Vienna University of Technology
Inst. Information Systems
Favoritenstraße 9
1000 Vienna, Austria
e-mail: *musliu@dbai.tuwien.ac.at*

Yusuke Nojima
Osaka Prefecture University
Dep. Computer Science and Intelligent Systems,
Graduate School of Engineering
1-1 Gakuen-Cho, Sakai
599-8531 Osaka, Japan
e-mail: *nojima@cs.osakafu-u.ac.jp*

Stefano Nolfi
Consiglio Nazionale delle Ricerche (CNR-ISTC)
Inst. Cognitive Sciences and Technologies
Via S. Martino della Battaglia, 44
00185 Roma, Italy
e-mail: *stefano.nolfi@istc.cnr.it*

Una-May O'Reilly
Massachusetts Institute of Technology
Computer Science and Artificial Intelligence Lab.
32 Vassar St.
Cambridge, MA 02139, USA
e-mail: *unamay@csail.mit.edu*

Miguel Pagola
Universidad Pública de Navarra
Dep. Automática y Computación
31006 Pamplona (Navarra), Spain
e-mail: *miguel.pagola@unavarra.es*

Lynne Parker
University of Tennessee
Dep. Electrical Engineering and Computer Science
1520 Middle Drive
Knoxville, TN 37996, USA
e-mail: *leparker@utk.edu*

Kevin M. Passino
The Ohio State University
Dep. Electrical and Computer Engineering
2015 Neil Avenue
Columbus, OH 43210-1272, USA
e-mail: *passino@ece.osu.edu*

Martin Pelikan
1271 Lakeside Dr. #3123
Sunnyvale, CA 94085, USA
e-mail: *martin@martinpelikan.net*

Irina Perfilieva
University of Ostrava
Inst. Research and Applications of Fuzzy Modeling
30. dubna 22
70103 Ostrava, Czech Republic
e-mail: *Irina.Perfilieva@osu.cz*

Henry Prade
Université Paul Sabatier
IRIT – Equipe ADRIA
118 route de Narbonne
31062 Toulouse Cedex 9, France
e-mail: *prade@irit.fr*

Mike Preuss
WWU Münster
Inst. Wirtschaftsinformatik
Leonardo-Campus 3
48149 Münster, Germany
e-mail: *mike.preuss@tu-dortmund.de*

José C. Principe
University of Florida
Dep. Electrical and Computer Engineering
Gainesville, FL 32611, USA
e-mail: *principe@cnel.ufl.edu*

Domenico Quagliarella
CIRA, Italian Aerospace Research Center
Fluid Dynamics Lab.
Via Maiorise
81043 Capua (CE), Italy
e-mail: *d.quagliarella@cira.it*

Nicanor Quijano
Universidad de los Andes
Dep. Electrical and Electronics Engineering
Cra 1Este # 19A-40
111711 Bogotá, Colombia
e-mail: *nquijano@uniandes.edu.co*

Jaroslav Ramík
Silesian University in Opava
Dep. Informatics and Mathematics
University Sq. 1934/3
73340 Karviná, Czech Republic
e-mail: *ramik@opf.slu.cz*

Ismael Rodríguez Fdez
University of Santiago de Compostela
Research Centre for Information Technologies
15782 Santiago de Compostela, Spain
e-mail: *ismael.rodriguez@usc.es*

Franz Rothlauf
Johannes Gutenberg University Mainz
Gutenberg School of Management and Economics
Jakob Welder-Weg 9
55099 Mainz, Germany
e-mail: *rothlauf@uni-mainz.de*

Jonathan E. Rowe
University of Birmingham
School of Computer Science
Birmingham, B15 2TT, UK
e-mail: *J.E.Rowe@cs.bham.ac.uk*

Imre J. Rudas
Óbuda University
Dep. Applied Mathematics
Bécsi út 96/b
1034 Budapest, Hungary
e-mail: *rudas@uni-obuda.hu*

Günter Rudolph
Technische Universität Dortmund
Fak. Informatik
Otto-Hahn-Str. 14
44227 Dortmund, Germany
e-mail: *guenter.rudolph@cs.tu-dortmund.de*

Gabriele Sadowski
Technische Universität Dortmund
Bio- und Chemieingenieurwesen
Emil-Figge-Str. 70
44227 Dortmund, Germany
e-mail: *gabriele.sadowski@bci.tu-dortmund.de*

Marco Scarpa
Universita' di Messina
Dip. Ingegneria Civile, Informatica
Contrada di Dio — S. Agata
98166 Messina, Italy
e-mail: *mscarpag@unime.it*

Werner Schafhauser
XIMES
Hollandstraße 12/12
1020 Vienna, Austria
e-mail: *schafhauser@ximes.com*

Roberto Sepúlveda Cruz
Av. del Parque No. 131º
B.C. 22414, Mesa de Otay, Tijuana, Mexico
e-mail: *rsepulve@citedi.mx*

Jennie Si
Arizona State University
School of Electrical, Computer and
Energy Engineering
Tempe, AZ 85287-5706, USA
e-mail: *si@asu.edu*

Marco Signoretto
Katholieke Universiteit Leuven
Kasteelpark Arenberg 10
3001 Leuven, Belgium
e-mail: *marco.signoretto@esat.kuleuven.be*

Andrzej Skowron
University of Warsaw
Faculty of Mathematics,
Computer Science and Mechanics
Banacha 2
02-097 Warsaw, Poland
e-mail: *skowron@mimuw.edu.pl*

Igor Škrjanc
University of Ljubljana
Faculty of Electrical Engineering
Tržaška 25
1000 Ljubljana, Slovenia
e-mail: *igor.skrjanc@fe.uni-lj.si*

Roman Słowiński
Poznań University of Technology
Inst. Computing Science
Piotrowo 2
60-965 Poznań, Poland
e-mail: *roman.slowinski@cs.put.poznan.pl*

Guido Smits
Dow Benelux BV
Core R&D
Herbert H. Dowweg 5
4542 NM Hoek, The Netherlands
e-mail: *gfsmits@dow.com*

Ronen Sosnik
Holon Institute of Technology (H.I.T.)
Electrical, Electronics and Communication
Engineering
52 Golomb St.
5810201 Holon, Israel
e-mail: *ronens@hit.ac.il*

Alessandro Sperduti
University of Padova
Dep. Pure and Applied Mathematics
Via Trieste, 63
351 21 Padova, Italy
e-mail: *sperduti@math.unipd.it*

Kasper Støy
IT University of Copenhagen
Rued Langgaards Vej 7
2300 Copenhagen S, Denmark
e-mail: *ksty@itu.dk*

Harrison Stratton
Arizona State University & Barrow
Neurological Institute
Phoenix, AZ 85013, USA
e-mail: *Harrison.Stratton@asu.edu*

Thomas Stützle
Université libre de Bruxelles (ULB)
IIRIDIA, CP 194/6
Av. F. Roosevelt 50
1050 Brussels, Belgium
e-mail: *stuetzle@ulb.ac.be*

Dirk Sudholt
University of Sheffield
Dep. Computer Science
211 Portobello
Sheffield, S1 4DP, UK
e-mail: *d.sudholt@sheffield.ac.uk*

Ron Sun
Rensselaer Polytechnic Institute
Cognitive Science Dep.
110 Eighth Street, Carnegie 302A
Troy, NY 12180, USA
e-mail: *rsun@rpi.edu*

Johan A. K. Suykens
Katholieke Universiteit Leuven
Kasteelpark Arenberg 10
3001 Leuven, Belgium
e-mail: *johan.suykens@esat.kuleuven.be*

Roman W. Swiniarski (deceased)

El-Ghazali Talbi
University of Lille
Computer Science CRISTAL
Bat.M3 cité scientifique
59655 Villeneuve d'Ascq, France
e-mail: *el-ghazali.talbi@univ-lille1.fr*

Lothar Thiele
Swiss Federal Institute of Technology Zurich
Computer Engineering and Networks Lab.
Gloriastrasse 35
8092 Zurich, Switzerland
e-mail: *thiele@ethz.ch*

Peter Tino
University of Birmingham
School of Computer Science
Edgbaston
Birmingham, B15 2TT, UK
e-mail: *P.Tino@cs.bham.ac.uk*

Joan Torrens
University of the Balearic Islands
Dep. Mathematics and Computer Science
Crta. Valldemossa km. 7,5
07122 Palma de Mallorca, Spain
e-mail: *jts224@uib.es*

Vito Trianni
Consiglio Nazionale delle Ricerche
Ist. Scienze e Tecnologie della Cognizione
via San Martino della Battaglia 44
00185 Roma, Italy
e-mail: *vito.trianni@istc.cnr.it*

Enric Trillas
European Centre for Soft Computing
Fundamentals of Soft Computing
33600 Mieres, Spain
e-mail: *enric.trillas@softcomputing.es*

Fevrier Valdez
Tijuana Institute of Technology
Calzada del Tecnológico S/N, Tomas Aquino
B.C. 22414, Tijuana, Mexico
e-mail: *fevrier@tectijuana.mx*

Nele Verbiest
Ghent University
Dep. Applied Mathematics,
Computer Science and Statistics
Krijgslaan 281 (S9)
9000 Ghent, Belgium
e-mail: *nele.verbiest@ugent.be*

Thomas Villmann
University of Applied Sciences Mittweida
Dep. Mathematics, Natural and Computer
Sciences
Technikumplatz 17
09648 Mittweida, Germany
e-mail: *thomas.villmann@hs-mittweida.de*

Milan Vlach
Charles University
Theoretical Computer Science and
Mathematical Logic
Malostranské náměstí 25
118 00 Prague, Czech Republic
e-mail: *Milan.Vlach@mff.cuni.cz*

Ekaterina Vladislavleva
Evolved Analytics Europe BVBA
A. Coppenslaan 27
2300 Turnhout, Belgium
e-mail: *katya@evolved-analytics.com*

Tobias Wagner
TU Dortmund University
Dep. Mechanical Engineering
Baroper Str. 303
44227 Dortmund, Germany
e-mail: *wagner@isf.de*

Jun Wang
The Chinese University of Hong Kong
Dep. Mechanical & Automation Engineering
Shatin, New Territories
Hongkong, Hong Kong
e-mail: *jwang@mae.cuhk.edu.hk*

Simon Wessing
Technische Universität Dortmund
Fak. Informatik
Otto-Hahn-Str. 14
44227 Dortmund, Germany
e-mail: *simon.wessing@tu-dortmund.de*

Wei-Zhi Wu
Zhejiang Ocean University
School of Mathematics, Physics and
Information Science
No.1 Haida South Road, Lincheng District
316022 Zhoushan, Zhejiang, China
e-mail: *wuwz@zjou.edu.cn*

Lei Xu
The Chinese University of Hong Kong
Dep. Computer Science and Engineering
Shatin, New Territories
Hong Kong, Hong Kong
e-mail: *lxu@cse.cuhk.edu.hk*

JingTao Yao
University of Regina
Dep. Computer Science
3737 Wascana Parkway
Regina, Saskatchewan, S4S 0A2, Canada
e-mail: *jtyao@cs.uregina.ca*

Yiyu Yao
University of Regina
Dep. Computer Science
3737 Wascana Parkway
Regina, Saskatchewan, S4S 0A2, Canada
e-mail: *yyao@cs.uregina.ca*

Andreas Zabel
TU Dortmund University
Dep. Mechanical Engineering
Baroper Str. 303
44227 Dortmund, Germany
e-mail: *zabel@isf.de*

Sławomir Zadrożny
Polish Academy of Sciences
Systems Research Inst.
ul. Newelska 6
01-447 Warsaw, Poland
e-mail: *Slawomir.Zadrozny@ibspan.waw.pl*

Zhigang Zeng
Huazhong University of Science and Technology
Dep. Control Science and Engineering
No. 1037, Luoyu Road
430074 Wuhan, China
e-mail: *zgzeng@hust.edu.cn*

Yan Zhang
University of Regina
Dep. Computer Science
3737 Wascana Parkway
Regina, Saskatchewan, S4S 0A2, Canada
e-mail: *zhang83y@cs.uregina.ca*

Zhi-Hua Zhou
Nanjing University
National Key Lab. for Novel Software Technology
210023 Nanjing, China
e-mail: *zhouzh@nju.edu.cn*

Contents

Part B   Fuzzy Logic

Part D   Neural Networks

Part G   Hybrid Systems

List of Abbreviations

Symbols

1-D	one-dimensional
2-D	two-dimensional
3-CNF-SAT	three variables/clause-conjunctive normal form-satisfiability
3-D	three-dimensional

A

A2A	all-to-all
AaaS	analytics-as-a-service
AANN	auto-associative neural network
ABC	artificial bee colony
ACC	anterior cingulate cortex
ACO	ant colony optimization
ACP	active categorical perception
ACS	action-centered subsystem
ACS	ant colony system
ACT-R	adaptive control of thought-rational
AD	anomaly detection
ADC	analog digital converter
ADF	additively decomposable function
ADF	automatically defined function
ADGLIB	adaptive genetic algorithm optimization library
AER	address event representation
AFPGA	adaptive full POD genetic algorithm
AFSA	artificial fish swarm algorithm
AI	anomaly identification
AI	artificial intelligence
AIC	Akaike information criterion
AICOMP	comparable based AI model
AIGEN	generative AI model
ALCS	anticipatory learning classifier system
ALD	approximate linear dependency
ALM	asset–liability management
ALU	arithmetic logic unit
ALU	arithmetic unit
AM	amplitude modulation
amBOA	adaptive variant of mBOA
AMPGA	adaptive mixed-flow POD genetic algorithm
AMS	anticipated mean shift
AMT	active media technology
ANN	artificial neural network
ANOVA	analysis of variance
ANYA	Angelov–Yager
AP	alternating-position crossover
AP	automatic programming
APA	affine projection algorithm
API	application programming interface
APS	aggregation pheromone system
APSD	auto power spectral density
AR	approximate reasoning
AR	average ranking
ARD	automatic relevance determination
ARGOT	adaptive representation genetic optimization technique
ARMOGA	adaptive range MOGA
ASIC	application-specific integrated circuit
ASP	answer-set programming
ATP	adenosine triphosphate
AUC	area under curve
AUC	area under ROC curve
AVITEWRITE	adaptive vector integration to endpoint handwriting

B

BBB	blood brain barrier
BCI	brain–computer interface
BCO	bee colony optimization
BDAS	Berkeley data analytics stack
BER	bit error rate
BeRoSH	behavior-based multiple robot system with host for object manipulation
BFA	basic fuzzy algebra
BG	basal ganglia
BIC	Bayesian information criterion
BINCSP	binary constraint satisfaction problem
BioHEL	bioinformatics-oriented hierarchical evolutionary learning
BKS	Bandler–Kohout subproduct
BLB	bag of little bootstrap
BMA	Bayes model averaging
BMDA	bivariate marginal distribution algorithm
BMF	binary matrix factorization
BMI	brain–machine interface
BnB	branch and bound
BOA	Bayesian optimization algorithm
BP	bereitschafts potential

BP	back-propagation		CNGM	computational neuro-genetic modeling
BPTT	back-propagation through time		CNN	cellular neural network
BSB	base system builder		CNS	central nervous system
BSD	bipolar satisfaction degree		COA	center of area
BSS	blind source separation		COG	center of gravity
BYY	Bayesian Yin-Yang		COGIN	coverage-based genetic induction
			COP	cluster of processors

C

			COP	constrained optimization problem
c-granule	complex granule		CORE	Computing Research and Education
CA	cellular automata		cos	center of set
CA	classification accuracy		COW	cluster of workstations
CA	complete F-transform-based fusion algorithm		CP	constraint programming
			CP	contrapositive symmetry
CAD	computer-aided design		CP net	conditional preference network
CAE	contrastive auto-encoder		CPF	centralized Pareto front
CAM	computer-assisted manufacturing		CPG	central pattern generator
CART	classification analysis and regression tree		CPSD	cross power spectral density
CBLS	constraint-based local search		CPT	cummulative prospect theory
CBR	case-based reasoner		CPU	central processing unit
CBR	case-based reasoning		CR	commonsense reasoning
CC	coherence criterion		CR	control register
CCF	cross correlation function		CRA	chemical reaction algorithm
CCG	controlling crossed genes		CRI	compositional rule of inference
CD	contrastive divergence		CS	cell saving
CEA	cellular evolutionary algorithm		CS1	cognitive system
CEBOT	cellular robot		CSA	contractual service agreement
CF	collaborative filtering		CSA	cumulative step-size adaptation
CF	compact flash		CSM	covariate shift minimization
cf	convergence factor		CSP	common spatial pattern
CFD	computational fluid dynamics		CSP	constraint satisfaction problem
CFG	context-free grammar		CST	class-shape transformation
CFS	correlation feature selection		CST	corticospinal tract
CG	center of gravity		CTMC	continuous-time finite Markov chain
CG	Cohen–Grossberg		CUDA	compute unified device architecture
cGA	compact genetic algorithm		CW	computing with words
CGP	Cartesian GP		CW	control word
CI	computational intelligence		CWW	computing with words
CIP	cross information potential		CX	cycle crossover
CIS	Computational Intelligence Society			
CLB	configurable logic block			

D

			DA	dopamine
clk	clock		DACE	design and analysis of computer experiments
CLM	component level model			
CMA	cingulate motor area		DAE	denoising auto-encoder
CMA	covariance matrix adaptation		DAG	directed acyclic graph
CML	coupled map lattice		DAL	logic for data analysis
cMOEA	cellular MOEA		DB	database
CMOS	complementary metal-oxide-semiconductor		DBN	deep belief network
CNF	conjunctive normal form		dBOA	decision-graph BOA

DC	direct current
DC/AD	change/activate-deactivate
DCA	de-correlated component analysis
DE	differential evolution
dEA	distributed evolutionary algorithm
DENFIS	dynamic neuro-fuzzy inference system
deSNN	dynamic eSNN
DEUM	density estimation using Markov random fields algorithm
DEUM	distribution estimation using Markov random fields
DGA	direct genetic algorithm
DIC	deviance information criterion
DL	deep learning
DLPFC	dorsolateral prefrontal cortex
DLR	German Aerospace Center
DLS	dynamic local search
DM	displacement mutation operator
DM	decision maker
DMA	direct memory access
dMOEA	distributed MOEA
DNA	deoxyribonucleic acid
DNF	disjunctive normal form
DNN	deep neural network
DOE	design of experiment
DOF	degree of freedom
DP	dynamic programming
DPF	distributed Pareto front
DPLL	Davis–Putnam–Logemann–Loveland
DPR	dynamic partial reconfiguration
DRC	domain relational calculus
DREAM	distributed resource evolutionary algorithm machine
DRRS	dynamically reconfigurable robotic system
DRS	dominance resistant solution
DRSA	dominance-based rough set approach
DSA	data space adaptation
DSMGA	dependency-structure matrix genetic algorithm
DSP	digital signal processing
DSP	digital signal processor
DSS	decision support system
dtEDA	dependency-tree EDA
DTI	diffusion tensor imaging
DTLZ	Deb–Thiele–Laumanns–Zitzler
DTRS	decision-theoretic rough set
DW	data word

E

EA	evolutionary algorithm
EAPR	early access partial reconfiguration
EBNA	estimation of Bayesian network algorithm
EC	embodied cognition
EC	evolutionary computation
EC	evolutionary computing
ECGA	extended compact genetic algorithm
ECGP	extended compact genetic programming
ECJ	Java evolutionary computation
ECoG	electrocorticography
ECOS	evolving connectionist system
EDA	estimation of distribution algorithm
EDP	estimation of distribution programming
EEG	electroencephalogram
EEG	electroencephalography
EFRBS	evolutionary FRBS
EFuNN	evolving fuzzy neural network
EGA	equilibrium genetic algorithm
EGNA	estimation of Gaussian networks algorithm
EGO	efficient global optimization
EHBSA	edge histogram based sampling algorithm
EHM	edge histogram matrix
EI	expected improvement
EKM	enhanced KM
EKMANI	enhanced Karnik–Mendel algorithm with new initialization
ELSA	evolutionary local selection algorithm
EM	exchange mutation operator
EM	expectation maximization
EMG	electromyography
EMNA	estimation of multivariate normal algorithm
EMO	evolutionary multiobjective optimization
EMOA	evolutionary multiobjective algorithm
EMSE	excess mean square error
EODS	enhanced opposite directions searching
EP	evolutionary programming
EP	exchange property
EPTV	extended possibilistic truth value
ER	edge recombination
ERA	epigenetic robotics architecture
ERA	Excellence in Research for Australia
ERD	event-related desynchronization
ERM	empirical risk minimization
ERS	event-related synchronization
ES	embedding system
ES	evolution strategy

ESA	enhanced simple algorithm
ESN	echo state network
eSNN	evolving spiking neural network
ESOM	evolving self-organized map
ETS	evolving Takagi–Sugeno system
EV	extreme value
EvoStar	Main European Events on Evolutionary Computation
EvoWorkshops	European Workshops on Applications of Evolutionary Computation
EW-KRLS	exponentially weighted KRLS
EX-KRLS	extended kernel recursive least square

F

FA	factor analysis
FA	firefly algorithm
FA	fractional anisotropy
FA-DP	fitness assignment and diversity preservation
FATI	first aggregation then inference
FB-KRLS	fixed-budget KRLS
FCA	formal concept analysis
FCM	fuzzy c-means algorithm
FDA	factorized distribution algorithm
FDRC	fuzzy domain relational calculus
FDT	fuzzy decision tree
FEMO	fair evolutionary multi-objective optimizer
FGA	fuzzy generic algorithm
FIM	fuzzy inference mechanism
FIM	fuzzy instance based model
FIR	finite impulse response
FIS	fuzzy inference system
FIS1	type-1 fuzzy inference system
FIS2	type-2 fuzzy inference system
FITA	first inference then aggregation
FL	fuzzy logic
FLC	fuzzy logic controller
FlexCo	flexible coprocessor
FLP	fuzzy linear programming
FLS	fuzzy logic system
FM	fuzzy modeling
FMG	full multi-grid
FMM	finite mixture model
FMM	fuzzy mathematical morphology
fMRI	functional magneto-resonance imaging
FNN	fuzzy neural network
FNN	fuzzy nearest neighbor
FOM	full-order model

FOU	footprint of uncertainty
FPGA	field programmable gate array
FPGA	full POD genetic algorithm
FPID	fuzzy PID
FPM	fractal prediction machine
FPSO	fuzzy particle swarm optimization
FPU	floating point unit
FQL	fuzzy query language
FRB	fuzzy rule-based
FRBCS	fuzzy rule-based classification systems
FRBS	fuzzy rule-based system
FRC	fuzzy-rule based classifier
FRI	fuzzy relational inference
FS	fuzzy set
FS	fuzzy system
FSVM	fuzzy support vector machine
FTR	F-transform image compression
FURIA	unordered fuzzy rule induction algorithm
FX	foreign exchange

G

GA	general achievement
GA	genetic algorithm
GABA	gamma-aminobutyric acid
GACV	generalized approximate cross-validation
GAGRAD	genetic algorithm gradient
GAOT	genetic algorithm optimization toolbox
GC	granular computing
GCP	graph coloring problem
GE	grammatical evolution
GECCO	Genetic and Evolutionary Computation Conference
GEFRED	generalized fuzzy relational database
GFP	green fluorescent protein
GFS	genetic fuzzy system
GGA	grouping genetic algorithm
GGP	grammar-based genetic programming
GM	gray matter
GMP	generalized modus ponens
GMPE	grammar model-based program evolution
GP	genetic algorithm
GP	genetic programming
GP	Gaussian process
GPCR	g-protein coupled receptor
GPGPU	general-purpose GPU
GPi	globus pallidus
GPIO	general-purpose input/output interface
GPS	genetic pattern search
GPU	graphics processing unit

GPX	greedy partition crossover
GRASP	greedy randomized adaptive search procedure
GRBM	Gaussian RBM
GRN	gene/protein regulatory network
GRN	gene regulatory network
GT2	generalized T2FS
GWAS	genome-wide association studies
GWT	global workspace theory

H

H-PIPE	hierarchical probabilistic incremental program evolution
hBOA	hierarchical BOA
HCF	hyper-cube framework
HCwL	hill climbing with learning
HDL	hardware description language
HFB	higher frequency band
HH	Hodgkin–Huxley
HM	health management
HMM	hidden Markov model
HPC	high-performance computing
HS	Hilbert space
HSS	heuristic space search
HT	Hough transform
HW	hardware
HWICAP	hardware internal configuration access point

I

i.i.d.	independent, identically distributed
IASC	iterative algorithm with stop condition
IB	indicator based
IBEA	indicator-based evolutionary algorithm
iBOA	incremental Bayesian optimization algorithm
IC	intelligent controller
IC	interrupt controller
ICA	independent component analysis
ICAP	internal configuration access point
ICE	induced chromosome element exchanger
ICML	International Conference on Machine Learning
IDA	intelligent distribution agent
IDEA	iterated density estimation evolutionary algorithm
IE	inference engine

IEEE	Institute of Electrical and Electronics Engineers
IF	intuitionistic fuzzy
IFA	independent factor analysis
IFVS	interval valued fuzzy set
IG	iterated greedy
IGR	interactive granular computing
IHA	iterative heuristic algorithm
ILS	iterated local search
IN	interneuron
INCF	International Neuroinformatics Coordinating Facility
IO-HMM	input/output hidden Markov model
IOB	input/output block
IoT	internet of things
IP	identity principle
IP	inductive programming
IP	intellectual property
IPIF	intellectual property interface
IPL	inferior parietal lobe
IPOP	increasing population size
IRGC	interactive rough granular computing
IRLS	iteratively re-weighted least squares
IRSA	indiscernibility-based rough set approach
ISE	Integrated Synthesis Environment
ISI	inter-spike interval
ISM	insertion mutation operator
IT2	interval type-2
IT2FC	interval T2FC
IT2FS	interval T2FS
ITAE	integral time absolute error
ITL	information theoretic learning
IUMDA	incremental univariate marginal distribution algorithm
IVM	inversion mutation operator

J

JDEAL	Java distributed evolutionary algorithms library
JEGA	John Eddy genetic algorithm
JMAF	java multi-criteria and multi-attribute analysis framework

K

KAF	kernel adaptive filter
KAPA	kernel affine projection algorithm
KB	knowledge base
KDD	knowledge discovery and data mining

KGA	Kriging-driven genetic algorithm		LT	linguistic term
KKT	Karush–Kuhn–Tucker		LUT	look-up table
KL	Kullback–Leibler		LV	linguistic variable
KLMS	kernel least mean square		LVT	linguistic-variable-term
KM	Karnik–Mendel		LWPR	locally-weighted projection regression algorithm
KMC	kernel Maximum Correntropy		LZ	leading zero
KNN	k nearest neighbor			
KPCA	kernel principal component analysis			
KRLS	kernel recursive least square			
KUR	Kurswae			

M

			M1	motor cortex
			M2M	machine-to-machine
			MA	Markovian agent
L			MA	memetic algorithm
			MAE	mean of the absolute error
LAN	local network		MAFRA	Java mimetic algorithms framework
LASSO	least absolute shrinkage and selection operator		MAM	Markovian agent model
LB	logic block		MAMP	multiple algorithms, multiple problems
LCS	learning classifier system		MAMS	multiple algorithms and multiple problem instances
LDA	latent Dirichlet allocation		MAP	maximum a posteriori
LDA	linear discriminant analysis		MARS	multivariate adaptive regression splines
LDS	limited discrepancy search		MASP	multiple algorithms and one single problem
LED	light emitting diode		mBOA	mixed Bayesian optimization algorithm
LEM	learning from examples module		MCA	minor component analysis
LERS	learning from examples using rough sets		MCDA	multi-criteria decision analysis
LFA	local factor analysis		MCDA	multiple criteria decision aiding
LFB	lower frequency band		MCDM	multiple criteria decision-making
LFDA	learning FDA		MCS	maximum cardinality search
LFM	linguistic fuzzy modeling		MCS	meta-cognitive subsystem
LFP	local field potential		MDL	minimum description length
LGP	linear GP		MDP	Markov decision process
LHS	latin hypercube sampling		MDS	multidimensional scaling
LI	law of importation		MEG	magnetoencephalogram
LIFM	leaky integrate-and-fire		MEG	magnetoencephalography
LLE	liquid–liquid equilibrium		MEL	minimal epistemic logic
LMI	linear matrix inequalities		MF	membership function
LMS	least mean square		MG	Mackey–Glass
LNS	large neighborhood search		MG	morphological gradient
LO	leading one		MH	metaheuristic
LOCVAL	locational value		MIL	multi-instance learning
LOO	leave-one-out		MIMIC	mutual information maximizing input clustering
LOOCV	leave-one-out cross-validation		MIML	multi-instance, multi-label learning
LOTZ	leading ones trailing zeroes		MISO	multiple inputs-single output
LP	logic programming		MKL	multiple kernel learning
LQR	linear-quadratic regulator		ML	machine learning
LR	logistic regression		ML	maximum likelihood
LRP	lateralized readiness potential		MLEM2	modified LEM2 algorithm
LS	least square			
LS	local search			
LSM	liquid state machine			
LSTM	long short term memory			

MLP	multilayer perceptron
MLR	multiple linear regression
MLR	multi-response linear regression
MM	mathematical morphology
MMA	multimemetic algorithm
MMAS	MAX-MIN ant system
MMEA	model-based multiobjective evolutionary algorithm
MMLD	man–machine learning dilemma
MN-EDA	Markov network EDA
MNN	memristor-based neural network
MNN	modular neural network
MOAMO	multiobjectivization-assisted multimodal optimization
MOE	multiobjective evolutionary
MoE	mixture of experts
MOEA	multiobjective evolutionary algorithm
MOEA/D	multiobjective evolutionary algorithm based on decomposition
MOGA	multiobjective genetic algorithm
MoGFS	multiobjective genetic fuzzy system
MOM	mean of maxima
MOMGA	multi-objective messy GA
MOO	multi-objective optimization
MOP	many-objective optimization problem
MOP	multiobjective problem
MOP	multiobjective optimization problem
MOPSO	multiobjective particle swarm optimization
MOSAIC	modular selection and identification for control
MOSES	meta-optimizing semantic evolutionary search
MOT	movement time
MPE	mean percentage error
MPE	most probable explanation
MPGA	mixed-flow POD genetic algorithm
MPI	message passing interface
MPM	marginal product model
MPP	massively parallel machine
MPS	multiprocessor system
MR	maximum ranking
MRCP	movement-related cortical potentials
MRI	magnetic resonance imaging
mRMR	minimal-redundancy-maximal-relevance
mRNA	messenger RNA
MRNN	memristor-based recurrent neural network
MS	master/slave
MS	motivational subsystem
MSA	minor subspace analysis

MSE	mean square error
MSG	max-set of Gaussian landscape generator
msMOEA	master–slave MOEA
MT	medial temporal
MTFL	multi-task feature learning
MTL	multi-task learning
MV	maximum value
MWRA	minimum-weight rooted arborescence

N

NACS	non-action-centered subsystem
NASA	National Aeronautics and Space Administration
NC	neural computation
NC	novelty criterion
NC	numerical control
NCL	negative correlation learning
NDS	nonlinear dynamical systems
NEAT	neuro-evolution of augmenting topologies
NES	natural evolution strategy
NeuN	neuronal nuclei antibody
NFA	non-Gaussian factor analysis
NFI	neuro-fuzzy inference system
NFL	no free lunch
NHBSA	node histogram based sampling algorithm
NIL	nondeterministic information logic
NIPS	neural information processing system
NLMS	normalized LMS
NLPCA	nonlinear principal components
NMF	negative matrix and tensor factorization
NMF	nonnegative matrix factorization
NN	neural network
NOW	networks of workstation
NP	neutrality principle
NP	nondeterministic polynomial-time
NPV	net present value
NR	noise reduction
NS	negative slope
NSGA	nondominated sorting genetic algorithm
NSPSO	nondominated sorting particle swarm optimization
NURBS	nonuniform rational B-spline

O

ODE	ordinary differential equation
OEM	original equipment manufacturer

OKL	online kernel learning
OLAP	online analytical processing
OM	operational momentum
OMA	ordered modular average
OP	ordering property
OPB	on-chip peripheral bus
OPL	open programming language
OR	operations research
OR	operational research
OS	overshoot
OWA	ordered weighted average
OWMax	ordered weighted maximum
OX1	order crossover
OX2	order-based crossover

P

PAC	probably approximately correct
PAES	Pareto-archived evolution strategy
PAR	place and route
PBC	perception-based computing
PBIL	population-based incremental learning
PbO	programming by optimization
PC	probabilistic computing
PC-SAFT	perturbed chain statistical associating fluid theory
PCA	principal component analysis
PCVM	probabilistic classifier vector machine
PD	Parkinson disease
PD	proportional-differential
PDDL	planning domain definition language
PDE	partial differential equation
pdf	probability density function
PDGP	parallel and distributed GP
PEEL	program evolution with explicit learning
PERT	program evaluation and review technique
PESA	Pareto-envelope based selection algorithm
PET	positron emission tomography
PFC	Pareto front computation
PFC	prefrontal cortex
PFM	precise fuzzy modeling
PHM	prognostics and health management
PIC	peripheral interface controller
PID	proportional-integral-derivative
PII	probabilistic iterative improvement
PIPE	probabilistic incremental program evolution

PLA	programmable logic array
PLB	processor local bus
PLS	partial least square
pLSA	probabilistic latent semantic analysis
PLV	phase lock value
PM	parallel model
PMBGA	probabilistic model-building genetic algorithm
PMC	premotor cortex
PMI	partial mutual information
PMX	partially-mapped crossover
PN	pyramidal neuron
PNS	peripheral nervous system
POD	proper orthogonal decomposition
PoE	product of experts
POR	preference order relation
POS	position-based crossover
PP	parallel platform
PPSN	parallel problem solving in nature
PR	partial reconfiguration
PRAS	polynomial-time randomized approximation scheme
PRM	partially reconfigurable module
PRODIGY	program distribution estimation with grammar model
PRR	partially reconfigurable region
PS	pattern search
PSA	principal subspace analysis
PSCM	problem-space computational model
PSD	power spectral density
PSD	predictive sparse decomposition
PSEA	Pareto sorting evolutionary algorithm
PSNR	peak signal-to-noise ratio
PSO	particle swarm optimization
PSS	problem space search
PSTH	peri-stimulus-time histogram
PTT	pursuit-tracking task
PV	principal value
PVS	persistent vegetative state
PWM	pulse width modulation

Q

Q–Q	quantile–quantile
QAP	quadratic assignment problem
QeSNN	quantum-inspired eSNN
QIP	quadratic information potential
QKLMS	quantized KLMS
QP	quadratic programming

R

r.k.	reproducing kernel
RAF	representable aggregation function
RAM	random access memory
RANS	Reynolds-averaged Navier–Stokes
RB	rule base
RBF	radial basis function
RBM	Boltzmann machine
rBOA	real-coded BOA
RECCo	robust evolving cloud-based controller
RecNN	recursive neural network
REGAL	relational genetic algorithm learner
REML	restricted maximum likelihood estimator
RET	relevancy transformation
RFID	radio frequency identification
RFP	red fluorescent protein
RGB	red-green-blue
RGN	random generation number
RHT	randomized Hough transform
RII	randomized iterative improvement
RISC	reduced intstruction set computer
RKHS	reproducing kernel Hilbert space
RL	reinforcement learning
RLP	randomized linear programming
RLS	randomized local search
RLS	recursive least square
RM-MEDA	regularity model based multiobjective EDA
RMI	remote method invocation
RMSE	root-mean-square error
RMSEP	root-mean-square error of prediction
RMTL	regularized multi-task learning
RN	regularization network
RNA	ribonucleic acid
RNN	recurrent neural network
ROC	receiver operating characteristic
ROI	region of interest
ROM	read only memory
ROM	reduced-order model
ROS	robot operating system
RP	readiness potential
RPC	remote procedure call
RPCL	rival penalized competitive learning
RS	rough set
rst	reset
RT	reaction time
RT	real-time
RTL	register transfer logic
RTRL	real-time recurrent learning

RUL	remaining useful life
RWS	roulette wheel selection

S

S-ACO	simple ant colony optimization
S-bit	section-bit
S3VM	semi-supervised support vector machine
SA	simple F-transform-based fusion algorithm
SA	simulated annealing
SAE	sparse auto-encoder
SAMP	one single algorithm and multiple problems
SARSA	state-action-reward-state-action
SASP	one single algorithm and one single problem
SAT	satisfiability
SBF	subspace-based function
SBO	surrogate-based optimization
SBR	similarity based reasoning
SBS	sequential backward selection
SBSO	surrogate based shape optimization
SBX	simulated binary crossover
SC	soft computing
SC	surprise criterion
SCH	school
SCNG	sparse coding neural gas
SD	structured data
SDE	stochastic differential equation
SDPE	standard deviation percentage error
SEAL	simulated evolution and learning
SEMO	simple evolutionary multi-objective optimizer
SF	scaling factor
SFS	sequential forward selection
SG-GP	stochastic grammar-based genetic programming
SHCLVND	stochastic hill climbing with learning by vectors of normal distribution
SI	swarm intelligence
SIM	simple-inversion mutation operator
SISO	single input single output
SLAM	simultaneous localization and mapping
SLF	superior longitudinal fasciculus
SLS	stochastic local search
SM	scramble mutation operator
SM	surrogate model
SMA	supplementary motor area

SMO	sequential minimum optimization
SMP	symmetric multiprocessor
SMR	sensorimotor rhythm
sMRI	structural magnetic resonance imaging
SMS-EMOA	S-metric selection evolutionary multiobjective algorithm
SNARC	spatial–numerical association of response code
SNE	stochastic neighborhood embedding
SNP	single nucleotide polymorphism
SNR	signal-noise-ratio
SOC	self-organized criticality
SOFM	self-organized feature maps
SOFNN	self-organizing fuzzy neural network
SOGA	single-objective genetic algorithm
SOM	self-organizing map
SPAM	set preference algorithm for multiobjective optimization
SPAN	spike pattern association neuron
SPD	strictly positive definite
SPEA	strength Pareto evolutionary algorithm
SPOT	sequential parameter optimization toolbox
SPR	static partial reconfiguration
SQL	structured query language
SR	stochastic resonance
SR	symbolic regression
SRD	standard reference dataset
SRF	strength raw fitness
SRM	spike response model
SRM	structural risk minimization
SRN	simple recurrent network
SRT	serial reaction time
SSM	state–space model
SSOCF	subset size-oriented common features
SSSP	single-source shortest path problem
StdGP	standard GP
STDP	spike-timing dependent plasticity
STDP	spike-timing dependent learning
STGP	strongly typed GP
SU	single unit
SURE-REACH	sensorimotor, unsupervised, redundancy-resolving control architecture
SUS	stochastic universal sampling
SVaR	simplified value at risk
SVC	support vector classification
SVD	singular value decomposition
SVM	support vector machine
SW	software
SW-KRLS	sliding window KRLS

T

T1	type-1
T1FC	type-1 fuzzy controller
T1FS	type-1 fuzzy set
T2	type-2
T2FC	type-2 fuzzy controller
T2FS	type-2 fuzzy set
T2IC	type-2 intelligent controller
T2MF	type-2 membership function
TAG3P	tree adjoining grammar-guided genetic programming
TD	temporal difference
TDNN	time delay neural network
TET	total experiment time
TFA	temporal factor analysis
TGBF	truncated generalized Bell function
TN	thalamus
TOGA	target objective genetic algorithm
TR	type reducer
TRC	tuple relational calculus
TS	tabu search
TS	time saving
TSK	Takagi–Sugeno–Kang
TSP	traveling salesman problem
TTGA	trainable threshold gate array
TWNFI	transductive weighted neuro-fuzzy inference system

U

UART	universal asynchronous receiver/transmitter
UAV	unmanned aerial vehicle
UCF	user constraint file
UCS	supervised classifier system
UCX	uniform cycle crossover
UMDA	univariate marginal distribution algorithm
UML	universal modeling language
UPMOPSO	user-preference multiobjective PSO
US EPA	United States Environmental Protection Agency
UW	underwriter

V

VB	variational Bayes
VC	Vapnik–Chervonenkis
VC	variable consistency
VCR	variance ratio criterion
VEGA	vector-evaluated GA

VHDL	VHSIC hardware description language
VHS	virtual heading system
VHSIC	very high speed integrated circuit
VLNS	very large neighborhood search
VLPFC	ventrolateral prefrontal cortex
VLSI	very large scale integration
VND	variable neighborhood descent
VNS	variable neighborhood search
VPRSM	variable precision rough set model
VQ	vector quantization
VQRS	vaguely quantified rough set

W

W2T	wisdom web of things
WAN	wide area network
WC	Wilson–Cowan
WEP	weight error power
WFG	walking fish group
WisTech	Wisdom Technology
WM	white matter

WM	working memory
WSN	wireless sensor network
WT	Wu–Tan
WTA	winner-take-all
WWKNN	weighted-weighted nearest neighbor

X

XACS	x-anticipatory classifier system
XB	Xie-Beni cluster validity index
XCS	X classifier system
XCSF	XCS for function approximation
xNES	exponential natural evolution strategy
XPS	Xilinx platform studio
XSG	Xilinx system generator

Z

ZCS	zeroth level classifier system
ZDT	Zitzler–Deb–Thiele
ZEN	Zonal Euler–Navier–Stokes

1. Introduction

Janusz Kacprzyk, Witold Pedrycz

This *Springer Handbook of Computational Intelligence* is a result of a broad project that has been launched by us to respond to an urgent need of a wide scientific and scholarly community for a comprehensive reference source on Computational Intelligence, a field of science that has for some decades enjoyed a growing popularity both in terms of the theory and methodology as well as numerous applications. As it is always the case in such situations, after some time once an area has reached maturity, and there is to some extent a consent in the community as to which paradigms, and tools and techniques may be useful and promising, the time will come when some

state of the art exposition, exemplified by this Springer Handbook, would be welcome. We think that *this* is the right moment.

The first and most important question that can be posed by many people, notably those who work in more traditional and relatively well-defined fields of science, is what *Computational Intelligence* is. There are many definitions that try to capture the very essence of that field, emphasize different aspects, and – by necessity – somehow reflect the individual research interests, preferences, prospective application areas, etc.

However, it seems that in recent years there has has been a wider and wider consent as to what basically Computational Intelligence is. Let us start with a citation coming from the Constitution of the IEEE (Institute of Electrical and Electronics Engineers) CIS (Computational Intelligence Society) – Article I, Section 5:

The Field of Interest of the Society is the theory, design, application, and development of biologically and linguistically motivated computational paradigms emphasizing neural networks, connectionist systems, genetic algorithms, evolutionary programming, fuzzy systems, and hybrid intelligent systems in which these paradigms are contained.

It seems that this is extremely up to the point, and we have basically followed this general philosophy in the composition of the Springer Handbook.

Let us first extend a little bit the above essential land comprehensive description of what computational intelligence is interested in, what deals with, which tools and techniques it uses, etc. Computational Intelligence is a broad and diverse collection of nature inspired computational methodologies and approaches, and tools and techniques that are meant to be used to model and solve complex real-world problems in various areas of science and technology in which the traditional approaches based on strict and well-defined tools and techniques, exemplified by *hard* mathematical modeling, optimization, control theory, stochastic analyses, etc., are either not feasible or not efficient. Of course, the term *nature inspired* should be meant in a broader sense of being *biologically inspired*, *socially inspired*, etc.

Those complex problems that are of interest of computational intelligence are often what may be called ill-posed which may make their exact solution, using the traditional hard tools and techniques, impossible.

However, we all know that such problems are quite effectively and efficiently solved in real life by human being, or – more generally – by living species. One can easily come to a conclusion that one should develop new tools, maybe less *precise* and not so well mathematically founded, that would provide a solution, maybe not optimal but good enough. This is exactly what Computational Intelligence is meant to provide.

Briefly speaking, it is most often considered that computational intelligence includes as its main components fuzzy logic, (artificial) neural networks, and evolutionary computation. Of course, these main elements should be properly meant. For instance, one should understand that fuzzy logic is to be complemented by rough set theory or multivalued logic, neural networks should be more generally meant as including all kinds of connectionist systems and also learning systems, and evolutionary computation should be viewed as the area including swarm intelligence, artificial immune systems, bacterial algorithms, etc. One can also add in this context many other approaches like the Dempster–Shafer theory, chaos theory, etc.

1.1 Details of the Contents

The afore-mentioned view of the very essence of computational intelligence has been followed by us when dividing the handbook into parts, which correspond to the particular fields that constitute the area of Computational Intelligence, and also in the selection of field editors, and then their selection of proper authors. We have been very fortunate to be able to attract as the field editors and authors of chapters the best people in the respective fields.

1.1.1 Part A Foundations

For obvious reasons, we start the *Springer Handbook of Computational Intelligence* with *Part A, Foundations*, which deliver a constructive survey of some carefully chosen topics that are of importance for virtually all ensuing parts of the handbook. This part, edited by Professors Radko Mesiar and Bernard De Baets, involved foundational works on multivalued logics, possibility theory, aggregation functions, measure-based integrals, the essence of extensions of fuzzy sets, *F*-transforms, mathematical programming, and games under imprecision and fuzziness. It is easy to see that the contributions cover topics that are of profound relevance.

1.1.2 Part B Fuzzy Logic

Part B, Fuzzy Logic, edited by Professors Luis Magdalena and Enrique Herrera-Viedma, attempts to present the most relevant elements and issues related to a vast area of fuzzy logic. First of all, a comprehensive account of foundations of fuzzy sets theory has been provided, emphasizing both theoretical and application oriented aspects. This has been followed by a state-of-the-art presentation of the concept, properties, and applications of fuzzy relations, including a brief historical perspective and future challenges. A similar account of the past, present, and future of an extremely important concept of fuzzy implications has then been authoritatively covered.

Then, various issues related to fuzzy systems modeling have been presented; notably the concept and properties of fuzzy-rule-based systems which are the core of fuzzy modeling, and the problem of interpretability of fuzzy systems. Fuzzy clustering, which is one of the most widely used tools and techniques, encountered in virtually all problems related to data analysis, modeling, control, etc., is then exposed, with focus on the past, present, and future challenges. Then, issues related to Zadeh's seminal idea of computing with words have been thoroughly studied from many perspectives, in particular a logical and algebraic one.

Since fuzzy (logic) control is undoubtedly the most vigorously reported industrial application of fuzzy logic, this subject has been presented in detail, both for the *conventional* fuzzy sets and their extensions, especially type 2 and interval type-2 fuzzy sets. Applications of fuzzy logic in autonomous robotics have been presented.

An account of fundamental issues and solutions related to the use of fuzzy logic in database and information management has then been given.

1.1.3 Part C Rough Sets

Part C, Rough Sets, edited by Professors Roman Słowiński and Yiyu Yao, starts with a comprehensive, rigorous, yet readable presentation of foundations of rough sets, followed by a similar exposition focused on the use of rough sets to decision making, aiding, and support. Then, rule induction is considered as a tool for modeling, decision making, and data analysis.

A number of important extensions of the basic concept of a rough set have then been presented, including the concept of a probabilistic rough set and a generalized rough set, along with a lucid exposition of their properties and possible applications.

A crucial problem of a fuzzy-rough hybridization is then discussed, followed by a more general exposition of rough systems.

1.1.4 Part D Neural Networks

Part D, Neural Networks, edited by Professors Cesare Alippi and Marios Polycarpou, starts with a general presentation of artificial neural network models, followed by presentations of some mode specific types exemplified by deep and modular neural networks.

Much attention in this part is devoted to machine learning, starting from a very general overview of the area and main tools and techniques employed, theoretical methods in machine learning, probabilistic modeling in machine learning, kernel methods in machine learning, etc.

An important problem area called neurodynamics is the subject of a next state-of-the-art survey, followed by a review of basic aspects, models, and challenges of computational neuroscience considered basically from the point of view of biophysical modeling of neural systems. Cognitive architectures, notably for agent-based systems, and a related problem of computational models of cognitive and motor control have then been presented in much detail.

Advanced issues involved in the so-called embodied intelligence, and neuroengineering, and neuromorphic engineering, emerging as promising paradigms for modeling and problem solving, have then been dealt with.

Evolving connectionist systems, which constitute a novel, very promising architectures for the modeling of various processes and systems have been surveyed. An important part is on real-world applications of machine learning completes this important part.

1.1.5 Part E Evolutionary Computation

Part E, Evolutionary Computation, edited by Professors Frank Neumann, Carsten Witt, Peter Merz, Car-

los A. Coello Coello, Oliver Schütze, Thomas Bartz-Beielstein, Jörn Mehnen, and Günther Raidl, concerns the third fundamental element of what is traditionally being considered to be the core of Computational Intelligence.

First, comprehensive surveys of genetic algorithms, genetic programming, evolution strategies, parallel evolutionary algorithms are presented, which are readable and constructive so that a large audience might find them useful and – to some extent – ready to use. Some more general topics like the estimation of distribution algorithms, indicator-based selection, etc., are also discussed.

An important problem, from a theoretical and practical point of view, of learning classifier systems is presented in depth.

Multiobjective evolutionary algorithms, which constitute one of the most important group, both from the theoretical and applied points of view, are discussed in detail, followed by an account of parallel multiobjective evolutionary algorithms, and then a more general analysis of many multiobjective problems.

Considerable attention has also been paid to a presentation of hybrid evolutionary algorithms, such as memetic algorithms, which have emerged as a very promising tool for solving many real-world problems in a multitude of areas of science and technology. Moreover, parallel evolutionary combinatorial optimization has been presented.

Search operators, which are crucial in all kinds of evolutionary algorithms, have been prudently analyzed. This analysis was followed by a thorough analysis of various issues involved in stochastic local search algorithms.

An interesting survey of various technological and industrial applications in mechanical engineering and design has been provided. Then, an account of the use of evolutionary combinatorial optimization in bioinformatics is given.

An analysis of a synergistic integration of metaheuristics, notably evolutionary computation, and constraint satisfaction, constraint programming, graph coloring, tree decomposition, and similar relevant problems completes the part.

1.1.6 Part F Swarm Intelligence

Part F, Swarm Intelligence, edited by Professors Christian Blum and Roderich Gross, starts with a concise yet comprehensive introduction to swarm intelligence in optimization and robotics, two fields of science in which this type of metaheuristics has been considered, and demonstrated to be powerful and useful.

Then, a preference-based multiobjective particle swarm optimization model is covered as a good tool for airfoil design. An ant colony optimization model for the minimum-weight rooted arborescence problem, which may be a good model for many diverse problem in computer science, decision analysis, data analysis, etc. is discussed.

An intelligent swarm of Markovian agents is the topic of a thorough analysis, which highlights the power and universality of this model. Moreover, a probabilistic modeling of swarm systems is surveyed.

Then some explicitly nature inspired algorithms based on how some species behave are presented, notably, a honey bee social foraging algorithm for resource allocation. Collective behavior modes and reconfigurability are discussed in swarm robotics, complemented by an exposition of problems and solutions related to the collective manipulation and construction.

1.1.7 Part G Hybrid Systems

Part G, Hybrid Systems, edited by Professors Oscar Castillo and Patricia Melin, starts with papers on various types of controllers developed with the aid of computational intelligence tools and techniques that are employed in a highly synergistic way.

First, an interesting and visionary study of robust evolving cloud-based controller which combines new conceptual ideas with novel computing architecture is provided. Then evolving embedded fuzzy controllers as well as the bio-inspired optimization of type-2 fuzzy controllers are surveyed. New hybrid modeling and solution tools are then presented; notably multiobjective genetic fuzzy systems. The use of modular neural networks and type-2 fuzzy logic is shown to be effective and efficient in various pattern recognition problems.

A novel idea of using chemical algorithms for the optimization of interval type-2 and type-1 fuzzy controllers for autonomous mobile robots is shown.

Finally, the implementation of bio-inspired optimization methods on graphic processing units is presented and its efficiency is emphasized.

1.2 Conclusions and Acknowledgments

To summarize, the coverage of topics in the particular parts has certainly provided a comprehensive, rigorous yet readable state-of-the-art survey of main research directions, developments and challenges in Computational Intelligence.

Both more advanced readers, who may look for details, and novice readers, who may look for some more general and readable introduction that could be employed in their later works, would certainly find this handbook useful.

Part A Foundations

Part A Foundations

Ed. by Bernard De Baets, Radko Mesiar

2. Many-Valued and Fuzzy Logics

Siegfried Gottwald

In this chapter, we consider particular classes of infinite-valued propositional logics which are strongly related to t-norms as conjunction connectives and to the real unit interval as a set of their truth degrees, and which have their implication connectives determined via an adjointness condition.

Such systems have in the last 10 years been of considerable interest, and the topic of important results. They generalize well-known systems of infinite-valued logic, and form a link to as different areas as, e.g., linear logic and fuzzy set theory.

We survey the most important ones of these systems, always explaining suitable algebraic semantics and adequate formal calculi, but also mentioning complexity issues.

Finally, we mention a type of extension which allows for graded notions of provability and entailment.

Classical *two-valued* logic is characterized by two basic principles. The *principle of extensionality* states that the truth value of any compound sentence depends only on the truth values of the components. The *Principle of bivalence*, also known as *tertium non datur*, states that any sentence is either true or false, nothing else is possible. Intuitively, a sentence is understood here as a formulation which has a truth value, i.e., which is true or false. In everyday language this excludes formulations like questions, requests, and commands. Nevertheless, this explanation sounds like a kind of circular formulation. So formally one first fixes a certain formalized language, and then lays down formal criteria of what should count as a well-formed formula of this language, and particularly what should count as a sentence.

The principle of bivalence excludes hence *self-contradictory* formulations which are true as well as false, and it also excludes so-called truth value gaps, i.e., formulations which are neither true nor false.

Based on the understanding that a sentence is a formulation which has a truth value, many-valued logic generalizes the understanding of what a truth value is and hence allows for *more* values as only the two classical values *true* and *false*. To indicate this generalization, we speak here of *truth degrees* in the case of many-valued logics. And this allows the additional convention to have the name truth value reserved for the values *true* and *false* in their standard understanding in the sense of classical logic.

There are many possibilities to choose particular sets of truth degrees. Thus there are quite different systems of many-valued logic. However, each partic-ular one of these systems $\mathbb{L}$ has a fixed set of truth degrees $\mathcal{W}_{\mathbb{L}}$. Furthermore, each such system has its set $\mathcal{D}_{\mathbb{L}} \subset \mathcal{W}_{\mathbb{L}}$ of designated truth degrees: formulas of the corresponding formalized language are *logically valid* iff they always have a designated truth degree.

Instead of the principle of bivalence each system $\mathbb{L}$ of many-valued logic satisfy a *principle of multivalence* in the sense that any sentence has to have exactly one truth degree out of $\mathcal{W}_{\mathbb{L}}$. And the *principle of extensionality* now states that the truth degree of any compound sentence depends only on the truth degrees of the components.

Fuzzy logics are particular infinite-valued logics which have, at least in their most simple forms, the real unit interval $[0, 1]$ as their truth degree sets, and which have the degree 1 as their only designated truth degree.

2.1 Basic Many-Valued Logics

If one looks systematically for many-valued logics which have been designed for quite different applications, one finds four main types of systems:

- The Łukasiewicz logics L_κ as explained in [2.1];
- The Gödel logics G_κ from [2.2];
- The product logic $\varPi$ studied in [2.3];
- The Post logics P_m for $2 \leq m \in \mathbb{N}$ from [2.4].

The first two types of many-valued logics each offer a uniformly defined family of systems which differ in their sets of truth degrees and comprise finitely valued logics for each one of the truth degree sets $\mathcal{W}_n = \{0, \frac{1}{n-1}, \frac{2}{n-1}, \ldots, 1\}$, $n \geq 2$, together with an infinite-valued system with truth degree set $\mathcal{W}_\infty = [0, 1]$. Common reference to the finite-valued and the infinite-valued cases is formally indicated by choosing $\kappa \in \{n \in \mathbb{N} \mid n \geq 2\} \cup \{\infty\}$ as an index.

For the fourth type an infinite-valued version is lacking.

In their original presentations, these logics look rather different, regarding their propositional parts. For the first-order extensions, however, there is a unique strategy: one adds a universal and an existential quantifier such that quantified formulas get, respectively, as their truth degrees the infimum and the supremum of all the particular cases in the range of the quantifiers.

As a reference for these and also other many-valued logics in general, the reader may consult [2.5].

Our primary interest here is in the infinite-valued versions of these logics. These ones have the closest connections to the fuzzy logics discussed later on. Therefore, we further on write simply L instead of L_∞, and G instead of G_∞.

For simplicity of notation, later on we often will use the same symbol for a connective and its truth degree function. It shall always become clear from the context what actually is meant.

2.1.1 The Gödel Logics

The simplest ones of these logics are the *Gödel logics* G_κ which have a conjunction $\wedge$ and a disjunction $\vee$ defined by the minimum and the maximum, respectively, of the truth degrees of the constituents

$$u \wedge v = \min\{u, v\}, \qquad u \vee v = \max\{u, v\} . \quad (2.1)$$

These Gödel logics have also a negation $\approx$ and an implication $\rightarrow_\mathsf{G}$ defined by the truth degree functions

$$\approx u = \begin{cases} 1, & \text{if } u = 0 ; \\ 0, & \text{if } u > 0 . \end{cases} \qquad u \rightarrow_\mathsf{G} v = \begin{cases} 1, & \text{if } u \leq v ; \\ v, & \text{if } u > v . \end{cases}$$
$$(2.2)$$

The systems differ in their truth degree sets: for each $2 \leq \kappa \leq \infty$ the truth degree set of G_κ is $\mathcal{W}_\kappa$.

As shown by *Dummett* [2.6], the infinite-valued propositional Gödel logic G has an *adequate axiomatization* which is provided by an adequate axiomatization of the intuitionistic propositional logic enriched with the additional axiom schema

$$(\varphi \to \psi) \vee (\psi \to \varphi) \,. \tag{2.3}$$

Later on, in Sect. 2.5, we will recognize another axiomatization because G is a particular *t*-norm-based logic.

2.1.2 The Łukasiewicz Logics

The *Łukasiewicz logics* L_κ, again with $2 \le \kappa \le \infty$, have originally been designed in [2.1] with only two primitive connectives, an implication $\to_L$ and a negation $\neg$ characterized by the truth degree functions

$$\neg u = 1 - u, \qquad u \to_L v = \min\{1, 1 - u + v\} \,. \tag{2.4}$$

The systems differ in their truth degree sets: for each $2 \le \kappa \le \infty$ the truth degree set of L_κ is $\mathcal{W}_\kappa$.

However, it is possible to define further connectives from these primitive ones. With

$$\varphi \,\&\, \psi =_{df} \neg(\varphi \to_L \neg\psi),$$
$$\varphi \veebar \psi =_{df} \neg\varphi \to_L \psi \tag{2.5}$$

one gets a (strong) conjunction and a (strong) disjunction with truth degree functions

$$u \,\&\, v = \max\{u + v - 1, 0\},$$
$$u \veebar v = \min\{u + v, 1\} \,, \tag{2.6}$$

usually called the *Łukasiewicz (arithmetical) conjunction* and the *Łukasiewicz (arithmetical) disjunction*. It should be mentioned that these connectives are linked together via a De Morgan's law using the standard negation of this system

$$\neg(u \,\&\, v) = \neg u \veebar \neg v \,. \tag{2.7}$$

With the additional definitions

$$\varphi \wedge \psi =_{df} \varphi \,\&\, (\varphi \to_L \psi)$$
$$\varphi \vee \psi =_{df} (\varphi \to_L \psi) \to_L \psi \tag{2.8}$$

one gets another (weak) conjunction $\wedge$ with truth degree function min, and a further (weak) disjunction $\vee$

with max as truth degree function, i.e., one has the conjunction and the disjunction of the Gödel logics also available.

The infinite-valued propositional Łukasiewicz logic L, with implication and negation as primitive connectives, has an adequate axiomatization consisting of the axiom schemata:

$(L_\infty 1)$ $\varphi \to_L (\psi \to_L \varphi)$,
$(L_\infty 2)$ $(\varphi \to_L \psi) \to_L ((\psi \to_L \chi) \to_L (\varphi \to_L \chi))$,
$(L_\infty 3)$ $(\neg\psi \to_L \neg\varphi) \to_L (\varphi \to_L \psi)$,
$(L_\infty 4)$ $((\varphi \to_L \psi) \to_L \psi) \to_L ((\psi \to_L \varphi) \to_L \varphi)$

together with the rule of detachment as the only inference rule.

Later on, in Sect. 2.5, we will recognize another axiomatization because L is a particular *t*-norm-based logic.

2.1.3 The Product Logic

The *product logic* Π, in detail explained in [2.3], has the real unit interval as truth degree set, has a fundamental conjunction $\odot$ with the ordinary product of reals as its truth degree function, and has an implication $\to_\Pi$ with the truth degree function

$$u \to_\Pi v = \begin{cases} 1, & \text{if } u \le v \,; \\ \dfrac{u}{v}, & \text{if } u < v \,. \end{cases} \tag{2.9}$$

Additionally, it has a truth degree constant $\bar{0}$ to denote the truth degree zero.

In this context, a negation and a further conjunction are defined as

$$\approx \varphi =_{df} \varphi \to_\Pi \bar{0},$$
$$\varphi \wedge \psi =_{df} \varphi \odot (\varphi \to_\Pi \psi) \,. \tag{2.10}$$

Routine calculations show that both connectives coincide with the corresponding ones of the infinite-valued Gödel logic G. And also the disjunction $\vee$ of this Gödel logic becomes available, now via the definition

$$\varphi \vee \psi =_{df} ((\varphi \to_\Pi \psi) \to_\Pi \psi)$$
$$\wedge ((\psi \to_\Pi \varphi) \to_\Pi \varphi) \,. \tag{2.11}$$

There is, however, no natural way to combine with this (infinite valued) product logic a whole family of finite-valued systems by simply restricting the set of truth degrees to some $\mathcal{W}_m$ as in the previous two cases:

besides W_2 no such set is closed under the ordinary product, and for W_2 the product coincides, e.g., with the minimum operation.

Later on, in Sect. 2.4, we will recognize an adequate axiomatization because also the product logic Π is a particular t-norm-based logic. Contrary to the previous cases of G and L, however, there is no essentially different axiomatization known as this later one.

2.1.4 The Post Logics

The Post system P_m for $m \geq 2$ has truth degree set W_m. These propositional systems have been originally formulated uniformly in negation and disjunction as basic connectives with the following truth degree functions

$$\approx u = \begin{cases} 1, & \text{for } u = 0 , \\ u - \dfrac{1}{m-1}, & \text{for } u \neq 0 , \end{cases}$$

$$u \vee v = \max\{u, v\} .$$

Contrary to the previous systems, the definition of negation here does not seem to be given in a uniform way independent of the number of truth degrees. However, it is always just a cyclic permutation of all the truth degrees (in their natural order).

For the sets of designated truth degrees, a canonical choice does not exist; already Post [2.4] has discussed the possibility that there may be chosen truth degrees different from 1 as designated ones. Nevertheless, $\mathcal{D}^P = \{1\}$ is a kind of standard choice.

The set of basic connectives of each one of the Post systems P_m is functionally complete, i. e., allows to represent every possible truth degree function (over W_m). Therefore, each one of the Post systems P_m, with $\mathcal{D}_P = \{1\}$ as the set of designated truth degrees, covers its corresponding Łukasiewicz system with the same set of truth degrees – in the sense that the set of L_m-tautologies is a subset of the set of P_m-tautologies, and that this set of P_m-tautologies does not contain any formula φ whose Łukasiewicz negation $\neg\varphi$ is L_m-satisfiable. And the same holds true for the corresponding m-valued Gödel system G_m.

If one enriches all the finitely many-valued (propositional) Łukasiewicz systems L_m with truth degree constants for all their truth degrees, then these enriched systems L_m^* become functionally complete. And this means that the extended m-valued Łukasiewicz systems L_m^* and the m-valued Post logics become interdefinable (for each fixed number m of truth degrees). Hence there is in principle no essential difference between both

types of (finitely valued) systems: all what can be expressed in the *Post world* can also be expressed in the (extended) *Łukasiewicz world*, and vice versa.

We omit to discuss adequate axiomatizations because these Post logics will not be of particular interest later on in this chapter. The interested reader might consult [2.5].

2.1.5 Algebraic Semantics

All these previously discussed many-valued logics have been introduced by their standard semantics.

Besides these standard semantics, all these many-valued logics have also *algebraic semantics* determined by suitable classes $\mathcal{K}$ of truth degree structures. The situation is similar here to the case of classical logic: the logically valid formulas in classical logic are also just all those formulas which are valid in all Boolean algebras.

Of course, these structures have the same signature as the language $\mathcal{L}$ of the corresponding logic, and they have to have – in the case that one discusses the corresponding first order logics – suprema and infima for all those subsets which may appear as value sets of formulas. Particularly, hence, they have to be (partially) ordered, or at least preordered.

For each formula φ of the language $\mathcal{L}$ of the corresponding logic, for each such (generalized truth degree) structure $\mathbf{A}$, and for each evaluation e which maps the set of atomic formulas of $\mathcal{L}$ into the carrier of $\mathbf{A}$, one has to define a value $\mathsf{Val}(\varphi, e)$, and finally one has to define what it means that such a formula φ is *valid in* $\mathbf{A}$. Then a formula φ is *logically valid* w.r.t. this class $\mathcal{K}$ iff φ is valid in all structures from $\mathcal{K}$.

Gödel and Łukasiewicz Logics

It is remarkable that for both these types of many-valued logics corresponding algebraic semantics have mainly been developed for the infinite-valued systems, and have been considered in the context of completeness proofs.

For the infinite-valued Gödel logic G such a class of structures is, according to the completeness proof given by *Dummett* [2.6], the class of all *Heyting algebras*, i. e., of all relatively pseudo-complemented lattices, which satisfy the prelinearity condition

$$(u \rightarrowtail v) \cup (v \rightarrowtail u) = \mathbf{1} . \tag{2.12}$$

Here $\cup$ is the lattice join and $\rightarrowtail$ the relative pseudo-complement.

For the infinite-valued Łukasiewicz logic L the corresponding class of structures is the class of all *MV-algebras*, first introduced again within a completeness proof by *Chang* [2.7], and more recently extensively studied in [2.8].

It is interesting to recognize that all these structures – prelinear Heyting algebras, MV-algebras, and product algebras – are Abelian lattice-ordered semigroups with an additional *residuation* operation.

For the finite-valued logics from both families, separately developed algebraic semantics did not yet find considerable interest.

Product Logic

The product logic, as introduced in [2.3], was from the very beginning designed as a logic which had, in parallel, a standard semantics – provided by the real unit interval and by a product-based conjunction as a fundamental connective – as well as an algebraic semantics, formed by the class of all product algebras – introduced in [2.3] again within a completeness proof.

We shall not explain more details here because this whole approach proved to become paradigmatic for the development of *t*-norm-based infinite-valued logics, a topic which shall be discussed later on, starting with Sect. 2.3.

Post Logics

Contrary to the situation for the Łukasiewicz and the Gödel systems, for the Post systems in their original form there exist only very few syntactically oriented studies toward constituting or investigating logical calculi for these systems. Instead, for the Post systems one mainly was interested in the corresponding algebraic structures, which were suitable to form an algebraic semantics, and investigated such structures earlier, and in more detail, as similar structures for the Łukasiewicz and the Gödel systems. *Rosenbloom* in a paper [2.9] of 1942 was the first one to do this. His algebraic structures shall here be called P-algebras for short – but not be considered in detail: the interested reader might, e.g., consult [2.5].

One of the main reasons for the difficulty and complexity of the defining conditions of P-algebras is the fact that the Post systems as well as the P-algebras have only two primitive notions, their connectives resp. their basic operations, but have maximal expressive power in the sense of being functionally complete. That this choice of the primitive notions really is the main obstacle toward a simplification became clear as *Epstein* [2.10] in 1960 changed these basic operations and found a much simpler class of definitionally equivalent algebras, now called Post algebras.

What are not covered by these basic considerations are possible infinite-valued generalizations of these logical calculi, or of these Post algebras. Approaches toward this problem started, e.g., with papers on generalizations of the notion of Post algebras like [2.11–13]. The most influential paper, however, which also discussed the corresponding logical systems was the paper [2.14] of Rasiowa in which Post algebras of the order $\omega + 1$ and the corresponding systems of infinitely many-valued (first-order) logic have been introduced. The algebraic theory of these Post algebras of the order $\omega + 1$ is partly given in [2.14].

Another such infinitely many-valued generalization of the standard Post systems is discussed, e.g., in [2.15, 16], Post algebras of the order $\omega + \omega^*$.

The Post algebras of finite or infinite order and the systems of many-valued logic related with them seem to be of particular importance for investigations in computer science, which rely on many-valued logic as a toolbox, because these Post systems are functionally complete and well suited to study the representability of truth degree functions on the basis of some predetermined set of basic truth degree functions, as determined, e.g., by available electronic components, cf. [2.17] for a still good introduction.

2.2 Fuzzy Sets

A fuzzy set A is usually a fuzzy subset of a given set $\mathbb{X}$ and characterized by its membership function $\mu_A : \mathbb{X} \mapsto [0, 1]$. The set $\mathbb{X}$ is often called the *universe of discourse*. This notation derives from [2.18]. So these fuzzy sets are (possibly) first-level objects of a cumulative hierarchy, with the elements of $\mathbb{X}$ as urelements. But the usual applications do not need higher level fuzzy sets. And also for our discussion of the background logic such higher level fuzzy sets do not matter.

2.2.1 Set Algebra for Fuzzy Sets

Mathematically, it is customary to identify such a fuzzy subset of $\mathbb{X}$ with its membership function. Accordingly,

$\mu_A(a)$ and $A(a)$ both are used to denote the membership degree of the object $a \in \mathbb{X}$ w.r.t. the fuzzy set A. For any binary operation $*$ between membership degrees the pointwise approach means to define from it a binary operation $\circledast$ for fuzzy sets such that the fuzzy set $A \circledast B$ is characterized by

$$A \circledast B(x) = A(x) * B(x) \quad \text{for all } x \in \mathbb{X} . \tag{2.13}$$

Hence, Zadeh's *standard intersection* $A \cap B$ and *union* $A \cup B$ are characterized by

$$\begin{aligned} A \cap B(x) &= \min\{A(x), B(x)\} , \\ A \cup B(x) &= \max\{A(x), B(x)\} . \end{aligned} \tag{2.14}$$

Additionally, again following the first proposal from [2.18], one usually also defines the *complement* $\complement_{\mathbb{X}} A$ of a fuzzy set A by the condition

$$\complement_{\mathbb{X}} A(x) = 1 - A(x) \quad \text{for all } x \in \mathbb{X} . \tag{2.15}$$

However, in [2.18] also other versions of such binary operations had been mentioned: an algebraic product AB and an algebraic sum $A + B$ defined through

$$\begin{aligned} AB(x) &= \min\{A(x) \cdot B(x)\} , \\ A + B(x) &= \min\{A(x) + B(x), 1\} \end{aligned} \tag{2.16}$$

as well as an absolute difference $|A - B|$ defined by

$$|A - B|(x) = |A(x) - B(x)| . \tag{2.17}$$

It is interesting to notice, and shall be explained in more detail in Sect. 2.2.2, that the operations (2.16) can be seen as generalized kinds of intersection and union operations, respectively. However, these operations are *not idempotent*: one has in general $AA \neq A$ as well as $A + A \neq A$.

2.2.2 Fuzzy Sets and Many-Valued Logic

It is well known that there is a strong parallelism between the standard set algebraic operations of intersection, union, and complementation, and classical logic, namely, the operations of conjunction, disjunction, and negation, determined by their truth value functions $\mathrm{et}, \mathrm{vel}, \mathrm{non}$, respectively. So one usually defines, e.g., the intersection $M \cap N$ of sets M, N by the condition

$$x \in M \cap N \iff x \in M \wedge x \in N , \tag{2.18}$$

with $\wedge$ for the conjunction operation of classical logic here.

In more abstract terms, the idea here is that these operations are defined in such a way that the power set algebra $\mathcal{P}(M)$ of any set M is (isomorphic to) the direct product $W^M = \prod_{a \in M} W$ of the Boolean algebra $W = (\{1, 0\}, \mathrm{et}, \mathrm{vel}, \mathrm{non})$ of truth values of classical logic.

A similar relationship can be recognized between the set algebra of fuzzy sets and suitable many-valued logics. It is simply necessary to consider the set of membership degrees for the fuzzy sets as set of truth degrees for a corresponding many-valued logic. So one can consider the operations (2.14) as intersection and union related to the Gödel logic G, or also related to the Łukasiewicz logic L. Similarly, the complementation (2.15) is related to the negation operation of the Łukasiewicz logics, and the *algebraic* operations in (2.16) are an intersection operation with respect to product logic, and a union operations with respect to (the strong disjunction of) Łukasiewicz logic. Even the operation (2.17) can be defined via Łukasiewicz logic: one gets immediately via the corresponding truth degree functions

$$|u - v| = \neg((u \to_{\mathsf{L}} v) \,\&\, (v \to_{\mathsf{L}} u)) . \tag{2.19}$$

In more abstract terms, again, the set algebraic operations with respect to a particular $[0, 1]$-valued logic $\mathbb{L}$ should be defined in such a way that the class $\mathcal{F}(\mathbb{X}) = [0, 1]^{\mathbb{X}}$ of fuzzy subsets of a universe of discourse $\mathbb{X}$ is (isomorphic to) the direct product $W^{\mathbb{X}} = \prod_{a \in \mathbb{X}} W$ of the truth degree algebra $W = ([0, 1], \dots)$ of this particular $[0, 1]$-valued logic.

2.2.3 *t*-Norms and *t*-Conorms

For the previously mentioned nonidempotent intersection operation, i.e., the algebraic product from (2.16), and for further similar possibilities the mathematically oriented part of the fuzzy community reached, mainly in the first half of the 1980s, a consensus that such generalized intersection operations should be defined via (2.13) from a triangular norm $*$. Such *triangular norms – t-norms* for short – had first been considered in the context of probabilistic metric spaces to get a suitable version of a triangle inequality, cf. e.g. [2.19], and found since independent interest in different contexts, cf. [2.20, 21]. They are *isotonic, associative*, and *commutative* binary operations in the unit interval which have 1 as their *neutral element*. This means that they make the unit interval an *ordered monoid*.

The class of all *t*-norms is, however, very large and not yet really well understood. So the question appears to restrict to suitable subclasses, e.g., to the continuous *t*-norms or to the left-continuous ones. (For a *t*-norm *T* left-continuity means that all the unary functions T_a with $T_a(x) = T(a, x)$ for each $a \in [0, 1]$ are left-continuous. For continuity the conditions (i) that *T* is continuous as a binary function and (ii) that all the T_a are continuous coincide [2.5, 20].) Standard examples for continuous *t*-norms are the min-operation in [0, 1], also called *Gödel t-norm* T_G, the arithmetic product in [0, 1], also called *product t-norm* T_P, and the *Łukasiewicz t-norm* $T_L : (u, v) \mapsto \max\{u + v - 1, 0\}$ which is the truth degree function of the strong conjunction in the Łukasiewicz many-valued systems. And a standard example for a left-continuous *t*-norm which is not continuous is the *nilpotent minimum* T_{NM} defined as

$$T_{NM}(u, v) = \begin{cases} \min\{u, v\}, & \text{if } u + v > 1 \\ 0 & \text{otherwise .} \end{cases} \quad (2.20)$$

These examples for continuous *t*-norms are even characteristic in the sense that each continuous *t*-norm is an ordinal sum of isomorphic versions of T_L, T_P, T_G, cf. [2.20] and also [2.5].

To explain what is meant by an isomorphic version of some *t*-norm, one has to start from an *order automorphism f* of the unit interval, i. e., from a continuous 1–1 onto map $f : [0, 1] \rightarrow [0, 1]$ with $f(0) = 0$ and $f(1) = 1$. Is now *T* a *t*-norm and $T^* : [0, 1]^2 \rightarrow [0, 1]$ defined by

$$T^*(x, y) = f^{-1}(T(f(x), f(y))), \quad (2.21)$$

which equivalently means

$$f(T^*(x, y)) = T(f(x), f(y)), \quad (2.22)$$

then T^* is again a *t*-norm and called an *isomorphic version* of *T*, and T, T^* are *isomorphic t-norms*.

Parallel with *t*-norms one often also considers *t*-conorms: these are isotonic, associative, and commutative binary operations in the unit interval which have 0 as their neutral element. For the set algebra of fuzzy sets they define (possibly) nonidempotent unions, and for the background logics they constitute (possibly) nonidempotent disjunctions.

There is a natural 1–1 duality between *t*-norms and *t*-conorms. By

$$1 - S(u, v) = T(1 - u, 1 - v) \quad (2.23)$$

one determines a *t*-conorm *S* for any *t*-norm *T*, and conversely determines a *t*-norm *T* for any *t*-conorm *S*. This relationship connects, e.g. the truth degree function $(u, v) \mapsto \max\{u + v - 1, 0\}$ of the Łukasiewicz strong conjunction with that one of the corresponding strong disjunction $(u, v) \mapsto \min\{u + v, 1\}$.

Obviously (2.23) constitutes, for the background logic, a de Morgan connection between suitably chosen conjunctions and disjunctions – as long as the function $u \mapsto 1 - u$ acts as the truth degree function of a negation. And indeed this is the truth degree function of the negation of the Łukasiewicz systems, which was already used in the definition (2.15) of the complement of a fuzzy set.

Summing up, one has for the background logic idempotent *weak* connectives for conjunction and disjunction, determined by the minimum and the maximum operation in [0, 1]. Furthermore, one is interested to have (possibly) nonidempotent *strong* connectives for conjunction and disjunction, determined by a *t*-norm and a *t*-conorm, usually one the dual of the other according to (2.23).

2.3 *t*-Norm-Based Logics

From the point of view of many-valued logic, a *t*-norm is a suitable candidate for a truth degree function of some generalized conjunction connective. Accepting this, one is essentially concerned with systems of many-valued logic with infinite truth degree set [0, 1]. And additionally one prefers to consider such systems which have the truth degree 1 as the only designated truth degree. (This means, e.g., that a formula of the language of such a system counts as logically valid just in case it always assumes this designated truth degree 1. This

notion, as well as the other notions from many-valued logic are explained in detail, e.g., in [2.5].)

2.3.1 Basic Ideas

Such a system of many-valued logic is called *t-norm based* (on some particular *t*-norm *T*) iff all the other connectives of it have associated truth degree functions which are defined from this *t*-norm *T*, using possibly some truth degree constants. Usually one considers together with the conjunction connective & with the truth degree function T an implication connective $\rightarrow$ with the truth degree function I_T characterized by

$$I_T(u, v) =_{\mathrm{df}} \sup\{z \mid T(u, z) \leq v\}, \qquad (2.24)$$

the so-called *R-implication* connected with T, and a *standard negation connective* $\neg$ with truth degree function $\mathbf{n}_T$, given as

$$\mathbf{n}_T(u) =_{\mathrm{df}} I_T(u, 0). \qquad (2.25)$$

As shall be explained in the Sect. 2.3.2, the definition (2.24) determines a reasonable implication function just in the case that the *t*-norm T is left continuous. Here *reasonable* essentially means that $\rightarrow_T$ satisfies a suitable version of the rule of detachment.

In more technical terms it means that for left continuous *t*-norms T condition (2.24) defines a *residuation operator* I_T, previously sometimes also called φ-operator, cf. [2.22]. And it means also, under this assumption of left continuity of T, that condition (2.24) is equivalent to the *adjointness condition*

$$T(u, w) \leq v \iff w \leq I_T(u, v), \qquad (2.26)$$

i. e., that the operations T and I_T form an *adjoint pair*.

Forced by these results one usually restricts, in this logical context, the considerations to left continuous – or even to continuous – *t*-norms.

But together with this restriction of the *t*-norms, a generalization of the possible truth degree sets sometimes is useful: one may accept each subset of the unit interval $[0, 1]$ as a truth degree set which is closed under the particular *t*-norm T and its residuum.

The restriction to *continuous t*-norms enables even the definition of the operations max and min, which make $[0, 1]$ into an (linearly) ordered lattice. On the one hand, one has from straightforward calculations that always

$$\min\{u, v\} = T(u, I_T(u, v)), \qquad (2.27)$$

and on the other hand one gets always [2.23, 24]

$$\max\{u, v\} = \min\{I_T(I_T(u, v), v), I_T(I_T(v, u), u)\}. \qquad (2.28)$$

It is a routine matter to check that the infinite-valued Gödel logic G, the infinite-valued Łukasiewicz logic L,

and also the product logic Π all are *t*-norm-based logics in the present sense.

The systems of fuzzy logic we discuss here are also sometimes called *R-fuzzy logics*, stressing the fact that our implication connectives $\rightarrow$ have as truth degree functions I_T the residuation operations, characterized by (2.24) or (2.26). Besides these R-fuzzy logics one occasionally, e.g., in [2.25, 26], discusses so-called *S-fuzzy logics* which are also based on some *t*-norm, but additionally take the Łukasiewicz negation $\mathbf{n}_L(u) = 1 - u$ or also some other negation function, sometimes together with a further *t*-conorm, as a basic connective.

These S-fuzzy logics define their implication connective like material implication might be defined in classical logic. However, these logics lose, in general, the rule of detachment as a sound rule of inference if they have the degree 1 as the only designated truth degree – or they allow all positive reals from $(0, 1]$ as designated truth degrees.

For a complete development of such *t*-norm-based logics one needs adequate axiomatizations. This seems to be, however, a difficult goal – essentially because of its dependency from the particular choice of the *t*-norms which determine these logics. Therefore, the first successful approaches intended to axiomatize common parts of a whole class of such logics. This will be discuss later in Sect. 2.4.

2.3.2 Left and Full Continuity of *t*-Norms

As had been mentioned in the previous section, the adjointness condition is an algebraic equivalent of the analytical notion of left continuity. This will be proved here.

Definition 2.1
A *t*-norm T is left continuous (continuous) iff all the unary functions $T_a : x \mapsto T(x, a)$ for $a \in [0, 1]$ are left continuous (continuous).

This definition of continuity for *t*-norms via their unary parametrizations coincides with the usual definition of continuity for a binary function, cf. [2.20].

Proposition 2.1
A *t*-norm T is left continuous iff T and its *R*-implication I_T form an adjoint pair.

Proofs are given, e.g., in [2.5, 20, 22].

It is interesting, and important later on, to also notice that the continuity of a *t*-norm has an algebraic equivalent.

Proposition 2.2

A *t*-norm T is continuous iff T and I_T satisfy the equation

$$T(a, I_T(a, b)) = \min\{a, b\}. \qquad (2.29)$$

Proof: Assume first that T is continuous. Then one has for $a \leq b \in [0, 1]$ immediately $T(a, I_T(a, b)) = T(a, 1) = a = \min\{a, b\}$. And one has for $b < a$

$$T(a, I_T(a, b)) = T(a, \max\{z \mid T(a, z) \leq b\})$$
$$= \max\{T(a, z) \mid T(a, z) \leq b\} \leq b \qquad (2.30)$$

already by the left continuity of T. Continuity of T furthermore gives from $0 = T(a, 0) \leq b < a = T(a, 1)$ the existence of some $c \in [0, 1]$ with $b = T(a, c)$, and thus $T(a, I_T(a, b)) = b = \min\{a, b\}$ by (2.30).

Assume conversely (2.29). Then the adjointness condition forces T to be left continuous. Hence for the continuity of T one has to show that T is also right continuous.

Suppose that this is not the case. Then there exist $a, b \in [0, 1]$, and also a decreasing sequence $(x_i)_{i \geq 0}$ with $\lim_{i \to \infty} x_i = b$ such that $T(a, b) \neq \inf_i T(a, x_i)$, i.e., such that $T(a, b) < \inf_i T(a, x_i)$. Consider now some d with $T(a, b) < d < \inf_i T(a, x_i) \leq a$. Then there does not exist some $c \in [0, 1]$ with $d = T(a, c)$, because otherwise one would have $d = T(a, c) > T(a, b)$, hence $c > b$ and thus $\inf_i T(a, x_i) \leq T(a, c) = d$ from the fact that $b = \lim_{i \to \infty} x_i$ and there thus exists some integer k with $x_k \leq c$. This means that the lack of right continuity for T contradicts condition (2.29). ∎

2.3.3 Extracting an Algebraic Framework

For the problem of adequate axiomatization of (classes of) *t*-norm-based systems of many-valued logic there is an important difference to the standard approach toward semantically based systems of many-valued logic: here there is no single, *standard* semantical matrix for the general approach.

The most appropriate way out of this situation seems to be: to find some suitable class(es) of algebraic structures which can be used to characterize these logical systems, and which preferably should be algebraic varieties, i.e., equationally definable.

From an algebraic point of view, the following conditions seem to be structurally important for *t*-norms:

- $\langle [0, 1], T, 1 \rangle$ is a commutative semigroup with a neutral element, i.e., a commutative monoid,
- $\leq$ is a (lattice) ordering in $[0, 1]$ which has 0 as universal lower bound and 1 as universal upper bound,
- Both structures *fit together*: T is nondecreasing w.r.t. this lattice ordering.

Thus it seems reasonable to consider commutative lattice-ordered monoids as the truth degree structures for the *t*-norm-based systems.

In general, however, commutative lattice-ordered monoids may have different elements as the universal upper bound of the lattice and as the neutral element of the monoid. This is not the case for the *t*-norm-based systems, they make $[0, 1]$ into an *integral* commutative lattice-ordered monoid as truth degree structure, namely, one in which the universal upper bound of the lattice ordering and the neutral element of the monoidal structure coincide.

Furthermore, one also likes to have the *t*-norm T combined with another operation, its *R*-implication operator, which forms together with T an adjoint pair: i.e., the commutative lattice-ordered monoid formed by the truth degree structure has also to be a *residuated* one.

Summing up, hence, we are going to consider *residuated lattices*, i.e., algebraic structures $\langle L, \cap, \cup, *, \rightarrowtail, 0, 1 \rangle$ such that L is a lattice under $\cap, \cup$ with the universal lower bound 0 and the universal upper bound 1, and a commutative lattice-ordered monoid under $*$ with neutral element 1, and such that the operations $*$ and $\rightarrowtail$ form an adjoint pair, i.e., satisfy

$$x * z \leq y \iff z \leq (x \rightarrowtail y). \qquad (2.31)$$

In this framework one additionally introduces, following the understanding of the negation connective given in (2.25), a further operation $-$ by

$$-x =_{df} x \rightarrowtail 0. \qquad (2.32)$$

Definition 2.2

A lattice-ordered monoid $\langle L, *, 1, \leq \rangle$ is *divisible* iff for all $a, b \in L$ with $a \leq b$ there exists some $c \in L$ with $a = b * c$.

For linearly ordered residuated lattices, one has another nice and useful characterization of divisibility.

Proposition 2.3
A linearly ordered residuated lattice $\langle L, \cap, \cup, *, \rightarrowtail, 0, 1 \rangle$ is divisible, i. e., corresponds to a divisible lattice-ordered monoid $\langle L, *, 1, \leqslant \rangle$, iff one has $a \cap b = a * (a \rightarrowtail b)$ for all $a, b \in L$. (Of course, $\leqslant$ here is the lattice ordering of the lattice $\langle L, \cap, \cup \rangle$.)

Proof: We first show that one has in each residuated lattice

$$a * (a \rightarrowtail b) = b \quad \Leftrightarrow \quad \exists x (a * x = b) \qquad (2.33)$$

for all $a, b \in L$. Of course, in the case $a * (a \rightarrowtail b) = b$ there exists an x such that $a * x = b$. So suppose $a * c = b$ for some $c \in L$. If one then would have $a * (a \rightarrowtail b) \neq b$, this would mean $a * (a \rightarrowtail b) < b = a * c$ because one always has $a * (a \rightarrowtail b) \leq b$ by the adjointness condition, and this hence would mean $c \not\leq a \rightarrowtail b$ (because otherwise $c \leq a \rightarrowtail b$ and hence $b = a * c \leq a * (a \rightarrowtail b)$ would be the case) and therefore also $a * c = c * a \not\leq b$ by the adjointness condition, a contradiction. Thus (2.33) is established.

Supposing now the divisibility of $\langle L, \cap, \cup, *, \rightarrowtail, 0, 1 \rangle$, then one has for all $b \leq a \in L$ from the existence of an x such that $(b = a * x)$ immediately $a * (a \rightarrowtail b) = b = a \cap b$. Otherwise one has $a \leq b$ by the linearity of the ordering and hence $a \rightarrowtail b = 1$ from the adjointness property, thus $a * (a \rightarrowtail b) = a * 1 = a = a \cap b$.

Assuming on the other hand that one always has $a \cap b = a * (a \rightarrowtail b)$; furthermore, for all $a \leq b \in L$ from $a = a \cap b = b \cap a$ one gets the equation $a = b * (b \rightarrowtail a)$, and hence there is an x such that $a = b * x$. ∎

Using this result, we can restate Proposition 2.2 in the following way.

Corollary 2.1
A t-algebra $[0, 1]_T = \langle [0, 1], \min, \max, T, I_T, 0, 1 \rangle$ is divisible iff the t-norm T is continuous.

A further restriction is suitable w.r.t. the class of residuated lattices because each t-algebra $[0, 1]_T$ is linearly ordered, and thus makes particularly the wff $(\varphi \rightarrow \psi) \vee (\psi \rightarrow \varphi)$ valid. Following *Hájek* [2.23, 24], one calls *BL-algebras* those divisible residuated lattices which also satisfy the *prelinearity* condition (2.12).

Definition 2.3
A structure $\mathbf{L} = \langle L, \vee, \wedge, *, \rightarrowtail, \mathbf{0}, \mathbf{1} \rangle$ is a *BL-algebra* iff:

i) $(L, \vee, \wedge, \mathbf{0}, \mathbf{1})$ is a bounded lattice with lattice ordering $\leqslant$,
ii) $(L, *, \mathbf{1}, \leqslant)$ is a lattice-ordered Abelian monoid,
iii) The operations $*$ and $\rightarrowtail$ satisfy the *adjointness* condition

$$x * y \leqslant z \quad \Longleftrightarrow \quad x \leqslant y \rightarrowtail z, \qquad (2.34)$$

iv) the *prelinearity* condition (2.12) is satisfied,
v) the *divisibility* condition is satisfied, i. e., one has always

$$x * (x \rightarrowtail y) = x \wedge y. \qquad (2.35)$$

It is interesting to notice that the prelinearity condition (2.12) can equivalently be characterized in another form, which will become important later on.

Proposition 2.4
In residuated lattices there are equivalent

(i) $(x \rightarrowtail y) \cup (y \rightarrowtail x) = 1$,
(ii) $((x \rightarrowtail y) \rightarrowtail z) * ((y \rightarrowtail x) \rightarrowtail z) \leqslant z$.

The proof is by routine calculations, cf., e.g., [2.5, 23].

2.4 Particular Fuzzy Logics

Now we shall discuss the core systems of t-norm-based logics. Of course, it would be preferable to be able to axiomatize each single t-norm-based logic directly. However, actually there is no way to do so. Hence other approaches have been developed. The core idea is first to develop systems which cover large parts which are common to all those t-norm-based logics.

The first successful approach came from *Hájek* who presented 1998 in the seminal monograph [2.23] the logic BL of all continuous t-norms, i. e., the common part of all the t-norm-based logics which are determined by a continuous t-norm. Inspired by this work a short time later *Esteva* and *Godo* [2.27] introduced 2001 the logic MTL of all left-continuous t-norms.

These logics are characterized by algebraic semantics: BL by the class of all t-algebras with a continuous t-norm, and MTL by the class of all t-algebras with a left-continuous t-norm. All those t-algebras are particular cases of residuated lattices.

It should be noticed, however, that already in 1996 *Höhle* [2.28] introduced the monoidal logic ML characterized by the class of all residuated lattices as their algebraic semantics.

And it should also be mentioned that, in the case of logics which are determined by an algebraic semantics, the problem of their adequate axiomatization becomes particularly well manageable, if the algebraic semantics is given as a variety of algebraic structures, i.e., as an equationally definable class of algebraic structures.

2.4.1 The Logic BL of All Continuous t-Norms

The class of t-algebras (with a continuous t-norm or not) is not a variety: it is not closed under direct products because each t-algebra is linearly ordered. Hence one may expect that it would be helpful for the development of a logic of continuous t-norms to extend the class of all divisible t-norm algebras in a moderate way to get a variety.

And indeed this idea works: it was developed by *Hájek* and in detail explained in [2.23].

The core point is that one considers instead of the divisible t-algebras $[0, 1]_T$, which are linearly ordered integral monoids, lattice-ordered integral monoids which satisfy the condition (2.35), which have an additional residuation operation connected with the semigroup operation via an adjointness condition (2.26), and which also satisfy the *prelinearity condition*

$$(x \rightarrowtail y) \vee (y \rightarrowtail x) = \mathbf{1}, \tag{2.36}$$

or equivalently

$$((x \rightarrowtail y) \rightarrowtail z) \rightarrowtail (((y \rightarrowtail x) \rightarrowtail z) \rightarrowtail z) = \mathbf{1}. \tag{2.37}$$

The axiomatization of *Hájek* [2.23] for the *basic t-norm logic* BL (in [2.5] denoted BTL), i.e., for the class of all well-formed formulas which are valid in all BL-algebras, is given in a language $\mathcal{L}_T$ which has as basic vocabulary the connectives $\rightarrow$, & and the truth degree constant $\overline{0}$, taken in each BL-algebra $\langle L, \cap, \cup, *, \rightarrowtail, \mathbf{0}, \mathbf{1} \rangle$ as the operations $\rightarrowtail$, $*$ and the element $\mathbf{0}$.

This t-norm-based logic BL has the following axiom schemata:

(Ax$_{\text{BL}}$1) $(\varphi \rightarrow \psi) \rightarrow ((\psi \rightarrow \chi) \rightarrow (\varphi \rightarrow \chi))$,
(Ax$_{\text{BL}}$2) $\varphi \& \psi \rightarrow \varphi$,
(Ax$_{\text{BL}}$3) $\varphi \& \psi \rightarrow \psi \& \varphi$,
(Ax$_{\text{BL}}$4) $(\varphi \rightarrow (\psi \rightarrow \chi)) \rightarrow (\varphi \& \psi \rightarrow \chi)$,
(Ax$_{\text{BL}}$5) $(\varphi \& \psi \rightarrow \chi) \rightarrow (\varphi \rightarrow (\psi \rightarrow \chi))$,
(Ax$_{\text{BL}}$6) $\varphi \& (\varphi \rightarrow \psi) \rightarrow \psi \& (\psi \rightarrow \varphi)$,
(Ax$_{\text{BL}}$7) $((\varphi \rightarrow \psi) \rightarrow \chi) \rightarrow (((\psi \rightarrow \varphi) \rightarrow \chi) \rightarrow \chi)$,
(Ax$_{\text{BL}}$8) $\overline{0} \rightarrow \varphi$,

and has as its (only) inference rule the rule of detachment.

Starting from the primitive connectives $\rightarrow$, &, and the truth degree constant $\overline{0}$, the language $\mathcal{L}_T$ of BL is extended by definitions of further connectives

$$\varphi \wedge \psi =_{\text{df}} \varphi \& (\varphi \rightarrow \psi), \tag{2.38}$$

$$\varphi \vee \psi =_{\text{df}} ((\varphi \rightarrow \psi) \rightarrow \psi)$$
$$\wedge ((\psi \rightarrow \varphi) \rightarrow \varphi), \tag{2.39}$$

$$\neg \varphi =_{\text{df}} \varphi \rightarrow \overline{0}, \tag{2.40}$$

where φ, ψ are formulas of the language of that system.

Calculations (in BL-algebras) show that the additional connectives $\wedge$, $\vee$ just have the lattice operations $\cap$, $\cup$ as their truth degree functions.

The system BL is an implicative logic in the sense of *Rasiowa* [2.29]. So one gets a general soundness and completeness result.

Theorem 2.1 General Completeness
A formula φ of the language $\mathcal{L}_T$ is derivable within the axiomatic system BL iff φ is valid in all BL-algebras.

However, it is shown in [2.23] that already the class of all BL-chains, i.e., of all linearly ordered BL-algebras, provides an adequate algebraic semantics.

Theorem 2.2 General Chain Completeness
A formula φ of $\mathcal{L}_T$ is derivable within the axiomatic system BL iff φ is valid in all BL-chains.

But even more is provable and leads back to the starting point of the whole approach: the theorems of BL are just those formulas which hold true w.r.t. all divisible t-algebras. This was, extending preliminary results from [2.24], finally proved in [2.30].

Theorem 2.3 Standard Completeness
The class of all formulas which are provable in the sys-

tem BL coincides with the class of all formulas which are logically valid in all *t*-algebras with a continuous *t*-norm.

The main steps in the proof are to show (i) that each BL-algebra is a subdirect product of subdirectly irreducible BL-chains, i. e., of linearly ordered BL-algebras which are not subdirect products of other BL-chains, and (ii) that each subdirectly irreducible BL-chain can be embedded into the ordinal sum of some BL-chains which are either trivial one-element BL-chains, or linearly ordered MV-algebras, or linearly ordered product algebras, such that (iii) each such ordinal summand is locally embedable into a *t*-norm-based residuated lattice with a continuous *t*-norm, cf. [2.24, 30] and again [2.5].

This is a lot more of algebraic machinery as necessary for the proof of the General Completeness Theorem 2.1 and thus offers a further indication that the extension of the class of divisible *t*-algebras to the class of BL-algebras made the development of the intended logical system easier. But even more can be seen from this proof: *the class of BL-algebras is the smallest variety which contains all the divisible t-algebras*, i. e., all the *t*-algebras determined by a continuous *t*-norm. And the algebraic reason for this is that each variety may be generated from its subdirectly irreducible elements, cf. again [2.31, 32].

Yet another generalization of Theorem 2.1 deserves to be mentioned. To state it, let us call *schematic extension* of BL every extension which consists in an addition of axiom schemata to the axiom schemata of BL. And let us denote such an extension by BL + *C*. And call BL(*C*)-algebra each BL-algebra **A** which makes **A**-valid all formulas of *C*, i. e., which is a model of *C*.

Then one can prove, as done in [2.23], an even more general completeness result.

Theorem 2.4 Strong General Completeness
For each set *C* of axiom schemata and any formula φ of $\mathcal{L}_T$ there are equivalent:

i) φ is derivable within BL + *C*;
ii) φ is valid in all BL(*C*)-algebras;
iii) φ is valid in all BL(*C*)-chains.

For the standard semantics this result holds true only in a restricted form: one has to restrict the consideration to *finite* sets *C* of axiom schemata, i. e., to finite theories. For the Łukasiewicz logic L, which is

the extension of BL by the schema $\neg\neg\varphi \rightarrow \varphi$ of double negation, this has already been shown in [2.23]. And for arbitrary continuous *t*-norms this follows from results of *Hanikóva* [2.33, 34].

Theorem 2.5 Strong Standard Completeness
For each finite set *C* of axiom schemata and any formula φ of $\mathcal{L}_T$ there are equivalent:

i) φ is derivable within BL + *C*;
ii) φ is valid in all *t*-algebras which are models of *C*.

2.4.2 The Logic MTL of All Left Continuous *t*-Norms

The guess of *Esteva* and *Godo* [2.27] has been that one should arrive at the logic of left continuous *t*-norms if one starts from the logic of continuous *t*-norms and deletes the continuity condition, i. e., the divisibility condition (2.35).

The algebraic approach needs only a small modification: in the definition of the BL-algebras one has simply to delete the divisibility condition. The resulting algebraic structures have been called *MTL-algebras*. They again form a variety.

Following this idea, one has to modify the previous axiom system in a suitable way. And one has to delete the definition (2.38) of the connective $\wedge$, because this definition (together with suitable axioms) essentially codes the divisibility condition. The definition (2.39) of the connective $\vee$ remains unchanged.

As a result one now considers a new system MTL of mathematical fuzzy logic, characterized semantically by the class of all MTL-algebras. It is connected with the axiom system:

$(\text{Ax}_{\text{MTL}}1)$ $(\varphi \rightarrow \psi) \rightarrow ((\psi \rightarrow \chi) \rightarrow (\varphi \rightarrow \chi))$,
$(\text{Ax}_{\text{MTL}}2)$ $\varphi \mathbin{\&} \psi \rightarrow \varphi$,
$(\text{Ax}_{\text{MTL}}3)$ $\varphi \mathbin{\&} \psi \rightarrow \psi \mathbin{\&} \varphi$,
$(\text{Ax}_{\text{MTL}}4)$ $(\varphi \rightarrow (\psi \rightarrow \chi)) \rightarrow (\varphi \mathbin{\&} \psi \rightarrow \chi)$,
$(\text{Ax}_{\text{MTL}}5)$ $(\varphi \mathbin{\&} \psi \rightarrow \chi) \rightarrow (\varphi \rightarrow (\psi \rightarrow \chi))$,
$(\text{Ax}_{\text{MTL}}6)$ $\varphi \wedge \psi \rightarrow \varphi$,
$(\text{Ax}_{\text{MTL}}7)$ $\varphi \wedge \psi \rightarrow \psi \wedge \varphi$,
$(\text{Ax}_{\text{MTL}}8)$ $\varphi \mathbin{\&} (\varphi \rightarrow \psi) \rightarrow \varphi \wedge \psi$,
$(\text{Ax}_{\text{MTL}}9)$ $\overline{0} \rightarrow \varphi$,
$(\text{Ax}_{\text{MTL}}10)$ $((\varphi \rightarrow \psi) \rightarrow \chi) \rightarrow (((\psi \rightarrow \varphi) \rightarrow \chi) \rightarrow \chi)$,

together with the rule of detachment (w.r.t. the implication connective $\rightarrow$) as (the only) inference rule.

Again, the system MTL is an implicative logic in the sense of *Rasiowa* [2.29], giving a general soundness and completeness result as for the previous system BL. Proofs of these results were given in [2.27].

Theorem 2.6 General Completeness
A formula φ of the language $\mathcal{L}_T$ is derivable within the system MTL iff φ is valid in all MTL-algebras.

Furthermore it is shown, in [2.27], that again already the class of all MTL-chains provides an adequate algebraic semantics.

Theorem 2.7 General Chain Completeness
A formula φ of $\mathcal{L}_T$ is derivable within the axiomatic system MTL iff φ is valid in all MTL-chains.

And again, similar as for the BL-case, even more is provable: the system MTL characterizes just these formulas which hold true w.r.t. all those t-norm-based logics which are determined by a left continuous t-norm, cf. [2.35].

Theorem 2.8 Standard Completeness
The class of formulas which are provable in the system MTL coincides with the class of formulas which are logically valid in all t-algebras with a left continuous t-norm.

This result again means, as the similar one for the logic of continuous t-norms, that the variety of all MTL-algebras is the smallest variety which contains all t-algebras with a left continuous t-norm.

Also for MTL an extended completeness theorem similar to Theorem 2.4 holds true. (The notions MTL + C and MTL(C)-algebra are used similar to the BL case.)

Theorem 2.9 Strong General Completeness
For each set C of axiom schemata and any formula φ of $\mathcal{L}_T$ the following are equivalent:

i) φ is derivable within the system MTL + C;
ii) φ is valid in all MTL(C)-algebras;
iii) φ is valid in all MTL(C)-chains.

For much more information on completeness matters for different systems of fuzzy logic the reader may consult [2.36].

2.4.3 Extensions of MTL

Because of the fact that the BL-algebras are the divisible MTL-algebras, one gets another adequate axiomatization of the basic t-norm logic BL.

Proposition 2.5

$$\mathsf{BL} = \mathsf{MTL} + \{\varphi \wedge \psi \to \varphi \,\&\, (\varphi \to \psi)\}\,.$$

Proof: Routine calculations in MTL-algebras give $x * (x \rightarrowtail y) \leq x$ and $x * (x \rightarrowtail y) \leq y$, and hence the inequality $x * (x \rightarrowtail y) \leq x \cap y$. In those MTL-algebras which are models of

$$\varphi \wedge \psi \to \varphi \,\&\, (\varphi \to \psi)\,. \tag{2.41}$$

also the converse inequality holds true, hence even $x * (x \rightarrowtail y) = x \cap y$. Thus the class of models of (2.41) is the class of all BL-algebras. So the result follows from the Completeness Theorem 2.1. ∎

Proposition 2.6

$$\mathsf{L} = \mathsf{BL} + \{\neg\neg\varphi \to \varphi\}$$

Proof: BL-algebras which also satisfy the equation $(x \rightarrowtail 0) \rightarrowtail 0 = x$ can be shown to be MV-algebras, cf. [2.5, 37]. And each MV-algebra is also a BL-algebra. Hence $\mathsf{BL} + \{\neg\neg\varphi \to \varphi\}$ is characterized by the class of all MV-algebras, so it is L according to Sect. 2.4.1. (There is also a syntactic proof available given in [2.23].) ∎

Proposition 2.7

$$\varPi = \mathsf{BL} + \{\varphi \wedge \neg\varphi \to 0,$$
$$\neg\neg\chi \to ((\varphi \,\&\, \chi \to \psi \,\&\, \chi) \to (\varphi \to \psi))\}\,.$$

Proof: This is essentially the original characterization of the product logic $\varPi$ as given in [2.3]. ∎

Proposition 2.8

$$\mathsf{G} = \mathsf{BL} + \{\varphi \to \varphi \,\&\, \varphi\}\,.$$

Proof: The prelinear Heyting algebras are just those BL-algebras for which the semigroup operation $*$ coincides with the lattice meet: $* = \wedge$, cf. [2.5, 38], and each Heyting algebra is also a BL-algebra. So the result follows again via Sect. 2.4.1. ∎

Similar remarks apply to further extensions of MTL.

2.4.4 Logics of Particular *t*-Norms

It is easy to recognize that two isomorphic (left continuous) *t*-norms T_1, T_2 determine the same *t*-norm-based logic: any order automorphism of $[0, 1]$ which transforms T_1 into T_2 according to (2.21) is an isomorphism between the *t*-algebras $[0, 1]_{T_1}$ and $[0, 1]_{T_2}$.

A continuous *t*-norm T is called *archimedean* iff $T(x, x) < x$ holds true for all $0 < x < 1$. And a *t*-norm T has *zero divisors* iff there exist $0 < a, b$ such that $T(a, b) = 0$.

It is well known that each continuous archimedean *t*-norm with zero divisors is isomorphic to the Łukasiewicz *t*-norm T_L, cf. [2.5, 20].

Proposition 2.9

Each *t*-norm-based logic, which is determined by a continuous archimedean *t*-norm with zero divisors, has the same axiomatizations as the infinite-valued Łukasiewicz logic L.

Furthermore, a continuous *t*-norm T is called *strict* iff it is strictly monotonous, i.e., satisfies for all $z \neq 0$

$$x < y \quad \Longleftrightarrow \quad T(x, z) < T(y, z) .$$

Again it is well known that each strict continuous *t*-norm is isomorphic to the product *t*-norm T_P, cf. [2.5, 20].

Proposition 2.10

Each *t*-norm-based logic which is determined by a strict continuous *t*-norm has the same axiomatizations as the infinite-valued product logic Π.

But there is a general solution of the axiomatization problem of those *t*-norm-based logics which are determined by a continuous *t*-norm.

In [2.39], *Esteva* et al. study the variety $\mathbb{BL}$ of all BL-algebras. They prove that each of its subvarieties which is generated by a single *T*-algebra over $[0, 1]$, *T* a continuous *t*-norm, is finitely axiomatizable. Additionally, they provide an algorithm to determine these finitely many axioms.

So the following main result is reached:

Theorem 2.10

Each *t*-norm-based fuzzy logic $\mathcal{L}_T$ determined by a continuous *t*-norm T is a finite axiomatic extension of the basic fuzzy logic BL.

For left continuous *t*-norms a similar result is lacking.

2.4.5 Extensions to First-Order Logics

The extensions of these propositional logics to first-order ones follows the standard lines of approach: one has to start from a first-order language ($\mathcal{L}$ with the two standard quantifiers $\forall, \exists$) and a suitable residuated lattice **A**, and has to define **A**-interpretations **M** by fixing a nonempty domain $M = |\mathbf{M}|$ and by assigning to each predicate symbol of $\mathcal{L}$ an **A**-valued relation in M (of suitable arity) and to each constant an element from (the support of) **A**.

Usually one supposes that the first-order language $\mathcal{L}$ has only predicate symbols and no function symbols. The insertion of function symbols proves to be a delicate matter, essentially because it is not completely clear what the basic properties of the identity predicate should be. The core problem is whether such an identity relation should be a crisp one or should be really graded. The paper [2.40] also surveys these problems of the identity relation.

The satisfaction relation is defined in the standard way. The quantifiers $\forall$ and $\exists$ are interpreted as taking the infimum or supremum, respectively, of all the values of the relevant instances.

To be sure that this approach works well one has either to suppose that the underlying lattices of the interpretations are complete lattices, or at least that all the necessary infima and suprema do exist in these lattices. Interpretations over lattices which satisfy this last mentioned condition are called *safe* by *Hájek* [2.23].

For the logic BL of continuous *t*-norms, *Hájek* [2.23] added the axioms:

($\forall$1) $(\forall x)\varphi(x) \to \varphi(t)$, where t is substitutable for x in φ,

($\exists$1) $\varphi(t) \to (\exists x)\varphi(x)$, where t is substitutable for x in φ,

($\forall$2) $(\forall x)(\chi \to \varphi) \to (\chi \to (\forall x)\varphi)$, where x is not free in χ,

($\exists$2) $(\forall x)(\varphi \rightarrow \chi) \rightarrow ((\exists x)\varphi \rightarrow \chi)$, where x is not free in χ,

($\forall$3) $(\forall x)(\chi \vee \varphi) \rightarrow \chi \vee (\forall x)\varphi$, where x is not free in χ

and the rule of generalization to the propositional calculus yielding the system BL$\forall$.

Then he was able to prove the following completeness theorem.

Theorem 2.11 General Chain Completeness

A first-order formula φ is BL$\forall$-provable iff it is valid in all safe interpretations over BL-chains.

This result can be extended to a lot of other first-order fuzzy logics, e.g., to MTL$\forall$.

We will not discuss further completeness results here but refer to the extended survey [2.40]. But it should be mentioned that, as suprema are not always maxima and infima not always minima, the truth degree of an existentially/universally quantified formula may not be the maximum/minimum of the truth degrees of the instances. It is, however, interesting to have conditions which characterize models in which the truth degrees of each existentially/universally quantified formula is witnessed as the truth degree of an instance. *Cintula* and *Hájek* [2.41] study this problem.

The topic of first-order fuzzy logics with identity deserves some attention. The core problem is, as in any many-valued logic, whether the identity symbol should be interpreted by the standard, i.e., *two-valued* identity relation, or whether one should allow for graded identity relations inside the interpretations.

Direct translations of the identity axioms of classical first-order logic into, e.g., the language of the Łukasiewicz systems force that the interpretation of the identity symbol has to be the standard identity relation,

cf. [2.42]. Similarly, for a wide class of first-order fuzzy logics the addition of the axioms:

Id1 $x \approx y \vee \neg x \approx y$
Id2 $x \approx x$
Id3 $x \approx y \rightarrow (\varphi(x, \vec{z}) \rightarrow \varphi(y, \vec{z}))$, for y substitutable for x in φ

forces that the identity symbol $\approx$ can only be understood as meaning standard identity. A general completeness theorem like Theorem 2.11 remains valid in this case too, cf. [2.40].

For the case of the Łukasiewicz logics, however, a slight modification of the standard identity axioms – particularly of the Leibniz schema, as given in [2.43], allows for graded identity relations, cf. also [2.5]. For fuzzy logics, in general, similarity relations, i.e., graded equivalence relations offer such an approach [2.23, 44]. For the restricted case of Horn formulas an approach is offered by *Bělohlávek* and *Vychodil* [2.45, 46]. They consider a first-order language with function symbols and the identity symbol $\approx$ as the only predicate symbol. Their models for sets of Horn formulas therefore have to be algebraic structures with graded identity relations. However, the aim of these authors is not to develop an identity logic, they mainly are interested to use the approach to characterize classes of algebraic structures with graded identity relations and to find *fuzzified* versions of results from universal algebra.

These authors even consider fuzzy sets of Horn formulas, i.e., they work in a Pavelka-style fuzzy logic as explained later in Sect. 2.6.1. But because this type of approach can be mirrored in standard fuzzy logics with sufficiently many truth degree constants, (Sect. 2.6.1) this approach is already discussed here.

2.5 Some Generalizations

The standard approach toward *t*-norm-based logics, as explained in Sects. 2.4.1 and 2.4.2, has been modified in various ways. The main background ideas are the extension or the modification of the expressive power of these logical systems.

2.5.1 Adding a Projection Operator

A first, quite fundamental addition to the standard vocabulary of the languages of *t*-norm-based systems was proposed in [2.47]: a unary propositional operator $\triangle$,

also known as *Baaz' Delta*, which has for *t*-algebras the semantics

$$\triangle(x) = 1 \quad \text{for } x = 1 \, ,$$
$$\triangle(x) = 0 \quad \text{for } x \neq 1 \, . \tag{2.42}$$

This unary connective can be added to the systems BL and MTL via the additional axioms

($\triangle$1) $\triangle\varphi \vee \neg\triangle\varphi \, ,$

($\triangle$2) $\triangle(\varphi \vee \varphi) \rightarrow (\triangle\varphi \vee \triangle\psi) \, ,$

(Δ3) $\triangle\varphi \to \varphi$,

(Δ4) $\triangle\varphi \to \triangle\triangle\varphi$,

(Δ5) $\triangle(\varphi \to \psi) \to (\triangle\varphi \to \triangle\psi)$.

This addition leaves all the essential theoretical results, like correctness and completeness theorems, valid: of course w.r.t. suitably expanded algebraic structures.

2.5.2 Adding an Idempotent Negation

A second stream of papers discusses the addition of an idempotent negation, i.e., a negation which satisfies the double negation law, for those cases where the standard negation of the *t*-norm-based system is not idempotent. This is, e.g., the case for the product logic which, as explained at the end of Sect. 2.1.3, has the Gödel negation (2.2) as its standard negation. By the way, it should be noticed that (routine calculations show that) this nonidempotent Gödel negation is the standard negation of all those *t*-norm algebras with a *t*-norm $\otimes$ which does not have zero-divisors. A very general approach is given in [2.48], and a more particular axiomatization problem discussed in [2.49].

2.5.3 Logics with Additional Strong Conjunctions

A third stream of papers, partly related to the previously mentioned one, is devoted to the problem of a unified treatment of different, usually two, *t*-norms and their related connectives within one logical system. Here the focus is on the join of the systems based on the Łukasiewicz *t*-norm and on the product *t*-norm. The great advantage of this unification is that the Łukasiewicz *t*-norm essentially allows to treat the addition, as may be seen from the truth degree function (2.6) of the Łukasiewicz (arithmetical) disjunction, and that the product *t*-norm adds the treatment of the usual product: and this means that the elementary arithmetic (in the unit interval) can be discussed in this combined system. This combined system has been considered in two strongly related forms, denoted by $Ł\Pi$ and $Ł\Pi\frac{1}{2}$. The distinction between both systems is that $Ł\Pi$ has both *t*-norms & and $\odot$ and their related (residual) implications and negations among their basic connectives, and that $Ł\Pi\frac{1}{2}$ adds a truth degree constant for the truth degree $\frac{1}{2}$. These two systems are discussed in detail in [2.50–54].

2.5.4 Logics Without a Truth Degree Constant

A fourth stream of papers intends to weaken the systems BL and MTL in such a way that one deletes the explicit reference to the truth degree constant $\overline{0}$ and considers the *falsity free* fragments of the previous systems. From the algebraic point of view their characteristic structures become the *hoops* which in general are defined as algebraic structures $H = \langle H, *, \Rightarrow, \mathbf{1}\rangle$ such that $\langle H, *, \mathbf{1}\rangle$ is a commutative monoid and that the further binary operation $\Rightarrow$ satisfies the equations

$$x \Rightarrow x = \mathbf{1} ,$$
$$x * (x \Rightarrow y) = y * (y \Rightarrow x) ,$$
$$(x * y) \Rightarrow z = x \Rightarrow (y \Rightarrow z) .$$

The definition

$$x \sqsubseteq y =_{\mathrm{df}} x \Rightarrow x = \mathbf{1}$$

provides an ordering $\sqsubseteq$ with the universal upper bound $\mathbf{1}$ which makes $\langle H, *, \mathbf{1}\rangle$ an ordered monoid, and which has the additional property that the operations $*, \Rightarrow$ become an adjoint pair w.r.t. this ordering.

In particular, hoops with the additional property

$$x \Rightarrow (y \Rightarrow z) \sqsubseteq (y \Rightarrow (x \Rightarrow z)) \Rightarrow z$$

can in a natural way be generated from *t*-algebras with continuous *t*-norms, as has been shown in [2.55]. So one has a kind of competing generalization of *t*-algebras. And for this kind of algebraic semantics, one can find adequate axiomatizations for the corresponding hoop logics quite similar to the approaches of Sects. 2.4.1 and 2.4.2. The details have been developed in [2.56].

2.5.5 Logics with a Noncommutative Strong Conjunction

And a fifth stream discusses the generalization of the algebraic semantics from the case of commutative lattice-ordered monoids with residuation to the case of noncommutative lattice-ordered semigroups. In this context, one tries to define noncommutative BL-algebras or noncommutative MTL-algebras, and similarly defines noncommutative *t*-norms, also called pseudo-*t*-norms. And these considerations become combined with the design of an adequate axiomatization, with similar results as in Sects. 2.4.1 and 2.4.2. Important papers on this topic are [2.57–63].

And finally it should be mentioned that *Hájek* [2.64] even gives a common generalization of all of these generalized fuzzy logics, thus giving up divisibility, the falsity constant, and commutativity. The corresponding algebras are called *fleas* (or *flea algebras*), and the logic is the *flea logic* FIL. There are examples of fleas on $(0, 1]$ not satisfying divisibility, nor commutativity, and having no least element.

2.6 Extensions with Graded Notions of Inference

The systems of *t*-norm-based logics discussed up to now have been designed to formalize the logical background for fuzzy sets, and they have degrees of truth for their formulas. But they all have crisp notions of consequence, i. e., of entailment and of provability.

Having in mind that fuzzy logics, also in their form as formalized logical systems, should be a (mathematical) tool for approximate reasoning makes it desirable that they should be able to deal with graded inferences too. This means inferences which start from *fuzzy sets of formulas*, and offer consequence hulls which again are fuzzy sets of formulas.

2.6.1 Pavelka-Style Approaches

This problem was first treated by *Pavelka* [2.65–67]. The basic monograph elaborating this approach is [2.44]. Accordingly, such approaches are sometimes called *Pavelka-style*, but they have – with emphasis on the syntactic side of the matter – also been coined approaches with *evaluated syntax*. Here we will call them *GI-approaches*.

Such an approach with graded inferences has to deal with fuzzy sets $\Sigma^{\approx}$ of formulas, i. e., besides formulas φ also their membership degrees $\Sigma^{\approx}(\varphi)$ in $\Sigma^{\approx}$. And these membership degrees are just the truth degrees. We may assume that these degrees again form a residuated lattice $\mathbf{L} = \langle L, \cap, \cup, *, \rightarrowtail, 0, 1 \rangle$. Thus we (slightly) generalize the standard notion of fuzzy set (with membership degrees from the real unit interval). Therefore, the appropriate language has the same logical connectives as in the previous considerations.

A GI-approach is an easy matter as long as the entailment relationship is considered. An evaluation e is a *model* of a fuzzy set $\Sigma^{\approx}$ of formulas iff

$$\Sigma^{\approx}(\varphi) \leqslant e(\varphi) \tag{2.43}$$

holds for each formula φ. This immediately yields that the semantic consequence hull of $\Sigma^{\approx}$ should be char-

acterized by the membership degrees

$$C^{\text{sem}}(\Sigma^{\approx})(\psi) = \bigwedge \{ e(\psi) \mid e \text{ model of } \Sigma^{\approx} \} \tag{2.44}$$

for each formula ψ.

For a syntactic characterization of this entailment relation, it is necessary to have some calculus $\mathbb{K}$ which treats formulas of the language together with truth degrees. So the language of this calculus has to extend the language of the basic logical system by having also symbols for the truth degrees. We indicate these symbols by overlined letters like $\overline{a}, \overline{c}$, and realize the common treatment of formulas and truth degrees by considering *evaluated formulas*, i. e., ordered pairs $(\overline{a}, \varphi)$ consisting of a truth degree symbol and a formula. This transforms each fuzzy set $\Sigma^{\approx}$ of formulas into a (usual) set of evaluated formulas, again denoted by $\Sigma^{\approx}$.

So $\mathbb{K}$ has to allow to derive evaluated formulas from sets of evaluated formulas, using suitable axioms and rules of inference. These axioms are usually only formulas φ which, however, are used in the derivations as the corresponding evaluated formulas $(\overline{1}, \varphi)$. The rules of inference have to deal with evaluated formulas.

Each $\mathbb{K}$-derivation of an evaluated formula $(\overline{a}, \varphi)$ counts as a derivation of φ *to the degree* $a \in L$. The *provability degree* of φ from $\Sigma^{\approx}$ in $\mathbb{K}$ is the supremum over all these degrees. The syntactic consequence hull of $\Sigma^{\approx}$ is the fuzzy set $C_{\mathbb{K}}^{\text{syn}}$ of formulas characterized by the membership function

$$C_{\mathbb{K}}^{\text{syn}}(\Sigma^{\approx})(\psi)$$
$$= \bigvee \{ a \in L \mid \mathbb{K} \text{ derives } (\overline{a}, \psi) \text{ out of } \Sigma^{\approx} \} \tag{2.45}$$

for each formula ψ.

Despite the fact that $\mathbb{K}$ is a standard calculus for evaluated formulas, this is – for infinite truth degree structures – an *infinitary* notion of provability for usual formulas.

For the infinite-valued Łukasiewicz logic L, this machinery works particularly well because it needs in an essential way the continuity of the residuation operation. The corresponding calculus $\mathbb{K}_L$ has as axioms any axiom system of the infinite-valued Łukasiewicz logic L which provides together with the rule of detachment an adequate axiomatization of L, but $\mathbb{K}_L$ replaces this standard rule of detachment by the generalized form

$$\frac{(\bar{a}, \varphi) \qquad (\bar{c}, \varphi \to \psi)}{(\overline{a * c}, \psi)} \qquad (2.46)$$

for evaluated formulas.

The soundness result then says that the $\mathbb{K}_L$-provability of an evaluated formula $(\bar{a}, \varphi)$ means that $a \leq e(\varphi)$ holds for every valuation e. And this just means that the formula $\bar{a} \to \varphi$ is valid; however, as a formula of an *extended* propositional language which has all the truth degree constants among its vocabulary. Of course, for this extended language the evaluations e have to satisfy $e(\bar{a}) = a$ for each $a \in [0, 1]$.

The soundness and completeness results for $\mathbb{K}_L$ say that a *strong completeness theorem* holds true giving

$$C^{\text{sem}}(\Sigma^{\approx})(\psi) = C^{\text{syn}}_{\mathbb{K}_L}(\Sigma^{\approx})(\psi) \qquad (2.47)$$

for each formula ψ and each fuzzy set $\Sigma^{\approx}$ of formulas.

If one takes the previously mentioned turn and extends the standard language of propositional L by truth degree constants for all degrees $a \in [0, 1]$, and if one reads each evaluated formula $(\bar{a}, \varphi)$ as the formula $\bar{a} \to \varphi$, then a slight modification $\mathbb{K}_L^+$ of the former calculus $\mathbb{K}_L$ again provides an adequate axiomatization: one has to add the *bookkeeping axioms*

$$(\bar{a} \,\&\, \bar{c}) \equiv \overline{a * c},$$
$$(\bar{a} \to \bar{c}) \equiv \overline{a \rightarrowtail_L c},$$

as explained, e.g., in [2.44]. And if one is interested to have evaluated formulas together with the extension of the language by truth degree constants, one has also to add the *logical constant introduction rule*

$$\frac{(\bar{a}, \varphi)}{\bar{a} \to \varphi} \,.$$

However, even a stronger result is available which refers only to a notion of derivability over a countable language. The completeness result (2.47), for $\mathbb{K}_L^+$ instead of $\mathbb{K}_L$, becomes already provable if one adds truth degree constants only for all the *rationals* in $[0, 1]$, as

was shown in [2.23]. And this extension of L, known as Rational Pavelka Logic, is even a conservative one, cf. [2.68], i.e., $\mathbb{K}_L^+$ proves only such *constant-free* formulas of the language with rational constants which are already provable in the standard infinite-valued Łukasiewicz logic L.

So the GI-approach with graded notion of provability and entailment can suitably be mirrored inside standard fuzzy logics with sufficiently many truth degree constants.

For more details the reader may also consult, e.g., [2.23, 44, 69].

2.6.2 A Lattice Theoretic Approach

For completeness, we also mention a much more abstract approach toward fuzzy logics with graded notions of entailment as the previously explained one for the *t*-norm-based fuzzy logics is.

The background for this generalization by *Gerla*, in detail explained in [2.70], is that (already) in systems of classical logic the syntactic as well as the semantic consequence relations, i.e., the provability as well as the entailment relations, are closure operators within the set of formulas. This is a fundamental observation made by *Tarski* [2.71] already in 1930. And the same holds true for the Pavelka style extensions of Sect. 2.6.1 and the operators C^{sem} and C^{syn} introduced in (2.44) and (2.45), respectively: they are generalized closure operators.

The context, chosen in [2.70], is that of *L*-fuzzy sets, with $\mathbf{L} = \langle L, \leq \rangle$ an arbitrary complete lattice. A *closure operator in* $\mathbf{L}$ is a mapping $J : L \to L$ satisfying for arbitrary $x, y \in L$ the well-known conditions

$$
\begin{array}{ll}
x \leqslant J(x) & \text{(increasingness)}, \\
x \leqslant y \Rightarrow J(x) \leqslant J(y) & \text{(isotonicity)}, \\
J(J(x)) = J(x) & \text{(idempotency)}.
\end{array}
$$

And a *closure system* in $\mathbf{L}$ is a subclass $C \subseteq L$ which is closed under arbitrary lattice meets.

For fuzzy logic such closure operators and closure systems are considered in the lattice $\mathcal{F}_L(\mathbb{F})$ of all fuzzy subsets of the set $\mathbb{F}$ of formulas of some suitable formalized language.

An *abstract fuzzy deduction system* now is an ordered pair $\mathcal{D} = (\mathcal{F}_L(\mathbb{F}), D)$ determined by a closure operator D in the lattice $\mathcal{F}_L(\mathbb{F})$. And the *fuzzy theories* T of such an abstract fuzzy deduction system, also called $\mathcal{D}$-theories, are the fixed points of D: $T = D(T)$, i.e., the deductively closed fuzzy sets of formulas.

A rather abstract setting is also chosen for the semantics of such an abstract fuzzy deduction system: an *abstract fuzzy semantics* $\mathcal{M}$ is nothing but a class of elements of the lattice $\mathcal{F}_L(\mathbb{F})$, i.e., a class of fuzzy sets of formulas. These fuzzy sets of formulas are called *models*. The only restriction is that the universal set over $\mathbb{F}$, i.e., the fuzzy subset of $\mathbb{F}$ which has always membership degree one, is not allowed as a model. The background idea here is that, for each standard interpretation $\mathfrak{A}$ (in the sense of many-valued logic – including an evaluation of the individual variables) for the formulas of $\mathbb{F}$, a model M is determined as the fuzzy set which has for each formula $\varphi \in \mathbb{F}$ the truth degree of φ in A as membership degree. Accordingly, the satisfaction relation $\models_{\mathcal{M}}$ coincides with inclusion: for models $M \in \mathcal{M}$ and fuzzy sets Σ of formulas one has

$$M \models_{\mathcal{M}} \Sigma \quad \leftrightarrow \quad \Sigma \subseteq M \,. \tag{2.48}$$

In this setting, one has a semantic and a syntactic consequence operator, both being closure operators, i.e., one has for each fuzzy set Σ of formulas from $\mathbb{F}$ a semantic as well as a syntactic consequence hull, given by

$$C^{sem}(\Sigma) = \bigcap \{M \in \mathcal{M} \mid M \models_{\mathcal{M}} \Sigma\} \,,$$
$$C^{syn}(\Sigma) = D(\Sigma) \,. \tag{2.49}$$

Similar to the classical case one has $C^{sem}(M) = M$ for each model $M \in \mathcal{M}$, i.e., each such model provides a C^{sem}-theory.

However, a general completeness theorem is not available. What one needs instead, in search for a completeness result, that are specifications which restrict the full generality of this approach, and lead mainly back to situations which have been discussed in the previous sections.

2.7 Some Complexity Results

Each (left-continuous) t-norm T determines four important sets of formulas:

- 1TAUT (T): The set of all 1-tautologies.
- posTAUT (T): The set of all positive tautologies.
- 1SAT (T): The set of 1-satisfiable formulas.
- posSAT (T): The set of all positively satisfiable formulas.

Here a 1-tautology is a formula valid in $[0, 1]_T$, i.e., having for each evaluation of propositional variables by elements of $[0, 1]$ the value 1 in $[0, 1]_T$. And a positive tautology is a formula which has for each evaluation a positive value in $[0, 1]_T$. Similarly 1-satisfiability means to have the $[0, 1]_T$ value 1 for some evaluation, and positive satisfiability means to have for some evaluation a positive $[0, 1]_T$-value.

In the same way, one defines analogous sets corresponding to sets of t-norms; in particular, with BL referring to the set of all continuous t-norms, one defines

1TAUT(BL)

$$= \bigcap \{1TAUT(T) \mid T \text{ a continuous } t\text{-norm}\} \,,$$

posTAUT(BL)

$$= \bigcap \{posTAUT(T) \mid T \text{ a continuous } t\text{-norm}\} \,,$$

and similarly for the satisfiability cases.

There are interesting results on the computational complexity of these sets. So it was, already in [2.23], shown that if the t-norm T is T_L, or T_G, or T_P, then $1TAUT(T)$ and $posTAUT(T)$ are co-NP-complete, and $1SAT(T)$ and $posSAT(T)$ are NP-complete. This result was partly strengthened in [2.34] yielding that $1TAUT(T)$ is co-NP-complete for each continuous t-norm T.

The corresponding results have been proved in [2.72, 73] for the logic BL of continuous t-norms.

Theorem 2.12
1TAUT(BL) and *posTAUT(BL)* are co-NP-complete, and *1SAT(BL)* as well as *posSAT(BL)* are NP-complete.

Furthermore, there are several results on equality or inequality among the sets involved [2.23, 73]. So one has, e.g.,

$$1SAT(G) = posSAT(G) = 1SAT(\varPi) = posSAT(\varPi) \,,$$

but also

$$1SAT(L) \neq posSAT(L)$$

and

$$posSAT(BL) = posSAT(L) \,.$$

For the 1-tautologicity the papers [2.74–76] contain interesting results relating the property of a formula being a 1-tautology of one of the logics L, G, Π to the property of being a 1-tautology of finitely many finite-valued logics of estimated complexity, and similar results for 1TAUT(BL). For example, φ is in 1TAUT(L) if and only if it is a 1-tautology of the finitely valued Łukasiewicz logic L_m for m being $2^{\#(\varphi)}$ where $\#(\varphi)$ is the number of occurrences of variables in φ.

Remind that for predicate logics $\mathbb{L}_{\mathcal{K}}\forall$, the general models are safe interpretations over any *linearly ordered* $\mathbb{L}_{\mathcal{K}}$-algebra, and the standard models – for t-norm-based logics – are interpretations over any t-algebra which is also an $\mathbb{L}_{\mathcal{K}}$-algebra.

Definition 2.4
Let φ be a closed formula of the language of $\mathbb{L}_{\mathcal{K}}\forall$:

1. φ is a *general* $\mathbb{L}_{\mathcal{K}}\forall$-*tautology* if φ is valid in each safe interpretation over any *linearly ordered* $\mathbb{L}_{\mathcal{K}}$-algebra; *genTAUT*($\mathbb{L}_{\mathcal{K}}\forall$) is the set of all general $\mathbb{L}_{\mathcal{K}}\forall$-tautologies.
2. φ is a *standard* $\mathbb{L}_{\mathcal{K}}\forall$-*tautology* if φ is valid in each safe interpretation over any *standard* $\mathbb{L}_{\mathcal{K}}$-algebra; *stTAUT*($\mathbb{L}_{\mathcal{K}}\forall$) is the set of all standard $\mathbb{L}_{\mathcal{K}}\forall$-tautologies.
3. φ is $\mathbb{L}_{\mathcal{K}}\forall$-*satisfiable* if φ is valid in some safe interpretation over some *linearly ordered* $\mathbb{L}_{\mathcal{K}}$-algebra; *genSAT*($\mathbb{L}_{\mathcal{K}}\forall$) is the set of all $\mathbb{L}_{\mathcal{K}}\forall$-satisfiable sentences.
4. φ is *standardly* $\mathbb{L}_{\mathcal{K}}\forall$-*satisfiable* if φ is valid in some interpretation over some *standard* $\mathbb{L}_{\mathcal{K}}$-algebra; *stSAT*($\mathbb{L}_{\mathcal{K}}\forall$) is the set of all standardly $\mathbb{L}_{\mathcal{K}}\forall$-satisfiable formulas.

It was already shown in [2.23] that if $\mathbb{L}$ is the logic BL or one of its specifications L, G, Π, then genTAUT($\mathbb{L}\forall$) is Σ_1-complete and genSAT($\mathbb{L}\forall$) is Π_1-complete. And this result has been extended in [2.77] to any t-norm-based logic $\mathbb{L}_T\forall$ for a continuous t-norm T.

For standard semantics the situation is different. Already *Ragaz* [2.78, 79] proved that the set stTAUT(L$\forall$) of standard tautologies of the infinite-valued Łukasiewicz logic is Π_2-complete. Generalizing this, *Hájek* [2.23] also showed that stTAUT(G$\forall$) = genTAUT(G$\forall$) and that therefore the set stTAUT(G$\forall$) of standard tautologies of the infinite-valued Gödel logic is Σ_1-complete.

These results have been considerably extended by *Montagna* [2.80] yielding the following facts.

Theorem 2.13
1. For each set $\mathcal{K}$ of continuous t-norms containing a t-norm different from T_G, the set *stTAUT*($\mathbb{L}_{\mathcal{K}}\forall$) is Π_2-hard.
2. If $\mathcal{K}$ is a nonempty set of continuous t-norms containing a t-norm which has, in its ordinal sum representation, a product component or a nonextremal (this means being neither first nor last summand) Łukasiewicz component then *stTAUT*($\mathbb{L}_{\mathcal{K}}\forall$) is not arithmetical.

The arithmetical complexity of the set stTAUT($\mathbb{L}_T\forall$) remains undetermined if T is, e.g., one of the t-norms L $\oplus$ L, G $\oplus$ L, L $\oplus$ G, or L $\oplus$ G $\oplus$ L, with $\oplus$ denoting the ordinal sum operation.

For standard satisfiability *Hájek* [2.23] proved that the sets stSAT(G$\forall$) and stSAT(L$\forall$) are Π_1-complete. He also proved, in [2.81], that the set stSAT($\Pi\forall$) is not arithmetical, and gave in [2.82] also the following result.

Theorem 2.14
If $\otimes$ is a continuous t-norm whose first ordinal sum component is G, or L, then *stSAT*($\mathbb{L}_{\otimes}\forall$) is Π_1-complete.

The reason is that one has, under these assumptions,

$$\mathsf{stSAT}(\mathbb{L}_{\otimes}\forall) = \mathsf{stSAT}(\mathsf{G}\forall) \, ,$$

as well as

$$\mathsf{stSAT}(\mathbb{L}_{\otimes}\forall) = \mathsf{stSAT}(\mathsf{L}\forall) \, .$$

Montagna [2.83] added for the product logic Π the more general.

Theorem 2.15
If $\mathcal{K}$ is a nonempty set of continuous t-norms containing T_P, or a t-norm whose first ordinal summand is T_P, then *stSAT*($\mathbb{L}_{\mathcal{K}}\forall$) is not arithmetical.

The complexity of stSAT($\mathbb{L}_T\forall$) for continuous t-norms T which do not have a first component in their ordinal sum representation is an open problem.

More complexity results are surveyed in [2.77] and more recently in [2.84, 85].

2.8 Concluding Remarks

The reader who is interested in further results, or in more details, might consult the recent *Handbook of Mathematical Fuzzy Logic* [2.86]. This Handbook surveys the whole field of mathematical fuzzy logics and offers the most actual state of the art in this field. For the wider topic of many-valued logics, [2.5] is still the best reference.

There is one approach, however, which is not discussed here and only shortly mentioned in [2.86]: a version of Church-style type theory based on suitable mathematical fuzzy logics, called *fuzzy type theory*, and

developed by *Novák* [2.87]. This fuzzy-type theory is particularly used for linguistic modeling. A good survey is [2.88].

There are further topics also often treated under the heading *fuzzy logic* which have been omitted in this survey. Here the focus was on what is called mathematical fuzzy logic or also *fuzzy logic in narrow sense*. Those other topics mainly are classified as *fuzzy logic in wider sense*. They include topics like fuzzy implications, approximate, and commonsense reasoning – and some of them are discussed elsewhere in this Handbook.

References

2.1   J. Łukasiewicz, A. Tarski: Untersuchungen über den Aussagenkalkül, c.r. Séances Soc. Sci. Lett. Vars. cl. III **23**, 30–50 (1930)

2.2   K. Gödel: Zum intuitionistischen Aussagenkalkül, Anz. Akad. Wiss. Wien: Math.-naturwiss. Kl. **69**, 65–66 (1932)

2.3   P. Hájek, L. Godo, F. Esteva: A complete many-valued logic with product-conjunction, Arch. Math. Log. **35**, 191–208 (1996)

2.4   E.L. Post: Introduction to a general theory of elementary propositions, Am. J. Math. **43**, 163–185 (1921)

2.5   S. Gottwald: *A Treatise on Many-Valued Logics*, Studies in Logic and Computation, Vol. 9 (Research Studies, Baldock 2001)

2.6   M. Dummett: A propositional calculus with denumerable matrix, J. Symb. Log. **24**, 97–106 (1959)

2.7   C.C. Chang: Algebraic analysis of many valued logics, Trans. Am. Math. Soc. **88**, 476–490 (1958)

2.8   R. Cignoli, I.M.L. D'Ottaviano, D. Mundici: *Algebraic Foundations of Many-Valued Reasoning*, Trends in Logic – Studia Logica Library, Vol. 7 (Kluwer, Dordrecht 2000)

2.9   P.C. Rosenbloom: Post algebras. I. postulates and general theory, Am. J. Math. **64**, 167–188 (1942)

2.10   G. Epstein: The lattice theory of post algebras, Trans. Am. Math. Soc. **95**, 300–317 (1960)

2.11   N. Cat-Ho: *Generalized Post Algebras and Their Applications to Some Infinitary Many-Valued Logics*, Diss. Math., Vol. 57 (PWN, Warsaw 1973)

2.12   P. Dwinger: Generalized post algebras, Bull. Acad. Polon. Sci. Sér. Sci. Math. Astronom. Phys. **16**, 559–563 (1968)

2.13   P. Dwinger: A survey of the theory of post algebras and their generalizations. In: *Modern Uses of Multiple-Valued Logic*, ed. by J.M. Dunn, G. Epstein (Reidel, Dordrecht 1977) pp. 53–75

2.14   H. Rasiowa: On generalised post algebras of order ω^+ and ω^+-valued predicate calculi, Bull. Acad.

Polon. Sci. Sér. Sci. Math. Astron. Phys. **21**, 209–219 (1973)

2.15   G. Epstein, H. Rasiowa: Theory and uses of post algebras of order $\omega + \omega^*$. Part I, 20th Int. Symp. Multiple-Valued Log., Charlotte/NC 1990 (IEEE Computer Society, New York 1990) pp. 42–47

2.16   G. Epstein, H. Rasiowa: Theory and uses of post algebras of order $\omega + \omega^*$. Part II, 21st Int. Symp. Multiple-Valued Log., Victoria/B.C., 1991 (IEEE Computer Society, New York 1991) pp. 248–254

2.17   D.C. Rine (Ed.): *Computer Science and Multiple Valued Logic*, 2nd edn. (North-Holland, Amsterdam 1984)

2.18   L.A. Zadeh: Fuzzy sets, Inf. Control **8**, 338–353 (1965)

2.19   B. Schweizer, A. Sklar: *Probabilistic Metric Spaces* (North-Holland, Amsterdam 1983)

2.20   E.P. Klement, R. Mesiar, E. Pap: *Triangular Norms* (Kluwer, Dordrecht 2000)

2.21   C. Alsina, M.J. Frank, B. Schweizer: *Associative functions. Triangular Norms and Copulas* (World Scientific, Hackensack 2006)

2.22   S. Gottwald: *Fuzzy Sets and Fuzzy Logic: Foundations of Application – From a Mathematical Point of View*. Artificial Intelligence (Verlag Vieweg, Wiesbaden, and Tecnea, Toulouse 1993)

2.23   P. Hájek: *Metamathematics of Fuzzy Logic*, Trends in Logic, Vol. 4 (Kluwer, Dordrecht 1998)

2.24   P. Hájek: Basic fuzzy logic and BL-algebras, Soft Comput. **2**, 124–128 (1998)

2.25   D. Butnariu, E.P. Klement, S. Zafrany: On triangular norm-based propositional fuzzy logics, Fuzzy Sets Syst. **69**, 241–255 (1995)

2.26   J. Hekrdla, E.P. Klement, M. Navara: Two approaches to fuzzy propositional logics, J. Multiple-Valued Log, Soft Comput. **9**, 343–360 (2003)

2.27   F. Esteva, L. Godo: Monoidal t-norm based logic: Toward a logic for left-continuous t-norms, Fuzzy Sets Syst. **124**, 271–288 (2001)

2.28 U. Höhle: On the fundamentals of fuzzy set theory, J. Math. Anal. Appl. **201**, 786–826 (1996)

2.29 H. Rasiowa: *An Algebraic Approach to Non-Classical Logics* (North-Holland/PWN, Amsterdam/Warsaw 1974)

2.30 R. Cignoli, F. Esteva, L. Godo, A. Torrens: Basic fuzzy logic is the logic of continuous t-norms and their residua, Soft Comput. **4**, 106–112 (2000)

2.31 S. Burris, H.P. Sankappanavar: *A Course in Universal Algebra* (Springer, New York 1981)

2.32 K. Denecke, S.L. Wismath: *Universal Algebra and Applications in Theoretical Computer Science* (Chapman Hall/CRC, Boca Raton 2002)

2.33 Z. Haniková: Standard algebras for fuzzy propositional calculi, Fuzzy Sets Syst. **124**, 309–320 (2001)

2.34 Z. Haniková: A note on the complexity of propositional logics of individual t-algebras, Neural Netw. World **12**, 453–460 (2002)

2.35 S. Jenei, F. Montagna: A proof of standard completeness for Esteva and Godo's logic MTL, Stud. Log. **70**, 183–192 (2002)

2.36 P. Cintula, F. Esteva, J. Gispert, L. Godo, F. Montagna, C. Noguera: Distinguished algebraic semantics for t-norm based fuzzy logics: Methods and algebraic equivalencies, Ann. Pure Appl. Log. **160**, 53–81 (2009)

2.37 U. Höhle: Presheaves over GL-monoids. In: *Non-Classical Logics and Their Applications to Fuzzy Subsets*, Theory and Decision Library, Series B, Vol. 32, ed. by U. Höhle, E.P. Klement (Kluwer, Dordrecht 1995) pp. 127–157

2.38 U. Höhle: Commutative, residuated l-monoids. In: *Non-Classical Logics and Their Applications to Fuzzy Subsets*, Theory and Decision Library, Series B, Vol. 32, ed. by U. Höhle, E.P. Klement (Kluwer, Dordrecht 1995) pp. 53–106

2.39 F. Esteva, L. Godo, F. Montagna: Equational characterization of the subvarieties of BL generated by t-norm algebras, Stud. Log. **76**, 161–200 (2004)

2.40 P. Cintula, P. Hájek: Triangular norm based predicate fuzzy logics, Fuzzy Sets Syst. **161**, 311–346 (2010)

2.41 P. Cintula, P. Hájek: On theories and models in fuzzy predicate logics, J. Symb. Log. **71**, 863–880 (2006)

2.42 H. Thiele: Theorie der endlichwertigen Łukasiewiczschen Prädikatenkalküle der ersten Stufe, Z. Math. Log. Grundl. Math. **4**, 108–142 (1958)

2.43 S. Gottwald: A generalized Łukasiewicz-style identity logic. In: *Mathematical Logic and Formal Systems*, Lecture Notes in Pure and Applied Mathematics, Vol. 94, ed. by L.P. de Alcantara (Marcel Dekker, New York 1985) pp. 183–195

2.44 V. Novák, I. Perfilieva, J. Močkoř: *Mathematical Principles of Fuzzy Logic* (Kluwer, Boston 1999)

2.45 R. Bělohlávek, V. Vychodil: Fuzzy Horn logic. I. Proof theory, Arch. Math. Log. **45**, 3–51 (2006)

2.46 R. Bělohlávek, V. Vychodil: Fuzzy Horn logic. II. Implicationally defined classes, Arch. Math. Log. **45**, 149–177 (2006)

2.47 M. Baaz: Infinite-valued Gödel logics with 0-1 projections and relativizations. In: *Gödel '96*, Lecture Notes in Logic, Vol. 6, ed. by P. Hájek (Springer, New York 1996) pp. 23–33

2.48 F. Esteva, L. Godo, P. Hájek, M. Navara: Residuated fuzzy logic with an involutive negation, Arch. Math. Log. **39**, 103–124 (2000)

2.49 S. Gottwald, S. Jenei: A new axiomatization for involutive monoidal t-norm based logic, Fuzzy Sets Syst. **124**, 303–307 (2001)

2.50 P. Cintula: The $Ł\Pi$ and $Ł\Pi\frac{1}{2}$ propositional and predicate logics, Fuzzy Sets Syst. **124**, 289–302 (2001)

2.51 P. Cintula: An alternative approach to the $Ł\Pi$ logic, Neural Netw. World **124**, 561–575 (2001)

2.52 P. Cintula: Advances in $Ł\Pi$ and $Ł\Pi\frac{1}{2}$ logics, Arch. Math. Log. **42**, 449–468 (2003)

2.53 F. Esteva, L. Godo: Putting together Łukasiewicz and product logics, Mathw. Soft Comput. **6**, 219–234 (1999)

2.54 F. Esteva, L. Godo, F. Montagna: The $Ł\Pi$ and $Ł\Pi\frac{1}{2}$ logics: Two complete fuzzy systems joining Łukasiewicz and product logics, Arch. Math. Log. **40**, 39–67 (2001)

2.55 P. Agliano, I.M.A. Ferreirim, F. Montagna: Basic hoops: An algebraic study of continuous t-norms, Stud. Log. **87**, 73–98 (2007)

2.56 F. Esteva, L. Godo, P. Hájek, F. Montagna: Hoops and fuzzy logic, J. Log. Comput. **13**, 531–555 (2003)

2.57 A. di Nola, G. Georgescu, A. Iorgulescu: Pseudo-BL algebras. I and II, J. Multiple-Valued Log. **8**, 671–750 (2002)

2.58 P. Flondor, G. Georgescu, A. Iorgulescu: Pseudo t-norms and pseudo-BL algebras, Soft Comput. **5**, 355–371 (2001)

2.59 P. Hájek: Embedding standard BL-algebras into non-commutative pseudo-BL-algebras, Tatra Mt. Math. Publ. **27**, 125–130 (2003)

2.60 P. Hájek: Fuzzy logics with non-commutative conjunctions, J. Log. Comput. **13**, 469–479 (2003)

2.61 P. Hájek: Observations on non-commutative fuzzy logics, Soft Comput. **8**, 28–43 (2003)

2.62 S. Jenei, F. Montagna: A proof of standard completeness for non-commutative monoidal t-norm logic, Neural Netw. World **13**, 481–488 (2003)

2.63 J. Kühr: Pseudo-BL algebras and PRl-monoids, Math. Bohem. **128**, 199–208 (2003)

2.64 P. Hájek: Fleas and fuzzy logic, J. Multiple-Valued Log. Soft Comput. **11**, 137–152 (2005)

2.65 J. Pavelka: On fuzzy logic. Part I, Z. Math. Log. Grundl. Math. **25**, 45–52 (1979)

2.66 J. Pavelka: On fuzzy logic. Part II, Z. Math. Log. Grundl. Math. **25**, 119–134 (1979)

2.67 J. Pavelka: On fuzzy logic. Part III, Z. Math. Log. Grundl. Math. **25**, 447–464 (1979)

2.68 P. Hájek, J. Paris, J. Shepherdson: Rational Pavelka predicate logic is a conservative extension of Łukasiewicz predicate logic, J. Symb. Log. **65**, 669–682 (2000)

2.69 E. Turunen: Well-defined fuzzy sentential logic, Math. Log. Quart. **41**, 236–248 (1995)

2.70 G. Gerla: *Fuzzy logic*, Mathematical Tools for Approximate Reasoning, Trends in Logic, Vol. 11 (Kluwer, Dordrecht 2001)

2.71 A. Tarski: Fundamentale Begriffe der Methodologie der deduktiven Wissenschaften, Monatsh. Math. Phys. **37**, 361–404 (1930)

2.72 M. Baaz, P. Hájek, F. Montagna, H. Veith: Complexity of *t*-tautologies, Ann. Pure Appl. Log. **113**, 3–11 (2002)

2.73 P. Hájek: Basic fuzzy logic and BL-algebras II, Soft Comput. **7**, 179–183 (2003)

2.74 S. Aguzzoli, A. Ciabattoni: Finiteness in infinite-valued Łukasiewicz logic, J. Log. Lang. Inf. **9**, 5–29 (2000)

2.75 S. Aguzzoli, B. Gerla: Finite-valued reductions of infinite-valued logics, Arch. Math. Log. **41**, 361–399 (2002)

2.76 S. Aguzzoli, B. Gerla: On countermodels in basic logic, Neural Netw. World **12**, 407–420 (2002)

2.77 P. Hájek: Arithmetical complexity of fuzzy predicate logics – A survey, Soft Comput. **9**(12), 935–941 (2005)

2.78 M. Ragaz: Arithmetische Klassifikation von Formelmengen der unendlichwertigen Logik, Ph.D. Thesis (Abteilung Mathematik der Eidgenössischen Technischen Hochschule Zürich, Zürich 1981)

2.79 M. Ragaz: Die Unentscheidbarkeit der einstelligen unendlichwertigen Prädikatenlogik, Arch. Math. Log. Grundlagenforsch. **23**, 129–139 (1983)

2.80 F. Montanga: On the predicate logics of continuous t-norm BL-algebras, Arch. Math. Log. **44**, 97–114 (2005)

2.81 P. Hájek: Fuzzy logic and arithmetical hierarchy III, Stud. Log. **68**, 129–142 (2001)

2.82 P. Hájek: Fuzzy logic and arithmetical hierarchy IV, First-Order Logic Revisited, Proc. Conf. FOL75 – 75 Years of First-Order Logic, Berlin, ed. by V. Hendricks (Logos, Berlin 2004) pp. 107–115

2.83 F. Montagna: Three complexity problems in quantified fuzzy logic, Stud. Log. **68**, 143–152 (2001)

2.84 Z. Haniková: Computational complexity of propositional fuzzy logics. In: *Handbook of Mathematical Fuzzy Logic*, Studies in Logic, Vol. 2, ed. by P. Cintula, P. Hájek, C. Noguera (College, London 2011) pp. 793–851

2.85 P. Hájek, F. Montagna, C. Noguera: Computational complexity of first-order fuzzy logics. In: *Handbook of Mathematical Fuzzy Logic*, Studies in Logic, ed. by P. Cintula, P. Hájek, C. Noguera (College Publ., London 2011) pp. 853–908

2.86 P. Cintula, P. Hájek, C. Noguera (Eds.): *Handbook of Mathematical Fuzzy Logic*, Studies in Logic, Vol. 37 (College Publ., London 2011)

2.87 V. Novák: On fuzzy type theory, Fuzzy Sets Syst. **149**(2), 235–273 (2005)

2.88 V. Novák: Reasoning about mathematical fuzzy logic and its future, Fuzzy Sets Syst. **192**, 25–44 (2012)

3. Possibility Theory and Its Applications: Where Do We Stand?

Didier Dubois, Henry Prade

This chapter provides an overview of possibility theory, emphasizing its historical roots and its recent developments. Possibility theory lies at the crossroads between fuzzy sets, probability, and nonmonotonic reasoning. Possibility theory can be cast either in an ordinal or in a numerical setting. Qualitative possibility theory is closely related to belief revision theory, and commonsense reasoning with exception-tainted knowledge in artificial intelligence. Possibilistic logic provides a rich representation setting, which enables the handling of lower bounds of possibility theory measures, while remaining close to classical logic. Qualitative possibility theory has been axiomatically justified in a decision-theoretic framework in the style of Savage, thus providing a foundation for qualitative decision theory. Quantitative possibility theory is the simplest framework for statistical reasoning with imprecise probabilities. As such, it has close connections with random set theory and confidence intervals, and can provide a tool for uncertainty propagation with limited statistical or subjective information.

Possibility theory is an uncertainty theory devoted to the handling of incomplete information. To a large extent, it is comparable to probability theory because it is based on set functions. It differs from the latter by the use of a pair of dual set functions (possibility and necessity measures) instead of only one. Besides, it is not additive and makes sense on ordinal structures. The name *Theory of Possibility* was coined by *Zadeh* [3.1], who was inspired by a paper by *Gaines* and *Kohout* [3.2]. In Zadeh's view, possibility distributions were meant to provide a graded semantics to natural language statements; on this basis, possibility degrees can be attached to other statements, as well as dual necessity degrees expressing graded certainty. However, possibility and necessity measures can also be the basis of a full-fledged representation of partial belief that parallels probability, without compulsory reference to linguistic information [3.3, 4]. It can be seen either as a coarse, nonnumerical version of probability theory, or a framework for reasoning with extreme probabilities, or yet a simple approach to reasoning with imprecise probabilities [3.5].

Besides, possibility distributions can also be interpreted as representations of preference, thus standing for a counterpart to a utility function. In this case, possibility degrees estimate degrees of feasibility of alternative choices, while necessity measures can represent priorities [3.6]. The possibility theory framework is also bipolar [3.7] because distributions may either restrict the possible states of the world (negative information pointing out the impossible), or model sets of actually observed possibilities (positive information pointing out the possible). Negative information refers to pieces of knowledge that are supposedly correct and act as constraints. Possibility and necessity measures rely on negative information.

Positive information refers to reports of actually observed states, or to sets of preferred choices. They induce two other set functions: guaranteed possibility measures and its dual, that are decreasing w.r.t. set inclusion [3.8].

After reviewing pioneering contributions to possibility theory, we recall its basic concepts namely the four set functions at work in possibility theory. Then we present the two main directions along which possibility theory has developed: the qualitative and quantitative settings. Both approaches share the same basic *maxitivity* axiom. They differ when it comes to conditioning, and to independence notions. We point out the connections with a coarse numerical integer-valued approach to belief representation, proposed by *Spohn* [3.9], now known as ranking theory [3.10].

In each setting, we discuss current and prospective lines of research. In the qualitative approach, we review the connections between possibility theory and modal logic, possibilistic logic and its applications to non-monotonic reasoning, logic programming and the like, possibilistic counterparts of Bayesian belief networks, the framework of soft constraints and the possibilistic approach to qualitative decision theory, and more recent investigations in formal concept analysis and learning. On the quantitative side, we review quantitative possibilistic networks, the connections between possibility theory, belief functions and imprecise probabilities, the connections with non-Bayesian statistics, and the application of quantitative possibility to risk analysis.

3.1 Historical Background

Zadeh was not the first scientist to speak about formalising notions of possibility. The modalities *possible* and *necessary* have been used in philosophy at least since the Middle Ages in Europe, based on Aristotle's and Theophrastus' works [3.11]. More recently these notions became the building blocks of modal logics that emerged at the beginning of the 20th century from the works of C.I. Lewis (see *Cresswell* [3.12]). In this approach, possibility and necessity are all-or-nothing notions, and handled at the syntactic level. More recently, and independently from Zadeh's view, the notion of possibility, as opposed to probability, was central in the works of one economist, and in those of two philosophers.

3.1.1 G.L.S. Shackle

A graded notion of possibility was introduced as a full-fledged approach to uncertainty and decision in 1940–1970 by the English economist *Shackle* [3.13], who called *degree of potential surprise* of an event its degree of impossibility, that is, retrospectively, the degree of necessity of the opposite event. Shackle's notion of possibility is basically epistemic, it is a *character of the chooser's particular state of knowledge in his present*. Impossibility is understood as disbelief. Potential surprise is valued on a disbelief scale, namely a positive interval of the form $[0, y^*]$, where y^* denotes the absolute rejection of the event to which it is assigned. In case everything is possible, all mutually exclusive hypotheses have zero surprise. At least one elementary hypothesis must carry zero potential surprise. The degree of surprise of an event, a set of elementary hypotheses, is the degree of surprise of its least surprising realization. Shackle also introduces a notion of conditional possibility, whereby the degree of surprise of a conjunction of two events A and B is equal to the maximum of the degree of surprise of A, and of the degree of surprise of B, should A prove true. The disbelief notion introduced later by *Spohn* [3.9, 10] employs the same type of convention as potential surprise, but uses the set of natural integers as a disbelief scale; his conditioning rule uses the subtraction of natural integers.

3.1.2 D. Lewis

In his 1973 book [3.14], the philosopher *David Lewis* considers a graded notion of possibility in the form of a relation between possible worlds he calls *comparative possibility*. He connects this concept of possibility to a notion of similarity between possible worlds. This asymmetric notion of similarity is also comparative, and is meant to express statements of the form: *a world j is at least as similar to world i as world k is*. Comparative similarity of j and k with respect to i is interpreted as the comparative possibility of j with respect to k viewed from world i. Such relations are assumed to be complete pre-orderings and are instrumental in defining the truth conditions of counterfactual statements (of the form *If I were rich, I would buy a big boat*). Comparative possibility relations $\geq_\Pi$ obey the key axiom: for all events A, B, C

$$A \geq_\Pi B \text{ implies } C \cup A \geq_\Pi C \cup B.$$

This axiom was later independently proposed by the first author [3.15] in an attempt to derive a possibilistic counterpart to comparative probabilities. Inde-

pendently, the connection between numerical possibility degrees and similarity was investigated by *Sudkamp* [3.16].

3.1.3 L.J. Cohen

A framework very similar to the one of Shackle was proposed by the philosopher *Cohen* [3.17] who considered the problem of legal reasoning. He introduced so-called *Baconian probabilities* understood as degrees of provability. The idea is that it is hard to prove someone guilty at the court of law by means of pure statistical arguments. The basic feature of degrees of provability is that a hypothesis and its negation cannot both be provable together to any extent (the contrary being a case for inconsistency). Such degrees of provability coincide with what is known as necessity measures.

3.1.4 L.A. Zadeh

In his seminal paper [3.1], *Zadeh* proposed an interpretation of membership functions of fuzzy sets as possibility distributions encoding flexible constraints induced by natural language statements. *Zadeh* tentatively articulated the relationship between possibility and probability, noticing that what is probable must preliminarily be possible. However, the view of possibility degrees developed in his paper refers to the idea of graded feasibility (degrees of ease, as in the example of *how many eggs can Hans eat for his breakfast*) rather than to the epistemic notion of plausibility laid bare by Shackle. Nevertheless, the key axiom of *maxitivity* for possibility measures is highlighted. In the two subsequent articles [3.18, 19], *Zadeh* acknowledged the connection between possibility theory, belief functions and upper/lower probabilities, and proposed their extensions to fuzzy events and fuzzy information granules.

3.2 Basic Notions of Possibility Theory

The basic building blocks of possibility theory originate in *Zadeh*'s paper [3.1] and were first extensively described in the authors' book [3.20], then further on in [3.3, 21]. More recent accounts are in [3.4, 5]. In this section, possibility theory is envisaged as a stand-alone theory of uncertainty.

3.2.1 Possibility Distributions

Let S be a set of states of affairs (or descriptions thereof), or states for short. This set can be the domain of an attribute (numerical or categorical), the Cartesian product of attribute domains, the set of interpretations of a propositional language etc.. A possibility distribution is a mapping π from S to a totally ordered scale L, with top denoted by 1 and bottom by 0. In the finite case $L = \{1 = \lambda_1 > \cdots \lambda_n > \lambda_{n+1} = 0\}$. The possibility scale can be the unit interval as suggested by Zadeh, or generally any finite chain, or even the set of nonnegative integers. It is often assumed that L is equipped with an order-reversing map denoted by $\lambda \in L \mapsto 1 - \lambda$.

The function π represents the state of knowledge of an agent (about the actual state of affairs), also called an *epistemic state* distinguishing what is plausible from what is less plausible, what is the normal course of things from what is not, what is surprising from what is expected. It represents a flexible restriction on what is the actual state with the following conventions (similar to probability, but opposite to Shackle's potential surprise scale (If $L = \mathbb{N}$, the conventions are opposite: 0 means possible and ∞ means impossible.)):

- $\pi(s) = 0$ means that state s is rejected as impossible;
- $\pi(s) = 1$ means that state s is totally possible (= plausible).

The larger $\pi(s)$, the more possible, i.e., plausible the state s is. Formally, the mapping π is the membership function of a fuzzy set [3.1], where membership grades are interpreted in terms of plausibility. If the universe S is exhaustive, at least one of the elements of S should be the actual world, so that $\exists s, \pi(s) = 1$ (normalization). This condition expresses the consistency of the epistemic state described by π.

Distinct values may simultaneously have a degree of possibility equal to 1. In the Boolean case, π is just the characteristic function of a subset $E \subseteq S$ of mutually exclusive states (a disjunctive set [3.22]), ruling out all those states considered as impossible. Possibility theory is thus a (fuzzy) set-based representation of incomplete information.

3.2.2 Specificity

A possibility distribution π is said to be at least as specific as another π' if and only if for each state of affairs s: $\pi(s) \leq \pi'(s)$ [3.23]. Then, π is at least as restrictive and informative as π', since it rules out at least as many states with at least as much strength. In the possibilistic framework, extreme forms of partial knowledge can be captured, namely:

- *Complete knowledge*: for some $s_0, \pi(s_0) = 1$ and $\pi(s) = 0, \forall s \neq s_0$ (only s_0 is possible)

- *Complete ignorance*: $\pi(s) = 1, \forall s \in S$ (all states are possible).

Possibility theory is driven by the *principle of minimal specificity*. It states that *any hypothesis not known to be impossible cannot be ruled out*. It is a minimal commitment, cautious information principle. Basically, we must always try to maximize possibility degrees, taking constraints into account.

Given a piece of information in the form x *is* F, where F is a fuzzy set restricting the values of the ill-known quantity x, it leads to represent the knowledge by the inequality $\pi \leq \mu_F$, the membership function of F. The minimal specificity principle enforces the possibility distribution $\pi = \mu_F$, if no other piece of knowledge is available. Generally there may be impossible values of x due to other piece(s) of information. Thus, given several pieces of knowledge of the form x *is* F_i, for $i = 1, \ldots, n$, each of them translates into the constraint $\pi \leq \mu_{F_i}$; hence, several constraints lead to the inequality $\pi \leq \min_{i=1}^{n} \mu_{F_i}$ and on behalf of the minimal specificity principle, to the possibility distribution

$$\pi = \min_{i=1}^{n} \pi_i \,,$$

where π_i is induced by the information item x *is* F_i. It justifies the use of the minimum operation for combining information items. It is noticeable that this way of combining pieces of information fully agrees with classical logic, since a classical logic base is equivalent to the logical conjunction of the logical formulas that belong to the base, and its models is obtained by intersecting the sets of models of its formulas. Indeed, in propositional logic, asserting a proposition ϕ amounts to declaring that any interpretation (state) that makes ϕ false is impossible, as being incompatible with the state of knowledge.

3.2.3 Possibility and Necessity Functions

Given a simple query of the form *does event A occur?* (is the corresponding proposition ϕ true?) where A is a subset of states, the response to the query can be obtained by computing degrees of possibility and necessity, respectively (if the possibility scale $L = [0, 1]$)

$$\Pi(A) = \sup_{s \in A} \pi(s); \quad N(A) = \inf_{s \notin A} 1 - \pi(s) \,.$$

$\Pi(A)$ evaluates to what extent A is consistent with π, while $N(A)$ evaluates to what extent A is certainly implied by π. The possibility–necessity duality is expressed by $N(A) = 1 - \Pi(A^c)$, where A^c is the complement of A. Generally, $\Pi(S) = N(S) = 1$ and $\Pi(\emptyset) = N(\emptyset) = 0$ (since π is normalized to 1). In the Boolean case, the possibility distribution comes down to the disjunctive (epistemic) set $E \subseteq S$ [3.3, 24]:

- $\Pi(A) = 1$ if $A \cap E \neq \emptyset$, and 0 otherwise: function Π checks whether proposition A is logically consistent with the available information or not.
- $N(A) = 1$ if $E \subseteq A$, and 0 otherwise: function N checks whether proposition A is logically entailed by the available information or not.

More generally, possibility and necessity measures represent degrees of plausibility and belief, respectively, in agreement with other uncertainty theories (see Sect. 3.4). Possibility measures satisfy the basic *maxitivity* property $\Pi(A \cup B) = \max(\Pi(A), \Pi(B))$. Necessity measures satisfy an axiom dual to that of possibility measures, namely $N(A \cap B) = \min(N(A), N(B))$. On infinite spaces, these axioms must hold for infinite families of sets. As a consequence, of the normalization of π, $\min(N(A), N(A^c)) = 0$ and $\max(\Pi(A), \Pi(A^c)) = 1$, where A^c is the complement of A, or equivalently $\Pi(A) = 1$ whenever $N(A) > 0$, which totally fits the intuition behind this formalism, namely that something somewhat certain should be fully possible, i. e., consistent with the available information.

3.2.4 Certainty Qualification

Human knowledge is often expressed in a declarative way using statements to which belief degrees are attached. Certainty-qualified pieces of uncertain information of the form A *is certain to degree* α can then be modeled by the constraint $N(A) \geq \alpha$. It represents a family of possible epistemic states π that obey this constraints. The least specific possibility distribution among them exists and is defined by [3.3]

$$\pi_{(A,\alpha)}(s) = \begin{cases} 1 & \text{if } s \in A, \\ 1 - \alpha & \text{otherwise}. \end{cases} \qquad (3.1)$$

If $\alpha = 1$, we get the characteristic function of A. If $\alpha = 0$, we get total ignorance. This possibility distribution is a key building block to construct possibility distributions from several pieces of uncertain knowledge. Indeed, e.g., in the finite case, any possibility distribution can be viewed as a collection of nested certainty-qualified statements. Let $E_i = \{s : \pi(s) \geq \lambda_i \in L\}$ be the

λ_i-cut of π. Then it is easy to check that $\pi(s) = \min_{i:s \notin E_i} 1 - N(E_i)$ (with the convention $\min_\emptyset = 1$).

We can also consider possibility-qualified statements of the form $\Pi(A) \geq \beta$; however, the least specific epistemic state compatible with this constraint expresses total ignorance.

3.2.5 Joint Possibility Distributions

Possibility distributions over Cartesian products of attribute domains $S_1 \times \cdots \times S_m$ are called joint possibility distributions $\pi(s_1, \ldots, s_n)$. The projection $\pi_k^\downarrow$ of the joint possibility distribution π onto S_k is defined as

$$\pi_k^\downarrow(s_k) = \Pi(S_1 \times \cdots S_{k-1} \times \{s_k\} \times \cdots S_{k+1} \times S_m)$$
$$= \sup_{s_i \in S_i, i \neq k} \pi(s_1, \ldots, s_n).$$

Clearly, $\pi(s_1, \ldots, s_n) \leq \min_{k=1}^m \pi_k^\downarrow(s_k)$ that is, a joint possibility distribution is at least as specific as the Cartesian product of its projections. When the equality holds, $\pi(s_1, \ldots, s_n)$ is called *separable*.

3.2.6 Conditioning

Notions of conditioning exist in possibility theory. Conditional possibility can be defined similarly to probability theory using a Bayesian-like equation of the form [3.3]

$$\Pi(B \cap A) = \Pi(B \mid A) \star \Pi(A).$$

where $\Pi(A) > 0$ and $\star$ is a *t*-norm (A nondecreasing Abelian semigroup operation on the unit interval having identity 1 and absorbing element 0 [3.25].); moreover $N(B \mid A) = 1 - \Pi(B^c \mid A)$. The above equation makes little sense for necessity measures, as it becomes trivial when $N(A) = 0$, that is under lack of certainty, while in the above definition, the equation becomes problematic only if $\Pi(A) = 0$, which is natural as then A is considered impossible. If operation $\star$ is the minimum, the equation $\Pi(B \cap A) = \min(\Pi(B \mid A), \Pi(A))$ fails to characterize $\Pi(B \mid A)$, and we must resort to the minimal specificity principle to come up with the qualitative conditioning rule [3.3]

$$\Pi(B \mid A) = \begin{cases} 1 & \text{if } \Pi(B \cap A) = \Pi(A) > 0, \\ \Pi(B \cap A) & \text{otherwise}. \end{cases}$$

$$(3.2)$$

It is clear that $N(B \mid A) > 0$ if and only if $\Pi(B \cap A) > \Pi(B^c \cap A)$. Moreover, if $\Pi(B \mid A) > \Pi(B)$ then $\Pi(B \mid A) = 1$, which points out the limited expressiveness of this qualitative notion (no gradual positive reinforcement of possibility). However, it is possible to have that $N(B) > 0, N(B^c \mid A_1) > 0, N(B \mid A_1 \cap A_2) > 0$ (i. e., oscillating beliefs). Extensive works on conditional possibility, especially qualitative, handling the case $\Pi(A) = 0$, have been recently carried out by *Coletti* and *Vantaggi* [3.26, 27] in the spirit of De Finetti's approach to subjective probabilities defined in terms of conditional measures and allowing for conditioning on impossible events.

In the numerical setting, due to the need of preserving for $\Pi(B \mid A)$ continuity properties of Π, we must choose $\star = $ product, so that

$$\Pi(B \mid A) = \frac{\Pi(B \cap A)}{\Pi(A)}$$

which makes possibilistic and probabilistic conditionings very similar [3.28] (now, gradual positive reinforcement of possibility is allowed). But there is yet another definition of numerical possibilistic conditioning, not based on the above equation as seen later in this chapter.

3.2.7 Independence

There are also several variants of possibilistic independence between events. Let us mention here the two basic approaches:

- *Unrelatedness*: $\Pi(A \cap B) = \min(\Pi(A), \Pi(B))$. When it does not hold, it indicates an epistemic form of mutual exclusion between A and B. It is symmetric but sensitive to negation. When it holds for all pairs made of A, B and their complements, it is an epistemic version of logical independence related to separability.
- *Causal independence*: $\Pi(B \mid A) = \Pi(B)$. This notion is different from the former one and stronger. It is a form of directed epistemic independence whereby learning A does not affect the plausibility of B. It is neither symmetric nor insensitive to negation: for instance, it is not equivalent to $N(B \mid A) = N(B)$.

Generally, independence in possibility theory is neither symmetric, nor insensitive to negation. For Boolean variables, independence between events is not equivalent to independence between variables. But since the possibility scale can be qualitative or quantitative, and there are several forms of conditioning, there are also various possible forms of independence. For studies of various notions and their properties see [3.29–32]. More discussions and references appear in [3.4].

3.2.8 Fuzzy Interval Analysis

An important example of a possibility distribution is a fuzzy interval [3.3, 20]. A fuzzy interval is a fuzzy set of reals whose membership function is unimodal and upper-semi continuous. Its α-cuts are closed intervals. The calculus of fuzzy intervals is an extension of interval arithmetics based on a possibilistic counterpart of a computation of random variable. To compute the addition of two fuzzy intervals A and B one has to compute the membership function of $A \oplus B$ as the degree of possibility $\mu_{A \oplus B}(z) = \Pi(\{(x, y) : x + y = z\})$, based on the possibility distribution $\min(\mu_A(x), \mu_B(y))$. There is a large literature on possibilistic interval analysis; see [3.33] for a survey of 20th-century references.

3.2.9 Guaranteed Possibility

Possibility distributions originally represent negative information in the sense that their role is essentially to rule out impossible states. More recently, [3.34, 35] another type of possibility distribution has been considered where the information has a positive nature, namely it points out actually possible states, such as observed cases, examples of solutions, etc. Positively-flavored possibility distributions will be denoted by δ and serve as evidential support functions. The conventions for interpreting them contrast with usual possibility distributions:

- $\delta(s) = 1$ means that state s is actually possible because of a high evidential support (for instance, s is a case that has been actually observed);
- $\delta(s) = 0$ means that state s has not been observed (yet: potential impossibility).

Note that $\pi(s) = 1$ indicates *potential* possibility, while $\delta(s) = 1$ conveys more information. In contrast, $\delta(s) = 0$ expresses ignorance.

A measure of *guaranteed possibility* can be defined, that differs from functions Π and N [3.34, 35]

$$\Delta(A) = \inf_{s \in A} \delta(s) \,.$$

It estimates to what extent *all* states in A are actually possible according to evidence. $\Delta(A)$ can be used as a degree of evidential support for A. Of course, this function possesses a conjugate ∇ such that $\nabla(A) = 1 - \Delta(A^c) = \sup_{s \notin A} 1 - \delta(s)$. Function $\nabla(A)$ evaluates the degree of potential necessity of A, as it is 1 only if some state s outside A is potentially impossible.

Uncertain statements of the form A *is possible to degree β* often mean that any realization of A is possible to degree β (e.g., *it is possible that the museum is open this afternoon*). They can then be modeled by a constraint of the form $\Delta(A) \geq \beta$. It corresponds to the idea of observed evidence.

This type of information is better exploited by assuming an informational principle opposite to the one of minimal specificity, namely, *any situation not yet observed is tentatively considered as impossible*. This is similar to the closed-world assumption. The most specific distribution $\delta_{(A,\beta)}$ in agreement with $\Delta(A) \geq \beta$ is

$$\delta_{(A,\beta)}(s) = \begin{cases} \beta & \text{if } s \in A , \\ 0 & \text{otherwise} . \end{cases}$$

Note that while possibility distributions induced from certainty qualified pieces of knowledge combine conjunctively, by discarding possible states, evidential support distributions induced by possibility-qualified pieces of evidence combine disjunctively, by accumulating possible states. Given several pieces of knowledge of the form *x is F_i is possible*, for $i = 1, \ldots, n$, each of them translates into the constraint $\delta \geq \mu_{F_i}$; hence, several constraints lead to the inequality $\delta \geq \max_{i=1}^n \mu_{F_i}$ and on behalf of another minimal commitment principle based on maximal specificity, we get the possibility distribution

$$\delta = \max_{i=1}^n \pi_i ,$$

where δ_i is induced by the information item *x is F_i is possible*. It justifies the use of the maximum operation for combining evidential support functions. Acquiring pieces of possibility-qualified evidence leads to updating $\delta_{(A,\beta)}$ into some wider distribution $\delta > \delta_{(A,\beta)}$. Any possibility distribution can be represented as a collection of nested possibility-qualified statements of the form $(E_i, \Delta(E_i))$, with $E_i = \{s : \delta(s) \geq \lambda_i\}$, since $\delta(s) = \max_{i:s \in E_i} \Delta(E_i)$, dually to the case of certainty-qualified statements.

3.2.10 Bipolar Possibility Theory

A bipolar representation of information using pairs (δ, π) may provide a natural interpretation of interval-valued fuzzy sets [3.8]. Although positive and negative information are represented in separate and different ways via δ and π functions, respectively, there is a coherence condition that should hold between positive and negative information. Indeed, observed information should not be impossible. Likewise, in terms of preferences, solutions that are preferred to some extent should not be unfeasible. This leads to enforce the coherence constraint $\delta \leq \pi$ between the two representations.

This condition should be maintained when new information arrives and is combined with the previous one. This does not go for free since degrees $\delta(s)$ tend to increase while degrees $\pi(s)$ tend to decrease due to the disjunctive and conjunctive processes that, respectively, govern their combination. Maintaining this coherence requires a revision process that works as follows. If the current information state is represented by the pair (δ, π), receiving a new positive (resp. negative) piece of information represented by δ^{new} (resp. π^{new}) to be enforced, leads to revising (δ, π) into $(\max(\delta, \delta^{\text{new}}), \pi^{\text{rev}})$ (resp. into $(\delta^{\text{rev}}, \min(\pi, \pi^{\text{new}}))$), using, respectively,

$$\pi^{\text{rev}} = \max(\pi, \delta^{\text{new}}) ; \tag{3.3}$$
$$\delta^{\text{rev}} = \min(\pi^{\text{new}}, \delta) . \tag{3.4}$$

It is important to note that when both positive and negative pieces of information are collected, there are two options:

- Either priority is given to positive information over negative information: it means that (past) positive information cannot be ruled out by (future) negative information. This may be found natural when very reliable observations (represented by δ) contradict tentative knowledge (represented by π). Then revising (δ, π) by $(\delta^{\text{new}}, \pi^{\text{new}})$ yields the new pair

$$(\delta^{\text{rev}}, \pi^{\text{rev}}) = (\max(\delta, \delta^{\text{new}}),$$
$$\max(\min(\pi, \pi^{\text{new}}), \max(\delta, \delta^{\text{new}})))$$

- Priority is given to negative information over positive information. It makes sense when handling preferences. Indeed, then, positive information may be viewed as wishes, while negative information reflects constraints. Then, revising (δ, π) by $(\delta^{\text{new}}, \pi^{\text{new}})$ would yield the new pair

$$(\delta^{\text{rev}}, \pi^{\text{rev}}) = (\min(\min(\pi, \pi^{\text{new}}), \max(\delta, \delta^{\text{new}})),$$
$$\min(\pi, \pi^{\text{new}})) .$$

It can be checked that the two latter revision rules generalize the two previous ones. With both revision options, it can be checked that if $\delta \leq \pi$ and $\delta^{\text{new}} \leq \pi^{\text{new}}$

hold, revising (δ, π) by $(\delta^{\text{new}}, \pi^{\text{new}})$ yields a new coherent pair. This revision process should not be confused with another one pertaining only to the negative part of the information, namely computing $\min(\pi, \pi^{\text{new}})$ may yield a possibility distribution that is not normalized, in the case of inconsistency. If such an inconsistency takes place, it should be resolved (by some appropriate renormalization) before one of the two above bipolar revision mechanisms can be applied.

3.3 Qualitative Possibility Theory

This section is restricted to the case of a finite state space S, typically S is the set of interpretations of a formal propositional language $\mathcal{L}$ based on a finite set of Boolean attributes $\mathcal{V}$. The usual connectives $\wedge$ (conjunction), $\vee$ (disjunction), and $\neg$ (negation) are used. The possibility scale is then taken as a finite chain, or the unit interval understood as an ordinal scale, or even just a complete preordering of states. At the other end, one may use the set of natural integers (viewed as an impossibility scale) equipped with addition, which comes down to a countable subset of the unit interval, equipped with the product t-norm, instrumental for conditioning. However, the qualitative nature of the latter setting is questionable, even if authors using it do not consider it as genuinely quantitative.

3.3.1 Possibility Theory and Modal Logic

In this section, the possibility scale is Boolean ($L = \{0, 1\}$) and a possibility distribution reduces to a subset of states E, for instance the models of a set of formulas K representing the beliefs of an agent in propositional logic. The presence of a proposition p in K can be modeled by $N([p]) = 1$, or $\Pi([\neg p]) = 0$ where $[p]$ is the set of interpretations of p; more generally the degrees of possibility and necessity can be defined by [3.36]:

- $N([p]) = \Pi([p]) = 1$ if and only if $K \models p$ (the agent believes p)
- $N([\neg p]) = \Pi([\neg p]) = 0$ if and only if $K \models \neg p$ (the agent believes $\neg p$)
- $N([p]) = 0$ and $\Pi([p]) = 1$ if and only if $K \not\models p$ and $K \not\models \neg p$ (the agent is unsure about p)

However, in propositional logic, it cannot be syntactically expressed that $N([p]) = 0$ nor $\Pi([p]) = 1$. To

do so, a modal language is needed [3.12], that prefixes propositions with modalities such as necessary ($\square$) and possible ($\lozenge$). Then $\square p$ encodes $N([p]) = 1$ (instead of $p \in K$ in classical logic), $\lozenge p$ encodes $\Pi([p]) = 1$. Only a very simple modal language $\mathcal{L}^{\square}$ is needed that encapsulates the propositional language $\mathcal{L}$. Atoms of this logic are of the form $\square p$, where p is any propositional formula. Well-formed formulas in this logic are obtained by applying standard conjunction and negation to these atoms

$$\mathcal{L}^{\square} = \square p, p \in \mathcal{L} \mid \neg \phi \mid \phi \wedge \psi .$$

The well-known conjugateness between possibility and necessity reads: $\lozenge p = \neg \square \neg p$. Maxitivity and minitivity axioms of possibility and necessity measure, respectively, read $\lozenge(p \vee q) = \lozenge p \vee \lozenge q$ and $\square(p \wedge q) = \square p \wedge \square q$ and are well known to hold in regular modal logics, and the consistency of the epistemic state is ensured by axiom $D: \square p \rightarrow \lozenge p$. This is the minimal epistemic logic (MEL) [3.37] needed to account for possibility theory. It corresponds to a small fragment of the logic KD without modality nesting and without objective formulas ($\mathcal{L}^{\square} \cap \mathcal{L} = \emptyset$). Models of such modal formulas are epistemic states: for instance, E is a model of $\square p$ means that $E \subseteq [p]$ [3.37, 38]. This logic is sound and complete with respect to this semantics, and enables propositions whose truth status is explicitly unknown to be reasoned about.

3.3.2 Comparative Possibility

A plausibility ordering is a complete preorder of states denoted by $\geq_\pi$, which induces a well-ordered partition $\{E_1, \ldots, E_n\}$ of S. It is the comparative counterpart of a possibility distribution π, i.e., $s \geq_\pi s'$ if and only if $\pi(s) \geq \pi(s')$. Indeed it is more natural to expect that an agent will supply ordinal rather than numerical information about his beliefs. By convention, E_1 contains the most normal states of fact, E_n the least plausible, or most surprising ones. Denoting by $\max(A)$ any most plausible state $s_0 \in A$, ordinal counterparts of possibility and necessity measures [3.15] are then defined as follows: $\{s\} \geq_\Pi \emptyset$ for all $s \in S$ and

$A \geq_\Pi B$ if and only if $\max(A) \geq_\pi \max(B)$

$A \geq_N B$ if and only if $\max(B^c) \geq_\pi \max(A^c)$.

Possibility relations $\geq_\Pi$ were proposed by *Lewis* [3.14] and they satisfy his characteristic property

$A \geq_\Pi B$ implies $C \cup A \geq_\Pi C \cup B$,

while necessity relations can also be defined as $A \geq_N B$ if and only if $B^c \geq_\Pi A^c$, and they satisfy a similar axiom

$A \geq_N B$ implies $C \cap A \geq_N C \cap B$.

The latter coincides with epistemic entrenchment relations in the sense of belief revision theory [3.39] (provided that $A >_\Pi \emptyset$, if $A \neq \emptyset$). Conditioning a possibility relation $\geq_\Pi$ by a nonimpossible event $C >_\Pi \emptyset$ means deriving a relation $\geq_\Pi^C$ such that

$A \geq_\Pi^C B$ if and only if $A \cap C \geq_\Pi B \cap C$.

These results show that possibility theory is implicitly at work in the principal axiomatic approach to belief revision [3.40], and that conditional possibility obeys its main postulates [3.41]. The notion of independence for comparative possibility theory was studied by *Dubois* et al. [3.31], for independence between events, and *Ben Amor* et al. [3.32] between variables.

3.3.3 Possibility Theory and Nonmonotonic Inference

Suppose S is equipped with a plausibility ordering. The main idea behind qualitative possibility theory is that the state of the world is always believed to be as normal as possible, neglecting less normal states. $A \geq_\Pi B$ really means that there is a normal state where A holds that is at least as normal as any normal state where B holds. The dual case $A \geq_N B$ is intuitively understood as A *is at least as certain as* B, in the sense that there are states where B fails to hold that are at least as normal as the most normal state where A does not hold. In

particular, the events accepted as true are those which are true in all the most plausible states, namely the ones such that $A >_N \emptyset$. These assumptions lead us to interpret the plausible inference $A \mid\approx B$ of a proposition B from another A, under a state of knowledge $\geq_\Pi$ as follows: *B should be true in all the most normal states were A is true*, which means $B >_\Pi^A B^c$ in terms of ordinal conditioning, that is, $A \cap B$ is more plausible than $A \cap B^c$. $A \mid\approx B$ also means that the agent considers B as an accepted belief in the context A.

This kind of inference is nonmonotonic in the sense that $A \mid\approx B$ does not always imply $A \cap C \mid\approx B$ for any additional information C. This is similar to the fact that a conditional probability $P(B \mid A \cap C)$ may be low even if $P(B \mid A)$ is high. The properties of the consequence relation $\mid\approx$ are now well understood, and are precisely the ones laid bare by *Lehmann* and *Magidor* [3.42] for their so-called *rational inference*. Monotonicity is only partially restored: $A \mid\approx B$ implies $A \cap C \mid\approx B$ provided that $A \mid\approx C^c$ does not hold (i. e., that states were A is true do not typically violate C). This property is called *rational monotony*, and, along with some more standard ones (like closure under conjunction), characterizes default possibilistic inference $\mid\approx$. In fact, the set $\{B, A \mid\approx B\}$ of accepted beliefs in the context A is deductively closed, which corresponds to the idea that the agent reasons with accepted beliefs in each context as if they were true, until some event occurs that modifies this context. This closure property is enough to justify a possibilistic approach [3.43] and adding the rational monotonicity property ensures the existence of a single possibility relation generating the consequence relation $\mid\approx$ [3.44].

Plausibility orderings can be generated by a set of *if-then* rules tainted with unspecified exceptions. This set forms a knowledge base supplied by an agent. Each rule *if A then B* is modeled by a constraint of the form $A \cap B >_\Pi A \cap B^c$ on possibility relations. There exists a single minimally specific element in the set of possibility relations satisfying all constraints induced by rules (unless the latter are inconsistent). It corresponds to the most compact plausibility ranking of states induced by the rules [3.44]. This ranking can be computed by an algorithm originally proposed by *Pearl* [3.45].

Qualitative possibility theory has been studied from the point of view of cognitive psychology. Experimental results [3.46] suggest that there are situations where people reason about uncertainty using the rules or possibility theory, rather than with those of probability theory, namely people jump to plausible conclusions based on assuming the current world is normal.

3.3.4 Possibilistic Logic

Qualitative possibility relations can be represented by (and only by) possibility measures ranging on any totally ordered set L (especially a finite one) [3.15]. This absolute representation on an ordinal scale is slightly more expressive than the purely relational one. For instance, one can express that a proposition is fully plausible ($\Pi(A) = 1$), while using a possibility relation, one can only say that it is among the most plausible ones. When the finite set S is large and generated by a propositional language, qualitative possibility distributions can be efficiently encoded in possibilistic logic [3.47–49].

A possibilistic logic base K is a set of pairs (p_i, α_i), where p_i is an expression in classical (propositional or first-order) logic and $\alpha_i > 0$ is a element of the value scale L. This pair encodes the constraint $N(p_i) \geq \alpha_i$ where $N(p_i)$ is the degree of necessity of the set of models of p_i. Each prioritized formula (p_i, α_i) has a fuzzy set of models (via certainty qualification described in Sect. 3.2) and the fuzzy intersection of the fuzzy sets of models of all prioritized formulas in K yields the associated plausibility ordering on S encoded by a possibility distribution π_K. Namely, an interpretation s is all the less possible as it falsifies formulas with higher weights, i.e.,

$$\pi_K(s) = 1 \text{ if } s \models p_i, \forall (p_i, \alpha_i) \in K, \tag{3.5}$$

$$\pi_K(s) = 1 - \max\{\alpha_i : (p_i, \alpha_i) \in K, s \not\models p_i\}$$
$$\text{otherwise}. \tag{3.6}$$

This distribution is obtained by applying the minimal specificity principle, since it is the largest one that satisfies the constraints $N(p_i) \geq \alpha_i$. If the classical logic base $\{p_i : (p_i, \alpha_i) \in K\}$ is inconsistent, π_K is not normalized, and a level of inconsistency equal to $\mathrm{inc}(K) = 1 - \max \pi_K$ can be attached to the base K. However, the set of formulas $\{p_i : (p_i, \alpha_i) \in K, \alpha_i > \mathrm{inc}(K)\}$ is always consistent.

Syntactic deduction from a set of prioritized clauses is achieved by refutation using an extension of the standard resolution rule, whereby $(p \vee q, \min(\alpha, \beta))$ can be derived from $(p \vee r, \alpha)$ and $(q \vee \neg r, \beta)$. This rule, which evaluates the validity of an inferred proposition by the validity of the weakest premiss, goes back to Theophrastus, a disciple of Aristotle. Another way of presenting inference in possibilistic logic relies on the fact that $K \vdash (p, \alpha)$ if and only if $K_\alpha = \{p_i : (p_i, \alpha_i) \in K, \alpha_i \geq \alpha\} \vdash p$ in the sense of classical logic. In particular, $\mathrm{inc}(K) = \max\{\alpha : K_\alpha \vdash \bot\}$. Inference in possi-

bilistic logic can use this extended resolution rule and proceeds by refutation since $K \vdash (p, \alpha)$ if and only if $\mathrm{inc}(\{(\neg p, 1)\} \cup K) \geq \alpha$. Computational inference methods in possibilistic logic are surveyed in [3.50].

Possibilistic logic is an inconsistency-tolerant extension of propositional logic that provides a natural semantic setting for mechanizing nonmonotonic reasoning [3.51], with a computational complexity close to that of propositional logic. Namely, once a possibility distribution on models is generated by a set of *if-then* rules $p_i \rightarrow q_i$ (as explained in Sect. 3.3.3 and modeled here using qualitative conditioning as $N(q_i \mid p_i) > 0$), weights $\alpha_i = N(\neg p_i \vee q_i)$ can be computed, and the corresponding possibilistic base built [3.51]. See [3.52] for an efficient method involving compilation.

Variants of possibilistic logic have been proposed in later works. A partially ordered extension of possibilistic logic has been proposed, whose semantic counterpart consists of partially ordered models [3.53]. Another approach for handling partial orderings between weights is to encode formulas with partially constrained weights in a possibilistic-like many-sorted propositional logic [3.54]. Namely, a formula (p, α) is rewritten as a classical two-sorted clause $p \vee ab_\alpha$, where ab_α means *the situation is α-abnormal*, and thus the clause expresses that p *is true or the situation is abnormal*, while more generally $(p, \min(\alpha, \beta))$ is rewritten as the clause $p \vee ab_\alpha \vee ab_\beta$. Then a known constraint between unknown weights such as $\alpha \geq \beta$ is translated into a clause $\neg ab_\alpha \vee ab_\beta$. In this way, a possibilistic logic base, where only partial information about the relative ordering between the weights is available under the form of constraints, can be handled as a set of classical logic formulas that involve symbolic weights.

An efficient inference process has been proposed using the notion of forgetting variables. This approach provides a technique for compiling a standard possibilistic knowledge bases in order to process inference in polynomial time [3.55]. Let us also mention quasi-possibilistic logic [3.56], an extension of possibilistic logic based on the so-called *quasi-classical logic*, a paraconsistent logic whose inference mechanism is close to classical inference (except that it is not allowed to infer $p \vee q$ from p). This approach copes with inconsistency between formulas having the same weight. Other types of possibilistic logic can also handle constraints of the form $\Pi(\phi) \geq \alpha$, or $\Delta(\phi) \geq \alpha$ [3.49].

There is a major difference between possibilistic logic and weighted many-valued logics [3.57]. Namely, in the latter, a weight $\tau \in L$ attached to a (many valued, thus nonclassical) formula p acts as a truth-value threshold, and (p, τ) in a fuzzy knowledge base expresses the Boolean requirement that the truth value of p should be at least equal to τ for (p, τ) to be valid. So in such fuzzy logics, while truth of p is many-valued, the validity of a weighted formula is two-valued. On the contrary, in possibilistic logic, truth is two-valued (since p is Boolean), but the validity of a possibilistic formula (p, α) is many-valued. In particular, it is possible to cast possibilistic logic inside a many-valued logic. The idea is to consider many-valued atomic sentences ϕ of the form (p, α), where p is a formula in classical logic. Then, one can define well-formed formulas such as $\phi \lor \psi$, $\phi \land \psi$, or yet $\phi \rightarrow \psi$, where the external connectives linking ϕ and ψ are those of the chosen many-valued logic. From this point of view, possibilistic logic can be viewed as a fragment of a many-valued logic that uses only one external connective: conjunction interpreted as minimum. This approach involving a Boolean algebra embedded in a nonclassical one has been proposed by *Boldrin* and *Sossai* [3.58] with a view to augment possibilistic logic with fusion modes cast at the object level. It is also possible to replace classical logic by a many-valued logic inside possibilistic logic. For instance, possibilistic logic has been extended to Gödel many-valued logic [3.59]. A similar technique has been used by *Hájek* et al. to extend possibilistic logic to a many-valued modal setting [3.60].

Lehmke [3.61] has cast fuzzy logics and possibilistic logic inside the same framework, considering weighted many-valued formulas of the form (p, θ), where p is a many-valued formula with truth set T, and θ is a *label* defined as a monotone mapping from the truth-set T to a *validity* set L (a set of possibility degrees). T and L are supposed to be complete lattices, and the set of labels has properties that make it a fuzzy extension of a filter. Labels encompass *fuzzy truth-values* in the sense of *Zadeh* [3.62], such as *very true*, *more or less true* that express uncertainty about (many-valued) truth in a graded way.

Rather than expressing statements such as *it is half-true that John is tall*, which presupposes a state of complete knowledge about John's height, one may be interested in handling states of incomplete knowledge, namely assertions of the form *all we know is that John is tall*. One way to do it is to introduce fuzzy constants in a possibilistic first-ordered logic. *Dubois*, *Prade*, and *Sandri* [3.63] have noticed that an imprecise restriction on the scope of an existential quantifier can be handled in the following way. From the two premises $\forall x \in A, \neg p(x, y) \lor q(x, y)$, and $\exists x \in B, p(x, a)$, where a is a constant, we can conclude that $\exists x \in B, q(x, a)$ pro-

vided that $B \subseteq A$. Thus, letting $p(B, a)$ stand for $\exists x \in B, p(x, a)$, one can write

$$\forall x \in A, \neg p(x, y) \lor q(x, y), p(B, a) \vdash q(B, a)$$

if $B \subseteq A$, B being an imprecise constant. Letting A and B be fuzzy sets, the following pattern can be validated in possibilistic logic

$$(\neg p(x, y) \lor q(x, y), \min(\mu_A(x), \alpha)), (p(B, a), \beta) \\ \vdash (q(B, a), \min(N_B(A), \alpha, \beta)),$$

where $N_B(A) = \inf_t \max(\mu_A(t), 1 - \mu_B(t))$ is the necessity measure of the fuzzy event A based on fuzzy information B. Note that A, which appears in the weight slot of the first possibilistic formula plays the role of a fuzzy predicate, since the formula expresses that *the more x is A, the more certain (up to level α) if p is true for (x, y), q is true for them as well.*

Alsinet and *Godo* [3.64, 65] have applied possibilistic logic to logic programming that allows for fuzzy constants [3.65, 66]. They have developed programming environments based on possibility theory. In particular, the above inference pattern can be strengthened, replacing B by its cut B_β in the expression of $N_B(A)$ and extended to a sound resolution rule. They have further developed possibilistic logic programming with similarity reasoning [3.67] and more recently argumentation [3.68, 69].

Lastly, in order to improve the knowledge representation power of the answer-set programming (ASP) paradigm, the stable model semantics has been extended by taking into account a certainty level, expressed in terms of necessity measure, on each rule of a normal logic program. It leads to the definition of a possibilistic stable model for weighted answer-set programming [3.70]. *Bauters* et al. [3.71] introduce a characterization of answer sets of classical and possibilistic ASP programs in terms of possibilistic logic where an ASP program specifies a set of constraints on possibility distributions.

3.3.5 Ranking Function Theory

A theory that parallels possibility theory to a large extent and that has been designed for handling issues in belief revision, nonmonotonic reasoning and causation, just like qualitative possibility theory is the one of ranking functions by *Spohn* [3.9, 10, 72]. The main difference is that it is not really a qualitative theory as it uses the set of integers including ∞ (denoted by $\mathbb{N}^+$)

as a value scale. Hence, it is more expressive than qualitative possibility theory, but it is applied to the same problems.

Formally [3.10], a ranking function is a mapping $\kappa : 2^S \to \mathbb{N}^+$ such that:

- $\kappa(\{s\}) = 0$ for some $s \in S$;
- $\kappa(A) = \min_{s \in A} \kappa(\{s\})$;
- $\kappa(\emptyset) = \infty$.

It is immediate to verify that the set function $\Pi(A) = 2^{-\kappa(A)}$ is a possibility measure. So a ranking function is an integer-valued measure of impossibility (disbelief). The function $\beta(A) = \kappa(A^c)$ is an integer-valued necessity measure used by Spohn for measuring belief, and it is clear that the rescaled necessity measure is $N(A) = 1 - 2^{-\beta(A)}$. Interestingly, ranking functions also bear close connection to probability theory [3.72], viewing $\kappa(A)$ as the exponent of an infinitesimal probability, of the form $P(A) = \epsilon^{\kappa(A)}$. Indeed the order of magnitude of $P(A \cup B)$ is then $\epsilon^{\min(\kappa(A),\kappa(B))}$. Integers also come up naturally if we consider Hamming distances between models in the Boolean logic context, if for instance, the degree of possibility of an interpretation is a function of its Hamming distance to the closest model of a classical knowledge base.

Spohn [3.9] also introduces conditioning concepts, especially:

- The so-called A-part of κ, which is a conditioning operation by event A defined by $\kappa(B \mid A) = \kappa(B \cap A) - \kappa(B)$;
- The (A, n)-conditionalization of κ, $\kappa(\cdot \mid (A \to n))$ which is a revision operation by an uncertain input enforcing $\kappa'(A^c) = n$, and defined by

$$\kappa(s \mid (A \to n)) = \begin{cases} \kappa(s \mid A) & \text{if } s \in A \\ n + \kappa(s \mid A^c) & \text{otherwise} . \end{cases}$$
(3.7)

This operation makes A more believed than A^c by n steps, namely,

$$\beta(A \mid (A \to n)) = 0 ; \quad \beta(A^c \mid (A \to n = n)) .$$

It is easy to see that the conditioning of ranking functions comes down to the product-based conditioning of numerical possibility measures, and to the infinitesimal counterpart of usual Bayesian conditioning of probabilities. The other conditioning rule can be obtained by means of Jeffrey's rule of conditioning [3.73] $P(B \mid (A, \alpha)) = \alpha P(B \mid A) + (1 - \alpha) P(B \mid A^c)$

by a constraint of the form $P(A) = \alpha$. Both qualitative and quantitative counterparts of this revision rule in possibility theory have been studied in detail [3.74, 75]. In fact, ranking function theory is formally encompassed by numerical possibility theory. Moreover, there is no fusion rule in Spohn theory, while fusion is one of the main applications of possibility theory (see Sect. 3.5).

3.3.6 Possibilistic Belief Networks

Another compact representation of qualitative possibility distributions is the possibilistic directed graph, which uses the same conventions as Bayesian nets, but relies on conditional possibility [3.76]. The qualitative approach is based on a symmetric notion of qualitative independence $\Pi(B \cap A) = \min(\Pi(A), \Pi(B))$ that is weaker than the causal-like condition $\Pi(B \mid A) = \Pi(B)$ [3.31]. Like joint probability distributions, joint possibility distributions can be decomposed into a conjunction of conditional possibility distributions (using minimum or product) in a way similar to Bayes nets [3.76]. A joint possibility distribution associated with variables $X_1, \ldots, X_n$, decomposed by the chain rule

$$\pi(X_1, , X_n) = \min(\pi(X_n \mid X_1, \ldots, X_{n-1}),$$
$$\ldots, \pi(X_2 \mid X_1), \pi(X_1)) .$$

Such a decomposition can be simplified by assuming conditional independence relations between variables, as reflected by the structure of the graph. The form of independence between variables at work here is conditional noninteractivity: Two variables X and Y are independent in the context Z, if for each instance (x, y, z) of (X, Y, Z) we have: $\pi(x, y \mid z) = \min(\pi(x \mid z), \pi(y \mid z))$.

Ben Amor and *Benferhat* [3.77] investigate the properties of qualitative independence that enable local inferences to be performed in possibilistic nets. Uncertainty propagation algorithms suitable for possibilistic graphical structures have been studied in [3.78]. It is also possible to propagate uncertainty in nondirected decompositions of joint possibility measures as done quite early by *Borgelt* et al. [3.79]. Counterparts of product-based numerical possibilistic nets using ranking functions exist as well [3.10]. Qualitative possibilistic counterparts of decision trees and influence diagrams for decision trees have been recently investigated [3.80, 81]. Compilation techniques for inference in possibilistic networks have been devised [3.82]. Finally, the study of possibilistic networks from the standpoint of causal reasoning has been investigated, using the concept of intervention, that comes down to enforcing the values of some variables so as to lay bare their influence on other ones [3.83, 84].

3.3.7 Fuzzy Rule–Based and Case–Based Approximate Reasoning

A typology of fuzzy rules has been devised in the setting of possibility theory, distinguishing rules whose purpose is to propagate uncertainty through reasoning steps, from rules whose main purpose is similarity-based interpolation [3.85], depending on the choice of a many-valued implication connective that models a rule. The bipolar view of information based on (δ, π) pairs sheds new light on the debate between conjunctive and implicative representation of rules [3.86]. Representing a rule as a material implication focuses on counterexamples to rules, while using a conjunction between antecedent and consequent points out examples of the rule and highlights its positive content. Traditionally in fuzzy control and modeling, the latter representation is adopted, while the former is the logical tradition. Introducing fuzzy implicative rules in modeling accounts for constraints or landmark points the model should comply with (as opposed to observed data) [3.87]. The bipolar view of rules in terms of examples and counterexamples may turn out to be very useful when extracting fuzzy rules from data [3.88].

Fuzzy rules have been applied to case-based reasoning (CBR). In general, CBR relies on the following implicit principle: *similar situations may lead to similar outcomes*. Thus, a similarity relation S between problem descriptions or situations, and a similarity measure T between outcomes are needed. This implicit CBR principle can be expressed in the framework of fuzzy rules as: "the more similar (in the sense of S) are the attribute values describing two situations, the more possible the similarity (in the sense of T) of the values of the corresponding outcome attributes." Given a situation s_0 associated to an unknown outcome t_0 and a current case (s, t), this principle enables us to conclude on the possibility of t_0 being equal to a value similar to t [3.89]. This acknowledges the fact that, often in practice, a database may contain cases that are rather similar with respect to the problem description attributes, but which may be distinct with respect to outcome attribute(s). This emphasizes that case-based reasoning can only lead to cautious conclusions.

The possibility rule *the more similar s and s_0, the more possible t and t_0 are similar*, is modeled in terms of a guaranteed possibility measure [3.90]. This leads

to enforce the inequality $\Delta_0(T(t, \cdot)) \geq \mu_S(s, s_0)$, which expresses that the guaranteed possibility that t_0 belongs to a high degree to the fuzzy set of values that are T-similar to t, is lower bounded by the S-similarity of s and s_0. Then the fuzzy set F of possible values t' for t_0 with respect to case (s, t) is given by

$$F_{t_0}(t') = \min(\mu_T(t, t'), \mu_S(s, s_0)),$$

since the maximally specific distribution such that $\Delta(A) \geq \alpha$ is $\delta = \min(\mu_A, \alpha)$. What is obtained is the fuzzy set $T(t, .)$ of values t' that are T-similar to t, whose possibility level is truncated at the global degree $\mu_S(s, s_0)$ of similarity of s and s_0. The max-based aggregation of the various contributions obtained from the comparison with each case (s, t) in the memory M of cases acknowledges the fact that each new comparison may suggest new possible values for t_0 and agrees with the positive nature of the information in the repository of cases. Thus, we obtain the following fuzzy set Es_0 of the possible values t' for t_0

$$Es_0(t') = \max_{(s,t) \in M} \min(S(s, s_0), T(t, t')).$$

This latter expression can be put in parallel with the evaluation of a flexible query [3.91]. This approach has been generalized to imprecisely or fuzzily described situations, and has been related to other approaches to instance-based prediction [3.92, 93].

3.3.8 Preference Representation

Possibility theory also offers a framework for preference modeling in constraint-directed reasoning. Both prioritized and soft constraints can be captured by possibility distributions expressing degrees of feasibility rather than plausibility [3.6]. Possibility theory offers a natural setting for fuzzy optimization whose aim is to balance the levels of satisfaction of multiple fuzzy constraints (instead of minimizing an overall cost) [3.94]. In such problems, some possibility distributions represent soft constraints on decision variables, other ones can represent incomplete knowledge about uncontrollable state variables. Qualitative decision criteria are particularly adapted to the handling of uncertainty in this setting. Possibility distributions can also model ill-known constraint coefficients in linear and nonlinear programming, thus leading to variants of chance-constrained programming [3.95].

Optimal solutions of fuzzy constraint-based problems maximize the satisfaction of the most violated constraint, which does not ensure the Pareto dominance of all such solutions. More demanding optimality notions have been defined, by canceling equally satisfied constraints (the so-called *discrimin* ordering) or using a leximin criterion [3.94, 96, 97].

Besides, the possibilistic logic setting provides a compact representation framework for preferences, where possibilistic logic formulas represent prioritized constraints on Boolean domains. This approach has been compared to qualitative conditional preference networks (CP nets), based on a systematic ceteris paribus assumption (preferential independence between decision variables). CP nets induce partial orders of solutions rather than complete preorders, as possibilistic logic does [3.98]. Possibilistic networks can also model preference on the values of variables, conditional to the value of other ones, and offer an alternative to conditional preference networks [3.98].

Bipolar possibility theory has been applied to preference problems where it can be distinguished between imperative constraints (modeled by propositions with a degree of necessity), and nonimperative wishes (modeled by propositions with a degree of guaranteed possibility level) [3.99]. Another kind of bipolar approach to qualitative multifactorial evaluation based on possibility theory, is when comparing objects in terms of their pros and cons where the decision maker focuses on the most important assets or defects. Such qualitative multifactorial bipolar decision criteria have been defined, axiomatized [3.100], and empirically tested [3.101]. They are qualitative counterparts of cumulative prospect theory criteria of *Kahneman* and *Tverski* [3.102].

Two issues in preference modeling based on possibility theory in a logic format are as follows:

- Preference statements of the form $\Pi(p) > \Pi(q)$ provide an incomplete description of a preference relation. One question is then how to complete this description by default. The principle of minimal specificity then means that a solution not explicitly rejected is satisfactory by default. The dual maximal specificity principle, says that a solution not supported is rejected by default. It is not always clear which principle is the most natural.
- A statement according to which it is better to satisfy a formula p than a formula q can in fact be interpreted in several ways. For instance, it may mean that the best solution satisfying p is better that the best solution satisfying q, which reads $\Pi(p) > \Pi(q)$ and can be encoded in possibilistic logic under minimal specificity assumption; a stronger statement is that the worst solution satisfying p is

better that the best solution satisfying q, which reads $\Delta(p) > \Pi(q)$. Other possibilities are $\Delta(p) > \Delta(q)$, and $\Pi(p) > \Delta(q)$. This question is studied in some detail by *Kaci* [3.103].

3.3.9 Decision–Theoretic Foundations

Zadeh [3.1] hinted that *since our intuition concerning the behavior of possibilities is not very reliable*, our understanding of them

> *would be enhanced by the development of an axiomatic approach to the definition of subjective possibilities in the spirit of axiomatic approaches to the definition of subjective probabilities.*

Decision-theoretic justifications of qualitative possibility were devised, in the style of Von Neumann and Morgenstern, and *Savage* [3.104] more than 15 years ago [3.105, 106].

On top of the set of states, assume there is a set X of consequences of decisions. A decision, or act, is modeled as a mapping f from S to X assigning to each state S its consequence $f(s)$. The axiomatic approach consists in proposing properties of a preference relation $\succeq$ between acts so that a representation of this relation by means of a preference functional $W(f)$ is ensured, that is, act f is as good as act g (denoted by $f \succeq g$) if and only if $W(f) \geq W(g)$. $W(f)$ depends on the agent's knowledge about the state of affairs, here supposed to be a possibility distribution π on S, and the agent's goal, modeled by a utility function u on X. Both the utility function and the possibility distribution map to the same finite chain L. A pessimistic criterion $W_\pi^-(f)$ is of the form

$$W_\pi^-(f) = \min_{s \in S} \max(n(\pi(s)), u(f(s))),$$

where n is the order-reversing map of L. $n(\pi(s))$ is the degree of certainty that the state is not s (hence the degree of surprise of observing s), $u(f(s))$ the utility of choosing act f in state s. $W_\pi^-(f)$ is all the higher as all states are either very surprising or have high utility. This criterion is actually a prioritized extension of the Wald maximin criterion. The latter is recovered if $\pi(s) = 1$ (top of L) $\forall s \in S$. According to the pessimistic criterion, acts are chosen according to their worst consequences, restricted to the most plausible states $S^* = \{s, \pi(s) \geq n(W_\pi^-(f))\}$. The optimistic counterpart of this criterion is

$$W_\pi^+(f) = \max_{s \in S} \min(\pi(s), u(f(s))).$$

$W_\pi^+(f)$ is all the higher as there is a very plausible state with high utility. The optimistic criterion was first proposed by *Yager* [3.107] and the pessimistic criterion by *Whalen* [3.108]. See *Dubois* et al. [3.109] for the resolution of decision problems under uncertainty using the above criterion, and cast in the possibilistic logic framework. Such criteria can be refined by the classical expected utility criterion [3.110].

These optimistic and pessimistic possibilistic criteria are particular cases of a more general criterion based on the Sugeno integral [3.111] specialized to possibility and necessity of fuzzy events [3.1, 20]

$$S_{\gamma,u}(f) = \max_{\lambda \in L} \min(\lambda, \gamma(F_\lambda)),$$

where $F_\lambda = \{s \in S, u(f(s)) \geq \lambda\}$, γ is a monotonic set function that reflects the decision-maker attitude in front of uncertainty: $\gamma(A)$ is the degree of confidence in event A. If $\gamma = \Pi$, then $S_{\Pi,u}(f) = W_\pi^+(f)$. Similarly, if $\gamma = N$, then $S_{N,u}(f) = W_\pi^-(f)$.

For any acts f, g, and any event A, let fAg denote an act consisting of choosing f if A occurs and g if its complement occurs. Let $f \wedge g$ (resp. $f \vee g$) be the act whose results yield the worst (resp. best) consequence of the two acts in each state. Constant acts are those whose consequence is fixed regardless of the state. A result in [3.112, 113] provides an act-driven axiomatization of these criteria, and enforces possibility theory as a *rational* representation of uncertainty for a finite state space S:

Theorem 3.1

Suppose the preference relation $\succeq$ on acts obeys the following properties:

1. $(X^S, \succeq)$ is a complete preorder.
2. There are two acts such that $f \succ g$.

3. $\forall A, \forall g$ and h constant, $\forall f, g \succeq h$ implies $gAf \succeq hAf$.
4. If f is constant, $f \succ h$ and $g \succ h$ imply $f \wedge g \succ h$.
5. If f is constant, $h \succ f$ and $h \succ g$ imply $h \succ f \vee g$.

Then there exists a finite chain L, an L-valued monotonic set function γ on S and an L-valued utility function u, such that $\succeq$ is representable by a Sugeno integral of $u(f)$ with respect to γ. Moreover, γ is a necessity (resp. possibility) measure as soon as property (4) (resp. (5)) holds for all acts. The preference functional is then $W_\pi^-(f)$ (resp. $W_\pi^+(f)$).

Axioms (4 and 5) contradict expected utility theory. They become reasonable if the value scale is finite, decisions are one-shot (no compensation) and provided that there is a big step between any level in the qualitative value scale and the adjacent ones. In other words, the preference pattern $f \succ h$ always means that f is significantly preferred to h, to the point of considering the value of h negligible in front of the value of f. The above result provides decision-theoretic foundations of possibility theory, whose axioms can thus be tested from observing the choice behavior of agents. See [3.114] for another approach to comparative possibility relations, more closely relying on Savage axioms, but giving up any comparability between utility and plausibility levels. The drawback of these and other qualitative decision criteria is their lack of discrimination power [3.115]. To overcome it, refinements of possibilistic criteria were recently proposed, based on lexicographic schemes. These refined criteria turn out to be by a classical (but big-stepped) expected utility criterion [3.110], and Sugeno integral can be refined by a Choquet integral [3.116]. For extension of this qualitative decision-making framework to multiple-stage decision, see [3.117].

3.4 Quantitative Possibility Theory

The phrase *quantitative possibility* refers to the case when possibility degrees range in the unit interval, and are considered in connection with belief function and imprecise probability theory. Quantitative possibility theory is the natural setting for a reconciliation between probability and fuzzy sets. In that case, a precise articulation between possibility and probability theories is useful to provide an interpretation to possibility and necessity degrees. Several such interpretations can

be consistently devised: a degree of possibility can be viewed as an upper probability bound [3.118], and a possibility distribution can be viewed as a likelihood function [3.119]. A possibility measure is also a special case of a Shafer plausibility function [3.120]. Following a very different approach, possibility theory can account for probability distributions with extreme values, infinitesimal [3.72] or having big steps [3.121]. There are finally close connections between possibility theory and idempotent analysis [3.122]. The theory of large deviations in probability theory [3.123] also handles set functions that look like possibility measures [3.124]. Here we focus on the role of possibility theory in the theory of imprecise probability.

3.4.1 Possibility as Upper Probability

Let π be a possibility distribution where $\pi(s) \in [0, 1]$. Let $\mathbf{P}(\pi)$ be the set of probability measures P such that $P \leq \Pi$, i.e., $\forall A \subseteq S, P(A) \leq \Pi(A)$. Then the possibility measure Π coincides with the upper probability function P^* such that $P^*(A) = \sup\{P(A), P \in \mathbf{P}(\pi)\}$ while the necessity measure N is the lower probability function P_* such that $P_*(A) = \inf\{P(A), P \in \mathbf{P}(\pi)\}$; see [3.118, 125] for details. P and π are said to be consistent if $P \in \mathbf{P}(\pi)$. The connection between possibility measures and imprecise probabilistic reasoning is especially promising for the efficient representation of nonparametric families of probability functions, and it makes sense even in the scope of modeling linguistic information [3.126].

A possibility measure can be computed from nested confidence subsets $\{A_1, A_2, \ldots, A_m\}$ where $A_i \subset A_{i+1}, i = 1, \ldots, m-1$. Each confidence subset A_i is attached a positive confidence level λ_i interpreted as a lower bound of $P(A_i)$, hence a necessity degree. It is viewed as a certainty qualified statement that generates a possibility distribution π_i according to Sect. 3.2. The corresponding possibility distribution is

$$\pi(s) = \min_{i=1,\ldots,m} \pi_i(s)$$

$$= \begin{cases} 1 & \text{if } u \in A_1 \\ 1 - \lambda_{j-1} & \text{if } j = \max\{i : s \notin A_i\} > 1 \end{cases}.$$

The information modeled by π can also be viewed as a nested random set $\{(A_i, v_i), i = 1, \ldots, m\}$, where $v_i = \lambda_i - \lambda_{i-1}$. This framework allows for imprecision (reflected by the size of the A_i's) and uncertainty (the v_i's). And v_i is the probability that the agent only knows that A_i contains the actual state (it is not $P(A_i)$). The random set view of possibility theory is well adapted to the idea of imprecise statistical data, as developed in [3.127, 128]. Namely, given a bunch of imprecise (not necessarily nested) observations (called focal sets), π supplies an approximate representation of the data, as $\pi(s) = \sum_{i:s \in A_i} v_i$.

In the continuous case, a fuzzy interval M can be viewed as a nested set of α-cuts, which are intervals $M_\alpha = \{x : \mu_M(x) \geq \alpha, \forall \alpha > 0\}$. In the continuous case,

note that the degree of necessity is $N(M_\alpha) = 1 - \alpha$, and the corresponding probability set $\mathbf{P}(\mu_M) = \{P : P(M_\alpha) \geq 1 - \alpha, \forall \alpha > 0\}$. Representing uncertainty by the family of pairs $\{(M_\alpha, 1 - \alpha) : \forall \alpha > 0\}$ is very similar to the basic approach of info-gap theory [3.129].

The set $\mathbf{P}(\pi)$ contains many probability distributions, arguably too many. *Neumaier* [3.130] has recently proposed a related framework, in a different terminology, for representing smaller subsets of probability measures using two possibility distributions instead of one. He basically uses a pair of distributions (δ, π) (in the sense of Sect. 3.2) of distributions, he calls *cloud*, where δ is a guaranteed possibility distribution (in our terminology) such that $\pi \geq \delta$. A cloud models the (generally nonempty) set $\mathbf{P}(\pi) \cap \mathbf{P}(1 - \delta)$, viewing $1 - \delta$ as a standard possibility distribution. The precise connections between possibility distributions, clouds and other simple representations of numerical uncertainty is studied in [3.131].

3.4.2 Conditioning

There are two kinds of conditioning that can be envisaged upon the arrival of new information E. The first method presupposes that the new information alters the possibility distribution π by declaring all states outside E impossible. The conditional measure $\pi(. \mid E)$ is such that $\Pi(B \mid E) \cdot \Pi(E) = \Pi(B \cap E)$. This is formally Dempster rule of conditioning of belief functions, specialized to possibility measures. The conditional possibility distribution representing the weighted set of confidence intervals is

$$\pi(s \mid E) = \begin{cases} \dfrac{\pi(s)}{\Pi(E)}, & \text{if } s \in E \\ 0 & \text{otherwise} . \end{cases}$$

De Baets et al. [3.28] provide a mathematical justification of this notion in an infinite setting, as opposed to the min-based conditioning of qualitative possibility theory. Indeed, the maxitivity axiom extended to the infinite setting is not preserved by the min-based conditioning. The product-based conditioning leads to a notion of independence of the form $\Pi(B \cap E) = \Pi(B) \cdot \Pi(E)$ whose properties are very similar to the ones of probabilistic independence [3.30].

Another form of conditioning [3.132, 133], more in line with the Bayesian tradition, considers that the possibility distribution π encodes imprecise statistical information, and event E only reflects a feature of the current situation, not of the state in general. Then the value $\Pi(B \| E) = \sup\{P(B \mid E), P(E) > 0, P \le \Pi\}$ is the result of performing a sensitivity analysis of the usual conditional probability over $\mathbf{P}(\pi)$ [3.134]. Interestingly, the resulting set function is again a possibility measure, with distribution

$$\pi(s \| E) = \begin{cases} \max\left(\pi(s), \dfrac{\pi(s)}{\pi(s) + N(E)}\right), & \text{if } s \in E \\ 0 & \text{otherwise}. \end{cases}$$

It is generally less specific than π on E, as clear from the above expression, and becomes noninformative when $N(E) = 0$ (i.e., if there is no information about E). This is because $\pi(\cdot \| E)$ is obtained from the focusing of the generic information π over the reference class E. On the contrary, $\pi(\cdot \mid E)$ operates a revision process on π due to additional knowledge asserting that states outside E are impossible. See *De Cooman* [3.133] for a detailed study of this form of conditioning.

3.4.3 Probability–Possibility Transformations

The problem of transforming a possibility distribution into a probability distribution and conversely is meaningful in the scope of uncertainty combination with heterogeneous sources (some supplying statistical data, other linguistic data, for instance). It is useful to cast all pieces of information in the same framework. The basic requirement is to respect the consistency principle $\Pi \ge P$. The problem is then either to pick a probability measure in $\mathbf{P}(\pi)$, or to construct a possibility measure dominating P.

There are two basic approaches to possibility/probability transformations, which both respect a form of probability–possibility consistency. One, due to *Klir* [3.135, 136] is based on a principle of information invariance, the other [3.137] is based on optimizing information content. Klir assumes that possibilistic and probabilistic information measures are commensurate. Namely, the choice between possibility and probability is then a mere matter of translation between languages *neither of which is weaker or stronger than the other* (quoting *Klir* and *Parviz* [3.138]). It suggests that entropy and imprecision capture the same facet of uncertainty, albeit in different guises. The other approach, recalled here, considers that going from possibility to probability leads to increase the precision of the considered representation (as we go from

a family of nested sets to a random element), while going the other way around means a loss of specificity.

From Possibility to Probability
The most basic example of transformation from possibility to probability is the Laplace principle of insufficient reason claiming that what is equally possible should be considered as equally probable. A generalized Laplacean indifference principle is then adopted in the general case of a possibility distribution π: the weights v_i bearing on the sets A_i from the nested family of levels cuts of π are uniformly distributed on the elements of these cuts A_i. Let P_i be the uniform probability measure on A_i. The resulting probability measure is $P = \sum_{i=1,\dots,m} v_i \cdot P_i$. This transformation, already proposed in 1982 [3.139] comes down to selecting the center of gravity of the set $\mathbf{P}(\pi)$ of probability distributions dominated by π. This transformation also coincides with Smets' pignistic transformation [3.140] and with the Shapley value of the *unamimity game* (another name of the necessity measure) in game theory. The rationale behind this transformation is to minimize arbitrariness by preserving the symmetry properties of the representation. This transformation from possibility to probability is one-to-one. Note that the definition of this transformation does not use the nestedness property of cuts of the possibility distribution. It applies all the same to nonnested random sets (or belief functions) defined by pairs $\{(A_i, v_i), i = 1, \dots, m\}$, where v_i are nonnegative reals such that $\sum_{i=1,\dots,m} v_i = 1$.

From Objective Probability to Possibility
From probability to possibility, the rationale of the transformation is not the same according to whether the probability distribution we start with is subjective or objective [3.106]. In the case of a statistically induced probability distribution, the rationale is to preserve as much information as possible. This is in line with the handling of Δ-qualified pieces of information representing observed evidence, considered in Sect. 3.2; hence we select as the result of the transformation of a probability measure P, the most specific possibility measure in the set of those dominating P [3.137]. This most specific element is generally unique if P induces a linear ordering on S. Suppose S is a finite set. The idea is to let $\Pi(A) = P(A)$, for these sets A having minimal probability among other sets having the same cardinality as A. If $p_1 > p_2 > \cdots > p_n$, then $\Pi(A) = P(A)$ for sets A of the form $\{s_i, \dots, s_n\}$, and the possibility distribution is defined

as $\pi_P(s_i) = \sum_{j=i,\dots,m} p_j$, with $p_j = P(\{s_j\})$. Note that π_P is a kind of cumulative distribution of P, already known as a Lorentz curve in the mathematical literature [3.141]. If there are equiprobable elements, the unicity of the transformation is preserved if equipossibility of the corresponding elements is enforced. In this case it is a bijective transformation as well. Recently, this transformation was used to prove a rather surprising agreement between probabilistic indeterminateness as measured by Shannon entropy, and possibilistic nonspecificity. Namely it is possible to compare probability measures on finite sets in terms of their relative *peakedness* (a concept adapted from *Birnbaum* [3.142]) by comparing the relative specificity of their possibilistic transforms. Namely let P and Q be two probability measures on S and π_P, π_Q the possibility distributions induced by our transformation. It can be proved that if $\pi_P \geq \pi_Q$ (i.e., P is less peaked than Q) then the Shannon entropy of P is higher than the one of Q [3.143]. This result give some grounds to the intuitions developed by *Klir* [3.135], without assuming any commensurability between entropy and specificity indices.

Possibility Distributions Induced by Prediction Intervals

In the continuous case, moving from objective probability to possibility means adopting a representation of uncertainty in terms of prediction intervals around the mode viewed as the *most frequent value*. Extracting a prediction interval from a probability distribution or devising a probabilistic inequality can be viewed as moving from a probabilistic to a possibilistic representation. Namely suppose a nonatomic probability measure P on the real line, with unimodal density ϕ, and suppose one wishes to represent it by an interval I with a prescribed level of confidence $P(I) = \gamma$ of hitting it. The most natural choice is the most precise interval ensuring this level of confidence. It can be proved that this interval is of the form of a cut of the density, i.e., $I_\gamma = \{s, \phi(s) \geq \theta\}$ for some threshold θ. Moving the degree of confidence from 0 to 1 yields a nested family of prediction intervals that form a possibility distribution π consistent with P, the most specific one actually, having the same support and the same mode as P and defined by [3.137]

$$\pi(\inf I_\gamma) = \pi(\sup I_\gamma) = 1 - \gamma = 1 - P(I_\gamma) \,.$$

This kind of transformation again yields a kind of cumulative distribution according to the ordering in-

duced by the density ϕ. Similar constructs can be found in the statistical literature (*Birnbaum* [3.142]). More recently *Mauris* et al. [3.144] noticed that starting from any family of nested sets around some characteristic point (the mean, the median,...), the above equation yields a possibility measure dominating P. Well-known inequalities of probability theory, such as those of Chebyshev and Camp-Meidel, can also be viewed as possibilistic approximations of probability functions. It turns out that for symmetric unimodal densities, each side of the optimal possibilistic transform is a convex function. Given such a probability density on a bounded interval $[a, b]$, the triangular fuzzy number whose core is the mode of ϕ and the support is $[a, b]$ is thus a possibility distribution dominating P regardless of its shape (and the tightest such distribution). These results justify the use of symmetric triangular fuzzy numbers as fuzzy counterparts to uniform probability distributions. They provide much tighter probability bounds than Chebyshev and Camp-Meidel inequalities for symmetric densities with bounded support. This setting is adapted to the modeling of sensor measurements [3.145]. These results are extended to more general distributions by *Baudrit* et al. [3.146], and provide a tool for representing poor probabilistic information. More recently, *Mauris* [3.147] unifies, by means of possibility theory, many old techniques independently developed in statistics for one-point estimation, relying on the idea of dispersion of an empirical distribution. The efficiency of different estimators can be compared by means of fuzzy set inclusion applied to optimal possibility transforms of probability distributions. This unified approach does not presuppose a finite variance.

Subjective Possibility Distributions

The case of a subjective probability distribution is different. Indeed, the probability function is then supplied by an agent who is in some sense forced to express beliefs in this form due to rationality constraints, and the setting of exchangeable bets. However his actual knowledge may be far from justifying the use of a single well-defined probability distribution. For instance in case of total ignorance about some value, apart from its belonging to an interval, the framework of exchangeable bets enforces a uniform probability distribution, on behalf of the principle of insufficient reason. Based on the setting of exchangeable bets, it is possible to define a subjectivist view of numerical possibility theory, that differs from the proposal of *Walley* [3.134]. The approach developed by *Dubois* et al. [3.148] relies on the assumption that when an agent constructs a probability measure by assigning prices to lotteries, this probability measure is actually induced by a belief function representing the agent's actual state of knowledge. We assume that going from an underlying belief function to an elicited probability measure is achieved by means of the above mentioned pignistic transformation, changing focal sets into uniform probability distributions. The task is to reconstruct this underlying belief function under a minimal commitment assumption. In the paper [3.148], we pose and solve the problem of finding the least informative belief function having a given pignistic probability. We prove that it is unique and consonant, thus induced by a possibility distribution. The obtained possibility distribution can be defined as the converse of the pignistic transformation (which is one-to-one for possibility distributions). It is subjective in the same sense as in the subjectivist school in probability theory. However, it is the least biased representation of the agent's state of knowledge compatible with the observed betting behavior. In particular, it is less specific than the one constructed from the prediction intervals of an objective probability. This transformation was first proposed in [3.149] for objective probability, interpreting the empirical necessity of an event as summing the excess of probabilities of realizations of this event with respect to the probability of the most likely realization of the opposite event.

Possibility Theory and Defuzzification

Possibilistic mean values can be defined using Choquet integrals with respect to possibility and necessity measures [3.133, 150], and come close to defuzzification methods [3.151]. Interpreting a fuzzy interval M, associated with a possibility distribution μ_M, as a family of

probabilities, upper and lower mean values $E^*(M)$ and $E_*(M)$, can be defined as [3.152]

$$E_*(M) = \int_0^1 \inf M_\alpha \, d\alpha; \quad E^*(M) = \int_0^1 \sup M_\alpha \, d\alpha \, ,$$

where M_α is the α-cut of M.

Then the mean interval $E(M) = [E_*(M), E^*(M)]$ of M is the interval containing the mean values of all random variables consistent with M, that is $E(M) = \{E(P) \mid P \in \mathbf{P}(\mu_M)\}$, where $E(P)$ represents the expected value associated with the probability measure P. That the *mean value* of a fuzzy interval is an interval seems to be intuitively satisfactory. Particularly the mean interval of a (regular) interval $[a, b]$ is this interval itself. The upper and lower mean values are linear with respect to the addition of fuzzy numbers. Define the addition $M + N$ as the fuzzy interval whose cuts are $M_\alpha + N_\alpha = \{s + t, s \in M_\alpha, t \in N_\alpha\}$ defined according to the rules of interval analysis. Then $E(M + N) = E(M) + E(N)$, and similarly for the scalar multiplication $E(aM) = aE(M)$, where aM has membership grades of the form $\mu_M(s/a)$ for $a \neq 0$. In view of this property, it seems that the most natural defuzzication method is the middle point $\hat{E}(M)$ of the mean interval (originally proposed by *Yager* [3.153]). Other defuzzification techniques do not generally possess this kind of linearity property. $\hat{E}(M)$ has a natural interpretation in terms of simulation of a fuzzy variable [3.154], and is the mean value of the pignistic transformation of M. Indeed it is the mean value of the empirical probability distribution obtained by the random process defined by picking an element α in the unit interval at random, and then an element s in the cut M_α at random.

3.5 Some Applications

Possibility theory has not been the main framework for engineering applications of fuzzy sets in the past. However, on the basis of its connections to symbolic artificial intelligence, to decision theory and to imprecise statistics, we consider that it has significant potential for further applied developments in a number of areas, including some where fuzzy sets are not yet always accepted. Only some directions are pointed out here.

3.5.1 Uncertain Database Querying and Preference Queries

The evaluation of a flexible query in the face of incomplete or fuzzy information amounts to computing the possibility and the necessity of the fuzzy event expressing the gradual satisfaction of the query [3.155]. This evaluation, known as fuzzy pattern matching [3.156, 157], corresponds to the extent to which fuzzy sets (representing the query) overlap, or include the possibility distributions (representing the available information). Such an evaluation procedure has been extended to symbolic labels that are no longer represented by possibility distributions, but which belong to possibilistic ontologies where approximate similarity and subsumption between labels are estimated in terms of possibility and necessity degrees, respectively [3.158]. These approaches presuppose a total lack of dependencies between ill-known attributes. A more general approach based on possible world semantics has been envisaged [3.159]. However, as for the probabilistic counterpart of this latter view, evaluating queries has a high computational cost [3.160]. This is why it has been proposed to only use certainty qualified values (or disjunctions of values), as in possibilistic logic, rather than general possibility distributions, for representing attribute values pervaded with uncertainty. It has been shown that it leads to a tractable extension of relational algebra operations [3.161, 162].

Besides, possibility theory is not only useful for representing qualitative uncertainty, but it may also be of interest for representing preferences, and as such may be applied to the handling of preferences queries [3.163]. Thus, requirements of the form *A and preferably B* (i.e., it is more satisfactory to have *A* and *B* than *A* alone), or *A or at least B* can be expressed using appropriate priority orderings, as in possibilistic logic [3.164]. Lastly, in bipolar queries [3.165–167], flexible constraints that are more or less compulsory are distinguished from additional wishes that are optional, as for instance in the request find the apartments that are cheap and maybe near the train station. Indeed, negative preferences express what is (more or less, or completely) impossible or undesirable, and by complementation state flexible constraints restricting the possible or acceptable values. Positive preferences are not compulsory, but rather express wishes; they state what attribute values would be really satisfactory.

3.5.2 Description Logics

Description logics (initially named terminological logics) are tractable fragments of first-order logic representation languages that handle notions of concepts, roles and instances, referring at the semantic level to the respective notions of set, binary relations, membership, and cardinality. They are useful for describing ontologies that consist in hierarchies of concepts in a particular domain, for the semantic web. Two ideas that, respectively, come from fuzzy sets and possibil-

ity theory, and that may be combined, may be used for extending the expressive power of description logics. On one hand, vague concepts can be approximated in practice by pairs of nested sets corresponding to the cores and the supports of fuzzy sets, thus sorting out the typical elements, in a way that agrees with fuzzy set operations and inclusions. On the other hand, a possibilistic treatment of uncertainty and exceptions can be performed on top of a description logic in a possibilistic logic style [3.168]. In both cases, the underlying principle is to remain as close as possible to classical logic for preserving computational efficiency as much as possible. Thus, formal expressions such as $(P \sqsupseteq_{\alpha}^{X} Q, \beta)$ intend to mean that *it is certain at least at level β that the degree of subsumption of concept P in Q is at least α*, in the sense of some *X*-implication (e.g., Gödel, or Kleene–Dienes implication). In particular, it can be expressed that *typical Ps are Qs*, or that *typical Ps are typical Qs*, or that an instance is typical of a concept. Such ideas have been developed by *Qi* et al. [3.169] toward implemented systems in connection with web research.

3.5.3 Information Fusion

Possibility theory offers a simple, flexible framework for information fusion that can handle incompleteness and conflict. For instance, intervals or fuzzy intervals can be merged, coming from several sources. The basic fusion modes are the conjunctive and disjunctive modes, presupposing, respectively, that all sources of information are reliable and that at least one is [3.170, 171]. In the conjunctive mode, the use of the minimum operation avoids assuming sources are independent. If they are, the product rule can be applied, whereby low plausibility degrees reinforce toward impossibility. Quite often, the results of a conjunctive aggregation are subnormalized, this indicating a conflict. Then, it is common to apply a renormalization step that makes this mode of combination brittle in case of strong conflict, and anyway the more numerous the sources the more conflicting they become. Weighted average of possibility degrees can be used but it does not preserve the properties of possibility measure. The use of the disjunctive mode is more cautious: it avoids the conflict at the expense of losing information. When many sources are involved the result becomes totally uninformative.

To cope with this problem, some ad hoc adaptive combination rules have been proposed that focus on maximal subsets of sources that are either fully consistent or not completely inconsistent [3.170]. This scheme has been further improved by *Oussalah* et al. [3.172]. *Oussalah* [3.173] has proposed a number of postulates a possibilistic fusion rule should satisfy. Another approach is to merge the set of cuts of the possibility distributions based on the maximal consistent subsets of sources (consistent subsets of cuts are merged using conjunction, and the results are merged disjunctively). The result is then a belief function [3.174]. Another option is to make a guess on the number of reliable sources and merge information inside consistent subsets of sources having this cardinality.

Possibilistic information fusion can be performed syntactically on more compact representations such as possibilistic logic bases [3.175] (the merging of possibilistic networks [3.176] has also been recently considered). The latter type of fusion may be of interest both from a computational and from representational point of view. Still it is important to make sure that the syntactic operations are counterparts of semantic ones. Fusion should be performed both at the semantic and at the syntactic levels equivalently. For instance, the conjunctive merging of two possibility distributions corresponds to the mere union of the possibilistic bases that represent them. More details for other operations can be found in [3.175, 177], and in the bipolar case in [3.99]. This line of research is pursued by *Qi* et al. [3.178]. They also proposed an approach to measuring conflict between possibilistic knowledge bases [3.179].

The distance-based approach [3.180] that applies to the fusion of classical logic bases can be embedded in the possibilistic fusion setting as well [3.177]. The distance between an interpretation s and each classical base K is usually defined as $d(s,K) = \min\{H(s,s^*) : s^* \models K\}$ where $H(s,s^*)$ is the Hamming distance that evaluates the number of literals with different signs in s and s^*). It is then easy to encode the distance $d(s,K)$ into a possibilistic knowledge base (interpreting possibility as Hamming-distance-based similarity to the models of K, i.e., $\pi(s) = a^{d(s,K)}, a \in (0,1)$). The result of the possibilistic fusion is a possibilistic knowledge base, the highest weight layer of which is the classical database that is searched for, provided that the distance merging operation is suitably translated to a possibilistic merging operation.

A similar problem exists in belief revision where an epistemic state, represented either by a possibility distribution or by a possibilistic logic base, is revised by an input information p [3.181]. Revision can be viewed as prioritized fusion, using for instance conditioning,

or other operations, depending if in the revised epistemic state one wants to enforce $N(p) = 1$, or $N(p) > 0$ only, or if we are dealing with an uncertain input (p, α). Then, the uncertain input may be understood as enforcing $N(p) \geq \alpha$ in any case, or as taking it into account only if it is sufficiently certain w.r.t. the current epistemic state.

3.5.4 Temporal Reasoning and Scheduling

Temporal reasoning may refer to time intervals or to time points. When handling time intervals, the basic building block is the one provided by Allen relations between time intervals. There are 13 relations that describe the possible relative locations of two intervals. For instance, given the two intervals $A = [a, a']$ and $B = [b, b']$, A is before (resp. after) B means $a' < b$ (resp. $b' < a$), A meets (resp. is met by) B means $a' = b$ (resp. $b' = a$), A overlaps (resp. is overlapped by) B iff $b > a$ and $a' > b$ and $b' > a'$ (resp. $a > b$ and $b' > a$, and $a' > b'$). The introduction of fuzzy features in temporal reasoning can be related to two different issues:

- First, it can be motivated by the need of a gradual, linguistic-like description of temporal relations even in the face of complete information. Then an extension of Allen relational calculus has been proposed, which is based on fuzzy comparators expressing linguistic tolerance, which are used in place of the exact relations $>, ='$, and $<$. Fuzzy Allen relations are thus defined from three fuzzy relations between dates that can be, for instance *approximately equal, clearly greater*, and *clearly smaller*, where, e.g., the extent to which x is approximately equal to y is the degree of membership of $x - y$ to some fuzzy set expressing something like *small* [3.182, 183].
- Second, the possibilistic handling of fuzzy or incomplete information leads to pervade classical Allen relations, and more generally fuzzy Allen relations, with uncertainty. Then patterns for propagating uncertainty and composing the different (fuzzy) Allen relations in a possibilistic way have been laid bare [3.184, 185].

Besides, the handling of temporal reasoning in terms of relations between time points can also be extended in case of uncertain information [3.186]. Uncertain relations between temporal points are represented by means of possibility distributions over the three basic relations $>, ='$, and $<$. Operations for computing inverse relations, for composing relations, for combining relations coming from different sources and pertaining to the same temporal points, or for handling negation, have been defined. This shows that possibilistic temporal uncertainty can be handled in the setting of point algebra. The possibilistic approach can then be favorably compared with a probabilistic approach previously proposed (first, the approach can be purely qualitative, thus avoiding the necessity of quantifying uncertainty if information is poor, and second, it is capable of modeling ignorance in a nonbiased way). Possibilistic logic has also been extended to a timed version where time intervals where a proposition is more or less certainly true is attached to classical propositional formulas [3.187].

Applications of possibility theory-based decision-making can be found in scheduling. One issue is to handle fuzzy due dates of jobs using the calculus of fuzzy constraints [3.188]. Another issue is to handle uncertainty in task durations in basic scheduling problems such as program evaluation and review technique (PERT) networks. A large literature exists on this topic [3.189, 190] where the role of fuzzy sets is not always very clear. Convincing solutions on this problem start with the works of *Chanas* and *Zielinski* [3.191, 192], where the problem is posed in terms of projecting a joint possibility theory on quantities of interest (earliest finishing times, or slack times) and where tasks can be possibly or certainly critical. A full solution applying Boolean possibility theory to interval uncertainty of tasks durations is described in [3.193], and its fuzzy extension in [3.194]. Other scheduling problems are solved in the same possibilistic framework by *Kasperski* and colleagues [3.195, 196], as well as more general optimization problems [3.197, 198].

3.5.5 Risk Analysis

The aim of risk analysis studies is to perform uncertainty propagation under poor data and without independence assumptions (see the papers in the special issue [3.199]). Finding the potential of possibilistic representations in computing conservative bounds for such probabilistic calculations is certainly a major challenge [3.200]. An important research direction is the comparison between fuzzy interval analysis [3.33] and random variable calculations with

a view to unifying them [3.201]. Methods for joint propagation of possibilistic and probabilistic information have been devised [3.202], based on casting both in a random set setting [3.203]; the case of probabilistic models with fuzzy interval parameters has also been dealt with [3.204]. The active area of fuzzy random variables is also connected to this question [3.205].

3.5.6 Machine Learning

Applications of possibility theory to learning have started to be investigated rather recently in different directions. For instance, taking advantage of the proximity between reinforcement learning and partially observed Markov decision processes, a possibilistic counterpart of reinforcement learning has been proposed after developing the possibilistic version of the latter [3.206]. Besides, by looking for big-stepped probability distributions, defined by discrete exponential distributions, one can mine data bases for discovering default rules [3.207]. Big-stepped probabilities mimick possibility measures in the sense that $P(A) > P(B)$ if and only if $\max_{s \in A} p(s) > \max_{s \in B} p(s)$. The version space approach to learning presents interesting similarities with the binary bipolar possibilistic representation setting, thinking of examples as positive information and of counterexamples as negative information [3.208]. The general bipolar setting, where intermediary degrees of possibility are allowed, provides a basis for extending version space approach in a graded way, where examples and counter examples can be weighted according to their importance. The graded version space approach agrees with the possibilistic extension of inductive logic programming [3.209]. Indeed, where the background knowledge may be associated with certainty levels, the examples may be more or less important to cover, and the set of rules that is learnt may be stratified in order to have a better management of exceptions in multiple-class classification problems, in agreement with the possibilistic approach to nonmonotonic reasoning.

Other applications of possibility theory can be found in fields such as data analysis [3.79, 210, 211], diagnosis [3.212, 213], belief revision [3.181], argumentation [3.68, 214, 215], etc.

3.6 Some Current Research Lines

A number of ongoing works deal with new research lines where possibility theory is central. In the following, we outline a few of those:

- *Formal concept analysis*: Formal concept analysis (FCA) studies Boolean data tables relating objects and attributes. The key issue of FCA is to extract so-called concepts from such tables. A concept is a maximal set of objects sharing a maximal number of attributes. The enumeration of such concepts can be carried out via a Galois connection between objects and attributes, and this Galois connection uses operators similar to the Δ function of possibility theory. Based on this analogy, other correspondences can be laid bare using the three other set functions of possibility theory [3.216, 217]. In particular, one of these correspondences detects independent subtables [3.22]. This approach can be systematized to fuzzy or uncertain versions of formal concept analysis.
- *Generalized possibilistic logic*: Possibilistic logic, in its basic version, attaches degrees of necessity to formulas, which turn them into graded modal formulas of the necessity kind. However only conjunction of weighted formulas are allowed. Yet very early we noticed that it makes sense to extend the language toward handing constraints on the degree of possibility of a formula. This requires allowing for negation and disjunctions of necessity-qualified proposition. This extension, still under study [3.218], puts together the KD modal logic and basic possibilistic logic. Recently it has been shown that nonmonotonic logic programming languages can be translated into generalized possibilistic logic, making the meaning of negation by default in rules much more transparent [3.219]. This move from basic to generalized possibilistic logic also enables further extensions to the multiagent and the multisource case [3.220] to be considered. Besides, it has been recently shown that a Sugeno integral can also be represented in terms of possibilistic logic, which enables us to lay bare the logical description of an aggregation process [3.221].

- *Qualitative capacities and possibility measures*: While a numerical possibility measure is equivalent to a convex set of probability measures, it turns out that in the qualitative setting, a monotone set function can be represented by means of a family of possibility measures [3.222, 223]. This line of research enables qualitative counterparts of results in the study of Choquet capacities in the numerical settings to be established. Especially, a monotone set function can be seen as the counterpart of a belief function, and various concepts of evidence theory can be adapted to this setting [3.224]. Sugeno integral can be viewed as a lower possibilistic expectation in the sense of Sect. 3.3.9 [3.223]. These results enable the structure of qualitative monotonic set functions to be laid bare, with possible connection with neighborhood semantics of nonregular modal logics [3.225].
- *Regression and kriging*: Fuzzy regression analysis is seldom envisaged from the point of view of possibility theory. One exception is the possibilistic regression initiated by *Tanaka* and *Guo* [3.211], where the idea is to approximate precise or set-valued data in the sense of inclusion by means of a set-valued or fuzzy set-valued linear function obtained by making the linear coefficients of a linear function fuzzy. The alternative approach is the fuzzy least squares of *Diamond* [3.226] where fuzzy data are interpreted as functions and a crisp distance between fuzzy sets is often used. However, in this approach, fuzzy data are questionably seen as objective entities [3.227]. The introduction of possibility theory in regression analysis of fuzzy data comes down to an epistemic view of fuzzy data whereby one tries to construct the envelope of all linear regression results that could have been obtained, had the data been precise [3.228]. This view has been applied to the kriging problem in geostatistics [3.229]. Another use of possibility theory consists in exploiting possibility–probability transforms to develop a form of quantile regression on crisp data [3.230], yielding a fuzzy function that is much more faithful to the data set than what a fuzzified linear function can offer.

Part A | 3

References

3.1   L.A. Zadeh: Fuzzy sets as a basis for a theory of possibility, Fuzzy Set. Syst. **1**, 3–28 (1978)

3.2   B.R. Gaines, L. Kohout: Possible automata, Proc. Int. Symp. Multiple-Valued Logics (Bloomington, Indiana 1975) pp. 183–196

3.3   D. Dubois, H. Prade: *Possibility Theory* (Plenum, New York 1988)

3.4   D. Dubois, H.T. Nguyen, H. Prade: Fuzzy sets and probability: Misunderstandings, bridges and gaps. In: *Fundamentals of Fuzzy Sets*, ed. by D. Dubois, H. Prade (Kluwer, Boston 2000) pp. 343–438, see also the bibliography in http://www.scholarpedia.org/article/Possibility_theory)

3.5   D. Dubois, H. Prade: Possibility theory: Qualitative and quantitative aspects. In: *Handbook of Defeasible Reasoning and Uncertainty Management Systems*, Vol. 1, ed. by D.M. Gabbay, P. Smets (Kluwer, Dordrecht 1998) pp. 169–226

3.6   D. Dubois, H. Fargier, H. Prade: Possibility theory in constraint satisfaction problems: Handling priority, preference and uncertainty, Appl. Intell. **6**, 287–309 (1996)

3.7   S. Benferhat, D. Dubois, S. Kaci, H. Prade: Modeling positive and negative information in possibility theory, Int. J. Intell. Syst. **23**, 1094–1118 (2008)

3.8   D. Dubois, H. Prade: An overview of the asymmetric bipolar representation of positive and negative information in possibility theory, Fuzzy Set. Syst. **160**, 1355–1366 (2009)

3.9   W. Spohn: Ordinal conditional functions: A dynamic theory of epistemic states. In: *Causation in Decision, Belief Change, and Statistics*, Vol. 2, ed. by W.L. Harper, B. Skyrms (Kluwer, Dordrecht 1988) pp. 105–134

3.10   W. Spohn: *The Laws of Belief: Ranking Theory and its Philosophical Applications* (Oxford Univ. Press, Oxford 2012)

3.11   I.M. Bocheński: *La Logique de Théophraste* (Librairie de l'Université de Fribourg en Suisse, Fribourg 1947)

3.12   B.F. Chellas: *Modal Logic, an Introduction* (Cambridge Univ. Press, Cambridge 1980)

3.13   G.L.S. Shackle: *Decision, Order and Time in Human Affairs*, 2nd edn. (Cambridge Univ. Press, Cambridge 1961)

3.14   D.L. Lewis: *Counterfactuals* (Basil Blackwell, Oxford 1973)

3.15   D. Dubois: Belief structures, possibility theory and decomposable measures on finite sets, Comput. Artif. Intell. **5**, 403–416 (1986)

3.16   T. Sudkamp: Similarity and the measurement of possibility, Actes Rencontres Francophones sur la Logique Floue et ses Applications (Cepadues Editions, Toulouse 2002) pp. 13–26

3.17   L.J. Cohen: *The Probable and the Provable* (Clarendon, Oxford 1977)

3.18   L.A. Zadeh: Fuzzy sets and information granularity. In: *Advances in Fuzzy Set Theory and Applications*, ed. by M.M. Gupta, R. Ragade, R.R. Yager (Amsterdam, North-Holland 1979) pp. 3–18

3.19   L.A. Zadeh: Possibility theory and soft data analysis. In: *Mathematical Frontiers of Social and Policy Sciences*, ed. by L. Cobb, R. Thrall (Westview, Boulder 1982) pp. 69–129

3.20   D. Dubois, H. Prade: *Fuzzy Sets and Systems: Theory and Applications* (Academic Press, New York 1980)

3.21   G.J. Klir, T. Folger: *Fuzzy Sets, Uncertainty and Information* (Prentice Hall, Englewood Cliffs 1988)

3.22   D. Dubois, H. Prade: Possibility theory and formal concept analysis: Characterizing independent subcontexts, Fuzzy Set. Syst. **196**, 4–16 (2012)

3.23   R.R. Yager: An introduction to applications of possibility theory, Hum. Syst. Manag. **3**, 246–269 (1983)

3.24   R.R. Yager: A foundation for a theory of possibility, Cybern. Syst. **10**(1–3), 177–204 (1980)

3.25   E.P. Klement, R. Mesiar, E. Pap: *Triangular Norms* (Kluwer, Dordrecht 2000)

3.26   G. Coletti, B. Vantaggi: Comparative models ruled by possibility and necessity: A conditional world, Int. J. Approx. Reason. **45**(2), 341–363 (2007)

3.27   G. Coletti, B. Vantaggi: T-conditional possibilities: Coherence and inference, Fuzzy Set. Syst. **160**(3), 306–324 (2009)

3.28   B. De Baets, E. Tsiporkova, R. Mesiar: Conditioning in possibility with strict order norms, Fuzzy Set. Syst. **106**, 221–229 (1999)

3.29   G. De Cooman: Possibility theory. Part I: Measure- and integral-theoretic groundwork; Part II: Conditional possibility; Part III: Possibilistic independence, Int. J. Gen. Syst. **25**, 291–371 (1997)

3.30   L.M. De Campos, J.F. Huete: Independence concepts in possibility theory, Fuzzy Set. Syst. **103**, 487–506 (1999)

3.31   D. Dubois, L. del Farinas Cerro, A. Herzig, H. Prade: Qualitative relevance and independence: A roadmap, Proc. 15th Int. Jt. Conf. Artif. Intell. Nagoya (1997) pp. 62–67

3.32   N. Ben Amor, K. Mellouli, S. Benferhat, D. Dubois, H. Prade: A theoretical framework for possibilistic independence in a weakly ordered setting, Int. J. Uncertain. Fuzziness Knowl.-Based Syst. **10**, 117–155 (2002)

3.33   D. Dubois, E. Kerre, R. Mesiar, H. Prade: Fuzzy interval analysis. In: *Fundamentals of Fuzzy Sets*, ed. by D. Dubois, H. Prade (Kluwer, Boston 2000) pp. 483–581

3.34   D. Dubois, H. Prade: Possibility theory as a basis for preference propagation in automated reasoning, Proc. 1st IEEE Int. Conf. Fuzzy Systems (FUZZ-IEEE'92), San Diego (1992) pp. 821–832

3.35   D. Dubois, P. Hájek, H. Prade: Knowledge-driven versus data-driven logics, J. Log. Lang. Inform. **9**, 65–89 (2000)

3.36   D. Dubois, H. Prade: Possibility theory, probability theory and multiple-valued logics: A clarification, Ann. Math. Artif. Intell. **32**, 35–66 (2001)

3.37   M. Banerjee, D. Dubois: A simple modal logic for reasoning about revealed beliefs, Lect. Notes Artif. Intell. **5590**, 805–816 (2009)

3.38   M. Banerjee, D. Dubois: A simple logic for reasoning about incomplete knowledge, Int. J. Approx. Reason. **55**, 639–653 (2014)

3.39   D. Dubois, H. Prade: Epistemic entrenchment and possibilistic logic, Artif. Intell. **50**, 223–239 (1991)

3.40   P. Gärdenfors: *Knowledge in Flux* (MIT Press, Cambridge 1988)

3.41   D. Dubois, H. Prade: Belief change and possibility theory. In: *Belief Revision*, ed. by P. Gärdenfors (Cambridge Univ. Press, Cambridge 1992) pp. 142–182

3.42   D. Lehmann, M. Magidor: What does a conditional knowledge base entail?, Artif. Intell. **55**, 1–60 (1992)

3.43   D. Dubois, H. Fargier, H. Prade: Ordinal and probabilistic representations of acceptance, J. Artif. Intell. Res. **22**, 23–56 (2004)

3.44   S. Benferhat, D. Dubois, H. Prade: Nonmonotonic reasoning, conditional objects and possibility theory, Artif. Intell. **92**, 259–276 (1997)

3.45   J. Pearl: System Z: A natural ordering of defaults with tractable applications to default reasoning, Proc. 3rd Conf. Theor. Aspects Reason. About Knowl. (Morgan Kaufmann, San Francisco 1990) pp. 121–135

3.46   E. Raufaste, R. Da Silva Neves, C. Mariné: Testing the descriptive validity of possibility theory in human judgements of uncertainty, Artif. Intell. **148**, 197–218 (2003)

3.47   H. Farreny, H. Prade: Default and inexact reasoning with possibility degrees, IEEE Trans. Syst. Man Cybern. **16**(2), 270–276 (1986)

3.48   D. Dubois, J. Lang, H. Prade: Possibilistic logic. In: *Handbook of Logic in AI and Logic Programming*, Vol. 3, ed. by D.M. Gabbay (Oxford Univ. Press, Oxford 1994) pp. 439–513

3.49   D. Dubois, H. Prade: Possibilistic logic: A retrospective and prospective view, Fuzzy Set. Syst. **144**, 3–23 (2004)

3.50   J. Lang: Possibilistic logic: Complexity and algorithms. In: *Algorithms for Uncertainty and Defeasible Reasoning*, (Kluwer, Dordrecht 2001) pp. 179–220

3.51   S. Benferhat, D. Dubois, H. Prade: Practical handling of exception-tainted rules and independence information in possibilistic logic, Appl. Intell. **9**, 101–127 (1998)

3.52   S. Benferhat, S. Yahi, H. Drias: A new default theories compilation for MSP-entailment, J. Autom. Reason. **45**(1), 39–59 (2010)

3.53   S. Benferhat, S. Lagrue, O. Papini: Reasoning with partially ordered information in a possibilistic logic framework, Fuzzy Set. Syst. **144**, 25–41 (2004)

3.54   S. Benferhat, H. Prade: Encoding formulas with partially constrained weights in a possibilistic-like many-sorted propositional logic, Proc. 9th Int. Jt. Conf. Artif. Intell. (IJCAI'05) (2005) pp. 1281–1286

3.55   S. Benferhat, H. Prade: Compiling possibilistic knowledge bases, Proc. 17th Eur. Conf. Artif. Intell. (Riva del Garda, Italy 2006)

3.56   D. Dubois, S. Konieczny, H. Prade: Quasi-possibilistic logic and its measures of information and conflict, Fundam. Inform. **57**, 101–125 (2003)

3.57   D. Dubois, F. Esteva, L. Godo, H. Prade: Fuzzy-set based logics – An history-oriented presentation of their main developments. In: *Handbook of the History of Logic, the Many-Valued and Nonmonotonic Turn in Logic*, Vol. 8, ed. by D.M. Gabbay, J. Woods (Elsevier, Amsterdam 2007) pp. 325–449

3.58   L. Boldrin, C. Sossai: Local possibilistic logic, J. Appl. Non-Class. Log. **7**, 309–333 (1997)

3.59   P. Dellunde, L. Godo, E. Marchioni: Extending possibilistic logic over Gödel logic, Int. J. Approx. Reason. **52**, 63–75 (2011)

3.60   P. Hájek, D. Harmancová, R. Verbrugge: A qualitative fuzzy possibilistic logic, Int. J. Approx. Reason. **12**(1), 1–19 (1995)

3.61   S. Lehmke: Logics which Allow Degrees of Truth and Degrees of Validity, Ph.D. Thesis (Universität Dortmund, Germany 2001)

3.62   L.A. Zadeh: Fuzzy logic and approximate reasoning (In memory of Grigore Moisil), Synthese **30**, 407–428 (1975)

3.63   D. Dubois, H. Prade, S. Sandri: A. Possibilistic logic with fuzzy constants and fuzzily restricted quantifiers. In: *Logic Programming and Soft Computing*, ed. by T.P. Martin, F. Arcelli-Fontana (Research Studies Press Ltd., Baldock, England 1998) pp. 69–90

3.64   T. Alsinet, L. Godo: Towards an automated deduction system for first-order possibilistic logic programming with fuzzy constants, Int. J. Intell. Syst. **17**, 887–924 (2002)

3.65   T. Alsinet: Logic Programming with Fuzzy Unification and Imprecise Constants: Possibilistic Semantics and Automated Deduction, Ph.D. Thesis (Technical University of Catalunya, Barcelona 2001)

3.66   T. Alsinet, L. Godo, S. Sandri: Two formalisms of extended possibilistic logic programming with context-dependent fuzzy unification: A comparative description, Electr. Notes Theor. Comput. Sci. **66**(5), 1–21 (2002)

3.67   T. Alsinet, L. Godo: Adding similarity-based reasoning capabilities to a Horn fragment of possibilistic logic with fuzzy constants, Fuzzy Set. Syst. **144**, 43–65 (2004)

3.68   T. Alsinet, C. Chesñevar, L. Godo, S. Sandri, G. Simari: Formalizing argumentative reasoning in a possibilistic logic programming setting with fuzzy unification, Int. J. Approx. Reason. **48**, 711–729 (2008)

3.69   T. Alsinet, C. Chesñevar, L. Godo, G. Simari: A logic programming framework for possibilistic argumentation: Formalization and logical properties, Fuzzy Set. Syst. **159**(10), 1208–1228 (2008)

3.70   P. Nicolas, L. Garcia, I. Stéphan, C. Lefèvre: Possibilistic uncertainty handling for answer set programming, Ann. Math. Artif. Intell. **47**(1/2), 139–181 (2006)

3.71   K. Bauters, S. Schockaert, M. De Cock, D. Vermeir: Possibilistic answer set programming revisited, Proc. 26th Conf. Uncertainty in Artif. Intell. (UAI'10), Catalina Island (2010) pp. 48–55

3.72   W. Spohn: A general, nonprobabilistic theory of inductive reasoning. In: *Uncertainty in Artificial Intelligence*, Vol. 4, ed. by R.D. Shachter (North Holland, Amsterdam 1990) pp. 149–158

3.73   R. Jeffrey: *The Logic of Decision*, 2nd edn. (Chicago Univ. Press, Chicago 1983)

3.74   D. Dubois, H. Prade: A synthetic view of belief revision with uncertain inputs in the framework of possibility theory, Int. J. Approx. Reason. **17**(2–3), 295–324 (1997)

3.75   S. Benferhat, D. Dubois, H. Prade, M.-A. Williams: A framework for iterated belief revision using possibilistic counterparts to Jeffrey's rule, Fundam. Inform. **99**(2), 147–168 (2010)

3.76   S. Benferhat, D. Dubois, L. Garcia, H. Prade: On the transformation between possibilistic logic bases and possibilistic causal networks, Int. J. Approx. Reason. **29**, 135–173 (2002)

3.77   N. Ben Amor, S. Benferhat: Graphoid properties of qualitative possibilistic independence relations, Int. J. Uncertain. Fuzziness Knowl.-Based Syst. **13**, 59–97 (2005)

3.78   N. Ben Amor, S. Benferhat, K. Mellouli: Anytime propagation algorithm for min-based possibilistic graphs, Soft Comput. **8**, 50–161 (2003)

3.79   C. Borgelt, J. Gebhardt, R. Kruse: Possibilistic graphical models. In: *Computational Intelligence in Data Mining*, ed. by G.D. Riccia (Springer, Wien 2000) pp. 51–68

3.80   W. Guezguez, N. Ben Amor, K. Mellouli: Qualitative possibilistic influence diagrams based on qualitative possibilistic utilities, Eur. J. Oper. Res. **195**, 223–238 (2009)

3.81   H. Fargier, N. Ben Amor, W. Guezguez: On the complexity of decision making in possibilistic decision trees, Proc. UAI (2011) pp. 203–210

3.82   R. Ayachi, N. Ben Amor, S. Benferhat: Experimental comparative study of compilation-based inference in Bayesian and possibilitic networks, Lect. Notes Comput. Sci. **6857**, 155–163 (2011)

3.83   S. Benferhat: Interventions and belief change in possibilistic graphical models, Artif. Intell. **174**(2), 177–189 (2010)

3.84   S. Benferhat, S. Smaoui: Inferring interventions in product-based possibilistic causal networks, Fuzzy Set. Syst. **169**(1), 26–50 (2011)

3.85   D. Dubois, H. Prade: What are fuzzy rules and how to use them, Fuzzy Set. Syst. **84**, 169–185 (1996)

3.86   D. Dubois, H. Prade, L. Ughetto: A new perspective on reasoning with fuzzy rules, Int. J. Intell. Syst. **18**, 541–567 (2003)

3.87   S. Galichet, D. Dubois, H. Prade: Imprecise specification of ill-known functions using gradual rules, Int. J. Approx. Reason. **35**, 205–222 (2004)

3.88   D. Dubois, E. Huellermeier, H. Prade: A systematic approach to the assessment of fuzzy association rules, Data Min. Knowl. Discov. **13**, 167–192 (2006)

3.89   D. Dubois, F. Esteva, P. Garcia, L. Godo, R. de Lopez Mantaras, H. Prade: Fuzzy set modelling in cased-based reasoning, Int. J. Intell. Syst. **13**(4), 345–373 (1998)

3.90   D. Dubois, E. Huellermeier, H. Prade: Fuzzy set-based methods in instance-based reasoning, IEEE Trans. Fuzzy Syst. **10**, 322–332 (2002)

3.91   D. Dubois, E. Huellermeier, H. Prade: Fuzzy methods for case-based recommendation and decision support, J. Intell. Inf. Syst. **27**, 95–115 (2006)

3.92   E. Huellermeier, D. Dubois, H. Prade: Model adaptation in possibilistic instance-based reasoning, IEEE Trans. Fuzzy Syst. **10**, 333–339 (2002)

3.93   E. Huellermeier: *Case-Based Approximate Reasoning* (Springer, Berlin 2007)

3.94   D. Dubois, P. Fortemps: Computing improved optimal solutions to max−min flexible constraint satisfaction problems, Eur. J. Oper. Res. **118**, 95–126 (1999)

3.95   M. Inuiguchi, H. Ichihashi, Y. Kume: Modality constrained programming problems: A unified approach to fuzzy mathematical programming problems in the setting of possibility theory, Inf. Sci. **67**, 93–126 (1993)

3.96   D. Dubois, H. Fargier, H. Prade: Refinements of the maximin approach to decision-making in fuzzy environment, Fuzzy Set. Syst. **81**, 103–122 (1996)

3.97   D. Dubois, P. Fortemps: Selecting preferred solutions in the minimax approach to dynamic programming problems under flexible constraints, Eur. J. Oper. Res. **160**, 582–598 (2005)

3.98   S. Kaci, H. Prade: Mastering the processing of preferences by using symbolic priorities in possibilistic logic, Proc. 18th Eur. Conf. Artif. Intell. (ECAI'08), Patras (2008) pp. 376–380

3.99   S. Benferhat, D. Dubois, S. Kaci, H. Prade: Bipolar possibility theory in preference modeling: Representation, fusion and optimal solutions, Inf. Fusion **7**, 135–150 (2006)

3.100   D. Dubois, H. Fargier, J.-F. Bonnefon: On the qualitative comparison of decisions having positive and negative features, J. Artif. Intell. Res. **32**, 385–417 (2008)

3.101   J.-F. Bonnefon, D. Dubois, H. Fargier, S. Leblois: Qualitative heuristics for balancing the pros and the cons, Theor. Decis. **65**, 71–95 (2008)

3.102   A. Tversky, D. Kahneman: Advances in prospect theory: Cumulative representation of uncertainty, J. Risk Uncertain. **5**, 297–323 (1992)

3.103   S. Kaci: *Working with Preferences: Less Is More* (Springer, Berlin 2011)

3.104   L.J. Savage: *The Foundations of Statistics* (Dover, New York 1972)

3.105   D. Dubois, L. Godo, H. Prade, A. Zapico: On the possibilistic decision model: From decision under uncertainty to case-based decision, Int. J. Uncertain. Fuzziness Knowl.-Based Syst. **7**, 631–670 (1999)

3.106   D. Dubois, H. Prade, P. Smets: New semantics for quantitative possibility theory, Lect. Notes Artif. Intell. **2143**, 410–421 (2001)

3.107   R.R. Yager: Possibilistic decision making, IEEE Trans. Syst. Man Cybern. **9**, 388–392 (1979)

3.108   T. Whalen: Decision making under uncertainty with various assumptions about available information, IEEE Trans. Syst. Man Cybern. **14**, 888–900 (1984)

3.109   D. Dubois, D. Le Berre, H. Prade, R. Sabbadin: Using possibilistic logic for modeling qualitative decision: ATMS-based algorithms, Fundam. Inform. **37**, 1–30 (1999)

3.110   H. Fargier, R. Sabbadin: Qualitative decision under uncertainty: Back to expected utility, Artif. Intell. **164**, 245–280 (2005)

3.111   M. Grabisch, T. Murofushi, M. Sugeno (Eds.): *Fuzzy Measures and Integrals – Theory and Applications* (Physica, Heidelberg 2000) pp. 314–322

3.112   D. Dubois, H. Prade, R. Sabbadin: Qualitative decision theory with Sugeno integrals. In: *Fuzzy Measures and Integrals – Theory and Applications*, (Physica, Heidelberg 2000) pp. 314–322

3.113   D. Dubois, H. Prade, R. Sabbadin: Decision-theoretic foundations of possibility theory, Eur. J. Oper. Res. **128**, 459–478 (2001)

3.114   D. Dubois, H. Fargier, P. Perny, H. Prade: Qualitative decision theory with preference relations and comparative uncertainty: An axiomatic approach, Artif. Intell. **148**, 219–260 (2003)

3.115   D. Dubois, H. Fargier, H. Prade, R. Sabbadin: A survey of qualitative decision rules under uncertainty. In: *Decision-making Process-Concepts and Methods*, ed. by D. Bouyssou, D. Dubois, M. Pirlot, H. Prade (Wiley, London 2009) pp. 435–473

3.116   D. Dubois, H. Fargier: Making discrete sugeno integrals more discriminant, Int. J. Approx. Reason. **50**(6), 880–898 (2009)

3.117   R. Sabbadin, H. Fargier, J. Lang: Towards qualitative approaches to multi-stage decision making, Int. J. Approx. Reason. **19**(3–4), 441–471 (1998)

3.118   D. Dubois, H. Prade: When upper probabilities are possibility measures, Fuzzy Set. Syst. **49**, 65–74 (1992)

3.119   D. Dubois, S. Moral, H. Prade: A semantics for possibility theory based on likelihoods, J. Math. Anal. Appl. **205**, 359–380 (1997)

3.120   G. Shafer: Belief functions and possibility measures. In: *Analysis of Fuzzy Information, Vol. I: Mathematics and Logic*, ed. by J.C. Bezdek (CRC, Boca Raton 1987) pp. 51–84

3.121   S. Benferhat, D. Dubois, H. Prade: Possibilistic and standard probabilistic semantics of conditional knowledge bases, J. Log. Comput. **9**, 873–895 (1999)

3.122   V. Maslov: *Méthodes Opérationelles* (Mir Publications, Moscow 1987)

3.123   A. Puhalskii: *Large Deviations and Idempotent Probability* (Chapman Hall, London 2001)

3.124   H.T. Nguyen, B. Bouchon-Meunier: Random sets and large deviations principle as a foundation for possibility measures, Soft Comput. **8**, 61–70 (2003)

3.125   G. De Cooman, D. Aeyels: Supremum-preserving upper probabilities, Inf. Sci. **118**, 173–212 (1999)

3.126   P. Walley, G. De Cooman: A behavioural model for linguistic uncertainty, Inf. Sci. **134**, 1–37 (1999)

3.127   J. Gebhardt, R. Kruse: The context model, Int. J. Approx. Reason. **9**, 283–314 (1993)

3.128   C. Joslyn: Measurement of possibilistic histograms from interval data, Int. J. Gen. Syst. **26**, 9–33 (1997)

3.129   Y. Ben-Haim: *Info-Gap Decision Theory: Decisions Under Severe Uncertainty*, 2nd edn. (Academic Press, London 2006)

3.130   A. Neumaier: Clouds, fuzzy sets and probability intervals, Reliab. Comput. **10**, 249–272 (2004)

3.131   S. Destercke, D. Dubois, E. Chojnacki: Unifying practical uncertainty representations Part I: Generalized p-boxes, Int. J. Approx. Reason. **49**, 649–663 (2008); Part II: Clouds, Int. J. Approx. Reason. **49**, 664–677 (2008)

3.132   D. Dubois, H. Prade: Bayesian conditioning in possibility theory, Fuzzy Set. Syst. **92**, 223–240 (1997)

3.133   G. De Cooman: Integration and conditioning in numerical possibility theory, Ann. Math. Artif. Intell. **32**, 87–123 (2001)

3.134   P. Walley: *Statistical Reasoning with Imprecise Probabilities* (Chapman Hall, London 1991)

3.135   G.J. Klir: A principle of uncertainty and information invariance, Int. J. Gen. Syst. **17**, 249–275 (1990)

3.136   J.F. Geer, G.J. Klir: A mathematical analysis of information-preserving transformations between probabilistic and possibilistic formulations of uncertainty, Int. J. Gen. Syst. **20**, 143–176 (1992)

3.137   D. Dubois, H. Prade, S. Sandri: On possibility/probability transformations. In: *Fuzzy Logic: State of the Art*, ed. by R. Lowen, M. Roubens (Kluwer, Dordrecht 1993) pp. 103–112

3.138   G.J. Klir, B.B. Parviz: Probability-possibility transformations: A comparison, Int. J. Gen. Syst. **21**, 291–310 (1992)

3.139   D. Dubois, H. Prade: On several representations of an uncertain body of evidence. In: *Fuzzy Information and Decision Processes*, ed. by

M. Gupta, E. Sanchez (North-Holland, Amsterdam 1982) pp. 167–181

3.140   P. Smets: Constructing the pignistic probability function in a context of uncertainty. In: *Uncertainty in Artificial Intelligence*, Vol. 5, ed. by M. Henrion (North-Holland, Amsterdam 1990) pp. 29–39

3.141   A.W. Marshall, I. Olkin: *Inequalities: Theory of Majorization and Its Applications* (Academic, New York 1979)

3.142   Z.W. Birnbaum: On random variables with comparable peakedness, Ann. Math. Stat. **19**, 76–81 (1948)

3.143   D. Dubois, E. Huellermeier: Comparing probability measures using possibility theory: A notion of relative peakedness, Inter. J. Approx. Reason. **45**, 364–385 (2007)

3.144   D. Dubois, L. Foulloy, G. Mauris, H. Prade: Probability-possibility transformations, triangular fuzzy sets, and probabilistic inequalities, Reliab. Comput. **10**, 273–297 (2004)

3.145   G. Mauris, V. Lasserre, L. Foulloy: Fuzzy modeling of measurement data acquired from physical sensors, IEEE Trans. Meas. Instrum. **49**, 1201–1205 (2000)

3.146   C. Baudrit, D. Dubois: Practical representations of incomplete probabilistic knowledge, Comput. Stat. Data Anal. **51**, 86–108 (2006)

3.147   G. Mauris: Possibility distributions: A unified representation of usual direct-probability-based parameter estimation methods, Int. J. Approx. Reason. **52**, 1232–1242 (2011)

3.148   D. Dubois, H. Prade, P. Smets: A definition of subjective possibility, Int. J. Approx. Reason. **48**, 352–364 (2008)

3.149   D. Dubois, H. Prade: Unfair coins and necessity measures: A possibilistic interpretation of histograms, Fuzzy Set. Syst. **10**(1), 15–20 (1983)

3.150   D. Dubois, H. Prade: Evidence measures based on fuzzy information, Automatica **21**, 547–562 (1985)

3.151   W. Van Leekwijck, E.E. Kerre: Defuzzification: Criteria and classification, Fuzzy Set. Syst. **108**, 303–314 (2001)

3.152   D. Dubois, H. Prade: The mean value of a fuzzy number, Fuzzy Set. Syst. **24**, 279–300 (1987)

3.153   R.R. Yager: A procedure for ordering fuzzy subsets of the unit interval, Inf. Sci. **24**, 143–161 (1981)

3.154   S. Chanas, M. Nowakowski: Single value simulation of fuzzy variable, Fuzzy Set. Syst. **25**, 43–57 (1988)

3.155   P. Bosc, H. Prade: An introduction to the fuzzy set and possibility theory-based treatment of soft queries and uncertain of imprecise databases. In: *Uncertainty Management in Information Systems*, ed. by P. Smets, A. Motro (Kluwer, Dordrecht 1997) pp. 285–324

3.156   M. Cayrol, H. Farreny, H. Prade: Fuzzy pattern-matching, Kybernetes **11**(2), 103–116 (1982)

3.157   D. Dubois, H. Prade, C. Testemale: Weighted fuzzy pattern matching, Fuzzy Set. Syst. **28**, 313–331 (1988)

3.158   Y. Loiseau, H. Prade, M. Boughanem: Qualitative pattern matching with linguistic terms, AI Commun. **17**(1), 25–34 (2004)

3.159   P. Bosc, O. Pivert: Modeling and querying uncertain relational databases: A survey of approaches based on the possible worlds semantics, Int. J. Uncertain. Fuzziness Knowl.-Based Syst. **18**(5), 565–603 (2010)

3.160   P. Bosc, O. Pivert: About projection-selection-join queries addressed to possibilistic relational databases, IEEE Trans. Fuzzy Syst. **13**, 124–139 (2005)

3.161   P. Bosc, O. Pivert, H. Prade: A model based on possibilistic certainty levels for incomplete databases, Proc. 3rd Int. Conf. Scalable Uncertainty Management (SUM 2009) (Springer, Washington 2009) pp. 80–94

3.162   P. Bosc, O. Pivert, H. Prade: An uncertain database model and a query algebra based on possibilistic certainty, Proc. 2nd Int. Conf. Soft Comput. and Pattern Recognition (SoCPaR10), ed. by T.P. Martin (IEEE, Paris 2010) pp. 63–68

3.163   A. HadjAli, S. Kaci, H. Prade: Database preference queries – A possibilistic logic approach with symbolic priorities, Ann. Math. Artif. Intell. **63**, 357–383 (2011)

3.164   P. Bosc, O. Pivert, H. Prade: A possibilistic logic view of preference queries to an uncertain database, Proc. 19th IEEE Int. Conf. Fuzzy Syst. (FUZZ-IEEE10), Barcelona (2010)

3.165   D. Dubois, H. Prade: Handling bipolar queries in fuzzy information processing. In: *Fuzzy Information Processing in Databases*, Vol. 1, ed. by J. Galindo (Information Science Reference, Hershey 2008) pp. 97–114

3.166   S. Zadrozny, J. Kacprzyk: Bipolar queries – An aggregation operator focused perspective, Fuzzy Set. Syst. **196**, 69–81 (2012)

3.167   P. Bosc, O. Pivert: On a fuzzy bipolar relational algebra, Inf. Sci. **219**, 1–16 (2013)

3.168   D. Dubois, J. Mengin, H. Prade: Possibilistic uncertainty and fuzzy features in description logic. A preliminary discussion. In: *Fuzzy Logic and the Semantic Web*, ed. by E. Sanchez (Elsevier, Amsterdam 2005)

3.169   G. Qi, Q. Ji, J.Z. Pan, J. Du: Extending description logics with uncertainty reasoning in possibilistic logic, Int. J. Intell. Syst. **26**(4), 353–381 (2011)

3.170   D. Dubois, H. Prade: Possibility theory and data fusion in poorly informed environments, Control Eng. Pract. **2**(5), 811–823 (1994)

3.171   D. Dubois, H. Prade, R.R. Yager: Merging fuzzy information. In: *Fuzzy Sets in Approximate Reasoning and Information Systems*, The Handbooks of Fuzzy Sets Series, ed. by J. Bezdek, D. Dubois, H. Prade (Kluwer, Boston 1999) pp. 335–401

3.172   M. Oussalah, H. Maaref, C. Barret: From adaptive to progressive combination of possibility distributions, Fuzzy Set. Syst. **139**(3), 559–582 (2003)

3.173   M. Oussalah: Study of some algebraic properties of adaptive combination rules, Fuzzy Set. Syst. **114**(3), 391–409 (2000)

3.174 S. Destercke, D. Dubois, E. Chojnacki: Possibilistic information fusion using maximum coherent subsets, IEEE Trans. Fuzzy Syst. **17**, 79–92 (2009)

3.175 S. Benferhat, D. Dubois, H. Prade: From semantic to syntactic approaches to information combination in possibilistic logic. In: *Aggregation and Fusion of Imperfect Information*, ed. by B. Bouchon-Meunier (Physica-Verlag, Heidelberg 1998) pp. 141–161

3.176 S. Benferhat: Merging possibilistic networks, Proc. 17th Eur. Conf. Artif. Intell. (Riva del Garda, Italy 2006)

3.177 S. Benferhat, D. Dubois, S. Kaci, H. Prade: Possibilistic merging and distance-based fusion of propositional information, Ann. Math. Artif. Intell. **34**(1–3), 217–252 (2002)

3.178 G. Qi, W. Liu, D.H. Glass, D.A. Bell: A split-combination approach to merging knowledge bases in possibilistic logic, Ann. Math. Artif. Intell. **48**(1/2), 45–84 (2006)

3.179 G. Qi, W. Liu, D.A. Bell: Measuring conflict and agreement between two prioritized knowledge bases in possibilistic logic, Fuzzy Set. Syst. **161**(14), 1906–1925 (2010)

3.180 S. Konieczny, J. Lang, P. Marquis: Distance based merging: A general framework and some complexity results, Proc. Int. Conf. Principles of Knowledge Representation and Reasoning (2002) pp. 97–108

3.181 S. Benferhat, D. Dubois, H. Prade, M.-A. Williams: A practical approach to revising prioritized knowledge bases, Stud. Log. **70**, 105–130 (2002)

3.182 S. Barro, R. Marín, J. Mira, A.R. Patón: A model and a language for the fuzzy representation and handling of time, Fuzzy Set. Syst. **61**, 153175 (1994)

3.183 D. Dubois, H. Prade: Processing fuzzy temporal knowledge, IEEE Trans. Syst. Man Cybern. **19**(4), 729–744 (1989)

3.184 D. Dubois, A. Hadj Ali, H. Prade: Fuzziness and uncertainty in temporal reasoning, J. Univer. Comput. Sci. **9**(9), 1168–1194 (2003)

3.185 M.A. Cárdenas Viedma, R. Marín, I. Navarrete: Fuzzy temporal constraint logic: A valid resolution principle, Fuzzy Set. Syst. **117**(2), 231–250 (2001)

3.186 D. Dubois, A. Hadj Ali, H. Prade: A possibility theory-based approach to the handling of uncertain relations between temporal points, Int. J. Intell. Syst. **22**, 157–179 (2007)

3.187 D. Dubois, J. Lang, H. Prade: Timed possibilistic logic, Fundam. Inform. **15**, 211–234 (1991)

3.188 D. Dubois, H. Fargier, H. Prade: Fuzzy constraints in job-shop scheduling, J. Intell. Manuf. **6**, 215–234 (1995)

3.189 R. Slowinski, M. Hapke (Eds.): *Scheduling under Fuzziness* (Physica, Heidelberg 2000)

3.190 D. Dubois, H. Fargier, P. Fortemps: Fuzzy scheduling: Modelling flexible constraints vs. coping with incomplete knowledge, Eur. J. Oper. Res. **147**, 231–252 (2003)

3.191 S. Chanas, P. Zielinski: Critical path analysis in the network with fuzzy activity times, Fuzzy Set. Syst. **122**, 195–204 (2001)

3.192 S. Chanas, D. Dubois, P. Zielinski: Necessary criticality in the network with imprecise activity times, IEEE Trans. Man Mach. Cybern. **32**, 393–407 (2002)

3.193 J. Fortin, P. Zielinski, D. Dubois, H. Fargier: Criticality analysis of activity networks under interval uncertainty, J. Sched. **13**, 609–627 (2010)

3.194 D. Dubois, J. Fortin, P. Zielinski: Interval PERT and its fuzzy extension. In: *Production Engineering and Management Under Fuzziness*, ed. by C. Kahraman, M. Yavuz (Springer, Berlin 2010) pp. 171–199

3.195 S. Chanas, A. Kasperski: Possible and necessary optimality of solutions in the single machine scheduling problem with fuzzy parameters, Fuzzy Set. Syst. **142**, 359–371 (2004)

3.196 A. Kasperski: A possibilistic approach to sequencing problems with fuzzy parameters, Fuzzy Set. Syst. **150**, 77–86 (2005)

3.197 J. Fortin, A. Kasperski, P. Zielinski: Some methods for evaluating the optimality of elements in matroids with ill-known weights, Fuzzy Set. Syst. **16**, 1341–1354 (2009)

3.198 A. Kasperski, P. Zielenski: Possibilistic bottleneck combinatorial optimization problems with ill-known weights, Int. J. Approx. Reason. **52**, 1298–1311 (2011)

3.199 J. C. Helton, W. L. Oberkampf (Eds.): Alternative representations of uncertainty, Reliab. Eng. Syst. Saf. **85**(1–3), (2004)

3.200 D. Guyonnet, B. Bourgine, D. Dubois, H. Fargier, B. Côme, J.-P. Chilès: Hybrid approach for addressing uncertainty in risk assessments, J. Environ. Eng. **129**, 68–78 (2003)

3.201 D. Dubois, H. Prade: Random sets and fuzzy interval analysis, Fuzzy Set. Syst. **42**, 87–101 (1991)

3.202 C. Baudrit, D. Guyonnet, D. Dubois: Joint propagation and exploitation of probabilistic and possibilistic information in risk assessment, IEEE Trans. Fuzzy Syst. **14**, 593–608 (2006)

3.203 C. Baudrit, I. Couso, D. Dubois: Joint propagation of probability and possibility in risk analysis: Towards a formal framework, Inter. J. Approx. Reason. **45**, 82–105 (2007)

3.204 C. Baudrit, D. Dubois, N. Perrot: Representing parametric probabilistic models tainted with imprecision, Fuzzy Set. Syst. **159**, 1913–1928 (2008)

3.205 M. Gil (Ed.): Fuzzy random variables, Inf. Sci. **133**, (2001) Special Issue

3.206 R. Sabbadin: Towards possibilistic reinforcement learning algorithms, FUZZ-IEEE **2001**, 404–407 (2001)

3.207 S. Benferhat, D. Dubois, S. Lagrue, H. Prade: A big-stepped probability approach for discovering default rules, Int. J. Uncertain. Fuzziness Knowl.-Based Syst. **11**(Supplement), 1–14 (2003)

Part A | 3

3.208   H. Prade, M. Serrurier: Bipolar version space learn-
         ing, Int. J. Intell. Syst. **23**, 1135–1152 (2008)

3.209   H. Prade, M. Serrurier: Introducing possibilistic
         logic in ILP for dealing with exceptions, Artif. In-
         tell. **171**, 939–950 (2007)

3.210   O. Wolkenhauer: *Possibility Theory with Applica-
         tions to Data Analysis* (Research Studies Press,
         Chichester 1998)

3.211   H. Tanaka, P.J. Guo: *Possibilistic Data Analysis for
         Operations Research* (Physica, Heidelberg 1999)

3.212   D. Cayrac, D. Dubois, H. Prade: Handling uncer-
         tainty with possibility theory and fuzzy sets in
         a satellite fault diagnosis application, IEEE Trans.
         Fuzzy Syst. **4**, 251–269 (1996)

3.213   S. Boverie: Online diagnosis of engine dyno test
         benches: A possibilistic approach, Proc. 15th. Eur.
         Conf. Artif. Intell. (IOS, Lyon, Amsterdam 2002)
         pp. 658–662

3.214   L. Amgoud, H. Prade: Reaching agreement through
         argumentation: A possibilistic approach, Proc. 9th
         Int. Conf. Principles of Knowledge Representation
         and Reasoning (KR'04), Whistler, BC, Canada (AAAI,
         Palo Alto 2004) pp. 175–182

3.215   L. Amgoud, H. Prade: Using arguments for mak-
         ing and explaining decisions, Artif. Intell. **173**(3–4),
         413–436 (2009)

3.216   D. Dubois, F. de Dupin Saint-Cyr, H. Prade:
         A possibility-theoretic view of formal concept
         analysis, Fundam. Inform. **75**, 195–213 (2007)

3.217   Y. Djouadi, H. Prade: Possibility-theoretic exten-
         sion of derivation operators in formal concept
         analysis over fuzzy lattices, Fuzzy Optim. Decis.
         Mak. **10**, 287–309 (2011)

3.218   D. Dubois, H. Prade: Generalized possibilistic logic,
         Lect. Notes Comput. Sci. **6929**, 428–432 (2011)

3.219   D. Dubois, H. Prade, S. Schockaert: Rules and meta-
         rules in the framework of possibility theory and
         possibilistic logic, Sci. Iran. **18**(3), 566–573 (2011)

3.220   D. Dubois, H. Prade: Toward multiple-agent exten-
         sions of possibilistic logic, Proc. IEEE Int. Conf. on
         Fuzzy Syst. (FUZZ-IEEE'07), London (2007) pp. 187–
         192

3.221   D. Dubois, H. Prade, A. Rico: A possibilistic logic
         view of Sugeno integrals. In: Proc. Eurofuse Work-
         shop on Fuzzy Methods for Knowledge-Based Sys-
         tems (EUROFUSE 2011), Advances in Intelligent and
         Soft Computing, Vol. 107, ed. by P. Melo-Pinto,
         P. Couto, C. Serôdio, J. Fodor, B. De Baets (Springer,
         Berlin 2011) pp. 19–30

3.222   G. Banon: Constructive decomposition of fuzzy
         measures in terms of possibility and necessity
         measures, Proc. VIth IFSA World Congress, Vol. I (São
         Paulo, Brazil 1995) pp. 217–220

3.223   D. Dubois: Fuzzy measures on finite scales as fami-
         lies of possibility measures, Proc. 7th Conf. Eur. Soc.
         Fuzzy Logic Technol. (EUSFLAT'11) (Atlantis, Annecy
         2011) pp. 822–829

3.224   H. Prade, A. Rico: Possibilistic evidence, Lect. Notes
         Artif. Intell. **6717**, 713–724 (2011)

3.225   D. Dubois, H. Prade, A. Rico: Qualitative capacities
         as imprecise possibilities, Lect. Notes Comput. Sci.
         7958, 169–180 (2011)

3.226   P. Diamond: Fuzzy least squares, Inf. Sci. **46**, 141–157
         (1988)

3.227   K. Loquin, D. Dubois: Kriging and epistemic uncer-
         tainty: A critical discussion. In: *Methods for Han-
         dling Imperfect Spatial Information*, Vol. 256, ed.
         by R. Jeansoulin, O. Papini, H. Prade, S. Schockaert
         (Springer, Berlin 2010) pp. 269–305

3.228   D. Dubois, H. Prade: Gradualness, uncertainty and
         bipolarity: Making sense of fuzzy sets, Fuzzy Set.
         Syst. **192**, 3–24 (2012)

3.229   K. Loquin, D. Dubois: A fuzzy interval analysis ap-
         proach to kriging with ill-known variogram and
         data, Soft Comput. **16**(5), 769–784 (2012)

3.230   H. Prade, M. Serrurier: Maximum-likelihood
         principle for possibility distributions viewed as
         families of probabilities, Proc. IEEE Int. Conf.
         Fuzzy Syst. (FUZZ-IEEE'11), Taipei (2011) pp. 2987–
         2993

4. Aggregation Functions on [0,1]

Radko Mesiar, Anna Kolesárová, Magda Komorníková

After a brief presentation of the history of aggregation, we recall the concept of aggregation functions on $[0, 1]$ and on a general interval $I \subseteq [-\infty, \infty]$. We give a list of basic examples as well as some peculiar examples of aggregation functions. After discussing the classification of aggregation functions on $[0, 1]$ and presenting the prototypical examples for each introduced class, we also recall several construction methods for aggregation functions, including optimization methods, extension methods, constructions based on given

aggregation functions, and introduction of weights. Finally, a remark on aggregation of more general inputs, such as intervals, distribution functions, or fuzzy sets, is added.

Aggregation (fusion, joining) of several input values into one, in some sense the most informative value, is a basic processing method in any field dealing with quantitative information. We only recall mathematics, physics, economy, sociology or finance, among others.

Basic arithmetical operations of addition and multiplication on $[0, \infty]$ are typical examples of aggregation functions. As another example let us recall integration and its application to geometry allowing us to compute areas, surfaces, volumes, etc.

4.1 Historical and Introductory Remarks

Just in the field of integration one can find the first historical traces of aggregation known in the written form. Recall the Moscow *mathematical* papyrus and its problem no. 14, dating back to 1850 BC, concerning the computation of the volume of a pyramidal frustum [4.1], or the exhaustive method allowing to compute several types of areas proposed by Eudoxus of Cnidos around 370 BC [4.2]. The roots of a recent penalty-based method of constructing aggregation functions [4.3] can be found in books of Appolonius of Perga (living in the period about 262–190 BC) who (motivated by the center of gravity problems) proposed an approach leading to the centroid, i. e., to the arithmetic mean, minimizing the sum of squares of the Euclidean distances of the given n points from an unknown but fixed one. Generalization of the Appolonius

of Perga method based on a general norm is known as the Fréchet mean, or also as the *Karcher* mean, and it was deeply discussed in [4.4].

Another type of mean, the Heronian mean of two nonnegative numbers x and y is given by the formula

$$\mathrm{He}(x, y) = \frac{1}{3} \left(x + \sqrt{xy} + y \right) . \tag{4.1}$$

It is named after Hero of Alexandria (10–70 AD) who used this aggregation function for finding the volume of a conical or pyramidal frustum. He showed that this volume is equal to the product of the height of the frustum and the Heronian mean of areas of parallel bases.

Another interesting historical example can be found in multivalued logic. Already Aristotle (384–322 BC) was a classical logician who did not fully accept the law of excluded middle, but he did not create a system of multivalued logic to explain this isolated remark (in the work De Interpretatione, chapter IX). Systems of multivalued logics considering 3, n (finitely many), and later also infinitely many truth degrees were introduced by *Łukasiewicz* [4.5], *Post* [4.6], *Gödel* [4.7], respectively, and in each of these systems the aggregation of truth values was considered (conjunction, disjunction).

Though several particular aggregation functions (or classes of aggregation functions) were discussed in many earlier works (we only recall means discussed around 1930 by *Kolmogorov* [4.8] and *Nagumo* [4.9], or later by *Aczél* [4.10], triangular norms and copulas studied by *Schweizer* and *Sklar* in 1960s of the previous century and summarized in [4.11]), an independent theory of aggregation can be dated only about 20 years back and the roots of its axiomatization can be found in [4.12–14]. Probably the first monograph devoted purely to aggregation is the monograph by *Calvo* et al. [4.15]. As a basic literature for any scientist interested in aggregation we recommend the monographs [4.16–18].

In this chapter, not only we summarize some earlier, but also some recent results concerning aggregation, including classification, construction methods, and several examples. We will deal with inputs and outputs from the unit interval $[0, 1]$. Note that though, in general, we can consider an arbitrary interval $I \subseteq [-\infty, \infty]$, there is no loss of generality (up to the isomorphism) when restricting our considerations to $I = [0, 1]$. As an example, consider the aggregation of nonnegative inputs, i.e., fix $I = [0, \infty[$. Then any aggregation function A on $[0, \infty[$ can be seen as an isomorphic transform of some aggregation function B on $[0, 1]$, restricted to $[0, 1[$ and satisfying two constraints:

i) $B(\mathbf{x}) = 1$ if and only if $\mathbf{x} = (1, \dots, 1)$,
ii) $\sup \{B(\mathbf{x}) \mid \mathbf{x} \in [0, 1[^n\} = 1, n \in \mathbb{N}$.

Note that any increasing bijection $\varphi : [0, 1[\to [0, \infty[$ can be applied as the considered isomorphism. For more details about aggregation on a general interval $I \subseteq [-\infty, \infty]$ refer to [4.17].

We can consider either aggregation functions with a fixed number $n \in \mathbb{N}$, $n \geq 2$, of inputs or extended aggregation functions defined for any number $n \in \mathbb{N}$ of inputs. The number n is called the arity of the aggregation function.

Definition 4.1

For a fixed $n \in \mathbb{N}$, $n \geq 2$, a function $A : [0, 1]^n \to [0, 1]$ is called an (*n*-ary) aggregation function whenever it is increasing in each variable and satisfies the boundary conditions

$$A(0, \dots, 0) = 0 \quad \text{and} \quad A(1, \dots, 1) = 1 \, .$$

A mapping $A : \bigcup_{n \in \mathbb{N}} [0, 1]^n \to [0, 1]$ is called an extended aggregation function whenever $A(x) = x$ for each $x \in [0, 1]$, and for each $n \in \mathbb{N}$, $n \geq 2$, $A \mid [0, 1]^n$ is an *n*-ary aggregation function.

The framework of extended aggregation functions is rather general, not relating different arities, and thus some additional constraints are often considered, such as associativity, decomposability, neutral element, etc.

The Heronian mean He given in (4.1) is an example of a binary aggregation function. Prototypical examples of extended aggregation functions on $[0, 1]$ are:

● The smallest extended aggregation function A_s given by

$$A_s(\mathbf{x}) = \begin{cases} 1 & \text{if } \mathbf{x} = (1, \dots, 1) \\ 0 & \text{else} \, . \end{cases}$$

● The greatest extended aggregation function A_g given by

$$A_g(\mathbf{x}) = \begin{cases} 0 & \text{if } \mathbf{x} = (0, \dots, 0) \\ 1 & \text{else} \, . \end{cases}$$

● The arithmetic mean M given by

$$M(x_1, \dots, x_n) = \frac{1}{n} \sum_{i=1}^{n} x_i \, .$$

● The geometric mean G given by

$$G(x_1, \dots, x_n) = \left(\prod_{i=1}^{n} x_i \right)^{\frac{1}{n}} \, .$$

● The product Π given by

$$\Pi(x_1, \dots, x_n) = \prod_{i=1}^{n} x_i \, .$$

● The minimum Min given by

$$\text{Min}(x_1, \dots, x_n) = \min \{x_1, \dots, x_n\} \, .$$

- The maximum Max given by

$$\mathrm{Max}(x_1,\ldots,x_n) = \max\{x_1,\ldots,x_n\}\ .$$

- The truncated sum S_L (also known as the Łukasiewicz t-conorm) given by

$$S_L(x_1,\ldots,x_n) = \min\left\{1, \sum_{i=1}^{n} x_i\right\}\ ,$$

- The 3-Π-operator E introduced in [4.19] and given by

$$E(x_1,\ldots,x_n) = \frac{\prod_{i=1}^{n} x_i}{\prod_{i=1}^{n} x_i + \prod_{i=1}^{n}(1-x_i)}\ ,$$

with some convention covering the case $\frac{0}{0}$,
- The Pascal weighted arithmetic mean W_P given by

$$W_P(x_1,\ldots,x_n) = \frac{1}{2^{n-1}} \sum_{i=1}^{n} \binom{n-1}{i-1} x_i\ .$$

As distinguished examples of n-ary aggregation functions for a fixed arity $n \geq 2$, recall the projections P_i and order statistics OS_i, $i = 1,\ldots,n$, given by

$$P_i(x_1,\ldots,x_n) = x_i$$

and

$$OS_i(x_1,\ldots,x_n) = x_{\sigma(i)}\ ,$$

where σ is an arbitrary permutation of $(1,\ldots,n)$ such that $x_{\sigma(1)} \leq x_{\sigma(2)} \leq \cdots \leq x_{\sigma(n)}$. Observe that the first projection $P_F = P_1$ and the last projection $P_L = P_n$ can be seen as instances of extended aggregation functions P_F and P_L, respectively. On the other hand, for any fixed $n \geq 2$, OS_1 is just Min $| [0,1]^n$ and $OS_n = $ Max $| [0,1]^n$.

As a peculiar example of an extended aggregation function we can introduce the mapping $V: \bigcup_{n\in\mathbb{N}} [0,1]^n \to [0,1]$ given by

$$V(x_1,\ldots,x_n) = \min\left\{\left(\sum_{i=1}^{n} x_i^n\right)^{\frac{1}{n}}, 1\right\}\ . \tag{4.2}$$

4.2 Classification of Aggregation Functions

Let us denote by $\mathcal{A}$ the class of all extended aggregation functions, and by $\mathcal{A}_n$ (for a fixed $n \geq 2$) the class of all n-ary aggregation functions. Several classifications of n-ary aggregation functions can be straightforwardly extended to the class $\mathcal{A}$. The basic classification proposed by *Dubois* and *Prade* [4.20] distinguishes (both for n-ary and extended aggregation functions):

- *Conjunctive* aggregation functions,
  $C = \{A \in \mathcal{A} \mid A \leq \mathrm{Min}\}$,
- *Disjunctive* aggregation functions,
  $\mathcal{D} = \{A \in \mathcal{A} \mid A \geq \mathrm{Max}\}$,
- *Averaging* aggregation functions,
  $\mathcal{A}v = \{A \in \mathcal{A} \mid \mathrm{Min} \leq A \leq \mathrm{Max}\}$,
- *Mixed* aggregation functions,
  $\mathcal{M} = \mathcal{A} \setminus (C \cup \mathcal{D} \cup \mathcal{A}v)$.

Considering purely averaging aggregation functions $\mathcal{A}v^p = \mathcal{A}v \setminus \{\mathrm{Min}, \mathrm{Max}\}$, we can see that the set $\{C, \mathcal{D}, \mathcal{A}v^p, \mathcal{M}\}$ forms a partition of $\mathcal{A}$. Note that the classes $\mathcal{A}, C, \mathcal{D}, \mathcal{A}v, \mathcal{A}v^p$ are convex, which is not the case of the class $\mathcal{M}$. For the previously introduced ex-

amples it holds:

- $M, G, W_P, P_F, P_L \in \mathcal{A}v^p$,
- $\Pi \in C$,
- $S_L, V \in \mathcal{D}$,
- $E \in \mathcal{M}$.

Observe that n-ary aggregation functions P_i and OS_i, $i = 1,\ldots,n$, are averaging, so are their convex sums, i. e., weighted arithmetic means

$$W = \sum_{i=1}^{n} w_i P_i\ ,$$

and ordered weighted averages (OWA operators) [4.21],

$$\mathrm{OWA} = \sum_{i=1}^{n} w_i OS_i\ ,$$

with $w_i \geq 0$ and $\sum_{i=1}^{n} w_i = 1$. The binary Heronian mean He given in (4.1) is a convex combination of

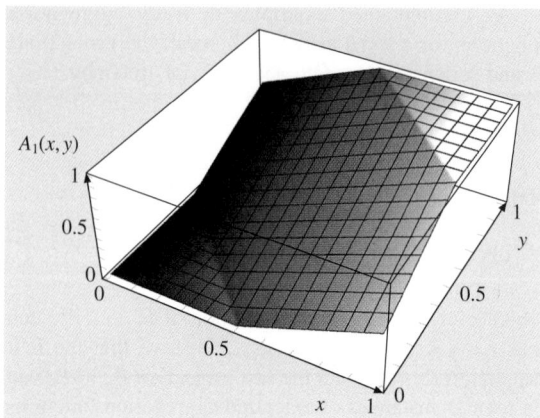

Fig. 4.1 3D plot of the aggregation function A_1 defined by (4.3)

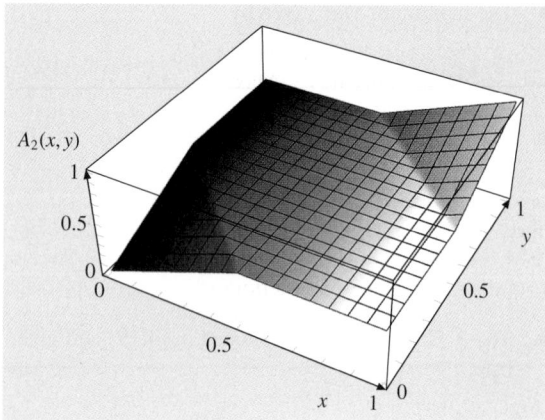

Fig. 4.2 3D plot of the aggregation function A_2 defined by (4.4)

the arithmetic mean M and the geometric mean G, $He = \frac{2}{3}M + \frac{1}{3}G$, and thus it is also averaging.

Consider two binary aggregation functions $A_1, A_2 :$ $[0, 1]^2 \rightarrow [0, 1]$ given by

$$A_1(x, y) = \text{Med}(0, 1, x + y - 0.5) \qquad (4.3)$$

and

$$A_2(x, y) = \text{Med}(x + y, 0.5, x + y - 1), \qquad (4.4)$$

where Med is the standard median operator. Then $A_1, A_2 \in \mathcal{M}$ but $\frac{1}{2}A_1 + \frac{1}{2}A_2 = M \in \mathcal{A}v$. The 3D plots of aggregation functions A_1, A_2 and M are depicted in Figs. 4.1–4.3.

More refined classifications of n-ary aggregation functions are related to order statistics OS_i, $i = 1, \ldots, n$. The conjunctive classification [4.22] deals with the partition of the class $\mathcal{A}_n$ given by $\{C_1, \ldots, C_n, R_C\}$, where the class of i-conjunctive aggregation functions, $i = 1, \ldots, n$, is defined by

$$C_i = \{A \in \mathcal{A}_n \mid \min\{\text{card}\{j \mid x_j \geq A(\boldsymbol{x})\}$$
$$\mid x \in [0, 1]^n\} = i\}$$
$$= \{A \in \mathcal{A}_n \mid A \leq \text{OS}_{n-i+1} \text{ but not } A \leq \text{OS}_{n-i}\},$$

where formally $\text{OS}_0 \equiv 0$.

In other words, A is i-conjunctive if and only if the aggregated value $A(\boldsymbol{x})$ is dominated by at least i input values independently of $\boldsymbol{x} \in [0, 1]^n$, but not by $(i + 1)$ values, in general.

Clearly, the classes $C_1, \ldots, C_n$ are pairwise disjoint and the remaining aggregation functions are members of the class $R_C = \mathcal{A}_n \setminus \bigcup_{i=1}^{n} C_i$. If we come back to the above-mentioned basic classification of aggregation functions (applied to $\mathcal{A}_n$), we obtain $C = C_n$ and $\mathcal{W}_C = \bigcup_{i=1}^{n-1} C_i = \mathcal{A}v \setminus \{\text{Min}\}$. The class $\mathcal{W}_C$ is called *weakly conjunctive* [4.22].

Similarly, we have a disjunctive type of classification of $\mathcal{A}_n$ related to the partition $\{\mathcal{D}_1, \ldots, \mathcal{D}_n, R_{\mathcal{D}}\}$, with

$$\mathcal{D}_i = \{A \in \mathcal{A}_n \mid A \geq \text{OS}_i \text{ but not } A \geq \text{OS}_{i+1}\},$$
$$i = 1, \ldots, n.$$

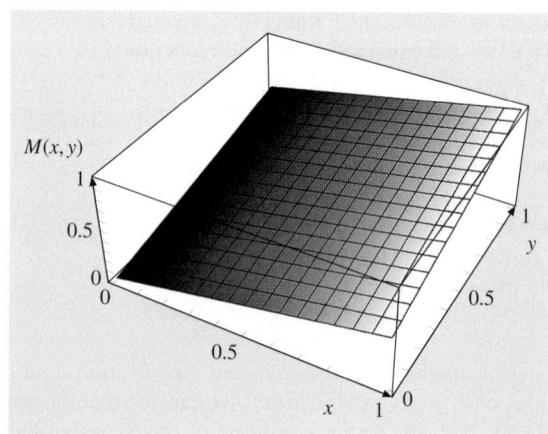

Fig. 4.3 3D plot of the aggregation function $\frac{1}{2}A_1 + \frac{1}{2}A_2$ $= M$

Then $\mathcal{D}_n = \mathcal{D}$ and for the class of *weakly disjunctive* aggregation functions $\mathcal{W}_\mathcal{D} = \bigcup_{i=1}^{n-1} \mathcal{D}_i$ we have $\mathcal{W}_\mathcal{D} = \mathcal{A}v \setminus \{\text{Max}\}$. Hence $\mathcal{W}_C \bigcap \mathcal{W}_\mathcal{D} = \mathcal{A}v^p$, and $A \in \bigcup_{i=1}^{n} C_i$ if and only if $A \le \text{Max}$, while $A \in \bigcup_{i=1}^{n} \mathcal{D}_i$ if and only if $A \ge \text{Min}$.

Note that the conjunctive and disjunctive classifications can be applied to aggregation functions defined on posets, too [4.22], and that this approach to the classification of aggregation functions on [0, 1] was already proposed by *Marichal* in [4.23] as *i*-tolerant and *i*-intolerant aggregation functions (Marichal's approach based on order statistics is applicable when considering chains only).

Observe that this approach to classification has no direct extension to extended aggregation functions. On the other hand, we have the next classification valid for extended aggregation functions only. We distinguish:

- *Dimension decreasing* aggregation functions forming the class $\mathcal{A}_\searrow$, satisfying $A(x_1,\dots,x_n,x_{n+1}) \le A(x_1,\dots,x_n)$ for any $n \in \mathbb{N}$, $x_1,\dots,x_{n+1} \in [0,1]$, but violating the equality, in general.
- *Dimension increasing* aggregation functions forming the class $\mathcal{A}_\nearrow$, satisfying $A(x_1,\dots,x_n,x_{n+1}) \ge A(x_1,\dots,x_n)$ for any $n \in \mathbb{N}$, $x_1,\dots,x_{n+1} \in [0,1]$, but violating the equality, in general.
- *Dimension averaging* aggregation functions forming the class $\overleftrightarrow{\mathcal{A}}$, satisfying $A(x_1,\dots,x_n,0) \le A(x_1,\dots,x_n) \le A(x_1,\dots,x_n,1)$ for any $n \in \mathbb{N}$, $x_1,\dots,x_n \in [0,1]$, and attaining strict inequalities for at least one $\boldsymbol{x} \in [0,1]^n$.

Evidently, the classes $\mathcal{A}_\searrow, \mathcal{A}_\nearrow$, and $\overleftrightarrow{\mathcal{A}}$ are disjoint and they, together with their reminder $\mathcal{A} \setminus (\mathcal{A}_\searrow \cup \mathcal{A}_\nearrow \cup \overleftrightarrow{\mathcal{A}})$, form a partition of $\mathcal{A}$. Let us note that each associative conjunctive aggregation function is dimension decreasing, and thus, $\Pi, \text{Min} \in \mathcal{A}_\searrow$. Similarly, each associative disjunctive aggregation function is dimension increasing, so, $S_L, \text{Max} \in \mathcal{A}_\nearrow$.

Recently, *Yager* has introduced extended aggregation functions with the self-identity property [4.24] characterized by the equality

$$A(x_1,\dots,x_n,A(x_1,\dots,x_n)) = A(x_1,\dots,x_n)$$

for any $n \in \mathbb{N}$ and $x_1,\dots,x_n \in [0,1]$ (e.g., the arithmetic mean M or the geometric mean G satisfy this

property). Evidently, each such aggregation function satisfies

$$A(x_1,\dots,x_n,0) \le A(x_1,\dots,x_n,x_{n+1})$$
$$\le A(x_1,\dots,x_n,1)$$

for all $n \in \mathbb{N}, x_1,\dots,x_{n+1} \in [0,1]$ and thus, if the strict inequalities are attained for some $n \in \mathbb{N}$ and

$$x_1,\dots,x_{n+1} \in [0,1],$$

A belongs to $\overleftrightarrow{\mathcal{A}}$. So, for example, $M, G \in \overleftrightarrow{\mathcal{A}}$. The extended aggregation function V (4.2) also belongs to $\overleftrightarrow{\mathcal{A}}$. On the other hand, the first projection P_F does not belong to

$$\mathcal{A}_\searrow \cup \mathcal{A}_\nearrow \cup \overleftrightarrow{\mathcal{A}},$$

and the last projection P_L belongs to $\overleftrightarrow{\mathcal{A}}$. Recall that if $A \in \mathcal{A}_\searrow$, it is also said to have the *downward attitude property* [4.24]. Similarly, the upward attitude property introduced in [4.24] corresponds to the class $\mathcal{A}_\nearrow$. Dimension increasing aggregation functions were also considered in [4.25].

Let us return to the basic classification of aggregation functions and recall several distinguished types of aggregation functions belonging to the classes $C, \mathcal{D}, \mathcal{A}v^p$, and $\mathcal{M}$:

- *Conjunctive aggregation functions*: Triangular norms [4.26, 27], copulas [4.27, 28], quasi-copulas [4.29, 30], and semicopulas [4.31].
- *Disjunctive aggregation functions*: Triangular conorms [4.26, 27], dual copulas [4.28].
- *Averaging aggregation functions*: (Weighted) quasi-arithmetic means [4.10], idempotent uninorms [4.32], integrals based on capacities, including the Choquet and Sugeno integrals [4.18, 33–36], also covering OWA [4.21], ordered weighted maximum (OWMax) [4.37] and ordered modular average (OMA) [4.38] operators, as well as lattice polynomials [4.39].
- *Mixed aggregation functions*: nonidempotent uninorms [4.40], gamma-operators [4.41], special convex sums in fuzzy linear programming [4.42].

For more details concerning these aggregation functions see [4.17] or references given above.

4.3 Properties and Construction Methods

Properties of aggregation functions are mostly related to the field of their application, such as multicriteria decision aid, multivalued logics, or probability theory, for example. Besides the standard analytical properties of functions, such as continuity, the Lipschitz property, and (perhaps adapted) algebraic properties, such as symmetry, associativity, bisymmetry, neutral element, annihilator, cancellativity, or idempotency [4.17, Chapter 2], the above-mentioned applied fields have brought into aggregation theory properties as decomposability, conjunctivity, or n-increasingness. Each of the mentioned properties can be introduced for n-ary aggregation functions (excepting decomposability), and thus also for extended aggregation functions. However, in the case of extended aggregation functions, some properties can be introduced in a stronger form, involving different arities in a single formula.

For example, the (weak) idempotency of $A \in \mathcal{A}$ means the idempotency of each $A \mid [0, 1]^n$, which means that for each $n \in \mathbb{N}$ and

$$x \in [0, 1] , \quad A(\underbrace{x, \ldots, x}_{n\text{-times}}) = x .$$

Note that an extended aggregation function A is idempotent if and only if it is averaging, i.e., $A \in \mathcal{A}v$. The strong idempotency [4.15] of an extended aggregation function $A \in \mathcal{A}$ means that

$$A(\underbrace{\boldsymbol{x}, \ldots, \boldsymbol{x}}_{k\text{-times}}) = A(\boldsymbol{x})$$

for each $k \in \mathbb{N}$ and $\boldsymbol{x} \in \bigcup_{n\in\mathbb{N}} [0, 1]^n$. For example, the extended aggregation function W_{P} is idempotent but not strongly idempotent.

Similarly, $e \in [0, 1]$ is a (weak) neutral element of an extended aggregation function $A \in \mathcal{A}$ if and only if for each $n \geq 2$ and $\boldsymbol{x} \in [0, 1]^n$ such that $x_j = e$ for $j \neq i$ it holds $A(\boldsymbol{x}) = x_i$. On the other hand, e is a strong neutral element of an extended aggregation function $A \in \mathcal{A}$ if and only if for any $n \geq 2, \boldsymbol{x} \in [0, 1]^n$ with $x_i = e$, it holds

$$A(x_1, \ldots, x_{i-1}, e, x_{i+1}, \ldots, x_n)$$
$$= A(x_1, \ldots, x_{i-1}, x_{i+1}, \ldots, x_n) .$$

Obviously, if e is a strong neutral element of $A \in \mathcal{A}$ then it is also a (weak) neutral element of A. As an example, consider the extended copula $D \in \mathcal{A}$ given by

$$D(x_1, \ldots, x_n) = x_1 \cdot \min \{x_2, \ldots, x_n\} .$$

Obviously, $e = 1$ is a weak neutral element of D. However $D\left(1, \frac{1}{2}, \frac{1}{2}\right) = \frac{1}{2} \neq \frac{1}{4} = D\left(\frac{1}{2}, \frac{1}{2}\right)$, i.e., $e = 1$ is not a strong neutral element of D. For a deeper discussion and exemplification of properties of aggregation functions we recommend [4.17].

Aggregation functions in many fields are constrained by the required properties – axioms in each considered field. As a typical example recall multivalued logics (fuzzy logics) with truth values domain $[0, 1]$, where conjunction is modeled by means of triangular norms [4.26, 43, 44]. Recall that a binary aggregation function $T : [0, 1]^2 \rightarrow [0, 1]$ is called a triangular norm (t-norm for short) whenever it is symmetric, associative and $e = 1$ is its neutral element. Due to associativity, there is a genuine extension of a t-norm T into an extended aggregation function (we will also use the same notation T in this case). Then $e = 1$ is a strong neutral element for the extended T. However, without some additional properties we still cannot determine a t-norm convenient for our purposes. Requiring, for example, the idempotency of T, we obtain that the only solution is $T = \mathrm{Min}$, the strongest triangular norm. Considering continuous triangular norms satisfying the diagonal inequalities $0 < T(x, x) < x$ for all $x \in]0, 1[$, we can show that T is isomorphic to the product Π, i.e., there is an automorphism $\varphi : [0, 1] \rightarrow [0, 1]$ such that $T(x, y) = \varphi^{-1} (\Pi (\varphi(x), \varphi(y)))$, and in the extended form, $T(x_1, \ldots, x_n) = \varphi^{-1} (\Pi (\varphi(x_1), \ldots, \varphi(x_n)))$. For more details and several other results we recommend [4.26].

As another example consider probability theory, namely the relationship between the joint distribution function F_Z of a random vector $Z = (X_1, \ldots, X_n)$, and the corresponding marginal one-dimensional distribution functions $F_{X_1}, \ldots, F_{X_n}$. By the *Sklar* theorem [4.45], for all $(x_1, \ldots, x_n) \in \overline{\mathbb{R}}$ we have

$$F_Z(x_1, \ldots, x_n) = C (F_{X_1}(x_1), \ldots, F_{X_n}(x_n))$$

for some n-ary aggregation function C. Obviously, constrained by the basic properties of probabilities, C should possess a neutral element $e = 1$ and annihilator (zero element) $a = 0$, and the function C should be n-increasing (i.e., probability $P (Z \in [u_1, v_1] \times \cdots \times [u_n, v_n]) \geq 0$ for any n-dimensional box $[u_1, v_1] \times \cdots \times [u_n, v_n]$), which yields an axiomatic definition of copulas. More details for interested readers can be found in [4.28]. Considering some additional constraints, we obtain special subclasses of copulas. For example, if we fix $n = 2$ and consider the stability of copulas with respect to positive powers, i.e., the property $C(x^\lambda, y^\lambda) = (C(x,y))^\lambda$ for each $\lambda \in]0, \infty[$ and each $(x,y) \in [0,1]^2$, then we obtain *extreme value copulas* (EV copulas) [4.46, 47]. Recall that a copula $C : [0,1]^2 \rightarrow [0,1]$ is an EV copula if and only if there is a convex function $d : [0,1] \rightarrow [0,1]$ such that for each $t \in [0,1]$, $\max\{t, 1-t\} \le d(t) \le 1$ and for all $(x,y) \in]0,1[^2$,

$$C(x,y) = (xy)^{d\left(\frac{\log x}{\log xy}\right)}$$

(observe that on $[0,1]^2 \backslash]0,1[^2$ for each copula it holds $C(x,y) = \min\{x,y\}$).

Our third example comes from economics. In multicriteria decision problems, we often meet the requirement of the comonotone additivity of the considered n-ary (extended) aggregation function A, i.e., we expect that $A(\boldsymbol{x} + \boldsymbol{y}) = A(\boldsymbol{x}) + A(\boldsymbol{y})$ for all $\boldsymbol{x}, \boldsymbol{y} \in [0,1]^n$ such that $\boldsymbol{x} + \boldsymbol{y} \in [0,1]^n$ and $(x_i - x_j)(y_i - y_j) \ge 0$ for any $i,j \in \{1, \dots, n\}$. The comonotonicity of $\boldsymbol{x}$ and $\boldsymbol{y}$ means that the ordering on $\{1, \dots, n\}$ induced by $\boldsymbol{x}$ is not contradictory to that one induced by $\boldsymbol{y}$. Due to *Schmeidler* [4.48], we know that then A is necessarily the Choquet integral based on the fuzzy measure $m : 2^{\{1, \dots, n\}} \rightarrow [0,1]$, $m(E) = A(1_E)$, given by (4.6).

The axiomatic approach to aggregation characterizes some special classes of aggregation functions. Another important look at aggregation involves construction methods. We can roughly divide them into the next four groups:

- Optimization methods,
- Extension methods,
- Constructions based on the given aggregation functions,
- Introduction of weights.

An exhaustive overview of construction methods for aggregation functions can be found in [4.17, Chapter 6]. Here we briefly recall the most distinguished ones.

A typical *optimization* method is the penalty-based approach proposed in [4.49] and generalized in [4.3], where dissimilarity functions were introduced, see also [4.50].

Definition 4.2
A function $D : [0,1]^2 \rightarrow [0, \infty[$ given by

$$D(x,y) = K(f(x) - f(y)) ,$$

where $f : [0,1] \rightarrow \mathbb{R}$ is a continuous strictly monotone function and $K : \mathbb{R} \rightarrow [0, \infty[$ is a convex function

attaining the unique minimum $K(0) = 0$, is called a *dissimilarity* function.

Theorem 4.1
Let $D : [0,1]^2 \rightarrow [0, \infty]$ be a dissimilarity function. Then for any $n \in \mathbb{N}$, $x_1, \dots, x_n \in [0,1]$, the function $h : [0,1] \rightarrow \mathbb{R}$ given by $h(t) = \sum_{i=1}^{n} D(x_i, t)$ attains its minimal value exactly on a closed interval $[a,b]$ and the formula

$$A(x_1, \dots, x_n) = \frac{a+b}{2}$$

defines a strongly idempotent symmetric extended aggregation function A on $[0,1]$.

Construction given in Theorem 4.1 covers:

- the arithmetic mean $(D(x,y) = (x-y)^2)$,
- quasi-arithmetic means $(D(x,y) = (f(x) - f(y))^2)$,
- the median $(D(x,y) = |x-y|)$,

among others. This method is a generalization of the Appolonius of Perga method. Note that in general, a function D need not be symmetric, i.e., K need not be an even function (compare with the symmetry of metrics). As a typical example, let us recall the dissimilarity function $D_c : [0,1]^2 \rightarrow [0, \infty]$, $c \in]0, \infty[$, given by

$$D_c(x,y) = \begin{cases} x-y & \text{if } x \ge y , \\ c(y-x) & \text{if } x < y , \end{cases}$$

yielding by means of Theorem 4.1 the α-quantile of a sample $(x_1, \dots, x_n)$ with $\alpha = \frac{1}{1+c}$.

As a possible generalization of Theorem 4.1, one can consider different dissimilarity functions D_i (which violates the symmetry of the constructed aggregation function A). Consider, for example, $D_1(x,y) = |x - y|$ and $D_2(x,y) = \dots = D_n(x,y) = \dots = (x-y)^2$. Then the minimization of the sum $\sum_{i=1}^{n} D_i(x_i, t)$ results in the extended aggregation function $A : \bigcup_{n \in \mathbb{N}} [0,1]^n \rightarrow [0,1]$ given by

$$A(x_1, \dots, x_n) =$$
$$\text{Med}\left(x_1, M(x_2, \dots, x_n) - 0.5, M(x_2, \dots, x_n) + 0.5\right) ,$$
$$(4.5)$$

whenever $n > 1$.

Some other generalizations based on a generalized approach to dissimilarity (penalty) functions can be found in [4.16].

Extension methods are based on a partial information that is available about an aggregation function. As
a typical example, we recall integral-based aggregation
functions. Suppose that for a fixed arity n the values of
an aggregation function A are known at Boolean inputs
only, i.e., we know $A \mid \{0,1\}^n$ only. Identifying subsets of the space $X = \{1,\ldots,n\}$ with the corresponding
characteristic functions, we get the set function $m:$
$2^X \to [0,1]$ given by $m(E) = A(1_E)$. Obviously, m is
monotone, i.e., $m(E_1) \le m(E_2)$ whenever $E_1 \subseteq E_2 \subseteq X$,
and $m(\emptyset) = A(0,\ldots,0) = 0$, $m(X) = A(1,\ldots,1) = 1$.
Note that m is often called a *fuzzy measure* [4.51, 52] or
a *capacity* [4.17].

Among several integral-based extension methods
we recall:

- The *Choquet* integral [4.53], $Ch_m : [0,1]^n \to [0,1]$,

$$Ch_m(x) = \sum_{i=1}^{n} x_{\sigma(i)} \cdot (m(E_{\sigma,i}) - m(E_{\sigma,i+1})) ,$$

(4.6)

where $\sigma : X \to X$ is a permutation such that $x_{\sigma(1)} \le$
$x_{\sigma(2)} \le \cdots \le x_{\sigma(n)}$, $E_{\sigma,i} = \{\sigma(i),\ldots,\sigma(n)\}$ for $i =$
$1,\ldots,n$, and $E_{\sigma,n+1} = \emptyset$. Note that the Choquet in-
tegral can be seen as a weighted arithmetic mean
with the weights dependent on the ordinal structure
of the input vector x. If the capacity m is additive,
i.e., $m(E) = \sum_{i\in E} m(\{i\})$, then

$$Ch_m(x) = \sum_{i=1}^{n} w_i x_i ,$$

where for the weights it holds $w_i = m(\{i\})$, $i \in X$
(hence $\sum_{i=1}^{n} w_i = 1$).
- The *Sugeno* integral [4.51], $Su_m : [0,1]^n \to [0,1]$,

$$Su_m(x) = \max \{\min \{x_{\sigma(i)}, m(E_{\sigma,i})\} \mid i \in X\} .$$

If m is maxitive, i.e., $m(E) = \max \{m(\{i\}) \mid i \in E\}$,
then we recognize the weighted maximum
$Su_m(x) = \max \{\min \{x_i, v_i\} \mid i \in X\}$, with weights
$v_i = m(\{i\})$ (hence $\max \{v_i \mid i \in X\} = 1$).
- The copula-based integral [4.34], $I_{C,m} : [0,1]^n \to$
$[0,1]$, where $C : [0,1]^2 \to [0,1]$ is a binary copula,

$$I_{C,m}(x) = \sum_{i=1}^{n} \left(C\left(x_{\sigma(i)}, m(E_{\sigma,i})\right) \right.$$
$$\left. -C\left(x_{\sigma(i)}, m(E_{\sigma,i+1})\right)\right) .$$

This integral covers the Choquet integral if C is
equal to the product copula Π, $I_{\Pi,m} = Ch_m$, as well
as the Sugeno integral in the case of the greatest
copula Min, $I_{Min,m} = Su_m$. Observe that if the capacity m is symmetric, i.e., $m(E) = v_{card E}$, where
$0 = v_0 \le v_1 \le \cdots \le v_n = 1$, then $I_{C,m}$ turns to OMA
operator introduced in [4.38]. Its special instances
are the OWA operators [4.21] based on the Choquet
integral,

$$OWA(x) = \sum_{i=1}^{n} x_{\sigma(i)} \cdot w_i ,$$

with $w_i = v_i - v_{i-1}$, and the OWMax operator [4.37],

$$OWMax(x) = \max \{\min \{x_{\sigma(i)}, v_i\} \mid i \in X\} .$$

For better understanding, fix $n = 2$, i.e., consider
$X = \{1,2\}$. Then $m(\{1\}) = a$ and $m(\{2\}) = b$ are any
constants from $[0,1]$, and $m(\emptyset) = 0$, $m(X) = 1$ due to
the boundary conditions. The following equalities hold:

- $Ch_m(x,y) = \begin{cases} ax+(1-a)y & \text{if } x \ge y, \\ (1-b)x+by & \text{else}, \end{cases}$
- $Su_m(x,y) = \max \{\min \{a,x\}, \min \{b,y\}, \min \{x,y\}\}$,
- $I_{C,m}(x,y) = \begin{cases} C(x,a)+y-C(y,a) & \text{if } x \ge y, \\ C(y,b)+x-C(x,b) & \text{else}. \end{cases}$

The considered capacity m is symmetric if and only
if $a = b$, and then:

- $Ch_m(x,y) = OWA(x,y) = (1-a) \cdot \min \{x,y\} + a \cdot$
$\max \{x,y\}$,
- $Su_m(x,y) = OWMax(x,y) = Med(x,a,y)$ is the so-
called a-median [4.54, 55],
- $I_{C,m}(x,y) = OMA(x,y) = f_1(\min \{x,y\}) +$
$f_2(\max \{x,y\})$, where $f_1, f_2 : [0,1] \to [0,1]$ are
given by $f_1(t) = t - C(t,a)$ and $f_2(t) = C(t,a)$.

For more details concerning integral-based
constructions of aggregation functions we recommend [4.34, 36, 56] or [4.34] by *Klement, Mesiar*, and
Pap.

Another kind of extension methods exploiting
capacities is based on the Möbius transform. Recall that for a capacity $m : 2^X \to [0,1]$, its Möbius
transform $\mu : 2^X \to \mathbb{R}$ is given by

$$\mu(E) = \sum_{L \subseteq E} (-1)^{card(E\setminus L)} m(L) .$$

Theorem 4.2

[4.57] Let $C : [0, 1]^n \to [0, 1]$ be an n-ary copula, and $m : 2^X \to [0, 1]$ a capacity. Then the function $A_{C,m} : [0, 1]^n \to [0, 1]$ given by

$$A_{C,m}(\mathbf{x}) = \sum_{E \subseteq X} \mu(E) \cdot C(\mathbf{x} \vee 1_{E^c})$$

is an aggregation function.

Special instances of Theorem 4.2 are the *Lovász* extension [4.58] corresponding to the strongest copula Min ($A_{\mathrm{Min},m} = I_{\Pi,m} = \mathrm{Ch}_m$ is just the Choquet integral), and the Owen extension [4.59] corresponding to the product copula Π ($A_{\Pi,m}(\mathbf{x}) = \sum_{E \subseteq X} \left(\mu(E) \prod_{i \in E} x_i \right)$).

Several extension methods were introduced for binary copulas, for example, in the case when only the information about their diagonal section $\delta_C : [0, 1] \to [0, 1]$, $\delta_C(x) = C(x, x)$ is available. If $\delta : [0, 1] \to [0, 1]$ is any increasing 2-Lipschitz function such that $\delta(0) = 0, \delta(1) = 1$, and $\delta(x) \leq x$ for each $x \in [0, 1]$, then the formula

$$D(x, y) = \min \left\{ x, y, \frac{\delta(x) + \delta(y)}{2} \right\}, \quad (x, y) \in [0, 1]^2 ,$$

defines a binary copula with $\delta_D = \delta$. Note that D is the greatest symmetric copula with the given diagonal section. Among numerous papers dealing with such types of extensions we recommend the overview paper [4.60]. Similarly, one can extend horizontal or vertical sections to copulas [4.61]. An overview of extension methods for triangular norms can be found in [4.26].

The third group of construction methods involves methods creating new aggregation functions from the given ones. These methods are applied either to aggregation functions with a fixed arity n, or to extended aggregation functions. Some of them can be applied to any kind of aggregation functions. As a typical example, recall transformation of aggregation functions by means of an automorphism $\varphi : [0, 1] \to [0, 1]$ (i.e., an isomorphic transformation) given by

$$A_\varphi(x_1, \ldots, x_n) = \varphi^{-1} \left(A \left(\varphi(x_1), \ldots, \varphi(x_n) \right) \right) .$$
$$(4.7)$$

Transformation (4.7) preserves all algebraic properties as well as the classification of aggregation functions. However, some analytical properties can be broken, for example, the Lipschitz property or n-increasingness. Some special classes of aggregation functions can be

characterized by a unique member and its isomorphic transforms. Consider, for example, triangular norms. Then strict triangular norms are isomorphic to the product t-norm Π, nilpotent t-norms are isomorphic to the Łukasiewicz t-norm T_L. Similarly, quasi-arithmetic means with no annihilator are isomorphic to the arithmetic mean M. The only n-ary aggregation functions invariant under isomorphic transformations are the lattice polynomials [4.62], i.e., the Choquet integrals with respect to $\{0, 1\}$-valued capacities. So, for $n = 2$, only Min, Max, P_F and P_L are invariant under isomorphic transformations. There are several generalizations of (4.7). One can consider, for example, decreasing bijections $\eta : [0, 1] \to [0, 1]$ and define A_η via (4.7). This type of transformations reverses the conjunctivity of aggregation function into disjunctivity, and vice versa. It preserves the existence of a neutral element (annihilator), however, if e is a neutral element of A (a is an annihilator of A) then $\eta^{-1}(e)$ is a neutral element of A_η ($\eta^{-1}(a)$ is an annihilator of A_η). If η is involutive, i. e., if $\eta \circ \eta = \mathrm{id}_{[0,1]}$, then $(A_\eta)_\eta = A$, so there is a duality between A and A_η. The most applied duality is based on the standard (or Zadeh's) negation $\eta : [0, 1] \to [0, 1]$ given by $\eta(x) = 1 - x$. In that case, we use the notation $A^d = A_\eta$ and $A^d(x_1, \ldots, x_n) = 1 - A(1 - x_1, \ldots, 1 - x_n)$. As a distinguished example recall the class of triangular conorms which are just the dual aggregation functions to triangular norms, i. e., S is a triangular conorm [4.26] if and only if there is a triangular norm T such that $S = T^d$.

Further generalizations of (4.7) consider different automorphisms $\varphi, \varphi_1, \ldots, \varphi_n : [0, 1] \to [0, 1]$,

$$A_{\varphi, \varphi_1, \ldots, \varphi_n}(x_1, \ldots, x_n)$$
$$= \varphi \left(A \left(\varphi_1(x_1), \ldots, \varphi_n(x_n) \right) \right) . \quad (4.8)$$

Moreover, it is enough to suppose that $\varphi_1, \ldots, \varphi_n$ are monotone (not necessarily strictly) and satisfy $\varphi_i(0) = 0, \varphi_i(1) = 1, i = 1, \ldots, n$, as in such case it also holds that for any aggregation function A, $A_{\varphi, \varphi_1, \ldots, \varphi_n}$ given by (4.8) is an aggregation function.

Another construction well known from functional theory is linked to the composition of functions. We have two kinds of composition methods. In the first one, considering a k-ary aggregation function $B : [0, 1]^k \to [0, 1]$, we can choose arbitrary k aggregation functions $C_1, \ldots, C_k$ (either all of them are extended aggregation functions, or all of them are n-ary aggregation functions for some fixed $n > 1$), and then we can introduce a new aggregation function A (either extended, with the convention $A(x) = x$, $x \in [0, 1]$; or n-ary) such that

$$A(\mathbf{x}) = B\left(C_1(\mathbf{x}), \dots, C_k(\mathbf{x})\right) . \tag{4.9}$$

As a typical example of construction (4.9), consider B to be a weighted arithmetic mean W, $W(x_1, \dots, x_n) = \sum_{i=1}^{n} w_i x_i$. Then

$$A(\mathbf{x}) = \sum_{i=1}^{k} w_i \cdot C_i(\mathbf{x}) ,$$

i.e., A is a convex combination of aggregation functions $C_1, \dots, C_k$.

The second method is based on a partition of the space of coordinates $\{1, \dots, n\}$ into subspaces

$$\{1, \dots, n_1\}, \{n_1 + 1, \dots, n_1 + n_2\}, \dots,$$
$$\{n_1 + \cdots + n_{k-1} + 1, n\} .$$

Then, considering a k-ary aggregation function B: $[0, 1]^k \to [0, 1]$ and aggregation functions $C_i : [0, 1]^{n_i} \to [0, 1]$, $i = 1, \dots, k$, we can define a composite aggregation function $A : [0, 1]^n \to [0, 1]$ by

$$A(x_1, \dots, x_n)$$
$$= B\left(C_1(x_1, \dots, x_{n_1}), C_2(x_{n_1+1}, \dots, x_{n_1+n_2}),\right.$$
$$\left. \dots, C_k(x_{n_1+\cdots+n_{k-1}}, \dots, x_n)\right) . \tag{4.10}$$

This method can be generalized by considering an arbitrary partition of $\{1, \dots, n\}$ into $\{I_1, \dots, I_k\}$. As an example, consider the n-ary copula $C : [0, 1]^n \to [0, 1]$ defined for a fixed partition $\{I_1, \dots, I_k\}$ of $\{1, \dots, n\}$ by

$$C(x_1, \dots, x_n) = \prod_{i=1}^{k} \min\left\{x_j \mid j \in I_i\right\} .$$

For more details, see [4.63].

The third group containing constructions based on some given aggregation functions can be seen as a group of patchwork methods. As typical examples, we can recall several types of ordinal sums. Besides the well-known Min-based ordinal sums for conjunctive aggregation functions (especially for triangular norms and copulas) [4.26, 64], W-ordinal sums for copulas (or quasi-copulas) [4.65], as well as g-ordinal sums for copulas [4.66], we recall one kind of ordinal sums introduced in [4.67] which is applicable to arbitrary aggregation functions.

Theorem 4.3

Let $f : [0, 1] \to [-\infty, \infty]$ be a continuous strictly monotone function, and let $0 = a_0 < a_1 < \cdots < a_k = 1$ be a given sequence of real constants. Then for any system $(A_i)_{i=1}^{k}$ of n-ary (extended) aggregation functions the function $A : [0, 1]^n \to [0, 1]$ ($A : \cup_{n \in \mathbb{N}} [0, 1]^n \to [0, 1]$) given by

$$A(\mathbf{x}) = f^{-1}\left(\left(\sum_{j=1}^{k} f(a_{j-i} + (a_j - a_{j-1})A_j(\mathbf{x}^{(j)})\right)\right.$$
$$\left. - \sum_{j=1}^{k-1} f(a_j)\right) , \tag{4.11}$$

where

$$\mathbf{x}^{(j)} = \left(x_1^{(j)}, \dots, x_n^{(j)}\right)$$

and

$$x_i^{(j)} = \max\left\{0, \min\left\{1, \frac{x_i - a_{j-1}}{a_j - a_{j-1}}\right\}\right\} ,$$

is an n-ary (extended) aggregation function.

Observe that if all A_i's are triangular norms (copulas, quasi-copulas, triangular conorms, continuous aggregation functions, idempotent aggregation functions, symmetric aggregation functions) then so is the newly constructed aggregation function A.

The fourth group contains construction methods allowing one to introduce weights into the aggregation procedure. The *quantitative* look at weights can be seen as the corresponding *repetition* of inputs, and the weights roughly correspond to the occurrence of single input arguments. For example, when considering a strongly idempotent (symmetric) aggregation function constructed by means of a dissimilarity function D (see Theorem 4.1) and weights $w_1, \dots, w_n$ (at least one of them should be positive, and all of them are nonnegative), we look for minimizers of the sum $\sum_{i=1}^{n} w_i D(x_i, t)$. For example, if $D(x, y) = (x-y)^2$, then we obtain the weighted arithmetic mean

$$W(x_1, \dots, x_n) = \frac{\sum_{i=1}^{n} w_i x_i}{\sum_{i=1}^{n} w_i} .$$

This approach can also be introduced in the case when different dissimilarity functions are applied. As an example, consider the aggregation function $A : [0, 1]^n \to$

[0, 1] given by (4.5). We look for minimizers of the expression $w_1|x_1 - t| + \sum_{i=2}^{n} w_i(x_i - t)^2$ and the resulting weighted aggregation function $A_{\mathbf{w}} : [0, 1]^n \to [0, 1]$ is given by

$$A_{\mathbf{w}}(x_1, \ldots, x_n)$$
$$= \text{Med}\left(x_1, M(x_2, \ldots, x_n) - \frac{w_1}{2\sum_{i=2}^{n} w_i},\right.$$
$$\left. M(x_2, \ldots, x_n) + \frac{w_1}{2\sum_{i=2}^{n} w_i}\right).$$

Considering the integer weights $\mathbf{w} = (w_1, \ldots, w_n)$, for an extended aggregation function A which is symmetric and strongly idempotent, we obtain the weighted aggregation function $A_{\mathbf{w}} : [0, 1]^n \to [0, 1]$ given by

$$A_{\mathbf{w}}(x_1, \ldots, x_n)$$
$$= A\left(\underbrace{x_1, \ldots, x_1}_{w_1\text{-times}}, \underbrace{x_2, \ldots, x_2}_{w_2\text{-times}}, \ldots, \underbrace{x_n, \ldots, x_n}_{w_n\text{-times}}\right).$$

The strong idempotency of A also allows one to introduce rational weights into aggregation. Observe that

for each $k \in \mathbb{N}$, the weights $k \cdot \mathbf{w}$ result in the same weighted aggregation function as when considering the weights $\mathbf{w}$ only. For general weights the limit approach described in [4.17, Proposition 6.27] should be applied.

The *qualitative* approach to weights considers a transformation of inputs $x_1, \ldots, x_n$ accordingly to the considered weights (importances) $w_1, \ldots, w_n \in [0, 1]$, with constraint that at least once it holds $w_i = 1$. This approach is applied when we consider an extended aggregation function A with a strong neutral element $e \in [0, 1]$. Then the weighted aggregation function $A_{\mathbf{w}} : [0, 1]^n \to [0, 1]$ is given by

$$A_{\mathbf{w}}(x_1, \ldots, x_n) = A\left(h(w_1, x_1), \ldots, h(w_n, x_n)\right),$$

where $h : [0, 1]^2 \to [0, 1]$ is a relevancy transformation (RET) operator [4.24, 68] satisfying $h(0, x) = e$, $h(1, x) = x$, which is increasing in the second coordinate as well as in the first coordinate for all $x \geq e$, while $h(\cdot, x)$ is decreasing for all $x \leq e$. As an example, consider the RET operator h given by

$$h(w, x) = wx + (1 - w)e.$$

For more details, we recommend [4.17, Chapter 6].

4.4 Concluding Remarks

As already mentioned, all introduced results (sometimes for special types of aggregation functions only) can be straightforwardly extended to any interval $I \subseteq [-\infty, \infty]$. Moreover, one can aggregate more general objects than real numbers. For example, a quite expanding field concerns interval mathematics. The aggregation of interval inputs can be done coordinate-wise,

$$A([x_1, y_1], \ldots, [x_n, y_n])$$
$$= [A_1(x_1, \ldots, x_n), A_2(y_1, \ldots, y_n)],$$

where A_1, A_2 are an arbitrary couple of classical aggregation functions such that $A_1 \leq A_2$ (mostly $A_1 = A_2$ is considered). However, there are also more sophisticated approaches [4.69].

Already in 1942, *Menger* [4.43] introduced the aggregation of distribution functions whose supports are contained in $[0, \infty]$ (distance functions), which led not only to the concept of triangular norms [4.44], but also to triangle functions directly aggregating

such distribution functions [4.70]. Some triangle functions are derived from special aggregation functions (triangular norms), some of them have more complex background (as a distinguished example recall the standard convolution of distribution functions). For an overview and details we recommend [4.71, 72].

In 1965, *Zadeh* [4.73] introduced fuzzy sets. Their aggregation, in particular union and intersection, is again built by means of special aggregation functions on [0, 1], namely by means of triangular conorms and triangular norms [4.26]. Triangular norms also play an important role in the *Zadeh* extension principle [4.74–76] allowing to extend standard aggregation functions acting on real inputs to the generalized aggregation functions acting on fuzzy inputs. As a typical example recall the arithmetic of fuzzy numbers [4.77]. In some special fuzzy logics also uninorms have found the application in modeling conjunctions. Among recent generalizations of fuzzy set theory recall the type 2-fuzzy sets, including interval-valued fuzzy sets, or

Part A | 4.4

n-fuzzy sets. In all these fields, a deep study of aggregation functions is one of the major theoretical tasks to build a sound background.

Observe that all mentioned particular domains are covered by the aggregation on posets, where up to now only some particular general results are known [4.22, 78]. We expect an enormous growth of interest in this field, as it can be seen, for example, in its special subdomain dealing with computing and aggregation with words [4.79–81].

References

4.1   R.C. Archibald: Mathematics before the Greeks Science, Science **71**(1831), 109–121 (1930)

4.2   D. Smith: *History of Mathematics* (Dover, New York 1958)

4.3   T. Calvo, R. Mesiar, R.R. Yager: Quantitative weights and aggregation, IEEE Trans. Fuzzy Syst. **12**(1), 62–69 (2004)

4.4   H. Karcher: Riemannian center of mass and mollifier smoothing, Commun. Pure Appl. Math. **30**(5), 509–541 (1977), published online: 13 October 2006

4.5   J. Łukasiewicz: O logice trójwartosciowej (in Polish), Ruch Filoz. **5**, 170–171 (1920); English translation: On three–valued logic. In: *Selected Works by Jan Lukasiewicz*, ed. by L. Borkowski (North-Holland, Amsterdam 1970) pp. 87–88

4.6   E.L. Post: Introduction to a general theory of elementary propositions, Am. J. Math. **43**, 163–185 (1921)

4.7   K. Gödel: Zum intuitionistischen Aussagenkalkül, Anz. Akad. Wiss. Wien **69**, 65–66 (1939)

4.8   A.N. Kolmogoroff: Sur la notion de la moyenne, Accad. Naz. Lincei Mem. Cl. Sci. Fis. Mat. Nat. Sez. **12**, 388–391 (1930)

4.9   M. Nagumo: Über eine Klasse der Mittelwerte, Jpn. J. Math. **6**, 71–79 (1930)

4.10   J. Aczél: *Lectures on Functional Equations and Their Applications* (Academic, New York 1966)

4.11   B. Schweizer, A. Sklar: *Probabilistic Metric Spaces*, Ser. Probab. Appl. Math, Vol. 5 (North-Holland, New York 1983)

4.12   G.J. Klir, T.A. Folger: *Fuzzy Sets, Uncertainty, and Information* (Prentice-Hall, Hemel Hempstead 1988)

4.13   A. Kolesárová, M. Komorníková: Triangular norm–based iterative aggregation and compensatory operators, Fuzzy Sets Syst. **104**, 109–120 (1999)

4.14   R. Mesiar, M. Komorníková: Triangular norm–based aggregation of evidence under fuzziness. In: *Studies in Fuzziness and Soft Computing, Aggregation and Fusion of Imperfect Information*, Vol. 12, ed. by B. Bouchon-Meunier (Physica, Heidelberg 1998) pp. 11–35

4.15   T. Calvo, A. Kolesárová, M. Komorníková, R. Mesiar: *A Review of Aggregation Operators* (Univ. of Alcalá Press, Alcalá de Henares, Madrid 2001)

4.16   G. Beliakov, A. Pradera, T. Calvo: *Aggregation Functions: A Guide for Practitioners* (Springer, Berlin 2007)

4.17   M. Grabisch, J.-L. Marichal, R. Mesiar, E. Pap: *Aggregation Functions*, Encyclopedia of Mathematics and Its Applications, Vol. 127 (Cambridge Univ. Press, Cambridge 2009)

4.18   Y. Narukawa (Ed.): *Modeling Decisions: Information Fusion and Aggregation Operators*, Cognitive Technologies (Springer, Berlin, Heidelberg 2007)

4.19   R.R. Yager, D.P. Filev: *Essentials of Fuzzy Modelling and Control* (Wiley, New York 1994)

4.20   D. Dubois, H. Prade: On the use of aggregation operations in information fusion processes, Fuzzy Sets Syst. **142**, 143–161 (2004)

4.21   R.R. Yager: On ordered weighted averaging aggregation operators in multicriteria decision making, IEEE Trans. Syst. Man. Cybern. **18**, 183–190 (1988)

4.22   M. Komorníková, R. Mesiar: Aggregation functions on bounded partially ordered sets and their classification, Fuzzy Sets Syst. **175**(1), 48–56 (2011)

4.23   J.-L. Marichal: *k*-intolerant capacities and Choquet integrals, Eur. J. Oper. Res. **177**(3), 1453–1468 (2007)

4.24   R.R. Yager: Aggregation operators and fuzzy systems modeling, Fuzzy Sets Syst. **67**(2), 129–146 (1995)

4.25   M. Gagolewski, P. Grzegorzewski: Arity–monotonic extended aggregation operators, Commun. Comput. Inform. Sci. **80**, 693–702 (2010)

4.26   E.P. Klement, R. Mesiar, E. Pap: *Triangular Norms* (Kluwer, Dordrecht 2000)

4.27   C. Alsina, M.J. Frank, B. Schweizer: *Associative Functions, Triangular Norms and Copulas* (World Scientific, Hackensack 2006)

4.28   R.B. Nelsen: *An Introduction to Copulas*, Lecture Notes in Statistics, Vol. 139, 2nd edn. (Springer, New York 2006)

4.29   C. Alsina, R.B. Nelsen, B. Schweizer: On the characterization of a class of binary operations on distribution functions, Stat. Probab. Lett. **17**(2), 85–89 (1993)

4.30   C. Genest, J.J. Quesada Molina, J.A. Rodriguez Lallena, C. Sempi: A characterization of quasi-copulas, J. Multivar. Anal. **69**, 193–205 (1999)

4.31   B. Bassano, F. Spizzichino: Relations among univariate aging, bivariate aging and dependence for exchangeable lifetimes, J. Multivar. Anal. **93**, 313–339 (2005)

4.32   B. De Baets: Idempotent uninorms, Eur. J. Oper. Res. **180**, 631–642 (1999)

4.33   D. Denneberg: *Non-Additive Measure and Integral* (Kluwer, Dordrecht 1994)

4.34   E.P. Klement, R. Mesiar, E. Pap: A universal integral as common frame for Choquet and Sugeno integral, IEEE Trans. Fuzzy Syst. **18**, 178–187 (2010)

4.35   E. Pap: *Null-Additive Set Functions* (Kluwer, Dordrecht 1995)

4.36   Z. Wang, G.J. Klir: *Generalized Measure Theory* (Springer, New York 2009)

4.37   D. Dubois, H. Prade: A review of fuzzy set aggregation connectives, Inform. Sci. **36**, 85–121 (1985)

4.38   R. Mesiar, A. Mesiarová-Zemánková: The ordered modular averages, IEEE Trans. Fuzzy Syst. **19**, 42–50 (2011)

4.39   M. Couceiro, J.-L. Marichal: Representations and characterizations of polynomial functions on chains, J. Multiple-Valued Log. Soft Comput. **16**(1–2), 65–86 (2010)

4.40   J.C. Fodor, R.R. Yager, A. Rybalov: Structure of uninorms, Int. J. Uncertain. Fuzziness Knowledge-Based Syst. **5**, 411–427 (1997)

4.41   H.J. Zimmermann, P. Zysno: Latent connectives in human decision making, Fuzzy Sets Syst. **4**, 37–51 (1980)

4.42   M.K. Luhandjula: Compensatory operators in fuzzy linear programming with multiple objectives, Fuzzy Sets Syst. **8**(3), 245–252 (1982)

4.43   K. Menger: Statistical metrics, Proc. Natl. Acad. Sci. **28**, 535–537 (1942)

4.44   B. Schweizer, A. Sklar: Statistical metric spaces, Pac. J. Math. **10**(1), 313–334 (1960)

4.45   A. Sklar: *Fonctions de répartition à n dimensions et leurs marges*, Vol. 8 (Institut de Statistique, LUniversité de Paris, Paris 1959) pp. 229–231

4.46   J. Galambos: *The Asymptotic Theory of Extreme Order Statistics*, 2nd edn. (Krieger, Melbourne 1987)

4.47   J.A. Tawn: Bivariate extreme value theory: Models and estimation, Biometrika **75**, 397–415 (1988)

4.48   D. Schmeidler: Integral representation without additivity, Proc. Am. Math. Soc. **97**(2), 255–261 (1986)

4.49   R.R. Yager: Fusion od ordinal information using weighted median aggregation, Int. J. Approx. Reason. **18**, 35–52 (1998)

4.50   T. Calvo, G. Beliakov: Aggregation functions based on penalties, Fuzzy Sets Syst. **161**, 1420–1436 (2010)

4.51   M. Sugeno: Theory of fuzzy integrals and applications, Ph.D. Thesis (Tokyo Inst. of Technology, Tokyo 1974)

4.52   Z. Wang, G.J. Klir: *Fuzzy Measure Theory* (Plenum, New York 1992)

4.53   G. Choquet: Theory of capacities, Ann. Inst. Fourier **5**(54), 131–295 (1953)

4.54   J.C. Fodor: An extension of Fung-Fu's theorem, Int. J. Uncertain. Fuziness Knowledge-Based Syst. **4**, 235–243 (1996)

4.55   L.W. Fung, K.S. Fu: An axiomatic approach to rational decision making in a fuzzy environment. In: *Fuzzy Sets and Their Applications to Cognitive and Decision Processes*, ed. by L.A. Zadeh, K.S. Fu, K. Tanaka, M. Shimura (Academic, New York 1975) pp. 227–256

4.56   M. Grabisch, T. Murofushi, M. Sugeno (Eds.): *Fuzzy Measures and Integrals. Theory and Applications* (Physica, Heidelberg 2000)

4.57   A. Kolesárová, A. Stupňanová, J. Beganová: Aggregation-based extensions of fuzzy measures, Fuzzy Sets Syst. **194**, 1–14 (2012)

4.58   L. Lovász: Submodular functions and convexity. In: *Mathematical Programming: The State of the Art*, ed. by A. Bachem, M. Grotschel, B. Korte (Springer, Berlin, Heidelberg 1983) pp. 235–257

4.59   G. Owen: Multilinear extensions of games, Manag. Sci. **18**, 64–79 (1972)

4.60   F. Durante, A. Kolesárová, R. Mesiar, C. Sempi: Copulas with given diagonal sections: novel constructions and applications, Int. J. Uncertain. Fuzziness Knowlege-Based Syst. **15**(4), 397–410 (2007)

4.61   F. Durante, A. Kolesárová, R. Mesiar, C. Sempi: Copulas with given values on a horizontal and a vertical section, Kybernetika **43**(2), 209–220 (2007)

4.62   S. Ovchinnikov, A. Dukhovny: Integral representation of invariant functionals, J. Math. Anal. Appl. **244**, 228–232 (2000)

4.63   R. Mesiar, V. Jágr: *d*-dimensional dependence functions and Archimax copulas, Fuzzy Sets Syst. **228**, 78–87 (2013)

4.64   R. Mesiar, C. Sempi: Ordinal sums and idempotents of copulas, Aequ. Math. **79**(1–2), 39–52 (2010)

4.65   R. Mesiar, J. Szolgay: *W*-ordinal sums of copulas and quasi-copulas, Proc. MAGIA 2004 Conf. Kočovce (2004) pp. 78–83

4.66   R. Mesiar, V. Jágr, M. Juráňová, M. Komorníková: Univariate conditioning of copulas, Kybernetika **44**(6), 807–816 (2008)

4.67   R. Mesiar, B. De Baets: New construction methods for aggregation operators, IPMU'2000 Int. Conf. Madrid (Springer, Berlin, Heidelberg 2000) pp. 701–706

4.68   M. Šabo, A. Kolesárová, Š. Varga: RET operators generated by triangular norms and copulas, Int. J. Uncertain. Fuzziness Knowledge-Based Syst. **9**, 169–181 (2001)

4.69   G. Deschrijver, E.E. Kerre: Aggregation operators in interval-valued fuzzy and atanassov's intuitionistic fuzzy set theory. In: *Fuzzy Sets and Their Extensions: Representation, Aggregation and Models, Studies in Fuzziness and Soft Computing*, ed. by H. Bustince, F. Herrera, J. Montesa (Springer, Berlin, Heidelberg 2008) pp. 183–203

4.70   A.N. Šerstnev: On a probabilistic generalization of metric spaces, Kazan. Gos. Univ. Učen. Zap. **124**, 3–11 (1964)

4.71   S. Saminger-Platz, C. Sempi: A primer on triangle functions I, Aequ. Math. **76**(3), 201–240 (2008)

4.72   S. Saminger-Platz, C. Sempi: A primer on triangle functions II, Aequ. Math. **80**(3), 239–268 (2010)

4.73   L.A. Zadeh: Fuzzy sets, Inform. Control **8**, 338–353 (1965)

Part A | 4

4.74   L.A. Zadeh: The concept of a linguistic variable and its application to approximate reasoning, Part I, Inform. Sci. **8**, 199–251 (1976)

4.75   L.A. Zadeh: The concept of a linguistic variable and its application to approximate reasoning, Part II, Inform. Sci. **8**, 301–357 (1975)

4.76   L.A. Zadeh: The concept of a linguistic variable and its application to approximate reasoning, Part III, Inform. Sci. **9**, 43–80 (1976)

4.77   D. Dubois, E.E. Kerre, R. Mesiar, H. Prade: Fuzzy interval analysis. In: *Fundamentals of Fuzzy Sets*, The Handbook of Fuzzy Sets Series, ed. by D. Dubois, H. Prade (Kluwer, Boston 2000) pp. 483–582

4.78   G. De Cooman, E.E. Kerre: Order norms on bounded partially ordered sets, J. Fuzzy Math. **2**, 281–310 (1994)

4.79   F. Herrera, S. Alonso, F. Chiclana, E. Herrera-Viedma: Computing with words in decision making: Foundations, trends and prospects, Fuzzy Optim. Decis. Mak. **8**(4), 337–364 (2009)

4.80   L.A. Zadeh: *Computing with Words – Principal Concepts and Ideas*, Stud. Fuzziness Soft Comput, Vol. 277 (Springer, Berlin, Heidelberg 2012)

4.81   L.A. Zadeh (Ed.): *Computing with Words in Information/Intelligent System 1: Foundations*, Stud. Fuzziness Soft Comput, Vol. 33 (Springer, Berlin, Heidelberg 2012) p. 1

5. Monotone Measures-Based Integrals

Erich P. Klement, Radko Mesiar

The theory of classical measures and integral reflects the genuine property of several quantities in standard physics and/or geometry, namely the σ-additivity. Though monotone measure not assuming σ-additivity appeared naturally in models extending the classical ones (for example, inner and outer measures, where the related integral was considered by Vitali already in 1925), their intensive research was initiated in the past 40 years by the computer science applications in areas reflecting human decisions, such as economy, psychology, multicriteria decision support, etc. In this chapter, we summarize basic types of monotone measures together with the basic monotone measures-based integrals, including the Choquet and Sugeno integrals, and we introduce the concept of universal integrals proposed by Klement et al. to give a common roof for all mentioned

integrals. Benvenuti's integrals linked to semicopulas are shown to be a special class of universal integrals. Up to several other integrals, we also introduce decomposition integrals due to Even and Lehrer, and show which decomposition integrals are inside the framework of universal integrals.

Before Cauchy, there was no definition of the integral in the actual sense of the word *definition*, though the integration was already well established and in many areas applied method. Recall that constructive approaches to integration can be traced as far back as the ancient Egypt around 1850 BC; the *Moscow Mathematical Papyrus* (Problem 14) contains a formula of a frustum of a square pyramid [5.1]. The first documented systematic technique, capable of determining integrals, is the method of exhaustion of the ancient Greek astronomer Eudoxus of Cnidos (ca. 370 BC) [5.2] who tried to find areas and volumes by approximating them by a (large) number of shapes for which the area or volume was known. This method was further developed by Archimedes in third-century BC who calculated the area of parabolas and gave an approximation to the area of a circle. Similar methods were independently developed in China around third-century AD by Liu Hui, who used it to find the area of the circle. This

was further developed in the fifth century by the Chinese mathematicians Zu Chongzhi and Zu Geng to find the volume of a sphere. In the same century, the Indian mathematician Aryabhata used a similar method in order to find the circumference of a circle. More than 1000 years later, *Johannes Kepler* invented the *Kepler'sche Fassregel* [5.3] (also known as *Simpson rule*) in order to compute the (approximative) volume of (wine) barrels.

Based on the fundamental work of *Isaac Newton* and *Gottfried Wilhelm Leibniz* in the 18th century (see [5.4, 5]), the first indubitable access to integration was given by *Bernhard Riemann* in his Habilitation Thesis at the University of Göttingen [5.6]. Note that Riemann has generalized the Cauchy definition of integral defined for continuous real functions (of one variable) defined on a closed interval $[a, b]$.

Among several other developments of the integration theory, recall the Lebesgue approach covering measurable functions defined on a measurable space and general σ-additive measures. Here we recall the final words of H. Lebesgue from his lecture held at a conference in Copenhagen on May 8, 1926, entitled *The Development of the Notion of the Integral* (for the full text see [5.7]):

> ... *if you will, that a generalization made not for the vain pleasure of generalizing, but rather for the solution of problems previously posed, is always a fruitful generalization. The diverse applications which have already taken the concepts which we have just examine prove this super-abundantly.*

All till now mentioned approaches to integration are related to measurable spaces, measurable real functions and (σ-)additive real-valued measures. Though there are many generalizations and modifications concerning the range and domain of considered functions and measures (and thus integrals), in this chapter we will stay in the above-mentioned framework, with the only exception that the (σ-)additivity of measures is relaxed into their monotonicity, thus covering many natural generalizations of (σ-)additive measures, such as outer or inner measures, lower or upper envelopes of systems of measures, etc.

Maybe the first approach to integration not dealing with the additivity was due to *Vitali* [5.8]. Vitali was looking for integration with respect to lower/upper measures and his approach is completely covered by the later, more general, approach of *Choquet* [5.9], see Sect. 5.1. Note that the Choquet integral is a generalization of the Lebesgue integral in the sense

that they coincide whenever the considered measure is σ-additive (i.e., when the Lebesgue integral is meaningful).

A completely different approach, influenced by the starting development of fuzzy set theory [5.10], is due to *Sugeno* [5.11]. Sugeno even called his integral as fuzzy integral (and considered set functions as fuzzy measures), though there is no fuzziness in this concept (Sect. 5.1). Later, several approaches generalizing or modifying the above-mentioned integrals were introduced. In this chapter, we give a brief overview of these integrals, i.e., integrals based on monotone measures. In the next section, some preliminaries and basic notions are recalled, as well as the Choquet and Sugeno integrals. Section 5.2 brings a generalization of both Choquet and Sugeno integrals, now known as the Benvenuti integral. In Sect. 5.3, universal integrals as a rather general framework for monotone measures-based integral is given and discussed, including copula-based integrals, among others. In Sect. 5.4, we bring some integrals not giving back the underlying measure. Finally, some possible applications are indicated and some concluding remarks are added. Note that we will not discuss integrals defined only for some special subclasses of monotone measures, such as pseudoadditive integrals [5.12, 13] or t-conorms-based integrals of *Weber* [5.14]. Moreover, we restrict our considerations to normed measures satisfying $m(X) = 1$, and to functions with range contained in [0, 1]. This is done for the sake of higher transparentness and the generalizations for $m(X) \in {]}0, \infty]$ and functions with different ranges will be covered by the relevant quotations only.

5.1 Preliminaries, Choquet, and Sugeno Integrals

For a fixed measurable space $(X, \mathcal{A})$, where $\mathcal{A}$ is a σ-algebra of subsets of the universe X, we denote by $\mathcal{F}_{(X,\mathcal{A})}$ the set of all $\mathcal{A}$-measurable functions $f : X \to [0, 1]$, and by $\mathcal{M}_{(X,\mathcal{A})}$ the set of all monotone measures $m : \mathcal{A} \to [0, 1]$ (i.e., $m(\emptyset) = 0$, $m(X) = 1$ and $m(A) \leq m(B)$ whenever $A \subset B \subset X$). Note that functions f from $\mathcal{F}_{(X,\mathcal{A})}$ can be seen as membership functions of fuzzy events on $(X, \mathcal{A})$, and that monotone measures are in different references also called fuzzy measures, capacities, pre-measures, etc. Moreover, if X is finite, we will always consider $\mathcal{A} = 2^X$ only. In such case, any monotone measure $m \in \mathcal{M}_{(X,\mathcal{A})}$ is determined by $2^{|X|} - 2$ weights from [0, 1] (measures of proper subsets of X) constraint by the monotonicity condition only, and to

each monotone measure $m : \mathcal{A} \to [0, 1]$ we can assign its Möbius transform $M_m : \mathcal{A} \to \mathbb{R}$ given by

$$M_m(A) = \sum_{B \subseteq A} (-1)^{|A \setminus B|} \cdot m(B). \tag{5.1}$$

Then

$$m(A) = \sum_{B \subseteq A} M_m(B). \tag{5.2}$$

Moreover, dual monotone measure $m^d : \mathcal{A} \to [0, 1]$ is given by $m^d(A) = 1 - m(A^c)$.

Among several distinguished subclasses of monotone measures from $\mathcal{M}_{(X,\mathcal{A})}$ we recall these classes, supposing the finiteness of X:

- *Additive* measures, $m(A \cup B) = m(A) + m(B)$ whenever $A \cap B = \emptyset$;
- *Maxitive* measures, $m(A \cup B) = m(A) \vee m(B)$ (these measures are called also possibility measures [5.15, 16]);
- *k-additive* measures, $M_m(A) = 0$ whenever $|A| > k$ (hence additive measures are 1-additive);
- *Belief* measures, $M_m(A) \geq 0$ for all $A \subset X$;
- *Plausibility* measures, m^d is a belief measure;
- *Symmetric* measures, $M_m(A)$ depends on $|A|$ only.

For more details on monotone measures, we recommend [5.17–19] and [5.20].

Concerning the functions, for any $c \in [0, 1], A \in \mathcal{A}$ we define a basic function $b(c, A) : X \to [0, 1]$ by

$$b(c, A)(x) = \begin{cases} c & \text{if } x \in A, \\ 0 & \text{else}. \end{cases}$$

Obviously, basic functions can be related to the characteristic functions, $1_A = b(1, A)$ and $b(x, A) = c \cdot 1_A$. However, as we are considering more general types of multiplication as the standard product, in general, we prefer not to depend in our consideration on the standard product.

The first integral introduced for monotone measures was proposed by *Choquet* [5.9] in 1953.

Definition 5.1

For a fixed monotone measure $m \in \mathcal{M}_{(X,\mathcal{A})}$, a functional $\mathrm{Ch}_m : \mathcal{F}_{(X,\mathcal{A})} \to [0, 1]$ given by

$$\mathrm{Ch}_m(f) = \int_0^1 m(f \geq t) \mathrm{d}t \tag{5.3}$$

is called the *Choquet* integral (with respect to m), where the right-hand side of (5.3) is the classical Riemann integral.

Note that the Choquet integral is well defined because of the monotonicity of m. Observe that if m is σ-additive, i. e., if it is a probability measure on $(X, \mathcal{A})$, then the function $h : [0, 1] \to [0, 1]$ given by $h(t) = m(f \geq t)$ is the standard survival function of the random variable f, and then $\mathrm{Ch}_m(f) = \int_0^1 h(t)\mathrm{d}t = \int_X f \, \mathrm{d}m$ is the standard expectation of f (i. e., Lebesgue integral of f with respect to m).

Due to *Schmeidler* [5.21, 22], we have the following axiomatization of the Choquet integral.

Theorem 5.1

A functional $I : \mathcal{F}_{(X,\mathcal{A})} \to [0, 1], I(1_X) = 1$, is the Choquet integral with respect to monotone measure $m \in \mathcal{M}_{(X,\mathcal{A})}$ given by $m(A) = I(1_A)$ if and only if I is comonotone additive, i. e., if $I(f + g) = I(f) + I(g)$ for all $f, g \in \mathcal{F}_{(X,\mathcal{A})}$ such that $f + g \in \mathcal{F}_{(X,\mathcal{A})}$ and f and g are comonotone, $(f(x) - f(y)) \cdot (g(x) - g(y)) \geq 0$ for any $x, y \in X$.

We recall some properties of the Choquet integral.

It is evident that the Choquet integral Ch_m is an increasing functional, $\mathrm{Ch}_m(f) \leq \mathrm{Ch}_m(g)$ for any $m \in \mathcal{M}_{(X,\mathcal{A})}, f, g \in \mathcal{F}_{(X,\mathcal{A})}$ such that $f \leq g$. Moreover, for each $A \in \mathcal{A}$ it holds $\mathrm{Ch}_m(b(c, A)) = c \cdot m(A)$, and especially $\mathrm{Ch}_m(1_A) = m(A)$.

Remark 5.1

i) Due to results of *Šipoš* [5.23], see also [5.24], the comonotone additivity of the functional I in Theorem 5.1, which implies its positive homogeneity, $I(cf) = c \cdot I(f)$ for all $c > 0$ and $f \in \mathcal{F}_{(X,\mathcal{A})}$ such that $cf \in \mathcal{F}_{(X,\mathcal{A})}$ can be replaced by the positive homogeneity of I and its horizontal additivity, i. e.,

$$I(f) = I(f \wedge a) + I(f - f \wedge a)$$

for all $f \in \mathcal{F}_{(X,\mathcal{A})}$ and $a \in [0, 1]$.

ii) Choquet integral $\mathrm{Ch}_m : \mathcal{F}_{(X,\mathcal{A})} \to [0, 1]$ is continuous from below,

$$\lim_{n \to \infty} \mathrm{Ch}_m(f_n) = \mathrm{Ch}_m(f)$$

whenever for $(f_n)_{n \in \mathbb{N}} \in \mathcal{F}_{(X,\mathcal{A})}^{\mathbb{N}}$ we have $f_n \leq f_{n+1}$ for all $n \in \mathbb{N}$ and $f = \lim_{n \to \infty} f_n$, if and only if m is continuous from below,

$$\lim_{n \to \infty} m(A_n) = m(A)$$

whenever for $(A_n)_{n \in \mathbb{N}} \in \mathcal{A}^{\mathbb{N}}$ we have $A_n \subset A_{n+1}$ for all $n \in \mathbb{N}$ and $A = \bigcup_{n \in \mathbb{N}} A_n$.

iii) Choquet integral $\mathrm{Ch}_m : \mathcal{F}_{(X,\mathcal{A})} \to [0, 1]$ is subadditive (superadditive),

$$I(f + g) \leq I(f) + I(g) \quad (I(f + g) \geq I(f) + I(g))$$

for all $f, g, f + g \in \mathcal{F}_{(X,\mathcal{A})}$, if and only if m is sub-modular (supermodular),

$$m(A \cup B) + m(A \cap B) \leq m(A) + m(B) ,$$
$$(m(A \cup B) + m(A \cap B) \geq m(A) + m(B))$$

for all $A, B \in \mathcal{A}$.

iv) For any $m \in \mathcal{M}_{(X,\mathcal{A})}$ and $f \in \mathcal{F}_{(X,\mathcal{A})}$ it holds

$$\mathrm{Ch}_{m^d}(f) = 1 - \mathrm{Ch}_m(1-f) ,$$

i.e., in the framework of aggregation functions [5.25] the dual to a Choquet integral (with respect to a monotone measure m) is again the Choquet integral (with respect to the dual monotone measure m^d).

For the proofs and more details about the above results on Choquet integral, we recommend [5.18, 19, 24, 26].

Restricting our considerations to finite universes, we have also the next evaluation formula due to *Chateauneuf* and *Jaffray* [5.27]

$$\mathrm{Ch}_m(f) = \sum_{A \subseteq X} M_m(A) \cdot \min (f(x) \,|\, x \in A) . \tag{5.4}$$

In the Dempster–Shafer theory of evidence [5.28, 29], belief measures are considered, and then the Möbius transform $M_m : 2^X \setminus \{\emptyset\} \to [0, 1]$ of a belief measure m is called a *basic probability assignment*. Evidently, M_m can be seen as a probability measure (of singletons) on the finite space $2^X \setminus \{\emptyset\}$ (with cardinality $2^{|X|} - 1$), and defining a function $F : 2^X \setminus \{\emptyset\} \to [0, 1]$ by $F(A) = \min (f(x)|x \in A)$, the formula (5.4) can be seen as the Lebesgue integral of F with respect to M_m (i.e., it is the standard expectation of variable F)

$$\mathrm{Ch}_m(f) = \sum_{A \in 2^X \setminus \{\emptyset\}} F(A) \cdot M_m(A) .$$

Another genuine relationship of Choquet and Lebesgue integrals in the framework of the Dempster–Shafer theory is based on the fact that each belief measure m can be seen as a lower envelope of the class of dominating probability measures, i.e., for each $A \subseteq X$ (X is finite)

$$m(A) = \inf \{P(A)|P \geq m\} .$$

Then

$$\mathrm{Ch}_m(f) = \inf \left\{ \int_X f \, \mathrm{d}P | P \geq m \right\} .$$

Similarly, for the related plausibility measure m^d, it holds

$$\mathrm{Ch}_{m^d}(f) = \sup \left\{ \int_X f \, \mathrm{d}P | P \leq m^d \right\}$$

$$= \sup \left\{ \int_X f \, \mathrm{d}P | P \geq m \right\} .$$

For interested readers, we recommend the collection [5.30].

In general, for any monotone measure $m \in \mathcal{M}_{(X,\mathcal{A})}$ and any measurable (continuous from below) function $f \in \mathcal{F}_{(X,\mathcal{A})}$ there is a probability measure $P_{m,f}$ on $(X, \mathcal{A})$ so that

$$\mathrm{Ch}_m(f) = \int_X f \, \mathrm{d}P_{m,f} , \tag{5.5}$$

see, e.g., [5.24, Theorem 2.6], where the right-hand side of (5.5) is the standard Lebesgue integral. Moreover, if $f, g \in \mathcal{F}_{(X,\mathcal{A})}$ are comonotone, one can find unique probability measure P allowing to express the Choquet integral of f and g with respect to m as the Lebesgue integral of f and g with respect to P, respectively. As an immediate consequence of (5.5), Jensen's inequality for Choquet integral can be shown to be valid. Similarly, if f and g are comonotone, based on the above observations, one can prove the Minkowski and Chebyshev inequality. For more details, see [5.31].

For $k \in \mathbb{N}$, consider a probability measure P on the product space $(X, \mathcal{A})^k$, and define a set function $m : \mathcal{A} \to [0, 1]$ by $m(A) = P(A^k)$. Then $m \in \mathcal{M}_{(X,\mathcal{A})}$ is a k-additive monotone measure (and belief measure, as well), and for all $f \in \mathcal{F}_{(X,\mathcal{A})}$ it holds

$$\mathrm{Ch}_m(f) = \int_{X^k} F \, \mathrm{d}P , \tag{5.6}$$

where $F : X^k \to [0, 1]$ is given by $F(x_1, \ldots, x_k) = \min (f(x_1), \ldots, f(x_k))$. For more details see [5.32].

The Sugeno integral (in the original sources called fuzzy integral) was introduced by *Sugeno* in 1972 in Japanese in [5.33] and in English in 1974 in [5.11]. Inspired by the fuzzy set theory introduced by *Zadeh* [5.10], Sugeno has proposed a way how to formalize human subjectivity in spirit similar to the randomness but based only on ordinal scales. His concept is not fuzzy, though both fuzzy set theory and Sugeno's integral theory exploit the same aggregation functions (sup and inf), and considering functions $f \in \mathcal{F}_{(X,\mathcal{A})}$ as membership functions of fuzzy subsets of X, the corresponding Sugeno integral can be seen as a version of expectation of fuzzy sets.

Definition 5.2

For a fixed monotone measure $m \in \mathcal{M}_{(X,\mathcal{A})}$, a functional $\mathrm{Su}_m : \mathcal{F}_{(X,\mathcal{A})} \to [0,1]$ given by

$$\mathrm{Su}_m(f) = \sup \{ \min(t, m(f \geq t)) \,|\, t \in [0,1] \} \qquad (5.7)$$

is called the *Sugeno integral* (with respect to m).

There is an equivalent formula for the Sugeno integral, compare ([5.11, Definition 3.1]),

$$\mathrm{Su}_m(f) = \sup \{ \min(m(A), \inf \{ f(x) | x \in A \}) \,|\, A \in \mathcal{A} \} , \qquad (5.8)$$

which in the case of finite X (and using also lattice notation $\sup = \vee$, $\min = \wedge$) can be rewritten as

$$\mathrm{Su}_m(f) = \bigvee_{A \subset X} \left(M_m^{\vee}(A) \wedge \min(f(x) | x \in A) \right) , \qquad (5.9)$$

showing the striking similarity with the evaluation formula (5.4) for the Choquet integral. Here the set function $M_m^{\vee} : 2^X \setminus \{\emptyset\} \to [0,1]$ is the so-called possibilistic Möbius transform introduced by *Mesiar* in [5.34] and given by

$$M_m^{\vee}(A) = \begin{cases} 0 & \text{if } m(A) = m(B) \text{ for some } B \subsetneq A , \\ m(A) & \text{else} . \end{cases}$$

Sugeno integral has properties similar to the Choquet integral. Indeed, it is nondecreasing functional such that $\mathrm{Su}_m(b(c,A)) = c \wedge m(A)$, and in particular $\mathrm{Su}_m(1_A) = m(A)$. Moreover, Su_m is comonotone maxitive, i.e., $\mathrm{Su}_m(f \vee g) = \mathrm{Su}_m(f) \vee \mathrm{Su}_m(g)$ for any comonotone $f, g \in \mathcal{F}_{(X,\mathcal{A})}$, and min-homogeneous, $\mathrm{Su}_m(c \wedge f) = c \wedge \mathrm{Su}_m(f)$. We have the next axiomatization of the Sugeno integral due to *Marichal* [5.35] (compare with Theorem 5.1 for the Choquet integral).

Theorem 5.2

A functional $I : \mathcal{F}_{(X,\mathcal{A})} \to [0,1]$, $I(1_X) = 1$, is the Sugeno integral with respect to monotone measure $m \in \mathcal{M}_{(X,\mathcal{A})}$ given by $m(A) = I(1_A)$ if and only if I is comonotone maxitive and min-homogeneous.

For alternative axiomatizations see [5.24].

Choquet and Sugeno integrals with respect to a monotone measure m may differ not more than $\frac{1}{4}$, i.e., for all $f \in \mathcal{F}_{(X,\mathcal{A})}$ it holds

$$|\mathrm{Ch}_m(f) - \mathrm{Su}_m(f)| \leq \frac{1}{4} .$$

Moreover, $\mathrm{Ch}_m(f) = \mathrm{Su}_m(f)$ for all $f \in \mathcal{F}_{(X,\mathcal{A})}$ if and only if $m(A) \in \{0,1\}$ for all $A \in \mathcal{A}$, and then

$$\mathrm{Ch}_m(f) = \mathrm{Su}_m(f) = \sup \{ \inf \{ f(x) | x \in A \} \,|\, m(A) = 1 \} ,$$

which in case X is finite turns out to be a lattice polynomial.

Note that if X has cardinality n and $m(A) \in \{0,1\}$ for all $A \subseteq X$, then $\mathrm{Ch}_m = \mathrm{Su}_m : [0,1]^n \to [0,1]$ are the only n-ary continuous aggregation functions invariant under each automorphism $\phi : [0,1] \to [0,1]$, i.e., $\phi \circ \mathrm{Ch}_m(f) = \mathrm{Ch}_m(f \circ \phi)$ for each $f[0,1]^n$ (for $f = (a_1, \ldots, a_n), f \circ \phi = (\phi(a_1), \ldots, \phi(a_n))$). For more details see [5.36].

Example 5.1

i) Let $X = \{1,2,3\}$ and define $m : 2^X \to [0,1]$ by

$$m(A) = \begin{cases} 0 & \text{if card } A \leq 1 , \\ 1 & \text{otherwise} . \end{cases}$$

Then, for each $f = (x,y,z) \in [0,1]^3$,

$$\mathrm{Ch}_m(f) = \mathrm{Su}_m(f) = (x \wedge y) \vee (x \wedge z) \vee (y \wedge z)$$
$$= \mathrm{med}(x,y,z)$$

brings the classical median.

ii) Let $X = \{1,2\}$ and define $m : 2^X \to [0,1]$ by $m(A) = \frac{\mathrm{card}\,A}{2}$. Then, for each $f = (x,y) \in [0,1]^2$,

$$\mathrm{Ch}_m(f) = \frac{x+y}{2}$$

(i.e., Ch_m is the standard arithmetic mean), while

$$\mathrm{Su}_m(f) = (x \wedge y) \vee \left((x \vee y) \wedge \frac{1}{2} \right) .$$

For $f_1 = \left(\frac{1}{2}, 1 \right)$, $\mathrm{Ch}_m(f_1) = \frac{3}{4}$ and $\mathrm{Su}_m(f_1) = \frac{1}{2}$. For $f_2 = \left(0, \frac{1}{2} \right)$, $\mathrm{Ch}_m(f_2) = \frac{1}{4}$ and $\mathrm{Su}_m(f_2) = \frac{1}{2}$.

Part A | 5.1

In general,

$$|\mathrm{Ch}_m(f) - \mathrm{Su}_m(f)| = \frac{1}{2}\,(|x-y| \wedge |x+y-1|) \le \frac{1}{4}\,.$$

iii) Let $X = [0,1]$, $\mathcal{A} = \mathcal{B}([0,1])$ and let $m : \mathcal{A} \to [0,1]$ be given by $m(A) = \lambda^p(A)$, where $p \in \,]0, \infty[$ is a fixed constant and $\lambda : \mathcal{A} \to [0,1]$ is the standard Lebesgue measure. For any Lebesgue measure preserving function $f : X \to [0,1]$, such as $f(x) = x$, $f(x) = 1 - x$, or $f(x) = |2x-1|$, we have

$$\mathrm{Ch}_m(f) = \int\limits_0^1 m(f \ge t)\,\mathrm{d}t = \int\limits_0^1 (1-t)^p \mathrm{d}t = \frac{1}{p+1}$$

and

$$\mathrm{Su}_m(f) = \sup\{\min(t, (1-t)^p)\,|t \in [0,1]\} = c_p\,,$$

where c_p is the unique solution of the equation $t = (1-t)^p$, $t \in \,]0,1[$. Hence,

if $p = 1$, $\mathrm{Ch}_m(f) = \mathrm{Su}_m(f) = \dfrac{1}{2}$;

if $p = 2$, $\mathrm{Ch}_m(f) = \dfrac{1}{3}$

and $\mathrm{Su}_m(f) = \dfrac{3 - \sqrt{5}}{2} \doteq 0.382$;

if $p = 3$, $\mathrm{Ch}_m(f) = \dfrac{2}{3}$

and $\mathrm{Su}_m(f) = \dfrac{\sqrt{5}-1}{2} \doteq 0.618$.

5.2 Benvenuti Integral

Comparing Theorems 5.1 and 5.2, we see a striking similarity in the axiomatic characterization of the Choquet and Sugeno integrals. This similarity was generalized under a common roof by *Benvenuti* et al. [5.24], calling there introduced integral *general fuzzy integral*. This integral is now also known as Benvenuti integral (compare [5.25]).

Choquet integral is linked to the standard arithmetic operations $+$ and $\cdot$ on $[0, \infty]$, while the Sugeno integral deals with lattice operations $\wedge$ and $\vee$ on $[0,1]$. To generalize these two couples of operations, pseudoaddition $\oplus$ and pseudomultiplication $\odot$ was introduced in [5.24].

Definition 5.3
Let $u \in [1, \infty]$ be a fixed constant. An operation $\oplus : [0, u]^2 \to [0, u]$ is called a *pseudoaddition* on $[0, u]$ whenever it is associative, nondecreasing in both components, 0 is its neutral element, and $\oplus$ is continuous.

Observe that the structure $([0, u], \oplus)$ with $\oplus$ a pseudoaddition on $[0, u]$ is just an *I*-semigroup of *Mostert* and *Shields* [5.37] and hence $\oplus$ is also commutative. Moreover, considering the principles of Galois connections, we can introduce a *pseudodifference* $\ominus$ related to $\oplus$ satisfying, for all $a, b, c \in [0, u]$, $(a \ominus b) \le c$ if and only if $a \le b \ominus c$.

It is not difficult to see the link to the pseudodifference considered already by *Weber* [5.14].

Lemma 5.1
Let $\oplus : [0, u]^2 \to [0, u]$ be a given pseudoaddition on $[0, u]$. The related pseudo difference $\ominus : [0, u]^2 \to [0, u]$ is given by

$$a \ominus b = \inf\{c \in [0, u]\,|\,b \oplus c \ge a\}\,.$$

Considering the standard addition $+$ on $[0, \infty]$, and $a \ge b$, then the corresponding (pseudo-) difference is the standard difference $a - b$. On the other hand, $\vee$ is a pseudoaddition on any interval $[0, u]$, and its corresponding pseudodifference $\ominus_\vee$ is given by

$$a \ominus_\vee b = \begin{cases} 0 & \text{if } a \le b\,, \\ a & \text{otherwise}\,. \end{cases}$$

Due to [5.37], each pseudoaddition $\oplus$ on $[0, u]$ can be represented as an ordinal sum,

$$a \oplus b = \begin{cases} g_k^{-1}(g_k(\beta_k) \wedge (g_k(a) + g_k(b))) \\ \qquad \text{if } (a, b) \in \,]\alpha_k, \beta_k[^2\,, \\ a \vee b \qquad \text{otherwise}\,, \end{cases}$$

where $(]\alpha_k, \beta_k[)_{k \in \mathcal{K}}$ is a disjoint system of open subintervals of $[0, u]$, and $g_k : [\alpha_k, \beta_k] \to [0, \infty]$ is a continuous strictly increasing function such that $g_k(\alpha_k) = 0$, $k \in \mathcal{K}$ ($\mathcal{K}$ can be also empty). Two extremal cases correspond to $\oplus = \vee$ (when $\mathcal{K}$ is empty) and

Archimedean pseudoaddition $\oplus$ on $[0, u]$ generated by $g : [0, u] \to [0, \infty]$ (when $\mathcal{K}$ is singleton, say $\mathcal{K} = \{1\}$, and $\alpha_1 = 0$, $\beta_1 = u$),

$$a \oplus b = g^{-1}\left(g(u) \wedge (g(a) + g(b))\right) .$$

Then g is called an *additive generator* of $\oplus$ and it is unique up to a positive multiplicative constant.

Note that if g is a bijection, i. e., $g(u) = \infty$, then $a \oplus b = g^{-1}(g(a) + g(b))$ and $\oplus$ is called a *strict pseudoaddition*.

For a fixed pseudoaddition $\oplus$ on $[0, u]$, *Benvenuti* et al. [5.24] have introduced a $\oplus$-fitting pseudomultiplication $\odot$.

Definition 5.4

Fix $u, v \in [1, \infty]$ and let $\oplus$ be a given pseudoaddition on $[0, u]$. A mapping $\odot : [0, u] \times [0, v] \to [0, u]$ is called a $\oplus$-*fitting pseudomultiplication* whenever it is nondecreasing in both components, 0 is its annihilator, i. e., $0 \odot b = a \odot 0 = 0$ for all $a \in [0, u]$, $b \in [0, v]$, it is left distributive over $\oplus$, i. e., $(a \oplus b) \odot c = (a \odot c) \oplus (b \odot c)$ for all $a, b \in [0, u]$, $c \in [0, v]$, and it is lower semicontinuous, i. e.,

$$\left(\vee_{n \in \mathbb{N}} a_n\right) \odot \left(\vee_{m \in \mathbb{N}} b_m\right) = \vee_{n, m \in \mathbb{N}} (a_n \odot b_m) .$$

The left distributivity of a pseudomultiplication $\odot$ over $\vee$ simply means the nondecreasingness of $\odot$ in the first coordinate, and thus there are several kinds of $\vee$-fitting pseudomultiplication $\odot$. On the other hand, this is a rather restrictive constraint when $\oplus$ is Archimedean, i. e., generated by an additive generator $g : [0, u] \to [0, \infty]$.

Proposition 5.1

Let $\oplus : [0, u]^2 \to [0, u]$ be an Archimedean pseudoaddition generated by an additive generator $g : [0, u] \to [0, \infty]$. A mapping $\odot : [0, u] \times [0, v] \to [0, u]$ is a $\oplus$-fitting pseudomultiplication if and only if there is a lower semicontinuous nondecreasing function $h : [0, v] \to [0, \infty]$ such that $h(w) = 0$ for some $w \in [0, v]$, and $g(u) \cdot h(a) \geq g(u)$ for all $a \in \,]w, v]$, so that

$$a \odot b = g^{-1}\left(g(u) \wedge g(a) \cdot h(b)\right) .$$

In particular, if $\oplus$ is a strict pseudoaddition, then $h : [0, v] \to [0, \infty]$ is a lower semicontinuous nondecreasing function, satisfying $h(0) = 0$, and $a \odot b = g^{-1}(g(a) \cdot h(b))$.

Definition 5.5

Let $u, v \in [1, \infty]$ be fixed given constants and let $\oplus : [0, u]^2 \to [0, u]$ be a given pseudoaddition, and $\odot : [0, u] \times [0, v] \to [0, u]$ be a given $\oplus$-fitting pseudomultiplication such that $1 \odot 1 \leq 1$. For a fixed monotone measure $m \in \mathcal{M}_{(X, \mathcal{A})}$, a functional $\mathcal{B}_m^{\oplus, \odot} : \mathcal{F}_{(X, \mathcal{A})} \to [0, 1]$ given by

$$\mathcal{B}_m^{\oplus, \odot}(f) = \sup \left\{ \bigoplus_{i=1}^{n} (a_i \odot m(A_i)) \,|\, n \in \mathbb{N}, \right.$$

$$\left. \bigoplus_{i=1}^{n} b(a_i, A_i) \leq f, (A_i)_{i=1}^{n} \text{ is a chain} \right\}$$

is called Benvenuti integral (with respect to m, based on $\oplus$ and $\odot$).

Observe that if $s \in \mathcal{F}_{(X, \mathcal{A})}$ is a simple function, range $s = \{b_1, \ldots, b_n\}$, $b_1 < b_2 < \cdots < b_n$, then

$$\mathcal{B}_m^{\oplus, \odot}(s) = \bigoplus_{i=1}^{n} \left((b_i \ominus b_{i-1}) \odot m(s \geq b_i)\right) ,$$

with the convention $b_0 = 0$. Then for any $f \in \mathcal{F}_{(X, \mathcal{A})}$,

$$\mathcal{B}_m^{\oplus, \odot}(f)$$
$$= \sup \left\{ \mathcal{B}_m^{\oplus, \odot}(s) \,|\, s \in \mathcal{F}_{(X, \mathcal{A})} \text{ is simple, } s \leq f \right\} .$$

Evidently,

$$\mathcal{B}_m^{\oplus, \odot}(b(a, A)) = a \odot m(A)$$

and hence

$$\mathcal{B}_m^{\oplus, \odot}(1_A)) = m(A)$$

for all $m \in \mathcal{M}_{(X, \mathcal{A})}$, $A \in \mathcal{A}$ only if $1 \odot b = b$ for all $b \in [0, 1]$.

If $\oplus$ is a strict pseudoaddition on $[0, u]$ generated by an additive generator g, this means that $\odot$ restricted to $[0, 1]^2$ is given by

$$a \odot b = g^{-1}\left(\frac{g(a) \cdot g(b)}{g(1)}\right) .$$

If $\oplus$ is a nonstrict pseudoaddition, then there is no $\oplus$-fitting pseudomultiplication $\odot$ such that $1 \odot b = b$ for all $b \in [0, 1]$.

Note that for the standard arithmetic operations $+$ and $\cdot$ on $[0, \infty]$, $\mathcal{B}_m^{+, \cdot} = \mathrm{Ch}_m$, i. e., the Choquet integral is recovered. Similarly, $\mathcal{B}_m^{\vee, \wedge} = \mathrm{Su}_m$.

Example 5.2

i) Let $u = v = 1$, $\oplus = \vee$ and $\odot : [0,1]^2 \to [0,1]$ be given by $a \odot b = a^p \cdot b^q$, $p, q \in]0, \infty[$. Then $\mathcal{B}_m^{\oplus, \odot}(f)) = \sup \{t^p \cdot (m(f \geq t))^q \,|\, t \in [0,1]\}$ for any $m \in \mathcal{M}_{(X,\mathcal{A})}$ and $f \in \mathcal{F}_{(X,\mathcal{A})}$, and $\mathcal{B}_m^{\oplus, \odot}(1_A) = (m(A))^q$. Note that if $p = q = 1$, the Shilkret integral $\mathrm{Sh}_m = \mathcal{B}_m^{\oplus, \odot}$ is recovered, see [5.19, 38]. In general, $\mathcal{B}_m^{\oplus, \odot}(f) = \mathrm{Sh}_{m^q}(f^p)$ for any $f \in \mathcal{F}_{(X,\mathcal{A})}$.

ii) For a strict pseudoaddition $\oplus$ on $[0, u]$ and a $\oplus$-fitting pseudomultiplication $\odot$ on $[0, u] \times [0, v]$, see Proposition 5.1, the constraint $1 \odot 1 \leq 1$ means $h(b) \leq 1$, and then $\mathcal{B}_m^{\oplus, \odot}(f) = g^{-1}\left(\mathrm{Ch}_{h(m)}\left(g(f)\right)\right)$, i.e., $\mathcal{B}_m^{\oplus, \odot}$ is obtained as a transformation of the Choquet integral.

For more details, we recommend the original source [5.24], but also [5.25, 39].

Remark 5.2

When considering $u = 1$, a pseudoaddition $\oplus$ on $[0, 1]$ becomes a (continuous) *t*-conorm. Integrals based on *t*-conorms closely related to Benvenuti integrals were discussed by *Murofushi* and *Sugeno* [5.40], resulting to the two classes of *t*-conorm based integrals. Those based on the smallest *t*-conorm $\vee$ coincide with Benvenuti integral based on $\vee$, with stronger requirements on the corresponding $\vee$-fitting pseudomultiplication $\odot$. The second one, based on continuous Archimedean *t*-conorms, is a special transform of the Choquet integral, compare Example 5.2 ii),

$$MS_m(f) = k\left(\mathrm{Ch}_{h(m)}(g(f))\right),$$

with appropriately chosen functions $k, h, g : [0,1] \to [0, \infty]$. Note that the Murofushi–Sugeno integral covers also the integral of *Weber* [5.14] based on strict *t*-conorms. Another closely related approach to integration, fixing $u = v = \infty$, can be found in [5.41], where Choquet-like integrals were introduced and discussed. For more details on these types of integrals we refer to [5.42, 43].

5.3 Universal Integrals

The concept of universal integrals on $[0, \infty]$ was proposed and discussed in [5.44]. As already mentioned, we will restrict our considerations to the interval $[0, 1]$.

Definition 5.6

Let S be the class of all measurable spaces. A mapping

$$I : \bigcup_{(X,\mathcal{A}) \in S} \left(\mathcal{M}_{(X,\mathcal{A})} \times \mathcal{F}_{(X,\mathcal{A})}\right) \to [0,1]$$

is called a universal integral whenever it satisfies

UI1 I is nondecreasing in both components;

UI2 there is a semicopula $\otimes : [0,1]^2 \to [0,1]$ (i.e., $\otimes$ is nondecreasing in both components and $1 \otimes a = a \otimes 1$ for all $a \in [0,1]$) such that $I(m, b(a, E)) = a \otimes m(E)$ for all $a \in [0,1]$, any $(X, \mathcal{A}) \in S$, $m \in \mathcal{M}_{(X,\mathcal{A})}$ and $E \in \mathcal{A}$;

UI3 $I(m_1, f_1) = I(m_2, f_2)$ whenever $(m_i, f_i) \in (X_i, \mathcal{A}_i)$, $i = 1, 2$, and $m_1(f_1 \geq t) = m_2(f_2 \geq t)$ for all $t \in [0, 1]$.

Observe that the axiom (UI1) reflects the standard monotonicity of integrals. On the other hand, (UI2) expresses the fact that an integral of a basic function $b(a, E)$ with respect to a monotone measure m depends on the values a and $m(E)$ only, independently of the considered measurable space $(X, \mathcal{A})$ and a monotone measure $m \in \mathcal{M}_{(X,\mathcal{A})}$ (compare the truth values principle in the propositional logics). Finally, (UI3) generalizes the well-known fact from the probability theory that two random variables (defined possibly on two different probability spaces) have the same expectation whenever their distribution functions coincide (in fact, for a probability measure P, $P(f \geq t)$ defines a survival function which is complementary to the related distribution function).

There are several construction methods for universal integrals. First of all, for any given semicopula $\otimes : [0,1]^2 \to [0,1]$, one can introduce the smallest universal integral $I_\otimes$ and the greatest universal integral $I^\otimes$ related to $\otimes$ through (UI2):

$$I_\otimes(m, f) = \sup\{t \otimes m(f \geq t) \,|\, t \in [0,1]\}$$

and

$$I^\otimes(m, f) = \mathrm{essup}_m(f) \otimes m(\mathrm{supp}\, f),$$

where

$$\mathrm{essup}_m(f) = \sup\{t \in [0,1] | m(f \geq t) > 0\}$$

and

$$\mathrm{supp}\, f = \{x \in X | f(x) > 0\}\ .$$

Observe that $I_\wedge(m,\cdot) = \mathrm{Su}_m$ is the Sugeno integral, $I_\Pi(m,\cdot) = \mathrm{Sh}_m$ is the Shilkret integral (Π denotes the product semicopula), while I_T with T a strict t-norm is an integral introduced by *Weber* in [5.45].

Considering the Benvenuti integral based on a pseudo-addition $\oplus$ on $[0,u]$ and a $\oplus$-fitting pseudomultiplication $\odot : [0,u] \times [0,v] \to [0,u]$, $u, v, \in [1,\infty]$, such that $\otimes = \odot|[0,1]^2$ is a semicopula, one get a universal integral given by

$$I^{\oplus,\otimes}(m,f) = \mathcal{B}_m^{\oplus,\odot}(f)\ .$$

Note that $I^{+,\cdot}(m,f) = \mathrm{Ch}_m(f)$ and $I^{\vee,\wedge}(m,f) = \mathrm{Su}_m(f)$.

As an important class of universal integrals we introduce copula-based integrals. Recall that a semicopula $C : [0,1]^2 \to [0,1]$ is called a copula [5.46] whenever it is supermodular, i.e., for any $x, y \in [0,1]^2$ it holds

$$C(x \vee y) + C(x \wedge y) \geq C(x) + C(y)\ .$$

Note that there is a one-to-one correspondence between copulas and probability measures on Borel subsets of $[0,1]^2$ with uniformly distributed margins, this relation is stated by the equality

$$P_C([0,a] \times [0,b]) = C(a,b),\ (a,b) \in [0,1]^2\ .$$

The next result is extracted from [5.44], also compare [5.47, 48].

Proposition 5.2
Let $C : [0,1]^2 \to [0,1]$ be a fixed copula. Then the mapping

$$K_C : \bigcup_{(X,\mathcal{A}) \in \mathcal{S}} \left(\mathcal{M}_{(X,\mathcal{A})} \times \mathcal{F}_{(X,\mathcal{A})}\right) \to [0,1]$$

given by

$$K_C(m,f) = P_C\left(\{(u,v) \in [0,1]^2 | v \leq m(f \geq u)\}\right)$$

is a universal integral (with C being the corresponding semicopula).

Note that for the product copula Π, $K_\Pi(m,\cdot) = \mathrm{Ch}_m$ is the Choquet integral, while for the greatest copula $\wedge = \mathrm{Min}$, $K_\wedge(m,\cdot) = \mathrm{Su}_m$ is the Sugeno integral. For the smallest copula $W : [0,1]^2 \to [0,1]$ given by

$$W(a,b) = \max(0, a + b - 1)\ ,$$

K_W was called *opposite Sugeno integral* in [5.49] and it is given by

$$K_W(m,f) = \lambda\left(\{t \in [0,1] | m(f \geq t) \geq 1 - t\}\right)\ ,$$

where λ is the standard Lebesgue measure on Borel subsets of $[0,1]$.

Remark 5.3
The class of universal integrals is convex, i.e., for I_1, I_2 universal integrals and a constant $c \in [0,1]$, also

$$I = cI_1 + (1-c)I_2$$

is a universal integral (related to the semicopula $\odot = c \cdot \odot_1 + (1-c) \cdot \odot_2$).

Though the class of semicopulas is also convex, for the weakest universal integrals we can ensure only the inequality

$$I_{c \cdot \odot_1 + (1-c) \cdot \odot_2} \leq cI_{\odot_1} + (1-c)I_{\odot_2}\ .$$

On the other hand, for the convex class of copulas it holds

$$K_{cC_1 + (1-c)C_2} = cK_{C_1} + (1-c)K_{C_2}\ ,$$

i.e., the class of copula-based integrals is convex.

5.4 General Integrals Which Are Not Universal

There are several integrals defined on any measurable space $(X, \mathcal{A})$, for any monotone measure $m \in \mathcal{M}_{(X,\mathcal{A})}$ and any function $f \in \mathcal{F}_{(X,\mathcal{A})}$ which are not universal. We recall two of them based on the standard arithmetic operations $+$ and $\cdot$.

Definition 5.7
A mapping

$$G: \bigcup_{(X,\mathcal{A}) \in S} \left(\mathcal{M}_{(X,\mathcal{A})} \times \mathcal{F}_{(X,\mathcal{A})}\right) \to [0, \infty]$$

given by

$$G(m,f) = \sup \left\{ \sum_{i=1}^{n} a_i \cdot m(A_i) \mid n \in \mathbb{N}, a_1, \dots, a_n \geq 0, \right.$$

$$\sum_{i=1}^{n} b(a_i, A_i) \leq f \text{ and } (A_i)_{i=1}^{n}$$

$$\left. \text{is a disjoint subsystem of } \mathcal{A} \right\}$$

is called a PAN-integral.

Note that this integral was introduced by *Yang* [5.50], see also [5.51] in more general setting on $[0, \infty]$ involving operations $\oplus$ and $\odot$. Due to the results of [5.52], each PAN-integral on $[0, 1]$ is either a transformation of integral given in Definition 5.7, $I(m,f) = g^{-1}(G(g(m), g(f)))$ for some automorphism $g: [0, 1] \to [0, 1]$, or if $\oplus = \vee$, it is a special instant of integrals $I_\odot$ discussed in Sect. 5.3. Also observe that a deep discussion on PAN-integral G can be found in [5.53].

PAN-integral allows one to recognize the underflying monotone measure m only if m is superadditive. Moreover, as a major defect of this integral we recall that it does not exclude the equality of integrals based on two different monotone measures, i.e., there are monotone measures $m_1, m_2 \in \mathcal{M}_{(X,\mathcal{A})}, m_1 \neq m_2$, such that $G(m_1,f) = G(m_2,f)$ for all $f \in \mathcal{F}_{(X,\mathcal{A})}$. Note that PAN-integral coincide with the Lebesgue integral whenever m is σ-additive. A similar situation is linked to the concave integral introduced by *Lehrer* [5.54], see also [5.55].

Definition 5.8
A mapping

$$L: \bigcup_{(X,\mathcal{A}) \in S} \left(\mathcal{M}_{(X,\mathcal{A})} \times \mathcal{F}_{(X,\mathcal{A})}\right) \to [0, \infty]$$

given by

$$L(m,f) = \sup \left\{ \sum_{i=1}^{n} a_i \cdot m(A_i) \mid n \in \mathbb{N}, \right.$$

$$\left. a_1, \dots, a_n \geq 0, \sum_{i=1}^{n} b(a_i, A_i) \leq f \right\}$$

is called a concave integral.

Observe that this integral is concave in the sense that for each $m \in \mathcal{M}_{(X,\mathcal{A})}, f, g \in \mathcal{F}_{(X,\mathcal{A})}$ and $c \in [0, 1]$,

$$L(m, cf + (1-c)g) \geq cL(m,f) + (1-c)L(m,g).$$

Concave integral coincides with the Choquet integral whenever m is supermodular. However, also here $L(m_1,f) = L(m_2,f)$ may hold for all $f \in \mathcal{F}_{(X,\mathcal{A})}$ for some monotone measures $m_1, m_2 \in \mathcal{M}_{(X,\mathcal{A})}, m_1 \neq m_2$. Finally, recall that it trivially holds

$$L(m,f) \geq G(m,f) \text{ and } L(m,f) \geq \mathrm{Ch}_m(f)$$

for all $m \in \mathcal{M}_{(X,\mathcal{A})}$ and $f \in \mathcal{F}_{(X,\mathcal{A})}$.

Example 5.3
i) Consider $X = [0, 1]$, $\mathcal{A} = \mathcal{B}([0, 1])$ and λ the standard Lebesgue measure on $\mathcal{A}$. Let $m = \lambda^p, p \in]0, 1[$. Then for any $f \in \mathcal{F}_{(X,\mathcal{A})}$ with nonvanishing support (i. e., $m(f > 0) > 0$) it holds

$$G(m,f) = L(m,f) = +\infty.$$

On the other hand, for $m = \lambda^2$ (observe that m is supermodular, and thus also superadditive) we get, considering $f = \mathrm{id}_X$,

$$G(m,f) = \frac{2}{13} \text{ while } L(m,f) = \mathrm{Ch}_m(f) = \frac{1}{3}.$$

ii) For $X = \{1, 2, 3\}$ and $\mathcal{A} = 2^X$, let $m_a: \mathcal{A} \to \mathbb{R}$ be given by $m_a(\emptyset) = 0, m_a(A) = 0.1$ if $\mathrm{card}\, A = 1$,

$m_a(A) = a$ if card $A = 2$ and $m_a(X) = 1$. Evidently, $m_a \in \mathcal{M}_{(X,\mathcal{A})}$ if and only if $a \in [0.1, 1]$. Let $f \in \mathcal{F}_{(X,\mathcal{A})}$ be given by $f(1) = \frac{1}{3}, f(2) = \frac{2}{3}, f(3) = 1$. Then

$$G(m_a, f) = \sup \left\{ \frac{1}{3} \cdot 1, \frac{1}{3} \cdot 0.1 + \frac{2}{3} \cdot a \right\}$$

$$= \begin{cases} \frac{1}{3} & \text{if } a \in [0.1, 0.45], \\ \frac{0.1 + 2a}{3} & \text{if } a \in]0.45, 1], \end{cases}$$

and

$$L(m_a, f) = \sup \left\{ \frac{1}{3} \cdot 1 + \frac{1}{3} \cdot a + \frac{1}{3} \cdot 0.1, \frac{1}{3} \cdot 1 \right.$$

$$\left. + \frac{1}{3} \cdot 0.1 + \frac{2}{3} \cdot 0.1, \frac{1}{3} \cdot a + \frac{2}{3} \cdot a \right\}$$

$$= \begin{cases} \frac{1 \cdot 3}{3} & \text{if } a \in [0.1, 0.2[, \\ \frac{1.1 + a}{3} & \text{if } a \in [0.2, 0.55], \\ a & \text{if } a \in]0.55, 1]. \end{cases}$$

Moreover,

$$\mathrm{Ch}_{m_a}(f) = \frac{1.1 + a}{3}.$$

Observe that m_a is supermodular if and only if $a \in [0.2, 0.55]$ and then

$$L(m_a, f) = \mathrm{Ch}_{m_a}(f) = \frac{1.1 + a}{3}.$$

iii) For X finite and $m \in \mathcal{M}_{(X,\mathcal{A})}$ such that $m(A) \in \{0, 1\}$ for all $A \subseteq X$, all universal integrals coincide, independently of the underlying semicopula $\otimes$, $I_m(f) = \sup \{\min (f(x)|x \in A) \,|m(A) = 1\}$. However, this does not hold for PAN-integral $G(m, \cdot)$ neither for the concave integral $L(m, \cdot)$. Consider as an example the greatest monotone measure $m^* \in \mathcal{M}_{(X, 2^X)}$ given by

$$m^*(A) = \begin{cases} 0 & \text{if } A = \emptyset, \\ 1 & \text{else}. \end{cases}$$

Then for any universal integral I it holds $I(m^*, f) = \max (f(x)|x \in X)$, but $G(m^*, f) = L(m^*, f) = \sum_{x \in X} f(x)$.

iv) The only monotone measures $m \in \mathcal{M}_{(X, 2^X)}$, X finite, such that all universal integrals as well as the

PAN and concave integrals coincide, are so-called unanimity measures

$$m_B, \ B \subseteq X, \ B \neq \emptyset, \ m_B(A) = \begin{cases} 1 & \text{if } B \subseteq A, \\ 0 & \text{else}. \end{cases}$$

Then

$$I(m_B, f) = G(m_B, f) = L(m_B, f)$$
$$= \min (f(x)|x \in B).$$

Recently, a new concept of decomposition integrals was proposed in [5.56], unifying the PAN integral G, the concave integral L, the Choquet integral Ch, and the Shilkret integral Sh.

Definition 5.9
Let $(X, \mathcal{A})$ be a measurable space and let $\mathcal{H}$ be a system of some finite subsystems (i. e., of collections) from $\mathcal{A}$. Then the mapping

$$D_{\mathcal{H}} : \mathcal{M}_{(X,\mathcal{A})} \times \mathcal{F}_{(X,\mathcal{A})} \to [0, \infty]$$

given by

$$D_{\mathcal{H}}(m, f) = \sup \left\{ \sum_{i \in I} a_i \cdot m(A_i)|a_i \geq 0, i \in I, \right.$$

$$\left. \sum_{i \in I} b(a_i, A_i) \leq f, \ (A_i)_{i \in I} \in \mathcal{H} \right\}$$

is called a $\mathcal{H}$-decomposition integral.

Consider the next decomposition systems

$\mathcal{H}^{(n)} = \{(A_i)_{i=1}^n$ is a chain in $\mathcal{A}\}$, $n \in \mathbb{N}$;
$\mathcal{H}_G = \{(A_i)_{i \in I}$ is a finite measurable partition of $X\}$;
$\mathcal{H}_L = \mathcal{A}$;
$\mathcal{H}_{\mathrm{Ch}} = \{\mathcal{B}|\mathcal{B}$ is a finite chain in $\mathcal{A}\}$.

Then

$$D_{\mathcal{H}^{(1)}}(m, \cdot) = Sh_m;$$
$$D_{\mathcal{H}_G} = G;$$
$$D_{\mathcal{H}_L} = L;$$
$$D_{\mathcal{H}_{\mathrm{Ch}}}(m, \cdot) = \mathrm{Ch}_m.$$

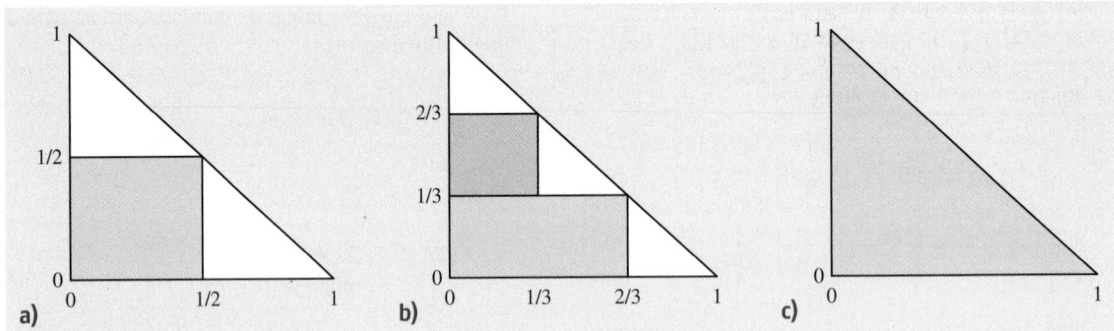

Fig. 5.1a-c The function $\lambda (\mathrm{id}_X \geq t)$ with *shaded areas* expressing the corresponding integrals $D_{\mathcal{H}^{(1)}} (\lambda, \mathrm{id}_X)$ **(a)**, $D_{\mathcal{H}^{(2)}} (\lambda, \mathrm{id}_X)$ **(b)**, $\mathrm{Ch}_\lambda (\mathrm{id}_X)$ **(c)**

Further, the only decomposable integrals which are also universal integrals are the Choquet integral and $\mathcal{H}^{(n)}$-decomposition integrals $D_{\mathcal{H}^{(n)}}$ and they satisfy

$$\mathrm{Sh} = D_{\mathcal{H}^{(1)}} \leq D_{\mathcal{H}^{(2)}} \leq \cdots \leq D_{\mathcal{H}^{(n)}} \leq \cdots \leq \mathrm{Ch} \,.$$

Observe that if X is finite, card $X = n$, then $D_{\mathcal{H}^{(n)}} = \mathrm{Ch}$ and that

$$\mathrm{Ch} = \lim_{n \to \infty} D_{\mathcal{H}^{(n)}} = \sup \{ D_{\mathcal{H}^{(n)}} | n \in \mathbb{N} \} \,.$$

For more details and further discussion about decomposition integrals, we recommend [5.56–58].

Example 5.4
Using the notation from Example 5.3 i), it holds

$$D_{\mathcal{H}^{(n)}} (\lambda, \mathrm{id}_X) = \frac{n}{2(n + 1)}$$

and

$$\lim_{n \to \infty} \frac{n}{2(n + 1)} = \frac{1}{2} = \mathrm{Ch}_\lambda (\mathrm{id}_X) \,.$$

For better understanding, see Fig. 5.1 with the graph of the function $\lambda (\mathrm{id}_X \geq t)$ and with shaded areas expressing the corresponding integrals.

5.5 Concluding Remarks, Application Fields

We have recalled and discussed several kinds of integrals defined on any measurable space for any monotone measure and any nonnegative measurable functions, restricting our considerations to the unit interval $[0, 1]$. There are several possible extensions of these integrals to the bipolar scale $[-1, 1]$, i. e., for integrating functions with range in $[-1, 1]$. Recall only the case of the Choquet integral with bipolar extensions of different kinds, such as:

- *Asymmetric* Choquet integral,

$$\mathrm{Ch}_m^{\mathrm{as}}(f) = \mathrm{Ch}_m(f^+) - \mathrm{Ch}_{m^d}(f^-) \,,$$

where $f^+ : X \to [0, 1]$ is given by $f^+(x) = \max (0, f(x))$, $f^- : X \to [0, 1]$ is given by $f^-(x) = \max (0, -f(x))$, and $m^d : \mathcal{A} \to [0, 1]$ is a monotone measure dual to m. For more details see [5.18, 19, 26];

- *Symmetric* (Šipoš) Choquet integral,

$$\mathrm{Ch}_m^{\mathrm{sym}}(f) = \mathrm{Ch}_m(f^+) - \mathrm{Ch}_m(f^-) \,,$$

see [5.18, 19, 23, 26];
- In the case when X is finite, two another extensions called a *balanced* Choquet integral [5.59] and a *merging* Choquet integral [5.60] reflecting (partial) compensation of positive and negative inputs were also introduced and discussed. Further generalizations yield the background of cummulative prospect theory CPT (Cummulative Prospect Theory) of *Tversky* and *Kahneman* [5.61, 62], however, then two monotone measures are considered,

$$\mathrm{Ch}_{m_1, m_2} (f) = \mathrm{Ch}_{m_1} (f^+) - \mathrm{Ch}_{m_2} (f^-) \,.$$

Observe that economical applications of CPT have resulted into Nobel Prize for Tversky and Kahneman in 2002.

Some of introduced integrals were introduced because of solving some practical problems. For example, concave integral of *Lehrer* [5.54] is a solution of an optimization problem looking for a maximal global performance.

Among many fields where integrals discussed in this chapter are an important tool, we recall decision making under multiple criteria, multiobjective optimization, multiperson decision making, pattern recognition and classification, image analysis, etc. For more details, we recommend [5.25, Appendix B] or [5.19].

References

5.1 R.C. Archibald: Mathematics before the Greeks, Science **71**(1831), 109–121 (1930)

5.2 D. Smith: *History of Mathematics* (Dover Publications, New York 1958)

5.3 J. Kepler: *Nova Stereometria Doliorum Vinariorum* (Linz 1615)

5.4 G.W. Leibniz: Nova methodus pro maximis et minimis (New method for maximums and minimums; 1684). In: *A Source Book in Mathematics*, ed. by D.J. Struik (Harvard Univ. Press, Cambridge 1969) p. 271

5.5 I. Newton: Principia (1687) (S. Chandrasekhar: *Newtons Principia for the Common Reader*, Oxford Univ. Press, Oxford, 1995)

5.6 G.F.B. Riemann: On the Hypotheses Which Underlie Geometry, Habilitation Thesis (Universität Göttingen, Göttingen 1854), published first in Proc. R. Philos. Soc. Göttingen **13**, 87–132 (1854) in German

5.7 S.B. Chae: *Lebesgue Integration* (Marcel Dekker, Inc., New York 1980)

5.8 G. Vitali: Sulla definizione di integrale delle funzioni di una variabile, Ann. Mat. Pura Appl. IV **2**, 111–121 (1925)

5.9 G. Choquet: Theory of capacities, Ann. Inst. Fourier (Grenoble) **5**, 131–292 (1953)

5.10 L.A. Zadeh: Fuzzy sets, Inform. Control **8**, 338–353 (1965)

5.11 M. Sugeno: Theory of Fuzzy Integrals and Applications, Ph.D. Thesis (Tokyo Inst. of Technology, Tokyo 1974)

5.12 M. Sugeno, T. Murofushi: Pseudo–additive measures and integrals, J. Math. Anal. Appl. **122**, 197–222 (1987)

5.13 E. Pap: Integral generated by decomposable measure, Univ. u Novom Sadu Zb. Rad. Prirod.-Mat. Fak. Ser. Mat. **20**(1), 135–144 (1990)

5.14 S. Weber: ⊥–decomposable measures and integrals for Archimedean t–conorms ⊥, J. Math. Anal. Appl. **101**, 114–138 (1984)

5.15 L.A. Zadeh: Fuzzy sets as a basis for a theory of possibility, Fuzzy Sets Syst. **1**, 3–28 (1978)

5.16 D. Dubois, H. Prade: *Fuzzy Sets and Systems, Theory and Applications* (Academic, New York 1980)

5.17 M. Grabisch, T. Murofushi, M. Sugeno (Eds.): *Fuzzy Measures and Integrals, Theory and Applications* (Physica, Heidelberg 2000)

5.18 E. Pap: *Null–Additive Set Functions* (Kluwer, Dordrecht 1995)

5.19 Z. Wang, G.J. Klir: *Generalized Measure Theory* (Springer, New York 2009)

5.20 E. Pap (Ed.): *Handbook of Measure Theory* (Elsevier, Amsterdam 2002)

5.21 D. Schmeidler: Integral Representation without additivity, Proc. Am. Math. Soc. **97**(2), 255–261 (1986)

5.22 D. Schmeidler: Subjective probability and expected utility without additivity, Econometrica **57**, 571–587 (1989)

5.23 J. Šipoš: Integral with respect to a pre-measure, Math. Slov. **29**, 141–155 (1979)

5.24 P. Benvenuti, R. Mesiar, D. Vivona: Monotone set functions-based integrals. In: *Handbook of Measure Theory*, ed. by E. Pap (Elsevier, Amsterdam 2002) pp. 1329–1379

5.25 M. Grabisch, J.-L. Marichal, R. Mesiar, E. Pap: *Aggregation Functions*, Encyclopedia of Mathematics and Its Applications, Vol. 127 (Cambridge Univ. Press, Cambridge 2009)

5.26 D. Denneberg: *Non–Additive Measure and Integral* (Kluwer, Dordrecht 1994)

5.27 A. Chateauneuf, J.-Y. Jaffray: Some characterizations of lower probabilities and other monotone capacities through the use of Möbius inversion, Math. Soc. Sci. **17**, 263–283 (1989)

5.28 A.P. Dempster: Upper and lower probabilities induced by a multi-valued mapping, Ann. Math. Stat. **38**, 325–339 (1967)

5.29 G. Shafer: *A Mathematical Theory of Evidence* (Princeton Univ. Press, Princeton, NJ 1976)

5.30 R.R. Yager, L. Liu: Classic works of the Dempster-Shafer theory of belief functions. In: *Studies in Fuzziness and Soft Computing*, ed. by R.R. Yager, L. Liu (Springer, Berlin 2008)

5.31 R. Mesiar, J. Li, E. Pap: The Choquet integral as Lebesgue integral and related inequalities, Kybernetika **46**(6), 1098–1107 (2010)

5.32 R. Mesiar: k-order additive fuzzy measures, Int. J. Uncertain. Fuzziness Knowl.-Based Syst. **7**(6), 561–568 (1999)

5.33 M. Sugeno: Fuzzy measure and fuzzy integral, Trans. Soc. Instrum. Control Eng. **8**, 95–102 (1972)

5.34 R. Mesiar: k-order Pan-discrete fuzzy measures, Proc. IFSA'97 **1**, 488–490 (1997)

5.35 J.L. Marichal: An axiomatic approach of the discrete Sugeno integral as a tool to aggregate interacting criteria in a qualitative framework, IEEE Trans. Fuzzy Syst. **9**(1), 164–172 (2001)

5.36   S. Ovchinnikov, A. Dukhovny: Integral representation of invariant functionals, J. Math. Anal. Appl. **244**, 228–232 (2000)

5.37   P.S. Mostert, A.L. Shield: On the structure of semigroups on a compact manifold with boundary, Ann. Math. **65**, 117–143 (1957)

5.38   N. Shilkret: Maxitive measures and integration, Indag. Math. **33**, 109–116 (1971)

5.39   W. Sander, J. Siedekum: Multiplication, distributivity and fuzzy-integral II & III, Kybernetika **41**(4), 497–518 (2005)

5.40   T. Murofushi, M. Sugeno: Fuzzy *t*-conorm integrals with respect to fuzzy measures: generalizations of Sugeno integral and Choquet integral, Fuzzy Sets Syst. **42**, 51–57 (1991)

5.41   R. Mesiar: Choquet–like integrals, J. Math. Anal. Appl. **194**, 477–488 (1995)

5.42   E. Pap: Pseudo-convolution and its applications. In: *Fuzzy Measures and Integrals, Theory and Applications*, ed. by M. Grabisch, T. Murofushi, M. Sugeno (Physica, Heidelberg 2000) pp. 171–204

5.43   W. Sander, J. Siedekum: Multiplication, distributivity and fuzzy-integral I, Kybernetika **41**(3), 397–422 (2005)

5.44   E.P. Klement, R. Mesiar, E. Pap: A universal integral as common frame for Choquet and Sugeno integral, IEEE Trans. Fuzzy Syst. **18**, 178–187 (2010)

5.45   S. Weber: Two integrals and some modified version – critical remarks, Fuzzy Sets Syst. **20**, 97–105 (1986)

5.46   R.B. Nelsen: *An Introduction to Copulas*, Lecture Notes in Statistics, Vol. 139, 2nd edn. (Springer, New York 2006)

5.47   H. Imaoka: Comparison between three fuzzy integrals. In: *Fuzzy Measures and Integrals, Theory and Applications*, ed. by M. Grabisch, T. Murofushi, M. Sugeno (Physica, Heidelberg 2000) pp. 273–286

5.48   E.P. Klement, R. Mesiar, E. Pap: Measure-based aggregation operators, Fuzzy Sets Syst. **142**(1), 3–14 (2004)

5.49   H. Imaoka: On a subjective evaluation model by a generalized fuzzy integral, Int. J. Uncertain. Fuzziness Knowl.–Based Syst. **5**, 517–529 (1997)

5.50   Q. Yang: The pan-integral on fuzzy measure space, Fuzzy Math. **3**, 107–114 (1985), in Chinese

5.51   Z. Wang, G.J. Klir: *Fuzzy Measure Theory* (Plenum, New York 1992)

5.52   R. Mesiar, J. Rybárik: Pan-operations structure, Fuzzy Sets Syst. **74**, 365–369 (1995)

5.53   Q. Zhang, R. Mesiar, J. Li, P. Struk: Generalized Lebesgue integral, Int. J. Approx. Reason. **52**(3), 427–443 (2011)

5.54   E. Lehrer: A new integral for capacities, Econ. Theory **39**, 157–176 (2009)

5.55   E. Lehrer, R. Teper: The concave integral over large spaces, Fuzzy Sets Syst. **159**, 2130–2144 (2008)

5.56   Y. Even, E. Lehrer: Decomposition-integral: unifying Choquet and the concave integrals, Econ. Theory **56**, 33–58 (2014)

5.57   R. Mesiar, A. Stupňanová: Decomposition integrals, Int. J. Approx. Reason. **54**(8), 1252–1259 (2013)

5.58   A. Stupňanová: Decomposition integrals, Comm. Comput. Info. Sci. **300**, 542–548 (2012)

5.59   A. Mesiarová-Zemánková, R. Mesiar, K. Ahmad: The balancing Choquet integral, Fuzzy Sets Syst. **161**(7), 2243–2255 (2010)

5.60   R. Mesiar, A. Mesiarová-Zemánková, K. Ahmad: Discrete Choquet integral and some of its symmetric extensions, Fuzzy Sets Syst. **184**(1), 148–155 (2011)

5.61   A. Tversky, D. Kahneman: Advances in prospect theory: Cumulative representation of uncertainty, J. Risk Uncertain. **5**, 297–323 (1992)

5.62   A. Tversky, D. Kahneman: Rational choice and the framing of decisions, J. Bus. **59**(278), 251–278 (1986)

6. The Origin of Fuzzy Extensions

Humberto Bustince, Edurne Barrenechea, Javier Fernández, Miguel Pagola, Javier Montero

Many different kinds of sets have been defined within the framework of fuzzy sets. This paper focusses on those fuzzy set extensions that address the difficulties that experts find in order to build the membership values. In particular, we analyze type-2 fuzzy sets, interval-valued fuzzy sets, Atanassov's intuitionistic fuzzy sets, or bipolar sets of type-2 and Atanassov's interval-valued fuzzy sets. After stating a general approach to these extensions, we remark some structural problems in the extension problem and stress some applications for which the results obtained with extensions are better than those obtained with Zadeh's fuzzy sets.

Many different types of fuzzy sets have appeared in the literature since *Zadeh* introduced the concept of fuzzy set (or type-1 fuzzy set) [6.1]. Roughly speaking, the basic characteristics of all those definitions are the following:

i)   They are particular instances of the *L*-fuzzy sets defined by *Goguen* [6.2].
ii) They arise from theoretical problems and are very efficient to solve such theoretical problems.
iii) The specific characteristics of the new definitions do not use to play a formal role, quite often becoming an easy adaptation of Zadeh's fuzzy sets.
iv) It is not always shown to what extent the new proposal implies a practical advantage when compared to Zadeh's fuzzy sets.

The last point gives rise to a key criticism when additional information is needed for the management of a new kind of fuzzy sets, but the improvement we obtain in practice cannot be justified by the effort required to obtain such an information. But more important than that is the previous criticism (iii), about the difficulty of building the best family of sets for the application we are considering. Surprisingly, this key issue has not captured the attention of too many researchers.

In this paper, we shall focuss on those sets conceived to address the problem stated by Zadeh in 1971 in order to address the difficulty of finding the membership degree of each element (we shall refer to these sets as *extensions of the fuzzy sets*), and then we shall point out applications that can be found in the literature in which the use of some extensions provides better results than the use of type-1 fuzzy sets, according to the comparison carried out in the papers where this improvement is shown. Once the definition of extension of fuzzy sets has been introduced, we shall describe some of its properties and remark the structural problems of the different types of these extensions. Among those extensions we shall consider type-2 fuzzy sets, interval-valued fuzzy sets, Atanassov's intuitionistic fuzzy sets or type-2 bipolar fuzzy sets and Atanassov's interval-valued fuzzy sets.

We have organized this chapter as follows. In Sect. 6.2 we start recalling the reasons that led Zadeh to introduce fuzzy sets. We also remind the basic notions in Brouwer's intuitionistic theory to later justify the terminological problems linked to the sets defined by Atanassov. In Sect. 6.3 we present the origin of the extensions of fuzzy sets as well as the definitions. Section 6.4 is devoted to type-2 fuzzy sets. We stress the problems related to the definition of the basic operations and the terminology. In Sect. 6.5 we analyze a particular case of the previous sets, namely, interval-valued fuzzy sets. We present their properties and different construction methods, depending on the application that we are dealing with. We also refer to the papers in which it is shown that the results that we obtain with these sets are better than those obtained with other techniques. In Sects. 6.6 and 6.7 we describe the sets defined by Atanassov. Section 6.8 explains the links between the considered extensions. In Sect. 6.9 we exhibit some other definitions of fuzzy sets in the literature that do not fall into the scope of our notion of extension. We finish with some conclusions and references.

6.1 Considerations Prior to the Concept of Extension of Fuzzy Sets

In classical logic, propositions can only be either true or false. Aristotle formulated the basic principles of this logic: the noncontradiction principle (a statement cannot be true and false at the same time) and the middle-excluded principle (every statement is either true or false).

It is easy to note that there are many situations for which more than two truth values are needed. This fact led C.S. Peirce to say that Aristotle's formulation is the *simplest hypothesis* we can work with. In fact, meanwhile human knowledge representation is based upon concepts [6.3], and these concepts are not crisp in nature, we should not expect that human beings use binary logic so often in their daily life.

Everyday situations such as taste, meaning of adjectives, etc., can only be studied precisely if gradings more complex than true or false are considered. Even very widely used mathematical models can lead to paradoxes. For instance, quite often we are forced to establish arbitrary cuts in order to make reality fit our binary model.

These considerations led to propose different logical formulations which allowed for more than two truth values, like Brouwer's intuitionistic logic (partially caught by the so-called intuitionistic propositional calculus modeled by Heyting algebras), multivalued logics presented by Lukasiewicz, or Zadeh's fuzzy logic (which replaces the set $\{0, 1\}$ by the set $[0, 1]$, for example.

6.1.1 Brouwer's Intuitionistic Logic

In 1907, the Dutch mathematician L.E.J. Brouwer (1881–1966) introduced the intuitionistic logic. Between the precursors of intuitionistic logic, we can include Kronecker, Poincare, Borel, or Weyl.

For intuitionistic researchers, the objects of study in Mathematics are just some intuitions of the mind and the constructions that can be made with them. Hence, the intuitionistic mathematics only handles built objects and only recognizes the properties assigned to these objects in their construction. In particular, the negation of the impossibility of a fact is not a construction of such a fact, and so both the double negation principle and the reduction ad absurdum method are not acceptable for the intuitionist. In the same way, it may happen that it is impossible to build both a fact and its negation, so also the middle-excluded principle is excluded by intuitionism.

In 1930 Heyting, a Brouwer's disciple, went one step ahead and defined a propositional calculus in terms of axioms and rules in Hilbert's style. This calculus is known as intuitionistic propositional calculus (intuitionistic logic). For several decades, the research in intuitionism was almost stopped. But it has reappeared with strength in the logic of categories and topos [6.4, 5]. In this sense, the studies by *Takeuti* and *Titani* in 1984 [6.6] on intuitionistic fuzzy logic and intuitionistic fuzzy set theory are of special interest for us. In [6.7], it is settled that

> *Takeuti and Titani's intuitionitic fuzzy logic is simply an extension of intuitionistic logic, i.e., all formulas provable in the intuitionistic logic are provable in their logic. They give a sequent calculus which extends Heyting intuitionistic logic, an extension that does not collapse to classical logic and keeps the flavor of intuitionism.*

6.1.2 Lukasiewicz's Multivalued Logics

In 1920s, Jan Lukasiewicz (1878–1956) along with Lesniewski founded a school of logic in Warsaw that became one of the most important mathematical teams in the world, and among whose members was Alfred Tarski.

Lukasiewicz's idea consists in distributing the truth values uniformly on the $[0, 1]$ interval: if n values are considered, they should be $0, \frac{1}{n-1}, \frac{2}{n-1}, \dots, \frac{n-2}{n-1}, 1$; if they are infinite, we should take $Q \cap [0, 1]$. Negation is defined as $n(x) = 1 - x$, and the following operation is also defined: $x \oplus y = \min(1, x + y)$.

6.1.3 Zadeh's Fuzzy Logic. First Generalization by Goguen

Consistently to Lukasiewicz's studies, *Zadeh* [6.1] introduced fuzzy logic in his 1965 paper, *Fuzzy Sets*. Born in Azerbaijan in 1921, he moved to the University of California at Berkeley in 1959. His ideas on fuzzy sets were soon applied to different areas such as artificial intelligence, natural language, decision making, expert systems, neural networks, control theory, etc.

In mathematics, every subset of a given referential universe U can be identified with its *characteristic function* f; that is, the function $f: U \rightarrow \{0, 1\}$ which takes the value 1 if the element belongs to the considered subset and 0 in other case. In contrast, a fuzzy set is a mapping from the universe U to $[0, 1]$; that is,

---

Definition 6.1

A fuzzy set (or type-1 fuzzy set) A over a referential set U is an object

$$A = \{(u_i, \mu_A(u_i)) | u_i \in U\},$$

where $\mu_A: U \rightarrow [0, 1]$.

$\mu_A(u_i)$ represents the degree of membership of the element $u_i \in U$ to the set A. The elements for which $\mu_A(u_i) = 1$ belong to the set A; those for which $\mu_A(u_i) = 0$ do not belong to A and there are elements with a greater or smaller degree of membership to A depending on $\mu_A(u_i)$.

We are going to denote by $FS(U)$ the class of fuzzy sets defined over U; that is, $FS(U) \equiv [0, 1]^U$. The membership degree of an element $u_i \in U$ to the fuzzy set A is usually denoted by $A(u_i)$ instead of $\mu_A(u_i)$.

From the classical definition of union and intersection for crisp sets, Zadeh proposes the following definitions:

$$\begin{aligned} A \cup B(u_i) &= \max(A(u_i), B(u_i)), \\ A \cap B(u_i) &= \min(A(u_i), B(u_i)). \end{aligned} \tag{6.1}$$

A key concept in the following developments is that of lattice. We review now its definition, that can be found for instance in [6.8].

Recall that an order relationship over a set L is a relation $\leq_L$ such that

i) $x \leq_L x$ for all $x \in L$ (reflexivity);
ii) if $x \leq_L y$ and $y \leq_L z$ then $x \leq_L z$ for any $x, y, z \in L$ (transitivity);
iii) if $x \leq_L y$ and $y \leq_L x$, then $x = y$, for any $x, y \in L$ (antisymmetry).

If $\leq_L$ is an order relationship over L then $(L, \leq_L)$ is called a partially ordered set. Now, in order to define a lattice we need first to introduce the following definition.

Definition 6.2

Let $(L, \leq_L)$ be a partially ordered set and $A \subset L$ (in the sense of the usual set theory). The greatest lower bound of A (if it exists) is the element $x_{\inf} \in L$ such that:

i) $x_{\inf} \leq_L z$ for all $z \in A$ and
ii) for any $y \in L$ such that $y \leq_L z$ for all $z \in A$ it follows that $y \leq_L x_{\inf}$.

Analogously, the least upper bound of A (if it exists) is the element $x_{\sup} \in L$ such that

i) $z \leq_L x_{\sup}$ for all $z \in A$ and
ii) for any $y \in L$ such that $z \leq_L y$ for all $z \in A$ it follows that $x_{\sup} \leq_L y$.

Now we can introduce the notion of lattice.

Definition 6.3

A lattice is a partially ordered set $(L, \leq_L)$ such that any two elements $x, y \in L$ have the greatest lower bound or meet, denoted by $x \wedge y$ and the lowest upper bound or join, denoted by $x \vee y$. A lattice L is called complete if any subset of L has the lowest upper bound and the greatest lower bound.

Given a lattice $(L \leq_L)$, we will call supremum of L and denote by 1_L the lowest upper bound of L (if it exists). Analogously, we will call the infimum of L and denote by 0_L the greatest lower bound of L. In case both 1_L and 0_L exist, L is called a bounded lattice.

Observe that if we know how the join and meet operations are defined for any two elements of a set L, we can recover the ordering $\leq_L$ just by defining for any $x, y \in L$

$x \leq_L y$ if and only if $x \wedge y = x$

if and only if $x \vee y = y$

Taking into account (6.1) and Definition 6.3, it is easy to prove the following theorem.

Theorem 6.1
$(FS(U), \cup, \cap)$ is a complete lattice.

From Theorem 6.1 and the concept of lattice, we can define the following partial order relation: For $A, B \in FS(U)$

$A \leq_{FS} B$ if and only if $A(u_i) \leq B(u_i)$

for every $u_i \in U$.

The first criticism to fuzzy sets theory arises from this order relation $\leq_{FS}$. Since Zadeh presented fuzzy sets to represent uncertainty, it comes out that $\leq_{FS}$ is a crisp relation. Note that the following may happen: Let U be a referential set with 1000 elements and let A and B be two fuzzy sets over U such that for every element except for one $A(u_i) \leq B(u_i)$. Then, from the previous relation, A is not less than B. This fact led *Willmott* [6.9], *Bandler* and *Kohout* [6.10] and others to consider the concept of inclusion measure. These measures have been widely used in fuzzy morphologic mathematics [6.11], in image processing [6.12], etc.

It is easy to see that with the operations defined in (6.1) and the standard negation, $n(x) = 1 - x$ for all $x \in [0, 1]$, neither the noncontradiction principle nor the middle excluded principle hold. Nowadays, operations in (6.1) are given in terms of *t*-norms and *t*-conorms [6.13–16].

Definition 6.1 can be clearly extended to consider mappings valued over any kind of set. In particular, for our future developments and following Goguen's work [6.2], it is interesting to consider the case of mappings that take values over a lattice L. In this case, we speak of L-fuzzy sets.

Taking into account Definition 6.3 Goguen presents the concept of L-fuzzy set as follows:

Definition 6.4

Let $(L, \vee, \wedge)$ be a lattice. An L-fuzzy set over the referential set U is a mapping

$A: U \to L$.

For a given lattice L, we will denote by L-$FS(U)$, the space of L-fuzzy sets over the referential U. That is, L-$FS(U) \equiv L^U$.

Union and intersection of L-fuzzy sets can be easily defined as follows.

Definition 6.5

Let L be a lattice, and let $\vee$ and $\wedge$ be its join and meet operators respectively. Then intersection and union are defined, respectively, by:

i)

$$\cap_L : L\text{-}FS(U) \times L\text{-}FS(U) \to L\text{-}FS(U) \text{ given by}$$
$$\cap_L(A, B)(u_i) = A(u_i) \wedge B(u_i) .$$

In order to recover the usual notation for fuzzy sets, we will write $\cap_L(A, B)$ as $A \cap_L B$;

ii)

$$\cup_L : L\text{-}FS(U) \times L\text{-}FS(U) \to L\text{-}FS(U) \text{ given by}$$
$$\cup_L(A, B)(u_i) = A(u_i) \vee B(u_i) .$$

In order to recover the usual notation for fuzzy sets, we will write $\cup_L(A, B)$ as $A \cup_L B$.

We can state the following result for L-fuzzy sets.

Proposition 6.1

Let L be a bounded lattice with a supremum given by 1_L and an infimum given by 0_L. Let $\vee$ and $\wedge$ be the join and meet operators of L, respectively. Then, the set

$(L\text{-}FS(U), \leq_{L\text{-}FS(U)})$ is a bounded lattice, where the order is defined as

$$A \leq_{L\text{-}FS(U)} B \text{ if and only if } A \cup_L B = B$$

or equivalently

$$A \leq_{L\text{-}FS(U)} B \text{ if and only if } A \cap_L B = A .$$

That is

$$A \leq_{L\text{-}FS(U)} B \text{ if and only if } A(u_i) \vee B(u_i) = B(u_i)$$
for all $u_i \in U$

or equivalently

$$A \leq_{L\text{-}FS(U)} B \text{ if and only if } A(u_i) \wedge B(u_i) = A(u_i)$$
for all $u_i \in U .$

The supremum of this lattice is given by

$$1_{L\text{-}FS(U)} : U \to L ,$$
$$u_i \to 1_L$$

and the infimum is given by

$$0_{L\text{-}FS(U)} : U \to L$$
$$u_i \to 0_L .$$

6.2 Origin of the Extensions

In 1971, *Zadeh* in his paper [6.17] settled that the construction of the fuzzy sets, that is, the determination of the membership degree of each element to the set, is the biggest problem for using fuzzy sets theory in applications. This fact led him to introduce the concept of type-2 fuzzy set.

Later, in December 11, 2008, in the *bisc-group* mail list Zadeh proposes the following definitions.

Definition 6.6

Fuzzy logic is a precise system of reasoning, deduction, and computation in which the objects of discourse and analysis are associated with information which is, or is allowed to be, imperfect.

Definition 6.7

Imperfect information is defined as information which in one or more respects is imprecise, uncertain, vague, incomplete, partially true, or partially possible.

On the same date and place, Zadeh made the following remarks:

1. In fuzzy logic everything is or is allowed to be a matter of degree. Degrees are allowed to be fuzzy.
2. Fuzzy logic is not a replacement for bivalent logic or bivalent-logic-based probability theory. Fuzzy logic adds to bivalent logic and bivalent-logic-based probability theory a wide range of concepts and techniques for dealing with imperfect information.
3. Fuzzy logic is designed to address problems in reasoning, deduction, and computation with imperfect information which are beyond the reach of traditional methods based on bivalent logic and bivalent-logic-based probability theory.
4. In fuzzy logic the writing instrument is a spray pen (Fig. 6.1) with a precisely known adjustable spray pattern. In bivalent logic the writing instrument is a ballpoint pen.

Part A | 6.2

5. The importance of fuzzy logic derives from the fact that in much of the real-world imperfect information is the norm rather than exception.

All these considerations justify the use of fuzzy sets theory whenever objects are linked to soft concepts, those that do not show clear boundaries. Of course, applications might require tools other than fuzzy [6.18]. In any case, if we decide to use fuzzy sets and it is hard for us to build the characteristic functions of the involved sets, then we must use set representations that take into account these difficulties, and focus on those fuzzy sets that we call *extensions*.

So the origin of the concept of extension of fuzzy sets is directly associated with the idea of building fuzzy sets that allow us to represent objects that are described through imperfect information, and that also allow us to represent the lack of knowledge or uncertainty associated with the membership degrees that are given by the experts.

It is clear that working with extensions implies that we need to use more information than in the basic model of Zadeh. As already pointed out, in order to justify the use of these extensions in practice, the results obtained with them must be better than those obtained with usual fuzzy sets.

6.3 Type-2 Fuzzy Sets

The idea of taking into account the experts' uncertainty when they build the membership degrees of the elements to a given fuzzy sets led *Zadeh* to present in 1971 the notion of type-2 fuzzy set [6.17] as follows: A type-2 fuzzy set is a fuzzy set over a referential set U for which the membership degrees of the elements are given by fuzzy sets defined over the referential set $[0, 1]$.

The mathematical formalization of this concept was made in 1976 by *Mizumoto* and *Tanaka* in [6.19] and in 1979 by *Dubois* and *Prade* in [6.20] as follows:

Definition 6.8
A type-2 fuzzy set is a mapping $A: U \to FS([0, 1])$.

In Fig. 6.2 we show an example of type-2 fuzzy set.
We denote by $T2FS(U)$ the set of all type-2 fuzzy sets over U. That is

$$T2FS(U) \equiv (FS([0, 1]))^U .$$

6.3.1 Type-2 Fuzzy Sets as a Lattice

From Definition 6.8, the following result is obvious.

Corollary 6.1
Type-2 fuzzy sets are a particular type of Goguen's L-fuzzy sets.

Taking into account Corollary 6.1, it is clear that we can define the following operations over type-2 fuzzy sets [6.21].

Definition 6.9
The operations of union $\cup_{T2}$ and intersection $\cap_{T2}$ of $A, B \in T2FS(U)$ (in the sense of lattices) are defined, respectively, as

$$\cup_{T2}(A, B): U \to FS([0, 1]) \text{ given by}$$
$$A \cup_{T2} B(u_i) = A(u_i) \cup B(u_i)$$

and

$$\cap_{T2}(A, B): U \to FS([0, 1]) ,$$
$$A \cap_{T2} B(u_i) = A(u_i) \cap B(u_i) .$$

Proposition 6.2
The set $(T2FS(U), \cup_{T2}, \cap_{T2})$ is a bounded lattice with respect to the order

$$A \leq_{T2FS(U)} B \text{ if and only if } A \cup_{T2} B = B$$

or equivalently

$$A \leq_{T2FS(U)} B \text{ if and only if } A \cap_{T2} B = A .$$

That is

$$A \leq_{T2FS(U)} B \text{ if and only if } A(u_i) \cup B(u_i) = B(u_i)$$
$$\text{for all } u_i \in U$$

or equivalently

$$A \leq_{T2FS(U)} B \text{ if and only if } A(u_i) \cap B(u_i) = A(u_i)$$
$$\text{for all } u_i \in U .$$

The supremum of this lattice is given by $1_{T2FS(U)}$: $U \to FS(U)$ where, for every $u_i \in U$, $1_{T2FS(U)}(u_i)$ is

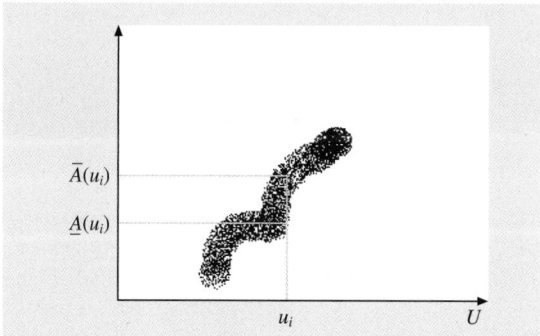

Fig. 6.1 The writing instrument is a spray pen

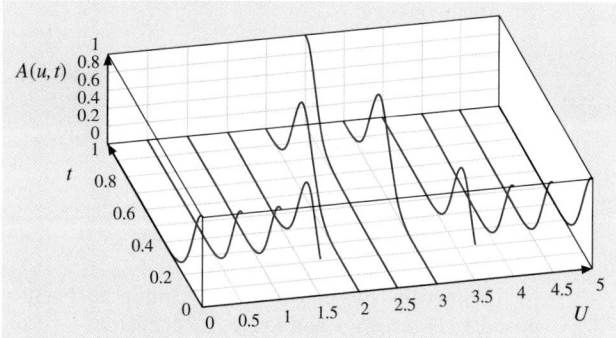

Fig. 6.2 Example of a type-2 fuzzy set

the fuzzy sets that assigns to every $t \in [0, 1]$ membership equal to 1. The infimum is given by $0_{T2FS(U)}: U \to FS(U)$ where, for every $u_i \in U$, $0_{T2FS(U)}(u_i)$ is the fuzzy sets that assigns to every $t \in [0, 1]$ membership equal to 0.

6.3.2 Remarks on the Notation

Mizumoto and *Tanaka* in 1976 [6.19] and *Mendel* and *John* in 2000 [6.22] used the following notation:

$$\int_{u \in U} \int_{t \in J_u} \frac{A(u, t)}{(u, t)}, \qquad J_u \subset [0, 1],$$

where J_u is the primary membership of $u \in U$ and, for each fixed $u = u_0$, the fuzzy set $\int_{t \in J_{u_0}} A(u_0, t)/t$ is the secondary membership of u_0.

From our point of view, this notation is not the most appropriate one, so now we try to introduce a more clarifying notation. Observe that a type-2 fuzzy set assigns to an element in the referential U a mapping $A(u): [0, 1] \to [0, 1]$. To represent fuzzy sets (or type-1 fuzzy sets) defined by a mapping A it is quite usual the notation

$$\{(u_i, A(u_i)) \mid u \in U\}. \tag{6.2}$$

In this type-1 case, $A(u)$ is a real number in $[0, 1]$ for every $u_i \in U$. In the case of type-2 fuzzy sets, if we imitate this notation, we formally lead to $\{(u_i, A(u_i)) \mid u_i \in U\}$. But now for each $u_i \in U$, we have that $A(u_i)$ is not a real number but a mapping (a type-1 fuzzy set)

$$A(u): [0, 1] \to [0, 1],$$
$$t \to A(u)(t).$$

Taking into account these considerations *Harding* et al. [6.21] and *Aisbett* et al. [6.23] suggested the following notation for a type-2 fuzzy set A:

$$A = \{(u_i, (t, A(u_i)(t)) \mid u_i \in U, t \in [0, 1]\}.$$

But an easier one to use one could be the following.

Definition 6.10
Let $A: U \to FS([0, 1])$ be a type-2 fuzzy set. Then A is denoted as

$$\{(u_i, A(u_i, t)) \mid u_i \in U, t \in [0, 1]\}.$$

where $A(u_i, \cdot): [0, 1] \to [0, 1]$ is defined as $A(u_i, t) = A(u_i)(t)$.

6.3.3 A First Definition of Operations Between Type-2 Fuzzy Sets: Lattice–Based Approach

With Definition 6.10, if we have two type-2 fuzzy sets

$$A = \{(u_i, (A(u_i, t)) \mid u_i \in U, t \in [0, 1]\}$$

and

$$B = \{(u_i, (B(u_i, t)) \mid u_i \in U, t \in [0, 1]\}$$

we have (Fig. 6.3)

$$A \cup_{T2FS} B = \{(u_i, A \cup B(u_i, t)) \mid u_i \in U, t \in [0, 1]\},$$

where, for each $u_i \in U$ and each $t \in [0, 1]$, we have

$$A \cup B(u_i, t) = \max(A(u_i, t), B(u_i, t))$$
$$= \max(A(u_i)(t), B(u_i)(t)) \tag{6.3}$$

Analogously,

$$A \cap_{T2FS} B = \{(u_i, A \cap B(u_i, t)) \mid u_i \in U, \ t \in [0, 1]\},$$

where, for each $u_i \in U$ and each $t \in [0, 1]$, we have

$$A \cap B(u_i, t) = \min(A(u_i, t), B(u_i, t))$$
$$= \min(A(u_i)(t), B(u_i)(t)) \qquad (6.4)$$

Observe that this notation is very similar to that proposed by *Deschrijver* and *Kerre* [6.24, 25].

6.3.4 Problems with the Lattice–Based Definitions. Operations Based on Zadeh's Extension Principle

Although meaningful from a mathematical point of view, as pointed out by *Dubois* and *Prade* in [6.26], from these definitions we do not recover the usual ones for fuzzy sets. To see it, just consider a finite referential set $U = \{u_1, u_2, u_3\}$ with three elements, and consider the following two fuzzy sets over U. We use the notation of (6.2) for the sake of brevity.

$$A = \left\{ \left(u_1, \frac{1}{2}\right), \left(u_2, \frac{1}{3}\right), (u_3, 1) \right\}$$

and

$$B = \left\{ \left(u_1, \frac{1}{4}\right), \left(u_2, \frac{1}{2}\right), \left(u_3, \frac{1}{7}\right) \right\}.$$

Then we have, for instance,

$$A \cup B = \left\{ \left(u_1, \frac{1}{2}\right), \left(u_2, \frac{1}{2}\right), (u_3, 1) \right\}$$

On the other hand, we can also see A and B as type-2 fuzzy sets, that we denote by A_2 and B_{T2}, respectively, just taking

$$A_{T2}(u_1)(t) = \begin{cases} 1 & \text{if } t = \frac{1}{2} \\ 0 & \text{in other case} \end{cases},$$

$$A_{T2}(u_2)(t) = \begin{cases} 1 & \text{if } t = \frac{1}{3} \\ 0 & \text{in other case} \end{cases},$$

$$A_{T2}(u_3)(t) = \begin{cases} 1 & \text{if } t = 1 \\ 0 & \text{in other case} \end{cases},$$

and

$$B_{T2}(u_1)(t) = \begin{cases} 1 & \text{if } t = \frac{1}{4} \\ 0 & \text{in other case} \end{cases},$$

$$B_{T2}(u_2)(t) = \begin{cases} 1 & \text{if } t = \frac{1}{2} \\ 0 & \text{in other case} \end{cases},$$

$$B_{T2}(u_3)(t) = \begin{cases} 1 & \text{if } t = \frac{1}{7} \\ 0 & \text{in other case} \end{cases}.$$

Then we have

$$A_{T2} \cup_{T2FS} B_{T2}(u_1)(t) = \begin{cases} 1 & \text{if } t = \frac{1}{4} \text{ or } t = \frac{1}{2} \\ 0 & \text{in other case} \end{cases},$$

$$A_{T2} \cup_{T2FS} B_{T2}(u_2)(t) = \begin{cases} 1 & \text{if } t = \frac{1}{2} \text{ or } t = \frac{1}{3} \\ 0 & \text{in other case} \end{cases},$$

and

$$A_{T2} \cup_{T2FS} B_{T2}(u_3)(t) = \begin{cases} 1 & \text{if } t = \frac{1}{7} \text{ or } t = 1 \\ 0 & \text{in other case} \end{cases},$$

which does not coincide with our previous result. Moreover, observe that we do not even recover a fuzzy set but a *true* type-2 fuzzy set.

In order to solve this problem, several authors [6.19, 22, 26] proposed the following definitions of the operations of union and intersection.

6.3.5 Second Definition of the Operations: Zadeh's Extension Principle Approach

Definition 6.11
Given two type-2 fuzzy sets

$$A = \{(u_i, A(u_i, t)) \mid u_i \in U, \ t \in [0, 1]\}$$

and

$$B = \{(u_i, B(u_i, t)) \mid u_i \in U, \ t \in [0, 1]\}$$

we can define (Fig. 6.4)

$$A \sqcap B = \{(u_i, A \sqcap B(u_i, t)) \mid u_i \in U, \ t \in [0, 1]\}$$

with

$$A \sqcap B(u_i, t) = \sup_{\min(z, w) = t} \min(A(u_i, z), B(u_i, w))$$

and

$$A \sqcup B = \{(u_i, A \sqcup B(u_i, t)) \mid u_i \in U , \ t \in [0,1]\}$$

with

$$A \sqcup B(u_i, t) = \sup_{\max(z,w)=t} \min(A(u_i, z), B(u_i, w)) .$$

For instance, let us recover our previous example. Consider the type-2 fuzzy sets A_{T2} and B_{T2}. Then we have that

$$A_{T2} \sqcup B_{T2}(u_1, t) =$$
$$\begin{cases} 0 & \text{if } t \notin \{\frac{1}{4}, \frac{1}{2}\} \\ \sup_{\max(z,w)=t} (\min(A_{T2}(u_1, z), B_{T2}(u_1, w))) . \\ & \text{in other case} \end{cases}$$

But if $t = \frac{1}{4}$, then, as $\frac{1}{2} > \frac{1}{4}$ and since $A_{T2}(u_1, z) = 0$ for all $z \leq \frac{1}{4}$, it follows that $\min(A_{T2}(u_1, z), B_{T2}(u_1, w)) = 0$ whenever $\max(z, w) = \frac{1}{4}$. Finally, if $t = \frac{1}{2}$, then $\min(A_{T2}(u_1, \frac{1}{2}), B_{T2}(u_1, \frac{1}{4})) = 1$, so we finally arrive at

$$A_{T2} \sqcup B_{T2}(u_1, t) = \begin{cases} 0 & \text{if } t \neq \frac{1}{2} \\ 1 & \text{if } t = \frac{1}{2} \end{cases} .$$

Since for u_2 and u_3 the same arguments work, we see that we indeed recover the fuzzy case. In particular, with respect to these new operations, we have the following result [6.21].

___

Proposition 6.3
Let U be a referential set. $(T2FS(U), \sqcup, \sqcap)$ is not a lattice.

___

In fact, the problem is that the absorption laws

$$A \sqcap (A \sqcup B) = A$$

and

$$A \sqcup (A \sqcap B) = A$$

do not hold. Nevertheless, it is also possible to provide a positive result [6.21].

___

Proposition 6.4
Let U be a referential set. Then for any $A, B, C \in T2FS(U)$ the following properties hold:

i) $A \sqcup A = A$ and $A \sqcap A = A$;
ii) $A \sqcup B = B \sqcup A$ and $A \sqcap B = B \sqcap A$;
iii) $A \sqcup (B \sqcup C) = (A \sqcup B) \sqcup C$.

That is, $(T2FS(U), \sqcup, \sqcap)$ is a bisemilattice.

___

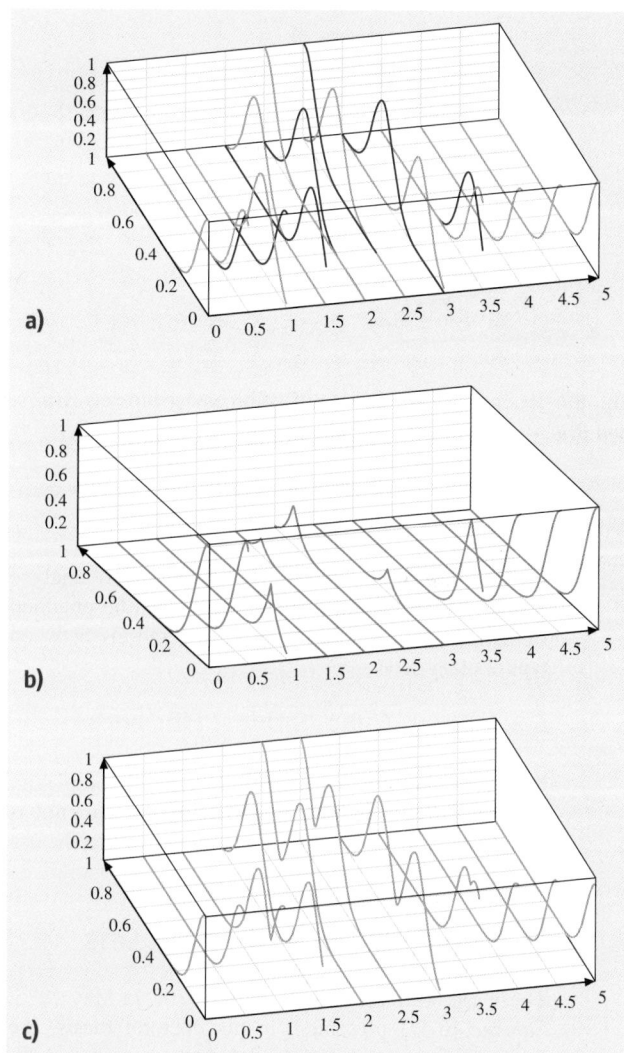

Fig. 6.3a–c Two different type 2 fuzzy sets (**a**) $A \cup_{T2FS} B$ (**b**) $A \cap_{T2FS} B$ (**c**)

___

Remark 6.1
We should remark the following:

1. If we work with the operations defined in Eqs. (6.3) and (6.4), and consider fuzzy sets as particular instances of type-2 fuzzy sets, then we do not recover the classical operations defined by Zadeh.
2. If we use the operations in Definition 6.11, then we recover Zadeh's classical operations for fuzzy sets, but we do not have a lattice structure. This fact makes that the use of type-2 fuzzy sets in many

Part A | 6.3

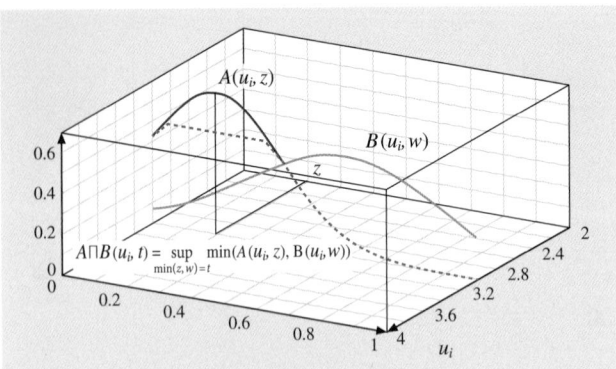

Fig. 6.4 Example of intersection of two membership sets $A(u_i, t)$ and $B(u_i, t)$. *Green line* is the set obtained

applications, such as decision making, is very complicate.

Obviously, an interesting problem is to analyze which further conclusions and results can be obtained from this new formulation of the operations between type-2 fuzzy sets.

6.3.6 About Computational Efficiency

Note also that although the computational complexity and the efficiency in time of type-2 fuzzy sets are not as high as used to be a few years ago, it is clear that the use

of these kinds of sets introduces additional complexity in any given problem. For this reason, many times the possible improvement of results is not as big as replacing type-1 fuzzy sets by type-2 fuzzy sets in many applications.

On the other hand, we can also define type-3 fuzzy sets as those fuzzy sets whose membership of each element is given by a type-2 fuzzy set [6.27]. Even more, it is possible to define recursively type-n fuzzy sets as those fuzzy sets whose membership values are type-$(n-1)$ fuzzy sets. The computational efficiency of these sets decreases as the complexity level of the building increases. From a theoretical point of view, we consider that it is necessary to carry out a complete analysis of type-n fuzzy sets structures and operations. But up to now no applications has been developed on the basis of a type-n fuzzy sets.

6.3.7 Applications

It is worth to mention the works by *Mendel* in computing with words and perceptual computing [6.28–31], of *Hagras* [6.32, 33], of *Sepulveda* et al. [6.34] in control, of *Xia* et al. in mobiles [6.35] and of *Wang* in neural networks [6.36]. We will see in the next section that the advantage of using these kinds of sets versus usual fuzzy sets has been shown only for a particular type of them, namely, the so-called interval-valued fuzzy sets.

6.4 Interval-Valued Fuzzy Sets

These sets were introduced in the 1970s. In May 1975, *Sambuc* [6.37] presented, in his doctoral thesis, the concept of an interval-valued fuzzy set named a Φ-fuzzy set. That same year, *Jahn* [6.38] wrote about these sets. One year later, *Grattan-Guinness* [6.39] established a definition of an interval-valued membership function. In that decade interval-valued fuzzy sets appeared in the literature in various guises and it was not until the 1980s, with the work of *Gorzalczany* and *Türksen* [6.40–45], that the importance of these sets, as well as their name, was definitely established.

Let us denote by $L([0, 1])$ the set of all closed subintervals in $[0, 1]$, that is,

$$L([0, 1]) = \left\{ \mathbf{x} = \left[\underline{x}, \overline{x}\right] \mid \left(\underline{x}, \overline{x}\right) \in [0, 1]^2 \right. \tag{6.5}$$
$$\left. \text{and } \underline{x} \le \overline{x} \right\} .$$

Definition 6.12
An interval-valued fuzzy set (or interval type-2 fuzzy set) A on the universe $U \ne \emptyset$ is a mapping

$$A: U \to L([0, 1]) ,$$

such that the membership degree of $u \in U$ is given by $A(u) = [\underline{A}(u), \overline{A}(u)] \in L([0, 1])$, where $\underline{A}: U \to [0, 1]$ and $\overline{A}: U \to [0, 1]$ are mappings defining the lower and the upper bounds of the membership interval $A(u)$, respectively (Fig. 6.5).

From Definition 6.12, it is clear that for these sets the membership degree of each element $u_i \in U$ to A is given by a closed subinterval in $[0, 1]$; that is, $A(u_i) = [\underline{A}(u_i), \overline{A}(u_i)]$. Obviously, if for every $u_i \in U$, we have $\underline{A}(u_i) = \overline{A}(u_i)$, then the considered set is a fuzzy set. So

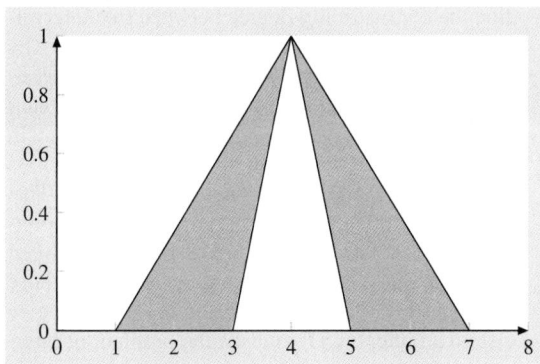

Fig. 6.5 Example of interval valued fuzzy set

fuzzy sets are particular cases of interval-valued fuzzy sets.

In 1989, *Deng* [6.46] presented the concept of Grey sets. Later Dubois proved that these are also interval-valued fuzzy sets.

We denote by $IVFS(U)$ the class of all interval-valued fuzzy sets over U; that is, $IVFS(U) \equiv L([0, 1])^U$. From Zadeh's definitions of union and intersections, Sambuc proposed the following definition:

Definition 6.13
Given $A, B \in IVFS(U)$.

$$A \cup_{L([0,1])} B(u_i) = [\max(\underline{A}(u_i), \underline{B}(u_i)),$$
$$\max(\overline{A}(u_i), \overline{B}(u_i))]$$
$$A \cap_{L([0,1])} B(u_i) = [\min(\underline{A}(u_i), \underline{B}(u_i)),$$
$$\min(\overline{A}(u_i), \overline{B}(u_i))]$$

These operations can be generalized by the use of the widely analyzed concepts of IV t-conorm and IV t-norm [6.47–49].

Corollary 6.2
Interval valued fuzzy sets are a particular case of L-fuzzy sets.

Proof: Just note that $L([0, 1])$ with the operations in Definition 6.13 is a lattice. ∎

Proposition 6.5
The set $(IVFS(U), \cup_{L([0,1])}, \cap_{L([0,1])})$ is a bounded lattice, where the order is defined as

$$A \leq_{IVFS(U)} B \text{ if and only if } A \cup_{L([0,1])} B = B$$

or equivalently

$$A \leq_{IVFS(U)} B \text{ if and only if } A \cap_{L([0,1])} B = A .$$

That is

$$A \leq_{IVFS(U)} B \text{ if and only if}$$
$$\max(\underline{A}(u_i), \underline{B}(u_i)) = \underline{B}(u_i) \text{ and}$$
$$\max(\overline{A}(u_i), \overline{B}(u_i)) = \overline{B}(u_i)$$

for all $u_i \in U$, or equivalently

$$A \leq_{IVFS(U)} B \text{ if and only if}$$
$$\min(\underline{A}(u_i), \underline{B}(u_i)) = \underline{A}(u_i) \text{ and}$$
$$\min(\overline{A}(u_i), \overline{B}(u_i)) = \overline{A}(u_i)$$

for all $u_i \in U$.

From Proposition 6.5, we deduce that the order $A \leq_{IVFS(U)} B$ if and only if $\underline{A}(u_i) \leq \underline{B}(u_i)$ and $\overline{A}(u_i) \leq \overline{B}(u_i)$ for all $u_i \in U$ is not linear. The use of these sets in decision making has led several authors to consider the problem of defining total orders between intervals [6.50]. In this sense, in [6.51] a construction method for such orders by means of aggregation functions can be found.

6.4.1 Two Interpretations of Interval-Valued Fuzzy Sets

From our point of view, interval-valued fuzzy sets can be understood in two different ways [6.52]:

1. The membership degree of an element to the set is a value that belongs to the considered interval. The interval representation is used since we cannot say precisely which that number is. For this reason, we provide bounds for that number. We think this is the correct interpretation for these sets.
2. The membership degree of each element is the whole closed subinterval provided as membership. From a mathematical point of view, this interpretation is very interesting, but, in our opinion, it is very difficult to understand it in the applied field. Moreover, in this case, we find the following paradox [6.53]:
For fuzzy sets and with the standard negation it holds that $\min(A(u_i), 1 - A(u_i)) \leq 0.5$ for all $u_i \in U$. But for interval-valued fuzzy sets, if we use the stan-

Part A | 6.4

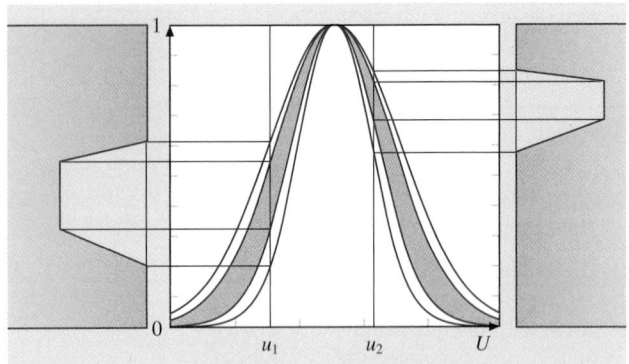

Fig. 6.6 Construction of type-2 fuzzy sets from interval-valued fuzzy sets

dard negation $N(A(u_i)) = [1 - \overline{A}(u_i), 1 - \underline{A}(u_i)]$, we have that there is no equivalent bound for

$$\min \left[\underline{A}(u_i), \overline{A}(u_i)\right], \left[1 - \overline{A}(u_i), 1 - \underline{A}(u_i)\right] .$$

6.4.2 Shadowed Sets Are a Particular Case of Interval-Valued Fuzzy Sets

The so-called shadow sets were suggested by *Pedrycz* [6.54] and developed later together with *Vukovic* [6.55, 56]. A shadowed set B induced by a given fuzzy set A defined in U is an interval-valued fuzzy set in U that maps elements of U into 0,1 and the unit interval $[0, 1]$, i.e., B is a mapping $B: U \to \{0, 1, [0, 1]\}$, where 0, 1, $[0, 1]$ denote complete exclusion from B, complete inclusion in B and complete ignorance, respectively. Shadow sets are isomorphic with a three-valued logic.

6.4.3 Interval-Valued Fuzzy Sets Are a Particular Case of Type-2 Fuzzy Sets

In 1995, *Klir* and *Yuan* proved in [6.27] that from an interval-valued fuzzy set, we can build a type-2 fuzzy set as pointed out in Fig. 6.6.
    Later in 2007 *Deschrijver* and *Kerre* [6.24, 25] and *Mendel* [6.57], proved that interval-valued fuzzy sets are particular cases of type-2 fuzzy sets.

6.4.4 Some Problems with Interval-Valued Fuzzy Sets

1. Taking into account the definition of interval-valued fuzzy sets, we follow *Gorzalczany* [6.41] and de-

fine the compatibility degree between two interval-valued fuzzy sets as an element in $L([0, 1])$. The other information measures [6.58–62] (interval-valued entropy, interval-valued similarity, etc.) should also be given by an interval. However, in most of the works about these measures, the results are given by a number, and not by an interval. This consideration leads us to settle that, from a theoretical point of view, we should distinguish between two different types of information measures: those which give rise to a number and those which give rise to an interval. Obviously, the problem of interpreting both types of measures arises. Moreover, if the result of the measure is an interval, we should consider its amplitude as a measure of the lack of knowledge [6.63] linked to the considered measure.

2. In [6.57], *Mendel* writes:

    It turns out that an interval type-2 fuzzy set is the same as an interval-valued fuzzy set for which there is a very extensive literature. These two seemingly different kinds of fuzzy sets were historically approached from very different starting points, which as we shall explain next has turned out to be a very good thing.

    Nonetheless, we consider that interval-valued fuzzy sets are a particular case of interval type-2 fuzzy sets and therefore they are not the same thing.

3. Due to the current characteristics of computers, we can say that the computational cost of working with these sets is not much higher than the cost of working with type-1 fuzzy sets [6.64].

4. We have already said that the commonly used order is not linear. This is a problem for some applications, such as decision making. In [6.65], it is shown that the choice of the order should depend on the considered application. Often experts do not have enough information to choose a total order. This is a big problem since the choice of the order influences strongly the final outcome.

6.4.5 Applications

We can say that there already exist applications of interval-valued fuzzy sets that provide results which are better than those obtained with fuzzy sets. For instance:

1. *In classification problems.* Specifically, in [6.66–69] a methodology to enhance the performance of fuzzy rule-based classification systems (FRBCSs)

is presented. The methodology used in these papers has the following structure:

1) An initial FRBCS is generated by using a fuzzy rule learning algorithm.
2) The linguistic labels of the learned fuzzy rules are modeled with interval-valued fuzzy sets in order to take into account the ignorance degree associated with the assignment of a number as the membership degree of the elements to the sets. These sets are constructed starting from the fuzzy sets used in the learning process and their shape is determined by the value of one or two parameters.
3) The fuzzy reasoning method is extended so as to take into account the ignorance represented by the interval-valued fuzzy sets throughout the inference process.
4) The values of the system's parameters, for instance the ones determining the shape of the interval-valued fuzzy sets, are tuned applying evolutionary algorithms. See [6.66–69] for details about the specific features of each proposal.

The methodology allows us to statistically outperforming the performance of the following approaches:

a) In [6.66], the performance of the initial FRBCS generated by the *Chi* et al. algorithm [6.70] and the fuzzy hybrid genetics-based machine learning method [6.71] are outperformed. In addition, the results of the GAGRAD (genetic algorithm gradient) approach [6.72] are notably improved.
b) A new tuning approach is defined in [6.67], where the results obtained by the tuning of the lateral position of the linguistic labels ([6.73]) and the performance provided by the tuning approach based on the linguistic 3-tuples representation [6.74] are outperformed.
c) Fuzzy decision trees (FDTs) are used as the learning method in [6.68]. In this contribution, numerous decision trees are enhanced, including crisp decision trees, FDTs, and FDTs constructed using genetic algorithms. For instance, the well-known C4.5 decision tree ([6.75]) or the fuzzy decision tree proposed by *Janikow* [6.76] is outperformed.
d) The proposal presented in [6.69] is the most remarkable one, since it allows outperforming two state-of-the-art fuzzy classifiers, namely, the FARC-HD method [6.77] and the unordered fuzzy rule induction algorithm (FURIA) [6.78].

Furthermore, the performance of the fuzzy counterpart of the presented approach is outperformed as well.

2. *Image processing.* In [6.63, 79–85], it has been shown that if we use interval-valued fuzzy sets to represent those areas of an image for which the experts have problems to build the fuzzy membership degrees, then edges, segmentation, etc., are much better.
3. In some decision-making problems, it has also been shown that the results obtained with interval-valued fuzzy sets are better than the ones obtained with fuzzy sets [6.86]. They have also been used in Web problems [6.87], pattern recognition [6.88], medicine [6.89], etc., see also [6.90, 91].

Construction of Interval–Valued Fuzzy Sets

In many cases, it is easier for experts to give the membership degrees by means of numbers instead of by means of intervals. In this case it may happen that the obtained results are not the best ones. If this is so, we should build intervals from the numerical values provided by the experts. For this reason, we study methods to build intervals from real numbers. For any such methods, we require the following:

i) The numerical value provided by the expert should be interior to the considered interval. We require this property since we assume that the membership degree for the expert is a number but he or she is not able to fix it exactly so he or she provides two bounds for it.
ii) The amplitude of the built interval is going to represent the degree of ignorance of the expert to fix the numerical value he or she has provided us.

The previous considerations have led us to define in [6.63] the concept of ignorance degree G_i associated with the value given by an expert. In such definition, it is settled that if the degree of membership given by the expert is equal to 0 or 1, then the ignorance is equal to 0, since the expert is sure of the fact that the element belongs or does not belong to the considered set. However, if the provided membership degree is equal to 0.5, then ignorance is maximal, since the expert does not know at all whether the element belongs or not to the set. Different considerations and construction methods for such ignorance functions using overlap functions can be found in [6.92].

Taking into account the previous argumentation, in Fig. 6.7 we show the schema of construction of an interval from a membership degree μ given by the expert

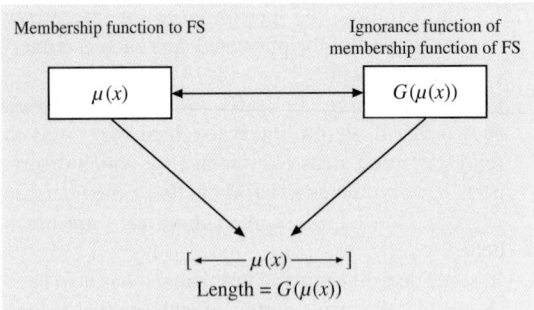

Fig. 6.7 Construction with ignorance functions

and from an ignorance function G_i chosen for the considered problem [6.63]:

There exist other methods for constructing interval-valued fuzzy sets. The choice of the method depends on the application we are working in. One of the most used methods in magnetic resonance image processing (for fuzzy theory) is the following: several doctors are asked for building, for an specific region of an image, a fuzzy set representing that region. At the end, we will have several fuzzy sets, and with them we build an interval-valued fuzzy set as follows. For each element's membership, we take as lower bound the minimum of the values provided by the doctors, and as the upper bound, the maximum. This method has shown itself very useful in particular images [6.83]. In Fig. 6.8, we represent the proposed construction.

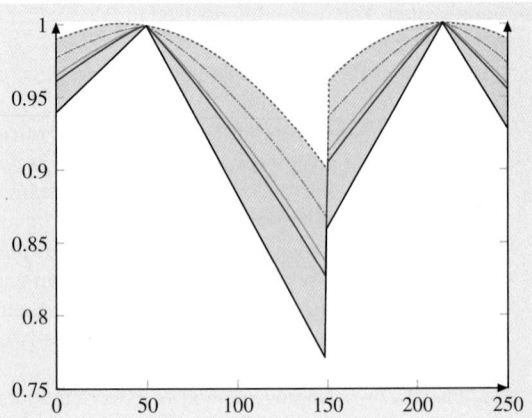

Fig. 6.8 Construction with different experts

In [6.63], it is shown that for some specific ultrasound images, if we use fuzzy theory to obtain the objects in the image, results are worse than if we use interval-valued fuzzy sets using the method proposed by *Tizhoosh* in [6.84]. Such method consists of the following (see Fig. 6.9): from the numerical membership degree μ_A given by the expert and from a numerical coefficient $\alpha > 1$, associated with the doubt of the expert when he or she constructs μ_A we generate the membership interval

$$\left[\mu_{A_t}^{\alpha}(q), \mu_{A_t}^{\frac{1}{\alpha}}(q) \right].$$

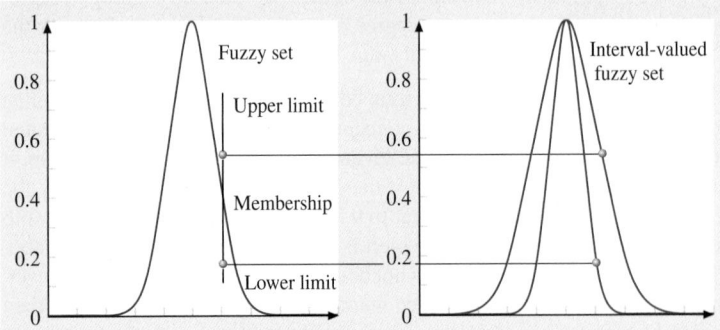

Fig. 6.9 Tizhoosh's construction

6.5 Atanasssov's Intuitionistic Fuzzy Sets or Bipolar Fuzzy Sets of Type 2 or IF Fuzzy Sets

In 1983, *Atanassov* presented his definition of intuitionistic fuzzy set [6.93]. This paper was written in Bulgarian, and in 1986 he presented his ideas in English in [6.94].

Definition 6.14
An intuitionistic fuzzy set over U is an expression A given by

$$A = \{(u_i, \mu_A(u_i), \nu_A(u_i)) | u_i \in U\},$$

where $\mu_A : U \longrightarrow [0, 1]$

$$\nu_A : U \longrightarrow [0, 1]$$

such that $0 \leq \mu_A(u_i) + \nu_A(u_i) \leq 1$ for every $u_i \in U$.

Atanassov also introduced the following two essential characteristics of these sets:

1. The complementary of

$$A = \{(u_i, \mu_A(u_i), \nu_A(u_i)) | u_i \in U\}$$

   is

$$A_c = \{(u_i, \nu_A(u_i), \mu_A(u_i)) | u_i \in U\}.$$

2. For each $u_i \in U$, the intuitionistic or hesitance index of such element in the considered set A is given by

$$\pi_A(u_i) = 1 - \mu_A(u_i) - \nu_A(u_i).$$

$\pi_A(u_i)$ is a measure of the hesitance of the expert to assign a numerical value to $\mu_A(u_i)$ and $\nu_A(u_i)$. For this reason, we consider that these sets are an extension of fuzzy sets. It is clear that if for each $u_i \in U$ we take $\nu_A(u_i) = 1 - \mu_A(u_i)$, then the considered set A is a fuzzy set in Zadeh's sense. So fuzzy sets are a particular case of those defined by Atanassov.

In 1993, *Gau* and *Buehre* [6.95] introduced the concept of vague set and later in 1994 it was shown that these are the same as those introduced by Atanassov in 1983 [6.96].

We denote by $A - IFS(U)$ the class of all intuitionistic sets (in the sense of Atanassov) defined over the referential U. Atanassov also gave the following definition:

Definition 6.15
Given $A, B \in A - IFS(U)$.

$$A \cup_{A-IFS} B = \{(u_i, \max(\mu_A(u_i), \mu_B(u_i)),$$
$$\min(\nu_A(u_i), \nu_B(u_i))) | u_i \in U\}$$
$$A \cap_{A-IFS} B = \{(u_i, \min(\mu_A(u_i), \mu_B(u_i)),$$
$$\max(\nu_A(u_i), \nu_B(u_i))) | u_i \in U\}.$$

Definitions of connectives for Atanassov's sets in terms of *t*-norms, etc. can be found in [6.49, 97].

Corollary 6.3
Atanassov's intuitionistic fuzzy sets are a particular case of *L*-fuzzy sets.

Proof: Just note that $L = \{(x_1, x_2) | x_1 + x_2 \leq 1$ with $x_1, x_2 \in [0, 1]\}$ with the operations in Definition 6.15 is a lattice. ∎

Proposition 6.6
The set $(A - IFS(U), \cup_{A-IFS}, \cap_{A-IFS})$ is a bounded lattice, where the order is defined as

$$A \leq_{A-IFS} B \text{ if and only if } A \cup_{A-IFS} B = B$$

or equivalently

$$A \leq_{A-IFS} B \text{ if and only if } A \cap_{A-IFS} B = A.$$

From Proposition 6.6, we see that the order

$$A \leq_{A-IFS} B \text{ if and only if } \mu_A(u_i) \leq \mu_B(u_i) \text{ and}$$
$$\nu_A(u_i) \geq \nu_B(u_i) \text{ for all } u_i \in U$$

is not linear. Different methods to get linear orders for these sets can be found in [6.50, 51].

6.5.1 Relation Between Interval-Valued Fuzzy Sets and Atanassov's Intuitionistic Fuzzy Sets: Two Different Concepts

In 1989, *Atanassov* and *Gargov* [6.98] and later *Deschrijver* and *Kerre* [6.24] proved that from an interval-valued fuzzy set we can build an intuitionistic fuzzy set and vice-versa.

Theorem 6.2
The mapping

$$\Phi: IVFS(U) \rightarrow A - IFS(U) ,$$
$$A \rightarrow A',$$

where $A' = \{(u_i, \underline{A}(u_i), 1 - \overline{A}(u_i)) | u_i \in U\}$, is a bijection.

Theorem 6.2 shows that interval-valued fuzzy sets and Atanassov's intuitionistic fuzzy sets, are equivalent from a mathematical point of view. But, as pointed out in [6.52], the absence of a structural component in their description might explain this result, since from a conceptual point of view they very different models:

a) The representation of the membership of an element to a set using an interval means that the expert doubts about the exact value of such membership, so such an expert provides two bounds, and we never consider the representation of the nonmembership to a set.

b) By means of the intuitionistic index we, represent the hesitance of the expert in simultaneously building the membership and the nonmembership degrees.

From an applied point of view, the conceptual difference between both concepts has also been clearly displayed in [6.99]. On page 204 of this paper, Ye adapts an example by *Herrera* and *Herrera-Viedma* appeared in 2000 [6.100]. Ye's example runs as follows: n experts are asked about a money investment in four different companies. Ye considers that the membership to the set that represents each company is given by the number of experts who would invest their money in that company (normalized by n), and the nonmembership is given by the number of experts who would not invest their money in that company. Clearly, the intuitionistic index corresponds to the experts that do not provide either a positive or a negative answer about investing in that company. In this way, Ye proves that:

1. The results obtained with this representation are more realistic than those obtained in [6.100] using Zadeh's fuzzy sets.
2. In the considered problem, the interval interpretation does not make much sense besides its use as a mathematical tool.

6.5.2 Some Problems with the Intuitionistic Sets Defined by Atanassov

Besides the missed structural component pointed out in [6.52]:

1. In these sets, each element has two associated values. For this reason, we consider that the information measures as entropy [6.59, 61], similarity [6.101, 102], etc. should also be given by two numerical values. That is, in our opinion, we should distinguish between those measures that provide a single number and those others that provide two numbers. This fact is discussed in [6.103] where the two concepts of entropy given in [6.59] and [6.61] are jointly used to represent the uncertainty linked to Atanassov's intuitionistic fuzzy set. So we think that it is necessary to carry out a conceptual revision of the definitions of similarity, dissimilarity, entropy, comparability, etc., given for these sets. Even more since nowadays working with two numbers instead of a single one does not imply a much larger computational cost.

2. As in the case of interval-valued fuzzy sets, in many applications, there is a problem to choose the most appropriate linear order associated with that application [6.50, 51]. We should remark that the chosen order directly influences the final outcome, so it is necessary to study the conditions that determine the choice of one order or another [6.65].

6.5.3 Applications

Extensions have shown themselves very useful in problems of decision making [6.99, 104–108]. In general, they work very well in problems for which we have to represent the difference between the positive and the negative representation of something [6.109], in particular in cognitive psychology and medicine [6.110]. Also in image processing they have been used often, as in [6.111, 112]. We should remark that the mathematical equivalence between these sets and interval-valued fuzzy sets makes that in many applications in which interval-valued fuzzy sets are useful, so are Atanassov's intuitionistic fuzzy sets [6.113].

6.5.4 The Problem of the Name

From Sect. 6.1.1, it is clear that the term *intuitionistic* was used in 1907 by Brouwer, in 1930 by Heyting, etc. So, 75 years before Atanassov used it, it already had a specific meaning in logic. Moreover, one year after Atanassov first used it in Bulgarian, Takeuti and Titani (1984) presented a set representation for Heyting ideas, using the expression *intuitionistic fuzzy sets*. From our point of view, this means that in fact the correct terminology is that of Takeuti and Titani. Nevertheless, all these facts have originated a serious notation problem in the literature about the subject.

In 2005, in order to solve these problems, *Dubois* et al. published a paper [6.7] on the subject and, they proposed to replace the name intuitionistic fuzzy sets by *bipolar fuzzy sets*, justifying this change. Later, *Atanassov* has answered in [6.114], where he defends the reasons he had to choose the name intuitionistic and states a clear fact: the sets he defined are much more cited and used than those defined by Takeuti and Titani, so in his opinion the name must not change.

In *Dubois* and *Prade*'s works about bipolarity types [6.115, 116], these authors stated that Atanassov's sets are included in the type-2 bipolar sets, so they call these sets fuzzy bipolar sets of type-2.

But we must say that nine years before Dubois et al.'s paper about the notation, *Zhang* in [6.117, 118] used the word bipolar in connection with the fuzzy sets theory and presented the concept of bipolar-valued set.

All these considerations have led some authors to propose the name *Atanassov's intuitionistic fuzzy sets*. However, Atanassov himself disagrees with this notation and asserts that his notation must be hold; that is, intuitionistic fuzzy sets. Other authors use the name IF-sets (intuitionistic fuzzy) [6.119].

In any case, only time will fix the appropriate names.

6.6 Atanassov's Interval-Valued Intuitionistic Fuzzy Sets

In 1989, *Atanassov* and *Gargov* presented the following definition [6.98]:

Definition 6.16
An Atanassov's interval-valued intuitionistic fuzzy set over U is an expression A given by

$$A = \{(u_i, M_A(u_i), N_A(u_i)) | u_i \in U\},$$

where $M_A : U \longrightarrow L([0, 1])$,

$$N_A : U \longrightarrow L([0, 1])$$

such that $0 \le \overline{M_A}(u_i) + \overline{N_A}(u_i) \le 1$ for every $u_i \in U$.

In this definition, authors adapt Atanassov's intuitionistic sets to Zadeh's ideas on the problem of building the membership degrees of the elements to the fuzzy set. Moreover, if for every $u_i \in U$, we have that $\underline{M_A}(u_i) = \overline{M_A}(u_i)$ and $\underline{N_A}(u_i) = \overline{N_A}(u_i)$, then we recover an Atanassov's intuitionistic fuzzy set, so the latter are a particular case of Atanassov's interval-valued intuitionistic fuzzy sets. As in the case of Atanassov's intuitionistic fuzzy sets, the complementary of a set is obtained by interchanging the membership and nonmembership intervals.

We represent by $A - IVIFS(U)$ the class of all Atanassov's interval-valued intuitionistic fuzzy sets over a referential set U.

Definition 6.17
Given $A, B \in A - IVIFS(U)$.

$$A \cup_{A-IVIFS} B = \{(u_i, A \cup_{A-IVIFS} B(u_i)) | u_i \in U\}$$
where $A \cup_{A-IVIFS} B(u_i)$
$$= \left[\left(\max\left(\underline{M_A}(u_i), \underline{M_B}(u_i)\right), \max\left(\overline{M_A}(u_i), \overline{M_B}(u_i)\right)\right)\right]$$
$$\left[\min\left(\underline{N_A}(u_i), \underline{N_B}(u_i)\right), \min\left(\overline{N_A}(u_i), \overline{N_B}(u_i)\right)\right],$$
$$A \cap_{A-IVIFS} B = \{(u_i, A \cap_{A-IVIFS} B(u_i)) | u_i \in U\}$$
where $A \cap_{A-IVIFS} B(u_i)$
$$= \left(\left[\min\left(\underline{M_A}(u_i), \underline{M_B}(u_i)\right), \min\left(\overline{M_A}(u_i), \overline{M_B}(u_i)\right)\right]\right.$$
$$\left.\left[\max\left(\underline{N_A}(u_i), \underline{N_B}(u_i)\right), \max\left(\overline{N_A}(u_i), \overline{N_B}(u_i)\right)\right]\right),$$

Corollary 6.4
Atanassov's interval-valued intuitionistic fuzzy sets are a particular case of *L*-fuzzy sets.

Proof: Just note that $LL([0, 1]) = \{(\mathbf{x}, \mathbf{y}) \in L([0, 1])^2 | \bar{x} + \bar{y}\}$ with the operations in Definition 6.17 is a lattice. ∎

Proposition 6.7
The set $(A - IVIFS(U), \cup_{A-IVIFS}, \cap_{A-IVIFS})$ is a bounded lattice, where the order is defined as

$$A \le_{A-IVIFS} B \text{ if and only if } A \cup_{A-IVIFS} B = B$$

Part A | 6.6

or equivalently

$$A \leq_{A-IVIFS} B \text{ if and only if } A \cap_{A-IVIFS} B = A .$$

Note that $A \leq_{A-IVIFS} B$ if and only if $M_A(u_i) \leq_{IVFS} M_B(u_i)$ and $N_A(u_i) \geq_{IVFS} N_B(u_i)$ for all $u_i \in U$; that is, $A \leq_{A-IVIFS} B$ if and only if $\underline{M_A}(u_i) \leq \underline{M_B}(u_i), \overline{M_A}(u_i) \leq \overline{M_B}(u_i), \underline{N_A}(u_i) \geq \underline{N_B}(u_i)$, and $\overline{N_A}(u_i) \geq \overline{N_B}(u_i)$ for all $u_i \in U$ is not linear.

We make the following remarks regarding these extensions:

1. It is necessary to study two different types of information measures: those whose outcome is a single number [6.120] and those whose outcomes are two

intervals in [0, 1] [6.120]. It is necessary a study of both types.

2. Nowadays, there are many works using these sets [6.121–123]. However none of them displays an example where the results obtained with these sets are better than those obtained with fuzzy sets or other techniques. As it happened until recent years with interval-valued fuzzy sets, it is necessary to find an application that provides better results using these extensions rather than using other sets. To do so, we should compare the results with those obtained with other techniques, which is something that it is not done for the moment in the papers that make use of these sets. From the moment, most of the studies are just theoretical [6.124–126].

6.7 Links Between the Extensions of Fuzzy Sets

Taking into account the study carried out in previous sections, we can describe the following links between the different extensions.

1. $FS \subset IVFS \equiv \text{Grey Sets} \equiv A - IFS \equiv$
    $\text{Vague sets} \subset A - IVIFS \subset L - FS .$

2. If we consider the operations in Definition 6.11, we have the sequence of inclusions:

$$FS \subset IVFS \equiv \text{Grey Sets} \equiv A - IFS$$
$$\equiv \text{Vague sets} \subset T2FS \subset L - FS .$$

6.8 Other Types of Sets

In this section, we present the definition of other types of sets that have arisen from the idea of Zadeh's fuzzy set. However, for us none of them should be considered an extension of a fuzzy set, since we do not represent with them the degree of ignorance or uncertainty of the expert.

6.8.1 Probabilistic Sets

These sets were introduced in 1981 by *Hirota* [6.127].

Definition 6.18
Let (Ω, B, P) be a probability space and let $B(0, 1)$ denote the family of Borel sets in [0, 1]. A probabilistic set A over the universe U is a function

$$A: U \times \Omega \to ([0, 1], B(0, 1)) ,$$

where $A(u_i, \cdot)$ is measurable for each $u_i \in U$.

6.8.2 Fuzzy Multisets and *n*-Dimensional Fuzzy Sets

The idea of multiset was given by *Yager* in 1986 [6.128] and later developed by *Miyamoto* [6.129]. In these multilevel sets, several degrees of membership are assigned to each element.

Definition 6.19
Let U be a nonempty set and $n \in N^+$. A fuzzy multiset A over U is given by

$$A = \{(u_i, \mu_{A_1}(u_i), \mu_{A_2}(u_i), \dots, \mu_{A_n}(u_i))|u_i \in U\} ,$$

where $\mu_{A_i}: U \to [0, 1]$ is called the *i*th membership degree of A.

If in Definition 6.19 we require that: $\mu_{A_1} \leq \mu_{A_2} \leq \cdots \leq \mu_{A_n}$ we have an *n-Dimensional fuzzy set* [6.130, 131]. Nevertheless, it is worth to point out the relation of these families of fuzzy set with the classification model proposed in [6.132], and the particular model proposed in [6.133], where fuzzy preference intensity was arranged according to the basic preference attitudes.

6.8.3 Bipolar Valued Set or Bipolar Set

In 1996, *Zhang* presented the concept of bipolar set as follows [6.117]:

Definition 6.20
A bipolar-valued set or a bipolar set on U is an object

$$A = \{(u_i, \varphi^+(u_i), \varphi^-(u_i))|u_i \in U\}$$

with $\varphi^+: U \to [0,1]$, $\varphi^-: U \to [-1,0]$.

In these sets, the value $\varphi^-(u_i)$ must be understood as how much the environment of the problem opposes to the fulfillment of $\varphi^+(u_i)$. Nowadays interesting studies exist about these sets [6.134–138].

6.8.4 Neutrosophic Sets or Symmetric Bipolar Sets

These sets were first studied by Smarandache in 2002 [6.139]. They arise from Atanassov's intuitionistic fuzzy sets ignoring the restriction on the sum of the membership and the nonmembership degrees.

Definition 6.21
A neutrosophic set or symmetric bipolar set on U is an object

$$A = \{(u_i, \mu_A(u_i), \nu_A(u_i))|u_i \in U\},$$

with $\mu_A: U \to [0,1]$, $\nu_A: U \to [0,1]$.

6.8.5 Hesitant Sets

These sets were introduced by *Torra* and *Narukawa* in 2009 to deal with decision-making problems [6.140, 141].

Definition 6.22
Let $\wp([0,1])$ be the set of all subsets of the unit interval and U be a nonempty set. Let $\mu_A: U \to \wp([0,1])$, then a *hesitant fuzzy set* (HFS in short) A defined over U is given by

$$A = \{(u_i, \mu_A(u_i))|u_i \in U\}. \tag{6.6}$$

6.8.6 Fuzzy Soft Sets

Based on the definition of soft set [6.142], *Maji* et al. present the following definition [6.143].

Definition 6.23
A pair (F,A) is called a fuzzy soft set over U, where F is a mapping given by $F: A \to FP(U)$.

Where $FP(U)$ denotes the set of all fuzzy subsets of U.

6.8.7 Fuzzy Rough Sets

From the concept of rough set given by *Pawlak* in [6.144], *Dubois* and *Prade* in 1990 proposed the following definition [6.145]. *From different point of views these sets could be considered as an extension of fuzzy sets in our sense*, besides these sets are being exhaustively studied, for this reason we consider that these sets need another chapter.

Definition 6.24
Let U be a referential set and R be a fuzzy similarity relation on U. Take $A \in \mathcal{F}S(U)$. A fuzzy rough set over U is a pair $(R \downarrow A, R \uparrow A) \in \mathcal{F}S(U) \times \mathcal{F}S(U)$, where

- $R \downarrow A: U \to [0,1]$ is given by
  $R \downarrow A(u) = \inf_{v \in U} \max(1 - R(v,u), A(v))$
- $R \uparrow A: U \to [0,1]$ is given by
  $R \uparrow A(u) = \sup_{v \in U} \min(R(v,u), A(v))$.

Part A | 6.8

6.9 Conclusions

In this chapter, we have reviewed the main types of fuzzy sets defined since 1965. We have classified these sets in two groups: those that take into account the problem of building the membership functions, which we have included in the so-called extensions of fuzzy sets, and those that appear as an answer to such a key issue.

We have introduced the definitions and first properties of the extensions, that is, type-2 fuzzy sets, interval-valued fuzzy sets; Atanassov's intuitionistic fuzzy sets or type-2 bipolar fuzzy sets, and Atanassov's interval-valued fuzzy sets. We have described the properties and problems linked to type-2 fuzzy sets, and we have presented several construction methods for interval-valued fuzzy sets, depending on the application. We have also referred to some papers where it is shown that the use of interval-valued fuzzy sets improves the results obtained with fuzzy sets.

In general, we have stated the main problem in fuzzy sets extensions, namely, to find applications for which the results obtained with these sets are better than those obtained with other techniques. This has only been proved, up to now, for interval-valued fuzzy sets. We think that the great defy for some sets that are initially justified as a theoretical need is to prove their practical usefulness.

References

6.1 L.A. Zadeh: Fuzzy sets, Inf. Control **8**, 338–353 (1965)

6.2 J.A. Goguen: L-fuzzy sets, J. Math. Anal. Appl. **18**, 145–174 (1967)

6.3 J.T. Cacioppo, W.L. Gardner, C.G. Berntson: Beyond bipolar conceptualizations and measures: The case of attitudes and evaluative space, Pers. Soc. Psychol. Rev. **1**, 3–25 (1997)

6.4 R. Goldblatt: *Topoi: The Categorial Analysis of Logic* (North-Holland, Amsterdam 1979)

6.5 S.M. Lane, I. Moerfijk: *Sheaves in Geometry and Logic* (Springer, New York 1992)

6.6 G. Takeuti, S. Titani: Intuitionistic fuzzy logic and intuitionistic fuzzy set theory, J. Symb. Log. **49**, 851–866 (1984)

6.7 D. Dubois, S. Gottwald, P. Hajek, J. Kacprzyk, H. Prade: Terminological difficulties in fuzzy set theory – The case of intuitionistic fuzzy sets, Fuzzy Sets Syst. **156**(3), 485–491 (2005)

6.8 G. Birkhoff: *Lattice Theory* (American Mathematical Society, Providence 1973)

6.9 R. Willmott: *Mean Measures in Fuzzy Power-Set Theory*, Report No. FRP-6 (Dep. Math., Univ. Essex, Colchester 1979)

6.10 W. Bandler, L. Kohout: Fuzzy power sets, fuzzy implication operators, Fuzzy Sets Syst. **4**, 13–30 (1980)

6.11 B. De Baets, E.E. Kerre, M. Gupta: The fundamentals of fuzzy mathematical morphology – part 1: Basic concepts, Int. J. Gen. Syst. **23**(2), 155–171 (1995)

6.12 L.K. Huang, M.J. Wang: Image thresholding by minimizing the measure of fuzziness, Pattern Recognit. **29**(1), 41–51 (1995)

6.13 H. Bustince, J. Montero, E. Barrenechea, M. Pagola: Semiautoduality in a restricted family of aggregation operators, Fuzzy Sets Syst. **158**(12), 1360–1377 (2007)

6.14 T. Calvo, A. Kolesárová, M. Komorníková, R. Mesiar: Aggregation operators: Properties, classes and construction methods. In: *Aggregation Operators New Trends and Applications*, ed. by T. Calvo, G. Mayor, R. Mesiar (Physica, Heidelberg 2002) pp. 3–104

6.15 J. Fodor, M. Roubens: *Fuzzy preference modelling and multicriteria decision support*, Theory and Decision Library (Kluwer, Dordrecht 1994)

6.16 E.P. Klement, R. Mesiar, E. Pap: *Triangular norms, trends in logic*, Studia Logica Library (Kluwer, Dordrecht 2000)

6.17 L.A. Zadeh: Quantitative fuzzy semantics, Inf. Sci. **3**, 159–176 (1971)

6.18 E.E. Kerre: A first view on the alternatives of fuzzy sets theory. In: *Computational Intelligence in Theory and Practice*, ed. by B. Reusch, K.-H. Temme (Physica, Heidelberg 2001) pp. 55–72

6.19 M. Mizumoto, K. Tanaka: Some properties of fuzzy sets of type 2, Inf. Control **31**, 312–340 (1976)

6.20 D. Dubois, H. Prade: Operations in a fuzzy-valued logic, Inf. Control **43**(2), 224–254 (1979)

6.21 J. Harding, C. Walker, E. Walker: The variety generated by the truth value algebra of type-2 fuzzy sets, Fuzzy Sets Syst. **161**, 735–749 (2010)

6.22 J.M. Mendel, R.I. John: Type-2 fuzzy sets made simple, IEEE Trans. Fuzzy Syst. **10**, 117–127 (2002)

6.23 J. Aisbett, J.T. Rickard, D.G. Morgenthaler: Type-2 fuzzy sets as functions on spaces, IEEE Trans. Fuzzy Syst. **18**(4), 841–844 (2010)

6.24 G. Deschrijver, E.E. Kerre: On the position of intuitionistic fuzzy set theory in the framework of theories modelling imprecision, Inf. Sci. **177**, 1860–1866 (2007)

6.25 G. Deschrijver, E.E. Kerre: On the relationship between some extensions of fuzzy set theory, Fuzzy Sets Syst. **133**, 227–235 (2003)

6.26   D. Dubois, H. Prade: *Fuzzy Sets and Systems: Theory and Applications* (Academic, New York 1980)

6.27   G.J. Klir, B. Yuan: *Fuzzy Sets and Fuzzy Logic: Theory and Applications* (Prentice-Hall, New Jersey 1995)

6.28   J.M. Mendel: Type-2 fuzzy sets for computing with words, IEEE Int. Conf. Granul. Comput., Atlanta (2006), GA 8-8

6.29   J.M. Mendel: Computing with words and its relationships with fuzzistics, Inf. Sci. **177**(4), 988–1006 (2007)

6.30   J.M. Mendel: Historical reflections on perceptual computing, Proc. 8th Int. FLINS Conf. (FLINS'08) (World Scientific, Singapore 2008) pp. 181–187

6.31   J.M. Mendel: Computing with words: Zadeh, Turing, Popper and Occam, IEEE Comput. Intell. Mag. **2**, 10–17 (2007)

6.32   H. Hagras: Type-2 FLCs: A new generation of fuzzy controllers, IEEE Comput. Intell. Mag. **2**, 30–43 (2007)

6.33   H. Hagras: A hierarchical type-2 fuzzy logic control architecture for autonomous mobile robots, IEEE Trans. Fuzzy Syst. **12**, 524–539 (2004)

6.34   R. Sepulveda, O. Castillo, P. Melin, A. Rodriguez-Diaz, O. Montiel: Experimental study of intelligent controllers under uncertainty using type-1 and type-2 fuzzy logic, Inf. Sci. **177**, 2023–2048 (2007)

6.35   X.S. Xia, Q.L. Liang: Crosslayer design for mobile ad hoc networks using interval type-2 fuzzy logic systems, Int. J. Uncertain. Fuzziness Knowl. Syst. **16**(3), 391–408 (2008)

6.36   C.H. Wang, C.S. Cheng, T.T. Lee: Dynamical optimal training for interval type-2 fuzzy neural network (T2FNN), IEEE Trans. Syst. Man Cybern. B **34**(3), 14621477 (2004)

6.37   R. Sambuc: Fonction Φ-Flous, Application a l'aide au Diagnostic en Pathologie Thyroidienne, These de Doctorat en Medicine (Univ. Marseille, Marseille 1975)

6.38   K.U. Jahn: Intervall-wertige Mengen, Math. Nachr. **68**, 115–132 (1975)

6.39   I. Grattan-Guinness: Fuzzy membership mapped onto interval and many-valued quantities, Z. Math. Log. Grundl. Math. **22**, 149–160 (1976)

6.40   A. Dziech, M.B. Gorzalczany: Decision making in signal transmission problems with interval-valued fuzzy sets, Fuzzy Sets Syst. **23**(2), 191–203 (1987)

6.41   M.B. Gorzalczany: A method of inference in approximate reasoning based on interval-valued fuzzy sets, Fuzzy Sets Syst. **21**, 1–17 (1987)

6.42   M.B. Gorzalczany: An interval-valued fuzzy inference method. Some basic properties, Fuzzy Sets Syst. **31**(2), 243–251 (1989)

6.43   I.B. Türksen: Interval valued fuzzy sets based on normal forms, Fuzzy Sets Syst. **20**(2), 191–210 (1986)

6.44   I.B. Türksen, Z. Zhong: An approximate analogical reasoning schema based on similarity measures and interval-valued fuzzy sets, Fuzzy Sets Syst. **34**, 323–346 (1990)

6.45   I.B. Türksen, D.D. Yao: Representation of connectives in fuzzy reasoning: The view through normal forms, IEEE Trans. Syst. Man Cybern. **14**, 191–210 (1984)

6.46   J.L. Deng: Introduction to grey system theory, J. Grey Syst. **1**, 1–24 (1989)

6.47   H. Bustince: Indicator of inclusion grade for interval-valued fuzzy sets. Application to approximate reasoning based on interval-valued fuzzy sets, Int. J. Approx. Reason. **23**(3), 137–209 (2000)

6.48   G. Deschrijver: The Archimedean property for t-norms in interval-valued fuzzy set theory, Fuzzy Sets Syst. **157**(17), 2311–2327 (2006)

6.49   G. Deschrijver, C. Cornelis, E.E. Kerre: On the representation of intuitionistic fuzzy t-norms and t-conorms, IEEE Trans. Fuzzy Syst. **12**(1), 45–61 (2004)

6.50   Z. Xu, R.R. Yager: Some geometric aggregation operators based on intuitionistic fuzzy sets, Int. J. Gen. Syst. **35**, 417–433 (2006)

6.51   H. Bustince, J. Fernandez, A. Kolesárová, M. Mesiar: Generation of linear orders for intervals by means of aggregation functions, Fuzzy Sets Syst. **220**, 69–77 (2013)

6.52   J. Montero, D. Gomez, H. Bustince: On the relevance of some families of fuzzy sets, Fuzzy Sets Syst. **158**(22), 2429–2442 (2007)

6.53   H. Bustince, F. Herrera, J. Montero: *Fuzzy Sets and Their Extensions: Representation Aggregation and Models* (Springer, Berlin 2007)

6.54   W. Pedrycz: Shadowed sets: Representing and processing fuzzy sets, IEEE Trans. Syst. Man Cybern. B **28**, 103–109 (1998)

6.55   W. Pedrycz, G. Vukovich: Investigating a relevance off uzzy mappings, IEEE Trans. Syst. Man Cybern. B **30**, 249–262 (2000)

6.56   W. Pedrycz, G. Vukovich: Granular computing with shadowed sets, Int. J. Intell. Syst. **17**, 173–197 (2002)

6.57   J.M. Mendel: Advances in type-2 fuzzy sets and systems, Inf. Sci. **177**, 84–110 (2007)

6.58   H. Bustince, J. Montero, M. Pagola, E. Barrenechea, D. Gomez: A survey of interval-valued fuzzy sets. In: *Handbook of Granular Computing*, ed. by W. Pedrycz (Wiley, New York 2008)

6.59   P. Burillo, H. Bustince: Entropy on intuitionistic fuzzy sets and on interval-valued fuzzy sets, Fuzzy Sets Syst. **78**, 305–316 (1996)

6.60   A. Jurio, M. Pagola, D. Paternain, C. Lopez-Molina, P. Melo-Pinto: Interval-valued restricted equivalence functions applied on clustering techniques, Proc. Int. Fuzzy Syst. Assoc. World Congr. Eur. Soc. Fuzzy Log. Technol. Conf. (2009) pp. 831–836

6.61   E. Szmidt, J. Kacprzyk: Entropy for intuitionistic fuzzy sets, Fuzzy Sets Syst. **118**(3), 467–477 (2001)

6.62   H. Rezaei, M. Mukaidono: New similarity measures of intuitionistic fuzzy sets, J. Adv. Comput. Intell. Inf. **11**(2), 202–209 (2007)

6.63   H. Bustince, M. Pagola, E. Barrenechea, J. Fernandez, P. Melo-Pinto, P. Couto, H.R. Tizhoosh, J. Montero: Ignorance functions. An application to the calculation of the threshold in prostate ultrasound images, Fuzzy Sets Syst. **161**(1), 20–36 (2010)

6.64   D. Wu: Approaches for reducing the computational cost of interval type-2 fuzzy logic systems: Overview and comparisons, IEEE Trans. Fuzzy Syst. **21**(1), 80–99 (2013)

6.65   H. Bustince, M. Galar, B. Bedregal, A. Kolesárová, R. Mesiar: A new approach to interval-valued Choquet integrals and the problem of ordering in interval-valued fuzzy set applications, IEEE Trans. Fuzzy Syst. **21**(6), 1150–1162 (2013)

6.66   J. Sanz, H. Bustince, F. Herrera: Improving the performance of fuzzy rule-based classification systems with interval-valued fuzzy sets and genetic amplitude tuning, Inf. Sci. **180**(19), 3674–3685 (2010)

6.67   J. Sanz, A. Fernandez, H. Bustince, F. Herrera: A genetic tuning to improve the performance of fuzzy rule-based classification systems with interval-valued fuzzy sets: Degree of ignorance and lateral position, Int. J. Approx. Reason. **52**(6), 751–766 (2011)

6.68   J. Sanz, A. Fernandez, H. Bustince, F. Herrera: IIVFDT: Ignorance functions based interval-valued fuzzy decision tree with genetic tuning, Int. J. Uncertain. Fuzziness Knowl.-Based Syst. **20**(Suppl. 2), 1–30 (2012)

6.69   J. Sanz, A. Fernandez, H. Bustince, F. Herrera: IV-TURS: A linguistic fuzzy rule-based classification system based on a new interval-valued fuzzy reasoning method with tuning and rule selection, IEEE Trans. Fuzzy Syst. **21**(3), 399–411 (2013)

6.70   Z. Chi, H. Yan, T. Pham: *Fuzzy Algorithms with Applications to Image Processing and Pattern Recognition* (World Scientific, Singapore 1996)

6.71   H. Ishibuchi, T. Yamamoto, T. Nakashima: Hybridization of fuzzy GBML approaches for pattern classification problems, IEEE Trans. Syst. Man Cybern. B **35**(2), 359–365 (2005)

6.72   J. Dombi, Z. Gera: Rule based fuzzy classification using squashing functions, J. Intell. Fuzzy Syst. **19**(1), 3–8 (2008)

6.73   R. Alcalá, J. Alacalá-Fdez, F. Herrera: A proposal for the genetic lateral tuning of linguistic fuzzy systems and its interaction with rule selection, IEEE Trans. Fuzzy Syst. **15**(4), 616–635 (2007)

6.74   R. Alcalá, J. Alacalá-Fdez, M. Graco, F. Herrera: Rule base reduction and genetic tuning of fuzzy systems based on the linguistic 3-tuples representation, Soft Comput. **11**(5), 401–419 (2007)

6.75   J. Quinlan: *C4.5: Programs for Machine Learning* (Morgan Kaufmann, San Mateo 1993)

6.76   C.Z. Janikow: Fuzzy decision trees: Issues and methods, IEEE Trans. Syst. Man Cybern. B **28**(1), 1–14 (1998)

6.77   J. Alacalá-Fdez, R. Alcalá, F. Herrera: A fuzzy association rule-based classification model for high-dimensional problems with genetic rule selection and lateral tuning, IEEE Trans. Fuzzy Syst. **19**(5), 857–872 (2011)

6.78   J. Hühn, E. Hüllermeier: FURIA: An algorithm for unordered fuzzy rule induction, Data Min. Knowl. Discov. **19**(3), 293–319 (2009)

6.79   E. Barrenechea, H. Bustince, B. De Baets, C. Lopez-Molina: Construction of interval-valued fuzzy relations with application to the generation of fuzzy edge images, IEEE Trans. Fuzzy Syst. **19**(5), 819–830 (2011)

6.80   H. Bustince, P.M. Barrenechea, J. Fernandez, J. Sanz: "Image thresholding using type II fuzzy sets." Importance of this method, Pattern Recognit. **43**(9), 3188–3192 (2010)

6.81   H. Bustince, E. Barrenechea, M. Pagola, J. Fernandez: Interval-valued fuzzy sets constructed from matrices: Application to edge detection, Fuzzy Sets Syst. **60**(13), 1819–1840 (2009)

6.82   M. Galar, F. Fernandez, G. Beliakov, H. Bustince: Interval-valued fuzzy sets applied to stereo matching of color images, IEEE Trans. Image Process. **20**, 1949–1961 (2011)

6.83   M. Pagola, C. Lopez-Molina, J. Fernandez, E. Barrenechea, H. Bustince: Interval type-2 fuzzy sets constructed from several membership functions. Application to the fuzzy thresholding algorithm, IEEE Trans. Fuzzy Syst. **21**(2), 230–244 (2013)

6.84   H.R. Tizhoosh: Image thresholding using type-2 fuzzy sets, Pattern Recognit. **38**, 2363–2372 (2005)

6.85   M.E. Yuksel, M. Borlu: Accurate segmentation of dermoscopic images by image thresholding based on type-2 fuzzy logic, IEEE Trans. Fuzzy Syst. **17**(4), 976–982 (2009)

6.86   C. Shyi-Ming, W. Hui-Yu: Evaluating students answer scripts based on interval-valued fuzzy grade sheets, Expert Syst. Appl. **36**(6), 9839–9846 (2009)

6.87   F. Liu, H. Geng, Y.-Q. Zhang: Interactive fuzzy interval reasoning for smart web shopping, Appl. Soft Comput. **5**(4), 433–439 (2005)

6.88   C. Byung-In, C.-H.R. Frank: Interval type-2 fuzzy membership function generation methods for pattern recognition, Inf. Sci. **179**(13), 2102–2122 (2009)

6.89   H.M. Choi, G.S. Min, J.Y. Ahn: A medical diagnosis based on interval-valued fuzzy sets, Biomed. Eng. Appl. Basis Commun. **24**(4), 349–354 (2012)

6.90   J.M. Mendel, H. Wu: Type-2 fuzzistics for symmetric interval type-2 fuzzy sets: Part 1: Forward problems, IEEE Trans. Fuzzy Syst. **14**(6), 781–792 (2006)

6.91   D. Wu, J.M. Mendel: A vector similarity measure for linguistic approximation: Interval type-2 and type-1 fuzzy sets, Inf. Sci. **178**(2), 381–402 (2008)

6.92   A. Jurio, H. Bustince, M. Pagola, A. Pradera, R.R. Yager: Some properties of overlap and grouping functions and their application to image thresholding, Fuzzy Sets Syst. **229**, 69–90 (2013)

6.93   K.T. Atanassov: Intuitionistic fuzzy sets, VII ITKRs Session, Central Sci.-Tech. Libr. Bulg. Acad. Sci., Sofia (1983) pp. 1684–1697, (in Bulgarian)

6.94   K.T. Atanassov: Intuitionistic fuzzy sets, Fuzzy Sets Syst. **20**, 87–96 (1986)

6.95   W.L. Gau, D.J. Buehrer: Vague sets, IEEE Trans. Syst. Man Cybern. **23**(2), 610–614 (1993)

6.96   H. Bustince, P. Burillo: Vague sets are intuitionistic fuzzy sets, Fuzzy Sets Syst. **79**(3), 403–405 (1996)

6.97   H. Bustince, E. Barrenechea, P. Pagola: Generation of interval-valued fuzzy and Atanassov's intuitionistic fuzzy connectives from fuzzy connectives and from K_α operators: Laws for conjunctions and disjunctions, amplitude, Int. J. Intell. Syst. **32**(6), 680–714 (2008)

6.98   K.T. Atanassov, G. Gargov: Interval valued intuitionistic fuzzy sets, Fuzzy Sets Syst. **31**(3), 343–349 (1989)

6.99   J. Ye: Fuzzy decision-making method based on the weighted correlation coefficient under intuitionistic fuzzy environment, Eur. J. Oper. Res. **205**(1), 202–204 (2010)

6.100 F. Herrera, E. Herrera-Viedma: Linguistic decision analysis: Steps for solving decision problems under linguistic information, Fuzzy Sets Syst. **115**, 67–82 (2000)

6.101 L. Baccour, A.M. Alimi, R.I. John: Similarity measures for intuitionistic fuzzy sets: State of the art, J. Intell. Fuzzy Syst. **24**(1), 37–49 (2013)

6.102 E. Szmidt, J. Kacprzyk, P. Bujnowski: Measuring the amount of knowledge for Atanassovs intuitionistic fuzzy sets, Lect. Notes Comput. Sci. **6857**, 17–24 (2011)

6.103 N.R. Pal, H. Bustince, M. Pagola, U.K. Mukherjee, D.P. Goswami, G. Beliakov: Uncertainties with Atanassov's intuitionistic fuzzy sets: fuzziness and lack of knowledge, Inf. Sci. **228**, 61–74 (2013)

6.104 U. Dudziak, B. Pekala: Equivalent bipolar fuzzy relations, Fuzzy Sets Syst. **161**(2), 234–253 (2010)

6.105 Z. Xu: Approaches to multiple attribute group decision making based on intuitionistic fuzzy power aggregation operators, Knowl. Syst. **24**(6), 749–760 (2011)

6.106 Z. Xu, H. Hu: Projection models for intuitionistic fuzzy multiple attribute decision making, Int. J. Inf. Technol. Decis. Mak. **9**(2), 257–280 (2010)

6.107 Z. Xu: Priority weights derived from intuitionistic multiplicative preference relations in decision making, IEEE Trans. Fuzzy Syst. **21**(4), 642–654 (2013)

6.108 X. Zhang, Z. Xu: A new method for ranking intuitionistic fuzzy values and its application in multi-attribute decision making, Fuzzy Optim. Decis. Mak. **11**(2), 135–146 (2012)

6.109 T. Chen: Multi-criteria decision-making methods with optimism and pessimism based on Atanassov's intuitionistic fuzzy sets, Int. J. Syst. Sci. **43**(5), 920–938 (2012)

6.110 S.K. Biswas, A.R. Roy: An application of intuitionistic fuzzy sets in medical diagnosis, Fuzzy Sets Syst. **117**, 209–213 (2001)

6.111 I. Bloch: Lattices of fuzzy sets and bipolar fuzzy sets, and mathematical morphology, Inf. Sci. **181**(10), 2002–2015 (2011)

6.112 P. Melo-Pinto, P. Couto, H. Bustince, E. Barrenechea, M. Pagola, F. Fernandez: Image segmentation using Atanassov's intuitionistic fuzzy sets, Expert Syst. Appl. **40**(1), 15–26 (2013)

6.113 P. Couto, A. Jurio, A. Varejao, M. Pagola, H. Bustince, P. Melo-Pinto: An IVFS-based image segmentation methodology for rat gait analysis, Soft Comput. **15**(10), 1937–1944 (2011)

6.114 K.T. Atanassov: Answer to D. Dubois, S. Gottwald, P. Hajek, J. Kacprzyk and H. Prade's paper Terminological difficulties in fuzzy set theory – The case of Intuitionistic fuzzy sets, Fuzzy Sets Syst. **156**(3), 496–499 (2005)

6.115 D. Dubois, H. Prade: An introduction to bipolar representations of information and preference, Int. J. Intell. Syst. **23**, 866–877 (2008)

6.116 D. Dubois, H. Prade: An overview of the asymmetric bipolar representation of positive and negative information in possibility theory, Fuzzy Sets Syst. **160**(10), 1355–1366 (2009)

6.117 W.R. Zhang: NPN fuzzy sets and NPN qualitative algebra: a computational framework for bipolar cognitive modeling and multiagent decision analysis, IEEE Trans. Syst. Man Cybern. B **26**(4), 561–574 (1996)

6.118 W.R. Zhang: Bipolar logic and bipolar fuzzy partial orderings for clustering and coordination, Proc. 6th Joint Conf. Inf. Sci. (2002) pp. 85–88

6.119 P. Grzegorzewski: On some basic concepts in probability of IF-events, Inf. Sci. **232**, 411–418 (2013)

6.120 H. Bustince, P. Burillo: Correlation of interval-valued intuitionistic fuzzy sets, Fuzzy Sets Syst. **74**(2), 237–244 (1995)

6.121 J. Wu, F. Chiclana: Non-dominance and attitudinal prioritisation methods for intuitionistic and interval-valued intuitionistic fuzzy preference relations, Expert Syst. Appl. **39**(18), 13409–13416 (2012)

6.122 Z. Xu, Q. Chen: A multi-criteria decision making procedure based on interval-valued intuitionistic fuzzy bonferroni means, J. Syst. Sci. Syst. Eng. **20**(2), 217–228 (2011)

6.123 J. Ye: Multicriteria decision-making method using the Dice similarity measure based on the reduct intuitionistic fuzzy sets of interval-valued intuitionistic fuzzy sets, Appl. Math. Model. **36**(9), 4466–4472 (2012)

6.124 A. Aygunoglu, B.P. Varol, V. Cetkin, H. Aygun: Interval-valued intuitionistic fuzzy subgroups based on interval-valued double t-norm, Neural Comput. Appl. **21**(1), S207–S214 (2012)

6.125 M. Fanyong, Z. Qiang, C. Hao: Approaches to multiple-criteria group decision making based on interval-valued intuitionistic fuzzy Choquet integral with respect to the generalized lambda-Shapley index, Knowl. Syst. **37**, 237–249 (2013)

6.126 W. Wang, X. Liu, Y. Qin: Interval-valued intuitionistic fuzzy aggregation operators 14, J. Syst. Eng. Electron. **23**(4), 574–580 (2012)

Part A | 6

6.127 K. Hirota: Concepts of probabilistic sets, Fuzzy Sets Syst. **5**, 31–46 (1981)

6.128 R.R. Yager: On the theory of bags, Int. J. Gen. Syst. **13**, 23–37 (1986)

6.129 S. Miyamoto: Multisets and fuzzy multisets. In: *Soft Computing and Human-Centered Machines*, ed. by Z.-Q. Liu, S. Miyamoto (Springer, Berlin 2000) pp. 9–33

6.130 Y. Shang, X. Yuan, E.S. Lee: The n-dimensional fuzzy sets and Zadeh fuzzy sets based on the finite valued fuzzy sets, Comput. Math. Appl. **60**, 442–463 (2010)

6.131 B. Bedregal, G. Beliakov, H. Bustince, T. Calvo, R. Mesiar, D. Paternain: A class of fuzzy multisets with a fixed number of memberships, Inf. Sci. **189**, 1–17 (2012)

6.132 A. Amo, J. Montero, G. Biging, V. Cutello: Fuzzy classification systems, Eur. J. Oper. Res. **156**, 459–507 (2004)

6.133 J. Montero: Arrow's theorem under fuzzy rationality, Behav. Sci. **32**, 267–273 (1987)

6.134 A. Mesiarová, J. Lazaro: Bipolar Aggregation operators, Proc. AGOP2003, Al-calá de Henares (2003) pp. 119–123

6.135 A. Mesiarová-Zemánková, R. Mesiar, K. Ahmad: The balancing Choquet integral, Fuzzy Sets Syst. **161**(17), 2243–2255 (2010)

6.136 A. Mesiarová-Zemánková, K. Ahmad: Multi-polar Choquet integral, Fuzzy Sets Syst. **220**, 1–20 (2013)

6.137 W.R. Zhang, L. Zhang: YinYang bipolar logic and bipolar fuzzy logic, Inf. Sci. **165**(3/4), 265–287 (2004)

6.138 W.R. Zhang: YinYang Bipolar T-norms and T-conorms as granular neurological operators, Proc. IEEE Int. Conf. Granul. Comput., Atlanta (2006) pp. 91–96

6.139 F. Smarandache: A unifying field in logics: Neutrosophic logic, Multiple-Valued Logic **8**(3), 385–438 (2002)

6.140 V. Torra: Hesitant fuzzy sets, Int. J. Intell. Syst. **25**, 529539 (2010)

6.141 V. Torra, Y. Narukawa: On hesitant fuzzy sets and decision, Proc. Conf. Fuzzy Syst. (FUZZ IEEE) (2009) pp. 1378–1382

6.142 D. Molodtsov: Soft set theory. First results, Comput. Math. Appl. **37**, 19–31 (1999)

6.143 P.K. Maji, R. Biswas, R. Roy: Fuzzy soft sets, J. Fuzzy Math. **9**(3), 589–602 (2001)

6.144 Z. Pawlak: Rough sets, Int. J. Comput. Inf. Sci. **11**, 341–356 (1982)

6.145 D. Dubois, H. Prade: Rough fuzzy-sets and fuzzy rough sets, Int. J. Gen. Syst. **17**(3), 191–209 (1990)

7. F-Transform

F-Transform

Irina Perfilieva

The theory of the *F*-transform is presented and discussed from the perspective of the latest developments and applications. Various fuzzy partitions are considered. The definition of the *F*-transform is given with respect to a generalized fuzzy partition, and the main properties of the *F*-transform are listed. The applications to image processing, namely image compression, fusion and edge detection, are discussed with sufficient technical details.

7.1 Fuzzy Modeling

Fuzzy modeling is still regarded as a modern technique with a nonclassical background. The goal of this chapter is to bridge standard mathematical methods and methods for the construction of fuzzy approximation models. We will present the theory of the *fuzzy transform* (the *F-transform*), which was introduced in [7.1] for the purpose of encompassing both classical (usually, integral) transforms and approximation models based on fuzzy IF–THEN rules (*fuzzy approximation models*). We start with an informal characterization of integral transforms, and from this discussion, we examine the similarities and differences among integral transforms, the *F*-transform, and fuzzy approximation models. An integral transform is performed using some kernel. The kernel is represented by a function of two variables and can be understood as a *collection of local factors* or closeness areas around elements of an original space. Each factor is then assigned an aver-

age value of a transforming object (usually, a function). Consequently, the transformed object is a new function defined on a space of *local factors*. The *F*-transform can be implicitly characterized by a *discrete* kernel that is associated with a finite collection of fuzzy subsets (local factors or closeness areas around chosen *nodes*) of an original space. We say that this collection establishes a *fuzzy partition* of the space. Then, similar to integral transforms, the *F*-transform assigns an average value of a transforming object to each fuzzy subset from the fuzzy partition of the space. Consequently, the *F*-transformed object is a finite vector of average values.

Similar to the *F*-transform, a fuzzy approximation model can also be implicitly characterized by a discrete kernel that establishes a fuzzy partition of an original space. Each element of the established fuzzy partition is a fuzzy set in the IF part (antecedent) of the respective fuzzy IF–THEN rule. The rule characterizes a correspondence between an antecedent and an average value of a transforming object (singleton model) or a fuzzy subset of a space of object values (fuzzy set model).

To emphasize the differences among integral transforms, the F-transform, and fuzzy approximation models, we note that the last two are actually finite collections of local descriptions of a considered object. Each collection produces a global description of the considered object in the form of the direct F-transform or the system of fuzzy IF–THEN rules.

The idea of producing collections of local descriptions by fuzzy IF–THEN rules originates from the early works of *Zadeh* [7.2–5] and from the *Takagi–Sugeno* [7.6] approximation models.

Similar to the conventional integral transforms (the Fourier and Laplace transforms, for example), the F-transform performs a transformation of an original universe of functions into a universe of their *skeleton*

models (vectors of F-transform components) for which further computations are easier (see, e.g., an application to the initial value problem with fuzzy initial conditions [7.7]). In this respect, the F-transform can be as useful in applications as traditional transforms (see applications to image compression [7.8, 9] and time series processing [7.10–14], for example). Moreover, sometimes the F-transform can be more efficient than its counterparts; see the details below.

The structure of this chapter is as follows. In Sect. 7.2, we consider various fuzzy partitions: uniform and with and without the Ruspini condition, among others; in Sect. 7.3, definitions of the F-transforms (direct and inverse) and their main properties are considered; in Sect. 7.4, the discrete F-transform is defined; in Sect. 7.5, the direct and inverse F-transform of a function of two variables is introduced; in Sect. 7.6, a higher degree F-transform is considered; in Sect. 7.7, applications of the F-transform and F^1-transform to image processing are discussed.

7.2 Fuzzy Partitions

In this section, we present a short overview of various fuzzy partitions of a universe in which transforming objects (functions) are defined. As we learned from Sect. 7.1, a fuzzy partition is a finite collection of fuzzy subsets of the universe that determines a discrete kernel and thus a respective transform. Therefore, we have as many F-transforms as fuzzy partitions.

7.2.1 Fuzzy Partition with the Ruspini Condition

The *fuzzy partition with the Ruspini condition* (7.1) (simply, *Ruspini partition*) was introduced in [7.1]. This condition implies normality of the respective fuzzy partition, i.e., the *partition-of-unity*. It then leads to a simplified version of the inverse F-transform. In later publications [7.15, 16], the Ruspini condition was weakened to obtain an additional degree of freedom and a better approximation by the inverse F-transform.

Definition 7.1

Let $x_1 < \cdots < x_n$ be fixed nodes within $[a, b]$ such that $x_1 = a, x_n = b$ and $n \geq 2$. We say that the fuzzy sets $A_1, \ldots, A_n$, identified with their membership functions defined on $[a, b]$, establish a Ruspini partition of

$[a, b]$ if they fulfill the following conditions for $k = 1, \ldots, n$:

1. $A_k : [a, b] \to [0, 1]$, $A_k(x_k) = 1$
2. $A_k(x) = 0$ if $x \notin (x_{k-1}, x_{k+1})$, where for uniformity of notation, we set $x_0 = a$ and $x_{n+1} = b$
3. $A_k(x)$ is continuous
4. $A_k(x)$, for $k = 2, \ldots, n$, strictly increases on $[x_{k-1}, x_k]$ and $A_k(x)$ for $k = 1, \ldots, n-1$, strictly decreases on $[x_k, x_{k+1}]$
5. for all $x \in [a, b]$,

$$\sum_{k=1}^{n} A_k(x) = 1 . \tag{7.1}$$

The condition (7.1) is known as the Ruspini condition. The membership functions $A_1, \ldots, A_n$ are called the *basic functions*. A point $x \in [a, b]$ is *covered* by the basic function A_k if $A_k(x) > 0$.

The shape of the basic functions is not predetermined and therefore, it can be chosen according to additional requirements (e.g., smoothness). Let us give examples of various fuzzy partitions with the Ruspini condition. In Fig. 7.1, two such partitions with triangular and cosine basic functions are shown. The following formulas represent generic fuzzy partitions with the

Ruspini condition and triangular functions

$$A_1(x) = \begin{cases} 1 - \dfrac{(x-x_1)}{h_1}, & x \in [x_1, x_2], \\ 0, & \text{otherwise}, \end{cases}$$

$$A_k(x) = \begin{cases} \dfrac{(x-x_{k-1})}{h_{k-1}}, & x \in [x_{k-1}, x_k], \\ 1 - \dfrac{(x-x_k)}{h_k}, & x \in [x_k, x_{k+1}], \\ 0, & \text{otherwise}, \end{cases}$$

$$A_n(x) = \begin{cases} \dfrac{(x-x_{n-1})}{h_{n-1}}, & x \in [x_{n-1}, x_n], \\ 0, & \text{otherwise}, \end{cases}$$

where $k = 2, \ldots n-1$ and $h_k = x_{k+1} - x_k$.

We say that a Ruspini partition of $[a,b]$ is *h-uniform* if its nodes $x_1, \ldots, x_n$, where $n \geq 3$, are equidistant, i.e., $x_k = a + h(k-1)$, for $k = 1, \ldots, n$, where $h = (b-a)/(n-1)$, and the two additional properties are met:

6. $A_k(x_k - x) = A_k(x_k + x)$, for all $x \in [0, h]$, $k = 2, \ldots, n-1$,
7. $A_k(x) = A_{k-1}(x-h)$, for all $k = 2, \ldots, n-1$ and $x \in [x_k, x_{k+1}]$, and $A_{k+1}(x) = A_k(x-h)$, for all $k = 2, \ldots, n-1$ and $x \in [x_k, x_{k+1}]$.

7.2.2 Fuzzy Partitions with the Generalized Ruspini Condition

Fuzzy partitions with the generalized Ruspini condition were introduced in [7.15]. The generalization consists in replacing *partition-of-unity* (7.1) by *fuzzy r-partition* (7.2). This type of partition was investigated in [7.15, 17], where the focus was on smoothing or filtering data using the inverse F-transform. The following definition is taken from [7.15].

Definition 7.2

Let $r \geq 1$ and $n \geq 2$ be fixed integers such that $r \leq n$. Let $a = x_1 < \cdots < x_n = b$ be nodes within $[a, b]$, and let $x_{1-r} < \cdots < x_0 < a$ and $b < x_{n+1} < \cdots < x_{n+r}$ be nodes outside of $[a, b]$. A fuzzy r-partition of $[a, b]$ is a family of $n + 2r - 2$ continuous, normal, convex fuzzy sets

$$A_{2-r}^{(r)}, \ldots, A_1^{(r)}, \ldots, A_n^{(r)}, \ldots, A_{n+r-1}^{(r)}$$

such that the following conditions are fulfilled:

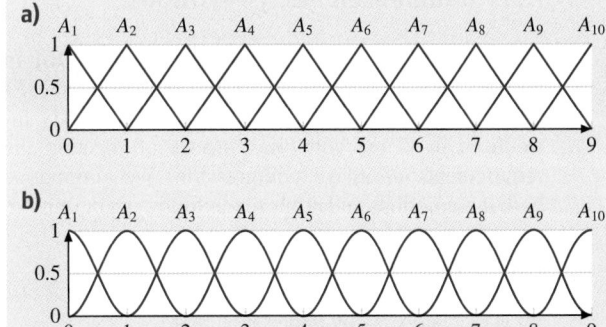

Fig. 7.1a,b Two Ruspini partitions with triangular (**a**) and cosine basic functions (**b**)

1. For $k = 1, \ldots, n$, $A_k^{(r)}$ is a continuous function on $[a, b]$ such that $A_k^{(r)}(x_k) = 1$ and $A_k^{(r)}(x) = 0$ for $x \notin [\max(x_{k-r}, a), \min(x_{k+r}, b)]$
2. For $k = 1, \ldots, n$, $A_k^{(r)}$ is increasing on $[\max(x_{k-r}, a), x_k]$ and decreasing on $[x_k, \min(x_{k+r}, b)]$
3. For $k = -r+2, \ldots, 0$, $A_k^{(r)}$ is decreasing on $[\max(x_k, a), x_{k+r}]$
4. For $k = n+1, \ldots, n+r-1$, $A_k^{(r)}$ is increasing on $[x_{k-r}, \min(x_k, b)]$
5. For all $x \in [a, b]$, the following *partition-of-r* condition holds

$$\sum_{k=-r+2}^{n+r-1} A_k^{(r)}(x) = r. \tag{7.2}$$

If $r = 1$, then a fuzzy r-partition in the sense of Definition 7.2 becomes the standard fuzzy partition in the sense of Definition 7.1, i.e., the *partition-of-unity*. In Fig. 7.2, the fuzzy 2-partition with triangular basic functions is shown.

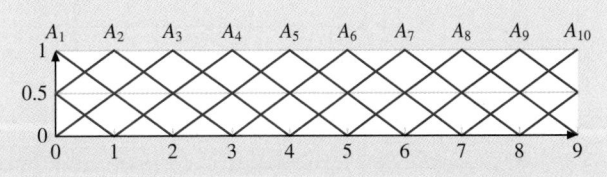

Fig. 7.2 An example of a fuzzy 2-partition with triangular basic functions

7.2.3 Generalized Fuzzy Partitions

A *generalized fuzzy partition* appeared in [7.16] in connection with the notion of the higher degree *F*-transform. Its even weaker version was implicitly introduced in [7.18] with the purpose of meeting the requirements of image compression. We summarize both these notions and propose the following definition.

Definition 7.3

Let $[a, b]$ be an interval on $\mathbb{R}$, $n \geq 2$, and let $x_0, x_1, \ldots, x_n, x_{n+1}$ be nodes such that

$$a = x_0 \leq x_1 < \cdots < x_n \leq x_{n+1} = b .$$

We say that the fuzzy sets

$$A_1, \ldots, A_n : [a, b] \to [0, 1]$$

constitute a *generalized fuzzy partition* of $[a, b]$ if for every $k = 1, \ldots, n$ there exist $h'_k, h''_k \geq 0$ such that

$$h'_k + h''_k > 0 , [x_k - h'_k, x_k + h''_k] \subseteq [a, b]$$

and the following three conditions are fulfilled:

1. (locality) – $A_k(x) > 0$ if $x \in (x_k - h'_k, x_k + h''_k)$, and $A_k(x) = 0$ if $x \in [a, b] \setminus [x_k - h'_k, x_k + h''_k]$
2. (continuity) – A_k is continuous on $[x_k - h'_k, x_k + h''_k]$
3. (covering) – for $x \in [a, b]$, $\sum_{k=1}^{n} A_k(x) > 0$.

It is important to remark that by conditions of *locality* and *continuity*,

$$\int_a^b A_k(x)\mathrm{d}x > 0 .$$

An (h, h')-*uniform* generalized fuzzy partition of $[a, b]$ is defined for equidistant nodes

$$x_k = a + h(k - 1) , k = 1, \ldots, n ,$$

where $h = (b-a)/(n-1)$, $h' > h/2$ and two additional properties are satisfied:

4. $A_k(x) = A_{k-1}(x - h)$ for all $k = 2, \ldots, n-1$ and $x \in [x_k, x_{k+1}]$, and $A_{k+1}(x) = A_k(x - h)$ for all $k = 2, \ldots, n-1$ and $x \in [x_k, x_{k+1}]$.
5. $h'_1 = h''_n = 0$, $h''_1 = h'_2 = \cdots = h''_{n-1} = h'_n = h'$ and for all $k = 2, \ldots, n-1$ and all $x \in [0, h']$, $A_k(x_k - x) = A_k(x_k + x)$.

An (h, h')-uniform generalized fuzzy partition of $[a, b]$ can also be defined using the *generating function* $A_0 : [-1, 1] \to [0, 1]$, which is assumed to be *even*, continuous, and positive everywhere except for on boundaries, where it vanishes. (The function $A_0 : [-1, 1] \to \mathbb{R}$ is even if for all $x \in [0, 1]$, $A_0(-x) = A_0(x)$.) Then, basic functions A_k of an (h, h')-uniform generalized fuzzy partition are shifted copies of A_0 in the sense that

$$A_1(x) = \begin{cases} A_0\left(\dfrac{x - x_1}{h'}\right), & x \in [x_1, x_1 + h'] , \\ 0, & \text{otherwise} , \end{cases}$$

and for $k = 2, \ldots, n-1$,

$$A_k(x) = \begin{cases} A_0\left(\dfrac{x - x_k}{h'}\right), & x \in [x_k - h', x_k + h'] , \\ 0, & \text{otherwise} , \end{cases}$$

$$A_n(x) = \begin{cases} A_0\left(\dfrac{x - x_n}{h'}\right), & x \in [x_n - h', x_n] , \\ 0, & \text{otherwise} . \end{cases} \tag{7.3}$$

As an example, we note that the function $A_0(x) = 1 - |x|$ is a generating function for all uniform triangular partitions. The difference between them is in parameters h

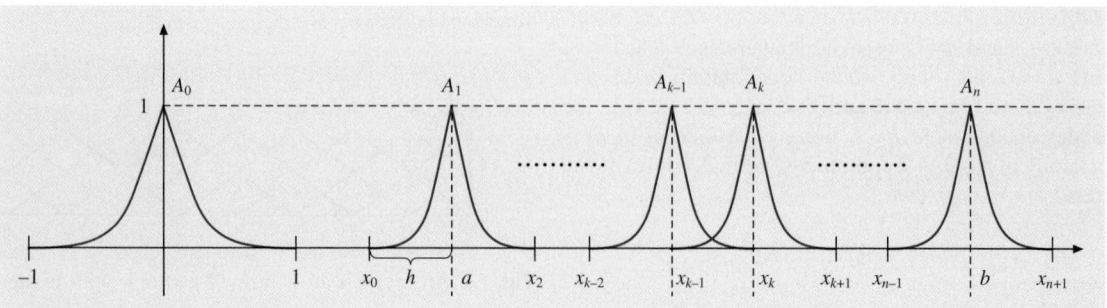

Fig. 7.3 Generating function A_0 of an h-uniform generalized fuzzy partition (after [7.19])

and h'. An (h, h)-uniform generalized fuzzy partition is simply called an h-uniform one (Fig. 7.3).

see [7.20]. In this case, every basic function has a generic representation in terms of a kernel $\varphi : \mathbb{R}^+ \to \mathbb{R}$ such that

$$A_k(x) = \varphi(\|x - x_k\|_2), \quad k = 1, \ldots, n.$$

Remark 7.1
A generalized fuzzy partition can also be considered in connection with radial membership functions;

7.3 Fuzzy Transform

The F-transform establishes a correspondence between a set of continuous functions on an interval of real numbers and the set of n-dimensional (real) vectors. Each component of the resulting vector is a weighted local mean of a corresponding function over an area covered by a corresponding basic function. The vector of the F-transform components is a simplified representation of an original function that can be used instead of the original function in many applications. Among them, let us mention applications to image compression [7.8, 9], image fusion, image reduction, time series processing [7.10–14], and the initial value problem with fuzzy initial conditions [7.7].

7.3.1 Direct F-Transform

In this section, we give the definition of the F-transform according to [7.1] and recall the main properties of it. We assume that the universe is an interval $[a, b]$ and $x_1 < \cdots < x_n$ are fixed nodes from $[a, b]$ such that $x_1 = a$, $x_n = b$ and $n \geq 2$. Let us formally extend the set of nodes by $x_0 = a$ and $x_{n+1} = b$. Let $A_1, \ldots, A_n$ be the basic functions that form a fuzzy partition of $[a, b]$ according to Definition 7.3. Let $C([a, b])$ be the set of continuous functions on the interval $[a, b]$. The following definition introduces the fuzzy transform of a function $f \in C([a, b])$.

Definition 7.4
Let $A_1, \ldots, A_n$ be the basic functions that form a generalized fuzzy partition of $[a, b]$ and f be any function from $C([a, b])$. We say that the n-tuple of real numbers $\mathbf{F}[f] = (F_1, \ldots, F_n)$ given by

$$F_k = \frac{\int_a^b f(x) A_k(x) \mathrm{d}x}{\int_a^b A_k(x) \mathrm{d}x}, \quad k = 1, \ldots, n, \quad (7.4)$$

is the (integral) F-transform of f with respect to $A_1, \ldots, A_n$.

The elements $F_1, \ldots, F_n$ are called the *components of the F-transform*. If $A_1, \ldots, A_n$ is an h-uniform Ruspini partition, then (7.4) may be simplified as follows,

$$F_1 = \frac{2}{h} \int_{x_1}^{x_2} f(x) A_1(x) \mathrm{d}x,$$

$$F_n = \frac{2}{h} \int_{x_{n-1}}^{x_n} f(x) A_n(x) \mathrm{d}x,$$

$$F_k = \frac{1}{h} \int_{x_{k-1}}^{x_{k+1}} f(x) A_k(x) \mathrm{d}x, \quad k = 2, \ldots, n-1. \quad (7.5)$$

The following is a list of some properties of the F-transform of f with respect to a generalized fuzzy partition of $[a, b]$:

(a) If for all $x \in [a, b], f(x) = C$, then
$F_k = C, k = 1, \ldots, n$.
(b) If $f = \alpha g + \beta h$, then
$\mathbf{F}[f] = \alpha \mathbf{F}[g] + \beta \mathbf{F}[h]$.
(c) If $[c, d] = \{f(x) \mid x \in [a, b]\}$, then
$F_k = \min_{[c,d]} \int_a^b (f(x) - y)^2 A_k(x) \mathrm{d}x, \quad k = 1, \ldots, n$.
(d) If f is twice continuously differentiable on $[a, b]$, then $F_k = f(x_k) + O(h^2), k = 1, \ldots, n$. (This is true for an h-uniform Ruspini partition of $[a, b]$ only. A similar estimation of the F-transform component F_k as a linear combination of $f(x_k - r + 1), \ldots, f(x_k), \ldots, f(x_k + r - 1)$ can be established for a fuzzy r-partition [7.15].)
(e) If a generalized fuzzy partition is (h, h')-uniform, then for each $k = 1, \ldots, n-1$,

$$|f(t) - F_k| \leq 2\omega(\tilde{h}, f),$$

$$|f(t) - F_{k+1}| \leq 2\omega(\tilde{h}, f),$$

where $\tilde{h} = \max(h, h'), t \in [x_k, x_k + \tilde{h}]$, and

$$\omega(\tilde{h}, f) = \max_{|\delta| \leq \tilde{h}} \max_{x \in [a, b - \delta]} |f(x + \delta) - f(x)|.$$

$$(7.6)$$

(f)

$$\int_a^b f(x)dx = h\left(\frac{F_1}{2} + \frac{F_n}{2} + \sum_{k=2}^{n-1} F_k\right).$$

(This is true for an *h*-uniform Ruspini partition of $[a,b]$ only.)

7.3.2 Inverse *F*-Transform

It is clear that an original nonconstant function f cannot be precisely reconstructed from its F-transform $\mathbf{F}[f]$ because we lose information when passing from f to $\mathbf{F}[f]$. However, the inverse F-transform $\hat{f}$ that can be reconstructed (using the inversion formula (7.7)) approximates f in such a way that universal convergence can be established.

Definition 7.5

Let $A_1, \ldots, A_n$ be the basic functions that form a generalized fuzzy partition of $[a, b]$ and f be a function from $C([a,b])$. Let $\mathbf{F}[f] = (F_1, \ldots, F_n)$ be the F-transform of f with respect to $A_1, \ldots, A_n$. Then, the function $\hat{f} : [a, b] \to \mathbb{R}$ represented by

$$\hat{f}(x) = \frac{\sum_{k=1}^n F_k A_k(x)}{\sum_{k=1}^n A_k(x)}, \tag{7.7}$$

is called *the inverse F-transform.*

Remark 7.2

If a fuzzy partition of $[a, b]$ fulfills the generalized Ruspini condition (7.2) with $r \geq 1$, then the inversion formula (7.7) can be simplified to

$$\hat{f}(x) = \frac{1}{r} \sum_{k=1}^n F_k A_k(x)$$

or to (in the case of the Ruspini partition for which $r = 1$)

$$\hat{f}(x) = \sum_{k=1}^n F_k A_k(x).$$

The following theorem demonstrates that the inverse F-transform $\hat{f}$ can approximate a continuous function f with arbitrary precision. Thus, it explains why the F-transform has convincing applications in various fields, including image and time series processing, and data mining [7.21]. In Fig. 7.4, we illustrate

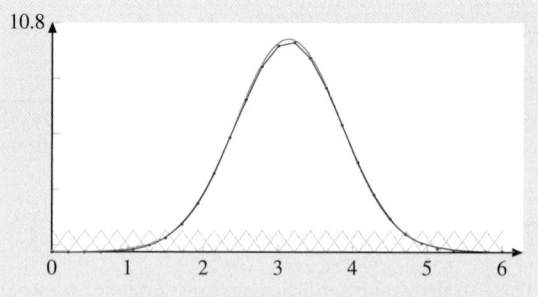

Fig. 7.4 The function $f(x) = 10e^{-(x-\pi)^2}$ (*gray*) and its inverse *F*-transform (*brown*) with respect to the uniform Ruspini partition of $[0, 6]$ by 29 triangular-shaped basic functions. The *F*-transform components are marked by *small circles*

how the inverse F-transform approximates the function $10e^{-(x-\pi)^2}$.

Theorem 7.1

Let f be a continuous function on $[a, b]$. Then, for any $\varepsilon > 0$, there exist n_ε and a generalized fuzzy partition $A_1, \ldots, A_{n_\varepsilon}$ of $[a, b]$ such that for all $x \in [a, b]$,

$$eq8|f(x) - \hat{f}_\varepsilon(x)| \leq \varepsilon, \tag{7.8}$$

where $\hat{f}_\varepsilon$ is the inverse F-transform of f with respect to the fuzzy partition $A_1, \ldots, A_{n_\varepsilon}$.

From Theorem 7.2, which is given below, we learn that for a pointwise approximation (as in Theorem 7.1), it is sufficient to compute the F-transform with respect to the simplest triangular fuzzy partition. Therefore, almost all applications of the F-transform are based on this type of partition.

Theorem 7.2

Let f be any continuous function on $[a, b]$, and let $A'_1, \ldots, A'_n$ and $A''_1, \ldots, A''_n$, for $n \geq 3$, be the basic functions that form different (h, h')-uniform generalized fuzzy partitions of $[a, b]$. Let $\hat{f}'$ and $\hat{f}''$ be the two inverse F-transforms of f with respect to different sets of basic functions $A'_1, \ldots, A'_n$ or $A''_1, \ldots, A''_n$. Then, for arbitrary $x \in [a, b]$,

$$|\hat{f}'(x) - \hat{f}''(x)| \leq 4\omega(\tilde{h}, f),$$

where $h = \frac{b-a}{n-1}$, $\tilde{h} = \max(h, h')$ and $\omega(\tilde{h}, f)$ is the modulus of continuity (7.6) of f on the interval $[a, b]$.

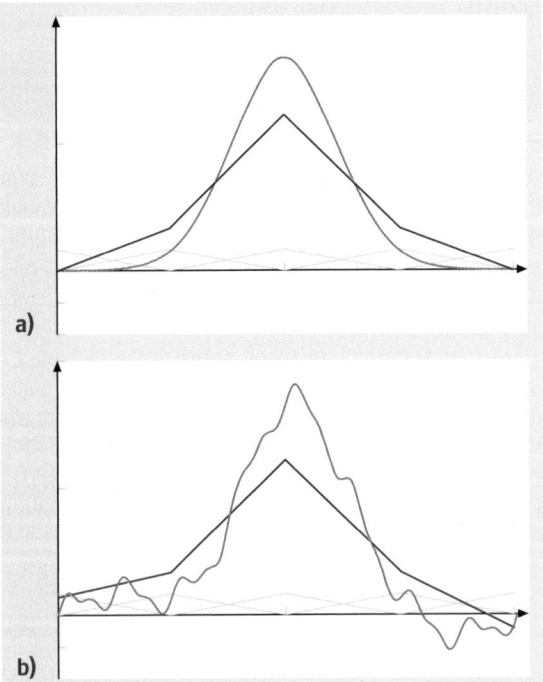

Fig. 7.5 (a) Function $f(x) = 10e^{-(x-\pi)^2}$ (*gray*) and its inverse F-transform (*brown*) with respect to the Ruspini partition given by the triangular-shaped basic functions $A_1, \ldots, A_5$ (*gray*). **(b)** Noisy function $f + s$ (*gray*), where $s(x) = \sin(2x) + 0.6\sin(8x) + 0.3\sin(16x)$, and its inverse F-transform (*brown*) with respect to the same fuzzy partition. Both inverse F-transforms $\hat{f}$ and $\widehat{f+s}$ are equal on $[x_2, x_4]$

The proofs of Theorems 7.1 and 7.2 can be obtained from the respective proofs in [7.22, Theorems 2 and 3] after some necessary changes caused by the usage of the generalized fuzzy partition.

Below, we list some properties of the inverse F-transform $\hat{f}$ of f that were considered and proved in [7.1, 15, 23]. If not specially mentioned, it is assumed that the F-transform is computed with respect to a generalized fuzzy partition of $[a, b]$:

(a) If for all $x \in [a, b]$, $f(x) = C$, then $\hat{f}(x) = C$
(b) If $f = \alpha g + \beta h$, then $\hat{f} = \alpha \hat{g} + \beta \hat{h}$
(c) $\int_a^b f(x)dx = \int_a^b \hat{f}(x)dx$ (This is true for the fuzzy r-partition ($r \geq 1$) of $[a, b]$ only.)
(d) Let $A_1, \ldots, A_n$ be an h-uniform Ruspini partition of $[a, b]$, where $h = (b - a)/(n - 1)$ and $n > 3$. Let $s : [a, b] \to \mathbb{R}$ be a continuous function such that one of the following two conditions are fulfilled:

   (i) s is $2h$-periodical and for all $x \in [0, h]$, $s(x_k - x) = -s(x_k + x)$, where $k = 2, \ldots, n-1$
   (ii) s is h-periodical and $\int_{x_{k-1}}^{x_k} s(x)dx = 0$, where $k = 2, \ldots, n-1$.

Then, for $x \in [x_2, x_{n-1}]$,

$$\hat{f} = \widehat{f+s}.$$

The last property is known as *noise removal*. This phrase implies that both functions f (non-noisy) and $f + s$ (noisy) have the same inverse F-transform. The noise is represented by s and characterized by conditions (i) or (ii). We illustrate this property in Fig. 7.5.

7.4 Discrete F-Transform

The discrete case of the F-transform, for which an original function f is defined (may be computed) on a finite set $P = \{p_1, \ldots, p_l\} \subseteq [a, b]$, was introduced in [7.1]. We will adapt the mentioned definition to the case of a generalized fuzzy partition of $[a, b]$.

We assume that the domain P of the function f is *sufficiently dense with respect to the fixed partition*, i.e.,

$$(\forall k)(\exists j)A_k(p_j) > 0.$$

Then, the (discrete) F-transform of f is defined as follows.

Definition 7.6

Let $A_1, \ldots, A_n$, for $n > 2$, be the basic functions that form a generalized fuzzy partition of $[a, b]$, and let func-

tion f be defined on the set $P = \{p_1, \ldots, p_l\} \subseteq [a, b]$, which is sufficiently dense with respect to the partition. We say that the n-tuple of real numbers $(F_1, \ldots, F_n)$ is the discrete F-transform of f with respect to $A_1, \ldots, A_n$ if

$$F_k = \frac{\sum_{j=1}^l f(p_j)A_k(p_j)}{\sum_{j=1}^l A_k(p_j)}. \tag{7.9}$$

It is not difficult to demonstrate that the components of the discrete F-transform have similar properties to those listed in Sect. 7.3.1.

In the discrete case, we define the inverse F-transform on the same set P on which the original function is defined.

Definition 7.7

Let $A_1, \ldots, A_n$, for $n > 2$, be the basic functions that form a generalized fuzzy partition of $[a, b]$, and let function f be defined on the set $P = \{p_1, \ldots, p_l\} \subseteq [a, b]$, which is sufficiently dense with respect to the partition. Moreover, let $\mathbf{F}[f] = (F_1, \ldots, F_n)$ be the discrete F-transform of f w.r.t. $A_1, \ldots, A_n$. Then, the function $\hat{f} : P \to \mathbb{R}$ represented by

$$\hat{f}(p_j) = \frac{\sum_{k=1}^n F_k A_k(p_j)}{\sum_{k=1}^n A_k(p_j)} \tag{7.10}$$

is *the inverse discrete F-transform* of f.

Remark 7.3

If a fuzzy partition of $[a, b]$ fulfills the generalized Ruspini condition (7.2) with $r \geq 1$, i. e., for all $p_j \in P$, $\sum_{k=1}^n A_k(p_j) = r$, then the inversion formula (7.10) can be simplified to

$$\hat{f}(p_j) = \frac{1}{r} \sum_{k=1}^n F_k A_k(p_j)$$

or (in the case of Ruspini partition, i. e., $r = 1$) to

$$\hat{f}(p_j) = \sum_{k=1}^n F_k A_k(p_j) .$$

Analogous to Theorem 7.1, we can show that the inverse discrete F-transform $\hat{f}$ can approximate the original discrete function f on P with arbitrary precision [7.1]. Moreover, the properties (a)–(c) that are listed in Sect. 7.3.2 have valid discrete analogies.

An interesting comparison between the discrete F-transform and the least-square approximation was made in [7.20]. It was demonstrated that the discrete F-transform is invariant with respect to the interpolating and least-squares approximation of the set $\{(p_j, f(p_j)) \mid j = 1, \ldots, l\}$. This means that the best approximation of f on P in the form of $\sum_{i=1}^n \alpha_i A_i$, where $n \leq l$, has the same direct discrete F-transform as the original f.

7.5 F-Transforms of Functions of Two Variables

The direct and inverse F-transform of a function of two (and more) variables is a direct generalization of the case of one variable. We introduce it briefly and refer to [7.1] for more details.

Suppose that the universe is a rectangle $[a, b] \times [c, d] \subseteq \mathbb{R} \times \mathbb{R}$ and that $x_1 < \cdots < x_n$ are the fixed nodes of $[a, b]$ and $y_1 < \cdots < y_m$ are the fixed nodes of $[c, d]$ such that $x_1 = a$, $x_n = b$, $y_1 = c$, $x_m = d$ and $n, m \geq 2$. Let us formally extend the set of nodes by setting $x_0 = a$, $y_0 = c$, $x_{n+1} = b$, and $y_{m+1} = d$. Assume that $A_1, \ldots, A_n$ are the basic functions that form a generalized fuzzy partition of $[a, b]$ and $B_1, \ldots, B_m$ are basic functions that form a generalized fuzzy partition of $[c, d]$. Then, the rectangle $[a, b] \times [c, d]$ is partitioned into fuzzy sets $A_k \times B_l$ with the membership functions $(A_k \times B_l)(x, y) = A_k(x)B_l(y)$, $k = 1, \ldots, n, l = 1, \ldots, m$. Let $C([a, b] \times [c, d])$ be the set of continuous functions of two variables on the domain and $f \in C([a, b] \times [c, d])$.

Definition 7.8

Let $A_1, \ldots, A_n$ be the basic functions that form a generalized fuzzy partition of $[a, b]$ and $B_1, \ldots, B_m$ be the basic functions that form a generalized fuzzy partition of $[c, d]$. Let f be any function from $C([a, b] \times [c, d])$. We say that the $n \times m$-matrix of real numbers $\mathbf{F}[f] = (F_{kl})_{n \times m}$ is the (integral) F-transform of f with respect

to $A_1, \ldots, A_n$ and $B_1, \ldots, B_m$ if for each $k = 1, \ldots, n$, $l = 1, \ldots, m$,

$$F_{kl} = \frac{\int_c^d \int_a^b f(x, y) A_k(x) B_l(y) dx dy}{\int_c^d \int_a^b A_k(x) B_l(y) dx dy} . \tag{7.11}$$

The components F_{kl} (7.11) have properties (adapted to the case of two variables) similar to those listed in Sect. 7.3.1. For example, the property (e) has the following form (we assume that $A_1, \ldots, A_n$ form an h_1-uniform Ruspini partition of $[a, b]$ and $B_1, \ldots, B_m$ form an h_2-uniform Ruspini partition of $[c, d]$)

$$\int_c^d \int_a^b f(x, y) dx dy$$
$$= \frac{h_1 h_2}{4} (F_{11} + F_{1m} + F_{n1} + F_{nm})$$
$$+ \frac{h_1 h_2}{2} \left(\sum_{k=2}^{n-1} F_{k1} + \sum_{k=2}^{n-1} F_{km} + \sum_{l=2}^{m-1} F_{1l} + \sum_{l=2}^{m-1} F_{nl} \right)$$
$$+ h_1 h_2 \sum_{k=2}^{n-1} \sum_{l=2}^{m-1} F_{kl} .$$

In the discrete case, when an original function f is known only at points $(p_i, q_j) \in [a, b] \times [c, d]$, where $i =$

$1,\ldots,N$ and $j=1,\ldots,M$, the (discrete) F-transform of f can be introduced in a manner analogous to the case of a function of one variable. This case is important for applications of the F-transform to image processing [7.8, 9, 18, 24–26].

Definition 7.9
Let a function f be given at nodes $(p_i, q_j) \in [a,b] \times [c,d]$, for which $i=1,\ldots,N$ and $j=1,\ldots,M$, and $A_1,\ldots,A_n$ and $B_1,\ldots,B_m$, where $n < N$ and $m < M$, be the basic functions that form generalized fuzzy partitions of $[a,b]$ and $[c,d]$, respectively. Suppose that sets P and Q of these nodes are sufficiently dense with respect to the chosen partitions. We say that the $n \times m$-matrix of real numbers $\mathbf{F}[f] = (F_{kl})_{nm}$ is the discrete F-transform of f with respect to $A_1,\ldots,A_n$ and $B_1,\ldots,B_m$ if

$$F_{kl} = \frac{\sum_{j=1}^{M}\sum_{i=1}^{N} f(p_i,q_j)A_k(p_i)B_l(q_j)}{\sum_{j=1}^{M}\sum_{i=1}^{N} A_k(p_i)B_l(q_j)} \qquad (7.12)$$

holds for all $k=1,\ldots,n$, $l=1,\ldots,m$.

7.6 F^1-Transform

In [7.16], a higher degree F-transform was introduced for the purpose for advanced applications in time series and image processing [7.26, 27]. In this section, we give a description of the F^1-transform, which has working applications, and refer to [7.16] for the F^m-transform for which $m > 1$.

Throughout this section, we assume that $A_1,\ldots,A_n$, $n > 2$ is an h-uniform generalized fuzzy partition of $[a,b]$ such that there exists a generating function $A_0 : [-1,1] \to [0,1]$ such that for all $k=1,\ldots,n$, A_k is defined by (7.3) (the illustration is in Fig. 7.3).

Let k be a fixed integer from $\{1,\ldots,n\}$, and let $L_2(A_k)$ be a normed space of square-integrable functions $f : [x_{k-1}, x_{k+1}] \to \mathbb{R}$, where the norm $\|f\|_k$ is given by

$$\|f\|_k = \sqrt{\frac{\int_{x_{k-1}}^{x_{k+1}} f^2(x)A_k(x)dx}{\int_{x_{k-1}}^{x_{k+1}} A_k(x)dx}}.$$

By $L_2(A_1,\ldots,A_n)$ we denote a set of functions $f : [a,b] \to \mathbb{R}$ such that for all $k=1,\ldots,n$, $f|_{[x_{k-1},x_{k+1}]} \in L_2(A_k)$, where $f|_{[x_{k-1},x_{k+1}]}$ is the restriction of f on $[x_{k-1},x_{k+1}]$.

For any function f from $L_2(A_1,\ldots,A_n)$ we define the F^1-transform of f with respect to $A_1,\ldots,A_n$ as the

The inverse F-transform of a function of two variables is a simple extension of (7.7). It will be given below for the continuous version of a function.

Definition 7.10
Let $A_1,\ldots,A_n$ and $B_1,\ldots,B_m$ be the basic functions that form generalized fuzzy partitions of $[a,b]$ and $[c,d]$, respectively. Let f be a function from $C([a,b] \times [c,d])$ and $\mathbf{F}[f]$ be the F-transform of f with respect to $A_1,\ldots,A_n$ and $B_1,\ldots,B_m$. Then, the function $\hat{f} : [a,b] \times [c,d] \to \mathbb{R}$ represented by

$$\hat{f}(x,y) = \frac{\sum_{k=1}^{n}\sum_{l=1}^{m} F_{kl}A_k(x)B_l(y)}{\sum_{k=1}^{n}\sum_{l=1}^{m} A_k(x)B_l(y)} \qquad (7.13)$$

is called the the inverse F-transform.

Similar to the case of a function of one variable, we can prove that the inverse F-transform $\hat{f}$ can approximate the original continuous function f with arbitrary precision, and the (adapted) properties (a)–(c), which are listed in Sect. 7.3.2, are fulfilled.

vector of linear functions

$$\mathbf{F}^1[f] = (c_{1,0}+c_{1,1}(x-x_1),\ldots,c_{n,0}+c_{n,1}(x-x_n)), \qquad (7.14)$$

where for every $k=1,\ldots,n$,

$$c_{k,0} = \frac{\int_{x_{k-1}}^{x_{k+1}} f(x)A_k(x)dx}{hs_0},$$

$$c_{k,1} = \frac{\int_{x_{k-1}}^{x_{k+1}} f(x)(x-x_k)A_k(x)dx}{\int_{x_{k-1}}^{x_{k+1}} (x-x_k)^2 A_k(x)dx}, \qquad (7.15)$$

and

$$s_0 = \int_{-1}^{1} A_0(x)dx.$$

The kth component of the vector $\mathbf{F}^1[f]$ is denoted by $F_k^1[f]$.

The following is a list of properties of the F^1-transform of f with respect to a generalized fuzzy partition of $[a,b]$. They are particular cases of the properties of the F^m-transform proved in [7.16]:

(a) Let F_k and $c_{k,0}+c_{k,1}(x-x_k)$, for $k=1,\ldots,n$, be respective kth components of $\mathbf{F}^1[f]$ and $\mathbf{F}[f]$. Then, $F_k = c_{k,0}$.

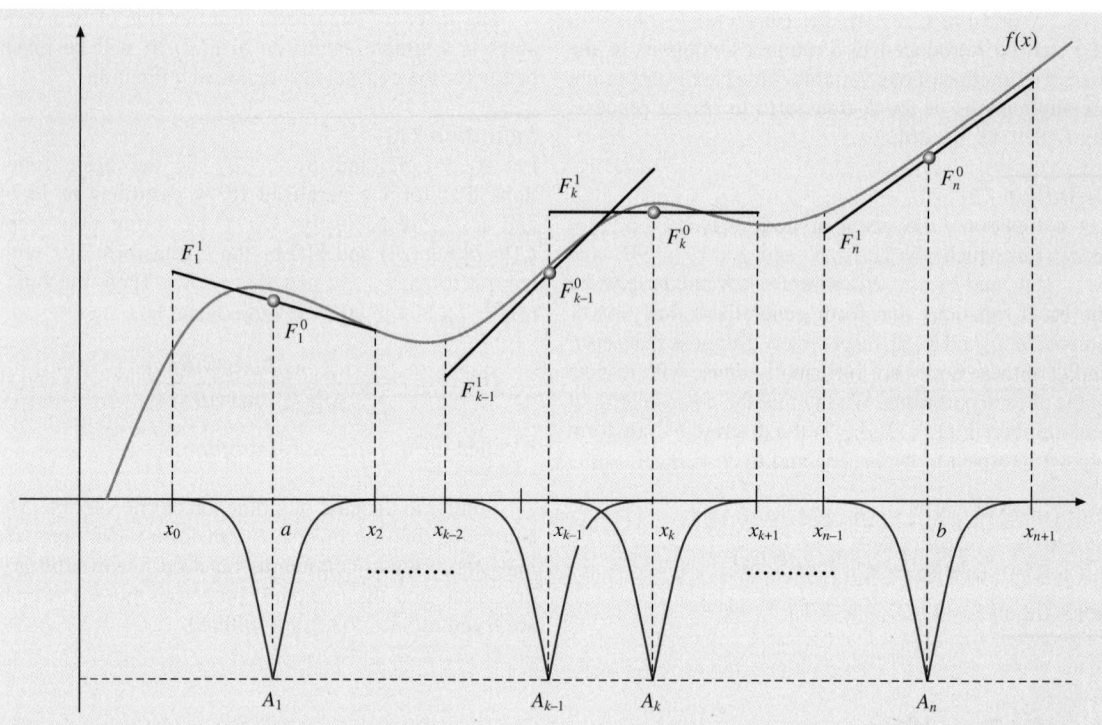

Fig. 7.6 Function f, its F^1-transform components $F_1^1, \ldots, F_k^1, \ldots, F_n^1$ (*linear segments*) and its F^0-transform components $F_1^0, \ldots, F_k^0, \ldots, F_n^0$ (*star nodes*) (after [7.16])

(b) If for all $x \in [a, b]$, $f(x) = d + cx$, then all the components of F^1-transform of $d + cx$ are equal to $(d + cx_k) + c(x - x_k)$, $k = 1, \ldots, n$.

(c) If $f = \alpha g + \beta h$, then $\mathbf{F}^1[f] = \alpha \mathbf{F}^1[g] + \beta \mathbf{F}^1[h]$.

(d) $c_{k,0} + c_{k,1}(x - x_k) = \min \|f(x) - (d + c(x - x_k))\|_k$, $k = 1, \ldots, n$, where min is considered over the set of functions of the form $(d + c(x - x_k))$.

(e) If f is four times continuously differentiable on $[a, b]$, then

$$c_{k,0} = f(x_k) + O(h^2),$$
$$c_{k,1} = f'(x_k) + O(h), \quad k = 1, \ldots, n.$$

In Fig. 7.6, we show a schematic representation of the F^1-transform components of a generic function f.

Finally, we give simplified expressions of F^1-transform components with respect to an h-uniform triangular fuzzy partition [7.16]

$$c_{k,0} = \frac{\int_{x_{k-1}}^{x_{k+1}} f(x) A_k(x) \mathrm{d}x}{h}, \tag{7.16}$$

$$c_{k,1} = \frac{12 \int_{x_{k-1}}^{x_{k+1}} f(x)(x - x_k) A_k(x) \mathrm{d}x}{h^3}, \tag{7.17}$$

where $k = 1, \ldots, n$.

7.7 Applications

In this section, we consider applications of the F-transform and F^1-transform to image processing.

7.7.1 Image Compression and Reconstruction

A method of lossy image compression and reconstruction using fuzzy relations was proposed in [7.19]. The dominant idea was a choice of suitable granulation (represented by a fuzzy relation) of an image domain. We will refer to this method as FEQ. F-transform image compression (FTR) is based on the same idea of granulation but connects it with fuzzy partitions [7.1, 9]. In the cited papers, two approaches were proposed: a uniform fuzzy partition of the entire domain [7.1] and a two-step partition [7.9] in which initially the entire domain is partitioned into blocks and second, each block is uniformly partitioned into fuzzy sets. Both approaches were compared with JPEG and other compression techniques (including FEQ) [7.9], and the conclusion was that the F-transform-based method is slightly worse than JPEG but better than FEQ. Two further improvements of the F-transform-based compression have been proposed in [7.18, 28], where an advantage over JPEG was achieved in many cases.

In this section, after reiterating the principles of image compression and reconstruction using the F-transform and its inverse, we explain how a proper choice of a fuzzy partition improves the quality of the reconstructed image. A detailed elaboration and comparison with other existing techniques is in [7.18, 28] and will be presented in subsequent papers.

Principles of Image Compression Using the F-Transform

Let a grayscale image of size $N \times M$ pixels be represented by a function of two variables $u : N \times M \to [0, 1]$. The value $u(i, j)$ represents the intensity range of each pixel in the gray scale. The problem of image compression is to reduce the image's size to save space or transmission time. A desirable size $n \times m$ (where $n < N$ and $m < M$) of a compressed image can be obtained from the *compression ratio*, $\rho = nm/(NM)$. If a compression method is lossy (JPEG, FEQ, and the F-transform, for example), then the respective reconstruction $\hat{u}$ to a full size image is compared with the original image using the two quality indices PSNR (peak signal-to-noise ratio) and RMSE (root-mean-square error), where

$$\text{PSNR} = 20 \ln \frac{255}{\text{RMSE}} ,$$

and

$$\text{RMSE} = \sqrt{\frac{\sum_{i=1}^{N} \sum_{j=1}^{M} [u(i,j) - \hat{u}(i,j)]^2}{NM}} .$$

Simple F-Transform Compression

In [7.1], we proposed representing a compressed image by the $n \times m$ matrix $\mathbf{U}$ of F-transform components

$$\mathbf{U} = \begin{pmatrix} U_{11} & \cdots & U_{1m} \\ \vdots & \vdots & \vdots \\ U_{n1} & \cdots & u_{nm} \end{pmatrix} ,$$

computed over uniform fuzzy partitions (usually, triangular) $A_1, \ldots, A_n$ and $B_1, \ldots, B_m$ of the entire domains

$[1, N]$ and $[1, M]$, respectively

$$U_{kl} = \frac{\sum_{j=1}^{M} \sum_{i=1}^{N} u(i,j) A_k(i) B_l(j)}{\sum_{j=1}^{M} \sum_{i=1}^{N} A_k(i) B_l(j)} ,$$

$$k = 1, \ldots, n ; \quad l = 1, \ldots, m .$$

We proposed reconstructing $\mathbf{U}$ to a full-size image using the inverse F-transform of u such that

$$\hat{u}(i,j) = \sum_{k=1}^{n} \sum_{l=1}^{m} U_{kl} A_k(i) B_l(j) .$$

This method does not take advantage of any property of the original image and therefore, its quality is not very high. Let us illustrate it on the image *Cameraman* taken from the Corel Gallery. In Fig. 7.7, we show the original image and its reconstruction using the simple F-transform compression described above. The compression ratio is $\rho = 0.25$, and PSNR $= 25.422$ (compare with PSNR $= 38.8$ for JPEG with a similar compression ratio).

F-Transform Compression with Block Decomposition

This F-transform-based compression [7.9] was inspired by the JPEG method in which, at first, the entire domain was decomposed into blocks and then, each block was compressed according to a compression ratio. In [7.9], the same principle is used. In the first step, a decomposition into blocks of the same size is performed, where the size (chosen experimentally) is such that a certain quality of approximation by the inverse F-transform should be guaranteed (Theorem 7.1). Each block is then uniformly partitioned into cosine-shaped fuzzy sets and compressed by the simple F-transform method according to a compression ratio. In comparison with the simple F-transform compression, this method considers the peculiarities of the original images when making

a)   b)

Fig. 7.7a,b Original image *Cameraman* (**a**) and its reconstruction after applying the simple F-transform compression (**b**) with PSNR $= 25.422$

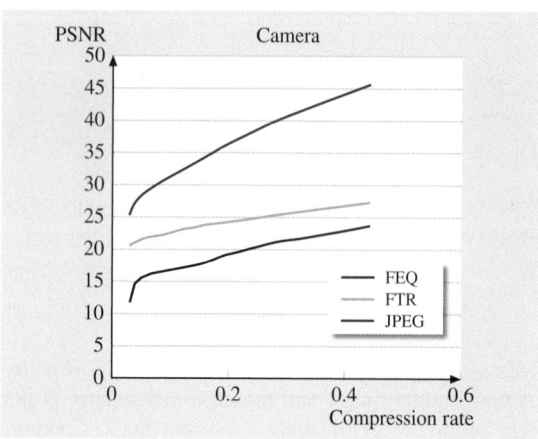

Fig. 7.8 The PSNR values of the image *Cameraman* compressed using three methods: FEQ, the *F*-transform with block decomposition, and JPEG (after [7.29])

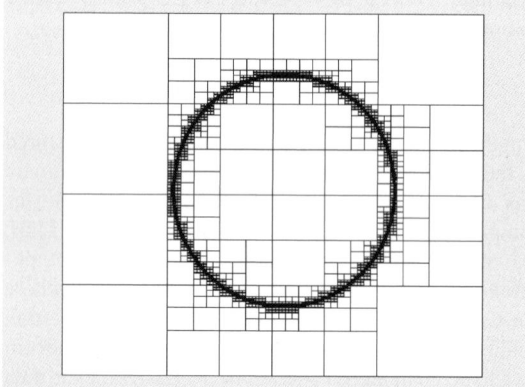

Fig. 7.9 The quad tree algorithm and the generalized fuzzy partition on its base

Fig. 7.10a,b Two reconstructions of the image *Cameraman* after applying the advanced *F*-transform compression (the ratio is 0.188) with the histogram restoring (**a**) and without it (**b**). The PSNR values are 29 (**a**) and 30 (**b**)

the block decomposition. In Fig. 7.8, we show the PSNR quality measure of the image *Cameraman* compressed using three methods: FEQ, *F*-transform with block decomposition and JPEG. It is easily observed that the JPEG method is still better than the *F*-transform with block decomposition, whereas the latter is better than FEQ. However, for the particular image *Cameraman* and the compression ratio $\rho = 0.25$, the value of PSNR of the *F*-transform with block decomposition is similar to that of the simple *F*-transform compression: 25.0676 versus 25.422, respectively. This means that the uniform partition, even when applied to both steps independently, is not effective with respect to the quality estimated by PSNR. In the next subsection, we propose an *F*-transform compression method [7.18] that is almost nonlossy and is based on a nonuniform generalized partition adapted to each particular image.

Advanced Image Compression

If we analyze the properties of the *F*-transform (Sect. 7.3.1), then it is immediate from (a) that the more the function behaves like a constant, the better is the approximation quality of the inverse *F*-transform. Thus, the following recommendation regarding the choice of a proper generalized fuzzy partition can be made:

● A generalized fuzzy partition of the domain $[1, N] \times [1, M]$ into fuzzy sets $A_k \times B_l$, where $k = 1, \ldots, n$ and $l = 1, \ldots, m$, should guarantee that the difference between extremal values of the image over each $A_k \times B_l$ is not greater than $\varepsilon > 0$ or (if the preceding condition cannot be fulfilled) the area of $A_k \times B_l$ is not greater than $\delta > 0$.

There are several algorithms that can produce a generalized fuzzy partition with the mentioned property. In [7.18], we used the quad tree algorithm for this purpose; see the illustration in Fig. 7.9.

Let us add that the advanced image compression algorithm [7.18] uses the following two tricks to increase the quality of the reconstructed image:

● Preserve sharp edges
● Restore the histogram of the original image.

Figure 7.10 shows how the histogram restoration influences the quality of the reconstructed image. In Fig. 7.11, we see that the PSNR values of the advanced *F*-transform and the JPEG are almost equal.

7.7.2 Image Fusion

Image fusion aims to integrate complementary distorted multisensor, multitemporal, and/or multiview scenes into one new image that contains the *best* parts of each scene. Thus, the primary problem in image fusion is to find the least distorted scene for every pixel.

A local focus measure is traditionally used for the selection of an undistorted scene. The scene that maximizes the focus measure is selected. Usually, the focus measure is a measure of high-frequency occurrences in the image spectrum. This measure is used when a source of distortion is connected with blurring, which suppresses high frequencies in an image. In this case, it is desirable that a focus measure decreases with an increase in blurring.

There are various fusion methodologies that are currently in use. They can be classified according to the primary technique: aggregation operators [7.22], fuzzy methods [7.30], optimization methods (e.g., neural networks and genetic algorithms [7.29]), and multiscale decomposition methods based on various transforms (e.g., discrete wavelet transforms; [7.31]).

The *F*-transform approach to image fusion was proposed in [7.32, 33]. The primary idea is a combination of (at least) two fusion operators, both of which are based on the *F*-transform. The first fusion operator is applied to the *F*-transform components of scenes and is based on a robust partition of the scene domain. The second fusion operator is applied to the residuals of scenes with respect to

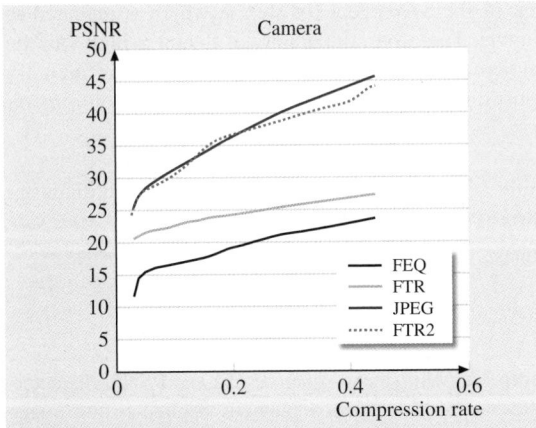

Part A | 7.7

Fig. 7.11 The PSNR values of the image *Cameraman* compressed using four methods: FEQ, the *F*-transform with block decomposition, the advanced *F*-transform, and JPEG

inverse *F*-transforms with fused components and is based on a finer partition of the same domain. Although this approach is not explicitly based on focus measures, it uses the fusion operator, which is able to choose an undistorted scene among the available blurred scenes.

Principles of Image Fusion Using the *F*-Transform

In this subsection, we present a short overview of the two methods of fusion that were proposed in [7.32, 33] and introduce a new method [7.34] that is a weighted combination of those two. We will demonstrate that the new method is computationally more effective than the first two.

The *F*-transform fusion is based on a certain decomposition of an image. We assume that the image u is a discrete real function $u = u(x, y)$ defined on the $N \times M$ array of pixels $P = \{(i, j) \mid i = 1, \ldots, N, j = 1, \ldots, M\}$ such that $u : P \to \mathbb{R}$. Moreover, let fuzzy sets $A_1, \ldots, A_n$ and $B_1, \ldots, B_m$, where $2 \leq n \leq N, 2 \leq m \leq M$, establish uniform Ruspini partitions of $[1, N]$ and $[1, M]$, respectively. We begin with the following representation of u on P,

$$u(x, y) = u_{nm}(x, y) + e(x, y) , \qquad (7.18)$$

$$e(x, y) = u(x, y) - u_{nm}(x, y) , \qquad (7.19)$$

where u_{nm} is the inverse *F*-transform of u and e is the respective first difference. If we replace e in (7.18) by its inverse *F*-transform e_{NM} with respect to the finest partition of $[1, N] \times [1, M]$, the above representation can then be rewritten as follows,

$$u(x, y) = u_{nm}(x, y) + e_{NM}(x, y) . \qquad (7.20)$$

We call (7.20) a *one-level decomposition* of u on P. If u is smooth, then the function e_{NM} is small (this claim follows from the property (e) in Sect. 7.3.1), and we can stop at this level. In the opposite case, we continue with the decomposition of the first difference e in (7.18). We decompose e into its inverse *F*-transform $e_{n'm'}$ (with respect to a finer fuzzy partition of $[1, N] \times [1, M]$ with $n' : n < n' \leq N$ and $m' : m < m' \leq M$ basic functions) and the second difference e'. Thus, we obtain the *second-level decomposition* of u on P

$$u(x, y) = u_{nm}(x, y) + e_{n'm'}(x, y) + e'(x, y) ,$$

$$e'(x, y) = e(x, y) - e_{n'm'}(x, y) .$$

In the same manner, we can obtain a *higher level decomposition* of u on P

$$u(x,y) = u_{n_1 m_1}(x,y) + e^{(1)}_{n_2 m_2}(x,y) + \cdots$$
$$+ e^{(k-2)}_{n_{k-1} m_{k-1}}(x,y) + e^{(k-1)}(x,y), \quad (7.21)$$

where

$$0 < n_1 \leq n_2 \leq \cdots \leq n_{k-1} \leq N,$$
$$0 < m_1 \leq m_2 \leq \cdots \leq m_{k-1} \leq M,$$
$$e^{(1)}(x,y) = u(x,y) - u_{n_1 m_1}(x,y),$$
$$e^{(i)}(x,y) = e^{(i-1)}(x,y) - e^{(i-1)}_{n_i m_i}(x,y),$$
$$i = 2, \ldots, k-1. \quad (7.22)$$

Three Algorithms for Image Fusion
In [7.33], we proposed two algorithms:

1. The *simple F*-transform-based fusion algorithm (SA) and
2. The *complete F*-transform-based fusion algorithm (CA).

The principal role in the fusion algorithms CA and SA is played by the *fusion operator* $\kappa : \mathbb{R}^K \to \mathbb{R}$, which is defined as follows:

$$\kappa(x_1, \ldots, x_K) = x_p, \text{ if } |x_p| = \max(|x_1|, \ldots, |x_K|). \quad (7.23)$$

The Simple F-Transform-Based Fusion Algorithm
In this subsection, we present a *block* description of the SA without technical details, which can be found in [7.33]. We assume that $K \geq 2$ input (channel) images $c_1, \ldots, c_K$ with various types of degradation are given. Our aim is to recognize undistorted parts in the given images and to fuse them into one image. The algorithm is based on the decompositions given in (7.20), which are applied to each channel image:

1. Choose values n and m such that $2 \leq n \leq N, 2 \leq m \leq M$ and create a fuzzy partition of $[1, N] \times [1, M]$ by fuzzy sets $A_k \times B_l$, where $k = 1, \ldots, n$ and $l = 1, \ldots, m$.
2. Decompose the input images $c_1, \ldots, c_K$ into inverse *F*-transforms and error functions according to the one-level decomposition (7.20).
3. Apply the fusion operator (7.23) to the respective *F*-transform components of $c_1, \ldots, c_K$, and obtain the fused *F*-transform components of a new image.

4. Apply the fusion operator to the respective *F*-transform components of the error functions e_i, $i = 1, \ldots, K$, and obtain the fused *F*-transform components of a new error function.
5. Reconstruct the fused image from the inverse *F*-transforms with the fused components of the new image and the fused components of the new error function.

The SA-based fusion is very efficient if we can guess values n and m that characterize a proper fuzzy partition. Usually, this is performed manually according to the user's skills. The dependence on fuzzy partition parameters can be considered as a primary shortcoming of this otherwise effective algorithm. Two recommendations follow from our experience:

- For complex images (with many small details), higher values of n and m yield better results.
- If a triangular shape for a basic function is chosen, than the generic choice of n and m is such that the corresponding values of n_p and m_p are equal to 3 (recall that n_p is the number of points that are covered by every full basic function A_k).

The Complete F-Transform-Based Fusion Algorithm
The CA-based fusion does not depend on the choice of only one fuzzy partition (as in the case of the SA) because it runs through a sequence (7.22) of increasing values of n and m. The algorithm is based on the decomposition presented in (7.21), which is applied to each channel image. The description of the CA is similar to that of the SA except for step 4, which is repeated in a cycle. Therefore, the quality of fusion is high, but the implementation of the CA is rather slow and memory consuming, especially for large images. For an illustration, Fig. 7.12, Tables 7.1 and 7.2.

Table 7.1 Basic characteristics of the three algorithms applied to the image *Balls*

Image	Resolution	Time (s)			Memory (MB)		
		CA	SA	ESA	CA	SA	ESA
Balls	1600 × 1200	340	1.2	36	270	58	152

Table 7.2 MSE (mean-square error) and PSNR characteristics of the three fusion methods applied to the image *Balls*

Image set	MSE			PSNR		
	CA	SA	ESA	CA	SA	ESA
Balls	1.28	6.03	0.86	48.91	43.81	52.57

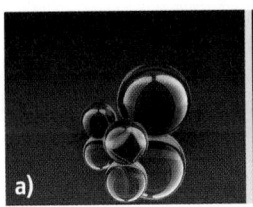 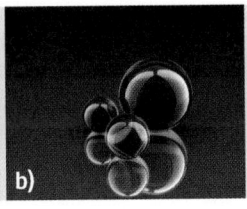 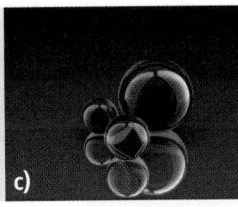

Fig. 7.12a-c The SA (**a**), CA (**b**) and ESA (**c**) fusions of the image *Balls*. The ESA fusion has the best quality (Table 7.2)

Enhanced Simple Fusion Algorithm

In [7.34], we proposed an algorithm that is as fast as the SA and as efficient as the CA. We aimed at achieving the following goals:

- Avoid running through a long sequence of possible partitions (as in the case of CA).
- Automatically adjust the parameters of the fusion algorithm according to the level of blurring and the location of blurred areas in input images.

The algorithm adds another run of the *F*-transform over the first difference (7.18). The explanation is as follows: the first run of the *F*-transform is aimed at edge detection in each input image, whereas the second run propagates only sharp edges (and their local areas) to the fused image. We refer to this algorithm as to *enhanced simple algorithm* (ESA) and give its informal description:

for all input (channel) images **do**
  Compute the *inverse F-transform*
  Compute the *first absolute difference* between the original image and the inverse *F*-transform of it
  Compute the *second absolute difference* between the first one and its inverse *F*-transform and set them as the pixel weights
end for
for all pixels in an image **do**
  Compute the value of *sow* – the sum of the weights over all input images
  for all input images **do**
    Compute the value of *wr* – the ratio between the weight of a current pixel and *sow*
  end for
  Compute the fused value of a pixel in the resulting image as a weighted (by *wr*) sum of input image values
end for

The primary advantages of the ESA are:

- *Time* – the execution time is smaller than for the CA (Table 7.1).

- *Quality* – the quality of the ESA fusion is better than that of the SA and for particular cases (Table 7.2), it is better than that of the CA.

Because of space limitations, we present only one illustration of the *F*-transform fusion performed using the three algorithms, SA, CA, and ESA. We chose the image *Balls* with geometric figures to demonstrate that our fusion methods are able to reconstruct edges. In Fig. 7.13, two (channel) inputs of the image *Balls* are given, and in Fig. 7.12, three fusions of the same image are demonstrated.

In Table 7.1, we demonstrate that the complexity (measured by the execution time or by the memory used) of the ESA is greater than the complexity of the SA and less than the complexity of the CA.

In Table 7.2, we demonstrate that for the particular image *Balls*, the quality of fusion (measured by the values of MSE and PSNR) of the ESA result is better (the MSE value is smaller) than the quality of the SA result and even than the quality of the CA result.

7.7.3 F^1-Transform Edge Detector

Edge detection is inevitable in image processing. In particular, it is a first step in feature extraction and image segmentation. We focused on the Canny edge detector [7.35], which is widely used in computer vision. It was developed to ensure three basic criteria: good detection, good localization, and minimal response. In

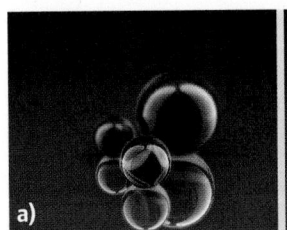 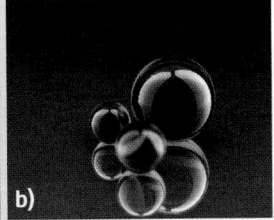

Fig. 7.13a,b Two inputs for the image *Balls*. The *central ball* is blurred in (**a**), and conversely, it is the only sharp ball in (**b**)

Fig. 7.14a–d Original images (**a,c**) and their F^1-transform edges (**b,d**)

these aspects, the Canny detector can be considered an *optimal* edge detector. In [7.26], we proposed using the F^1-transform with the purpose of simplifying the first two steps of the Canny algorithm. Below, we provide the details of our proposal.

The Canny algorithm is a multistep procedure for detecting edges as the local maxima of the gradient magnitude. The first step, performed using a Gaussian filter, is image smoothing and filtering noise. The second step is computation of a gradient of the image function to find the local maxima of the gradient magnitude and the gradient's direction at each point. This step is performed using a convolution of the original image with directional masks (edge detection operators, such as those of *Roberts, Prewitt*, and *Sobel*, are some examples of these filters). The next step is called nonmaximum suppression [7.36], and it selects those points whose gradient magnitudes are maximal in the corresponding gradient direction. The final step is tracing edges and hysteresis thresholding, which leads to preserving the continuity of edges.

In our experiment, we removed the first two steps in the Canny algorithm and replaced them by computation of approximate gradient values using the F^1-transform. The reason is that the F^1-transform (similar to the ordinary F-transform) filters out noise when computing approximate values of the first partial derivatives given by (7.15). We assume that the image is represented by a discrete function $u : P \rightarrow \mathbb{R}$ of two variables, where $P = \{(i,j) \mid i = 1, \ldots, N, j = 1, \ldots, M\}$ is an $N \times M$ array of pixels, and the fuzzy sets $A_1, \ldots, A_n$ and $B_1, \ldots, B_m$ establish a uniform triangular fuzzy partition of $[1, N]$ and $[1, M]$, respectively.

Let $x_1, \ldots, x_n \in [1, N]$ and $y_1, \ldots, y_m \in [1, M]$ be the h_x and h_y-equidistant nodes of $[1, N]$ and $[1, M]$, respectively.

According to property (e) in Sect. 7.6, the coefficients $c_{k,1}$ of the linear polynomials of the F^1-transform components are approximate values of the first partial derivatives of the image function at nodes (x_k, y_l) (for simplicity, we assume $k = 2, \ldots, n-1$ and $l = 2, \ldots, m$), where by (7.17) and (7.5) the following hold,

$$c_{k,1}(y_l) = \frac{12}{h_x^3 h_y} \sum_{i=1}^{N} \sum_{j=1}^{M} u(i,j)(i - x_k) A_k(i) B_l(j),$$

$$c_{l,1}(x_k) = \frac{12}{h_x h_y^3} \sum_{i=1}^{N} \sum_{j=1}^{M} u(i,j)(j - y_l) A_k(i) B_l(j).$$

Then, we can write approximations of the first partial derivatives as the respective inverse F-transforms

$$G_x(i,j) \approx \sum_{k=1}^{n} \sum_{l=1}^{m} c_{k,1}(y_l) A_k(i) B_l(j),$$

and

$$G_y(i,j) \approx \sum_{k=1}^{n} \sum_{l=1}^{m} c_{l,1}(x_k) A_k(i) B_l(j).$$

All other steps of the Canny algorithm – namely, finding the local maxima of the gradient magnitude and its direction, nonmaximum suppression, tracing edges through the image and hysteresis thresholding – are the same as in the original procedure.

In the two examples in Fig. 7.14, we demonstrate the results of the F^1-transform edge detector on images chosen from the dataset available at ftp://figment.csee.usf.edu/pub/ROC/edge_comparison_dataset.tar.gz.

We observe that many thin edges/lines are detected as well as their connectedness and smoothness. Moreover, the following properties are retained:

- Smoothness of circular lines
- Concentricness circles
- Smoothness of sharp connections.

7.8 Conclusions

In this chapter, the theory of the *F*-transform has been discussed from the perspective of the latest developments and applications. The importance of a proper choice of fuzzy partition has been stressed. Various fuzzy partitions have been considered, including the most general partition (currently known). The definition of the *F*-transform has been adapted to the generalized fuzzy partition, and the main properties of the *F*-transform have been re-established. The applications to image processing, namely image compression, fusion and edge detection, have been discussed with sufficient technical details.

References

7.1 I. Perfilieva: Fuzzy transforms: Theory and applications, Fuzzy Sets Syst. **157**, 993–1023 (2006)

7.2 L.A. Zadeh: Outline of a new approach to the analysis of complex systems and decision processes, IEEE Trans. Syst. Man Cybern. **SMC-3**, 28–44 (1973)

7.3 L.A. Zadeh: The concept of a linguistic variable and its application to approximate reasoning, Part I, Inf. Sci. **8**, 199–257 (1975)

7.4 L.A. Zadeh: The concept of a linguistic variable and its application to approximate reasoning, Part II, Inf. Sci. **8**, 301–357 (1975)

7.5 L.A. Zadeh: The concept of a linguistic variable and its application to approximate reasoning, Part III, Inf. Sci. **9**, 43–80 (1975)

7.6 T. Takagi, M. Sugeno: Fuzzy identification of systems and its application to modeling and control, IEEE Trans. Syst. Man Cybern. **15**, 116–132 (1985)

7.7 I. Perfilieva, H. De Meyer, B. De Baets, D. Pisková: Cauchy problem with fuzzy initial condition and its approximate solution with the help of fuzzy transform, Proc. WCCI 2008, IEEE Int. Conf. Fuzzy Syst., Hong Kong (2008) pp. 2285–2290

7.8 I. Perfilieva: Fuzzy transforms and their applications to image compression, Lect. Notes Artif. Intell. **3849**, 19–31 (2006)

7.9 F. Di Martino, V. Loia, I. Perfilieva, S. Sessa: An image coding/decoding method based on direct and inverse fuzzy transforms, Int. J. Approx. Reason. **48**, 110–131 (2008)

7.10 V. Novák, M. Štěpnička, A. Dvořák, I. Perfilieva, V. Pavliska, L. Vavřičková: Analysis of seasonal time series using fuzzy approach, Int. J. Gen. Syst. **39**, 305–328 (2010)

7.11 I. Perfilieva, V. Novák, V. Pavliska, A. Dvořák, M. Štěpnička: Analysis and prediction of time series using fuzzy transform, Proc. WCCI 2008, IEEE Int. Conf. Neural Netw., Hong Kong (2008) pp. 3875–3879

7.12 F. Di Martino, V. Loia, S. Sessa: Fuzzy transforms method in prediction data analysis, Fuzzy Sets Syst. **180**, 146–163 (2011)

7.13 M. Štěpnička, A. Dvořák, V. Pavliska, L. Vavřičková: A linguistic approach to time series modeling with the help of *F*-transform, Fuzzy Sets Syst. **180**, 164–184 (2011)

7.14 L. Troiano, P. Kriplani: Supporting trading strategies by inverse fuzzy transform, Fuzzy Sets Syst. **180**, 121–145 (2011)

7.15 L. Stefanini: F-transform with parametric generalized fuzzy partitions, Fuzzy Sets Syst. **180**, 98–120 (2011)

7.16 I. Perfilieva, M. Daňková, B. Bede: Towards F-transform of a higher degree, Fuzzy Sets Syst. **180**, 3–19 (2011)

7.17 M. Holčapek, T. Tichý: A smoothing filter based on fuzzy transform, Fuzzy Sets Syst. **180**, 69–97 (2011)

7.18 P. Hurtik, I. Perfilieva: Image compression methodology based on fuzzy transform, Int. Jt. Conf. CISIS'12-ICEUTE'12-SOCO'12 Special Sessions, Adv. Intell. Soft Comput., Vol. 189 (Springer, Berlin, Heidelberg 2013) pp. 525–532

7.19 K. Hirota, W. Pedrycz: Fuzzy relational compression, IEEE Trans. Syst. Man Cybern. **29**, 407–415 (1999)

7.20 G. Patanè: Fuzzy transform and least-squares approximation: Analogies, differences, and generalizations, Fuzzy Sets Syst. **180**, 41–54 (2011)

7.21 I. Perfilieva, V. Novák, A. Dvořák: Fuzzy transform in the analysis of data, Int. J. Approx. Reason. **48**, 36–46 (2008)

7.22 R.S. Blum: Robust image fusion using a statistical signal processing approach, Inf. Fusion **6**, 119–128 (2005)

7.23 I. Perfilieva, R. Valášek: Fuzzy transforms in removing noise. In: *Computational Intelligence, Theory and Applications*, ed. by B. Reusch (Springer, Heidelberg 2005) pp. 225–234

7.24 F. Di Martino, V. Loia, S. Sessa: A segmentation method for images compressed by fuzzy transforms, Fuzzy Sets Syst. **161**, 56–74 (2010)

7.25 I. Perfilieva, M. Daňková: Image fusion on the basis of fuzzy transforms, Proc. 8th Int. FLINS Conf., Madrid (2008) pp. 471–476

7.26 I. Perfilieva, P. Hodáková, P. Hurtik: F^1-transform edge detector inspired by Canny's algorithm. In: *Advances on Computational Intelligence (IPMU2012)*, ed. by S. Greco, B. Bouchon-Meunier, G. Coletti, M. Fedrizzi, B. Matarazzo, R.R. Yager (Catania, Italy 2012) pp. 230–239

7.27   V. Novák, I. Perfilieva, V. Pavliska: The use of higher-order F-transform in time series analysis, World Congr. IFSA 2011 AFSS 2011, Surabaya (2011) pp. 2211–2216

7.28   I. Perfilieva, B. De Baets: Fuzzy transform of monotonous functions, Inf. Sci. **180**, 3304–3315 (2010)

7.29   A. Mumtaz, A. Masjid: Genetic algorithms and its applications to image fusion, IEEE Int. Conf. Emerg. Technol., Rawalpindi (2008) pp. 6–10

7.30   R. Ranjan, H. Singh, T. Meitzler, G.R. Gerhart: Iterative image fusion technique using fuzzy and neuro fuzzy logic and applications, Proc. IEEE Fuzzy Inf. Process. Soc., Detroit (2005) pp. 706–710

7.31   G. Piella: A general framework for multiresolution image fusion: From pixels to regions, Inf. Fusion **4**, 259–280 (2003)

7.32   I. Perfilieva, M. Daňková, H.P.M. Vajgl: The use of f-transform for image fusion algorithms, Proc. Int. Conf. Soft Comput. Pattern Recognit. (SoCPaR2010), Cergy Pontoise (2010) pp. 472–477

7.33   I. Perfilieva, M. Daňková, P. Hodáková, M. Vajgl: F-transform based image fusion. In: *Image Fusion*, ed. by O. Ukimura (InTech, Rijeka 2011), pp. 3–22, available online from http://www.intechopen.com/books/image-fusion/f-transform-based-image-fusion

7.34   M. Vajgl, I. Perfilieva, P. Hodáková: Advanced F-transform-based image fusion, Adv. Fuzzy Syst. **2012**, 125086 (2012)

7.35   J. Canny: A computational approach to edge detection, IEEE Trans. Pattern Anal. Mach. Intell. **PAMI-8**(6), 679–698 (1986)

7.36   A. Rosenfeld, M. Thurston: Edge and curve detection for visual scene analysis, IEEE Trans. Comput. **C-20**(5), 562–569 (1971)

8. Fuzzy Linear Programming and Duality

Jaroslav Ramík, Milan Vlach

The chapter is concerned with linear programming problems whose input data may be fuzzy while the values of variables are always real numbers. We propose a rather general approach to these types of problems, and present recent results for problems in which the notions of feasibility and optimality are based on the fuzzy relations of possibility and necessity. Special attention is devoted to the weak and strong duality.

Formulation of an abstract model applicable to a complex decision problem usually involves a tradeoff between the accuracy of the problem description and the tractability of the resulting model. One of the widespread models of decision problems is based on the assumption of linearity of constraints and optimization criteria, in spite of the fact that, in most instances of real decision problems, not all constraints and optimization criteria are linear. Fortunately, in many such cases, solutions of decision problems obtained through linear programming are exact or numerically tractable approximations. Given the practical relevance of linear programming, it is not surprising that attempts to extend linear programming theory to problems involving fuzzy data have been appearing since the early days of fuzzy sets. To obtain a meaningful extension of linear programming to problems involving fuzzy data, one has to specify a suitable class of permitted fuzzy numbers, introduce fundamental arithmetic operations with such fuzzy numbers, define inequalities between fuzzy numbers, and clarify the meaning of feasibility and optimality. Because this can be done in many different ways, we can hardly expect a unique extension that would be so clean and clear like the theory of linear programming without fuzzy data. Instead, there exist several variants of the theory for fuzzy linear programming, the results of which resemble in various degrees some of the useful results established in the conventional linear programming.

Certainly, the most influential papers for the early development of optimization theory for problems with fuzzy data were papers written by *Bellman* and *Zadeh* [8.1], and *Zimmermann* [8.2]. As pointed out in a recent paper by *Dubois* [8.3], fuzzy optimization that is based on the Bellman and Zadeh, and Zimmermann ideas comes down to max–min bottleneck optimization. Thus, strictly speaking, the fuzzy linear programming problems are not necessarily linear in the standard sense.

Throughout the chapter, we assume that some or all of the input data defining the problem may be fuzzy while the values of variables are always real numbers. For problems with fuzzy decision variables, see e.g. [8.4]. Moreover, we not always satisfy the requirement of the symmetric model of [8.1, 2] which demands that the constraints and criteria are to be treated in the same way. In general, we take into consideration the fact that in many situations, in practice, the degree of feasibility may be essentially different from the degree of optimality attainment.

The structure of the chapter is briefly described as follows. In the next section, we first recall the basic results of the conventional linear programming, especially the results on duality. As a canonical problem we consider the problem of the form: Given real numbers $b_1, b_2, \ldots, b_m, c_1, c_2, \ldots, c_n, a_{11}, a_{12}, \ldots, a_{mn}$,

$$\text{maximize} \quad c_1 x_1 + c_2 x_2 + \cdots + c_n x_n$$
$$\text{subject to} \quad a_{i1} x_1 + a_{i2} x_2 + \cdots + a_{in} x_n \leq b_i \,,$$
$$i = 1, 2, \ldots, m \,,$$
$$x_j \geq 0 \,, \quad j = 1, 2, \ldots, n \,.$$

Then we review the basic notions and terminology of fuzzy set theory, which we need for precise formulation and description of results of linear programming problems involving fuzzy data. After these necessary preliminaries, in Sect. 3, we introduce and study fuzzy linear programming problems. We focus attention on analogous canonical form, namely on the following

problem

$$\text{maximize} \quad \tilde{c}_1 x_1 \tilde{+} \cdots \tilde{+} \tilde{c}_n x_n$$
$$\text{subject to} \quad \tilde{a}_{i1} x_1 \tilde{+} \cdots \tilde{+} \tilde{a}_{in} x_n \; \tilde{P} \tilde{b}_i \,,$$
$$i = 1, 2, \ldots, m \,,$$
$$x_j \geq 0 \,, \quad j = 1, 2, \ldots, n \,,$$

where $\tilde{c}_j, \tilde{a}_{ij}$, and $\tilde{b}_i$ are fuzzy quantities and the meanings of *subject to* and *maximize* are based on the standard possibility and necessity relations introduced in [8.5]. The final section is devoted to duality theory for fuzzy linear programming problems. First, we recall some of the early approaches that are based on the ideas of *Bellman* and *Zadeh* [8.1], and *Zimmermann* [8.2]. Then we present recent results of *Ramík* [8.6, 7].

8.1 Preliminaries

8.1.1 Linear Programming

Linear programming is concerned with optimization problems whose objective functions are linear in the unknowns and whose constraints are linear inequalities or linear equalities in the unknowns. The form of a linear programming problem may differ from one problem to another but, fortunately, there are several standard forms to which any linear programming problem can be transformed. We shall use the following canonical form.

Given real numbers $b_1, b_2, \ldots, b_m, \; c_1, c_2, \ldots, c_n$, $a_{11}, a_{12}, \ldots, a_{mn}$,

$$\text{maximize} \quad c_1 x_1 + c_2 x_2 + \cdots + c_n x_n \tag{8.1}$$
$$\text{subject to} \quad a_{i1} x_1 + a_{i2} x_2 + \cdots + a_{in} x_n \leq b_i \,,$$
$$i = 1, 2, \ldots, m \,, \tag{8.2}$$
$$x_j \geq 0 \,,$$
$$j = 1, 2, \ldots, n \,. \tag{8.3}$$

The set of all n-tuples $(x_1, x_2, \ldots, x_n)$ of real numbers that simultaneously satisfy inequalities (8.2) and (8.3) is called the *feasible region* of problem (8.1)–(8.3) and the elements of feasible region are called *feasible solutions*. A feasible solution $\hat{x}$ such that no other feasible solution x satisfies

$$c_1 x_1 + c_2 x_2 + \cdots + c_n x_n > c_1 \hat{x}_1 + c_2 \hat{x}_2 + \cdots + c_n \hat{x}_n$$

is called an *optimal solution* of (8.1)–(8.3), and the set of all optimal solutions is called the *optimal region*.

Using the same data $b_1, b_2, \ldots, b_m, \; c_1, c_2, \ldots, c_n$, $a_{11}, a_{12}, \ldots, a_{mn}$, we can associate with problem (8.1)–(8.3) another linear programming problem, namely, the problem

$$\text{minimize} \quad y_1 b_1 + y_2 b_2 + \cdots + y_m b_m \tag{8.4}$$
$$\text{subject to} \quad y_1 a_{1j} + y_2 a_{2j} + \cdots + y_m a_{mj} \geq c_j \,,$$
$$j = 1, 2, \ldots, n \,, \tag{8.5}$$
$$y_i \geq 0 \,,$$
$$i = 1, 2, \ldots, m \,. \tag{8.6}$$

Analogously to the case of maximization, we say that the set of all m-tuples $(y_1, y_2, \ldots, y_m)$ of real numbers that simultaneously satisfy inequalities (8.5) and (8.6) is the feasible region of problem (8.4)–(8.6), and that an element $\hat{y}$ of the feasible region such that no other element y of the feasible region satisfies

$$b_1 y_1 + b_2 y_2 + \cdots + b_m y_m < b_1 \hat{y}_1 + b_2 \hat{y}_2 + \cdots + b_m \hat{y}_m$$

is an optimal solution of (8.4)–(8.6).

The problem (8.1)–(8.3) is then called the *primal problem* and the associated problem (8.4)–(8.6) is called the *dual problem* to (8.1)–(8.3). However, this terminology is relative because if we rewrite the dual problem into the form of the equivalent primal problem and again construct the corresponding dual, then we obtain a linear programming problem which is equivalent to the original primal problem. In other words, the dual to the dual is the primal. Consequently, it is just the matter of convenience which of these problems is taken as the primal problem.

The main theoretical results on linear programming are concerned with mutual relationship between the primal problem and its dual problem. They can be summarized as follows, see also [8.8, 9].

Let $\mathbb{R}^n$ and $\mathbb{R}^n_+$ denote the set of real n-vectors and real nonnegative n-vectors equipped by the usual euclidean distance. For $n = 1$, we simplify the notation to $\mathbb{R}$ and $\mathbb{R}_+$. The scalar product of vectors x and y from $\mathbb{R}^n$ is denoted by xy:

1. If x is a feasible solution of the primal problem and if y is a feasible solution of the dual problem, then $cx \leq yb$.
2. If $\bar{x}$ is a feasible solution of the primal problem, and if $\bar{y}$ is a feasible solution of the dual problem, and if $c\bar{x} = \bar{y}b$, then $\bar{x}$ is optimal for the primal problem and $\bar{y}$ is optimal for the dual problem.
3. If the feasible region of the primal problem is nonempty and the objective function $x \mapsto cx$ is not bounded above on it, then the feasible region of the dual problem is empty.
4. If the feasible region of the dual problem is nonempty and the objective function $y \mapsto yb$ is not bounded below on it, then the feasible region of the primal problem is empty.

It turns out that the following deeper results concerning mutual relation between the primal and dual problems hold:

5. If either of the problem (8.1)–(8.3) or (8.4)–(8.6) has an optimal solution, so does the other, and the corresponding values of the objective functions are equal.
6. If both problems (8.1)–(8.3) and (8.4)–(8.6) have feasible solutions, then both of them have optimal solutions and the corresponding optimal values are equal.
7. A necessary and sufficient condition that feasible solutions x and y of the primal and dual problems are optimal is that

$$x_j > 0 \Rightarrow yA^j = c_j, \quad 1 \leq j \leq n,$$
$$x_j = 0 \Leftarrow yA^j > c_j, \quad 1 \leq j \leq n,$$
$$y_i > 0 \Rightarrow A_i x = b_i, \quad 1 \leq i \leq m,$$
$$y_i = 0 \Leftarrow A_i x < b_i, \quad 1 \leq i \leq m,$$

where A^j and A_i stand for the j-th column and i-th row of $A = \{a_{ij}\}$, respectively.

It is also well known that the essential duality results of linear programming can be expressed as a saddle-point property of the Lagrangian function, see [8.10]:

8. Let $L: \mathbb{R}^n_+ \times \mathbb{R}^m_+ \to \mathbb{R}$ be the *Lagrangian function* for the primal problem (8.1)–(8.3), that is, $L(x, y) = cx + y(b - Ax)$. The necessary and sufficient condition that $\bar{x} \in \mathbb{R}^n_+$ be an optimal solution of the primal problem (8.1)–(8.3) and $\bar{y} \in \mathbb{R}^m_+$ be an optimal solution of the dual problem (8.4)–(8.6) is that $(\bar{x}, \bar{y})$ be a saddle point of L; that is, for all $x \in \mathbb{R}^n_+$ and $y \in \mathbb{R}^m_+$,

$$L(x, \bar{y}) \leq L(\bar{x}, \bar{y}) \leq L(\bar{x}, y) . \tag{8.7}$$

8.1.2 Sets and Fuzzy Sets

A well-known fact about subsets of a given set is that their properties and their mutual relations can be studied by means of their characteristic functions. However, these two notions are different, and the notion of characteristic function of a subset of a set is more complicated than that of a subset of a set. Indeed, because the characteristic function χ_A of a subset A of a fixed given set X is a mapping from X into the set $\{0, 1\}$, we not only need the underlying set X and its subset A but also one additional set; in particular, the set $\{0, 1\}$. In addition, we also need the notion of an ordered pair and the notion of the Cartesian product of sets because functions are specially structured binary relations; in this case, special subsets of $X \times \{0, 1\}$.

The phrases *the membership function of a fuzzy set* ... or *the fuzzy set defined by membership function* ... (and similar ones), which are very common in the fuzzy set literature, clearly indicate that a *fuzzy set* and its *membership function* are different mathematical objects. If we introduce fuzzy sets by means of their membership functions, that is, by replacing the range $\{0, 1\}$ of characteristic functions with the unit interval $[0, 1]$ of real numbers ordered by the standard ordering $\leq$, then we are tacitly assuming that the membership functions of fuzzy sets on X are related to fuzzy sets on X in an analogous way as the characteristic functions of subsets of X are related to subsets of X. What are those objects that we call fuzzy sets on X in set-theoretic terms? Obviously, they are more complex than just subsets of X because the class of functions mapping X into the lattice $([0, 1], \leq)$ is much richer than the class of functions mapping X into $\{0, 1\}$. We follow the opinion that fuzzy sets are special-nested families of subsets of a set, see [8.11].

Definition 8.1

A *fuzzy subset* of a nonempty set X (or a *fuzzy set on X*) is a family $\{A_\alpha\}_{\alpha\in[0,1]}$ of subsets of X such that $A_0 = X, A_\beta \subset A_\alpha$ whenever $0 \leq \alpha \leq \beta \leq 1$, and $A_\beta = \cap_{0\leq\alpha<\beta}A_\alpha$ whenever $0 < \beta \leq 1$.

Definition 8.2

If $A = \{A_\alpha\}_{\alpha\in[0,1]}$ is a fuzzy subset of X, then the *membership function of A* is the function μ_A from X into the unit interval $[0,1]$ defined by $\mu_A(x) = \sup\{\alpha : x \in A_\alpha\}$.

Remark 8.1

It is worth noting that by defining a fuzzy subset of a set X as a special family of subsets of X, we can easily avoid certain troublesome phrases. For example, we are used to say *a subset A of X* and not *a subset* $\chi_A : X \to \{0,1\}$ *of a set X*. Similarly, it is more natural to say *a fuzzy subset A of X* than to say *a fuzzy subset* $\mu_A : X \to [0,1]$ *in X*. Moreover, if a fuzzy set on X would be defined as a function μ from X to $[0,1]$, then we would obtain statements like *fuzzy set μ is function μ*, or *a fuzzy set μ is convex if and only if μ is quasiconcave*.

Let A be a subset of a set X and let $\{A_\alpha\}_{\alpha\in[0,1]}$ be the family of subsets of X defined by $A_0 = X$ and $A_\alpha = A$ for each positive α from $[0,1]$. It can easily be seen that this family is a fuzzy set on X and that its membership function is equal to the characteristic function of A; see [8.12, 13] for details. This one-to-one correspondence between the characteristic functions of subsets of X and the membership functions of certain fuzzy sets on X provides an embedding of the set of subsets of X into the set of fuzzy sets on X. Consequently, we can view subsets of X as special fuzzy sets on X. When we need to distinguish the latter from the other fuzzy sets on X, we call them the *crisp fuzzy sets* on X. Moreover, we can also view the elements of X as a special fuzzy sets on X by additionally employing the one-to-one correspondence that assigns to each element x of X the singleton $\{x\}$. When we need to distinguish an element $x \in X$ from the crisp fuzzy sets on X corresponding to $\{x\}$, we write $k(x)$ for the latter.

We denote the collection of all fuzzy sets on X by $\mathcal{F}(X)$. When A is from $\mathcal{F}(X)$ and μ_A is the membership function of A, then we use the following terminol-

ogy. The value $\mu_A(x)$ is called the *membership degree* of x in A. The set $\{x \in X : \mu_A(x) = 1\}$ is called the *core* of A. If the core of A is nonempty, then A is said to be *normalized*. The *complement* of A is the fuzzy set $c(A)$ on X whose membership function is $\mu_{c(A)}(x) = 1 - \mu_A(x)$. For each $\alpha \in [0,1]$, the set $\{x \in X \mid \mu_A(x) \geq \alpha\}$ is called the *α-cut of A* and is denoted by $[A]_\alpha$. If X is a nonempty subset of a real finite-dimensional normed space, then a fuzzy set A in X is called *closed, bounded, compact,* or *convex* if the α-cut $[A]_\alpha$ is a closed, bounded, compact or convex subset of X for every $\alpha \in (0,1]$, respectively.

Following the terminology of [8.7], we say that a fuzzy subset A of $\mathbb{R}$ is a *fuzzy quantity* whenever A is normal, compact, and its membership function μ_A is semistrictly quasiconcave in the following sense: The membership function μ_A of A is *semistrictly quasiconcave* on $\mathbb{R}$ if there exist $a,b,c,d \in \mathbb{R}$, $-\infty < a \leq b \leq c \leq d < +\infty$, such that

$$\mu_A(t) = 0 \quad \text{if } t < a \text{ or } t > d,$$
$$\mu_A \quad \text{is strictly increasing on the interval } [a,b],$$
$$\mu_A(t) = 1 \quad \text{if } b \leq t \leq c,$$
$$\mu_A \quad \text{is strictly decreasing on the interval } [c,d].$$

The set of all fuzzy quantities is denoted by $\mathcal{F}_0(\mathbb{R})$. Note that $\mathcal{F}_0(\mathbb{R})$ contains well-known classes of fuzzy numbers: crisp (real) numbers, crisp intervals, triangular fuzzy numbers, trapezoidal, and bell-shaped fuzzy numbers etc. However, $\mathcal{F}_0(\mathbb{R})$ does not contain fuzzy sets with *stair-like* membership functions.

Recall that the binary relations on X are subsets of the Cartesian product $X \times X$ and that the fuzzy sets on $X \times X$ are called the *fuzzy binary relation on X*, or simply *fuzzy relation* on X. Because the binary relations on X are subsets of $X \times X$, we can view them as special fuzzy relations on X; namely, as those fuzzy relations on X whose membership functions are equal to the characteristic functions of the corresponding binary relations. Again, we call them *crisp*. Since the membership functions of fuzzy sets provide a mathematical tool for introducing grades in the notion of set membership, the fuzzy relations on X can be used for introducing grades in comparison of elements of X. However, if we need to compare not only elements of X but also fuzzy sets on X, then we need binary relations and fuzzy binary relations on the set of fuzzy sets on X, that is, on $\mathcal{F}(X) \times \mathcal{F}(X)$.

Let R be a fuzzy relation on X and let Q be a fuzzy relation on $\mathcal{F}(X)$, that is, R belongs to $\mathcal{F}(X \times X)$ and Q belongs to $\mathcal{F}(\mathcal{F}(X) \times \mathcal{F}(X))$. We say that Q is a *fuzzy extension* (or briefly an *extension*) of R from X to $\mathcal{F}(X)$ if, for each pair x and y in X,

$$\mu_Q(k(x), k(y)) = \mu_R(x, y) . \qquad (8.8)$$

8.2 Fuzzy Linear Programming

As mentioned in the beginning, we can hardly expect that some unique extension of the conventional linear programming to problems with fuzzy data can be established which would be so clean and clear like the theory of the conventional linear programming in finite-dimensional spaces. This can also be easily seen from the current literature where we can find a number of different extensions, the results of which resemble in various degrees some of the useful results established in the conventional linear programming.

When dealing with problems that arise from the canonical linear programming problem (8.1)–(8.3) by permitting the input data c_j, a_{ij}, and b_i in (8.1)–(8.3) to be fuzzy quantities, we distinguish the fuzzy quantities from real numbers by writing the tilde above the corresponding symbol. Thus, we write $\tilde{c}_j$, $\tilde{a}_{ij}$, and $\tilde{b}_i$ and consequently $\mu_{\tilde{c}_j} : \mathbb{R} \to [0, 1]$, $\mu_{\tilde{a}_{ij}} : \mathbb{R} \to [0, 1]$ and $\mu_{\tilde{b}_i} : \mathbb{R} \to [0, 1]$, respectively, for $i \in \mathcal{M} = \{1, 2, \ldots, m\}$ and $j \in \mathcal{N} = \{1, 2, \ldots, n\}$. When the tilde is omitted, it signifies that the corresponding data or values of variables are considered to be real numbers. Notice that if $\tilde{c}_j$ and $\tilde{a}_{ij}$ are fuzzy quantities, then, for every $(x_1, x_2, \ldots, x_n)$ from $\mathbb{R}^n$, the fuzzy subsets $\tilde{c}_1 x_1 \,\tilde{+}\, \cdots \,\tilde{+}\, \tilde{c}_n x_n$ and $\tilde{a}_{i1} x_1 \,\tilde{+}\, \cdots \,\tilde{+}\, \tilde{a}_{in} x_n$ of $\mathbb{R}$ defined by the extension principle are again fuzzy quantities. Also notice that it is possible to consider the conventional linear programming problems as special cases of such fuzzy problems because the real numbers can be identified with crisp fuzzy quantities.

As the canonical fuzzy counterpart of the canonical linear programming problem (8.1)–(8.3), we consider the problem

$$\text{maximize} \quad \tilde{c}_1 x_1 \,\tilde{+}\, \cdots \,\tilde{+}\, \tilde{c}_n x_n$$
$$\text{subject to} \quad (\tilde{a}_{i1} x_1 \,\tilde{+}\, \cdots \,\tilde{+}\, \tilde{a}_{in} x_n) \, \tilde{P}_i \, \tilde{b}_i \,, \quad i \in \mathcal{M} \,,$$
$$x_j \geq 0 \,, \quad j \in \mathcal{N}, \quad (8.9)$$

where, for each $i \in \mathcal{M}$, the fuzzy quantities $\tilde{a}_{i1} x_1 \,\tilde{+}\, \cdots \,\tilde{+}\, \tilde{a}_{in} x_n$ and $\tilde{b}_i$ from $\mathcal{F}_0(\mathbb{R})$ are compared by a fuzzy relation $\tilde{P}_i$ on $\mathcal{F}_0(\mathbb{R})$, and where the meanings of *subject to* and *maximize*, that is, the meanings of feasibility and optimality, remain to be specified.

Because the set of the conventional binary relations on X can be embedded into the set of fuzzy relation on X, we also obtained from (8.8) extensions of conventional binary relations on X to fuzzy relations on $\mathcal{F}(X)$.

Primarily, we shall study the case in which all $\tilde{P}_i$ appearing in the constraints of problem (8.9) are the same. Namely, let $\tilde{P}$ be a fuzzy relation on $\mathcal{F}_0(\mathbb{R})$ and let us assume that $\tilde{P}_i = \tilde{P}$ for all $i \in \mathcal{M}$. Then (8.9) simplifies to

$$\text{maximize} \quad \tilde{c}_1 x_1 \,\tilde{+}\, \cdots \,\tilde{+}\, \tilde{c}_n x_n$$
$$\text{subject to} \quad (\tilde{a}_{i1} x_1 \,\tilde{+}\, \cdots \,\tilde{+}\, \tilde{a}_{in} x_n) \, \tilde{P} \, \tilde{b}_i \,, \quad i \in \mathcal{M} \,,$$
$$x_j \geq 0 \,, \quad j \in \mathcal{N}, \quad (8.10)$$

where the meaning of *feasibility* and *optimality* are specified as follows.

- *Feasibility*: Let β be a positive number from $[0, 1]$. By a β-*feasible region* of problem (8.10) we understand the β-cut of the fuzzy subset $\tilde{X}$ of $\mathbb{R}^n$ whose membership function $\mu_{\tilde{X}}$ is given by

$$\mu_{\tilde{X}}(x) = \begin{cases} \displaystyle\min_{1 \leq i \leq m} \mu_{\tilde{P}}(\tilde{a}_{i1} x_1 \,\tilde{+}\, \cdots \,\tilde{+}\, \tilde{a}_{in} x_n, \tilde{b}_i) \\ \qquad\qquad \text{if } x_j \geq 0 \text{ for all } j \in \mathcal{N}, \\ 0 \qquad\qquad \text{otherwise} . \end{cases}$$
$$(8.11)$$

The elements of β-feasible region are called β-*feasible solutions* of problem (8.10), and $\tilde{X}$ defined by (8.11) is called the *feasible region* of problem (8.10). It is worth mentioning that when the data in (8.10) are crisp, then $\tilde{X}$ become the feasible region of the canonical linear programming problem (8.1)–(8.3).

- *Optimality*: When specifying the meaning of optimization, we have to take into account that the set of fuzzy values of the objective function is not linearly ordered, and that the relation for making comparison of elements of this set may be independent of that used in the notion of feasibility. We propose to use the notion of α-efficient (α-nondominated) solution of the fuzzy linear programming (FLP) problem. (Some other approaches can be found in the literature; for example, see [8.6].)

First, we observe that a feasible solution $\hat{x}$ of nonfuzzy problem (8.1)–(8.3) is optimal exactly when there is no other feasible solution x such that $cx > c\bar{x}$. This suggests the introduction of a suitable fuzzy extensions of $>$. Let $\tilde{Q}$ be a fuzzy relation on $\mathbb{R}$ and let $\alpha \in (0, 1]$. If $\tilde{a}$ and $\tilde{b}$ are fuzzy quantities, then we write

$$\tilde{a}\,\tilde{Q}_\alpha\,\tilde{b}\,, \text{ if } \mu_{\tilde{Q}}(\tilde{a}, \tilde{b}) \geq \alpha \tag{8.12}$$

and call $\tilde{Q}_\alpha$ the *α-relation on $\mathbb{R}$ associated to $\tilde{Q}$*. We also write

$$\tilde{a}\,\tilde{Q}^*_\alpha\,\tilde{b}\,, \text{ if } (\tilde{a}\,\tilde{Q}_\alpha\,\tilde{b} \text{ and } \mu_{\tilde{Q}}(\tilde{b}, \tilde{a}) < \alpha), \tag{8.13}$$

and call $\tilde{Q}^*_\alpha$ the *strict α-relation on $\mathbb{R}$ associated to $\tilde{Q}$*. Now let α and β be positive numbers from $[0, 1]$. We say that a β-feasible solution $\hat{x}$ of (8.10) is *(α, β)-maximal solution* of (8.10) if there is no β-feasible solution x of (8.10) different from $\hat{x}$ such that

$$\tilde{c}_1\hat{x}_1 \,\tilde{+}\, \tilde{c}_2\hat{x}_2 \,\tilde{+}\, \cdots \,\tilde{+}\, \tilde{c}_n\hat{x}_n\tilde{Q}^*_\alpha\tilde{c}_1 x_1$$
$$\tilde{+}\, \tilde{c}_2 x_2 \,\tilde{+}\, \cdots \,\tilde{+}\, \tilde{c}_n x_n\,. \tag{8.14}$$

Remark 8.2
Note that $\tilde{Q}_\alpha$ and $\tilde{Q}^*_\alpha$ are binary relations on the set of fuzzy quantities $\mathcal{F}_0(\mathbb{R})$ that are constructed from fuzzy relation $\tilde{Q}$ at the level $\alpha \in (0, 1]$, and that relation $\tilde{Q}^*_\alpha$ is the strict relation associated with the relation $\tilde{Q}_\alpha$. Also notice that if $\tilde{a}$ and $\tilde{b}$ are crisp fuzzy numbers corresponding to real numbers a and b, respectively, and $\tilde{Q}$ is a fuzzy extension of relation $\leq$, then $a\,\tilde{Q}_\alpha\,b$ holds if and only if $a \leq b$ does. Then, for $\alpha \in (0, 1)$, $a\,\tilde{Q}^*_\alpha\,b$ if and only if $a < b$.

Significance and usefulness of duality results for linear programming problems with fuzzy data depend crucially on the choice of fuzzy relations $\tilde{P}$ and $\tilde{Q}$ appearing in the definition of feasibility and optimality. In what follows, we use the natural extensions of binary relations $\leq$ and $\geq$ on $\mathbb{R}$ to fuzzy relations on $\mathcal{F}(\mathbb{R})$ that are based on the possibility and necessity relations Pos and Nec defined on $\mathcal{F}(\mathbb{R})$ by

$$\mu_{\text{Pos}}(\tilde{a}, \tilde{b})$$
$$= \sup\{\min(\mu_{\tilde{a}}(x), \mu_{\tilde{b}}(y), \mu_R(x, y))|x, y \in \mathbb{R}\}, \tag{8.15}$$

$$\mu_{\text{Nec}}(\tilde{a}, \tilde{b})$$
$$= \inf\{\max(1 - \mu_{\tilde{a}}(x), 1 - \mu_{\tilde{b}}(y),$$
$$\mu_R(x, y))|x, y \in \mathbb{R}\}\,. \tag{8.16}$$

Equivalently, we write $\tilde{a} \preceq^{\text{Pos}} \tilde{b}$ and $\tilde{a} \prec^{\text{Nec}} \tilde{b}$, instead of $\mu_{\text{Pos}}(\tilde{a}, \tilde{b})$ and $\mu_{\text{Nec}}(\tilde{a}, \tilde{b})$, respectively, and by $\tilde{a} \succeq^{\text{Pos}} \tilde{b}$ we mean $\tilde{b} \preceq^{\text{Pos}} \tilde{a}$.

The proofs of the following propositions can be found in [8.7].

Proposition 8.1
Let $\tilde{a}, \tilde{b} \in \mathcal{F}_0(\mathbb{R})$ be fuzzy quantities. Then, for each $\alpha \in (0, 1)$, we have

$$\mu_{\text{Pos}}(\tilde{a}, \tilde{b}) \geq \alpha \text{ iff } \inf[\tilde{a}]_\alpha \leq \sup[\tilde{b}]_\alpha\,, \tag{8.17}$$
$$\mu_{\text{Nec}}(\tilde{a}, \tilde{b}) \geq \alpha \text{ iff } \sup[\tilde{a}]_{1-\alpha} \leq \inf[\tilde{b}]_{1-\alpha}\,. \tag{8.18}$$

Let $\tilde{d} \in \mathcal{F}_0(\mathbb{R})$ be a fuzzy quantity, let $\beta \in [0, 1]$, and let $\tilde{d}^{\text{L}}(\beta)$ and $\tilde{d}^{\text{R}}(\beta)$ be defined by

$$\tilde{d}^{\text{L}}(\beta) = \inf\{t|t \in [\tilde{d}]_\beta\} = \inf[\tilde{d}]_\beta,$$
$$\tilde{d}^{\text{R}}(\beta) = \sup\{t|t \in [\tilde{d}]_\beta\} = \sup[\tilde{d}]_\beta. \tag{8.19}$$

Proposition 8.2
i) Let $\tilde{P} = \preceq^{\text{Pos}}$ and let $\beta \in [0, 1]$. A vector $x = (x_1, \ldots, x_n)$ is a β-feasible solution of the FLP problem (8.10) if and only if it is a nonnegative solution of the system of inequalities

$$\sum_{j \in \mathcal{N}} \tilde{a}^{\text{L}}_{ij}(\beta)x_j \leq \tilde{b}^{\text{R}}_i(\beta)\,, \quad i \in \mathcal{M}\,.$$

ii) Let $\tilde{P} = \prec^{\text{Nec}}$. A vector $x = (x_1, \ldots, x_n)$ is a β-feasible solution of the FLP problem (8.10) if and only if it is a nonnegative solution of the system of inequalities

$$\sum_{j \in \mathcal{N}} \tilde{a}^{\text{R}}_{ij}(1 - \beta)x_j \leq \tilde{b}^{\text{L}}_i(1 - \beta)\,, \quad i \in \mathcal{M}\,.$$

The following proposition is a simple consequence of the above results applied to the particular fuzzy relations $\tilde{P} = \preceq^{\text{Pos}}$ and $\tilde{P} = \prec^{\text{Nec}}$.

Proposition 8.3
Let $\tilde{a}$ and $\tilde{b}$ be fuzzy quantities, $\alpha \in (0, 1]$.
i) Let $\tilde{P} = \preceq^{\text{Pos}}$ be a fuzzy relation on $\mathbb{R}$ defined by (8.15). Then

$$\begin{aligned}\tilde{a}\,\tilde{P}_\alpha\,\tilde{b} &\text{ iff } \tilde{a}^{\text{L}}(\alpha) \leq \tilde{b}^{\text{R}}(\alpha),\\ \tilde{a}\,\tilde{P}^*_\alpha\,\tilde{b} &\text{ iff } \tilde{a}^{\text{R}}(\alpha) < \tilde{b}^{\text{L}}(\alpha).\end{aligned} \tag{8.20}$$

ii) Let $\tilde{P} = \prec^{\text{Nec}}$ be a fuzzy relation on $\mathbb{R}$ defined by (8.16). Then

$$\tilde{a}\,\tilde{P}_\alpha\,\tilde{b} \quad \text{iff } \tilde{a}^R(1-\alpha) \leq \tilde{b}^L(1-\alpha),$$
$$\tilde{a}\,\tilde{P}^*_\alpha\,\tilde{b} \quad \text{iff } \tilde{a}^R(1-\alpha) \leq \tilde{b}^L(1-\alpha) \text{ and}$$
$$\tilde{a}^L(1-\alpha) < \tilde{b}^R(1-\alpha).$$

$$(8.21)$$

As to the *optimal* solution of FLP problem, we obtain the following result, see also [8.7].

Proposition 8.4

Let $\alpha, \beta \in (0,1)$ and let $\tilde{X}$ be a feasible region of the FLP problem (8.10) with $\tilde{P} = \preceq^{\text{Pos}}$. Let c_j be such that

$\tilde{c}_j^L(\alpha) \leq c_j \leq \tilde{c}_j^R(\alpha)$ for all $j \in \mathcal{N}$. If $x^* = (x_1^*, \ldots, x_n^*)$ is an optimal solution of the LP problem

$$\text{maximize} \quad \sum_{j \in \mathcal{N}} c_j x_j$$
$$\text{subject to} \quad \sum_{j \in \mathcal{N}} \tilde{a}_{ij}^L(\beta) x_j \leq \tilde{b}_i^R(\beta), \quad i \in \mathcal{M},$$
$$x_j \geq 0, \quad j \in \mathcal{N},$$

$$(8.22)$$

then x^* is an (α, β)-maximal solution of the FLP problem (8.10).

8.3 Duality in Fuzzy Linear Programming

8.3.1 Early Approaches

Dual Pairs of Rödder and Zimmermann

One of the early approaches to duality in linear programming problems involving fuzziness is due to *Rödder* and *Zimmermann* [8.14]. To be able to state the problems considered by Rödder and Zimmermann concisely, we first observe that conditions (8.7) bring up the pair of optimization problems

$$\text{maximize} \min_{y \geq 0} L(x,y) \quad \text{subject to } x \in \mathbb{R}^n_+, \quad (8.23)$$

$$\text{minimize} \max_{x \geq 0} L(x,y) \quad \text{subject to } y \in \mathbb{R}^m_+. \quad (8.24)$$

Let μ and μ' be the real-valued functions on $\mathbb{R}^n_+$ and $\mathbb{R}^m_+$, respectively, and let $\{v_x\}_{x \in \mathbb{R}^n_+}$ and $\{v'_y\}_{y \in \mathbb{R}^m_+}$ be families of real-valued functions on $\mathbb{R}^m_+$ and $\mathbb{R}^n_+$, respectively. Furthermore, let φ_y and ψ_x be real-valued functions on $\mathbb{R}^n_+$ and $\mathbb{R}^m_+$ defined by

$$\varphi_y(x) = \min(\mu(x), v_x(y)), \quad (8.25)$$
$$\psi_x(y) = \min(\mu'(y), v'_y(x)). \quad (8.26)$$

Now let us consider the following pair of families of optimization problems

Family $\{P_y\}$: Given $y \in \mathbb{R}^m_+$,

$$\text{maximize} \quad \varphi_y(x) \quad \text{subject to } x \in \mathbb{R}^n_+.$$

Family $\{D_x\}$: Given $x \in \mathbb{R}^n_+$,

$$\text{maximize} \quad \psi_x(y) \quad \text{subject to } y \in \mathbb{R}^m_+.$$

Motivated and supported by economic interpretation, *Rödder* and *Zimmermann* [8.14] propose to specify

functions μ and μ' and families $\{v_x\}$ and $\{v'_y\}$ as follows: Given an $m \times n$ matrix A, $m \times 1$ vector b, $1 \times n$ vector c, and real numbers γ and δ, define the functions μ, μ', v_x and v'_y by

$$\mu(x) = \min(1, 1 - (\gamma - cx)),$$
$$\mu'(y) = \min(1, 1 - (yb - \delta)); \quad (8.27)$$
$$v_x(y) = \max(0, y(b - Ax)),$$
$$v'_y(x) = \max(0, (yA - c)x). \quad (8.28)$$

Strictly speaking, we do not obtain a duality scheme as conceived by Kuhn because there is no relationship between the numbers γ and δ. Indeed, if the family $\{P_y\}_{y \geq 0}$ is considered to be the primal problem, then we have the situation in which the primal problem is completely specified by data A, b, c, and γ. However, these data are not sufficient for specification of family $\{D_x\}_{x \geq 0}$ because the definition of $\{D_x\}_{x \geq 0}$ requires knowledge of δ. Thus, from the point of view that the dual problem is to be constructed only on the basis of the primal problem data, every choice of δ determines a certain family dual to $\{P_y\}_{y \geq 0}$. In this sense, we could say that every choice of δ gives a duality, the δ-*duality*. Analogously, if the primal problem is $\{D_x\}_{x \geq 0}$, then every choice of γ determines some family $\{P_y\}_{y \geq 0}$ dual to $\{D_x\}_{x \geq 0}$, and we obtain the γ-*duality*. In other words, for every γ, δ, we obtain (γ, δ)-duality. It is worth noticing that families $\{P_y\}$ and $\{D_x\}$ consist of uncountably many linear optimization problems. Moreover, every problem of each of these families may have uncountably many optimal solutions. Consequently, the solution of the problem given by family $\{P_y\}_{y \geq 0}$ is the family $\{X(y)\}_{y \geq 0}$ of subsets of $\mathbb{R}^n_+$ where $X(y)$ is the set of maximizers of φ_y over $\mathbb{R}^n_+$. Analogously, the family

$\{Y(x)\}_{x\geq 0}$ of maximizers of ψ_x over $\mathbb{R}^m_+$ is the solution of problem given by family $\{D_x\}_{x\geq 0}$. Rödder and Zimmermann propose to replace the families $\{P_y\}$ and $\{D_x\}$ by the families $\{P'_y\}$ and $\{D'_x\}$ of problems defined as follows

- Family $\{P'_u\}$: For every $u \geq 0$,

  maximize λ

  subject to $\lambda \leq 1 + cx - \gamma$

  $\lambda \leq u(b - Ax)$

  $x \geq 0$, (8.29)

- Family $\{D'_x\}$: For every $x \geq 0$,

  minimize η

  subject to $\eta \geq ub - \delta - 1$

  $\eta \geq (c - uA)x$

  $u \geq 0$. (8.30)

They call these families of optimization problems the *fuzzy dual pair* and claim that the families $\{P_y\}$ and $\{D_x\}$ become families $\{P'_y\}$ and $\{D'_x\}$ when μ, μ', ν_x and ν'_y are defined by (8.27) and (8.28). To see that this claim cannot be substantiated, it suffices to observe that the value of function φ_y cannot be greater than 1, whereas the value of λ is not bounded above whenever A and b are such that both cx and $-yAx$ are positive for some $x \in \mathbb{R}^n_+$.

To obtain a valid conversion, one needs to add the inequalities $\lambda \leq 1$ and $\eta \geq -1$ to the constraints. Thus, it seems that more suitable choice of functions ν_x and ν'_y in the Rödder and Zimmermann duality scheme would be

$$\nu_x(y) = \min(1, 1 + y(b - Ax)), \quad (8.31)$$

$$\nu'_y(x) = \min(1, 1 + (yA - c)x). \quad (8.32)$$

Another objection to the Rödder and Zimmermann model arises from the fact that the duality results for the proposed fuzzy dual pair do not reduce to the standard duality results for the crisp scenario, that is, for $\lambda = 1, \eta = -1$. Again an easy remedy is to work with ν_x and ν'_y defined by (8.31) and (8.32) instead of ν_x and ν'_y from (8.28). Similar approaches can be found in [8.15, 16].

Dual Pairs of Bector and Chandra

In contrast to the usual practice, in the Rödder and Zimmermann model, the range of membership functions μ

and μ' is $(-\infty, 1]$, and the range of membership functions ν_x and ν'_u is $[0, \infty)$ or $[1, \infty)$ instead of usual $[0, 1]$. *Bector* and *Chandra* [8.17] proposed to replace the relations $\leq$ and $\geq$ appearing in the dual pair of linear programming problems by suitable fuzzy relations on $\mathbb{R}$. In particular, the inequality $\leq$ appearing in the i-th constraint of the primal problem (8.1)–(8.3) is replaced by the fuzzy relation $\preceq_i$ whose membership function $\mu_{\preceq_i}: \mathbb{R} \times \mathbb{R} \to [0, 1]$ is defined by

$$\mu_{\preceq_i}(\alpha, \beta) = \begin{cases} 1 & \text{if } \alpha \leq \beta \\ 1 - \frac{\alpha - \beta}{p_i} & \text{if } \beta < \alpha \leq \beta + p_i , \\ 0 & \text{if } \beta + p_i < \alpha \end{cases}$$

where p_i is a positive number. Analogously, the inequality $\geq$ appearing in the j-th constraint of the dual problem (8.4)–(8.6) is replaced by the fuzzy relation $\succeq_j$ with the membership function

$$\mu_{\succeq_j}(\alpha, \beta) = \begin{cases} 1 & \text{if } \alpha \geq \beta \\ 1 - \frac{\beta - \alpha}{q_j} & \text{if } \beta > \alpha \geq \beta - q_j , \\ 0 & \text{if } \beta - q_j > \alpha \end{cases}$$

where q_j is a positive number. The degree of satisfaction with which $x \in \mathbb{R}^n$ fulfills the i-th fuzzy constraint $A_i x \preceq_i b_i$ of the primal problem is expressed by the fuzzy subset of $\mathbb{R}^n$ whose membership function μ_i is defined by $\mu_i(x) = \mu_{\preceq_i}(A_i x, b_i)$, and the degree of satisfaction with which $y \in \mathbb{R}^m$ fulfills the j-th fuzzy constraint $yA^j \succeq_j c_j$ of the dual problem is expressed by the fuzzy subset of $\mathbb{R}^m$ whose membership function μ_j is defined by $\mu_j(y) = \mu_{\succeq_j}(yA^j, c_j)$.

Similarly, we can express the degree of satisfaction with a prescribed aspiration level γ of the objective function value cx by the fuzzy subset of $\mathbb{R}^n$ given by $\mu_0(x) = \mu_{\succeq_0}(cx, \gamma)$ where, for the tolerance given by a positive number p_0, the membership function $\mu_{\succeq_0}$ is defined by

$$\mu_{\succeq_0}(\alpha, \beta) = \begin{cases} 1 & \text{if } \alpha \geq \beta \\ 1 - \frac{\beta - \alpha}{p_0} & \text{if } \beta > \alpha \geq \beta - p_0 \\ 0 & \text{if } \beta - p_0 > \alpha . \end{cases}$$

Analogously, for the degree of satisfaction with the aspiration level δ and tolerance q_0 in the dual problem, we have $\mu_0(y) = \mu_{\preceq_0}(\delta, yb)$ where

$$\mu_{\preceq_0}(\alpha, \beta) = \begin{cases} 1 & \text{if } \alpha \leq \beta \\ 1 - \frac{\alpha - \beta}{q_0} & \text{if } \beta < \alpha \leq \beta + q_0 \\ 0 & \text{if } \beta + q_0 < \alpha . \end{cases}$$

This leads to the following pair of linear programming problems.

Given positive numbers $p_0, p_1, \ldots, p_m$, and a real number γ, maximize λ subject to

$$(\lambda - 1)p_0 \leq cx - \gamma\,,$$
$$(\lambda - 1)p_i \leq b_i - A_i x\,, \quad 1 \leq i \leq m\,,$$
$$0 \leq \lambda \leq 1\,, \quad x \geq 0\,. \tag{8.33}$$

Given positive numbers $q_0, q_1, \ldots, q_n$, and a real number δ, minimize $-\eta$ subject to

$$(\eta - 1)q_0 \leq \delta - yb\,,$$
$$(\eta - 1)q_j \leq yA^j - c_j\,, \quad 1 \leq j \leq n\,,$$
$$0 \leq \eta \leq 1\,, \quad y \geq 0\,. \tag{8.34}$$

Bector and Chandra call this pair the *modified fuzzy pair of primal dual linear programming problems*, and they show that if x, λ, and u, η are feasible solutions of the corresponding problems, then

$$(\lambda - 1)\sum_{i=1}^m u_i p_i + (\eta - 1)\sum_{j=1}^n q_j x_j$$
$$\leq \sum_{i=1}^m u_i b_i - \sum_{j=1}^n c_j x_j\,,$$
$$(\lambda - 1)p_0 + (\eta - 1)q_0$$
$$\leq \sum_{j=1}^n c_j x_j - \sum_{i=1}^m u_i b_i + (\delta - \gamma)\,.$$

It follows that, for the crisp scenario $\lambda = 1$ and $\eta = 1$, we have

$$\sum_{j=1}^n c_j x_j \leq \sum_{i=1}^m u_i b_i \leq \sum_{j=1}^n c_j x_j + (\delta - \gamma)\,.$$

Moreover, for $\gamma \leq \delta$, feasible solutions $\bar{x}, \bar{\lambda}$ and $\bar{u}, \bar{\eta}$ are optimal if

- $(\bar{\lambda} - 1)\sum_{i=1}^m \bar{u}_i p_i + (\bar{\eta} - 1)\sum_{j=1}^n q_j \bar{x}_j$
  $= \sum_{i=1}^m \bar{u}_i b_i - \sum_{j=1}^n c_j \bar{x}_j$,
- $(\bar{\lambda} - 1)p_0 + (\bar{\eta} - 1)q_0$
  $= \sum_{j=1}^n c_j \bar{x}_j - \sum_{i=1}^m \bar{u}_i b_i + (\delta - \gamma)$.

Again we see that the dual problem is not stated by using only the data available in the primal problem. Indeed, if problem (8.33) is considered to be the primal problem, then to state its dual problem one needs

additional information; namely, a number δ and numbers $q_0, q_1, \ldots, q_n$; if problem (8.34) is considered to be primal, then one needs a number γ and numbers $p_0, p_1, \ldots, p_m$.

Dual Pairs of Verdegay
Verdegay's approach to duality in fuzzy linear problems presented in [8.18] is based on two natural ideas: (i) Solutions to problems involving fuzziness should be fuzzy; (ii) the dual problem to a problem with fuzziness only in constraints should involve fuzziness only in the objective.

The primal problem considered in [8.19] has the form

$$\text{maximize} \quad cx$$
$$\text{subject to} \quad A_i x \preceq_i b_i\,, \quad i = 1, 2, \ldots, m$$
$$x \geq 0\,, \tag{8.35}$$

where the valued relation $\preceq_i$ in the ith constraint is the same as in the previous section, that is,

$$\mu^i_{\preceq}(\alpha, \beta) = \begin{cases} 1 & \text{if} \quad \alpha \leq \beta \\ 1 - \frac{\alpha - \beta}{p_i} & \text{if} \quad \beta < \alpha \leq \beta + p_i \\ 0 & \text{if} \quad \beta + p_i < \alpha\,. \end{cases}$$

The fuzzy solution of problem (8.35) is given by the fuzzy subset of $\mathbb{R}^n$ whose each γ-cut, $0 < \gamma \leq 1$, is the solution set of the problem

$$\text{maximize} \quad cx$$
$$\text{subject to} \quad \mu^i_{\preceq}(A_i x, b_i) \geq \gamma\,, \quad i = 1, 2, \ldots, m$$
$$x \geq 0\,. \tag{8.36}$$

Consequently, we obtain the following problem of parametric linear programming.
For $0 < \gamma \leq 1$,

$$\text{maximize} \quad cx$$
$$\text{subject to} \quad A_i x \leq b_i + (1 - \gamma)p_i\,, \quad i = 1, 2, \ldots, m$$
$$x \geq 0\,. \tag{8.37}$$

Consider now the ordinary dual problem to (8.37), that is, for $0 < \gamma \leq 1$,

$$\text{minimize} \quad \sum_{i=1}^m u_i(b_i + (1 - \gamma)p_i)$$
$$\text{subject to} \quad uA \geq c$$
$$u \geq 0\,. \tag{8.38}$$

This suggests to introduce variables $y_1, y_2, \ldots, y_m$ by

$$y_i = b_i + \delta p_i, \quad i = 1, 2, \ldots, m$$

with $\delta = 1 - \gamma$, and consider the family of problems: Given $0 < \delta \le 1$ and $u \ge 0$ with $uA \ge c$,

$$\text{minimize } \sum_{i=1}^{m} u_i y_i$$

$$\text{subject to } y_i \ge b_i + \delta p_i, \quad i = 1, 2, \ldots, m. \quad (8.39)$$

Consequently, in terms of the membership functions $\mu_{\le}^i$, we obtain the family of problems: Given $0 < \delta \le 1$ and $u \ge 0$ with $uA \ge c$,

$$\text{minimize } \sum_{i=1}^{m} u_i y_i$$

$$\text{subject to } \mu_{\le}^i(y_i, b_i) \le 1 - \delta, \ i = 1, 2, \ldots, m. \quad (8.40)$$

8.3.2 More General Approach

In this section, we return to the general canonical FLP problem, that is, to the problem

$$\text{maximize } \tilde{c}_1 x_1 \tilde{+} \cdots \tilde{+} \tilde{c}_n x_n$$

$$\text{subject to } (\tilde{a}_{i1} x_1 \tilde{+} \cdots \tilde{+} \tilde{a}_{in} x_n) \, \tilde{P} \, \tilde{b}_i, \quad i \in \mathcal{M},$$

$$x_j \ge 0, \quad j \in \mathcal{N}. \quad (8.41)$$

We will call it the *primal FLP problem* and denote it by $\mathfrak{P}$. The feasible region of $\mathfrak{P}$ is introduced by (8.11) and the meaning of (α, β)-maximal solution is explained in (8.14) and (8.22).

To introduce the dual problem to problem $\mathfrak{P}$, we first define a suitable notion of duality for fuzzy relations on $\mathcal{F}(X)$. Let Φ and Ψ be mappings from $\mathcal{F}(X \times X)$ into $\mathcal{F}(\mathcal{F}(X) \times \mathcal{F}(X))$, and let Θ be a nonempty subset of $\mathcal{F}(X \times X)$. We say that mapping Φ is *dual* to mapping Ψ on Θ, if

$$\Phi(c(P)) = c(\Psi(P)) \quad (8.42)$$

for each $P \in \Theta$. Moreover, if P is in Θ and Φ is *dual* to Ψ on Θ, then we say that the fuzzy relation $\Phi(P)$ on $\mathcal{F}(X)$ is *dual* to fuzzy relation $\Psi(P)$.

The *dual FLP problem* (denoted by $\mathfrak{D}$) to problem $\mathfrak{P}$ is formulated as

$$\text{minimize } \tilde{b}_1 y_1 \tilde{+} \cdots \tilde{+} \tilde{b}_m y_m$$

$$\text{subject to } \tilde{c}_j \, \tilde{Q} \, (\tilde{a}_{1j} y_1 \tilde{+} \cdots \tilde{+} \tilde{a}_{mj} y_m), \quad j \in \mathcal{N},$$

$$y_i \ge 0, \quad i \in \mathcal{M}. \quad (8.43)$$

Here, $\tilde{P}$ and $\tilde{Q}$ are dual fuzzy relations to each other, particularly $\tilde{P} = \preceq^{\text{Pos}}$, $\tilde{Q} = \prec^{\text{Nec}}$, or, $\tilde{P} = \prec^{\text{Nec}}$, $\tilde{Q} = \preceq^{\text{Pos}}$. In problem $\mathfrak{P}$, *maximization* is considered with respect to fuzzy relation $\tilde{P}$, in problem $\mathfrak{D}$, *minimization* is considered with respect to fuzzy relation $\tilde{Q}$. The pair of FLP problems $\mathfrak{P}$ and $\mathfrak{D}$, that is, (8.41) and (8.43), is called the *primal-dual pair of FLP problems*. Now, we introduce a concept of feasible region of problem $\mathfrak{D}$, which is a modification of the feasible region of primal problem $\mathfrak{P}$, see also [8.20].

Let $\mu_{\tilde{a}_{ij}}$ and $\mu_{\tilde{c}_j}$, $i \in \mathcal{M}, j \in \mathcal{N}$, be the membership functions of fuzzy quantities $\tilde{a}_{ij}$ and $\tilde{c}_j$, respectively. Let $\tilde{P}$ be a fuzzy extension of a binary relation P on $\mathbb{R}$. A fuzzy set $\tilde{Y}$, whose membership function $\mu_{\tilde{Y}}$ is defined for all $y \in \mathbb{R}^m$ by

$$\mu_{\tilde{Y}}(y) = \begin{cases} \min\{\mu_{\tilde{P}}(\tilde{c}_1, \tilde{a}_{11} y_1 \tilde{+} \cdots \tilde{+} \tilde{a}_{m1} y_m), \\ \quad \ldots, \mu_{\tilde{P}}(\tilde{c}_n, \tilde{a}_{1n} y_1 \tilde{+} \cdots \tilde{+} \tilde{a}_{mn} y_m)\} \\ \quad \text{if } y_i \ge 0 \text{ for all } i \in \mathcal{M}, \\ 0 \quad \text{otherwise}, \end{cases}$$

$$(8.44)$$

is called a *fuzzy set of feasible region* or shortly *feasible region* of dual FLP problem (8.43). Moreover, if $\beta \in (0, 1]$, then the vectors belonging to $[\tilde{Y}]_\beta$ are called β-*feasible solutions* of problem (8.43).

By the parallel way, we define an *optimal solution* of the dual FLP problem $\mathfrak{D}$.

Let $\tilde{c}_j$, $\tilde{a}_{ij}$, and $\tilde{b}_i$, $i \in \mathcal{M}, j \in \mathcal{N}$, be fuzzy quantities on $\mathbb{R}$. Let $\tilde{Q}$ be a fuzzy relation on $\mathcal{F}(\mathbb{R})$ that is a fuzzy extension of the usual binary relation $\le$ on $\mathbb{R}$, and let $\alpha, \beta \in (0, 1]$. A β-feasible solution of (8.43) $y \in [\tilde{Y}]_\beta$ is called the (α, β)-*minimal solution of* (8.43) if there is no $y' \in [\tilde{Y}]_\beta$, $y' \ne y$, such that

$$\tilde{b}_1 y_1' \tilde{+} \tilde{b}_2 y_2' \tilde{+} \cdots \tilde{+} \tilde{b}_m y_m' \, \tilde{Q}_\alpha^* \, \tilde{b}_1 y_1 \tilde{+} \tilde{b}_2 y_2$$

$$\tilde{+} \cdots \tilde{+} \tilde{b}_m x_m. \quad (8.45)$$

where $\tilde{Q}_\alpha^*$ is the strict α-relation on $\mathbb{R}$ associated to $\tilde{Q}$.

Let P be the usual binary operation $\le$ on $\mathbb{R}$. Now, we shall investigate FLP problems (8.36) and (8.43) with pairs of dual fuzzy relations in the constraints, particularly $\tilde{P} = \preceq^{\text{Pos}}$, $\tilde{Q} = \prec^{\text{Nec}}$, or, $\tilde{P} = \prec^{\text{Nec}}$, $\tilde{Q} = \preceq^{\text{Pos}}$. The values of objective functions $\tilde{z}$ and $\tilde{w}$ are *maximized* and *minimized* with respect to fuzzy relation $\tilde{P}$ and $\tilde{Q}$, respectively.

The feasible region of the primal FLP problem $\mathfrak{P}$ is denoted by $\tilde{X}$, the feasible region of the dual FLP problem $\mathfrak{D}$ by $\tilde{Y}$. Clearly, $\tilde{X}$ is a fuzzy subset of $\mathbb{R}^n$, $\tilde{Y}$ is a fuzzy subset of $\mathbb{R}^m$.

Note that in the crisp case, that is, when the parameters $\tilde{c}_j$, $\tilde{a}_{ij}$, and $\tilde{b}_i$ are crisp fuzzy quantities, then by (8.15) and (8.16), the relations $\preceq^{\mathrm{Pos}}$ and $\prec^{\mathrm{Nec}}$ coincide with $\leq$. Hence, $\mathfrak{P}$ and $\mathfrak{D}$ forms a primal–dual pair of linear programming problems in the classical sense. The following proposition is a useful modification of Proposition 8.4 and gives a sufficient conditions for y^* to be an (α, β)-minimal solution of the FLP problem (8.43).

Proposition 8.5

Let $\tilde{c}_j$, $\tilde{a}_{ij}$ and $\tilde{b}_i$ be fuzzy quantities for all $i \in \mathcal{M}$ and $j \in \mathcal{N}$, $\alpha, \beta \in (0, 1)$. Let $\tilde{Y}$ be a feasible region of the FLP problem (8.43) with $\tilde{P} = \preceq^{\mathrm{Pos}}$. Let b_i be such that $\tilde{b}_i^{\mathrm{L}}(\alpha) \leq b_i \leq \tilde{b}_i^{\mathrm{R}}(\alpha)$ for all $i \in \mathcal{M}$. If $y^* = (y_1^*, \ldots, y_m^*)$ is an optimal solution of the LP problem

$$\text{minimize} \sum_{i \in \mathcal{M}} b_i y_i$$

$$\text{subject to} \sum_{i \in \mathcal{M}} \tilde{a}_{ij}^{\mathrm{R}}(\beta) y_i \geq \tilde{c}_j^{\mathrm{L}}(\beta), \quad j \in \mathcal{N},$$

$$y_i \geq 0, \quad i \in \mathcal{M}, \quad (8.46)$$

then y^* is a (α, β)-minimal solution of the FLP problem (8.43).

Dual Pairs of Ramík

When presenting duality theorems obtained by Ramík in [8.21] (see also [8.7]), we always present two versions: i) for fuzzy relation $\preceq^{\mathrm{Pos}}$ and ii) for fuzzy relation $\prec^{\mathrm{Nec}}$. In order to prove duality results we assume that the level of satisfaction α of the objective function is equal to the level of satisfaction β of the constraints. Otherwise, the duality theorems in our formulation do not hold. The proofs of the following theorems can be found in [8.7].

Theorem 8.1 First Weak Duality Theorem

Let $\tilde{c}_j$, $\tilde{a}_{ij}$ and $\tilde{b}_i$ be fuzzy quantities, $i \in \mathcal{M}$ and $j \in \mathcal{N}$, $\alpha \in (0, 1)$.

i) Let $\tilde{X}$ be a feasible region of the FLP problem (8.36) with $\tilde{P} = \preceq^{\mathrm{Pos}}$, and $\tilde{Y}$ be a feasible region of the FLP problem (8.43) with $\tilde{Q} = \prec^{\mathrm{Nec}}$. If a vector $x = (x_1, \ldots, x_n) \geq 0$ belongs to $[\tilde{X}]_\alpha$ and $y = (y_1, \ldots, y_m) \geq 0$ belongs to $[\tilde{Y}]_{1-\alpha}$, then

$$\sum_{j \in \mathcal{N}} \tilde{c}_j^{\mathrm{R}}(\alpha) x_j \leq \sum_{i \in \mathcal{M}} \tilde{b}_i^{\mathrm{R}}(\alpha) y_i. \quad (8.47)$$

ii) Let $\tilde{X}$ be a feasible region of the FLP problem (8.36) with $\tilde{P} = \prec^{\mathrm{Nec}}$, $\tilde{Y}$ be a feasible region of the FLP problem (8.43) with $\tilde{Q} = \preceq^{\mathrm{Pos}}$. If a vector $x = (x_1, \ldots, x_n) \geq 0$ belongs to $[\tilde{X}]_{1-\alpha}$ and $y = (y_1, \ldots, y_m) \geq 0$ belongs to $[\tilde{Y}]_\alpha$, then

$$\sum_{j \in \mathcal{N}} \tilde{c}_j^{\mathrm{L}}(\alpha) x_j \leq \sum_{i \in \mathcal{M}} \tilde{b}_i^{\mathrm{L}}(\alpha) y_i. \quad (8.48)$$

Theorem 8.2 Second Weak Duality Theorem

Let $\tilde{c}_j$, $\tilde{a}_{ij}$ and $\tilde{b}_i$ be fuzzy quantities for all $i \in \mathcal{M}$ and $j \in \mathcal{N}$, $\alpha \in (0, 1)$.

i) Let $\tilde{X}$ be a feasible region of the FLP problem (8.36) with $\tilde{P} = \preceq^{\mathrm{Pos}}$, $\tilde{Y}$ be a feasible region of the FLP problem (8.43) with $\tilde{Q} = \prec^{\mathrm{Nec}}$. If for some $x = (x_1, \ldots, x_n) \geq 0$ belonging to $[\tilde{X}]_\alpha$ and $y = (y_1, \ldots, y_m) \geq 0$ belonging to $[\tilde{Y}]_{1-\alpha}$ it holds

$$\sum_{j \in \mathcal{N}} \tilde{c}_j^{\mathrm{R}}(\alpha) x_j = \sum_{i \in \mathcal{M}} \tilde{b}_i^{\mathrm{R}}(\alpha) y_i, \quad (8.49)$$

then x is an (α, α)-maximal solution of the FLP problem (8.36) and y is an $(1 - \alpha, 1 - \alpha)$-minimal solution of the FLP problem $\mathfrak{D}$, (8.43).

ii) Let $\tilde{X}$ be a feasible region of the FLP problem (8.36) with $\tilde{P} = \prec^{\mathrm{Nec}}$, $\tilde{Y}$ be a feasible region of the FLP problem (8.43) with $\tilde{Q} = \preceq^{\mathrm{Pos}}$. If for some $x = (x_1, \ldots, x_n) \geq 0$ belonging to $[\tilde{X}]_{1-\alpha}$ and $y = (y_1, \ldots, y_m) \geq 0$ belonging to $[\tilde{Y}]_\alpha$ it holds

$$\sum_{j \in \mathcal{N}} \tilde{c}_j^{\mathrm{L}}(\alpha) x_j = \sum_{i \in \mathcal{M}} \tilde{b}_i^{\mathrm{L}}(\alpha) y_i, \quad (8.50)$$

then x is an $(1 - \alpha, 1 - \alpha)$-maximal solution of the FLP problem $\mathfrak{P}$, (8.36) and y is an (α, α)-minimal solution of the FLP problem $\mathfrak{D}$, (8.43).

Remark 8.3

In the crisp case, Theorems 8.1 and 8.2 are the standard linear programming weak duality theorems.

Remark 8.4

Let $\alpha \geq 0.5$. Then $[\tilde{X}]_\alpha \subset [\tilde{X}]_{1-\alpha}$, $[\tilde{Y}]_\alpha \subset [\tilde{Y}]_{1-\alpha}$, hence in the first weak duality theorem we can change the assumptions as follows: $x \in [\tilde{X}]_\alpha$ and $y \in [\tilde{Y}]_\alpha$.

However, the statements of the theorem remain unchanged. The same holds for the second weak duality theorem.

Finally, let us direct our attention to the *strong duality*. Motivated by the pairs of Propositions 8.4 and 8.5, in Theorem 8.2, we consider a pair of dual LP problems corresponding to FLP problems (8.36) and (8.43) with fuzzy relations $\tilde{P} = \preceq^{\mathrm{Pos}}, \tilde{Q} = \prec^{\mathrm{Nec}}, \alpha = \beta$

$$\text{maximize} \sum_{j \in \mathcal{N}} \tilde{c}_j^{\mathrm{R}}(\alpha) x_j$$

(P1) subject to $\sum_{j \in \mathcal{N}} \tilde{a}_{ij}^{\mathrm{L}}(\alpha) x_j \leq \tilde{b}_i^{\mathrm{R}}(\alpha)$, $i \in \mathcal{M}$,

$$x_j \geq 0, \quad j \in \mathcal{N}, \tag{8.51}$$

$$\text{minimize} \sum_{i \in \mathcal{M}} \tilde{b}_i^{\mathrm{R}}(\alpha) y_i$$

(D1) subject to $\sum_{i \in \mathcal{M}} \tilde{a}_{ij}^{\mathrm{L}}(\alpha) y_i \geq \tilde{c}_j^{\mathrm{R}}(\alpha)$, $j \in \mathcal{N}$,

$$y_i \geq 0, \quad i \in \mathcal{M}. \tag{8.52}$$

Moreover, we consider a pair of dual LP problems with fuzzy relations $\tilde{P} = \prec^{\mathrm{Nec}}, \tilde{P}^D = \preceq^{\mathrm{Pos}}$

$$\text{maximize} \sum_{j \in \mathcal{N}} \tilde{c}_j^{\mathrm{L}}(\alpha) x_j$$

(P2) subject to $\sum_{j \in \mathcal{N}} \tilde{a}_{ij}^{\mathrm{R}}(\alpha) x_j \leq \tilde{b}_i^{\mathrm{L}}(\alpha)$, $i \in \mathcal{M}$,

$$x_j \geq 0, \quad j \in \mathcal{N}, \tag{8.53}$$

$$\text{minimize} \sum_{i \in \mathcal{M}} \tilde{b}_i^{\mathrm{L}}(\alpha) y_i$$

(D2) subject to $\sum_{i \in \mathcal{M}} \tilde{a}_{ij}^{\mathrm{R}}(\alpha) y_i \geq \tilde{c}_j^{\mathrm{L}}(\alpha)$, $j \in \mathcal{N}$,

$$y_i \geq 0, \quad i \in \mathcal{M}. \tag{8.54}$$

Notice that (P1) and (D1) are classical dual linear programming problems and the same holds for (P2) and (D2).

Theorem 8.3 Strong Duality Theorem
Let $\tilde{c}_j$, $\tilde{a}_{ij}$, and $\tilde{b}_i$ be fuzzy quantities for all $i \in \mathcal{M}$ and $j \in \mathcal{N}$.

i) Let $\tilde{X}$ be a feasible region of the FLP problem (8.36) with $\tilde{P} = \preceq^{\mathrm{Pos}}$, $\tilde{Y}$ be a feasible region of the FLP problem (8.43) with $\tilde{Q} = \prec^{\mathrm{Nec}}$. If for some $\alpha \in (0,1)$, $[\tilde{X}]_\alpha$ and $[\tilde{Y}]_{1-\alpha}$ are nonempty, then there exists x^* – an (α, α)-maximal solution of the FLP problem $\mathfrak{P}$, and there exists y^* – an $(1-\alpha, 1-\alpha)$-minimal solution of the FLP problem $\mathfrak{D}$ such that

$$\sum_{j \in \mathcal{N}} \tilde{c}_j^{\mathrm{R}}(\alpha) x_j^* = \sum_{i \in \mathcal{M}} \tilde{b}_i^{\mathrm{R}}(\alpha) y_i^* . \tag{8.55}$$

ii) Let $\tilde{X}$ be a feasible region of the FLP problem (8.36) with $\tilde{P} = \prec^{\mathrm{Nec}}$, $\tilde{Y}$ be a feasible region of the FLP problem (8.43) with $\tilde{Q} = \preceq^{\mathrm{Pos}}$. If for some $\alpha \in (0,1)$, $[\tilde{X}]_{1-\alpha}$ and $[\tilde{Y}]_\alpha$ are nonempty, then there exists x^* – an $(1-\alpha, 1-\alpha)$-maximal solution of the FLP problem $\mathfrak{P}$, and y^* – an (α, α)-minimal solution of the FLP problem $\mathfrak{D}$ such that

$$\sum_{j \in \mathcal{N}} \tilde{c}_j^{\mathrm{L}}(\alpha) x_j^* = \sum_{i \in \mathcal{M}} \tilde{b}_i^{\mathrm{L}}(\alpha) y_i^* . \tag{8.56}$$

Remark 8.5
In the crisp case, Theorem 8.3 is the standard linear programming (strong) duality theorem.

Remark 8.6
Let $\alpha \geq 0.5$. Then $[\tilde{X}]_\alpha \subset [\tilde{X}]_{1-\alpha}$, $[\tilde{Y}]_\alpha \subset [\tilde{Y}]_{1-\alpha}$, hence in the strong duality theorem, we can assume $x \in [\tilde{X}]_\alpha$ and $y \in [\tilde{Y}]_\alpha$. Evidently, the statement of the theorem remains unchanged.

Remark 8.7
Theorem 8.3 provides only the existence of the (α, α)-maximal solution (or $(1-\alpha, 1-\alpha)$-maximal solution) of the FLP problem $\mathfrak{P}$, and $(1-\alpha, 1-\alpha)$-minimal solution ((α, α)-minimal solution) of the FLP problem $\mathfrak{D}$ such that (8.55) or (8.56) holds. However, the proof of the theorem gives also the method for finding the solutions by solving linear programming problems (P1) and (D1).

8.4 Conclusion

The leading idea of this chapter is based on the fact that, in many cases, the solutions of decision problems obtained through linear programming are numerically tractable approximations of the original nonlinear problems. Because of the practical relevance of linear programming and taking into account a vast literature on this subject, we extended linear programming theory to problems involving fuzzy data. To obtain a meaningful extension of linear programming to problems involving fuzzy data, we specified a suitable class of permitted fuzzy values called fuzzy quantities or fuzzy numbers, introduced fundamental arithmetic operations with such fuzzy numbers, defined inequalities between fuzzy numbers, and clarified the meaning of feasibility and optimality concepts. On the other hand, we did not deal with linear programming problems involving fuzzy variables. In the literature, it has been done in many different ways, here we focused on such variants of the theory for fuzzy linear programming the results of which resemble in various degrees some of the useful results established in the conventional linear programming. The final and main section of this work has been devoted to duality theory for fuzzy linear programming problems. We recalled some of the early approaches that are based on the ideas of Bellman and Zadeh, and Zimmermann, and then we presented our own recent results.

References

8.1   R.E. Bellman, L.E. Zadeh: Decision making in a fuzzy environment, Manag. Sci. **17**, B141–B164 (1970)

8.2   H.-J. Zimmermann: Fuzzy programming and linear programming with several objective functions, Fuzzy Sets Syst. **1**, 45–55 (1978)

8.3   D. Dubois: The role of fuzzy sets in decision sciences: Old techniques and new directions, Fuzzy Sets Syst. **184**, 3–28 (2011)

8.4   C. Stanciulescu, P. Fortemps, M. Install, V. Wertz: Multiobjective fuzzy linear programming problems with fuzzy decision variables, Eur. J. Oper. Res. **149**, 654–675 (2003)

8.5   D. Dubois, H. Prade: Ranking fuzzy numbers in the setting of possibility theory, Inf. Sci. **30**, 183–224 (1983)

8.6   J. Ramík: Duality in fuzzy linear programming: Some new concepts and results, Fuzzy Optim. Decis. Mak. **4**, 25–39 (2005)

8.7   J. Ramík: Duality in fuzzy linear programming with possibility and necessity relations, Fuzzy Sets Syst. **157**, 1283–1302 (2006)

8.8   A.L. Soyster: A duality theory for convex programming with set-inclusive constraints, Oper. Res. **22**, 892–898 (1974)

8.9   D.J. Thuente: Duality theory for generalized linear programs with computational methods, Oper. Res. **28**, 1005–1011 (1980)

8.10 H.W. Kuhn: Nonlinear programming – A historical view, SIAM-AMS **9**, 1–26 (1976)

8.11 J. Ramík, M. Vlach: *Generalized Concavity in Optimization and Decision Making* (Kluwer, Dordrecht 2001)

8.12 D.A. Ralescu: A generalization of the representation theorem, Fuzzy Sets Syst. **51**, 309–311 (1992)

8.13 J. Ramík, M. Vlach: A non-controversial definition of fuzzy sets, Lect. Notes Comput. Sci. **3135**, 201–207 (2004)

8.14 W. Rödder, H.-J. Zimmermann: *Duality in Fuzzy Linear Programming, Extremal Methods and System Analysis* (Springer, New York 1980) pp. 415–429

8.15 H. Rommelfanger, R. Slowinski: Fuzzy linear programming with single or multiple objective functions, Handb. Fuzzy Sets Ser. **1**, 179–213 (1998)

8.16 H.-C. Wu: Duality theory in fuzzy linear programming problems with fuzzy coefficients, Fuzzy Optim. Decis. Mak. **2**, 61–73 (2003)

8.17 C.R. Bector, C. Chandra: On duality in linear programming under fuzzy environment, Fuzzy Sets Syst. **125**, 317–325 (2002)

8.18 J.L. Verdegay: A dual approach to solve the fuzzy linear programming problem, Fuzzy Sets Syst. **14**, 131–141 (1984)

8.19 M. Inuiguchi, H. Ichihashi, Y. Kume: Some properties of extended fuzzy preference relations using modalities, Inf. Sci. **61**, 187–209 (1992)

8.20 M. Inuiguchi, J. Ramík, T. Tanino, M. Vlach: Satisficing solutions and duality in interval and fuzzy linear programming, Fuzzy Sets Syst. **135**, 151–177 (2003)

8.21 H. Hamacher, H. Lieberling, H.-J. Zimmermann: Sensitivity analysis in fuzzy linear programming, Fuzzy Sets Syst. **1**, 269–281 (1978)

9. Basic Solutions of Fuzzy Coalitional Games

Tomáš Kroupa, Milan Vlach

This chapter is concerned with basic concepts of solution for coalitional games with fuzzy coalitions in the case of finitely many players and transferable utility. The focus is on those solutions which preoccupy the main part of cooperative game theory (the core and the Shapley value). A detailed discussion or just the comprehensive overview of current trends in fuzzy games is beyond the reach of this chapter. Nevertheless, we mention current developments and briefly discuss other solution concepts.

The theory of cooperative games builds and analyses mathematical models of situations in which players can form coalitions and make binding agreements on how to share results achieved by these coalitions. One of the basic models of cooperative games is a *cooperative game in coalitional form* (briefly a *coalitional game* or a *game*). Following *Osborne* and *Rubinstein* [9.1] we assume that the data specifying a coalitional game are composed of:

- A nonempty set Ω (the set of players) and a nonempty set X (the set of consequences),
- A mapping V that assigns to every subset S of Ω a subset $V(S)$ of X, and
- A family $\{\succ_i\}_{i\in\Omega}$ of binary relations on X (players' preference relations).

The set Ω of all players is usually referred to as the *grand coalition*, subsets of Ω are called *coalitions*, and the mapping V is called the *characteristic function* (or *coalition function*) of the game.

This definition provides a rather general framework for analyzing many classes of coalitional games. The games of this type are usually called coalitional games *without side payments* or *without transferable payoff* (or *utility*). Obviously, for many purposes, this framework is too general because it neither specifies some useful structure of the set of consequences nor properties of preference relations. At the same time, this framework is also too restrictive because of requiring that the domain of the characteristic function must be the system of all subsets of the player set.

In this chapter, we are mainly concerned with coalitional games in which the number of players is finite. The number of players will be denoted by n and, without loss of generality, the players will be named by integers $1, 2, \ldots, n$. In other words, we set $\Omega = N$ where $N = \{1, 2, \ldots, n\}$. Moreover, we assume that the sets $V(S)$ of consequences are subsets of the n-dimensional real linear space $\mathbb{R}^n$, and that each player i prefers $(x_1, \ldots, x_n)$ to $(y_1, \ldots, y_n)$ if and only if $x_i > y_i$. Furthermore, we significantly restrict the generality by considering only the so-called coalitional games with transferable payoff or utility. This class of games is a subclass of games without transferable utility that is characterized by the property: for each coalition S, there exists a real number $v(S)$ such that

$$V(S) = \left\{ x \in \mathbb{R}^n : \sum_{i\in S} x_i \le v(S) \text{ and } x_j = 0 \text{ if } j \notin S \right\}.$$

Evidently, each such game can be identified with the corresponding real-valued function v defined on the system of all subsets of N.

In coalitional games, whether with transferable or nontransferable utility, each player has only two alternatives of participation in a nonempty coalition: full participation or no participation. This assumption is too restrictive in many situations, and there has been a need for models that give players the possibility of participation in some or all intermediate levels between these two extreme involvements.

The first mathematical models in the form of coalitional games in which the players are permitted to participate in a coalition not only fully or not at all but also partially were proposed by *Butnariu* [9.2] and *Aubin* [9.3]. Aubin notices that the idea of partial participation in a coalition was used already in the *Shapley–Shubik* paper on market games [9.4]. In these models, the subsets of N no longer represent every possible coalition. Instead, a notion of a coalition has to be introduced that makes it possible to represent the partial membership degrees.

It has become customary to assume that a membership degree of player $i \in N$ is determined by a number a_i in the unit interval $I = [0, 1]$, and to call the resulting vector $a = (a_1, \ldots, a_n) \in I^n$ a *fuzzy coalition*. The

n-dimensional cube I^n is thus identified with the set of all fuzzy coalitions. Every subset S of N, that is, every coalition S, can be viewed as an n-vector from $\{0, 1\}^n$ whose ith components is 1 when $i \in S$ and 0 when $i \notin S$. These special fuzzy coalitions are often called *crisp* coalitions. Hence, we may think of the set of all fuzzy coalitions I^n as the convex closure of the set $\{0, 1\}^n$ of all crisp coalitions. This leads to the notion of an n-player coalitional game with fuzzy coalitions and transferable utility (briefly a fuzzy game) as a bounded function $v: I^n \to \mathbb{R}$ satisfying $v(0) = 0$.

It turns out that most classes of coalitional games with transferable utility and most solution concepts have natural counterparts in the theory of fuzzy games with transferable utility. Therefore, in what follows, we start with the classical case (Sect. 9.1) and then deal with the fuzzy case (Sect. 9.2). Taking into account that, in comparison with the classical case, the theory of fuzzy games is relatively less developed, we focus attention on two well-established solution concepts of fuzzy games: the core and the Shapley value.

9.1 Coalitional Games with Transferable Utility

We know from the beginning of this chapter that from the mathematical point of view, every n-player coalitional game with transferable utility can be identified with a real-valued function v defined on the system of all subsets of the set $N = \{1, 2, \ldots, n\}$. For convenience, we assume that always $v(\emptyset) = 0$.

It is customary to interpret the value $v(S)$ of the characteristic function v at coalition S as the worth of coalition S or the total payoff that coalition S will be able to distribute among its members, provided exactly the coalition S forms. However, equally well, the number $v(S)$ may represent the total cost of reaching some common goal of coalition S that must be shared by the members of S; or some other quantity, depending on the application field. In conformity with the players preferences stated previously, we usually assume that $v(S)$ represents the total payoff that S can distribute among its members.

Since the preferences are fixed, we denote the game given through N and v by (N, v), or simply v, and the collection of all games with fixed N by G_N. The sum $v + w$ of games from G_N defined by $(v + w)(S) = v(S) + w(S)$ for each coalition S is again a game from G_N. Moreover, if multiplication of $v \in G_N$ by a real number α is defined by $(\alpha v)(S) = \alpha v(S)$ for each coali-

tion S, then αv also belongs to G_N. An important and well-known fact is that G_N endowed with these two algebraic operations is a real linear space.

Example 9.1 Simple games
If the range of a game v is the two-element set $\{0, 1\}$ only, then the game can be viewed as a model of a voting system where each coalition $A \subseteq N$ is either *winning* ($v(A) = 1$) or *loosing* ($v(A) = 0$). Then it is natural to assume that the game also satisfies *monotonicity*; that is, if coalition A is winning and B is a coalition with $A \subseteq B$, then B is also winning. It is also natural to consider only games with at least one winning coalition. Thus, we define a *simple game* [9.5, Section 2.2.3] to be a $\{0, 1\}$-valued coalitional game v such that the grand coalition is winning and $v(A) \leq v(B)$, whenever $A \subseteq B$ for each $A, B \subseteq N$.

We say that a game v is *superadditive* if $v(A \cup B) \geq v(A) + v(B)$, for every disjoint pair of coalitions $A, B \subseteq N$. Consequently, in a superadditive game, it may be advantageous for members of disjoint coalitions A and B to form coalition $A \cup B$ because every pair of disjoint coalitions can obtain jointly at least as much as they could have obtained separately. Consequently, it is advantageous to form the largest possible coalitions, that is, the grand coalition.

The strengthening of the property of superadditivity is the assumption of nondecreasing marginal contribution of a player to each coalition with respect to coalition inclusion: a game v is said to be *convex* whenever

$$v(A \cup \{i\}) - v(A) \leq v(B \cup \{i\}) - v(B)$$

for each $i \in N$ and every $A \subseteq B \subseteq N \setminus \{i\}$. It can be directly checked that convexity of v is equivalent to

$$v(A \cup B) + v(A \cap B) \geq v(A) + v(B)$$
$$\text{for every } A, B \subseteq N .$$

Example 9.2

Let B be a nonempty coalition in a simple game (N, v). Then the game v_B given by

$$v_B(A) = \begin{cases} 1, & A \supseteq B, \\ 0, & \text{otherwise} , \end{cases} \quad A \subseteq N ,$$

is a convex simple game.

Example 9.3 Bankruptcy game [9.6]

Let $e > 0$ be the total value of assets held in a bankruptcy estate of a debtor and let N be the set of all creditors. Furthermore, let $d_i > 0$ be the debt to creditor $i \in N$. Assume that $e \leq \sum_{i \in N} d_i$. The *bankruptcy game* is then the game such that, for every $A \subseteq N$,

$$v(A) = \max \left(0, e - \sum_{i \in N \setminus A} d_i \right) .$$

It can be shown that the bankruptcy game is convex.

There is a variety of solution concepts for coalitional games with n players. Some, like the core, stable set or bargaining set, may consist of sets of real n-vectors, while others offer as a solution of a game a single real n-vector.

9.1.1 The Core

Let v be an n-player coalitional game with transferable utility. The *core* of v is the set of all efficient payoff vectors $x \in \mathbb{R}^n$ upon which no coalition can improve,

that is,

$$C(v) = \left\{ x \in \mathbb{R}^n \,\middle|\, \sum_{i \in N} x_i = v(N) \right.$$
$$\left. \text{and } \sum_{i \in A} x_i \geq v(A) \text{ for each } A \subseteq N \right\} . \quad (9.1)$$

The Bondareva–Shapley theorem [9.5, Theorem 3.1.4] gives a necessary and sufficient condition for the core nonemptiness in terms of the so-called balanced systems. It is easy to see that the core of every game is a (possibly empty) convex polytope. Moreover, the core of a convex game is always nonempty and its vertices can be explicitly characterized [9.7].

9.1.2 The Shapley Value

Let $f = (f_1, f_2, \ldots, f_n)$ be a mapping that assigns to every game v from some collection of games from G_N a real n-vector $f(v) = (f_1(v), f_2(v), \ldots, f_n(v))$. Following the basic interpretation of values of a characteristic functions as the total payoff, we can interpret the values of components of such a function as payoffs to individual players in game v.

Let $\mathcal{A}$ be a nonempty collection of games from G_N. A *solution function on* $\mathcal{A}$ is a mapping f from $\mathcal{A}$ into the n-dimensional real linear space $\mathbb{R}^n$. If the domain $\mathcal{A}$ of f is not explicitly specified, then it is assumed to be G_N. The collection of such mappings is too broad to contain only the mappings that lead to sensible solution concepts. Hence, to obtain reasonable solution concepts we have to require that the solution functions have some reasonable properties. One of the natural properties in many contexts is the following property of efficiency.

Property 9.1 Efficiency

A solution function f on a subset $\mathcal{A}$ of G_N is *efficient on* $\mathcal{A}$ if $f_1(v) + f_2(v) + \cdots + f_n(v) = v(N)$ for every game v from $\mathcal{A}$.

This property can be interpreted as a combination of the requirements of the *feasibility* defined by $f_1(v) + f_2(v) + \cdots + f_n(v) \leq v(N)$ and *collective rationality* defined by $f_1(v) + f_2(v) + \cdots + f_n(v) \geq v(N)$.

In addition to satisfying the efficiency condition, solution functions are required to satisfy a number of other desirable properties. To introduce some of them, we need further definitions.

Player i from N is a *null* player in game v if $v(S \cup \{i\}) = v(S)$ for every coalition S that does not contain player i; that is, participation of a null player in a coalition does not contribute anything to the coalition in question.

Player i from N is a *dummy* player in game v if $v(S \cup \{i\}) = v(S) + v(\{i\})$ for every coalition S that does not contain player i; that is, a dummy player contributes to every coalition the same amount, his or her value of the characteristic function.

Players i and j from N are *interchangeable* in game v if $v(S \cup \{i\}) = v(S \cup \{j\})$ for every coalition S that contains neither player i nor player j. In other words, two players are interchangeable if they can replace each other in every coalition that contain one of them.

Property 9.2 Null player
A solution function f satisfies the *null player property* if $f_i(v) = 0$ whenever $v \in G_N$ and i is a null player in v.

Property 9.3 Dummy player
A solution function f satisfies the *dummy player property* if $f_i(v) = v(\{i\})$ whenever $v \in G_N$ and i is a dummy player in v.

Property 9.4 Equal treatment
A solution function f satisfies the *equal treatment property* if $f_i(v) = f_j(v)$ for every $v \in G_N$ and every pair of players i, j that are interchangeable in v.

These three properties are quite reasonable and attractive, especially from the point of fairness and impartiality: a player who contributes nothing should get nothing; a player who contributes the same amount to every coalition cannot expect to get anything else than he or she contributed; and two players who contribute the same to each coalition should be treated equally by the solution function.

The next property reflects the natural requirement that the solution function should be independent of the players' names. Let v be a game from G_N and $\pi : N \to N$ be a permutation of N, and let the image of coalition S under π be denoted by $\pi(S)$. It is obvious that, for every $v \in G_N$, the function πv defined on G_N by $(\pi v)(S) = v(\pi(S))$ is again a game from G_N. Apparently, the game πv differs from game v only in players' names; they are interchanged by the permutation π.

Property 9.5 Anonymity
A solution function f is said to be *anonymous* if, for every permutation π of N, we have $f_i(\pi v) = f_{\pi(i)}(v)$ for every game $v \in G_N$ and every player $i \in N$.

When a game consists of two independent games played separately by the same players or if a game is split into a sum of games, then it is natural to require the following property of additivity.

Property 9.6 Additivity
A solution function f on G_N is said to be *additive* if $f(u + v) = f(u) + f(v)$ for every pair of games u and v from G_N.

The requirement of additivity differs from the previous conditions in one important aspect. It involves two different games that may or may not be mutually dependent. In contrast, the dummy player and equal treatment properties involve only one game, and the anonymity property involves only those games which are completely determined by a single game.

Remark 9.1
The terminology introduced in the literature for various properties of players and solution functions is not completely standardized. For example, some authors use the term *dummy player* and *symmetric players* (or *substitutes*), for what we call null player and interchangeable players, respectively. Moreover, the term *symmetry* is sometimes used for our equal treatment and sometimes for our anonymity.

One of the most studied and most influential single-valued solution concept for coalitional games with transferable utility is the Shapley solution function or briefly the Shapley value, proposed by *Shapley* in 1953 [9.8]. The simplest way of introducing the Shapley value is to define it explicitly by the following well-known formula for calculation of its components.

Definition 9.1
The Shapley value on a subset $\mathcal{A}$ of G_N is a solution function φ on $\mathcal{A}$ whose components $\varphi_1(v), \varphi_2(v), \ldots, \varphi_n(v)$ at game $v \in \mathcal{A}$ are defined by

$$\varphi_i(v) = \sum_{S : i \in S} \frac{(s-1)!(n-s)!}{n!} [v(S) - v(S \setminus \{i\})] ,$$

(9.2)

where the sum is meant over all coalitions S containing player i, and s generically stands for the number of players in coalition S.

To clarify the basic idea behind this definition, we first recall the notion of players' marginal contributions to coalitions.

Definition 9.2

For each player i and each coalition S, a *marginal contribution* of player i to coalition S in game v from G_N is the number $m_i^v(S)$ defined by

$$m_i^v(S) = \begin{cases} v(S) - v(S \setminus \{i\}) & \text{if} \quad i \in S \\ v(S \cup \{i\}) - v(S) & \text{if} \quad i \notin S \end{cases}.$$

Now imagine a procedure for dividing the total payoff $v(N)$ among the members of N in which the players enter a room in some prescribed order and each player receives his or her marginal contribution as payoff to the coalition of players already being in the room. Suppose that the prescribed order is $(\pi(1), \pi(2), \ldots, \pi(n))$ where $\pi: N \to N$ is a fixed permutation of N. Then the procedure under consideration determines the payoffs to individual players as follows: before the first player $\pi(1)$ entered the room, there was the empty coalition waiting in the room. After player $\pi(1)$ enters, the coalition in the room becomes $\{\pi(1)\}$ and the player receives $v(\{\pi(1)\}) - v(\emptyset)$. Similarly, before the second player $\pi(2)$ entered the room, there was coalition $\{\pi(1)\}$ waiting in the room. After player $\pi(2)$ enters, coalition $\{\pi(1), \pi(2)\}$ is formed in the room and player $\pi(2)$ receives $v(\{\pi(1), \pi(2)\}) - v(\{\pi(1)\})$. This continues till the last player $\pi(n)$ enters and receives $v(N) - v(N \setminus \{\pi(n)\})$.

Let S_i^π denote the coalition of players preceding player i in the order given by $(\pi(1), \pi(2), \ldots, \pi(n))$; that is, $S_i^\pi = \{\pi(1), \pi(2), \ldots, \pi(j-1)\}$ where j is the uniquely determined member of N such that $i = \pi(j)$. Because there are $n!$ possible orders, the arithmetical average of the marginal contributions of player i taken over all possible orderings is equal to the number $(1/n!) \sum m_i^v(S_i^\pi)$ where the sum is understood over all permutations π of N. This number is exactly the i-th component of the Shapley value. Therefore, in addition to the equality (9.2) we also have the equality

$$\varphi_i(v) = \frac{1}{n!} \sum_\pi m_i^v\left(S_i^\pi\right) \tag{9.3}$$

for computing the components of the Shapley value.

In addition to satisfying the condition of efficiency, the Shapley value has a number of other useful properties. In particular, it satisfies all properties 9.2–9.6. Remarkably, no other solution function on G_N satisfies the properties of null player, equal treatment, and additivity at the same time.

Theorem 9.1 Shapley

For each N, there exists a unique solution function on G_N satisfying the properties of efficiency, null player, equal treatment, and additivity; this solution function is the Shapley value introduced by Definition 9.1.

The standard proof of this basic result follows from the following facts:

- The collection $\{u_T : T \neq \emptyset, T \subseteq N\}$ of *unanimity games* defined by

$$u_T(S) = \begin{cases} 1 & \text{if } T \subseteq S \\ 0 & \text{otherwise,} \end{cases} \tag{9.4}$$

  form a base of the linear space G_N.
- The null player and equal treatment properties guarantee that φ is determined uniquely on multiples of unanimity games.
- The property of additivity (combined with the fact that the unanimity games form a basis) makes it possible to extend φ in a unique way to the whole space G_N.

In addition to the original axiomatization by Shapley, there exist several equally beautiful alternative axiomatizations of the Shapley value that do not use the property of additivity [9.9, 10].

9.1.3 Probabilistic Values

Let us fix some player i and, for every coalition S that does not contain player i, denote by $\alpha_i(S)$ the number $s!(n-s-1)!/n!$. The family $\{\alpha_i(S) : S \subseteq N \setminus \{i\}\}$ is a probability distribution over the set of coalitions not containing player i. Because the i-th component of the Shapley value can be computed by

$$\varphi_i(v) = \sum_{S \subseteq N \setminus \{i\}} \alpha_i(S)[v(S \cup \{i\}) - v(S)] \,,$$

we see that the i-th component of the Shapley value is the expected marginal contribution of player i with respect to the probability measure $\{\alpha_i(S) : S \subseteq N \setminus \{i\}\}$ and that the Shapley value belongs to the following class of solution functions:

Definition 9.3

A solution function f on a subset $\mathcal{A}$ of G_N is called *probabilistic on* $\mathcal{A}$ if, for each player i, there exists a probability distribution $\{p_i(S) : S \subseteq N \setminus \{i\}\}$ on the collection of coalitions not containing i such that

$$f_i(v) = \sum_{S \subseteq N \setminus \{i\}} p_i(S)[v(S \cup \{i\}) - v(S)] \qquad (9.5)$$

for every $v \in \mathcal{A}$.

The family of probabilistic solution functions embraces an enormous number of functions [9.11]. The efficient probabilistic solution functions are often called *quasivalues*, and the anonymous probabilistic solution functions are called *semivalues*. Since the Shapley value is anonymous and efficient on G_N, we know that it is both a quasivalue and a semivalue on G_N. Moreover, the Shapley value is the only probabilistic solution function with these properties.

Theorem 9.2 Weber
If N has at least three elements, then the Shapley value is the unique probabilistic solution function on G_N, that is, anonymous and efficient.

Another widely known probabilistic solution function is the function proposed originally only for voting games by *Banzhaf* [9.12].

Definition 9.4
The *Banzhaf value* on G_N is a solution function ψ on G_N whose components at game v are defined by

$$\psi_i(v) = \sum_{S \subseteq N \setminus \{i\}} \frac{1}{2^{n-1}} [v(S \cup \{i\}) - v(S)]. \qquad (9.6)$$

Again, by simple computation, we can verify that the Banzhaf solution is a probabilistic solution function. Consequently, the i-th component of the Banzhaf solution is the expected marginal contribution of player i with respect to the probability measure $\{\beta_i(S) : S \subseteq N \setminus \{i\}\}$, where $\beta_i(S) = 1/2^{n-1}$ for each subset S of $N \setminus \{i\}$. From the probabilistic point of view, the Banzhaf solution is based on the assumption that each player i is equally likely to join any subcoalition of $N \setminus \{i\}$. On the other hand, the Shapley value is based on the assumption that the coalition the player i enters is equally likely to be of any size s, $0 \le s \le n-1$, and that all coalitions of this size are equally likely.

9.2 Coalitional Games with Fuzzy Coalitions

Since the publication of *Aubin*'s seminal paper [9.3], cooperative scenarios allowing for players' fractional membership degrees in coalitions have been studied. In such situations, the subsets of N no longer model every possible coalition. Instead, a notion of coalitions has to be introduced that makes it possible to represent the partial membership degrees. It has become customary to assume that a membership degree of player $i \in N$ is determined by a number a_i in the unit interval $I = [0, 1]$, and to call the resulting vector $a = (a_1, \ldots, a_n) \in I^n$ a *fuzzy coalition*. (The choice of I^n is not the only possible choice, see [9.13] or the discussion in [9.14].) The n-dimensional cube I^n is thus identified with the set of all fuzzy coalitions. Every subset A of N, that is, every classical coalition, can be viewed as a vector $\mathbb{1}_A \in \{0, 1\}^n$ with coordinates

$$(\mathbb{1}_A)_i = \begin{cases} 1 & \text{if } i \in A, \\ 0 & \text{otherwise}. \end{cases}$$

These special fuzzy coalitions are also called *crisp* coalitions. When $A = \{i\}$ is a singleton, we write simply $\mathbb{1}_i$ in place of $\mathbb{1}_{\{i\}}$. Hence, we may think

of the set of all fuzzy coalitions I^n as the convex closure of the set $\{0, 1\}^n$ of all crisp coalitions; see [9.14] for further explanation of this convexification process.

Several definitions of fuzzy games appear in the literature [9.3, 15]. We adopt the one used by *Azrieli* and *Lehrer* [9.13]. However, note that the authors of [9.13] use a slightly more general definition, since they consider a fuzzy coalition a to be any nonnegative real vector such that $a \le q$, where $q \in \mathbb{R}^n$ is a given nonnegative vector.

Definition 9.5
An n-player *game (with fuzzy coalitions and transferable utility)* is a bounded function $v : I^n \to \mathbb{R}$ satisfying $v(\mathbb{1}_\emptyset) = 0$.

If we want to emphasize the dependence of Definition 9.5 on the number n of players, then we write (I^n, v) in place of v. Further, by $\bar{v}$ we denote the restriction of v to all crisp coalitions

$$\bar{v}(A) = v(\mathbb{1}_A), \quad A \subseteq N. \qquad (9.7)$$

Hence, every game with fuzzy coalitions v induces a classical coalition game $\bar{v}$ with transferable utility.

Most solution concepts of the cooperative game theory have been generalized to games with fuzzy coalitions. A *payoff vector* is any vector x with n real coordinates, $x = (x_1, \ldots, x_n) \in \mathbb{R}^n$. In a particular game with fuzzy coalitions (I^n, v), each player $i \in N$ obtains the amount of utility x_i as a result of his cooperative activity. Consequently, a fuzzy coalition $a \in I^n$ gains the amount

$$\langle a, x \rangle = \sum_{i=1}^{n} a_i x_i \, ,$$

which is just the weighted average of the players' payoffs x with respect to their participation levels in the fuzzy coalition a. By a *feasible payoff* in game (I^n, v), we understand a payoff vector x with $\langle \mathbb{1}_N, x \rangle \leq v(\mathbb{1}_N)$.

The following general definition captures most solution concepts for games with fuzzy coalitions.

Definition 9.6

Let Γ_N be a class of all games with fuzzy coalitions (I^n, v) and let Λ_N be its nonempty subclass. A *solution* on Λ_N is a function σ that associates with each game (I^n, v) in Λ_N a subset $\sigma(I^n, v)$ of the set

$$\{x \in \mathbb{R}^n | \langle \mathbb{1}_N, x \rangle \leq v(\mathbb{1}_N)\}$$

of all feasible payoffs in game (I^n, v).

The choice of σ is governed by all thinkable rules of economic rationality. Every solution σ is thus determined by a system of restrictions on the set of all feasible payoff vectors in the game. For example, we may formulate a set of axioms for σ to satisfy or single out inequalities making the payoffs in $\sigma(I^n, v)$ stable, in some sense.

9.2.1 Multivalued Solutions

Core

The core is a solution concept σ defined on the whole class of games with fuzzy coalitions Γ_N. We present the definition that appeared in [9.3].

Definition 9.7

Let $N = \{1, \ldots, n\}$ be the set of all players and $v \in \Gamma_N$. The *core* of v is a set

$$C(v) = \{x \in \mathbb{R}^n | \langle \mathbb{1}_N, x \rangle = v(\mathbb{1}_N), \ \langle a, x \rangle \geq v(a) \, ,$$
$$\text{for every } a \in I^n\} \, .$$

$$(9.8)$$

In words, the core of v is the set of all payoff vectors x such that no coalition $a \in I^n$ is better off when accepting any other payoff vector $y \notin C(v)$. This is a consequence of the two conditions in (9.8): *Pareto efficiency* $\langle \mathbb{1}_N, x \rangle = v(\mathbb{1}_N)$ requires that the profit of the grand coalition is distributed among all the players in N and *coalitional rationality* $\langle a, x \rangle \geq v(a)$ means that no coalition $a \in I^n$ accepts less than is its profit $v(a)$.

Observe that the core $C(v)$ of a game with fuzzy coalitions v is an intersection of uncountably many halfspaces $\langle a, x \rangle \geq v(a)$ with the affine hyperplane $\langle \mathbb{1}_N, x \rangle = v(\mathbb{1}_N)$. This implies that the core is a possibly empty compact convex subset of $\mathbb{R}^n$, since $C(v)$ is included in the core (9.1) of a classical coalition game $\bar{v}$ given by (9.7). In this way, we may think of the Aubin's core $C(v)$ as a refinement of the classical core (9.1).

A payoff x in the core $C(v)$ must meet uncountably many restrictions represented by all coalitions I^n. This raises several questions:

1. When is $C(v)$ nonempty/empty?
2. When is $C(v)$ reducible to the intersection of finitely many sets only?
3. For every $a \in I^n$, is there a core element $x \in C(v)$ giving coalition a exactly its worth $v(a)$?
4. Is there an allocation rule for assigning payoffs in $C(v)$ to fuzzy coalitions?

Azrieli and *Lehrer* formulated a necessary and a sufficient condition for the core nonemptiness [9.13], thus generalizing the well-known Bondareva–Shapley theorem for classical coalition games. We will need an additional notion in order to state their result. The *strong superadditive cover* of a game $v \in \Gamma_N$ is a game $\hat{v} \in \Gamma_N$ such that, for every $a \in I^n$,

$$\hat{v}(a) = \sup \left\{ \sum_{k=1}^{\ell} \lambda_k v(a^k) \middle| \ell \in \mathbb{N}, \ a^i \leq a, \ \lambda_i \geq 0, \right.$$

$$\left. \sum_{k=1}^{\ell} \lambda_k a^k = a, \quad i = 1, \ldots, \ell \right\} \, .$$

The nonemptiness of $C(v)$ depends on value of $\hat{v}$ at one point only.

Theorem 9.3 Azrieli and Lehrer [9.13]

Let $v \in \Gamma_N$. The core $C(v)$ is nonempty if and only if $v(\mathbb{1}_N) = \hat{v}(\mathbb{1}_N)$.

The above theorem answers Question 1. Nevertheless, it may be difficult to check the condition $v(\mathbb{1}_N) =$

$\hat{v}(\mathbb{1}_N)$. Can we simplify this task for some classes of games? In particular, can we show that the shape of the core is simpler on some class of games? This leads naturally to Question 2. *Branzei* et al. [9.16] showed that the class of games for which this holds true is the class of convex games. We say that a game $v \in \Gamma_N$ is *convex*, whenever the inequality

$$v(a+c) - v(a) \leq v(b+c) - v(b) \qquad (9.9)$$

is satisfied for every $a, b, c \in I^n$ such that $b + c \in I^n$ and $a \leq b$. A word of caution is in order here: in general, as shown in [9.13], the convexity of the game $v \in \Gamma_N$ does not imply and is not implied by the convexity of v as an n-place real function. The game-theoretic convexity captures the economic principle of nondecreasing marginal utility. Interestingly, this property makes it possible to simplify the structure of $C(v)$.

Theorem 9.4 Branzei et al. [9.12]
Let $v \in \Gamma_N$ be a convex game. Then $C(v) \neq \emptyset$ and, moreover, $C(v)$ coincides with the core $C(\bar{v})$ of the classical coalition game $\bar{v}$.

The previous theorem, which solves Question 2, provides in fact the complete characterization of core on the class of convex games with fuzzy coalitions. Indeed, since the game $\bar{v}$ is convex, we can use the result of *Shapley* [9.7] to describe the shape of $C(v) = C(\bar{v})$.

The point 3 motivates the following definition. A game $v \in \Gamma_N$ is said to be *exact* whenever for every $a \in I^n$, there exists $x \in C(v)$ such that $\langle a, x \rangle = v(a)$. The class of exact games can be explicitly described [9.13].

Theorem 9.5
Let $v \in \Gamma_N$. Then the following properties are equivalent:

i) v is exact;
ii) $v(a) = \min\{\langle a, x \rangle | x \in C(v)\}$;
iii) v is simultaneously
   a) a concave, positively homogeneous function on I^n, and
   b) $v(\lambda a + (1-\lambda)\mathbb{1}_N) = \lambda v(a) + (1-\lambda)v(\mathbb{1}_N)$, for every $a \in I^n$ and every $0 \leq \lambda \leq 1$.

The second equivalent property enables us to generate many examples of exact games – it is enough to take the minimum of a family of linear functions, each of which coincides at point $\mathbb{1}_N$.

Question 4 amounts to asking for the existence of allocation rules in the sense of *Lehrer* [9.17] or dynamic procedures for approximating the core elements by *Wu* [9.18]. A bargaining procedure for recovering the elements of the Aubin's core $C(v)$ is discussed in [9.19], where the authors present the so-called Cimmino-style bargaining scheme. For a game $v \in \Gamma_N$ and some *initial payoff* $x^0 \in \mathbb{R}^n$, the goal is to recover a sequence of payoffs converging to a core element, provided that $C(v) \neq \emptyset$. We consider a probability measure that captures the bargaining power of coalitions $a \in I^n$: a *coalitional assessment* is any complete probability measure ν on I^n. In what follows, we will require that ν is such that, for every Lebesgue measurable set $A \subseteq I^n$,

$$\nu(A) > 0, \quad \text{whenever } A \text{ is open or } \mathbb{1}_N \in A. \quad (9.10)$$

Let $x \in \mathbb{R}^n$ be an arbitrary payoff and $a \in I^n$. We denote

$$C_a(v) = \begin{cases} \{y \in \mathbb{R}^n | \langle a, y \rangle \geq v(a)\} & a \in I^n \setminus \{\mathbb{1}_N\}, \\ \{y \in \mathbb{R}^n | \langle \mathbb{1}_N, y \rangle = v(\mathbb{1}_N)\} & a = \mathbb{1}_N. \end{cases}$$

What happens when payoff x is accepted by a, that is, $x \in C_a(v)$? Then coalition a has no incentive to bargain for another payoff. On the contrary, if $x \notin C_a(v)$, then a may seek the payoff $P_a x \in C_a(v)$ such that $P_a x$ is the closest to x in some sense. Specifically, we will assume that $P_a x$ minimizes the Euclidean distance of x from set $C_a(v)$. This yields the formula

$$P_a x$$
$$= \arg\min_{y \in C_a(v)} \|y - x\|$$
$$= \begin{cases} x + \dfrac{\max\{0, v(a) - \langle a, x \rangle\}}{\|a\|^2} a & a \in I^n \setminus \{\mathbb{1}_\emptyset, \mathbb{1}_N\}, \\ x + \dfrac{v(\mathbb{1}_N) - \langle \mathbb{1}_N, x \rangle}{n} \mathbb{1}_N & a = \mathbb{1}_N, \\ x & a = \mathbb{1}_\emptyset, \end{cases}$$

where $\|\cdot\|$ is the Euclidean norm. After all coalitions $a \in I^n$ have raised their requests on the new payoff $P_a x$, we will average their demands with respect to the coalitional assessment ν in order to obtain a new proposal payoff vector Px. Hence, Px is computed as

$$Px = \int_{I^n} P_a x \, d\nu(a).$$

The integral on the right-hand side is well defined, whenever v is Lebesgue measurable. The amalgamated projection operator P is the main tool in the *Cimmino-style bargaining procedure*: an initial payoff x^0 is arbitrary, and we put $x^k = Px^{k-1}$, for each $k = 1, 2, \ldots$

Theorem 9.6

Let $v \in \Gamma_N$ be a continuous game with fuzzy coalitions and let v be a coalitional assessment satisfying (9.10):

1. If the sequence $(x^k)_{k \in \mathbb{N}}$ generated by the Cimmino procedure is bounded and

$$\lim_{k \to \infty} \int_{[0,1]^n} \| x^k - P_a x^k \| \, dv(a) = 0, \quad (9.11)$$

then $C(v) \neq \emptyset$ and $\lim_{k \to \infty} x^k \in C(v)$.

2. If the sequence $(x^k)_{k \in \mathbb{N}}$ is unbounded or (9.11) does not hold, then $C(v) = \emptyset$.

The interested reader is invited to consult [9.19] for further details and numerical experiments.

9.2.2 Single-Valued Solutions

Shapley value

Aubin defined Shapley value on spaces of games with fuzzy coalitions possessing nice analytical properties [9.3, 14, Chapter 13.4]. Specifically, let a function $v : \mathbb{R}^n \to \mathbb{R}$ be positively homogeneous and Lipschitz in the neighborhood of $\mathbb{1}_N$. Such functions are termed *generalized sharing games with side payments* by *Aubin* [9.14, Chap. 13.4]. The restriction of v onto the cube I^n is clearly a game with fuzzy coalitions and therefore we would not make any distinction between v and its restriction to I^n. In addition, assume that function v is continuously differentiable at $\mathbb{1}_N$ and denote by G_N^1 the class of all such games with fuzzy coalitions. Hence, we may put

$$\sigma(v) = \nabla v(\mathbb{1}_N), \quad v \in G_N^1. \quad (9.12)$$

Each coordinate $\sigma_i(v)$ of the gradient vector $\sigma(v)$ captures the marginal contribution of player $i \in N$ to the grand coalition $\mathbb{1}_N$. As pointed out by Aubin, the gradient measures the roles of the players as pivots in game v. Moreover, the operator σ given by (9.12) can be considered as a generalized Shapley value on the class of games G_N^1 (cf. Theorem 9.1): *Aubin* proved [9.14, Chapter 13.4] that the operator defined by (9.12) satisfies

$$\langle \mathbb{1}_N, \sigma(v) \rangle = v(\mathbb{1}_N),$$

for every game $v \in G_N^1$, and

$$\sigma_i(\pi v) = \sigma_{\pi(i)}(v),$$

for every player $i \in N$ and every permutation π of N. Moreover, σ fulfills a certain variant of the Dummy Property.

When defining a value on games with fuzzy coalitions, many other authors [9.15, 20] proceed in the following way: a classical cooperative game is extended from the set of all crisp coalitions to the set of all fuzzy coalitions. The main issue is to decide on the nature of this extension procedure and to check that the extended game with fuzzy coalitions inherits all or at least some properties of the function that is extended (such as superadditivity or convexity). Clearly, there are as many choices for the extension as there are possible interpolations of a real function on $\{0, 1\}^n$ to the cube $[0, 1]^n$.

Tsurumi et al. [9.20] used the Choquet integral as an extension. Specifically, for every $a \in I^n$, let $V_a = \{a_i | a_i > 0, i \in N\}$ and let $n_a = |V_a|$. Without loss of generality, we may assume that the elements of V_a are ordered and write them as $b_1 < \cdots < b_{n_a}$. Further, put $[a]_y = \{i \in N | a_i \geq y\}$, for each $a \in I^n$ and for each $y \in [0, 1]$.

Definition 9.8

A game with fuzzy coalitions (I^n, v) is a *game with Choquet integral form* whenever

$$v(a) = \sum_{i=1}^{n_a} \bar{v}([a]_{b_i})(b_i - b_{i-1}), \quad a \in I^n,$$

where $b_0 = 0$. Let Γ_N^C be the class of all games with Choquet integral form.

In the above definition the function v is the so-called Choquet integral [9.21] of a with respect to the restriction $\bar{v}$ of v to all crisp coalitions. It was shown that every game $v \in \Gamma_N^C$ is monotone whenever $\bar{v}$ is monotone [9.20, Lemma 2] and that v is a continuous function on I^n [9.20, Theorem 2]. The authors define a mapping

$$f : \Gamma_N^C \to ([0, \infty)^n)^{I^n},$$

which is called a *Shapley function*, by the following assignment

$$f_i(v)(a) = \sum_{i=1}^{n_a} f_i^0(\bar{v})([a]_{b_i})(b_i - b_{i-1}),$$

$$i \in N, \; v \in \Gamma_N^C, \; a \in I^n,$$

where

$$f_i^0(\bar{v})(B)$$

$$= \sum_{\substack{A \subseteq B \\ i \in A}} \frac{(|A|-1)!(|B|-|A|)!)}{|B|!} (\bar{v}(A) - \bar{v}(A \setminus \{i\})),$$

$$B \subseteq N,$$

whenever $i \in B$, and $f_i^0(\bar{v})(B) = 0$, otherwise. Observe that $f_i(v)(a)$ is the Choquet integral of a with respect to $f_i^0(v)$ and that $f_i^0(\bar{v})(B)$ is the Shapley value of $\bar{v}$ with the grand coalition N replaced with the coalition B. Before we show that the Shapley function has some expected properties, we prepare the following definitions. Let $a \in I^n$ and $i, j \in N$. For each $b \in I^n$ with $b \leq a$, define a vector b_{ij}^a whose coordinates are

$$\left(b_{ij}^a\right)_k = \begin{cases} b_i \wedge a_j & k = i, \\ b_j \wedge a_i & k = j, \\ b_k & \text{otherwise}, \end{cases} \quad k \in N.$$

For an arbitrary $b \in I^n$, put

$$\left(b_{ij}'[a]\right)_k = \begin{cases} b_j & k = i, \\ b_i & k = j, \\ b_k & \text{otherwise}, \end{cases} \quad k \in N.$$

Clearly, we have both $b_{ij}^a \leq a$ and $b_{ij}'[a] \leq a$. The following theorem is proved in [9.20].

Theorem 9.7
The operator $f : \Gamma_N^C \to ([0, \infty)^n)^{I^n}$ has the following properties:

1. If $v \in \Gamma_N^C$ and $a \in I^n$, then

   $$\sum_{i \in N} f_i(v)(a) = v(a) \quad \text{and} \quad f_j(v)(a) = 0,$$

   for every $j \in N$ such that $a_j = 0$.
2. If $v \in \Gamma_N^C$, $a \in I^n$, and $b \in I^n$ such that $v(b \wedge c) = v(b)$, for every $c \in I^n$ with $c \leq a$, then

   $$f_i(v)(a) = f_i(v)(b) \quad \text{for every } i \in N.$$

3. If $v \in \Gamma_N^C$, $a \in I^n$, a_{ij}^a is such that $v(a_{ij}^a \wedge c) = v(b)$, for every $c \in I^n$ with $c \leq a$, and $v(b) = v(b_{ij}')$ for every $b \in I^n$ with $b \leq a_{ij}^a$, then

   $$f_i(v)(a) = f_j(v)(a).$$

4. If $v, w \in \Gamma_N^C$, then $v + w \in \Gamma_N^C$, and

   $$f_i(v + w)(a) = f_i(v)(a) + f_i(w)(a),$$

   for every $i \in N$ and every $a \in I^n$.

The previous theorem thus says that the Shapley function f on the class of games Γ_N^C has the properties analogous to the Shapley value: efficiency, the carrier property, symmetry, and additivity.

Butnariu and *Kroupa* [9.15] studied a value operator on the class of fuzzy games (I^n, v) satisfying

$$v(a) = \sum_{t \in [0,1]} \psi(t) v(a^t), \quad a \in I^n,$$

where $\psi : [0, 1] \to \mathbb{R}$ fulfills

$$(\psi(t) = 0 \text{ iff } t = 0) \text{ and } \psi(1) = 1$$

and

$$a^t = \{i \in N | a_i = t\}.$$

The class of such fuzzy games is denoted by Γ_N^ψ. The so-called Shapley mapping function can be axiomatized on Γ_N^ψ [9.15, Axioms 1–3]: it turns out that there is only one Shapley mapping $\Phi : \Gamma_N^\psi \to (\mathbb{R}^n)^{I^n}$ [9.15, Theorem 1].

Theorem 9.8
There exists a unique Shapley mapping $\Phi : \Gamma_N^\psi \to (\mathbb{R}^n)^{I^n}$ and it is given by the following formula:

$$\Phi_i(v)(a) =$$

$$\begin{cases} \psi(r) \sum_{S \in \mathcal{P}_i(a^r)} \frac{(|S|-1)!(|a^r|-|S|)!}{|a^r|!} (v(S) - v(S \setminus \{i\})), \\ \qquad \text{if } a_i = r > 0, \\ 0, \quad \text{otherwise}, \end{cases}$$

where

$$\mathcal{P}_i(a^r) = \{R \subseteq N | i \in R \text{ and } R \subseteq a^r\}.$$

The expected total allocation of player $i \in N$ is then obtained as

$$\hat{\Phi}_i(v) = \int_{I^n} \Phi_i(v)(a) \, da,$$

provided that the above Lebesgue integral exists. The operator $\hat{\Phi} = (\hat{\Phi}_1, \ldots, \hat{\Phi}_n)$ is called the *cumulative value* of v. If the weight function ψ is bounded and Lebesgue integrable, then [9.15, Theorem 2] shows that the cumulative value is well-defined and its coordinates are

$$\hat{\Phi}_i(v) = v(\mathbb{1}_i) \int_0^1 \psi(t)\, dt$$

for each $i \in N$.

Owen's approach to classical Shapley value [9.22] cannot be, strictly speaking, classified as an attempt to define a Shapley-style value on some class of games with fuzzy coalitions, but we mention his construction for the sake of completeness. The idea is to extend a game $v \in G_N$ with crisp coalitions from its domain $\{\mathbb{1}_A | A \subseteq N\}$ to the whole unit cube I^n by way of multilinear interpolation. The resulting *multilinear extension* $\tilde{v}$ can be described explicitly as the function

$$\tilde{v}(a) = \sum_{A \subseteq N} \left[\prod_{i \in A} a_i \prod_{i \notin A} (1 - a_i) \right] v(A),$$
$$a = (a_1, \ldots, a_n) \in I^n. \tag{9.13}$$

Function $\tilde{v}$ is linear in each of its variables separately and $v(A) = \tilde{v}(\mathbb{1}_A)$, for each $A \subseteq N$. The usual formula (9.13) for the Shapley value $\phi(v)$ of v now takes the following *diagonal form* [9.22]

$$\phi_i(v) = \int_0^1 \frac{\partial \tilde{v}}{\partial x_i}(t, \ldots, t)\, dt. \tag{9.14}$$

Hence, $\phi_i(v)$ is completely determined by the behavior of the function $\tilde{v}$ in the neighborhood of the diagonal in I^n. The formula (9.14) is important from the computational point of view: its use in connection with statistical techniques can enhance computations with the Shapley value – see [9.23, Chap. XII.4] for further details.

Since the space of games with crisp coalitions is finite dimensional unlike the space of games with fuzzy coalitions, there is no general approach to the Shapley value of fuzzy games. Even a direct comparison of the cumulative value introduced above with the Shapley function on the space of games Γ_N^C of *Tsurumi* et al. [9.20] is hardly possible since the domains of Shapley operators are essentially different. The selection of the right space of games and an appropriate solution thus vary from one application to another.

9.3 Final Remarks

We presented results concerning basic concepts of solution for coalitional games with fuzzy coalitions and finitely many players in the case of transferable utility. We concentrated on those solutions which preoccupy the main part of cooperative game theory (the core and the Shapley value). A detailed discussion or just the comprehensive overview of the current trends in fuzzy games is beyond the reach of this chapter. Nevertheless, in this section we mention current developments and briefly discuss other solution concepts. The reader should always consult the relevant reference for the specification of the concepts used by the cited authors; for example, we can find at least two definitions of a convex fuzzy game:

1. *Azrieli* and *Lehrer* [9.13] and [9.16] use the definition (9.9) employed herein;
2. *Tsurumi* et al. [9.20] call a game with fuzzy coalitions v *convex* whenever

$$v(a \vee b) + v(a \wedge b) \ge v(a) + v(b)$$

holds true for every $a, b \in I^n$.

Shellshear [9.24] employs the concavification of the fuzzy game – the strong supperadditive cover – in order to show [9.24, Theorem 4.4] that the strong supperadditive cover has a stable core if and only if the original game has a stable core. Further, he investigates important properties of the concavification and its superdifferential; new necessary and sufficient conditions for core stability are given in [9.23, Chap. XII.4].

Yang et al. [9.25] introduced the concept of bargaining sets for games with fuzzy coalitions; they prove that the bargaining set coincides with the Aubin core whenever a game is continuous and convex. *Liu* and *Liu* [9.26] extended the results from [9.25] in order to overcome some weakness of the previously used fuzzy bargaining sets. The concept of the classical Mas-Colells bargaining set was also generalized and the authors proved existence theorems for such fuzzy bargaining sets. Moreover, both Aumann and Maschler and Mas-Colell fuzzy bargaining sets of a continuous convex cooperative fuzzy game coincide with its Aubin core.

A fuzzy game is represented as a convex program in [9.27]. It is shown that the optimum of the program determines the optimal coalitions as well as the optimal rewards for the players. Further, this framework seems to unify a number of existing representations of solutions: the core, the least core, and the nucleolus.

Wu [9.28] investigates various types of cores based on the dominance among payoff vectors and the con-

cepts of the true payoff and quasi-payoff of a fuzzy coalition.

Interpretational difficulties related to fuzzy games are pointed out by *Mareš* and *Vlach* in [9.29]. The authors propose an alternative model for a fuzzy coalition – a collection of crisp coalitions – and discuss some of its consequences.

References

9.1   M.J. Osborne, A. Rubinstein: *A Course in Game Theory* (MIT Press, Cambridge 1994)

9.2   D. Butnariu: Fuzzy games: A description of the concept, Fuzzy Sets Syst. **1**, 181–192 (1978)

9.3   J.-P. Aubin: Coeur et valeur des jeux flous à paiements latéraux, Comptes Rendus de l'Académie des Sciences Série A **279**, 891–894 (1974)

9.4   L.S. Shapley, M. Shubik: On Market Games, J. Econ. Theory **1**, 9–25 (1969)

9.5   B. Peleg, P. Sudhölter: *Introduction to the Theory of Cooperative Games*, Theory and Decision Library: Series C. Game Theory, Vol. 34, 2nd edn. (Springer, Berlin 2007)

9.6   R.J. Aumann, M. Maschler: Game theoretic analysis of a bankruptcy problem from the Talmud, J. Econ. Theory **36**(2), 195–213 (1985)

9.7   L.S. Shapley: Cores of convex games, Int. J. Game Theory **1**, 11–26 (1972)

9.8   L.S. Shapley: A value for *n*-person games. In: *Contributions to the Theory of Games. Vol. II*, Annals of Mathematics Studies, Vol. 28, ed. by H.W. Kuhn, A.W. Tucker (Princeton Univ. Press, Princeton 1953) pp. 307–317

9.9   H.P. Young: Monotonic solutions of cooperative games, Int. J. Game Theory **14**, 65–72 (1985)

9.10   S. Hart, A. Mas-Colell: Potential, value and consistency, Econometrica **57**, 589–614 (1989)

9.11   R. Weber: Probabilistic values for games. In: *The Shapley Value*, ed. by A.E. Roth (Cambridge Univ. Press, Cambridge 1988) pp. 101–120

9.12   J.F. Banzhaf III: Weighted voting does not work: A mathematical analysis, Rutgers Law Rev. **19**, 317–343 (1965)

9.13   Y. Azrieli, E. Lehrer: On some families of cooperative fuzzy games, Int. J. Game Theory **36**(1), 1–15 (2007)

9.14   J.-P. Aubin: *Optima and Equilibria*, Graduate Texts in Mathematics, Vol. 140, 2nd edn. (Springer, Berlin 1998)

9.15   D. Butnariu, T. Kroupa: Shapley mappings and the cumulative value for *n*-person games with

fuzzy coalitions, Eur. J. Oper. Res. **186**(1), 288–299 (2008)

9.16   R. Branzei, D. Dimitrov, S. Tijs: *Models in Cooperative Game Theory*, Lecture Notes in Economics and Mathematical Systems, Vol. 556 (Springer, Berlin 2005)

9.17   E. Lehrer: Allocation processes in cooperative games, Int. J. Game Theory **31**(3), 341–351 (2003)

9.18   L.S.Y. Wu: A dynamic theory for the class of games with nonempty cores, SIAM J. Appl. Math. **32**(2), 328–338 (1977)

9.19   D. Butnariu, T. Kroupa: Enlarged cores and bargaining schemes in games with fuzzy coalitions, Fuzzy Sets Syst. **5**(160), 635–643 (2009)

9.20   M. Tsurumi, T. Tanino, M. Inuiguchi: A Shapley function on a class of cooperative fuzzy games, Eur. J. Oper. Res. **129**(3), 596–618 (2001)

9.21   D. Denneberg: *Non-Additive Measure and Integral*, Theory and Decision Library B. Mathematical and Statistical Methods Series, Vol. 27 (Kluwer, Dordrecht 1994)

9.22   G. Owen: Multilinear extensions of games, Manag. Sci. **18**, P64–P79 (1971)

9.23   G. Owen: *Game Theory*, 3rd edn. (Academic, San Diego 1995)

9.24   E. Shellshear: A note on characterizing core stability with fuzzy games, Int. Game Theory Rev. **13**(01), 105–118 (2011)

9.25   W. Yang, J. Liu, X. Liu: Aubin cores and bargaining sets for convex cooperative fuzzy games, Int. J. Game Theory **40**(3), 467–479 (2011)

9.26   J. Liu, X. Liu: Fuzzy extensions of bargaining sets and their existence in cooperative fuzzy games, Fuzzy Sets Syst. **188**(1), 88–101 (2012)

9.27   M. Keyzer, C. van Wesenbeeck: Optimal coalition formation and surplus distribution: Two sides of one coin, Eur. J. Oper. Res. **215**(3), 604–615 (2011)

9.28   H.-C. Wu: Proper cores and dominance cores of fuzzy games, Fuzzy Optim. Decis. Mak. **11**(1), 47–72 (2012)

9.29   M. Mareš, M. Vlach: Disjointness of fuzzy coalitions, Kybernetika **44**(3), 416–429 (2008)

Part B

Part B Fuzzy Logic

Ed. by Enrique Herrera Viedma, Luis Magdalena

10. Basics of Fuzzy Sets

János C. Fodor, Imre J. Rudas

In this chapter we summarize basic knowledge on fuzzy logics and fuzzy sets. After a short historical overview of ideas strongly connected to and preceding the notion of fuzzy logics and fuzzy sets, we outline links between many-valued and fuzzy logics. Then fuzzy subsets of a universe are introduced. Interpretations of unary and binary connectives in fuzzy logics as appropriate functions (operations) on the unit interval are central to the approach. Fundamental knowledge on these function classes is presented then, including results on triangular norms and conorms, as well as on impli-

cations. Our concluding remarks suggest further reading, beyond the basics.

In everyday life we use and process vague, imprecise linguistic terms like *young*, *hot*, or *around midnight*. Classical mathematics is unable and inadequate to provide models that can express the complex semantics of such terms. Fuzzy sets, introduced by *Zadeh* [10.1] on the basis of his observation that

> more often than not, the classes of objects encountered in the real physical world do not have precisely defined criteria or membership,

are appropriate for modeling the semantics of vague linguistic terms. Fuzzy sets offer a framework to deal with predicates whose satisfaction is a matter of degree.

Some forerunners discussed ideas or formal definitions for describing vague predicates or classes with imprecise boundaries, very close to the basic notions introduced by *Zadeh* [10.1]. We should mention *Peirce* [10.2], *Russel* [10.3], *Łukasiewicz* [10.4], *Black* [10.5], *Weyl* [10.6], *Kaplan* and *Schott* [10.7]. The mathematician *Karl Menger* was the first (in 1951) who used the term *ensemble flou* (the French counterpart for fuzzy set) in the title of a paper in French [10.8]. In addition, Menger's work on probabilistic metric spaces also led to the introduction of so-called *triangular norms and conorms*, extensively studied by

Schweizer and *Sklar* [10.9], and which later have turned out to be basic operators for fuzzy sets [10.10]. For more historical facts we refer to [10.11].

We want to emphasize that Zadeh's motivations and background were quite different from those of the above-mentioned authors. He introduced the concept of a fuzzy set completely independently of their proposals in order to provide a tool for representing and reasoning with the available information in a manner similar to the way humans express knowledge and summarize data.

This Chapter is organized as follows. In the next section we briefly recall some notions from classical mathematics and its underlying two-valued (Boolean) logic. We extend this material, and in Sect. 10.3 we introduce key terms related to fuzzy sets. Sect. 10.4 contains the core knowledge on interpretations of connectives in fuzzy logic and fuzzy set-theoretic operations. This includes fundamentals of negations, triangular norms and conorms, together with the most important parametric families and particular operations. Fuzzy implications are also handled in a similar way. Concluding remarks are given at the end, including several suggested literature items for further reading.

10.1 Classical Mathematics and Logic

Classical mathematics is based on two-valued logic, in which the set of truth values consists of two elements: $\{0, 1\}$. There are two basic binary operations $\wedge$ (AND), $\vee$ (OR), and the unary complement $\neg$ (NOT). All other logical operations, e.g. the *implication* $\rightarrow$, the *logical equivalence* $\leftrightarrow$, and the *exclusive or* XOR, can be constructed from the three basic operations $\wedge, \vee, \neg$.

A *proposition* is either an atomic propositional variable $p_1, p_2, \ldots$, or a compound expression $(p \wedge q)$, $(p \vee q)$, or $\neg p$, where p and q are propositions. A proposition is either *true* (with truth value 1) or *false* (with truth value 0), but not both.

A *set* A is a collection of objects in a given universe X, where, for each possible object x from X, it either belongs to the set A (in symbols: $x \in A$) or not ($x \notin A$). A set A is a *subset* of B if all objects in A are in B as well (in symbols: $A \subseteq B$). We write $A \subset B$ if $A \subseteq B$ and there is at least one element in B which is not in A. The set of all subsets of X is denoted by $\mathcal{P}(X)$. The empty set, which does not contain any object, is denoted by $\emptyset$.

We consider three fundamental operations on sets. The *intersection* of two sets A and B, denoted by $A \cap B$, is the set of objects from X which belong both to A and B. The *union* of two sets A and B, denoted by $A \cup B$, is the set of objects from X which belong at least to one of the sets A and B. The *complement* of a set A, denoted by A^c, is the set of objects from X which do not belong to A.

A function $\chi_A : X \rightarrow \{0, 1\}$ is called the *characteristic function* of the set A if

$$\chi_A(x) = \begin{cases} 1 & \text{if } x \in A \\ 0 & \text{if } x \notin A \end{cases} .$$

The characteristic function discriminates between *members* and *nonmembers* of the set A. With the help of characteristic functions, set operations can be expressed as follows

$$\chi_{A \cap B}(x) = \chi_A(x) \wedge \chi_B(x) ,$$
$$\chi_{A \cup B}(x) = \chi_A(x) \vee \chi_B(x) ,$$
$$\chi_{A^c}(x) = \neg \chi_A(x) .$$

10.2 Fuzzy Logic, Membership Functions, and Fuzzy Sets

The idea behind *fuzzy logic* is to replace the set of truth values $\{0, 1\}$ by the entire unit interval $[0, 1]$. Then a *fuzzy set* on a universe X is represented by a function which maps each element $x \in X$ to a *degree of membership* from the unit interval $[0, 1]$. Larger values indicate higher degrees of membership.

For several decades, many-valued logic was considered as a pure mathematical topic. The introduction of fuzzy sets [10.1] produced a new impact to the investigation of many-valued logics. Informally speaking, fuzzy logic is understood as an extension of many-valued logics, with an ultimate goal of providing foundations for *approximate reasoning* with imprecise propositions using fuzzy set theory as the principal tool.

A many-valued propositional logic in which the class of truth values is modelled by the unit interval $[0, 1]$, and which forms an extension of the classical Boolean logic, i. e., the two-valued logic with truth values $\{0, 1\}$, is quite often called a *fuzzy logic* [10.10]. For sake of simplicity, it is assumed that all fuzzy logics have the same syntax, they may differ only by their semantics.

A *membership function* μ_A is a mapping from the universal set X to the unit interval, i. e., $\mu_A : X \rightarrow [0, 1]$. Membership functions are direct generalizations of characteristic functions. In a logical setting, the degree of membership $\mu_A(x)$ can also be seen as the truth value of the statement x *is element of A*.

Notice that a membership grade can have three meanings:

- *Degree of similarity.* The membership grade $\mu_A(x)$ represents the degree of proximity of x from prototype elements of A.
- *Degree of preference.* A represents a set of more or less preferred objects, and $\mu_A(x)$ represents an intensity of preference in favor of object x.
- *Degree of uncertainty.* The degree $\mu_A(x)$ can be viewed as the degree of plausibility that a parameter p has value x, given that all that is known about it is that p *is A*.

These three semantics of fuzzy sets appear in the works of Zadeh and he was the first to propose each of them.

A *fuzzy set* on X (or a *fuzzy subset* of X) is defined as the collection of the ordered pairs of elements of X and their membership grades. Practically, a fuzzy set A on X is identified with the membership function μ_A. The family of all fuzzy subsets of X is denoted by $\mathcal{F}(X)$. Classical subsets of X are special fuzzy subsets on X, and are called *crisp sets*. Note that one may represent membership grades not only by the unit interval but also by a (partially or completely) ordered set.

Given two fuzzy subsets A and B of X, we say that A is *equal* to B (in symbols $A = B$) if $\mu_A = \mu_B$, and that A is a *subset* of B (in symbols $A \subseteq B$) if $\mu_A \le \mu_B$.

As it is emphasized in [10.11], membership functions express a *vertical view* of fuzzy sets. Another view is to consider a fuzzy set as a nested family of classical sets, by using the notion of α-*cuts*. For any $\alpha \in [0, 1]$ we can introduce the α-cut A_α of a fuzzy set A. By definition, A_α is the crisp subset of X that contains all the elements of X that have a membership grade greater than or equal to the specified value α. More formally, $A_\alpha = \{x \in X | \mu_A(x) \ge \alpha\}$, $\alpha \in [0, 1]$.

In the sequel, membership functions and fuzzy sets will be denoted by the same symbol: we write simply $A(x)$ instead of $\mu_A(x)$ for $A \in \mathcal{F}(X)$ and $x \in X$.

10.3 Connectives in Fuzzy Logic

In order to generalize the classical set-theoretical operations like intersection, union and complement, it is quite natural to use interpretations of logic connectives $\land$, $\lor$ and $\neg$, respectively. Indeed, the values $(A \cap B)(x)$, $(A \cup B)(x)$ and $A^c(x)$ describe the truth values of the statements x *is element of A **AND** x is element of B*, x *is element of A **OR** x is element of B*, and x *is **NOT** element of A*, respectively.

We introduce appropriate classes of functions $N : [0, 1] \to [0, 1]$, $T : [0, 1]^2 \to [0, 1]$ and $S : [0, 1]^2 \to [0, 1]$ in order to interpret logic operations $\neg$, $\land$ and $\lor$, respectively, on the evaluation set $[0, 1]$. In addition, fuzzy implications are also introduced later on.

Then, the *complement of a fuzzy set A*, the *intersection and union of fuzzy sets A, B* are specified by the functions N, T and S, respectively, such that

$$A_N^c(x) = N(A(x)),$$
$$(A \cap_T B)(x) = T(A(x), B(x)),$$
$$(A \cup_S B)(x) = S(A(x), B(x)),$$ (10.1)

where $x \in X$, $A, B \in \mathcal{F}(X)$. Therefore, desired properties of fuzzy set-theoretic (or equivalently, logic) operations can be obtained through the corresponding properties of the above functions N, T and S.

10.3.1 Negations

Starting with the negation $\neg$, it is clear that its interpretation should map 1 to 0 and 0 to 1, in order to be an extension of the interpretation of the classical two-valued negation. Another natural property is that the interpretation of the negation $\neg$ be a non-increasing function. To simplify notations, and since there is no confusion possible, an interpretation of the negation $\neg$ will also be called a negation.

Definition 10.1

A decreasing function $N : [0, 1] \to [0, 1]$ with $N(0) = 1$, $N(1) = 0$ is called a *negation*. A strictly decreasing continuous negation is called a *strict negation*. A strict negation N is said to be a *strong negation* if N is also involutive: $N(N(x)) = x$ holds for all $x \in [0, 1]$.

Since a strict negation N is a strictly increasing and continuous function, its inverse N^{-1} is also a strict negation, generally different from N. Obviously, we have $N^{-1} = N$ if and only if N is involutive: $N(N(x)) = x$ holds for all $x \in [0, 1]$. This means that the graph of the function N is symmetric with respect to the line $\{(x, y) \mid x = y\}$.

Another important property of a strict negation N is that there exists a unique value $0 < \nu < 1$ such that $N(\nu) = \nu$. Then we also have $N^{-1}(\nu) = \nu$.

A negation which is neither strong nor strict is the *Gödel* (or intuitionistic) *negation* given by

$$N_{\mathbf{G}}(x) = \begin{cases} 1 & \text{if } x = 0 \\ 0 & \text{if } x \in \,]0, 1] \end{cases}.$$

By duality, we can define the *dual Gödel* negation as follows

$$N_{\mathbf{dG}}(x) = \begin{cases} 1 & \text{if } x \in [0, 1[\\ 0 & \text{if } x = 1 \end{cases}.$$

It is easy to see that for any negation N we have

$$N_{\mathbf{G}} \le N \le N_{\mathbf{dG}}.$$

A strict but not strong negation can be given by $N(x) = 1 - x^3$.

A parametric family of strong negations is defined as follows (see [10.12] under the name λ-complement)

$$N_\lambda(x) = \frac{1-x}{1+\lambda x}, \quad \lambda > -1 .$$

The *standard negation* N_s is defined simply as $N_s(x) = 1 - x, \quad x \in [0, 1]$. This is the most frequently used negation, which is obviously a strong negation. It plays a key role in the representation of strong negations presented in the following theorem. In this chapter we call a continuous, strictly increasing function $\varphi : [0, 1] \to [0, 1]$ with $\varphi(0) = 0$, $\varphi(1) = 1$ an *automorphism* of the unit interval.

Theorem 10.1

A function $N : [0, 1] \to [0, 1]$ is a strong negation if and only if there exists an automorphism φ of the unit interval such that [10.13]

$$N(x) = \varphi^{-1}(1 - \varphi(x)), \quad x \in [0, 1] . \tag{10.2}$$

In this case N_φ denotes N in (10.2) and is called a φ-*transform* of the standard negation. If the complement of fuzzy sets on X is defined by N_φ, we use the short notation A_φ^c for $A \in \mathcal{F}(X)$, instead of writing $A_{N_\varphi}^c$.

10.3.2 Triangular Norms and Conorms

It is assumed that the conjunction $\wedge$, which is always in the tuple of connectives, is interpreted by a t-norm, which, in a canonical way, is a generalization of the interpretation of the conjunction in Boolean logic. In a logical sense, a t-conorm is an ideal candidate for the interpretation of the disjunction $\vee$, since it is a canonical extension of the interpretation of the two-valued disjunction. This is clear from the following definition.

Definition 10.2

A *triangular norm* (shortly: a *t-norm*) is a function $T : [0, 1]^2 \to [0, 1]$ which is associative, commutative and increasing, and satisfies the boundary condition $T(1, x) = x$ for all $x \in [0, 1]$.

A *triangular conorm* (shortly: a *t-conorm*) is a function $S : [0, 1]^2 \to [0, 1]$ which is associative, commutative and increasing, with boundary condition $S(0, x) = x$ for all $x \in [0, 1]$.

The class of t-norms (with slightly different axioms) was introduced in the theory of statistical (probabilistic) metric spaces as a tool for generalizing the classical triangular inequality by *Menger* [10.14] (see also *Schweizer* and *Sklar* [10.9], *Alsina* et al. [10.15], and *Klement* et al. [10.10]).

Notice that continuity of a t-norm and a t-conorm is not taken for granted. Even more: conditions in Definition 10.2 do not even imply that all t-norms, as two-place functions, are measurable (see [10.16] for a counter-example).

However, the definition implies the following properties

$$T(x, y) \leq \min(x, y) ,$$
$$S(x, y) \geq \max(x, y) \quad (x, y \in [0, 1]) ,$$

and

$$T(0, y) = 0, \quad S(1, y) = 1 \quad \text{for all} \quad y \in [0, 1] .$$

The smallest t-norm is the *drastic product* T_D given by

$$T_D(x, y) = \begin{cases} 0 & \text{if } (x, y) \in [0, 1[\\ \min(x, y) & \text{otherwise.} \end{cases}$$

The greatest (and the only idempotent) t-norm is obviously $T_M = \min$, the *minimum t-norm*. Thus, for any t-norm T we have

$$T_D \leq T \leq T_M .$$

The smallest and greatest t-norms (T_D and T_M) together with the *product t-norm* $T_P(x, y) = xy$, and the *Łukasiewicz t-norm* T_L given by

$$T_L(x, y) = \max(0, x + y - 1)$$

are called *basic t-norms*.

The first known left-continuous and not continuous t-norm is the *nilpotent minimum* [10.17] defined by

$$T_{nM}(x, y) = \begin{cases} 0 & \text{if } x + y \leq 1 \\ \min(x, y) & \text{otherwise.} \end{cases}$$

A t-conorm S is called the dual to the t-norm T if $S(x, y) = 1 - T(1 - x, 1 - y)$ holds for all $x, y \in [0, 1]$.

The t-conorm $S_M = \max$ is the smallest t-conorm and it is dual to the greatest t-norm $\min$. The dual to the drastic product is the t-conorm S_D given by

$$S_D(x, y) = \begin{cases} 1 & \text{if } (x, y) \in]0, 1]^2 \\ \max(x, y) & \text{otherwise.} \end{cases}$$

For each t-conorm S, we have $S_M \leq S \leq S_D$.

The dual t-conorm to the product $T_\mathbf{P}$ is called the *probabilistic sum* and it is denoted by $S_\mathbf{P}$, with $S_\mathbf{P}(x, y) = x + y - xy$.

The Łukasiewicz t-conorm $S_\mathbf{L}$, called also the *bounded sum*, is given by $S_\mathbf{L}(x, y) = \min(1, x + y)$.

From algebraic point of view, a function $T: [0, 1]^2 \to [0, 1]$ is a t-norm if and only if $([0, 1], T, \le)$ is a fully ordered commutative semigroup with neutral element 1 and annihilator 0. Similarly, a function $S: [0, 1]^2 \to [0, 1]$ is a t-conorm if and only if $([0, 1], S, \le)$ is a fully ordered commutative semigroup with neutral element 0 and annihilator 1.

Clearly, for every t-norm T and strict negation N, the operation S defined by

$$S(x, y) = N^{-1}(T(N(x), N(y))), \quad x, y \in [0, 1] \tag{10.3}$$

is a t-conorm. In addition, if N is a strong negation then $N^{-1} = N$, and we have for $x, y \in [0, 1]$ that $T(x, y) = N(S(N(x), N(y)))$. In this case S and T are called *N-duals*. In case of the standard negation (i.e., when $N = N_s$) we simply speak about *duals*. Obviously, equality (10.3) expresses the De Morgan's law.

Definition 10.3

A triplet (T, S, N) is called a *De Morgan triplet* if and only if T is a t-norm, S is a t-conorm, N is a strong negation and they satisfy (10.3).

It is worth noting that, given a De Morgan triplet (T, S, N), the tuple $([0, 1], T, S, N, 0, 1)$ can never be a Boolean algebra [10.18]: in order to satisfy distributivity we must have $T = \min$ and $S = \max$, in which case it is impossible to have both $T(x, N(x)) = 0$ and $S(x, N(x)) = 1$ for all $x \in [0, 1]$. Depending on the operations used, one can, however, obtain rather general and useful structures such as, for instance, De Morgan algebras, residuated lattices, l-monoids, Girard algebras, MV-algebras, see [10.10].

There are several examples of De Morgan triplets. We list in Table 10.1 those ones that are related to the examples above. Let φ be an automorphism of the unit interval, N_φ the corresponding strong negation, and $x, y \in [0, 1]$.

A function $K: [0, 1]^2 \to [0, 1]$ will often be called a *binary operation* on $[0, 1]$. For an automorphism φ of $[0, 1]$, the *φ-transform* K_φ of such a K is defined by $K_\varphi(x, y) = \varphi^{-1}(K(\varphi(x), \varphi(y)))$, $x, y \in [0, 1]$. Thus, Table 10.1 contains φ-transforms of some *fundamental* t-norms and t-conorms.

Continuous Archimedean t-Norms and t-Conorms

A broad class of problems consists of the representation of multi-place functions in general by composition of *simpler* functions and functions of fewer variables (see *Ling* [10.19] for a brief survey), such as

$$K(x, y) = g(f(x) + f(y)),$$

where K is a two-place function and f, g are real functions. In that general framework, the representation of (two-place) associative functions by appropriate one-place functions is a particular problem. It was *Abel* who first obtained such a representation in 1826 [10.20], by assuming also commutativity, strict monotonicity and differentiability. Since Abel's result, a lot of contributions have been made to representations of associative functions (and generally speaking, of abstract semigroups).

For any $x \in [0, 1]$, any $n \in \mathbb{N}$, and for any associative binary operation K on $[0,1]$, denote $x_K^{(n)}$ the n-th power of x defined by

$$x_K^{(1)} = x,$$

and

$$x_K^{(n)} = K(\underbrace{x, \ldots, x}_{n\text{-times}}) \text{ for } n \ge 2.$$

Definition 10.4

A t-norm T (resp. a t-conorm S) is said to be:

a) *Continuous* if T (resp. S) as a function is continuous on the unit interval;
b) *Archimedean* if for each $(x, y) \in]0, 1[^2$ there is an $n \in \mathbb{N}$ such that $x_T^{(n)} < y$ (resp. $x_S^{(n)} > y$).

Note that the definition of the Archimedean property is borrowed from the theory of semigroups.

Table 10.1 Some N_φ-dual triangular norms and conorms

T	S
$\min(x, y)$	$\max(x, y)$
$\varphi^{-1}(\varphi(x)\varphi(y))$	$\varphi^{-1}(\varphi(x) + \varphi(y) - \varphi(x)\varphi(y))$
$\varphi^{-1}(\max(\varphi(x) + \varphi(y) - 1, 0))$	$\varphi^{-1}(\min(\varphi(x) + \varphi(y), 1))$
$\begin{cases} 0 & \text{if } \varphi(x) + \varphi(y) \le 1 \\ \min(x, y) & \text{otherwise} \end{cases}$	$\begin{cases} \max(x, y) & \text{if } \varphi(x) + \varphi(y) < 1 \\ 1 & \text{otherwise} \end{cases}$

We state here the representation theorem of *continuous Archimedean t-norms and t-conorms* attributed very often to *Ling* [10.19]. In fact, her main theorem can be deduced from previously known results on topological semigroups, see [10.21–23]. Nevertheless, the advantage of Ling's approach is twofold: treating two different cases in a unified manner and establishing elementary proofs.

Theorem 10.2

A t-norm T is continuous and Archimedean if and only if there exists a strictly decreasing and continuous function $t : [0, 1] \to [0, \infty]$ with $t(1) = 0$ such that

$$T(x, y) = t^{(-1)}(t(x) + t(y)) \qquad (x, y \in [0, 1]) ,$$

(10.4)

where $t^{(-1)}$ is the pseudoinverse of t defined by

$$t^{(-1)}(x) = \begin{cases} t^{-1}(x) & \text{if } x \le t(0) \\ 0 & \text{otherwise.} \end{cases}$$

Moreover, representation (10.4) is unique up to a positive multiplicative constant. [10.19]

We say that T *is generated by t* if T has representation (10.4). In this case t is said to be an *additive generator* of T.

Theorem 10.3

A t-conorm S is continuous and Archimedean if and only if there exists a strictly increasing and continuous function $s : [0, 1] \to [0, \infty]$ with $s(0) = 0$ such that

$$S(x, y) = s^{(-1)}(s(x) + s(y)) \qquad (x, y \in [0, 1]) ,$$

(10.5)

where $s^{(-1)}$ is the pseudoinverse of s defined by

$$s^{(-1)}(x) = \begin{cases} s^{-1}(x) & \text{if } x \le s(1) \\ 1 & \text{otherwise.} \end{cases}$$

Moreover, representation (10.5) is unique up to a positive multiplicative constant. [10.19]

We say that a continuous Archimedean t-conorm S *is generated by s* if S has representation (10.5). In this case s is said to be an *additive generator* of S.

Remark that *Aczél* published the representation of strictly increasing, continuous and associative two-place functions on open or half-open real intervals [10.24–26]. This was the starting point to be generalized by *Ling* [10.19].

Definition 10.5

We say that a t-norm T has *zero divisors* if there exist $x, y \in]0, 1[$ such that $T(x, y) = 0$. T is said to be *positive* if $x, y > 0$ imply $T(x, y) > 0$. A t-norm T or a t-conorm S is called *strict* if it is a continuous and strictly increasing function in each place on $]0, 1[^2$. T is called *nilpotent* if it is continuous and Archimedean with zero divisors. Triangular conorms which are duals of nilpotent t-norms are also called *nilpotent*.

The representation theorem of t-norms (resp. t-conorms) does not indicate any condition on the value of a generator function at 0 (resp. at 1). On the basis of this value, one can classify continuous Archimedean t-norms (resp. t-conorms) as it is stated in the following theorem.

Theorem 10.4

Let T be a continuous Archimedean t-norm with additive generator t, and S be a continuous Archimedean t-conorm with additive generator s. Then:

a) T is nilpotent if and only if $t(0) < +\infty$;
b) T is strict if and only if $t(0) = \lim_{x \to 0} t(x) = +\infty$;
c) S is nilpotent if and only if $s(1) < +\infty$;
d) S is strict if and only if $s(1) = \lim_{x \to 1} s(x) = +\infty$.

Using the general representation theorem of continuous Archimedean t-norms, we can give another form of representation for a class of continuous t-norms with zero divisors. More exactly, for continuous t-norms T such that $T(x, N(x)) = 0$ holds with a *strict negation N* for all $x \in [0, 1]$. Such t-norms are Archimedean, as it was proved in [10.27]. The following theorem is established after [10.28].

Theorem 10.5

A continuous t-norm T is such that $T(x, N(x)) = 0$ holds for all $x \in [0, 1]$ with a strict negation N if and only if there exists an automorphism φ of the unit interval such that for all $x, y \in [0, 1]$ we have

$$T(x, y) = \varphi^{-1}(\max\{\varphi(x) + \varphi(y) - 1, 0\})$$

(10.6)

and

$$N(x) \le \varphi^{-1}(1 - \varphi(x)) .$$

As a consequence, we obtain that any nilpotent t-norm is isomorphic to (i. e., is a φ-transform of) the Łukasiewicz t-norm $T_L(x, y) = \max(x + y - 1, 0)$. Similarly, any strict t-norm T is isomorphic to the algebraic product: there is an automorphism φ of $[0, 1]$ such that $T(x, y) = \varphi^{-1}(\varphi(x)\varphi(y))$, for all $x, y \in [0, 1]$.

Similar statements can be proved for t-conorms, see [10.29] for more details. For instance, any strict t-conorm is isomorphic to the probabilistic sum S_P, and any nilpotent t-conorm is isomorphic to the bounded sum S_L.

Continuous t-Norms and t-Conorms

Suppose that $\{[\alpha_m, \beta_m]\}_{m \in M}$ is a countable family of non-overlapping, closed, proper subintervals of $[0,1]$. With each $[\alpha_m, \beta_m]$ associate a continuous Archimedean t-norm T_m. Let T be a function defined on $[0, 1]^2$ by

$$T(x, y)$$
$$= \begin{cases} \alpha_m + (\beta_m - \alpha_m)\, T_m\left(\dfrac{x - \alpha_m}{\beta_m - \alpha_m}, \dfrac{y - \alpha_m}{\beta_m - \alpha_m}\right) \\ \qquad \text{if } (x, y) \in [\alpha_m, \beta_m]^2 \\ \min(x, y) \qquad \text{otherwise .} \end{cases}$$
$$(10.7)$$

Then T is a continuous t-norm. In this case T is called the *ordinal sum* of $\{([\alpha_m, \beta_m], T_m)\}_{m \in M}$ and each T_m is called a *summand*.

Similar construction works for t-conorms S. Just replace T_m with a continuous Archimedean t-conorm S_m, and min with max in (10.7). Thus defined S is a continuous t-conorm, called the *ordinal sum* of $\{([\alpha_m, \beta_m], S_m)\}_{m \in M}$, where each S_m is called a *summand*.

Assume now that T is a continuous t-norm. Then, T is either the minimum, or T is Archimedean, or there exist a family $\{([\alpha_m, \beta_m], T_m)\}_{m \in M}$ with continuous Archimedean summands T_m such that T is equal to the ordinal sum of this family, see [10.19, 22]. It has also been proved there that a continuous t-conorm S is either the maximum, or Archimedean, or there exist a family $\{([\alpha_m, \beta_m], S_m)\}_{m \in M}$ with continuous Archimedean summands S_m such that S is the ordinal sum of this family.

Parametric Families of Triangular Norms

We close this subsection by giving taste of the wide variety of parametric t-norm families. For a comprehensive list with detailed properties please look in [10.10].

Frank t-norms $\{T_\lambda^F\}_{\lambda \in [0, \infty]}$. Let $\lambda > 0, \lambda \neq 1$ be a real number. Define a continuous Archimedean t-norm T_λ^F in the following way

$$T_\lambda^F(x, y) = \log_\lambda \left(1 + \frac{(\lambda^x - 1)(\lambda^y - 1)}{\lambda - 1}\right)$$
$$\times (x, y \in [0, 1]) .$$

We can extend this definition for $\lambda = 0, \lambda = 1$ and $\lambda = \infty$ by taking the appropriate limits. Thus we get T_0^F, T_1^F and T_∞^F as follows

$$T_0^F(x, y) = \lim_{\lambda \to 0} T_\lambda^F(x, y) = \min\{x, y\},$$
$$T_1^F(x, y) = \lim_{\lambda \to 1} T_\lambda^F(x, y) = xy,$$
$$T_\infty^F(x, y) = \lim_{\lambda \to \infty} T_\lambda^F(x, y) = \max\{x + y - 1, 0\}.$$

Each T_λ^F is a strict t-norm for $\lambda \in]0, \infty[$. The corresponding additive generators t_λ^F are given by

$$t_\lambda^F(x) = \begin{cases} -\log x & \text{if } \lambda = 1 \\ -\log \frac{\lambda^x - 1}{\lambda - 1} & \text{if } \lambda \in]0, \infty[, \lambda \neq 1 \end{cases} .$$

The family $\{T_\lambda^F\}_{\lambda \in [0, \infty]}$ is called the *Frank family* of t-norms (see *Frank* [10.30]). Note that members of this family are decreasing functions of the parameter λ (see e.g. [10.31]).

The De Morgan law enables us to define the Frank family of t-conorms $\{S_\lambda^F\}_{\lambda \in [0, \infty]}$ by

$$S_\lambda^F(x, y) = 1 - T_\lambda^F(1 - x, 1 - y)$$

for any $\lambda \in [0, \infty]$.

In [10.30] one can find the following interesting characterization of these parametric families.

Theorem 10.6

A t-norm T and a t-conorm S satisfy the functional equation

$$T(x, y) + S(x, y) = x + y \qquad (x, y \in [0, 1]) \qquad (10.8)$$

if and only if

a) there is a number $\lambda \in [0, \infty]$ such that $T = T_\lambda^F$ and $S = S_\lambda^F$, or

b) T is representable as an ordinal sum of t-norms, each of which is a member of the family $\{T_\lambda^F\}$, $0 < \lambda \leq \infty$, and S is obtained from T via (10.8). [10.30]

Hamacher t-norms $\{T_\lambda^H\}_{\lambda\in[0,\infty]}$. Let us define three parameterized families of t-norms, t-conorms and strong negations, respectively, as follows.

$$T_\lambda^H(x,y) = \frac{xy}{\lambda+(1-\lambda)(x+y-xy)}, \quad \lambda \geq 0 ,$$

$$S_\beta^F(x,y) = \frac{x+y+(\beta-1)xy}{1+\beta xy}, \quad \beta \geq -1 ,$$

$$N_\gamma(x) = \frac{1-x}{1+\gamma x}, \quad \gamma > -1 .$$

Hamacher proved the following characterization theorem [10.32].

Theorem 10.7
(T,S,N) is a De Morgan triplet such that

$$T(x,y) = T(x,z) \Longrightarrow y = z ,$$
$$S(x,y) = S(x,z) \Longrightarrow y = z ,$$
$$\forall z \leq x \;\; \exists y,y' \text{ such that } T(x,y) = z ,$$
$$S(z,y') = x$$

and T and S are rational functions if and only if there are numbers $\lambda \geq 0, \beta \geq -1$ and $\gamma > -1$ such that $\lambda = \frac{1+\beta}{1+\gamma}$ and $T = T_\lambda^F, S = S_\beta^F$ and $N = N_\gamma$.

Remark that another characterization of the Hamacher family of t-norms with positive parameter has been obtained in [10.33] as solutions of a functional equation.

Dombi t-norms $\{T_\lambda^D\}_{\lambda\in[0,\infty]}$. The formula for this t-norm family is given by

$$T_\lambda^D(x,y) = \begin{cases} T_D(x,y) & \text{if } \lambda = 0 \\ T_M(x,y) & \text{if } \lambda = \infty \\ \frac{1}{1+\left(\left(\frac{1-x}{x}\right)^\lambda+\left(\frac{1-y}{y}\right)^\lambda\right)^{\frac{1}{\lambda}}} & \text{if } \lambda \in]0,\infty[\end{cases} .$$

Essential properties of these t-norms and other well-known families can be found in [10.31].

10.3.3 Fuzzy Implications

Turning to the interpretation of the implication $\rightarrow$ in fuzzy logics, it becomes apparent that there are several logical formulae which, in the Boolean two-valued logic, are equivalent to the implication, but give rise to different interpretations when replacing $\{0,1\}$ by

the unit interval $[0,1]$. As in the case of the negation, we call an interpretation of the implication $\rightarrow$ also an implication (or sometimes, a fuzzy implication). A comprehensive study of fuzzy implications can be found in the book [10.34].

In a very broad sense, any function $I : [0,1]^2 \rightarrow [0,1]$ which is decreasing/increasing and preserves the values of the crisp implication on $\{0,1\}$ is considered as a fuzzy implication.

Definition 10.6
A function $I : [0,1]^2 \rightarrow [0,1]$ is called a *fuzzy implication* if and only if it satisfies the following conditions:

I1. $I(0,0) = I(0,1) = I(1,1) = 1; I(1,0) = 0$.
I2. If $x \leq z$ then $I(x,y) \geq I(z,y)$ for all $y \in [0,1]$.
I3. If $y \leq t$ then $I(x,y) \leq I(x,t)$ for all $x \in [0,1]$.

The reason behind I1 is obvious, while a fuzzy implication is required to be decreasing/increasing (i. e., I2 and I3 should be satisfied) because it measures that the consequent is more true than the antecedent [10.35].

Clearly, a fuzzy implication I has the following properties (as a consequence of the definition):

I4. $I(0,x) = 1$ for all $x \in [0,1]$.
I5. $I(x,1) = 1$ for all $x \in [0,1]$.

Note that originally we defined a fuzzy implication in a slightly different form [10.29, Definition 1.15], which is equivalent to Definition 10.6.

Further properties may be required for a fuzzy implication that can be important also in some applications:

I6. $I(1,x) = x$ for all $x \in [0,1]$. [10.36]
I7. $I(x,I(y,z)) = I(y,I(x,z))$ for all $x,y,z \in [0,1]$. [10.36]
I8. $x \leq y$ if and only if $I(x,y) = 1$ for all $x,y \in [0,1]$. [10.37]
I9. $N(x) = I(x,0)$ is a strong negation $(x \in [0,1])$.
I10. $I(x,y) \geq y$ for all $x,y \in [0,1]$. [10.38]
I11. $I(x,x) = 1$ for all $x \in [0,1]$. [10.39]
I12. $I(x,y) = I(N(y),N(x))$ with a strong negation N, for all $x,y \in [0,1]$.
I13. I is a continuous function.

Property I6 yields that tautology cannot justify anything. Condition I7 is called the exchange principle, and is based on the following equivalence:

if P_1 then (if P_2 then P_3) $\Longleftrightarrow$ *if $(P_1$ AND $P_2)$ then P_3* .

I8 expresses that implication defines an ordering, I9 reflects that $P \to Q = \neg P$ if Q is false. I10 is the numerical counterpart of $P \to (Q \to P)$. I11 is called the identity principle and it yields that $P \to P$ is always true. I12, the contraposition law (or in other words, the contrapositive symmetry), expresses a relationship between modus ponens and modus tollens, see [10.35]. In general, this is a strong condition, see [10.17]. I13 prevents implication from reacting in a chaotic way to a small change of the truth value of either the antecedent or the consequent. This is also a fairly restrictive condition.

Fuzzy Implications Defined by t-Norms, t-Conorms and Negations

To be consistent, implications and conjunctions (or implications and disjunctions) cannot be studied independently. Thus, we introduce two particular classes of fuzzy implications based on t-norms, t-conorms and negations. These were identified in [10.36, 40–44].

For a left-continuous t-norm T, its T-residuum [10.10] I_T generalizes the Boolean implication, we prefer the name R-implication for I_T (see the next definition and the Remark after that).

Another way to introduce an implication (called S-implication) which is an extension of the Boolean implication is to exploit the fact that, in a two-valued logic, the formulae $p \to q$ and $\neg p \vee q$ are equivalent.

Definition 10.7

Suppose (T, S, N) is a De Morgan triplet.

An R-implication I_T associated with the t-norm T is defined by

$$I_T(x, y) = \sup\{z | T(x, z) \leq y\} \quad (x, y \in [0, 1]) . \tag{10.9}$$

An S-implication $I_{S,N}$ associated with the t-conorm S and the strong negation N is defined by

$$I_{S,N}(x, y) = S(N(x), y) \quad (x, y \in [0, 1]) . \tag{10.10}$$

It is easy to see that both I_T and $I_{S,N}$ satisfy properties I1–I3 for any t-norm T, t-conorm S and strong negation N, thus they are fuzzy implications. Note that if T is a continuous Archimedean t-norm with additive generator t then

$$I_T(x, y) = t^{-1}(\max\{t(y) - t(x), 0\}) \quad (x, y \in [0, 1]) .$$

Let us emphasize an important link between R-implications defined by left-continuous t-norms, and residuums in lattice-ordered monoids.

Assume that L is a non-empty set, $(L, \preceq)$ is a lattice and $(L, *)$ is a semigroup with neutral element. We introduce some definitions, for more details see [10.45].

i) The triplet $(L, *, \preceq)$ is called a *lattice-ordered monoid* (or an *l-monoid*) if for all $x, y, z \in L$ we have:
   LM1) $x * (y \vee z) = (x * y) \vee (x * z)$,
   LM2) $(x \vee y) * z = (x * z) \vee (y * z)$.
ii) An l-monoid $(L, *, \preceq)$ is said to be *commutative* if the semigroup $(L, *)$ is commutative.
iii) A commutative l-monoid $(L, *, \preceq)$ is called a *commutative, residuated l-monoid* if there exists a further binary operation $\to *$ on L, i.e., a function $\to * : L^2 \to L$ (the *-residuum), such that for all $x, y, z \in L$ we have (R) $x * y \preceq z$ if and only if $x \preceq y \to * z$.
iv) An l-monoid $(L, *, \preceq)$ is called *integral* if there is a greatest element in the lattice $(L, \preceq)$ (often called the universal upper bound) which coincides with the neutral element of the semigroup $(L, *)$.

It is evident that $([0, 1], T, \leq)$ is a commutative integral l-monoid if and only if the function $T : [0, 1]^2 \to [0, 1]$ is a t-norm. It turns out that the left-continuity of a t-norm can be characterized by the fact that the corresponding l-monoid is residuated. In this case the T-residuum I_T is given by (10.9), see [10.10]. Because of its interpretation in $[0, 1]$-valued logics, the T-residuum is also called a *residual implication* (or briefly, an *R-implication*).

Given a left-continuous t-norm T, the R-implication I_T is left-continuous in its first and right-continuous in its second argument, and it is continuous if and only if the underlying t-norm is nilpotent [10.29, 46].

For the sake of completeness we mention a third type of connectives used in quantum logic and called QL-implication defined as follows

$$I_{T,S,N}(x, y) = S(N(x), T(x, y)) \quad (x, y \in [0, 1]) . \tag{10.11}$$

For the idea behind QL-implications, see [10.47]. In general, I_T, S, N violates property I2, so it is not a fuzzy implication in the sense of Definition 10.6. Conditions under that I2 is satisfied can be found in [10.17].

Negations Defined by Implications

As we have seen, several types of negations can be introduced in fuzzy logic. The link between fuzzy implications and negations can be expressed by requiring that the function N defined by

$$N(x) = I(x, 0) \quad \text{for all } x \in [0, 1],$$

be a negation, where I is a fuzzy implication. This is motivated by the corresponding classical rule.

Suppose that $I : [0, 1]^2 \to [0, 1]$ is a function satisfying I3, I7, I8 and define $N(x) = I(x, 0)$ ($x \in [0, 1]$). Then (a) N is a negation; (b) $x \leq N(N(x))$ for all $x \in [0, 1]$;

(c) $N(N(N(x))) = N(x)$ for all $x \in [0, 1]$. If, in addition, N is continuous then it is involutive [10.29]. Thus, under the above conditions, N cannot be a noninvolutive strict negation: it is either discontinuous or a strong negation. If N is continuous then I fulfils I12 with this N.

For a positive t-norm T (like min or the algebraic product), the negation obtained via its R-implication is not continuous at all. In fact, in this case we have that

$$I_T(x, 0) = \begin{cases} 1 & \text{if } x = 0 \\ 0 & \text{if } x > 0 \end{cases} \quad (x \in [0, 1]).$$

10.4 Concluding Remarks

The study of fuzzy implications, triangular norms and their extensions is a never ending story. During such research, fundamental new properties and classes have been discovered, essential results have been proved and applied to diverse problem classes. These are beyond the goal of the present chapter. Nevertheless, we name just a few directions.

Firstly, we mention *uninorms* [10.48, 49], a joint extension of both t-norms and t-conorms, with neutral element being an arbitrary number between 0 and 1. Their study includes the Frank functional equation [10.50], their residual operators [10.51], different extensions [10.52, 53], characterizing their mathematical properties [10.54] and important subclasses such as idempotent [10.55] and representable uninorms

[10.56]. It turns out that some special uninorms have already been hidden, without using the name uninorm, in the classical expert system MYCIN [10.57].

Secondly, some recent papers on fuzzy implications are briefly listed, in which the interested reader can find further references. After the book [10.34] was published, several important contributions have been made by the authors themselves, like [10.58]. Some *algebraic properties* of fuzzy implications such as distributivity [10.59] and contrapositive symmetry [10.60], the law of importation or the exchange principle [10.61], typically in the form of functional equations, have also been studied intensively. New construction methods have also been introduced and deeply studied [10.62–65].

References

10.1   L. Zadeh: Fuzzy sets, Inf. Control **8**, 338–353 (1965)
10.2   C. Hartshorne, P. Weiss (Eds.): *Principles of Philosophy*, Collected Papers of Charles Sanders Peirce (Harvard University Press, Cambridge 1931)
10.3   B. Russell: Vagueness, Austr. J. Philos. **1**, 84–92 (1923)
10.4   J. Łukasiewicz: Philosophical remarks on many-valued systems of propositional logic. In: *Selected Works*, Studies in Logic and the Foundations of Mathematics, ed. by L. Borkowski (North-Holland, Amsterdam 1970) pp. 153–179
10.5   M. Black: Vagueness, Philos. Sci. **4**, 427–455 (1937)
10.6   H. Weyl: The ghost of modality. In: *Philosophical Essays in Memory of Edmund Husserl*, ed. by M. Farber (Cambridge, Cambridge 1940) pp. 278–303

10.7   A. Kaplan, H.F. Schott: A calculus for empirical classes, Methods **III**, 165–188 (1951)
10.8   K. Menger: Ensembles flous et fonctions aleatoires, C. R. Acad. Sci. Paris **232**, 2001–2003 (1951)
10.9   B. Schweizer, A. Sklar: *Probabilistic Metric Spaces* (North-Holland, Amsterdam 1983)
10.10   E.P. Klement, R. Mesiar, E. Pap: *Triangular Norms* (Kluwer, Dordrecht 2000)
10.11   D. Dubois, W. Ostasiewicz, H. Prade: Fuzzy sets: History and basic notions. In: *Fundamentals of Fuzzy Sets*, (Kluwer, Dordrecht 2000), Chap. 1, p. 21–124
10.12   M. Sugeno: Fuzzy measures and fuzzy integrals: A survey. In: *Fuzzy Automata and Decision Processes*, ed. by G.N. Saridis, M.M. Gupta, B.R. Gaines (North-Holland, Amsterdam 1977) pp. 89–102

10.13   E. Trillas: Sobre funciones de negación en la teorí a de conjuntos difusos, Stochastica **III**, 47–60 (1979)

10.14   K. Menger: Statistical metric spaces, Proc. Natl. Acad. Sci. USA **28**, 535–537 (1942)

10.15   C. Alsina, M.J. Frank, B. Schweizer: *Associative Functions: Triangular Norms and Copulas* (Word Scientific, Hoboken 2006)

10.16   E.P. Klement: Operations on fuzzy sets – An axiomatix approach, Inf. Sci. **27**, 221–232 (1982)

10.17   J.C. Fodor: Contrapositive symmetry of fuzzy implications, Fuzzy Sets Syst. **69**, 141–156 (1995)

10.18   R. Sikorski: *Boolean Algebras* (Springer, Berlin 1964)

10.19   C.H. Ling: Representation of associative functions, Publ. Math. Debr. **12**, 189–212 (1965)

10.20   N.H. Abel: Untersuchung der Fuctionen zweier unabhängig veränderlichen Grössen x und y wie $f(x, y)$, welche die Eigenschaft haben, dass $f(z. f(x, y))$ eine symmetrische Function von x, y und z ist, J. Reine Angew. Math. **1**, 11–15 (1826)

10.21   W.M. Faucett: Compact semigroups irreducibly connected between two idempotents, Proc. Am. Math. Soc. **6**, 741–747 (1955)

10.22   P.S. Mostert, A.L. Shields: On the structure of semigroups on a compact manifold with boundary, Annu. Math. **65**, 117–143 (1957)

10.23   A.B. Paalman-de Mirinda: *Topological Semigroups*, Technical Report (Mathematisch Centrum, Amsterdam 1964)

10.24   J. Aczél: Über eine Klasse von Funktionalgleichungen, Comment. Math. Helv. **54**, 247–256 (1948)

10.25   J. Aczél: Sur les opérations définies pour des nombres réels, Bull. Soc. Math. Fr. **76**, 59–64 (1949)

10.26   J. Aczél: *Lectures on Functional Equations and their Applications* (Academic, New York 1966)

10.27   S. Ovchinnikov, M. Roubens: On fuzzy strict preference, indifference and incomparability relations, Fuzzy Sets Syst. **47**, 313–318 (1992)

10.28   S. Ovchinnikov, M. Roubens: On strict preference relations, Fuzzy Sets Syst. **43**, 319–326 (1991)

10.29   J. Fodor, M. Roubens: *Fuzzy Preference Modelling and Multicriteria Decision Support* (Kluwer, Dordrecht 1994)

10.30   M.J. Frank: On the simultaneous associativity of $F(x, y)$ and $x + y - F(x, y)$, Aeq. Math. **19**, 194–226 (1979)

10.31   E.P. Klement, R. Mesiar, E. Pap: A characterization of the ordering of continuous t-norms, Fuzzy Sets Syst. **86**, 189–195 (1997)

10.32   H. Hamacher: *Über logische Aggrationen nichtbinär explizierter Entscheidungskriterien; Ein axiomatischer Beitrag zur normativen Entscheidungstheorie* (Fischer, Frankfurt 1978)

10.33   J.C. Fodor, T. Keresztfalvi: Characterization of the Hamacher family of t-norms, Fuzzy Sets Syst. **65**, 51–58 (1994)

10.34   M. Baczyński, B. Jayaram: *Fuzzy Implications* (Springer, Berlin 2008)

10.35   P. Smets, P. Magrez: Implication in fuzzy logic, Int. J. Approx. Reason. **1**, 327–347 (1987)

10.36   E. Trillas, L. Valverde: On some functionally expressable implications for fuzzy set theory, Proc. 3rd Int. Seminar on Fuzzy Set Theory (Johannes Kepler Universität, Linz 1981) pp. 173–190

10.37   B.R. Gaines: Foundations of fuzzy reasoning, Int. J. Man-Mach. Stud. **8**, 623–668 (1976)

10.38   R.R. Yager: An approach to inference in approximate reasoning, Int. J. Man-Mach. Stud. **13**, 323–338 (1980)

10.39   W. Bandler, L.J. Kohout: Fuzzy power sets and fuzzy implication operators, Fuzzy Sets Syst. **4**, 13–30 (1980)

10.40   E. Trillas, L. Valverde: On implication and indistinguishability in the setting of fuzzy logic. In: *Management Decision Support Systems using Fuzzy Sets and Possibility Theory*, ed. by J. Kacprzyk, R.R. Yager (Verlag TÜV Rheinland, Köln 1985) pp. 198–212

10.41   H. Prade: Modèles mathématiques de l'imprécis et de l'incertain en vue d'applications au raisonnement naturel, Ph.D. Thesis (Université P. Sabatier, Toulouse 1982)

10.42   S. Weber: A general concept of fuzzy connectives, negations and implications based on t-norms and t-conorms, Fuzzy Sets Syst. **11**, 115–134 (1983)

10.43   D. Dubois, H. Prade: Fuzzy logics and the generalized modus ponens revisited, Int. J. Cybern. Syst. **15**, 293–331 (1984)

10.44   D. Dubois, H. Prade: Fuzzy set-theoretic differences and inclusions and their use in the analysis of fuzzy equations, Control Cybern. **13**, 129–145 (1984)

10.45   G. Birkhoff: *Lattice Theory*, Collected Publications, Vol. 25 (Am. Math. Soc., Providence 1967)

10.46   U. Bodenhofer: *A Similarity-Based Generalization of Fuzzy Orderings*, Schriften der Johannes-Kepler-Universität Linz, Vol. 26 (Universitätsverlag Rudolf Trauner, Linz 1999)

10.47   D. Dubois, H. Prade: Fuzzy sets in approximate reasoning, part 1: Inference with possibility distributions, Fuzzy Sets Syst. **40**, 143–202 (1991)

10.48   R.R. Yager, A. Rybalov: Uninorm aggregation operators, Fuzzy Sets Sys. **80**, 111–120 (1996)

10.49   J.C. Fodor, R.R. Yager, A. Rybalov: Structure of uninorms, Int. J. Uncertain. Fuzziness Knowl.-Based Syst. **5**(4), 411–427 (1997)

10.50   T. Calvo, B. De Baets, J. Fodor: The functional equations of frank and alsina for uninorms and nullnorms, Fuzzy Sets Syst. **120**, 385–394 (2001)

10.51   B. De Baets, J. Fodor: Residual operators of uninorms, Soft Comput. **3**, 89–100 (1999)

10.52   M. Mas, G. Mayor, J. Torrens: T-operators and uninorms on a finite totally ordered set, Int. J. Intell. Syst. **14**, 909–922 (1999)

10.53   M. Mas, M. Monserrat, J. Torrens: On left and right uninorms, Int. J. Uncertain. Fuzziness Knowl.-Based Syst. **9**, 491–507 (2001)

10.54   M. Mas, G. Mayor, J. Torrens: The modularity condition for uninorms and t-operators, Fuzzy Sets Syst. **126**, 207–218 (2002)

10.55   B. De Baets: Idempotent uninorms, Eur. J. Oper. Res. **118**, 631–642 (1999)

10.56   J. Fodor, B. De Baets: A single-point characterization of representable uninorms, Fuzzy Sets Syst. **202**, 89–99 (2012)

10.57   B. De Baets, J.C. Fodor: Van melle?s combining function in mycin is a representable uninorm: An alternative proof, Fuzzy Sets Syst. **104**, 133–136 (1999)

10.58   B. Jayaram, M. Baczyński, R. Mesiar: R-implications and the exchange principle: The case of border continuous t-norms, Fuzzy Sets Syst. **224**, 93–105 (2013)

10.59   M. Baczyński: On two distributivity equations for fuzzy implications and continuous, archimedean t-norms and t-conorms, Fuzzy Sets Syst. **211**, 34–54 (2013)

10.60   M. Baczyński, F. Qin: Some remarks on the distributive equation of fuzzy implication and the contrapositive symmetry for continuous, archimedean t-norms, Int. J. Approx. Reason. **54**, 290–296 (2013)

10.61   S. Massanet, J. Torrens: The law of importation versus the exchange principle on fuzzy implications, Fuzzy Sets Syst. **168**, 47–69 (2011)

10.62   S. Massanet, J. Torrens: On a new class of fuzzy implications: h-implications and generalizations, Inf. Sci. **181**, 2111–2127 (2011)

10.63   S. Massanet, J. Torrens: On some properties of threshold generated implications, Fuzzy Sets Syst. **205**, 30–49 (2012)

10.64   S. Massanet, J. Torrens: Threshold generation method of construction of a new implication from two given ones, Fuzzy Sets Syst. **205**, 50–75 (2012)

10.65   S. Massanet, J. Torrens: On the vertical threshold generation method of fuzzy implication and its properties, Fuzzy Sets Syst. **206**, 32–52 (2013)

11. Fuzzy Relations: Past, Present, and Future

Susana Montes, Ignacio Montes, Tania Iglesias

Relations are used in many branches of mathematics to model concepts like *is lower than*, *is equal to*, etc. Initially, only crisp relations were considered, but in the last years, fuzzy relations have been revealed as a very useful tool in psychology, engineering, medicine, economics or any mathematically based field. A first approach to the concept of fuzzy relations is given in this chapter. Thus, operations among fuzzy relations are defined in general. When considering the particular case of fuzzy binary relations, their main properties are studied. Also, some particular cases of fuzzy binary relations are considered and related among them. Of course, this chapter is just a starting point to study in detail more specialized literature.

The notion of relation plays a central role in various fields of mathematics. As a consequence of that, it is a very important concept in all engineering, science, and mathematically based fields.

Crisp or classical relations show a problem; they do not allow to express partial levels of relationship among two elements. This is a problem in many practical situations since not always an element is *clearly* related to another one. The valued theory arises with the aim of allowing to assign degrees to the relations between alternatives. As it is well known, according to fuzzy set theory [11.1], the connection established among two alternatives admits different degrees of intensity and that intensity is represented by a value in the interval $[0, 1]$. The idea of working with values different from 0 and 1 to express the relationship between two elements was already considered by Łukasiewicz in the 1920s

when he introduced his three-valued logic, and later by *Luce* [11.2] or *Menger* [11.3], but it was *Zadeh* [11.4] who formally defined the concept of a fuzzy (or multi-valued) relation.

In the history of fuzzy mathematics, fuzzy relations were early considered to be useful in various applications: fuzzy modeling, fuzzy diagnosis, and fuzzy control. They also have applications in fields such as psychology, medicine, economics, and sociology. For this reason, they have been extensively investigated. For a contemporary general approach to fuzzy relations one should look at *Bělohláveks* book [11.5], and also to other general publications, as for instance the books by *Klir* and *Yuan* [11.6] and *Turunen* [11.7].

Since this chapter is entirely devoted to fuzzy relations, our aim is to give a detailed introduction to them for a nonexpert reader.

11.1 Fuzzy Relations

Assume that X and Y are two given sets. A *fuzzy relation* R is a mapping from the Cartesian product $X \times Y$ to the interval $[0, 1]$. Therefore, R is basically a fuzzy set in the universe $X \times Y$. This means that, for any $x \in X$ and any $y \in Y$, the value $R(x, y)$ measures the strength with which R connects x with y. If $R(x, y)$ is close to 1, x is related to y by R. If $R(x, y)$ is a value close to 0, then it hardly connects x with y and so on.

This definition can be extended to the Cartesian products of more than two sets and then they are called *n-ary fuzzy relations*. Note that fuzzy sets may be viewed as degenerate, 1-ary fuzzy relations.

Example 11.1

If we consider the case $X = Y = [0, 3]$, we could define the fuzzy relation *approximately equal to* as follows

$$R(x, y) = e^{-|x-y|}, \quad \forall (x, y) \in [0, 3] \times [0, 3]$$

which is represented in Fig. 11.1.

The *domain* of a fuzzy relation R is a fuzzy set on X, whose membership function is given by

$$\text{dom} \, R(x) = \sup_{y \in Y} R(x, y)$$

and the *range* is given by

$$\text{ran} \, R(x) = \sup_{x \in X} R(x, y) \, .$$

When X and Y are finite sets, we can consider a matrix representation for any fuzzy relation. The entry on the line x and column y of the associated matrix is the value $R(x, y)$.

Example 11.2

Let us consider the set X formed by three papers, $X = \{p_1, p_2, p_3\}$, and let Y be a set formed by five different topics $Y = \{t_1, t_2, t_3, t_4, t_5\}$. The fuzzy relation R measuring the degree of relationship of any paper with any topic is defined by

R	t_1	t_2	t_3	t_4	t_5
p_1	1.0	0.7	0.9	0.4	0.2
p_2	0.5	0.8	1.0	0.3	0.9
p_3	0.7	0.5	0.8	0.3	0.8

and its domain is given by the fuzzy subset of X

$$\text{dom} \, R = \{(p_1, 1), (p_2, 1), (p_3, 0.8)\} \, ,$$

and its range by

$$\text{ran} \, R = \{(t_1, 1), (t_2, 0.8), (t_3, 1), (t_4, 0.4), (t_5, 0.9)\} \, .$$

11.1.1 Operations on Fuzzy Relations

All concepts and operations applicable to fuzzy sets are applicable to fuzzy relations as well. Thus, for any fuzzy relations R and Q on the Cartesian product $X \times Y$, we have

- Given a *t*-norm T (for a complete study about *t*-norms and *t*-conorms, we refer to [11.8]), the *T-intersection* (or just intersection if there is not ambiguity) of R and Q is the fuzzy relation on $X \times Y$ defined by

$$R \cap_T Q(x, y) = T(R(x, y), Q(x, y)) \, ,$$
$$\forall (x, y) \in X \times Y \, .$$

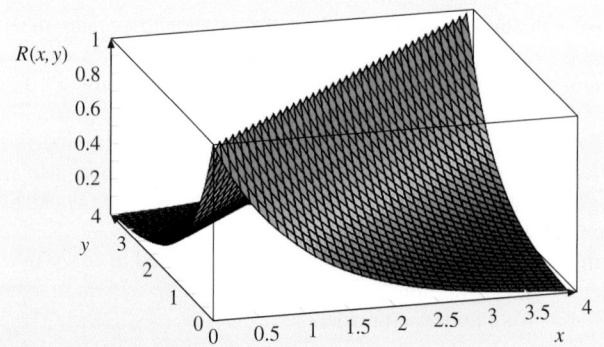

Fig. 11.1 Membership function of the fuzzy relation R (*approximately equal to*) introduced in Example 11.1

Initially, T was considered to be the minimum operator.

● Given a t-conorm S, the *S-union* of R and Q is the fuzzy relation on $X \times Y$ defined by

$$R \cup_S Q(x, y) = S(R(x, y), Q(x, y)),$$
$$\forall(x, y) \in X \times Y.$$

At the initial proposal, the maximum t-conorm was considered.

● The *transpose* or *inverse* of the fuzzy relation R, denoted as in the classical case by R^{-1}, is the fuzzy relation that satisfies

$$R^{-1}(x, y) = R(y, x), \quad \forall(x, y) \in X \times Y.$$

● The *complement* of a fuzzy relation is not unique. It depends on the negator n we choose. The n-complement of the fuzzy relation R, denoted by R^c, is the fuzzy relation defined by

$$R^c(x, y) = n(R(x, y)), \quad \forall(x, y) \in X \times Y.$$

Although the definition is given for any negator, the most widely used one is the standard negator ($n(x) = 1 - x$). In this case, it is called the *standard complement* and is defined by $R^c(x, y) = 1 - R(x, y)$.

● The *dual* of the fuzzy relation R is defined and denoted as in the classical case. The fuzzy relation R^d is the complement of the transpose of R

$$R^d(x, y) = n(R(y, x)), \quad \forall(x, y) \in X \times Y.$$

That is, $R^d = (R^{-1})^c$.

● We say that R is *contained* in Q, and we denote it by $R \subseteq Q$, if and only if for all $(x, y) \in X \times Y$ the inequality $R(x, y) \leq Q(x, y)$ holds.

● R and Q are said to be *equal* if and only if for all $(x, y) \in X \times Y$ we have the inequality $R(x, y) = Q(x, y)$, that is, $R \subseteq Q$ and $Q \subseteq R$.

11.1.2 Specific Operations on Fuzzy Relations

In the previous items, we are only considering that fuzzy relations can be seen as fuzzy sets on $X \times Y$ and we have adapted the corresponding definitions. However, fuzzy relations involve additional concepts and operations. The most important are: compositions, projections and cylindrical extensions, among others.

● Let R and Q two fuzzy relations on $X \times Y$ and $Y \times Z$, respectively. The *T-composition* of R and Q is the

fuzzy relation defined by

$$R \circ_T Q(x, y) = \sup_{y \in Y} T(R(x, y), Q(y, z)),$$
$$\forall(x, z) \in X \times Z.$$

Due to associativity and nondecreasingness of the t-norms, the following result can be easily proven.

Proposition 11.1

Let R, Q, and P be the three fuzzy relations on $X \times Y$, $Y \times Z$, $Z \times U$, respectively. Then:

i) $R \circ_T (Q \circ_T P) = (R \circ_T Q) \circ_T P$,
ii) If R' is another fuzzy relation on $X \times Y$ such that $R \subseteq R'$, then $R \circ_T Q \subseteq R' \circ_T Q$.

● Let R be a fuzzy relation on $X \times Y$. We can project R with respect X and Y as follows:

$$R_X(x) = \sup_{y \in Y} R(x, y), \quad \forall x \in X, \text{ and}$$
$$R_Y(y) = \sup_{x \in X} R(x, y), \quad \forall y \in Y$$

where R_X and R_Y denote the *projected relation of R* to X and Y, respectively.

It is clear that the projection to X coincides with the domain of R and the projection to Y with the range. The definition given for 2-ary fuzzy relations can be generalized to n-ary relations. Thus, if R is a fuzzy relation on $X_1 \times X_2 \times \cdots \times X_n$, the projected relation of R to the subspace $X_{i_1} \times X_{i_2} \times \cdots \times X_{i_k}$ is defined by

$$R_{X_{i_1} \times X_{i_2} \times \cdots \times X_{i_k}}(x_{i_1}, x_{i_2}, \ldots, x_{i_k})$$
$$= \sup_{x_{j_1}, x_{j_2}, \ldots, x_{j_m}} R(x_1, x_2, \ldots, x_n),$$

where $X_{j_1}, X_{j_2}, \ldots, X_{j_m}$ represent the omitted dimensions and $X_{i_1}, X_{i_2}, \ldots, X_{i_k}$ the remained ones. That is

$$\{i_1, i_2, \ldots, i_k\} \cup \{j_1, j_2, \ldots, j_m\} = \{1, 2, \ldots, n\}$$

and

$$\{i_1, i_2, \ldots, i_k\} \cap \{j_1, j_2, \ldots, j_m\} = \emptyset.$$

● Another operation on relations, which is in some sense an inverse to the projection, is called a *cylindrical extension*. If A is a fuzzy subset of X, then its cylindrical extension to Y is the fuzzy relation defined by

$$\text{cyl} A(x, y) = A(x), \quad \forall(x, y) \in X \times Y.$$

11.2 Cut Relations

Any fuzzy relation R has an associated family of crisp relations $\{R_\alpha | \alpha \in [0, 1]\}$, called cut relations, which are defined by

$$R_\alpha = \{(x, y) \in X \times Y | R(x, y) \geq \alpha\}.$$

It is clear that they are just the α-cuts of R, considered as a fuzzy set. Thus, it is immediate that they form a chain (nested family) of relations, that is,

$$\emptyset \subseteq R_{\alpha_m} \subseteq R_{\alpha_{m-1}} \subseteq \cdots \subseteq R_{\alpha_1} \subseteq X \times Y$$

if $0 \leq \alpha_1 \leq \ldots \alpha_{m-1} \leq \alpha_m \leq 1$.

Moreover, it is possible to represent a fuzzy relation by means of its cuts relations, since

$$R(x, y) = \sup_{\alpha \in [0,1]} \min(\alpha, R(x, y)),$$
$$\forall (x, y) \in X \times Y,$$

which is denoted by

$$R = \sup_{\alpha \in [0,1]} \alpha R_\alpha.$$

11.3 Fuzzy Binary Relations

In the particular case $X = Y$, fuzzy relations are called *fuzzy binary relations* or *valued binary relations* and they have specific and interesting properties. A detailed proof of the results presented here can be found in [11.9].

The first specific characteristic of fuzzy binary relations is that, apart from the matrix representation, they admit a graph representation if X is finite. In this *directed graph*, X is the set of nodes (vertices) and R is the set of arcs (edges). The arc from x to y exists if and only if x and y are related in some sense ($R(x, y) > 0$). A number on each arc represents the membership degree of this elements to R.

Example 11.3

If we consider $X = \{x, y, z, t\}$, the fuzzy binary relation

R	x	y	z	t
x	1.0	0.4	0.2	0.0
y	0.6	0.9	0.0	0.0
z	0.0	0.0	0.0	0.0
t	0.0	0.0	0.8	0.0

can also be represented by the graph in Fig. 11.2.

In the following, we will list some basic properties of fuzzy binary relations. Usually, these properties are translations of the equivalent for the particular case of (crisp) binary relations.

11.3.1 Reflexivity

The most used definition of reflexivity for fuzzy binary relations was given by *Zadeh* in 1971 [11.4]. Thus,

a fuzzy binary relation R on X is said to be *reflexive* iff

$$R(x, x) = 1, \quad \forall x \in X.$$

This means that every vertex in the graph originates a simple loop.

Other less restrictive definitions have also been considered in the literature. Thus, we say that R is ϵ-*reflexive* [11.10], with $\epsilon \in (0, 1]$, iff

$$R(x, x) \geq \epsilon, \quad \forall x \in X$$

and *weakly reflexive* [11.10] iff

$$R(x, x) \geq R(x, y), \quad \forall x, y \in X.$$

Of course, in the particular case of crisp binary relations, all of them are the usual definition of reflexivity. Moreover, if R is reflexive, it is ϵ-reflexive for any

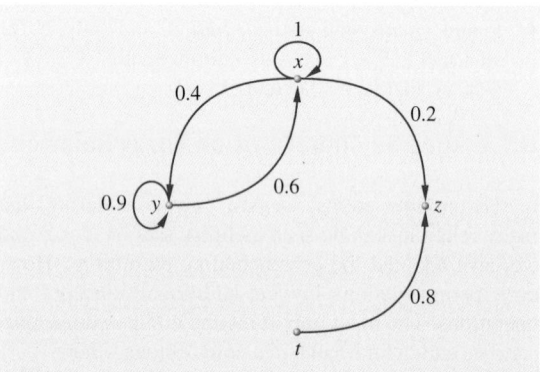

Fig. 11.2 Directed graph associated to a fuzzy binary relation

$\epsilon \in (0, 1]$ and weakly reflexive. The remaining implications are not true in general.

A cutworthy study of this property is given in the following proposition.

Proposition 11.2

Let R be a fuzzy binary relation on X. If R is reflexive, then its associated cut relations R_α, with $\alpha \in (0, 1]$, are reflexive.

The cutworthy property is not fulfilled, in general, by ϵ-reflexive or weakly reflexive fuzzy binary relations.

11.3.2 Irreflexivity

A fuzzy binary relation that is *irreflexive*, or *antireflexive*, is a fuzzy binary relation where no element is related in any degree to itself. Formally

$$R(x, x) = 0, \quad \forall x \in X.$$

This means that there is not any vertex in the graph originating a simple loop.

Analogous to the case of reflexivity, we can consider some generalizations of this concept:

- ϵ-Irreflexive, with $\epsilon \in [0, 1)$: $R(x, x) \leq \epsilon, \forall x \in X$
- Weakly irreflexive: $R(x, x) \leq R(x, y), \forall x, y \in X$.

The behavior of cut relations is similar to the previous case.

Proposition 11.3

Let R be a fuzzy binary relation on X. If R is irreflexive, then its associated cut relations R_α, with $\alpha \in (0, 1]$, are irreflexive.

Again this condition is not fulfilled for ϵ-reflexivity or weak reflexivity.

11.3.3 Symmetry

For symmetry, there is not any change with the classical definition for crisp relations. Thus, a fuzzy binary relation R on X is said to be *symmetric* if

$$R(x, y) = R(y, x), \quad \forall x, y \in X.$$

This is equivalent to require that R and its inverse are equal

$$R = R^{-1}.$$

Of course, if R is symmetric, so are their associated cut relations for any $\alpha \in (0, 1]$.

11.3.4 Antisymmetry

In the crisp case, a binary relation R is antisymmetric iff xRy and yRx which implies that $x = y$. This is equivalent to require that $x \neq y$ implies that $(x, y) \notin R \cap R^{-1}$. Thus, the intersection can be used in order to define antisymmetry. In the fuzzy case, the intersection will be defined, as usual, by means of a t-norm T. The definition will be directly related to the t-norm, and therefore the used t-norm should appear in the name of the property. Thus, a fuzzy binary relation R on X is said to be T-antisymmetric if

$$x \neq y \Longrightarrow T(R(x, y), R(y, x)) = 0$$

or, equivalently

$$R \cap_T R^{-1}(x, y) = 0, \quad \forall x \neq y.$$

It is immediate that if T and T' are t-norms such that $T' \leq T$, then the T-antisymmetry of a fuzzy relation implies its T'-antisymmetry.

In 1971, *Zadeh* [11.4] proposed to use the minimum t-norm for this aim, and he called it *perfect antisymmetry* or just antisymmetry. In that case, its cut relations are antisymmetric, for any $\alpha \in (0, 1]$. This is also true for T-antisymmetry for a positive t-norm $(x, y > 0 \Rightarrow T(x, y) > 0)$, since in this case, T-antisymmetry and perfect antisymmetry are equivalent. However, the cutworthy property is not fulfilled, in general, for any other T-norm.

Clearly, perfect antisymmetry implies the T-antisymmetry for any t-norm T. However, perfect antisymmetry can be too restrictive in many cases, since it excludes relations where $R(x, y)$ and $R(y, x)$ are almost zero. If we consider t-norms with zero divisors, this problem is solved. In this way, the case when $R(x, y)$ and $R(y, x)$ are high is avoided, but the equality to zero is not required. For instance, if we consider the Łukasiewicz t-norm $(T(x, y) = \max(x + y - 1, 0))$, T-antisymmetry is equivalent to require that

$$R(x, y) + R(y, x) \leq 1, \quad \forall x, y \in X \text{ such that } x \neq y.$$

11.3.5 Asymmetry

Asymmetry is a stronger condition, since it is not only required for pairs of elements (x, y) such that $x \neq y$, but also for any pair of elements in $X \times X$. Thus, the T-*asymmetry* of a fuzzy binary relation R on X is defined by

$$T(R(x, y), R(y, x)) = 0, \quad \forall x, y \in X$$

or, equivalently,

$$R \cap_T R^{-1} = \emptyset.$$

Clearly, T-asymmetry implies T'-asymmetry if $T' \leq T$ and then, classic asymmetry ($T = \min$) implies T-asymmetry for any t-norm T. Usually it is called just asymmetry. In particular, asymmetry is equivalent to the T-asymmetry if T is positive. In that case, its associated cut relations are crisp asymmetric relations for any $\alpha \in (0, 1]$.

It is possible to relate asymmetry and irreflexivity as follows:

Proposition 11.4

Let R be a T-asymmetric fuzzy binary relation on X. The following statements hold:

i) R is irreflexive if and only if T is a positive t-norm;
ii) R is ϵ-irreflexive for $\epsilon < 1$ if and only if T has zero divisors and ϵ belongs to the interval $(0, \sup\{x \in [0, 1] | T(x, x) = 0\}]$.

11.3.6 Transitivity

The pairwise comparison of possible alternatives is a first step in many approaches to decision making. If this first step lacks coherence, the whole decision process might become meaningless. A popular criterion for coherence is the transitivity of the involved relations, expressing that the strength of the link between two alternatives cannot be weaker than the strength of any chain involving another alternative [11.11].

The usual definition of transitivity for fuzzy relations is related to a t-norm and it is a generalization of the proposal given by *Zadeh* in 1971 [11.4]. Thus, a fuzzy binary relation R on X is said to be T-*transitive* if

$$T(R(x, y), R(y, z)) \leq R(x, z)$$

for all $x, y, z \in X$.

As it happened for the concepts of T-asymmetry and T-antisymmetry, T-transitivity is not unique as it

happened for classical relations. When T is the minimum t-norm that definition can also be expressed as

$$R(x, z) \geq \max_{y \in X}(\min(R(x, y), R(y, z))), \quad \forall x, z \in X.$$

Then, it is sometimes called *max-min-transitivity*. This coincides with the initial definition proposed by Zadeh.

The T-transitivity is a natural way of extending the original definition by Zadeh, specially after t-norms and t-conorms began to be used in the 1970s by different authors to generalize the intersection and the union. However, many other types of transitivity were defined. From the least restrictive ones as the *minimal transitivity* [11.12] defined by

$$R(x, y) = 1 \text{ and } R(y, z) = 1 \Longrightarrow R(x, z) = 1$$

or the *preference sensitive transitivity* [11.13], also called *quasitransitivity* [11.12], defined by

$$R(x, y) > 0 \quad \text{and} \quad R(y, z) > 0 \Longrightarrow R(x, z) > 0.$$

The *weak* and *parametric transitivities* [11.14] defined, respectively, by

$$R(x, y) > R(y, x) \text{ and } R(y, z) > R(z, y)$$
$$\Longrightarrow R(x, z) > R(z, x)$$

and by

$$R(x, y) > \theta > R(y, x) \text{ and } R(y, z) > \theta > R(z, y)$$
$$\Longrightarrow R(x, z) > \theta > R(z, x)$$

where θ is a fixed value in the interval $[0, 1)$.

Or the *weighted mean transitivity* [11.15] stating that the inequalities $R(x, y) > 0$ and $R(y, z) > 0$ require the existence of some $\theta \in (0, 1)$ such that

$$R(x, z) \geq \theta \cdot \max(R(x, y), R(y, z))$$
$$+ (1 - \theta) \cdot \min(R(x, y), R(y, z))$$

among others.

T-norms offer a way of defining transitivity for fuzzy relations, but it is known that these operators are too restrictive in some cases. If we have a look at the properties an operator defining transitivity must satisfy, associativity is only necessary if we try to extend the definition to more than three elements. Concerning the commutativity and the boundary condition, a much weaker condition is sufficient to generalize the classical definition: commutativity and boundary condition on $\{0, 1\}^2$. Thus, recent studies about transitivity for fuzzy binary relations are not restricted to t-norms, but a much more general definition of transitivity is considered, the one obtained by considering only the necessary conditions [11.16]. Thus, we consider a *conjunctor*, that is an increasing binary operator $f : [0, 1]^2 \rightarrow [0, 1]$ which coincides with the Boolean conjunction on $\{0, 1\}^2$. Recall that this definition preserves the concept of conjunction for classical relations. It is also clear that the notion of conjunctor is much more general than the one of t-norm. Neither associativity nor commutativity are required. Note that conjunctors are even not required to have neutral element 1.

Given a conjunctor f, we can define the f-*transitivity* of a fuzzy relation R in the same way as it is defined for t-norms

$$f(R(x, y), R(y, z)) \leq R(x, z), \quad \forall x, y, z \in X.$$

Since conjunctors are a much wider family of operators, this definition includes more types of transitivity than the definition given just for t-norms. It is trivial that if we restrict this definition to t-norms, we get the classical definition of T-transitivity.

The definition for conjunctors is a too general notion in a particular case: if we consider a reflexive relation R f-transitive, where f is a conjunctor, that conjunctor must be bounded by the minimum. That is, only conjunctors smaller than or equal to minimum can define the transitivity of reflexive fuzzy relations.

11.3.7 Negative Transitivity

Another important property is negative transitivity, which is a dual property of transitivity. Thus, given a t-conorm S, a fuzzy binary relation R on X is said to be *negatively S-transitive* if

$$R(x, z) \leq S(R(x, y), R(y, z))$$

for all $x, y, z \in X$.

If T is a t-norm and S is its dual t-conorm, R is T-transitive if and only if its dual R^d is negatively S-transitive.

Clearly, if R is negatively S-transitive, then it is negatively S'-transitive for any other t-conorm S' such that $S \leq S'$. In particular, the negative transitivity of the maximum implies the negatively S-transitivity for any t-conorm S.

11.3.8 Semitransitivity

In the classical case, a crisp relation R is semitransitive if xRy and yRz implies that there exists $t \in X$ such that xRt or tRz.

If we consider t-norms and t-conorms to generalize AND and OR, respectively, we obtain that a fuzzy binary relation R on X is T-S-*semitransitive* if

$$T(R(x, y), R(y, z)) \leq S(R(x, t), R(t, z))$$

for every $x, y, z, t \in X$.

It is clear that T-S-semitransitivity implies T'-S'-semitransitivity of any t-norm T' such that $T' \leq T$ and any t-conorm S' such that $S \leq S'$. Thus, the classical semitransitivity with the minimum t-norm and the maximum t-conorm implies the T-S-transitivity for any t-norm T and any t-conorm S.

As a consequence of the definition, for any T-S-semitransitive fuzzy relation R we have that:

- If R is reflexive, then it is negatively S-transitive
- If R is irreflexive, then it is T-transitive.

Moreover, we can easily prove the following propositions [11.9].

Proposition 11.5
If R is T-transitive and negatively S-transitive, then R is T-S-semitransitive.

Proposition 11.6
Suppose that T is a continuous t-norm in the De Morgan triple (T, S, n). If R is T-asymmetric and negatively S-transitive then R is T-S-semitransitive.

11.3.9 Completeness

In the crisp set theory, the concept of completeness is clear, the relation R defined on X is complete if every two elements are related by R, that is, if at least xRy or yRx for any pair of values x different from y in X. And it is still equivalent, for reflexive relations, to the concept of strong completeness (xRy or yRx, $\forall x, y \in X$). It is logical that in the setting of classical relations, the completeness is equivalent to the absence of incomparability. There is no pair of elements that cannot be compared since they are related at least by R or R^{-1}.

When we try to generalize this concept to fuzzy relations, the problem arises when trying to fuzzify the notion *related at least by R or R^{-1}*. In the classical case, it is clear that x and y are related by R if and only if $R(x,y) = 1$ or $R(y,x) = 1$. This could be a first way to define the concept of completeness for fuzzy relations.

Thus, a fuzzy binary relation R defined on X is *strongly complete* if

$$\max(R(x,y), R(y,x)) = 1, \quad \forall x, y \in X.$$

Perny and Roy [11.17] call it just complete.

But this condition could be considered too restrictive for fuzzy relations. Consider, for example, the case in which both $R(x,y) = 0.95$ and $R(y,x) = 0.95$. By the definition given above (strong completeness), R is not complete but it is clear that x and y are related by R. Taking into account this type of situations, other less restrictive completeness conditions were proposed.

Among the most employed in the literature, we find the one known as weak completeness. A fuzzy relation R defined on X is *weakly complete* if

$$R(x,y) + R(y,x) \geq 1, \quad \forall x, y \in X.$$

This condition is called *connectedness* in [11.13, 18] while in other works [11.14] this name makes reference to the strong completeness.

It is clear that this definition is much less restrictive than the one called strong completeness. Strongly complete relations are a particular type of weakly complete relations.

If we take a careful look at these two definitions, we can express them by means of a t-conorm. On the one hand, we can quickly identify the maximum t-conorm as the operator that relates R and its transpose in the first definition. On the other hand, since $R(x,y) + R(y,x) \geq 1$ is equivalent to $\min(R(x,y) + R(y,x), 1) = 1$, then the weakly completeness relates R and R^{-1} by means of the Łukasiewicz t-conorm.

These two conditions are special cases of what is called S-completeness. A fuzzy relation R defined on X is called *S-complete* [11.19], where S is a t-conorm, if

$$S(R(x,y), R(y,x)) = 1, \quad \forall x, y \in X.$$

It is immediate that this is equivalent to require that

$$R \cup_S R^{-1} = X \times X.$$

Remark that the previous definition corresponds, according to [11.19], to *strong S-completeness*, while the concept of S-completeness only requires the equality

$S(R(x,y), R(y,x)) = 1$ for pairs of different elements, $x \neq y$.

A direct consequence of the definition is that for any two t-conorms S and S', such that $S \leq S'$, S-completeness (respectively, strong S-completeness) implies S'-completeness (respectively, strong S'-completeness). It is easy to check that strong completeness can be identified not only with the S-completeness defined by the maximum t-conorm, but also with any t-conorm of which the dual t-norm has no zero divisors. The S-completeness of a fuzzy relation R is equivalent to the T-antisymmetry of its dual R^d [11.19], where T is the dual t-norm of S by means of a strong negation n.

The behavior of cut relations are the same as in the previous properties. Thus, if R is a max-complete (resp. strongly max-complete) fuzzy binary relation, then its associated cut relations R_α are complete (resp. strongly complete) crisp binary relations, for any $\alpha \in (0,1]$.

As it happens in the crisp case, we can relate completeness and reflexivity. In this case, we obtain different results depending on the chosen t-norm.

Proposition 11.7
Let R be a strongly S-complete fuzzy binary relation on X:

1. R is reflexive if and only if the dual t-norm associated to S is a positive t-norm.
2. R is ϵ-reflexive with $\epsilon < 1$ if, and only if, S is a nilpotent t-conorm and ϵ belongs to the interval $[\inf\{x \in [0,1] | S(x,x) = 1\}, 1)$.

Table 11.1 Some properties of fuzzy binary relations

Property	Definition
Reflexivity	$R(x,x) = 1, \quad \forall x \in X$
Irreflexivity	$R(x,x) = 0, \quad \forall x \in X$
Symmetry	$R(x,y) = R(y,x), \quad \forall x, y \in X$
T-antisymmetry	$T(R(x,y), R(y,x)) = 0, \quad \forall x, y \in X, x \neq y$
T-asymmetry	$T(R(x,y), R(y,x)) = 0, \quad \forall x, y \in X$
T-transitivity	$T(R(x,y), R(y,z)) \leq R(x,z), \quad \forall x, y, z \in X$
Negative S-transitivity	$R(x,z) \leq S(R(x,y), R(y,z)), \quad \forall x, y, z \in X$
T-S-semitransitivity	$T(R(x,y), R(y,z)) \leq S(R(x,t), R(t,z)), \quad \forall x, y, z, t \in X$
S-completeness	$S(R(x,y), R(y,x)) = 1, \quad \forall x, y \in X$
Strong completeness	$\max(R(x,y), R(y,x)) = 1, \quad \forall x, y \in X$
Weak completeness	$R(x,y) + R(y,x) \geq 1, \quad \forall x, y \in X$
T-linearity	$n_T(R(x,y)) \leq R(y,x), \quad \forall x, y \in X$

Completeness also plays an important role in order to relate transitivity and negative transitivity [11.9]. Thus, given a De Morgan triple (T, S, n), with T a continuous t-norm, and a strongly S-complete fuzzy binary relation R, the T-transitivity of R implies:

- Its negatively S-transitivity
- The T-transitivity of R^d.

11.3.10 Linearity

S-completeness is also very related to the concept of T-linearity. Given a t-norm T, a fuzzy relation R defined on X is called T-*linear* [11.20] if

$$n_T(R(x, y)) \leq R(y, x), \quad \forall x, y \in X,$$

where n_T stands for the negator $n_T(x) = \sup\{z \in [0, 1] | T(x, z) = 0\}$.

S-completeness is equivalent to T-linearity whenever T is nilpotent and S is the dual t-conorm of T by using n_T, that is, $S(x, y) = n_T(T(n_T(x), n_T(y)))$.

As the Łukasiewicz t-norm T_L is in particular a nilpotent t-norm, the weak completeness, that is S_L-completeness, is equivalent to T_L-linearity.

We summarize the properties and definitions we have introduced in this section in Table 11.1.

11.4 Particular Cases of Fuzzy Binary Relations

In this section, we deal with some particular cases of fuzzy binary relations, which are very important in several fields and they generalize classic concepts.

11.4.1 Similarity Relation

The notion of *similarity* is essentially a generalization of the notion of equivalence.

More concretely, a *T-indistinguishability relation R* is a fuzzy binary relation which is reflexive, symmetric, and T-transitive. Sometimes it is also called *fuzzy equivalence relation* or *equality relation*. $R(x, y)$ is interpreted as the degree of indistinguishability (or similarity) between x and y.

In this definition, reflexivity expresses the fact that every object is completely indistinguishable from itself. Symmetry says that the degree in which x and y are indistinguishable is the same as the degree in which y and x are indistinguishable. For transitivity, as it depends on a t-norm, we have a more flexible property. In particular, when we use the product t-norm, we obtain the so-called *possibility relations* introduced by *Menger* in [11.3]; if we choose the Łukasiewicz t-norm, we obtain the relations called *likeness* introduced by *Ruspini* [11.21]; while for the minimum t-norm we obtain *similarity relations* [11.22].

When the transitivity is not required, the relation R is said to be a *proximity relation*.

11.4.2 Fuzzy Order

Next, we make a quick overview on some different fuzzy ordering relations, focusing on the properties they shall satisfy. Consider a fuzzy binary relation R on the set X. R is called:

- *Partial T-preorder* or *T-quasiorder* if R is reflexive and T-transitive;
- *Total T-preorder* or *linear T-quasiorder* if R is strongly complete and T-transitive;
- *Partial T-order* if R is antisymmetric and T-transitive;
- *Strict partial T-order* if R is asymmetric and T-transitive;
- *Total T-order* or *linear T-order* if R is a complete partial T-order, that is, R is antisymmetric, T-transitive and complete;
- *Strict total T-order* if R is a complete strict partial T-order, that is, R is asymmetric, T-transitive and complete.

As in the previous concepts, when T is the t-norm of the minimum, we call R simply total preorder, total order, etc.

The previous definitions are summarized in Table 11.2.

Table 11.2 Fuzzy binary relations by properties

	Reflexivity	Symmetry	Antisymmetry	Asymmetry	Transitivity
Preorder	Yes				Yes
Order	Yes	No	Yes		Yes
Strict order	Yes	No	Yes	Yes	Yes
Proximity	Yes	Yes	No	No	
T-indistinguishability	Yes	Yes	No	No	Yes

11.5 Present and Future of Fuzzy Relations

In this chapter, we have tried to give to the reader a first approach to the concept of fuzzy relations. Of course this is just a starting point. Due to the current development of the topics related to fuzzy relations, a researcher interested in this notion should study in detail more specialized materials.

Here we have presented the definition of classic fuzzy relations, which take values in the interval [0, 1].

However, lattice-valued fuzzy relations are considered as a interesting tool in some areas. In that case, the relation R is a map from $X \times Y$ in a complete lattice L. Some approaches and interesting references for this lattice-valued relations are in [11.23].

A particular interesting case is when the lattice is a chain of linguistic labels. Some approaches in this direction can be found, for instance, in [11.24–26].

References

11.1 L.A. Zadeh: Fuzzy sets, Inf. Control **8**, 338–353 (1965)
11.2 R.D. Luce: *Individual Choice Behavior* (Wiley, New York 1959)
11.3 K. Menger: Probabilistic theories of relations, Proc. Natl. Acad. Sci. USA **37**, 178–180 (1951)
11.4 L.A. Zadeh: Similarity relations and fuzzy ordering, Inf. Sci. **3**, 177–200 (1971)
11.5 R. Bělohlávek: *Fuzzy Relational System* (Kluwer, Dordrecht 2002)
11.6 G.J. Klir, B. Yuan: *Fuzzy Sets and Fuzzy Logic* (Prentice Hall, Upper Saddle River 1995)
11.7 E. Turunen: *Mathematics Behind Fuzzy Logic* (Physica, Heidelberg 1999)
11.8 E.P. Klement, R. Mesiar, E. Pap: *Triangular Norms* (Kluwer, Dordrecht 2000)
11.9 J. Fodor, M. Roubens: *Fuzzy Preference Modelling and Multicriteria Decision Support* (Kluwer, Dordrecht 1994)
11.10 R.T. Yeh: Toward an algebraic theory of fuzzy relational systems, Proc. Int. Congr. Cybern. Namur (1973) pp. 205–223
11.11 D. Dubois, H. Prade: *Fuzzy Sets and Systems Theory and Applications* (Academic, New York 1980)
11.12 P.C. Fishburn: Binary choice probabilities: On the varieties of stochstic transitivity, J. Math. Psychol. **10**, 327–352 (1973)
11.13 C. Alsina: On a family of connectives for fuzzy sets, Fuzzy Sets Syst. **16**, 231–235 (1985)
11.14 J. Fodor: A new look at fuzzy connectives, Fuzzy Sets Syst. **57**, 141–148 (1993)
11.15 J. Azcél, F.S. Roberts, Z. Rosenbaum: On scientific laws without dimensional constants, J. Math. Anal. Appl. **119**, 389–416 (1986)
11.16 S. Díaz, B. De Baets, S. Montes: General results on the decomposition of transitive fuzzy relations, Fuzzy Optim. Decis. Mak. **9**, 1–29 (2010)
11.17 P. Perny, B. Roy: The use of fuzzy outranking relations in preference modelling, Fuzzy Sets Syst. **49**, 33–53 (1992)
11.18 J. Fodor: Traces of fuzzy binary relatios, Fuzzy Sets Syst. **50**, 331–341 (1992)
11.19 J.J. Buckley: Ranking alternatives using fuzzy numbers, Fuzzy Sets Syst. **15**, 21–31 (1985)
11.20 J. Azcél: *Lectures on Functional Equations and Applications* (Academic, New York 1966)
11.21 E. Ruspini: Recent Developments. In: *Fuzzy Clustering*, (Pergamon, Oxford 1982) pp. 133–147
11.22 J. Fodor, M. Roubens: Aggregation of strict preference relations. in MCDM procedures. In: *Fuzzy Approach to Reasoning and Decision-Making*, ed. by V. Novák, J. Ramík, M. Mareš, M. Černý, J. Nekola (Academia, Prague 1992) pp. 163–171
11.23 J. Jiménez, S. Montes, B. Seselja, A. Tepavcevic: Lattice-valued approach to closed sets under fuzzy relations: Theory and applications, Comput. Math. Appl. **62**(10), 3729–3740 (2011)
11.24 L. Martínez, L.G. Pérez, M.I. Barroco: Filling incomplete linguistic preference relations up by priorizating experts' opinions, Proc. EUROFUSE09 (2009) pp. 3–8

11.25 R.M. Rodríguez, L. Martínez, F. Herrera: A group
decision making model dealing with comparative
linguistic expressions based on hesitant fuzzy linguistic term set, Inf. Sci. **241**, 28–42 (2013)

11.26 J.M. Tapia-García, M.J. del Moral, M.A. Martínez,
E. Herrera-Viedma: A consensus model for group
decision making problems with linguistic interval
fuzzy preference relations, Expert Syst. Appl. **39**,
10022–10030 (2012)

Part B | 11

12. Fuzzy Implications: Past, Present, and Future

Michał Baczynski, Balasubramaniam Jayaram, Sebastia Massanet, Joan Torrens

Fuzzy implications are a generalization of the classical two-valued implication to the multi-valued setting. They play a very important role both in the theory and applications, as can be seen from their use in, among others, multivalued mathematical logic, approximate reasoning, fuzzy control, image processing, and data analysis. The goal of this chapter is to present the evolution of fuzzy implications from their beginnings to the current days. From the theoretical point of view, we present the basic facts, as well as the main topics and lines of research around fuzzy implications. We also devote a specific section to state and recall a list of main application fields where fuzzy implications are employed, as well as another one to the main open problems on the topic.

Part B | 12

Fuzzy logic connectives play a fundamental role in the theory of fuzzy sets and fuzzy logic. The basic fuzzy connectives that perform the role of generalized *And*, *Or*, and *Not* are *t*-norms, *t*-conorms, and negations, respectively, whereas fuzzy conditionals are usually managed through fuzzy implications. Fuzzy implications play a very important role both in theory and applications, as can be seen from their use in, among others, multivalued mathematical logic, approximate reasoning, fuzzy control, image processing, and data analysis. Thus, it is hardly surprising that many researchers have devoted their efforts to the study of implication functions. This interest has become more evident in the last decade when many works have appeared and have led to some surveys [12.1, 2] and even some research monographs entirely devoted to this topic [12.3, 4]. Thus, most of the known results and applications of fuzzy implications until the publication date were collected in [12.3], and very recently the edited volume [12.4] has been published complimenting the earlier monograph

with the most recent lines of investigation on fuzzy implications.

In this regard, we have decided to devote this chapter, as the title suggests, to present the evolution of fuzzy implications from their beginnings to the present time. The idea is not to focus on a list of results already collected in other works, but unraveling the relations and highlighting the importance in the development and progress that fuzzy implications have experienced along the time. From the theoretical point of view we present the basic facts, as well as the main topics and lines of research around fuzzy implications, recalling in most of the cases where the corresponding results can be found, instead of listing them. Of course, we also devote a specific section to state and recall a list of the main application fields where fuzzy implications are employed. A final section looks ahead to the future by listing some of the main open-problem-solutions of which are certain to enrich the existing literature on the topic.

12.1 Fuzzy Implications: Examples, Properties, and Classes

Fuzzy implications are a generalization of the classical implication to fuzzy logic. It is a well-established fact that fuzzy concepts have to generalize the corresponding crisp one, and consequently fuzzy implications restricted to $\{0,1\}^2$ must coincide with the classical implication. Currently, the most accepted definition of a fuzzy implication is the following one.

Definition 12.1 [12.3, Definition 1.1.1]
A function $I:[0,1]^2 \to [0,1]$ is called a *fuzzy implication* if it satisfies the following conditions:

(I1) $I(x,z) \geq I(y,z)$ when $x \leq y$, for all $z \in [0,1]$
(I2) $I(x,y) \leq I(x,z)$ when $y \leq z$, for all $x \in [0,1]$
(I3) $I(0,0) = I(1,1) = 1$ and $I(1,0) = 0$.

This definition is flexible enough to allow uncountably many fuzzy implications. This great repertoire of fuzzy implications allows a researcher to pick out, depending on the context, that fuzzy implication which satisfies some desired additional properties. Many additional properties, all of them arising from tautologies in classical logic, have been postulated in many works. The most important of them are collected below:

- *(NP)*: The *left neutrality principle*,

$$I(1,y) = y, \quad y \in [0,1].$$

- *(EP)*: The *exchange principle*,

$$I(x,I(y,z)) = I(y,I(x,z)), \quad x,y,z \in [0,1].$$

- *(OP)*: The *ordering property*,

$$x \leq y \iff I(x,y) = 1, \quad x,y \in [0,1].$$

- *(IP)*: The *identity principle*,

$$I(x,x) = 1, \quad x \in [0,1].$$

- *(CP(N))*: The *contrapositive symmetry* with respect to a fuzzy negation N,

$$I(x,y) = I(N(y),N(x)), \quad x,y \in [0,1].$$

Given a fuzzy implication I, its *natural negation* is defined as $N_I(x) = I(x,0)$ for all $x \in [0,1]$. This function is always a fuzzy negation. For the definitions of basic fuzzy logic connectives like fuzzy negations, t-norms and t-conorms please see [12.5]. Moreover, N_I can be continuous, strict, or strong and these are also additional properties usually required of a fuzzy implication I.

Table 12.1 lists the most well-known fuzzy implications along with the additional properties they satisfy [12.3, Chap.1]. In addition, the following

Table 12.1 Basic fuzzy implications and the additional properties they satisfy where N_C, N_{D_1}, and N_{D_2} stand for the classical, the least and the greatest fuzzy negations, respectively

Name	Formula	(NP)	(EP)	(IP)	(OP)	(CP(N))	N_I
Łukasiewicz	$I_{LK}(x,y) = \min\{1, 1-x+y\}$	✓	✓	✓	✓	N_C	N_C
Gödel	$I_{GD}(x,y) = \begin{cases}1 & \text{if } x \leq y \\ y & \text{if } x > y\end{cases}$	✓	✓	✓	✓	X	N_{D_1}
Reichenbach	$I_{RC}(x,y) = 1-x+xy$	✓	✓	X	X	N_C	N_C
Kleene–Dienes	$I_{KD}(x,y) = \max\{1-x, y\}$	✓	✓	X	X	N_C	N_C
Goguen	$I_{GG}(x,y) = \begin{cases}1 & \text{if } x \leq y \\ \frac{y}{x} & \text{if } x > y\end{cases}$	✓	✓	✓	✓	X	N_{D_1}
Rescher	$I_{RS}(x,y) = \begin{cases}1 & \text{if } x \leq y \\ 0 & \text{if } x > y\end{cases}$	X	X	✓	✓	N_C	N_{D_1}
Yager	$I_{YG}(x,y) = \begin{cases}1 & \text{if } (x,y) = (0,0) \\ y^x & \text{if } (x,y) \neq (0,0)\end{cases}$	✓	✓	X	X	X	N_{D_1}
Weber	$I_{WB}(x,y) = \begin{cases}1 & \text{if } x < 1 \\ y & \text{if } x = 1\end{cases}$	✓	✓	✓	X	X	N_{D_2}
Fodor	$I_{FD}(x,y) = \begin{cases}1 & \text{if } x \leq y \\ \max\{1-x,y\} & \text{if } x > y\end{cases}$	✓	✓	✓	✓	N_C	N_C

two implications

$$I_0(x, y) = \begin{cases} 1, & \text{if } x = 0 \text{ or } y = 1 \,, \\ 0, & \text{if } x > 0 \text{ and } y < 1 \,, \end{cases}$$

$$I_1(x, y) = \begin{cases} 1, & \text{if } x < 1 \text{ or } y > 0 \,, \\ 0, & \text{if } x = 1 \text{ and } y = 0 \,, \end{cases}$$

are the least and the greatest fuzzy implications, respectively, of the family of all fuzzy implications.

Beyond these examples of fuzzy implications, several families of these operations have been proposed and deeply studied. There exist basically two strategies in order to define classes of fuzzy implications. The most usual strategy is based on some combinations of aggregation functions. In this way, t-norms and t-conorms [12.5] were the first classes of aggregation functions used to generate fuzzy implications. Thus, the following are the three most important classes of fuzzy implications of this type:

1) (S, N)-implications defined as

$$I_{S,N}(x, y) = S(N(x), y) \,, \quad x, y \in [0, 1] \,,$$

where S is a t-conorm and N a fuzzy negation. They are the immediate generalization of the classical boolean material implication $p \to q \equiv \neg p \lor q$. If N is involutive, they are called strong or S-implications.

2) *Residual* or *R-implications* defined by

$$I_T(x, y) = \sup\{z \in [0, 1] \mid T(x, z) \leq y\} \,, \quad x, y \in [0, 1] \,,$$

where T is a t-norm. When they are obtained from left-continuous t-norms, they come from residuated lattices based on the residuation property

$$T(x, y) \leq z \Leftrightarrow I(x, z) \geq y \,, \quad \text{for all } x, y, z \in [0, 1] \,.$$

3) *QL-operations* defined by

$$I_{T,S,N}(x, y) = S(N(x), T(x, y)) \,, \quad x, y \in [0, 1] \,,$$

where S is a t-conorm, T is a t-norm and N is a fuzzy negation. Their origin is the quantum mechanic logic.

Note that R- and (S, N)-implications are always implications in the sense of Definition 12.1, whereas QL-operations are not implications in general (they are called QL-implications when they actually are).

A characterization of those QL-operations which are also implications is still open (Sect. 12.4), but a common necessary condition is $S(N(x), x) = 1$ for all $x \in [0, 1]$. Yet another class of fuzzy implications is that of Dishkant or D-operations [12.6] which are the contraposition of QL-operations with respect to a strong fuzzy negation.

These initial classes were successfully generalized considering more general classes of aggregation functions, mainly uninorms, generating new classes of fuzzy implications with interesting properties. In this way, (U, N), RU-implications and QLU-operations have been deeply analyzed [12.3, Chap. 5], [12.6].

A second approach to obtain fuzzy implications is based on the direct use of unary monotonic functions. In this way, the most important families are Yager's f- and g-generated fuzzy implications which can be seen as implications generated from additive generators of continuous Archimedean t-norms and t-conorms, respectively [12.3, Chap. 3]:

1) *Yager's f-generated implications* are defined as

$$I_f(x, y) = f^{-1}(x \cdot f(y)) \,, \quad x, y \in [0, 1] \,,$$

with the understanding $0 \cdot \infty = 0$, where $f : [0, 1] \to [0, \infty]$ is a strictly decreasing and continuous function with $f(1) = 0$.

2) *Yager's g-generated implications* are defined as

$$I_g(x, y) = g^{-1}\left(\min\left\{\frac{1}{x} \cdot g(y), g(1)\right\}\right) \,,$$

$$x, y \in [0, 1] \,,$$

with the understanding $\frac{1}{0} = \infty$ and $\infty \cdot 0 = \infty$ where $g : [0, 1] \to [0, \infty]$, is a strictly increasing and continuous function with $g(0) = 0$.

The above classes give rise to fuzzy implications with different additional properties which are collected in Table 12.2. All the results referred in Table 12.2 are from [12.3, Chaps. 2 and 3].

One of the main topics in this field is the characterization of each of these families of fuzzy implications through algebraic properties. This is an essential step in order to understand the behavior of these families. The available characterization results of the above families of implications are collected below.

Theorem 12.1 [12.3, Theorem 2.4.10]
For a function $I : [0, 1]^2 \to [0, 1]$ the following statements are equivalent:

Part B | 12.1

Table 12.2 Classes of fuzzy implications and the additional properties they satisfy

Class / Properties	(NP)	(EP)	(IP)	(OP)	(CP(N))	N_I
(S, N)-imp.	✓	✓	Thm. 2.4.17	Thm. 2.4.19	Prop. 2.4.3	N
R-imp. with l-c. T	✓	✓	✓	✓	Prop. 2.5.28	N_T
QL-imp.	✓	Thm. 2.6.19	Sect. 2.6.3	Sect. 2.6.4	Sect. 2.6.5	N
f-gen.	✓	✓	×	×	Thm. 3.1.7	Prop. 3.1.6
g-gen.	✓	✓	Thm. 3.2.8	Thm. 3.2.9	×	N_{D_1}

i) I is an (S, N)-implication with a continuous (strict, strong) fuzzy negation N.

ii) I satisfies (I1), (EP), and N_I is a continuous (strict, strong) fuzzy negation.

Moreover, in this case the representation $I(x, y) = S(N(x), y)$ is unique with $N = N_I$ and $S(x, y) = I(\mathfrak{R}_N(x), y)$ (for the definition of $\mathfrak{R}_N$ see [12.3, Lemma 1.4.10]).

Theorem 12.2 [12.3, Theorem 2.5.17]
For a function $I: [0, 1]^2 \to [0, 1]$ the following statements are equivalent:

i) I is an R-implication generated from a left-continuous t-norm.

ii) I satisfies (I2), (EP), (OP) and it is right continuous with respect to the second variable.

Moreover, the representation

$$I(x, y) = \max\{t \in [0, 1] | T(x, t) \le y\}$$

is unique with

$$T(x, y) = \min\{t \in [0, 1] | I(x, t) \ge y\} .$$

As already said, it is still an open question when QL-operations are fuzzy implications. However, in the continuous case, when S and N are the φ-conjugates of the Łukasiewicz t-conorm $S_{\mathbf{LK}}$ and the classical negation $N_{\mathbf{C}}$, respectively, for some order automorphism φ on the unit interval, the QL-operation has the following expression

$$I_{T,S,N}(x, y) = I_{\varphi, T}(x, y)$$
$$= \varphi^{-1}(1 - \varphi(x) + \varphi(T(x, y))) ,$$
$$x, y \in [0, 1] ,$$

and we have the following characterization result.

Theorem 12.3 [12.3, Theorem 2.6.12]
For a QL-operation $I_{\varphi, T}$, where T is a t-norm and φ is an automorphism on the unit interval, the following statements are equivalent:

i) $I_{\varphi, T}$ is a QL-implication.

ii) $T_{\varphi^{-1}}$ satisfies the Lipschitz condition, i. e.,

$$|T_{\varphi^{-1}}(x_1, y_1) - T_{\varphi^{-1}}(x_2, y_2)|$$
$$\le |x_1 - x_2| + |y_1 - y_2| , \quad x_1, x_2, y_1, y_2 \in [0, 1] .$$

In addition, (U, N)-implications are characterized in [12.3, Theorem 5.3.12] and more recently, Yager's f and g-generated [12.7] and RU-implications [12.8] have been also characterized. Finally, due to its importance in many results, we recall the characterization of the family of the conjugates of the Łukasiewicz implication.

Theorem 12.4 [12.3, Theorem 7.5.1]
For a function $I: [0, 1]^2 \to [0, 1]$ the following statements are equivalent:

i) I is continuous and satisfies both (EP) and (OP).

ii) I is a φ-conjugate with the Łukasiewicz implication $I_{\mathbf{LK}}$, i. e., there exists an automorphism φ on the unit interval, which is uniquely determined, such that I has the form

$$I(x, y) = (I_{\mathbf{LK}})_\varphi(x, y)$$
$$= \varphi^{-1}(\min\{1 - \varphi(x) + \varphi(y), 1\}) ,$$
$$x, y \in [0, 1] .$$

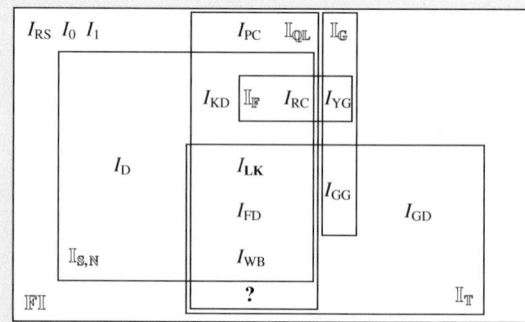

Fig. 12.1 Intersections between the main classes of fuzzy implications

For the conjugates of the other basic implications in Table 12.1, see the characterization results in [12.3, Sect. 7.5].

The great number of classes of fuzzy implications induces the study of the intersection between the different classes which brings out both the unity that exists among this diversity of classes and where the basic implications from Table 12.1 are located. The intersections among the main classes of fuzzy implications were studied in [12.3, Chap. 4] and are graphically displayed in Fig. 12.1 (note that $\mathbb{FI}$, $\mathbb{I}_{\mathrm{S,N}}$, $\mathbb{I}_{\mathrm{T}}$, $\mathbb{I}_{\mathrm{QL}}$, $\mathbb{I}_{\mathrm{F}}$ and $\mathbb{I}_{\mathrm{G}}$ denote the families of all fuzzy implications, (S,N)-implications, R-implications, QL-implications, Yager's f-generated implications and Yager's g-generated impli-

cations, respectively). In this figure, we have included the fuzzy implications of Table 12.1 and the following fuzzy implications which are examples of implications lying in some intersection between some families

$$I_{g^{\lambda}}(x,y) = \min\left\{1, \frac{y}{x^{\frac{1}{\lambda}}}\right\},$$

$$I_{\mathrm{D}}(x,y) = \begin{cases} 1, & \text{if } x = 0, \\ y, & \text{if } x > 0, \end{cases}$$

$$I_{\mathrm{PC}}(x,y) = 1 - (\max\{x(x+xy^2-2y),0\})^{\frac{1}{2}}.$$

Also note that it is still an open problem to prove if $(\mathbb{I}_{\mathrm{QL}} \cap \mathbb{I}_{\mathrm{T}}) \setminus \mathbb{I}_{\mathrm{S,N}} = \emptyset$.

12.2 Current Research on Fuzzy Implications

In the previous sections, we have seen some functional equations, namely, the exchange property (EP), the contrapositive symmetry (CP(N)) and the like. In this section, we deal with a few functional equations (or inequalities) involving fuzzy implications. These equations, once again, arise as the generalizations of the corresponding tautologies in classical logic involving boolean implications.

12.2.1 Functional Equations and Properties

A study of such equations stems from their applicability. The need for a plethora of fuzzy implications possessing various properties is quite obvious. On the one hand, they allow us to clearly classify and characterize different fuzzy implications, while on the other hand, they make themselves appealing to different applications. Thus, the functional equations presented in this section are chosen to reflect this dichotomy.

Distributivity over other Fuzzy Logic Operations
The distributivity of fuzzy implications over different fuzzy logic connectives, like t-norms, t-conorms, and uninorms is reduced to four equations

$$I(x, C_1(y,z)) = C_2(I(x,y), I(x,z)),\tag{12.1}$$

$$I(x, D_1(y,z)) = D_2(I(x,y), I(x,z)),\tag{12.2}$$

$$I(C(x,y), z) = D(I(x,z), I(y,z)),\tag{12.3}$$

$$I(D(x,y), z) = C(I(x,z), I(y,z)),\tag{12.4}$$

satisfied for all $x,y,z \in [0,1]$, where I is some generalization of classical implication, C, C_1, C_2 are some

generalizations of classical conjunction and D, D_1, D_2 are some generalizations of classical disjunction.

All the above equations can be investigated in two different ways. On the one hand, one can assume that function I belongs to some known class of fuzzy implications and investigate the connectives C_i, D_i that satisfy (12.1)–(12.4), as is done in the following works, for e.g., *Trillas* and *Alsina* [12.9], *Balasubramaniam* and *Rao* [12.10], *Ruiz-Aguilera* and *Torrens* [12.11, 12] and *Massanet* and *Torrens* [12.13]. On the other hand, one can assume that the connectives C_i, D_i come from the known classes of functions and investigate the fuzzy implications I that satisfy (12.1)–(12.4). See the works of *Baczyński* [12.14, 15], *Baczyński* and *Jayaram* [12.16], *Baczyński* and *Qin* [12.17, 18] for such an approach.

The above distributive equations play an important role in reducing the complexity of fuzzy systems, since the number of rules directly affects the computational duration of the overall application (we will discuss this problem again in Sect. 12.3.2).

Law of Importation
One of the desirable properties of a fuzzy implication is the law of importation as given below

$$I(x, I(y,z)) = I(T(x,y), z), \quad x,y,z \in [0,1],\tag{12.5}$$

where T is a t-norm (or, in general, some conjunction). It generalizes the classical tautology $(p \wedge q) \to r \equiv (p \to (q \to r))$ into fuzzy logic context. This equation has been investigated for many different families of fuzzy implications (for results connected with main classes see [12.3, Sect. 7.3]). Fuzzy implications satisfying (12.5) have been found extremely useful in fuzzy relational inference mechanisms, since one can obtain an equivalent hierarchical scheme which significantly decreases the computational complexity of the system without compromising on the approximation capability of the inference scheme. For more on this, we refer the readers to the following works [12.19, 20]. Related with (12.5) is its equivalence with (EP) that has been an open problem till the recent paper [12.21], where it is proved that (12.5) is stronger than (EP) and equivalent when N_I is continuous.

T-Conditionality or Modus Ponens

Another property investigated in the scientific literature, which is of great practical importance (see also Sect. 12.3.1), is the so-called T-conditionality, defined in the following way. If I is a fuzzy implication and T is a t-norm, then I is called an *MP-fuzzy implication* for T, if

$$T(x, I(x, y)) \leq y, \quad x, y \in [0, 1]. \tag{12.6}$$

Investigations of (12.6) have been done for the three main families of fuzzy implications, namely, (S, N)-, R-, and QL-implications [12.3, Sect. 7.4].

Nonsaturating Fuzzy Implications

Investigations connected with subsethood measures (see Sect. 12.3.3) and constructing strong equality functions by aggregation of implication functions by the formula $\Psi(x, y) = M(I(x, y), I(y, x))$, where M is some symmetric function, have led researchers to consider under which properties a fuzzy implication I satisfies the following conditions:

(P1) $I(x, y) = 1$ if and only if $x = 0$ or $y = 1$;
(P2) $I(x, y) = 0$ if and only if $x = 1$ and $y = 0$.

In [12.22], the authors considered the possible relationships between these two properties and the properties usually required of implication operations. Moreover, they developed different construction methods of strong equality indexes using fuzzy implications that satisfy these two additional properties.

Special Fuzzy Implications

Special implications were introduced by *Hájek* and *Kohout* [12.23] in their investigations on some statistics on marginals. The authors further have shown that they are related to special GUHA-implicative quantifiers (see, for instance, [12.24–26]). Thus, special fuzzy implications are related to data mining. In their quest to obtain some many-valued connectives as extremal values of some statistics on contingency tables with fixed marginals, they especially focussed on special homogenous implicational quantifiers and showed that:

Each special implicational quantifier determines a special implication. Conversely, each special implication is given by a special implicational quantifier.

Definition 12.2
A fuzzy implication I is said to be special, if for any $\varepsilon > 0$ and for all $x, y \in [0, 1]$ such that $x + \varepsilon, y + \varepsilon \in [0, 1]$ the following condition is satisfied

$$I(x, y) \leq I(x + \varepsilon, y + \varepsilon). \tag{12.7}$$

Recently, *Jayaram* and *Mesiar* [12.27] have investigated the above functional equation. Their study shows that among the main classes of fuzzy implications, no f-implication is a special implication, while the Goguen implication I_{GG} is the only special g-implication. Based on the available results, they have conjectured that the (S, N)-implications that are special also turn out to be R-implications. However, in the case of R-implications (generated from any t-norm) they have obtained the following result.

Theorem 12.5 [12.27, Theorem 4.6]
Let T be any t-norm and I_T be the R-implication obtained from T. Then the following statements are equivalent:

i) I_T satisfies (12.7).
ii) T satisfies the 1-Lipschitz condition.
iii) T has an ordinal sum representation $(\langle e_\alpha, a_\alpha, T_\alpha \rangle)_{\alpha \in A}$ where each t-norm T_α, $\alpha \in A$ is generated by a convex additive generator (for the definition of ordinal sum, see [12.5]).

Having shown that the families of (S, N)-, f-, and g-implications do not lead to any new special implications, *Jayaram* and *Mesiar* [12.27] turned to the most natural question: *Are there any other special implications, than those that could be obtained as residuals of t-norms?* This led them to propose some interesting constructions of fuzzy implications which were also special – one such construction is given in Definition 12.4 in Sect. 12.2.2.

12.2.2 New Classes and Generalizations

Another current research line on fuzzy implications is devoted to the study of new classes and generalizations of the already known families. The research in this direction has been extensively developed in recent years. Among many generalizations of already known classes of implications that have been dealt with in the literature, we highlight the following ones.

Generalizations of R-implications

The family of residual implications is one of the most commonly selected families for generalization. As already mentioned in Sect. 12.1, the RU-implications were the first generalization obtained via residuation from uninorms instead of from t-norms. In the same line, many other families of aggregation functions have been used to derive residual implications:

1. Copulas, quasi-copulas, and semicopulas were used in [12.28]. The main results in this work relate to the axiomatic characterizations of those functions I that are the residual implications of left-continuous commutative semicopulas, the residuals of quasi-copulas, and the residuals of associative copulas. For details on these characterizations, that involve up to ten different axioms, see [12.28].
2. Representable aggregation functions (RAFs) were used in [12.29]. These are aggregation functions constructed from additive generators of continuous Archimedean t-conorms and strong negations. The interest in the residual implications obtained from them lies in the fact that they are always continuous and in many cases they also satisfy the modus ponens with a nilpotent t-conorm. In particular, residual implications that depend only on a strong negation N are deduced from the general method just by considering specific generators of continuous Archimedean t-conorms.
3. A more general situation is studied in [12.30] where residual implications derived from binary functions $F: [0,1]^2 \rightarrow [0,1]$ are studied. In this case, the paper deals with the minimal conditions that F must satisfy in order to obtain an implication by residuation. The same is done in order to obtain residual implications satisfying each one of the most usual properties.
4. It is well known that residual implications derived from continuous Archimedean t-norms can be expressed directly from the additive generator of the t-norm. A generalization of this idea is

presented in [12.31], where strictly decreasing functions $f: [0,1] \rightarrow [0,+\infty]$ with $f(1) = 0$ are used to derive implications as follows

$$I(x,y) = \begin{cases} 1, & \text{if } x \leq y, \\ f^{(-1)}(f(y^+) - f(x)), & \text{if } x > y, \end{cases}$$

where $f(y^+) = \lim_{y \rightarrow y+} f(y)$ and $f(1^+) = f(1)$. Properties of these implications are studied and many new examples are also derived in [12.31].

Generalizations of (S, N)-Implications

Once again a first generalization of this class of implications has been done using uninorms leading to the (U, N)-implications mentioned in Sect. 12.1, but recently many other aggregation functions were also employed.

This is the case for instance in [12.32], where the authors make use of TS-functions obtained from a t-norm T, a t-conorm S and a continuous, strictly monotone function $f: [0,1] \rightarrow [-\infty, +\infty]$ through the expression

$$TS_{\lambda,f}(x,y) = f^{-1}((1-\lambda)f(T(x,y)) + \lambda f(S(x,y)))$$

for $x, y \in [0,1]$, where $\lambda \in (0,1)$. Operators defined by $I(x,y) = TS_{\lambda,f}(N(x),y)$ are studied in [12.32] giving the conditions under which they are fuzzy implications.

Another approach is based on the use of dual representable aggregation functions G, that are simply the N-dual of RAFs, introduced earlier. In this case, the corresponding (G, N)-operator is always a fuzzy implication and several examples and properties of this class can be found in [12.33]. See also [12.34] where it is proven that they satisfy (EP) (or (12.5)) if and only if G is in fact a nilpotent t-conorm.

Generalizations of Yager's Implications

In this case, the generalizations usually deal with the possibility of varying the generator used in the definition of the implication. A first step in this line was taken in [12.35] by considering multiplicative generators of t-conorms, but it was proven in [12.36] that this new class is included in the family of all (S, N)-implications obtained from t-conorms and continuous fuzzy negations.

Another approach was given in [12.37] introducing (f, g)-implications. In this case, the idea is to generalize f-generated Yager's implications by substituting the factor x by $g(x)$ where $g: [0,1] \rightarrow [0,1]$ is an increasing function satisfying $g(0) = 0$ and $g(1) = 1$.

In the same direction, a generalization of f- and g-generated Yager's implications based on aggregation operators is presented and studied in [12.38], where the implications are constructed by replacing the product t-norm in Yager's implications by any aggregation function.

Finally, *h-implications* were introduced in [12.39] and are constructed from additive generators of representable uninorms as follows.

Definition 12.3 ([12.39])
Let $h: [0, 1] \to [-\infty, \infty]$ be a strictly increasing and continuous function with $h(0) = -\infty$, $h(e) = 0$ for an $e \in (0, 1)$ and $h(1) = +\infty$. The function $I^h: [0, 1]^2 \to [0, 1]$ defined by

$$I^h(x, y) = \begin{cases} 1, & \text{if } x = 0, \\ h^{-1}(x \cdot h(y)), & \text{if } x > 0 \text{ and } y \le e, \\ h^{-1}\left(\frac{1}{x} \cdot h(y)\right), & \text{if } x > 0 \text{ and } y > e, \end{cases}$$

is called an *h-implication*.

This kind of implications maintains several properties of those satisfied by Yager's implications, like (EP) and (12.5) with the product t-norm, but at the same time they satisfy other interesting ones. For more details on this kind of implications, as well as some generalizations of them, see [12.39].

12.2.3 New Construction Methods

In this section, we recall some construction methods of fuzzy implications. The relevance of these methods is based on their capability of preserving the additional properties satisfied by the initial implication(s). First, note that some of them were already collected in [12.3, Chaps. 6 and 7], like:

- The φ-conjugation of a fuzzy implication I

$$I_\varphi(x, y) = \varphi^{-1}(I(\varphi(x), \varphi(y))), \quad x, y \in [0, 1],$$

  where φ is an order automorphism on $[0, 1]$.
- The min and max operations from two given fuzzy implications

$$(I \vee J)(x, y) = \max\{I(x, y), J(x, y)\}, \, x, y \in [0, 1],$$
$$(I \wedge J)(x, y) = \min\{I(x, y), J(x, y)\}, \, x, y \in [0, 1].$$

- The convex combinations of two fuzzy implications, where $\lambda \in [0, 1]$

$$I_{I,J}^\lambda(x, y) = \lambda \cdot I(x, y) + (1 - \lambda) \cdot J(x, y),$$
$$x, y \in [0, 1].$$

- The N-reciprocation of a fuzzy implication I

$$I_N(x, y) = I(N(y), N(x)), \quad x, y \in [0, 1],$$

  where N is a fuzzy negation.
- The upper, lower, and medium contrapositivization of a fuzzy implication I defined, respectively, as

$$I_N^u(x, y) = \max\{I(x, y), I_N(x, y)\}$$
$$= (I \vee I_N)(x, y),$$
$$I_N^l(x, y) = \min\{I(x, y), I_N(x, y)\}$$
$$= (I \wedge I_N)(x, y),$$
$$I_N^m(x, y) = \min\{I(x, y) \vee N(x), I_N(x, y) \vee y\},$$

  where N is a fuzzy negation and $x, y \in [0, 1]$. Please note that the lower (upper) contrapositivization is based on applying the min (max) method to a fuzzy implication I and its N-reciprocal.

It should be emphasized that the first major work to explore contrapositivization in detail, in its own right, was that of *Fodor* [12.40], where he discusses the contrapositive symmetry of fuzzy implications for the three main families, namely, S-, R-, and QL-implications. In fact, during this study Fodor discovered the nilpotent minimum t-norm T_{nM}, which is by far the first left-continuous but noncontinuous t-norm known in the literature. This study had a major impact on the development of left-continuous t-norms with strong natural negation, for instance, see the early works of *Jenei* [12.41, and references therein].

The above fact clearly illustrates how the study of functional equations involving fuzzy implications have also had interesting spin-offs and have immensely benefited other areas and topics in fuzzy logic connectives.

Among the new construction methods proposed in the recent literature, we can roughly divide them into the following categories.

Implications Generated from Negations
The first method was introduced by *Jayaram* and *Mesiar* in [12.42], while they were studying special implications (see Definition 12.2). From this study, they introduced the neutral special implications with a given negation and they studied the main properties of this new class.

Definition 12.4 [12.42]

Let N be a fuzzy negation such that $N \leq N_C$. Then the function $I_{[N]}: [0, 1]^2 \to [0, 1]$ given by

$$I_{[N]}(x, y) = \begin{cases} 1, & \text{if } x \leq y, \\ y + \dfrac{N(x - y)(1 - x)}{1 - x + y}, & \text{if } x > y, \end{cases}$$

with the understanding $\frac{0}{0} = 0$, is called the *neutral special implication generated from N*.

The second method of generation of fuzzy implications from fuzzy negations was introduced in [12.43].

Definition 12.5 [12.43]

Let N be a fuzzy negation. The function $I^{[N]}: [0, 1]^2 \to [0, 1]$ is defined by

$$I^{[N]}(x, y) = \begin{cases} 1, & \text{if } x \leq y, \\ \dfrac{(1 - N(x))y}{x} + N(x), & \text{if } x > y. \end{cases}$$

Again, several properties of these new implications can be derived, specially when the following classes of fuzzy negations are considered

$$N_A(x) = \begin{cases} 1, & \text{if } x \in A, \\ 0, & \text{if } x \notin A, \end{cases}$$

$$N_{A,\beta}(x) = \begin{cases} 1, & \text{if } x \in A, \\ \dfrac{1 - x}{1 + \beta x}, & \text{if } x \notin A, \end{cases}$$

where $A = [0, \alpha)$ with $\alpha \in (0, 1)$ or $A = [0, \alpha]$ with $\alpha \in [0, 1]$. Note that $N_{\{0\}} = N_{D_1}$ and $N_{\{0\},\beta}$ is the Sugeno class of negations. Note also that $I^{[N]}$ can be expressed as $I^{[N]}(x, y) = S_P(N(x), I_{GG}(x, y))$ for all $x, y \in [0, 1]$. From this observation, replacing S_P for any t-conorm S and I_{GG} for any implication I, the function

$$I^{[N,S,I]}(x, y) = S(N(x), I(x, y)), \quad x, y \in [0, 1],$$

is always a fuzzy implication.

Implications Constructed from Two Given Implications

In this section, we present methods that generate a fuzzy implication from two given ones.

The first method is based on an adequate scaling of the second variable of the two initial implications and it is called the *threshold generation method* [12.44].

Definition 12.6 [12.44]

Let I_1 and I_2 be two fuzzy implications and $e \in (0, 1)$. The function $I_{I_1 - I_2}: [0, 1]^2 \to [0, 1]$ defined by

$$I_{I_1 - I_2}(x, y) = \begin{cases} 1, & \text{if } x = 0, \\ e \cdot I_1\left(x, \dfrac{y}{e}\right), & \text{if } x > 0 \text{ and } y \leq e, \\ e + (1 - e) \cdot I_2\left(x, \dfrac{y - e}{1 - e}\right), & \\ & \text{if } x > 0 \text{ and } y > e, \end{cases}$$

is called the *e-threshold generated implication from I_1 and I_2*.

This method allows for a certain degree of control over the rate of increase in the second variable of the generated implication. Furthermore, the importance of this method derives from the fact that it allows us to characterize h-implications as the threshold generated implications of an f-generated and a g-generated implication [12.13, Theorem 2 and Remark 30]. Further, in contrast to many other generation methods of fuzzy implications from two given ones, it preserves (EP) and (12.5) if the initial implications possess them. Moreover, for an $e \in (0, 1)$, the e-threshold generated implications can be characterized as those implications that satisfy $I(x, e) = e$ for all $x > 0$.

The threshold generation method given above is based on splitting the domain of the implication with a horizontal line and then scaling the two initial implications in order to be well defined in those two regions. An alternate but analogous method can be proposed by using a vertical line instead of a horizontal line. This is the idea behind the *vertical threshold generation* method of fuzzy implications. This method does not preserve as many properties as the horizontal threshold method, but some results can still be proven. In particular, they are characterized as those fuzzy implications such that $I(e, y) = e$ for all $y < 1$ [12.45].

The following two construction methods were presented in [12.46]. Given two implications I, J, the following operations are introduced

$$(I \nabla J)(x, y) = I(J(y, x), J(x, y)),$$
$$(I \otimes J)(x, y) = I(x, J(x, y)),$$

for all $x, y \in [0, 1]$. The properties of these new operations as well as the structure of the set of all implications $\mathbb{FI}$ equipped with each one of these operations is studied in [12.46].

Other Construction Methods

In addition to the above methods, we would like to recall the following interesting method based on conditional probability and conditional distribution functions presented by *Grzegorzewski* in [12.47].

Definition 12.7 [12.47, 48]
The function $I_C: [0,1]^2 \to [0,1]$ given by

$$I_C(x,y) = \begin{cases} 1, & \text{if } x = 0, \\ \dfrac{C(x,y)}{x}, & \text{if } x > 0, \end{cases}$$

where C is a copula, is called a *probabilistic implication* based on copula C.

Conditions on copula C ensuring that the corresponding I_C is an implication, as well as properties of these implications are detailed in [12.48]. The main interest on this kind of implications lies in the fact that they are a powerful link between probability theory and fuzzy implications theory that can be useful in approximate reasoning. Moreover, results on these probabilistic implications can also be useful for examining and interpreting the behavior of some stochastic events. Some early results in this direction have appeared in [12.49, 50], where some generalizations of the previous idea are considered. In particular in [12.51], *survival implications* based on the probability that a given object will survive a fixed time into a population are studied. In this case, the survival implications are defined by

$$I_C^*(x,y) = \begin{cases} 1, & \text{if } x = 0, \\ \dfrac{x + y - 1 + C(1-x, 1-y)}{x}, & \text{if } x > 0, \end{cases}$$

where C is again a copula.

Finally, we only briefly mention that there exist other construction methods. For instance, *Massanet* and *Torrens* [12.13, 44, 45] have proposed methods of constructing implications derived from a given implication I and a fuzzy negation N as part of their study on some properties of horizontal and vertical threshold generated implications.

12.2.4 Fuzzy Implications in Nonclassical Settings

When we deal with uncertainty through fuzzy sets and fuzzy logic the natural framework is the unit interval $[0,1]$ and hence the logical connectives to be used are interpreted as operators on this interval. However, there are many different tools that have been proposed for managing uncertainty. In this context, some extensions of fuzzy logic and fuzzy sets have also been developed. One can list at least the following extensions: interval-valued fuzzy sets, Atanassov intuitionistic fuzzy sets (that are equivalent to the interval-valued approach, [12.52]), interval-valued intuitionistic fuzzy sets, type-2 fuzzy sets, fuzzy multisets, n-dimensional fuzzy sets, and hesitant fuzzy sets.

For all these extensions, the usual logical connectives like fuzzy conjunctions and fuzzy disjunctions need to be studied to develop a comprehensive theory, and especially fuzzy implications in order to make inferences in each one of these extensions. Due to space constraints, we only recall some aspects of interval-valued (or intuitionistic) fuzzy implications and the references where they can be found.

Interval-Valued Approach

A good compilation of the known results related to fuzzy implications (and other operations) in the interval-valued framework, can be found in [12.53] or [12.54] wherein, interval-valued or intuitionistic (S, N)- and R-implications are developed and some of their properties are presented. Works that deal with the construction of these classes of interval-valued implications can also be found in the literature. For instance, in [12.55] a construction method for the residual implication associated with a representable t-norm (constructed from two standard t-norms T_1 and T_2 with $T_1 \leq T_2$) is presented. Similarly, (S, N)- and R-implications generated from:

i) Aggregation functions and a standard fuzzy negation are presented in [12.56].
ii) Some classes of interval-valued aggregation functions based on t-norms and t-conorms are dealt with in [12.57].
iii) The so-called K_α-operators have been proposed in [12.58].

Discrete Approach

Note that all the above mentioned tools are mainly used in the management of imprecise quantitative information. However, experts deal with many problems where qualitative information is usually expressed through linguistic terms. Qualitative information is often interpreted to take values in a totally ordered finite scale like

$$\{\textit{Extremely Bad, Very Bad, Bad, Fair,} \\ \textit{Good, Very Good, Extremely Good}\}. \tag{12.8}$$

In these cases, the representative finite chain $L_n = \{0, 1, \ldots, n\}$ is usually considered to model these linguistic hedges and several researchers have developed an extensive study of operations on L_n, usually called *discrete operations*. This approach allows avoiding numerical interpretations and consequently, the fuzzification and defuzzification steps become unnecessary. In this framework, the smoothness condition is usually considered as the discrete counterpart of continuity. In fact, in the discrete framework this property is equivalent to the divisibility property as well as to the Lipschitz condition. In this way, smooth discrete t-norms and t-conorms were studied and characterized in [12.59] and also discrete fuzzy implications derived from them have been introduced.

As in the case of $[0, 1]$, the four most usual ways to construct discrete implications from t-norms and t-conorms on L_n are (S, N)-, R-, QL-, and D-implications. The first two classes derived from smooth t-norms and t-conorms and the only strong negation on L_n (given by $N_0(x) = n - x$) were studied in [12.60]. In the smooth case, it is proven that the intersection between (S, N)- and R-implications contains only the Łukasiewicz implication [12.60, Proposition 10]. Further, the nonsmooth case has also been investigated showing a parameterized family of nonsmooth t-norms T for which the corresponding R-implication coincides with the (S, N)-implication derived from the N_0-dual of T. The case of discrete QL- and D-operators is studied in [12.61], where characterization results on when such operators are in fact implications are given and, moreover, it is proven that both these classes coincide in the smooth case.

However, the modeling of linguistic information is limited because the information provided by experts for each variable must be expressed by a simple linguistic term. In most cases, this is a problem for experts because their opinion does not agree with a concrete term. On the contrary, experts' values are usually expressions like *better than Good, between Fair and Very Good*, or other even more complex expressions.

To avoid the limitation above, an approach has recently appeared trying to increase the flexibility of the elicitation of linguistic information. This approach deals with the possibility of extending monotonic operations on L_n to operations on the set of *discrete fuzzy numbers* whose support is a subinterval of L_n, usually denoted by $\mathcal{A}_1^{L_n}$. The idea lies in the fact that any discrete fuzzy number $A \in \mathcal{A}_1^{L_n}$ can be considered (identifying the scale $\mathcal{L}$ given in (12.8) with the chain L_6) as an assignment of a $[0, 1]$-value to each term in our linguistic scale. As an example, the above mentioned expression *between Fair and Very Good* can be performed, for instance, by a discrete fuzzy number $A \in \mathcal{A}_1^{L_6}$, with support given by the subinterval

[Fair, Very Good] = {Fair, Good, Very Good} ,

(that corresponds to the subinterval $[3, 5]$ in L_6). The values of A in its support should be described by experts, allowing in this way a complete flexibility of the qualitative valuation. Usual operations like t-norms, t-conorms, strong negations, aggregation functions, and also fuzzy implications have been introduced in this framework. The case of (S, N)-, QL- and D-implications can be found in [12.62, 63] and the case of R-implications in [12.64].

12.3 Fuzzy Implications in Applications

So far, we have discussed the theoretical aspects of fuzzy implications, namely, analytical and algebraic. In this section, we discuss their applicational value which shows a wide spectrum of areas wherein they are employed and how the gamut of properties that a fuzzy implication possesses plays an important role in its employability.

12.3.1 FL_n-Fuzzy Logic in the Narrow Sense

Boolean implications are employed in inference schemas like *modus ponens, modus tollens*, etc., where the reasoning is done with statements or propositions whose truth-values are two valued. Fuzzy implications play a similar role in the generalizations of the above inference schemas, where reasoning is done with fuzzy statements whose truth-value lies in $[0, 1]$ instead of $\{0, 1\}$.

Fuzzy Propositions

An expression of the form **x** *is* **A** where A is a fuzzy set on an appropriate domain U, with reference to the context, is termed as a *Fuzzy Statement* or a *Fuzzy Proposition*. (The above two interpretations bear a close resemblance to the *Adjunctive* and *Connective* interpretations as given in [12.65, pp. 331], though they are originally given for a binary operator. For other views and interpretation of the above statement, see, for instance, *Bezdek* et al., [12.66].)

Let it be given that **x** *is* **A** and also that x assumes the precise value, let us say, $x = u$, where $u \in U$, the domain of A. Then the truth value of the above fuzzy statement is obtained as follows

$$t(\mathbf{x} \; is \; \mathbf{A}) = A(u) \,,$$

i. e., the truth value of the above fuzzy statement, given that x is precisely known, is equal to the degree to which u – the value x assumes – is itself compatible with the fuzzy set A. Thus greater the membership degree of u in the concept A, higher is the truth value of the fuzzy statement.

Consider the statement *John is Tall* and that x – the height of John – is precisely given to be $5'10'' \in U$. Now, $A(5'10'')$ gives the membership degree of $5'10''$ in the concept $A = Tall$, which can be interpreted as how much *John* belongs to the set of all *Tall* men, or equivalently, how much *John is Tall* is true, which is nothing but the truth-value $t(John \; is \; Tall)$.

Fuzzy Conditionals or Fuzzy *IF–THEN* Rules

A fuzzy statement of the type discussed above $\widetilde{x} \; is \; \mathbf{A}$ can be interpreted in yet another way, namely, as a linguistic statement, i. e., as an assignment of a fuzzy set to a variable.

Let $A: U \to [0, 1]$ be a fuzzy set on a suitable domain U. Then A can be taken to represent a concept. A *linguistic variable* of U is a symbol $\widetilde{x}$ that can assume or be assigned any fuzzy subset of U. Then a linguistic statement $\widetilde{x} \; is \; A$ is interpreted as the linguistic variable $\widetilde{x}$ taking the *linguistic value A*.

For example, let U denote the set of all values in degrees centigrade. If the linguistic variable $\widetilde{x}$ denotes *Temperature*, then it can assume the following linguistic values A, namely, *high, more or less high, medium, cool, very cold*, etc. Each of the linguistic values (say $A = $ cool) is represented by a fuzzy set on the domain U of the linguistic variable $\widetilde{x}$, i. e., $A: U \to [0, 1]$.

The shape of the graph of the function represents the concept (say high temperature). The concept of high temperature is itself again context dependent. For example, high temperature (fever) for a human being is different from the high temperature in a blast furnace, and accordingly the domain of the linguistic variable is selected.

A fuzzy *IF–THEN* rule is of the form

$$\text{IF } \widetilde{x} \text{ is } A \text{ THEN } \widetilde{y} \text{ is } B \,, \tag{12.9}$$

where A, B are linguistic expressions/values assumed by the linguistic variables $\widetilde{x}, \widetilde{y}$. For example,

IF $\widetilde{x}$ (temperature) is A (high)
THEN $\widetilde{y}$ (pressure) is B (low).

Generalized Modus Ponens

Let α, β be two fuzzy propositions as given above and let $\alpha \longrightarrow \beta$ be the fuzzy conditional which is a fuzzy IF–THEN rule as above. In classical logic, one uses rules of deduction, like modus ponens and modus tollens to deduce new knowledge from a given set of propositions. For instance, modus ponens states that $\alpha \wedge (\alpha \longrightarrow \beta) \vdash \beta$.

In fuzzy logic, since we deal with fuzzy propositions whose truth values vary over the entire $[0, 1]$ interval we employ fuzzy logic operations. Typically $\wedge$ is interpreted as a *t*-norm T and for the $\longrightarrow$ a fuzzy implication is used.

Unlike with classical propositions, when we deal with fuzzy propositions it is not always given that from $\alpha \wedge (\alpha \longrightarrow \beta)$ one obtains β. This type of deduction is known as *generalized modus ponens* (GMP) and the study of pairs of operators $(\wedge, \longrightarrow)$, or alternately, a *t*-norm and fuzzy implication (T, I), that can be employed in GMP becomes important. It can be shown that this property translates to studying pairs (T, I) that satisfy the functional equation $T(x, I(x, y)) \leq y$ for $x, y \in [0, 1]$, which is nothing but T-conditionality as dealt with in Sect. 12.2.1.

Proof by Contradiction

In classical logic, many a time one proves a statement of the form $\alpha \longrightarrow \beta$ by proving its contrapositive, i. e., $\neg \beta \longrightarrow \neg \alpha$. However, in the setting of fuzzy logic, often the negation $\neg$ used is noninvolutive, i. e., $\neg\neg\alpha \neq \alpha$.

For instance, when the underlying fuzzy logic operations come from the Gödel residuated lattice $([0, 1], T_{\mathbf{M}}, I_{\mathbf{GD}}, \wedge, \vee)$, the natural negation of the fuzzy implication $I_{\mathbf{GD}}$ is not involutive and $I_{\mathbf{GD}}$ is not contrapositive w.r.t. any fuzzy negation. This led to the study of contrapositivization of fuzzy implications which was begun by *Fodor* [12.40] and is dealt with in Sect. 12.2.3 above.

12.3.2 Approximate Reasoning

One of the best known application areas of fuzzy logic is *approximate reasoning* (AR), wherein from imprecise inputs and fuzzy premises or rules we obtain, often, imprecise conclusions [12.67]. AR with fuzzy sets encompasses a wide variety of inference schemes and have been readily embraced in many fields, especially among others: decision making, expert systems, and control. Fuzzy implications play a vital role in many of these inference mechanisms, a brief discussion of which is presented below.

Inference Mechanisms in AR

Let us be given a set of n fuzzy IF–THEN rules of the form given in (12.10)

$$\text{If } \widetilde{x} \text{ is } A_i \text{ Then } \widetilde{y} \text{ is } B_i , \quad i = 1, 2, \ldots, n , \quad (12.10)$$

where A_i, B_i are fuzzy sets on input and output domains. Now, given a fuzzy input, i.e., a fuzzy proposition or a statement of the form $\widetilde{x}$ is A', the role of an inference mechanism is to obtain a fuzzy output B' that satisfies some desirable properties [12.68, 69].

Note that, if we denote the fuzzy rules as $A_i \longrightarrow B_i$, $i = 1, 2, \ldots, n$, as is typically done, then these are exactly the fuzzy conditionals discussed above in Sect. 12.3.1. Further, if we denote the input as A' then an inference mechanism implements the generalized modus ponens by *composing* the fuzzy input A' with all the rules $A_i \longrightarrow B_i$ to obtain the fuzzy output B'.

There are two established ways to accomplish the above, namely, *fuzzy relational inference* (FRI) and *similarity based reasoning* (SBR). Fuzzy implications play a major role in both the types of inference mechanisms as detailed below.

Fuzzy Relational Inference (FRI)

In a fuzzy relational inference, all the rules $A_i \longrightarrow B_i$ are combined into a single fuzzy relation R and the output B' is obtained as an image of the input A' composed with R.

A fuzzy IF–THEN rule base of the form (12.10) is modeled as a fuzzy relation $\hat{R}(x, y) : X \times Y \to [0, 1]$ as follows

$$\hat{R}(x, y) = \wedge_{i=1}^{n} (A_i(x) \to B_i(y))$$
$$= \wedge_{i=1}^{n} (I(A_i(x), B_i(y))) , \quad (12.11)$$

which reflects the conditional nature of the rules and where I is usually a fuzzy implication. Then given a fact $\widetilde{x}$ is A', the inferred output B' is obtained either as:

i) sup-T composition, as in the compositional rule of inference (CRI) of *Zadeh* [12.70], or
ii) An inf-I composition, as in the Bandler–Kohout subproduct (BKS) [12.71],

of $A'(x)$ and $\hat{R}(x, y)$, i.e.,

$$B'(y) = A'(x) \overset{T}{\circ} \hat{R}(x, y) = \sup_{x \in X} T(A'(x), \hat{R}(x, y)) , \quad (12.12)$$

$$B'(y) = A'(x) \overset{I}{\triangleleft} \hat{R}(x, y) = \inf_{x \in X} I(A'(x), \hat{R}(x, y)) , \quad (12.13)$$

where T can be any t-norm and I is any fuzzy implication.

It is clear from (12.12) and (12.13) that the important role fuzzy implications and their properties play in the goodness of an inference scheme. In the following subsection, we present a few issues where this role is highlighted.

Issues in FRI

While the rule base is an example of a single input single output (SISO) case, in practice we need multi-input single-output (MISO) rules of the form given below, with m input domains $X_j, j = 1, 2, \ldots, m$,

$$R_i : \text{ IF } \widetilde{x}_1 \text{ is } A_{i1} \text{ AND } \widetilde{x}_2 \text{ is } A_{i2} \text{ AND}$$
$$\ldots \text{ AND } \widetilde{x}_n \text{ is } A_{in} \text{ THEN } \widetilde{y} \text{ is } B_i .$$

While MISO rule bases are of great practical necessity, they spring up some new issues when they are employed in FRIs.

Combinatorial Explosion of Rules and Distributivity of Fuzzy Implications

Let there be k_j fuzzy sets defined on each of the domains $X_j, j = 1, 2, \ldots, m$. Then in a complete MISO rule base, we will have $n = k_1 \times k_2 \times \cdots k_m$ number of rules. Clearly, as m or k_j increases n increases and we have a combinatorial explosion of rules.

In a seminal work on studying this issue, *Combs* and *Andrews* [12.72] proposed an equivalent transformation of the CRI to mitigate the computational cost. The authors showed that the distributivity of fuzzy implications over t-norms play a major role in this transformation. This was further studied by *Balasubramaniam* and *Rao* [12.10] and its use in SBR was also demonstrated later by *Jayaram* [12.73].

Computational Complexity, Hierarchical Systems, and the Law of Importation

Let us consider an MISO rule base. From (12.11), it is clear that the relation $\hat{R}$ obtained is a multidimensional matrix, with $\hat{R}: X_1 \times X_2 \times \cdots \times X_m \times Y \to [0, 1]$. In fact, when one uses the First-Infer-Then-Aggregate mechanism in an FRI, either CRI or BKS, one needs to store n such m-dimensional matrices. Further, the input A' is also an m-dimensional matrix and the computation of the output gets costlier.

To overcome this, *Jayaram* [12.19] proposed an alternate hierarchical inference scheme which can be shown to be equivalent both in the CRI [12.19] and BKS [12.20] setting, when the underlying operators are such that the t-norm T and the fuzzy implication I satisfy the law of importation (12.5).

12.3.3 Fuzzy Subsethood Measures

Inclusion or subsethood of sets is an important concept. The first such definition of inclusion of a fuzzy set A over X in another fuzzy set B, was given by *Zadeh* [12.74] as follows

$$A \subset_Z B \Longleftrightarrow A(x) \le B(x) \,,$$
$$\text{for all } x \in X \,.$$

Note that this definition was more or less *crisp*, since an A was either contained in B or not. A more general notion of degree of inclusion was missing in the above definition. Subsequently many fuzzy subsethood measures, denoted (usually) *Inc*, were proposed.

Axiomatic Studies on Fuzzy Subsethood Measures

From the isomorphism that exists between classical set theory and classical logic, we know that $A \subseteq B$ is equivalent to $\chi_A \Longrightarrow \chi_B$, where χ_X is the characteristic function of the set X. Thus, early fuzzy subsethood measures also mimicked this equivalence by defining them based on fuzzy implications. Many researchers, in particular, *Sinha* and *Dougherty* [12.75], *Kitainik* [12.76], *Bandler* and *Kohout* [12.77] proposed sets of axioms for an *Inc* to satisfy.

It is easy to see that all of the above axiomatic approaches, eventually lead to employing implications as the underlying operators to define the corresponding *Inc*

measure, as given below

$$Inc_{\mathbf{SD}}(A, B) = \inf_{x \in X} \min\left(1, \lambda(A(x)) + \lambda(1 - B(x))\right) \,,$$
$$Inc_{\mathbf{K}}(A, B) = \inf_{x \in X} \varphi(I_{\mathbf{KD}}(B(x), A(x)) \,,$$
$$1 - I_{\mathbf{KD}}(A(x), B(x))) \,,$$
$$Inc_{\mathbf{BK}}(A, B) = \inf_{x \in X} (I(A(x), B(x))) \,,$$

where $\lambda: [0, 1] \to [0, 1]$ is a decreasing function with some additional properties, $\varphi: \mathcal{A} \to [0, 1]$ a function with additional properties where $\mathcal{A} = \{(x, y) \in [0, 1]^2 | x \ge y\}$ and I is any fuzzy implication.

From the above formulae the important position a fuzzy implication I holds in measuring fuzzy subsethood is apparent. Note that the *Inc* measure is used extensively in similarity based reasoning (SBR) and in fuzzy mathematical morphology (FMM) which are discussed below.

12.3.4 Fuzzy Control

While Sect. 12.3.2 dealt with FRIs which are largely used in the context of decision making and expert systems, in this section we deal with another type of fuzzy inference mechanism (FIM) that is used in fuzzy control, where the approximation properties of the FIM are important.

Similarity-Based Reasoning (SBR)

Let us once again consider a fuzzy IF–THEN rule base of the form (12.10) and a fuzzy input A'. In an SBR inference scheme, the following steps are employed to produce the output:

- *Matching*: The input A' is matched against each of the antecedents A_i of the rules (12.10) using a matching function M to obtain the corresponding similarity values $s_i = M(A', A_i) \in [0, 1]$ for $i = 1, 2, \dots, n$.
- *Modification*: Each of the similarity values s_i is used to modify the corresponding consequent B_i of the rule (12.10) using a modification J to obtain the modified output $B'_i = J(s_i, B_i)$.
- *Aggregation*: Finally all the modified outputs B'_i are aggregated to obtain an overall output $B = G(B'_1, \dots, B'_n)$.

In notations, we can write the above as

$$B'(y) = G_{i=1}^n \left(J(M(A', A_i), B_i(y))\right) \,, \quad y \in Y \,.$$
$$(12.14)$$

Fuzzy Implications and Matching Functions

Clearly, since $A, A_i' \in \mathcal{F}(X)$, we see that the matching function $M : \mathcal{F}(X) \times \mathcal{F}(X) \to [0, 1]$. Typically, a fuzzy subsethood measure Inc is employed as an M. While there exist M that are not based on fuzzy implications, it is seen that those that are based on fuzzy implications often satisfy many of the desirable properties required on the matching function M in different contexts, for instance, when the SBR is required to be interpolative, monotonic or for the SBR to possess good approximation properties. For more on this topic, see the works of Jayaram [12.73] or *Mandal* and *Jayaram* [12.78].

Fuzzy Implications and Modification Functions

From (12.14), it is clear that the modification function J can be seen simply as a binary function on $[0, 1]$. While any fuzzy logic operation could be used for J, fuzzy implications are preferred either due to their properties or due to the conditional nature of the underlying rules. For instance, when $J = I$ a fuzzy implication, if the original output B_i is normal then the modified output B_i' is also normal, which is usually not the case when one uses, say, a t-norm. In fact, different properties of I like (OP), (IP) and the nature of its natural negation N_I all play a role in the reasonableness of the final output of an SBR.

In real-life systems, the input and output domains X, Y are subsets of $\mathbb{R}$. Now, let the consequents B_i be of bounded support, i.e., $\{y \in Y \subset \mathbb{R}|B_i(y) > 0\} = [a, b] \subsetneq Y$ for some finite $a, b \in \mathbb{R}$. When an I whose N_I is not the Gödel least negation $N_{\mathbf{D1}}$ is employed, the support of B_i' becomes larger and in the case N_I is involutive then the support of the modified output sets B_i' become the whole of the set Y. This often makes the modified output sets B_i' to be nonconvex (and of larger support) and makes it difficult to apply standard defuzzification methods. For more on these see the works of *Štěpnička* and *De Baets* [12.79]. The above discussion brings out an interesting aspect of fuzzy implications. While fuzzy implications I whose N_I are strong are to be preferred in the setting of fuzzy logic FL_n for inferencing as noted in Sect. 12.3.1 above, an I with an N_I that is not even continuous is to be preferred in inference mechanisms used in fuzzy control.

By the *core* of a fuzzy set B on Y, we mean the set $\{y \in Y|B(y) = 1\}$. Now, an I which possesses (OP) or (IP) is preferred in an SBR to ensure there is an overlap between the cores of the modified outputs B_i' – a property that is so important to ensure *coherence* in the system [12.80] and that, once again, standard defuzzification methods can be applied.

12.3.5 Fuzzy Mathematical Morphology

Consider a 2D binary image $\mathcal{P}$, i.e., the value at a pixel is either 0 or 1. $\mathcal{P}$ can be seen as a function from $X \subset \mathbb{R}^2 \to \{0, 1\}$ or just a classical subset $X \subset \mathbb{R}^2$. Mathematical morphology (MM) is a set-theoretic method for the extraction of shape information from a scene. Here, a $Y \subset \mathbb{R}^2$ – which can be seen as another image Q and often referred to as the structuring element – is used to transform the original image $\mathcal{P}$ by some well-defined local operators termed *Dilation* and *Erosion* as defined below

$$D(\mathcal{P}, Q) = \{v \in \mathbb{R}^2|\mathcal{A}_v(Q) \cap \mathcal{P} \neq \emptyset\}, \tag{12.15}$$

$$E(\mathcal{P}, Q) = \{v \in \mathbb{R}^2|\mathcal{A}_v(Q) \subseteq \mathcal{P}\}, \tag{12.16}$$

where $\mathcal{A}_v(Q) = \{u \in \mathbb{R}^2|u - v \in Q\}$ is the translation of Q by $v \in \mathbb{R}^2$.

FMM is the extension of MM to gray-level images by using fuzzy sets and possibility theory. Note that a gray-level image $\mathcal{P}$ can be interpreted as a fuzzy set $X \subset \mathbb{R}^2 \to [0, 1]$ where the pixel value is interpreted as its membership degree to the original data set. This fuzzified image is then processed via morphological operators that are extensions of the boolean ones.

In the literature, one finds two approaches to this extension:

i) As a formal translation of crisp equations using t-norms and negations, by employing a fuzzy intersection for $\cap$ in (12.15) and a fuzzy subsethood measure Inc for $\subseteq$ in (12.16), and
ii) Using adjunction and residual implications.

While the first approach is based on the duality between dilation and erosion, the second approach stems more from an algebraic setting.

De Baets [12.81, 82] took the second approach, and defined the *fuzzy dilation* and *erosion* as follows

$$\tilde{D}(\mathcal{P}, Q)(y) = \sup_{x \in \mathcal{A}_v(Y) \cap X} [C(\mathcal{P}(x - y), Q(x))],$$

$$\tilde{E}(\mathcal{P}, Q)(y) = \inf_{x \in \mathcal{A}_v(Y)} [I(\mathcal{P}(x - y), Q(x))],$$

where C is any fuzzy conjunction and I is a fuzzy implication.

When the pair of operations (C, I) satisfy the adjunction property, or equivalently, I is a residual implication obtained from C, then many interesting aspects emerge. Firstly, it can be shown that *opening* and *closing* operations, which are some morphological operations obtained from the defined $\tilde{D}, \tilde{E}$ turn out to be idempotent, which is highly desirable [12.83]. Secondly, it can be shown, as was done by *Nachtegael* and *Kerre* [12.84], that this approach is more general and many other approaches become a specific case of it. Thirdly, recently, *Bloch* [12.85] showed that both the above approaches based on duality and adjunction are equivalent under some rather general and mild conditions, but those that often lead to highly desirable settings.

Recently, the approach initiated by De Baets has been enlarged by considering uninorms instead of *t*-norms and their residual implications with good results in edge detection, as well as in noise reduction [12.86, 87].

12.4 Future of Fuzzy Implications

Since the publication of [12.2, 3], the peak of interest in fuzzy implications has led to a rapid progress in attempts to solve open problems in this topic. Specially, in [12.3], many open problems were presented covering all the subtopics of this field: characterizations, intersections, additional properties, etc. Many of these problems have been already solved and the solutions have been collected in [12.88]. However, there still remain many open problems involving fuzzy implications. Thus, in this section, we will list some of them whose choice has been dictated either based on the importance of the problem or the significance of the solution.

The first subset corresponds to open problems dealing with the satisfaction of particular additional properties of fuzzy implications. The first one deals with the law of importation (LI). Recently, some works have dealt with this property and its equivalence to the exchange principle and from them, some new characterizations of (S, N)- and R-implications based on (12.5) have been proposed, see [12.21]. However, some questions are still open. Firstly, (12.5) with a *t*-norm (or a more general conjunction) and (EP) are equivalent when N_I is a continuous negation, but the equivalence in general is not fully determined.

Problem 12.1
Characterize all the cases when (LI) and (EP) are equivalent.

Secondly, it is not yet known which fuzzy implications satisfy (LI) when the conjunction operation is fixed.

Problem 12.2
Given a conjunction C (usually a *t*-norm or a conjunctive uninorm), characterize all fuzzy implications I that satisfy (LI) with this conjunction C. For instance, which implications I satisfy the following functional equation

$$I(xy, z) = I(x, I(y, z))$$

that comes from (LI) with $T = T_P$?

Another problem now concerning only the exchange principle follows.

Problem 12.3
Give a necessary condition on a nonborder continuous *t*-norm T for the corresponding I_T to satisfy (EP).

It should be mentioned that some related work on the above problem appeared in [12.89].

Some other open problems with respect to the satisfaction of particular additional properties are based on the preservation of these properties from some initial fuzzy implications to the generated one using some construction methods like max, min, or the convex combination method.

Problem 12.4
Characterize all fuzzy implications I, J such that $I \vee J$, $I \wedge J$ and K^λ satisfy (EP) or (LI), where $\lambda \in [0, 1]$.

The above problem is also related to the following one:

Problem 12.5
Characterize the convex closures of the following families of fuzzy implications: (S, N)-, R- and Yager's f- and g-generated implications.

Another open problem which has immense applicational value is the satisfaction of the T-conditionality by the Yager's families of fuzzy implications.

Problem 12.6
Characterize Yager's f-generated and g-generated implications satisfying the T-conditionality property with some *t*-norm T.

The following two open problems are related to the characterization of some particular classes of fuzzy implications.

Problem 12.7

What is the characterization of (S, N)-implications generated from noncontinuous negations?

Problem 12.8

Characterize triples (T, S, N) such that the corresponding QL-operation $I_{T,S,N}$ satisfies (I1).

Finally, a fruitful topic where many open problems are still to be solved is the study of the intersections among the classes of fuzzy implications (Fig. 12.1).

Problem 12.9

i) Is there a fuzzy implication I, other than the Weber implication $I_{\mathbf{WB}}$, which is both an (S, N)-implication and an R-implication which is obtained from a nonborder continuous t-norm and cannot be obtained as the residual of any other left-continuous t-norm?

ii) If the answer to the above question is affirmative, characterize the above nonempty intersection.

Problem 12.10

i) Characterize the nonempty intersection between (S, N)-implications and QL-implications, i.e., $\mathbb{I}_{S,N} \cap \mathbb{I}_{QL}$.

ii) Is the Weber implication $I_{\mathbf{WB}}$ the only QL-implication that is also an R-implication obtained from a nonleft continuous t-norm? If not, give other examples from the above intersection and hence, characterize the nonempty intersection between R-implications and QL-implications.

iii) Prove or disprove by giving an example: that there is no fuzzy implication which is both a QL- and an R-implication, but it is not an (S, N)-implication, i.e., $(\mathbb{I}_{QL} \cap \mathbb{I}_T) \setminus \mathbb{I}_{S,N} = \emptyset$.

References

12.1   M. Baczyński, B. Jayaram: (S, N)- and R-implications: A state-of-the-art survey, Fuzzy Sets Syst. **159**(14), 1836–1859 (2008)

12.2   M. Mas, M. Monserrat, J. Torrens, E. Trillas: A survey on fuzzy implication functions, IEEE Trans. Fuzzy Syst. **15**(6), 1107–1121 (2007)

12.3   M. Baczyński, B. Jayaram: *Fuzzy Implications*, Studies in Fuzziness and Soft Computing, Vol. 231 (Springer, Berlin, Heidelberg 2008)

12.4   M. Baczyński, G. Beliakov, H. Bustince, A. Pradera (Eds.): *Advances in Fuzzy Implication Functions*, Studies in Fuzziness and Soft Computing, Vol. 300 (Springer, Berlin, Heidelberg 2013)

12.5   E.P. Klement, R. Mesiar, E. Pap: *Triangular norms* (Kluwer, Dordrecht 2000)

12.6   M. Mas, M. Monserrat, J. Torrens: Two types of implications derived from uninorms, Fuzzy Sets Syst. **158**(23), 2612–2626 (2007)

12.7   S. Massanet, J. Torrens: On the characterization of Yager's implications, Inf. Sci. **201**, 1–18 (2012)

12.8   I. Aguiló, J. Suñer, J. Torrens: A characterization of residual implications derived from left-continuous uninorms, Inf. Sci. **180**(20), 3992–4005 (2010)

12.9   E. Trillas, C. Alsina: On the law $[(p \wedge q) \rightarrow r] = [(p \rightarrow r) \vee (q \rightarrow r)]$ in fuzzy logic, IEEE Trans. Fuzzy Syst. **10**(1), 84–88 (2002)

12.10   J. Balasubramaniam, C.J.M. Rao: On the distributivity of implication operators over T and S norms, IEEE Trans. Fuzzy Syst. **12**(2), 194–198 (2004)

12.11   D. Ruiz-Aguilera, J. Torrens: Distributivity of strong implications over conjunctive and disjunctive uninorms, Kybernetika **42**(3), 319–336 (2006)

12.12   D. Ruiz-Aguilera, J. Torrens: Distributivity of residual implications over conjunctive and disjunctive uninorms, Fuzzy Sets Syst. **158**(1), 23–37 (2007)

12.13   S. Massanet, J. Torrens: On some properties of threshold generated implications, Fuzzy Sets Syst. **205**(16), 30–49 (2012)

12.14   M. Baczyński: On the distributivity of fuzzy implications over continuous and Archimedean triangular conorms, Fuzzy Sets Syst. **161**(10), 1406–1419 (2010)

12.15   M. Baczyński: On the distributivity of fuzzy implications over representable uninorms, Fuzzy Sets Syst. **161**(17), 2256–2275 (2010)

12.16   M. Baczyński, B. Jayaram: On the distributivity of fuzzy implications over nilpotent or strict triangular conorms, IEEE Trans. Fuzzy Syst. **17**(3), 590–603 (2009)

12.17   F. Qin, M. Baczyński, A. Xie: Distributive equations of implications based on continuous triangular norms (I), IEEE Trans. Fuzzy Syst. **20**(1), 153–167 (2012)

12.18   M. Baczyński, F. Qin: Some remarks on the distributive equation of fuzzy implication and the contrapositive symmetry for continuous, Archimedean t-norms, Int. J. Approx. Reason. **54**(2), 290–296 (2012)

12.19   B. Jayaram: On the law of importation $(x \wedge y) \rightarrow z \equiv (x \rightarrow (y \rightarrow z))$ in fuzzy logic, IEEE Trans. Fuzzy Syst. **16**(1), 130–144 (2008)

12.20 M. Štěpnička, B. Jayaram: On the suitability of the Bandler–Kohout subproduct as an inference mechanism, IEEE Trans. Fuzzy Syst. **18**(2), 285–298 (2010)

12.21 S. Massanet, J. Torrens: The law of importation versus the exchange principle on fuzzy implications, Fuzzy Sets Syst. **168**(1), 47–69 (2011)

12.22 H. Bustince, J. Fernandez, J. Sanz, M. Baczyński, R. Mesiar: Construction of strong equality index from implication operators, Fuzzy Sets Syst. **211**(16), 15–33 (2013)

12.23 P. Hájek, L. Kohout: Fuzzy implications and generalized quantifiers, Int. J. Uncertain. Fuzziness Knowl. Syst. **4**(3), 225–233 (1996)

12.24 P. Hájek, M.H. Chytil: The GUHA method of automatic hypotheses determination, Computing **1**(4), 293–308 (1966)

12.25 P. Hájek, T. Havránek: The GUHA method-its aims and techniques, Int. J. Man-Mach. Stud. **10**(1), 3–22 (1977)

12.26 P. Hájek, T. Havránek: *Mechanizing Hypothesis Formation: Mathematical Foundations for a General Theory* (Springer, Heidelberg 1978)

12.27 B. Jayaram, R. Mesiar: On special fuzzy implications, Fuzzy Sets Syst. **160**(14), 2063–2085 (2009)

12.28 F. Durante, E. Klement, R. Mesiar, C. Sempi: Conjunctors and their residual implicators: Characterizations and construction methods, Mediterr. J. Math. **4**(3), 343–356 (2007)

12.29 M. Carbonell, J. Torrens: Continuous *R*-implications generated from representable aggregation functions, Fuzzy Sets Syst. **161**(17), 2276–2289 (2010)

12.30 Y. Ouyang: On fuzzy implications determined by aggregation operators, Inf. Sci. **193**, 153–162 (2012)

12.31 V. Biba, D. Hliněná: Generated fuzzy implications and known classes of implications, Acta Univ. M. Belii Ser. Math. **16**, 25–34 (2010)

12.32 H. Bustince, J. Fernandez, A. Pradera, G. Beliakov: On (*TS, N*)-fuzzy implications, Proc. AGOP 2011, Benevento, ed. by B. De Baets, R. Mesiar, L. Troiano (2011) pp. 93–98

12.33 I. Aguiló, M. Carbonell, J. Suñer, J. Torrens: Dual representable aggregation functions and their derived *S*-implications, Lect. Notes Comput. Sci. **6178**, 408–417 (2010)

12.34 S. Massanet, J. Torrens: An overview of construction methods of fuzzy implications. In: *Advances in Fuzzy Implication Functions*, Studies in Fuzziness and Soft Computing, Vol. 300, ed. by M. Baczyński, G. Beliakov, H. Bustince, A. Pradera (Springer, Berlin, Heidelberg 2013) pp. 1–30

12.35 J. Balasubramaniam: Yager's new class of implications J_f and some classical tautologies, Inf. Sci. **177**(3), 930–946 (2007)

12.36 M. Baczyński, B. Jayaram: Yager's classes of fuzzy implications: Some properties and intersections, Kybernetika **43**(2), 157–182 (2007)

12.37 A. Xie, H. Liu: A generalization of Yager's f-generated implications, Int. J. Approx. Reason. **54**(1), 35–46 (2013)

12.38 S. Massanet, J. Torrens: On a generalization of Yager's implications, Commun. Comput. Inf. Sci. Ser. **298**, 315–324 (2012)

12.39 S. Massanet, J. Torrens: On a new class of fuzzy implications: *h*-implications and generalizations, Inf. Sci. **181**(11), 2111–2127 (2011)

12.40 J.C. Fodor: Contrapositive symmetry of fuzzy implications, Fuzzy Sets Syst. **69**(2), 141–156 (1995)

12.41 S. Jenei: New family of triangular norms via contrapositive symmetrization of residuated implications, Fuzzy Sets Syst. **110**(2), 157–174 (2000)

12.42 B. Jayaram, R. Mesiar: I-Fuzzy equivalence relations and I-fuzzy partitions, Inf. Sci. **179**(9), 1278–1297 (2009)

12.43 Y. Shi, B.V. Gasse, D. Ruan, E.E. Kerre: On dependencies and independencies of fuzzy implication axioms, Fuzzy Sets Syst. **161**(10), 1388–1405 (2010)

12.44 S. Massanet, J. Torrens: Threshold generation method of construction of a new implication from two given ones, Fuzzy Sets Syst. **205**, 50–75 (2012)

12.45 S. Massanet, J. Torrens: On the vertical threshold generation method of fuzzy implication and its properties, Fuzzy Sets Syst. **226**, 32–52 (2013)

12.46 N.R. Vemuri, B. Jayaram: Fuzzy implications: Novel generation process and the consequent algebras, Commun. Comput. Inf. Sci. Ser. **298**, 365–374 (2012)

12.47 P. Grzegorzewski: Probabilistic implications, Proc. EUSFLAT-LFA 2011, ed. by S. Galichet, J. Montero, G. Mauris (Aix-les-Bains, France 2011) pp. 254–258

12.48 P. Grzegorzewski: Probabilistic implications, Fuzzy Sets Syst. **226**, 53–66 (2013)

12.49 P. Grzegorzewski: On the properties of probabilistic implications. In: *Eurofuse 2011, Advances in Intelligent and Soft Computing*, Vol. 107, ed. by P. Melo-Pinto, P. Couto, C. Serôdio, J. Fodor, B. De Baets (Springer, Berlin, Heidelberg 2012) pp. 67–78

12.50 A. Dolati, J. Fernández Sánchez, M. Úbeda-Flores: A copula-based family of fuzzy implication operators, Fuzzy Sets Syst. **211**(16), 55–61 (2013)

12.51 P. Grzegorzewski: Survival implications, Commun. Comput. Inf. Sci. Ser. **298**, 335–344 (2012)

12.52 G. Deschrijver, E. Kerre: On the relation between some extensions of fuzzy set theory, Fuzzy Sets Syst. **133**(2), 227–235 (2003)

12.53 G. Deschrijver, E. Kerre: Triangular norms and related operators in L^*-fuzzy set theory. In: *Logical, Algebraic, Analytic, Probabilistic Aspects of Triangular Norms*, ed. by E. Klement, R. Mesiar (Elsevier, Amsterdam 2005) pp. 231–259

12.54 G. Deschrijver: Implication functions in interval-valued fuzzy set theory. In: *Advances in Fuzzy Implication Functions*, Studies in Fuzziness and Soft Computing, Vol. 300, ed. by M. Baczyński, G. Beliakov, H. Bustince, A. Pradera (Springer, Berlin, Heidelberg 2013) pp. 73–99

12.55 C. Alcalde, A. Burusco, R. Fuentes-González: A constructive method for the definition of interval-valued fuzzy implication operators, Fuzzy Sets Syst. **153**(2), 211–227 (2005)

12.56 H. Bustince, E. Barrenechea, V. Mohedano: Intuitionistic fuzzy implication operators-an expression and main properties, Int. J. Uncertain. Fuzziness Knowl. Syst. **12**(3), 387–406 (2004)

12.57 G. Deschrijver, E. Kerre: Implicators based on binary aggregation operators in interval-valued fuzzy set theory, Fuzzy Sets Syst. **153**(2), 229–248 (2005)

12.58 R. Reiser, B. Bedregal: *K*-operators: An approach to the generation of interval-valued fuzzy implications from fuzzy implications and vice versa, Inf. Sci. **257**, 286–300 (2013)

12.59 G. Mayor, J. Torrens: Triangular norms in discrete settings. In: *Logical, Algebraic, Analytic, and Probabilistic Aspects of Triangular Norms*, ed. by E.P. Klement, R. Mesiar (Elsevier, Amsterdam 2005) pp. 189–230

12.60 M. Mas, M. Monserrat, J. Torrens: *S*-implications and *R*-implications on a finite chain, Kybernetika **40**(1), 3–20 (2004)

12.61 M. Mas, M. Monserrat, J. Torrens: On two types of discrete implications, Int. J. Approx. Reason. **40**(3), 262–279 (2005)

12.62 J. Casasnovas, J. Riera: *S*-implications in the set of discrete fuzzy numbers, Proc. IEEE-WCCI 2010, Barcelona (2010), pp. 2741–2747

12.63 J.V. Riera, J. Torrens: Fuzzy implications defined on the set of discrete fuzzy numbers, Proc. EUSFLAT-LFA 2011 (2011) pp. 259–266

12.64 J.V. Riera, J. Torrens: Residual implications in the set of discrete fuzzy numbers, Inf. Sci. **247**, 131–143 (2013)

12.65 P. Smets, P. Magrez: Implication in fuzzy logic, Int. J. Approx. Reason. **1**(4), 327–347 (1987)

12.66 J.C. Bezdek, D. Dubois, H. Prade: *Fuzzy Sets in Approximate Reasoning and Information Systems* (Kluwer, Dordrecht 1999)

12.67 D. Driankov, H. Hellendoorn, M. Reinfrank: *An Introduction to Fuzzy Control*, 2nd edn. (Springer, London 1996)

12.68 D. Dubois, H. Prade: Fuzzy sets in approximate reasoning, Part 1: Inference with possibility distributions, Fuzzy Sets Syst. **40**(1), 143–202 (1991)

12.69 G.J. Klir, B. Yuan: *Fuzzy sets and fuzzy logic-theory and applications* (Prentice Hall, Hoboken 1995)

12.70 L.A. Zadeh: Outline of a new approach to the analysis of complex systems and decision processes, IEEE Trans. Syst. Man Cybern. **3**(1), 28–44 (1973)

12.71 W. Bandler, L.J. Kohout: Semantics of implication operators and fuzzy relational products, Int. J. Man-Mach. Stud. **12**(1), 89–116 (1980)

12.72 W.E. Combs, J.E. Andrews: Combinatorial rule explosion eliminated by a fuzzy rule configuration, IEEE Trans. Fuzzy Syst. **6**(1), 1–11 (1998)

12.73 B. Jayaram: Rule reduction for efficient inferencing in similarity based reasoning, Int. J. Approx. Reason. **48**(1), 156–173 (2008)

12.74 L.A. Zadeh: Fuzzy sets, Inf. Control **8**(3), 338–353 (1965)

12.75 D. Sinha, E.R. Dougherty: Fuzzification of set inclusion: Theory and applications, Fuzzy Sets Syst. **55**(1), 15–42 (1991)

12.76 L. Kitainik: Fuzzy inclusions and fuzzy dichotomous decision procedures. In: *Optimization models using fuzzy sets and possibility theory*, ed. by J. Kacprzyk, S. Orlovski (Reidel, Dordrecht 1987) pp. 154–170

12.77 W. Bandler, L. Kohout: Fuzzy power sets and fuzzy implication operators, Fuzzy Sets Syst. **4**(1), 13–30 (1980)

12.78 S. Mandal, B. Jayaram: Approximation capability of SISO SBR fuzzy systems based on fuzzy implications, Proc. AGOP 2011, ed. by B. De Baets, R. Mesiar, L. Troiano (University of Sannio, Benevento 2011) pp. 105–110

12.79 M. Štěpnička, B. De Baets: Monotonicity of implicative fuzzy models, Proc. FUZZ-IEEE, 2010 Barcelona (2010), pp. 1–7

12.80 D. Dubois, H. Prade, L. Ughetto: Checking the coherence and redundancy of fuzzy knowledge bases, IEEE Trans. Fuzzy Syst. **5**(3), 398–417 (1997)

12.81 B. De Baets: Idempotent closing and opening operations in fuzzy mathematical morphology, Proc. ISUMA-NAFIPS'95, Maryland (1995), pp. 228–233

12.82 B. De Baets: Fuzzy morphology: A logical approach. In: *Uncertainty Analysis in Engineering and Science: Fuzzy Logic, Statistics, Neural Network Approach*, ed. by B.M. Ayyub, M.M. Gupta (Kluwer, Dordrecht 1997) pp. 53–68

12.83 J. Serra: *Image Analysis and Mathematical Morphology* (Academic, London, New York 1988)

12.84 M. Nachtegael, E.E. Kerre: Connections between binary, gray-scale and fuzzy mathematical morphologies original, Fuzzy Sets Syst. **124**(1), 73–85 (2001)

12.85 I. Bloch: Duality vs. adjunction for fuzzy mathematical morphology and general form of fuzzy erosions and dilations, Fuzzy Sets Syst. **160**(13), 1858–1867 (2009)

12.86 M. González-Hidalgo, A. Mir Torres, D. Ruiz-Aguilera, J. Torrens Sastre: Edge-images using a uninorm-based fuzzy mathematical morphology: Opening and closing. In: *Advances in Computational Vision and Medical Image Processing, Computational Methods in Applied Sciences*, Vol. 13, ed. by J. Tavares, N. Jorge (Springer, Berlin, Heidelberg 2009) pp. 137–157

12.87 M. González-Hidalgo, A. Mir Torres, J. Torrens Sastre: Noisy image edge detection using a uninorm fuzzy morphological gradient, Proc. ISDA 2009 (IEEE Computer Society, Los Alamitos 2009) pp. 1335–1340

12.88 M. Baczyński, B. Jayaram: Fuzzy implications: Some recently solved problems. In: *Advances in Fuzzy*

Implication Functions, Studies in Fuzziness and Soft Computing, Vol. 300, ed. by M. Baczyński, G. Beliakov, H. Bustince, A. Pradera (Springer, Berlin, Heidelberg 2013) pp. 177–204

12.89   B. Jayaram, M. Baczyński, R. Mesiar: *R*-implications and the exchange principle: The case of border continuous *t*-norms, Fuzzy Sets Syst. **224**, 93–105 (2013)

13. Fuzzy Rule-Based Systems

Luis Magdalena

Fuzzy rule-based systems are one of the most important areas of application of fuzzy sets and fuzzy logic. Constituting an extension of classical rule-based systems, these have been successfully applied to a wide range of problems in different domains for which uncertainty and vagueness emerge in multiple ways. In a broad sense, fuzzy rule-based systems are rule-based systems, where fuzzy sets and fuzzy logic are used as tools for representing different forms of knowledge about the problem at hand, as well as for modeling the interactions and relationships existing between its variables. The use of fuzzy statements as one of the main constituents of the rules allows capturing and handling the potential uncertainty of the represented knowledge. On the other hand, thanks to the use of fuzzy logic, inference methods have become more robust and flexible. This chapter will mainly analyze what is a fuzzy rule-based system (from both conceptual and structural points of view), how is it built, and how can be used. The analysis will start by considering the two main conceptual components of these systems, knowledge, and reasoning, and how they are represented. Then, a review of the main structural approaches to fuzzy rule-based systems will be considered. Hierarchical fuzzy systems will also be analyzed. Once defined the components, struc-

ture and approaches to those systems, the question of design will be considered. Finally, some conclusions will be presented.

From the point of view of applications, one of the most important areas of fuzzy sets theory is that of fuzzy rule-based systems (FRBSs). These kind of systems constitute an extension of classical rule-based systems, considering *IF–THEN* rules whose antecedents and consequents are composed of fuzzy logic (FL) statements, instead of classical logic ones.

Conventional approaches to knowledge representation are based on bivalent logic, which has associated a serious shortcoming: the inability to reason in situations of uncertainty and imprecision. As a consequence, conventional approaches do not provide an adequate framework for this mode of reasoning familiar to humans, and most commonsense reasoning falls into this category.

In a broad sense, an FRBS is a rule-based system where fuzzy sets and FL are used as tools for representing different forms of knowledge about the problem at hand, as well as for modeling the interactions and relationships existing between its variables.

The use of fuzzy statements as one of the main constituents of the rules, allows capturing and handling the potential uncertainty of the represented knowledge. On the other hand, thanks to the use of fuzzy logic, inference methods have become more robust and flexible.

Due to these properties, FRBSs have been successfully applied to a wide range of problems in different domains for which uncertainty and vagueness emerge in multiple ways [13.1–5].

The analysis of FRBSs will start by considering the two main conceptual components of these systems, knowledge and reasoning, and how they are represented. Then, a review of the main structural approaches to FRBSs will be considered. Hierarchical fuzzy systems would probably match in this previous section, but being possible to combine the hierarchical approach with any of the structural models defined there, it seems better to consider it independently. Once defined the components, structure, and approaches to those systems, the question of design will be considered. Finally, some conclusions will be presented. It is important to notice that this chapter will concentrate on the general aspects related to FRBSs without deepening in the foundations of FL which are widely considered in previous chapters.

13.1 Components of a Fuzzy Rule-Based System

Knowledge representation in FRBSs is enhanced with the use of linguistic variables and their linguistic values, that are defined by context-dependent fuzzy sets whose meanings are specified by gradual membership functions [13.6–8]. On the other hand, FL inference methods such as generalized Modus Ponens, generalized Modus Tollens, etc., form the basis for approximate reasoning [13.9]. Hence, FL provides a unique computational framework for inference in rule-based systems. This idea implies the presence of two clearly different concepts in FRBSs: knowledge and reasoning. This clear separation between knowledge and reasoning (the *knowledge base* (KB) and processing structure shown in Fig. 13.1) is the key aspect of knowledge-based systems, so that from this point of view, FRBSs can be considered as a type of knowledge-based system.

The first implementation of an FRBS dealing with real inputs and outputs was proposed by *Mamdani* [13.10], who considering the ideas published just a few months before by *Zadeh* [13.9] was able to augment his initial formulation allowing the application of fuzzy systems (FSs) to a control problem, so creating the first fuzzy control application. These kinds of FSs are also referred to as FRBSs with fuzzifier and defuzzifier or, more commonly, as *fuzzy logic controllers* (FLCs), as proposed by the author in his pioneering paper [13.11], or Mamdani FRBSs. From the beginning, the term FLC became popular since control systems design constituted the main application of Mamdani FRBSs. At present, control is only one more of the many application areas of FRBSs.

The generic structure of a Mamdani FRBS is shown in Fig. 13.1. The KB stores the available knowledge about the problem in the form of fuzzy *IF–THEN* rules. The processing structure, by means of these rules, puts into effect the inference process on the system inputs.

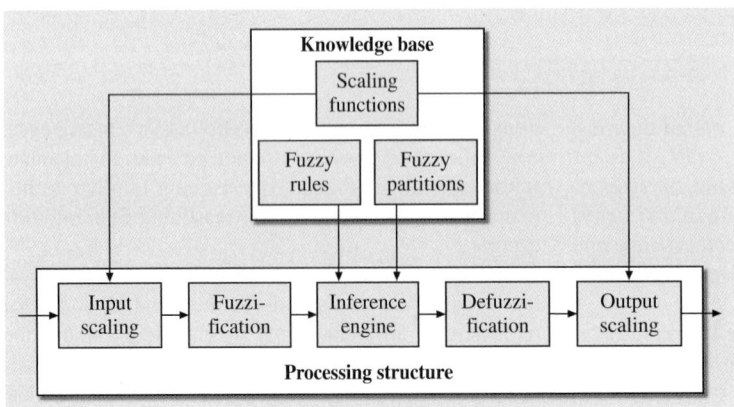

Fig. 13.1 General structure of a Mamdani FRBS

The fulfillment of rule antecedent gives rise to the execution of its consequent, i. e., one output is produced. The overall process includes several steps. The input and output scalings produce domain adaptations. Fuzzification interface establishes a mapping between crisp values in the input domain U, and fuzzy sets defined on the same universe of discourse. On the other hand, the defuzzification interface performs the opposite operation by defining a mapping between fuzzy sets defined in the output domain V and crisp values defined in the same universe. The central step of the process is inference.

The next two subsections analyze in depth the two main components of an FRBS, the KB and the processing structure, considering the case of a Mamdani FRBS.

13.1.1 Knowledge Base

The KB of an FRBS serves as the repository of the problem-specific knowledge – that models the relationship between input and output of the underlying system – upon which the inference process reasons to obtain from an observed input, an associated output.

This knowledge is represented in the form of rules, and the most common rule structure in Mamdani FRBSs involves the use of linguistic variables [13.6–8]. Hence, when dealing with multiple inputs-single output (MISO) systems, these *linguistic* rules possess the following form

$$IF\ X_1\ is\ LT_1\ and\ \dots\ and\ X_n\ is\ LT_n$$
$$THEN\ Y\ is\ LT_o\ , \tag{13.1}$$

with X_i and Y being, respectively, the input and output linguistic variables, and with LT_i being linguistic terms associated with these variables.

Note that the KB contains two different information levels, i. e., the linguistic variables (providing fuzzy rule semantics in the form of fuzzy partitions) and the linguistic rules representing the expert knowledge. Apart from that, a third component, scaling functions, is added in many FRBSs to act as an interfacing component for domain adaptation between the external world and the universes of discourse used at the level of the fuzzy partitions. This conceptual distinction drives to the three separate entities that constitute the KB:

- The *fuzzy partitions* (also called Frames of Cognition) describe the sets of linguistic terms associated

with each variable and considered in the linguistic rules, and the membership functions defining the semantics of these linguistic terms. Each linguistic variable involved in the problem will have associated a fuzzy partition of its domain. Figure 13.2 shows a fuzzy partition using triangular membership functions. This structure provides a natural framework to include expert knowledge in the form of fuzzy rules. The fuzzy partition shown in the figure uses five linguistic terms {*very small, small, medium, large,* and *very large*}, (represented as VS, S, M, L, and VL, respectively) with the interval $[l, r]$ being its domain (Universe of discourse). The figure also shows the membership function associated to each of these five terms.

- A *rule base* (RB) is comprised of a collection of linguistic rules (as the one shown in (13.1)) that are joined by the *also* operator. In other words, multiple rules can fire simultaneously for the same input.
- Moreover, the KB also comprises the *scaling functions* or scaling factors that are used to transform between the universe of discourse in which the fuzzy sets are defined from/to the domain of the system input and output variables.

It is important to note that the RB can present several structures. The usual one is the list of rules, although a decision table (also called rule matrix) becomes an equivalent and more compact representation for the same set of linguistic rules when only a few input variables (usually one or two) are considered by the FRBS.

Let us consider an FRBS where two input variables (x_1 and x_2) and a single output variable (y) are involved, with the following term sets associated: {*small, medium, large*}, {*short, medium, long*} and {*bad, medium, good*}, respectively. The following RB

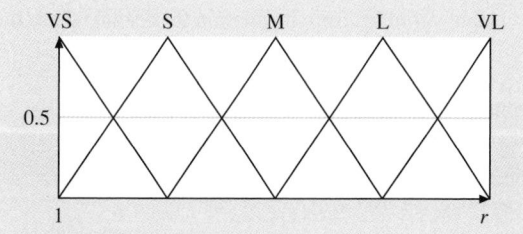

Fig. 13.2 Example of a fuzzy partition

composed of five linguistic rules

R_1: IF X_1 is small and X_2 is short THEN Y is bad, also

R_2: IF X_1 is small and X_2 is medium THEN Y is bad, also

R_3: IF X_1 is medium and X_2 is short THEN Y is medium, also

R_4: IF X_1 is large and X_2 is medium THEN Y is medium, also

R_5: IF X_1 is large and X_2 is long THEN Y is good ,
(13.2)

can be represented by the decision table shown in Table 13.1.

Before concluding this section, we should notice two aspects. On one hand, the structure of a linguistic rule may be more generic if a connective other than the *and* operator is used to aggregate the terms in the rule antecedent. However, it has been demonstrated that the above rule structure is generic enough to subsume other possible rule representations [13.12]. The above rules are therefore commonly used throughout the literature due to their simplicity and generality. On the other hand, linguistic rules are not the only option and rules with a different structure can be considered, as we shall see in Sect. 13.2.

13.1.2 Processing Structure

The functioning of FRBSs has been described as the interaction of knowledge and reasoning. Once briefly considered the knowledge component, this section will analyze the reasoning (processing) structure. The processing structure of a Mamdani FRBS is composed of the following five components:

- The *input scaling* that transforms the values of the input variables from its domain to the one where the input fuzzy partitions are defined.

Table 13.1 Example of a decision table

x_2	x_1		
	small	*medium*	*large*
short	bad	medium	
medium	bad		medium
long			good

- A *fuzzification interface* that transforms the crisp input data into fuzzy values that serve as the input to the fuzzy reasoning process.
- An *inference engine* that infers from the fuzzy inputs to several resulting output fuzzy sets according to the information stored in the KB.
- A *defuzzification interface* that converts the fuzzy sets obtained from the inference process into a crisp value.
- The *output scaling* that transforms the defuzzified value from the domain of the output fuzzy partitions to that of the output variables, constituting the global output of the FRBS.

In the following, the five elements will be briefly described.

The Input/Output Scaling
Input/output scaling maps (applying the corresponding scaling functions or factors contained in the KB) the input/output variables to/from the universes of discourse over which the corresponding linguistic variables were defined.

This mapping can be performed with different functions ranging from a simple scaling factor to linear and nonlinear functions.

The initial idea for scaling was the use of *scaling factors* with a tuning purpose [13.13], giving a certain adaptation capability to the fuzzy system.

Additional degrees of freedom could be obtained by using a more complex scaling function. A second option is the use of *linear scaling* with a function of the form

$$f(x) = \lambda \cdot x + \nu ,$$
(13.3)

where the scaling factor λ enlarges or reduces the operating range, which in turn decreases or increases the sensitivity of the system in respect to that input variable, or the corresponding gain in the case of an output variable. The parameter ν shifts the operating range and plays the role of an offset for the corresponding variable.

Finally, it is possible to use more complex mappings generating *nonlinear scaling*. A common nonlinear scaling function is

$$f(x) = \text{sign}(x) \cdot |x|^\alpha .$$
(13.4)

This nonlinear scaling increases ($\alpha > 1$) or decreases ($\alpha < 1$) the relative sensitivity in the region closer to the central point of the interval and has the opposite effect when moving far from the central point [13.14].

The Fuzzification Interface

The fuzzification interface enables Mamdani FRBSs to handle crisp input values. Fuzzification establishes a mapping from crisp input values to fuzzy sets defined in the universe of discourse of those inputs. The membership function of the fuzzy set A' defined over the universe of discourse U associated to a crisp input value x_0 is computed as

$$\mu_{A'} = F(x_0), \tag{13.5}$$

in which F is a fuzzification operator.

The most common choice for the fuzzification operator F is the *point wise* or *singleton fuzzification*, where A' is built as a singleton with support x_0, i. e., it presents the following membership function:

$$\mu_{A'}(x) = \begin{cases} 1, & \text{if } x = x_0 \\ 0, & \text{otherwise}. \end{cases} \tag{13.6}$$

Nonsingleton options [13.15] are also possible and have been considered in some cases as a tool to represent the imprecision of measurements.

The Inference System

The inference system is the component that derives the fuzzy outputs from the input fuzzy sets according to the relation defined through the fuzzy rules. The usual fuzzy inference scheme employs the generalized Modus Ponens, an extension to the classical Modus Ponens [13.9]

$$\begin{aligned} &\textit{IF X is A THEN Y is B} \\ &\textit{X is A'} \\ &\qquad \textit{Y is B'}. \end{aligned} \tag{13.7}$$

In this expression, *IF X is A THEN Y is B* describes a conditional statement that in this case is a fuzzy conditional statement, since A and B are fuzzy sets, and X and Y are linguistic variables. A fuzzy conditional statement like this one represents a fuzzy relation between A and B defined in $U \times V$. This fuzzy relation is expressed again by a fuzzy set (R) whose membership function $\mu_R(x, y)$ is given by

$$\mu_R(x, y) = I(\mu_A(x), \mu_B(y)), \quad \forall x \in U, y \in V, \tag{13.8}$$

in which $\mu_A(x)$ and $\mu_B(y)$ are the membership functions of the fuzzy sets A and B, and I is a fuzzy implication operator that models the existing fuzzy relation.

Going back to (13.7), the result of applying generalized Modus Ponens is obtaining the fuzzy set B' (through its membership function) by means of the *compositional rule of inference* [13.9]:

If R is a fuzzy relation defined in U and V, and A' is a fuzzy set defined in U, then the fuzzy set B', induced by A', is obtained from the composition of R and A',

that is

$$B' = A' \circ R. \tag{13.9}$$

Now it is needed to compute the fuzzy set B' from A' and R. According to the definition of composition (*T-composition*) given in the chapter devoted to fuzzy relations, the result will be

$$\mu_{B'}(y) = \sup_{x \in U} T(\mu_{A'}(x), \mu_R(x, y)), \tag{13.10}$$

where T is a triangular norm (t-norm). The concept and properties of t-norms have been previously introduced in the chapter devoted to fuzzy sets.

Given now an input value $X = x_0$, obtaining A' in accordance with (13.6) (where $\mu_{A'}(x) = 0 \ \forall x \neq x_0$), and considering the properties of t-norms ($T(1, a) = a$, $T(0, a) = 0$), the previous expression is reduced to

$$\begin{aligned} \mu_{B'}(y) &= T(\mu_{A'}(x_0), \mu_R(x_0, y)) \\ &= T(1, \mu_R(x_0, y)) = \mu_R(x_0, y). \end{aligned} \tag{13.11}$$

The only additional point to arrive to the final value of $\mu_{B'}(y)$ is the definition of R, the fuzzy relation representing the Implication. This is a somehow controversial question. Since the very first applications of FRBSs [13.10, 11] this relation has been implemented with the minimum (product has been also a common choice). If we analyze the definition of fuzzy implication given in the corresponding chapter, it is clear that the minimum does not satisfy all the conditions to be a fuzzy implication, so, why is it used? It can be said that initially it was a short of heuristic decision, which demonstrated really good results being accepted and reproduced in all subsequent applications. Further analysis can offer different explanations to this choice [13.16–18].

In any case, assuming the minimum as the representation for R, (13.11) produces the following final result:

$$\mu_{B'}(y) = \min(\mu_A(x_0), \mu_B(y)). \tag{13.12}$$

Considering now an n-dimensional input space, the inference will establish a mapping between fuzzy sets defined in the Cartesian product ($U = U_1 \times U_2 \times \cdots \times U_n$) of the universes of discourse of the input variables $X_1, \ldots, X_n$, and fuzzy sets defined in V, being the universe of discourse of the output variable Y. Therefore, when applied to the ith rule of the RB, defined as

$$R_i : IF\ X_1\ is\ A_{i1}\ and\ \ldots\ and\ X_n\ is\ A_{in}\ THEN\ Y\ is\ B_i\ ,$$
$$(13.13)$$

considering an input value $x_0 = (x_1, \ldots, x_n)$, the output fuzzy set B' will be obtained by replacing $\mu_A(x_0)$ in (13.12) with,

$$\mu_{A_i}(x_0) = T(\mu_{A_{i1}}(x_1), \ldots, \mu_{A_{in}}(x_n))\ ,$$

where T is a fuzzy conjunctive operator (a t-norm).

The Defuzzification Interface

The inference process in Mamdani-type FRBSs operates at the level of individual rules. Thus, the application of the compositional rule of inference to the current input, using the m rules in the KB, generates m output fuzzy sets B_i'. The defuzzification interface has to aggregate the information provided by the m individual outputs and obtain a crisp output value from the aggregated set. This task can be done in two different ways [13.1, 12, 19]: *Mode A-FATI* (first aggregate, then infer) and *Mode B-FITA* (first infer, then aggregate).

Mamdani originally suggested the mode A-FATI in his first conception of FLCs [13.10]. In the last few years, the Mode B-FITA is becoming more popular [13.19–21], in particular, in real-time applications which demand a fast response time.

Mode A-FATI: First Aggregate, then Infer. In this case, the defuzzification interface operates as follows:

- Aggregate the individual fuzzy sets B_i' into an overall fuzzy set B' by means of a fuzzy aggregation operator G (usually named as the *also* operator):

$$\mu_{B'}(y) = G\left\{\mu_{B_1'}(y), \mu_{B_2'}(y), \ldots, \mu_{B_m'}(y)\right\}\ .$$
$$(13.14)$$

- Employ a defuzzification method, D, transforming the fuzzy set B' into a crisp output value y_0:

$$y_0 = D(\mu_{B'}(y))\ .$$
$$(13.15)$$

Usually, the aggregation operator G is implemented by the maximum (a t-conorm), and the defuzzifier D is the *center of gravity* (CG) or the *mean of maxima* (MOM), whose expressions are as follows:

- CG:

$$y_0 = \frac{\int_Y y \cdot \mu_{B'}(y)\mathrm{d}y}{\int_Y \mu_{B'}(y)\mathrm{d}y}\ .$$
$$(13.16)$$

- MOM:

$$y_{\inf} = \inf\{z | \mu_{B'}(z) = \max_y \mu_{B'}(y)\}$$
$$y_{\sup} = \sup\{z | \mu_{B'}(z) = \max_y \mu_{B'}(y)\}$$
$$y_0 = \frac{y_{\inf} + y_{\sup}}{2}\ .$$
$$(13.17)$$

Mode B-FITA: First Infer, then Aggregate. In this second approach, the contribution of each fuzzy set is considered separately and the final crisp value is obtained by means of an averaging or selection operation applied to the set of crisp values derived from each of the individual fuzzy sets B_i'.

The most common choice is either the CG or the *maximum value* (MV), then weighted by the matching degree. Its expression is shown as follows:

$$y_0 = \frac{\sum_{i=1}^m h_i \cdot y_i}{\sum_{i=1}^m h_i}\ ,$$
$$(13.18)$$

with y_i being the CG or the MV of the fuzzy set B_i', inferred from rule R_i, and $h_i = \mu_{A_i}(x_0)$ being the matching between the system input x_0 and the antecedent (premise) of rule i.

Hence, this approach avoids aggregating the rule outputs to generate the final fuzzy set B', reducing the computational burden compared to mode A-FATI defuzzification.

This defuzzification mode constitutes a different approach to the notion of the *also* operator, and it is directly related to the idea of interpolation and the approach of Takagi–Sugeno–Kang (TSK) fuzzy systems, as can be seen by comparing (13.18) and (13.25).

13.2 Types of Fuzzy Rule-Based Systems

As discussed earlier, the first proposal of an FRBS was that of Mamdani, and this kind of system has been considered as the basis for the general description of previous section. This section will focus on the different structures that can be considered when building an FRBS.

13.2.1 Linguistic Fuzzy Rule-Based Systems

This approach corresponds to the original *Mamdani* FRBS [13.10, 11], being the main tool to develop Linguistic models, and is the approach that has been mainly considered to this point in the chapter.

A Mamdani FRBS provides a natural framework to include expert knowledge in the form of linguistic rules. This knowledge can be easily combined with rules which are automatically generated from data sets that describe the relation between system input and output. In addition, this knowledge is highly interpretable. The fuzzy rules are composed of input and output variables, which take values from their term sets having a meaning (a semantics) associated with each linguistic term. Therefore, each rule is a description of a condition-action statement that exhibits a clear interpretation to a human – for this reason, these kinds of systems are usually called *linguistic* or *descriptive Mamdani FRBSs*. This property makes Mamdani FRBSs appropriate for applications in which the emphasis lies on model interpretability, such as fuzzy control [13.20, 22, 23] and linguistic modeling [13.4, 21].

13.2.2 Variants of Mamdani Fuzzy Rule-Based Systems

Although Mamdani FRBSs possess several advantages, they also come with some drawbacks. One of the problems, especially in linguistic modeling applications, is their limited accuracy in some complex problems, which is due to the structure of the linguistic rules. [13.24] and [13.25] analyzed these limitations concluding that the structure of the fuzzy linguistic *IF–THEN* rule is subject to certain restrictions because of the use of linguistic variables:

- There is a lack of flexibility in the FRBS due to the rigid partitioning of the input and output spaces.

- When the input variables are mutually dependent, it becomes difficult to find a proper fuzzy partition of the input space.

- The homogeneous partition of the input and output space becomes inefficient and does not scale well as the dimensionality and complexity of the input–output mapping increases.

- The size of the KB increases rapidly with the number of variables and linguistic terms in the system. This problem is known as the *course of dimensionality*. In order to obtain an accurate FRBS, a fine level of granularity is needed, which requires additional linguistic terms. This increase in granularity causes the number of rules to grow, which complicates the interpretability of the system by a human. Moreover, in the vast majority of cases, it is possible to obtain an equivalent FRBS that achieves the same accuracy with a fewer number of rules whose fuzzy sets are not restricted to a fixed input space partition.

Both variants of linguistic Mamdani FRBSs described in this section attempt to solve the said problems by making the linguistic rule structure more flexible.

DNF Mamdani Fuzzy Rule-Based Systems

The first extension to Mamdani FRBSs aims at a different rule structure, the so-called *disjunctive normal form (DNF) fuzzy rule*, which has the following form [13.26, 27]:

$$IF\ X_1\ is\ \widetilde{A}_1\ and\ \dots\ and\ X_n\ is\ \widetilde{A}_n$$
$$THEN\ Y\ is\ B\,, \tag{13.19}$$

where each input variable X_i takes as its value a set of linguistic terms $\widetilde{A}_i$, whose members are joined by a disjunctive operator, while the output variable remains a usual linguistic variable with a single label associated. Thus, the complete syntax for the antecedent of the rule is

$$X_1\ is\ \widetilde{A}_1 = \{A_{11}\ or\ \dots\ or\ A_{1l_1}\}\ and\ \dots$$
$$and\ X_n\ is\ \widetilde{A}_n = \{A_{n1}\ or\ \dots\ or\ A_{nl_n}\}\,. \tag{13.20}$$

An example of this kind of rule is shown as follows. Let us suppose we have three input variables, X_1, X_2, and X_3, and one output variable, Y, such that the linguistic term sets D_i $(i = 1, 2, 3)$ and F, associated with each variable, are

$$D_1 = \{A_{11}, A_{12}, A_{13}\}$$
$$D_2 = \{A_{21}, A_{22}, A_{23}, A_{24}, A_{25}\}$$
$$D_3 = \{A_{31}, A_{32}\} \quad F = \{B_1, B_2, B_3\} . \tag{13.21}$$

In this case, an example of DNF rule will be

$$IF\ X_1\ is\ \{A_{11}\ or\ A_{12}\}\ and\ X_2\ is\ \{A_{23}\ or\ A_{24}\}$$
$$and\ X_3\ is\ \{A_{31}\ or\ A_{32}\}\ THEN\ Y\ is\ B_2 . \tag{13.22}$$

This expression contains an additional *connective* different than the *and* considered in all previous rules. The *or* connective is computed through a t-conorm, the maximum being the most commonly used.

The main advantage of this rule structure is its ability to integrate in a single expression (a single DNF rule) the information corresponding to several elemental rules (the rules commonly used in Mamdani FRBSs). In this example, (13.22) corresponds to 8 ($2 \times 2 \times 2$) rules of the equivalent system expressed as (13.1). This property produces a certain level of compression of the rule base, being quite helpful when the number of input variables increases, alleviating the effect of the course of dimensionality.

Approximate Mamdani-Type Fuzzy Rule-Based Systems

While the previous DNF fuzzy rule structure does not involve an important loss in the linguistic Mamdani

a) Descriptive Knowledge base

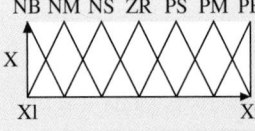

R1: If X is NB then Y is NB   R5: If X is PS then Y is PS
R2: If X is NM then Y is NM   R6: If X is PM then Y is PM
R3: If X is NS then Y is NS   R7: If X is PB then Y is PB
R4: If X is ZR then Y is ZR

b) Approximate fuzzy rule base

R1: If X is △ then Y is ◺
R2: If X is ◺ then Y is △
R3: If X is ◿ then Y is △
R4: If X is △ then Y is △

Fig. 13.3a,b Comparison between a descriptive KB and an approximate fuzzy rule base

FRBS interpretability, the point of departure for the second extension is to obtain an FRBS which achieves a better accuracy at the cost of reduced interpretability. These systems are called *approximate Mamdani-type FRBSs* [13.1, 25, 28–30], in opposition to the previous *descriptive* or *linguistic Mamdani FRBSs*.

The structure of an approximate FRBS is similar to that of a descriptive one shown in Fig. 13.1. The difference is that in this case, the rules do not refer in their definition to predefined fuzzy partitions of the linguistic variables. In an approximate FRBS, each rule defines its own fuzzy sets instead of using a linguistic label pointing to a particular fuzzy set of the partition of the underlying linguistic variable. Thus, an approximate fuzzy rule has the following form:

$$IF\ X_1\ is\ A_1\ and\ \ldots\ and\ X_n\ is\ A_n\ THEN\ Y\ is\ B . \tag{13.23}$$

The major difference with respect to the rule structure considered in linguistic Mamdani FRBSs is the fact that the input variables X_i and the output one Y are fuzzy variables instead of linguistic variables and, thus, A_i and B are not linguistic terms (LT_i) as they were in (13.1), but independently defined fuzzy sets that elude an intuitive linguistic interpretation. In other words, rules of approximate nature are *semantic free*, whereas descriptive rules operate in the context formulated by means of the linguistic terms semantics.

Therefore, approximate FRBSs do not relay on fuzzy partitions defining a semantic context in the form of linguistic terms. The fuzzy partitions are somehow integrated into the fuzzy rule base in which each rule subsumes the definition of its underlying input and output fuzzy sets, as shown in Fig. 13.3(b).

Approximate FRBSs demonstrate some specific advantages over linguistic FRBSs making them particularly useful for certain types of applications [13.25]:

- The major advantage of the approximate approach is that each rule employs its own distinct fuzzy sets resulting in additional degrees of freedom and an increase in expressiveness. It means that the tuning of a certain fuzzy set in a rule will have no effect on other rules, while changing a fuzzy set of a fuzzy partition in a descriptive model affects all rules considering the corresponding linguistic label.

- Another important advantage is that the number of rules can be adapted to the complexity of the problem. Simple input–output relationships are modeled with a few rules, but still more rules can be added as the complexity of the problem increases. Therefore, approximate FRBSs constitute a potential remedy to the course of dimensionality that emerges when scaling to multidimensional systems.

These properties enable approximate FRBSs to achieve a better accuracy than linguistic FRBS in complex problem domains. However, despite their benefits, they also come with some drawbacks:

- Their main drawback compared to the descriptive FRBS is the degradation in terms of interpretability of the RB as the fuzzy variables no longer share a unique linguistic interpretation. Still, unlike other kinds of approximate models such as neural networks that store knowledge implicitly, the knowledge in an approximate FRBS remains explicit as the system behavior is described by local rules. Therefore, approximate FRBSs can be considered as a compromise between the apparent interpretability of descriptive FRBSs and the type of black-box behavior, typical for nondescriptive, implicit models.
- The capability to approximate a set of training data accurately can lead to over-fitting and therefore to a poor generalization capability to cope with previously unseen input data.

According to their properties, fuzzy modeling [13.1] constitutes the major application of approximate FRBSs, as model accuracy is more relevant than description ability. Approximate FRBSs are usually not the first choice for linguistic modeling and fuzzy control problems. Hence, descriptive and approximate FRBSs are considered as complementary rather than competitive approaches. Depending on the problem domain and requirements on the obtained model, one should use one or the other approach. Approximate FRBSs are recommendable in case one wants to trade interpretability for improved accuracy.

13.2.3 Takagi–Sugeno–Kang Fuzzy Rule-Based Systems

Instead of working with linguistic rules of the kind introduced in the previous section, *Sugeno* et al. [13.31, 32] proposed a new model based on rules whose antecedent is composed of linguistic variables and the consequent is represented by a function of the input variables. The most common form of this kind of rules is the one in which the consequent expression consti-

tutes a linear combination of the variables involved in the antecedent

IF X_1 is A_1 and ... and X_n is A_n

$$\text{THEN } Y = p_0 + p_1 \cdot X_1 + \cdots + p_n \cdot X_n , \qquad (13.24)$$

where X_i are the input variables, Y is the output variable, and $p = (p_0, p_1, \ldots, p_n)$ is a vector of real parameters. Regarding A_i, they are either a direct specification of a fuzzy set (thus X_i being fuzzy variables) or a linguistic label that points to a particular member of a fuzzy partition of a linguistic variable. These rules, and consequently the systems using them, are usually called *TSK fuzzy rules*, in reference to the names of their first proponents.

The output of a TSK FRBS, using a KB composed of m rules, is obtained as a weighted sum of the individual outputs provided by each rule, Y_i, $i = 1, \ldots, m$, as follows:

$$\frac{\sum_{i=1}^{m} h_i \cdot Y_i}{\sum_{i=1}^{m} h_i} , \qquad (13.25)$$

in which $h_i = T(A_{i1}(x_1), \ldots, A_{in}(x_n))$ is the matching degree between the antecedent part of the ith rule and the current inputs to the system, $x_0 = (x_1, \ldots, x_n)$. T stands for a conjunctive operator modeled by a t-norm. Therefore, to design the inference engine of TSK FRBSs, the designer only selects this conjunctive operator T, with the most common choices being the minimum and the product. As a consequence, TSK systems do not need defuzzification, being their outputs real numbers.

This type of FRBS divides the input space in several fuzzy subspaces and defines a linear input–output relationship in each one of these subspaces [13.31]. In the inference process, these partial relationships are combined in the said way for obtaining the global input–output relationship, taking into account the dominance of the partial relationships in their respective areas of application and the conflicts emerging in the overlapping zones. As a result, the overall system performs as a sort of interpolation of the local models represented by each individual rule.

TSK FRBSs have been successfully applied to a large variety of practical problems. The main advantage of these systems is that they present a set of compact system equations that allows the parameters p_i to be estimated by means of classical regression methods, which facilitates the design process. However, the main drawback associated with TSK FRBSs is the form of the rule consequents, which does not provide a natural framework for representing expert knowledge that is afflicted with uncertainty. Still, it becomes possible to integrate expert knowledge in these FRBSs by slightly modifying the rule consequent: for each linguistic rule with consequent *Y is B*, provided by an expert, its consequent is substituted by $Y = p_0$, with p_0 standing for the modal point of the fuzzy set associated with the label *B*. These kinds of rules are usually called *simplified TSK rules* or *zero-order TSK rules*.

However, TSK FRBSs are more difficult to interpret than Mamdani FRBSs due to two different reasons:

- The structure of the rule consequents is difficult to be understood by human experts, except for zero-order TSK.
- Their overall output simultaneously depends on the activation of the rule antecedents and on the evaluation of the function defining rule consequent, that depends itself on the crisp inputs as well, rather than being constant.

TSK FRBSs are used in fuzzy modeling [13.4, 31] as well as control problems [13.31, 33].

As with Mamdani FRBSs, it is also possible to built descriptive as well as approximate TSK systems.

13.2.4 Singleton Fuzzy Rule–Based Systems

The singleton FRBS, where the rule consequent takes a single real-valued number, may be considered as a particular case of the linguistic FRBS (the consequent is a fuzzy set where the membership function is one for a specific value and zero for the remaining ones) or of the TSK-type FRBS (the previously described zero-order TSK systems).

Its rule structure is the following

> *IF X_1 is A_1 and ... and X_n is A_n*
> *THEN Y is y_0* .   (13.26)

Since the single consequent seems to be more easily interpretable than a polynomial function, the singleton FRBS may be used to develop linguistic fuzzy models. Nevertheless, compared with the linguistic FRBS, the fact of having a different consequent value for each rule (no global semantic is used for the output variable) worsens the interpretability.

13.2.5 Fuzzy Rule–Based Classifiers

Previous sections have implicitly considered FRBSs working with inputs and, what is more important, outputs which are real variables. These kinds of fuzzy systems show an interpolative behavior where the overall output is a combination of the individual outputs of the fired rules. This interpolative behavior is explicit in TSK models but it is also present in Mamdani systems. This situation gives FRBSs a sort of smooth output, generating soft transitions between rules, and being one of the significant properties of FRBSs.

A completely different situation is that of having a problem where the output takes values from a finite list of possible values representing categories or classes. Under those circumstances, the interpolative approach of previously defined aggregation and defuzzification methods, makes no sense. As a consequence, some additional comments will be added to highlight the main characteristics of fuzzy rule-based classifiers (FRBCs), and the differences with other FRBSs.

A fuzzy rule-based classifier is an automatic classification system that uses fuzzy rules as knowledge representation tool. Therefore, the fuzzy classification rule structure is as follows

> *IF X_1 is A_1 and ... and X_n is A_n*
> *THEN Y is C* ,   (13.27)

with *Y* being a categorical variable, so *C* being a class label. The processing structure is similar to that previously described in what concerns to the evaluation of matching degree between each rule's antecedent and current input, i.e., for each rule R_i we obtain $h_i = T(A_{i1}(x_1), \dots, A_{in}(x_n))$. Once obtained h_i, the winner rule criteria could be applied so that the overall output is assigned with the consequent of the rule achieving the highest matching degree (highest value of h_i). More elaborated evaluations as voting are also possible.

Other alternative representations that include a certainty degree or weight for each rule have also been considered [13.34]. In this case, the previously described rule will also include a rule weight w_i that weights the matching degree during the inference process. The effect will be that the winning rule will be that achieving the highest value of $h_i \cdot w_i$, or in the case of voting schemes, the influence of the vote of the rule will be proportional to this value.

13.2.6 Type-2 Fuzzy Rule–Based Systems

The idea of extending fuzzy sets by allowing membership functions to include some kind of uncertainty was already mentioned by *Zadeh* in early papers [13.6–8]. The idea, that was not really exploited for a long period, has achieved now a significant presence in the literature with the proposal of Type-2 fuzzy systems and Interval type-2 fuzzy systems [13.35]. The main concept is that the membership degree is not a value but a fuzzy set or an interval, respectively. The effect is obtaining additional degrees of freedom being available in the design process, but increasing the complexity of the processing structure that requires now a type-reduction step added to the overall process described in previous section. As the complexity of the type reduction process is much lower for Interval type-2 fuzzy systems than in the general case of type-2 fuzzy systems, interval approaches are the most widely considered now in the area of Type-2 fuzzy sets.

13.2.7 Fuzzy Systems with Implicative Rules

Rule-based systems mentioned to this point consider rules that, having the form *if X is A then Y is B*, model the inference through a t-norm, usually minimum or product (Sect. 13.1.2). With this interpretation, rules are described as conjunctive rules, representing joint sets of possible input and output values. As mentioned in the chapter devoted to fuzzy control, these rules should be seen not as logical implications but rather as input–output associations.

That kind of rule is the one commonly used in real applications to the date. However, different authors have pointed out that the same rule will have a completely different meaning when modeled in terms of material implications (the approach for Boolean *if–then* statements in propositional logic) [13.18]. As a result, in addition to the common interpretation of fuzzy rules that is widely considered in the literature, some authors are exploring the modeling of fuzzy rules (with exactly the same structure previously mentioned) by means of *material implications* [13.36]. Even being in a quite preliminary stage of development, it is interesting to mention this ideas since it constitutes a completely different interpretation of FRBSs, offering so new possibilities to the field.

13.3 Hierarchical Fuzzy Rule-Based Systems

The knowledge structure of FRBSs offers different options to introduce hierarchical structures. Rules, partitions, or variables can be distributed at different levels according to their specificity, granularity, relevance, etc. This section will introduce different approaches to hierarchical FRBSs.

It would be possible to consider hierarchical fuzzy systems as a different type of FRBS, so including it in previous section, or as a design option to build *simpler* FRBSs, being then included as part of the next section. Including it in previous section could be a little bit confusing since it is possible to combine the hierarchical approach with several of the structural models defined there, it seems better to consider it independently devoting a section to analyze them.

The definition of hierarchical fuzzy systems as a method to solve problems with a higher level of complexity than those usually focused on with FRBSs, has produced some good results. In most of the cases, the underlying idea is to cope with the complexity of a problem by applying some kind of decomposition that generates a hierarchy of lower complexity systems [13.37].

Several methods to establish hierarchies in fuzzy controllers have been proposed. These methods could be grouped according to the way they structure the inference process, and the knowledge applied.

A first approach defines the hierarchy as a prioritization of rules in such a way that rules with a different level of specificity receive a different priority, having higher priority those rules being more specific [13.38, 39]. With this kind of hierarchy, a generic rule is applied only when no suitable specific rule is available. In this case, the hierarchy is the effect of a particular implication mechanism applying the rules by taking into account its priority. This methodology defines the hierarchy (the decomposition) at the level of rules. The rules are grouped into prioritized levels to design a hierarchical fuzzy controller.

Another option is that of considering a hierarchy of fuzzy partitions with different granularity [13.40]. From that point, an FRBS is structured in layers, where each layer contains fuzzy partitions with a different granularity, as well as the rules using those fuzzy partitions. Usually, every partition in a certain layer has the same number of fuzzy terms. In this case, rules at different layers have different granularity, being somehow related to the idea of specificity of the previous paragraph. It is even possible to generate a multilevel grid-like partition where only for some specific regions of the input space (usually those regions showing poor performance) a higher granularity is considered [13.41], with a similar approach to that already considered in some neuro-fuzzy systems [13.42].

A completely different point of view is that of introducing the decomposition at the level of variables. In this case, the input space is decomposed into subspaces of lower dimensionality, and each input variable is only considered at a certain level of the hierarchy. The result is a cascade structure of FRBSs where, in addition to a subset of the input variables, the output of each level is considered as one of the inputs to the following level [13.43]. As a result, the system is decomposed into a finite number of reduced-order subsystems, eliminating the need for a large-sized inference engine. This decomposition is usually stated as a way to maintain under control the problems generated by the so-called course of dimensionality, the exponential growth of the number of rules related to the number of variables of the system.

The number of rules of an FRBS with n input variables and l linguistic terms per variable, will be l^n. In this approach to hierarchical FRBSs, the variables (and rules) are divided into different levels in such a way that those considered the most influential variables are chosen as input variables at the first level, the next most important variables are chosen as input variables at the

second level, and so on. The output variable of each level is introduced as input variable at the following level.

With that structure, the rules at first level of the FRBS have a similar structure to any Mamdani FRBS, i.e., that describe by (13.1), but at k-th level ($k > 1$), rules include the output of the previous level as input

$$IF\ X_{N_k+1}\ is\ LT_{N_k+1}\ and\ \dots\ and\ X_{N_k+n_k}\ is\ LT_{N_k+n_k}$$
$$and\ O_{k-1}\ is\ LT_{Ok-1}\ THEN\ O_k\ is\ LT_{Ok}\ ,\quad(13.28)$$

where the value N_k determines the input variables considered in previous levels

$$N_k = \sum_{t=1}^{k-1} n_t\ ,\quad(13.29)$$

with n_t being the number of system variables applied at level t. Variable O_k represent the output of the k level of the hierarchy. All outputs are intermediate variables except for the output of the last level that will be Y (the overall output of the system).

With this structure it is shown [13.43] that the number of rules in a complete rule base could be reduced to a linear function of the number of variables, while in a conventional FRBS it was an exponential function of the number of variables.

13.4 Fuzzy Rule–Based Systems Design

Once defined the components and functioning of an FRBS, it is time to consider its design, i.e., how to built an FRBS to solve a certain problem while showing some specific properties. The present section will focus on this question.

An FRBS can be characterized according to its structure and its behavior. When referring to its structure, we can consider questions as the dimension of the system (number of variables, fuzzy sets, rules, etc.) as well as other aspects related to properties of its components (distinguishability of the fuzzy sets, redundancy of the fuzzy rules, etc.). On the other hand, the characterization related to the behavior mostly analyzes properties considering the input–output relation defined by the FRBS. In this area, we can include questions as stability or accuracy. Finally, there is a third question that simultaneously involves structure and behavior. This question is interpretability, a central aspect

in fuzzy systems design that is considered in an independent chapter.

13.4.1 FRBS Properties

All the structural properties to be mentioned are related to properties of the KB, and basically cover characteristics related to the individual fuzzy sets, the fuzzy partitions related to each input and output variable, the fuzzy rules, and the rule set as a whole.

The elemental components of the KB are fuzzy sets. At this level, we have several questions to be analyzed as normality, convexity, or differentiability of fuzzy sets; all of them being related to the properties of the membership function ($\mu_A(x)$) defining the fuzzy set (A). In most applications the considered fuzzy sets adopt predefined shapes as triangular, trapezoidal, Gaussian, or bell; the fuzzy sets are then defined by only changing some parameters of these *parameterized functions*. In summary, most fuzzy sets considered in FRBSs are normal and convex sets belonging to one of two possible families: piecewise linear functions and differentiable functions. Piecewise linear functions are basically triangular and trapezoidal functions offering a reduced complexity from the processing point of view. On the other hand, differentiable functions are mainly Gaussian, bell, and sigmoidal functions being better adapted to some kind of differential learning approaches as those used in neuro-fuzzy systems, but adding complexity from the processing point of view.

Once individual fuzzy sets have been considered, the following level is that of fuzzy partitions related to each variable. The main characteristics of a fuzzy partition are cardinality, coverage, and distinguishability. Cardinality corresponds to the number of fuzzy sets that compose the fuzzy partitions. In most cases, this number ranges from 3 to 9, with 9 being an upper limit commonly accepted after the ideas of *Miller* [13.44]. The larger the number of fuzzy sets in the partition, the most difficult the design and interpretation of the FRBS. Coverage corresponds to the minimum membership degree with which any value of the variable (x), through its universe of discourse (U), will be assigned to at least a fuzzy set (A_i) in the partition. Coverage is then defined as

$$\min_{x \in U} \max_{i=1...n} \mu_{A_i}(x) , \qquad (13.30)$$

being n the cardinality of the partition. As an example, the fuzzy partition in Fig. 13.2 has a coverage of 0.5. Finally, distinguishability of fuzzy sets is related to the level of overlapping of their membership functions, being analyzed with different expressions.

On the basis of the fuzzy sets and fuzzy partitions, the fuzzy rules are built. The first structural question regarding fuzzy rules is the type of fuzzy rule to be considered: Mamdani, TSK, descriptive or approximate, DNF, etc. If we consider now the interaction between the different fuzzy rules of a fuzzy system, questions as knowledge consistency or redundancy appear, i.e., does a fuzzy system include pieces of knowledge (usually rules) providing contradictory (or redundant) information for a specific situation. Finally, when considering the rule base as a whole, completeness and complexity are to be considered. Completeness refers to the fact that any potential input value will fire at least one rule.

Considering now behavioral properties, the most widely analyzed are stability and accuracy. It is also possible to take into account other properties as continuity or robustness, but we will concentrate in those having the larger presence in the literature. Behavioral properties are related to the overall system, i.e., to the processing structure as well as to the KB.

Stability is a key aspect of dynamical systems analysis, and plays a central role in control theory. FRBSs are nonlinear dynamical systems, and after its early application to control problems, the absence of a formal stability analysis was seriously criticized. As a consequence, the stability question received significant attention from the very beginning, at present being a widely studied problem [13.45] for both Mamdani and TSK fuzzy systems, considering the use of different approaches as Lyapunov's methods, Popov criterion or norm-based analysis among others.

Another question with a continuous presence in the literature is that of accuracy and the somehow related concept of universal approximation. The idea of fuzzy systems as universal approximators states that, given any continuous real-valued function on a compact subset of R^n, we can, at least in theory, find an FRBS that approximates this function to any degree. This property has been established for different types of fuzzy systems [13.46–48]. On this basis, the idea of building fuzzy models with an unbounded level of accuracy can be considered. In any case, it is important to notice that previous papers proof the existence of such a model but assuming at the same time an unbounded complexity, i.e., the number of fuzzy sets and rules involved in the fuzzy system will usually grow as the accuracy improves. That means that improving accuracy is possible but always with a cost related either to the complexity of the system or to the relaxation of some of its properties (usually interpretability).

13.4.2 Designing FRBSs

Given a modeling, classification, or control problem to be solved, and assumed it will be focused on through an FRBS, there are several steps in the process of design. The first decision is the choice between the different types of systems mentioned in Sect. 13.2, particularly Mamdani and TSK approaches. They offer different characteristics related to questions as their accuracy and interpretability, as well as different methods for the derivation of its KB.

Once chosen a type of FRBS, its design implies the construction of its processing structure as well as the derivation of its KB. Even considering that there are several options to modify the processing structure of the system (Sect. 13.1.2), most designers consider a standard inference engine and concentrate on the knowledge extraction problem.

Going now to the knowledge extraction problem, some of its parts are common to any modeling process (being fuzzy or not). Questions as the selection of the input and output variables and the determination of the range of those variables are generic to any modeling approach. The specific aspects related to the fuzzy environment are the definition of the fuzzy sets or the fuzzy partition related to each of those variables, and the derivation of a suitable set of fuzzy rules. These two components can be jointly derived in a single process, or sequentially performed by considering first the design of the fuzzy partition associated with each variable and then the fuzzy rules. The design process can be based on two main sources of information: expert knowledge and experimental data.

If we first consider the definition of fuzzy sets and fuzzy partitions, quite different approaches [13.49] can be applied. Even the idea of simply generating a uni-formly distributed strong fuzzy partition of a certain cardinality is widely considered.

Going now to rules, Mamdani FRBSs are particularly adapted to expert knowledge extraction, and knowledge elicitation for that kind of system has been widely considered in the literature. In any case, there is not a standard methodology for fuzzy knowledge extraction from experts and at present most practical works consider either a direct data-driven approach, or the integration of expert and data-driven knowledge extraction [13.50].

When considering data-driven knowledge extraction, there is an almost endless list of approaches. Some options are the use of ad-hoc methods based on data covering measures (as [13.46]), the generation of fuzzy decisions trees [13.51], the use of clustering techniques [13.52], and the use of hybrid systems where genetic fuzzy systems [13.53] and neuro fuzzy systems [13.54] represent the most widely considered approaches to fuzzy systems design. Some of those techniques produce both the partitions (or fuzzy sets) and the rules in a single process.

13.5 Conclusions

Fuzzy rule-based systems constitute a tool for representing knowledge and reasoning on it. Jointly with fuzzy clustering techniques, FRBSs are probably the developments of fuzzy sets theory leading to the larger number of applications. These systems, being a kind of rule-based system, can be analyzed as knowledge-based systems showing a structure with two main components: knowledge and processing. The processing structure relays on many concepts presented in previous chapters as fuzzy implications, connectives, relations and so on. In addition, some new concepts as fuzzification and defuzzification are required when constructing a fuzzy rule-based system. But the central concept of fuzzy rule-based systems are fuzzy rules. Different types of fuzzy rules have been considered, particularly those having a fuzzy (or not) consequent, producing different types of FRBS. In addition, new formulations are being considered, e.g., implicative rules. Eventually, the representation capabilities of fuzzy sets have been considered as too limited to represent some specific kinds of knowledge or information, and some extended types of fuzzy sets have been defined. Type-2 fuzzy sets are an example of extension of fuzzy sets.

Having been said that FRBSs are knowledge-based systems, and as a consequence, its design involves, apart from aspects related to the processing structure, the elicitation of a suitable KB properly describing the way to solve the problem under consideration. Even considering the large number of problems solved using FRBSs, there is not a clear design methodology defining a well-established design protocol. In addition, two completely different sources of knowledge, requiring different extraction approaches, have been considered when building FRBSs: expert knowledge and data. Many expert and data-driven knowledge extraction techniques and methods are described in the literature and can be considered. Connected to this question, as part of the process to provide automatic knowledge extraction capabilities to FRBSs, many hybrid approaches have been proposed, genetic fuzzy systems and neuro-fuzzy systems being the most widely considered.

In summary, FRBSs are a powerful tool to solve real world problems, but many theoretical aspects and design questions remain open for further investigation.

References

13.1 A. Bardossy, L. Duckstein: *Fuzzy Rule-Based Modeling with Application to Geophysical, Biological and Engineering Systems* (CRC, Boca Raton 1995)

13.2 Z. Chi, H. Yan, T. Pham: *Fuzzy Algorithms: With Applications to Image Processing and Pattern Recognition* (World Scientific, Singapore 1996)

13.3 K. Hirota: *Industrial Applications of Fuzzy Technology* (Springer, Berlin, Heidelberg 1993)

13.4 W. Pedrycz: *Fuzzy Modelling: Paradigms and Practice* (Kluwer Academic, Dordrecht 1996)

13.5 R.R. Yager, L.A. Zadeh: *An Introduction to Fuzzy Logic Applications in Intelligent Systems* (Kluwer Academic, Dordrecht 1992)

13.6 L.A. Zadeh: The concept of a linguistic variable and its applications to approximate reasoning – Part I, Inf. Sci. **8**(3), 199–249 (1975)

13.7 L.A. Zadeh: The concept of a linguistic variable and its applications to approximate reasoning – Part II, Inf. Sci. **8**(4), 301–357 (1975)

13.8 L.A. Zadeh: The concept of a linguistic variable and its applications to approximate reasoning – Part III, Inf. Sci. **9**(1), 43–80 (1975)

13.9 L.A. Zadeh: Outline of a new approach to the analysis of complex systems and decision processes, IEEE Trans. Syst. Man Cybern. **3**, 28–44 (1973)

13.10 E.H. Mamdani: Applications of fuzzy algorithm for control of simple dynamic plant, Proc. IEE **121**(12), 1585–1588 (1974)

13.11 E.H. Mamdani, S. Assilian: An experiment in linguistic synthesis with a fuzzy logic controller, Int. J. Man-Mach. Stud. **7**, 1–13 (1975)

13.12 L.X. Wang: *Adaptive Fuzzy Systems and Control: Design and Analysis* (Prentice Hall, Englewood Cliffs 1994)

13.13 T.J. Procyk, E.H. Mamdani: A linguistic self-organizing process controller, Automatica **15**(1), 15–30 (1979)

13.14 L. Magdalena: Adapting the gain of an FLC with genetic algorithms, Int. J. Approx. Reas. **17**(4), 327–349 (1997)

13.15 G.C. Mouzouris, J.M. Mendel: Nonsingleton fuzzy logic systems: Theory and application, IEEE Trans. Fuzzy Syst. **5**, 56–71 (1997)

13.16 F. Klawonn, R. Kruse: Equality relations as a basis for fuzzy control, Fuzzy Sets Syst. **54**(2), 147–156 (1993)

13.17 J.M. Mendel: Fuzzy logic systems for engineering: A tutorial, Proc. IEEE **83**(3), 345–377 (1995)

13.18 D. Dubois, H. Prade: What are fuzzy rules and how to use them, Fuzzy Sets Syst. **84**, 169–185 (1996)

13.19 O. Cordón, F. Herrera, A. Peregrín: Applicability of the fuzzy operators in the design of fuzzy logic controllers, Fuzzy Sets Syst. **86**, 15–41 (1997)

13.20 D. Driankov, H. Hellendoorn, M. Reinfrank: *An Introduction to Fuzzy Control* (Springer, Berlin, Heidelberg 1993)

13.21 M. Sugeno, T. Yasukawa: A fuzzy-logic-based approach to qualitative modeling, IEEE Trans. Fuzzy Syst. **1**(1), 7–31 (1993)

13.22 C.C. Lee: Fuzzy logic in control systems: Fuzzy logic controller – Part I, IEEE Trans. Syst. Man Cybern. **20**(2), 404–418 (1990)

13.23 C.C. Lee: Fuzzy logic in control systems: Fuzzy logic controller – Part II, IEEE Trans. Syst. Man Cybern. **20**(2), 419–435 (1990)

13.24 A. Bastian: How to handle the flexibility of linguistic variables with applications, Int. J. Uncertain. Fuzziness Knowl.-Based Syst. **3**(4), 463–484 (1994)

13.25 B. Carse, T.C. Fogarty, A. Munro: Evolving fuzzy rule based controllers using genetic algorithms, Fuzzy Sets Syst. **80**, 273–294 (1996)

13.26 A. González, R. Pèrez, J.L. Verdegay: Learning the structure of a fuzzy rule: A genetic approach, Fuzzy Syst. Artif. Intell. **3**(1), 57–70 (1994)

13.27 L. Magdalena, F. Monasterio: A fuzzy logic controller with learning through the evolution of its knowledge base, Int. J. Approx. Reas. **16**(3/4), 335–358 (1997)

13.28 R. Alcalá, J. Casillas, O. Cordón, F. Herrera: Building fuzzy graphs: Features and taxonomy of learning for non-grid-oriented fuzzy rule-based systems, J. Intell. Fuzzy Syst. **11**(3/4), 99–119 (2001)

13.29 O. Cordón, F. Herrera: A three-stage evolutionary process for learning descriptive and approximate fuzzy logic controller knowledge bases from examples, Int. J. Approx. Reas. **17**(4), 369–407 (1997)

13.30 L. Koczy: Fuzzy if … then rule models and their transformation into one another, IEEE Trans. Syst. Man Cybern. **26**(5), 621–637 (1996)

13.31 T. Takagi, M. Sugeno: Fuzzy identification of systems and its application to modeling and control, IEEE Trans. Syst. Man Cybern. **15**(1), 116–132 (1985)

13.32 M. Sugeno, G.T. Kang: Structure identification of fuzzy model, Fuzzy Sets Syst. **28**(1), 15–33 (1988)

13.33 R. Palm, D. Driankov, H. Hellendoorn: *Model Based Fuzzy Control* (Springer, Berlin, Heidelberg 1997)

13.34 H. Ishibuchi, T. Nakashima: Effect of rule weights in fuzzy rule-based classification systems, IEEE Trans. Fuzzy Syst. **9**, 506–515 (2001)

13.35 J.M. Mendel: Type-2 fuzzy sets and systems: An overview, Comput. Intell. Mag. IEEE **2**(1), 20–29 (2007)

13.36 H. Jones, B. Charnomordic, D. Dubois, S. Guillaume: Practical inference with systems of gradual implicative rules, IEEE Trans. Fuzzy Syst. **17**, 61–78 (2009)

13.37 V. Torra: A review of the construction of hierarchical fuzzy systems, Int. J. Intell. Syst. **17**(5), 531–543 (2002)

13.38 R.R. Yager: On a hierarchical structure for fuzzy modeling and control, IEEE Trans. Syst. Man Cybern. **23**(4), 1189–1197 (1993)

Part B | 13

13.39   R.R. Yager: On the construction of hierarchical fuzzy systems models, IEEE Trans. Syst. Man Cybern. C **28**(1), 55–66 (1998)

13.40   O. Cordón, F. Herrera, I. Zwir: Linguistic modeling by hierarchical systems of linguistic rules, IEEE Trans. Fuzzy Syst. **10**, 2–20 (2002)

13.41   E. D'Andrea, B. Lazzerini: A hierarchical approach to multi-class fuzzy classifiers, Exp. Syst. Appl. **40**(9), 3828–3840 (2013)

13.42   H. Takagi, N. Suzuki, T. Koda, Y. Kojima: Neural networks designed on approximate reasoning architecture and their applications, IEEE Trans. Neural Netw. **3**(5), 752–760 (1992)

13.43   G.V.S. Raju, J. Zhou, R.A. Kisner: Hierarchical fuzzy control, Int. J. Control **54**(5), 1201–1216 (1991)

13.44   G.A. Miller: The magical number seven, plus or minus two: Some limits on our capacity for processing information, Psychol. Rev. **63**, 81–97 (1956)

13.45   K. Michels, F. Klawonn, R. Kruse, A. Nürnberger: *Fuzzy Control: Fundamentals, Stability and Design of Fuzzy Controllers* (Springer, Berlin, Heidelberg 2006)

13.46   L.-X. Wang, J.M. Mendel: Fuzzy basis functions, universal approximation, and orthogonal least-squares learning, IEEE Trans. Neural Netw. **3**(5), 807–813 (1992)

13.47   B. Kosko: Fuzzy systems as universal approximators, IEEE Trans. Comput. **43**(11), 1329–1333 (1994)

13.48   J.L. Castro: Fuzzy logic controllers are universal approximators, IEEE Trans. Syst. Man Cybern. **25**(4), 629–635 (1995)

13.49   R. Krishnapuram: Membership function elicitation and learning. In: *Handbook of Fuzzy Computation*, ed. by E.H. Ruspini, P.P. Bonissone, W. Pedrycz (IOP Publ., Bristol 1998) pp. 349–368

13.50   J.M. Alonso, L. Magdalena: HILK++: An interpretability-guided fuzzy modeling methodology for learning readable and comprehensible fuzzy rule-based classifiers, Soft Comput. **15**(10), 1959–1980 (2011)

13.51   N.R. Pal, S. Chakraborty: Fuzzy rule extraction from id3-type decision trees for real data, IEEE Trans. Syst. Man Cybern. B **31**(5), 745–754 (2001)

13.52   M. Delgado, A.F. Gómez-Skarmeta, F. Martín: A fuzzy clustering-based rapid prototyping for fuzzy rule-based modeling, IEEE Trans. Fuzzy Syst. **5**, 223–233 (1997)

13.53   O. Cordón, F. Herrera, F. Hoffmann, L. Magdalena: *Genetic Fuzzy Systems: Evolutionary Tuning and Learning of Fuzzy Knowledge Bases* (World Scientific, Singapore 2001)

13.54   D.D. Nauck, A. Nürnberger: Neuro-fuzzy systems: A short historical review. In: *Computational Intelligence in Intelligent Data Analysis*, ed. by C. Moewes, A. Nürnberger (Springer, Berlin, Heidelberg 2013) pp. 91–109

14. Interpretability of Fuzzy Systems: Current Research Trends and Prospects

Jose M. Alonso, Ciro Castiello, Corrado Mencar

Fuzzy systems are universally acknowledged as valuable tools to model complex phenomena while preserving a readable form of knowledge representation. The resort to natural language for expressing the terms involved in fuzzy rules, in fact, is a key factor to conjugate mathematical formalism and logical inference with human-centered interpretability. That makes fuzzy systems specifically suitable in every real-world context where people are in charge of crucial decisions. This is because the self-explanatory nature of fuzzy rules profitably supports expert assessments. Additionally, as far as interpretability is investigated, it appears that (a) the simple adoption of fuzzy sets in modeling is not enough to ensure interpretability; (b) fuzzy knowledge representation must confront the problem of preserving the overall system accuracy, thus yielding a trade-off which is frequently debated. Such issues have attracted a growing interest in the research community and became to assume a central role in the current literature panorama of computational intelligence. This chapter gives an overview of the topics related to fuzzy system interpretability, facing the ambitious goal of proposing some answers to a number of open challenging questions: What is interpretability? Why interpretability is worth considering? How to ensure interpretability, and how to assess (quantify) it? Finally, how to design interpretable fuzzy models?

The objective of this chapter is to provide some answers for the questions posed above. Section 14.1 deals with the challenging task of setting a proper definition of interpretability. Section 14.2 introduces the main constraints and criteria that can be adopted to ensure interpretability when designing interpretable fuzzy systems. Section 14.3 gives a brief overview of the soundest indexes for

assessing interpretability. Section 14.4 presents the most popular approaches for designing fuzzy systems endowed with a good interpretability-accuracy trade-off. Section 14.5 enumerates some application fields where interpretability is a main concern. Section 14.6 sketches a number of challenging tasks which should be addressed in the near future. Finally, some conclusions are drawn in Sect. 14.7.

The key factor for the success of fuzzy logic stands in the ability of modeling and processing *perceptions* instead of measurements [14.1]. In most cases, such perceptions are expressed in natural language. Thus, fuzzy logic acts as a mathematical underpinning for modeling and processing perceptions described in natural language.

Historically, it has been acknowledged that fuzzy systems are endowed with the capability to conjugate a complex behavior and a simple description in terms of linguistic rules. In many cases, the compilation of fuzzy systems has been accomplished *manually*; with human knowledge purposely injected in fuzzy rules in order to model the desired behavior (the rules could be eventually tuned to improve the system accuracy). In addition, the great success of fuzzy logic led to the development of many algorithms aimed at acquiring knowledge from data (expressing it in terms of fuzzy rules). This made the automatic design of fuzzy systems (through data-driven design techniques) feasible. Moreover, theoretical studies proved the universal approximation capabilities of such systems [14.2].

The adoption of data-driven design techniques is a common practice nowadays. Nevertheless, while fuzzy sets can be generally used to model perceptions, some of them do not lead to a straight interpretation in natural language. In consequence, the adoption of accuracy-driven algorithms for acquiring knowledge from data often results in unintelligible models. In those cases, the fundamental plus of fuzzy logic is lost and the derived models are comparable to other measurement-based models (like neural networks) in terms of knowledge interpretability.

In a nutshell, interpretability is not granted by the adoption of fuzzy logic which represents a necessary yet not a sufficient requirement for modeling and processing perceptions. However, interpretability is a quality that is not easy to define and quantify. Several open and challenging questions arise while considering interpretability in fuzzy modeling: *What* is interpretability? *Why* interpretability is worth considering? How to *ensure* interpretability? How to *assess* (quantify) interpretability? How to *design* interpretable fuzzy models? And so on.

14.1 The Quest for Interpretability

Answering the question *What is interpretability?* is not straightforward. Defining interpretability is a challenging task since it deals with the analysis of the relation occurring between two heterogeneous entities: a model of the system to be designed (usually formalized through a mathematical definition) and a human user (meant not as a passive beneficiary of a system's outcome, but as an active reader and interpreter of the model's working engine). In this sense, interpretability is a quality which is inherent in the model and yet it refers to an act performed by the user who is willing to grasp and explain the meaning of the model.

To pave the way for the definition of such a relation, a common ground must be settled. This could be represented by a number of fundamental properties to be incorporated into a model, so that its formal description becomes compatible with the user's knowledge representation. In this way, the human user may interface the mathematical model resting on concepts that appear to be suitable to deal with it. The quest for interpretability, therefore, calls for the identification of several features. Among them, resorting to an appropriate framework for knowledge representation is a crucial element and the adoption of a fuzzy inference engine based on fuzzy

rules is straightforward to approach the linguistic-based formulation of concepts which is typical of the human abstract thought.

A distinguishing feature of a fuzzy rule-based model is the double level of knowledge representation. The lower level of representation is constituted by the formal definition of the fuzzy sets in terms of their membership functions, as well as the aggregation functions used for inference. This level of representation defines the *semantics* of a fuzzy rule-based model as it determines the behavior of the model, i.e. the input/output mapping for which it is responsible.

On the higher level of representation, knowledge is represented in the form of rules. They define a formal structure where linguistic variables are involved and reciprocally connected by some formal operators, such as *AND*, *THEN*, and so on. Linguistic variables correspond to the inputs and outputs of the model. The (symbolic) values they assume are related to linguistic terms which, in turn, are mapped to the fuzzy sets defined in the lower level of representation. The formal operators are likewise mapped to the aggregation functions. This mapping provides the interpretative transition that is quite common in the mathematical context: a formal structure is assigned semantics by mapping symbols (linguistic terms and operators) to objects (fuzzy sets and aggregation functions).

In principle, the mapping of linguistic terms to fuzzy sets is arbitrary. It just suffices that identical linguistic terms are mapped to identical fuzzy sets. Of course, this is not completely true for formal operators (e.g., *t*-norms, implications, etc.). The corresponding aggregation functions should satisfy a number of constraints; however some flexibility is possible. Nevertheless, the mere use of symbols in the high level of knowledge representation implies the establishment of a number of semiotic relations that are fundamental for the quest of interpretability of a fuzzy model. In particular, linguistic terms – as usually picked from natural language – must be fully meaningful for the expected reader since they denote concepts, i. e. mental representations that allow people to draw appropriate inferences about the entities they encounter.

Concepts and fuzzy sets, therefore, are both denoted by linguistic terms. Additionally, concepts and fuzzy sets play a similar role: the former (being part of the human knowledge) contribute to determine the behavior of a person; the latter (being the basic elements of a fuzzy rule base) contribute to determine the behavior of a system to be modeled. As a consequence, concepts and fuzzy sets are implicitly connected by means of common linguistic terms they are related to, which refer to object classes in the real world. The key essence of interpretability is therefore the property of *cointension* [14.3] between fuzzy sets and concepts, consisting in the possibility of referring to similar classes of objects: such a possibility is assured by the use of common linguistic terms.

Semantic cointension is a key issue when dealing with interpretability of fuzzy systems. It has been introduced and centered on the role of fuzzy sets, but it can be easily extended to refer to some more complex structures, such as fuzzy rules or the whole fuzzy models. In this regard, a crisp assertion about the importance of cointension pronounced at the level of the whole model is given by the *Michalski*'s *Comprehensibility Postulate* [14.4]:

The results of computer induction should be symbolic descriptions of given entities, semantically and structurally similar to those a human expert might produce observing the same entities. Components of these descriptions should be comprehensible as single chunks of information, directly interpretable in natural language, and should relate quantitative and qualitative concepts in an integrated fashion.

It should be observed that the above postulate has been formulated in the general area of machine learning. Nevertheless, the assertion made by Michalski has important consequences in the specific area of fuzzy modeling (FM) too. According to the Comprehensibility Postulate, results of computer induction should be described symbolically. Symbols are necessary to communicate information and knowledge; hence, pure numerical methods, such as neural networks, are not suited for meeting interpretability unless an interpretability-oriented postprocessing of the resulting knowledge is performed.

The key point of the Michalski's postulate is the human centrality of the results of a computer induction process. The importance of the human component implicitly suggests a novel aspect to be taken into account in the quest for interpretability. Actually, the semantic cointension is related to one facet of the interpretability process, which can be referred to as *comprehensibility* of the content and behavior of a fuzzy model. In other words, cointension concerns the semantic interpretation performed by a user determined to comprehend such a model. On the other hand, when we turn to consider the cognitive capabilities of human brains and their intrinsic limitations, then a different facet of the interpretability process can be defined in terms of *readability* of the bulk of information conveyed by a fuzzy model. In that case, simplicity is required to perform the interpretation process because of the limited ability to store information in the human brain's short-term memory [14.5]. Therefore, structural measures concerning the complexity of a rule base affect the cognitive efforts of a user determined to read and interpret a fuzzy model.

Comprehensibility and readability represent two facets of a common issue and both of them are to be considered while assessing the interpretability process. In particular, this distinction should be acknowledged when criteria are specifically designed to provide a quantitative definition of interpretability.

14.1.1 Why Is Interpretability So Important?

A great number of inductive modeling techniques are currently available to acquire knowledge from data. Many of these techniques provide predictive models that are very accurate and flexible enough to be applied in a wide range of applications. Nevertheless, the resulting models are usually considered as black boxes, i. e. models whose behavior cannot be easily explained in terms of the model structure. On the other hand, the use of fuzzy rule-based models is a matter of design choice: whenever interpretability is a key factor, fuzzy rule-based models should be naturally preferred. It is worth noting that interpretability is a distinguishing feature of fuzzy rule-based models. Several reasons justify a choice inclined toward interpretability. They include but are not limited to:

- *Integration*: In an interpretable fuzzy rule-based model the acquired knowledge can be easily verified and related to the domain knowledge of a human expert. In particular, it is easy to verify if the acquired knowledge expresses new and interesting relations about the data; also, the acquired knowledge can be refined and integrated with expert knowledge.
- *Interaction*: The use of natural language as a mean for knowledge communication enables the possibility of interaction between the user and the model. Interactivity is meant to explore the acquired knowledge. In practice, it can be done at symbolical level (by adding new rules or modifying existing ones) and/or at numerical level (by modifying the fuzzy sets denoted by linguistic terms; or by adding new linguistic terms denoting new fuzzy sets).
- *Validation*: The acquired knowledge can be easily validated against common-sense knowledge and domain-specific knowledge. This capability enables the detection of semantic inconsistencies that may have different causes (misleading data involved in the inductive process, local minimum where the inductive process may have been trapped, data overfitting, etc.). This kind of anomaly detection is important to drive the inductive process toward a qualitative improvement of the acquired knowledge.
- *Trust*: The most important reason to adopt interpretable fuzzy models is their inherent ability to convince end users about the reliability of a model (especially those users not concerned with knowledge acquisition techniques). An interpretable fuzzy rule-based model is endowed with the capability of explaining its inference process so that users may be confident on how it produces its outcomes. This is particularly important in such domains as medical diagnosis, where a human expert is the ultimate responsible for a decision.

14.1.2 A Historical Review

It has been long time since *Zadeh's* seminal work on fuzzy sets [14.6] and nowadays there are lots of fruitful research lines related to fuzzy logic [14.7]. Hence, we can state that fuzzy sets and systems have become the subjects of a mature research field counting several works both theoretical and applied in their scope. Figure 14.1 shows the distribution of publications per year regarding interpretability issues. Three main phases can be identified taking into account the historical evolution of FM.

From 1965 to 1990

During this initial period, interpretability emerged naturally as the main advantage of fuzzy systems. Researchers concentrated on building fuzzy models mainly working with expert knowledge and a few simple linguistic variables [14.8–10] and linguistic rules usually referred to as *Mamdani* rules [14.11]. As a result, those designed fuzzy models were characterized by their high interpretability. Moreover, interpretability is assumed as an intrinsic property of fuzzy systems. Therefore, there are only a few publications regarding interpretability issues. Note that the first proposal of a fuzzy rule-based system (FRBS) was presented by Mamdani who was able to augment Zadeh's initial formulation allowing the application of fuzzy systems to a control problem. These kinds of fuzzy systems are also referred to as *fuzzy logic controllers*, as proposed by the author in his pioneering paper. In addition, Mamdani-type FRBSs soon became the main tool to develop linguistic models. Of course, many other rule formats were arising and gaining importance. In addition to Mamdani FRBSs, probably the most famous FRBSs are those proposed by *Takagi* and *Sugeno* [14.12], the popular TSK fuzzy systems, where the conclusion is a function of the input values. Due to their current popularity, in the following we will use the term *fuzzy system* to denote Mamdani-type FRBSs and their subsequent extensions.

From 1990 to 2000

In the second period the focus was set on accuracy. Researchers realized that expert knowledge was not enough to deal with complex systems. Thus, they explored the use of fuzzy machine learning techniques to automatically extract knowledge from data [14.13, 14]. Accordingly, those designed fuzzy models became composed of extremely complicated fuzzy rules with high accuracy but at the cost of disregarding inter-

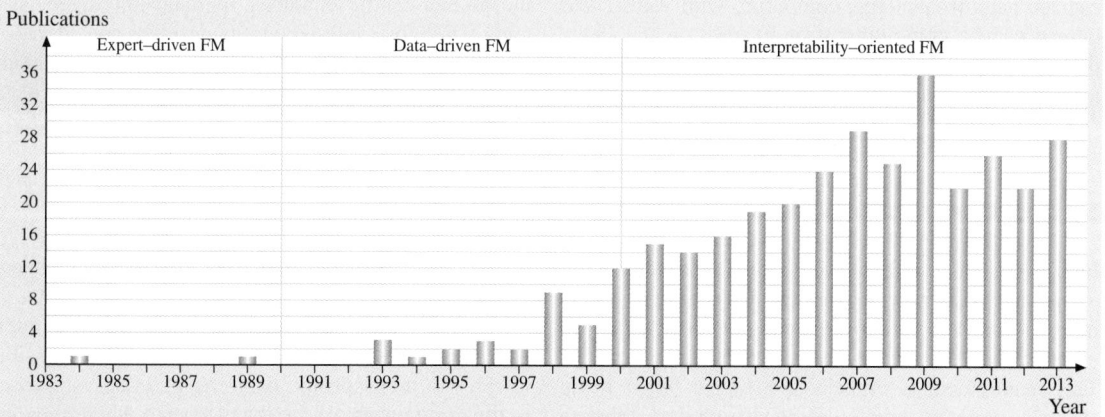

Fig. 14.1 Publications per year related to interpretability issues

pretability as a side effect. Obviously, automatically generated rules were rarely as readable as desired. Along this period some researchers started claiming that fuzzy models are not interpretable per se. Interpretability is a matter of careful design. Thus, interpretability issues must be deeply analyzed and seriously discussed. Although the amount of publications related to interpretability issues is still small in this period, please pay attention to the fact that publications begin to grow exponentially at the end of this second phase.

From 2000 to 2013

After the two previous periods, researchers realized that both expert-driven (from 1965 to 1990) and data-driven (from 1990 to 2000) design approaches have their own advantages and drawbacks, but they are somehow complementary. For instance, expert knowledge is general and easy to interpret but hard to formalize. On the contrary, knowledge derived from data can be extracted automatically but it becomes quite specific and its interpretation is usually hard [14.15]. Moreover, researchers were aware of the need of taking into account simultaneously interpretability and accuracy during the design of fuzzy models. As a result, during this third phase the main challenge was how to combine expert knowledge and knowledge extracted from data, with the aim of designing compact and robust systems with a good interpretability–accuracy trade-off. When considering both interpretability and accuracy in FM, two main strategies turn up naturally [14.16]: *linguistic fuzzy modeling* (LFM) and *precise fuzzy modeling* (PFM). On the one hand, in LFM, designers first focus on the interpretability of the model, and then they try to improve its accuracy [14.17]. On the other hand, in PFM, design-

ers first build a fuzzy model maximizing its accuracy, and then they try to improve its interpretability [14.18]. As an alternative, since accuracy and interpretability represent conflicting goals by nature, multiobjective fuzzy modeling strategies (considering accuracy and interpretability as objectives) have become very popular [14.19, 20].

At the same time, there has been a great effort for formalizing interpretability issues. As a result, the number of publications has grown. Researchers have actively looked for the right definition of interpretability. In addition, several interpretability constraints have been identified. Moreover, interpretability assessment has become a hot research topic. In fact, several interpretability indexes (able to guide the FM design process) have been defined. Nevertheless, a universal index widely admitted is still missing. Hence, further research on interpretability issues is demanded.

Unfortunately, although the number of publications was growing exponentially until 2009, later it started decreasing. We would like to emphasize the impact of the two pioneer books [14.17, 18] edited in 2003. They contributed to make the fuzzy community aware of the need to take into account again interpretability as a main research concern. It is worth noting that the first formal definition of interpretability (in the fuzzy literature) was included in [14.18]. It was given by *Bodenhofer* and *Bauer* [14.21] who established an axiomatic treatment of interpretability at the level of linguistic variables.

We encourage the fuzzy community to keep paying attention to interpretability issues because there is still a lot of research to be done. Interpretability must be the central point on system modeling. In fact, some of the hottest and most recent research topics like pre-

Part B | 14.1

cisiated natural language, computing with words, and human centric computing strongly rely on the interpretability of the designed models. The challenge is to better exploit fuzzy logic techniques for improving the human-centric character of many intelligent systems. Therefore, interpretability deserves consideration as a main research concern and the number of publications should grow again in the next years.

14.2 Interpretability Constraints and Criteria

Interpretability is a quality of fuzzy systems that is not immediate to quantify. Nevertheless, a quantitative definition is required both for assessing the interpretability of a fuzzy system and for designing new fuzzy systems. This requirement is especially stringent when fuzzy systems are automatically designed from data, through some knowledge extraction procedure.

A common approach for defining interpretability is based on the adoption of a number of constraints and criteria that, taken as a whole, provide for a definition of interpretability. This approach is inherent to the subjective nature of interpretability, because the validity of some conditions/criteria is not universally acknowledged and may depend on the application context.

In the literature, a large number of interpretability constraints and criteria can be found. Some of them are widely accepted, while others are controversial. The nature of these constraints and criteria is also diverse. Some are neatly defined as a mathematical condition, others have a fuzzy character and their satisfaction is a matter of degree. This section is addressed to give a brief yet homogeneous outline of the best known interpretability constraints and criteria. The reader is referred to the specialized literature for deeper insights on this topic [14.22, 23].

Several ways are available to categorize interpretability constraints and criteria. It could be possible to refer to their specific nature (e.g., crisp vs. fuzzy), to the components of the fuzzy system where they are applied, or to the description level of the fuzzy system itself. Here, as depicted in Fig. 14.2, we choose a hierarchical organization that starts from the most basic components of a fuzzy system, namely the involved fuzzy sets, and goes on toward more complex levels, such as fuzzy partitions, fuzzy rules, up to considering the model as a whole.

14.2.1 Constraints and Criteria for Fuzzy Sets

Fuzzy sets are the basic elements of fuzzy systems and their role is to express elementary yet imprecise concepts that can be denoted by linguistic labels. Here we assume that fuzzy sets are defined on a universe of discourse represented by a closed interval of the real line (this is the case of most fuzzy systems, especially those acquired from data). Thus, fuzzy sets are the building blocks to translate a numerical domain in a linguistically quantified domain that can be used to communicate knowledge.

Generally speaking, single fuzzy sets are employed to express elementary concepts and, through the use of connectives, are combined to represent more complex concepts. However, not all fuzzy sets can be related to elementary concepts, since the membership function of a fuzzy set may be very awkward but still legitimate from a mathematical viewpoint. Actually, a subclass of fuzzy sets should be considered, so that its members can be easily associated with elementary concepts and tagged by the corresponding linguistic labels. Fuzzy sets of this subclass must verify a number of basic interpretability constraints, including:

- *Normality*: At least one element of the universe of discourse is a prototype for the fuzzy set, i.e. it is characterized by a full membership degree.

High-level	**Fuzzy rule bases**	→ Compactness → Average firing rules → Logical view → Completeness → Locality
	Fuzzy rules	→ Description length → Granular output
Abstraction levels	**Fuzzy partitions**	→ Justifiable number of elements → Distinguishability → Coverage → Relation preservation → Prototypes on special elements
Low-level	**Fuzzy sets**	→ Normality → Continuity → Convexity

Fig. 14.2 Interpretability constraints and criteria in different abstraction levels

A normal fuzzy set represents a concept that fully qualifies at least one element of the universe of discourse, i. e. the concept has at least one example that fulfills it. On the other hand, a subnormal fuzzy set is usually a consequence of a partial contradiction (it is easy to show that the degree of inclusion of a subnormal fuzzy set in the empty set is nonzero).

- *Continuity*: The membership function is continuous on the universe of discourse. As a matter of fact, most concepts that can be naturally represented through fuzzy sets derive from a perceptual act, which comes from external stimuli that usually vary in continuity. Therefore, continuous fuzzy sets are better in accordance with the perceptive nature of the represented concepts.
- *Convexity*: In a convex fuzzy set, given three elements linearly placed on the axis related to the universe of discourse, the degree of membership of the middle element is always greater than or equal to the minimum membership degree of the side elements [14.24]. This constraint encodes the rule that if a property is satisfied by two elements, then it is also satisfied by an element settled between them.

14.2.2 Constraints and Criteria for Fuzzy Partitions

The key success factor of fuzzy logic in modeling is the ability of expressing knowledge *linguistically*. Technically, this is realized by linguistic variables, i. e. variables that assume symbolic values called linguistic terms. The peculiarity of linguistic variables with respect to classical symbolic approaches is the interpretation of linguistic terms as fuzzy sets. The collection of fuzzy sets used as interpretation of the linguistic terms of a linguistic variable forms a fuzzy partition of the universe of discourse.

To understand the role of a fuzzy partition, we should consider that it is meant to define a relation among fuzzy sets. Such a relation must be co-intensive with the one connecting the elementary concepts represented by the fuzzy sets involved in the fuzzy partition. That is the reason why the design of fuzzy partitions is so crucial for the overall interpretability of a fuzzy system. The most critical interpretability constraints for fuzzy partitions are:

- *Justifiable number of elements*: The number of fuzzy sets included in a linguistic variable must be small enough so that they can be easily remembered

and recalled by users. Psychological studies suggest at most nine fuzzy sets or even less [14.5, 25]. Usually, three to five fuzzy sets are convenient choices to set the partition cardinality.

- *Distinguishability*: Since fuzzy sets are denoted by distinct linguistic terms, they should refer to well-distinguished concepts. Therefore, fuzzy sets in a partition should be well separated, although some overlapping is admissible because usually perception-based concepts are not completely disjoint. Several alternatives are available to quantify distinguishability, including similarity and possibility [14.26].
- *Coverage*: Distinguishable fuzzy sets are necessary, but if they are too much separated they risk to under-represent some subset of the universe of discourse. The coverage constraint requires that each element of the universe of discourse must belong to at least one fuzzy set of the partition with a membership degree not less than a threshold [14.22]. This requirement involves that each element of the universe of discourse has some quality that is well represented in the fuzzy partition. On the other hand, the lack of coverage is a signal of incompleteness of the fuzzy partition that may hamper the overall comprehensibility of the system's knowledge. Coverage and distinguishability are somewhat conflicting requirements that are usually balanced by fuzzy partitions that enforce the intersection of adjacent fuzzy sets to elements whose maximum membership degree is equal to a threshold (usually the value of this threshold is set to 0.5).
- *Relation preservation*: The concepts that are represented by the fuzzy sets in a fuzzy partition are usually cross related. The most immediate relation which can be conceived among concepts is related to the order (e.g., *Low* preceding *Medium*, preceding *High*, and so on). Relations of this type must be preserved by the corresponding fuzzy sets in the fuzzy partition [14.27].
- *Prototypes on special elements*: In many problems, some elements of the universe of discourse have some special meaning. A common case is the meaning of the bounds of the universe of discourse, which usually represent some extreme qualities (e.g., *Very Large* or *Very Small*). Other examples are possible, which could be aside from the bounds of the universe of discourse being, instead, more problem-specific (e.g., prototypes could be conceived for the icing point of water, the typical

human body temperature, etc.). In all these cases, the prototypes of some fuzzy sets of the partition must coincide with such special elements.

14.2.3 Constraints and Criteria for Fuzzy Rules

In most cases, a fuzzy system is defined over a multidimensional universe of discourse that can be split into many one-dimensional universes of discourse, each of them associated with a linguistic variable. A subset of these linguistic variables is used to represent the input of a system, while the remaining variables (usually only one variable) are used to represent the output. The input/output behavior is expressed in terms of rules. Each rule prescribes a linguistic output value when the input matches the rule condition (also called rule premise), usually expressed as a logical combination of soft constraints. A soft constraint is a linguistic proposition (specification) that ties a linguistic variable to a linguistic term (e.g., *Temperature is High*). Furthermore, the soft constraints combined in a rule condition may involve different linguistic variables (e.g., *Temperature is High AND Pressure is Low*).

A fuzzy rule is a unit of knowledge that has the twofold role of determining the system behavior and communicating this behavior in a linguistic form. The latter feature urges to adopt a number of interpretability constraints which are to be added up to the constraints required for fuzzy sets and fuzzy partitions. Some of the most general interpretability constraints and criteria for fuzzy rules are as follows:

- *Description length*: The description length of a fuzzy rule is the sum of the number of soft constraints occurring in the condition and in the consequent of the rule (it is usually known as *total rule length*). In most cases, only one linguistic variable is represented in a rule consequent, therefore the description length of a fuzzy rule is directly related to the complexity of the condition. A small number of soft constraints in a rule implies both high readability and semantic generality; hence, short rules should be preferred in fuzzy systems.
- *Granular outputs*: The main strength of fuzzy systems is their ability to represent and process imprecision in both data and knowledge. Imprecision is part of fuzzy inference, therefore the inferred output of a fuzzy system should carry information about the imprecision of its knowledge. This can be accomplished by using fuzzy sets as outputs. De-

fuzzification collapses fuzzy sets into single scalars; it should be therefore used only when strictly necessary and in those situations where outputs are not the object of user interpretation.

14.2.4 Constraints and Criteria for Fuzzy Rule Bases

As previously stated, the interpretability of a rule base taken as a whole has two facets: (1) a structural facet (*readability*), which is mainly related to the easiness of reading the rules; (2) a semantic facet (*comprehensibility*), which is related to the information conveyed to the users who are willing to understand the system behavior. The following interpretability constraints and criteria are commonly defined to ensure the structural and semantic interpretability of fuzzy rule bases.

- *Compactness*: A compact rule base is defined by a small number of rules. This is a typical structural constraint that advocates for simple representation of knowledge in order to allow easy reading and understanding. Nevertheless, a small number of rules usually involves low accuracy; it is therefore very common to balance compactness and accuracy in a trade-off that mainly depends on user needs.
- *Average firing rules*: When an input is applied to a fuzzy system, the rules whose conditions are verified to a degree greater than zero are *firing*, i.e. they contribute to the inference of the output. On an average, the number of firing rules should be as small as possible, so that users are able to understand the contributions of the rules in determining the output.
- *Logical view*: Fuzzy rules resemble logical propositions when their linguistic description is considered. Since linguistic description is the main mean for communicating knowledge, it is necessary that logical laws are applicable to fuzzy rules; otherwise, the system behavior may result counter intuitive. Therefore, the validity of some basic laws of the propositional logic (like *Modus Ponens*) and the truth-preserving operations (e.g., application of distributivity, De Morgan laws, etc.) should also be verified for fuzzy rules.
- *Completeness*: The behavior of a fuzzy system is well defined for all inputs in the universe of discourse; however, when the maximum firing strength determined by an input is too small, it is not easy to justify the behavior of the system in terms of the

activated rules. It is therefore required that for each possible input at least one rule is activated with a firing strength greater than a threshold value (usually set to 0.5) [14.22].

- *Locality*: Each rule should define a local model, i.e. a fuzzy region in the universe of discourse where the behavior of the system is mainly due to the rule and only marginally by other rules that are simultaneously activated [14.28]. This requirement is necessary to avoid that the final output of the system is a consequence of an interpolative behavior of different rules that are simultaneously activated with high firing strengths. On the other hand, a moderate overlapping of local models is admissible in order to enable a smooth transition from a local model to another when the input

values gradually shift from one fuzzy region to another.

In summary, a number of interpretable constraints and criteria apply to all levels of a fuzzy system. This section highlights only the constraints that are general enough to be applied independently on the modeling problem; however, several problem-specific constraints are also reported in the literature (e.g., attribute correlation). Sometimes interpretability constraints are conflicting (as exemplified by the dichotomy distinguishability versus coverage) and, in many cases, they conflict with the overall accuracy of the system. A balance is therefore required, asking in its turn for a way to assess interpretability in a qualitative but also quantitative way. This is the main subject of the next section.

14.3 Interpretability Assessment

The interpretability constraints and criteria presented in previous section belong to two main classes: (1) structural constraints and criteria referring to the static description of a fuzzy model in terms of the elements that compose it; (2) semantic constraints and criteria quantifying interpretability by looking at the behavior of the fuzzy system. Whilst structural constraints address the *readability* of a fuzzy model, semantic constraints focus on its *comprehensibility*.

Of course, interpretability assessment must regard both global (description readability) and local (inference comprehensibility) points of view. It must also take into account both structural and semantic issues when considering all components (fuzzy sets, fuzzy partitions, linguistic partitions, linguistic propositions, fuzzy rules, fuzzy operators, etc.) of the fuzzy system under study.

Thus, assessing interpretability represents a challenging task mainly because the analysis of interpretability is extremely subjective. In fact, it clearly depends on the feeling and background (knowledge, experience, etc.) of the person who is in charge of making the evaluation. Even though having subjective indexes would be really appreciated for personalization purposes, looking for a universal metric widely admitted also makes the definition of objective indexes mandatory. Hence, it is necessary to consider both objective and subjective indexes. On the one hand, objective indexes are aimed at making feasible fair comparisons among different fuzzy models designed for solving

the same problem. On the other hand, subjective indexes are thought for guiding the design of customized fuzzy models, thus making easier to take into account users' preferences and expectations during the design process.

The rest of this section gives an overview on the most popular interpretability indexes which turn out from the specialized literature. Firstly, *Zhou* and *Gan* [14.29] established a two-level taxonomy regarding interpretability issues. They distinguished between low-level (also called fuzzy set level) and high-level (or fuzzy rule level). This taxonomy was extended by *Alonso* et al. [14.30] who introduced a conceptual framework for characterizing interpretability. They considered both fuzzy partitions and fuzzy rules at several abstraction levels. Moreover, in [14.31] *Mencar* et al. remarked the need to distinguish between readability (related to structural issues) and comprehensibility (related to semantic issues). Later, *Gacto* et al. [14.32] proposed a double axis taxonomy regarding semantic and structural properties of fuzzy systems, at both partition and rule base levels. Accordingly, they pointed out four groups of indexes. Below, we briefly introduce the two most sounded indexes inside each group (they are summarized in Fig. 14.3):

G1. *Structural-based interpretability at fuzzy partition level*:
- *Number of features*.
- *Number of membership functions*.

	Fuzzy partition level	Fuzzy rule base level
Structural–based interpretability	G1 Number of features Number of membership functions	G2 Number of rules Number of conditions
Semantic–based interpretability	G3 Context–adaptation–based index GM3M index	G4 Semantic–cointension–based index Co–firing–based–comprehensibility index

Fig. 14.3 Interpretability indexes considered in this work

G2. *Structural-based interpretability at fuzzy rule base level*:

- *Number of rules*. This index is the most widely used [14.30].
- *Number of conditions*. This index corresponds to the previously mentioned *total rule length* which was coined by *Ishibuchi* et al. [14.33].

G3. *Semantic-based interpretability at fuzzy partition level*:

- *Context-adaptation-based index* [14.34]. This index was introduced by Botta et al. with the aim of guiding the so-called context adaptation approach for multiobjective evolutionary design of fuzzy rule-based systems. It is actually an interpretability index based on fuzzy ordering relations.
- *GM3M index* [14.35]. Gacto et al. proposed an index defined as the geometric mean of three single metrics. The first metric computes the displacement of the tuned membership functions with respect to the initial ones. The second metric evaluates the changes in the shapes of membership functions in terms of lateral amplitude rate. The third metric measures the area similarity. This index was used to preserve the semantic interpretability of fuzzy partitions along multiobjective evolutionary rule selection and tuning processes aimed at designing fuzzy models with a good interpretability-accuracy trade-off.

G4. *Semantic-based interpretability at fuzzy rule base level*:

- *Semantic-cointension-based index* [14.36]. This index exploits the cointension concept coined by *Zadeh* [14.3]. In short, two different concepts referring almost to the same entities are taken as cointensive. Thus, a fuzzy system is deemed as comprehensible only when the explicit semantics (defined by fuzzy sets attached to linguistic terms as well as fuzzy operators) embedded in the fuzzy model is cointensive with the implicit semantics inferred by the user while reading the linguistic representation of the rules. In the case of classification problems, semantic cointension can be evaluated through a *logical view* approach, which evaluates the degree of fulfillment of a number of logical laws exhibited by a given fuzzy rule base [14.31]. The idea mainly relies on the assumption that linguistic propositions resemble logical propositions, for which a number of basic logical laws are expected to hold.

- *Co-firing-based comprehensibility index* [14.37]. It measures the complexity of understanding the fuzzy inference process in terms of information related to co-firing rules, i.e. rules firing simultaneously with a given input vector. This index emerges in relation with a novel approach for fuzzy system comprehensibility analysis, based on visual representations of the fuzzy rule-based inference process. Such representations are called fuzzy inference-grams (fingrams) [14.38, 39]. Given a fuzzy rule base, a fingram plots it graphically as a social network made of nodes representing fuzzy rules and edges connecting nodes in terms of rule interaction at the inference level. Edge weights are computed by paying attention to the number of co-firing rules. Thus, looking carefully at all the information provided by a fingram it becomes easy and intuitive understanding the structure and behavior of the fuzzy rule base it represents.

Notice that, most published interpretability indexes only deal with structural issues, so they correspond to groups G1 and G2. Indexes belonging to these groups are mainly quantitative. They essentially analyze the structural complexity of a fuzzy model by counting the number of elements (membership functions, rules, etc.) it contains. As a result, these indexes can be deemed as objective ones. Although these indexes are usually quite simple (that is the reason why we have just listed them above), they are by far the most popular ones. On the contrary, only a few interpretability indexes are able to assess the comprehensibility of a fuzzy model dealing with semantic issues (they belong to groups G3 and G4). This is mainly due to the fact that these indexes must take into account not only quantitative but also qualitative aspects of the modeled fuzzy system. They are inherently subjective and therefore not easy to for-malize (that is the reason why we have provided more details above). Anyway, the interested reader is referred to the cited papers for further information. Moreover, a much more exhaustive list of indexes can be found in [14.32].

Even though there has been a great effort in the last years to propose new interpretability indexes, a universal index is still missing. Hence, defining such an index remains a challenging task. Anyway, we would like to highlight the need to address another encouraging challenge which is a careful design of interpretable fuzzy systems guided by one or more of the already existing interpretability indexes.

14.4 Designing Interpretable Fuzzy Systems

Linguistic (Mamdani-type) fuzzy systems are widely known as a powerful tool to develop linguistic models [14.11]. They are made up of two main components:

- The *inference engine*, that is the component of the fuzzy system in charge of the fuzzy processing tasks.
- The *knowledge base* (KB), that is the component of the fuzzy system that stores the knowledge about the problem being solved. It is composed of:
  - The *fuzzy partitions*, describing the linguistic terms along with the corresponding membership functions defining their semantics, and
  - The *fuzzy rule base*, constituted by a collection of linguistic rules with the following structure

$$\text{IF } X_1 \text{ is } A_1 \text{ and } \ldots \text{ and } X_n \text{ is } A_n$$
$$\text{THEN } Y_1 \text{ is } B_1 \text{ and } \ldots \text{ and } Y_m \text{ is } B_m$$

with X_i and Y_j being input and output linguistic variables, respectively, and A_i and B_j being linguistic terms defined by the corresponding fuzzy partitions. This structure provides a natural framework to include expert knowledge in the form of linguistic fuzzy rules. In addition to expert knowledge, induced knowledge automatically extracted from experimental data (describing the relation between system input and output) can also be easily formalized in the same rule base. Expert and induced knowledge are complementary. Furthermore, they are represented in a highly interpretable structure. The fuzzy rules are composed of input and output linguistic variables which take values from their term sets having a meaning associated with each linguistic label. As a result, each rule is a description of a condition-action statement that offers a clear interpretation to a human.

The accuracy of a fuzzy system directly depends on two aspects, the composition of the KB (fuzzy partitions and fuzzy rules) and the way in which it implements the fuzzy inference process. Therefore, the design process of a fuzzy system includes two main tasks which are going to be further explained in the following subsections, regarding both interpretability and accuracy:

- *Generation of the KB* in order to formulate and describe the knowledge that is specific to the problem domain.
- *Conception of the inference engine*, that is the choice of the different fuzzy operators that are employed by the inference process.

Mamdani-type fuzzy systems favor interpretability. Therefore, they are usually considered when looking for interpretable fuzzy systems. However, it is important to remark that they are not interpretable per se. Notice that designing interpretable fuzzy systems is a matter of careful design.

14.4.1 Design Strategies for the Generation of a KB Regarding the Interpretability-Accuracy Trade-Off

The two main objectives to be addressed in the FM field are the *interpretability* and *accuracy*. Of course, the ideal aim would be to satisfy both objectives to a high degree but, since they represent conflicting goals, it is generally not possible. Regardless of the approach, a common scheme is found in the existing literature:

- Firstly, the main objective (interpretability or accuracy) is tackled defining a specific model structure to be used, thus setting the FM approach.
- Then, the modeling components (model structure and/or modeling process) are improved by means of different mechanisms to achieve the desired ratio between interpretability and accuracy.

This procedure resulted in four different possibilities:

1. LFM with improved interpretability,
2. LFM with improved accuracy,
3. PFM with improved interpretability, and
4. PFM with improved accuracy.

Option (1) gives priority to interpretability. Although a fuzzy system designed by LFM uses a model structure with high descriptive power, it has some problems (curse of dimensionality, excessive number of input variables or fuzzy rules, garbled fuzzy sets, etc.) that make it not as interpretable as desired. In consequence, there is a need of interpretability improvements to restore the pursued balance.

On the contrary, option (4) considers accuracy as the main concern. However, obtaining more accuracy in PFM does not pay attention to the interpretability of the model. Thus, this approach goes away from the aim of this chapter. It acts close to black-box techniques, so it does not follow the original objective of FM (not taking profit from the advantages that distinguish it from other modeling techniques).

Finally, the two remaining options, (2) and (3), propose improvement mechanisms to compensate for the initial imbalance in the quest for the best trade-off between interpretability and accuracy. In summary, three main approaches exist depending on how the two objectives are optimized (sequentially or at once):

- First interpretability then accuracy (*LFM with improved accuracy*).
- First accuracy then interpretability (*PFM with improved interpretability*).
- Multiobjective design. Both objectives are optimized at the same time.

The rest of this section provides additional details related to each of these approaches.

First Interpretability Then Accuracy

LFM has some inflexibility due to the use of linguistic variables with global semantics that establishes a general meaning of the used fuzzy sets [14.40]:

1. There is a lack of flexibility in the fuzzy system because of the rigid partitioning of the input and output spaces.
2. When the system input variables are dependent, it is very hard to find out right fuzzy partitions of the input spaces.
3. The usual homogeneous partitioning of the input and output spaces does not scale to high-dimensional spaces. It yields to the well-known curse of dimensionality problem that is characteristic of fuzzy systems.
4. The size of the KB directly depends on the number of variables and linguistic terms in the model. The derivation of an accurate linguistic fuzzy system usually requires a big number of linguistic terms. Unfortunately, this fact causes the number of rules to rise significantly, which may cause the system to lose the capability of being readable by human beings. Of course, in most cases it would be possible to obtain an equivalent fuzzy system with a much smaller number of rules by renouncing to that kind of rigidly partitioned input space.

However, it is possible to make some considerations to face the disadvantages enumerated above. Basically, two ways of improving the accuracy in LFM can be considered by performing the improvement in:

- The *model structure*, slightly changing the rule structure to make it more flexible, or in
- The *modeling process*, extending the model design to other components beyond the rule base, such as the fuzzy partitions, or even considering more sophisticated derivations of it.

Note that, the so-called strong fuzzy partitions are widely used because they satisfy most of the interpretability constraints introduced in Sect. 14.2.2. The design of fuzzy partitions may be integrated within the whole derivation process of a fuzzy system with different schemata:

- *Preliminary design*. It involves extracting fuzzy partitions automatically by induction (usually performed by nonsupervised clustering techniques) from the available dataset.

- *Embedded design.* Following a meta-learning process, this approach first derives different fuzzy partitions and then samples its efficacy running an embedded basic learning method of the entire KB [14.41].
- *Simultaneous design.* The process of designing fuzzy partitions is developed together with the derivation of other components such as the fuzzy rule base [14.42].
- *A posteriori design.* This approach involves tuning of the previously defined fuzzy partitions once the remaining components have been obtained. Usually, the tuning process changes the membership function shapes with the aim of improving the accuracy of the linguistic model [14.43]. Nevertheless, sometimes it also takes care of getting better interpretability (e.g., merging membership functions [14.44]).

It is also possible to opt for using more sophisticated rule base learning methods while the fuzzy partitions and the model structure are kept unaltered. Usually, all these improvements have the final goal of enhancing the *interpolative reasoning* the fuzzy system develops. For instance, the COR (cooperative rules) method follows the primary objective of inducing a better cooperation among linguistic rules [14.45].

As an alternative, other authors advocate the extension of the usual linguistic model structure to make it more flexible. As *Zadeh* highlighted in [14.46], a way to do so without losing the description ability to a high degree is to use linguistic hedges (also called *linguistic modifiers* in a wider sense). In addition, the rule structure can be extended through the definition of double-consequent rules, weighted rules, rules with exceptions, hierarchical rule bases, etc.

First Accuracy Then Interpretability

The birth of more flexible fuzzy systems such as TSK or approximate ones (allowing the FM to achieve higher accuracy) entailed the eruption of PFM. Nevertheless, the modeling tasks with these kinds of fuzzy systems increasingly resembled black-box processes. Consequently, nowadays several researchers share the idea of rescuing the seminal intent of FM, i.e. to preserve the good interpretability advantages offered by fuzzy systems. This fact is usually attained by reducing the complexity of the model [14.47]. Furthermore, there are

approaches aimed at improving the local description of TSK-type fuzzy rules:

- *Merging/removing fuzzy sets in precise fuzzy systems.* The interpretability of TSK-type fuzzy systems may be improved by removing those fuzzy sets that, after an automatic adaptation and/or acquisition, do not contribute significantly to the model behavior. Two aspects must be considered:
  - *Redundancy.* It refers to the coexistence of similar fuzzy sets representing compatible concepts. In consequence, models become more complex and difficult to understand (the distinguishability constraint is not satisfied).
  - *Irrelevancy.* It arises when fuzzy sets with a constant membership degree equal to 1, or close to it, are used. These kinds of fuzzy sets do not furnish relevant information.

  The use of similarity measures between fuzzy sets the has been proposed to automatically detect these undesired fuzzy sets [14.48]. Through first merging/removing fuzzy sets and then merging fuzzy rules, the precise fuzzy model goes through an interpretability improvement process that makes it less complex (more compact) and more easily interpretable (more transparent).
- *Ordering/selecting TSK-type fuzzy rules.* An efficient way to improve the interpretability in FM is to select a subset of significant fuzzy rules that represent in a more compact way the system to be modeled. Moreover, as a side effect this selection of important rules reduces the possible redundancy existing in the fuzzy rule base, thus improving the generalization capability of the system, i.e., its accuracy. For instance, resorting to orthogonal transformations [14.49] is one of the most successful approaches in this sense.
- *Exploiting the local description of TSK-type fuzzy rules.* TSK-type fuzzy systems are usually considered as the combination of simple models (the rules) that describe local behaviors of the system to be modeled. Hence, insofar as each fuzzy rule is either forced to have a smoother consequent polynomial function or to develop an isolated action, the interpretability will be improved:
  - *Smoothing the consequent polynomial function* [14.50]. Through imposing several constraints to the weights involved in the polynomial function of each rule consequent then a convex combination of the input variables is

performed. This contributes to a better understanding of the model.

- *Isolating the fuzzy rule actions* [14.47]. The description of each fuzzy rule is improved when the overlapping between adjacent input fuzzy sets is reduced. Note that the performance region of a rule is more clearly defined by avoiding that other rules have high firing degree in the same area.

Multiobjective Design

Since interpretability and accuracy are widely recognized as conflicting goals, the use of multiobjective evolutionary (MOE) strategies is becoming more and more popular in the quest for the best interpretability-accuracy trade-off [14.19, 51]. *Ducange* and *Marcelloni* [14.52] proposed the following taxonomy of multiobjective evolutionary fuzzy systems:

- *MOE Tuning.* Given an already defined fuzzy system, its main parameters (typically membership function parameters but also fuzzy inference parameters) are refined through MOE strategies [14.53, 54].
- *MOE Learning.* The components of a fuzzy system KB, the both fuzzy partitions forming the database (DB) and fuzzy rules forming the rule-base (RB), are automatically generated from experimental data.
  - *MOE DB Learning.* The most relevant variables are identified and the optimum membership function parameters are defined from scratch. It usually wraps a RB heuristic-based learning process [14.55].
  - *MOE RB Selection.* Starting from an initial RB, a set of nondominated RBs is generated by selecting subsets of rules exhibiting different trade-offs between interpretability and accuracy [14.56]. In some works [14.35, 57], MOE RB selection and MOE tuning are carried out together.
  - *MOE RB Learning.* The entire set of fuzzy rules is fully defined from scratch. In this approach, uniformly distributed fuzzy partitions are usually considered [14.58].
  - *MOE KB Learning.* Simultaneous evolutionary learning of all KB components (DB and RB). Concurrent learning of fuzzy partitions and fuzzy rules proved to be a powerful tool in the quest for a good balance between interpretability and accuracy [14.59].

It is worthy to note that for the sake of clarity we have only cited some of the most relevant papers in the field of MOE fuzzy systems. For further details, the interested reader is referred to [14.51, 52] where a much more exhaustive review of related works is carried out.

14.4.2 Design Decisions at Fuzzy Processing Level

Although there are studies analyzing the behavior of the existing fuzzy operators for different purposes, unfortunately this question has not been considered yet as a whole from the interpretability point of view. Keeping in mind the interpretability requirement, the implementation the of the inference engine must address the following careful design choices:

- *Select the right conjunctive operator to be used in the antecedent of the rule.* Different operators (belonging to the *t*-norm family) are available to make this choice [14.60].
- *Select the operator to be used in the fuzzy implication of IF-THEN rules.* Mamdani proposed to use the minimum operator as the *t*-norm for implication. Since then, various other *t*-norms have been suggested as implication operator [14.60], for instance the algebraic product. Other important family of implication operators are the fuzzy implication functions [14.61], one of the most usual being the Lukasiewicz's one. Less common implication operators such as force-implications [14.62], *t*-conorms and operators not belonging to any of the most known implication operator families [14.63, 64] have been considered too.
- *Choose the right inference mechanism.* Two main strategies are available:
  - *FATI (First Aggregation Then Inference).* All antecedents of the rules are aggregated to form a multidimensional fuzzy relation. Via the composition principle the output fuzzy set is derived. This strategy is preferred when dealing with implicative rules [14.65].
  - *FITA (First Inference Then Aggregation).* The output of each rule is first inferred, and then all individual fuzzy outputs are aggregated. This is the common approach when working with the usual conjunctive rules. This strategy has become by far the most popular, especially in case of real-time applications. The choice for an output aggregation method (in some cases this is called the *also* operator) is closely related to

the considered implication operator since it has to be related to the interpretation of the rules (which is connected to the kind of implication).
- *Choose the most suitable defuzzification interface operation mode.* There are different options being the most widely used the center of area, also called center of gravity, and the mean of maxima. Even though most methods are based on geometrical or statistical interpretations, there are also parametric methods, adaptive methods including human knowledge, and even evolutionary adaptive methods [14.66].

14.5 Interpretable Fuzzy Systems in the Real World

Interpretable fuzzy systems have an immediate impact on real-world applications. In particular, their usefulness is appreciable in all application areas that put humans at the center of computing. Interpretable fuzzy systems, in fact, conjugate knowledge acquisition capabilities with the ability of communicating knowledge in a human-understandable way.

Several application areas can take advantage from the use of interpretable fuzzy systems. In the following, some of them are briefly outlined, along with a few notes on specific applications and potentialities.

- *Environment*: Environmental issues are often challenging because of the complex dynamics, the high number of variables and the consequent uncertainty characterizing the behavior of subjects under study. Computational intelligence techniques come into play when tolerance for imprecision can be exploited to design convenient models that are suitable to understand phenomena and take decisions. Interpretable fuzzy systems show a clear advantage over black-box systems in providing knowledge that is capable of explaining complex and nonlinear relationships by using linguistic models. Real-world environmental applications of interpretable fuzzy systems include: harmful bioaerosol detection [14.67]; modeling habitat suitability in river management [14.68]; modeling pesticide loss caused by meteorological factors in agriculture [14.69], and so on.
- *Finance*: This is a sector where human-computer cooperation is very tight. Cooperation is carried out in different ways, including the use of computers to provide business intelligence for decision support in financial operations. In many cases financial decisions are ultimately made by experts, who can benefit from automated analyses of big masses of data flowing daily in markets. To this pursuit, Computational intelligence approaches are spreading among the tools used by financial experts in their decisions, including interpretable fuzzy systems for stock return predictions [14.70], exchange rate forecasting [14.71], portfolio risk monitoring [14.72], etc.
- *Industry*: Industrial applications could take advantage from interpretable fuzzy systems when there is the need of explaining the behavior of complex systems and phenomena, like in fault detection [14.73]. Also, control plans for systems and processes can be designed with the help of fuzzy systems. In such cases, a common practice is to start with an initial expert knowledge (used to design rules which are usually highly interpretable) that is then tuned to increase the accuracy of the controller. However, any unconstrained tuning could destroy the original interpretability of the knowledge base, whilst, by taking into account interpretability, the possibility of revising and modifying the controller (or the process manager) can be enhanced [14.74].
- *Medicine and Health-care*: As a matter of fact, in almost all medical contexts intelligent systems can be invaluable decision support tools, but people are the ultimate actors in any decision process. As a consequence, people need to rely on intelligent systems, whose reliability can be enhanced if their outcomes may be explained in terms that are comprehensible by human users. Interpretable fuzzy systems could play a key role in this area because of the possibility of acquiring knowledge from data and communicating it to users. In the literature, several approaches have been proposed to apply interpretable fuzzy systems in different medical problems, like assisted diagnosis [14.75], prognosis prediction [14.76], patient subgroup discovery [14.77], etc.
- *Robotics*: The complexity of robot behavior modeling can be tackled by an integrated approach where a first modeling stage is carried out by combining human expert and empirical knowledge acquired from experimental trials. This integrated approach requires that the final knowledge base is provided to experts for further maintenance: this task could be done effectively only if the acquired knowledge

is interpretable by the user. Some concrete applications of this approach can be found in robot localization systems [14.78] and motion analysis [14.79, 80].

- *Society*: The focus of intelligent systems for social issues has noticeably increased in recent years. For

reasons that are common to all the previous application areas, interpretable fuzzy systems have been applied in a wide variety of scopes, including quality of service improvement [14.81], data mining with privacy preservation [14.82], social network analysis [14.37], and so on.

14.6 Future Research Trends on Interpretable Fuzzy Systems

Research on interpretable fuzzy systems is open in several directions. Future trends involve both theoretical and methodological aspects of interpretability. In the following, some trends are outlined amongst the possible lines of research development [14.7].

- *Interpretability definition*: The blurred nature of interpretability requires continuous investigations on possible definitions that enable a computable treatment of this quality in fuzzy systems. This requirement casts the research on interpretable fuzzy systems toward cross-disciplinary investigations. For instance, this research line includes investigations on computable definitions of some conceptual qualities, like *vagueness* (which has to be distinguished from imprecision and fuzziness). Also, the problem of interpretability of fuzzy systems can be intended as a particular instance of the more general problem of communication between granular worlds [14.83], where many aspects of interpretability could be treated in a more abstract way.
- *Interpretability assessment*: A prominent objective is the adoption of a common framework for characterizing and assessing interpretability with the aim of avoiding misleading notations. Within such a framework, novel metrics could be devised, especially for assessing subjective aspects of interpretability, and integrated with objective interpretability measures to define more significant interpretability indexes.
- *Design of interpretable fuzzy models*: A current research trend in designing interpretable fuzzy models makes use of multiobjective genetic algorithms in

order to deal with the conflicting design objectives of accuracy and interpretability. The effectiveness and usefulness of these approaches, especially those concerning advanced schemes, have to be verified against a number of indexes, including indexes that integrate subjective measures. This verification process is particularly required when tackling high-dimensional problems. In this case, the combination of linguistic and graphical approaches could be a promising approach for descriptive and exploratory analysis of interpretable fuzzy systems.

- *Representation of fuzzy systems*: For very complex problems the use of novel forms of representation (different from the classical rule based) may help in representing complex relationship in comprehensible ways thus yielding a valid aid in designing interpretable fuzzy systems. For instance, a multilevel representation could enhance the interpretability of fuzzy systems by providing different granularity levels for knowledge representation. On the one hand, the highest granulation levels give a coarse (yet immediately comprehensible) description of knowledge, while lower levels provide for more detailed knowledge.

As a final remark, it is worth observing that interpretability is one aspect of the multifaceted problem of *human-centered* design of fuzzy systems [14.84]. Other facets include acceptability (e.g., according to ethical rules), interestingness of fuzzy rules, applicability (e.g., with respect to law), etc. Many of them are not yet in the research mainstream but they clearly represent promising future trends.

14.7 Conclusions

Interpretability is an indispensable requirement for designing fuzzy systems, yet it cannot be assumed to hold by the simple fact of using fuzzy sets for modeling. Interpretability must be encoded in some computational

methods in order to drive the design of fuzzy systems, as well as to assess the interpretability of existing models. The study of interpretability issues started about two decades ago and led to a number of theoretical and methodological results of paramount value in fuzzy modeling. Nevertheless, research is still open both in depth – through new ways of encoding and assessing interpretability – and in breadth, by integrating interpretability in the more general realm of human centered computing.

References

14.1   L.A. Zadeh: From computing with numbers to computing with words–from manipulation of measurements to manipulation of perceptions, IEEE Trans. Circuits Syst. I: Fundam. Theory Appl. **45**(1), 105–119 (1999)

14.2   L.-X. Wang, J.M. Mendel: Fuzzy basis functions, universal approximation, and orthogonal least squares learning, IEEE Trans. Neural Netw. **3**, 807–814 (1992)

14.3   L.A. Zadeh: Is there a need for fuzzy logic?, Inf. Sci. **178**(13), 2751–2779 (2008)

14.4   R.S. Michalski: A theory and methodology of inductive learning, Artificial Intell. **20**(2), 111–161 (1983)

14.5   G.A. Miller: The magical number seven, plus or minus two: Some limits on our capacity for processing information, Psychol. Rev. **63**, 81–97 (1956)

14.6   L.A. Zadeh: Fuzzy sets, Inf. Control **8**, 338–353 (1965)

14.7   J.M. Alonso, L. Magdalena: Editorial: Special issue on interpretable fuzzy systems, Inf. Sci. **181**(20), 4331–4339 (2011)

14.8   L.A. Zadeh: The concept of a linguistic variable and its application to approximate reasoning. Part I, Inf. Sci. **8**, 199–249 (1975)

14.9   L.A. Zadeh: The concept of a linguistic variable and its application to approximate reasoning. Part II, Inf. Sci. **8**, 301–357 (1975)

14.10   L.A. Zadeh: The concept of a linguistic variable and its application to approximate reasoning. Part III, Inf. Sci. **9**, 43–80 (1975)

14.11   E.H. Mamdani: Application of fuzzy logic to approximate reasoning using linguistic synthesis, IEEE Trans. Comput. **26**(12), 1182–1191 (1977)

14.12   T. Takagi, M. Sugeno: Fuzzy identification of systems and its applications to modelling and control, IEEE Trans. Syst. Man Cybern. B Cybern. **15**, 116–132 (1985)

14.13   E. Hüllermeier: Fuzzy methods in machine learning and data mining: Status and prospects, Fuzzy Sets Syst. **156**(3), 387–406 (2005)

14.14   E. Hüllermeier: Fuzzy sets in machine learning and data mining, Appl. Soft Comput. **11**(2), 1493–1505 (2011)

14.15   S. Guillaume: Designing fuzzy inference systems from data: An interpretability-oriented review, IEEE Trans. Fuzzy Syst. **9**(3), 426–443 (2001)

14.16   R. Alcalá, J. Alcalá-Fdez, J. Casillas, O. Cordón, F. Herrera: Hybrid learning models to get the interpretability-accuracy trade-off in fuzzy modeling, Soft Comput. **10**(9), 717–734 (2006)

14.17   J. Casillas, O. Cordón, F. Herrera, L. Magdalena (Eds.): *Accuracy Improvements in Linguistic Fuzzy Modeling*, Studies in Fuzziness and Soft Computing, Vol. 129 (Springer, Berlin, Heidelberg 2003)

14.18   J. Casillas, O. Cordón, F. Herrera, L. Magdalena (Eds.): *Interpretability Issues in fuzzy modeling*, Studies in Fuzziness and Soft Computing, Vol. 128 (Springer, Berlin, Heidelberg 2003)

14.19   O. Cordón: A historical review of evolutionary learning methods for Mamdani-type fuzzy rule-based systems: Designing interpretable genetic fuzzy systems, Int. J. Approx. Reason. **52**, 894–913 (2011)

14.20   F. Herrera: Genetic fuzzy systems: Taxonomy, current research trends and prospects, Evol. Intell. **1**, 27–46 (2008)

14.21   U. Bodenhofer, P. Bauer: A formal model of interpretability of linguistic variables, Stud. Fuzzin. Soft Comput. **128**, 524–545 (2003)

14.22   C. Mencar, A.M. Fanelli: Interpretability constraints for fuzzy information granulation, Inf. Sci. **178**(24), 4585–4618 (2008)

14.23   J. de Valente Oliveira: Semantic constraints for membership function optimization, IEEE Trans. Syst. Man Cybern. A **29**(1), 128–138 (1999)

14.24   W. Pedrycz, F. Gomide: *An Introduction to Fuzzy Sets. Analysis and Design* (MIT Press, Cambridge 1998)

14.25   T.L. Saaty, M.S. Ozdemir: Why the magic number seven plus or minus two, Math. Comput. Model. **38**(3–4), 233–244 (2003)

14.26   C. Mencar, G. Castellano, A.M. Fanelli: Distinguishability quantification of fuzzy sets, Inf. Sci. **177**(1), 130–149 (2007)

14.27   U. Bodenhofer, P. Bauer: Interpretability of linguistic variables: A formal account, Kybernetika **41**(2), 227–248 (2005)

14.28   A. Riid, E. Rüstern: *Transparent Fuzzy Systems in Modelling and Control*, Studies in Fuzziness and Soft Computing, Vol. 128 (Springer, Berlin, Heidelberg 2003) pp. 452–476

14.29   S.-M. Zhou, J.Q. Gan: Low-level interpretability and high-level interpretability: A unified view of data-driven interpretable fuzzy system modelling, Fuzzy Sets Syst. **159**(23), 3091–3131 (2008)

14.30   J.M. Alonso, L. Magdalena, G. González-Rodríguez: Looking for a good fuzzy system interpretability index: An experimental approach, Int. J. Approx. Reason. **51**(1), 115–134 (2009)

14.31   C. Mencar, C. Castiello, R. Cannone, A.M. Fanelli: Design of fuzzy rule-based classifiers with semantic cointension, Inf. Sci. **181**(20), 4361–4377 (2011)

14.32 M.J. Gacto, R. Alcalá, F. Herrera: Interpretability of linguistic fuzzy rule-based systems: An overview of interpretability measures, Inf. Sci. **181**(20), 4340–4360 (2011)

14.33 H. Ishibuchi, T. Nakashima, T. Murata: Three-objective genetics-based machine learning for linguistic rule extraction, Inf. Sci. **136**(1–4), 109–133 (2001)

14.34 A. Botta, B. Lazzerini, F. Marcelloni, D.C. Stefanescu: Context adaptation of fuzzy systems through a multi-objective evolutionary approach based on a novel interpretability index, Soft Comput. **13**(5), 437–449 (2009)

14.35 M.J. Gacto, R. Alcalá, F. Herrera: Integration of an index to preserve the semantic interpretability in the multiobjective evolutionary rule selection and tuning of linguistic fuzzy systems, IEEE Trans. Fuzzy Syst. **18**(3), 515–531 (2010)

14.36 C. Mencar, C. Castiello, R. Cannone, A.M. Fanelli: Interpretability assessment of fuzzy knowledge bases: A cointension based approach, Int. J. Approx. Reason. **52**(4), 501–518 (2011)

14.37 J.M. Alonso, D.P. Pancho, O. Cordón, A. Quirin, L. Magdalena: Social network analysis of co-fired fuzzy rules. In: *Soft Computing: State of the Art Theory and Novel Applications*, ed. by R.R. Yager, A.M. Abbasov, M. Reformat, S.N. Shahbazova (Springer, Berlin, Heidelberg 2013) pp. 113–128

14.38 D.P. Pancho, J.M. Alonso, O. Cordón, A. Quirin, L. Magdalena: FINGRAMS: Visual representations of fuzzy rule-based inference for expert analysis of comprehensibility, IEEE Trans. Fuzzy Syst. **21**(6), 1133–1149 (2013)

14.39 D.P. Pancho, J.M. Alonso, L. Magdalena: Quest for interpretability-accuracy trade-off supported by fingrams into the fuzzy modeling tool GUAJE, Int. J. Comput. Intell. Syst. **6**(1), 46–60 (2013)

14.40 A. Bastian: How to handle the flexibility of linguistic variables with applications, Int. J. Uncertain. Fuzzin. and Knowl.-Based Syst. **2**(4), 463–484 (1994)

14.41 O. Cordón, F. Herrera, P. Villar: Generating the knowledge base of a fuzzy rule-based system by the genetic learning of the data base, IEEE Trans. Fuzzy Syst. **9**(4), 667–674 (2001)

14.42 A. Homaifar, E. McCormick: Simultaneous design of membership functions and rule sets for fuzzy controllers using genetic algorithms, IEEE Trans. Fuzzy Syst. **3**(2), 129–139 (1995)

14.43 B.-D. Liu, C.-Y. Chen, J.-Y. Tsao: Design of adaptive fuzzy logic controller based on linguistic-hedge concepts and genetic algorithms, IEEE Trans. Syst. Man Cybern. B Cybern. **31**(1), 32–53 (2001)

14.44 J. Espinosa, J. Vandewalle: Constructing fuzzy models with linguistic integrity from numerical data-AFRELI algorithm, IEEE Trans. Fuzzy Syst. **8**(5), 591–600 (2000)

14.45 J. Casillas, O. Cordón, F. Herrera: COR: A methodology to improve ad hoc data-driven linguistic rule learning methods by inducing cooperation among rules, IEEE Trans. Syst. Man Cybern. B Cybern. **32**(4), 526–537 (2002)

14.46 L.A. Zadeh: Outline of a new approach to the analysis of complex systems and decision processes, IEEE Trans. Syst. Man. Cybern. **3**(1), 28–44 (1973)

14.47 A. Riid, E. Rüstern: Identification of transparent, compact, accurate and reliable linguistic fuzzy models, Inf. Sci. **181**(20), 4378–4393 (2011)

14.48 M. Setnes, R. Babuška, U. Kaymak, H.R. van Nauta Lemke: Similarity measures in fuzzy rule base simplification, IEEE Trans. Syst. Man Cybern. B Cybern. **28**(3), 376–386 (1998)

14.49 P.A. Mastorocostas, J.B. Theocharis, V.S. Petridis: A constrained orthogonal least-squares method for generating TSK fuzzy models: Application to short-term load forecasting, Fuzzy Sets Syst. **118**(2), 215–233 (2001)

14.50 A. Fiordaliso: A constrained Takagi-Sugeno fuzzy system that allows for better interpretation and analysis, Fuzzy Sets Syst. **118**(2), 307–318 (2001)

14.51 M. Fazzolari, R. Alcalá, Y. Nojima, H. Ishibuchi, F. Herrera: A review of the application of multi-objective evolutionary fuzzy systems: Current status and further directions, IEEE Trans. Fuzzy Syst. **21**(1), 45–65 (2013)

14.52 P. Ducange, F. Marcelloni: Multi-objective evolutionary fuzzy systems, Lect. Notes Artif. Intell. **6857**, 83–90 (2011)

14.53 J. Alcalá-Fdez, F. Herrera, F. Márquez, A. Peregrín: Increasing fuzzy rules cooperation based on evolutionary adaptive inference systems, Int. J. Intell. Syst. **22**(4), 1035–1064 (2007)

14.54 P. Fazendeiro, J. De Valente Oliveira, W. Pedrycz: A multiobjective design of a patient and anaesthetist-friendly neuromuscular blockade controller, IEEE Trans. Bio-Med. Eng. **54**(9), 1667 (2007)

14.55 R. Alcalá, M.J. Gacto, F. Herrera: A fast and scalable multi-objective genetic fuzzy system for linguistic fuzzy modeling in high-dimensional regression problems, IEEE Trans. Fuzzy Syst. **19**(4), 666–681 (2011)

14.56 H. Ishibuchi, T. Murata, I.B. Türksen: Single-objective and two-objective genetic algorithms for selecting linguistic rules for pattern classification problems, Fuzzy Sets Syst. **89**(2), 135–150 (1997)

14.57 R. Alcalá, Y. Nojima, F. Herrera, H. Ishibuchi: Multiobjective genetic fuzzy rule selection of single granularity-based fuzzy classification rules and its interaction with the lateral tuning of membership functions, Soft Comput. **15**(12), 2303–2318 (2011)

14.58 J. Casillas, P. Martínez, A.D. Benítez: Learning consistent, complete and compact sets of fuzzy rules in conjunctive normal form for regression problems, Soft Comput. **13**(5), 451–465 (2009)

14.59 M. Antonelli, P. Ducange, B. Lazzerini, F. Marcelloni: Learning concurrently data and rule bases of Mamdani fuzzy rule-based systems by exploiting a novel interpretability index, Soft Comput. **15**(10), 1981–1998 (2011)

14.60 M.M. Gupta, J. Qi: Design of fuzzy logic controllers based on generalized T-operators, Fuzzy Sets Syst. **40**(3), 473–489 (1991)

14.61 E. Trillas, L. Valverde: On implication and indistinguishability in the setting of fuzzy logic. In: *Management Decision Support Systems Using Fuzzy Logic and Possibility Theory*, ed. by J. Kacpryzk, R.R. Yager (Verlag TÜV Rheinland, Köln 1985) pp. 198–212

14.62 C. Dujet, N. Vincent: Force implication: A new approach to human reasoning, Fuzzy Sets Syst. **69**(1), 53–63 (1995)

14.63 J. Kiszka, M. Kochanska, D. Sliwinska: The influence of some fuzzy implication operators on the accuracy of a fuzzy model – Part I, Fuzzy Sets Syst. **15**, 111–128 (1985)

14.64 J. Kiszka, M. Kochanska, D. Sliwinska: The influence of some fuzzy implication operators on the accuracy of a fuzzy model – Part II, Fuzzy Sets Syst. **15**, 223–240 (1985)

14.65 H. Jones, B. Charnomordic, D. Dubois, S. Guillaume: Practical inference with systems of gradual implicative rules, IEEE Trans. Fuzzy Syst. **17**(1), 61–78 (2009)

14.66 O. Cordón, F. Herrera, F.A. Márquez, A. Peregrín: A study on the evolutionary adaptive defuzzification methods in fuzzy modeling, Int. J. Hybrid Int. Syst. **1**(1), 36–48 (2004)

14.67 P. Pulkkinen, J. Hytonen, H. Koivisto: Developing a bioaerosol detector using hybrid genetic fuzzy systems, Eng. Appl. Artif. Intell. **21**(8), 1330–1346 (2008)

14.68 E. Van Broekhoven, V. Adriaenssens, B. de Baets: Interpretability-preserving genetic optimization of linguistic terms in fuzzy models for fuzzy ordered classification: An ecological case study, Int. J. Approx. Reason. **44**(1), 65–90 (2007)

14.69 S. Guillaume, B. Charnomordic: Interpretable fuzzy inference systems for cooperation of expert knowledge and data in agricultural applications using FisPro, IEEE Int. Conf. Fuzzy Syst., Barcelona (2010) pp. 2019–2026

14.70 A. Kumar: Interpretability and mean-square error performance of fuzzy inference systems for data mining, Intell. Syst. Account. Finance Manag. **13**(4), 185–196 (2005)

14.71 F. Cheong: A hierarchical fuzzy system with high input dimensions for forecasting foreign exchange rates, Int. J. Artif. Intell. Soft Comput. **1**(1), 15–24 (2008)

14.72 A. Ghandar, Z. Michalewicz, R. Zurbruegg: Enhancing profitability through interpretability in algorithmic trading with a multiobjective evolutionary fuzzy system, Lect. Notes Comput. Sci. **7492**, 42–51 (2012)

14.73 S. Altug, M.-Y. Chow, H.J. Trussell: Heuristic constraints enforcement for training of and rule extraction from a fuzzy/neural architecture. Part II: Implementation and application, IEEE Trans. Fuzzy Syst. **7**(2), 151–159 (1999)

14.74 A. Riid, E. Rüstern: Interpretability of fuzzy systems and its application to process control, IEEE Int. Conf. Fuzzy Syst., London (2007) pp. 1–6

14.75 I. Gadaras, L. Mikhailov: An interpretable fuzzy rule-based classification methodology for medical diagnosis, Artif. Intell. Med. **47**(1), 25–41 (2009)

14.76 J.M. Alonso, C. Castiello, M. Lucarelli, C. Mencar: Modelling interpretable fuzzy rule-based classifiers for medical decision support. In: *Medical Applications of Intelligent Data Analysis: Research Advancements*, ed. by R. Magdalena, E. Soria, J. Guerrero, J. Gómez-Sanchis, A.J. Serrano (IGI Global, Hershey 2012) pp. 254–271

14.77 C.J. Carmona, P. Gonzalez, M.J. del Jesus, M. Navio-Acosta, L. Jimenez-Trevino: Evolutionary fuzzy rule extraction for subgroup discovery in a psychiatric emergency department, Soft Comput. **15**(12), 2435–2448 (2011)

14.78 J.M. Alonso, M. Ocaña, N. Hernandez, F. Herranz, A. Llamazares, M.A. Sotelo, L.M. Bergasa, L. Magdalena: Enhanced WiFi localization system based on soft computing techniques to deal with small-scale variations in wireless sensors, Appl. Soft Comput. **11**(8), 4677–4691 (2011)

14.79 J.M. Alonso, L. Magdalena, S. Guillaume, M.A. Sotelo, L.M. Bergasa, M. Ocaña, R. Flores: Knowledge-based intelligent diagnosis of ground robot collision with non detectable obstacles, J. Int. Robot. Syst. **48**(4), 539–566 (2007)

14.80 M. Mucientes, J. Casillas: Quick design of fuzzy controllers with good interpretability in mobile robotics, IEEE Trans. Fuzzy Syst. **15**(4), 636–651 (2007)

14.81 F. Barrientos, G. Sainz: Interpretable knowledge extraction from emergency call data based on fuzzy unsupervised decision tree, Knowl.-Based Syst. **25**(1), 77–87 (2011)

14.82 L. Troiano, L.J. Rodríguez-Muñiz, J. Ranilla, I. Díaz: Interpretability of fuzzy association rules as means of discovering threats to privacy, Int. J. Comput. Math. **89**(3), 325–333 (2012)

14.83 A. Bargiela, W. Pedrycz: *Granular Computing: An Introduction* (Kluwer Academic Publishers, Boston, Dordrecht, London 2003)

14.84 A. Bargiela, W. Pedrycz: *Human-Centric Information Processing Through Granular Modelling*, Studies in Computational Intelligence, Vol. 182 (Springer, Berlin, Heidelberg 2009)

Part B | 14

15. Fuzzy Clustering – Basic Ideas and Overview

Sadaaki Miyamoto

This chapter overviews basic formulations as well as recent studies in fuzzy clustering. A major part is devoted to the discussion of fuzzy c-means and their variations. Recent topics such as kernel-based fuzzy c-means and clustering with semi-supervision are mentioned. Moreover, fuzzy hierarchical clustering is overviewed and fundamental theorem is given.

15.1 Fuzzy Clustering

Data clustering is an old subject [15.1–4] but recently more researchers are developing different techniques and application fields are enlarging. Fuzzy clustering [15.5–10] is also popular in a variety of fuzzy systems. This chapter reviews basic ideas of fuzzy clustering, and provides a brief overview of recent studies. First, we consider the most popular method in fuzzy clustering, i.e., fuzzy c-means. There are many variations, extensions, and applications of fuzzy c-means, some of which are described here. Recent studies on kernel-based methods and clustering with semisupervision are also discussed in relation to fuzzy c-means.

Moreover, another fuzzy clustering is briefly mentioned which uses the transitive closure of fuzzy relations [15.10]. This method is shown to be equivalent to the well-known methods of the single linkage of agglomerative hierarchical clustering [15.11].

15.2 Fuzzy c-Means

We begin with basic notations and then introduces the method of fuzzy c-means by *Dunn* [15.5, 6] and *Bezdek* [15.7, 8].

15.2.1 Notations

The set of objects for clustering is denoted by $X = \{x_1, \ldots, x_N\}$ where each objects is a point of p-dimensional Euclidean space $\mathbf{R}^p$: $x_k = (x_k^1, \ldots, x_k^p)$, $k = 1, \ldots, N$. Clusters are denoted either by G_i or simply by i when no confusion arises. Clustering uses a similarity or dissimilarity measure. In this section, a dissimilarity measure denoted by $D(x, y)$, $x, y \in \mathbf{R}^p$, is used.

Although we have different choices for dissimilarity measure, a standard measure is the squared Euclidean distance

$$D(x, y) = \|x - y\|^2 = \sum_{j=1}^{p} (x^j - y^j)^2 . \tag{15.1}$$

In fuzzy c-means and related methods, the number of clusters, denoted by c is assumed to be given beforehand. The membership of object x_k to cluster i is assumed to be given by u_{ki}. Moreover, the collection of all memberships is denoted by matrix $U = (u_{ki})$. It is natural to assume that $u_{ki} \in [0, 1]$ for all $1 \leq i \leq c$

and $1 \le k \le N$, and, moreover, $\sum_{j=1}^{c} u_{kj} = 1$, for all $1 \le k \le N$.

The method of fuzzy c-means also uses a center for a cluster, which is denoted by $v_i = (v_i^1, \ldots, v_i^p) \in \mathbf{R}^p$ for cluster i. For the ease of reference, all cluster centers are summarized into $V = (v_1, \ldots, v_c)$.

Basic K-Means Algorithm

Many studies of clustering handles K-means [15.12] as a standard method.

Algorithm 15.1 KM: Basic K-means algorithm

KM0:   Generate randomly c cluster centers.
KM1:   Allocate each object x_k $(k = 1, \ldots, N)$ to the cluster of the nearest center.
KM2:   Calculate new cluster centers as the centroid (the center of gravity). If all cluster centers are convergent, stop. Otherwise, go to KM1.
End KM.

Note that the centroid of a cluster G_i is given by $v_i = \frac{1}{|G_i|} \sum_{x_k \in G_i} x_k$, where $|G_i|$ is the number of objects in G_i.

15.2.2 Fuzzy c-Means Algorithm

It should first be noted that the basic idea of fuzzy c-means is an alternative optimization of an objective function proposed by *Dunn* [15.5, 6] and *Bezdek* [15.7, 8]

$$J(U, V) = \sum_{i=1}^{c} \sum_{k=1}^{N} (u_{ki})^m D(x_k, v_i) \quad (m > 1) ,$$

$$(15.2)$$

where $D(x_k, v_i)$ is the squared Euclidean distance (15.1).

Using this objective function, the following alternative optimization is carried out.

Algorithm 15.2 FCM: Fuzzy c-means algorithm

FCM0:   Generate randomly initial fuzzy clusters. Let the solutions be $(\bar{U}, \bar{V})$
FCM1:   Minimize $J(U, \bar{V})$ with respect to U. Let the optimal solution be a new $\bar{U}$.
FCM2:   Minimize $J(\bar{U}, V)$ with respect to V. Let the optimal solution be a new $\bar{V}$.

FCM3:   If the solution $(\bar{U}, \bar{V})$ is convergent, stop. Else go to FCM1.
End FCM.

Note that optimization with respect to U is with the constraint

$$u_{ki} \in [0, 1], \quad \forall 1 \le i \le c, 1 \le k \le N ,$$

$$\sum_{j=1}^{c} u_{kj} = 1, \quad \forall 1 \le k \le N , \tag{15.3}$$

while optimization with respect to V is without any constraint.

It is not difficult to have the optimal solutions as follows

$$\bar{u}_{ki} = \frac{\frac{1}{D(x_k, \bar{v}_i)^{\frac{1}{m-1}}}}{\sum_{j=1}^{c} \frac{1}{D(x_k, \bar{v}_j)^{\frac{1}{m-1}}}} , \tag{15.4}$$

$$\bar{v}_i = \frac{\sum_{k=1}^{N} (\bar{u}_{ki})^m x_k}{\sum_{k=1}^{N} (\bar{u}_{ki})^m} . \tag{15.5}$$

The derivations are omitted; the readers should refer to [15.8] or other textbooks.

Note also that (15.4) appears ill-defined when $x_k = v_i$. In such a case, we use

$$\bar{u}_{ki} = \frac{1}{1 + \sum_{j \ne i} \frac{1}{D(x_k, \bar{v}_j)^{\frac{1}{m-1}}}} , \tag{15.6}$$

which has the same value as (15.4) without a singular point.

Moreover, we write these equations without the use of bars like

$$u_{ki} = \frac{\frac{1}{D(x_k, v_i)^{\frac{1}{m-1}}}}{\sum_{j=1}^{c} \frac{1}{D(x_k, v_j)^{\frac{1}{m-1}}}}$$

$$v_i = \frac{\sum_{k=1}^{N} (u_{ki})^m x_k}{\sum_{k=1}^{N} (u_{ki})^m} ,$$

for simplicity and without any confusion.

15.2.3 A Natural Classifier

These solutions lead us to the following natural fuzzy classifier with a given set of cluster centers V

$$U_i(x; V) = \frac{1}{1 + \sum_{j \ne i} \frac{1}{D(x, \bar{v}_j)^{\frac{1}{m-1}}}} . \tag{15.7}$$

There is nothing strange in (15.7), since $U_i(x; V)$ has been derived from u_{ki} simply by replacing object x_k by the variable x.

This replacement appears rather trivial and it also appears that $U_i(x; V)$ has no further information than u_{ki}. On the contrary, this function is important if we wish to observe theoretical properties of fuzzy c-means.

The following propositions are not difficult to prove and hence the proofs are omitted [15.13]. In particular, the first proposition is trivial.

Proposition 15.1
$U_i(x_k; V) = u_{ki}$, i. e., the fuzzy classifier interpolates the membership value u_{ki}.

Proposition 15.2
When $|x|$ go to infinity, $U_i(x; V)$, $i = 1, \ldots, c$, approaches the same value of $1/c$

$$\lim_{\|x\| \to \infty} U_i(x; V) = \frac{1}{c}.$$

Proposition 15.3
The maximum value of $U_i(x; V)$, $i = 1, \ldots, c$, is at $x = v_i$

$$\max_{x \in \mathbf{R}^p} U_i(x; V) = U_i(v_i, V) = 1.$$

The significance of the function $U_i(x; V)$ is shown in these propositions. An object x_k is a fixed point, while x is a variable. Without such a variable, we cannot observe theoretical properties of fuzzy c-means.

15.2.4 Variations of Fuzzy c-Means

Many variations of fuzzy c-means have been studied, among which we first mention fuzzy c-varieties [15.8], fuzzy c-regressions [15.14], and the method of *Gustafson* and *Kessel* [15.15] to take clusterwise covariance into account. Note that these are relatively old variations and they all are based on variations of objective functions including the change of $D(x, v)$.

In this section, we use the additional symbols $\langle x, y \rangle = x^\top y = y^\top x$, which is the standard scalar product of the Euclidean space $\mathbf{R}^p$. Moreover, we introduce

$$D(x, v; S) = (x - v)^\top S^{-1}(x - v),$$

which is the squared Mahalanobis distance.

Fuzzy c-Varieties
Let us first consider a q-dimensional subspace

$$\text{span}\{s_1, \ldots, s_q\} = \{a_1 s_1 + \cdots + a_q s_q : \\ -\infty < a_j < +\infty, \\ k = 1, \ldots, q\},$$

where $s_1, \ldots, s_q$ is a set of orthonormal set of vectors with $q < p$. s_0 is a given vector of $\mathbf{R}^p$. A linear variety L in $\mathbf{R}^p$ is represented by

$$\mathcal{L} = \{l = s_0 + a_1 s_1 + \cdots + a_q s_q : \\ -\infty < a_j < +\infty, k = 1, \ldots, q\}.$$

Let

$$P(x, l) = \arg\max_{l \in \mathcal{L}} \langle x - s_0, l - s_0 \rangle$$

be the projection of x onto L. We then define

$$D(x_k, l_i) = \|x_k - s_0\|^2 - P(x_k, l_i)^2.$$

We consider the objective function for fuzzy c-varieties

$$J(U, L) = \sum_{i=1}^{c} \sum_{k=1}^{N} (u_{ki})^m D(x_k, l_i) \quad (m > 1), \quad (15.8)$$

where $L = (l_1, \ldots, l_c)$.

The derivation of the solutions is omitted here, but the solutions are as follows:

$$u_{ki} = \frac{\frac{1}{D(x_k, l_i)^{\frac{1}{m-1}}}}{\sum_{j=1}^{c} \frac{1}{D(x_k, l_j)^{\frac{1}{m-1}}}}, \quad (15.9)$$

$$\bar{s}_0^{(i)} = \frac{\sum_{k=1}^{N} (u_{ki})^m x_k}{\sum_{k=1}^{N} (u_{ki})^m}, \quad (15.10)$$

while $s_j^{(i)}$ $(j = 1, \ldots, q)$ is the normalized eigenvector corresponding to the q maximum eigenvalues of the matrix

$$A_i = \sum_{k=1}^{N} (u_{ki})^m \left(x_k - s_0^{(i)}\right) \left(x_k - s_0^{(i)}\right)^\top. \quad (15.11)$$

Note that the superscript $s_j^{(i)}$ shows those vectors for cluster i.

Therefore, alternative optimization of (15.8) is done by calculating (15.9), (15.10), and the eigenvectors for (15.11) repeatedly until convergence.

Fuzzy *c*-Regression Models

In this section, we assume that $x = (x^1, \ldots, x^p)$ is a p-dimensional independent variable, while y is a scalar-valued dependent variable. Hence data set $\{(x_1, y_1), \ldots, (x_N, y_N)\}$ is handled. We consider c regression models

$$y = \sum_{j=1}^{p} \beta_i^j x^j + \beta_i^{p+1}, \quad i = 1, \ldots, c.$$

Hence, the squared error is taken to be the dissimilarity

$$D((x_k, y_k), B_i) = \left(y_k - \sum_{j=1}^{p} \beta_i^j x_k^j - \beta_i^{p+1} \right)^2$$

and an objective function

$$J(U, B) = \sum_{i=1}^{c} \sum_{k=1}^{N} (u_{ki})^m D((x_k, y_k), B_i) \quad (m > 1),$$

(15.12)

is considered, where $B_i = (\beta_i^1, \ldots, \beta_i^{p+1})$ and $B = (B_1, \ldots, B_c)$. To express the solutions in a compact manner, we introduce two vectors

$$z_k = (x_k^1, \ldots, x_k^p, 1), \quad \hat{\beta}_i = \left(\beta_i^1, \ldots, \beta_i^{p+1} \right).$$

Then we have

$$u_{ki} = \frac{\frac{1}{D((x_k, y_k), B_i)^{\frac{1}{m-1}}}}{\sum_{j=1}^{c} \frac{1}{D((x_k, y_k), B_i)^{\frac{1}{m-1}}}},$$

(15.13)

$$\hat{\beta}_i = \left(\sum_{k=1}^{N} (u_{ki})^m z_k z_k^\top \right)^{-1} \sum_{k=1}^{N} (u_{ki})^m y_k z_k.$$

(15.14)

Thus the alternative optimization of $J(U, B)$ is to calculate (15.13) and (15.14) iteratively until convergence.

The Method of Gustafson and Kessel

The method of Gustafson and Kessel enables us to incorporate clusterwise covariance variables denoted by $S_1, \ldots, S_c$. We consider

$$J(U, V, S) = \sum_{i=1}^{c} \sum_{k=1}^{N} (u_{ki})^m D(x_k, v_i; S_i) \quad (m > 1),$$

(15.15)

where a simplified symbol $S = (S_1, \ldots, S_c)$ and the clusterwise squared Mahalanobis distance $D(x_k, v_i; S_i)$ is used. Note also that S_i is with the constraint

$$|S_i| = \rho_i \quad (\rho_i > 0)$$

(15.16)

where ρ_i is a fixed parameter and $|S_i|$ is the determinant of S_i. We assume, for simplicity, $\rho_i = 1$ [15.16].

The solutions are as follows

$$u_{ki} = \left\{ \sum_{j=1}^{c} \left(\frac{D(x_k, v_i; S_i)}{D(x_k, v_j; S_j)} \right)^{\frac{1}{m-1}} \right\}^{-1},$$

(15.17)

$$v_i = \frac{\sum_{k=1}^{N} (u_{ki})^m x_k}{\sum_{k=1}^{N} (u_{ki})^m},$$

(15.18)

$$S_i = \frac{1}{|\hat{S}_i|^{\frac{1}{p}}} \sum_{k=1}^{N} (u_{ki})^m (x_k - v_i)(x_k - v_i)^\top.$$

(15.19)

where $\hat{S}_i = \sum_{k=1}^{N} (u_{ki})^m (x_k - v_i)(x_k - v_i)^\top$.

Since three types of variables are used for the method of Gustafson and Kessel, the alternative optimization iteratively calculates (15.17–15.19) until convergence.

15.2.5 Possibilistic Clustering

The possibilistic clustering [15.17, 18] proposed by *Krishnapuram* and *Keller* does not use the constraint (15.3) in the alternative optimization algorithm FCM. Rather, the optimization with respect to U is without any constraint. To handle $\arg\min_U J(U, \bar{V})$ of (15.2) without a constraiint leads to the trivial solution of $U = O$ (the zero matrix), and hence they proposed a modified objective function

$$J_{\text{pos}}(U, V) = \sum_{i=1}^{c} \sum_{k=1}^{N} (u_{ki})^m D(x_k, v_i)$$

$$+ \sum_{i=1}^{c} \eta_i \sum_{k=1}^{N} (1 - u_{ki})^m \quad (m > 1),$$

(15.20)

where $\eta_i \ (i = 1, \ldots, c)$ is a positive constant.

We easily have the optimal U

$$u_{ki} = \frac{1}{1 + \left(\frac{D(x_k, v_i)}{\eta_i} \right)^{\frac{1}{m-1}}},$$

(15.21)

while optimal V is given by (15.5).

The natural fuzzy classifier derived from the possibilistic clustering is

$$U_i(x; v_i) = \cfrac{1}{1 + \left(\cfrac{D(x,v_i)}{\eta_i}\right)^{\frac{1}{m-1}}} , \qquad i = 1, \ldots, c .$$

(15.22)

This classifier has the following properties:

___

Proposition 15.4
$U_i(x_k; v_i) = u_{ki}$ by (15.21), i. e., the possibilistic classifier interpolates the membership value u_{ki}.

___

Proposition 15.5
When $|x|$ go to infinity, $U_i(x; v_i)$ $(i = 1, \ldots, c)$ approaches zero

$$\lim_{\|x\| \to \infty} U_i(x; v_i) = 0 .$$

___

Proposition 15.6
The maximum value of $U_i(x; v_i)$ $(i = 1, \ldots, c)$ is at $x = v_i$

$$\max_{x \in \mathbf{R}^p} U_i(x; v_i) = U_i(v_i, v_i) = 1 .$$

___

15.2.6 Kernel-Based Fuzzy c-Means

The support vector machines [15.19, 20] is now one of the most popular methods of supervised classification. Since positive definite kernels are frequently used in support vector machines, the study of kernels has also been done by many researchers (e.g., [15.21]). The positive definite kernels can also be used for fuzzy c-means, as we see in this section.

The reason why we use kernels for clustering is that essentially the K-means and fuzzy c-means have linear boundaries between clusters.

Note that the K-means classifier uses the nearest center allocation rule for a given $x \in \mathbf{R}^p$

$$x \to G_i \iff i = \arg \min_{1 \le j \le c} D(x, v_i) ,$$

and hence the rules generates the Voronoi region [15.22] with the centers $v_1, \ldots, v_c$ which has piecewise linear boundaries.

For fuzzy c-means, the classifiers are fuzzy but if we introduce simplification of the rules by crisp reallocation [15.13] by

$$x \to G_i \iff i = \arg \max_{1 \le j \le c} U_i(x; V) .$$

Then we again have the Voronoi regions with the centers $v_1, \ldots, v_c$.

The introduction of the covariance variables enables the cluster boundaries to be quadratic, but more flexible nonlinear boundaries cannot be obtained.

In order to have clusters with nonlinear boundaries, we can use positive-definite kernels. Kernels are introduced by using a high-dimensional mapping $\Phi : \mathbf{R}^p \to H$, where H is a Hilbert space with the inner product $\langle \cdot, \cdot \rangle_H$ and the norm $\| \cdot \|_H$.

Given objects $x_1, \ldots x_N$, we consider its images by the mapping Φ: $\Phi(x_1), \ldots, \Phi(x_N)$. Note that the method of kernels does not assume that an explicit form of $\Phi(x_1), \ldots, \Phi(x_N)$ is known, but their inner product $\langle \Phi(x_i), \Phi(x_j) \rangle_H$ is assumed to be known. Specifically, a positive-definite function $K(x, y)$ is given and we assume

$$K(x, y) = \langle \Phi(x_i), \Phi(x_j) \rangle_H .$$

This assumption seems abstract, but if we are given an actual kernel function, the method becomes simple, for example, a well-known kernel is the Gaussian kernel

$$K(x, y) = \exp(-C\|x - y\|^2) .$$

Then what we handle is $\langle \Phi(x_i), \Phi(x_j) \rangle_H = \exp(-C\|x - y\|^2)$.

We now proceed to consider kernel-based fuzzy c-means [15.23]. The objective function uses $\Phi(x_1), \ldots, \Phi(x_N)$ and cluster centers $w_1, \ldots, w_c$ of H

$$J(U, V) = \sum_{i=1}^{c} \sum_{k=1}^{N} (u_{ki})^m \|\Phi(x_k) - w_i\|_H^2 \quad (m > 1) ,$$

(15.23)

where $W = (w_1, \ldots, w_c)$. We have

$$u_{ki} = \cfrac{\cfrac{1}{\|\Phi(x_k) - w_i\|_H^{\frac{2}{m-1}}}}{\sum_{j=1}^{c} \cfrac{1}{\|\Phi(x_k) - w_i\|_H^{\frac{2}{m-1}}}}$$

(15.24)

$$w_i = \frac{\sum_{k=1}^{N} (u_{ki})^m \Phi(x_k)}{\sum_{k=1}^{N} (u_{ki})^m} .$$

(15.25)

Note, however, that the explicit form of $\Phi(x_k)$ and hence w_i is not available. Therefore, we eliminate w_i from the iterative calculation. Thus, the updating w_i is replaced by the update of

$$D_H(x_k, w_i) = \|\Phi(x_k) - w_i\|^2 .$$

We then have

$$D_H(x_k, w_i) = K(x_k, x_k)$$

$$-\frac{2}{\sum_{k=1}^{N}(u_{ki})^m}\sum_{j=1}^{N}(u_{ji})^m K(x_j, x_k)$$

$$+\frac{1}{\left(\sum_{k=1}^{N}(u_{ki})^m\right)^2}$$

$$\times \sum_{j=1}^{N}\sum_{\ell=1}^{N}(u_{ji}u_{\ell i})^m K(x_j, x_\ell) .$$

$$(15.26)$$

Using (15.26), we calculate

$$u_{ki} = \frac{\frac{1}{D_H(x_k, w_i)^{\frac{1}{m-1}}}}{\sum_{j=1}^{c}\frac{1}{D_H(x_k, w_i)^{\frac{1}{m-1}}}} .$$

$$(15.27)$$

Thus the alternative optimization repeats (15.26) and (15.27) until convergence.

Fuzzy classifiers of kernel-based fuzzy c-means can also be derived [15.13]. We omit the details and show the function in the following

$$D(x; v_i) = K(x, x) - \frac{2}{\sum_{k=1}^{N}(u_{ki})^m}\sum_{j=1}^{N}(u_{ji})^m K(x, x_j)$$

$$+\frac{1}{\left(\sum_{k=1}^{N}(u_{ki})^m\right)^2}$$

$$\times \sum_{j=1}^{N}\sum_{\ell=1}^{N}(u_{ji}u_{\ell i})^m K(x_i, x_\ell) ,$$

$$(15.28)$$

$$U_i(x) = \left[\sum_{j=1}^{c}\left(\frac{D(x; v_i)}{D(x; v_j)}\right)^{\frac{1}{m-1}}\right]^{-1} .$$

$$(15.29)$$

A Simple Numerical Example

A well-known and simple example to see how the kernel method work to produce clusters with nonlinear boundaries is given in Fig. 15.1. There is a circular cluster inside another group of objects of a ring shape. We call it *ring around circle* data.

Figure 15.2 shows the shape of a fuzzy classifier (15.29) with $m = 2$ and $c = 2$ obtained from the *ring around circle* data. Thus the ring and the circle inside the ring are perfectly separated.

15.2.7 Clustering with Semi-Supervision

Recently many studies consider semisupervised learning (e.g., [15.24, 25]). In this section we briefly overview literature in fuzzy clustering with semi-supervision.

We begin with two classes of semisupervised learning after *Zhu* and *Goldberg* [15.25]. They defined *semisupervised classification* that has a set of labeled samples and another set of unlabeled samples. Another class is called constrained clustering which has two sets of *must-links* $ML = \{(x, y), \ldots\}$ and *cannot-links* $CL = \{(z, w), \ldots\}$. Two objects x and y in the must-link set has to be allocated in the same cluster, while z and w in the cannot-link set has to be allocated to different clusters.

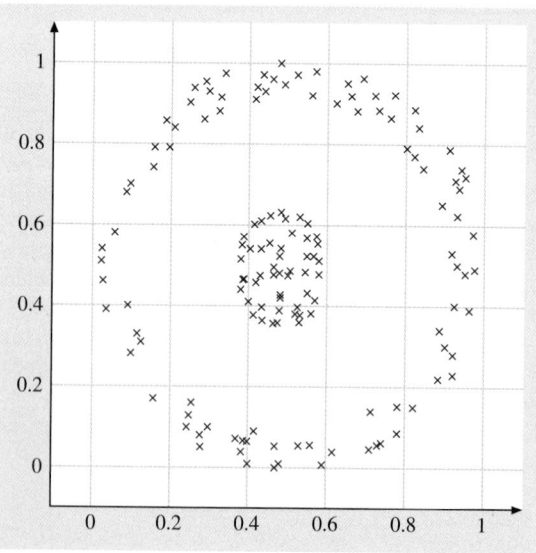

Fig. 15.1 An example of a circle and a ring around the circle

Let us briefly mention two studies in the first class of semisupervised classification. *Bouchachia* and *Pedrycz* [15.26] used the following objective function that has an additional term

$$J(U, V) = \sum_{i=1}^{c} \sum_{k=1}^{N} \{(u_{ki})^m D(x_k, v_i)$$
$$+ \alpha(u_{ki} - \tilde{u}_{ki})^m D(x_k, v_i)\} .$$

where $\tilde{u}_{ki}$ is a given membership showing semisupervision.

Miyamoto [15.27] proved that an objective function with entropy term [15.13, 28–31] can generalize the EM solution of the mixture of Gaussian densities with semisupervision [15.25, p. 27].

Another class of constrained clustering [15.32, 33] has also been studied using a modified objective function [15.34] with additional terms of the must-link and cannot-link

$$J(U, V) = \sum_{i=1}^{c} \sum_{k=1}^{N} (u_{ki})^2 D(x_k, v_i)$$
$$+ \alpha \sum_{k=1}^{N} \left(\sum_{(x_k, x_j) \in ML} \sum_{i \neq l} u_{ki} u_{jl} + \sum_{(x_k, x_j) \in CL} \sum_{i=1}^{c} u_{ki} u_{jk} \right) .$$

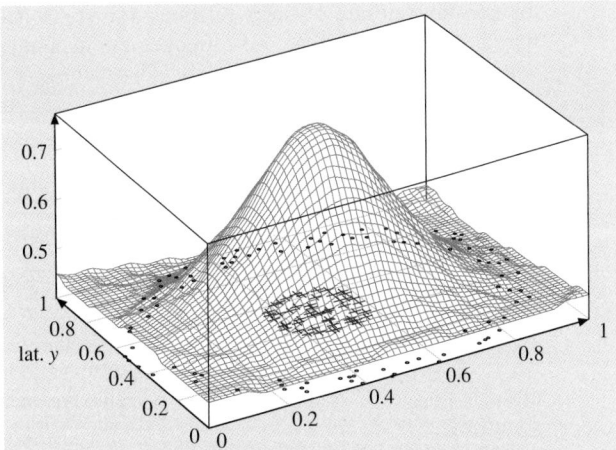

Fig. 15.2 Two clusters and a fuzzy classifier from the *ring around circle* data; fuzzy c-means with $m = 2$ is used

To summarize, the method of fuzzy c-means with semisupervision including constrained clustering has not yet gained wide popularity, due to the limited number of studies comparing with those in another field of machine learning where many papers and a number of books have been published [15.24, 25, 32]. Hence more results can be expected in this area of studies.

15.3 Hierarchical Fuzzy Clustering

There is still another method of fuzzy clustering that is very different from the above fuzzy c-means which is related to the single linkage in agglomerative hierarchical clustering.

In this section, we assume that objects $X = \{x_1, \ldots, x_N\}$ are not necessarily in an Euclidean space. Rather, a relation $S(x, y)$ satisfying reflexivity and symmetry

$$S(x, x) = 1 , \quad \forall x \in X , \tag{15.30}$$
$$S(x, y) = S(y, x) , \quad \forall x, y \in X \tag{15.31}$$

is assumed, where a larger value of $S(x, y)$ means that x and y are more similar.

We then describe the general algorithm of agglomerative hierarchical clustering as follows:

Algorithm 15.3 AHC: Algorithm of Agglomerative Hierarchical Clustering

AHC1:   Let initial clusters be individual objects $G_i = \{x_i\}, i = 1, \ldots, N$ and put $K = N$.

AHC2:   Find pair of clusters of maximum similarity

$$(G_p, G_q) = \arg \max_{i,j} S(G_i, G_j) .$$

Merge $G_r = G_p \cup G_q$. $K = K - 1$ and if $K = 1$, stop.

AHC3:   Update $S(G_r, G')$ for all other clusters G'. Go to AHC1.

End AHC.

The updating step of AHC3 admits different choices of similarity between clusters, among which the single linkage methods uses

$$S(G, G') = \max_{x \in G, y \in G'} S(x, y) .$$

Although there are other choices, discussion in this section is focused upon the single linkage.

On the other hand, studies in the 1970s including *Zadeh's* [15.10] proposed hierarchical clustering using the transitive closure of fuzzy relations $S(x, y)$. To define the transitive closure, we introduce the max–min composition

$$(S \circ T)(x, z) = \max_{y \in X} \min\{S(x, y), T(y, z)\},$$

where S and T are the fuzzy relations of X. Using the max-min composition, we can define the transitive closure S^* of S

$$S^*(x, y) = \max\{S(x, y), S^2(x, y), S^3(x, y), \ldots\},$$

where $S^2 = S \circ S$ and $S^k = S \circ S^{k-1}$. It also is not difficult to see $S^* = S^{N-1}$ when S is reflexive and symmetric.

When S is reflexive and symmetric, the transitive closure S^* is also reflexive and symmetric, and moreover transitive

$$S^*(x, y) \geq \min\{S^*(x, z), S^*(z, y)\}, \quad \forall z \in X.$$

If a fuzzy relation is reflexive, symmetric, and transitive, then it is called a fuzzy equivalence relation: it has a property that every α-cut is a crisp equivalence relation

$$[S^*]_\alpha(x, x) = 1, \quad \forall x \in X, \tag{15.32}$$

$$[S^*]_\alpha(x, y) = [S^*]_\alpha(y, x), \quad \forall x, y \in X, \tag{15.33}$$

$$[S^*]_\alpha(x, y) = 1, \ [S^*]_\alpha(y, z) = 1$$
$$\Rightarrow [S^*]_\alpha(x, z) = 1, \tag{15.34}$$

where $[S^*]_\alpha(x, y)$ is the α-cut of $S^*(x, y)$

$$[S^*]_\alpha(x, y) = 1 \iff S^*(x, y) \geq \alpha;$$
$$[S^*]_\alpha(x, y) = 0 \iff S^*(x, y) < \alpha.$$

Thus each α-cut of S^* induces an equivalence class of X, and moreover if α decreases, the equivalence class

becomes coarser, and therefore S^* defines a hierarchical clusters.

We now can describe a fundamental theorem on fuzzy hierarchical clustering.

Theorem 15.1 Miyamoto [15.11]

Given a set of objects $X = \{x_1, \ldots, x_N\}$ and a similarity measure $S(x, y)$ for all $x, y \in X$, the following four methods give the same hierarchical clusters:

1. Clusters by the single linkage,
2. Clusters by the transitive closure S^*,
3. Clusters as vertices of connected components of fuzzy graph with vertices X and edges $X \times X$ with membership values $S(x, y)$, and
4. Clusters generated from the maximum spanning tree of the network with vertices X and edges $X \times X$ with weight $S(x, y)$.

The above theorem needs some more explanations. Connected components of a fuzzy graph means the family of those connected components of all α-cuts of the fuzzy graph. Since connected components grows with decreasing α, those sets of vertices form hierarchical clusters. The minimum spanning tree is well known, but the maximum spanning tree is used instead. The way in which hierarchical clusters are generated is the same as the connected components of the fuzzy graph.

Although this theorem shows the importance of fuzzy hierarchical clustering, it appears that no new results that are useful in applications are included in this theorem. *Miyamoto* [15.35] showed, however, that other methods of DBSCAN [15.36] and *Wishart's* mode analysis [15.37] have close relations to the above results, and he discusses the possibility of further application of this theorem, e.g., to nonsymmetric similarity measure.

15.4 Conclusion

We overviewed fuzzy c-means and related studies. Kernel-based clustering algorithm and clustering with semisupervision were also discussed. Moreover, fuzzy hierarchical clustering which is based on fuzzy graphs and very different from fuzzy c-means was briefly reviewed.

New methods and algorithms based on the idea of fuzzy c-means are still being developed, as the fundamental idea has enough potential to produce many new techniques. On the other hand, fuzzy hierarchical clustering is rarely mentioned in the literature. However, there are possibilities for having new theory, methods, and applications in this area, as the fundamental mathematical structure is well established.

Many important studies were not mentioned in this overview, for example, *Ruspini*'s method [15.38] and cluster validation measures are important. Readers may read books on fuzzy clustering [15.8, 9, 13, 16, 39] for details of fuzzy clustering. Also, *Miyamoto* [15.11] can still be used for studying the fundamental theorem on fuzzy hierarchical clustering.

References

15.1   R.O. Duda, P.E. Hart: *Pattern Classification and Scene Analysis* (John Wiley, New York 1973)

15.2   B.S. Everitt: *Cluster Analysis*, 3rd edn. (Arnold, London 1993)

15.3   A.K. Jain, R.C. Dubes: *Algorithms for Clustering Data* (Prentice Hall, Englewood Cliffs, NJ 1988)

15.4   L. Kaufman, P.J. Rousseeuw: *Finding Groups in Data: An Introduction to Cluster Analysis* (Wiley, New York 1990)

15.5   J.C. Dunn: A fuzzy relative of the ISODATA process and its use in detecting compact well-separated clusters, J. Cybern. **3**, 32–57 (1974)

15.6   J.C. Dunn: Well-separated clusters and optimal fuzzy partitions, J. Cybern. **4**, 95–104 (1974)

15.7   J.C. Bezdek: Fuzzy Mathematics in Pattern Classification, Ph.D. Thesis (Cornell Univ., Ithaca 1973)

15.8   J.C. Bezdek: *Pattern Recognition with Fuzzy Objective Function Algorithms* (Plenum, New York 1981)

15.9   J.C. Bezdek, J. Keller, R. Krishnapuram, N.R. Pal: *Fuzzy Models and Algorithms for Pattern Recognition and Image Processing* (Kluwer, Boston 1999)

15.10   L.A. Zadeh: Similarity relations and fuzzy orderings, Inf. Sci. **3**, 177–200 (1971)

15.11   S. Miyamoto: *Fuzzy Sets in Information Retrieval and Cluster Analysis* (Kluwer, Dordrecht 1990)

15.12   J.B. MacQueen: Some methods of classification and analysis of multivariate observations, Proc. 5th Berkeley Symp. Math. Stat. Prob. (Univ. of California Press, Berkeley 1967) pp. 281–297

15.13   S. Miyamoto, H. Ichihashi, K. Honda: *Algorithms for Fuzzy Clustering* (Springer, Berlin 2008)

15.14   R.J. Hathaway, J.C. Bezdek: Switching regression models and fuzzy clustering, IEEE Trans. Fuzzy Syst. **1**(3), 195–204 (1993)

15.15   E.E. Gustafson, W.C. Kessel: Fuzzy clustering with a fuzzy covariance matrix, IEEE CDC (1979) pp. 761–766

15.16   F. Höppner, F. Klawonn, R. Kruse, T. Runkler: *Fuzzy Cluster Analysis* (Wiley, New York 1999)

15.17   R. Krishnapuram, J.M. Keller: A possibilistic approach to clustering, IEEE Trans. Fuzzy Syst. **1**, 98–110 (1993)

15.18   R.N. Davé, R. Krishnapuram: Robust clustering methods: A unified view, IEEE Trans. Fuzzy Syst. **5**, 270–293 (1997)

15.19   V.N. Vapnik: *Statistical Learning Theory* (Wiley, New York 1998)

15.20   V.N. Vapnik: *The Nature of Statistical Learning Theory*, 2nd edn. (Springer, New York 2000)

15.21   B. Schölkopf, A.J. Smola: *Learning with Kernels* (MIT Press, Cambridge 2002)

15.22   T. Kohonen: *Self-Organizing Maps*, 2nd edn. (Springer, Berlin 1997)

15.23   S. Miyamoto, D. Suizu: Fuzzy c-means clustering using kernel functions in support vector machines, J. Adv. Comput. Intell. Intell. Inf. **7**(1), 25–30 (2003)

15.24   O. Chapelle, B. Schölkopf, A. Zien (Eds.): *Semi-Supervised Learning* (MIT Press, Cambridge 2006)

15.25   X. Zhu, A.B. Goldberg: *Introduction to Semi-Supervised Learning* (Morgan Claypool, San Rafael 2009)

15.26   A. Bouchachia, W. Pedrycz: A semi-supervised clustering algorithm for data exploration, IFSA 2003, Lect. Notes Artif. Intell. **2715**, 328–337 (2003)

15.27   S. Miyamoto: An overview of hierarchical and non-hierarchical algorithms of clustering for semi-supervised classification, LNAI **7647**, 1–10 (2012)

15.28   R.-P. Li, M. Mukaidono: A maximum entropy approach to fuzzy clustering, Proc. 4th IEEE Int. Conf. Fuzzy Syst. (FUZZ-IEEE/IFES'95) (1995) pp. 2227–2232

15.29   S. Miyamoto, M. Mukaidono: Fuzzy c-means as a regularization and maximum entropy approach, Proc. 7th Int. Fuzzy Syst. Assoc. World Congr. IFSA'97, Vol. 2 (1997) pp. 86–92

15.30   H. Ichihashi, K. Honda, N. Tani: Gaussian mixture PDF approximation and fuzzy c-means clustering with entropy regularization, Proc. 4th Asian Fuzzy Syst. Symp., Vol. 1 (2000) pp. 217–221

15.31   H. Ichihashi, K. Miyagishi, K. Honda: Fuzzy c-means clustering with regularization by K-L information, Proc. 10th IEEE Int. Conf. Fuzzy Syst., Vol. 2 (2001) pp. 924–927

15.32   S. Basu, I. Davidson, K.L. Wagstaff: *Constrained Clustering* (CRC, Boca Raton 2009)

15.33   N. Shental, A. Bar-Hillel, T. Hertz, D. Weinshall: Computing Gaussian mixture models with EM using equivalence constraints, Advances in Neural Information Processing Systems 16, ed. by S. Thrun, L.K. Saul, B. Schölkopf (MIT Press, Cambridge 2004)

15.34   N. Wang, X. Li, X. Luo: Semi-supervised kernel-based fuzzy c-means with pairwise constraints, Proc. WCCI 2008 (2008) pp. 1099–1103

15.35   S. Miyamoto: Statistical and non-statistical models in clustering: An introduction and recent topics,

Analysis and Modelling of Complex Data in Behavioural and Social Sciences, JCS−CLADAG 12, ed. by A. Okada, D. Vicari, G. Ragozini (2012) pp. 3−6

15.36   M. Ester, H.-P. Kriegel, J. Sander, X.W. Xu: A density-based algorithm for discovering clusters in large spatial databases with noise, Proc. 2nd Int. Conf. Knowl. Discov. Data Min. (KDD-96) (AAAI Press, Menlo Park 1996)

15.37   D. Wishart: Mode analysis: A generalization of nearest neighbour which reduces chaining effects, Numer. Taxn., Proc. Colloq., ed. by A.J. Cole (1968) pp. 283−311

15.38   E.H. Ruspini: A new approach to clustering, Inf. Control **15**, 22−32 (1969)

15.39   D. Dumitrescu, B. Lazzerini, L.C. Jain: *Fuzzy Sets and Their Application to Clustering and Training* (CRC Press, Boca Raton 2000)

16. An Algebraic Model of Reasoning to Support Zadeh's CWW

Enric Trillas

In the very wide setting of a Basic Fuzzy Algebra, a formal algebraic model for Commonsense Reasoning is presented with fuzzy and crisp sets including, in particular, the usual case of the Standard Algebras of Fuzzy Sets. The aim with which the model is constructed is that of, first, adding to Zadeh's *Computing with Words* a wide perspective of ordinary reasoning in agreement with some basic characteristics of it, and second, presenting an operational ground on which linguistic terms can be represented, and schemes of inference posed. Additionally, the chapter also tries to express the author's belief that reasoning deserves to be studied like an Experimental Science.

16.1 A View on Reasoning

Thinking is a yet not scientifically well-known natural and complex neurophysiological phenomenon that, shown by people and given thanks to their brains, is mostly and significantly externalized in some observable physical ways, like it is the case of talking by means of uttered or written words. Only recently the functioning of the brain's systems started to be studied with the current methods of experimental science, and made some knowledge on its internal working possible.

Talking acquires full development with a typically social human manifestation called telling with, at least, its two modalities of discourse and narrative that, either in different oral, spatial hand's signs, or written forms, not only support telling but, together with abstraction, could be considered among the highest expressions of brain's capability of thinking, surely reinforced during evolution by the physical possibilities of the humans to tackle and to consider the possible usefulness of ob-

jects. Telling can be roughly described as consisting in chains of sentences organized with some purpose.

Thinking and telling are but names for abstract concepts covering the totality of those human actions designated by *to think*, *to tell*, and *to discuss*, of which only the last two can be directly observed by a layperson. With telling and discussing not only reasoning is shown, but also abstraction is conveyed. In this sense, it can be said that telling and discussing cannot exist without reasoning, and that they are intermingled in some inextricable form that allows us to guess for foreseeing what will come in the future, to imagine what could happen in it, and to express it by words. At this point, the human capability of conjecturing appears as something fundamental [16.1].

Foreseeing and imagining resulted essential for the growing and expanding of mankind on Earth, and are assisted by the human capabilities of guessing (or conjecturing), and refuting, for saying nothing on those emotions that so often drive human reasoning toward *creative thinking*. Of course, as thinking comprises more features than reasoning, like they are the case of feelings and imagining with sounds, images, etc., both concepts should not be confused.

This Introduction only refers generically and just in a co-lateral form, to telling as supporting discourse and narrative in natural language, and for whose understanding the context-dependent and purpose-driven concept of the meaning of statements [16.2], their semantics, is essential. This chapter mainly deals with an algebraic analysis of the reasoning that, generated by the physical processes in the brain thinking consists in, is externalized by means of the language of signs, oral or written expressions, figures, etc. It pretends nothing else than to be a first trial toward a possible and more general algebra of reasoning than are Boolean algebras, orthomodular lattices, and standard algebras of fuzzy sets, the algebraic structures in which classical propositional calculus, quantum physics' reasoning, and approximate reasoning are, respectively, represented and formally presented [16.3, 4].

Reasoning is considered the manifestation of rationality [16.5], a concept that comes from the Latin word *ratio* (namely, referred to comparing statements), and from very old, allowed to believe in a clear cut existing among the living species and under which the human one is the unique that is rational, that can reason. Reasoning is also, at its turn, an abstract concept referring to several ways for obtaining conclusions from a previous knowledge, or information, or evidence, given by statements that are called the premises; it is sometimes said that the premises are the reasons for the conclusions, or the reasons that support them. Conclusions are also statements, and without a previous knowledge neither reasoning, nor understanding, is possible.

Apparent processes of reasoning is what, to some extent, is observable and can be submitted to a *Menger*'s kind of *exact thinking* [16.6], by analyzing them in general enough algebraic terms. This is actually the final goal of this chapter, whose aim is to be placed at the ground of *Zadeh*'s *Computing with Words* (CWW) [16.7], helping to adopt in it the point of view of ordinary reasoning and not only that of the deductive one, and for a viewing of CWW close to the mathematical modeling of natural language and ordinary reasoning.

Note

Due, in a large part, to the author's lack of knowledge, there are many topics appearing as manifestations of thinking and, up to some extent, matching with reasoning, that cannot even be slightly taken into account in a paper that is, in itself, of a very limited scope, and only contains generic reflections in its nontechnical Sects. 16.1–16.4. Among these topics, there are some that, as far as the author knows, are not yet submitted to a systematic scientific study. It is the case, for instance, of what could be called *sudden direct action*, or *action under pressing*, as well as those concerning thinking and reasoning in both the *beaux arts*, and the music's creation [16.8].

Those topics can still deserve some scientific and subsequent philosophical reflections. In some of them like it is, for instance, the case in modern painting, it appears a yet mysterious kind of play between actual or virtual situations, and where the same objects of reasoning can be seen as unfinished, but not as not-finished [16.9]. That is, in which, the antonym seems to play a role different from that of the negate [16.10].

Being those topics yet open to more analysis, perhaps some of the methodologies of analogical or case-based reasoning could suggest some ways for an exploration of them in terms of fuzzy sets [16.11], and in view of a possible computational mechanization of some of their aspects.

16.2 Models

The pair telling–reasoning can be completed to a triplet of philosophically essential concepts with that of *modeling* that, facilitated by abstraction, is one among the best ways people have for capturing the basics of the phenomena appearing in some reality by not only taking some perspective and distance with them, but, after observation, recognizing their more basic treats. Thanks to *models*, not only the terms or words employed in the linguistic description of a phenomenon can be well enough understood by bounding its meaning in a formal frame, but only thanks to mathematical models the use of the safest type of reasoning, formal deduction, can be used for its study. In addition and currently, models are often useful formalisms through which the possibility of coping with human reasoning by means of computers can be done. This chapter essentially presents a new mathematical (algebraic) model for reasoning that is neither directly based on *truth*, but on *contradiction*, nor confuses reasoning and deducing, a confusion CWW should not fall in, since people mainly reason in nonformal and nondeductive ways.

If Experimental Science can be roughly described as an art for building plausible specific models of some reality with the aim of intellectually capturing it, in no case they (namely, the algebraic ones) are established forever but do change with time, something that shows they cannot be always confused with a time-cut of the reality tried to be modeled. On the other side, good models do warrant, at least, to preserve what already hold in them once new evidence on the corresponding reality is known.

Even if models are not to be confused with what they represent, same as an architect's mockup that should not be confused with a building following

from it, mathematical and computational modeling are among the greatest human acquisitions coming from abstraction and safe reasoning. Actually, a good deal of what characterizes the current civilization derives from models, and mathematical ones are usually considered at the very top of rationality since they had proven to allow for a good comprehension of several and important realities previously recognized as actually existing.

Models are a good help for a better understanding of the reality they model, and mathematical ones show the so-called *unreasonable effectiveness of mathematics* [16.12] for the understanding of reality thanks, in a good part, to formal deduction, the safest form of reasoning they allow to use. If it can metaphorically be said that if reality is in color, a model of this reality is a simplification of it in black and white; of course and up to some extent, modeling is an art with whose simplifications it should not be avoided what is essential for the description of the corresponding reality.

16.3 Reasoning

To give a definition of reasoning is very difficult, if not impossible since, at the end, it is a family of natural processes generated in the brain. A first operational question that can be posed is, *What is reasoning for?* of which, in a first approach, can be just said that reasoning is actually intermingled with the will of people to ask and to answer questions, to influence, to inform, to teach, to convince, or just to communicate with other people. Something usually managed by means of reciprocal telling, or dialogue, or conversation, between people.

In a second and perhaps complementary approach, it should also be said that reasoning also serves for satisfying human's will of searching new ideas that are not immediately seen in the evidence, and that can help for a further exploring of the reality to which the premises refer to. The human will for communicating and influencing, for foreseeing and for exploring, are made possible by means of reasoning that, in this perspective, seems to be a capability acquired thanks to the brain complexity once it, and through the senses, is in contact with the world and can try to understand what is and what happens in it by means of the neuronal/synaptic representations reached in the brain thanks, in part, to the external receptors of the human nervous system.

To answer a more scientifically sensitive and less psychological question, *What does reasoning appear to be?* let us place ourselves in two different, although overlapping, points of view: those from the premises and those from the conclusions. From the first, reasoning can try to confirm, to explain, to enlarge, or to refute, the information conveyed by the premises. From the second, and in confrontation with the context of the premises, conclusions can be classified in either necessary, or contingent. At its turn, the last can be either explanations or speculations that is, conclusions trying to foresee, either backward or forward and perhaps by *jumps* from the premises and without clear rules for it, respectively, something new and currently unknown, but that eventually can be *suggested* by the premises and, very often, reached thanks to some additional background knowledge. All this is what conducts to the typically human capability called *creativity*, many times obtained through either hypotheses (backward case), or speculations (forward jumping case). It even helps to take rational, pondered decisions, that are among the essential characteristics shown by the *intelligence* attributed to people.

For what concerns necessary conclusions, they are not usually for capturing something radically new since, in general, what is not included, or hidden, in the premises, but is external to them, is changing. From necessary premises not only necessary conclusions follow under some rules of inference, but also contingent ones that, in addition, cannot be always deployed from the premises under some well-known precise rules. Necessary conclusions are surely useful for arriving at a better understanding of what is strictly described by the premises, and for deploying what they contain, as it happens in formal sciences, and also to show what cannot be the case if the given premises do not perfectly reflect the reality. In this sense, refutation of either a part of the evidence, or of hypotheses, etc., is actually important [16.13].

16.3.1 A Remark on the Mathematical Reasoning

In the case of mathematics, where the only certified knowledge is furnished by the theorems proved by formal deduction, the former statement *necessary conclusions are not for capturing something radically new*, could be actually surprising, especially if it could mean that nothing *new* can be deductively deployed from some supposedly necessary premises. For instance, from the Peano's axioms defining the set $\mathbb{N}$ of natural numbers, a big amount of new and fertile concepts and theorems are deductively deployed and successfully applied to fields outside mathematics.

To quote a case: The not easily captured – for nonmathematicians – abstract concept of the real number is constructed after that of rational number (whose set is $\mathbb{Q}$), that comes from an equivalence between pairs of integers (set $\mathbb{Z}$), after defining the concept of the integer number from another equivalence between pairs of natural numbers and that, at the end and with a jump from pairs to some infinite sequences of rational numbers, makes of each equivalence's class of such sequences a real number (set $\mathbb{R}$). Not only real numbers are very useful in maths and outside them, but many more concepts arise after the real number is constructed in such a classificatory way, like, for instance, the two classifications of $\mathbb{R}$ in rational/irrational, and in algebraic/transcendental, from which a big amount of useful certified knowledge arises. Some hints on how all that happened are the following.

The just described process for knowing through classifying came along a large period of time in which the same concept of number suffered changes. For instance, irrational numbers were not seen as actual numbers by some old Greek thinkers for whom numbers were only the rational ones. Also, the natural number concept came from the counting of objects in the real world, the integer from indexing units in scales above and below some point, and the rational ones from systematically fractioning segments. Even for a long time the number 0 was neither known, nor latter on considered as an actual number, and the interest in irrational numbers grown from the necessity of managing expressions composed by roots of rational numbers, essentially in the solution of polynomial equations, as well as from the relevance of some *rare* numbers like π and e. Since along this process, math was involved in many practical problems, those mathematicians who finally constructed the real numbers in just pure mathematical terms and under the deductive procedures characterizing math, and including in it integers, rational, and irrationals, were strongly influenced by that large history. A hint on this influence is shown, for instance, by the names assigned to some classes of natural numbers: prime, quadratic, cubic, friend, etc., not to speak of the concept of a complex number coming from the so-called imaginary numbers.

The above-mentioned equivalences allowing to pass from $\mathbb{N}$ to $\mathbb{Z}$, $\mathbb{Z}$ to $\mathbb{Q}$, and $\mathbb{Q}$ to $\mathbb{R}$, are classifications in some sets that, at each case, are derived from the former one and that, once well formally constructed, and hence being of an increasingly abstract character and named with more or less common names, all of them are based, at the end, on the Peano's definition of $\mathbb{N}$. This definition makes $\mathbb{N}$ a very intriguing set, which curious and often surprising certified properties generated one of the more complex branches of maths, Number Theory, in which study many concepts of mathematical analysis and probability theory are used. Let us just remember the sophisticated proof that changed the old Fermat's conjecture into a theorem, that is, into certified mathematical knowledge. Before deductively proven, mathematical conjectures, and in particular those in Number Theory, are but speculations well based on many positive instances.

Even accepting that math is freely created by mathematicians from some accepted minimal number of noncontradictory and independent axioms, and that the only way of certifying its knowledge is by deductive proof, it should not be forgotten that to imagine what at each case can be deduced from the premises, and that is often done by analogy with a previously solved, or at least considered, case, is a sample of commonsense reasoning. In addition, the *beauty* mathematicians attribute to their results not only play an important role in the development of math, but also show how mathematicians do reason same as cultured people do. A nice example of the fact that mathematicians also do reason through commonsense reasoning, is shown by the typical statement *How beautiful this (supposed) concept or theorem is*, claimed before a deductive proof fails and avoids to accept it as mathematical knowledge.

Mathematicians such as philosophers, detectives, writers, businessmen, scientists, physicians, etc., reason thanks to their brains and with the experiential background knowledge stored in it. They are moved by curiosity, supported in imagination and conjecturing, and with the will of reaching new knowledge in their respective field. In addition, a high level of creativity seems to be a remarkable characteristic of great mathematicians. All that originated *Wigner*'s famous *unreasonable effectiveness of mathematics in the natural sciences* [16.12].

A way for obtaining a *positive confirmation against reality* of a reasoning, can be searched for through those conclusions showing a good level of agreement with the reality to which the information refers to, and a *negative* one through the refutations, or conclusions contradicting either the premises, or what necessarily follows from them. With respect to explanations, or hypotheses, from them not only necessarily should follow the premises, but also all that necessarily follows from them. Of course, all that can be qualified as necessary should not be only as *safe* as the premises could be, but obtained by means of precise rules allowing anybody to totally reproduce the processes going from the premises to the conclusions. In this sense, perhaps it could be better said that necessary conclusions are *safe* in the context of the premises, but that contingent conclusions are *unsafe*, and that the first should be obtained in a way showing that they are just reproducibly deployed from the premises. It seems clear enough that confirmation with reality should be searched, in general, by means of some theoretic or experimental testing of the conclusions against the reality to which they refer to.

With respect to premises and to ensure its safety, its set should be bound to show some internal consistency, like it is for instance to neither contain contradictory pairs of them, nor self-contradictory ones, since in such a case it does not seem acceptable that the premises can jointly convey information that could be taken as admissible in the model. The same happens with the set of necessary conclusions directly deployed from the premises, since in this case the existence of contradictory conclusions will delete its necessity. The case of contingent conclusions is different, as it comes from everyday experience, since the existence of contradictory explanations or contradictory speculations is not only not surprising at all, but it is sometimes the case of having contradictory hypotheses or speculations for or from the same phenomenon.

All that, if only expressed by words or linguistic terms, cannot facilitate by itself a clear, distinct, and complete comprehension on the subject of reasoning with fertility enough to go further. To increase the comprehension of the machinery of reasoning is for what a modeling of it in black and white [16.14] can be a good help toward a better understanding of what it is, or surrounds it, like mathematical models offer in experimental sciences from, at least, Newton's time. Of course, to establish a mathematical model for reasoning it is indeed necessary not only to be acquainted both with what reasoning is, and its different modalities, but to have a suitable frame of representation for all the involved linguistic terms, including the concept of contradiction.

The concept of representation in a suitable *formal frame* is essential for establishing models and, jointly with the use of the deductive (safe) reasoning the formal frame makes possible, is what not only marks an important frontier between Science and Philosophy, but also helps us to show the *unreasonable effectiveness of mathematics* in Science and Technology. For instance, like the set of rational numbers is a good enough formal frame for the shop's bill, the three-dimensional real space is a formal frame for 3D Euclidean geometry, the four-dimensional Riemann space is that for Relativity theory, and the infinite-dimensional Hilbert space is the frame for quantum physics. Mathematical models add to the study of reality the gift consisting in the possibility of systematically applying to its analysis the safest form of reasoning, that is, formal deduction.

When the linguistic terms translating the corresponding concepts are precise and, at least in principle, all the information that eventually can be needed is supposed to be available, like it is the case in the classical propositional calculus, the frame of Boolean algebras seems to be well suitable for representing these terms [16.3]. If the linguistic terms designating the involved basic concepts are precise but not all the information is always available, like it happens, for instance, in quantum physics, weaker structures than Boolean algebras, like orthomodular lattices are, could be taken into account. If there are involved essentially imprecise linguistic terms, like it happens in commonsense reasoning, then the so-called algebras of fuzzy sets seem to be suitable once the linguistic terms are well enough designed by fuzzy sets [16.15, 16].

The problem of selecting a convenient frame of representation for ordinary reasoning, defined by a minimal number of axioms, is a crucial point that should be established in agreement with both the methodological principle the Occam's Razor states by *Not introduce more entities than those strictly necessary*, and with the Menger's addenda, *Nor less than those with which some interesting results can be obtained* [16.17]. Such methodological principle and addenda are, of course, taken into account in what follows since, on the contrary, almost nothing could be added to which philosophers said. Two important features that *Menger*'s *exact thinking* [16.6] offers through mathematical models are that it is always clear in it under which presuppositions the obtained results can hold, that deductive (safe) reasoning can be extensively used through a mathematical symbolism translating the basic treats of the subject, and that what is not yet included in it has a possibility of, at least, being clearly situated outside the model and, perhaps, latter on included in a new and larger model. This is what, at the end, happened with the old Euclid's Geometry and Linear Algebra.

16.3.2 A Remark on Medical Reasoning

The field of medicine [16.18], that is full of imprecise technical concepts, is one in which the ordinary reasoning used in it, deserves a careful consideration. In particular, it is important to know if a medical concept can or cannot be specified by a classical set, in the negative case, since it is not possible to conduct the corresponding reasoning in the classical Boolean frame. For instance, if two concepts D and B, are interpreted as fuzzy sets, the statement $((D$ and $B)$ or $(D$ and not $B))$ is only equivalent to D in some standard algebras of fuzzy sets in which no law of duality holds [16.19].

In the typically clinical reasoning for diagnosing, there are many technical concepts that cannot be considered precise by being clearly subjected to degrees. It is hence the concept of *observed diabetes* that can be submitted to a (empirical) Sorites' process [16.20] to conclude that it cannot be represented by a classical set. For instance, a patient with $100 \, \text{mg/dl}$ of glucose in blood does not suffer diabetes, as well as with 101, 102, ..., and up to $120 \, \text{mg/dl}$, in which moment the patient could be diagnosed with diabetes. Nevertheless, since the crisp mark 120 is not liable in all cases by being a somehow changing experimental threshold, it is better to frame the diagnose in the setting of fuzzy logic and by also taking into account the weight, the kind of job, the age, as well as the usual alimentation of the patient. That is, the physicians do reason on the basis of the complex imprecise predicate that could be written by: (Diabetes/Patient) (p) $\sim$ glucose in blood (p) & weight (p) & job (p) & age (p) & alimentation (p), at its turn composed by *elemental* predicates, some of which are also imprecise.

The physicians cannot reason by only taking into account the amount of glucose in blood, and under the typical schemes of classical reasoning. Once the current medical concepts are translated into fuzzy terms, it is necessary to follow the reasoning under those schemes allowed in a suitable algebra of fuzzy sets [16.19], and once it is designed accordingly with the context where the concepts are inscribed in.

In addition, since what the processes conducting to diagnose try to find is a good enough hypothesis matching with the symptoms of the presumed illness, it is relevant for the researchers on medical reasoning to know about inductive abduction and speculation [16.18], and, mainly, for what respects to applying CWW. For all that the *laws* holding in the framework taken for representing the involved fuzzy terms is relevant.

16.4 Reasoning and Logic

Let us stop for a while at the question *Which is the relationship between reasoning and logic?* requiring to first stop at the concept of logic, classically understood as *the formal study of the laws of reasoning* and that, in addition and modernly, is basically understood by restricting reasoning to deduction. Today's logic is indeed the study of systems allowing the safest type of reasoning under which, and from a consistent set of premises, a consistent set of necessary conclusions, or logical

consequences, is derived in a step-by-step ruled process that can be fully reproduced by another person who masters the use of rules. Hence, nowadays logic consists in the study of the so-called deductive systems, and it mainly avoids other types of reasoning like those by analogy, by abduction, and by induction. After *Tarski* formalized deductive systems by means of consequence operators [16.21], *a logic* is defined, in mathematical terms, as a pair consisting in a set of statements and a consequence operator that can be applied to some (selected) subset of these statements, and once all that is represented in a formal frame.

Nevertheless, it seems that in commonsense, everyday, or ordinary reasoning, people only make deductions in, at most, a 25% of the cases [16.22], and that many conclusions are just reached either by the help of analogy from a precedent and similar case, or just by speculating at each case accordingly with some rules of thumb. In addition, some properties and schemes of reasoning that were classically considered like *laws of reasoning*, today cannot be seen as universally valid as it is with the distributive law in the reasoning of quantum physics, and with the several schemes of reasoning with fuzzy sets studied in [16.4] in the line of analyzing what is sometimes known as the preservation of the *Aristotelian form*.

Consequently, the analysis of more than 75% of nondeductive reasoning processes should be considered of an upmost importance for a more complete study of reasoning. In some sense, such study began with the work of *Peirce* [16.23] for understanding scientific

thinking, and latter on was continued with the studies on nonmonotonic reasoning in the field of Artificial Intelligence for helping to mechanize some ordinary ways of reasoning [16.24, 25]. Since *Computing with Words* deals with ordinary reasoning in natural language, it seems obvious that the nondeductive ways of reasoning should be not only in the back of CWW, but also taken into account in it.

Toward a formal, algebraic, study of such 75% is mainly devoted to this chapter that, essentially, can be seen as a trial for enlarging logic from formal deductive to everyday reasoning with, perhaps, a kind of returning to Middle Age's logic, as it can be considered from the Occam's saying [16.26] that *demonstration is the noblest part of logic*, reflecting that logic was seen in that time as more than the study of deduction, even if it is considered a crucial form of reasoning. In the model presented in this chapter, based on *conjectures* (a term coming from [16.27], deduction, as a modality of reasoning, plays a central role both in a weak and a strong form, respectively, corresponding to the formal, and to the ordinary ways of reasoning.

16.5 A Possible Scheme for an Algebraic Model of Commonsense Reasoning

It does not seem that the term *deduction* can refer to the same concept in formal and in ordinary reasoning, since in the first it appears more strict than in the second where, for instance, the conclusions (consequences) are not necessarily admissible like the premises. Think, for instance, on what basis philosophers consider is a deduction, and on what mathematicians refer to by a proof. Anyway, in both cases it should try to reflect a safe enough kind of reasoning, in the sense of attributing to the conclusions no less confidence than that attributed to the items of initial information. These items should also be admissible in the sense of being as safe as possible knowledge on some subject. The *good quality* of initial information is actually important in any process of reasoning.

Mathematics is considered the paradigm of deductive reasoning, but it does not mean (as it is remarked in Sect. 16.3) that mathematicians only reason deductively since when they search for something new they do reason like other people do, often by doing jumps from the initial information to unwarranted conclusions [16.22]. What is not at all accepted in mathematics are contradictions, and nondeductive proofs.

If for mathematicians the mathematical model, based on the admissibility of its axioms, is their reality, then for applied scientists or for engineers a model is just a representation of some reality. For instance, no engineer confuses the actual working of a machine with a dynamic model of it and, when launching a rocket, it is well known that the so-called *nominal* trajectory (computed from a mathematical model), is not exactly coincidental with the actual one, and the performance of the rocket's propulsion system is measured by taking into account the difference between these trajectories. In commonsense reasoning (CR) the situation yet shows currently sensible differences with these cases.

Some characteristics separating commonsense from formal deductive reasoning are as follows:

a) CR does not consist in a single type of reasoning, but in several. A reasoning in CR can be schematized by $P \vdash q$, where P is a set of admissible items of information, q is a conclusion under the considered type of reasoning, and the symbol $\vdash$ reflects the corresponding reasoning's process. Only in the case of a deductive reasoning these processes are done under a strict regulation.

Alternatively, if AC(P) reflects the set of attainable conclusions, the scheme can be changed to $q \in$ AC(P).

b) Often the items of information from which CR starts are expressed in natural language, with precise and imprecise linguistic terms, numbers, functions, pictures, etc. In addition and also often, such items of information are partial and/or partially liable with respect to the reality they are concerned with. In what follows, it will be supposed that no ambiguous terms are contained in these items, and that they are expressed by linguistic statements.

c) Often CR lacks monotony. That is, when the number of initial information items increases, then either the number of conclusive items decreases (anti-monotony), or there is no law for its variation (non monotony). Deduction is always monotonic, that is, no less conclusive items are obtained when the number of items of initial information grows.

d) It is typical of CR to *jump* from the initial information to some conclusions, that is, that no step-by-step/element-after-element way can be followed. Jumping is never the case in formal deduction, where the conclusions should be deployed from the initial items of information in a strict step-by-step manner and under previously known rules, even if the current *reasoner* avoids some trivial steps that, nevertheless, always can be easily recovered.

e) In CR, people try to obtain either explanations, or refutations, or what is hidden in the given information, or new ideas lucubrate from what it is supposedly known (the initial information). Only the third of these kinds of conclusions are typical of deduction.

f) A minimal limitation in CR is that of keeping some kind of *consistency* among the given items of initial information, like it is not containing two contradictory such items, and also between them and the conclusions.

Let us denote by P an accepted set of *items of admissible initial information* (premises), and by AC(P) the attainable conclusions under one of the last four types of CR in (e). It will be supposed that P is in some designated family F of sets able to consistently describe something, and that AC(P) is in a larger family C such that F ⊂ C. Hence, AC can be seen as a mapping AC: F → C. The sets in F are supposed to contain items of

admissible information, but this is not the case for those sets in C − F. Then:

a) To do an analysis of CR in mathematical terms, an algebraic frame for representing all that is involved in CR should be selected in a way of not introducing more objects and laws than those strictly necessary at each case like they are, for instance, a symbolic representation of the linguistic connectives *and, or, not,* and *If/Then.* The symbols that will be used in this chapter are $., +, ',$ and $\leq$, of which the first two are binary operations, the third is a unary operation, and the fourth is a binary relation.

b) Basic in all kind of reasoning is the concept of *consistency* even if it is not a unique way of seeing it. Three possible definitions of consistency are as follows:

- Consistency is identified with noncontradiction: If p, then it is never not-q, symbolically represented by $p \not\leq q'$.
- Consistency is identified with joint noncontradiction: If (p and q), then it is never not-(p and q), symbolically represented by $p.q \not\leq (p.q)'$.
- Consistency is identified with incompatibility: It is never (p and q), symbolically represented by $p.q = 0$, provided there exist a symbol 0 like in set theory is the empty set Ø.

At each case a suitable definition of consistency should be chosen accordingly with the corresponding context but, in what follows is just taken the first one. Of course, no other and less formal ways of seeing at consistency should be excluded and, in any case, the concept of consistency between pairs of elements should be extended to sets of premises and sets of conclusions to make them consisting, respectively, in admissible premises and attainable conclusions.

Notice that in a Boolean algebra, since it is $p.q = 0 \Leftrightarrow p \leq q' \Leftrightarrow p.q \leq (p.q)'$, the former definitions are equivalent, and, for this reason in the reasoning with precise linguistic terms there is no discussion for what refers to the concept of consistency.

c) p is *contradictory* with q, provided *If p, then not-q,* and p is *self-contradictory* when *If p, then not-p.* Notice that in ortholattices the only self-contradictory element is $p = 0$, and that with fuzzy sets endowed with the negation 1-id, the self-contradictory fuzzy sets are those such that $A \leq A' = 1 - A \Leftrightarrow A(x) \leq \frac{1}{2}$, for all $x \in X$.

d) If $P \subset AC(P)$, for all P in F, it is said that AC is *extensive*, and $AC(P)$ necessarily contains some items of admissible information.

e) If $P \subset Q$, both in F, then if $AC(P) \subset AC(Q)$, it is said that AC is *monotonic*. If $AC(Q) \subset AC(P)$, *antimonotonic*, and if AC is neither monotonic, nor antimonotonic, it is said that AC is *nonmonotonic* and there can exist cases in which, being $P \subset Q$, $AC(P)$ and $AC(Q)$ are not comparable under the set-inclusion $\subset$.

f) Provided q represents a statement, and q' represents its negation not-q: If q is in $AC(P)$, then q' is not in $AC(P)$, or q' is in $AC(P)^c$, it is said that AC is *consistent in* P. AC is just *consistent* if it is consistent in all P in F.

g) AC is said to be a *closure*, if for all $P \in F$ it is $AC(P) \in F$, and $AC(AC(P)) = AC^2(P) = AC(P)$.

h) $AC(P) \in C - F$ means that not all the elements in $AC(P)$ show the characteristics that make admissible those items in F.

Main Definitions

1) A mapping $AC : F \rightarrow C$ is said to be a weak-deduction operator [16.28] if it is monotonic, and consistent under a suitable definition.

2) A mapping $AC : F \rightarrow F$ is said to be a strong-deduction operator, or a Tarski's logical consequence operator [16.21], if it is a weak deduction one that is also extensive, and is a closure.

If AC is a weak-deduction operator, the elements in $AC(P)$ are called weak consequences of P. If AC is a Tarski's operator, the elements in $AC(P)$ are the strong or logical consequences of P. Since logicians universally consider that Tarski's operators translate the characteristics of formal deductive systems, or formal deduction, it will be here considered that weak-deduction operators translate those of (some kind of) commonsense deduction.

Remarks 16.1

a) Notice that, in the model, F represents the family of those sets whose elements are accepted as items of admissible initial information, and that such *admissibility*, once translated into a suitable definition of consistency, should be defined at each case. Hence, each time F should be conveniently chosen. For instance, a possible definition is: F is the family of those P for which there are no p and q in it and such that *If p, then not-p*.

b) At each case, it should be defined to which ground-set W the sets F and C are included in, and W should be endowed with operations able to represent all that is necessary for the formalization of CR. For instance, it will be supposed that there is a binary relation $\leq$ in W such that $p \leq q$ translates into W the linguistic statement *If p, then q*. Analogously, and in the same vein in which $'$ represents the linguistic *not*, there should be binary operations ., and +, representing, respectively, the linguistics *and*, and *or*.

c) Basic in CR is the idea of *conjecture*, and that of *refutation*. Once a weak or strong consequence operator AC is adopted, the refutations of P could be defined as those elements $r \in W$, such that $r' \in AC(P)$, that is those whose negation is deducible from P. At its turn, the conjectures from P can be defined as those elements $q \in W$ such that $q' \notin AC(P)$, that is, those whose negation is not deducible from P. In this sense, the conjectures are the elements that are not (deductive) refutations of P. Both concepts could be more precisely named AC-refutations, and AC-conjectures.

Since both precise and imprecise linguistic terms are usually managed in CR, in what follows W will be the set of all fuzzy sets in a universe of discourse X, that is, $W = [0, 1]^X$, the set of all functions $A : X \rightarrow [0, 1]$. This set will be endowed with the algebraic structure of a Basic Fuzzy Algebra, where the restriction of its operations to $\{0, 1\}^X$ makes this set a Boolean algebra, isomorphic to the power set 2^X endowed with the classical set-operations of intersection, union, and complement of subsets. Crisp sets allow us to represent precise linguistic terms as it is stated by the axiom of Specification in naïve set theory [16.29], an axiom that cannot be immediately extended to imprecise predicates since, for instance, they are not always represented by a single fuzzy set.

Definition 16.1 [16.1]

If .and + are binary operations and $'$ is a unary one, then $([0, 1]^X, ., +, ')$ is a Basic Fuzzy Algebra (BFA) provided it holds,

1. $A_0.A = A.A_0 = A_0, A_1.A = A.A_1 = A, A_0 + A = A + A_0 = A, A + A_1 = A_1 + A = A_1$

2. If $A \leq B$, then $C.A \leq C.B, A.C \leq B.C, C + A \leq C + B, A + C \leq B + C$, and $B' \leq A'$.

3. $A'_0 = A_1$, and $A'_1 = A_0$.

4. If $A, B \in \{0, 1\}^X$, then $A.B = \min(A, B)$, $A + B = \max(A, B)$, and $A' = 1 - A$,
where A_0 is the function $A_0(x) = 0$, A_1 is the function $A_1(x) = 1$, and $1 - A$ is $(1 - A)(x) = 1 - A(x)$, for all $x \in X$. Obviously, in 2^X, A_0 represents the empty set $\emptyset$, A_1 the ground set X, and $1 - A$ the complement A^c of A.
Of course, it is $A \leq B$ if and only if $A(x) \leq B(x)$, for all x in X, a partial order that with crisp sets reduces to $A \subset B$.

Notice that:

a) The *formal connectives*., $+$, and $'$, in a BFA are neither presumed to be functionally expressible, nor associative, nor commutative, nor distributive, nor dual, etc.

b) Only if . $= \min$, and $+ = \max$, the BFA is a lattice that, if the negation $'$ is a strong one ($A'' = A$, for all A), is a De Morgan–Kleene algebra. Hence, no BFA is a Boolean algebra, and not even an ortholattice.

c) It is not difficult to prove [16.28] that it is always $A.B \leq \min(A, B) \leq \max(A, B) \leq A + B$. Of course, the standard algebras of fuzzy sets are particular BFAs.

d) It is also easy to prove that:
   d.1) In a BFA with $+ = \max$, it holds the first law of semiduality: $A' + B' \leq (A.B)'$, regardless of which are . and $'$.
   d.2) In a BFA with . $= \min$, it holds the second law of semiduality: $(A + B)' \leq A'.B'$, regardless of which are $+$ and $'$.
   d.3) Regardless of $'$, in a BFA with . $= \min$, and $+ = \max$, both semiduality laws hold.

e) Obviously, all standard algebras of fuzzy sets [16.30] (those in which . is decomposed by a continuous t-norm, $+$ by a continuous t-conorm, and $'$ by a strong negation function) are BFAs.

Since BFAs are defined by just a few axioms in principle only allowing very simple calculations, what can be proven in their framework has a very general validity that is not modified by the addition of new independent axioms. One of the weaknesses of Boolean and De Morgan algebras, as well as of orthomodular lattices, for representing CR, just lies in the big amount of laws they enjoy and make them too rigid to afford the flexibility natural language and CR show in front of any artificially constructed language, and of formal reasoning.

Remarks 16.2

1. For what concerns the representation of a linguistic predicate L in a universe of discourse X by a fuzzy set, it is of an actual interest to reflect on what can mean the values $A_L(x)$, for $x \in X$ and $A_L : X \to [0, 1]$, the membership function of a fuzzy set labeled L. Just the expression *fuzzy set labeled* L, forces that the membership function A_L should translate something closely related to L, namely , to the meaning of L in X. It is not clear at all that all predicates can be represented by a function taking its values in the totally ordered unit interval of the real line: It should be added to the involved predicate the possibility of some numerical quantification of its meaning.

To well linguistically manage a numerically quantifiable predicate L in X it should, at least, be recognized when it is, or it is not, the case that x *is less P than* y, a linguistic (empirical and perceptively captured) relationship that can be translated into a binary relation $x \leq_L y$, with $\leq_L \subset X \times X$. This relation reflects how the *amount of L* varies on X, and once the pair $(X, \leq_L)$ is known, a measure of the extent up to which each $x \in X$, is L, is a mapping $M_L : X \to [0, 1]$ such that $x \leq_L y \Rightarrow M_L(x) \leq M_L(y)$, and those elements x such that $M_L(x) = 1$, if existing, can be called the prototypes of L in X. Analogously, those y such that $M_L(y) = 0$ can be called the antiprototypes of L in X. Obviously, and in the same vein that there is not a single probability measuring a random event, there is not always a single measure M_L.

If the use of the predicate is precise in X, all their elements x should be prototypes, or antiprototypes, that is $M_L(x)$ is in $\{0, 1\}$. When for some x it is $0 < M_L(x) < 1$, it is said that the use of L is imprecise in X. Once a triplet $(X, \leq_L, M_L)$ is known, it is a quantity that can be understood as reflecting the meaning of L in X [16.2]. Calling M_L an *ideal membership function* of the fuzzy set labeled L, it can be said that it exists when the meaning of L in X is a quantity.

Notice that each measure M_L defines a new binary relation given by $x \leq_{ML} y \Leftrightarrow M_L(x) \leq M_L(y)$, obviously verifying $\leq_L \subset \leq_{ML}$, that is, the new relation is larger than the former that is directly drawn from the perceived linguistic behavior of L in X.

It is said that M_L *perfectly reflects* L whenever $\leq_L = \leq_{ML}$, but, since the second is always a linear relation – for all x, y in X, it is either $M_L(x) \leq M_L(y)$, or $M_L(x) > M_L(y)$ –, and $\leq_L$ is not usually so, not always can be the case that M_L perfectly reflects L. This is one of the reasons for which, being $\leq_L$ often difficult to be completely known – for instance, if X is not finite – the designer just arrives to a function A_L (the membership function of the fuzzy set labeled L) that is not usually the ideal membership function M_L but an approximation of it, obtained through the data on L that are available to the designer. Of course, a good design is reached when it can be supposed that the value Sup $\{x \in X; /M_L(x) - A_L(x)/\}$ is minimized.

From all that it comes the importance of carefully designing [16.15, 16] the membership functions with which a fuzzy system is represented. Analogous comments can be made for what concerns the connectives., $+$, $'$, and the *axioms* they verify, to reach an election of the BFA ($[0,1]^X$, ., $+$, $'$) well linked to the currently considered problem.

2. The suitability of the wide structure of BFAs for representing CR comes from, for instance, the fact that the linguistic conjunction *and* is not always commutative specially when *time* intervenes, the laws of duality are not always valid when dealing with statements in natural language, the connectives' decomposability (or functional expressibility), is not always guaranteed, the distributive laws between . and $+$ not always hold, etc.

3. When the linguistic terms are represented in a set W endowed with a partial order $\leq$, and with operations . of conjunction, $+$ of disjunction, and $'$ of negation, respectively, representing the linguistic connectives If/Then, and, or, not, if (W, ., $+$) is a lattice (for all that concerns lattices, see [16.31]) at least the following five points do hold:

   I. It is $A \leq B \Leftrightarrow A.B = A \Leftrightarrow A + B = B$: The conditional statement *If A, then B* should be equivalent to the statements *A and B coincides with A*, and *A or B coincides with B*.

   II. If *and* is represented by the lattice's conjunction ., $A.B$ is the greatest lower bound of both A and B: It should be known the set of all that is below A (C is below A means $C \leq A$), the set of all that is below B, the intersection of these two sets, and that $A.B$ is the greatest element in this last set (respect to the partial order $\leq$).

   III. Analogously, for the case of *or*, if represented by the lattice's disjunction $+$, there should be known the sets of elements in W that are greater than A, those that are greater than B, their inter-

section, and that $A + B$ is the lowest element in this last set.

   IV. $A.B = B.A$: The meaning of the statements *A and B*, and *B and A* cannot be different.

   V. $A + B = B + A$: The meaning of the statements *A or B*, and *B or A* cannot be different.

All this shows that contrary to what usually happens in both CR and the applications, where all the previous information is not only costly in searching for, in money, and almost impossible to collect completely, a lot of structural information on the reasoning's context should be necessarily known for establishing the model. Something that is typical in formal sciences, but that in the case of CR, and also in many applications, produces some scepticism for the possibility of always taking (W, ., $+$) as a lattice.

To count with a representation's lattice for CR is but something to be considered rare or, at least, limited to some cases as it can be that of representing a formal-like type of reasoning with precise linguistic terms where the former five points are usually accepted. These are, for instance, the cases in Boolean algebras with $A \to B = A' + B = 1 \Leftrightarrow A \leq B$, and orthomodular lattices with $A \to B = A' + A.B = 1 \Leftrightarrow A \leq B$, with the respective implication operators $\to$ translating the corresponding linguistic If/Then.

With the standard algebras of fuzzy sets, a lattice is only reached when the connectives are given by the greatest *t*-norm min, and the lowest *t*-conorm max [16.30]. In this case, the *implication functions* with which it is $A \to B = A_1 \Leftrightarrow A \leq B$, are the T-residuated ones [16.30], functionally expressible through the numerical functions $J_T(a,b) = \text{Sup}\{r \in [0,1]; T(a,r) \leq b\}$, where T is a left-continuous *t*-norm, and that generalize the Boolean material conditional $A \to B = A' + B$ since, in a complete Boolean algebra, it is $A' + B = \text{Sup}\{C; A.C \leq B\}$. The fuzzy implications given by $(A \to B)(x,y) = J_T(A(x), B(y))$ enjoy many of the typical properties of Boolean algebras with the material conditional, and with them the standard algebra with min, max, and the strong negation 1-id, enjoys, among these algebras, the biggest amount of Boolean laws and makes of it a very particular algebra to be used for extensive use in CR.

Nevertheless, it should be remembered that what concerns CR, when time intervenes not always can coincide the meanings of the statements *A and B*, and B and A, as it is the case *He sneezed and came to bed*, and *He came to bed and sneezed*.

4. The BFA's structure, based on $[0, 1]^X$, can yet be made more abstract. It simply requires to consider, instead of $[0, 1]^X$, once pointwise ordered by "$A \leq B \Leftrightarrow A(x) \leq B(x)$, for all x in X, a poset $(L, \leq)$, with minimum 0 and maximum 1, endowed with two binary operations . and $+$, and a unary one, containing a subset L_0 ($\{0, 1\} \subseteq L_0 \subseteq L$) that, with the restrictions of the three operations ., $+$, and ', is a Boolean algebra, and verifying anal-

ogous laws to the former in 1 to 4. These algebraic structures are called [16.32] Formal Basic Flexible Algebras, and they are a shell comprising ortholattices, De Morgan algebras, BFAs and, of course, orthomodular lattices, Boolean algebras, and standard algebras of fuzzy sets, as particular cases. By taking (W, ., $+$, ') as a Formal Flexible Algebra, what follows can be generalized, with a few restrictions, to such abstract and general shell.

16.6 Weak and Strong Deduction: Refutations and Conjectures in a BFA (with a Few Restrictions)

Let ($[0, 1]^X$, ., $+$, ') be a BFA whose negation ' is a weak one, that is, restricted to verify the law $A \leq A''$ for all $A \in [0, 1]^X$, and whose conjunction . is associative, and commutative. No other properties are presumed and, hence, what follows contains by large the case with a standard algebra of fuzzy sets, and what can be obviously restricted to the Boolean algebra of crisp sets in $\{0, 1\}^X$.

Let us consider as the former family F of *admissible* premises, the F(.) comprising the finite sets (of premises) $P = \{A_1, \ldots, A_n\}$, such that their *conjunction* $A_P = A_1 \ldots A_n$ is not self-contradictory, that is, $A_P \nleq A_P'$. Of course, $A_P \nleq A_P'$ implies $A_i \nleq A_j'$, for all A_i, A_j in P, and, obviously, it should be $A_P \neq A_0$. Hence, sets $P \in F(.)$ neither contain contradictory premises, nor the *empty set* A_0. Notice that associativity and commutativity of the conjunction are presumed just to warrant a nonambiguous definition of A_P, and the restriction on the negation ' is just to allow some step in a proof. Under these conditions, the operator defined [16.28] by

$$C.(P) = \{B \in [0, 1]^X; A_P \leq B\},$$

translating into the BFA the statement *If A_1 and A_2 and ... and A_n, then B*, or *B follows from A_P in the order $\leq$*, verifies the following:

a) Usually, C.(P) is not in F(.), for instance, it is not always finite. Hence, in general it has no sense to reapply the operator C. to C.(P). That is, the operator $C.^2$ cannot be usually defined, and less again to make C. a closure.

b) Since $A_P \leq A_i$, $1 \leq i \leq n$, it is $P \subset C.(P) : C.$ is extensive.

c) If $P \subset Q$, with $Q = P \cup \{A_{n+1}, \ldots, A_m\}$, and since $A_P \leq A_1 \ldots A_n.A_{n+1} \ldots A_m$, then it follows $C.(P) \subset C.(Q) : C.$ is monotonic.

d) If $B \in C.(P)$, it is not $B' \in C.(P) : A_P \leq B$ and $A_P \leq B' \Rightarrow B \leq B'' \leq A_P'$, and it follows the absurd $A_P \leq A_P' : C.$ is consistent in all $P \in F(.)$.

e) Obviously, $A_1 \in C.(P)$, but $A_0 \notin C.(P)$. Hence, $C.(P) \neq \emptyset$. Analogously, $C.(P)$ cannot coincide with the full set $[0, 1]^X$ since it will imply the absurd $A_P = A_0$.

Consequently, all operators C. are consistent and extensive weak-deduction operators.

In addition,

f) If $C.(P)$ is not finite, then it is obviously Inf $C.(P) = A_P$.

g) No contradictory elements are in $C.(P)$: If $B, C \in C.(P)$ and it were $B \leq C'$, from $A_P \leq B$, $A_P \leq C$, it follows $B' \leq A_P'$ and $C' \leq A_P'$, and the absurd $A_P \leq B \leq C' \leq A_P'$.

h) If $._1 \leq ._2$, that is, the operation $._1$ is weaker than the operation $._2$, it is obviously $A_1._1 \ldots ._1 A_n \leq A_1._2 \ldots ._2 A_n$, and hence $C._2(P) \subset C._1(P)$. That is, the bigger the operation ., the smaller the set C. (P). Consequently, if it can be selected the operation min, it is Cmin (P) $\subset$ C.(P), for all operations., and all $P \in F(min)$: Cmin is the smallest among the operators C. .

Notice, that it is always F(min) $\subset$ F(.): the family with min is the smallest among those of admissible premises.

i) Provided $C.(P)$ is a finite set, and since C. is extensive, it is $C.(P) = P \cup \{A_{n+1}, \ldots, A_m\}$, and then $A_{C.(P)} = A_1 \ldots A_n.A_{n+1} \ldots A_m = A_P.A_{n+1}$

$\ldots A_m \leq A_P$. Thus, if $. = \min$, $A_{C.(P)} = \min(A_P, \min(A_1, \ldots, A_m)) = A_P$, that means $C\min(P) \in F(\min)$, and has sense to reapply Cmin. Since $P \subset C\min(P) \Rightarrow C\min(P) \subset C\min(C\min(P))$, and, if $A \in C\min(C\min(P))$, then $A_{C\min(P)} = A_P \leq A$, it is $A \in C\min(P)$, and from $C\min(C\min(P)) \subset C\min(P)$, it finally follows $C\min^2(P) = C\min(P)$. In conclusion, provided all the involved sets Cmin (P), with $P \in F(\min)$, were finite, it is $C\min : F(\min) \to F(\min)$, and Cmin will be a strong, or Tarski's, consequence operator.

i) Provided the family F of sets of premises is made free of only containing finite sets, since it always exist Inf and, bounding P to verify $\text{Inf } P \not\leq (\text{Inf } P)'$, for all $P \in F$, the operator

$$C_\infty(P) = \{B \in [0, 1]^X ; \text{Inf } P \leq B\},$$

obviously verifying Inf $C_\infty(P) = \text{Inf } P$, is not only extensive, consistent, and monotonic, but it is also a closure: From $C_\infty(P) \in F$, and $P \subset C_\infty(P)$, it follows $C_\infty(P) \subset C_\infty(C_\infty(P))$, but if $B \in C_\infty(C_\infty(P))$, that is, Inf $C_\infty(P) \leq B$, from Inf $C_\infty(P) = \text{Inf } P$, follows $B \in C_\infty(P)$. Finally, $C_\infty(C_\infty(P)) = C_\infty(P)$. C_∞, restricted to finite sets is just Cmin, and restricted to all the crisp sets in $\{0, 1\}^X$, is the consequence operator on which classical propositional calculus is developed [16.3].

Definition 16.2

Given a weak-deduction operator C. [16.28]:

1. The set of C.-refutations of P, is Ref.(P) = $\{B \in [0, 1]^X ; B' \in C.(P)\}$.
2. The set of C.-conjectures of P, is Conj.(P) = $\{B \in [0, 1]^X ; B' \in C.(P)^c\}$.

Namely, refutations are those fuzzy sets whose negation is weakly deducible from the premises, and conjectures those whose negation is not weakly deducible from them. Obviously, Conj.(P) = Ref.(P)^c.

Notice that it immediately follows that all operators Ref. are consistent, and monotonic, but not extensive, and that all operators Conj. are extensive, antimonotonic, not consistent, and consequently it cannot be stated that Conj.(P) is always in F(.). It is Conj.: $F(.) \to [0, 1]^{X.}$, and not all conjectures can be taken as items of admissible information.

It is also immediate that Ref.(P) $\cup$ Conj.(P) = $[0, 1]^X$, and Ref.(P) $\cap$ Conj.(P) = $\emptyset$, that is, both sets

constitute a partition of the set of all fuzzy sets in X, and Conj.(P) = Ref.(P)^c. Hence, the conjectures are those fuzzy sets that are nonrefutable in front of the information furnished by P.

Since, C min(P) $\subset$ C.(P), it follows that:

● Ref min(P) $\subset$ Ref.(P): Ref min is the smallest among refutation operators.
● Conj.(P) $\subset$ Conj min(P): Conj min is the biggest among conjecture operators. Namely,

$$P \subset C \min(P) \subset C.(P)$$
$$\subset \text{Conj.}(P) \subset \text{Conj min}(P),$$

a chain of inclusions showing that both weak and strong consequences are but a particular type of conjectures. Consequently, in the model deducing is but one of the forms of conjecturing as it is asked for in [16.27].

Remarks 16.3

1. Only if it is $. = \min$, it holds: Ref min (C min(P)) = Ref min(P), and Conj min (C min(P)) = Conj min(P), showing that strong consequences neither allow to obtain more refutations, nor more conjectures.
2. Since $A_0 \in \text{Ref.}(P)$, it is Ref.(P) $\neq \emptyset$. Nevertheless, Ref.(P) cannot coincide with the full set $[0, 1]^X$, since it will imply $A = A_0$.
3. The sets Conj.(P) cannot be empty since it will imply C.(P) = $[0, 1]^X$. On the other side, it is Conj.(P) = $[0, 1]^X \Leftrightarrow$ C.(P) = $\emptyset$. Hence, it is always,

$$\emptyset \not\subseteq \text{Conj.}(P) \not\subseteq [0, 1]^X.$$

In this model, the empty set A_0 cannot be taken for either conjecturing, or deducing, or refuting.

4. The particularization of the concept of conjecture to crisp sets, that is, to the fuzzy sets in $\{0, 1\}^X \subset [0, 1]^X$, reduces to take as F(min) the set of those crisp sets that are nonempty, since with crisp sets it is $A \subset A^c \Leftrightarrow A = \emptyset$, and thus Conj.(P) is the set of those $B \subset X$, such that $A \cap B \neq \emptyset$, since with crisp sets it is $A \subset B^c \Leftrightarrow A \cap B = \emptyset$.
With classical sets, that is, in Boolean algebras, there is no distinction between contradiction, and incompatibility.

5. After the classical definition: B is decidable $\Leftrightarrow$ either B is provable, or not B is provable, it can be defined the set of .-weakly decidable elements for

P by $C.(\text{P}) \cup \text{Ref.}(\text{P})$, and the set of strongly de-cidable elements for P by $C\min(\text{P}) \cup \text{Ref}\min(\text{P})$. Obviously, and for all operations ., strongly decid-able elements are .-weakly decidable ones, but not reciprocally [16.32].

The nonstrongly decidable elements are those fuzzy sets in the set

$$(C\min(\text{P})\text{URef}\min(\text{P}))^c$$
$$= C\min(\text{P})^c \cap \text{Ref}\min(\text{P})^c$$
$$= C\min(\text{P})^c \cap \text{Conj}\min(\text{P}),$$

that is, they are the conjectures that are not strong consequences. Analogously, given P, the .-weakly nondecidable elements are those fuzzy sets that are .-weak conjectures, but not .-weak consequences. Consequently, to obtain a classification of the dif-ference sets $\text{Conj.}(\text{P}) - C.(\text{P})$, and $\text{Conj}\min(\text{P}) - C\min(\text{P})$, is actually important.

6. Since Ref. is monotonic and consistent, it could be alternatively taken for defining conjectures [16.28] in the parallel form

$$\text{Conj} * .(\text{P}) = \{B \in 0, 1^X; B' \notin \text{Ref.}(\text{P})\} = C.(\text{P})^c,$$

where conjectures appear as just those fuzzy sets that are not .-weak consequences instead of those that are not refutations. Under this new definition, it is $\text{Ref.}(\text{P}) \subset \text{Conj} * .(\text{P})$. Notice that .-refutations B can be defined without directly referring to $C.$, by

$$\text{Ref.}(\text{P}) = \{B \in [0, 1]^X; A_P \le B'\}.$$

7. With all that, a tentative formal definition for a model of (Commonsense) Reasoning (CR) could

be essayed by posing *à la Popper* [16.13],

$$\text{CR.}(\text{P}) = \text{Ref.}(\text{P}) \cup \text{Conj.}(\text{P}),$$

once the operations., and $'$ are selected, and $C.(\text{P})$ defined with, at least, $A \ne A_0$. Anyway, this defi-nition gives nothing else than $\text{CR.}(\text{P}) = \text{Conj.}(\text{P}) \cup \text{Conj.}(\text{P})^c = \text{Ref.}(\text{P}) \cup \text{Ref.}(\text{P})^c$, with which the for-mal model for CR appears as nothing else than either *conjecturing*, or *refuting* once .-weak con-sequences are taken as the basic concept toward formalizing CR.

Thus, a new concept of strong-CR, can be intro-duced by

$$\text{CR}\min(\text{P}) = \text{Ref}\min(\text{P}) \cup \text{Conj}\min(\text{P}),$$

and in both cases deduction (weak and strong, re-spectively), is a type of conjecturing [16.27].

8. The imposed commutativity and associativity of the conjunction can be avoided by previously fix-ing an algorithm to define A_P. For instance, if P contains four premises, then the algorithm's steps can be the following: 1) Select an order for the premises and call them $A_1, \dots, A_4$. 2) Define A_P by $A_1.(A_2.(A_3.A_4))$. Then, of course, all that has been formerly said depends on the way chosen to de-fine A_P.

For what refers to the restriction on the negation $'$, notice that it is already verified in the cases (usual in fuzzy logic) where it is strong: $A'' = A$, for all $A \in [0, 1]^X$, as it is the case in the standard algebras of fuzzy sets.

16.7 Toward a Classification of Conjectures

Provided it is $\text{Conj.}(\text{P}) - C.(\text{P}) \ne \emptyset$, it is clear that the left-hand difference set is equal to

$$\{B \in \text{Conj.}(\text{P}); B < A_P\} \cup \{B \in \text{Conj.}(\text{P});$$
$$A_P \text{ is not } \le -\text{comparable with } B\}.$$

That is,

$$\text{Conj.}(\text{P}) = C.(\text{P}) \cup \{B \in [0, 1]^X;$$
$$A_P \not\le B' \& A_0 < B < A_P\}$$
$$\cup \{B \in [0, 1]^X; A_P \not\le B' \& A_P \textbf{ nc } B\},$$

with the symbol **nc** shortening *not $\le$-comparable with*.

Notice that the second set in this union contains the fuzzy sets B being neither empty, nor contradictory with A_P, and for which *If B, then A_P*. Consequently, it can be said that these fuzzy sets B *explain* A_P, or P, and, of course, they also *explain* any .-weak consequence of P: If $A_P \le C$, it follows $B < C$. Let us denote this set of conjectures by $\text{Hyp.}(\text{P})$, and call it the set of explicative conjectures or, for short, hypotheses for P. If $C.(\text{P}) \in F(.)$, it is clear that $\text{Hyp.}(C.(\text{P})) = \text{Hyp.}(\text{P})$, as it also happens with the strongest conjunction $. = \min$.

For what concerns the third set in the last union, call it Sp.(P), it is decomposable in the disjoint union

$$\{B \in [0,1]^X; A_P < B' \,\&\, A_P \textbf{ nc } B\} \cup \{B \in [0,1]^X;$$
$$A_P \textbf{ nc } B' \,\&\, A_P \textbf{nc} B\},$$

whose elements will be called speculative conjectures or, for short, speculations. Let us, respectively, denote by Sp.$_1$(P) and Sp.$_2$(P), the first and the second set in the decomposition. The elements in Sp.$_i$(P) will be called type-i speculations ($i = 1, 2$).

It should be pointed out that the symbol **nc** shows the *jumps* cited before, and that these jumps affect both types of speculations but, specially, those in the type-2. In the case in which $B \in$ Sp.$_1$(P), since $A_P > B$ is equivalent to $A'_P < B$, B could be captured by going forward from A'_P, but if $B \in$ Sp.$_2$(P) a jump from either A_P, or from A'_P, is necessary to reach B.

It should also be pointed out that weak consequences are reached by *moving* forward from A_P, that hypotheses are reached by *moving* backward from A_P, but that for speculations a jump forward from either A_P, or A'_P, is required. It is clear, in addition, that Conj.(P) = Conj.($\{A_P\}$), C.(P) = C.($\{A_P\}$), Hyp.(P) = Hyp.($\{A_P\}$), and Sp.(P) = Sp.($\{A_P\}$), since $\{A_P\} \in$ F.(.), and in this sense, A_P can be seen as the résumé of the information conveyed by P.

With all that, the set of conjectures from P is completely classified by the disjoint union

$$\text{Conj.(P)} = C.(P) \cup \text{Hyp.(P)} \cup \text{Sp.}_1(P) \cup \text{Sp.}_2(P),$$

since all the intersections $C.(P) \cap \text{Hyp.(P)}, \ldots,$ Sp.$_1$(P) $\cap$ Sp.$_2$(P), are empty. Hence, conjectures are either weak consequences, or hypotheses, or type-1 speculations, or type-2 speculations. Consequently, and once given $P \in$ F.(.), all the fuzzy sets in $[0,1]^X$ are classified in refutations, .-weak consequences, hypotheses, speculations of type-1 and of type-2, with strong consequences being a part of the weak ones. It should be pointed out that, in some particular cases, the sets Hyp.(P), or Sp.(P), can be empty.

Note

It can be said that the above algebraic model for CR contains strong and weak deduction, abduction (the search for hypotheses), and also speculative reasoning.

What happens with hypotheses and speculations for what relates to monotony? Since given two enchained sets of premises $P \subset Q$, both in F.(.), the conjunction

of their premises, call them respectively A_P and B_Q, obviously verify $B_Q \le A_P$, it is Hyp.(Q) $\subset$ Hyp.(P): the operator Hyp. : F(.) $\to [0,1]^X$, is antimonotonic. Notice that Hyp. is not a consistent operator and, consequently, it cannot be supposed that Hyp.(P) always contains admissible information. It is risky to take a hypothesis as a new premise.

Analogously, Sp. is also a mapping F(.) $\to [0,1]^X$, that is nonconsistent, and, as it is easy to see by means of simple examples with crisp sets, it is neither monotonic, nor antimonotonic, nor elements in Sp.(P) can be always taken as admissible information. That is, since there is no law for the *growing* of Sp. with the *growing* of the premises, it can be said that speculations are peculiar among conjectures. Such peculiarity is somewhat clarified by what follows: If $S \in Sp.$(P):

a) If S is such that $A_P.S \ne A_0$, it is $A_P.S < A_P$ since if it were $A_P.S = A_P$, then follows $A_P \le S$, or the absurd $S \in C.$(P). From $A_0 < A_P.S < A_P$, and provided $A_P.S$ is a conjecture, it follows $A_P.S \in$ Hyp.(P).

b) Since $A_P \le A_P + S$, it is always $A_P + S \in C.$(P).

c) Provided the law of semiduality $B' + C' \le (B.C)'$, holds in the BFA ($[0,1]^X, ., +,'$), as it happens with $+ =$ max, and since $A_P \le A_P + S' \le (A'_P.S)'$, it follows that $A'_P.S$ is a refutation.

Hence, by means of speculations S, hypotheses $A_P.S$ can be obtained provided $A_P.S \ne A_0$ is a conjecture, $A_P + S$ is always a weak consequence, and with semiduality it is $A'_P.S$ a refutation. In this sense, speculations can serve as a tool for deducing, for abducing, and for refuting: They are *auxiliary conjectures* for either deploying what is hidden in the premises, or for refuting the premises, or to explain them. For this reason, to speculate is an important type of nonruled reasoning, and whose mastering should be encouraged to be learned.

A Remark on Heuristics

Although the concept of *heuristics* is not yet formalized, from last paragraphs, and under a few and soft constraints, speculations can be seen as auxiliary conjectures that intermediate for advancing reasoning. Since to reach a speculation S a jump from A_P should be taken, since there are no direct and step-by-step links between A_P and S, the process to arrive by the intermediary of S to either a consequence $A_P + S$, or a hypothesis $A_P.S$, or a refutation $A'_P.S$, is a typically heuristic one, perhaps obtained at each case by some nonstep-by-step rule of thumb. For instance, a hypothesis like $H = A_P$. S can be reached after some *heuristic path* conducting to S by some jumping from A_P that, additionally, can be done in several and different steps.

At this respect, the formal characterization of those hypotheses that are reducible to the form $A_P.C$ for some fuzzy set $C \in [0, 1]^X$ is yet an open problem. Since in the case of orthomodular lattices and, of course, of Boolean algebras, it was proven that all hypotheses are reducible [16.1], but that in the case of nonorthomodular ortholattices nonreducible ones should exist, given P it can be analogously supposed the existence in $[0, 1]^X$ of nonreducible hypotheses. Consequently, if existing, such hypotheses can be seen as isolated ones that cannot be reached by the intermediary of a speculation, but only directly by a particular heuristic backward-track from A_P.

For what concerns hypotheses, and except in formal deductive reasoning where, in principle, they can be either safely accepted, or refused through deduction, in CR a crucial point is to know how a hypothesis can be deductively or inductively refuted [16.28, 33]. As it is well known, the idea of refuting a hypothesis is central in scientific research [16.13], and it can be formalized in the current model as follows [16.32].

Let us suppose that there is a doubt between which one of the two statements: $H \in \text{Hyp.}(P)$, and $H \notin \text{Hyp.}(P)$, is valid but knowing that $H \nleq H'$, that the presumed hypothesis is not self-contradictory:

a) Provided the first statement is valid, since $H < A_P$ and $A_P \leq C$ imply $H < C$, it is $C.(P) \subset C.(\{H\})$. Thus, if there is $D \in C.(P)$ such that $D \notin C.(\{H\})$, H cannot be a hypothesis for P: To *weak-deductively* refute H as a hypothesis for P, it suffices to find a weak consequence of P that is not a weak consequence of $\{H\}$.

   Of course, classical (strong) deductive refutation corresponds to the case in which it is possible to take $. = \min$.

b) In addition, it is $C.(P) \subset \text{Conj.}(\{H\})$. Indeed, $A_P \leq B$, and $H < A_P$, do not imply $B \leq H'$ since, in this case it follows $H \leq H'$, it is $B \notin \text{Conj.}(P)$. Thus, it is $B \nleq H'$, or $B \in \text{Conj.}(\{H\})$. Consequently, to *weak-inductively* refute H as a hypothesis for P it suffices to find a weak consequence of P that is not conjecturable from $\{H\}$.

   All that can be, *mutatis mutandis,* repeated with the strongest conjunction $. = \min$, and the concept of the strong-inductive refutation of H is reached.

16.8 Last Remarks

In former papers of the author [16.28, 32, 34], the concept of a conjecture was formalized in the settings of ortholattices, and De Morgan algebras. In the first, and since $A_P \leq A'_P$ implies $A_P = 0$, it suffices to take $A_P \neq 0$. What lacked was the case of the standard algebras of fuzzy sets with a t-norm and a *t*-conorm different, respectively, of min and max, a case that is subsumed in what is presented in [16.28], and now is completed in this chapter.

With only crisp sets, the résumé A_P of the information conveyed by the premises in P, obviously verifies $A_P \leq B' \Leftrightarrow A_P.B = A_0 \Leftrightarrow A_P.B \leq (A_P.B)'$, but this chain of equivalences fails to hold if A_P or B are proper fuzzy sets. Consequently, and in addition to $C.(P) = \{B; A_P \leq B\}$, it can be also considered the two operators [16.32],

$$C.^1(P) = \{B; A_P.B' = A_0\}, \quad \text{and}$$
$$C.^2(P) = \{B; A_P.B' \leq (A_P.B')'\},$$

with which the corresponding conjecture operators could also be defined by

$$\text{Conj.}^i(P) = \{B \in [0, 1]^X; B' \notin C.^i(P)\}, \quad i = 1, 2,$$

by taking consistency in the other two forms cited in Sect. 16.5.

Both operators $C.^i$ are not extensive, but monotonic, at least $C^1_{\min}$ is consistent if the negation is functionally expressible, it is unknown which other $C.^i$ ($i = 1, 2$) is consistent, $C.^1(P) \subset C.^2(P)$, and it is not actually clear if in both cases it is, or it is not, $C.^i(P) \subset \text{Conj.}^i(P)$ except when $. = \min$ [16.28].

Consequently, the door is open to consider alternative definitions for the concept of a conjecture depending on the way of defining when an element $B \in [0, 1]^X$ is *consistent* with the résumé A_P. Anyway, what it seems actually difficult is how to imagine a kind of nonconsistent deduction.

It is well known that in CR conclusions are often obtained by some kind of *analogy* or *similitude* with a previously considered case. Without trying to completely *formalize* analogical reasoning, let us introduce some ideas that, eventually, could conduct toward such formalization.

Define *B is analogous to A*, if it exists a family K of mappings

$$\sigma : [0, 1]^X \to [0, 1]^X \,,$$

such that $B = \sigma \circ A$, for $\sigma \in K$. Namely, B is K-analogous to A. At its turn the set of fuzzy sets that are *analogous or similar* to those in P can be defined by

$$K(A_P) = \{B \in [0, 1]^X \,;$$
$$B = \sigma \circ A_P, \sigma \in K\} \,.$$

Then:

1) If id $\leq \sigma \Rightarrow A_P \leq \sigma \circ A_P \Rightarrow \sigma \circ A_P \in$ C.(P)
2) If $\sigma < id \Rightarrow \sigma \circ A_P < A_P \Rightarrow \sigma \circ A_P \in$ Hyp.(P)
3) If id **nc** $\sigma \Rightarrow \sigma \circ A_P$ **nc** $A_P \Rightarrow \sigma \circ A_P \in$ Sp.(P)
4) If id $\leq \sigma' \Rightarrow A_P \leq \sigma' \circ A_P = (\sigma \circ A_P)' \Rightarrow (\sigma \circ A_P)' \in$ Ref.(P).

Depending on the possible ordering of the pairs (id, σ), and (id, σ'), with id the identity mapping in $[0, 1]^X$, that is, id$(A) = A$, for all $A \in [0, 1]^X$, the *fuzzy sets analogous to* A_P are either consequences or hypotheses, speculations, or refutations. Notice that in point [16.4] it is id σ' equivalent to $\sigma \leq$ id$'$, that is, $\sigma \circ A \leq A'$ for all fuzzy set A.

Of course, for a further study of analogical types of reasoning it lacks to submit transformations σ to some restricting properties, surely depending on the concrete case under consideration.

16.9 Conclusions

The establishment of a general framework for the several types of reasoning comprised in Commonsense Reasoning seems to be of an upmost importance. At least, it should be so for fuzzy researchers in the new field of Computing with Words (CWW) that, with a calculus able to simulate reasoning, tries to deal with sentences and arguments in natural language more complex than those considered in today's current fuzzy logic.

In the way toward a full development of CWW covering as many scenarios as possible of the people's ways of reasoning, it seems relevant not to forget the big amount of nondeductive reasoning people commonly do. This chapter just offers a wide framework to jointly consider the four modalities of deduction, abduction, speculation, and refutation, typical of both ordinary and also specialized reasoning, where deduction and refutation are *deductive* modalities of reasoning, but where abduction and speculation can be considered its *inductive* modalities [16.35, 36].

There remain some unended questions that concern the proposed model and, among them, can be posed the two following ones:

- The finding of rules in the *Mill's* style [16.37], for obtaining hypotheses and speculations from the premises. These rules could conduct to obtain computer programs or algorithms able, in some cases,

to find either hypotheses or speculations, and can be also useful for clarifying the concept of a *heuristics*.

- The study of what happens with the conjectures once new consistent information supplied by new premises, is added to the initial set P [16.38]. It should not to be forgot that ordinary reasoning is rarely made from a *static* initial set of items of information, but that the information comes in a kind of *flux* under which conjectures can vary of number and of character.

Like speculations facilitate heuristics for finding consequences, hypotheses, and refutations, analogical reasoning could constitute a *good trick* for obtaining conjectures and refutations, on the base of some earlier solved similar problems. It should be pointed out that what is not yet clear enough is how to compute the degree of liability an analogical conclusion could deserve.

What is not addressed in this chapter is the theoretic and practical important problem of which is the best hypothesis or speculation to be selected at each particular case. This question seems linked with the translation into the conclusions of some numerical *weights of confidence* previously attributed to the premises, and that depend on the case into consideration. Because in this chapter such weights are neither considered for premises, nor for conclusions, the presented model can be qualified as a *crisp* one, but not yet as a *fuzzy* one, since it fails taking into account the level of liability of the sentences represented in fuzzy terms. The semantics of the linguistic terms is here confined, through its most careful possible design [16.15, 16], to the contextual and purpose driven meaning of the involved fuzzy sets and fuzzy connectives, as well as to the possible linguistic interpretation of the accepted conclusions, but what is not yet taken into account is their degree of liability.

By viewing formal theories as abstract constructions that could help us to reach a better understanding of a subject inscribed in some reality, this chapter represents a *formal theory* of reasoning with precise and imprecise linguistic terms represented by fuzzy sets and already presented in [16.28]. It consists in a way of formalizing conjectures and refutations in the wide mathematical setting of BFAs whose axioms comprise the particular instances of ortholattices, De Morgan algebras, standard algebras of fuzzy sets, and, of course, Boolean algebras. As is shown by two of the forms allowing to define what weak deduction could be, depending on the kind of consistency chosen, and that, with crisp sets, also collapse in the classical case, this formalization cannot be yet considered as definitive, but open to further study.

Nevertheless, this new theory should not be confused with the actual human reasoning, and only through some work of an experimental character on CR it will be possible to clarify which degree of agreement with the reality of reasoning either the selected way for defining weak deduction, or a different one, does show. Provided this kind of work could be done, the observational appearance of some observed regularities in CR, or *observed laws* of the actual ordinary reasoning, and that can be *predicted* by some *invariants* in the model, is a very important topic for future research.

Anyway, to advance toward a kind of Experimental Science of CR, there are today no answers to some crucial questions like, for instance:

- Which regularities exist in natural language and in CR that, reflected by some invariants in the model, can be submitted to experimentation?

- How observation and experiments on CR could be systematically projected and programmed?
- Which quantities linked to such realities can exist in the model, and that could show numerical effects, measurable through experiments?

One of the reasons for such ignorance could come from both the small number of axioms the BFA-model have, and from the too big number of axioms the standard algebras of fuzzy sets show, for saying nothing on the Boolean and De Morgan algebras. When the number of axioms is too small, it seems difficult to think in a way to find invariants, and if there are too many axioms then the big number of relationships among them seems to show too much theoretical invariants, like it is, for instance, the indistinguishability between contradiction and incompatibility in Boolean algebras.

That ignorance can also come from the lack of numerical quantities associated with some important properties concerning both natural language and CR, like it can be, for instance, a degree of associativity in the case the model is just a BFA, or a degree of liability of speculations for allowing to reach refutations for some reality described by a set P of premises. It can also be the lack of a standard degree of similarity (like the one studied in [16.11]) to numerically reflect how much analogous to a possible hypothesis for P can be some known statement in order to be taken itself as the hypothesis.

Without some numerical quantities reflecting important characteristics of, for instance, words, it is difficult to think on how to project experiments for detecting some regularities or invariants without which it seems very difficult to conduct a deep scientific study of CR. Possible instances of such quantities are the measures of both the specificity [16.39], and the fuzziness [16.40] of imprecise predicates, and a experiment to be projected could be, perhaps, an analysis of how their values jointly vary in some context.

Actually and currently, an Experimental Science of CR is just a dream. Would this chapter also serve as a humble bell's peal toward such goal!

References

16.1   E. Trillas, A. Pradera, A. Alvarez: On the reducibility of hypotheses and consequences, Inf. Sci. **179**(23), 3957–3963 (2009)

16.2   I. García-Honrado, E. Trillas: An essay on the linguistic roots of fuzzy sets, Inf. Sci. **181**, 4061–4074 (2011)

16.3   E. Trillas, I. García-Honrado: Hacia un replanteamiento del cálculo proposicional clásico?, Agora **32**(1), 7–25 (2013)

16.4   E. Trillas, C. Alsina, E. Renedo: On some classical schemes of reasoning in fuzzy logic, New Math. Nat. Comput. **7**(3), 433–451 (2011)

16.5   A. Pradera, E. Trillas: A reflection on rational-ity, guessing, and measuring, IPMU, Annecy (2002) pp. 777–784

16.6   K. Menger: *Morality, Decision and Social Organization* (Reidel, Dordrecht 1974)

16.7   L.A. Zadeh: *Computing with Words. Principal Concepts and Ideas* (Springer, Berlin, Heidelberg 2012)

16.8   F.H. Rauscher, G.L. Shaw, K.N. Ky: Listening to Mozart enhances spatial-temporal reasoning, Neurosci. Lett. **185**, 44–47 (1995)

16.9   J. Berger: *The Success and Failure of Picasso* (Pantheon, New York 1989)

16.10   E. Trillas, C. Moraga, S. Guadarrama, S. Cubillo, E. Castiñeira: Computing with antonyms, Stud. Fuzziness Soft Comput. **217**, 133–153 (2007)

16.11   E. Castiñeira, S. Cubillo, E. Trillas: On a similarity ratio, Proc. EUSFLAT-ESTYLF (1999) pp. 239–242

16.12   E. Wigner: The unreasonable effectiveness of mathematics in the natural sciences, Commun. Pure Appl. Math. **13**(1), 1–14 (1960)

16.13   K. Popper: *Conjectures and Refutations* (Harper Row, New York 1968)

16.14   E. Trillas: Reasoning: In black & white?, Proc. NAFIPS (2012)

16.15   E. Trillas, S. Guadarrama: Fuzzy representations need a careful design, Int. J. Gen. Syst. **39**(3), 329–346 (2010)

16.16   E. Trillas, C. Moraga: Reasons for a careful design of fuzzy sets, Proc. EUSFLAT (2013), Forthcoming

16.17   K. Menger: A counterpart of Occam's Razor, Synthese **13**(4), 331–349 (1961)

16.18   A.C. Masquelet: *Le Raisonnement Médical* (PUF/Clarendon, Oxford 2006), in French

16.19   E. Trillas, C. Alsina, E. Renedo: On some schemes of reasoning in fuzzy logic, New Math. Nat. Comput. **7**(3), 433–451 (2011)

16.20   E. Trillas, L.A. Urtubey: Towards the dissolution of the Sorites paradox, Appl. Soft Comput. **11**(2), 1506–1510 (2011)

16.21   A. Tarski: *Logic, Semantics, Metamathematics* (Hackett, Cambridge 1956)

16.22   J.F. Sowa: E-mail to P. Werbos, 2011, in BISC-GROUP

16.23   C.S. Peirce: Deduction, induction, and hypothesis, Popul. Sci. Mon. **13**, 470–482 (1878)

16.24   J.F. Sowa: *Knowledge Representation* (Brooks/Cole, Farmington Hills 2000)

16.25   R. Reiter: Nonmonotonic reasoning, Annu. Rev. Comput. Sci. **2**, 147–186 (1987)

16.26   W. Ockham: *Summa Logica* (Parker, London 2012), in Spanish and Latin

16.27   W. Whewell: *Novum Organum Renovatum (Second Part of the Philosophy of Inductive Sciences)* (Parker, London 1858)

16.28   E. Trillas: A model for *Crisp Reasoning* with fuzzy sets, Int. J. Intell. Syst. **27**, 859–872 (2012)

16.29   P.R. Halmos: *Naïve Set Theory* (Van Nostrand, Amsterdam 1960)

16.30   A. Pradera, E. Trillas, E. Renedo: An overview on the construction of fuzzy set theories, New Math. Nat. Comput. **1**(3), 329–358 (2005)

16.31   G. Birkhoff: *Lattice Theory* (American Mathematical Society, New York 1967)

16.32   I. García-Honrado, E. Trillas: On an attempt to formalize guessing. In: *Soft Computing in Humanities and Social Sciences*, ed. by R. Seising, V. Sanz (Springer, Berlin, Heidelberg 2012) pp. 237–255

16.33   E. Trillas, S. Cubillo, E. Castiñeira: On conjectures in orthocomplemented lattices, Artif. Intell. **117**(2), 255–275 (2000)

16.34   E. Trillas, D. Sánchez: Conjectures in De Morgan algebras, Proc. NAFIPS (2012)

16.35   S. Baker: *Induction and Hypotheses* (Cornell Univ. Press, Ithaca 1957)

16.36   B. Bossanquet: *Logic or the Morphology of Knowledge*, Vol. I (Clarendon, Oxford 1911)

16.37   J.S. Mill: *Sir William Hamilton's Philosophy and the Principal Philosophical Questions Discussed in His Writings* (Longmans Green, London 1889)

16.38   I. García-Honrado, A.R. de Soto, E. Trillas: Some (Unended) queries on conjecturing, Proc. 1st World Conf. Soft Comput. (2011) pp. 152–157

16.39   L. Garmedia, R.R. Yager, E. Trillas, A. Salvador: Measures of fuzzy sets under T-indistinguishabilities, IEEE Trans. Fuzzy Syst. **14**(4), 568–572 (2006)

16.40   A. De Luca, S. Termini: A definition of a nonprobabilistic entropy in the setting of fuzzy sets theory, Inf. Control **20**, 301–312 (1972)

17. Fuzzy Control

Christian Moewes, Ralf Mikut, Rudolf Kruse

Fuzzy control is by far the most successful field of applied fuzzy logic. This chapter discusses human-inspired concepts of fuzzy control. After a short introduction to classical control engineering, three types of very well known fuzzy control concepts are presented: Mamdani-Assilian, Takagi-Sugeno and fuzzy logic-based controllers. Then three real-world fuzzy control applications are discussed. The chapter ends with a conclusion and a future perspective.

17.1 Knowledge-Driven Control

With no doubt, the biggest achievement of fuzzy logic with respect to industrial and commercial applications has been obtained by *fuzzy control*. Since its first practical use for *a simple dynamic plant* by *Mamdani* [17.1], over attention-getting applications such as the automatic train operation in Sendai, Japan [17.2], fuzzy control systems have become indispensable in the industry today (Until today more than 60 000 patents have been filed worldwide using the words *fuzzy* and *control* according to [17.3]). A wide range of real-world applications have also been described by *Hirota* [17.4], *Terano* and *Sugeno* [17.5], *Precup* and *Hellendoorn* [17.6].

Simply speaking, fuzzy control is a kind of defining a nonlinear table-based controller. Every entry in such a table can be seen as partial knowledge about the specified input–output behavior [17.7]. However, knowledge does not have to exist for every input–output combination. Thus, the transition function of a fuzzy controller is a typical nonlinear interpolation between defined regions of this *knowledge*. The knowledge is commonly

stored as *imprecise* rules consisting of imprecise terms such as *small*, *big*, *cold*, or *warm*. Consequently, these rules lead to an *imprecisely defined* transition function that is eventually *defuzzified* if a crisp decision is needed.

This procedure is sometimes advantageous when compared to classical control systems – especially for control problems that are usually solved intuitively by human beings, but not by computing machines, e.g., parking a car, riding the bike, boiling an egg [17.8]. This might be also a reason why fuzzy control did not originate from control engineering researchers. It had rather been inspired by *Zadeh* [17.9] who proposed rule-based systems for handling complex systems using fuzzy sets – a concept which he introduced 8 years before [17.10].

The focus of this chapter is a profound discussion of such human-inspired concepts of fuzzy control. Other fuzzy control approaches based on a fuzzification of well-known methods of the classical control theory (fuzzy PID control, fuzzy adaptive control, stability of fuzzy systems, fuzzy sliding mode, fuzzy observer, etc.) are only described briefly. For these topics, the reader is referred to other textbooks, e.g., *Tanaka* and *Wang* [17.11].

But before we formally present the concepts of fuzzy control, let us give a brief introduction of classical control engineering in Sect. 17.2. Then, in detail, we cover fuzzy control in Sect. 17.3, including the most well-known approaches of Mamdani and Assilian, Takagi and Sugeno, and *truly* fuzzy-logic-based controllers in the Sects. 17.3.1, 17.3.2, and 17.3.3, respectively. We also talk about their advantages and limitations. We discuss some more recent industrial applications in Sect. 17.4 and automatic learning strategies in Sect. 17.5. Finally, we conclude our presentation of fuzzy control in Sect. 17.6.

17.2 Classical Control Engineering

To introduce the problem of controlling a process, let us consider a technical system for which we dictate a desired behavior [17.12]. Generally speaking, we wish to reach a desired set value for a time-dependent *output variable* of the process. This output is influenced by a variable that we can influence, i.e., the *control variable*. Last but not least – to deal with unexpected influences – let a time-dependent *disturbance variable* be given that manipulates the output, too.

Then the current control value is typically specified by mainly two components, i.e., the present measurement values of the output variable ξ, the variation of the output $\Delta\xi = \frac{d\xi}{dt}$, and further variables which we do not specify here. We refer to n input variables $\xi_1 \in X_1, \ldots, \xi_n \in X_n$ of the controller (e.g., computed from the measured output variable of the process and its desired values) and one control variable $\eta \in Y$. Formally, the solution of a control problem is a desired control function $\varphi : X_1 \times \cdots \times X_n \to Y$ which sets a suitable control value $y = \varphi(\vec{x})$ for every input tuple $\vec{x} = (x^{(1)}, x^{(2)}, \ldots, x^{(n)}) \in X_1 \times \cdots \times X_n$. Controllers with multiple outputs are often handled as independent controllers with one output each.

In classical control, we can determine φ using different techniques. The most popular one for practical applications is the use of simple standard controllers such as the so-called *PID controller*. This controller uses three parameters to compute a weighted sum of proportional, integral, and derivative (PID) components of the error between the output variable and desired values to compute the control variable. In many relevant cases, a good control performance of the closed-loop feedback system with controller and process can be reached by simple tuning heuristics for these parameters. This strategy is successful for many processes that can be described by linear differential equations for the overall behavior or at least near all relevant setpoints. More advanced controllers are used in cases with nonlinear process behavior, time-variant process changes, or complicated process dynamics. They require both a mathematical process model based on a set of differential or difference equations and a fitness function to quantify the performance. Here, many different strategies exist starting from a setpoint-dependent adaptation of the PID parameters, additional feedforward components to react to setpoint changes or known disturbances, the estimation of internal process states by observers in state-space controllers, the online estimation of unknown process parameters in adaptive controllers, the use of inverted process models as controllers, or robust controllers that can handle bounded parameter changes. For all these controllers, many elaborate design and analysis techniques exist that guarantee an optimal behavior based on a known process model, see e.g., *Åström* and *Wittenmark* [17.13], *Goodwin* et al. [17.14].

However, it might be mathematically intractable or even impossible to define the exact differential equations for the process and the controller φ. For such cases, classical control theory cannot be applied at all. For instance, consider the decision process of human beings and compare it to formal mathematical equations. Many of us have the great ability to control diverse processes without knowing about higher mathematics at all – just think of a preschool child operating a bike or juggling a European football with its foot or even head.

17.3 Using Fuzzy Rules for Control

Probably the simplest way to obtain a human *control* behavior for a given process is to find out – by asking direct questions for instance – how a person would react given a specific situation. One alternative is to observe the process to be controlled, e.g., using sensors, and then discover substantive information in these signals. Both approaches can be seen as *knowledge-based* analysis which eventually provides us with a set of *linguistic rules*. We assume that these if–then rules are able to properly control the given process.

Let us briefly outline the operating principle of a fuzzy controller based on such if–then rules. Each rule consists of an antecedent and consequent part. The former relates to an imprecise description of the crisp measured input, whereas the latter defines a suitable fuzzy output for the input. In order to enable a computing machine to use such linguistic rules, mathematical terms of the linguistic expressions used in the rules need to be properly defined. Once a control input is present, more than one rule might (partly) fulfill the present concepts. Thus, there is a need for suitable accumulation methods for these instantiated rules to eventually compute one fuzzy output value. From this value, a crisp output value can be obtained if necessary.

This knowledge-based model of a fuzzy controller is conceptually shown in Fig. 17.1. The *fuzzification interface* operates on the current input value $\vec{x}_0$. Here too, $\vec{x}_0$ might be mapped to a desired space if it is necessary – one may want to normalize the input to the unit interval first. The fuzzification interface eventually translates $\vec{x}_0$ into a linguistic term described by a fuzzy set. The *knowledge base* is *head* of the controller, it serves as *database*. Here, every essential piece of information is stored, i.e., the variable ranges, domain transformations, and the definition of the fuzzy sets with their corresponding linguistic terms. Furthermore, it comprises the *rule base* that is required to linguistically describe and control the process. The *decision logic* computes the fuzzy output value of the given measurement value by taking into account the knowledge base. Last but not least, the *defuzzification interface* computes a crisp output value from the fuzzy one.

Two well-known and similar approaches have led to the tremendous use and the success of fuzzy control. The Mamdani–Assilian and the Takagi–Sugeno approaches are motivated intuitively in Sects. 17.3.1 and 17.3.2, respectively. They have in common that their interpretation of a linguistic *rule* diverges from

mathematical implications. Both types of controllers rather *associate* an input specified as an antecedent part with the given output given as a consequent part. A mathematically formal approach to fuzzy control as discussed in Sect. 17.3.3 leads to completely different computations. As it turns out in practice, controllers based on any kind of logical implications are usually too restrictive to suitably control a given process.

17.3.1 Mamdani–Assilian Control

Just one year after the publication of *Zadeh* [17.9], Ebrahim *Abe* Mamdani and his student Sedrak Assilian were the first who successfully controlled a simple process using fuzzy rules [17.1, 15]. They developed a fuzzy algorithm to control a steam engine based on human expert knowledge in an application-driven way. That is why today we refer to their approach as nowadays *Mamdani–Assilian control*.

The expert knowledge of a Mamdani–Assilian controller needs to be expressed by linguistic rules. Therefore, for every set X_i of given values for an input, we define suitable linguistic terms that summarize or partition this input by fuzzy sets. Let us consider the first set X_1 for which we define p_1 fuzzy sets $\mu_1^{(1)}, \ldots, \mu_{p_1}^{(1)} \in \mathcal{F}(X_\infty)$. Each of these fuzzy sets is mapped to one preferable linguistic term. Thus, X_1 is partitioned by these fuzzy sets. To ensure a better interpretability of each fuzzy set, we recommend to use just unimodal membership functions. In doing so, every fuzzy set can be seen as imprecisely defined value or interval. We furthermore urge to choose disjoint fuzzy sets for every

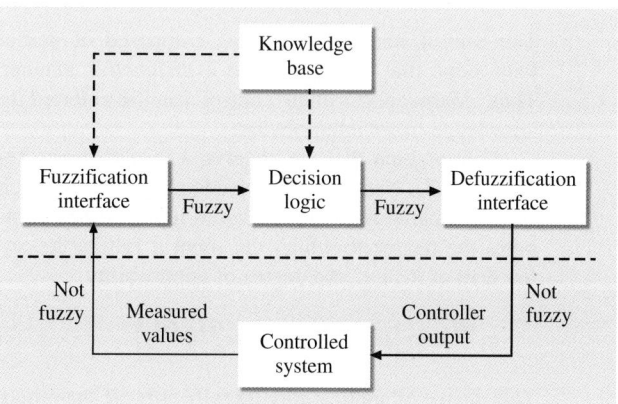

Fig. 17.1 Architecture of a fuzzy controller

partition, i. e., the fuzzy sets shall satisfy

$$i \neq j \Rightarrow \sup_{x \in X_1} \left\{ \min \left\{ \mu_i^{(1)}(x), \mu_j^{(1)}(x) \right\} \right\} \leq 0.5 .$$

When X_1 is partitioned into p_1 fuzzy sets $\mu_1^{(1)}, \ldots, \mu_{p_1}^{(1)}$, we can continue to partition the remaining sets $X_2, \ldots, X_n$ and Y in the same way. The eventual outcome of this procedure (i. e., the linguistic terms associated with the fuzzy sets for each variable) establishes the database in our knowledge base.

The rule base of any Mamdani–Assilian controller is specified by rules of the form

if ξ_1 is $A^{(1)}$ and ... and ξ_n is $A^{(n)}$ then η is B

(17.1)

where $A^{(1)}, \ldots, A^{(n)}$ and B symbolize linguistic terms which correspond to the fuzzy sets $\mu^{(1)}, \ldots, \mu^{(n)}$ and μ, respectively, according to the fuzzy partitions of $X_1 \times \cdots \times X_n$ and Y. Thus, the rule base consists of k control rules

$$R_r : \text{if } \xi_1 \text{ is } A_{i_{1,r}}^{(1)} \text{ and } \ldots \text{ and } \xi_n \text{ is } A_{i_{n,r}}^{(n)}$$
$$\text{then } \eta \text{ is } B_{i_r}, \quad r = 1, \ldots, k .$$

We again underline that these rules are not interpreted logically as mathematical implications. They rather specify the function $\eta = \varphi(\xi_1, \ldots, \xi_n)$ piecewise using the existing associations between the known input–output tuples, i. e.,

$$\eta \approx \begin{cases} B_{i_1} & \text{if } \xi_1 \approx A_{i_{1,1}}^{(1)} \text{ and } \ldots \text{ and } \xi_n \approx A_{i_{n,1}}^{(n)} , \\ \vdots & \vdots \\ B_{i_k} & \text{if } \xi_1 \approx A_{i_{1,k}}^{(1)} \text{ and } \ldots \text{ and } \xi_n \approx A_{i_{n,k}}^{(n)} . \end{cases}$$

The control function φ is thus composed of partial knowledge that we connect in a *disjunctive* manner. Thus, Mamdani–Assilian control can be referred to knowledge-based interpolation.

Now assume that we observe a measurement $\vec{x} \in X_1 \times \cdots \times X_n$. Naturally, the decision logic applies each rule R_r separately to the measured input. It then computes the degree to which the input $\vec{x}$ fulfills the antecedent of R_r, i. e., the degree of applicability

$$\alpha_r \stackrel{\text{def}}{=} \min \left\{ \mu_{i_{1,r}}^{(1)}(x^{(1)}), \ldots, \mu_{i_{n,r}}^{(n)}(x^{(n)}) \right\} . \quad (17.2)$$

This degree of applicability literally *cuts off* the output fuzzy set μ_{i_r} of the rule R_r at the level α_r which leads

to the rule's output fuzzy set

$$\mu_{\vec{x}}^{o(R_r)}(y) = \min \{ \alpha_r, \mu_{i_r}(y) \} . \quad (17.3)$$

When the decision logic does that for all α_r for $r = 1, \ldots, k$, then it unifies all output fuzzy sets $\mu_{\vec{x}}^{o(R_r)}$ by using a t-conorm. The standard Mamdani–Assilian controller uses the maximum as t-norm. Thus, the ultimate output fuzzy set

$$\mu_{\vec{x}}^{o}(y) = \max_{r=1,\ldots,k} \{ \min \{ \alpha_r, \mu_{i_r}(y) \} \} . \quad (17.4)$$

The whole process of evaluating Mamdani–Assilian rules is depicted in Fig. 17.2.

Of course, from a fuzzy set-theoretic and interpretational point of view, it suffices to keep $\mu_{\vec{x}}^{o}(y)$ as the final output *value*. However, fuzzy controllers are used in real-control application where a crisp control value is for sure needed, e.g., to increase the electric current of a hotplate when boiling an egg. That is why the fuzzy control output $\mu_{\vec{x}}^{o}$ is processed in the defuzzification interface. Depending on the implemented method of fuzzifying $\mu_{\vec{x}}^{o}$, a real value is ultimately obtained. Three methods are used in the literature extensively, i. e., the max criterion method, the mean of maxima (MOM) method, and the center of gravity (COG) method.

The max criterion method simply returns an arbitrary value $y \in Y$ such that $\mu_{\vec{x}}^{o}(y)$ obtains a maximum membership degree. However, this arbitrary value picked at random typically results in a nondeterministic control behavior. That is usually undesired as the interpretability and repeatability of already produced outcomes get lost. The MOM method returns the mean value of the set of elements $y \in Y$ that have maximal membership degrees in the output fuzzy set. Using this approach, it might happen that the defuzzified control value η may not even belong to the points leading to maximal membership degrees. Just consider for instance a bimodal normal fuzzy set as a fuzzy output. That is why MOM may lead to control actions that you would not await. Finally, the COG method returns the value at the center of gravity of the fuzzy output area $\mu_{\vec{x}}^{o}$, i. e.,

$$\eta = \left(\int_{y \in Y} \mu_{\vec{x}}^{o}(y) \cdot y \, dy \right) \bigg/ \left(\int_{y \in Y} \mu_{\vec{x}}^{o}(y) \, dy \right) .$$

(17.5)

Usually, the COG method is taken to defuzzify the fuzzy output as it typically leads to a smooth control

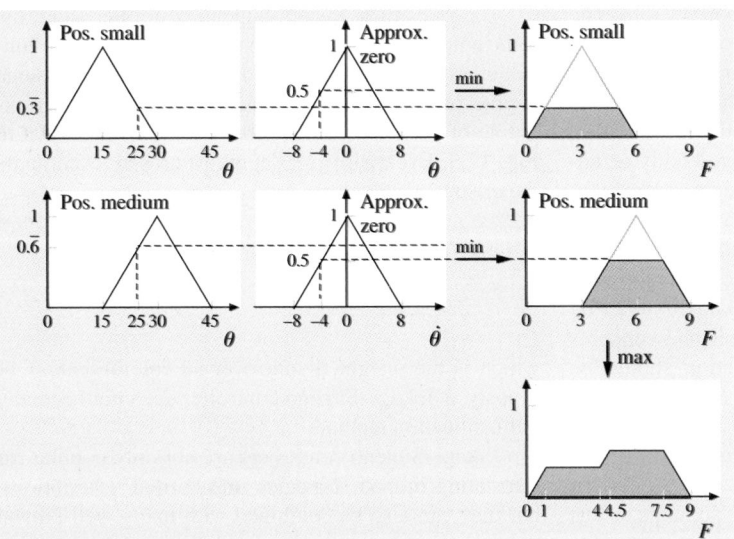

Fig. 17.2 The Mamdani–Assilian rule evaluation. Here we assume that the control process is described by two input variables $\theta, \dot{\theta}$ and one output F. Let the controller be specified by two Mamdani–Assilian rules. Here the input tuple $(25, -4)$ leads to the fuzzy output shown on the lower right side. If a crisp output is needed, any defuzzification method can eventually be applied

function. Nevertheless, it is possible to obtain an unreasonable output, too. We refer the gentle reader to *Kruse* et al. [17.16] for a richer treatment of defuzzification methods.

Coming back to the theoretical background of Mamdani–Assilian control, let us analyze its type of linguistic rules. Having a look at (17.3) again, we see that the minimum is used to serve as fuzzy implication. But the minimum does not fulfill all truth value combinations of the propositional logic's implication. To see this, let us consider $p \to q$ and assume that p is false. Then $p \to q$ is true – regardless of the truth value of q. The minimum of 0 and q, however, is 0. This logical flaw could be seen as inconsistency of the standard Mamdani–Assilian controller. On the contrary, it actually turns it into a very powerful technique. Created to solve a simple practical problem, we might speak from a heuristic method instead. When we do not see Mamdani–Assilian rules as logical implications but rather as *association* [17.17], then the controller is even theoretically sound: Every rule R_r associates an output fuzzy set B_{i_r} with n input fuzzy sets $A_{\tilde{\xi}_j, r}^{(j)}$ for $j = 1, \ldots, n$. Consequently, we *must* use a fuzzy conjunction, e.g., the minimum t-norm.

Mamdani and Assilian's heuristics can be obtained by the extension principle, too [17.18, 19]. If the fuzzy relation R that relates to the input $x^{(j)}$ and the output y satisfies a couple of extensionality properties, then Mamdani and Assilian's approach can also be obtained. Therefore, let E and E' be two similarity relations defined on the domains X and Y of x and y,

respectively. The extensionality of R on $X \times Y$ indicates that

$$\forall x \in X : \forall y, y' \in Y : R(x, y) \otimes E'(y, y') \leq R(x, y'),$$
$$\forall x, x' \in X : \forall y \in Y : R(x, y) \otimes E(x, x') \leq R(x', y).$$
$$(17.6)$$

Thus, if $(x, y) \in R$, then x is related to the neighborhood y. The same holds for y in relation to x. Then $A_r^{(j)}(x) = E(x, a_r^{(j)})$ and $B_r(x) = E'(y, b_r)$ can be regarded as fuzzy sets of values in the proximity of $a_r^{(j)}$ and b_r, respectively. Hence $\forall r = 1, \ldots, k$: $R(a_r^{(1)}, \ldots, a_r^{(p)}, b_r) = 1$. Applying this type of control definition to real-world problems, a practitioner must specify sensible similarity relations E_j and E' for each input ξ_j and output η, respectively. Eventually, using the extension principle for R, we obtain

$$R(x^{(1)}, \ldots, x^{(p)}, y) \geq \max_{r=1,\ldots,k}$$
$$\bigotimes \left(A_r^{(1)}(x^{(1)}), \ldots, A_r^{(p)}(x^{(p)}), A_r(y) \right).$$

In addition, if we use the minimum t-norm for $\otimes$, then we get exactly the approach of Mamdani–Assilian. *Boixader* and *Jacas* [17.20], *Klawonn* and *Castro* [17.21] show that indistinguishability or similarity is the connection between the extensionality property and fuzzy equivalence relations.

In practical applications, different t-norms and t-conorms as product instead of minimum and bounded sum instead of maximum play an important role. The reason is a stepwise multilinear interpolation behavior between the rules that cause a smoother function $y = \varphi(\vec{x})$ compared to minimum and maximum operators.

Depending on the input and output variables, Mamdani–Assilian controllers can intentionally or unintentionally copy concepts from classical control theory. As an example, a controller with inputs as proportional, integral and derivative errors between the output variable and desired values in combination with a symmetrical rule base has a similar behavior to a PID controller. Such a reinvention of established concepts with a more complicated implementation should be avoided though.

17.3.2 Takagi–Sugeno Control

Partitioning both the input and output domains seems to be reasonable from an interpretational point of view. Nevertheless, one might face control applications where a sufficient approximation quality can only be achieved using many linguistic terms for each dimension. This, in turn, will increase the potential number of rules which most probably worsen the ability to interpret the rule base of the fuzzy controller. For such control processes, we recommend to neglect the concept of partitioning the output domain and instead define functions that locally approximate the control behavior.

A controller that uses rules like this is called the Takagi–Sugeno controller [17.22]. The Takagi–Sugeno rules R_r for $r = 1, \ldots, k$ are typically defined as

$$R_r : \text{if } \xi_1 \text{ is } A_{i_1,r}^{(1)} \text{ and } \ldots \text{ and } \xi_n \text{ is } A_{i_n,r}^{(n)}$$
$$\text{then } \eta = f_r(\xi_1, \ldots, \xi_n) .$$

Most commonly, *linear* functions f_r can be found in many controllers, i. e.,

$$f_r(\vec{x}) = a_r^{(0)} + \sum_{i=1}^{n} a_r^{(i)} x^{(i)} .$$

The rules of a Takagi–Sugeno controller share the same antecedent parts with the Mamdani–Assilian controller, so does the decision logic computes the same degree of applicability α_r, i. e., using (17.2). An example of such a controller for two inputs is shown in Fig. 17.3. Eventually, all degrees are used to compute the crisp control value

$$\eta = \frac{\sum_{r=1}^{k} \alpha_r \cdot f_r(\vec{x})}{\sum_{r=1}^{k} \alpha_r}$$

which is the weighted sum over all rule outputs. Obviously, a Takagi–Sugeno controller does not need any defuzzification method.

Takagi–Sugeno controllers are not only popular for translating human strategies into formal descriptions, but they can also be combined with many well-known methods from classical control theory. For instance, many locally valid linear models of the process can be aggregated into one nonlinear model in the form of Takagi–Sugeno rules. In a next step, the desired behavior of the resulting closed-loop system is formulated. Stability can be defined in the strictest form as guaranteed convergence to the desired setpoint from any initial state, including robustness against bounded parameter uncertainties of the process model. Mathematical methods as Lyapunov functions in the form of linear matrix equations are now applied to design the Takagi–Sugeno fuzzy controller. For details about these concepts, we refer to *Feng* [17.23]. The main advantage of this strategy is the guaranteed performance, whereas many disadvantages come into play, too. This approach requires a model and sophisticated mathematical methods. Also it usually leads to a limited performance due to conservative design results and the limited interpretability. That is also why many recent papers propose iterative improvements, e.g., by proposing ways to handle other types of uncertainties such as varying time delays [17.24], and by reducing the conservativeness of the solutions [17.25].

R_1 : if ξ_1 is	[graph: 3, 9]		then $\eta_1 = 1 \cdot \xi_1 + 0.5 \cdot \xi_2 + 1$
R_2 : if ξ_1 is	[graph: 3, 9]	and ξ_2 is [graph: 4, 13]	then $\eta_2 = -0.1 \cdot \xi_1 + 4 \cdot \xi_2 + 1.2$
R_3 : if ξ_1 is	[graph: 3, 9 11, 18]	and ξ_2 is [graph: 4, 13]	then $\eta_3 = 0.9 \cdot \xi_1 + 0.7 \cdot \xi_2 + 9$
R_4 : kf ξ_1 is	[graph: 11, 18]	and ξ_2 is [graph: 4, 13]	then $\eta_4 = 0.2 \cdot \xi_1 + 0.1 \cdot \xi_2 + 0.2$

Fig. 17.3 A Takagi–Sugeno controller for two inputs and one output described by four rules. If a certain clause x_j is $A_{i_{j,r}}^{(j)}$ in any rule R_r is missing, then the corresponding membership function $\mu_{i_{j,r}}(x_j) \equiv 1$ for all linguistic values $i_{j,r}$. In this example, consider for instance x_2 in rule R_1. Thus, $\mu_{i_{2,1}}(x_2) \equiv 1$ for all $i_{2,1}$

17.3.3 Fuzzy Logic–Based Controller

Both controllers that have been introduced so far interpret every linguistic rule as an association of an n-dimensional fuzzy input point with a fuzzy output. Thus, we can interpret the set of fuzzy rules as setpoints of the control system. Recall, however, that this has nothing to do with logic inference since not all rules need to be activated by a given input.

When all rules are evaluated in a conjunctive manner, we can regard each fuzzy rule as a fuzzy constrain on a fuzzy input–output relation. The inference operation of such a controller is identical to approximate reasoning. Note that classical reasoning uses inference rules (so-called tautologies) to deductively infer crisp conclusions from *crisp* propositions. A generalization of classical reasoning is approximate reasoning applied to *fuzzy* propositions. *Zadeh* [17.9] proposed the first approaches to handle fuzzy sets in approximate reasoning. The gentle reader is referred to further details explained in *Zadeh* [17.26, 27]. The basic idea is to represent incomplete knowledge as possibility distributions.

Possibility theory [17.28] has been proposed to study and model imperfect descriptions of an existing element x_0 in a set $A \subseteq X$. It can be seen as a counterpart to probability theory. To formally define a possibility distribution $\Pi : 2^X \to [0, 1]$ we need the following axioms that seem to have similarities to the well-known Kolmogorov axioms:

$$\Pi(\emptyset) = 0 \,,$$
$$\Pi(A) \leq \Pi(B) \text{ if } A \subseteq B \text{ and}$$
$$\Pi(A \cup B) = \max\{\Pi(A), \Pi(B)\} \text{ for all } A, B \subset X \,.$$

The expression $\Pi(A) = 1$ includes that $x_0 \in A$ is unconditional possible. If $\Pi(A) = 0$, then it is impossible that $x_0 \in A$. *Zadeh* [17.29] models uncertainty about x_0 by the possibility measure $\Pi : 2^\Omega \to [0, 1]$, $\Pi(A) = \sup\{\mu(x) \mid x \in A\}$ when a fuzzy set $\mu : x \to [0, 1]$ is the only known description of x_0. Then the possibility measure is given by the possibility degrees of singletons, i. e., $\Pi(\{x\}) = \mu(x)$.

Now, consider only one-dimensional input and output spaces. Then we must specify a suitable two-dimensional possibility distribution. Let the rule

$$R : \text{if } \xi \text{ is } A \text{ then } \eta \text{ is } B$$

associate the input fuzzy set μ_A with the output fuzzy set μ_B. We can express this rule by a possibility distribution

$$\pi_{X,Y}(x, y) = I(\mu_A(x), \mu_B(y))$$

where I is an implication of any multivalued logic. So, we can compute the output by the composition of the input and the rule base, i. e., $\mu_B = \mu_A \circ \pi_{X,Y}$ where the fuzzy rules are expressed by the fuzzy relation $\pi_{X,Y}$ defined on $X \times Y$. The composition of a fuzzy set μ with a fuzzy relation π is defined by

$$\mu \circ \pi : Y \to [0, 1], \quad y \mapsto \sup_{x \in X} \{\min\{\mu(x), \pi(x, y)\}\} \,.$$

We can easily see that this is a fuzzification of the standard composition $\circ$ of two crisp sets $M \subseteq X$ and $R \subseteq X \times Y$, i. e.,

$$M \circ R \stackrel{\text{def}}{=} \{y \in Y \mid \exists x \in X : (x \in M \wedge (x, y) \in R)\} \subseteq Y \,.$$

The challenge in fuzzy control applications using relational equations is to search for a fuzzy relation π that satisfies all equations $\mu_{B_r} = \mu_{A_r} \circ \pi$ for every rule R_r with $r = 1, \ldots, k$. If multiple inputs $X_1, \ldots, X_n$ are used, then μ_A is defined on the product space $X = X_1 \times \cdots \times X_n$ as in (17.2). The fuzzy relation π can be found by determining the *Gödel relation* for every given relational equation, i. e.,

$$(x, y) \in \pi_{X,Y}^G \iff (x \in \mu_A \to y \in \mu_B) \,.$$

Here the implication arrow $\to$ represents the Gödel implication

$$a \to b = \begin{cases} 1 & \text{if } a \leq b, \\ b & \text{if } a > b. \end{cases}$$

So, actually a linguistic rule expresses the gradual rule in terms of *the more μ_A, the more μ_B*. Hence it constrains the fuzzy relation π by the inequality

$$\min(\mu_A(x), \pi(x, y)) \leq \mu_B(y)$$

for all $(x, y) \in X \times Y$. The Gödel implication is theoretically not the only way to represent π. *Dubois* and *Prade* [17.30, 31] give a variety of good conclusions, however, not to take another but the Gödel implication.

If the system of relational equations $\mu_{B_r} = \mu_{A_r} \circ \pi$ for $r = 1, \ldots, k$ is solvable, then the intersection of all rule's Gödel relations

$$\pi^G = \bigcap_{r=1}^{k} \pi_r^G(\mu_{A_r}(x), \mu_{B_r}(y))$$

is a solution with ∩ being the minimum *t*-norm. Due to the mathematical properties of the Gödel implication, the Gödel relation π^G is the greatest solution in terms of elementwise membership degrees.

To conclude this type of controller, we recall that the relation

$$\Pi(\{(x,y)\}) \stackrel{\text{def}}{=} \pi(x,y)$$

approximates if it is *possible* to assign the output value *y* to the input tuple *x*. Besides the overall *conjunctive* nature of the rules does *softly constrain* the control function φ. It might thus happen in practical applications that these constraints lead to contradictions if very narrow output fuzzy sets are assigned to overlapping input fuzzy sets. In such a case, the controller's output will be the empty fuzzy set which corresponds to no solution. One way to overcome this problem is to specify both narrow input fuzzy sets and broader output fuzzy sets. This procedure, however, limits the expressiveness and thus applicability of fuzzy logic-based controllers.

17.4 A Glance at Some Industrial Applications

Shortly after big success stories in the 1980s, mainly in Japan [17.2], many real-world control applications have been greatly solved using the Mamdani–Assilian approach all around the world. So did the research group of the paper's third author initiate the development of some automobile controllers.

We want to discuss two of these control processes that have been developed with Volkswagen AG (VW), i.e., the engine idle speed control [17.18] and the shift-point determination of an automatic transmission [17.32]. Both of these very successful Mamdani–Assilian controllers can nowadays be still found in VW automobiles. The idle speed controller is based on similarity relations which facilitates to interpret the control function as interpolation of a point-wise imprecisely known function. The shift-point determination continuously adapts the gearshift schedule between two extremes, i.e., economic and sporting. This controller determines a so-called sport factor and individually adapts the gearshift movements of the driver.

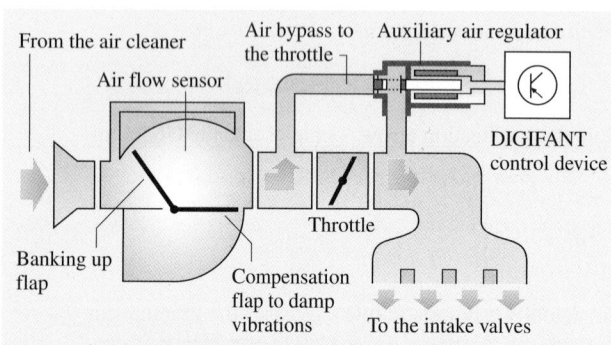

Fig. 17.4 Principle of the engine idle speed control

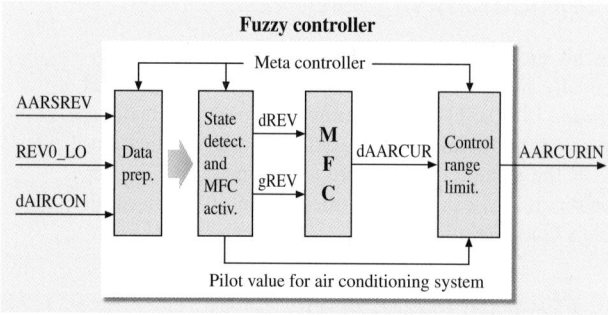

Fig. 17.5 Structure of the fuzzy controller

17.4.1 Engine Idle Speed Control

This controller shall adjust the idle speed of a spark ignition engine. Usually, a volumetric control is used to control the spark ignition engine. The principle is shown in Fig. 17.4. Here an auxiliary air regulator differs the cross-section of a bypass to the throttle.

The controller's task is to adjust the auxiliary air regulator's pulse width. In the case of a rapid fall of the number of revolutions, the controller shall drive the auxiliary air regulator to broaden the bypass cross-section. This increase of the air flow rate is measured by an air flow sensor which serves as controller signal. Then a new amount for the fuel injection has to be determined, and with a higher air flow rate the engine yields more torque. This, in turn, leads to a higher number of revolutions which could be decreased correspondingly by narrowing the bypass cross-section.

The ultimate goal is to reduce both fuel consumption and pollutant emissions. It is straightforward to achieve this goal by slowing down the idle speed. On the contrary, some automobile facilities, e.g., the air-conditioning system, are very often switched on and off which forces the number of revolutions to drop. So a very flexible controller is needed to adjust this process properly. *Schröder* et al. [17.32] even point out other problems of this control application.

As it turned out, the engineers who defined the similarity relations to model indistinguishability/similarity of two control states did not experience any big difficulties. Remember that the control expert must define a set of k input–output tuples $\left((x_r^{(1)}, \ldots, x_r^{(p)}), y_r\right)$. So, for each $r = 1, \ldots, k$ the output value y_r seems appropriate for the input $(x_r^{(1)}, \ldots, x_r^{(p)})$. Like that the control expert specifies a partial control function φ_0. According to (17.6), we directly obtain a Mamdani–Assilian controller by determining the extensional hull of φ_0 given the similarity relations. We thus obtain the rules from the partial control function φ_0 as

$R_r :$ if ξ_1 is approximately $x_r^{(1)}$ and $\ldots$

and ξ_p is approximately $x_r^{(p)}$

then η is approximately y_r .

Klawonn et al. [17.18] explain the more detailed theory of this approach.

Eventually, only two input variables are needed to control the engine idle speed controller, i. e.:

1. The deviation dREV [rpm] of the number of revolutions to the set value.
2. The gradient gREV [rpm] of the number of revolutions between two ignitions.

There exists just one output variable which is the change of current dAARCUR for the auxiliary air regulator. The final controller is shown in Fig. 17.5.

The control rules of the engine idle speed controller have been found from idle speed experiments. The partial control function $\varphi_0 : X_{(\mathrm{dREV})} \times X_{(\mathrm{gREV})} \rightarrow Y_{(\mathrm{dAARCUR})}$ is depicted in the upper half of Tab. 17.1.

The fuzzy controller has been defined by a similarity relation and the partial control mapping φ_0. With the center of area (COA) method, it yields a control surface as shown in Fig. 17.6. The function values here are evaluated in a grid of equally sampled input points. The respective Mamdani–Assilian controller has been found by relating each point of φ_0 to a linguistic term, e.g., negative big (nb), negative medium (nm), negative small (ns), and approximately zero (az). The resulting fuzzy partitions of dREV, gREV, and dAARCUR are displayed in Figs. 17.7, 17.8, 17.9, respectively. So, we obtain linguistic rules from φ_0 like

if dREV is A and gREV is B then dAARCUR is C .

The complete set of rules is given in the lower part of Tab. 17.1.

Klawonn et al. [17.18], *Schröder* et al. [17.32] show that this Mamdani–Assilian controller leads to a very smooth and thus better control behavior when com-

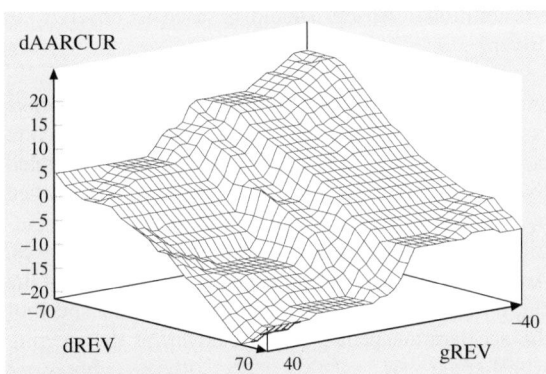

Fig. 17.6 Performance characteristics

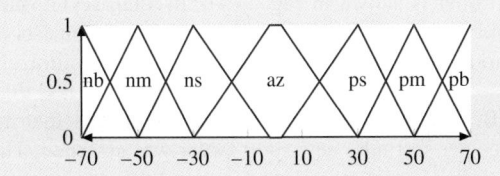

Fig. 17.7 Deviation dREV of the number of revolutions

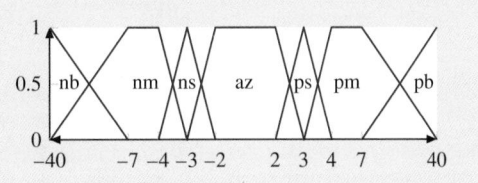

Fig. 17.8 Gradient gREV of the number of revolutions

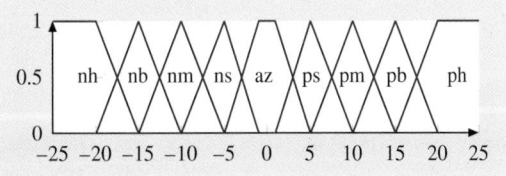

Fig. 17.9 Change of current dAARCUR for the auxiliary air regulator

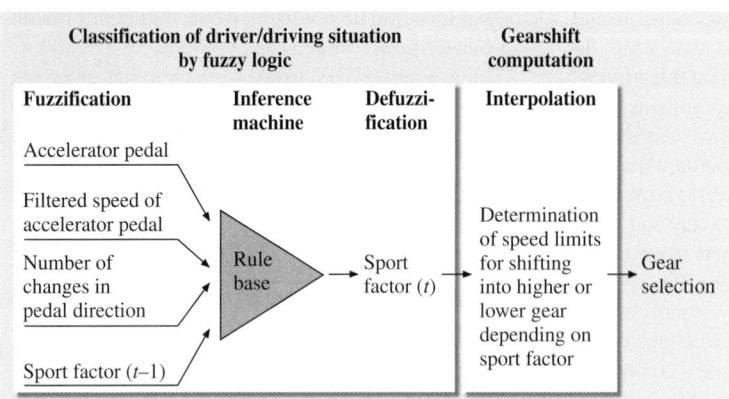

Fig. 17.10 Flowing shift-point determination with fuzzy logic

pared to classical controllers. Still, for this application, it has been much simpler to define a fuzzy controller than a one based on higher mathematics. Moreover, they found that this fuzzy controller reaches the desired setpoint precisely and fast. Last but not least, even increasing the load slowly does not change the control behavior significantly. So it is nearly impossible to experience any vibration, even after drastic load changes.

17.4.2 Flowing Shift–Point Determination

Conventional automatic transmissions select gears based on the so-called gearshift diagrams. Here, the gearshift simply depends on the accelerator position

Table 17.1 The partial control mapping φ_0 (*upper table*) and its corresponding fuzzy rule base (*lower table*)

		gREV						
		−40	−6	−3	0	3	6	40
	−70	20	15	15	10	10	5	5
	−50	20	15	10	10	5	5	0
	−30	15	10	5	5	5	0	0
dREV	0	5	5	0	0	0	−10	−5
	30	0	0	0	−5	−5	−10	−15
	50	0	−5	−5	−10	−15	−15	−20
	70	−5	−5	−10	−15	−15	−15	−15

		gREV						
		nb	nm	ns	az	ps	pm	pb
	nb	ph	pb	pb	pm	pm	ps	ps
	nm	ph	pb	pm	pm	ps	ps	az
	ns	pb	pm	ps	ps	az	az	az
dREV	az	ps	ps	az	az	az	nm	ns
	ps	az	az	az	ns	ns	nm	nb
	pm	az	ns	ns	nm	nb	nb	nh
	pb	ns	ns	nm	nb	nb	nb	nb

and the velocity. A lagging between the up and down shift avoids oscillating gearshift when the velocity varies slightly, e.g., during stop-and-go traffic. For instance, if the driver kicks gas with half throttle, the gearshift will start with the first gear. For a standardized behavior, a fixed diagram works well. Until 1994, the VW gear box had two different types of gearshift diagrams, i.e., economic *ECO* and sporting *SPORT*. An economic gearshift diagram switches gears at a low number of revolutions to reduce the fuel consumption. A sporting one leads to gearshifts at a higher number of revolutions. Since 1991 it was a research issue at VW to develop an individual adaption of shift-points. No additional sensors should be used to observe the driver.

The idea was that the car *observes* the driver [17.32] and classifies him or her into calm, normal, sportive (assigning a sport factor $\in [0, 1]$), or nervous (to calm down the driver). A test car from VW was operated by many different drivers. These people were classified by a human expert (passenger). Simultaneously, 14 attributes were continuously measured during test drives. Among them were variables such as the velocity of the car, the position of the acceleration pedal, the speed of the acceleration pedal, the kick down, or the steering wheel angle.

The final Mamdani controller was based on four input variables and one output. The basic structure of the controller is shown in Fig. 17.10. In total, seven rules could be identified at which the antecedent consists of up to four clauses. The program was highly optimized: It used 24 byte RAM and 702 byte ROM, i.e., less than 1 KB. The runtime was 80 ms which means that 12 times per second a new sport factor was assigned. The controller is in series since January 1995. It shows an excellent performance.

17.5 Automatic Learning of Fuzzy Controllers

The automatic generation of linguistic rules plays an important role in many applications, e.g., classification [17.33–36], regression [17.37–39], and image processing [17.40, 41]. Since fuzzy controllers are based on linguistic rules, automatic ways to tune and learn them have been developed for control applications as well [17.18, 19, 42–44].

How can a computer learn fuzzy rules from data to explain or support decisions like people do? We think that the fuzzy analysis of data can answer this question sufficiently [17.45]. The easiest and most common way is to use *fuzzy clustering* which automatically determines fuzzy sets from data.

Before we talk about the generation of linguistic rules from fuzzy clustering, however, let us briefly list some of the very diverse methods of fuzzy data analysis. *Grid-based* approaches define fixed fuzzy partitions for every variable. Every cell in that multidimensional grid may correspond to one rule [17.39]. Most well known are *hybrid methods* to induce fuzzy rules. Therefore, a fuzzy system is combined with computational intelligence techniques. For instance, *evolutionary algorithms* are used for guided searching the space of possible rule bases [17.46] or fuzzifying and thus summarizing a crisp set of rules [17.47]. *Neuro-fuzzy systems* use learning methods of artificial neural network (e.g., backpropagation) to tune the parameters of a network that can be directly understood as a fuzzy system [17.48]. Standard rule generation methods have been fuzzified as well (e.g., separate-and-conquer rule learning [17.49], decision trees [17.50], and support vector machines [17.51, 52]).

Using fuzzy clustering to learn fuzzy rules from data, we only refer to the standard fuzzy c-means algorithm (FCM) [17.53, 54]. Consider the input space $X \subset \mathbb{R}^n$ and the output space $Y \subset \mathbb{R}$. We observe m patterns $(\vec{x}_j, y_j) \in S \subseteq X \times Y$ where $j = 1, \ldots, m$. Running FCM on that dataset S leads to c cluster prototypes

$$\vec{c}_i = \left(c_i^{(1)}, \ldots, c_i^{(n)}, c_i^{(y)} \right)$$

with $i = 1, \ldots, c$ that can be seen as concatenation of both the input values $c_i^{(j)}, j = 1, \ldots, n$ and the output value $c_i^{(y)}$. Thus, every prototype represents one linguistic rule

$$R_i : \text{if } x \text{ is close to } \left(c_i^{(1)}, \ldots, c_i^{(n)} \right)$$
$$\text{then } y \text{ is close to } c_i^{(y)}.$$

Using the membership degrees U, we can rewrite these rules as

$$R_i : \text{if } \vec{u}_i^{\vec{x}}(\vec{x}) \text{ then } u_i^y(y). \tag{17.7}$$

The only problem is that FCM returns the membership degrees $\vec{u}_i(\vec{x}, y)$ of the product space $X \times Y$. To obtain rules like (17.7), we must *project* $\vec{u}_i$ onto $\vec{u}_i^{\vec{x}}$ and u_i^y. If $\vec{x}$ and y are restricted to $[\vec{x}_{\min}, \vec{x}_{\max}]$ and $[y_{\min}, y_{\max}]$, respectively, the projections are given by

$$\vec{u}_i^{\vec{x}}(\vec{x}) = \sup_{y \in [y_{\min}, y_{\max}]} \vec{u}_i(\vec{x}, y),$$
$$u_i^y(y) = \sup_{x \in [\vec{x}_{\min}, \vec{x}_{\max}]} \vec{u}_i(\vec{x}, y).$$

We can also project $\vec{u}_i$ onto each single input variable $X_1, \ldots, X_n$ by

$$u_{ik}(x^{(k)}) = \sup_{\vec{x}^{(\neg k)} \in [\vec{x}_{\min}^{(\neg k)}, \vec{x}_{\max}^{(\neg k)}]} \vec{u}_i^{\vec{x}}(\vec{x})$$

for $k = 1, \ldots, n$ where $\vec{x}^{(\neg k)} \stackrel{\text{def}}{=} (x^{(1)}, \ldots, x^{(k-1)}, x^{(k+1)}, \ldots, x^{(n)})$. We may thus write (17.7) in the form of a *Mamdani–Assilian rule* (17.1) as

$$R_i : \text{if } \bigwedge_{k=1}^{n} u_{ik}(x^{(k)}) \text{ then } u_i^y(y). \tag{17.8}$$

For one rule, the output value of an unseen input $\vec{x} \in \mathbb{R}^n$ will be equivalent to (17.2) if the minimum t-norm is used as conjunction $\wedge$. The overall output of the complete rule base is given by a disjunction $\vee$ of all rule outputs (cf. (17.4) if $\vee$ is the t-conorm maximum).

A crisp output can then again be computed by defuzzification, e.g., using the COG method (17.5). Since this computation is rather costly, the output membership functions u_i^y are commonly replaced by singletons, i.e.,

$$u_i^y(y) = \begin{cases} 1 & \text{if } y = c_i^{(y)}, \\ 0 & \text{otherwise.} \end{cases}$$

Since each rule consequently comprise the component $c_i^{(y)}$ of the cluster prototype, we can rewrite (17.8) as the *Sugeno-Yasukawa rule* [17.55]

$$R_i : \text{if } \bigwedge_{k=1}^{n} u_{ik}(x^{(k)}) \text{ then } y = c_i^{(y)}.$$

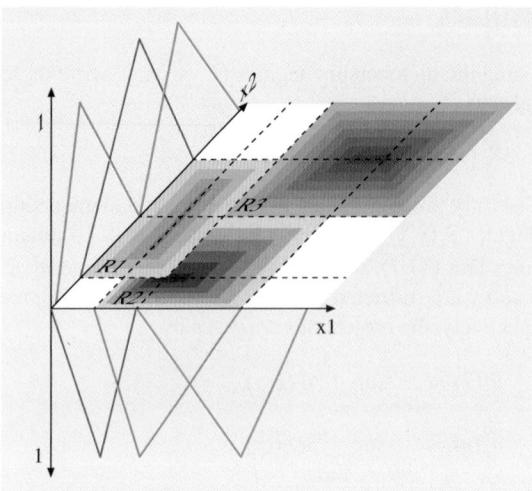

Fig. 17.11 Fuzzy rules and induced imprecise areas

These rules strongly resemble the neurons of an radial basis function (RBF) network. This will become clear if every membership function is Gaussian, i. e.,

$$\vec{u}_i^{\vec{x}}(\vec{x}) = \exp\left(\frac{\vec{x} - \boldsymbol{\mu}_i}{\boldsymbol{\sigma}_i}\right)^2 ,$$

and if there are normalized, i. e.,

$$\sum_{i=1}^{c} \vec{u}_i^{\vec{x}}(\vec{x}) = 1 \quad \text{for all} \quad \vec{x} \in \mathbb{R}^n .$$

Note that this link is used in neuro-fuzzy systems for both training fuzzy rules with backpropagation and initializing RBF networks with fuzzy rules [17.34].

17.5.1 Transfer Passenger Analysis Based on FCM

In this section, we present another real-world control problem that deals with the control of passenger movements and flows in terminal areas on an airport's land side. Especially during mass events such as world championships, concerts, or in the peak season in touristic areas, the capacities of passenger airports reach their upper limits. The conflict-free allocation, e.g., using intelligent destination boards and signpost, can increase the safety and security of passengers and airport employees. Thus, it is rational to study intelligent control approaches to allocate passengers from their arrival terminal to their departure terminal.

To evaluate different controllers, the German Aerospace Center (DLR) implemented a macroscopic

passenger flow model that simulates passenger movements. Here, probabilistic distributions are used to describe passenger movements in terminal areas. The approach of *Keller* and *Kruse* [17.56] constructs a fuzzy rule base using FCM to describe the transfer passenger amount between aircraft. These rules can be used as control feedback for the macroscopic simulation.

The following attributes of passengers are used to for analysis:

- The maximal amount of passengers in a certain aircraft (depending on the type of the aircraft)
- The distance between the airport of departure and the airport of destination (in three categories: short-, medium-, and long-haul)
- The time of departure
- The percentage of transfer passengers in the aircraft.

The number of clusters is determined by validity measures [17.41, 57] evaluating the whole partition of all data. The clustering algorithm is run for a varying number of clusters. The validity of the resulting partitions is then compared by different measures.

An example of resulting fuzzy clusters is given in Fig. 17.11. Here, every fuzzy cluster corresponds to one fuzzy rule. The color intensity indicates the firing strength of a specific rule. The imprecise areas are the fuzzy clusters where the color intensity indicates the membership degree. The tips of the fuzzy partitions are obtained in every domain by projections of the multidimensional cluster centers (as explained before in Sect. 17.5).

The fuzzy rules obtained by FCM are simplified through several steps. First, similar fuzzy sets are combined to one fuzzy set. Fuzzy sets similar to the universal fuzzy set are removed. Fuzzy rules with the same input clauses are either combined if they also share the same output clauses or else they are removed from the rule base. Eventually, FCM and the rule-simplifying process yield five rules.

Among them are the two following rules. If an aircraft with a relatively small amount of maximal passengers (80−200) has a short- or medium-haul destination departing late at night, then usually this flight has a high amount of transfer passengers (80−90%). If a flight with a medium-haul destination and a small aircraft (about 150 passengers) starts about noon, then it carries a relatively high amount of transfer passengers (ca. 70%). We refer the gentle reader to *Keller* and *Kruse* [17.56] for further details about this real-world control application.

17.6 Conclusions

In this chapter, we introduced fuzzy control – a human-inspired way to control a nonlinear process as an imprecisely defined function. We talked about classical control engineering and its limitations which also motivates the need for a human knowledge-based approach of control. This knowledge is typically represented as either Mamdani–Assilian rules or Takagi–Sugeno rules. We presented both types of fuzzy controllers, and also discussed the shortcomings of logic-based controllers, although they are mathematically well defined. We thoroughly presented two successful industrial applications of fuzzy control. We also stressed the necessity for automatic learning and tuning algorithms. We mentioned the most known approaches briefly and rule induction from fuzzy clustering in de-

tail. We also showed a real-world control application where such rules are used in the feedback loop of a simulation.

How will the future of fuzzy control look alike? Even in university lectures for control engineers, fuzzy control has become part of the curriculum years ago. With the drastically growing number of control systems – becoming more and more complex – a new generation of well-educated engineers and scientists will strengthen the presence of fuzzy controllers for real-world applications. In the far future, the use of data analysis techniques and algorithms will most probably drive evolving fuzzy controllers that are even able to react in nonstationary environments.

References

17.1 E.H. Mamdani: Application of fuzzy algorithms for the control of a simple dynamic plant, Proc. IEEE **121**(12), 1585–1588 (1974)

17.2 S. Yasunobu, S. Miyamoto: *Automatic Train Operation System by Predictive Fuzzy Control* (North-Holland, Amsterdam 1985) pp. 1–18

17.3 Google patents: http://patents.google.com/, last accessed on August 22, 2013

17.4 K. Hirota (Ed.): *Industrial Applications of Fuzzy Technology* (Springer, Tokio 1993)

17.5 T. Terano, M. Sugeno: *Applied Fuzzy Systems* (Academic, Boston 1994)

17.6 R.-E. Precup, H. Hellendoorn: A survey on industrial applications of fuzzy control, Comput. Ind. **62**(3), 213–226 (2011)

17.7 C. Moewes, R. Kruse: Fuzzy control for knowledge-based interpolation. In: *Combining Experimentation and Theory: A Hommage to Abe Mamdani*, Studies in Fuzziness and Soft Computing, Vol. 271, ed. by E. Trillas, P.P. Bonissone, L. Magdalena, J. Kacprzyk (Springer, Berlin, Heidelberg 2012) pp. 91–101

17.8 P. Podržaj, M. Jenko: A fuzzy logic-controlled thermal process for simultaneous pasteurization and cooking of soft-boiled eggs, Chemom. Intell. Lab. Syst. **102**(1), 1–7 (2010)

17.9 L.A. Zadeh: Outline of a new approach to the analysis of complex systems and decision processes, IEEE Trans. Syst. Man Cybern. **3**(1), 28–44 (1973)

17.10 L.A. Zadeh: Fuzzy sets, Inf. Control **8**(3), 338–353 (1965)

17.11 K. Tanaka, H.O. Wang: *Fuzzy Control Systems Design and Analysis: A Linear Matrix Inequality Approach* (Wiley, New York 2001)

17.12 K. Michels, F. Klawonn, R. Kruse, A. Nürnberger: *Fuzzy Control: Fundamentals, Stability and Design of Fuzzy Controllers*, Studies in Fuzziness and Soft Computing, Vol. 200 (Springer, Berlin, Heidelberg 2006)

17.13 K.J. Åström, B. Wittenmark: *Adaptive Control* (Courier Dover, Mineola 2008)

17.14 G.C. Goodwin, S.F. Graebe, M.E. Salgado: *Control System Design*, Vol. 240 (Prentice Hall, Upper Saddle River 2001)

17.15 E.H. Mamdani, S. Assilian: An experiment in linguistic synthesis with a fuzzy logic controller, Int. J. Man-Mach. Stud. **7**(1), 1–13 (1975)

17.16 R. Kruse, J. Gebhardt, F. Klawonn: *Foundations of Fuzzy Systems* (Wiley, Chichester 1994)

17.17 O. Cordón, M.J. del Jesus, F. Herrera: A proposal on reasoning methods in fuzzy rule-based classification systems, Int. J. Approx. Reason. **20**(1), 21–45 (1999)

17.18 F. Klawonn, J. Gebhardt, R. Kruse: Fuzzy control on the basis of equality relations with an example from idle speed control, IEEE Trans. Fuzzy Syst. **3**(3), 336–350 (1995)

17.19 F. Klawonn, R. Kruse: Equality relations as a basis for fuzzy control, Fuzzy Sets Syst. **54**(2), 147–156 (1993)

17.20 D. Boixader, J. Jacas: Extensionality based approximate reasoning, Int. J. Approx. Reason. **19**(3/4), 221–230 (1998)

17.21 F. Klawonn, J.L. Castro: Similarity in fuzzy reasoning, Mathw. Soft Comput. **2**(3), 197–228 (1995)

17.22 T. Takagi, M. Sugeno: Fuzzy identification of systems and its applications to modeling and control, IEEE Trans. Syst. Man Cybern. **15**(1), 116–132 (1985)

17.23 G. Feng: A survey on analysis and design of model-based fuzzy control systems, IEEE Trans. Fuzzy Syst. **14**(5), 676–697 (2006)

17.24 L. Wu, X. Su, P. Shi, J. Qiu: A new approach to stability analysis and stabilization of discrete-time ts fuzzy time-varying delay systems, IEEE Trans. Syst. Man Cybern. B: Cybern. **41**(1), 273–286 (2011)

17.25 K. Tanaka, H. Yoshida, H. Ohtake, H.O. Wang: A sum-of-squares approach to modeling and control of nonlinear dynamical systems with polynomial fuzzy systems, IEEE Trans. Fuzzy Syst. **17**(4), 911–922 (2009)

17.26 L.A. Zadeh: A theory of approximate reasoning, Proc. 9th Mach. Intell. Workshop, ed. by J.E. Hayes, D. Michie, L.I. Mikulich (Wiley, New York 1979) pp. 149–194

17.27 L.A. Zadeh: The role of fuzzy logic in the management of uncertainty in expert systems, Fuzzy Sets Syst. **11**(1/3), 197–198 (1983)

17.28 D. Dubois, H. Prade: *Possibility Theory: An Approach to Computerized Processing of Uncertainty* (Plenum Press, New York 1988)

17.29 L.A. Zadeh: Fuzzy sets as a basis for a theory of possibility, Fuzzy Sets Syst. **1**(1), 3–28 (1978)

17.30 D. Dubois, H. Prade: The generalized modus ponens under sup-min composition – A theoretical study. In: *Approximate Reasoning in Expert Systems*, ed. by M.M. Gupta, A. Kandel, W. Bandler, J.B. Kiszka (North-Holland, Amsterdam 1985) pp. 217–232

17.31 D. Dubois, H. Prade: Possibility theory as a basis for preference propagation in automated reasoning, 1992 IEEE Int. Conf. Fuzzy Syst. (IEEE, New York 1992) pp. 821–832

17.32 M. Schröder, R. Petersen, F. Klawonn, R. Kruse: Two paradigms of automotive fuzzy logic applications. In: *Applications of Fuzzy Logic: Towards High Machine Intelligence Quotient Systems*, Environmental and Intelligent Manufacturing Systems Series, Vol. 9, ed. by M. Jamshidi, A. Titli, L. Zadeh, S. Boverie (Prentice Hall, Upper Saddle River 1997) pp. 153–174

17.33 L.I. Kuncheva: *Fuzzy Classifier Design*, Studies in Fuzziness and Soft Computing, Vol. 49 (Physica, Heidelberg, New York 2000)

17.34 D. Nauck, R. Kruse: A neuro-fuzzy method to learn fuzzy classification rules from data, Fuzzy Sets Syst. **89**(3), 277–288 (1997)

17.35 R. Mikut, J. Jäkel, L. Gröll: Interpretability issues in data-based learning of fuzzy systems, Fuzzy Sets Syst. **150**(2), 179–197 (2005)

17.36 R. Mikut, O. Burmeister, L. Gröll, M. Reischl: Takagi-Sugeno-Kang fuzzy classifiers for a special class of time-varying systems, IEEE Trans. Fuzzy Syst. **16**(4), 1038–1049 (2008)

17.37 J.A. Dickerson, B. Kosko: Fuzzy function approximation with ellipsoidal rules, IEEE Trans. Syst. Man Cybern. B: Cybern. **26**(4), 542–560 (1996)

17.38 D. Nauck, R. Kruse: Neuro-fuzzy systems for function approximation, Fuzzy Sets Syst. **101**(2), 261–271 (1999)

17.39 L. Wang, J.M. Mendel: Generating fuzzy rules by learning from examples, IEEE Trans. Syst. Man Cybern. **22**(6), 1414–1427 (1992)

17.40 J.C. Bezdek, J. Keller, R. Krisnapuram, N.R. Pal: *Fuzzy Models and Algorithms for Pattern Recognition and Image Processing*, The Handbooks of Fuzzy Sets, Vol. 4 (Kluwer, Norwell 1999)

17.41 F. Höppner, F. Klawonn, R. Kruse, T. Runkler: *Fuzzy Cluster Analysis: Methods for Classification, Data Analysis and Image Recognition* (Wiley, New York 1999)

17.42 F. Klawonn, R. Kruse: Automatic generation of fuzzy controllers by fuzzy clustering, 1995 IEEE Int. Conf. Syst. Man Cybern.: Intell. Syst. 21st Century, Vol. 3 (IEEE, Vancouver 1995) pp. 2040–2045

17.43 F. Klawonn, R. Kruse: Constructing a fuzzy controller from data, Fuzzy Sets Syst. **85**(2), 177–193 (1997)

17.44 Z.-W. Woo, H.-Y. Chung, J.-J. Lin: A PID type fuzzy controller with self-tuning scaling factors, Fuzzy Sets Syst. **115**(2), 321–326 (2000)

17.45 R. Kruse, P. Held, C. Moewes: On fuzzy data analysis, Stud. Fuzzin. Soft Comput. **298**, 351–356 (2013)

17.46 O. Cordón, F. Gomide, F. Herrera, F. Hoffmann, L. Magdalena: Ten years of genetic fuzzy systems: Current framework and new trends, Fuzzy Sets Syst. **141**(1), 5–31 (2004)

17.47 C. Moewes, R. Kruse: Evolutionary fuzzy rules for ordinal binary classification with monotonicity constraints. In: *Soft Computing: State of the Art Theory and Novel Applications*, Studies in Fuzziness and Soft Computing, Vol. 291, ed. by R.R. Yager, A.M. Abbasov, M.Z. Reformat, S.N. Shahbazova (Springer, Berlin, Heidelberg 2013) pp. 105–112

17.48 D. Nauck, F. Klawonn, R. Kruse: *Foundations of Neuro-Fuzzy Systems* (Wiley, New York 1997)

17.49 J.C. Hühn, E. Hüllermeier: FR3: A fuzzy rule learner for inducing reliable classifiers, IEEE Trans. Fuzzy Syst. **17**(1), 138–149 (2009)

17.50 C. Olaru, L. Wehenkel: A complete fuzzy decision tree technique, Fuzzy Sets Syst. **138**(2), 221–254 (2003)

17.51 C. Moewes, R. Kruse: Unification of fuzzy SVMs and rule extraction methods through imprecise domain knowledge, Proc. Int. Conf. Inf. Process. Manag. Uncertain. Knowl.-Based Syst. (IPMU-08), ed. by J.L. Verdegay, L. Magdalena, M. Ojeda-Aciego (Torremolinos, Málaga 2008) pp. 1527–1534

17.52 C. Moewes, R. Kruse: On the usefulness of fuzzy SVMs and the extraction of fuzzy rules from SVMs, Proc. 7th Conf. Eur. Soc. Fuzzy Logic Technol. (EUSFLAT-2011) and LFA-2011, Vol. 17, ed. by S. Galichet, J. Montero, G. Mauris (Atlantis, Amsterdam, Paris 2011) pp. 943–948

17.53 J.C. Bezdek: Fuzzy Mathematics in Pattern Classification, Ph.D. Thesis (Cornell University, Itheca 1973)

17.54 J.C. Bezdek: *Pattern Recognition with Fuzzy Objective Function Algorithms* (Kluwer, Norwell 1981)

17.55 M. Sugeno, T. Yasukawa: A fuzzy-logic-based approach to qualitative modeling, IEEE Trans. Fuzzy Syst. **1**(1), 7–31 (1993)

17.56 A. Keller, R. Kruse: Fuzzy rule generation for transfer passenger analysis, Proc. 1st Int. Conf. Fuzzy Syst. Knowl. Discovery (FSDK'02), ed. by L. Wang, S.K. Halgamuge, X. Yao (Orchid Country Club, Singapore 2002) pp. 667–671

17.57 R. Kruse, C. Döring, M. Lesot: Fundamentals of fuzzy clustering. In: *Advances in Fuzzy Clustering and Its Applications*, ed. by J.V. de Oliveira, W. Pedrycz (Wiley, Chichester 2007) pp. 3–30

18. Interval Type-2 Fuzzy PID Controllers

Tufan Kumbasar, Hani Hagras

The aim of this chapter is to present a general overview about interval type-2 fuzzy PID (proportional-integral-derivative) controller structures. We will focus on the standard double input direct action type fuzzy PID controller structures and their present design methods. It has been shown in various works that the type-1 fuzzy PID controllers, using crisp type-1 fuzzy sets, might not be able to fully handle the high levels of uncertainties associated with control applications while the type-2 fuzzy PID controller using type-2 fuzzy sets might be able to handle such uncertainties to produce a better control performance. Thus, we will classify and examine the handled fuzzy PID controllers within two groups with respect to the fuzzy sets they employ, namely type-1 and interval type-2 fuzzy sets. We will present and examine the controller structures of the direct action type-1 fuzzy PID and interval type-2 fuzzy PID controllers on

a generic, a symmetrical 3×3 rule base. We will present general information about the type-1 fuzzy PID and interval type-2 fuzzy PID controllers tuning parameters and design strategies. Finally, we will present a simulation study to evaluate the control performance of the type-1 fuzzy PID and interval type-2 fuzzy PID on a first-order plus time-delay benchmark process.

18.1 Fuzzy Control Background

It is a known fact that the conventional PID controllers are the most popular controllers used in industry due to their simple structure and cost efficiency [18.1, 2]. However, the PID controller being linear is not suited for strongly nonlinear and uncertain systems. Thus, fuzzy logic controllers (FLCs) are extensively used as an alternative to PID control in processes where the system dynamics is either very complex or exhibit highly nonlinear characteristics. FLCs have achieved a huge success in real-world control applications since it does not require the process model and the controller can be constructed based on the human operator's control expertise.

In the fuzzy control literature, fuzzy PID controllers (FPID) are often mentioned as an alternative to the conventional PID controllers since they are analogous to the conventional PID controllers from the input–output

relationship point of view [18.3–6]. The FPID controllers can be classified into three major categories as direct action type, fuzzy gain scheduling type, and hybrid type fuzzy PID controllers [18.6]. The direct action type can also be classified into three categories according to the number of inputs as single input, double input, and triple input direct action FPID controllers [18.6]. In the literature, researchers mainly focused on and analyzed double input direct action FPID controllers [18.6–10]. Numerous techniques have been developed in the literature for analyzing and designing a wide variety of FPID control systems. After the pioneer study by *Qiao* and *Mizumoto* [18.7], the main research was focused on type-1 fuzzy PID controllers (T1-FPID); however, a growing number of techniques have been developed for interval type-2 fuzzy PID controllers (IT2-FPID) controllers, recently.

It has been demonstrated that type-2 fuzzy logic systems are much more powerful tools than ordinary (type-1) fuzzy logic systems to represent highly nonlinear and/or uncertain systems. As a consequence, type-2 fuzzy logic systems have been applied in various areas especially in control system design. The internal structure of the interval type-2 fuzzy logic controllers (IT2-FLC) is similar to the type-1 counterpart. However, the major difference is that at least one of the input fuzzy sets (FSs) is an interval type-2 fuzzy set (IT2-FS) [18.11]. Thus, a type reducer is needed to convert type-2 sets into a type-1 fuzzy set before a defuzzification procedure can be performed [18.12]. Generally, interval type-2 fuzzy logic systems achieve better control performance because of the additional degree of freedom provided by the footprint of uncertainty (FOU) in their membership functions [18.13]. Consequently, IT2-FLCs have attracted much research interest, especially in control applications, since they are a much more pow-

erful to handle uncertainties and nonlinearities [18.14]. Thus, several applications employed successfully IT2-FLCs such as pH control [18.13], liquid-level process control [18.15], autonomous mobile robots [18.14, 16, 17], and bioreactor control [18.18].

In this chapter, we will focus on the most commonly used double input direct action type FPID controller structures. We will first present the general structure of the FPID controller and then classify the FPID controllers within two groups with respect to the fuzzy sets they employ, namely type-1 and interval type-2 fuzzy sets. Thus, we will present and examine the structures of the T1-FPID and T2-FPID controllers on a generic, a symmetrical 3×3 rule base. We will present detailed information about their internal structures and design strategies presented in the literature. Finally, we will evaluate the control performance of the T1-FPID and IT2-FPID on a first-order plus time delay benchmark process.

18.2 The General Fuzzy PID Controller Structure

In this section, we present the general structure of the two input direct action type FPID controllers formed using a fuzzy PD controller with an integrator and a summation unit at the output [18.7–10]. The standard FPID controller is constructed by choosing the inputs to be error (e) and derivative of error (Δe) as shown and the output is the control signal (u) as illustrated in Fig. 18.1. The output of the FPID is defined as

$$u = \alpha U + \beta \int U \, \mathrm{d}t \,, \tag{18.1}$$

where U is the output of the fuzzy inference system.

The design parameters of the FPID controller structure can be summarized within two groups, structural parameters and tuning parameters [18.6]. The structural parameters include input/output variables to fuzzy in-

ference, fuzzy linguistic sets, type of membership functions, fuzzy rules, and the inference mechanism, i.e., the fuzzy logic controller. In the handled FPID structure, the FLC is constructed as a set of heuristic control rules, and the control signal is directly deduced from the knowledge base and the fuzzy inference as done in diagonal rule base generation approaches [18.7–10]. More detailed information about the internal structure of the FPID will be presented in the following subsections. Usually the structural parameters of the FPID controller structure are determined during an off-line design.

The tuning parameters include input/output scaling factors (SFs) and parameters of membership functions (MFs). As can be seen from Fig. 18.1, the handled FPID controller structure has two input and two output scaling factors [18.7–10]. The input SFs K_e (for error (e)) and K_d (for the change of error (Δe)) normalize the inputs to the common interval $[-1, 1]$ in which the membership functions of the inputs are defined (thus $e(t)$ and $\Delta e(t)$)) are converted after normalization into E and ΔE. While the FLC output (U) is mapped onto the respective actual output (u) domain by output SFs α and β. Usually, the tuning parameters can be calculated during offline design process as well as online adjustments of the controller to enhance the process performance [18.9, 10].

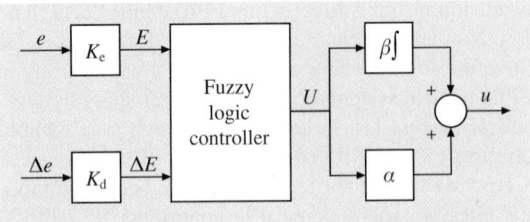

Fig. 18.1 Illustration of the FPID controller structure

In the following subsections, we will examine the structures of the T1-FPID and IT2-FPID controllers. We will present detailed information about the T1-FPID and IT2-FPID internal structures, design parameters, and tuning strategies. Finally, we will present a comparative simulation results to show the superiority of the interval type-2 fuzzy PID controller compared to its type-1 counterparts.

18.2.1 Type-1 Fuzzy PID Controllers

In this subsection, we will start by presenting the internal structure of the handled T1-FPID controller and then we will present brief information about the design strategies for the T1-FPID controller structure in the literature.

The Internal Structure of the Type-1 Fuzzy PID Controller

In the handled T1-FPID structure, a symmetrical 3×3 rule base is used as shown in Table 18.1. The rule structure of the type-1 fuzzy logic controller (T1-FLC) is as follows

$$R_m: \text{If } E \text{ is } A_{1k} \text{ and } \Delta E \text{ is } A_{2l} \text{ then } U \text{ is } G_m , \quad (18.2)$$

where E (normalized error) and ΔE (normalized change of error) are the inputs, U is the output of FLC, G_n is the consequent crisp set ($f = 1 \ldots F = 9$), and F is the number of rules. Here, A_{1k} and A_{2l} represent the type-1 membership functions (T1-MFs) ($k = 1, 2, K = 3$; $l = 1, 2, L = 3$), K, and L are the number of MFs that cover the universe of discourse of the inputs E and ΔE, respectively. In this chapter, we will employ three triangular type T1-MFs for each input domain (E and ΔE) and denote them as N (negative), Z (zero), and P (positive). The T1-MFs of the T1-FLC are defined with the three parameters (l_{ij}, c_{ij}, r_{ij}; $i = 1, I = 2$; $j = 1, 2, J = 3$), as shown in Fig. 18.2a. Here, I is the total number of the inputs ($I = 2$) and J ($J = K = L = 3$) is the total number of MFs. The outputs of the FLC are

Table 18.1 The rule base of the FPID controller

$E / \Delta E$	N	Z	P
N	N	NM	Z
Z	NM	Z	PM
P	Z	PM	P

defined with five crisp singleton consequents (negative (N) $= y_N$, negative medium (NM) $= y_{NM}$, zero (Z) $= y_Z$, positive medium (PM) $= y_{PM}$, positive (P) $= y_P$) as illustrated in Fig. 18.2b. The implemented T1-FLCs use the product implication and the center of sets defuzzification method. Thus, the output (U) of the T1-FLC is defined as

$$U = \frac{\sum_{m=1}^{M} f_m G_m}{\sum_{m=1}^{M} f_m} , \quad (18.3)$$

where f_m is the total firing strength for the m-th rule is defined as

$$f_m = \mu_{A_{1k}} * \mu_{A_{2l}} . \quad (18.4)$$

Here, $*$ represents the product implication (the t-norm) and $\mu_{A_{1k}}$ and $\mu_{A_{2l}}$ are the membership grades of the A_{1k} and A_{2l} T1-FMs, respectively.

Type-1 Fuzzy PID Design Strategies

In the design of the handled T1-FPID controller structure with a 3×3 rule base, the parameters to be determined are the scaling factors and the parameters of the antecedent and consequent membership functions. The antecedent MFs of the T1-FPID controller that are labeled as the N and P are defined for each input with two parameters each which are c_{i1}, r_{i1} (for N) and l_{i3}, c_{i3} (for P) ($i = 1, 2$), respectively, while the linguistic label Z is defined with three parameters which are l_{i2}, c_{i2}, r_{i2}, ($i = 1, 2$). Hence for two inputs, the total number of the antecedent membership function parameters to be designed for the T1-FPID is then $2 \times 7 = 14$. Moreover, five output consequent parameters (y_N, y_{NM}, y_Z, y_{PM}, y_P) have to be determined. Thus,

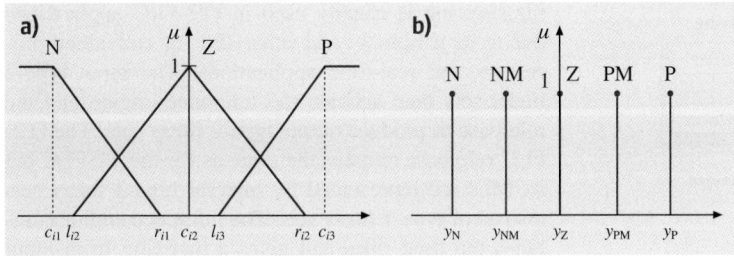

Fig. 18.2a,b Illustration of the (**a**) antecedent MFs (**b**) consequent MFs of the T1-FLC

in total 19 MF parameters have to be tuned. Besides the input and output scaling factors (K_e, K_d, α and β) of the T1-FPID controller must also be determined. Thus, there are 19 MF and four SF parameters ($19 + 4 = 23$) that have to be tuned for the handled T1-FPID controller.

In the fuzzy control literature, one method for the T1-FPID controller design is by employing evolutionary algorithms [18.19, 20]. Besides *Ahn* and *Truong* [18.21] used a robust extended Kalman filter to tune the membership functions of the fuzzy controller during the system operation process to improve the control performance in an online manner. Moreover, various heuristic and nonheuristic scaling factor tuning algorithms have been presented in the case of the systems that own nonlinearities, parameter changes, modeling errors, disturbances [18.7, 9, 10, 22–25].

18.2.2 Interval Type-2 Fuzzy PID Controllers

In this subsection, we will start by presenting the internal structure of the handled IT2-FPID controller and then we will present brief information about the design strategies for the IT2-FPID controller structure in the literature.

The Internal Structure of the Interval Type-2 Fuzzy PID Controller

In the handled IT2-FPID structures, the same 3×3 rule base presented for T1-FPID controller is used which is presented in Table 18.1. The internal structure of the IT2-FPID is similar to the type-1 counterpart. However, the major differences are that IT2-FLCs employ IT2-FSs (rather than type-1 fuzzy sets) and the IT2-FLCs process interval type-2 fuzzy sets (IT2-FSs) and thus the IT2-FLC has the extra type-reduction process [18.12, 14].

Type-2 fuzzy sets are the generalized forms of type-1 fuzzy sets. A type-2 fuzzy set ($\tilde{A}$) is characterized by

a type-2 membership function $\mu_{\tilde{A}}(x, u)$, i.e.,

$$\tilde{A} = \{((x, u), \mu_{\tilde{A}}(x, u)) \mid \forall x \in X,$$
$$\forall u \in J_x \subseteq [0, 1]\} \quad (18.5)$$

in which $0 \leq \mu_{\tilde{A}}(x, u) \leq 1$.

For a continuous universe of discourse, $\tilde{A}$ can be also expressed as

$$\tilde{A} = \int_{x \in X} \int_{u \in J_x} \mu_{\tilde{A}}(x, u)/(x, u), \quad J_x \subseteq [0, 1], \quad (18.6)$$

where $\iint$ denotes union over all admissible x and u [18.12, 14]. J_x is referred to as the primary membership of x, while $\mu_{\tilde{A}}(x, u)$ is a type-1 fuzzy set known as the secondary set. The uncertainty in the primary membership of a type-2 fuzzy set $\tilde{A}$ is defined by a region named footprint of uncertainty (FOU). The FOU can be described in terms of an upper membership function ($\overline{\mu}_{\tilde{A}}$) and a lower membership function ($\underline{\mu}_{\tilde{A}}$). The primary membership is called J_x, and its associated possible secondary membership functions can be trapezoidal, interval, etc. When the interval secondary membership function is employed an IT2-FS (such as the ones shown in Fig. 18.4a) is obtained [18.12, 14]. In other words, when $\mu_{\tilde{A}}(x, u) = 1$ for $\forall u \in J_x \subseteq [0, 1]$, an IT2-FS is constructed.

The internal structure of the IT2-FLC is given in Fig. 18.3. Similar to a T1-FLC, an IT2-FLC includes fuzzifier, rule-base, inference engine, and substitutes the defuzzifier by the output processor comprising a type reducer and a defuzzifier. The IT2-FLC uses interval type-2 fuzzy sets (such as the ones shown in Fig. 18.4a) to represent the inputs and/or outputs of the FLC. In the interval type-2 fuzzy sets all the third dimension values are equal to one. The use of IT2-FLC helps to simplify the computation (as opposed to the general type-2 FLC where the third dimension of the type-2 fuzzy sets can take any shape).

The IT2-FLC works as follows: the crisp inputs are first fuzzified into input type-2 fuzzy sets; singleton fuzzification is usually used in IT2-FLC applications due to its simplicity and suitability for embedded processors and real-time applications. The input type-2 fuzzy sets then activate the inference engine and the rule base to produce output type-2 fuzzy sets. The IT2-FLC rule base remains the same as for the T1-FLC but its MFs are represented by interval type-2 fuzzy sets instead of type-1 fuzzy sets. The inference engine combines the fired rules and gives a mapping from input

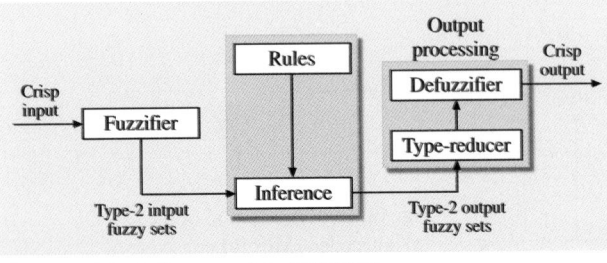

Fig. 18.3 Block diagram of the IT2-FLC

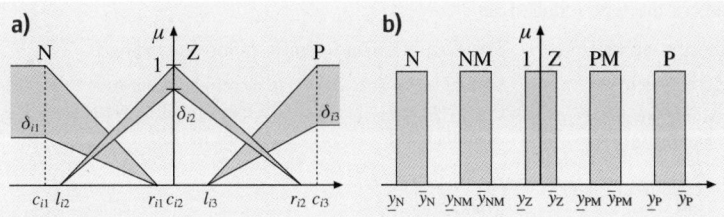

Fig. 18.4a,b Illustration of the (**a**) antecedent MFs (**b**) consequent MFs of the IT2-FLC

type-2 fuzzy sets to output type-2 fuzzy sets. The type-2 fuzzy output sets of the inference engine are then processed by the type reducer which combines the output sets and performs a centroid calculation which leads to type-1 fuzzy sets called the type-reduced sets. There are different types of type-reduction methods. In this paper, we will be using the center of sets type reduction as it has reasonable computational complexity that lies between the computationally expensive centroid type-reduction and the simple height and modified height type-reductions which have problems when only one rule fires [18.12]. After the type-reduction process, the type-reduced sets are defuzzified (by taking the average of the type-reduced sets) to obtain crisp outputs. More information about the type-2 fuzzy logic systems and their benefits can be found in [18.11, 12, 14].

The rule structure of the interval type-2 fuzzy logic controller is as follows

$$R_m: \text{If } E \text{ is } \tilde{A}_{1k} \text{ and } \Delta E \text{ is } \tilde{A}_{2l} \text{ then } U \text{ is } \tilde{G}_m \ , \quad (18.7)$$

where E (normalized error) and ΔE (normalized change of error) are the inputs, U is the output of IT2-FLC, $\tilde{G}_m$ is the consequent interval set ($\tilde{G}_m = [\underline{g}_m, \overline{g}_m]$, $m = 1, \dots, M = 9$) and M is the number of rules. The antecedents of the IT2-FLC are defined interval type-2 membership functions (IT2-MFs) ($\tilde{A}_{1k}, \tilde{A}_{2l}$) for the inputs E and ΔE, respectively, which can be simply obtained by extending/blurring the T1-MFs (A_{1k}, A_{2l}) of the T1-FLC. Here, the IT2-MF is defined with four parameters ($l_{ij}, c_{ij}, r_{ij}, \delta_{ij}; i = 1, 2, j = 1, 2, 3$), as shown in Fig. 18.4a. Since the input IT2-FS is described in terms of an upper membership function ($\overline{\mu}_{\tilde{A}}$) and a lower membership function ($\underline{\mu}_{\tilde{A}}$), the total firing strength for the mth rule is

$$\tilde{f}_m = \left[\overline{f}_m \underline{f}_m \right] \ , \quad (18.8)$$

where $\tilde{f}_m$ is the total firing interval and is defined as

$$\underline{f}_m = \underline{\mu}_{\tilde{A}_{1k}} * \underline{\mu}_{\tilde{A}_{2l}} \quad (18.9)$$

$$\overline{f}_m = \overline{\mu}_{\tilde{A}_{1k}} * \overline{\mu}_{\tilde{A}_{2l}} \ . \quad (18.10)$$

Here, $*$ represents the product implication (the t-norm) and $\underline{\mu}_{\tilde{A}_{1k}}, \underline{\mu}_{\tilde{A}_{2l}}$ and $\overline{\mu}_{\tilde{A}_{1k}}, \overline{\mu}_{\tilde{A}_{2l}}$ are the lower and upper membership grades of the $\tilde{A}_{1k}$ and $\tilde{A}_{2l}$ IT2-FMs, respectively.

The consequent membership functions of the IT2-FLC are defined with five interval consequents and label them as negative (N) = $[\underline{y}_N \overline{y}_N]$, negative medium (NM) = $[\underline{y}_{NM} \overline{y}_{NM}]$, zero (Z) = $[\underline{y}_Z \overline{y}_Z]$, positive medium (PM) = $[\underline{y}_{PM} \overline{y}_{PM}]$, and positive (P) = $[\underline{y}_P \overline{y}_P]$ as shown in Fig. 18.4b.

The implemented IT2-FLC uses the center of sets type reduction method [18.12]. It has been demonstrated that the defuzzified output of an IT2-FLC can be calculated as

$$U = \frac{U_l + U_r}{2} \ , \quad (18.11)$$

where U_l and U_r are the left- and right-end points, respectively, of the type reduced set, are defined as follows

$$U_l = \frac{\sum_{m=1}^{L} \overline{f}_m G_m + \sum_{L+1}^{M=9} \underline{f}_m G_m}{\sum_{m=1}^{L} \overline{f}_m + \sum_{L+1}^{M=9} \underline{f}_m} \ , \quad (18.12)$$

$$U_r = \frac{\sum_{m=1}^{R} \underline{f}_m G_m + \sum_{R+1}^{M=9} \overline{f}_m G_m}{\sum_{m=1}^{R} \underline{f}_m + \sum_{R+1}^{M=9} \overline{f}_m} \ . \quad (18.13)$$

The typed reduced set can be calculated by using the iterative Karnik and Mendel method (KM), which is given in Table 18.2 [18.26].

Interval Type-2 Fuzzy PID Design Strategies

In the interval type-2 fuzzy PID control design strategy, the scaling factors, the parameters of the antecedent, and consequent membership functions of the IT2-FPIDs have to be determined. The antecedent IT2-FSs of IT2-FPID controller that are labeled as the P and N for each input are defined with three parameters each which are ($c_{i1}, r_{i1}, \delta_{i1}$), and ($l_{i3}, c_{i3}, \delta_{i3}, i = 1, 2$), respectively, while the IT2-FS labeled as Z is defined with

Table 18.2 Calculation of the two end points of the type reduced set

Steps	The Karnik Mendel algorithm for computing U_l	The Karnik Mendel algorithm for computing U_r
1.	Sort $\underline{g}_m$ ($m = 1, \ldots, M$) in increasing order such that $\underline{g}_1 \leq \underline{g}_2 \leq \cdots \leq \underline{g}_M$. Match the corresponding weights $\bar{f}_M \underline{f}_M$ (with their index corresponds to the renumbered $\underline{g}_m$)	Sort $\bar{g}_m$ ($m = 1, 2, \ldots, M$) in increasing order such that $\bar{g}_1 \leq \bar{g}_2 \leq \cdots \leq \bar{g}_M$. Match the corresponding weights $\bar{f}_m \underline{f}_m$ (with their index corresponds to the renumbered $\bar{g}_m$)
2.	Initialize f_m by setting $$f_m = \frac{\bar{f}_m + \underline{f}_m}{2}, \quad m = 1, 2, \ldots M$$ Compute $$U = \frac{\sum_{m=1}^{M} f_m \underline{g}_m}{\sum_{m=1}^{M} f_m}$$	Initialize f_m by setting $$f_m = \frac{\bar{f}_m + \underline{f}_m}{2} \quad m = 1, 2, \ldots, M$$ Compute $$U = \frac{\sum_{m=1}^{M} f_m \underline{g}_m}{\sum_{m=1}^{M} f_m}$$
3.	Find the switch point L ($1 \leq L \leq M - 1$) such that $$\underline{g}_m \leq U \leq \underline{g}_{m+1}$$	Find the switch point R ($1 \leq L \leq M - 1$) such that $$\bar{g}_m \leq U \leq \bar{g}_{m+1}$$
4.	Set $f_m = \begin{cases} \bar{f}_m & m \leq L \\ \underline{f}_m & m > L \end{cases}$ and compute $$U' = \frac{\sum_{m=1}^{M} f_m \underline{g}_m}{\sum_{m=1}^{M} f_m}$$	Set $f_m = \begin{cases} \underline{f}_m & m \leq R \\ \bar{f}_m & m > R \end{cases}$ and compute $$U' = \frac{\sum_{m=1}^{M} f_m \bar{g}_m}{\sum_{m=1}^{M} f_m}$$
5.	Check if $U = U'$. If not go to step 3 and set $U' = U$. If yes stop and set $U_l = U'$.	Check if $U = U'$. If not go to step 3 and set $U' = U$. If yes stop and set $U_r = U'$.

four parameters ($l_{i2}, c_{i2}, r_{i2}, \delta_{i2}, i = 1, 2$). Consequently, for the two inputs the total numbers to be designed is $2 \times 10 = 20$. Moreover, 10 output consequent parameters ($\underline{y}_N, \bar{y}_N, \underline{y}_{NM}, \bar{y}_{NM}, \underline{y}_Z, \bar{y}_Z, \underline{y}_{PM}, \bar{y}_{PM}, \underline{y}_P, \bar{y}_P$) have to be determined. Hence, in total 30 MF parameters have to be tuned. Besides, the input and output scaling factors (K_e, K_d, α, and β) of the controller must also be determined. Thus, there are 30 MF and four SF parameters ($30 + 4 = 34$) have to be tuned for the handled IT2-FPID controller. It is obvious that, the IT2-FPID has 11 more tuning parameters, i. e., extra degrees of freedom, than the T1-FPID controller structure (23 parameters).

The systematic design of type-2 fuzzy controllers is a challenging problem since the output cannot be presented in a closed form due to the KM-type reduction method. To overcome this bottleneck, alternative type reduction algorithms which are closed-form approximations to the original KM-type reduction algorithm have been proposed and employed in controller design [18.27, 28]. However, the main difficulty is to tune the relatively big number of parameters of the

IT2-FPID controller structure. Thus, several studies have employed various techniques for the design problem including genetic algorithms [18.15, 29], particle swarm optimization [18.30], and ant colony optimization [18.31].

In practical point of view, the IT2-FPID controller design problem can be simply solved by blurring/extending MFs of an existing T1-FPID controller [18.16, 28, 32]. Moreover, each rule consequent can also be chosen as a crisp number ($\underline{g}_m = \bar{g}_m$) to reduce the number of parameters to be determined. It is also common to set the consequent parameters and the scaling factors to the same value of a predesigned T1-FPID controller. Thus, in the IT2-FPID design only the antecedent membership parameters have to be tuned [18.28]. This design approach will reduce the parameters to be tuned from 34 to 20 since only the parameters of the antecedent membership functions must be designed. The design of the antecedent MFs can be solved by extensively trial and error procedures or employing evolutionary algorithms [18.33].

18.3 Simulation Studies

In this section, we will compare the performances of the IT2-FPID controller with the T1-FPID controller for the following first-order plus-time delay process

$$G(s) = \frac{K}{\tau s + 1} e^{-Ls}, \tag{18.14}$$

where K is the gain, L is the time delay, and τ is the time constant of the process. The nominal system parameters are $K = 1, L = 1$ and $\tau = 1$.

We will first design a T1-FPID controller and then extend the type-1 fuzzy controller to design an IT2-FPID controller structure since type-2 fuzzy logic theory is a generalization and extension of type-1 fuzzy logic theory. We will characterize each input domain (E and ΔE) of the T1-FPID controllers with three uniformly distributed symmetrical triangular MFs. The parameters of the MFs are tabulated in Table 18.3. We will

set the consequent parameters as $y_N = -1.0$ (negative), $y_{NM} = -0.75$ (negative medium), $y_Z = 0.0$ (zero), $y_{PM} = 0.75$ (positive medium), $y_P = 1.0$ (positive) to obtain a standard diagonal rule base. Then, the scaling factors of the T1-FPID have been chosen such that to obtain a fast and satisfactory output response for

Table 18.3 The antecedent MF parameters of the T1-FPID and IT2-FPID controllers

		T1-FPID			IT2-FPID			
		l	*c*	*r*	*l*	*c*	*r*	*δ*
E	N		−1.0	0.0		−1.0	0.0	0.2
	Z	−1.0	0.0	1.0	−1.0	0.0	1.0	0.9
	P	0.0	1.0		0.0	1.0		0.2
ΔE	N		−1.0	0.0		−1.0	0.0	0.2
	Z	−1.0	0.0	1.0	−1.0	0.0	1.0	0.9
	P	0.0	1.0		0.0	1.0		0.2

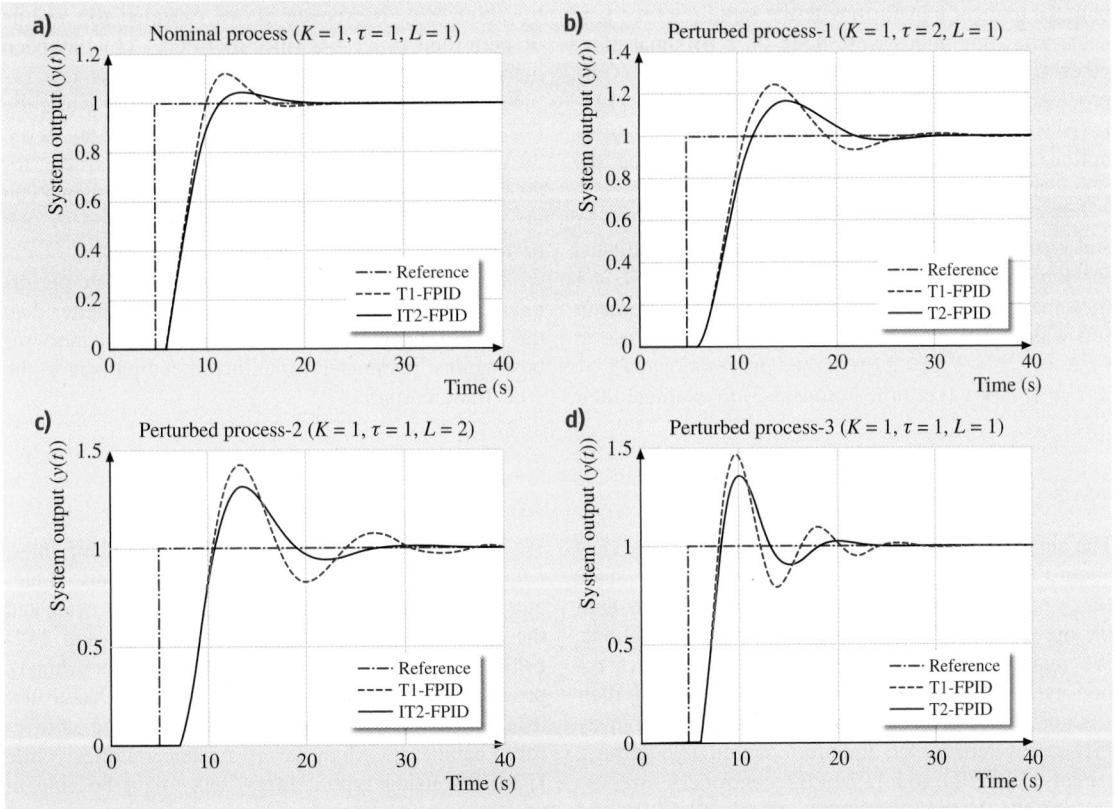

Fig. 18.5a–d Illustration of the step responses: (**a**) nominal process (**b**) perturbed process-1 (**c**) perturbed process-2 (**d**) perturbed process-3

Table 18.4 Control performance comparison of the FPID controllers

	Nominal process ($K=1, \tau=1, L=1$)			Perturbed process-1 ($K=1, \tau=2, L=1$)			Perturbed process-2 ($K=1, \tau=1, L=2$)			Perturbed process-3 ($K=2, \tau=1, L=1$)		
	OS	T_s	ITAE	OS	T_s	ITAE	OS	T_s	ITAE	OS	T_s	ITAE
T1-FPID	12	9.8	27.87	25	18.6	53.38	43	23.8	80.83	47	15.5	43.74
IT2-FPID	4	6.0	26.63	16	14.8	47.83	30	18.2	59.13	36	11.9	32.17

a unit step input. The scaling factors are set as $K_e = 1$, $K_d = 0.1$, $\alpha = 0.1$, and $\beta = 0.5$.

As it has been asserted, the IT2-FPID controller design will be accomplished by only blurring/extending antecedent MFs of the T1-FPID controller. Thus, we will set the consequent parameters and the scaling factors as the same values of its type-1 counterpart. The antecedent MFs of the IT2-FPID parameters are presented in Table 18.3. This setting will give the opportunity to illustrate how the extra degrees of freedom provided by FOUs affect the control system performances.

In the simulation studies, both FPID controllers are implemented as the discrete-time versions obtained with the bilinear transform with the sampling time $t_s = 0.1$ s. The simulations were done on a personal computer with an Intel Pentium Dual Core T2370 1.73 GHz processor, 2.99 GB RAM, and software package MATLAB/Simulink 7.4.0. Note that the simulation solver option is chosen as ode5 (Dormand-prince) and the step size is fixed at a value of 0.1 s.

The unit step response performances of the type-1 and type-2 fuzzy PID control systems are investigated for the nominal parameter set $K = 1$, $\tau = 1$, $L = 1$ (nominal process) and for three perturbed parameter sets which are $K = 1, \tau = 2, L = 1$ (perturbed process-1), $K = 1, \tau = 1, L = 2$ (perturbed process-2), and $K = 2, \tau = 1, L = 1$ (perturbed process-3) to examine their

robustness against parameter variations. In this context, we will consider three performance measures namely, the settling time (T_s), the overshoot (OS%), and the integral time absolute error (ITAE).

The system performances of the nominal and perturbed systems are illustrated in Fig. 18.5 and the performance measures are given in Table 18.4. As it can be clearly seen in Fig. 18.5, the IT2-FPID controller produces superior control performance in comparison to its type-1 controller counterpart. For instance, if we examine the results for nominal process, as compared to T1-FPID, the IT2-FPID control structure reduces the overshoot by about 66%; it also decreases the settling time by about 39% and the total IAE value by about 8%. Moreover, if we examine the results of perturbed process-2 (the time delay (L) has been increased 100%) it can be clearly seen that the T1-FPID control system response is oscillating while the IT2-FPI was able to reduces the overshoot by about 30%, the settling time by about 24% and the total ITAE value by about 27%. Similar comments can be made for presented other two perturbed system performances.

It can be concluded that the transient state performance of the IT2-FPID control structure is better than the T1-FPID controllers while it appears to be more robust against parameter variations in comparison to the type-1 counterpart.

18.4 Conclusion

The aim of this chapter is to present a general overview about FPID controller structures in the literature since fuzzy sets are recognized as a powerful tool to handle the faced uncertainties within control applications. We mainly focused on the double input direct action type fuzzy PID controller structures and their state-of-the-art design methods. We classified the fuzzy PID controllers in the literature within two groups, namely T1-FPID and IT2-FPID controllers. We examined the internal structures of the T1-FPID and IT2-FPID on a generic, a diagonal 3×3 rule base.

We presented detailed information about their internal structures, design parameters, and tuning strategies presented in the literature. Finally, we evaluated the control performance of the T1-FPID and IT2-FPID on a first-order plus time delay benchmark process. We illustrated that the T1-FPID controller using crisp type-1 fuzzy sets might not be able to fully handle the high levels of uncertainties while IT2-FPID using type-2 fuzzy sets might be able to handle such uncertainties to produce a better control performance.

References

18.1 S. Skogestad: Simple analytic rules for model reduction and PID controller tuning, J. Process Control **13**(4), 291–309 (2003)

18.2 M. Zhuang, D.P. Atherton: Automatic tuning of optimum PID controllers, Control Theory Appl., IEE Proc. D **140**(3), 216–224 (1993)

18.3 S. Galichet, L. Foulloy: Fuzzy controllers: Synthesis and equivalences, IEEE Trans. Fuzzy Syst. **3**, 140–148 (1995)

18.4 B.S. Moon: Equivalence between fuzzy logic controllers and PI controllers for single input systems, Fuzzy Sets Syst. **69**, 105–113 (1995)

18.5 T.T. Huang, H.Y. Chung, J.J. Lin: A fuzzy PID controller being like parameter varying PID, IEEE Int. Fuzzy Syst. Conf. Proc. **1**, 269–275 (1999)

18.6 B. Hu, G.K.I. Mann, R.G. Gasine: A systematic study of fuzzy PID controllers – Function-based evaluation approach, IEEE Trans. Fuzzy Syst. **9**(5), 699–711 (2001)

18.7 W.Z. Qiao, M. Mizumoto: PID type fuzzy controller and parameters adaptive method, Fuzzy Sets Syst. **78**, 23–35 (1996)

18.8 H.X. Li, H.B. Gatland: Conventional fuzzy control and its enhancement, IEEE Trans. Syst. Man Cybern. Part B **26**(5), 791–797 (1996)

18.9 M. Guzelkaya, I. Eksin, E. Yesil: Self-tuning of PID-type fuzzy logic controller coefficients via relative rate observer, Eng. Appl. Artif. Intell. **16**, 227–236 (2003)

18.10 X.-G. Duan, H.-X. Li, H. Deng: Effective tuning method for fuzzy PID with internal model control, Ind. Eng. Chem. Res. **47**, 8317–8323 (2008)

18.11 N.N. Karnik, J.M. Mendel, Q. Liang: Type-2 fuzzy logic systems, IEEE Trans. Fuzzy Syst. **7**, 643–658 (1999)

18.12 Q. Liang, J.M. Mendel: Interval type-2 fuzzy logic systems: Theory and design, IEEE Trans. Fuzzy Syst. **8**(5), 535–550 (2000)

18.13 T. Kumbasar, I. Eksin, M. Guzelkaya, E. Yesil: Type-2 fuzzy model based controller design for neutralization processes, ISA Transactions **51**(2), 277–287 (2012)

18.14 H. Hagras: A hierarchical type-2 fuzzy logic control architecture for autonomous mobile robots, IEEE Trans. Fuzzy Syst. **12**(4), 524–539 (2004)

18.15 D. Wu, W.W. Tan: Genetic learning and performance evaluation of internal type-2 fuzzy logic controllers, Eng. Appl. Artif. Intell. **19**, 829–841 (2006)

18.16 M. Galluzzo, B. Cosenza, A. Matharu: Control of a nonlinear continuous bioreactor with bifurcation by a type-2 fuzzy logic controller, Comput. Chem. Eng. **32**(12), 2986–2993 (2008)

18.17 J.S. Martínez, J. Mulot, F. Harel, D. Hissel, M.C. Péra, R.I. John, M. Amiet: Experimental validation of a type-2 fuzzy logic controller for energy management in hybrid electrical vehicles, Eng. Appl. Artif. Intell. **26**(7), 1772–1779 (2013)

18.18 C. Lynch, H. Hagras: Developing type-2 fuzzy logic controllers for handling the uncertainties in marine diesel engine speed control, J. Comput. Intell. Res. **4**(4), 402–422 (2009)

18.19 Y.T. Juang, Y.T. Chang, C.P. Huang: Design of fuzzy PID controllers using modified triangular membership functions, Inf. Sci. **178**, 1325–1333 (2008)

18.20 G. Fang, N.M. Kwok, Q. Ha: Automatic membership function tuning using the particle swarm optimization, 2008 IEEE Pasific-Asia Workshop Comput. Intell. Indust. Appl. (2008) pp. 324–328

18.21 K.K. Ahn, D.Q. Truong: Online tuning fuzzy PID controller using robust extended Kalman filter, J. Process Control **19**, 1011–1023 (2009)

18.22 R.K. Mudi, N.R. Pal: A robust self-tuning scheme for PI- and PD-type fuzzy controllers, IEEE Trans. Fuzzy Syst. **7**(1), 2–16 (1999)

18.23 Z.W. Woo, H.Y. Chung, J.J. Lin: A PID-type fuzzy controller with self-tuning scaling factors, Fuzzy Sets Syst. **115**, 321–326 (2000)

18.24 S. Bhatttacharya, A. Chatterjee, S. Munshi: A new self-tuned PID-type fuzzy controller as a combination of two-term controllers, ISA Transactions **43**, 413–426 (2004)

18.25 O. Karasakal, M. Guzelkaya, I. Eksin, E. Yesil: An error-based on-line rule weight adjustment method for fuzzy PID controllers, Expert Syst. Appl. **38**(8), 10124–10132 (2011)

18.26 H. Wu, J. Mendel: Enhanced karnik-mendel algorithms, IEEE Trans. Fuzzy Syst. **17**(4), 923–934 (2009)

18.27 M. Biglarbegian, W.W. Melek, J.M. Mendel: On the stability of interval type-2 TSK fuzzy logic control systems, IEEE Trans. Syst. Man. Cybern. Part B **4**(3), 798–818 (2010)

18.28 D. Wu: On the fundamental differences between type-1 and interval type-2 fuzzy logic controllers, IEEE Trans. Fuzzy Syst. **10**(5), 832–848 (2012)

18.29 R. Martínez, O. Castillo, L.T. Aguilar: Optimization of interval type-2 fuzzy logic controllers for a perturbed autonomous wheeled mobile robot using genetic algorithms, Inf. Sci. **179**(13), 2158–2174 (2009)

18.30 S.-K. Oh, H.-J. Jang, W. Pedrycz: A comparative experimental study of type-1/type-2 fuzzy cascade controller based on genetic algorithms and particle swarm optimization, Expert Syst. Appl. **38**(9), 11217–11229 (2011)

18.31 O. Castillo, R. Martínez, P. Melin, F. Valdez, J. Soria: Comparative study of bio-inspired algorithms applied to the optimization of type-1 and type-2 fuzzy controllers for an autonomous mobile robot, Inf. Sci. **19**(2), 19–38 (2012)

18.32 O. Linda, M. Manic: Comparative analysis of type-1 and type-2 fuzzy control in context of learning behaviors for mobile robotics, IECON 2010–36th Annu. Conf. IEEE Ind. Electron. Soc. (2010) pp. 1092–1098

18.33   O. Castillo, P. Melin: A review on the design and optimization of interval type-2 fuzzy controllers, Appl. Soft Comput. **12**, 1267–1278 (2012)

19. Soft Computing in Database and Information Management

Guy De Tré, Sławomir Zadrożny

Information is often imperfect. The sources of this imperfection include imprecision, uncertainty, incompleteness, and ambiguity. Soft computing techniques allow for coping more efficiently with such kinds of imperfection when handling data in information systems. In this chapter, we give an overview of selected soft computing techniques for database management. The chapter is subdivided in two parts which deal with the soft computing techniques, respectively, for information modelling and querying. A considerable part of the chapter is related to the issue of bipolarity of preferences and data representation which is among the important recent research trends.

19.1 Challenges for Modern Information Systems

Database systems nowadays form a basic component of almost every information system and their role is getting more and more important. Almost every person or company keeps track of a large amount of digital data and this amount is still growing everyday. Many ICT managers declare that *big data* is a point of attention for the coming years. Despite the fact that the concept of *big data* is not clearly defined, one can generally agree that it refers to the increasing need for efficiently storing and handling large amounts of information.

However, it is easy to observe that information is not always available in a perfect form. Just consider the fact that human beings communicate most of the time using vague terms hereby reflecting the fact that they do not know exact, precise values with certainty. In general, imperfection of information might be due to the imprecision, uncertainty, incompleteness, or ambiguity.

Our life and society have changed in such a way that we simply cannot neglect or discard imperfect information anymore. To be competitive, companies need to cope with all information that is available. Efficiently storing and handling imperfect information without introducing errors or causing data loss is therefore considered as one of the main challenges for information management in this century [19.1].

Soft computing offers formalisms and techniques for coping with imperfect data in a mathematically sound way [19.2]. The earliest research activities in this area dates back to the early eighties of the previous century. In this chapter, we present an overview of selected results of the research on soft computing techniques aimed at improving database modelling and database access in the presence of imperfect information. The scope of the chapter is further limited to database access techniques that are based on querying and specifying and handling user preferences in query formulations. Other techniques, not dealt with in this chapter, include:

- Self-query auto completion systems that help users in formulating queries by exploiting past queries, as used in recommendation systems [19.3].
- Navigational querying systems that allow intelligent navigation through the database [19.4].
- Cooperative querying systems that support *indirect* answers such as summaries, conditional answers, and contextual background information for (empty) results [19.5].

The remainder of the chapter is organized as follows. In Sect. 19.2, we give some preliminaries on (relational) databases, which are used as a basis for illustrating the described techniques. The next two Sects. 19.3 and 19.4 form the core of the chapter. In Sect. 19.3, an overview of soft computing techniques for the modelling and handling of imperfect data in databases is presented. Whereas, in Sect. 19.4 the main trends in soft computing techniques for flexible database querying are discussed. Both, querying of regular databases and querying of databases containing imperfect data are handled. The conclusions of the chapter are stated in Sect. 19.5.

19.2 Some Preliminaries

In order to review and discuss main contributions to the research area of soft computing in database and information management, we have to introduce the terminology and notation related to the basics of database management and fuzzy logic.

19.2.1 Relational Databases

The techniques presented in this work will be described as general as possible, so that they are in fact applicable to multiple database models. However, due to its popularity and mathematical foundations, the relational database model has been used as the original formal framework for many of these techniques. For that reason, we opt for using the relational model as underlying database model throughout the chapter.

A *relational database* can in an abstract sense be seen as a collection of *relations* or, informally, of *tables* which represent them. Informally speaking, the columns of a table represent its characteristics, whereas its rows reflect its content [19.6]. From a formal point of view, each relation R is defined via its relation schema [19.7]

$$R(A_1 : D_1, \ldots, A_n : D_n) , \qquad (19.1)$$

where $A_i : D_i, i = 1, \ldots, n$ are the *attributes* (columns) of the relation. Each attribute $A_i : D_i$ is characterized by its name A_i and its associated data type (domain) D_i, to be denoted also as dom_{A_i}. The data type D_i determines the allowed values for the attribute and the basic operators that can be applied on them. Each relation (table) represents a set of real world entities, each of them being modeled by a *tuple* (row) of the relation. Relations schemas are the basic components of a database schema. In this way, a table contains data describing a part of the real world being modeled by the *database schema*.

The most interesting operation on a database, from this chapter's perspective, is the retrieval of data satisfying certain conditions. Usually, to retrieve data, a user forms a *query* specifying these conditions (criteria). The conditions then reflect the user's preferences with respect to the information he or she is looking for. The retrieval process may be meant as the calculation of a matching degree for each tuple of relevant relation(s). Classically, a row either matches the query or not, i.e., the concept of matching is binary. In the context of soft computing, flexible criteria, soft aggregation, and soft ranking techniques can be used, so that tuple matching becomes a matter of a degree.

Usually two general formal approaches to the querying are assumed: the *relational algebra* and the *relational calculus*. The former has a procedural character: a query consists here of a sequence of operations on relations that finally yield requested data. These operations comprise five basic ones: union ($\cup$), difference ($\backslash$), projection (π), selection (σ), and cross product ($\times$) that may be combined to obtain some derived operations such as, e.g., intersection ($\cap$), division ($\div$), and join ($\bowtie$). The latter approach, known in two flavours as the tuple relational calculus (TRC) or the domain relational calculus (DRC), is of a more declarative nature. Here a query just describes what kind of data is requested, but how it is to be retrieved from a database is left to the database management system. The exact form of queries is not of an utmost importance for our considerations. However, some reported research in this area employs directly the de-facto standard querying language for relational databases, i. e. SQL (structured query language) [19.7, 8]. Thus, we will also sometimes refer to the SELECT-FROM-WHERE instruction of this language and more specifically consider its WHERE clause, where query conditions are specified.

19.2.2 Fuzzy Set Theory and Possibility Theory

We will use the following concepts and notation concerning fuzzy set theory [19.9]. A *fuzzy set* F in the universe U is characterized by a membership function

$$\mu_F : U \to [0, 1] : u \mapsto \mu_F(u) . \tag{19.2}$$

For each element $u \in U$, $\mu_F(u)$ denotes the membership grade or extent to which u belongs to F. The origins of membership functions may be different and depending on that they have different semantics [19.10]. With their traditional interpretation as *degrees of similarity*, membership grades allow it to appropriately represent vague concepts, like *tall man*, *expensive book*, and *large garden*, taking into account the gradual characteristics of such a concept. Membership grades can also express *degrees of preference*, hereby expressing that several values apply to a different extent. For example, the languages one speaks can be expressed by a fuzzy set $\{(English, 1), (French, 0.7), (Spanish, 0.2)\}$, and then the membership degrees represent skill levels attained in a given language. A fuzzy set can also be interpreted as a possibility distribution, in which case its membership grades denote *degrees of uncertainty*. Then, it can be used to represent, e.g., the uncertainty about the actual value of a variable, like *the height of a man*, *the price of a book* and *the size of a garden* [19.11, 12]. This interpretation is related to the concept of the *disjunctive fuzzy set*.

Possibility distributions are denoted by π. The notation π_X is often used to indicate that the distribution concerns the value of a variable X,

$$\pi_X : U \to [0, 1] : u \mapsto \pi_X(u) , \tag{19.3}$$

where X takes its value from a universe U.

Possibility and *necessity measures* can provide for the quantification of such an uncertainty. These mea-

sures are denoted by Π and N, respectively, i. e.,

$$\Pi : \tilde{\wp}(U) \to [0, 1] : A \mapsto \Pi(A) \quad \text{and}$$
$$N : \tilde{\wp}(U) \to [0, 1] : A \mapsto N(A) , \tag{19.4}$$

where the *fuzzy* power set $\tilde{\wp}(U)$ stands for the family of fuzzy sets defined over U. Assuming that all we know about the value of a variable X is a possibility distribution π_X, these measures, for a given fuzzy set F, assess to what extent, respectively, this set is consistent (Π) and its complement is inconsistent (N) with our knowledge on the value of X. More precisely, if π_X is the underlying possibility distribution, then

$$\Pi_X(F) = \sup_{u \in U} \min(\pi_X(u), \mu_F(u)) , \tag{19.5}$$

$$N_X(F) = \inf_{u \in U} \max(1 - \pi_X(u), \mu_F(u)) . \tag{19.6}$$

Sometimes the interval $[N_X(F), \Pi_X(F)]$ is used as an estimate of the consistency of F with the actual value of X. The possibility (necessity) that two variables X and Y, whose values are given by possibility distributions, π_X and π_Y, are in relation θ – e.g., equality – is computed as follows. The joint possibility distribution, π_{XY}, of X and Y on $U \times U$ (assuming noninteractivity of the variables) is given by

$$\pi_{XY}(u, w) = \min(\pi_X(u), \pi_Y(w)) . \tag{19.7}$$

The relation θ can be fuzzy and represented by a fuzzy set $F \in \tilde{\wp}(U \times U)$ such that $\mu_F(u, w)$ expresses to what extent u is in relation θ with w. The possibility (resp. necessity) measure associated with π_{XY} will be denoted by Π_{XY} (resp. N_{XY}). Then, we calculate the measures of possibility and necessity that the values of the variables are in relation θ as follows

$$\Pi(X \theta Y) = \Pi_{XY}(F)$$
$$= \sup_{u, w \in U} \min(\pi_X(u), \pi_Y(w), \mu_F(u, w)) , \tag{19.8}$$

$$N(X \theta Y) = N_{XY}(F) = \inf_{u, w \in U} \max(1 - \pi_X(u),$$
$$1 - \pi_Y(w), \mu_F(u, w)) . \tag{19.9}$$

Knowing the possibility distributions of two variables X and Y, one may also be interested on how these distributions are similar to each other. Obviously, (19.8)–(19.9) provide some assessment of this similarity, but other indices of similarity are also applicable.

Part B | 19.2

Table 19.1 Special EPTVs

$\tilde{\imath}^*(p)$	Interpretation
$(T, 1)$	p is true
$(F, 1)$	p is false
$(T, 1), (F, 1)$	p is unknown
$(\perp, 1)$	p is inapplicable
$(T, 1), (F, 1), (\perp, 1)$	Information about p is not available

This leads to a distinction between representation-based and value-based comparisons of possibility distributions [19.13]. We will discuss this later on in Sect. 19.4.2.

An important class of possibility distributions are *extended possibilistic truth values* (EPTV) [19.14]. An EPTV is defined as a possibility distribution (a disjunctive fuzzy set) in the universe $I^* = \{T, F, \perp\}$ that consists of the three truth values T (true), F (false) and $\perp$ (undefined). The set of all EPTVs is denoted as $\tilde{\wp}(I^*)$. They are meant to represent uncertainty as to the truth value of a proposition $p \in P$, where P denotes the universe of all propositions, in particular in the context of database querying. Thus, a valuation $\tilde{\imath}^*$ is assumed such that

$$\tilde{\imath}^* : P \to \tilde{\wp}(I^*) : p \mapsto \tilde{\imath}^*(p) \,. \tag{19.10}$$

In general the EPTV $\tilde{\imath}^*(p)$ representing (the knowledge of) the truth of a proposition $p \in P$ has the following format

$$\tilde{\imath}^*(p) = [(T, \mu_{\tilde{\imath}^*(p)}(T)), (F, \mu_{\tilde{\imath}^*(p)}(F)),$$
$$(\perp, \mu_{\tilde{\imath}^*(p)}(\perp))] \,. \tag{19.11}$$

Hereby, $\mu_{\tilde{\imath}^*(p)}(T)$, $\mu_{\tilde{\imath}^*(p)}(F)$ and $\mu_{\tilde{\imath}^*(p)}(\perp)$, respectively denote the possibility that p is true, false, or undefined. The latter value is also covering cases where p is not applicable or not supplied. EPTVs extend the approach based on the possibility distributions defined on just the set $\{T, F\}$ with an explicit facility to deal with the inapplicability of information as can for example occur with the evaluation of query conditions. In Table 19.1, some special cases of EPTVs are presented: These cases are verified as follows:

- If it is completely possible that the proposition is true and no other truth values are possible, then it means that the proposition is known to be true.
- If it is completely possible that the proposition is false and no other truth values are possible, then it means that the proposition is known to be false.
- If it is completely possible that the proposition is true, it is completely possible that the proposition is false and it is not possible that the proposition is inapplicable, then it means that the proposition is applicable, but its truth value is unknown. This EPTV will be called in short *unknown*.
- If it is completely possible that the proposition is inapplicable and no other truth values are possible, then it means that the proposition is inapplicable.
- If all truth values are completely possible, then this means that no information about the truth of the proposition or its applicability is available. The proposition might be inapplicable, but might also be true or false. This EPTV will be called in short *unavailable*.

19.3 Soft Computing in Information Modeling

Soft computing techniques make it possible to grasp imperfect information about a modeled part of the world and represent it directly in a database. If fuzzy set theory [19.9] or possibility theory [19.12] are used to model imperfect data in a database, the database is called a *fuzzy database*. Other approaches include those that are based on rough set theory [19.15] and on probability theory [19.16] and resulting databases are then called *rough databases* and *probabilistic databases*, respectively.

In what follows, we give an overview of the most important soft computing techniques for modeling imperfect information which are based on fuzzy set theory

and possibility theory. Hereby, we distinguish between basic techniques (Sect. 19.3.1) and more advanced techniques (Sect. 19.3.2). As explained in the preliminaries, we use the relational database model [19.6] as the framework for our descriptions. This is also motivated by the fact that initial research in this area has been done on the relational database model and this model is nowadays still the standard for database modeling. Soft-computing-related research on other database models like the (E)ER model, the object-relational model, the XML-model, and object-oriented database models exists. Overviews can, among others, be found in [19.17–22].

19.3.1 Modeling of Imperfect Information – Basic Approaches

In view of the correct handling of information, it is of utmost importance that the available information that has to be stored in a database is modeled as adequate as possible so as to avoid the information loss. The most straightforward application of fuzzy logic to the classical relational data model is by assuming that the relations in a database themselves are also fuzzy [19.23]. Each tuple of a relation (table) is associated with a membership degree. This approach is often neglected because the interpretation of the membership degree is unclear. On the other hand, it is worth noticing that fuzzy queries, as will be discussed in Sect. 19.4, in fact produce fuzzy relations. So, we will come back to this issue when discussing fuzzy queries in Sect. 19.4.

Most of the research on modeling imperfect information in databases using soft computing techniques is devoted to a proper representation and processing of an attribute value. Such a value, in general, may not be known perfectly due to many different reasons [19.24]. For example, due to the imprecision, as when the painting is dated to the *beginning of the fourteenth century*; or due to the unreliability, as when the source of information is not fully reliable; or due to the ambiguity, as when the provided value may have different meanings; or due to the inconsistency, as when there are multiple different values provided by different sources; or due to the incompleteness, as when when the value is completely missing or given as a set of possible alternatives (e.g., the picture was painted by Rubens or van Dyck). These various forms of information imperfection are not totally unconnected as well as may occur together. From the viewpoint of data representation they may be primarily seen as yielding uncertainty as to the actual value of an attribute and as such may be properly accounted for by a possibility distribution.

It is worth noticing that then the assignment of the value to an attribute may be identified with a Zadeh's [19.25] linguistic expression X *is* A, where X is a linguistic variable corresponding to the attribute while A is a (disjunctive) fuzzy set representing imperfect information on its value. Then, various combined forms of information imperfection may be represented by appropriate qualified linguistic expressions such as, e.g., X *is* A *with certainty at least* α [19.26]. Such qualified linguistic expressions may be in turn transformed into a X *is* B expression where a fuzzy set B is a function of A and other possible parameters of a qualified expression, like α in the previous example. Thus, the basic linguistic expression X *is* A indeed plays a fundamental role in the representation of the imperfect information.

The work of *Prade* and *Testemale* [19.27] is the most representative for the approaches to imperfect information modeling in a database based on the possibility theory. Other works in this vein include [19.27–32]. On the other hand, *Buckles* and *Petry* [19.33] as well as *Anvari* and *Rose* [19.34] assume the representation of attributes' values using sets of alternatives which may be treated as a simple binary possibility distribution. However, their motivation is different as they assume that domain elements are similar/indistinguishable to some extent and due to that it may be difficult to determine a precise value of an attribute. We will first briefly describe the approach of Prade and Testemale and, then, the model of Buckles and Petry.

Possibilistic Approach

In the possibilistic approach, disjunctive fuzzy sets are used to represent the *imprecisely* known value of an attribute A. Hence, such a fuzzy set is interpreted as a possibility distribution π_A and is defined on the domain dom_A of the attribute. The (degree of) possibility that the actual value of A is a particular element x of the domain of this attribute, $x \in dom_A$, equals $\pi_A(x)$. Every domain value $x \in dom_A$ with $\pi_A(x) \neq 0$ is thus a candidate for being the actual value of A. Together all candidate values and their associated possibility degrees reflect what is actually known about the attribute value. Thus, if the value of an attribute is not known precisely then a set of values may be specified (represented by the support of the fuzzy set used) and, moreover, particular elements of this set may be indicated as a more or less plausible values of the attribute in question.

A typical scenario in which such an imprecise value has to be stored in a database is when the value of an attribute is expressed using a linguistic term. For example, assume that the value of a painting is not known precisely, but the painting is known to be *very valuable*. Then, this information might be represented by the possibility distribution

$$\pi_{\text{value}}(x) = \mu_{\text{very\_valuable}}(x)$$

$$= \begin{cases} 0 & \text{if } x < 10M \\ \dfrac{x - 10}{10} & \text{if } 10M \leq x \leq 20M \\ 1 & \text{if } x > 20M. \end{cases}$$

The term *uncertainty* is in information management often used to refer to situations where one has to cope with several (distinct) candidate attribute values coming from different information sources. For example, one information source can specify that the phone number of a person is *X*, while another source can specify that it is *Y*. This kind of uncertainty can also be handled with the possibilistic approach as described above. In that case, a possibility distribution π_A over the domain of the attribute is used to model different options for the attribute's actual value. However, in general possibility distributions are less informative than probability distributions. They only inform the user about the relative likeliness of different options. Probability distributions provide more information and led to the so-called probabilistic databases [19.35–39]; cf. also [19.40].

Imprecision at the one hand and uncertainty at the other hand are orthogonal concepts: they can occur at the same time, as already mentioned earlier. For example, it might be uncertain whether the value of a painting is *3M*, *around 2M*, or *much cheaper*, where the latter two options are imprecise descriptions. Using the regular possibilistic modeling approach in such a case would yield in a single possibility distribution over the domain of values and would result in a loss of information on how the original three options were specified. Level-2 fuzzy sets, which are fuzzy sets defined over a domain of fuzzy sets [19.41, 42], can help to avoid this information loss [19.43].

In traditional databases, missing information is mostly handled by means of a pseudovalue, called a *null value* [19.44, 45]. In fact, information may be missing for many different reasons: the data may exist but be unknown (e.g., the salary of an employee may be unknown); the data may not exist nor apply (e.g., an unemployed person earns no salary) [19.46]. For the handling of nonapplicability a special pseudovalue is still required, but the case of unknown information can be adequately handled by using possibility theory. Indeed, as studied in [19.47], in the so-called *extended possibilistic approach*, the domain dom_A of an attribute *A* can be extended with an extra value $\perp_A$ that is interpreted as *regular value not applicable*. Missing information can then be adequately modeled by considering the following three special possibility distributions π_{UNK}, $\pi_{\text{N/A}}$, and π_{UNA}:

- Unknown value

$$\pi_{\text{UNK}}(x) = \begin{cases} 1, & \text{if } x \in dom_A \setminus \{\perp_A\} \\ 0, & \text{if } x = \perp_A \end{cases}$$

- Value not applicable

$$\pi_{\text{N/A}}(x) = \begin{cases} 0, & \text{if } x \in dom_A \setminus \{\perp_A\} \\ 1, & \text{if } x = \perp_t \end{cases}$$

- No information available

$$\pi_{\text{UNA}}(x) = 1, \forall x \in dom_A .$$

Similarity–Based Approach

The basic idea behind this approach [19.33] is that while specifying the value of a database attribute one may consider similar values as also being applicable. Thus, in general, the value of an attribute *A* is assumed to be a subset of its domain dom_A. Moreover, the domain dom_A is associated with a similarity relation S_A quantifying this similarity for each pair of elements $x, y \in dom_A$. The values $S_A(x, y)$ taken by S_A are in the unit interval [0, 1], where 0 corresponds to *totally different* and 1 to *totally similar*. Hence, S_A is a fuzzy binary relation that associates a membership grade to each pair of domain values. This relation is assumed to be reflexive, symmetric and satisfying some form of transitivity. This requirements have been found too restrictive and some approaches based on a weaker structure have been proposed in [19.48], where the proximity relation is used and all attractive properties of the original approach are preserved. Among these properties, the most important is the proper adaptation of the redundancy concept and of the relational algebra operations.

It has been quickly recognized that the rough sets theory [19.15] offers effective tools to deal with and analyze the indistinguishability/equivalence relation and the similarity-based approaches evolved into the rough-sets-based database model [19.49, 50].

There are also a number of hybrid models proposed in the literature. *Takahashi* [19.51] has proposed a model for a fuzzy relational database assuming possibility distributions as attribute values. Moreover, in his model fuzzy sets are used as tuples' truth values. For example, a tuple *t*, accompanied by such a truth qualification, may express that *It is quite true that the paintings origin is the beginning of the fifteenth century*.

Medina et al. [19.52] proposed a fuzzy database model called GEFRED (generalized fuzzy relational database) in an attempt to integrate both approaches: the possibilistic and similarity based one. The data are stored as generalized fuzzy relations that extend the relations of the relational model by allowing imprecise information and a compatibility degree associated with each attribute value.

19.3.2 Modeling of Imperfect Information – Selected Advanced Approaches

In this section, we will focus on some extensions to the possibility-based approach described in Sect. 19.3.1. As argued earlier, a disjunctive fuzzy set may very well represent the situation when an attribute A of a tuple for sure takes exactly one value from the domain dom_A (as it should also due to the classical relational model) but we do not know exactly which one. The complete ignorance is then modeled by the set dom_A. However, very often we can distinguish the elements of dom_A with respect to their plausibility as the actual value of the attribute. This information, based on some evidence, is represented by the membership function of a disjunctive fuzzy set which is further identified with a possibility distribution. However, the characteristic of the available evidence may be difficult to express using regular fuzzy sets. In the literature dealing with data representation there are first attempts to cover such cases using some extensions to the concept of the fuzzy set. We will now briefly review them.

Imprecise Membership Degrees

Prade and *Testemale* in their original approach [19.31] assume that the membership degrees of a mentioned disjunctive fuzzy set are known precisely. On the other hand, one can argue that they may be also known only in an imprecise way. It may be the case, in particular, when the value of an attribute is originally specified using a linguistic term. For example, if a painting is dated to the beginning of the fifteenth century then assigning, e.g., the degree of 0.6 to the year 1440 may be challenging for an expert who is to define the representation of this linguistic term. He or she may be much more comfortable stating that it is something, e.g., between 0.5 and 0.7. Some precision is lost and a *second level* uncertainty is then implied but it may better reflect the evidence actually available.

Type 2 fuzzy sets [19.25] make it possible to model the data in the case described earlier. In particular, their simplest form, the *interval valued fuzzy sets* (IFVSs), may be here of interest. In the case of interval-valued fuzzy sets [19.25] a membership degree is represented as an interval, as in the example given earlier. Thus an interval-valued fuzzy set X over a universe of discourse U is defined by two functions

$$\mu_X^l, \mu_X^u : U \to [0,1],$$

such that

$$0 \le \mu_X^l(x) \le \mu_X^u(x) \le 1, \quad \forall\, x \in U, \tag{19.12}$$

and may be denoted by

$$X = [< x, \mu_X^l(x), \mu_X^u(x) > |(x \in U) \wedge$$
$$(0 \le \mu_X^l(x) \le \mu_X^u(x) \le 1)]. \tag{19.13}$$

Constraint (19.12) reflects that $\mu_A^l(x)$ and $\mu_A^u(x)$ are, respectively, interpreted as a lower and an upper bound on the actual degree of membership of x in X.

Basically, the representation of information using (disjunctive) IVFS is conceptually identical with the original approach of Prade and Testemale while it provides some more flexibility in defining the meaning of linguistic terms. Some preliminary discussion on their use may be found in [19.53].

Bipolarity of Information

Bipolarity is related to the existence of the positive and negative information [19.54–58]. It manifests itself, in particular, when people are making judgments about some alternatives and take into account their positive and negative sides. From this point of view, bipolarity of information may play an important role in database querying and is discussed from this perspective in Sect. 19.4.1. Here we will briefly discuss the role of bipolarity in data representation. We will mostly follow in this respect the work of *Dubois* and *Prade* [19.58].

The value of an attribute may be not known precisely but some information on it may be available in the form of both *positive* and *negative* statements. In some situations positive information is provided, stating what values are possible, satisfactory, permitted, desired, or considered as being acceptable. In other situations, negative statements express what values are impossible, rejected or forbidden.

Different types of bipolarity can be distinguished [19.58, 59]:

- *Type I, symmetric univariate bipolarity*: positive and negative information are considered as being exact complements of each other as in, e.g., the probabillity theory; for instance, if the probability that a given painting is painted in the eighteenth century is stated to be 0.7 (positive information), then the probability that this painting was not painted in the eighteenth century equals 0.3 (negative information); this simple form of bipolarity is well supported by traditional information systems; this bipolarity is quantified on a *bipolar univariate scale* such as the intervals $[0, 1]$ or $[-1, 1]$;

- *Type II, symmetric bivariate bipolarity*: another, more flexible approach is to consider positive and negative information as being dual concepts, measured along two different scales but based on the same piece of evidence. The dependency between them is modeled by means of some duality relation. This kind of bipolarity is, among others, used in *Atanassov's* intuitionistic fuzzy sets [19.60] where each element of a set is assigned both a membership and nonmembership degree which do not have to sum up to 1 but their sum cannot exceed 1. For example, it could be stated that Rubens is a good candidate to be an author of a given painting to a degree 0.6 (due to some positive information) while at the same time he is not a good candidate to a degree 0.2 (due to some negative information); this bipolarity is quantified on a *unipolar bivariate scale* composed of two unipolar scales such as, e.g., two intervals [0, 1];
- *Type III, asymmetric/heterogeneous bipolarity*: in the most general case, positive and negative information is provided by two separate bodies of evidence, which are to some extent independent of each other and are of a different nature. A constraint to guarantee that the information does not contain contradictions can exist but beside of that both statements are independent of each other and hence giving rise to the notion of the heterogeneous bipolarity; this bipolarity is quantified as in the case of Type II bipolarity.

Type III bipolarity is of special interest from the point of view of data representation. *Dubois* and *Prade* [19.58] argue that in this context the heterogeneity of bipolarity is related to to the different nature of two bodies of evidence available. Namely, the negative information corresponds to the *knowledge* which puts some general constraints on the feasible values of an attribute. On the other hand, the positive information corresponds to *data*, i.e., observed cases which justify plausibility of a given element as the candidate for the value of an attribute.

This type of bipolarity is proposed to be represented for a tuple t by two separate possibility distributions, $\delta_{A(t)}$ and $\pi_{A(t)}$, defined on the domain of an attribute, dom_A [19.58] (and earlier works cited therein). A possibility distribution $\pi_{A(t)}$, as previously, represents the compatibility of particular elements $x \in dom_A$ with the available information on the value of the attribute A for a tuple t. This compatibility is quantified on a unipolar negative scale identified with the interval [0, 1]. The extreme values of π_A, i.e., $\pi_{A(t)}(x) = 1$ and $\pi_{A(t)}(x) = 0$ are meant to represent, respectively, that x is *potentially fully possible* to be the value of A at t (1 is a *neutral* element on this unipolar scale) and that x is totally *impossible* to be the value of A at t (0 is an extreme negative element on this unipolar scale). On the other hand, a possibility distribution $\delta_{A(t)}$ expresses the degree of support for an element $x \in dom_A$ to be the value of A at t provided by some evidence. In this case, $\delta_{A(t)}(x) = 0$ denotes the lack of such a support but is meant as just a neutral assessment while $\delta_{A(t)}(x) = 1$ denotes full support (1 is an extreme positive element on this scale).

For example, when the exact dating of a painting is unknown, one can be convinced it has been painted in some time range (e.g., related to the time period its author lived in) and also there may be some evidence supporting a particular period of time (e.g., due to the fact that other very similar paintings of a given author are known from this period). Thus, the former is a negative information, excluding some period of time while the latter is a positive information supporting given period.

Thus, $\delta_{A(t)}$ and $\pi_{A(t)}$ are said to represent, respectively, the set of *guaranteed/actually possible* and the set of *potentially possible* values of an attribute A for the tuple t. These possibility distributions are related by a consistency constraint: $\pi_{A(t)}(x) \geq \delta_{A(t)}(x)$ as x have to be first nonexcluded before it may be somehow supported by the evidence.

For a given attribute A and tuple t, $\pi_{A(t)}$ is based on the set of nonimpossible values $NI_{A(t)}$ while $\delta_{A(t)}$ relates to the set of actually possible values $G_{A(t)}$. The querying of such a *bipolar database* may be defined in terms of these sets [19.58].

19.4 Soft Computing in Querying

The research on soft computing in querying has already a long history. It has been inspired by the success of fuzzy logic in modeling natural language propositions. The use of such propositions in queries, in turn, seems to be very natural for human users of any information system, notably the database management system. Later on, the interest in fuzzy querying has been reinforced by the omnipresence of network-based applications, related to buzzwords of modern information technology, such as e-commerce, e-government, etc. These applications evidently call for a flexible querying capability when users are looking for some goods, hotel accommodations, etc., that may be best described using natural language terms such as cheap, large, close to the airport, etc. Another amplification of the interest in fuzzy querying comes from developments in the area of data warehousing and data mining related applications. For example, a combination of fuzzy querying and data mining interfaces [19.61, 62] or fuzzy logic and the OLAP (online analytical processing) technology [19.63] may lead to new, effective and more efficient solutions in this area. More recently, *big data* challenges can be seen as driving forces for research in soft querying. Indeed, efficiently querying huge quantities of heterogeneous structured and unstructured data is one of the prerequisites for efficiently handling *big data*.

19.4.1 Flexible Querying of Regular Databases

As a starting point, we consider a simplified form of database queries on a classical crisp relational database. Hereby a query is assumed to consist of a combination of conditions that are to be met by the data sought. Introducing flexibility is done by specifying fuzzy preferences. This can be done inside the query conditions via flexible search criterion and allows to express that some values are more desirable than others in a gradual way. Query conditions are allowed to contain natural language terms. Another option is to specify fuzzy preferences at the level of the aggregation. By assigning grades of importance to (groups of) conditions it can be indicated that the satisfaction of some query conditions is more desirable than the satisfaction of others.

Basic Approaches
One of the pioneering approaches in recognizing the power of fuzzy set theory for information retrieval purposes in general is [19.64]. The research on the application of soft computing in database querying research proper dates back to an early work of *Tahani* [19.65], proposing the modeling of linguistic terms in queries using elements of fuzzy logic. An important enhancement of this basic approach consisted in considering flexible aggregation operators [19.10, 66–68]. Another line of research focused on embedding fuzzy constructs in the syntax of the standard SQL [19.21, 69–74].

Fuzzy Preferences Inside Query Conditions. *Tahani* [19.65] proposed to use imprecise terms typical for natural language such as, e.g., *high*, *young* etc., to form conditions of an SQL-like querying language for relational databases. These imprecise linguistic terms are modeled using fuzzy sets defined in attributes domains. The binary satisfaction of a classical query is replaced with the *matching degree* defined in a straightforward way. Namely, a tuple t matches a simple (elementary) condition $A = l$, where A is an attribute (e.g., *price*) and l is a linguistic term (e.g., *high*) to a degree $\gamma(A = l, t)$ such that

$$\gamma(A = l, t) = \mu_l(A(t)) \, , \tag{19.14}$$

where $A(t)$ is the value of the attribute A at the tuple t and $\mu_l(\cdot)$ is the membership function of the fuzzy set representing the linguistic term l. The matching degree for compound conditions, e.g., *price = high AND (date = beginning-of-17-century OR origin = south-europe)* is obtained by a proper interpretation of the fuzzy logical connectives. For example

$$\gamma[(A_1 = l_1) \text{ AND } (A_2 = l_2), t] = \min[\mu_{l_1}(A_1(t)), \mu_{l_2}(A_2(t))] \, . \tag{19.15}$$

The relational algebra has been very early adapted for the purposes of fuzzy flexible querying of regular relational databases. The division operator attracted a special attention and its many *fuzzy* variants has been proposed, among other by *Yager* [19.75], *Dubois* and *Prade* [19.76], *Galindo* et al. [19.77], and *Bosc* et al. [19.78, 79]. *Takahashi* [19.80] was among the first authors to propose a fuzzy version of the relational calculus. His fuzzy query language (FQL) was meant as a fuzzy extension of the domain relational calculus (DRC).

Fuzzy Preferences Between Query Conditions. Often, a query is composed of several conditions of varying importance for the user. For example, a customer of a real-estate agency may be looking for a cheap apartment in a specific district of a city and located not higher that a given floor. However, for he or she the low price may be much more *important* than the two other features. It may be difficult to express such preferences in a traditional query language. On the other hand, it is very natural for flexible fuzzy querying approaches due to the assumed gradual character of the matching degree as well as due to the existence of sophisticated preference modeling techniques developed by fuzzy logic community. Thus, most approaches make it possible to assign to a condition an importance weight, usually represented by a number from the [0, 1] interval.

The impact of a weight can be modeled by first matching the condition as if there is no weight and only then modifying the resulting matching degree in accordance with the weight. A modification function that strengthens the match of more important conditions and weakens the match of less important conditions is used for this purpose.

The evaluation of a whole query against a tuple may be seen as an aggregation of the matching degrees of elementary conditions comprising the query against this tuple. Thus, an aggregation operator is involved which, in the case of a simple conjunction or disjunction of the elementary conditions is usually assumed to take the form of the minimum and maximum operator, respectively. If weights are assigned to the elementary conditions connected using conjunction or disjunction then, first, the matching degrees of these conditions are modified using the weights [19.81] and then they are aggregated as usual.

On the other hand, some special aggregation operators may be explicitly used in the query and then they guide the aggregation process. *Kacprzyk* et al. [19.66, 67] were first to propose the use in queries of an aggregation operator in the form of a *linguistic quantifier* [19.82]. Thus, the user may require, e.g., *most* of the elementary conditions to be fulfilled instead of all of them (what is required when the conjunction of the conditions is used) or instead of just one of them (what is required in the case of the disjunction. For example, the user may define paintings of his interest as those meeting *most* of the following conditions: *not expensive, painted in Italy, painted not later than in seventeenth century, accompanied by an attractive insurance offer* etc. The overall matching degree of a query involving a linguistic quantifier may be computed using any of the approaches used to model these quantifiers. In [19.66, 67], *Zadeh's* original approach is used [19.82] while in [19.83] *Yager's* approach based on the OWA operators is adopted [19.84]. Further studies on modeling sophisticated aggregation operators, notably linguistic quantifiers, in the flexible fuzzy queries include the papers by *Bosc* et al. [19.85, 86], *Galindo* et al. [19.21] and *Vila* et al. [19.87].

A recent book by *Bosc* and *Pivert* [19.88] contains a comprehensive survey of the sophisticated flexible database querying techniques.

Bipolar Queries

An important novel line of research concerning advanced querying of databases addresses the issue of the bipolar nature of users preferences. Some psychological studies (e.g., sources cited by [19.56]) show that while expressing his or her preferences a human being is separately considering *positive* and *negative* aspects of a given option. Thus, to account for this phenomenon, a query should be seen as a combination of two types of conditions: the satisfaction of one of them makes a piece of data desired while the satisfaction of the second makes it to be rejected. Such a query will be referred to as the *bipolar query* and will be denoted as a pair of conditions (C, P), where C, for convenience, denotes the complement of the negative condition and P denotes the positive condition. The relations between these two types of conditions may be analyzed from different viewpoints, and the conditions itself may be expressed in various ways. In Sect. 19.3.2, we have already introduced the concept of bipolarity in the context of data representation. Now, we will briefly survey different approaches to modeling the *bipolarity* with a special emphasis on the context of database querying.

Models of Bipolarity. Various scales may be used to express bipolarity of preferences. Basically, two models based on: a *bipolar univariate* scale and a *unipolar bivariate* scale [19.89] are usually considered. The former assumes one scale with three main levels of negative, neutral, and positive preference degrees, respectively. These degrees are gradually changing from one end of the scale to another accounting for some intermediate levels. In the second model, two scales are used which separately account for the positive and negative preference degrees. Often, the intervals $[-1, 1]$ and $[0, 1]$ are used to represent the scales in the respective models of bipolarity.

From the point of view of database querying, the first model may be seen as assuming that the user assesses both positive and negative aspects of a given piece of data (an attribute value or a tuple) and is in a position to come up with an overall scalar evaluation. This is convenient with respect to the ordering of the tuples in the answer to a query.

The second model is more general and makes it possible for the user to separately express his or her evaluation of positive and negative aspects of a given piece of data. This may be convenient if the user cannot, or is not willing to, combine his or her evaluations of positive and negative features of data. Obviously it requires some special means to order the query answer dataset with respect to a pair of evaluations.

Levels at Which the Bipolarity May be Expressed. Bipolar evaluations may concern the domain of an attribute or the whole set of tuples. This is a distinction of a practical importance, in particular if the elicitation of user preferences is considered.

In the former case, the user is supposed to be willing and in a position to partition the domains of selected attributes into (fuzzy) subsets of elements with positive, negative, and neutral evaluations. For example, the domain of the *price* attribute, characterizing paintings offered during an auction at a gallery, may be in the context of a given query subjectively partitioned using fuzzy sets representing the terms *cheap* (positive evaluation), *expensive* (negative evaluation), and some elements with a neutral evaluation.

In the case of bipolar evaluations at the tuples level a similar partitioning is assumed but concerning the whole set of tuples (here, representing the paintings). Usually, this partition will be defined again by (fuzzy) sets, this time defined with reference to possibly many attributes, i.e., defined on the cross product of the domains of several attributes. For example, the user may identify as negative these paintings which satisfy a compound condition *expensive and modern*. Thus, the evaluations in this case have a *comprehensive* character and concern the whole tuples, taking implicitly into account a possibly complex weighting scheme of particular attributes and their interrelations.

Referring to the models of bipolarity, it seems slightly more natural for the bipolar evaluations expressed on the level of the domain of an attribute to use a bipolar univariate scale while the evaluations on the level of the whole set of tuples would rather adopt a unipolar bivariate scale.

A General Interpretation of Bipolarity in the Context of Database Querying. In the most general interpretation, we do not assume anything more about the relation between positive and negative conditions. Thus, we have two conditions and each tuple is evaluated against them yielding a pair of matching degrees. Then an important question is how to order data in an answer to such a query. Basically, while doing that we should take into account the very nature of both matching degrees, i.e., the fact that they correspond to the positive and negative conditions. The situation here is somehow similar to that of decision making under risk. Namely, in the latter context a decision

maker who is risk-averse may not accept actions leading with some nonzero probability to a loss. On the other hand, a risk-prone decision maker may ignore the risk of an even serious loss as long as there are prospects for a high gain. Similar considerations may apply in the case of bipolar queries. Some users may be more concerned about negative aspects and will reject a tuple with a nonzero matching degree of the negative condition. Some other users may be more oriented on the satisfaction of the positive conditions and may be ready to accept the fact that given piece of data satisfies to some extent the negative conditions too. Thus, the bipolar query should be evaluated in a database in a way strongly dependent on the attitude of the user. In the extreme cases, the above-mentioned *risk-averse* and *risk-prone* attitudes would be represented by lexicographic orders. In the former case, the lexicographic ordering would be first nondecreasing with respect to the negative condition matching degree and then nonincreasing with respect to the positive condition matching degree. The less extreme attitudes of the users may be represented by various aggregation operators producing a scalar overall matching degree of a bipolar query.

An approach to a comprehensive treatment of so generally meant bipolar queries has been proposed by *Matthé* and *De Tré* [19.90], and further developed in [19.91]. In this approach, a pair of matching degrees of the positive and negative conditions is referred to as a *bipolar satisfaction degree* (BSD). The respective matching degrees are denoted as s and d, and called the *satisfaction degree* and the *dissatisfaction degree*, respectively. The ranking of data retrieved against a bipolar query in this approach may be obtained in various ways. One of the options is based on the difference $s - d$ of the two matching degrees. In this case, a *risk-neutral* attitude of the user is modeled: he or she does not favor neither positive nor negative evaluation.

The Required/Desired Semantics. Most of the research on bipolar queries has been so far focused on a special interpretation of the positive and negative conditions. Namely, the data items sought have to satisfy the complement of the latter condition, i.e., the condition denoted earlier as C, unconditionally while the former condition, i.e., the condition denoted as P, is of somehow secondary importance. For example, a painting one is looking for should be from seventeenth century and, *if possible* should be painted by one of the famous Flemish painters. The C condition is here *painted in the seventeenth century* (the original negative condition is of course *painted not in the seventeenth century*) while the positive condition P is *painted by one of the famous Flemish painters*. Thus, the condition C is *required* to be satisfied condition while the condition P may be referred to as a *desired* condition. Anyway, we still have two matching degrees, of conditions C and P, and the assumed relation between them determines the way the tuples should be ordered in the answer to a query.

The simplest approach is to use the desired condition's matching degree just to order the data items which satisfy the required condition. However, if the required condition is fuzzy, i.e., may be satisfied to a degree, it is not obvious what should it mean that it *is satisfied*. Some authors [19.56, 92] propose to adopt the *risk-averse* model of the user and use the corresponding lexicographic order with the primary account for the satisfaction of the condition C. This interpretation is predominant in the literature.

Another approach consists in employing an aggregation operator, which combines the degrees of matching of conditions C and P in such a way so that the *possibility* of satisfying both conditions C and P is explicitly taken into account, i.e., the focus is on a proper interpretation of the following expression which is identified with the bipolar query (C, P)

$$C \text{ and possibly } P. \tag{19.16}$$

Aggregation operators of this type have been studied in the literature under different names and in various contexts. In the framework of database querying it were *Lacroix* and *Lavency* [19.93] who first proposed it. It has been proposed independently in the context of default reasoning by *Yager* [19.94] and by *Dubois* and *Prade* [19.95]. The concept of this operator was also used by *Bordogna* and *Pasi* [19.96] in the context of textual information retrieval. Recently, a more general concept of a *query with preferences* and a corresponding new relational algebra operator, *winnow*, were introduced by *Chomicki* [19.97].

Zadrożny and *Kacprzyk* [19.98, 99] proposed a direct *fuzzification* of the concept of the *and possibly* operator, implicit in the work of *Lacroix* and *Lavency* [19.93]. In their approach, the essence of the *and possibly* operator modeling consists in taking into account the whole database (set of tuples) while combining the required and desired conditions matching degrees. Namely:

● If there is a tuple which satisfies both conditions then and only then it is actually *possible* to satisfy

both of them and each tuple have to meet both of them, i.e., the *and possibly* turns into a regular conjunction $C \wedge P$,

● If there is no such tuple then it is *not possible* to satisfy both conditions and the desired one can be disregarded, i.e., the query reduces to C.

These are however two extreme cases and actually it may be the case that the two conditions may be simultaneously satisfied to *a degree*. Then, the (C, P) query may be also matched to a degree which is identified with the truth of the following formula

$$C(t) \text{ and possibly } P(t)$$
$$\equiv C(t) \wedge \exists s(C(s) \wedge P(s)) \Rightarrow P(t) \tag{19.17}$$

This formula has been proposed by *Lacroix* and *Lavency* [19.93] for the crisp case. Its fuzzy counterpart [19.98–100] requires to choose a proper interpretation of the logical connectives, and may take the following form

$$C(t) \text{ and possibly } P(t)$$
$$\equiv \min \left\{ C(t), \max \left[1 - \max_{s \in \Gamma} \min(C(s), P(s)), P(t) \right] \right\} \tag{19.18}$$

where Γ denotes the whole set of tuples being queried.

Formula (19.16) is derived from (19.17) using the classical fuzzy interpretation of the logical connectives via the max and min operators. *Zadrożny* and *Kacprzyk* [19.100–102] studied the properties of the counterparts of (19.18) obtained using a broader class of the operators modeling logical connectives.

It is worth noting that if the *required/preferred* semantics is assumed and the bipolar evaluations are expressed at the level of an attribute domain then it is reasonable to impose some consistency conditions on the form of both fuzzy sets representing condition C and P. Namely, it may be argued that a domain element should be first acceptable, i.e., should satisfy the required condition C, before it may be desired, i.e., satisfy the condition P. Such consistency conditions between fuzzy sets C and P may be conveniently expressed using the concepts of twofold fuzzy sets or *Atanassov* intuitionistic fuzzy sets/interval-valued fuzzy sets, referred to earlier in Sect. 19.3.2. For an in-depth discussion of such consistency conditions the reader is referred to [19.56, 92, 103].

The growing interest in modeling bipolarity of user preferences in queries resulted recently in some further studies and interpretations of the *and possibly* operator as well as in the concept of new similar operators such as the *or at least* operator. For more details, the reader is referred to [19.104, 105].

19.4.2 Flexible Querying of Fuzzy Databases

Possibilistic Approach

The possibilistic approach to data modeling is based on the sound foundations of a well-developed theory, i.e., the possibility theory. Thus, the standard relational algebra operations have their counterparts in an algebra for retrieving information from a fuzzy possibilistic relational database, proposed by *Prade* and *Testemale* [19.31]. Let us consider the selection operation σ. In the classical relational algebra it is an unary operation which for a given relation R returns another relation $\sigma(R)$, comprising these tuples of R which satisfy a condition c (such a condition is a kind of a parameter of the selection operations). In the possibilistic approach, the selection operation has to be redefined so as to make it compatible with the assumed data representation. To this end, two types of elementary conditions are considered:

(i) $A \, \theta \, a$, where A is an attribute, θ is a comparison operator (fuzzy or not) and a is a constant (fuzzy or not);
(ii) $A_i \, \theta \, A_j$, where A_i and A_j are attributes.

In general, an exact value of an attribute is unknown and, thus, the matching degree is defined as the possibility and necessity of the match between this value and the constant (case (i) above) or the value of another attribute (case (ii) above). Hence, the formulas (19.17)–(19.18) are used to compute the possibility and necessity in the following way.

In case (i), the possibility distribution $\pi_{A(t)}(\cdot)$ representing the value of the attribute A at a tuple t is used to compute the possibility measure of a set F, crisp or fuzzy, of elements of dom_A being in the relation θ with elements representing the constant a. The membership function of the set F is

$$\mu_F(x) = \sup_{y \in dom_A} \min(\mu_\theta(x, y), \mu_a(y)), \quad x \in dom_A ,$$
(19.19)

where $\mu_a(\cdot)$ is the membership function of the constant a and $\mu_\theta(\cdot)$ represents the fuzzy comparison operator (fuzzy relation) θ. Then, the pair $(\Pi_{A(t)}(F), N_{A(t)}(F))$

represents the membership of a tuple t to the relation being the result of the selection operator.

In case (ii), the joint possibility distribution $\pi_{(A_i(t), A_j(t))}(\cdot)$ is used to compute the possibility measure of a subset F of the Cartesian product of domains of A_i and A_j comprising the pairs of elements being in relation θ. The membership function of the set F is defined as follows

$$\mu_F(x, y) = \mu_\theta(x, y), \quad x \in dom_{A_i}, \ y \in dom_{A_j} .$$
(19.20)

Then, the pair $(\Pi_{(A_i(t), A_j(t))}(F), N_{(A_i(t), A_j(t))}(F))$ represents the membership degree of a tuple t to the relation being the result of the selection operator. If the attributes A_i and A_j are *noninteractive* [19.27] then the computing of the possibility measure is simplified. Namely

$$\pi_{(A_i(t), A_j(t))}(x, y) = \min(\pi_{A_i(t)}(x), \pi_{A_j(t)}(y)) .$$
(19.21)

It is worth noting that *Prade* and *Testemale* [19.27], in fact, consider the answer to a query as composed of two fuzzy sets of tuples:

- Those which necessarily match the query; the membership function degree for each tuple of this set is defined by $N_{A(t)}(F)$);
- Those which possibly match the query; the membership function degree for each tuple of this set is defined by $\Pi_{A(t)}(F)$).

Prade and *Testemale* [19.27] consider also the case when the selection operation is used with a compound condition C, i.e., $C = C_1 \wedge C_2$ or $C = C_1 \vee C_2$ or $C = \neg C_1$. Due to the fact that, in general, a calculus of uncertainty, exemplified by the possibility theory, cannot be truth functional [19.106], it is not enough to compute the possibility and necessity measures for elementary conditions using possibility distributions representing values of particular attributes and then combine them using an appropriate operator. In order to secure effective computing of the result of the selection operator it is thus assumed that the attributes referred to in elementary conditions are noninteractive. In such a case, truth-functional combination of the obtained possibility and necessity measures is justified.

Dubois and *Prade* [19.58] propose a technique of querying bipolar data using bipolar queries. A tuple may be classified to many categories with respect to an answer to such a query.

In the *extended possibilistic approach* [19.107] the matching degree of an elementary condition against a tuple t is expressed by an EPTV (cf., page 298). This EPTV represents the extent to which it is (un)certain that t belongs to the result of a flexible query. Let us consider a query condition of the form A *is* l, where A denotes an attribute and l denotes a fuzzy set representing a linguistic term used, in a query such as, e.g., *low* in *Price is low* in a query. Then, the EPTV representing the matching degree will be computed as

$$\mu_{\tilde{T}^*(A \text{ is } l)}(T) = \sup_{x \in dom_A} \min(\pi_A(x), \mu_l(x)), \quad (19.22)$$

$$\mu_{\tilde{T}^*(A \text{ is } l)}(F) = \sup_{x \in dom_A - \{\perp\}} \min(\pi_A(x), 1 - \mu_l(x)), \quad (19.23)$$

$$\mu_{\tilde{T}^*(A \text{ is } l)}(\perp) = \min(\pi_A(\perp), 1 - \mu_l(\perp)), \quad (19.24)$$

where $\pi_A(\cdot)$ denotes the possibility distribution representing the value of the attribute A (to simplify the notaion we omit here a reference to a tuple t). In the case of a compound query condition, the resulting EPTV can be obtained by aggregating the EPTVs computed for the elementary conditions. Hereby, generalizations of the logical connectives of the conjunction ($\wedge$), disjunction ($\vee$), negation ($\neg$), implication ($\rightarrow$), and equivalence ($\leftrightarrow$) can be applied according to [19.14, 108].

Baldwin et al. [19.109] have implemented a system for querying a possibilistic relational database using semantic unification and the evidential support logic rule to combine matching degrees of the elementary conditions. The queries are composed of one or more conditions, the importance of each condition, a *filtering* function (similar to the notion of quantifier) and a threshold. The particularity of their work is the process, semantic unification, used for matching the fuzzy values of the criteria with the possibility distributions representing the values of the attributes. As a result, one obtains an interval $[n, p]$, where, similarly to the previous case, n (necessity) is the certain degree of matching and p (possibility) is the maximum possible degree of matching. However, this time the calculations are based on the mass assignments theory developed by Baldwin. In this approach, an interactive iterative process of the querying is postulated.

Bosc and *Pivert* [19.110] proposed another type of a query against a possibilistic database. Namely, the user may be interested in finding tuples which have a specific features of the possibility distribution representing the value of an attribute. Thus, the condition of such a query does not refer to the value of an attribute

itself but to the characteristics of its possibility distribution. This new type of queries may be illustrated with the following examples:

I. Find tuples such that all the values $a_1, a_2, \ldots, a_n$ are possible for an attribute A.
II. Find tuples such that more than n values are possible to a degree higher than λ for an attribute A.
III. Find tuples where for attribute A the value a_1 is more possible than the value a_2.
IV. Find tuples where for attribute A only one value is completely possible.

The matching degree for such queries is computed in a fairly straightforward way. For the query of type I. it may be computed as $\min(\pi_A(a_1), \pi_A(a_2), \ldots, \pi_A(a_n))$.

The reader is referred for more details to the following sources on fuzzy querying in the possibilistic setting [19.28, 29, 32, 111].

Similarity-Based Approach

The research on querying in similarity-based fuzzy databases is best presented in a series of papers by *Petry* et al. [19.112–114]. A complete set of operations of the relational algebra has been defined for the similarity relation-based model. These operations result from their classical counterparts by the replacement of the concept of equality of two domain values with the concept of their similarity. The conditions of queries are composed of crisp predicates as in a regular query language. Additionally, a set of level thresholds may be submitted as a part of the query. A threshold may be specified for each attribute appearing in query's condition. Such a threshold indicates what degree of similarity of two values from the domain of a given attribute justifies to consider them equal. The concept of the threshold level also plays a central role in the definition of the redundancy in this database model and thus is important for the relational algebra operations as they are usually assumed to be followed by the reduction of redundant tuples. In this model, two tuples are redundant if the values of all corresponding attributes are similar (to a level higher than a selected degree) rather than equal, as it is the case in the traditional relational data model.

Hybrid models, mentioned earlier, are usually accompanied by their own querying schemes. For example, the GEFRED model is equipped with a generalized fuzzy relational algebra. *Galindo* et al. [19.115] have extended the GEFRED model with a fuzzy domain relational calculus (FDRC) and in [19.116] the fuzzy quantifiers have been included.

19.5 Conclusions

In this chapter, we have presented an overview of selected contributions in the areas of data representation and querying. We have focused on approaches rooted in the relational data model. In the literature, many approaches have been also proposed for, e.g., fuzzy object oriented models or fuzzy spatial information modeling which are not covered here due to the lack of space. We have compiled an extensive list of references which will hopefully help the reader to study particular approaches in detail.

References

19.1   H.F. Korth, A. Silberschatz: Database research faces the information explosion, Communication ACM **40**(2), 139–142 (1997)

19.2   L.A. Zadeh: Fuzzy logic, neural networks, and soft computing, Communication ACM **37**(3), 77–84 (1994)

19.3   D. Kastrinakis, Y. Tzitzikas: Advancing search query autocompletion services with more and better suggestions, Lect. Notes Comput. Sci. **6189**, 35–49 (2010)

19.4   R. Ozcan, I.S. Altingovde, O. Ulusoy: Exploiting navigational queries for result presentation and caching in Web search engines, J. Am. Soc. Inf. Sci. Technol. **62**(4), 714–726 (2011)

19.5   T. Gaasterland, P. Godfrey, J. Minker: An overview of cooperative answering, J. Intell. Inf. Syst. **1**, 123–157 (1992)

19.6   E.F. Codd: A relational model of data for large shared data banks, Communication ACM **13**(6), 377–387 (1970)

19.7   R. Elmasri, S. Navathe: *Fundamentals of Database Systems*, 6th edn. (Addison Wesley, Boston 2011)

19.8   ISO/IEC 9075-1:2011: Information technology – Database languages – SQL – Part 1: Framework (SQL/Framework) (2011)

19.9   L.A. Zadeh: Fuzzy sets, Inf. Control **8**(3), 338–353 (1965)

19.10   D. Dubois, H. Prade: The three semantics of fuzzy sets, Fuzzy Sets Syst. **90**(2), 141–150 (1997)

19.11   L.A. Zadeh: Fuzzy sets as a basis for a theory of possibility, Fuzzy Sets Syst. **1**(1), 3–28 (1978)

19.12   D. Dubois, H. Prade: *Possibility Theory* (Plenum, New York 1988)

19.13   P. Bosc, L. Duval, O. Pivert: Value-based and representation-based querying of possibilistic databases. In: *Recent Research Issues on Fuzzy Databases*, ed. by G. Bordogna, G. Pasi (Physica, Heidelberg 2000) pp. 3–27

19.14   G. De Tré: Extended possibilistic truth values, Int. J. Intell. Syst. **17**, 427–446 (2002)

19.15   Z. Pawlak: Rough sets, Int. J. Parallel Program. **11**(5), 341–356 (1982)

19.16   A.N. Kolmogorov: *Foundations of the Theory of Probability*, 2nd edn. (Chelsea, New York 1956)

19.17   R. De Caluwe (Ed.): *Fuzzy and Uncertain Object-oriented Databases: Concepts and Models* (World Scientific, Signapore 1997)

19.18   A. Yazici, R. George: *Fuzzy Database Modeling* (Physica, Heidelberg 1999)

19.19   G. Bordogna, G. Pasi: Linguistic aggregation operators of selection criteria in fuzzy informational retrieval, Int. J. Intell. Syst. **10**(2), 233–248 (1995)

19.20   Z. Ma (Ed.): *Advances in Fuzzy Object-Oriented Databases: Modeling and Applications* (Idea Group, Hershey 2005)

19.21   J. Galindo, A. Urrutia, M. Piattini (Eds.): *Fuzzy Databases: Modeling, Design and Implementation* (Idea Group, Hershey 2006)

19.22   J. Galindo (Ed.): *Handbook of Research on Fuzzy Information Processing in Databases* (Idea Group, Hershey 2008)

19.23   K.V.S.V.N. Raju, A.K. Majumdar: The study of joins in fuzzy relational databases, Fuzzy Sets Syst. **21**(1), 19–34 (1987)

19.24   P. Smets: Imperfect Information: Imprecision and uncertainty. In: *Uncertainty Management in Information Systems: From Needs to Solution*, ed. by A. Motro, P. Smets (Kluwer, Boston 1996) pp. 225–254

19.25   L.A. Zadeh: The concept of a linguistic variable and its application to approximate reasoning, Part I, Inf. Sci. **8**(3), 43–80 (1975)

19.26   D. Dubois, H. Prade: Fuzzy sets in approximate reasoning, Part 1: Inference with possibility distributions, Fuzzy Sets Syst. **40**, 143–202 (1991)

19.27   H. Prade, C. Testemale: Representation of soft constraints and fuzzy attribute values by means of possibility distributions in databases. In: *Analysis of Fuzzy Information*, ed. by J.C. Bezdek (Taylor Francis, Boca Raton 1987) pp. 213–229

19.28   M. Umano: FREEDOM-0: A fuzzy database system. In: *Fuzzy Information and Decision Processes*, ed. by M.M. Gupta, E. Sanchez (Elsevier, Amsterdam 1982) pp. 339–347

19.29   M. Zemankova-Leech, A. Kandel: *Fuzzy relational Data Bases: A Key to Expert Systems* (Verlag TÜV Rheinland, Cologne 1984)

19.30   M. Zemankova-Leech, A. Kandel: Implementing imprecision in information systems, Inf. Sci. **37**(1-3), 107–141 (1985)

19.31 H. Prade, C. Testemale: Generalizing database relational algebra for the treatment of incomplete or uncertain information and vague queries, Inf. Sci. **34**, 115–143 (1984)

19.32 M. Umano, S. Fukami: Fuzzy relational algebra for possibility-distribution-fuzzy-relational model of fuzzy data, J. Intell. Inf. Syst. **3**, 7–27 (1994)

19.33 B.P. Buckles, F.E. Petry: A fuzzy representation of data for relational databases, Fuzzy Sets Syst. **7**, 213–226 (1982)

19.34 M. Anvari, G.F. Rose: Fuzzy relational databases. In: *Analysis of Fuzzy Information*, ed. by J.C. Bezdek (Taylor Francis, Boca Raton 1987) pp. 203–212

19.35 E. Wong: A statistical approach to incomplete information in database systems, ACM Trans. Database Syst. **7**, 470–488 (1982)

19.36 M. Pittarelli: An algebra for probabilistic databases, IEEE Trans. Knowl. Data Eng. **6**, 293–303 (1994)

19.37 O. Benjelloun, A. Das Sarma, C. Hayworth, J. Widom: An introduction to ULDBs and the trio system, IEEE Data Eng. Bull. **29**(1), 5–16 (2006)

19.38 E. Michelakis, D.Z. Wang, M.N. Garofalakis, J.M. Hellerstein: Granularity conscious modeling for probabilistic databases, Proc. ICDM DUNE (2007) pp. 501–506

19.39 L. Antova, T. Jansen, C. Koch, D. Olteanu: Fast and simple relational processing of uncertain data, Proc. 24th Int. Conf. Data Eng. (2008) pp. 983–992

19.40 P. Bosc, O. Pivert: Modeling and querying uncertain relational databases: A survey of approaches based on the possible world semantics, Int. J. Uncertain. Fuzziness Knowl. Syst. **18**(5), 565–603 (2010)

19.41 L.A. Zadeh: Quantitative fuzzy semantics, Inf. Sci. **3**(2), 177–200 (1971)

19.42 S. Gottwald: Set theory for fuzzy sets of a higher level, Fuzzy Sets Syst. **2**(2), 125–151 (1979)

19.43 G. De Tré, R. De Caluwe: Level-2 fuzzy sets and their usefulness in object-oriented database modelling, Fuzzy Sets Syst. **140**, 29–49 (2003)

19.44 Y. Vassiliou: Null values in data base management: A denotational semantics approach, Proc. SIGMOD Conf. (1979) pp. 162–169

19.45 C. Zaniolo: Database relations with null values, J. Comput. Syst. Sci. **28**(1), 142–166 (1984)

19.46 E.F. Codd: Missing information (applicable and inapplicable) in relational databases, ACM SIGMOD Rec. **15**(4), 53–78 (1986)

19.47 G. De Tré, R. De Caluwe, H. Prade: Null values in fuzzy databases, J. Intell. Inf. Syst. **30**, 93–114 (2008)

19.48 S. Shenoi, A. Melton: Proximity relations in the fuzzy relational database model, Fuzzy Sets Syst. **31**(3), 285–296 (1989)

19.49 T. Beaubouef, F.E. Petry, B.P. Buckles: Extension of the relational database and its algebra with rough set techniques, Comput. Intell. **11**, 233–245 (1995)

19.50 T. Beaubouef, F.E. Petry, G. Arora: Information-theoretic measures of uncertainty for rough sets and rough relational databases, Inf. Sci. **109**, 185–195 (1998)

19.51 Y. Takahashi: Fuzzy database query languages and their relational completeness theorem, IEEE Trans. Knowl. Data Eng. **5**, 122–125 (1993)

19.52 J.M. Medina, O. Pons, M.A. Vila: GEFRED. A generalized model of fuzzy relational databases, Inf. Sci. **76**(1/2), 87–109 (1994)

19.53 S. Zadrożny, G. De Tré, J. Kacprzyk: On some approaches to possibilistic bipolar data modeling in databases. In: *Advances in Fuzzy Sets, Intuitionistic Fuzzy Sets. Generalized Nets and Related Topics, Vol. II: Applications*, ed. by K.T. Atanassov, O. Hryniewicz, J. Kacprzyk, M. Krawczak, Z. Nahorski, E. Szmidt, S. Zadrożny (EXIT, Warsaw 2008) pp. 197–220

19.54 S. Benferhat, D. Dubois, S. Kaci, H. Prade: Bipolar possibilistic representations, Proc. 18th Conf. Uncertain. Artif. Intell. (2002) pp. 45–52

19.55 D. Dubois, H. Fargier: Qualitative decision-making with bipolar information, Proc. 10th Int. Conf. Princ. Knowl. Represent. Reason. (2006) pp. 175–186

19.56 D. Dubois, H. Prade: Handling bipolar queries in fuzzy information processing. In: *Handbook of Research on Fuzzy Information Processing in Databases*, ed. by J. Galindo (Information Science Reference, Hershey 2008) pp. 97–114

19.57 D. Dubois, H. Prade (eds.): Int. J. Intell. Syst. **23**(8/10) 863–1152 (2008), Special issues on bipolar representations of information and preference (Part 1A, 1B, 2)

19.58 D. Dubois, H. Prade: An overview of the asymmetric bipolar representation of positive and negative information in possibility theory, Fuzzy Sets Syst. **160**(10), 1355–1366 (2009)

19.59 D. Dubois, H. Prade: An introduction to bipolar representations of information and preference, Int. J. Intell. Syst. **23**(8), 866–877 (2008)

19.60 K.T. Atanassov: *On Intuitionistic Fuzzy Sets Theory* (Springer, Berlin, Heidelberg, 2012)

19.61 J. Kacprzyk, S. Zadrożny: Linguistic database summaries and their protoforms: Towards natural language based knowledge discovery tools, Inf. Sci. **173**(4), 281–304 (2005)

19.62 J. Kacprzyk, S. Zadrożny: Computing with words is an implementable paradigm: Fuzzy queries, linguistic data summaries, and natural-language generation, IEEE Trans. Fuzzy Syst. **18**(3), 461–472 (2010)

19.63 A. Laurent: Querying fuzzy multidimensional databases: Unary operators and their properties, Int. J. Uncertain. Fuzziness Knowl. Syst. **11**, 31–46 (2003)

19.64 T. Radecki: Mathematical model of information retrieval based on the theory of fuzzy sets, Inf. Process. Manag. **13**(2), 109–116 (1977)

19.65 V. Tahani: A conceptual framework for fuzzy query processing – A step toward very intelligent database systems, Inf. Process. Manag. **13**(5), 289–303 (1977)

19.66 J. Kacprzyk, A. Ziółkowski: Database queries with fuzzy linguistic quantifiers, IEEE Trans. Syst. Man Cybern. **16**(3), 474–479 (1986)

19.67 J. Kacprzyk, S. Zadrożny, A. Ziółkowski: FQUERY III+: A *human-consistent* database querying system based on fuzzy logic with linguistic quantifiers, Inf. Syst. **14**(6), 443–453 (1989)

19.68 P. Bosc, O. Pivert: An approach for a hierarchical aggregation of fuzzy predicates, Proc. 2nd IEEE Int. Conf. Fuzzy Syst. (1993) pp. 1231–1236

19.69 P. Bosc, O. Pivert: Some approaches for relational databases flexible querying, J. Intell. Inf. Syst. **1**(3/4), 323–354 (1992)

19.70 P. Bosc, O. Pivert: Fuzzy querying in conventional databases. In: *Fuzzy Logic for the Management of Uncertainty*, ed. by L.A. Zadeh, J. Kacprzyk (Wiley, New York 1992) pp. 645–671

19.71 P. Bosc, O. Pivert: SQLf: A relational database language for fuzzy querying, IEEE Trans. Fuzzy Syst. **3**, 1–17 (1995)

19.72 J. Kacprzyk, S. Zadrożny: FQUERY for Access: Fuzzy querying for windows-based DBMS. In: *Fuzziness in Database Management Systems*, ed. by P. Bosc, J. Kacprzyk (Physica, Heidelberg 1995) pp. 415–433

19.73 J. Galindo, J.M. Medina, O. Pons, J.C. Cubero: A server for Fuzzy SQL queries, Lect. Notes Artif. Intell. **1495**, 164–174 (1998)

19.74 J. Kacprzyk, S. Zadrożny: Computing with words in intelligent database querying: Standalone and internet-based applications, Inf. Sci. **134**(1–4), 71–109 (2001)

19.75 R.R. Yager: Fuzzy quotient operators for fuzzy relational databases, Proc. Int. Fuzzy Eng. Symp. (1991) pp. 289–296

19.76 D. Dubois, H. Prade: Semantics of quotient operators in fuzzy relational databases, Fuzzy Sets Syst. **78**, 89–93 (1996)

19.77 J. Galindo, J.M. Medina, J.C. Cubero, M.T. Garcia: Relaxing the universal quantifier of the division in fuzzy relational databases, Int. J. Intell. Syst. **16**(6), 713–742 (2001)

19.78 P. Bosc, O. Pivert, D. Rocacher: Tolerant division queries and possibilistic database querying, Fuzzy Sets Syst. **160**(15), 2120–2140 (2009)

19.79 P. Bosc, O. Pivert: On diverse approaches to bipolar division operators, Int. J. Intell. Syst. **26**(10), 911–929 (2011)

19.80 Y. Takahashi: A fuzzy query language for relational databases. In: *Fuzziness in Database Management Systems*, ed. by P. Bosc, J. Kacprzyk (Physica, Heidelberg 1995) pp. 365–384

19.81 D. Dubois, H. Prade: Using fuzzy sets in flexible querying: Why and how? In: *Flexible Query Answering Systems*, ed. by T. Andreasen, H. Christiansen, H.L. Larsen (Kluwer, Boston 1997) pp. 45–60

19.82 L.A. Zadeh: A computational approach to fuzzy quantifiers in natural languages, Comput. Math. Appl. **9**, 149–184 (1983)

19.83 S. Zadrożny, J. Kacprzyk: Issues in the practical use of the OWA operators in fuzzy querying, J. Intell. Inf. Syst. **33**(3), 307–325 (2009)

19.84 R.R. Yager: Interpreting linguistically quantified propositions, Int. J. Intell. Syst. **9**, 541–569 (1994)

19.85 P. Bosc, O. Pivert, L. Lietard: Aggregate operators in database flexible querying, Proc. IEEE Int. Conf. Fuzzy Syst. (2001) pp. 1231–1234

19.86 P. Bosc, L. Lietard, O. Pivert: Sugeno fuzzy integral as a basis for the interpretation of flexible queries involving monotonic aggregates, Inf. Process. Manag. **39**(2), 287–306 (2003)

19.87 M.A. Vila, J.-C. Cubero, J.-M. Medina, O. Pons: Using OWA operator in flexible query processing. In: *The Ordered Weighted Averaging Operators: Theory and Applications*, ed. by R.R. Yager, J. Kacprzyk (Kluwer, Boston 1997) pp. 258–274

19.88 O. Pivert, P. Bosc: *Fuzzy Preference Queries to Relational Database* (Imperial College, London 2012)

19.89 M. Grabisch, S. Greco, M. Pirlot: Bipolar and bivariate models in multicriteria decision analysis: Descriptive and constructive approaches, Int. J. Intell. Syst. **23**, 930–969 (2008)

19.90 T. Matthé, G. De Tré: Bipolar query satisfaction using satisfaction and dissatisfaction degrees: Bipolar satisfaction degrees, Proc. ACM Symp. Appl. Comput. (2009) pp. 1699–1703

19.91 T. Matthé, G. De Tré, S. Zadrożny, J. Kacprzyk, A. Bronselaer: Bipolar database querying using bipolar satisfaction degrees, Int. J. Intell. Syst. **26**(10), 890–910 (2011)

19.92 D. Dubois, H. Prade: Bipolarity in flexible querying, Proc. 5th Int. Conf. Flex. Query Answ. Syst. (2002) pp. 174–182

19.93 M. Lacroix, P. Lavency: Preferences: Putting more knowledge into queries, Proc. 13 Int. Conf. Very Large Databases (1987) pp. 217–225

19.94 R.R. Yager: Using approximate reasoning to represent default knowledge, Artif. Intell. **31**(1), 99–112 (1987)

19.95 D. Dubois, H. Prade: Default reasoning and possibility theory, Artif. Intell. **35**(2), 243–257 (1988)

19.96 G. Bordogna, G. Pasi: Linguistic aggregation operators of selection criteria in fuzzy information retrieval, Int. J. Intell. Syst. **10**(2), 233–248 (1995)

19.97 J. Chomicki: Preference formulas in relational queries, ACM Trans. Database Syst. **28**(4), 427–466 (2003)

19.98 S. Zadrożny: Bipolar queries revisited, Proc. Model. Decis. Artif. Intell. (2005) pp. 387–398

19.99   S. Zadrożny, J. Kacprzyk: Bipolar queries and queries with preferences, Proc. 17th Int. Conf. Database Expert Syst. Appl. (2006) pp. 415–419

19.100   S. Zadrożny, J. Kacprzyk: Bipolar queries: An aggregation operator focused perspective, Fuzzy Sets Syst. **196**, 69–81 (2012)

19.101   S. Zadrożny, J. Kacprzyk: Bipolar queries using various interpretations of logical connectives, Proc. IFSA Congr. (2007) pp. 181–190

19.102   S. Zadrożny, J. Kacprzyk: Bipolar queries: An approach and its various interpretations, Proc. IFSA/EUSFLAT Conf. (2009) pp. 1288–1293

19.103   G. De Tré, S. Zadrożny, A. Bronselaer: Handling bipolarity in elementary queries to possibilistic databases, IEEE Trans. Fuzzy Syst. **18**(3), 599–612 (2010)

19.104   D. Dubois, H. Prade: Modeling *and if possible* and *or at least*: Different forms of bipolarity in flexible querying. In: *Flexible Approaches in Data, Information and Knowledge Management*, ed. by O. Pivert, S. Zadrożny (Springer, Berlin, Heidelberg 2014)

19.105   L. Liétard, N. Tamani, D. Rocacher: Fuzzy bipolar conditions of type *or else*, Proc. FUZZ-IEEE 2011 (2011) pp. 2546–2551

19.106   D. Dubois, H. Prade: Gradualness, uncertainty and bipolarity: Making sense of fuzzy sets, Fuzzy Sets Syst. **192**, 3–24 (2012)

19.107   G. De Tré, R. De Caluwe: Modelling uncertainty in multimedia database systems: An extended possibilistic approach, Int. J. Uncertain. Fuzziness Knowl. Syst. **11**(1), 5–22 (2003)

19.108   G. De Tré, B. De Baets: Aggregating constraint satisfaction degrees expressed by possibilistic truth values, IEEE Trans. Fuzzy Syst. **11**(3), 361–368 (2003)

19.109   J.F. Baldwin, M.R. Coyne, T.P. Martin: Querying a database with fuzzy attribute values by iterative updating of the selection criteria, Proc. IJCAI'93 Workshop Fuzzy Logic Artif. Intell. (1993) pp. 62–76

19.110   P. Bosc, O. Pivert: On representation-based querying of databases containing ill-known values, Proc. 10th Int. Symp. Found. Intell. Syst. (1997) pp. 477–486

19.111   P. Bosc, O. Pivert: Possibilistic databases and generalized yes/no queries, Proc. 15th Int. Conf. Database Expert Syst. Appl. (2004) pp. 912–916

19.112   B.P. Buckles, F.E. Petry: Query languages for fuzzy databases. In: *Management Decision Support Systems Using Fuzzy Sets and Possibility Theory*, ed. by J. Kacprzyk, R. Yager (Verlag TÜV Rheiland, Cologne 1985) pp. 241–251

19.113   B.P. Buckles, F.E. Petry, H.S. Sachar: A domain calculus for fuzzy relational databases, Fuzzy Sets Syst. **29**, 327–340 (1989)

19.114   F.E. Petry: *Fuzzy Databases: Principles and Applications* (Kluwer, Boston 1996)

19.115   J. Galindo, J.M. Medina, M.C. Aranda: Querying fuzzy relational databases through fuzzy domain calculus, Int. J. Intell. Syst. **14**, 375–411 (1999)

19.116   J. Galindo, J.M. Medina, J.C. Cubero, M.T. Garcia: Fuzzy quantifiers in fuzzy domain calculus, Proc. Int. Conf. Inf. Process. Manag. Uncertain. Knowl. Syst. (2000) pp. 1697–1704

20. Application of Fuzzy Techniques to Autonomous Robots

Ismael Rodríguez Fdez, Manuel Mucientes, Alberto Bugarín Diz

The application of fuzzy techniques in robotics has become widespread in the last years and in different fields of robotics, such as behavior design, coordination of behavior, perception, localization, etc. The significance of the contributions was high until the end of the 1990s, where the main aim in robotics was the implementation of basic behaviors. In the last years, the focus in robotics moved to building robots that operate autonomously in real environments; the actual impact of fuzzy techniques in the robotics community is not as deep as it was in the early stages of robotics or as it is in other application areas (e.g., medicine, processes industry ...). In spite of this, new emerging areas in robotics such as human–robot interaction, or well-established ones, such as perception, are good examples of new potential realms of applications where (hybridized) fuzzy approaches will surely be capable of exhibiting their capacity to deal with such complex and dynamic scenarios.

Part B | 20.1

20.1 Robotics and Fuzzy Logic

Although many other classical definitions could be stated, an autonomous robot may be defined as a machine that collects data from the environment through sensors, processes these data taking into account its previous knowledge of the world, and acts according to a goal. This definition is general and covers the different types of robots available today: indoor wheeled robots, autonomous cars, unmanned aerial vehicles (UAVs), autonomous underwater vehicles, robotic arms, humanoid robots, robotic heads, etc.

Between the early 1990s and today robotics has evolved a lot. From our point of view, it is possible to distinguish three stages in robotics research in the last years. At the beginning, the focus was on endowing robots with a number of simple behaviors to solve basic tasks like wall-following, obstacle avoidance,

entering rooms, etc. Later, the objective in robotics moved to building truly autonomous robots, which required the mapping of the environment, the localization of the robot, and navigation or motion planning. Although these topics are still open, the focus of current robotics is starting to move to a third stage where much higher level and integrated capabilities, such as advanced perception, learning of complex behaviors, or human–robot interaction, are involved.

Within this context, fuzzy logic has been widely used in robotics for several purposes. The main advantage of using fuzzy logic in robotics is its ability to manage the uncertainty due to sensors, actuators, and also in the knowledge about the world. Until the end of the 1990s, the contributions of fuzzy logic were mainly in three fields [20.1]: behaviors, coordination

Table 20.1 Distribution of the 98 publications considered in the respective rankings: Thomson–Reuters Web of Science (after [20.6]; JCR 2012-WoK) for journals and Microsoft Academic Search (MAS 2013; after [20.5] for conferences

Quartile	Journals no.papers	%papers	Conferences no. papers	%papers
Q1	68	81	10	72
Q2	10	12	1	7
Q3	2	2	1	7
Q4	1	1	0	0
No ranking	3	4	2	14
Total	84	100	14	100

of behaviors, and perception. The design of behaviors for solving specific and simple tasks in robotics has been undoubtedly the most successful application of fuzzy logic in robotics. As examples of these behaviors we have wall-following, navigation, trajectory tracking, moving objects tracking, etc. Also, the selection and/or combination of these basic behaviors has been solved with fuzzy logic [20.2–4]. Finally, fuzzy techniques have contributed to perception in two lines: i) for the preprocessing of sensor data, prior to their use as input to the behavior and ii) for modeling the uncertainty both in occupancy and feature-based maps.

In this chapter we describe and analyze the contributions of fuzzy techniques to robotics in the period 2003–2013. A number of 98 references related to the topic *fuzzy and robotics* have finally been selected for being categorized and described with the aim to focus on recent papers describing uses of fuzzy-based or hybrid methods in relevant tasks. Both basic behaviors and also high-level tasks of the robotics area as it is understood nowadays were considered. Table 20.1 describes the relevance of the papers considered in terms of their position in the well-known rankings Thomson–Reuters Web of Knowledge for journals and Microsoft Academic Search (MAS 2013) [20.5] for conferences. We decided to consider the MAS 2013 ranking since other conference rankings such as CORE-ERA (Computing Research and Education-Excellence in Research for Australia) were not up to date at the moment and do not extensively cover the robotics area. It can be seen that a vast majority of the papers are ranked in the Q1 of the respective lists (81% for journals and 72% for conferences).

All the references were revised and classified according to the robotics area they mainly addressed and were included into one of the 12 categories described in Sects. 20.2–20.13. Also the fuzzy technique they use (together with its hybridization with other soft computing techniques if this is the case) was annotated in order to assess which are the most active areas of soft computing in the field of robotics and also to evaluate their actual impact in the field.

The results of this revision, classification, and analysis are presented in the sections that follow.

20.2 Wall-Following

When autonomous robots navigate within indoor environments (e.g., industrial or civil buildings), they have to be endowed with a number of basic capabilities that allow the robot to perform specific tasks during operation. These basic capabilities are usually referred to in the robotics literature as behaviors. Some examples of usual robot behaviors are the ability to move along corridors, to follow walls at a given distance, to turn corners, and to cross open areas in rooms. Wall-following behavior is one of the most relevant ones and has been very widely dealt with in the robotics literature, since it is one of the basic behaviors to be executed when the robot is exploring an unknown area, or when it is moving between two points in a map.

The characteristic that makes a fuzzy controller useful for the implementation of this and other behaviors is the ability that fuzzy controllers have to cope with noisy inputs. This noise appears when the sensors of the robot detect the surrounding environment and is an inherent feature to the whole field of robotic sensors.

The importance of wall-following behavior for a car-like mobile robot was pointed out in [20.7]. In this work, a fuzzy logic control system was used in order to implement human-like driving skills by an autonomous mobile robot. Four different sensor-based behaviors were merged in order to synthesize the concepts of the maneuvers needed. These behaviors were: wall-following, corner control, garage-parking, and parallel-parking. A description of the design and implementation of a velocity controller for wall-following can be found in [20.8, 9]. In [20.8] fuzzy temporal rules were used in order to filter the sensorial noise of a Nomad

200 mobile robot and to endow the rule base with high expressiveness. The use of these types of rules has noticeably improved the robustness and reliability of the system.

Wall-following behavior has also been used as a testing benchmark for automatic learning of fuzzy controllers. In [20.10] an evolutionary algorithm was used to automatically learn the fuzzy controller, taking into account the tradeoff between complexity and accuracy. Continuing this work, in [20.11] the focus was on reducing the expert knowledge demanded for designing the controller. No restrictions are placed either on the number of linguistic labels or on the values that define the membership functions. Finally, in [20.12] a simple but effective learning methodology was presented. This methodology was proposed in order to not only generate fuzzy controllers with good behavior in simulated experiments but also to use them directly in the real robot, with no further post-processing or implementing a tuning stage.

More recent studies have addressed the use of type-2 fuzzy logic. These proposals are mostly motivated by the claim that type-1 fuzzy sets cannot handle the high levels of uncertainty that are usually present in real-world applications to the same extent that type-2 fuzzy sets can.

As for type-1 fuzzy logic controllers (FLCs), this behavior has been used for testing the automatic learning of type-2 FLCs. In [20.13] a genetic algorithm was used for tuning the type-2 fuzzy membership functions of a previously defined controller. A more complex learning scheme was presented in [20.14]. The antecedent is learned using a type-2 fuzzy clustering based on examples and without expert knowledge. The actions in the consequent part of rules is selected

from a set of defined control actions through a hybrid method composed of a reinforcement learning algorithm and ant colony optimization. Both works used a real Pioneer robot to demonstrate the viability of the controllers.

In order to show the advantages of using type-2 fuzzy logic, a comparative analysis of type-1 and interval type-2 fuzzy controllers was presented in [20.15]. A particle swarm optimization algorithm was used to optimize a type-1 FLC. Next, the interval type-2 fuzzy controller was constructed, blurring the membership functions. The results obtained by a real mobile robot showed that the interval type-2 fuzzy controller can cope better with dynamic uncertainties in the sensory inputs due to the softening and smoothing of the output control surface.

However, these works only focus on interval type-2 fuzzy logic. The high computational complexities associated with general type-2 fuzzy logic systems (FLSs) have, until recently, prevented their application to real-world control problems. In [20.16] this problem was addressed by introducing a complete representation framework, which is referred to as zSlices-based general type-2 fuzzy systems. As a proof-of-concept application, this framework was implemented for a mobile robot, which operates in a real-world outdoor environment using the wall-following behavior. In this case, the proposed approach outperformed type-1 and interval type-2 fuzzy controllers in terms of errors in the distance to the wall. For this behavior, type-2 approaches exhibit a better performance when compared to type-1 approaches. Nevertheless, from a robotics point of view, type-2 proposals still did not outperform in general other fuzzy approaches and have a very limited impact on this area.

20.3 Navigation

Robot navigation consists of a series of actions, which are summarized in the ability of the robot to go from a starting point to a goal without a planned route. Navigation is one of the main issues that a mobile robot must solve in order to operate.

In this behavior the ability to work in dynamic environments whose structure is unknown, with great uncertainty, with moving objects or objects that may change their position is of great importance. The capacity to work under these conditions is the best motivation to use fuzzy logic in this particular task.

Some work in this area has been done in simulated environments. In [20.17] a multi-sensor fusion technique was used to integrate all types of sensors and combine them to obtain information about the environment. In this way, the environment can be perceived comprehensively, and the ability of the FLC is improved. In [20.18, 19] the navigation behavior was studied for a set of robots that navigate in the same environment. Both works used a Petri net to negotiate the priority of the robots. Also, in [20.19] different FLCs were compared. Each controller used a different number of labels for each variable as well as a different shape of the fuzzy sets. It was concluded that utilizing a Gaussian membership function is better for navigation in environments with a high number of moving objects.

Moreover, there are several works where this behavior has been successfully implemented in a real robot. A hardware implementation of a FLC for navigation in mobile robots was presented in [20.20]. The design methodology allows to transform a FLC into a system that is suitable for easy implementation on a digital signal processor (DSP). This methodology was tested with good results in a ROMEO 4R car-like vehicle for a parking problem. Moreover, in [20.21] the design of a new fuzzy logic-based navigation algorithm for autonomous robots was illustrated. It effectively achieves correct environment modeling, and processes noisy and uncertain sensory data on a low-cost Khepera robot.

The ability to avoid dead-end paths was studied in [20.22, 23]. In these works the minimum risk approach was used to avoid local minima. A novel path-searching behavior was developed to recommend the local direction with minimum risk, where the risk was modeled using fuzzy logic. Another approximation to solve this problem was presented in [20.24]. While the fuzzy logic body of the algorithm performs the main tasks of obstacle avoidance and target seeking, an actual/virtual target switching strategy solves the problem of dead-ends on the way to the target.

In [20.25] a new approach was proposed that employs a fuzzy discrete event system to implement the behavior coordinator mechanism that selects relevant behaviors at a particular moment to produce an appropriate system response. The possible transition from one state to another when an event occurs was modeled using fuzzy sets.

In addition to the use of conventional FLCs, different approaches have been used in the last decade. In [20.26] a novel reactive type-2 fuzzy logic architecture was used. Type-2 fuzzy logic was used for both implementing the basic navigation behaviors and also the strategies for their coordination. The proposed architecture was implemented in a robot and successfully tested in indoor and outdoor environments.

In the same way as several learning approaches have been used for wall-following behavior, different works focused on automatically learning navigation skills. In [20.27] a novel fuzzy Q-learning approach was presented, where the weights of the fuzzy rules of the controller were learned through a reinforcement algorithm.

In [20.28] a neuro-fuzzy network that is able to add rules to the rule base was presented. The criteria for adding rules was based on a performance evaluation within a genetic algorithm that explores the new situations to add. A comparison of three different neuro-fuzzy approaches with classical fuzzy controllers can be found in [20.29]. It is shown that neuro-fuzzy approaches perform better and that the best results were obtained by an optimization made by a genetic algorithm for both Mamdani and Takagi–Sugeno approaches.

Although mobile wheeled robots are the most common area of application for navigation, other types of robots also use this behavior. One of the most impressive types of robot that implement the navigation behavior are the unmanned aerial vehicles (UAVs). In [20.30] two fuzzy controllers (one for altitude and the other for latitude–longitude) were combined in order to control the navigation of a small UAV. In [20.31] the design of a Takagi–Sugeno controller for an unmanned helicopter was presented. The controller proposed is a fuzzy gain-scheduler used for stable and robust altitude, roll, pitch, and yaw control. Testing in both papers was performed in simulated environments to show the results obtained by the controllers and, therefore, real testing on real UAVs was not reported.

Another type of robot that demands navigation capabilities are robotic manipulators. Simulated manipulators were used in [20.32, 33]. The strategy followed in [20.32] was to use a fuzzy inference process to tune the gain of a sliding-mode control and the weights of a neural network controller in the presence of disturbance or big tracking errors. It was shown that the combination of these controllers can guarantee stability. In [20.33] a genetic algorithm was presented in order to optimize the controllers of two robotic manipulators working on the same environment.

Other examples of applications of navigation are: a robotic fish motion control algorithm [20.34] endowed with an orientation control system based on a FLC; the data transmission latency or data loss considered in [20.35] for internet-based teleoperation of robots (when data transmission fails, the robot automatically moves and protects itself using a fuzzy controller optimized using a co-evolutionary algorithm); and the stabilization of a unicycle mobile robot described in [20.36], where a type-2 FLC was used, and computer simulations confirmed its good performance in different navigation problems.

20.4 Trajectory Tracking

Tracking refers to the ability of a robot to follow a predetermined series of movements or a predefined path. Tracking is a relevant behavior in the industrial field, where robots usually have to perform a repeated pattern or trajectory with high precision.

Developing accurate analytical models for such systems and hence reliable controllers based on such models is extremely difficult and in general unfeasible even for not very complex trajectories. Applying fuzzy control strategies for such systems seems appropriate, since with these systems the nonlinear system identification methodologies can be exploited with the help of inherent knowledge. Furthermore, suitable stability conditions for such controllers to guarantee global asymptotic stability can be determined.

In [20.37] a control structure that makes possible the integration of a kinematic controller and an adaptive fuzzy controller was developed for mobile robots. A highly robust and flexible system that automatically follows a sequence of discrete way-points was presented in [20.38]. In [20.39] the combination of a velocity controller with a simple fuzzy system that limits on-line the advancing speed of the vehicle to allow it to follow an assigned path in compliance with the kinematic constraints was presented.

More recent studies in tracking for mobile robots have focused on more advanced and complex systems. The work in [20.40] focused on the design of a dynamic Petri recurrent fuzzy neural network. This network structure was applied to the path-tracking control of a non-holonomic mobile robot for verifying its validity. Also, in [20.41] the tracking control of a mobile robot with uncertainties in the robot kinematics, the robot dynamics, and the wheel actuator dynamics was investigated. A robust adaptive controller was proposed for back-stepping a FLC. Finally, [20.42] proposed a complete control law comprising an evolutionary programming-based kinematic control and an adaptive fuzzy sliding-mode dynamic control.

Although mobile robots have a great utility in industry, the robotic manipulators are the most used type of robot in this sector. One main issue is to develop controllers that deal with uncertainties. [20.43] addressed trajectory tracking problems of robotic manipulators using a fuzzy rule controller designed to deal with uncertainty. In [20.44] two adaptive fuzzy systems are employed to approximate the nonlinear and uncertain terms occurring in the robot arm and the joint motor dynamics. Other adaptive controllers for motion control of multi-link robot manipulators can be found in [20.45–47].

Other works focused on the ability to adapt the fuzzy control to different demands or requisites over time with different approaches. In [20.48] a fuzzy sliding mode controller was proposed for robotic manipulators. The membership functions of the control gain are updated on-line and, therefore, the controller is not a conventional fuzzy controller but an adaptive one. In [20.49] a design method that constructs the fuzzy rule base from a conventional proportional-integral-derivative (PID) controller in an incremental way using recursive feedback was proposed. A direct fuzzy control system for the regulation of robot manipulators was presented in [20.50]. The bounds of the applied torques in this case are adjusted by means of the output membership functions parameters in such a way that the maximum torque demanded by the controller always ranges between the limits given by the manufacturer.

A different perspective of robotic manipulators, but also a very common one, is the scheme of different systems that need to communicate. In [20.51] a new observer-controller structure for robot manipulators with model uncertainty using only the position measurements is proposed. In this method, adaptive fuzzy logic is used to approximate the nonlinear and uncertain robot dynamics in both the observer and the controller.

The effects of network-induced delay and data packet dropout for a class of nonlinear networked control systems for a flexible arm was investigated in [20.52]. The non-linear networked control systems were approximated by linear networked Takagi–Sugeno fuzzy models. An iterative algorithm for constructing the fuzzy model was proposed. Also, in [20.53], the delay transmission of a signal through an internet and wireless module was studied.

As well as for the case of mobile robots, in the recent years of research in robotic manipulators both the automatic design and the learning of different parts of the controllers have been addressed. For example, in [20.54] a novel tracking control design for robotic systems using fuzzy wavelet networks was presented. Fuzzy wavelet networks were used to estimate unknown functions and, therefore, to solve the problem of demanding prior knowledge of the controlled plant.

Intelligent control approaches such as neural networks for the approximation of nonlinear systems have also received considerable attention. They are very effective in coping with structured parametric uncertainty and unstructured disturbance by using their powerful learning capability. Thus, neuro-fuzzy network controllers are the usual choice for robot manipulators.

In [20.55] the position control of modular and reconfigurable robots was addressed. A neuro-fuzzy control architecture was used for tuning the gains inside the PID controller. An improvement was achieved with respect to classic controllers in terms of error of the trajectory that is tracked. Another neuro-fuzzy robust tracking control law was implemented in [20.56]. In this work, the controller guaranteed transient and asymptotic performance.

Other different approaches of neuro-fuzzy controllers have also been developed. In [20.57] a stable discrete-time adaptive tracking controller using a neuro-fuzzy dynamic-inversion for a robotic manipulator was presented. The dynamics of the manipulator were approximated by a dynamic Takagi–Sugeno fuzzy model. With the aim of improving the robustness of the controller, in [20.58] a novel parameter adjustment scheme using a neuro-fuzzy inference system architecture was presented.

More real environment applications of neuro-fuzzy networks approaches have also been discussed in the literature [20.59]. In [20.60, 61] a robust neural-fuzzy-network control was investigated for the joint position control of an n-link robot manipulator for periodic motion, with the aim of achieving high-precision position tracking. In [20.62] an approximate Takagi–

Sugeno type neuro-fuzzy state–space model for a flexible robotic arm was presented. The model was trained using a particle swarm optimization technique.

In the last years, more advanced learning algorithms have been used. In [20.63] an algorithm to learn the path-following behavior for multi-link mobile robots was presented. In this approach the learning complexity of the path-following behavior is reduced, as long paths are divided into a set of small motion primitives that can reach almost every point in the neighborhood. In [20.64], a robot manipulator was controlled by a FLC, where the parameters of the Gaussian membership functions were optimized with particle swarm optimization. Also, in [20.65] the authors described the application of ant colony optimization and particle swarm optimization to the optimization of the membership function parameters of a FLC. The aim in this case was to find the optimal trajectory tracking controller for an autonomous wheeled mobile robot.

Interval type-2 FLCs have been described in several case studies to handle uncertainties. However, one of the main issues in adopting such systems on a larger scale is the lack of a systematic design methodology. [20.66] presented a novel design methodology of interval type-2 Takagi–Sugeno–Kang (TSK) FLCs for modular and reconfigurable robot manipulators for tracking purposes with uncertain dynamic parameters. Moreover, [20.67] provided a problem-driven design methodology together with a systematic assessment of the performance quality and uncertainty robustness of interval type-2 FLCs. The method was evaluated on the problem of position control of a delta parallel robot.

20.5 Moving Target Tracking

Service robots have to be endowed with the capacity of working in dynamic environments with high uncertainty. Typical environments with these characteristics are airports, hallways of buildings, corridors of hospitals, domestic environments, etc. One of the most important factors when working in real environments are the moving objects in the surrounding of the robot. The knowledge of the position, speed, and heading of the moving objects is fundamental for the execution of tasks like localization, route planning, interaction with humans, or obstacle avoidance.

In [20.68] a module that allows the mobile robot to localize the target precisely in the environment was pre-

sented. Both direction and distance to the target were measured using infrared sensors. Also, a fuzzy target tracking control unit was proposed. This control unit comprises a behavior network for each action of the tracking control and a gate network for combining the information of the infrared sensors. It was shown that the proposed control scheme is, indeed, effective and feasible through some simulated and real examples of the behavior.

In [20.69, 70] a pattern classifier system for the detection of moving objects using laser range finders data was presented. An evolutionary algorithm was used to learn the classifier system based on the quantified fuzzy temporal rules model. These quantified fuzzy temporal rules are able to analyze the persistence of the fulfillment of a condition in a temporal reference by using fuzzy quantifiers. Moreover, in [20.71] the authors presented a deep experimental study on the performance of different evolutionary fuzzy systems for moving object following. Several environments with different degrees of complexity and a real environment were used in order to show the applicability of the methodologies presented.

20.6 Perception

Perception is an essential part of any robotic system, since it is the functionality through which the robot incorporates information from the environment. Several sensors have been typically used in mobile robots over the years. Ultrasound, laser range finders, acoustic signals, or cameras are some examples of the most used sensors. The information obtained by these sensors can be used in order to solve various problems such as object recognition, collision avoidance, navigation, or some particular objects.

The perception of landmarks is a very useful and meaningful strategy for helping in localization or navigation. Furthermore, it is a quite common strategy to detect known landmarks that are present in the environments instead of adding artificial landmarks (e.g., visual beacons, radio frequency identification (RFID) labels, ...) in order to preserve environments with the less possible external intervention or manipulation. Within this context, one of the most commonly used landmarks in indoor environments are doors, since they indicate relevant points of interaction and also for their static nature. In [20.72] fuzzy temporal rules were used for detecting doors using the information obtained from ultrasound sensors. This paradigm was used to model the temporal variations of the sensor signals together with the model of the necessary knowledge for detection. A different approach for door detection using computer vision was presented in [20.73]. Doors are found in gray-level images by detecting the borders and are distinguished from other similar shapes using a fuzzy system designed using expert knowledge. Also, a tuning mechanism based on a genetic algorithm was used to improve the performance of the system according to the particularities of the environment in which it is going to be employed.

In robotic soccer games perception also plays a principal role. The work presented in [20.74] describes a type-2 FLC to accurately track a mobile object, in this case a ball, from a robot agent. Both players and ball positions must be tracked using a low-computational cost image processing algorithm. The fuzzy controller aims to overcome the uncertainty added by this image processing.

20.7 Planning

For more complex behaviors in robotics, systems must be able to achieve certain higher-level goals. In order to do that, the robot needs to make choices that maximize the utility or value of the available alternatives. The process by which this objective is solved is called planning. When the robot is not the only actor (as is usual), it must check periodically if the environment matches with the predictions made and change its plan accordingly.

Path planning is a typical task that is needed in most mobile robots that work in unknown environments. The problem consists in determining the path to be followed by the robot in order to reach a goal. In the case when not only the path, but also the movements of the robot at each instant are determined, planning is referred to as motion planning. In [20.75] a two-layered goal-oriented motion planning strategy using fuzzy logic was developed for a Koala mobile robot navigating in an unknown environment. The information about the global goal and the long-range sensorial data are used by the first layer of the planner to produce an intermediate goal in a favorable direction. The second layer of the planner takes this sub-goal and guides the robot to reach it while avoiding collisions using short-range sensorial data.

Other path planning approaches that use fuzzy logic can be found in more recent years. A cooperative control in a multi-agent architecture was applied in [20.76] in order to implement high cognitive capabilities like planning. The agents provided basic behaviors (such as moving to a point) sharing the robot resources and negotiating when conflicts arose. A new proposal to solve path planning was also proposed in [20.77]. It was based in an ant colony optimization to find the best route with a fuzzy cost function. In [20.78] fuzzy logic was used in order to discretize the environment in relation to a soccer robot. Then, a multi-objective evolutionary algorithm was designed in order to optimize the actions needed in order to reach the ball.

Some research has also been reported in the field of robot soccer, for high-level planning of team be-

havior. In [20.79] an extensive fuzzy behavior-based architecture was proposed. The behavior-based architecture decomposes the complex multi-robotic system into smaller modules of roles, behaviors, and actions. Each individual behavior was implemented using a FLC. The same approach was used for coordinating the various behaviors and select the most appropriate role for each robot. Continuing this work, in [20.80] an evolutionary algorithm approach was used to optimize each FLC for each layer of the architecture.

20.8 SLAM

Simultaneous localization and mapping (SLAM) is a field of robotics that has as its main objective the construction of a map of the environment while at the same time keeping track of the current location of the robot inside the map that is being built. Mapping consists of integrating the information gathered with the robot's sensors into a given representation. In contrast to this, localization is the problem of estimating where the robot is placed on a map. In practice, these two problems cannot be solved independently of each other. Before a robot can answer the question of how the environment looks like given a set of observations, it needs to know from which locations these observations have been made. At the same time, it is hard to estimate the current position of a robot (or any vehicle) without a map.

Different fuzzy approximations have been used in order to help to solve the SLAM problem. In [20.81] the development of a new neuro-fuzzy-based adaptive Kalman filtering algorithm was proposed. The neuro-fuzzy-based supervision for the Kalman filtering algorithm is carried out with the aim of reducing the mismatch between the theoretical and the actual covariance of the innovation sequences. To do that, it attempts to estimate the elements of the covariance matrix at

each sampling instant when a measurement update step is carried out. Also, a fuzzy adaptive extended information filtering scheme was used in [20.82] for ultrasonic localization and pose tracking of an autonomous mobile robot. The scheme was presented in order to improve the estimation accuracy and robustness for the proposed localization system with a system having a lack of information and noise.

A novel hybrid method for integrating fuzzy logic and genetic algorithms (genetic fuzzy systems, GFSs) to solve the SLAM problem was presented in [20.83]. The core of the proposed SLAM algorithm searches for the most probable map such that the associated poses provide the robot with the best localization information. Prior knowledge about the problem domain was transferred to the genetic algorithm in order to speed up convergence. Fuzzy logic is employed to serve this purpose and allows the algorithm to conduct the search starting from a potential region of the pose space. The underlying fuzzy mapping rules infer the uncertainty in the location of the robot after executing a motion command and generate a sample-based prediction of its current position. The robustness of the proposed algorithm has been shown in different indoor experiments using a Pioneer 3AT mobile robot.

20.9 Cooperation

A multi-agent system is a system composed of multiple interacting intelligent agents within an environment. When the agents are robots, these systems lead to a more challenging task because of their implicit real-world environment, which is presumably difficult to

model. There are two fundamental needs for multi-agent approaches. On one hand, some problems can be naturally too complex or impossible to be accomplished by a single robot. On the other hand, there can be benefits for using several simple robots because they are cheaper and more fault tolerant than having a single complex robot. Also, multi-agent systems can be helpful for social and life science problems.

An illustrative example of this type of system is shown in [20.84], where a mobile sensor network approach composed by robots that cooperate was presented. The objective of the sensor network was the localization of hazardous contaminants in an unknown large-scale area. The robots have a swarm controller that controls the behavior for the localization for each robot, whose actions are based on a fuzzy logic control system that is identical for all robots.

Control cooperation of robots has been of great interest in the last years. Fuzzy logic controllers have obtained good performance in some simulation experiments. The idea of applying fuzzy controllers comes from the fact that soft computing techniques have proved to be efficient for poorly defined system optimization and multi-agent coordination. In [20.85] a multi-agent control system was proposed, based on a fuzzy inference system for a group of two wheeled mobile robots executing a common task. An application of this control system is the control of robotic formations moving on the plane such as a group of guard robots taking care of an area and dealing with potential intruders. The use of fuzzy logic in this work allows easy expression of rules, and the multi-agent structure supports separation of team and individual knowledge. Another example can be found in [20.86]. In this work, a collision free target tracking problem of a multi-agent robot system was presented. Game theory provides an effective tool to solve this problem. To enhance robustness, a fuzzy controller tunes the cost function weights directly for the game theoretic solution and helps to achieve a prescribed value of cost function components.

20.10 Legged Robots

Wheeled robots have dominated the state of the art of mobile robots. However, in the last decade, there has been an interest to find alternatives for those environments in which wheeled robots are not able to operate. When the terrain is variable and unprepared, adding legs to robots might be a solution. Legged robots can navigate on and adapt to any kind of surfaces (such as rough, rocky, sandy, and steep terrains) and step over obstacles; they can adapt. Moreover, legged robots help the exploration of human and animal locomotion.

One of the main differences between legged and wheeled robots is that legged robots require the system to generate an appropriate gait to move, whereas wheels just need to roll. To clarify this, gait is the movement pattern of limbs in animals and humans used for locomotion over a variety of surfaces. The same concept is used to design the pattern of movement of robots on different surfaces. In [20.87], the learning of a biped gait was solved using reinforcement learning. The aim of this work was to improve the learning rate through the incorporation of expert knowledge using fuzzy logic. This fuzzy logic was incorporated in the reinforcement system through neuro-fuzzy architectures. Moreover, fuzzy rule-based feedback is incorporated instead of numerical reinforcement signals. A different approximation was carried out in [20.88], where two different genetic algorithm approaches were used in order to improve the performance of a FLC designed to model the gait generation problem of a biped robot. In both works, computer simulations were done in order to compare the different approaches of the control systems in terms of stability.

The work in [20.89] focused on the design of a leg for a quadrupedal galloping machine. For that, two intelligent strategies, a fuzzy and an heuristic controller, were developed for verification on a one-legged system. The fuzzy controller consists of a fuzzy rule base with an adaptation mechanism that modifies the rule output centers to correct velocity. These techniques were successfully implemented for operating one leg at speeds necessary for a dynamic gallop. It was shown that the fuzzy controller outperformed the heuristic controller without relying on a model of the system.

Finally for the gait problems, in [20.90] a fuzzy logic vertical ground reaction force controller was developed for a robotic cadaveric gait, which altered tendon forces in real time and iteratively adjusted the robotic trajectory in order to track a target reaction. This controller was validated using a novel dynamic cadaveric gait simulator. The fuzzy logic rule-based controller was able to track the target with a very low tracking error, demonstrating its ability to accurately control this type of robot.

Besides the gait problem, the biped robotic system contains a great deal of uncertainties associated with the mechanism dynamics and environment parameters. In [20.91] it was suggested that type-2 fuzzy logic control systems could be a better way to deal with the uncertainty in a robotic system. A novel type-2 fuzzy switching control system was proposed for biped robots, which includes a type-2 fuzzy modeling algorithm. As in the previous work, simulated experiments were used in order to compare the performance of the controller proposed with other dynamical intelligent control methods.

In [20.92] a fuzzy controller, consisting of a fuzzy prefilter (designed by a genetic algorithm) in the feed-forward loop and a PID-like fuzzy controller in the feed-back loop, was proposed for foot trajectory tracking control of a hydraulically actuated hexapod robot. A COMET-III real robot was used in this work and the experimental results exhibit that the proposed controller manifests better foot trajectory tracking performance compared to an optimal classical controller like the state feedback linear-quadratic regulator (LQR) controller.

20.11 Exoskeletons and Rehabilitation Robots

The latest advances in assistive robotics has had a great impact in different fields. For instance, in military applications, these technologies can allow soldiers to carry a higher payload and walk further without requiring more effort or producing fatigue. However, the field where the impact of this type of robotics is the greatest is healthcare. The aging of the population will be one of the main problems in the near future, since more people are going to need some type of assistance on a daily basis. This dependency suffered by the elderly can be partially resolved with the use of robotic systems that help people with a lack of mobility or strength. Robotic systems for assistance can provide total or partial movement to these people. Moreover, rehabilitation using these systems can make regaining movement-related functions easier and faster.

Some studies were developed in recent years that use fuzzy techniques on the rehabilitation of upper-limb motion (shoulder joint motion and elbow joint motion). The principal reason to use fuzzy logic is to deal with complicated, ill-defined, and dynamic processes, which are intrinsically difficult to being modelled mathematically. Moreover, fuzzy logic control incorporates human knowledge and experience directly without relying on a detailed model of the control system.

In [20.93] an exoskeleton and its fuzzy control system to assist the human upper-limb motion of physically weak persons was presented. The proposed robot automatically assists human motion mainly based on electromyogram signals on the skin surface. In a later work [20.94], the authors introduced a hierarchical neuro-fuzzy controller for a robotic exoskeleton where the angles of the elbow and shoulder are modeled using fuzzy sets. Additionally, fuzzy sets are used in order to set a trigger in the activity of the muscles. In order to solve the same problem, a hybrid position/force fuzzy logic control system was presented in [20.95]. The objective of this work was to assist the subject in performing both passive and active movements along the designed trajectories with specified loads.

More recent studies have been published under the paradigm of fuzzy sliding mode control. In [20.96] a novel adaptive self-organizing fuzzy sliding mode control for the control of a 3-degree-of-freedom (DOF) rehabilitation robot was presented. An interesting characteristic of this approach is the ability to establish and regulate the fuzzy rule base dynamically. Going a step further along that same line, in [20.97] an adaptive self-organizing fuzzy sliding mode control robot was proposed for a 2-DOF rehabilitation robot.

For comparison and performance measurement purposes, one common practice in order to examine the effectiveness of the proposed exoskeleton in motion assistance is to use human subjects who perform different cooperative motions of the elbow and shoulder [20.93]. Different performance measures can be used to show the correctness of the proposed systems. In [20.94] the angles obtained by the exoskeleton were compared, while in [20.95] the results were shown in terms of force and stability. Finally, in [20.96, 97], the performance of the robotic rehabilitation system was measured in terms of response of the system to movements and tracking errors.

20.12 Emotional Robots

Future robots need a transparent interface that regular people can interpret, such as an emotional human-like face. Moreover, such robots must exhibit behaviors that are perceived as believable and life like. In general terms, the use of fuzzy techniques in this field did not have a great impact. However, the fuzzy approach can not only simplify the design task, but also enrich the interaction between humans and robots.

An application of fuzzy logic for this type of robot can be found in [20.98]. In this research it was proposed to use fuzzy logic for effectively building the whole behavior system of face emotion expression robots. It was shown how these behaviors could be constructed by a fuzzy architecture that not only seems more realistic but can also be easily implemented.

In [20.73] a fuzzy system that establishes a level of possibility about the degree of interest that the people around the robot may have in interacting with it was presented. Firstly, a method to detect and track persons using stereo vision was proposed. Then, the interest of

each person was computed using fuzzy logic by analyzing its position and its level of attention to the robot. The level of attention is estimated by analyzing whether or not the person is looking at the robot.

A more recent work of video-based emotion recognition was presented in [20.99]. In this work, a fuzzy rule-based approach was used for emotion recognition from facial expressions. The fuzzy classification itself analyzes the deformation of a face separately in each image. In contrast to most existing approaches, also blended emotions with varying intensities as proposed by psychologists can be handled. Other work that was based on physiological measures was presented in [20.100]. In this work, a fuzzy inference engine was developed to estimate human responses. The authors demonstrated in a later work [20.101] that a hidden Markov model is able to achieve better classification results than the previously reported fuzzy inference engine.

20.13 Fuzzy Modeling

Some extensions to fuzzy logic have been developed in the last decade in the field of mobile robotics. In [20.102] a probabilistic type-2 FLS was proposed for modeling and control. [20.89] focused on the design of a leg for a quadrupedal galloping machine. For that, two intelligent strategies (fuzzy and heuristic controllers) were developed for verification on a one-legged system. The fuzzy controller consists of a fuzzy rule base with an adaption mechanism that modifies the rule output centers to correct velocity errors. These techniques were successfully implemented for operating one leg at the speeds demanded for a dynamic gallop. It was shown that the fuzzy controller outperformed the heuristic controller without relying on a model of the systems. This proposal aims to solve the lack of capability of FLSs to handle various uncertainties identified by this work in practical applications. Two examples were used to validate the probabilistic fuzzy model: a function approximation and a robotic application. The robotic application was successfully implemented for the control of a simulated biped robot.

A novel representation of robot kinematics was proposed in [20.103, 104] in order to merge qualitative and quantitative reasoning. Fuzzy reasoning is good at communicating with sensing and control level sub-

systems by means of fuzzification and defuzzification methods. It has powerful reasoning strategies utilizing compiled knowledge through conditional statements so as to easily handle mathematical and engineering systems. Fuzzy reasoning also provides a means for handling uncertainty in a natural way, making it robust in significantly noisy environments. However, in this work a lack of ability in fuzzy reasoning alone to deal with qualitative inference about complex systems was pointed out.

It is argued that qualitative reasoning can compensate this drawback. Qualitative reasoning has the advantage of operating at the conceptual modeling level, reasoning symbolically with models that retain the mathematical structure of the problem rather than the input/output representation of rule bases. Moreover, the computational cause–effect relations contained in qualitative models facilitate analyzing and explaining the behavior of a structural model. The kinematics of a PUMA 560 robot is modelled for the trajectory tracking task. Thus demonstrating the ability of fuzzy reasoning. Simulation results demonstrated that the proposed method effectively provides a two-way connection for robot representations used for both numerical and symbolic robotic tasks.

20.14 Comments and Conclusions

The significance of the contributions of fuzzy logic to robotics is quite different in the three stages that we have identified. In the first stage, the objective was to endow the robot with a set of simple behaviors to solve basic tasks like wall-following, obstacle avoidance, moving object tracking, trajectory tracking, etc. Fuzzy logic significantly contributed in this stage, not only with the design of behaviors, but also with the coordination or fusion among them (more than 75% of the papers considered in this chapter deal with these topics). In the second stage, the focus moved to implementing autonomous robots that are able to operate in real environments (museums, hospitals, homes, . . .), which should, therefore, be able to generate a map of the environment, localize in the map, and also navigate between different positions (motion planning). The first two tasks were joined under the SLAM field, which has been one of the most important topics in robotics in the recent years. Motion planning has been another relevant field, that experimented a great improvement with the use of heuristic search algorithms and the inclusion of kinematic constraints in the planning. The contributions of fuzzy logic to this second wave have been marginal; SLAM techniques are dominated by probabilistic and optimization approaches, while the best motion planning proposals rely on heuristic search processes and probabilistic approaches to manage the uncertainty.

We have assessed which is the actual impact of the recently reported research on fuzzy-based approaches to robotics research and applications from a quantitative/qualitative point of view. In order to have an estimation we have considered the two journals with the highest impact factor in the 2012 Thomson–Reuters Web of Knowledge (the *International Journal of Robotics Research* IJRR and *IEEE Transactions on Robotics*, IEEE-TR) and looked for papers that included the term *fuzzy* in the Abstract in the period considered (2003–2013). We found that only one paper fulfilled such conditions in IJRR and only seven papers in IEEE-TR. These results indicate that the actual impact and diffusion of fuzzy approaches in the relevant robotics arena is very limited. A vast majority of research and application results of fuzzy approaches in robotics are, therefore, published and presented in papers and conferences related to soft computing. In fact,

only 22 out of the 98 papers considered in this chapter (i. e., 22%) were published in robotics-related forums. Among these almost all papers (20 out of 22, 91%) described FLC of the Mamdani type for different tasks, which suggests that FLC is without a doubt the area with the highest impact in papers and conferences of the most genuine robotics area (i. e., out of the soft computing related publications).

Furthermore, FLC is the most active area of research and applications, since 66% of the papers considered in this chapter describe Mamdani-based fuzzy controllers for all the behaviors and high-level tasks considered. Other hybrid methodologies such as neuro-fuzzy networks or fuzzy-based ones such as type-2 fuzzy sets and Takagi–Sugeno rules follow at a large distance (12%, 9%, and 7%, respectively), but with almost no impact in robotic-centered publications.

Although these topics are still open, nowadays the focus in robotics is moving to other higher-level fields (third stage), like perception, learning of complex behaviors, or human–robot interaction. Perception requirements are not just the construction of occupancy or feature-based maps, but the recognition of objects, the classification of objects, the identification of actions, etc.; in summary, scene understanding. Moreover, perception has to combine different sources of information, with visual and volumetric data being the two main sources. Contributions from fuzzy techniques are still few in number, but from our point of view, they can contribute to this topic and will surely do so in those cases where high-level reasoning is required, and also in the description of scenes.

From the point of view of human–robot interaction, there are two directions in which fuzzy logic may also contribute significantly. The first one is the interpretation of the emotional state of the people interacting with the robot. This interpretation uses several information sources (visual, acoustic, etc.), requires expert knowledge to build the classification rules, and the kind of uncertainty of the data could be adequately modeled with fuzzy sets. The second direction is the expressiveness of the robot, which will be fundamental for social robotics. Again, and for the same reasons as for the previous topic, fuzzy logic approaches may generate significant contributions in the field.

References

20.1   A. Saffiotti: The uses of fuzzy logic in autonomous robot navigation, Soft Comput. **1**(4), 180–197 (1997)

20.2   E.H. Ruspini: Fuzzy logic in the flakey robot, Proc. Int. Conf. Fuzzy Log. Neural Netw., Iizuka (1990) pp. 767–770

20.3   A. Saffiotti, E.H. Ruspini, K. Konolige: Blending reactivity and goal-directedness in a fuzzy controller, IEEE Int. Conf. Fuzzy Syst., San Francisco (1993) pp. 134–139

20.4   A. Saffiotti, K. Konolige, E.H. Ruspini: A multi-valued-logic approach to integrating planning and control, Artif. Intell. **76**(1), 481–526 (1995)

20.5   Microsoft, Inc: http://academic.research.microsoft.com/

20.6   Thomson Reuters: http://thomsonreuters.com/ thomson-reuters-web-of-science/

20.7   T.S. Li, S.-J. Chang, Y.-X. Chen: Implementation of human-like driving skills by autonomous fuzzy behavior control on an FPGA-based car-like mobile robot, IEEE Trans. Ind. Electron. **50**(5), 867–880 (2003)

20.8   M. Mucientes, R. Iglesias, C.V. Regueiro, A. Bugarín, S. Barro: A fuzzy temporal rule-based velocity controller for mobile robotics, Fuzzy Sets Syst. **134**(1), 83–99 (2003)

20.9   V.M. Peri, D. Simon: Fuzzy logic control for an autonomous robot, Proc. Annu. Meet. N. Am. Fuzzy Inf. Process. Soc. (2005) pp. 337–342

20.10   M. Mucientes, D.L. Moreno, A. Bugarín, S. Barro: Evolutionary learning of a fuzzy controller for wall-following behavior in mobile robotics, Soft Comput. **10**(10), 881–889 (2006)

20.11   M. Mucientes, D.L. Moreno, A. Bugarn, S. Barro: Design of a fuzzy controller in mobile robotics using genetic algorithms, Appl. Soft Comput. **7**(2), 540–546 (2007)

20.12   M. Mucientes, R. Alcalá, J. Alcalá-Fdez, J. Casillas: Learning weighted linguistic rules to control an autonomous robot, Int. J. Intell. Syst. **24**, 226–251 (2009)

20.13   C. Wagner, H. Hagras: A genetic algorithm based architecture for evolving type-2 fuzzy logic controllers for real world autonomous mobile robots, Proc. IEEE Int. Conf. Fuzzy Syst. (2007) pp. 1–6

20.14   C.-F. Juang, C.-H. Hsu: Reinforcement ant optimized fuzzy controller for mobile-robot wall-following control, IEEE Trans. Ind. Electron. **56**(10), 3931–3940 (2009)

20.15   O. Linda, M. Manic: Comparative analysis of type-1 and type-2 fuzzy control in context of learning behaviors for mobile robotics, Proc. 36th Annu. Conf. IEEE Ind. Electron. Soc. (2010) pp. 1092–1098

20.16   C. Wagner, H. Hagras: Toward general type-2 fuzzy logic systems based on zSlices, IEEE Trans. Fuzzy Syst. **18**(4), 637–660 (2010)

20.17   A. Zhu, S.X. Yang: A fuzzy logic approach to reactive navigation of behavior-based mobile robots, Proc. IEEE Int. Conf. Robot. Autom. (ICRA) '04, Vol. 5 (2004) pp. 5045–5050

20.18   D.R. Parhi: Navigation of mobile robots using a fuzzy logic controller, J. Int. Robot. Syst. **42**(3), 253–273 (2005)

20.19   S.K. Pradhan, D.R. Parhi, A.K. Panda: Fuzzy logic techniques for navigation of several mobile robots, Appl. Soft Comput. **9**(1), 290–304 (2009)

20.20   I. Baturone, F.J. Moreno-Velo, V. Blanco, J. Ferruz: Design of embedded DSP-based fuzzy controllers for autonomous mobile robots, IEEE Trans. Ind. Electron. **55**(2), 928–936 (2008)

20.21   F. Cupertino, V. Giordano, D. Naso, L. Delfine: Fuzzy control of a mobile robot, Robot. Autom. Mag. IEEE **13**(4), 74–81 (2006)

20.22   M. Wang, J.N.-K. Liu: Fuzzy logic based robot path planning in unknown environment, Proc. Int. Conf. Mach. Learn. Cybern., Vol. 2 (2005) pp. 813–818

20.23   M. Wang, J.N.K. Liu: Fuzzy logic-based real-time robot navigation in unknown environment with dead ends, Robot. Auton. Syst. **56**(7), 625–643 (2008)

20.24   O.R.E. Motlagh, T.S. Hong, N. Ismail: Development of a new minimum avoidance system for a behavior-based mobile robot, Fuzzy Sets Syst. **160**(13), 1929–1946 (2009)

20.25   R. Huq, G.K.I. Mann, R.G. Gosine: Behavior-modulation technique in mobile robotics using fuzzy discrete event system, IEEE Trans. Robotics **22**(5), 903–916 (2006)

20.26   H.A. Hagras: A hierarchical type-2 fuzzy logic control architecture for autonomous mobile robots, IEEE Trans. Fuzzy Syst. **12**(4), 524–539 (2004)

20.27   P. Ritthipravat, T. Maneewarn, D. Laowattana, J. Wyatt: A modified approach to fuzzy Q learning for mobile robots, Proc. IEEE Int. Conf. Syst., Man Cybern., Vol. 3 (2004) pp. 2350–2356

20.28   L.-H. Chen, C.-H. Chiang: New approach to intelligent control systems with self-exploring process, IEEE Trans. Syst. Man Cybern. B **33**(1), 56–66 (2003)

20.29   N.B. Hui, V. Mahendar, D.K. Pratihar: Time-optimal, collision-free navigation of a car-like mobile robot using neuro-fuzzy approaches, Fuzzy Sets Syst. **157**(16), 2171–2204 (2006)

20.30   L. Doitsidis, K.P. Valavanis, N.C. Tsourveloudis, M. Kontitsis: A framework for fuzzy logic based UAV navigation and control, Proc. IEEE Int. Conf. Robot. Autom. (ICRA) '04, Vol. 4 (2004) pp. 4041–4046

20.31   B. Kadmiry, D. Driankov: A fuzzy gain-scheduler for the attitude control of an unmanned helicopter, IEEE Trans. Fuzzy Syst. **12**(4), 502–515 (2004)

Part B | 20

20.32 H. Hu, P.-Y. Woo: Fuzzy supervisory sliding-mode and neural-network control for robotic manipulators, IEEE Trans. Ind. Electron. **53**(3), 929–940 (2006)

20.33 E.A. Merchan-Cruz, A.S. Morris: Fuzzy-GA-based trajectory planner for robot manipulators sharing a common workspace, IEEE Trans. Robot. **22**(4), 613–624 (2006)

20.34 J. Yu, M. Tan, S. Wang, E. Chen: Development of a biomimetic robotic fish and its control algorithm, IEEE Trans. Syst. Man Cybern. B **34**(4), 1798–1810 (2004)

20.35 K.-B. Sim, K.-S. Byun, F. Harashima: Internet-based teleoperation of an intelligent robot with optimal two-layer fuzzy controller, IEEE Trans. Ind. Electron. **53**(4), 1362–1372 (2006)

20.36 R. Martínez, O. Castillo, L.T. Aguilar: Optimization of interval type-2 fuzzy logic controllers for a perturbed autonomous wheeled mobile robot using genetic algorithms, Inf. Sci. **179**(13), 2158–2174 (2009)

20.37 T. Das, I.N. Kar: Design and implementation of an adaptive fuzzy logic-based controller for wheeled mobile robots, IEEE Trans. Control Syst. Technol. **14**(3), 501–510 (2006)

20.38 E. Maalouf, M. Saad, H. Saliah: A higher level path tracking controller for a four-wheel differentially steered mobile robot, Robot. Auton. Syst. **54**(1), 23–33 (2006)

20.39 G. Antonelli, S. Chiaverini, G. Fusco: A fuzzy-logic-based approach for mobile robot path tracking, IEEE Trans. Fuzzy Syst. **15**(2), 211–221 (2007)

20.40 R.-J. Wai, C.-M. Liu: Design of dynamic petri recurrent fuzzy neural network and its application to path-tracking control of nonholonomic mobile robot, IEEE Trans. Ind. Electron. **56**(7), 2667–2683 (2009)

20.41 Z.-G. Hou, A.-M. Zou, L. Cheng, M. Tan: Adaptive control of an electrically driven nonholonomic mobile robot via backstepping and fuzzy approach, IEEE Trans. Control Syst. Technol. **17**(4), 803–815 (2009)

20.42 C.-Y. Chen, T.-H.S. Li, Y.-C. Yeh: EP-based kinematic control and adaptive fuzzy sliding-mode dynamic control for wheeled mobile robots, Inf. Sci. **179**(1), 180–195 (2009)

20.43 Z. Song, J. Yi, D. Zhao, X. Li: A computed torque controller for uncertain robotic manipulator systems: Fuzzy approach, Fuzzy Sets Syst. **154**(2), 208–226 (2005)

20.44 J.P. Hwang, E. Kim: Robust tracking control of an electrically driven robot: Adaptive fuzzy logic approach, IEEE Trans. Fuzzy Syst. **14**(2), 232–247 (2006)

20.45 S. Purwar, I.N. Kar, A.N. Jha: Adaptive control of robot manipulators using fuzzy logic systems under actuator constraints, Fuzzy Sets Syst. **152**(3), 651–664 (2005)

20.46 C.-S. Chiu: Mixed feedforward/feedback based adaptive fuzzy control for a class of mimo nonlinear systems, IEEE Trans. Fuzzy Syst. **14**(6), 716–727 (2006)

20.47 N. Goléa, A. Goléa, K. Barra, T. Bouktir: Observer-based adaptive control of robot manipulators: Fuzzy systems approach, Appl. Soft Comput. **8**(1), 778–787 (2008)

20.48 Y. Guo, P.-Y. Woo: An adaptive fuzzy sliding mode controller for robotic manipulators, IEEE Trans. Syst. Man Cybern. A **33**(2), 149–159 (2003)

20.49 Y.L. Sun, M.J. Er: Hybrid fuzzy control of robotics systems, IEEE Trans. Fuzzy Syst. **12**(6), 755–765 (2004)

20.50 V. Santibañez, R. Kelly, M.A. Llama: A novel global asymptotic stable set-point fuzzy controller with bounded torques for robot manipulators, IEEE Trans. Fuzzy Syst. **13**(3), 362–372 (2005)

20.51 E. Kim: Output feedback tracking control of robot manipulators with model uncertainty via adaptive fuzzy logic, IEEE Trans. Fuzzy Syst. **12**(3), 368–378 (2004)

20.52 X. Jiang, Q.-L. Han: On designing fuzzy controllers for a class of nonlinear networked control systems, IEEE Trans. Fuzzy Syst. **16**(4), 1050–1060 (2008)

20.53 C.-L. Hwang, L.-J. Chang, Y.-S. Yu: Network-based fuzzy decentralized sliding-mode control for car-like mobile robots, IEEE Trans. Ind. Electron. **54**(1), 574–585 (2007)

20.54 C.-K. Lin: Nonsingular terminal sliding mode control of robot manipulators using fuzzy wavelet networks, IEEE Trans. Fuzzy Syst. **14**(6), 849–859 (2006)

20.55 W.W. Melek, A.A. Goldenberg: Neurofuzzy control of modular and reconfigurable robots, IEEE/ASME Trans. Mechatron. **8**(3), 381–389 (2003)

20.56 Y.-C. Chang: Intelligent robust control for uncertain nonlinear time-varying systems and its application to robotic systems, IEEE Trans. Syst. Man Cybern. B **35**(6), 1108–1119 (2005)

20.57 F. Sun, L. Li, H.-X. Li, H. Liu: Neuro-fuzzy dynamic-inversion-based adaptive control for robotic manipulators–discrete time case, IEEE Trans. Ind. Electron. **54**(3), 1342–1351 (2007)

20.58 M.O. Efe: Fractional fuzzy adaptive sliding-mode control of a 2-DOF direct-drive robot arm, IEEE Trans. Syst. Man Cybern. B **38**(6), 1561–1570 (2008)

20.59 C.-S. Chen: Dynamic structure neural-fuzzy networks for robust adaptive control of robot manipulators, IEEE Trans. Ind. Electron. **55**(9), 3402–3414 (2008)

20.60 R.-J. Wai, P.-C. Chen: Robust neural-fuzzy-network control for robot manipulator including actuator dynamics, IEEE Trans. Ind. Electron. **53**(4), 1328–1349 (2006)

20.61 R.-J. Wai, Z.-W. Yang: Adaptive fuzzy neural network control design via a T-S fuzzy model for a robot manipulator including actuator dynamics, IEEE Trans. Syst. Man Cybern. B **38**(5), 1326–1346 (2008)

20.62 A. Chatterjee, R. Chatterjee, F. Matsuno, T. Endo: Augmented stable fuzzy control for flexible robotic arm using LMI approach and neuro-fuzzy state space modeling, IEEE Trans. Ind. Electron. **55**(3), 1256–1270 (2008)

20.63 F.J. Marín, J. Casillas, M. Mucientes, A.A. Transeth, S.A. Fjerdingen, I. Schjølberg: Learning intelligent controllers for path-following skills on snake-like robots, LNAI. Intell. Robot. Appl. Proc. 4th Int. Conf., II ICIRA (2011), 525–535 (2011)

20.64 Z. Bingül, O. Karahan: A fuzzy logic controller tuned with PSO for 2 DOF robot trajectory control, Expert Syst. Appl. **38**(1), 1017–1031 (2011)

20.65 O. Castillo, R. Martínez-Marroquín, P. Melin, F. Valdez, J. Soria: Comparative study of bio-inspired algorithms applied to the optimization of type-1 and type-2 fuzzy controllers for an autonomous mobile robot, Inf. Sci. **192**, 19–38 (2012)

20.66 M. Biglarbegian, W.W. Melek, J.M. Mendel: Design of novel interval type-2 fuzzy controllers for modular and reconfigurable robots: Theory and experiments, IEEE Trans. Ind. Electron. **58**(4), 1371–1384 (2011)

20.67 O. Linda, M. Manic: Uncertainty-robust design of interval type-2 fuzzy logic controller for delta parallel robot, IEEE Trans. Ind. Inf. **7**(4), 661–670 (2011)

20.68 T.S. Li, S.-J. Chang, W. Tong: Fuzzy target tracking control of autonomous mobile robots by using infrared sensors, IEEE Trans. Fuzzy Syst. **12**(4), 491–501 (2004)

20.69 M. Mucientes, A. Bugarín: People detection with quantified fuzzy temporal rules, Proc. IEEE Int. Conf. Fuzzy Syst. (2007) pp. 1149–1154

20.70 M. Mucientes, A. Bugarín: People detection through quantified fuzzy temporal rules, Pattern Recognit. **43**(4), 1441–1453 (2010)

20.71 M. Mucientes, J. Alcalá-Fdez, R. Alcalá, J. Casillas: A case study for learning behaviors in mobile robotics by evolutionary fuzzy systems, Expert Syst. Appl. **37**, 1471–1493 (2010)

20.72 P. Carinena, C.V. Regueiro, A. Otero, A.J. Bugarin, S. Barro: Landmark detection in mobile robotics using fuzzy temporal rules, IEEE Trans. Fuzzy Syst. **12**(4), 423–435 (2004)

20.73 R.M. Muñoz-Salinas, E. Aguirre, M. García-Silvente, A. González: A fuzzy system for visual detection of interest in human-robot interaction, Proc. 2nd Int. Conf. Mach. Intell. (ACIDCA-ICMI2005) (2005) pp. 574–581

20.74 J. Figueroa, J. Posada, J. Soriano, M. Melgarejo, S. Rojas: A type-2 fuzzy controller for tracking mobile objects in the context of robotic soccer games, Proc. 14th IEEE Int. Conf. Fuzzy Syst. FUZZ '05 (2005) pp. 359–364

20.75 X. Yang, M. Moallem, R.V. Patel: A layered goal-oriented fuzzy motion planning strategy for mobile robot navigation, IEEE Trans. Syst. Man Cybern. B **35**(6), 1214–1224 (2005)

20.76 B. Innocenti, B. López, J. Salvi: A multi-agent architecture with cooperative fuzzy control for a mobile robot, Robot. Auton. Syst. **55**(12), 881–891 (2007)

20.77 M.A. Garcia, O. Montiel, O. Castillo, R. Sepúlveda, P. Melin: Path planning for autonomous mobile robot navigation with ant colony optimization and fuzzy cost function evaluation, Appl. Soft Comput. **9**(3), 1102–1110 (2009)

20.78 J.-H. Kim, Y.-H. Kim, S.-H. Choi, I.-W. Park: Evolutionary multi-objective optimization in robot soccer system for education, IEEE Comput. Intell. Mag. **4**(1), 31–41 (2009)

20.79 P. Vadakkepat, O.C. Miin, X. Peng, T.-H. Lee: Fuzzy behavior-based control of mobile robots, IEEE Trans. Fuzzy Syst. **12**(4), 559–565 (2004)

20.80 P. Vadakkepat, X. Peng, B.K. Quek, T.H. Lee: Evolution of fuzzy behaviors for multi-robotic system, Robot. Auton. Syst. **55**(2), 146–161 (2007)

20.81 A. Chatterjee, F. Matsuno: A neuro-fuzzy assisted extended Kalman filter-based approach for simultaneous localization and mapping (SLAM) problems, IEEE Trans. Fuzzy Syst. **15**(5), 984–997 (2007)

20.82 H.-H. Lin, C.-C. Tsai, J.-C. Hsu: Ultrasonic localization and pose tracking of an autonomous mobile robot via fuzzy adaptive extended information filtering, IEEE Trans. Instrum. Meas. **57**(9), 2024–2034 (2008)

20.83 M. Begum, G.K.I. Mann, R.G. Gosine: Integrated fuzzy logic and genetic algorithmic approach for simultaneous localization and mapping of mobile robots, Appl. Soft Comput. **8**(1), 150–165 (2008)

20.84 X. Cui, T. Hardin, R.K. Ragade, A.S. Elmaghraby: A swarm-based fuzzy logic control mobile sensor network for hazardous contaminants localization, Proc. IEEE Int. Conf. Mob. Ad-hoc Sens. Syst. (2004) pp. 194–203

20.85 D.H.V. Sincák: Multi-robot control system for pursuit-evasion problem, J. Electr. Eng. **60**(3), 143–148 (2009)

20.86 I. Harmati, K. Skrzypczyk: Robot team coordination for target tracking using fuzzy logic controller in game theoretic framework, Robot. Auton. Syst. **57**(1), 75–86 (2009)

20.87 C. Zhou, Q. Meng: Dynamic balance of a biped robot using fuzzy reinforcement learning agents, Fuzzy Sets Syst. **134**(1), 169–187 (2003)

20.88 R.K. Jha, B. Singh, D.K. Pratihar: On-line stable gait generation of a two-legged robot using a genetic-fuzzy system, Robot. Auton. Syst. **53**(1), 15–35 (2005)

20.89 J.G. Nichol, S.P.N. Singh, K.J. Waldron, L.R. Palmer, D.E. Orin: System design of a quadrupedal galloping machine, Int. J. Robot. Res. **23**(10/11), 1013–1027 (2004)

Part B | 20

20.90   P.M. Aubin, E. Whittaker, W.R. Ledoux: A robotic cadaveric gait simulator with fuzzy logic vertical ground reaction force control, IEEE Trans. Robot. **28**(1), 246–255 (2012)

20.91   Z. Liu, Y. Zhang, Y. Wang: A type-2 fuzzy switching control system for biped robots, IEEE Trans. Syst. Man Cybern. C **37**(6), 1202–1213 (2007)

20.92   R.K. Barai, K. Nonami: Optimal two-degree-of-freedom fuzzy control for locomotion control of a hydraulically actuated hexapod robot, Inf. Sci. **177**(8), 1892–1915 (2007)

20.93   K. Kiguchi, T. Tanaka, K. Watanabe, T. Fukuda: Exoskeleton for human upper-limb motion support, Proc. IEEE Int. Conf. Robot. Autom. (ICRA) '03, Vol. 2 (2003) pp. 2206–2211

20.94   K. Kiguchi, T. Tanaka, T. Fukuda: Neuro-fuzzy control of a robotic exoskeleton with EMG signals, IEEE Trans. Fuzzy Syst. **12**(4), 481–490 (2004)

20.95   M.-S. Ju, C.-C.K. Lin, D.-H. Lin, I.-S. Hwang, S.-M. Chen: A rehabilitation robot with force-position hybrid fuzzy controller: Hybrid fuzzy control of rehabilitation robot, IEEE Trans. Neural Syst. Rehabil. Eng. **13**(3), 349–358 (2005)

20.96   M.-K. Chang, T.-H. Yuan: Experimental implementations of adaptive self-organizing fuzzy sliding mode control to a 3-DOF rehabilitation robot, Int. J. Innov. Comput. Inf. Control **5**(10), 3391–3404 (2009)

20.97   M.-K. Chang: An adaptive self-organizing fuzzy sliding mode controller for a 2-DOF rehabilitation robot actuated by pneumatic muscle actuators, Control Eng. Pract. **18**(1), 13–22 (2010)

20.98   H. Mobahi, S. Ansari: Fuzzy perception, emotion and expression for interactive robots, Proc. IEEE Int. Conf. Syst., Man Cybern., Vol. 4 (2003) pp. 3918–3923

20.99   N. Esau, E. Wetzel, L. Kleinjohann, B. Kleinjohann: Real-time facial expression recognition using a fuzzy emotion model, Proc. IEEE Int. Conf. Fuzzy Syst., FUZZ-IEEE (2007) pp. 1–6

20.100   D. Kulic, E. Croft: Anxiety detection during human-robot interaction, Proc. IEEE/RSJ Int. Conf. Intell. Robot. Syst. (IROS) (2005) pp. 616–621

20.101   D. Kulic, E.A. Croft: Affective state estimation for human–robot interaction, IEEE Trans. Robot. **23**(5), 991–1000 (2007)

20.102   Z. Liu, H.-X. Li: A probabilistic fuzzy logic system for modeling and control, IEEE Trans. Fuzzy Syst. **13**(6), 848–859 (2005)

20.103   H. Liu: A fuzzy qualitative framework for connecting robot qualitative and quantitative representations, IEEE Trans. Fuzzy Syst. **16**(6), 1522–1530 (2008)

20.104   H. Liu, D.J. Brown, G.M. Coghill: Fuzzy qualitative robot kinematics, IEEE Trans. Fuzzy Syst. **16**(3), 808–822 (2008)

Part C Rough Sets

Ed. by Roman Słowiński, Yiyu Yao

21. Foundations of Rough Sets

Andrzej Skowron, Andrzej Jankowski, Roman W. Swiniarski (deceased)

The rough set (RS) approach was proposed by Pawlak as a tool to deal with imperfect knowledge. Over the years the approach has attracted attention of many researchers and practitioners all over the world, who have contributed essentially to its development and applications. This chapter discusses the RS foundations from rudiments to challenges.

21.1 Rough Sets: Comments on Development

The rough set (RS) approach was proposed by *Zdzisław Pawlak* in 1982 [21.1, 2] as a tool for dealing with imperfect knowledge, in particular with vague concepts. Many applications of methods based on rough set theory alone or in combination with other approaches have been developed. This chapter discusses the RS foundations from rudiments to challenges.

In the development of rough set theory and applications, one can distinguish three main stages. While the first period was based on the assumption that objects are perceived by means of partial information represented by attributes, in the second period it was assumed that information about the approximated concepts is also partial. Approximation spaces and searching strategies for relevant approximation spaces were recognized as the basic tools for rough sets. Important achievements both in theory and applications were obtained. Nowadays, a new period for rough sets is emerging, which is also briefly characterized in this chapter.

The rough set approach seems to be of fundamental importance in artificial intelligence AI and cognitive sciences, especially in machine learning, data mining, knowledge discovery from databases, pattern recognition, decision support systems, expert systems, intelligent systems, multiagent systems, adaptive systems, autonomous systems, inductive reasoning, commonsense reasoning, adaptive judgment, conflict analysis.

Rough sets have established relationships with many other approaches such as fuzzy set theory, granular computing (GC), evidence theory, formal concept analysis, (approximate) Boolean reasoning, multicriteria decision analysis, statistical methods, decision theory, and matroids. Despite the overlap with many other theories rough set theory may be considered as an independent discipline in its own right. There are reports on many hybrid methods obtained by combining rough sets with other approaches such as soft computing (fuzzy sets, neural networks, genetic algorithms), statistics, natural computing, mereology, principal component analysis, singular value decomposition, or support vector machines.

The main advantage of rough set theory in data analysis is that it does not necessarily need any preliminary or additional information about data like probability distributions in statistics, basic probability assignments in evidence theory, a grade of membership, or the value of possibility in fuzzy set theory.

One can observe the following advantages about the rough set approach:

i) Introduction of efficient algorithms for finding hidden patterns in data.
ii) Determination of optimal sets of data (data reduction); evaluation of the significance of data.
iii) Generation of sets of decision rules from data.
iv) Easy-to-understand formulation.
v) Straightforward interpretation of results obtained.
vi) Suitability of many of its algorithms for parallel processing.

Due to space limitations, many important research topics in rough set theory such as various logics related to rough sets and many advanced algebraic properties of rough sets are only mentioned briefly in this chapter.

From the same reason, we herein restrict the references on rough sets to the basic papers by *Zdzisław Pawlak* (such as [21.1, 2]), some survey papers [21.3–5], and some books including long lists of references to papers on rough sets. The basic ideas of rough set theory and its extensions as well as many interesting applications can be found in a number of books, issues of *Transactions on Rough Sets*, special issues of other journals, numerous proceedings of international conferences, and tutorials [21.3, 6, 7]. The reader is referred to the cited books and papers, references therein, as well as to web pages [21.8, 9].

The chapter is structured as follows. In Sect. 21.2 we discuss some basic issues related to vagueness and vague concepts. The rough set philosophy is outlined in Sect. 21.3. The basic concepts for rough sets such as indiscernibility and approximation are presented in Sect. 21.4. Decision systems and rules are covered in Sect. 21.5. The basic information about dependencies is included in Sect. 21.6. Attribute reduction belonging to one of the basic problems of rough sets is discussed in Sect. 21.7. Rough membership function as a tool for measuring degrees of inclusion of sets is presented in Sect. 21.8. The role of discernibility and Boolean reasoning for solving problems related to rough sets is briefly explained in Sect. 21.9. In Sect. 21.10 a short discussion on rough sets and induction is included. Several generalizations of the approach proposed by Pawlak are discussed in Sect. 21.11. In this section some emerging research directions related to rough sets are also outlined. In Sect. 21.12 some comments about logics based on rough sets are included. The role of adaptive judgment is emphasized.

21.2 Vague Concepts

Mathematics requires that all mathematical notions (including sets) must be exact, otherwise precise reasoning would be impossible. However, philosophers [21.10], and recently computer scientists as well as other researchers, have become interested in *vague* (imprecise) concepts. Moreover, in the twentieth century one can observe the drift paradigms in modern science from dealing with precise concepts to vague concepts, especially in the case of complex systems (e.g., in economy, biology, psychology, sociology, and quantum mechanics).

In classical set theory, a set is uniquely determined by its elements. In other words, this means that every element must be uniquely classified as belonging to the set or not. That is to say the notion of a set is a *crisp* (precise) one. For example, the set of odd numbers is crisp because every integer is either odd or even.

In contrast to odd numbers, the notion of a beautiful painting is vague, because we are unable to classify uniquely all paintings into two classes: *beautiful* and *not beautiful*. With some paintings it cannot be decided whether they are beautiful or not and thus they remain in the doubtful area. Thus, *beauty* is not a precise but a vague concept.

Almost all concepts that we use in natural language are vague. Therefore, common sense reasoning based on natural language must be based on vague concepts and not on classical logic. An interesting discussion of this issue can be found in [21.11]. The idea of vagueness can be traced back to the ancient Greek philosopher Eubulides of Megara (ca. 400 BC) who first formulated the so-called *sorites* (heap) and *falakros* (bald man) paradoxes [21.10]. There is a huge literature on issues related to vagueness and vague concepts in philosophy [21.10].

Vagueness is often associated with the boundary region approach (i. e., existence of objects which cannot be uniquely classified relative to a set or its complement), which was first formulated in 1893 by the father of modern logic, the German logician, *Gottlob Frege* (1848–1925) ([21.12]). According to Frege the concept must have a sharp boundary. To the concept without a sharp boundary there would correspond an area that would not have any sharp boundary line all around. This means that mathematics must use crisp, not vague concepts, otherwise it would be impossible to reason precisely.

One should also note that vagueness also relates to insufficient specificity, as the result of a lack of feasible searching methods for sets of features adequately describing concepts. A discussion on vague (imprecise) concepts in philosophy includes their following characteristic features [21.10]: (i) the presence of borderline cases, (ii) boundary regions of vague concepts are not crisp, (iii) vague concepts are susceptible to sorites paradoxes. In the sequel we discuss the first two issues in the RS framework. The reader can find a discussion on the application of the RS approach to the third item in [21.11].

21.3 Rough Set Philosophy

Rough set philosophy is founded on the assumption that with every object of the universe of discourse we associate some information (data, knowledge). For example, if objects are patients suffering from a certain disease, symptoms of the disease form information about the patients. Objects characterized by the same information are indiscernible (similar) in view of the available information about them.

The *indiscernibility relation* generated in this way is the mathematical basis of rough set theory. This understanding of indiscernibility is related to the idea of *Gottfried Wilhelm Leibniz* that objects are indiscernible if and only if all available functionals take identical values on them (Leibniz's law of indiscernibility: the identity of indiscernibles) [21.13]. However, in the rough set approach indiscernibility is defined relative to a given set of functionals (attributes).

Any set of all indiscernible (similar) objects is called an elementary set and forms a basic granule (atom) of knowledge about the universe. Any union of some elementary sets is referred to as a *crisp* (precise) set. If a set is not crisp, then it is called *rough* (imprecise, vague). Consequently, each rough set has *borderline cases* (*boundary-line*), i.e., objects which cannot be classified with certainty as members of either the set or its complement. Obviously, crisp sets have no borderline elements at all. This means that borderline cases cannot be properly classified by employing available knowledge.

Thus, the assumption that objects can be *seen* only through the information available about them leads to the view that knowledge has granular structure. Due to the granularity of knowledge, some objects of interest cannot be discerned and appear as the same (or similar). As a consequence, vague concepts in contrast to precise concepts, cannot be characterized in terms of information about their elements. Therefore, in the proposed approach, we assume that any vague concept is replaced by a pair of precise concepts – called the lower and the upper approximation of the vague concept. The lower approximation consists of all objects which definitely belong to the concept and the upper approximation contains all objects which possibly belong to the concept. The difference between the upper and the lower approximation constitutes the boundary region of the vague concept. Approximation operations are the basic operations in rough set theory. Properties of the boundary region (expressed, e.g., by the rough membership function) are important in the rough set methods.

Hence, rough set theory expresses vagueness not by means of membership, but by employing a boundary region of a set. If the boundary region of a set is empty it means that the set is crisp, otherwise the set is rough (inexact). A nonempty boundary region of a set means that our knowledge about the set is not sufficient to define the set precisely.

Rough set theory it is not an alternative to classical set theory but it is embedded in it. Rough set theory can be viewed as a specific implementation of Frege's idea of vagueness, i.e., imprecision in this approach is expressed by a boundary region of a set.

21.4 Indiscernibility and Approximation

The starting point of rough set theory is the indiscernibility relation, which is generated by information about objects of interest (Sect. 21.1). The indiscernibility relation expresses the fact that due to a lack of information (or knowledge) we are unable to discern some objects by employing available information (or knowledge).

This means that, in general, we are unable to deal with each particular object but we have to consider granules (clusters) of indiscernible objects as a fundamental basis for our theory.

From a practical point of view, it is better to define basic concepts of this theory in terms of data. Therefore, we will start our considerations from a data set called an *information system*. An information system can be represented by a data table containing rows labeled by objects of interest and columns labeled by attributes and entries of the table are attribute values. For example, a data table can describe a set of patients in a hospital. The patients can be characterized by some attributes, like *age, sex, blood pressure, body temperature*, etc. With every attribute a set of its values is associated, e.g., values of the attribute *age* can be *young, middle*, and *old*. Attribute values can also be numerical. In data analysis the basic problem that we are interested in is to find patterns in data, i.e., to find a relationship between some set of attributes, e.g., we might be interested whether *blood pressure* depends on *age and sex*.

More formally, suppose we are given a pair $\mathcal{A} = (U, A)$ of nonempty, finite sets U and A, where U is the *universe* of *objects*, and an $A-a$ set consisting of *attributes*, i.e., functions $a: U \longrightarrow V_a$, where V_a is the set of values of attribute a, called the *domain* of a. The pair $\mathcal{A} = (U, A)$ is called an *information system*. Any information system can be represented by a data table with rows labeled by objects and columns labeled by attributes. Any pair (x, a), where $x \in U$ and $a \in A$ defines the table entry consisting of the value $a(x)$. Note that in statistics or machine learning such a data table is called a sample [21.14].

Any subset B of A determines a binary relation $\mathrm{IND}(B)$ on U, called an *indiscernibility relation*, defined by

$$x\mathrm{IND}(B)y \text{ if and only if} \atop a(x) = a(y) \text{ for every } a \in B, \tag{21.1}$$

where $a(x)$ denotes the value of attribute a for the object x.

Obviously, $\mathrm{IND}(B)$ is an equivalence relation. The family of all equivalence classes of $\mathrm{IND}(B)$, i.e., the partition determined by B, will be denoted by $U/\mathrm{IND}(B)$, or simply U/B; the equivalence class of $\mathrm{IND}(B)$, i.e., the block of the partition U/B, containing x will be denoted by $B(x)$ (other notation used: $[x]_B$ or $[x]_{\mathrm{IND}(B)}$). Thus in view of the data we are unable, in general, to observe individual objects but we are forced to reason only about the accessible granules of knowledge.

If $(x, y) \in \mathrm{IND}(B)$ we will say that x and y are *B-indiscernible*. Equivalence classes of the relation $\mathrm{IND}(B)$ (or blocks of the partition U/B) are referred to as *B-elementary sets* or *B-elementary granules*. In the rough set approach the elementary sets are the basic building blocks (concepts) of our knowledge about reality. The unions of *B-elementary sets* are called *B-definable sets*. Let us note that in applications we consider only some subsets of the family of definable sets, e.g., defined by conjunction of descriptors only. This is due to the computational complexity of the searching problem for relevant definable sets in the whole family of definable sets.

For $B \subseteq A$, we denote by $\mathrm{Inf}_B(x)$ the *B-signature* of $x \in U$, i.e., the set $\{(a, a(x)): a \in B\}$. Let $\mathrm{INF}(B) = \{\mathrm{Inf}_B(x): x \in U\}$. Then for any objects $x, y \in U$ the following equivalence holds: $x\mathrm{IND}(B)y$ if and only if $\mathrm{Inf}_B(x) = \mathrm{Inf}_B(y)$.

The indiscernibility relation is used to define the approximations of concepts. We define the following two operations on sets $X \subseteq U$

$$B_*(X) = \{x \in U: B(x) \subseteq X\}, \tag{21.2}$$
$$B^*(X) = \{x \in U: B(x) \cap X \neq \emptyset\}, \tag{21.3}$$

assigning to every subset X of the universe U two sets $B_*(X)$ and $B^*(X)$ called the *B-lower* and the *B-upper approximation* of X, respectively. The set

$$\mathrm{BN}_B(X) = B^*(X) - B_*(X), \tag{21.4}$$

will be referred to as the *B-boundary region* of X.

From the definition we obtain the following interpretation: (i) the *lower approximation* of a set X with respect to B is the set of all objects, which can for *certain* be classified to X using B (are *certainly* in X in view of B), (ii) the *upper approximation* of a set X with respect to B is the set of all objects which can *possibly* be classified to X using B (are *possibly* in X in view of B), (iii) the *boundary region* of a set X with respect to B is the set of all objects, which can be classified neither to X nor to not-X using B.

Due to the granularity of knowledge, rough sets cannot be characterized by using available knowledge. The definition of approximations is clearly depicted in Fig. 21.1.

The approximations have the following properties

$$B_*(X) \subseteq X \subseteq B^*(X) \,,$$
$$B_*(\emptyset) = B^*(\emptyset) = \emptyset \,, B_*(U) = B^*(U) = U \,,$$
$$B^*(X \cup Y) = B^*(X) \cup B^*(Y) \,,$$
$$B_*(X \cap Y) = B_*(X) \cap B_*(Y) \,,$$
$$X \subseteq Y \text{ implies } B_*(X) \subseteq B_*(Y)$$
$$\text{and } B^*(X) \subseteq B^*(Y) \,,$$
$$B_*(X \cup Y) \supseteq B_*(X) \cup B_*(Y) \,,$$
$$B^*(X \cap Y) \subseteq B^*(X) \cap B^*(Y) \,,$$
$$B_*(-X) = -B^*(X) \,,$$
$$B^*(-X) = -B_*(X) \,,$$
$$B_*(B_*(X)) = B^*(B_*(X)) = B_*(X) \,,$$
$$B^*(B^*(X)) = B_*(B^*(X)) = B^*(X) \,. \tag{21.5}$$

Let us note that the inclusions (for union and intersection) in (21.5) cannot, in general, be substituted by the equalities. This has some important algorithmic and logical consequences.

Now we are ready to give the definition of rough sets. If the boundary region of X is the empty set, i. e., $\mathrm{BN}_B(X) = \emptyset$, then the set X is *crisp* (*exact*) with respect to B; in the opposite case, i. e., if $\mathrm{BN}_B(X) \neq \emptyset$, the set X is referred to as *rough* (*inexact*) with respect to B. Thus any rough set, in contrast to a crisp set, has a nonempty boundary region. This is the idea of vagueness proposed by Frege.

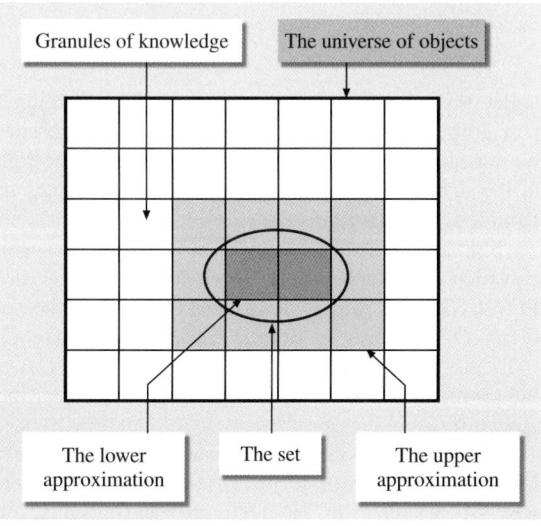

Fig. 21.1 A rough set

Let us observe that the definition of rough sets refers to data (knowledge), and is *subjective*, in contrast to the definition of classical sets, which is in some sense an *objective* one.

A rough set can also be characterized numerically by the following coefficient

$$\alpha_B(X) = \frac{\mathrm{card}(B_*(X))}{\mathrm{card}(B^*(X))} \,, \tag{21.6}$$

called the *accuracy of approximation*, where $X \neq \emptyset$ and $\mathrm{card}(X)$ denotes the cardinality of X. Obviously $0 \leq \alpha_B(X) \leq 1$. If $\alpha_B(X) = 1$ then X is *crisp* with respect to B (X is *precise* with respect to B), and otherwise, if $\alpha_B(X) < 1$ then X is *rough* with respect to B (X is *vague* with respect to B). The accuracy of approximation can be used to measure the quality of approximation of decision classes on the universe U. One can use another measure of accuracy defined by $1 - \alpha_B(X)$ or by

$$1 - \frac{\mathrm{card}(\mathrm{BN}_B(X))}{\mathrm{card}(U)} \,.$$

Some other measures of approximation accuracy are also used, e.g., based on entropy or some more specific properties of boundary regions. The choice of a relevant accuracy of approximation depends on a particular data set. Observe that the accuracy of approximation of X can be tuned by B. Another approach to the accuracy of approximation can be based on the variable precision rough set model (VPRSM).

In [21.10], it is stressed that boundaries of vague concepts are not crisp. In the definition presented in this chapter, the notion of boundary region is defined as a crisp set $\mathrm{BN}_B(X)$. However, let us observe that this definition is relative to the subjective knowledge expressed by attributes from B. Different sources of information may use different sets of attributes for concept approximation. Hence, the boundary region can change when we consider these different views. Another reason for boundary change may be related to incomplete information about concepts. They are known only on samples of objects. Hence, when new objects appear again the boundary region may change. From the discussion in the literature it follows that vague concepts cannot be approximated with satisfactory quality by *static* constructs such as induced membership inclusion functions, approximations, or models derived, e.g., from a sample. An understanding of vague concepts can be only realized in a process in which the induced models adaptively match the concepts in dynamically changing environments. This conclusion seems to have important consequences for the further development of rough set theory in combination with fuzzy sets and other soft computing paradigms for adaptive approximate reasoning.

21.5 Decision Systems and Decision Rules

In this section, we discuss the decision rules (constructed over a selected set B of features or a family of sets of features), which are used in inducing classification algorithms (classifiers), making it possible to classify unseen objects to decision classes. Parameters which are tuned in searching for a classifier with high quality are its description size (defined, e.g., by used decision rules) and its quality of classification (measured, e.g., by the number of misclassified objects on a given set of objects). By selecting a proper balance between the accuracy of classification and the description size one can search for classifier with a high quality of classification also on testing objects. This approach is based on the minimum description length principle (MDL) [21.15].

In an information system $\mathcal{A} = (U, A)$ we sometimes distinguish a partition of A into two disjoint classes $C, D \subseteq A$ of attributes, called *condition* and *decision (action)* attributes, respectively. The tuple $\mathcal{A} = (U, C, D)$ is called a *decision system* (or a *decision table*).

Let $V = \bigcup \{V_a : a \in C\} \cup \bigcup \{V_d \mid d \in D\}$. Atomic formulae over $B \subseteq C \cup D$ and V are expressions $a = v$ called *descriptors (selectors) over B and V*, where $a \in B$ and $v \in V_a$. The set of formulae over B and V, denoted by $\mathcal{F}(B, V)$, is the least set containing all atomic formulae over B and V and closed under the propositional connectives $\wedge$ (conjunction), $\vee$ (disjunction) and $\neg$ (negation). By $\|\varphi\|_{\mathcal{A}}$ we denote the meaning of $\varphi \in \mathcal{F}(B, V)$ in the decision system $\mathcal{A}$, which is the set of all objects in U with the property φ. These sets are defined by

$$\|a = v\|_{\mathcal{A}} = \{x \in U \mid a(x) = v\},$$
$$\|\varphi \wedge \varphi'\|_{\mathcal{A}} = \|\varphi\|_{\mathcal{A}} \cap \|\varphi'\|_{\mathcal{A}};$$
$$\|\varphi \vee \varphi'\|_{\mathcal{A}} = \|\varphi\|_{\mathcal{A}} \cup \|\varphi'\|_{\mathcal{A}};$$
$$\|\neg\varphi\|_{\mathcal{A}} = U - \|\varphi\|_{\mathcal{A}}.$$

The formulae from $\mathcal{F}(C, V)$, $\mathcal{F}(D, V)$ are called *condition formulae of $\mathcal{A}$* and *decision formulae of $\mathcal{A}$*, respectively.

Any object $x \in U$ belongs to the *decision class* $\|\bigwedge_{d \in D} d = d(x)\|_{\mathcal{A}}$ of $\mathcal{A}$. All decision classes of $\mathcal{A}$ create a partition U/D of the universe U.

A *decision rule* for $\mathcal{A}$ is any expression of the form $\varphi \Rightarrow \psi$, where $\varphi \in \mathcal{F}(C, V)$, $\psi \in \mathcal{F}(D, V)$, and $\|\varphi\|_{\mathcal{A}} \neq \emptyset$. Formulae φ and ψ are referred to as the *predecessor* and the *successor* of decision rule $\varphi \Rightarrow \psi$. Decision rules are often called *IF ... THEN ...* rules. Such rules are used in machine learning.

Decision rule $\varphi \Rightarrow \psi$ is *true* in $\mathcal{A}$ if and only if $\|\varphi\|_{\mathcal{A}} \subseteq \|\psi\|_{\mathcal{A}}$. Otherwise, one can measure its *truth degree* by introducing some inclusion measure of $\|\varphi\|_{\mathcal{A}}$ in $\|\psi\|_{\mathcal{A}}$. Let us denote by $\text{card}_{\mathcal{A}}(\varphi)$ (or $\text{card}(\varphi)$, if this does not lead to confusion) the number of objects from U that satisfies formula φ, i.e., the cardinality of $\|\varphi\|_{\mathcal{A}}$. According to *Łukasiewicz* [21.16], one can assign to formula φ the value

$$\frac{\text{card}(\varphi)}{\text{card}(U)},$$

and to the implication $\varphi \Rightarrow \psi$ the fractional value

$$\frac{\text{card}(\varphi \wedge \psi)}{\text{card}(\varphi)},$$

under the assumption that $\|\varphi\|_{\mathcal{A}} \neq \emptyset$. The fractional part proposed by Łukasiewicz was adapted much later by machine learning and data mining community, e.g., in the definitions of the accuracy of decision rules or confidence of association rules.

For any decision system $\mathcal{A} = (U, C, D)$ one can consider a *generalized decision* function $\delta_A: U \longrightarrow \text{POW}(\text{INF}(D))$, where for any $x \in U$, $\delta_A(x)$ is the set of all D-signatures of objects from U which are C-indiscernible with x, $A = C \cup D$ and $\text{POW}(\text{INF}(D))$ is the powerset of the set $\text{INF}(D)$ of all possible decision signatures.

The decision system $\mathcal{A}$ is called *consistent (deterministic)*, if $\text{card}(\delta_A(x)) = 1$, for any $x \in U$. Otherwise $\mathcal{A}$ is said to be *inconsistent (nondeterministic)*. Hence, a decision system is inconsistent if it consists of some objects with different decisions but that are indiscernible with respect to condition attributes. Any set consisting of all objects with the same generalized decision value is called a *generalized decision class*.

Now, one can consider certain (possible) rules for decision classes defined by the lower (upper) approximations of such generalized decision classes of $\mathcal{A}$. This approach can be extended by using the relationships of rough sets with the evidence theory (Dempster–Shafer theory) by considering rules relative to decision classes defined by the lower approximations of unions of decision classes of $\mathcal{A}$.

Numerous methods have been developed for the generation of different types of decision rules, and the reader is referred to the literature on rough sets for details. Usually, one is searching for decision rules that are (semi) optimal with respect to some optimization criteria describing the quality of decision rules in concept approximations.

21.6 Dependencies

Another important issue in data analysis is discovering dependencies between attributes in a given decision system $\mathcal{A} = (U, C, D)$. Intuitively, a set of attributes D depends totally on a set of attributes C, denoted $C \Rightarrow D$, if the values of attributes from C uniquely determine the values of attributes from D. In other words, D depends totally on C, if there exists a functional dependency between values of C and D.

D can depend partially on C. Formally such a dependency can be defined in the following way. We will say that D *depends on* C to a *degree* k ($0 \leq k \leq 1$), denoted by $C \Rightarrow_k D$, if

$$k = \gamma(C, D) = \frac{\mathrm{card}(\mathrm{POS}_C(D))}{\mathrm{card}(U)}, \tag{21.7}$$

where

$$\mathrm{POS}_C(D) = \bigcup_{X \in U/D} C_*(X), \tag{21.8}$$

which is called a *positive region* of the partition U/D with respect to C, is the set of all elements of U that can

be uniquely classified to blocks of the partition U/D, by means of C.

If $k = 1$, we say that D *depends totally* on C, and if $k < 1$, we say that D *depends partially* (to *degree* k) on C. If $k = 0$, then the *positive region* of the partition U/D with respect to C is empty.

The coefficient k expresses the ratio of all elements of the universe, which can be properly classified to blocks of the partition U/D, employing attributes C and is called the *degree of the dependency*.

It can be easily seen that if D depends totally on C, then $\mathrm{IND}(C) \subseteq \mathrm{IND}(D)$. This means that the partition generated by C is finer than the partition generated by D.

Summing up: D is *totally* (*partially*) dependent on C, if *all* (*some*) elements of the universe U can be uniquely classified to blocks of the partition U/D, employing C. Observe that (21.7) defines only one of the possible measures of dependency between attributes. Note that one can consider dependencies between arbitrary subsets of attributes in the same way. One also can compare the dependency discussed in this section with dependencies considered in databases.

21.7 Reduction of Attributes

We often face the question as to whether we can remove some data from a data table and still preserve its basic properties, that is – whether a table contains some superfluous data. Let us express this idea more precisely.

Let $C, D \subseteq A$ be sets of condition and decision attributes, respectively. We will say that $C' \subseteq C$ is a *D-reduct* (reduct with *respect* to D) of C, if C' is a minimal

subset of C such that

$$\gamma(C, D) = \gamma(C', D). \tag{21.9}$$

The intersection of all D-reducts is called a *D-core* (core with *respect* to D). Because the core is the intersection of all reducts, it is included in every reduct, i. e., each element of the core belongs to some reduct.

Thus, in a sense, the core is the most important subset of attributes, since none of its elements can be removed without affecting the classification power of attributes. Certainly, the geometry of reducts can be more comlex. For example, the core can be empty but there can exist a partition of reducts into a few sets with nonempty intersection.

Many other kinds of reducts and their approximations have been discussed in the literature. They are defined relative to different quality measures. For example, if one changes the condition (21.9) to $\partial_A(x) = \partial_B(x)$, (where $A = C \cup D$ and $B = C' \cup D$), then the defined reducts preserve the generalized decision. Other kinds of reducts preserve, e.g., (i) the distance between attribute value vectors for any two objects, if this distance is greater than a given threshold, (ii) the distance between entropy distributions between any two objects, if this distance exceeds a given. Yet another kind of reducts is defined by the so-called reducts relative to object used for the generation of decision rules.

Reducts are used for building data models. Choosing a particular reduct or a set of reducts has an impact on the model size as well as on its quality in describing a given data set. The model size together with the model quality are two basic components tuned in selecting relevant data models. This is known as the minimum length principle. Selection of relevant kinds of reducts is an important step in building data models. It turns out that the different kinds of reducts can be efficiently computed using heuristics based, e.g., on the Boolean reasoning approach.

Let us note that analogously to the information flow [21.17] one can consider different theories over information or decision systems representing different views on knowledge encoded in the systems. In particular, this approach was used for inducing concurrent models from data tables. For more details the reader is referred to the books cited at the beginning of the chapter.

21.8 Rough Membership

Let us observe that rough sets can be also defined employing the rough membership function (21.10) instead of approximation. That is, consider

$$\mu_X^B : U \to [0, 1] \,,$$

defined by

$$\mu_X^B(x) = \frac{\text{card}(B(x) \cap X)}{\text{card}(B(x))} \,, \qquad (21.10)$$

where $x \in X \subseteq U$. The value $\mu_X^B(x)$ can be interpreted as the degree that x belongs to X in view of knowledge about x expressed by B or the degree to which the elementary granule $B(x)$ is included in the set X. This means that the definition reflects a subjective knowledge about elements of the universe, in contrast to the classical definition of a set related to objective knowledge.

Rough membership function can also be interpreted as the conditional probability that x belongs to X given B. One may refer to Bayes' theorem as the origin of this function. This interpretation was used by several researchers in the rough set community.

One can observe that the rough membership function has the following properties:

1) $\mu_X^B(x) = 1$ iff $x \in B_*(X)$,
2) $\mu_X^B(x) = 0$ iff $x \in U - B^*(X)$,
3) $0 < \mu_X^B(x) < 1$ iff $x \in \text{BN}_B(X)$,
4) $\mu_{U-X}^B(x) = 1 - \mu_X^B(x)$ for any $x \in U$,
5) $\mu_{X \cup Y}^B(x) \geq \max(\mu_X^B(x), \, \mu_Y^B(x))$ for any $x \in U$,
6) $\mu_{X \cap Y}^B(x) \leq \min(\mu_X^B(x), \, \mu_Y^B(x))$ for any $x \in U$.

From the properties it follows that the rough membership differs essentially from the fuzzy membership [21.18], for properties 5) and 6) show that the membership for union and intersection of sets, in general, cannot be computed – as in the case of fuzzy sets – from their constituents' membership. Thus formally rough membership is different from fuzzy membership. Moreover, the rough membership function depends on available knowledge (represented by attributes from B). Besides, the rough membership function, in contrast to the fuzzy membership function, has a probabilistic flavor.

Let us also mention that rough set theory, in contrast to fuzzy set theory, clearly distinguishes two very important concepts, vagueness and uncertainty, very often confused in the AI literature. Vagueness is the property of concepts. Vague concepts can be approximated using the rough set approach. Uncertainty is the property of elements of a set or a set itself (e.g., only examples and/or counterexamples of elements of a considered set are given). Uncertainty of elements of a set can be expressed by the rough membership function.

Both fuzzy and rough set theory represent two different approaches to vagueness. Fuzzy set theory addresses *gradualness* of knowledge, expressed by the fuzzy membership, whereas rough set theory addresses *granularity* of knowledge, expressed by the indiscernibility relation. One can also cope with knowledge gradualness using the rough membership. A nice illustration of this difference was given by Dider Dubois and Henri Prade in their example related to image processing, where fuzzy set theory refers to gradualness of gray level, whereas rough set theory is about the size of pixels.

Consequently, these theories do not compete with each other but are rather complementary. In particular, the rough set approach provides tools for approximate construction of fuzzy membership functions. The rough-fuzzy hybridization approach has proved to be successful in many applications. An interesting discussion of fuzzy and rough set theory in the approach to vagueness can be found in [21.11].

21.9 Discernibility and Boolean Reasoning

The discernibility relations are closely related to indiscernibility relations and belong to the most important relations considered in rough set theory. Tools for discovering and classifying patterns are based on *reasoning schemes* rooted in various paradigms. Such patterns can be extracted from data by means of methods based, e.g., on discernibility and Boolean reasoning.

The ability to discern between perceived objects is important for constructing many entities like reducts, decision rules, or decision algorithms. In the standard approach the discernibility relation $DIS(B) \subseteq U \times U$ is defined by $x DIS(B) y$ if and only if $non(x IND(B) y)$, i.e., $B(x) \cap B(y) = \varnothing$. However, this is, in general, not the case for generalized approximation spaces.

The idea of Boolean reasoning is based on the construction for a given problem P of a corresponding Boolean function f_P with the following property: the solutions for the problem P can be decoded from prime implicants of the Boolean function f_P [21.19–21]. Let us mention that to solve real-life problems it is necessary to deal with very large Boolean functions.

A successful methodology based on the discernibility of objects and Boolean reasoning has been developed for computing many important ingredients for applications. These applications include generation of reducts and their approximations, decision rules, association rules, discretization of real-valued attributes, symbolic value grouping, searching for new features defined by oblique hyperplanes or higher-order surfaces, pattern extraction from data, as well as conflict resolution or negotiation [21.4, 6].

Most of the problems related to the generation of the above-mentioned entities are NP-complete or NP-hard. However, it was possible to develop efficient heuristics returning suboptimal solutions of the problems. The results of experiments on many data sets are very promising. They show very good quality of solutions generated by the heuristics in comparison with other methods reported in the literature (e.g., with respect to the classification quality of unseen objects). Moreover, they are very efficient from the point of view of the time necessary to compute the solution. Many of these methods are based on discernibility matrices. However, it is possible to compute the necessary information about these matrices without their explicit construction (i.e., by sorting or hashing original data).

It is important to note that the methodology makes it possible to construct heuristics with a very important *approximation property*, which can be formulated as follows: expressions, called *approximate implicants*, generated by heuristics that are *close* to prime implicants define approximate solutions for the problem.

Mining large data sets is one of the biggest challenges in knowledge discovery and data mining (KDD). In many practical applications, there is a need for data mining algorithms running on terminals of a client–server database system where the only access to database (located in the server) is enabled by queries in structured query language (SQL).

Let us consider two illustrative examples of problems for large data sets: (i) searching for short reducts, (ii) searching for best partitions defined by cuts on continuous attributes. In both cases, the traditional implementations of rough sets and Boolean reasoning-based methods are characterized by a high computational cost. The critical factor for the time complexity of algorithms solving the discussed problems is the number of data access operations. Fortunately some efficient modifications of the original algorithms were proposed by relying on concurrent retrieval of higher-level statistics, which are sufficient for the heuristic search of reducts and partitions [21.4, 6]. The rough set approach was also applied in the development of other scalable big data processing techniques (e.g., [21.22]).

21.10 Rough Sets and Induction

The rough set approach is strongly related to inductive reasoning (e.g., in rough set-based methods for inducing classifiers or clusters [21.6]). The general idea for inducing classifiers is as follows. From a given decision table a set of granules in the form of decision rules is induced together with arguments *for* and *against* each decision rule and decision class. For any new object with known signature one can select rules matching this object. Note that the left-hand sides of decision rules are described by formulae that make it possible to check for new objects if they satisfy them assuming that the signatures of these objects are known. In this way, one can consider two semantics of formulae: on a sample of objects U and on its extension $U^* \supseteq U$. Definitely, one should consider a risk related to such generalization, e.g., in the decision rule induction. Next, a conflict resolution should be applied to resolve conflicts between matched rules by new object voting for different decisions. In the rough set approach, the process of inducing classifiers can be considered as the process of inducing approximations of concepts over

extensions of approximation spaces (defined over samples of objects represented by decision systems). The whole procedure can be generalized for the case of approximation of more complex information granules. It is worthwhile mentioning that approaches for inducing approximate reasoning schemes have also been developed.

A typical approach in machine learning is based on inducing classifiers from samples of objects. These classifiers are used for prediction decisions on objects unseen so far, if only the signatures of these objects are available. This approach can be called global, i.e., leading to decision extension from a given sample of objects on the whole universe of objects. This global approach has some drawbacks (see the Epilog in [21.23]). Instead of this one can try to use transduction [21.23], semi-supervised learning, induced local models relative to new objects, or adaptive learning strategies. However, we are still far away from fully understanding the discovery processes behind such generalization strategies [21.24].

21.11 Rough Set–Based Generalizations

The original approach by Pawlak was based on indiscernibility defined by equivalence relations. Any such indiscernibility relation defines a partition of the universe of objects. Over the years many generalizations of this approach were introduced, many of which are based on coverings rather than partitions. In particular, one can consider the similarity (tolerance)-based rough set approach, binary relation based rough sets, neighborhood and covering rough sets, the dominance-based rough set approach, hybridization of rough sets and fuzzy sets, and many others.

One should note that dealing with coverings requires solving several new algorithmic problems such as the selection of family of definable sets or resolving problems with the selection of the relevant definition of the approximation of sets among many possible ones. One should also note that for a given problem (e.g., a

classification problem) one should discover the relevant covering for the target classification task. In the literature there are numerous papers dedicated to theoretical aspects of the covering rough set approach. However, still much more work should be done on rather hard algorithmic issues, e.g., for the relevant covering discovery.

Another issue to be solved is related to inclusion measures. Parameters of such measures are tuned in inducing of the high quality approximations. Usually, this is done on the basis of the minimum description length principle. In particular, approximation spaces with rough inclusion measures have been investigated. This approach was further extended to the rough mereological approach. More general cases of approximation spaces with rough inclusion have also been discussed in the literature, including approximation spaces in GC.

Finally, it is worthwhile mentioning the approach for ontology approximation used in hierarchical learning of complex vague concepts [21.6].

In this section, we discuss in more detail some issues related to the above-mentioned generalizations. Several generalizations of the classical rough set approach based on approximation spaces defined as pairs of the form (U, R), where R is the equivalence relation (called indiscernibility relation) on the set U, have been reported in the literature. They are related to different views on important components used in the definition of rough sets. In the definition of rough sets different kinds of structural sets that are examples of information granules are used. From mathematical point of view, one may treat them as sets defined over the hierarchy of the powerset of objects. Among them are the following ones:

- Elementary granules (neighborhoods) of objects (e.g., similarity, tolerance, dominance neighborhoods, fuzzy neighborhoods, rough-fuzzy neighborhoods, fuzzy rough neighborhoods, families of neighborhoods).
- Granules defined by accessible information about objects (e.g., only partial information on the signature of objects may be accessible).
- Methods for the definition of higher-order information granules (e.g., defined by the left-hand sides of induced decision rules or clusters of similar information granules).
- Inclusion measures making it possible to define the degrees of inclusion and/or closeness between information granules (e.g., the degrees of inclusion granules defined by accessible information about objects into elementary granules).
- Aggregation methods of inclusion or/and closeness degrees.
- Methods for the definition of approximation operations, including strategies for extension of approximations from samples of objects to larger sets of objects.
- Algebraic structures of approximation spaces.

Let us consider some examples of generalizations of the rough set approach proposed by Pawlak in 1982.

A generalized approximation space [21.25] can be defined by a tuple $\mathcal{A}S = (U, I, \nu)$, where I is the *uncertainty function* defined on U with values in the powerset POW(U) of U ($I(x)$ is the *neighborhood* of x) and ν is the *inclusion function* defined on the Cartesian product POW$(U) \times$ POW(U) with values in the interval $[0, 1]$

measuring the degree of inclusion of sets. The lower and upper approximation operations can be defined in $\mathcal{A}S$ by

$$\mathrm{LOW}(\mathcal{A}S, X) = \{x \in U : \nu(I(x), X) = 1\}, \quad (21.11)$$
$$\mathrm{UPP}(\mathcal{A}S, X) = \{x \in U : \nu(I(x), X) > 0\}. \quad (21.12)$$

In the case considered by *Pawlak* [21.2], $I(x)$ is equal to the equivalence class $B(x)$ of the indiscernibility relation $\mathrm{IND}(B)$; in the case of the tolerance (or similarity) relation $T \subseteq U \times U$ we take $I(x) = [x]_T = \{y \in U : xTy\}$, i.e., $I(x)$ is equal to the tolerance class of T defined by x.

The standard rough inclusion relation ν_{SRI} is defined for $X, Y \subseteq U$ by

$$\nu_{\mathrm{SRI}}(X, Y) = \begin{cases} \dfrac{\mathrm{card}(X \cap Y)}{\mathrm{card}(X)}, & \text{if } X \neq \emptyset, \\ 1, & \text{otherwise}. \end{cases} \quad (21.13)$$

For applications it is important to have some constructive definitions of I and ν.

One can consider another way to define $I(x)$. Usually together with $\mathcal{A}S$ we consider some set $\mathcal{F}$ of formulae describing sets of objects in the universe U of $\mathcal{A}S$ defined by semantics $\|\cdot\|_{\mathcal{A}S}$, i.e., $\|\alpha\|_{\mathcal{A}S} \subseteq U$ for any $\alpha \in \mathcal{F}$. If $\mathcal{A}S = (U, A)$ then we will also write $\|\alpha\|_U$ instead of $\|\alpha\|_{\mathcal{A}S}$. Now, one can take the set

$$N_{\mathcal{F}}(x) = \{\alpha \in \mathcal{F} : x \in \|\alpha\|_{\mathcal{A}S}\}, \quad (21.14)$$

and $I(x) = \{\|\alpha\|_{\mathcal{A}S} : \alpha \in N_{\mathcal{F}}(x)\}$. Hence, more general uncertainty functions with values in POW(POW(U)) can be defined and in consequence different definitions of approximations are considered. For example, one can consider the following definitions of approximation operations in this approximation space $\mathcal{A}S$

$$\mathrm{LOW}(\mathcal{A}S_\circ, X)$$
$$= \{x \in U : \nu(Y, X) = 1 \text{ for some } Y \in I(x)\}, \quad (21.15)$$
$$\mathrm{UPP}(\mathcal{A}S_\circ, X)$$
$$= \{x \in U : \nu(Y, X) > 0 \text{ for any } Y \in I(x)\}. \quad (21.16)$$

There are also different forms of rough inclusion functions. Let us consider two examples. In the first example of a rough inclusion function, a threshold $t \in (0, 0.5)$ is used to relax the degree of inclusion of sets.

The rough inclusion function ν_t is defined by

$$\nu_t(X, Y)$$
$$= \begin{cases} 1 & \text{if } \nu_{\text{SRI}}(X, Y) \geq 1 - t, \\ \dfrac{\nu_{\text{SRI}}(X, Y) - t}{1 - 2t} & \text{if } t \leq \nu_{\text{SRI}}(X, Y) < 1 - t, \\ 0 & \text{if } \nu_{\text{SRI}}(X, Y) \leq t. \end{cases}$$

(21.17)

One can obtain approximations considered in the variable precision rough set approach (VPRSM) by substituting in (21.12) and (21.13) the rough inclusion function ν_t defined by (21.17) instead of ν, assuming that Y is a decision class and $I(x) = B(x)$ for any object x, where B is a given set of attributes. Another example of application of the standard inclusion was developed by using probabilistic decision functions. The rough inclusion relation can be also used for function approximation and relation approximation [21.25].

The approach based on inclusion functions has been generalized to the *rough mereological approach* [21.26]. The inclusion relation $x\mu_r y$ with the intended meaning *x is a part of y to a degree at least r* has been taken as the basic notion of the rough mereology being a generalization of the *Leśniewski* mereology [21.27].

Usually families of approximation spaces labeled by some parameters are considered. By tuning such parameters according to chosen criteria (e.g., minimal description length) one can search for the optimal approximation space for concept approximation.

Our knowledge about the approximated concepts is often partial and uncertain. For example, concept approximation should be constructed from examples and counterexamples of objects for the concepts [21.14]. Hence, concept approximations constructed from a given sample of objects are extended, using inductive reasoning, on objects not yet observed. The rough set approach for dealing with concept approximation under such partial knowledge is now well developed.

Searching strategies for relevant approximation spaces are crucial for real-life applications. They include the discovery of uncertainty functions, inclusion measures, as well as selection of methods for approximations of decision classes and strategies for inductive extension of approximations from samples on larger sets of objects.

Approximations of concepts should be constructed under dynamically changing environments. This leads to a more complex situation where the boundary regions are not crisp sets, which is consistent with the postulate of the higher-order vagueness considered by philosophers [21.10]. Different aspects of vagueness in the rough set framework have been discussed.

It is worthwhile mentioning that a rough set approach to the approximation of compound concepts has been developed. For such concepts, it is hardly possible to expect that they can be approximated with the high quality by the traditional methods [21.23, 28]. The approach is based on hierarchical learning and ontology approximation. Approximation methods of concepts in distributed environments have been developed. The reader may find surveys of algorithmic methods for concept approximation based on rough sets and Boolean reasoning in the literature.

In several papers, the problem of ontology approximation was discussed together with possible applications to approximation of compound concepts or to knowledge transfer. In any ontology [21.29] (vague) concepts and local dependencies between them are specified. Global dependencies can be derived from local dependencies. Such derivations can be used as hints in searching for relevant compound patterns (information granules) in approximation of more compound concepts from the ontology. The ontology approximation problem is one of the fundamental problems related to approximate reasoning in distributed environments. One should construct (in a given language that is different from the language in which the ontology is specified) not only approximations of concepts from ontology, but also vague dependencies specified in the ontology. It is worthwhile mentioning that an ontology approximation should be induced on the basis of incomplete information about concepts and dependencies specified in the ontology. Information granule calculi based on rough sets have been proposed as tools making it possible to solve this problem. Vague dependencies have vague concepts in premises and conclusions.

The approach to approximation of vague dependencies based only on degrees of closeness of concepts from dependencies and their approximations (classifiers) is not satisfactory for approximate reasoning. Hence, more advanced approach should be developed. Approximation of any vague dependency is a method which for any object allows us to compute the arguments *for* and *against* its membership to the dependency conclusion on the basis of analogous arguments relative to the dependency premises. Any argument is a compound information granule (compound pattern). Arguments are fused by local schemes (production rules) discovered from data. Further fusions are possible through composition of local schemes, called approximate reasoning schemes (AR) [21.30]. To estimate the degree to which (at least) an object belongs to concepts from ontology the arguments *for* and *against* those concepts are collected and next a conflict resolution strategy is applied to them to predict the degree.

Several generalizations of the rough set approach introduced by Pawlak in 1982 are discussed in this handbook in more detail:

- The similarity (tolerance)-based rough set approach (Chap. 25)
- Binary relation based rough sets (Chap. 25)
- Neighborhood and covering rough sets (Chap. 25)
- The dominance-based rough set approach (Chap. 22)
- The probabilistic rough set approach and its probabilistic extension called the variable consistency dominance-based rough set approaches (Chap. 24)
- Parameterized rough sets based on Bayesian confirmation measures (Chap. 22)
- Stochastic rough set approaches (Chap. 22)
- Generalizations of rough set approximation operations (Chap. 25)
- Hybridization of rough sets and fuzzy sets (Chap. 26)
- Rough sets on abstract algebraic structures (e.g., lattices) (Chap. 25).

There are some other well-established or emerging domains not covered in the chapter where some generalizations of rough sets are proposed as the basic tools, often in combination with other existing approaches. Among them are rough sets based on [21.6]:

i)   Incomplete information and/or decision systems
ii) Nondeterministic information and/or decision systems
iii) The rough set model on two universes
iv) Dynamic information and/or decision systems
v)   Dynamic networks of information and/or decision systems.

Moreover, rough sets play a crucial role in the development of granular computing (GC) [21.31]. The extension to interactive granular computing (IGR) [21.32] requires generalization of basic concepts such as information and decision systems, as well as methods of inducing hierarchical structures of information and decision systems.

Let us note that making progress in understanding interactive computations is one of the key problems in developing high quality intelligent systems working in complex environments [21.33]. The current research projects aim at developing foundations of IGC based on the rough set approach in combination with other soft computing approaches, in particular with fuzzy sets. The approach is called interactive rough granular computing (IRGC). In IRGC computations are based on interactions of complex granules (c-granules, for short). Any c-granule consists of a physical part and a mental part that are linked in a special way [21.32]. IRGC is treated as the basis for (see [21.6] and references in this book):

i)   Wistech Technology, in particular for approximate reasoning, called adaptive judgment about properties of interactive computations
ii) Context induction
iii) Reasoning about changes
iv) Process mining (this research was inspired by [21.34])
v)   Perception-based computing (PBC)
vi) Risk management in computational systems [21.32].

Interactive computations based on c-granules seem to create a good background, e.g., for modeling computations in Active Media Technology (AMT) and Wisdom Web of Things (W2T). We plan to investigate their role for foundations of natural computing too. Let us also mention that the interactive computations based on c-granules are quite different in nature than Turing computations. Hence, we plan to investigate relationships of interactive computability based on c-granules and Turing computability.

21.12 Rough Sets and Logic

The father of contemporary logic was the German mathematician Gottlob Frege (1848–1925). He thought that mathematics should not be based on the notion of set but on the notions of logic. He created the first axiomatized logical system but it was not understood by the logicians of those days. During the first three decades of the twentieth century there was a rapid development in logic, bolstered to a great extent by Polish logicians, in particular by Alfred Tarski, Stanisław Leśniewski, Jan Łukasiewicz, and next by Andrzej Mostowski and Helena Rasiowa. The development of computers and their applications stimulated logical research and widened their scope.

When we speak about logic, we generally mean *deductive logic*. It gives us tools designed for deriving true propositions from other true propositions. Deductive reasoning always leads to true conclusions from true premises. The theory of deduction has well-established, generally accepted theoretical foundations. Deductive reasoning is the main tool used in mathematical reasoning.

Rough set theory has contributed to some extent to various kinds of deductive reasoning. Particularly, various kinds of logics based on the rough set approach have been investigated; rough set methodology has contributed essentially to modal logics, many valued logic, intuitionistic logic, and others (see, e.g., references in the book [21.6] and in articles [21.3, 4]). A summary of this research can be found in [21.35, 36] and the interested reader is advised to consult these volumes.

In natural sciences (e.g., in physics) *inductive reasoning* is of primary importance. The characteristic feature of such reasoning is that it does not begin from axioms (expressing general knowledge about the reality) like in deductive logic, but some partial knowledge (examples) about the universe of interest are the starting point of this type of reasoning, which are generalized next and they constitute the knowledge about a wider reality than the initial one. In contrast to deductive reasoning, inductive reasoning does not lead to true conclusions but only to probable (possible) ones. Also, in contrast to the logic of deduction, the logic of induction does not have uniform, generally accepted, theoretical foundations as yet, although many important and interesting results have been obtained, e.g., concerning statistical and computational learning and others.

Verification of the validity of hypotheses in the logic of induction is based on experiment rather than the formal reasoning of the logic of deduction. Physics is the best illustration of this fact. The research on modern inductive logic has a several centuries' long history. It is worthwhile mentioning here the outstanding English philosophers Francis Bacon (1561–1626) and *John Stuart Mill* (1806–1873) [21.37].

The creation of computers and their innovative applications essentially contributed to the rapid growth of interest in inductive reasoning. This domain is developing very dynamically thanks to computer science. Machine learning, knowledge discovery, reasoning from data, expert systems, and others are examples of new directions in inductive reasoning. Rough set theory is very well suited as a theoretical basis for inductive reasoning. Basic concepts of this theory fit very well to represent and analyze knowledge acquired from examples, which can be next used as a starting point for generalization. Besides, in fact, rough set theory has been successfully applied in many domains to find patterns in data (data mining) and acquire knowledge from examples (learning from examples). Thus, rough set theory seems to be another candidate as a mathematical foundation of inductive reasoning.

The most interesting from a computer science point of view is *common sense* reasoning. We use this kind of reasoning in our everyday lives, and we face examples of such kind of reasoning in newspapers, radio, TV, etc., in political, economics, etc., and in debates and discussions.

The starting point for such reasoning is the knowledge possessed by a specific group of people (*common knowledge*) concerning some subject and intuitive methods of deriving conclusions from it. Here we do not have the possibility to resolve the dispute by means of methods given by deductive logic (reasoning) or by inductive logic (experiment). So the best known methods for solving the dilemma are voting, negotiations, or even war. See, e.g., Gulliver's Travels [21.38], where the hatred between Tramecksan (High-Heels) and Slamecksan (Low-Heels) or disputes between Big-Endians and Small-Endians could not be resolved without a war. These methods do not reveal the truth or falsity of the thesis under consideration at all. Of course, such methods are not acceptable in mathematics or physics. Nobody is going to solve the truth of Fermat's theorem or Newton's laws by voting, negotiations, or declare a war.

Reasoning of this kind is the least studied from the theoretical point of view and its structure is not sufficiently understood, in spite of many interesting theoretical research in this domain [21.39]. The meaning of commonsense reasoning, considering its scope and significance for some domains, is fundamental, and rough set theory can also play an important role in it, but more fundamental research must be done to this end. In particular, the rough truth introduced and studied in [21.40] seems to be important for investigating commonsense reasoning in the rough set framework.

Let us consider a simple example. In the decision system considered we assume $U =$ Birds is a set of birds that are described by some condition attributes from a set A. The decision attribute is a binary attribute Flies with possible values *yes* if the given bird flies and *no*, otherwise. Then, we define the set of abnormal birds by $Ab_A(\text{Birds}) = A_*(\{x \in \text{Birds}: \text{Flies}(x) = no\})$. Hence, we have, $Ab_A(\text{Birds}) = \text{Birds} - A^*(\{x \in \text{Birds}: \text{Flies}(x) = yes\})$ and $\text{Birds} - Ab_A(\text{Birds}) = A^*(\{x \in \text{Birds}: \text{Flies}(x) = yes\})$. This means that for normal birds it is consistent, with knowledge represented by A, to assume that they can fly, i.e., it is possible that they can fly. One can optimize $Ab_A(\text{Birds})$ using A to obtain minimal boundary region in the approximation of $\{x \in \text{Birds}: \text{Flies}(x) = no\}$.

It is worthwhile mentioning that in [21.41] an approach was presented that combines the rough sets with nonmonotonic reasoning. Some basic concepts are distinguished, which can be approximated on the basis of sensor measurements and more complex concepts that are approximated using so-called transducers defined by first-order theories constructed over approximated concepts. Another approach to commonsense reasoning was developed in a number of papers. The approach is based on an ontological framework for approximation. In this approach, approximations are constructed for concepts and dependencies between the concepts represented in a given ontology, expressed, e.g., in natural language. Still another approach combining rough sets with logic programming has been developed. Let us also note that *Pawlak* proposed a new approach to conflict analysis [21.42]. The approach was next extended in the rough set framework.

To recapitulate, let us consider the following characteristics of the three above-mentioned kinds of reasoning:

a) *Deductive*
   i)   *Reasoning methods*: axioms and rules of inference
   ii) *Applications*: mathematics
   iii) *Theoretical foundations*: complete theory
   iv) *Conclusions*: true conclusions from true premisses
   v)   *Hypotheses verification*: formal proof
b) *Inductive*
   i)   *Reasoning methods*: generalization from examples
   ii) *Applications*: natural sciences (physics)
   iii) *Theoretical foundations*: lack of generally accepted theory

iv) *Conclusions*: not true but probable (possible)
v)   *Hypotheses verification*: empirical experiment
c) *Common sense*
   i)   *Reasoning methods*: reasoning method based on common sense knowledge with intuitive rules of inference expressed in natural language
   ii) *Applications*: everyday life, humanities
   iii) *Theoretical foundations*: lack of generally accepted theory
   iv) *Conclusions*: obtained by mixture of deductive and inductive reasoning based on concepts expressed in natural language, e.g., with application of different inductive strategies for conflict resolution (such as voting, negotiations, cooperation, war) based on human behavioral patterns
   v)   *Hypotheses verification*: human behavior.

There are numerous issues related to approximate reasoning under uncertainty. These issues are discussed in books on granular computing, rough mereology, and the computational complexity of algorithmic problems related to these issues. For more details, the reader is referred to the following books [21.26, 31, 43, 44].

Finally, we would like to stress that still much more work should be done to develop approximate reasoning about complex vague concepts to make progress in the development of intelligent systems. According to *Leslie Valiant* [21.45] (who is the 2011 winner of the ACM Turing Award, for his fundamental contributions to the development of computational learning theory and to the broader theory of computer science):

A fundamental question for artificial intelligence is to characterize the computational building blocks that are necessary for cognition. A specific challenge is to build on the success of machine learning so as to cover broader issues in intelligence ... This requires, in particular a reconciliation between two contradictory characteristics – the apparent logical nature of reasoning and the statistical nature of learning.

It is worthwhile presenting two more views. The first one by *Lotfi A. Zadeh*, the founder of fuzzy sets and the computing with words (CW) paradigm [21.46]:

Manipulation of perceptions plays a key role in human recognition, decision and execution processes. As a methodology, computing with words provides a foundation for a computational theory of per-

ceptions – a theory which may have an important bearing on how humans make and machines might make – perception-based rational decisions in an environment of imprecision, uncertainty and partial truth. ... computing with words, or CW for short, is a methodology in which the objects of computation are words and propositions drawn from a natural language.

The other view is that of *Judea Pearl* [21.47] (the 2011 winner of the ACM Turing Award, the highest distinction in computer science, for fundamental contributions to artificial intelligence through the development of a calculus for probabilistic and causal reasoning):

Traditional statistics is strong in devising ways of describing data and inferring distributional parameters from sample. Causal inference requires two additional ingredients: a science-friendly language for articulating causal knowledge, and a mathematical machinery for processing that knowledge, combining it with data and drawing new causal conclusions about a phenomenon.

The question arises about the logic relevant for the above-mentioned tasks. First, let us observe that the satisfiability relations in the IRGC framework can be treated as tools for constructing new granules. In fact, for a given satisfiability relation one can define the semantics of formulae related to this relation, i. e., which are the candidates for the new relevant granules. We would like to emphasize one a very important feature. The relevant satisfiability relation for the considered problems is not given but it should be induced (discovered) from partial information given by information or decision systems. For real-life problems it is often necessary to discover a hierarchy of satisfiability relations before we obtain the relevant target one. Granules constructed on different levels of this hierarchy finally lead to relevant ones for approximation of complex vague concepts related to complex granules expressed using natural language.

The reasoning making it possible to derive relevant c-granules for solutions of the target tasks is called *adaptive judgment*. *Intuitive judgment* and *ra-*

tional judgment are distinguished as different kinds of judgment [21.48]. Deduction and induction as well as abduction or analogy-based reasoning are involved in adaptive judgment. Among the tasks for adaptive judgment are the following ones, which support reasoning under uncertainty toward: searching for relevant approximation spaces, discovery of new features, selection of relevant features, rule induction, discovery of inclusion measures, strategies for conflict resolution, adaptation of measures based on the minimum description length principle, reasoning about changes, perception (action and sensory) attributes' selection by agent control, adaptation of quality measures over computations relative to agents, adaptation of object structures, discovery of relevant contexts, strategies for knowledge representation and interaction with knowledge bases, ontology acquisition and approximation, learning in dialog of inclusion measures between granules from different languages (e.g., the formal language of the system and the user's natural language), strategies for adaptation of existing models, strategies for development and evolution of communication language among agents in distributed environments, strategies for risk management in distributed computational systems. Definitely, in the language used by agents for dealing with adaptive judgment (i. e., intuitive and rational) some deductive systems known from logic may be applied for reasoning about knowledge relative to closed worlds. This may happen, e.g., if the agent languages are based on classical mathematical logic. However, if we move to interactions in open worlds, then new specific rules or patterns relative to a given agent or group of agents in such worlds should be discovered. The process of inducing such rules or patterns is influenced by uncertainty because they are induced by agents under uncertain and/or imperfect knowledge about the environment.

The concepts discussed, such as interactive computation and adaptive judgment, are among the basic concepts in Wisdom Technology (WisTech) [21.49, 50]. Let us mention here the WisTech meta-equation

$$\begin{aligned} \text{WISDOM} = {}& \text{INTERACTIONS} \\ & + \text{ADAPTIVE JUDGMENT} \\ & + \text{KNOWLEDGE} \,. \end{aligned} \tag{21.18}$$

21.13 Conclusions

In the chapter, we have discussed some basic issues and methods related to rough sets together with some generalizations, including those related to relationships of rough sets with inductive reasoning. We have also listed some current research directions based on interactive rough granular computing. For more details, the reader is referred to the literature cited at the beginning of this chapter (see also [21.9]).

References

21.1   Z. Pawlak: *Rough Sets: Theoretical Aspects of Reasoning about Data*, Theory and Decision Library D, Vol. 9 (Kluwer, Dordrecht 1991)

21.2   Z. Pawlak: Rough sets, Int. J. Comp. Inform. Sci. **11**, 341–356 (1982)

21.3   Z. Pawlak, A. Skowron: Rudiments of rough sets, Inform. Sci. **177**(1), 3–27 (2007)

21.4   Z. Pawlak, A. Skowron: Rough Sets: Some Extensions, Inform. Sci. **177**(1), 28–40 (2007)

21.5   Z. Pawlak, A. Skowron: Rough sets and Boolean reasoning, Inform. Sci. **177**(1), 41–73 (2007)

21.6   A. Skowron, Z. Suraj (Eds.): *Rough Sets and Intelligent Systems. Professor Zdzisław Pawlak in Memoriam*, Intelligent Systems Reference Library, Vol. 42/43 (Springer, Berlin, Heidelberg 2013)

21.7   I. Chikalov, V. Lozin, I. Lozina, M. Moshkov, H.S. Nguyen, A. Skowron, B. Zielosko: *Three Approaches to Data Analysis. Test Theory, Rough Sets and Logical Analysis of Data*, Intelligent Systems Reference Library, Vol. 41 (Springer, Berlin, Heidelberg 2012)

21.8   Ch. Cornelis (Ed.): International Rough Sets Society, online available from: http://www.roughsets.org

21.9   Z. Suraj: Rough Set Database System, online available from: http://www.rsds.univ.rzeszow.pl

21.10   R. Keefe: *Theories of Vagueness*, Cambridge Studies in Philosophy (Cambridge Univ. Press, Cambridge 2000)

21.11   S. Read: *Thinking about Logic: An Introduction to the Philosophy of Logic* (Oxford Univ. Press, Oxford 1994)

21.12   G. Frege: *Grundgesetze der Arithmetik*, Vol. 2 (Verlag von Hermann Pohle, Jena 1903)

21.13   G.W. Leibniz: Discourse on Metaphysics. In: *Philosophical Essays (1686)*, ed. by R. Ariew, D. Garber (Hackett, Indianapolis 1989) pp. 35–68

21.14   T. Hastie, R. Tibshirani, J.H. Friedman: *The Elements of Statistical Learning: Data Mining, Inference, and Prediction* (Springer, Berlin, Heidelberg 2001)

21.15   J. Rissanen: Modeling by shortes data description, Automatica **14**, 465–471 (1978)

21.16   J. Łukasiewicz: Die logischen Grundlagen der Wahrscheinlichkeitsrechnung. In: Jan Łukasiewicz – Selected Works, ed. by L. Borkowski (North Holland/Polish Scientific Publishers, Amsterdam, Warsaw 1970) pp. 16–63

21.17   J. Barwise, J. Seligman: *Information Flow: The Logic of Distributed Systems* (Cambridge Univ. Press, Cambridge 1997)

21.18   L.A. Zadeh: Fuzzy sets, Inform. Control **8**, 338–353 (1965)

21.19   G. Boole: *The Mathematical Analysis of Logic* (G. Bell, London 1847), Reprinted by Philosophical Library, New York 1948

21.20   G. Boole: *An Investigation of the Laws of Thought* (Walton, London 1954)

21.21   F.M. Brown: *Boolean Reasoning* (Kluwer, Dordrecht 1990)

21.22   D. Slezak: Infobright, online available from: http://www.infobright.com/

21.23   V. Vapnik: *Statistical Learning Theory* (Wiley, New York 1998)

21.24   R.S. Michalski: A theory and methodology of inductive learning, Artif. Intell. **20**, 111–161 (1983)

21.25   J. Stepaniuk: *Rough-Granular Computing in Knowledge Discovery and Data Mining* (Springer, Berlin, Heidelberg 2008)

21.26   L. Polkowski: *Approximate Reasoning by Parts. An Introduction to Rough Mereology*, Intelligent Systems Reference Library, Vol. 20 (Springer, Berlin, Heidelberg 2011)

21.27   S. Leśniewski: Grungzüge eines neuen Systems der Grundlagen der Mathematik, Fundam. Math. **14**, 1–81 (1929)

21.28   L. Breiman: Statistical modeling: The two cultures, Stat. Sci. **16**(3), 199–231 (2001)

21.29   S. Staab, R. Studer: *Handbook on Ontologies*, International Handbooks on Information Systems (Springer, Berlin 2004)

21.30   S.K. Pal, L. Polkowski, A. Skowron: *Rough-Neural Computing: Techniques for Computing with Words*, Cognitive Technologies (Springer, Berlin, Heidelberg 2004)

21.31   W. Pedrycz, S. Skowron, V. Kreinovich (Eds.): *Handbook of Granular Computing* (Wiley, Hoboken 2008)

21.32   A. Jankowski: *Practical Issues of Complex Systems Engineering: Wisdom Technology Approach* (Springer, Berlin, Heidelberg 2015), in preparation

21.33   D. Goldin, S. Smolka, P. Wegner: *Interactive Computation: The New Paradigm* (Springer, Berlin, Heidelberg 2006)

21.34   Z. Pawlak: Concurrent versus sequential – the rough sets perspective, Bulletin EATCS **48**, 178–190 (1992)

21.35   L. Polkowski: *Rough Sets: Mathematical Foundations*, Advances in Soft Computing (Physica, Berlin, Heidelberg 2002)

21.36   M. Chakraborty, P. Pagliani: *A Geometry of Approximation: Rough Set Theory – Logic, Algebra and Topology of Conceptual Patterns* (Springer, Berlin, Heidelberg 2008)

21.37   D.M. Gabbay, S. Hartmann, J. Woods (Eds.): *Inductive Logic*, Handbook of the History of Logic, Vol. 10 (Elsevier, Amsterdam 2011)

21.38   J. Swift: *Gulliver's Travels into Several Remote Nations of the World* (anonymous publisher, London 1726)

21.39   D.M. Gabbay, C.J. Hogger, J.A. Robinson: *Non-monotonic Reasoning and Uncertain Reasoning*, Handbook of Logic in Artificial Intelligence and Logic Programming, Vol. 3 (Clarendon, Oxford 1994)

21.40   Z. Pawlak: Rough logic, Bull. Pol. Ac.: Tech. **35**(5/6), 253–258 (1987)

21.41   P. Doherty, W. Łukaszewicz, A. Skowron, A. Szałas: *Knowledge Engineering: A Rough Set Approach*, Studies in Fizziness and Soft Computing, Vol. 202 (Springer, Berlin, Heidelberg 2006)

21.42   Z. Pawlak: An inquiry into anatomy of conflicts, Inform. Sci. **109**, 65–78 (1998)

21.43   M.J. Moshkov, M. Piliszczuk, B. Zielosko: *Partial Covers, Reducts and Decision Rules in Rough Sets – Theory and Applications*, Studies in Computational Intelligence, Vol. 145 (Springer, Berlin, Heidelberg 2008)

21.44   P. Delimata, M.J. Moshkov, A. Skowron, Z. Suraj: *Inhibitory Rules in Data Analysis: A Rough Set Approach*, Studies in Computational Intelligence, Vol. 163 (Springer, Berlin, Heidelberg 2009)

21.45   Web page of Professor Leslie Valiant, online available from: http://people.seas.harvard.edu/~valiant/researchinterests.htm

21.46   L.A. Zadeh: From computing with numbers to computing with words – From manipulation of measurements to manipulation of perceptions, IEEE Trans. Circuits Syst. **45**, 105–119 (1999)

21.47   J. Pearl: Causal inference in statistics: An overview, Stat. Surv. **3**, 96–146 (2009)

21.48   D. Kahneman: Maps of Bounded Rationality: Psychology for behavioral economics, Am. Econ. Rev. **93**, 1449–1475 (2002)

21.49   A. Jankowski, A. Skowron: A wistech paradigm for intelligent systems, Lect. Notes Comput. Sci. **4374**, 94–132 (2007)

21.50   A. Jankowski, A. Skowron: Logic for artificial intelligence: The Rasiowa–Pawlak school perspective. In: *Andrzej Mostowski and Foundational Studies*, ed. by A. Ehrenfeucht, V. Marek, M. Srebrny (IOS, Amsterdam 2008) pp. 106–143

22. Rough Set Methodology for Decision Aiding

Roman Słowiński, Salvatore Greco, Benedetto Matarazzo

Since its conception, the dominance-based rough set approach (DRSA) has been adapted to a large variety of decision problems. In this chapter we outline the rough set methodology designed for multi-attribute decision aiding. DRSA was proposed as an extension of the Pawlak concept of rough sets in order to deal with ordinal data. We focus on decision problems where all attributes describing objects of a decision problem have ordered value sets (scales). Such attributes are called criteria, and thus the problems are called multi-criteria decision problems. Criteria are real-valued functions of gain or cost type, depending on whether a greater value is better or worse, respectively. In these problems, we also assume the presence of a well defined decision maker (DM) (single of group DM) concerned by multi-criteria classification, choice, and ranking.

Ordinal data are typically encountered in multi-attribute decision problems, where a set of objects (also called actions, acts, solutions, etc.) evaluated by a set of attributes (also called criteria, variables, features, etc.) raises one of the following questions: (i) how to assign the objects to some ordered classes (*ordinal classification*), (ii) how to choose the best subset of objects (*choice* or its particular case – *optimization*), or (iii) how to rank the objects from the best to the worst (*ranking*). The answer to all of these questions involves an aggregation of the multi-attribute evaluation of objects, which takes into account a law relating the evaluation with the classification, or choice, or ranking decision. This law has to be discovered by inductive learning from data describing the considered decision situation. In the case of decision problems that correspond to some physical phenomena, this law is a model of cause–effect relationships, and in the case of human decision making, this law is a decision maker's preference model. In DRSA, these models have the form of a set of *if ..., then ...* decision rules. In the case of multi-attribute classification the syntax of rules is: *if evaluation of object a is better (or worse) than given values of some attributes, then a belongs to at* *least (at most) a given class*, and in the case of multi-attribute choice or ranking: *if object a is preferred to object b in at least (at most) given degrees with respect to some attributes, then a is preferred to b in at least (at most) a given degree.* These models are used to work out a recommendation concerning unseen objects in the context of one of the three problem statements.

22.1 Data Inconsistency as a Reason for Using Rough Sets

The data describing a given decision situation include either observations of DM's past decisions in the same decision context, or examples of decisions consciously elicited by the DM on the demand of an analyst. These data hides the value system of the DM, and thus they are called *preference information*. This way of preference information elicitation is called *indirect*, in opposition to *direct* elicitation when the DM is supposed to provide information leading directly to the definition of all preference model parameters, like weights and discrimination thresholds of criteria, trade-off rates, etc. [22.1].

Past decisions or decision examples may, however, be *inconsistent* with the *dominance principle* commonly accepted for multi-criteria decision problems. Decisions are inconsistent with the dominance principle if:

- In case of ordinal classification: object *a* has been assigned to a worse decision class than object *b*, although *a* is at least as good as *b* on all the considered criteria, i. e., *a* dominates *b*.
- In the case of choice and ranking: a pair of objects (a, b) has been assigned a degree of preference worse than pair (c, d), although differences of evaluations between *a* and *b* on all the considered criteria is at least as strong as the respective differences of evaluations between *c* and *d*, i. e., pair (a, b) dominates pair (c, d).

Thus, in order to build a preference model from partly inconsistent preference information, we had the idea to structure this data using the concept of a *rough set* introduced by *Pawlak* [22.2, 3]. Originally, however, Pawlak's understanding of inconsistency was different to the above inconsistency with the dominance principle. The original rough set philosophy (Chap. 21) is based on the assumption that with every object of the universe *U* there is associated a certain amount of

information (data, knowledge). This information can be expressed by means of a number of attributes that describe the objects. Objects which have the same description are said to be indiscernible (or similar) with respect to the available information. The *indiscernibility relation* thus generated constitutes the mathematical basis of rough set theory. It induces a partition of the universe into blocks of indiscernible objects, called elementary sets, which can be used to build knowledge about a real or abstract world. The use of the indiscernibility relation results in information *granulation*.

Any subset *X* of the universe may be expressed in terms of these blocks either precisely (as a union of elementary sets) or approximately. In the latter case, the subset *X* may be characterized by two ordinary sets, called the *lower* and *upper approximations*. A rough set is defined by means of these two approximations, which coincide in the case of an ordinary set. The lower approximation of *X* is composed of all the elementary sets included in *X* (whose elements, therefore, certainly belong to *X*), while the upper approximation of *X* consists of all the elementary sets which have a non-empty intersection with *X* (whose elements, therefore, may belong to *X*). The difference between the upper and lower approximations constitutes the boundary region of the rough set, whose elements cannot be characterized with certainty as belonging or not to *X* (by using the available information). The information about objects from the boundary region is, therefore, inconsistent or ambiguous. The cardinality of the boundary region states, moreover, the extent to which it is possible to express *X* in exact terms, on the basis of the available information. For this reason, this cardinality may be used as a measure of vagueness of the information about *X*.

Some important characteristics of the rough set approach make it a particularly interesting tool in a variety of problems and concrete applications. For example, it is possible to deal with both quantitative and qualitative input data, and inconsistencies need not be removed prior to the analysis. In terms of the output information, it is possible to acquire *a posteriori* information regarding the relevance of particular attributes and their subsets to the quality of approximation considered within the problem at hand. Moreover, the lower and upper approximations of a partition of *U* into decision classes prepare the ground for inducing *certain* and *possible* knowledge patterns in the form of *if … then …* decision rules.

Several attempts have been made to employ rough set theory for decision aiding [22.4, 5]. The Indiscernibility-based Rough Set Approach (IRSA) is not able, however, to handle inconsistencies with respect to the dominance principle.

22.1.1 From Indiscernibility-Based Rough Sets to Dominance-Based Rough Sets

An extension of IRSA which deals with inconsistencies with respect to the dominance principle, which are typical for preference data, was proposed by *Greco* et al. in [22.6–8]. This extension is the dominance-based rough set approach (DRSA), which is mainly based on the substitution of the indiscernibility relation by a dominance relation in the rough approximation of decision classes. An important consequence of this fact is the possibility of inferring (from observations of past decisions or from exemplary decisions) the DM's

preference model in terms of decision rules which are logical statements of the type *if …, then …*. The separation of *certain* and *uncertain* knowledge about the DM's preferences is carried out by the distinction of different kinds of decision rules, depending upon whether they are induced from lower approximations of decision classes or from the difference between upper and lower approximations (composed of inconsistent examples). Such a preference model is more general than the classical functional models considered within multi-attribute utility theory or the relational models considered, for example, in outranking methods [22.9–11].

This chapter is based on previous publications of the authors, in particular, on [22.12–14]. In the next section, we explain the need for replacing the indiscernibility relation by the dominance relation in the definition of rough sets when reasoning about ordinal data. This leads us to Sect. 22.3, where DRSA is presented with respect to multi-criteria ordinal classification. This section also includes two special versions of DRSA: variable consistency DRSA (VC DRSA) and stochastic DRSA. Section 22.4 presents DRSA with respect to multi-criteria choice and ranking. Section 22.5 characterizes some relevant extensions of DRSA, and Sect. 22.6 presents applications of DRSA to some operational research problems. Section 22.7 summarizes the features of DRSA applied to multi-criteria decision problems and concludes the chapter.

22.2 The Need for Replacing the Indiscernibility Relation by the Dominance Relation when Reasoning About Ordinal Data

When trying to apply the rough set concept based on indiscernibility to reasoning about preference ordered data, it has been noted that IRSA ignores not only the preference order in the value sets of attributes but also the monotonic relationship between evaluations of objects on such attributes (called criteria) and the preference ordered value of decision (classification decision or degree of preference) [22.6, 15–17].

In order to explain the importance of the above monotonic relationship for data describing multi-criteria decision problems, let us consider the example of a data set concerning pupils' achievements in a high school. Suppose that among the criteria used for evaluation of the pupils there are results in *Mathematics* (*Math*) and *Physics* (*Ph*). There is also a *General*

Achievement (*GA*) result, which is considered as a classification decision. The value sets of all three criteria are composed of three values: *bad*, *medium*, and *good*. The preference order of these values is obvious: *good* is better than *medium* and *bad*, and *medium* is better than *bad*. The three values *bad*, *medium*, and *good* can be number-coded as 1, 2, and 3, respectively, making a gain-type criterion scale. One can also notice a *semantic correlation* between the two criteria and the classification decision, which means that an improvement in one criterion should not worsen the classification decision, while the other criterion value is unchanged. Precisely, an improvement of a pupil's score in *Math* or *Ph*, with other criterion value unchanged, should not worsen the pupil's general achievement (*GA*), but rather improve it. In general terms, this requirement is concordant with the dominance principle defined in Sect. 22.1.

This semantic correlation is also called a monotonicity constraint, and thus, an alternative name of the classification problem with semantic correlation between evaluation criteria and classification decision is *ordinal classification with monotonicity constraints*.

Two questions naturally follow the consideration of this example:

- What classification rules can be drawn from the pupils' data set?
- How does the semantic correlation influence the classification rules?

The answer to the first question is: monotonic *if ..., then ...* decision rules. Each decision rule is characterized by a *condition profile* and a *decision profile*, corresponding to vectors of threshold values on evaluation criteria and on classification decision, respectively. The answer to the second question is that condition and decision profiles of a decision rule should observe the dominance principle (monotonicity constraint) if the rule has at least one pair of semantically correlated criteria spanned over the condition and decision part. We say that one profile *dominates* another if the values of criteria of the first profile are not worse than the values of criteria of the second profile.

Let us explain the dominance principle with respect to decision rules on the pupils' example. Suppose that two rules induced from the pupils' data set relate *Math* and *Ph* on the condition side, with *GA* on the decision side:

- rule #1: if *Math* = *medium* and *Ph* = *medium*, then *GA* = *good*,
- rule #2: if *Math* = *good* and *Ph* = *medium*, then *GA* = *medium*.

The two rules do not observe the dominance principle because the condition profile of rule #2 dominates the condition profile of rule #1, while the decision profile of rule #2 is dominated by the decision profile of rule #1. Thus, in the sense of the dominance principle, the two rules are inconsistent, i. e., they are wrong.

One could say that the above rules are true because they are supported by examples of pupils from the analyzed data set, but this would mean that the examples are also inconsistent. The *inconsistency* may come from many sources. Examples include:

- Missing attributes (regular ones or criteria) in the description of objects. Maybe the data set does not

include such attributes as the *opinion of the pupil's tutor* expressed only verbally during an assessment of the pupil's *GA* by a school assessment committee.

- Unstable preferences of decision makers. Maybe the members of the school assessment committee changed their view on the influence of *Math* on *GA* during the assessment.

Handling these inconsistencies is of crucial importance for data structuring prior to induction of decision rules. They cannot be simply considered as noise or error to be eliminated from data, or amalgamated with consistent data by some averaging operators. They should be identified and presented as uncertain rules.

If the semantic correlation was ignored in prior knowledge, then the handling of the above-mentioned inconsistencies would be impossible. Indeed, there would be nothing wrong with rules #1 and #2. They would be supported by different examples discerned by the attributes considered.

It has been acknowledged by many authors that *rough set theory* provides an excellent framework for dealing with inconsistencies in knowledge discovery [22.3, 18–24]. These authors show that the paradigm of rough set theory is that of *granular computing*, because the main concept of the theory (rough approximation of a set) is built up of blocks of objects which are indiscernible by a given set of attributes, called *granules of knowledge*. In the space of regular attributes, the indiscernibility granules are bounded sets. Decision rules induced from indiscernibility-based rough approximations are also built up of such granules.

It appears, however, as demonstrated by the above pupils' example, that rough sets and decision rules built up of indiscernibility granules are not able to handle inconsistency with respect to the dominance principle. For this reason, we have proposed an extension of the granular computing paradigm that enables us to take into account prior knowledge about multi-criteria evaluation with monotonicity constraints. The combination of the new granules with the idea of rough approximation is the *DRSA* approach [22.6, 8, 12–16, 25–27].

In the following, we present the concept of granules, which permit us to handle prior knowledge about multi-criteria evaluation with monotonicity constraints when inducing decision rules.

Let U be a finite set of objects (universe) and let Q be a finite set of attributes divided into a set C of *condition attributes* and a set D of *decision attributes*, where $C \cap D = \emptyset$. Also, let X_q be the set of possible evaluations of considered objects with respect to attribute $q \in Q$, so that

$$X_C = \prod_{q=1}^{|C|} X_q \quad \text{and} \quad X_D = \prod_{q=1}^{|D|} X_q$$

are attribute spaces corresponding to sets of condition and decision attributes, respectively. The elements of X_C and X_D can be interpreted as possible evaluations of objects on attributes from set $C = \{1, \ldots, |C|\}$ and from set $D = \{1, \ldots, |D|\}$, respectively. In the following, with a slight abuse of notation, we shall denote the value of object $x \in U$ on attribute $q \in Q$ by x_q.

Suppose, for simplicity, that all condition attributes in C and all decision attributes in D are criteria, and that C and D are semantically correlated.

Let $\succeq_q$ be a weak preference relation on U, representing a preference on the set of objects with respect to criterion $q \in \{C \cup D\}$. Now, $x_q \succeq y_q$ means x_q *is at least as good as* y_q *with respect to criterion q*. On the one hand, we say that *x dominates y* with respect to $P \subseteq C$ (shortly, *x P-dominates y*) in the condition attribute space X_P (denoted by xD_Py) if $x_q \succeq y_q$ for all $q \in P$. Assuming, without loss of generality, that the domains of the criteria are number-coded (i.e., $X_q \subseteq \mathbb{R}$ for any $q \in C$) and that they are ordered so that the preference increases with the value (gain-type), we can say that xD_Py is equivalent to $x_q \geq y_q$ for all $q \in P$, $P \subseteq C$. Observe that for each $x \in X_P$, xD_Px, i.e., P-dominance D_P is reflexive. Moreover, for any $x, y, z \in X_P$, xD_Py and yD_Pz imply xD_Pz, i.e., P-dominance D_P is a transitive relation. Being a reflexive and transitive relation, P-dominance D_P is a partial preorder. On the other hand, the analogous definition holds in the decision attribute space X_R, $R \subseteq D$, where $x_q \geq y_q$ for all $q \in R$ will be denoted by xD_Ry.

The dominance relations xD_Py and xD_Ry ($P \subseteq C$ and $R \subseteq D$) are directional statements where x is a subject and y is a referent.

If $x \in X_P$ is the referent, then one can define a set of objects $y \in X_P$ dominating x, called the *P-dominating set* (denoted by $D_P^+(x)$) and defined as $D_P^+(x) = \{y \in U : yD_Px\}$. If $x \in X_P$ is the subject, then one can define a set of objects $y \in X_P$ dominated by x, called the *P-dominated set* (denoted by $D_P^-(x)$) and defined as $D_P^-(x) = \{y \in U : xD_Py\}$.

P-dominating sets $D_P^+(x)$ and P-dominated sets $D_P^-(x)$ correspond to *positive* and *negative dominance cones* in X_P, with the origin x.

With respect to the decision attribute space X_R (where $R \subseteq D$), the R-dominance relation enables us to define the following sets

$$Cl_R^{\geq x} = \{y \in U : yD_Rx\}, \quad Cl_R^{\leq x} = \{y \in U : xD_Ry\}.$$

$Cl_{t_q} = \{x \in X_D : x_q = t_q\}$ is a decision class with respect to $q \in D$. $Cl_R^{\geq x}$ is called the *upward union* of classes, and $Cl_R^{\leq x}$ is the *downward union* of classes. If $y \in Cl_R^{\geq x}$, then y belongs to class Cl_{t_q}, $x_q = t_q$, or better, on each decision attribute $q \in R$. On the other hand, if $y \in Cl_R^{\leq x}$, then y belongs to class Cl_{t_q}, $x_q = t_q$, or worse, on each decision attribute $q \in R$. The downward and upward unions of classes correspond to the *positive* and *negative dominance cones* in X_R, respectively.

In this case, the granules of knowledge are open sets in X_P and X_R defined by dominance cones $D_P^+(x)$, $D_P^-(x)$ ($P \subseteq C$) and $Cl_R^{\geq x}$, $Cl_R^{\leq x}$ ($R \subseteq D$), respectively. Then, classification rules to be induced from data are functions representing granules $Cl_R^{\geq x}$, $Cl_R^{\leq x}$ by granules $D_P^+(x)$, $D_P^-(x)$, respectively, in the condition attribute space X_P, for any $P \subseteq C$ and $R \subseteq D$ and for any $x \in X_P$.

22.3 The Dominance-based Rough Set Approach to Multi-Criteria Classification

22.3.1 Granular Computing with Dominance Cones

When inducing classification rules, a set D of decision attributes is, usually, a singleton, $D = \{d\}$. Let us make this assumption for further presentation, although it is not necessary for DRSA. The decision attribute d makes a partition of U into a finite number of classes, $Cl = \{Cl_t, t = 1, \ldots, n\}$. Each object $x \in U$

belongs to one and only one class, $Cl_t \in Cl$. The upward and downward unions of classes boil down to, respectively,

$$Cl_t^{\geq} = \bigcup_{s \geq t} Cl_s,$$

$$Cl_t^{\leq} = \bigcup_{s \leq t} Cl_s,$$

where $t = 1, \ldots, n$. Notice that for $t = 2, \ldots, n$ we have $Cl_t^\geq = U - Cl_{t-1}^\leq$, i.e., all the objects not belonging to class Cl_t or better, belong to class Cl_{t-1} or worse.

Let us explain how the rough set concept has been generalized in DRSA, so as to enable granular computing with dominance cones.

Given a set of criteria, $P \subseteq C$, the inclusion of an object $x \in U$ to the upward union of classes $Cl_t^\geq$, $t = 2, \ldots, n$, is *inconsistent with the dominance principle* if one of the following conditions holds:

- x belongs to class Cl_t or better but it is P-dominated by an object y belonging to a class worse than Cl_t, i.e., $x \in Cl_t^\geq$ but $D_P^+(x) \cap Cl_{t-1}^\leq \neq \emptyset$;
- x belongs to a worse class than Cl_t but it P-dominates an object y belonging to class Cl_t or better, i.e., $x \notin Cl_t^\geq$ but $D_P^-(x) \cap Cl_t^\geq \neq \emptyset$.

If, given a set of criteria $P \subseteq C$, the inclusion of $x \in U$ to $Cl_t^\geq$, where $t = 2, \ldots, n$, is inconsistent with the dominance principle, we say that x belongs to $Cl_t^\geq$ with *some ambiguity*. Thus, x belongs to $Cl_t^\geq$ *without any ambiguity* with respect to $P \subseteq C$, if $x \in Cl_t^\geq$ and there is no inconsistency with the dominance principle. This means that all objects P-dominating x belong to $Cl_t^\geq$, i.e., $D_P^+(x) \subseteq Cl_t^\geq$. Geometrically, this corresponds to the inclusion of the complete set of objects contained in the positive dominance cone originating in x, in the positive dominance cone $Cl_t^\geq$ originating in Cl_t.

Furthermore, *x possibly belongs to $Cl_t^\geq$* with respect to $P \subseteq C$ if one of the following conditions holds:

- According to decision attribute d, x belongs to $Cl_t^\geq$.
- According to decision attribute d, x does not belong to $Cl_t^\geq$, but it is inconsistent in the sense of the dominance principle with an object y belonging to $Cl_t^\geq$.

In terms of ambiguity, x possibly belongs to $Cl_t^\geq$ with respect to $P \subseteq C$, if x belongs to $Cl_t^\geq$ with or without any ambiguity. Due to the reflexivity of the P-dominance relation D_P, the above conditions can be summarized as follows: x *possibly belongs* to class Cl_t or better, with respect to $P \subseteq C$, if among the objects P-dominated by x there is an object y belonging to class Cl_t or better, i.e.,

$$D_P^-(x) \cap Cl_t^\geq \neq \emptyset .$$

Geometrically, this corresponds to the non-empty intersection of the set of objects contained in the negative dominance cone originating in x, with the positive dominance cone $Cl_t^\geq$ originating in Cl_t.

For $P \subseteq C$, the set of all objects belonging to $Cl_t^\geq$ without any ambiguity constitutes the *P-lower approximation* of $Cl_t^\geq$, denoted by $\underline{P}(Cl_t^\geq)$, and the set of all objects that possibly belong to $Cl_t^\geq$ constitutes the *P-upper approximation* of $Cl_t^\geq$, denoted by $\overline{P}(Cl_t^\geq)$. More formally

$$\underline{P}(Cl_t^\geq) = \{x \in U : D_P^+(x) \subseteq Cl_t^\geq\} ,$$
$$\overline{P}(Cl_t^\geq) = \{x \in U : D_P^-(x) \cap Cl_t^\geq \neq \emptyset\} ,$$

where $t = 1, \ldots, n$. Analogously, one can define the *P-lower approximation* and the *P-upper approximation* of $Cl_t^\leq$

$$\underline{P}(Cl_t^\leq) = \{x \in U : D_P^-(x) \subseteq Cl_t^\leq\} ,$$
$$\overline{P}(Cl_t^\leq) = \{x \in U : D_P^+(x) \cap Cl_t^\leq \neq \emptyset\} ,$$

where $t = 1, \ldots, n$.

The P-lower and P-upper approximations of $Cl_t^\geq$, $t = 1, \ldots, n$, can also be expressed in terms of unions of positive dominance cones as follows

$$\underline{P}(Cl_t^\geq) = \bigcup_{D_P^+(x) \subseteq Cl_t^\geq} D_P^+(x) ,$$
$$\overline{P}(Cl_t^\geq) = \bigcup_{x \in Cl_t^\geq} D_P^+(x) .$$

Analogously, the P-lower and P-upper approximations of $Cl_t^\leq$, $t = 1, \ldots, n$, can be expressed in terms of unions of negative dominance cones as follows

$$\underline{P}(Cl_t^\leq) = \bigcup_{D_P^-(x) \subseteq Cl_t^\leq} D_P^-(x) ,$$
$$\overline{P}(Cl_t^\leq) = \bigcup_{x \in Cl_t^\leq} D_P^-(x) .$$

The P-lower and P-upper approximations so defined satisfy the following *inclusion properties* for each $t \in \{1, \ldots, n\}$ and for all $P \subseteq C$

$$\underline{P}(Cl_t^\geq) \subseteq Cl_t^\geq \subseteq \overline{P}(Cl_t^\geq) ,$$
$$\underline{P}(Cl_t^\leq) \subseteq Cl_t^\leq \subseteq \overline{P}(Cl_t^\leq) .$$

All the objects belonging to $Cl_t^\geq$ and $Cl_t^\leq$ with some ambiguity constitute the *P-boundary* of $Cl_t^\geq$ and $Cl_t^\leq$, denoted by $Bn_P(Cl_t^\geq)$ and $Bn_P(Cl_t^\leq)$, respectively. They can be represented, in terms of upper and lower approximations, as follows

$$Bn_P(Cl_t^\geq) = \overline{P}(Cl_t^\geq) - \underline{P}(Cl_t^\geq),$$
$$Bn_P(Cl_t^\leq) = \overline{P}(Cl_t^\leq) - \underline{P}(Cl_t^\leq),$$

where $t = 1, \ldots, n$. The P-lower and P-upper approximations of the unions of classes $Cl_t^\geq$ and $Cl_t^\leq$ have an important *complementarity property*. It says that if object x belongs without any ambiguity to class Cl_t or better, then it is impossible that it could belong to class Cl_{t-1} or worse, i. e.,

$$\underline{P}(Cl_t^\geq) = U - \overline{P}(Cl_{t-1}^\leq), \quad t = 2, \ldots, n.$$

Due to the complementarity property, $Bn_P(Cl_t^\geq) = Bn_P(Cl_{t-1}^\leq)$, for $t = 2, \ldots, n$, which means that if x belongs with ambiguity to class Cl_t or better, then it also belongs with ambiguity to class Cl_{t-1} or worse.

Considering application of the lower and the upper approximations based on dominance D_P, $P \subseteq C$, to any set $X \subseteq U$, instead of the unions of classes $Cl_t^\geq$ and $Cl_t^\leq$, one obtains upward lower and upper approximations $\underline{P}^\geq(X)$ and $\overline{P}^\geq(X)$, as well as downward lower and upper approximations $\underline{P}^\leq(X)$ and $\overline{P}^\leq(X)$, as follows

$$\underline{P}^\geq(X) = \{x \in U : D_P^+(x) \subseteq X\},$$
$$\overline{P}^\geq(X) = \{x \in U : D_P^-(x) \cap X \neq \emptyset\},$$
$$\underline{P}^\leq(X) = \{x \in U : D_P^-(x) \subseteq X\},$$
$$\overline{P}^\leq(X) = \{x \in U : D_P^+(x) \cap X \neq \emptyset\}.$$

From the definition of rough approximations $\underline{P}^\geq(X), \overline{P}^\geq(X), \underline{P}^\leq(X)$ and $\overline{P}^\leq(X)$, we can also obtain the following properties of the P-lower and P-upper approximations [22.28, 29]:

1. $\underline{P}^\geq(\emptyset) = \overline{P}^\geq(\emptyset) = \underline{P}^\leq(\emptyset) = \overline{P}^\leq(\emptyset) = \emptyset$,
   $\underline{P}^\geq(U) = \overline{P}^\geq(U) = \underline{P}^\leq(U) = \overline{P}^\leq(U) = U$,

2. $\overline{P}^\geq(X \cup Y) = \overline{P}^\geq(X) \cup \overline{P}^\geq(Y)$,
   $\overline{P}^\leq(X \cup Y) = \overline{P}^\leq(X) \cup \overline{P}^\leq(Y)$,

3. $\underline{P}^\geq(X \cap Y) = \underline{P}^\geq(X) \cap \underline{P}^\geq(Y)$,
   $\underline{P}^\leq(X \cap Y) = \underline{P}^\leq(X) \cap \underline{P}^\leq(Y)$,

4. $X \subseteq Y \Rightarrow \overline{P}^\geq(X) \subseteq \overline{P}^\geq(Y)$,
   $X \subseteq Y \Rightarrow \overline{P}^\leq(X) \subseteq \overline{P}^\leq(Y)$,

5. $X \subseteq Y \Rightarrow \underline{P}^\geq(X) \subseteq \underline{P}^\geq(Y)$,
   $X \subseteq Y \Rightarrow \underline{P}^\leq(X) \subseteq \underline{P}^\leq(Y)$,

6. $\underline{P}^\geq(X \cup Y) \supseteq \underline{P}^\geq(X) \cup \underline{P}^\geq(Y)$,
   $\underline{P}^\leq(X \cup Y) \supseteq \underline{P}^\leq(X) \cup \underline{P}^\leq(Y)$,

7. $\overline{P}^\geq(X \cap Y) \subseteq \overline{P}^\geq(X) \cap \overline{P}^\geq(Y)$,
   $\overline{P}^\leq(X \cap Y) \subseteq \overline{P}^\leq(X) \cap \overline{P}^\leq(Y)$,

8. $\underline{P}^\geq(\underline{P}^\geq(X)) = \overline{P}^\geq(\underline{P}^\geq(X)) = \underline{P}^\geq(X)$,
   $\underline{P}^\leq(\underline{P}^\leq(X)) = \overline{P}^\leq(\underline{P}^\leq(X)) = \underline{P}^\leq(X)$,

9. $\overline{P}^\geq(\overline{P}^\geq(X)) = \underline{P}^\geq(\overline{P}^\geq(X)) = \overline{P}^\geq(X)$,
   $\overline{P}^\leq(\overline{P}^\leq(X)) = \underline{P}^\leq(\overline{P}^\leq(X)) = \overline{P}^\leq(X)$.

From the knowledge discovery point of view, P-lower approximations of unions of classes represent *certain knowledge* provided by criteria from $P \subseteq C$, while P-upper approximations represent *possible knowledge* and the P-boundaries contain *doubtful knowledge* provided by the criteria from $P \subseteq C$.

22.3.2 Variable Consistency Dominance-Based Rough Set Approach (VC-DRSA)

The above definitions of rough approximations are based on a strict application of the dominance principle. However, when defining non-ambiguous objects, it is reasonable to accept a limited proportion of negative examples, particularly for large data tables. This relaxed version of DRSA is called the variable consistency dominance-based rough set approach (VC-DRSA) model [22.30].

For any $P \subseteq C$, we say that $x \in U$ belongs to $Cl_t^\geq$ *with no ambiguity at consistency level* $l \in (0, 1]$, if $x \in Cl_t^\geq$ and at least $l * 100\%$ of all objects $y \in U$ dominating x with respect to P also belong to $Cl_t^\geq$, i. e.,

$$\frac{|D_P^+(x) \cap Cl_t^\geq|}{|D_P^+(x)|} \geq l.$$

The term $|D_P^+(x) \cap Cl_t^\geq| / |D_P^+(x)|$ is called *rough membership* and can be interpreted as conditional probability $\Pr(y \in Cl_t^\geq \mid y \in D_P^+(x))$. The level l is called the *consistency level* because it controls the degree of consistency between objects qualified as belonging to $Cl_t^\geq$ without any ambiguity. In other words, if $l < 1$, then at most $(1 - l) * 100\%$ of all objects $y \in U$ dominating x with respect to P do not belong to $Cl_t^\geq$ and thus contradict the inclusion of x in $Cl_t^\geq$.

Analogously, for any $P \subseteq C$ we say that $x \in U$ belongs to $Cl_t^\leq$ *with no ambiguity at consistency level* $l \in (0, 1]$, if $x \in Cl_t^\leq$ and at least $l * 100\%$ of all the objects $y \in U$ dominated by x with respect to P also belong to $Cl_t^{\leq}$, i.e.,

$$\frac{|D_P^-(x) \cap Cl_t^{\leq}|}{|D_P^-(x)|} \geq l \,.$$

The rough membership $|D_P^-(x) \cap Cl_t^{\leq}|/|D_P^-(x)|$ can be interpreted as conditional probability $\Pr(y \in Cl_t^{\leq} \mid y \in D_P^-(x))$. Thus, for any $P \subseteq C$, each object $x \in U$ is either ambiguous or non-ambiguous at consistency level l with respect to the upward union $Cl_t^{\geq}$ ($t = 2, \ldots, n$) or with respect to the downward union $Cl_t^{\leq}$ ($t = 1, \ldots, n-1$).

The concept of non-ambiguous objects at some consistency level l naturally leads to the definition of P-lower approximations of the unions of classes $Cl_t^{\geq}$ and $Cl_t^{\leq}$, which can be formally presented as follows

$$\underline{P}^l(Cl_t^{\geq}) = \left\{ x \in Cl_t^{\geq} : \frac{|D_P^+(x) \cap Cl_t^{\geq}|}{|D_P^+(x)|} \geq l \right\} \,,$$

$$\underline{P}^l(Cl_t^{\leq}) = \left\{ x \in Cl_t^{\leq} : \frac{|D_P^-(x) \cap Cl_t^{\leq}|}{|D_P^-(x)|} \geq l \right\} \,.$$

Given $P \subseteq C$ and consistency level l, we can define the P-upper approximations of $Cl_t^{\geq}$ and $Cl_t^{\leq}$, denoted by $\overline{P}^l(Cl_t^{\geq})$ and $\overline{P}^l(Cl_t^{\leq})$, respectively, by complementation of $\underline{P}^l(Cl_{t-1}^{\leq})$ and $\underline{P}^l(Cl_{t+1}^{\geq})$ with respect to U as follows

$$\overline{P}^l(Cl_t^{\geq}) = U - \underline{P}^l(Cl_{t-1}^{\leq}) \,, t = 2, \ldots, n,$$

$$\overline{P}^l(Cl_t^{\leq}) = U - \underline{P}^l(Cl_{t+1}^{\geq}) \,, t = 1, \ldots, n-1.$$

$\overline{P}^l(Cl_t^{\geq})$ can be interpreted as the set of all the objects belonging to $Cl_t^{\geq}$, which are *possibly ambiguous* at consistency level l. Analogously, $\overline{P}^l(Cl_t^{\leq})$ can be interpreted as the set of all the objects belonging to $Cl_t^{\leq}$, which are *possibly ambiguous* at consistency level l. The P-boundaries (P-doubtful regions) of $Cl_t^{\geq}$ and $Cl_t^{\leq}$ are defined as

$$Bn_P(Cl_t^{\geq}) = \overline{P}^l(Cl_t^{\geq}) - \underline{P}^l(Cl_t^{\geq}) \,,$$

$$Bn_P(Cl_t^{\leq}) = \overline{P}^l(Cl_t^{\leq}) - \underline{P}^l(Cl_t^{\leq}) \,,$$

where $t = 1, \ldots, n$. The VC-DRSA model provides some degree of flexibility in assigning objects to lower and upper approximations of the unions of decision classes. It can easily be demonstrated that for $0 < l' < l \leq 1$ and $t = 2, \ldots, n$,

$$\underline{P}^l(Cl_t^{\geq}) \subseteq \underline{P}^{l'}(Cl_t^{\geq}) \quad \text{and} \quad \overline{P}^{l'}(Cl_t^{\geq}) \subseteq \overline{P}^l(Cl_t^{\geq}) \,.$$

The VC-DRSA model was inspired by Ziarko's model of the *variable precision* rough set approach [22.31]. However, there is a significant difference in the definition of rough approximations because $\underline{P}^l(Cl_t^{\geq})$ and $\overline{P}^l(Cl_t^{\geq})$ are composed of non-ambiguous and ambiguous objects at the consistency level l, respectively, while Ziarko's $\underline{P}^l(Cl_t)$ and $\overline{P}^l(Cl_t)$ are composed of P-indiscernibility sets such that at least $l *$ 100% of these sets are included in Cl_t or have a non-empty intersection with Cl_t, respectively. If one would like to use Ziarko's definition of variable precision rough approximations in the context of multiple-criteria classification, then the P-indiscernibility sets should be substituted by P-dominating sets $D_P^+(x)$. However, then the notion of ambiguity that naturally leads to the general definition of rough approximations [22.21] loses its meaning. Moreover, a bad side effect of the direct use of Ziarko's definition is that a lower approximation $\underline{P}^l(Cl_t^{\geq})$ may include objects y assigned to Cl_h, where h is much less than t, if y belongs to $D_P^+(x)$, which was included in $\underline{P}^l(Cl_t^{\geq})$. When the decision classes are preference ordered, it is reasonable to expect that objects assigned to far worse classes than the considered union are not counted to the lower approximation of this union.

The VC-DRSA model presented above has been generalized in [22.32, 33]. The generalized model applies two types of consistency measures in the definition of lower approximations:

- Gain-type consistency measures $f_{\geq t}^P(x), f_{\leq t}^P(x)$

$$\underline{P}^{\alpha_{\geq t}}(Cl_t^{\geq}) = \{ x \in Cl_t^{\geq} : f_{\geq t}^P(x) \geq \alpha_{\geq t} \} \,,$$

$$\underline{P}^{\alpha_{\leq t}}(Cl_t^{\leq}) = \{ x \in Cl_t^{\leq} : f_{\leq t}^P(x) \geq \alpha_{\leq t} \} \,,$$

- Cost-type consistency measures $g_{\geq t}^P(x), g_{\leq t}^P(x)$

$$\underline{P}^{\beta_{\geq t}}(Cl_t^{\geq}) = \{ x \in Cl_t^{\geq} : g_{\geq t}^P(x) \geq \beta_{\geq t} \} \,,$$

$$\underline{P}^{\beta_{\leq t}}(Cl_t^{\leq}) = \{ x \in Cl_t^{\leq} : g_{\leq t}^P(x) \geq \beta_{\leq t} \} \,,$$

where $\alpha_{\geq t}$, $\alpha_{\leq t}$, $\beta_{\geq t}$, $\beta_{\leq t}$, are threshold values on the consistency measures that condition the inclusion of object x in the P-lower approximation of $Cl_t^{\geq}$, or

$Cl_t^{\leq}$. Here are the consistency measures considered in [22.33]: for all $x \in U$ and $P \subseteq C$

$$\mu_{\geq t}^{P}(x) = \frac{|D_P^+(x) \cap Cl_t^{\geq}|}{|D_P^+(x)|},$$

$$\mu_{\leq t}^{P}(x) = \frac{|D_P^-(x) \cap Cl_t^{\leq}|}{|D_P^-(x)|},$$

$$\overline{\mu}_{\geq t}^{P}(x) = \max_{\substack{R \subseteq P, \\ z \in D_R^-(x) \cap Cl_t^{\geq}}} \frac{|D_R^+(z) \cap Cl_t^{\geq}|}{|D_R^+(z)|},$$

$$\overline{\mu}_{\leq t}^{P}(x) = \max_{\substack{R \subseteq P, \\ z \in D_R^+(x) \cap Cl_t^{\leq}}} \frac{|D_R^-(z) \cap Cl_t^{\leq}|}{|D_R^-(z)|},$$

$$B_{\geq t}^{P}(x) = \frac{|D_P^+(x) \cap Cl_t^{\geq}| \, |Cl_{t-1}^{\leq}|}{|D_P^+(x) \cap Cl_{t-1}^{\leq}| \, |Cl_t^{\geq}|}, \quad t = 2, \dots, m,$$

$$B_{\leq t}^{P}(x) = \frac{|D_P^-(x) \cap Cl_t^{\leq}| \, |Cl_{t+1}^{\geq}|}{|D_P^-(x) \cap Cl_{t+1}^{\geq}| \, |Cl_t^{\leq}|},$$

$$t = 1, \dots, m-1,$$

$$\varepsilon_{\geq t}^{P}(x) = \frac{|D_P^+(x) \cap Cl_{t-1}^{\leq}|}{|Cl_{t-1}^{\leq}|}, \quad t = 2, \dots, m,$$

$$\varepsilon_{\leq t}^{P}(x) = \frac{|D_P^-(x) \cap Cl_{t+1}^{\geq}|}{|Cl_{t+1}^{\geq}|}, \quad t = 1, \dots, m-1,$$

$$\varepsilon_{\geq t}^{'P}(x) = \frac{|D_P^+(x) \cap Cl_{t-1}^{\leq}|}{|Cl_t^{\geq}|}, \quad t = 2, \dots, m,$$

$$\varepsilon_{\leq t}^{'P}(x) = \frac{|D_P^-(x) \cap Cl_{t+1}^{\geq}|}{|Cl_t^{\leq}|}, \quad t = 1, \dots, m-1,$$

$$\varepsilon_{\geq t}^{*P}(x) = \max_{r \leq t} \varepsilon_{\geq r}^{P}(x),$$

$$\varepsilon_{\leq t}^{*P}(x) = \max_{r \geq t} \varepsilon_{\leq r}^{P}(x),$$

with

$$\mu_{\geq t}^{P}(x), \ \mu_{\leq t}^{P}(x), \ \overline{\mu}_{\geq t}^{P}(x),$$
$$\overline{\mu}_{\leq t}^{P}(x), \ B_{\geq t}^{P}(x), \ B_{\leq t}^{P}(x)$$

being gain-type consistency measures and

$$\varepsilon_{\geq t}^{P}(x), \ \varepsilon_{\leq t}^{P}(x), \ \varepsilon_{\geq t}^{'P}(x), \ \varepsilon_{\leq t}^{'P}(x), \ \varepsilon_{\geq t}^{*P}(x), \ \varepsilon_{\leq t}^{*P}(x)$$

being cost-type consistency measures.

To be concordant with the rough set philosophy, consistency measures should enjoy some monotonicity properties (Table 22.1). A consistency measure is monotonic if it does not decrease (or does not increase) when:

Table 22.1 Monotonicity properties of consistency measures (after [22.33])

Consistency measure	(m1)	(m2)	(m3)	(m4)
$\mu_{\geq t}^{P}(x), \mu_{\leq t}^{P}(x)$ (rough membership)	no	yes	yes	no
$\overline{\mu}_{\geq t}^{P}(x), \overline{\mu}_{\leq t}^{P}(x)$	yes	yes	yes	yes
$B_{\geq t}^{P}(x), B_{\leq t}^{P}(x)$ (Bayesian)	no	no	no	no
$\varepsilon_{\geq t}^{P}(x), \varepsilon_{\leq t}^{P}(x)$	yes	yes	no	yes
$\varepsilon_{\geq t}^{*P}(x), \varepsilon_{\leq t}^{*P}(x)$	yes	yes	yes	yes
$\varepsilon_{\geq t}^{'P}(x), \varepsilon_{\leq t}^{'P}(x)$	yes	yes	yes	yes

(m1) The set of attributes is growing.
(m2) The set of objects is growing.
(m3) The union of ordered classes is growing.
(m4) x improves its evaluation, so that it dominates more objects.

For every $P \subseteq C$, the objects being consistent in the sense of the dominance principle with all upward and downward unions of classes are called *P-correctly classified*. For every $P \subseteq C$, the *quality of approximation of classification Cl* by the set of criteria P is defined as the ratio between the number of *P*-correctly classified objects and the number of all the objects in the decision table. Since the objects which are *P*-correctly classified are those that do not belong to any *P*-boundary of unions $Cl_t^{\geq}$ and $Cl_t^{\leq}$, $t = 1, \dots, n$, the quality of approximation of classification Cl by the set of criteria P, can be written as

$$\gamma_P(Cl) = \frac{\left| \left(U - \left(\bigcup_{t \in \{1, \dots, n\}} Bn_P(Cl_t^{\geq}) \right) \right. \right.}{|U|}$$
$$\cup \frac{\left. \left(\bigcup_{t \in \{1, \dots, n\}} Bn_P(Cl_t^{\leq}) \right) \right) \right|}{|U|}$$
$$= \frac{\left| \left(U - \left(\bigcup_{t \in \{1, \dots, n\}} Bn_P(Cl_t^{\geq}) \right) \right) \right|}{|U|}.$$

$\gamma_P(Cl)$ can be seen as a measure of the quality of knowledge that can be extracted from the decision table, where P is the set of criteria and Cl is the classification considered.

Each minimal subset $P \subseteq C$, such that $\gamma_P(Cl) = \gamma_C(Cl)$, is called a *reduct* of Cl and is denoted by RED_{Cl}. Note that a decision table can have more than one reduct. The intersection of all reducts is called the *core* and is denoted by $CORE_{Cl}$. Criteria from $CORE_{Cl}$ cannot be removed from the decision table without deteriorating the knowledge to be discovered. This means that in set C there are three categories of criteria:

- *Indispensable* criteria included in the core.
- *Exchangeable* criteria included in some reducts but not in the core.
- *Redundant* criteria that are neither indispensable nor exchangeable, thus not included in any reduct.

Note that reducts are minimal subsets of criteria conveying the relevant knowledge contained in the decision table. This knowledge is relevant for the explanation of patterns in a given decision table but not necessarily for prediction.

It has been shown in [22.34] that the quality of classification satisfies properties of set functions called *fuzzy measures*. For this reason, we can use the quality of classification for the calculation of indices that measure the relevance of particular attributes and/or criteria, in addition to the strength of interactions between them. The useful indices are: the value index and interaction indices of Shapley and Banzhaf; the interaction indices of Murofushi-Soneda and Roubens; and the Möbius representation [22.15]. All these indices can help to assess the interaction between the criteria considered and can help us to choose the best reduct.

22.3.3 Stochastic Dominance–based Rough Set Approach

From a probabilistic point of view, the assignment of object x_i to *at least* class t can be made with probability $\Pr(y_i \geq t \mid x_i)$, where y_i is the classification decision for x_i, $t = 1, \ldots, n$. This probability is supposed to satisfy the usual axioms of probability

$$\Pr(y_i \geq 1 \mid x_i) = 1 \,,$$
$$\Pr(y_i \leq t \mid x_i) = 1 - \Pr(y_i \geq t+1 \mid x_i) \,, \text{ and}$$
$$\Pr(y_i \geq t \mid x_i) \leq \Pr(y_i \geq t' \mid x_i) \quad \text{for } t \geq t' \,.$$

These probabilities are unknown but can be estimated from data.

For each class $t = 2, \ldots, n$, we have a binary problem of estimating the conditional probabilities $\Pr(y_i \geq t \mid x_i) = 1$, $\Pr(y_i < t \mid x_i)$. It can be solved by *isotonic regression* [22.35]. Let $y_{it} = 1$ if $y_i \geq t$, otherwise $y_{it} = 0$. Let also p_{it} be the estimate of the probability $\Pr(y_i \geq t \mid x_i)$. Then, choose estimates p_{it}^* which minimize the squared distance to the class assignment y_{it}, subject to the monotonicity constraints

$$\text{Minimize } \sum_{i=1}^{|U|} (y_{it} - p_{it})^2$$
$$\text{subject to } p_{it} \geq p_{jt} \text{ for all } x_i, x_j \in U \text{ such that } x_i \succeq x_j \,,$$

where $x_i \succeq x_j$ means that x_i dominates x_j.

Then, stochastic α-lower approximations for classes *at least* t and *at most* $t-1$ can be defined as

$$\underline{P}^\alpha(Cl_t^{\geq}) = \{x_i \in U : \Pr(y_i \geq t \mid x_i) \geq \alpha\} \,,$$
$$\underline{P}^\alpha(Cl_{t-1}^{\leq}) = \{x_i \in U : \Pr(y_i < t \mid x_i) \geq \alpha\} \,.$$

Replacing the unknown probabilities $\Pr(y_i \geq t \mid x_i)$ and $\Pr(y_i < t \mid x_i)$ by their estimates p_{it}^* and $1 - p_{it}^*$ obtained from isotonic regression, we obtain

$$\underline{P}^\alpha(Cl_t^{\geq}) = \{x_i \in U : p_{it}^* \geq \alpha\} \,,$$
$$\underline{P}^\alpha(Cl_{t-1}^{\leq}) = \{x_i \in U : p_{it}^* \leq 1 - \alpha\} \,,$$

where parameter $\alpha \in [0.5, 1]$ controls the allowed amount of inconsistency.

Solving isotonic regression requires $O(|U|^4)$ time, but a good heuristic needs only $O(|U|^2)$. In fact, as shown in [22.35], we do not really need to know the probability estimates to obtain stochastic lower approximations. We only need to know for which object x_i, $p_{it}^* \geq \alpha$ and for which x_i, $p_{it}^* \leq 1 - \alpha$. This can be found by solving a linear programming (reassignment) problem.

As before, $y_{it} = 1$ if $y_i \geq t$, otherwise $y_{it} = 0$. Let d_{it} be the decision variable which determines a new class assignment for object x_i. Then, reassign objects from union of classes indicated by y_{it} to the union of classes indicated by d_{it}^*, such that the new class assignments are consistent with the dominance principle, where d_{it}^* results from solving the following linear programming problem

$$\text{Minimize } \sum_{i=1}^{|U|} w_{y_{it}} |y_{it} - d_{it}|$$
$$\text{subject to } d_{it} \geq d_{jt} \quad \text{for all } x_i, x_j \in U$$
$$\text{such that } x_i \succeq x_j \,,$$

where $w_{y_{it}}$ are some positive weights and $x_i \succeq x_j$ means that x_i dominates x_j.

Due to unimodularity of the constraint matrix, the optimal solution of this linear programming problem is always integer, i.e., $d_{it}^* \in \{0, 1\}$. For all objects consistent with the dominance principle, $d_{it}^* = y_{it}$. If we set $w_0 = \alpha$ and $w_1 = \alpha - 1$, then the optimal solution d_{it}^* satisfies: $d_{it}^* = 1 \Leftrightarrow p_{it}^* \geq \alpha$. If we set $w_0 = 1 - \alpha$ and $w_1 = \alpha$, then the optimal solution d_{it}^* satisfies: $d_{it}^* = 0 \Leftrightarrow p_{it}^* \leq 1 - \alpha$.

For each $t = 2, \ldots, n$, solving the reassignment problem twice, we can obtain the lower approximations $\underline{P}^\alpha(Cl_t^{\geq})$, $\underline{P}^\alpha(Cl_{t-1}^{\leq})$, without knowing the probability estimates!

22.3.4 Induction of Decision Rules

Using the terms of knowledge discovery, the dominance-based rough approximations of upward and downward unions of classes are applied on the data set in the pre-processing stage. In result of this stage, the data are structured in a way that facilitates induction of *if ..., then ...* decision rules with a guaranteed consistency level. For a given upward or downward union of classes, $Cl_t^{\geq}$ or $Cl_s^{\leq}$, the decision rules induced under the hypothesis that objects belonging to $\underline{P}(Cl_t^{\geq})$ or $\underline{P}(Cl_s^{\leq})$ are positive and all the others are negative, suggests an assignment to *class Cl_t or better*, or to *class Cl_s or worse*, respectively. On the other hand, the decision rules induced under a hypothesis that objects belonging to the intersection $\overline{P}(Cl_s^{\leq}) \cap \overline{P}(Cl_t^{\geq})$ are positive and all the others are negative, suggest an assignment to some classes between Cl_s and $Cl_t (s < t)$.

In the case of preference ordered data it is meaningful to consider the following five types of decision rules:

1. *Certain $D_{\geq}$-decision rules.* These provide lower profile descriptions for objects belonging to $Cl_t^{\geq}$ without ambiguity:
   if $x_{q1} \succeq_{q1} r_{q1}$ *and* $x_{q2} \succeq_{q2} r_{q2}$ *and* ... $x_{qp} \succeq_{qp} r_{qp}$, *then* $x \in Cl_t^{\geq}$, where for each $w_q, z_q \in X_q$, $w_q \succeq_q z_q$ means w_q *is at least as good as* z_q.
2. *Possible $D_{\geq}$-decision rules.* Such rules provide lower profile descriptions for objects belonging to $Cl_t^{\geq}$ with or without any ambiguity:
   if $x_{q1} \succeq_{q1} r_{q1}$ *and* $x_{q2} \succeq_{q2} r_{q2}$ *and* ... $x_{qp} \succeq_{qp} r_{qp}$, *then* x possibly belongs to $Cl_t^{\geq}$.
3. *Certain $D_{\leq}$-decision rules.* These give upper profile descriptions for objects belonging to $Cl_t^{\leq}$ without ambiguity:
   if $x_{q1} \preceq_{q1} r_{q1}$ *and* $x_{q2} \preceq_{q2} r_{q2}$ *and* ... $x_{qp} \preceq_{qp} r_{qp}$, *then* $x \in Cl_t^{\leq}$, where for each $w_q, z_q \in X_q$, $w_q \preceq_q z_q$ means w_q *is at most as good as* z_q.
4. *Possible $D_{\leq}$-decision rules.* These provide upper profile descriptions for objects belonging to $Cl_t^{\leq}$ with or without any ambiguity:
   if $x_{q1} \preceq_{q1} r_{q1}$ *and* $x_{q2} \preceq_{q2} r_{q2}$ *and* ... $x_{qp} \preceq_{qp} r_{qp}$, *then* x possibly belongs to $Cl_t^{\leq}$.
5. *Approximate $D_{\geq \leq}$-decision rules.* These represent simultaneously lower and upper profile descriptions for objects belonging to $Cl_s \cup Cl_{s+1} \cup \cdots \cup Cl_t$ without the possibility of discerning the actual class:
   if $x_{q1} \succeq_{q1} r_{q1}$ *and* ... $x_{qk} \succeq_{qk} r_{qk}$ *and* $x_{qk+1} \preceq_{qk+1} r_{qk+1}$ *and* ... $x_{qp} \preceq_{qp} r_{qp}$, *then* $x \in Cl_s \cup Cl_{s+1} \cup \cdots \cup Cl_t$.

In the left-hand side of a $D_{\geq \leq}$-decision rule we can have $x_q \succeq_q r_q$ and $x_q \preceq_q r'_q$, where $r_q \leq r'_q$, for the same $q \in C$. Moreover, if $r_q = r'_q$, the two conditions boil down to $x_q \sim_q r_q$, where for each $w_q, z_q \in X_q$, $w_q \sim_q z_q$ means w_q *is indifferent to* z_q.

A rule is *minimal* if there is no other rule with a left-hand side that has at least the same weakness (which means that it uses a subset of elementary conditions and/or weaker elementary conditions) and which has a right-hand side that has at least the same strength (which means a $D_{\geq}$- or a $D_{\leq}$-decision rule assigning objects to the same union or sub-union of classes, or a $D_{\geq \leq}$-decision rule assigning objects to the same or larger set of classes).

Rules of type 1) and 3) represent certain knowledge extracted from the decision table, while the rules of type 2) and 4) represent possible knowledge. Rules of type 5) represent doubtful knowledge.

Rules of type 1) and 3) are *exact* if they do not cover negative examples; they are *probabilistic*, otherwise. In the latter case, each rule is characterized by a confidence ratio, representing the probability that an object matching the left-hand side of the rule also matches its right-hand side.

A set of decision rules is *complete* if it is able to cover all objects from the decision table in such a way that consistent objects are re-classified to their original classes and inconsistent objects are classified to clusters of classes that refer to this inconsistency. Each set of decision rules that is complete and non-redundant is called *minimal*. Note that an exclusion of any rule from this set makes it non-complete.

In the case of VC-DRSA, the decision rules are probabilistic because they are induced from the P-lower approximations whose composition is controlled by the user-specified consistency level l. Consequently, the value of confidence α for the rule should be constrained from the bottom. It is reasonable to require that the smallest accepted confidence level of the rule should not be lower than the currently used consistency level l. Indeed, in the worst case, some objects from the P-lower approximation may create a rule using all the criteria from P, thus giving a confidence $\alpha \geq l$.

Observe that the syntax of decision rules induced from dominance-based rough approximations uses the concept of dominance cones: each condition profile is a dominance cone in X_C, and each decision profile is a dominance cone in X_D. In both cases the cone is positive for $D_{\geq}$-rules and negative for $D_{\leq}$-rules.

Also note that dominance cones that correspond to condition profiles can originate in any point of X_C, without the risk of being too specific. Thus, in contrast to granular computing based on the indiscernibility (or similarity) relation, in the case of granular computing based on dominance, the condition attribute space X_C need not be discretized [22.28, 36, 37].

Procedures for induction of rules from dominance-based rough approximations have been proposed in [22.38, 39]. A publicly available computer implementation of one of these procedures is called JMAF (java multi-criteria and multi-attribute analysis framework) [22.40, 41].

The utility of decision rules is threefold: they *explain* (summarize) decisions made on objects from the dataset, they can be used to *make decisions* with respect to new (unseen) objects which are matching conditions of some rules, and they permit to *build up a strategy of intervention* [22.42]. The attractiveness of particular decision rules can be measured in many different ways; however, the most convincing measures are Bayesian confirmation measures enjoying a special monotonicity property, as reported in [22.43, 44].

In [22.45], a new methodology for the induction of monotonic decision trees from dominance-based rough approximations of preference ordered decision classes was proposed.

It is finally worth noting that several algebraic models have been proposed for DRSA [22.29, 46, 47] – the algebraic structures are based on bipolar disjoint representation (positive and negative) of the interior and the exterior of a concept. These algebra models give elegant representations of the basic properties of dominance-based rough sets. Moreover, a topology for DRSA in a bitopological space was proposed in [22.48].

22.3.5 Rule-based Classification Algorithms

We will now comment upon the application of decision rules to some objects described by criteria from C. When applying $D_\geq$-decision rules to an object x, it is possible that x either matches the left hand side of at least one decision rule or it does not. In the case of at least one such match, it is reasonable to conclude that x belongs to class Cl_t, because it is the lowest class of the upward union $Cl_t^\geq$ which results from intersection of all the right hand sides of the rules covering x. More precisely, if x matches the left-hand side of rules $\rho_1, \rho_2, \ldots, \rho_m$, having right-hand sides $x \in C_{t1}^\geq$, $x \in Cl_{t2}^\geq, \ldots, x \in Cl_{tm}^\geq$, then x is assigned to class Cl_t, where $t = \max\{t1, t2, \ldots, tm\}$. In the case of no matching, we can conclude that x belongs to Cl_1, i.e., to the worst class, since no rule with a right-hand side suggesting a better classification of x covers this object.

Analogously, when applying $D_\leq$-decision rules to the object x, we can conclude that x belongs either to class Cl_z (because it is the highest class of the downward union $Cl_t^\leq$ resulting from the intersection of all the right-hand sides of the rules covering x), or to class Cl_n, i.e., to the best class, when x is not covered by any rule. More precisely, if x matches the left-hand side of rules $\rho_1, \rho_2, \ldots, \rho_m$, having right-hand sides $x \in C_{t1}^\leq$, $x \in Cl_{t2}^\leq, \ldots, x \in Cl_{tm}^\leq$, then x is assigned to class Cl_t, where $t = \min\{t1, t2, \ldots, tm\}$. In the case of no matching, it is concluded that x belongs to the best class Cl_n because no rule with a right-hand side suggesting a worse classification of x covers this object. Finally, when applying $D_{\geq\leq}$-decision rules to x, it is possible to conclude that x belongs to the union of all the classes suggested in the right-hand side of the rules covering x.

A new classification algorithm was proposed in [22.49]. Let $\varphi_1 \to \psi_1, \ldots, \varphi_k \to \psi_k$, be the rules matching object x. Then, $R_t(x) = \{j: Cl_t \in \psi_j, j = 1, \ldots, k\}$ denotes the set of rules matching x, which recommend assignment of object x to a union including class Cl_t, and $R_{\neg t}(x) = \{j: Cl_t \notin \psi_j, j = 1, \ldots, k\}$ denotes the set of rules matching x, which do not recommend assignment of object x to a union including class Cl_t. $\|\varphi_j\|$, $\|\psi_j\|$ are sets of objects with property φ_j and ψ_j, respectively, $j = 1, \ldots, k$. For a classified object x, one has to calculate the score for each candidate class

$$score(Cl_t, x) = score^+(Cl_t, x) - score^-(Cl_t, x),$$

where

$$score^+(Cl_t, x) = \frac{\left|\bigcup_{j\in R_t(x)}(\|\varphi_j\| \cap Cl_t)\right|^2}{\left|\bigcup_{j\in R_t(x)}\|\varphi_j\|\right| \times |Cl_t|}$$

and

$$score^-(Cl_t, x)$$
$$= \frac{\left|\bigcup_{j\in R_{\neg t}(x)}(\|\varphi_j\| \cap \|\psi_j\|)\right|^2}{\left|\bigcup_{j\in R_{\neg t}(x)}\|\varphi_j\|\right| \times \left|\bigcup_{j\in R_{\neg t}(x)}\|\psi_j\|\right|}.$$

$score^+(Cl_t, x)$ and $score^-(Cl_t, x)$ can be interpreted in terms of conditional probability as a product of confidence and coverage of the matching rules

$$score^+(Cl_t, x) = \Pr(\{\varphi_j : j \in R_t(x)\} | Cl_t)$$
$$\times \Pr(Cl_t | \{\varphi_j : j \in R_t(x)\})\,,$$
$$score^-(Cl_t, x) = \Pr(\{\varphi_j : j \in R_{\neg t}(x)\} | \neg Cl_t)$$
$$\times \Pr(\neg Cl_t | \{\varphi_j : j \in R_{\neg t}(x)\})\,.$$

The recommendation of the univocal classification $x \to Cl_t$ is such that

$$Cl_t = \arg \max_{t \in \{1,\dots,n\}} [score(Cl_t, x)]\,.$$

Examples illustrating the application of DRSA to multi-criteria classification in a didactic way can be found in [22.12–14, 50].

22.4 The Dominance-based Rough Set Approach to Multi-Criteria Choice and Ranking

22.4.1 Differences with Respect to Multi-Criteria Classification

One of the very first extensions of DRSA concerned preference ordered data representing pairwise comparisons (i. e., binary relations) between objects on both, condition and decision attributes [22.7, 8, 25, 51]. Note that while classification is based on the absolute evaluation of objects, choice and ranking refer to pairwise comparisons of objects. In this case, the decision rules to be discovered from the data characterize a comprehensive binary relation on the set of objects. If this relation is a preference relation and if, among the condition attributes, there are some criteria which are semantically correlated with the comprehensive preference relation, then the data set (serving as the learning sample) can be considered as preference information provided by a DM in a multi-criteria choice or ranking problem. In consequence, the comprehensive preference relation characterized by the decision rules discovered from this data set can be considered as a *preference model* of the DM. It may be used to explain the decision policy of the DM and to recommend an optimal choice or preference ranking with respect to new objects.

Let us consider a finite set A of objects evaluated by a finite set C of criteria. The optimal choice (or the preference ranking) in set A is semantically correlated with the criteria from set C. The preference information concerning the multi-criteria choice or ranking problem is a data set in the form of a pairwise comparison table which includes pairs of some *reference objects* from a subset $B \subseteq A \times A$. This is described by preference relations on particular criteria and a comprehensive preference relation. One such example is a weak preference relation called the *outranking relation*. By using DRSA for the analysis of the pairwise

comparison table, we can obtain a rough approximation of the *outranking relation* by a dominance relation. The decision rules induced from the rough approximation are then applied to the complete set A of the objects associated with the choice or ranking. As a result, one obtains a four-valued outranking relation on this set. In order to obtain a recommendation, it is advisable to use an exploitation procedure based on the net flow score of the objects. We present this methodology in more detail below.

22.4.2 The Pairwise Comparison Table as Input Preference Information

Given a multi-criteria choice or ranking problem, a DM can express the preferences by pairwise comparisons of the reference objects. In the following, xSy denotes the presence, while $xS^c y$ denotes the absence of the outranking relation for a pair of objects $(x, y) \in A \times A$. Relation xSy reads *object x is at least as good as object y*.

For each pair of reference objects $(x, y) \in B \subseteq A \times A$, the DM can select one of the three following possibilities:

1. Object x is as good as y, i. e., xSy.
2. Object x is worse than y, i. e., $xS^c y$.
3. The two objects are incomparable at the present stage.

A pairwise comparison table, denoted by S_{PCT}, is then created on the basis of this information. The first m columns correspond to the criteria from set C. The last, i. e., the $(m + 1)$-th column, represents the comprehensive binary preference relation S or S^c. The rows correspond to the pairs from B. For each pair in S_{PCT}, a difference between criterion values is put in the corresponding column. If the DM judges that two objects are incomparable, then the corresponding pair does not appear in S_{PCT}.

We will define S_{PCT} more formally. For any criterion $g_i \in C$, let T_i be a finite set of binary relations defined on A on the basis of the evaluations of objects from A with respect to the considered criterion g_i, such that for every $(x, y) \in A \times A$ exactly one binary relation $t \in T_i$ is verified. More precisely, given the domain V_i of $g_i \in C$, if $v'_i, v''_i \in V_i$ are the respective evaluations of $x, y \in A$ by means of g_i and $(x, y) \in t$, with $t \in T_i$, then for each $w, z \in A$ having the same evaluations v'_i, v''_i by means of g_i, $(w, z) \in t$. Furthermore, let T_d be a set of binary relations defined on set A (comprehensive pairwise comparisons) such that at most one binary relation $t \in T_d$ is verified for every $(x, y) \in A \times A$.

The *pairwise comparison table* is defined as the data table $S_{PCT} = \langle B, C \cup \{d\}, T_G \cup T_d, f \rangle$, where $B \subseteq A \times A$ is a non-empty *set of exemplary pairwise comparisons of reference objects*, $T_G = \bigcup_{g_i \in G} T_i$, d is a decision corresponding to the comprehensive pairwise comparison (comprehensive preference relation), and $f : B \times (C \cup \{d\}) \to T_G \cup T_d$ is a total function such that $f[(x, y), q] \in T_i$ for every $(x, y) \in A \times A$ and for each $g_i \in C$, and $f[(x, y), q] \in T_d$ for every $(x, y) \in B$. It follows that for any pair of reference objects $(x, y) \in B$ there is verified one and only one binary relation $t \in T_d$. Thus, T_d induces a partition of B. In fact, the data table S_{PCT} can be seen as a decision table, since the set C of considered criteria and the decision d are distinguished.

We consider a pairwise comparison table where the set T_d is composed of two binary relations defined on A:

- x outranks y (denoted by xSy or $(x, y) \in S$), where $(x, y) \in B$,
- x does not outrank y (denoted by $xS^c y$ or $(x, y) \in S^c$), where $(x, y) \in B$, and $S \cup S^c = B$.

Observe that the binary relation S is reflexive, but not necessarily transitive or complete.

22.4.3 Rough Approximation of Preference Relations

In the following, we will distinguish between two types of evaluation scales of criteria: *cardinal* and *ordinal*. Let C^N be the set of criteria expressing preferences on a cardinal scale, and let C^O be the set of criteria expressing preferences on an ordinal scale, such that $C^N \cup C^O = C$ and $C^N \cap C^O = \emptyset$. Moreover, for each $P \subseteq C$, we denote by P^O the subset of P composed of criteria expressing preferences on an ordinal scale, i. e., $P^O = P \cap C^O$, and by P^N we denote the subset of P composed of criteria expressing preferences on a cardinal scale, i. e., $P^N = P \cap C^N$. Of course, for each $P \subseteq C$, we have $P = P^N \cup P^O$ and $P^N \cap P^O = \emptyset$.

The meaning of the two scales is such that in the case of the cardinal scale we can specify the intensity of the preference for a given difference of evaluations, while in the case of the ordinal scale, this is not possible and we can only establish an order of evaluations.

Multi-Graded Dominance

We assume that the pairwise comparisons of reference objects on cardinal criteria from set C^N can be represented in terms of *graded preference relations* (for example, *very weak preference, weak preference, strict preference, strong preference,* and *very strong preference*), denoted by P_q^h: for each $q \in C^N$ and for every $(x, y) \in A \times A$, $T_i = \{P_i^h, h \in H_i\}$, where H_i is a particular subset of the relative integers and:

- $xP_i^h y$, $h > 0$, means that object x is preferred to object y by degree h with respect to criterion g_i.
- $xP_i^h y$, $h < 0$, means that object x is not preferred to object y by degree h with respect to criterion g_i.
- $xP_i^0 y$ means that object x is similar (asymmetrically indifferent) to object y with respect to criterion g_i.

Within the preference context, the similarity relation P_i^0, even if not symmetric, resembles the indifference relation. Thus, in this case, we call this similarity relation *asymmetric indifference*. Of course, for each $g_i \in C$ and for every $(x, y) \in A \times A$,

$$[xP_i^h y, h > 0] \Rightarrow [yP_i^k x, k \leq 0],$$
$$[xP_i^h y, h < 0] \Rightarrow [yP_i^k x, k \geq 0].$$

Let $P = P^N$ and $P^O = \emptyset$. Given $P \subseteq C$ ($P \neq \emptyset$), $(x, y), (w, z) \in A \times A$, the pair of objects (x, y) is said to dominate (w, z) with respect to criteria from P (denoted by $(x, y)D_P(w, z)$), if x is preferred to y at least as strongly as w is preferred to z with respect to each $g_i \in P$. More precisely, *at least as strongly as* means *by at least the same degree*, i. e., $h \geq k$, where $h, k \in H_i$, $xP_i^h y$, and $wP_i^k z$, for each $g_i \in P$.

Let $D_{\{i\}}$ be the dominance relation confined to the single criterion $g_i \in P$. The binary relation $D_{\{i\}}$ is reflexive ($(x, y)D_{\{i\}}(x, y)$ for every $(x, y) \in A \times A$), transitive ($(x, y)D_{\{i\}}(w, z)$ and $(w, z)D_{\{i\}}(u, v)$ imply $(x, y)D_{\{i\}}(u, v)$ for every $(x, y), (w, z), (u, v) \in A \times A$), and complete ($(x, y)D_{\{i\}}(w, z)$ and/or $(w, z)D_{\{i\}}(x, y)$

for all $(x, y), (w, z) \in A \times A$. Therefore, $D_{\{i\}}$ is a complete preorder on $A \times A$. Since the intersection of complete preorders is a partial preorder, and $D_P = \bigcap_{g_i \in P} D_{\{i\}}$, $P \subseteq C$, the dominance relation D_P is a partial preorder on $A \times A$.

Let $R \subseteq P \subseteq C$ and $(x, y), (u, v) \in A \times A$; then the following implication holds

$$(x, y)D_P(u, v) \implies (x, y)D_R(u, v) .$$

Given $P \subseteq C$ and $(x, y) \in A \times A$, we define the following:

- A set of pairs of objects dominating (x, y), called the *P-dominating set*, denoted by $D_P^+(x, y)$ and defined as $\{(w, z) \in A \times A : (w, z)D_P(x, y)\}$.
- A set of pairs of objects dominated by (x, y), called the *P-dominated set*, denoted by $D_P^-(x, y)$ and defined as $\{(w, z) \in A \times A : (x, y)D_P(w, z)\}$.

The P-dominating sets and the P-dominated sets defined on B for all pairs of reference objects from B are *granules of knowledge* that can be used to express P-lower and P-upper approximations of the comprehensive outranking relations S and S^c, respectively,

$$\underline{P}(S) = \left\{ (x, y) \in B : D_P^+(x, y) \subseteq S \right\} ,$$

$$\overline{P}(S) = \bigcup_{(x,y) \in S} D_P^+(x, y) .$$

$$\underline{P}(S^c) = \{ (x, y) \in B : D_P^-(x, y) \subseteq S^c \} ,$$

$$\overline{P}(S^c) = \bigcup_{(x,y) \in S^c} D_P^-(x, y) .$$

It was proved in [22.7] that

$$\underline{P}(S) \subseteq S \subseteq \overline{P}(S), \quad \underline{P}(S^c) \subseteq S^c \subseteq \overline{P}(S^c) .$$

Furthermore, the following complementarity properties hold

$$\underline{P}(S) = B - \overline{P}(S^c), \quad \overline{P}(S) = B - \underline{P}(S^c) ,$$

$$\underline{P}(S^c) = B - \overline{P}(S), \quad \overline{P}(S^c) = B - \underline{P}(S) .$$

The P-boundaries (P-doubtful regions) of S and S^c are defined as

$$Bn_P(S) = \overline{P}(S) - \underline{P}(S) ,$$

$$Bn_P(S^c) = \overline{P}(S^c) - \underline{P}(S^c) .$$

From the above it follows that $Bn_P(S) = Bn_P(S^c)$.

The concepts of the quality of approximation, reducts, and core can also be extended to the approximation of the outranking relation by the multi-graded dominance relations.

In particular, the coefficient

$$\gamma_P = \frac{|\underline{P}(S) \cup \underline{P}(S^c)|}{|B|}$$

defines the *quality of approximation of S and S^c* by $P \subseteq C$. It expresses the ratio of all pairs of reference objects $(x, y) \in B$ correctly assigned to S and S^c by the set P of criteria to all the pairs of objects contained in B. Each minimal subset $P \subseteq C$, such that $\gamma_P = \gamma_C$, is called a *reduct* of C (denoted by $RED_{S_{\text{PCT}}}$). Note that S_{PCT} can have more than one reduct. The intersection of all B-reducts is called the *core* (denoted by $CORE_{S_{\text{PCT}}}$).

It is also possible to use the variable consistency model on S_{PCT} [22.52], if one is aware that some of the pairs in the positive or negative dominance sets belong to the opposite relation, while at least $l * 100\%$ of pairs belong to the correct one. Then the definition of the lower approximations of S and S^c boils down to

$$\underline{P}(S) = \left\{ (x, y) \in B : \frac{\left| D_P^+(x, y) \cap S \right|}{\left| D_P^+(x, y) \right|} \geq l \right\} ,$$

$$\underline{P}(S^c) = \left\{ (x, y) \in B : \frac{|D_P^-(x, y) \cap S^c|}{|D_P^-(x, y)|} \geq l \right\} .$$

Dominance Without Degrees of Preference

The degree of graded preference considered above is defined on a cardinal scale of the strength of preference. However, in many real world problems, the existence of such a quantitative scale is rather questionable. This is the case with ordinal scales of criteria. In this case, the dominance relation is defined directly on evaluations $g_i(x)$ for all objects $x \in A$. Let us explain this latter case in more detail.

Let $P = P^O$ and $P^N = \emptyset$, then, given $(x, y), (w, z) \in A \times A$, the pair (x, y) is said to dominate the pair (w, z) with respect to criteria from P (denoted by $(x, y)D_P(w, z)$), if for each $g_i \in P$, $g_i(x) \geq g_i(w)$ and $g_i(z) \geq g_i(y)$.

Let $D_{\{i\}}$ be the dominance relation confined to the single criterion $g_i \in P^O$. The binary relation $D_{\{i\}}$ is reflexive, transitive, but non-complete (it is possible that *not* $(x, y)D_{\{i\}}(w, z)$ and *not* $(w, z)D_{\{i\}}(x, y)$ for some $(x, y), (w, z) \in A \times A$). Therefore, $D_{\{i\}}$ is a partial preorder. Since the intersection of partial preorders is also a partial preorder and $D_P = \bigcap_{g_i \in P} D_{\{i\}}$, $P = P^O$, the dominance relation D_P is a partial preorder.

If some criteria from $P \subseteq C$ express preferences on a quantitative or a numerical non-quantitative scale and others on an ordinal scale, i.e., if $P^N \neq \emptyset$ and $P^O \neq \emptyset$, then, given $(x, y), (w, z) \in A \times A$, the pair (x, y) is said to dominate the pair (w, z) with respect to criteria from P, if (x, y) dominates (w, z) with respect to both P^N and P^O. Since the dominance relation with respect to P^N is a partial preorder on $A \times A$ (because it is a multi-graded dominance) and the dominance with respect to P^O is also a partial preorder on $A \times A$ (as explained above), then the dominance D_P, being the intersection of these two dominance relations, is a partial preorder. In consequence, all the concepts introduced in the previous section can be restored using this specific definition of dominance.

22.4.4 Induction of Decision Rules from Rough Approximations of Preference Relations

Using the rough approximations of preference relations S and S^c defined in Sect. 22.4.3, it is possible to induce a generalized description of the preference information contained in a given S_{PCT} in terms of suitable decision rules. The syntax of these rules involves the concept of *upward cumulated preferences* (denoted by $P_i^{\geq h}$) and *downward cumulated preferences* (denoted by $P_i^{\leq h}$), with the following interpretation:

- $xP_i^{\geq h}y$ means
  x is preferred to y with respect to g_i by at least degree h;
- $xP_i^{\leq h}y$ means
  x is preferred to y with respect to g_i by at most degree h.

The exact definition of the cumulated preferences, for each $(x, y) \in A \times A$, $g_i \in C^N$, and $h \in H_i$, can be represented as follows:

- $xP_i^{\geq h}y$ if $xP_i^k y$, where $k \in H_i$ and $k \geq h$;
- $xP_i^{\leq h}y$ if $xP_i^k y$, where $k \in H_i$ and $k \leq h$.

Let also $G_i = \{g_i(x), x \in A\}$, $g_i \in C^O$. The decision rules have then the following syntax:

1. $D_{\geq}$-decision rules:

If $xP_{i1}^{\geq h(i1)}y$ and $\ldots$ $xP_{ie}^{\geq h(ie)}y$ and $g_{ie+1}(x) \geq r_{ie+1}$ and $g_{ie+1}(y) \leq s_{ie+1}$ and $\ldots$ $g_{ip}(x) \geq r_{ip}$ and $g_{ip}(y) \leq s_{ip}$, then xSy,

where

$$P = \{g_{i1}, \ldots, g_{ip}\} \subseteq C,$$
$$P^N = \{g_{i1}, \ldots, g_{ie}\},$$
$$P^O = \{g_{ie+1}, \ldots, g_{ip}\},$$
$$(h(i1), \ldots, h(ie)) \in H_{i1} \times \cdots \times H_{ie}$$

and

$$(r_{ie+1}, \ldots, r_{ip}), (s_{ie+1}, \ldots, s_{ip}) \in G_{ie+1} \times \cdots \times G_{ip}.$$

These rules are supported by pairs of objects from the P-lower approximation of S only.

2. $D_{\leq}$-decision rules:

If $xP_{i1}^{\leq h(i1)}y$ and $\ldots$ $xP_{ie}^{\leq h(ie)}y$ and $g_{ie+1}(x) \leq r_{ie+1}$ and $g_{ie+1}(y) \geq s_{ie+1}$ and $\ldots$ $g_{ip}(x) \leq r_{ip}$ and $g_{ip}(y) \geq s_{ip}$, then $xS^c y$,

where

$$P = \{g_{i1}, \ldots, g_{ip}\} \subseteq C,$$
$$P^N = \{g_{i1}, \ldots, g_{ie}\},$$
$$P^O = \{g_{ie+1}, \ldots, g_{ip}\},$$
$$(h(i1), \ldots, h(ie)) \in H_{i1} \times \cdots \times H_{ie}$$

and

$$(r_{ie+1}, \ldots, r_{ip}), (s_{ie+1}, \ldots, s_{ip}) \in G_{ie+1} \times \cdots \times G_{ip}.$$

These rules are supported by pairs of objects from the P-lower approximation of S^c only.

3. $D_{\geq \leq}$-decision rules:

If $xP_{i1}^{\geq h(i1)}y$ and $\ldots$ $xP_{ie}^{\geq h(ie)}y$ and $xP_{ie+1}^{\leq h(ie+1)}y$ $\ldots xP_{if}^{\leq h(if)}y$ and $g_{if+1}(x) \geq r_{if+1}$ and $g_{if+1}(y) \leq s_{if+1}$ and $\ldots$ $g_{ig}(x) \geq r_{ig}$ and $g_{ig}(y) \leq s_{ig}$ and $g_{ig+1}(x) \leq r_{ig+1}$ and $g_{ig+1}(y) \geq s_{ig+1}$ and $\ldots$ $g_{ip}(x) \leq r_{ip}$ and $g_{ip}(y) \geq s_{ip}$, then xSy or $xS^c y$,

where

$$O' = \{g_{i1}, \ldots, g_{ie}\} \subseteq C,$$
$$O'' = \{g_{ie+1}, \ldots, g_{if}\}\} \subseteq C,$$
$$P^N = O' \cup O'',$$

O', and O'' are not necessarily disjoint,

$$P^O = \{g_{if+1}, \ldots, g_{ip}\},$$
$$(h(i1), \ldots, h(if)) \in H_{i1} \times \cdots \times H_{if},$$
$$(r_{if+1}, \ldots, r_{ip}), (s_{if+1}, \ldots, s_{ip})$$
$$\in G_{if+1} \times \cdots \times G_{ip}.$$

These rules are supported by pairs of objects from the P-boundary of S and S^c only.

22.4.5 Application of Decision Rules to Multi-Criteria Choice and Ranking

The decision rules induced from a given S_{PCT} describe the comprehensive preference relations S and S^c either exactly ($D_{\geq}$- and $D_{\leq}$-decision rules) or approximately ($D_{\geq\leq}$-decision rules). A set of these rules covering all pairs of S_{PCT} represents a preference model of the DM who gave the preference information in terms of pairwise comparison of reference objects. The application of these decision rules on a new subset $M \subseteq A$ of objects induces a specific preference structure on M.

In fact, any pair of objects $(u, v) \in M \times M$ can match the decision rules in one of four ways:

- At least one $D_{\geq}$-decision rule and neither $D_{\leq}$ nor $D_{\geq\leq}$-decision rules.
- At least one $D_{\leq}$-decision rule and neither $D_{\geq}$ nor $D_{\geq\leq}$-decision rules.
- At least one $D_{\geq}$-decision rule and at least one $D_{\leq}$-decision rule, or at least one $D_{\geq\leq}$-decision rule, or at least one $D_{\geq\leq}$-decision rule and at least one $D_{\geq}$ and/or at least one $D_{\leq}$-decision rule.
- No decision rule.

These four ways correspond to the following four situations of outranking, respectively:

- uSv and *not* $uS^c v$, i.e., *true* outranking (denoted by $uS^T v$).
- $uS^c v$ and *not* uSv, i.e., *false* outranking (denoted by $uS^F v$).
- uSv and $uS^c v$, i.e., *contradictory* outranking (denoted by $uS^K v$).
- *not* uSv and *not* $uS^c v$, i.e., *unknown* outranking (denoted by $uS^U v$).

The above four situations, which together constitute the so-called *four-valued outranking* [22.53], have been introduced to underline the presence and absence of *positive* and *negative* reasons for the outranking. Moreover, they make it possible to distinguish contradictory situations from unknown ones.

A final *recommendation* (optimal choice or ranking) can be obtained upon suitable exploitation of this structure, i.e., of the presence and the absence of outranking S and S^c on M. A possible exploitation procedure consists of calculating a specific score, called the net flow score, for each object $x \in M$

$$S_{nf}(x) = S^{++}(x) - S^{+-}(x) + S^{-+}(x) - S^{-}(x),$$

where

- $S^{++}(x) = |\{y \in M: \text{there is at least one decision rule which affirms } xSy\}|$;
- $S^{+-}(x) = |\{y \in M: \text{there is at least one decision rule which affirms } ySx\}|$;
- $S^{-+}(x) = |\{y \in M: \text{there is at least one decision rule which affirms } yS^c x\}|$;
- $S^{-}(x) = |\{y \in M: \text{there is at least one decision rule which affirms } xS^c y\}|$.

The recommendation in ranking problems consists of the total preorder determined by $S_{nf}(x)$ on M. In choice problems, it consists of the object(s) $x^* \in M$ such that

$$S_{nf}(x^*) = \max_{x \in M} \{S_{nf}(x)\}.$$

The above procedure has been characterized with reference to a number of desirable properties in [22.53, 54]. A computer implementation of the whole approach, called jRank (ranking generator using DRSA) is publicly available [22.55].

Recently, *Fortemps* et al. [22.56] extended DRSA to multi-criteria choice and ranking on multi-graded preference relations, instead of uni-graded relations S and S^c.

It is also worth mentioning that there is a machine learning approach to multi-criteria choice and ranking using ensembles of decision rules. The approach presented by *Dembczyński* et al. [22.57] makes a bridge between stochastic methods of preference learning and DRSA for choice and ranking. Examples illustrating the application of DRSA to multi-criteria choice and ranking in a didactic way can be found in [22.12–14, 50, 54].

Part C | 22.4

22.5 Important Extensions of DRSA

The existing literature describes many extensions of DRSA that make it a useful tool for other practical applications. These extensions are:

- DRSA to decision under risk and uncertainty [22.58];
- DRSA to decision under uncertainty and time preference [22.59];
- DRSA handling missing data [22.60, 61];
- DRSA for imprecise object evaluations and assignments [22.62];
- Dominance-based approach to induction of association rules [22.63];
- Fuzzy-rough hybridization of DRSA [22.8, 64–67];
- DRSA as a way of operator-free fuzzy-rough hybridization [22.28, 67, 68];
- DRSA to granular computing [22.36, 37];
- DRSA to case-based reasoning [22.69, 70];
- DRSA for hierarchical structure of evaluation criteria [22.71];
- DRSA to decision involving multiple decision makers [22.72, 73];
- DRSA to interactive multi-objective optimization [22.74];
- DRSA to interactive evolutionary multi-objective optimization under risk and uncertainty [22.75, 76].

It is worth stressing that dealing with ordinal data and monotonicity constraints also makes sense in general classification problems, where the notion of preference has no meaning. Even when the ordering seems irrelevant, the presence or the absence of a property have an ordinal interpretation. If two properties are related, one of the two: the presence or the absence of one property should make more (or less) probable the presence of the other property. A formal proof showing that the IRSA is a particular case of the DRSA was given in [22.28]. With this in mind, DRSA can be seen as a general framework for analysis of classification data. Although it was designed for ordinal classification problems with monotonicity constraints, DRSA can be used to solve a general classification problem where no additional information about ordering is taken into account.

The idea behind this claim is the following [22.77]. We assume, without loss of generality, that the value sets of all regular attributes are number-coded. While this is natural for numerical attributes, categorical attributes must get numerical codes for categories. In this way, the value sets of all regular attributes become ordered (as all sets of numbers are ordered). Now, to analyze a non-ordinal classification problem using DRSA, we transform the decision table such that each regular attribute is cloned (doubled). It is assumed that the value set of each original attribute is ordered with respect to increasing preference (gain-type criterion), and the value set of its clone is ordered with respect to decreasing preference (cost-type criterion). Using DRSA, for each $t \in \{1, \ldots, n\}$, we approximate two sets of objects from the decision table: class Cl_t and its complement $\neg Cl_t$. Obviously, we can calculate dominance-based rough approximations of the two sets. Moreover, they can serve to induce *if ..., then ...* decision rules recommending assignment to class Cl_t or to its complement $\neg Cl_t$. In this way, we reformulated the original non-ordinal classification problem to an ordinal classification problem with monotonicity constraints. Due to cloning of attributes with opposite preference orders, we can have rules that cover a subspace in the condition space, which is bounded from the top and from the bottom – this leads (without discretization) to more synthetic rules than those resulting from the IRSA.

22.6 DRSA to Operational Research Problems

DRSA is also a useful instrument in the toolbox of operational research (OR). DRSA has been adapted to solve the following OR problems:

- Interactive multi-objective optimization [22.74] applied to OR problems, such as portfolio management, project scheduling, and production planning.

- Interactive evolutionary multi-objective optimization under risk and uncertainty [22.75, 76].
- Decision under uncertainty and time preference [22.59], which is useful for dealing with many OR problems where uncertainty of outcomes and their distribution over time play a fundamental role, such as portfolio selection [22.78], scheduling with

time-resource interactions, and inventory management.
- Global investment risk analysis on partially missing data [22.79].

- Explanation of recommendations following from robust ordinal regression applied to multi-criteria ranking problems in terms of rules [22.80].

22.7 Concluding Remarks on DRSA Applied to Multi-Criteria Decision Problems

Let us point out the main features of the methodology described:

- The input data set describing a given decision situation is the preference information elicited by the DM in terms of exemplary decisions (class assignments or pairwise comparisons of some objects).
- The rough set analysis of preference information using DRSA supplies some useful elements of knowledge about the decision situation. These are: the relevance of particular criteria, information about their interaction, minimal subsets of criteria (reducts) conveying important knowledge contained in the exemplary decisions and the set of the non-reducible criteria (core).
- The methodology presented is based on elementary concepts and mathematical tools (sets and set operations, binary relations), without recourse to any complex algebraic or analytical structures; the main idea is very natural and the key concept of dominance is rational and objective.
- DRSA structures the input data prior to induction of decision rules. The structuring takes into account inconsistencies of the preference information with respect to the dominance principle. Due to the structuring the induced decision rules are certain or possible, depending whether they are induced from lower or upper approximations (of unions of classes or preference relations), respectively.
- The preference model induced from the rough approximations defined on the preference information is expressed in the natural and comprehensible language of *if ..., then ...* decision rules, fulfilling the postulate of transparency and interpretability of preference models in decision aiding; each decision rule can be clearly identified with those parts of the preference information (decision examples) which support the rule; the rules inform the DM about the relationships between conditions and decisions; in this way, the rules permit traceability of the decision aiding process and give understandable justifications for the decision to be made, so that the resulting preference model constituted for the DM is a *glass box* rather than a *black box*. Finally, the decision rule preference model is more general than all existing models of conjoint measurement due to its capacity of handling inconsistent preferences [22.9–11, 81].

- Apart from their clear meaning, the decision rules are characterized by some interestingness measures, among which Bayesian confirmation measures appear to be the most appropriate, as shown in the studies [22.43, 44].
- The decision rules do not convert ordinal information into numeric information but keep the ordinal character of input data due to the syntax proposed.
- Heterogeneous information (qualitative and quantitative, ordered and non-ordered) and scales of preference (ordinal, cardinal) can be processed within DRSA, while classical methods consider only quantitative ordered evaluations (with rare exceptions).
- No prior discretization of the quantitative domains of criteria is necessary.

References

22.1   J. Figueira, S. Greco, M. Ehrgott (Eds.): *Multiple Criteria Decision Analysis: State of the Art Surveys* (Springer, Berlin 2005)

22.2   Z. Pawlak: Rough sets, Int. J. Comput. Inf. Sci. **11**, 341–356 (1982)

22.3   Z. Pawlak: *Rough sets: Theoretical Aspects of Reasoning about Data* (Kluwer, Dordrecht 1991)

22.4   Z. Pawlak, R. Słowiński: Rough set approach to multi-attribute decision analysis, Eur. J. Oper. Res. **72**, 443–459 (1994)

22.5   R. Słowiński: Rough set learning of preferential attitude in multi-criteria decision making, Lect. Notes Artif. Intell. **689**, 642–651 (1993)

22.6   S. Greco, B. Matarazzo, R. Słowiński: A new rough set approach to evaluation of bankruptcy risked. In: *Operational Tools in the Management of Financial Risk*C. Zopounidis (Kluwer, Dordrecht 1998) pp. 121–136

22.7   S. Greco, B. Matarazzo, R. Słowiński: Rough approximation of a preference relation by dominance relations, Eur. J. Oper. Res. **117**, 63–83 (1999)

22.8   S. Greco, B. Matarazzo, R. Słowiński: The use of rough sets and fuzzy sets in MCDM. In: *Multicriteria Decision Making*, International Series in Opearations Research & Management Science, Vol. 21, ed. by T. Gal, T. Stewart, T. Hanne (Kluwer, Dordrecht 1999) pp. 397–455

22.9   S. Greco, B. Matarazzo, R. Słowiński: Preference representation by means of conjoint measurement & decision rule model. In: *Aiding Decisions with Multiple Criteria-Essays*, International Series in Operations Reasearch & Management Science, Vol. 44, ed. by D. Bouyssou, E. Jacquet-Lagrèze, P. Perny, R. Słowiński, D. Vanderpooten, P. Vincke (Kluwer, Dordrecht 2002) pp. 263–313

22.10   S. Greco, B. Matarazzo, R. Słowiński: Axiomatic characterization of a general utility function and its particular cases in terms of conjoint measurement and rough-set decision rules, Eur. J. Oper. Res. **158**, 271–292 (2004)

22.11   R. Słowiński, S. Greco, B. Matarazzo: Axiomatization of utility, outranking and decision-rule preference models for multiple-criteria classification problems under partial inconsistency with the dominance principle, Control Cybern. **31**, 1005–1035 (2002)

22.12   R. Słowiński, S. Greco, B. Matarazzo: Rough-set-based decision support. In: *Search Methodologies: Introductory Tutorials in Optimization and Decision Support Techniques*, 2nd edn., ed. by E.K. Burke, G. Kendall (Springer, New York 2014) pp. 557–609

22.13   R. Słowiński, S. Greco, B. Matarazzo: Rough Sets in Decision Making. In: *Encyclopedia of Complexity and Systems Science*, ed. by R.A. Meyers (Springer, New York 2009) pp. 7753–7786

22.14   R. Słowiński, S. Greco, B. Matarazzo: Rough set and rule-based multicriteria decision aiding, Pesquisa Oper. **32**(2), 213–269 (2012)

22.15   S. Greco, B. Matarazzo, R. Słowiński: Rough sets theory for multicriteria decision analysis, Eur. J. Oper. Res. **129**, 1–47 (2001)

22.16   R. Słowiński, S. Greco, B. Matarazzo: Rough set analysis of preference-ordered data, Lect. Notes Artif. Intell. **2475**, 44–59 (2002)

22.17   R. Słowiński, J. Stefanowski, S. Greco, B. Matarazzo: Rough sets based processing of inconsistent information in decision analysis, Control Cybern. **29**, 379–404 (2000)

22.18   Z. Pawlak, J.W. Grzymala-Busse, R. Słowiński, W. Ziarko: Rough sets, Communications ACM **38**, 89–95 (1995)

22.19   L. Polkowski: *Rough Sets: Mathematical Foundations* (Physica, Heidelberg 2002)

22.20   R. Słowiński (Ed.): *Intelligent Decision Support: Handbook of Applications and Advances of the Rough Sets Theory* (Kluwer, Dordrecht 1992)

22.21   R. Słowiński, D. Vanderpooten: A generalised definition of rough approximations, IEEE Trans. Knowl. Data Eng. **12**, 331–336 (2000)

22.22   R. Słowiński, C. Zopounidis: Application of the rough set approach to evaluation of bankruptcy risk, Intell. Syst. Account. Financ. Manag. **4**, 27–41 (1995)

22.23   W. Ziarko: Rough sets as a methodology for data mining. In: *Rough Sets in Knowledge Discovery*, (Physica, Heidelberg 1998) pp. 554–576

22.24   R. Słowiński: A generalization of the indiscernibility relation for rough set analysis of quantitative information, Riv. Mat. Sci. Econ. Soc. **15**, 65–78 (1992)

22.25   S. Greco, B. Matarazzo, R. Słowiński: Extension of the rough set approach to multicriteria decision support, INFOR **38**, 161–196 (2000)

22.26   S. Greco, B. Matarazzo, R. Słowiński: Rough sets methodology for sorting problems in presence of multiple attributes and criteria, Eur. J. Oper. Res. **138**, 247–259 (2002)

22.27   S. Greco, B. Matarazzo, R. Słowiński: Determining task and methods Calssification: Multicriteria classification. In: *Handbook of Data Mining and Knowledge Discovery*, ed. by W. Kloesgen, J. Zytkow (Oxford Univ. Press, Oxford 2002), 318–328

22.28   S. Greco, B. Matarazzo, R. Słowiński: Dominance-based rough set approach as a proper way of handling graduality in rough set theory, Lect. Notes Comput. Sci. **4400**, 36–52 (2007)

22.29   S. Greco, B. Matarazzo, R. Słowiński: The bipolar complemented de Morgan Brouwer-Zadeh distributive lattice as an algebraic structure for the dominance-based rough set approach, Fundam. Inf. **115**, 25–56 (2012)

22.30   S. Greco, B. Matarazzo, R. Słowiński, J. Stefanowski: Variable consistency model of dominance-based rough set approach, Lect. Notes Artif. Intell. **2005**, 170–181 (2001)

22.31   W. Ziarko: Variable precision rough sets model, J. Comput. Syst. Sci. **46**, 39–59 (1993)

22.32   S. Greco, B. Matarazzo, R. Słowiński: Parameterized rough set model using rough membership and Bayesian confirmation measures, Int. J. Approx. Reason. **49**, 285–300 (2008)

22.33   J. Błaszczyński, S. Greco, R. Słowiński, M. Szeląg: Monotonic variable consistency rough set approaches, Int. J. Approx. Reason. **50**(7), 979–999 (2009)

22.34   S. Greco, B. Matarazzo, R. Słowiński: Assessment of a value of information using rough sets and fuzzy measures. In: *Fuzzy Sets and Their Applications*, ed.

by J. Chocjan, J. Leski (Silesian Univ. Technol. Press, Gliwice 2001) pp. 185–193

22.35 W. Kotłowski, K. Dembczyński, S. Greco, R. Słowiński: Stochastic dominance-based rough set model for ordinal classification, Inf. Sci. **178**(21), 4019–4037 (2008)

22.36 S. Greco, B. Matarazzo, R. Słowiński: Granular computing for reasoning about ordered data: The dominance-based rough set approach. In: *Handbook of Granular Computing*, ed. by W. Pedrycz, A. Skowron, V. Kreinovich (Wiley, Chichester 2008) pp. 347–373

22.37 S. Greco, B. Matarazzo, R. Słowiński: Granular computing and data mining for ordered data – The dominance-based roughset approach. In: *Encyclopedia of Complexity and Systems Science*, ed. by R.A. Meyers (Springer, New York 2009) pp. 4283–4305

22.38 S. Greco, B. Matarazzo, R. Słowiński, J. Stefanowski: An algorithm for induction of decision rules consistent with dominance principle, Lect. Notes Artif. Intell. **2005**, 304–313 (2001)

22.39 J. Błaszczyński, R. Słowiński, M. Szeląg: Sequential covering rule induction algorithm for variable consistency rough set approaches, Inf. Sci. **181**, 987–1002 (2011)

22.40 J. Błaszczyński, S. Greco, B. Matarazzo, R. Słowiński, M. Szeląg: jMAF – Dominance-based rough set data analysis framework. In: *Rough Sets and Intelligent Systems*, Intelligent Systems Reference Library, Vol. 42, ed. by A. Skowron, Z. Suraj (Springer, Berlin 2013) pp. 185–209

22.41 J. Błaszczyński, S. Greco, B. Matarazzo, R. Słowiński, M. Szeląg: jMAF (java multi-criteria and multi-attribute analysis framework), 2013, available at: http://www.cs.put.poznan.pl/jBlaszczynski/Site/jRS.html

22.42 S. Greco, B. Matarazzo, N. Pappalardo, R. Słowiński: Measuring expected effects of interventions based on decision rules, J. Exp. Theor. Artif. Intell. **17**(1/2), 103–118 (2005)

22.43 S. Greco, Z. Pawlak, R. Słowiński: Can Bayesian confirmation measures be useful for rough set decision rules?, Eng. Appl. Artif. Intell. **17**(4), 345–361 (2004)

22.44 S. Greco, R. Słowiński, I. Szczęch: Properties of rule interestingness measures and alternative approaches to normalization of measures, Inf. Sci. **216**, 1–16 (2012)

22.45 S. Giove, S. Greco, B. Matarazzo, R. Słowiński: Variable consistency monotonic decision trees, Lect. Notes Artif. Intell. **2475**, 247–254 (2002)

22.46 S. Greco, B. Matarazzo, R. Słowiński: Algebra and topology for dominance-based rough set approach. In: *Advances in Intelligent Information Systems, Studies in Computational Intelligence*, (Springer, Berlin 2010) pp. 43–78

22.47 S. Greco, B. Matarazzo, R. Słowiński: Dominance-based rough set approach to granular computing.

22.48 In: *Novel Developments in Granular Computing*, ed. by J. Yao (Hershey, New York 2010) pp. 439–527

22.48 S. Greco, B. Matarazzo, R. Słowiński: On topological dominance-based rough set approach, Lect. Notes Comput. Sci. **6190**, 21–45 (2010)

22.49 J. Błaszczyński, S. Greco, R. Słowiński: Multi-criteria classification – A new scheme for application of dominance-based decision rules, Eur. J. Oper. Res. **181**(3), 1030–1044 (2007)

22.50 S. Greco, B. Matarazzo, R. Słowiński: Decision rule approach. In: *Multiple Criteria Decision Analysis: State of the Art Surveys*, (Springer, New York 2005) pp. 507–562

22.51 S. Greco, B. Matarazzo, R. Słowiński: Rule-based decision support in multicriteria choice and ranking, Lect. Notes Artif. Intell. **2143**, 29–47 (2001)

22.52 R. Słowiński, S. Greco, B. Matarazzo: Mining decision-rule preference model from rough approximation of preference relation, Proc. 26th IEEE Annu. Int. Conf. Comput. Softw. Appl., Oxford (2002) pp. 1129–1134

22.53 S. Greco, B. Matarazzo, R. Słowiński, A. Tsoukias: Exploitation of a rough approximation of the outranking relation in multicriteria choice and ranking. In: *Trends in Multicriteria Decision Making*, Lecture Notes in Economics and Mathematical Systems, Vol. 465, ed. by T.J. Stewart, R.C. van den Honert (Springer, Berlin 1998) pp. 45–60

22.54 M. Szeląg, S. Greco, R. Słowiński: Rule-based approach to multicriteria ranking. In: *Multicriteria Decision Aid and Artificial Intelligence: Links, Theory and Applications*, ed. by M. Doumpos, E. Grigoroudis (Wiley-Blackwell, London 2013) pp. 127–160

22.55 M. Szelag, S. Greco, R. Slowinski: jRank (ranking generator using DRSA), 2013, available at: http://www.cs.put.poznan.pl/mszelag/Software/jRank/jRank.html

22.56 P. Fortemps, S. Greco, R. Słowiński: Multicriteria decision support using rules that represent rough-graded preference relations, Eur. J. Oper. Res. **188**(1), 206–223 (2008)

22.57 K. Dembczyński, W. Kotłowski, R. Słowiński, M. Szeląg: Learning of rule ensembles for multiple attribute ranking problems. In: *Preference Learning*, ed. by J. Fürnkranz, E. Hüllermeier (Springer, Berlin 2010) pp. 217–247

22.58 S. Greco, B. Matarazzo, R. Słowiński: Rough set approach to decisions under risk, Lect. Notes Artif. Intell. **2005**, 160–169 (2001)

22.59 S. Greco, B. Matarazzo, R. Słowiński: Dominance-based rough set approach to decision under uncertainty and time preference, Ann. Oper. Res. **176**, 41–75 (2010)

22.60 S. Greco, B. Matarazzo, R. Słowiński: Handling missing values in rough set analysis of multi-attribute and multi-criteria decision problems, Lect. Notes Artif. Intell. **1711**, 146–157 (1999)

22.61 S. Greco, B. Matarazzo, R. Słowiński: Dealing with missing data in rough set analysis of multi-

Part C | 22

attribute and multi-criteria decision problems. In: *Decision Making: Recent Developments and Worldwide Applications*, ed. by S.H. Zanakis, G. Doukidis, C. Zopounidis (Kluwer, Dordrecht 2000) pp. 295–316

22.62   K. Dembczyński, S. Greco, R. Słowiński: Rough set approach to multiple criteria classification with imprecise evaluations and assignments, Eur. J. Oper. Res. **198**(2), 626–636 (2009)

22.63   S. Greco, B. Matarazzo, R. Słowiński, J. Stefanowski: Mining association rules in preference-ordered data, Lect. Notes Artif. Intell. **2366**, 442–450 (2002)

22.64   S. Greco, B. Matarazzo, R. Słowiński: Rough set processing of vague information using fuzzy similarity relations. In: *Finite Versus Infinite – Contributions to an Eternal Dilemma*, ed. by C.S. Calude, G. Paun (Springer, Berlin 2000) pp. 149–173

22.65   S. Greco, B. Matarazzo, R. Słowiński: Fuzzy extension of the rough set approach to multicriteria and multiattribute sorting. In: *Preferences and Decisions under Incomplete Knowledge*, ed. by J. Fodor, B. De Baets, P. Perny (Physica, Heidelberg 2000) pp. 131–151

22.66   S. Greco, M. Inuiguchi, R. Słowiński: Dominance-based rough set approach using possibility and necessity measures, Lect. Notes Artif. Intell. **2475**, 85–92 (2002)

22.67   S. Greco, M. Inuiguchi, R. Słowiński: A new proposal for fuzzy rough approximations and gradual decision rule representation, Lect. Notes Comput. Sci. **3135**, 319–342 (2003)

22.68   S. Greco, M. Inuiguchi, R. Słowiński: Fuzzy rough sets and multiple-premise gradual decision rules, Int. J. Approx. Reason. **41**, 179–211 (2005)

22.69   S. Greco, B. Matarazzo, R. Słowiński: Case-based reasoning using gradual rules induced from dominance-based rough approximations, Lect. Notes Artif. Intell. **5009**, 268–275 (2008)

22.70   M. Szeląg, S. Greco, J. Błaszczyński, R. Słowiński: Case-based reasoning using dominance-based decision rules, Lect. Notes Artif. Intell. **6954**, 404–413 (2011)

22.71   K. Dembczyński, S. Greco, R. Słowiński: Methodology of rough-set-based classification and sorting with hierarchical structure of attributes and criteria, Control Cybern. **31**, 891–920 (2002)

22.72   S. Greco, B. Matarazzo, R. Słowiński: Dominance-based rough set approach to decision involving multiple decision makers, Lect. Notes Comput. Sci. **4259**, 306–317 (2006)

22.73   S. Greco, B. Matarazzo, R. Słowiński: Dominance-based rough set approach on pairwise comparison tables to decision involving multiple decision makers, Lect. Notes Comput. Sci. **6954**, 126–135 (2011)

22.74   S. Greco, B. Matarazzo, R. Słowiński: Dominance-based rough set approach to interactive multiobjective optimization, Lect. Notes Comput. Sci. **5252**, 121–156 (2008)

22.75   S. Greco, B. Matarazzo, R. Słowiński: Dominance-based rough set approach to interactive evolutionary multiobjective optimization. In: *Preferences and Decisions: Models and Applications, Studies in Fuzziness*, (Springer, Berlin 2010) pp. 225–260

22.76   S. Greco, B. Matarazzo, R. Słowiński: Interactive evolutionary multiobjective optimization using dominance-based rough set approach, Proc. IEEE World Congr. Comput. Intell. 2010 (WCCI 2010), Barcelona, Spain (2010) pp. 3026–3033

22.77   J. Błaszczyński, S. Greco, R. Słowiński: Inductive discovery of laws using monotonic rules, Eng. Appl. Artif. Intell. **25**(2), 284–294 (2012)

22.78   S. Greco, B. Matarazzo, R. Słowiński: Beyond Markowitz with multiple criteria decision aiding, J. Bus. Econ. **83**(1), 29–60 (2013)

22.79   S. Greco, B. Matarazzo, R. Słowiński, S. Zanakis: Global investing risk: A case study of knowledge assessment via rough sets, Ann. Oper. Res. **185**, 105–138 (2011)

22.80   S. Greco, R. Słowiński, P. Zielniewicz: Putting dominance-based rough set approach and robust ordinal regression together, Decis. Support Syst. **54**, 891–903 (2013)

22.81   S. Greco, B. Matarazzo, R. Słowiński: Conjoint measurement and rough set approach for multicriteria sorting problems in presence of ordinal criteria. In: *A-MCD-A: Aide Multi-Critère à la Décision – Multiple Criteria Decision Aiding*, (European Commission, Ispra 2001) pp. 117–144

23. Rule Induction from Rough Approximations

Jerzy W. Grzymala-Busse

Rule induction is an important technique in data mining or machine learning. Knowledge is frequently expressed by rules in many areas of artificial intelligence (AI), including rule-based expert systems. In this chapter we discuss only *supervised learning* in which all cases of the input data set are pre-classified by an expert.

23.1 Complete and Consistent Data

Our basic assumption is that the data sets are presented as decision tables. An example of a decision table is presented in Table 23.1. Rows of the decision table represent *cases* and columns represent *variables*. The set of all cases is denoted by U. In Table 23.1, $U = \{1, 2, 3, 4, 5, 6, 7, 8\}$. Some variables are called *attributes* while one selected variable is called a *decision* and is denoted by d. The set of all attributes will be denoted by A. In Table 23.1, $A = \{Wind, Humidity, Temperature\}$ and $d = Trip$. For an attribute a and case x, $a(x)$ denotes the value of the attribute a for case x. For example, $Wind(1) = low$.

Let B be a subset of the set A of all attributes. Complete data sets are characterized by the indiscernibility relation $IND(B)$ [23.1, 2] defined as follows: for any $x, y \in U$,

$$(x, y) \in IND(B) \text{ if and only if } a(x) = a(y)$$
$$\text{for any } a \in B . \quad (23.1)$$

Obviously, $IND(B)$ is an equivalence relation. The equivalence class of $IND(B)$ containing $x \in U$ will be denoted by $[x]_B$ and called a *B-elementary* set. A-elementary sets will be called *elementary*. Any union

of B-elementary sets will be called a *B-definable* set. By analogy, the A-definable set will be called definable. The elementary sets of the partition $\{d\}^*$ are called *concepts*. In Table 23.1, the concepts are $\{1, 2, 3\}$, $\{4, 5\}$, and $\{6, 7, 8\}$. The set of all equivalence classes $[x]_B$, where $x \in U$, is a partition on U denoted by B^*. For Table 23.1, $A^* = \{\{1\}, \{2\}, \{3\}, \{4\}, \{5\}, \{6\}, \{7\}, \{8\}\}$. All members of A^* are elementary sets.

We will quote some definitions from [23.3]. A rule r is an expression of the following form

$$(a_1, v_1) \& (a_2, v_2) \& \ldots \& (a_k, v_k) \rightarrow (d, w) , \quad (23.2)$$

Table 23.1 A complete and consistent decision table

	Attributes			Decision
Case	**Wind**	**Humidity**	**Temperature**	**Trip**
1	low	low	medium	yes
2	low	low	low	yes
3	low	medium	medium	yes
4	low	medium	high	maybe
5	medium	low	medium	maybe
6	medium	high	low	no
7	high	high	high	no
8	medium	high	high	no

where $a_1, a_2, \ldots, a_k$ are distinct attributes, d is a decision, $v_1, v_2, \ldots, v_k$ are respective attribute values, and w is a decision value.

A case x is *covered* by a rule r if and only if any attribute–value pair of r is satisfied by the corresponding value of x. For example, case 1 from Table 23.1 is covered by the following rule r:

$$(Wind, low) \,\&\, (Humidity, low) \,\to\, (Trip, yes) \,.$$

The concept C defined by rule r is *indicated* by r. The above rule r indicates concept $\{1, 2, 3\}$.

A rule r is *consistent* with the data set if and only if for any case x covered by r, x is a member of the concept indicated by r. The above rule is consistent with the data set represented by Table 23.1. A rule set R is consistent with the data set if and only if for any $r \in R$, r is consistent with the data set. The rule set containing the above rule is consistent with the data set represented by Table 23.1.

We say that a concept C is *completely* covered by a rule set R if and only if for every case x from C there exists a rule r from R such that r covers x. For example, the single rule

$$(Wind, low) \,\to\, (Trip, yes)$$

completely covers the concept $\{1, 2, 3\}$. On the other hand, this rule is not consistent with the data set represented by Table 23.1. A rule set R is *complete* for a data set if and only if every concept from the data set is completely covered by R.

In this chapter we will discuss how to induce rule sets that are complete and consistent with the data set.

23.1.1 Global Coverings

The simplest approach to rule induction is based on finding the smallest subset B of the set A of all attributes that is sufficient to be used in a rule set. Such reducing of the attribute set is one of the main and frequently used techniques in rough set theory [23.1, 2, 4]. This approach is also called a *feature selection*. In Table 23.1 the attribute *Humidity* is redundant (irrelevant). The remaining two attributes (*Wind* and *Temperature*) distinguish all eight cases. Let us make it more precise using the fundamental definitions of rough set theory [23.1, 2, 4].

For a decision d we say that $\{d\}$ depends on B if and only if $B^* \leq \{d\}^*$, i.e., for any elementary set X in B there exists a concept C from $\{d\}^*$ such that $X \subseteq C$.

note that for partitions π and τ on U, if for any $X \in \pi$ there exists $Y \in \tau$ such that $X \subseteq Y$, then we say that π is smaller than or equal to τ and denote it by $\pi \leq \tau$. A *global covering* (or *relative reduct*) of $\{d\}$ is a subset B of A such that $\{d\}$ depends on B and B is minimal in A. The algorithm to compute a single global covering is presented below.

Algorithm 23.1 Algorithm to compute a single global covering

1: (**input**: the set A of all attributes,
   partition $\{d\}^*$ on U;
   output: a single global covering R);
2: **begin**
3: compute partition A^*;
4: $P := A$;
5: $R := \emptyset$;
6:   **if** $A^* \leq \{d\}^*$
7:     **then**
8:     **begin**
9:     **for** each attribute a in A **do**
10:       **begin**
11:         $Q := P - \{a\}$;
12:         compute partition Q^*;
13:         **if** $Q^* \leq \{d\}^*$
14:         **then** $P := Q$
15:       **end** {for}
16:       $R := P$
17:     **end** {then}
18: **end** {algorithm}.

Let us use this algorithm for Table 23.1. First,

$$A^* = \{\{1\}, \{2\}, \{3\}, \{4\}, \{5\}, \{6\}, \{7\}, \{8\}\}$$
$$\leq \{Trip\}^* \,.$$

Initially,

$$P = A \text{ and } Q = P - Wind \,,$$
$$Q = \{Humidity, Temperature\} \,,$$

and then we compute Q^*, where

$$Q^* = \{\{1, 5\}, \{2\}, \{3\}, \{4\}, \{6\}, \{7, 8\}\} \,.$$

We find that $Q^* \not\leq \{Trip\}^*$. Thus, $P = A$. Next, we try to delete *Humidity* from P. We obtain $Q = \{Wind, Temperature\}$ and then we compute Q^*, where $Q^* = \{\{1, 3\}, \{2\}, \{4\}, \{5\}, \{6\}, \{7, 8\}\}$. This time $Q^* \leq \{Trip\}^*$, so $P = \{Wind, Temperature\}$.

We still need to check $Q = P - \{Temperature\}$, $Q = \{Wind\}$ and $Q^* = \{\{1,2,3,4\}, \{5,6,8\}, \{7\}\}$, and $Q^* \not\leq \{Trip\}^*$. Thus $R = \{Wind, Temperature\}$ is a global covering.

For a given global covering rules are induced by examining cases of the data set. Initially, such a rule contains all attributes from the global covering with the corresponding attribute values, then a *dropping conditions* technique is used; we try to drop one condition (attribute–value pair) at a time, starting from the leftmost condition, checking whether the rule is still consistent with the data set, then we try to drop the next condition, and so on. For example,

$(Wind, low)$ & $(Temperature, medium) \rightarrow (Trip, yes)$

is our first candidate for a rule. If we are going to drop the first condition, the above rule will be reduced to

$(Temperature, medium) \rightarrow (Trip, yes)$.

However, this rule covers the case 5, so it is not consistent with the data set represented by Table 23.1. By dropping the second condition from the initial rule we obtain

$(Wind, low) \rightarrow (Trip, yes)$,

but this rule is not consistent with the data represented by Table 23.1 either, since it covers case 4, so we conclude that the initial rule is the simplest possible. This rule covers two cases: 1 and 3.

It is not difficult to check that the rule

$(Wind, low)$ & $(Temperature, low) \rightarrow (Trip, yes)$

is as simple as possible and that it covers only case 2. Thus, the above two rules consistently and completely cover the concept $\{1,2,3\}$.

The above algorithm is implemented as LEM1 (Learning from Examples Module, version 1). It is a component of the data mining system LERS (Learning from Examples Using Rough Sets). A similar system was described in [23.5].

23.1.2 Local Coverings

The LEM1 algorithm is based on calculus on partitions on the entire universe U. Another approach to rule induction, based on attribute–value pairs, is presented in

the LEM2 algorithm (Learning from Examples Module, version 2), another component of LERS. We will quote a few definitions from [23.6, 7].

For an attribute–value pair $(a, v) = t$, a *block* of t, denoted by $[t]$, is a set of all cases from U such that for attribute a have value v, i.e.,

$$[(a,v)] = \{x \mid a(x) = v\} . \tag{23.3}$$

Let T be a set of attribute–value pairs. The block of T, denoted by $[T]$, is the following set

$$\bigcap_{t \in T} [t] . \tag{23.4}$$

Let B be a subset of U. Set B *depends* on a set T of attribute–value pairs $t = (a, v)$ if and only if $[T]$ is nonempty and

$$[T] \subseteq B . \tag{23.5}$$

Set T is a *minimal complex* of B if and only if B depends on T and no proper subset T' of T exists such that B depends on T'. Let $\mathcal{T}$ be a nonempty collection of nonempty sets of attribute–value pairs. Then $\mathcal{T}$ is a *local covering* of B if and only if the following conditions are satisfied:

1. each member T of $\mathcal{T}$ is a minimal complex of B,
2. $\bigcup_{t \in \mathcal{T}} [T] = B$, and $\mathcal{T}$ is minimal, i.e., $\mathcal{T}$ has the smallest possible number of members.

An algorithm for finding a single local covering, called LEM2, is presented below. For a set X, $|X|$ denotes the cardinality of X.

Algorithm 23.2 LEM2

1: (**input**: a set B,
   output: a single local covering $\mathcal{T}$ of set B);
2: **begin**
3: $G := B$;
4: $\mathcal{T} := \emptyset$;
5: **while** $G \neq \emptyset$
6:   **begin**
7:     $T := \emptyset$;
8:     $T(G) := \{t | [t] \cap G \neq \emptyset\}$;
9:     **while** $T = \emptyset$ **or** $[T] \not\subseteq B$
10:      **begin**
11:        select a pair $t \in T(G)$
12:        such that $||t] \cap G|$ is
13:        maximum; if a tie

```
14:        occurs, select a pair
15:        t ∈ T(G) with the
16:        smallest cardinality of [t];
17:        if another tie occurs,
18:        select first pair;
19:        T := T ∪ {t} ;
20:        G := [t] ∩ G ;
21:        T(G) := {t|[t] ∩ G ≠ ∅};
22:        T(G) := T(G) − T ;
23:     end {while}
24:   for each t ∈ T do
25:      if [T − {t}] ⊆ B
26:         then T := T − {t};
27:   𝒯 := 𝒯 ∪ {T};
28:   G := B − ∪_{T∈𝒯}[T];
29: end {while};
30: for each T ∈ 𝒯 do
31:    if ∪_{S∈𝒯−{T}}[S] = B
32:       then 𝒯 := 𝒯 − {T};
33: end {procedure}.
```

We will trace the LEM2 algorithm applied to the following input set $\{1,2,3\} = [(Trip, yes)]$. The tracing of LEM2 is presented in the Tables 23.2 and 23.3. The corresponding comments are:

1. The set $G = \{1,2,3\}$. The best attribute-value pair t, with the largest cardinality of the intersection of $[t]$ and G (presented in the third column of Table 23.2) is $(Wind, low)$. The corresponding entry in the third column of Table 23.2 is bulleted. However, $[(Wind, low)] = \{1,2,3,4\} \nsubseteq \{1,2,3\} = B$, hence we need to look for the next t.
2. The set G is the same, $G = \{1,2,3\}$. There are four attribute–value pairs with $\|[t \cap G] = 2$. Two of them have the same cardinality as $[t]$, so we

select the first (top) pair, $(Humidity, low)$. This time $\{1,2,3,4\} \cap \{1,2,5\} = \{1,2\} \subseteq \{1,2,3\}$, so $\{(Wind, low), (Humidity, low)\}$ is the first element T of $\mathcal{T}$.

3. The new set $G = B − [T] = \{1,2,3\} − \{1,2\} = \{3\}$. The pair $[(Humidity, medium)]$ has the smallest cardinality of $[t]$, so it is the best choice. However, $[(Humidity, medium)] = \{3,4\} \nsubseteq \{1,2,3\}$, hence we need to look for the next t.

4. The pair $[(Temperature, medium)]$ is the best choice, and $\{3,4\} \cap \{1,3,5\} = \{3\} \subseteq \{1,2,3\}$, so $\{(Humidity, medium), (Temperature, medium)\}$ is the second element T of $\mathcal{T}$.

Thus,

$$\mathcal{T} = \{\{(Wind, low), (Humidity, low)\},$$
$$\{(Humidity, medium), (Temperature, medium)\}\} .$$

Therefore, the LEM2 algorithm induces the following rule set

$$(Wind, low) \,\&\, (Humidity, low)$$
$$\rightarrow (Trip, yes)$$
$$(Humidity, medium) \,\&\, (Temperature, medium)$$
$$\rightarrow (Trip, yes) .$$

Rules induced from local coverings differ from rules induced from global coverings. In many cases the former are simpler than the latter. For example, for Table 23.1 and the concept $[(Trip, no)]$, the LEM2 algorithm would induce just one rule that covers all three cases

$$(Humidity, high) \rightarrow (Trip, no) .$$

Table 23.2 Computing a local covering for the concept $[(Trip, yes)]$, part I

$(a,v) = t$	$[(a,v)]$	$\{1,2,3\}$	$\{1,2,3\}$
$(Wind, low)$	$\{1,2,3,4\}$	$\{1,2,3\}$ •	$\{1,3\}$
$(Wind, medium)$	$\{5,6,8\}$	−	−
$(Wind, high)$	$\{7\}$	−	−
$(Humidity, low)$	$\{1,2,5\}$	$\{1,2\}$	$\{1,2\}$ •
$(Humidity, medium)$	$\{3,4\}$	$\{3\}$	$\{3\}$
$(Humidity, high)$	$\{6,7,8\}$	−	−
$(Temperature, low)$	$\{2,6\}$	$\{2\}$	$\{1,3\}$
$(Temperature, medium)$	$\{1,3,5\}$	$\{1,3\}$	$\{1,3\}$
$(Temperature, high)$	$\{4,7,8\}$	−	−
Comments		1	2

Table 23.3 Computing a local covering for the concept $[(Trip, yes)]$, part II

$(a,v) = t$	$[(a,v)]$	$\{3\}$	$\{3\}$
$(Wind, low)$	$\{1,2,3,4\}$	$\{3\}$	$\{3\}$
$(Wind, medium)$	$\{5,6,8\}$	−	−
$(Wind, high)$	$\{7\}$	−	−
$(Humidity, low)$	$\{1,2,5\}$	−	−
$(Humidity, medium)$	$\{3,4\}$	$\{3\}$ •	−
$(Humidity, high)$	$\{6,7,8\}$	−	−
$(Temperature, low)$	$\{2,6\}$	−	−
$(Temperature, medium)$	$\{1,3,5\}$	$\{3\}$	$\{3\}$ •
$(Temperature, high)$	$\{4,7,8\}$	−	−
Comments		3	4

On the other hand, the attribute *Humidity* is not included in the global covering. The rules induced from the global covering are

$$(\textit{Temperature}, \textit{high}) \rightarrow (\textit{Trip}, \textit{no}).$$

$$(\textit{Wind}, \textit{medium}) \;\&\; (\textit{Temperature}, \textit{low})$$
$$\rightarrow (\textit{Trip}, \textit{no}).$$

23.1.3 Classification

Rule sets, induced from data sets, are used most frequently to classify new, unseen cases. A *classification system* has two inputs: a rule set and a data set containing new cases and it classifies every case as being a member of some concept. A classification system used in LERS is a modification of the well-known bucket brigade algorithm [23.7–9].

The decision of to which concept a case belongs is made on the basis of three factors: *strength*, *specificity*, and *support*. These factors are defined as follows: *strength* is the total number of cases correctly classified by the rule during training. *Specificity* is the total number of attribute–value pairs on the left-hand side of the rule. The matching rules with a larger number of attribute–value pairs are considered more specific. The third factor, *support*, is defined as the sum of products of strength and specificity for all matching rules indicating the same concept. The concept C for which the support, i.e., the following expression

$$\sum_{\substack{\text{matching rules } r \text{ describing } C}} \text{Strength}(r) *$$
$$\text{Specificity}(r) \qquad (23.6)$$

is the largest is the winner, and the case is classified as being a member of C.

In the classification system of LERS, if complete matching is impossible, all partially matching rules are identified. These are rules with at least one attribute–value pair matching the corresponding attribute–value pair of a case. For any partially matching rule r, the additional factor, called *Matching_factor* (r), is computed. Matching_factor (r) is defined as the ratio of the number of matched attribute–value pairs of r with a case to the total number of attribute–value pairs of r. In partial matching, the concept C for which the following expression

$$\sum_{\substack{\text{partially matching} \\ \text{rules } r \text{ describing } C}} \text{Matching\_factor}(r) *$$
$$\text{Strength}(r) *$$
$$\text{Specificity}(r) . \qquad (23.7)$$

is the largest is the winner and the case is classified as being a member of C.

Since the classification system is a part of the LERS data mining system, rules induced by any component of LERS, such as LEM1 or LEM2, are presented in the LERS format, in which every rule is associated with three numbers: the total number of attribute–value pairs on the left-hand side of the rule (i.e., specificity), the total number of cases correctly classified by the rule during training (i.e., strength), and the total number of training cases matching the left-hand side of the rule, i.e., the rule domain size.

23.2 Inconsistent Data

Frequently data sets contain conflicting cases, i.e., cases with the same attribute values but from different concepts. An example of such a data set is presented in Table 23.4. Cases 4 and 5 have the same values for all three attributes, yet their decision values are different (they belong to different concepts). Similarly, cases 7 and 8 also conflict. Rough set theory handles inconsistent data by introducing lower and upper approximations for every concept [23.1, 2].

There exists a very simple test for consistency: $A^* \leq \{d\}^*$. If this condition is false, the corresponding data set is not consistent. For Table 23.4, $A^* = \{\{1\}, \{2\}, \{3\}, \{4, 5\}, \{6, 7, 8\}, \{9\}, \{10\}\}$, and $\{d\}^* = \{\{1, 2, 3, 4\}, \{5, 6, 7\}, \{8, 9, 10\}\}$, so $A^* \not\leq \{d\}^*$.

Let B be a subset of the set A of all attributes. For inconsistent data sets, in general, a concept X is not a definable set. However, set X may be approximated by two B-definable sets; the first one is called a *B-lower approximation* of X, denoted by $\underline{B}X$ and defined as follows

$$\{x \in U | [x]_B \subseteq X\}. \qquad (23.8)$$

The second set is called a *B-upper approximation* of X, denoted by $\overline{B}X$ and defined as follows

$$\{x \in U | [x]_B \cap X \neq \emptyset\}. \qquad (23.9)$$

In (23.8) and (23.9) lower and upper approximations are constructed from singletons x; we say that we are using the so-called *first method*. The *B*-lower approximation of *X* is the largest *B*-definable set contained in *X*. The *B*-upper approximation of *X* is the smallest *B*-definable set containing *X*.

As was observed in [23.2], for complete decision tables we may use a *second method* to define the *B*-lower approximation of *X*, by the following formula

$$\cup\{[x]_B | x \in U, [x]_B \subseteq X\}, \qquad (23.10)$$

while the *B*-upper approximation of *x* may be defined, using the second method, by

$$\cup\{[x]_B | x \in U, [x]_B \cap X \neq \emptyset\}. \qquad (23.11)$$

Obviously, both (23.8) and (23.10) define the same set. Similarly, (23.9) and (23.11) also define the same set. For Table 23.4,

$$\underline{A}\{1, 2, 3, 4\} = \{1, 2, 3\}$$

and

$$\overline{A}\{1, 2, 3, 4\} = \{1, 2, 3, 4, 5\}.$$

It is well known that for any $B \subseteq A$ and $X \subseteq U$,

$$\underline{B}X \subseteq X \subseteq \overline{B}X, \qquad (23.12)$$

hence any case *x* from $\underline{B}X$ is *certainly* a member of *X*, while any member *x* of $\overline{B}X$ is *possibly* a member of *X*. This observation is used in the LERS data mining system. If an input data set is inconsistent, LERS computes lower and upper approximations for any concept and then induces *certain* rules from the lower approximation and *possible* rules from the upper approximation. For example, if we want to induce certain and possible

rule sets for the concept [(*Trip*, *yes*)] from Table 23.4, we need to consider the following two data sets, presented in Tables 23.5 and 23.6.

Table 23.5 was obtained from Table 23.4 by assigning the value *yes* of the decision *Trip* to all cases from the lower approximation of [(*Trip*, *yes*)] and by replacing all remaining values of *Trip* by a special value, say *SPECIAL*. Similarly, Table 23.6 was obtained from Table 23.4 by assigning the value *yes* of the decision *Trip* to all cases from the upper approximation of [(*Trip*, *yes*)] and by replacing all remaining values of *Trip* by the value *SPECIAL*. Obviously, both tables 23.5 and 23.6 are consistent. Therefore, we may use the LEM1 or LEM2 algorithms to induce rules from Tables 23.5 and 23.6. The rule set induced by the LEM2 algorithm from Table 23.5 is:

- 2, 2, 2

  (*Wind*, *low*) &

  (*Humidity*, *low*) → (*Trip*, *yes*) ,

Table 23.5 A new data set for inducing certain rules for the concept [(*Trip*, *yes*)]

	Attributes			Decision
Case	Wind	Humidity	Temperature	Trip
1	low	low	medium	yes
2	low	low	low	yes
3	low	medium	medium	yes
4	low	medium	high	SPECIAL
5	low	medium	high	SPECIAL
6	medium	low	medium	SPECIAL
7	medium	low	medium	SPECIAL
8	medium	low	medium	SPECIAL
9	high	high	high	SPECIAL
10	medium	high	high	SPECIAL

Table 23.4 An inconsistent decision table

	Attributes			Decision
Case	Wind	Humidity	Temperature	Trip
1	low	low	medium	yes
2	low	low	low	yes
3	low	medium	medium	yes
4	low	medium	high	yes
5	low	medium	high	maybe
6	medium	low	medium	maybe
7	medium	low	medium	maybe
8	medium	low	medium	no
9	high	high	high	no
10	medium	high	high	no

Table 23.6 A new data set for inducing possible rules for the concept [(*Trip*, *yes*)]

	Attributes			Decision
Case	Wind	Humidity	Temperature	Trip
1	low	low	medium	yes
2	low	low	low	yes
3	low	medium	medium	yes
4	low	medium	high	yes
5	low	medium	high	yes
6	medium	low	medium	SPECIAL
7	medium	low	medium	SPECIAL
8	medium	low	medium	SPECIAL
9	high	high	high	SPECIAL
10	medium	high	high	SPECIAL

- 2, 1, 1

  (*Humidity*, *medium*) &
  (*Temperature*, *medium*) → (*Trip*, *yes*) ,

- 1, 4, 4

  (*Temperature*, *high*) → (*Trip*, *SPECIAL*) ,

- 1, 4, 4

  (*Wind*, *medium*) → (*Trip*, *SPECIAL*) ,

where all rules are presented in the LERS format, see
Sect. 23.1.3.

Obviously, only rules with (*Trip*, *yes*) on the right-hand side are informative; the remaining rules, with (*Trip*, *SPECIAL*) on the right-hand side should be ignored. These two rules are *certain*. The only informative rule induced by the LEM2 algorithm from Table 23.6 is:

- 1, 4, 5

  (*Wind*, *low*) → (*Trip*, *yes*) .

This rule is *possible*.

23.3 Decision Table with Numerical Attributes

An example of a data set with numerical attributes is presented in Table 23.7.

In rule induction from numerical data a preliminary step called *discretization* [23.10–12] is usually conducted. During discretization a domain of the numerical attribute is divided into intervals defined by cut-points (left and right delimiters of intervals). Such an interval, delimited by two cut-points, c and d, will be denoted by $c \ldots d$. In this chapter we will discuss how to do both processes concurrently: rule induction and discretization. First we need to check whether our data set is consistent. Note that numerical data are, in general, consistent, but inconsistent numerical data are possible. For inconsistent numerical data we need to compute lower and upper approximations and the induce certain and possible rule sets. In the data set from Table 23.7, $A^* = \{\{1\}, \{2\}, \{3\}, \{4\}, \{5\}, \{6\}, \{7\}, \{8\}\}$, $\{d\}^* = \{\{1, 2, 3\}, \{4, 5\}, \{6, 7, 8\}\}$, so $A^* \leq \{d\}^*$, and the data set is consistent.

A modified LEM2 algorithm for rule induction, called MLEM2 [23.13], does not need any preliminary discretization of numerical attributes. The domain of

any numerical attribute is sorted first. Then potential cut-points are selected as averages of any two consecutive values of the sorted list. For each cut-point c the MLEM2 algorithm creates two blocks, the first block contains all cases for which values of the numerical attribute are smaller than c, the second block contains the remaining cases (with values of the numerical attribute larger than c). Once all such blocks have been computed, rule induction in MLEM2 is conducted the same way as in LEM2. We will illustrate rule induction

Table 23.8 Computing a local covering for the concept [(*Trip*, *yes*)], part I

$(a, v) = t$	$[(a, v)]$	$\{1, 2, 3\}$	$\{1, 2, 3\}$
(*Wind*, 4..6)	$\{1, 3\}$	$\{1, 3\}$	$\{1, 3\}$ •
(*Wind*, 6..30)	$\{2, 4, 5, 6, 7, 8\}$	$\{2\}$	$\{2\}$
(*Wind*, 4..10)	$\{1, 2, 3, 4\}$	$\{1, 2, 3\}$ •	—
(*Wind*, 10..30)	$\{5, 6, 7, 8\}$	—	—
(*Wind*, 4..14)	$\{1, 2, 3, 4, 5, 8\}$	$\{1, 2, 3\}$	—
(*Wind*, 14..30)	$\{6, 7\}$	—	—
(*Wind*, 4..23)	$\{1, 2, 3, 4, 5, 6, 8\}$	$\{1, 2, 3\}$	—
(*Wind*, 23..30)	$\{7\}$	—	—
(*Humidity*, low)	$\{1, 2, 5\}$	$\{1, 2\}$	$\{1, 2\}$
(*Humidity*, medium)	$\{3, 4\}$	$\{3\}$	$\{3\}$
(*Humidity*, high)	$\{6, 7, 8\}$	—	—
(*Temperature*, low)	$\{2, 6\}$	$\{2\}$	$\{1, 3\}$
(*Temperature*, medium)	$\{1, 3, 5\}$	$\{1, 3\}$	$\{1, 3\}$
(*Temperature*, high)	$\{4, 7, 8\}$	—	—
Comments		1	2

Table 23.7 A data set with numerical attributes

	Attributes			Decision
Case	Wind	Humidity	Temperature	Trip
1	4	low	medium	yes
2	8	low	low	yes
3	4	medium	medium	yes
4	8	medium	high	maybe
5	12	low	medium	maybe
6	16	high	low	no
7	30	high	high	no
8	12	high	high	no

Table 23.9 Computing a local covering for the concept [(*Trip, yes*)], part II

(a, v) = t	[(a, v)]	{2}	{2}
(*Wind*, 4..6)	{1, 3}	–	–
(*Wind*, 6..30)	{2, 4, 5, 6, 7, 8}	{2}	{2}
(*Wind*, 4..10)	{1, 2, 3, 4}	{2}	{2}
(*Wind*, 10..30)	{5, 6, 7, 8}	–	–
(*Wind*, 4..14)	{1, 2, 3, 4, 5, 8}	{2}	{2}
(*Wind*, 14..30)	{6, 7}	–	–
(*Wind*, 4..23)	{1, 2, 3, 4, 5, 6, 8}	{2}	{2}
(*Wind*, 23..30)	{7}	–	–
(*Humidity*, low)	{1, 2, 5}	{2}	{2} •
(*Humidity*, medium)	{3, 4}	–	–
(*Humidity*, high)	{6, 7, 8}	–	–
(*Temperature*, low)	{2, 6}	{2} •	–
(*Temperature*, medium)	{1, 3, 5}	–	–
(*Temperature*, high)	{4, 7, 8}	–	–
Comments		3	4

from Table 23.7 using the MLEM2 rule induction algorithm. The MLEM2 algorithm is shown in Tables 23.8 and 23.9. The corresponding comments are

1. The set $G = \{1, 2, 3\}$. The best attribute–value pair t, with the largest cardinality of the intersection of $[t]$ and G (presented in the third column of Table 23.8) is (*Wind*, 4..10). The corresponding entry in the third column of Table 23.8 is bulleted. However,

$$[(Wind, 4..10)] = \{1, 2, 3, 4\} \not\subseteq \{1, 2, 3\} = B,$$

hence we need to look for the next t.

2. Set G is the same, $G = \{1, 2, 3\}$. There are dashes for rows (*Wind*, 4..14) and (*Wind*, 4..23) since the corresponding intervals contain 4..10. There are four attribute–value pairs with $||t \cap G| = 2$. The best attribute–value pair, with the smallest cardinality of $[t]$, is (*Wind*, 4..6). This time

$$\{1, 2, 3, 4\} \cap \{1, 3\} = \{1, 3\} \subseteq \{1, 2, 3\}.$$

Obviously, the common part of both intervals is 4..6, so $\{(Wind, 4..6)\}$ is the first element T of $\mathcal{T}$.

3. The new set $G = B - [T] = \{1, 2, 3\} - \{1, 3\} = \{2\}$. The pair [(*Temperature, low*)] has the smallest cardinality of $[t]$, so it is the best choice. However, [(*Temperature, low*)] $= \{2, 6\} \not\subseteq \{1, 2, 3\}$, hence we need to look for the next t.

4. The pair [(*Humidity, low*)] is the best choice, and

$$\{3, 4\} \cap \{1, 3, 5\} = \{3\} \subseteq \{1, 2, 3\},$$

so $\{[(Temperature, low), (Humidity, low)]\}$ is the second element T of $\mathcal{T}$.

As a result,

$$\mathcal{T} = \{\{(Wind, 4..6)\}, \{(Temperature, low), (Humidity, low)\}\}.$$

In other words, the MLEM2 algorithm induces the following rule set for Table 23.7:

- 1, 2, 2

    (*Wind*, 4..6) → (*Trip, yes*),

- 2, 1, 1

    (*Temperature, low*) & (*Humidity, low*) → (*Trip, yes*).

23.4 Incomplete Data

Real-life data are frequently incomplete. In this section we will consider incompleteness in the form of missing attribute values. We will distinguish three types of missing attribute values:

- *Lost values*, denoted by ?, where the original values existed, but are currently unavailable, since these values have been, for example, erased or the operator forgot to input them. In rule induction we

will induce rules from existing, specified attribute values.

- *Do not care conditions*, denoted by *, where the original values are mysterious. For example, data were collected in a form of the interview, some questions were considered to be irrelevant or were embarrassing. Let us say that in an interview associated with the diagnosis of a disease, there is a question about eye color. For some people such a question is irrelevant. In rule induction we are assuming that the attribute value is any value from the attribute domain.

- *Attribute-concept value*, denoted by −. This interpretation is a special case of the *do not care* condition: it is restricted to attribute values typical for the concept to which the case belongs. For example, typical values of temperature for patients sick with flu are: high and very-high, for a patient the temperature value is missing, but we know that this patient is sick with flu, if using the attribute-concept interpretation, we will assume that possible temperature values are: high and very-high.

We will assume that for any case at least one attribute value is specified (i. e., is not missing) and that all decision values are specified. An example of a decision table with missing attribute values is presented in Table 23.10.

The definition of consistent data from Sect. 23.2 cannot be applied to data with missing attribute values, since for such data the standard definition of the indiscernibility relation must be extended. Moreover, it is well known that the standard definitions of lower and upper approximations are not applicable to data with missing attribute values. In Sect. 23.4.1 we will discuss three generalizations of the standard approximations: singleton, subset, and concept.

23.4.1 Singleton, Subset, and Concept Approximations

For incomplete data the definition of a block of an attribute-value pair is modified [23.14]:

- If for an attribute a there exists a case x such that $a(x) = ?$, i. e., the corresponding value is lost, then

Table 23.10 An incomplete decision table

	Attributes			Decision
Case	Wind	Humidity	Temperature	Trip
1	low	low	medium	yes
2	?	low	*	yes
3	*	medium	medium	yes
4	low	?	high	maybe
5	medium	−	medium	maybe
6	*	high	low	no
7	−	high	*	no
8	medium	high	high	no

the case x should not be included in any blocks $[(a, v)]$ for all values v of attribute a.

- If for an attribute a there exists a case x such that the corresponding value is a *do not care* condition, i. e., $a(x) = *$, then the case x should be included in blocks $[(a, v)]$ for all specified values v of attribute a.
- If for an attribute a there exists a case x such that the corresponding value is an attribute–concept value, i. e., $a(x) = -$, then the corresponding case x should be included in blocks $[(a, v)]$ for all specified values $v \in V(x, a)$ of attribute a, where

$$V(x, a) = \{a(y) \mid a(y) \text{ is specified}, \\ y \in U, \ d(y) = d(x)\}. \quad (23.13)$$

For Table 23.10,

$$V(5, Humidity) = \emptyset \text{ and}$$
$$V(7, Wind) = \{medium\},$$

so the blocks of attribute–value pairs are

$$[(Wind, low)] = \{1, 3, 4, 6\},$$
$$[(Wind, medium)] = \{3, 5, 6, 7, 8\},$$
$$[(Humidity, low)] = \{1, 2\},$$
$$[(Humidity, medium)] = \{3\},$$
$$[(Humidity, high)] = \{6, 7, 8\},$$
$$[(Temperature, low)] = \{2, 6, 7\},$$
$$[(Temperature, medium)] = \{1, 2, 3, 5, 7\},$$
$$[(Temperature, high)] = \{2, 4, 7, 8\}.$$

For a case $x \in U$, the *characteristic set* $K_B(x)$ is defined as the intersection of the sets $K(x, a)$, for all $a \in B$, where the set $K(x, a)$ is defined in the following way:

- If $a(x)$ is specified, then $K(x, a)$ is the block $[(a, a(x)]$ of attribute a and its value $a(x)$.
- If $a(x) = ?$ or $a(x) = *$ then the set $K(x, a) = U$.
- If $a(x) = -$, then the corresponding set $K(x, a)$ is equal to the union of all blocks of attribute–value pairs (a, v), where $v \in V(x, a)$ if $V(x, a)$ is nonempty. If $V(x, a)$ is empty, $K(x, a) = U$.

Part C | 23.4

For Table 23.10 and $B = A$,

$$K_A(1) = \{1,3,4,6,7\} \cap \{1,2\} \cap \{1,2,3,5,7\}$$
$$= \{1\},$$
$$K_A(2) = U \cap \{1,2\} \cap U = \{1,2\},$$
$$K_A(3) = U \cap \{3\} \cap \{1,2,3,5,7\} = \{3\},$$
$$K_A(4) = \{1,3,4,6\} \cap U \cap \{1,2,3,5,7\} = \{4\},$$
$$K_A(5) = \{3,5,6,7,8\} \cap U \cap (\{1,2,3,5,7\})$$
$$= \{3,5,7\},$$
$$K_A(6) = U \cap \{6,7,8\} \cap \{2,6,7\} = \{6,7\},$$
$$K_A(7) = \{3,5,6,7,8\} \cap \{6,7,8\} \cap U = \{6,7,8\},$$
$$K_A(8) = \{3,5,6,7,8\} \cap \{6,7,8\} \cap \{2,4,7,8\}$$
$$= \{7,8\}.$$

The characteristic set $K_B(x)$ may be interpreted as the set of cases that are indistinguishable from x using all attributes from B and using a given interpretation of missing attribute values. For completely specified data sets (i.e., data sets with no missing attribute values), characteristic sets are reduced to elementary sets. The *characteristic relation* $R(B)$ is a relation on U defined for $x, y \in U$ as follows

$$(x,y) \in R(B) \text{ if and only if } y \in K_B(x). \quad (23.14)$$

The characteristic relation $R(B)$ is reflexive but – in general – does not need to be symmetric or transitive. Obviously, the characteristic relation $R(B)$ is known if we know characteristic sets $K_B(x)$ for all $x \in U$ and vice versa. In our example, $R(A) = \{(1,1),(2,1),(2,2),(3,3),(4,4),(5,3),(5,5),(6,6),$ $(6,7),(7,6),(7,7),(7,8),(8,7),(8,8)\}$. For a complete decision table, the characteristic relation $R(B)$ is reduced to the indiscernibility relation [23.2].

Definability for completely specified decision tables should be modified to fit into incomplete decision tables. For incomplete decision tables, a union of some intersections of attribute–value pair blocks, where such attributes are members of B and are distinct, will be called B-*locally definable* sets. A union of characteristic sets $K_B(x)$, where $x \in X \subseteq U$ will be called a B-*globally definable* set. Any set X that is B-globally definable is B-locally definable; the converse is not true.

For example, the set $\{2\}$ is A-locally definable since $\{2\} = [(Humidity, low)] \cap [(Temperature, high)]$. However, the set $\{2\}$ is not A-globally definable. On the other hand, the set $\{5\} =$ is not even locally definable since all blocks of attribute–value pairs containing case 5 contain

also the case 7 as well. Obviously, if a set is not B-locally definable then it cannot be expressed by rule sets using attributes from B. Thus we should induce rules from sets that are at least A-locally definable.

For incomplete decision tables lower and upper approximations may be defined in a few different ways. We suggest three different definitions of lower and upper approximations for incomplete decision tables, following [23.14–16]. Let X be a concept, a subset of U, let B be a subset of the set A of all attributes, and let $R(B)$ be the characteristic relation of the incomplete decision. Our first definition uses an idea similar to the first method in Sect. 23.2, and is based on constructing both approximations from single elements of the set U. We will call these approximations *singleton*. A singleton B-lower approximation of X is defined as follows

$$\underline{B}X = \{x \in U \mid K_B(x) \subseteq X\}. \quad (23.15)$$

A singleton B-upper approximation of X is

$$\overline{B}X = \{x \in U \mid K_B(x) \cap X \neq \emptyset\}. \quad (23.16)$$

In our example of the decision table presented in Table 23.10, the singleton A-lower and A-upper approximations of the concept: $\{1,2,3\}$ are:

$$\underline{A}\{1,2,3\} = \{1,2,3\}, \quad (23.17)$$
$$\overline{A}\{1,2,3\} = \{1,2,3,5\}. \quad (23.18)$$

We may easily observe that the set $\{1,2,3,5\} = (\overline{A}\{1,2,3\})$ is not A-locally definable since in all blocks of attribute–value pairs cases 5 and 7 are inseparable. Thus, as it was observed in, e.g., [23.14–16], singleton approximations should not be used, theoretically, for rule induction.

The second method of defining lower and upper approximations for complete decision tables uses another idea: lower and upper approximations are unions of elementary sets, subsets of U. Therefore, we may define lower and upper approximations for incomplete decision tables by analogy with the second method in Sect. 23.2, using characteristic sets instead of elementary sets. There are two ways to do this. Using the first way, a *subset* B-lower approximation of X is defined as follows

$$\underline{B}X = \cup\{K_B(x) \mid x \in U, K_B(x) \subseteq X\}. \quad (23.19)$$

A *subset* B-upper approximation of X is

$$\overline{B}X = \cup\{K_B(x) \mid x \in U, K_B(x) \cap X \neq \emptyset\}. \quad (23.20)$$

For any concept X, singleton B-lower and B-upper approximations of X are subsets of the subset B-lower and B-upper approximations of X, respectively [23.16], because the characteristic relation $R(B)$ is reflexive. For the decision table presented in Table 23.10, the subset A-lower and A-upper approximations are

$$\underline{A}\{1,2,3\} = \{1,2,3\},$$
$$\overline{A}\{1,2,3\} = \{1,2,3,5,7\}.$$

The second possibility is to modify the subset definition of lower and upper approximation by replacing the universe U from the subset definition by a concept X. A *concept B-lower approximation* of the concept X is defined as follows

$$\underline{B}X = \cup\{K_B(x) \mid x \in X, K_B(x) \subseteq X\}. \quad (23.21)$$

Obviously, the subset B-lower approximation of X is the same set as the concept B-lower approximation of X. A *concept B-upper approximation* of the concept X is defined as follows

$$\overline{B}X = \cup\{K_B(x) \mid x \in X, K_B(x) \cap X \neq \emptyset\}$$
$$= \cup\{K_B(x) \mid x \in X\}. \quad (23.22)$$

The concept upper approximations were defined in [23.17] and [23.18] as well. The concept B-upper approximation of X is a subset of the subset B-upper approximation of X [23.16]. For the decision table presented in Table 23.10, the concept A-upper approximations is

$$\overline{A}\{1,2,3\} = \{1,2,3\}.$$

Note that for complete decision tables, all three definitions of lower and upper approximations, singleton, subset, and concept, are reduced to the same standard definition of lower and upper approximations, respectively.

23.4.2 Modified LEM2 Algorithm

The same MLEM2 rule induction from Sect. 23.3 may be used for rule induction from incomplete data; the only difference is a different definition of blocks of attribute–value pairs. Let us apply the MLEM2 algorithm to the data set from Table 23.10. First, we need to make a decision as to what kind of approximations we are going to use: singleton, subset, or concept. In our example, we use concept approximation. For Table 23.10,

$$\underline{A}\{1,2,3\} = \overline{A}\{1,2,3\} = \{1,2,3\},$$

we will trace the MLEM2 algorithm applied to the set $\{1,2,3\}$; this way our certain rule set, for the concept $[(Trip, yes)]$, is at the same time certain and possible. The tracing of LEM2 is presented in the Tables 23.11.

The corresponding comments are:

1. The set $G = \{1,2,3\}$. The best attribute–value pair t, with the largest cardinality of the intersection of $[t]$ and G (presented in the third column of Table 23.11) is $(Temperature, medium)$. The corresponding entry in the third column of Table 23.11 is bulleted. However,

$$[(Temperature, medium)]$$
$$= \{1,2,3,5,7\} \nsubseteq \{1,2,3\} = B,$$

hence we need to look for the next t.

2. Set G is the same, $G = \{1,2,3\}$. There are two attribute–value pairs with $|[t \cap G| = 2$. One of them, $(Humidity, low)$ has the smallest cardinality of $[t]$, so we select it. This time

$$\{1,2,3,5,7\} \cap \{1,2\} = \{1,2\} \subseteq \{1,2,3\}.$$

However, $(Temperature, medium)$ is redundant, since $[(Humidity, low)] \subseteq \{1,2,3\}$, hence $\{(Humidity, low)\}$ is the first element T of $\mathcal{T}$.

3. The new set $G = B - [T] = \{1,2,3\} - \{1,2\} = \{3\}$. The pair $[(Humidity, medium)$ has the smallest cardinality of $[t]$, so it is the best choice. Additionally, $[(Humidity, medium)] = \{3\} \subseteq \{1,2,3\}$, hence we are done, the set $T = \{(Humidity, medium)\}$.

Table 23.11 Computing a rule set for the concept $[(Trip, yes)]$, Table 23.10

$(a, v) = t$	$[(a,v)]$	$\{1,2,3\}$	$\{1,2,3\}$	$\{3\}$
$(Wind, low)$	$\{1,3,4,6\}$	$\{1,3\}$	$\{1,3\}$	$\{3\}$
$(Wind, medium)$	$\{3,5,6,7,8\}$	$\{3\}$	$\{3\}$	$\{3\}$
$(Humidity, low)$	$\{1,2\}$	$\{1,2\}$	$\{1,2\}$ •	$-$
$(Humidity, medium)$	$\{3\}$	$\{3\}$	$\{3\}$	$\{3\}$ •
$(Humidity, high)$	$\{6,7,8\}$	$-$	$-$	$-$
$(Temperature, low)$	$\{2,6,7\}$	$\{2\}$	$-$	$-$
$(Temperature, medium)$	$\{1,2,3,5,7\}$	$\{1,2,3\}$•	$-$	$\{3\}$
$(Temperature, high)$	$\{2,4,7,8\}$	$\{2\}$	$-$	$-$
Comments		1	2	3

Part C | 23.4

Therefore, $\mathcal{T} = \{\{(Humidity, low)\}, \{(Humidity, medium)\}\}$. The MLEM2 algorithm induces the following rule set for Table 23.10:

- 1, 2, 2

$$(Humidity, low) \rightarrow (Trip, yes),$$

- 1, 1, 1

$$(Humidity, medium) \rightarrow (Trip, yes).$$

23.4.3 Probabilistic Approximations

In this section we are going to generalize singleton, subset, and concept approximations from Sect. 23.4.1 to corresponding approximations that are defined using an additional parameter (or threshold), denoted by α, and interpreted as a probability. A generalization of standard approximations, called *probabilistic approximations*, has been studied in many papers [23.19–26].

Let B be a subset of the attribute set A and X be a subset of U.

A B-singleton probabilistic approximation of X with the threshold α, $0 < \alpha \le 1$, denoted by $\mathrm{appr}_{\alpha,B}^{\mathrm{singleton}}(X)$, is defined as follows

$$\{x \mid x \in U,\ Pr(X \mid K_B(x)) \ge \alpha\},$$

where

$$Pr(X \mid K_B(x)) = \frac{|X \cap K_B(x)|}{|K_B(x)|}$$

is the conditional probability of X given $K_B(x)$ and $|Y|$ denotes the cardinality of set Y.

A B-subset probabilistic approximation of the set X with the threshold α, $0 < \alpha \le 1$, denoted by $\mathrm{appr}_{\alpha,B}^{\mathrm{subset}}(X)$, is defined as follows

$$\cup\{K_B(x) \mid x \in U,\ Pr(X \mid K_B(x)) \ge \alpha\}.$$

A B-concept probabilistic approximation of the set X with the threshold α, $0 < \alpha \le 1$, denoted by $\mathrm{appr}_{\alpha,B}^{\mathrm{concept}}(X)$, is defined as follows

$$\cup\{K_B(x) \mid x \in X,\ Pr(X \mid K_B(x)) \ge \alpha\}.$$

For simplicity, if $B = A$, an A-singleton, B-subset, and B-concept probabilistic approximations will be called singleton, subset, and concept prob-

abilistic approximations, and will be denoted by $\mathrm{appr}_{\alpha}^{\mathrm{singleton}}(X)$, $\mathrm{appr}_{\alpha}^{\mathrm{subset}}(X)$, and $\mathrm{appr}_{\alpha}^{\mathrm{concept}}(X)$, respectively.

Obviously, for the concept X, the probabilistic approximation of a given type (singleton, subset, or concept) of X computed for the threshold equal to the smallest positive conditional probability $Pr(X \mid [x])$ is equal to the standard upper approximation of X of the same type. Additionally, the probabilistic approximation of a given type of X computed for the threshold equal to 1 is equal to the standard lower approximation of X of the same type.

For the data set from Table 23.12, the set of blocks of attribute–value pairs is

$$[(Wind, low)] = \{1, 3, 5\},$$
$$[(Wind, high)] = \{4, 6, 7, 8\},$$
$$[(Humidity, low)] = \{1, 2, 3, 5\},$$
$$[(Humidity, high)] = \{1, 4, 6, 7, 8\},$$
$$[(Temperature, low)] = \{1, 2, 5, 6\},$$
$$[(Temperature, high)] = \{1, 4, 6, 7, 8\}.$$

The corresponding characteristic sets are

$$K_A(1) = K_A(3) = \{1, 3, 5\},$$
$$K_A(2) = \{1, 2, 5\},$$
$$K_A(4) = \{4, 6, 8\},$$
$$K_A(5) = \{1, 5\},$$
$$K_A(6) = K_A(8) = \{4, 6, 8\},$$
$$K_A(7) = \{4, 6, 7, 8\}.$$

Conditional probabilities of the concept $\{1, 2, 3, 4\}$ given a characteristic set $K_A(x)$ are presented in Table 23.13.

Table 23.12 An incomplete decision table

Case	Attributes			Decision
	Wind	Humidity	Temperature	Trip
1	low	low	*	yes
2	?	low	low	yes
3	low	low	?	yes
4	high	high	high	yes
5	low	*	low	no
6	high	high	*	no
7	high	?	high	no
8	high	high	high	no

Table 23.13 Conditional probabilities

$K_A(x)$	{1, 2, 5}	{1, 3, 5}	{1, 5}	{4, 6, 8}	{4, 6, 7, 8}
$Pr(\{1,2,4,6\} \mid K_A(x))$	0.667	0.667	0.5	0.333	0.25

For Table 23.13, all probabilistic approximations (singleton, subset, and concept) are

$$\text{appr}_{0.25}^{\text{singleton}}(\{1,2,3,4\}) = U,$$

$$\text{appr}_{0.333}^{\text{singleton}}(\{1,2,3,4\}) = \{1,2,3,4,5,6,8\},$$

$$\text{appr}_{0.5}^{\text{singleton}}(\{1,2,3,4\}) = \{1,2,3,5\},$$

$$\text{appr}_{0.667}^{\text{singleton}}(\{1,2,3,4\}) = \{1,2,3\},$$

$$\text{appr}_{1}^{\text{singleton}}(\{1,2,3,4\}) = \emptyset,$$

$$\text{appr}_{0.25}^{\text{subset}}(\{1,2,3,4\}) = U,$$

$$\text{appr}_{0.333}^{\text{subset}}(\{1,2,3,4\}) = \{1,2,3,4,5,6,8\},$$

$$\text{appr}_{0.5}^{\text{subset}}(\{1,2,3,3\}) = \{1,2,3,5\},$$

$$\text{appr}_{0.667}^{\text{subset}}(\{1,2,3,4\}) = \{1,2,3,5\},$$

$$\text{appr}_{1}^{\text{subset}}(\{1,2,3,4\}) = \emptyset,$$

$$\text{appr}_{0.25}^{\text{concept}}(\{1,2,3,4\}) = \{1,2,3,4,5,6,8\},$$

$$\text{appr}_{0.333}^{\text{concept}}(\{1,2,3,4\}) = \{1,2,3,4,5,6,8\},$$

$$\text{appr}_{0.5}^{\text{concept}}(\{1,2,3,4\}) = \{1,2,3,5\},$$

$$\text{appr}_{0.667}^{\text{concept}}(\{1,2,3,4\}) = \{1,2,3,5\},$$

$$\text{appr}_{1}^{\text{concept}}(\{1,2,3,4\}) = \emptyset.$$

For rule induction from probabilistic approximations of the given concept a technique similar to the

Table 23.14 A modified decision table

	Attributes			Decision
Case	Wind	Humidity	Temperature	Trip
1	low	low	*	yes
2	?	low	low	yes
3	low	low	?	yes
4	high	high	high	SPECIAL
5	low	*	low	no
6	high	high	*	SPECIAL
7	high	?	high	SPECIAL
8	high	high	high	SPECIAL

one in Sect. 23.2 may be used. For any concept and the probabilistic approximation of the concept we will create a new decision table. Let us illustrate this idea with inducing a rule set for the concept [(Trip, yes)] from Table 23.12 using concept probabilistic approximation with $\alpha = 0.5$. The corresponding modified decision table is presented in Table 23.14.

In the data set presented in Table 23.14, all values of Trip are copied from Table 23.12 for all cases from

$$\text{appr}_{0.5}^{\text{concept}}(\{1,2,3,4\}) = \{1,2,3,5\},$$

while for all remaining cases values of Trip are replaced by the SPECIAL value. The MLEM2 rule induction algorithm, using concept upper approximation should be used with the corresponding type of upper approximation (singleton, subset, and concept). In our example, the MLEM2 rule induction algorithm, using concept upper approximation, induces the following rule set from Table 23.14:

- 1, 3, 4

  $(Humidity, low) \rightarrow (Trip, yes),$

- 1, 4, 4

  $(Wind, high) \rightarrow (Trip, SPECIAL),$

- 2, 1, 2

  $(Wind, low) \& (Temperature, low) \rightarrow (Trip, no).$

The only rules that are useful should have (Trip, yes) on the right-hand side. Thus, the only rule that survives is:

- 1, 3, 4

  $(Humidity, low) \rightarrow (Trip, yes).$

23.5 Conclusions

Investigation of rule induction methods is subject to intensive research activity. New versions of rule induction algorithms based on probabilistic approximations have been explored [23.27, 28]. Novel rule induction algorithms in which computation of proba-bilistic approximations is done in parallel with rule induction were recently developed and experimentally tested [23.29]. The LEM2 algorithm was implemented in a bagged version [23.30], using ideas of ensemble learning.

References

23.1   Z. Pawlak: Rough sets, Int. J. Comput. Inf. Sci. **11**, 341–356 (1982)

23.2   Z. Pawlak: *Rough Sets. Theoretical Aspects of Reasoning about Data* (Kluwer Academic, Boston 1991)

23.3   J.W. Grzymala-Busse: Rule induction. In: *Data Mining and Knowledge Discovery Handbook, Second Edition*, ed. by O. Maimon, L. Rokach (Springer, Berlin, Heidelberg 2010) pp. 249–265

23.4   Z. Pawlak, J.W. Grzymala-Busse, R. Slowinski, W. Ziarko: Rough sets, Commun. ACM **38**, 89–95 (1995)

23.5   J.G. Bazan, M.S. Szczuka, A. Wojna, M. Wojnarski: On the evolution of rough set exploration system. In: *Rough Sets and Current Trends in Computing*, ed. by S. Tsumoto, R. Słowiński, J. Komorowski, J.W. Grzymala-Busse (Springer, Berlin, Heidelberg 2004) pp. 592–601

23.6   J.W. Grzymala-Busse: LERS – A system for learning from examples based on rough sets. In: *Intelligent Decision Support. Handbook of Applications and Advances of the Rough Set Theory*, ed. by R. Slowinski (Kluwer Academic, Boston 1992) pp. 3–18

23.7   J. Stefanowski: *Algorithms of Decision Rule Induction in Data Mining* (Poznan University of Technology Press, Poznan 2001)

23.8   L.B. Booker, D.E. Goldberg, J.F. Holland: Classifier systems and genetic algorithms. In: *Machine Learning. Paradigms and Methods*, ed. by J.G. Carbonell (MIT Press, Cambridge 1990) pp. 235–282

23.9   J.H. Holland, K.J. Holyoak, R.E. Nisbett, P.R. Thagard: *Induction. Processes of Inference, Learning, and Discovery* (MIT Press, Cambridge 1986)

23.10   M.R. Chmielewski, J.W. Grzymala-Busse: Global discretization of continuous attributes as preprocessing for machine learning, Int. J. Approx. Reason. **15**(4), 319–331 (1996)

23.11   J.W. Grzymala-Busse: Discretization of numerical attributes. In: *Handbook of Data Mining and Knowledge Discovery*, ed. by W. Kloesgen, J. Zytkow (Oxford Univ. Press, Oxford 2002) pp. 218–225

23.12   J.W. Grzymala-Busse: Mining numerical data – A rough set approach, Trans. Rough Sets **11**, 1–13 (2010)

23.13   J.W. Grzymala-Busse: MLEM2: A new algorithm for rule induction from imperfect data, Proc. 9th Int. Conf. Inform. Proc. Manag. Uncertain. Knowl.-Based Syst. (2002) pp. 243–250

23.14   J.W. Grzymala-Busse: Data with missing attribute values: Generalization of indiscernibility relation and rule induction, Trans. Rough Sets **1**, 78–95 (2004)

23.15   J.W. Grzymala-Busse: Rough set strategies to data with missing attribute values, Proc. Workshop Found. New Dir. Data Min. (2003) pp. 56–63

23.16   J.W. Grzymala-Busse: Characteristic relations for incomplete data: A generalization of the indiscernibility relation. In: *Rough Sets and Current Trends in Computing*, ed. by S. Tsumoto, R. Słowiński, J. Komorowski, J.W. Grzymala-Busse (Springer, Berlin, Heidelberg 2004) pp. 244–253

23.17   T.Y. Lin: Topological and fuzzy rough sets. In: *Intelligent Decision Support. Handbook of Applications and Advances of the Rough Sets Theory*, ed. by R. Slowinski (Kluwer Academic, Boston 1992) pp. 287–304

23.18   R. Slowinski, D. Vanderpooten: A generalized definition of rough approximations based on similarity, IEEE Trans. Knowl. Data Eng. **12**, 331–336 (2000)

23.19   J.W. Grzymala-Busse, W. Ziarko: Data mining based on rough sets. In: *Data Mining: Opportunities and Challenges*, ed. by J. Wang (Idea Group, Hershey 2003) pp. 142–173

23.20   J.W. Grzymala-Busse, Y. Yao: Probabilistic rule induction with the LERS data mining system, Int. J. Intell. Syst. **26**, 518–539 (2011)

23.21   Z. Pawlak, A. Skowron: Rough sets: Some extensions, Inf. Sci. **177**, 28–40 (2007)

23.22   Z. Pawlak, S.K.M. Wong, W. Ziarko: Rough sets: probabilistic versus deterministic approach, Int. J. Man-Mach. Stud. **29**, 81–95 (1988)

23.23   Y.Y. Yao: Probabilistic rough set approximations, Int. J. Approx. Reason. **49**, 255–271 (2008)

23.24   Y.Y. Yao, S.K.M. Wong: A decision theoretic framework for approximate concepts, Int. J. Man-Mach. Stud. **37**, 793–809 (1992)

23.25   W. Ziarko: Variable precision rough set model, J. Comput. Syst. Sci. **46**(1), 39–59 (1993)

23.26   W. Ziarko: Probabilistic approach to rough sets, Int. J. Approx. Reason. **49**, 272–284 (2008)

23.27   P.G. Clark, J.W. Grzymala-Busse: Experiments on probabilistic approximations, IEEE Int. Conf. Granul. Comput. (2011) pp. 144–149

23.28   P.G. Clark, J.W. Grzymala-Busse, M. Kuehnhausen: Local probabilistic approximations for incomplete data, Lect. Notes Comput. Sci. **7661**, 93–98 (2012)

23.29   J.W. Grzymala-Busse, W. Rzasa: A local version of the MLEM2 algorithm for rule induction, Fundam. Inform. **100**, 99–116 (2010)

23.30   C. Cohagan, J.W. Grzymala-Busse, Z.S. Hippe: Experiments on mining inconsistent data with bagging and the MLEM2 rule induction algorithm, Int. J. Granul. Comput. Rough Sets Intell. Syst. **2**, 257–271 (2012)

24. Probabilistic Rough Sets

Yiyu Yao, Salvatore Greco, Roman Słowiński

As quantitative generalizations of Pawlak rough sets, probabilistic rough sets consider degrees of overlap between equivalence classes and the set. An equivalence class is put into the lower approximation if the conditional probability of the set, given the equivalence class, is equal to or above one threshold; an equivalence class is put into the upper approximation if the conditional probability is above another threshold hold. We review a basic model of probabilistic rough sets (i.e., decision-theoretic rough set model) and variations. We present the main results of probabilistic rough sets by focusing on three issues: (a) interpretation and calculation of the required thresholds, (b) estimation of the required conditional probabilities, and (c) interpretation and applications of probabilistic rough set approximations.

Part C | 24

24.1 Motivation for Studying Probabilistic Rough Sets

Rough set theory [24.1, 2] provides a simple and elegant method for analyzing data represented in a tabular form called an information table. The rows of the table represent a finite set of objects, the columns represent a finite set of attributes, and each cell represents the value of an object on the corresponding attribute. With a limited number of attributes, we may only be able to describe some subsets of objects precisely [24.3, 4]. Those subsets that can be precisely described are called definable sets, and all other subsets are called undefinable sets. A fundamental notion of rough set theory is the approximation of a subset of objects by a pair of definable sets from below and above, or equivalently, by three pairwise disjoint positive, negative, and boundary regions [24.4].

Pawlak rough set approximations are characterized by a zero tolerance of errors. That is, an object in the lower approximation certainly belongs to set and an object in the complement of the upper approximation certainly does not belong to the set. This has motivated the introduction of many different generalizations of rough sets. By introducing certain levels of errors, probabilistic rough sets [24.5, 6] are quantitative generalizations of the qualitative Pawlak rough sets. Although several specific models of probabilistic rough sets had been considered by some authors [24.7–10], a more general model, called decision-theoretic rough set (DTRS)

model, was first proposed by *Yao* et al. [24.11, 12] based on the well-established Bayesian decision theory. Other probabilistic models include variable precision rough sets [24.13, 14], Bayesian rough sets [24.15–18], parameterized rough sets [24.19, 20], game-theoretic rough sets [24.21, 22], variable-consistency-indiscernibility-based and dominance-based rough sets [24.23, 24], stochastic dominance-based rough sets [24.25], naive Bayesian rough sets [24.26], information-theoretic rough sets [24.27], confirmation-theoretic rough sets [24.28], and many different types of probabilistic rough set approximations [24.29, 30].

In this chapter, we present a basic model of probabilistic rough sets and a brief review of other probabilistic rough set models. We examine in particular three fundamental issues, namely, the interpretation and computation of the pair of thresholds, the estimation of probability, and an application of three regions. We also show how a probabilistic approach can be applied when information related to some order representing the extent to which some property related to considered attributes has to be taken into account. This situation is handled by the well-known rough set extension called dominance-based rough set approach [24.31–34]. A full understanding of these issues will greatly increase the chance of success when applying probabilistic rough sets in real-world applications.

24.2 Pawlak Rough Sets

We present a semantically meaningful definition of rough set approximations and a simple method for constructing rough set approximations.

24.2.1 Rough Set Approximations

In rough set theory, a finite set of objects is described by using a finite set of attributes in a tabular form, called an information table [24.2]. Formally, an information table can be expressed as

$$S = (U, AT, \{V_a \mid a \in AT\}, \{I_a \mid a \in AT\}),$$

where

U is a finite nonempty set of objects called universe ,

AT is a finite nonempty set of attributes ,

V_a is a nonempty set of values for $a \in AT$,

$I_a : U \rightarrow V_a$ is an information function .

The information table provides all available information about the set of objects, based on which we can perform tasks of analysis and inference.

In an information table, one can introduce a description language, as suggested by *Marek* and *Pawlak* [24.3], to formally describe objects. We consider a language DL that is recursively defined as follows

(1) $(a = v) \in DL$, where $a \in AT, v \in V_a$,

(2) if $p, q \in DL$, then $\neg p, p \wedge q, p \vee q \in DL$.

Formulas defined by (1) are called atomic formulas. The satisfiability of formula p by an object x, written

$x \models p$, is defined as follows

(i) $x \models (a = v)$, iff $I_a(x) = v$,
(ii) $x \models \neg p$, iff $\neg(x \models p)$,
(iii) $x \models p \wedge q$, iff $x \models p$ and $x \models q$,
(iv) $x \models p \vee q$, iff $x \models p$ or $x \models q$.

If p is a formula, the set $m(p) \subseteq U$ defined by

$$m(p) = \{x \in U \mid x \models p\} \qquad (24.1)$$

is called the meaning set of p. That is, the meaning set $m(p)$ consists of all those objects that satisfy the formula p.

With the introduction of a description language, we can formally describe an important characteristics of an information table, namely, some subsets of objects are definable or describable while others are not. A subset of objects $X \subseteq U$ is called a definable set [24.3, 4] if there exists a formula p such that

$$X = m(p), \qquad (24.2)$$

otherwise, X is called an undefinable set. The formula p is called a description of X. Let $\mathrm{DEF}(U) \subseteq 2^U$ denote the family of all definable sets, where 2^U is the power set of U. By definition, $\mathrm{DEF}(U)$ contains the empty set $\emptyset$, the entire universe U and is closed under set complement, intersection, and union. In other words, $\mathrm{DEF}(U)$ is a sub-Boolean algebra of the power set 2^U.

For any subset of objects $X \subseteq U$, may be either definable or undefinable, we define the following pair of lower and upper approximations

$\underline{apr}(X) = $ the largest definable set contained by X

$$= \bigcup \{G \in \mathrm{DEF}(U) \mid G \subseteq X\},$$

$\overline{apr}(X) = $ the smallest definable set containing X

$$= \bigcap \{G \in \mathrm{DEF}(U) \mid X \subseteq G\}. \qquad (24.3)$$

By definition, it follows that $\underline{apr}(X) \subseteq X \subseteq \overline{apr}(X)$ for any $X \subseteq U$, and $\underline{apr}(X) = X = \overline{apr}(X)$ if and only if $X \in \mathrm{DEF}(U)$. The definition is semantically meaningful in the sense that it clearly explains the motivation for introducing rough set approximations and provides an interpretation of the approximations. However, one cannot use this definition to construct rough set approximations easily.

24.2.2 Construction of Rough Set Approximations

A simple method for constructing rough set approximations is through an equivalence relation. For an attribute $a \in AT$, the information function I_a maps an object in U to a value of V_a, that is, $I_a(x) \in V_a$. For an attribute $a \in AT$, we can define an equivalence relation E_a as follows: for $x, y \in U$

$$xE_a y \iff I_a(x) = I_a(y). \qquad (24.4)$$

The equivalence class containing x is denoted by $[x]_a$. Similarly, for a subset of attributes $A \subseteq AT$, we define an equivalence relation E_A

$$xE_A y \iff \forall a \in A(I_a(x) = I_a(y)). \qquad (24.5)$$

The equivalence class containing x is denoted by $[x]_A$. By definition, it follows that, for $a \in AT$ and $A \subseteq AT$,

$$E_{\{a\}} = E_a, \qquad [x]_{\{a\}} = [x]_a,$$
$$E_A = \bigcap_{a \in A} E_a, \qquad [x]_A = \bigcap_{a \in A} [x]_a. \qquad (24.6)$$

That is, we can construct the equivalence relation induced by a subset of attributes A by using equivalence relations induced by individual attributes in A.

Consider the equivalence relation $E_A \subseteq U \times U$ induced by a subset of attributes $A \subseteq AT$. The equivalence relation E_A induces a partition U/E_A of U, i.e., a family of nonempty and pairwise disjoint subsets whose union is the universe. For an object $x \in U$, its equivalence class is given by

$$[x]_A = \{y \in U \mid xE_A y\}. \qquad (24.7)$$

By taking the union of a family of equivalence classes, one can construct an atomic sub-Boolean $B(U/E_A)$ of 2^U with U/E_A as the set of atoms

$$B(U/E_A) = \{\bigcup F \mid F \subseteq U/E_A\}. \qquad (24.8)$$

That is, $B(U/E_A)$ contains the empty set $\emptyset$, the whole set U, and is closed with respect to set complement, intersection, and union. The three notions of equivalence relation E, the partition U/E_A, and atomic Boolean algebra $B(U/E_A)$ uniquely determine each other. We can therefore use E_A, U/E_A, and $B(U/E_A)$ interchangeably. The pair $apr = (U, E_A)$, equivalently, the pair $apr = (U, U/E_A)$ or the pair $apr = (U, B(U/E_A))$, is called an approximation space. Although three different representations are equivalent, each of them provides a different hint when we generalize rough sets. The pair $apr = (U, E_A)$ is useful for generalizing rough sets using a nonequivalence relation [24.35]. The partition U/E_A may be viewed as a granulation of the universe U and the pair $apr = (U, U/E_A)$ relates rough sets and granular computing [24.36]. The pair $apr = (U, B(U/E_A))$ leads to a subsystem-based formulation and generalizations [24.37].

For a subset of attributes $A \subseteq AT$, if we restrict the formulas of DL by using only attributes in A, we obtain a sublanguage $DL(A) \subseteq DL$. It can be proved that the family of all definable sets $DEF_A(U)$ defined by $DL(A)$ is exactly the sub-Boolean algebra $B(U/E_A)$. With respect to a subset of attributes $A \subseteq AT$, each object x is described by a logic formula

$$\bigwedge_{a \in A} a = I_a(x) , \tag{24.9}$$

where $I_a(x) \in V_a$ and the atomic formula $a = I_a(x)$ indicate that the value of an object on attribute a is $I_a(x)$. The equivalence class containing x, namely, $[x]_{E_A}$, is the set of those objects that satisfy the formula $\wedge_{a \in A} a = I_a(x)$. The formula can be viewed as a description of objects that are equivalent to x with respect to A, including x itself.

Based on the equivalence of $DEF_A(U)$ and $B(U/E_A)$, we can equivalently define rough set approximations by using the equivalence classes $[x]_A$. For simplicity, we also simply write $[x]$ when no confusion arises.

For a subset of objects $X \subseteq U$, the pair of lower and upper approximations can be equivalently defined by

$$apr(X) = \{x \in U \mid [x] \subseteq X\} ,$$
$$\overline{apr}(X) = \{x \in U \mid [x] \cap X \neq \emptyset\} . \tag{24.10}$$

Construction of rough set approximation by this definition is much easier. Alternatively, one can also define three pairwise disjoint positive, negative, and boundary regions [24.38]

$$POS(X) = \{x \in U \mid [x] \subseteq X\} ,$$
$$NEG(X) = \{x \in U \mid [x] \cap X = \emptyset\} ,$$
$$BND(X) = \{x \in U \mid [x] \not\subseteq X \wedge [x] \cap X \neq \emptyset\} . \tag{24.11}$$

The pair of approximations and three regions determines each other as follows

$$POS(X) = apr(X) ,$$
$$NEG(X) = (\overline{apr}(X))^c ,$$
$$BND(X) = \overline{apr}(X) - apr(X) , \tag{24.12}$$

and

$$apr(X) = POS(X) ,$$
$$\overline{apr}(X) = POS(X) \cup BND(X) , \tag{24.13}$$

where $(\cdot)^c$ denotes the complement of a set. Each representation provides a distinctive interpretation of rough set approximations. We will use the three-region approximation in the rest of this chapter, due to its close connections to three-way decisions.

24.3 A Basic Model of Probabilistic Rough Sets

Decision-theoretic rough set (DTRS) model proposed by *Yao* et al. [24.11, 12] gives rises to a general form of probabilistic rough set approximations by using a pair of thresholds on conditional probabilities. The results enable us to formulate a basic model of probabilistic rough sets. However, we introduce the model in a way that is different from DTRS. We first interpret Pawlak rough sets in terms of probability and the two extreme value of probability (i. e., 1 and 0) and then generalize 1 and 0 into a pair of thresholds (α, β) with $0 \leq \beta < \alpha \leq 1$.

The Pawlak rough sets consider only qualitative relationship between an equivalence class and a set, namely, an equivalence is a subset of the set or has

a nonempty intersection with the set. This qualitative nature becomes clearer with a probabilistic interpretation [24.6]. Suppose $Pr(X|[x])$ denotes the conditional probability that an object is in X given that the object is in $[x]$. The conditions for defining rough set three regions can be equivalently expressed as

$$[x] \subseteq X \iff Pr(X|[x]) \geq 1 ;$$
$$[x] \cap X = \emptyset \iff Pr(X|[x]) \leq 0 ;$$
$$[x] \not\subseteq X \wedge [x] \cap X \neq \emptyset \iff 0 < Pr(X|[x]) < 1 . \tag{24.14}$$

Although a probability can never be greater than 1 or less than 0, we purposely use the conditions ≥ 1 and

≤ 0 whose intended meaning will become clearer later. By those conditions, Pawlak three regions can be equivalently expressed as

$$\text{POS}(X) = \{x \in U \mid Pr(X|[x]) \geq 1\},$$
$$\text{NEG}(X) = \{x \in U \mid Pr(X|[x]) \leq 0\},$$
$$\text{BND}(X) = \{x \in U \mid 0 < Pr(X|[x]) < 1\}. \quad (24.15)$$

They show that Pawlak rough sets only use the two extreme values, i.e., 1 and 0, of probability.

It is natural to generalize Pawlak rough sets by replacing 1 and 0 with some other values in the unit interval $[0, 1]$. Given a pair of thresholds α, β with $0 \leq \beta < \alpha \leq 1$, the main results of probabilistic rough sets are the (α, β)-probabilistic regions defined by

$$\text{POS}_{(\alpha,\beta)}(X) = \{x \in U \mid Pr(X|[x]) \geq \alpha\},$$
$$\text{NEG}_{(\alpha,\beta)}(X) = \{x \in U \mid Pr(X|[x]) \leq \beta\},$$
$$\text{BND}_{(\alpha,\beta)}(X) = \{x \in U \mid \beta < Pr(X|[x]) < \alpha\}.$$
$$(24.16)$$

The Pawlak rough set model is a special case in which $\alpha = 1$ and $\beta = 0$. In the case when $0 < \beta = \alpha < 1$, the three regions are given by

$$\text{POS}_{(\alpha,\alpha)}(X) = \{x \in U \mid Pr(X|[x]) > \alpha\},$$
$$\text{NEG}_{(\alpha,\alpha)}(X) = \{x \in U \mid Pr(X|[x]) < \alpha\},$$
$$\text{BND}_{(\alpha,\alpha)}(X) = \{x \in U \mid Pr(X|[x]) = \alpha\}. \quad (24.17)$$

It may be commented that this special case is perhaps more of mathematical interest, rather than practical applications. We use this particular definition in order to establish connection to existing studies. As will be shown in subsequent discussions, when $Pr(X|[x]) = \alpha = \beta$, the costs of assigning objects in $[x]$ to the positive, boundary, and negative regions, respectively, are the same. In fact, one may simply define two regions by assigning objects in the boundary region into either the positive or boundary region.

The main results of the basic model of probabilistic rough sets were first proposed by *Yao* et al. [24.11, 12] in a DTRS model, based on Bayesian decision theory. The DTRS model covers all specific models introduced before it. The interpretation of Pawlak rough sets in terms of conditional probability, i.e., the model characterized by $\alpha = 1$ and $\beta = 0$, was first given by *Wong* and *Ziarko* [24.10]. A 0.5-model, characterized by $\alpha = \beta = 0.5$, was introduced by *Wong* and *Ziarko* [24.8] and *Pawlak* et al. [24.7], in which the positive region is defined by probability greater than 0.5, the negative by probability less than 0.5, and the boundary by probability equal to 0.5. A model characterized by $\alpha > 0.5$ and $\beta = 0.5$ was suggested by *Wong* and *Ziarko* [24.9]. Most recent developments on decision-theoretic rough sets can be found in a book edited by *Li* et al. [24.39] and papers [24.21, 40–50] in a journal special issue edited by *Yao* et al. [24.51].

24.4 Variants of Probabilistic Rough Sets

Since the introduction of decision-theoretic rough set model, several new models have been proposed and investigated. They offer related but different directions in generalizing Pawlak rough sets by incorporating probabilistic information.

24.4.1 Variable Precision Rough Sets

The first version of variable precision rough sets was introduced by *Ziarko* [24.14], in which the standard set inclusion $[x] \subseteq X$ is generalized into a graded set inclusion $s([x], X)$ called a measure of the relative degree of misclassification of $[x]$ with respect to X. A particular measure suggested by Ziarko is given by

$$s([x], X) = 1 - \frac{|[x] \cap X|}{|[x]|}, \quad (24.18)$$

where $|\cdot|$ denotes the cardinality of a set. By introducing a threshold $0 \leq z < 0.5$, one can define three regions as follows

$$\text{VPOS}_z(X) = \{x \in U \mid s([x], X) \leq z\},$$
$$\text{VNEG}_z(X) = \{x \in U \mid s([x], X) \geq 1 - z\},$$
$$\text{VBND}_z(X) = \{x \in U \mid z < s([x], X) < 1 - z\}. \quad (24.19)$$

A more generalized version using a pair of thresholds was late introduced by *Katzberg* and *Ziarko* [24.13] as follows: for $0 \leq l < u \leq 1$,

$$\text{VPOS}_{(l,u)}(X) = \{x \in U \mid s([x], X) \leq l\},$$
$$\text{VNEG}_{(l,u)}(X) = \{x \in U \mid s([x], X) \geq u\},$$
$$\text{VBND}_{(l,u)}(X) = \{x \in U \mid l < s([x], X) < u\}. \quad (24.20)$$

The one-threshold model may be considered as a special case of the two-threshold model with $l = z$ and $u = 1 - z$.

One may interpret the ratio in (24.18) as an estimation of the conditional probability $Pr(X|[x])$, namely

$$s([x], X) = 1 - \frac{|[x] \cap X|}{|[x]|} = 1 - Pr(X|[x]) . \quad (24.21)$$

By setting $\alpha = 1 - l$ and $\beta = 1 - u$, we immediately have

$$\begin{aligned}
\mathrm{POS}_{(1-l,1-u)}(X) &= \{x \in U \mid Pr(X|[x]) \geq 1 - l\} \\
&= \{x \in U \mid s([x], X) \leq l\} \\
&= \mathrm{VPOS}_{(l,u)}(X) , \\
\mathrm{NEG}_{(1-l,1-u)}(X) &= \{x \in U \mid Pr(X|[x]) \geq 1 - u\} \\
&= \{x \in U \mid s([x], X) \geq u\} \\
&= \mathrm{VNEG}_{(l,u)}(X) , \\
\mathrm{BND}_{(1-l,1-u)}(X) &= \{x \in U \mid 1 - l < Pr(X|[x]) < 1 - u\} \\
&= \{x \in U \mid l < s([x], X) < u\} \\
&= \mathrm{VBND}_{(l,u)}(X) . \quad (24.22)
\end{aligned}$$

It follows that, when the particular set-inclusion measure defined by (24.18) is used, the variable precision rough sets are coincident with the decision-theoretic rough sets.

Variable precision rough sets provide an alternative direction in generalizing Pawlak rough sets by considering a graded set-inclusion relation, which is not necessarily restricted to a probabilistic interpretation. If we use other set-inclusion measures, we will obtain other types of quantitative rough sets [24.38, 52]. Unfortunately, subsequent developments lose this crucial feature in an attempt to unify variable precision rough sets into probabilistic rough sets [24.53].

24.4.2 Parameterized Rough Sets

Parameterized rough sets, proposed by *Greco* et al. [24.19, 20], generalize probabilistic rough sets by introducing a Bayesian confirmation measure and a pair of thresholds on the confirmation measure, in addition to a pair of thresholds on conditional probability. According to *Fitelson* [24.54], *measures of confirmation* quantify the degree to which a piece of evidence E provides *evidence for or against* or *support for or against* a hypothesis H.

A measure of confirmation of a piece of evidence E with respect to a hypothesis H is denoted by $c(E, H)$. A confirmation measure $c(E, H)$ is required to satisfy the following minimal property:

$$c(E, H) = \begin{cases} > 0 & \text{if } Pr(H|E) > Pr(H) \\ = 0 & \text{if } Pr(H|E) = Pr(H) \\ < 0 & \text{if } Pr(H|E) < Pr(H). \end{cases} \quad (i)$$

Two well-known Bayesian confirmation measures are [24.55]

$$\begin{aligned}
c_d([x], X) &= Pr(X|[x]) - Pr(X) , \\
c_r([x], X) &= \frac{Pr(X|[x])}{Pr(X)} . \quad (24.23)
\end{aligned}$$

These measures have a probabilistic interpretation. The parameterized rough sets can be therefore viewed as a different formulation of probabilistic rough sets.

A first discussion about relationships between confirmation measures and rough sets were proposed by *Greco* et al. [24.56]. Other contributions related to the properties of confirmation measures with special attention to application to rough sets are given in [24.57].

Given a pair of thresholds (s, t) with $t < s$, three (α, β, s, t)-parameterized regions are defined in [24.19, 20]

$$\begin{aligned}
\mathrm{PPOS}_{(\alpha,\beta,s,t)}(X) &= \{x \in U \mid Pr(X|[x]) \geq \alpha \\
&\quad \land c([x], X) \geq s\} , \\
\mathrm{PNEG}_{(\alpha,\beta,s,t)}(X) &= \{x \in U \mid Pr(X|[x]) \leq \beta \\
&\quad \land c([x], X) \leq t\} , \\
\mathrm{PBND}_{(\alpha,\beta,s,t)}(X) &= \{x \in U \mid (Pr(X|[x]) > \beta \\
&\quad \lor c([x], X) > t) \\
&\quad \land (Pr(X|[x]) < \alpha \\
&\quad \lor c([x], X) < s)\} . \quad (24.24)
\end{aligned}$$

There exist many Bayesian confirmation measures, which makes the model of parameterized rough sets more flexible. On the other hand, due to lack of a general agreement on a Bayesian confirmation measure, choosing an appropriate confirmation measure for a particular application may not be an easy task.

24.4.3 Confirmation-Theoretic Rough Sets

Although many Bayesian confirmation measures are related to the conditional probability $Pr(X|[x])$, *Zhou* and *Yao* [24.26] argued that the conditional probability $Pr(X|[x])$ and a Bayesian confirmation measure have very different semantics and should be used for different purposes. For example, the conditional probability $Pr(X|[x])$ gives us an absolute degree of confidence in classifying objects from $[x]$ as belonging to X. On the other hand, a Bayesian measure, for example, c_d or c_r, normally reflects a change of confidence in X before and after knowing $[x]$. Thus, a Bayesian confirmation measures is useful to weigh the strength of evidence $[x]$ with respect to the hypothesis X. A mixture of conditional probability and confirmation measure in the parameterized rough sets may cause a semantic difficulty in interpreting the three regions.

To resolve this difficulty, *Zhou* and *Yao* [24.28] suggested a separation of the parameterized model into two models. One is the conventional probabilistic model and the other is a confirmation-theoretic model. For a Bayesian confirmation measure $c([x], X)$ and a pair of thresholds (s, t) with $t < s$, three confirmation regions are defined by

$$\text{CPOS}_{(s,t)}(X) = \{[x] \in U/R \mid c([x], X) \geq s\},$$
$$\text{CNEG}_{(s,t)}(X) = \{[x] \in U/R \mid c([x], X) \leq t\},$$
$$\text{CBND}_{(s,t)}(X) = \{[x] \in U/R \mid t < c([x], X) < s\}.$$
$$(24.25)$$

For the case with $s = t$, we define

$$\text{CPOS}_{(s,s)}(X) = \{[x] \in U/R \mid c([x], X) > s\},$$
$$\text{CNEG}_{(s,s)}(X) = \{[x] \in U/R \mid c([x], X) < s\},$$
$$\text{CBND}_{(s,s)}(X) = \{[x] \in U/R \mid c([x], X) = s\}.$$
$$(24.26)$$

In the definition, each equivalence class may be viewed as a piece of evidence. Thus, the partition U/E, instead of the universe, is divided into three regions. An equivalence class in the positive region supports X to a degree at least s, an equivalence class in the negative region supports X to a degree at most t and may be viewed as against X, and an equivalence class in the boundary region is interpreted as neutral toward X.

24.4.4 Bayesian Rough Sets

Bayesian rough sets were proposed by *Ślęzak* and *Ziarko* [24.15, 16] as a probabilistic model in which the required pair of thresholds is interpreted using the a pri-

ori probability $Pr(X)$. They introduced Bayesian rough sets and variable precision Bayesian rough sets.

For the Bayesian rough sets, the three regions are defined by

$$\text{BPOS}(X) = \{x \in U \mid Pr(X|[x]) > Pr(X)\},$$
$$\text{BNEG}(X) = \{x \in U \mid Pr(X|[x]) < Pr(X)\},$$
$$\text{BBND}(X) = \{x \in U \mid Pr(X|[x]) = Pr(X)\}.\ (24.27)$$

Bayesian rough sets can be viewed as a special case of the decision-theoretic rough sets when $\alpha = \beta = Pr(X)$. Semantically, they are very different, however. In contrast to decision-theoretic rough sets, for a set with a higher a priori probability $Pr(X)$, many equivalence classes may not be put into the positive region in the Bayesian rough set model, as the condition $Pr(X|[x]) > Pr(X)$ may not hold. For example, the positive region of the entire universe is always empty, namely, $\text{BPOS}(U) = \emptyset$. This leads to a difficulty in interpreting the positive region as a lower approximation of a set.

The difficulty with Bayesian rough sets stems from the fact that they are in fact a special model of confirmation-theoretic rough sets, which is suitable for classifying pieces of evidence (i. e., equivalence classes), but is inappropriate for approximating a set. Recall that one Bayesian confirmation measure is given by $c_d([x], X) = Pr(X|[x]) - Pr(X)$. Therefore, Bayesian rough sets can be expressed as confirmation-theoretic rough sets as follows,

$$\begin{aligned}
\text{BPOS}(X) &= \{x \in U \mid Pr(X|[x]) > Pr(X)\}, \\
&= \{x \in U \mid c_d([x], X) > 0\}, \\
&= \bigcup \text{CPOS}_{(0,0)}(X), \\
\text{BNEG}(X) &= \{x \in U \mid Pr(X|[x]) < Pr(X)\}, \\
&= \{x \in U \mid c_d([x], X) < 0\}, \\
&= \bigcup \text{CNEG}_{(0,0)}(X), \\
\text{BBND}(X) &= \{x \in U \mid Pr(X|[x]) = Pr(X)\} \\
&= \{x \in U \mid c_d([x], X) = 0\}, \\
&= \bigcup \text{CBND}_{(0,0)}(X).
\end{aligned}$$
$$(24.28)$$

That is, the Bayesian rough sets are a model of confirmation-theoretic rough sets characterized by the Bayesian confirmation measure c_d with a pair of thresholds $s = t = 0$. *Ślęzak* and *Ziarko* [24.17] showed that

Bayesian rough sets can also be interpreted by using other Bayesian confirmation measures.

The three regions of the variable precision Bayesian rough sets are defined as follows [24.16]: for $\epsilon \in [0, 1)$

$$
\begin{aligned}
\text{VBPOS}_\epsilon(X) &= \{x \in U \mid Pr(X|[x]) \\
&\geq 1 - \epsilon(1 - Pr(X))\} , \\
\text{VBNEG}_\epsilon(X) &= \{x \in U \mid Pr(X|[x]) \leq \epsilon Pr(X)\} , \\
\text{VBBND}_\epsilon(X) &= \{x \in U \mid \epsilon Pr(X) < Pr(X|[x]) \\
&< 1 - \epsilon(1 - Pr(X))\} . \quad (24.29)
\end{aligned}
$$

Consider the Bayesian confirmation measure

$$
c_r([x], X) = Pr(X|[x])/Pr(X) .
$$

For the condition of the positive region, when $Pr(X^c) \neq 0$ we have

$$
Pr(X|[x]) \geq 1 - \epsilon(1 - Pr(X)) \iff c_r([x], X^c) \leq \epsilon . \tag{24.30}
$$

Similarly, for the condition defining the negative region, when $Pr(X) \neq 0$ we have

$$
Pr(X|[x]) \leq \epsilon Pr(X) \iff c_r([x], X) \leq \epsilon . \tag{24.31}
$$

That is, $[x]$ is put into the positive region if it confirms X^c to a degree less than or equal to ϵ and is put into the negative region if it confirms X to a degree less than or equal to ϵ. In this way, we get a confirmation-theoretic interpretation of variable precision Bayesian rough sets.

Unlike the confirmation-theoretic model defined by (24.25), the positive region of variable precision Bayesian rough sets is defined based on the confirmation of the complement of X and negative region is defined based on the confirmation of X. This definition is a bit awkward to interpret. Generally speaking, it may be more natural to define the positive region by those equivalence classes that confirm X to at least a certain degree. This suggests that one can redefine variable precision Bayesian rough sets by using the framework of confirmation-theoretic rough sets. Moreover, one can use a pair of thresholds instead of one threshold.

24.5 Three Fundamental Issues of Probabilistic Rough Sets

For practical applications of probabilistic rough sets, one must consider at least the following three fundamental issues [24.58, 59]:

- Interpretation and determination of the required pair of thresholds,
- Estimation of the required conditional probabilities, and
- Interpretation and applications of three probabilistic regions.

For each of the three issues, this section reviews one example of the possible methods.

24.5.1 Decision–Theoretic Rough Set Model: Determining the Thresholds

A decision-theoretic model formulates the construction of rough set approximations as a Bayesian decision problem with a set of two states and a set of three actions [24.11, 12]. The set of states is given by $\Omega = \{X, X^c\}$ indicating that an element is in X and not in X, respectively. For simplicity, we use the same symbol to denote both a subset X and the corresponding state. Corresponding to the three regions, the set of actions

is given by $\mathcal{A} = \{a_P, a_B, a_N\}$, denoting the actions in classifying an object x, namely, deciding $x \in \text{POS}(X)$, deciding $x \in \text{BND}(X)$, and deciding $x \in \text{NEG}(X)$, respectively. The losses regarding the actions for different states are given by the 3×2 matrix

	X (P)	X^c (N)
a_P	λ_{PP}	λ_{PN}
a_B	λ_{BP}	λ_{BN}
a_N	λ_{NP}	λ_{NN}

In the matrix, λ_{PP}, λ_{BP}, and λ_{NP} denote the losses incurred for taking actions a_P, a_B, and a_N, respectively, when an object belongs to X, and λ_{PN}, λ_{BN} and λ_{NN} denote the losses incurred for taking the same actions when the object does not belong to X.

The expected losses associated with taking different actions for objects in $[x]$ can be expressed as

$$
\begin{aligned}
R(a_P|[x]) &= \lambda_{PP} Pr(X|[x]) + \lambda_{PN} Pr(X^c|[x]) , \\
R(a_B|[x]) &= \lambda_{BP} Pr(X|[x]) + \lambda_{BN} Pr(X^c|[x]) , \\
R(a_N|[x]) &= \lambda_{NP} Pr(X|[x]) + \lambda_{NN} Pr(X^c|[x]) . \\
& \quad (24.32)
\end{aligned}
$$

The Bayesian decision procedure suggests the following minimum-risk decision rules

(P) If $R(a_P|[x]) \leq R(a_B|[x])$
  and $R(a_P|[x]) \leq R(a_N|[x])$, decide $x \in \mathrm{POS}(X)$;
(B) If $R(a_B|[x]) \leq R(a_P|[x])$
  and $R(a_B|[x]) \leq R(a_N|[x])$, decide $x \in \mathrm{BND}(X)$;
(N) If $R(a_N|[x]) \leq R(a_P|[x])$
  and $R(a_N|[x]) \leq R(a_B|[x])$, decide $x \in \mathrm{NEG}(X)$.

In order to make sure that the three regions are mutually disjoint, tie-breaking criteria should be added when two or three actions have the same risk. We use the following ordering for breaking a tie: a_P, a_N, a_B.

Consider a special class of loss functions with

(c0) $\lambda_{PP} \leq \lambda_{BP} < \lambda_{NP}, \quad \lambda_{NN} \leq \lambda_{BN} < \lambda_{PN}$.
$$\tag{24.33}$$

That is, the loss of classifying an object x belonging to X into the positive region $\mathrm{POS}(X)$ is less than or equal to the loss of classifying x into the boundary region $\mathrm{BND}(X)$, and both of these losses are strictly less than the loss of classifying x into the negative region $\mathrm{NEG}(X)$. The reverse order of losses is used for classifying an object not in X. With the condition (c0) and the equation $Pr(X|[x]) + Pr(X^c|[x]) = 1$, we can express the decision rules (P)–(N) in the following simplified form (for a detailed derivation, see references [24.58])

(P) If $Pr(X|[x]) \geq \alpha$
  and $Pr(X|[x]) \geq \gamma$, decide $x \in \mathrm{POS}(X)$;
(B) If $Pr(X|[x]) \leq \alpha$
  and $Pr(X|[x]) \geq \beta$, decide $x \in \mathrm{BND}(X)$;
(N) If $Pr(X|[x]) \leq \beta$
  and $Pr(X|[x]) \leq \gamma$, decide $x \in \mathrm{NEG}(X)$,

where

$$
\begin{aligned}
\alpha &= \frac{(\lambda_{PN} - \lambda_{BN})}{(\lambda_{PN} - \lambda_{BN}) + (\lambda_{BP} - \lambda_{PP})}, \\
\beta &= \frac{(\lambda_{BN} - \lambda_{NN})}{(\lambda_{BN} - \lambda_{NN}) + (\lambda_{NP} - \lambda_{BP})}, \\
\gamma &= \frac{(\lambda_{PN} - \lambda_{NN})}{(\lambda_{PN} - \lambda_{NN}) + (\lambda_{NP} - \lambda_{PP})}.
\end{aligned}
\tag{24.34}
$$

Each rule is defined by two out of the three parameters. By setting $\alpha > \beta$, namely

$$
\begin{aligned}
&\frac{(\lambda_{PN} - \lambda_{BN})}{(\lambda_{PN} - \lambda_{BN}) + (\lambda_{BP} - \lambda_{PP})} \\
&> \frac{(\lambda_{BN} - \lambda_{NN})}{(\lambda_{BN} - \lambda_{NN}) + (\lambda_{NP} - \lambda_{BP})},
\end{aligned}
\tag{24.35}
$$

we obtain the following condition on the loss function [24.58]

(c1) $\dfrac{\lambda_{NP} - \lambda_{BP}}{\lambda_{BN} - \lambda_{NN}} > \dfrac{\lambda_{BP} - \lambda_{PP}}{\lambda_{PN} - \lambda_{BN}}$.
$$\tag{24.36}$$

The condition (c1) implies that $1 \geq \alpha > \gamma > \beta \geq 0$. In this case, after tie-breaking, we have the simplified rules [24.58]

(P) If $Pr(X|[x]) \geq \alpha$, decide $x \in \mathrm{POS}(X)$;
(B) If $\beta < Pr(X|[x]) < \alpha$, decide $x \in \mathrm{BND}(X)$;
(N) If $Pr(X|[x]) \leq \beta$, decide $x \in \mathrm{NEG}(X)$.

The parameter γ is no longer needed. Each object can be put into one and only one region by using rules (P), (B), and (N). The (α, β)-probabilistic positive, negative and boundary regions are given, respectively, by

$$
\begin{aligned}
\mathrm{POS}_{(\alpha,\beta)}(X) &= \{x \in U \mid Pr(X|[x]) \geq \alpha\}, \\
\mathrm{BND}_{(\alpha,\beta)}(X) &= \{x \in U \mid \beta < Pr(X|[x]) < \alpha\}, \\
\mathrm{NEG}_{(\alpha,\beta)}(X) &= \{x \in U \mid Pr(X|[x]) \leq \beta\}. \quad (24.37)
\end{aligned}
$$

The formulation provides a solid theoretical basis and a practical interpretation of the probabilistic rough sets. The threshold parameters are systematically calculated from a loss function.

In the development of decision-theoretic rough sets, we assume that a loss function is given by experts in a particular application. There are studies on other types of loss functions and their acquisition [24.60]. Several other proposals have also been made regarding the interpretation and computation of the thresholds, including game-theoretic rough sets [24.21, 22], information-theoretic rough sets [24.27], and an optimization-based framework [24.43, 61, 62].

24.5.2 Naive Bayesian Rough Set Model: Estimating the Conditional Probability

Naive Bayesian rough set model was proposed by *Yao* and *Zhou* [24.59] as a practical method for estimating the conditional probability. First, we perform the logit transformation of the conditional probability

$$\text{logit}(Pr(X|[x])) = \log \frac{Pr(X|[x])}{1 - Pr(X|[x])}$$
$$= \log \frac{Pr(X|[x])}{Pr(X^c|[x])} , \qquad (24.38)$$

which is a monotonically increasing transformation of $Pr(X|[x])$. Then, we apply the Bayes' theorem

$$Pr(X|[x]) = \frac{Pr([x]|X)Pr(X)}{Pr([x])} , \qquad (24.39)$$

to infer the a posteriori probability $Pr(X|[x])$ from the likelihood $Pr([x]|X)$ of $[x]$ with respect to X and the a priori probability $Pr(X)$. Similarly, for X^c we also have

$$Pr(X^c|[x]) = \frac{Pr([x]|X^c)Pr(X^c)}{Pr([x])} . \qquad (24.40)$$

By substituting results of (24.39) and (24.40) into (24.38), we immediately have

$$\text{logit}(Pr(X|[x])) = \log O(X|[x])$$
$$= \log \frac{Pr(X|[x])}{Pr(X^c|[x])}$$
$$= \log \frac{Pr([x]|X)}{Pr([x]|X^c)} \cdot \frac{Pr(X)}{Pr(X^c)}$$
$$= \log \frac{Pr([x]|X)}{Pr([x]|X^c)} + \log O(X) , \qquad (24.41)$$

where $O(X|[x])$ and $O(X)$ are the a posterior and the a prior odds, respectively, and $Pr([x]|X)/Pr([x]|X^c)$ is the likelihood ratio.

A threshold value on the probability can be expressed as another threshold value on logarithm of the likelihood ratio. For the positive region, we have

$$Pr(X|[x]) \geq \alpha$$
$$\Longleftrightarrow \log \frac{Pr(X|[x])}{Pr(X^c|[x])} \geq \log \frac{\alpha}{1 - \alpha}$$
$$\Longleftrightarrow \log \left(\frac{Pr([x]|X)}{Pr([x]|X^c)} \cdot \frac{Pr(X)}{Pr(X^c)} \right) \geq \log \frac{\alpha}{1 - \alpha}$$
$$\Longleftrightarrow \log \frac{Pr([x]|X)}{Pr([x]|X^c)} \geq \log \frac{Pr(X^c)}{Pr(X)} + \log \frac{\alpha}{1 - \alpha}$$
$$= \alpha' . \qquad (24.42)$$

Similar expressions can be obtained for the negative and boundary regions. The three regions can now be written as

$$\text{POS}_{(\alpha,\beta)}(X) = \left\{ x \in U \mid \log \frac{Pr([x]|X)}{Pr([x]|X^c)} \geq \alpha' \right\} ,$$
$$\text{BND}_{(\alpha,\beta)}(X) = \left\{ x \in U \mid \beta' < \log \frac{Pr([x]|X)}{Pr([x]|X^c)} < \alpha' \right\} ,$$
$$\text{NEG}_{(\alpha,\beta)}(X) = \left\{ x \in U \mid \log \frac{Pr([x]|X)}{Pr([x]|X^c)} \leq \beta' \right\} ,$$
$$\qquad (24.43)$$

where

$$\alpha' = \log \frac{Pr(X^c)}{Pr(X)} + \log \frac{\alpha}{1 - \alpha} ,$$
$$\beta' = \log \frac{Pr(X^c)}{Pr(X)} + \log \frac{\beta}{1 - \beta} . \qquad (24.44)$$

With the transformation, we need to estimate the likelihoods that are relatively easier to obtain.

Suppose that an equivalence relation E_A is defined by using a subset of attributes $A \subseteq AT$. In the naive Bayesian rough set model, we estimate the likelihood ratio $Pr([x]_A|X)/Pr([x]_A|X^c)$ through the likelihoods $Pr([x]_a|X)$ and $Pr([x]_a|X^c)$ defined by individual attributes, as the latter can be estimated more accurately. For this purpose, based on the results in (24.6), we make the following naive conditional independence assumptions

$$Pr([x]_A|X) = Pr \left(\bigcap_{a \in A} [x]_a | X \right) = \prod_{a \in A} Pr([x]_a|X) ,$$
$$Pr([x]_A|X^c) = Pr \left(\bigcap_{a \in A} [x]_a | X^c \right) = \prod_{a \in A} Pr([x]_a|X^c) .$$
$$\qquad (24.45)$$

By inserting them into (24.42) and assuming that $[x]$ is defined by a subset of attributes $A \subseteq AT$, namely, $[x] = [x]_A$, we have

$$\log \frac{Pr([x]_A|X)}{Pr([x]_A|X^c)} \geq \alpha'$$
$$\Longleftrightarrow \log \frac{\prod_{a \in A} Pr([x]_a|X)}{\prod_{a \in A} Pr([x]_a|X^c)} \geq \alpha'$$
$$\Longleftrightarrow \sum_{a \in A} \log \frac{Pr([x]_a|X)}{Pr([x]_a|X^c)} \geq \alpha' . \qquad (24.46)$$

Similar conditions can be derived for negative and boundary regions. Finally, the three regions can be defined as

$$\text{POS}_{(\alpha,\beta)}(X) = \left\{ x \in U \mid \sum_{a \in A} \log \frac{Pr([x]_a|X)}{Pr([x]_a|X^c)} \geq \alpha' \right\},$$

$$\text{BND}_{(\alpha,\beta)}(X) = \left\{ x \in U \mid \beta' \right.$$

$$\left. < \sum_{a \in A} \log \frac{Pr([x]_a|X)}{Pr([x]_a|X^c)} < \alpha' \right\},$$

$$\text{NEG}_{(\alpha,\beta)}(C) = \left\{ x \in U \mid \sum_{a \in A} \log \frac{Pr([x]_a|X)}{Pr([x]_a|X^c)} \leq \beta' \right\},$$

$$\tag{24.47}$$

where

$$\alpha' = \log \frac{Pr(X^c)}{Pr(X)} + \log \frac{\alpha}{1 - \alpha},$$

$$\beta' = \log \frac{Pr(X^c)}{Pr(X)} + \log \frac{\beta}{1 - \beta}. \tag{24.48}$$

We obtain a model in which we only need to estimate likelihoods of equivalence classes induced by individual attributes.

The likelihoods $Pr([x]_a|X)$ and $Pr([x]_a|X^c)$ may be simply estimated based on the following frequencies

$$Pr([x]_a|X) = \frac{|[x]_a \cap X|}{|X|},$$

$$Pr([x]_a|X^c) = \frac{|[x]_a \cap X^c|}{|X^c|},$$

where $[x]_a = \{y \in U \mid I_a(y) = I_a(x)\}$. An equivalence class defined by a single attribute is usually large in comparison with an equivalence classes defined by a subset of attributes. Probability estimation based on the former may be more accurate than based on the latter.

Naive Bayesian rough sets provide only one of possible ways to estimate the conditional probability. Other estimation methods include logistic regress [24.46] and the maximum likelihood estimators [24.63].

24.5.3 Three-Way Decisions: Interpreting the Three Regions

A theory of three-way decisions [24.64] is motivated by the needs for interpreting the three regions [24.65–67]

and moves beyond rough sets. The main results of three-way decisions can be found in two recent books edited by *Jia* et al. [24.68] and *Liu* et al. [24.69], respectively. We present an interpretation of rough set three regions based on the framework of three-way decisions.

In an information table, with respect to a subset of attributes $A \subseteq AT$, an object x induces a logic formula

$$\bigwedge_{a \in A} a = I_a(x), \tag{24.49}$$

where $I_a(x) \in V_a$ and the atomic formula $a = I_a(x)$ indicates that the value of an object on attribute a is $I_a(x)$. An object y satisfies the formula if $I_a(y) = I_a(x)$ for all $a \in A$, that is

$$\left(y \models \bigwedge_{a \in A} a = I_a(x) \right) \Longleftrightarrow \forall a \in A \ (I_a(y) = I_a(x)). \tag{24.50}$$

With these notations, we are ready to interpret rough set in three regions.

From the three regions, we can construct three classes of rules for classifying an object, called the positive, negative, and boundary rules [24.58, 66, 67]. They are expressed in the following forms, for $y \in U$:

- Positive rule induced by an equivalence class $[x] \subseteq \text{POS}_{(\alpha,\beta)}(X)$

$$\text{if } y \models \bigwedge_{a \in A} a = I_a(x), \ \text{accept } y \in X$$

- Negative rule induced by an equivalence class $[x] \subseteq \text{NEG}_{(\alpha,\beta)}(X)$

$$\text{if } y \models \bigwedge_{a \in A} a = I_a(x), \ \text{reject } y \in X$$

- Boundary rule induced by an equivalence class $[x] \subseteq \text{BND}_{(\alpha,\beta)}(X)$

$$\text{if } y \models \bigwedge_{a \in A} a = I_a(x), \ \text{neither accept}$$

$$\text{nor reject } y \in X.$$

The three types of rules have very different semantic interpretations as defined by their respective decisions. A positive rule allows us to *accept* an object y to be a member of X, because y has a higher probability of being in X due to the facts that $y \in [x]_A$ and $Pr(X|[x]_A) \geq \alpha$. A negative rule enables us to *reject* an object y to be a member of X, because y has lower probability of being in X due to the facts that $y \in [x]_A$ and $Pr(X|[x]_A) \leq \beta$. When the probability of y being in X is neither high nor low, a boundary rule makes a noncommitment decision. Although we explicitly give the class of boundary rules for convenience and completeness, we do not really need this class, once we have both classes of positive and negative rules. Whenever we can not accept nor reject an object to be a member of X, we choose a noncommitment decision.

Both actions of acceptance and rejection as associated with errors and costs. The error rate of a positive rule is given by $1 - Pr(X|[x])$, which, by definition of the three regions, is at or below $1 - \alpha$. The error rate of negative rule is given by $Pr(X|[x])$ and is at or below β. It becomes clear that the introduction of a noncommitment decision is to ensure both a low level of acceptance error and a low level of rejection error. According to the 3×2 table in Sect. 24.5.1, the cost a positive rule is $\lambda_{PP}Pr(X|[x]_A) + \lambda_{PN}(1 - Pr(X|[x]_A))$ and is bounded above by $\alpha\lambda_{PP} + (1-\alpha)\lambda_{PN}$. The cost a negative rule is $\lambda_{NP}Pr(X|[x]_A) + \lambda_{NN}(1 - Pr(X|[x]_A))$ and is bounded above by $\beta\lambda_{NP} + (1-\beta)\lambda_{NN}$. From view of cost, a noncommitment decision is preferred if its cost is less than an action of acceptance or rejection.

24.6 Dominance–Based Rough Set Approaches

Very often value sets V_a of some attributes $a \in AT$ are ordered in the sense that it is meaningful to consider a binary relation $\succsim_a$ on V_a such that for $x, y \in U$, $I_a(x) \succsim_a I_a(y)$ means that x possesses some property related to attribute a at least as much as y. In this case, it is natural to consider $\succsim_a$ as complete preorder on V_a, i.e., a transitive and strongly complete binary relation on V_a (let us remember that strong completeness means that for all $v_a, u_a \in V_a$ we have $v_a \succsim_a u_a$ or $u_a \succsim_a v_a$ and that this implies the reflexivity of $\succsim_a$). Observe that the binary relation $\succsim_a^U$ on U defined as $x \succsim_a^U y$ if $I_a(x) \succsim_a I_a(y)$ for all $x, y \in U$ is a complete preorder. The first type of properties considered in this perspective were preferences encountered in Multiple Criteria Decision Aiding (MCDA) (for a comprehensive collection of state of the art surveys see [24.70]), where for $x, y \in U$, $I_a(x) \succsim I_a(y)$ means x is at least as good as y with respect to attribute a that in this case is called *criterion*. If there are attributes $a \in AT$ related to some complete preorder $\succsim_a$, then the indiscernibility relation is unable to produce granules in U taking into account the order generated by $\succsim_a$. To do so, the indiscernibility relation has to be substituted by a new binary relation on U that, using a term coming from MCDA, is called *dominance relation*. Suppose, for simplicity, all attributes a from AT are criteria related to corresponding complete preorders $\succsim_a$.

We say that x *dominates* y with respect to $A \subseteq AT$ (shortly, x *A-dominates* y) denoted by $x \succsim_A^U y$, if $I_a(x) \succsim_a I(y)$ for all $a \in A$. Since $\succsim_a^U$ is a complete preorder on U for each $a \in AT$, $\succsim_A$ is a partial preoder on U, i.e. $\succsim_A$ is a reflexive and transitive binary relation on U.

For any $x \in U$ and for each nonempty $A \subseteq AT$, we can define a positive and a negative cone of dominance, denoted by $D_A^+(x)$ and $D_A^-(x)$, respectively,

$$D_A^+(x) = \{y \in U \mid y \succsim_A^U x\} \,,$$
$$D_A^-(x) = \{y \in U \mid x \succsim_A^U y\} \,. \tag{24.51}$$

For simplicity, we also simply write $D^+(x)$ and $D^-(x)$ when no confusion arises.

Let us explain how the rough set concept has been generalized to the dominance-based rough set approach (DRSA) in order to enable granular computing with dominance cones (for more details, see Chap. 22, and [24.31–34, 71–74]).

For any $X \subseteq U$ we define upward lower and upper approximations $apr^+(X)$ and $\overline{apr}^+(X)$, as well as downward lower and upper approximations $apr^-(X)$ and $\overline{apr}^-(X)$, as follows

$$apr^+(X) = \{x \in U \mid D^+(x) \subseteq X\} \,,$$
$$\overline{apr}^+(X) = \{x \in U \mid D^-(x) \cap X \neq \varnothing\} \,,$$
$$apr^-(X) = \{x \in U \mid D^-(x) \subseteq X\} \,,$$
$$\overline{apr}^-(X) = \{x \in U \mid D^+(x) \cap X \neq \varnothing\} \,. \tag{24.52}$$

For any $X \subseteq U$, using cones of dominance $D^+(x)$ and $D^-(x)$, we can define three upward pairwise disjoint positive, negative and boundary regions

$$POS^+(X) = \{x \in U \mid D^+(x) \subseteq X\} \,,$$
$$NEG^+(X) = \{x \in U \mid D^-(x) \cap X = \varnothing\} \,,$$
$$BND^+(X) = \{x \in U \mid D^+(x) \nsubseteq X$$
$$\text{and } D^-(x) \cap X \neq \varnothing\} \,. \tag{24.53}$$

Analogously, for any $X \subseteq U$, we can define three downward pairwise disjoint positive, negative, and boundary regions

$$POS^-(X) = \{x \in U \mid D^-(x) \subseteq X\},$$

$$NEG^-(X) = \{x \in U \mid D^+(x) \cap X = \emptyset\},$$

$$BND^-(X) = \{x \in U \mid D^-(x) \not\subseteq X$$
$$\text{and } D^+(x) \cap X \neq \emptyset\}. \quad (24.54)$$

Observe that the following complementarity properties hold: For all $X \subseteq U$

$$POS^+(X) = NEG^-(U - X),$$

$$POS^-(X) = NEG^+(U - X),$$

$$BND^+(X) = BND^-(U - X),$$

$$BND^-(X) = BND^+(U - X). \quad (24.55)$$

For all $X \subseteq U$, the pair of upward approximations and three upward regions determine each others as follows

$$POS^+(X) = \underline{apr}^+(X),$$

$$NEG^+(X) = (\overline{apr}^+(X))^c,$$

$$BND^+(X) = \overline{apr}^+(X) - \underline{apr}^+(X), \quad (24.56)$$

and

$$\underline{apr}^+(X) = POS^+(X),$$

$$\overline{apr}^+(X) = POS^+(X) \cup BND(X). \quad (24.57)$$

Analogously, for all $X \subseteq U$, the pair of downward approximations and three downward regions determine each others as follows

$$POS^-(X) = \underline{apr}^-(X),$$

$$NEG^-(X) = (\overline{apr}^-(X))^c,$$

$$BND^-(X) = \overline{apr}^-(X) - \underline{apr}^-(X), \quad (24.58)$$

and

$$\underline{apr}^-(X) = POS^-(X),$$

$$\overline{apr}^-(X) = POS^-(X) \cup BND(X). \quad (24.59)$$

24.7 A Basic Model of Dominance–Based Probabilistic Rough Sets

DRSA considers only qualitative relationship between positive and negative cones $D^+(x)$ and $D^-(x)$, and a set X, namely, a positive or negative cone is a subset of the set or has a nonempty intersection with the set. This qualitative nature becomes clearer with a probabilistic interpretation. Suppose $Pr(X|D^+(x))$ denotes the conditional probability that an object is in X, given that the object is in $D^+(x)$, as well as $Pr(X|D^-(x))$ denotes the conditional probability that an object is in X, given that the object is in $D^-(x)$. The conditions for defining rough set three upward regions can be equivalently expressed as

$$D^+(x) \subseteq X \iff Pr(X|D^+(x)) \geq 1;$$
$$D^-(x) \cap X = \emptyset \iff Pr(X|D^-(x)) \leq 0;$$
$$D^+(x) \not\subseteq X \land D^-(x) \cap X \neq \emptyset$$
$$\iff Pr(X|D^+(x)) < 1$$
$$\land Pr(X|D^-(x)) > 0. \quad (24.60)$$

Analogously, the conditions for defining rough set three upward regions can be equivalently expressed as

$$D^-(x) \subseteq X \iff Pr(X|D^-(x)) \geq 1;$$
$$D^+(x) \cap X = \emptyset \iff Pr(X|D^+(x)) \leq 0;$$
$$D^-(x) \not\subseteq X \land D^+(x) \cap X \neq \emptyset$$
$$\iff Pr(X|D^-(x)) < 1 \land Pr(X|D^+(x)) > 0. \quad (24.61)$$

By those conditions, DRSA upward and downward three regions can be equivalently expressed as

$$POS^+(X) = \{x \in U \mid Pr(X|D^+(x)) \geq 1\},$$

$$NEG^+(X) = \{x \in U \mid Pr(X|D^-(x)) \leq 0\},$$

$$BND^+(X) = \{x \in U \mid Pr(X|D^+(x)) < 1$$
$$\land Pr(X|D^-(x)) > 0\},$$

$$POS^-(X) = \{x \in U \mid Pr(X|D^-(x)) \geq 1\},$$

$$NEG^-(X) = \{x \in U \mid Pr(X|D^+(x)) \leq 0\},$$

$$BND^-(X) = \{x \in U \mid Pr(X|D^-(x)) < 1$$
$$\land Pr(X|D^+(x)) > 0\}. \quad (24.62)$$

Observe that DRSA approximations use only the two extreme values, i.e., 1 and 0, of probability.

Part C | 24.7

It is natural to generalize DRSA approximations by replacing 1 and 0 with some other values in the unit interval $[0, 1]$. Given a pair of thresholds α, β with $0 \leq \beta < \alpha \leq 1$, the main results of probabilistic DRSA are the (α, β)-probabilistic regions defined by

$$\text{POS}^+_{(\alpha, \beta)}(X) = \{x \in U \mid Pr(X|D^+(x)) \geq \alpha\},$$

$$\text{NEG}^+_{(\alpha, \beta)}(X) = \{x \in U \mid Pr(X|D^-(x)) \leq \beta\},$$

$$\text{BND}^+_{(\alpha, \beta)}(X) = \{x \in U \mid Pr(X|D^+(x)) < \alpha$$
$$\wedge Pr(X|D^-(x)) > \beta\},$$

$$\text{POS}^-_{(\alpha, \beta)}(X) = \{x \in U \mid Pr(X|D^-(x)) \geq \alpha\},$$

$$\text{NEG}^-_{(\alpha, \beta)}(X) = \{x \in U \mid Pr(X|D^+(x)) \leq \beta\},$$

$$\text{BND}^-_{(\alpha, \beta)}(X) = \{x \in U \mid Pr(X|D^-(x)) < \alpha$$
$$\wedge Pr(X|D^+(x)) > \beta\}. \quad (24.63)$$

The DRSA rough set model is a special case in which $\alpha = 1$ and $\beta = 0$. In the case when $0 < \beta = \alpha < 1$, the three regions are given by

$$\text{POS}^+_{(\alpha, \alpha)}(X) = \{x \in U \mid Pr(X|D^+(x)) \geq \alpha\},$$

$$\text{NEG}^+_{(\alpha, \alpha)}(X) = \{x \in U \mid Pr(X|D^-(x)) \leq \alpha\},$$

$$\text{BND}^+_{(\alpha, \alpha)}(X) = \{x \in U \mid Pr(X|D^+(x)) < \alpha$$
$$\wedge Pr(X|D^-(x)) > \alpha\},$$

$$\text{POS}^-_{(\alpha, \alpha)}(X) = \{x \in U \mid Pr(X|D^-(x)) \geq \alpha\},$$

$$\text{NEG}^-_{(\alpha, \alpha)}(X) = \{x \in U \mid Pr(X|D^+(x)) \leq \alpha\},$$

$$\text{BND}^-_{(\alpha, \alpha)}(X) = \{x \in U \mid Pr(X|D^-(x)) < \alpha$$
$$\wedge Pr(X|D^+(x)) > \alpha\}. \quad (24.64)$$

24.8 Variants of Probabilistic Dominance–Based Rough Set Approach

Several models generalizing dominance-based rough sets by incorporating probabilistic information can be considered.

24.8.1 Variable Consistency Dominance–Based Rough Sets

In a first version of variable consistency dominance-based rough sets [24.23] (see also [24.24]) the standard set inclusions $D^+(x) \subseteq X$ and $D^-(x) \subseteq X$ can be generalized into graded set inclusion $s^+(D^+(x), X)$ and $s^-(D^-(x), X)$ called measure of the relative upward and downward degree of misclassification of $D^+(x)$ and $D^-(x)$ with respect to X, respectively. A particular upward and downward measure is given by

$$s^+(D^+(x), X) = 1 - \frac{|D^+(x) \cap X|}{|D^+(x)|},$$

$$s^-(D^-(x), X) = 1 - \frac{|D^-(x) \cap X|}{|D^-(x)|}.$$

$$(24.65)$$

By introducing a threshold $0 \leq z < 0.5$, one can define three upward and downward regions as follows

$$\text{VPOS}^+_z(X) = \{x \in U \mid s^+(D^+(x), X) \leq z\},$$

$$\text{VNEG}^+_z(X) = \{x \in U \mid s^-(D^-(x), X) \geq 1 - z\},$$

$$\text{VBND}^+_z(X) = \{x \in U \mid s^+(D^+(x), X) > z$$
$$\wedge s^-(D^-(x), X) < 1 - z\},$$

$$\text{VPOS}^-_z(X) = \{x \in U \mid s^-(D^-(x), X) \leq z\},$$

$$\text{VNEG}^-_z(X) = \{x \in U \mid s^+(D^+(x), X) \geq 1 - z\},$$

$$\text{VBND}^-_z(X) = \{x \in U \mid s^-(D^-(x), X) > z$$
$$\wedge s^+(D^+(x), X) < 1 - z\}. \quad (24.66)$$

A more generalized version using a pair of thresholds can be defined as follows: for $0 \leq l < u \leq 1$,

$$\text{VPOS}^+_{(l,u)}(X) = \{x \in U \mid s^+(D^+(x), X) \leq l\},$$

$$\text{VNEG}^+_{(l,u)}(X) = \{x \in U \mid s^-(D^-(x), X) \geq u\},$$

$$\text{VBND}^+_{(l,u)}(X) = \{x \in U \mid s^+(D^+(x), X) > l$$
$$\wedge s^-(D^-(x), X) < u\},$$

$$\text{VPOS}^-_{(l,u)}(X) = \{x \in U \mid s^-(D^-(x), X) \leq l\},$$

$$\text{VNEG}^-_{(l,u)}(X) = \{x \in U \mid s^+(D^+(x), X) \geq u\},$$

$$\text{VBND}^-_{(l,u)}(X) = \{x \in U \mid s^-(D^-(x), X) > l$$
$$\wedge s^+(D^+(x), X) < u\}. \quad (24.67)$$

The one-threshold model may be considered as a special case of the two-threshold model with $l = z$ and $u = 1 - z$.

One may interpret the ratio in (24.65) as an estimation of the conditional probability $Pr(X|D^+(x))$ and $Pr(X|D^-(x))$, namely,

$$s^+(Pr(X|D^+(x)), X) = 1 - \frac{|D^+(x) \cap X|}{|D^+(x)|}$$

$$= 1 - Pr(X|D^+(x)),$$

$$s^-(Pr(X|D^-(x)), X) = 1 - \frac{|D^-(x) \cap X|}{|D^-(x)|}$$

$$= 1 - Pr(X|D^-(x)).$$

(24.68)

By setting $\alpha = 1 - l$ and $\beta = 1 - u$, we immediately get

$$POS^+_{A,(1-l,1-u)}(X) = \{x \in U \mid Pr(X|D^+(x)) \geq 1 - l\}$$

$$= \{x \in U \mid s^+(D^+(x), X) \leq l\}$$

$$= VPOS^+_{(l,u)}(X),$$

$$NEG^+_{(1-l,1-u)}(X) = \{x \in U \mid Pr(X|D^-(x)) \leq 1 - u\}$$

$$= \{x \in U \mid s^-(D^-(x), X) \geq u\}$$

$$= VNEG^+_{(l,u)}(X),$$

$$BND^+_{(1-l,1-u)}(X) = \{x \in U \mid Pr(X|D^+(x)) < 1 - l$$
$$\wedge Pr(X|D^-(x)) > 1 - u\}$$

$$= \{x \in U \mid s^+(D^+(x), X) > l$$
$$\wedge s^-(D^-(x), X) < u\}$$

$$= VBND^+_{(l,u)}(X),$$

$$POS^-_{(1-l,1-u)}(X) = \{x \in U \mid Pr(X|D^-(x)) \geq 1 - l\}$$

$$= \{x \in U \mid s^-(D^-(x), X) \leq l\}$$

$$= VPOS^-_{(l,u)}(X),$$

$$NEG^-_{(1-l,1-u)}(X) = \{x \in U \mid Pr(X|D^+(x)) \leq 1 - u\}$$

$$= \{x \in U \mid s^+(D^+(x), X) \geq u\}$$

$$= VNEG^+_{(l,u)}(X),$$

$$BND^+_{(1-l,1-u)}(X) = \{x \in U \mid Pr(X|D^-(x)) < 1 - l$$
$$\wedge Pr(X|D^+(x)) > 1 - u\}$$

$$= \{x \in U \mid s^-(D^-(x), X) > l$$
$$\wedge s^+(D^+(x), X) < u\}$$

$$= VBND^-_{(l,u)}(X).$$

(24.69)

24.8.2 Parameterized Dominance–Based Rough Sets

Parameterized rough sets based on dominance [24.24] generalize variable consistency DRSA by introducing a Bayesian confirmation measure and a pair of thresholds on the confirmation measure, in addition to a pair of thresholds on conditional probability. Let $c^+(D^+(x), X)$ and $c^-(D^-(x), X)$ denote a Bayesian upward and downward confirmation measure, respectively, that indicate the degree to which positive or negative cones $D^+(x)$ and $D^-(x)$ confirm the hypothesis X. The upward and downward Bayesian confirmation measures corresponding to those ones introduced in Sect. 24.4.3 are

$$c_d^+(D^+(x), X) = Pr(X|D^+(x)) - Pr(X),$$

$$c_d^-(D^-(x), X) = Pr(X|D^-(x)) - Pr(X),$$

$$c_r^+(D^+(x), X) = \frac{Pr(X|D^+(x))}{Pr(X)},$$

$$c_r^-(D^+(x), X) = \frac{Pr(X|D^-(x))}{Pr(X)}.$$

(24.70)

Given a pair of thresholds (s, t) with $t < s$, three (α, β, s, t)-parameterized regions can be defined as follows

$$PPOS^+_{(\alpha,\beta,s,t)}(X) = \{x \in U \mid Pr(X|D^+(x)) \geq \alpha$$
$$\wedge c^+(D^+(x), X) \geq s\},$$

$$PNEG^+_{(\alpha,\beta,s,t)}(X) = \{x \in U \mid Pr(X|D^-(x)) \leq \beta$$
$$\wedge c^-(D^-(x), X) \leq t\},$$

$$PBND^+_{(\alpha,\beta,s,t)}(X) = \{x \in U \mid (Pr(X|D^+(x)) < \alpha$$
$$\vee c^+(D^+(x), X) < s)$$
$$\wedge (Pr(X|D^-(x)) > \beta$$
$$\vee c^-(D^-(x), X) > t)\}$$

$$PPOS^-_{(\alpha,\beta,s,t)}(X) = \{x \in U \mid Pr(X|D^-(x)) \geq \alpha$$
$$\wedge c^-(D^-(x), X) \geq s\},$$

$$PNEG^-_{(\alpha,\beta,s,t)}(X) = \{x \in U \mid Pr(X|D^+(x)) \leq \beta$$
$$\wedge c^+(D^+(x), X) \leq t\},$$

$$PBND^-_{(\alpha,\beta,s,t)}(X) = \{x \in U \mid (Pr(X|D^-(x)) < \alpha$$
$$\vee c^-(D^-(x), X) < s)$$
$$\wedge (Pr(X|D^+(x)) > \beta$$
$$\vee c^+(D^+(x), X) > t)\}.$$ (24.71)

Let us remember that a family of consistency measures, called gain-type consistency measures, and inconsistency measures, called cost-type consistency measures, larger than confirmation measures, and the related dominance-based rough sets have been considered in [24.24]. For any $x \in U$ and $X \subseteq U$, for a consistency measure $m_c(x, X)$, x can be assigned to the positive region of X if $m_c(x, X) \geq \alpha$, with α being a proper threshold, while for an inconsistency measure $m_{ic}(x, X)$, x can be assigned to the positive region of X if $m_c(x, X) \leq \alpha$. A consistency measure $m_c(x, X)$ or an inconsistency measure $m_{ic}(x, X)$ are monotonic (Sect. 22.3.2 in Chap. 22) if they do not deteriorate when:

(m1) The set of attributes is growing,
(m2) The set of objects is growing,
(m3) x improves its evaluation, so that it dominates more objects.

Among the considered consistency and inconsistency measures, one that can be considered very interesting because it enjoys all the considered monotonity properties (m1)–(m3) while maintaining a reasonably easy formulation is the inconsistency measures ε' which is expressed as follows:

● In the case of dominance-based upward approximation

$$\varepsilon'^{+}(x, X) = \frac{|D^{+}(x) \cap (U - X)|}{|X|}$$

● In the case of dominance-based downward approximation

$$\varepsilon'^{-}(x, X) = \frac{|D^{-}(x) \cap (U - X)|}{|X|} .$$

Observe that as explained in [24.24], consistency and inconsistency measures can be properly reformulated in order to be used in indiscernibility-based rough sets. For example, inconsistency measure ε' in case of indicernibility-based rough sets becomes

$$\varepsilon'^{+}(x, X) = \frac{|[x] \cap (U - X)|}{|X|} .$$

24.8.3 Confirmation–Theoretic Dominance–Based Rough Sets

A separation of the parameterized model into two models within DRSA can be constructed as follows. One is the conventional probabilistic model and the other is a confirmation-theoretic model. For an upward and a downward Bayesian confirmation measure ($c^{+}(D^{+}(x), X)$ and $c^{-}(D^{-}(x), X)$), and a pair of

thresholds (s, t) with $t < s$, three confirmation regions are defined by

$$\text{CPOS}^{+}_{(s,t)}(X) = \{x \in U \mid c^{+}(D^{+}(x), X) \geq s\} ,$$
$$\text{CNEG}^{+}_{(s,t)}(X) = \{x \in U \mid c^{-}(D^{-}(x), X) \leq t\} ,$$
$$\text{CBND}^{+}_{(s,t)}(X) = \{x \in U \mid c^{+}(D^{+}(x), X) < s$$
$$\wedge c^{-}(D^{-}(x), X) > t\} ,$$
$$\text{CPOS}^{-}_{(s,t)}(X) = \{x \in U \mid c^{+}(D^{-}(x), X) \geq s\} ,$$
$$\text{CNEG}^{-}_{(s,t)}(X) = \{x \in U \mid c^{-}(D^{+}(x), X) \leq t\} ,$$
$$\text{CBND}^{-}_{(s,t)}(X) = \{x \in U \mid c^{-}(D^{-}(x), X) < s$$
$$\wedge c^{+}(D^{+}(x), X) > t\} . \quad (24.72)$$

For the case with $s = t$, we define

$$\text{CPOS}^{+}_{(s,s)}(X) = \{x \in U \mid c^{+}(D^{+}(x), X) \geq s\} ,$$
$$\text{CNEG}^{+}_{(s,s)}(X) = \{x \in U \mid c^{-}(D^{-}(x), X) \leq s\} ,$$
$$\text{CBND}^{+}_{(s,s)}(X) = \{x \in U \mid c^{+}(D^{+}(x), X) < s$$
$$\wedge c^{-}(D^{-}(x), X) > s\} ,$$
$$\text{CPOS}^{-}_{(s,s)}(X) = \{x \in U \mid c^{+}(D^{+}(x), X) \geq s\} ,$$
$$\text{CNEG}^{-}_{(s,s)}(X) = \{x \in U \mid c^{-}(D^{-}(x), X) \leq s\} ,$$
$$\text{CBND}^{-}_{(s,s)}(X) = \{x \in U \mid c^{+}(D^{+}(x), X) < s$$
$$\wedge c^{+}(D^{+}(x), X) > s\} . \quad (24.73)$$

24.8.4 Bayesian Dominance–Based Rough Sets

Bayesian DRSA model in which the required pair of thresholds is interpreted using a priori probability $Pr(X)$ can be defined as an extension of the Bayesian DRSA and variable consistency Bayesian DRSA, as explained below.

For the Bayesian DRSA, the three upward and downward regions are defined by

$$\text{BPOS}^{+}(X) = \{x \in U \mid Pr(X|D^{+}(x)) > Pr(X)\} ,$$
$$\text{BNEG}^{+}(X) = \{x \in U \mid Pr(X|D^{-}(x)) < Pr(X)\} ,$$
$$\text{BBND}^{+}(X) = \{x \in U \mid Pr(X|D^{+}(x)) \leq Pr(X)$$
$$\wedge Pr(X|D^{-}(x)) \geq Pr(X)\} ,$$
$$\text{BPOS}^{-}(X) = \{x \in U \mid Pr(X|D^{-}(x)) > Pr(X)\} ,$$
$$\text{BNEG}^{-}(X) = \{x \in U \mid Pr(X|D^{+}(x)) < Pr(X)\} ,$$
$$\text{BBND}^{-}(X) = \{x \in U \mid Pr(X|D^{-}(x)) \leq Pr(X)$$
$$\wedge Pr(X|D^{+}(x)) \geq Pr(X)\} . \quad (24.74)$$

Bayesian dominance-based rough sets can be viewed as a special case of the decision-theoretic DRSA when $\alpha = \beta = Pr(X)$.

Recalling the upward and downward DRSA, Bayesian confirmation measures

$$c_d^+(D^+(x), X) = Pr(X|D^+(x)) - Pr(X)$$

and

$$c_d^-(D^-(x), X) = Pr(X|D^-(x)) - Pr(X) ,$$

Bayesian dominance-based rough sets can be expressed as confirmation-theoretic dominance-based rough sets as follows

$$\text{BPOS}^+(X) = \{x \in U \mid Pr(X|D^+(x)) > Pr(X)\} ,$$
$$= \{x \in U \mid c_d^+(D^+(x), X) > 0\} ,$$
$$\text{BNEG}^+(X) = \{x \in U \mid Pr(X|D^-(x)) < Pr(X)\} ,$$
$$= \{x \in U \mid c_d^-(D^-(x), X) < 0\} ,$$
$$\text{BBND}(X)^+ = \{x \in U \mid Pr(X|D^+(x)) \leq Pr(X)$$
$$\wedge Pr(X|D^-(x)) \geq Pr(X)\}$$
$$= \{x \in U \mid c_d^+(D^+(x), X) \leq 0$$
$$\wedge c_d^-(D^-(x), X) \geq 0\} ,$$

$$\text{BPOS}^-(X) = \{x \in U \mid Pr(X|D^-(x)) > Pr(X)\} ,$$
$$= \{x \in U \mid c_d^-(D^-(x), X) > 0\} ,$$
$$\text{BNEG}^+(X) = \{x \in U \mid Pr(X|D^-(x)) < Pr(X)\} ,$$
$$= \{x \in U \mid c_d^+(D^+(x), X) < 0\} ,$$
$$\text{BBND}(X)^+ = \{x \in U \mid Pr(X|D^-(x)) \leq Pr(X)$$
$$\wedge Pr(X|D^+(x)) \geq Pr(X)\}$$
$$= \{x \in U \mid c_d^-(D^-(x), X) \leq 0$$
$$\wedge c_d^+(D^+(x), X) \geq 0\} . \qquad (24.75)$$

That is, the Bayesian rough sets are models of confirmation-theoretic rough sets characterized by the upward and downward Bayesian confirmation measures c_d^+ and c_d^- with a pair of thresholds $s = t = 0$.

The three upward and downward regions of the variable precision Bayesian rough sets are defined as follows: for $\epsilon \in [0, 1)$,

$$\text{VBPOS}_\epsilon^+(X) = \{x \in U \mid Pr(X|[x])$$
$$\geq 1 - \epsilon(1 - Pr(X))\} ,$$
$$\text{VBNEG}_\epsilon(X) = \{x \in U \mid Pr(X|[x]) \leq \epsilon Pr(X)\} ,$$
$$\text{VBBND}_\epsilon(X) = \{x \in U \mid \epsilon Pr(X) < Pr(X|[x])$$
$$< 1 - \epsilon(1 - Pr(X))\} . \qquad (24.76)$$

24.9 Three Fundamental Issues of Probabilistic Dominance–Based Rough Sets

Also for probabilistic dominance-based rough sets, one must consider the three fundamental issues of interpretation and determination of the required pair of thresholds, estimation of the required conditional probabilities, and interpretation and applications of three probabilistic regions.

These three issues are considered in this section with respect to dominance-based rough sets.

24.9.1 Decision–Theoretic Dominance–Based Rough Set Model: Determining the Thresholds

Following [24.75], a decision-theoretic model formulates the construction of dominance-based rough set approximations as a Bayesian decision problem with a set of two states $\Omega = \{X, X^c\}$, indicating that an element is in X and not in X, respectively. In the case of up-ward dominance-based rough sets, we consider a set of three actions $\mathcal{A}^+ = \{a_P^+, a_B^+, a_N^+\}$, with a_P^+ deciding $x \in \text{POS}^+(X)$, a_B^+ deciding $x \in \text{BND}^+(X)$, and a_N^+ deciding $x \in \text{NEG}^+(X)$, respectively. In case of downward dominance-based rough sets, we consider a set of three actions $\mathcal{A}^- = \{a_P^-, a_B^-, a_N^-\}$, with a_P^- deciding $x \in \text{POS}^-(X)$, a_B^- deciding $x \in \text{BND}^-(X)$, and a_N^- deciding $x \in \text{NEG}^-(X)$, respectively. The losses regarding the actions for different states are given by the 6×2 matrix

	X (P)	X^c (N)
a_P^+	λ_{PP}^+	λ_{PN}^+
a_B^+	λ_{BP}^+	λ_{BN}^+
a_N^+	λ_{NP}^+	λ_{NN}^+
a_P^-	λ_{PP}^-	λ_{PN}^-
a_B^-	λ_{BP}^-	λ_{BN}^-
a_N^-	λ_{NP}^-	λ_{NN}^-

Part C | 24.9

In the matrix:

- In the case that upward dominance-based rough approximations are considered, λ_{PP}^+, λ_{BP}^+, and λ_{NP}^+ denote the losses incurred for taking actions a_P^+, a_B^+, and a_N^+, respectively, when an object belongs to X, and λ_{PN}^+, λ_{BN}^+ and λ_{NN}^+ denote the losses incurred for taking the same actions when the object does not belong to X,
- In the case that downward dominance-based rough approximations are considered, λ_{PP}^-, λ_{BP}^-, and λ_{NP}^- denote the losses incurred for taking actions a_P^-, a_B^-, and a_N^-, respectively, when an object belongs to X, and λ_{PN}^-, λ_{BN}^- and λ_{NN}^- denote the losses incurred for taking the same actions when the object does not belong to X.

In the case that upward dominance-based rough approximations are considered, the expected losses associated with taking different actions for objects in $D^+(x)$ can be expressed as

$$R(a_P^+|D^+(x)) = \lambda_{PP}^+ Pr(X|D^+(x))$$
$$+ \lambda_{PN}^+ Pr(X^c|D^+(x)),$$
$$R(a_B^+|D^+(x)) = \lambda_{BP}^+ Pr(X|D^+(x))$$
$$+ \lambda_{BN}^+ Pr(X^c|D^+(x)),$$
$$R(a_N^+|D^+(x)) = \lambda_{NP}^+ Pr(X|D^+(x))$$
$$+ \lambda_{NN}^+ Pr(X^c|D^+(x)). \quad (24.77)$$

In the case that downward dominance-based rough approximations are considered, the expected losses associated with taking different actions for objects in $D^-(x)$ can be expressed as

$$R(a_P^-|D^-(x)) = \lambda_{PP}^- Pr(X|D^-(x))$$
$$+ \lambda_{PN}^- Pr(X^c|D^-(x)),$$
$$R(a_B^-|D^-(x)) = \lambda_{BP}^- Pr(X|D^-(x))$$
$$+ \lambda_{BN}^- Pr(X^c|D^-(x)),$$
$$R(a_N^-|D^-(x)) = \lambda_{NP}^- Pr(X|D^-(x))$$
$$+ \lambda_{NN}^- Pr(X^c|D^-(x)). \quad (24.78)$$

In the case that upward dominance-based rough approximations are considered, the Bayesian decision procedure suggests the following minimum-risk decision rules

(P$^+$) If $R(a_P^+|D^+(x)) \leq R(a_B^+|D^+(x))$
and $R(a_P^+|[x]) \leq R(a_N^+|D^+(x))$,
decide $x \in \text{POS}^+(X)$;

(B$^+$) If $R(a_B^+|D^+(x)) \leq R(a_P^+|D^+(x))$
and $R(a_B^+|[x]) \leq R(a_N^+|D^+(x))$,
decide $x \in \text{BND}^+(X)$;

(N$^+$) If $R(a_N^+|D^+(x)) \leq R(a_P^+|D^+(x))$
and $R(a_N^+|[x]) \leq R(a_B^+|D^+(x))$,
decide $x \in \text{NEG}^+(X)$.

In the case that downward dominance-based rough approximations are considered, the Bayesian decision procedure suggests the following minimum-risk decision rules

(P$^-$) If $R(a_P^-|D^-(x)) \leq R(a_B^-|D^-(x))$
and $R(a_P^-|[x]) \leq R(a_N^-|D^-(x))$,
decide $x \in \text{POS}^-(X)$;

(B$^-$) If $R(a_B^-|D^-(x)) \leq R(a_P^-|D^-(x))$
and $R(a_B^-|[x]) \leq R(a_N^-|D^-(x))$,
decide $x \in \text{BND}^-(X)$;

(N$^-$) If $R(a_N^-|D^-(x)) \leq R(a_P^-|D^-(x))$
and $R(a_N^-|[x]) \leq R(a_B^-|D^-(x))$,
decide $x \in \text{NEG}^-(X)$.

Also in the case that dominance-based rough approximations are considered, when two or three actions have the same risk, one can use the same ordering for breaking a tie used in case indiscernibility-based rough approximations are used: a_P^+, a_N^+, a_B^+ in case upward rough approximations are considered, and a_P^-, a_N^-, a_B^- in case downward rough approximations are considered.

Analogously to Sect. 24.5.1, let us consider the special class of loss functions with

(c0$^+$). $\lambda_{PP}^+ \leq \lambda_{BP}^+ < \lambda_{NP}^+$, $\quad \lambda_{NN}^+ \leq \lambda_{BN}^+ < \lambda_{PN}^+$,
(c0$^-$). $\lambda_{PP}^- \leq \lambda_{BP}^- < \lambda_{NP}^-$, $\quad \lambda_{NN}^- \leq \lambda_{BN}^- < \lambda_{PN}^-$.
$$(24.79)$$

With the conditions (c0$^+$) and (c0$^-$), and the equations

$$Pr(X|D^+(x)) + Pr(X^c|D^+(x)) = 1$$

and

$$Pr(X|D^-(x)) + Pr(X^c|D^-(x)) = 1,$$

we can express the decision rules (P$^+$)–(N$^+$) and (P$^-$)–(N$^-$) in the following simplified form

 (P$^+$) If $Pr(X|D^+(x)) \geq \alpha^+$,

   and $Pr(X|D^+(x)) \geq \gamma^+$,

   decide $x \in \text{POS}^+(X)$;

 (B$^+$) If $Pr(X|D^+(x)) \leq \alpha^+$

   and $Pr(X|D^+(x)) \geq \beta^+$,

   decide $x \in \text{BND}^+(X)$;

 (N$^+$) If $Pr(X|D^+(x)) \leq \beta^+$

   and $Pr(X|D^+(x)) \leq \gamma^+$,

   decide $x \in \text{NEG}^+(X)$;

 (P$^-$) If $Pr(X|D^-(x)) \geq \alpha^-$

   and $Pr(X|D^-(x)) \geq \gamma^+$,

   decide $x \in \text{POS}^-(X)$;

 (B$^-$) If $Pr(X|D^-(x)) \leq \alpha^-$

   and $Pr(X|D^-(x)) \geq \beta^+$,

   decide $x \in \text{BND}^-(X)$;

 (N$^-$) If $Pr(X|D^-(x)) \leq \beta^-$

   and $Pr(X|D^-(x)) \leq \gamma^+$,

   decide $x \in \text{NEG}^-(X)$.

where

$$\alpha^+ = \frac{(\lambda^+_{PN} - \lambda^+_{BN})}{(\lambda^+_{PN} - \lambda^+_{BN}) + (\lambda^+_{BP} - \lambda^+_{PP})},$$

$$\beta^+ = \frac{(\lambda^+_{BN} - \lambda^+_{NN})}{(\lambda^+_{BN} - \lambda^+_{NN}) + (\lambda^+_{NP} - \lambda^+_{BP})},$$

$$\gamma^+ = \frac{(\lambda^+_{PN} - \lambda^+_{NN})}{(\lambda^+_{PN} - \lambda^+_{NN}) + (\lambda^+_{NP} - \lambda^+_{PP})},$$

$$\alpha^- = \frac{(\lambda^-_{PN} - \lambda^-_{BN})}{(\lambda^-_{PN} - \lambda^-_{BN}) + (\lambda^-_{BP} - \lambda^-_{PP})},$$

$$\beta^- = \frac{(\lambda^-_{BN} - \lambda^-_{NN})}{(\lambda^-_{BN} - \lambda^-_{NN}) + (\lambda^-_{NP} - \lambda^-_{BP})},$$

$$\gamma^- = \frac{(\lambda^-_{PN} - \lambda^-_{NN})}{(\lambda^-_{PN} - \lambda^-_{NN}) + (\lambda^-_{NP} - \lambda^-_{PP})}. \tag{24.80}$$

By setting $\alpha^+ > \beta^+$ and $\alpha^- > \beta^-$, we obtain that $1 \geq \alpha^+ > \gamma^+ > \beta^+ \geq 0$ and $1 \geq \alpha^- > \gamma^- > \beta^- \geq 0$, that, after tie breaking, we give the following simplified rules

 (P$^+$) If $Pr(X|D^+(x)) \geq \alpha^+$,

   decide $x \in \text{POS}^+(X)$;

 (B$^+$) If $\beta^+ < Pr(X|D^+(x)) < \alpha^+$,

   decide $x \in \text{BND}^+(X)$;

 (N$^+$) If $Pr(X|D^+(x)) \leq \beta^+$,

   decide $x \in \text{NEG}^+(X)$,

 (P$^-$) If $Pr(X|D^-(x)) \geq \alpha^-$,

   decide $x \in \text{POS}^-(X)$;

 (B$^-$) If $\beta^- < Pr(X|D^-(x)) < \alpha^-$,

   decide $x \in \text{BND}^-(X)$;

 (N$^-$) If $Pr(X|D^-(x)) \leq \beta^-$,

   decide $x \in \text{NEG}^-(X)$,

so that the parameters γ^+ and γ^- are no longer needed. Each object can be put into one and only one upward region, and one and only one downward region by using rules (P$^+$), (B$^+$) and (N$^+$), and (P$^-$), (B$^-$) and (N$^-$), respectively. The upward (α^+, β^+)-probabilistic positive, negative, and boundary regions and downward (α^-, β^-)-probabilistic positive, negative, and boundary regions are given, respectively, by

$$\text{POS}^+_{(\alpha^+, \beta^+)}(X) = \{x \in U \mid Pr(X|D^+(x)) \geq \alpha^+\},$$

$$\text{BND}^+_{(\alpha^+, \beta^+)}(X) = \{x \in U \mid \beta^+ < Pr(X|D^+(x))$$
$$< \alpha^+\},$$

$$\text{NEG}^+_{(\alpha^+, \beta^+)}(X) = \{x \in U \mid Pr(X|D^+(x)) \leq \beta\},$$

$$\text{POS}^-_{(\alpha^-, \beta^-)}(X) = \{x \in U \mid Pr(X|D^-(x)) \geq \alpha^-\},$$

$$\text{BND}^-_{(\alpha^-, \beta^-)}(X) = \{x \in U \mid \beta^- < Pr(X|D^-(x))$$
$$< \alpha^-\},$$

$$\text{NEG}_{(\alpha^-, \beta^-)}(X) = \{x \in U \mid Pr(X|D^-(x)) \leq \beta^-\}. \tag{24.81}$$

An alternative decision theoretic model for dominance-based rough sets taking into account in the cost function the conditional probabilities $P(X|D^+(x))$ and $P(X^c|D^-(x))$ for upward rough approximations, as well as the conditional probabilities $P(X|D^-(x))$ and $P(X^c|D^+(x))$ for downward rough approximations, can be defined as follows.

In the case that upward dominance-based rough approximations are considered, the expected losses associated with taking different actions for objects in

$D^+(x)$ can be expressed as

$$R(a_P^+|D^+(x), D^-(x)) = \lambda_{PP}^+ Pr(X|D^+(x))$$
$$+ \lambda_{PN}^+ Pr(X^c|D^-(x)),$$
$$R(a_B^+|D^+(x), D^-(x)) = \lambda_{BP}^+ Pr(X|D^+(x))$$
$$+ \lambda_{BN}^+ Pr(X^c|D^-(x)),$$
$$R(a_N^+|D^+(x), D^-(x)) = \lambda_{NP}^+ Pr(X|D^+(x))$$
$$+ \lambda_{NN}^+ Pr(X^c|D^-(x)).$$
$$(24.82)$$

In the case that downward dominance-based rough approximations are considered, the expected losses associated with taking different actions for objects in $D^-(x)$ can be expressed as

$$R(a_P^-|D^+(x), D^-(x)) = \lambda_{PP}^- Pr(X|D^-(x))$$
$$+ \lambda_{PN}^- Pr(X^c|D^+(x)),$$
$$R(a_B^-|D^+(x), D^-(x)) = \lambda_{BP}^- Pr(X|D^-(x))$$
$$+ \lambda_{BN}^- Pr(X^c|D^+(x)),$$
$$R(a_N^-|D^+(x), D^-(x)) = \lambda_{NP}^- Pr(X|D^-(x))$$
$$+ \lambda_{NN}^- Pr(X^c|D^+(x)).$$
$$(24.83)$$

24.9.2 Stochastic Dominance–Based Rough Set Approach: Estimating the Conditional Probability

Naive Bayesian rough set model presented for rough sets based on indiscernibility in Sect. 24.5.2 can be extended quite straightforwardly to rough sets based on dominance. Thus, in this section, we present a different approach to estimate probabilities for rough approximations: stochastic rough set approach [24.25] (see also Sect. 22.3.3 in Chap. 22). It can be applied also to rough sets based on indiscernibility, but here we present this approach taking into consideration rough sets based on dominance. In the following, we shall consider upward dominance-based approximations of a given $X \subseteq U$. However, the same approach can be used for downward dominance-based approximations. From a probabilistic point of view, the assignment of object x to $X \subseteq U$ can be made with probability $Pr(X|D^+(x))$ and $Pr(X|D^-(x))$. This probability is supposed to satisfy the usual axioms of probability

$$Pr(U|D^+(x)) = 1,$$

$$Pr(U - X|D^+(x)) = 1 - Pr(X|D^+(x)),$$
$$Pr(U|D^-(x)) = 1,$$
$$Pr(U - X|D^-(x)) = 1 - Pr(X|D^-(x)).$$

Moreover, this probability has to satisfy an axiom related to the choice of the rough upward approximation, i.e., the positive monotonic relationships one expects between membership in $X \subseteq U$ and possession of the properties related to attributes from AT, i.e., the dominance relation $\succsim$: for any $x, y \in U$ such that $x \succsim y$

(i) $Pr(X|D^+(x)) \geq Pr(X|D^+(y))$,
(ii) $Pr(U - X|D^-(x)) \leq Pr(U - X|D^-(y))$.

Condition (i) says that if objects x possesses properties related to attributes from AT at least as object y, i.e., $x \succsim y$, then the probability that x belongs to X has to be not smaller than the probability that y belongs to X. Analogously, Condition (ii) says that since $x \succsim y$, then the probability that x does not belong to X should not be greater than the probability that y does not belong to X. Observe that (ii) can be written also as

(ii) $Pr(X|D^-(x)) \geq Pr(X|D^-(y))$.

These probabilities are unknown but can be estimated from data. For each $X \subseteq U$, we have a binary problem of estimating the conditional probabilities $Pr(X|D^+(x)) = 1 - Pr(U - X|D^+(x))$ and the conditional probabilities $Pr(X|D^-(x)) = 1 - Pr(U - X|D^-(x))$. It can be solved by *isotonic regression* [24.25]. For $X \subseteq U$ and for any $x \in U$, let $y(x, X) = 1$ if $x \in X$, otherwise $y(x, X) = 0$. Then one can choose estimates $Pr^*(X|D^+(x))$ and $Pr^*(X|D^-(x))$ with $Pr^*(X|D^+(x))$ and $Pr^*(X|D^-(x))$ which minimize the squared distance to the class assignment $y(x, X)$, subject to the monotonicity constraints related to the dominance relation $\succsim$ on the attributes from AT (see also Sect. 22.3.3 in Chap. 22)

Minimize

$$\sum_{x \in U} (y(x, X) - Pr(X|D^+(x)))^2$$
$$+ (y(x, X) - Pr(X|D^-(x)))^2$$

subject to

$$Pr(X|D^+(x)) \geq Pr(X|D^+(z)) \text{ and}$$
$$Pr(X|D^-(x)) \geq Pr(X|D^-(z)) \text{ if } x \succsim z,$$

for all $x, z \in U$.

Then, stochastic α-lower approximations of $X \subseteq U$ can be defined as

$$\underline{P}^\alpha(X) = \{x \in U : Pr(X|D^+(x)) \geq \alpha\},$$
$$\underline{P}^\alpha(U-X) = \{x \in U : Pr(U-X|D^-(x)) \geq \alpha\}.$$

Replacing the unknown probabilities

$$Pr(X|D^+(x))$$

and

$$Pr(U-X|D^-(x))$$

by their estimates

$$Pr^*(X|D^+(x))$$

and

$$Pr^*(U-X|D^-(x))$$

obtained from isotonic regression, we get

$$\underline{P}^\alpha(X) = \left\{x \in U : Pr^*(X|D^+(x)) \geq \alpha\right\},$$
$$\underline{P}^\alpha(U-X) = \left\{x \in U : Pr^*(U-X|D^-(x)) \geq \alpha\right\},$$

where parameter $\alpha \in [0.5, 1]$ controls the allowed amount of inconsistency.

Solving isotonic regression requires $O(|U|^4)$ time, but a good heuristic needs only $O(|U|^2)$.

In fact, as shown in [24.25] and recalled in Sect. 22.3.3 in Chap. 22, we do not really need to know the probability estimates to obtain stochastic lower approximations. We only need to know for which object $x \in U$, $Pr^*(X|D^+(x)) \geq \alpha$ and for which $x \in U$, $Pr^*(U-X|D^-(x)) \geq \alpha$ (i.e., $Pr^*(X|D^-(x)) \leq 1 - \alpha$). This can be found by solving a linear programming (reassignment) problem.

As before, $y(x, X) = 1$ if $x \in X$, otherwise $y(x, X) = 0$. Let $d(x, X)$ be the decision variable which determines a new class assignment for object x. Then, reassign objects to X if $d^*(x, X) = 1$, and to $U - X$ if $d^*(x, X) = 0$, such that the new class assignments are consistent with the dominance principle, where $d^*(x, X)$ results from solving the following linear programming problem

$$\text{Minimize} \sum_{x \in U} w_{y(x,X)} |y(x, X) - d(x, X)|$$

$$\text{subject to } d(x, X) \geq d(z, X) \text{ if } x \succsim z$$
$$\text{for all } x, z \in U$$

where w_1 and w_0 are arbitrary positive weights.

Due to unimodularity of the constraint matrix, the optimal solution of this linear programming problem is always integer, i.e., $d^*(x, X) \in \{0, 1\}$. For all objects consistent with the dominance principle, $d^*(x, X) = y(x, X)$. If we set $w_0 = \alpha$ and $w_1 = \alpha - 1$, then the optimal solution $d^*(x, X)$ satisfies: $d^*(x, X) = 1 \Leftrightarrow Pr^*(X|D^+(x)) \geq \alpha$. If we set $w_0 = 1 - \alpha$ and $w_1 = \alpha$, then the optimal solution $d^*(x, X)$ satisfies: $d^*(x, X) = 0 \Leftrightarrow Pr^*(X|D^-(x)) \leq 1 - \alpha$.

Solving the reassignment problem twice, we can obtain the lower approximations $\underline{P}^\alpha(X)$, $\underline{P}^\alpha(U-X)$, without knowing the probability estimates.

24.9.3 Three-Way Decisions: Interpreting the Three Regions in the Case of Dominance-Based Rough Sets

In this section, we present an interpretation of dominance-based rough set three regions taking into consideration the framework of three-way decisions.

In an information table, with respect to a subset of attributes $A \subseteq AT$, an object x induces logic formulae

$$\bigwedge_{a \in A} I_a(x) \succsim_a v_a, \tag{24.84}$$

$$\bigwedge_{a \in A} v_a \succsim_a I_a(x), \tag{24.85}$$

where $I_a(x), v_a \in V_a$ and

- The atomic formula $v_a \succsim_a I_a(x)$ indicates that object x taking value $I_a(x)$ on attribute a possess a property related to a not more than any object y taking value $I_a(y) = v_a$ on attribute a.
- The atomic formula $I_a(x) \succsim_a v_a$ indicates that object x taking value $I_a(x)$ on attribute a possess a property related to a not less than any object y taking value $I_a(y) = v_a$ on attribute a.

Thus, an object y satisfies the formula

$$\bigwedge_{a \in A} I_a(x) \succsim_a v_a, \text{ if } I_a(y) \succsim v_a \text{ for all } a \in A,$$

that is,

$$\left(y \models \bigwedge_{a \in A} I_a(x) \succsim_a v_a \iff \forall a \in A, \ (I_a(y) \succsim_a v_a)\right). \tag{24.86}$$

Analogously, an object y satisfies the formula

$$\bigwedge_{a \in A} v_a \succsim_a I_a(x) \,, \text{ if } v_a \succsim I_a(y) \text{ for all } a \in A \,,$$

that is,

$$\left(y \models \bigwedge_{a \in A} v_a \succsim_a I_a(x) \iff \forall a \in A, \ (v_a \succsim_a I_a(y)) \right).$$

$$(24.87)$$

With these notations, we are ready to interpret upward and downward dominance-based rough set three regions.

From the upward and downward three regions, we can construct three classes of rules for classifying an object, called the upward and downward positive, negative, and boundary rules.

They are expressed in the following forms: for $y \in U$,

● Positive rule induced by an upward cone

$$D^+(x) \subseteq \text{POS}^+_{(\alpha,\beta)}(X):$$

if $y \models \bigwedge_{a \in A} I_a(x) \succsim_a v_a$, accept $y \in X$,

● Negative rule induced by the complement of an upward cone

$$U - D^+(x) \subseteq \text{NEG}^+_{(\alpha,\beta)}(X):$$

if $y \models \neg \bigwedge_{a \in A} I_a(x) \succsim_a v_a$, reject $y \in X$,

● Boundary rule induced by an upward cone $D^+(x)$ and its complement $U - D^+(x)$ such that

$$D^+(x) \nsubseteq \text{POS}^+_{(\alpha,\beta)}(X)$$

and $(U - D^+(x)) \nsubseteq \text{NEG}^+_{(\alpha,\beta)}(X):$

if $y \models \bigwedge_{a \in A} I_a(x) \succsim_a v_a \wedge \neg \bigwedge_{a \in A} I_a(x) \succsim_a u_a,$

neither accept nor reject $y \in X$,

● Positive rule induced by an downward cone

$$D^-(x) \subseteq \text{POS}^-_{(\alpha,\beta)}(X):$$

if $y \models \bigwedge_{a \in A} v_a \succsim_a I_a(x)$, accept $y \in X$,

● Negative rule induced by the complement of a downward cone

$$U - D^-(x) \subseteq \text{NEG}^-_{(\alpha,\beta)}(X):$$

if $y \models \neg \bigwedge_{a \in A} v_a \succsim_a I_a(x)$, reject $y \in X$,

● Boundary rule induced by a downward cone $D^-(x)$ and its complement $U - D^-(x)$ such that

$$D^-(x) \nsubseteq \text{POS}^-_{(\alpha,\beta)}(X)$$

and $U - D^-(x) \nsubseteq \text{NEG}^-_{(\alpha,\beta)}(X):$

if $y \models \bigwedge_{a \in A} v_a \succsim_a I_a(x) \wedge \neg \bigwedge_{a \in A} u_a \succsim_a I_a(x),$

neither accept nor reject $y \in X$.

The three types of rules have a semantic interpretations analogous to those induced by probabilistic rough sets based on indiscernibility presented in Sect. 24.5.3. Let us consider the rules related to POS^+ and NEG^+. A positive rule allows us to *accept* an object y to be a member of X, because y has a higher probability of being in X due to the facts that $y \in D^+(x)$ and $Pr(X|D^+(x)) \geq \alpha^+$. A negative rule enables us to *reject* an object y to be a member of X, because y has lower probability of being in X due to the facts that $y \in D^+(x)$ and $Pr(X|D^+(x)) \leq \beta^+$. When the probability of y being in X is neither high nor low, a boundary rule makes a noncommitment decision.

The error rate of a positive rule is given by $1 - Pr(X|D^+(x))$, which, by definition of the three regions, is at or below $1 - \alpha^+$. The error rate of negative rule is given by $Pr(X|D^+(x))$ and is at or below β^+. The cost of a positive rule is $\lambda^+_{PP} Pr(X|D^+(x)) + \lambda^+_{PN}(1 - Pr(X|D^+(x)))$ and is bounded above by $\alpha^+ \lambda^+_{PP} + (1 - \alpha^+)\lambda^+_{PN}$. The cost of a negative rule is $\lambda^+_{NP} Pr(X|D^+(x)) + \lambda^+_{NN}(1 - Pr(X|D^+(x)))$ and is bounded above by $\beta^+ \lambda^+_{NP} + (1 - \beta^+)\lambda^+_{NN}$.

24.10 Conclusions

A basic probabilistic rough set model is formulated by using a pair of thresholds on conditional probabilities, which leads to flexibility and robustness when performing classification or decision-making tasks. Three theories are the supporting pillars of probabilistic rough sets. Bayesian decision theory enables us to determine and interpret the required thresholds by using more operable notions such as loss, cost, risk, etc. Bayesian inference ensures us to estimate the conditional probability accurately. A theory of three-way decisions allows us to make a wise decision in the presence of incomplete or insufficient information.

Other probabilistic rough set models have also been described. We have shown how a probabilistic approach can be applied when information related to some order. The order concerns degrees in which an object has some properties related to considered attributes. This kind of order can be handled by the well-known rough set extension called dominance-based rough set approach.

One may expect a continuous growth of interest in probabilistic approaches to rough sets. An important task is to examine fully, in the light of three fundamental issues concerning the basic model, the semantics of each model, in order to identify its limitations and appropriate areas of applications.

References

24.1   Z. Pawlak: Rough set, Int. J. Inf. Comput. Sci. **11**, 341–356 (1982)
24.2   Z. Pawlak: *Rough Sets: Theoretical Aspects of Reasoning About Data* (Kluwer, Dordrecht 1991)
24.3   W. Marek, Z. Pawlak: Information storage and retrieval systems: mathematical foundations, Theor. Comput. Sci. **1**, 331–354 (1976)
24.4   Y.Y. Yao: A note on definability and approximations. In: *Transactions on Rough Sets VII*, Lecture Notes in Computer Science, Vol. 4400, ed. by J.F. Peters, A. Skowron, V.W. Marek, E. Orlowska, R. Słowiński, W. Ziarko (Springer, Heidelberg 2007) pp. 274–282
24.5   Y.Y. Yao: Probabilistic approaches to rough sets, Expert Syst. **20**, 287–297 (2003)
24.6   Y.Y. Yao: Probabilistic rough set approximations, Int. J. Approx. Reason. **49**, 255–271 (2008)
24.7   Z. Pawlak, S.K.M. Wong, W. Ziarko: Rough sets: Probabilistic versus deterministic approach, Int. J. Man-Mach. Stud. **29**, 81–95 (1988)
24.8   S. K. M. Wong, W. Ziarko: A probabilistic model of approximate classification and decision rules with uncertainty in inductive learning, Technical Report CS-85-23 (Department of Computer Science, University of Regina 1985)
24.9   S.K.M. Wong, W. Ziarko: INFER – an adaptive decision support system based on the probabilistic approximate classifications, Proc. 6th Int. Workshop on Expert Syst. Their Appl., Vol. 1 (1986) pp. 713–726
24.10   S.K.M. Wong, W. Ziarko: Comparison of the probabilistic approximate classification and the fuzzy set model, Fuzzy Sets Syst. **21**(3), 357–362 (1987)
24.11   Y.Y. Yao, S.K.M. Wong: A decision theoretic framework for approximating concepts, Int. J. Man-Mach. Stud. **37**, 793–809 (1992)
24.12   Y.Y. Yao, S.K.M. Wong, P. Lingras: A decision-theoretic rough set model. In: *Methodologies for Intelligent Systems*, Vol. 5, ed. by Z.W. Ras, M. Ze-

mankova, M.L. Emrich (North-Holland, New York 1990) pp. 17–24
24.13   J.D. Katzberg, W. Ziarko: Variable precision rough sets with asymmetric bounds. In: *Rough Sets, Fuzzy Sets and Knowledge Discovery*, ed. by W. Ziarko (Springer, Heidelberg 1994) pp. 167–177
24.14   W. Ziarko: Variable precision rough set model, J. Comput. Syst. Sci. **46**, 39–59 (1993)
24.15   D. Ślęzak, W. Ziarko: Bayesian rough set model, Proc. Found. Data Min. (FDM 2002) (2002) pp. 131–135
24.16   D. Ślęzak, W. Ziarko: Variable precision Bayesian rough set model, Rough Sets, Fuzzy Sets, Data Mining and Granular Comput. (RSFGrC 2013), Lect. Notes Comput. Sci. (Lect. Notes Artif. Intel.), Vol. 2639, ed. by G.Y. Wang, Q. Liu, Y.Y. Yao, A. Skowron (Springer, Heidelberg 2003) pp. 312–315
24.17   D. Ślęzak, W. Ziarko: The investigation of the Bayesian rough set model, Int. J. Approx. Reason. **40**, 81–91 (2005)
24.18   H.Y. Zhang, J. Zhou, D.Q. Miao, C. Gao: Bayesian rough set model: a further investigation, Int. J. Approx. Reason. **53**, 541–557 (2012)
24.19   S. Greco, B. Matarazzo, R. Słowiński: Rough membership and Bayesian confirmation measures for parameterized rough sets. In: *Rough Sets, Fuzzy Sets, Data Mining and Granular Computing*, Lecture Notes in Computer Science, Vol. 3641, ed. by D. Ślęzak, G.Y. Wang, M. Szczuka, I. Duntsch, Y.Y. Yao (Springer, Heidelberg 2005) pp. 314–324
24.20   S. Greco, B. Matarazzo, R. Słowiński: Parameterized rough set model using rough membership and Bayesian confirmation measures, Int. J. Approx. Reason. **49**, 285–300 (2008)
24.21   N. Azam, J.T. Yao: Analyzing uncertainties of probabilistic rough set regions with game-theoretic rough sets, Int. J. Approx. Reason. **55**, 142–155 (2014)

Part C | 24

24.22   J.P. Herbert, J.T. Yao: Game-theoretic rough sets, Fundam. Inf. **108**, 267–286 (2011)

24.23   S. Greco, B. Matarazzo, R. Słowiński, J. Stefanowski: Variable consistency model of dominance-based rough set approach. In: *Rough Sets and Current Trends in Computing*, Lecture Notes in Computer Science, Vol. 2005, ed. by W. Ziarko, Y.Y. Yao (Springer, Heidelberg 2001) pp. 170–181

24.24   J. Błaszczyński, S. Greco, R. Słowiński, M. Szelag: Monotonic variable consistency rough set approaches, Int. J. Approx. Reason. **50**, 979–999 (2009)

24.25   W. Kotłowski, K. Dembczyński, S. Greco, R. Słowiński: Stochastic dominance-based rough set model for ordinal classification, Inf. Sci. **178**, 4019–4037 (2008)

24.26   B. Zhou, Y.Y. Yao: Feature selection based on confirmation-theoretic rough sets. In: *Rough Sets and Current Trends in Computing*, Lecture Notes in Computer Science, Vol. 8536, ed. by C. Cornelis, M. Kryszkiewicz, D. Ślęzak, E.M. Ruiz, R. Bello, L. Shang (Springer, Heidelberg 2014) pp. 181–188

24.27   X.F. Deng, Y.Y. Yao: An information-theoretic interpretation of thresholds in probabilistic rough sets. In: *Rough Sets and Knowledge Technology*, Lecture Notes in Computer Science, Vol. 7414, ed. by T.R. Li, H.S. Nguyen, G.Y. Wang, J. Grzymala-Busse, R. Janicki (Springer, Heidelberg 2012) pp. 369–378

24.28   B. Zhou, Y.Y. Yao: Comparison of two models of probabilistic rough sets. In: *Rough Sets and Knowledge Technology*, Lecture Notes in Computer Science, Vol. 8171, ed. by P. Lingras, M. Wolski, C. Cornelis, S. Mitra, P. Wasilewski (Springer, Heidelberg 2013) pp. 121–132

24.29   J.W. Grzymala-Busse: Generalized parameterized approximations. In: *Rough Sets and Knowledge Technology*, Lecture Notes in Computer Science, Vol. 6954, ed. by J.T. Yao, S. Ramanna, G.Y. Wang, Z. Suraj (Springer, Heidelberg 2011) pp. 36–145

24.30   J.W. Grzymala-Busse: Generalized probabilistic approximations. In: *Transactions on Rough Sets*, Lecture Notes in Computer Science, Vol. 7736, ed. by J.F. Peters, A. Skowron, S. Ramanna, Z. Suraj, X. Wang (Springer, Heidelberg 2013) pp. 1–16

24.31   S. Greco, B. Matarazzo, R. Słowiński: Rough sets theory for multicriteria decision analysis, Eur. J. Oper. Res. **129**, 1–47 (2001)

24.32   S. Greco, B. Matarazzo, R. Słowiński: Decision rule approach. In: *Multiple Criteria Decision Analysis: State of the Art Surveys*, ed. by J.R. Figueira, S. Greco, M. Ehrgott (Springer, Berlin 2005) pp. 507–562

24.33   R. Słowiński, S. Greco, B. Matarazzo: Rough sets in decision making. In: *Encyclopedia of Complexity and Systems Science*, ed. by R.A. Meyers (Springer, New York 2009) pp. 7753–7786

24.34   R. Słowiński, S. Greco, B. Matarazzo: Rough set and rule-based multicriteria decision aiding, Pesqui. Oper. **32**, 213–269 (2012)

24.35   Y.Y. Yao: Relational interpretations of neighborhood operators and rough set approximation operators, Inf. Sci. **111**, 239–259 (1998)

24.36   Y.Y. Yao: Information granulation and rough set approximation, Int. J. Intell. Syst. **16**, 87–104 (2001)

24.37   Y.Y. Yao, Y.H. Chen: Subsystem based generalizations of rough set approximations. In: *Foundations of Intelligent Systems*, Lecture Notes in Computer Science, Vol. 3488, ed. by M.S. Hacid, N.V. Murray, Z.W. Raś, S. Tsumoto (Springer, Heidelberg 2005) pp. 210–218

24.38   Y.Y. Yao, X.F. Deng: Quantitative rough sets based on subsethood measures, Inf. Sci. **267**, 702–715 (2014)

24.39   H.X. Li, X.Z. Zhou, T.R. Li, G.Y. Wang, D.Q. Miao, Y.Y. Yao: *Decision-Theoretic Rough Set Theory and Recent Progress* (Science Press, Beijing 2011)

24.40   H. Yu, G.Z. Liu, Y.G. Wang: An automatic method to determine the number of clusters using decision-theoretic rough set, Int. J. Approx. Reason. **55**, 101–115 (2014)

24.41   F. Li, M. Ye, D.X. Chen: An extension to rough c-means clustering based on decision-theoretic rough sets model, Int. J. Approx. Reason. **55**, 116–129 (2014)

24.42   J. Li, T.P.X. Yang: An axiomatic characterization of probabilistic rough sets, Int. J. Approx. Reason. **55**, 130–141 (2014)

24.43   X.Y. Jia, Z.M. Tang, W.H. Liao, L. Shang: On an optimization representation of decision-theoretic rough set model, Int. J. Approx. Reason. **55**, 156–166 (2014)

24.44   F. Min, Q.H. Hu, W. Zhu: Feature selection with test cost constraint, Int. J. Approx. Reason. **55**, 167–179 (2014)

24.45   J.W. Grzymala-Busse, G.P. Clark, M. Kuehnhausen: Generalized probabilistic approximations of incomplete data, Int. J. Approx. Reason. **55**, 180–196 (2014)

24.46   D. Liu, T.R. Li, D.C. Liang: Incorporating logistic regression to decision-theoretic rough sets for classifications, Int. J. Approx. Reason. **55**, 197–210 (2014)

24.47   B. Zhou: Multi-class decision-theoretic rough sets, Int. J. Approx. Reason. **55**, 211–224 (2014)

24.48   H.Y. Qian, H. Zhang, L.Y. Sang, Y.J. Liang: Multigranulation decision-theoretic rough sets, Int. J. Approx. Reason. **55**, 225–237 (2014)

24.49   P. Lingras, M. Chen, Q.D. Miao: Qualitative and quantitative combinations of crisp and rough clustering schemes using dominance relations, Int. J. Approx. Reason. **55**, 238–258 (2014)

24.50   W.M. Shao, Y. Leung, Z.W. Wu: Rule acquisition and complexity reduction in formal decision contexts, Int. J. Approx. Reason. **55**, 259–274 (2014)

24.51   J.T. Yao, X.X. Li, G. Peters: Decision-theoretic rough sets and beyond, Int. J. Approx. Reason. **55**, 9–100 (2014)

24.52   X.Y. Zhang, D.Q. Miao: Two basic double-quantitative rough set models of precision and grade and their investigation using granular computing, Int. J. Approx. Reason. **54**, 1130–1148 (2013)

24.53 W. Ziarko: Probabilistic approach to rough sets, Int. J. Approx. Reason. **49**, 272–284 (2008)

24.54 B. Fitelson: Studies in Bayesian Confirmation Theory, Ph.D. Thesis (University of Wisconsin, Madison 2001)

24.55 R. Festa: Bayesian confirmation. In: *Experience, Reality, and Scientific Explanation*, ed. by M. Galavotti, A. Pagnini (Kluwer, Dordrecht 1999) pp. 55–87

24.56 S. Greco, Z. Pawlak, R. Słowiński: Can Bayesian confirmation measures be useful for rough set decision rules?, Eng. Appl. Artif. Intell. **17**, 345–361 (2004)

24.57 S. Greco, R. Słowiński, I. Szczęch: Properties of rule interestingness measures and alternative approaches to normalization of measures, Inf. Sci. **216**, 1–16 (2012)

24.58 Y.Y. Yao: Two semantic issues in a probabilistic rough set model, Fundam. Inf. **108**, 249–265 (2011)

24.59 Y.Y. Yao, B. Zhou: Naive Bayesian rough sets. In: *Rough Sets and Knowledge Technology*, Lecture Notes in Computer Science, Vol. 6401, ed. by J. Yu, S. Greco, P. Lingras, G.Y. Wang, A. Skowron (Springer, Heidelberg 2010) pp. 719–726

24.60 D.C. Liang, D. Liu, W. Pedrycz, P. Hu: Triangular fuzzy decision-theoretic rough sets, Int. J. Approx. Reason. **54**, 1087–1106 (2013)

24.61 H.X. Li, X.Z. Zhou: Risk decision making based on decision-theoretic rough set: a three-way view decision model, Int. J. Comput. Intell. Syst. **4**, 1–11 (2011)

24.62 D. Liu, T.R. Li, D. Ruan: Probabilistic model criteria with decision-theoretic rough sets, Inf. Sci. **181**, 3709–3722 (2011)

24.63 K. Dembczyński, S. Greco, W. Kotłowski, R. Słowiński: Statistical model for rough set approach to multicriteria classification. In: *Knowledge Discoveery in Databases*, Lecture Notes in Computer Science, Vol. 4702, ed. by J.N. Kok, J. Koronacki, R. de Lopez Mantaras, S. Matwin, D. Mladenic, A. Skowron (Springer, Heidelberg 2007) pp. 164–175

24.64 Y.Y. Yao: An outline of a theory of three-way decisions. In: *Rough Sets and Current Trends in Computing*, Lecture Notes in Computer Science,

Vol. 7413, ed. by J.T. Yao, Y. Yang, R. Słowiński, S. Greco, H.X. Li, S. Mitra, L. Polkowski (Springer, Heidelberg 2012) pp. 1–17

24.65 Y.Y. Yao: Three-way decision: an interpretation of rules in rough set theory. In: *Rough Sets and Knowledge Technology*, Lecture Notes in Computer Science, Vol. 5589, ed. by P. Wen, Y.F. Li, L. Polkowski, Y.Y. Yao, S. Tsumoto, G.Y. Wang (Springer, Heidelberg 2009) pp. 642–649

24.66 Y.Y. Yao: Three-way decisions with probabilistic rough sets, Inf. Sci. **180**, 341–353 (2010)

24.67 Y.Y. Yao: The superiority of three-way decisions in probabilistic rough set models, Inf. Sci. **181**, 1080–1096 (2011)

24.68 X.Y. Jia, L. Shang, X.Z. Zhou, J.Y. Liang, D.Q. Miao, G.Y. Wang, T.R. Li, Y.P. Zhang: *Theory of Three-Way Decisions and Application* (Nanjing Univ. Press, Nanjing 2012)

24.69 D. Liu, T.R. Li, D.Q. Miao, G.Y. Wang, J.Y. Liang: *Three-Way Decisions and Granular Computing* (Science Press, Beijing 2013)

24.70 J.R. Figueira, S. Greco, M. Ehrgott: *Multiple Criteria Decision Analysis: State of the Art Surveys* (Springer, Berlin 2005)

24.71 S. Greco, B. Matarazzo, R. Słowiński: A new rough set approach to evaluation of bankruptcy risk. In: *Rough Fuzzy and Fuzzy Rough Sets*, ed. by C. Zopounidis (Kluwer, Dordrecht 1998) pp. 121–136

24.72 S. Greco, B. Matarazzo, R. Słowiński: The use of rough sets and fuzzy sets in MCDM. In: *Multicriteria Decision Making*, Int. Ser. Opear. Res. Manage. Sci., Vol. 21, ed. by T. Gal, T. Stewart, T. Hanne (Kluwer, Dordrecht 1999) pp. 397–455

24.73 S. Greco, B. Matarazzo, R. Słowiński: Extension of the rough set approach to multicriteria decision support, INFOR **38**, 161–196 (2000)

24.74 S. Greco, B. Matarazzo, R. Słowiński: Rough sets methodology for sorting problems in presence of multiple attributes and criteria, Eur. J. Oper. Res. **138**, 247–259 (2002)

24.75 S. Greco, R. Słowiński, Y. Yao: Bayesian decision theory for dominance-based rough set approach. In: *Rough Sets and Knowledge Technology*, Lecture Notes in Computer Science, Vol. 4481, ed. by J.T. Yao, P. Lingras, W.Z. Wu, M. Szczuka, N. Cercone (Springer, Heidelberg 2007) pp. 134–141

25. Generalized Rough Sets

JingTao Yao, Davide Ciucci, Yan Zhang

This chapter reviews three formulations of rough set theory, i. e., element-based definition, granule-based definition, and subsystem-based definition. These formulations are adopted to generalize rough sets from three directions. The first direction is to use an arbitrary binary relation to generalize the equivalence relation in the element-based definition. The second is to use a covering to generalize the partition in the granule-based definition, and the third to use a subsystem to generalize the Boolean algebra in the subsystem-based definition. In addition, we provide some insights into the theoretical aspects of these generalizations, mainly with respect to relations with non-classical logic and topology theory.

In the *Pawlak* rough set model, the relationships of objects are defined by equivalence relations [25.1, 2]. In addition, we may obtain two other equivalent structures: the partition, induced by the equivalence relations, and an atomic Boolean algebra, formed by the equivalence classes as its set of atoms [25.2, 3]. In other words, we have three equivalent formulations of rough sets, namely, the equivalence relation-based formulation, the partition-based formulation, and the Boolean algebra-based formulation [25.4]. The approximation operators $\underline{apr}$ and $\overline{apr}$ are defined by an equivalence relation E, a partition U/E, and Boolean algebra $B(U/E)$, respectively [25.3, 5]. Although mathematically equivalent, these three formulations give different insights into the theory. More interestingly, when rough sets are generalized, the three formulations are no longer equivalent and thus

give new directions for the exploration of rough sets.

This chapter aims to explore these different generalizations. The discussion is organized in two parts. In the first part, we review and summarize relation-based, covering-based, and subsystem-based rough sets, based on several articles by *Yao* [25.4, 6, 7]. In the second part, we will give some insight into the theoretical aspects of these generalizations, mainly with respect to relations with nonclassical logic (modal and many-valued) and topology theory. It is to be noted that this second part partially overlaps with the first one, however, the scopes are different. Indeed, whereas the first part explains the models and their genesis, the second one is only devoted to some theoretical aspects. As such, the second part can be skipped by readers who are not so interested in fine details but may still have a clear view of the whole landscape of these kinds of generalized rough sets.

25.1 Definition and Approximations of the Models

In this section we discuss three equivalent formulations, namely, the equivalence relation-based formulation, the partition-based formulation, and the Boolean algebra-based formulation.

25.1.1 A Framework for Generalizing Rough Sets

For a systematic study on the generalization of Pawlak rough sets, *Yao* provided a framework to classifying commonly used definitions of rough set approximations into three types: the element-based definition, the granule-based definition, and the subsystem-based definition [25.5]. He argued that these types offer three directions for generalizing rough set models. We adapt this framework in the following discussion.

Suppose the universe U is a finite and nonempty set and let $E \subseteq U \times U$ be an equivalence relation on U. The equivalence class containing x is denoted as

$$[x]_E = \{y | y \in U, xEy\} .$$

The family of all equivalence classes is known as a quotient set denoted by

$$U/E = \{[x]_E | x \in U\} .$$

U/E defines a partition of U. A family of all definable sets form $B(U/E)$. A family of all definable sets can be obtained from U/E by adding the empty set $\emptyset$ and making itself closed under set union. A family of all definable sets is a subsystem of 2^U, that is, $B(U/E) \subseteq 2^U$ [25.1]. The standard rough set theory deals with the approximation of any subset of U in terms of definable subsets in $B(U/E)$. From different representations of an equivalence relation, three definitions of Pawlak rough set approximations can be obtained as follows:

- Element-based definitions [25.3, 5]

$$\underline{\mathrm{apr}}(A) = \{x | x \in U, [x]_E \subseteq A\}$$
$$= \{x | x \in U, \forall y \in U[xEy \Rightarrow y \in A]\} ,$$
$$\overline{\mathrm{apr}}(A) = \{x | x \in U, [x]_E \cap A \neq \emptyset\}$$
$$= \{x | x \in U, \exists y \in U[xEy \wedge y \in A]\} . \quad (25.1)$$

- Granule-based definitions [25.3, 5, 8]

$$\underline{\mathrm{apr}}(A) = \bigcup\{[x]_E | [x]_E \in U/E, [x]_E \subseteq A\}$$
$$= \bigcup\{X | X \in U/E, X \subseteq A\} ,$$
$$\overline{\mathrm{apr}}(A) = \bigcup\{[x]_E | [x]_E \in U/E, [x]_E \cap A \neq \emptyset\}$$
$$= \bigcup\{X | X \in U/E, X \cap A \neq \emptyset\} . \quad (25.2)$$

- Subsystem-based definition [25.3, 5, 8]

$$\underline{\mathrm{apr}}(A) = \bigcup\{X | X \in B(U/E), X \subseteq A\} ,$$
$$\overline{\mathrm{apr}}(A) = \bigcap\{X | X \in B(U/E), A \subseteq X\} . \quad (25.3)$$

The three equivalent definitions offer different interpretations of rough set approximations [25.5]. According to the element-based definition, an element x is in the lower approximation $\underline{\mathrm{apr}}(A)$ of a set A if all of its equivalent elements are in A; the element is in the upper approximation $\overline{\mathrm{apr}}(A)$ if at least one of its equivalent elements is in A [25.5]. According to the granule-based definition, $\underline{\mathrm{apr}}(A)$ is the union of equivalence classes that are subsets of A; $\overline{\mathrm{apr}}(A)$ is the union of equivalence classes that have a nonempty intersection with A [25.5]. According to the subsystem-based definition, $\underline{\mathrm{apr}}(A)$ is the largest definable set in the subsystem $B(U/E)$ that is contained in A; $\overline{\mathrm{apr}}(A)$ is the smallest definable set in the subsystem $B(U/E)$ that contains A [25.5].

Figure 25.1, adapted from *Yao* and *Yao* [25.4], shows three directions in generalized rough set models. In the Pawlak model, the definitions of approximation operators based on the equivalence relation, partition and Boolean algebra $B(U/E)$ are equivalent. The symbol $\Leftrightarrow$ is used to show a one-to-one two-way construction process. However, the generalized definitions of approximation operators using arbitrary binary relations, coverings, and subsystems are not equivalent.

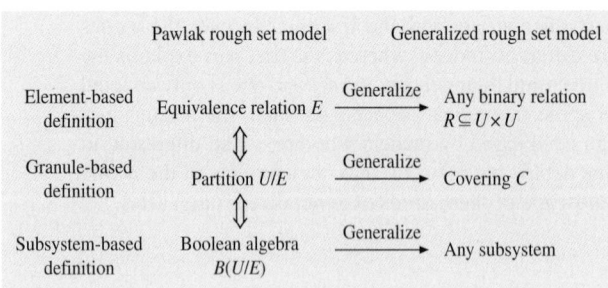

Fig. 25.1 Different formulations of approximation operators

In other words, the corresponding subsystem or covering may not be found based on an arbitrary binary relation. With each formulation, various definitions of approximation operators can be examined. One may consider an arbitrary binary relation in generalizing the equivalence relation in the element-based definition, a covering in generalizing the partition in the granule-based definition, and other subsystems in generalizing the Boolean algebra in the subsystem-based definition [25.4, 5].

25.1.2 Binary Relation-Based Rough Sets

In the development of the theory of rough sets, approximation operators are typically defined by using equivalence relations which are reflexive, symmetric, and transitive [25.2]. The Pawlak rough set model can be extended by using any arbitrary binary relation to replace the equivalence relation. *Wybraniec-Skardowska* introduced different rough set models based on various types of binary relations [25.9]. *Pawlak* pointed out that any type of relations may be assumed on the universe for the development of a rough set theory [25.10]. *Yao* et al. extended conventional rough set models by considering various types of relations by drawing results from modal logics [25.11]. Similarly to defining different types of modal logic systems, different rough set models were defined by using classes of binary relations satisfying various sets of properties, formed by serial, reflective, symmetric, transitive, and Euclidean relations, and their combinations. *Slowinski* and *Vanderpooten* considered a special case in which a reflexive (not necessarily symmetric and transitive) similarity relation was used [25.12]. *Greco* et al. examined a fuzzy rough approximation based on fuzzy similarity relations [25.13]. *Guan* and *Wang* investigated the relationships among 12 different basic definitions of approximations and suggested the suitable generalized definitions of approximations for each class of generalized indiscernibility relations [25.14].

A binary relation R may be conveniently represented by a mapping $n\colon U \to 2^U$, i.e., n is a neighborhood operator and $n(x)$ consists of all R-related elements of x. In the element-based definition, the equivalence class $[x]_E$ can be viewed as a neighborhood of x consisting of objects equivalent to x. In general, one may consider any type neighborhood of x, consisting of objects related to x, to form more general approximation operators. By extending (25.1), we can define lower and upper approximation operators as follows [25.15]

$$\underline{\mathrm{apr}}_n(A) = \{x \mid x \in U, n(x) \subseteq A\}$$
$$= \{x \mid x \in U, \forall y \in U(y \in n(x) \Rightarrow y \in A)\},$$
$$\overline{\mathrm{apr}}_n(A) = \{x \mid x \in U, n(x) \cap A \neq \emptyset\}$$
$$= \{x \mid x \in U, \exists y(y \in n(x) \wedge y \in A)\}.$$

$$(25.4)$$

The set $\underline{\mathrm{apr}}_n(A)$ consists of elements whose R-related elements are all in A, and $\overline{\mathrm{apr}}_n(A)$ consists of elements such that at least one of whose R-related elements is in A. The lower and upper approximation operators $\underline{\mathrm{apr}}_n$ and $\overline{\mathrm{apr}}_n$ pair are a generalized rough set of A induced by the binary relation R.

A neighborhood operator can be defined by using a binary relation [25.6, 12, 16]. Suppose $R \subseteq U \times U$ is a binary relation on the universe U. A successor neighborhood operator $R\cdot\colon U \to 2^U$ can be defined as

$$xR\cdot = \{y \mid y \in U, xRy\}.$$

Conversely, a binary relation can be constructed from its successor neighborhood as

$$xRy \Leftrightarrow y \in xR\cdot.$$

Generalized approximations by a neighborhood operator can be equivalently formulated by using a binary relation [25.4]. This formulation connects generalized approximation operators with the necessity and possibility operators in modal logic [25.6]. There are many types of generalized approximation operators defined by neighborhood operators that are induced by a binary relation or a family of binary relations [25.3, 15–19].

For an arbitrary relation, generalized rough set operators do not necessarily satisfy all the properties in the Pawlak rough set model. Nevertheless, the following properties hold in rough set models induced by any binary relation [25.3, 6, 20]

(L1) $\underline{\mathrm{apr}}(A) = (\overline{\mathrm{apr}}(A^c))$,

(L2) $\underline{\mathrm{apr}}(U) = U$,

(L3) $\underline{\mathrm{apr}}(A \cap B) = \underline{\mathrm{apr}}(A) \cap \underline{\mathrm{apr}}(B)$,

(L4) $\underline{\mathrm{apr}}(A \cup B) \supseteq \underline{\mathrm{apr}}(A) \cup \underline{\mathrm{apr}}(B)$,

(L5) $A \subseteq B \Rightarrow \underline{\mathrm{apr}}(A) \subseteq \underline{\mathrm{apr}}(B)$,

(K) $\underline{\mathrm{apr}}(A^c \cup B) \subseteq (\underline{\mathrm{apr}}(A))^c \cup \underline{\mathrm{apr}}(B)$,

(U1) $\overline{\mathrm{apr}}(A) = (\underline{\mathrm{apr}}(A^c))$,

$(U2)$ $\overline{\mathrm{apr}}(\emptyset) = \emptyset$,

$(U3)$ $\overline{\mathrm{apr}}(A \cup B) = \overline{\mathrm{apr}}(A) \cup \overline{\mathrm{apr}}(B)$,

$(U4)$ $\overline{\mathrm{apr}}(A \cap B) \subseteq \overline{\mathrm{apr}}(A) \cap \overline{\mathrm{apr}}(B)$,

$(U5)$ $A \subseteq B \Rightarrow \overline{\mathrm{apr}}(A) \subseteq \overline{\mathrm{apr}}(B)$.

A relation R is a serial relation if for all $x \in U$ there exists a $y \in U$ such that xRy; a relation is a reflexive relation if for all $x \in U$ the relationship xRx holds; a relation is symmetric relation if for all $x, y \in U$, xRy implies yRx holds; a relation is transitive relation if for three elements $x, y, z \in U$, xRy and yRZ imply xRz; a relation is Euclidean when for all $x, y, z \in U$, xRy and xRz imply yRz [25.6, 15]. By using mapping n, we can express equivalently the conditions on a binary relation as follows [25.6, 15, 21]:

Serial	$x \in U, n(x) \neq \emptyset$
Reflexive	$x \in U, x \in n(x)$
Symmetric	$x, y \in U, x \in n(y) \Rightarrow y \in n(x)$
Transitive	$x, y \in U, y \in n(x) \Rightarrow n(y) \subseteq n(x)$
Euclidean	$x, y \in U, y \in n(x) \Rightarrow n(x) \subseteq n(y)$.

Different binary relations have different properties. The five properties of a binary relation, namely, the serial, reflexive, symmetric, transitive, and Euclidean properties, induce five properties for the approximation operators [25.6, 20, 21]. We use the same labeling system as in modal logic to label these properties [25.6]:

Serial	Property (D)
	$\mathrm{apr}(A) \subseteq \overline{\mathrm{apr}}(A)$ holds
Reflexive	Property (T)
	$\mathrm{apr}(A) \subseteq A$ holds
Symmetric	Property (B)
	$A \subseteq \mathrm{apr}(\overline{\mathrm{apr}}(A))$ holds
Transitive	Property (4)
	$\mathrm{apr}(A) \subseteq \mathrm{apr}(\mathrm{apr}(A))$ holds
Euclidean:	Property (5)
	$\overline{\mathrm{apr}}(A) \subseteq \mathrm{apr}(\overline{\mathrm{apr}}(A))$ holds.

By combining these properties, one can construct more rough set models [25.6, 20]. Other than the above mentioned properties, (K) denotes the property that any binary relation holds, i.e., no special property is required. We use a series of property labels, i.e., (K), (D), (T), (B), (4), (5), to represent the rough set models built on relations with these properties. For example, the KTB rough set model is built on a compatibility relation R, i.e., with reflexive and symmetric properties. In such a model, properties (K), (D), (T) and (B) hold, however, properties (4) and (5) do not hold. Property (D) does not explicitly appear in this label because (D) can

be obtained from (T). If R is reflexive, symmetric, and transitive, i.e., R is an equivalence relation, we obtain the Pawlak rough set model [25.6, 20]. The approximation operators satisfies all properties (D), (T), (B), (4), and (5).

Figure 25.2 summarizes the relationships between these models [25.6, 20, 21]. The label of the model indicates the characterization properties of that model. A line connecting two models indicates that the model on the upper level is also a model on the lower level. For example, a KT5 model is a KT4 model, as KT5 connects down to KT4. It should be noted that the lines that can be derived by transitivity are not explicitly shown. The model K may be considered as the basic model because it does not require any special property on the binary relation. All other models are built on top of the model K and it can be regarded as the weakest model. The model KT5, i.e., the Pawlak rough set model, is the strongest model.

With the element-based definition, we can obtain binary relation-based rough set models by generalizing the equivalence relation to binary relations. Different binary relations can induce different rough set models with different properties, as was discussed above. This generalization not only deepens our understanding of rough sets, but also enriches the rough set theory.

25.1.3 Covering–Based Rough Sets

A covering of a universe is a family of subsets of the universe such that their union is the universe. By allowing nonempty overlap of two subsets, a covering is a generalized mathematical structure of a partition [25.22]. These subsets in a covering or a partition can be considered as granules based on the concepts

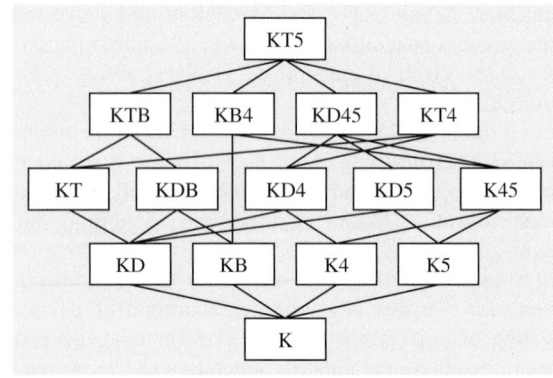

Fig. 25.2 Rough set models (after [25.6])

in granular computing [25.23]. By generalizing the partition to covering in granule-based approximation definitions, we form a more general definition and we call this approach a granule-based definition. In this section, we mainly investigate covering-based rough sets.

Zakowski proposed the notion of covering-based rough set approximations in 1983 [25.24]. Lower and upper approximation operators are defined by a straightforward generalization of the rough set definition proposed by Pawlak. However, the generalized approximation operators are not dual to each other with respect to set complements [25.15, 25]. *Pomykala* studied two pairs of dual approximation operators [25.25]. The lower approximation operator in one pair is the same as the *Zakowski* lower approximation operator, and the upper approximation operator in the other pair is same as the *Zakowski* upper approximation operator [25.25]. *Pomykala* also suggested and examined additional pairs of dual approximation operators that are induced by a covering. Furthermore, he considered coverings produced by tolerance relations in an incomplete information table [25.26].

Instead of using duality, *Wybraniec-Skardowska* studied pairs of approximation operators linked together by a different type of relations [25.9]. Given an upper approximation operator, the corresponding lower approximation operator is defined based on the upper approximations of singleton subsets. Several such pairs of approximation operators were studied based on a covering and a tolerance relation defined by a covering, including some of those used by *Zakowski* [25.24] and *Pomykala* [25.25]. *Yao* investigated dual approximation operators by using coverings induced by the predecessor and/or successor neighborhoods of serial or inverse serial binary relations. The two pairs of dual approximation operators introduced by *Pomykala* were examined and the conditions for their equivalence to those obtained from a binary relation were given [25.5, 15]. *Couso* and *Dubois* proposed a loose pair and a tight pair [25.27]. They presented an interesting investigation of the two pairs of approximation operators within the context of incomplete information. The two pairs of operators were shown to be related to the family of approximation operators produced by all partitions consistent with a covering induced by an ill-known attribute function in an incomplete information table. *Restrepo* et al. investigated different relationships between commonly used operators using concepts of duality and other properties [25.28]. They also showed that a pair of lower operators and an upper approxi-

mation operator can be dual and adjoint at the same time.

By using the minimum neighborhood of an object (i. e., the intersection of subsets in the minimal description of the object), *Wang* et al. introduced a pair of dual approximation operators [25.29]. The same pair of approximation operators was also used and examined by *Xu* and *Wang* [25.30] and *Xu* and *Zhang* [25.31]. *Zhu*'s team systematically studied five types of approximation operators [25.19, 32–36]. The lower approximation operator is the Zakowski lower approximation operator, and the upper approximation operators are different. They investigated properties of these operators and their relationships and provided set of axioms for characterizing these operators. *Liu* examined covering-based rough sets from constructive and axiomatic approaches [25.37]. The relationships among four types of covering-based rough sets and the topologies induced by different covering approximations were discussed. *Zhang* and *Luo* investigated relationships between relation-based rough sets and covering-based rough sets [25.38]. They also presented some sufficient and necessary conditions for different types of covering-based rough sets to be equal.

We will elaborate on how to obtain rough sets by generalizing a partition to a covering, as well as duality, loose pairs, and tight pairs of approximation operators. Let $\mathbb{C}$ be a covering of the universe U. By replacing a partition U/E with a covering $\mathbb{C}$ and equivalence classes with subsets in $\mathbb{C}$ in the granule-based definition, a pair of approximation operators can be obtained [25.24]. However, they are not a pair of dual operators [25.25]. To overcome this problem, *Yao* suggested that one can generalize one of them and define the other by duality [25.7, 15]. The granule-based definition can be generalized in two ways, i. e., (1) the lower approximation operator is extended from partition to covering and the upper approximation operator is redefined by duality, (2) the upper approximation operator is extended from partition to covering and the lower approximation operator is redefined by duality [25.4, 7]. The results are two pairs of dual approximation operators [25.4]

$$\underline{\mathrm{apr}}'(A) = \bigcup \{X | X \in \mathbb{C}, X \subseteq A\}$$
$$= \{x | x \in U, \exists X \in \mathbb{C}[x \in X, X \subseteq A]\},$$
$$\overline{\mathrm{apr}}'(A) = (\underline{\mathrm{apr}}'(A^c))^c$$
$$= \{x | x \in U, \forall X \in \mathbb{C}[x \in X \Rightarrow X \cap A \neq \emptyset]\}.$$
$$(25.5)$$

Part C | 25.1

and

$$\underline{\text{apr}}''(A) = (\overline{\text{apr}}''(A^c))^c$$
$$= \{x | x \in U, \forall X \in \mathbb{C}[x \in X \Rightarrow X \subseteq A]\},$$
$$\overline{\text{apr}}''(A) = \bigcup\{X | X \in \mathbb{C}, X \cap A \neq \emptyset\}$$
$$= \{x | x \in U, \exists X \in \mathbb{C}[x \in X, X \cap A \neq \emptyset]\}. \tag{25.6}$$

We may define two pairs of dual approximation operators for each covering. Both of pairs are consistent with the Pawlak definition. The following relationships hold for the above approximation operators [25.4, 25]

$$\underline{\text{apr}}''(A) \subseteq \underline{\text{apr}}'(A) \subseteq A \subseteq \overline{\text{apr}}'(A) \subseteq \overline{\text{apr}}''(A). \tag{25.7}$$

Therefore, the pair $(\underline{\text{apr}}'(A), \overline{\text{apr}}'(A))$ is called a pair of tighter approximation and the pair $(\underline{\text{apr}}''(A), \overline{\text{apr}}''(A))$ is called a looser approximation [25.25]. Furthermore, any approximation produced by other authors are bounded by

$$(\underline{\text{apr}}'(A), \overline{\text{apr}}'(A))$$

and

$$(\underline{\text{apr}}''(A), \overline{\text{apr}}''(A))$$

if

$$\underline{\text{apr}}(A) \subseteq A \subseteq \overline{\text{apr}}(A).$$

In addition to this fundamental generalization of a rough set to a covering, more than 20 different approximation pairs have been defined [25.4, 39]. Their properties were studied in [25.39], where the inclusion relationship occurring among two sets and their approximations was considered. All these approaches were categorized in a recent study [25.4].

We recall some notions that will be useful in Sect. 25.2.2 when dealing with the topological characterization of approximations.

Definition 25.1 [25.39]

Let $\mathbb{C}$ be a covering on a universe U and $x \in U$. We define:

- The neighborhood of x: $\gamma(x) = \cap\{C \in \mathbb{C} : x \in C\}$.
- The friends of x: $\delta(x) = \cup\{C \in \mathbb{C} : x \in C\}$.
- The partition generated by a covering: $\Pi_{\mathbb{C}}(x) = \{y \in U : \forall C \in \mathbb{C}, (x \in C \leftrightarrow y \in C)\}$.

Using the above operators, the following approximation pairs are introduced

$$L_1(A) = \cup\{\delta(x) : \delta(x) \subseteq A\},$$
$$U_1(A) = L_1(A^c)^c, \tag{25.8}$$
$$L_2(A) = \cup\{C \subseteq A\},$$
$$U_2(A) = L_2(A^c)^c = \cap\{C^c : C \in \mathbb{C}, C \cap A = \emptyset\}, \tag{25.9}$$
$$L_3(A) = L_2(A),$$
$$U_3(A) = \cup\{C : C \cap A \neq \emptyset\} \setminus L_3(A^c), \tag{25.10}$$
$$L_4(A) = \cup\{\Pi_{\mathbb{C}}(x) \subseteq A\},$$
$$U_4(A) = \cup\{\Pi_{\mathbb{C}}(x) \cap A \neq \emptyset\}, \tag{25.11}$$
$$L_5(A) = \{x \in U : \gamma(x) \subseteq A\},$$
$$U_5(A) = \{x \in U : \gamma(x) \cap A \neq \emptyset\}, \tag{25.12}$$
$$L_6(A) = U_6(A^c)^c,$$
$$U_6(A) = \cup\{\gamma(x) : x \in A\}. \tag{25.13}$$

These approximation pairs have been introduced and studied in several papers: approximation pair 1 (25.8) can be found in [25.40], approximation pair 2 (25.9) in [25.5, 40–43], approximation pair 3 (25.10) in [25.42], approximation pair 4 (25.11) in many papers starting from [25.44], approximation pair 5 (25.12) in [25.41, 45], and approximation pair 6 (25.13) in [25.45, 46]. As we will discuss, they all show nice topological properties.

By simply replacing a partition with a covering, a generalized mathematical structure of a partition, in the granule-based definition, we form new rough sets. The lower and upper approximation operators are not necessary dual. We may redefine one of them to obtain the dual approximation operators. There are two types of approximation, a tight pair and a loose pair. The two pairs provide the boundary when new approximation operators are introduced.

25.1.4 Subsystem–Based Rough Sets

In the Pawlak rough set model, the same subsystem is used to define lower and upper approximation operators. When generalizing the subsystem-based definition, two subsystems may be used, one for the lower approximation, which is closed under union, and the other for the upper approximation, which is closed under intersection [25.4, 7, 47]. To ensure duality of approximation operators, the two subsystems should be dual systems with respect to set complement [25.4]. Given a closure system $\overline{S}$, its dual system $\underline{S}$ can be constructed as $\underline{S} = \{\sim X | X \in \overline{S}\}$. The system $\overline{S}$ contains the universe U and is closed under set intersection. The system $\underline{S}$ contains the empty set $\emptyset$ and is closed under set union [25.4]. A pair of lower and upper approximation operators with respect to S is defined as [25.4]

$$\underline{apr}(A) = \bigcup \{X | X \in \underline{S}, X \subseteq A\},$$
$$\overline{apr}(A) = \bigcap \{X | X \in \overline{S}, A \subseteq X\}. \tag{25.14}$$

In the Pawlak rough set model, the two systems $\underline{S}$ and $\overline{S}$ are the same, namely, $\underline{S} = \overline{S}$, which is closed under set complement, union, and intersection. That is, it is a Boolean algebra. The subsystem-based definition provides a way to approximate any set in 2^U by a pair of sets in $\underline{S}$ and $\overline{S}$, respectively [25.4]. The subsystem-based definition can be generalized by using different mathematical structures, such as topological spaces [25.7, 47, 48], closure systems [25.7, 47], lattices [25.7, 49], and posets [25.7, 50].

For an arbitrary topological space, the family of open sets is different from the family of closed sets. Let $(U, O(U))$ be a topological space, where $O(U) \subseteq 2^U$ is a family of subsets of U called open sets. The family of (topological) open sets contains $\emptyset$ and U. The family of open sets is closed under union and finite intersection. The family of all (topological) closed sets $C(U) = \{\neg X | X \in O(U)\}$ contains $\emptyset$ and U, and is closed under intersection and finite union. A pair of generalized approximation operators can be defined by replacing $B(U/E)$ with $O(U)$ for the lower approximation operator, and $B(U/E)$ with $C(U)$ for the upper approximation operator [25.5]. The definitions of approximation operators are [25.5, 7, 50]

$$\underline{apr}(A) = \bigcup \{X | X \in O(U), X \subseteq A\},$$
$$\overline{apr}(A) = \bigcap \{X | X \in C(U), A \subseteq X\}. \tag{25.15}$$

The rough set model can be generalized by using closure systems. A family of subsets of U, $C(U)$, is called a closure system if it contains U and is closed under intersection. By collecting the complements of members of $C(U)$, we can obtain another system $O(U) = \{\neg X | X \in C(U)\}$, which contains the empty set $\emptyset$ and is closed under union. In this case, a pair of approximation operators in a closure system can be defined by replacing $B(U/E)$ with $O(U)$ for the lower approximation operator, and $B(U/E)$ with $C(U)$ for the upper approximation operator. The definitions of

approximation operators are [25.5, 7]

$$\underline{apr}(A) = \bigcup \{X | X \in O(U), X \subseteq A\},$$
$$\overline{apr}(A) = \bigcap \{X | X \in C(U), A \subseteq X\}. \tag{25.16}$$

The power set of the universe is a special lattice. Suppose $(\mathcal{B}, \neg, \wedge, \vee, 0, 1)$ is a finite Boolean algebra and $(\mathcal{B}_0, \neg, \wedge, \vee, 0, 1)$ is a sub-Boolean algebra. One may approximate an element of $\mathcal{B}$ by using elements of $\mathcal{B}_0$ [25.7]

$$\underline{apr}(x) = \bigvee \{y | y \in \mathcal{B}_0, y \leq x\},$$
$$\overline{apr}(x) = \bigwedge \{y | y \in \mathcal{B}_0, x \leq y\}. \tag{25.17}$$

We consider a more generalized definition in which the Boolean algebra $\mathcal{B}$ is replaced by a completely distributive lattice [25.51], and one subsystem is used. A subsystem $O(\mathcal{B})$ of $\mathcal{B}$ satisfies the following axioms [25.7]:

(O1) $0 \in O(\mathcal{B}), 1 \in O(\mathcal{B})$;
(O2) for any subsystem $\mathcal{D} \subseteq O(\mathcal{B})$, if there exists a least upper bound $LUB(\mathcal{D}) = \bigvee \mathcal{D}$, it belongs to $O(\mathcal{B})$;
(O3) $O(\mathcal{B})$ is closed under finite meets.

Elements of $O(\mathcal{B})$ are referred to as inner definable elements. The complement of an inner definable element is called an outer definable element. The set of outer definable elements $C(\mathcal{B}) = \{\neg x | x \in O(\mathcal{B})\}$ is characterized by the following axioms [25.7]:

(C1) $0 \in C(\mathcal{B}), 1 \in C(\mathcal{B})$;
(C2) for any subsystem $\mathcal{D} \subseteq C(\mathcal{B})$, if there exists a greatest lower bound $GLB(\mathcal{D}) = \bigwedge \mathcal{D}$, it belongs to $C(\mathcal{B})$;
(C3) $C(\mathcal{B})$ is closed under finite joins.

From the sets of inner and outer definable elements, we define the following approximation operators [25.7]

$$\underline{apr}(x) = \bigvee \{y | y \in O(\mathcal{B}), y \leq x\},$$
$$\overline{apr}(x) = \bigwedge \{y | y \in C(\mathcal{B}), x \leq y\}. \tag{25.18}$$

Let $(L, \leq, 0, 1)$ be a bounded lattice. Suppose $O(L)$ is a subset of L such that it contains 0 and is closed under join, and $C(L)$ a subset of L such that it contains 1 and is closed under meets. They are complete lattices, although the meet of $O(L)$ and the join of $C(L)$ may be different from those of L. Based on these two systems, we can define two other approximation operators as follows [25.7]

$$\underline{\text{apr}}(x) = \bigvee\{y|y \in O(L), y \leq x\},$$
$$\overline{\text{apr}}(x) = \bigwedge\{y|y \in C(L), x \leq y\}. \quad (25.19)$$

The operator $\overline{\text{apr}}$ is a closure operator [25.7]. $C(L)$ corresponds to the closure system in the set-theoretic framework. However, since a lattice may not be complemented, we must explicitly consider both $O(L)$ and $C(L)$. That is, the system $(L, O(L), C(L))$, or equivalently the system $(L, \text{apr}, \overline{\text{apr}})$, is used for the generalization of Pawlak approximation operators.

The subsystem-based formulation provides an important interpretation of rough set theory. It allows us to study rough set theory in the contexts of many algebraic systems [25.47]. This naturally leads to the generalization of rough set approximations. With the subsystem-based definition, we examine the generalized approximation operators by using topological space, closure systems, lattices, and posets in this subsection.

25.2 Theoretical Approaches

In this section, we further develop the previously outlined links with modal logics and topology.

25.2.1 Logical Setting

The *minimal modal system K* [25.52] is at the basis of any modal logic. Its language $\mathcal{L}$ is the usual one of propositional logic plus necessity $\square$ and possibility $\Diamond$. That is, $\mathcal{L} = a \in \mathcal{V}|\neg\alpha|\alpha \wedge \beta|\square(\alpha)$, where $\mathcal{V} = \{a, b, c, \ldots\}$ is the set of propositional variables and $\neg, \wedge, \square$ are the negation, conjunction, and necessity connectives. As usual, other connectives can be derived: disjunction $\alpha \vee \beta$ stands for $(\alpha' \wedge \beta')'$, implication $\alpha \rightarrow \beta$ stands for $\alpha' \vee \beta$, and possibility is $\Diamond\alpha = \neg\square(\neg\alpha)$.

The axioms are those of Boolean logic plus the axioms to characterize the modal connectives:

(B1) $\alpha \rightarrow (\beta \rightarrow \alpha)$
(B2) $(\beta \rightarrow (\alpha \rightarrow \mu)) \rightarrow ((\beta \rightarrow \alpha) \rightarrow (\beta \rightarrow \gamma))$
(B3) $(\alpha' \rightarrow \beta') \rightarrow (\beta \rightarrow \alpha)$
(K) $\square(\alpha \rightarrow \beta) \rightarrow (\square\alpha \rightarrow \square\beta)$.

The rules are modus ponens: *If $\vdash \alpha$ and $\vdash \alpha \rightarrow \beta$ then $\vdash \beta$* and necessitation: *If $\vdash \alpha$ then $\vdash \square\alpha$.*

In our context, the semantics is given through a model $M = (X, \mathcal{R}, v)$, where (X, R) is an approximation space (that is, a universe with a binary relation) and v is the interpretation that given a variable returns a subset of elements of the universe: $v(a) \subseteq X$. Using the standard modal logic terminology, X is the set of possible worlds, R the accessibility relation, and $v(a)$ represents the set of possible worlds where a holds. The interpretation v can recursively be extended to any formula α as

$$v(\neg\alpha) = v(\alpha)^c,$$
$$v(\alpha_1 \wedge \alpha_2) = v(\alpha_1) \cap v(\alpha_2),$$
$$v(\alpha_1 \vee \alpha_2) = v(\alpha_1) \cup v(\alpha_2),$$

and modal operators are mapped to lower and upper approximations according to Definition 25.4

$$v(\square\alpha) = \underline{\text{apr}}_n(v(\alpha)),$$
$$v(\Diamond\alpha) = \overline{\text{apr}}_n(v(\alpha)).$$

It is well known from modal logic [25.52] that, once the basic axioms (B1)–(B3) and (K) are fixed, then a different modal axiom according to Table 25.1 corresponds to any relation property.

Clearly, these axioms reflect the properties on rough approximations given in Sect. 25.1.2 and can be used to generate all the logics given in Fig. 25.2.

Other kinds of generalized rough sets models have been studied in the literature under the framework of modal logic. In particular, nondeterministic information

Table 25.1 Correspondence between modal axioms and relation properties

Name	Axiom	Property
T	$\square\alpha \rightarrow \alpha$	Reflexive
4	$\square\alpha \rightarrow \square\square(\alpha)$	Transitive
5	$\Diamond\alpha \rightarrow \square(\Diamond(\alpha))$	Euclidean
D	$\square\alpha \rightarrow \Diamond\alpha$	Serial
B	$\alpha \rightarrow \square\Diamond\alpha$	Symmetric

logic (NIL) [25.53] is defined to capture those information tables in which more than one value can correspond to each pair (object, attribute). For instance, if we have a feature *color* then it is allowed that to each object can be assigned more than one color. Given this extended definition, several new relations can be introduced; these where studied in the *Orlowska–Pawlak* seminal paper [25.53] and some subsequent studies [25.54–56]. Some of these relations are:

- *Similarity (connection)*
  xSy iff $f(x,a) \cap f(y,a) \neq \emptyset$ for all $a \in A$.
- *Inclusion*
  $x\mathit{I}y$ iff $f(x,a) \subseteq f(y,a)$ for all $a \in A$.
- *Indiscernibility*
  $x \, Ind \, y$ iff $f(x,a) = f(y,a)$ for all $a \in A$.
- *Weak indiscernibility*
  $x \, \mathcal{W}_i$ iff $f(x,a) = f(y,a)$ for some $a \in A$.
- *Weak similarity*
  $x \, \mathcal{W}_s$ iff $f(x,a) \cap f(y,a) \neq \emptyset$ for some $a \in A$.
- *Complementarity*
  $x \, Com \, y$ iff $f(x,a) = VAL_a \setminus f(y,a)$.

We also mention the logic for data analysis (DAL) [25.57], which is meant to deal with approximation spaces with more than one equivalence relation (X, R_i).

Besides modal logic, in standard rough set theory based on one equivalence relation, several authors have dealt with a many-valued logic approach [25.58, 59], also with some criticism from the point of view of the interpretation of results [25.60]. On the other hand, there have been only a few attempts to link generalized rough sets and many-valued logic. One of the reasons is the intrinsic difficulty that arises when trying to define intersection and union of rough sets (in an algebraic context defining a lattice and not only a poset) without imposing some restrictions (see, for example, [25.61, 62]).

A recent work [25.63] deals with many-valued logic in coverings and in particular the apr'', $\overline{apr}''$ approximations defined in (25.6). The novelty of the approach in the introduction of a *subordination relation* among objects

$$x \preceq y \quad \text{iff} \quad \forall C \in C(y \in C \implies x \in C),$$

which is strictly linked to the notion of neighborhood. Indeed,

$$x \preceq y \quad \text{iff} \quad x \in \gamma(y).$$

We also remark that a similar preorder relation defined by a topology is used in the bitopological approach to dominance-based rough sets [25.64]. This link could bring new insight into the many-valued approach to covering-based rough sets.

The syntax of the logic in [25.63] consists of two types of variables: *object variables* $x, y, \ldots$ and *set variables* $A, B, \ldots$ Atomic formulae are $x \preceq y$ and $x \in A$ (where A can be a set variable or a composition of set variables) and compound formulae are obtained with the usual logical connectives $\neg, \wedge, \vee$. The axioms are given in the form of sequent calculus, and the interpretation mapping I is given with respect to a covering C in atomic formulae as

$$I(x \preceq y) = \begin{cases} t & \text{if } v(x) \preceq v(y) \\ f & \text{otherwise} \end{cases},$$

where v maps each object variable to an object in the universe U and

$$I(x \in A) = \begin{cases} t & \text{if } v(x) \in \underline{apr}''(w(A)) \\ f & \text{if } v(x) \in \underline{apr}''(w(A)^c) \\ u & \text{otherwise} \end{cases},$$

where w maps each set variable and set formula to a subset of objects and u is a third truth value representing the unknown. The interpretation extends to compound formulae by truth functional application of Kleene three-valued logic. The logic is proven to be sound but complete only with respect to the sublanguage of atomic formulae.

We remark that this logic suffers from the problems of using three-valued logic to capture an epistemic notion such as is the case of Kleene-valued logics with respect to uncertainty. For instance, even if we are not sure if an element $x \in A$, we can undoubtedly say that $(x \in A)$ or $\neg(x \in A)$ (*tertium non datur*). On the contrary, with the above interpretation we can obtain that $I((x \in A) \wedge \neg(x \in A)) = u$, whenever x is in the boundary of A, that is $I(x \in A) = u$.

25.2.2 Topology

We saw in Sect. 25.1.4 that the subsystem approach can be generalized by the help of topological notions. Here, we further develop this topic and show which covering-based approximations have a topological behavior.

Let us consider a lattice structure and define on it a notion of closure [25.65, 66].

Definition 25.2
Given a lattice $\mathcal{L}$, a map $c : \mathcal{L} \mapsto \mathcal{L}$ is a *closure operator* if for all $x, y \in \mathcal{L}$:

(C1op) $x \leq c(x)$
(C2op) If $x \leq y$ then $c(x) \leq c(y)$
(C3op) $c(c(x)) = c(x)$

The map c is a *topological closure* if in addition (C1op)–(C3op) satisfies

(C4op) $c(a) \vee c(b) = c(a \vee b)$.

The map c is an *Alexandroff closure* if in addition (C1op)–(C3op) satisfies

(C5op) $\vee_j c(a_j) = c(\vee_j a_j)$.

Of course, any Alexandroff closure is a topological one and on a finite universe, the two notions coincide.

On a complemented lattice, an interior operator is defined by duality as $i(x) = c(x')'$ and properties dual to (C1op)–(C5op) hold. On the other hand, if the lattice is not complemented, an interior operator must be explicitly defined, as discussed in Sect. 25.1.4.

To the above algebraic definition of a closure operator there corresponds an equivalent one based on closed sets as we saw in Sect. 25.1.4. More precisely:

Definition 25.3
Let $\mathcal{L}$ be a lattice and $C \subseteq \mathcal{L}$ a subset of elements which is closed under arbitrary intersections, that is, axioms (C1)–(C3) are satisfied. Then, a *closure oper-*

ator satisfying properties (C1op–C3op) is defined as $c(a) = \wedge \{u \in C : a \leq u\}$.

A *topological closure* is such that the union of a finite family of closed elements is closed, i. e., $(\vee_{i \in I} c_i) \in C$ with I a finite set of indexes and an *Alexandroff topology* if closed under arbitrary union.

Now, if the subsystem rough sets are naturally based on a topological ground, also covering rough sets can be classified with respect to topological properties. First of all, let us consider the approximations $\text{apr}'(A), \overline{\text{apr}}'(A)$ defined in (25.5). They are an interior and a closure operator, respectively. On the other hand, approximation $\overline{\text{apr}}''(A)$ in (25.6) is not a closure, since in general, it does not satisfy condition (C3).

Moreover, let us consider a covering $\mathbb{C}(X)$ of a universe and the neighborhood of an element $x \in X$ with respect to $\mathbb{C}(X)$ defined as $\gamma(x)$ in Definition 25.1. It is well known that an Alexandroff closure operator is induced as the map $c_\gamma : \mathcal{P}(X) \mapsto \mathcal{P}(X)$ defined as $c_\gamma(A) = \cup\{\gamma(a) : a \in A\}$, which correspond to the upper approximation U_6 in (25.13), and consequently the dual operator L_6 is an interior operator.

More generally, all the upper approximations U_1–U_6 are closure operators. In particular U_4–U_6 are also topological closures, and since duality holds with respect to all approximation pairs but (L_3, U_3) and since $L_3 = L_2$, then all lower approximations are interior operators. This result can be easily established by checking that the properties satisfied by the approximations include those of Definition 25.2 (see Table 25.1 in [25.39]).

25.3 Conclusion

Three equivalent approaches to Pawlak rough sets can be given based on an equivalence relation, a partition of the universe, or a Boolean algebra. These different views generate three different possible generalizations of the classical model: binary relation, covering, and subsystem-based rough sets. We have reviewed these models and given the definitions of rough approximations in the different contexts. It can be seen that different models show interesting mathematical properties. In particular, binary relations-based rough sets have their roots in modal logic, whereas covering and subsystem-based rough sets are linked to topology.

Nowadays, generalized rough sets are continuously defined and we can encounter, for instance, more than 20 definitions of approximations based on coverings [25.4]. There is, however, a lack of interpretation in this collection. Efforts should be made to understand the meaning and usefulness of the already defined approximations. This should also be considered when defining new approximations. Besides an intrinsic theoretical interest, a logical approach could also be useful in this direction. Indeed, if in the case of binary relation-based rough sets we have a clear logical framework, the same cannot be said about covering and subsystem-based rough sets, where only few results are known.

References

25.1   Z. Pawlak: Rough sets, Int. J. Parallel Program. **11**(5), 341–356 (1982)

25.2   Z. Pawlak: *Rough Sets: Theoretical Aspects of Reasoning About Data* (Kluwer, Dordrecht 1991)

25.3   Y.Y. Yao: Two views of the theory of rough sets in finite universes, Int. J. Approx. Reason. **15**(4), 291–317 (1996)

25.4   Y.Y. Yao, B.X. Yao: Covering based rough set approximations, Inf. Sci. **200**, 91–107 (2012)

25.5   Y.Y. Yao: On generalizing rough set theory, Proc. Int. Conf. Rough Sets Fuzzy Sets Data Min. Granul. Comput. (Springer, Berlin Heidelberg 2003) pp. 44–51

25.6   Y.Y. Yao, T.Y. Lin: Generalization of rough sets using modal logic, Intell. Autom. Soft Comput. **2**(2), 103–120 (1996)

25.7   Y.Y. Yao: On generalizing Pawlak approximation operators, Proc. Int. Conf. Rough Sets Curr. Trends Comput. (Springer, Berlin Heidelberg 1998) pp. 298–307

25.8   Y.Y. Yao, T. Wang: On rough relations: An alternative formulation, Proc. Int. Conf. New Dir. Rough Sets Data Min. and Granul.-Soft Comput. (Springer, Berlin Heidelberg 1999) pp. 82–90

25.9   U. Wybraniec-Skardowska: On a generalization of approximation space, Bull. Pol. Acad. Sci. Math. **37**(1-6), 51–62 (1989)

25.10   Z. Pawlak: Hard and soft sets, Rough Sets, Proc. Int. Workshop Rough Sets Knowl. Discov. (Springer, London 1994) pp. 130–135

25.11   Y.Y. Yao, X. Li, T.Y. Lin, Q. Liu: Representation and classification of rough set models, Proc. Int. Workshop Rough Sets Soft Comput. (SCS: Society for Computer Simulation, San Diego 1995) pp. 44–47

25.12   R. Slowinski, D. Vanderpooten: A generalized definition of rough approximations based on similarity, Knowl. Data Eng., IEEE Trans. **12**(2), 331–336 (2000)

25.13   S. Greco, B. Matarazzo, R. Slowinski: Fuzzy similarity relation as a basis for rough approximations, Proc. Int. Conf. Rough Sets Curr. Trends Comput. (Springer, Berlin Heidelberg 1998) pp. 283–289

25.14   L.H. Guan, G.Y. Wang: Generalized approximations defined by non-equivalence relations, Inf. Sci. **193**, 163–179 (2012)

25.15   Y.Y. Yao: Relational interpretations of neighborhood operators and rough set approximation operators, Inf. Sci. **111**(1), 239–259 (1998)

25.16   W.Z. Wu, W.X. Zhang: Neighborhood operator systems and approximations, Inf. Sci. **144**(1), 201–217 (2002)

25.17   H.M. Abu-Donia: Comparison between different kinds of approximations by using a family of binary relations, Knowl.-Based Syst. **21**(8), 911–919 (2008)

25.18   Y.Y. Yao: Generalized rough set models. In: *Rough Sets in Knowledge Discovery*, ed. by L. Polkowski, A. Skowron (Physica, Heidelberg 1998) pp. 286–318

25.19   W. Zhu: Relationship between generalized rough sets based on binary relation and covering, Inf. Sci. **179**(3), 210–225 (2009)

25.20   Y.Y. Yao, S.K.M. Wang, T.Y. Lin: A review of rough set models. In: *Rough Sets and Data Mining: Analysis for Imprecise Data*, ed. by L. Polkowski, A. Skowron (Kluwer, Boston 1997) pp. 47–75

25.21   Y.Y. Yao: Constructive and algebraic methods of the theory of rough sets, J. Inf. Sci. **109**(1), 21–47 (1998)

25.22   J.T. Yao, Y.Y. Yao: Induction of classification rules by granular computing, Proc. Int. Conf. Rough Sets Curr. Trends Comput. (Springer, Berlin Heidelberg 2002) pp. 331–338

25.23   J.T. Yao, A.V. Vasilakos, W. Pedrycz: Granular computing: Perspectives and challenges, IEEE Trans. Cybern. **43**(6), 1977–1989 (2013)

25.24   W. Zakowski: Approximations in the space (u, Π), Demonstr. Math. **16**(40), 761–769 (1983)

25.25   J.A. Pomykala: Approximation operations in approximation space, Bull. Pol. Acad. Sci. Math. **35**, 653–662 (1987)

25.26   J.A. Pomykała: On definability in the nondeterministic information system, Bull. Pol. Acad. Sci.: Math. **36**(3/4), 193–210 (1988)

25.27   I. Couso, D. Dubois: Rough sets, coverings and incomplete information, Fundam. Inf. **108**(3), 223–247 (2011)

25.28   M. Restrepo, C. Cornelis, J. Gómez: Duality, conjugacy and adjointness of approximation operators in covering-based rough sets, Int. J. Approx. Reason. **55**(1), 469–485 (2014)

25.29   J. Wang, D. Dai, Z. Zhou: Fuzzy covering generalized rough sets, J. Zhoukou Teach. Coll. **21**(2), 20–22 (2004), in Chinese

25.30   Z. Xu, Q. Wang: On the properties of covering rough sets model, J. Henan Norm. Univ. **33**(1), 130–132 (2005), in Chinese

25.31   W.H. Xu, W.X. Zhang: Measuring roughness of generalized rough sets induced by a covering, Fuzzy Sets Syst. **158**(22), 2443–2455 (2007)

25.32   W. Zhu, F.Y. Wang: Some results on covering generalized rough sets, Pattern Recogn. Artif. Intell. **15**(1), 6–13 (2002)

25.33   W. Zhu, F.Y. Wang: Reduction and axiomization of covering generalized rough sets, Inf. Sci. **152**, 217–230 (2003)

25.34   W. Zhu: Properties of the second type of covering-based rough sets, Proc. Int. Web Intell. Intell. Agent Technol. (IEEE, Piscataway 2006) pp. 494–497

25.35   W. Zhu, F.Y. Wang: A new type of covering rough set, Proc. Int. Conf. Intell. Syst. (IEEE, Piscataway 2006) pp. 444–449

25.36   W. Zhu, F.Y. Wang: On three types of covering-based rough sets, IEEE Trans. Knowl. Data Eng. **19**(8), 1131–1144 (2007)

Part C | 25

25.37 G.L. Liu: The relationship among different covering approximations, Inf. Sci. **250**, 178–183 (2013)

25.38 Y.-L. Zhang, M.-K. Luo: Relationships between covering-based rough sets and relation-based rough sets, Inf. Sci. **225**, 55–72 (2012)

25.39 P. Samanta, M.K. Chakraborty: Generalized rough sets and implication lattices, Trans. Rough Sets **14**, 183–201 (2011)

25.40 J.A. Pomykala: *Approximation, Similarity and Rough Construction*, ILLC Prepublication Series for Computation and Complexity Theory, Vol. 93 (Univ. Amsterdam, Amsterdam 1993)

25.41 T.-J. Li: Rough approximation operators in covering approximation spaces, RSCTC2006 Proc. (Springer, Berlin Heidelberg 2006) pp. 174–182

25.42 D. Slezak, P. Wasilewski: Granular sets – Foundations and case study of tolerance spaces, RSFD-GrC2007 Proc. (Springer, Berlin Heidelberg 2007) pp. 435–442

25.43 G. Cattaneo, D. Ciucci: Lattices with interior and closure operators and abstract approximation spaces, Trans. Rough Sets **10**, 67–116 (2009)

25.44 Z. Bonikowski: A certain conception of the calculus of rough sets, Notre Dame J. Formal Log. **33**(3), 412–421 (1992)

25.45 K.Y. Qin, Y. Gao, Z. Pei: On covering rough sets, RSKT2007 Proc. (Springer, Berlin Heidelberg 2007) pp. 34–41

25.46 W. Zhu: Topological approaches to covering rough sets, Inf. Sci. **177**(6), 1499–1508 (2007)

25.47 Y.Y. Yao, Y.H. Chen: Subsystem based generalizations of rough set approximations, Proc. Int. Conf. Found. Intell. Syst. (Springer, Berlin Heidelberg 2005) pp. 210–218

25.48 A. Wiweger: On topological rough sets, Bull. Pol. Acad. Sci. Math. **37**, 89–93 (1989)

25.49 J. Järvinen: On the structure of rough approximations, Fundam. Inf. **53**(2), 135–153 (2002)

25.50 G. Cattaneo: Abstract approximation spaces for rough theories, Rough Sets Knowl. Discov. **1**, 59–98 (1998)

25.51 M. Gehrke, E. Walker: On the structure of rough sets, Bull. Pol. Acad. Sci. Math. **40**, 235–245 (1992)

25.52 B.F. Chellas: *Modal Logic: An Introduction* (Cambridge Univ. Press, Cambridge 1988)

25.53 E. Orlowska, Z. Pawlak: Representation of non-deterministic information, Theor. Comput. Sci. **29**, 27–39 (1984)

25.54 D. Vakarelov: A model logic for similarity relations in Pawlak knowledge representation systems, Fundam. Inf. **15**(1), 61–79 (1991)

25.55 D. Vakarelov: Modal logics for knowledge representation systems, Theor. Comput. Sci. **90**(2), 433–456 (1991)

25.56 P. Balbiani, D. Vakarelov: A modal logic for indiscernibility and complementarity in information systems, Fundam. Inf. **50**(3/4), 243–263 (2002)

25.57 F. del Cerro, L.E. Orlowska: DAL – A logic for data analysis, Theor. Comput. Sci. **36**, 251–264 (1985)

25.58 M. Banerjee, K. Chakraborty: Algebras from rough sets. In: *Rough-Neural Computing: Techniques for Computing with Words*, ed. by S.K. Pal, A. Skowron, L. Polkowski (Springer, Berlin Heidelberg 2004) pp. 157–188

25.59 M. Banerjee, M.A. Khan: Propositional logics from rough set theory, Trans. Rough Sets **6**, 1–25 (2007)

25.60 D. Ciucci, D. Dubois: Truth-functionality, rough sets and three-valued logics, Proc. ISMVL (IEEE, Piscataway 2010) pp. 98–103

25.61 Z. Bonikowski, E. Bryniarski, U. Wybraniec-Skardowska: Extensions and intentions in the rough set theory, Inf. Sci. **107**(1–4), 149–167 (1998)

25.62 G. Cattaneo, D. Ciucci: On the lattice structure of preclusive rough sets, IEEE Int. Conf. Fuzzy Syst., Piscataway (2004)

25.63 B. Konikowska: Three-valued logic for reasoning about covering-based rough sets. In: *Rough Sets and Intelligent Systems – Professor Z. Pawlak in Memoriam*, Intelligent Systems Reference Library, Vol. 42, ed. by A. Skowron, Z. Suraj (Springer, Berlin Heidelberg 2013) pp. 439–461

25.64 S. Greco, B. Matarazzo, R. Słowiński: Algebra and topology for dominance-based rough set approach. In: *Advances in Intelligent Information Systems*, Studies in Computational Intelligence, Vol. 265, ed. by Z.W. Ras, L.-S. Tsay (Springer, Berlin Heidelberg 2010) pp. 43–78

25.65 B.A. Davey, H.A. Priestley: *Introduction to Lattices and Order* (Cambridge Univ. Press, Cambridge 1990)

25.66 N. Caspard, B. Monjardet: The lattices of closure systems, closure operators, and implicational systems on a finite set: A survey, Discret. Appl. Math. **127**(2), 241–269 (2003)

26. Fuzzy-Rough Hybridization

Masahiro Inuiguchi, Wei-Zhi Wu, Chris Cornelis, Nele Verbiest

Fuzzy sets and rough sets are known as uncertainty models. They are proposed to treat different aspects of uncertainty. Therefore, it is natural to combine them to build more powerful mathematical tools for treating problems under uncertainty. In this chapter, we describe the state-of-the-art in the combinations of fuzzy and rough sets dividing into three parts.

In the first part, we describe two kinds of models of fuzzy rough sets: one is classification-oriented model and the other is approximation-oriented model. We describe the fundamental properties and show the relations of those models. Moreover, because those models use logical connectives such as conjunction and implication functions, the selection of logical connectives can sometimes be a question. Then we propose a logical connective-free model of fuzzy rough sets.

In the second part, we develop a generalized fuzzy rough set model. We first introduce general types of belief structures and their induced dual pairs of belief and plausibility functions in the fuzzy environment. We then build relationships between belief and plausibility functions in the Dempster–Shafer theory of evidence and the lower and upper approximations in rough set theory in various situations. We also provide the potential applications of the main results to intelligent information systems.

In the third part, we give an overview of the practical applications of fuzzy rough sets. The main focus will be on the machine-learning domain. In

particular, we review fuzzy-rough approaches for attribute selection, instance selection, classification, and prediction.

26.1 Introduction to Fuzzy-Rough Hybridization

Rough set approaches [26.1, 2] have been successfully applied to various fields related to data analysis, knowledge discovery, decision analysis, and so on. In order to expand the application area and to develop its theory

further, rough sets have been generalized under various settings. There are two different generalizations. One relaxes the precision so that the sizes of lower and upper approximations are controlled by a precision parameter.

This generalized rough set is called a variable precision rough set. The other generalizes the approximation space, i.e., the structure of background knowledge. Many researchers generalized an equivalence relation which is often referred to as an indiscernibility relation to a general binary relation or a family. Many other researchers [26.3–23] generalized an equivalence relation to a fuzzy binary relation or a family of fuzzy sets.

In this chapter, we describe the generalizations of rough sets in the latter sense. More precisely, we concentrate on the fuzzy generalizations of rough set approaches called fuzzy rough hybridizations. Fuzzy rough sets were originally proposed by *Nakamura* [26.3] and by *Dubois* and *Prade* [26.4, 5]. The fundamental properties of fuzzy rough sets have been investigated by *Dubois* and *Prade* [26.4, 5] and *Radzikowska* and *Kerre* [26.9]. In those studies, an equivalence relation of approximation space in the original rough sets is generalized to a fuzzy equivalence relation. *Greco* et al. [26.7] proposed fuzzy rough sets under a fuzzy dominance relation. Those fuzzy rough sets are based on possibility and necessity measures directly. Moreover, this type of fuzzy rough sets is defined under more generalized settings [26.11, 15] and different types of fuzzy rough sets were proposed based on certainty qualifications by *Inuiguchi* and *Tanino* [26.10, 12] and also based on modifier functions by *Greco* et al. [26.24, 25]. The fuzzy rough set model can be used to deal with attribute reduction in information systems with fuzzy decision while the fuzzy rough set model can be employed in reasoning and knowledge acquisition with decision tables with real-valued conditional attributes or quantitative data (see, for example, [26.26–36]).

In the first part of this chapter, we introduce three models of fuzzy rough sets. Those fuzzy sets are classified into two groups, i.e., classification-oriented fuzzy rough set models and approximation-oriented fuzzy rough set models proposed by *Inuiguchi* [26.37] originally in the crisp settings. In the classification-oriented models, we are interested in a set to which objects belong. We evaluate each object whether its membership to a set X is consistent with all information we have at hand or not. The positive region of X is defined by collecting all objects whose memberships to X are consistent with whole information. The possible region of X is defined by collecting all objects whose memberships to X are conceivable from some part of information but not consistent with all information. Then the fuzzy rough set of X is defined by a pair of the positive and possible regions of X. On

the contrary, in approximation-oriented models, we are interested in the approximations of a set by using elementary sets of a family. We approximate a set X by unions of the elementary sets and by intersections of the complementary sets of the elementary sets. The lower and upper approximations are defined by the inner and outer approximations of X, respectively. A rough set of X is defined by a pair of the lower and upper approximations. We describe that one of the three models belongs to the group of classification-oriented models and the remaining two models belong to the group of approximation-oriented models.

Another important method used to deal with uncertainty in intelligent systems is the Dempster–Shafer theory of evidence [26.38]. Shafer's belief and plausibility functions are constructed under the assumption that the focal elements in the belief structure are all crisp. In some situations, it seems to be quite natural that the evidence mass may be assigned to a fuzzy subset of the universe of discourse. In fact, combining the Dempster–Shafer theory and fuzzy set theory has been suggested to be a way to deal with different kinds of uncertain information in intelligent systems in a number of studies. It is demonstrated that the lower and upper approximation operators in rough set theory have strong relationship with the belief and plausibility functions in the Dempster–Shafer theory of evidence [26.21, 23, 39–44]. The Dempster–Shafer theory of evidence may be used to analyze knowledge acquisition in information systems (see, for example, [26.45–49]).

In the second part of this chapter, we will explore the relationships between belief and plausibility functions in the Dempster–Shafer theory of evidence and the lower and upper approximations in rough set theory with their potential applications to intelligent information systems.

Both fuzzy set and rough set theories have fostered broad research communities and have been applied in a wide range of settings. More recently, this has also extended to the hybrid fuzzy rough set models. The third part of this chapter tries to give a sample of those applications, which are in particular numerous for machine learning but which also cover many other fields, like image processing, decision making, and information retrieval.

Note that we do not consider applications that simply involve a joint application of fuzzy sets and rough sets, like for instance a rough classifier that induces fuzzy rules. Rather, we focus on applications that specifically involve one of the fuzzy rough set models discussed in the previous sections.

This chapter is organized as follows. In the next section, three models of fuzzy rough sets are explained dividing into two groups. In Sect. 26.3, we introduce generalized fuzzy belief structures with application in fuzzy information systems. In Sect. 26.4, we give an overview of the practical applications of fuzzy rough sets focusing on the machine-learning domain.

26.2 Classification- Versus Approximation-Oriented Fuzzy Rough Set Models

In this section, we review three kinds of fuzzy rough sets from classification-oriented and approximation-oriented points of view. Focusing on the membership of an object to a set X under the indiscernibility relation, the classical rough set defined by a pair of lower and upper approximations of a set X can be seen as a classifier of objects into three disjoint regions: positive, negative, and boundary regions of a set X. Namely, the lower approximation defines the positive region, the complement of the upper approximation defines the negative region and the difference between upper and lower approximations defines the boundary region. On the other hand, focusing on the approximations of X by means of elementary sets of the partition, the rough set of X defines the inner and outer approximations of X. Namely, the lower approximation defines the inner approximation and the upper approximation defines the outer approximation. Those two different views of rough sets give different definitions of rough sets in the generalized settings (see *Inuiguchi* [26.50]). In this section, we describe fuzzy rough sets in a generalized setting from those points of view and show the fundamental properties, differences, and similarities.

26.2.1 Classification-Oriented Fuzzy Rough Sets

Definitions in Crisp Setting
In this subsection, we define fuzzy rough sets under the interpretation of rough sets as classification of objects into positive, negative, and boundary regions of a set and describe their properties. As the introduction, we first describe the definitions of positive and possible regions of a set in the crisp setting. Let U be a set of all objects. Assume that we do not know objects which fit with a particular concept C but we have pieces of information that tell some objects fit with C and that the other objects do not fit. Let $X \subseteq U$ be the set of objects which are supposed to fit with C in the information and $U - X$ the set of objects which are sup-

posed not to fit with C in the information. On the other hand, there is knowledge about C expressed by a binary relation $P \subseteq U \times U$. Under the binary relation P, we presume y *fits with* C from facts $(y, x) \in P$ and x *fits with* C.

Under this circumstance, we investigate credible members of X and plausible members of X. Objects whose membership to X is consistent with the knowledge can be understood as credible members of X, while objects whose membership to X is presumable from the information and the knowledge can be understood as plausible members. For convenience, we define $P(x) = \{y \in U \mid (y, x) \in P\}$ which is the set of objects whose membership to X is presumed from the fact $x \in X$. Therefore, if $x \in X$ satisfies $\forall y \in P(x)$, $y \in X$ or simply, $P(x) \subseteq X$, x can be considered a credible member of X. Thus, the set of credible members of X is defined by

$$P_*(X) = \{x \in X \mid P(x) \subseteq X\} \\ = X \cap \{x \in U \mid P(x) \subseteq X\}. \tag{26.1}$$

On the other hand, we may presume $x \in X$ if $x \in X$ or $\exists y \in X$, $x \in P(y)$ under the information and the knowledge. Then the set of plausible members of X can be defined by

$$P^*(X) = X \cup \{x \in U \mid \exists y \in X, \ x \in P(y) \neq \emptyset\}. \tag{26.2}$$

$P_*(X)$ is called the positive region of X and $P^*(X)$ is called the possible region of X. Moreover, we do not assume the reflexivity of P, i.e., $\forall x \in U, (x, x) \in P$. This is why we take the intersection with X in the definition of $P_*(X)$ and the union with X in the definition of $P^*(X)$. Those intersection and union can be dropped when P is reflexive.

When there is knowledge about C expressed by a binary relation $Q \subseteq U \times U$ instead of P. Under the binary relation Q, we presume y *does not fit with* C from facts

$(y, x) \in Q$ and x *does not fit with* C. In this case, we directly obtain positive and possible regions of $U - X$, respectively, by

$$Q_*(U - X) = \{x \in U - X \mid Q(x) \subseteq U - X\}$$
$$= (U - X) \cap \{x \in U \mid Q(x) \subseteq U - X\}, \tag{26.3}$$

$$Q^*(U - X) = (U - X) \cup \{x \in U \mid \exists y \in U - X,$$
$$x \in Q(y) \neq \emptyset\}. \tag{26.4}$$

Because an object that is not a member of $Q_*(U - X)$ can be seen as a plausible member of X and an object which is not a member of $Q^*(U - X)$ can be seen as a credible member of X, we may define positive and possible regions of X by

$$\bar{Q}_*(X) = U - Q^*(U - X),$$
$$\bar{Q}^*(X) = U - Q_*(U - X). \tag{26.5}$$

Inuiguchi [26.50] investigated the properties of those positive and possible regions.

Definitions in Fuzzy Setting and Their Properties

We now extend those definitions of positive and possible regions into the fuzzy setting. First, we assume a fuzzy set $X \subseteq U$ and a fuzzy binary relation $P \subseteq U \times U$ are given. Their membership functions $\mu_X(x)$ and $\mu_P(y, x)$ show the membership degree of $x \in U$ to a fuzzy set X and the degree to what extent we presume that y is a member of X from the fact x is a member of a fuzzy set X, where $\mu_X : U \to [0, 1]$ and $\mu_P : U \times U \to [0, 1]$. We define $P(x)$ by its membership function $\mu_{P(x)}(y) = \mu_P(y, x)$.

To define the positive region under this circumstance, we should consider the consistency degree of the information that x is a member of X to membership degree $\mu_X(x)$ with the knowledge P. This can be measured by the truth value of statement $y \in P(x)$ implies $y \in X$ under fuzzy sets $P(x)$ and X. The truth value of this statement can be defined by a necessity measure $\inf_{y \in U} I(\mu_{P(x)}(y), \mu_X(y))$ with an implication function $I : [0, 1] \times [0, 1] \to [0, 1]$ such that $I(0, 0) = I(0, 1) = I(1, 1) = 1$, $I(1, 0) = 0$, $I(\cdot, a)$ is decreasing for any $a \in [0, 1]$ and $I(a, \cdot)$ is increasing for any $a \in [0, 1]$. Therefore, in the analogy to (26.1), the membership function of the positive region $P_*(X)$ of X can be

defined by

$$\mu_{P_*(X)}(x) = \min\left(\mu_X(x), \inf_{y \in U} I(\mu_{P(x)}(y), \mu_X(y))\right)$$
$$= \min\left(\mu_X(x), \inf_{y \in U} I(\mu_P(y, x), \mu_X(y))\right), \tag{26.6}$$

where we note the intersection $C \cap D$ of two fuzzy sets $C, D \subseteq U$ is normally defined by $\mu_{C \cap D}(x) = \min(\mu_C(x), \mu_D(x))$, $\forall x \in U$. $\mu_{C \cap D}$, μ_C and μ_D are membership functions of $C \cap D$, C and D. However, some researchers use t-norms [26.51] instead of the min operation. A t-norm t is a conjunction function $t : [0, 1] \times [0, 1] \to [0, 1]$ such that (t1) $\forall a \in [0, 1]$, $t(a, 1) = t(1, a) = a$ (boundary condition), (t2) $\forall a, b \in [0, 1]$, $t(a, b) = t(b, a)$ (commutativity) and (t3) $\forall a, b, c \in [0, 1]$, $t(a, t(b, c)) = t(t(a, b), c)$ (associativity).

Now let us define the possible region when X and P are a fuzzy set and a fuzzy binary relation, respectively. To do this, we should define the truth value of statement *there exists* $y \in X$ *such that* $x \in P(y)$ under fuzzy sets X and $P(x)$. The truth value of this statement can be obtained by a possibility measure $\sup_{y \in U} T(\mu_{P(y)}(x), \mu_X(y))$ with a conjunction function $T : [0, 1] \times [0, 1] \to [0, 1]$ such that $T(1, 1) = 1$, $T(0, 0) = T(0, 1) = T(1, 0) = 0$ and T is increasing in both arguments. Therefore, in the analogy to (26.2), the membership function of the possible region $P^*(X)$ of X can be defined by

$$\mu_{P^*(X)}(x) = \max\left(\mu_X(x), \sup_{y \in U} T(\mu_{P(y)}(x), \mu_X(y))\right)$$
$$= \max\left(\mu_X(x), \sup_{y \in U} T(\mu_P(x, y), \mu_X(y))\right), \tag{26.7}$$

where we note the union $C \cup D$ of two fuzzy sets $C, D \subseteq U$ is normally defined by $\mu_{C \cup D}(x) = \max(\mu_C(x), \mu_D(x))$, $\forall x \in U$. $\mu_{C \cup D}$ is a membership functions of $C \cup D$. However, some researchers use t-conorms [26.51] instead of the max operation. A t-conorm s is a function $s : [0, 1] \times [0, 1] \to [0, 1]$ such that (s1) $\forall a \in [0, 1]$, $s(a, 0) = s(0, a) = a$ (boundary condition), (s2) $\forall a, b \in [0, 1]$, $s(a, b) = s(b, a)$ (commutativity), (s3) $\forall a, b, c \in [0, 1] s(a, s(b, c)) = s(s(a, b), c)$ (associativity). and (s4) $\forall a, b, c, d$ such that $a \geq c$ and $b \geq d$; $s(a, b) \geq s(c, d)$ (monotonicity).

Note that we do not assume the reflexivity of P, i.e., $\mu_P(x, x) = 1$, $\forall x \in U$ so that we take the minimum between μ_X and $\inf_{y \in U} I(\mu_{P(x)}(y), \mu_X(y))$ in Eq. (26.6) and the maximum between μ_X and $\sup_{y \in U} T(\mu_{P(y)}(x), \mu_X(y))$ in (26.7). When P is reflexive, $I(1, a) \leq a$ and $T(1, a) \geq a$ for all $a \in [0, 1]$, we have

$$\mu_{P_*(X)}(x) = \inf_{y \in U} I(\mu_{P(x)}(y), \mu_X(y)) ,$$
$$\mu_{P^*(X)}(x) = \sup_{y \in U} T(\mu_{P(y)}(x), \mu_X(y)) . \qquad (26.8)$$

Those definitions of lower and upper approximations have been proposed by *Dubois* and *Prade* [26.4, 5] and *Radzikowska* and *Kerre* [26.9]. They assumed the reflexivity of P and $I(1, a) = T(1, a) = a$, for all $a \in [0, 1]$. Moreover, the definitions of (26.8) are used even when P is not reflexive and neither I nor T satisfy the boundary conditions $I(1, a) = T(1, a) = a$, for all $a \in [0, 1]$ [26.15, 52]. In such generalized situation, we may loose the inclusiveness of $P_*(X)$ in X and that of X in $P^*(X)$ for $P_*(X)$ and $P^*(X)$ defined by (26.8). The definitions of $P_*(X)$ and $P^*(X)$ by (26.6) and (26.7) obtained from the interpretations of positive and possible regions of X satisfy the inclusiveness of $P_*(X)$ in X and that of X in $P^*(X)$ even in the generalized situation.

Using the positive region $P_*(X)$ and the possible region $P^*(X)$, we can define a fuzzy rough set of X as a pair $(P_*(X), P^*(X))$. We can call such fuzzy rough sets as classification-oriented fuzzy rough sets under a positively extensive relation P of X (for short CP-fuzzy rough sets). Note that the relation P depends on the meaning of a set X. Thus, we cannot always define the CP-rough set of $U - X$ by the same relation P.

To define a CP-rough set of $U - X$, we should introduce another fuzzy relation $Q \subseteq U \times U$ such that $\mu_{Q(x)}(y) = \mu_Q(y, x)$ represents the degree to what extent we presume an object y as a member of $U - X$ from the fact x is a member of $U - X$, where $\mu_Q: U \times U \to [0, 1]$ is a membership function of a fuzzy relation Q. In the same way, we define positive and possible regions of $U - X$ under fuzzy relation Q by the following membership functions

$$\mu_{Q_*(U-X)}(x)$$
$$= \min\left(n(\mu_X(x)), \inf_{y \in U} I(\mu_Q(y, x), n(\mu_X(y)))\right), \qquad (26.9)$$

$$\mu_{Q^*(U-X)}(x)$$
$$= \max\left(n(\mu_X(x)), \sup_{y \in U} T(\mu_Q(x, y), n(\mu_X(y)))\right), \qquad (26.10)$$

where $U - X$ is defined by a membership function $n(\mu_X(\cdot))$ and $n: [0, 1] \to [0, 1]$ is a strong negation which is a decreasing function such that $n(n(a)) = a, a \in [0, 1]$ (involutive). The involution implies the continuity of n.

Using $Q_*(X)$ and $Q^*(X)$, in analogy to (26.5), we can define the positive region $\bar{Q}_*(X)$ and the possible region $\bar{Q}^*(X)$ of X by the following membership functions

$$\mu_{\bar{Q}_*(X)}(x)$$
$$= \min\left(\mu_X(x), \inf_{y \in U} n(T(\mu_Q(x, y), n(\mu_X(y))))\right), \qquad (26.11)$$

$$\mu_{\bar{Q}^*(X)}(x)$$
$$= \max\left(\mu_X(x), \sup_{y \in U} n(I(\mu_Q(y, x), n(\mu_X(y))))\right). \qquad (26.12)$$

We can define a fuzzy rough set of X as a pair $(\bar{Q}_*(X), \bar{Q}^*(X))$ with the positive region $\bar{Q}_*(X)$ and the possible region $\bar{Q}^*(X)$. We can call this type of rough sets as classification-oriented fuzzy rough sets under a negatively extensive relation Q of X (for short CN-fuzzy rough sets).

Let us discuss the properties of CP- and CN-fuzzy rough sets. By definition, we have

$$P_*(X) \subseteq X \subseteq P^*(X) ,$$
$$\bar{Q}_*(X) \subseteq X \subseteq \bar{Q}^*(X) , \qquad (26.13)$$
$$P_*(\emptyset) = P^*(\emptyset) = \bar{Q}_*(\emptyset) = \bar{Q}^*(\emptyset) = \emptyset , \qquad (26.14)$$
$$P_*(U) = P^*(U) = \bar{Q}_*(U) = \bar{Q}^*(U) = U , \qquad (26.15)$$

$$P_*(X \cap Y) = P_*(X) \cap P_*(Y) ,$$
$$P^*(X \cup Y) = P^*(X) \cup P^*(Y) , \qquad (26.16)$$
$$\bar{Q}_*(X \cap Y) = \bar{Q}_*(X) \cap \bar{Q}_*(Y) ,$$
$$\bar{Q}^*(X \cup Y) = \bar{Q}^*(X) \cup \bar{Q}^*(Y) , \qquad (26.17)$$
$$X \subseteq Y \text{ implies } P_*(X) \subseteq P_*(Y) ,$$
$$X \subseteq Y \text{ implies } P^*(X) \subseteq P^*(Y) , \qquad (26.18)$$
$$X \subseteq Y \text{ implies } \bar{Q}_*(X) \subseteq \bar{Q}_*(Y) ,$$
$$X \subseteq Y \text{ implies } \bar{Q}^*(X) \subseteq \bar{Q}^*(Y) , \qquad (26.19)$$
$$P_*(X \cup Y) \supseteq P_*(X) \cup P_*(Y) ,$$
$$P^*(X \cap Y) \subseteq P^*(X) \cap P^*(Y) , \qquad (26.20)$$
$$\bar{Q}_*(X \cup Y) \supseteq \bar{Q}_*(X) \cup \bar{Q}_*(Y) ,$$
$$\bar{Q}^*(X \cap Y) \subseteq \bar{Q}^*(X) \cap \bar{Q}^*(Y) , \qquad (26.21)$$

Part C | 26.2

where the inclusion relation between two fuzzy sets A and B is defined by $\mu_A(x) \leq \mu_B(x)$, for all $x \in U$.

The properties satisfied under some conditions are listed as follows (see *Inuiguchi* [26.37]):

⟨1⟩ When $I(a,b) = n(T(a,n(b)))$, for all $a,b \in [0,1]$ and Q is the converse of P, i.e., $\mu_Q(x,y) = \mu_P(y,x)$, for all $x,y \in U$, we have

$$P_*(X) = U - Q^*(U-X) = \bar{Q}_*(X) \,, \qquad (26.22)$$

$$P^*(X) = U - Q_*(U-X) = \bar{Q}^*(X) \,. \qquad (26.23)$$

⟨2⟩ When $T(a,I(a,b)) \leq b$ holds for all $a,b \in [0,1]$, we have

$$X \supseteq P^*(P_*(X)) \supseteq P_*(X) \supseteq P_*(P_*(X)) \,, \qquad (26.24)$$

$$X \subseteq \bar{Q}_*(\bar{Q}^*(X)) \subseteq \bar{Q}^*(X) \subseteq \bar{Q}^*(\bar{Q}^*(X)) \,. \qquad (26.25)$$

⟨3⟩ When $I(a,T(a,b)) \geq b$ holds for all $a,b \in [0,1]$, we have

$$X \subseteq P_*(P^*(X)) \subseteq P^*(X) \subseteq P^*(P^*(X)) \,, \qquad (26.26)$$

$$X \supseteq \bar{Q}^*(\bar{Q}_*(X)) \supseteq \bar{Q}_*(X) \supseteq \bar{Q}_*(\bar{Q}_*(X)) \,. \qquad (26.27)$$

⟨4⟩ Let P and Q be T'-transitive. The following assertions are valid:
  (a) When I is upper semicontinuous and satisfies $I(a,I(b,c)) = I(T'(b,a),c)$ for all $a,b,c \in [0,1]$, we have

$$P_*(P_*(X)) = P_*(X) \,, \qquad \bar{Q}^*(\bar{Q}^*(X)) = \bar{Q}^*(X) \,. \qquad (26.28)$$

  (b) When $T = T'$ is lower semicontinuous and satisfies $T(a,T(b,c)) = T(T(a,b),c)$ for all $a,b,c \in [0,1]$ (associativity), we have

$$P^*(P^*(X)) = P^*(X) \,, \qquad \bar{Q}_*(\bar{Q}_*(X)) = \bar{Q}_*(X) \,. \qquad (26.29)$$

⟨5⟩ When P and Q are reflexive and T-transitive, the following assertions are valid:
  (a) If $I(a,\cdot)$ is upper semicontinuous, $I(1,a) \leq a$, and $T = \xi[I]$ is associative, then we have

$$P^*(P_*(X)) = P_*(X) \,, \qquad \bar{Q}_*(\bar{Q}^*(X)) = \bar{Q}^*(X) \,. \qquad (26.30)$$

(b) If $I(a,b) = n(\xi[I](a,n(b)))$ and the conditions of (a) are satisfied, then we have

$$P_*(P^*(X)) = P^*(X) \,, \qquad \bar{Q}^*(\bar{Q}_*(X)) = \bar{Q}_*(X) \,. \qquad (26.31)$$

Here a fuzzy relation P is said to be T'-transitive, if and only if P satisfies $\mu_P(x,z) \geq T'(\mu_P(x,y),\mu_P(y,z))$ for all $x,y,z \in U$ and for a conjunction function T'. We can generate a function $\xi[I]: [0,1] \times [0,1] \to [0,1]$ by $\xi[I](a,b) = \inf\{s \in [0,1] \mid I(a,s) \geq b\}$ when a function $I: [0,1] \times [0,1] \to [0,1]$ is given. $\xi[I]$ is a conjunction function when I satisfies $I(1,a) < 1$ for all $a \in [0,1)$.

Concerning to the assumption of ⟨1⟩, it is known that a function I' defined by $I'(a,b) = n(T(a,n(b)))$ is an implication function and that a function T' defined by $T'(a,b) = n(I(a,n(b)))$ is a conjunction function (see, for example, *Inuiguchi* and *Sakawa* [26.51, 53]). The assumption of ⟨2⟩ corresponds to modus ponens, i.e., A and $(A \to B)$ implies B. Therefore, it is a natural assumption. However, this cannot hold for any implication and conjunction functions. For example, consider functions $T(a,b) = \min(a,b)$ and $I(a,b) = \max(1-a,b)$ which are often used in possibility theory. $T(a,I(a,b)) \leq b$ does not always hold. On the other hand, the assumption holds for any T and I such that $T(a,b) \leq \min(a,b)$ for all $a,b \in [0,1]$ and $I(a,b) \leq b$ for all $a,b \in [0,1]$ satisfying $a > b$. Thus, a t-norm T and a residual implication I of a t-norm satisfies the assumption, i.e., I is defined by $I(a,b) = \sup\{s \in [0,1] \mid t(a,s) \leq b\}$, for $a,b \in [0,1]$. The assumption of ⟨3⟩ is dual with that of ⟨2⟩. Namely, for any implication function I, there exists a conjunction function T' such that $I(a,b) = n(T'(a,n(b)))$, and for any conjunction function T, there exists an implication function I' such that $T(a,b) = n(I'(a,n(b)))$. Using T' and I', the assumption $I(a,T(a,b)) \geq b$ is equivalent to $T'(a,I'(a,b)) \leq b$ which is the same as the assumption of ⟨2⟩.

The assumption of ⟨3⟩ is satisfied with I and T such that $I(a,b) \geq \max(n(a),b)$ for all $a,b \in [0,1]$ and $T(a,b) \geq b$ for all $a,b \in [0,1]$ satisfying $a > n(b)$. The assumption of ⟨4⟩-(a) is satisfied with residual implication functions of lower semicontinuous t-norms T' and S-implication functions with respect to lower semicontinuous t-norms T', where an S-implication function I with respect to the t-norm T' is defined by $I(a,b) = n(T'(a,n(b)))$, $a,b \in [0,1]$ with a strong negation n. The assumption of ⟨4⟩-(b) is satisfied with lower semicontinuous t-norms T. These assumptions are satisfied with a lot of famous implication and conjunction functions.

26.2.2 Approximation-Oriented Fuzzy Rough Sets

Definitions in Crisp Setting

In this section, we define fuzzy rough sets under the interpretation of rough sets as approximation of sets and describe their properties. We first describe the definitions of lower and upper approximations in the crisp setting. We assume a family of subsets in U, $\mathcal{F} = \{F_i \mid i = 1, 2, \ldots, p\}$ is given. Each elementary set F_i is a meaningful set of objects such as a set of objects satisfying some properties. F_is can be seen as information granules with which we would like to express a set of objects. Given a set $X \subseteq U$, an understated expression of X, or in other words, an inner approximation of X by means of unions of F_is is obtained by

$$\mathcal{F}_*^{\cup}(X) = \bigcup \{F_i \in \mathcal{F} \mid F_i \subseteq X\}. \qquad (26.32)$$

On the other hand, an overstated expression of X, or in other words, an outer approximation of X by means of unions of F_is is obtained by

$$\mathcal{F}_{\cup}^*(X) = \bigcap \left\{ \bigcup_{i \in J} F_i \;\middle|\; \bigcup_{i \in J} F_i \supseteq X, \right.$$
$$\left. J \subseteq \{1, 2, \ldots, p, \circ\} \right\}, \qquad (26.33)$$

where we define $F_\circ = U$. We add $F_\circ$ considering cases where there is no $J \subseteq \{1, 2, \ldots, p\}$ such that $\bigcup_{i \in J} F_i \supseteq X$. In such cases, we obtain $\mathcal{F}_{\cup}^*(X) = U$ owing to the existence of $F_\circ = U$. $\mathcal{F}_*^{\cup}(X)$ and $\mathcal{F}_{\cup}^*(X)$ are called lower and upper approximations of X, respectively.

Applying those approximations to $U - X$, we obtain $\mathcal{F}_*^{\cup}(U - X)$ and $\mathcal{F}_{\cup}^*(U - X)$. From those, we obtain

$$\mathcal{F}_*^{\cap}(X) = U - \mathcal{F}_{\cup}^*(U - X)$$
$$= \bigcup \left\{ U - \bigcup_{i \in J} F_i \;\middle|\; \bigcup_{i \in J} F_i \supseteq U - X, \right.$$
$$\left. J \subseteq \{1, 2, \ldots, p, \circ\} \right\}, \qquad (26.34)$$

$$\mathcal{F}_{\cap}^*(X) = U - \mathcal{F}_*^{\cup}(U - X)$$
$$= \bigcap \{U - F_i \mid F_i \subseteq U - X,$$
$$i \in \{1, 2, \ldots, p, \bullet\}\}, \qquad (26.35)$$

where we define $F_\bullet = \emptyset$. We note that $\mathcal{F}_*^{\cap}(X)$ and $\mathcal{F}_{\cap}^*(X)$ are not always the same as $\mathcal{F}_*^{\cup}(X)$ and $\mathcal{F}_{\cup}^*(X)$,

respectively. The properties of those lower and upper approximations are studied by *Inuiguchi* [26.50].

Definitions by Certainty Qualifications in Fuzzy Setting

We extend those lower and upper approximations to cases where $\mathcal{F}$ is a family of fuzzy sets in U and X is a fuzzy set in U. To do this, we extend the intersection, union, complement, and the inclusion relation into the fuzzy setting. The intersection, union, and complement are defined by the min operation, the max operation and a strong negation n, i. e., $C \cap D$, $C \cup D$ and $U - C$ for fuzzy sets C and D are defined by membership functions $\mu_{C \cap D}(x) = \min(\mu_C(x), \mu_D(x))$, $\forall x \in C$, $\mu_{C \cup D}(x) = \max(\mu_C(x), \mu_D(x))$, $\forall x \in C$, $\mu_{U-C}(x) = n(\mu_C(x))$, $\forall x \in C$, respectively. The inclusion relation $C \subseteq D$ is extended to inclusion relation with degree $Inc(C, D) = \inf_x I(\mu_C(x), \mu_D(x))$, where I is an implication function.

First let us define a lower approximation by extending (26.32). In (26.32), before applying the union, we collect F_i such that $F_i \subseteq X$. This procedure cannot be extended simply into the fuzzy setting, because the inclusion relation has a degree showing to what extent the inclusion holds in the fuzzy setting. Namely, each F_i has a degree $q_i = Inc(F_i, X)$. This means that X includes F_i to a degree q_i. Therefore, by using F_i, X is expressed as a fuzzy set including F_i to a degree q_i. In other words, X is a fuzzy set Y satisfying

$$Inc(F_i, Y) = \inf_x I(\mu_{F_i}(x), \mu_Y(x)) = q_i. \qquad (26.36)$$

We note that there exists a solution satisfying (26.36) because q_i is defined by $Inc(F_i, X)$. There can be many solutions Y satisfying (26.36) and the intersection and union of those solutions can be seen as inner and outer approximations of X by F_i. Because we are now extending (26.32) and interested in the lower approximation, we consider the intersection of fuzzy sets including F_i to a degree q_i. Let us consider

$$Inc(F_i, Y) = \inf_x I(\mu_{F_i}(x), \mu_Y(x)) \geq q_i, \qquad (26.37)$$

instead of (26.36). Equation (26.37) is called a converse-certainty qualification [26.10] (or possibility-qualification). Because $I(a, \cdot)$ is increasing for any $a \in [0, 1]$ for an implication function I and (26.36) has a solution, the intersection of solutions of (26.36) is the same as the intersection of solutions of (26.37). Moreover, because I is upper semicontinuous, we obtain the intersection of solutions of (26.37) as the smallest solution $\check{Y}$ of (26.37) defined by the following membership function

$$\mu_{\check{Y}}(x) = \inf\{s \in [0, 1] \mid I(\mu_{F_i}(x), s) \geq q_i\}$$
$$= \xi[I](\mu_{F_i}(x), q_i) . \qquad (26.38)$$

We have $\check{Y} \subseteq X$. Because we have many $F_i \in \mathcal{F}$, the lower approximation $\mathcal{F}_*^{\xi}(X)$ of X is defined by the following membership function

$$\mu_{\mathcal{F}_*^{\xi}(X)}(x)$$

$$= \sup_{F \in \mathcal{F}} \xi[I] \left(\mu_F(x), \inf_{y \in U} I(\mu_F(y), \mu_X(y)) \right), \quad (26.39)$$

where $\mathcal{F}$ can have infinitely many elementary fuzzy sets F.

Because (26.32) is extended to (26.39), (26.35) is extended to the following equation in the sense that $\mathcal{F}_{\xi}^*(X) = U - F_*^{\xi}(U - X)$

$$\mu_{\mathcal{F}_{\xi}^*(X)}(x)$$

$$= \inf_{F \in \mathcal{F}} n \left(\xi[I] \left(\mu_F(x), \inf_{y \in U} I(\mu_F(y), n(\mu_X(y))) \right) \right),$$

$$(26.40)$$

where $\mu_{\mathcal{F}_{\xi}^*(X)}$ is the membership function of the upper approximation $\mathcal{F}_{\xi}^*(X)$ of X.

Now let us consider the extension of (26.33). In this case, before applying the intersection, we collect $\bigcup_{i \in J} F_i$ such that $\bigcup_{i \in J} F_i \supseteq X$. In the fuzzy setting, each $\bigcup_{i \in J} F_i$ has a degree $r_J = Inc(X, \bigcup_{i \in J} F_i)$. This means that X is included in $\bigcup_{i \in J} F_i$ to a degree r_J. Therefore, by using F_i, $i \in J$, X is expressed as a fuzzy set included in $\bigcup_{i \in J} F_i$ to a degree r_J. In other words, X is a fuzzy set Y satisfying

$$Inc \left(Y, \bigcup_{i \in J} F_i \right) = \inf_x I \left(\mu_Y(x), \max_{i \in J} \mu_{F_i}(x) \right)$$

$$= r_J . \qquad (26.41)$$

We note that there exists a solution satisfying (26.41) because r_J is defined by $Inc(X, \bigcup_{i \in J} F_i)$. There can be many solutions Y satisfying (26.41) and the intersection and union of those solutions can be seen as inner and outer approximations of X by $\bigcup_{i \in J} F_i$. Because we are now extending (26.33) and interested in the upper approximation, we consider the union of fuzzy sets in-

cluding $\bigcup_{i \in J} F_i$ to a degree r_J. Let us consider

$$Inc \left(Y, \bigcup_{i \in J} F_i \right) = \inf_x I \left(\mu_Y(x), \max_{i \in J} \mu_{F_i}(x) \right)$$

$$\geq r_J , \qquad (26.42)$$

instead of (26.41). Equation (26.42) is called a certainty qualification [26.10, 54]. Because $I(\cdot, a)$ is decreasing for any $a \in [0, 1]$ for an implication function I and (26.41) has a solution, the union of solutions of (26.41) is the same as the union of solutions of (26.42). Moreover, because I is upper semicontinuous, we obtain the union of solutions of (26.42) as the greatest solution $\hat{Y}$ of (26.42) defined by the following membership function

$$\mu_{\hat{Y}}(x) = \sup \left\{ s \in [0, 1] \mid I \left(s, \max_{i \in J} \mu_{F_i}(x) \right) \geq q_i \right\}$$

$$= \sigma[I] \left(r_J, \max_{i \in J} \mu_{F_i}(x) \right),$$

$$(26.43)$$

where we define $\sigma[I](a, b) = \sup\{s \in [0, 1] \mid I(s, b) \geq a\}$ for $a, b \in [0, 1]$. We have $X \subseteq \hat{Y}$. Because we have many $\bigcup_{i \in J} F_i$, the upper approximation $\mathcal{F}_{\sigma}^*(X)$ of X is defined by the following membership function

$$\mu_{\mathcal{F}_{\sigma}^*(X)}(x)$$

$$= \inf_{\mathcal{T} \subseteq \mathcal{F}} \sigma[I] \left(\inf_{y \in U} I \left(\mu_X(y), \sup_{F \in \mathcal{T}} \mu_F(y) \right), \right.$$

$$\left. \sup_{F \in \mathcal{T}} \mu_F(x) \right), \qquad (26.44)$$

where $\mathcal{F}$ can have infinitely many elementary fuzzy sets F. We note that $\sigma[I]$ becomes an implication function.

Because (26.33) is extended to (26.44), (26.34) is extended to the following equation in the sense that $\mathcal{F}_*^{\sigma}(X) = U - F_{\sigma}^*(U - X)$

$$\mu_{\mathcal{F}_*^{\sigma}(X)}(x)$$

$$= \sup_{\mathcal{T} \subseteq \mathcal{F}} n \left(\sigma[I] \left(\inf_{y \in U} I \left(n(\mu_X(y)), \sup_{F \in \mathcal{T}} \mu_F(y) \right), \right. \right.$$

$$\left. \left. \sup_{F \in \mathcal{T}} \mu_F(x) \right) \right),$$

$$(26.45)$$

where $\mu_{\mathcal{F}_*^{\sigma}(X)}$ is the membership function of the upper approximation $\mathcal{F}_*^{\sigma}(X)$ of X.

These four approximations were originally proposed by *Inuiguchi* and *Tanino* [26.10]. They selected a pair $(\mathcal{F}_*^\xi(X), \mathcal{F}_\xi^*(X))$ to define a rough set of X. However, in connection with the crisp case, *Inuiguchi* [26.37] selected pairs $(\mathcal{F}_*^\xi(X), \mathcal{F}_\sigma^*(X))$ and $(\mathcal{F}_*^\sigma(X), \mathcal{F}_\xi^*(X))$ for the definitions of rough sets of X. In this chapter, a pair $(\mathcal{F}_*^\xi(X), \mathcal{F}_\xi^*(X))$ is called a ξ-fuzzy rough set and a pair $(\mathcal{F}_*^\sigma(X), \mathcal{F}_\sigma^*(X))$ a σ-fuzzy rough set.

Properties

First, we show properties about the representations of lower and upper approximations defined by (26.39), (26.40), (26.44), and (26.45). We have the following equalities (see *Inuiguchi* [26.37])

$$\mu_{\mathcal{F}_*^\xi(X)}(x) = \sup\{\xi[I](\mu_F(x), h) \mid F \in \mathcal{F}, h \in [0, 1]$$

such that

$$\xi[I](\mu_F(y), h) \le \mu_X(y), \forall y \in U\}, \tag{26.46}$$

$$\mu_{\mathcal{F}_*^\sigma(X)}(x) = \sup\left\{n\left(\sigma[I]\left(h, \sup_{F \in \mathcal{T}} \mu_F(x)\right)\right)\right|$$

$$\mathcal{T} \subseteq \mathcal{F}, h \in [0, 1]$$

such that

$$\sigma[I]\left(h, \sup_{F \in \mathcal{T}} \mu_F(y)\right) \ge n(\mu_X(y)), \forall y \in U\right\}, \tag{26.47}$$

$$\mu_{\mathcal{F}_\xi^*(X)}(x) = \inf\{n(\xi[I](\mu_F(x), h)) \mid$$

$$F \in \mathcal{F}, h \in [0, 1]$$

such that

$$\xi[I](\mu_F(y), h) \le n(\mu_X(y)), \forall y \in U\}, \tag{26.48}$$

$$\mu_{\mathcal{F}_\sigma^*(X)}(x)$$

$$= \inf\left\{\sigma[I]\left(h, \sup_{F \in \mathcal{T}} \mu_F(x)\right) \mid \mathcal{T} \subseteq \mathcal{F}, h \in [0, 1]$$

such that

$$\sigma[I]\left(h, \sup_{F \in \mathcal{T}} \mu_F(y)\right) \ge \mu_X(y), \forall y \in U\right\}. \tag{26.49}$$

Using these equations, the following properties can be easily obtained:

$$\mathcal{F}_*^\xi(X) \subseteq X \subseteq \mathcal{F}_\xi^*(X),$$

$$\mathcal{F}_*^\sigma(X) \subseteq X \subseteq \mathcal{F}_\sigma^*(X), \tag{26.50}$$

$$\mathcal{F}_*^\xi(\emptyset) = \mathcal{F}_*^\sigma(\emptyset) = \emptyset,$$

$$\mathcal{F}_\xi^*(U) = \mathcal{F}_\sigma^*(U) = U, \tag{26.51}$$

$$X \subseteq Y \text{ implies } \mathcal{F}_*^\xi(X) \subseteq \mathcal{F}_*^\xi(Y),$$

$$X \subseteq Y \text{ implies } \mathcal{F}_*^\sigma(X) \subseteq \mathcal{F}_*^\sigma(Y), \tag{26.52}$$

$$X \subseteq Y \text{ implies } \mathcal{F}_\xi^*(X) \subseteq \mathcal{F}_\xi^*(Y),$$

$$X \subseteq Y \text{ implies } \mathcal{F}_\sigma^*(X) \subseteq \mathcal{F}_\sigma^*(Y), \tag{26.53}$$

$$\mathcal{F}_*^\xi(X \cup Y) \supseteq \mathcal{F}_*^\xi(X) \cup \mathcal{F}_*^\xi(Y),$$

$$\mathcal{F}_*^\sigma(X \cup Y) \supseteq \mathcal{F}_*^\sigma(X) \cup \mathcal{F}_*^\sigma(Y), \tag{26.54}$$

$$\mathcal{F}_\xi^*(X \cap Y) \subseteq \mathcal{F}_\xi^*(X) \cap \mathcal{F}_\xi^*(Y),$$

$$\mathcal{F}_\sigma^*(X \cap Y) \subseteq \mathcal{F}_\sigma^*(X) \cap \mathcal{F}_\sigma^*(Y), \tag{26.55}$$

$$\mathcal{F}_*^\xi(U - X) = U - \mathcal{F}_\xi^*(X),$$

$$\mathcal{F}_*^\sigma(U - X) = U - \mathcal{F}_\sigma^*(X). \tag{26.56}$$

Furthermore, we can prove the following properties (see *Inuiguchi* [26.37]):

⟨7⟩ The following assertions are valid:
(a) If $a > 0$, $b < 1$ imply $I(a, b) < 1$ and

$$\inf_{x \in U} \sup_{F \in \mathcal{F}} \mu_F(x) > 0,$$

then we have

$$\mathcal{F}_*^\xi(U) = U \text{ and } \mathcal{F}_\xi^*(\emptyset) = \emptyset.$$

(b) If $b < 1$ implies $I(1, b) < 1$ and

$$\inf_{x \in U} \sup_{F \in \mathcal{F}} \mu_F(x) = 1,$$

then we have

$$\mathcal{F}_*^\xi(U) = U \text{ and } \mathcal{F}_\xi^*(\emptyset) = \emptyset.$$

(c) If $a > 0$, $b < 1$ imply $I(a, b) < 1$ and $\forall x \in U$, $\exists F \in \mathcal{F}$ such that $\mu_F(x) < 1$, then we have $\mathcal{F}_*^\sigma(U) = U$ and $\mathcal{F}_\sigma^*(\emptyset) = \emptyset$.
(d) If $a > 0$ implies $I(a, 0) < 1$ and $\forall x \in U$, $\exists F \in \mathcal{F}$ such that $\mu_F(x) = 0$, then we have $\mathcal{F}_*^\sigma(U) = U$ and $\mathcal{F}_\sigma^*(\emptyset) = \emptyset$.

⟨8⟩ We have

$$\mathcal{F}_*^\sigma(X \cap Y) = \mathcal{F}_*^\sigma(X) \cap \mathcal{F}_*^\sigma(Y),$$

$$\mathcal{F}_\sigma^*(X \cup Y) = \mathcal{F}_\sigma^*(X) \cup \mathcal{F}_\sigma^*(Y). \tag{26.57}$$

Moreover, if $\forall a \in [0, 1]$, $I(a, a) = 1$ and $\forall F_i, F_j \in \mathcal{F}$, $F_i \ne F_j$, $F_i \cap F_j = \emptyset$, we have

$$\mathcal{F}_*^\xi(X \cap Y) = \mathcal{F}_*^\xi(X) \cap \mathcal{F}_*^\xi(Y),$$

$$\mathcal{F}_\xi^*(X \cup Y) = \mathcal{F}_\xi^*(X) \cup \mathcal{F}_\xi^*(Y). \tag{26.58}$$

⟨9⟩ We have

$$\mathcal{F}_*^\xi(\mathcal{F}_*^\xi(X)) = \mathcal{F}_*^\xi(X) ,$$
$$\mathcal{F}_*^\sigma(\mathcal{F}_*^\sigma(X)) = \mathcal{F}_*^\sigma(X) , \qquad (26.59)$$
$$\mathcal{F}_\xi^*(\mathcal{F}_\xi^*(X)) = \mathcal{F}_\xi^*(X) ,$$
$$\mathcal{F}_\sigma^*(\mathcal{F}_\sigma^*(X)) = \mathcal{F}_\sigma^*(X) . \qquad (26.60)$$

Inuiguchi and *Tanino* [26.10] first proposed this type of fuzzy rough sets. They demonstrated the advantage in approximation when P is reflexive and symmetric, I is Dienes implication, and T is minimum operation. *Inuiguchi* and *Tanino* [26.55] showed that by selection of a necessity measure expressible various inclusion situations, the approximations become better, i.e., the differences between lower and upper approximations satisfying (26.58) become smaller. Moreover, *Inuiguchi* and *Tanino* [26.56] applied these fuzzy rough sets to function approximation.

26.2.3 Relations Between Two Kinds of Fuzzy Rough Sets

Under the given fuzzy relations P and Q described in Sect. 26.2.1, we discuss the relations between two kinds of fuzzy rough sets. Families of fuzzy sets are defined by $\mathcal{P} = \{P(x) \mid x \in U\}$ and $\mathcal{Q} = \{Q(x) \mid x \in U\}$. We have the following assertions:

⟨10⟩ When P and Q are reflexive, $I(1, a) = a$, we have

$$P_*(X) \subseteq \mathcal{P}_*^\xi(X) , \quad \mathcal{Q}_\xi^*(X) \subseteq \bar{Q}^*(X) . \qquad (26.61)$$

⟨11⟩ When P and Q are reflexive, X is a crisp set, $a \leq b$ if and only if $I(a, b) = 1$ and $T(a, 1) = a$ for all $a \in [0, 1]$, we have

$$\mathcal{P}_\sigma^*(X) \subseteq P^*(X) , \quad \bar{Q}_*(X) \subseteq \mathcal{Q}_*^\sigma(X) . \qquad (26.62)$$

⟨12⟩ When P and Q are T-transitive, the following assertions are valid:
(a) When $T = \xi[I]$ is associative, we have

$$\mathcal{P}_*^\xi(X) \subseteq P_*(X) , \quad \bar{Q}^*(X) \subseteq \mathcal{Q}_\xi^*(X) . \qquad (26.63)$$

(b) When $T = \xi[\sigma[I]]$ and $\sigma[I](a, \sigma[I](b, c)) = \sigma[I](b, \sigma[I](a, c))$ for all $a, b, c \in [0, 1]$, we have

$$P^*(X) \subseteq \mathcal{P}_\sigma^*(X) , \quad \mathcal{Q}_*^\sigma(X) \subseteq \bar{Q}_*(X) . \qquad (26.64)$$

Here we define

$$\zeta[T](a, b) = \sup\{s \in [0, 1] \mid T(a, s) \leq b\} .$$

This functional ζ can produce an implication function from a conjunction function T. Note that $\zeta[\xi[I]] = I$ and $\xi[\zeta[T]] = I$ for upper semicontinuous I and lower semicontinuous T (see *Inuiguchi* and *Sakawa* [26.53]).

26.2.4 The Other Approximation-Oriented Fuzzy Rough Sets

Greco et al. [26.24, 25] proposed fuzzy rough sets corresponding to a gradual rule [26.57], *the more an object is in G, the more it is in X* with fuzzy sets G and X. Corresponding to this gradual rule, we may define the lower approximation $G_*^+(X)$ of X and the upper approximation $G_+^*(X)$ of X, respectively, by the following membership functions

$$\mu_{G_*^+(X)}(x) = \inf\{\mu_X(z) \mid z \in U, \ \mu_G(z) \geq \mu_G(x)\} , \qquad (26.65)$$

$$\mu_{G_+^*(X)}(x) = \sup\{\mu_X(z) \mid z \in U, \ \mu_G(z) \leq \mu_G(x)\} . \qquad (26.66)$$

When we have a gradual rule, *the less an object is in G, the more it is in X*, we define the lower approximation $G_*^-(X)$ of X and the upper approximation $G_-^*(X)$ of X, respectively, by the following membership functions

$$\mu_{G_*^-(X)}(x) = \inf\{\mu_X(z) \mid z \in U, \ \mu_G(z) \leq \mu_G(x)\} , \qquad (26.67)$$

$$\mu_{G_-^*(X)}(x) = \sup\{\mu_X(z) \mid z \in U, \ \mu_G(z) \geq \mu_G(x)\} . \qquad (26.68)$$

Moreover, when a complex gradual rule, *the more an object is in G^+ and the less it is in G^-, the more it is in X* is given, the lower approximation $G_*^\pm(X)$ and upper approximation $G_\pm^*(X)$ are defined, respectively, by the following equations

$$\mu_{G_*^\pm(X)}(x) = \inf\{\mu_X(z) \mid z \in U, \ \mu_G^+(z) \geq \mu_G^+(x),$$
$$\mu_G^-(z) \leq \mu_G^-(x)\} , \qquad (26.69)$$

$$\mu_{G_\pm^*(X)}(x) = \sup\{\mu_X(z) \mid z \in U, \ \mu_G^+(z) \leq \mu_G^+(x),$$
$$\mu_G^-(z) \geq \mu_G^-(x)\} , \qquad (26.70)$$

where we define $G = \{G^+, G^-\}$.

The fuzzy rough sets are defined by pairs of those lower and upper approximations. This approach is advantageous in (i) no logical connectives such as implication function, conjunction function, etc., are used and (ii) the fuzzy rough sets correspond to gradual rules (see *Greco* et al. [26.24, 25]). However, we need a background knowledge about the monotone properties between G (or $\mathcal{G}$) and X.

This approach can be seen from a viewpoint of modifier functions of fuzzy sets. A modifier function φ is generally a function from $[0, 1]$ to $[0, 1]$ [26.58]. Functions defined by $\varphi_1(x) = x^2$, $\varphi_2(x) = \sqrt{x}$ and $\varphi_3(x) = 1 - x$ are known as modifier functions corresponding to modifying words *very, more, or less* and *not*. Namely, given a fuzzy set A, we may define fuzzy sets *very A*, *more or less A* and *not-A* by the following membership functions

$$\mu_{\text{very } A}(x) = (\mu_A(x))^2 ,$$

$$\mu_{\text{more or less } A}(x) = \sqrt{\mu_A(x)} ,$$

$$\mu_{\text{not-}A}(x) = 1 - \mu_A(x) . \tag{26.71}$$

Such modifier functions are often used in approximate/fuzzy reasoning [26.59, 60], especially in the indirect method of fuzzy reasoning which is called also, truth value space method.

Namely, we may define the lower approximation $\Phi_*(X)$ of X and the upper approximation $\Phi^*(X)$ of X by means of a fuzzy set G by the following membership functions

$$\mu_{\Phi_*(X)}(x) = \varphi_*^{G \to X}(\mu_G(x)) ,$$

$$\mu_{\Phi^*(X)}(x) = \varphi_{G \to X}^*(\mu_G(x)) , \tag{26.72}$$

where $\Phi = \{\varphi_*^{G \to X}, \varphi_{G \to X}^*\}$ and modifier functions $\varphi_*^{G \to X}$ and $\varphi_{G \to X}^*$ are selected to satisfy

$$\varphi_*^{G \to X}(\mu_G(x)) \le \mu_X(x) ,$$

$$\varphi_{G \to X}^*(\mu_G(x)) \ge \mu_X(x) , \quad \forall x \in U . \tag{26.73}$$

Indeed,

$$\xi[I](\cdot, \inf_{y \in U} I(\mu_G(y), \mu_X(y)))$$

and

$$\sigma[I](\inf_{y \in U} I(\mu_X(y), \mu_G(y)), \cdot)$$

are modifier functions satisfying (26.73) and these are used to define $\mathcal{F}_*^\xi(X)$ and $\mathcal{F}_\sigma^*(X)$ in (26.39) and (26.44), respectively. We note that we consider multiple fuzzy sets $G = F \in \mathcal{F}$ in (26.39) and apply the union because we have

$$\xi[I](\mu_G, \inf_{y \in U} I(\mu_G(y), \mu_X(x)) \le \mu_X(x) ,$$

$$\forall x \in U \text{ for all } G \in \mathcal{F} .$$

Similarly we consider multiple fuzzy sets G defined by $\mu_G(x) = \sup_{F \in \mathcal{T} \subseteq \mathcal{F}} \mu_F(x)$, $x \in U$ in (26.44) and we apply the intersection because we have $\sigma[I](\inf_{y \in U} I(\mu_G(y), \mu_X(y)), \mu_G(x)) \ge \mu_X(x)$, $\forall x \in U$ for all those fuzzy sets G.

In the definitions of (26.65)–(26.68), the following modifier functions are used, respectively

$$\varphi_*^+(\alpha) = \sup\{\psi_*^+(\beta) \mid \beta \in [0, \alpha]\} ,$$

$$\varphi_+^*(\alpha) = \inf\{\psi_+^*(\beta) \mid \beta \in [\alpha, 1]\} , \tag{26.74}$$

$$\varphi_*^-(\alpha) = \sup\{\psi_*^-(\beta) \mid \beta \in [\alpha, 1]\} ,$$

$$\varphi_-^*(\alpha) = \inf\{\psi_-^*(\beta) \mid \beta \in [0, \alpha]\} , \tag{26.75}$$

where we define

$$\psi_*^+(\alpha) = \inf\{\mu_X(z) \mid z \in U, \ \mu_G(z) \ge \alpha\} , \tag{26.76}$$

$$\psi_+^*(\alpha) = \sup\{\mu_X(z) \mid z \in U, \ \mu_G(z) \le \alpha\} , \tag{26.77}$$

$$\psi_*^-(\alpha) = \inf\{\mu_X(z) \mid z \in U, \ \mu_G(z) \le \alpha\} , \tag{26.78}$$

$$\psi_-^*(\alpha) = \sup\{\mu_X(z) \mid z \in U, \ \mu_G(z) \ge \alpha\} , \tag{26.79}$$

with $\inf \emptyset = 0$ and $\sup \emptyset = 1$. We note that φ_*^+ and φ_+^* are monotonically increasing which φ_*^- and φ_-^* are monotonically decreasing. These monotonicities are imposed in order to fit the supposed gradual rules. However, such monotonicities do not hold for functions ψ_*^+, ψ_+^*, ψ_*^- and ψ_-^*. In the cases of (26.69) and (26.70), we should extend the modifier function to a generalized modifier function which is a function from $[0, 1] \times [0, 1]$ to $[0, 1]$ because we have two fuzzy sets in the premise of the corresponding gradual rule. The associated generalized modifier functions with (26.69) and (26.70) are obtained as

$$\varphi_*^\pm(\alpha_1, \alpha_2)$$
$$= \sup\{\psi_*^\pm(\beta_1, \beta_2) \mid \beta_1 \in [0, \alpha_1], \ \beta_2 \in [\alpha_2, 1]\} , \tag{26.80}$$

$$\varphi_\pm^*(\alpha_1, \alpha_2)$$
$$= \sup\{\psi_\pm^*(\beta_1, \beta_2) \mid \beta_1 \in [\alpha_1, 1], \ \beta_2 \in [0, \alpha_2]\} , \tag{26.81}$$

where we define

$$\psi_*^\pm(\beta_1, \beta_2)$$
$$= \inf\{\mu_X(z) \mid z \in U, \ \mu_G^+(z) \ge \beta_1, \mu_G^-(z) \le \beta_2\} , \tag{26.82}$$

$$\psi_\pm^*(\beta_1, \beta_2)$$
$$= \sup\{\mu_X(z) \mid z \in U, \ \mu_G^+(z) \le \beta_1, \mu_G^-(z) \ge \beta_2\} , \tag{26.83}$$

with $\inf \emptyset = 0$ and $\sup \emptyset = 1$. We note $\varphi_*^{\pm}$ and $\varphi_{\pm}^*$ are monotonically increasing in the first argument and monotonically decreasing in the second argument.

Moreover, when we do not have any background knowledge about the relation between G and X which is expressed by a gradual rule. We may define the lower approximation $G_*(X)$ and the upper approximation $G^*(X)$ by the following membership functions

$$\mu_{G_*(X)}(x) = \inf\{\mu_X(z) \mid z \in U, \ \mu_G(z) = \mu_G(x)\}, \tag{26.84}$$

$$\mu_{G^*(X)}(x) = \sup\{\mu_X(z) \mid z \in U, \ \mu_G(z) = \mu_G(x)\}. \tag{26.85}$$

The modifier functions associate with these approximations are obtained as

$$\varphi_*(\alpha) = \inf\{\mu_X(z) \mid z \in U, \ \mu_G(z) = \alpha\}, \tag{26.86}$$

$$\varphi^*(\alpha) = \sup\{\mu_X(z) \mid z \in U, \ \mu_G(z) = \alpha\}, \tag{26.87}$$

where we define $\inf \emptyset = 0$ and $\sup \emptyset = 1$. Equation (26.87) is same as the inverse truth qualification [26.59, 60] of X based on G.

We describe the properties of the approximations defined by (26.65) to (26.68). However. the other approximations defined by (26.69), (26.70), (26.84), and (26.85) have the similar results. We have the following properties for the approximations defined by (26.65) to (26.68) (see *Greco* et al. [26.25] for a part of these properties):

$$G_*^+(X) \subseteq X \subseteq G_+^*(X),$$

$$G_*^-(X) \subseteq X \subseteq G_-^*(X), \tag{26.88}$$

$$G_*^+(\emptyset) = G_+^*(\emptyset) = G_*^-(\emptyset) = G_-^*(\emptyset) = \emptyset, \tag{26.89}$$

$$G_*^+(U) = G_+^*(U) = G_*^-(U) = G_-^*(U) = U, \tag{26.90}$$

$$G_*^+(X \cap Y) = G_*^+(X) \cap G_*^+(Y),$$

$$G_*^-(X \cap Y) = G_*^-(X) \cap G_*^-(Y), \tag{26.91}$$

$$G_+^*(X \cup Y) = G_+^*(X) \cup G_+^*(Y),$$

$$G_-^*(X \cup Y) = G_-^*(X) \cup G_-^*(Y), \tag{26.92}$$

$$X \subseteq Y \text{ implies } G_*^+(X) \subseteq G_*^+(Y),$$

$$X \subseteq Y \text{ implies } G_+^*(X) \subseteq G_+^*(Y), \tag{26.93}$$

$$X \subseteq Y \text{ implies } G_*^-(X) \subseteq G_*^-(Y),$$

$$X \subseteq Y \text{ implies } G_-^*(X) \subseteq G_-^*(Y), \tag{26.94}$$

$$G_*^+(X \cup Y) \supseteq G_*^+(X) \cup G_*^+(Y),$$

$$G_+^*(X \cap Y) \subseteq G_+^*(X) \cap G_+^*(Y), \tag{26.95}$$

$$G_*^-(X \cup Y) \supseteq G_*^-(X) \cup G_*^-(Y),$$

$$G_-^*(X \cap Y) \subseteq G_-^*(X) \cap G_-^*(Y), \tag{26.96}$$

$$G_*^+(U \backslash X) = U \backslash G_-^*(X) = U \backslash (U \backslash G)_+^*(X)$$
$$= (U \backslash G)_*^-(U \backslash X), \tag{26.97}$$

$$G_*^-(U \backslash X) = U \backslash G_+^*(X) = U \backslash (U \backslash G)_-^*(X)$$
$$= (U \backslash G)_*^+(U \backslash X), \tag{26.98}$$

$$G_*^+(G_*^+(X)) = G_+^*(G_*^+(X)) = G_*^+(X),$$

$$G_+^*(G_+^*(X)) = G_+^*(G_+^*(X)) = G_+^*(X), \tag{26.99}$$

$$G_*^-(G_*^-(X)) = G_-^*(G_*^-(X)) = G_*^-(X),$$

$$G_-^*(G_-^*(X)) = G_-^*(G_-^*(X)) = G_-^*(X), \tag{26.100}$$

where $U \backslash X$ is a fuzzy set defined by its membership function $\mu_{U \backslash X}(x) = N(\mu_X(x))$, $\forall x \in U$ with a strictly decreasing function $N: [0, 1] \to [0, 1]$. We found that all fundamental properties [26.2] of the classical rough set are preserved.

26.2.5 Remarks

Three types of fuzzy rough set models have been described, divided into two groups: classification-oriented and approximation-oriented models. The classification-oriented fuzzy rough set models are much more investigated by many researchers. However, the approximation-oriented fuzzy rough set models would be more important because they are associated with rules. While approximation-oriented fuzzy rough set models based on modifiers are strongly related to the gradual rules, approximation-oriented fuzzy rough set models based on certainty qualification have relations to uncertain generation rule (uncertain qualification rule: certainty rule and possibility rule) [26.54], i. e., a rule such as *the more an object is in A, the more certain (possible) it is in B*. While approximation-oriented fuzzy rough set models based on modifiers need a modifier function for each granule G, the approximation-oriented fuzzy rough set models based on certainty qualification need only a degree of inclusion for each granule F. Therefore, the latter may work well for data compression such as image compression, speech compression, and so on.

26.3 Generalized Fuzzy Belief Structures with Application in Fuzzy Information Systems

In rough set theory there exists a pair of approximation operators, the lower and upper approximations, whereas in the Dempster–Shafer theory of evidence there exists a dual pair of uncertainty measures, the belief and plausibility functions. In this section, general types of belief structures and their induced dual pairs of belief and plausibility functions are first introduced. Relationships between belief and plausibility functions in the Dempster–Shafer theory of evidence and the lower and upper approximations in rough set theory are then established. It is shown that the probabilities of lower and upper approximations induced from an approximation space yield a dual pair of belief and plausibility functions. And for any belief structure there must exist a probability approximation space such that the belief and plausibility functions defined by the given belief structure are, respectively, the lower and upper probabilities induced by the approximation space. The pair of lower and upper approximations of a set capture the non-numeric aspect of uncertainty of the set which can be interpreted as the qualitative representation of the set, whereas the pair of the belief and plausibility measures of the set capture the numeric aspect of uncertainty of the set which can be treated as the quantitative characterization of the set. Finally, the potential applications of the main results to intelligent information systems are explored.

26.3.1 Belief Structures and Belief Functions

In this section, we recall some basic notions related to belief structures with their induced belief and plausibility functions.

Belief and Plausibility Functions Derived from a Crisp Belief Structure

Throughout this section, U will be a nonempty set called the universe of discourse. The class of all subsets (respectively, fuzzy subsets) of U will be denoted by $\mathcal{P}(U)$ (respectively, by $\mathcal{F}(U)$). For any $A \in \mathcal{F}(U)$, the complement of A will be denoted by $\approx A$, i.e.,

$$(\approx A)(x) = 1 - A(x) \text{ for all } x \in U .$$

The basic representational structure in the Dempster–Shafer theory of evidence is a belief structure.

Definition 26.1

Let U be a nonempty set which may be infinite, a set function $m: \mathcal{P}(U) \to [0, 1]$ is referred to as a crisp basic probability assignment if it satisfies axioms (M1) and (M2)

$$(\text{M1}) \ m(\emptyset) = 0, \quad (\text{M2}) \sum_{X \subseteq U} m(X) = 1 .$$

A set $X \in \mathcal{P}(U)$ with nonzero basic probability assignment is referred to as a focal element of m. We denote by $\mathcal{M}$ the family of all focal elements of m. The pair $(\mathcal{M}, m)$ is called a crisp belief structure on U.

Lemma 26.1

Let $(\mathcal{M}, m)$ be a crisp belief structure on U. Then the focal elements of m constitute a countable set.

Proof: For any $n \in \{1, 2, \ldots\}$, let

$$H_n = \{A \in \mathcal{M} | m(A) > 1/n\} .$$

By axiom (M2) we can see that for each $n \in \{1, 2, \ldots\}$, H_n is a finite set. Since $\mathcal{M} = \bigcup_{n=1}^{\infty} H_n$, we conclude that $\mathcal{M}$ is countable. ∎

Associated with each belief structure, a pair of belief and plausibility functions can be defined.

Definition 26.2

Let $(\mathcal{M}, m)$ be a crisp belief structure on U. A set function Bel: $\mathcal{P}(U) \to [0, 1]$ is called a CC-belief function on U if

$$\text{Bel}(X) = \sum_{M \subseteq X} m(M), \quad \forall X \in \mathcal{P}(U) . \tag{26.101}$$

A set function Pl: $\mathcal{P}(U) \to [0, 1]$ is called a CC-plausibility function on U if

$$\text{Pl}(X) = \sum_{M \cap X \neq \emptyset} m(M), \quad \forall X \in \mathcal{P}(U) . \tag{26.102}$$

Remark 26.1

Since $\mathcal{M}$ is a countable set, the change of convergence may not change the values of the infinite (countable) sums in (26.101) and (26.102). Therefore, Definition 26.2 is reasonable.

The CC-belief function and CC-plausibility function based on the same belief structure are connected by the dual property

$$Pl(X) = 1 - Bel(\approx X), \quad \forall X \in \mathcal{P}(U) \qquad (26.103)$$

and moreover,

$$Bel(X) \leq Pl(X), \quad \forall X \in \mathcal{P}(U). \qquad (26.104)$$

When U is finite, a CC-belief function can be equivalently defined as a monotone *Choquet* capacity [26.61] on U which satisfies the following properties [26.38]:

(MC1) $Bel(\emptyset) = 0$,
(MC2) $Bel(U) = 1$,
(MC3) for all $X_i \in \mathcal{P}(U), i = 1, 2, \ldots, k$,

$$Bel\left(\bigcup_{i=1}^{k} X_i\right) \geq$$

$$\sum_{\emptyset \neq J \subseteq \{1,2,\ldots,k\}} (-1)^{|J|+1} Bel\left(\bigcap_{i \in J} X_i\right).$$

$$(26.105)$$

Similarly, a CC-plausibility function can be equivalently defined as an alternating Choquet capacity on U which satisfies the following properties:

(AC1) $Pl(\emptyset) = 0$,
(AC2) $Pl(U) = 1$,
(AC3) for all $X_i \in \mathcal{P}(U), i = 1, 2, \ldots, k$,

$$Pl\left(\bigcap_{i=1}^{k} X_i\right) \leq$$

$$\sum_{\emptyset \neq J \subseteq \{1,2,\ldots,k\}} (-1)^{|J|+1} Pl\left(\bigcup_{i \in J} X_i\right). \qquad (26.106)$$

A monotone Choquet capacity is a belief function in which the basic probability assignment can be calculated by using the Möbius transform

$$m(X) = \sum_{Y \subseteq X} (-1)^{|X \setminus Y|} Bel(Y), X \in \mathcal{P}(U). \qquad (26.107)$$

A crisp belief structure can also be induced by a dual pair of fuzzy belief and plausibility functions.

Definition 26.3
Let $(\mathcal{M}, m)$ be a crisp belief structure on U. A fuzzy set function Bel: $\mathcal{F}(U) \to [0, 1]$ is called a CF-belief function on U if

$$Bel(X) = \sum_{A \in \mathcal{M}} m(A) N_A(X), \quad \forall X \in \mathcal{F}(U).$$

$$(26.108)$$

A fuzzy set function Pl: $\mathcal{F}(U) \to [0, 1]$ is called a CF-plausibility function on U if

$$Pl(X) = \sum_{A \in \mathcal{M}} m(A) \Pi_A(X), \quad \forall X \in \mathcal{F}(U), \qquad (26.109)$$

where $N_A: \mathcal{F}(U) \to [0, 1]$ and $\Pi_A: \mathcal{F}(U) \to [0, 1]$ are, respectively, the necessity measure and the possibility measure determined by the crisp set A defined as follows

$$N_A(X) = \bigwedge_{u \in A} X(u), \quad X \in \mathcal{F}(U), \qquad (26.110)$$

$$\Pi_A(X) = \bigvee_{u \in A} X(u), \quad X \in \mathcal{F}(U). \qquad (26.111)$$

Belief and Plausibility Functions Derived from a Fuzzy Belief Structure

Definition 26.4
Let U be a nonempty set which may be infinite. A set function $m: \mathcal{F}(U) \to [0, 1]$ is referred to as a fuzzy basic probability assignment, if it satisfies axioms (FM1) and (FM2)

(FM1) $m(\emptyset) = 0$,

(FM2) $\sum_{X \in \mathcal{F}(U)} m(X) = 1$.

A fuzzy set $X \in \mathcal{F}(U)$ with $m(X) > 0$ is referred to as a focal element of m. We denote by $\mathcal{M}$ the family of all focal elements of m. The pair $(\mathcal{M}, m)$ is called a fuzzy belief structure.

Lemma 26.2
[26.62] Let $(\mathcal{M}, m)$ be a fuzzy belief structure on W. Then the focal elements of m constitute a countable set.

In the discussion to follow, all the focal elements are supposed to be normal, i. e., for any $A \in \mathcal{M}$, there exists an $x \in U$ such that $A(x) = 1$. Associated with the fuzzy belief structure $(\mathcal{M}, m)$, two pairs of fuzzy belief and plausibility functions can be derived.

Definition 26.5

Let U be a nonempty set which may be infinite, and $(\mathcal{M}, m)$ a fuzzy belief structure on U. A crisp set function Bel: $\mathcal{P}(U) \rightarrow [0, 1]$ is referred to as a FC-belief function on U if

$$\text{Bel}(X) = \sum_{A \in \mathcal{M}} m(A) N_A(X), \quad \forall X \in \mathcal{P}(U).$$

$$(26.112)$$

A crisp set function Pl: $\mathcal{P}(U) \rightarrow [0, 1]$ is called a FC-plausibility function on U if

$$\text{Pl}(X) = \sum_{A \in \mathcal{M}} m(A) \Pi_A(X), \quad \forall X \in \mathcal{P}(U), \quad (26.113)$$

where $N_A: \mathcal{P}(U) \rightarrow [0, 1]$ and $\Pi_A: \mathcal{P}(U) \rightarrow [0, 1]$ are, respectively, the necessity measure and the possibility measure determined by the fuzzy set A defined as follows

$$N_A(X) = \bigwedge_{u \notin X} (1 - A(u)), X \in \mathcal{P}(U) \quad (26.114)$$

$$\Pi_A(X) = \bigvee_{u \in X} A(u), X \in \mathcal{P}(U). \quad (26.115)$$

Definition 26.6

Let U be a nonempty set which may be infinite, and $(\mathcal{M}, m)$ a fuzzy belief structure on U. A fuzzy set function Bel: $\mathcal{F}(U) \rightarrow [0, 1]$ is referred to as a FF-belief function on U if

$$\text{Bel}(X) = \sum_{A \in \mathcal{M}} m(A) N_A(X), \quad \forall X \in \mathcal{F}(U).$$

$$(26.116)$$

A fuzzy set function Pl: $\mathcal{F}(U) \rightarrow [0, 1]$ is called a FF-plausibility function on U if

$$\text{Pl}(X) = \sum_{A \in \mathcal{M}} m(A) \Pi_A(X), \quad \forall X \in \mathcal{F}(U). \quad (26.117)$$

Where $N_A: \mathcal{F}(U) \rightarrow [0, 1]$ and $\Pi_A: \mathcal{F}(U) \rightarrow [0, 1]$ are, respectively, the necessity measure and the possibility measure determined by the fuzzy set A defined as follows

$$N_A(X) = \bigwedge_{u \in U} (X(u) \vee (1 - A(u))), X \in \mathcal{F}(U),$$

$$(26.118)$$

$$\Pi_A(X) = \bigvee_{u \in U} (X(u) \wedge A(u)), X \in \mathcal{F}(U). \quad (26.119)$$

It can be proved that the belief and plausibility functions derived from the same fuzzy belief structure $(\mathcal{M}, m)$ are dual, that is,

$$\text{Bel}(X) = 1 - \text{Pl}(\approx X), \quad \forall X \in \mathcal{F}(U). \quad (26.120)$$

And

$$\text{Bel}(X) \leq \text{Pl}(X), \quad \forall X \in \mathcal{F}(U). \quad (26.121)$$

Moreover, Bel is a fuzzy monotone Choquet capacity of infinite order on U which satisfies axioms (FMC1)–(FMC3),

(FMC1) Bel($\emptyset$) = 0,
(FMC2) Bel(U) = 1,
(FMC3) For $X_i \in \mathcal{F}(U), i = 1, 2, \ldots, n, n \in \mathbf{N}$,

$$\text{Bel}\left(\bigcup_{i=1}^{n} X_i\right) \geq$$

$$\sum_{\emptyset \neq J \subseteq \{1,2,\ldots,n\}} (-1)^{|J|+1} \text{Bel}\left(\bigcap_{j \in J} X_j\right).$$

$$(26.122)$$

And Pl is a fuzzy alternating Choquet capacity of infinite order on U which obeys axioms (FAC1)–(FAC3),

(FAC1) Pl($\emptyset$) = 0,
(FAC2) Pl(U) = 1,
(FAC3) For $X_i \in \mathcal{F}(U), i = 1, 2, \ldots, n, n \in \mathbf{N}$,

$$\text{Pl}\left(\bigcap_{i=1}^{n} X_i\right) \leq \sum_{\emptyset \neq J \subseteq \{1,2,\ldots,n\}} (-1)^{|J|+1} \text{Pl}\left(\bigcup_{i \in J} X_i\right).$$

$$(26.123)$$

26.3.2 Belief Structures of Rough Approximations

In this section, we show relationships between various belief and plausibility functions in Dempster–Shafer theory of evidence and the lower and upper approximations in rough set theory with potential applications.

Belief Functions
versus Crisp Rough Approximations

Definition 26.7

Let U and W be two nonempty universes of discourse. A subset $R \in \mathcal{P}(U \times W)$ is referred to as a binary relation from U to W. The relation R is referred to as serial if for any $x \in U$ there exists $y \in W$ such that $(x, y) \in R$. If $U = W$, $R \in \mathcal{P}(U \times U)$ is called a binary relation on U, $R \in \mathcal{P}(U \times U)$ is referred to as reflexive if $(x, x) \in R$ for all $x \in U$; R is referred to as symmetric if $(x, y) \in R$ implies $(y, x) \in R$ for all $x, y \in U$; R is referred to as transitive if for any $x, y, z \in U$, $(x, y) \in R$ and $(y, z) \in R$ imply $(x, z) \in R$; R is referred to as Euclidean if for any $x, y, z \in U$, $(x, y) \in R$ and $(x, z) \in R$ imply $(y, z) \in R$; R is referred to as an equivalence relation if R is reflexive, symmetric and transitive.

Assume that R is an arbitrary binary relation from U to W. One can define a set-valued mapping $R_s \colon U \to \mathcal{P}(W)$ by

$$R_s(x) = \{y \in W | (x, y) \in R\}, \quad x \in U . \qquad (26.124)$$

$R_s(x)$ is called the successor neighborhood of x with respect to R [26.63]. Obviously, any set-valued mapping F from U to W defines a binary relation from U to W by setting $R = \{(x, y) \in U \times W | y \in F(x)\}$. For $A \in \mathcal{P}(W)$, let $j(A) = R_s^{-1}(A)$ be the counter-image of A under the set-valued mapping R_s, i.e.,

$$j(A) = \begin{cases} R_s^{-1}(A) = \{u \in U | R_s(u) = A\}, \\ \quad \text{if } A \in \{R_s(x) | x \in U\} , \\ \emptyset, \quad \text{otherwise} . \end{cases} \qquad (26.125)$$

Then it is well known that j satisfies the properties (J1) and (J2)

(J1) $A \neq B \Longrightarrow j(A) \cap j(B) = \emptyset$,

(J2) $\displaystyle\bigcup_{A \in \mathcal{P}(W)} j(A) = U$.

Definition 26.8

If R is an arbitrary relation from U to W, then the triple (U, W, R) is referred to as a generalized approximation space. For any set $A \subseteq W$, a pair of lower and upper approximations, $\underline{R}(A)$ and $\overline{R}(A)$, are, respectively, defined by

$$\underline{R}(A) = \{x \in U | R_s(x) \subseteq A\},$$
$$\overline{R}(A) = \{x \in U | R_s(x) \cap A \neq \emptyset\} . \qquad (26.126)$$

The pair $(\underline{R}(A), \overline{R}(A))$ is referred to as a generalized crisp rough set and $\underline{R}$ and $\overline{R} \colon \mathcal{P}(W) \to \mathcal{P}(U)$ are called the lower and upper generalized approximation operators, respectively.

If U is countable set, P a normalized probability measure on U, i.e., $P(\{x\}) > 0$ for all $x \in U$, and R an arbitrary relation from U to W, then $((U, P), W, R)$ is referred to as a probability approximation space.

Theorem 26.1 [26.43]

Assume that $((U, P), W, R)$ is a serial probability approximation space, for $X \in \mathcal{P}(W)$, define

$$m(X) = P(j(X)),$$
$$\mathrm{Bel}(X) = P(\underline{R}(X)),$$
$$\mathrm{Pl}(X) = P(\overline{R}(X)) . \qquad (26.127)$$

Then $m \colon \mathcal{P}(W) \to [0, 1]$ is a basic probability assignment on W and $\mathrm{Bel} \colon \mathcal{P}(W) \to [0, 1]$ and $\mathrm{Pl} \colon \mathcal{P}(W) \to [0, 1]$ are, respectively, the CC-belief and CC-plausibility functions on W.

Conversely, for any crisp belief structure $(\mathcal{M}, m)$ on W which may be infinite. If $\mathrm{Bel} \colon \mathcal{P}(W) \to [0, 1]$ and $\mathrm{Pl} \colon \mathcal{P}(W) \to [0, 1]$ are, respectively, the CC-belief and CC-plausibility functions defined in Definition 26.2, then there exists a countable set U, a serial relation R from U to W, and a normalized probability measure P on U such that

$$\mathrm{Bel}(X) = P(\underline{R}(X)),$$
$$\mathrm{Pl}(X) = P(\overline{R}(X)), \quad \forall X \in \mathcal{P}(W) . \qquad (26.128)$$

The notion of information systems (sometimes called data tables, attribute-value systems, knowledge representation systems etc.) provides a convenient tool for the representation of objects in terms of their attribute values.

An information system is a pair (U, A), where $U = \{x_1, x_2, \ldots, x_n\}$ is a nonempty, finite set of objects called the universe of discourse and $A = \{a_1, a_2, \ldots, a_m\}$ is a nonempty, finite set of attributes, such that $a \colon U \to V_a$ for any $a \in A$, where V_a is called the domain of a.

Each nonempty subset $B \subseteq A$ determines an indiscernibility relation as follows

$$R_B = \{(x, y) \in U \times U | a(x) = a(y), \forall a \in B\} . \qquad (26.129)$$

Since R_B is an equivalence relation on U, it forms a partition $U/R_B = \{[x]_B | x \in U\}$ of U, where $[x]_B$ denotes the equivalence class determined by x with respect to (w.r.t.) B, i.e., $[x]_B = \{y \in U | (x, y) \in R_B\}$.

Let (U, A) be an information system, $B \subseteq A$, for any $X \subseteq U$, denote

$$\underline{R_B}(X) = \{x \in U | [x]_B \subseteq X\} ,$$
$$\overline{R_B}(X) = \{x \in U | [x]_B \cap X \neq \emptyset\} , \qquad (26.130)$$

where $\underline{R_B}(X)$ and $\overline{R_B}(X)$ are, respectively, referred to as the lower and upper approximations of X w.r.t. (U, R_B), the knowledge generated by B. Objects in $\underline{R_B}(X)$ can be certainty classified as elements of X on the basis of knowledge in (U, R_B), whereas objects in $\overline{R_B}(X)$ can only be classified possibly as elements of X on the basis of knowledge in (U, R_B)).

For $B \subseteq A$ and $X \subseteq U$, denote $\mathrm{Bel}_B(X) = P(\underline{R_B}(X))$ and $\mathrm{Pl}_B(X) = P(\overline{R_B}(X))$, where $P(Y) = |Y|/|U|$ and $|Y|$ is the cardinality of a set Y. Then Bel_B and Pl_B are CC-belief function and CC-plausibility function on U, respectively, and the corresponding mass distribution is

$$m_B(Y) = \begin{cases} P(Y), & \text{if } Y \in U/R_B , \\ 0, & \text{otherwise} . \end{cases}$$

A decision system (sometimes called decision table) is a pair $(U, C \cup \{d\})$ where (U, C) is an information system, and d is a distinguished attribute called the decision; in this case C is called the conditional attribute set, d is a map $d : U \rightarrow V_d$ of the universe U into the value set V_d, we assume, without any loss of generality, that $V_d = \{1, 2, \ldots, r\}$. Define

$$R_d = \{(x, y) \in U \times U | d(x) = d(y)\} . \tag{26.131}$$

Then we obtain the partition $U/R_d = \{D_1, D_2, \ldots, D_r\}$ of U into decision classes, where $D_j = \{x \in U | d(x) = j\}, j \leq r$. If $R_C \subseteq R_d$, then the decision system $(U, C \cup \{d\})$ is consistent, otherwise it is inconsistent. One can acquire certainty decision rules from consistent decision systems and uncertainty decision rules from inconsistent decision systems.

Belief Functions versus Rough Fuzzy Approximations

Definition 26.9
Let (U, W, R) be a generalized approximation space, for a fuzzy set $A \in \mathcal{F}(W)$, the lower and upper approximations of A, $\underline{RF}(A)$ and $\overline{RF}(A)$, with respect to the approximation space (U, W, R) are fuzzy sets of U whose membership functions, for each $x \in U$, are defined, respectively, by

$$\overline{RF}(A)(x) = \bigvee_{y \in R_s(x)} A(y), \qquad x \in U , \tag{26.132}$$

$$\underline{RF}(A)(x) = \bigwedge_{y \in R_s(x)} A(y), \qquad x \in U . \tag{26.133}$$

The pair $(\underline{RF}(A), \overline{RF}(A))$ is referred to as a generalized rough fuzzy set, and $\underline{RF}$ and $\overline{RF} : \mathcal{F}(W) \rightarrow \mathcal{F}(U)$ are

referred to as lower and upper generalized rough fuzzy approximation operators, respectively.

In the discussion to follow, we always assume that $(U, \mathcal{A}, P)$ is a probability space, i.e., U is a nonempty set, $\mathcal{A} \subseteq \mathcal{P}(U)$ a σ-algebra on U, and P a probability measure on U.

Definition 26.10
A fuzzy set $A \in \mathcal{F}(U)$ is said to be measurable w.r.t. $(U, \mathcal{A})$ if $A : U \rightarrow [0, 1]$ is a measurable function w.r.t. $\mathcal{A} - \mathcal{B}([0, 1])$, where $\mathcal{B}([0, 1])$ is the family of Borel sets on $[0, 1]$. We denote by $\mathcal{F}(U, \mathcal{A})$ the family of all measurable fuzzy sets of U w.r.t. $\mathcal{A} - \mathcal{B}([0, 1])$.

For any measurable fuzzy set $A \in \mathcal{F}(U, \mathcal{A})$, since $A_\alpha \in \mathcal{A}$ for all $\alpha \in [0, 1]$, A_α is a measurable set on the probability space $(U, \mathcal{A}, P)$ and then $P(A_\alpha) \in [0, 1]$. Note that $f(\alpha) = P(A_\alpha)$ is monotone decreasing and left continuous, it can be seen that $f(\alpha)$ is integrable, we denote the integrand as $\int_0^1 P(A_\alpha) d\alpha$.

Definition 26.11
If a fuzzy set A is measurable w.r.t. $(U, \mathcal{A})$, and P is a probability measure on $(U, \mathcal{A})$. Denote

$$P(A) = \int_0^1 P(A_\alpha) d\alpha , \tag{26.134}$$

$P(A)$ is called the probability of A.

For a singleton set $\{x\}$, we will write $P(x)$ instead of $P(\{x\})$ for short.

Proposition 26.1
[26.21, 64] The fuzzy probability measure P in Definition 26.11 satisfies the following properties:

(1) $P(A) \in [0, 1]$ and $P(A) + P(\approx A) = 1$, for all $A \in \mathcal{F}(U, \mathcal{A})$.
(2) P is countably additive, i.e., for $A_i \in \mathcal{F}(U, \mathcal{A})$, $i = 1, 2, \ldots, A_i \cap A_j = \emptyset, \forall i \neq j$, then

$$P\left(\bigcup_{i=1}^{\infty} A_i\right) = \sum_{i=1}^{\infty} P(A_i) . \tag{26.135}$$

(3) $A, B \in \mathcal{F}(U, \mathcal{A}), A \subseteq B \Longrightarrow P(A) \leq P(B)$.

(4) If $U = \{u_i | i = 1, 2, \ldots\}$ is an infinite countable set and $\mathcal{A} = \mathcal{P}(U)$, then for all $A \in \mathcal{F}(U)$,

$$P(A) = \int_0^1 P(A_\alpha) d\alpha = \sum_{x \in U} A(x) P(x) . \quad (26.136)$$

(5) If U is a finite set with $|U| = n$, $\mathcal{A} = \mathcal{P}(U)$, and $P(u) = 1/n$, then $P(A) = \int_0^1 P(A_\alpha) d\alpha = |A|/n$ for all $A \in \mathcal{P}(U)$.

Theorem 26.2

Assume that $((U, P), W, R)$ is a serial probability approximation space, for $X \in \mathcal{F}(W)$, define

$$m(X) = P(j(X)) ,$$
$$\mathrm{Bel}(X) = P(\underline{RF}(X)) ,$$
$$\mathrm{Pl}(X) = P(\overline{RF}(X)) . \quad (26.137)$$

Then $m \colon \mathcal{P}(W) \to [0, 1]$ is a basic probability assignment on W and $\mathrm{Bel} \colon \mathcal{F}(W) \to [0, 1]$ and $\mathrm{Pl} \colon \mathcal{F}(W) \to [0, 1]$ are, respectively, the CF-belief and CF-plausibility functions on W.

Conversely, for any crisp belief structure $(\mathcal{M}, m)$ on W which may be infinite. If $\mathrm{Bel} \colon \mathcal{F}(W) \to [0, 1]$ and $\mathrm{Pl} \colon \mathcal{F}(W) \to [0, 1]$ are, respectively, the CF- belief and CF-plausibility functions defined in Definition 26.3, then there exists a countable set U, a serial relation R from U to W, and a normalized probability measure P on U such that

$$\mathrm{Bel}(X) = P(\underline{RF}(X)) ,$$
$$\mathrm{Pl}(X) = P(\overline{RF}(X)) , \quad \forall X \in \mathcal{F}(W). \quad (26.138)$$

For a decision table $(U, C \cup \{d\})$, where $V_d = \{d_1, d_2, \ldots, d_r\}$, d is called a fuzzy decision if, for each $x \in U$, $d(x)$ is a fuzzy subset of V_d, i. e., $d \colon U \to \mathcal{F}(V_d)$, with no lose of generality, we represent d as follows

$$d(x_i) = d_{i1}/d_1 + d_{i2}/d_2 + \cdots + d_{ir}/d_r, i$$
$$= 1, 2, \ldots, n, \quad (26.139)$$

where $d_{ij} \in [0, 1]$. In this case, $(U, C \cup \{d\})$ is called an information system with fuzzy decision. For the fuzzy decision d, we define a fuzzy indiscernibility binary relation R_d on U as follows: For $i, k = 1, 2, \ldots, n$

$$R_d(x_i, x_k) = \min\{1 - |d_{ij} - d_{kj}| | j = 1, 2, \ldots, r\} . \quad (26.140)$$

Then, we obtain a fuzzy similarity class $S_d(x)$ of $x \in U$ in the system $(U, C \cup \{d\})$ as follows

$$S_d(x)(y) = R_d(x, y), \quad y \in U . \quad (26.141)$$

Since $S_d(x)(x) = R_d(x, x) = 1$, we see that $S_d(x) \colon U \to [0, 1]$ is a normalized fuzzy set of U. Denote by U/R_d the fuzzy similarity classes induced by the fuzzy decision d, i. e.

$$U/R_d = \{S_d(x) | x \in U\} . \quad (26.142)$$

For $B \subseteq C$ and $X \in \mathcal{F}(U)$, we define the lower and upper approximations of X w.r.t. (U, R_B) as follows

$$\underline{RF_B}(X)(x) = \bigwedge_{y \in S_B(x)} X(y), \quad x \in U ,$$
$$\overline{RF_B}(X)(x) = \bigvee_{y \in S_B(x)} X(y), \quad x \in U .$$
$$\quad (26.143)$$

Theorem 26.3

Let $(U, C \cup \{d\})$ be an information system with fuzzy decision. For $B \subseteq C$ and $X \in \mathcal{F}(U)$, if $\underline{RF_B}(X)$ and $\overline{RF_B}(X)$ are, respectively, the lower and upper approximations of X w.r.t. (U, R_B) defined by Definition 26.9, denote

$$\mathrm{Bel}_B(X) = P(\underline{RF_B}(X)) ,$$
$$\mathrm{Pl}_B(X) = P(\overline{RF_B}(X)) , \quad (26.144)$$

where $P(X) = \sum_{x \in U} X(x)/|U|$ for $X \in \mathcal{F}(U)$, then $\mathrm{Bel}_B \colon \mathcal{F}(U) \to [0, 1]$ and $\mathrm{Pl}_B \colon \mathcal{F}(U) \to [0, 1]$ are, respectively, a CF-belief function and a CF-plausibility function on U, and the corresponding basic probability assignment m_B is

$$m_B(Y) = \begin{cases} P(Y) = |Y|/|U|, & \text{if } Y \in U/R_B , \\ 0, & \text{otherwise} . \end{cases}$$
$$\quad (26.145)$$

Belief Functions versus Fuzzy Rough Approximations

Definition 26.12

Let U and W be two nonempty universes of discourse. A fuzzy subset $R \in \mathcal{F}(U \times W)$ is referred to as a binary relation from U to W, $R(x, y)$ is the degree of relation between x and y, where $(x, y) \in U \times W$. The fuzzy relation R is referred to as serial if for each $x \in U$, $\bigvee_{y \in W} R(x, y) = 1$. If $U = W$, $R \in \mathcal{F}(U \times U)$ is called a fuzzy binary relation on U, R is referred to as a reflexive fuzzy relation if $R(x, x) = 1$ for all $x \in U$; R is referred to as a symmetric fuzzy relation if $R(x, y) = R(y, x)$ for all $x, y \in U$; R is referred to as a transitive fuzzy relation if $R(x, z) \geq \vee_{y \in U}(R(x, y) \wedge R(y, z))$ for all $x, z \in U$; R is referred to as an equivalence fuzzy relation if it is reflexive, symmetric, and transitive.

Definition 26.13

Let U and W be two nonempty universes of discourse and R a fuzzy relation from U to W. The triple (U, W, R) is called a generalized fuzzy approximation space. For any set $A \in \mathcal{F}(W)$, the lower and upper approximations of A, $\underline{FR}(A)$ and $\overline{FR}(A)$, with respect to the approximation space (U, W, R) are fuzzy sets of U whose membership functions, for each $x \in U$, are defined, respectively, by

$$\overline{FR}(A)(x) = \bigvee_{y \in W} [R(x, y) \wedge A(y)], \quad x \in U,$$

$$\underline{FR}(A)(x) = \bigwedge_{y \in W} [(1 - R(x, y)) \vee A(y)], \quad x \in U.$$

(26.146)

The pair $(\underline{FR}(A), \overline{FR}(A))$ is referred to as a generalized fuzzy rough set, and $\underline{FR}$ and $\overline{FR}$: $\mathcal{F}(W) \to \mathcal{F}(U)$ are referred to as lower and upper generalized fuzzy rough approximation operators, respectively.

Theorem 26.4

Let (U, W, R) be a serial fuzzy approximation space in which U is a countable set and P a probability measure on U. If $\underline{FR}$ and $\overline{FR}$ are the fuzzy rough approximation operators defined in Definition 26.13, denote

$$Bel(X) = P(\underline{FR}(X)),$$
$$Pl(X) = P(\overline{FR}(X)), \quad X \in \mathcal{F}(W).$$

(26.147)

Then Bel: $\mathcal{F}(W) \to [0, 1]$ and Pl: $\mathcal{F}(W) \to [0, 1]$ are, respectively, FF-fuzzy belief and FF-plausibility functions on W.

Conversely, if $(\mathcal{M}, m)$ is a fuzzy belief structure on W, Bel: $\mathcal{F}(W) \to [0, 1]$ and Pl: $\mathcal{F}(W) \to [0, 1]$ are the pair of FF-fuzzy belief function and FF-plausibility function defined in Definition 26.6, then there exists a countable set U, a serial fuzzy relation R from U to W, and a probability measure P on U such that for all

$X \in \mathcal{F}(W)$,

$$Bel(X) = P(\underline{FR}(X)) = \sum_{x \in U} \underline{FR}(X)(x)P(x),$$

(26.148)

$$Pl(X) = P(\overline{FR}(X)) = \sum_{x \in U} \overline{FR}(X)(x)P(x).$$

(26.149)

A pair (U, A) is called a fuzzy information system if each $a \in A$ is a fuzzy attribute, i. e., for each $x \in U$, $a(x)$ is a fuzzy subset of V_d, that is, $a: U \to \mathcal{F}(V_a)$. Similar to (26.140), we can define a reflexive fuzzy binary relation R_a on U, and consequently, for any attribute subset $B \subseteq A$ one can define a reflexive fuzzy relation R_B as follows

$$R_B = \bigcap_{a \in B} R_a.$$

(26.150)

For $X \in \mathcal{F}(U)$, denote

$$Bel_B(X) = P(\underline{FR_B}(X)), \quad Pl_B(X) = P(\overline{FR_B}(X)),$$

(26.151)

where $P(X) = \sum_{x \in U} X(x)/|U|$ for $X \in \mathcal{F}(U)$. Then, according to Theorem 26.4, Bel_B: $\mathcal{F}(U) \to [0, 1]$ and Pl_B: $\mathcal{F}(U) \to [0, 1]$ are respectively, FF-fuzzy belief function and FF-plausibility function on U. More specifically, if X in (26.151) is crisp subset of U, then Bel_B: $\mathcal{P}(U) \to [0, 1]$ and Pl_B: $\mathcal{P}(U) \to [0, 1]$ defined by (26.151) are, respectively, FC-fuzzy belief functions and FC-plausibility functions on U. Based on these observations, we believe that FF-fuzzy belief functions and FF-plausibility functions can be used to analyze uncertainty fuzzy information systems with fuzzy decision and whereas FC-fuzzy belief functions and FC-plausibility functions can be employed to deal with knowledge discovery in fuzzy information systems with crisp decision.

26.3.3 Conclusion of This Section

The lower and upper approximations of a set capture the non-numeric aspect of uncertainty of the set which can be interpreted as the qualitative representation of the set, whereas the pair of the belief and plausibility measures of the set characterize the numeric aspect of uncertainty of the set which can be treated as the quantitative characterization of the set. In this section, we have introduced some generalized belief and plausibility and belief functions on the Dempster–Shafer theory of evidence. We have shown that the fuzzy belief and plausibility functions can be interpreted as the lower and upper approximations in rough set theory. That is, the belief and plausibility functions in the Dempster–Shafer theory of evidence can be represented as the probabilities of lower and upper approximations in rough set theory; thus, rough set theory may be regarded as the basis of the Dempster–Shafer theory of evidence. Also the Dempster–Shafer theory of evidence in the fuzzy environment provides a potentially useful tool for reasoning and knowledge acquisition in fuzzy systems and fuzzy decision systems.

26.4 Applications of Fuzzy Rough Sets

Both fuzzy set and rough set theories have fostered broad research communities and have been applied in a wide range of settings. More recently, this has also extended to the hybrid fuzzy rough set models. This section tries to give a sample of those applications, which are in particular numerous for machine learning but which also cover many other fields, like image processing, decision making, and information retrieval.

Note that we do not consider applications that simply involve a joint application of fuzzy sets and rough sets, like for instance a rough classifier that induces fuzzy rules. Rather, we focus on applications that specifically involve one of the fuzzy rough set models discussed in the previous sections.

26.4.1 Applications in Machine Learning

Feature Selection

The most prominent application of classical rough set theory is undoubtedly semantics-preserving data dimensionality reduction: the removal of attributes (features) from information systems (An information system (U, A) consists of a nonempty set U of objects which are described by a set of attributes A.) without sacrificing the ability to discern between different objects. A minimal attribute subset $B \subseteq A$ that maintains objects' discernibility is called a *reduct*. For classification tasks, it is sufficient to be able to discern between objects belonging to different classes, in which case a *decision reduct*, also called *relative reduct*, is sought.

The traditional rough set model sets forth a crisp notion of discernibility, where two objects are either discernible or not w.r.t. a set of attributes B based on their values for all attributes in B. To be able to handle numerical data, discretization is required. Fuzzy-rough feature selection avoids this external preprocessing step by incorporating graded indiscernibility between objects directly into the data reduction process. On the other hand, by the use of fuzzy partitions, such that objects can belong to different classes to varying degrees, a more flexible data representation is obtained.

Chronologically, the oldest proposal to apply fuzzy rough sets to feature selection is due to *Kuncheva* [26.26] in 1992. However, rather than using Dubois and Prade's definition, she proposed her own notion of a fuzzy rough set based on an inclusion measure. Based on this, she defined a quality measure for evaluating attribute subsets w.r.t. their ability to approximate a predetermined fuzzy partition on the data, and illustrated its usefulness on a medical data set.

Jensen and *Shen* [26.27, 29] were the first to propose a reduction method that generalizes the classical rough set positive region and dependency function. In particular, the dependency degree γ_B, with $B \subseteq A$, is used to guide a hill-climbing search in which, starting from $B = \emptyset$, in each step an attribute a is added such that $\gamma_{B \cup \{a\}}$ is maximal. The search ends when there is no further increase in the measure. This is the Quick Reduct algorithm. In [26.65] they replaced this simple greedy search heuristic by a more complex one based on ant colony optimization.

Hu et al. [26.66] formally defined the notions of reduct and decision reduct in the fuzzy-rough case, referring to the invariance of the fuzzy partition induced by the data, and of the fuzzy positive region, respectively. They also showed that minimal subsets that are invariant w.r.t. (conditional) entropy are (decision) reducts.

Tsang et al. [26.67] proposed a method based on the discernibility matrix and function to find all decision reducts where invariance of the fuzzy positive region defined using Dubois and Prade's definition is imposed, and proved its correctness. In [26.68], an extension of this method is defined that finds all decision reducts where the approximations are defined using a lower semicontinuous t-norm $\mathcal{T}$ and its R-implicator. The particular case using Łukasiewicz connectives was studied in [26.69]. Later, *Zhao* and *Tsang* [26.31] studied relationships that exist between different kinds of decision reducts, defined using different types of fuzzy connectives.

In [26.70], *Jensen* and *Shen* introduced three different quality measures for evaluating attribute subsets: the first one is a revised version of their previously defined degree of dependency, the second one is based on the fuzzy boundary region, and the third one on the satisfaction of the clauses of the fuzzy discernibility function. On the other hand, in [26.71], *Cornelis* et al. proposed the definition of fuzzy $\mathcal{M}$-decision reducts, where $\mathcal{M}$ is an increasing, [0, 1]-valued quality measure. They studied two measures based on the fuzzy positive region and two more based on the fuzzy discernibility function, and applied them to classification and regression problems.

In [26.33], *Chen* and *Zhao* studied the concept of local reduction: instead of looking for a global reduction, where the whole positive region is considered as an invariant, they focus on subsets of decision classes and identify the conditional attributes that provide minimal descriptions for them.

Over the past few years, there has also been considerable interest in the application of noise-tolerant fuzzy rough set models to feature selection, where the aim is to make the reduction more robust in the presence of noisy or erroneous data. For instance, *Hu* et al. [26.72] defined fuzzy rough sets as an extension of variable precision rough sets, and used a corresponding notion of positive region to guide a greedy search algorithm. In [26.73], *Cornelis* and *Jensen* evaluated the vaguely quantified rough set (VQRS) approach to feature selection. They found that because the model does not satisfy monotonicity w.r.t. the fuzzy relation R, adding more attributes does not always lead to an expansion of the fuzzy positive region, and the hill-climbing search sometimes runs into troubles. Furthermore, in [26.74] *Hu* et al., inspired by the idea of soft margin support vector machines, introduced soft fuzzy rough sets and applied them to feature selection.

He et al. [26.75] consider the problem of fuzzy-rough feature selection for decision systems with fuzzy decisions, that is, where the decision attribute is characterized by a fuzzy T-similarity relation instead of a crisp one. This is the case of regression problems. They give an algorithm for finding all decision reducts and another one for finding a single reduction.

The relatively high complexity of fuzzy-rough feature selection algorithms somewhat limits is applicability to large datasets. In view of this, *Chen* et al. [26.76] propose a fast algorithm to obtain one reduct, based on a procedure to find the minimal elements of the discernibility matrix of [26.67]. The algorithm is compared w.r.t. execution time with the proposals in [26.70] and [26.67], and turns out to be a lot faster. On the other hand, *Qian* et al. [26.77] implement an efficient version of feature selection using the model of *Hu* et al. [26.72].

The use of kernel functions as fuzzy similarity relations in feature selection algorithms has also sparked researchers' interest. In particular, *Du* et al. [26.78] apply fuzzy-rough feature selection with kernelized fuzzy rough sets to yawn detection, while *Chen* et al. [26.79] propose parameterized attribute reduction with Gaussian kernel-based fuzzy rough sets. *He* and *Wu* [26.80] develop a new method to compute membership for fuzzy support vector machines (FSVMs) by using a Gaussian kernel-based fuzzy rough set, and employ a technique of attribute reduction using Gaussian kernel-based fuzzy rough sets to perform feature selection for FSVMs.

Finally, *Derrac* et al. [26.81] combine fuzzy-rough feature selection with evolutionary instance selection.

Instance Selection

Instance selection can be seen as the orthogonal task to feature selection: here the goal is to reduce an information system (U, A) by removing objects from U. The first work on instance selection using fuzzy rough set theory was presented in [26.82]. The main idea is that instances for which the fuzzy rough lower approximation membership is lower than a certain threshold are removed. This idea was improved in [26.83], where the selection threshold is optimized. This method has been applied in combination with evolutionary feature selection in [26.84] and for imbalanced classification problems in [26.85, 86], in combination with resampling methods.

Classification

Fuzzy rough sets have been widely used for classification purposes, either by means of rule induction or by plugging them into existing classifiers like nearest neighbor classifiers, decision trees, and support vector machines (SVM).

The earliest work on rule induction using fuzzy rough set theory can be found in [26.25]. In this paper, the authors propose a fuzzy rough framework to induce fuzzy decision rules that does not use any fuzzy logical connectives. Later, in [26.30], an approach that generates rules from data using fuzzy reducts was presented, with a fuzzy rough feature selection preprocessing step. In [26.87], the authors noticed that using feature selection as a preprocessing step often leads to too specific rules, and proposed an algorithm for simultaneous feature selection and rule induction. In [26.88, 89], a rule-based classifier is built using the so-called consistency degree as a critical value to keep the discernibility information invariant in the rule-induction process. Another approach to fuzzy rough rule induction can be found in [26.90], where rules are found from training data with hierarchical and quantitative attribute values. The most recent work can be found in [26.91], where fuzzy equivalence relations are used to model different types of attributes in order to obtain small rule sets from hybrid data, and in [26.92] where a harmony search algorithm is proposed to generate emerging rule sets.

In [26.93], the K nearest neighbor method was improved using fuzzy set theory. So far, three different fuzzy-rough-based approaches have been used to improve this fuzzy nearest neighbor (FNN) classifier. In [26.94], the author introduces a fuzzy rough ownership function and plugs it into the FNN algorithm. In [26.95–98], the extent to which the nearest neighbors belong to the fuzzy lower and upper approximations of a certain class are used to predict the class of the target instance, these techniques are applied in [26.99] for mammographic risk analysis. Finally, in [26.100], the FNN algorithm is improved using the fuzzy rough positive regions as weights for the nearest neighbors.

During the last decade, several authors have worked on fuzzy rough improvements of decision trees. The common idea of these methods is that during the construction phase of the decision tree, the feature significances are measured using fuzzy rough techniques [26.101–104]. In [26.105–107], the kernel functions of the SVM are redefined using fuzzy rough sets, to take into account the inconsistency between conditional attributes and the decision class. In [26.80], this approach is combined with fuzzy rough feature selection. In [26.108], SVMs are reformulated by plugging in the fuzzy rough memberships of all training samples into the constraints of the SVMs.

Clustering

Many authors have worked on clustering methods that use both fuzzy set theory and rough set theory, but to the best of our knowledge, only two approaches use fuzzy rough sets for clustering. In [26.109], fuzzy rough sets are used to measure the intracluster similarity, in order to estimate the optimal number of clusters. In [26.110], a fuzzy rough measure is used to measure the similarity between genes in microarray analysis, in order to generate clusters such that genes within a cluster are highly

correlated to the sample categories, while those in different clusters are as dissimilar as possible.

Neural Networks

There are many approaches to incorporate fuzzy rough set theory in neural networks. One option is to use fuzzy rough set theory to reduce the problem that samples in the same input clusters can have different classes. The resulting fuzzy rough neural networks are designed such that they work as fuzzy rough membership functions [26.111–114]. A related approach is to use fuzzy rough set theory to find the importance of each subset of information sources of subnetworks [26.115]. Other approaches use fuzzy rough set theory to measure the importance of each feature in the input layer of the neural network [26.116–118].

26.4.2 Other Applications

Image Processing

Fuzzy rough sets have been used in several domains of image processing. They are especially suitable for these tasks because they can capture both indiscernibility and vagueness, which are two important aspects of image processing.

In [26.119, 120], fuzzy-rough-based image segmenting methods are proposed and applied in a traditional Chinese medicine tongue image segmentation experiment. Often, fuzzy rough attribute reduction methods are proposed for image processing problems, as in [26.121] or in [26.122], where the methods are applied for face recognition. In [26.123], a method for edge detection is proposed by building a hierarchy of rough-fuzzy sets to exploit the uncertainty and vagueness at different image resolutions. Another aspect of image processing is texture segmentation, this problem is tackled in [26.124] using rough-fuzzy sets. In [26.125], the authors solve the image classification problem using a nearest neighbor clustering algorithm based on fuzzy rough set theory, and apply their algorithm to hand gesture recognition. In [26.126], a combined approach of neural network classification systems with a fuzzy rough sets based feature reduction method is presented. In [26.127], fuzzy rough feature reduction techniques are applied to a large-scale Mars McMurdo panorama image.

Decision Making

Fuzzy rough set theory has many applications in decision making. In [26.128], the authors calculate the fuzzy rough memberships of software components in previous projects and decide based on these values which ones to reuse in a new program. In [26.129, 130], a multiobjective decision-making model based on fuzzy rough set theory is used to solve the inventory problem. In [26.131], variable precision fuzzy rough sets are used to develop a decision making model, that is applied for IT offshore outsourcing risk evaluation. Another approach can be found in [26.132] where the decision corresponds to the decision corresponding with the instance with maximal sum of lower and upper soft fuzzy rough approximation. Recent work can be found in [26.133], where a fuzzy rough set model over two universes is defined to develop a general decision-making framework in an uncertainty environment for solving a medical diagnosis problem.

Information Retrieval, Data Mining, and the Web

Fuzzy rough sets have been used to model imprecision and vagueness in databases. In [26.134], the au-thors develop a fuzzy rough relational database, while in [26.135], a fuzzy rough extension of a rough object classifier for relational database mining is studied. In [26.136], fuzzy rough set theory is used to mine from incomplete datasets, while in [26.137], fuzzy rough sets are incorporated in mining agents for predicting stock prices. More recently, fuzzy rough sets have been applied to identify imprecision in temporal database models [26.138, 139].

In [26.140, 141], fuzzy rough set theory is used to approximate document queries. In the context of the semantic web, a lot of work has been done on fuzzy rough description logics. The first paper on this topic can be found in [26.142], where a fuzzy rough ontology was proposed. Later, in [26.143], the authors propose a fuzzy rough extension of the descriptive logic $\mathcal{SHIN}$. A fuzzy rough extension of the descriptive logic $\mathcal{ALC}$ can be found in [26.144]. In [26.145, 146], an improved and more general approach is presented.

References

26.1 Z. Pawlak: Rough sets, Int. J. Comput. Inf. Sci. **11**, 341–356 (1982)

26.2 Z. Pawlak: *Rough Sets: Theoretical Aspects of Reasoning About Data* (Kluwer, Boston 1991)

26.3 A. Nakamura: Fuzzy rough sets, Notes Mult.-Valued Log. Jpn. **9**, 1–8 (1988)

26.4 D. Dubois, H. Prade: Rough fuzzy sets and fuzzy rough sets, Int. J. Gen. Syst. **17**, 191–209 (1990)

26.5 D. Dubois, H. Prade: Putting rough sets and fuzzy sets together. In: *Intelligent Decision Support*, ed. by R. Słowiński (Kluwer, Boston 1992) pp. 203–232

26.6 N.N. Morsi, M.M. Yakout: Axiomatics for fuzzy rough sets, Fuzzy Sets Syst. **100**, 327–342 (1998)

26.7 S. Greco, B. Matarazzo, R. Słowiński: The use of rough sets and fuzzy sets in MCDM. In: *Multicriteria Decision Making*, ed. by T. Gál, T.J. Steward, T. Hanne (Kluwer, Boston 1999) pp. 397–455

26.8 D. Boixader, J. Jacas, J. Recasens: Upper and lower approximations of fuzzy sets, Int. J. Gen. Syst. **29**, 555–568 (2000)

26.9 A.M. Radzikowska, E.E. Kerre: A comparative study of fuzzy rough set, Fuzzy Sets Syst. **126**, 137–155 (2002)

26.10 M. Inuiguchi, T. Tanino: New fuzzy rough sets based on certainty qualification. In: *Rough-Neural Computing*, ed. by K. Pal, L. Polkowski, A. Skowron (Springer, Berlin, Heidelberg 2003) pp. 278–296

26.11 W.-Z. Wu, J.-S. Mi, W.-X. Zhang: Generalized fuzzy rough sets, Inf. Sci. **151**, 263–282 (2003)

26.12 M. Inuiguchi: Generalization of rough sets: From crisp to fuzzy cases, Lect. Notes Artif. Intell. **3066**, 26–37 (2004)

26.13 A.M. Radzikowska, E.E. Kerre: Fuzzy rough sets based on residuated lattices, Lect. Notes Comput. Sci. **3135**, 278–296 (2004)

26.14 W.-Z. Wu, W.-X. Zhang: Constructive and ax-iomatic approaches of fuzzy approximation operators, Inf. Sci. **159**, 233–254 (2004)

26.15 J.-S. Mi, W.-X. Zhang: An axiomatic characterization of a fuzzy generalization of rough sets, Inf. Sci. **160**, 235–249 (2004)

26.16 W.-Z. Wu, Y. Leung, J.-S. Mi: On characterizations of $(\mathcal{I}, \mathcal{T})$-fuzzy rough approximation operators, Fuzzy Sets Syst. **15**, 76–102 (2005)

26.17 D.S. Yeung, D.G. Chen, E.C.C. Tsang, J.W.T. Lee, X.Z. Wang: On the generalization of fuzzy rough sets, IEEE Trans. Fuzzy Syst. **13**, 343–361 (2005)

26.18 M. DeCock, C. Cornelis, E.E. Kerre: Fuzzy rough sets: The forgotten step, IEEE Trans. Fuzzy Syst. **15**, 121–130 (2007)

26.19 T.J. Li, W.X. Zhang: Rough fuzzy approximations on two universes of discourse, Inf. Sci. **178**, 892–906 (2008)

26.20 J.-S. Mi, Y. Leung, H.-Y. Zhao, T. Feng: Generalized fuzzy rough sets determined by a triangular norm, Inf. Sci. **178**, 3203–3213 (2008)

26.21   W.-Z. Wu, Y. Leung, J.-S. Mi: On generalized fuzzy belief functions in infinite spaces, IEEE Trans. Fuzzy Syst. **17**, 385–397 (2009)

26.22   X.D. Liu, W. Pedrycz, T.Y. Chai, M.L. Song: The development of fuzzy rough sets with the use of structures and algebras of axiomatic fuzzy sets, IEEE Trans. Knowl. Data Eng. **21**, 443–462 (2009)

26.23   W.-Z. Wu: On some mathematical structures of T-fuzzy rough set algebras in infinite universes of discourse, Fundam. Inf. **108**, 337–369 (2011)

26.24   S. Greco, M. Inuiguchi, R. Słowiński: Rough sets and gradual decision rules, Lect. Notes Artif. Intell. **2639**, 156–164 (2003)

26.25   S. Greco, M. Inuiguchi, R. Słowiński: Fuzzy rough sets and multiple-premise gradual decision rules, Int. J. Approx. Reason. **41**(2), 179–211 (2006)

26.26   L.I. Kuncheva: Fuzzy rough sets: Application to feature selection, Fuzzy Sets Syst. **51**, 147–153 (1992)

26.27   R. Jensen, Q. Shen: Fuzzy-rough attributes reduction with application to web categorization, Fuzzy Sets Syst. **141**, 469–485 (2004)

26.28   R. Jensen, Q. Shen: Semantics-preserving dimensionality reduction: Rough and fuzzy-rough based approaches, IEEE Trans. Knowl. Data Eng. **16**, 1457–1471 (2004)

26.29   R. Jensen, Q. Shen: Fuzzy-rough sets assisted attribute selection, IEEE Trans. Fuzzy Syst. **15**, 73–89 (2007)

26.30   X.Z. Wang, E.C.C. Tsang, S.Y. Zhao, D.G. Chen, D.S. Yeung: Learning fuzzy rules from fuzzy samples based on rough set technique, Fuzzy Sets Syst **177**, 4493–4514 (2007)

26.31   S.Y. Zhao, E.C.C. Tsang: On fuzzy approximation operators in attribute reduction with fuzzy rough sets, Inf. Sci. **178**, 3163–3176 (2008)

26.32   S.Y. Zhao, E.C.C. Tsang, D.G. Chen: The model of fuzzy variable precision rough sets, IEEE Trans. Fuzzy Syst. **17**, 451–467 (2009)

26.33   D.G. Chen, S.Y. Zhao: Local reduction of decision system with fuzzy rough sets, Fuzzy Sets Syst. **161**, 1871–1883 (2010)

26.34   Q.H. Hu, L. Zhang, D.G. Chen, W. Pedrycz, D.R. Yu: Gaussian kernel based fuzzy rough sets: Model, uncertainty measures and applications, Int. J. Approx. Reason. **51**, 453–471 (2010)

26.35   Q.H. Hu, D.R. Yu, W. Pedrycz, D.G. Chen: Kernelized fuzzy rough sets and their applications, IEEE Trans. Knowl. Data Eng. **23**, 1649–1667 (2011)

26.36   Q.H. Hu, L. Zhang, S. An, D. Zhang, D.R. Yu: On robust fuzzy rough set models, IEEE Trans. Fuzzy Syst. **20**, 636–651 (2012)

26.37   M. Inuiguchi: Classification- versus approximation-oriented fuzzy rough sets, Proc. Inf. Process. Manag. Uncertain. Knowl.-Based Syst. (2004), CD-ROM

26.38   G. Shafer: *A Mathematical Theory of Evidence* (Princeton Univ. Press, Princeton 1976)

26.39   A. Skowron: The relationship between rough set theory and evidence theory, Bull. Polish Acad. Sci. Math. **37**, 87–90 (1989)

26.40   A. Skowron: The rough sets theory and evidence theory, Fundam. Inf. **13**, 245–262 (1990)

26.41   A. Skowron, J. Grzymala-Busse: From rough set theory to evidence theory. In: *Advance in the Dempster-Shafer Theory of Evidence*, ed. by R.R. Yager, M. Fedrizzi, J. Kacprzyk (Wiley, New York 1994) pp. 193–236

26.42   W.-Z. Wu, Y. Leung, W.-X. Zhang: Connections between rough set theory and Dempster-Shafer theory of evidence, Int. J. Gen. Syst. **31**, 405–430 (2002)

26.43   W.-Z. Wu, J.-S. Mi: Some mathematical structures of generalized rough sets in infinite universes of discourse, Lect. Notes Comput. Sci. **6499**, 175–206 (2011)

26.44   Y.Y. Yao, P.J. Lingras: Interpretations of belief functions in the theory of rough sets, Inf. Sci. **104**, 81–106 (1998)

26.45   P.J. Lingras, Y.Y. Yao: Data mining using extensions of the rough set model, J. Am. Soc. Inf. Sci. **49**, 415–422 (1998)

26.46   W.-Z. Wu: Attribute reduction based on evidence theory in incomplete decision systems, Inf. Sci. **178**, 1355–1371 (2008)

26.47   W.-Z. Wu: Knowledge reduction in random incomplete decision tables via evidence theory, Fundam. Inf. **115**, 203–218 (2012)

26.48   W.-Z. Wu, M. Zhang, H.-Z. Li, J.-S. Mi: Knowledge reduction in random information systems via Dempster-Shafer theory of evidence, Inf. Sci. **174**, 143–164 (2005)

26.49   M. Zhang, L.D. Xu, W.-X. Zhang, H.-Z. Li: A rough set approach to knowledge reduction based on inclusion degree and evidence reasoning theory, Expert Syst. **20**, 298–304 (2003)

26.50   M. Inuiguchi: Generalization of rough sets and rule extraction, Lect. Notes Comput. Sci. **3100**, 96–119 (2004)

26.51   E.P. Klement, R. Mesiar, E. Pap: *Triangular Norms* (Kluwer, Boston 2000)

26.52   W. Wu, J. Mi, W. Zhang: Generalized fuzzy rough sets, Inf. Sci. **151**, 263–282 (2003)

26.53   M. Inuiguchi, M. Sakawa: On the closure of generation processes of implication functions from a conjunction function. In: *Proc. 4th Int. Conf. Soft Comput.* 1996) pp. 327–330

26.54   D. Dubois, H. Prade: Fuzzy sets in approximate reasoning, Part 1: Inference with possibility distributions, Fuzzy Sets Syst. **40**, 143–202 (1991)

26.55   M. Inuiguchi, T. Tanino: A new class of necessity measures and fuzzy rough sets based on certainty qualifications, Lect. Notes Comput. Sci. **2005**, 261–268 (2001)

26.56   M. Inuiguchi, T. Tanino: Function approximation by fuzzy rough sets. In: *Intelligent Systems for Information Processing: From Representa-*

tion to Applications, ed. by B. Bouchon-Meunier, L. Foulloy, R.R. Yager (Elsevier, Amsterdam 2003) pp. 93–104

26.57 D. Dubois, H. Prade: Gradual inference rules in approximate reasoning, Inf. Sci. **61**, 103–122 (1992)

26.58 L.A. Zadeh: A fuzzy set-theoretic interpretation of linguistic hedge, J. Cybern. **2**, 4–34 (1974)

26.59 J.F. Baldwin: A new approach to approximate reasoning using a fuzzy logic, Fuzzy Sets Syst. **2**(4), 309–325 (1979)

26.60 Y. Tsukamoto: An approach to fuzzy reasoning method. In: *Advances in Fuzzy Set Theory and Applications*, ed. by M.M. Gupta, R.K. Ragade, R.R. Yager (North-Holland, New-York 1979) pp. 137–149

26.61 G. Choquet: Theory of capacities, Ann. l'institut Fourier **5**, 131–295 (1954)

26.62 L. Biacino: Fuzzy subsethood and belief functions of fuzzy events, Fuzzy Sets Syst. **158**, 38–49 (2007)

26.63 Y.Y. Yao: Generalized rough set model. In: *Rough Sets in Knowledge Discovery 1. Methodology and Applications*, ed. by L. Polkowski, A. Skowron (Physica, Heidelberg 1998) pp. 286–318

26.64 D.G. Chen, W.X. Yang, F.C. Li: Measures of general fuzzy rough sets on a probabilistic space, Inf. Sci. **178**, 3177–3187 (2006)

26.65 R. Jensen, Q. Shen: Fuzzy-rough data reduction with ant colony optimization, Fuzzy Sets Syst. **149**(1), 5–20 (2005)

26.66 Q. Hu, D. Yu, Z. Xie: Information-preserving hybrid data reduction based on fuzzy-rough techniques, Pattern Recogn. Lett. **27**(5), 414–423 (2006)

26.67 E.C.C. Tsang, D.G. Chen, D.S. Yeungm, X.Z. Wang, J.W.T. Lee: Attributes reduction using fuzzy rough sets, IEEE Trans. Fuzzy Syst. **16**(5), 1130–1141 (2008)

26.68 D. Chen, E. Tsang, S. Zhao: Attribute reduction based on fuzzy rough sets, Lect. Notes Comput. Sci. **4585**, 73–89 (2007)

26.69 D. Chen, E. Tsang, S. Zhao: An approach of attributes reduction based on fuzzy tl-rough sets, Proc. IEEE Int. Conf. Syst. Man Cybern. (2007) pp. 486–491

26.70 R. Jensen, Q. Shen: New approaches to fuzzy-rough feature selectio, IEEE Trans. Fuzzy Syst. **17**(4), 824–838 (2009)

26.71 C. Cornelis, G.H. Martin, R. Jensen, D. Slezak: Feature selection with fuzzy decision reducts, Inf. Sci. **180**(2), 209–224 (2010)

26.72 Q. Hu, X.Z. Xie, D.R. Yu: Hybrid attribute reduction based on a novel fuzzy-rough model and information granulation, Pattern Recogn. **40**(12), 3509–3521 (2007)

26.73 C. Cornelis, R. Jensen: A noise-tolerant approach to fuzzy-rough feature selection, Proc. IEEE Int. Conf. Fuzzy Syst. (2008) pp. 1598–1605

26.74 Q. Hu, S.A. An, D.R. Yu: Soft fuzzy rough sets for robust feature evaluation and selection, Inf. Sci. **180**(22), 4384–4440 (2010)

26.75 Q. He, C.X. Wu, D.G. Chen, S.Y. Zhao: Fuzzy rough set based attribute reduction for information systems with fuzzy decisions, Knowl.-Based Syst. **24**(5), 689–696 (2011)

26.76 D.G. Chen, L. Zhang, S.Y. Zhao, Q.H. Hu, P.F. Zhu: A novel algorithm for finding reducts with fuzzy rough sets, IEEE Trans. Fuzzy Syst. **20**(2), 385–389 (2012)

26.77 Y.H. Qian, C. Li, J.Y. Liang: An efficient fuzzy-rough attribute reduction approach, Lect. Notes Artif. Intell. **6954**, 63–70 (2011)

26.78 Y. Du, Q. Hu, D.G. Chen, P.J. Ma: Kernelized fuzzy rough sets based yawn detection for driver fatigue monitoring, Fundam. Inf. **111**(1), 65–79 (2011)

26.79 D.G. Chen, Q.H. Hu, Y.P. Yang: Parameterized attribute reduction with Gaussian kernel based fuzzy rough sets, Inf. Sci. **181**(23), 5169–5179 (2011)

26.80 Q. He, C.X. Wu: Membership evaluation and feature selection for fuzzy support vector machine based on fuzzy rough sets, Soft Comput. **15**(6), 1105–1114 (2011)

26.81 J. Derrac, C. Cornelis, S. Garcia, F. Herrera: Enhancing evolutionary instance selection algorithms by means of fuzzy rough set based feature selection, Inf. Sci. **186**(1), 73–92 (2012)

26.82 R. Jensen, C. Cornelis: Fuzzy-rough instance selection, Proc. IEEE Int. Conf. Fuzzy Syst. (2010) pp. 1–7

26.83 N. Verbiest, C. Cornelis, F. Herrera: Granularity-based instance selection, Proc. 20th Ann. Belg.-Dutch Conf. Mach. Learn. (2011) pp. 101–103

26.84 J. Derrac, N. Verbiest, S. Garcia, C. Cornelis, F. Herrera: On the use of evolutionary feature selection for improving fuzzy rough set based prototype selection, Soft Comput. **17**(2), 223–238 (2013)

26.85 E. Ramentol, N. Verbiest, R. Bello, Y. Caballero, C. Cornelis, F. Herrera: Smote-frst: A new resampling method using fuzzy rough set theory, Proc. 10th Int. FLINS Conf. Uncertain. Model. Knowl. Eng. Decis. Mak. (2012) pp. 800–805

26.86 N. Verbiest, E. Ramentol, C. Cornelis, F. Herrera: Improving smote with fuzzy rough prototype selection to detect noise in imbalanced classification data, Proc. 13th Ibero-Am. Conf. Artif. Intell. (2012) pp. 169–178

26.87 R. Jensen, C. Cornelis, Q. Shen: Hybrid fuzzy-rough rule induction and feature selection, Proc. IEEE Int. Conf. Fuzzy Syst. (2009) pp. 1151–1156

26.88 E. Tsang, S.Y. Zhao, J. Lee: Rule induction based on fuzzy rough sets, Proc. Int. Conf. Mach. Learn. Cybern. (2007) pp. 3028–3033

26.89 S. Zhao, E. Tsang, D. Chen, X. Wang: Building a rule-based classifier – a fuzzy-rough set approach, IEEE Trans. Knowl. Data Eng. **22**, 624–638 (2010)

Part C | 26

26.90   T.P. Hong, Y.L. Liou, S.L. Wang: Fuzzy rough sets with hierarchical quantitative attributes, Expert Syst. Appl. **36**(3), 6790–6799 (2009)

26.91   Y. Liu, Q. Zhou, E. Rakus-Andersson, G. Bai: A fuzzy-rough sets based compact rule induction method for classifying hybrid data, Lect. Notes Comput. Sci. **7414**, 63–70 (2012)

26.92   R. Diao, Q. Shen: A harmony search based approach to hybrid fuzzy-rough rule induction, Proc. 21st Int. Conf. Fuzzy Syst. (2012) pp. 1549–1556

26.93   J.M. Keller, M.R. Gray, J.R. Givens: A fuzzy k-nearest neighbor algorithm, IEEE Trans. Syst. Man Cybern. **15**, 580–585 (1985)

26.94   M. Sarkar: Fuzzy-rough nearest neighbor algorithms in classification, Fuzzy Sets Syst. **158**, 2134–2152 (2007)

26.95   R. Jensen, C. Cornelis: A new approach to fuzzy-rough nearest neighbour classification, Lect. Notes Comput. Sci. **5306**, 310–319 (2008)

26.96   R. Jensen, C. Cornelis: Fuzzy-rough nearest neighbour classification and prediction, Theor. Comput. Sci. **412**, 5871–5884 (2011)

26.97   Y. Qu, C. Shang, Q. Shen, N.M. Parthalain, W. Wu: Kernel-based fuzzy-rough nearest neighbour classification, IEEE Int. Conf. Fuzzy Syst. (2011) pp. 1523–1529

26.98   H. Bian, L. Mazlack: Fuzzy-rough nearest-neighbor classification approach, 22nd Int. Conf. North Am. Fuzzy Inf. Process. Soc. (2003) pp. 500–505

26.99   M.N. Parthalain, R. Jensen, Q. Shen, R. Zwiggelaar: Fuzzy-rough approaches for mammographic risk analysis, Intell. Data Anal. **13**, 225–244 (2010)

26.100   N. Verbiest, C. Cornelis, R. Jensen: Fuzzy rough positive region-based nearest neighbour classification, Proc. 20th Int. Conf. Fuzzy Syst. (2012) pp. 1961–1967

26.101   R. Jensen, Q. Shen: Fuzzy-rough feature significance for decision trees, Proc. 2005 UK Workshop Comput. Intell. (2005) pp. 89–96

26.102   R. Bhatt, M. Gopal: FRCT: Fuzzy-rough classification trees, Pattern Anal. Appl. **11**, 73–88 (2008)

26.103   M. Elashiri, H. Hefny, A.A. Elwahab: Induction of fuzzy decision trees based on fuzzy rough set techniques, Proc. Int. Conf. Comput. Eng. Syst. (2011) pp. 134–139

26.104   J. Zhai: Fuzzy decision tree based on fuzzy-rough technique, Soft Comput. **15**, 1087–1096 (2011)

26.105   D. Chen, Q. He, X. Wang: Frsvms: Fuzzy rough set based support vector machines, Fuzzy Sets Syst. **161**, 596–607 (2010)

26.106   Z. Zhang, D. Chen, Q. He, H. Wang: Least squares support vector machines based on fuzzy rough set, IEEE Int. Conf. Syst. Man Cybern. (2010) pp. 3834–3838

26.107   Z. Xue, W. Liu: A fuzzy rough support vector regression machine, 9th Int. Conf. Fuzzy Syst. Knowl. Discov. (2012) pp. 840–844

26.108   D. Chen, S. Kwong, Q. He, H. Wang: Geometrical interpretation and applications of membership functions with fuzzy rough sets, Fuzzy Sets Syst. **193**, 122–135 (2012)

26.109   F. Li, F. Min, Q. Liu: Intra-cluster similarity index based on fuzzy rough sets for fuzzy c-means algorithm, Lect. Notes Comput. Sci. **5009**, 316–323 (2008)

26.110   P. Maji: Fuzzy rough supervised attribute clustering algorithm and classification of microarray data, IEEE Trans. Syst. Man Cybern., Part B: Cybern. **41**, 222–233 (2011)

26.111   M. Sarkar, B. Yegnanarayana: Fuzzy-rough neural networks for vowel classification, IEEE Int. Conf. Syst. Man Cybern., Vol. 5 (1998) pp. 4160–4165

26.112   J.Y. Zhao, Z. Zhang: Fuzzy rough neural network and its application to feature selection, Fourth Int. Workshop Adv. Comput. Intell. (2011) pp. 684–687

26.113   D. Zhang, Y. Wang: Fuzzy-rough neural network and its application to vowel recognition, 45th IEEE Conf. Control Decis. (2006) pp. 221–224

26.114   M. JianXu, L. Caiping, W. Yaonan: Remote sensing images classification using fuzzy-rough neural network, IEEE Fifth Int. Conf. Bio-Inspir. Comput. Theor. Appl. (2010) pp. 761–765

26.115   M. Sarkar, B. Yegnanarayana: Application of fuzzy-rough sets in modular neural networks, IEEE Joint World Congr. Comput. Intell. Neural Netw. (1998) pp. 741–746

26.116   A. Ganivada, P. Sankar: A novel fuzzy rough granular neural network for classification, Int. J. Comput. Intell. Syst. **4**, 1042–1051 (2011)

26.117   M. Sarkar, B. Yegnanarayana: Rough-fuzzy set theoretic approach to evaluate the importance of input features in classification, Int. Conf. Neural Netw. (1997) pp. 1590–1595

26.118   A. Ganivada, S.S. Ray, S.K. Pal: Fuzzy rough granular self-organizing map and fuzzy rough entropy, Theor. Comput. Sci. **466**, 37–63 (2012)

26.119   L. Jiangping, P. Baochang, W. Yuke: Tongue image segmentation based on fuzzy rough sets, Proc. Int. Conf. Environ. Sci. Inf. Appl. Technol. (2009) pp. 367–369

26.120   L. Jiangping, W. Yuke: A shortest path algorithm of image segmentation based on fuzzy-rough grid, Proc. Int. Conf. Comput. Intell. Softw. Eng. (2009) pp. 1–4

26.121   A. Petrosino, A. Ferone: Rough fuzzy set-based image compression, Fuzzy Sets Syst. **160**, 1485–1506 (2009)

26.122   L. Zhou, W. Li, Y. Wu: Face recognition based on fuzzy rough set reduction, Proc. Int. Conf. Hybrid Inf. Technol. (2006) pp. 642–646

26.123   A. Petrosino, G. Salvi: Rough fuzzy set based scale space transforms and their use in image analysis, Int. J. Approx. Reason. **41**, 212–228 (2006)

26.124   A. Petrosino, M. Ceccarelli: Unsupervised texture discrimination based on rough fuzzy sets and parallel hierarchical clustering, Proc. IEEE Int. Conf. Pattern Recogn. (2000) pp. 1100–1103

26.125 X. Wang, J. Yang, X. Teng, N. Peng: Fuzzy-rough set based nearest neighbor clustering classification algorithm, Proc. 2nd Int. Conf. Fuzzy Syst. Knowl. Discov. (2005) pp. 370–373

26.126 C. Shang, Q. Shen: Aiding neural network based image classification with fuzzy-rough feature selection, Proc. IEEE Int. Conf. Fuzzy Syst. (2008) pp. 976–982

26.127 S. Changjing, D. Barnes, S. Qiang: Effective feature selection for mars mcmurdo terrain image classification, Proc. Int. Conf. Intell. Syst., Des. Appl. (2009) pp. 1419–1424

26.128 D.V. Rao, V.V.S. Sarma: A rough–fuzzy approach for retrieval of candidate components for software reuse, Pattern Recogn. Lett. **24**, 875–886 (2003)

26.129 G. Cong, J. Zhang, T. Huazhong, K. Lai: A variable precision fuzzy rough group decision-making model for it offshore outsourcing risk evaluation, J. Glob. Inf. Manag. **16**, 18–34 (2008)

26.130 J. Xu, L. Zhao: A multi-objective decision-making model with fuzzy rough coefficients and its application to the inventory problem, Inf. Sci. **180**, 679–696 (2010)

26.131 J. Xu, L. Zhao: A class of fuzzy rough expected value multi-objective decision making model and its application to inventory problems, Comput. Math. Appl. **56**(8), 2107–2119 (2008)

26.132 B. Sun, W. Ma: Soft fuzzy rough sets and its application in decision making, Artif. Intell. Rev. **41**(1), 67–80 (2014)

26.133 B. Suna, W. Ma, Q. Liu: An approach to decision making based on intuitionistic fuzzy rough sets over two universes, J. Oper. Res. Soc. **64**(7), 1079–1089 (2012)

26.134 T. Beaubouef, F. Petry: Fuzzy rough set techniques for uncertainty processing in a relational database, Int. J. Intell. Syst. **15**(5), 389–424 (2000)

26.135 R.R. Hashemi, F.F. Choobineh: A fuzzy rough sets classifier for database mining, Int. J. Smart Eng. Syst. Des. **4**, 107–114 (2002)

26.136 T.P. Hong, L.H. Tseng, B.C. Chien: Mining from incomplete quantitative data by fuzzy rough sets, Expert Syst. Appl. **37**, 2644–2653 (2010)

26.137 Y.F. Wang: Mining stock price using fuzzy rough set system, Expert Syst. Appl. **24**, 13–23 (2003)

26.138 A. Burney, N. Mahmood, Z. Abbas: Advances in fuzzy rough set theory for temporal databases, Proc. 11th WSEAS Int. Conf. Artif. Intell. Knowl. Eng. Data Bases (2012) pp. 237–242

26.139 A. Burney, Z. Abbas, N. Mahmood, Q. Arifeen: Application of fuzzy rough temporal approach in patient data management (frt-pdm), Int. J. Comput. **6**, 149–157 (2012)

26.140 P. Srinivasan, M. Ruiz, D.H. Kraft, J. Chen: Vocabulary mining for information retrieval: Rough sets and fuzzy sets, Inf. Process. Manag. **37**, 15–38 (2001)

26.141 M. DeCock, C. Cornelis: Fuzzy rough set based web query expansion, Proc. Rough Sets Soft Comput. Intell. Agent Web Technol., Int. Workshop (2005) pp. 9–16

26.142 L. Dey, M. Abulaish, R. Goyal, K. Shubham: A rough-fuzzy ontology generation framework and its application to bio-medical text processing, Proc. 5th Atl. Web Intell. Conf. (2007) pp. 74–79

26.143 Y. Jiang, J. Wang, P. Deng, S. Tang: Reasoning within expressive fuzzy rough description logics, Fuzzy Sets Syst. **160**, 3403–3424 (2009)

26.144 F. Bobillo, U. Straccia: Generalized fuzzy rough description logics, Inf. Sci. **189**, 43–62 (2012)

26.145 Y. Jiang, Y. Tang, J. Wang, S. Tang: Reasoning within intuitionistic fuzzy rough description logics, Inf. Sci. **179**, 2362–2378 (2009)

26.146 F. Bobillo, U. Straccia: Supporting fuzzy rough sets in fuzzy description logics, Lect. Notes Comput. Sci. **5590**, 676–687 (2009)

Part D

Neural N

Part D Neural Networks

Ed. by Cesare Alippi, Marios Polycarpou

27. Artificial Neural Network Models

Peter Tino, Lubica Benuskova, Alessandro Sperduti

We outline the main models and developments in the broad field of artificial neural networks (ANN). A brief introduction to biological neurons motivates the initial formal neuron model – the perceptron. We then study how such formal neurons can be generalized and connected in network structures. Starting with the biologically motivated layered structure of ANN (feed-forward ANN), the networks are then generalized to include feedback loops (recurrent ANN) and even more abstract generalized forms of feedback connections (recursive neuronal networks) enabling processing of structured data, such as sequences, trees, and graphs. We also introduce ANN models capable of forming topographic lower-dimensional maps of data (self-organizing maps). For each ANN type we outline the basic principles of training the corresponding ANN models on an appropriate data collection.

The human brain is arguably one of the most exciting products of evolution on Earth. It is also the most powerful information processing tool so far. Learning based on examples and parallel signal processing lead to emergent macro-scale behavior of neural networks in the brain, which cannot be easily linked to the behavior of individual micro-scale components (neurons). In this chapter, we will introduce *artificial neural network* (ANN) models motivated by the brain that can learn in the presence of a *teacher*. During the course of learning the teacher specifies what the right responses to input examples should be. In addition, we will also mention ANNs that can learn without a teacher, based on principles of self-organization.

To set the context, we will begin by introducing basic neurobiology. We will then describe the *perceptron* model, which, even though rather old and simple, is an important building block of more complex *feed-forward ANN* models. Such models can be used to approximate complex non-linear functions or to learn a variety of association tasks. The feed-forward models are capable of processing patterns without temporal association. In the presence of temporal dependencies, e.g., when learning to predict future elements of a time series (with certain prediction horizon), the feed-forward ANN needs to be extended with a memory mechanism to account for temporal structure in the data. This will naturally lead us to *recurrent neural network* models (RNN), which besides feed-forward connections also contain feedback loops to preserve, in the form of the information processing state, information about the past. RNN can be further extended to *recursive ANNs* (RecNN), which can process structured data such as trees and acyclic graphs.

27.1 Biological Neurons

It is estimated that there are about 10^{12} neural cells (*neurons*) in the human brain. Two-thirds of the neurons form a 4–6 mm thick cortex that is assumed to be the center of cognitive processes. Within each neuron com-

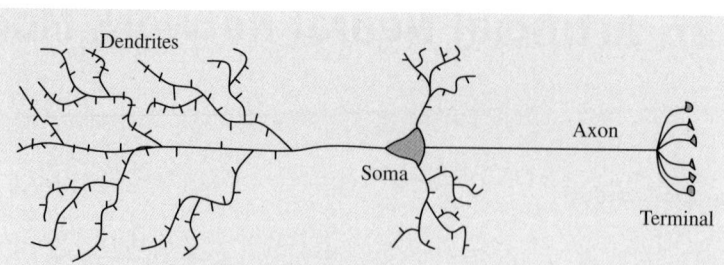

Fig. 27.1 Schematic illustration of the basic information processing structure of the biological neuron

plex biological processes take place, ensuring that it can process signals from other neurons, as well as send its own signals to them. The signals are of electro-chemical nature. In a simplified way, signals between the neurons can be represented by real numbers quantifying the *intensity* of the incoming or outgoing signals. The point of signal transmission from one neuron to the other is called the *synapse*. Within synapse the incoming signal can be reinforced or damped. This is represented by the *weight* of the synapse. A single neuron can have up to $10^3 - 10^5$ such *points of entry* (synapses). The input to the neuron is organized along *dendrites* and the *soma* (Fig. 27.1). Thousands of dendrites form a rich tree-like structure on which most synapses reside.

Signals from other neurons can be either excitatory (positive) or inhibitory (negative), relayed via excitatory or inhibitory synapses. When the sum of the positive and negative contributions (signals) from other neurons, weighted by the synaptic weights, becomes greater than a certain excitation threshold, the neuron will generate an electric spike that will be transmitted over the output channel called the *axon*. At the end of

axon, there are thousands of output branches whose terminals form synapses on other neurons in the network. Typically, as a result of input excitation, the neuron can generate a series of spikes of some average frequency – about $1 - 10^2$ Hz. The frequency is proportional to the overall stimulation of the neuron.

The first principle of information coding and representation in the brain is *redundancy*. It means that each piece of information is processed by a redundant set of neurons, so that in the case of partial brain damage the information is not lost completely. As a result, and crucially – in contrast to conventional computer architectures, gradually increasing damage to the computing substrate (neurons plus their interconnection structure) will only result in gradually decreasing processing capabilities (*graceful degradation*). Furthermore, it is important what set of neurons participate in coding a particular piece of information (*distributed representation*). Each neuron can participate in coding of many pieces of information, in conjunction with other neurons. The information is thus associated with patterns of distributed activity on sets of neurons.

27.2 Perceptron

The perceptron is a simple neuron model that takes input signals (patterns) coded as (real) *input* vectors $\bar{x} = (x_1, x_2, \ldots, x_{n+1})$ through the associated (real) vector of synaptic *weights* $\bar{w} = (w_1, w_2, \ldots, w_{n+1})$. The output o is determined by

$$o = f(\text{net}) = f(\bar{w} \cdot \bar{x}) =$$

$$f\left(\sum_{j=1}^{n+1} w_j x_j\right) = f\left(\sum_{j=1}^{n} w_j x_j - \theta\right), \quad (27.1)$$

where net denotes the weighted sum of inputs, (i. e., dot product of weight and input vectors), and f is the *activation function*. By convention, if there are n inputs to

the perceptron, the input $(n+1)$ will be fixed to -1 and the associated weight to $w_{n+1} = \theta$, which is the value of the *excitation threshold*.

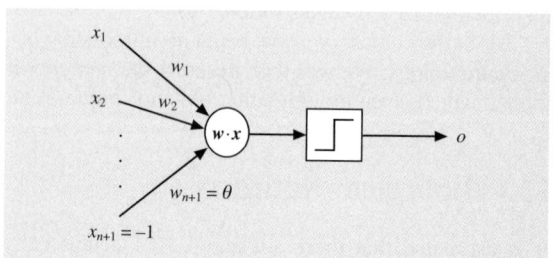

Fig. 27.2 Schematic illustration of the perceptron model

In 1958 *Rosenblatt* [27.1] introduced a discrete perceptron model with a bipolar activation function (Fig. 27.2)

$$f(\text{net}) = sign(\text{net})$$

$$= \begin{cases} +1 & \text{if net} \geq 0 \Leftrightarrow \sum_{j=1}^{n} w_j x_j \geq \theta \\ -1 & \text{if net} < 0 \Leftrightarrow \sum_{j=1}^{n} w_j x_j < \theta . \end{cases} \tag{27.2}$$

The *boundary* equation

$$\sum_{j=1}^{n} w_j x_j - \theta = 0, \tag{27.3}$$

parameterizes a hyperplane in n-dimensional space with normal vector $\bar{w}$.

The perceptron can classify input patterns into two classes, if the classes can indeed be separated by an $(n-1)$-dimensional hyperplane (27.3). In other words, the perceptron can deal with linearly-*separable problems* only, such as logical functions AND or OR. XOR, on the other hand, is not linearly separable (Fig. 27.3). Rosenblatt showed that there is a simple training rule that will find the separating hyperplane, provided that the patterns are linearly separable.

As we shall see, a general rule for training many ANN models (not only the perceptron) can be formulated as follows: the weight vector $\bar{w}$ is changed proportionally to the product of the input vector and a *learning signal s*. The learning signal s is a function of $\bar{w}, \bar{x}$, and possibly a teacher feedback d

$$s = s(\bar{w}, \bar{x}, d) \quad \text{or} \quad s = s(\bar{w}, \bar{x}) . \tag{27.4}$$

In the former case, we talk about *supervised learning* (with direct guidance from a teacher); the latter case is known as *unsupervised learning*. The update of the j-th weight can be written as

$$w_j(t+1) = w_j(t) + \Delta w_j(t) = w_j(t) + \alpha s(t) x_j(t) . \tag{27.5}$$

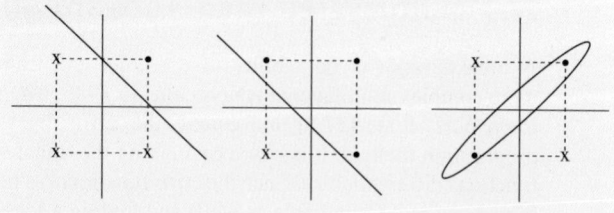

Fig. 27.3 Linearly separable and non-separable problems

The positive constant $0 < \alpha \leq 1$ is called the learning rate.

In the case of the perceptron, the learning signal is the disproportion (difference) between the desired (target) and the actual (produced by the model) response, $s = d - o = \delta$. The update rule is known as the δ (delta) rule

$$\Delta w_j = \alpha(d - o)x_j . \tag{27.6}$$

The same rule can, of course, be used to update the activation threshold $w_{n+1} = \theta$.

Consider a training set

$$A_{\text{train}} = \{(\bar{x}^1, d^1)(\bar{x}^2, d^2) \dots (\bar{x}^p, d^p) \dots (\bar{x}^P, d^P)\}$$

consisting of P (input,target) couples. The perceptron training algorithm can be formally written as:

- *Step* 1: Set $\alpha \in (0, 1)$. Initialize the weights randomly from $(-1, 1)$. Set the counters to $k = 1, p = 1$ (k indexes sweep through A_{train}, p indexes individual training patterns).
- *Step* 2: Consider input $\bar{x}^p$, calculate the output $o = sign(\sum_{j=1}^{n+1} w_j x_j^p)$.
- *Step* 3: Weight update: $w_j \leftarrow w_j + \alpha(d^p - o^p)x_j^p$, for $j = 1, \dots, n + 1$.
- *Step* 4: If $p < P$, set $p \leftarrow p + 1$, go to step 2. Otherwise go to step 5.
- *Step* 5: Fix the weights and calculate the cumulative error E on A_{train}.
- *Step* 6: If $E = 0$, finish training. Otherwise, set $p = 1$, $k = k + 1$ and go to step 2. A new training epoch starts.

27.3 Multilayered Feed-Forward ANN Models

A breakthrough in our ability to construct and train
more complex multilayered ANNs came in 1986, when
Rumelhart et al. [27.2] introduced the error backpropagation method. It is based on making the transfer
functions differentiable (hence the error functional to be
minimized is differentiable as well) and finding a local
minimum of the error functional by the gradient-based
steepest descent method.

We will show derivation of the back-propagation
algorithm for two-layer feed-forward ANN as demonstrated, e.g., in [27.3]. Of course, the same principles
can be applied to a feed-forward ANN architecture with
any (finite) number of layers. In feed-forward ANNs
neurons are organized in layers. There are no connections among neurons within the same layer; connections
only exist between successive layers. Each neuron from
layer l has connections to each neuron in layer $l+1$.

As has already been mentioned, the *activation functions* need to differentiable and are usually of the
sigmoid *shape*. The most common activation functions
are

● *Unipolar sigmoid*:

$$f(\text{net}) = \frac{1}{1 + \exp(-\lambda \text{net})} \qquad (27.7)$$

● *Bipolar sigmoid* (hyperbolic tangent):

$$f(\text{net}) = \frac{2}{1 + \exp(-\lambda \text{net})} - 1 . \qquad (27.8)$$

The constant $\lambda > 0$ determines steepness of the sigmoid curve and it is commonly set to 1. In the limit
$\lambda \to \infty$ the bipolar sigmoid tends to the sign function
(used in the perceptron) and the unipolar sigmoid tends
to the step function.

Consider the single-layer ANN in Fig. 27.4. The
output and input vectors are $\bar{y} = (y_1, \ldots, y_j, \ldots, y_J)$
and $\bar{o} = (o_1, \ldots, o_k, \ldots, o_K)$, respectively, where $o_k =
f(\text{net}_k)$ and

$$\text{net}_k = \sum_{j=1}^{J} w_{kj} y_j . \qquad (27.9)$$

Set $y_J = -1$ and $w_{kJ} = \theta_k$, a threshold for $k = 1, \ldots, K$
output neurons. The desired output is $\bar{d} = (d_1, \ldots, d_k,
\ldots, d_K)$.

After training, we would like, for all training patterns $p = 1, \ldots, P$ from A_{train}, the model output to

closely resemble the desired values (target). The training problem is transformed to an optimization one by
defining the error function

$$E_p = \frac{1}{2} \sum_{k=1}^{K} (d_{pk} - o_{pk})^2 , \qquad (27.10)$$

where p is the training point index. E_p is the sum of
squares of errors on the output neurons. During learning we seek to find the weight setting that minimizes
E_p. This will be done using the gradient-based steepest
descent on E_p,

$$\Delta w_{kj} = -\alpha \frac{\partial E_p}{\partial w_{kj}} = -\alpha \frac{\partial E_p}{\partial(\text{net}_k)} \frac{\partial(\text{net}_k)}{\partial w_{kj}} = \alpha \delta_{ok} y_j ,$$
$$(27.11)$$

where α is a positive learning rate. Note that
$-\partial E_p/\partial(\text{net}_k) = \delta_{ok}$, which is the generalized training
signal on the k-th output neuron. The partial derivative
$\partial(\text{net}_k)/\partial w_{kj}$ is equal to y_j (27.9). Furthermore,

$$\delta_{ok} = -\frac{\partial E_p}{\partial(\text{net}_k)} = -\frac{\partial E_p}{\partial o_k} \frac{\partial o_k}{\partial(\text{net}_k)} = (d_{pk} - o_{pk})f'_k ,$$
$$(27.12)$$

where f'_k denotes the derivative of the activation function with respect to net_k. For the unipolar sigmoid
(27.7), we have $f'_k = o_k(1 - o_k)$. For the bipolar sigmoid
(27.8), $f'_k = (1/2)(1 - o_k^2)$. The rule for updating the j-th
weight of the k-th output neuron reads as

$$\Delta w_{kj} = \alpha(d_{pk} - o_{pk})f'_k y_j , \qquad (27.13)$$

where $(d_{pk} - o_{pk})f'_k = \delta_{ok}$ is *generalized error signal*
flowing back through all connections ending in the k-
the output neuron. Note that if we put $f'_k = 1$, we would
obtain the perceptron learning rule (27.6).

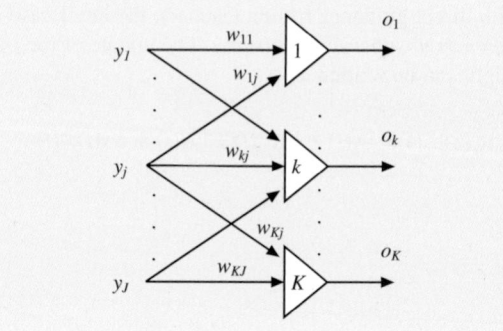

Fig. 27.4 A single-layer ANN

We will now extend the network with another layer, called the hidden layer (Fig. 27.5).

Input to the network is identical with the input vector $\bar{x} = (x_1, \ldots, x_i, \ldots, x_I)$ for the hidden layer. The output neurons process as inputs the outputs $\bar{y} = (y_1, \ldots, y_j, \ldots, y_J)$, $y_j = f(\text{net}_j)$ from the hidden layer. Hence,

$$\text{net}_j = \sum_{i=1}^{I} v_{ji} x_i . \tag{27.14}$$

As before, the last (in this case the I-th) input is fixed to -1. Recall that the same holds for the output of the J-th hidden neuron. Activation thresholds for hidden neurons are $v_{jI} = \theta_j$, for $j = 1, \ldots, J$.

Equations (27.11)–(27.13) describe modification of weights from the hidden to the output layer. We will now show how to modify weights from the input to the hidden layer. We would still like to minimize E_p (27.10) through the steepest descent.

The hidden weight v_{ji} will be modified as follows

$$\Delta v_{ji} = -\alpha \frac{\partial E_p}{\partial v_{ji}} = -\alpha \frac{\partial E_p}{\partial (\text{net}_j)} \frac{\partial (\text{net}_j)}{\partial v_{ji}} = \alpha \delta_{yj} x_i . \tag{27.15}$$

Again, $-\partial E_p / \partial (\text{net}_j) = \delta_{yj}$ is the generalized training signal on the j-th hidden neuron that should flow on the input weights. As before, $\partial (\text{net}_j) / \partial v_{ji} = x_i$ (27.14). Furthermore,

$$\delta_{yj} = -\frac{\partial E_p}{\partial (\text{net}_j)} = -\frac{\partial E_p}{\partial y_j} \frac{\partial y_j}{\partial (\text{net}_j)} = -\frac{\partial E_p}{\partial y_j} f'_j , \tag{27.16}$$

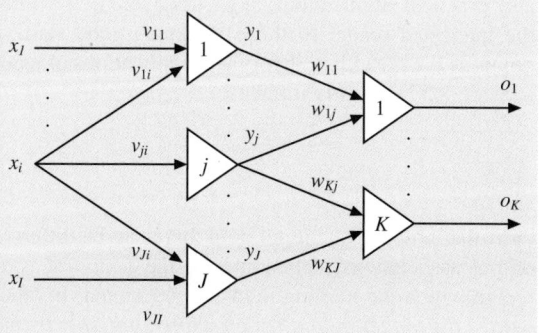

$$\text{Fig. 27.5}\ \text{A two-layer feed-forward ANN}$$

where f'_j is the derivative of the activation function in the hidden layer with respect to net_j,

$$\frac{\partial E_p}{\partial y_j} = -\sum_{k=1}^{K} (d_{pk} - o_{pk}) \frac{\partial f(\text{net}_k)}{\partial y_j}$$

$$= -\sum_{k=1}^{K} (d_{pk} - o_{pk}) \frac{\partial f(\text{net}_k)}{\partial (\text{net}_k)} \frac{\partial (\text{net}_k)}{\partial y_j} . \tag{27.17}$$

Since f'_k is the derivative of the output neuron sigmoid with respect to net_k and $\partial (\text{net}_k) / \partial y_j = w_{kj}$ (27.9), we have

$$\frac{\partial E_p}{\partial y_j} = -\sum_{k=1}^{K} (d_{pk} - o_{pk}) f'_k w_{kj} = -\sum_{k=1}^{K} \delta_{ok} w_{kj} . \tag{27.18}$$

Plugging this to (27.16) we obtain

$$\delta_{yj} = \left(\sum_{k=1}^{K} \delta_{ok} w_{kj} \right) f'_j . \tag{27.19}$$

Finally, the weights from the input to the hidden layer are modified as follows

$$\Delta v_{ji} = \alpha \left(\sum_{k=1}^{K} \delta_{ok} w_{kj} \right) f'_j x_i . \tag{27.20}$$

Consider now the general case of m hidden layers. For the n-th hidden layer we have

$$\Delta v_{ji}^n = \alpha \delta_{yj}^n x_i^{n-1} , \tag{27.21}$$

where

$$\delta_{yj}^n = \left(\sum_{k=1}^{K} \delta_{ok}^{n+1} w_{kj}^{n+1} \right) (f_j^n)' , \tag{27.22}$$

and $(f_j^n)'$ is the derivative of the activation function of the n-layer with respect to net_j^n.

Often, the learning speed can be improved by using the so-called momentum term

$$\Delta w_{kj}(t) \leftarrow \Delta w_{kj}(t) + \mu \Delta w_{kj}(t-1) ,$$
$$\Delta v_{ji}(t) \leftarrow \Delta v_{ji}(t) + \mu \Delta v_{ji}(t-1) , \tag{27.23}$$

where $\mu \in (0, 1)$ is the momentum rate.

Consider a training set

$$A_{\text{train}} = \{(\bar{x}^1, \bar{d}^1)(\bar{x}^2, \bar{d}^2) \ldots (\bar{x}^p, \bar{d}^p) \ldots (\bar{x}^P, \bar{d}^P)\} .$$

The back-propagation algorithm for training feed-forward ANNs can be summarized as follows:

- *Step* 1: Set $\alpha \in (0, 1)$. Randomly initialize weights to *small* values, e.g., in the interval $(-0.5, 0.5)$. Counters and the error are initialized as follows: $k = 1, p = 1, E = 0$. E denotes the accumulated error across training patterns

$$E = \sum_{p=1}^{P} E_p , \qquad (27.24)$$

where E_p is given in (27.10). Set a tolerance threshold ε for the error. The threshold will be used to stop the training process.
- *Step* 2: Apply input $\bar{x}^p$ and compute the corresponding $\bar{y}^p$ and $\bar{o}^p$.
- *Step* 3: For every output neuron, calculate δ_{ok} (27.12), for hidden neuron determine δ_{yj} (27.19).
- *Step* 4: Modify the weights $w_{kj} \leftarrow w_{kj} + \alpha \delta_{ok} y_j$ and $v_{ji} \leftarrow v_{ji} + \alpha \delta_{yj} x_i$.
- *Step* 5: If $p < P$, set $p = p + 1$ and go to step 2. Otherwise go to step 6.
- *Step* 6: Fixing the weights, calculate E. If $E < \varepsilon$, stop training, otherwise permute elements of A_{train}, set $E = 0, p = 1, k = k + 1$, and go to step 2.

Consider a feed-forward ANN with fixed weights and single output unit. It can be considered a real-valued function G on I-dimensional vectorial inputs,

$$G(\bar{x}) = f\left(\sum_{j=1}^{J} w_j f\left(\sum_{i=1}^{I} v_{ji} x_i\right)\right) .$$

There has been a series of results showing that such a parameterized function class is *sufficiently rich* in the space of *reasonable* functions (see, e.g., [27.4]). For example, for any smooth function F over a compact domain and a precision threshold ε, for sufficiently large number J of hidden units there is a weight setting so that G is not further away from F than ε (in L-2 norm).

When training a feed-forward ANN a key decision must be made about how complex the model should be. In other words, how many hidden units J one should use. If J is too small, the model will be too rigid (high

bias) and it will not be able to sufficiently adapt to the data. However, under different samples from the same data generating process, the resulting trained models will vary relatively little (low variance). On the other hand, if J is too high, the model will be too complex, modeling even such irrelevant features of the data such as output noise. The particular data will be interpolated exactly (low bias), but the variability of fitted models under different training samples from the same process will be immense. It is, therefore, important to set J to an *optimal* value, reflecting the complexity of the data generating process. This is usually achieved by splitting the data into three disjoint sets – *training, validation*, and *test sets*. Models with different numbers of hidden units are trained on the training set, their performance is then checked on a held-out validation set. The *optimal* number of hidden units is selected based on the (smallest) validation error. Finally, the test set is used for independent comparison of selected models from different model classes.

If the data set is not large enough, one can perform such a *model selection* using k-fold *cross-validation*. The data for model construction (this data would be considered training and validation sets in the scenario above) is split into k disjoint folds. One fold is selected as the validation fold, the other $k - 1$ will be used for training. This is repeated k times, yielding k estimates of the validation error. The validation error is then calculated as the mean of those k estimates.

We have described data-based methods for model selection. Other alternatives are available. For example, by turning an ANN into a probabilistic model (e.g., by including an appropriate output noise model), under some prior assumptions on weights (e.g., a-priori small weights are preferred), one can perform Bayesian model selection (through *model evidence*) [27.5].

There are several seminal books on feed-forward ANNs with well-documented theoretical foundations and practical applications, e.g., [27.3, 6, 7]. We refer the interested reader to those books as good starting points as the breadth of theory and applications of feed-forward ANNs is truly immense.

27.4 Recurrent ANN Models

Consider a situation where the associations in the training set we would like to learn are of the following (abstract) form: $a \to \alpha$, $b \to \beta$, $b \to \alpha$, $b \to \gamma$, $c \to \alpha$, $c \to \gamma$, $d \to \alpha$, etc., where the Latin and Greek letters stand for input and output vectors, respectively. It is

clear that now for one input item there can be different output associations, depending on the *temporal context* in which the training items are presented. In other words, *the model output is determined not only by the input, but also by the history of presented items so far.*

Obviously, the feed-forward ANN model described in the previous section cannot be used in such cases and the model must be further extended so that the temporal context is properly represented.

The architecturally simplest solution is provided by the so-called *time delay neural network* (TDNN) (Fig. 27.6). The input *window into the past* has a finite length D. If the output is an estimate of the next item of the input time series, such a network realizes a nonlinear autoregressive model of order D.

If we are lucky, even such a simple solution can be sufficient to capture the temporal structure hidden in the data. An advantage of the TDNN architecture is that some training methods developed for feed-forward networks can be readily used. A disadvantage of TDNN networks is that fixing a finite order D may not be adequate for modeling the temporal structure of the data generating source. TDNN enables the feed-forward ANN to see, besides the current input at time t, the other inputs from the past up to time $t - D$. Of course, during the training, it is now imperative to preserve the order of training items in the training set. TDNN has been successfully applied in many fields where spatial-temporal structures are naturally present, such as robotics, speech recognition, etc. [27.8, 9].

In order to extend the ANN architecture so that the *variable* (even unbounded) length of input window can be flexibly considered, we need a different way of capturing the temporal context. This is achieved through the so-called *state space formulation*. In this case, we will need to change our outlook on training. The new architectures of this type are known as *recurrent neural networks* (*RNN*).

As in feed-forward ANNs, there are connections between the successive layers. In addition, and in contrast to feed-forward ANNs, connections between neurons of the same layer are allowed, but subject to a *time de-*lay. It also may be possible to have connections from a higher-level layer to a lower layer, again subject to a time delay. In many cases it is, however, more convenient to introduce an additional fictional *context layer* that contains delayed activations of neurons from the selected layer(s) and represent the resulting RNN architecture as a feed-forward architecture with some fixed one-to-one delayed connections. As an example, consider the so-called *simple recurrent network* (SRN) of Elman [27.10] shown in Fig. 27.7. The output of SRN at time t is given by

$$o_k^{(t)} = f\left(\sum_{j=1}^{J} m_{kj} y_j^{(t)}\right),$$

$$y_j^{(t)} = f\left(\sum_{i=1}^{J} w_{ji} y_i^{(t-1)} + \sum_{i=1}^{I} v_{ji} x_i^{(t)}\right). \qquad (27.25)$$

The hidden layer constitutes the state of the input-driven dynamical system whose role it is to represent the relevant (with respect to the output) information

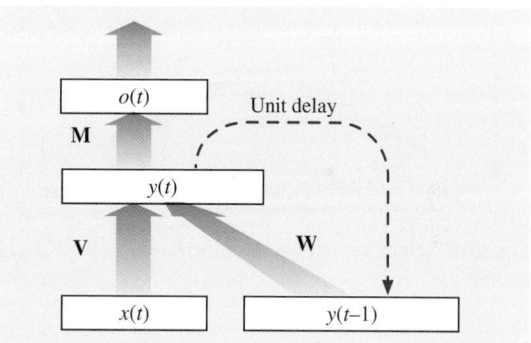

Fig. 27.7 Schematic depiction of the SRN architecture

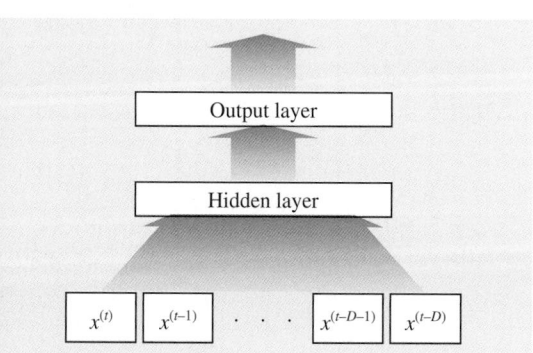

Fig. 27.6 TDNN of order D

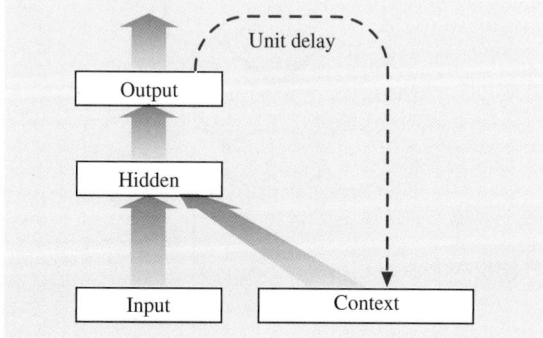

Fig. 27.8 Schematic depiction of the Jordan's RNN architecture

about the input history seen so far. The state (as in generic state space model) is updated recursively.

Many variations on such architectures with time-delayed feedback loops exist. For example, *Jordan* [27.11] suggested to feed back the outputs as the relevant temporal context, or *Bengio* et al. [27.12] mixed the temporal context representations of SRN and the Jordan network into a single architecture. Schematic representations of these architectures are shown in Figs. 27.8 and 27.9.

Training in such architectures is more complex than training of feed-forward ANNs. The principal problem is that changes in weights propagate in time and this needs to be explicitly represented in the update rules. We will briefly mention two approaches to training RNNs, namely *back-propagation through time* (BPTT) [27.13] and *real-time recurrent learning* (RTRL) [27.14]. We will demonstrate BPTT on a clas-

sification task, where the label of the input sequence is known only after T time steps (i.e., after T input items have been processed). The RNN is unfolded in time to form a feed-forward network with T hidden layers. Figure 27.10 shows a simple two-neuron RNN and Fig. 27.11 represents its unfolded form for $T = 2$ time steps.

The first input comes at time $t = 1$ and the last at $t = T$. Activities of context units are initialized at the beginning of each sequence to some fixed numbers. The unfolded network is then trained as a feed-forward network with T hidden layers. At the end of the sequence, the model output is determined as

$$o_k^{(T)} = f\left(\sum_{j=1}^{J} m_{kj}^{(T)} y_j^{(T)}\right),$$

$$y_j^{(t)} = f\left(\sum_{i=1}^{J} w_{ji}^{(t)} y_i^{(t-1)} + \sum_{i=1}^{I} v_{ji}^{(t)} x_i^{(t)}\right). \quad (27.26)$$

Having the model output enables us to compute the error

$$E(T) = \frac{1}{2}\sum_{k=1}^{K}\left(d_k^{(T)} - o_k^{(T)}\right)^2. \quad (27.27)$$

The hidden-to-output weights are modified according to

$$\Delta m_{kj}^{(T)} = -\alpha\frac{\partial E(T)}{\partial m_{kj}} = \alpha\delta_k^{(T)} y_j^{(T)}, \quad (27.28)$$

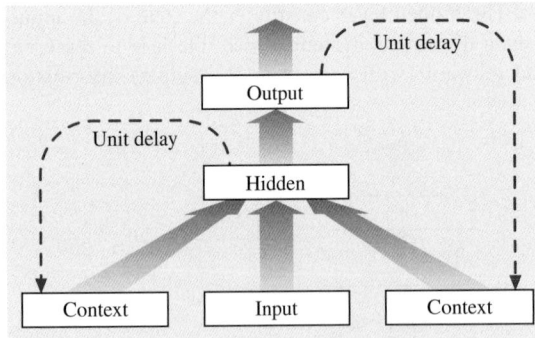

Fig. 27.9 Schematic depiction of the Bengio's RNN architecture

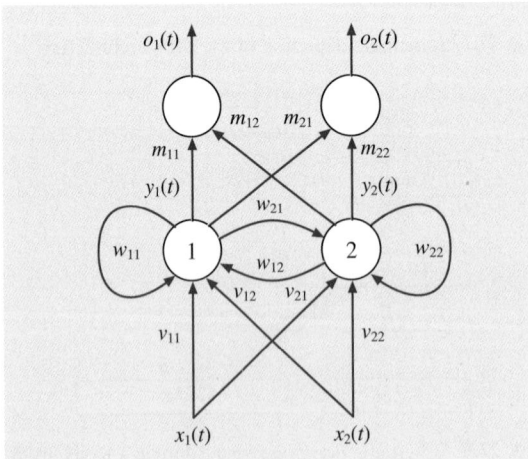

Fig. 27.10 A two-neuron SRN

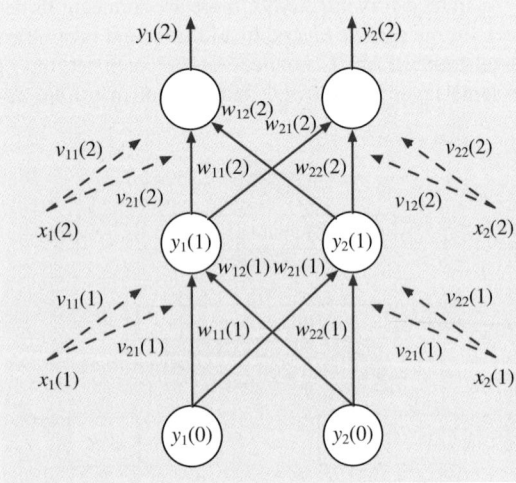

Fig. 27.11 Two-neuron SRN unfolded in time for $T = 2$

where

$$\delta_k^{(T)} = \left(d_k^{(T)} - o_k^{(T)} \right) f' \left(\text{net}_k^{(T)} \right) . \tag{27.29}$$

The other weight updates are calculated as follows

$$\Delta w_{hj}^{(T)} = -\alpha \frac{\partial E(T)}{\partial w_{hj}} = \alpha \delta_h^{(T)} y_j^{(T-1)} ;$$

$$\delta_h^{(T)} = \left(\sum_{k=1}^{K} \delta_k^{(T)} m_{kh}^{(T)} \right) f' \left(\text{net}_h^{(T)} \right) \tag{27.30}$$

$$\Delta v_{ji}^{(T)} = -\alpha \frac{\partial E(T)}{\partial v_{ji}} = \alpha \delta_j^{(T)} x_i^{(T)} ;$$

$$\delta_j^{(T)} = \left(\sum_{k=1}^{K} \delta_k^{(T)} m_{kj}^{(T)} \right) f' \left(\text{net}_j^{(T)} \right) \tag{27.31}$$

$$\Delta w_{hj}^{(T-1)} = \alpha \delta_h^{(T-1)} y_j^{(T-2)} ;$$

$$\delta_h^{(T-1)} = \left(\sum_{j=1}^{J} \delta_j^{(T-1)} w_{jh}^{(T-1)} \right) f' \left(\text{net}_h^{(T-1)} \right) \tag{27.32}$$

$$\Delta v_{ji}^{(T-1)} = \alpha \delta_j^{(T-1)} x_i^{(T-1)} ;$$

$$\delta_j^{(T-1)} = \left(\sum_{h=1}^{J} \delta_h^{(T-1)} w_{hj}^{(T-1)} \right) f' \left(\text{net}_j^{(T-1)} \right) , \tag{27.33}$$

etc. The final weight updates are the averages of the T partial weight update suggestions calculated on the unfolded network

$$\Delta w_{hj} = \frac{\sum_{t=1}^{T} \Delta w_{hj}^{(t)}}{T} \quad \text{and} \quad \Delta v_{ji} = \frac{\sum_{t=1}^{T} \Delta v_{ji}^{(t)}}{T} . \tag{27.34}$$

For every new training sequence (of possibly different length T) the network is unfolded to the desired length and the weight update process is repeated. In some cases (e.g., continual prediction on time series), it is necessary to set the maximum unfolding length L that will be used in every update step. Of course, in such cases we can lose vital information from the past. This problem is eliminated in the RTRL methodology.

Consider again the SRN architecture in Fig. 27.6. In RTRL the weights are updated *on-line*, i. e., at every

time step t

$$\Delta w_{kj}^{(t)} = -\alpha \frac{\partial E^{(t)}}{\partial w_{kj}^{(t)}} ,$$

$$\Delta v_{ji}^{(t)} = -\alpha \frac{\partial E^{(t)}}{\partial v_{ji}^{(t)}} ,$$

$$\Delta m_{jl}^{(t)} = -\alpha \frac{\partial E^{(t)}}{\partial m_{jl}^{(t)}} . \tag{27.35}$$

The updates of hidden-to-output weights are straightforward

$$\Delta m_{kj}^{(t)} = \alpha \delta_k^{(t)} y_j^{(t)} = \alpha \left(d_k^{(t)} - o_k^{(t)} \right) f_k' \left(\text{net}_k^{(t)} \right) y_j^{(t)} . \tag{27.36}$$

For the other weights we have

$$\Delta v_{ji}^{(t)} = \alpha \sum_{k=1}^{K} \left(\delta_k^{(t)} \sum_{h=1}^{J} w_{kh} \frac{\partial y_h^{(t)}}{\partial v_{ji}} \right) ,$$

$$\Delta w_{ji}^{(t)} = \alpha \sum_{k=1}^{K} \left(\delta_k^{(t)} \sum_{h=1}^{J} w_{kh} \frac{\partial y_h^{(t)}}{\partial w_{ji}} \right) , \tag{27.37}$$

where

$$\frac{\partial y_h^{(t)}}{\partial v_{ji}} = f' \left(\text{net}_h^{(t)} \right) \left(x_i^{(t)} \delta_{jh}^{\text{Kron.}} + \sum_{l=1}^{J} w_{hl} \frac{\partial y_l^{(t-1)}}{\partial v_{ji}} \right) \tag{27.38}$$

$$\frac{\partial y_h^{(t)}}{\partial w_{ji}} = f' \left(\text{net}_h^{(t)} \right) \left(x_i^{(t)} \delta_{jh}^{\text{Kron.}} + \sum_{l=1}^{J} w_{hl} \frac{\partial y_l^{(t-1)}}{\partial w_{ji}} \right) , \tag{27.39}$$

and $\delta_{jh}^{\text{Kron.}}$ is the Kronecker delta ($\delta_{jh}^{\text{Kron.}} = 1$, if $j = h$; $\delta_{jh}^{\text{Kron.}} = 0$ otherwise). The partial derivatives required for the weight updates can be recursively updated using (27.37)–(27.39). To initialize training, the partial derivatives at $t = 0$ are usually set to 0.

There is a well-known problem associated with gradient-based parameter fitting in recurrent networks (and, in fact, in any parameterized state space models of similar form) [27.15]. In order to *latch* an important piece of past information for future use, the state-transition dynamics (27.25) should have an attractive set.

However, in the neighborhood of such an attractive set, the derivatives of the dynamic state-transition map vanish. Vanishingly small derivatives cannot be reliably propagated back through time in order to form a useful latching set. This is known as the *information latching problem*. Several suggestions for dealing with information latching problem have been made, e.g., [27.16]. The most prominent include *long short term memory* (LSTM) RNN [27.17] and *reservoir computation* models [27.18].

LSTM models operate with a specially designed formal neuron model that contains so-called *gate* units. The gates determine when the input is *significant* (in terms of the task given) to be remembered, whether the neuron should continue to remember the value, and when the value should be output. The LSTM architecture is especially suitable for situations where there are long time intervals of unknown size between *important* events. LSTM models have been shown to provide superior results over traditional RNNs in a variety of applications (e.g., [27.19, 20]).

Reservoir computation models try to avoid the information latching problem by fixing the state-transition part of the RNN. Only linear readout from the state activations (hidden recurrent layer) producing the output is fit to the data. The state space with the as-sociated dynamic state transition structure is called the *reservoir*. The main idea is that the reservoir should be sufficiently complex so as to capture a large number of potentially useful features of the input stream that can be then exploited by the simple readout.

The reservoir computing models differ in how the fixed reservoir is constructed and what form the readout takes. For example, *echo state networks* (ESN) [27.21] have fixed RNN dynamics (27.25), but with a lin-ear hidden-to-output layer map. *Liquid state machines* (LSM) [27.22] also have (mostly) linear readout, but the reservoirs are realized through the dynamics of a set of coupled spiking neuron models. *Fractal prediction machines* (FPM) [27.23] are reservoir RNN models for processing discrete sequences. The reservoir dynamics is driven by an affine iterative function system and the readout is constructed as a collection of multinomial distributions. Reservoir models have been successfully applied in many practical applications with competitive results, e.g., [27.21, 24, 25].

Several books that are solely dedicated to RNNs have appeared, e.g., [27.26–28] and they contain a much deeper elaboration on theory and practice of RNNs than we were able to provide here.

27.5 Radial Basis Function ANN Models

In this section we will introduce another implemen-tation of the idea of feed-forward ANN. The activa-tions of hidden neurons are again determined by the *closeness* of inputs $\bar{x} = (x_1, x_2, \ldots, x_n)$ to weights $\bar{c} = (c_1, c_2, \ldots, c_n)$. Whereas in the feed-forward ANN in Sect. 27.3, the closeness is determined by the dot-product of $\bar{x}$ and $\bar{c}$, followed by the sigmoid activation function, in *radial basis function* (RBF) *networks* the closeness is determined by the squared Euclidean dis-tance of $\bar{x}$ and $\bar{c}$, transferred through the inverse expo-nential. The output of the j-th hidden unit with input weight vector $\bar{c}_j$ is given by

$$\varphi_j(\bar{x}) = \exp\left(-\frac{\|\bar{x} - \bar{c}_j\|^2}{\sigma_j^2}\right), \tag{27.40}$$

where σ_j is the *activation strength* parameter of the j-th hidden unit and determines the width of the spherical (un-normalized) Gaussian. The output neurons are usu-ally linear (for regression tasks)

$$o_k(\bar{x}) = \sum_{j=1}^{J} w_{kj} \varphi_j(\bar{x}). \tag{27.41}$$

The RBF network in this form can be simply viewed as a form of kernel regression. The J functions φ_j form a set of J linearly independent basis functions (e.g., if all the centers $\bar{c}_j$ are different) whose span (the set of all their linear combinations) forms a lin-ear subspace of functions that are realizable by the given RBF architecture (with given centers $\bar{c}_j$ and kernel widths σ_j).

For the training of RBF networks, it important that the basis functions $\varphi_j(\bar{x})$ cover the structure of the in-puts space *faithfully*. Given a set of training inputs $\bar{x}^p$ from $A_{\text{train}} = \{(\bar{x}^1, \bar{d}^1)(\bar{x}^2, \bar{d}^2) \ldots (\bar{x}^p, \bar{d}^p) \ldots (\bar{x}^P, \bar{d}^P)\}$, many RBF-ANN training algorithms determine the cen-ters $\bar{c}_j$ and widths σ_j based on the inputs $\{\bar{x}^1, \bar{x}^2, \ldots, \bar{x}^P\}$ only. One can employ different clustering algo-rithms, e.g., *k-means* [27.29], which attempts to position the centers among the training inputs so that the overall sum of (Euclidean) distances be-tween the centers and the inputs they represent (i.e., the inputs falling in their respective Voronoi com-partments – the set of inputs for which the cur-rent center is the closest among all the centers) is minimized:

- *Step* 1: Set J, the number of hidden units. The optimum value of J can be obtained through a model selection method, e.g., cross-validation.
- *Step* 2: Randomly select J training inputs that will form the initial positions of the J centers $\bar{c}_j$.
- *Step* 3: At time step t:
  a) Pick a training input $\bar{x}(t)$ and find the center $\bar{c}(t)$ closest to it.
  b) Shift the center $\bar{c}(t)$ towards $\bar{x}(t)$

$$\bar{c}(t) \leftarrow \bar{c}(t) + \rho(t)(\bar{x}(t) - \bar{c}(t)) ,$$
$$\text{where } 0 \leq \rho(t) \leq 1 . \qquad (27.42)$$

The learning rate $\rho(t)$ usually decreases in time towards zero. The training is stopped once the centers settle in their positions and move only slightly (some norm of weight updates is below a certain threshold). Since k-means is guaranteed to find only locally optimal solutions, it is worth re-initializing the centers and re-running the algorithm several times, keeping the solution with the lowest quantization error.

Once the centers are in their positions, it is easy to determine the RBF widths, and once this is done, the output weights can be solved using methods of linear regression.

Of course, it is more optimal to position the centers with respect to *both* the inputs and target outputs in the training set. This can be formulated, e.g., as a gradient descent optimization. Furthermore, covering of the input space with spherical Gaussian kernels may not be optimal, and algorithms have been developed for learning of general covariance structures. A comprehensive review of RBF networks can be found, e.g., in [27.30].

Recently, it was shown that if enough hidden units are used, their centers can be set randomly at very little cost, and determination of the only remaining free parameters – output weights – can be done cheaply and in a closed form through linear regression. Such architectures, known as *extreme learning machines* [27.31] have shown surprisingly high performance levels. The idea of extreme learning machines can be considered as being analogous to the idea of reservoir computation, but in the *static* setting. Of course, extreme learning machines can be built using other implementations of feed-forward ANNs, such as the sigmoid networks of Sect. 27.3.

27.6 Self-Organizing Maps

In this section we will introduce ANN models that learn without any signal from a teacher, i.e., learning is based solely on training inputs – there are no output targets. The ANN architecture designed to operate in this setting was introduced by *Kohonen* under the name *self-organizing map* (SOM) [27.32]. This model is motivated by organization of neuron sensitivities in the brain cortex.

In Fig. 27.12a we show schematic illustration of one of the principal organizations of biological neural networks. In the bottom layer (grid) there are receptors representing the inputs. Every element of the inputs (each receptor) has forward connections to all neurons in the upper layer representing the cortex. The neurons are organized spatially on a grid. Outputs of the neurons represent activation of the SOM network. The neurons, besides receiving connections from the input receptors, have a lateral interconnection structure among themselves, with connections that can be excitatory, or inhibitory. In Fig. 27.12b we show a formal SOM architecture – neurons spatially organized on a grid receive inputs (elements of input vectors) through connections with synaptic weights.

A particular feature of the SOM is that it can map the training set on the neuron grid in a manner that preserves the training set's topology – two input patterns *close* in the input space will activate neurons most that are close on the SOM grid. Such *topological mapping* of inputs (feature mapping) has been observed in biological neural networks [27.32] (e.g., visual maps, orientation maps of visual contrasts, or auditory maps, frequency maps of acoustic stimuli).

Teuvo Kohonen presented one possible realization of the Hebb rule that is used to train SOM. Input

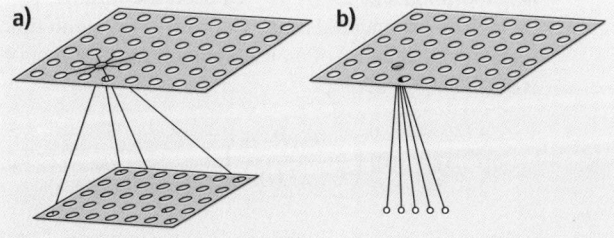

Fig. 27.12a,b Schematic representation of the SOM ANN architectures

weights of the neurons are initialized as small random numbers. Consider a training set of inputs, $A_{\text{train}} = \{\bar{x}_p\}_{p=1}^P$ and linear neurons

$$o_i = \sum_{j=1}^m w_{ij} x_j = \bar{w}_i \bar{x} \,, \tag{27.43}$$

where m is the input dimension and $i = 1, \ldots, n$. Training inputs are presented in random order. At each training step, we find the (winner) neuron with the weight vector *most similar* to the current input $\bar{x}$. The measure of similarity can be based on the dot product, i. e., the index of the winner neuron is $i^* = \arg \max(\bar{w}_i^T \bar{x})$, or the (Euclidean) distance $i^* = \arg \min_i \|\bar{x} - \bar{w}_i\|$. After identifying the winner the learning continues by adapting the winner's weights *along with the weights all its neighbors on the neuron grid*. This will ensure that nearby neurons on the grid will eventually represent similar inputs in the input space. This is moderated by a *neighborhood function $h(i^*, i)$* that, given a winner neuron index i^*, quantifies how many other neurons on the grid should be adapted

$$\bar{w}_i (t+1) = \bar{w}_i (t) + \alpha (t) \cdot h \left(i^*, i\right) \cdot (\bar{x} (t) - \bar{w}_i (t)) \,. \tag{27.44}$$

The learning rate $\alpha(t) \in (0, 1)$ decays in time as $1/t$, or $\exp(-kt)$, where k is a positive time scale constant. This ensures convergence of the training process. The simplest form of the neighborhood function operates with rectangular neighborhoods,

$$h(i^*, i) = \begin{cases} 1, & \text{if } d_M(i^*, i) \le \lambda (t) \\ 0, & \text{otherwise} \,, \end{cases} \tag{27.45}$$

where $d_M(i^*, i)$ represents the $2\lambda(t)$ (*Manhattan*) distance between neurons i^* and i on the map grid. The neighborhood size $2\lambda (t)$ should decrease in time, e.g., through an exponential decay as or $\exp(-qt)$, with time scale $q > 0$. Another often used neighborhood function is the Gaussian kernel

$$h(i^*, i) = \exp \left(-\frac{d_E^2 (i^*, i)}{\lambda^2 (t)}\right) \,, \tag{27.46}$$

where $d_E(i^*, i)$ is the Euclidean distance between i^* and i on the grid, i. e., $d_E(i^*, i) = \|\bar{r}_{i^*} - \bar{r}_i\|$, where $\bar{r}_i$ is the co-ordinate vector of the i-th neuron on the grid SOM.

Training of SOM networks can be summarized as follows:

- *Step* 1: Set α_0, λ_0 and t_{max} (maximum number of iterations). Randomly (e.g., with uniform distribution) generate the synaptic weights (e.g., from $(-0.5, 0.5)$). Initialize the counters: $t = 0$, $p = 1$; t indexes time steps (iterations) and p is the input pattern index.
- *Step* 2: Take input $\bar{x}^p$ and find the corresponding winner neuron.
- *Step* 3: Update the weights of the winner and its topological neighbors on the grid (as determined by the neighborhood function). Increment t.
- *Step* 4: Update α and λ.
- *Step* 5: If $p < P$, set $p \leftarrow p + 1$, go to step 2 (we can also use randomized selection), otherwise go to step 6.
- *Step* 6: If $t = t_{\text{max}}$, finish the training process. Otherwise set $p = 1$ and go to step 2. A new training epoch begins.

The SOM network can be used as a tool for nonlinear data visualization (grid dimensions 1, 2, or 3). In general, SOM implements constrained vector quantization, where the codebook vectors (vector quantization centers) cannot move freely in the data space during adaptation, but are constrained to lie on a lower dimensional manifold Ψ in the data space. The dimensionality of Ψ is equal to the dimensionality of the neural grid. The neural grid can be viewed as a discretized version of the local co-ordinate system Υ (e.g., computer screen) and the weight vectors in the data space (connected by the neighborhood structure on the neuron grid) as its image in the data space. In this interpretation, the neuron positions on the grid represent co-ordinate functions (in the sense of differential geometry) mapping elements of the manifold Ψ to the coordinate system Υ. Hence, the SOM algorithm can also be viewed as one particular implementation of manifold learning.

There have been numerous successful applications of SOM in a wide variety of applications, e.g., in image processing, computer vision, robotics, bioinformatics, process analysis, and telecommunications. A good survey of SOM applications can be found, e.g., in [27.33]. SOMs have also been extended to temporal domains, mostly by the introduction of additional feedback connections, e.g., [27.34–37]. Such models can be used for topographic mapping or constrained clustering of time series data.

27.7 Recursive Neural Networks

In many application domains, data are naturally organized in structured form, where each data item is composed of several components related to each other in a non-trivial way, and the specific nature of the task to be performed is strictly related not only to the information stored at each component, but also to the structure connecting the components. Examples of structured data are parse trees obtained by parsing sentences in natural language, and the molecular graph describing a chemical compound.

Recursive neural networks (RecNN) [27.38, 39] are neural network models that are able to directly process structured data, such as trees and graphs. For the sake of presentation, here we focus on positional trees. Positional trees are trees for which each child has an associated index, its position, with respect to the siblings. Let us understand how RecNN is able to process a tree by analogy with what happens when unfolding a RNN when processing a sequence, which can be understood as a special case of tree where each node v possesses a single child.

In Fig. 27.13 (top) we show the unfolding in time of a sequence when considering a graphical model (*re-cursive network*) representing, for a generic node v, the functional dependencies among the input information x_v, the state variable (hidden node) y_v, and the output variable o_v. The operator q^{-1} represents the shift operator in time (unit time delay), i.e., $q^{-1}y_t = y_{t-1}$, which applied to node v in our framework returns the child of node v. At the bottom of Fig. 27.13 we have reported the unfolding of a binary tree, where the recursive network uses a generalization of the shift operator, which given an index i and a variable associated to a vertex v returns the variable associated to the i-th child of v, i.e., $q_i^{-1}y_v = y_{\mathrm{ch}_i[v]}$. So, while in RNN the network is *unfolded in time*, in RecNN the network is *unfolded on the structure*. The result of unfolding, in both cases, is the *encoding network*. The encoding network for the sequence specifies how the components implementing the different parts of the recurrent network (e.g., each node of the recurrent network could be instantiated by a layer of neurons or by a full feed-forward neural network with hidden units) need to be interconnected. In the case of the tree, the encoding network has the same semantics: this time, however, a set of parameters (weights) for each child should be considered, leading to a net-

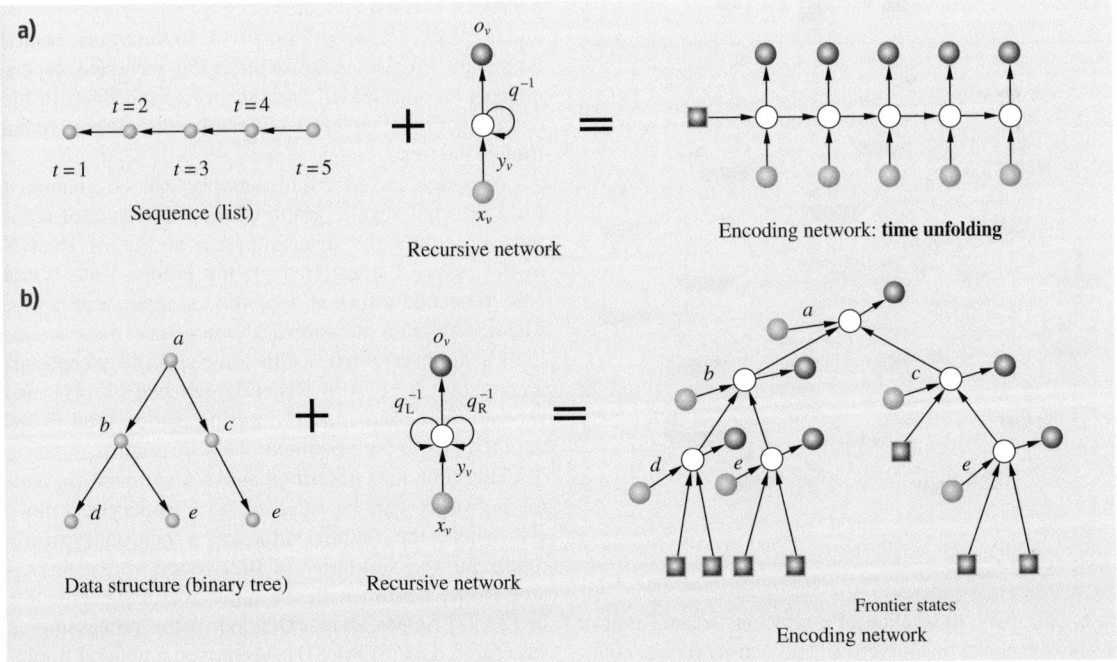

Fig. 27.13a,b Generation of the encoding network (**a**) for a sequence and (**b**) a tree. Initial states are represented by squared nodes

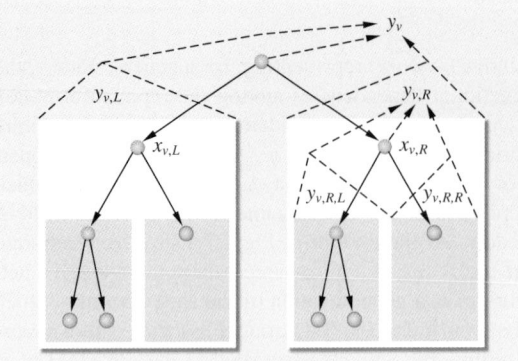

Fig. 27.14 The causality style of computation induced by the use of recursive networks is made explicit by using nested boxes to represent the recursive dependencies of the hidden variable associated to the root of the tree ◄

work that, given a node v, can be described by the equations

$$o_k^{(v)} = f\left(\sum_{j=1}^{J} m_{kj} y_j^{(v)}\right),$$

$$y_j^{(v)} = f\left(\sum_{s=1}^{d}\sum_{i=1}^{J} w_{ji}^s y_i^{(\text{ch}_s[v])} + \sum_{i=1}^{I} v_{ji} x_i^{(v)}\right),$$

where d is the maximum number of children an input node can have, and weights w_{ji}^s are indexed on the s-th child. Note that it is not difficult to generalize *all the learning algorithms* devised for RNN to these extended equations.

It should be remarked that recursive networks clearly introduce a causal style of computation, i.e., the computation of the hidden and output variables for a vertex v only depends on the information attached to v and the hidden variables of the children of v. This dependence is satisfied recursively by all v's descendants and is clearly shown in Fig. 27.14. In the figure, nested boxes are used to make explicit the recursive dependencies among hidden variables that contribute to the determination of the hidden variable y_v associated to the root of the tree.

Although an encoding network can be generated for a directed acyclic graph (DAG), this style of computation limits the discriminative ability of RecNN to the class of trees. In fact, the hidden state is not able to encode information about the *parents* of nodes. The introduction of *contextual processing*, however, allows us to discriminate, with some specific exceptions, among DAGs [27.40]. Recently, *Micheli* [27.41] also showed how contextual processing can be used to extend RecNN to the treatment of *cyclic* graphs.

The same idea described above for supervised neural networks can be adapted to unsupervised models, where the output value of a neuron typically represents the similarity of the weight vector associated to the neuron with the input vector. Specifically, in [27.37] SOMs were extended to the processing of structured data (SOM-SD). Moreover, a general framework for self-organized processing of structured data

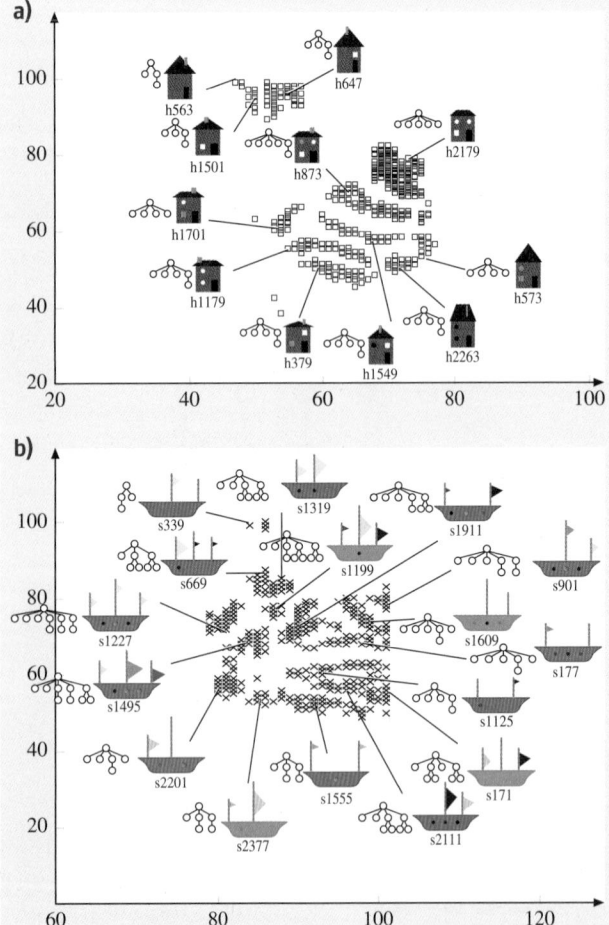

Fig. 27.15a,b Schematic illustration of a principal organization of biological self-organizing neural network (**a**) and its formal counterpart SOM ANN architecture (**b**)

was proposed in [27.42]. The key concepts introduced are:

i)   The explicit definition of a representation space R equipped with a similarity measure $d_R(\cdot, \cdot)$ to evaluate the similarity between two hidden states.

ii) The introduction of a general representation function, denoted $rep(\cdot)$, which transforms the activation of the map for a given input into an hidden state representation.

In these models, each node v of the input structure is represented by a tuple $[\bar{x}_v, \bar{r}_{v_1}, \ldots, \bar{r}_{v_d}]$, where $\bar{x}_v$ is a real-valued vectorial encoding of the information attached to vertex v, and $\bar{r}_{v_i}$ are real-valued vectorial representations of hidden states returned by the $rep(\cdot)$ function when processing the activation of the map for the i-th neighbor of v. Each neuron n_j in the map is associated to a weight vector $[\bar{w}_j, \bar{c}_j^1, \ldots, \bar{c}_j^d]$. The computation of the winner neuron is based on the joint contribution of the similarity measures $d_x(\cdot, \cdot)$ for the input information, and $d_R(\cdot, \cdot)$ for the hidden states, i.e., the internal representations. Some parts of a SOM-SD map trained on DAGs representing visual patterns are shown in Fig. 27.15. Even in this case the style of computation is causal, ruling out the treatment of undirected and/or cyclic graphs. In order to cope with general graphs, recently a new model, named GraphSOM [27.43], was proposed.

27.8 Conclusion

The field of artificial neural networks (ANN) has grown enormously in the past 60 years. There are many journals and international conferences specifically devoted to neural computation and neural network related models and learning machines. The field has gone a long way from its beginning in the form of simple threshold units existing in isolation (e.g., the perceptron, Sect. 27.2) or connected in circuits. Since then we have learnt how to generalize such networks as parameterized differentiable models of various sorts that can be fit to data (*trained*), usually by transforming the learning task into an optimization one.

ANN models have found numerous successful practical applications in many diverse areas of science and engineering, such as astronomy, biology, finance, geology, etc. In fact, even though basic feed-forward ANN architectures were introduced long time ago, they continue to surprise us with successful applications, most recently in the form of *deep networks* [27.44]. For example, a form of deep ANN recently achieved the best performance on a well-known benchmark problem – the recognition of handwritten digits [27.45]. This is quite remarkable, since such a simple ANN architecture trained in a purely data driven fashion was able to outperform the current state-of-art techniques, formulated in more sophisticated frameworks and possibly incorporating domain knowledge.

ANN models have been formulated to operate in supervised (e.g., feed-forward ANN, Sect. 27.3; RBF networks, Sect. 27.5), unsupervised (e.g., SOM models, Sect. 27.6), semi-supervised, and reinforcement learning scenarios and have been generalized to process inputs that are much more general than simple vector data of fixed dimensionality (e.g., the recurrent and recursive networks discussed in Sects. 27.4 and 27.7). Of course, we were not able to cover all important developments in the field of ANNs. We can only hope that we have sufficiently motivated the interested reader with the variety of modeling possibilities based on the idea of interconnected networks of formal neurons, so that he/she will further consult some of the many (much more comprehensive) monographs on the topic, e.g., [27.3, 6, 7].

We believe that ANN models will continue to play an important role in modern computational intelligence. Especially the inclusion of ANN-like models in the field of probabilistic modeling can provide techniques that incorporate both explanatory model-based and data-driven approaches, while preserving a much fuller modeling capability through operating with full distributions, instead of simple point estimates.

References

27.1 F. Rosenblatt: The perceptron, a probabilistic model for information storage and organization in the brain, Psychol. Rev. **62**, 386–408 (1958)

27.2 D.E. Rumelhart, G.E. Hinton, R.J. Williams: Learning internal representations by error propagation. In: *Parallel Distributed Processing: Explorations in the Microstructure of Cognition. Vol. 1 Foundations*, ed. by D.E. Rumelhart, J.L. McClelland (MIT Press/Bradford Books, Cambridge 1986) pp. 318–363

27.3 J. Zurada: *Introduction to Artificial Neural Systems* (West Publ., St. Paul 1992)

27.4 K. Hornik, M. Stinchcombe, H. White: Multilayer feedforward networks are universal approximators, Neural Netw. **2**, 359–366 (1989)

27.5 D.J.C. MacKay: Bayesian interpolation, Neural Comput. **4**(3), 415–447 (1992)

27.6 S. Haykin: *Neural Networks and Learning Machines* (Prentice Hall, Upper Saddle River 2009)

27.7 C. Bishop: *Neural Networks for Pattern Recognition* (Oxford Univ. Press, Oxford 1995)

27.8 T. Sejnowski, C. Rosenberg: Parallel networks that learn to pronounce English text, Complex Syst. **1**, 145–168 (1987)

27.9 A. Weibel: Modular construction of time-delay neural networks for speech recognition, Neural Comput. **1**, 39–46 (1989)

27.10 J.L. Elman: Finding structure in time, Cogn. Sci. **14**, 179–211 (1990)

27.11 M.I. Jordan: Serial order: A parallel distributed processing approach. In: *Advances in Connectionist Theory*, ed. by J.L. Elman, D.E. Rumelhart (Erlbaum, Hillsdale 1989)

27.12 Y. Bengio, R. Cardin, R. DeMori: Speaker independent speech recognition with neural networks and speech knowledge. In: *Advances in Neural Information Processing Systems II*, ed. by D.S. Touretzky (Morgan Kaufmann, San Mateo 1990) pp. 218–225

27.13 P.J. Werbos: Generalization of backpropagation with application to a recurrent gas market model, Neural Netw. **1**(4), 339–356 (1988)

27.14 R.J. Williams, D. Zipser: A learning algorithm for continually running fully recurrent neural networks, Neural Comput. **1**(2), 270–280 (1989)

27.15 Y. Bengio, P. Simard, P. Frasconi: Learning long-term dependencies with gradient descent is difficult, IEEE Trans. Neural Netw. **5**(2), 157–166 (1994)

27.16 T. Lin, B.G. Horne, P. Tino, C.L. Giles: Learning long-temr dependencies with NARX recurrent neural networks, IEEE Trans. Neural Netw. **7**(6), 1329–1338 (1996)

27.17 S. Hochreiter, J. Schmidhuber: Long short-term memory, Neural Comput. **9**(8), 1735–1780 (1997)

27.18 M. Lukosevicius, H. Jaeger: *Overview of Reservoir Recipes*, Technical Report, Vol. 11 (School of Engineering and Science, Jacobs University, Bremen 2007)

27.19 A. Graves, M. Liwicki, S. Fernandez, R. Bertolami, H. Bunke, J. Schmidhuber: A novel connectionist system for improved unconstrained handwriting recognition, IEEE Trans. Pattern Anal. Mach. Intell. **31**, 5 (2009)

27.20 S. Hochreiter, M. Heusel, K. Obermayer: Fast model-based protein homology detection without alignment, Bioinformatics **23**(14), 1728–1736 (2007)

27.21 H. Jaeger, H. Hass: Harnessing nonlinearity: predicting chaotic systems and saving energy in wireless telecommunication, Science **304**, 78–80 (2004)

27.22 W. Maass, T. Natschlager, H. Markram: Real-time computing without stable states: A new framework for neural computation based on perturbations, Neural Comput. **14**(11), 2531–2560 (2002)

27.23 P. Tino, G. Dorffner: Predicting the future of discrete sequences from fractal representations of the past, Mach. Learn. **45**(2), 187–218 (2001)

27.24 M.H. Tong, A. Bicket, E. Christiansen, G. Cottrell: Learning grammatical structure with echo state network, Neural Netw. **20**, 424–432 (2007)

27.25 K. Ishii, T. van der Zant, V. Becanovic, P. Ploger: Identification of motion with echo state network, Proc. OCEANS 2004 MTS/IEEE-TECHNO-OCEAN Conf., Vol. 3 (2004) pp. 1205–1210

27.26 L. Medsker, L.C. Jain: *Recurrent Neural Networks: Design and Applications* (CRC, Boca Raton 1999)

27.27 J. Kolen, S.C. Kremer: *A Field Guide to Dynamical Recurrent Networks* (IEEE, New York 2001)

27.28 D. Mandic, J. Chambers: *Recurrent Neural Networks for Prediction: Learning Algorithms, Architectures and Stability* (Wiley, New York 2001)

27.29 J.B. MacQueen: Some models for classification and analysis if multivariate observations, Proc. 5th Berkeley Symp. Math. Stat. Probab. (Univ. California Press, Oakland 1967) pp. 281–297

27.30 M.D. Buhmann: *Radial Basis Functions: Theory and Implementations* (Cambridge Univ. Press, Cambridge 2003)

27.31 G.-B. Huang, Q.-Y. Zhu, C.-K. Siew: Extreme learning machine: theory and applications, Neurocomputing **70**, 489–501 (2006)

27.32 T. Kohonen: *Self-Organizing Maps*, Springer Series in Information Sciences, Vol. 30 (Springer, Berlin, Heidelberg 2001)

27.33 T. Kohonen, E. Oja, O. Simula, A. Visa, J. Kangas: Engineering applications of the self-organizing map, Proc. IEEE **84**(10), 1358–1384 (1996)

27.34 T. Koskela, M. Varsta, J. Heikkonen, K. Kaski: Recurrent SOM with local linear models in time series prediction, 6th Eur. Symp. Artif. Neural Netw. (1998) pp. 167–172

27.35 T. Voegtlin: Recursive self-organizing maps, Neural Netw. **15**(8/9), 979–992 (2002)

27.36 M. Strickert, B. Hammer: Merge som for temporal data, Neurocomputing **64**, 39–72 (2005)

27.37   M. Hagenbuchner, A. Sperduti, A. Tsoi: Self-organizing map for adaptive processing of structured data, IEEE Trans. Neural Netw. **14**(3), 491–505 (2003)

27.38   A. Sperduti, A. Starita: Supervised neural networks for the classification of structures, IEEE Trans. Neural Netw. **8**(3), 714–735 (1997)

27.39   P. Frasconi, M. Gori, A. Sperduti: A general framework for adaptive processing of data structures, IEEE Trans. Neural Netw. **9**(5), 768–786 (1998)

27.40   B. Hammer, A. Micheli, A. Sperduti: Universal approximation capability of cascade correlation for structures, Neural Comput. **17**(5), 1109–1159 (2005)

27.41   A. Micheli: Neural network for graphs: A contextual constructive approach, IEEE Trans. Neural Netw. **20**(3), 498–511 (2009)

27.42   B. Hammer, A. Micheli, A. Sperduti, M. Strickert: A general framework for unsupervised processing of structured data, Neurocomputing **57**, 3–35 (2004)

27.43   M. Hagenbuchner, A. Sperduti, A.-C. Tsoi: Graph self-organizing maps for cyclic and unbounded graphs, Neurocomputing **72**(7–9), 1419–1430 (2009)

27.44   Y. Bengio, Y. LeCun: Greedy Layer-Wise Training of Deep Network. In: *Advances in Neural Information Processing Systems 19*, ed. by B. Schölkopf, J. Platt, T. Hofmann (MIT Press, Cambridge 2006) pp. 153–160

27.45   D.C. Ciresan, U. Meier, L.M. Gambardella, J. Schmidhuber: Deep big simple neural nets for handwritten digit recognition, Neural Comput. **22**(12), 3207–3220 (2010)

28. Deep and Modular Neural Networks

Ke Chen

In this chapter, we focus on two important areas in neural computation, i.e., deep and modular neural networks, given the fact that both deep and modular neural networks are among the most powerful machine learning and pattern recognition techniques for complex AI problem solving. We begin by providing a general overview of deep and modular neural networks to describe the general motivation behind such neural architectures and fundamental requirements imposed by complex AI problems. Next, we describe background and motivation, methodologies, major building blocks, and the state-of-the-art hybrid learning strategy in context of deep neural architectures. Then, we describe background and motivation, taxonomy, and learning algorithms pertaining to various typical modular neural networks in a wide context. Furthermore, we also examine relevant

issues and discuss open problems in deep and modular neural network research areas.

28.1 Overview

The human brain is a generic effective and efficient system that solves complex and difficult problems and generates the trait of intelligence and creation. Neural computation has been inspired by brain-related research in different disciplines, e.g., biology and neuroscience, on various levels ranging from a simple single-neuron to complex neuronal structure and organization [28.1]. Among many discoveries in brain-related sciences, two of the most important properties are modularity and hierarchy of neuronal organization in the human brain.

Neuroscientific research has revealed that the central nervous system (CNS) in the human brain is a distributed, massively parallel, and self-organizing modular system [28.1–3]. The CNS is composed of several regions such as the spinal cord, medulla oblongata, pons, midbrain, diencephalon, cerebellum, and the two cerebral hemispheres. Each such region forms a functional module and all regions are interconnected with other parts of the brain [28.1]. In particular, the cerebral cortex consists of several regions attributed to main perceptual and cognitive tasks, where modularity emerges in two different aspects: i.e., structural and functional modularity. Structural modularity is observable from the fact that there are sparse connections between different neuronal groups but neurons are often densely connected within a neuronal group, while functional modularity is evident from different response patterns produced by neural modules for different perceptual and cognitive tasks. Modularity evidence in the human brain strongly suggests that domain-specific modules are required by specific tasks and different modules can cooperate for high level, complex tasks, which primarily motivates the modular neural network (MNN) development in neural computation (NC) [28.4, 5].

Apart from modularity, the human brain also exhibits a functional and structural hierarchy given the fact that information processing in the human brain is done in a hierarchical way. Previous studies [28.6, 7] suggested that there are different cortical visual areas that lead to hierarchical information representations to carry out highly complicated visual tasks, e.g., object recognition. In general, hierarchical information processing enables the human brain to accomplish complex perceptual and cognitive tasks in an effective and extremely efficient way, which mainly inspires the study of deep neural networks (DNNs) of multiple layers in NC.

In general, both DNNs and MNNs can be categorized into biologically plausible [28.8] and artificial

models [28.9] in NC. The main difference between biologically plausible and artificial models lies their methodologies that a biologically plausible model often takes both structural and functional resemblance to its biological counterpart into account, while an artificial model simply works towards modeling the functionality of a biological system without considering those bio-mimetic factors. Due to the limited space, in this chapter we merely focus on artificial DNNs and MNNs. Readers interested in biologically plausible models are referred to the literature, e.g., [28.4], for useful information.

28.2 Deep Neural Networks

In this section, we overview main deep neural network (DNN) techniques with an emphasis on the latest progress. We first review background and motivation for DNN development. Then we describe major building blocks and relevant learning algorithms for constructing different DNNs. Next, we present a hybrid learning strategy in the context of NC. Finally, we examine relevant issues related to DNNs.

28.2.1 Background and Motivation

The study of NC dates back to the 1940s when McCullod and Pitts modeled a neuron mathematically. After that NC was an active area in AI studies until *Minsky* and *Papert* published their influential book, Perceptron [28.10], in 1969. In the book, they formally proved the limited capacities of the single-layer perceptron and further concluded that there is a slim chance to expand its capacities with its multi-layer version, which significantly slowed down NC research until the back-propagation (BP) algorithm was invented (or reinvented) to solve the learning problem in a multi-layer perceptron (MLP) [28.11].

In theory, the BP algorithm enables one to train an MLP of many hidden layers to form a powerful DNN. Such an attractive technique has aroused tremendous enthusiasm in applying DNNs in different fields [28.9]. Apart from a few exceptions, e.g., [28.12], researchers soon found that an MLP of more than two hidden layers often failed [28.13] due to the well-known fact that MLP learning involves an extremely difficult non-convex optimization problem, and the gradient-based local search used in the BP algorithm easily gets stuck in an unwanted local minimum. As a result, most re-

searchers gradually gave up deep architectures and devoted their attention to shallow learning architectures of theoretical justification, e.g., the formal but non-constructive proof that an MLP of single hidden layer may be a universal function approximator [28.14] and a support vector machine (SVM) [28.15], instead. It has been shown that shallow architectures often work well with support of effective feature extraction techniques (but these are often handcrafted). However, recent theoretic justification suggests that learning models of insufficient depth have a fundamental weakness as they cannot efficiently represent the very complicated functions often required in complex AI tasks [28.16, 17].

To solve complex the non-convex optimization problem encountered in DNN learning, *Hinton* and his colleagues made a breakthrough by coming up with a hybrid learning strategy in 2006 [28.18]. The novel learning strategy combines unsupervised and supervised learning paradigms where a layer-wise greedy unsupervised learning is first used to construct an initial DNN with chosen building blocks (such an initial DNN alone can also be used for different purposes, e.g., unsupervised feature learning [28.19]), and supervised learning is then fulfilled based on the pre-trained DNN. Their seminal work led to an emerging machine learning (ML) area, *deep learning*. As a result, different building blocks and learning algorithms have been developed to construct various DNNs. Both theoretical justification and empirical evidence suggest that the hybrid learning strategy [28.18] greatly facilitates learning of DNNs [28.17].

Since 2006, DNNs trained with the hybrid learning strategy have been successfully applied in different and complex AI tasks, such as pattern recognition [28.20–23], various computer vision tasks [28.24–26], audio classification and speech information processing [28.27–31], information retrieval [28.32–34], natural language processing [28.35–37], and robotics [28.38]. Thus, DNNs have become one of the most promising ML and NC techniques to tackle challenging AI problems [28.39].

28.2.2 Building Blocks and Learning Algorithms

In general, a building block is composed of two parametric models, *encoder* and *decoder*, as illustrated in Fig. 28.1. An encoder transforms a raw input or a low-level representation x into a high-level and abstract representation $h(x)$, while a decoder generates an output $\hat{x}$, a reconstructed version of x, from $h(x)$. The learning building block is a self-supervised learning task that minimizes an elaborate *reconstruction cost function* to find appropriate parameters in encoder and decoder. Thus, the distinction between two building blocks of different types lies in their encoder and decoder mechanisms and reconstruction cost functions (as well as optimization algorithms used for parameter estimation). Below we describe different building blocks and their learning algorithms in terms of the generic architecture shown in Fig. 28.1.

Auto-Encoders

The auto-encoder [28.40] and its variants are simple building blocks used to build an MLP of many layers. It is carried out by an MLP of one hidden layer. As depicted in Fig. 28.2, the input and the hidden layers constitute an encoder to generate a M-dimensional representation $h(x) = (h_1(x), \ldots, h_M(x))^\mathsf{T}$ (hereinafter, we use the notation $h(x) = (h_m(x))_{m=1}^M$ to indicate a vector–element relationship for simplifying the presentation) for a given input $x = (x_n)_{n=1}^N$ in N-dimensional space

$$h(x) = f(Wx + b_h) \, ,$$

where W is a connection weight matrix between the input and the hidden layers, b_h is the bias vector for all hidden neurons, and $f(\cdot)$ is a transfer function, e.g., the sigmoid function [28.9]. Let $f(u) = (f(u_k))_{k=1}^K$ be a collective notation for output of all K neurons in a layer. Accordingly, the hidden and the output layers form a decoder that yields a reconstructed version $\hat{x} = (\hat{x}_n)_{n=1}^N$

$$\hat{x} = f(W^\mathsf{T} h(x) + b_o) \, ,$$

where W^T is the transpose of the weight matrix W and b_o is the bias vector for all output neurons. Note that the auto-encoder can be viewed as a special case of auto-associator when the same weights are tied to be used in connections between different layers, which will be clearly seen in the learning algorithm later on. Doing so avoids an unwanted solution when an over-complete representation, i.e., $M > N$, is required [28.22].

Further studies [28.41] suggest that the auto-encoder is unlikely to lead to the discovery of a more useful representation than the input despite the fact that a representation should encode much of the information conveyed in the input whenever the auto-encoder produces a good reconstruction of its input. As a result, a variant named the denoising auto-encoder (DAE) was proposed to capture stable structures underlying the distribution of its observed input. The basic idea is as follows: instead of learning the auto-encoder from the intact input, the DAE will be trained to recover the

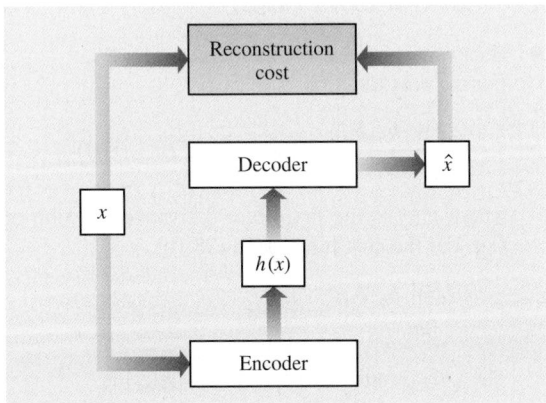

Fig. 28.1 Schematic diagram of a generic building block architecture

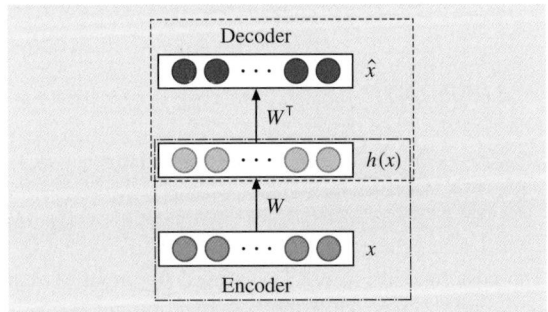

Fig. 28.2 Auto-encoder architecture

Part D | 28.2

original input from its distorted version of partial destruction [28.41]. As is illustrated in Fig. 28.3, the DAE leads to a more useful representation $h(\tilde{x})$ by restoring the corrupted input $\tilde{x}$ to a reconstructed version $\hat{x}$ as close to the clean input x as possible. Thus, the encoder yields a representation as

$$h(\tilde{x}) = f(W\tilde{x} + b_h) ,$$

and the decoder produces a restored version $\hat{x}$ via the representation $h(\tilde{x})$

$$\hat{x} = f(W^T h(\tilde{x}) + b_o) .$$

To produce a corrupted input, we need to distort a clean input by corrupting it with appropriate noise. Depending on the attribute nature of input, there are three kinds of noise used in the corruption process: i.e., the isotropic *Gaussian noise*, $N(0, \sigma^2 I)$, *masking noise* (by setting some randomly chosen elements of x to zero) and *salt-and-pepper* noise (by flipping some randomly chosen elements' values of x to the maximum or the minimum of a given range). Normally, Gaussian noise is used for input of real or continuous values, while and masking and salt-and-pepper noise is applied to input of discrete values, e.g., pixel intensities of gray images. It is worth stating that the variance σ^2 in Gaussian noise and the number of randomly chosen elements in masking and salt-and-pepper noise are hyper-parameters that affect DAE learning. By corrupting a clean input with the chosen noise, we achieve an example, $(\tilde{x}, x)$, for self-supervised learning.

Given a training set of T examples $\{(x_t, x_t)\}_{t=1}^T$ (auto-encoder) or $\{(\tilde{x}_t, x_t)\}_{t=1}^T$ (DAE) two reconstruction cost functions are commonly used for learning auto-encoders as follows

$$L(W, b_h, b_o) = \frac{1}{2T} \sum_{t=1}^{T} \sum_{n=1}^{N} (x_{tn} - \hat{x}_{tn})^2 , \quad (28.1a)$$

$$L(W, b_h, b_o)$$
$$= -\frac{1}{T} \sum_{t=1}^{T} \sum_{n=1}^{N} (x_{tn} \log \hat{x}_{tn} + (1 - x_{tn}) \log(1 - \hat{x}_{tn})) . \quad (28.1b)$$

The cost function in (28.1a) is used for input of real or discrete values, while the cost function in (28.1b) is employed especially for input of binary values.

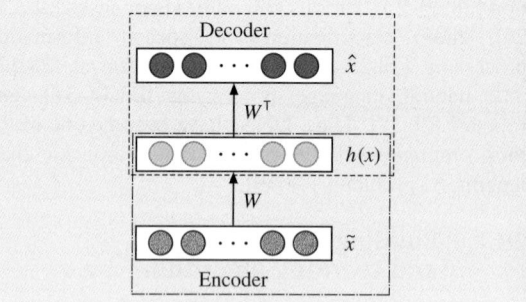

Fig. 28.3 Denoising auto-encoder architecture

To minimize reconstruction functions in (28.1), application of the stochastic gradient descent algorithm [28.12] leads to a generic learning algorithm for training the auto-encoder and its variant, summarized as follows:

Auto-Encoder Learning Algorithm. Given a training set of T examples, $\{(z_t, x_t)\}_{t=1}^T$ where $z_t = x_t$ for the auto-encoder or $z_t = \tilde{x}_t$ for the DAE, and a transfer function, $f(\cdot)$, randomly initialize all parameters, W, b_h and b_o, in auto-encoders and pre-set a learning rate ϵ. Furthermore, the training set is randomly divided into several batches of T_B examples, $\{(z_t, x_t)\}_{t=1}^{T_B}$, and then parameters are updated based on each batch:

● Forward computation
For the input z_t ($t = 1, \cdots, T_B$), output of the hidden layer is

$$h(z_t) = f(u_h(z_t)) , \quad u_h(z_t) = Wz_t + b_h .$$

And output of the output layer is

$$\hat{x}_t = f(u_o(z_t)) , \quad u_o(z_t) = W^T h(z_t) + b_o .$$

● Backward gradient computation
For the cost function in (28.1a),

$$\frac{\partial L(W, b_h, b_o)}{\partial u_o(z_t)} = \left((\hat{x}_{tn} - x_{tn}) f'(u_{o,n}(z_t))\right)_{n=1}^N ,$$

where $f'(\cdot)$ is the first-order derivative function of $f(\cdot)$. For the cost function in (28.1b),

$$\frac{\partial L(W, b_h, b_o)}{\partial u_o(z_t)} = \hat{x}_t - x_t .$$

Then, the gradient with respect to $h(z_t)$ is

$$\frac{\partial L(W, b_h, b_o)}{\partial h(z_t)} = W \frac{\partial L(W, b_h, b_o)}{\partial u_o(z_t)} .$$

Applying the chain rule achieves the gradient with respect to $\boldsymbol{u}_h(z_t)$ as

$$\frac{\partial L(W, \boldsymbol{b}_h, \boldsymbol{b}_o)}{\partial \boldsymbol{u}_h(z_t)} = \left(f'(u_{h,m}(z_t)) \frac{\partial L(W, \boldsymbol{b}_h, \boldsymbol{b}_o)}{\partial h_m(z_t)} \right)_{m=1}^M .$$

Gradients with respect to biases are

$$\frac{\partial L(W, \boldsymbol{b}_h, \boldsymbol{b}_o)}{\partial \boldsymbol{b}_o} = \frac{\partial L(W, \boldsymbol{b}_h, \boldsymbol{b}_o)}{\partial \boldsymbol{u}_o(z_t)} ,$$

and

$$\frac{\partial L(W, \boldsymbol{b}_h, \boldsymbol{b}_o)}{\partial \boldsymbol{b}_h} = \frac{\partial L(W, \boldsymbol{b}_h, \boldsymbol{b}_o)}{\partial \boldsymbol{u}_h(z_t)} .$$

- Parameter update
  Applying the gradient descent method and tied weights leads to update rules

$$W \leftarrow W - \frac{\epsilon}{T_B} \sum_{t=1}^{T_B} \left[\frac{\partial L(W, \boldsymbol{b}_h, \boldsymbol{b}_o)}{\partial \boldsymbol{u}_h(z_t)} [z_t]^{\mathsf{T}} \right.$$
$$\left. + \boldsymbol{h}(z_t) \left(\frac{\partial L(W, \boldsymbol{b}_h, \boldsymbol{b}_o)}{\partial \boldsymbol{u}_o(z_t)} \right)^{\mathsf{T}} \right] ,$$

$$\boldsymbol{b}_o \leftarrow \boldsymbol{b}_o - \frac{\epsilon}{T_B} \sum_{t=1}^{T_B} \frac{\partial L(W, \boldsymbol{b}_h, \boldsymbol{b}_o)}{\partial \boldsymbol{u}_o(z_t)} ,$$

and

$$\boldsymbol{b}_h \leftarrow \boldsymbol{b}_h - \frac{\epsilon}{T_B} \sum_{t=1}^{T_B} \frac{\partial L(W, \boldsymbol{b}_h, \boldsymbol{b}_o)}{\partial \boldsymbol{u}_h(z_t)} .$$

The above three steps repeat for all batches, which leads to a training epoch. The learning algorithm runs iteratively until a termination condition is met (typically based on a cross-validation procedure [28.12]).

The Restricted Boltzmann Machine

Strictly speaking, the restricted Boltzmann machine (RBM) [28.42] is an energy-based generative model, a simplified version of the generic Boltzmann machine. As illustrated in Fig. 28.4, an RBM can be viewed as a probabilistic NN of two layers, i.e., *visible* and *hidden* layers, with bi-directional connections. Unlike the Boltzmann machine, there are no lateral connections among neurons in the same layer in an RBM. With the bottom-up connections from the visible to the

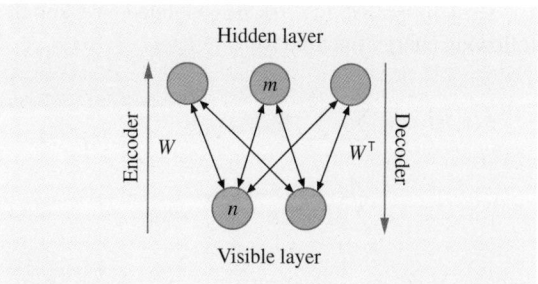

Fig. 28.4 Restricted Boltzmann machine (RBM) architecture

hidden layer, RBM forms an encoder that yields a probabilistic representation $\boldsymbol{h} = (h_m)_{m=1}^M$ for input data $\boldsymbol{v} = (v_n)_{n=1}^N$

$$P(\boldsymbol{h}|\boldsymbol{v}) = \prod_{m=1}^M P(h_m|\boldsymbol{v}) ,$$

$$P(h_m|\boldsymbol{v}) = \phi \left(\sum_{n=1}^N W_{mn} v_n + b_{h,m} \right) , \qquad (28.2)$$

where W_{mn} is the connection weight between the visible neuron n and the hidden neuron m, and $b_{h,m}$ is the bias of the hidden neuron m. $\phi(u) = \frac{1}{1+e^{-u}}$ is the sigmoid transfer function. As h_m is assumed to take a binary value, i.e., $h_m \in \{0, 1\}$, $P(h_m|\boldsymbol{v})$ is interpreted as the probability of $h_m = 1$. Accordingly, RBM performs a probabilistic decoder via the top-down connections from the hidden to the visible layer to reconstruct an input with the probability

$$P(\boldsymbol{v}|\boldsymbol{h}) = \prod_{m=1}^M P(v_n|\boldsymbol{h}) ,$$

$$P(v_n|\boldsymbol{h}) = \phi \left(\sum_{m=1}^N W_{nm} h_m + b_{v,n} \right) , \qquad (28.3)$$

where W_{nm} is the connection weight between the hidden neuron m and the visible neuron n, and $b_{v,n}$ is the bias of visible neuron n. Like connection weights in auto-encoders, bi-directional connection weights are tied, i.e., $W_{mn} = W_{nm}$, as shown in Fig. 28.4. By learning a parametric model of the data distribution $P(\boldsymbol{v})$ derived from the joint probability $P(\boldsymbol{v}, \boldsymbol{h})$ for a given data set, RBM yields a probabilistic representation that tends to reconstruct any data subject to $P(\boldsymbol{v})$.

The joint probability $P(v, h)$ is defined based on the following energy function for $v_n \in \{0, 1\}$

$$E(v, h) = -\sum_{m=1}^{M}\sum_{n=1}^{N} W_{mn} h_m v_n$$

$$-\sum_{m=1}^{M} h_m b_{h,m} - \sum_{n=1}^{N} v_n b_{v,m} . \quad (28.4)$$

As a result, the joint probability is subject to the Boltzmann distribution

$$P(v, h) = \frac{e^{-E(v,h)}}{\sum_v \sum_h e^{-E(v,h)}} . \quad (28.5)$$

Thus, we achieve the data probability by marginalizing the joint probability as follows

$$P(v) = \sum_h P(v, h) = \sum_h P(v|h)P(h) . $$

In order to achieve the most likely reconstruction, we need to maximize the log-likelihood of $P(v)$. Therefore, the reconstruction cost function of an RBM is its negative log-likelihood function

$$L(W, b_h, b_v) = -\log P(v) = -\log \sum_h P(v|h)P(h) . $$
$$(28.6)$$

From (28.5) and (28.6), it is observed that the direct use of a gradient descent method for optimal parameters often leads to intractable computation due to the fact that the exponential number of possible hidden-layer configurations needs to be summed over in (28.5) and then used in (28.6). Fortunately, an approximation algorithm named *contrastive divergence (CD)* has been proposed to solve this problem [28.42]. The key ideas behind the CD algorithm are (i) using Gibbs sampling based on the conditional distributions in (28.2) and (28.3), and (ii) running only a few iterations of Gibbs sampling by treating the data x input to an RBM as the initial state, i.e., $v^0 = x$, of the Markov chain at the visible layer. Many studies have suggested that only the use of one iteration of the Markov chain in the CD algorithm works well for building up a deep belief network (DBN) in practice [28.17, 18, 22], and hence the algorithm is dubbed *CD-1* in this situation, a special case of the *CD-k* algorithm that executes k iterations of the Markov chain in the Gibbs sampling process.

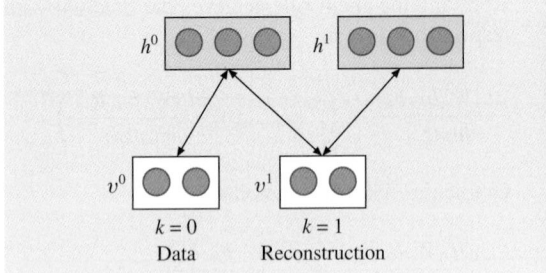

Fig. 28.5 Gibbs sampling process in the *CD-1* algorithm

Figure 28.5 illustrates a Gibbs sampling process used in the *CD-1* algorithm as follows

i) Estimating probabilities $P(h_m^0|v^0)$, for $m = 1, \cdots, M$, with the encoder defined in (28.2) and then forming a realization of h^0 by sampling with these probabilities.

ii) Applying the decoder defined in (28.3) to estimate probabilities $P(v_n^1|h^0)$, for $n = 1, \cdots, N$, and then producing a reconstruction of v^1 via sampling.

iii) With the reconstruction, estimating probabilities $P(h_m^1|v^1)$, for $m = 1, \cdots, M$, with the encoder.

With the above Gibbing sampling procedure, the *CD-1* algorithm is summarized as follows:

Algorithm 28.1 RBM CD−1 Learning Algorithm
Given a training set of T instances, $\{x_t\}_{t=1}^{T}$, randomly initialize all parameters, W, b_h and b_v, in an RBM and pre-set a learning rate ϵ:

- Positive phase
  - Present an instance to the visible layer, i.e., $v^0 = x_t$.
  - Estimate probabilities with the encoder: $\hat{P}(h^0|v^0) = (P(h_m^0|v^0))_{m=1}^{M}$ by using (28.2).
- Negative phase
  - Form a realization of h^0 by sampling with probabilities $\hat{P}(h^0|v^0)$.
  - With the realization of h^0, apply the decoder to estimate probabilities: $\hat{P}(v^1|h^0) = (P(v_n^0|h^0))_{n=1}^{N}$ by using (28.3), and then produce a reconstruction v^1 via sampling based on $\hat{P}(v^1|h^0)$.
  - With the encoder and the reconstruction, estimate probabilities: $\hat{P}(h^1|v^1) = (P(h_m^1|v^1))_{m=1}^{M}$ by using (28.2).
- Parameter update
  Based on Gibbs sampling results in the positive and

the negative phases, parameters are updated as follows:

$$W \leftarrow W + \epsilon \left(\hat{P}(h^0|v^0)(v^0)^\mathsf{T} - \hat{P}(h^1|v^1)(v^1)^\mathsf{T} \right),$$

$$b_h \leftarrow b_h + \epsilon \left(\hat{P}(h^0|v^0) - \hat{P}(h^1|v^1) \right),$$

and

$$b_v \leftarrow b_v + \epsilon \left(v^0 - v^1 \right).$$

The above three steps repeat for all instances in the given training set, which leads to a training epoch. The learning algorithm runs iteratively until it converges.

Predictive Sparse Decomposition

Predictive sparse decomposition (PSD) [28.43] is a building block obtained by combining sparse coding [28.44] and auto-encoder ideas. In a PSD building block, the encoder is specified by

$$h(x_t) = G\tanh(W_E x_t + b_h), \tag{28.7}$$

where W_E is the $M{\times}N$ connection matrix between input and hidden neurons in the encoder and $G = \mathrm{diag}(g_{mm})$ is an $M{\times}M$ learnable diagonal gain matrix for an M-dimensional representation of an N-dimensional input, x_t, b_h are biases of hidden neurons, and $\tanh(\cdot)$ is the hyperbolic tangent transfer function [28.9]. Accordingly, the decoder is implemented by a linear mapping used in the sparse coding [28.44]

$$\hat{x}_t = W_D h(x_t), \tag{28.8}$$

where W_D is an $N{\times}M$ connection matrix between hidden and output neurons in the decoder, and each column of W_D always needs to be normalized to a unit vector to avoid trivial solutions [28.44].

Given a training set of T instances, $\{x_t\}_{t=1}^T$, the PSD cost function is defined as

$$L_{\mathrm{PSD}}\left(G, W_E, W_D, b_h; h^*(x_t)\right)$$
$$= \sum_{t=1}^T \|W_D h^*(x_t) - x\|_2^2 + \alpha \|h^*(x_t)\|_1$$
$$+ \beta \|h^*(x_t) - h(x_t)\|_2^2, \tag{28.9}$$

where $h^*(x_t)$ is the optimal sparse hidden representation of x_t while $h(x_t)$ is the output of the encoder in (28.7) based on the current parameter values. In (28.9), α and β are two hyper-parameters to control regularization strengths, and $\|\cdot\|_1$ and $\|\cdot\|_2$ are $\mathcal{L}_1$ and $\mathcal{L}_2$ norm, respectively. Intuitively, in the multi-objective

cost function defined in (28.9), the first term specifies reconstruction errors, the second term refers to the magnitude of non-sparse representations, and the last term drives the encoder towards yielding the optimal representation.

For learning a PSD building block, the cost function in (28.9) needs to be optimized simultaneously with respect to the hidden representation and all the parameters. As a result, a learning algorithm of two alternate steps has been proposed to solve this problem [28.43] as follows:

Algorithm 28.2 PSD Learning algorithm
Given a training set of T instances $\{x_t\}_{t=1}^T$ randomly initialize all the parameters, W_E, W_D, G, b_h, and the optimal sparse representation $\{h^*(x_t)\}_{t=1}^T$ in a PSD building block and pre-set hyper-parameters α and β as well as learning rates ϵ_i ($i = 1, \cdots, 4$):

- Optimal representation update
  In this step, the gradient descent method is applied to find the optimal sparse representation based on the current parameter values of the encoder and the decoder, which leads to the following update rule

$$h^*(x_t) \leftarrow h^*(x_t) - \epsilon_1 \big[\alpha \mathrm{sign}(h^*(x_t)) \\ + \beta(h^*(x_t) - h(x_t)) \\ + (W_D)^\mathsf{T}(W_D h^*(x_t) - x_t) \big],$$

  where $\mathrm{sign}(\cdot)$ is the sign function; $\mathrm{sign}(u) = \pm 1$ if $u \gtrless 0$ and $\mathrm{sign}(u) = 0$ if $u = 0$.
- Parameter update
  In this step, $h^*(x_t)$ achieved in the above step is fixed. Then the gradient descent method is applied to the cost function (28.9) with respect to all encoder and decoder parameters, which results in the following update rules

$$W_E \leftarrow W_E - \epsilon_2 g(x_t)(x_t)^\mathsf{T},$$
$$b_h \leftarrow b_h - \epsilon_3 g(x_t).$$

Here $g(x_t)$ is obtained by

$$g(x_t) = \big[g_{mm}^{-1}(g_{mm}^2 - h_m^2(x_t)) \\ (h_m^*(x_t) - h_m(x_t)) \big]_{m=1}^M.$$
$$G \leftarrow G - \epsilon_2 \mathrm{diag}\left[\left(h_m^*(x_t) - h_m(x_t) \right) \\ \times \tanh\left(\sum_{n=1}^N [W_E]_{mn} x_{tn} + b_{v,m} \right) \right],$$

and

$$W_D \leftarrow W_D - \epsilon_4 \left[W_D \boldsymbol{h}^*(\boldsymbol{x}_t) - \boldsymbol{x}_t \right] \left[\boldsymbol{h}^*(\boldsymbol{x}_t) \right]^{\mathsf{T}} .$$

Normalize each column of W_D such that

$$\| [W_D]_{\cdot n} \|_2^2 = 1 \text{ for } n = 1, \cdots, N .$$

The above two steps repeat for all the instances in the given training set, which leads to a training epoch. The learning algorithm runs iteratively until it converges.

Other Building Blocks

While the auto-encoders and the RBM are building blocks widely used to construct DNNs, there are other building blocks that are either derived from existing building blocks for performance improvement or are developed with an alternative principle. Such building blocks include regularized auto-encoders and RBM variants. Due to the limited space, we briefly overview them below.

Recently, a number of auto-encoder variants have been developed by adding a regularization term to the standard reconstruction cost function in (28.1) and hence are dubbed regularized auto-encoders. The contrastive auto-encoder (CAE) is a typical regularized version of the auto-encoder with the introduction of the norm of the Jacobian matrix of the encoder evaluated at each training example $\boldsymbol{x}_t$ into the standard reconstruction cost function [28.45]

$$L_{\text{CAE}}(W, \boldsymbol{b}_h, \boldsymbol{b}_o) = L(W, \boldsymbol{b}_h, \boldsymbol{b}_o) + \alpha \sum_{t=1}^{\mathsf{T}} \| J(\boldsymbol{x}_t) \|_F^2 ,$$
(28.10)

where α is a trade-off parameter to control the regularization strength and $\| J(\boldsymbol{x}_t) \|_F^2$ is the Frobenius norm of the Jacobian matrix of the encoder and is calculated as follows

$$\| J(\boldsymbol{x}_t) \|_F^2 = \sum_{m=1}^{M} \sum_{n=1}^{N} \left[f'(u_{h,m}(\boldsymbol{x}_t)) \right]^2 W_{mn}^2 .$$

Here, $f'(\cdot)$ is the first-order derivative of a transfer function $f(\cdot)$, and $f'[u_{h,m}(\boldsymbol{x}_t)] = h_m(\boldsymbol{x}_t)[1 - h_m(\boldsymbol{x}_t)]$ when $f(\cdot)$ is the sigmoid function [28.9]. It is straightforward to apply the stochastic gradient method [28.12]

to the CAE cost function in (28.10) to derive a learning algorithm used for training a CAE. Furthermore, an improved version of CAE was also proposed by penalizing additional higher order derivatives [28.46]. The sparse auto-encoder (SAE) is another class of regularized auto-encoders. The basic idea underlying SAEs is the introduction of a sparse regularization term working on either hidden neuron biases, e.g., [28.47], or their outputs, e.g., [28.48], into the standard reconstruction cost function. Different forms of sparsity penalties, e.g., $\mathcal{L}_1$ norm and student-t, are employed for regularization, and the learning algorithm is derived by applying the coordinate descent optimization procedure to a new reconstruction cost function [28.47, 48].

The RBM described above works only for an input of binary values. When an input has real values, a variant named Gaussian RBM (GRBM) [28.49], has been proposed with the following energy function

$$\begin{aligned} E(\boldsymbol{v}, \boldsymbol{h}) = -\sum_{m=1}^{M} \sum_{n=1}^{N} W_{mn} h_m \frac{v_n}{\sigma_n} \\ - \sum_{m=1}^{M} h_m b_{h,m} - \sum_{n=1}^{N} \frac{(v_n - b_{v,n})^2}{2\sigma_n^2} , \end{aligned}$$
(28.11)

where σ_n is the standard deviation of the Gaussian noise for the visible neuron n. In the *CD* learning algorithm, the update rule for the hidden neurons remains the same except that each v_n is substituted by $\frac{v_n}{\sigma_n}$, and the update rule for all visible neurons needs to use reconstructions $\boldsymbol{v}$, produced by sampling from a Gaussian distribution with mean $\sigma_n \sum_{m=1}^{M} W_{nm} h_m + b_{v,n}$ and variance σ_n^2 for $n = 1, \cdots, N$. In addition, an improved GRBM was also proposed by introducing an alternative parameterization of the energy function in (28.11) and incorporating it into the *CD* algorithm [28.50]. Other RBM variants will be discussed later on, as they often play a different role from being used to construct a DNN.

28.2.3 Hybrid Learning Strategy

Based on the building blocks described in Sect. 28.2.2, we describe a systematic approach to establishing a feed-forward DNN for supervised and semi-supervised learning. This approach employs a hybrid learning strategy that combines unsupervised and supervised learning paradigms to overcome the optimization difficulty in training DNNs. The hybrid learning strategy [28.18, 40] first applies layer-wise unsupervised learning to set up a DNN and initialize parameters with input data only and then uses a global supervised learning algorithm with teachers' information to train all the parameters in the initialized DNN for a given task.

Layer-Wise Unsupervised Learning

In the hybrid learning strategy, unsupervised learning is a layer-wise greedy learning process that constructs a DNN with a chosen building block and initializes parameters in a layer-by-layer way.

Suppose we want to establish a DNN of K ($K>1$) hidden layers and denote output of hidden layer k as $\boldsymbol{h}_k(\boldsymbol{x})$ ($k = 1, \cdots, K$) for a given input $\boldsymbol{x}$ and output of the output layer as $\boldsymbol{o}(\boldsymbol{x})$, respectively. To facilitate the presentation, we stipulate $\boldsymbol{h}_0(\boldsymbol{x}) = \boldsymbol{x}$. Then, the generic layer-wise greedy learning procedure can be summarized as follows:

Algorithm 28.3 Layer-wise greedy learning procedure

Given a training set of T instances $\{\boldsymbol{x}_t\}_{t=1}^T$, randomly initialize its parameters in a chosen building block and pre-set all hyper-parameters required for learning such a building block:

- Train a building block for hidden layer k
  - Set the number of neurons required by hidden layer k to be the dimension of the hidden representation in the chosen building block.
  - Use the training data set $\{\boldsymbol{h}_{k-1}(\boldsymbol{x}_t)\}_{t=1}^T$ train the building block to achieve its optimal parameters.
- Construct a DNN up to hidden layer k
  With the trained building block in the above step, discard its decoder part, including all associated parameters, and stack its hidden layer on the existing DNN with connection weights of the encoder and biases of hidden neurons achieved in the above step (the input layer $\boldsymbol{h}_0(\boldsymbol{x}) = \boldsymbol{x}$ is viewed as the starting architecture of a DNN).

The above steps are repeated for $k = 1, \cdots, K$. Then, the output layer $\boldsymbol{o}(\boldsymbol{x})$ is stacked onto hidden layer K with randomly initialized connection weights so as to finalize the initial DNN construction and its parameter initialization.

Figure 28.6 illustrated two typical instances for constructing an initial DNN via the layer-wise greedy learning procedure described above. Figure 28.6a

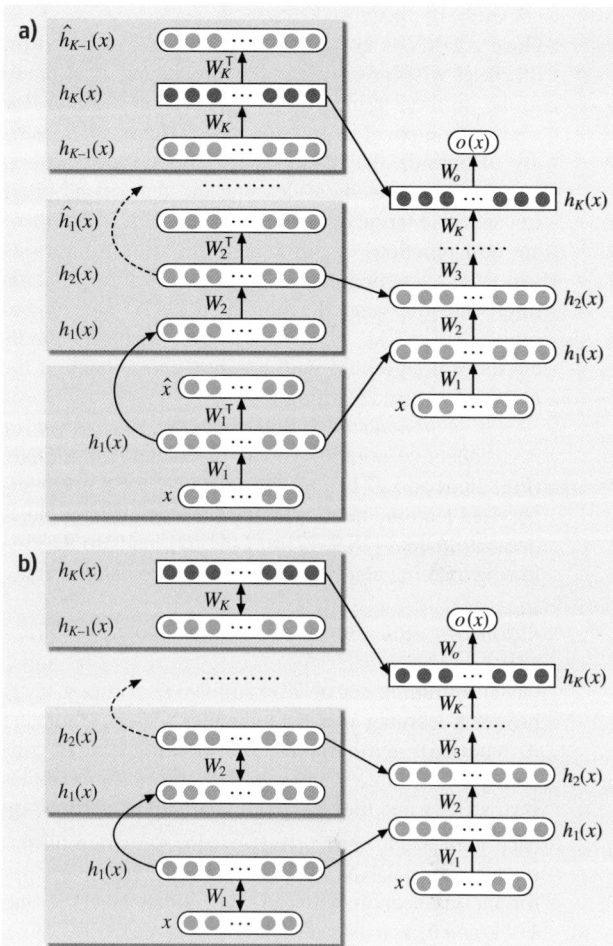

Fig. 28.6a,b Construction of a DNN with a building block via layer-wise greedy learning. (**a**) Auto-encoder or its variants. (**b**) RBM or its variants

shows a schematic diagram of the layer-wise greedy learning process with the auto-encoder or its variants; to construct the hidden layer k, the output layer and its associated parameters W_k^T and $\boldsymbol{b}_{o,k}$ are removed and the remaining part is stacked onto hidden layer $k-1$, and W_o is a randomly initialized weight matrix for the connection between the hidden layer K and the output layer. When a DNN is constructed with the RBM or its variants, all backward connection weights in the decoder are abandoned after training and only the hidden layer with those forward connection weights and biases of hidden neurons are used to construct the DNN, as depicted in Fig. 28.6b.

Global Supervised Learning

Once a DNN is constructed and initialized based on the layer-wise greedy learning procedure, it is ready to be further trained in a supervision fashion for a classification or regression task. There are a variety of optimization methods for supervised learning, e.g., stochastic gradient descent and the second-order Levenberg–Marquadt methods [28.9, 12]. Also there are cost functions of different forms used for various supervised learning tasks and regularization towards improving the generalization of a DNN. Due to the limited space, we only review the stochastic gradient descent algorithm with a generic cost function for global supervised learning.

For a generic cost function $L(\Theta, \mathcal{D})$, where Θ is a collective notation of all parameters in a DNN (Fig. 28.6) and $\mathcal{D}$ is a training data set for a given supervised learning task, applying the stochastic gradient descent method [28.9, 12] to $L(\Theta, \mathcal{D})$ leads to the following learning algorithm for fine-tuning parameters.

Algorithm 28.4 Global supervised learning algorithm

Given a training set of T examples $\mathcal{D} = \{(x_t, y_t)\}_{t=1}^{T}$ pre-set a learning rate ϵ (and other hyper-parameters if required). Furthermore, the training set is randomly divided into many mini-batches of T_B examples $\{(x_t, y_t)\}_{t=1}^{T_B}$ and then parameters are updated based on each mini-batch. $\Theta = (\{W_k\}_{k=1}^{K+1}, \{b_k\}_{k=1}^{K+1})$ are all parameters in a DNN, where W_k is the weight matrix for the connection between the hidden layers k and $k-1$, and b_k is biases of neurons in layer k (Fig. 28.6). Here, input and output layers are stipulated as layers 0 and $K+1$, respectively, i. e., $h_0(x_t) = x_t$, $W_o = W_{K+1}$, $b_o = b_{K+1}$ and $o(x_t) = h_{K+1}(x_t)$:

- Forward computation
  Given the input x_t, for $k = 1, \cdots, K+1$, the output of layer k is

$$h_k(x_t) = f(u_k(x_t)), \quad u_k(x_t) = W_k h_{k-1}(x_t) + b_k .$$

- Backward gradient computation
  Given a cost function on each mini-batch, $L_B(\Theta, \mathcal{D})$ calculate gradients at the output layer, i. e.,

$$\frac{\partial L_B(\Theta, \mathcal{D})}{\partial h_{K+1}(x_t)} = \frac{\partial L_B(\Theta, \mathcal{D})}{\partial o(x_t)} ,$$

$$\frac{\partial L_B(\Theta, \mathcal{D})}{\partial u_{K+1}(x_t)} = \left(\frac{\partial L_B(\Theta, \mathcal{D})}{\partial h_{(K+1)j}(x_t)} f'(u_{(K+1)j}(x_t)) \right)_{j=1}^{|o|} ,$$

where $f'(\cdot)$ is the first-order derivative of the transfer function $f(\cdot)$.
For all hidden layers, i. e., $k = K, \cdots, 1$, applying the chain rule leads to

$$\frac{\partial L_B(\Theta, \mathcal{D})}{\partial u_k(x_t)} = \left(\frac{\partial L_B(\Theta, \mathcal{D})}{\partial h_{kj}(x_t)} f'(u_{kj}(x_t)) \right)_{j=1}^{|h_k|} ,$$

and

$$\frac{\partial L_B(\Theta, \mathcal{D})}{\partial h_k(x_t)} = \left(W_{k+1}^{(i)} \right)^{\mathsf{T}} \frac{\partial L_B(\Theta, \mathcal{D})}{\partial u_{k+1}(x_t)} .$$

For $k = K+1, \cdots, 1$, gradients with respect to biases of layer k are

$$\frac{\partial L_B(\Theta, \mathcal{D})}{\partial b_k} = \frac{\partial L_B(\Theta, \mathcal{D})}{\partial u_k(x_t)} .$$

- Parameter update
  Applying the gradient descent method results in the following update rules:
  For $k = K+1, \cdots, 1$

$$W_k \leftarrow W_k - \frac{\epsilon}{T_B} \sum_{t=1}^{T_B} \frac{\partial L_B(\Theta, \mathcal{D})}{\partial u_k(x_t)} (h_{k-1}(x_t))^{\mathsf{T}} ,$$

$$b_k \leftarrow b_k - \frac{\epsilon}{T_B} \sum_{t=1}^{T_B} \frac{\partial L_B(\Theta, \mathcal{D})}{\partial u_k(x_t)} .$$

The above three steps repeat for all mini-batches, which leads to a training epoch. The learning algorithm runs iteratively until a termination condition is met (typically based on a cross-validation procedure [28.12]).

For the above learning algorithm, the BP algorithm [28.11] is a special case when the transfer function is the sigmoid function, i. e., $f(u) = \phi(u)$, and the cost function is the mean square error (MSE) function, i. e., for each mini-batch

$$L_B(\Theta, \mathcal{D}) = \frac{1}{2T_B} \sum_{t=1}^{T_B} \| o(x_t) - y_t \|_2^2 .$$

Thus, we have

$$\phi'(u) = \phi(u)(1 - \phi(u)) ,$$

$$\frac{\partial L_B(\Theta, \mathcal{D})}{\partial o(x_t)} = \frac{1}{T_B} \sum_{t=1}^{T_B} (o(x_t) - y_t) ,$$

and

$$\frac{\partial L_B(\Theta, \mathcal{D})}{\partial \boldsymbol{u}_{K+1}(\boldsymbol{x}_t)}$$

$$= \left[\frac{1}{T_B} \sum_{t=1}^{T_B} (o_j(\boldsymbol{x}_t) - y_{tj}) o_j(\boldsymbol{x}_t) (1 - o_j(\boldsymbol{x}_t)) \right]_{j=1}^{|o|} .$$

28.2.4 Relevant Issues

In the literature, the hybrid learning strategy described in Sect. 28.2.3 is often called the semi-supervised learning strategy [28.17, 39]. Nevertheless, semi-supervised learning implies the situation that there are few labeled examples but many unlabeled instances in a training set. Indeed, such a strategy works well in a situation where both unlabeled and labeled data in a training set are used for layer-wise greedy learning, and only labeled data are used for fine-tuning in global supervised learning. However, other studies, e.g., [28.28, 29, 51], also show that this strategy can considerably improve the generalization of a DNN even though there are abundant labeled examples in a training data set. Hence we would rather name it hybrid learning. On the other hand, our review focuses on only primary supervised learning tasks in the context of NC. In a wider context, the unsupervised learning process itself develops a novel approach to automatic feature discovery/extraction via learning, which is an emerging ML area named representation learning [28.39]. In such a context, some DNNs can perform a generative model. For instance, the DBN [28.18] is a RBM-based DNN by retaining both forward and backward connections during layer-wise greedy learning. To be a generative model, the DBN needs an alternative learning algorithm, e.g., the wake–sleep algorithm [28.18], for global unsupervised learning. In general, the global unsupervised learning for a generative DNN is still a challenging problem.

While the hybrid learning strategy has been successfully applied to many complex AI tasks, in general, it is still not entirely clear why such a strategy works well empirically. A recent study attempted to provide some justification of the role played by layer-wise greedy learning for supervised learning [28.51]. The findings can be summarized as follows: such an unsupervised learning process brings about a regularization effect that initializes DNN parameters towards the basin of attraction corresponding to a *good* local minimum, which facilitates global supervised learning in terms of generalization [28.51]. In general, a deeper understanding of such a learning strategy will be required in the future.

On the other hand, a successful story was recently reported [28.52] where no unsupervised pre-training was used in the DNN learning for a non-trivial task; which poses another open problem as to when and where such a learning strategy must be employed for training a DNN to yield a satisfactory generalization performance.

Recent studies also suggest that the use of artificially distorted or deformed training data and unsupervised *front-ends* can considerably improve the performance of DNNs regardless of the hybrid learning strategy. As DNN learning is of the data-driven nature, augmenting training data with known input deformation amounts to the use of more representative examples conveying intrinsic variations underlying a class of data in learning. For example, speech corrupted by some known channel noise and deformed images by using affine transformation and adding noise have significantly improved the DNN performance in various speech and visual information processing tasks [28.22, 28, 29, 51, 52]. On the other hand, the generic building blocks reviewed in Sect. 28.2.2 can be extended to be specialist *front-ends* by exploiting intrinsic data structures. For instance, the RBM has several variants, e.g., [28.53–55], to capture covariance and other statistical information underlying an image. After unsupervised learning, such *front-ends* generate powerful representations that greatly facilitate further DNN learning in visual information processing.

While our review focuses on only fully connected feed-forward DNNs, there are alternative and more effective DNN architectures for specific tasks. Convolutional DNNs [28.12] make use of topological locality constraints underlying images to form more effective locally connected DNN architecture. Furthermore, various pooling techniques [28.56] used in convolutional DNNs facilitate learning invariant and robust features. With appropriate building blocks, e.g., the PSD reviewed in Sect. 28.2.2, convolutional DNNs work very well with the hybrid learning strategy [28.43, 57]. In addition, novel DNN architectures need to be developed by exploring the nature of a specific problem, e.g., a regularized Siamese DNN was recently developed for generic speaker-specific information extraction [28.28, 29]. As a result, novel DNN architecture development and model selection are among important DNN research topics.

Finally, theoretical justification of deep learning and the hybrid learning strategy, along with other developed recently techniques, e.g., parallel graphics processing unit (GPU) computing, enable researchers to develop large-scale DNNs to tackle very complex real world problems. While some theoretic justification has been provided in the literature, e.g., [28.16, 17, 39], to show strengths in their potential capacity and efficient representational schemes of DNNs, more and more successful applications of DNNs, particulary working with the hybrid learning strategy, lend evidence to support the argument that DNNs are one of the most promising learning systems for dealing with complex and large-scale real world problems. For example, such evidence can be found from one of the latest developments in a DNN application to computer vision where it is demonstrated that applying a DNN of nine layers constructed with the SAE building block via layer-wise greedy learning results in the favorable performance in object recognition of over 22 000 categories [28.58].

28.3 Modular Neural Networks

In this section, we review main modular neural networks (MNN) and their learning algorithms with our own taxonomy. We first review background and motivation for MNN research and present our MNN taxonomy. Then we describe major MNN architectures and relevant learning algorithms. Finally, we examine relevant issues related to MNNs in a boarder context.

28.3.1 Background and Motivation

Soon after neural network (NN) research resurged in the middle of the 1980s, MNN studies emerged; they have become an important area in NC since then. There are a variety of motivations that inspire MNN researches, e.g., biological, psychological, computational, and implementation motivations [28.4, 5, 9]. Here, we only describe the background and motivation of MNN researches from learning and computational perspectives.

From the learning perspective, MNNs have several advantages over monolithic NNs. First of all, MNNs adopt an alternative methodology for learning, so that complex problem can be solved based an ensemble of simple NNs, which might avoid/alleviate the complex optimization problems encountered in monolithic NN learning without decreasing the learning capacity. Next, modularity enables MNNs to use a priori knowledge flexibly and facilitates knowledge integration and update in learning. To a great extent, MNNs are immune to temporal and spatial *cross-talk*, a problem faced by monolithic NNs during learning [28.9]. Finally, theoretical justification and abundant empirical evidence show that an MNN often yields a better generalization than its component networks [28.5, 59]. From the computational perspective, modularization in MNNs leads to more efficient and robust computation, given the fact that MNNs often do not suffer from a high coupling burden in a monolithic NN and hence tend to have a lower overall structural complexity in tackling the same problem [28.5]. This main computational merit makes MNNs scalable and extensible to large-scale MNN implementation.

There are two highly influential principles that are often used in artificial MNN development; i. e., *divide-and-conquer* and *diversity-promotion*. The divide-and-conquer principle refers to a generic methodology that tackles a complex and difficult problem by dividing it into several relatively simple and easy subproblems, whose solutions can be combined seamlessly to yield a final solution. On the other hand, theoretical justification [28.60, 61] and abundant empirical studies [28.62] suggest that apart from the condition that component networks need to reach some certain accuracy, the success of MNNs are largely attributed to diversity among them. Hence, the promotion of diversity in MNNs becomes critical in their design and development. To understand motivations and ideas underlying different MNNs, we believe that it is crucial to examine how two principles are applied in their development.

There are different taxonomies of MNNs [28.4, 5, 9]. In this chapter, we present an alternative taxonomy that highlights the interaction among component networks in an MNN during learning. As a result, there is a dichotomy between *tightly* and *loosely coupled models* in MNNs. In a tightly coupled MNN, all component networks are jointly trained in a dependent way by taking their interaction into account during a single learning stage, and hence all parameters of different networks (and combination mechanisms if there are any) need to be updated simultaneously by minimizing a cost function defined at the global level. In contrast, training of a loosely coupled MNN often undergoes multiple stages in a hierarchical or sequential way where learning undertaken in different stages may be either correlated or uncorrelated via different strategies. We believe that such a taxonomy facilitates not only understanding different MNNs especially from a learning perspective but also relating MNNs to generic ensemble learning in a broader context.

28.3.2 Tightly Coupled Models

There are two typical tightly coupled MNNs: the mixture of experts (MoE) [28.63, 64] and MNNs trained via negative correlation learning (NCL) [28.65].

Mixture of Experts

The MoE [28.63, 64] refers to a class of MNNs that dynamically partition input space to facilitate learning in a complex and non-stationary environment. By applying the divide-and-conquer principle, a soft-competition idea was proposed to develop the MoE architecture. That is, at every input data point, multiple expert networks compete to take on a given supervised learning task. Instead of winner-take-all, all expert networks may work together but the winner expert plays a more important role than the losers.

The MoE architecture is composed of N expert networks and a gating network, as illustrated in Fig. 28.7. The n-th expert network produces an output vector, $o_n(x)$, for an input, x. The gating network receives the vector x as input and produces N scalar outputs that form a partition of the input space at each point x. For the input x, the gating network outputs N linear combination coefficients used to verdict the importance of all expert networks for a given supervised learning task. The final output of MoE is a convex weighted sum of all the output yielded by N expert networks. Although NNs of different types can be used as expert networks, a class of generalized linear NNs are often employed where such an NN is linear with a single output non-

linearity [28.64]. As a result, output of the n-th expert network is a generalized linear function of the input x

$$o_n(x) = f(W_n x),$$

where W_n is a parameter matrix, a collective notation for both connection weights and biases, and $f(\cdot)$ is a nonlinear transfer function. The gating network is also a generalized linear model, and its n-th output $g(x, v_n)$ is the softmax function of $v_n^T x$

$$g(x, v_n) = \frac{e^{v_n^T x}}{\sum_{k=1}^N e^{v_k^T x}},$$

where v_n is the n-th column of the parameter matrix V in the gating network and is responsible for the linear combination coefficient regarding the expert network n. The overall output of the MoE is the weighted sum resulted from the soft-competition at the point x

$$o(x) = \sum_{n=1}^N g(x, v_n) o_n(x).$$

There is a natural probabilistic interpretation of the MoE [28.64]. For a training example (x, y), the values of $g(x, V) = (g(x, v_n))_{n=1}^N$ are interpreted as the multinomial probabilities associated with the decision that terminates in a regressive process that maps x to y. Once a decision has been made that leads to a choice of regressive process n, the output y is chosen from a probability distribution $P(y|x, W_n)$. Hence, the overall probability of generating y from x is the mixture of the probabilities of generating y from each component distribution and the mixing proportions are subject to a multinomial distribution

$$P(y|x, \Theta) = \sum_{n=1}^N g(x, v_n) P(y|x, W_n), \qquad (28.12)$$

where Θ is a collective notation of all the parameters in the MoE, including both expert and gating network parameters. For different learning tasks, specific component distribution models are required. For example, the probabilistic component model should be a Gaussian distribution for a regression task, while a Bernoulli distribution and multinomial distributions are required for binary and multi-class classification tasks, respectively. In general, MoE is viewed as a conditional mixture model for supervised learning, a non-trivial extension of finite mixture model for unsupervised learning.

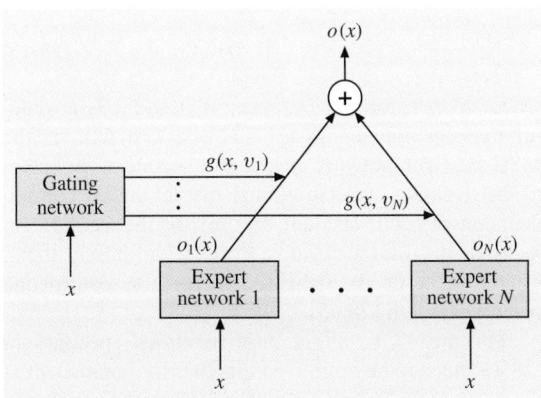

Fig. 28.7 Architecture of the mixture of experts

By means of the above probabilistic interpretation, learning in the MoE is treated as a maximum likelihood problem defined based on the model in (28.12). An expectation-maximization (EM) algorithm was proposed to update parameters in the MoE [28.64]. It is summarized as follows:

Algorithm 28.5 EM algorithm for MoE learning

Given a training set of T examples $\mathcal{D} = \{(\boldsymbol{x}_t, \boldsymbol{y}_t)\}_{t=1}^{\mathsf{T}}$ pre-set the number of expert networks, N, and randomly initialize all the parameters $\Theta = \{V, (W_n)_{n=1}^N\}$ in the MoE:

- E-step
  For each example, $(\boldsymbol{x}_t, \boldsymbol{y}_t)$ in $\mathcal{D}$, estimate posterior probabilities, $\boldsymbol{h}_t = (h_{nt})_{n=1}^N$, with the current parameter values, $\hat{V}$ and $\{\hat{W}_n\}_{n=1}^N$

$$h_{nt} = \frac{g(\boldsymbol{x}_t, \hat{\boldsymbol{v}}_n) P(\boldsymbol{y}_t | \boldsymbol{x}_t, \hat{W}_n)}{\sum_{k=1}^N g(\boldsymbol{x}_t, \hat{\boldsymbol{v}}_k) P(\boldsymbol{y}_t | \boldsymbol{x}_t, \hat{W}_k)} .$$

- M-step
  - For expert network n $(n = 1, \cdots, n)$, solve the maximization problems

$$\hat{W}_n = \arg\max_{W_n} \sum_{t=1}^{\mathsf{T}} h_{nt} \log P(\boldsymbol{y}_t | \boldsymbol{x}_t, W_n) ,$$

with all examples in $\mathcal{D}$ and posterior probabilities $\{\boldsymbol{h}_t\}_{t=1}^{\mathsf{T}}$ achieved in the E-step.
  - For the gating network, solve the maximization problem

$$\hat{V} = \arg\max_V \sum_{t=1}^{\mathsf{T}} \sum_{n=1}^N h_{nt} \log g(\boldsymbol{x}_t, \boldsymbol{v}_n) ,$$

with training examples, $\{(\boldsymbol{x}_t, \boldsymbol{h}_t)\}_{t=1}^{\mathsf{T}}$, derived from posterior probabilities $\{\boldsymbol{h}_t\}_{t=1}^{\mathsf{T}}$.

Repeat the E-step and the M-step alternately until the EM algorithm converges.

To solve optimization problems in the M-step, the iteratively re-weighted least squares (IRLS) algorithm was proposed [28.64]. Although the IRLS algorithm has the strength to solve maximum likelihood problems arising from MoE learning, it might result in some instable performance due to its incorrect assumption on multi-class classification [28.66]. As learning in the gating network is a multi-class classification

task in essence, the problem always exists if the IRLS algorithm is used in the EM algorithm. Fortunately, improved algorithms were proposed to remedy this problem in the EM learning [28.66]. In summary, numerous MoE variants and extensions have been developed in the past 20 years [28.59], and the MoE architecture turns out to be one of the most successful MNNs.

Negative Correlation Learning

The NCL [28.65] is a learning algorithm to establish an MNN consisting of diverse neural networks (NNs) by promoting the diversity among component networks during learning. The NCL development was clearly inspired by the bias-variance analysis of generalization errors [28.60, 61]. As a result, the NCL encourages cooperation among component networks via interaction during learning to manage the bias-variance trade-off.

In the NCL, an unsupervised penalty term is introduced to the MSE cost function for each component NN so that the error diversity among component networks is explicitly managed via training towards negative correlation. Suppose that an NN ensemble $F(\boldsymbol{x}, \Theta)$ is established by simply taking the average of N neural networks $f(\boldsymbol{x}, W_n)$ $(n = 1, \cdots, N)$, where W_n denotes all the parameters in the n-th component network and $\Theta = \{W_n\}_{n=1}^N$. Given a training set $\mathcal{D} = \{(\boldsymbol{x}_t, \boldsymbol{y}_t)\}_{t=1}^{\mathsf{T}}$ the NCL cost function for the n-th component network [28.65] is defined as follows

$$
\begin{aligned}
L(\mathcal{D}, W_n) = &\frac{1}{2T} \sum_{t=1}^{\mathsf{T}} \| f(\boldsymbol{x}_t, W_n) - \boldsymbol{y}_t \|_2^2 \\
&- \frac{\lambda}{2T} \sum_{t=1}^{\mathsf{T}} \| f(\boldsymbol{x}_t, W_n) - F(\boldsymbol{x}_t, \Theta) \|_2^2 ,
\end{aligned}
\tag{28.13}
$$

where $F(\boldsymbol{x}_t, \Theta) = \frac{1}{N} \sum_{n=1}^N f(\boldsymbol{x}_t, W_n)$ and λ is a trade-off hyper-parameter. In (28.13), the first term is the MSE cost for network n and the second term refers to the negative correlation cost. By taking all component networks into account, minimizing the second term leads to maximum negative correlation among them. Therefore, λ needs to be set properly to control the penalty strength [28.65].

For the NCL, all N cost functions specified in (28.13) need to be optimized together for parameter estimation. Based on the stochastic descent method, the generic NCL algorithm is summarized as follows:

Algorithm 28.6 Negative correlation learning algorithm

Given a training set of T examples $\mathcal{D} = \{(\boldsymbol{x}_t, \boldsymbol{y}_t)\}_{t=1}^{\mathsf{T}}$ pre-set the number of component networks, N, and learning rate, ϵ, as well as randomly initialize all the parameters $\Theta = \{W_n\}_{n=1}^{N}$ in component networks:

- Output computation
  For each example, $(\boldsymbol{x}_t, \boldsymbol{y}_t)$ in $\mathcal{D}$, calculate output of each component network $f(\boldsymbol{x}_t, W_n)$ and that of the NN ensemble by

$$F(\boldsymbol{x}_t, \Theta) = \frac{1}{N} \sum_{n=1}^{N} f(\boldsymbol{x}_t, W_n) \,.$$

- Gradient computation
  For component network n $(n = 1, \cdots, N)$, calculate the gradient of the NCL cost function in (28.13) with respect to the parameters based on all training examples in $\mathcal{D}$

$$\frac{\partial L(\mathcal{D}, W_n)}{\partial W_n} = \frac{1}{T} \sum_{t=1}^{T} \left\{ \|f(\boldsymbol{x}_t, W_n) - \boldsymbol{y}_t\|_2 \right. $$
$$\left. - \frac{\lambda(N-1)}{N} \|f(\boldsymbol{x}_t, W_n) - F(\boldsymbol{x}_t, \Theta)\|_2 \right\} \frac{\partial f(\boldsymbol{x}_t, W_n)}{\partial W_n} \,.$$

- Parameter update
  For component network n $(n = 1, \cdots, N)$, update the parameters

$$W_n \leftarrow W_n - \epsilon \frac{\partial L(\mathcal{D}, W_n)}{\partial W_n} \,.$$

Repeat the above three steps until a pre-specified termination condition is met.

While the NCL was originally proposed based on the MSE cost function, the NCL idea can be extended to other cost functions without difficulty. Hence, applying appropriate optimization techniques on alternative cost functions leads to NCL algorithms of different forms accordingly.

28.3.3 Loosely Coupled Models

In a loosely coupled model, component networks are trained independently or there is no direct interaction among them during learning. There are a variety of MNNs that can be classified as loosely coupled models. Below we review several typical loosely coupled MNNs.

Neural Network Ensemble

An neural network ensemble here refers to a committee machine where a number of NNs trained independently but their outputs are somehow combined to reach a consensus as a final solution. The development of NN ensembles is explicitly motivated by the diversity-promotion principle [28.67, 68].

Intuitively, errors made by component NNs can be corrected by taking their diversity or mutual complement into account. For example, three NNs, NN_i ($i = 1, 2, 3$), trained on the same data set have different yet imperfect performance on test data. Given three test points, $\boldsymbol{x}_t$, ($t = 1, 2, 3$), NN_1 yields the correct output for $\boldsymbol{x}_2$ and $\boldsymbol{x}_3$ but does not for $\boldsymbol{x}_1$, NN_2 yields the correct output for $\boldsymbol{x}_1$, and $\boldsymbol{x}_3$ but does not for $\boldsymbol{x}_2$ and NN_3 yields the correct output for $\boldsymbol{x}_1$ and $\boldsymbol{x}_2$ but does not for $\boldsymbol{x}_3$, respectively. In such circumstances, an error made by one NN can be corrected by other two NNs with a majority vote so that the ensemble can outperform any component NNs. Formally, there is a variety of theoretical justification [28.60, 61] for NN ensembles. For example, it has been proven for regression that the NN ensemble performance is never inferior to the average performance of all component NNs [28.60]. Furthermore, a theoretical bias-variance analysis [28.61] suggests that the promotion of diversity can improve the performance of NN ensembles provided that there is an adequate trade-off between bias and variance. In general, there are two non-trivial issues in constructing NN ensembles; i. e., creating diverse component NNs and ensembling strategies.

Depending on the nature of a given problem [28.5, 62], there are several methodologies for creating diverse component NNs. First of all, a NN learning process itself can be exploited. For instance, learning in a monolithic NN often needs to solve a complex non-convex optimization problem [28.9]. Hence, a local-search-based learning algorithm, e.g., BP [28.11], may end up with various solutions corresponding to local minima due to random initialization. In addition, model selection is required to find out an appropriate NN structure for a given problem. Such properties can be exploited to create component networks in a homogeneous NN ensemble [28.67]. Next, NNs of different types trained on the same data may also yield different performance and hence are candidates in a heterogeneous NN ensemble [28.5]. Finally, exploration/exploitation of input space and different representations is an alternative methodology for creating different component NNs. Instead of training an NN on the input space, NNs can be trained on different input subspaces achieved by a partitioning method, e.g., random partitioning [28.69], which results in a subspace NN ensemble. Whenever raw data can be characterized by different representations, NNs trained on different feature sets would constitute a multi-view NN ensemble [28.70].

Ensembling strategies are required for different tasks. For regression, some optimal fusion rules have been developed for NN ensembles, e.g., [28.68], which are supported by theoretical justification, e.g., [28.60, 61]. For classification, ensembling strategies are more complicated but have been well-studied in a wider context, named combination of multiple classifiers. As is shown in Fig. 28.8, ensembling strategies are generally divided into two categories: learnable and non-learnable. Learnable strategies use a parametric model to learn an optimal fusion rule, while non-learnable strategies fulfil the combination by directly using the statistics of all competent network outputs along with simple measures. As depicted in Fig. 28.8, there are six main non-learnable fusion rules: sum, product, min, max, median, and majority vote; details of such non-learnable rules can be found in [28.71]. Below, we focus on the main learnable ensembling strategies in terms of classification.

In general, learnable ensembling strategies are viewed as an application of the stacked generalization principle [28.72]. In light of stacked generalization, all component NNs serve as level 0 generalizers, and a learnable ensembling strategy carried out by a combination mechanism would perform a level 1 generalizer working on the output space of all component NNs to improve the overall generalization. In this sense, such a combination mechanism is trained on a validation data set that is different from the training data set used in component NN learning. As is shown in Fig. 28.8, combination mechanisms have been developed from different perspectives, i.e., input-dependent and input-independent.

An input-dependent mechanism combines component NNs based on test data; i.e., given two inputs, x_1 and x_2; there is the property: $c(x_1|\Theta) \neq c(x_2|\Theta)$ if $x_1 \neq x_2$, where $c(x|\Theta) = (c_n(x|\Theta))_{n=1}^{N}$ is an input-dependent mechanism used to combine an ensemble of N component NNs and Θ collectively denotes all learnable parameters in this parametric model. As a result, output of an NN ensemble with the input-dependent ensembling strategy is of the following form

$$o(x) = \Omega \left(o_1(x), \cdots, o_N(x) \mid c(x|\Theta) \right),$$

where $o_n(x)$ is output of the n-th component NN for $n = 1, \cdots, N$ and Ω indicates a method on how to apply $c(x|\Theta)$ to component NNs. For example, Ω may be a linear combination scheme such that

$$o(x) = \sum_{n=1}^{N} c_n(x|\Theta) o_n(x). \tag{28.14}$$

As listed in Fig. 28.8, soft-competition and associative switch are two commonly used input-dependent combination mechanisms. The soft-competition mechanism can be regarded as a special case of the MoE described earlier when all expert networks were trained on a data set independently in advance. In this case, the gating network plays the role of the combination mechanism

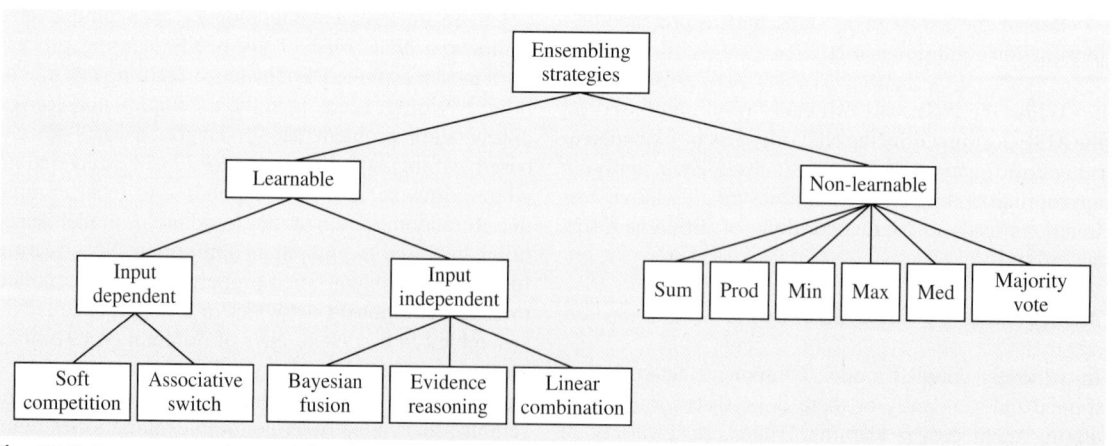

Fig. 28.8 A taxonomy of ensembling strategies

by deciding the importance of component NNs via soft-competition. Although various learning models may be used as such a gating network, a RBF-like (radial basis function) parametric model [28.73] trained on the EM algorithm has been widely used for this purpose. Unlike a soft-competition mechanism that produces the continuous-value weight vector $c(x)$ used in (28.14), the associative switch [28.74] adopts a winner-take-all strategy, i.e., $\sum_{n=1}^{N} c_n(x|\Theta) = 1$ and $c_n(x|\Theta) \in \{0, 1\}$. Thus, an associative switch yields a specific code for a given input so that the output of the best performed component NN can be selected as the final output of the NN ensemble according to (28.14). The associative switch learning is a multi-class classification problem, and an MLP is often used to carry it out [28.74]. Although an input-dependent ensembling strategy is applicable to most NN ensembles, it is difficult to apply it to multi-view NN ensembles, since different representations need to be considered simultaneously in training a combination mechanism. Fortunately, such issues have been explored in a wider context on how to use different representations simultaneously for ML [28.70, 75–79] so that both soft-competition and associative switch mechanisms can be extended to multi-view NN ensembles.

In contrast, an input-independent mechanism combines component NNs based on the dependence of their outputs without considering input data directly. Given two inputs x_1 and x_2, and $x_1 \neq x_2$, the same $c(\Theta)$ may be applied to outputs of component NNs, where $c(\Theta) = (c_n(\Theta))_{n=1}^{N}$ is an input-independent combination mechanism used to combine an ensemble of N component NNs. Several input-independent mechanisms have been developed [28.62], which often fall into one of three categories, i.e., Bayesian fusion, evidence reasoning, and a linear combination scheme, as shown in Fig. 28.8. Bayesian fusion [28.80] refers to a class of combination schemes that use the information collected from errors made by component NNs on a validation set in order to find out the optimal output of the maximum a posteriori probability, $C^* = \arg\max_{1 \leq l \leq L} P(C_l|o_1(x), \cdots, o_N(x), \Theta)$, via Bayesian reasoning, where C_l is the label for the l-th class in a classification task of L classes, and Θ here encodes the information gathered, e.g., a confusion matrix achieved during learning [28.80]. Similarly, evidence reasoning mechanisms make use of alternative reasoning theories [28.80], e.g., the Dempster–Shafer theory, to yield the best output for NN ensembles via an evidence reasoning process that works on all outputs of component NNs in an ensemble. Finally, linear com-bination schemes of different forms are also popular as input-independent combination mechanisms [28.62]. For instance, the work presented in [28.68] exemplifies how to achieve optimal linear combination weights in a linear combination scheme.

Constructive Modularization Learning

Efforts have also been made towards constructive modularization learning for a given supervised learning task. In such work, the divide-and-conquer principle is explicitly applied in order to develop a constructive learning strategy for modularization. The basic idea behind such methods is to divide a difficult and complex problem into a number of subproblems that are easily solvable by NNs of proper capacities, matching the requirements of the subproblems, and then the solutions to subproblems are combined seamlessly to form a solution to the original problem. On the other hand, constructive modularization learning may alleviate the model selection problem encountered by a monolithic NN. As NNs of simple and even different architectures may be used to solve subproblems, empirical studies suggest that an MNN generated via constructive modularization learning is often insensitive to component NN architectures and hence is less likely to suffer from overall overfitting or underfitting [28.81]. Below we describe two constructive modularization learning strategies [28.81–83] for exemplification.

The partitioning-based strategy [28.81, 82] performs the constructive modularization learning by applying the divide-and-conquer principle explicitly. For a given supervised learning task, the strategy consists of two learning stages: *dividing* and *conquering*. In the dividing stage, it first recursively partitions the input space into overlapping subspaces, which facilitates dealing with various uncertainties, by taking into supervision information into account until the nature of each subproblem defined in generated subspaces

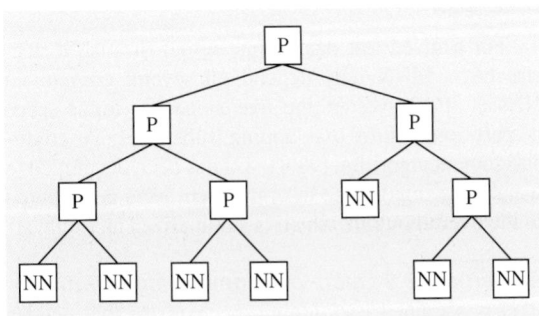

Fig. 28.9 A *self-generated* tree-structured MNN

matches the capacity of one pre-selected NN. In the conquering stage, an NN works on a given input subspace to complete the corresponding learning subtask. As a result, a tree-structured MNN is *self-generated*, where a learnable partitioning mechanism P, is situated at intermediate levels and NNs works at leaves of the tree, as illustrated in Fig. 28.9. To enable the partition-based constructive modularization learning, two generic algorithms have been proposed, i. e., growing and credit-assignment algorithms [28.81, 82] as summarized below.

Algorithm 28.7 Growing algorithm

Given a training set $\mathcal{D}$, set $X \leftarrow \mathcal{D}$. Randomly initialize parameters in all component NNs in a given repository and pre-set hyper-parameters in a learnable partitioning mechanism and compatibility criteria, respectively:

- Compatibility test
  For a training (sub)set X, apply the compatibility criteria to X to examine whether the learning task defined on X matches the capacity of a component NN in the repository.
- Partitioning space
  If none in the repository can solve the problem defined on X, then train the partitioning mechanism on the current X to partition it into two overlapped X_l and X_r. Set $X \leftarrow X_l$, then go to the *compatibility test* step. Set $X \leftarrow X_r$, then go to the *compatibility test* step.
  Otherwise, go to the *subproblem solving* step.
- Subproblem solving
  Train this NN on X with an appropriate learning algorithm. The trained NN resides at the current leaf node.

The growing algorithm expands a tree-structured MNNs until learning problems defined on all partitioned subspaces are solvable with NNs in the repository.

For a given test data point, output of such a tree-structured MNN may depend on several component NNs at the leaves of the tree since the input space is partitioned into overlapping subspaces. A credit-assignment algorithm [28.81, 82] has been developed to weight the importance of component NNs contributed to the overall output, which is summarized as follows:

Algorithm 28.8 Credit–assignment algorithm

$P(x)$ is a trained partitioning mechanisms that resides at a nonterminal node and partitions the current input

(sub)space into two subspaces with an overlapping defined by $-\tau \leq P(x) \leq \tau (\tau > 0)$. $C_L(\cdot)$, and $C_R(\cdot)$ are two credit assignment functions for two subspaces, respectively. For a test data point $\tilde{x}$:

- Initialization
  Set $\alpha(\tilde{x}) \leftarrow 1$ and `Pointer` $\leftarrow$ `Root`.
- Credit assignment
  As a recursive credit propagation process to assign credits to all the component NNs at leaf nodes that $\tilde{x}$ can reach, $CR[\alpha(\tilde{x}),$ `Pointer`$]$ consists of three steps:
  - If `Pointer` points to a leaf node, then output $\alpha(\tilde{x})$ and stop.
  - If $P(\tilde{x}) \leq \tau$, $\alpha(\tilde{x}) \leftarrow \alpha(\tilde{x}) \times C_L(P(\tilde{x}))$ and invoke $CR[\alpha(\tilde{x}),$ `Pointer.Leftchild`$]$.
  - If $P(\tilde{x}) \geq -\tau$, $\alpha(\tilde{x}) \leftarrow \alpha(\tilde{x}) \times C_R(P(\tilde{x}))$ and invoke $CR[\alpha(\tilde{x}),$ `Pointer.Rightchild`$]$.

Thus, the output of a *self-generated* MNN is

$$o(\tilde{x}) = \sum_{n \in \mathcal{N}} \alpha_n(\tilde{x}) \times o_n(\tilde{x}) \, ,$$

where $\mathcal{N}$ denotes all the component NNs that $\tilde{x}$ can reach, and $\alpha_n(\tilde{x})$ and $o_n(\tilde{x})$ are the credit assigned and the output of the n-th component NN in $\mathcal{N}$ for $\tilde{x}$, respectively.

To implement such a strategy, hyper-planes placed with heuristics [28.81] or linear classifiers trained with the Fisher discriminative analysis [28.82] were first used as the partition mechanism and NNs such as MLP or RBF can be employed to solve subproblems. Accordingly, two piece-wise linear credit assignment functions [28.81, 82] were designed for the hyperplane partitioning mechanism, so that $C_L(x) + C_R(x) = 1$. Heuristic compatibility criteria were developed by considering learning errors and efficiency [28.81, 82]. By using the same constructive learning algorithms described above, an alternative implementation was also proposed by using the self-organization map as a partitioning mechanism and SVMs were used for subproblem solving [28.84]. Empirical studies suggest that the partitioning-based strategy leads to favorable results in various supervised learning tasks despite different implementations [28.81, 82, 84].

By applying the divide-and-conquer principle, *task decomposition* [28.83] is yet another constructive modularization learning strategy for classification. Unlike the partitioning-based strategy, the task decomposition strategy converts a multi-class classification task into a number of binary classification subtasks in a brute-force way and each binary classification subtask is expected to be fulfilled by a simple NN. If a subtask is too *difficult* to carry out by a given NN, the subtask is allowed to be further decomposed into simpler binary classification subtasks. For a multi-class classification task of M categories, the task decomposition strategy first exhaustively decomposes it into $\frac{1}{2}M(M-1)$ different primary binary subtasks where each subtask merely concerns classification between two different classes without taking remaining $M-2$ classes into account, which differs from the commonly used one-against-rest decomposition method. In general, the original multi-class classification task may be decomposed into more binary subtasks if some primary subtasks are too *difficult*. Once the decomposition is completed, all the subtasks are undertaken by pre-selected simple NNs, e.g., MLP of one hidden layer, in parallel. For a final solution to the original problem, three non-learnable operations, *min*, *max*, and *inv*, were proposed to combine individual binary classification results achieved by all the component NNs. By applying three operations properly, all the component NNs are integrated together to form a min-max MNN [28.83].

28.3.4 Relevant Issues

In general, studies of MNNs closely relate to several areas in different disciplines, e.g., ML and statistics. We here examine several important issues related to MNNs in a wider context.

As described above, a tightly coupled MNN leads to an optimal solution to a given supervised learning problem. The MoE is rooted in the finite mixture model (FMM) studied in probability and statistics and becomes a non-trivial extension to conditional models where each expert is a parametric conditional probabilistic model and the mixture coefficients also depend on input [28.64]. While the MoE has been well studied for 20 years [28.59] in different disciplines, there still exist some open problems in general, e.g., model selection, global optimal solution, and convergence of its learning algorithms for arbitrary component models. Different from the FMM, the product of experts (PoE) [28.42] was also proposed to combine a number of experts (parametric probabilistic models) by taking their product and normalizing the result into account. The PoE has been argued to have some advantages over the MoE [28.42] but has so far merely been developed in the context of unsupervised learning. As a result, extending the PoE to

conditional models for supervised learning would be a non-trivial topic in tightly coupled MNN studies. On the other hand, the NCL [28.65] directly applies the bias-variance analysis [28.60, 61] to construction of an MNN. This implies that MNNs could be also built up via alternative loss functions that properly promote diversities among component MNNs during learning.

Almost all existing NN ensemble methods are now included in ensemble learning [28.85], which is an important area in ML, or the multiple classifier system [28.62] in the context of pattern recognition. In statistical ensemble learning, generic frameworks, e.g., boosting [28.86] and bootstrapping [28.87], were developed to construct ensemble learners where any learning models including NNs may be used as component learners. Hence, most of common issues raised for ensemble learning are applicable to NN ensembles. Nevertheless, ensemble learning researches suggest that behaviors of component learners may considerably affect the stability and overall performance of ensemble learning. As exemplified in [28.88], properties of different NN ensembles are worth investigating from both theoretical and application perspectives.

While constructive modularization learning provides an alternative way of model selection, it is generally a less developed area in MNNs, and existing methods are subject to limitation due to a lack of theoretical justification and underpinning techniques. For example, a critical issue in the partitioning-based strategy [28.81, 82] is how to measure the nature of a subproblem to decide if any further partitioning is required and the appropriateness of a pre-selected NN to a subproblem in terms of its capacity. In previous studies [28.81, 82], a number of heuristic and simple criteria were proposed based on learning errors and efficiency. Although such heuristic criteria work practically, there is no theoretical justification. As a result, more sophisticated compatibility criteria need to be developed for such a constructive learning strategy based on the latest ML development, e.g., manifold and adaptive kernel learning. Fortunately, the partitioning-based strategy has inspired the latest developments in ML [28.89]. In general, constructive modularization learning is still a non-trivial topic in MNN research.

Finally, it is worth stating that our MNN review here only focuses on supervised learning due to the limited space. Most MNNs described above may be extended to other learning paradigms, e.g., semi-supervised and unsupervised learning. More details on such topics are available in the literature, e.g., [28.90, 91].

28.4 Concluding Remarks

In this chapter, we have reviewed two important areas, DNNs and MNNs, in NC. While we have presented several sophisticated techniques that are ready for applications, we have discussed several challenging problems in both deep and modular neural network research as well. Apart from other non-trivial issues discussed in the chapter, it is worth emphasizing that it is still an open problem to develop large-scale DNNs and MNNs and integrate them for modeling highly intelligent behaviors, although some progress has been made recently [28.58]. In a wider context, DNNs and MNNs are closely related to two active areas, deep learning and ensemble learning, in ML. We anticipate that motivation and methodologies from different perspectives will mutually benefit each other and lead to effective solutions to common challenging problems in the NC and ML communities.

References

28.1 E.R. Kandel, J.H. Schwartz, T.M. Jessell: *Principle of Neural Science*, 4th edn. (McGraw-Hill, New York 2000)

28.2 G.M. Edelman: *Neural Darwinism: Theory of Neural Group Selection* (Basic Books, New York 1987)

28.3 J.A. Fodor: *The Modularity of Mind* (MIT Press, Cambridge 1983)

28.4 F. Azam: Biologically inspired modular neural networks, Ph.D. Thesis (School of Electrical and Computer Engineering, Virginia Polytechnic Institute and State University, Blacksburg 2000)

28.5 G. Auda, M. Kamel: Modular neural networks: A survey, Int. J. Neural Syst. **9**(2), 129–151 (1999)

28.6 D.C. Van Essen, C.H. Anderson, D.J. Fellman: Information processing in the primate visual system, Science **255**, 419–423 (1992)

28.7 J.H. Kaas: Why does the brain have so many visual areas?, J. Cogn. Neurosci. **1**(2), 121–135 (1989)

28.8 G. Bugmann: Biologically plausible neural computation, Biosystems **40**(1), 11–19 (1997)

28.9 S. Haykin: *Neural Networks and Learning Machines*, 3rd edn. (Prentice Hall, New York 2009)

28.10 M. Minsky, S. Papert: *Perceptrons* (MIT Press, Cambridge 1969)

28.11 D.E. Rumelhurt, G.E. Hinton, R.J. Williams: Learning internal representations by error propagation, Nature **323**, 533–536 (1986)

28.12 Y. LeCun, L. Bottou, Y. Bengio, P. Haffner: Gradient based learning applied to document recognition, Proc. IEEE **86**(9), 2278–2324 (1998)

28.13 G. Tesauro: Practical issues in temporal difference learning, Mach. Learn. **8**(2), 257–277 (1992)

28.14 G. Cybenko: Approximations by superpositions of sigmoidal functions, Math. Control Signals Syst. **2**(4), 302–314 (1989)

28.15 N. Cristianini, J. Shawe-Taylor: *An Introduction to Support Vector Machines and Other Kernel-Based Learning Methods* (Cambridge University Press, Cambridge 2000)

28.16 Y. Bengio, Y. LeCun: Scaling learning algorithms towards AI. In: *Large-Scale Kernel Machines*, ed. by L. Bottou, O. Chapelle, D. DeCoste, J. Weston (MIT Press, Cambridge 2006), Chap. 14

28.17 Y. Bengio: Learning deep architectures for AI, Found. Trends Mach. Learn. **2**(1), 1–127 (2009)

28.18 G.E. Hinton, S. Osindero, Y. Teh: A fast learning algorithm for deep belief nets, Neural Comput. **18**(9), 1527–1554 (2006)

28.19 Y. Bengio: Deep learning of representations for unsupervised and transfer learning, JMLR: Workshop Conf. Proc., Vol. 7 (2011) pp. 1–20

28.20 H. Larochelle, D. Erhan, A. Courville, J. Bergstra, Y. Bengio: An empirical evaluation of deep architectures on problems with many factors of variation, Proc. Int. Conf. Mach. Learn. (ICML) (2007) pp. 473–480

28.21 R. Salakhutdinov, G.E. Hinton: Learning a nonlinear embedding by preserving class neighbourhood structure, Proc. Int. Conf. Artif. Intell. Stat. (AISTATS) (2007)

28.22 H. Larochelle, Y. Bengio, J. Louradour, P. Lamblin: Exploring strategies for training deep neural networks, J. Mach. Learn. Res. **10**(1), 1–40 (2009)

28.23 W.K. Wong, M. Sun: Deep learning regularized Fisher mappings, IEEE Trans. Neural Netw. **22**(10), 1668–1675 (2011)

28.24 S. Osindero, G.E. Hinton: Modeling image patches with a directed hierarchy of Markov random field, Adv. Neural Inf. Process. Syst. (NIPS) (2007) pp. 1121–1128

28.25 I. Levner: Data driven object segmentation, Ph.D. Thesis (Department of Computer Science, University of Alberta, Edmonton 2008)

28.26 H. Mobahi, R. Collobert, J. Weston: Deep learning from temporal coherence in video, Proc. Int. Conf. Mach. Learn. (ICML) (2009) pp. 737–744

28.27 H. Lee, Y. Largman, P. Pham, A. Ng: Unsupervised feature learning for audio classification using convolutional deep belief networks, Adv. Neural Inf. Process. Syst. (NIPS) (2009)

28.28 K. Chen, A. Salman: Learning speaker-specific characteristics with a deep neural architecture, IEEE Trans. Neural Netw. **22**(11), 1744–1756 (2011)

28.29   K. Chen, A. Salman: Extracting speaker-specific information with a regularized Siamese deep network, Adv. Neural Inf. Process. Syst. (NIPS) (2011)

28.30   A. Mohamed, G.E. Dahl, G.E. Hinton: Acoustic modeling using deep belief networks, IEEE Trans. Audio Speech Lang. Process. **20**(1), 14–22 (2012)

28.31   G.E. Dahl, D. Yu, L. Deng, A. Acero: Context-dependent pre-trained deep neural networks for large-vocabulary speech recognition, IEEE Trans. Audio Speech Lang. Process. **20**(1), 30–42 (2012)

28.32   R. Salakhutdinov, G.E. Hinton: Semantic hashing, Proc. SIGIR Workshop Inf. Retr. Appl. Graph. Model. (2007)

28.33   M. Ranzato, M. Szummer: Semi-supervised learning of compact document representations with deep networks, Proc. Int. Conf. Mach. Learn. (ICML) (2008)

28.34   A. Torralba, R. Fergus, Y. Weiss: Small codes and large databases for recognition, Proc. Int. Conf. Comput. Vis. Pattern Recogn. (CVPR) (2008) pp. 1–8

28.35   R. Collobert, J. Weston: A unified architecture for natural language processing: Deep neural networks with multitask learning, Proc. Int. Conf. Mach. Learn. (ICML) (2008)

28.36   A. Mnih, G.E. Hinton: A scalable hierarchical distributed language model, Adv. Neural Inf. Process. Syst. (NIPS) (2008)

28.37   J. Weston, F. Ratle, R. Collobert: Deep learning via semi-supervised embedding, Proc. Int. Conf. Mach. Learn. (ICML) (2008)

28.38   R. Hadsell, A. Erkan, P. Sermanet, M. Scoffier, U. Muller, Y. LeCun: Deep belief net learning in a long-range vision system for autonomous off-road driving, Proc. Intell. Robots Syst. (IROS) (2008) pp. 628–633

28.39   Y. Bengio, A. Courville, P. Vincent: Representation learning: A review and new perspectives, IEEE Trans. Pattern Anal. Mach. Intell. **35**(8), 1798–1827 (2013)

28.40   Y. Bengio, P. Lamblin, D. Popovici, H. Larochelle: Greedy layer-wise training of deep networks, Adv. Neural Inf. Process. Syst. (NIPS) (2006)

28.41   P. Vincent, H. Larochelle, I. Lajoie, Y. Bengio, P.A. Manzagol: Stacked denoising autoencoders: Learning useful representations in a deep network with a local denoising criterion, J. Mach. Learn. Res. **11**, 3371–3408 (2010)

28.42   G.E. Hinton: Training products of experts by minimizing contrastive divergence, Neural Comput. **14**(10), 1771–1800 (2002)

28.43   K. Kavukcuoglu, M. Ranzato, Y. LeCun: Fast inference in sparse coding algorithms with applications to object recognition. CoRR, arXiv:1010.3467 (2010)

28.44   B.A. Olshausen, D.J. Field: Sparse coding with an overcomplete basis set: A strategy employed by V1?, Vis. Res. **37**, 3311–3325 (1997)

28.45   S. Rifai, P. Vincent, X. Muller, X. Glorot, Y. Bengio: Contracting auto-encoders: Explicit invariance during feature extraction, Proc. Int. Conf. Mach. Learn. (ICML) (2011)

28.46   S. Rifai, G. Mesnil, P. Vincent, X. Muller, Y. Bengio, Y. Dauphin, X. Glorot: Higher order contractive auto-encoder, Proc. Eur. Conf. Mach. Learn. (ECML) (2011)

28.47   M. Ranzato, C. Poultney, S. Chopra, Y. LeCun: Efficient learning of sparse representations with an energy based model, Adv. Neural Inf. Process. Syst. (NIPS) (2006)

28.48   M. Ranzato, Y. Boureau, Y. LeCun: Sparse feature learning for deep belief networks, Adv. Neural Inf. Process. Syst. (NIPS) (2007)

28.49   G.E. Hinton, R. Salakhutdinov: Reducing the dimensionality of data with neural networks, Science **313**, 504–507 (2006)

28.50   K. Cho, A. Ilin, T. Raiko: Improved learning of Gaussian-Bernoulli restricted Boltzmann machines, Proc. Int. Conf. Artif. Neural Netw. (ICANN) (2011)

28.51   D. Erhan, Y. Bengio, A. Courville, P.A. Manzagol, P. Vincent, S. Bengio: Why does unsupervised pre-training help deep learning?, J. Mach. Learn. Res. **11**, 625–660 (2010)

28.52   D.C. Ciresan, U. Meier, L.M. Gambardella, J. Schmidhuber: Deep big simple neural nets for handwritten digit recognition, Neural Comput. **22**(1), 1–14 (2010)

28.53   M. Ranzato, A. Krizhevsky, G.E. Hinton: Factored 3-way restricted Boltzmann machines for modeling natural images, Proc. Int. Conf. Artif. Intell. Stat. (AISTATS) (2010) pp. 621–628

28.54   M. Ranzato, V. Mnih, G.E. Hinton: Generating more realistic images using gated MRF's, Adv. Neural Inf. Process. Syst. (NIPS) (2010)

28.55   A. Courville, J. Bergstra, Y. Bengio: Unsupervised models of images by spike-and-slab RBMs, Proc. Int. Conf. Mach. Learn. (ICML) (2011)

28.56   H. Lee, R. Grosse, R. Ranganath, A.Y. Ng: Unsupervised learning of hierarchical representations with convolutional deep belief networks, Commun. ACM **54**(10), 95–103 (2011)

28.57   D. Hau, K. Chen: Exploring hierarchical speech representations with a deep convolutional neural network, Proc. U.K. Workshop Comput. Intell. (UKCI) (2011)

28.58   Q. Le, M. Ranzato, R. Monga, M. Devin, K. Chen, G.S. Corrado, J. Dean, A.Y. Ng: Building high-level features using large scale unsupervised learning, Proc. Int. Conf. Mach. Learn. (ICML) (2012)

28.59   S.E. Yuksel, J.N. Wilson, P.D. Gader: Twenty years of mixture of experts, IEEE Trans. Neural Netw. Learn. Syst. **23**(8), 1177–1193 (2012)

28.60   A. Krogh, J. Vedelsby: Neural network ensembles, cross validation, and active learning, Adv. Neural Inf. Process. Syst. (NIPS) (1995)

28.61   N. Ueda, R. Nakano: Generalization error of ensemble estimators, Proc. Int. Conf. Neural Netw. (ICNN) (1996) pp. 90–95

28.62   L.I. Kuncheva: *Combining Pattern Classifiers* (Wiley-Interscience, Hoboken 2004)

28.63   R.A. Jacobs, M.I. Jordan, S. Nowlan, G.E. Hinton: Adaptive mixture of local experts, Neural Comput. **3**(1), 79–87 (1991)

28.64   M.I. Jordan, R.A. Jacobs: Hierarchical mixture of experts and the EM algorithm, Neural Comput. **6**(2), 181–214 (1994)

28.65   Y. Liu, X. Yao: Simultaneous training of negatively correlated neural networks in an ensemble, IEEE Trans. Syst. Man Cybern. B **29**(6), 716–725 (1999)

28.66   K. Chen, L. Xu, H.S. Chi: Improved learning algorithms for mixture of experts in multi-class classification, Neural Netw. **12**(9), 1229–1252 (1999)

28.67   L.K. Hansen, P. Salamon: Neural network ensembles, IEEE Trans. Pattern Anal. Mach. Intell. **12**(10), 993–1001 (1990)

28.68   M.P. Perrone, L.N. Cooper: Ensemble methods for hybrid neural networks. In: *Artificial Neural Networks for Speech and Vision*, ed. by R.J. Mammone (Chapman-Hall, New York 1993) pp. 126–142

28.69   T.K. Ho: The random subspace method for constructing decision forests, IEEE Trans. Pattern Anal. Mach. Intell. **20**(8), 823–844 (1998)

28.70   K. Chen, L. Wang, H.S. Chi: Methods of combining multiple classifiers with different feature sets and their applications to text-independent speaker identification, Int. J. Pattern Recogn. Artif. Intell. **11**(3), 417–445 (1997)

28.71   J. Kittler, M. Hatef, R.P.W. Duin, J. Matas: On combining classifiers, IEEE Trans. Pattern Anal. Mach. Intell. **20**(3), 226–239 (1998)

28.72   D.H. Wolpert: Stacked generalization, Neural Netw. **2**(3), 241–259 (1992)

28.73   L. Xu, M.I. Jordan, G.E. Hinton: An alternative model for mixtures of experts, Adv. Neural Inf. Process. Syst. (NIPS) (1995)

28.74   L. Xu, A. Krzyzak, C.Y. Suen: Associative switch for combining multiple classifiers, J. Artif. Neural Netw. **1**(1), 77–100 (1994)

28.75   K. Chen: A connectionist method for pattern classification with diverse feature sets, Pattern Recogn. Lett. **19**(7), 545–558 (1998)

28.76   K. Chen, H.S. Chi: A method of combining multiple probabilistic classifiers through soft competition on different feature sets, Neurocomputing **20**(1–3), 227–252 (1998)

28.77   K. Chen: On the use of different representations for speaker modeling, IEEE Trans. Syst. Man Cybern. C **35**(3), 328–346 (2005)

28.78   Y. Yang, K. Chen: Temporal data clustering via weighted clustering ensemble with different representations, IEEE Trans. Knowl. Data Eng. **23**(2), 307–320 (2011)

28.79   Y. Yang, K. Chen: Time series clustering via RPCL ensemble networks with different representations, IEEE Trans. Syst. Man Cybern. C **41**(2), 190–199 (2011)

28.80   L. Xu, A. Krzyzak, C.Y. Suen: Methods of combining multiple classifiers and their applications to handwriting recognition, IEEE Trans. Syst. Man Cybern. **22**(3), 418–435 (1992)

28.81   K. Chen, X. Yu, H.S. Chi: Combining linear discriminant functions with neural networks for supervised learning, Neural Comput. Appl. **6**(1), 19–41 (1997)

28.82   K. Chen, L.P. Yang, X. Yu, H.S. Chi: A self-generating modular neural network architecture for supervised learning, Neurocomputing **16**(1), 33–48 (1997)

28.83   B.L. Lu, M. Ito: Task decomposition and module combination based on class relations: A modular neural network for pattern classification, IEEE Trans. Neural Netw. Learn. Syst. **10**(5), 1244–1256 (1999)

28.84   L. Cao: Support vector machines experts for time series forecasting, Neurocomputing **51**(3), 321–339 (2003)

28.85   T.G. Dietterich: Ensemble learning. In: *Handbook of Brain Theory and Neural Networks*, ed. by M.A. Arbib (MIT Press, Cambridge 2002) pp. 405–408

28.86   Y. Freund, R.E. Schapire: Experiments with a new boosting algorithm, Proc. Int. Conf. Mach. Learn. (ICML) (1996) pp. 148–156

28.87   L. Breiman: Bagging predictors, Mach. Learn. **24**(2), 123–140 (1996)

28.88   H. Schwenk, Y. Bengio: Boosting neural networks, Neural Comput. **12**(8), 1869–1887 (2000)

28.89   J. Wang, V. Saligrama: Local supervised learning through space partitioning, Adv. Neural Inf. Process. Syst. (NIPS) (2012)

28.90   X.J. Zhu: *Semi-supervised learning literature survey, Technical Report, School of Computer Science* (University of Wisconsin, Madison 2008)

28.91   J. Ghosh, A. Acharya: Cluster ensembles, WIREs Data Min. Knowl. Discov. **1**(2), 305–315 (2011)

29. Machine Learning

James T. Kwok, Zhi-Hua Zhou, Lei Xu

This tutorial provides a brief overview of a number of important tools that form the crux of the modern machine learning toolbox. These tools can be used for supervised learning, unsupervised learning, reinforcement learning and their numerous variants developed over the years. Because of the lack of space, this survey is not intended to be comprehensive. Interested readers are referred to conference proceedings such as Neural Information Processing Systems (NIPS) and the International Conference on Machine Learning (ICML) for the most recent advances.

29.1 Overview

Machine learning represents one of the most prolific developments in modern artificial intelligence. It provides a new generation of computational techniques and tools that support understanding and extraction of useful knowledge from complicated data sets. So what is machine learning? *Simon* [29.1] defined machine learning as:

changes in the system that are adaptive in the sense that they enable the system to do the same task or tasks drawn from the same population more effectively the next time.

Hence, fundamentally, the emphasis of machine learning is on the system's ability to adapt or change. Typically, this is in response to some form of experience provided to the system. After learning or adaptation, the system is expected to have better future performance on the same or a related task.

Over the past decades, machine learning has grown from a few toy applications to being almost everywhere. It is now being applied to numerous real-world applications. For example, the control of autonomous robots that can navigate on their own, the filtering of spam from mailboxes, the recognition of characters from handwriting, the recognition of speech on mobile devices, the detection of faces in digital cameras, and so on. Indeed, one can find applications of machine learning from everyday consumer products to advanced information systems in corporations. Studies of machine learning may be overviewed from either the perspective of learning intelligent systems or that of a machine learning toolbox. For the former, learning is considered as a process of an intelligent system for coordinately solving three levels of inverse problems, namely problem solving for making pattern recognition and various other tasks, parameter learning for estimating unknown parameters in the system, and model selection for shaping system configuration with an appropriate scale or complexity to describe regularities underlying a finite size of samples. Different learning approaches are featured by differences in one or more of three ingredients, namely as a learner that has an appropriate system configuration, a theory that guides learning, and an algorithm or dynamic procedure that implements learning. Examples of studies from this prospect are recently overviewed in [29.2, 3], and will not be further addressed in this chapter. Instead, this chapter aims at a tutorial on studies of machine learning from the second prospect, that is, on those important collections pooled in the machine learning toolbox for decades. Actually, the current prosperity of machine learning comes from not only further developments of the classical statistical modeling and neural network learning, but also emerging achievements of machine learning and data mining in recent decades. Due to limited space, the focus of this tutorial will be particularly placed on those advancements made in the last two decades or so.

Classically, there are three basic learning paradigms, namely, supervised learning (Sect. 29.2), unsupervised learning (Sect. 29.3), and reinforcement learning (Sect. 29.4). In supervised learning, the learner is provided with a set of inputs together with the corresponding desired outputs. This is similar to the familiar human learning process for pattern recognition, in which a teacher provides examples to teach children to recognize different objects (say, for example, animals). Such a pattern recognition task is featured by data with each input sample associated with a label, namely labeled data. In the current literature on machine learning, the term labeled data is even generally used to refer data with each input associated with an output beyond simply a label, which is also adopted in this chapter. Section 29.2 provides not only a tutorial on basic issues of supervised learning but

also an overview on a number of interesting topics developed in recent years. The coverage of this section is not complete, e.g., it does not cover the supervised learning studies in the literature on neural networks. Interested readers are referred to a number of survey papers, e.g., especially those on multi-layer perceptron and radial basis functions [29.4, 5].

Unlike supervised learning, the tasks of unsupervised learning are featured by data that consist of only inputs, namely, the data is unlabeled and there is no longer the presence of a teacher. Unsupervised learning aims at finding certain dependence structure underlying data via optimizing a learning principle. Considering different types of structures, studies include not only classic topics of data clustering, subspace, and topological maps, but also emerging topics of learning latent factor models, hidden state–space models, and hierarchical structures. Section 29.3 also consists of two parts. The first part provides a tutorial on three classic topics, while the second part makes an overview on emerging topics. Extensive studies have been made on unsupervised learning for many decades. Instead of seeking a complete coverage, this section focuses on a tutorial on fundamentals and an overview on interesting developments of recent years, mainly based on a more systematic overview [29.6]. Further, readers are referred to several recent survey papers, e.g., [29.7] for an overview on 50 years of studies beyond k-means for data clustering, [29.8, 9] for subspace and manifold learning, and [29.10] for topological maps.

The third paradigm is reinforcement learning. Upon observing the current environment and obtaining some input (if any), the learner makes an action and changes to a new environment, receiving an evaluation (award or punish) value about the action. A learning process makes a series of actions with the received total award maximized. Different to unsupervised learning, the learner gets a guidance from an external evaluation. Also unlike supervised learning in which the teacher clearly specifies the output that corresponds to an input, in reinforcement learning the learner is only provided with an evaluative value about the action made. Section 29.4 starts at giving a tutorial on basic issues of reinforcement learning, especially temporal difference TD learning and Q-learning, plus improvements on the Q-learning with the help of some unsupervised learning methods.

Besides these three basic learning paradigms, many more variants have been developed in recent years because of the advances in machine learning. Some of these will also be described in this tutorial. They are often a hybrid of the previous learning paradigms. A very popular variant is semi-supervised learning (Sect. 29.5), which uses both labeled data (as in supervised learning) and unlabeled data (as in unsupervised learning) for training. This is advantageous as labeled data typically are expensive and involve tedious human effort, while a large amount of unlabeled data can often be obtained in an inexpensive manner (e.g., simply downloadable from the web). Another hybrid of supervised learning and unsupervised learning is discriminative clustering. Here, one adopts a cost function originally used for supervised learning as a clustering criterion. A well-known example in this category is called maximum margin clustering [29.11–13], which tries to maximize the margin (used as a criterion in constructing the highly successful supervised learning model: support vector machine) between clusters.

Moreover, instead of just constructing one learner from training data, one can construct a set of learners and combine them to solve the problem. This approach, known as *ensemble learning* [29.14], has become very popular in recent years and will be discussed in more detail in Sect. 29.6. Finally, before learning can proceed, the data need to be appropriately represented by a set of features. In many real-world data sets, there are often a large number of features, many of which are abundant or irrelevant. Feature selection and extraction aim at automatically extracting the good features and removing the bad ones, and this will be covered in Sect. 29.7.

29.2 Supervised Learning

A supervised learner is provided with some labeled data (often called *training data*). This consists of a set of *training samples*, each of which is an input together with the corresponding desired outputs. Hence, the first step in machine learning is to collect these training samples. Moreover, as each training sample needs to be represented in a form amendable by the computer algorithm, one has to define a set of *features*. As an example, consider the task of recognizing handwritten characters on an envelope. To construct the training samples, obviously one first has to collect a number of envelopes with handwritten characters on. Then, the characters on each envelope have to be separated from each other. This can be performed either manually or automatically by some image segmentation algorithm. Afterwards, each character is a block of pixels (typically rectangular). A simple feature representation will be to use the intensities of these raw pixels. Each input is represented as a vector of feature values, and this vector is called the *feature vector*. Obviously, it is important to have a good set of features to work with. The presence of bad features may confuse the learning algorithm and makes learning more difficult. For example, in the context of character recognition, the color of the ink is not relevant to the identity of the character and so can be considered as a bad feature. Depending on the domain knowledge, more sophisticated features can be manually defined. It is desirable that good features can be automatically extracted and bad features automatically removed. More details on these feature selection/extraction algorithms will be covered in Sect. 29.7. Finally, each character on the envelope has to be manually labeled.

In practice, as the real-world data are often dirty, a significant amount of time may have to be spent on data pre-processing in order to create the training data. There are many forms of *dirty* data. For example, it can be incomplete in that certain attribute values (e.g., occupation) may be lacking; it can contain outliers or errors (e.g., the *salary* is negative); parts of it may be inconsistent (e.g., the customer's age is 42 but his/her birthday is *03/07/2012*); it may also be redundant in that there are duplicate records or unnecessary attributes. All these problems may be due to faulty or careless data collection, human/hardware/software problems, errors in data transmission; or that the data may have come from a number of different data sources. In all cases, data pre-processing can have a significant impact on the resultant machine learning system, as no quality data implies no quality learning results!

29.2.1 Classification and Regression

The two main goals in supervised learning are (i) classification, which aims at assigning the input pattern to different categories (also called classes or labels); and (ii) regression, which aims at predicting a real value or vector associated with the input. The basic idea and the training/testing procedures in regression are similar to those in classification. Hence, we will mainly focus on the classification problem in the sequel.

The simplest case for classification is binary classification, in which there are only two classes. Examples include classifying an email as spam or non-spam; and classifying an image as face or non-face. For each sample, the supervised learner examines the feature values in that sample and predicts the class that the sample belongs to. Essentially, the supervised learner partitions the whole feature space (the space of all possible feature value combinations) into two regions, one for each class. The boundary is called the *decision boundary*. A wide variety of models can be used to construct this decision boundary. A simple example is the linear classifier, which creates a linear boundary. Depending on the task, the linear classifier may be too simple to differentiate the two classes. Then, one can also use a more complicated decision boundary, such as a quadratic surface, leading to a quadratic classifier. In machine learning, a large number of various models that are capable of producing nonlinear decision boundaries have been proposed. The most popular ones include the decision tree classifier, nearest neighbor classifiers, neural network classifiers, Bayesian classifiers, and support vector machines. Each of these models has some parameters that have to be adapted to the particular data set. For example, the parameters of a linear classifier include the weight on each feature (which controls the slope of the linear boundary) and a bias (which controls the offset). To estimate or train these parameters, one has to provide a training set, where the i-th training pattern (x_i, y_i) consists of an input x_i and the corresponding target output label y_i (for regression problems, this y_i is a real value or vector). The greater the amount of training data, intuitively the more accurate the learned model. However, since the training data in supervised learning are labeled, obtaining these output labels typically involve expensive and tedious human effort. Hence, recent machine learning algorithms also try to utilize data that are unlabeled, leading to the development of semi-supervised learning algorithms in Sect. 29.5.

Given the model, different strategies can be used to learn the model parameters so that it fits the training set (i.e., train the model). Parameter estimation and feature selection (Sect. 29.7) can sometimes be performed together. However, note that there is the danger of *overfitting*, which occurs when the model performs better than other models on the training data, but worse on the entire data distribution as it has captured the trends of the noise underlying the data. Often this happens when the model is excessively complex, such as when it has a lot more parameters than can be re-

liably estimated from the limited number of training patterns. To combat overfitting, one can constrain the model's freedom during training by adding a regularizer or Bayesian to the parameters or model beforehand. Alternatively, one can stop the learning procedure before convergence (*early stopping*) or remove part of the model when training is complete (*pruning*). If there are noisy training samples that significantly deviate from the underlying input–output trend, one can also perform outlier detection to first remove these outlying samples.

There are two general approaches to train the model parameters. The first approach treats the model as a generative model that defines how the data are generated (typically by using a probabilistic model). One can then maximize the likelihood by varying the parameters, or to maximize the posterior probability of the parameters given the training data. Alternatively, one can take a discriminative approach that directly considers how the output is related to the input. The parameters can be obtained by *empirical risk minimization*, which seeks the parameters which best fit the training data. The risk is dependent on the loss function, which measures the difference between the prediction and the target output. Let y_i be the target output for sample i, and $\hat{y}_i$ be the predicted output from the supervised learner. For classification problems, commonly used loss functions include the logistic loss $\ln(1 + \exp(-y_i\hat{y}_i))$ and the hinge loss $\max(0, 1 - y_i\hat{y}_i)$; and for regression problems, the most common loss function is the square loss $(y_i - \hat{y}_i)^2$. However, in order to combat overfitting, it is better to perform *regularized risk minimization* instead of empirical risk minimization. Regularized risk consists of two components. The first component is the loss as in empirical risk minimization. The second component is a regularizer, which helps to control the model complexity and prevents overfitting. Various regularizers have been proposed. Let $w = [w_1, w_2, \ldots, w_d]^T$ be the vector of parameters. A popular regularizer is the ℓ_2-norm of w, i.e.,

$$\|w\|_2^2 = \sum_{i=1}^{d} w_i^2 \,.$$

This leads to ridge regression when the linear model is used, and is commonly called *weight decay* in the neural networks literature. Instead of using the ℓ_2-norm, one can use the ℓ_0-norm $\|w\|_0$, which counts the number of nonzero w_i in the model. However, this is nonconvex and the associated optimization is more difficult. A common way to alleviate this problem is by using the

ℓ_1-norm

$$\|w\|_1 = \sum_{i=1}^{d} |w_i| \,,$$

which is still convex (as for the ℓ_2-norm) but can still lead to a sparse parameter solution (as for the ℓ_0-norm). When used with the square loss on the linear model, this leads to the well-known lasso model.

Once trained, the classifier can be used to predict the label of an unseen test sample. The underlying assumption is that this test sample comes from the same distribution as that of the training samples. In this case, we expect the trained classifier to be able to generalize well to this new sample. This can also be formally described by generalization error bounds in computational learning theory.

There are multiple ways to measure the performance of a trained classifier. An obvious performance evaluation criterion is classification accuracy, which is the fraction of test samples that are correctly classified (by comparing the prediction obtained from the classifier and the true class output of the test sample). As mentioned above, because of the issue of overfitting, it can be misleading to simply gauge classification accuracy on the training set. Instead, one can measure classification accuracy on a separate validation set (which is used as a proxy for the underlying data distribution), or use cross-validation. Moreover, sometimes, when the sample sizes of the two classes differ significantly, this accuracy may again be misleading, as one may attain an apparently high accuracy by simply predicting the test sample to belong to the majority class. In these cases, other measures such as precision, recall, and F-measure may be more useful. Moreover, while the classifier's accuracy is often an important criterion, other aspects may also be important, such as the training and testing of computational complexities (including both time and space), user-friendliness (e.g., is the trained model considered as a black-box or can it be easily conveyed and explained to the users), etc.

While binary classification assumes the presence of only two output classes, many real-world applications have more than two (say, K), leading to a multi-class classification problem. There are two common approaches to reduce a multi-class classification problem to binary classification problems, namely, the one-vs-rest (also called one-vs-all) approach and the one-vs-one approach. In the one-vs-rest approach, K binary classifiers are constructed, each one separating the samples belonging to the i-th class from those that do not.

On prediction, the test sample is sent to all the K classifiers, and its label corresponds to the classifier with the highest output. In the one-vs-one approach, one binary classifier is built for each pair of outputs (e.g., outputs i and j), and each classifier tries to discriminate samples belonging to the i class from those belonging to the j class. Thus, there are a total of $\frac{1}{2}K(K-1)$ classifiers. On prediction, the test sample is again sent to all the binary classifiers, and the class that receives the largest number of votes is output.

29.2.2 Other Variants of Supervised Learning

Multi-Label Classification

While an instance can only belong to one and only one class in multi-class classification, an instance in multi-label classification can belong to multiple classes. Many real-world applications involve multi-label classification. For example, in text categorization, a document can belong to more than one category, such as government and health; in bioinformatics, a gene may be associated with more than one function, such as metabolism, transcription, and protein synthesis; and in image classification, an image may belong to multiple semantic categories, such as beach and urban. Note that the number of labels associated with an unseen instance is unknown and can also vary from instance to instance. Hence, this makes the multi-label classification problem more complicated than the multi-class classification problem. In the special case where the number of labels associated with each instance is always equal to one, obviously multi-label classification reduces to multi-class classification.

In general, multi-label classification algorithms can be divided into two categories: *problem transformation* and *algorithm adaptation* [29.15]. Problem transformation methods transform a multi-label classification problem into one or more single-label classification problems. The basic approach (called *binary relevance*) simply decomposes a multi-label problem with K labels into K binary classification problems, one for each label. In other words, the i-th classifier is a binary classifier that tries to decide whether the sample belongs to the i-th class. However, since this considers the labels independently, any possible correlations among labels will be ignored, leading to inferior performance in problems with highly correlated labels. More refined variants thus take the label correlation into account during training, a similar idea that is also exploited in multi-task learning (Sect. 29.2.2). On the other hand, algorithm adaptation methods extend a specific learning algorithm for multi-label classification. The specific extension is thus tailor-made for each individual learning algorithm and less general. Example learning algorithms that have been extended in this way include boosting, decision trees, ensemble methods, neural networks, support vector machines, genetic algorithms, and the nearest-neighbor classifier. Recent surveys on the progress of multi-label classification and its use in different applications can be found in [29.15, 16].

In many applications, the labels are often organized in a hierarchy, either in the form of a tree (such as documents in Wikipedia) or as a directed acyclic graph (such as gene ontology). An instance is associated with a label only if it is also associated with the label's parent(s) in the hierarchy. Recently, progress has also been made in multi-label classification in these structured label hierarchies [29.17–19].

Multi-Instance Learning

In multi-instance learning (MIL), the training set is composed of many *bags* each containing multiple instances, where a positive bag contains at least one positive instance, whereas a negative bag contains only negative instances; labels of the training bags are known, but labels of the instances are unknown. The task is to make predictions for labels of unseen bags. The multi-instance learning framework is illustrated in Fig. 29.1. Notice that the instances are described by the same feature set, rather than different feature sets.

The MIL learning framework originated from the study of drug activity [29.20], where a molecule with multiple low-energy shapes is known to be useful to make a drug, whereas it is unknown which shape is crucial. Later, many real tasks are found to be natural multi-instance learning problems. For example, in image retrieval if we regard each image patch as an instance, then the fact that an user is interested (or not interested) in an image implies that there are at least one patch (or no patch) that contains his/her interesting objects.

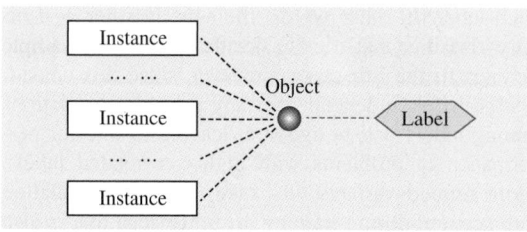

Fig. 29.1 Illustration of multi-instance learning

Most MIL methods attempt to adapt single-instance supervised learning algorithms to the multi-instance representation by shifting their focus from discrimination on instances to discrimination on bags; there are also methods that try to adapt the multi-instance representation to single-instance algorithms by representation transformation [29.21]. Recently, it has been recognized that the instances in the bags should not be treated independently [29.22]; otherwise MIL is a special case of semi-supervised learning [29.23].

In addition to classification, multi-instance regression and clustering have been studied, and different versions of generalized multi-instance learning have been defined [29.24, 25]. To deal with complicated data objects that are associated with multiple labels simultaneously, a new framework, multi-instance, multi-label learning (MIML) [29.26], was developed recently.

Notice that the original MIL assumption implies that there exists a *key instance* in a positive bag; later, some other assumptions were introduced [29.27]. For example, some methods assumed that there is no key instance and every instance contributes to the bag label.

Multi-View Learning

In many real tasks there is more than one feature set. For example, a video film can be described by audio features, image features, etc.; a web page can be described by features characterizing its own content, features characterizing its linked pages, etc. A classical routine is to take these features together and represent each instance using a concatenated feature vector. The different feature sets, however, usually convey information from different channels, and therefore, it may be better to consider the difference explicitly. This motivates multi-view learning, where each feature set is called a *view*.

Each instance in multi-view learning is represented by multiple feature vectors each in a different, usually non-overlapping feature set. Multi-view learning methods in supervised learning setting are closely related to studies of *information fusion, combining classifiers* [29.28–30], and *ensemble methods* [29.14]. A popular representative is to construct a model from each view, and then combine their predictions using *voting* or *averaging*. The models are often assigned with different weights, reflecting their different strength, reliability, and/or importance.

Multi-views make great sense when unlabeled data are considered. For example, it has been proved that when there are *sufficient and redundant views* (that is, each view contains sufficient information for constructing a good model, and the two views are conditionally independent given the class label), *co-training* is able to boost the performance of any initial weak learner to an arbitrary performance using unlabeled data [29.31]. Later, it was found that such a process is beneficial even when the two views satisfy weaker assumptions, such as weak dependence, expansion, or large diversity [29.32–34], and when there are really sufficient and redundant views, even semi-supervised learning with a single labeled example is possible [29.35]. Moreover, in *active learning* where the learner actively selects some unlabeled instances to query their labels from an *oracle* (such as a human expert), it has been proved that multi-view learning enables exponential improvement of sample complexity in a setting close to real tasks [29.36], whereas previously it was believed that only polynomial improvement is possible.

Multi-Task Learning

Many real-world problems involve the learning of a number of similar asks. Consider the simple example of learning to recognize the numeric digits 0–9. One can build ten separate classifiers, one for each digit. However, apparently these ten classifiers share some common features, e.g., many of the digits consist of loops and strokes. Hence, the ability to detect these higher latent features is of common interest to all these classifiers, and learning all these tasks together will allow them to borrow strength from each other. Moreover, when the number of training examples is rare for each task, most single-task learning methods may fail. By learning them together, better generalization performance can be obtained by harnessing the intrinsic task relationships. Consequently, this leads to the development of *multi-task learning* (MTL) [29.37]. These different tasks have different output spaces and can also have different input features. But it is also quite often that these different tasks share the same set of input features. In this case, the problem is similar to multi-label classification (Sect. 29.2.2).

A popular MTL approach is regularized multi-task learning (RMTL) [29.38, 39]. It assumes that the tasks are highly related, and encourages the parameters of all the tasks to be close. More specifically, let there be T tasks and denote the parameter associated with the t-th task by w_t. RMTL assumes that all the w_t's are close to some shared task $\bar{w}$, and that the w_t's differ by each other only in a term Δw_t's as $w_t = \bar{w} + \Delta w_t$. Hence, $\bar{w}$ represents the component that is shared by all the tasks, and thus can benefit from learning all the tasks

together; while Δw_t is the component that is specific to each individual task, and can be used to capture the individual variations. Alternatively, other MTL methods, such as multi-task feature learning (MTFL) [29.40], assumes that all the tasks lie in a shared low-dimensional space.

Moreover, tasks are supposed to form several clusters rather than from the same group. If such a task clustering structure is known, then a simple remedy is to constrain task sharing to be just within the same cluster [29.39, 41]. More generally, all the tasks are related in different degrees, which can be represented by a network of task relationships [29.42]. In this case, MTL can also be performed. In practice, however, such an explicit knowledge of task clusters/network may not be readily available.

A number of efforts have made towards identifying task relationships simultaneously during parameter learning, e.g., learning a low-dimensional subspace shared by most of the tasks [29.43], finding the correlations between tasks [29.44], and inferring the clustering structure [29.45, 46], as well as integrating low-rank and group-sparse structures for robust multi-task learning [29.47].

Transfer Learning

As discussed in Sect. 29.1, traditionally, *machine learning* is defined as:

> changes in the system that are adaptive in the sense that they enable the system to do the same task or tasks drawn from the same population more effectively the next time.

However, recently, there has been increasing interest in adapting a classifier/regressor trained in one task for use in another. This so-called *transfer learning* is particularly crucial when the target application is in short supply of labeled data. For example, it is very expensive to calibrate a WiFi localization model in a large-scale environment. To reduce re-calibration effort, we might want to adapt the localization model trained in one time period (source domain) for a new time period (target domain), or to adapt the localization model trained on a mobile device (source domain) for a new mobile device (target domain). However, the WiFi signal strength is a function of time, device, and other dynamic factors. Thus, transfer learning is used to adapt the distributions of WiFi data collected over time or across devices.

In general, transfer learning addresses the problem of how to utilize plentiful labeled data in a source domain to solve related but different problems in a target domain, even when the training and testing problems have different distributions or features. The success to transfer learning from one context to another context depends on how similar the learning task is to the transferred task. There are two main approaches to transfer learning. The first approach tries to learn a common set of features from both domains, which can then be used for knowledge transfer [29.48–50]. Intuitively, a good feature representation should be able to reduce the difference in distributions between domains as much as possible, while at the same time preserving important (geometric or statistical) properties of the original data. With a good feature representation, we can apply standard machine learning methods to train classifiers or regression models in the source domain for use in the target domain. The second approach to transfer learning is based on instances [29.51–53]. It tries to learn different weights on the source examples for better adaptation in the target domain. For example, in the *kernel mean matching* algorithm [29.52], instances in a reproducing kernel Hilbert space are re-weighted based on the theory of *maximum mean discrepancy*.

Cost-Sensitive Learning

In many real tasks, the costs of making different types of mistakes are usually unequal. In such situations, maximizing the *accuracy* (or equivalently, minimizing the number of mistakes) may not provide the optimal decision. For example, two instances that each cost 10 dollars are less important than one instance that costs 50 dollars. Cost-sensitive learning methods attempt to minimize the *total cost* by reducing serious mistakes through sacrificing minor mistakes.

There are two types of misclassification costs, i. e., *example-dependent* or *class-dependent* cost. The former assumes that every example has its own misclassification cost, whereas the latter assumes that every class has its own misclassification cost. To obtain example-dependent cost is usually much more difficult in real practice, and therefore, most studies focus on class-dependent cost.

The essence of most cost-sensitive learning methods is *rescaling* (or *rebalance*), which tries to rebalance the classes such that the influence of each class in the learning process is in proportion to its costs. Suppose the cost of misclassifying the i-th class to the j-th class is $cost_{ij}$. For binary classification, it can be derived from the Bayes risk theory that the optimal rescaling ratio of the i-th class against the j-th class is $\tau_{ij} = \frac{cost_{ij}}{cost_{ji}}$ [29.54]. For multi-class problems, however, there is no direct solution to obtain the optimal rescaling ratios [29.55], and one may want to decompose a multi-class problem to a series of binary problems to solve.

Rescaling can be implemented in different ways, e.g., re-weighting or re-sampling the training examples of different classes, or even moving the decision threshold directly towards the cheaper class. It can be easily incorporated into existing supervised learning algorithms. For example, for support vector machines, the corresponding optimization problem can be written as

$$\min_{w,b,\xi} \quad \frac{1}{2}\|w\|_{\mathcal{H}}^2 + C\sum_{i=1}^{m} cost(x_i)\xi_i$$
$$\text{s.t.} \quad y_i(w^T\phi(x_i) + b) \geq 1 - \xi_i$$
$$\xi_i \geq 0 \quad i = 1, \ldots, m, \tag{29.1}$$

where ϕ is the feature induced from a kernel function and $cost(x_i)$ is the cost for misclassifying x_i. It can be found that the only difference with the classical support vector machine is the insertion of $cost(x_i)$.

It is often difficult to know precise costs in real practice, and some recent studies have tried to address this issue [29.56]. Notice that a learning process may involve various costs such as the *test cost*, *teacher cost*, *intervention cost*, etc. [29.57], and these costs can also be considered in cost-sensitive learning. Last but not least, it should be noted that the variants introduced in this section already go beyond the classic paradigm of supervised learning. Many of them are integrated with unsupervised learning. Some further issues will be also addressed in the following sections.

29.3 Unsupervised Learning

Given a set $X_N = \{x_t\}_{t=1}^{N}$ of unlabeled data samples, unsupervised learning aims at finding a certain dependence structure underlying data X_N with help of a learning principle. The simplest one is the structure

of merely a point μ in a vector space as illustrated in Fig. 29.2a(A(2)). It represents each sample x_t with an error measure $\varepsilon_t = \|x_t - \mu\|^2$. The best μ may be obtained under a learning principle, e.g., minimizing the following error

$$E_2 = \sum_{t=1}^{N} \varepsilon_t \,, \tag{29.2}$$

which results in that μ is simply the mean of the samples.

Efforts made have been far from a simple point structure. As illustrated in Fig. 29.2, these efforts are roughly grouped into several closely related streams. One consists of those listed in Fig. 29.2a(A), featured by increasing the dimensionality of the modeling structure from a single point to a line, plane, and subspace. The second stream consists of those listed in Fig. 29.2a(B), with multiple structures replacing its counterparts listed in Fig. 29.2a(A). The third stream consists of those listed in Fig. 29.2b(C), based on matrix/graph representation of underlying dependence structures. Moreover, another stream is featured with underlying dependencies in tree structures, such as temporal modeling, hierarchical learning, and causal tree structuring, as illustrated in Fig. 29.2b(D). This section will provide a tutorial on the basic structures listed in Fig. 29.2. Also, an overview will be made of a number of emerging topics, mainly coming from a recent systematic overview [29.6].

Additionally, there is also a stream of studies that not only consist of unsupervised learning as a major ingredient but also include features of supervised learning and reinforcement learning, some of which are referred to under the term semi-supervised learning, while others are referred to under the names of semi-unsupervised learning, hybrid learning, mixture of experts, etc. Among them, semi-supervised learning has become a well-adopted name in the literature of machine learning and will be further introduced in Sect. 29.5. Moreover, readers are further referred to Sect. 4.3 of [29.58] and [29.59] for a general formulation called semi-blind learning.

29.3.1 Principal Subspaces and Independent Factor Analysis

When a point μ is replaced with a line structure as illustrated in Fig. 29.2a(A(3)), e.g., represented by a vector $w_\mu = w - \mu$ of a unit length, we consider the error ε_t as the shortest distance from x_t to the line. Then, minimizing E by (29.2) results in that w_μ is the principal component direction of the sample set X_N, that is, we

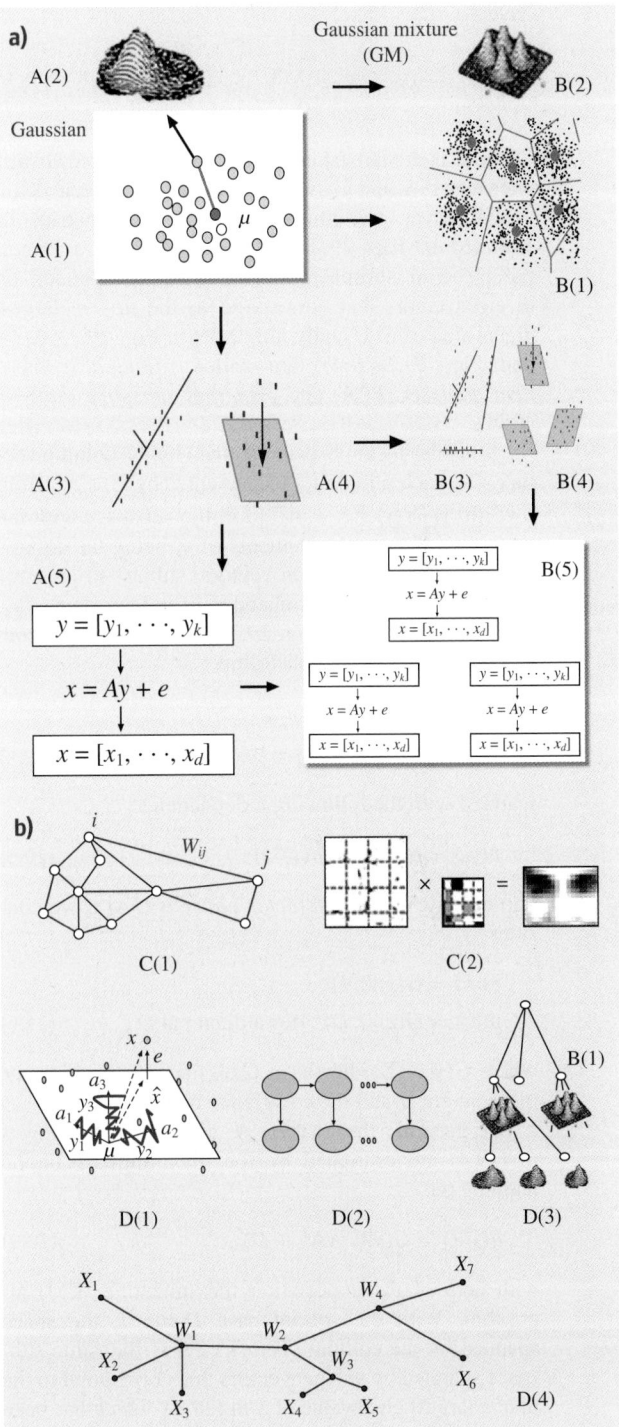

Fig. 29.2a,b Four streams of unsupervised learning studies featured by types of underlying dependence structures

Part D | 29.3

have

$$Sw_\mu = \lambda w_\mu, \quad S = \frac{1}{N}\sum_{t=1}^{N}(x_t - \mu)(x_t - \mu)^T, \quad (29.3)$$

where λ is the largest eigenvalue of the sample covariance matrix S, and w_μ is the corresponding eigenvector.

Moreover, we consider a plane or subspace illustrated in Fig. 29.2a(A(4)), resulting in a principal plane or subspace, i.e., a subspace spanned by m eigenvectors that correspond to the first m largest eigenvalue of S. Usually, the tasks in Fig. 29.2a(A(3)) and Fig. 29.2a(A(4)) are called *principal component analysis (PCA)* and *principal subspace analysis (PSA)*.

Considering a subspace spanned by the column vectors of A, each sample x_t is represented by $\hat{x}_t = Ay_t$ from a vector $y_t = [y_t^{(1)}, \ldots, y_t^{(m)}]^T$ in the subspace with the mutually independent elements of y_t being the coordinates along these column vectors, subject to an error $e_t = x_t - \hat{x}_t$ that is uncorrelated to or independent of y_t. Thus, as illustrated in Fig. 29.2a(A(5)), x_t comes from an underlying subspace as follows

$$x_t = \hat{x}_t + e_t = Ay_t + e_t,$$
$$Ee_ty_t^T = 0 \text{ or } p(e_t|y_t) = p(e_t), \quad (29.4)$$

featured with the following independence

$$p(y_t) = p(y_t^{(1)}) \cdots p(y_t^{(m)}). \quad (29.5)$$

Particularly, it is called *factor analysis (FA)* if we consider

$$p(y_t) = G(y_t|0, I),$$
$$p(e_t) = G(e_t|0, D) \quad \text{for a diagonal } D, \quad (29.6)$$

where $G(x|\mu, \Sigma)$ denotes a Gaussian density with the mean vector μ and the covariance matrix Σ.

In general, the matrix $\mathbf{A}$ and other unknowns in (29.4) are estimated by the maximum likelihood learning on

$$q(x|\theta) = G(x|0, \mathbf{A}\mathbf{A}^T + D), \quad (29.7)$$

with help of the expectation maximization (EM) algorithm. With the special case $D = \sigma^2 I$, the space spanned by the column vectors of A is the same subspace spanned by m eigenvectors that correspond to the first m largest eigenvalue of S in (29.3), which has been a well-known fact since Anderson's work in 1959. In the last two decades, there has been a renewed interest

in the machine learning literature under the new name of probabilistic PCA.

Classically, the principal subspace is obtained via computing the eigenvectors of the sample covariance Σ. However, Σ is usually poor in accuracy when the sample size N is small while the dimensionality of x_t is high. Alternatively, Oja's rule and variants thereof have been proposed to learn the eigenvectors adaptively per sample without directly computing Σ. Also, extensions have been made on adaptive robust PCA learning on data with outliers and on the adaptive principle curve as the line in Fig. 29.2a(A(3)) extended to a curve.

A hyperplane has two dual representations. One is spanned by several one-dimensional unit vectors, while the other is represented by a unit length normal vector w that is orthogonal to this subspace. In the latter case, minimizing E results in that w_μ is still a solution of (29.3) but with λ becoming the smallest eigenvalue instead of the largest one. Accordingly, the problem is called *minor component analysis (MCA)*. In general, an m-dimensional subspace in R^d may also be represented by the spanning vectors of a $d-m$ complementary subspace, for which minimizing E results in $d-m$ eigenvectors that correspond to the first m smallest eigenvalues of S. The problem is called *minor subspace analysis (MSA)*. When $m > d/2$, the minor subspace needs fewer free parameters than the principal subspace does. In a dual subspace pattern classification, each class is represented by either a principal subspace or a minor subspace. Because they are different from PCA and PSA, MCA and MSA are more prone to noises. For further details about these topics the interested reader is referred to Sect. 3.2.1 of [29.58] and a recent overview [29.60]. In the following, we add brief summaries on three typical methods.

PCA versus ICA

Independent component analysis (ICA) has been widely studied in the past two decades. The key point is to seek a linear mapping $y_t = Wx_t$ such that the components $y_t^{(1)}, \ldots, y_t^{(m)}$ of y_t become mutually independent, as shown in (29.5), as an extension of PCA by which the components $[y_t^{(1)}, \ldots, y_t^{(m)}]$ of $y_t = Wx_t$ become mutually de-correlated when the rows of W consist of the eigenvectors of the first m largest eigenvalue of S in (29.3). Strictly speaking, this is inexact since the counterpart of ICA should be called de-correlated component analysis (DCA), with independence among $y_t^{(1)}, \ldots, y_t^{(m)}$ in the second order of statistics. PCA is one extreme case of DCA that chooses those de-correlated components with the first m largest variances/eigenvalues, while MCA is another extreme case that chooses those with the first m smallest variances/eigenvalues. Correspondingly, the extended counterpart of PCA/MCA should be the principal/minor ICA that chooses the independent components with the first m largest/smallest variances. Readers are referred to Sect. 2.4 of [29.61] for further details. Several adaptive learning algorithms have been developed for implementing ICA, but their implementation cannot be guaranteed (29.5). Theoretically, theorems have been proved that such a guarantee can be reached as long as one bit of information is provided to each component $y_t^{(j)}$ [29.62].

FA-a, FA-b, and Model Selection

In the literature of statistics and machine learning, the model (29.4) with (29.6) is conventionally referred to as FA. Actually, we have $Ay = \tilde{A}\tilde{y}$ with $\tilde{A} = A\phi^{-1}, \tilde{y} = \phi y$ for any unknown nonsingular matrix. Among the different choices to handle this indeterminacy, the standard one, shortly denoted by FA-a, imposes (29.6) on y, which reduces an indeterminacy of a general nonsingular ϕ to an orthonormal matrix. One other choice is given by (29.4) with

$$A^T A = I, \ p(y_t) = G(y_t|0, \Lambda) \text{ for a diagonal } \Lambda, \tag{29.8}$$

shortly denoted by FA-b. We have $\hat{x}_t = Ay = A\Lambda\Lambda^{-1}y = A\Lambda\phi^T\phi\Lambda^{-1}y = \tilde{A}\tilde{y}$, with $\tilde{A} = A\Lambda\phi^T$, $\tilde{y} = \phi\Lambda^{-1}y$, and $\phi^T\phi = I$, i.e., its $\hat{x}_t$ is equivalent to the one by FA-a for a given m with an invertible Λ^{-1}. In other words, FA-a and FA-b are equivalent for a learning principle based on $e_t = x_t - \hat{x}_t$, e.g., minimizing E by (29.2) or maximizing the likelihood on x_t. Moreover, FA-a and FA-b are still equivalent when model selection is used for determining an appropriate value m by one of the classic model selection criteria to be introduced later in (29.14). However, FA-a and FA-b become considerably different to the Bayesian Yin-Yang (BYY) harmony learning in Sect. 3.2.1 of [29.58] and also to automatic model selection in general. Empirically, experiments show that not only BYY harmony learning but also the variational Bayes method perform considerably better on FA-b than on FA-a [29.63].

Non-Gaussian FA

Both FA-a and FA-b still suffer an indeterminacy of an $m \times m$ orthonormal matrix ϕ, which can be further removed when at most one of the components $y_t^{(1)}, \ldots, y_t^{(m)}$ is Gaussian. Accordingly, (29.4) with non-Gaussian components $p(y_t^{(j)})$ in (29.5) is called *non-Gaussian FA* (NFA). It is also referred to as *independent FA* (IFA), although NFA sounds better, since the concept of IFA covers not only NFA but also FA-a and FA-b. One useful special case of NFA is called binary FA when y_t is a binary vector. Moreover, in the degenerated case $e_t = 0$, obtaining A of $x_t = Ay_t$ subject to (29.5) is equivalent to getting $W = A^{-1}$ such that $Wx_t = y_t$ to satisfy (29.5). For this reason, NFA with $e_t \neq 0$ is also sometimes referred to as noisy ICA. Strictly speaking, the map $x_t \to y_t$ towards (29.5), being an inverse of NFA, should be nonlinear instead of a linear $y_t = Wx_t$. Maximum likelihood learning is implemented with help of the EM algorithm, which was developed in the middle to end of the 1990s for BFA/NFA, respectively. Also, learning algorithms have been proposed for implementing BYY harmony learning with automatic model selection on m. Recently, both BFA and NFA were used for transcription regulatory networks in gene analysis; for further details the reader is referred to the overview [29.60] and especially its *Roadmap*.

29.3.2 Multi-Model-Based Learning: Data Clustering, Object Detection, and Local Regression

The task of data clustering is partitioning a set of samples into several clusters such that samples within a sample cluster are similar while samples from different clusters should be as different as possible. An indicator matrix $\mathbf{P} = [p_{\ell,t}]$ with $PP^T = I$ is used to represent one possible partition of a sample set $X_N = \{x_t\}_{t=1}^N$ into one of $\ell = 1, \ldots, k$ clusters, i.e., $p_{\ell,t} = 1$ if x_t belongs to the ℓ-th cluster, otherwise $p_{\ell,t} = 1$. For multi-model-based clustering, each cluster is modeled by one structure, with $p_{\ell,t}$ obtained by a competition of using the structure of each cluster to represent a sample x_t. The structure of each cluster could be one of the ones listed on the left-hand side of Fig. 29.2; multiple clusters are thus represented by multiple structures listed on the right-hand side of Fig. 29.2, which feature the basic topics of the second stream of studies.

We still start from the simplest point structure illustrated in Fig. 29.2a(A(1)), extending to the structure of multi-points illustrated in Fig. 29.2a(B(1)). With data already divided into k clusters, it is easy to obtain the mean μ_j of each cluster. Given $\{\mu_j\}_{j=1}^k$ fixed, it is also easy to divide X_N into k clusters by

$$p_{\ell,t} = \begin{cases} 1, & \ell = \arg\min_j \varepsilon_{j,t} , \\ 0, & \text{otherwise} , \end{cases} \quad (29.9)$$

where $\varepsilon_{j,t} = \|x_t - \mu_j\|^2$, i.e., x_t is assigned to the ℓ-th cluster if $p_{\ell,t} = 1$. The key idea of the k-means algorithm is alternatively getting $p_{\ell,t}$ and computing μ_j from an initialization. Although it aims at minimizing E_2 by 29.2 with $\varepsilon_t = \sum_{\ell=1}^k p_{\ell,t}\varepsilon_{j,t}$, k-means typically results in a local minimum of E_2, depending on the initialization.

Merely using the mean μ_j is not good for describing a cluster beyond a ball shape. Instead, it is extended to considering the Gaussian illustrated in Fig. 29.2a(A(2)) and thus its counterpart in Fig. 29.2a(B(2)), i.e., the following Gaussian mixture

$$q(x|\theta) = \sum_{j=1}^k \alpha_j G(x|\mu_j, \Sigma_j) . \quad (29.10)$$

K-means can be extended to getting $p_{\ell,t}$ by (29.11) with

$$\varepsilon_{j,t} = -\ln[\alpha_j G(x|\mu_j, \Sigma_j)] . \quad (29.11)$$

and computing each Gaussian by

$$\alpha_\ell^* = \frac{\sum_t p_{\ell,t}}{N},$$

$$\mu_\ell^* = \frac{1}{N\alpha_\ell^*} \sum_t p_{\ell,t} x_t ,$$

$$\Sigma_\ell^* = \frac{1}{N\alpha_\ell^*} \sum_t p_{\ell,t}(x_t - \mu_\ell^*)(x_t - \mu_\ell^*)^{\mathrm{T}} , \quad (29.12)$$

which actually performs a type of elliptic clustering. Instead of getting $p_{\ell,t}$ by (29.9), we compute

$$p_{\ell,t} = q(\ell|x_t, \theta^*), \ q(\ell|x_t, \theta) = e^{-\varepsilon_{j,t}} / \sum_{i=1}^k e^{-\varepsilon_{i,t}} . \quad (29.13)$$

Actually, alternatively iterating (29.13) and (29.12) is the well-known EM algorithm for carrying out maximum likelihood learning on the Gaussian mixture.

Another important topic is to determine an appropriate k (model selection), i.e. how many clusters are needed. Classic model selection seeks a best $k^* =$ $\arg\min_k J(k)$ with a criterion $J(k)$ in a format as follows

$$J(k) = -L(k, \theta^*) + \omega(k, N), \ \theta^* = \arg\max_\theta L(k, \theta) , \quad (29.14)$$

where $L(k, \theta)$ is the likelihood function of $q(x|\theta)$, and $\omega(k, N) > 0$ increases with k and decreases with N. One typical example is called the Bayesian information criterion (BIC) or minimum description length (MDL). To obtain k^* one needs to enumerate a set of k values and estimate θ^* for each k value, which incurs an extensive computation and is thus difficult to scale up to a large number of clusters.

Alternatively, *automatic model selection* aims at obtaining k^* during learning θ^* by a mechanism or principle that is different from the maximum likelihood. This learning drives away an extra cluster via a certain indicator $\rho_j \to 0$, e.g., $\rho_j = \alpha_j$ or $\rho_j = \alpha_j Tr[\Sigma_j]$. One early effort is rival penalized competitive learning (RPCL). RPCL learning does not implement (29.12) by either (29.9) or (29.13), with $p_{\ell,t}$ given as follows

$$p_{\ell,t} = \begin{cases} 1, & \ell^* = \arg\min_j \varepsilon_{j,t} , \\ -\gamma, & \ell = \arg\min_{\ell \neq \ell^*} \varepsilon_{j,t} , \\ 0, & \text{otherwise} , \end{cases} \quad (29.15)$$

by which learning is made on a cluster when $p_{\ell,t} = 1$, and penalizing or de-learning is made on a cluster when $p_{\ell,t} = -\gamma$, with a heuristic penalizing strength of roughly $\gamma \approx 0.005 \approx 0.05$.

The BYY harmony learning gets rid of the difficulty of finding an appropriate penalizing strength, with both parameter learning and model selection made under the *Ying Yang* best harmony principle. The algorithm obtained still implements (29.12) and replaces $p_{\ell,t}$ by (29.12) with

$$p_{\ell,t} = q(\ell|x_t, \theta^*)(1 + \Delta\pi_{\ell,t}) ,$$

$$\Delta\pi_{\ell,t} = \sum_j q(j|x_t, \theta^*)\varepsilon_{j,t} - \varepsilon_{\ell,t} . \quad (29.16)$$

where $\Delta\pi_{\ell,t} > 0$ means that the j-th component is better than the average of all the components for describing the sample x_t. We further update the j-th component in (29.12) to enhance the description. If $0 > \Delta\pi_{\ell,t} > -1$, i.e., the fitness by the j-th component to x_t is below the average but still not too far away, updating of the j-th component remains the same trend as in (29.12) but with reduced strength. Moreover, when $-1 > \Delta\pi_{\ell,t}$, the updating on the j-th component reverses direction to become de-learning, similar to updating the rival in RPCL learning.

RPCL learning, which was proposed in 1992, and BYY harmony learning, which was developed in 1995, are similar in nature to the popular sparse learning method, which was developed in 1995 [29.64, 65], and prior-based automatic model selection approaches [29.66–68], that is, extra parts in a model are removed as some parameters are pushed towards zero. Without any priors on the parameters, these prior-based approaches degenerate to maximum likelihood learning, while RPCL learning and further improved by incorporating appropriate priors.

For further details about automatic model selection, prior-aided learning, and model selection criteria the reader is referred to Sect. 2.2 of [29.58, 69] and [29.70] for recent overviews. Also, readers are referred to Sect. 2.1 and Table 1 of [29.58] for a tutorial on several algorithms for learning Gaussian mixture, including the ones introduced above. In the following, only three typical ones are briefly summarized.

Local Subspaces and Local Factor Analysis

As illustrated in Fig. 29.2a(B(3)–(5)), the structure of multi-points illustrated in Fig. 29.2a(B(1)) can be extended into multiple subspaces and FA models. Still, we can obtain $p_{\ell,t}$ by (29.9), (29.15), and (29.13) with $\varepsilon_{j,t}$ given by either (29.11) with $\Sigma_j = A_j A_j^T + D_j$ or simply the shortest square distance from x_t to the j-th subspace. Given data divided into k clusters, we may estimate the subspace or FA of each cluster as introduced in Sect. 29.3.1, which leads to extensions of the k-means algorithm, the EM algorithm, and the BYY harmony learning algorithm for learning FAs or subspaces that locate at different $\{\mu_j\}_{j=1}^k$. Moreover, readers are referred to [29.71] and [29.2] for learning local FAs with both the number k and the dimensions $\{m_j\}$ determined automatically during BYY harmony learning.

Object Detection and Pattern-Based Clustering

The structures in Fig. 29.3a(B(3),(4)) are applicable to the tasks of detecting lines and subspaces among image data, which are topics that are widely studied in the literature of pattern recognition and handled by the well-known Hough transform (HT) and randomized HT (RHT) [29.70]. Extensions can be made to detect multiple objects such as circles, ellipses, lines and other shapes, as well as so-called pattern based clustering, still obtaining $p_{\ell,t}$ by (29.9), (29.15), and (29.13) but with $\varepsilon_{j,t}$ being the shortest square distance from x_t

to each shape. However, it is no longer possible to use (29.12) for updating the parameters θ_j of each shape. Instead, learning is done by

$$\theta_\ell^{\text{new}} = \theta_\ell^{\text{old}} + \eta p_{\ell,t} \nabla_{\theta_\ell} \varepsilon_{\ell,t} , \qquad (29.17)$$

where $\eta > 0$ is a learning step size; for further details the reader is referred to [29.69, 70] and [29.8].

Mixture of Experts, RBF Networks, SBF Functions

Let each Gaussian to be associated with a function $f(x|\phi_j)$ for a mapping $x \to z$, we consider the task of learning

$$q(z_t|x_t, \psi) = \sum_{j=1}^k q(\ell|x_t, \theta) G[z_t|f_j(x_t, \phi_j), \Gamma_j] , \qquad (29.18)$$

from a set $D_N = \{x_t, z_t\}_{t=1}^N$ of labeled data. The above $q(z_t|x_t, \psi)$ is actually the alternative mixture of experts [29.72], featured by a combination of unsupervised learning for the Gaussian mixture by (29.10) and supervised learning for every $f(x|\phi_j)$. For a regression task, typically we consider $f(x|\phi_j) = w_j^T x + c_j$ with $E[z|x] = \sum_{j=1}^k q(j|x, \theta) f(x, \phi_j)$ implementing a type of piecewise linear regression. In implementation, we still obtain $p_{\ell,t}$ by (29.9), (29.15), and (29.13) but with the following $\varepsilon_{j,t}$

$$\varepsilon_{j,t} = -\ln[\alpha_j G(x|\mu_j, \Sigma_j) G(z_t|f_j(x_t, \phi_j)] , \qquad (29.19)$$

and then compute each Gaussian by (29.12), as well as update $G(z_t|f_j(x_t, \phi_j), \Gamma_j)$. When $\alpha_j = |\Sigma_j| / \sum_{i=1}^k |\Sigma_i|$, it becomes equivalent to an extended normalized radial basis function (RBF) network and a normalized RBF network simply with $w_j = 0$. Moreover, letting each subspace be associated with $f(x|\phi_j)$ will lead to subspace-based functions (SBFs). For further details readers are referred to [29.5] and Sect. 7 of [29.69].

29.3.3 Matrix-Based Learning: Similarity Graph, Nonnegative Matrix Factorization, and Manifold Learning

We proceed to the third stream, featured with graph/matrix structures. We start with the sample similarity graph, with each node for a sample and each edge attached with a similarity measure between two samples, as illustrated in Fig. 29.2b(C(1)). Such a graph is also equivalently represented by a symmetric matrix $\mathbf{W} = [w_{ij}]$.

One similarity measure is simply the inner product $w_{ij} = x_i^T x_j$ of two samples. Given a data matrix $\mathbf{X} = [x_1, \ldots, x_N]$, we simply have $\mathbf{W} = \mathbf{X}^T\mathbf{X}$. We seek an indicator matrix $\mathbf{P} = [p_{\ell,t}]$ that divides $X_N = \{x_t\}_{t=1}^N$ into k clusters, with help of the following maximization

$$\max_{HH^T=I, H\geq 0} \text{Tr}[\mathbf{H}^T\mathbf{W}\mathbf{H}],$$
$$\mathbf{H} = \text{diag}[n_1^{-0.5}, \ldots, n_k^{-0.5}]P, \qquad (29.20)$$

where $\mathbf{H} \geq 0$ is a nonnegative matrix with each element $h_{ij} \geq 0$, and n_ℓ is the number of samples in the ℓ-th cluster. It can be shown that this problem is equivalent to minimizing E_2 by (29.2) with $\varepsilon_t = \sum_{\ell=1}^k p_{\ell,t} \|x_t - \mu_j\|^2$, i.e., the same target that k-means aims at.

Computationally, (29.20) is a typical intractable binary quadratic programming problem, for which various approximate methods are proposed. The most simple one is dropping the constraint $\mathbf{H} \geq 0$ to do a PCA analysis about the matrix $\mathbf{W}$. That is, the columns of $\mathbf{H}$ consist of the k eigenvectors of $\mathbf{W}$ that correspond to the first k largest eigenvalues. Then, each element of the matrix $\text{diag}[n_1^{0.5}, \ldots, n_k^{0.5}]\mathbf{H}$ is chopped into 1 or 0 by a rule of thumb.

Another similarity measure is $w_{ij} = \exp(-\|x_i - x_j\|^2)$, based on which we consider dividing the nodes of a graph into balanced two sets A, B such the total sum of w_{ij} associated with edges connecting the two sets becomes as small as possible. Using a vector $f = [f_1, \ldots, f_N]^T$ with $f_t = 1$ if $x_t \in A$ and $f_t = -1$ if $x_t \in B$, the problem is formulated as follows

$$\min_f f^T L f, \; s.t. \; [1, \ldots, 1]f = 0,$$
$$L = D - W, \; D = \text{diag}[w_{11}, \ldots, w_{NN}], \qquad (29.21)$$

where L is the graph Laplacian. Again, it is an intractable combinatorial problem and needs to consider some approximation. A typical one is given as follows

$$\min_f \frac{f^T L f}{f^T f}, \; s.t. \; [1, \ldots, 1]f = 0. \qquad (29.22)$$

Its solution f is the eigenvector of L that corresponds the second smallest eigenvalue. Moreover, this idea has been extended to cutting a graph into multiple clusters, which leads to approximately finding $\mathbf{H}$ with its

columns being the eigenvectors of $\tilde{\mathbf{W}} = D^{-0.5}WD^{-0.5}$, corresponding to the first k largest eigenvalues.

Moreover, the above studies are closely related to nonnegative matrix factorization (NMF) problems [29.73]. For example, the above problem can be equivalently expressed as factorization $\tilde{\mathbf{W}} \approx \mathbf{H}^T\mathbf{H}$ by

$$\min_{HH^T=I, H\geq 0} \|\tilde{\mathbf{W}} - \mathbf{H}^T\mathbf{H}\|^2. \qquad (29.23)$$

More generally, the NMF problem considers that

$$\mathbf{X} \approx \mathbf{FH}, \; \mathbf{X} \geq 0, \; \mathbf{F} \geq 0, \mathbf{H} \geq 0, \qquad (29.24)$$

as illustrated in Fig. 29.2b(C(1)). One typical method is to iterate the following multiplicative update rule

$$\mathbf{H}_{ij}^{\text{new}} = \mathbf{H}_{ij}^{\text{old}} \frac{(\mathbf{F}^T\mathbf{X})_{ij}}{(\mathbf{F}^T F\mathbf{H})_{ij}}, \; \mathbf{F}_{ij}^{\text{new}} = \mathbf{F}_{ij}^{\text{old}} \frac{(\mathbf{XH}^T)_{ij}}{(\mathbf{FHH}^T)_{ij}}, \qquad (29.25)$$

which guarantees nonnegativity and is supposed to converge to a local solution of the following minimization

$$\min_{HH^T=I, H\geq 0, F\geq 0} \|\mathbf{X} - \mathbf{FH}\|^2. \qquad (29.26)$$

Particularly, if we also impose the constraint $\mathbf{F}^T\mathbf{F} = \mathbf{I}$, the resulted $\mathbf{H}$ divides the columns of $\mathbf{X}$ into k clusters, while the resulted $\mathbf{F}$ also divides the rows of $\mathbf{X}$ into k clusters, and is thus called *bi-clustering* [29.74].

Several NMF learning algorithms have been developed in the literature. In [29.75], a binary matrix factorization (BMF) algorithm was developed under BYY harmony learning for clustering proteins that share similar interactions, featured with the nature of automatically determining the cluster number, while this number has to be pre-given for most existing BMF algorithms.

In the past decade, the similarity graph and especially the graph Laplacian L have also taken important roles in another popular topic called *manifold learning* [29.76, 77]. Considering a mapping $Y \approx WX$, a locality preserving projection is made to minimize the sum of each distance between two mapped points on the graph, subject to a unity L_2 norm of this projection WX.

Alternatively, we may also regard that X is generated via $X = AY + E$ such that the topological dependence among Y is preserved by considering

$$q(Y) \propto e^{-\frac{1}{2}\text{Tr}[YLY^T\mathbf{\Lambda}^{-1}]}, \qquad (29.27)$$

where Λ is a positive diagonal matrix. Learning is implemented by BYY harmony learning, during which automatic model selection is made via updating $q(Y)$ to drive some diagonal elements of Λ towards zeros. For further details readers are referred to the end part of Sect. 5 in [29.58].

29.3.4 Tree-Based Learning: Temporal Ladder, Hierarchical Mixture, and Causal Tree

Unsupervised learning also includes learning temporal and hierarchical underlying dependence structures, as illustrated in Fig. 29.2b(D). Instead of directly modeling temporal dependence underlying data $X = [x_1, \ldots, x_N]$, its structure is typically represented in a hidden space, while non-temporal or spatial dependence is represented by a relation from the hidden space to the space where X is observed, in the ladder structure illustrated in Fig. 29.2b(D(1)).

One typical example is the classic hidden Markov model (HMM). Its hidden space is featured by a discrete variable that jumps between a set of discrete values or states $\{s_j\}$, with temporal dependence described by the jumping probabilities between the states, typically considering $p(s_j|s_i)$ of jumping from one state s_i to another s_j. The relation from the hidden space to the space of X is described by $p(x_t|s_i)$ for the probability that the value of x_t is emitted from the state s_i. Classically, the values of x_t are also a set of labels. The task is learning from $X = [x_1, \ldots, x_N]$ two probability matrices $\mathbf{Q} = [p(s_j|s_i)]$ and $\mathbf{E} = [p(x_t|s_i)]$. Given the number of states, learning is typically implemented to maximize the likelihood $p(X|\mathbf{Q}, \mathbf{E})$ by the well-known Baum–Welch algorithm.

Another example is the classic state–space model (SSM), which has been widely studied in the literature of control theory and signal processing since the 1960s; this has also been called a linear dynamical system with considered with renewed interest since the beginning of the 2000s. As illustrated in Fig. 29.2b(D(2)), its hidden space is featured by an m-dimensional subspace and temporal dependence is described by one first-order vector autoregressive model as follows

$$y_t = By_{t-1} + \varepsilon_t, \; Ey_{t-1}\varepsilon_t^{\mathrm{T}} = 0,$$
$$\varepsilon_t \approx G(\varepsilon_t|0, \Lambda), \; \Lambda \text{ is diagonal}, \qquad (29.28)$$

while the spatial dependence is represented by a relation between the coordinates of the state–space and the coordinates of the space of X, e.g., typically by (29.4).

Though the EM algorithm has also been suggested for learning the SSM parameters, the performance is usually unsatisfactory because an SSM is generally not identifiable due to an indeterminacy of any unknown nonsingular matrix, similar to what was discussed previously with respect to the FA in (29.5). Favorably, it has been shown that the indeterminacy of not only any unknown nonsingular matrix but also an unknown orthonormal matrix is usually removed by additionally requiring a diagonal matrix B, which leads to temporal factor analysis (TFA).

TFA is an extension of the FA by (29.4) with (29.6) replaced by (29.28). As introduced in Sect. 29.3.1, the FA is generalized into NFA when (29.6) becomes (29.5) with each $p(y_t^{(j)})$ being non-Gaussian. The NFA with a real vector y_t can be further extended into a temporal NFA when (29.5) is also extended by (29.28) with

$$p(\varepsilon_t) = p(\varepsilon_t^{(1)}) \cdots p(\varepsilon_t^{(m)}).$$

Moreover, the BFA (i. e., NFA with a binary vector y_t) can be extended into a temporal BFA. Also, TFA has been extended into an integration of several TFA models coordinated by an HMM. For further details readers are referred to Sect. 5.2 of [29.58] for a recent overview on TFA and its extensions.

A ladder is merely a special type of tree structure. Hierarchical modeling is one other type of tree structure, as illustrated in Fig. 29.2b(D(3)). Again, the EM algorithm has been extended to implement learning on a hierarchical or tree mixture of Gaussians [29.78]. Also, a learning algorithm is available for implementing BYY harmony learning with tree configuration determined during learning. A learning algorithm for a three-level hierarchical mixture of Gaussians is shown in Fig. 12 of [29.3], featured by a hierarchical learning flow circling from bottom up as one step and then top down as the other step. Similar to (29.16), where there is a term of Δ featuring the difference of BYY harmony learning from EM learning, there is also such a Δ term on each level of hierarchy. If these Δ terms are set to be zero, the algorithm degenerates back to the EM algorithm. For further details readers are referred to Sect. 5.1 in [29.3] and especially equation (55) therein.

Many applications consider several sets of samples. Each set is known to come from one model or pattern class. Typically, one does unsupervised learning on each set of samples by a hierarchical mixture of Gaussians, and then integrates individual hierarchical models in a supervised way to form a classifier.

Alternatively, we may put together all the individual hierarchical mixtures with each as a branch of one higher level root of a tree, and then do learning as shown in Fig. 12 of [29.3]. The BYY harmony learning algorithm (including the EM algorithm as its degenerated case) for learning a two-level hierarchical mixture of Gaussians is shown in Sect. 5.3 of [29.58], and especially Fig. 11 therein. This type of learning can be regarded as semi-supervised learning in the sense that each sample has two teaching labels. One is known, indicating which individual hierarchy x_t comes from, while the other is unknown to be determined, indicating which Gaussian component x_t comes from. Even generally, this type of learning provides a general formulation that involves the multi-label classification of Sect. 29.2.2 (especially labels with a hierarchy).

There are also real applications that consider a combination of ladder structures and hierarchical structures. For example, what is widely used in speech processing is an HMM model with each hidden state associated with a two-level hierarchical Gaussian mixture as illustrated in Fig. 11 of [29.58]. Also, extensions are made with each Gaussian mixture replaced by a mixture of local subspaces or FA or NFA models. For further details readers are referred to Sect. 5.3 and Fig. 14 of [29.3]. Another example is considering a two-level hierarchical model with both HMM for modeling nonstationary temporal dependence and TFA for modeling stationary temporal dependence. For further details readers are referred to Sect. 5.2.2 of [29.58].

Another typical tree structure, as illustrated in Fig.29.2b(D(4)), is a learning probabilistic tree, i.e., a joint distribution of a set of variables on a tree with one node per variable. The most well-known study is structuring such tree models for a given set of bi-valued variables, as done by *Pearl* in 1986 [29.79]. Following this, one study in 1987 [29.80] extends this to construct tree representations of continuous variables. It has been proved that the tree can be structured from the correlations observed between pairs of variables if the visible variables are governed by a tree decomposable joint normal distribution. Moreover, the conditions for tree decomposable normal distribution are less restrictive than those of bi-valued variables.

Nowadays, many advances have been made along this line. Some of the basic results, e.g., (29.15) and (29.17) in [29.80], has become a widely used technique in network construction for detecting whether an edge describes a direct link or a duplicated indirect link. For example, considering the association between two nodes i, j linked to a third node w with the correlation coefficients ρ_{iw} and ρ_{jw}, we can remove the link i, j if its correlation coefficient ρ_{ij} fails to satisfy, i.e.,

$$\rho_{ij} > \rho_{iw}\rho_{wj} . \tag{29.29}$$

Otherwise we may either choose to keep the link i, j, or let three nodes to be linked to a newly added node and then remove all the original links among the three nodes.

29.4 Reinforcement Learning

Differently to unsupervised learning, reinforcement learning gets guidance from external evaluation. Also, unlike supervised learning in which the teacher clearly specifies the output that corresponds to an input, reinforcement learning is only provided with an evaluative value about the action made. Furthermore, reinforcement learning is featured by a dynamic process in discrete time steps. At each step, upon observing the current environment and getting some input (if any), the learner makes an action and moves to a new state, receiving an award or punish value about the action. The aim is to maximize the total award received.

This section provides a brief tutorial on the basic issues of reinforcement learning, especially TD learning and Q-learning. Then, improvements on Q-learning are proposed by replacing its built-in winner-take-all competition mechanism with some unsupervised learning

methods. For further reading readers are referred to tutorials and reviews in [29.6, 81, 82].

29.4.1 Markov Decision Processes

Reinforcement learning is closely related to Markov decision processes (MDP), which consist of a series of states $s_0, s_1, \ldots, s_t, s_{t+1}, \ldots$. At a state s_t, an action $a_t = \pi(s_t)$ is selected from the set A of actions according to a policy π, which makes the environment move to a new state s_{t+1}, and the reward r_{t+1} associated with the transition (s_t, a_t, s_{t+1}) is received. The goal is to collect as much reward as possible, that is, to maximize the total reward or return

$$R = \sum_{t=0}^{N-1} r_{t+1} ,$$

where N denotes the random time when a terminal state is reached. In the case of nonepisodic problems the return $R = \sum_{t=0}^{\infty} \gamma^t r_{t+1}$ is considered by a discount-factor $0 \leq \gamma \leq 1$.

Given an initial distribution based on which the initial state is sampled at random, we can assign the expected return $E[R|\pi]$ to policy π. Since the actions are selected according to π, the task is to specify an algorithm that can be used to find a policy π to maximize $E[R]$. Suppose we know the state transition probability $p_a(s'|s) = P(s_{t+1} = s'|s_t = s)$ and the corresponding reward $r_{t+1} = R_a(s'|s)$, the standard family of algorithms to calculate this optimal policy is featured by iterating the following two steps

(1) $\pi(s) = \arg \max_a V^a(s)$,

(2) $V^{\pi}(s) = \sum_{s'} p_{\pi(s)}(s'|s)[R_{\pi(s)}(s'|s) + \gamma V(s')]$,

$$(29.30)$$

with $V^{\pi}(s)$ estimating $E[R|s, \pi]$. The iteration can be made in one of several variants as follows:

● Doing step (1) once and then repeating step (2) several times or until it converges. Then step (1) is done once again, and so on.
● Doing step (2) by solving a set of linear equations.
● Substituting the calculation of $\pi(s)$ into the calculation of $V^*(s) = \max_{\pi} V^{\pi}(s)$, resulting in a combined step

$$V^*(s) = \max_a \left\{ \sum_{s'} p_a(s'|s)[R_a(s'|s) + \gamma V^*(s')] \right\},$$

$$(29.31)$$

which is called backward induction and is iterated for all states until it converges to what is called the Bellman equation.
● Preferentially applying the steps to states that are in some way of importance.

Under some mild regularity conditions, all the implementations will reach a policy that achieves these optimal values of $V^*(s) = \max_{\pi} V^{\pi}(s)$ and thus also maximizes the expected return $E[V^{\pi}(s)]$, where s is a state that is randomly sampled from the underlying distribution.

In the implementation of MDPs we need to know the probability $p_a(s'|s)$ per action a. Reinforcement learning avoids obtaining this $p_a(s'|s)$ with the help of stochastic approximation. The two most popular examples are temporal difference (TD) learning and Q-learning, respectively. The name TD derives from its use of differences in predictions over successive time steps to drive learning, while the name Q comes from its use of a function that calculates the quality of a state-action combination.

29.4.2 TD Learning and Q-Learning

TD learning aims at predicting a measure of the total amount of reward expected over the future. At time t, we seek an estimate $\hat{r}_t$ of $R_t = \sum_{i=1}^{\infty} \gamma^{i-1} r_{t+i}$ with $0 \leq \gamma < 1$. Each estimate is a prediction because it involves future values of r. We can write $\hat{r}_t = \Pi(s_t)$, where Π is a prediction function. The prediction at any given time step is updated to bring it closer to the prediction of the same quantity at the next step, based on the error correction $\delta_{t+1} = R_t - \Pi_t(s_t)$. To obtain R_t exactly requires waiting for the arrival of all the future values of r. Instead, we use $R_t = r_{t+1} + \gamma R_{t+1}$ with $\Pi_t(s_{t+1})$ as an estimate of R_{t+1} available at step t, that is, we have

$$\delta_{t+1} = r_{t+1} + \gamma \Pi_t(s_{t+1}) - \Pi_t(s_t), \qquad (29.32)$$

which is termed the temporal difference error (or TD error).

The simplest TD algorithm updates the prediction function Π_t at step t into to a new prediction function Π_{t+1} as follows

$$\Pi_{t+1}(x) = \begin{cases} \Pi_t(x) + \eta \delta_{t+1} & \text{if } x = s_t \\ \Pi_t(x) & \text{otherwise}, \end{cases} \qquad (29.33)$$

where η is a learning step size and x denotes any possible input signal. The simplest format is a prediction function implemented as a lookup table. Suppose that s_t takes only a finite number of values and that there is an entry in a lookup table to store a prediction for each of these values. At step t, the state s_t moves to the next s_{t+1} based on the current status of the table, e.g., the table entry for s_{t+1} is the largest across the table, or s_{t+1} is selected according to a fixed policy. When r_{t+1} is observed, only the table entry for s_t changes from its current value of $\hat{r}_t = \Pi_t(s_t)$ to $\Pi_t(s_t) + \eta \delta_{t+1}$.

The algorithm uses a prediction of a later quantity $\Pi_t(s_{t+1})$ to update a prediction of an earlier quantity $\Pi_t(s_t)$. As learning proceeds, later predictions tend to become accurate sooner than earlier ones, resulting in an overall error reduction. This depends on whether an input sequence has sufficient regularity to make predicting possible. When s_t comes from the states of a Markov chain, on which the r values are given by a function of these states, a prediction function may exist that accurately gives the expected value of the quantity R_t for each t.

Another view of the TD algorithm is that it operates to maintain the following consistency condition

$$\Pi(s_t) = r_{t+1} + \gamma \Pi(s_{t+1}) , \qquad (29.34)$$

which must be satisfied by correct predictions. By the theory of *Markov* decision processes, any function that satisfies $R_t = r_{t+1} + \gamma R_{t+1}$ for all t must actually give the correct predictions. The TD error indicates how far the current prediction function deviates from this condition, and the algorithm acts to reduce this error towards this condition. Actually, $\Pi_t(s_t) + \eta \delta_{t+1} = (1 - \eta)\Pi_t(s_t) + \eta[r_{t+1} + \gamma \Pi_t(s_{t+1})]$ is a type of stochastic approximation to the value function in (29.30), without directly requiring to know the probability $p_a(s'|s)$.

Alternatively, Q-learning calculates the quality of a state-action combination, i. e., estimating $Q(s_t, a_t)$ of R_t conditionally on the action a_t at s_t. The implementation of Q-learning consists of

$$a_t = \arg\max_{a \in A} Q_t(s_t, a) ,$$

$$Q_{t+1}(x, a) = \begin{cases} Q_t(x, a) + \eta \delta_{t+1}(a) & \text{if } x = s_t\, a = a_t \\ Q_t(x, a) & \text{otherwise,} \end{cases}$$

$$\delta_{t+1}(a) = r(s_t, a) + \gamma \max_{\ell} Q_t(s_{t+1}, \ell) - Q_t(s_t, a) .$$

$$(29.35)$$

At s_t, an action $a_t \in A$ is obtained in an easy computation, and then makes a move to a new state s_{t+1}. Receiving the reward $r_{t+1} = r(s_t, a_t)$ associated with the transition (s_t, a_t, s_{t+1}), only the table entry for s_t and a_t is updated.

The format of δ_{t+1} is similar to the one in (29.32) with the prediction $\Pi_t(s_t)$ replaced by $Q(s_t, a_t)$ and $\Pi_t(s_{t+1})$ replaced by $\max_a Q_t(s_{t+1}, a)$. Alternatively, we may select a_{t+1} by a fixed policy and then obtain δ_{t+1} with $\max_a Q_t(s_{t+1}, a)$ replaced by $Q_t(s_{t+1}, a_{t+1})$, which leads to a variant of the Q-learning rule called state-action-reward-state-action (SARSA). Under some mild regularity conditions, similarly to TD learning, both Q-learning and SARSA converge to prediction functions that make optimal action choices.

Both TD learning and Q-learning have variants and extensions. In the following, we briefly summarize two typical streams:

- In (29.33) and (29.35), only the table entry for s_t is modified, though r_{t+1} provides useful information for learning earlier predictions as well. Under the name of eligibility traces, an exponentially decaying memory trace is provided on a number of previous input signals so that each new observation can update the parameters related to these signals.
- In addition to a lookup table, the prediction function can be replaced by a more advanced prediction function. It could be a linear or nonlinear regression function $F_t(\omega_\tau, \theta)$ with input signals $\omega_\tau = \{x_\tau^{(1)}, \ldots, x_\tau^{(m)}\}$. Each $x_\tau^{(j)}$ could be either a state or an action or even one additional feature around a state in one eligibility trace, where τ can be different from t. Then, learning adjusts θ to reduce the error δ_{t+1} or $\delta_{t+1}(a_t)$.

29.4.3 Improving Q-Learning by Unsupervised Methods

Examining the Q-learning by (29.35), we observe that it shares some common features with the multi-model-based learning introduced in Sect. 29.3. For a set A of finite many actions, we use the index $\ell = 1, \ldots, k$ to denote each action. Obtaining a_t in (29.35) is equivalent to obtaining $p_{\ell, t}$ in (29.9) with $\varepsilon_{j,t} = -Q_t(s_t, j)$, that is, a selection is made by winner-take-all (WTA) competition. Then, updating $Q_t(s_t, a)$ in (29.35) can be rewritten as follows

$$Q_{t+1}(s_t, \ell) = Q_t(s_t, \ell) + \eta\, p_{\ell,t} \delta_{t+1}(\ell),$$
$$Q_{t+1}(s, \ell) = Q_t(s, \ell), \text{ for } s \neq s_t , \qquad (29.36)$$

which is similar to the general updating rule by (29.17), with $p_{\ell,t}$ selecting which column of Q table to update. This motivates the following improvements on Q-learning, motivated by the multi-model-based learning methods in Sect. 29.3.2.

First, the WTA selection of the above $p_{\ell,t}$ can be replaced by an estimation of the posteriori probabilities as follows

$$p_{\ell,t} = q(\ell|s_t), \quad q(\ell|s) = \mathrm{e}^{Q_t(s,\ell)} \Big/ \sum_{j=1}^{k} \mathrm{e}^{Q_t(s,j)} .$$

$$(29.37)$$

Putting this into (29.38), we improve the weak points incurred from a WTA competition by updating all the columns of the Q table with the weights by $p_{\ell,t}$, as a counterpart of (29.12) of the well-known EM algorithm for learning a finite mixture.

Second, $\delta_{t+1}(a)$ in (29.35) uses $\max_a Q_t(s_{t+1}, a)$ as a prediction of the Q-value at s_{t+1}, also by a WTA competition that gives an optimistic choice. Alternatively, we can use the following more reliable one

$$\delta_{t+1}(s_t, a_t) = r(s_t, a_t) + \gamma \Delta \pi_{\ell,t}(s_{t+1}),$$

$$\Delta \pi_{\ell,t}(s) = \sum_j q(j|s) Q_t(s,j) - Q_t(s_t, a_t), \quad (29.38)$$

where $q(j|s_{t+1})$ is given by (29.37) with $s = s_{t+1}$ instead of $s = s_t$, to obtain $\Delta \pi_{\ell,t}(s_{t+1})$ with $q(\ell|s)$.

Third, instead of $p_{\ell,t}$ given by (29.37), we may also use a counterpart of (29.16) to implement Q-learning with help of BYY harmony learning. That is, we consider

$$p_{\ell,t} = q(\ell|s_t)[1 + \Delta \pi_{\ell,t}(s_t)], \quad (29.39)$$

by which an action is encouraged when its value is higher than the average of all actions, while an action is discouraged when its value is below the average but still not too far away, and then is repelled when its value is far below this average.

Moreover, we may simplify the above $p_{\ell,t}$ by focusing on a few of major actions, e.g., the winning action $a_t = \arg\max_{a \in A} Q_t(s_t, a)$ and its rival action similar to rival penalized competitive learning (RPCL) by $p_{\ell,t}$ given as follows

$$p_{\ell,t} = \begin{cases} 1, & \ell^* = \arg\max_j Q_t(s_t, j), \\ -\gamma, & \ell = \arg\max_{\ell \neq \ell^*} Q_t(s_t, j), \\ 0, & \text{otherwise}, \end{cases} \quad (29.40)$$

i.e., the winning action is encouraged while its rival is repelled.

BYY harmony learning and RPCL learning lead to discriminative Q-learning by which actions at each state become more discriminative and thus easier to be selected. As a result, confusing branches in a searching tree will be pruned away. Moreover, we may discard one extra action if we observe that its corresponding

$$\alpha_\ell^{\text{new}} = (1 - \eta)\alpha_\ell^{\text{new}} + \eta q(\ell|s_t), \quad (29.41)$$

is pushed to zero. Actually, this is the nature of automatic model selection, which controls the complexity of function $Q(s, j)$.

29.5 Semi-Supervised Learning

In many real tasks it is easy to obtain a large amount of unlabeled training data but labeling them is expensive because of the requirement of great human effort and expertise or high execution cost. Semi-supervised learning [29.83–86] attempts to exploit unlabeled data to help improve the learning performance without assuming human intervention. In situations where the unlabeled data are exactly the test data, it is also called *transductive learning* [29.87].

Figure 29.3 illustrates why unlabeled data (*gray points*) can be helpful. It can be seen that although both classification boundaries are consistent with labeled data, the boundary obtained by considering unlabeled data is better in generalization. One reason is that the unlabeled data can disclose some information about data distribution which is helpful for model construction.

There are two popular assumptions connecting the distribution information disclosed by unlabeled data with label information. The *cluster assumption* assumes

that data with similar inputs have similar class labels; the *manifold assumption* assumes that data live in a low-dimensional manifold, whereas unlabeled data can help to identify that manifold. The latter can be regarded as a generalization of the former because it is usually assumed that the cluster structure of the data will be more easily found in the lower-dimensional

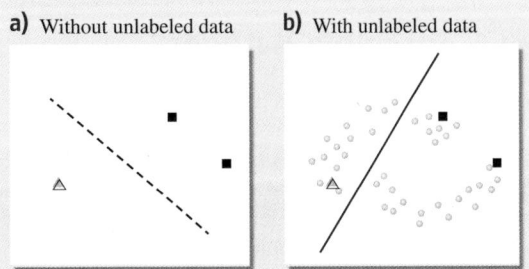

Fig. 29.3a,b Illustration of the usefulness of unlabeled data

manifold. These assumptions are closely related to *low-density separation*, which specifies that the boundary should not go across high-density regions in the instance space.

Many semi-supervised learning methods have been developed. Roughly speaking, they can be categorized into four categories. In *generative methods*, both labeled and unlabeled data are assumed to be generated by the same model, and thus, the unlabeled data can be exploited to model the label estimation or parameter estimation process. For example, if we assume the data come from a mixture model with T components, i. e.,

$$f(x|\theta) = \sum_{t=1}^{T} \alpha_t f(x|\theta_t) , \tag{29.42}$$

where α_t is mixing coefficient and $\theta = \{\theta_t\}$ are the model parameters, then label c_i can be determined by the mixture component m_i and the instance x_i according to the *maximum* a posteriori criterion

$$\arg\max_k \sum_j P(c_i = k|m_i = j, x_i) P(m_i = j|x_i) , \tag{29.43}$$

where estimating $P(c_i = k|m_i = j, x_i)$ requires label information, but unlabeled data can be used to help estimate $P(m_i = j|x_i)$, and hence improve the learning performance. Actually, the posteriori probability is equivalently given by a *mixture of experts* that will be further addressed next in Sect. 29.6.1.

In *semi-supervised support vector machines* (S3VM), unlabeled data are used directly to help adjust the decision boundary, as illustrated in Fig. 29.3. Given l labeled examples and u unlabeled instances, the goal is usually accomplished by minimizing an objective

$$\frac{1}{2}\|w\|_{\mathcal{H}}^2 + C_1 \sum_{i=1}^{l} \ell[y_i, f(x_i)] + C_2 \sum_{j=1}^{u} \ell[\hat{y}_j, f(x_j)] , \tag{29.44}$$

where the first term is structural risk, the second term is empirical risk on the labeled data (x_i, y_i), the third term is empirical risk on the unlabeled instances x_j $(j = 1, \ldots, u)$ and the estimated outputs $\hat{y}_j$, whereas C_1/C_2 balance the contribution of labeled/unlabeled data.

Graph-based methods construct a graph whose nodes are the training instances (both labeled and unlabeled), and the edges between nodes reflect a certain relation, such as similarity, between the corresponding examples. Then, the learning process is accomplished by propagating label information on the graph.

Disagreement-based methods generate multiple learners and exploit the disagreements among the learners, where unlabeled data serve as a kind of platform for information exchange; if one learner is much more confident on a disagreed unlabeled instance than other learner(s), then it will *teach* other(s) by assigning a predicted *pseudo-label* to the instance. A representative of this category is *co-training* [29.31], which constructs two learners from two different views, and thus is closely related to multi-view learning.

In addition to classification, semi-supervised regression, dimension reduction, clustering, etc., have also been well studied. It is worth mentioning that exploiting unlabeled data does not always improve the performance, and sometimes the performance may be even worse than using only the labeled data. Some recent studies have tried to address this issue under the name of *safe* semi-supervised learning [29.88].

29.6 Ensemble Methods

29.6.1 Basic Concepts

Ordinary learning methods try to construct one learner from training data, whereas ensemble methods [29.14] try to construct a set of learners and combine them to solve the problem. Such kinds of learning methods are also called *committee-based learning*, *meta-learning*, or *multiple classifier systems*, although ensemble methods have also been found to be helpful in clustering [29.14, 89, 90] and various tasks other than classification.

Figure 29.4 shows a common ensemble architecture. An ensemble contains a number of *base learners*, or *individual learners*, *component learners*, or *weak learners* because the main purpose of ensemble methods is to generate strong learners by combining learners whose generalization performances are not strong. Base learners can be generated by a *base learning algorithm*, such as a decision tree algorithm, a neural network algorithm, etc., and such ensembles are called *homogeneous* ensembles because they contain homogeneous base learners. An ensemble can also be *heterogeneous* if multiple types of base learners are included.

The generalization ability of an ensemble is often much stronger than that of base learners. Roughly, there are three threads of studies that lead to the state-of-the art of ensemble methods. The *combining classifiers* thread was mostly studied in the pattern recognition community, where researchers usually focused on the design of powerful combining rules to obtain a strong combined classifier [29.28, 29]. The *mixture of experts* thread generally considered a divide-and-conquer strategy, trying to learn a mixture of parametric models jointly [29.91]. Equation (29.43) is actually a mixture of experts for classification, with $P(c_i = k | m_i = j, x_i)$ being the individual expert and $P(m_i = j | x_i)$ the gating net, especially for the one given in [29.72] where the gating net is given by the posteriori of $\alpha_t f(x | \theta_t)$ in (29.42). The *ensembles of weak learners* thread often works on weak learners and tries to design powerful algorithms to boost performance from weak to strong. Readers are referred to [29.30] for a recent survey on *combining classifiers* and *mixture-of-experts* as well as their relations, and to Sect. 5 of [29.86] for a brief overview on all the three threads.

Generally, an ensemble is built in two steps; that is, generating the base learners and then combining them. It is worth noting that the computational cost of constructing an ensemble is often not much larger than creating a single learner. This is because when using a single learner, one usually has to generate multiple versions of the learner for model selection or parameter tuning; this is comparable to generating multiple base learners in ensembles, whereas the computational cost for combining base learners is often small because most combining rules are simple.

The term *boosting* refers to a family of algorithms originated in [29.92], with AdaBoost [29.93] as its rep-resentative. This kind of algorithm is usually provably able to convert weak learners that are just slightly better than random guess to strong learners that have nearly perfect performance.

Algorithm 29.1 shows the pseudo-code of AdaBoost. Roughly speaking, the basic idea of boosting is to let later learners try to correct the mistakes made by earlier learners, and this is accomplished by deriving in each round a new data distribution which makes the earlier mistakes more evident. The base learners should be able to learn with specific distributions; this is usually accomplished by re-weighting or re-sampling the training examples according to the data distribution in each round. Such a learning process is very similar to residual minimization, and it has a close relation to additive models, inspiring an interpretation that AdaBoost is a stagewise estimation procedure for fitting an additive logistic regression model with an exponential loss [29.94]. Notice that AdaBoost was designed for binary classification, but it has many variants for multi-class problems [29.93, 95, 96].

It has been proved [29.93] that the generalization error of AdaBoost is upper bounded by

$$\epsilon \le \epsilon_D + \tilde{O}\left(\sqrt{\frac{dT}{m}}\right),\tag{29.45}$$

with probability at least $1 - \delta$, where ϵ_D is the error on the training sample D, d is the VC-dimension of base learners, m is the number of training samples, and $\tilde{O}(\cdot)$ is used instead of $O(\cdot)$ to hide logarithmic terms and constant factors. This generalization bound implies that the complexity d of base learners and the number T of learning rounds need to be constrained; otherwise AdaBoost will overfit. Empirical studies, however, show that AdaBoost often seems resistant to overfitting; that is, the test error often tends to decrease even after the training error reaches zero.

Algorithm 29.1 The AdaBoost Algorithm
Input: data set $D = \{(x_1, y_1), (x_2, y_2), \ldots, (x_m, y_m)\}$;
     Base learning algorithm $\mathfrak{L}$; number of learning rounds T.

Process:
1: $\mathcal{D}_1(x) = 1/m$. % initialize the weight distribution
2: **for** $t = 1, \ldots, T$:
3:    $h_t = \mathfrak{L}(D, \mathcal{D}_t)$;   % train a classifier h_t from D under distribution $\mathcal{D}_t$
4:    $\epsilon_t = P_{x \approx \mathcal{D}_t}(h_t(x) \ne f(x))$; % evaluate the error of h_t
5:   **if** $\epsilon_t > 0.5$ **then break**

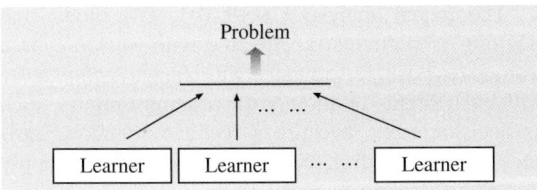

Fig. 29.4 A common ensemble architecture

6:   $\alpha_t = \frac{1}{2}\ln\left(\frac{1-\epsilon_t}{\epsilon_t}\right)$;   % determine the weight of h_t

7:   $\mathcal{D}_{t+1}(x) = \frac{\mathcal{D}_t(x)\exp(-\alpha_t f(x)h_t(x))}{Z_t}$ % update
the distribution, where Z_t is
% a normalization factor which enables $\mathcal{D}_{t+1}$
to be a distribution

8: **end**

Output: $H(x) = \text{sign}\left(\sum_{t=1}^{T}\alpha_t h_t(x)\right)$

29.6.2 Boosting

For binary classification, formally, $f(x) \in \{-1,+1\}$, the margin of the classifier h on the instance x is defined as $f(x)h(x)$, and similarly, the margin of the ensemble $H(x) = \sum_{t=1}^{T}\alpha_t h_t(x)$ is $f(x)H(x) = \sum_{t=1}^{T}\alpha_t f(x)h_t(x)$, whereas the *normalized margin* is

$$f(x)H(x) = \frac{\sum_{t=1}^{T}\alpha_t f(x)h_t(x)}{\sum_{t=1}^{T}\alpha_t}, \qquad (29.46)$$

where α_t are the weights of base learners. Given any threshold $\theta > 0$ of margin over the training sample D, it was proved in [29.97] that the generalization error of the ensemble is bounded with probability at least $1 - \delta$ by

$$\epsilon \leq 2^T \prod_{t=1}^{T}\sqrt{\epsilon_t^{1-\theta}(1-\epsilon_t)^{1+\theta}}$$

$$+ \tilde{O}\left(\sqrt{\frac{d}{m\theta^2} + \ln\frac{1}{\delta}}\right), \qquad (29.47)$$

where ϵ_t is the training error of the base learner h_t. This bound implies that when other variables are fixed, the larger the margin over the training set, the smaller the generalization error. Thus, [29.97] argued that AdaBoost tends to be resistant to overfitting because it can increase the ensemble margin even after the training error reaches zero.

This margin-based explanation seems reasonable; however, it was later questioned [29.98] by the fact that (29.47) depends heavily on the minimum margin, whereas there are counterexamples where an algorithm is able to produce uniformly larger minimum margins than AdaBoost, but the generalization performance drastically decreases. From then on, there was much debate about whether the margin-based explanation holds; more details can be found in [29.14].

One drawback of AdaBoost lies in the fact that it is very sensitive to noise. Great efforts have been devoted to address this issue [29.99, 100]. For example, RobustBoost [29.101] tries to improve the noise tolerance ability by boosting the normalized classification margin, which was believed to be closely related to the generalization error.

29.6.3 Bagging

In contrast to *sequential ensemble methods* such as boosting where the base learners are generated in a sequential style to exploit the *dependence* between the base learners, bagging [29.99] is a kind of *parallel ensemble method* where the base learners are generated in parallel, attempting to exploit the *independence* between the base learners.

The name bagging came from the abbreviation of *Bootstrap AGGregatING*. Algorithm 29.2 shows its pseudo-code, where $\mathbb{I}$ is the indicator function. Bagging applies *bootstrap sampling* [29.102] to obtain multiple different data subsets for training the base learners. Given m training examples, a set of m training examples is generated by sampling with replacement; some original examples appear more than once, whereas some do not present. By applying the process T times, T such sets are obtained, and then each set is used for training a base learner. Bagging can deal with binary as well as multi-class problems by using *majority voting* to combine base learners; it can also be applied to regression by using *averaging* for combination.

Algorithm 29.2 The Bagging Algorithm

Input: Data set $D = \{(x_1, y_1), (x_2, y_2), \ldots, (x_m, y_m)\}$;
Base learning algorithm $\mathfrak{L}$;
Number of base learners T.

Process:
1: **for** $t = 1, \ldots, T$:
2:   $h_t = \mathfrak{L}(D, \mathcal{D}_{bs})$ % $\mathcal{D}_{bs}$ is bootstrap distribution
3: **end**

Output: $H(x) = \arg\max_{y \in \mathcal{Y}} \sum_{t=1}^{T} \mathbb{I}(h_t(x) = y)$

Theoretical analysis [29.99, 103, 104] shows that Bagging is particularly effective with *unstable* base learners (whose performance will change significantly with even slight variation of the training sample), such as decision trees, because it has a tremendous variance reduction effect, whereas it is not wise to apply Bagging to stable learners, such as nearest neighbor classifiers.

A prominent extension of Bagging is the random forest method [29.105], which has been successfully deployed in many real tasks. Random forest incorporates a randomized feature selection process in constructing the individual decision trees. For each individual decision tree, at each step of split selection, it randomly selects a feature subset, and then executes conventional split selection within the feature subset. The recommended size of feature subsets is the logarithm of the number of all features [29.105].

29.6.4 Stacking

Stacking [29.106–108] trains a *meta-learner* (or *second-level learner*), to combine the individual learners (or *first-level learners*). First-level learners are often generated by different learning algorithms, and therefore, stacked ensembles are often heterogeneous, although it is also possible to construct homogeneous stacked ensembles. Also, one similar approach was proposed in IJCNN1991 [29.109], with *meta-learner* referred to by a different name called *associative switch*, which is learned from examples for combining multiple classifiers.

Stacking can be viewed as a generalized framework of many ensemble methods, and can also be regarded as a specific combination method, i. e., *combining by learning*. It uses the original training examples to construct the first-level learners, and then generates a new data set to train the meta-learner, where the first-level learners' outputs are used as input features whereas the original labels are used as labels. Notice that there will be a high risk of overfitting if the exact data that are used to train the first-level learners are also used to generate the new data set for the meta-learner. Hence, it is recommended to exclude the training examples for the first-level learners from the data that are used for the meta-learner, and a cross-validation procedure is usually used.

It is crucial to consider the types of features for the new training data, and the types of learning algorithms for the meta-learner [29.106]. It has been suggested [29.110] to use class probabilities instead of crisp class labels as features for the new data, and to use multi-response linear regression (MLR) for the meta-learner. It has also been suggested [29.111] to use different sets of features for the linear regression problems in MLR.

If stacking (and many other ensemble methods) is simply viewed as assigning weights to combine different models, then it is closely related to *Bayes model averaging* (BMA), which assigns weights to models based on posterior probabilities. In theory, if the correct data generation model is in consideration and if the noise level is low, BMA is never worse and often better than stacking. In practice, however, BMA rarely performs better than stacking, because the correct data generalization model is usually unknown, whereas BMA is quite sensitive to model approximation error [29.112].

29.6.5 Diversity

If the base learners are independent, an amazing combination effect will occur. Taking binary classification, for an example, suppose each base learner has an independent generalization error ϵ and T learners are combined via majority voting. Then, the ensemble makes an error only when at least half of its base learners make errors. Thus, by *Hoeffding inequality*, the generalization error of the ensemble is

$$\sum_{k=0}^{\lfloor T/2 \rfloor} \binom{T}{k} (1-\epsilon)^k \epsilon^{T-k} \le \exp\left(-\frac{1}{2} T (2\epsilon - 1)^2\right) ,$$

(29.48)

which implies that the generalization error decreases exponentially to the ensemble size T, and ultimately approaches zero as T approaches infinity.

It is practically impossible to obtain really independent base learners, but it is generally accepted that to construct a good ensemble, the base learners should be as accurate as possible, and as *diverse* as possible. This has also been confirmed by *error-ambiguity decomposition* and *bias-variance-covariance decomposition* [29.113–115]. Generating diverse base learners, however, is not easy, because these learners are generated from the same training data for the same learning problem, and thus they are usually highly correlated. Actually, we need to require that the base learners must not be very poor; otherwise their combination may even worsen the performance.

Usually, combining only accurate learners is often worse than combining some accurate ones together with some relatively weak ones, because the complementarity is more important than pure accuracy. Notice that it is possible to do some selection to construct a smaller but stronger ensemble after obtaining all base learners [29.116], possibly because this way makes it easier to trade off between individual performance and diversity.

Unfortunately, there is not yet a clear understanding about diversity although it is crucial for ensemble methods. Many efforts have been devoted to designing diversity measures, however, none of them is well-accepted [29.14, 117]. In practice, heuristics are usually employed to generate diversity, and popular strategies include manipulating data samples, input features, learning parameters, and output representations [29.14].

29.7 Feature Selection and Extraction

Real-world data are often high-dimensional and contain many spurious features. For example, in face recognition, an image of size $m \times n$ is often represented as a vector in R^{mn}, which can be very high-dimensional for typical values of m and n. Similarly, biological databases such as microarray data can have thousands or even tens of thousands of genes as features. Such a large number of features can easily lead to the curse of dimensionality and severe overfitting. A simple approach is to manually remove irrelevant features from the data. However, this may not be feasible in practice. Hence, automatic dimensionality reduction techniques, in the form of either feature selection or feature extraction, play a fundamental role in many machine learning problems.

Feature selection selects only a relevant subset of features for use with the model. In feature selection, the features may be scored either individually or as a subset. Not only can feature selection improve the generalization performance of the resultant classifier, the use of fewer features is also less computationally expensive and thus implies faster testing. Moreover, it can eliminate the need to collect a large number of irrelevant and redundant features, and thus reduces cost. The discovery of a small set of highly predictive variables also enhances our understanding of the underlying physical, biological, or natural processes, beyond just the building of accurate *black-box* predictors.

Feature selection and extraction has been a classic topic in the literature of pattern recognition for several decades; many results obtained before the 1980s are systematically summarized in [29.118]. Reviews on further studies in the recent three decadse are referred to [29.119–121]. Roughly, feature selection methods can be classified into three main paradigms: filters, wrappers, and the embedded approach [29.120]. Filters score the usefulness of the feature subset obtained as a pre-processing step. Commonly used scores include mutual information and the inter/intra class distance. This filtering step is performed independently of the classifier and is typically least computationally expensive among the three paradigms. Wrappers, on the other hand, score the feature subsets according to their prediction performance when used with the classifier. In other words, the classifier is trained on each of the candidate feature subsets, and the one with the best score is then selected. However, as the number of candidate feature subsets can be very large, this approach is computationally expensive, though it is also expected to perform better than filters. Both filters and wrappers rely on search strategies to guide the search for the *best* feature subset. While a large number of search strategies can be used, one is often limited to the computationally simple greedy strategies: (i) forward, in which features are added to the candidate set one by one; or (ii) backward, in which one starts with the full feature set and deletes features one by one. Finally, embedded methods combine feature selection with the classifier to create a sparse model. For example, one can use the ℓ_1 regularizer which shrinks the coefficients of the useless features to zero, essentially removing them from the model. Another popular algorithm is called recursive feature elimination [29.122] for use with support vector machines. It repeatedly constructs a model and then removes those features with low weights. Empirically, embedded methods are often more efficient than filters and wrappers [29.120].

While most feature selection methods are supervised, there are also recent works on feature selection in the unsupervised learning setting. However, unsupervised feature selection is much more difficult due to the lack of label information to guide the search for relevant features. Most unsupervised feature selection methods are based on the filter approach [29.123–125], though there are also some studies on wrappers [29.126] and embedded approaches [29.124, 127–129].

Recently, feature selection in multi-task learning has been receiving increasing attention. Recall that the ℓ_1 regularizer is commonly used to induce feature selection in single-task learning; this is extended to the mixed norms in MTL. Specifically, let $W = [w_1, w_2, \dots, w_T]$, where $w_t \in R^d$ is the parameter associated with the t-th task. To enforce joint sparsity across the T tasks, the $\ell_{\infty,1}$ norm of W is used as the regularizer, i. e., $\|W\|_{\infty,1} = \sum_{j=1}^{d} \max_{1 \le i \le T} |W_{ji}|$ [29.130]. In other words, one uses an ℓ_∞ norm on the rows of the W to combine the contributions of each row (feature) from all the tasks, and then combine the features by using the ℓ_1 norm, which, because of its sparsity-encouraging property, leads to only a few nonzero rows of W.

Instead of only selecting a subset from the existing set of features, feature extraction aims at extracting a set of new features from the original features. This can be viewed as performing dimensionality reduction that maps the original features to a new lower-dimensional feature space, while ensuring that the overall structure of the data points remains intact. The unsupervised methods previously introduced in Sect. 29.3.1 can all be used for feature extraction. The classic ones consist of *principal component analysis* (PSA) and *principal subspace analysis* (PSA), and their complementary counterparts *minor component analysis* (MCA) and *minor subspace analysis* (MSA), as well as the closely related *factor analysis* (FA), while independent component analysis (ICA) and *non-Gaussian factor analysis* (NFA) are further developments of PCA and FA, respectively. Another popular further development of PCA is kernel principal component analysis (KPCA) [29.131].

Moreover, feature extraction and unsupervised learning are coordinately conducted in many learning tasks, such as local factor analysis (LFA) in Sect. 29.3.2, nonnegative matrix factorization (NMF) and manifold learning in Sect. 29.3.3, temporal and hierarchical learning in Sect. 29.3.4, as well as other latent factor featured methods. Furthermore, the use of supervised information can lead to even better discriminative features for classification problems. Linear discriminant analysis (LDA) is the most classic example, which results in the Bayes optimal transform direction in the special case that the two classes are normally distributed with the same covariance. Learning multiple layer perceptron or neural networks can be regarded as nonlinear extensions of LDA, with hidden units extracting optimal features for supervised classification and regression.

Part D | 29

References

29.1   H. Simon: Why should machines learn? In: *Machine Learning. An Artificial Intelligence Approach*, ed. by I.R. Anderson, R.S. Michalski, J.G. Carbonell, T.M. Mitchell (Tioga Publ., Palo Alto 1983)

29.2   L. Xu: Bayesian Ying Yang learning, Scholarpedia **2**(3), 1809 (2007)

29.3   L. Xu: Bayesian Ying-Yang system, best harmony learning, and five action circling, Front. Electr. Electr. Eng. China **5**(3), 281–328 (2010)

29.4   L. Xu, S. Klasa, A. Yuille: Recent advances on techniques static feed-forward networks with supervised learning, Int. J. Neural Syst. **3**(3), 253–290 (1992)

29.5   L. Xu: Learning algorithms for RBF functions and subspace based functions. In: *Handbook of Research on Machine Learning, Applications and Trends: Algorithms, Methods and Techniques*, ed. by E. Olivas, J.D.M. Guerrero, M.M. Sober, J.R.M. Benedito, A.J.S. López (Inform. Sci. Ref., Hershey 2009) pp. 60–94

29.6   L. Xu: Several streams of progresses on unsupervised learning: A tutorial overview, Appl. Inf. **1** (2013)

29.7   A. Jain: Data clustering: 50 years beyond k-means, Pattern Recognit. Lett. **31**, 651–666 (2010)

29.8   H. Kriegel, P. Kroger, A. Zimek: Clustering high-dimensional data: A survey on subspace clustering, pattern-based clustering, and correlation clustering, ACM Trans. Knowl. Discov. Data **3**(1), 1 (2009)

29.9   H. Yin: Advances in adaptive nonlinear manifolds and dimensionality reduction, Front. Electr. Electr. Eng. China **6**(1), 72–85 (2011)

29.10   T.T. Kohonen Honkela: Kohonen network, Scholarpedia **2**(1), 1568 (2007)

29.11   L. Xu, J. Neufeld, B. Larson, D. Schuurmans: Maximum margin clustering, Adv. Neural Inf. Process. Syst. (2004) pp. 1537–1544

29.12   K. Zhang, I. Tsang, J. Kwok: Maximum margin clustering made practical, IEEE Trans. Neural Netw. **20**(4), 583–596 (2009)

29.13   Y.-F. Li, I. Tsang, J. Kwok, Z.-H. Zhou: Tighter and convex maximum margin clustering, Proc. 12th Int. Conf. Artif. Intell. Stat. (2009)

29.14   Z.-H. Zhou: *Ensemble Methods: Foundations and Algorithms* (Taylor Francis, Boca Raton 2012)

29.15   G. Tsoumakas, I. Katakis, I. Vlahavas: Mining multi-label data. In: *Data Mining and Knowledge Discovery Handbook*, 2nd edn., ed. by O. Maimon, L. Rokach (Springer, Berlin, Heidelberg 2010)

29.16   C. Silla, A. Freitas: A survey of hierarchical classification across different application domains, Data Min. Knowl. Discov. **22**(1/2), 31–72 (2010)

29.17   W. Bi, J. Kwok: Multi-label classification on tree- and DAG-structured hierarchies, Proc. 28th Int. Conf. Mach. Learn. (2011)

29.18 W. Bi, J. Kwok: Hierarchical multilabel classification with minimum Bayes risk, Proc. Int. Conf. Data Min. (2012)

29.19 W. Bi, J. Kwok: Mandatory leaf node prediction in hierarchical multilabel classification, Adv. Neural Inf. Process. Syst. (2012)

29.20 T.G. Dietterich, R.H. Lathrop, T. Lozano-Pérez: Solving the multiple-instance problem with axis-parallel rectangles, Artif. Intell. **89**(1–2), 31–71 (1997)

29.21 Z.-H.M.-L. Zhou Zhang: Solving multi-instance problems with classifier ensemble based on constructive clustering, Knowl. Inf. Syst. **11**(2), 155–170 (2007)

29.22 Z.-H. Zhou, Y.-Y. Sun, Y.-F. Li: Multi-instance learning by treating instances as non-i.i.d. samples, Proc. 26th Int. Conf. Mach. Learn. (2009) pp. 1249–1256

29.23 Z.-H. Zhou, J.-M. Xu: On the relation between multi-instance learning and semi-supervised learning, Proc. 24th Int. Conf. Mach. Learn. (2007) pp. 1167–1174

29.24 N. Weidmann, E. Frank, B. Pfahringer: A two-level learning method for generalized multi-instance problem, Proc. 14th Eur. Conf. Mach. Learn. (2003) pp. 468–479

29.25 S.D. Scott, J. Zhang, J. Brown: On generalized multiple-instance learning, Int. J. Comput. Intell. Appl. **5**(1), 21–35 (2005)

29.26 Z.-H. Zhou, M.-L. Zhang, S.-J. Huang, Y.-F. Li: Multi-instance multi-label learning, Artif. Intell. **176**(1), 2291–2320 (2012)

29.27 J. Foulds, E. Frank: A review of multi-instance learning assumptions, Knowl. Eng. Rev. **25**(1), 1–25 (2010)

29.28 L. Xu, A. Krzyzak, C. Suen: Several methods for combining multiple classifiers and their applications in handwritten character recognition, IEEE Trans. Syst. Man Cybern. SMC **22**(3), 418–435 (1992)

29.29 J. Kittler, M. Hatef, R. Duin, J. Matas: On combining classifiers, IEEE Trans. Pattern Anal. Mach. Intell. **20**(3), 226–239 (1998)

29.30 L. Xu, S.I. Amari: Combining classifiers and learning mixture-of-experts. In: *Encyclopedia of Artificial Intelligence*, ed. by J. Dopioco, J. Dorado, A. Pazos (Inform. Sci. Ref., Hershey 2008) pp. 318–326

29.31 A. Blum, T. Mitchell: Combining labeled and unlabeled data with co-training, Proc. 11th Annu. Conf. Comput. Learn. Theory (1998) pp. 92–100

29.32 S. Abney: Bootstrapping, Proc. 40th Annu. Meet. Assoc. Comput. Linguist. (2002) pp. 360–367

29.33 M.-F. Balcan, A. Blum, K. Yang: Co-training and expansion: Towards bridging theory and practice, Adv. Neural Inf. Process. Syst. (2005) pp. 89–96

29.34 W. Wang, Z.-H. Zhou: A new analysis of co-training, Proc. 27th Int. Conf. Mach. Learn. (2010) pp. 1135–1142

29.35 Z.-H. Zhou, D.-C. Zhan, Q. Yang: Semi-supervised learning with very few labeled training examples, Proc. 22nd AAAI Conf. Artif. Intell. (2007) pp. 675–680

29.36 W. Wang, Z.-H. Zhou: Multi-view active learning in the non-realizable case, Adv. Neural Inf. Process. Syst. (2010) pp. 2388–2396

29.37 R. Caruana: Multitask learning, Mach. Learn. **28**(1), 41–75 (1997)

29.38 T. Evgeniou, M. Pontil: Regularized multi-task learning, Proc. 10th Int. Conf. Know. Discov. Data Min. (2004) pp. 109–117

29.39 T. Evgeniou, C.A. Micchelli, M. Pontil: Learning multiple tasks with kernel methods, J. Mach. Learn. Res. **6**, 615–637 (2005)

29.40 A. Argyriou, T. Evgeniou, M. Pontil: Multi-task feature learning, Adv. Neural Inf. Process. Syst. (2007) pp. 41–48

29.41 A. Argyriou, T. Evgeniou, M. Pontil: Convex multi-task feature learning, Mach. Learn. **73**(3), 243–272 (2008)

29.42 T. Kato, H. Kashima, M. Sugiyama, K. Asai: Multi-task learning via conic programming, Adv. Neural Inf. Process. Syst. (2007) pp. 737–744

29.43 R. Ando, T. Zhang: A framework for learning predictive structures from multiple tasks and unlabeled data, J. Mach. Learn. Res. **6**, 1817–1853 (2005)

29.44 Y. Zhang, D.-Y. Yeung: A convex formulation for learning task relationships in multi-task learning, Proc. 24th Conf. Uncertain. Artif. Intell. (2010) pp. 733–742

29.45 L. Jacob, F. Bach, J. Vert: Clustered multi-task learning: A convex formulation, Adv. Neural Inf. Process. Syst. (2008) pp. 745–752

29.46 L.J. Zhong Kwok: Convex multitask learning with flexible task clusters, Proc. 29th Int. Conf. Mach. Learn. (2012)

29.47 J. Chen, J. Zhou, J. Ye: Integrating low-rank and group-sparse structures for robust multi-task learning, Proc. 17th Int. Conf. Knowl. Discov. Data Min. (2011) pp. 42–50

29.48 S. Pan, J. Kwok, Q. Yang, J. Pan: Adaptive localization in A dynamic WiFi environment through multi-view learning, Proc. 22nd AAAI Conf. Artif. Intell. (2007) pp. 1108–1113

29.49 S. Pan, J. Kwok, Q. Yang: Transfer learning via dimensionality reduction, Proc. 23rd AAAI Conf. Artif. Intell. (2008)

29.50 S. Pan, I. Tsang, J. Kwok, Q. Yang: Domain adaptation via transfer component analysis, IEEE Trans. Neural Netw. **22**(2), 199–210 (2011)

29.51 W. Dai, Q. Yang, G. Xue, Y. Yu: Boosting for transfer learning, Proc. 24th Int. Conf. Mach. Learn. (2007) pp. 193–200

29.52 J. Huang, A. Smola, A. Gretton, K. Borgwardt, B. Schölkopf: Correcting sample selection bias by unlabeled data, Adv. Neural Inf. Process. Syst. (2007) pp. 601–608

29.53   M. Sugiyama, S. Nakajima, H. Kashima, P.V. Buenau, M. Kawanabe: Direct importance estimation with model selection and its application to covariate shift adaptation, Adv. Neural Inf. Process. Syst. (2008)

29.54   C. Elkan: The foundations of cost-sensitive learning, Proc. 17th Int. Jt. Conf. Artif. Intell. (2001) pp. 973–978

29.55   Z.-H. Zhou, X.-Y. Liu: On multi-class cost-sensitive learning, Proc. 21st Natl. Conf. Artif. Intell. (2006) pp. 567–572

29.56   X.-Y. Liu, Z.-H. Zhou: Learning with cost intervals, Proc. 16th Int. Conf. Knowl. Discov. Data Min. (2010) pp. 403–412

29.57   P.D. Turney: Types of cost in inductive concept learning, Proc. 17th Int. Conf. Mach. Learn. (2000) pp. 15–21

29.58   L. Xu: On essential topics of BYY harmony learning: Current status, challenging issues, and gene analysis applications, Front. Electr. Elect. Eng. China 7(1), 147–196 (2012)

29.59   L. Xu: Semi-blind bilinear matrix system, BYY harmony learning, and gene analysis applications, Proc. 6th Int. Conf. New Trends Inf. Sci. Serv. Sci. Data Min. (2012) pp. 661–666

29.60   L. Xu: Independent subspaces. In: Encyclopedia of Artificial Intelligence, ed. by J. Dopioco, J. Dorado, A. Pazos (Inform. Sci. Ref., Hershey 2008) pp. 903–912

29.61   L. Xu: Independent component analysis and extensions with noise and time: A Bayesian Ying-Yang learning perspective, Neural Inf. Process. Lett. Rev. 1(1), 1–52 (2003)

29.62   L. Xu: One-bit-matching ICA theorem, convex-concave programming, and distribution approximation for combinatorics, Neural Comput. 19, 546–569 (2007)

29.63   S. Tu, L. Xu: Parameterizations make different model selections: Empirical findings from factor analysis, Front. Electr. Electr. Eng. China 6(2), 256–274 (2011)

29.64   P. Williams: Bayesian regularization and pruning using A Laplace prior, Neural Comput. 7(1), 117–143 (1995)

29.65   R. Tibshirani: Regression shrinkage and selection via the lasso, J. R. Stat. Soc. Ser. B: Methodol. 58(1), 267–288 (1996)

29.66   M. Figueiredo, A. Jain: Unsupervised learning of finite mixture models, IEEE Trans. Pattern Anal. Mach. Intell. 24(3), 381–396 (2002)

29.67   C. McGrory, D. Titterington: Variational approximations in Bayesian model selection for finite mixture distributions, Comput. Stat. Data Anal. 51(11), 5352–5367 (2007)

29.68   A. Corduneanu, C. Bishop: Variational Bayesian model selection for mixture distributions, Proc. 8th Int. Conf. Artif. Intell. Stat. (2001) pp. 27–34

29.69   L. Xu: Rival penalized competitive learning, Scholarpedia 2(8), 1810 (2007)

29.70   L. Xu: A unified perspective and new results on RHT computing, mixture based learning, and multi-learner based problem solving, Pattern Recognit. 40(8), 2129–2153 (2007)

29.71   L. Xu: BYY harmony learning, structural RPCL, and topological self-organizing on mixture models, Neural Netw. 8–9, 1125–1151 (2002)

29.72   L. Xu, M. Jordan, G. Hinton: An alternative model for mixtures of experts, Adv. Neural Inf. Process. Syst. (1995) pp. 633–640

29.73   D. Lee, H. Seung: Learning the parts of objects by non-negative matrix factorization, Nature 401(6755), 788–791 (1999)

29.74   S. Madeira: A. Oliveira, Biclustering algorithms for biological data analysis: A survey, IEEE Trans. Comput. Biol. Bioinform. 1(1), 25–45 (2004)

29.75   S. Tu, R. Chen, L. Xu: A binary matrix factorization algorithm for protein complex prediction, Proteome Sci. 9(Suppl 1), S18 (2011)

29.76   X. He, P. Niyogi: Locality preserving projections, Adv. Neural Inf. Process. Syst. (2003) pp. 152–160

29.77   X. He, B. Lin: Tangent space learning and generalization, Front. Electr. Electr. Eng. China 6(1), 27–42 (2011)

29.78   M.M. Meila Jordan: Learning with mixtures of trees, J. Mach. Learn. Res. 1, 1–48 (2000)

29.79   J. Pearl: Fusion, propagation and structuring in belief networks, Artif. Intell. 29(3), 241–288 (1986), Sep.

29.80   L. Xu, J. Pearl: Structuring causal tree models with continuous variables, Proc. 3rd Annu. Conf. Uncertain. Artif. Intell. (1987) pp. 170–179

29.81   A. Barto: Temporal difference learning, Scholarpedia 2(11), 1604 (2007)

29.82   F. Woergoetter, B. Porr: Reinforcement learning, Scholarpedia 3(3), 1448 (2008)

29.83   O. Chapelle, B. Schölkopf, A. Zien: Semi-Supervised Learning (MIT, Cambridge 2006)

29.84   X. Zhu: Semi-supervised learning literature survey (Univ. of Wisconsin, Madison 2008)

29.85   Z.-H. Zhou, M. Li: Semi-supervised learning by disagreement, Knowl. Inform. Syst. 24(3), 415–439 (2010)

29.86   Z.-H. Zhou: When semi-supervised learning meets ensemble learning, Front. Electr. Electr. Eng. China 6(1), 6–16 (2011)

29.87   V.N. Vapnik: Statistical Learning Theory (Wiley, New York 1998)

29.88   Y.-F. Li, Z.-H. Zhou: Towards making unlabeled data never hurt, Proc. 28th Int. Conf. Mach. Learn. (2011) pp. 1081–1088

29.89   A. Fred, A.K. Jain: Data clustering using evidence accumulation, Proc. 16th Int. Conf. Pattern Recognit. (2002) pp. 276–280

29.90   A. Strehl, J. Ghosh: Cluster ensembles – A knowledge reuse framework for combining multiple partitions, J. Mach. Learn. Res. 3, 583–617 (2002)

Part D | 29

29.91   R. Jacobs, M. Jordan, S. Nowlan, G. Hinton: Adaptive mixtures of local experts, Neural Comput. **3**, 79–87 (1991)

29.92   R.E. Schapire: The strength of weak learnability, Mach. Learn. **5**(2), 197–227 (1990)

29.93   Y. Freund, R.E. Schapire: A decision-theoretic generalization of on-line learning and an application to boosting, J. Comput. Syst. Sci. **55**(1), 119–139 (1997)

29.94   J. Friedman, T. Hastie, R. Tibshirani: Additive logistic regression: A statistical view of boosting (with discussions), Ann. Stat. **28**(2), 337–407 (2000)

29.95   R.E. Schapire, Y. Singer: Improved boosting algorithms using confidence-rated predictions, Mach. Learn. **37**(3), 297–336 (1999)

29.96   J. Zhu, S. Rosset, H. Zou, T. Hastie: Multi-class AdaBoost, Stat. Interface **2**, 349–360 (2009)

29.97   R.E. Schapire, Y. Freund, P. Bartlett, W.S. Lee: Boosting the margin: A new explanation for the effectiveness of voting methods, Ann. Stat. **26**(5), 1651–1686 (1998)

29.98   L. Breiman: Prediction games and arcing algorithms, Neural Comput. **11**(7), 1493–1517 (1999)

29.99   L. Breiman: Bagging predictors, Mach. Learn. **24**(2), 123–140 (1996)

29.100   C. Domingo, O. Watanabe: Madaboost: A modification of AdaBoost, Proc. 13th Annu. Conf. Comput. Learn. Theory (2000) pp. 180–189

29.101   Y. Freund: An adaptive version of the boost by majority algorithm, Mach. Learn. **43**(3), 293–318 (2001)

29.102   B. Efron, R. Tibshirani: *An Introduction to the Bootstrap* (Chapman Hall, New York 1993)

29.103   A. Buja, W. Stuetzle: Observations on bagging, Stat. Sin. **16**(2), 323–351 (2006)

29.104   J.H.P. Friedman Hall: On bagging and nonlinear estimation, J. Stat. Plan. Inference **137**(3), 669–683 (2007)

29.105   L. Breiman: Random forests, Mach. Learn. **45**(1), 5–32 (2001)

29.106   D.H. Wolpert: Stacked generalization, Neural Netw. **5**(2), 241–260 (1992)

29.107   L. Breiman: Stacked regressions, Mach. Learn. **24**(1), 49–64 (1996)

29.108   P. Smyth, D. Wolpert: Stacked density estimation, Adv. Neural Inf. Process. Syst. (1998) pp. 668–674

29.109   L. Xu, A. Krzyzak, C. Sun: Associative switch for combining multiple classifiers, Int. Jt. Conf. Neural Netw. (1991) pp. 43–48

29.110   K.M. Ting, I.H. Witten: Issues in stacked generalization, J. Artif. Intell. Res. **10**, 271–289 (1999)

29.111   A.K. Seewald: How to make stacking better and faster while also taking care of an unknown weakness, Proc. 19th Int. Conf. Mach. Learn. (2002) pp. 554–561

29.112   B. Clarke: Comparing Bayes model averaging and stacking when model approximation error cannot be ignored, J. Mach. Learn. Res. **4**, 683–712 (2003)

29.113   A. Krogh, J. Vedelsby: Neural network ensembles cross validation, and active learning, Adv. Neural Inf. Process. Syst. (1995) pp. 231–238

29.114   N.R. Ueda Nakano: Generalization error of ensemble estimators, Proc. IEEE Int. Conf. Neural Netw. (1996) pp. 90–95

29.115   G. Brown, J.L. Wyatt, P. Tino: Managing diversity in regression ensembles, J. Mach. Learn. Res. **6**, 1621–1650 (2005)

29.116   Z.-H. Zhou, J. Wu, W. Tang: Ensembling neural networks: Many could be better than all, Artif. Intell. **137**(1–2), 239–263 (2002)

29.117   L.I. Kuncheva, C.J. Whitaker: Measures of diversity in classifier ensembles and their relationship with the ensemble accuracy, Mach. Learn. **51**(2), 181–207 (2003)

29.118   P. Devijver, J. Kittler: *Pattern Recognition: A Statistical Approach* (Prentice Hall, New York 1982)

29.119   Y. Saeys, I. Inza, P. Larraaga: A review of feature selection techniques in bioinformatics, Bioinformatics **19**(23), 2507–2517 (2007)

29.120   I. Guyon, A. Elisseeff: An introduction to variable and feature selection, J. Mach. Learn. Res. **3**, 1157–1182 (2003)

29.121   A. Jain, R. Duin, J. Mao: Statistical pattern recognition: A review, IEEE Trans. Pattern Anal. Mach. Intell. **22**, 1 (2000)

29.122   I. Guyon, J. Weston, S. Barnhill, V. Vapnik: Gene selection for cancer classification using support vector machines, Mach. Learn. **46**(1–3), 389–422 (2002)

29.123   M. Dash, K. Choi, P. Scheuermann, H. Liu: Feature selection for clustering – A filter solution, Proc. 2nd Int. Conf. Data Min. (2002) pp. 115–122, Dec.

29.124   M. Law, M. Figueiredo, A. Jain: Simultaneous feature selection and clustering using mixture models, IEEE Trans. Pattern Anal. Mach. Intell. **26**(9), 1154–1166 (2004)

29.125   P. Mitra, C. Murthy, S.K. Pal: Unsupervised feature selection using feature similarity, IEEE Trans. Pattern Anal. Mach. Intell. **24**(3), 301–312 (2002)

29.126   V. Roth: The generalized LASSO, IEEE Trans. Neural Netw. **15**(1), 16–28 (2004)

29.127   C. Constantinopoulos, M. Titsias, A. Likas: Bayesian feature and model selection for Gaussian mixture models, IEEE Trans. Pattern Anal. Mach. Intell. **28**(6), 1013–1018 (2006)

29.128   J. Dy, C. Brodley: Feature selection for unsupervised learning, J. Mach. Learn. Res. **5**, 845–889 (2004)

29.129   B. Zhao, J. Kwok, F. Wang, C. Zhang: Unsupervised maximum margin feature selection with manifold regularization, Proc. Int. Conf. Comput. Vis. Pattern Recognit. (2009)

29.130   B. Turlach, W. Venables, S. Wright: Simultaneous variable selection, Technometrics **27**, 349–363 (2005)

29.131   B. Schölkopf, A. Smola: *Learning with Kernels* (MIT Press, Cambridge 2002)

30. Theoretical Methods in Machine Learning

Badong Chen, Weifeng Liu, José C. Principe

The problem of optimization in machine learning is well established but it entails several approximations. The theory of Hilbert spaces, which is principled and well established, helps solve the representation problem in machine learning by providing a rich (universal) class of functions where the optimization can be conducted. Working with functions is cumbersome, but for the class of reproducing kernel Hilbert spaces (RKHSs) it is still manageable provided the algorithm is restricted to inner products. The best example is the support vector machine (SVM), which is a batch mode algorithm that uses a very efficient (supralinear) optimization procedure. However, the problem of SVMs is that they display large memory and computational complexity. For the large-scale data limit, SVMs are restrictive because for fast operation the Gram matrix, which increases with the square of the number of samples, must fit in computer memory. The computation in this best-case scenario is also proportional to number of samples square. This is not specific to the SVM algorithm and is shared by kernel regression. There are also other relevant data processing scenarios such as streaming data (also called a time series) where the size of the data is unbounded and potentially nonstationary, therefore batch mode is not directly applicable and brings added difficulties.

Online learning in kernel space is more efficient in many practical large scale data applications. As the training data are sequentially presented to the learning system, online kernel learning, in general, requires much less memory and computational bandwidth. The drawback is that online algorithms only converge weakly (in mean square) to the optimal solution, i. e., they only have guaranteed convergence within a ball of radius ε around the optimum (ε is controlled by the user). But because the theoretical optimal ML solution has many approximations, this is one more approximation

that is worth exploring practically. The most important recent advance in this field is the development of the kernel adaptive filters (KAFs). The KAF algorithms are developed in reproducing kernel Hilbert space (RKHS), by using the linear structure of this space to implement well-established linear adaptive algorithms (e.g., LMS, RLS, APA, etc.) and to obtain nonlinear filters in the original input space. The main goal of this chapter is to bring closer to readers, from both machine learning and signal processing communities, these new online learning techniques. In this chapter, we focus mainly on the kernel least mean square (KLMS), kernel recursive least squares (KRLSs), and the kernel affine projection algorithms (KAPAs). The derivation of the algorithms and some key aspects, such as the mean-square convergence and the sparsification of the solutions, are discussed. Several illustration examples are also presented to demonstrate the learning performance.

30.1 Background Overview

The general goal of machine learning is to build a model from data with the goal of extracting useful structure contained in the data. More specifically, machine learning can be defined as a process by which the topology and the free parameters of a neural network (i. e., the learning machine) are adapted through a process of stimulation by the environment in which the network is embedded [30.1]. There is a wide variety of machine learning algorithms. Based on the desired response of the algorithm or the type of input available during training, machine learning algorithms can be divided into several categories: *supervised learning, unsupervised learning, semi-supervised learning, reinforcement learning*, and so on [30.2]. In this chapter, however, we focus mainly on supervised learning, and in particular, on the regression tasks. The goal of supervised learning is, in general, to infer a function that maps inputs to desired outputs or labels that should have the generalization property, that is, it should perform well on unseen data instances.

In supervised machine learning problems, we assume that the data pairs $\{x_i, z_i\}$ collected from real-world experiments are stochastic and drawn independently from an unknown probability distribution $P(X, Z)$ that represents the underlying phenomenon we wish to model. The optimization problem is normally formulated in terms of the expected risk $R(f)$ defined as $R(f) = \int L(f(x), z)dP(x, z)$, where a loss function $L(f(x), z)$ translates the goal of the analysis, and f belongs to a functional space. The optimization problem is to find the minimal expected risk $R(f)$ among all possible functions, i. e. $f^* = \min_f R(f)$. Unfortunately, we cannot work with arbitrary functions in our model, so we restrict f to a mapper class F, and very likely $f^* \notin F$. For instance, if our mapper is linear, then the functional class is the linear set which is small, albeit important, and so we will approximate f^* by the closest linear function, committing sometimes a large error. But even if the mapper is a multilayer perceptron with fixed topology, the same problem exists although the error will likely be smaller. The best solution is therefore $f_F^* = \min_{f \in F} R(f)$ and it represents the first source of implementation error experimenters face. But this is not the only problem. We also normally do not know $P(X, Z)$ in advance (indeed in machine learning this is normally the goal of the analysis). Therefore, we resort to the law of large numbers and approximate the expected risk by $\hat{R}_N(f) = 1/N \sum_i L(f(x_i), z_i)$ which we call the empirical risk. Therefore, our optimization

goal becomes $f_N^* = \min_{f \in F} \hat{R}_N(f)$, which is normally achieved by optimization algorithms. The difference between the optimal solution $R(f^*)$ and the solution achieved $R_N(f_N^*)$ with the finite number of samples and the chosen mapper can be written as

$$
\hat{R}_N(f_N^*) - R(f^*)
$$
$$
= \left[R(f_F^*) - R(f^*) \right] + \left[\hat{R}_N(f_N^*) - R(f_F^*) \right] ,
$$

where the first term is the *approximation error* while the second term is the *estimation error*. The optimization itself is also subject to constraints as we can imagine. The major compromise is how to treat the two terms. Statisticians favor algorithms to decrease as fast as possible the second term (estimation error), while optimization experts concentrate on supra-linear algorithms to minimize the first term (approximation error). But in large-scale data problems, one major consideration is the optimization time under these optimal assumptions, which can become prohibitively large. This paper among others [30.3] defines a third error ρ called the *optimization error* to approximate the practical optimal solution $R_N(f_N^*)$ by $R_N(\tilde{f}_N)$, provided one can find $R_N(\tilde{f}_N)$ simply with algorithms that are $O(N)$ in time and memory usage. Basically, the final solution of $R_N(\tilde{f}_N)$ will exist in a neighborhood of the optimal solution of radius ρ.

Let us now explain how the powerful mathematical tool called the RKHS has been widely utilized in the areas of machine learning [30.4, 5]. It is well known that the probability of linearly *shattering* data tends to one with the increase in dimensionality of the data space. However, the main bottleneck of this technique was the large number of free parameters of the high-dimensional classifiers, which results in two difficult issues: expensive computation and the need to regularize the solutions. The RKHS (also kernel space or feature space) provides a nice way to simplify the computation. The dimension of an RKHS can be very high (even infinite), but by the *kernel trick* the calculation in RKHS can still be done efficiently in the original input space if the algorithms can be expressed in terms of the inner products. *Vapnik* proposed a robust regularizer in support vector machine (SVM), which promoted the application of RKHS in pattern recognition [30.4, 5].

Kernel-based learning algorithms have been successfully applied in batch settings (say SVM). The batch kernel learning algorithms, however, usually require significant memory and computational burden due to the necessity of retaining all the training data and calculating a large Gram matrix. In many practical situations, the online kernel learning (OKL) is more efficient. Since the training data are sequentially (one by one) presented to the learning system, OKL in general requires much less memory and computational cost. Another key advantage of OKL algorithms is that they can easily deal with nonstationary (time varying) environments (i. e., where the data statistics change over time).

Traditional linear adaptive filtering algorithms like the least mean square (LMS) and recursive least squares (RLSs) are the most well known and simplest online learning algorithms, especially in signal processing community [30.6–8]. In recent years, many researchers devoted to use the RKHS to design the optimal nonlinear adaptive filters, namely, the kernel adaptive filters (KAF) [30.9]. The KAF algorithms are developed in RKHS, by using the linear structure (inner product) of this space to implement the well-established linear adaptive algorithms and to obtain (by *kernel trick*) nonlinear filters in the original input space. Up to now, there have been many KAF algorithms. Typical examples include the KLMS [30.10], kernel affine projection algorithms (KAPA) [30.11], kernel recursive least squares (KRLS) [30.12], and the extended kernel recursive least squares (EX-KRLSs) [30.13]. If the kernel is a Gaussian, these nonlinear filters build a radial basis function (RBF) network with a growing structure, where centers are placed at the projected samples and the weights are directly related to the errors at each sample.

The main bottleneck of KAF algorithms (and many other OKL algorithms) is their growing structure. This drawback will result in increasing memory and computational requirements, especially in continuous adaptation situation where the number of centers grows unbounded. In order to make the KAF algorithms practically useful, it is crucial to find a way to curb the network growth and to obtain a compact representation. Some *sparsification* rules can be applied to address this issue [30.9]. According to these sparsification rules,

the new input is accepted as a new center (i. e., inserted into the *center dictionary*) only if it is judged as an *important* input under a certain criterion. Popular sparsification criteria include the novelty criterion (NC) [30.14], coherence criterion (CC) [30.15], approximate linear dependency (ALD) criterion [30.12], surprise criterion (SC) [30.16], and so on. In addition, the *quantization* approach can also be used to sparsify the solution and produce a compact network with desirable accuracy [30.17].

Besides the RKHS, fundamental concepts and principles from information theory can also be applied in the areas of signal processing and machine learning. For example, information theoretic descriptors like entropy and divergence have been widely used as similarity metrics and optimization criteria in *information theoretic learning* (ITL) [30.18]. These descriptors are particularly useful for nonlinear and non-Gaussian situations since they capture the information content and higher order statistics of signals rather than simply their energy (i. e., second-order statistics like variance and correlation). Recent studies show that the ITL is closely related to RKHS. The quantity of *correntropy* in ITL is in essence a correlation measure in RKHS [30.19]. Many ITL costs can also be formulated in an RKHS induced by the Mercer kernel function defined as the *cross information potential* (CIP) [30.20]. The popular *quadratic information potential* (QIP) can be expressed as a squared norm in this RKHS. The estimators of information theoretic quantities can also be reinterpreted in RKHS. For example, the nonparametric kernel estimator of the QIP can be expressed as a squared norm of the mean vector of the data in kernel space [30.20].

The focus of the present chapter is mainly on a large family of online kernel learning algorithms, the kernel adaptive filters. Several basic learning algorithms are introduced. Some key aspects about these algorithms are discussed, and several illustration examples are presented. Although our focus is on the kernel adaptive filtering, the basic ideas will be applicable to many other online learning methods.

30.2 Reproducing Kernel Hilbert Spaces

A Hilbert space is a linear, complete, and normed space equipped with an inner product. A reproducing kernel Hilbert space is a special Hilbert space associated with a kernel κ such that it reproduces (via an inner product) each function f in the space. Let X be a set (usually a compact subset of $\mathbb{R}^d$) and $\kappa(x, y)$ be a real-valued bivariate function on $X \times X$. Then the function $\kappa(x, y)$ is said to be *nonnegative definite* if for any finite point set $\{x_i \in X\}_{i=1}^N$ and for any real number set $\{\alpha_i \in \mathbb{R}\}_{i=1}^N$,

$$\sum_{i=1}^N \sum_{j=1}^N \alpha_i \alpha_j \kappa(x_i, x_j) \geq 0 . \tag{30.1}$$

If the above inequality is strict for all nonzero vectors $\boldsymbol{\alpha} = [\alpha_1, \dots, \alpha_N]^T$, the function $\kappa(x, y)$ is said to be *strictly positive definite* (SPD). The following theorem shows that any symmetric and nonnegative definite bivariate function $\kappa(x, y)$ is a *reproducing kernel*.

Theorem 30.1 (Moore–Aronszajn [30.21, 22])
Any symmetric, nonnegative definite function $\kappa(x, y)$ defines implicitly a Hilbert space $\mathcal{H}_k$ that consists of functions on X such that:

1) $\forall x \in X, \kappa(., x) \in \mathcal{H}_k$,
2) $\forall x \in X, \forall f \in \mathcal{H}_k, f(x) = \langle f, \kappa(., x) \rangle_{\mathcal{H}_k}$,
   where $\langle ., . \rangle_{\mathcal{H}_k}$ denotes the inner product in $\mathcal{H}_k$.

Property (2) is the so-called *reproducing property*. In this case, we call $\kappa(x, y)$ a *reproducing kernel*, and $\mathcal{H}_k$ an RKHS defined by the reproducing kernel $\kappa(x, y)$. Usually, the space X is also called the *input space*. Property 1) indicates that each point in the input space X can be mapped onto a function in a potentially much higher dimensional RKHS $\mathcal{H}_k$. The nonlinear mapping from the input space to RKHS is defined as $\Phi(x) = \kappa(., x)$. In particular, we have

$$\langle \Phi(x), \Phi(y) \rangle_{\mathcal{H}_k} = \langle \kappa(., x), \kappa(., y) \rangle_{\mathcal{H}_k}$$
$$= \kappa(x, y) . \tag{30.2}$$

Thus the inner products in high-dimensional RKHS can be simply calculated via kernel evaluation. This is normally called the *kernel trick*. Note that the RKHS is defined by the selected kernel function, and the similarity between functions in the RKHS is also defined by the kernel since it defines the inner product of functions.

The next theorem guarantees the existence of a nonlinear mapping between the input space and a high-dimensional feature space (a vector space in which the training data are embedded).

Theorem 30.2 (Mercer's [30.23])
Let $\kappa \in L_\infty(X \times X)$ be a symmetric bivariate kernel function. If κ is the kernel of a positive integral oper-

ator in $L_2(X)$, and X is a compact subset of $\mathbb{R}^d$, then

$$\forall \psi \in L_2(X): \int_X \kappa(x, y)\psi(x)\psi(y)\mathrm{d}x\mathrm{d}y \geq 0 . \tag{30.3}$$

Let $\varphi_i \in L_2(X)$ be the normalized orthogonal eigenfunctions of the integral operator, and λ_i the corresponding positive eigenvalues. Then

$$\kappa(x, y) = \sum_{i=1}^M \lambda_i \varphi_i(x)\varphi_i(y) , \tag{30.4}$$

where $M \leq \infty$. Since the eigenvalues are positive, one can readily construct a nonlinear mapping φ from the input space X to a feature space $\mathbb{F}$

$$\varphi: X \to \mathbb{F} ,$$
$$\varphi(x) = \left[\sqrt{\lambda_1}\varphi_1(x), \sqrt{\lambda_2}\varphi_2(x), \dots \right]^T . \tag{30.5}$$

The dimension of $\mathbb{F}$ is M, i.e., the number of the positive eigenvalues (which can be infinite in the strictly positive definite kernel case).

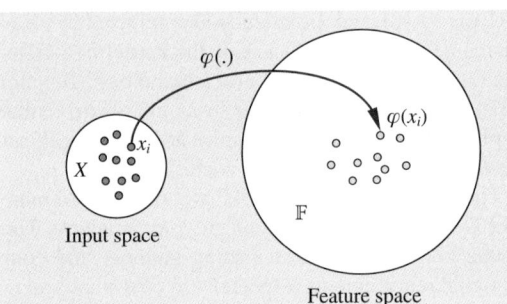

Fig. 30.1 Nonlinear map $\varphi(\cdot)$ between the input space and the feature space

Table 30.1 Some well-known kernels defined over $X \times X$ $X \subset \mathbb{R}^d (c > 0, \sigma > 0, p \in \mathbb{N})$

Kernels	Expressions
Polynomial	$\kappa(x, y) = (c + x^T y)^p$
Exponential	$\kappa(x, y) = \exp(x^T y / 2\sigma^2)$
Sigmoid	$\kappa(x, y) = \tanh(x^T y / \sigma + c)$
Gaussian	$\kappa(x, y) = \exp(-\|x - y\|^2 / 2\sigma^2)$
Laplacian	$\kappa(x, y) = \exp(-\|x - y\| / 2\sigma^2)$
Cosine	$\kappa(x, y) = \exp(\angle(x, y))$
Multiquadratic	$\kappa(x, y) = \sqrt{\|x - y\|^2 + c}$
Inverse multiquadratic	$\kappa(x, y) = 1 / \sqrt{\|x - y\|^2 + c}$

The feature space $\mathbb{F}$ is *isometric–isomorphic* to the RKHS $\mathcal{H}_k$ induced by the kernel. This can be easily recognized by identifying $\boldsymbol{\varphi}(x) = \Phi(x) = \kappa(., x)$. In general, we do not distinguish these two spaces if no confusion arises.

Now the basic idea of kernel-based learning algorithms can be simply described as follows: Via a nonlinear mapping $\boldsymbol{\varphi} \colon \mathcal{X} \to \mathbb{F}$ (or $\Phi \colon \mathcal{X} \to \mathcal{H}_k$), the data $\{x_i \in \mathcal{X}\}_{i=1}^{N}$ are mapped into a high dimensional (usually $M \gg d$) feature space $\mathbb{F}$ with a linear structure (Fig. 30.1). Then a learning problem in $\mathcal{X}$ is solved in $\mathbb{F}$ instead, by working with $\{\boldsymbol{\varphi}(x_i) \in \mathbb{F}\}_{i=1}^{N}$. As long as an algorithm can be formulated in terms of the inner products in $\mathbb{F}$, all the operations can be done in the input space (via kernel evalua-tions). Because $\mathbb{F}$ is high dimensional, a simple linear learning algorithm (preferably one expressed solely in terms of inner products) in $\mathbb{F}$ can solve arbitrarily nonlinear problems in the input space, provided that $\mathbb{F}$ is rich enough to represent the mapping (the feature space can be *universal* if it is infinite dimensional).

The kernel function κ is a crucial factor in all kernel methods because it defines the similarity between data points. Some well-known kernels are listed in Table 30.1. Among these kernels, the Gaussian kernel is most popular and is, in general, a default choice due to its universal approximating capability (Gaussian kernel is strictly positive definite), desirable smoothness and numerical stability.

30.3 Online Learning with Kernel Adaptive Filters

In this section, we discuss several important online kernel learning algorithms, i.e., the kernel adaptive filtering algorithms. Suppose our goal is to learn a continuous input–output mapping $f \colon \mathbb{U} \to \mathbb{D}$ based on a sequence of input–output examples (the so called training data) $\{\boldsymbol{u}(i), d(i)\}, i = 1, 2, \ldots$, where $\mathbb{U} \subset \mathbb{R}^m$ is the input domain, $\mathbb{D} \subset \mathbb{R}$ is the desired output space. This supervised learning problem can be solved online (sequentially) using an adaptive filter. Figure 30.2 shows a general scheme of an adaptive filter. Usually, an adaptive filter consists of three elements:

1) The input–output training data.
2) The structure (or topology) of the filter, with a set of unknown parameters (or weights) $\boldsymbol{w}$.
3) An optimization criterion J (or cost function).

An adaptive filtering algorithm will adjust the filter parameters so as to minimize the disparity (measured by the cost function) between the filtering and desired outputs. The filter topology can be a simple linear structure (e.g., the FIR filter) or any nonlinear network structure (e.g., MLPs, RBF, etc.). The cost function is, in general, the mean square error (MSE) or the least-squares (LSs) cost. The adaptive filtering algorithm is usually a gradient-based algorithm.

The great appeal of developing adaptive filters in RKHS is to utilize the linear structure of this space to implement well-established linear adaptive filtering algorithms and to achieve nonlinear filters in the input space. Compared with other nonlinear adaptive filters, the KAFs have several desirable features:

1) If choosing a universal kernel (e.g., Gaussian kernel), they are universal approximators.
2) Under MSE criterion, the performance surface is still quadratic so gradient descent learning does not suffer from local minima.
3) If pruning the redundant features, they have moderate complexity in terms of computation and memory.

Table 30.2 gives the comparison of different adaptive filters [30.9].

30.3.1 Kernel Least Mean Square (KLMS) Algorithm

Among the family of the KAF, the KLMS is the simplest, which is derived by directly mapping the linear

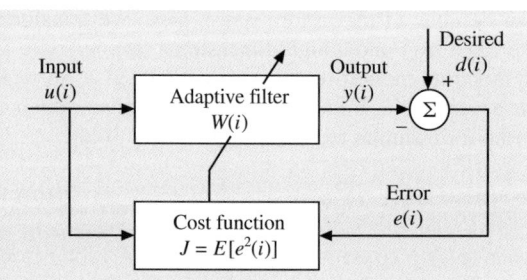

Fig. 30.2 General configuration of an adaptive filter

Table 30.2 Comparison of different adaptive filters

Adaptive filters	Modeling capacity	Convexity	Complexity
Linear adaptive filters	Linear only	Yes	Very simple
Hammerstein, Weiner models	Limited nonlinearity	No	Simple
Volterra, Wiener series	Universal	Yes	Very high
Time-lagged neural networks	Universal	No	Modest
Recurrent neural networks	Universal	No	High
Kernel adaptive filters	Universal	Yes	Modest
Recursive Bayesian filters	Universal	No	Very high

least mean square (LMS) algorithm into RKHS [30.10]. Before proceeding, we simply discuss the well-known LMS algorithm.

LMS Algorithm

Usually, the LMS algorithm assumes a linear finite impulse response (FIR) filter (or transversal filter), whose output, at i iteration, is simply a linear combination of the input

$$y(i) = w(i-1)^T u(i) , \tag{30.6}$$

where $w(i-1)$ denotes the estimated weight vector at $(i-1)$ iteration. With the above linear model, the LMS algorithm can be given as follows

$$\begin{cases} w(0) = 0 , \\ e(i) = d(i) - w(i-1)^T u(i) , \\ w(i) = w(i-1) + \eta e(i) u(i) , \end{cases} \tag{30.7}$$

where $e(i) = d(i) - y(i)$ is the prediction error, and $\eta > 0$ is the step size. The LMS algorithm is in essence a stochastic gradient-based algorithm under the instantaneous MSE cost $J(i) = e^2(i)/2$. In fact, the weight update equation of the LMS can be simply derived as

$$\begin{aligned} w(i) &= w(i-1) - \eta \frac{\partial}{\partial w(i-1)} \left(\frac{1}{2} e^2(i) \right) \\ &= w(i-1) - \eta e(i) \frac{\partial}{\partial w(i-1)} (d(i) \\ &\quad - w(i-1)^T u(i)) \\ &= w(i-1) + \eta e(i) u(i) . \end{aligned} \tag{30.8}$$

The LMS algorithm has been widely applied in adaptive signal processing due to its simplicity and efficiency [30.6–8]. The robustness of the LMS has been proven in [30.24], and it has been shown that a single realization of the LMS is optimal in the H_∞ sense. The step size η is a crucial parameter and has significant influence on the learning performance. It controls

the compromise between convergence speed and misadjustment. In practice, the selection of step size should guarantee the stability and convergence rate of the algorithm.

The LMS algorithm is sensitive to the input power. In order to guarantee the stability and improve the performance, one often uses the normalized LMS (NLMS) algorithm, which is a variant of the LMS algorithm, where the step-size parameter is normalized by the input power, that is

$$w(i) = w(i-1) + \frac{\eta}{\|u(i)\|^2} e(i) u(i) . \tag{30.9}$$

KLMS Algorithm

The LMS algorithm assumes a linear FIR filter, and hence the performance will become very poor if the unknown mapping is highly nonlinear. To overcome this limitation, we are motivated to formulate a *similar* algorithm in a high-dimensional feature space (or equivalent RKHS), which is capable of learning arbitrary nonlinear mapping. This is the motivation of the development of the kernel adaptive filtering algorithms.

Let us come back to the previous nonlinear learning problem, i.e., learning a continuous arbitrary input–output mapping f based on a sequence of input–output examples $\{u(i), d(i)\}, i = 1, 2, \ldots$ Online learning finds sequentially an estimate of f such that f_i (the estimate at iteration i) is updated based on the last estimate f_{i-1} and the current example $\{u(i), d(i)\}$. This recursive process can be done in the feature space. First, we transform the input $u(i)$ into a high-dimensional feature space $\mathbb{F}$ by a kernel-induced nonlinear mapping $\varphi(\cdot)$. Second, we assume a linear model in the feature space, which is in the form similar to the linear model in (30.6)

$$y(i) = \Omega(i-1)^T \varphi(i) , \tag{30.10}$$

where $\varphi(i) = \varphi(u(i))$ is the mapped *feature vector* from the input space to the feature space, $\Omega(i-1)$ denotes a high-dimensional weight vector in feature space.

Third, we develop a *linear* adaptive filtering algorithm based on the model (30.10) and the transformed training data $\{\varphi(i), d(i)\}$, $i = 1, 2, \ldots$ If we can formulate this *linear* adaptive algorithm in terms of the inner products, we will obtain a nonlinear adaptive algorithm in input space, namely the kernel adaptive filtering algorithm.

Performing the LMS algorithm on the model (30.10) with new example sequence $\{\varphi(i), d(i)\}$ yields the KLMS algorithm [30.10]

$$\begin{cases} \Omega(0) = \mathbf{0}, \\ e(i) = d(i) - \Omega(i-1)^T \varphi(i), \\ \Omega(i) = \Omega(i-1) + \eta e(i) \varphi(i). \end{cases} \quad (30.11)$$

The KLMS is very similar to the LMS algorithm, except for the dimensionality (or richness) of the projection space. By identifying $\varphi(u) = \kappa(u, .)$, one can easily obtain the learning rule in the input space

$$\begin{cases} f_0 = 0, \\ e(i) = d(i) - f_{i-1}(u(i)), \\ f_i = f_{i-1} + \eta e(i) \kappa(u(i), .). \end{cases} \quad (30.12)$$

The KLMS can be viewed as the solution of the following regularized least squares problem

$$\min_{f_i \in \mathcal{H}_k} (y(i) - f_i(u(i)))^2 + \frac{1-\eta}{\eta} \|f_i - f_{i-1}\|_{\mathcal{H}_k}^2. \quad (30.13)$$

The above formula can be rewritten as

$$\min_{\Delta f_i \in \mathcal{H}_k} (e(i) - \Delta f_i(u(i)))^2 + \frac{1-\eta}{\eta} \|\Delta f_i\|_{\mathcal{H}_k}^2, \quad (30.14)$$

where $\Delta f_i = f_i - f_{i-1}$. From (30.14), we observe:

1) The learning of KLMS at iteration i is equivalent to solving a regularized least squares problem, in which the previous estimate f_{i-1} is frozen, and only the adjustment term Δf_i is solved.
2) In this least squares problem, there is only one training example involved. i. e., $\{u(i), e(i)\}$.
3) The regularization factor is directly related to the step size via $\gamma = (1 - \eta)/\eta$.

It has been proven in [30.10] that the KLMS has *self-regularization* property, i. e., the step size plays a similar role as the regularization parameter.

Given an input u, the output of the KLMS filter, at iteration i, will be

$$f_i(u) = \eta \sum_{j=1}^{i} e(j) \kappa(u(j), u). \quad (30.15)$$

If the kernel is a radial kernel (e.g., Gaussian kernel), the KLMS creates a growing RBF network by allocating a new kernel unit for every new example with input $u(i)$ as the center and $\eta e(i)$ as the coefficient. The network topology of the KLMS filter is shown in Fig. 30.3. The procedure of KLMS is summarized in Algorithm 30.1.

It is also straightforward to derive the normalized KLMS algorithm. The weight update equation of normalized KLMS will be

$$\begin{aligned} \Omega(i) &= \Omega(i-1) + \frac{\eta}{\|\varphi(i)\|^2} e(i) \varphi(i) \\ &= \Omega(i-1) + \frac{\eta}{\kappa(u(i), u(i))} e(i) \varphi(i). \end{aligned} \quad (30.16)$$

If the kernel function is the Gaussian kernel (Table 30.1), we have $\kappa(u(i), u(i)) \equiv 1$. In this case, the KLMS is automatically normalized.

Algorithm 30.1 Kernel Least Mean Square Algorithm

Initialization:
  Choose kernel κ and step size η
  $a_1 = \eta d(1), C(1) = \{u(1)\}, f_1 = a_1 \kappa(u(1), .)$
Computation:
  while $\{u(i), d(i)\}(i > 1)$ available **do**

1) Compute the filter output: $f_{i-1}(u(i)) = \sum_{j=1}^{i-1} a_j \kappa(u(i), u(j))$
2) Compute the error: $e(i) = d(i) - f_{i-1}(u(i))$
3) Store the new center: $C(i) = \{C(i-1), u(i)\}$

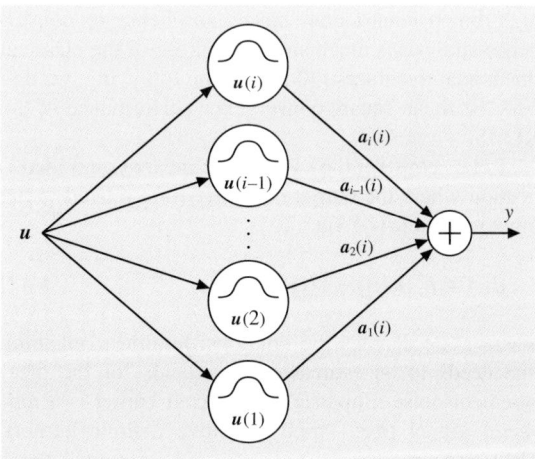

Fig. 30.3 The network topology of the KLMS filter

4) Compute and store the new coefficient: $a_i = \eta e(i)$

end while

(where a denotes the coefficient vector and $C(i)$ denotes the dictionary at i iteration)

The KLMS is a simple kernel learning algorithm, which requires $O(i)$ operations per iteration. The role of the step size η in KLMS remains, in principle, the same as the step size in traditional LMS. Specifically, it controls the compromise between convergence speed and misadjustment. The step-size parameter in KLMS is also directly related to the *optimization error* introduced in [30.3].

In KLMS, the kernel is usually chosen to be a Gaussian kernel. The kernel size (or kernel bandwidth) σ in the Gaussian kernel is a crucial parameter that controls the degree of smoothing and consequently has significant influence on the learning performance. In practice, the kernel size can be set manually, or estimated by rule-based methods (e.g., Silverman's rule [30.25]), or determined automatically using cross-validation.

Mean Square Convergence Performance
The mean square convergence analysis is very important for adaptive filters. For linear adaptive filters, much research has been done in this area and significant results have been achieved. For nonlinear adaptive filters, the mean square convergence analysis is, in general, rather complicated and little studied. The mean square convergence analysis of the KLMS is, however, relatively tractable since it is a simple linear algorithm in high-dimensional feature space, and hence its convergence analysis is much similar to those of the classical linear adaptive filters [30.26]. In the following, we discuss the mean square convergence performance of the KLMS.

Let us consider the case of nonlinear system identification where the output data $\{d(i)\}$ are related to the input vectors $\{u(i)\}$ via

$$d(i) = f^*(u(i)) + v(i), \tag{30.17}$$

where $f^*(\cdot)$ denotes the unknown nonlinear mapping that needs to be estimated, $v(i)$ stands for the measurement noise. Suppose the selected kernel is a *universal* kernel (i.e., strictly positive definite kernel). Then, according to the *universal approximation* prop-

erty [30.27], there is a weight vector $\Omega^* \in \mathbb{F}$ such that

$$d(i) = \Omega^{*T}\varphi(i) + v(i). \tag{30.18}$$

The prediction error $e(i)$ can thus be expressed as

$$e(i) = \tilde{\Omega}(i-1)^T\varphi(i) + v(i) = e_a(i) + v(i), \tag{30.19}$$

where $\tilde{\Omega}(i-1) = \Omega^* - \Omega(i-1)$ is the *weight error vector* in $\mathbb{F}$, $e_a(i) \triangleq \tilde{\Omega}(i-1)^T\varphi(i)$ is the a priori error at iteration i.

Subtracting Ω^* from both sides of the weight update equation $\Omega(i) = \Omega(i-1) + \eta e(i)\varphi(i)$, we get

$$\tilde{\Omega}(i) = \tilde{\Omega}(i-1) - \eta e(i)\varphi(i). \tag{30.20}$$

Define the a posteriori error $e_p(i) \triangleq \tilde{\Omega}(i)^T\varphi(i)$. Then we have

$$e_p(i) = e_a(i) + (\tilde{\Omega}(i)^T - \tilde{\Omega}(i-1)^T)\varphi(i). \tag{30.21}$$

By incorporating (30.20),

$$\begin{aligned}e_p(i) &= e_a(i) - \eta e(i)\varphi(i)^T\varphi(i) \\ &= e_a(i) - \eta e(i)\kappa(u(i), u(i)).\end{aligned} \tag{30.22}$$

Combining (30.20) and (30.22), and eliminating the prediction error $e(i)$, yields

$$\tilde{\Omega}(i) = \tilde{\Omega}(i-1) + \frac{(e_p(i) - e_a(i))\varphi(i)}{\kappa(u(i), u(i))}. \tag{30.23}$$

Squaring both sides of (30.23), and after some straightforward manipulations, we obtain

$$\begin{aligned}&\|\tilde{\Omega}(i)\|^2 + \frac{e_a^2(i)}{\kappa(u(i), u(i))} \\ &= \|\tilde{\Omega}(i-1)\|^2 + \frac{e_p^2(i)}{\kappa(u(i), u(i))},\end{aligned} \tag{30.24}$$

where $\|\tilde{\Omega}(i)\|^2 = \tilde{\Omega}(i)^T\tilde{\Omega}(i)$ is the *weight error power* (WEP) in feature space $\mathbb{F}$. Further, taking expectations of both sides of (30.24) yields

$$\begin{aligned}&E\left[\|\tilde{\Omega}(i)\|^2\right] + E\left[\frac{e_a^2(i)}{\kappa(u(i), u(i))}\right] \\ &= E\left[\|\tilde{\Omega}(i-1)\|^2\right] + E\left[\frac{e_p^2(i)}{\kappa(u(i), u(i))}\right].\end{aligned} \tag{30.25}$$

The above equation is referred to as the *energy conservation relation* in feature space [30.26], which shows how the WEP in feature space evolves in time. The expression of this fundamental relation is in the form similar to those of the energy conservation relation for classical linear adaptive filters [30.28–30]. In fact, this is not surprising, since the KLMS is a linear (but high-dimensional) adaptive algorithm in feature space.

Substituting $e_p(i) = e_a(i) - \eta e(i)\kappa(\boldsymbol{u}(i),\boldsymbol{u}(i))$ into the energy conservation relation (30.25) yields

$$E\left[\|\tilde{\boldsymbol{\Omega}}(i)\|^2\right] = E\left[\|\tilde{\boldsymbol{\Omega}}(i-1)\|^2\right] - 2\eta E[e_a(i)e(i)]$$
$$+ \eta^2 E[e(i)^2 \kappa(\boldsymbol{u}(i),\boldsymbol{u}(i))]\,. \tag{30.26}$$

When choosing Gaussian kernel, we have $\kappa(\boldsymbol{u}(i), \boldsymbol{u}(i)) \equiv 1$, and hence

$$E\left[\|\tilde{\boldsymbol{\Omega}}(i)\|^2\right] = E\left[\|\tilde{\boldsymbol{\Omega}}(i-1)\|^2\right]$$
$$- 2\eta E[e_a(i)e(i)] + \eta^2 E[e^2(i)]\,. \tag{30.27}$$

Gaussian kernel is a normalized and *shift-invariant* kernel, which makes the analysis much simpler. Since Gaussian kernel is also a default kernel in KLMS, in the following, we will focus on Gaussian kernel and use (30.27) to analyze the mean square convergence behavior of the KLMS. It is straightforward to generalize the discussion to arbitrary shift-invariant kernels. In the following, we give an assumption that will be used in the analysis.

Assumption 30.1 A1

The noise $v(i)$ is zero-mean, independent, identically distributed (i.i.d.), and independent of the a priori estimation error $e_a(i)$.

The above assumption is commonly used in convergence analysis for classical linear adaptive filtering algorithms [30.8]. A sufficient condition for the independence between $v(i)$ and $e_a(i)$ is the independence between $v(i)$ and the input sequence $\{\boldsymbol{u}(i)\}$.

Combining (30.27) and assumption A1, we have

$$E\left[\|\tilde{\boldsymbol{\Omega}}(i)\|^2\right] = E\left[\|\tilde{\boldsymbol{\Omega}}(i-1)\|^2\right] - 2\eta E\left[e_a^2(i)\right]$$
$$+ \eta^2 \left(E\left[e_a^2(i)\right] + \xi_v^2\right)\,, \tag{30.28}$$

where ξ_v^2 denotes the noise power (variance). It is worth noting that Eq. (30.28) depends on the noise $v(i)$ through ξ_v^2 only.

A Sufficient Condition for Mean Square Convergence. From (30.28), one can easily derive

$$E\left[\|\tilde{\boldsymbol{\Omega}}(i)\|^2\right] \le E\left[\|\tilde{\boldsymbol{\Omega}}(i-1)\|^2\right]$$
$$\Leftrightarrow -2\eta E\left[e_a^2(i)\right] + \eta^2 \left(E\left[e_a^2(i)\right] + \xi_v^2\right) \le 0$$
$$\Leftrightarrow \eta \le \frac{2E\left[e_a^2(i)\right]}{E\left[e_a^2(i)\right] + \xi_v^2}\,. \tag{30.29}$$

Thus, if we choose the step size such that $\forall i,\ \eta \le 2E[e_a^2(i)]/(E[e_a^2(i)] + \xi_v^2)$, the WEP in feature space will be monotonically decreasing (and hence convergent). This sufficient condition for the mean square convergence is, interestingly, identical to that of the normalized LMS algorithm. The essential reason for this is that the Gaussian kernel is a shift-invariant and normalized kernel ($\kappa(\boldsymbol{u}, \boldsymbol{u}) \equiv 1$). From (30.29), one can also observe that, when the noise power ξ_v^2 is very small, the upper bound on step size will be approximately equal to 2.0.

Steady-State Mean Square Performance. Take the limit of Eq. (30.28) as $i \to \infty$,

$$\lim_{i\to\infty} E\left[\|\tilde{\boldsymbol{\Omega}}(i)\|^2\right] = \lim_{i\to\infty} E\left[\|\tilde{\boldsymbol{\Omega}}(i-1)\|^2\right]$$
$$- 2\eta \lim_{i\to\infty} E\left[e_a^2(i)\right]$$
$$+ \eta^2 \left(\lim_{i\to\infty} E\left[e_a^2(i)\right] + \xi_v^2\right)\,. \tag{30.30}$$

If the WEP in feature space reaches a steady-state value, i.e., $\lim_{i\to\infty} E[\|\tilde{\boldsymbol{\Omega}}(i)\|^2] = \lim_{i\to\infty} E[\|\tilde{\boldsymbol{\Omega}}(i-1)\|^2]$, we have

$$-2\eta \lim_{i\to\infty} E\left[e_a^2(i)\right] + \eta^2 \left(\lim_{i\to\infty} E\left[e_a^2(i)\right] + \xi_v^2\right) = 0\,. \tag{30.31}$$

It follows that

$$\lim_{i\to\infty} E\left[e_a^2(i)\right] = \frac{\eta\xi_v^2}{2 - \eta}\,. \tag{30.32}$$

The a priori error power $E[e_a^2(i)]$ is also referred to as the *excess mean square error* (EMSE) in the adaptive filtering community. From (30.32), we see that the steady-state EMSE of KLMS depends only on the step

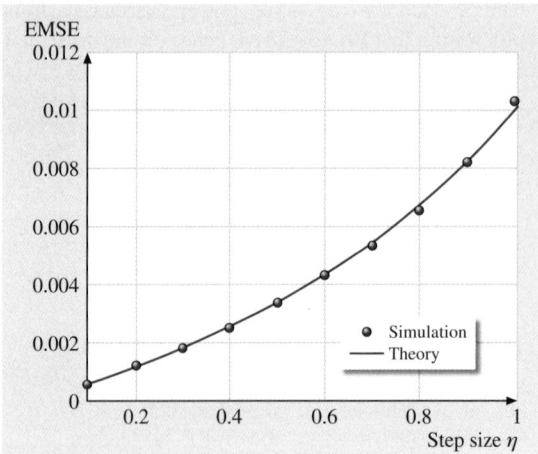

Fig. 30.4 Simulated and theoretical EMSE versus step size

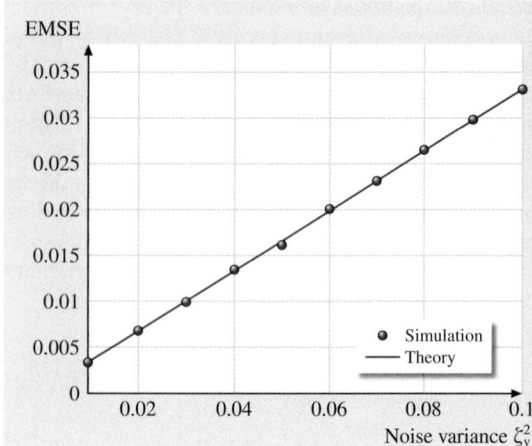

Fig. 30.5 Simulated and theoretical EMSE versus noise variance

size and noise variance, and is NOT related to the kernel size and the unknown nonlinear mapping. We should point out here that, although the kernel size does not affect the KLMS steady-state accuracy, it has crucial influence on the convergence rate. In most practical situations, the training data are finite and the algorithm can never reach the steady state. In these cases, the kernel size also has significant influence on the final accuracy (not the steady-state accuracy).

We present here a simple simulation example to verify the obtained theoretical results. Suppose that the training data are generated by the following nonlinear system [30.26]

$$d(i) = \sin(u(i)) + 0.5u(i-1) - 0.1u^2(i-2) + v(i) .$$
(30.33)

The input sequence $\{u(i)\}$ is assumed to be a white Gaussian process with variance 1.0, and $\{v(i)\}$ is a zero-mean white noise that is independent of $\{u(i)\}$. In the simulation, except mentioned otherwise, the step size is set at $\eta = 0.5$, the noise variance is $\xi_v^2 = 0.01$, and the kernel size is $\sigma = 1.0$. For different values of the step size, noise variance, and kernel size, the simulated and theoretical EMSE are illustrated in Figs. 30.4–30.6. Evidently, the experimental and theoretical results agree very well.

Network Size Control

The KLMS filter network grows rapidly with each new sample following a nonparametric approach. Due to finite resources one must cut the growing structure of the filter and constrain the network size (number of the

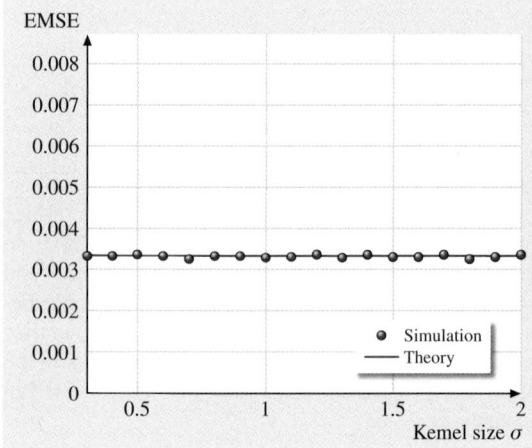

Fig. 30.6 Simulated and theoretical EMSE versus kernel size

centers). Some *sparsification* methods can be applied to cope with this issue. According to these methods, new samples are inserted into the dictionary, only if they satisfy a certain *sparsification criterion*. In the following, we briefly discuss several useful sparsification criteria.

Suppose at i iteration, the dictionary is $C(i) = \{c_1, c_1, \ldots, c_{m_i}\}$, and the coefficient vector $a(i) = \{a_1, a_2, \ldots, a_{m_i}\}$, where c_j is the jth center, a_j is the jth coefficient, and m_i the dictionary size (or network size) at i iteration. In this case, the learned mapping is

$$f_i(.) = \sum_{j=1}^{m_i} a_j(i)\kappa(c_j, u) .$$
(30.34)

When a new example $\{u(i+1), d(i+1)\}$ is presented, the learning system needs to decide whether $u(i+1)$ should be inserted into the dictionary. This decision procedure is, in general, based on some sparsification criterion.

Novelty Criterion. Platt's NC [30.14] first computes the distance of $u(i+1)$ to the present dictionary

$$\text{dis}_1 = \min_{c_j \in C(i)} \|u(i+1) - c_j\| . \tag{30.35}$$

If dis_1 is smaller than some preset threshold $\delta_1 (\delta_1 > 0)$, $u(i+1)$ will not be added into the dictionary. Otherwise, it computes the prediction error $e(i+1) = d(i+1) - f_i(u(i+1))$. Only if the magnitude of the prediction error is larger than another preset threshold $\delta_2 (\delta_2 > 0)$, $u(i+1)$ will be accepted as a new center. If the input domain $\mathbb{U}$ is a compact subset, the NC criterion always produces a dictionary with finite elements.

Coherence Criterion. According to the CC [30.15], the input $u(i+1)$ will be inserted into the dictionary if its coherence remains below a given threshold μ_0, that is

$$\max_{c_j \in C(i)} |\kappa(u(i+1), u(c_j))| \leq \mu_0 . \tag{30.36}$$

ALD Criterion. The ALD uses the distance of the new input to the linear span of the current dictionary in feature space, that is [30.12]

$$\text{dis}_2 = \min_{\forall b} \left\| \varphi(u(i+1)) - \sum_{c_j \in C(i)} b_j \varphi(c_j) \right\| . \tag{30.37}$$

ALD is computationally expensive especially when the dictionary size m_i is very large. In order to simplify the computation, one can use the following approximate distance

$$\text{dis}_3 = \min_{\forall b, \forall c_j \in C(i)} \left\| \varphi(u(i+1)) - b\varphi(c_j) \right\| . \tag{30.38}$$

Surprise Criterion. *Surprise* is a subjective information measure of an example $\{u, d\}$ with respect to a learning system $\mathcal{L}$, which is defined as the negative log likelihood of the example given the learning system's estimate on the data distribution [30.16]

$$S_{\mathcal{L}}(u, d) = -\ln p(u, d \mid \mathcal{L}) , \tag{30.39}$$

where $p(u, d \mid \mathcal{L})$ is the subjective probability of (u, d) hypothesized by $\mathcal{L}$. The surprise $S_{\mathcal{L}}(u, d)$ measures how *surprising* the exemplar is to the learning system. The surprise of the new example $\{u(i+1), d(i+1)\}$ is

$$\begin{aligned} S_{\mathcal{L}(i)}&(u(i+1), d(i+1)) \\ &= -\ln p(u(i+1), d(i+1) \mid \mathcal{L}(i)) , \end{aligned} \tag{30.40}$$

where $\mathcal{L}(i)$ denotes the present learning system. To simplify notation, one can write $S_{\mathcal{L}(i)}(u(i+1), d(i+1))$ as $S(i+1)$.

By the definition, if surprise $S(i+1)$ is large, the new example $\{u(i+1), d(i+1)\}$ contains something new for the system to learn or it is suspicious. Otherwise, if surprise $S(i+1)$ is very small, the new datum is well expected by the learning system $\mathcal{L}(i)$ and thus contains little information to be learned. Usually one can classify the new example into three categories

$$\begin{cases} \text{abnormal: } S(i+1) > T_1 , \\ \text{learnable: } T_1 \geq S(i+1) \geq T_2 , \\ \text{redundant: } S(i+1) < T_2 , \end{cases} \tag{30.41}$$

where T_1 and T_2 are threshold parameters. The choice of the thresholds and learning strategies defines the characteristics of the learning system. In general, a new center will be added only if the example is learnable, i.e., $T_1 \geq S(i+1) \geq T_2$.

Besides the aforementioned sparsification methods, there is another technique, called the *quantization approach*, to reduce the network size of KLMS. By quantization approach, the input space is quantized, if the quantization of the new input has already been assigned a center, no new center will be added, while the coefficient of that center will be updated. This new algorithm is called the quantized KLMS (QKLMS) algorithm [30.17]. The mapping update equation of the QKLMS can be simply expressed as

$$\begin{cases} f_0 = 0 , \\ e(i) = d(i) - f_{i-1}(u(i)) , \\ f_i = f_{i-1} + \eta e(i)\kappa(Q[u(i)], .) , \end{cases} \tag{30.42}$$

where $Q[.]$ is a quantization operator over input space. A simple online vector quantization (VQ) method has also been proposed in [30.17]. The QKLMS algorithm (with simple online VQ) is described in Algorithm 30.2.

Algorithm 30.2 Quantized Kernel Least Mean Square Algorithm

Initialization:
Choose kernel κ, step size η, quantization size ε
$a_1 = \eta d(1), C(1) = \{u(1)\}, f_1 = a_1 \kappa(u(1), .)$
Computation:
while $\{u(i), d(i)\} (i > 1)$ available **do**

1) Compute the prediction error

$$e(i) = d(i) - \sum_{j=1}^{size(C(i-1))} a_j(i-1)\kappa(C_j(i-1), u(i))$$

2) Compute the distance between $u(i)$ and $C(i-1)$

$$dis(u(i), C(i-1)) = \|u(i) - C_{j*}(i-1)\|$$

where

$$j^* = \underset{1 \le j \le size(C(i-1))}{argmin} \|u(i) - C_j(i-1)\|$$

3) **if** $dis(u(i), C(i-1)) \le \varepsilon$, **then**

$$C(i) = C(i-1), \quad a_{j*}(i) = a_{j*}(i-1) + \eta e(i)$$

else

$$C(i) = \{C(i-1), u(i)\}, \quad a(i) = [a(i-1), \eta e(i)]$$

end if

end while
(where $C_j(i-1)$ denotes the jth element of the dictionary $C(i-1)$).

Kernel Maximum Correntropy (KMC) Algorithm

Like most conventional adaptive filtering algorithms, the KLMS adopts the MSE as the optimality cost function. The MSE is mathematically tractable, computationally simple, and optimal for linear Gaussian systems. However, MSE may be a poor cost for nonlinear or/and non-Gaussian (e.g., heavy-tail distributions) situations, since it constraints only the second-order statistics. To cope with this problem, one may use a non-MSE cost, such as a higher order statistics, or an *information theoretic criterion* (entropy, correntropy, divergence, etc.). In particular, the *kernel maximum correntropy* (KMC) algorithm has been developed in [30.31], which is derived by applying the *maximum correntropy criterion* (MCC) to KLMS.

The correntropy defines a new correlation function between two random variables. Let X and Y be two random variables with the same dimensions, the cor-

rentropy is defined by [30.19]

$$V(X, Y) = E_{XY}[\kappa_{corr}(X, Y)]$$
$$= \int \kappa_{corr}(x, y) dF_{XY}(x, y), \tag{30.43}$$

where $\kappa_{corr}(., .)$ is a Mercer kernel (usually Gaussian kernel), and $F_{XY}(x, y)$ denotes the joint distribution function of X, Y. Since any Mercer kernel induces a nonlinear mapping $\varphi(\cdot)$ from the input space to a high-dimensional (possibly infinite) feature space, and the inner product of two points $\varphi(X)$ and $\varphi(Y)$ in feature space can be implicitly computed by using the Mercer kernel, so the correntropy (30.43) can alternatively be expressed as

$$V(X, Y) = E[\langle \varphi(X), \varphi(Y) \rangle], \tag{30.44}$$

where $\langle ., . \rangle$ denotes the inner product in the feature space induced by $\kappa_{corr}(., .)$. Clearly, correntropy is a generalized correlation function and it is also positive definite, i.e., it defines a new RKHS for inference. By a simple Taylor series expansion on the kernel, one can see that correntropy provides a number that is the sum of all the statistical moments expressed by the kernel. In many applications, this sum may be sufficient to quantify better than correlation the relationships of interest and it is much simpler to estimate than the higher order statistical moments. Therefore, it can be considered a new type of statistical descriptor and a new cost function for adaptive system training.

Under MCC criterion, the learning cost function is $V(d(i), y(i)) = E[\kappa_{corr}(d(i), y(i))]$. Dropping the expectation operator, one obtains the instantaneous cost function $\tilde{V}(d(i), y(i)) = \kappa_{corr}(d(i), y(i))$. Thus, a *stochastic gradient algorithm* in RKHS $\mathcal{H}_k$ (which is induced by κ, NOT by κ_{corr}) can be readily derived as follows

$$f_i = f_{i-1} + \eta \frac{\partial}{\partial f_{i-1}} \tilde{V}(d(i), y(i))$$
$$= f_{i-1} + \eta \frac{\partial}{\partial f_{i-1}} \kappa_{corr}(d(i), y(i))$$
$$= f_{i-1} + \eta \frac{\partial}{\partial y(i)} \kappa_{corr}(d(i), y(i)) \frac{\partial}{\partial f_{i-1}} y(i)$$
$$= f_{i-1} + \eta \frac{\partial}{\partial y(i)} \kappa_{corr}(d(i),$$
$$y(i)) \frac{\partial}{\partial f_{i-1}} \langle f_{i-1} \mid \kappa(u(i), .) \rangle_{\mathcal{H}_k}$$
$$= f_{i-1} + \eta \frac{\partial}{\partial y(i)} \kappa_{corr}(d(i), y(i)) \kappa(u(i), .), \tag{30.45}$$

where $\partial/\partial f_{i-1}$ denotes *Frechet's* differential. This algorithm is called the KMC algorithm. If $\kappa_{corr}(.,.)$ is a Gaussian kernel, i.e., $\kappa_{corr}(d(i), y(i)) = \exp(-(e(i)^2/2\sigma^2))$, then KMC (30.45) becomes

$$f_i = f_{i-1} + \eta \frac{\partial}{\partial y(i)} \exp\left(-\frac{e(i)^2}{2\sigma^2}\right)\kappa(\boldsymbol{u}(i), .)$$

$$= f_{i-1} + \frac{\eta}{\sigma^2} \exp\left(-\frac{e(i)^2}{2\sigma^2}\right) e(i)\kappa(\boldsymbol{u}(i), .) .$$

(30.46)

The algorithm of (30.46) is, in fact, a KLMS algorithm with step size $\eta' = \frac{\eta}{\sigma^2} \exp(-(e(i)^2/2\sigma^2))$.

To achieve a better performance, one should select a suitable kernel size for correntropy. Note that there may be two kernel sizes in KMC: the kernel size for the RKHS of filter and the kernel size for the cost function. Here we talk about the latter. A kernel size update rule has been proposed in [30.32], which is

$$\sigma(i+1) = \alpha\sigma(i) + (1-\alpha)\sqrt{\frac{\beta_G}{\beta_e}}\sigma_e^2 ,$$

(30.47)

where $\sigma(i)$ denotes the kernel size of correntropy at iteration i, $0 \le \alpha \le 1$ is a forgetting factor, β_G is the kurtosis of the Gaussian distribution (i.e., $\beta_G = 3$), and β_e and σ_e^2 are, respectively, the kurtosis and variance of the prediction error.

30.3.2 Kernel Recursive Least Squares (KRLS) Algorithm

The *recursive least squares* (RLS) is another popular algorithm in the traditional linear adaptive filtering literature, which recursively updates the estimated autocorrelation matrix of the input signal vector and the cross-correlation vector between the input vector and the desired response. The convergence rate of RLS is, in general, much faster than the LMS algorithm. This improvement in performance, however, is achieved at the expense of an increase in computational complexity. Similar to the LMS algorithm, the RLS algorithm can also be *kernelized*. Next, we will discuss the KRLS algorithm [30.12]. The derivation of KRLS is based on a least squares formulation in the feature space.

Based on a sequence of available examples (up to and including time $i-1$) $\{\boldsymbol{u}(j), d(j)\}_{j=1}^{i-1}$, the regularized least squares regression in $\mathcal{H}_k$ can be formulated as

$$\min_{f \in \mathcal{H}_k} \sum_{j=1}^{i-1} (d(j) - f(\boldsymbol{u}(j)))^2 + \gamma\|f\|_{\mathcal{H}_k}^2 ,$$

(30.48)

where $\gamma \ge 0$ is the regularization factor that controls the smoothness of the solution (to avoid overfitting). Note that in KLMS, the step size performs a similar role as the regularization factor (*self-regularization property*), and hence there is no need to add explicitly a regularization factor in KLMS.

By the representer theorem [30.33], the function f in $\mathcal{H}_k$ minimizing (30.48) can be expressed as a linear combination of the kernels in terms of the available data

$$f(.) = \sum_{j=1}^{i-1} \alpha_j \kappa(\boldsymbol{u}(j), .) .$$

(30.49)

The learning problem can also be defined as finding $\boldsymbol{\alpha} \in \mathbb{R}^{i-1}$ that minimizes

$$\min_{\boldsymbol{\alpha}(i-1) \in \mathbb{R}^{i-1}} \|\boldsymbol{d}(i-1) - \boldsymbol{K}(i-1)\boldsymbol{\alpha}(i-1)\|^2$$
$$+ \gamma\boldsymbol{\alpha}(i-1)^T\boldsymbol{K}(i-1)\boldsymbol{\alpha}(i-1) ,$$

(30.50)

where $\boldsymbol{\alpha}(i-1) = [\alpha_1, \dots, \alpha_{i-1}]^T$, $\boldsymbol{d}(i-1) = [d(1), \dots, d(i-1)]^T$, and $\boldsymbol{K}(i-1) \in \mathbb{R}^{(i-1)\times(i-1)}$ is the Gram matrix with elements $\boldsymbol{K}_{jk} = \kappa(\boldsymbol{u}(j), \boldsymbol{u}(k))$, $j, k = 1, 2, \dots, i-1$. The solution of (30.50) will be

$$\boldsymbol{\alpha}^* = (\gamma\boldsymbol{I} + \boldsymbol{K}(i-1))^{-1}\boldsymbol{d}(i-1) ,$$

(30.51)

where $\boldsymbol{I}$ denotes an identity matrix with appropriate dimension. Of course, the above least squares problem can alternatively be formulated in feature space $\mathbb{F}$

$$\min_{\boldsymbol{\Omega} \in \mathbb{F}} \sum_{j=1}^{i-1} (d(j) - \boldsymbol{\Omega}^T\boldsymbol{\varphi}(j))^2 + \gamma\|\boldsymbol{\Omega}\|^2$$

(30.52)

The solution of (30.52) can be derived as [30.9]

$$\boldsymbol{\Omega}^* = \boldsymbol{\Phi}(i-1)\boldsymbol{\alpha}^*$$
$$= \boldsymbol{\Phi}(i-1)(\gamma\boldsymbol{I} + \boldsymbol{K}(i-1))^{-1}\boldsymbol{d}(i-1) , \quad (30.53)$$

where $\boldsymbol{\Phi}(i-1) = [\boldsymbol{\varphi}(1), \dots, \boldsymbol{\varphi}(i-1)]$ (Hence the Gram matrix $\boldsymbol{K}$ can also be expressed as $\boldsymbol{K} = \boldsymbol{\Phi}^T\boldsymbol{\Phi}$). The KRLS algorithm will update this solution recursively as new data $(\boldsymbol{u}(i), d(i))$ become available.

When the new data $(\boldsymbol{u}(i), d(i))$ are available, the optimal solution of (30.53) becomes

$$\boldsymbol{\Omega}^* = \boldsymbol{\Phi}(i)(\gamma\boldsymbol{I} + \boldsymbol{K}(i))^{-1}\boldsymbol{d}(i) .$$

(30.54)

Denote

$$\mathbf{Q}(i) = (\gamma \mathbf{I} + \mathbf{K}(i))^{-1} = (\gamma \mathbf{I} + \mathbf{\Phi}(i)^T \mathbf{\Phi}(i))^{-1} . \tag{30.55}$$

It is easy to see

$$\mathbf{Q}(i)^{-1} = \begin{bmatrix} \mathbf{Q}(i-1)^{-1} & \mathbf{h}(i) \\ \mathbf{h}(i)^T & \gamma + \boldsymbol{\varphi}(i)^T \boldsymbol{\varphi}(i) \end{bmatrix}, \tag{30.56}$$

where $\mathbf{h}(i) = \mathbf{\Phi}(i-1)^T \boldsymbol{\varphi}(i)$. Using the block matrix inversion identity [30.9], one can derive

$$\mathbf{Q}(i) = r(i)^{-1} \begin{bmatrix} \mathbf{Q}(i-1)r(i) + z(i)z(i)^T & -z(i) \\ -z(i)^T & 1 \end{bmatrix}, \tag{30.57}$$

where $z(i) = \mathbf{Q}(i-1)\mathbf{h}(i)$, $r(i) = \gamma + \kappa(\mathbf{u}(i), \mathbf{u}(i)) - z(i)^T \mathbf{h}(i)$. Then the coefficient vector can be updated as

$$\begin{aligned}
\boldsymbol{\alpha}^*(i) &= \mathbf{Q}(i)\mathbf{d}(i) \\
&= r(i)^{-1} \begin{bmatrix} \mathbf{Q}(i-1)r(i) + z(i)z(i)^T & -z(i) \\ -z(i)^T & 1 \end{bmatrix} \\
&\quad \begin{bmatrix} \mathbf{d}(i-1) \\ d(i) \end{bmatrix} \\
&= \begin{bmatrix} \boldsymbol{\alpha}^*(i-1) - z(i)r(i)^{-1}e(i) \\ r(i)^{-1}e(i) \end{bmatrix},
\end{aligned} \tag{30.58}$$

where $e(i) = d(i) - \mathbf{h}(i)^T \boldsymbol{\alpha}^*(i-1)$ is the prediction error.

Now we have obtained a recursive algorithm to solve the kernel least squares problem, namely, the KRLS algorithm (see Algorithm 30.3). The computational cost of KRLS is $O(i^2)$ per iteration. The KRLS also produces a network with linear growth. All the previously mentioned *sparsification* or *quantization* approaches can still be applied to curb the network growth. Notice that the algorithm presented here is just the basic KRLS algorithm. There are many variants or extensions of KRLS, including the exponentially weighted KRLS (EW-KRLS) [30.9], sliding window KRLS (SW-KRLS) [30.34], fixed-budget KRLS (FB-KRLS) [30.35], extended KRLS (EX-KRLS) [30.13], and so on.

Algorithm 30.3 Kernel Recursive Least Squares Algorithm

Initialization:
   Set the regularization parameter $\gamma > 0$
   $C(1) = \{\mathbf{u}(1)\}$, $\mathbf{Q}(1) = [\kappa(\mathbf{u}(1), \mathbf{u}(1)) + \gamma]^{-1}$,
   $\boldsymbol{\alpha}^*(1) = \mathbf{Q}(1)d(1)$

Computation:
   while $\{\mathbf{u}(i), y(i)\}(i > 1)$ available **do**
      $\mathbf{h}(i) = [\kappa(\mathbf{u}(i), \mathbf{u}(1)), \ldots, \kappa(\mathbf{u}(i), \mathbf{u}(i-1))]^T$
      $z(i) = \mathbf{Q}(i-1)\mathbf{h}(i)$
      $r(i) = \gamma + \kappa(\mathbf{u}(i), \mathbf{u}(i)) - z(i)^T \mathbf{h}(i)$
      $\mathbf{Q}(i) = r(i)^{-1} \begin{bmatrix} \mathbf{Q}(i-1)r(i) + z(i)z(i)^T & -z(i) \\ -z(i)^T & 1 \end{bmatrix}$
      $e(i) = d(i) - \mathbf{h}(i)^T \boldsymbol{\alpha}^*(i-1)$
      $\boldsymbol{\alpha}^*(i) = \begin{bmatrix} \boldsymbol{\alpha}^*(i-1) - z(i)r(i)^{-1}e(i) \\ r(i)^{-1}e(i) \end{bmatrix}$

   end while
(where $C_j(i-1)$ denotes the *j*th element of the dictionary $C(i-1)$).

30.3.3 Kernel Affine Projection Algorithms (KAPA)

The KAPA algorithms are nonlinear extensions of the conventional *affine projection algorithms* (APAs) in kernel space, which include the KLMS and KRLS as special cases [30.9]. Before presenting the KAPA algorithms, we give a brief introduction of the APA algorithms.

Let d be a zero-mean scalar-valued random variable, and $\mathbf{u}$ be a zero-mean $m \times 1$ random variable with a positive-definite covariance matrix $\mathbf{R_u} = E[\mathbf{uu}^T]$. Denote r_{du} the cross-covariance vector $r_{du} = E[d\mathbf{u}]$. Then the weight vector $\mathbf{w}$ that solves

$$\min_{\mathbf{w} \in \mathbb{R}^m} E|d - \mathbf{w}^T \mathbf{u}|^2 + \lambda \|\mathbf{w}\|^2 , \tag{30.59}$$

is given by

$$\mathbf{w}^* = (\lambda \mathbf{I} + \mathbf{R_u})^{-1} r_{du} .$$

The solution $\mathbf{w}^*$ of (30.59) can also be recursively solved using a common gradient-based method

$$\begin{aligned}
\mathbf{w}(i) &= \mathbf{w}(i-1) + \eta[r_{du} - (\lambda \mathbf{I} + \mathbf{R_u})\mathbf{w}(i-1)] \\
&= (1 - \eta\lambda)\mathbf{w}(i-1) + \eta[r_{du} - \mathbf{R_u}\mathbf{w}(i-1)] ,
\end{aligned} \tag{30.60}$$

Table 30.3 Weight update equations of four KAPA algorithms

Algorithms	Update equations
KAPA-1	$\boldsymbol{\Omega}(i) = \boldsymbol{\Omega}(i-1) + \eta \boldsymbol{\Phi}(i)[\boldsymbol{d}(i) - \boldsymbol{\Phi}(i)^T \boldsymbol{\Omega}(i-1)]$
KAPA-2	$\boldsymbol{\Omega}(i) = \boldsymbol{\Omega}(i-1) + \eta \boldsymbol{\Phi}(i)(\varepsilon \mathbf{I} + \boldsymbol{\Phi}(i)^T \boldsymbol{\Phi}(i))^{-1}[\boldsymbol{d}(i) - \boldsymbol{\Phi}(i)^T \boldsymbol{\Omega}(i-1)]$
KAPA-3	$\boldsymbol{\Omega}(i) = (1 - \eta\lambda)\boldsymbol{\Omega}(i-1) + \eta \boldsymbol{\Phi}(i)[\boldsymbol{d}(i) - \boldsymbol{\Phi}(i)^T \boldsymbol{\Omega}(i-1)]$
KAPA-4	$\boldsymbol{\Omega}(i) = (1 - \eta)\boldsymbol{\Omega}(i-1) + \eta \boldsymbol{\Phi}(i)(\lambda \mathbf{I} + \boldsymbol{\Phi}(i)^T \boldsymbol{\Phi}(i))^{-1}\boldsymbol{d}(i)$

Table 30.4 Several kernel learning algorithms related to KAPA

Algorithms	Relation to KAPA
KLMS [30.10]	KAPA-1 ($L = 1$)
NKLMS [30.9]	KAPA-2 ($L = 1$)
NORMA [30.36]	KAPA-3 ($L = 1$)
Kernel Adaline [30.37]	KAPA-1 ($L = N$)
RA-RBF [30.38]	KAPA-3 ($\eta\lambda = 1, L = N$)
SW-KRLS [30.34]	KAPA-4 ($\eta = 1$)
RegNet [30.39]	KAPA-4 ($\eta = 1, L = N$)

or Newton's recursion (for the case $\lambda \neq 0$)

$$w(i) = w(i-1) + \eta(\lambda \boldsymbol{I} + R_u)^{-1}$$
$$\times [r_{du} - (\lambda \boldsymbol{I} + R_u)w(i-1)]$$
$$= (1 - \eta)w(i-1) + \eta(\lambda \boldsymbol{I} + R_u)^{-1}r_{du} . \tag{30.61}$$

If the regularization factor $\lambda = 0$, Newton's recursion should be

$$w(i) = w(i-1) + \eta(\varepsilon \boldsymbol{I} + R_u)^{-1}[r_{du} - R_u w(i-1)] , \tag{30.62}$$

where ε is a small positive number to avoid numerical instability.

Suppose we have access to the observations (training examples) of u and d: $\{u(i), d(i)\}$, $i = 1, 2, \ldots$ Then the APA algorithms can be easily derived by approximating R_u and r_{du} in Algorithms (30.60)–(30.62). Based on the L most recent observations, the covariance matrix R_u and the cross-covariance vector r_{du} can be simply approximated by

$$\begin{cases} \hat{R}_u = \dfrac{1}{L} U(i)U(i)^T , \\ \hat{r}_{du} = \dfrac{1}{L} U(i)d(i) , \end{cases} \tag{30.63}$$

where

$$U(i) = [u(i-L+1), \ldots, u(i)]_{m \times L} ,$$
$$d(i) = [d(i-L+1), \ldots, d(i)]^T .$$

Combining (30.60) and (30.63) yields

$$w(i) = (1 - \eta\lambda)w(i-1) + \eta U(i)[d(i) - U(i)^T w(i-1)] . \tag{30.64}$$

When $\lambda = 0$, (30.64) becomes

$$w(i) = w(i-1) + \eta U(i)[d(i) - U(i)^T w(i-1)] . \tag{30.65}$$

Similarly, combining (30.61) and (30.63), we have

$$w(i) = (1 - \eta)w(i-1)$$
$$+ \eta(\lambda \mathbf{I} + U(i)U(i)^T)^{-1}U(i)d(i)$$
$$= (1 - \eta)w(i-1)$$
$$+ \eta U(i)(\lambda \mathbf{I} + U(i)^T U(i))^{-1}d(i) , \tag{30.66}$$

where the second equation comes from *the matrix inverse lemma*.

Further, using the approximations of (30.63), algorithm (30.62) becomes

$$w(i) = w(i-1) + \eta(\varepsilon \mathbf{I} + U(i)U(i)^T)^{-1}$$
$$\times U(i)[d(i) - U(i)^T w(i-1)] . \tag{30.67}$$

Algorithm (30.67) is equivalent to (by the matrix inverse lemma)

$$w(i) = w(i-1) + \eta U(i)(\varepsilon \mathbf{I} + U(i)^T U(i))^{-1}$$
$$\times [d(i) - U(i)^T w(i-1)] . \tag{30.68}$$

Algorithms (30.65), (30.68), (30.64), and (30.66) are, respectively, referred to as the APA-1, APA-2, APA-3, and APA-4 algorithms.

Reformulating the above APA algorithms in feature space yields the KAPA algorithms [30.11], whose weight update equations are summarized in Table 30.3.

The KAPA algorithms are directly related to many other OKL algorithms [30.9]. Typical examples are presented in Table 30.4.

Part D | 30.3

30.4 Illustration Examples

30.4.1 Chaotic Time Series Prediction

Mackey–Glass (MG) Time Series

First, we consider the Mackey–Glass time series. The time sequence is generated (with a sampling period $T = 6\,\mathrm{s}$) from the following time-delay differential equation [30.9]

$$\frac{\mathrm{d}x(t)}{\mathrm{d}t} = -bx(t) + \frac{ax(t-\tau)}{1+x(t-\tau)^{10}}\,, \qquad (30.69)$$

where $b = 0.1$, $a = 0.2$, and $\tau = 30$. The 10 most recent values ($\boldsymbol{u}(i) = [x(i-10),\dots,x(i-1)]^T$) in the past are used as the input to predict the present value $x(i)$. A segment of 500 samples is used as the training data and another 100 as the testing data. The data are corrupted by additive Gaussian noise with zero mean and variance 0.0016. Figure 30.7 shows the learning curves of LMS and KLMS algorithms. Evidently, the KLMS converges

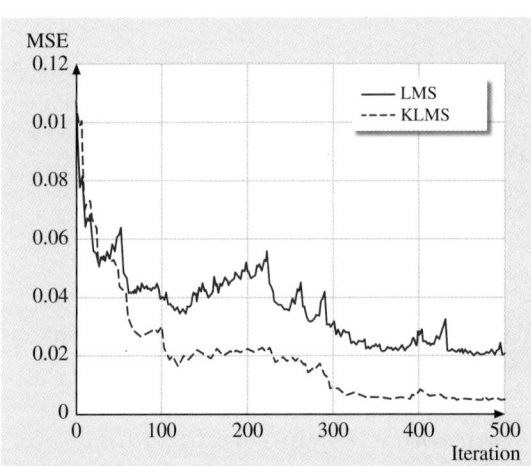

Fig. 30.7 Learning curves of LMS and KLMS in MG time series prediction (adopted from [30.9])

Table 30.5 Performance comparison among LMS, KLMS, and RN

Algorithms	Training MSE	Testing MSE
LMS	0.021 ± 0.002	0.026 ± 0.007
KLMS ($\eta = 0.1$)	0.0074 ± 0.0003	0.0069 ± 0.0008
KLMS ($\eta = 0.2$)	0.0054 ± 0.0004	0.0056 ± 0.0008
KLMS ($\eta = 0.6$)	0.0062 ± 0.0012	0.0058 ± 0.0017
RN ($\lambda = 0$)	0 ± 0	0.012 ± 0.004
RN ($\lambda = 1$)	0.0038 ± 0.0002	0.0039 ± 0.0008
RN ($\lambda = 10$)	0.011 ± 0.0001	0.010 ± 0.0003

to a much smaller value of the testing MSE. This is an expected result as the MG time series is a nonlinear system. In the simulation, the Gaussian kernel is used, and the kernel parameter ($a = 1/(2\sigma^2)$) is set at $a = 1.0$. The step sizes of the LMS and KLMS are both set at 0.2. Table 30.5 presents the performance comparison among LMS, KLMS with different step sizes, and regularization network (RN) with different regularization parameters. The performance of KLMS is much better than LMS and is comparable to RN with the best regularization. This is indeed surprising since RN is a batch mode kernel regression method while KLMS is a simple stochastic gradient algorithm in RKHS.

Lorenz Time Series

Next, we consider the Lorenz chaotic time series, generated from a nonlinear, three-dimensional dynamic system [30.17]

$$\begin{cases} \dfrac{\mathrm{d}x}{\mathrm{d}t} = -\beta x + yz\,, \\[2mm] \dfrac{\mathrm{d}y}{\mathrm{d}t} = \delta(z-y)\,, \\[2mm] \dfrac{\mathrm{d}z}{\mathrm{d}t} = -xy + \rho y - z\,, \end{cases} \qquad (30.70)$$

where the parameters are $\beta = 8/3$, $\delta = 10$, and $\rho = 28$. The sample data are obtained using the first-order approximation with step size 0.01. The state x is picked for short-term prediction task. The signal is preprocessed to be zero mean and unit variance (Fig. 30.8).

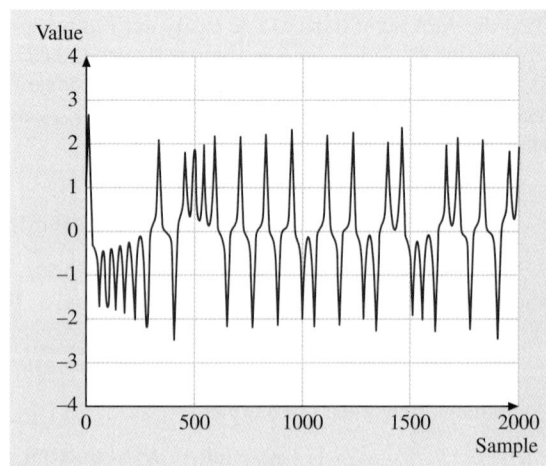

Fig. 30.8 A segment of the processed Lorenz time series

We use the previous five consecutive samples $u(i) = [x(i-5), \ldots, x(i-1)]^T$ to predict the current sample $x(i)$. The performances of QKLMS, KLMS-NC, KLMS-SC, and the standard KLMS are compared. Here, KLMS-NC and KLMS-SC denote the sparsified KLMS with, respectively, the novelty and surprise criterion. The Gaussian kernel with the kernel parameter $a = 1.0$ is used. The step sizes are all set at $\eta = 0.1$, and the other parameters are tuned such that all the algorithms except KLMS yield almost the same final network size (Fig. 30.9). Figure 30.10 shows the average learning curves over 100 simulation runs with different segments of the signal, where the testing MSE

is calculated based on 200 test data (the filter is fixed in the testing phase). Simulation results clearly indicate that the QKLMS exhibits much better performance, achieving almost the same testing MSE as the KLMS but with small network size.

30.4.2 Frequency Doubling

In frequency doubling, both the input and desired data for the learning system are sine waves with frequencies f_0 and $2f_0$, respectively (Fig. 30.11). In this example, 1500 samples are used as the training data and another 200 as the testing data. The data are cor-

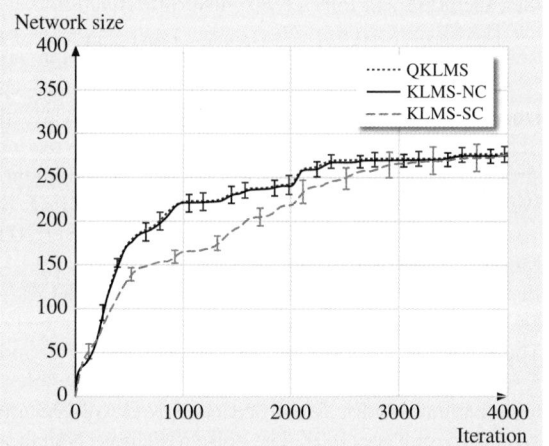

Fig. 30.9 Network sizes of QKLMS, KLMS-NC, and KLMS-SC in Lorenz time series prediction

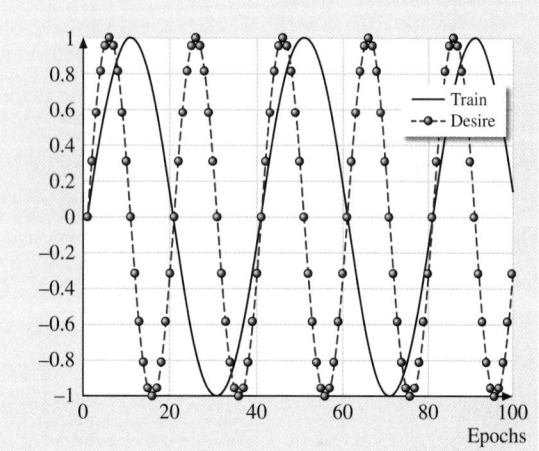

Fig. 30.11 Simulation data in frequency doubling (adopted from [30.31])

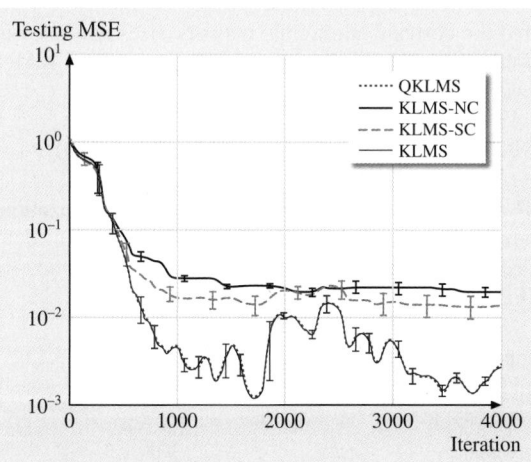

Fig. 30.10 Learning curves of QKLMS, KLMS-NC, KLMS-SC, and KLMS in Lorenz time series prediction

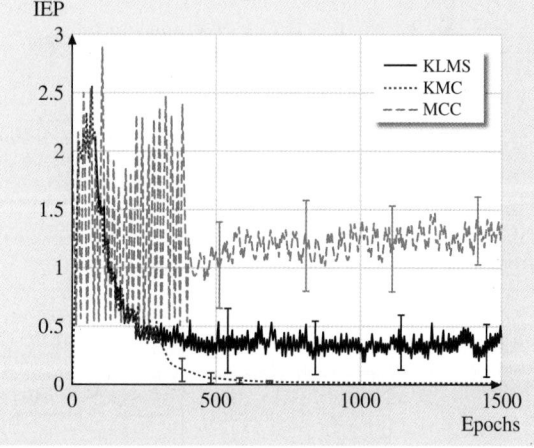

Fig. 30.12 Learning curves of KLMS, KMC, and MCC in frequency doubling (adopted from [30.31])

Part D | 30.4

rupted by an impulsive mixture Gaussian noise, with probability density function $p(x) \approx 0.9\mathcal{N}(0, 0.01) + 0.1\mathcal{N}(2, 0.01)$ [30.31]. Let the dimension of the input vector be 2. Figure 30.12 shows the average learning curves of KLMS, KMC, and MCC (adaptive FIR filter under MCC criterion). It is clear that the KMC algorithm outperforms both KLMS and MCC algorithms. Simulation results suggest that the KMC algorithm performs well under impulsive noise environment.

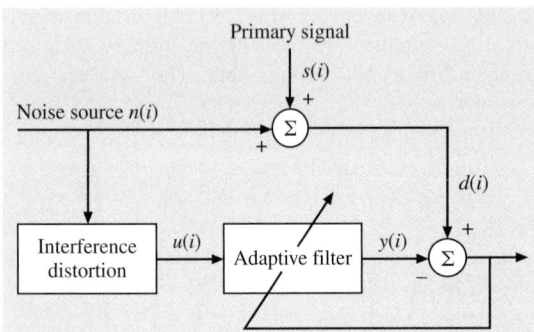

Fig. 30.13 Basic structure of the noise cancellation system

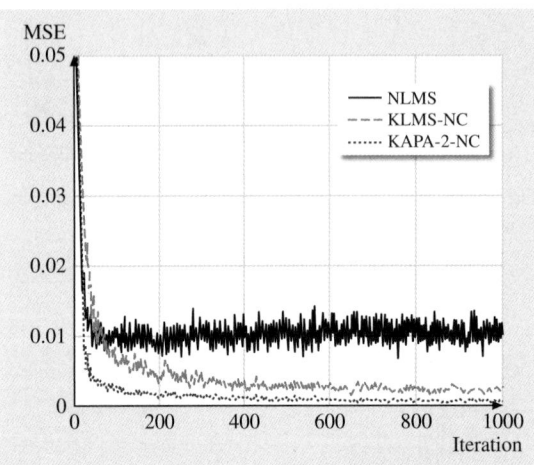

Fig. 30.14 Average learning curves of NLMS, KLMS-NC, and KAPA-2-NC in noise cancellation (adopted from [30.9])

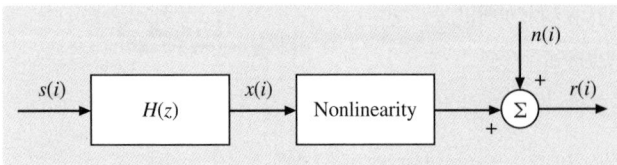

Fig. 30.15 Basic structure of the nonlinear channel

30.4.3 Noise Cancellation

Noise cancellation is very important in signal processing where an unknown interference has to be removed based on some reference measurement. Figure 30.13 shows the basic structure of a noise cancellation system. The goal of the noise cancellation is to use the reference measurement $u(i)$ as the input to the adaptive filter and to obtain the filter output $y(i)$ as an estimate of the unknown noise source $n(i)$, such that the noise can be subtracted from the noisy measurement $d(i)$ to improve the signal-to-noise ratio (SNR).

In this example, the noise source is assumed to be white and uniformly distributed over $[-0.5, 0.5]$. Further, the nonlinear interference distortion function is

$$u(i) = n(i) - 0.2u(i-1) - u(i-1)n(i-1) + 0.1n(i-1) + 0.4u(i-2) . \tag{30.71}$$

During the training phase the primary signal is assumed to be $s(i) = 0$, that is, the system simply tries to reconstruct the noise source from the reference measurement. We use the NLMS, KLMS-NC, KAPA-2-NC ($L = 10$) algorithms. The average learning curves over 200 Monte Carlo simulations are illustrated in Fig. 30.14. In the simulation, the step sizes for NLMS, KLMS-NC, and KAPA-2-NC are 0.2, 0.5, and 0.2, respectively. The Gaussian kernel is used for both KLMS-NC and KAPA-2-NC with kernel parameter $a = 1.0$. The tolerance parameters for KLMS-NC and KAPA-2-NC are $\delta_1 = 0.15$ and $\delta_2 = 0.01$. The noise reduction (NR) factor, defined as

$$10 \log_{10} \frac{E[n(i)^2]}{E[n(i) - y(i)]^2} ,$$

and the corresponding final network sizes are listed in Table 30.6. The performance improvement of KAPA over KLMS is obvious.

30.4.4 Nonlinear Channel Equalization

The final example is on nonlinear channel equalization, where the nonlinear channel consists of a serial connection of a linear filter $H(z)$ and a static nonlinearity (Fig. 30.15). The problem setting is as follows:

Table 30.6 Performance comparison of NLMS, KLMS, and KAPA-2 in noise cancellation

Algorithms	Network size	NR (dB)
NLMS	N/A	9.09 ± 0.45
KLMS-NC	407 ± 14	15.58 ± 0.48
KAPA-2-NC	370 ± 14	21.99 ± 0.80

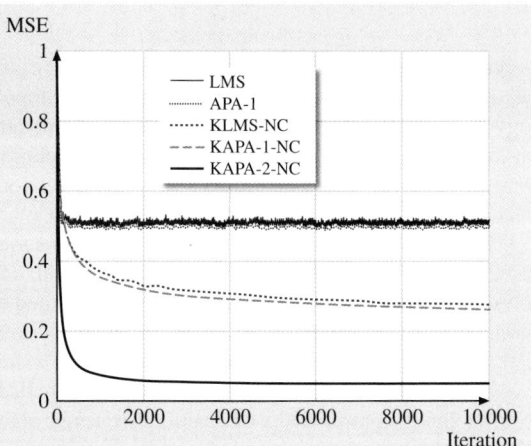

Fig. 30.16 Learning curves of LMS, APA-1, KLMS-NC, KAPA-1-NC, and KAPA-2-NC in nonlinear channel equalization (adopted from [30.9])

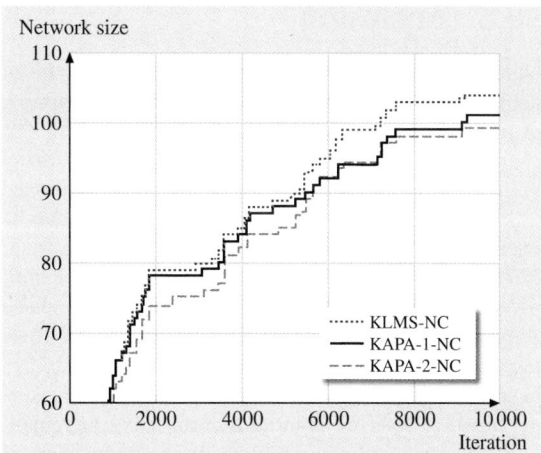

Fig. 30.17 Network sizes of KLMS-NC, KAPA-1-NC, and KAPA-2-NC over training in nonlinear channel equalization (adopted from [30.9])

A binary signal $\{s(1), s(2), \ldots, s(N)\}$ is fed into a nonlinear channel. At the receiver end of the channel, the signal is further corrupted by additive i.i.d. Gaussian noise, and then is observed as $\{r(1), r(2), \ldots, r(N)\}$. The objective of channel equalization is to learn an *inverse filter* that recovers the original signal with as low an error rate as possible. This problem can be formulated as a regression problem with input–output training data $\{(r(t+D), r(t+D-1), \ldots, r(t+D-l+1)), s(t)\}$, where l is the time embedding length, and D is the equalization time lag. In this example, the nonlinear channel is defined by $x(t) = s(t) + 0.5s(t-1)$, $r(t) = x(t) - 0.9x(t)^2 + n(t)$, where $n(t)$ is a white Gaussian noise with variance σ^2.

We compare the performance of LMS, APA-1, KLMS-NC, KAPA-1-NC ($L = 10$) and KAPA-2-NC ($L = 10$). The noise variance is assumed to be 0.01. $l = 3$, and $D = 2$ in the equalizer. For KLMS-NC, KAPA-1-NC and KAPA-2-NC, the Gaussian kernel with kernel parameter 1.0 is used, and the NC is employed with $\delta_1 = 0.26$, $\delta_1 = 0.08$. Figure 30.16 shows the average learning curves over 50 Monte Carlo simulations, where the MSE is calculated between the continuous output (i. e., before taking the hard decision) and the desired signal. Figure 30.17 plots the dynamic changes of the network sizes during the training. In addition, different noise variances are set. To

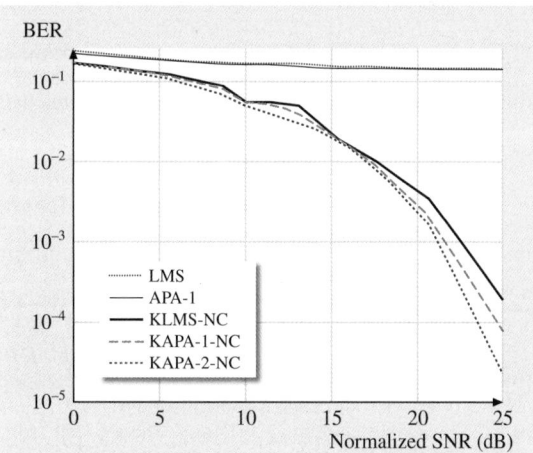

Fig. 30.18 Performance comparison of LMS, APA-1, KLMS-NC, KAPA-1-NC, and KAPA-2-NC with different SNR in nonlinear channel equalization (adopted from [30.9])

make the comparison fair, we tune the NC parameters (δ_1 and δ_2) to make the network size almost the same (around 100) in each scenario. The simulation results in terms of bit error rate (BER) are presented in Fig. 30.18, where the normalized SNR is defined as $10 \log_{10}(1/\sigma^2)$.

Part D | 30.4

30.5 Conclusion

Online learning has found its place in a wide range of applications, especially in situations where the number of training data is extremely large or the data statistics change fast over time. Recent studies suggest that many online learning algorithms can be efficiently extended to kernel space, provided that these algorithms can be expressed in terms of inner products, since the inner products in high-dimensional kernel space can be simply calculated using the kernel function in input space. At present, most of the well-known linear adaptive filtering algorithms, such as the LMS, RLS, and APA, have been *kernelized*. These new algorithms, namely the kernel adaptive filtering algorithms, can solve incrementally arbitrary nonlinear problems in the input space, if the kernel space is rich (high-dimensional) enough to represent the mapping. In general, KAFs naturally create a growing RBF network, learning the network topology and adapting the free parameters directly from data at the same time. However, by sparsifying the solution one can achieve a compact model with small network size even in continuous adaptation situations.

Illustration examples, including chaotic time series prediction, frequency doubling, noise cancellation, and nonlinear channel equalization, have been presented to demonstrate the performance and usefulness of the kernel adaptive filtering algorithms.

The use of the kernel trick and online sparsification techniques to develop other online learning algorithms (supervised, unsupervised, and reinforcement) is a promising direction for future research.

References

30.1   S. Haykin: *Neural Networks and Learning Machines*, 3rd edn. (Prentice Hall, Upper Saddle River 2009)

30.2   E. Alpaydin: *Introduction to Machine Learning* (MIT Press, Cambridge 2004)

30.3   L. Bottou, O. Bousquet: The tradeoffs of large-scale learning. In: *Optimization for Machine Learning*, ed. by S. Sra, S. Nowozin, S.J. Wright (MIT Press, Cambridge 2011) pp. 351–368

30.4   V. Vapnik: *The Nature of Statistical Learning Theory* (Springer, New York 1995)

30.5   B. Scholkopf, A.J. Smola: *Learning with Kernels, Support Vector Machines, Regularization, Optimization and Beyond* (MIT Press, Cambridge 2002)

30.6   B. Widrow, S.D. Stearns: *Adaptive Signal Processing* (Prentice-Hall, Englewood Cliffs 1985)

30.7   S. Haykin: *Adaptive Filtering Theory*, 3rd edn. (Prentice Hall, Upper Saddle River 1996)

30.8   A.H. Sayed: *Fundamentals of Adaptive Filtering* (Wiley, Hoboken 2003)

30.9   W. Liu, J.C. Principe, S. Haykin: *Kernel Adaptive Filtering: A Comprehensive Introduction* (Wiley, Hoboken 2010)

30.10 W. Liu, P. Pokharel, J. Principe: The kernel least mean square algorithm, IEEE Trans. Signal Process. **56**, 543–554 (2008)

30.11 W. Liu, J. Principe: Kernel affine projection algorithm, EURASIP J. Adv. Signal Process. **2008**, 784292 (2008)

30.12 Y. Engel, S. Mannor, R. Meir: The kernel recursive least-squares algorithm, IEEE Trans. Signal Process. **52**, 2275–2285 (2004)

30.13 W. Liu, Il Park, Y. Wang, J.C. Principe: Extended kernel recursive least squares algorithm, IEEE Trans. Signal Process. **57**, 3801–3814 (2009)

30.14 J. Platt: A resource-allocating network for function interpolation, Neural Comput. **3**, 213–225 (1991)

30.15 C. Richard, J.C.M. Bermudez, P. Honeine: Online prediction of time series data with kernels, IEEE Trans. Signal Process. **57**, 1058–1066 (2009)

30.16 W. Liu, Il Park, J.C. Principe: An information theoretic approach of designing sparse kernel adaptive filters, IEEE Trans. Neural Netw. **20**, 1950–1961 (2009)

30.17 B. Chen, S. Zhao, P. Zhu, J.C. Principe: Quantized kernel least mean square algorithm, IEEE Trans. Neural Netw. Learn. Syst. **23**(1), 22–32 (2012)

30.18 J.C. Principe: *Information Theoretic Learning: Renyi's Entropy and Kernel Perspectives* (Springer, New York 2010)

30.19 W. Liu, P.P. Pokharel, J.C. Principe: Correntropy: properties and applications in non-Gaussian signal processing, IEEE Trans. Signal Process. **55**(11), 5286–5298 (2007)

30.20 J.-W. Xu, A. Paiva, I. Park, J.C. Principe: A reproducing kernel Hilbert space framework for information-theoretic learning, IEEE Trans. Signal Process. **56**(12), 5891–5902 (2008)

30.21 E. Moore: On properly positive Hermitian matrices, Bull. Am. Math. Soc. **23**(59), 66–67 (1916)

30.22 N. Aronszajn: The theory of reproducing kernels and their applications, Cambr. Philos. Soc. Proc. **39**, 133–153 (1943)

30.23 J. Mercer: Functions of positive and negative type, and their connection with the theory of integral equations, Philos. Trans. R. Soc. Lond. **209**, 415–446 (1909)

30.24 B. Hassibi, A.H. Sayed, T. Kailath: The H_∞ optimality of the LMS algorithm, IEEE Trans. Signal Process. **44**, 267–280 (1996)

30.25 B.W. Silverman: *Density Estimation for Statistics and Data Analysis* (Chapman Hall/CRC, London 1986)

30.26 B. Chen, S. Zhao, P. Zhu, J.C. Principe: Mean square convergence analysis of the kernel least mean square algorithm, Signal Process. **92**, 2624–2632 (2012)

30.27 I. Steinwart: On the infuence of the kernel on the consistency of support vector machines, J. Mach. Learn. Res. **2**, 67–93 (2001)

30.28 N.R. Yousef, A.H. Sayed: A unified approach to the steady-state and tracking analysis of adaptive filters, IEEE Trans. Signal Process. **49**, 314–324 (2001)

30.29 T.Y. Al-Naffouri, A.H. Sayed: Adaptive filters with error nonlinearities: mean-square analysis and optimum design, EURASIP J. Appl. Signal Process. **4**, 192–205 (2001)

30.30 T.Y. Al-Naffouri, A.H. Sayed: Transient analysis of adaptive filters with error nonlinearities, IEEE Trans. Signal Process. **51**, 653–663 (2003)

30.31 S. Chen, B. Chen, J.C. Principe: Kernel adaptive filtering with maximum correntropy criterion, Proc. Int. Joint Conf. Neural Netw. (IJCNN) (2011) pp. 2012–2017

30.32 S. Zhao, B. Chen, J.C. Principe: An adaptive kernel width update for correntropy, Proc. Intern. Joint Conf. Neural Netw. (IJCNN) (2012), pp. 1–5

30.33 C.J.C. Burges: A tutorial on support vector machines for pattern recognition, Data Min. Knowl. Discov. **2**, 121–167 (1998)

30.34 S. Van Vaerenbergh, J. Via, I. Santamaria: A sliding window kernel RLS algorithm and its application to nonlinear channel identification, IEEE Int. Conf. Acoust., Speech, Signal Process. (ICASSP), Toulouse (2006)

30.35 S. Van Vaerenbergh, I. Santamaria, W. Liu, J.C. Principe: Fixed-budget kernel recursive least-squares, 2010 IEEE Int. Conf. Acoust, Speech Signal Process. (ICASSP), Dallas (2010) pp. 1882–1885

30.36 J. Kivinen, A. Smola, R.C. Williamson: Online learning with kernels, IEEE Trans. Signal Process. **52**(8), 2165–2176 (2004)

30.37 T.-T. Frieb, R.F. Harrison: A kernel-based ADALINE, Proc. Eur. Symp. Artif. Neural Netw. 1999 (1999) pp. 245–250

30.38 W. Liu, P.P. Pokharel, J.C. Principe: Recursively adapted radial basis function networks and its relationship to resource allocating networks and online kernel learning, 2007 IEEE Workshop Mach. Learn. Signal Process., Thessaloniki (2007) pp. 300–305

30.39 F. Girosi, M. Jones, T. Poggio: Regularization theory and neural networks architectures, Neural Comput. **7**, 219–269 (1995)

Part D | 30

31. Probabilistic Modeling in Machine Learning

Davide Bacciu, Paulo J.G. Lisboa, Alessandro Sperduti, Thomas Villmann

Probabilistic methods are the heart of machine learning. This chapter shows links between core principles of information theory and probabilistic methods, with a short overview of historical and current examples of unsupervised and inferential models. Probabilistic models are introduced as a powerful idiom to describe the world, using random variables as building blocks held together by probabilistic relationships. The chapter discusses how such probabilistic interactions can be mapped to directed and undirected graph structures, which are the Bayesian and Markov networks. We show how these networks are subsumed by the broader class of the probabilistic graphical models, a general framework that provides concepts and methodological tools to encode, manipulate and process probabilistic knowledge in a computationally efficient way. The chapter then introduces, in more detail, two topical methodologies that are central to probabilistic modeling in machine learning. First, it discusses latent variable models, a probabilistic approach to capture complex relationships between a large number of observable and measurable events (data, in general), under the assumption that these are generated by an unknown, nonobservable process. We show how the parameters of a probabilistic model involving such nonobservable information can be efficiently estimated using the concepts underlying the expectation–maximization algorithms. Second, the chapter introduces a notable example

Part D | 31.1

of latent variable model, that is of particular relevance for representing the time evolution of sequence data, that is the hidden Markov model. The chapter ends with a discussion on advanced approaches for modeling complex data-generating processes comprising nontrivial probabilistic interactions between latent variables and observed information.

31.1 Probabilistic and Information-Theoretic Methods

Information theory is closely connected to probability theory and statistics. In particular, the standard definition of information contained in a random variable X with a probability density function $P(X)$ is well known to be $I(X) = -\log(P(X))$, with the corresponding *Shannon entropy*, in differential form, given by the average information

$$H(P) = -\int P(X) \log(P(X)) \, dx . \tag{31.1}$$

One of the fundamental theorems of information theory, the second Gibbs theorem, states that the normal distribution achieves maximum entropy, hence maximal average information from all distributions with known variance. To show this in the univariate case, consider the normal distribution in the standard form

$$P(X) = \frac{1}{\sqrt{2\pi\sigma^2}} \exp\left(-\frac{(X-\mu)^2}{2\sigma^2}\right).$$

It is straightforward to show that for the natural logarithm

$$-\int P(X)\log(P(X))\mathrm{d}x = \frac{1}{2} + \log\left(\sqrt{2\pi\sigma^2}\right)$$

$$= -\int G(X)\log(P(X))\mathrm{d}x,$$

where $G(X)$ is any arbitrary density function with variance $\int G(X)(X-\mu)^2\mathrm{d}x = \sigma^2$. Therefore, the difference in average information between the two density functions necessarily observes the following

$$-\int P(X)\log(P(X))\mathrm{d}x + \int G(X)\log(G(X))\mathrm{d}x$$

$$= -\int G(X)\log(P(X))\mathrm{d}x + \int G(X)\log(G(X))\mathrm{d}x$$

$$= -\int G(X)\log\left(\frac{P(X)}{G(X)}\right)\mathrm{d}x$$

$$\geq \int G(X)\left(1 - \frac{P(X)}{G(X)}\right)\mathrm{d}x = 0$$

using Jensen's inequality $\log(x) \leq x-1$ and the normalization property $\int P(X) = \int Q(X) = 1$. This is a particular instance of Gibbs inequality and proves that the asymptotic distribution of the *central limit theorem* also maximizes entropy.

This led, in probability theory, to the definition of natural measures of dissimilarity closely related to the expectation of information difference, e.g., the *Kullback–Leibler* (KL) divergence [31.1]

$$D_{\mathrm{KL}}(P\|Q) = \int P(X)\log\left(\frac{P(X)}{Q(X)}\right)\mathrm{d}x, \qquad (31.2)$$

as generalized distances between probability distributions P and Q.

The KL divergence occurs frequently in machine learning, where the development of learning strategies links information theory with statistical and biologically motivated concepts. For instance, the perceptron model was established as a simple but mathematically tractable model of a biological neuron as the smallest information processing unit in brains [31.2]. Recognition that gradient descent provided a pragmatic but effective solution to the credit assignment problem, namely which values the hidden nodes should have, led to the multilayer perceptron as powerful computational tools for classification and regression. Initially *maximum likelihood* optimization was used for parameter estimation, following the tried and tested statistical concepts of normally distributed errors leading to a sum-of-squares loss function in regression and, for classification, the Bernoulli distribution for binary data and the so-called cross-entropy (31.2) for multinomial class assignments, the latter two likelihood functions measuring information divergence averaged over the true distribution given by the empirical class labels.

Information theoretic aspects (e.g., mutual information) were also considered in neural models in order to avoid overtraining [31.3], for instance in Boltzmann networks which directly mirror information principles in statistical mechanics [31.4]. Related approaches are used currently for deep learning models, where information principles drive the feature representations [31.5].

The correspondence between maximum entropy and maximum likelihood outlined above is just one aspect of the application of information-theoretic concepts in machine learning. The next section outlines further developments linked first to source identification through blind signal separation and matrix factorization methods. These concepts from signal processing identify important degrees of freedom that may be used as hidden variables in probabilistic models, discussed later in the chapter. Furthermore, the application of information-theoretic methods extends also to the automatic identification of prototypes for use in compact data representations that include *dictionaries* defined by methods such as vector quantization, typically with unsupervised approaches.

Supervised methods are introduced as probabilistic models, focusing first on discriminative methods. This indicates that the maximum likelihood approach is limited in its predictive power in generalization to out-of-sample data, because it allows models to be generated with very little bias but with considerable variance – for a more detailed discussion of this point refer [31.6]. What this means in practice is that flexible models such as neural networks are prone to overfitting unless the complexity of the model is controlled along with the extent to which the model fits the data. The latter is described by the likelihood, but the model complexity can be controlled in a number of different ways. In probabilistic models an efficient framework to maximize the generality of probabilistic inference models is to apply the maximum a posteriori (MAP) framework which optimizes the posterior probability of the model parameters given the data but also given prior distributions for the parameters, typically limiting their size by assuming a zero-centred normal distribution as the prior. This is the basis of the method of *automatic relevance determination*, explained in Sect. 31.2.

While discriminative models are efficient approximators for nonlinear response functions, both in regression and in the estimation of class conditional density functions, they are difficult to interpret and can generally be considered as black boxes, meaning that they are not readily interpreted to give insights about the data. A topical and widely used alternative approach is to model the joint distribution of the data directly. This is ideally done by factorization into subgraphs into which the multivariate structure of the data is broken-up using strict conditional independence requirements, as discussed in Sect. 31.2. Inference can then proceed using Bayes theorem introduced in (31.6).

An alternative approach to modeling the joint distribution of the covariates is to use the mutual correlation in the data to identify important degrees of freedom that may be *hidden* in the sense that they are not directly observed. This generates *latent variable* representations that naturally fit into the framework of probabilistic modeling. However, the introduction of additional variables also introduces complexity into the optimization process for estimating their values. This leads naturally to the introduction of *expectation maximization* (EM), a general approach of particular value for estimating mixture models, discussed in Sect. 31.3.

So far the modeling methodologies focus on snapshots of the data, without taking into consideration the time evolution of the covariates. To do this requires explicit parametrization, for which arguably the most widely used probabilistic approach is *hidden Markov models* (HMM). These models are build on the concepts of conditional independence, latent variables, and expectation maximization to model the time evolution of sequences of covariate measurements, in the last substantive Sect. 31.4

31.1.1 Information–Theoretic Methods

While the statistical properties of perceptrons are widely investigated [31.6], the more difficult problem of establishing statistical independence is becoming increasingly important and novel algorithms have been presented during the last decade [31.7]. Their applicability is enormous, ranging from variable selection, to *blind source separation* (BSS) and statistical causality. Frequently, the difficult question of statistical dependence in data is replaced by the easier consideration of estimation and application of data correlations for learning strategies. A recent approach tries to determine independence by generalized correlation functions [31.8]. In this context of decorrelation and independence, BSS and nonnegative matrix factorizations [31.9] of data channels are based on statistical deconvolution. A comprehensive overview for BSS, independent component analysis (ICA) and nonnegative matrix and tensor factorization (NMF) can be found in [31.10–12], respectively. Different aspects can be investigated, like ICA and BSS maximizing conditional probabilities [31.11]. A relevant connection exists between NMF and probabilistic graphical models comprising hidden variables [31.13], which is briefly discussed in Sect. 31.3.4.

Other recent approaches in this field incorporate information theoretic principles directly: *Pham* [31.14] investigated BSS based on mutual information, whereas [31.15] applied β-divergences. The *infomax* principle for ICA was considered in depth [31.16], as was the problem of learning overcomplete data representations and performing overcomplete noisy blind source separation, e.g., the sparse coding neural gas (SCNG) [31.17]. Recent results including modern divergences (generalized α-β-divergences) were recently published [31.18]. Obviously, information theoretic divergence measures like Rényi-divergences (belonging to the family of α-divergences) capture directly the statistical information contained in the data, as expressed by the probability density function [31.19, 20]. This property can be used for unsupervised model estimation for instance in vector quantization, when divergences are used as dissimilarity measure [31.21].

Information optimum vector quantization by prototypes is a widely investigated topic in clustering and data compression, based on the optimization of the γ-reconstruction error

$$E_{VQ}(\gamma) = \int \|\mathbf{v} - \mathbf{w}(\mathbf{v})\|_E^{\gamma} P(V = \mathbf{v}) \, d\mathbf{v} ,$$

where $P(V = \mathbf{v})$ is the data density of the vector data $\mathbf{v}$ and $\|\mathbf{v} - \mathbf{w}(\mathbf{v})\|_E$ is the Euclidean distance of the data vector and the prototype $\mathbf{w}(\mathbf{v})$ representing it. One of the key results concerning information theoretic principles for vector quantization is *Zador*'s magnification law [31.22]: if the data vectors **v** are given in q-dimensional Euclidean space, then the magnification law $\rho \sim P^\alpha$ holds. Here, $\rho(\mathbf{w})$ is the prototype density with the magnification factor

$$\alpha = \frac{q}{q + \gamma} \,.$$

This is the basic principle of vector quantization based on Euclidean distances. For different schemes like self-organizing maps, *Neural Gas* variants with slightly different magnification factors are obtained depending on the choice of neighborhood cooperation scheme applied during prototype adaptation [31.23–25]. Information optimum magnification for $\alpha = 1$ is equivalent to maximum mutual information [31.22]. Yet, it is possible to control the magnification for most of these algorithms by different strategies like localized or frequency sensitive competitive learning. For an overview, we refer to [31.23]. If the Euclidean distance is replaced by divergence measures, optimum magnification $\alpha = 1$ can also be achieved by maximum entropy learning [31.26], or by the utilization of correntropy [31.27]. Vector quantization algorithms directly derived from information theoretic principles based on Rényi entropies are intensively studied in [31.28], also highlighting its connection to graph clustering and Mercer kernel-based learning [31.29].

Other information theoretic vector quantizers optimize the mutual information between data and prototypes, or the respective KL divergence, instead of minimizing a reconstruction error [31.30]. Based on this principle, several data embedding, or dimensionality reduction techniques, have been developed as alternatives to multidimensional scaling. These approaches are frequently used to visualize data. Prominent examples are *stochastic neighborhood embedding* (SNE) [31.31] or variants thereof: for instance, t-SNE uses outlier-robust Student-t-distributions for data characterization instead of Gaussians [31.32]. The generalization to other divergences than KL can be found in [31.33].

Another role for information theory in machine learning is in *feature selection*. Removing irrelevant or redundant features not only leads to a simplification of the model and a reduced requirement for data acquisition, but it is also central for maximizing the generality of the model when it is applied to future data. Most feature selection approaches are supervised schemes, hence using class information or expected regression

values. Strategies to achieve this goal can be classical Bayesian inference schemes of which *automatic relevance determination* (ARD) is a good example (described further in Sect. 31.2), or statistical approaches based on mutual correlation or covariances [31.34, 35]. An alternative approach to feature selection is to use *mutual information*

$$I(X, Y) = D_{\mathrm{KL}}(J(X, Y) \| P(X)Q(Y))$$

between random variables X and Y with probability densities P and Q, respectively, and joint density J [31.36]. Here, the features are treated as random variables to be compared and mutual information measures the information loss resulting from removal of variables from the model. Learning classification together with feature weighting in vector quantization is known as *relevance learning* [31.37]. Recent developments to introduce sparseness according to information theoretic constraints are discussed in [31.38, 39].

Information-theoretic measures such as mutual information, can be explicitly estimated from data [31.40]. This is used in the context of vectorial data analysis to obtain consistent and reliable estimators with topographic maps or kernels [31.41]. Further applications of information theoretic learning also use Rényi entropy

$$H_a(P) = \frac{1}{1 - \alpha} \log \left(\int (P(X))^\alpha \, \mathrm{d}x \right)$$

as a cost function instead of the mean squared error, resorting, for computational efficiency, to Parzen estimators [31.42] or nearest neighbor entropy estimation models. For effective computation of an approximate of the mutual information $I(X, Y)$, the quadratic Rényi entropy $H_2(p)$ or the closely related information energy are common choices [31.43]. Parzen window-based estimators for some information theoretic cost functions have also been shown to be cost functions in a corresponding Mercer kernel space [31.44]. In particular, a classification rule based on an information theoretic criterion has been shown to correspond to a linear classifier in the kernel space. This leads to the formulation of the support vector machine (SVM) from information theory principles.

31.1.2 Probabilistic Models

Kernel models are known for having excellent discrimination performance, but they are typically not well calibrated. This is because they are designed to be efficient binary class allocation models rather than estimators of the posterior probability for membership of each class C. As an example, SVMs allocate inputs to classes on the basis on a binary-valued indicator variable that generally does not have a link function to a probability density estimate. This type of models is known as discriminative models, a well-known variant being Fisher's linear discriminant. As the name implies, the central model is linear in the covariates,

$$y = w^T x$$

optimizing, for binary classification, a discriminant function derived from the mean m_i and variance s_i of each class (i. e., $i = 1, 2$), namely

$$J(w) = \frac{(m_1 - m_2)^2}{s_1^2 + s_2^2}.$$

In general, given the two data cohorts, the covariance matrix of the data S has a strict decomposition into within- and between-class covariance matrices as $S = S_w + S_b$. For an overall data mean vector m and a total of N_j data points in each class, these matrices are given by

$$S = \sum_{i=1}^{N} \left((x_i - m)^T (x_i - m) \right),$$

$$S_w = \sum_{j=1}^{2} \sum_{i=1}^{N} \left((x_i - m_j)^T (x_i - m_j) \right),$$

$$S_b = \sum_{j=1}^{2} \left(N_j (m_j - m)^T (m_j - m) \right).$$

The solution to the optimization of $J(w)$ is

$$w \propto S_w^{-1} (m_2 - m_1),$$

where the inverse of the within-class covariance matrix S_w positions the discriminant hyperplane so as to minimize the overlap between the projections of the data points in each class onto the direction of the weight w. This illustrates the observation that, in general, this projection will not be calibrated with a probabilistic estimate such as the logit

$$\mathrm{logit}(P(C|X)) = \log \left(\frac{P(C|X)}{1 - P(C|X)} \right).$$

The correct calibration is found in a class of generalized linear models of the form

$$y(x) = f(w^T x + w_0),$$

where $f(\cdot)$ is known as the *activation function* in machine learning and its inverse is called a *link function* by statisticians [31.6]. Perhaps the best-known choice of activation is the sigmoid function, where the probabilistic model becomes logistic regression and the linear index $w^T x$ represents exactly the logit $(P(C|X))$. This is very widely used and a generally well-calibrated model, even when severe class imbalance is present.

It is often quoted that generalized linear models are limited by the discriminant forms determined by the linear scores, which must therefore be hyperplanes. However, this ignores the observation that, in most practical applications, suitable attribute representations are defined using domain knowledge, typically by binning variables into discrete states. This turns the probabilistic estimators into linear-in-the-parameter models with significant discrimination potential for nonlinearly separable data. In effect, if the link function is properly tuned to the noise structure of the data and in particular when there are larger numbers of independent covariates, well-designed generalized linear models are competitive with flexible machine learning models, the more so as the limitation of using a linear-in-the-parameters scoring index now works as a form of regularization limiting the complexity of the model. Moreover, the linear index provides a strong element of interpretability whose importance to application domain experts cannot be overestimated. Notwithstanding the power of machine learning, generalized linear models should always be used as benchmarks to set against nonlinear models.

An alternative to probabilistic linear models is the wide range of flexible direct estimators of $P(C|X)$ among which arguably the most widely used model remains the multilayer perceptron (MLP). Similarly to linear statistical models, however, it is important to note that the estimation of class conditional probabilities with an MLP is contingent on using a correct activation function at the output node together with a suitable choice of loss function, which must be one of the entropy functions outlined in the previous section. So, in binary classification, the log-likelihood function with a Bernoulli distribution should be used in conjunction with a sigmoid activation function. In the multinomial case, we would need an extension of the sigmoid function, the softmax activation, together with the cross-entropy as the loss function, since this is the correct measure of the divergence between the estimated and observed probability density functions. Similarly, for nonlinear regression, the activation function should be linear with the usual sum-of-squares error function, provided the inherent noise in the data can be assumed to be normally distributed with zero mean, since this is where the loss function is derived from. In the event where the noise variance, for instance, is dependent on the covariates, heteroscedastic noise models must be used to derive appropriate loss functions [31.6].

While the strength of neural networks is their universal approximation capability, in the sense of fitting any multivariate surface to an arbitrarily small error, this flexibility also makes them prone to overfitting, potentially resulting in data models with little bias but large variance, in direct contrast to generalized linear models. In both cases, it is necessary to control the complexity of the model and this is best done by adding a penalty term to enforce the principle of parsimony, colloquially known as Occam's razor (*lex parsimoniae*). Arguably, the most commonly used and effective scheme is to apply Bayes' theory at the level of fitting the model parameters, then to the regularization hyperparameters, and finally to model selection itself.

As we saw previously, the output of the MLP represents a direct estimate of the posterior probability of class membership $P(C|X)$. This approach can be generalized for the analysis of longitudinal data where each individual subject is follow up over a period of time starting with a defined recruitment point and ending either at the end of a defined observation period or when an event of interest is observed, whichever occurs first. This is often called *survival modeling* and is typically used to estimate event rates in the presence of censorship, e.g., where the outcome of interest, for instance recovery from an illness, is observed in some subjects for only part of the allowed period of follow-up due to other events taking over, such as another condition setting-in, which prevent the observer from ever knowing whether or not the subject would have recovered from the original illness, which is the event of interest. For discrete time, these models can be estimated using the standard MLP with an additional input node coding the time intervals. The output of the MLP again represents a conditional probability, but now the probability of the subject surviving each time interval given that the subject survived until the start of the time interval. This defines the hazard function $h_l(\mathbf{x}_i)$, for subject with covariate vector $\mathbf{x}_i$ and predictions over the lth discrete

time interval, which is given by

$$h_l(\mathbf{x}_i) = P(T \le t_l | T > t_{l-1}, \mathbf{x}_i) \,.$$

For a single event of interest, i.e., a single *risk* factor, the log-likelihood function exactly mirrors that used in binary classification, treating as independent the probability estimates for each of the N subjects and over the discrete time intervals where the subject was observed, i.e., up to the end of the follow-up period or until censorship. This leads to the following loss function

$$L_B = \prod_{i=1}^{N} \prod_{l=1}^{l_i} \left\{ h_l(\mathbf{x}_i)^{d_{il}} \left[1 - h_l(\mathbf{x}_i) \right]^{(1-d_{il})} \right\} \,, \quad (31.3)$$

where the binary indicator variable $d_{il} = 0$ if the event of interest was not observed for the subject during the specific time interval, and is 1 otherwise. This loss function is known as a partial likelihood, since it is measured only over time periods where the outcomes for each subject are observed, an approach that has been extended to the multinomial case to provide a rigorous treatment of censorship with flexible models in the context known as competing risks [31.45].

Application of the Bayesian regularization framework consists in maximizing the posterior probability for the model parameters w, given the data set $\mathcal{D}$, the regularization hyperparameters α and the choice of the model structure, e.g., selected covariates H, namely

$$P(w|\mathcal{D}, \alpha, H) = \frac{P(\mathcal{D}|w, \alpha, H)P(w|\alpha, H)}{P(\mathcal{D}|\alpha, H)} \,. \quad (31.4)$$

The first term on the right-hand side of Eq. (31.4) denotes the probability of the model fitting the data, represented by the exponential of the entropy term discussed in the introduction and defined for longitudinal data by (31.3), hence

$$P(\mathcal{D}|w, \alpha, H) = \mathrm{e}^{-L_B} \,.$$

The second term in (31.4) represents a *prior* distribution of the model parameters typically with a quadratic loss term corresponding to independent zero-mean univariate Gaussian distributions, sometimes called weight decay terms. A particularly efficient implementation of Bayesian regularization is to assign a separate weight decay term to each covariate, indexing the covariates by m of which there are N_α, with the N_m hidden nodes indexed by n. This allows each covariate to be separately turned on or off depending on how informative it is for fitting the observations about the outcome variable, a process known as *automatic relevance determination* (ARD) [31.4]. Expressed in full, this gives

$$P(w|\alpha, H) = \frac{e^{-G(w,\alpha)}}{Z_w(\alpha)}, \quad \text{where} \quad G(w, \alpha)$$

$$= \frac{1}{2}\sum_{m=1}^{N_\alpha} \alpha_m \sum_{n=1}^{N_m} w_{mn}^2 \quad \text{and} \quad Z_w = \prod_{m=1}^{N_\alpha} \left(\frac{2\pi}{\alpha_m}\right)^{\frac{N_m}{2}}.$$

In principle, the best values for the regularization hyperparameters, i.e., the weight decay parameters α, are those which minimize their *posterior* probability

$$P(\alpha|\mathcal{D}, H) = \frac{P(\mathcal{D}|\alpha, H)P(\alpha|H)}{P(\mathcal{D}|H)}.$$

However, the denominator of (31.4) cannot be obtained in closed form, so a Laplace approximation is typically around a stationary point in the loss function as a function of the weights. This amounts to a local Taylor expansion of

$$P(\mathcal{D}|\alpha, H) = \int P(\mathcal{D}|w, \alpha, H)P(w|\alpha, H)\mathrm{d}w$$

$$= \int \frac{e^{-S(w,\alpha)}}{Z_w(\alpha)}\mathrm{d}w,$$

where the linear term in the weights vanishes because of stationarity leading to

$$S^*(w, \alpha) \approx S(w^{\mathrm{MP}}, \alpha) + \frac{1}{2}(w - w^{\mathrm{MP}})^T A(w - w^{\mathrm{MP}}),$$

from which the posterior probability for the hyperparameter results

$$P(\alpha|\mathcal{D}, H) \propto \frac{e^{-S(w^{\mathrm{MP}},\alpha)}}{Z_W(\alpha)}(2\pi)^{\frac{N_w}{2}} \det(A)^{-1/2}.$$

In practice, what this means is that the log-odds ratio, given by the activation of the output node of the MLP can be assumed to have a univariate normal distribution whose variance is given by the Hessian of the matrix S with respect to the weights; g is the gradient of the activation a with respect to the weights, namely

$$P(a|X, \mathcal{D}) = \frac{1}{1}(2\pi s^2)^{1/2}e^{-\left(\frac{(a-a_{\mathrm{MP}})^2}{2s^2}\right)}$$

with a_{MP} denoting the most probable value of the activation function, i.e., the direct output of the MLP without marginalization, and

$$s^2(x) = g^T A^{-1}g.$$

The so-called marginalized estimate of the MLP output is now the posterior distribution integrated over the activation a. In the above expression, g is the gradient of the activation with respect to the network weights and A is the corresponding Hessian; hence the matrix of second partial derivatives. For binary classification and single-risks modeling, this is given by a neat analytical expression

$$h(x_i, l) = \int g(a)P(a|X_i = x_i, l, \mathcal{D})\mathrm{d}a$$

$$= g\left(\frac{a^{\mathrm{MP}}(x_i, l)}{\sqrt{1 + (\pi/8)g^T A^{-1}g}}\right) \tag{31.5}$$

with $g(\cdot)$ denoting the sigmoid function. This adjustment to the original MLP output, i.e., a^{MP}, shows the regularization process in operation: stationary points, where the weights are well defined, have small variance s^2 and therefore their value remains almost unchanged. Conversely, flat valleys in the loss function, where stationary points for the weights have broad Gaussian distributions, are penalized by reducing the value of the argument of the sigmoid function in (31.5) toward nil, reflecting an increase in uncertainty by shifting the MLP output toward the *don't know* threshold.

A probabilistic alternative to discriminative approaches consists of generative models, where Bayes' theorem is once again put into practice to estimate the *posterior* probability of class membership $P(C_k|X)$, from the *class conditional density* functions $P(X|C_k)$ and *prior* probabilities for the classes $P(C_k)$, that is

$$P(C_k|X) = \frac{P(X|C_k)P(C_k)}{\sum_k P(X|C_k P(C_k))}, \tag{31.6}$$

where classes are indexed by k and the sum-rule has been used to expand the denominator. Suitable models for the probability density functions (pdf) of the data given each class will depend on the nature of the data. However, it is straightforward to show for two classes that if the pdfs are normal distributions with equal variance, then the posterior probability will have exactly the functional form of the logistic regression model. This can be taken as an explanation in probabilistic terms of the potential limitations of this linear model, since different classes in practice tend to have distinct variances, even when that data sets for each class are approximately normally distributed. A natural extension of this approach is to use a mixture of Gaussian distributions. This is a very flexible model that can parameterize also multimodal density functions. In the interest of space, we refer the interested reader to a standard textbook [31.6].

The two approaches of discriminative and generative models may be combined by using generative models to build kernels. These kernels define similarity between two covariate vectors $\mathbf{x}$ and $\mathbf{x}'$ by correlation between the respective pdfs, with the values of the kernel function given by $k(\mathbf{x}, \mathbf{x}') = P(X = \mathbf{x}) \cdot P(X' = \mathbf{x}')$ for suitable choices of the probability functions. A ker-

nel so designed will naturally form a Gram matrix. Such kernels lead naturally to the use of latent variables

$$k(\mathbf{x}, \mathbf{x}') = \sum_i P(X = \mathbf{x} | Z = i) P(X' = \mathbf{x}' | Z = i)$$
$$\times P(Z = i) \,,$$

with weighting coefficients $P(Z)$ reflecting the strength of the latent variable Z indexed by i. An example of this approach in practice will be seen in the HMMs later in this chapter (see Sect. 31.4.2).

31.2 Graphical Models

In this section, we give a basic introduction to *graphical models*, a general framework for dealing with uncertainty in a computationally efficient way. Probabilistic models that we treat in the next sections belong to this framework. Here, we introduce the two main classes of graphical models, Bayesian and Markov networks, discussing different methods for performing probabilistic inference. Specific instances of learning within this framework, are given in the next two sections. For the sake of presentation, here we limit our presentation to discrete random variables; however, graphical models can be defined on continuous variables or mixed variables. The material covered in this section is based on [31.6, 46, 47].

A graphical model allows us to represent a family of joint probability distributions in terms of a directed or undirected graph, where nodes are associated with random variables, and edges represent some form of direct probabilistic interaction between variables. Being able to compactly represent the joint probability distribution of a set of random variables $X = \{X_1, \ldots, X_n\}$ is very important: any probabilistic query involving the variables $X_1, \ldots, X_n$ can be answered by knowing their joint probability distribution $P(X_1, \ldots, X_n)$. For example, assume variables to be discrete, and suppose we want to know the posterior probability of X_1 and X_2 given all the other variables, i.e., $P(X_1, X_2 | X_3, \ldots, X_n)$. We can easily answer this query by computing

$$P(X_1, X_2 | X_3, \ldots, X_n)$$
$$= \frac{P(X_1, \ldots, X_n)}{\sum_{\substack{X_1 \in dom(X_1) \\ X_2 \in dom(X_2)}} P(X_1 = x_1, X_2 = x_2, X_3, \ldots, X_n)} \,.$$

Unfortunately, storing the joint probability values associated with all the different assignments $x_1, \ldots, x_n$ is not feasible: if d_i is the size of $dom(X_1)$, all the different assignments are $\prod_{i=1}^{n} d_i$, i.e., an exponential number of entries. This situation, however, constitutes the worst case. In fact, in many application domains, independence properties allow us to factorize the joint distribution into compact parts which can be stored efficiently. Graphical models provide the *language* to compactly represent these factors, enabling in many cases inference and learning over a compact parameterization of the joint distribution as graphical manipulations.

Graphical models can be characterized according to the type of probabilistic interaction between variables they model. Directed graphs (*Bayesian networks*) are used to express causal relationships between random variables (i.e., *cause → effect* relationships), while undirected graphs (*Markov networks*) are better suited to express probabilistic constraints among subset of variables to which it is difficult to ascribe a directionality (graphical models containing both directed and undirected edges are possible; however, they will not be covered here). In both cases, the joint distribution is factorized according to the notion of *conditional independence*.

Definition 31.1 Conditional Independence
Let $\mathcal{X}$, $\mathcal{Y}$, $\mathcal{Z}$ be sets of random variables with $X_i \in \mathcal{X}$, $Y_i \in \mathcal{Y}$, $Z_i \in \mathcal{Z}$. $\mathcal{X}$ is conditionally independent of $\mathcal{Y}$ given $\mathcal{Z}$ (denoted as $\mathcal{X} \perp\!\!\!\perp \mathcal{Y} | \mathcal{Z}$) in a distribution P if, for all values $x_i \in dom(X_i)$, $y_i \in dom(Y_i)$, $z_i \in dom(Z_i)$

$$P(\mathcal{X} = \mathbf{x}, \mathcal{Y} = \mathbf{y} | \mathcal{Z} = \mathbf{z}) = P(\mathcal{X} = \mathbf{x} | \mathcal{Z} = \mathbf{z})$$
$$\times P(\mathcal{Y} = \mathbf{y} | \mathcal{Z} = \mathbf{z}) \,,$$

where $X = \mathbf{x}$ denotes $X_1 = x_1, \ldots, X_{n_X} = x_{n_X}$, $\mathcal{Y} = \mathbf{y}$ denotes $Y_1 = y_1, \ldots, Y_{n_Y} = y_{n_Y}$, $\mathcal{Z} = \mathbf{z}$ denotes $Z_1 = z_1, \ldots, Z_{n_Z} = z_{n_Z}$, and $n_X = |X|, n_Y = |\mathcal{Y}|, n_Z = |\mathcal{Z}|$.

It is not difficult to see that if $X \perp\!\!\!\perp \mathcal{Y}|\mathcal{Z}$, then it is also true that $P(X|\mathcal{Y}, \mathcal{Z}) = P(X|\mathcal{Z})$. In fact, using the product rule for probabilities, we have $P(X, \mathcal{Y}|\mathcal{Z}) = P(X|\mathcal{Y}, \mathcal{Z})P(\mathcal{Y}|\mathcal{Z})$.

In the following, we will discuss how conditional independence is used within Bayesian and Markov networks to factorize the joint distribution. Inference and learning will be discussed as well.

31.2.1 Bayesian Networks

Bayesian networks are directed acyclic graphs used to model causal relationships between random variables: an edge $X_1 \rightarrow X_2$ is used to express the fact that variable X_1 (*cause*) influences variable X_2 (*effect*). The combination of this interpretation in conjunction with the exploitation of conditional independence, where applicable, allows the efficient probabilistic modeling of many relevant application domains. In general, the product rule can be used to factorize the joint distribution of variables $X_1, X_2, X_3, \ldots, X_n$ as

$$P(X_1, X_2, X_3, \ldots, X_n) = \prod_{i=1}^{n} P(X_i|X_1, X_2, \ldots, X_{i-1}) .$$
$$(31.7)$$

The conditional independence relationships can be used to *simplify* the form of each factor in (31.7), i.e., by eliminating variables from the conditioning part, thus drastically reducing the number of probability values that need to be specified to define the factor. For example, if we assume that all the variables are Boolean, then the number of entries needed to define $P(X_n|X_1, X_2, \ldots, X_{n-1},)$ would be 2^{n-1}. If we consider a simple scenario in which the variable X_n is dependent only on X_{n-1}, the corresponding simplified factor becomes $P(X_n|X_1, X_2, \ldots, X_{n-1}) = P(X_n|X_{n-1})$, which only requires two entries.

The *naïve Bayes* model used in classification tasks can be understood as a Bayesian network, where the variable associated with the class label C is the cause and the variables $X_1, \ldots, X_n$ used to describe the attributes of the current input are the effects. The underlying conditional independence assumption is fairly simplistic, but allows a very parsimonious factorization of the joint distribution. By assuming that the class label does not depend on the attributes, and that the at-

tributes are conditionally independent with respect to each other given the class label, i.e., $\forall i, j\, P(X_i, X_j|C) = P(X_i|C)P(X_j|C)$, *naïve Bayes* factorizes the joint distribution as

$$P(C, X_1, X_2, X_3, \ldots, X_n) = P(C) \prod_{i=1}^{n} P(X_i|C) .$$

The details of this model are not discussed in this chapter, but a good didactic reference is [31.6].

In general, after simplification via conditional independence, factors are in the form $P(X_i|X_{j_1}, \ldots, X_{j_k})$, where $X_{j_1}, \ldots, X_{j_k}$ are denoted as *parents* of X_i, and the notation $\text{pa}(X_i)$ is used with the following meaning $\text{pa}(X_i) = \{X_{j_1}, \ldots, X_{j_k}\}$. The factor associated with variable X_i can thus be rewritten as $P(X_i|\text{pa}(X_i))$ and the joint distribution as

$$P(X_1, X_2, X_3, \ldots, X_n) = \prod_{i=1}^{n} P(X_i|\text{pa}(X_i)) . \quad (31.8)$$

The graphical representation of a Bayesian network is shown in Fig. 31.1. The graphical model includes one node for each involved variable. Moreover, a variable that is conditioned (effect) with respect to a parent one (cause) receives a directed edge from that variable. For example, in the Bayesian network represented in Fig. 31.1, we have $\text{pa}(X_7) = \{X_2, X_3\}$, i.e., the set constituted by the two nodes from which X_7 receives an edge. This means that the factor associated with X_7 is $P(X_7|X_2, X_3)$. In Fig. 31.1, we have reported one popular way to specify the parameters of $P(X_7|X_2, X_3)$ when the involved variables are discrete, i.e., the *conditional*

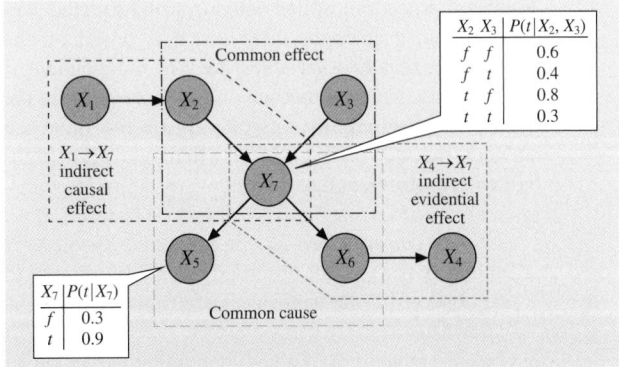

Fig. 31.1 An example of Bayesian network. Conditional probability tables are shown only for variables X_5 and X_7. Different types of probabilistic influence among variables are highlighted

probability distribution table (CPD table). The CPD of X_7 in Fig. 31.1, for instance, reports the probability of $X_7 = t$, given each possible assignment of values to its parents. The CPD table associated with X_5 is reported as well. By using the CPD tables associated to all nodes, the joint distribution can be rewritten as

$$P(X_1, \ldots, X_7) = P(X_1)P(X_3)P(X_2|X_1)P(X_7|X_2, X_3)$$
$$\times P(X_5|X_7)P(X_6|X_7)P(X_4|X_6) \ .$$

Note that different distributions can be obtained by using different values for the entries of the CPD tables. Thus, a Bayesian network is actually representing a family of distributions: all the distributions that are consistent with the conditional independence assumptions used to simplify the factors. In fact, up to now, we have discussed how starting from a *universal* decomposition of the joint distribution via the product rule (note that such decomposition is not unique as it depends on the presentation order assigned to the variables), a set of conditional independence assumptions can be used to simplify the factors, leading to the corresponding graphical representation given by the Bayesian network. An important question, however, is whether the topological structure of a Bayesian network allows for the direct identification of other (conditional) independence relationships, i. e., whether there exist other (conditional) independence relationships that *must* hold for *any* joint distribution P that is compatible with the structure of a specific Bayesian network (note that additional relationships may hold only for *some* specific distributions, i. e., some specific assignment of values to the entries of the CDP tables). As we will see later, the answer to this question is important to devise general-purpose inference algorithms on Bayesian networks. A general procedure, called *d-separation* (directed separation), can answer the question. It is based on the observation that two variables are not independent if one can influence the other via one or more paths in the graph. Let us exemplify this concept on the Bayesian network reported in Fig. 31.1, where we have highlighted four different basic cases:

1. *Indirect causal effect*: X_1 can influence X_7 via X_2 if and only if X_2 is not observed (a variable is said to be observed if the value assigned to that variable is known).
2. *Indirect evidential effect*: X_4 can influence X_7 via X_6 if and only if X_6 is not observed.
3. *Common cause*: X_5 can influence X_6 (and viceversa) via X_7 if and only if X_7 is not observed.

4. *Common effect*: X_2 can influence X_3 (and viceversa) if and only if either X_7 or one of X_7's descendants (in this case, X_5, X_6, X_4) is observed.

The topological structure encountered in the *common effect* is called *v-structure* and it plays a relevant role in the *d-separation* procedure. In general, it is clear from above that probabilistic influence does not follow edge direction. Thus, when considering a longer trail, e.g., the path from X_1 to X_4, we have to consider whether each part of the trail allows probabilistic influence to flow or not (according to the four basic cases described above).

Definition 31.2 Active Trail

Let $X_1, \ldots, X_k$ be a trail in a Bayesian network **G**, and $\mathcal{E}$ be a subset of observed variables in **G**. The trail $X_1, \ldots, X_k$ is active given $\mathcal{E}$ if:

- Whenever a *v*-structure $X_{i-1} \rightarrow X_i \leftarrow X_{i+1}$ does occur, X_i or one of its descendants belong to $\mathcal{E}$;
- No other node along the trail belongs to $\mathcal{E}$.

Of course, by definition, if $X_1 \in \mathcal{E}$ or $X_n \in \mathcal{E}$ the trail is not active. Examples of active/not active trails from the Bayesian network represented in Fig. 31.1 are: the trail X_1, X_2, X_7, X_6, X_4 is active given the set $\mathcal{E} = \{X_3, X_5\}$, while it is not active whenever either X_2 or X_7 or X_6 belongs to $\mathcal{E}$; on the other hand, the trail X_1, X_2, X_7, X_3 is active if $X_2 \notin \mathcal{E}$ and either X_7 or X_5 or X_6 or X_4 belongs to $\mathcal{E}$.

The Bayesian network represented in Fig. 31.1 does not allow more than one trail between any couple of nodes. In general, however, two nodes may have several trails connecting them and one node can influence the other one as long as there exist at least one active trail among them. This intuition is captured by the definition of the concept of *d-separation*.

Definition 31.3 d-Separation

Let $\mathcal{X}$, $\mathcal{Y}$, $\mathcal{Z}$ be nonintersecting sets of nodes of a Bayesian network. $\mathcal{X}$ and $\mathcal{Y}$ are *d*-separated given $\mathcal{Z}$ if there is no active trail between any node $X \in \mathcal{X}$ and $Y \in \mathcal{Y}$ given $\mathcal{Z}$.

The *d*-separation test can be used to precisely characterize the independence relationships which hold for probabilistic distributions that factorize according to the given Bayesian network.

In the following, we introduce another class of graphical models, i.e., Markov networks, which are described by undirected graphs.

31.2.2 Markov Networks

Directed edges in Bayesian networks are suited to describe causal relationships between random variables. In many cases, however, the probabilistic interaction between two variables is not directional. In these cases, it is natural to consider *undirected graphs*, i.e., *Markov networks*. An undirected edge between variables X and Y represents a probabilistic constraint between the two variables. On the other hand, if X and Y are not connected, then we can state a conditional independence assertion involving them if and only if there are no active trails connecting them in the graph. Note that, since edges are now undirected, a trail is not active if and only if any of the variables in the trail is observed. This leads us to discuss which kind of joint distribution factorization a Markov network does represent.

If we go back to the concept of active trail, it is clear that if we consider a subset S of fully connected nodes in the undirected graph, i.e., nodes in S are connected to each other, then any $X, Y \in S$ will be connected by so many trails involving nodes in $S \setminus \{X, Y\}$ that it is wise to consider a single factor ϕ_S involving all nodes in S. Technically, S is called a *clique*, and we are actually interested in *maximal cliques*, i.e., cliques which cannot be extended in size by considering another node of the graph. For example, the maximal cliques of the Markov network given in Fig. 31.2 are

$$c_1 = \{X_1, X_3, X_5\}, c_2 = \{X_1, X_2\},$$
$$c_3 = \{X_2, X_4\}, c_4 = \{X_3, X_4\}.$$

Note that, while $\{X_1, X_5\}$ is a clique, it is not maximal since we can add X_3 obtaining a larger clique.

A different factor can be associated with each maximal clique c_i. By using a global normalization constant for the joint distribution factorization, a factor associated with a clique c_i can be modeled by a *potential function* $\phi_{c_i}(\cdot)$, i.e., any nonnegative function (see Fig. 31.2 for involving Boolean variables). Thus, the factorization of the joint distribution for the example in Fig. 31.2 is

$$P(X_1, X_2, X_3, X_4, X_5) = \frac{1}{Z}\phi_{c_1}(X_1, X_3, X_5)\phi_{c_2}(X_1, X_2)$$
$$\times \phi_{c_3}(X_2, X_4)\phi_{c_4}(X_3, X_4),$$

where the normalization constant

$$Z = \sum_{\forall i, x_i \in X_i} \phi_{c_1}(X_1, X_3, X_5)\phi_{c_2}(X_1, X_2)$$
$$\times \phi_{c_3}(X_2, X_4)\phi_{c_4}(X_3, X_4)$$

is called the *partition function*. If with $\mathbf{x}$ we denote an assignment of values to the variables $X_1, \ldots, X_n$ and with $\mathbf{x}_{c_i}$ the corresponding assignments associated with variables in the clique c_i, the general formulas for a Markov network are

$$P(X_1, \ldots, X_n) = \frac{1}{Z}\prod_{\forall i, c_i} \phi_{c_i}(\mathbf{x}_{c_i}),$$

where

$$Z = \sum_{\mathbf{x}} \prod_{\forall i, c_i} \phi_{c_i}(\mathbf{x}_{c_i}).$$

If the potential functions are restricted to be strictly positive, then it is possible to find a precise correspondence between factorization and conditional independence. In fact, if we consider the set of all possible distributions defined over variables of a given Markov network, then the set of such distributions that are consistent with the conditional independence statements that can be derived by using the adapted concept of active trails and d-separation coincides with the set of distributions that can be expressed as a factorization of the form given above with respect to maximal cliques of the network (*Hammersley–Clifford* theorem).

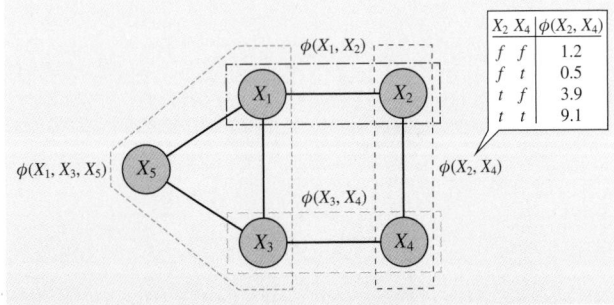

Fig. 31.2 An example of Markov network involving five variables. *Maximal cliques* and corresponding *potential functions* are highlighted. An example of potential function is given for clique $\{X_2, X_4\}$, where we have assumed that X_2 and X_4 are Boolean variables

For practical reasons, it is convenient to express a strictly positive potential function as a *Boltzmann distribution*, i.e.,

$$\phi_{c_i}(\mathbf{x}_{c_i}) = e^{-E(\mathbf{x}_{c_i})} \, ,$$

where $E(\mathbf{x}_{c_i})$ is called an *energy function*. Since the joint distribution is the product of potentials, the total energy is obtained by adding the energy functions of each of the maximal cliques. Energy functions are very useful since, in the absence of a specific probabilistic interpretation for the potential functions, assignments of values that have high probability can be given low energies, while less probable assignments will correspond to high energies.

Let us give an example of application of Markov networks: image de-noising. The task is to remove noise from a binary image $\mathcal{Y}$ where the pixels Y_i are -1 or $+1$. Each observed pixel Y_i is obtained by a noise-free image X with pixels X_i where, with some small probability, the sign of the pixel is flipped. Since neighboring pixels in the noise-free image are strongly correlated, as well as the two variables Y_i and X_i, due to the small flipping probability, we can use a Markov network like the one depicted in Fig. 31.3 to capture this knowledge. The total energy function encoding such prior knowledge would be

$$E(X, \mathcal{Y}) = -\beta \sum_{X_i, X_j \in X} X_i X_j - \eta \sum_{\substack{X_i \in X \\ Y_i \in \mathcal{Y}}} X_i Y_i \, ,$$

where all the maximal cliques are considered and couples of pixels with the same sign get lower energy values. Since we are interested in removing noise from

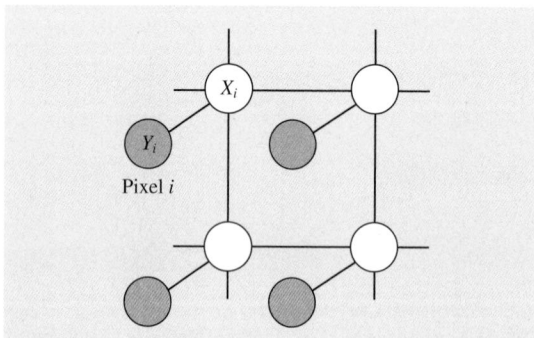

Fig. 31.3 A Markov network for image de-noising. Y_i is the binary variable representing the state of pixel i in the noisy observed image, while X_i refers to the noise-free image

the observed pixels Y_i, we add a bias toward pixel values that have one particular sign, by summing a term hX_i to the energy function for each pixel in the noise-free image

$$E(X, \mathcal{Y}) = h \sum_{X_i \in X} X_i - \beta \sum_{X_i, X_j \in X} X_i X_j - \eta \sum_{\substack{X_i \in X \\ Y_i \in \mathcal{Y}}} X_i Y_i \, .$$

Note that his operation is legal since it corresponds to multiplying the potential function, which are arbitrary nonnegative functions, by a nonnegative function.

The factorized joint distribution over $\mathcal{Y}$ and X is then defined as

$$P(X, \mathcal{Y}) = \frac{1}{Z} e^{-E(X, \mathcal{Y})} \, .$$

Probabilistic inference can now be performed by clamping the value of $\mathcal{Y}$ to the observed image, which implicitly corresponds to a conditional distribution $P(X|\mathcal{Y})$ over free images, and by computing the assignments to X that minimizes the total energy of the Markov model, i.e., the assignment of values to pixels of X with highest probability given the observed image $\mathcal{Y}$. The resulting assignment of values to X will return the (presumed) noise-free version of $\mathcal{Y}$.

In the following, we briefly present different approaches to perform probabilistic inference in Bayesian and Markov networks.

31.2.3 Inference

Performing probabilistic inference in a graphical model over a set of random variables X means being able to answer any probabilistic query involving X. Since a graphical model, either a Bayesian or a Markov network, describes a factorization of the joint distribution, any probabilistic query can be answered, so the problem reduces to find efficient procedures to perform inference. In the following, we report some of the most typical form of queries:

- *Conditional*: In this case, we are interested in computing $P(\mathcal{Y}|\mathcal{E} = \mathbf{e})$, where $\mathcal{Y}, \mathcal{E} \subset X$, with $\mathcal{Y} \cap \mathcal{E} = \emptyset$, where $\mathcal{Y}$ are the *query* variables and $\mathcal{E} = \{E_1, \dots, E_k\}$ are the *evidence* variables for which specific values $\mathbf{e} = \{e_1, \dots, e_k\}$ have been observed.
- *Most probable assignment*: Given evidence $\mathcal{E} = \mathbf{e}$, we are interested in computing the most likely assignment $\mathbf{y}^*$ to $\mathcal{Y} \subseteq X \setminus \mathcal{E}$. There are two main variants for this kind of query: *most probable explanation* (MPE) and *maximum a posteriori* (MAP).

A MPE query must solve the problem

$$\mathbf{y}^* = \arg\max_{\mathbf{y}} P(\mathcal{Y} = \mathbf{y}, \mathcal{E} = \mathbf{e}),$$

where $\mathcal{Y} = \mathcal{X} \setminus \mathcal{E}$, while a MAP query must solve the problem

$$\mathbf{y}^* = \arg\max_{\mathbf{y}} \sum_{\mathbf{z}} P(\mathcal{Y} = \mathbf{y}, \mathcal{Z} = \mathbf{z} | \mathcal{E} = \mathbf{e}),$$

where $\mathcal{Z} = \mathcal{X} \setminus \mathcal{E} \setminus \mathcal{Y}$.

From the point of view of inference, both directed and undirected networks can be treated in the same way. In fact, directed networks can be converted to undirected networks. This is done by observing that factors in directed networks can be understood as factors corresponding to cliques in an undirected graph obtained by mutually connecting all the parents of each node by new undirected edges and by dropping direction from the original directed edges. This procedure is known as *moralization* and the resulting undirected graph is the *moral* graph. By this means, all the variables involved in factors of the directed graph (e.g., CPTs) will be contained in corresponding cliques of the moral graph. Thus, we can focus on undirected graphs.

From a computational point of view, in the worst case, probabilistic inference is difficult: every type of probabilistic inference in graphical models is $\mathcal{NP}$-hard or harder. Specifically, the complexity of inference is related to a topological property of the graphical network called *treewidth*. Approximate inference methods have been devised to deal with such computational complexity. Unfortunately, approximate inference turns out to be hard, in the worst case. Nevertheless, if the *treewidth* of the graphical network is not too large (e.g., in polytrees), exact inference can be performed in a reasonable amount of time. Moreover, in many practical cases, approximate inference is efficient and adequate.

There are three major approaches to perform inference: *exact* algorithms, *sampling* algorithms, and *variational* algorithms. The former tries to compute the exact probabilities while avoiding repeated computations. The second approach aims to efficiently approximate probabilities by sampling, in a smart way, the universe of events. Finally, the third approach allows us to treat both exact and approximate inference within the same conceptual framework. In the following, we briefly sketch the main ideas underpinning these approaches.

Exact Algorithms

Let us illustrate one of the basic ideas of exact algorithms, i.e., *variable elimination*, by using the Markov network shown in Fig. 31.2, where we assume all variables to be Boolean. Suppose we are interested in computing the marginal probability $P(X_2)$. We can get it by summing the factorized joint distribution over the remaining variables

$$P(X_2) = \sum_{x_1} \sum_{x_3} \sum_{x_4} \sum_{x_5} \frac{1}{Z} \phi(X_1, X_3, X_5)$$
$$\times \phi(X_1, X_2)\phi(X_2, X_4)\phi(X_3, X_4).$$

Naïve computation of the above equation would require $O(2^5)$ operations, since each summand involves five Boolean variables. However, we can rearrange the summands in a smarter way

$$P(X_2) = \frac{1}{Z} \sum_{x_1} \phi(X_1, X_2) \sum_{x_4} \phi(X_2, X_4) \sum_{x_3} \phi(X_3, X_4)$$
$$\times \sum_{x_5} \phi(X_1, X_3, X_5)$$
$$= \frac{1}{Z} \sum_{x_1} \phi(X_1, X_2) \sum_{x_4} \phi(X_2, X_4)$$
$$\times \sum_{x_3} \phi(X_3, X_4) m_5(X_1, X_3)$$
$$= \frac{1}{Z} \sum_{x_1} \phi(X_1, X_2) \sum_{x_4} \phi(X_2, X_4) m_3(X_1, X_4)$$
$$= \frac{1}{Z} \sum_{x_1} \phi(X_1, X_2) m_4(X_1, X_2)$$
$$= \frac{1}{Z} m_1(X_2),$$

where the m_i terms are the intermediate factors obtained by summation on variable X_i. Note that Z can be computed by summing on variable X_2. Moreover, the total computational complexity reduces to $O(2^3)$ since no more than three variables occur together in any summand. In general, the maximal number of variables that occur in any summand is determined by the elimination order. Since many different elimination orders may be used, the lowest complexity is obtained by the order that minimizes this maximal number, which is related to the *treewidth* of the graph. Unfortunately, finding the optimal elimination order is $\mathcal{NP}$-hard.

One positive aspect of the elimination approach is that it also works for continuous variables since

$$\boxed{X_1, X_3, X_5} \xrightarrow{X_1, X_3} \boxed{X_1, X_3, X_4} \xrightarrow{X_1, X_4} \boxed{X_1, X_2, X_4} \xrightarrow{X_1, X_2} \boxed{X_1, \underline{X_2}}$$

Fig. 31.4 Example of cluster graph, where the direction of the flow of computation is shown under each edge, while the scope of the computed factor transmitted to the other node after variable elimination is shown over each edge

it is only based on the topology of the graph. However, the elimination procedure returns only a single marginal probability, while it is often of interest to compute more than one marginal probability. Luckily, we can generalize the idea to efficiently compute all the single marginals. Here we give some hints on how to do it. Consider the sequence of intermediate factors generated in the example above. They can be indexed by the variables in their scope, i.e., $\psi_{1,3,5}$ $= \phi(X_1, X_3, X_5)$, $\psi_{1,3,4} = \phi(X_3, X_4)m_5(X_1, X_3)$, $\psi_{1,2,4}$ $= \phi(X_2, X_4)m_3(X_1, X_4)$, $\psi_{1,2} = \phi(X_1, X_2)m_4(X_1, X_2)$. Graphically, we can represent them via a *cluster graph*, where each node is associated with a subset of variables (i.e., the scope of intermediate factors) and the undirected edges *support* the flow of computation of the elimination process. In our example, the cluster graph is shown in Fig. 31.4, where we have shown the direction of the flow of computation under each edge, and the scope of the computed factor *transmitted* to the other node after variable elimination over each edge. The variable X_2 in the rightmost node is underlined to remark that it is the *target* of the flow of computation. In general, since each edge is associated with a variable elimination, it is not difficult to realize that the cluster graph is in fact a tree (called *clique tree* or *junction tree*). This structure can also be used for computing other marginals. In order to see that, we have to observe that the scope of the rightmost node is a subset of the scope of the node at its left, so it can be merged with this last node; moreover, each *initial potential* must be associated with a node with consistent scope, e.g.,

$$\boxed{X_1, X_3, X_5} \overline{\underset{\phi(X_1, X_3, X_5)}{}} \xrightarrow{X_1, X_3} \boxed{X_1, X_3, X_4} \overline{\underset{\phi(X_3, X_4)}{}} \xrightarrow{X_1, X_4} \boxed{X_1, X_2, X_4} \underset{\phi(X_1, X_2)\phi(X_2, X_4)}{}$$

Now, suppose we want to compute $P(X_3)$ by eliminating all the other variables. We have to select a node which contains X_3 in its scope, e.g., the middle node. The flow of computation should now converge toward that node, as shown in

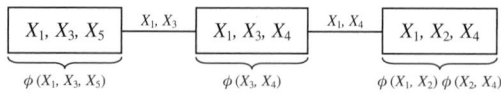

$$\boxed{X_1, X_3, X_5} \xrightarrow[m_5]{X_1, X_3} \boxed{X_1, \underline{X_3}, X_4} \xleftarrow[m_2]{X_1, X_4} \boxed{X_1, X_2, X_4}$$

Any elimination order consistent with the above flow will do the work, e.g., we first consider the leftmost node and eliminate X_5 by transmitting the *message*

$$m_5(X_1, X_3) = \sum_{x_5} \phi(X_1, X_3, X_5)$$

to the middle node. Then, we do the same for the rightmost node, by eliminating X_2 and transmitting the message

$$m_2(X_1, X_4) = \sum_{x_2} \phi(X_1, X_2)\phi(X_2, X_4) .$$

Finally, the middle node can merge the two received messages with the local potential obtaining

$$\phi(X_3, X_4)m_5(X_1, X_3)m_2(X_1, X_4) ,$$

which is an unnormalized version of the joint distribution $P(X_1, X_3, X_4)$. Marginal $P(X_3)$ can then be computed by summing out X_1 and X_4 and normalizing the result. Note that the same flow can be used to compute $P(X_1)$ and $P(X_4)$: in the first case, the final stage will sum out X_3 and X_4, while in the second case it will sum out X_1 and X_3.

In general, *all* the factors needed by *all* the nodes to compute the marginals of the variables in their scope, can be computed by a *sum-product message passing* scheme where, having selected an arbitrary node as *root*, messages are transmitted from the leaves up to the root and then back from the root to the leaves. If evidence is present, *restricted* potentials (i.e., potentials where evidence variables are bound to the observed values) are used. MEP and MAP queries can be answered by using a *max-sum algorithm*, which is a variation of the sum-product algorithm exploiting a *trellis* over all the values the variables can take. The message passing scheme sketched above can also be implemented using *division*, giving raise to the *Belief Update* algorithm.

Sampling Algorithms

The strategy adopted by sampling algorithms to perform (approximate) inference is to approximate the joint distribution via estimates computed on a set of *representative* instantiations of all, or some of, the variables of the graphical model. Unlike exact inference, some techniques are specialized for directed networks. For example, a simple approach to estimate the joint probability in a Bayesian network is *Forward Sampling*. It starts by considering any topological ordering of the variables, e.g., for the network in Fig. 31.1 the order $X_1, X_3, X_2, X_7, X_5, X_6, X_4$ will do the job. Then random samples are generated by following the order and by picking a value for each variable according to its distribution. Note that variables with conditional distributions will be considered only when specific values for their parents have already been generated, so that the conditional probability for those variables is fully specified. Once M full samples are generated in this way, the probability of a specific event $P(\mathcal{E} = \mathbf{e})$ is estimated as the fraction of samples where variables in $\mathcal{E}$ take values $\mathbf{e}$. If the query is of the form $P(\mathcal{Y}|\mathcal{E} = \mathbf{e})$, samples which are not consistent with the evidence are rejected (*rejection sampling*) and the remaining samples used to estimate the conditional distribution on variables $\mathcal{Y}$. With this approach, however, a large amount of generated samples are discarded.

An improvement on this aspect is given by the *likelihood weighting* algorithm, which is based on the observation that evidence variables can be forced to assume *only* the observed values in a sample as long as the sample is weighted by the likelihood of the evidence. This means that a weight is associated with each sample and the weight is given by the product of all the posterior probabilities corresponding to the observed values for the evidence variables, i.e.,

$$w_{\text{sample}} = \prod_{E_i \in \mathcal{E}} P(E_i = e_i | \text{pa}(E_i)) .$$

Estimates are then computed considering weighted samples. Likelihood weighting turns out to be a special case of a more general approach called *importance sampling* which aims at estimating the expectation of a function relative to some distribution.

Improved sampling methods, which can also be applied to Markov networks, are given by *Markov chain Monte Carlo* methods. Unlike the methods described so far, these methods generate a *sequence* of samples, in such a way that later samples are generated by distributions that provably approximate with increasing precision the target posterior probability (i.e., the query $P(\mathcal{Y}|\mathcal{E} = \mathbf{e})$).

The simpler method uses *Gibbs sampling*: an initial assignment of values for the unobserved variables

is generated from an initial distribution; subsequently, in turn, each unobserved variable is sampled using the posterior probability *given* the current sample for all other variables. This distribution can be computed efficiently by using only factors associated with the *Markov blanket*, i.e., the neighbors of the variable to be resampled in the Markov network (in Bayesian networks, the Markov blanket of a node is given by the set of its parents, its children and the parents of its children). Using the theory of *Markov chains* (discussed in Sect. 31.4.1), it is possible to show that, under some assumptions, the sequence of generated distributions converges to a stationary distribution, where the fraction of time in which a specific assignment of values to variables (sample) does occur in the sequence is exactly proportional to the posterior probability of that assignment.

A drawback of Gibbs sampling is that it uses only *local moves* (i.e., resampling of a single variable), leading to very slow convergence for assignments with low probability. More effective methods, based on the *Metropolis–Hastings* approach, enable for a broader range of moves. Further, more advanced approaches allow us to consider partial assignments in conjunction with a closed-form distribution for unassigned variables. Others use deterministic methods to explicitly search for high-probability assignments to approximate the joint distribution.

Variational Algorithms

Probabilistic inference can be formulated as a constrained optimization problem. This allows both to rediscover exact inference algorithms, such as the ones we have briefly discussed above, and to design approximated inference algorithms, by simplifying either the objective function to optimize and/or the admissible region for optimization. The possibility to devise theoretically founded approximation algorithms is particularly appealing in cases where the joint distribution is characterized by a factorization with associated large treewidth. Research in this area has been recently very active, yielding to several interesting results. Here we do not have the space for a proper technical treatment, so we try to give only a brief introduction to the main ideas.

Variational approaches are based on the idea of approximating an intractable probabilistic distribution with a simpler one, which allows for inference. This simpler distribution is selected from a family of tractable distributions, as the distribution that is the *best* approximation to the desired one. Can we define a measure of the quality of the approximation that can be used for the minimization process? A good measure is the KL-divergence introduced in (31.2). Let us denote a distribution that factorizes according to the graphical model G as

$$P_G(X) = \frac{1}{Z} \prod_{\forall i, c_i} \phi_{c_i}(\mathbf{x}_{c_i}) \qquad (31.9)$$

and let $Q(X)$ be a member of the tractable distributions we use to approximate $P_G(X)$. Then, a nice feature of KL-divergence is that it allows us to efficiently solve the optimization problem

$$\arg \min_{Q(X)} D_{KL}(Q(X) \| P_G(X))$$

without requiring to perform inference in $P_G(X)$. In fact, using the factorization of $P_G(X)$ in (31.9), it is not difficult to show that

$$D_{KL}(Q(X) \| P_G(X)) = \log Z - \sum_{\forall i, c_i} \mathbb{E}_Q[\log \phi_{c_i}]$$
$$+ \mathbb{E}_Q[\log Q(X)], \quad (31.10)$$

and, since $\log Z$ does not depend on $Q(X)$, minimizing $D_{KL}(Q(X) \| P_G(X))$ is equivalent to maximizing the *energy functional* term

$$\sum_{\forall i, c_i} \mathbb{E}_Q[\log \phi_{c_i}] - \mathbb{E}_Q[\log Q(X)] .$$

Following from the definition in (31.1), $H_Q(X) = -\mathbb{E}_Q[\log Q(X)]$ is the entropy of Q, while the first term in (31.10) is referred to as *energy term*.

Different variational methods correspond to different strategies for optimizing the energy functional. The name *variational* is used since all of them adopt the general strategy of reformulating the optimization problem by introducing new variational parameters to be used for optimization. In particular, each specific choice of values for the variational parameters expresses one member, i.e., $Q(X)$, of the family of tractable distributions we want to use. The optimization procedure searches the space of variational parameters to find the $Q^*(X)$ that best approximates $P_G(X)$. It is important to understand that the family of tractable distributions will actually corresponds to a set of constraints, involving the variational parameters that must be satisfied while maximizing the energy functional. By using *Lagrange multipliers* these constraints can be merged together with the energy functional, giving rise to a *Lagrangian* function that must be maximized. By taking the partial derivatives with respect to the variational parameters and the Lagrange multipliers, the solution to the optimization problem can be characterized by a set of *fixed-point equations*. These equations can then be used to straightforwardly devise an iterative solution.

Different variational methods work with different types of approximations. There are two main sources of approximation, which can be used singularly or in conjunction. One source is the energy functional, which can be substituted by a functional easy to manipulate while preserving a good degree of approximation. Another source of approximation are the constraints, i.e., the definition of the family of tractable distributions, which may not be fully consistent with the factorization represented by the graphical model (in this case, denoted as pseudo-distributions).

We do not have space here to give more details; however, it is worth to mention that while convergence proofs of several variational methods are available, it is not so common to find theoretical guarantees on the approximation error made by the specific method.

31.3 Latent Variable Models

Knowledge hindered in the complex relation between a large number of observable variables can be surfaced under the assumption that a simpler and unobservable process exists, which is responsible for generating the complex behavior of manifest data. Such an unobservable generative process can be modeled through the use of *latent variables*, as opposed to observable variables, that are not directly measurable, but can be inferred from observations and can explain the relation between manifest data. Intuitively, latent variables can be un-derstood as an attempt to model the unknown physical process generating the observations or as an abstraction providing a simplified representation of the manifest data, e.g., clusters.

Probabilistic models that attempt to explain observations in terms of latent variables are called *latent variable models*. In probabilistic terms, the simplification introduced by latent variables results in conditional independence assumptions, such that (subsets of) observable variables can be considered conditionally independent when their hidden explanation, i. e., the latent variable assignment, is given. Similarly to observed variables, latent variables can be discrete or continuous: their nature, together with that of the observations, determines different types of probabilistic models. Nevertheless, parameter estimation in the different latent variable models can be achieved through a general iterative principle, known as *expectation–maximization*.

31.3.1 Latent Space Representation

To understand the intuition at the basis of latent space representation, consider a joint distribution $P(X) = P(X_1, \ldots, X_N)$ defined over N joint observed random variables X_i. As discussed in Sect. 31.2.1, without any simplifying assumption, the number of free parameters of this simple model grows as $O(2^{N-1})$ for Boolean variables, which quickly becomes unmanageable for large N. One way to control the number of free parameters of a model, without taking too simplistic assumptions (e.g., X_i being i.i.d.), is to introduce a collection of *latent*, or hidden, variables $Z = \{Z_1, \ldots, Z_K\}$. The latent variables are unobserved but can be used to factorize the joint distribution $P(X)$ while allowing to capture (some of) the correlations between the $X = \{X_1, \ldots, X_N\}$ observed variables. More formally, latent variables are such that

$$P(X) = \int_z P(X|Z = z)P(Z = z)dz \,, \qquad (31.11)$$

that is the general formulation for the likelihood of a *latent variable model*. The details of the latent variable model, and the tractability of the integral in (31.11), are determined by the form of the conditional distribution $P(X|Z)$ and by the marginal probability $P(Z)$. A common approach in latent variable models is to assume that observed variables become conditionally independent given the latent variables, that is

$$P(X) = \int_z \prod_{i=1}^N P(X_i|Z = z)P(Z = z)dz \,. \qquad (31.12)$$

A basic assumption for this latent model to be effective, is that the conditional and marginal distributions should be more *tractable* than the joint distribution $P(X)$. For instance, in a simple scenario with discrete observations and latent variables, this entails that $K \ll N$. Not surprisingly, the same intuition is applied, in a deterministic context, for dimensionality reduction (cf. the number of projection directions in PCA) and clustering.

Different types of latent variable models are defined based on the nature of the latent and observed variables, as well as depending on the form of the conditional and marginal probabilities. In the following, we discuss two general classes of latent variable models with continuous and discrete hidden variables, which are *factor analysis* and *mixture models*, respectively.

31.3.2 Learning with Latent Variables: The Expectation–Maximization Algorithm

Learning, in a probabilistic setting, entails working with the model likelihood. In latent variable models, the likelihood in (31.11) might be difficult to treat due to the marginalization inside the logarithm, which can potentially couple all the model parameters. Despite the diversity of the models that can be designed, based on the general expression in (31.11), there exist a general principle to estimate their parameters.

The *expectation–maximization* (EM) algorithm [31.48] is a general iterative method for the maximization of the likelihood under latent variables. The key intuition of the EM algorithm is to define an alternative objective function where the parameter coupling introduced by the marginalization of the hidden variables is removed. The EM algorithm maximizes the marginal data likelihood $P(X|\theta)$, where θ are the model parameters, through a tractable lower bound defined by introducing a function of the latent variables, i. e., $Q(Z)$, into the data likelihood through marginalization. For notational simplicity, consider the case of discrete latent variables. For any nonzero distribution $Q(Z)$, it holds

$$\begin{aligned} \mathcal{L}(\theta) &= \log P(X|\theta) \\ &= \log \sum_z P(X, Z = z|\theta) \\ &= \log \sum_z Q(z) \frac{P(X, Z = z|\theta)}{Q(z)} \\ &\geq \sum_z Q(z) \log P(X, Z|\theta) \\ &\quad - \sum_z Q(z) \log Q(z) = \tilde{\mathcal{L}}(Q, \theta) \,, \qquad (31.13) \end{aligned}$$

where the lower bound $\tilde{\mathcal{L}}(Q, \theta) \leq \mathcal{L}(\theta)$ is obtained by the application of the Jensen inequality to the concave log function. The joint distribution $P(X, Z|\theta)$ is known as the *complete data likelihood*, where the term *complete* refers to the fact that the marginal data likelihood $P(X|\theta)$ is *completed* with the observations **z** for the latent variables.

The *Expectation–maximization* algorithm defines an alternate optimization process where the bound $\tilde{L}(Q, \theta)$ is maximized with respect to $Q(\cdot)$ and θ. In general, this is performed by two independent maximization steps that are repeated until convergence:

- Expectation (E) Step: For θ fixed, find the distribution $Q^{(t+1)}(\mathbf{z})$ that maximizes the bound $\tilde{L}(Q, \theta^{(t)})$;
- Maximization (M) Step: Given the distribution $Q(\mathbf{z})^{(t+1)}$, estimate the model parameters $\theta^{(t+1)}$ that maximize the bound $\tilde{L}(Q^{(t+1)}, \theta)$;

where the superscript denotes the estimate at time t. Clearly, the optimal solution for the *E-step* is attained when

$$Q^{(t+1)}(\mathbf{z}) = P(Z = \mathbf{z}|X, \theta^{(t)}) , \qquad (31.14)$$

that is when the lower bound in (31.13) becomes an equality. In practice, to explicitly evaluate the complete likelihood in $\tilde{L}(Q, \theta^{(t)})$, we would need to observe the **z** assignments. These are unknown, since latent variables are unobservable. However, given the marginalization of **z** in (31.13), we can substitute the unavailable **z** observations with their expected values, by considering them as another random variable. To this end, it suffices that the E-step computes the expected value of the *complete log-likelihood* $\log P(X, Z|\theta)$ with respect to Z. These observations provide the final form of the *classical EM* algorithm:

- *E*-step: Given the current estimate of the model parameters $\theta^{(t)}$, compute

$$Q^{(t+1)}(\theta|\theta^{(t)}) = E_{Z|X, \theta^{(t)}}[\log P(X, Z|\theta)] ;$$
$$(31.15)$$

- *M*-step: Find the new estimate of the model parameters

$$\theta^{(t+1)} = \arg\max_{\theta} Q^{(t+1)}(\theta|\theta^{(t)}) . \qquad (31.16)$$

In other words, the *E-step* estimates the value of the otherwise unobserved latent variables, while the *M-step* finds the parameters that maximize the current estimate of the log-likelihood. In practice, the *E-step* often reduces to estimating the expectation of Z as its posterior

$P(Z|X, \theta^{(t)})$, while the *M-step* uses these values as sufficient statistics to update the model parameters $\theta^{(t+1)}$. This alternate optimization is typically iterated until the log-likelihood does not change much between consecutive estimates, or when a number of maximum iterations is reached. Note that the two-step EM optimization process is prone to local optima. Hence, its convergence can be slow and, often, its solutions tend to be dependent on the initialization.

The EM algorithm assumes that we can calculate the expected value of the complete log-likelihood. However, there are cases in which the required summation is not computationally feasible (e.g., with infinite summations where the integral has no close-form solution): in this cases, the approximated inference methods described in Sect. 31.2.3 can be used to define nonexact EM algorithms. For instance, stochastic versions of the EM algorithm are obtained by approximating the infeasible summation using (e.g., Gibbs) sampling from the posterior distribution $P(Z|X, \theta)$. The classical EM algorithm is a ML method providing point estimates of the model parameters θ. The *variational Bayes* (VB) [31.6] method has been introduced to obtain a fully Bayesian solution that returns a posterior distribution of the parameters $P(\theta)$, instead of their point estimate. VB is based on an analytical approximation of the joint posterior of the latent variables and model parameters that yields to a generalization of the EM alternate optimization, where the maximization at the *M-step* is taken over possible distributions $Q(\theta)$, instead of on θ itself.

31.3.3 Linear Factor Analysis

Factor analysis (FA) is an example of a latent variable model for continuous hidden and manifest variables. In its simplest linear form, it is a classical statistical model widely used for generative dimensionality reduction. Similarly to its deterministic counterparts, e.g., PCA, it forms a low-dimensional embedding of a set of observations $\mathcal{D} = (\mathbf{x}_1, \dots, \mathbf{x_n})$, where each observation **x** is a D-dimensional vector of reals. FA finds a lower dimensional probabilistic representation of $\mathcal{D}$, by assuming that the features of each **x** are independently generated by K real-valued latent variables $Z = \{Z_1, \dots, Z_K\}$, with $K \ll D$ (see the associated graphical model in Fig. 31.5).

The FA model, assumes that observations are linked to the latent vectors through a linear model

$$\mathbf{x} = F\mathbf{z} + b + \epsilon , \qquad (31.17)$$

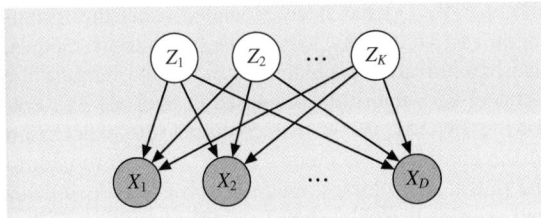

Fig. 31.5 Linear factor analysis: the observed D-dimensional variable X is related to the K latent variables $Z = \{Z_1, \ldots, Z_K\}$ through a linear mapping

where $\epsilon \sim \mathcal{N}(\epsilon|0, \Psi)$ is the Gaussian distributed noise with zero mean and covariance Ψ, b is a bias vector and F is the factor loading matrix. The latent variables are the *factors* and are generally assumed to be distributed as $Z \sim \mathcal{N}(\mathbf{z}|0, I_K) = P(Z)$, where I_K is the K-dimensional identity matrix. Under such Gaussian assumptions, and given the linear model in (31.17), the conditional distribution of the observations is

$$P(X = \mathbf{x}|Z = \mathbf{z}) = \mathcal{N}(\mathbf{x}|F\mathbf{z} + b, \Psi), \qquad (31.18)$$

which, inserted in (31.11), provides the distribution for the FA complete likelihood

$$P(X) = \int_{\mathbf{z}} P(X|Z)P(Z)d\mathbf{z} = \mathcal{N}(\mathbf{x}|b, FF^T + \Psi).$$
$$\qquad (31.19)$$

The form of the noise covariance Ψ determines the type of FA model: in general, this is chosen as a diagonal matrix with a vector of $(\psi_1, \ldots, \psi_D)$ values on the main diagonal. When the diagonal elements are all equal to a single value $\sigma^2 \in \mathbb{R}$, the FA reduces to the special case of the *Probabilistic PCA* [31.49].

Learning of the FA parameters $\theta = (\Psi, F)$ (b is usually set a priori to the mean of the data) is obtained by maximum likelihood estimation. The most popular approach to obtain such estimates is based on solving an eigen-decomposition problem. Given the nature of FA as a latent variable model, its θ parameters can also be estimated by applying EM to the logarithm of the complete likelihood in (31.19). The latter approach is, however, less used in general, given its slower convergence.

31.3.4 Mixture Models

The term *mixture models* identifies a large family of latent variable models comprising discrete hidden variables and generic manifest variables. A mixture model assumes that each observation is generated by a weighted contribution of a number of simple distributions, selected by the hidden variables. The simplest form of mixture model assumes that an observation is independently generated by a single mixture component. Widely popular elements of this family are the *Gaussian mixture model* for continuous observations and the *mixture of unigrams* for multinomial data. In the following, we discuss an example of more articulated generative processes comprising observations with mixed component memberships.

Probabilistic Latent Semantic Analysis

Probabilistic latent semantic analysis (pLSA) [31.50] has been introduced to model mixed membership observations, where a manifest sample is allowed to be generated by multiple latent variables. Its primary application is on documental analysis, where latent variables are interpreted as topics to be identified in a collection of documents. Intuitively, in the mixture of unigrams, each document is assigned to a unique topic and, as a consequence, all the words in a document are constrained to belong to a single topic. The pLSA model relaxes this assumption by allowing words in a document to belong to different topics, obtaining a multitopic representation for the documents in the collection.

The typical pLSA setting includes a dataset of multinomial samples, which are the documents $\mathcal{D} = \{d_1, \ldots, d_N\}$. Each document is an L-dimensional vector of word counts of length equal to the size of the reference dictionary. In other words, the ith observed sample is a vector $d_i = (w_1^i, \ldots, w_L^i)$, where w_j^i is the number of occurrences of the jth word of the vocabulary in the ith document. This data is typically summarized in a rectangular $L \times N$ integer matrix n, such that each row $n(\cdot, d_i)$ contains the word counts for document d_i. The variables identifying words and documents, i.e., W_j and D_i, are observed, in contrast with the set of topics $Z = \{Z_1, \ldots, Z_K\}$, which are the *latent variables*. In pLSA, every observation $n(w_j, d_i)$ is associated with a latent topic z_k by means of the hidden variable Z_k.

The fundamental probabilities associated with this model are $P(D = d_i)$, that is, the document probability, $P(W = w_j|Z = z_k)$, that is, the probability of word w_j conditioned on topic z_k, and $P(Z = z_k|D = d_i)$, that is the conditional probability of topic z_k given document d_i. Given the nature of the manifest and hidden variables, all probabilities involved in pLSA are multinomials. The pLSA defines a (quasi) generative model for the word/document co-occurrences whose generative process is described by Fig. 31.6, using *plate notation*. This is a concise representation for graphical models involving replications: rectangular plates denote replication of their content for a number of times given by term on the bottom right (e.g., N and L_d for the outer and inner plates in Fig. 31.6, respectively); each shaded circular item denotes an observed variable, while empty circles identify latent variables.

The conditional independence relationships in Fig. 31.6 allow us to factorize the joint word-topic distribution: by using the parent decomposition rule introduced in (31.8), it yields

$$P(W_j, D_i) = P(D_i)P(W_j|D_i)$$

$$= P(D_i) \sum_{k=1}^{K} P(Z_k|D_i)P(W_j|Z_k) \,, \quad (31.20)$$

that is the specific pLSA form of the general latent topic factorization in (31.12). The second equality in (31.20) is given by the marginalization of the latent topics Z_k and by the conditional independence assumption of the pLSA model, stating that word w_j and document d_i can be considered independent given the state of the latent variable Z_k. In other words, the word distribution of a document is modeled as a convex combination of K topic-specific distributions $P(W_j|Z_k)$. Such decomposition has a well-known characterization in terms of *Nonnegative matrix factorization* [31.13].

Estimation of the pLSA parameters $\theta = \{P(W_j|Z_k), P(Z_k|D_i)\}$ is obtained by maximization of the log-likelihood

$$\mathcal{L}(\theta) = \log \prod_{i=1}^{D} \prod_{j=1}^{W} P(W_j, D_i)^{n(w_j, d_i)} = \sum_{i=1}^{D} \sum_{j=1}^{W} n(w_j, d_i)$$

$$\times \log \left\{ P(D_i) \sum_{k=1}^{K} P(Z_k|D_i)P(W_j|Z_k) \right\}, \quad (31.21)$$

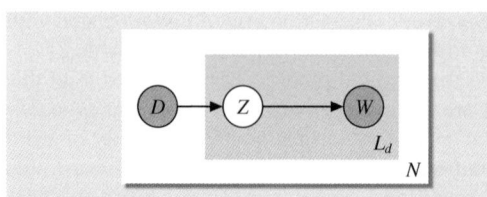

Fig. 31.6 Graphical model for the probabilistic latent semantic analysis: indices for the random variables D, Z, and W are omitted in the plate notation. The term L_d denotes replication for the L_d words present in the dth document

where $P(W_j, D_i)$ has been expanded using the formulation in (31.20). As with other latent topic models, this maximization problem can be solved through the iterative EM-algorithm discussed in Sect. 31.3.2. Following (31.15), the *E-step* computes the expectation of the complete likelihood $P(Z, W, D)$ with respect to the pLSA latent topics, assuming observed documents and words. It easily shows that the resulting E-step computes

$$P(Z_k|W_j, D_i) = \frac{P(Z_k|D_i)^{(t)} P(W_j|Z_k)^{(t)}}{\sum_{k'=1}^{K} P(Z_{k'}|D_i)^{(t)} P(W_j|Z_{k'})^{(t)}} \,, \quad (31.22)$$

that is the probability of the topic Z_k given word W_j in document D_i, estimated using the current values (at time t) of the model parameters $\theta^{(t)} = \{P(W_j|Z_k)^{(t)}, P(Z_k|D_i)^{(t)}\}$. Note that the decomposition on the right-hand side of Eq. (31.22) has been obtained by factorization of the posterior $P(Z_k|W_j, D_i)$ using the Bayes theorem.

The *M-step* equations (31.16) are obtained by differentiating the pLSA log-likelihood, extended with appropriate Lagrange multipliers for normalization, with respect to the $P(Z_k|D_i)$ and $P(W_j|Z_k)$ parameters. The resulting update equations are

$$P(Z_k|D_i)^{(t+1)} = \frac{\sum_{j=1}^{W} n(w_j, d_i)P(Z_k|W_j, D_i)}{\sum_{j=1}^{W} n(w_j, d_i)} \,, \quad (31.23)$$

$$P(W_j|Z_k)^{(t+1)} = \frac{\sum_{i=1}^{D} n(w_j, d_i)P(Z_k|W_j, D_i)}{\sum_{j=1}^{W} \sum_{i=1}^{D} n(w_j, d_i)P(Z_k|W_j, D_i)} \,. \quad (31.24)$$

The two-step optimization is iterated until a likelihood convergence criterion is met: often a validation set, or a *tempered* version of the EM are used in order to avoid model overfitting [31.50].

Advanced Topic Models

The pLSA was the first mixed membership model allowing a single observed sample to be generated by multiple latent topics at the same time. However, pLSA cannot be considered a fully generative model. In fact, the document-specific mixing weights for the topics are not sampled from a distribution, rather they are selected from $P(z_k|d_i)$ based on the index of document d_i. Hence, pLSA indexes only those documents that are in the training set $\mathcal{D}$ and cannot directly model the generative process of unseen test documents. In other words, the pLSA is basically assigning null probabilities to all inputs that are not in the training set. The folding-in heuristic has been proposed to opportunistically solve this limitation, by assigning latent variables in the test-data to their MAP values before computing the test-set perplexity. However, the folding-in approach has been shown to lead to overly optimistic estimates of the test-set log-likelihood [31.51].

The *latent Dirichlet allocation* (LDA) [31.52] has been proposed as a Bayesian approach to address such modeling limitation of pLSA. It extends pLSA by treating the multinomial weights $P(Z|D)$ as additional latent random variables, sampling them from a Dirichlet distribution, that is the conjugate prior of a multinomial distribution. Using conjugate distribution eases inference as it ensures that the posterior distribution has the same form of the prior. The latent variable decomposition of the LDA log-likelihood is

$$P(W = w|\phi, \alpha, \beta) = \int \sum_z P(W = w|Z = z, \phi)$$

$$\times P(Z = z|\theta)P(\theta|\alpha)P(\phi|\beta)d\theta ,$$
$$(31.25)$$

where $P(W|Z, \phi)$ is the multinomial word-topic distribution with parameters ϕ sampled from the Dirichlet distribution $P(\phi|\beta)$. The term $P(Z|\theta)$ is the topic distribution having θ as document-specific multinomial parameter being sampled from the Dirichlet $P(\theta|\alpha)$.

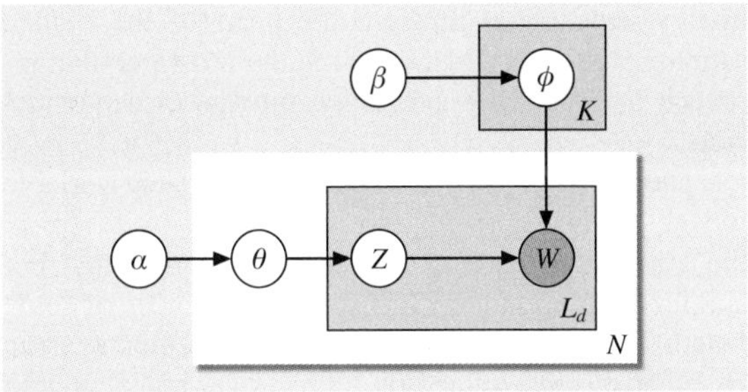

Fig. 31.7 Graphical model for the latent Dirichlet allocation

The terms α and β are the hyperparameters of the Dirichlet distribution, see Fig. 31.7 for the model plate notation. Direct EM inference is impossible for LDA, since the integral in (31.25) is intractable due to the couplings between the parameters within the topic marginalization. Again, approximate and stochastic Bayesian inference methods, such as those in Sect. 31.2.3, are used to fit the LDA parameters, including VB [31.52], expectation propagation [31.53], and Gibbs sampling [31.54].

The principles underlying pLSA and LDA have inspired the development of latent topic models that account for more articulated assumptions on the form of the hidden generative process. For instance, hierarchical LDA [31.55] proposes a generative process where observations are generated by a topic tree instead of being drawn from a flat topic collection. Further, specialized latent variable models have been developed for specific applications, such as author-topic analysis in scientific literature [31.56] and image understanding [31.57].

31.4 Markov Models

Time series and, more generally, sequences are a form of structured data that represents a list of observations for which a complete order can be defined, e.g., time in a temporal sequence. Let a sequence of length T be $\mathbf{y}_n = y_1, \ldots, y_T$, where the bold notation is used to denote the fact that $\mathbf{y}$ is a compound object (in practice, however, this is can be treated as a set of random variables). The term y_t is used to denote the tth observation with respect to the total order. Position t is often referred to as time when dealing with time-series data.

Two sequences are generally the results of independent trials, hence they can be considered i.i.d. samples. However, the elements composing a sequence fail to meet such i.i.d. property. Therefore, in principle, a prob-

abilistic model for $\mathbf{y}$ would be required to specify the joint distribution $P(Y_1, \ldots, Y_T)$. For discrete valued observations y_t, the joint distribution grows exponentially with the size of the observation domain. Clearly, this would make the use of the probabilistic model fairly impractical due to the exponential size of the parameter space. To reduce such parameterization, *Markov chains* make the simplifying assumption that an observation occurring at some position t of the sequence, only depends on a limited number of its predecessors with respect to the complete order. In a time series, this entails that an observation at the present time, only depends on the history of a limited number of past observations. Markov chains allow us to model such history dependence and are the heart of the *hidden Markov model* (HMM), which is the most popular approach to model the generative process of sequential data.

The HMM is a notable example of latent variable model: in the following, we provide an overview of the associated learning and inference problems. For simplicity, presentation focuses on sequences of finite length T and discrete time t. Sequence elements y_t can be either discrete valued or defined over reals, without major impact on the model. The section also discusses how the HMM causation assumption can be modified to give rise to alternative approaches, with interesting applications that overshoot simple sequence modeling.

31.4.1 Markov Chains

A Markov chain is a simple stochastic process for sequences. It assumes that an observation y_t at time (position) t only depends on a finite set of $L \geq 1$ predecessors in the sequence. The number of predecessor L influencing the new observation is the order of the Markov chain.

Definition 31.4 Markov Chain
An L-order Markov chain is a sequence of random variables $\mathbf{Y} = Y_1, \dots, Y_T$ such that for every $t \in \{1, \dots, T\}$, it holds

$$P(Y_t = y_t | Y_1, \dots, Y_{t-1}, Y_{t+1}, \dots, Y_T)$$
$$= P(Y_t = y_t | Y_{t-L}, \dots, Y_{t-1}) . \quad (31.26)$$

Following from the discussions in Sect. 31.2.1, (31.26) states that the L predecessors of Y_t define the set of its Bayesian parents $\mathrm{pa}(Y_t) = \{Y_{t-L}, \dots, Y_{t-1}\}$. For a first-order Markov chain, i.e., $L = 1$, (31.26) reduces to $P(Y_t = y_t | Y_{t-1} = y_{t-1})$. Such conditional independence assumption formally encodes the intuition that the current observation can be predicted from the sole knowledge of the preceding sample. The graphical model of a first-order Markov chain is shown in Fig. 31.8, whose joint distribution decomposes as

$$P(Y_1, \dots, Y_T) = P(Y_1)P(Y_2|Y_1), P(Y_3|Y_2)$$
$$\times \dots P(Y_T|Y_{T-1})$$
$$= P(Y_1) \prod_{t=2}^{T} P(Y_t|Y_{t-1}) . \quad (31.27)$$

The first element Y_1 has an empty conditioning part given that is has no predecessor. Its probability $P(Y_1)$

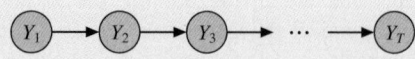

Fig. 31.8 Graphical model for a first-order Markov chain of length T, where $\mathrm{pa}(Y_t) = \{Y_{t-1}\}$

is referred to as *marginal* or *prior* probability, while the term $P(Y_t|Y_{t-1})$ is the *transition* probability.

A Markov chain is *stationary* or *homogeneous*, if the transition probability does not depend on the time (position) t. In other words, the parameterization of the Markov chain is such that

$$P(Y_t = y' | Y_{t-1} = y) = f(y', y) ,$$

where the transition distribution is a function $f(y', y)$ of the sole observations y, y'. An interesting stationary first-order Markov chain is that whose random variables take values from a finite alphabet of discrete symbols $i, j \in \{1, \dots, M\}$. In these chains, the transition probability

$$A_{ij} = P(Y_t = i | Y_{t-1} = j) \quad (31.28)$$

denotes the probability of occurrence of the ith symbol preceded by symbol j. For convenience, such probability is represented by the element A_{ij} of the $M \times M$ transition matrix $A = [A_{ij}]_{i,j=1}^{M}$. Similarly, the marginal distribution defines the elements

$$\pi_i = P(Y_1 = i) \quad (31.29)$$

of the $M \times 1$ initial state vector $\pi = [\pi_i]_{i=1}^{M}$. These Markov chains can be straightforwardly interpreted as state-transition systems, where each symbol i of the alphabet is a state and a state-transition arrow exists between states i and j having a nonzero A_{ij} entry in the transition matrix.

The Markov chains described by (31.28) and (31.29), despite their simplicity, have found wide application, e.g., in modeling of physical phenomena, economic time series, and information retrieval. Learning Markov chains requires fitting the M^2 parameters of the transition matrix plus an M-dimensional prior, where M is the size of the observation alphabet. Efficient methods exists to fit stationary first-order Markov chains by maximum likelihood (ML). By using the decomposition in (31.26), substituting the definitions in (31.28) and (31.29), the Markov chain log-likelihood for a generic sequence **y** writes

$$\mathcal{L}(\theta) = \log P(\mathbf{Y} = \mathbf{y}|\theta) = \log \prod_{i'=1}^{M} \pi_{i'}^{\delta(y_1=i')}$$

$$\times \prod_{t=2}^{T} \prod_{i,j=1}^{M} A_{ij}^{\delta(y_t=i,y_{t-1}=j)}, \qquad (31.30)$$

where $\theta = (A, \pi)$ are the model parameters and $\delta(y_t = i, y_{t-1} = j)$ is the indicator function. For instance, it equals 1 if a transition from $y_{t-1} = j$ to $y_t = i$ can be observed in the sequence and it is 0 otherwise. Similarly, $\delta(y_1 = i') = 1$ if and only if the first symbol of the sequence is i'. The final expression of the log-likelihood is obtained by taking the log into the products and adding appropriate Lagrange multipliers for normalization. The ML estimate is obtained by differentiating this final expression with respect to parameters A_{ij} and π_i, yielding

$$A_{ij} = \frac{\sum_{t=2}^{T} \delta(y_t = i, y_{t-1} = j)}{\sum_{t=2}^{T} \sum_{i=1}^{M} \delta(y_t = i, y_{t-1} = j)}, \qquad (31.31)$$

$$\pi_i = \frac{\delta(y_1 = i)}{\sum_{i=1}^{M} \delta(y_1 = i)}. \qquad (31.32)$$

Intuitively, the ML estimate corresponds to counting the number of transitions from symbol j to i across time (similarly for the initial state). Generalization to a set of N samples sequences $\mathbf{y}^n$ is straightforward: it suffices to count transitions both in time and across samples, and similarly for the initial symbols y_1^n.

31.4.2 Hidden Markov Models

Markov chains model sequential data assuming that sequence elements are generated by a fully observable stochastic process. In the discrete-state Markov chain, this requires each state of the process to correspond to an observable element of the sequence, i.e., en event. On the other hand, most real-world systems generate observable events that are correlated, but not coincident, with the state of the generating process. More importantly, the only available information can be the outcome of the stochastic process at each time, i.e., event y_t, while the state of the system remains unobservable, i.e., *hidden*. The HMM allows modeling more general stochastic processes where the state transition dynamics is disentangled from the observable information generated by the process. The state-transition

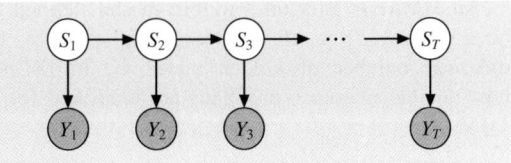

Fig. 31.9 A first-order HMM with hidden states S_t chosen on the discrete domain $\{1, \ldots, C\}$, for $t = 1 \ldots T$

dynamics is assumed to be nonobservable and is modeled by a Markov chain of discrete and finite latent variables, i.e., the *hidden states*. The observable information is then generated by such hidden states similarly to how latent variables generate observations in mixture models (see Sect. 31.3.4).

The graphical model of an HMM is exemplified in Fig. 31.9: the hidden states are latent variables S_t, while the sequence elements Y_t are observed.

The conditional dependence expressed by the arrow $S_t \rightarrow Y_t$ indicates that the observed element of the sequence at time t is generated by the corresponding hidden state S_t through the *emission distribution* $b_{s_t}(y_t) = P(Y_t = y_t|S_t = s_t)$. The unknown state-transition dynamics is modeled by the first-order Markov chain of discrete and finite hidden states S_t. By applying the Markovian decomposition in (31.27) to the hidden states chain, the joint distribution of the observed sequence $\mathbf{y} = y_1, \ldots, y_T$ and associated hidden states $\mathbf{s} = s_1, \ldots, s_T$ writes as

$$P(\mathbf{Y} = \mathbf{y}, \mathbf{S} = \mathbf{s}) = P(S_1) \prod_{t=2}^{T} P(S_t|S_{t-1})P(Y_t|S_t).$$

$$(31.33)$$

The actual parameterization of the probabilities in (31.33) depends on the form of the observation and hidden states variables. From Sect. 31.8, a stationary hidden states chain is known to be regulated by the $C \times C$ matrix of *state transitions* $A_{ij} = P(S_t = i|S_{t-1} = j)$ and by the C-dimensional vector of *initial state probabilities* $\pi_i = P(S_t = i)$, where i, j are drawn from $\{1, \ldots, C\}$. For discrete sequence observations $y_t \in \{1, M\}$, the emission distribution is an $M \times C$ emission matrix B such that its elements are

$$b_i(k) = B_{ki} = P(Y_t = k|S_t = i). \qquad (31.34)$$

For continuous observations y_t, the state assignment $S_t = i$ selects the ith emission distributions $b_i(y_t) = P(Y_t|S_t = i)$ from a mixture of C candidates.

An HMM is a latent variable model defined by the $\theta = (\pi, A, B)$ parameters and, implicitly, by the (unkown) number of hidden states C. In [31.58], three notable inference problems are identified for an HMM.

Definition 31.5 Evaluation Problem
Given a model θ and an observed sequence **y**, determine the likelihood $P(\mathbf{Y} = \mathbf{y}|\theta)$ of the sequence being generated by the model.

Definition 31.6 Learning Problem
Given a dataset of N observed sequences $\mathcal{D} = \{\mathbf{y}^1, \ldots, \mathbf{y}^N\}$ and the number of hidden states C, find the parameters π, A and B that maximize the probability of model $\theta = \{\pi, A, B\}$ having generated the sequences in $\mathcal{D}$.

Definition 31.7 Optimal States Problem
Given a model θ and an observed sequence **y**, find an optimal state assignment $\mathbf{s} = s_1^*, \ldots, s_T^*$ for the underlying hidden Markov chain.

These classical inference problems are addressed using efficient and numerically stable recursive algorithms that exploit message passing on the HMM junction tree (Sect. 31.2.3) to factorize the, otherwise hardly tractable, joint maximization problems. The underlying intuition is a recursive computation of intermediate probabilities (messages) that are passed forward and backward along the sequence (the junction tree, in practice) to accumulate evidence for solving the joint problem. A discussion of the key aspects of these solutions is provided in the following.

Evaluation
The evaluation problem refers to measuring how well a given HMM matches an observed sequences. Let the model be $\theta = (\pi, A, B)$ and the observed sequence $\mathbf{y} = y_1, \ldots, y_T$, the objective is to find $P(\mathbf{Y} = \mathbf{y}|\theta)$. To effectively compute this probability in the HMM assumption, it is needed to introduce the hidden states assignment corresponding to the observed sequence **y**. Following the general approach for latent variable models in Eq. (31.11), these are introduced through marginalization on the joint assignment $\mathbf{s} = s_1, \ldots, s_T$

$$P(\mathbf{Y}|\theta) = \sum_s P(\mathbf{Y}, \mathbf{S} = \mathbf{s}|\theta)$$

$$= \sum_{s_1, \ldots, s_T} P(S_1) \prod_{t=2}^{T} P(S_t|S_{t-1})P(Y_t|S_t),$$

$$(31.35)$$

where the joint probability $P(\mathbf{Y}, \mathbf{S}|\theta)$ has been factorized according to the HMM assumption in (31.33).

Direct computation of (31.35) is generally infeasible, as it would require $O(TC^T)$ operations. This probability can be efficiently computed, with $O(TC^2)$ operations, through accumulation of a recursive term that is computed by scanning the sequence from left to right. The procedure is known as *forward algorithm*: let $\mathbf{y}_{1:t}$ be the observed subsequence from position 1 to t, define the *forward probability* as

$$\alpha_t(i) = P(\mathbf{Y}_{1:t} = \mathbf{y}_{1:t}, S_t = i|\theta) \qquad (31.36)$$

that is the probability of observing a partial sequence up to position t and the underlying hidden process being in state i at time t. A recursive formulation of the $\alpha_t(i)$ term is obtained by introducing the hidden state S_{t-1} by marginalization, yielding

$$\alpha_t(i) = \sum_{j=1}^{C} P(\mathbf{Y}_{1:t} = \mathbf{y}_{1:t}, S_t = i, S_{t-1} = j|\theta)$$

$$= \sum_{j=1}^{C} P(Y_t = y_t|S_t = i, \theta)$$

$$\times P(S_t = i|S_{t-1} = j, \theta)$$

$$\times P(\mathbf{Y}_{1:t-1} = \mathbf{y}_{1:t-1}, S_{t-1} = j|\theta)$$

$$= b_i(y_t) \sum_{j=1}^{C} A_{ij}\alpha_{t-1}(j), \qquad (31.37)$$

where the second equality follows from the conditional independence assumptions of the model. Since, $\mathrm{pa}(S_t) = \{S_{t-1}\}$, the chain element S_t is completely determined by the hidden state at previous time S_{t-1}; similarly, emission Y_t is conditional independent from the rest, given the hidden state S_t.

The forward recursion scans the observed sequence from left to right and recursively computes the $\alpha_t(i)$ values in each position $t = 1, \ldots, T$ using (31.37). At each observed position t, the $\alpha_t(i)$ values are computed for each $i \in \{1, \ldots, C\}$, since the hidden states are not observed. The basis of the recursion is at $t = 1$, where the

(31.37) reduces to $\alpha_1(i) = b_i(y_1)\pi_i$, such that y_1 is the first element of the observed sequence. The likelihood of the full sequence $\mathbf{y} = \mathbf{y}_{1:T}$ is computed at the end of the forward recursion as

$$P(\mathbf{Y}|\theta) = \sum_{i=1}^{C} P(\mathbf{Y}_{1:T}, S_T = i|\theta) = \sum_{i=1}^{C} \alpha_T(i) .$$

(31.38)

Learning

Learning of an HMM $\theta = (\pi, A, B)$ amounts to finding the values of the parameters π, A and B that are most likely to have generated a dataset of observed i.i.d. sequences $\mathcal{D} = \{\mathbf{y}^1, \ldots, \mathbf{y}^N\}$. From the evaluation problem, we know how to measure the quality of the matching between a sequence $\mathbf{y}$ and a model θ using the likelihood $P(\mathbf{Y}|\theta)$. The HMM learning problem can be solved through ML estimation of θ parameters considering the hidden states as latent variables. As discussed in Sect. 31.3.2, this problem can be solved through application of the EM algorithm, whose HMM version is referred to as *Baum–Welch algorithm* [31.59], which is a form of *sum-product* inference algorithm introduced in Sect. 31.2.3. Marginalization of the hidden states as in (31.35), yields to the HMM log-likelihood on the dataset $\mathcal{D}$

$$\mathcal{L}(\theta) = \log \prod_{n=1}^{N} P(\mathbf{Y}^n|\theta)$$

$$= \log \prod_{n=1}^{N} \left\{ \sum_{s_1^n, \ldots, s_{T_n}^n} P(S_1^n) \right.$$

$$\left. \times \prod_{t=2}^{T_n} P(S_t^n|S_{t-1}^n) P(Y_t^n|S_t^n) \right\} ,$$

(31.39)

where overscript n refers to the nth sequence $\mathbf{y}^n$ and T_n is the corresponding length. The likelihood in (31.39) is intractable due to the nonobservable state assignment that introduces the marginalization term. Following the principles of the EM algorithm, we assume to know the unobserved state assignment, as in (31.30). This can be achieved by introducing indicator variables z_{ti}^n for the unknown assignment, such that $z_{ti}^n = 1$ if the chain is in state i at position t of the nth sequences, and it is 0 otherwise. Given this (assumed) knowledge about the

hidden state assignments, if is possible to write the corresponding *completed* likelihood

$$\mathcal{L}_c(\theta)$$

$$= \log \prod_{n=1}^{N} \left\{ \prod_{i=1}^{C} P(S_1^n = i)^{z_{1i}^n} \right.$$

$$\left. \times \prod_{t=2}^{T_n} \prod_{i,j=1}^{C} P(S_t^n = i|S_{t-1}^n = j)^{z_{ti}^n z_{(t-1)i}^n} P(Y_t^n|S_t^n = i)^{z_{ti}^n} \right\}$$

$$= \sum_{n=1}^{N} \left\{ \sum_{i=1}^{C} z_{1i}^n \log \pi_i + \sum_{t=2}^{T_n} \sum_{i,j=1}^{C} z_{ti}^n z_{(t-1)i}^n \right.$$

$$\left. \times \log A_{ij} + z_{ti}^n \log b_i(y_t^n) \right\} ,$$

(31.40)

where the latter equality introduces the parameters θ in place of the corresponding probabilities and brings the logarithms into the products.

The EM procedure is applied to the complete log-likelihood in (31.40). Following (31.15), the E-step computes the expected value of $\mathcal{L}_c(\theta)$ with respect to the distribution of the indicator variables $\mathcal{Z} = \{z_{ti}^n\}$, conditional on the observed sequences $\mathcal{D}$ and the current estimate of the parameters $\theta^{(k)}$. Given $\mathcal{L}_c(\theta)$ as in (31.40), taking its conditional expectation with respect to the hidden variables $\mathcal{Z}$, it yields to the following posterior probability:

$$E_{\mathcal{Z}|\mathbf{Y}, \theta^{(k)}}[z_{ti}] = P(S_t = i|\mathbf{y}) ,$$

(31.41)

where superscript n is omitted for notational simplicity. The estimation of this posterior is known as the *smoothing* problem. In the Baum–Welch algorithm, this is efficiently solved by a double recursion that exploits the following decomposition of the joint probability

$$P(S_t = i, \mathbf{y}) = P(S_t = i, \mathbf{Y}_{1:t}, \mathbf{Y}_{t+1:T})$$

$$= P(S_t = i, \mathbf{Y}_{1:t})$$

$$\times P(\mathbf{Y}_{t+1:T}|S_t = i) = \alpha_t(i)\beta_t(i) ,$$

(31.42)

where the observed contribution from the predecessors of t (i.e., $\mathbf{Y}_{1:t}$) is separated from that of its successors (i.e., $\mathbf{Y}_{t+1:T}$). The cancelations in (31.42) follow from the fact that S_t d-separates (see definition in Sect. 31.2.1) the elements of the two subsequences, i.e., $\mathbf{Y}_{1:t}$ and $\mathbf{Y}_{t+1:T}$.

The first term in (31.42) is the $\alpha_t(i)$ probability defined in (31.36), which can be computed through the *forward algorithm*. The $\beta_t(i)$ term can also be computed through a recursive procedure known as *backward algorithm*, due to the inverted direction with respect to the forward recursion. Consider the following recursive decomposition

$$
\begin{aligned}
\beta_{t-1}(j) &= P(\mathbf{Y}_{t:T}|S_{t-1}=j) \\
&= \sum_{i=1}^{C} P(\mathbf{Y}_{t:T}, S_t=i|S_{t-1}=j) \\
&= \sum_{i=1}^{C} P(Y_t|S_t=i)P(\mathbf{Y}_{(t+1):T}|S_t=i) \\
&\quad \times P(S_t=i|S_{t-1}=j) \\
&= \sum_{i=1}^{C} b_i(y_t)\beta_t(i)A_{ij} \,,
\end{aligned}
\tag{31.43}
$$

it can be computed for $2 \le t \le T$ by scanning the sequence backward, assuming $\beta_T(j)=1$ for each $j \in \{1,\dots,C\}$.

The final expression of the smoothed posterior in (31.41) is given by the joint $\alpha - \beta$ recursions, known as the *forward–backward algorithm*, that is

$$
\begin{aligned}
\gamma_t(i) &= P(S_t=i|\mathbf{Y}) = \frac{P(S_t=i, \mathbf{Y})}{P(\mathbf{Y})} \\
&= \frac{\alpha_t(i)\beta_t(i)}{\sum_{j=1}^{C} \alpha_t(j)\beta_t(j)} \,.
\end{aligned}
\tag{31.44}
$$

Note that the forward and backward recursions can be ran in parallel, since the values of α and β do not depend on each other. To complete the derivations of the sufficient statistics for the M-step, it is also necessary to estimate the joint posterior

$$
E_{Z|\mathbf{Y},\theta^{(k)}}[z_{ti}z_{(t-1)j}] = P(S_t=i, S_t=j|\mathbf{Y}) \,, \tag{31.45}
$$

which can be straightforwardly factorized into known probabilities along the lines of (31.42). It turns out that such joint posterior can be estimated using the $\alpha - \beta$ probabilities computed by the *forward–backward algorithm*, that is

$$
\begin{aligned}
\gamma_{t,t-1}(i,j) &= P(S_t=i, S_t=j|\mathbf{Y}) \\
&= \frac{\alpha_{t-1}(j)A_{ij}b_i(y_t)\beta_t(i)}{\sum_{m,l=1}^{C} \alpha_{t-1}(m)A_{lm}b_3(y_t)\beta_t(l)} \,.
\end{aligned}
\tag{31.46}
$$

Parameters $\theta = (\pi, A, B)$ are re-estimated at the M-step, with update equations that follow straightforwardly from the maximization problem in (31.16). It suffices to differentiate (31.40), extended with appropriate Lagrange multipliers to account for the sum-to-one constraints. Intuitively, the update equations can be straightforwardly written from the ML estimates for observable Markov chains in (31.31) and (31.32). It suffices to substitute the observed state counts, obtained through the indicator function $\delta(\cdot)$, with the virtual counts $\gamma(\cdot)$ estimated by (31.44) and (31.46) at the E-step. For the hidden state transition and initial state distributions this yields to

$$
A_{ij} = \frac{\sum_{n=1}^{N} \sum_{t=2}^{T^n} \gamma_{t,t-1}^n(i,j)}{\sum_{n=1}^{N} \sum_{t=2}^{T^n} \gamma_{t-1}^n(j)}
$$

$$
\text{and} \quad \pi_i = \sum_{n=1}^{N} \gamma_1^n(i) \,. \tag{31.47}
$$

The estimate of the parameters B depends on the form of the emission distribution: if the observed sequences take values k from a finite alphabet $\{1,\dots,M\}$, the corresponding multinomial emission in (31.34) is updated by

$$
B_{ki} = \sum_{n=1}^{N} \sum_{t=1}^{T_n} \gamma_t^n(i)\delta(y_t=k) \,, \tag{31.48}
$$

where $\delta(\cdot)$ is the indicator function counting the occurrences of the symbols k in the observed sequences. Real-valued sequences are modeled usually through Gaussian emissions, whose parameters are fit as usual through maximization of the complete log-likelihood.

Particular care must be taken to avoid numerical problems when implementing the *forward–backward algorithm*. Both recursions work with multiplications of small numbers: hence, the values of α and β can underflow for long sequences. To this end, it is advisable to perform them in log-space or to work with scaled versions of the α and β probabilities [31.60]. A sequential version of the smoothing algorithm exists [31.61] that directly computes the smoothed posterior $\gamma_t(i) = P(S_t=i|\mathbf{Y})$ through a γ-recursion that uses the α values generated by the forward algorithm.

Optimal State

Once a model θ has been trained, it can be interesting to determine the most likely hidden state assignment $\mathbf{s}^*$ that has generated an observed sequence $\mathbf{y}$. This inference problem, known also as decoding, has different solutions, since several optimal assignment exists, depending on the interpretation of what an optimal assignment is. For instance, the optimal hidden sequence can be the one maximizing the expected count of correct states. On the other hand, an optimal assignment might be the sequence of hidden states $\mathbf{s}^*$ with the maximum joint probability $P(\mathbf{Y} = \mathbf{y}, \mathbf{S} = \mathbf{s}^*)$.

The former optimality condition is solved by selecting, at each position t, the most likely state given by the sequence, i. e.,

$$s_t^* = \arg \max_{i=1,\ldots,C} P(S_t = i | \mathbf{Y}) . \qquad (31.49)$$

Clearly, this amounts to select the most likely state for each position independently, using the posterior computed by the Baum–Welch algorithm. Conversely, the latter optimality condition estimates the joint hidden state assignment

$$\mathbf{s}^* = \arg \max_{\mathbf{s}} P(\mathbf{Y}, \mathbf{S} = \mathbf{s}) . \qquad (31.50)$$

This is a complex inference problem that can be efficiently solved though a dynamic programming approach, known as the *Viterbi algorithm*. Note that the two optimality definitions generally lead to different solutions. For instance, the Viterbi solution is constrained to provide only state transitions allowed by the generating distribution, while this is not the case for the Baum, Welch solution, given that hidden states are selected independently.

The *Viterbi algorithm* is based on a backward recursion that exploits a factorization of the maximization problem in (31.50). Consider the restricted problem of determining the hidden state of the tail element T

$$\max_{s_T} P(\mathbf{Y}, S_T = s_T) = \max_{s_T} \prod_{t=1}^{T} P(Y_t | S_t) P(S_t | S_{t-1})$$

$$= \prod_{t=1}^{T-1} P(Y_t | S_t) P(S_t | S_{t-1}) \max_{s_T} P(Y_t | S_T) P(S_T | S_{T-1}) , \qquad (31.51)$$

where the joint probability factorizes according to the Markov chain assumption. We can isolate the maximization problem in the rightmost term

$$\epsilon_{T-1}(s_{T-1}) = \max_{s_T} P(Y_t | S_T = s_T)$$

$$\times P(S_T = s_T | S_{T-1} = s_{T-1}) , \qquad (31.52)$$

that is a message conveying information on the maximization of the tail element to the penultimate position. Substituting the definition of $\epsilon_{T-1}(s_{T-1})$ back

in (31.51) and adding the maximization with respect to s_{T-1}, suggests the recursive formulation of $\epsilon.(\cdot)$ for a generic position $t-1$, i. e.,

$$\epsilon_{t-1}(s_{t-1}) = \max_{s_t} P(Y_t | S_t = s_t)$$

$$\times P(S_t = s_t | S_{t-1} = s_{t-1}) \epsilon_t(s_t) , \qquad (31.53)$$

for $2 \leq t \leq T$, where $\epsilon_T(s_T) = 1$ is the basis of the recursion. At each step t of the backward recursion, the Viterbi algorithm computes the ϵ-message for each possible assignment of the hidden state of t and propagates it to the predecessor $t-1$. The recursion ends at the initial element of the sequence, where the initial optimal state is obtained as

$$s_1^* = \arg \max_s P(Y_t | S_1 = s) P(S_1 = s_1) \epsilon_1(s) . \qquad (31.54)$$

The assignment of the remaining hidden states is obtained by backtracking through the forward recursion

$$s_t^* = \arg \max_s P(Y_t | S_t = s)$$

$$\times P(S_t = s | S_{t-1} = s_{t-1}^*) \epsilon_t(s) . \qquad (31.55)$$

Note that the *Viterbi* algorithm is a special case of a *max-sum* inference algorithm introduced in Sect. 31.2.3.

31.4.3 Related Models

Higher Order Markov Models
Hidden Markov models serve as a starting point for the design on more complex Markov generative processes, besides the obvious extension to higher order hidden chains [31.62]. *Factorial HMMs* [31.63] generalize the original model by defining super states that are collections of K discrete hidden states, each being part of an independent Markov chain (see Fig. 31.10). This factorial model results in K hidden Markov chains running in parallel: at each time step, the emission depends on the K-dimensional super state, but each state variable is decoupled from those of the other chains and evolves according to its own dynamics. By this means, it is possible to efficiently encode the state dynamics of K objects evolving independently that interact to jointly determine the observation (e.g., K cars moving in the traffic and jointly determining traffic jams).

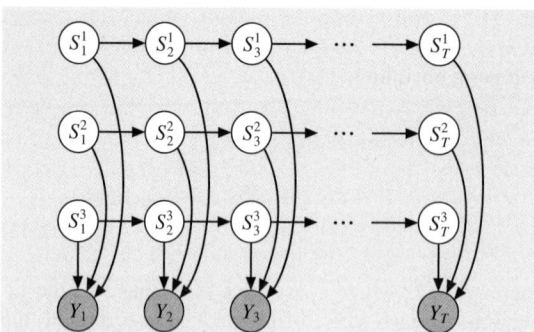

Fig. 31.10 Factorial HMM with $K = 3$ independent hidden Markov chains

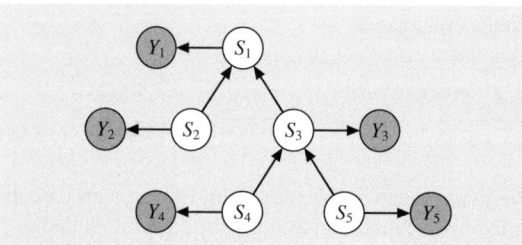

Fig. 31.11 A bottom-up hidden tree Markov model for a simple structure with five nodes: the generative process follows the direction of the *arrows*, i.e., from the leaves to the root ($t = 1$)

Nonhomogenous HMMs

Relaxation of the homogeneity assumption led to the *input/output hidden Markov model* (IO-HMM) [31.64] that allow modeling the causal dependence of the hidden generative process from an additional input sequence **x**. Basically, the IO-HMM enables nonhomoge- neous transition and emission distributions that are ex plicitly dependent (i. e., parameterized) on the currently observed label of the input sequence. An IO-HMM im plements a mapping, referred to as *transduction*, from an observed input sequence **x** into an output (target sequence **y**, realized by the input-conditional hidden process $P(\mathbf{Y}|\mathbf{X})$. Interesting applications of IO-HMM are in learning transformations between modalities in multimedia data [31.65], exploratory analysis of finan cial time series [31.66] and gene data analysis [31.67].

HMMs for Structured Data

Hidden tree Markov models represent the generative process of more complex, tree-structured information (see Fig. 31.11). Differently from the sequential do main, the direction of the generative process leads to different representational capabilities when dealing with trees. Top-down approaches [31.68] model all pos sible paths from the root to the leaves of the tree Bottom-up models [31.69] propose a generative process from the leaves to the root, where complex structures are generated by composition of simpler substructures Recently, an extension of the IO-HMM has been pro posed to learn transductions between trees [31.70].

Bayesian and Nonparametric Extensions

HMMs have been extended to allow a countably infinite number of hidden states through a Bayesian approach where state distributions are modeled by Dirichlet pro cesses [31.71]. Abstracting from the direction of the arrows in Fig. 31.9 leads to a discriminative proba bilistic model known as liner-chain *conditional random fields* [31.72], whose capability to model long term de pendences is widely used in natural language parsing and computer vision.

31.5 Conclusion and Further Reading

Graphical models have been discussed as an excel lent framework for probabilistic modeling of articulated processes that can be described by a *static* set of ran dom variables tied up by probabilistic relationships. Such relationships need not to be necessarily known, a-priori. Several approaches exists to infer them from data, i. e., to determine the presence of a correspond ing edge in the graphical model. However, the same approaches tend to *fix* the structure of the graphical model, once this is determined from the data. In other words, these graphical models represent a *static* picture of the process, where the set of random variables and associated relationships is held fixed from a point on ward. The nature of sequence data calls for the ability to model more *dynamic* phenomena. Processing of video information requires Markov networks that can unfold their structure across the video sequence. Even *classic* text analysis needs to account for novel generative dy namics, where texts are produced as dynamic streams instead of being static collections of words, e.g., con sider blog posts and associated comments, or the stream of social networks status updates. Therefore, the horizon of current research is pushing graphical models to more *dynamic* formulations where, on the one hand, the structure is allowed to change over time and, on the other hand, the model is allowed to dynamically self-tune the number of parameters that is most adequate to represent the process at each time. Following the intuitions underlying the HMM approach, dynamical graphical models are being proposed that are capable of unfolding their structure across time, to better model the dynamics of complex time-varying processes. At the same time, concepts from nonparametric Bayesian statics are being used to develop models where latent variables can be dynamically adjusted to sample from a virtually infinite set of events and where the very same structure of the latent space is adapted across time, i. e., through variable addition and pruning. Such a new class of dynamic graphical models introduces novel computational challenges associated with inference and representation of dynamic knowledge. The answers to this challenges can be partly found in the chapter, in the approximated inference methods discussed for static models and in the principles underlying the unfolding of Markov chains. Finally, it is worth to note that deep learning, described in Chap. 2, is an instance of graphical model where both nonlinearity and dynamic representations play an important role.

References

31.1 S. Kullback, R.A. Leibler: On information and sufficiency, Ann. Math. Stat. **22**, 79–86 (1951)

31.2 F. Rosenblatt: The perceptron: A probabilistic model for information storage and organization in the brain, Psychol. Rev. **65**, 386–408 (1958)

31.3 G. Deco, W. Finnoff, H.G. Zimmermann: Unsupervised mutual information criterion for elemination of overtraining in supervised mulilayer networks, Neural Comput. **7**, 86–107 (1995)

31.4 D.J.C. Mackay: *Information Theory, Inference and Learning Algorithms* (Cambridge Univ. Press, Cambridge 2003)

31.5 R. Salakhutdinov, G. Hinton: Using deep belief nets to learn covariance kernels for Gaussian processes, Adv. Neural Inf. Process. Syst. **20**, 1249–1256 (2008)

31.6 C.M. Bishop: *Pattern Recognition and Machine Learning* (Springer, New York 2006)

31.7 S. Seth, J.C. Principe: Variable selection: A statistical dependence perspective, Proc. Int. Conf. Mach. Learn. Appl. (ICMLA) (2010)

31.8 M. Rao, S. Seth, J. Xu, Y. Chen, H. Tagare, J.C. Principe: A test of independence based on a generalized correlation function, Signal Process. **91**, 15–27 (2011)

31.9 D.D. Lee, H.S. Seung: Learning the parts of objects by non-negative matrix factorization, Nature **401**(6755), 788–791 (1999)

31.10 P. Comon, C. Jutten: *Handbook of Blind Source Separation* (Academic, Oxford 2010)

31.11 A. Hyvärinen, J. Karhunen, E. Oja: *Independent Component Analysis* (Wiley, New York 2001)

31.12 A. Cichocki, R. Zdunek, A.H. Phan, S.-I. Amari: *Nonnegative Matrix Tensor Factorizations* (Wiley, Chichester 2009)

31.13 E. Gaussier, C. Goutte: Relation between plsa and nmf and implications, Proc. 28th Int. ACM Conf. Res. Dev. Inf. Retr. (SIGIR'05) (ACM, New York 2005) pp. 601–602

31.14 D.T. Pham: Mutual information approach to blind separation of stationary sources, IEEE Trans. Inf. Theory **48**, 1935–1946 (2002)

31.15 M. Minami, S. Eguchi: Robust blind source separation by beta divergence, Neural Comput. **14**, 1859–1886 (2002)

31.16 T.-W. Lee, M. Girolami, T.J. Sejnowski: Independent component analysis using an extended infomax algorithm for mixed sub-Gaussian and super-Gaussian sources, Neural Comput. **11**(2), 417–441 (1999)

31.17 K. Labusch, E. Barth, T. Martinetz: Sparse coding neural gas: Learning of overcomplete data representations, Neuro **72**(7–9), 1547–1555 (2009)

31.18 A. Cichocki, S. Cruces, S.-I. Amari: Generalized alpha-beta divergences and their application to robust nonnegative matrix factorization, Entropy **13**, 134–170 (2011)

31.19 I. Csiszár: Axiomatic characterization of information measures, Entropy **10**, 261–273 (2008)

31.20 F. Liese, I. Vajda: On divergences and informations in statistics and information theory, IEEE Trans. Inf. Theory **52**(10), 4394–4412 (2006)

31.21 T. Villmann, S. Haase: Divergence based vector quantization, Neural Comput. **23**(5), 1343–1392 (2011)

31.22 P.L. Zador: Asymptotic quantization error of continuous signals and the quantization dimension, IEEE Trans. Inf. Theory **28**, 149–159 (1982)

31.23 T. Villmann, J.-C. Claussen: Magnification control in self-organizing maps and neural gas, Neural Comput. **18**(2), 446–469 (2006)

31.24 B. Hammer, A. Hasenfuss, T. Villmann: Magnification control for batch neural gas, Neurocomputing **70**(7–9), 1225–1234 (2007)

31.25 E. Merényi, A. Jain, T. Villmann: Explicit magnification control of self-organizing maps for "forbidden" data, IEEE Trans. Neural Netw. **18**(3), 786–797 (2007)

31.26 T. Villmann, S. Haase: Magnification in divergence based neural maps, Proc. Int. Jt. Conf. Artif. Neural Netw. (IJCNN 2011), ed. by R. Mikkulainen (IEEE, Los Alamitos 2011) pp. 437–441

Part D | 31

31.27   R. Chalasani, J.C. Principe: Self organizing maps with the correntropy induced metric, Proc. Int. Jt. Conf. Artif. Neural Netw. (IJCNN 2010) (IEEE, Barcelona 2010) pp. 1–6

31.28   T. Lehn-Schiøler, A. Hegde, D. Erdogmus, J.C. Principe: Vector quantization using information theoretic concepts, Nat. Comput. **4**(1), 39–51 (2005)

31.29   R. Jenssen, D. Erdogmus, J.C. Principe, T. Eltoft: The Laplacian PDF distance: A cost function for clustering in a kernel feature space, Adv. Neural Inf. Process. Syst., Vol. 17 (MIT Press, Cambridge 2005) pp. 625–632

31.30   A. Hegde, D. Erdogmus, T. Lehn-Schiøler, Y.N. Rao, J.C. Principe: Vector quantization by density matching in the minimum Kullback-Leibler-divergence sense, Proc. Int. Jt. Conf. Artif. Neural Netw. (IJCNN), Budapest (IEEE, New York 2004) pp. 105–109

31.31   G.E. Hinton, S.T. Roweis: Stochastic neighbor embedding, Adv. Neural Inf. Process. Syst., Vol. 15 (MIT Press, Cambridge 2002) pp. 833–840

31.32   L. van der Maaten, G. Hinten: Visualizing data using t-SNE, J. Mach. Learn. Res. **9**, 2579–2605 (2008)

31.33   K. Bunte, S. Haase, M. Biehl, T. Villmann: Stochastic neighbor embedding (SNE) for dimension reduction and visualization using arbitrary divergences, Neurocomputing **90**(9), 23–45 (2012)

31.34   M. Strickert, F.-M. Schleif, U. Seiffert, T. Villmann: Derivatives of pearson correlation for gradient-based analysis of biomedical data, Intel. Artif. Rev. Iberoam. Intel. Artif. **37**, 37–44 (2008)

31.35   M. Strickert, B. Labitzke, A. Kolb, T. Villmann: Multi-spectral image characterization by partial generalized covariance, Proc. Eur. Symp. Artif. Neural Netw. (ESANN'2011), Louvain-La-Neuve, ed. by M. Verleysen (2011) pp. 105–110

31.36   V. Gómez-Verdejo, M. Verleysen, J. Fleury: Information-theoretic feature selection for functional data classification, Neurocomputing **72**(16–18), 3580–3589 (2009)

31.37   B. Hammer, T. Villmann: Generalized relevance learning vector quantization, Neural Netw. **15**(8/9), 1059–1068 (2002)

31.38   T. Villmann, M. Kästner: Sparse functional relevance learning in generalized learning vector quantization, Lect. Notes Comput. Sci. **6731**, 79–89 (2011)

31.39   M. Kästner, B. Hammer, M. Biehl, T. Villmann: Functional relevance learning in generalized learning vector quantization, Neurocomputing **90**(9), 85–95 (2012)

31.40   A. Kraskov, H. Stogbauer, P. Grassberger: Estimating mutual information, Phys. Rev. E **69**(6), 66–138 (2004)

31.41   Y.-I. Moon, B. Rajagopalan, U. Lall: Estimating mutual information by kernel density estimators, Phys. Rev. E **52**, 2318–2321 (1995)

31.42   J.C. Principe: *Information Theoretic Learning* (Springer, Heidelberg, 2010)

31.43   R. Andonie, A. Cataron: An information energy LVQ approach for feature ranking, Eur. Symp. Artif. Neural Netw. 2004, ed. by M. Verleysen (d-side, Evere 2004) pp. 471–476

31.44   R. Jenssen, D. Erdogmus, J.C. Principe, T. Eltoft: Some equivalences between kernel methods and information theoretic methods, J. VLSI Signal Process. **45**, 49–65 (2006)

31.45   P.J.G. Lisboa, T.A. Etchells, I.H. Jarman, C.T.C. Arsene, M.S.H. Aung, A. Eleuteri, A.F.G. Taktak, F. Ambrogi, P. Boracchi, E. Biganzoli: Partial logistic artificial neural network for competing risks regularized with automatic relevance determination, IEEE Trans. Neural Netw. **20**(9), 1403–1416 (2009)

31.46   M.I. Jordan: Graphical models, Stat. Sci. **19**, 140–155 (2004)

31.47   D. Koller, N. Friedman: *Probabilistic Graphical Models: Principles and Techniques – Adaptive Computation and Machine Learning* (MIT Press, Cambridge 2009)

31.48   A.P. Dempster, N.M. Laird, D.B. Rubin: Maximum likelihood from incomplete data via the EM algorithm, J. R. Stat. Soc. Ser. B **39**(1), 1–38 (1977)

31.49   M.E. Tipping, C.M. Bishop: Probabilistic principal component analysis, J. R. Stat. Soc. Ser. B **61**(3), 611–622 (1999)

31.50   T. Hofmann: Unsupervised learning by probabilistic latent semantic analysis, Mach. Learn. **42**(1/2), 177–196 (2001)

31.51   M. Welling, C. Chemudugunta, N. Sutter: Deterministic latent variable models and their pitfalls, SIAM Int. Conf. Data Min. (2008)

31.52   D.M. Blei, A.Y. Ng, M.I. Jordan: Latent Dirichlet allocation, J. Mach. Learn. Res. **3**, 993–1022 (2003)

31.53   T. Minka, J. Lafferty: Expectation propagation for the generative aspect model, Proc. Conf. Uncertain. AI (2002)

31.54   T. Griffiths, M. Steyvers: Finding scientific topics, Proc. Natl. Acad. Sci. USA **101**, 5228–5235 (2004)

31.55   M. Blei, D. Blei, T. Griffiths, J. Tenenbaum: Hierarchical topic models and the nested Chinese restaurant process, Adv. Neural Inf. Process. Syst., Vol. 16 (MIT Press, Cambridge 2004) p. 17

31.56   M. Rosen-Zvi, T. Griffiths, M. Steyvers, P. Smyth: The author-topic model for authors and documents, Proc. 20th Conf. Uncertain. Artif. Intell., UAI '04 (AUAI, Corvallis 2004) pp. 487–494

31.57   L.-J. Li, L. Fei-Fei: What, where and who? classifying events by scene and object recognition, IEEE 11th Int. Conf. Comput. Vis. (ICCV) 2007 (2007), pp. 1–8

31.58   L.R. Rabiner: A tutorial on hidden markov models and selected applications in speech recognition, Proc. IEEE **77**(2), 257–286 (1989)

31.59   L.E. Baum, T. Petrie: Statistical inference for probabilistic functions of finite state Markov chains, Ann. Math. Stat. **37**(6), 1554–1563 (1966)

31.60 S.E. Levinson, L.R. Rabiner, M.M. Sondhi: An introduction to the application of the theory of probabilistic functions of a Markov process to automatic speech recognition, Bell Syst. Tech. J. **62**(4), 1035–1074 (1983)

31.61 P.A. Devijver: Baum's forward-backward algorithm revisited, Pattern Recogn. Lett. **3**(6), 369–373 (1985)

31.62 M. Brand, N. Oliver, A. Pentland: Coupled hidden Markov models for complex action recognition, Computer Vision and Pattern Recognition, Proc., 1997 IEEE (1997) pp. 994–999

31.63 Z. Ghahramani, M.I. Jordan: Factorial hidden Markov models, Mach. Learn. **29**(2), 245–273 (1997)

31.64 Y. Bengio, P. Frasconi: Input-output HMMs for sequence processing, IEEE Trans. Neural Netw. **7**(5), 1231–1249 (1996)

31.65 Y. Li, H.Y. Shum: Learning dynamic audio-visual mapping with input-output hidden Markov models, IEEE Trans. Multimed. **8**(3), 542–549 (2006)

31.66 B. Knab, A. Schliep, B. Steckemetz, B. Wichern: Model-based clustering with hidden Markov models and its application to financial time-series data, Proc. GfKl 2002 Data Sci. Appl. Data Anal. (Springer, Berlin, Heidelberg 2003) pp. 561–569

31.67 M. Seifert, M. Strickert, A. Schliep, I. Grosse: Exploiting prior knowledge and gene distances in the analysis of tumor expression profiles with extended hidden Markov models, Bioinformatics **27**(12), 1645–1652 (2011)

31.68 M. Diligenti, P. Frasconi, M. Gori: Hidden tree markov models for document image classification, IEEE Trans. Pattern Anal. Mach. Intell. **25**(4), 519–523 (2003)

31.69 D. Bacciu, A. Micheli, A. Sperduti: Compositional generative mapping for tree-structured data – Part I: Bottom-up probabilistic modeling of trees, IEEE Trans. Neural Netw. Learn. Syst. **23**(12), 1987–2002 (2012)

31.70 D. Bacciu, A. Micheli, A. Sperduti: An input-output hidden Markov model for tree transductions, Neurocomputing **112**, 34–46 (2013)

31.71 M.J. Beal, Z. Ghahramani, C.E. Rasmussen: The infinite hidden Markov model, Adv. Neural Inf. Process. Syst. **14**, 577–584 (2002)

31.72 C. Sutton, A. McCallum: An introduction to conditional random fields for relational learning. In: *Introduction to Statistical Relational Learning*, ed. by L. Getoor, B. Taskar (MIT Press, Cambridge 2006) pp. 93–128

Part D | 31

32. Kernel Methods

Marco Signoretto, Johan A. K. Suykens

This chapter addresses the study of kernel methods, a class of techniques that play a major role in machine learning and nonparametric statistics. Among others, these methods include support vector machines (SVMs) and least squares SVMs, kernel principal component analysis, kernel Fisher discriminant analysis, and Gaussian processes. The use of kernel methods is systematic and properly motivated by statistical principles. In practical applications, kernel methods lead to flexible predictive models that often outperform competing approaches in terms of generalization performance. The core idea consists of mapping data into a high-dimensional space by means of a feature map. Since the feature map is normally chosen to be nonlinear, a linear model in the feature space corresponds to a nonlinear rule in the original domain. This fact suits many real world data analysis problems that often require nonlinear models to describe their structure.

In Sect. 32.1 we present historical notes and summarize the main ingredients of kernel methods. In Sect. 32.2 we present the core ideas of statistical learning and show how regularization can be employed to devise practical learning algorithms. In Sect. 32.3 we show a selection of techniques that are representative of a large class of kernel methods; these techniques – termed primal–dual methods – use Lagrange duality as the main mathematical tools. Section 32.4 discusses Gaussian processes, a class of kernel methods that uses a Bayesian approach to perform inference and learning. Section 32.5 recalls different approaches for the tuning of parameters. In Sect. 32.6 we review the mathematical properties of different yet equivalent notions of kernels and recall a number of specialized kernels for learning problems involving structured data. We conclude the chapter by presenting applications in Sect. 32.7.

Part D | 32

32.1 Background

This chapter addresses the study of kernel methods, a class of techniques that play a major role in machine learning and nonparametric statistics.

The development of kernel-based techniques [32.1, 2] has been an important activity within machine learning in the last two decades. In this period, a number of powerful kernel-based learning algorithms were proposed. Among others, these methods include support vector machines (SVMs) and least squares SVMs, kernel principal component analysis, kernel Fisher discriminant analysis, and Gaussian processes. The use of kernel methods is systematic and properly motivated by statistical principles. In practical applications, kernel methods lead to flexible predictive models that often outperform competing approaches in terms of generalization performance. The core idea consists of mapping data into a high-dimensional space by means of a feature map. Since the feature map is normally chosen to be nonlinear, a linear model in the feature space corresponds to a nonlinear rule in the original domain. This fact suits many real world data analysis problems that often require nonlinear models to describe their structure.

32.1.1 Summary of the Chapter

In the rest of this section we present historical notes and summarize the main ingredients of kernel methods. In Sect. 32.2 we present the core ideas of statistical learning and show how regularization can be employed to devise practical learning algorithms. In Sect. 32.3 we show a selection of techniques that are representative of a large class of kernel methods; these techniques – termed primal–dual methods – use Lagrange duality as the main mathematical tools. Section 32.4 discusses Gaussian processes, a class of kernel methods that uses a Bayesian approach to perform inference and learning. Section 32.5 recalls different approaches for the tuning of parameters. In Sect. 32.6 we review the mathematical properties of different yet equivalent notions of kernels and recall a number of specialized kernels for learning problems involving structured data. We conclude the chapter by presenting applications in Sect. 32.7. Additional information can be found in a number of existing tutorials on SVMs and kernel methods, including [32.3–7].

32.1.2 Historical Background

The study of the mathematical foundation of kernels can be traced back at least to the beginning of the nineteenth century in connection with a general theory of integral equations [32.8, 9]. According to [32.10] the theory of reproducing kernel Hilbert spaces (RKHS) was first applied to detection and estimation problems by Parzen [32.11]. Properties of (reproducing) kernels are thoroughly presented in [32.12]. A first systematic treatment in the domain of nonparametric statistics can be found in [32.13]. Modern mathematical reviews include [32.14, 15]. The first use of kernels in the context of machine learning is generally attributed to [32.16]. The linear support vector algorithm, which undoubtedly had a prominent role in the history of kernel methods, made its first appearance in Russia in the 1960s [32.17, 18], in the framework of the *statistical learning theory* developed by *Vapnik* and *Chervonenkis* [32.19, 20]. Later, the idea was developed in connection to kernels by *Vapnik* and co-workers at AT&T labs [32.21–24]. The novel approach was rooted on a solid theoretical foundation. Additionally, studies began to report state-of-the-art performances in a number of applications, which further stimulated research on kernel-based techniques.

32.1.3 The Main Ingredients

Before delving into the details we now present the general setting for statistical learning problems and then briefly review the main ingredients of a substantial part of kernel methods used in machine learning.

Setting for Statistical Learning

The setting of learning from examples comprises three components [32.25]:

1. A *generator* of input data. We shall assume that data can be represented as vectors of $\mathbb{R}^D$. These vectors are independently and identically distributed (i.i.d.) according to a fixed but unknown probability distribution $p(x)$.
2. A *supervisor* that, given input data x, returns an output value y according to a conditional distribution $p(y|x)$ also fixed and unknown. Note that the supervisor might or might not be present.

3. A *learning machine* (or *learning algorithm*) able to choose an hypothesis

$$f(x; \theta) . \qquad (32.1)$$

Note that the hypothesis f is a function of x and depends upon a parameter vector θ belonging to a set Θ. The corresponding *hypothesis space* is then

$$\mathcal{S} = \{f(x; \Theta) : \theta \in \Theta\} , \qquad (32.2)$$

which is one-to-one with the *parameter space* Θ.

When the supervisor is present the learning problem is called *supervised*. The goal is to find that hypothesis that best mimics the supervisor response. When the supervisor is not present, the learning problem is called *unsupervised*. In this case, the aim is to find an hypothesis that represents the best concise representation of the data produced by the generator.

In both cases we might be interested either in the whole domain or we might be concerned only with a specific subset of points.

Feature Mapping and Kernel Trick

Kernel methods are a special class of learning algorithms. Their main idea consists of mapping input points, generally represented as elements of $\mathbb{R}^D$, into an high-dimensional inner product space $\mathcal{F}$, called the *feature space*. The mapping is performed by means of a *feature map* ϕ

$$\phi : \mathbb{R}^D \to \mathcal{F} ,$$
$$x \mapsto \phi(x) . \qquad (32.3)$$

One then approaches the learning task of interest by finding a linear model in the features space according to training points $\phi(x_1), \dots, \phi(x_N) \in \mathcal{F}$. Since the feature map is normally chosen to be nonlinear, a linear model in the feature space corresponds to a nonlinear rule in $\mathbb{R}^D$. Alternative kernel methods differ in the way the linear model in the feature space is found. Nonetheless, a common feature across different techniques is the following. If the algorithm can be expressed solely in terms of inner products, one can restate the problem in terms of evaluations of a *kernel function*

$$k : \mathbb{R}^D \times \mathbb{R}^D \to \mathbb{R} ,$$
$$(x, y) \mapsto k(x, y) , \qquad (32.4)$$

by letting

$$k(x, y) = \phi(x)^\top \phi(y) . \qquad (32.5)$$

This fact, usually referred to as the *kernel trick*, is of particular interest for the cases where the feature space is infinite dimensional, which prevents direct computation in the feature space. In practice, one often starts with designing a *positive definite* kernel, which guarantees the existence of a feature map ϕ satisfying (32.4).

Primal–Dual Estimation Techniques

As we shall see, an important class of machine learning methods consists of *primal–dual* learning techniques [32.1, 2, 26]. In this case, one starts from a *primal model* representation of the type

$$f(x; w, b) = w^\top \phi(x) + b$$
$$= \sum_i w_i \phi_i(x) + b . \qquad (32.6)$$

With reference to (32.1) note that here we have the tuple $\theta = (w, b)$. The *primal problem* is then a mathematical optimization problem aimed at finding optimal $w \in \mathcal{F}$ and $b \in \mathbb{R}$. Notably, the right-hand side of (32.6) is affine in $\phi(x)$; however, since ϕ is in general a nonlinear mapping, f is a nonlinear function of x.

A first approach consists of solving the primal problem. The information content of training data is absorbed into the primal model's parameters during the procedure to find optimal parameters; the evaluation of the model (32.6) on new patterns (*out-of-sample extension*) does no longer require the use of training data; therefore, they can be discarded after training.

A second approach relies on Lagrangian duality arguments. In this case, the solution is represented in terms of dual variables $\alpha_1, \alpha_2, \dots, \alpha_N$ and solved in $\alpha, \in \mathbb{R}^N$ and $b \in \mathbb{R}$. The *dual model* representation is then

$$f(x; \alpha, b) = \sum_{n=1}^{N} \alpha_n k(x_n, x) + b \qquad (32.7)$$

and depends upon the training patterns $x_1, x_2, \dots, x_N \in \mathbb{R}^D$. The representation in (32.6) is usually called *parametric*, while (32.7) is the *nonparametric* representation [32.26].

Part D | 32.1

32.2 Foundations of Statistical Learning

In this section, we briefly recall the main nomenclature and give a basic introduction on statistical learning theory. Historically, statistical learning theory constituted the theoretical foundation upon which the main methods of support vector machines were grounded. The theory is similar in spirit to a number of alternative complexity criteria and bias-variance trade-off curves. Nowadays, it remains a powerful framework for the design of learning algorithms.

32.2.1 Supervised and Unsupervised Inductive Learning

We have already introduced the distinction between *supervised* and *unsupervised*. Three important learning tasks are found within this categorization: *regression*, *classification*, and *density estimation*. In *regression* the supervisor's response takes values in the real numbers. In *classification* the supervisor's output takes values in the discrete finite set of possible labels $\mathcal{Y}$. In particular, in the *binary classification* problem $\mathcal{Y}$ consists of two elements, e.g., $\mathcal{Y} = \{-1, 1\}$. *Density estimation* is an instance of unsupervised learning: there is no supervisor. The functional relation to be learned from examples is the probability density $p(x)$ (the generator). Supervised and unsupervised learning are concerned with estimating a function (an optimal hypothesis) over the whole input domain $\mathbb{R}^D$ based upon a finite set of training points. Therefore, they are *inductive* approaches aiming at the general picture.

32.2.2 Semi-Supervised and Transductive Learning

Semi-Supervised Inductive Learning
In supervised learning the N training data are i.i.d. pairs

$$\{(x_1, y_1), (x_2, y_2), \ldots, (x_N, y_N)\} \subset \mathbb{R}^D \times \mathcal{Y}, \quad (32.8)$$

each of which is assumed to be drawn according to

$$p(x, y) = p(y|x)p(x) . \quad (32.9)$$

There is yet another inductive approach, namely *semi-supervised learning*. In semi-supervised learning one has a set of labeled pairs (32.8), as in supervised learning, as well as a set of unlabeled data

$$\{x_{N+1}, x_{N+2}, \ldots, x_{N+T}\} \subset \mathbb{R}^D \quad (32.10)$$

i.i.d. from the generator $p(x)$, as in unsupervised learning. The purpose is the same as in supervised learning to find an approximation of the supervisor response. However this goal is achieved by a learning algorithm that takes into account the additional information coming from the unlabeled data. According to [32.27] semi-supervised learning was popularized for the first time in the mid-1970s although similar ideas appeared earlier. Alternative semi-supervised learning algorithms differ in the way they exploit the information from the unlabeled set. One popular idea is to assume that the (possibly high-dimensional) input data lie (roughly) on a low-dimensional manifold [32.28–31].

Transductive Learning
In induction one seeks for the general picture with the purpose of making an out-of-sample prediction. This is an ambitious goal that might be unmotivated in certain settings. What if all the (unlabeled) data are given in advance? Suppose that one is only interested in prediction at given (finitely many) points. It is expected that this less ambitious task results in simple inference problems. These ideas are reflected in the approach found in *transductive learning* formulations. As in semi-supervised learning in transductive learning one has training pairs (32.65) as well as test (unlabeled) data (32.10). However, differently from semi-supervised learning one is only interested in making predictions for the test data (32.10).

32.2.3 Bounds on Generalization Error

Transductive and inductive inference share the common goal of achieving the lowest possible error on test data. In contrast with induction, transduction assumes that input test data are given in advance and consist of a finite discrete set of patterns drawn from the same distribution as the training set. From this perspective, it is clear that both transductive and inductive learning are concerned with *generalization*. In turn, a powerful framework to study the problem of generalization is the *structural risk minimization* (SRM) principle.

Expected and Empirical Risk
The starting point is the definition of a *loss* $L(y, f(x; \theta))$, or discrepancy, between the response y of the supervisor to a given input x and the response $f(x; \theta)$ of the learning algorithm (that can be transductive or inductive). Formally, the generalization error can be defined

s the *expected risk*

$$R(\theta) = \int L(y, f(x; \theta)) p(x, y) \mathrm{d}x \mathrm{d}y . \qquad (32.11)$$

rom a mathematical perspective the goal of learning is the minimization of this quantity. However, $p(x, y)$ is nknown and one can rely only on the sample version f (32.11), namely the *empirical risk*

$$R_{\text{emp}}^N(\theta) = \sum_{n=1}^{N} L(y_n, f(x_n, \theta)) . \qquad (32.12)$$

A possible learning approach is based on *empirical risk minimization* (ERM) and encompasses maximum likelihood (ML) inference [32.25]. It consists of finding

$$\hat{\theta}_N := \arg \min_{\theta \in \Theta} R_{\text{emp}}^N(\theta) . \qquad (32.13)$$

Consistency

Definition 32.1
The ERM approach is said to be *consistent* if

$$R_{\text{emp}}^N(\hat{\theta}_N) \xrightarrow{N} \inf_{\theta \in \Theta} R(\theta) ,$$

$$R(\hat{\theta}_N) \xrightarrow{N} \inf_{\theta \in \Theta} R(\theta) ,$$

where $\xrightarrow{N}$ denotes convergence in probability for $N \to \infty$.

n words: the ERM is consistent if, as the number of raining patterns N increases, *both* the expected risk $R(\hat{\theta}_N)$ and the empirical risk $R_{\text{emp}}^N(\hat{\theta}_N)$ converge to the minimal possible risk $\min_{\theta \in \Theta} R(\theta)$, see Fig. 32.1 for n illustration.

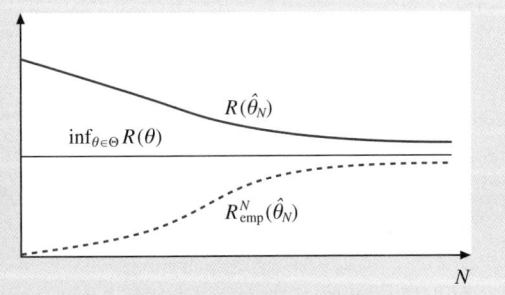

Fig. 32.1 Consistency of ERM

It was shown in [32.32] that the necessary and sufficient condition for consistency is that

$$P\left\{ \sup_{\theta \in \Theta} |R(\theta) - R_{\text{emp}}^N(\theta)| \geq \epsilon \right\} \xrightarrow{N} 0 , \quad \forall \epsilon > 0 . \qquad (32.14)$$

In turn, the necessary and sufficient conditions for (32.14) to hold true were established in 1968 by Vapnik [32.33, 34] and are based on *capacity factors*.

Capacity Factors
Consistency is one of the main theoretical questions in statistics. From a learning perspective, however, it does not address the most important aspect. The aspect that one should be mostly concerned with is how to control the generalization of a certain learning algorithm. Whereas consistency is an asymptotic result, we want to minimize the expected risk given that we have available only finitely many observations to train the learning algorithm. It turns out, however, that consistency is central to address also this aspect [32.25]. Additionally, a crucial role for answering this question is played by capacity factors that, roughly speaking, are all measures of how well the set of functions $\{f(x; \theta) : \theta \in \Theta\}$ can separate data. A more detailed description is given in the following (precise definitions and formulas can be found in [32.25, Chap. 2]). In general, the theory states that without restricting the set of admissible functions, the ERM is not consistent. The interested reader is referred to [32.25, 35].

VC Entropy. The first capacity factor (here and below VC is used as an abbreviation for Vapnik–Chervonenkis.) relates to the expected number of *equivalence classes* according to which the training patterns divide the set of functions $\{f(x; \theta) : \theta \in \Theta\}$ (an equivalence class is a subset of $\{f(x; \theta) : \theta \in \Theta\}$ consisting of functions that attribute the same labels to the input pattern in the training set). We denote the VC entropy by $\text{En}(p, N)$, where the symbols emphasize the dependence of the VC Entropy on the underlying joint probability p and the number of training patterns N. The condition

$$\lim_{N \to \infty} \frac{\text{En}(p, N)}{N} = 0$$

forms the necessary and sufficient condition for (32.14) to hold true *with respect to the fixed probability density p*.

Growth Function. It corresponds to the maximal number of *equivalence classes* with respect to all the possible training samples of cardinality N. As such, it is a distribution-independent version of the VC entropy obtained via a worst-case approach. We denote it by $\mathrm{Gr}(N)$. The condition

$$\lim_{N\to\infty} \frac{\ln \mathrm{Gr}(N)}{N} = 0$$

forms the necessary and sufficient condition for (32.14) to hold true *for all the probability densities p*.

VC Dimension. This is the cardinality of the largest set of points that the algorithm can shatter; we denote it by $\dim_{\mathrm{VC}}$. Note that $\dim_{\mathrm{VC}}$ is a property of $\{f(x;\theta) : \theta \in \Theta\}$, which neither depends on N nor on p. Roughly speaking it tells how flexible the set of functions is. A finite value of $\dim_{\mathrm{VC}}$ forms the necessary and sufficient condition for (32.14) to hold true *for all the probability densities p*.

The three capacities are related by the chain of inequalities [32.33, 34]

$$\mathrm{En}(p,N) \le \ln \mathrm{Gr}(N) \le \dim_{\mathrm{VC}}\left(\ln\frac{N}{\dim_{\mathrm{VC}}} + 1\right).$$
(32.15)

Finite-Sample Bounds

One of the key results of the theory developed by *Vapnik* and *Chervonenkis* is the following probabilistic bound. With probability $1 - \eta$ simultaneously for all $\theta \in \Theta$ it holds that [32.25]

$$R(\theta) \le R_{\mathrm{emp}}^N(\theta) + \sqrt{\frac{\mathrm{En}(p,2N) - \ln\eta}{N}}.$$
(32.16)

Note that the latter depends on p. The result says that, for a fixed set of functions $\{f(x;\theta) : \theta \in \Theta\}$, one can pick that $\theta \in \Theta$ that minimizes $R_{\mathrm{emp}}^N(\theta)$ and in this way obtain the best guarantee on $R(\theta)$. Now, taking into account (32.15) one can formulate the following bound based on the growth function

$$R(\theta) \le R_{\mathrm{emp}}^N(\theta) + \sqrt{\frac{\ln \mathrm{Gr}(2N) - \ln\eta}{N}}.$$
(32.17)

In the same way one has

$$R(\theta) \le R_{\mathrm{emp}}^N(\theta) + \sqrt{\frac{\dim_{\mathrm{VC}}\left(\ln\frac{2N}{\dim_{\mathrm{VC}}} + 1\right) - \ln\eta}{N}}.$$
(32.18)

Figure 32.2 illustrates the main idea.

Note that both (32.17) and (32.18) are distribution independent. Additionally (32.18) only depends upon the VC dimension (which, contrary to Gr, is in dependent from N). Unfortunately there is no free lunch: (32.17) is less tight than (32.16) and (32.18) is less tight than (32.17).

So far we gave a flavor of the theoretical framework in which the support vector algorithms were originally conceived. Recent research reinterpreted and significantly improved the error bounds using mathematical tools from approximation and learning theory, functional analysis, and statistics. The interested reader is referred to [32.36] and [32.37]. Although tighter bounds exist, the study of sharper bounds remains a challenge for future research. In fact, existing bounds are normally too loose to lead to practical model selection techniques, i.e., strategies for tuning the parameters that control the capacity of the model's class. Nonetheless, the theory provides important guidelines for the derivation of algorithms.

The Role of Transduction

It turns out that a key step in obtaining the bound (32.16) is based upon the *symmetrization lemma*

$$P\left\{\sup_\theta |R(\theta) - R_{\mathrm{emp}}^N(\theta)| \ge \epsilon\right\}$$
$$\le 2P\left\{\sup_\theta |R_{\mathrm{emp}_1}^{N_1}(\theta) - R_{\mathrm{emp}_2}^{N_2}(\theta)| \ge \frac{\epsilon}{2}\right\},$$
(32.19)

where $R_{\mathrm{emp}_1}^{N_1}$ and $R_{\mathrm{emp}_2}^{N_2}$ are constructed upon two different i.i.d. samples, precisely as in transduction. More specifically, (32.16) comes from upper-bounding the right-hand side of (32.19) [32.38]. More generally it is apparent that to obtain all bounds of this type the key element remains the symmetrization lemma [32.25].

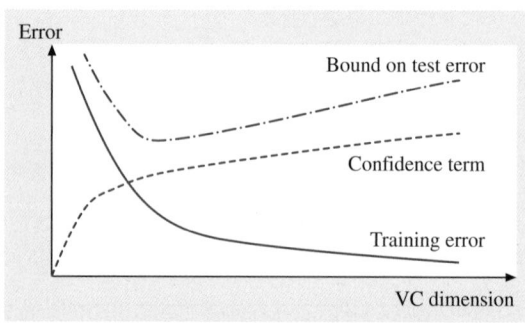

Fig. 32.2 Illustration of the generalization bound depending on capacity factors

Notably starting from the latter one can derive bounds explicitly designed for the transductive case where one of the two samples plays the role of the training set and the other of the test set. In light of this, *Vapnik* argues that transductive inference is a fundamental step in machine learning. Additionally, since the bounds for transduction are tighter than those for induction, the theory suggests that, whenever possible, transductive inference should be preferred over inductive inference. Practical algorithms can take advantage of this fact by implementing the *adaptive* version of the structural risk minimization (SRM) principle that we discuss next.

32.2.4 Structural Risk Minimization and Regularization

The Structural Risk Minimization Principle

The structure of the bounds above suggests that one should minimize the empirical risk while controlling some measure of complexity. The idea behind the SRM, introduced in the 1970s, is to construct nested subsets of functions

$$ S_1 \subset S_1 \subset \cdots \subset S_L = S = \{f(x, \theta) : \theta \in \Theta\} \, , $$
(32.20)

where each subset S_l has capacity h_l (VC entropy, growth function, or VC dimension) with $h_1 < h_2 < \cdots < h_l$ and S is the entire hypothesis space. Then one chooses an element of the nested subsets so that the second term in the right-hand side of the bounds is kept under control; within that subset one then picks that specific function that minimizes the empirical risk. As *Vapnik* points out in [32.38]:

> [...] to find a good solution using a finite (limited) number of training examples one has to construct a (smart) structure which reflects prior knowledge about the problem of interest.

In practice one can use the information coming from the unlabeled data to define a smart structure to improve the learning. In other words, the side information coming from unlabeled data can serve the purpose of devising a *data-dependent* set of functions. On top of this, one should use additional side information over the structure of the problem, whenever available. Indeed, using *informative representations* for the input data is also a way to construct a smart set of functions. In fact, representing the data in a suitable form implies a mapping from the input space to a more convenient set of features. We will discuss this aspect more extensively in Sect. 32.6.

Learning Through Regularization

So far we have addressed the theory but we have not talked about how to practically implement it. It is understood that the essential idea of SRM is to find the best trade-off between the empirical risk and some measure of complexity (the capacity) of the hypothesis space. This ensures that the left-hand side of VC bounds – the expected risk that we are interested in to achieve generalization – is minimized. In practice there are different ways to define the sets in the sequence (32.20). The generic set S_l could be the set of polynomials of degree l or a set of splines with l nodes. However, it is in connection to *regularization theory* that practical implementations of the SRM principle find their natural domain.

Tikhonov Theory

Regularization theory was introduced by *Andrey Tikhonov* [32.39–41] as a way to solve ill-posed problems. Ill-posed problems are problems that are not well posed in the sense of *Hadamard* [32.42]. Consider solving in f a linear operatorial equation of the type

$$ Af = b \, . $$
(32.21)

In the general case, f is an element of a Hilbert space, A is a compact operator, and b is an element of its range. Even if a solution exists, it is often observed that a slight perturbation of the right-hand side b causes large deviations in the solution f. *Tikhonov* proposed to solve this problem by minimizing a functional of the type

$$ \|Af - b\|^2 + \lambda \, \Gamma(f) \, , $$

where $\|\cdot\|$ is a suitable norm on the range of A, λ is some hyperparameter, and Γ is a regularization functional (sometimes called *stabilizer*). The theory of such an approach was developed by *Tikhonov* and *Ivanov*; in particular it was shown that there exists a strategy to choose λ depending on the accuracy of b that asymptotically leads to the desired solution $f^\star$. This was shown under the assumption that there exists $c^\star$ such that $f^\star \in \{f : \Gamma(f) \leq c^\star\}$.

According to *Vapnik* [32.20], the theory of Tikhonov regularization differs from statistical learning theory in a number of ways. To begin with Tikhonov regularization considers specific structures in the nested sequence (32.20) (depending on the way Γ is defined);

secondly it requires the solution to be in the hypothesis space; finally the theory developed by Tikhonov and Ivanov was not concerned with guarantees for a finite number of observations.

When f is an element of a reproducing kernel Hilbert space (RKHS) (Sect. 32.6), the theory is best known through the work of *Wahba* [32.13, 43].

SRM and Regularization in RKHSs

SVMs, and more generally, primal–dual learning algorithms, represent an important class of kernel methods. The primal–dual approach emphasizes the geometrical aspects of the problem and it is particularly insightful when (32.7) is used to define a discriminative rule arising in a classification problem. We will consider this class of learning algorithms in later sections. Alternatively, the setting of RKHSs provides a convenient way to define the sequence (32.20). When the hypothesis space S coincides with a RKHS of functions $\mathcal{H}$, a nested sequence can be constructed by bounding the norm in $\mathcal{H}$, used as a proxy for the complexity of models

$$S_l = \{ f \in \mathcal{H} : \|f\| \le a_l \} . \tag{32.22}$$

It turns out that there is a measure of capacity of S_l, which is an increasing function of a_l [32.44]. This capacity measure can be used to derive probabilistic bounds in line with (32.16), (32.17), and (32.18). In practice, instead of solving the *constrained problem*

$$\min_{f \in \mathcal{H}} R^N_{\mathrm{emp}}(f)$$

$$\text{subject to } \|f\| \le a_l \tag{32.23}$$

for any l, one normally solves the provably equivalent *penalized problem*

$$\min_{f \in \mathcal{H}} R^N_{\mathrm{emp}}(f) + \lambda_l \|f\|^2 \tag{32.24}$$

and pick the optimal λ_l appropriately. Note that in (32.23) and (32.24) we wrote $R^N_{\mathrm{emp}}(f)$ instead of $R^N_{\mathrm{emp}}(\theta)$, as before. In fact, in this case, the solution of the learning problem is found by formulating a *convex variational problem* where the function f itself plays the role of the optimization variable θ. In practice, however, the *representer theorem* [32.13, 43, 45, 46] shows that a representation of the optimal f only depends upon an expansion of kernel functions centered at the training patterns. This result leads to a representation for f in line with (32.7). More specifically it holds that $f(x) =$

$\sum_{n=1}^{N} \alpha_n k(x_n, x)$ where $\alpha \in \mathbb{R}^N$ is found solving a finite dimensional optimization problem. The latter is convex [32.47] provided that L in the empirical risk (32.12) is a convex loss function.

Abstract Penalized Empirical Risk Minimization Problems

The penalized empirical risk minimization problem was introduced in (32.24) in the setting of RKHS of functions. However, it shall be noted that it is a very general idea. Ultimately this can be related to the generality of (32.21). The latter can either refer to infinite dimensional problems or to a finite system of linear equations involving objects living in some finite dimensional space. Therefore, for the sake of generality, one can consider in place of (32.24) the problem

$$\min_{\theta \in \Theta} R^N_{\mathrm{emp}}(\theta) + \lambda \, \Gamma(\theta) , \tag{32.25}$$

where Θ — which is one-to-one with the hypothesis space — either coincides with some abstract vector space, or it is a subset of it; R^N_{emp} is the empirical risk and $\Gamma : \Theta \to \mathbb{R}$ is a suitable penalty function. This, in particular, includes the situations where θ is a vector, a matrix, or a higher-order tensor (i. e., a higher-order array generalizing the notion of matrices).

32.2.5 Types of Regularization

A penalty frequently used in practice is $\Gamma(\theta) = \|\theta\|^2$ where $\|\theta\|$ is the *Hilbertian norm* defined upon the space's inner product

$$\|\theta\|^2 = \langle \theta, \theta \rangle . \tag{32.26}$$

This choice leads to *ridge regression* [32.48, 49]. Note that, in this case, $\|\theta\| = 0$ if and only if θ is the zero vector of the space. A more general class of quadratic penalties is represented by *seminorms*. A seminorm is allowed to assign zero length to some nonzero vectors (in addition to the zero vector). They are commonly used in smoothing splines [32.13, 50] where the unknown is decomposed into an unpenalized parametric component and a penalized nonparametric part.

LASSO and Non-Hilbertian Norms

The methods that we present in the next sections are all instances of the problem class in (32.25); although this is not necessarily emphasized in the presentation, they all employ a simple quadratic penalty. This is central to relying on Lagrange duality theory [32.47, 51], which, in turn, constitutes the main technical tool for the derivation of a large class of kernel methods. However, it is important to mention that in the last decade much research effort has been expended on the design of alternative penalties (correspondingly, there has been increased interest in other notions of duality, such as *Fenchel duality*). This arises from the realization that using a certain penalty is also a way to convey prior knowledge. This fact is best understood within a Bayesian setting, in light of a *maximum a posteriori* (MAP) interpretation of (32.25), see, e.g., [32.44]. A penalty term based on the space's inner product has been replaced with various type of *non-Hilbertian norms*. These are norms that, contrary to (32.26), do not arise from inner products. LASSO (least absolute shrinkage and selection operator, [32.52]) is perhaps the most prominent example of such cases. In LASSO one considers linear functions

$$f(x; \theta) = \langle \theta, x \rangle = \sum_{d=1}^{D} \theta_d x_d$$

and uses the l_1 norm

$$\|\theta\|_1 = \sum_{d=1}^{D} |\theta_d| \tag{32.27}$$

to promote the sparsity of the parameter vector θ. Note that this corresponds to defining the structure (32.20) according to

$$S_l = \{f(x; \theta) : \|\theta\|_1 \le a_l\} . \tag{32.28}$$

Like ridge regression, LASSO is a continuous shrinkage method that achieves good prediction performance via a bias-variance trade-off. Since usually the estimated coefficient vector has many entries equal to zero, the approach has the further advantage over ridge regression of giving rise to interpretable models.

More recently, different structure-inducing penalties have been proposed as a promising alternative [32.53–55]. The general idea is to convey structural assumption on the problem, such as grouping or hierarchies over the set of input variables, by suitably crafting the penalty. In this way, the users are permitted to customize the regularization approach according to their subjective knowledge on the task. Correspondingly, as in (32.28), one (implicitly) forms a smart structure of

nested subsets of functions, in agreement with the SRM principle.

These ideas have been generalized to the case where Θ is infinite dimensional, in particular in the framework of the multiple kernel learning (MKL) problem. This was investigated both from a functional viewpoint [32.56, 57], and from the more pragmatic point of view of optimization [32.58–60].

Spectral Regularization

Yet another generalization of (32.25) arises in the context of *multitask learning* [32.61–63]. In this setting one approaches simultaneously different learning tasks under some common constraint(s). The general idea, sometimes also known as *collaborative filtering*, is that one can take advantage of shared features across tasks. In practical applications it was shown that one can significantly gain in terms of generalization performance from exploiting such prior knowledge. From the point of view of learning through regularization, a sensible approach is given in [32.64, 65]. Suppose one has T datasets, one for each task; the t-th dataset has N_t observations. Note that, in general, $N_1 \ne N_2 \ne \cdots \ne N_T$. In this setting one has to learn vectors θ_t, $t = 1, 2, \ldots, T$, one per task; the parameter space is, therefore, a space of matrices, $\Theta = \mathbb{R}^{F \times T}$ where F is, possibly, infinity. The idea translates into penalized empirical risk minimization problems of the type

$$\min_{\theta = [\theta_1, \theta_2, \cdots, \theta_T] \in \mathbb{R}^{F \times T}} \sum_{t=1}^{T} R_{\text{emp}}^{N_t}(\theta_t) + \lambda \|\theta\|_* , \tag{32.29}$$

where $\|\theta\|_*$ is the *nuclear norm*

$$\|\theta\|_* = \sum_{r=1}^{R} \sigma_r(\theta) \tag{32.30}$$

and $\sigma_1(\theta), \sigma_2(\theta), \cdots, \sigma_R(\theta)$ are the $R \le \min(T, F)$ nonzero singular values of the $F \times T$ matrix θ. Note that (32.30) corresponds to the l_1 norm of the vector of singular values. The definition also remains valid in the infinite dimensional case under some regularity assumptions [32.66]. The nuclear norm is the convex envelope of the rank function on the spectral-norm unit ball [32.67]; roughly speaking, it represents the best convex relaxation of the rank function.

The use of the nuclear norm in (32.29) is motivated by the assumption that the parameter vectors of related tasks should be approximately linearly dependent. This assumption is meaningful for a number of cases of interest. Other uses of the nuclear norm exist; ultimately, this is due to the fact that notions of rank are ubiquitous in the mathematical formulations stemming from real-life problems. As a consequence, the nuclear norm is a very versatile mathematical tool to impose structure on (seemingly) very diverse settings. This includes the identification of linear time-invariant systems [32.68, 69] and the analysis of nonstationary cointegrated systems [32.70]. Finally we mention that, in place of (32.30), one can consider *spectral penalties* [32.71, 72] that include the nuclear norm as a special case.

32.3 Primal–Dual Methods

The purpose of this section is to introduce the methods that have served as the archetypal approaches for a large class of kernel methods. In the process, we detail the Lagrange duality argument underlying general primal–dual techniques. We begin by giving a short overview of the formulations of SVMs introduced by Vapnik; successively, we discuss a number of modifications and extensions.

32.3.1 SVMs for Classification

Margin
The problem of pattern recognition amounts to finding the label $y \in \{-1, 1\}$ that corresponds to a generic input point $x \in \mathbb{R}^D$. This task can be approached by assigning a label $\hat{y}$ according to the model

$$\hat{y} = \text{sign}\left[w^\top \phi(x) + b\right], \tag{32.31}$$

where $w^\top \phi(x) + b$ is a hyperplane, found by a learning algorithm based on training data

$$\{(x_1, y_1), (x_2, y_2), \ldots, (x_N, y_N)\} \subset \mathbb{R}^D \times \{-1, 1\}. \tag{32.32}$$

The concept of feature map ϕ was presented in short in Sect. 32.1.3. Later we will discuss the role of ϕ in more detail. For now, it suffices to say that ϕ is expected to capture features that are important for the discrimination of points. In the simplest case, ϕ is the identity map, i.e., $\phi : x \mapsto x$. Note that $w^\top \phi(x) + b$ is a primal model of the type (32.6), with $w \in \mathcal{F}$ and $b \in \mathbb{R}$.

In general, one can see that there are several possible separating hyperplanes, see Fig. 32.3.

The solution picked by the support vector classification (SVC) algorithm is the one that separates the data with the maximal margin. More precisely, *Vapnik* considered a rescaling of the problem so that points closest to the separating hyperplane satisfy the normalizing condition

$$\left|w^\top \phi(x) + b\right| = 1. \tag{32.33}$$

The two hyperplanes $w^\top \phi(x) + b = 1$ and $w^\top \phi(x) + b = -1$ are called *canonical hyperplanes*, and the distance between them is called the *margin*, see Fig. 32.4.

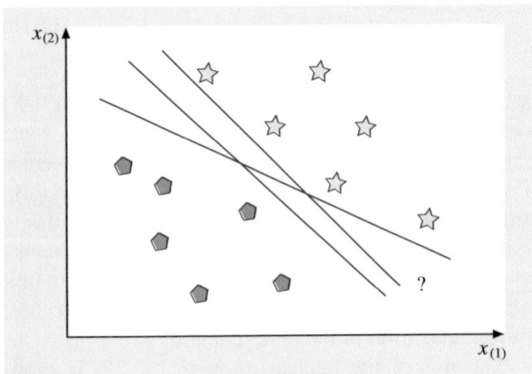

Fig. 32.3 Several possible separating hyperplanes exist

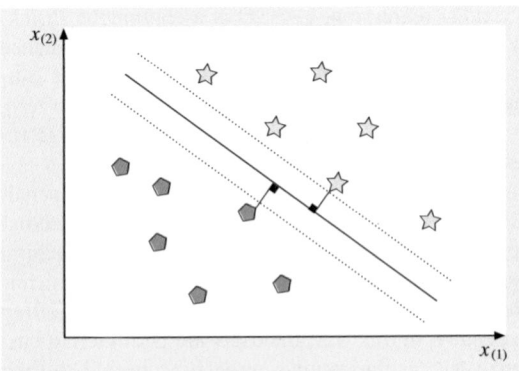

Fig. 32.4 SVC finds the solution that maximizes the margin

Assuming that the classification problem is *separable*, i.e., that there exists at least a hyperplane separating the training data (32.65), one obtains a canonical representation (w, b) satisfying

$$y_n \left(w^\top \phi(x_n) + b \right) \geq 1, \quad n = 1, \ldots, N. \quad (32.34)$$

Let us assume, without loss of generality, that $y_1 = 1$ and $y_2 = -1$. If the corresponding patterns x_1 and x_2 are among the closest points to the separating hyperplane, the scaling imposed by *Vapnik* implies

$$w^\top \phi(x_1) + b = 1,$$
$$w^\top \phi(x_2) + b = -1, \quad (32.35)$$

which, in turn, leads to

$$w^\top (\phi(x_1) - \phi(x_2)) = 2. \quad (32.36)$$

Now the normal vector to the separating hyperplane $w^\top \phi(x) + b$ is $(1/\|w\|)w$. The margin is equal to the component of the vector $\phi(x_1) - \phi(x_2)$ along $(1/\|w\|)w$, i.e., the projection

$$(1/\|w\|)w^\top (\phi(x_1) - \phi(x_2)).$$

Using (32.36) one obtains that the margin is equal to $2/\|w\|$; correspondingly, the distance between the points satisfying (32.33) and the separating hyperplane is $1/\|w\|$. By minimizing $\|w\|$, subject to the set of constraints (32.34), one obtains a *maximal margin classifier* that maximizes the margin between the two classes. This hyperplane, in turn, can be naturally envisioned as the simplest solution given the observed data.

Primal Problem

In practice, for computational reasons it is more convenient to minimize $\frac{1}{2}\|w\|^2 = \frac{1}{2}w^\top w$ rather than $\|w\|$. Additionally, it is in general unrealistic to assume that the classification problem is separable. In practical applications, one should try to find a set of features (in fact, a feature mapping from the input domain to a more convenient representation) that allow to separate the two classes as much as possible. Nonetheless, there might be no boundary that can perfectly separate the data; therefore one should tolerate misclassifications. Taking this requirement into account leads to the primal problem for the SVC algorithm [32.24]. This is the quadratic programming (QP) problem

$$\min_{w,b,\xi} \; J_P(w, \xi) = \frac{1}{2} w^\top w + c \sum_{n=1}^{N} \xi_n$$

subject to $y_n(w^\top \phi(x_n) + b) \geq 1 - \xi_n, \quad n = 1, \ldots, N$
$$\xi_n \geq 0, \quad n = 1, \ldots, N, \quad (32.37)$$

where $c > 0$ is a user-defined parameter. In this problem, one accounts for misclassifications by replacing the set of constraints (32.34), with the set of constraints

$$y_n \left(w^\top \phi(x_n) + b \right) \geq 1 - \xi_n, \quad n = 1, \ldots, N, \quad (32.38)$$

where $\xi_1, \xi_2, \ldots, \xi_N$ are positive slack variables. It is clear that for higher values of c one penalizes more the violations of the conditions in (32.34).

Dual Problem

The Lagrangian corresponding to (32.37) is

$$\mathcal{L}(w, b, \xi; \alpha, \nu) = J_P(w, \xi)$$
$$- \sum_{n=1}^{N} \alpha_n \left(y_n(w^\top \phi(x_n) + b) - 1 + \xi_n \right) - \sum_{n=1}^{N} \nu_n \xi_n, \quad (32.39)$$

with Lagrangian multipliers $\alpha_n \geq 0$, $\nu_n \geq 0$ for $n = 1, \ldots, N$. The solution is given by the saddle point of the Lagrangian

$$\max_{\alpha, \nu} \min_{w, b, \xi} \mathcal{L}(w, b, \xi; \alpha, \nu). \quad (32.40)$$

One obtains

$$\begin{cases} \dfrac{\partial \mathcal{L}}{\partial w} = 0 \rightarrow w = \sum_{n=1}^{N} \alpha_n y_n \phi(x_n), \\[2ex] \dfrac{\partial \mathcal{L}}{\partial b} = 0 \rightarrow \sum_{n=1}^{N} \alpha_n y_n = 0, \\[2ex] \dfrac{\partial \mathcal{L}}{\partial \xi_n} = 0 \rightarrow 0 \leq \alpha_n \leq c, \quad n = 1, \ldots, N. \end{cases} \quad (32.41)$$

The dual problem is then the QP problem

$$\max_{\alpha} \; J_D(\alpha)$$

subject to $\sum_{n=1}^{N} \alpha_n y_n = 0$
$$0 \leq \alpha_n \leq c, \quad n = 1, \ldots, N, \quad (32.42)$$

where

$$J_D(\alpha) = -\frac{1}{2} \sum_{m,n=1}^{N} y_m y_n k(x_m, x_n) \alpha_m \alpha_n + \sum_{n=1}^{N} \alpha_n \, , \tag{32.43}$$

and we used the kernel trick

$$k(x_m, x_n) = \phi(x_m)^\top \phi(x_n) \, , \quad m, n = 1, \dots, N \, . \tag{32.44}$$

The classifier based on the dual model representation is

$$\text{sign} \left[\sum_{n=1}^{N} \alpha_n y_n k(x, x_n) + b \right] \, , \tag{32.45}$$

where α_n are positive real numbers obtained solving (32.42) and b is obtained based upon Karush–Kuhn–Tucker (KKT) optimality conditions, i. e., the set of conditions that must be satisfied at the optimum of a constrained optimization problem. These are

$$\begin{cases} \frac{\partial \mathcal{L}}{\partial w} = 0 \to w = \sum_{n=1}^{N} \alpha_n y_n \phi(x_n) \, , \\[2mm] \frac{\partial \mathcal{L}}{\partial b} = 0 \to \sum_{n=1}^{N} \alpha_n y_n = 0 \, , \\[2mm] \frac{\partial \mathcal{L}}{\partial \xi_n} = 0 \to c - \alpha_n - v_n = 0 \, , \quad n = 1, \dots, N \\[2mm] \alpha_n \left(y_n (w^\top \phi(x_n) + b) - 1 + \xi_n \right) = 0 \, , \\[1mm] \qquad n = 1, \dots, N \, , \\[2mm] v_n \xi_n = 0 \, , \quad n = 1, \dots, N \, , \\[2mm] \alpha_n \geq 0 \, , \quad n = 1, \dots, N \, , \\[2mm] v_n \geq 0 \, , \quad n = 1, \dots, N \, . \end{cases} \tag{32.46}$$

From these equations it can be seen that, at optimum, we have

$$y_n (w^\top \phi(x_n) + b) - 1 = 0 \ \text{if} \ 0 < \alpha_n < c \, , \tag{32.47}$$

from which one can compute b.

SVC as a Penalized Empirical Risk Minimization Problem

The derivation so far followed the classical approach due to *Vapnik*; the main argument comes from geometrical insights on the pattern recognition problem.

Whenever the feature space is finite dimensional, one can approach learning either by solving the primal problem or by solving the dual one; when this is not the case one can still use the dual problem and rely on the dual representation obtained.

Before proceeding, we highlight a different, yet equivalent problem formulation. For the primal problem this reads

$$\min_{w,b} \sum_{n=1}^{N} \left[1 - y_n \left(w^\top \phi(x_n) + b \right) \right]_+ + \lambda \, w^\top w \, , \tag{32.48}$$

where we let $\lambda = 1/(2c)$ and we define $[\cdot]_+$ by

$$[a]_+ = \begin{cases} a, & \text{if } a \geq 0 \\ 0, & \text{otherwise} \, . \end{cases} \tag{32.49}$$

Problem (32.37) is an instance of (32.25) obtained by letting (note that $\mathcal{F} \times \mathbb{R}$ is naturally equipped with the inner product $\langle (w_1, b_1), (w_2, b_2) \rangle = w_1^\top w_2 + b_1 b_2$ and is a (finite-dimensional) Hilbert space (HS).) $\Theta = \mathcal{F} \times \mathbb{R}$ and taking as penalty the seminorm

$$\Gamma : (w, b) \mapsto w^\top w \, . \tag{32.50}$$

This shows that (32.37) is essentially a regularized empirical risk minimization problem that can be analyzed in the framework of the SRM principle presented in Sect. 32.2.

VC Bounds for Classification

In Sect. 32.2.3 we already discussed bounds on the generalization error in terms of capacity factors. In particular, (32.18) states a bound for the case of VC dimension. The larger this VC dimension the smaller the training error (empirical risk) can become but the confidence term (second term on the right-hand side of (32.18) will grow. The minimum of the sum of these two terms is then a good compromise solution. For SVM classifiers, *Vapnik* has shown that hyperplanes satisfying $\|w\| \leq a$ have a VC dimension h that is upper bounded by

$$h \leq \min([r^2 a^2], N) + 1 \, , \tag{32.51}$$

where $[\cdot]$ represents the integer part and r is the radius of the smallest ball containing the points $\phi(x_1), \phi(x_2), \dots, \phi(x_N)$ in the feature space $\mathcal{F}$.

Note that for each value of a there exists a corresponding value of λ in (32.48), correspondingly, a value of c in (32.37) or (32.42). Additionally, the radius r can also be computed by solving a QP problem, see, e.g., [32.26]. It follows that one could compute solutions corresponding to multiple values of the hyperparameters, find the corresponding empirical risk and radius and then pick the model corresponding to the least value of the right-hand side of the bound (32.18). As we have already remarked, however, the bound (32.18) is often too conservative. Sharper bounds and frameworks alternative to VC theory have been derived, see, e.g. [32.73]. In practice, however, the choice of parameters is often guided by data-driven model selection criteria, see Sect. 32.5.

Relative Margin and Data-Dependent Regularization

Although maximum margin classifiers have proved to be very effective, alternative notions of data separation have been proposed.

The authors of [32.74, 75], for instance, argue that maximum margin classifiers might be misled by direction of large variations. They propose a way to correct this drawback by measuring the margin not in an absolute sense but rather relative to the spread of data in any projection direction. Note that this can be seen as a way to conveniently craft the hypothesis space, an important aspect that we discussed in Sect. 32.2.

32.3.2 SVMs for Function Estimation

In addition to classification, the support vector methodology has also been introduced for linear and nonlinear function estimation problems [32.25]. For the general nonlinear case, output values are assigned according to the primal model

$$\hat{y} = w^\top \phi(x) + b. \tag{32.52}$$

In order to estimate the model's parameter w and b, from training data consisting of N input–output pairs, *Vapnik* proposed to evaluate the empirical risk according to

$$R_{\text{emp}} = \frac{1}{N} \sum_{n=1}^{N} \left| y_n - w^\top \phi(x_n) - b \right|_\epsilon \tag{32.53}$$

with the so called Vapnik's ϵ-insensitive loss function defined as

$$|y - f(x)|_\epsilon = \begin{cases} 0, & \text{if } |y - f(x)| \le \epsilon \\ |y - f(x)| - \epsilon, & \text{otherwise}. \end{cases} \tag{32.54}$$

The idea is illustrated in Fig. 32.5.

The corresponding primal optimization problem is the QP problem

$$\min_{w,b,\xi} J_P(w, \xi, \xi^*) = \frac{1}{2} w^\top w + c \sum_{n=1}^{N} \left(\xi_n + \xi_n^* \right)$$

subject to $y_n - w^\top \phi(x_n) - b \le \epsilon + \xi_n$,

$$n = 1, \ldots, N,$$

$$w^\top \phi(x_n) + b - y_n \le \epsilon + \xi_n^*,$$

$$n = 1, \ldots, N,$$

$$\xi_n, \quad \xi_n^* \ge 0, \, n = 1, \ldots, N, \tag{32.55}$$

where $c > 0$ is a user-defined parameter that determines the amount up to which deviations from the desired accuracy ϵ are tolerated. Following the same approach considered above for (32.37), one obtains the dual QP problem

$$\max_{\alpha, \alpha^*} J_D(\alpha_m, \alpha_m^*)$$

subject to $\sum_{n=1}^{N} (\alpha_m - \alpha_m^*) = 0$,

$$0 \le \alpha_n \le c, \quad n = 1, \ldots, N,$$

$$0 \le \alpha_n^* \le c, \quad n = 1, \ldots, N, \tag{32.56}$$

where

$$J_D(\alpha_m, \alpha_m^*) =$$

$$-\frac{1}{2} \sum_{m,n=1}^{N} (\alpha_m - \alpha_m^*)(\alpha_n - \alpha_n^*) k(x_m, x_n)$$

$$-\epsilon \sum_{n=1}^{N} (\alpha_n - \alpha_n^*) + \sum_{n=1}^{N} y_n (\alpha_n - \alpha_n^*). \tag{32.57}$$

Note that, whereas (32.37) and (32.42) have tuning parameter c, in (32.55) and (32.56) one has the additional parameter ϵ.

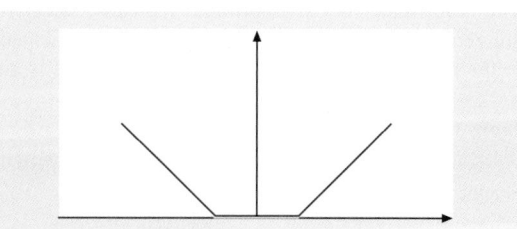

Fig. 32.5 ϵ-insensitive loss

Part D | 32.3

Before continuing, we note that a number of interesting modifications of the original SVR primal–dual formulations exist. In particular, [32.76] proposed the ν-tube support vector regression. In this method, the objective $J_P(w,\xi,\xi^*)$ in (32.55) is replaced by

$$J_P(w,\xi,\xi^*,\epsilon)$$
$$= \frac{1}{2}w^\top w + c\left(\nu\epsilon + \frac{1}{N}\sum_{n=1}^{N}\left(\xi_n+\xi_n^*\right)\right). \quad (32.58)$$

In the latter, ϵ is an optimization variable rather than a hyperparameter, as in (32.37); ν, on the other hand, is fixed by the user and controls the fraction of support vectors that is allowed outside the tube.

32.3.3 Main Features of SVMs

Here we briefly highlight the main features of support vector algorithms, making a direct comparison with classical neural networks.

Choice of Kernel
A number of possible kernels, such as the Gaussian radial basis function (RBF) kernel, can be chosen in (32.44). Some examples are included in Table 32.1.

In general, it is clear from (32.44) that a valid kernel function must preserve the fundamental properties of the inner-product. That is, for the equality to hold, the bivariate function $k: \mathbb{R}^D \times \mathbb{R}^D \to \mathbb{R}$ is required to be symmetric and positive definite. Note that this, in particular, imposes restriction on τ in the polynomial kernel. A more in depth discussion on kernels is postponed to Sect. 32.6.

Global Solution
(32.37) and its dual (32.42) are convex problems. This means that any local minimum must also be global. Therefore, even though SVCs share similarities with neural network schemes (see below), they do not suffer from the well-known issue of local minima.

Sparseness
The dual model is parsimonious: typically, many α's are zero at the solution with the nonzero ones located in the proximity of the decision boundary. This is also

Table 32.1 Some examples of kernel functions

kernel name	$k(x,y)$
Linear	$x^\top y$
Polynomial of degree $d>0$	$(\tau + x^\top y)^d$, for $\tau \geq 0$
Gaussian RBF	$\exp(-\|x-y\|^2/\sigma^2)$

desirable in all those setting were one requires fast online out-of-sample evaluations of models.

Neural Network Interpretation
Both primal (parametric) and dual (nonparametric) problems admit neural network representations [32.26], see Fig. 32.6. Note that in the dual problem the size of the QP problem is not influenced by the dimension D of the input space, nor it is influenced by the dimension of the feature space. Notably, in classical multilayer perceptrons one has to fix the number of hidden units in advance; in contrast, in SVMs the number of hidden units follows from the QP problem and corresponds to the number of support vectors.

SVM Solvers
The primal and dual formulations presented above are all QP problems. This means that one can rely on general purpose QP solvers for the training of models. Additionally, a number of specialized decomposition methods have been developed, including the sequential minimum optimization (SMO) algorithm [32.77]. Publicly available software packages such as libSVM [32.78, 79] and SVMlight [32.80] include implementations of efficient solvers.

32.3.4 The Class of Least–Squares SVMs

We discuss here the class of least-squares SVMs (LS-SVMs) obtained by simple modifications to the SVMs formulations. The arising problems relate to a number of existing methods and entail certain advantages with respect to the original SVM formulations.

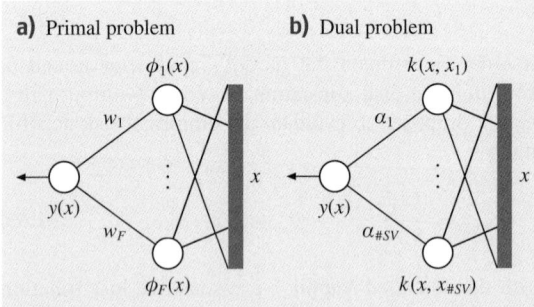

Fig. 32.6a,b Primal–dual network interpretations of SVMs [32.26]. (a) The number F of hidden units in the primal weight space corresponds to the dimensionality of the feature space. (b) The number # SV of hidden units in the dual weight space corresponds to the number of nonzero α's

LS-SVMs for Classification

We illustrate the idea with respect to the formulation for classification (LSs-SVC). The approach, originally proposed in [32.81], considers the primal problem

$$\min_{w,b,\xi} J_P(w,\epsilon) = \frac{1}{2} w^\top w + \frac{\gamma}{2} \sum_{n=1}^{N} \epsilon_n^2$$

subject to $y_n(w^\top \phi(x_n) + b) = 1 - \epsilon_n$,

$$n = 1, \dots, N.$$

(32.59)

This formulation simplifies the primal problem (32.37) in two ways. First, the inequality constraints are replaced by equality constraints; the 1's on the right-hand side in the constraints are regarded as target values rather than being treated as a threshold. An error ϵ_n is allowed so that misclassifications are tolerated in the case of overlapping distributions. Secondly, a squared loss function is taken for these error variables. The Lagrangian for (32.59) is

$$\mathcal{L}(w, b, \epsilon; \alpha, \nu)$$

$$= J_P(w, \epsilon) - \sum_{n=1}^{N} \alpha_n \left(y_n(w^\top \phi(x_n) + b) - 1 + \epsilon_n \right),$$

(32.60)

where α's are Lagrange multipliers that can be positive or negative since the problem now only entails equality constraints. The KKT conditions for optimality yield

$$\begin{cases} \dfrac{\partial \mathcal{L}}{\partial w} = 0 \rightarrow w = \sum_{n=1}^{N} \alpha_n y_n \phi(x_n), \\[2mm] \dfrac{\partial \mathcal{L}}{\partial b} = 0 \rightarrow \sum_{n=1}^{N} \alpha_n y_n = 0, \\[2mm] \dfrac{\partial \mathcal{L}}{\partial \epsilon_n} = 0 \rightarrow \gamma \epsilon_n = \alpha_n, \quad n = 1, \dots, N, \\[2mm] \dfrac{\partial \mathcal{L}}{\partial \alpha_n} = 0 \rightarrow y_n(w^\top \phi(x_n) + b) - 1 + \epsilon_n = 0, \\[2mm] \quad n = 1, \dots, N \end{cases}$$

(32.61)

By eliminating the primal variables w and b, one obtains the KKT system [32.82]

$$\begin{bmatrix} 0 & y^\top \\ y & \Omega + I/\gamma \end{bmatrix} \begin{bmatrix} b \\ \alpha \end{bmatrix} = \begin{bmatrix} 0 \\ 1_N \end{bmatrix},$$

(32.62)

where

$$y = [y_1, y_2, \dots, y_N]^\top$$

and

$$1_N = [1, 1, \dots, 1]^\top$$

are N-dimensional vectors and Ω is defined entry-wise by

$$(\Omega)_{mn} = y_m y_n \phi(x_m)^\top \phi(x_n) = y_m y_n k(x_m, x_n).$$

(32.63)

In the latter, we used the kernel trick introduced before. The dual model obtained corresponds to (32.45) where α and b are now obtained solving the linear system (32.62), rather than a more complex QP problem, as in (32.42). Notably, for LS-SVMs (and related methods) one can exploit a number of computational shortcuts related to spectral properties of the kernel matrix; for instance, one can compute solutions for different values of γ at the price of computing the solution of a single problem, which cannot be done for QPs [32.83–85].

The LS-SVC is easily extended to handle multiclass problems [32.26]. Extensive comparisons with alternative techniques (including SVC) for binary and multiclass classification are considered in [32.26, 86]. The results show that, in general, LS-SVC either outperforms or perform comparably to the alternative techniques. Interestingly, it is clear from the primal problem (32.61) that LS-SVC maximizes the margin while minimizing the within-class scattering from targets $\{+1, -1\}$. As such, LS-SVC is naturally related to Fisher discriminant analysis in the feature space [32.26]; see also [32.2, 87, 88].

Alternative Formulations

Besides classification, a primal–dual approach similar to the one introduced above has been considered for function estimation [32.26]; in this case too the dual model representation is obtained by solving a linear system of equations rather than a QP problem, as in SVR. This approach to function estimation is similar to a number of techniques, including smoothing splines [32.13], regularization networks [32.44, 89], kernel ridge regression [32.90], and Kriging [32.91]. LS-SVM solutions also share similarities with Gaussian processes [32.92, 93], which we discuss in more detail in the next section.

Other formulations have been considered within a primal–dual framework. These include principal component analysis [32.94], which we discuss next, spectral clustering [32.95], canonical correlation analysis [32.96], dimensionality reduction and data visualization [32.97], recurrent networks [32.98], and optimal control [32.99]; see also [32.100]. In all these cases the estimation problem of interest is conceived at the primal level as an optimization problem with equality constraints, rather than inequality constraints as in SVMs. The constraints relate to the model which is expressed in terms of the feature map. From the KKT optimality conditions one jointly finds the optimal model representation and the model estimate. As for the case of classification the dual model representation is expressed in terms of kernel functions.

Sparsity and Robustness

An important difference with SVMs, is that the dual model found via LS-SVMs depends upon all the training data. Reduction and pruning techniques have been used to achieve the sparse representation in a second stage [32.26, 101]. A different approach, which makes use of the primal–dual setting, leads to *fixed-size* techniques [32.26], which relate to the *Nyström method* proposed in [32.102] but lead to estimation in the primal setting. Optimized versions of fixed-size LS-SVMs are currently applicable to large data sets with millions of data points for training and tuning on a personal computer [32.103].

In LS-SVM the estimation of the support values is only optimal in the case of a Gaussian distribution of the error variables [32.26]; [32.101] shows how to obtain robust estimates for regression by applying a weighted version of LS-SVM. The approach is suitable in the case of outliers or non-Gaussian error distributions with heavy tails.

32.3.5 Kernel Principal Component Analysis

Principal component analysis (PCA) is one of the most important techniques in the class of unsupervised learning algorithms. PCA linearly transforms a number of possibly correlated variables into uncorrelated features called principal components. The transformation is performed to find directions of maximal variation. Often, few principal components can account for most of the structure in the original dataset. PCA is not suitable for discovering nonlinear relationships among the original variables. To overcome this limitation [32.104] originally proposed the idea of performing PCA in a feature space rather than in the input space.

Regardless of the space where the transformation is performed, there is a number of different ways to characterize the derivation of the PCA problem [32.105]. Ultimately, PCA analysis is readily performed by solving an eigenvalue problem. Here we consider a primal–dual formulation similar to the one introduced above for LS-SVC [32.106]. In this way the eigenvalue problem is seen to arise from optimality conditions. Notably the approach emphasizes the underlying model, which is important for finding the projection of out-of-sample points along the direction of maximal variation. The analysis assumes the knowledge of N training data pairs

$$\{x_1, x_2, \ldots, x_N\} \subset \mathbb{R}^D \tag{32.64}$$

i.i.d. according to the generator $p(x)$. The starting point is to define the generic *score variable* z as

$$z(x) = w^\top (\phi(x) - \hat{\mu}_\phi) . \tag{32.65}$$

The latter represents one projection of $\phi(x) - \mu_\phi$ into the target space. Note that we considered data centered in the feature space with

$$\hat{\mu}_\phi = \frac{1}{N} \sum_{n=1}^N \phi(x_n) \tag{32.66}$$

corresponding to the center of the empirical distribution. The primal problem consists of the following constrained formulation [32.94]

$$\max_{w,z} \quad -\frac{1}{2} w^\top w + \frac{\gamma}{2} \sum_{n=1}^N z_n^2$$

$$\text{subject to} \quad z_n = w^\top (\phi(x_n) - \hat{\mu}_\phi) , \quad n = 1, \ldots, N . \tag{32.67}$$

where $\gamma > 0$. The latter maximizes the empirical variance of z while keeping the norm of the corresponding parameter vector w small by the regularization term. One can also include a bias term, see [32.26] for a derivation.

The Lagrangian corresponding to (32.67) is

$$\mathcal{L}(w, z; \alpha) = -\frac{1}{2} w^\top w + \frac{\gamma}{2} \sum_{n=1}^N z_n^2$$

$$- \sum_{n=1}^N \alpha_n (z_n - w^\top (\phi(x_n) - \hat{\mu}_\phi)) , \tag{32.68}$$

with conditions for optimality given by

$$
\begin{cases}
\dfrac{\partial \mathcal{L}}{\partial w} = 0 \rightarrow w = \sum_{n=1}^{N} \alpha_n \left(\phi(x_n) - \mu_\phi \right), \\[2mm]
\dfrac{\partial \mathcal{L}}{\partial z_n} = 0 \rightarrow \alpha_n = \gamma z_n, \quad n = 1, \ldots, N, \\[2mm]
\dfrac{\partial \mathcal{L}}{\partial \alpha_n} = 0 \rightarrow z_n = w^\top \left(\phi(x_n) - \hat{\mu}_\phi \right), \\[2mm]
\quad n = 1, \ldots, N.
\end{cases}
\tag{32.69}
$$

By eliminating the primal variables w and z, one obtains for $n = 1, \ldots, N$

$$
\frac{1}{\gamma} \alpha_n - \sum_{m=1}^{N} \alpha_m \left(\phi(x_n) - \hat{\mu}_\phi \right) \left(\phi(x_m) - \hat{\mu}_\phi \right) = 0.
\tag{32.70}
$$

The latter is an eigenvalue decomposition that can be stated in matrix notation as

$$
\Omega_c \alpha = \lambda \alpha,
\tag{32.71}
$$

where $\lambda = 1/\gamma$ and Ω_c is the centered Gram matrix defined entry-wise by

$$
[\Omega_c]_{nm} = k(x_n, x_m) - \frac{1}{N} \sum_{l=1}^{N} k(x_m, x_l)
$$
$$
- \frac{1}{N} \sum_{l=1}^{N} k(x_n, x_l) + \frac{1}{N^2} \sum_{i=1}^{N} \sum_{j=1}^{N} k(x_j, x_i).
\tag{32.72}
$$

As before, one may choose any positive definite kernel; a typical choice corresponds to the Gaussian RBF kernel. By solving the eigenvalue problem (32.71) one finds N pairs of eigenvalues and eigenvectors

$$
(\lambda_m, \alpha^m), \quad m = 1, 2, \ldots, N.
$$

Correspondingly, one finds N score variables with dual representation

$$
z_m(x) = \sum_{n=1}^{N} \alpha_n^m \left(k(x_n, x) - \frac{1}{N} \sum_{l=1}^{N} k(x, x_l) \right.
$$
$$
\left. - \frac{1}{N} \sum_{l=1}^{N} k(x_n, x_l) + \frac{1}{N^2} \sum_{i=1}^{N} \sum_{j=1}^{N} k(x_j, x_i) \right),
\tag{32.73}
$$

in which α^m is the eigenvector associated to the eigenvalue λ_m. Note that all eigenvalues are positive and real because Ω_c is symmetric and positive semidefinite; the eigenvectors are mutually orthogonal, i.e., $(\alpha^l)^\top \alpha^m = 0$, for $l \neq m$. Note that when the feature map is nonlinear, the number of score variables associated to nonzero eigenvalues might exceed the dimensionality D of the input space. Typically, one selects then the minimal number of score variables that preserve a certain reconstruction accuracy, see [32.26, 104, 105].

Finally, observe that by the second optimality condition in (32.61), one has $z_l = \lambda_l \alpha^l$ for $l = 1, 2, \ldots, N$. From this, we obtain that the score variables are empirically uncorrelated. Indeed, we have for $l \neq m$

$$
\sum_{n=1}^{N} z_l(x_n) z_m(x_n)
$$
$$
= \sum_{n=1}^{N} \lambda_l \lambda_m (\alpha^l)^\top \alpha^m = 0.
\tag{32.74}
$$

32.4 Gaussian Processes

So far we have dealt with primal-dual kernel methods; regularization was motivated by the SRM principle, which achieves generalization by trading off empirical risk with the complexity of the model class. This is representative of a large number of procedures. However, it leaves out an important class of kernel-based probabilistic techniques that goes under the name of *Gaussian processes*. In Gaussian processes one uses a Bayesian approach to perform inference and learning. The main idea goes back at least to the work of *Wiener* [32.107] and *Kolmogorov* [32.108] on time-series analysis.

As a first step, one poses a probabilistic model which serves as a prior. This prior is updated in the light of training data so as to obtain a predictive distribution. The latter represents a spectrum of possible answers. In contrast, in the standard SVM/LS-SVM framework one obtains only point-wise estimates. The approach, however, is analytically tractable only for a limited number of cases of interests. In the following, we summarize the main ideas in the context of regression where tractability is ensured by Gaussian posteriors; the interested reader is referred to [32.109] for an in-depth review.

32.4.1 Definition

A real-valued stochastic process f is a Gaussian process (GP) if for every finite set of indices $x_1, x_2, \ldots, x_N$ in an index set X, the tuple

$$f_x = (f(x_1), f(x_2), \ldots, f(x_N))$$ (32.75)

is a multivariate Gaussian random variable taking values in $\mathbb{R}^N$. Note that the index set X represents the set of all possible inputs. This might be a countable set, such as $\mathbb{N}$ (e.g. a discrete time index) or, more commonly in machine learning, the Euclidean space $\mathbb{R}^D$. A GP f is fully specified by a *mean function* $m : X \to \mathbb{R}$ and a *covariance function* $k : X \times X \to \mathbb{R}$ defined by

$$m(x) = \mathbb{E}[f(x)] \,,$$ (32.76)
$$k(x, x') = \mathbb{E}[(f(x) - m(x))(f(x') - m(x'))] \,.$$ (32.77)

In light of this, one writes

$$f \sim \mathcal{GP}(m, k) \,.$$ (32.78)

Usually, for notational simplicity one takes the mean function to be zero, which we consider here; however, this need not to be the case.

Note that the specification of the covariance function implies a distribution over any finite collection of random variables obtained sampling the process f at given locations. Specifically, we can write for (32.75)

$$f_x \sim \mathcal{N}(0, K) \,,$$ (32.79)

which means that f_x follows a multivariate zero-mean Gaussian distribution with $N \times N$ covariance matrix K defined entry-wise by

$$[K]_{nm} = k(x_n, x_m) \,.$$

The typical use of a GP is in a regression context, which we consider next.

32.4.2 GPs for Regression

In regression one observes a dataset of input–output pairs (x_n, y_n), $n = 1, 2, \ldots, N$ and wants to make a prediction at one or more test points. In the following, we call y the vector obtained staking the target observations and denote by X the collection of input training patterns (32.64). In order to carry on the Bayesian inference, one needs a model for the generating process. It is generally assumed that function values are observed in noise, that is,

$$y_n = f(x_n) + \epsilon_n \,, \quad n = 1, 2, \ldots, N \,.$$ (32.80)

One further assumes that ϵ_n are i.i.d. zero-mean Gaussian random variables independent of the process f and with variance σ^2. Under these circumstances, the noisy observations are Gaussian with mean zero and covariance function

$$c(x_m, x_n) = \mathbb{E}[y_m y_n] = k(x_m, x_n) + \sigma^2 \delta_{nm} \,,$$ (32.81)

where the Kronecker delta function δ_{nm} is 1 if $n = m$ and 0 otherwise.

Suppose now that we are interested in the value f_* of the process at a single test point x_* (the approach that we discuss below extends to multiple test points in a straightforward manner). By relying on properties of Gaussian probabilities, we can readily write the joint distribution of the test function value and the noisy training observations. This reads

$$\begin{bmatrix} y \\ f_* \end{bmatrix} \sim \mathcal{N}\left(0, \begin{bmatrix} K + \sigma^2 I_N & k_x \\ k_x^\top & k_* \end{bmatrix}\right),$$ (32.82)

where I_N is the $N \times N$ identity matrix, $k_* = k(x_*, x_*)$. and finally

$$k_x = [k(x_1, x_*), k(x_2, x_*), \ldots, k(x_N, x_*)]^\top \,.$$ (32.83)

Prediction with Noisy Observations

Using the conditioning rule for multivariate Gaussian distributions, namely:

$$\begin{bmatrix} x \\ y \end{bmatrix} \sim \mathcal{N}\left(\begin{bmatrix} a \\ b \end{bmatrix}, \begin{bmatrix} A & C \\ C^\top & B \end{bmatrix}\right) \Rightarrow$$
$$y|x \sim \mathcal{N}\left(b + C^\top A^{-1}(x - a), B - C^\top A^{-1} C\right),$$

one arrives at the key predictive equation for GP regression

$$f_* | y, X, x_* \sim \mathcal{N}\left(m_*, \sigma_*^2\right),$$ (32.84)

where

$$m_* = k_x^\top (K + \sigma^2 I)^{-1} y,$$ (32.85)

$$\sigma_*^2 = k_* - k_x^\top (K + \sigma^2 I)^{-1} k_x.$$ (32.86)

Note that, by letting $\alpha = (K + \sigma^2 I)^{-1} y$, one obtains for the mean value

$$m_* = \sum_{n=1}^{N} \alpha_n k(x_n, x_*).$$ (32.87)

Up to the bias term b, the latter coincides with the typical dual model representation (32.7), in which x plays the role of a test point. Therefore, one can see that, in the framework of GPs, the covariance function plays the same role of the kernel function. The variance σ_*^2, on the other hand, is seen to be obtained from the prior covariance, by subtracting a positive term which accounts for the information about the process conveyed by training data.

Weight-Space View

We have presented GPs through the so-called *function space view* [32.109], which ultimately captures the distinctive nature of this class of methodologies. Here we illustrate a different view, which allows one to achieve three objectives: 1) it is seen that Bayesian linear models are a special instance of GPs; 2) the role of Bayes' rule is highlighted; 3) one obtains additional insight on the relationship with the feature map and kernel function used before within primal–dual techniques.

Bayesian Regression. The starting point is to characterize f as a parametric model involving a set of basis functions $\psi_1, \psi_2, \ldots, \psi_F$

$$f(x) = \sum_i w_i \psi_i(x) = w^\top \psi(x).$$ (32.88)

Note that F might be infinity. For the special case where ψ is the identity mapping, one recognizes in (32.80) the standard modeling assumptions for Bayesian linear regression analysis. Inference is based on the posterior distribution over the weights, computed by Bayes' rule

$$\text{posterior} = \frac{\text{likelihood} \times \text{prior}}{\text{marginal likelihood}},$$

$$p(w | y, X) = \frac{p(y | X, w) p(w)}{p(y | X)},$$ (32.89)

where the marginal likelihood (a.k.a. normalizing constant) is independent of the weights

$$p(y | X) = \int p(y | X, w) p(w) \mathrm{d}w.$$ (32.90)

Explicit Feature Space Formulation. To make a prediction for a test pattern x_* we average over all possible parameter values with weights corresponding to the posterior probability

$$p(f_* | y, X, x_*) = \int p(f_* | w, x_*) p(w | y, X) \mathrm{d}w.$$ (32.91)

One can see that computing the posterior $p(w | y, X)$ based upon the prior

$$p(w) = \mathcal{N}(0, \Sigma_p)$$ (32.92)

gives the predictive model

$$f_* | y, X, x_* \sim \mathcal{N}\left(\psi(x_*)^\top \Sigma_p \Psi A y, \right.$$

$$\left. \psi(x_*)^\top \Sigma_p \psi(x_*) - \psi(x_*)^\top \Sigma_p \Psi A \, \Psi^\top \Sigma_p \psi(x_*) \right),$$ (32.93)

where $A = (\Psi^\top \Sigma_p \Psi + \sigma^2 I_N)^{-1}$ and we denoted by

$$\Psi = [\psi(x_1), \psi(x_2), \ldots, \psi(x_N)]$$

the feature representation of the training patterns. It is not difficult to see that (32.93) is (32.84) in disguise. In particular, one has

$$k(x, y) = \psi(x)^\top \Sigma_p \psi(y).$$ (32.94)

The positive definiteness of Σ_p ensures the existence and uniqueness of the square root $\Sigma_p^{1/2}$. If we now define

$$\phi(x) = \Sigma_p^{1/2} \psi(x),$$ (32.95)

we retrieve the relationship in (32.44). We conclude that the kernel function considered in the previous sections can be interpreted as the covariance function of a GP.

32.4.3 Bayesian Decision Theory

Bayesian inference is particularly appealing when prediction is intended for supporting decisions. In this case, one requires a loss function $L(f_{\text{true}}, f_{\text{guess}})$ which specifies the penalty obtained by guessing f_{guess} when the true value is f_{true}. Note that the predictive distribution (32.84) or – equivalently – (32.93) was derived without reference to the loss function. This is a major difference with respect to the techniques developed within the framework of statistical learning. Indeed, in the non-Bayesian framework of penalized empirical risk minimization, prediction and loss are entangled; one tackles learning in a somewhat more direct way. In contrast, in the Bayesian setting there is a clear distinction between 1) the model that generated the data and 2) capturing the consequences of making guesses. In light of this, [32.109] advises one to beware of arguments like *a Gaussian likelihood implies a squared error loss*. In order to find the point prediction that incurs the minimal *expected* loss, one can define the merit function

$$R\left(f_{\text{guess}}|y, X, x_*\right)$$
$$= \int L(f_*, f_{\text{guess}})p(f_*|y, X, x_*)\mathrm{d}f_* . \tag{32.96}$$

Note that, since the true value f_{true} is unknown, the latter averages with respect to the model's opinion $p(f_*|y, X, x_*)$ on what the truth might be. The corresponding best guess is

$$f_{\text{opt}} = \arg\min_{f_{\text{guess}}} R(f_{\text{guess}}|y, X, x_*) . \tag{32.97}$$

Since $p(f_*|y, X, x_*)$ is Gaussian and hence symmetric, f_{opt} always coincides with the mean m_* whenever the loss is also symmetric. However, in many practical problems such as in critical safety applications, the loss can be asymmetrical. In these cases, one must solve the optimization problem in (32.97). Similar considerations hold for classification, see [32.109]. For an account on decision theory see [32.110].

32.5 Model Selection

Kernel-based models depend upon a number of parameters which are determined during training by numerical procedures. Still, one or more hyperparameters usually need to be tuned by the user. In SVC, for instance, one has to fix the value of c. The choice of the kernel function, and of the corresponding parameters, also needs to be properly addressed.

In general, performance measures used for model selection include k-fold cross-validation, leave-one-out (LOO) cross-validation, generalized approximate cross-validation (GACV), approximate span bounds, VC bounds, and radius-margin bounds. For discussions and comparisons see [32.111, 112]. Another approach found in the literature is kernel-target alignment [32.113].

32.5.1 Cross-Validation

In practice, model selection based on cross-validation is usually preferred over generalization error bounds. Criticism for cross-validation approaches is related to the high computational load involved; [32.114] presents an efficient methodology for hyperparameter tuning and model building using LS-SVMs. The

approach is based on the closed form LOO cross-validation computation for LS-SVMs, only requiring the same computational cost of one single LS-SVM training. Leave-one-out cross-validation-based estimates of performance, however, generally exhibit a relatively high variance and are, therefore, prone to over-fitting. To amend this, [32.115] proposed the use of Bayesian regularization at the second level of inference.

32.5.2 Bayesian Inference of Hyperparameters

Many authors have proposed a full Bayesian framework for kernel-based algorithms in the spirit of the methods developed by *MacKay* for classical MLPs [32.116–118]. In particular, [32.26] discusses the case of LS-SVMs. It is shown that, besides leading to tuning strategies, the approach allows us to take probabilistic interpretations of the outputs; [32.109] discusses the Bayesian model selection for GPs. In general, the Bayesian framework consists of multiple levels of inference. The parameters (i. e., with reference to the primal model, w and b) are inferred at level 1. Contrary to MLPs, this usually entails the solution of a convex op-mization problem or even solving a linear system, as n LS-SVMs and GPs. The regularization parameter(s) nd the kernel parameter(s) are inferred at higher levels.

The method progressively integrates out the parameters by using the evidence at a certain level of inference as the likelihood at the successive level.

2.6 More on Kernels

We have already seen that a kernel arising from an inner roduct can be interpreted as the covariance function f a Gaussian process. In this section we further study ne mathematical properties of different yet equivalent otions of kernels. In particular, we will discuss that ositive definite kernels are *reproducing*, in a sense that ve are about to clarify. We will then review a number f specialized kernels for learning problems involving tructured data.

2.6.1 Positive Definite Kernels

Denote by X a nonempty index set. A symmetric function $k : X \times X \to \mathbb{R}$ is a *positive definite kernel* if for ny $N \in \mathbb{N}$ and for any tuple $(x_1, x_2, \ldots, x_N) \in X^N$, the Gram matrix K defined entry-wise by $K_{nm} = k(x_n, x_m)$, atisfies (note that, by definition, a positive definite kernel satisfies $k(x, x) \geq 0$ for any $x \in X$)

$$\alpha^\top K \alpha = \sum_{n=1}^{N} \sum_{m=1}^{N} K_{nm} \alpha_n \alpha_m \geq 0 \ \forall \alpha \in \mathbb{R}^N .$$

n particular, suppose $\mathcal{F}$ is some Hilbert space (HS) for an elementary introduction see [32.66]) with inner product $\langle \cdot, \cdot \rangle$. Then for any function $\phi : X \to \mathcal{F}$ one has

$$\sum_{n=1}^{N} \sum_{m=1}^{N} \langle \phi(x_n), \phi(x_m) \rangle \alpha_n \alpha_m$$

$$= \sum_{n=1}^{N} \sum_{m=1}^{N} \langle \alpha_n \phi(x_n), \alpha_m \phi(x_m) \rangle$$

$$= \left\| \sum_{n=1}^{N} \alpha_n \phi(x_n) \right\|^2 \geq 0 . \tag{32.98}$$

From the first line one can then see that

$$k : (x, y) \mapsto \langle \phi(x), \phi(y) \rangle \tag{32.99}$$

s a positive definite kernel in the sense specified above. A continuous positive definite kernel k is often called Mercer kernel [32.36].

Note that in Sect. 32.1.3 we denoted $\langle \phi(x), \phi(y) \rangle$ by $\phi(x)^\top \phi(y)$, implicitly making the assumption that the feature space $\mathcal{F}$ is a finite dimensional Euclidean space. However, one can show that the feature space associated to certain positive definite kernels (such as the Gaussian RBF [32.119]) is an infinite dimensional HS; in turn, the inner product in such a space is commonly denoted as $\langle \cdot, \cdot \rangle$.

32.6.2 Reproducing Kernels

Evaluation Functional
Let $(\mathcal{H}, \langle \cdot, \cdot \rangle)$ be a HS of real-valued functions (the theory of RKHSs generally deals with complex-valued functions [32.12, 14]; here we stick to the real setting for simplicity) on X equipped with the norm $\|f\| = \sqrt{\langle f, f \rangle}$. For $x \in X$ we denote by L_x the evaluation functional

$$L_x : \mathcal{H} \to \mathbb{R} ,$$
$$f \mapsto f(x) . \tag{32.100}$$

L_x is said to be bounded if there exists $c > 0$ such that $|L_x f| = |f(x)| \leq c\|f\|$ for all $f \in \mathcal{H}$. By the Riesz representation theorem [32.120] if L_x is bounded then there exists a unique $\eta_x \in \mathcal{H}$ such that for any $f \in \mathcal{H}$

$$L_x f = \langle f, \eta_x \rangle . \tag{32.101}$$

Reproducing Kernel
A function

$$k : X \times X \to \mathbb{R} ,$$
$$(x, y) \mapsto k(x, y) \tag{32.102}$$

is said to be a *reproducing kernel* of $\mathcal{H}$ if and only if

$$\forall x \in X, k(\cdot, x) \in \mathcal{H} , \tag{32.103}$$
$$\forall x \in X, \forall f \in \mathcal{H} \ \langle f, k(\cdot, x) \rangle = f(x) . \tag{32.104}$$

Note that by $k(\cdot, x)$ we mean the function $k(\cdot, x) : t \mapsto k(t, x)$.

The definition of reproducing kernel (r.k.) implies that $k(\cdot, x) = \eta_x$, i.e., $k(\cdot, x)$ is the representer of the evaluation functional L_x; (32.104) goes under the name of *reproducing property*. From (32.103) and (32.104) it is clear that

$$k(x, y) = \langle k(\cdot, x), k(\cdot, y) \rangle, \quad \forall x, y \in X ; \qquad (32.105)$$

since $\langle \cdot, \cdot \rangle$ is symmetric, it follows that $k(x, y) = k(y, x)$. A HS of functions that possesses a reproducing kernel is called a *reproducing kernel Hilbert space* (RKHS).

Finally, notice that the reproducing kernel (r.k.) of a space of a HS of functions corresponds in a one-to-one manner with the definition of the inner product $\langle \cdot, \cdot \rangle$; changing the inner product implies a change in the reproducing kernel.

Basic Properties of RKHSs

Let $(G, \langle \cdot, \cdot \rangle)$ be a HS of functions. If, for any x, the evaluation functional L_x is bounded, then it is clear that G is a RKHS with reproducing kernel

$$k(x, y) = \langle \eta_x, \eta_y \rangle . \qquad (32.106)$$

Vice versa, if G admits a reproducing kernel k, then all evaluation functionals are bounded. Indeed, we have

$$|f(x)| = \langle f, k(\cdot, x) \rangle \leq \|f\| \|k(\cdot, x)\|$$
$$= \sqrt{k(x, x)} \|f\| , \qquad (32.107)$$

where we simply relied on the Cauchy–Schwarz inequality. Boundedness of evaluation functionals means that all the functions in the space are well defined for all x. Note that this is not the case, for instance, of the space of *square-integrable functions*.

It is not difficult to prove that, in a RKHS $\mathcal{H}$, the representation of a bounded linear functional A is simply $Ak(\cdot, x)$, i.e., it is obtained by applying A to the representer of L_x. As an example, take the functional evaluating the derivative of f at x

$$D_x : f \mapsto f'(x) .$$

If D_x is bounded on $\mathcal{H}$, then the property implies that

$$f'(x) = \langle f, k'(\cdot, x) \rangle, \quad \forall f \in \mathcal{H} ,$$

where $k'(\cdot, x)$ is the derivative of the function $k(\cdot, x)$.

32.6.3 Equivalence Between the Two Notions

Moore–Aronszajn Theorem

If we let

$$\phi : X \to \mathcal{H} ,$$
$$x \mapsto k(x, \cdot) , \qquad (32.108)$$

one can see that, in light of (32.98), the reproducing kernel k is also a positive definite kernel. The converse result, stating that a positive definite kernel is the reproducing kernel of a HS of functions $(\mathcal{H}, \langle \cdot, \cdot \rangle)$, is found in the Moore-Aronszajn theorem [32.12]. This completes the equivalence between positive definite kernel and reproducing kernels.

Feature Maps and the Mercer Theorem

Note that (32.108) is a first feature map associated to the kernel function k. Correspondingly, this shows that the RKHS $\mathcal{H}$ is a possible instance of the feature space. A different feature map can be given in view of the Mercer theorem [32.8, 121], which historically played a major role in the development of kernel methods. The theorem states that every positive definite kernel can be written as

$$k(x, y) = \sum_{i=1}^{\infty} \mu_i e_i(x) e_i(y) , \qquad (32.109)$$

where the series in the right-hand side of (32.109) converges absolutely and uniformly, $(e_i)_i$ is an orthonormal sequence of eigenfunctions, and $(\mu_i)_i$ is the corresponding sequence of nonnegative eigenvalues such that for some measure ν

$$\int k(x, y) e_i(y) d\nu(y) = \mu_i e_i(x) \forall x \in X , \qquad (32.110)$$

$$\int e_j(y) e_i(y) d\nu(y) = \delta_{ij} . \qquad (32.111)$$

The eigenfunctions $(e_i)_i$ belong to the RKHS $(\mathcal{H}, \langle \cdot, \cdot \rangle)$ associated to k. In fact, by (32.110) one has

$$e_i(x) = \frac{1}{\mu_i} \int k(x, y) e_i(y) d\nu(y) , \qquad (32.112)$$

and therefore e_i can be approximated by elements in the span of $(k_x)_{x \in X}$ [32.36]. One can further see that $(\sqrt{\mu_i} e_i)_i$ is an orthonormal basis for $\mathcal{H}$; indeed one has

$$\langle \sqrt{\mu_i}e_i, \sqrt{\mu_j}e_j \rangle$$

$$\stackrel{\text{by (110)}}{=} \left\langle \frac{1}{\sqrt{\mu_i}} \int k(\cdot,y)e_i(y)d\nu(y), \sqrt{\mu_j}e_j \right\rangle$$

$$= \int \frac{\sqrt{\mu_j}}{\sqrt{\mu_i}} \langle k(\cdot,y), e_j \rangle e_i(y)d\nu(y)$$

$$\stackrel{\text{by (104)}}{=} \int \frac{\sqrt{\mu_j}}{\sqrt{\mu_i}} e_j(y)e_i(y)d\nu(y) \stackrel{\text{by(111)}}{=} \frac{\sqrt{\mu_j}}{\sqrt{\mu_i}}\delta_{ij}\,.$$

$$\tag{32.113}$$

Note that we considered the general case in which the expansion (32.109) involves infinitely many terms. However, there are positive definite kernels (e.g., the polynomial kernel) for which only finitely many eigenvalues are nonzero.

In light of the Mercer theorem one can see that a different feature map is given by

$$\phi : \mathcal{X} \to \mathcal{F}\,,$$
$$x \mapsto (\sqrt{\mu_i}e_i(x))_i\,. \tag{32.114}$$

Note that ϕ maps x into an infinite dimensional vector with i-th entry $\phi_i(x) = \sqrt{\mu_i}e_i(x)$.

Connecting Functional and Parametric View

One can see now that $\sum_i w_i\phi_i(x)$ in the primal model (32.6) corresponds to the evaluation in x of a function f in a RKHS. To see this, we start from decomposing f according to the orthonormal basis $(\sqrt{\mu_i}e_i)_i$

$$f = \sum_i \langle f, \sqrt{\mu_i}e_i \rangle \sqrt{\mu_i}e_i$$
$$= \sum_i w_i \sqrt{\mu_i}e_i\,, \tag{32.115}$$

where we let $w_i = \langle f, \sqrt{\mu_i}e_i \rangle \sqrt{\mu_i}$. Now one has

$$L_x f = \langle f, k(\cdot,x) \rangle = \left\langle \sum_i w_i \sqrt{\mu_i}e_i, k(\cdot,x) \right\rangle$$

$$= \sum_i w_i \sqrt{\mu_i}\langle e_i, k(\cdot,x) \rangle = \sum_i w_i \sqrt{\mu_i}e_i(x)$$

$$= \sum_i w_i \phi_i(x)\,, \tag{32.116}$$

where we applied the reproducing property on e_i and used the definition of feature map (32.114). Addition-

ally, notice that one has

$$\|f\|^2 = \langle f, f \rangle = \left\langle \sum_i w_i \sqrt{\mu_i}e_i, \sum_j w_j \sqrt{\mu_j}e_j \right\rangle$$

$$= \sum_i \sum_j w_i w_j \langle \sqrt{\mu_i}e_i, \sqrt{\mu_j}e_j \rangle$$

$$\stackrel{\text{by(113)}}{=} \sum_i \sum_j w_i w_j \frac{\sqrt{\mu_j}}{\sqrt{\mu_i}}\delta_{ij}$$

$$= \sum_i w_i^2\,. \tag{32.117}$$

This shows that the penalty $w^\top w$, used within the primal problems of Sect. 32.3, can be connected to the squared norm of a function. The interested reader is referred to [32.14, 36] for additional properties of kernels.

32.6.4 Kernels for Structured Data

In applications where data are well represented by vectors in a Euclidean space one usually uses the Gaussian RBF kernel, which is *universal* [32.122]. Nonetheless, there exists an entire set of rules according to which one can design new kernels from elementary positive definite functions [32.1]. Although the idea of kernels has been around for a long time, it was only in the 1990s that the machine learning community started to realize that the index set $\mathcal{X}$ does not need to be (a subset of) some Euclidean space. This significantly improved the applicability of kernel-based algorithms to a broad range of data types, including sequence data, graphs and trees, and XML and HTML documents [32.123].

Probabilistic Kernels

One powerful approach consists of applying a kernel that brings generative models into a (possibly discriminative) kernel-based method [32.124, 125]. Generative models can deal naturally with missing data and in the case of hidden Markov models can handle sequences of varying length. A popular probabilistic similarity measure is the *Fisher kernel* [32.126, 127]. The key intuition behind this approach is that similarly structured objects should induce similar log-likelihood gradients in the parameters of a predefined class of generative models [32.126]. Different instances exist, depending on the generative model of interest, see also [32.1].

Graph Kernels and Dynamical Systems

Graphs can very naturally represent entities, their attributes, and their relationships to other entities; this makes them one of the most widely used tools for modeling structured data. Various type of kernels for graphs have been proposed, see [32.128, 129] and references therein. The approach can be extended to carry on recognitions and decisions for tasks involving dynamical systems; in fact, kernels on dynamical systems are related to graph kernels through the dynamics of random walks [32.128, 130].

Tensors and Kernels

Tensors are multidimensional arrays that represent higher-order generalizations of vectors and matrices. Tensor-based methods are often particularly effective in low signal-to-noise ratios and when the number of observations is small in comparison with the dimensionality of the data. They are used in domains ranging from neuroscience to vision and chemometrics, where tensors best capture the multiway nature of the data [32.131]. The authors of [32.132] proposed a family of kernels that exploit the algebraic structure of data tensors. The approach is related to a generalization of the singular value decomposition (SVD) to higher-order tensors [32.133]. The essential idea is to measure similarity based upon a Grassmannian distance of the subspaces spanned by matrix unfolding of data tensors. It can be shown that the approach leads to perfect separation of tensors generated by different sets of rank-1 atoms [32.132]. Within this framework, [32.134] proposed a kernel function for multichannel signals; the idea exploits the spectral information of tensors of fourth order cross-cumulants associated to each multichannel signal.

32.7 Applications

Kernel methods have been shown to be successful in many different applications. In this section we mention only a few examples.

32.7.1 Text Categorization

Recognition of objects and handwritten digits is studied in [32.135–137]; natural language text categorization is discussed in [32.138, 139]. The task consists of classifying documents based on their content. Attribute value representation of text is used to adequately represent the document text; typically, each distinct word in a document represents a feature with values corresponding to the number of occurrences.

32.7.2 Time–Series Analysis

The use of kernel methods for time-series prediction has been discussed in a number of papers [32.140–144], with applications ranging from electric load forecasting [32.145] to financial time series prediction [32.146]. Nonlinear system identification by LS-SVMs is discussed in [32.26] and references therein; [32.134] studies the problem of training a discriminative classifier given a set of labeled multivariate time series. Applications include brain decoding tasks based on magnetoencephalography (MEG) recordings.

32.7.3 Bioinformatics and Biomedical Applications

Gene expression analysis performed by SVMs is discussed in [32.147]. Applications in metabolomics, genetics, and proteomics are presented in the tutorial paper [32.148]; [32.149] discussed different techniques for the integration of side information in models based on gene expression data to improve the accuracy of diagnosis and prognosis in cancer; [32.150] provides an introduction to general data fusion problems using SVMs with application to computational biology problems. Detection of remote protein homologies by SVMs is discussed in [32.151], which combines discriminative methods with generative models. Bioengineering and bioinformatics applications can also be found in [32.152–154]. Survival analysis based on primal-dual techniques in discussed in [32.155, 156].

References

32.1   J. Shawe-Taylor, N. Cristianini: *Kernel Methods for Pattern Analysis* (Cambridge Univ. Press, Cambridge 2004)

32.2   B. Schölkopf, A.J. Smola: *Learning with Kernels: Support Vector Machines, Regularization, Optimization, Beyond* (MIT Press, Cambridge 2002)

32.3   A.J. Smola, B. Schölkopf: A tutorial on support vector regression, Stat. Comput. **14**(3), 199–222 (2004)

32.4   T. Hofmann, B. Schölkopf, A.J. Smola: Kernel methods in machine learning, Ann. Stat. **36**(3), 1171–1220 (2008)

32.5   K.R. Müller, S. Mika, G. Ratsch, K. Tsuda, B. Schölkopf: An introduction to kernel-based learning algorithms, IEEE Trans. Neural Netw. **12**(2), 181–201 (2001)

32.6   F. Jäkel, B. Schölkopf, F.A. Wichmann: A tutorial on kernel methods for categorization, J. Math. Psychol. **51**(6), 343–358 (2007)

32.7   C. Campbell: Kernel methods: A survey of current techniques, Neurocomputing **48**(1), 63–84 (2002)

32.8   J. Mercer: Functions of positive and negative type, and their connection with the theory of integral equations, Philos. Trans. R. Soc. A **209**, 415–446 (1909)

32.9   E.H. Moore: On properly positive Hermitian matrices, Bull. Am. Math. Soc. **23**(59), 66–67 (1916)

32.10   T. Kailath: RKHS approach to detection and estimation problems – I: Deterministic signals in Gaussian noise, IEEE Trans. Inf. Theory **17**(5), 530–549 (1971)

32.11   E. Parzen: An approach to time series analysis, Ann. Math. Stat. **32**, 951–989 (1961)

32.12   N. Aronszajn: Theory of reproducing kernels, Trans. Am. Math. Soc. **68**, 337–404 (1950)

32.13   G. Wahba: *Spline Models for Observational Data*, CBMS–NSF Regional Conference Series in Applied Mathematics, Vol. 59 (SIAM, Philadelphia 1990)

32.14   A. Berlinet, C. Thomas-Agnan: *Reproducing Kernel Hilbert Spaces in Probability and Statistics* (Springer, New York 2004)

32.15   S. Saitoh: *Integral Transforms, Reproducing Kernels and Their Applications*, Chapman Hall/CRC Research Notes in Mathematics, Vol. 369 (Longman, Harlow 1997)

32.16   M. Aizerman, E.M. Braverman, L.I. Rozonoer: Theoretical foundations of the potential function method in pattern recognition learning, Autom. Remote Control **25**, 821–837 (1964)

32.17   V. Vapnik: Pattern recognition using generalized portrait method, Autom. Remote Control **24**, 774–780 (1963)

32.18   V. Vapnik, A. Chervonenkis: A note on one class of perceptrons, Autom. Remote Control **25**(1), 112–120 (1964)

32.19   V. Vapnik, A. Chervonenkis: *Theory of Pattern Recognitition* (Nauka, Moscow 1974), in Russian, German Translation: W. Wapnik, A. Tscherwonenkis, *Theorie der Zeichenerkennung* (Akademie-Verlag, Berlin 1979)

32.20   V. Vapnik: *Estimation of Dependences Based on Empirical Data* (Springer, New York 1982)

32.21   B.E. Boser, I.M. Guyon, V.N. Vapnik: A training algorithm for optimal margin classifiers, Proc. 5th Ann. ACM Workshop Comput. Learn. Theory, ed. by D. Haussler (1992) pp. 44–152

32.22   I. Guyon, B. Boser, V. Vapnik: Automatic capacity tuning of very large VC-dimension classifiers, Adv. Neural Inf. Process. Syst. **5**, 147–155 (1993)

32.23   I. Guyon, V. Vapnik, B. Boser, L. Bottou, S.A. Solla: Structural risk minimization for character recognition, Adv. Neural Inf. Process. Syst. **4**, 471–479 (1992)

32.24   C. Cortes, V. Vapnik: Support vector networks, Mach. Learn. **20**, 273–297 (1995)

32.25   V. Vapnik: *The Nature of Statistical Learning Theory* (Springer, New York 1995)

32.26   J.A.K. Suykens, T. Van Gestel, J. De Brabanter, B. De Moor, J. Vandewalle: *Least squares support vector machines* (World Scientific, Singapore 2002)

32.27   O. Chapelle, B. Schölkopf, A. Zien: *Semi-Supervised Learning* (MIT Press, Cambridge 2006)

32.28   M. Belkin, P. Niyogi: Semi-supervised learning on Riemannian manifolds, Mach. Learn. **56**(1), 209–239 (2004)

32.29   M. Belkin, P. Niyogi, V. Sindhwani: Manifold regularization: A geometric framework for learning from labeled and unlabeled examples, J. Mach. Learn. Res. **7**, 2399–2434 (2006)

32.30   M. Belkin, P. Niyogi: Laplacian eigenmaps for dimensionality reduction and data representation, Neural Comput. **15**(6), 1373–1396 (2003)

32.31   V. Sindhwani, P. Niyogi, M. Belkin: Beyond the point cloud: From transductive to semi-supervised learning, Int. Conf. Mach. Learn. (ICML), Vol. 22 (2005) pp. 824–831

32.32   V. Vapnik, A. Chervonenkis: The necessary and sufficient conditions for consistency in the empirical risk minimization method, Pattern Recognit. Image Anal. **1**(3), 283–305 (1991)

32.33   V. Vapnik, A. Chervonenkis: Uniform convergence of frequencies of occurrence of events to their probabilities, Dokl. Akad. Nauk SSSR **181**, 915–918 (1968)

32.34   V. Vapnik, A. Chervonenkis: On the uniform convergence of relative frequencies of events to their probabilities, Theory Probab. Appl. **16**(2), 264–280 (1971)

32.35   O. Bousquet, S. Boucheron, G. Lugosi: Introduction to statistical learning theory, Lect. Notes Comput. Sci. **3176**, 169–207 (2004)

32.36   F. Cucker, D.X. Zhou: *Learning Theory: An Approximation Theory Viewpoint*, Cambridge Monographs on Applied and Computational Mathematics (Cambridge Univ. Press, New York 2007)

32.37   I. Steinwart, A. Christmann: *Support Vector Machines*, Information Science and Statistics (Springer, New York 2008)

32.38   V. Vapnik: Transductive inference and semi-supervised learning. In: *Semi-Supervised Learning*, ed. by O. Chapelle, B. Schölkopf, A. Zien (MIT Press, Cambridge 2006) pp. 453–472

32.39   A.N. Tikhonov: On the stability of inverse problems, Dokl. Akad. Nauk SSSR **39**, 195–198 (1943)

32.40   A.N. Tikhonov: Solution of incorrectly formulated problems and the regularization method, Sov. Math. Dokl. **5**, 1035 (1963)

32.41   A.N. Tikhonov, V.Y. Arsenin: *Solutions of Ill-posed Problems* (W.H. Winston, Washington 1977)

32.42   J. Hadamard: Sur les problèmes aux dérivées partielles et leur signification physique, Princet. Univ. Bull. **13**, 49–52 (1902)

32.43   G. Kimeldorf, G. Wahba: Some results on Tchebycheffian spline functions, J. Math. Anal. Appl. **33**, 82–95 (1971)

32.44   T. Evgeniou, M. Pontil, T. Poggio: Regularization networks and support vector machines, Adv. Comput. Math. **13**(1), 1–50 (2000)

32.45   B. Schölkopf, R. Herbrich, A.J. Smola: A generalized representer theorem, Proc. Ann. Conf. Comput. Learn. Theory (COLT) (2001) pp. 416–426

32.46   F. Dinuzzo, B. Schölkopf: The representer theorem for Hilbert spaces: A necessary and sufficient condition, Adv. Neural Inf. Process. Syst. **25**, 189–196 (2012)

32.47   S.P. Boyd, L. Vandenberghe: *Convex Optimization* (Cambridge Univ. Press, Cambridge 2004)

32.48   A.E. Hoerl, R.W. Kennard: Ridge regression: Biased estimation for nonorthogonal problems, Technometrics **12**(1), 55–67 (1970)

32.49   D.W. Marquardt: Generalized inverses, ridge regression, biased linear estimation, and nonlinear estimation, Technometrics **12**(3), 591–612 (1970)

32.50   C. Gu: *Smoothing Spline ANOVA Models* (Springer, New York 2002)

32.51   D.P. Bertsekas: *Nonlinear Programming* (Athena Scientific, Belmont 1995)

32.52   R. Tibshirani: Regression shrinkage and selection via the LASSO, J. R. Stat. Soc. Ser. B **58**(1), 267–288 (1996)

32.53   P. Zhao, G. Rocha, B. Yu: The composite absolute penalties family for grouped and hierarchical variable selection, Ann. Stat. **37**, 3468–3497 (2009)

32.54   R. Jenatton, J.Y. Audibert, F. Bach: Structured variable selection with sparsity-inducing norms, J. Mach. Learn. Res. **12**, 2777–2824 (2011)

32.55   M. Yuan, Y. Lin: Model selection and estimation in regression with grouped variables, J. R. Stat. Soc. Ser. B **68**(1), 49–67 (2006)

32.56   C.A. Micchelli, M. Pontil: Learning the Kernel Function via Regularization, J. Mach. Learn. Res. **6**, 1099–1125 (2005)

32.57   C.A. Micchelli, M. Pontil: Feature space perspectives for learning the kernel, Mach. Learn. **66**(2), 297–319 (2007)

32.58   F.R. Bach, G.R.G. Lanckriet, M.I. Jordan: Multiple kernel learning, conic duality, and the SMO algorithm, Proc. 21st Int. Conf. Mach. Learn. (ICML) (ACM, New York 2004)

32.59   G.R.G. Lanckriet, T. De Bie, N. Cristianini, M.I. Jordan, W.S. Noble: A statistical framework for genomic data fusion, Bioinformatics **20**(16), 2626–2635 (2004)

32.60   F.R. Bach, R. Thibaux, M.I. Jordan: Computing regularization paths for learning multiple kernels, Adv. Neural Inf. Process. Syst. **17**, 41–48 (2004)

32.61   J. Baxter: Theoretical models of learning to learn. In: *Learning to Learn*, ed. by L. Pratt, S. Thrun (Springer, New York 1997) pp. 71–94

32.62   R. Caruana: Multitask learning. In: *Learning to Learn*, ed. by S. Thrun, L. Pratt (Springer, New York 1998) pp. 95–133

32.63   S. Thrun: Life-long learning algorithms. In: *Learning to Learn*, ed. by S. Thrun, L. Pratt (Springer, New York 1998) pp. 181–209

32.64   A. Argyriou, T. Evgeniou, M. Pontil: Multi-task feature learning, Adv. Neural Inf. Process. Syst. **19**, 41–48 (2007)

32.65   A. Argyriou, T. Evgeniou, M. Pontil: Convex multi-task feature learning, Mach. Learn. **73**(3), 243–272 (2008)

32.66   L. Debnath, P. Mikusiński: *Hilbert Spaces with Application* (Elsevier, San Diego 2005)

32.67   M. Fazel: Matrix Rank Minimization with Application, Ph.D. Thesis (Stanford University, Stanford 2002)

32.68   Z. Liu, L. Vandenberghe: Semidefinite programming methods for system realization and identification, Proc. 48th IEEE Conf. Decis. Control (CDC) (2009) pp. 4676–4681

32.69   Z. Liu, L. Vandenberghe: Interior-point method for nuclear norm approximation with application to system identification, SIAM J. Matrix Anal. Appl. **31**(3), 1235–1256 (2009)

32.70   M. Signoretto, J.A.K. Suykens: Convex estimation of cointegrated var models by a nuclear norm penalty, Proc. 16th IFAC Symp. Syst. Identif. (SYSID) (2012)

32.71   A. Argyriou, C.A. Micchelli, M. Pontil: On spectral learning, J. Mach. Learn. Res. **11**, 935–953 (2010)

32.72   J. Abernethy, F. Bach, T. Evgeniou, J.P. Vert: A new approach to collaborative filtering: Operator estimation with spectral regularization, J. Mach. Learn. Res. **10**, 803–826 (2009)

32.73   P.L. Bartlett, S. Mendelson: Rademacher and Gaussian complexities: Risk bounds and structural results, J. Mach. Learn. Res. **3**, 463–482 (2003)

32.74   P.K. Shivaswamy, T. Jebara: Maximum relative margin and data-dependent regularization, J. Mach. Learn. Res. **11**, 747–788 (2010)

32.75   P.K. Shivaswamy, T. Jebara: Relative margin machines, Adv. Neural Inf. Process. Syst. **21**(1–8), 7 (2008)

32.76   B. Schölkopf, A.J. Smola, R.C. Williamson, P.L. Bartlett: New support vector algorithms, Neural Comput. **12**(5), 1207–1245 (2000)

32.77   J. Platt: Fast training of support vector machines using sequential minimal optimization. In: *Advances in Kernel Methods – Support Vector Learning*, ed. by B. Schölkopf, C.J.C. Burges, A.J. Smola (MIT Press, Cambridge 1999) pp. 185–208

32.78   C.C. Chang, C.J. Lin: LIBSVM: a library for support vector machines, ACM Trans. Intell. Syst. Technol. **2**(3), 27 (2011)

32.79   R.E. Fan, P.H. Chen, C.J. Lin: Working set selection using second order information for training support vector machines, J. Mach. Learn. Res. **6**, 1889–1918 (2005)

32.80   T. Joachims: Making large-scale SVM learning practical. In: Advance in Kernel Methods – Support Vector Learning, ed. by B. Schölkopf, C.J.C. Burges, A.J. Smola (MIT Press, Cambridge 1999) pp. 169–184

32.81   J.A.K. Suykens, J. Vandewalle: Least squares support vector machine classifiers, Neural Process. Lett. **9**(3), 293–300 (1999)

32.82   J. Nocedal, S.J. Wright: *Numerical Optimization* (Springer, New York 1999)

32.83   K. Pelckmans, J. De Brabanter, J.A.K. Suykens, B. De Moor: The differogram: Non-parametric noise variance estimation and its use for model selection, Neurocomputing **69**(1), 100–122 (2005)

32.84   K. Saadi, G.C. Cawley, N.L.C. Talbot: Fast exact leave-one-out cross-validation of least-square support vector machines, Eur. Symp. Artif. Neural Netw. (ESANN-2002) (2002)

32.85   R.M. Rifkin, R.A. Lippert: Notes on regularized least squares, Tech. Rep. MIT-CSAIL-TR-2007-025, CBCL-268 (2007)

32.86   T. Van Gestel, J.A.K. Suykens, B. Baesens, S. Viaene, J. Vanthienen, G. Dedene, B. De Moor, J. Vandewalle: Benchmarking least squares support vector machine classifiers, Mach. Learn. **54**(1), 5–32 (2004)

32.87   G. Baudat, F. Anouar: Generalized discriminant analysis using a kernel approach, Neural Comput. **12**(10), 2385–2404 (2000)

32.88   S. Mika, G. Rätsch, J. Weston, B. Schölkopf, K.R. Müllers: Fisher discriminant analysis with kernels, Proc. 1999 IEEE Signal Process. Soc. Workshop (1999) pp. 41–48

32.89   T. Poggio, F. Girosi: Networks for approximation and learning, Proc. IEEE **78**(9), 1481–1497 (1990)

32.90   C. Saunders, A. Gammerman, V. Vovk: Ridge regression learning algorithm in dual variables, Int. Conf. Mach. Learn. (ICML) (1998) pp. 515–521

32.91   N. Cressie: The origins of kriging, Math. Geol. **22**(3), 239–252 (1990)

32.92   D.J.C. MacKay: Introduction to Gaussian processes, NATO ASI Ser. F Comput. Syst. Sci. **168**, 133–166 (1998)

32.93   C.K.I. Williams, C.E. Rasmussen: Gaussian processes for regression, Advances in Neural Information Processing Systems, Vol.8 (MIT Press, Cambridge 1996) pp. 514–520

32.94   J.A.K. Suykens, T. Van Gestel, J. Vandewalle, B. De Moor: A support vector machine formulation to pca analysis and its kernel version, IEEE Trans. Neural Netw. **14**(2), 447–450 (2003)

32.95   C. Alzate, J.A.K. Suykens: Multiway spectral clustering with out-of-sample extensions through weighted kernel PCA, IEEE Trans. Pattern Anal. Mach. Intell. **32**(2), 335–347 (2010)

32.96   T. Van Gestel, J.A.K. Suykens, J. De Brabanter, B. De Moor, J. Vandewalle: Kernel canonical correlation analysis and least squares support vector machines, Lect. Notes Comput. Sci. **2130**, 384–389 (2001)

32.97   J.A.K. Suykens: Data visualization and dimensionality reduction using kernel maps with a reference point, IEEE Trans. Neural Netw. **19**(9), 1501–1517 (2008)

32.98   J.A.K. Suykens, J. Vandewalle: Recurrent least squares support vector machines, IEEE Trans. Circuits Syst. I: Fundam. Theory Appl. **47**(7), 1109–1114 (2000)

32.99   J.A.K. Suykens, J. Vandewalle, B. De Moor: Optimal control by least squares support vector machines, Neural Netw. **14**(1), 23–35 (2001)

32.100 J.A.K. Suykens, C. Alzate, K. Pelckmans: Primal and dual model representations in kernel-based learning, Stat. Surv. **4**, 148–183 (2010)

32.101 J.A.K. Suykens, J. De Brabanter, L. Lukas, J. Vandewalle: Weighted least squares support vector machines: Robustness and sparse approximation, Neurocomputing **48**(1), 85–105 (2002)

32.102 C.K.I. Williams, M. Seeger: Using the Nyström method to speed up kernel machines, Adv. Neural Inf. Process. Syst. **15**, 682–688 (2001)

32.103 K. De Brabanter, J. De Brabanter, J.A.K. Suykens, B. De Moor: Optimized fixed-size kernel models for large data sets, Comput. Stat. Data Anal. **54**(6), 1484–1504 (2010)

32.104 B. Schölkopf, A. Smola, K.-R. Müller: Nonlinear component analysis as a kernel eigenvalue problem, Neural Comput. **10**, 1299–1319 (1998)

32.105 I. Jolliffe: Principle Component Analysis. In: *Encyclopedia of Statistics in Behavioral Science*, (Wiley, Chichester 2005)

32.106 J.A.K. Suykens, T. Van Gestel, J. Vandewalle, B. De Moor: A support vector machine formulation to PCA analysis and its kernel version, IEEE Trans. Neural Netw. **14**(2), 447–450 (2003)

32.107   N. Weiner: *Extrapolation, Interpolation, Smoothing of Stationary Time Series with Engineering Applications* (MIT Press, Cambridge 1949)

32.108   A.N. Kolmogorov: Sur l'interpolation et extrapolation des suites stationnaires, CR Acad. Sci. **208**, 2043–2045 (1939)

32.109   C.E. Rasmussen, C.K.I. Williams: *Gaussian Processes for Machine Learning*, Vol. 1 (MIT Press, Cambridge 2006)

32.110   J.O. Berger: *Statistical Decision Theory and Bayesian Analysis* (Springer, New York 1985)

32.111   K. Duan, S.S. Keerthi, A.N. Poo: Evaluation of simple performance measures for tuning svm hyperparameters, Neurocomputing **51**, 41–59 (2003)

32.112   P.L. Bartlett, S. Boucheron, G. Lugosi: Model selection and error estimation, Mach. Learn. **48**(1), 85–113 (2002)

32.113   N. Shawe-Taylor, A. Kandola: On kernel target alignment, Adv. Neural Inf. Process. Syst. **14**(1), 367–373 (2002)

32.114   G.C. Cawley: Leave-one-out cross-validation based model selection criteria for weighted LS-SVMs, Int. Joint Conf. Neural Netw. (IJCNN) (2006) pp. 1661–1668

32.115   G.C. Cawley, N.L.C. Talbot: Preventing over-fitting during model selection via Bayesian regularisation of the hyper-parameters, J. Mach. Learn. Res. **8**, 841–861 (2007)

32.116   D.J.C. MacKay: Bayesian interpolation, Neural Comput. **4**, 415–447 (1992)

32.117   D.J.C. MacKay: The evidence framework applied to classification networks, Neural Comput. **4**(5), 720–736 (1992)

32.118   D.J.C. MacKay: Probable networks and plausible predictions – A review of practical Bayesian methods for supervised neural networks, Netw. Comput. Neural Syst. **6**(3), 469–505 (1995)

32.119   I. Steinwart, D. Hush, C. Scovel: An explicit description of the reproducing kernel Hilbert spaces of Gaussian RBF kernels, IEEE Trans. Inform. Theory **52**, 4635–4643 (2006)

32.120   J.B. Conway: *A Course in Functional Analysis* (Springer, New York 1990)

32.121   F. Riesz, B.S. Nagy: *Functional Analysis* (Frederick Ungar, New York 1955)

32.122   I. Steinwart: On the influence of the kernel on the consistency of support vector machines, J. Mach. Learn. Res. **2**, 67–93 (2002)

32.123   T. Gärtner: *Kernels for Structured Data*, Machine Perception and Artificial Intelligence, Vol. 72 (World Scientific, Singapore 2008)

32.124   D. Haussler: *Convolution kernels on discrete structures*, Tech. Rep. (UC Santa Cruz, Santa Cruz 1999)

32.125   T. Jebara, R. Kondor, A. Howard: Probability product kernels, J. Mach. Learn. Res. **5**, 819–844 (2004)

32.126   T.S. Jaakkola, D. Haussler: Exploiting generative models in discriminative classifiers, Adv. Neural Inf. Process. Syst. **11**, 487–493 (1999)

32.127   K. Tsuda, S. Akaho, M. Kawanabe, K.R. Müller: Asymptotic properties of the Fisher kernel, Neural Comput. **16**(1), 115–137 (2004)

32.128   S.V.N. Vishwanathan, N.N. Schraudolph, R. Kondor, K.M. Borgwardt: Graph kernels, J. Mach. Learn. Res. **11**, 1201–1242 (2010)

32.129   T. Gärtner, P. Flach, S. Wrobel: On graph kernels Hardness results and efficient alternatives, Lect. Notes Comput. Sci. **2777**, 129–143 (2003)

32.130   S.V.N. Vishwanathan, A.J. Smola, R. Vidal: Binet-Cauchy kernels on dynamical systems and its application to the analysis of dynamic scenes, Int. J. Comput. Vis. **73**(1), 95–119 (2007)

32.131   P.M. Kroonenberg: *Applied Multiway Data Analysis* (Wiley, Hoboken 2008)

32.132   M. Signoretto, L. De Lathauwer, J.A.K. Suykens: A kernel-based framework to tensorial data analysis, Neural Netw. **24**(8), 861–874 (2011)

32.133   L. De Lathauwer, B. De Moor, J. Vandewalle: A multilinear singular value decomposition, SIAM J. Matrix Anal. Appl. **21**(4), 1253–1278 (2000)

32.134   M. Signoretto, E. Olivetti, L. De Lathauwer, J.A.K. Suykens: Classification of multichannel signals with cumulant-based kernels, IEEE Trans. Signal Process. **60**(5), 2304–2314 (2012)

32.135   Y. LeCun, L.D. Jackel, L. Bottou, A. Brunot, C. Cortes, J.S. Denker, H. Drucker, I. Guyon, U.A. Muller, E. Sackinger, P. Simard, V. Vapnik: Comparison of learning algorithms for handwritten digit recognition, Int. Conf. Artif. Neural Netw. (ICANN) 2 (1995) pp. 53–60

32.136   D. Decoste, B. Schölkopf: Training invariant support vector machines, Mach. Learn. **46**(1), 161–190 (2002)

32.137   V. Blanz, B. Schölkopf, H. Bülthoff, C. Burges, V. Vapnik, T. Vetter: Comparison of view-based object recognition algorithms using realistic 3D models, Lect. Notes Comput. Sci. **1112**, 251–256 (1996)

32.138   T. Joachims: Text categorization with support vector machines: Learning with many relevant features, Lect. Notes Comput. Sci. **1398**, 137–142 (1998)

32.139   S. Dumais, J. Platt, D. Heckerman, M. Sahami: Inductive learning algorithms and representation for text categorization, Proc. 7th Int. Conf. Inf. Knowl. Manag. (1998) pp. 148–155

32.140   S. Mukherjee, E. Osuna, F. Girosi: Nonlinear prediction of chaotic time series using support vector machines, 1997 IEEE Workshop Neural Netw. Signal Process. VII (1997) pp. 511–520

32.141   D. Mattera, S. Haykin: Support vector machines for dynamic reconstruction of a chaotic system. In: *Advances in Kernel Methods*, ed. by B. Schölkopf, C.J.C. Burges, A.J. Smola (MIT Press, Cambridge 1999) pp. 211–241

32.142   K.R. Müller, A. Smola, G. Rätsch, B. Schölkopf, J. Kohlmorgen, V. Vapnik: Predicting time series with support vector machines, Lect. Notes Comput. Sci. **1327**, 999–1004 (1997)

32.143 M. Espinoza, J.A.K. Suykens, B. De Moor: Short term chaotic time series prediction using symmetric ls-svm regression, Proc. 2005 Int. Symp. Nonlinear Theory Appl. (NOLTA) (2005) pp. 606–609

32.144 M. Espinoza, T. Falck, J.A.K. Suykens, B. De Moor: Time series prediction using ls-svms, Eur. Symp. Time Ser. Prediction (ESTSP), Vol. 8 (2008) pp. 159–168

32.145 M. Espinoza, J.A.K. Suykens, R. Belmans, B. De Moor: Electric load forecasting, IEEE Control Syst. **27**(5), 43–57 (2007)

32.146 T. Van Gestel, J.A.K. Suykens, D.E. Baestaens, A. Lambrechts, G. Lanckriet, B. Vandaele, B. De Moor, J. Vandewalle: Financial time series prediction using least squares support vector machines within the evidence framework, IEEE Trans. Neural Netw. **12**(4), 809–821 (2001)

32.147 M.P.S. Brown, W.N. Grundy, D. Lin, N. Cristianini, C.W. Sugnet, T.S. Furey, M. Ares, D. Haussler: Knowledge-based analysis of microarray gene expression data by using support vector machines, Proc. Natl. Acad. Sci. USA **97**(1), 262–267 (2000)

32.148 J. Luts, F. Ojeda, R. Van de Plas, B. De Moor, S. Van Huffel, J.A.K. Suykens: A tutorial on support vector machine-based methods for classification problems in chemometrics, Anal. Chim. Acta **665**(2), 129 (2010)

32.149 A. Daemen, M. Signoretto, O. Gevaert, J.A.K. Suykens, B. De Moor: Improved microarray-based decision support with graph encoded interactome data, PLoS ONE **5**(4), 1–16 (2010)

32.150 S. Yu, L.C. Tranchevent, B. Moor, Y. Moreau: *Kernel-based Data Fusion for Machine Learning*, Studies in Computational Intelligence, Vol. 345 (Springer, Berlin 2011)

32.151 T. Jaakkola, M. Diekhans, D. Haussler: A discriminative framework for detecting remote protein homologies, J. Comput. Biol. **7**(1/2), 95–114 (2000)

32.152 C. Lu, T. Van Gestel, J.A.K. Suykens, S. Van Huffel, D. Timmerman, I. Vergote: Classification of ovarian tumors using Bayesian least squares support vector machines, Lect. Notes Artif. Intell. **2780**, 219–228 (2003)

32.153 F. Ojeda, M. Signoretto, R. Van de Plas, E. Waelkens, B. De Moor, J.A.K. Suykens: Semi-supervised learning of sparse linear models in mass spectral imaging, Pattern Recognit. Bioinform. (PRIB) (Nijmegen) (2010) pp. 325–334

32.154 D. Widjaja, C. Varon, A.C. Dorado, J.A.K. Suykens, S. Van Huffel: Application of kernel principal component analysis for single lead ECG-derived respiration, IEEE Trans. Biomed. Eng. **59**(4), 1169–1176 (2012)

32.155 V. Van Belle, K. Pelckmans, S. Van Huffel, J.A.K. Suykens: Support vector methods for survival analysis: A comparison between ranking and regression approaches, Artif. Intell. Med. **53**(2), 107–118 (2011)

32.156 V. Van Belle, K. Pelckmans, S. Van Huffel, J.A.K. Suykens: Improved performance on high-dimensional survival data by application of survival-SVM, Bioinformatics **27**(1), 87–94 (2011)

33. Neurodynamics

Robert Kozma, Jun Wang, Zhigang Zeng

This chapter introduces basic concepts, phenomena, and properties of neurodynamic systems. it consists of four sections with the first two on various neurodynamic behaviors of general neurodynamics and the last two on two types of specific neurodynamic systems. The neurodynamic behaviors discussed in the first two sections include attractivity, oscillation, synchronization, and chaos. The two specific neurodynamics systems are memrisitve neurodynamic systems and neurodynamic optimization systems.

33.1 Dynamics of Attractor and Analog Networks

An attractor, as a well-known mathematical object, is central to the field of nonlinear dynamical systems (NDS) theory, which is one of the indispensable conceptual underpinnings of complexity science. An attractor is a set towards which a variable moves according to the dictates of a nonlinear dynamical system, evolves over time, such that points get close enough to the attractor, and remain close even if they are slightly disturbed. To well appreciate what an attractor is, some corresponding NDS notions, such as phase or state space, phase portraits, basins of attractions, initial conditions, transients, bifurcations, chaos, and strange attractors are needed to tame some of the unruliness of complex systems.

Most of us have at least some inkling of what *nonlinear* means, which can be illustrated by the most well-known and vivid example of the *butterfly effect* of a chaotic system that is nonlinear. It has prompted

the use of the image of tiny air currents produced by a butterfly flapping its wing in Brazil, which are then amplified to the extent that they may influence the building up of a thunderhead in Kansas. Although no one can actually claim that there is such a linkage between Brazilian lepidopterological dynamics and climatology in the Midwest of the USA, it does serve to vividly portray nonlinearity in the extreme.

As the existence of both the nonlinearity and the capacity in passing through different regimes of stability and instability, the outcomes of the nonlinear dynamical system are unpredictable. These different regimes of a dynamical system are understood as different phases governed by different attractors, which means that the dynamics of each phase of a dynamical system are constrained within the circumscribed range allowable by that phase's attractors.

Part D | 33.1

33.1.1 Phase Space and Attractors

To better grasp the idea of phase space, a time series and phase portrait have been used to represent the data points. Time series display changes in the values of variables on the *y*-axis (or the *z*-axis), and time on the *x*-axis as in a time series chart, however, the phase portrait plots the variables against each other and leaves time as an implicit dimension not explicitly plotted. Attractors can be displayed by phase portraits as the long-term stable sets of points of the dynamical system. This means that the locations in the phase portrait towards which the system's dynamics are attracted after transient phenomena have died down. To illustrate phase space and attractors, two examples are employed.

Imagine a child on a swing and a parent pulling the swing back. This gives a good push to make the child move forward. When the child is not moving forward, he will move backward on the swing as shown in Fig. 33.1. The unpushed swing will come to rest as shown in the times series chart and phase space. The time series show an oscillation of the speed of the swing, which slows down and eventually stops, that is its flat lines. In phase space, the swing's speed is plotted against the distance of the swing from the central point called a fixed point attractor since it attracts the system's dynamics in the long run. The fixed point attractor in the center of Fig. 33.2 is equivalent to the flat line in Fig. 33.3. The fixed point attractor is another way to see and say that an unpushed swing will come to a state of rest in the long term. The curved lines with arrows spiraling down to the center point in Fig. 33.2 display what is called the basin of attraction for the unpushed swing. These basins of attraction represent various initial conditions for the unpushed swing, such as starting heights and initial velocities.

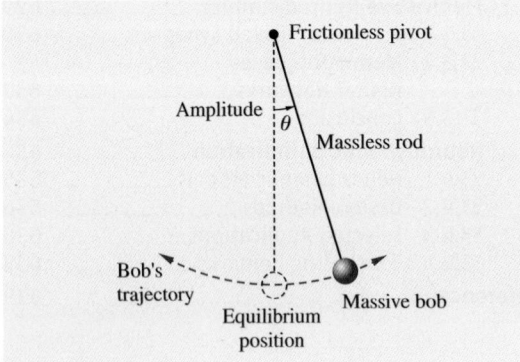

Fig. 33.1 Schematics of an unpushed swing

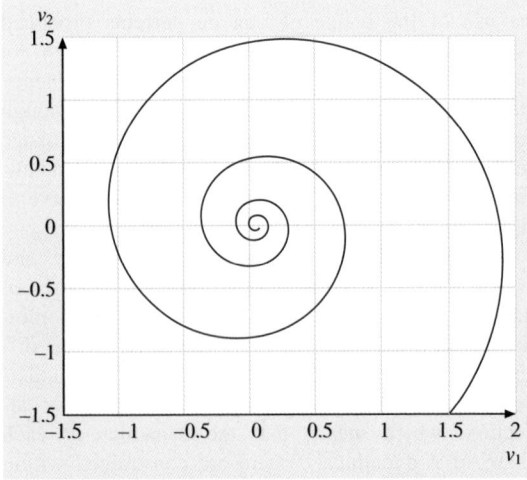

Fig. 33.2 Phase portrait and fixed point attractor of an unpushed swing

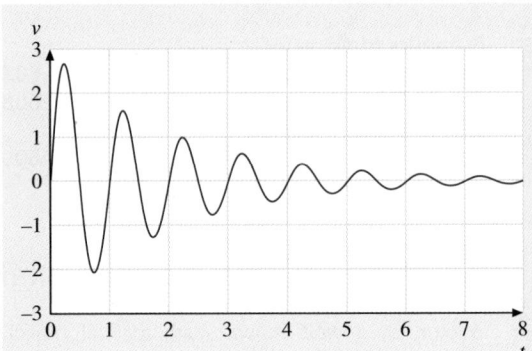

Fig. 33.3 Time series of the unpushed swing

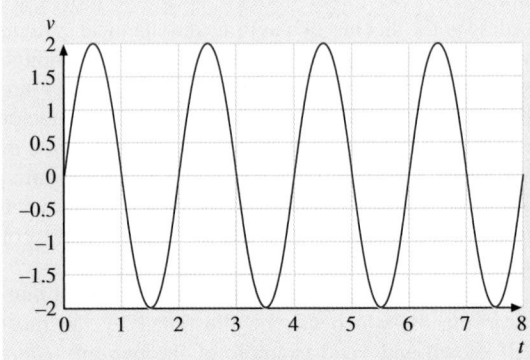

Fig. 33.4 Time series chart of the pushed swing

Now consider another type of a similar dynamical system, this time the swing is pushed each time it comes back to where the parent is standing. The time series chart of the pushed swing is shown in Fig. 33.4 as a continuing oscillation. This oscillation is around a zero value for y and is positive when the swing is going in one direction and negative when the swing is going in the other direction. As a phase space diagram, the states of variables against each other are shown in Fig. 33.5. The unbroken oval in Fig. 33.5 is a different kind of attractor from the fixed point one in Fig. 33.2. This attractor is well known as a limit cycle or periodic attractor of a pushed swing. It is called a limit cycle because it represents the cyclical behavior of the oscillations of the pushed swing as a limit to which the dynamical systems adheres under the sway of this attractor. It is periodic because the attractor oscillates around the same values, as the swing keeps going up and down until the s has a same heights from the lowest point. Such dynamical system can be called periodic for it has a repeating cycle or pattern.

By now, what we have learned about attractors can be summarized as follows: they are spatially displayed phase portraits of a dynamical system as it changes over the course of time, thus they represent the long-term dynamics of the system so that whatever the initial conditions represented as data points are, their trajectories in phase space fall within its basins of attraction, they are attracted to the attractor. In spite of wide usage in mathematics and science, as *Robinson* points out there is still no precise definition of an attractor, although many have been offered [33.2]. So he suggests thinking about an attractor as a phase portrait that attracts a large set of initial conditions and has some sort of minimality property, which is the smallest portrait in the phase space of the system. The attractor has the property of attracting the initial conditions after any initial transient behavior has died down. The minimality requirement implies the invariance or stability of the attractor. As a minimal object, the attractor cannot be split up into smaller subsets and retains its role as what dominates a dynamical system during a particular phase of its evolution.

33.1.2 Single Attractors of Dynamical Systems

Standard methods for the study of stability of dynamical systems with a unique attractor include the Lyapunov method, the Lasalles invariance principle, and the combination of thereof. Usually, given the properties of a (unique) attractor, we can realize a dynamical system with such an attractor.

Since the creation of the fundamental theorems of Lyapunov stability, many researchers have gone further and proved that most of the fundamental Lyapunov theories are reversible. Thus, from theory, this demonstrates that these theories are efficacious; i.e., there necessarily exists the corresponding Lyapunov function if the solution has some kind of stability. However, as for the construction of an appropriate V function for the determinant of stability, researchers are still interested. The difference between the existence and its construction is large. However, there is no general rule for the construction of the Lyapunov function. In some cases, different researchers have different methods for the construction of the Lyapunov function based on their experience and technique. Those, who can construct a good quality Lyapunov function, can get more useful information to demonstrate the effectiveness of their theories. Certainly, many successful Lyapunov functions have a practical background. For example, some equations inferred from the physical model have a clear physical meaning such as the mechanics guard system, in which the total sum of the kinetic energy and potential energy is the appropriate V function. The linear approximate method can be used; i.e., for the nonlinear differential equation, firstly find its corresponding linear differential equation's quadric form positive defined V function, then consider the nonlinear quality for the construction of a similar V function.

Grossberg proposed and studied additive neural networks because they add nonlinear contributions to the

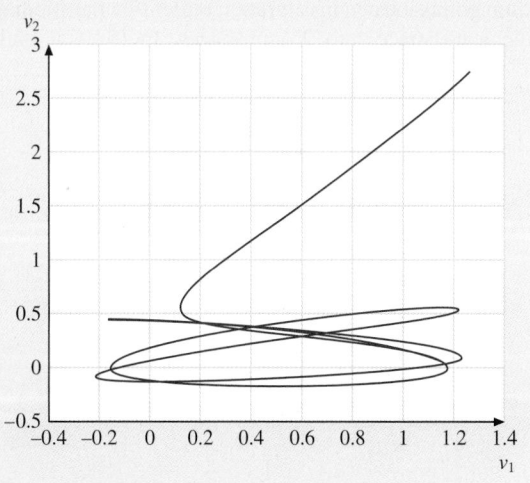

Fig. 33.5 Phase portrait and limit cycle attractor of a pushed swing (after [33.1])

neuron activity. The additive neural network has been used for many applications since the 1960s [33.3, 4], including the introduction of self-organizing maps. In the past decades, neural networks as a special kind of nonlinear systems have received considerable attention. The study of recurrent neural networks with their various generalizations has been an active research area [33.5–17]. The stability of recurrent neural networks is a prerequisite for almost all neural network applications. Stability analysis is primarily concerned with the existence and uniqueness of equilibrium points and global asymptotic stability, global exponential stability, and global robust stability of neural networks at equilibria. In recent years, the stability analysis of recurrent neural networks with time delays has received much attention [33.18, 19]. Single attractors of dynamical systems are shown in Fig. 33.6.

33.1.3 Multiple Attractors of Dynamical Systems

Multistable systems have attracted extensive interest in both modeling studies and neurobiological research in

recent years due to their feasibility to emulate and explain biological behavior [33.20–34]. Mathematically, multistability allows the system to have multiple fixed points and periodic orbits. As noted in [33.35], more than 25 years of experimental and theoretical work has indicated that the onset of oscillations in neurons and in neuron populations is characterized by multistability.

Multistability analysis is different from monostability analysis. In monostability analysis, the objective is to derive conditions that guarantee that each nonlinear system contains only one equilibrium point, and all the trajectories of the neural network converge to it. Whereas in multistability analysis, nonlinear systems are allowed to have multiple equilibrium points. Stable and unstable equilibrium points, and even periodic trajectories may co-exist in a multistable system.

The methods to study the stability of dynamical systems with a unique attractor include the Lyapunov method, the Lasalles invariance principle, and the combination of the two methods. One unique attractor can be realized by one dynamical system, but it is much more complicated for multiple attractors to be realized by one dynamical system or dynamical multisystems because of the compatibility, agreement, and behavior optimization among the systems. Generally, the usual global stability conditions are not adequately applicable to multistable systems. The latest results on multistability of neural networks can be found in [33.36–52]. It is shown in [33.45, 46] that the n-neuron recurrent neural networks with one step piecewise linear activation function can have 2^n locally exponentially stable equilibrium points located in saturation regions by partitioning the state space into 2^n subspaces. In [33.47], mul-

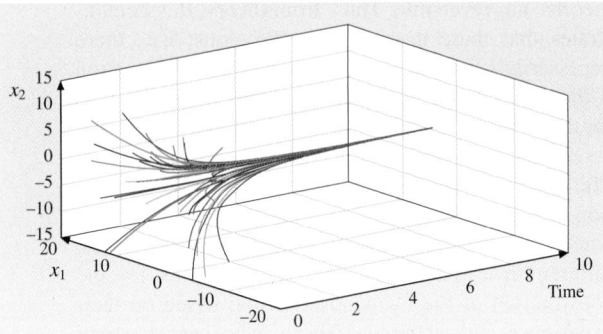

Fig. 33.6 Single attractors of dynamical systems

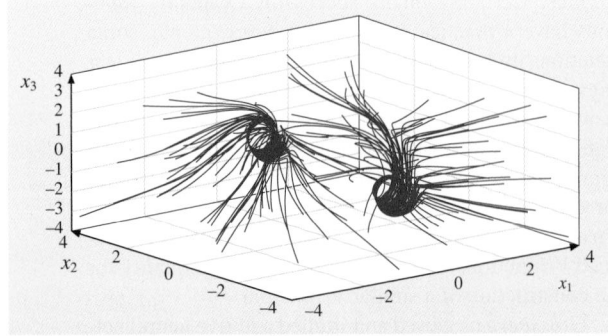

Fig. 33.7 Two limit cycle attractors of dynamical systems

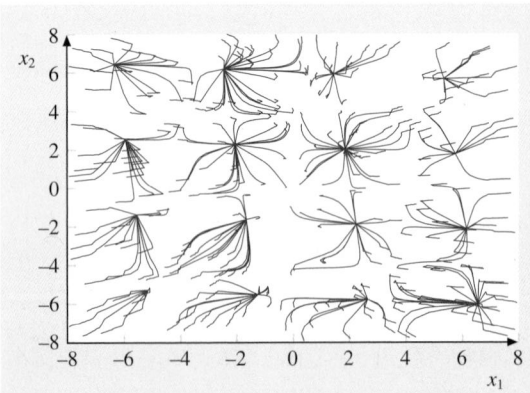

Fig. 33.8 2^4 equilibrium point attractors of dynamical systems

istability of almost periodic solutions of recurrently connected neural networks with delays is investigated. In [33.48], by constructing a Lyapunov functional and using matrix inequality techniques, a delay-dependent multistability criterion on recurrent neural networks is derived. In [33.49], the neural networks with a class of nondecreasing piecewise linear activation functions with $2r$ corner points are considered. It is proved that the n-neuron dynamical systems can have and only have $(2r + 1)^n$ equilibria under some conditions, of which $(r + 1)^n$ are locally exponentially stable and others are unstable. In [33.50], some multistability properties for a class of bidirectional associative memory recurrent neural networks with unsaturation piecewise linear transfer functions are studied based on local inhibition. In [33.51], for two classes of general activation functions, multistability of competitive neural networks

with time-varying and distributed delays is investigated by formulating parameter conditions and using inequality techniques. In [33.52], the existence of 2^n stable stationary solutions for general n-dimensional delayed neural networks with several classes of activation functions is presented through formulating parameter conditions motivated by a geometrical observation. Two limit cycle attractors and 2^4 equilibrium point attractors of dynamical systems are shown in Figs. 33.7 and 33.8, respectively.

33.1.4 Conclusion

In this section, we briefly introduced what attractors can be summarized as, and phase space and attractors. Furthermore, single-attractor and multiattractors of dynamical systems were also discussed.

33.2 Synchrony, Oscillations, and Chaos in Neural Networks

33.2.1 Synchronization

Biological Significance of Synchronization

Neurodynamics deals with dynamic changes of neural properties and behaviors in time and space at different levels of hierarchy in neural systems. The characteristic spiking dynamics of individual neurons is of fundamental importance. In large-scale systems, such as biological neural networks and brains with billions of neurons, the interaction among the connected neural components is crucial in determining collective properties. In particular, synchronization plays a critical role in higher cognition and consciousness experience [33.53–57]. Large-scale synchronization of neuronal activity arising from intrinsic asynchronous oscillations in local electrical circuitries neurons are at the root of cognition. Synchronization at the level of neural populations is characterized next.

There are various dynamic behaviors of potential interest for neural systems. In the simplest case, the system behavior converges to a fixed point, when all major variables remain unchanged. A more interesting dynamic behavior emerges when the system behavior periodically repeats itself at period T, which will be described first. Such periodic oscillations are common in neural networks and are often caused by the presence of inhibitory neurons and inhibitory neural populations. Another behavior emerges when the system

neither converges to a fixed point nor exhibits periodic oscillations, rather it maintains highly complex, chaotic dynamics. Chaos can be microscopic effect at the cellular level, or mesoscopic dynamics of neural populations or cortical regions. At the highest level of hierarchy, chaos can emerge as the result of large-scale, macroscopic effect across cortical areas in the brain.

Considering the temporal dynamics of a system of interacting neural units, limit cycle oscillations and chaotic dynamics are of importance. Synchronization in limit cycle oscillations is considered first, which illustrates the basic principles of synchronization. The extension to more complex (chaotic) dynamics is described in Sect. 33.2.3. Limit cycle dynamics is described as a cyclic repetition of the system's behavior at a given time period T. The cyclic repetition covers all characteristics of the system, e.g., microscopic currents, potentials, and dynamic variables; see, e.g., the Hodgkin–Huxley model of neurons [33.58]. Limit cycle oscillations can be described as a cyclic loop of the system trajectory in the space of all variables. The state of the system is given as a point on this trajectory at any given time instant. As time evolves, the point belonging to the system traverses along the trajectory. Due to the periodic nature of the movement, the points describing the system at time t and $t + T$ coincide fully. We can define a convenient reference system by selecting a center point of the trajectory and describe the motion as the vector pointing from the center to the actual state on the trajectory. This vector has an amplitude and phase in a suitable coordinate system, denoted as $\xi(t)$ and $\Phi(t)$, respectively. The evolution of the phase in an isolated oscillator with frequency ω_0 can be given as follows

$$\frac{d\Phi(t)}{dt} = \omega_0 . \tag{33.1}$$

Several types of synchronization can be defined. The strongest synchronization takes place when two (or multiple) units have identical behaviors. Considering limit cycle dynamics, strong synchronization means that the oscillation amplitude and phase are the same for all units. This means complete synchrony. An example of two periodic oscillators is given by the clocks shown in Fig. 33.9a–c [33.59]. Strong synchronization means that the two pendulums are connected with a rigid object forcing them move together. The lack of connection between the two pendulums means the absence of synchronization, i.e., they move completely independently. An intermediate level of synchrony may arise with weak coupling between the pendulums, such as a spring or a flexible band. Phase synchrony takes place when the amplitudes are not the same, but the

phases of the oscillations could still coincide. Figure 33.9b–d depicts the case of out-of-phase synchrony, when the phases of the two oscillators are exactly the opposite.

Amplitude Measures of Synchrony
Denote by $a_j(t)$ the time signal produced by the individual units (neurons); $j = 1, \ldots, N$, and the overall signal of interacting units (A) is determined as

$$A(t) = 1/N \sum_{j=1}^{N} a_j(t) . \tag{33.2}$$

The variance of time series $A(t)$ is given as follows

$$\sigma_A^2 = \left\langle A^2(t) \right\rangle - \left\langle A(t) \right\rangle^2 . \tag{33.3}$$

Here $\langle f(t) \rangle$ denotes time averaging over a give time window. After determining the variance of the individual channels $\sigma_{A_j}^2$ based on (33.3), the synchrony χ_N in the system with N components is defined as follows

$$\chi_N^2 = \frac{\sigma_A^2}{1/N \sum_{i=1}^{N} \sigma_{A_i}^2} . \tag{33.4}$$

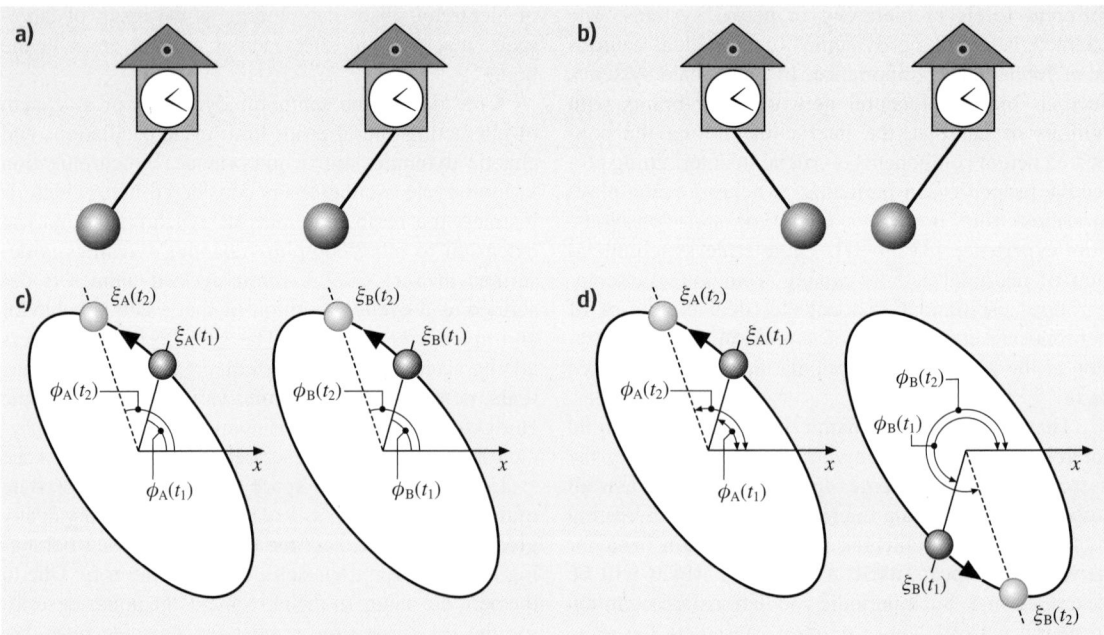

Fig. 33.9a–d Synchronization in pendulums, in phase and out of phase (after [33.59]). *Bottom plots*: Illustration of periodic trajectories, case of in-phase **(a–c)** and out-of-phase oscillations **(b–d)**

This synchrony measure has a nonzero value in synchronized and partially synchronized systems $0 < \chi_N < 1$, while $\chi_N = 0$ means the complete absence of synchrony in neural networks [33.60].

Fourier transform-based signal processing methods are very useful for the characterization of synchrony in time series, and they are widely used in neural network analysis. The Fourier transform makes important assumptions on the analyzed time series, including stationary or slowly changing statistical characteristics and ergodicity. In many applications these approximations are appropriate. In analyzing large-scale synchrony on brain signals, however, alternative methods are also justified. Relevant approaches include the Hilbert transform for rapidly changing brain signals [33.61, 62]. Here both Fourier and Hilbert-based methods are outlined and avenues for their applications in neural networks are indicated. Define the cross correlation function (CCF) between discretely sampled time series $x_i(t)$ and $x_i(t)$, $t = 1, \ldots, N$ as follows

$$\mathrm{CCF}_{ij}(\tau) = \frac{1}{T} \sum_{t=1}^{T-\tau} [x_i(t+\tau) - \langle x_i \rangle][x_j(t) - \langle x_j \rangle] .$$

(33.5)

Here $\langle x_i \rangle$ is the mean of the signal over period T, and it is assumed that $x_i(t)$ is normalized to unit variance. For completely correlated pairs of signals, the maximum of the cross correlation is 1, for uncorrelated signals it equals 0. The cross power spectral density $\mathrm{CPSD}_{ij}(\omega)$, cross spectrum for short, is defined as the Fourier transform of the cross correlation as follows: $\mathrm{CPSD}_{ij}(\omega) = \mathcal{F}(\mathrm{CCF}_{ij}(\tau))$. If $i = j$, i.e., the two channels coincide, then we talk about autocorrelation and auto power spectral density $\mathrm{APSD}_{ii}(\omega)$; for details of Fourier analysis, see [33.63]. Coherence γ^2 is defined by normalizing the cross spectrum by the autospectra

$$\gamma_{ij}^2(\omega) = \frac{|\mathrm{CPSD}_{ij}(\omega)|^2}{|\mathrm{APSD}_{ii}(\omega)||\mathrm{APSD}_{jj}(\omega)|} .$$

(33.6)

The coherence satisfies $0 \leq \gamma^2(\omega) \leq 1$ and it contains useful information on the frequency content of the synchronization between signals. If coherence is close to unity at some frequencies, it means that two signals are closely related or synchronized; a coherence near zero means the absence of synchrony at those frequencies. Coherence functions provide useful information on synchrony in brain signals at various frequency bands [33.64]. For other information-theoretical characterizations, including mutual information and entropy measures.

Phase Synchronization

If the components of the neural network are weakly interacting, the synchrony evaluated using the amplitude measure χ in (33.4) may be low. There can still be a meaningful synchronization effect in the system, based on phase measures. Phase synchronization is defined as the global entrainment of the phases [33.65], which means a collective adjustment of their rhythms due to their weak interaction. At the same time, in systems with phase synchronization the amplitudes need not be synchronized. Phase synchronization is often observed in complex chaotic systems and it has been identified in biological neural networks [33.61, 65].

In complex systems, the trajectory of the system in the phase space is often very convoluted. The approach described in (33.1), i.e., choosing a center point for the oscillating cycle in the phase space with natural frequency ω_0, can be nontrivial in chaotic systems. In such cases, the Hilbert transform-based approach can provide a useful tool for the characterization of phase synchrony. Hilbert analysis determines the analytic signal and its instantaneous frequency, which can be used to describe phase synchronization effects. Considering time series $s(t)$, its analytic signal $z(t)$ is defined as follows [33.62]

$$z(t) = s(t) + \mathrm{i}\hat{s}(t) = A(t)\mathrm{e}^{\mathrm{i}\Phi(t)} .$$

(33.7)

Here $A(t)$ is the analytic amplitude, $\Phi(t)$ is the analytic phase, while $\hat{s}(t)$ is the Hilbert transform of $s(t)$, given by

$$\hat{s}(t) = \frac{1}{\pi}\mathrm{PV} \int_{-\infty}^{+\infty} \frac{s(t)}{t - \tau} \mathrm{d}\tau ,$$

(33.8)

where PV stands for the principal value of the integral computed over the complex plane. The analytic signal and its instantaneous phase can be determined for an arbitrary broadband signal. However, the analytic signal has clear meaning only at a narrow frequency band, therefore, the bandpass filter should precede the evaluation of analytic signal in data with broad frequency content.

The Hilbert method of analytic signals is illustrated using actual local field potentials measured over rabbits with an array of chronically implanted intracranial electrodes [33.67]. The signals have been filtered in the theta band (3–7 Hz). An example of time series $s(t)$ is shown in Fig. 33.10a. The Hilbert transform $\hat{s}(t)$ is depicted in Fig. 33.10b in red, while blue color shows $s(t)$. Figure 33.10c shows the analytic

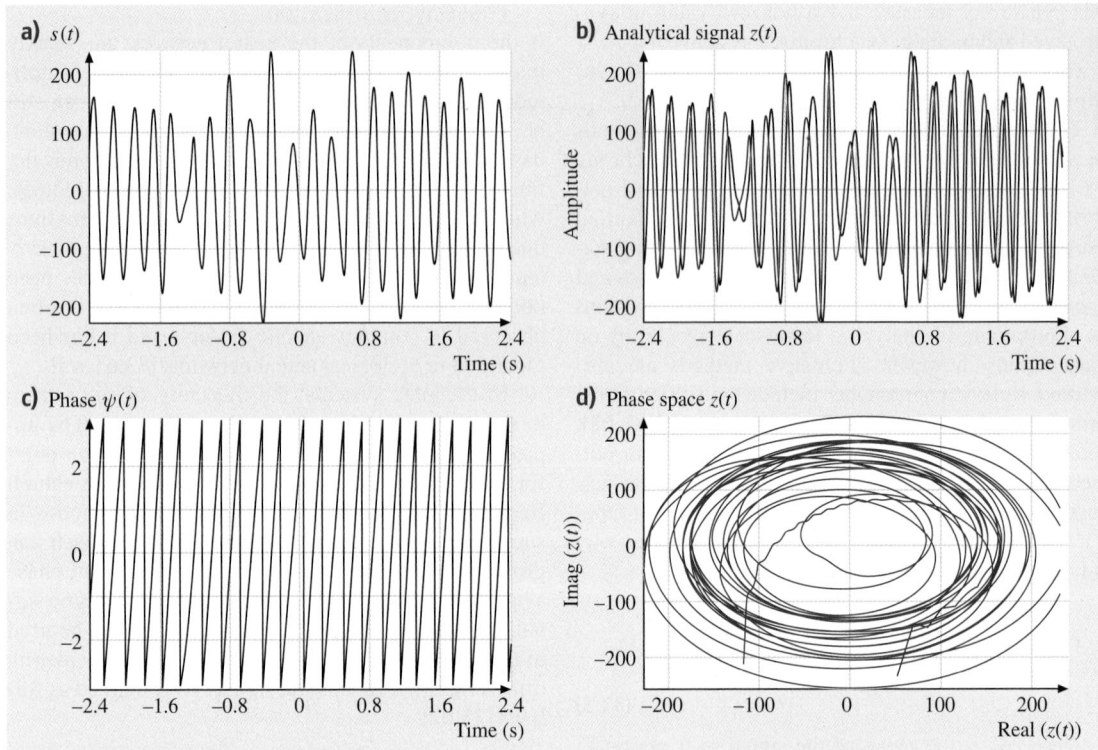

Fig. 33.10a–d Demonstration of the Hilbert analytic signal approach on electroencephalogram (EEG) signals (after [33.66]); (**a**) signal $s(t)$; (**b**) Hilbert transform $\hat{s}(t)$ (*red*) of signal $s(t)$ (*blue*); (**c**) instantaneous phase $\Phi(t)$; and analytic signal in complex plane $z(t)$

phase $\Phi(t)$, and Fig. 33.10d depicts the analytic $z(t)$ signal in the complex plane. Figure 33.11 shows the unwrapped instantaneous phase with bifurcating phase curves indicating desynchronization at specific time instances -1.3 s, -0.4 s, and 1 s. The plot on the right-hand side of Fig. 33.11 depicts the evolution of the instantaneous frequency in time. The frequency is around 5 Hz most of the time, indicating phase synchronization. However, it has very large dispersion at a few specific instances (desynchronization).

Synchronization between channels x and y can be measured using the phase lock value (PLV) defined as follows [33.61]

$$\mathrm{PLV}_{xy}(t) = \left| \frac{1}{T} \int_{t-T/2}^{t+T/2} \mathrm{e}^{\mathrm{i}[\Phi_x(\tau) - \Phi_y(\tau)]} \mathrm{d}\tau \right| . \qquad (33.9)$$

PLV ranges from 1 to 0, where 1 indicates complete phase locking. PLV defined in (33.9) determines an average value over a time window of length T. Note that

PLV is a function of t by applying the given sliding window. PLV is also the function of the frequency, which is being selected by the bandpass filter during the preprocessing phase. By changing the frequency band and time, the synchronization can be monitored at various conditions. This method has been applied productively in cognitive experiments [33.68].

Synchronization–Desynchronization Transitions

Transitions between neurodynamic regimes with and without synchronization have been observed and exploited for cognitive monitoring. The Haken–Kelso–Bunz (HKB) model is one of the prominent and elegant approaches providing a theoretical framework for synchrony switching, based on the observations related to bimanual coordination [33.69]. The HKB model invokes the concepts of metastability and multistability as fundamental properties of cognition. In the experiment, the subjects were instructed to follow the rhythm of a metronome with their index fingers in an anti-phase manner. It was observed that by increasing the metronome frequency, the subject spontaneously switched their anti-phase movement to in-phase at a certain oscillation frequency and maintained it thereon even if the metronome frequency was decreased again below the given threshold.

The following simple equation is introduced to describe the dynamics observed: $\mathrm{d}\Delta\Phi/\mathrm{d}t = -\sin(\Phi) - 2\varepsilon\sin(2\Phi)$. Here $\Delta\Phi = \phi_1 - \phi_2$ is the phase difference between the two finger movements, control parameter ε is related to the inverse of the introduced oscillatory frequency. The system dynamics is illustrated in Fig. 33.12 by the potential surface V, where stable fixed points correspond to local minima. For low oscillatory frequencies (high ε), there are stable equilibria at antiphase conditions. As the oscillatory frequency increases

(low ε) the dynamics transits to a state where only the in-phase equilibrium is stable.

Another practical example of synchrony-desynchrony transition in neural networks is given by image processing. An important basic task of neural networks is image segmentation, which is difficult to accomplish with excitatory nodes only. There is evidence that biological neural networks use inhibitory connections for completing basic pattern separation and integration tasks [33.70]. Synchrony between interacting neurons may indicate the recognition of an input. A typical neural network architecture implementing such a switch between synchronous and nonsynchronous states using local excitation and global inhibition is shown in Fig. 33.13. This system uses amplitude difference to

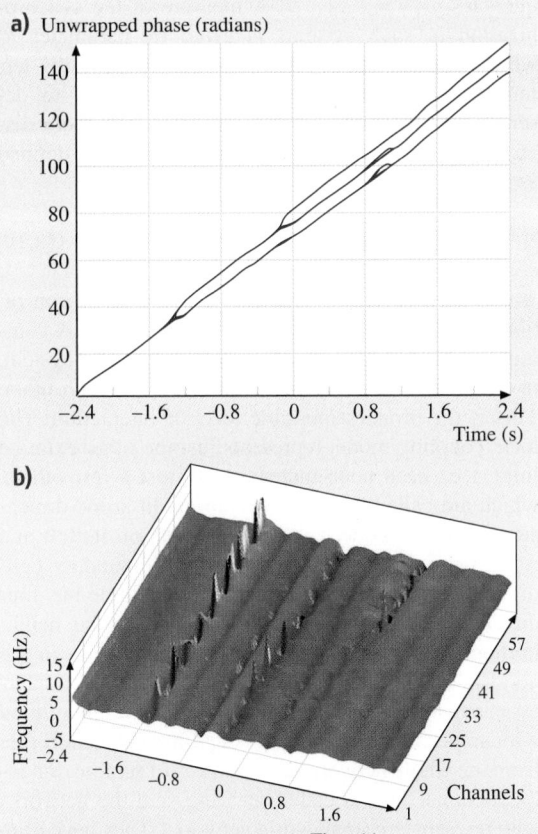

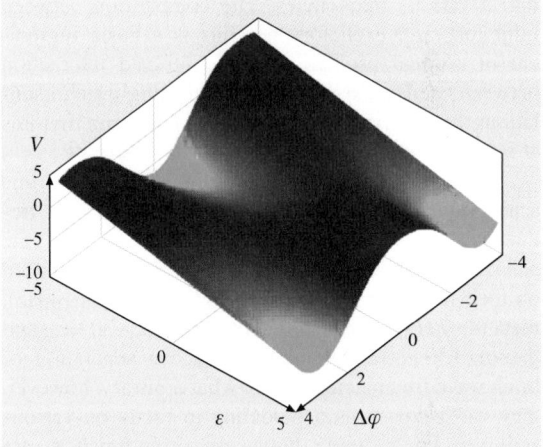

Fig. 33.12 Illustration of the potential surface (V) of the HKB system as a function of the phase difference in radians $\Delta\Phi$ and inverse frequency ε. The transition from antiphase to in-phase behavior is seen as the oscillation frequency increases (ε decreases)

Fig. 33.11a,b Illustration of instantaneous phases; (**a**) unwrapped phase with bifurcating phase curves indicating desynchronization at specific time instances -1.3 s, -0.4 s, and 1 s; (**b**) evolution of instantaneous frequency in time

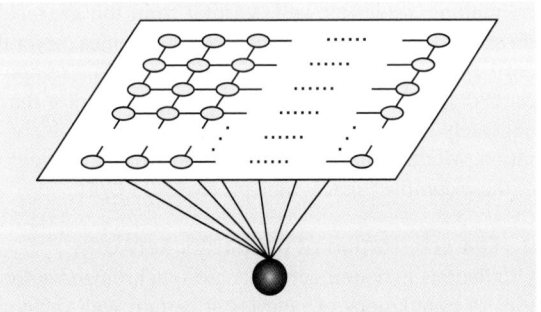

Fig. 33.13 Neural network with local excitation and a global inhibition node (*black*; after [33.70])

measure synchronization between neighboring neurons. Phase synchronization measures have been proposed as well to accomplish the segmentation and recognition tasks [33.71]. Phase synchronization provides a very useful tool for learning and control of the oscillations in weakly interacting neighborhoods.

33.2.2 Oscillations in Neural Networks

Oscillations in Brains

The interaction between opposing tendencies in physical and biological systems can lead to the onset of oscillations. Negative feedback between the system's components plays an important role in generating oscillations in electrical systems. Brains as large-scale bioelectrical networks consist of components oscillating at various frequencies. The competition between inhibitory and excitatory neurons is a basic ingredient of cortical oscillations. The intricate interaction between oscillators produces the amazingly rich oscillations that we experimentally observe as brain rhythms at multiple time scales [33.72, 73].

Oscillations occur in the brain at different time scales, starting from several milliseconds (high frequencies) to several seconds (low frequencies). One can distinguish between oscillatory components based on their frequency contents, including delta (1−4 Hz), theta (4−7 Hz), alpha (7−12 Hz), beta (12−30 Hz), and gamma (30−80 Hz) bands. The above separation of brain wave frequencies is somewhat arbitrary, however, they can be used as a guideline to focus on various activities. For example, higher cognitive functions are broadly assumed be manifested in oscillations in the higher beta and gamma bands.

Brain oscillations take place in time and space. A large part of cognitive activity happens in the cortex, which is a convoluted surface of the six-layer cortical sheet of gyri and sulci. The spatial activity is organized on multiple scales as well, starting from the neuronal level (μm), to granules (mm), cortical activities (several cm), and hemisphere-wide level (20 cm). The temporal and spatial scales are not independent, rather they delicately interact and modulate each other during cognition. Modern brain monitoring tools provide insight to these complex space–time processes [33.74].

Characterization of Oscillatory Networks

Oscillations in neural networks are synchronized activities of populations of neurons at certain well-defined frequencies. Neural systems are often modeled as the interaction of components which oscillate at specific,

well-defined frequencies. Oscillatory dynamics can correspond to either microscopic neurons, to mesoscopic populations of tens of thousands neurons, or to macroscopic neural populations including billions of neurons. Oscillations at the microscopic level have been thoroughly studied using spiking neuron models, such as the Hodgkin–Huxley equation (HH). Here we focus on populations of neurons, which have some natural oscillation frequencies. It is meaningful to assume that the natural frequencies are not identical due to the diverse properties of populations in the cortex. Interestingly, the diversity of oscillations at the microscopic and mesoscopic levels can give rise to large-scale synchronous dynamics at higher levels. Such emergent oscillatory dynamics is the primary subject of this section.

Consider N coupled oscillators with natural frequencies ω_j; $j = 1, \ldots, N$. A measure of the synchronization in such systems is given by parameter R, which is often called the *order parameter*. This terminology was introduced by *Haken* [33.75] to describe the emergence of macroscopic order from disorder. The time-varying order parameter $R(t)$ is defined as [33.76]

$$R(t) = |1/N \times \Sigma_{j=1}^{N} e^{i\theta_j(t)}| \, . \tag{33.10}$$

Order parameter R provides a useful synchronization measure for coupled oscillatory systems. A common approach is to consider a globally coupled system, in which all the components interact with each other. This is the broadest possible level of interaction. The local coupling model represents just the other extreme limit, i.e., each node interacts with just a few others, which are called its direct neighbors. In a one-dimensional array, a node has two neighbors on its left and right, respectively (assuming periodic boundary conditions). In a two-dimensional lattice, a node has four direct neighbors, and so on. The size of the neighborhood can be expanded, so the connectivity in the network becomes more dense. There is of special interest in networks that have a mostly regular neighborhood with some further neighbors added by a selection rule from the whole network. The addition of remote or nonlocal connections is called rewiring, and the networks with rewiring are small world networks. They have been extensively studied in network theory [33.76–78]. Figure 33.14 illustrates local (top left) and global coupling (bottom right), as well as intermediate coupling, with the bottom left plot giving an example of network with random rewiring.

The Kuramoto Model

The Kuramoto model [33.79] is a popular approach to describe oscillatory neural systems. It implements mean-field (global) coupling. The synchronization in this model allows an analytical solution, which helps to interpret the underlying dynamics in clear mathematical terms [33.76]. Let θ_j and ω_j denote the phase and the inherent frequency of the i-th oscillator. The oscillators are coupled by a nonlinear interaction term depending on their pair-wise phase differences. In the Kuramoto model, the following sinusoidal coupling term has been used to model neural systems

$$\frac{\mathrm{d}\theta_j}{\mathrm{d}t} = \omega_j + \frac{K}{N}\Sigma_{j=1}^N \sin(\theta_i - \theta_j), \quad j = 1,\ldots,N .$$

(33.11)

Here K denotes the coupling strength and $K = 0$ means no coupling. The system in (33.11) and its generalizations have been studied extensively since its first introduction by *Kuramoto* [33.79]. Kuramoto used Lorenztian initial distribution of phases θ defined as: $L(\theta) = \gamma/\{\pi(\gamma^2 + (\omega - \omega_0)^2)\}$. This leads to the asymptotic solution $N \to \inf$ and $t \to \inf$ for order parameter R in simple analytic terms

$$R = \sqrt{1 - (K_c/K)} \quad \text{if } K > K_c, R = 0 \text{ otherwise} .$$

(33.12)

Here K_c denotes the critical coupling strength given by $K_c = 2\gamma$. There is no synchronization between the oscillators if $K \le K_c$, and the synchronization becomes stronger as K increases at supercritical conditions $K > K_c$, see Fig. 33.15. Inputs can be used to control synchronization, i.e., a highly synchronized system can be (partially) desynchronized by input stimuli [33.80, 81]. Alternatively, input stimuli can induce large-scale synchrony in a system with low level of synchrony, as evidenced by cortical observations [33.82].

Neural Networks as Dynamical Systems

A dynamical system is defined by its equation of motion, which describes the location of the system as a function of time t

$$\frac{\mathrm{d}X(t,\lambda)}{\mathrm{d}t} = F(X), \quad X \in \mathbb{R}^n .$$

(33.13)

Here X is the state vector describing the state of the system in the n-dimensional Euclidean space $X = X(x_1,\ldots,x_n) \in \mathbb{R}^n$ and λ is the vector of system parameters. Proper initial conditions must be specified and it is assumed that $F(X)$ is a sufficiently smooth nonlinear function. In neural dynamics it is often assumed that the state space is a smooth manifold, and the goal is to study the evolution of the trajectory of $X(t)$ in the state space as time varies along the interval $[t_0, T]$.

The Cohen–Grossberg (CG) equation is a general formulation of the motion of a neural network as a dynamical system with distributed time delays in the presence of inputs. The CG model has been studied

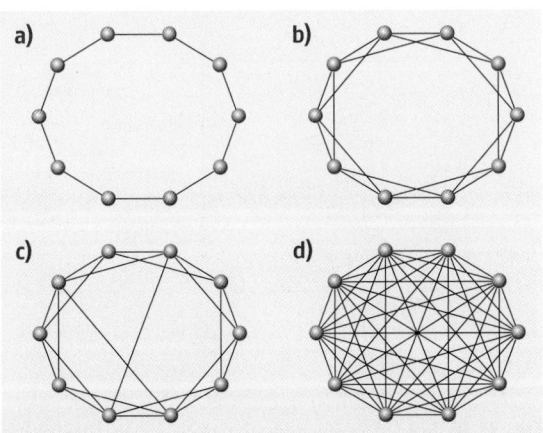

Fig. 33.14a–d Network architectures with various connectivity structures: (**a**) local, (**b**) and (**c**) are intermediate, and (**d**) global (mean-field) connectivity

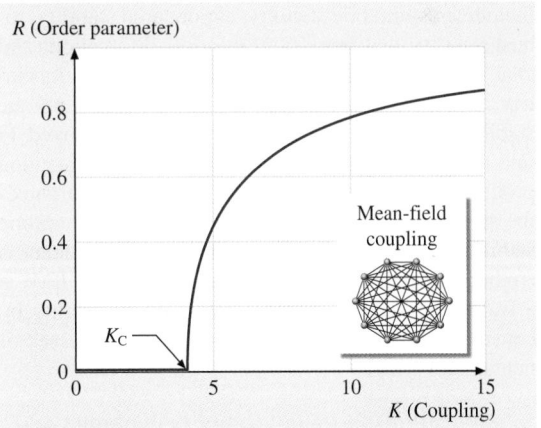

Fig. 33.15 Kuramoto model in the mean-field case. Dependence of order parameter R on the coupling strength K. Below a critical value K_c, the order parameter is 0, indicating the absence of synchrony; synchrony emerges for K above the critical value

thoroughly in the past decades and it served as a starting point for various other approaches. The general form of the CG model is [33.83]

$$\frac{\mathrm{d}z_i(t)}{\mathrm{d}t} = -a_i(x_i(t)) \left[b_i(x_i(t)) - \sum_{j=1}^{N} a_{ij} f_j(x_j(t)) \right.$$
$$\left. - \sum_{j=1}^{N} b_{ij} f_j(x_j(t - \tau_{ij})) + u_j \right],$$
$$i = 1, \dots, N. \tag{33.14}$$

Here $X(t) = [x_1(t), x_2(t), \dots, x_N(t)]^\mathsf{T}$ is the state vector describing a neural network with N neurons. Function $a_i(t)$ describes the amplification, $b_i(t)$ denotes a properly behaved function to guarantee that the solution remains bounded, $f_i(x)$ is the activation function, u_i denotes external input, a_{ij} and b_{ij} are components of the connection weight matrix and the delayed connection weight matrix, respectively, and τ_{ij} describes the time delays between neurons, $i, j = 1, \dots, n$. The solution of (33.14) can be determined after specifying suitable initial conditions.

There are various approaches to guarantee the stability of the CG equation as it approaches its equilibria under specific constraints. Global convergence assuming symmetry of the connectivity matrix has been shown [33.83]. The symmetric version of a simplified CG model has become popular as the Hopfield or Hopfield–Tank model [33.84]. Dynamical properties of CG equation have been studied extensively, including asymptotic stability, exponential stability, robust stability, and stability of periodic bifurcations and chaos. Symmetry requirements for the connectivity matrix have been relaxed, still guaranteeing asymptotic stability [33.85]. CG equations can be employed to find the optimum solutions of a nonlinear optimization problem when global asymptotic stability guarantees the stability of the solution [33.86]. Global asymptotic stability of the CG neural network with time delay is studied using linear matrix inequalities (LMI). LMI is a fruitful approach for global exponential stability by constructing Lyapunov functions for broad classes of neural networks.

Bifurcations in Neural Network Dynamics
Bifurcation theory studies the behavior of dynamical systems in the neighborhood of bifurcation points, i. e., at points when the topology of the state space abruptly changes with continuous variation of a system parameter. An example of the state space is given by the folded surface in Fig. 33.16, which illustrates a cusp bifurcation point. Here $\lambda = [a, b]$ is a two-dimensional parameter vector, $X \in \mathbb{R}^1$ [33.87]. As parameter b increases, the initially unfolded manifold undergoes a bifurcation through a cusp folding with three possible values of state vector X. This is an example of pitchfork bifurcation, when a stable equilibrium point bifurcates into one unstable and two stable equilibria. The projection to the $a - b$ plane shows the cusp bifurcation folding with multiple equilibria. The presence of multiple equilibria provides the conditions for the onset of oscillatory states in neural networks. The transition from fixed point to limit cycle dynamics can described by bifurcation theory.

Neural Networks with Inhibitory Feedback
Oscillations in neural networks are typically due to delayed, negative feedback between neural population. Mean-field models are described first, starting with Wilson–Cowan (WC) oscillators, which are capable of producing limit cycle oscillations. Next, a class of more general networks with excitatory–inhibitory feedback are described, which can generate unstable limit cycle oscillations.

The Wilson–Cowan model is based on statistical analysis of neural populations in the mean-field limit, i. e., assuming that all components of the system fully interact [33.88, 89]. In the brain it may describe a single cortical column in one of the sensory cortices, which in turn interacts with other columns to generate synchronous or asynchronous oscillations, depending on the cognitive state. In its simplest manifestation, the WC model has one excitatory and one inhibitory com-

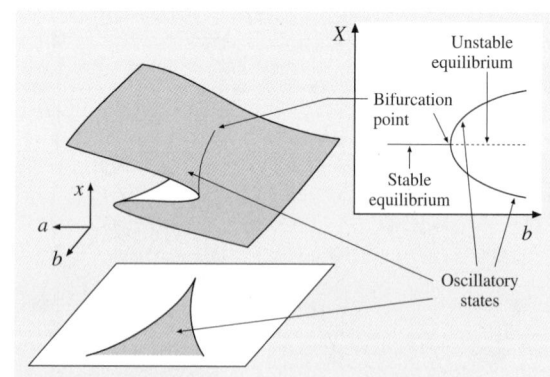

Fig. 33.16 Folded surface in the state space illustrating cusp bifurcation following (after [33.87]). By increasing parameter b, the stable equilibrium bifurcates to two stable and one unstable equilibria

ponent, with interaction weights denoted as w_{EE}, w_{EI}, w_{IE}, and w_{II}. Nonlinear function f stands for the standard sigmoid with rate constant a

$$\frac{dX_E}{dt} = -X_E + f(w_{EE}X_E + w_{IE}X_I + P_E) \,, \quad (33.15)$$

$$\frac{dX_I}{dt} = -X_I + f(w_{EI}X_E + w_{II}X_I + P_I) \,, \quad (33.16)$$

$$f(x) = 1/[1 + e^{-ax}] \,. \quad (33.17)$$

P_E and P_I describe the effect of input stimuli through the excitatory and inhibitory nodes, respectively. The inhibitory weights are negative, while the excitatory ones are positive. The WC system has been extensively studied with dynamical behaviors including fixed point and oscillatory regimes. In particular, for fixed weight values, it has been shown that the WC system undergoes a pitchfork bifurcation by changing P_E or P_I input levels. Figure 33.17 shows the schematics of the two-node system, as well as the illustration of the oscillatory states following the bifurcation with parameters $w_{EE} = 11.5$, $w_{II} = -2$, $w_{EI} = -w_{IE} = -10$, and input values $P_E = 0$ and $P_I = -4$, with rate constant $a = 1$. Stochastic versions of the Wilson–

Cowan oscillators have been extensively developed as well [33.90]. Coupled Wilson–Cowan oscillators have been used in learning models and have demonstrated applicability in a number of fields, including visual processing and pattern classification [33.91–93].

Oscillatory neural networks with interacting excitatory–inhibitory units have been developed in *Freeman* K sets [33.94]. That model uses an asymmetric sigmoid function $f(x)$ modeled based on neurophysiological activations and given as follows

$$f(x) = q\{1 - \exp(-[1/(q(e^x - 1))])\} \,. \quad (33.18)$$

Here q is a parameter specifying the slope and maximal asymptote of the sigmoid curve. The sigmoid has unit gain at zero, and has maximum gain at positive x values due to its asymmetry, see (33.18). This property provides the opportunity for self-sustained oscillations without input at a wide range of parameters. Two versions of the basic oscillatory units have been studied, either one excitatory and one inhibitory unit, or two excitatory and two inhibitory units. This is illustrated in Fig. 33.18. Stability conditions of the fixed point and limit cycle oscillations have been identified [33.95, 96]. The system with two E and two I units has the advantage that it avoids self-feedback, which is uncharacteristic in biological neural populations. Interestingly, the extended system has an operating regime with an unstable equilibrium without stable equilibria. This condition leads to an inherent instability in a dynamical regime when the system oscillates without input. Oscillations in the unstable region have been characterized and conditions for sustained unstable oscillations derived [33.96]. Simulations in the region confirmed the existence of limit cycles in the unstable regime with highly irregular oscillatory shapes of the cycle, see Fig. 33.18, upper plot. Regions with regular limit cycle oscillations and fixed point oscillations have been identified as well, see Fig. 33.18, middle and bottom [33.97].

Spatiotemporal Oscillations in Heterogeneous NNs

Neural networks describe the collective behavior of populations of neurons. It is of special interest to study populations with a large-number of components having complex, nonlinear interactions. Homogeneous populations of neurons allow mathematical modeling in mean-field approximation, leading to oscillatory models such as the Wilson–Cowan oscillators and *Freeman* KII sets. Field models with heterogeneous structure and dynamic

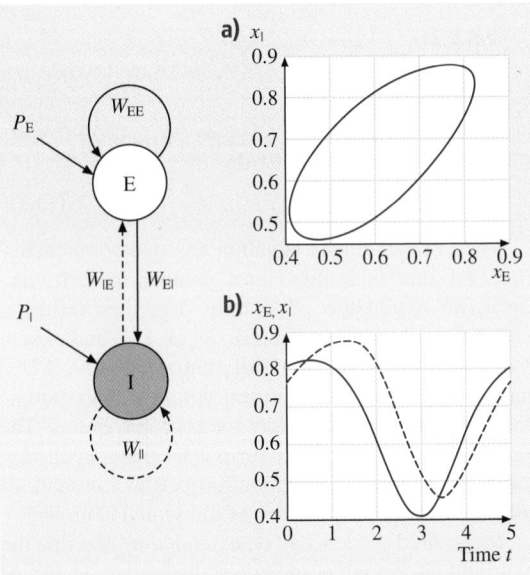

Fig. 33.17 Schematic diagram of the Wilson–Cowan oscillator with excitatory (E) and inhibitory (I) populations; *solid lines* show excitatory, *dashed* show inhibitory connections. The *right panels* show the trajectory in the phase space of $X_E - X_I$ and the time series of the oscillatory signals (after [33.90])

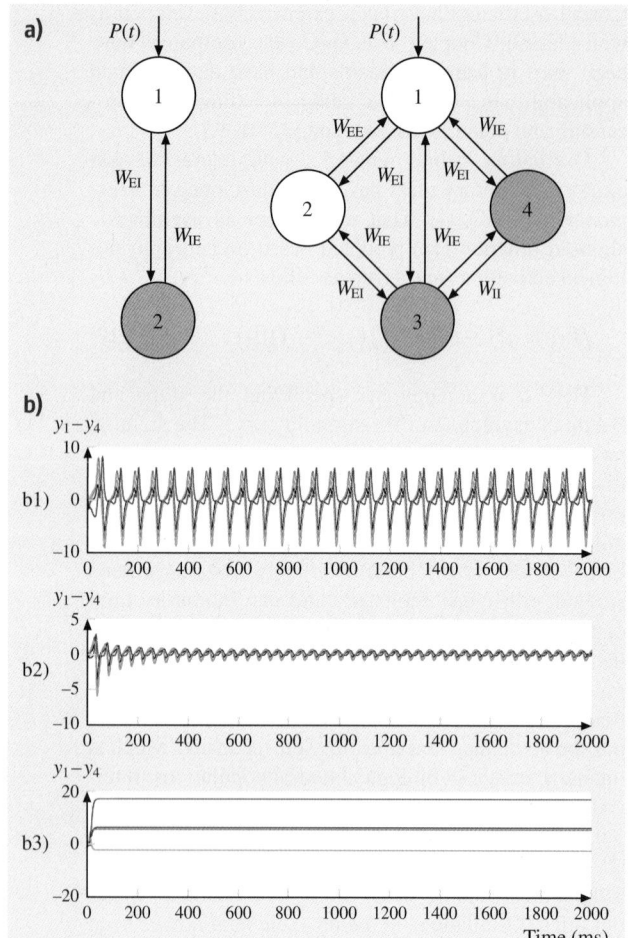

Fig. 33.18a,b Illustration of excitatory–inhibitory models. (a) *Left*: simplified model with one excitatory (E) and one inhibitory (I) node. *Right*: extended model with two E and two I nodes. (b) Simulations with the extended model with two E and two I nodes; $y_1 - y_4$ show the activations of the nodes; b1: limit cycle oscillations in the unstable regime; b2: oscillations in the stable limit cycle regime; b3: fixed point regime (after [33.97])

current, and spike timing. Each neuron is represented by a point in the state space given by the above coordinates comprising vector $X(t) \in \mathbb{R}^n$, and the evolution of a neuron is given with its trajectory the state space. Neuropils can include millions and billions of neurons; thus the phase space of the neurons contains a myriads of trajectories. Using the ensemble density approach of population modeling, the distribution of neurons in the state space at a given time t is described by a probability density function $p(X, t)$. The ensemble density approach models the evolution of the probability density in the state space [33.98]. One popular approach uses the Langevin formalism given next.

Field Theories of Neural Networks
Consider the stochastic process $X(t)$, which is described by the Langevin equation [33.99]

$$dX(t) = \mu(X(t))dt + \sigma(X(t))dW(t) . \qquad (33.19)$$

Here μ and σ denote the drift and variance, respectively, and $dW(t)$ is a Wiener process (Brown noise) with normally distributed increments. The probability density $p(X, t)$ of Langevin equation (33.19) satisfies the following form of the Fokker–Planck equation, after omitting higher-order terms

$$\frac{\partial p(X, t)}{\partial t} = -\sum_{i=1}^{n} \frac{\partial}{\partial x_i} [\mu_i(X) p(X, t)]$$

$$+ \sum_{i=1}^{n} \sum_{j=1}^{n} \frac{\partial^2}{\partial x_i \partial x_j} [D_{ij}(X) p(X, t)] .$$

$$(33.20)$$

The Fokker–Planck equation has two components. The first one is a flow term containing drift vector $\mu_i(X)$, while the other term describes diffusion with diffusion coefficient matrix $D_{ij}(X, t)$. The Fokker–Planck equation is a partial differential equation (PDE) that provides a deterministic description of macroscopic events resulting from random microscopic events. The mean-field approximation describes time-dependent, ensemble average population properties, instead of keeping track of the behavior of individual neurons.

Mean-field models can be extended to describe the evolution of neural populations distributed in physical space. Considering the cortical sheet as a de facto continuum of the highly convoluted neural tissue (the neuropil), field theories of brains are developed using partial differential equations in space and time. The corresponding PDEs are wave equations. Consider a simple one-dimensional model to describe the dynamics of

variables are of great interest as well, as they are the prerequisite of associative memory functions of neural networks.

A general mathematical formulation views the neuropil, the interconnected neural tissue of the cortex, as a dynamical system evolving in the phase space, see (33.13). Consider a population of spiking neurons each of which is modeled by a Hodgkin–Huxley equation. The state of a neuron at any time instant is determined by its depolarization potential, microscopic

the current density $\Phi(x, t)$ as a macroscopic variable. In the simple case of translational invariance of the connectivity function between arbitrary two points of the domain with exponential decay, the following form of the wave equation is obtained [33.100]

$$\frac{\partial^2 \Phi}{\partial t^2} + (\omega_0^2 - v^2 \Delta)\Phi + 2\omega_0 \frac{\partial \Phi}{\partial t}$$
$$= \left(\omega_0^2 + \omega_0 \frac{\partial}{\partial t} \right) S[\Phi(x, t) + P(x, t)] . \qquad (33.21)$$

Here $\Delta = \partial^2/x^2$ is the Laplacian in one dimension, $S(.)$ is a sigmoid transfer function for firing rates, $P(x, t)$ describes the effect of inputs; $\omega_0 = v/\sigma$, where v is the propagation velocity along lateral axons, and σ is the spatial relaxation constant of the applied exponential decay function [33.100]. The model can be extended to excitatory–inhibitory components. An example of simulations with a one-dimensional neural field model incorporating excitatory and inhibitory neurons is given in Fig. 33.19 [33.101]. The figure shows the propagation of two traveling pulses and the emergence of transient complex behavior ultimately leading to an elevated firing rate across the whole tissue [33.101]. For recent developments in brain field models, see [33.90, 102].

Coupled Map Lattices for NNs

Spatiotemporal dynamics in complex systems has been modeled using coupled map lattices (CML) [33.103]. CMLs use continuous state space and discrete time and space coordinates. In other words, CMLs are defined on (finite or infinite lattices) using discrete time iterations. Using periodic boundary conditions, the array can be

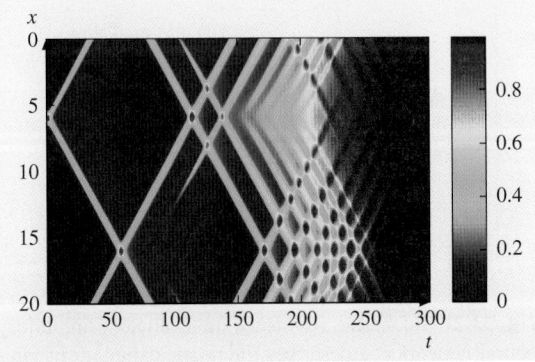

Fig. 33.19 Numerical simulations of a one-dimensional neural field model showing the interaction of two traveling pulses (after [33.101])

folded into a circle in one dimension, or into a torus for lattices of dimension 2 or higher. CML dynamics is described as follows

$$x_{n+1}(i) = (1-\varepsilon)f(x_n(i)) + \varepsilon \frac{1}{K} \sum_{k=-K/2}^{K/2} f(x_n(i+k)) ,$$
$$(33.22)$$

where $x_n(i)$ is the value of node i at iteration step n, $i = 1, \ldots, N$; N is the size of the lattice. Note that in (33.22) a periodic boundary condition applies. $f(.)$ is a nonlinear mapping function used in the iterations and ε is the coupling strength, $0 \leq \varepsilon \leq 1$. $\varepsilon = 0$ means no coupling, while $\varepsilon = 1$ is maximum coupling. The CML rule defined in (33.22) has two terms. The first term on the right-hand side is an iterative update of the i-th state, while the second term describes coupling between the units. Parameter K has a special role in coupled map lattices; it defines the size of the neighborhoods. $K = N$ describes mean-field coupling, while smaller K values belong to smaller neighborhoods. The geometry of the system is similar to the ones given in Fig. 33.14. The case of local neighborhood is the upper left diagram in Fig. 33.14, while mean-field coupling is the lower right diagram. Similar rules have been defined for higher-dimensional lattices.

CMLs exhibit very rich dynamic behavior, including fixed points, limit cycles, and chaos, depending on the choice of control parameters, ε, K, and function $f(.)$ [33.103, 104]. An example of the cubic sigmoid function

$$f(x, a) = ax^3 - ax + x$$

is shown in Fig. 33.20, together with the bifurcation diagram with respect to parameter a. By increasing the value of parameter a, the map exhibits bifurcations from fixed point to limit cycle, and ultimately to the chaotic regime.

Complex CML dynamics has been used to design dynamic associative memory systems. In CML, each memory is represented as a spatially coherent oscillation and is learnt by a correlational learning rule operating in limit cycle or chaotic regimes. In such systems, both the memory capacity and the basin volume for each memory are larger in CML than in the Hopfield model employing the same learning rule [33.105]. CML chaotic memories reduce the problem of spurious memories, but they are not immune to it. Spurious memories prevent the system from exploiting its memory capacity to the fullest extent.

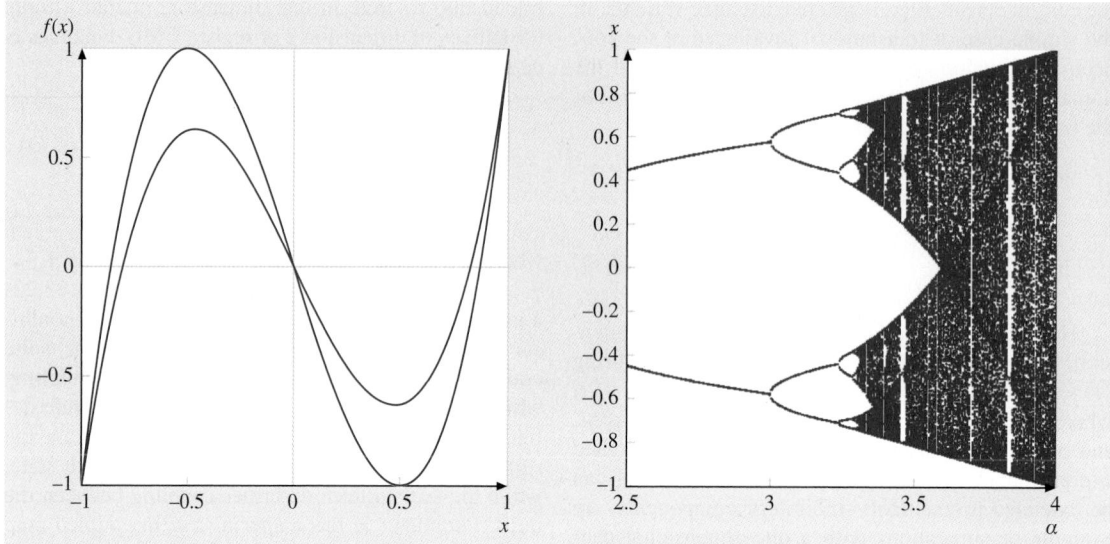

Fig. 33.20a,b Transfer function for CML: (**a**) shape of the cubic transfer function $f(x, a) = ax^3 - ax + x$; (**b**) bifurcation diagram over parameter a

Stochastic Resonance

Field models of brain networks develop deterministic PDEs (Fokker–Planck equation) for macroscopic properties based on a statistical description of the underlying stochastic dynamics of microscopic neurons. In another words, they are deterministic systems at the macroscopic level. Stochastic resonance (SR) deals with conditions when a bistable or multistable system exhibits strong oscillations under weak periodic perturbations in the presence of random noise [33.106]. In a typical SR situation, the weak periodic carrier wave is insufficient to cross the potential barrier between the equilibria of a multistable system. Additive noise enables the system to surmount the barrier and exhibit oscillations as it transits between the equilibria. SR is an example of processes when properly tuned random noise improves the performance of a nonlinear system and it is highly relevant to neural signal processing [33.107, 108].

A prominent example of SR in a neural network with excitatory and inhibitory units is described in [33.109]. In the model developed, the activation rate of excitatory and inhibitory neurons is described by μ_e and μ_i, respectively. The ratio $\alpha = \mu_e/\mu_i$ is an important parameter of the system. The investigated neural populations exhibit a range of dynamic behaviors, including convergence to fixed point, damped oscillations, and persistent oscillations. Figure 33.21 summarizes the main findings in the form of a phase diagram in the space of parameters α and *noise* level. The diagram contains three regions. Region I is at low noise levels and it corresponds to oscillations decaying to a fixed point at an exponential rate. Region II corresponds to high noise, when the neural activity exhibits damped oscillations as it approaches the steady state. Region III, however, demonstrates sustained oscillations for an intermediate level of noise. If a is above a critical value (see the tip of Region III),

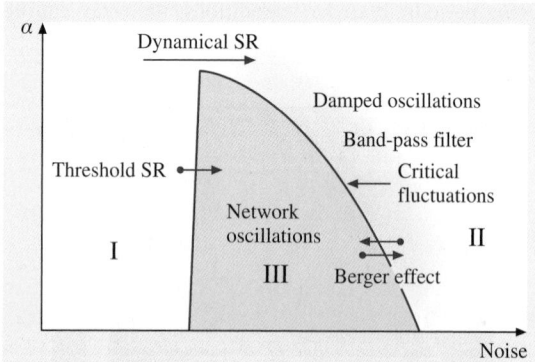

Fig. 33.21 Stochastic resonance in excitatory–inhibitory neural networks; α describes the relative strength of inhibition. Region I: fixed point dynamics. Region II: damped oscillatory regime. Region III: sustained periodic oscillations illustrating stochastic resonance (after [33.109])

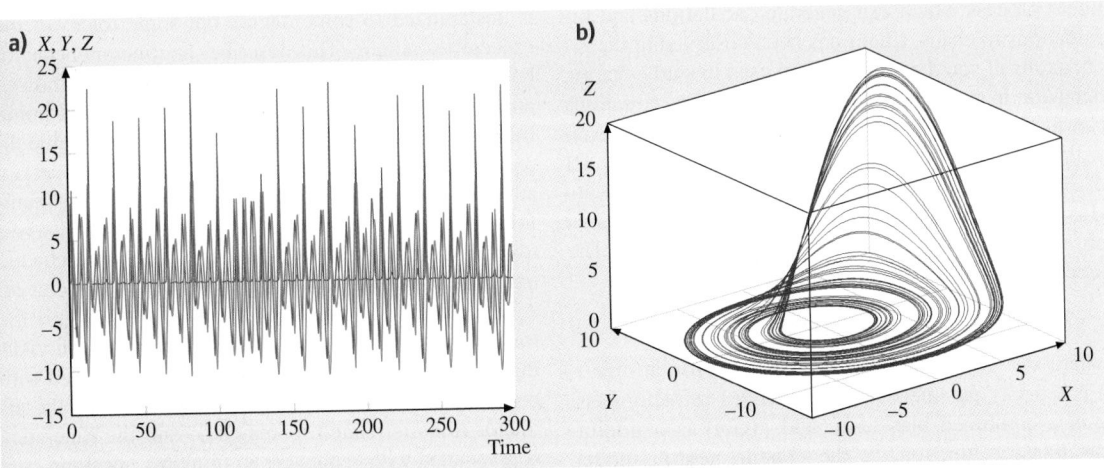

Fig. 33.22a,b Lorenz attractor in the chaotic regime; (**a**) time series of the variables X, Y, and Z; (**b**) butterfly-winged chaotic Lorenz attractor in the phase space spanned by variables X, Y, and Z

the activities in the steady state undergo a first-order phase transition at a critical noise level. The intensive oscillations in Region III at an intermediate noise level show that the output of the system (oscillations) can be enhanced by an optimally selected noise level.

The observed phase transitions may be triggered by neuronal avalanches, when the neural system is close to a critical state and the activation of a small number of neurons can generate an avalanche process of activation [33.110]. Neural avalanches have been described using self-organized criticality (SOC), which has been identified in neural systems [33.111]. There is much empirical evidence of the cortex conforming to the self-stabilized, scale-free dynamics with avalanches during the existence of some quasi-stable states [33.112, 113]. These avalanches maintain a metastable background state of activity.

Phase transitions have been studied in models with extended layers of excitatory and inhibitory neuron populations, respectively. A specific model uses random cellular neural networks to describe conditions with sustained oscillations [33.114]. The role of various control parameters has been studied, including noise level, inhibition, and rewiring. Rewiring describes long axonal connections to produce neural network architectures resembling connectivity patterns with short and long-range axons in the neuropil. By properly tuning the parameters, the system can reside in a fixed point regime in isolation, but it will switch to persistent oscillations under the influence of learnt input patterns [33.115].

33.2.3 Chaotic Neural Networks

Emergence of Chaos in Neural Systems

Neural networks as dynamical systems are described by the state vector $X(t)$ which obeys the equation of motion (33.13). Dynamical systems can exhibit fixed point, periodic, and chaotic behaviors. Fixed points and periodic oscillations, and transitions from one to the other through bifurcation dynamics has been described in Sect. 33.2.2. The trajectory of a chaotic system does not converge to a fixed point or limit cycle, rather it converges to a chaotic attractor. Chaotic attractors, or strange attractors, have the property that they define a fractal set in the state space, moreover, chaotic trajectories close to each other at some point, diverge from each other exponentially fast as time evolves [33.116, 117].

An example of the chaotic Lorenz attractor is shown in Fig. 33.22. The Lorenz attractor is defined by a system of three ordinary differential equations (ODEs) with nonlinear coupling, originally derived for the description of the motion of viscous flows [33.118]. The time series belonging to variables X, Y, Z are shown in Fig. 33.22a for parameters in the chaotic region, while the strange attractor is illustrated by the trajectory in the phase space, see Fig. 33.22b.

Chaotic Neuron Model

In chaotic neural networks the individual components exhibit chaotic behavior, and the goal is to study the order emerging from their interaction. Nerve membranes produce propagating action potentials in a highly nonlinear process which can generate oscillations and bifurcations to chaos. Chaos has been observed in the giant axons of squid and it has been used to study chaotic behavior in neurons. The Hodgkin–Huxley equations can model nonlinear dynamics in the squid giant axon with high accuracy [33.58]. The chaotic neuron model of *Aihara* et al. is an approximation of the Hodgkin–Huxley equation and it reproduces chaotic oscillations observed in the squid giant axon [33.119, 120]. The model uses the following simple iterative map

$$x(t+1) = kx(t) - \alpha f(x(t)) + a , \qquad (33.23)$$

where $x(t)$ is the state of the chaotic neuron at time t, k is a decay parameter, α characterizes refractoriness, a is a combined bias term, and $f(y(t))$ is a nonlinear transfer function. In the chaotic neuron model, the log sigmoid transfer function is used, see (33.17). Equation (33.23) combined with the sigmoid produces a piece-wise monotonous map, which generates chaos.

Chaotic neural networks composed of chaotic neurons generate spatio-temporal chaos and are able to retrieve previously learnt patterns as the chaotic trajectory traverses the state space. Chaotic neural networks are used in various information processing systems with abilities of parallel distributed processing [33.121–123]. Note that CMLs also consist of chaotic oscillators produced by a nonlinear local iterative map, like in chaotic neural networks. CMLs define a spatial relationship among their nodes to describe spatio-temporal fluctuations. A class of cellular neural networks combines the explicit spatial relationships similar to CMLs with detailed temporal dynamics using Cohen–Grossberg model [33.83] and it has been used successfully in neural network applications [33.124, 125].

Collective Chaos in Neural Networks

Chaos in neural networks can be an emergent macroscopic property stemming from the interaction of nonlinear neurons, which are not necessarily chaotic in isolation. Starting from the microscopic neural level up to the macroscopic level of cognition and consciousness, chaos plays an important role in neurodynamics [33.82, 126–129]. There are various routes to chaos in neural systems, including period-doubling bifurcations to chaos, chaotic intermittency, and collapse of a two-dimensional torus to chaos [33.130, 131].

Chaotic itinerancy is a special form of chaos, which is between ordered dynamics and fully developed chaos. Chaotic itinerancy describes the trajectory through high-dimensional state space of neural activity [33.132]. In chaotic itinerancy the chaotic system

is destabilized to some degree but some traces of the trajectories remain. This describes an itinerant behavior between the states of the system containing destabilized attractors or *attractor ruins*, which can be fixed point, limit cycle, torus, or strange attractor with unstable directions. Dynamical orbits are attracted to a certain attractor ruin, but they leave via an unstable manifold after a (short or long) stay around it and move toward another attractor ruin. This successive chaotic transition continues unless a strong input is received. A schematic diagram is shown in Fig. 33.23, where the trajectory of a chaotic itinerant system is shown visiting attractor ruins. Chaotic itinerancy is associated with perceptions and memories, the chaos between the attractor ruins is related to searches, and the itinerancy is associated with sequences in thinking, speaking, and writing.

Frustrated chaos is a dynamical system in a neural network with a global attractor structure when local connectivity patterns responsible for stable oscillatory behaviors become intertwined, leading to mutually competing attractors and unpredictable itinerancy between brief appearances of these attractors [33.133]. Similarly to chaotic itinerancy, frustrated chaos is related to destabilization of the dynamics and it generates itinerant, wavering oscillations between the orbits of the network, the trajectories of which have been stable with the original connectivity pattern. Frustrated chaos is shown to belong to the family of intermittency type of chaos [33.134, 135].

To characterize chaotic dynamics, tools of statistical time series analysis are useful. The studies may involve time and frequency domains. Time domain analysis

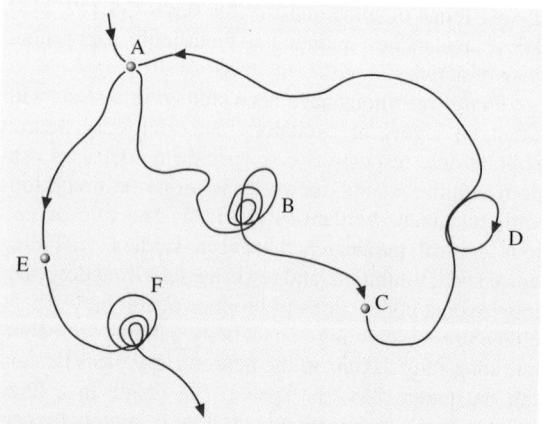

Fig. 33.23 Schematic illustration of itinerant chaos with a trajectory visiting attractor ruins (after [33.132])

includes attractor reconstruction, i. e., the attractor is depicted in the state space. Chaotic attractors have fractal dimensions, which can be evaluated using one of the available methods [33.136–138]. In the case of low-dimensional chaotic systems, the reconstruction can be illustrated using two or three-dimensional plots. An example of attractor reconstruction is given in Fig. 33.22 for the Lorenz system with three variables. Attractor reconstruction of a time series can be conducted using time-delay coordinates [33.139].

Lyapunov spectrum analysis is a key tool in identifying and describing chaotic systems. Lyapunov exponents measure the instability of orbits in different directions in the state space. It describes the rate of exponential divergence of trajectories that were once close to each other. The set of corresponding Lyapunov exponents constitutes the Lyapunov spectrum. The maximum Lyapunov exponent Λ^* is of crucial importance; as a positive leading Lyapunov exponent $\Lambda^* > 0$ is the hallmark of chaos. $X(t)$ describes the trajectory of the system in the phase space starting from $X(0)$ at time $t = 0$. Denote by $X_{\Delta x_0}(t)$ the perturbed trajectory starting from $[X(0) + \Delta x_0]$. The leading Lyapunov exponent can be determined using the following relationship [33.140]

$$\Lambda^* = \lim_{\substack{t \to \infty \\ \Delta x_0 \to 0}} t^{-1} \ln[|X_{\Delta x_0}(t) - X(t)|/|\Delta x_0|] \,,$$

(33.24)

where $\Lambda^* < 0$ corresponds to convergent behavior, $\Lambda^* = 0$ indicates periodic orbits, and $\Lambda^* > 0$ signifies chaos. For example, the Lorenz attractor has $\Lambda^* = 0.906$, indicating strong chaos (Fig. 33.24). Equation (33.24) measures the divergence for infinitesimal perturbations in the limit of infinite time series. In practical situations, especially for short time series, it is often difficult to distinguish weak chaos from random perturbations. One must be careful with conclusions about the presence of chaos when Λ^* has a value close to zero. Lyapunov exponents are widely used in brain monitoring using electroencephalogram (EEG) analysis, and various methods are available for characterization of normal and pathological brain conditions based on Lyapunov spectra [33.141, 142].

Fourier analysis conducts data processing in the frequency domain, see (33.5) and (33.6). For chaotic signals, the shape of the power spectra is of special interest. Power spectra often show $1/f^\alpha$ power law behavior in log–log coordinates, which is the indication of scale-free system and possibly chaos. Power-law scaling in systems at SOC is suggested by a linear decrease in log power with increasing log frequency [33.143]. Scaling properties of criticality facilitate the coexistence of spatially coherent cortical activity patterns for a duration ranging from a few milliseconds to a few seconds. Scale-free behavior characterizes chaotic brain activity both in time and frequency domains. For completeness, we mention the Hilbert space analysis as an alternative to Fourier methods. The analytic signal approach based on Hilbert analysis is widely used in brain monitoring.

Emergent Macroscopic Chaos in Neural Networks

Freeman's K model describes spatio-temporal brain chaos using a hierarchical approach. Low-level K sets were introduced in the 1970s, named in the honor of Aharon Kachalsky, an early pioneer of neural dynamics [33.82, 94]. K sets are multiscale models, describing an increasing complexity of structure and dynamics. K sets are mesoscopic models and represent an intermediate level between microscopic neurons and macroscopic brain structures. K-sets are topological specifications of the hierarchy of connectivity in neural populations in brains. K sets describe the spatial patterns of phase and amplitude of the oscillations generated by neural populations. They model observable fields of neural activity comprising electroencephalograms (EEGs), local field potentials (LFPs), and magnetoencephalograms (MEGs) [33.144]. K sets form a hierarchy for cell assemblies with components starting from K0 to KIV [33.145, 146].

K0 sets represent noninteractive collections of neurons forming cortical microcolumns; a K0 set models a neuron population of $\approx 10^3 - 10^4$ neurons. K0 models dendritic integration in average neurons and an asymmetric sigmoid static nonlinearity for axon transmission. The K0 set is governed by a point attractor with zero output and stays at equilibrium except when perturbed. In the original K-set models, K0s are described by a state-dependent, linear second-order ordinary differential equation (ODE) [33.94]

$$ab \, \mathrm{d}^2 X(t)/\mathrm{d}t^2 + (a + b) \, \mathrm{d}X(t)/\mathrm{d}t + P(t) = U(t) \,.$$

(33.25)

Here a and b are biologically determined time constants. $X(t)$ denotes the activation of the node as a function of time. $U(t)$ includes an asymmetric sigmoid function $Q(x)$, see (33.18), acting on the weighted sum of activation from neighboring nodes and any external input.

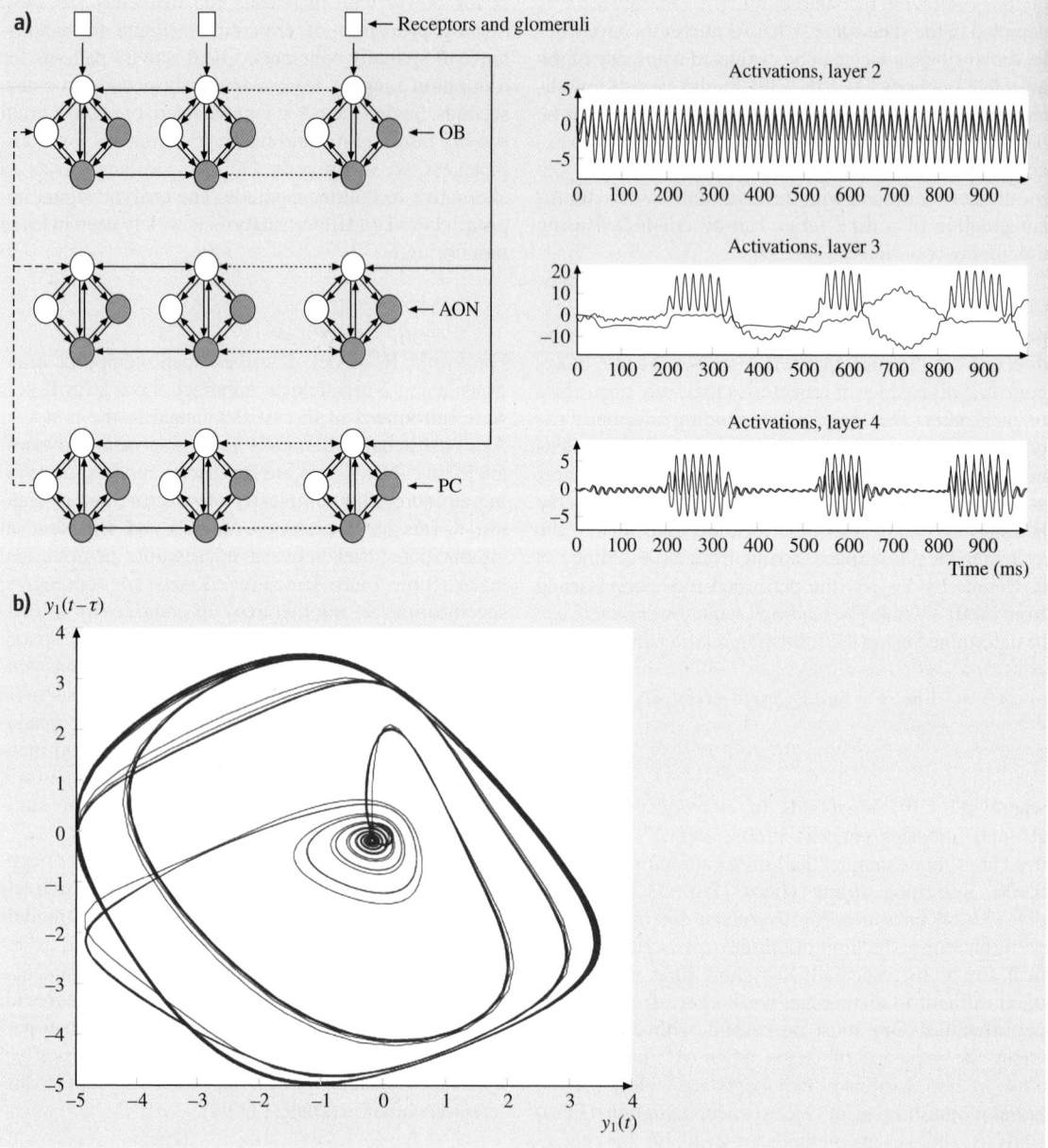

Fig. 33.24a–c KIII diagram and behaviors; (**a**) 3 double layer hierarchy of KIII and time series over each layer, exhibiting intermittent chaotic oscillations, (**b**) phase space reconstruction using delayed time coordinates

KI sets are made of interacting K0 sets, either excitatory or inhibitory with positive feedback. The dynamics of KI is described as convergence to a nonzero fixed point. If KI has sufficient functional connection density, then it is able to maintain a nonzero state of background activity by mutual excitation (or inhibition).

KI typically operates far from thermodynamic equilibrium. Neural interaction by stable mutual excitation (or mutual inhibition) is fundamental to understanding brain dynamics. KII sets consists of interacting excitatory and inhibitory KI sets with negative feedback. KII sets are responsible for the emergence of limit cycle oscillation due to the negative feedback between the neural populations. Transitions from point attractor to limit cycle attractor can be achieved through a suitable level of feedback gain or by input stimuli, see Fig. 33.18.

KIII sets made up of multiple interacting KII sets. Examples include the sensory cortices. KIII sets generate broadband, chaotic oscillations as background activity by combined negative and positive feedback among several KII populations with incommensurate frequencies. The increase in nonlinear feedback gain that is driven by input results in the destabilization of the background activity and leads to the emergence of a spatial amplitude modulation (AM) pattern in KIII. KIII sets are responsible for the embodiment of meaning in AM patterns of neural activity shaped by synaptic interactions that have been modified through learning in KIII layers. The KIII model is illustrated in Fig. 33.24 with three layers of excitatory–inhibitory nodes. In Fig. 33.24a the temporal dynamics is illustrated in each layer, while Fig. 33.24b shows the phase space reconstruction of the attractor. This is a chaotic behavior resembling the dynamics of the Lorenz attractor in Fig. 33.22. KIV sets are made up of interacting KIII units to model intentional neurodynamics of the limbic system. KIV exhibits global phase transitions, which are the manifestations of hemisphere-wide cooperation through intermittent large-scale synchronization. KIV is the domain of Gestalt formation and preafference through the convergence of external and internal sensory signals leading to intentional action [33.144, 146].

Properties of Collective Chaotic Neural Networks

KIII is an associative memory, encoding input data in spatio-temporal AM patterns [33.147, 148]. KIII chaotic memories have several advantages as compared to convergent recurrent networks:

1. They produce robust memories based on relatively few learning examples even in noisy environment.
2. The encoding capacity of a network with a given number of nodes is exponentially larger than their convergent counterparts.
3. They can recall the stored data very quickly, just as humans and animals can recognize a learnt pattern within a fraction of a second.

The recurrent Hopfield neural network can store an estimated $0.15N$ input patterns in stable attractors,

where N is the number of neurons [33.84]. Exact analysis by *Mceliece* et al. [33.149] shows that the memory capacity of the Hopfield network is $N/(4\log N)$. Various generalizations provide improvements over the initial memory gain [33.150, 151]. It is of interest to evaluate the memory capacity of the KIII memory. The memory capacity of chaotic networks which encode input into chaotic attractors is, in principle, exponentially increased with the number of nodes. However, the efficient recall of the stored memories is a serious challenge. The memory capacity of KIII as a chaotic associative memory device has been evaluated with noisy input patterns. The results are shown in Fig. 33.25, where the performance of Hopfield and KIII memories are compared; the top two plots are for Hopfield nets, while the lower two figures describe KIII results [33.152]. The light color shows recognition rate close to 100%, while the dark color means poor recognition approaching 0. The right-hand column has higher noise levels. The Hopfield network shows the well-known linear gain curve ≈ 0.15. The KIII model, on the other hand, has a drastically better performance. The boundary separating the correct and incorrect classification domains is superlinear; it has been fitted with as a fifth-order polynomial.

Cognitive Implications of Intermittent Brain Chaos

Developments in brain monitoring techniques provide increasingly detailed insights into spatio-temporal neurodynamics and neural correlates of large-scale cognitive processing [33.74, 153–155]. Brains as large-scale dynamical systems have a basal state, which is a high-dimensional chaotic attractor with a dynamic trajectory wandering broadly over the attractor landscape [33.82, 126]. Under the influence of external stimuli, cortical dynamics is destabilized and condenses intermittently to a lower-dimensional, more organized subspace. This is the act of perception when the subject identifies the stimulus with a meaning in the context of its previous experience. The system stays intermittently in the condensed, more coherent state, which gives rise to a spatio-temporal AM activity pattern corresponding to the stimulus in the given context. The AM pattern is meta-stable and it disintegrates as the system returns to the high-dimensional chaotic basal state (less synchrony) Brain dynamics is described as a sequence of phase transitions with intermittent synchronization-desynchronization effects. The rapid emergence of synchronization can be initiated by (Hebbian) neural assemblies that lock into synchronization

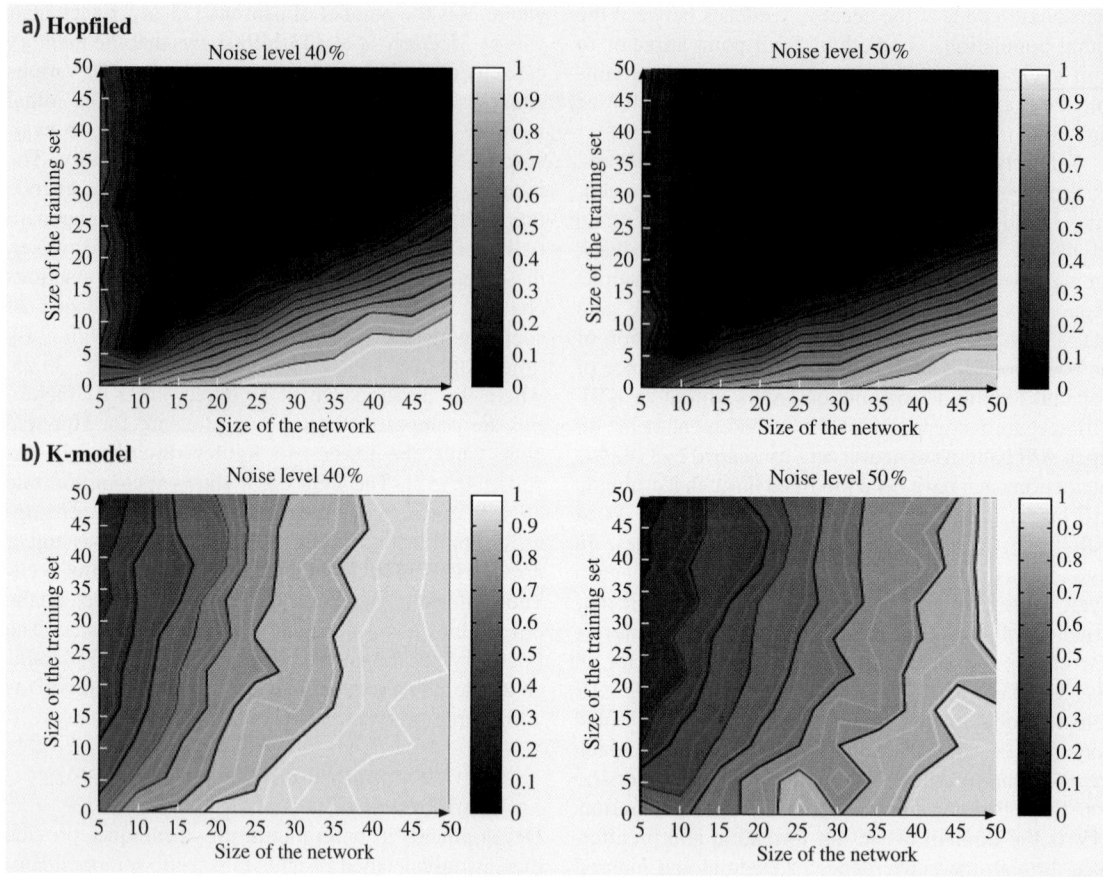

Fig. 33.25a,b Comparison of the memory capacity of (**a**) Hopfield and (**b**) KIII neural networks; the noise level is 40% (*left*); 50% (*right*); the lighter the color the higher the recall accuracy. Observe the linear gain for Hopfield networks and the superlinear (fifth-order) separation for KIII (after [33.152])

across widespread cortical and subcortical areas [33.82, 156, 157].

Intermittent oscillations in spatio-temporal neural dynamics are modeled by a neuropercolation approach. Neuropercolation is a family of probabilistic models based on the theory of probabilistic cellular automata on lattices and random graphs and it is motivated by structural and dynamical properties of neural populations. Neuropercolation constructs the hierarchy of interactive populations in networks as developed in *Freeman* K models [33.94, 144], but replace differential equations with probability distributions from the observed random networks that evolve in time [33.158]. Neuropercolation considers populations of cortical neurons which sustain their background state by mutual excitation, and their stability is guaranteed by the neural refractory periods. Neural populations transmit and re-

ceive signals from other populations by virtue of small-world effects [33.77, 159]. Tools of statistical physics and finite-size scaling theory are applied to describe critical behavior of the neuropil. Neuropercolation theory provides a mathematical approach to describe phase transitions and critical phenomena in large-scale, interactive cortical networks. The existence of phase transitions is proven in specific probabilistic cellular automata models [33.160, 161].

Simulations by neuropercolation models demonstrate the onset of large-scale synchronization-desynchronization behavior [33.162]. Figure 33.26 illustrates results of intermittent phase desynchronization for neuropercolation with excitatory and inhibitory populations. Three main regimes can be distinguished, separated by critical noise values $\varepsilon_1 > \varepsilon_0$. In Regime I $\varepsilon > \varepsilon_1$, Fig. 33.26a, the channels are not synchronous and

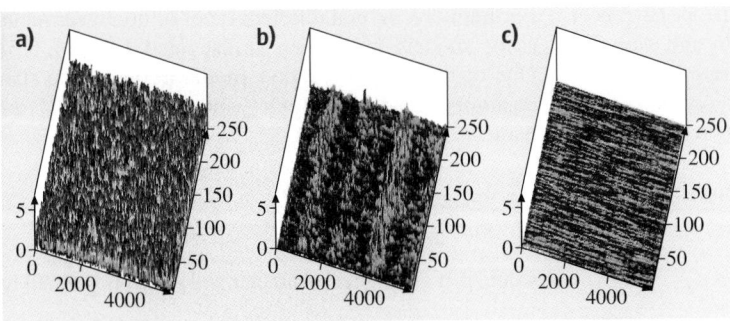

Fig. 33.26a–c Phase synchronization–desynchronization with excitatory–inhibitory connections in neuropercolation with 256 granule nodes; the z-axis shows the pairwise phase between the units. (**a**) No synchrony; (**b**) intermittent synchrony; (**c**) highly synchronized, frozen phase regime (after [33.162])

the phase values are distributed broadly. In Regime II $\varepsilon_1 > \varepsilon > \varepsilon_0$, Fig. 33.26b, the phase lags are drastically reduced indicating significant synchrony over extended time periods. Regime III is observed for high values of $\varepsilon_0 > \varepsilon$, when the channels demonstrate highly synchronized, frozen dynamics, see Fig. 33.26c. Similar transitions can be induced by the relative strength of inhibition, as well as by the fraction of rewiring across the network [33.114, 115, 163]. The probabilistic model of neural populations reproduces important properties of the spatio-temporal dynamics of cortices and is a promising approach for large-scale cognitive models.

33.3 Memristive Neurodynamics

Sequential processing of fetch, decode, and execution of instructions through the classical von Neumann digital computers has resulted in less efficient machines as their ecosystems have grown to be increasingly complex [33.164]. Though modern digital computers are fast and complex enough to emulate the brain functionality of animals like spiders, mice, and cats [33.165, 166], the associated energy dissipation in the system grows exponentially along the hierarchy of animal intelligence. For example, to perform certain cortical simulations at the cat scale even at an 83 times slower firing rate, the IBM team has to employ Blue Gene/P (BG/P), a super computer equipped with 147 456 CPUs and 144 TBs of main memory. On the other hand, the human brain contains more than 100 billion neurons and each neuron has more than 20 000 synapses [33.167]. Efficient circuit implementation of synapses, therefore, is very important to build a brain-like machine. One active branch of this research area is cellular neural networks (CNNs) [33.168, 169], where lots of multiplication circuits are utilized in a complementary metal-oxide-semiconductor (CMOS) chip. However, since shrinking the current transistor size is very difficult, introducing a more efficient approach is essential for further development of neural network implementations.

The memristor was first authorized by *Chua* as the fourth basic circuit element in electrical circuits in 1971 [33.170]. It is based on the nonlinear character-

istics of charge and flux. By supplying a voltage or current to the memristor, its resistance can be altered. In this way, the memristor remembers information. In that seminal work, Chua demonstrated that the memristance $M(q)$ relates the charge q and the flux φ in such a way that the resistance of the device will change with the applied electric field and time

$$M = \frac{d\varphi}{dq} \, . \tag{33.26}$$

The parameter M denotes the memristance of a charge controlled memristor, measured in ohms. Thus, the memristance M can be controlled by applying a voltage or current signal across the memristor. In other words, the memristor behaves like an ordinary resistor at any given instance of time, where its resistance depends on the complete history of the device [33.170].

Although the device was proposed nearly four decades ago, it was not until 2008 that researchers from HP Labs showed that the devices they had fabricated were indeed two-terminal memristors [33.171]. Figure 33.27 shows the I–V characteristics of a generic memristor, where memristance behavior is observed for TiO_2-based devices. A TiO_{2-x} layer with oxygen vacancies is placed on a perfect TiO_2 layer, and these layers are sandwiched between platinum electrodes. In metal oxide materials, the switching from R_{off} to R_{on} and vice versa occurs as a result of ion migration, due to the enormous electric fields applied across the nanoscale structures. These memristors have been fabricated using nanoimprint lithography and were successfully integrated on a CMOS substrate in [33.172]. Apart from these metal-oxide memristors, memristance has also been demonstrated using magnetic materials based on their magnetic domain wall motion and spin-torque induced magnetization switching in [33.173].

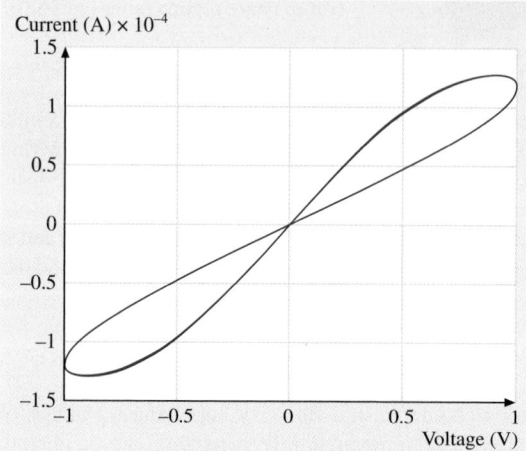

Fig. 33.27 Typical *I–V* characteristic of memristor (after [33.171]). The pinched hysteresis loop is due to the nonlinear relationship between the memristance current and voltage. The parameters of the memristor are $R_{on} = 100\,\Omega$, $R_{off} = 16\,K\Omega$, $R_{init} = 11\,k\Omega$, $D = 10\,nm$, $u_v = 10^{-10}\,cm^2\,s^{-1}\,V^{-1}$, $p = 10$ and $V_{in} = \sin(2\pi t)$. The memristor exhibits the feature of pinched hysteresis, which means that a lag occurs between the application and the removal of a field and its subsequent effect, just like the feature of neurons in the human brain

Furthermore, several different types of nonlinear memristor models have been investigated [33.174, 175]. One of them is the window model in which the state equation is multiplied by window function $F_p(\omega)$, namely

$$\frac{d\omega}{dt} = \mu_v \frac{R_{on}}{D} i(t) F_p(\omega) ,\qquad(33.27)$$

where p is an integer parameter and $F_p(\omega)$ is defined by

$$F_p(\omega) = 1 - \left(2\frac{\omega}{D} - 1\right)^{2p} ,\qquad(33.28)$$

which is shown in Fig. 33.28.

33.3.1 Memristor–Based Synapses

The design of simple weighting circuits for synaptic multiplication between arbitrary input signals and weights is extremely important in artificial neural systems. Some efforts have been made to build neuron-like analog neural networks [33.178–180]. However, this research has gained limited success so far because of the difficulty in implementing the synapses efficiently. Based on the memristor, a novel weighting circuit was proposed by *Kim* et al. [33.176, 181, 182] as shown in Fig. 33.29. The memristors provide a bridge-like switching for achieving either positive or negative weighting. Though several memristors are employed to emulate a synapse, the total area of the memristors is less than that of a single transistor. To compensate for the spatial nonuniformity and nonideal response of the memristor bridge synapse, a modified chip-in-the-loop learning scheme suitable

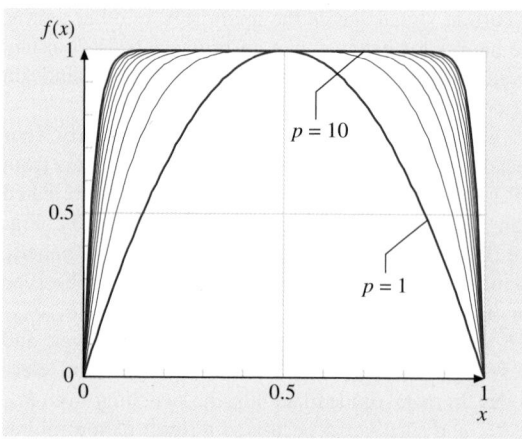

Fig. 33.28 Window function for different integer *p*

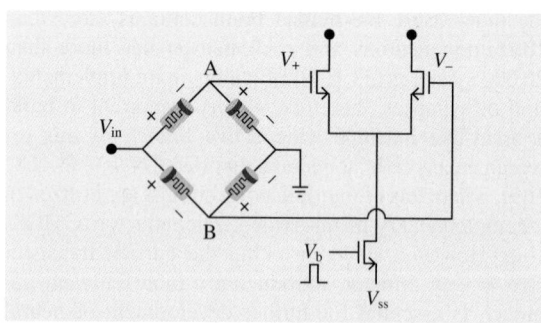

Fig. 33.29 Memristor bridge circuit. The synaptic weight is programmable by varying the input voltage. The weighting of the input signal is also performed in this circuit (after [33.176])

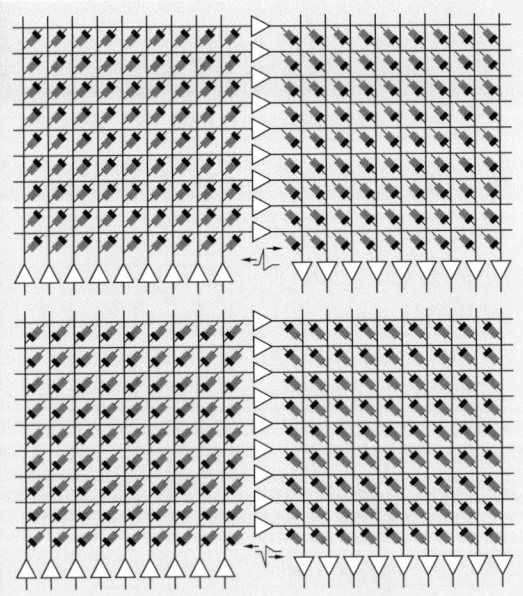

Fig. 33.30 Neuromorphic memristive computer equipped with STDP (after [33.177])

for the proposed neural network architecture is investigated [33.176]. In the proposed method, the initial learning is conducted by software, and the behavior of the software-trained network is learned via the hardware network by learning each of the single layered neurons of the network independently. The forward calculation of single layered neuron learning is implemented through circuit hardware and is followed by a weight updating phase assisted by a host computer. Unlike conventional chip-in-the-loop learning, the need for the readout of synaptic weights for calculating weight updates in each epoch is eliminated by virtue of the memristor bridge synapse and the proposed learning scheme.

On the other hand, spike-timing-dependent learning (STDP), which is a powerful learning paradigm for spiking neural systems because of its massive parallelism, potential scalability, and inherent defect, fault, and failure-tolerance, can be implemented by using a crossbar memristive array combined with neurons that asynchronously generate spikes of a given shape [33.177, 185]. Such spikes need to be sent back through the neurons to the input terminal as in Fig. 33.30. The shape of the spikes turns out to be very similar to the neural spikes observed in realistic biological neurons. The STDP learning function obtained by combining such neurons with memristors is exactly obtained from neurophysiological experiments on real synapses. Such nanoscale synapses can be combined with CMOS neurons which is possible to create neuromorphic hardware several orders of magnitude denser than in conventional CMOS. This method offers better control over power dissipation; fewer constraints on the design of memristive materials used for nanoscale synapses; greater freedom in learning algorithms than traditional design of synapses since the synaptic learning dynamics can be dynamically turned on or off; greater control over the precise form and timing of the STDP equations; the ability to implement a variety of other learning laws besides STDP; better circuit diversity since the approach allows different learning laws to be implemented in different areas of a single chip

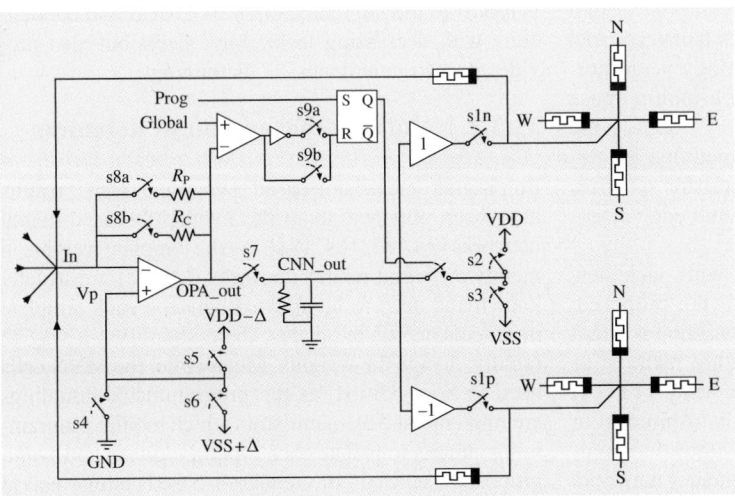

Fig. 33.31 Memristor-based cellular neural networks cell (after [33.183])

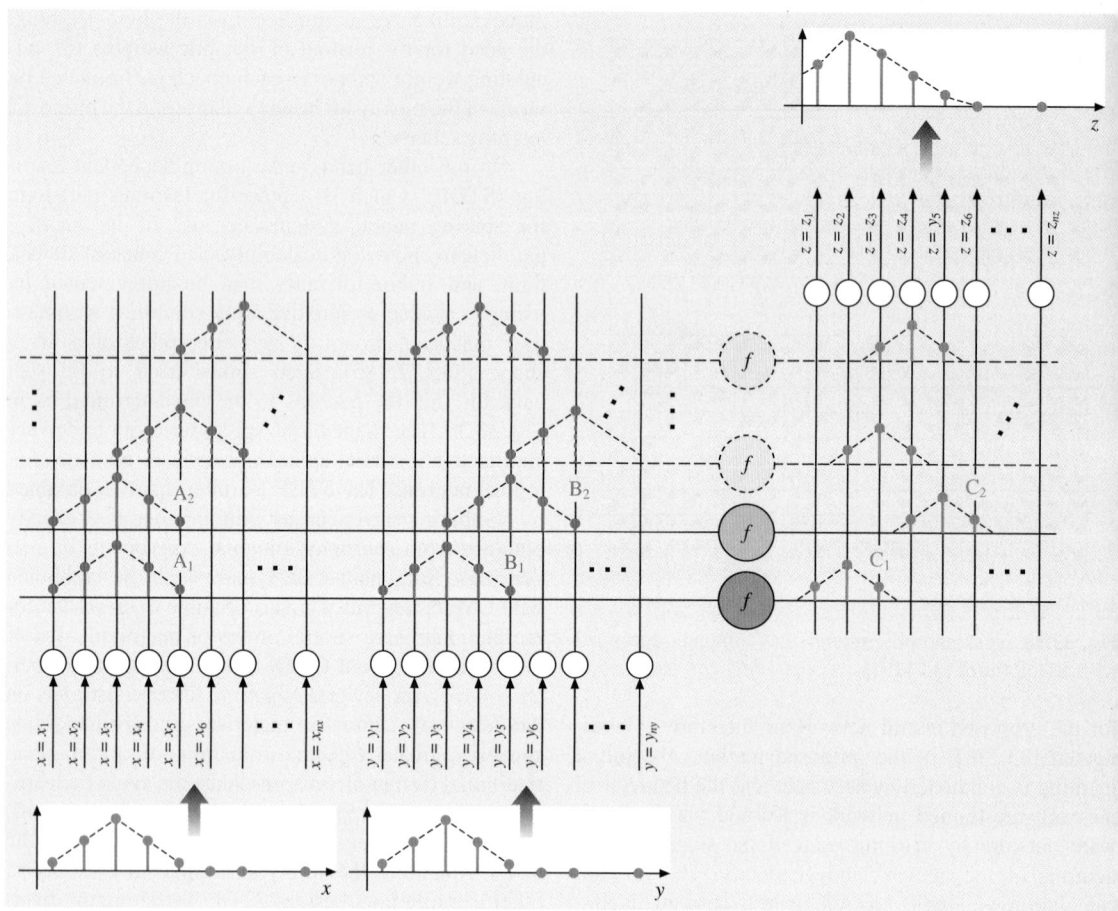

Fig. 33.32 Simple realization of MNN based on fuzzy concepts (after [33.184])

using the same memristive material for all synapses. Furthermore, an analog CMOS neuromorphic design utilizing STDP and memristor synapses is investigated for use in building a multipurpose analogy neuromorphic chip [33.186]. In order to obtain a multipurpose chip, a suitable architecture is established. Based on the technique of IBM 90 nm CMOS9RF, neurons are designed to interface with Verilog-A memristor synapses models to perform the XOR operation and edge detection function.

To make the neurons compatible with such new synapses, some novel training methods are proposed. For instance, *Manem* et al. proposed a variation-tolerant training method to efficiently reconfigure memristive synapses in a trainable threshold gate array (TTGA) system [33.187]. The training process is inspired from the gradient descent machine learning algorithm commonly used to train artificial threshold neural networks

known as perceptrons. The proposed training method is robust to the unpredictability of CMOS and nanocircuits with decreasing technology sizes, but also provides its own randomness in its training.

33.3.2 Memristor-Based Neural Networks

Employing memristor-based synapses, some results have been obtained about the memristor-based neural networks [33.183, 184, 188]. As the template weights in memristor-based neural networks (MNNs) are usually known and need to be updated between each template in a sequence of templates, there should be a way to rapidly change the weights. Meanwhile, the MNN cells need to be modified, as the programmable couplings are implemented by memristors which require programming circuits to isolate each other. *Lehtonen* and *Laiho* proposed a new cell of memristor-based cellular neural network that can be used to program the templates. For this purpose, a voltage *global* is input into the cell. This voltage is used to convey the weight of one connection into the cells [33.183]. The level of virtual ground and switches are controlled so that the memristor connected to a particular neighbor is biased above the programming threshold, until it reaches the desired resistance value.

Merrikh-Bayat et al. presented a new way to explain the relationships between logical circuits and artificial neural networks, logical circuits and fuzzy logic, and artificial neural networks and fuzzy inference systems, and proposed a new neuro-fuzzy computing system, which can effectively be implemented via the memristor-crossbar structure [33.184]. A simple realization of MNNs is shown in Figs. 33.32–33.34. Figure 33.32 shows that it is possible to interpret the working procedure of conventional artificial neural network ANN without changing its structure. In this figure, each row of the structure implements a simple fuzzy rule or minterm. Figure 33.33 shows how the activation function of neurons can be implemented when the activation function is modeled by a *t*-norm operator. Matrix multiplication is performed by vector circuit in Fig. 33.34. This circuit consists of a simple memristor crossbar where each of its rows is connected to the virtually grounded terminal of an operational amplifier that plays the role of a neuron with identity activation function. The advantages of the proposed system are twofold: first, its hardware can be directly trained using the Hebbian learning rule and without the need to perform any optimization; second, this system has a great ability to deal with a huge number of input-output training data without facing problems like overtraining.

Howard et al. proposed a spiking neuro-evolutionary system which implements memristors as plastic connections [33.188]. These memristors provide a learning architecture that may be beneficial to the evolutionary design process that exploits parameter self-adaptation and variable topologies, allow the number of neurons, connection weights, and interneural connectivity pattern to emerge. This approach allows the evolution of networks with appropriate complexity to emerge whilst exploiting the memristive properties of the connections to reduce learning time.

To investigate the dynamic behaviors of memristor-based neural networks, *Zeng* et al. proposed the memristor-based recurrent neural networks (MRNNs) [33.189, 190] shown in Fig. 33.35, where $x_i(.)$ is the

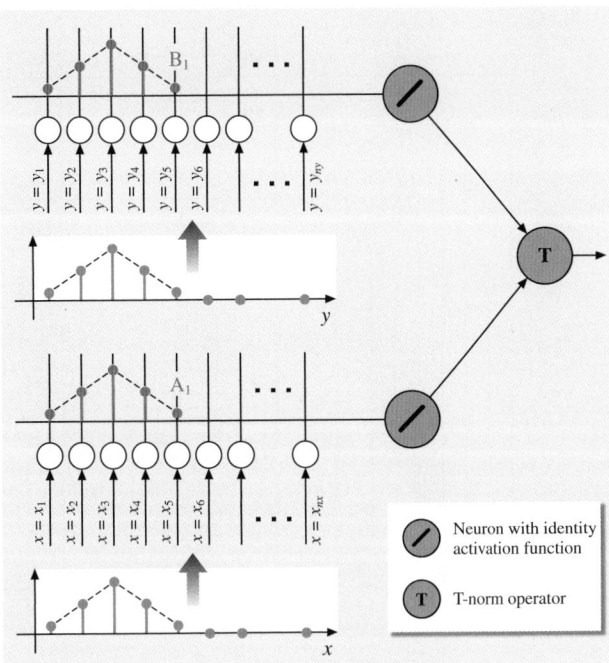

Fig. 33.33 Implementation of the activation function of neurons (after [33.184])

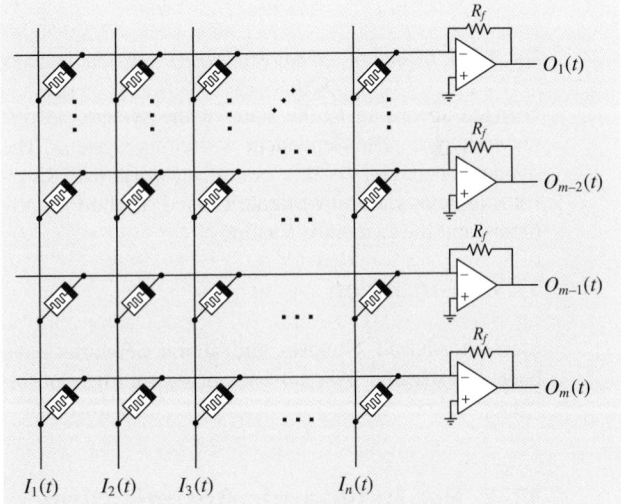

Fig. 33.34 Memristor crossbar-based circuit (after [33.184])

state of the i-th subsystem, $f_j(.)$ is the amplifier, M_{fij} is the connection memristor between the amplifier $f_j(.)$ and state $x_i(.)$, R_i and C_i are the resistor and capacitor, I_i is the external input, a_i, b_i are the outputs, $i, j = 1, 2, \ldots, n$. The parameters in this neural network are

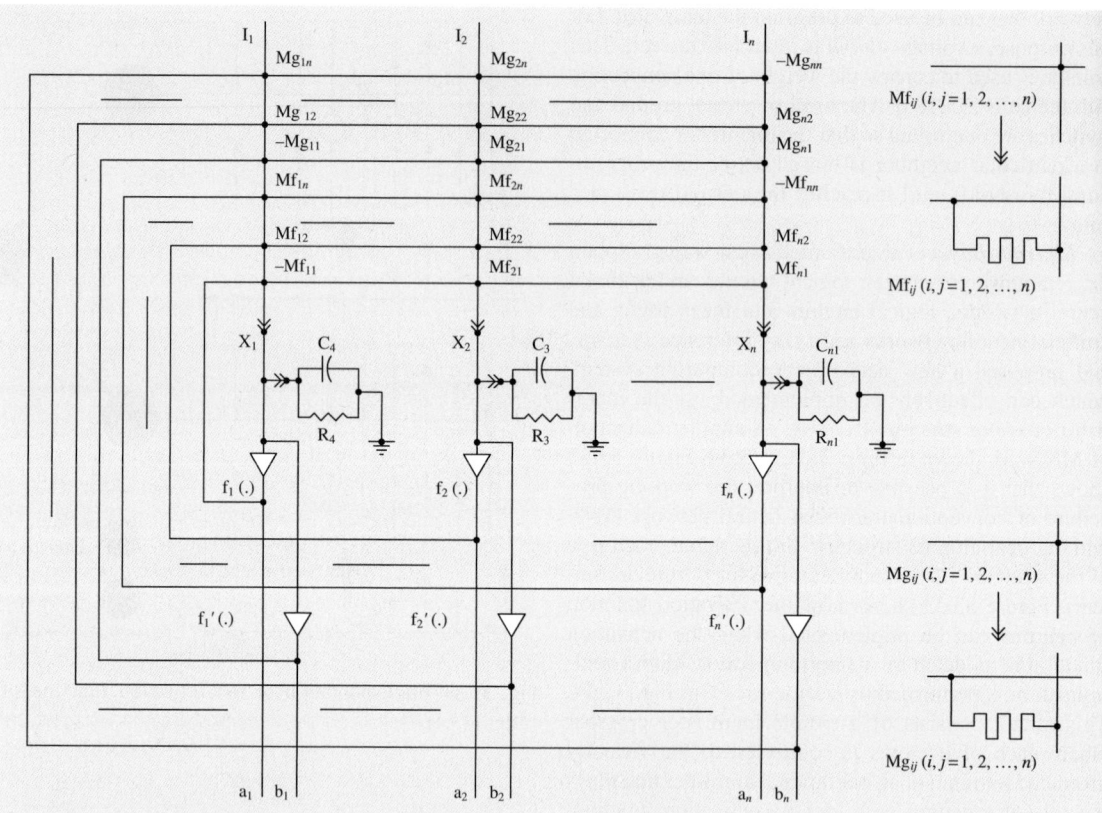

Fig. 33.35 Circuit of a memristor-based recurrent network (after [33.189])

changed according to the state of the system, so this network is a state-dependent switching system. The dynamic behavior of this neural network with time-varying delays was investigated based on the Filippov theory and the Lyapunov method.

33.3.3 Conclusion

Memristor-based synapses and neural networks have been investigated by many scientists for their possi-ble applications in analog, digital information process-ing, and memory and logic applications. However, the problem, of how to take advantage of the nonvolatile memory of memristors, nanoscale, low-power dissipa-tion, and so on to design a method to process and store the information, which needs learning and memory, into the synapses of the memristor-based neural networks at the dynamical mapping space by a more rational space-parting method, is still an open issue. Further investiga-tion is needed to shorten such a gap.

33.4 Neurodynamic Optimization

Optimization is omnipresent in nature and society, and an important tool for problem-solving in science, en-gineering, and commerce. Optimization problems arise in a wide variety of applications such as the design, planning, control, operation, and management of en-gineering systems. In many applications (e.g., online pattern recognition and in-chip signal processing in mo-bile devices), real-time optimization is necessary or desirable. For such applications, conventional optimiza-tion techniques may not be competent due to stringent requirements on computational time. It is computation-ally challenging when optimization procedures are to be performed in real time to optimize the performance of dynamical systems.

The brain is a profound dynamic system and its neurons are always active from birth to death. When a decision is to be made in the brain, many of its neurons are highly activated to gather information, search memory, compare differences, and make inferences and decisions. Recurrent neural networks are brain-like nonlinear dynamic system models and can be properly designed to imitate biological counterparts and serve as goal-seeking parallel computational models for solving optimization problems in a variety of settings. Neurodynamic optimization can be physically realized in designated hardware such as application-specific integrated circuits (ASICs) where optimization is carried out in a parallel and distributed manner, where the convergence rate of the optimization process is independent of the problem dimensionality. Because of the inherent nature of parallel and distributed information processing, neurodynamic optimization can handle large-scale problems. In addition, neurodynamic optimization may be used for optimizing dynamic systems in multiple time scales with parameter-controlled convergence rates. These salient features are particularly desirable for dynamic optimization in decentralized decision-making scenarios [33.191–194]. While population-based evolutionary approaches to optimization have emerged as prevailing heuristic and stochastic methods in recent years, neurodynamic optimization deserves great attention in its own right due to its close ties with optimization and dynamical systems theories, as well as its biological plausibility and circuit implementability with very large scale integration (VLSI) or optical technologies.

33.4.1 Neurodynamic Models

The past three decades witnessed the birth and growth of neurodynamic optimization. Although a couple of circuit-based optimization methods were developed earlier [33.195–197], it was perhaps *Hopfield* and *Tank* who spearheaded neurodynamic optimization research in the context of neural computation with their seminal work in the mid 1980s [33.198–200]. Since the inception, numerous neurodynamic optimization models in various forms of recurrent neural networks have been developed and analyzed, see [33.201–256], and the references therein. For example, *Tank* and *Hopfield* extended the continuous-time Hopfield network for linear programming and showed their experimental results with a circuit of operational amplifiers and other discrete components on a breadboard [33.200]. *Kennedy* and *Chua* developed a circuit-based recurrent neural network for nonlinear programming [33.201]. It is proven that the state of the neurodynamics is globally convergent and an equilibrium corresponding to an approximate optimal solution of the given optimization problems.

Over the years, neurodynamic optimization research has made significant progress with models with improved features for solving various optimization problems. Substantial improvements of neurodynamic optimization theory and models have been made in the following dimensions:

i) Solution quality: designed based on smooth penalty methods with a finite penalty parameter; the earliest neurodynamic optimization models can converge to approximate solutions only [33.200, 201]. Later on, better models designed based on other design principles can guarantee to state or output convergence to exact optimal solutions of solvable convex and pseudoconvex optimization problems with or without any conditions [33.204, 205, 208, 210], etc.

ii) Solvability scope: the solvability scope of neurodynamic optimization has been expanded from linear programming problems [33.200, 202, 208, 211, 212, 214–219, 223, 242, 244, 251], to quadratic programming problems [33.202–206, 210, 214, 217, 218, 220, 225, 226, 229, 233, 240–243, 247], to smooth convex programming problems with various constraints [33.201, 204, 205, 210, 214, 222, 224, 228, 230, 232, 234, 237, 245, 246, 257], to nonsmooth convex optimization problems [33.235, 248, 250–256], and recently to nonsmooth optimization with some nonconvex objective functions or constraints [33.239, 249, 254–256].

iii) Convergence property: the convergence property of neurodynamic optimization models has been extended from near-optimum convergence [33.200, 201], to conditional exact-optimum global convergence [33.205, 208, 210], to guaranteed global convergence [33.204, 205, 214–216, 218, 219, 222, 226–228, 230, 232, 234, 240, 243, 245, 247, 250, 253, 256, 257], to faster global exponential convergence [33.206, 224, 225, 228, 233, 237, 239, 241, 246, 254], to even more desirable finite-time convergence [33.235, 248, 249, 251, 252, 255], with increasing convergence rate.

iv) Model complexity: the neurodynamic optimization models for constrained optimization are essentially multilayer due to the introduction of instrumental variables for constraint handling (e.g., Lagrange multipliers or dual variables). The architectures of later neurodynamic optimization models for solving linearly constrained optimization problems have been reduced from multilayer structures to single-layer ones with decreasing model complexity to facilitate their implementation [33.243, 244, 251, 252, 254, 255].

Activation functions are a signature component of neural network models for quantifying the firing state activities of neurons. The activation functions in existing neurodynamic optimization models include smooth ones (e.g., sigmoid), as shown in Fig. 33.36a,b [33.200, 208–210], nonsmooth ones (e.g., piecewise-linear) as shown in Fig. 33.36c,d [33.203, 206], and even discontinuous ones as shown in Fig. 33.36e,f [33.243, 244, 251, 252, 254, 255].

33.4.2 Design Methods

The crux of neurodynamic optimization model design lies in the derivation of a convergent neurodynamic equation that prescribes the states of the neurodynamics. A properly derived neurodynamic equation can ensure that the states of neurodynamics reaches an equilibrium that satisfies the constraints and optimizes the objective function. Although the existing neurodynamic optimization models are highly diversified with many different features, the design methods or principles for determining their neurodynamic equations can be categorized as follows:

i) Penalty methods
ii) Lagrange methods
iii) Duality methods
iv) Optimality methods.

Penalty Methods

Consider the general constrained optimization problem

$$\text{minimize} \quad f(x)$$
$$\text{subject to} \quad g(x) \leq 0,$$
$$\qquad\qquad h(x) = 0,$$

where $x \in Re^n$ is the vector of decision variables, $f(x)$ is an objective function, $g(x) = [g_1(x), \dots, g_m(x)]^\mathsf{T}$ is a vector-valued function, and $h(x) = [h_1(x), \dots, h_p(x)]^\mathsf{T}$ a vector-valued function.

A penalty method starts with the formulation of a smooth or nonsmooth energy function based on a given objective function $f(x)$ and constraints $g(x)$ and $h(x)$. It plays an important role in neurodynamic optimization. Ideally, the minimum of a formulated energy function corresponds to the optimal solution of the original optimization problem. For constrained optimization, the minimum of the energy function has to satisfy a set of constraints. Most early approaches formulate an energy function by incorporating objective function and constraints through functional transformation and numerical weighting [33.198–201]. Functional transformation is usually used to convert constraints to a penalty function to penalize the violation of constraints; e.g., a smooth penalty function is as follows

$$p(x) = \frac{1}{2} \sum_{i=1}^{m} \{[-g_i(x)]^+\}^2 + \sum_{j=1}^{p} [h_j(x)]^2 \,,$$

where $[y]^+ = \max\{0, y\}$. Numerical weighting is often used to balance constraint satisfaction and objective optimization, e.g.,

$$E(x) = f(x) + wp(x) \,,$$

where w is a positive weight.

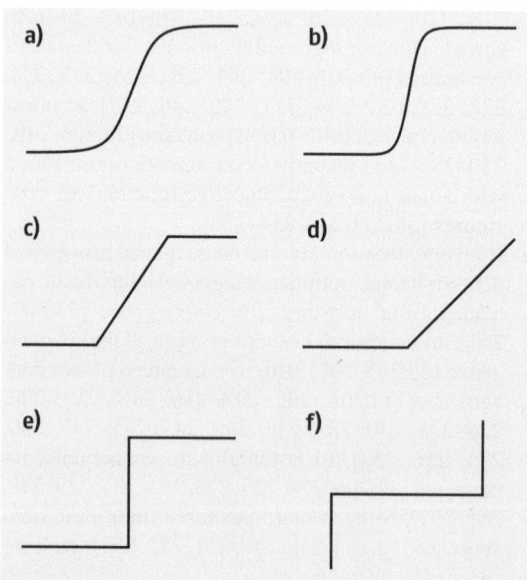

Fig. 33.36a–f Three classes of activation functions in neurodynamic optimization models: smooth in (**a**) and (**b**), nonsmooth in (**c**) and (**d**), and discontinuous in (**e**) and (**f**)

In smooth penalty methods, neurodynamic equations are usually derived as the negative gradient flow of the energy function in the form of a differential equation

$$\frac{dx(t)}{dt} \propto -\nabla E(x(t)) .$$

If the energy function is bounded below, the stability of the neurodynamics can be ensured. Nevertheless, the major limitation is that the neurodynamics designed using a smooth penalty method with any fixed finite penalty parameter can converge to an approximate optimal solution only, as a compromise between constraint satisfaction and objective optimization. One way to remedy the approximated limitation of smooth penalty design methods is to introduce a variable penalty parameter. For example, a time-varying decaying penalty parameter (called temperature) is used in deterministic annealing networks to achieve exact optimality with a slow cooling schedule [33.208, 210].

If the objective function or penalty function is nonsmooth, the gradient has to be replaced by a generalized gradient and the neurodynamics can be modeled using a differential inclusion [33.235, 248, 249, 251, 252, 255]. Two advantages of nonsmooth penalty methods over smooth ones are possible constraint satisfaction and objective optimization with some finite penalty parameters and finite-time convergence of the resulting neurodynamics. Needless to say, nonsmooth neurodynamics are much more difficult to analyze to guarantee their stability.

Lagrange Methods

A Lagrange method for designing a neurodynamic optimization model begins with the formulation of a Lagrange function (Lagrangian) instead of an energy function [33.204, 205]. A typical Lagrangian is defined as

$$L(x, \lambda, \mu) = f(x) + \sum_{i=1}^{m} \lambda_i g_i(x) + \sum_{j=1}^{p} \mu_j h_j(x) ,$$

where $\lambda = (\lambda_1, \ldots, \lambda_m)^\mathsf{T}$ and $\lambda = (\mu_1, \ldots, \mu_p)^\mathsf{T}$ are Lagrange multipliers, for inequality constraints $g(x)$ and equality constraints $h(x)$, respectively.

According to the saddle-point theorem, the optimal solution can be determined by minimizing the Lagrangian with respect to x and maximizing it with respect to λ and μ. Therefore, neurodynamic equations

can be derived in an augmented space

$$\epsilon \frac{dx(t)}{dt} = -\nabla_x L(x(t), \lambda(t), \mu(t)) ,$$

$$\epsilon \frac{d\lambda(t)}{dt} = -\nabla_\lambda L(x(t), \lambda(t), \mu(t)) ,$$

$$\epsilon \frac{d\mu(t)}{dt} = -\nabla_\mu L(x(t), \lambda(t), \mu(t)) ,$$

where ϵ is a positive time constant. The equilibrium of the Lagrangian neurodynamics satisfy the Lagrange necessary optimality conditions.

Duality Methods

For convex optimization, the objective functions of primal and dual problems reach the same value at their optima. In view of this duality property, the duality methods for designing neurodynamic optimization models begin with the formulation of an energy function consisting of a duality gap between the primal and dual problems and a constraint-based penalty function, e.g.,

$$E(x, y) = \frac{1}{2}(f(x) - f_d(y))^2 + p(x) + p_d(y) ,$$

where y is a vector of dual variables, $f_d(y)$ is the dual objective function to be maximized, $p(x)$ and $p_d(y)$ are, respectively, smooth penalty functions to penalize the violations of constraints of primal (original) and dual problems. The corresponding neurodynamic equation can be derived with guaranteed global stability as the negative gradient flow of the energy function similarly as in the aforementioned smooth penalty methods [33.216, 218, 222, 226, 258, 259]. Neurodynamic optimization models designed by using duality design methods can guarantee global convergence to the exact optimal solutions of convex optimization problems without any parametric condition.

In addition, using duality methods, dual networks and their simplified/improved versions can be designed for quadratic programming with reduced model complexity by mapping their global convergent optimal dual state variables to optimal primal solutions via linear or piecewise-linear output functions [33.240, 247, 260–263].

Optimality Methods

The neurodynamic equations of some recent models are derived based on optimality conditions (e.g., the

Karush–Kuhn–Tucker condition) and projection methods. Basically, the methods are to map the equilibrium of the designed neurodynamic optimization models to the equivalent equalities given by optimality conditions and projection equations (i. e., all equilibria essentially satisfy the optimality conditions) [33.225, 227, 228]. For several types of common geometric constraints (such as nonnegative constraints, bound constraints, and spherical constraints), some projection operators map the neuron state variables onto the convex feasible regions by using their activation functions and avoid the use of excessive dual variables as in the dual networks, and thus lower the model complexity. For neurodynamic optimization models designed using optimality methods, stability analysis is needed explicitly to ensure that the resulting neurodynamics are stable.

Once a neurodynamic equation has been derived and its stability is proven, the next step is to determine the architecture of the neural network in terms of the neurons and connections based on the derived neurodynamic equation. The last step is usually devoted to simulation or emulation to test the performance of the neural network numerically or physically. The simulation/emulation results may reveal additional properties or characteristics for further analysis or model redesign.

33.4.3 Selected Applications

Over the last few decades, neurodynamic optimization has been widely applied in many fields of science, engineering, and commerce, as highlighted in the following selected nine areas.

Scientific Computing
Neurodynamic optimization models ave been developed for solving linear equations and inequalities and computing inverse or pseudoinverse matrices [33.240, 264–268].

Network Routing
Neurodynamic optimization models have been developed or applied for shortest-path routing in networks modeled by using weighted directed graphs [33.258, 269–271].

Machine Learning
Neurodynamic optimization has been applied for support vector machine learning to take the advantages of its parallel computational power [33.272–274].

Data Processing
The data processing applications of neurodynami optimization include, but are not limited to, sort ing [33.275–277], winners-take-all selection [33.240 277, 278], data fusion [33.279], and data reconcilia tion [33.254].

Signal/Image Processing
The applications of neurodynamic optimization for sig nal and image processing include, but are not limite to, recursive least-squares adaptive filtering, overcom plete signal representations, time delay estimation, an image restoration and reconstruction [33.191, 203, 204, 280–283].

Communication Systems
The telecommunication applications of neurodynami optimization include beamforming [33.284, 285]) an simulations of DS-CDMA mobile communication sys tems [33.229].

Control Systems
Intelligent control applications of neurodynamic op timization include pole assignment for synthesizing linear control systems [33.286–289] and model predic tive control for linear/nonlinear systems [33.290–292].

Robotic Systems
The applications of neurodynamic optimization in in telligent robotic systems include real-time motion plan ning and control of kinematically redundant robot ma nipulators with torque minimization or obstacle avoid ance [33.259–263, 267, 293–298] and grasping force optimization for multifingered robotic hands [33.299].

Financial Engineering
Recently, neurodynamic optimization was also applied for real-time portfolio selection based on an equivalen probability measure to optimize the asset distribution in financial investments; [33.255, 300].

33.4.4 Concluding Remarks

Neurodynamic optimization provides a parallel distributed computational model for solving many optimization problems. For convex and convex-like optimization, neurodynamic optimization models are available with guaranteed optimality, expended applicability, improved convergence properties, and reduced model complexity. Neurodynamic optimization approaches have been demonstrated to be effective and efficient for many applications, especially those with real-time solution requirements.

The existing results can still be further improved to expand their solvability scope, increase their convergence rate, or reduce their model complexity. With the view that neurodynamic approaches to global optimization and discrete optimization are much more interesting and challenging, it is necessary to develop neurodynamic models for nonconvex optimization and combinatorial optimization. In addition, neurodynamic optimization approaches could be more widely applied for many other application areas in conjunction with conventional and evolutionary optimization approaches.

References

33.1 R. Abraham: *Dynamics: The Geometry of Behavior* (Aerial, Santa Cruz 1982)

33.2 J. Robinson: Attractor. In: *Encyclopedia of Nonlinear Science*, ed. by A. Scott (Routledge, New York 2005) pp. 26–28

33.3 S. Grossberg: Nonlinear difference–differential equations in prediction and learning theory, Proc. Natl. Acad. Sci. **58**, 1329–1334 (1967)

33.4 S. Grossberg: Global ratio limit theorems for some nonlinear functional differential equations I, Bull. Am. Math. Soc. **74**, 93–100 (1968)

33.5 H. Zhang, Z. Wang, D. Liu: Robust exponential stability of recurrent neural networks with multiple time-varying delays, IEEE Trans. Circuits Syst. II: Express Br. **54**, 730–734 (2007)

33.6 A.N. Michel, K. Wang, D. Liu, H. Ye: Qualitative limitations incurred in implementations of recurrent neural networks, IEEE Cont. Syst. Mag. **15**(3), 52–65 (1995)

33.7 H. Zhang, Z. Wang, D. Liu: Global asymptotic stability of recurrent neural networks with multiple time varying delays, IEEE Trans. Neural Netw. **19**(5), 855–873 (2008)

33.8 S. Hu, D. Liu: On the global output convergence of a class of recurrent neural networks with time-varying inputs, Neural Netw. **18**(2), 171–178 (2005)

33.9 D. Liu, S. Hu, J. Wang: Global output convergence of a class of continuous-time recurrent neural networks with time-varying thresholds, IEEE Trans. Circuits Syst. II: Express Br. **51**(4), 161–167 (2004)

33.10 H. Zhang, Z. Wang, D. Liu: Robust stability analysis for interval Cohen–Grossberg neural networks with unknown time varying delays, IEEE Trans. Neural Netw. **19**(11), 1942–1955 (2008)

33.11 M. Han, J. Fan, J. Wang: A dynamic feedforward neural network based on Gaussian particle swarm optimization and its application for predictive control, IEEE Trans. Neural Netw. **22**(9), 1457–1468 (2011)

33.12 S. Mehraeen, S. Jagannathan, M.L. Crow: Decentralized dynamic surface control of large-scale interconnected systems in strict-feedback form using neural networks with asymptotic stabilization, IEEE Trans. Neural Netw. **22**(11), 1709–1722 (2011)

33.13 Y. Zhang, T. Chai, H. Wang: A nonlinear control method based on anfis and multiple models for a class of SISO nonlinear systems and its application, IEEE Trans. Neural Netw. **22**(11), 1783–1795 (2011)

33.14 Y. Chen, W.X. Zheng: Stability and L_2 performance analysis of stochastic delayed neural networks, IEEE Trans. Neural Netw. **22**(10), 1662–1668 (2011)

33.15 M. Di Marco, M. Grazzini, L. Pancioni: Global robust stability criteria for interval delayed full-range cellular neural networks, IEEE Trans. Neural Netw. **22**(4), 666–671 (2011)

33.16 W.-H. Chen, W.X. Zheng: A new method for complete stability analysis of cellular neural networks with time delay, IEEE Trans. Neural Netw. **21**(7), 1126–1139 (2010)

33.17 H. Zhang, Z. Wang, D. Liu: Global asymptotic stability and robust stability of a general class of Cohen–Grossberg neural networks with mixed delays, IEEE Trans. Circuits Syst. I: Regul. Pap. **56**(3), 616–629 (2009)

33.18 X.X. Liao, J. Wang: Algebraic criteria for global exponential stability of cellular neural networks with multiple time delays, IEEE Trans. Circuits Syst. I **50**, 268–275 (2003)

33.19 Z.G. Zeng, J. Wang, X.X. Liao: Global exponential stability of a general class of recurrent neural networks with time-varying delays, IEEE Trans. Circuits Sys. I **50**(10), 1353–1358 (2003)

33.20 D. Angeli: Multistability in systems with counterclockwise input-output dynamics, IEEE Trans. Autom. Control **52**(4), 596–609 (2007)

33.21 D. Angeli: Systems with counterclockwise input-output dynamics, IEEE Trans. Autom. Control **51**(7), 1130–1143 (2006)

33.22 D. Angeli: Convergence in networks with counterclockwise neural dynamics, IEEE Trans. Neural Netw. **20**(5), 794–804 (2009)

33.23 J. Saez-Rodriguez, A. Hammerle-Fickinger, O. Dalal, S. Klamt, E.D. Gilles, C. Conradi: Multistability of signal transduction motifs, IET Syst. Biol. **2**(2), 80–93 (2008)

33.24 L. Chandrasekaran, V. Matveev, A. Bose: Multistability of clustered states in a globally inhibitory network, Phys. D **238**(3), 253–263 (2009)

33.25   B.K. Goswami: Control of multistate hopping intermittency, Phys. Rev. E **78**(6), 066208 (2008)

33.26   A. Rahman, M.K. Sanyal: The tunable bistable and multistable memory effect in polymer nanowires, Nanotechnology **19**(39), 395203 (2008)

33.27   K.C. Tan, H.J. Tang, W.N. Zhang: Qualitative analysis for recurrent neural networks with linear threshold transfer functions, IEEE Trans. Circuits Syst. I: Regul. Pap. **52**(5), 1003–1012 (2005)

33.28   H.J. Tang, K.C. Tan, E.J. Teoh: Dynamics analysis and analog associative memory of networks with LT neurons, IEEE Trans. Neural Netw. **17**(2), 409–418 (2006)

33.29   L. Zou, H.J. Tang, K.C. Tan, W.N. Zhang: Nontrivial global attractors in 2-D multistable attractor neural networks, IEEE Trans. Neural Netw. **20**(11), 1842–1851 (2009)

33.30   D. Liu, A.N. Michel: Sparsely interconnected neural networks for associative memories with applications to cellular neural networks, IEEE Trans. Circuits Syst. II: Analog Digit, Signal Process. **41**(4), 295–307 (1994)

33.31   M. Brucoli, L. Carnimeo, G. Grassi: Discrete-time cellular neural networks for associative memories with learning and forgetting capabilities, IEEE Trans. Circuits Syst. I: Fundam. Theory Appl. **42**(7), 396–399 (1995)

33.32   R. Perfetti: Dual-mode space-varying self-designing cellular neural networks for associative memory, IEEE Trans. Circuits Syst. I: Fundam. Theory Appl. **46**(10), 1281–1285 (1999)

33.33   G. Grassi: On discrete-time cellular neural networks for associative memories, IEEE Trans. Circuits Syst. I: Fundam. Theory Appl. **48**(1), 107–111 (2001)

33.34   L. Wang, X. Zou: Capacity of stable periodic solutions in discrete-time bidirectional associative memory neural networks, IEEE Trans. Circuits Syst. II: Express Br. **51**(6), 315–319 (2004)

33.35   J. Milton: Epilepsy: Multistability in a dynamic disease. In: *Self- Organized Biological Dynamics Nonlinear Control: Toward Understanding Complexity, Chaos, and Emergent Function in Living Systems*, ed. by J. Walleczek (Cambridge Univ. Press, Cambridge 2000) pp. 374–386

33.36   U. Feudel: Complex dynamics in multistable systems, Int. J. Bifurc. Chaos **18**(6), 1607–1626 (2008)

33.37   J. Hizanidis, R. Aust, E. Scholl: Delay-induced multistability near a global bifurcation, Int. J. Bifurc. Chaos **18**(6), 1759–1765 (2008)

33.38   G.G. Wells, C.V. Brown: Multistable liquid crystal waveplate, Appl. Phys. Lett. **91**(22), 223506 (2007)

33.39   G. Deco, D. Marti: Deterministic analysis of stochastic bifurcations in multi-stable neurodynamical systems, Biol. Cybern. **96**(5), 487–496 (2007)

33.40   J.D. Cao, G. Feng, Y.Y. Wang: Multistability and multiperiodicity of delayed Cohen–Grossberg neural networks with a general class of activation functions, Phys. D **237**(13), 1734–174■ (2008)

33.41   C.Y. Cheng, K.H. Lin, C.W. Shih: Multistability ■ recurrent neural networks, SIAM J. Appl. Math **66**(4), 1301–1320 (2006)

33.42   Z. Yi, K.K. Tan: Multistability of discrete-time re current neural networks with unsaturating piece wise linear activation functions, IEEE Trans. Neur■ Netw. **15**(2), 329–336 (2004)

33.43   Z. Yi, K.K. Tan, T.H. Lee: Multistability analysi■ for recurrent neural networks with unsaturatin■ piecewise linear transfer functions, Neural Com■ put. **15**(3), 639–662 (2003)

33.44   Z.G. Zeng, T.W. Huang, W.X. Zheng: Multistabili■ of recurrent neural networks with time-varyin■ delays and the piecewise linear activation func tion, IEEE Trans. Neural Netw. **21**(8), 1371–137■ (2010)

33.45   Z.G. Zeng, J. Wang, X.X. Liao: Stability analysis ■ delayed cellular neural networks described us ing cloning templates, IEEE Trans. Circuits Syst. ■ Regul. Pap. **51**(11), 2313–2324 (2004)

33.46   Z.G. Zeng, J. Wang: Multiperiodicity and expo nential attractivity evoked by periodic externa inputs in delayed cellular neural networks, Neu ral Comput. **18**(4), 848–870 (2006)

33.47   L.L. Wang, W.L. Lu, T.P. Chen: Multistability an■ new attraction basins of almost-periodic solu tions of delayed neural networks, IEEE Trans. Neu ral Netw. **20**(10), 1581–1593 (2009)

33.48   G. Huang, J.D. Cao: Delay-dependent multista bility in recurrent neural networks, Neural Netw **23**(2), 201–209 (2010)

33.49   L.L. Wang, W.L. Lu, T.P. Chen: Coexistence an■ local stability of multiple equilibria in neural net works with piecewise linear nondecreasing ac tivation functions, Neural Netw. **23**(2), 189–20■ (2010)

33.50   L. Zhang, Z. Yi, J.L. Yu, P.A. Heng: Some mul tistability properties of bidirectional associativ■ memory recurrent neural networks with unsat urating piecewise linear transfer functions, Neu rocomputing **72**(16–18), 3809–3817 (2009)

33.51   X.B. Nie, J.D. Cao: Multistability of competitiv■ neural networks with time-varying and dis tributed delays, Nonlinear Anal.: Real World Appl **10**(2), 928–942 (2009)

33.52   C.Y. Cheng, K.H. Lin, C.W. Shih: Multistability and convergence in delayed neural networks, Phys. ■ **225**(1), 61–74 (2007)

33.53   T.J. Sejnowski, C. Koch, P.S. Churchland: Computational neuroscience, Science **241**(4871), 1299 (1988)

33.54   G. Edelman: *Remembered Present: A Biologica■ Theory of Consciousness* (Basic Books, New Yor■ 1989)

33.55   W.J. Freeman: *Societies of Brains: A Study in th■ Neuroscience of Love and Hate* (Lawrence Erl baum, New York 1995)

33.56 R. Llinas, U. Ribary, D. Contreras, C. Pedroarena: The neuronal basis for consciousness, Philos. Trans. R. Soc. B **353**(1377), 1841 (1998)

33.57 F. Crick, C. Koch: A framework for consciousness, Nat. Neurosci. **6**(2), 119–126 (2003)

33.58 A.L. Hodgkin, A.F. Huxley: A quantitative description of membrane current and its application to conduction and excitation in nerve, J. Physiol. **117**(4), 500 (1952)

33.59 A. Pikovsky, M. Rosenblum: Synchronization, Scholarpedia **2**(12), 1459 (2007)

33.60 D. Golomb, A. Shedmi, R. Curtu, G.B. Ermentrout: Persistent synchronized bursting activity in cortical tissues with low magnesium concentration: A modeling study, J. Neurophysiol. **95**(2), 1049–1067 (2006)

33.61 M.L.V. Quyen, J. Foucher, J.-P. Lachaux, E. Rodriguez, A. Lutz, J. Martinerie, F.J. Varela: Comparison of Hilbert transform and wavelet methods for the analysis of neuronal synchrony, J. Neurosci. Methods **111**(2), 83–98 (2001)

33.62 W.J. Freeman, L.J. Rogers: Fine temporal resolution of analytic phase reveals episodic synchronization by state transitions in gamma EEGs, J. Neurophysiol. **87**(2), 937–945 (2002)

33.63 G.E.P. Box, G.M. Jenkins, G.C. Reinsel: *Ser. Probab. Stat*, Time Series Analysis: Forecasting and Control, Vol. 734 (Wiley, Hoboken 2008)

33.64 R.W. Thatcher, D.M. North, C.J. Biver: Development of cortical connections as measured by EEG coherence and phase delays, Hum. Brain Mapp. **29**(12), 1400–1415 (2007)

33.65 A. Pikovsky, M. Rosenblum, J. Kurths: *Synchronization: A Universal Concept in Nonlinear Sciences*, Vol. 12 (Cambridge Univ. Press, Cambridge 2003)

33.66 J. Rodriguez, R. Kozma: Phase synchronization in mesoscopic electroencephalogram arrays. In: *Intelligent Engineering Systems Through Artificial Neural Networks Series*, ed. by C. Dagli (ASME, New York 2007) pp. 9–14

33.67 J.M. Barrie, W.J. Freeman, M.D. Lenhart: Spatiotemporal analysis of prepyriform, visual, auditory, and somesthetic surface EEGs in trained rabbits, J. Neurophysiol. **76**(1), 520–539 (1996)

33.68 G. Dumas, M. Chavez, J. Nadel, J. Martinerie: Anatomical connectivity influences both intra- and inter-brain synchronizations, PloS ONE **7**(5), e36414 (2012)

33.69 J.A.S. Kelso: *Dynamic Patterns: The Self-Organization of Brain and Behavior* (MIT Press, Cambridge 1995)

33.70 S. Campbell, D. Wang: Synchronization and desynchronization in a network of locally coupled Wilson–Cowan oscillators, IEEE Trans. Neural Netw. **7**(3), 541–554 (1996)

33.71 H. Kurokawa, C.Y. Ho: A learning rule of the oscillatory neural networks for in-phase oscillation, IEICE Trans. Fundam. Electron. Commun. Comput. Sci. **80**(9), 1585–1594 (1997)

33.72 G. Buzsaki: *Rhythms of the Brain* (Oxford Univ. Press, New York 2009)

33.73 A.K. Engel, P. Fries, W. Singer: Dynamic predictions: Oscillations and synchrony in top-down processing, Nat. Rev. Neurosci. **2**(10), 704–716 (2001)

33.74 W.J. Freeman, R.Q. Quiroga: *Imaging Brain Function with EEG: Advanced Temporal and Spatial Analysis of Electroencephalographic Signals* (Springer, New York 2013)

33.75 H. Haken: Cooperative phenomena in systems far from thermal equilibrium and in nonphysical systems, Rev. Mod. Phys. **47**(1), 67 (1975)

33.76 S.H. Strogatz: Exploring complex networks, Nature **410**(6825), 268–276 (2001)

33.77 O. Sporns, D.R. Chialvo, M. Kaiser, C.C. Hilgetag: Organization, development and function of complex brain networks, Trends Cogn. Sci. **8**(9), 418–425 (2004)

33.78 B. Bollobás, R. Kozma, D. Miklos (Eds.): *Handbook of Large-Scale Random Networks*, Bolyai Soc. Math. Stud., Vol. 18 (Springer, Berlin, Heidelberg 2009)

33.79 Y. Kuramoto: Cooperative dynamics of oscillator community, Prog. Theor. Phys. Suppl. **79**, 223–240 (1984)

33.80 M.G. Rosenblum, A.S. Pikovsky: Controlling synchronization in an ensemble of globally coupled oscillators, Phys. Rev. Lett. **92**(11), 114102 (2004)

33.81 O.V. Popovych, P.A. Tass: Synchronization control of interacting oscillatory ensembles by mixed nonlinear delayed feedback, Phys. Rev. E **82**(2), 026204 (2010)

33.82 W.J. Freeman: The physiology of perception, Sci. Am. **264**, 78–85 (1991)

33.83 M.A. Cohen, S. Grossberg: Absolute stability of global pattern formation and parallel memory storage by competitive neural networks, IEEE Trans. Syst. Man Cybern. **13**(5), 815–826 (1983)

33.84 J.J. Hopfield, D.W. Tank: Computing with neural circuits – A model, Science **233**(4764), 625–633 (1986)

33.85 C.M. Marcus, R.M. Westervelt: Dynamics of iterated-map neural networks, Phys. Rev. A **40**(1), 501 (1989)

33.86 W. Yu, J. Cao, J. Wang: An LMI approach to global asymptotic stability of the delayed Cohen-Grossberg neural network via nonsmooth analysis, Neural Netw. **20**(7), 810–818 (2007)

33.87 F.C. Hoppensteadt, E.M. Izhikevich: *Weakly Connected Neural Networks*, Applied Mathematical Sciences, Vol. 126 (Springer, New York 1997)

33.88 H.R. Wilson, J.D. Cowan: Excitatory and inhibitory interactions in localized populations of model neurons, Biophys. J. **12**(1), 1–24 (1972)

33.89 H.R. Wilson, J.D. Cowan: A mathematical theory of the functional dynamics of cortical and thalamic nervous tissue, Biol. Cybern. **13**(2), 55–80 (1973)

33.90   P.C. Bressloff: Spatiotemporal dynamics of continuum neural fields, J. Phys. A: Math. Theor. **45**(3), 033001 (2011)

33.91   D. Wang: Object selection based on oscillatory correlation, Neural Netw. **12**(4), 579–592 (1999)

33.92   A. Renart, R. Moreno-Bote, X.-J. Wang, N. Parga: Mean-driven and fluctuation-driven persistent activity in recurrent networks, Neural Comput. **19**(1), 1–46 (2007)

33.93   M. Ursino, E. Magosso, C. Cuppini: Recognition of abstract objects via neural oscillators: interaction among topological organization, associative memory and gamma band synchronization, IEEE Trans. Neural Netw. **20**(2), 316–335 (2009)

33.94   W.J. Freeman: *Mass Action in the Nervous System* (Academic, New York 1975)

33.95   D. Xu, J. Principe: Dynamical analysis of neural oscillators in an olfactory cortex model, IEEE Trans. Neural Netw. **15**(5), 1053–1062 (2004)

33.96   R. Ilin, R. Kozma: Stability of coupled excitatory-inhibitory neural populations and application to control of multi-stable systems, Phys. Lett. A **360**(1), 66–83 (2006)

33.97   R. Ilin, R. Kozma: Control of multi-stable chaotic neural networks using input constraints, 2007. IJCNN 2007. Int. Jt. Conf. Neural Netw., Orlando (2007) pp. 2194–2199

33.98   G. Deco, V. K. Jirsa, P. A. Robinson, M. Breakspear, K. Friston: The dynamic brain: From spiking neurons to neural masses and cortical fields, PLoS Comput. Biol. **4**(8), e1000092 (2008)

33.99   L. Ingber: Generic mesoscopic neural networks based on statistical mechanics of neocortical interactions, Phys. Rev. A **45**(4), 2183–2186 (1992)

33.100   V.K. Jirsa, K.J. Jantzen, A. Fuchs, J.A. Scott Kelso: Spatiotemporal forward solution of the EEG and meg using network modeling, IEEE Trans. Med. Imaging **21**(5), 493–504 (2002)

33.101   S. Coombes, C. Laing: Delays in activity-based neural networks, Philos. Trans. R. Soc. A **367**(1891), 1117–1129 (2009)

33.102   V.K. Jirsa: Neural field dynamics with local and global connectivity and time delay, Philos. Trans. R. Soc. A **367**(1891), 1131–1143 (2009)

33.103   K. Kaneko: Clustering, coding, switching, hierarchical ordering, and control in a network of chaotic elements, Phys. D **41**(2), 137–172 (1990)

33.104   R. Kozma: Intermediate-range coupling generates low-dimensional attractors deeply in the chaotic region of one-dimensional lattices, Phys. Lett. A **244**(1), 85–91 (1998)

33.105   S. Ishii, M.-A. Sato: Associative memory based on parametrically coupled chaotic elements, Phys. D **121**(3), 344–366 (1998)

33.106   F. Moss, A. Bulsara, M.F. Schlesinger (Eds.): *The proceedings of the NATO Advanced Research Workshop: Stochastic Resonance in Physics and Biology* (Plenum Press, New York 1993)

33.107   S.N. Dorogovtsev, A.V. Goltsev, J.F.F. Mendes: Critical phenomena in complex networks, Rev. Mod. Phys. **80**(4), 1275 (2008)

33.108   M.D. McDonnell, L.M. Ward: The benefits of noise in neural systems: bridging theory and experiment, Nat. Rev. Neurosci. **12**(7), 415–426 (2011)

33.109   A.V. Goltsev, M.A. Lopes, K.-E. Lee, J.F.F. Mendes: Critical and resonance phenomena in neural networks, arXiv preprint arXiv:1211.5686 (2012)

33.110   P. Bak: *How Nature Works: The Science of Self-Organized Criticality* (Copernicus, New York 1996)

33.111   J.M. Beggs, D. Plenz: Neuronal avalanches in neocortical circuits, J. Neurosci. **23**(35), 11167–11177 (2003)

33.112   J.M. Beggs: The criticality hypothesis: How local cortical networks might optimize information processing, Philos. Trans. R. Soc. A **366**(1864) 329–343 (2008)

33.113   T. Petermann, T.C. Thiagarajan, M.A. Lebedev, M.A.L. Nicolelis, D.R. Chialvo, D. Plenz: Spontaneous cortical activity in awake monkeys composed of neuronal avalanches, Proc. Natl. Acad. Sci. **106**(37), 15921–15926 (2009)

33.114   M. Puljic, R. Kozma: Narrow-band oscillations in probabilistic cellular automata, Phys. Rev. E **78**(2), 026214 (2008)

33.115   R. Kozma, M. Puljic, W.J. Freeman: Thermodynamic model of criticality in the cortex based on EEG/ECoG data. In: *Criticality in Neural Systems*, ed. by D. Plenz, E. Niebur (Wiley, Hoboken 2014) pp. 153–176

33.116   J.-P. Eckmann, D. Ruelle: Ergodic theory of chaos and strange attractors, Rev. Mod. Phys. **57**(3), 617 (1985)

33.117   E. Ott, C. Grebogi, J.A. Yorke: Controlling chaos, Phys. Rev. Lett. **64**(11), 1196–1199 (1990)

33.118   E.N. Lorenz: Deterministic nonperiodic flow, J. Atmos. Sci. **20**(2), 130–141 (1963)

33.119   K. Aihara, T. Takabe, M. Toyoda: Chaotic neural networks, Phys. Lett. A **144**(6), 333–340 (1990)

33.120   K. Aihara, H. Suzuki: Theory of hybrid dynamical systems and its applications to biological and medical systems, Philos. Trans. R. Soc. A **368**(1930), 4893–4914 (2010)

33.121   G. Matsumoto, K. Aihara, Y. Hanyu, N. Takahashi, S. Yoshizawa, J.-I. Nagumo: Chaos and phase locking in normal squid axons, Phys. Lett. A **123**(4), 162–166 (1987)

33.122   K. Aihara: Chaos engineering and its application to parallel distributed processing with chaotic neural networks, Proc. IEEE **90**(5), 919–930 (2002)

33.123   L. Wang, S. Li, F. Tian, X. Fu: A noisy chaotic neural network for solving combinatorial optimization problems: Stochastic chaotic simulated annealing, IEEE Trans. Syst. Man Cybern. B **34**(5), 2119–2125 (2004)

33.124 Z. Zeng, J. Wang: Improved conditions for global exponential stability of recurrent neural networks with time-varying delays, IEEE Trans. Neural Netw. **17**(3), 623–635 (2006)

33.125 M.D. Marco, M. Grazzini, L. Pancioni: Global robust stability criteria for interval delayed full-range cellular neural networks, IEEE Trans. Neural Netw. **22**(4), 666–671 (2011)

33.126 C.A. Skarda, W.J. Freeman: How brains make chaos in order to make sense of the world, Behav. Brain Sci. **10**(2), 161–195 (1987)

33.127 H.D.I. Abarbanel, M.I. Rabinovich, A. Selverston, M.V. Bazhenov, R. Huerta, M.M. Sushchik, L.L. Rubchinskii: Synchronisation in neural networks, Phys.-Usp. **39**(4), 337–362 (1996)

33.128 H. Korn, P. Faure: Is there chaos in the brain? II. experimental evidence and related models, c.r. Biol. **326**(9), 787–840 (2003)

33.129 R. Kozma, W.J. Freeman: Intermittent spatio-temporal desynchronization and sequenced synchrony in ECoG signals, Chaos Interdiscip. J. Nonlinear Sci. **18**(3), 037131 (2008)

33.130 K. Kaneko: *Collapse of Tori and Genesis of Chaos in Dissipative Systems* (World Scientific Publ., Singapore 1986)

33.131 K. Aihara: Chaos in neural networks. In: *The Impact of Chaos on Science and Society*, ed. by C. Grebogi, J.A. Yorke (United Nations Publ., New York 1997) pp. 110–126

33.132 I. Tsuda: Toward an interpretation of dynamic neural activity in terms of chaotic dynamical systems, Behav. Brain Sci. **24**(5), 793–809 (2001)

33.133 H. Bersini, P. Sener: The connections between the frustrated chaos and the intermittency chaos in small Hopfield networks, Neural Netw. **15**(10), 1197–1204 (2002)

33.134 P. Berge, Y. Pomeau, C. Vidal: *Order in Chaos* (Herman, Paris and Wiley, New York 1984)

33.135 Y. Pomeau, P. Manneville: Intermittent transition to turbulence in dissipative dynamical systems, Commun. Math. Phys. **74**(2), 189–197 (1980)

33.136 T. Higuchi: Relationship between the fractal dimension and the power law index for a time series: A numerical investigation, Phys. D **46**(2), 254–264 (1990)

33.137 B. Mandelbrot: *Fractals and Chaos: The Mandelbrot Set and Beyond*, Vol. 3 (Springer, New York 2004)

33.138 K. Falconer: *Fractal Geometry: Mathematical Foundations and Applications* (Wiley, Hoboken 2003)

33.139 T. Sauer, J.A. Yorke, M. Casdagli: Embedology, J. Stat. Phys. **65**(3), 579–616 (1991)

33.140 A. Wolf, J.B. Swift, H.L. Swinney, J.A. Vastano: Determining Lyapunov exponents from a time series, Phys. D **16**(3), 285–317 (1985)

33.141 L.D. Iasemidis, J.C. Sackellares, H.P. Zaveri, W.J. Williams: Phase space topography and the Lyapunov exponent of electrocorticograms in partial seizures, Brain Topogr. **2**(3), 187–201 (1990)

33.142 S. Micheloyannis, N. Flitzanis, E. Papanikolaou, M. Bourkas, D. Terzakis, S. Arvanitis, C.J. Stam: Usefulness of non-linear EEG analysis, Acta Neurol. Scand. **97**(1), 13–19 (2009)

33.143 W.J. Freeman: A field-theoretic approach to understanding scale-free neocortical dynamics, Biol. Cybern. **92**(6), 350–359 (2005)

33.144 W.J. Freeman, H. Erwin: Freeman k-set, Scholarpedia **3**(2), 3238 (2008)

33.145 R. Kozma, W. Freeman: Basic principles of the KIV model and its application to the navigation problem, Integr. Neurosci. **2**(1), 125–145 (2003)

33.146 R. Kozma, W.J. Freeman: The KIV model of intentional dynamics and decision making, Neural Netw. **22**(3), 277–285 (2009)

33.147 H.-J. Chang, W.J. Freeman, B.C. Burke: Biologically modeled noise stabilizing neurodynamics for pattern recognition, Int. J. Bifurc. Chaos **8**(2), 321–345 (1998)

33.148 R. Kozma, J.W. Freeman: Chaotic resonance – methods and applications for robust classification of noisy and variable patterns, Int. J. Bifurc. Chaos **11**(6), 1607–1629 (2001)

33.149 R.J. McEliece, E.C. Posner, E. Rodemich, S. Venkatesh: The capacity of the Hopfield associative memory, IEEE Trans. Inf. Theory **33**(4), 461–482 (1987)

33.150 J. Ma: The asymptotic memory capacity of the generalized Hopfield network, Neural Netw. **12**(9), 1207–1212 (1999)

33.151 V. Gripon, C. Berrou: Sparse neural networks with large learning diversity, IEEE Trans. Neural Netw. **22**(7), 1087–1096 (2011)

33.152 I. Beliaev, R. Kozma: Studies on the memory capacity and robustness of chaotic dynamic neural networks, Int. Jt. Conf. Neural Netw., IEEE (2006) pp. 3991–3998

33.153 D.A. Leopold, N.K. Logothetis: Multistable phenomena: Changing views in perception, Trends Cogn. Sci. **3**(7), 254–264 (1999)

33.154 E.D. Lumer, K.J. Friston, G. Rees: Neural correlates of perceptual rivalry in the human brain, Science **280**(5371), 1930–1934 (1998)

33.155 G. Werner: Metastability, criticality and phase transitions in brain and its models, Biosystems **90**(2), 496–508 (2007)

33.156 W.J. Freeman: Understanding perception through neural *codes*, IEEE Trans. Biomed. Eng. **58**(7), 1884–1890 (2011)

33.157 R. Kozma, J.J. Davis, W.J. Freeman: Synchronized minima in ECoG power at frequencies between beta-gamma oscillations disclose cortical singularities in cognition, J. Neurosci. Neuroeng. **1**(1), 13–23 (2012)

33.158 R. Kozma: Neuropercolation, Scholarpedia **2**(8), 1360 (2007)

33.159   E. Bullmore, O. Sporns: Complex brain networks: Graph theoretical analysis of structural and functional systems, Nat. Rev. Neurosci. **10**(3), 186–198 (2009)

33.160   R. Kozma, M. Puljic, P. Balister, B. Bollobas, W. Freeman: Neuropercolation: A random cellular automata approach to spatio-temporal neurodynamics, Lect. Notes Comput. Sci. **3305**, 435–443 (2004)

33.161   P. Balister, B. Bollobás, R. Kozma: Large deviations for mean field models of probabilistic cellular automata, Random Struct. Algorithm. **29**(3), 399–415 (2006)

33.162   R. Kozma, M. Puljic, L. Perlovsky: Modeling goal-oriented decision making through cognitive phase transitions, New Math. Nat. Comput. **5**(1), 143–157 (2009)

33.163   M. Puljic, R. Kozma: Broad-band oscillations by probabilistic cellular automata, J. Cell. Autom. **5**(6), 491–507 (2010)

33.164   S. Jo, T. Chang, I. Ebong, B. Bhadviya, P. Mazumder, W. Lu: Nanoscale memristor device as synapse in neuromorphic systems, Nano Lett. **10**, 1297–1301 (2010)

33.165   L. Smith: *Handbook of Nature-Inspired and Innovative Computing: Integrating Classical Models with Emerging Technologies* (Springer, New York 2006) pp. 433–475

33.166   G. Indiveri, E. Chicca, R. Douglas: A VLSI array of low-power spiking neurons and bistable synapses with spike-timing dependent plasticity, IEEE Trans. Neural Netw. **17**, 211–221 (2006)

33.167   Editors of Scientific American: *The Scientific American Book of the Brain* (Scientifc American, New York 1999)

33.168   L. Chua, L. Yang: Cellular neural networks, Theory. IEEE Trans. Circuits Syst. **CAS-35**, 1257–1272 (1988)

33.169   C. Zheng, H. Zhang, Z. Wang: Improved robust stability criteria for delayed cellular neural networks via the LMI approach, IEEE Trans. Circuits Syst. II – Expr. Briefs **57**, 41–45 (2010)

33.170   L. Chua: Memristor - The missing circuit element, IEEE Trans. Circuits Theory **CT-18**, 507–519 (1971)

33.171   D. Strukov, G. Snider, D. Stewart, R. Williams: The missing memristor found, Nature **453**, 80–83 (2008)

33.172   Q. Xia, W. Robinett, M. Cumbie, N. Banerjee, T. Cardinali, J. Yang, W. Wu, X. Li, W. Tong, D. Strukov, G. Snider, G. Medeiros-Ribeiro, R. Williams: Memristor - CMOS hybrid integrated circuits for reconfigurable logic, Nano Lett. **9**, 3640–3645 (2009)

33.173   X. Wang, Y. Chen, H. Xi, H. Li, D. Dimitrov: Spintronic memristor through spin-torque-induced magnetization motion, IEEE Electron Device Lett. **30**, 294–297 (2009)

33.174   Y. Joglekar, S. Wolf: The elusive memristor: Properties of basic electrical circuits, Eur. J. Phys. **30**, 661–675 (2009)

33.175   M. Pickett, D. Strukov, J. Borghetti, J. Yang, G. Snider, D. Stewart, R. Williams: Switching dynamics in titanium dioxide memristive devices, J. Appl. Phys. **106**(6), 074508 (2009)

33.176   S. Adhikari, C. Yang, H. Kim, L. Chua: Memristor bridge synapse-based neural network and its learning, IEEE Trans. Neural Netw. Learn. Syst. **23**(9), 1426–1435 (2012)

33.177   B. Linares-Barranco, T. Serrano-Gotarredona: Exploiting memristance in adaptive spiking neuromorphic nanotechnology systems, 9th IEEE Conf. Nanotechnol., Genoa (2009) pp. 601–604

33.178   M. Holler, S. Tam, H. Castro, R. Benson: An electrically trainable artificial neural network (ETANN) with 10240 *Floating gate* synapsess, Int. J. Conf. Neural Netw., Washington (1989) pp. 191–196

33.179   H. Withagen: Implementing backpropagation with analog hardware, Proc. IEEE ICNN-94, Orlando (1994) pp. 2015–2017

33.180   S. Lindsey, T. Lindblad: Survey of neural network hardware invited paper, Proc. SPIE Appl. Sci. Artif. Neural Netw. Conf., Orlando (1995) pp. 1194–1205

33.181   H. Kim, M. Pd Sah, C. Yang, T. Roska, L. Chua: Neural synaptic weighting with a pulse-based memristor circuit, IEEE Trans. Circuits Syst. I **59**(1), 148–158 (2012)

33.182   H. Kim, M. Pd Sah, C. Yang, T. Roska, L. Chua: Memristor bridge synapses, Proc. IEEE **100**(6), 2061–2070 (2012)

33.183   E. Lehtonen, M. Laiho: CNN using memristors for neighborhood connections, 12th Int. Workshop Cell. Nanoscale Netw. Appl. (CNNA), Berkeley (2010)

33.184   F. Merrikh-Bayat, F. Merrikh-Bayat, S. Shouraki: The neuro-fuzzy computing system with the capacity of implementation on a memristor crossbar and optimization-free hardware training, IEEE Trans. Fuzzy Syst. **22**(5), 1272–1287 (2014)

33.185   G. Snider: Spike-timing-dependent learning in memristive nanodevices, IEEE Int. Symp. Nanoscale Archit., Anaheim (2008) pp. 85–92

33.186   I. Ebong, D. Deshpande, Y. Yilmaz, P. Mazumder: Multi-purpose neuro-architecture with memristors, 11th IEEE Conf. Nanotechnol., Portland, Oregon (2011) pp. 1194–1205

33.187   H. Manem, J. Rajendran, G. Rose: Stochastic gradient descent inspired training technique for a CMOS/Nano memristive trainable threshold gateway, IEEE Trans. Circuits Syst. I **59**(5), 1051–1060 (2012)

33.188   G. Howard, E. Gale, L. Bull, B. Costello, A. Adamatzky: Evolution of plastic learning in spiking networks via memristive connections, IEEE Trans. Evol. Comput. **16**(5), 711–729 (2012)

33.189   S. Wen, Z. Zeng, T. Huang: Exponential stability analysis of memristor-based recurrent neural networks with time-varying delays, Neurocomputing **97**(15), 233–240 (2012)

33.190 A. Wu, Z. Zeng: Dynamics behaviors of memristor-based recurrent neural networks with time-varying delays, Neural Netw. **36**, 1–10 (2012)

33.191 A. Cichocki, R. Unbehauen: *Neural Networks for Optimization and Signal Processing* (Wiley, New York 1993)

33.192 J. Wang: Recurrent neural networks for optimization. In: *Fuzzy Logic and Neural Network Handbook*, ed. by C.H. Chen (McGraw-Hill, New York 1996), pp. 4.1–4.35

33.193 Y. Xia, J. Wang: Recurrent neural networks for optimization: The state of the art. In: *Recurrent Neural Networks: Design and Applications*, ed. by L.R. Medsker, L.C. Jain (CRC, Boca Raton 1999), 13–45

33.194 Q. Liu, J. Wang: Recurrent neural networks with discontinuous activation functions for convex optimization. In: *Integration of Swarm Intelligence and Artifical Neutral Network*, ed. by S. Dehuri, S. Ghosh, S.B. Cho (World Scientific, Singapore 2011), 95–119

33.195 I.B. Pyne: Linear programming on an electronic analogue computer, Trans. Am. Inst. Elect. Eng. **75**(2), 139–143 (1956)

33.196 L.O. Chua, G. Lin: Nonlinear programming without computation, IEEE Trans. Circuits Syst. **31**(2), 182–189 (1984)

33.197 G. Wilson: Quadratic programming analogs, IEEE Trans. Circuits Syst. **33**(9), 907–911 (1986)

33.198 J.J. Hopfield, D.W. Tank: Neural computation of decisions in optimization problems, Biol. Cybern. **52**(3), 141–152 (1985)

33.199 J.J. Hopfield, D.W. Tank: Computing with neural circuits – a model, Science **233**(4764), 625–633 (1986)

33.200 D.W. Tank, J.J. Hopfield: Simple *neural* optimization networks: an A/D converter, signal decision circuit, and a linear programming circuit, IEEE Trans. Circuits Syst. **33**(5), 533–541 (1986)

33.201 M.P. Kennedy, L.O. Chua: Neural networks for nonlinear programming, IEEE Trans. Circuits Syst. **35**(5), 554–562 (1988)

33.202 A. Rodriguez-Vazquez, R. Dominguez-Castro, A. Rueda, J.L. Huertas, E. Sanchez-Sinencio: Nonlinear switch-capacitor *neural* networks for optimization problems, IEEE Trans. Circuits Syst. **37**(3), 384–398 (1990)

33.203 S. Sudharsanan, M. Sundareshan: Exponential stability and a systematic synthesis of a neural network for quadratic minimization, Neural Netw. **4**, 599–613 (1991)

33.204 S. Zhang, A.G. Constantinides: Lagrange programming neural network, IEEE Trans. Circuits Syst. **39**(7), 441–452 (1992)

33.205 S. Zhang, X. Zhu, L. Zou: Second-order neural nets for constrained optimization, IEEE Trans. Neural Netw. **3**(6), 1021–1024 (1992)

33.206 A. Bouzerdoum, T.R. Pattison: Neural network for quadratic optimization with bound constraints, IEEE Trans. Neural Netw. **4**(2), 293–304 (1993)

33.207 M. Ohlsson, C. Peterson, B. Soderberg: Neural networks for optimization problems with inequality constraints: The knapsack problem, Neural Comput. **5**, 331–339 (1993)

33.208 J. Wang: Analysis and design of a recurrent neural network for linear programming, IEEE Trans. Circuits Syst. I **40**(9), 613–618 (1993)

33.209 W.E. Lillo, M.H. Loh, S. Hui, S.H. Zak: On solving constrained optimization problems with neural networks: A penalty method approach, IEEE Trans. Neural Netw. **4**(6), 931–940 (1993)

33.210 J. Wang: A deterministic annealing neural network for convex programming, Neural Netw. **7**(4), 629–641 (1994)

33.211 S.H. Zak, V. Upatising, S. Hui: Solving linear programming problems with neural networks: A comparative study, IEEE Trans. Neural Netw. **6**, 94–104 (1995)

33.212 Y. Xia, J. Wang: Neural network for solving linear programming problems with bounded variables, IEEE Trans. Neural Netw. **6**(2), 515–519 (1995)

33.213 M. Vidyasagar: Minimum-seeking properties of analog neural networks with multilinear objective functions, IEEE Trans. Autom. Control **40**(8), 1359–1375 (1995)

33.214 M. Forti, A. Tesi: New conditions for global stability of neural networks with application to linear and quadratic programming problems, IEEE Trans. Circuits Syst. I **42**(7), 354–366 (1995)

33.215 A. Cichocki, R. Unbehauen, K. Weinzierl, R. Holzel: A new neural network for solving linear programming problems, Eur. J. Oper. Res. **93**, 244–256 (1996)

33.216 Y. Xia: A new neural network for solving linear programming problems and its application, IEEE Trans. Neural Netw. **7**(2), 525–529 (1996)

33.217 X. Wu, Y. Xia, J. Li, W.K. Chen: A high-performance neural network for solving linear and quadratic programming problems, IEEE Trans. Neural Netw. **7**(3), 1996 (1996)

33.218 Y. Xia: A new neural network for solving linear and quadratic programming problems, IEEE Trans. Neural Netw. **7**(6), 1544–1547 (1996)

33.219 Y. Xia: Neural network for solving extended linear programming problems, IEEE Trans. Neural Netw. **8**(3), 803–806 (1997)

33.220 M.J. Perez-Ilzarbe: Convergence analysis of a discrete-time recurrent neural network to perform quadratic real optimization with bound constraints, IEEE Trans. Neural Netw. **9**(6), 1344–1351 (1998)

33.221 M.C.M. Teixeira, S.H. Zak: Analog neural nonderivative optimizers, IEEE Trans. Neural Netw. **9**(4), 629–638 (1998)

33.222   Y. Xia, J. Wang: A general methodology for designing globally convergent optimization neural networks, IEEE Trans. Neural Netw. **9**(6), 1331–1343 (1998)

33.223   E. Chong, S. Hui, H. Zak: An analysis of a class of neural networks for solving linear programming problems, IEEE Trans. Autom. Control **44**(11), 1995–2006 (1999)

33.224   Y. Xia, J. Wang: Global exponential stability of recurrent neural networks for solving optimization and related problems, IEEE Trans. Neural Netw. **11**(4), 1017–1022 (2000)

33.225   X. Liang, J. Wang: A recurrent neural network for nonlinear optimization with a continuously differentiable objective function and bound constraints, IEEE Trans. Neural Netw. **11**(6), 1251–1262 (2000)

33.226   Y. Leung, K. Chen, Y. Jiao, X. Gao, K. Leung: A new gradient-based neural network for solving linear and quadratic programming problems, IEEE Trans. Neural Netw. **12**(5), 1074–1083 (2001)

33.227   X. Liang: A recurrent neural network for nonlinear continuously differentiable optimization over a compact convex subset, IEEE Trans. Neural Netw. **12**(6), 1487–1490 (2001)

33.228   Y. Xia, H. Leung, J. Wang: A projection neural network and its application to constrained optimization problems, IEEE Trans. Circuits Syst. I **49**(4), 447–458 (2002)

33.229   R. Fantacci, M. Forti, M. Marini, D. Tarchi, G. Vannuccini: A neural network for constrained optimization with application to CDMA communication systems, IEEE Trans. Circuits Syst. II **50**(8), 484–487 (2003)

33.230   Y. Leung, K. Chen, X. Gao: A high-performance feedback neural network for solving convex nonlinear programming problems, IEEE Trans. Neural Netw. **14**(6), 1469–1477 (2003)

33.231   Y. Xia, J. Wang: A general projection neural network for solving optimization and related problems, IEEE Trans. Neural Netw. **15**, 318–328 (2004)

33.232   X. Gao: A novel neural network for nonlinear convex programming, IEEE Trans. Neural Netw. **15**(3), 613–621 (2004)

33.233   X. Gao, L. Liao, W. Xue: A neural network for a class of convex quadratic minimax problems with constraints, IEEE Trans. Neural Netw. **15**(3), 622–628 (2004)

33.234   Y. Xia, J. Wang: A recurrent neural network for nonlinear convex optimization subject to nonlinear inequality constraints, IEEE Trans. Circuits Syst. I **51**(7), 1385–1394 (2004)

33.235   M. Forti, P. Nistri, M. Quincampoix: Generalized neural network for nonsmooth nonlinear programming problems, IEEE Trans. Circuits Syst. I **51**(9), 1741–1754 (2004)

33.236   Y. Xia, G. Feng, J. Wang: A recurrent neural network with exponential convergence for solving convex quadratic program and linear piecewise equations, Neural Netw. **17**(7), 1003–1015 (2004)

33.237   Y. Xia, J. Wang: Recurrent neural networks for solving nonlinear convex programs with linear constraints, IEEE Trans. Neural Netw. **16**(2), 379–386 (2005)

33.238   Q. Liu, J. Cao, Y. Xia: A delayed neural network for solving linear projection equations and its applications, IEEE Trans. Neural Netw. **16**(4), 834–84 (2005)

33.239   X. Hu, J. Wang: Solving pseudomonotone variational inequalities and pseudoconvex optimization problems using the projection neural network, IEEE Trans. Neural Netw. **17**(6), 1487–149 (2006)

33.240   S. Liu, J. Wang: A simplified dual neural networ for quadratic programming with its KWTA application, IEEE Trans. Neural Netw. **17**(6), 1500–151 (2006)

33.241   Y. Yang, J. Cao: Solving quadratic programming problems by delayed projection neural network, IEEE Trans. Neural Netw. **17**(6), 1630–1634 (2006)

33.242   X. Hu, J. Wang: Design of general projection neural network for solving monotone linear variational inequalities and linear and quadratic optimization problems, IEEE Trans. Syst. Man Cybern. B **37**(5), 1414–1421 (2007)

33.243   Q. Liu, J. Wang: A one-layer recurrent neural network with a discontinuous hard-limiting activation function for quadratic programming, IEEE Trans. Neural Netw. **19**(4), 558–570 (2008)

33.244   Q. Liu, J. Wang: A one-layer recurrent neural network with a discontinuous activation function for linear programming, Neural Comput. **20**(5), 1366–1383 (2008)

33.245   Y. Xia, G. Feng, J. Wang: A novel neural networl for solving nonlinear optimization problems with inequality constraints, IEEE Trans. Neural Netw **19**(8), 1340–1353 (2008)

33.246   M.P. Barbarosou, N.G. Maratos: A nonfeasible gradient projection recurrent neural networl for equality-constrained optimization problems IEEE Trans. Neural Netw. **19**(10), 1665–1677 (2008)

33.247   X. Hu, J. Wang: An improved dual neural network for solving a class of quadratic programming problems and its k-winners-take-all application IEEE Trans. Neural Netw. **19**(12), 2022–2031 (2008)

33.248   X. Xue, W. Bian: Subgradient-based neural networks for nonsmooth convex optimization problems, IEEE Trans. Circuits Syst. I **55**(8), 2378–239 (2008)

33.249   W. Bian, X. Xue: Subgradient-based neural networks for nonsmooth nonconvex optimization problems, IEEE Trans. Neural Netw. **20**(6), 1024–1038 (2009)

33.250   X. Hu, C. Sun, B. Zhang: Design of recurrent neural networks for solving constrained least absolute deviation problems, IEEE Trans. Neural Netw. **21**(7), 1073–1086 (2010)

3.251 Q. Liu, J. Wang: Finite-time convergent recurrent neural network with a hard-limiting activation function for constrained optimization with piecewise-linear objective functions, IEEE Trans. Neural Netw. **22**(4), 601–613 (2011)

3.252 Q. Liu, J. Wang: A one-layer recurrent neural network for constrained nonsmooth optimization, IEEE Trans. Syst. Man Cybern. **40**(5), 1323–1333 (2011)

3.253 L. Cheng, Z. Hou, Y. Lin, M. Tan, W.C. Zhang, F. Wu: Recurrent neural network for nonsmooth convex optimization problems with applications to the identification of genetic regulatory networks, IEEE Trans. Neural Netw. **22**(5), 714–726 (2011)

3.254 Z. Guo, Q. Liu, J. Wang: A one-layer recurrent neural network for pseudoconvex optimization subject to linear equality constraints, IEEE Trans. Neural Netw. **22**(12), 1892–1900 (2011)

3.255 Q. Liu, Z. Guo, J. Wang: A one-layer recurrent neural network for constrained pseudoconvex optimization and its application for dynamic portfolio optimization, Neural Netw. **26**(1), 99–109 (2012)

3.256 W. Bian, X. Chen: Smoothing neural network for constrained non-Lipschitz optimization with applications, IEEE Trans. Neural Netw. Learn. Syst. **23**(3), 399–411 (2012)

3.257 Y. Xia: An extended projection neural network for constrained optimization, Neural Comput. **16**(4), 863–883 (2004)

3.258 J. Wang, Y. Xia: Analysis and design of primal-dual assignment networks, IEEE Trans. Neural Netw. **9**(1), 183–194 (1998)

3.259 Y. Xia, G. Feng, J. Wang: A primal-dual neural network for online resolving constrained kinematic redundancy in robot motion control, IEEE Trans. Syst. Man Cybern. B **35**(1), 54–64 (2005)

3.260 Y. Xia, J. Wang: A dual neural network for kinematic control of redundant robot manipulators, IEEE Trans. Syst. Man Cybern. B **31**(1), 147–154 (2001)

3.261 Y. Zhang, J. Wang: A dual neural network for constrained joint torque optimization of kinematically redundant manipulators, IEEE Trans. Syst. Man Cybern. B **32**(5), 654–662 (2002)

3.262 Y. Zhang, J. Wang, Y. Xu: A dual neural network for bi-criteria kinematic control redundant manipulators, IEEE Trans. Robot. Autom. **18**(6), 923–931 (2002)

3.263 Y. Zhang, J. Wang, Y. Xia: A dual neural network for redundancy resolution of kinematically redundant manipulators subject to joint limits and joint velocity limits, IEEE Trans. Neural Netw. **14**(3), 658–667 (2003)

3.264 A. Cichocki, R. Unbehauen: Neural networks for solving systems of linear equations and related problems, IEEE Trans. Circuits Syst. I **39**(2), 124–138 (1992)

3.265 A. Cichocki, R. Unbehauen: Neural networks for solving systems of linear equations – part II: Minimax and least absolute value problems, IEEE Trans. Circuits Syst. II **39**(9), 619–633 (1992)

33.266 J. Wang: Recurrent neural networks for computing pseudoinverse of rank-deficient matrices, SIAM J. Sci. Comput. **18**(5), 1479–1493 (1997)

33.267 G.G. Lendaris, K. Mathia, R. Saeks: Linear Hopfield networks and constrained optimization, IEEE Trans. Syst. Man Cybern. B **29**(1), 114–118 (1999)

33.268 Y. Xia, J. Wang, D.L. Hung: Recurrent neural networks for solving linear inequalities and equations, IEEE Trans. Circuits Syst. I **46**(4), 452–462 (1999)

33.269 J. Wang: A recurrent neural network for solving the shortest path problem, IEEE Trans. Circuits Syst. I **43**(6), 482–486 (1996)

33.270 J. Wang: Primal and dual neural networks for shortest-path routing, IEEE Trans. Syst. Man Cybern. A **28**(6), 864–869 (1998)

33.271 Y. Xia, J. Wang: A discrete-time recurrent neural network for shortest-path routing, IEEE Trans. Autom. Control **45**(11), 2129–2134 (2000)

33.272 D. Anguita, A. Boni: Improved neural network for SVM learning, IEEE Trans. Neural Netw. **13**(5), 1243–1244 (2002)

33.273 Y. Xia, J. Wang: A one-layer recurrent neural network for support vector machine learning, IEEE Trans. Syst. Man Cybern. B **34**(2), 1261–1269 (2004)

33.274 L.V. Ferreira, E. Kaszkurewicz, A. Bhaya: Support vector classifiers via gradient systems with discontinuous right-hand sides, Neural Netw. **19**(10), 1612–1623 (2006)

33.275 J. Wang: Analysis and design of an analog sorting network, IEEE Trans. Neural Netw. **6**, 962–971 (1995)

33.276 B. Apolloni, I. Zoppis: Subsymbolically managing pieces of symbolical functions for sorting, IEEE Trans. Neural Netw. **10**(5), 1099–1122 (1999)

33.277 J. Wang: Analysis and design of k-winners-take-all model with a single state variable and Heaviside step activation function, IEEE Trans. Neural Netw. **21**(9), 1496–1506 (2010)

33.278 Q. Liu, J. Wang: Two k-winners-take-all networks with discontinuous activation functions, Neural Netw. **21**, 406–413 (2008)

33.279 Y. Xia, M. S. Kamel: Cooperative learning algorithms for data fusion using novel L1 estimation, IEEE Trans. Signal Process. **56**(3), 1083–1095 (2008)

33.280 B. Baykal, A.G. Constantinides: A neural approach to the underdetermined-order recursive least-squares adaptive filtering, Neural Netw. **10**(8), 1523–1531 (1997)

33.281 Y. Sun: Hopfield neural network based algorithms for image restoration and reconstruction – Part I: Algorithms and simulations, IEEE Trans. Signal Process. **49**(7), 2105–2118 (2000)

33.282 X.Z. Wang, J.Y. Cheung, Y.S. Xia, J.D.Z. Chen: Minimum fuel neural networks and their applications to overcomplete signal representations, IEEE Trans. Circuits Syst. I **47**(8), 1146–1159 (2000)

33.283 X.Z. Wang, J.Y. Cheung, Y.S. Xia, J.D.Z. Chen: Neural implementation of unconstrained minimum

L1-norm optimization–least absolute deviation model and its application to time delay estimation, IEEE Trans. Circuits Syst. II **47**(11), 1214–1226 (2000)

33.284   P.-R. Chang, W.-H. Yang, K.-K. Chan: A neural network approach to MVDR beamforming problem, IEEE Trans. Antennas Propag. **40**(3), 313–322 (1992)

33.285   Y. Xia, G.G. Feng: A neural network for robust LCMP beamforming, Signal Process. **86**(3), 2901–2912 (2006)

33.286   J. Wang, G. Wu: A multilayer recurrent neural network for on-line synthesis of minimum-norm linear feedback control systems via pole assignment, Automatica **32**(3), 435–442 (1996)

33.287   Y. Zhang, J. Wang: Global exponential stability of recurrent neural networks for synthesizing linear feedback control systems via pole assignment, IEEE Trans. Neural Netw. **13**(3), 633–644 (2002)

33.288   Y. Zhang, J. Wang: Recurrent neural networks for nonlinear output regulation, Automatica **37**(8), 1161–1173 (2001)

33.289   S. Hu, J. Wang: Multilayer recurrent neural networks for online robust pole assignment, IEEE Trans. Circuits Syst. I **50**(11), 1488–1494 (2003)

33.290   Y. Pan, J. Wang: Model predictive control of unknown nonlinear dynamical systems based on recurrent neural networks, IEEE Trans. Ind. Electron. **59**(8), 3089–3101 (2012)

33.291   Z. Yan, J. Wang: Model predictive control of nonlinear systems with unmodeled dynamics based on feedforward and recurrent neural networks, IEEE Trans. Ind. Inf. **8**(4), 746–756 (2012)

33.292   Z. Yan, J. Wang: Model predictive control of tracking of underactuated vessels based on recurrent neural networks, IEEE J. Ocean. Eng. **37**(4), 717–72 (2012)

33.293   J. Wang, Q. Hu, D. Jiang: A Lagrangian networ for kinematic control of redundant robot manip ulators, IEEE Trans. Neural Netw. **10**(5), 1123–11: (1999)

33.294   H. Ding, S.K. Tso: A fully neural-network-base planning scheme for torque minimization of re dundant manipulators, IEEE Trans. Ind. Electror **46**(1), 199–206 (1999)

33.295   H. Ding, J. Wang: Recurrent neural networks fe minimum infinity-norm kinematic control of re dundant manipulators, IEEE Trans. Syst. Man Cy bern. A **29**(3), 269–276 (1999)

33.296   W.S. Tang, J. Wang: Two recurrent neural networl for local joint torque optimization of kinemat ically redundant manipulators, IEEE Trans. Sys Man Cybern. B **30**(1), 120–128 (2000)

33.297   W.S. Tang, J. Wang: A recurrent neural networ for minimum infinity-norm kinematic control e redundant manipulators with an improved prob lem formulation and reduced architectural com plexity, IEEE Trans. Syst. Man Cybern. B **31**(1), 98 105 (2001)

33.298   Y. Zhang, J. Wang: Obstacle avoidance for kine matically redundant manipulators using a du neural network, IEEE Trans. Syst. Man Cybern. **4**(1), 752–759 (2004)

33.299   Y. Xia, J. Wang, L.-M. Fok: Grasping force opti mization of multi-fingered robotic hands usin a recurrent neural network, IEEE Trans. Robot. Au tom. **20**(3), 549–554 (2004)

33.300   Q. Liu, C. Dang, T. Huang: A one-layer recurrer neural network for real-time portfolio optimiza tion with probability criterion, IEEE Trans. Cyberr **43**(1), 14–23 (2013)

34. Computational Neuroscience – Biophysical Modeling of Neural Systems

Marrison Stratton, Jennie Si

Only within the past few decades have we had the tools capable of probing the brain to search for the fundamental components of cognition. Modern numerical techniques coupled with the fabrication of precise electronics have allowed us to identify the very substrates of our own minds. The pioneering work of Hodgkin and Huxley provided us with the first biologically validated mathematical model describing the flow of ions across the membranes of giant squid axon. This model demonstrated the fundamental principles underlying how the electrochemical potential difference, maintained across the neuronal membrane, can serve as a medium for signal transmission. This early model has been expanded and improved to include elements not originally described through collaboration between biologists, computer scientists, physicists and mathematicians. Multi-disciplinary efforts are required to understand this system that spans multiple orders of magnitude and involves diverse cellular signaling cascades. The massive amount of data published concerning specific functionality within neural networks is currently one of the major challenges faced in neuroscience. The diverse and sometimes disparate data collected across many laboratories must be collated into the same framework before we can transition to a general theory explaining the brain. Since this broad field would typically be the subject of its own textbook, here we will focus on the fundamental physical relationships that can be used to understand biological processes in the brain.

34.1 Anatomy and Physiology of the Nervous System

The animal brain is undoubtedly a unique organ that has evolved from humble beginnings starting with small groups of specialized cells in organisms long ago. The vast complexity found within the cortices of the mammalian brain is the result of selection across countless generations. The purpose of these early cells is in principle consistent with the function of our entire brain. Both serve to assess environmental variables in order to produce output that is situationally relevant. This production of appropriate behavioral responses is essential for an organism to successfully obtain resources in complex and often hostile surroundings. The morphology of the brain is species dependent. Its structure is functionally correlated to the necessary output a specific animal requires to survive in a particular environment. For example, the commonly used laboratory mouse has a brain structure that is coarsely similar to a human. However, the mouse possesses particularly enlarged olfactory bulbs situated in the front of the skull. This is in great contrast to the human olfactory centers that are considerably smaller, but perform the same function. This dramatic difference in the size of the olfactory bulb relates to differences in environmental variables that exist between lab mice and humans. In contrast to humans, mice live in environments where scent is a highway of information. Odorant molecules can provide crucial signals regarding changes to the environment that indicate such things as the approach of a predator or the presence of food. In contrast, humans have evolved a diverse set of methods for gathering food and avoiding predators that are highly dependent upon visual stimulation. As a consequence of this developmental variable, our brains have evolved to efficiently process visual stimuli with incredible speed and acuity [34.1]. This paradigm of form fits function exists throughout nature and allows experimentalists to take advantage of shared anatomical and physiological characteristics. By adjusting for differences in evolutionary history, we can confidently perform experiments on neurons from other animals. This data can then be extended and translated to gather information about the properties of our own nervous system. From this point on, unless otherwise specified, any reference to the nervous system refers to that of higher mammal species including rodents and primates, and humans.

34.1.1 Introduction to the Anatomy of the Nervous System

Cells of the nervous system arise from ectodermal embryonic tissue and generally develop into two distinct groups: the cells of the peripheral nervous system (PNS) and those of the central nervous system (CNS).

The brain and the spinal cord together constitute the CNS, with the nerves of the body (peripheral nerves and autonomic ganglion forming the PNS. The PNS refers to neurons and sensory organs located outside the blood brain barrier (BBB) created by the meninges which is a three-layer, dynamic, protective system isolating the CNS from the circulatory system. Additionally, the primary immune system does not extend into the brain, leaving it particularly vulnerable. Exploration of the PNS has formed the foundation of neuroscience research because these neurons tend to be large and easy to locate and remove for experimental examination. While many of the foundational principles of neuroscience were discovered within the PNS, the CNS has been the primary target of most recent neuroscience research. This is primarily due to the emergence of the frontal cortices as the substrate for conscious thought and action. The CNS contains many sub-regions that can be broadly separated into the spinal cord, the brainstem, cerebellar cortex, and cerebral cortex. Increasing complexity can be observed moving from spinal cord to frontal cortex, which demonstrates that the most forward structures are the most recently evolved. The recently evolved frontal cortex structures are of high interest to neuroscience researchers and are the subject of many computational investigations attempting to elicit their function [34.2].

The human brain is divided into four major regions (Fig. 34.1): the cortex, which includes the four lobes of the brain, the midbrain, the brainstem, and the cerebellum. The brainstem, midbrain, and cerebellum, also known as subcortical regions were the first to evolve and play various roles in the regulation of basic physiological function and relay of information to the cortex. The brainstem continues caudally as the spinal cord and contains numerous nuclei for the processing of information generated by spinal neurons. The midbrain is of particular importance concerning the integration and transmission of information from the spinal cord and brainstem to the cortex. The thalamus, part of the midbrain, is often called the gateway to the cerebral cortex as it is located centrally and retains projections to all parts of the cortex, and thus plays a pivotal role in the transmission of subcortical information to various association areas of the cortex. The cortex is divided into four lobes including frontal, parietal, temporal, and occipital. Each lobe contains areas of specialized function as well as areas of association. Incoming sensory information requires

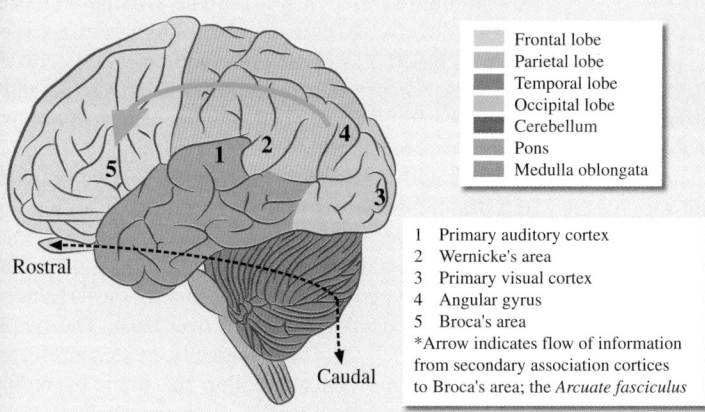

Fig. 34.1 A cartoon representation of the human brain with the major lobes colored according to the key *on the left*. The *blue arrow* highlights, as an example, a specific pathway that heavily contributes to the output of language by bringing together auditory and visual information in the parietal lobe. This information that contains associations between stimuli is then forwarded to the frontal cortex where it is integrated with the rest of the bodies' sensory information and will guide the output of speech sounds

Frontal lobe
Parietal lobe
Temporal lobe
Occipital lobe
Cerebellum
Pons
Medulla oblongata

1   Primary auditory cortex
2   Wernicke's area
3   Primary visual cortex
4   Angular gyrus
5   Broca's area
*Arrow indicates flow of information
from secondary association cortices
to Broca's area; the *Arcuate fasciculus*

Rostral

Caudal

context, hence information first travels to the areas of specialized function and subsequently traverses various association areas for integration before reaching its target of motor output [34.3]. An example of this flow of information within the cortex is depicted in Fig. 34.1, which details the progression of information from the primary auditory cortex for deciphering sound to Broca's area, which is involved in speech production. Figure 34.1 also demonstrates the general anatomy of the human brain, including the brainstem, cerebellum, and cortex.

Reviewed simply, the flow of information throughout this comprehensive system begins in sensory cells and ends in motor output via peripheral nerves. Sensory cells, such as mechanoreceptors for tactile sensation, transduce and forward signals to the spinal cord for first level processing. Sensory fibers run specifically along the dorsal surface of the spinal cord and ascend the entirety of the cord. Some synapse locally on the spinal cord and others project fully to the cortical surface of the brain. The information from sensory cells is first received in the proximal cortex, associations between stimuli are made, and this information is again relayed to the front of the brain where it is used to form a plan of action based on the specific input pattern. The front of the brain then begins a processing cascade that flows back toward the central sulcus of the brain and output motor neurons are triggered. These output signals descend the brain as a large bundle that go on to form the ventral surface of the spinal cord, where they either end on a local spinal neuron or are routed distally to a muscle. For a more complete discussion of neural anatomy and information processing, please refer to Kandel et al. [34.1].

34.1.2 Sensation – Environmental Signal Detection

Transduction of physical environmental stimuli into electrical and chemical signals is the common function shared by all sensory systems. This transduction provides baseline input to the brain from an array of sensors placed throughout the body in the skin, eyes, ears, mouth, and nose. These highly specialized cells respond only to the application of very precise external stimuli. While each receptor cell is specialized for a specific signal, there is much variation within each sensory system, allowing the most pertinent information to be extracted from the environment. An example of this variation exists within the eye, where there are two primary detector cell types: rods and cones [34.4]. The former is involved in the sensation of light and dark and the latter for the perception of colors. Even within the collection of cone cells there are further specializations that allow for the detection of various color wavelengths most specifically red, blue, and green. A similar pattern is observed in cells of the inner ear, where hair cells are housed in osseous cavities lined with membrane. This makes these cells specialized for the detection of vibrations in the air occurring at different frequencies. Sensory specialization is further mapped onto the brain where distinct regions of the cortex correspond to specific environmental stimuli or motor output patterns. Sensation is a vital part of cognition as the representation of the world we each possess is built upon our own unique sensory experiences. This literally means that we have shaped the surfaces of our cortices based on our experiences as individuals.

34.1.3 Associations – The Foundation of Cognition

The brain is an integrative structure capable of transforming sensory stimuli while forming associations between stimuli based on temporal and local parameters. Sensory integration allows organisms to relate pertinent information about environmental variables in real time based on successful behavioral patterns of the past. The mechanism driving these associative properties of the brain has been the subject of countless scientific endeavors yielding a basic understanding. Although progress has been made in understanding how the nervous system adapts to the environment, there are countless questions that emerge as new discoveries are made that require a constant revision of parameters. Kandel [34.5] was among the first to develop experiments capable of demonstrating the molecular and cellula r mechanisms underlying learning within a biological system.

Developing associations across a variety of sensory pathways is a major constituent of learning within animals. These associations are formed by altering cellular physiology based on input experience for a certain neuron or population of connected neurons. This cellular learning is the foundation for consciousness and there are many processes at the cellular level that contribute to learning in different ways. Some association mechanisms directly alter the number of synapses between neurons based on a history of communication between the cells, where other processes will affect the cell's DNA to accommodate a certain input pattern [34.6]. All of the mechanisms that influence learning in the brain have yet to be defined, which leaves a large opportunity for conjecture as to what constitutes learning and what does not. Modeling maintains a distinct advantage, as models of cellular communication and learning can be prototyped in silico to account for the large number of modifications occurring over time. The output of these models can then logically guide our search for learning mechanisms within the brain by outlining a possible path where learning mechanisms could be discovered. While we understand that association between neurons and glial cells likely form the basis of cognition, we have yet been able to recreate this phenomenon to completely explain its nature. This is partly due to the fact that the mammalian brain is so large and contains so many networks that detection and characterization of all changes occurring within the system simultaneously is extremely difficult. To resolve this issue, many neuroscientists have turned to using model organisms that provide a reduced set of neurons with which to experiment and demonstrate fundamental theories.

34.2 Cells and Signaling Among Cells

The brain is composed of two major cell types: neurons and glial cells. Both of these cell types are essential for the brain to function properly. Standard models place electrically excitable neurons in a signaling role with glial cells serving an indispensable support role although emerging evidence suggests glial cells could be more heavily involved with signaling than previously thought. The majority of research in computational neuroscience focuses on modeling neurons and their role in generating conscious behavior. This is largely due to the fact that their maintenance of a potential difference across the membrane serves as an efficient and robust communication pathway that can be modeled using established electrical dynamics. Unlike other cells of the body the neuron is a non-differentiating cell, which means it does not continuously undergo cellular mitosis or meiosis. This simple difference in the life cycle of these cells affords them an indispensable role in the animal nervous system – memory. Each cell in the body has an innate type of memory that begins by receiving messages from their environment at the membrane. These signals can sometimes propagate into the nucleus where alterations in DNA can occur, which will ultimately affect the function of the cell. Neurons have this type of DNA memory but also form connections with their neighboring cells or even cells that are located in other brain regions [34.7]. Because neurons do not divide these connections are not reset and can last for periods of time that are much longer than the life cycle of a typical cell. Connections between neurons are called synapses and form the fundamental communication element in the nervous system.

34.2.1 Neurons – Electrically Excitable Cells

Neurons are composed of a cell body referred to as the soma, which contains membrane-bound organelles that are found in most cells, in addition to one or more protoplasmic projections: an axon and dendrites. The soma contains the nucleus and is the central portion of the neuron. Among neurons, the most common pathway for signal transmission is from the axon terminal of one neuron to another's dendrites, which then relays the signal to the neuron's soma and on to the axon to be transmitted to another neuron. In particular, dendrites receive and conduct electrochemical signals from other neurons to the soma and play an integral role in determining the extent to which action potentials are generated. Dendrites are composed of many branches called dendritic trees and can create extensive and unique branched networks between neurons. The axon is the anatomical structure through which action potentials are transmitted away from the soma to other neurons or other types of cells such as muscles, ganglia, or glands [34.8]. Axons vary in length tremendously and bundle together to form large peripheral nerves that course from the spinal cord to the toe. They vary in composition depending upon their location in the PNS or the CNS and may be myelinated or unmyelinated. Myelin is a fatty, dielectric insulating layer that speeds signal conduction along the axon by forming discrete regions of low resistance and high conduction velocity. In between the discrete myelinated portions of axon, there are *nodes of Ranvier* to repeat the signal along the next segment of axon. Axons normally maintain an equal radius throughout their course and terminate at a synapse, where the electrochemical signal will be transmitted from the neuron to the target cell, which may be another neuron or another type of cell. A synapse is formed by the end(s) of one neuron's axon, called the axon terminal and the dendrites, axon, or soma of the receiving neuron. The synapse is fundamentally a transducer that converts electrical signals from a membrane potential wave to a chemical signal that modifies the state of a downstream cell [34.9].

Neurons are classified by the branching pattern and location of their dendrites and axons, their physiological function and location within the nervous system. Structural and functional classification, and the type of neurotransmitter released are relevant to modeling and will be reviewed. Structurally, neurons are classified as unipolar, pseudo-unipolar, bipolar, or multi-polar. Unipolar neurons contain one protoplasmic projection that divides distally into sensory and transmitting portions of an axon. Bipolar neurons retain two projections from the soma, one from which dendrites extend and the other from which the axon extends. The majority of neurons are multipolar neurons, which normally contain one long axon and many dendritic projections.

Figure 34.2a exemplifies bipolar, hippocampal neurons and Fig. 34.2b demonstrates the anatomy of a cortical, glutamatergic neuron. Functionally, neurons can be classified according to their electrophysiology and be described as tonic, phasic, or fast-spiking. However, it is more effective to describe the different firing patterns experienced by neurons, as most neurons exhibit variable firing patterns. Tonic firing involves continuous responses to stimuli and recurrent generation of

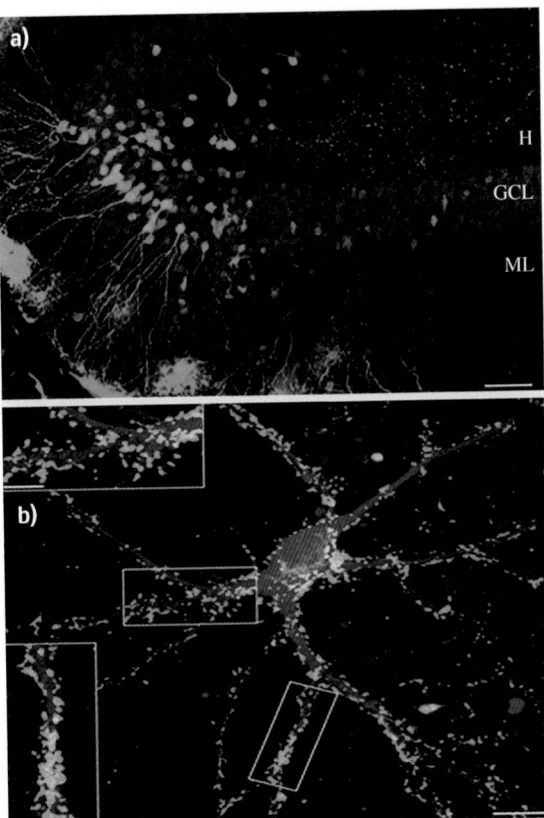

Fig. 34.2 (a) Fluorescently labeled hippocampal slice showing neuronal cell bodies stained in *blue* with neuronal nuclei antibody (NeuN). The GFP (green fluorescent protein) and RFP (red fluorescent protein) labeled neurons extending their processes are of dentate granule cell origin and have matured over the course of the experiment as this slice was imaged at postnatal day 53. **(b)** Here we see a neuron that has had its cell body stained with a *red* fluorescent marker and presynaptic glutamate receptors indicated in *green*. The sheer volume of presynaptic targets can be seen here along with two segments that have been selected and magnified to give a better view of how synapses cover the dendrite surface

action potentials. Tonic firing patterns are observed in large excitatory neurons during basal levels of activity to provide constant communication between elements of the network. Phasic firing patterns consist of bursts of action potentials, often in quick succession that has dramatic downstream effects. Bursting and phasic firing are highly studied phenomena that often signal a shift in steady state activity levels in the circuits where they are observed. Along with the branching patterns of dendrites and axons, neurons may be classified by the type of neurotransmitter released at synapses. The two dominant neurotransmitters in the brain are glutamate and gamma-aminobutyric acid (GABA), which generally mediate excitatory and inhibitory neurotransmission respectively [34.10].

34.2.2 Glial Cells – Supporting Neural Networks

Glial cells are fundamentally different from neurons in that they do not form synapses with other cells and generally do not maintain a membrane potential. Although they are not directly involved in signaling between neurons through synaptic means, these cells play a large role in the maintenance of synapses as well as signal integrity. There are four major types of glial cells in the brain: oligodendrocytes, astrocytes, ependymal cells, and micro-glial cells [34.11]. Oligodendrocytes secrete myelin, the insulating dielectric material covering axons, which facilitates signal transmission over longer distances. Ependymal cells line the ventricles of the brain and secrete the cerebral spinal fluid that bathes the brain and provides a route for expulsion of waste and the intake of nutrients. Micro-glial cells act as a type of immune system for the brain by digesting dead cells and collecting material that should not be present or could be damaging to cells of the brain. Astrocytes are abundant, star-shaped cells that generally surround neurons and provide nutrition and oxygen, and remove waste that could be toxic to a neuron if left to accumulate via cerebral spinal fluid. This astrocyte driven waste-removal system is essential for normal physiological function and structurally resembles the lymphatic system found in the rest of the body's tissues. Remarkably, new evidence suggests astrocytes may actually participate in synaptic communication, and that communication may be bi-directional [34.12]. Communication between astrocytes and neurons has yet to be fully characterized and provides an opportune target for new venture into neural network modeling. This new evidence of implicating astrocytes will shift our understanding of the brain as a communications structure and will open new questions that can be addressed using computational models. The current models of non-linear neural systems are already complex and must be altered to account for the evidence present in this new paradigm.

34.2.3 Transduction Proteins – Cellular Signaling Mechanisms

All cells of the body contain functional proteins embedded in their lipid bi-layer that are responsible for transducing environmental signals across the cellular membrane. The incredible variety of proteins present in the nervous system serves as mediators of cellular communication. Each protein will have its own unique structure depending on its function, and there are large groups of proteins that all share common properties such as the g-protein coupled receptors (GPCRs), ligand gated ion channels, passive ion channels, and a plethora of others. All of these will not be defined here as the classification and physiology of membrane proteins can be considered the subject of a whole field and are reviewed in detail by *Grillner* in [34.12]. The most important protein varieties for our consideration are those that control the flow of ions across the lipid bi-layer, which can be achieved in a number of ways. Some directly pass ions through a small pore in their center that is selective to certain ion species based on their electron distribution. Others use stored chemical energy to transfer a subset of ions outside of the cell while bringing others into the cell allowing a gradient to be established by using stored cellular energy to push ions against their potential energy gradient.

34.2.4 Electrochemical Potential Difference – Signaling Medium

Neurons carry information to their targets in the form of fluctuations of their membrane potential. Changes in the value of the membrane potential can trigger a variety of cellular signals including the opening of voltage gated proteins or the release of chemical neurotransmitters at a synapse [34.13]. As mentioned earlier, neurons generate their membrane potential by selectively transporting ions across the cellular membrane using energy, normally in the form of adenosine triphosphate (ATP). With the right combination of proteins neurons are able to shift their membrane potential in response to communication from other cells in a discrete fashion. This discrete wave along the membrane is known as an ac-ion potential and is only generated by a neuron once a certain level of activation has been attained. A single neuron in a network must receive communication from other cells, normally at its dendrites, which will cause the accumulation of positive ions inside the cell. Once the positive ions have accumulated to a certain critical level the cell will generate an action potential that flows down the axon to the cells targets. This action potential is the fundamental signaling unit within the nervous system and triggers the release of neurotransmitters when it arrives at a synapse. Once an action potential has been generated there is a period where the cell cannot create another wave, and this time is known as the absolute refractory period.

The potential energy across the membrane is a complex value that results from the transport of ions against their concentration gradient (Fig. 34.3). If these ions were left to freely diffuse the membrane potential would eventually deteriorate as each ion moved toward its own value of equilibrium potential. This equilibrium

potential, also known as the reversal potential or the Nernst potential, can be calculated using the following relationship for any ionic species x

$$E_x = \frac{RT}{zF} \ln \frac{[X]_o}{[X]_i} . \quad (34.1)$$

In the above equation, known as the Nernst equation, E_x represents the equilibrium potential for a certain ion species with $[X]_o$ and $[X]_i$ representing the external and internal concentrations of the ion, respectively [34.14]. Additionally, R is the Rydberg constant, T is the temperature in kelvin, z is the atomic number of the ion, and F is the Faraday constant followed by the natural logarithm of the concentration difference. At the reversal potential for an ionic species there is a net force of zero on ions in the systems and these particles will be at rest. From this relationship we can see that the energy driving the fluctuations in membrane potential originates from both electrical and chemical sources.

While the Nernst equation is capable of determining the reversal potential for a single ionic species, physiological systems often have many additional ions that participate in cellular signaling. The Nernst equation was expanded to form the Goldman–Hodgkin–Katz equation that yields the membrane potential in a resting system composed of multiple ionic species as shown below

$$u = \frac{RT}{F} \ln \frac{P_K \left[K^+\right]_o + P_{Na} \left[Na^+\right]_o + P_{Cl} \left[Cl^-\right]_i}{P_K \left[K^+\right]_i + P_{Na} \left[Na^+\right]_i + P_{Cl} \left[Cl^-\right]_o} . \quad (34.2)$$

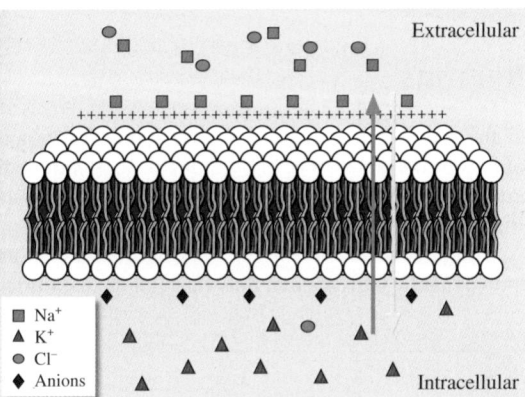

Fig. 34.3 The lipid bi-layer is composed of phosphate groups on the extra and intracellular faces of the membrane with a variety of lipids attached that essentially will self-assemble when dissolved in water at the appropriate concentration. The extra cellular surface has a net positive charge relative to the cytoplasmic side that is relatively negative. The membrane acts as a semi-permeable barrier that prevents the passage of molecules based on their attraction to water. Here we see that potassium ions have accumulated within the cell along with positive anions and sodium, and chloride can be found outside the cell due to the action of specific transporters embedded in the membrane. The asymmetric distribution of charge across the membrane causes a potential difference to exist in terms of both the electrical potential of the ions and their chemical nature to diffuse down their concentration gradient

Table 34.1 Here we see the concentration distribution of different salt species as they are found within a voltage clamp experiment using a segment of giant squid axon. These values will be different in each experimental preparation and should be considered carefully as small variations can have large implications for the firing patterns observed. Take note that here there are no positive anions present as the center of the giant squid axon is devoid of organelles unlike the inside of a mammalian neuron of the central nervous system that would likely contain many

Ion	Cytoplasmic concentration	Extracellular concentration	Equilibrium potential
K^+	400	20	-75
Na^+	50	440	$+55$
Cl^-	52	560	-60
Organic anions ($^-$)	385	None	None

Here u is the resting membrane potential of the cellular system under consideration and the constants are the same as shown above in the Nernst equation [34.15]. However, instead of relying solely upon the concentration of the particular ion inside and outside the cell here we can see that the number of ion channels, represented by the variable P, is also taken into account. Here P is the membrane's permeability to that particular ionic species in the unit cm/s. This equation is used to calculate the membrane potential of a particular physiological system with multiple internal and external salt species such as potassium, sodium, and chloride. Notice that the relationship between membrane potential and chloride concentration is the inverse due to its negative valence.

The simplest description for the maintenance of the membrane potential derives from the asymmetric distribution of ions across the cell's semi-permeable membrane. The concentrations displayed above were the first calculated for any neuron and were measured

from the axon of a giant squid. By carefully measuring the potential difference between the inside of the axon and the external solution Hodgkin and Huxley were able to determine the amount of current flowing across the axon's membrane under varying conditions [34.16]. The potential difference between the inside and outside of the cell is generated by manipulating the concentration of each ion with respect to its charge. Through careful observation of Table 34. above one can note that this resting condition will lead to a state where the cell is relatively negative on the inner membrane and positive along the outer wall of the membrane. It is this relative potential that varies along the surface of the neural membrane and it is what is responsible for carrying information along the length of the cell. These concentrations are considered bulk values and hardly ever deviate from these concentrations unless a period of sustained firing has occurred, where the cell can deplete this potential difference.

34.3 Modeling Biophysically Realistic Neurons

34.3.1 Electrical Properties of Neurons

When modeling a system one must first consider its physical dimensions and elements in order to construct a model that is true to reality and mathematically sound. By investigating the most fundamental structure of a neuron it is simple to see how this system can be easily related to that of an electronic capacitor. We have a system composed of two electrically conductive mediums, the extracellular fluid and the cytoplasm, which are separated by a dielectric layer that is also the phospholipid bi-layer. Therefore, from this description of a neuron we can generate the following relationship for the membrane potential

$$u = \frac{Q}{C}, \qquad (34.3)$$

where u represents the potential across the membrane and it is equal to the quotient of the charge along the membrane surface Q and the capacitance of the membrane itself C [34.17]. From this relationship it is clear that the membrane potential relies on the species and number of charges distributed across the membrane surface as well as the lipid constituents of the bi-layer. The capacitance of the membrane in a neuron is generally around $1\,\mu\mathrm{F/cm}^2$ but this can fluctuate depending

on the local ionic and lipid composition. Four major ionic currents are most often found in the cellular membrane and considered in a biologically realistic model (Fig. 34.4).

Similarly to electrical capacitors, neurons are reliant upon dynamic currents that flow through the membrane

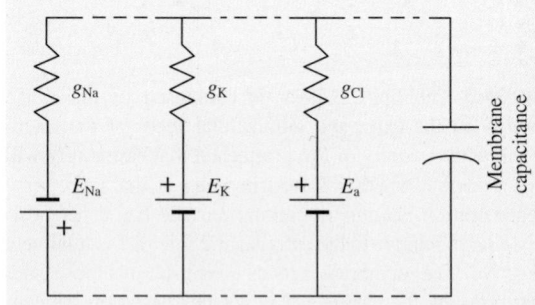

Fig. 34.4 A schematic equivalent circuit diagram of the four major ionic currents most often found in the cellular membrane. The resistors represent the varying conductivity of membrane channels for each ion and the batteries are each ion's respective concentration gradient. On the right we can see the membrane has an inherent capacitive current that acts to slow the spread of membrane currents and is manifested by the physical structure of the cell

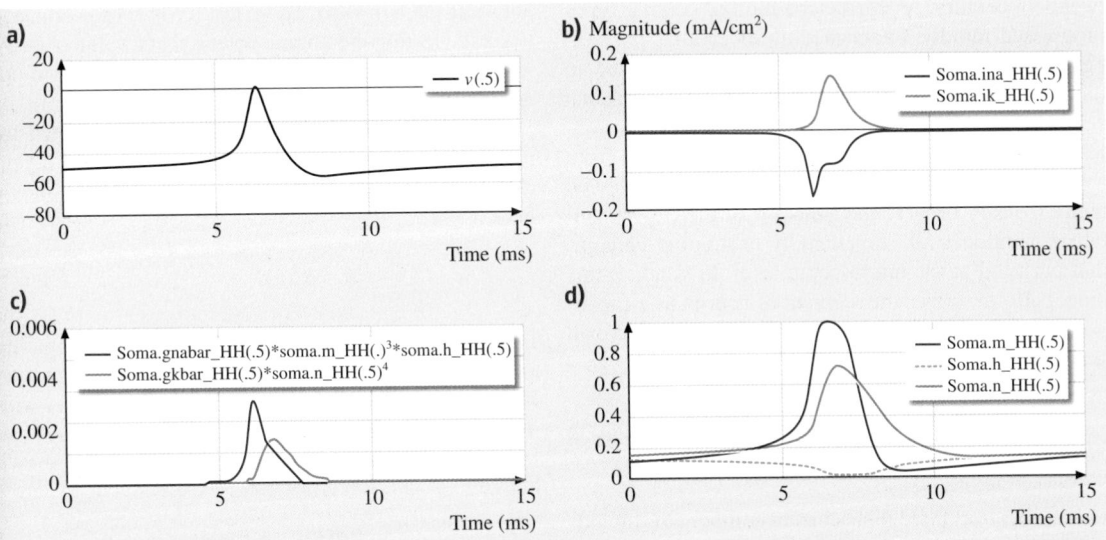

Fig. 34.5 (a) The membrane potential measured at the middle of a spherical single compartment model cell denoted by $v(.5)$. The cell has Hodgkin–Huxley type current dynamics. The y axis shows membrane potential at the center of the cell, $v(.5)$, and the x axis represents time in ms. **(b)** The *blue trace* shows the total magnitude, in mA/cm², of the outward potassium current within the Hodgkin–Huxley (HH) model. The *trace red* shows the total magnitude of the inward HH sodium current in the same units of mA/cm². **(c)** The *red trace* is the variable $g_{Na}+m^3h$ as shown in (34.5). The *blue trace* shows the potassium gating constant g_K+n^4 also from (34.5). **(d)** The state of each of the gating variables m, n, and h where the value 1 represents fully open and 0 represents fully closed or inactivated channels. This simulation was conducted within the NEURON simulation environment with a single compartment of area $29\,000\,\mu^2$, initialized at $-50\,\text{mV}$, with physiological concentrations of calcium, chloride, potassium, and sodium ions

into and out of the cell. One of the most important ions is potassium (K⁺), which conducts current based on the following relationship, $i_K = (\gamma_K \times u) - (\gamma_K \times E_K) = \gamma_K \times (u - E_K)$, where γ is the ionic conductance, u is the membrane potential, and E is the ionic reversal potential. The final term on the right-hand side of the equation is known as the electromotive force and can be calculated independently for each ionic current. This relationship shows how a neuron with both membrane potential and a potassium concentration gradient produces net potassium current. This potassium current can be generalized across the entire surface using this relationship, $g_K = N_K \times \gamma_K$, where g is the ionic conductance, N is the number of channels open at rest, and γ (gamma) is the permeability of an individual potassium channel. Determination of individualized channel conductance is performed in a lab setting using the patch clamp technique to isolate single ion channels. Once isolated, these channels can be tested using pharmacological techniques to determine their single channel conductivity [34.18]. These experiments are very sensitive and must be performed for each channel of interest

within the model as they cannot be represented without accurate biological data.

Describing a neuron as a capacitor yields interesting properties that we can infer about neurons from the large established set of knowledge regarding capacitance. The innate capacitive nature of the membrane actually affects the passage of current and, therefore, is relevant to our discussion here. When current is injected into our capacitive system it is inherently slowed based upon the time course of the current injection and the capacitance of the system represented by this relation $\Delta u = \frac{I_C \times \Delta t}{C}$. Therefore, the magnitude of the change in potential across the capacitor is relative to the duration of the current, presenting a natural latency in signal transmission that must be accounted for when modeling.

34.3.2 Early Empirical Models – Hodgkin and Huxley

Investigation into the structure of the brain began with improvements to the light microscope in the early

twentieth century. As characterization of cellular types progressed rapidly, understanding of cellular physiology lagged quite far behind. This is largely due to the fact that the technology to manipulate individual cells had yet to be invented. To go around this problem Hodgkin and Huxley used an axon harvested from a giant squid as their experimental system. This allowed them to easily observe the behavior of the experimental preparation while functionally examining changes that occurred at the microscopic level. In general, this model also describes the segment of neuron as a capacitor with x ionic currents and applied current as shown here

$$C\frac{du}{dt} = -\sum_x I_x(t) + I(t) . \tag{34.4}$$

This model uses three ionic current components including potassium, sodium, and a non-specific leak current that are described below

$$\sum_x I_x = g_{Na+} m^3 h(u - E_{Na+})$$
$$+ g_{K+} n^4 (u - E_{K+}) + g_L (u - E_L) . \tag{34.5}$$

Each of the terms above represents a specific ion with a gating variable that has been experimentally fitted to display the characteristics of each specific channel. A simulation of these parameters in a model cell expressing the standard Hodgkin–Huxley type channels is displayed in Fig. 34.5. In this case, the sodium current is determined using two different gating variables because it has two distinct phases, activation and inactivation. During the inactivation phase the conduction pore of the channel is blocked by a string of intracellular positive amino acids that literally plug the channel closed. This is distinct from potassium or leak channels that do not have an inactivation mechanism built into the protein. The differential form of each gating variable for the Hodgkin–Huxley model is shown below (Tables 34.2 and 34.3) in terms of the experimentally determined parameters α and β

$$m' = \alpha_m(u)(1-m) - \beta_m(u) ,$$
$$n' = \alpha_n(u)(1-n) - \beta_n(u)n ,$$
$$h' = \alpha_h(u)(1-h) - \beta_h(u) . \tag{34.6}$$

These values were obtained during initial laboratory experimentation by Hodgkin and Huxley using the gi-

ant squid axon in vitro. These results have been adjusted to set the resting membrane potential at a value of $0\,mV$ instead of the typical $-65\,mV$ as is seen in many modern interpretations.

This model has been expanded and interpreted for many systems outside the giant squid axon, and a general form for determining gating variables is shown here

$$\theta' = -\frac{1}{\tau_\theta(u)}[\theta - \theta_0(u)] . \tag{34.7}$$

In this differential form, Θ represents a particular gating variable of interest. When the membrane voltage u, is fixed at a certain value then Θ approaches $\theta_0(u)$ with a time constant represented by $\tau_\theta(u)$. These values can be calculated using the transformation equations shown below

$$\theta_0(u) = \frac{\alpha_\theta(u)}{[\alpha_\theta(u) + \beta_\theta(u)]} ,$$
$$\tau_\theta(u) = \frac{1}{[\alpha_\theta(u) + \beta_\theta(u)]} . \tag{34.8}$$

Table 34.2 Parameters associated with dynamics of ions within biological membranes, specifically of the giant squid. The first column shows the gating variables associated with each ionic species. The center column is the equilibrium potential that can be found for each ion within the system and is shown in millivolts. On the right the overall conductance of each ionic species within the model of action potential generation and membrane potential maintenance is shown

Ion species – gating variables (θ)	Reversal potential E_θ (mV)	Conductance g_θ (mS/cm^2)
Na – ($\theta = m, h$)	115	120
K – ($\theta = n$)	−12	36
L – no gating	10.6	0.3

Table 34.3 Gating variables with specific relationships that describe the particular function of each of the gating variables

Gating variable (θ)	α_θ (u/mV)	β_θ (u/mV)
n	$\dfrac{(0.1 - 0.01u)}{\left[e^{(1-0.1u)} - 1\right]}$	$0.125e^{\frac{-u}{80}}$
m	$\dfrac{(2.5 - 0.1u)}{\left[e^{(2.5-0.1u)} - 1\right]}$	$4e^{\frac{-u}{18}}$
h	$0.07e^{\frac{-u}{20}}$	$\dfrac{1}{\left[e^{(3-0.1u)} + 1\right]}$

34.3.3 Compartmental Modeling – Anatomical Reduction

Consider a simple case of a neuron with a spherical cell body. The time course of the potential change across the membrane can be described using the following

$$\Delta u(t) = I_m R_m \left(1 - e^{\frac{-t}{\tau}}\right). \tag{34.9}$$

In the above relationship I_m and R_m are the membrane current and resistance, respectively. The rightmost term contains e to the power of time divided by the membrane time constant τ.

Spatially, the current in our spherical cell body will decay along the length of the membrane according to the relationship below

$$\Delta u(x) = \Delta u_0 e^{-x/\omega}. \tag{34.10}$$

Above we can see that the spatial decay of potential relies upon the potential where the current was initially injected Δu_0, x is the distance from the site of the current injection, and ω is the membrane length constant. This constant is defined as, $\omega = \sqrt{r_m/r_a}$, where r_m is the membrane resistance and r_a is the axial resistance along the length of the compartment. The length constant is also dependent upon the radius of the segment with large axons conducting current more easily than smaller axons. In addition, each cell has individualized values of membrane and axial resistance that are based on the particular distribution of protein and cellular organelles in each experimental scenario.

34.3.4 Cable Equations – Connecting Compartments

One of the most convenient ways of modeling a neuron is by simplifying its neural structure. This can be done by approximating the shape of parts of the cell surface as cylinders (Fig. 34.6), since this is very close to the real shape of a neuronal process. By starting with a simple cylinder, we can consider this as the fundamental unit for computation where quantities regulating the system will be derived and repeated along the length of the model neuron [34.19]. The one-dimensional cable equation can effectively describe the propagation of current along a length of cylinder that does not branch as shown below

$$\frac{\partial V}{\partial T} + F(V) = \frac{\partial^2 V}{\partial X^2}. \tag{34.11}$$

In the above, V and F are both independent functions of time and space represented by T and X, respectively. Since most neurons are branched we can consider a neuron to be composed of many one-dimensional elements that can be arranged to form the branching structure. The boundary conditions of each cylinder are used to calculate specific values within each region of membrane. To derive this relationship one must first consider conservation of charge as shown here,

$$\sum i_a - \int_A i_m dA = 0,$$

where the leftmost term is the sum of axial currents entering the section and i_m represents the integrated transmembrane currents over the entire segment area. Expanding this description to include an electrode current source s we obtain the following relationship,

$$\sum i_a - \int_A i_m dA + \int_A i_s dA = 0.$$

The electrode current can be simplified to a point source of current due to the fact that most electrode current sources are much smaller than the cell itself.

To simplify our model we can split the neuron up into j compartments of size m and area A. All the properties of the compartment can be represented as the average at the middle of the compartment shown here, $i_{mj}A_j = \sum_k i_{akj}$. The current between compartments can be defined using Ohm's law where the voltage drop between compartments is divided by the axial resistance between compartmental centers $i_{akj} = (v_k - v_j)/r_{jk}$. Therefore, our compartment can then be described using the following relationship, $i_{mj}A_j = \sum_k (v_k - v_j)/r_{jk}$. The total membrane current can then be found using this expression,

$$i_{mj}A_j = c_j \frac{dv_j}{dt} + i_{ion}(v_j, t),$$

where c_j is the compartmental capacitance and $i_{ion}(v_j, t)$ is a function that captures the varying values of ion channel conductance in the membrane.

A set of branched cables can be constructed from individual segments to yield a set of differential equations that follow the form shown in (34.12) below

$$c_j frac dv_j dt + i_{ion}(v_j, t) = \sum_k \frac{(v_k - v_j)}{r_{jk}}. \tag{34.12}$$

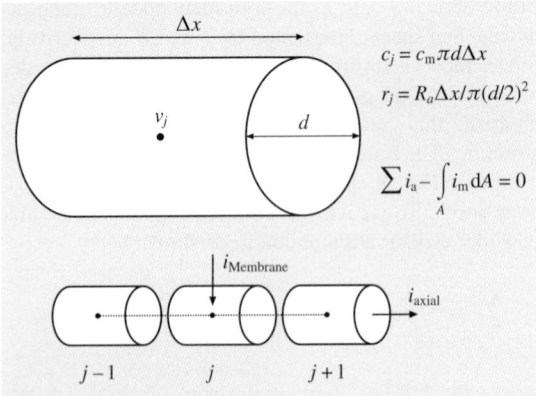

Fig. 34.6 Generalized diagram of a single compartment representing a segment of a neuron in a biophysical model using cable equations. Below, three compartments are shown to indicate axial and membrane current components that are represented by their summated valued at the center of each compartment [34.6]

In order for the equations above to be completely valid we must assume that the axial current flow between various compartments can be closely approximated by calculating the value of the current at the center of each compartment. This implies that the current can vary linearly across compartments with the compartment size chosen specifically to account for any spatial variations that may exist within the experimental system,

$$c_j \frac{dv_j}{dt} + i_{\text{ion}}(v_j, t) = \frac{v_{j-1} - v_j}{r_{j-1,k}} + \frac{v_{j+1} - v_j}{r_{j+1,k}}. \quad (34.13)$$

The above is a special case (34.14), which was discussed [34.5], where specific attention is paid to the axial current in adjacent compartments when a uniformly

distributed current passes through an initial neuronal segment with constant diameter [34.20]. Compartment have length Δx and diameter d. The capacitance can be written as $C_m \pi d \Delta x$ and the axial resistance of each compartment is defined as $R_a \Delta x / \pi \left(\frac{d}{2}\right)^2$, where C_m is the specific membrane capacitance and R_a is the specific cytoplasmic resistivity. Manipulation of (34.13) using the above consideration then yields the following (34.14)

$$C_m \frac{dv_j}{dt} + i_j(v_j, t) = \left(\frac{d}{4R_a}\right) \left(\frac{v_{j+1} - 2v_j + v_{j-1}}{\Delta x^2}\right). \quad (34.14)$$

Here the total ionic current specified above has been replaced with $i_j(v_j, t)$, a term that expands our consideration of injected ions by using a current density function. Now, if we consider the case where compartment size becomes infinitely small and the right-hand side then reduces to the second partial derivative of membrane potential with respect to the distance from the compartment of interest j. This reduction yields (34.15) below

$$C_m \frac{dv}{dt} + i(v, t) = \left(\frac{d}{4R_a}\right) \left(\frac{\partial^2 V}{\partial x^2}\right). \quad (34.15)$$

After multiplying both sides by the membrane resistance and with a simple application of Ohm's law we can see that $i R_m = v$ and, therefore, we find

$$C_m R_m \frac{dv}{dt} + v = \left(\frac{dR_m}{4R_a}\right) \left(\frac{\partial^2 v}{\partial x^2}\right). \quad (34.16)$$

This relationship can be scaled using the constants for time $\tau_m = R_m C_m$ and space $\omega = \sqrt{r_m/r_a}$, respectively.

34.4 Reducing Computational Complexity for Large Network Simulations

34.4.1 Reducing Computational Complexity – Large Scale Models

The Hodgkin–Huxley model [34.21] sets the foundation for mathematically modeling detailed temporal dynamics of how action potentials in neurons are initiated and propagated. The set of nonlinear ordinary differential equations of the form (34.4)–(34.6) were developed to describe the electrical characteristics of the squid giant axon. Given the many different mem-

brane currents that may be involved in the firing of an action potential, the Hodgkin–Huxley model represents the simplest possible representation of neuronal dynamics yet realistically captures the biophysical relationship between the voltage and time dependence of cell membrane currents. Even so, it is a daunting task to study a large neural network based on interconnected neurons each of which is modeled by Hodgkin–Huxley equations. The effort made in [34.22, 23] is a good illustration of the inherent challenges. Even on a single

euron level, recent studies have shown that it would
be easier to tune parameters in a less biophysically
realistic model under the general scope of *integrate-
and-fire*, or threshold models, which approximate the
pulse-like electrical activity as a threshold process. Or
in other words, such models are less sensitive in their
model parameters and thus provide more robust and accurate model fitting results given profiles of injected
current waveforms [34.24–26]. These threshold models
are easy to work with but they are phenomenological approaches to modeling true neural behavior. It
would not be possible to use these models to study
membrane voltage profile over a precise time course,
and it would be impossible to assess environmental
parameters such as temperature change, chemical environment change, pharmacological manipulations of
the ion channels and their impact on the membrane
dynamics.

Given the many challenges of mathematical modeling of realistic neurons and neuronal networks under
fine spatial and temporal resolution, great efforts have
been made on several fronts to advance the study
of neural network dynamic behaviors. Common to
all approaches, the role of time in neuronal activities
is emphasized and, thus, the models are usually described by nonlinear ordinary differential equations.
In the following, we will examine some of these dynamic models that are built on different premises and
considerations.

34.4.2 Firing Rate-Based Dynamic Models

The neural firing rate-based encoding scheme assumes
that information about environmental stimulus is contained in the firing rates of the neurons. Thus, the
specific spike times are under-represented. Sufficient
evidence points out that in most sensory systems, the
firing rate increases, generally non-linearly, with increasing stimulus intensity, and measurement of firing
rates has become a standard tool for describing the
properties of all types of sensory or cortical neurons,
partly due to the relative ease of measuring rates experimentally. However, this approach neglects all the
information possibly contained in the exact timing of
the spikes [34.1]. Maybe it is inefficient, but the rate
coding is robust and easy to measure and thus has been
used as a standard or basic tool for studying sensory or
cortical neuron characteristics in association with external stimuli or behaviors. The class of firing rate-based
dynamic models mainly takes into account two considerations. First, these models account for a population of

neurons in the model. As such, these models aim at simulating large-scale neural network behaviors. Second,
these models were motivated by associative memory
processes where the time reflects the memory recall
process.

Consider a population of neurons, and let $r_i(t)$ denote the mean firing rate of a target neuron i, and $r_j(t)$,
$j \in N_i = \{j/j$ is presynaptic to $i\}$, the mean firing rates
of all neurons presynaptic to neuron i. Let $h_i(t)$ be the
input to target neuron i, which is

$$h_i(t) = \sum_{j \in N_i} w_{ij}(t) r_j(t) . \tag{34.17}$$

Equation (34.17) takes into account all presynaptic neurons' contributions weighted by synaptic efficacy w_{ij}.
A representative class of the firing rate model as studied
extensively in the artificial neural networks community,
which was first popularized by *Hopfield* [34.27], can
then be described at the fixed-point of an associative
memory process as

$$r_i(t) = \Theta \left(\sum_{j \in N_i} w_{ij}(t) r_j(t) \right) . \tag{34.18}$$

In (34.18), $\Theta(.)$ is considered a gain function. Consequently the firing rate dynamics associated with this
associative memory process can be defined by introducing a time constant τ in the associative network as

$$\tau \frac{dr_i(t)}{dt} = -r_i(t) + \Theta \left(\sum_{j \in N_i} w_{ij}(t) r_j(t) \right) . \tag{34.19}$$

The firing rate model (34.19) also has another interesting interpretation where the mean firing rate $r_i(t)$ is
considered to be the spatial averaged neural potential
$F_i(t)$ of neuron i due to contributions from a local population of neurons $j \in N_i = \{j/j$ is presynaptic to $i\}$. As
such, (34.19) becomes

$$\tau \frac{dF_i(t)}{dt} = -F_i(t) + \Theta \left(\sum_{j \in N_i} w_{ij}(t) F_j(t) \right) . \tag{34.20}$$

The model described by (34.20) was used for the
analysis of a large neural network where slow neural
dynamics were assumed in order to describe spatially
homogeneous motoneurons [34.28].

34.4.3 Spike Response Model

The spike response model [34.29] uses response kernels to account for the integral effect of presynaptic action potentials. With two linear kernels and under a simple renewal assumption, it can be shown that the spike response model (SRM) is a generalization of the integrate-and-fire neuron. The spike response model describes the membrane potential $u_i(t)$ of neuron i as

$$u_i(t) = \eta\left(t - \hat{t}_i\right) + \int_0^\infty \kappa\left(t - \hat{t}_i, s\right) I^{\text{ext}}\left(t - s\right) \mathrm{d}s\,,$$

(34.21)

where η represents the typical form of an action potential, which includes both depolarization and repolarization, as well as the process of settling down to the resting potential; $\hat{t}_i$ stands for neuron i firing an action potential at that time. Also in the equation, the kernel $\kappa(t - \hat{t}_i, s)$ is a linear impulse response function of the membrane potential to a unit input current. Imagine it as a time course of an additive membrane potential to

the membrane potential of neuron i after its action potential fired at $\hat{t}_i$. The term $I^{\text{ext}}(t - s)$ accounts for all external driving currents.

According to the spike response model, an action potential is fired if the membrane potential $u_i(t)$ crosses the threshold $\theta(t - \hat{t}_i)$ from below, where it is noted that this threshold value is a function of $(t - \hat{t}_i)$. The consideration of using a dynamic threshold in a phenomenological neuron model was proven to be an important contributor to the success of using SRM for spike-time prediction under random conductance injection [34.30]. Actually, SRM outperformed a standard leaky integrate-and-fire model significantly when tested on the same experimental data. In [34.14], the authors performed an analytical reduction from the full conductance-based model to the spike response model for fast spiking neurons. After estimating the three parameters in the SRM, namely the kernels $(t - \hat{t}_i)$, $\kappa(t - \hat{t}_i, s)$, and the dynamic threshold $\theta(t - \hat{t}_i)$, the authors show that the full conductance-based model of a fast-spiking neuron model is well approximated by a single variable SRM model.

34.5 Conclusions

In this chapter we have provided an introduction to some established modeling approaches to studying biological neural systems. Motivations behind these models are twofold. First and foremost, biological realism is considered to be of the utmost importance. Given the complex nature of a biological neuron, reduced-order neuronal models that can be or have been validated by the Hodgkin–Huxley model or biological data have been developed. As discussed, these models only scratch the surface of providing an accurate and realistic account of a real neural system, not even a specific brain area or something capable of explaining a behavioral parameter completely and thoroughly. Nonetheless, this decade has probably seen the most progress in terms of computational modeling approaches to understanding the brain. The International Neuroinformatics

Coordinating Facility (INCF) was established about 10 years ago with a focus on coordinating and promoting neuroinformatics research activities. Their activities are based on maintenance of database and computational infrastructure to support neuroscience research and applications. The grand scale and ambitious neural modeling project of simulating a human brain in the next 10 years led by Henry Markram is another example of the urgency and timeliness of studying the brain by using advanced computing machinery. The wonders of new technology have certainly provided us with the tools critical to pondering some of the most challenging questions about the brain. An insurmountable amount of work has yet to be done to fill the huge gap between a single neuron and its model proposed by Hodgkin–Huxley and the true understanding of the human brain.

References

34.1   E.R. Kandel, J.H. Schwartz, T.M. Jessell, S.A. Siegelbaum, A.J. Hudspeth: *Principles of Neural Science*, 5th edn. (McGraw–Hill, New York 2013)

34.2   M.L. Hines, N.T. Carnevale: Neuron: A tool for neuroscientists, Neuroscientist **7**(2), 123–135 (2001)

34.3   N. Baumann, D. Pham-Dinh: Biology of oligo-
        dendrocyte and myelin in the mammalian central
        nervous system, Physiol. Rev. **81**(2), 871–927 (2001)
34.4   R. Brette, M. Rudolphy, T. Carnevale, M. Hines,
        D. Beeman, J.M. Bower, M. Diesmann, A. Mor-
        rison, P.H. Goodman, F.C. Harris Jr., M. Zirpe,
        T. Natschläger, D. Pecevski, A.B. Ermentrout,
        M. Djurfeldt, A. Cansner, O. Rochel, T. Vieville,
        E. Mulles, A.P. Davison, S. El Boustani, A. Destexhe,
        J. Harris, C. Frederick, B. Ermentrout: Simulation
        of networks of spiking neurons: A review of tools
        and strategies, J. Comput. Neurosci. **23**(3), 349–398
        (2007)
34.5   E.R. Kandel, J.H. Schwartz: Molecular biology of
        learning: Modulation of transmitter release, Sci-
        ence **218**(4571), 433–443 (1982)
34.6   N.T. Carnevale, M.L. Hines: *The NEURON Book* (Cam-
        bridge Univ. Press, Cambridge 2006)
34.7   J. Chen: A simulation study investigating the im-
        pact of dendritic morphology and synaptic topol-
        ogy on neuronal firing patterns, Neural Comput.
        22(4), 1086–1111 (2010)
34.8   J. Crank, A.B. Crowley: On an implicit scheme for the
        isotherm migration method along orthogonal flow
        lines in two dimensions, Int. J. Heat Mass Transf.
        22(10), 1331–1337 (1979)
34.9   R.J. Douglas, K.A.C. Martin: Recurrent neuronal cir-
        cuits in the neocortex, Curr. Biol. **17**(13), R496–500
        (2007)
34.10   W. Gerstner: Time structure of the activity in neural
        network models, Phys. Rev. E **51**(1), 738–758 (1995)
34.11   W. Gerstner, R. Naud: How good are neuron mod-
        els?, Science **326**(5951), 379–380 (2009)
34.12   S. Grillner: The motor infrastructure: From ion
        channels to neuronal networks, Nat. Rev. Neurosci.
        4(7), 573–586 (2003)
34.13   A.L. Hodgkin, A.F. Huxley: Propagation of electri-
        cal signals along giant nerve fibres, Proc. R. Soc. B
        140(899), 177–183 (1952)
34.14   R. Jolivet, T.J. Lewis, W. Gerstner: Generalized
        integrate-and-fire models of neuronal activity ap-
        proximate spike trains of a detailed model to a high
        degree of accuracy, J. Neurophysiol. **92**(2), 959–976
        (2004)
34.15   K.M. Stiefel, J.T. Sejnowski: Mapping function onto
        neuronal morphology, J. Neurophysiol. **98**(1), 513–
        526 (2007)

34.16   C.L. Kutscher: Chemical transmission in the mam-
        malian nervous system, Neurosci. Biobehav. Rev.
        2(2), 123–124 (1978)
34.17   L.F. Abbott: Modulation of function and gated
        learning in a network memory, Proc. Natl. Acad. Sci.
        USA **87**(23), 9241–9245 (1990)
34.18   S.B. Laughlin, T.J. Sejnowski: Communication in
        neuronal networks, Science **301**(5641), 1870–1874
        (2003)
34.19   P. Lledo, G. Gheusi, J. Vincent: Information pro-
        cessing in the mammalian olfactory system, Phys-
        iol. Rev. **85**(1), 281–317 (2005)
34.20   M.L. Hines, N.T. Carnevale: The NEURON simula-
        tion environment, Neural Comput. **9**(6), 1179–1209
        (1997)
34.21   A.L. Hodgkin, A.F. Huxley: A quantitative descrip-
        tion of membrane current and its application to
        conduction and excitation in nerve, Bull. Math.
        Biol. **52**(1), 25–71 (1990)
34.22   W.W. Lytton: Adapting a feedforward heteroas-
        sociative network to Hodgkin–Huxley dynamics,
        J. Comput. Neurosci. **5**(4), 353–364 (1998)
34.23   W.W. Lytton: Optimizing synaptic conductance cal-
        culation for network simulations, Neural Comput.
        8(3), 501–509 (1996)
34.24   M. Migliore, C. Cannia, W.W. Lytton, H. Markram,
        M.L. Hines: Parallel network simulations with NEU-
        RON, J. Comput. Neurosci. **21**(2), 119–129 (2006)
34.25   Y. Sun, D. Zhou, A.V. Rangan, D. Cai: Library-based
        numerical reduction of the Hodgkin–Huxley neu-
        ron for network simulation, J. Comput. Neurosci.
        27(3), 369–390 (2009)
34.26   X. Wang: Decision making in recurrent neuronal
        circuits, Neuron **60**(2), 215–234 (2008)
34.27   J.J. Hopfield: Neural networks and physical systems
        with emergent collective computational abilities,
        Proc. Nat. Acad. Sci. USA **79**(8), 2554–2558 (1982)
34.28   J.L. Feldman, J.D. Cowan: Large-scale activ-
        ity in neural nets I: Theory with applications
        to motoneuron pool, Biol. Cybern. **17**(1), 29–38
        (1975)
34.29   W. Gerstner, W. Kistler: *Spiking Neuron Models*
        (Cambridge Univ. Press, Cambridge 2002)
34.30   R. Jolivet, R. Kobayashi, A. Rauch, R. Naud, S. Shi-
        nomoto, W. Gerstner: A benchmark test for a quan-
        titative assessment of simple neuron models,
        J. Neurosci. Methods **169**, 417–424 (2008)

35. Computational Models of Cognitive and Motor Control

Ali A. Minai

Most of the earliest work in both experimental and theoretical/computational system neuroscience focused on sensory systems and the peripheral (spinal) control of movement. However, over the last three decades, attention has turned increasingly toward *higher* functions related to cognition, decision making and voluntary behavior. Experimental studies have shown that specific brain structures – the prefrontal cortex, the premotor and motor cortices, and the basal ganglia – play a central role in these functions, as does the dopamine system that signals reward during reinforcement learning. Because of the complexity of the issues involved and the difficulty of direct observation in deep brain structures, computational modeling has been crucial in elucidating the neural basis of cognitive control, decision making, reinforcement learning, working memory, and motor control. The resulting computational models are also very useful in engineering domains such as robotics, intelligent agents, and adaptive control. While it is impossible to encompass the totality of such modeling work, this chapter provides an overview of significant efforts in the last 20 years.

It also outlines many of the theoretical issues underlying this work, and discusses significant experimental results that motivated the computational models.

35.1 Overview

Mental function is usually divided into three parts: perception, cognition, and action – the so-called *sense-think-act cycle*. Though this view is no longer held dogmatically, it is useful as a structuring framework for discussing mental processes. Several decades of theory and experiment have elucidated an intricate, multiconnected functional architecture for the brain [35.1, 2] – a simplified version of which is shown in Fig. 35.1. While all regions and functions shown – and many not shown – are important, this figure provides a summary of the main brain regions involved in perception, cognition, and action. The highlighted blocks in Fig. 35.1 are

discussed in this chapter, which focuses mainly on the higher level mechanisms for the control of behavior.

The control of action (or behavior) is, in a real sense, the primary function of the nervous system. While such actions may be voluntary or involuntary, most of the interest in modeling has understandably focused on voluntary action. This chapter will follow this precedent.

It is conventional to divide the neural substrates of behavior into *higher* and *lower* levels. The latter involves the musculoskeletal apparatus of action (muscles, joints, etc.) and the neural networks of the spinal

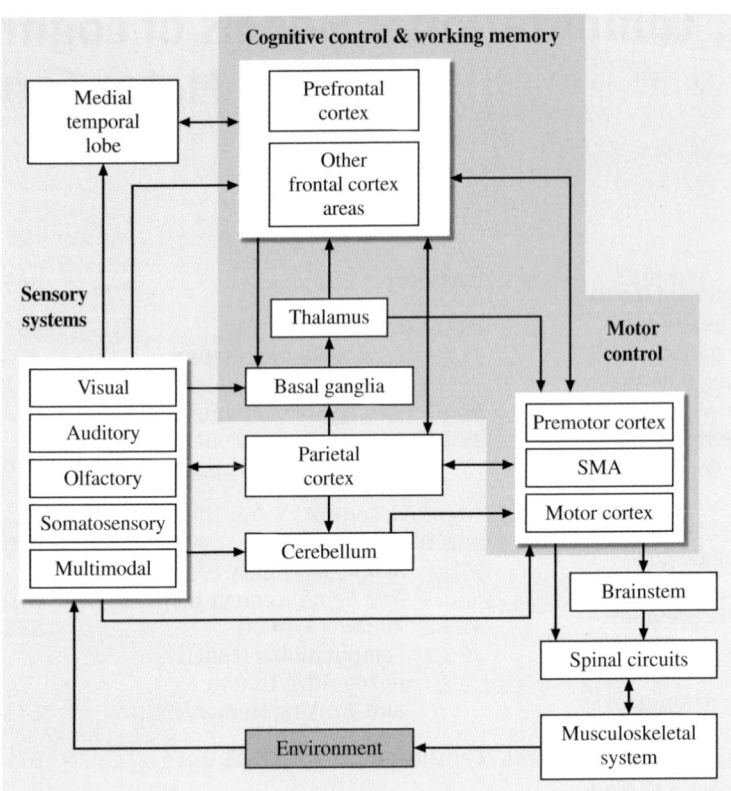

Fig. 35.1 A general schematic of primary signal flow in the nervous system. Many modulatory regions and connections, as well as several known connections, are not shown. The *shaded areas* indicate the components covered in this chapter

cord and brainstem. These systems are seen as representing the actuation component of the action system, which is controlled by the higher level system comprising cortical and subcortical structures. This division between a controller (the brain) and the plant (the body and spinal networks), which parallels the models used in robotics, has been criticized as arbitrary and unhelpful [35.3, 4], and there has recently been a shift of interest toward more embodied views of cognition [35.5, 6]. However, the conventional division is useful for organizing material covered in this chapter, which focuses primarily on the higher level systems, i. e., those above the spinal cord and the brainstem.

The higher level system can be divided further into a *cognitive control* component involving action selection, configuration of complex actions, and the learning of appropriate behaviors through experience, and a *motor control* component that generates the control signals for the lower level system to execute the selected action. The latter is usually identified with the motor cortex (M1), premotor cortex (PMC), and the supplementary motor area (SMA), while the former is seen as involv-

ing the prefrontal cortex (PFC), basal ganglia (BG), the anterior cingulate cortex (ACC) and other cortical and subcortical regions [35.7]. With regard to the generation of actions per se, an influential viewpoint for the higher level system is summarized by *Doya* [35.8]. It proposes that higher level control of action has three major loci: the cortex, the cerebellum, and the BG. Of these, the cortex – primarily the M1 – provides a self-organized repertoire of possible actions that, when triggered, generate movement by activating muscles via spinal networks, the cerebellum implements fine motor control configured through error-based supervised learning [35.9], and the BG provide the mechanisms for selecting among actions and learning appropriate ones through reinforcement learning [35.10–13]. The motor cortex and cerebellum can be seen primarily as motor control (though see [35.14]), whereas the BG falls into the domain of cognitive control and working memory (WM). The PFC is usually regarded as the locus for higher order choice representations, plans, goals, etc. [35.15–18], while the ACC is thought to be involved in conflict monitoring [35.19–21].

35.2 Motor Control

Given its experimental accessibility and direct relevance to robotics, motor control has been a primary area of interest for computational modeling [35.22–24]. Mathematical, albeit non-neural, theories of motor control were developed initially within the framework of dynamical systems. One of these directions led to models of action as an emergent phenomenon [35.3, 25–33] arising from interactions among preferred coordination modes [35.34]. This approach has continued to yield insights [35.29] and has been extended to multiactor situations as well [35.33, 35–37]. Another approach within the same framework is the *equilibrium point hypothesis* [35.38, 39], which explains motor control through the change in the equilibrium points of the musculoskeletal system in response to neural commands. Both these dynamical approaches have paid relatively less attention to the neural basis of motor control and focused more on the phenomenology of action in its context. Nevertheless, insights from these models are fundamental to the emerging synthesis of action as an embodied cognitive function [35.5, 6].

A closely related investigative tradition has been developed from the early studies of gaits and other rhythmic movements in cats, fish, and other animals [35.40–45], leading to computational models for *central pattern generators* (CPGs), which are neural networks that generate characteristic periodic activity patterns autonomously or in response to control signals [35.46]. It has been found that rhythmic movements can be explained well in terms of CPGs – located mainly in the spinal cord – acting upon the coordination modes inherent in the musculoskeletal system. The key insight to emerge from this work is that a wide range of useful movements can be generated by modulation of these CPGs by rather simple motor control signals from the brain, and feedback from sensory receptors can shape these movements further [35.43]. This idea was demonstrated in recent work by *Ijspeert* et al. [35.47] showing how the same simple CPG network could produce both swimming and walking movements in a robotic salamander model using a simple scalar control signal.

While rhythmic movements are obviously important, computational models of motor control are often motivated by the desire to build humanoid or biomorphic robots, and thus need to address a broader range of actions – especially aperiodic and/or voluntary movements. Most experimental work on aperiodic movement has focused on the paradigm of manual reaching [35.30, 48–64]. However, seminal work has also been done

with complex reflexes in frogs and cats [35.65–72], isometric tasks [35.73, 74], ball-catching [35.75], drawing and writing [35.60, 76–81], and postural control [35.71, 72, 82, 83].

A central issue in understanding motor control is the *degrees of freedom problem* [35.84] which arises from the immense redundancy of the system – especially in the context of multijoint control. For any desired movement – such as reaching for an object – there are an infinite number of control signal combinations from the brain to the muscles that will accomplish the task (see [35.85] for an excellent discussion). From a control viewpoint, this has usually been seen as a problem because it precludes the clear specification of an objective function for the controller. To the extent that they consider the generation of specific control signals for each action, most computational models of motor control can be seen as direct or indirect ways to address the degrees of freedom problem.

35.2.1 Cortical Representation of Movement

It has been known since the seminal work by *Penfield* and *Boldrey* [35.86] that the stimulation of specific locations in the M1 elicit motor responses in particular locations on the body. This has led to the notion of a motor homunculus – a map of the body on the M1. However, the issue of exactly what aspect of movement is encoded in response to individual neurons is far from settled. A crucial breakthrough came with the discovery of *population coding* by *Georgopoulos* et al. [35.49]. It was found that the activity of specific neurons in the hand area of the M1 corresponded to reaching movements in particular directions. While the tuning of individual cells was found to be rather broad (and had a sinusoidal profile), the joint activity of many such cells with different tuning directions coded the direction of movement with great precision, and could be decoded through neurally plausible estimation mechanisms. Since the initial discovery, population codes have been found in other regions of the cortex that are involved in movement [35.49, 53, 54, 60, 77–80, 87]. Population coding is now regarded as the primary basis of directional coding in the brain, and is the basis of most brain–machine interfaces (BMI) and brain-controlled prosthetics [35.88, 89]. Neural network models for population coding have been developed by several researchers [35.90–93], and population coding has come to be seen as a general neural representational strategy with application far beyond motor control [35.94]. Excellent reviews are provided in [35.95, 96]. Mathematical and computational models for Bayesian inference with population codes are discussed in [35.97, 98].

An active research issue in the cortical coding of movement is whether it occurs at the level of *kinematic variables*, such as direction and velocity, or in terms of *kinetic variables*, such as muscle forces and joint torques. From a cognitive viewpoint, a kinematic representation is obviously more useful, and population codes suggest that such representations are indeed present in the motor cortex [35.48, 53, 54, 60, 77–80, 99, 100] and PFC [35.15, 101]. However, movement must ultimately be constructed from the appropriate kinetic variables, i. e., by controlling the forces generated by specific muscles and the resulting joint torques. Studies have indicated that some neurons in the M1 are indeed tuned to muscle forces and joint torques [35.58, 59, 73, 99, 100, 102, 103]. This apparent multiplicity of cortical representations has generated significant debate among researchers [35.74]. One way to resolve this issue is to consider the kinetic and kinematic representations as dual representations related through the constraints of the musculoskeletal system. However, Shah et al. [35.104] have used a simple computational model to show that neural populations tuned to kinetic or kinematic variables can act jointly in motor control without the need for explicit coordinate transformations.

Graziano et al. [35.105] studied movements elicited by the sustained electrode stimulation of specific sites in the motor cortex of monkeys. They found that different sites led to specific complex, multijoint movements such as bringing the hand to the mouth or lifting the hand above the head regardless of the initial position. This raises the intriguing possibility that individual cells or groups of cells in the M1 encode goal-directed movements that can be triggered as units. The study also indicated that this encoding is not open-loop, but can compensate – at least to some degree – for variation or extraneous perturbations. The M1 and other related regions (e.g., the supplementary motor area and the PMC) appear to encode spatially organized maps of a few *canonical* complex movements that can be used as basis functions to construct other actions [35.105–107]. A neurocomputational model using self-organized feature maps has been proposed in [35.108] for the representation of such canonical movements.

In addition to rhythmic and reaching movements, there has also been significant work on the neural basis of sequential movements, with the finding that such neural codes for movement sequences exist in the supplementary motor area [35.109–111], cerebellum [35.112, 113], BG [35.112], and the PFC [35.101]. Coding for multiple goals in sequential reaching has been observed in the parietal cortex [35.114].

35.2.2 Synergy-based Representations

A rather different approach to studying the construction of movement uses the notion of motor primitives, often termed *synergies* [35.63, 115, 116]. Typically, these synergies are manifested in coordinated patterns of spatiotemporal activation over groups of muscles, implying a force field over posture space [35.117, 118]. Studies in frogs, cats, and humans have shown that a wide range of complex movements in an individual subject can be explained as the modulated superposition of a few synergies [35.63, 65–72, 115, 119, 120]. Given a set of n muscles, the n-dimensional time-varying vector of activities for the muscles during an action can be written as

$$\boldsymbol{m}^q(t) = \sum_{k=1}^{N} c_k^q \boldsymbol{g}_k \left(t - t_k^q\right), \tag{35.1}$$

where $\boldsymbol{g}_k(t)$ is a time-varying synergy function that takes only nonnegative values, c_k^q is the gain of the kth synergy used for action q, and t_k^q is the temporal offset with which the kth synergy is triggered for action q [35.69]. The key point is that a broad range of actions can be constructed by choosing different gains and offsets over the same set of synergies, which represent a set of hard-coded basis functions for the construction of movements [35.120, 121]. Even more interestingly, it appears that the synergies found empirically across different subjects of the same species are rather consistent [35.67, 72], possibly reflecting the inherent constraints of musculoskeletal anatomy. Various neural loci have been suggested for synergies, including the spinal cord [35.67, 107, 122], the motor cortex [35.123], and combinations of regions [35.85, 124].

Though synergies are found consistently in the analysis of experimental data, their actual existence in the neural substrate remains a topic for debate [35.125, 126]. However, the idea of constructing complex movements from motor primitives has found ready application in robotics [35.127–132], as discussed later in this chapter. A hierarchical neurocomputational model of motor synergies based on attractor networks has recently been proposed in [35.133, 134].

35.2.3 Computational Models of Motor Control

Motor control has been modeled computationally at many levels and in many ways, ranging from explicitly control-theoretic models through reinforcement-based models to models based on emergent dynamical patterns. This section provides a brief overview of these models.

As discussed above the M1, premotor cortex (PMC) and the supplementary motor area (SMA) are seen as providing self-organized *codes* for specific actions, including information on direction, velocity, force, low-level sequencing, etc., while the PFC provides higher level codes needed to construct more complex actions. These codes, comprising a *repertoire* of actions [35.10, 106], arise through self-organized learning of activity patterns in these cortical systems. The BG system is seen as the primary locus of selection among the actions in the cortical repertoire. The architecture of the system involving the cortex, BG, and the thalamus, and in particular the internal architecture of the BG [35.135], makes this system ideally suited to selectively disinhibiting specific cortical regions, presumably activating codes for specific actions [35.10, 136, 137]. The BG system also provides an ideal substrate for learning appropriate actions through a dopamine-mediated reinforcement learning mechanism [35.138–141].

Many of the influential early models of motor control were based on control-theoretic principles [35.142–144], using forward and inverse kinematic and dynamic models to generate control signals [35.55, 57, 145–150] – see [35.146] for an excellent introduction. These models have led to more sophisticated ones, such as MOSAIC (modular selection and identification for control) [35.151] and AVITEWRITE (adaptive vector integration to endpoint handwriting) [35.81]. The MOSAIC model is a mixture of experts, consisting of many parallel modules, each comprising three subsystems. These are: A forward model relating motor commands to predicted position, a responsibility predictor that estimates the applicability of the current module, and an inverse model that learns to generate control signals for desired movements. The system generates motor commands by combining the recommendations of the inverse models of all modules weighted by their applicability. Learning in the model is based on a variant of the EM algorithm. The model in [35.57] is a comprehensive neural model with both cortical and spinal components, and builds upon the earlier VITE model in [35.55]. The AVITEWRITE model [35.81], which is

a further extension of the VITE model, can generate the complex movement trajectories needed for writing by using a combination of pre-specified phenomenological motor primitives (synergies). A cerebellar model for the control of timing during reaches has been presented by *Barto* et al. [35.152].

The use of neural maps in models of motor control was pioneered in [35.153, 154]. These models used self-organized feature maps (SOFMs) [35.155] to learn visuomotor coordination. *Baraduc* et al. [35.156] presented a more detailed model that used multiple maps to first integrate posture and desired movement direction and then to transform this internal representation into a motor command. The maps in this and most subsequent models were based on earlier work by [35.90–93]. An excellent review of this approach is given in [35.94]. A more recent and comprehensive example of the map-based approach is the SURE-REACH (sensorimotor, unsupervised, redundancy-resolving control architecture) model in [35.157] which focuses on exploiting the redundancy inherent in motor control [35.84]. Unlike many of the other models, which use neutrally implausible error-based learning, SURE-REACH relies only on unsupervised and reinforcement learning. Maps are also the central feature of a general cognitive architecture called ERA (epigenetic robotics architecture) by *Morse* et al. [35.158].

Another successful approach to motor control models is based on the use of motor primitives, which are used as basis functions in the construction of diverse actions. This approach is inspired by the experimental observation of motor synergies as described above. However, most models based on primitives implement them nonneurally, as in the case of AVITEWRITE [35.81]. The most systematic model of motor primitives has been developed by *Schaal* et al. [35.129–132]. In this model, motor primitives are specified using differential equations, and are combined after weighting to produce different movements. Recently, *Matsubara* et al. [35.159] have shown how the primitives in this model can be learned systematically from demonstrations. *Drew* et al. [35.123] proposed a conceptual model for the construction of locomotion using motor primitives (synergies) and identified the characteristics of such primitives experimentally. A neural model of motor primitives based on hierarchical attractor networks has been proposed recently in [35.133, 134, 160], while *Neilson* and *Neilson* [35.85, 124] have proposed a model based on coordination among adaptive neural filters.

Motor control models based on primitives can be simpler than those based on trajectory tracking because the controller typically needs to choose only the weights (and possibly delays) for the primitives rather than specifying details of the trajectory (or forces). Among other things, this promises a potential solution to the degrees of freedom problem [35.84] since the coordination inherent in the definition of motor primitives reduces the effective degrees of freedom in the system. Another way to address the degrees of freedom problem is to use an optimal control approach with a specific objective function. Researchers have proposed objective functions such as minimum jerk [35.161], minimum torque [35.162], minimum acceleration [35.163], or minimum energy [35.85], but an especially interesting idea is to optimize the distribution of variability across the degrees of freedom in a task-dependent way [35.144, 164–167]. From this perspective, motor control trades off variability in task-irrelevant dimensions for greater accuracy in task-relevant ones. Thus, rather than specifying a trajectory, the controller focuses only on correcting consequential errors. This also explains the experimental observation that motor tasks achieve their goals with remarkable accuracy while using highly variable trajectories to achieve the same goal. *Trainin* et al. [35.168] have shown that the optimal control principle can be used to explain the observed neural coding of movements in

the cortex. *Biess* et al. [35.169] have proposed a detailed computational model for controlling an arm in three-dimensional space by separating the spatial and temporal components of control. This model is based on optimizing energy usage and jerk [35.161], but is not implemented at the neural level.

An alternative to these prescriptive and constructivist approaches to motor control is provided by models based on dynamical systems [35.3, 25–27, 29, 31–33]. The most important way in which these models diverge from the others is in their use of emergence as the central organizational principle of control. In this formulation, control programs, structures, primitives, etc., are not preconfigured in the brain–body system, but emerge under the influence of task and environmental constraints on the affordances of the system [35.33]. Thus, the dynamical systems view of motor control is fundamentally ecological [35.170], and like most ecological models, is specified in terms of low-dimensional state dynamics rather than high-dimensional neural processes. Interestingly, a correspondence can be made between the dynamical and optimal control models through the so-called *uncontrolled manifold* concept [35.31, 33, 39, 171]. In both models, the dimensions to be controlled and those that are left uncontrolled are decided by external constraints rather than internal prescription, as in classical models.

35.3 Cognitive Control and Working Memory

A lot of behavior – even in primates – is automatic, or almost so. This corresponds to actions (or internal behaviors) so thoroughly embedded in the sensorimotor substrate that they emerge effortlessly from it. In contrast, some tasks require significant cognitive effort for one or more reason, including:

1. An automatic behavior must be suppressed to allow the correct response to emerge, e.g., in the Stroop task [35.172].
2. Conflicts between incoming information and/or recalled behaviors must be resolved [35.19, 20].
3. More contextual information – e.g., social context – must be taken into account before acting.
4. Intermediate pieces of information need to be stored and recalled during the performance of the task, e.g., in sequential problem solving.

5. The timing of subtasks within the overall task is complex, e.g., in delayed-response tasks or other sequential tasks [35.173].

Roughly speaking, the first three fall under the heading of *cognitive control*, and the latter two of *working memory*. However, because of the functions are intimately linked, the terms are often subsumed into each other.

35.3.1 Action Selection and Reinforcement Learning

Action selection is arguably the central component of the cognitive control process. As the name implies, it involves selectively triggering an action from a repertoire of available ones. While action selection is a complex process involving many brain regions, a consensus has emerged that the BG system plays a central role in its mechanism [35.10, 12, 14]. The architecture of the BG system and the organization of its projections to and from the cortex [35.135, 174, 175] make it ideally suited to function as a state-dependent gating system for specific functional networks in the cortex. As shown in Fig. 35.2, the hypothesis is that the striatal layer of the BG system, receiving input from the cortex, acts as a pattern recognizer for the current cognitive state. Its activity inhibits specific parts of the globus pallidus (GPi), leading to disinhibition of specific neural assemblies in the cortex – presumably allowing the behavior/action encoded by those assemblies to proceed [35.10]. The associations between cortical activity patterns and behaviors are key to the functioning of the BG as an action selection system, and the configuration and modulation of these associations are thought to lie at the core of cognitive control. The neurotransmitter dopamine (DA) plays a key role here by serving as a reward signal [35.138–140] and modulating reinforcement learning [35.176, 177] in both the BG and the cortex [35.141, 178–180].

35.3.2 Working Memory

All nontrivial behaviors require task-specific information, including relevant domain knowledge and the relative timing of subtasks. These are usually grouped under the function of WM. An influential model of WM in [35.181] identifies three components in WM: (1) a *central executive*, responsible for attention, decision making, and timing; (2) a *phonological loop*, responsible for processing incoming auditory information, maintaining it in short-term memory, and rehearsing utterances; and (3) a *visuospatial sketchpad*, responsible for processing and remembering visual information, keeping track of *what* and *where* information, etc. An *episodic buffer* to manage relationships between the other three components is sometimes included [35.182]. Though already rather abstract, this model needs even more generalized interpretation in the context of many cognitive tasks that do not directly involve visual or auditory data. Working memory function is most closely identified with the PFC [35.183–185].

Almost all studies of WM consider only short-term memory, typically on the scale of a few seconds [35.186]. Indeed, one of the most significant – though lately controversial – results in WM research is the finding that only a small number of items can be *kept in mind* at any one time [35.187, 188]. However, most cognitive tasks require context-dependent repertoires of knowledge and behaviors to be enabled collectively over longer periods. For example, a player must continually think of chess moves and strategies over the course of a match lasting several hours. The configuration of context-dependent repertoires for extended periods has been termed *long-term working memory* [35.189].

35.3.3 Computational Models of Cognitive Control and Working Memory

Several computational models have been proposed for cognitive control, and most of them share common features. The issues addressed by the models include action selection, reinforcement learning of appropriate actions, decision making in choice tasks, task sequencing and timing, persistence and capacity in WM, task switching, sequence learning, and the configuration of context-appropriate workspaces. Most of the models discussed below are neural with a range of biological plausibility. A few important nonneural models are also mentioned.

A comprehensive model using spiking neurons and incorporating many biological features of the BG system has been presented in [35.13, 193]. This model focuses only on the BG and explicitly on the dynamics of dopamine modulation. A more abstract but broader model of cognitive control is the *agents of the mind* model in [35.14], which incorporates the cerebellum as well as the BG. In this model, the BG provide the action selection function while the cerebellum acts to refine and amplify the choices. A series of interrelated models have been developed by *O'Reilly, Frank* et al. [35.17, 179, 194–199]. All these models use the adaptive gating function of the BG in combination with the WM function of the prefrontal cortex to explain how executive function can arise without explicit top-down control – the so-called *homunculus* [35.196, 197]. A comprehensive review of these and other models of cognitive control is given in [35.200]. Models of goal-directed action mediated by the PFC have also been presented in [35.201] and [35.202]. *Reynolds* and *O'Reilly* et al. [35.203] have proposed a model for configuring hierarchically organized representations in the PFC via reinforcement learning. Computational models of cognitive control and working have also been used to explain mental pathologies such as schizophrenia [35.204].

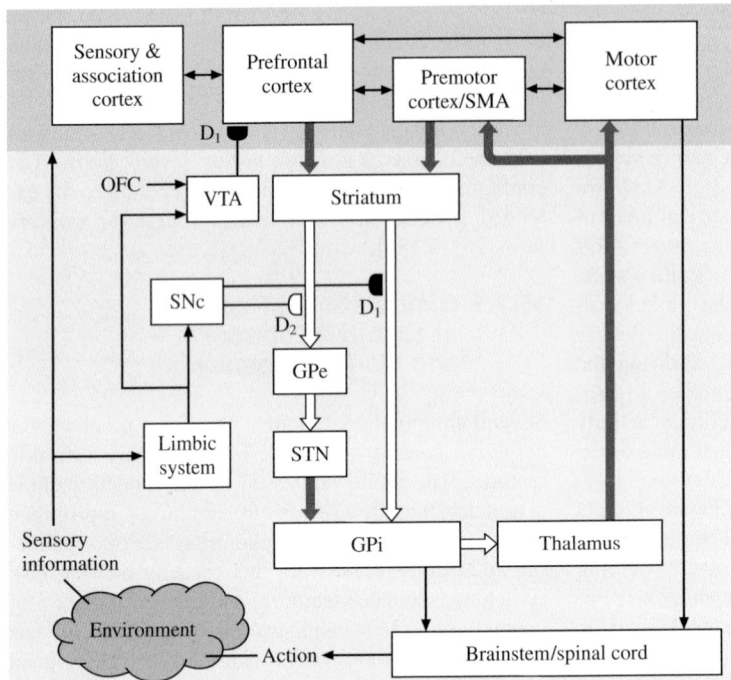

Fig. 35.2 The action selection and reinforcement learning substrate in the BG. *Wide filled arrows* indicate excitatory projections while *wide unfilled arrows* represent inhibitory projections. *Linear arrows* indicate generic excitatory and inhibitory connectivity between regions. The inverted D-shaped contacts indicate modulatory dopamine connections that are crucial to reinforcement learning. Abbreviations: SMA = supplementary motor area; SNc = substantia nigra pars compacta; VTA = ventral tegmental area; OFC = orbitofrontal cortex; GPe = globus pallidus (external nuclei); GPi = globus pallidus (internal nuclei); STN = subthalamic nucleus; D_1 = excitatory dopamine receptors; D_2 = inhibitory dopamine receptors. The primary neurons of GPi are inhibitory and active by default, thus keeping all motor plans in the motor and premotor cortices in check. The neurons of the striatum are also inhibitory but usually in an inactive *down* state (after [35.190]). Particular subgroups of striatal neurons are activated by specific patterns of cortical activity (after [35.136]), leading first to disinhibition of specific actions via the direct input from striatum to GPi, and then by re-inhibition via the input through STN. Thus the system gates the triggering of actions appropriate to current cognitive contexts in the cortex. The dopamine input from SNc projects a *reward* signal based on limbic system state, allowing desirable context-action pairs to be reinforced (after [35.191, 192]) – though other hypotheses also exist (after [35.14]). The dopamine input to PFC from the VTA also signals reward and other task-related contingencies

An important aspect of cognitive control is switching between tasks at various time-scales [35.205, 206]. *Imamizu* et al. [35.207] compared two computational models of task switching – a mixture-of-experts (MoE) model and MOSAIC – using brain imaging. They concluded that task switching in the PFC was more consistent with the MoE model and that in the parietal cortex and cerebellum with the MOSAIC model.

An influential abstract model of cognitive control is the *interactive activation model* in [35.208, 209]. In this model, learned behavioral schemata contend for activation based on task context and cognitive state. While this model captures many phenomenological aspects of behavior, it is not explicitly neural. *Botvinick* and *Plaut* [35.173] present an alternative neural model that relies on distributed neural representations and the dynamics of recurrent neural networks rather than explicit schemata and contention. *Dayan* et al. [35.210, 211] have proposed a neural model for implementing complex rule-based decision making where decisions are based on sequentially unfolding contexts. A partially neural model of behavior based on the CLARION cognitive model has been developed in [35.212].

Recently, *Grossberg* and *Pearson* [35.213] have presented a comprehensive model of WM called LIST PARSE. In this model, the term *working memory* is applied narrowly to the storage of temporally ordered items, i.e., lists, rather than more broadly to all short-term memory. Experimentally observed effects such as recency (better recall of late items in the list) and primacy (better recall of early items in the list) are explained by this model, which uses the concept of *competitive queuing* for sequences. This is based on the observation [35.101, 214] that multiple elements of a behavioral sequence are represented in the PFC as simultaneously active codes with activation levels representing the temporal order. Unlike the WM models discussed in the previous paragraph, the WM in LIST PARSE is embedded within a full cognitive control model with action selection, trajectory generation, etc. Many other neural models for chains of actions have also been proposed [35.214–224].

Higher level cognitive control is characterized by the need to fuse information from multiple sensory modalities and memory to make complex decisions. This has led to the idea of a *cognitive workspace*. In the *global workspace theory* (GWT) developed in [35.225], information from various sensory, episodic, semantic, and motivational sources comes together in a global workspace that forms brief, task-specific integrated representations that are broadcast to all subsystems for use in WM. This model has been implemented computationally in the *intelligent distribution agent* (IDA) model by *Franklin* et al. [35.226, 227]. A neurally implemented workspace model has been developed by *Dehaene* et al. [35.172, 228, 229] to explain human subjects' performance on effortful cognitive tasks (i.e., tasks that require suppression of automatic responses), and the basis of consciousness. The construction of cognitive workspaces is closely related to the idea of long-term working memory [35.189]. Unlike short-term working memory, there are few computational models for long-term working memory. Neural models seldom cover long periods, and implicitly assume that a chaining process through recurrent networks (e.g., [35.173]) can maintain internal attention. *Iyer* et al. [35.230, 231] have proposed an explicitly neurodynamical model of this function, where a stable but modulatable pattern of activity called a *graded attractor* is used to selectively bias parts of the cortex in the context-dependent fashion. An earlier model was proposed in [35.232] to serve a similar function in the hippocampal system.

Another class of models focuses primarily on single decisions within a task, and assume an underlying stochastic process [35.186, 233–235]. Typically, these models address two-choice short-term decisions made over a second or two [35.186]. The decision process begins with a starting point and accumulates information over time resulting in a diffusive (random walk) process. When the diffusion reaches one of two boundaries on either side of the starting point, the corresponding decision is made. This elegant approach can model such concrete issues as decision accuracy, decision time, and the distribution of decisions without any reference to the underlying neural mechanisms, which is both its chief strength and its primary weakness. Several connectionist models have also been developed based on paradigms similar to the diffusion approach [35.236–238]. The neural basis of such models has been discussed in detail in [35.239]. A population-coding neural model that makes Bayesian decisions based on cumulative evidence has been described by *Beck* et al. [35.98].

Reinforcement learning [35.176] is widely used in many engineering applications, but several models go beyond purely computational use and include details of the underlying brain regions and neurophysiology [35.141, 240]. Excellent reviews of such models are provided in [35.241–243]. Recently, models have also been proposed to show how dopamine-mediated learning could work with spiking neurons [35.244] and population codes [35.245].

Computational models that focus on working memory per se (i.e., not on the entire problem of cognitive control) have mainly considered how the requirement of selective temporal persistence can be met by biologically plausible neural networks [35.246, 247]. Since working memories must bridge over temporal durations (e.g., in remembering a cue over a delay period), there must be some neural mechanism to allow activity patterns to persist selectively in time. A natural candidate for this is attractor dynamics in recurrent neural networks [35.248, 249], where the recurrence allows some activity patterns to be stabilized by reverberation [35.250]. The neurophysiological basis of such persistent activity has been studied in [35.251]. A central feature in many models of WM is the role of dopamine in the PFC [35.252–254]. In particular, it is believed that dopamine sharpens the response of PFC neurons involved in WM [35.255] and allows for reliable storage of timing information in the presence of distractors [35.246]. The model in [35.246, 252] includes several biophysical details such as the effect of dopamine on different ion channels and its differential modulation of various receptors. More abstract neural models for WM have been proposed in [35.256] and [35.257].

A especially interesting type of attractor network uses the so-called *bump attractors* – spatially localized patterns of activity stabilized by local network connectivity and global competition [35.258]. Such a network has been used in a biologically plausible model of WM in the PFC in [35.259], which demonstrates that the memory is robust against distracting stimuli. A similar conclusion is drawn in [35.180] based on another bump attractor model of working memory. It shows that dopamine in the PFC can provide robustness against distractors, but robustness against internal noise is achieved only when dopamine in the BG locks the state of the striatum. Recently, *Mongillo* et al. [35.260] have proposed the novel hypothesis that the persistence of neural activity in WM may be due to calcium-mediated facilitation rather than reverberation through recurrent connectivity.

35.4 Conclusion

This chapter has attempted to provide an overview of neurocomputational models for cognitive control, WM, and motor control. Given the vast body of both experimental and computational research in these areas, the review is necessarily incomplete, though every attempt has been made to highlights the major issues, and to provide the reader with a rich array of references covering the breadth of each area.

The models described in this chapter relate to several other mental functions including sensorimotor integration, memory, semantic cognition, etc., as well as to areas of engineering such as robotics and agent systems. However, these links are largely excluded from the chapter – in part for brevity, but mainly because most of them are covered elsewhere in this Handbook.

References

35.1   J. Fuster: The cognit: A network model of cortical representation, Int. J. Psychophysiol. **60**, 125–132 (2006)

35.2   J. Fuster: *The Prefrontal Cortex* (Academic, London 2008)

35.3   M.T. Turvey: Coordination, Am. Psychol. **45**, 938–953 (1990)

35.4   D. Sternad, M.T. Turvey: Control parameters, equilibria, and coordination dynamics, Behav. Brain Sci. **18**, 780–783 (1996)

35.5   R. Pfeifer, M. Lungarella, F. Iida: Self-organization, embodiment, and biologically inspired robotics, Science **318**, 1088–1093 (2007)

35.6   A. Chemero: *Radical Embodied Cognitive Science* (MIT Press, Cambridge 2011)

35.7   J.C. Houk, S.P. Wise: Distributed modular architectures linking basal ganglia, cerebellum, and cerebral cortex: Their role in planning and controlling action, Cereb. Cortex **5**, 95–110 (2005)

35.8   K. Doya: What are the computations of the cerebellum, the basal ganglia and the cerebral cortex?, Neural Netw. **12**, 961–974 (1999)

35.9   M. Kawato, H. Gomi: A computational model of four regions of the cerebellum based on feedback-error learning, Biol. Cybern. **68**, 95–103 (1992)

35.10   A.M. Graybiel: Building action repertoires: Memory and learning functions of the basal ganglia, Curr. Opin. Neurobiol. **5**, 733–741 (1995)

35.11   A.M. Graybiel: The basal ganglia and cognitive pattern generators, Schizophr. Bull. **23**, 459–469 (1997)

35.12   A.M. Graybiel: The basal ganglia: Learning new tricks and loving it, Curr. Opin. Neurobiol. **15**, 638–644 (2005)

35.13   M.D. Humphries, R.D. Stewart, K.N. Gurney: A physiologically plausible model of action selection and oscillatory activity in the basal ganglia, J. Neurosci. **26**, 12921–12942 (2006)

35.14   J.C. Houk: Agents of the mind, Biol. Cybern. **92**, 427–437 (2005)

35.15   E. Hoshi, K. Shima, J. Tanji: Neuronal activity in the primate prefrontal cortex in the process of motor selection based on two behavioral rules, J. Neurophysiol. **83**, 2355–2373 (2000)

35.16   E.K. Miller, J.D. Cohen: An integrative theory of prefrontal cortex function, Annu. Rev. Neurosci. **4**, 167–202 (2001)

35.17   N.P. Rougier, D.C. Noelle, T.S. Braver, J.D. Cohen, R.C. O'Reilly: Prefrontal cortex and flexible cognitive control: Rules without symbols, PNAS **102**, 7338–7343 (2005)

35.18 J. Tanji, E. Hoshi: Role of the lateral prefrontal cortex in executive behavioral control, Physiol. Rev. **88**, 37–57 (2008)

35.19 M.M. Botvinick, J.D. Cohen, C.S. Carter: Conflict monitoring and anterior cingulate cortex: An update, Trends Cogn. Sci. **8**, 539–546 (2004)

35.20 M.M. Botvinick: Conflict monitoring and decision making: Reconciling two perspectives on anterior cingulate function, Cogn. Affect Behav. Neurosci. **7**, 356–366 (2008)

35.21 J.W. Brown, T.S. Braver: Learned predictions of error likelihood in the anterior cingulate cortex, Science **307**, 1118–1121 (2005)

35.22 J.S. Albus: New approach to manipulator control: The cerebellar model articulation controller (CMAC), J. Dyn. Sys. Meas. Control **97**, 220–227 (1975)

35.23 D. Marr: A theory of cerebellar cortex, J. Physiol. **202**, 437–470 (1969)

35.24 M.H. Dickinson, C.T. Farley, R.J. Full, M.A.R. Koehl, R. Kram, S. Lehman: How animals move: An integrative view, Science **288**, 100–106 (2000)

35.25 H. Haken, J.A.S. Kelso, H. Bunz: A theoretical model of phase transitions in human hand movements, Biol. Cybern. **51**, 347–356 (1985)

35.26 E. Saltzman, J.A.S. Kelso: Skilled actions: A task dynamic approach, Psychol. Rev. **82**, 225–260 (1987)

35.27 P.N. Kugler, M.T. Turvey: *Information, Natural Law, and the Self-Assembly of Rhythmic Movement* (Lawrence Erlbaum, Hillsdale 1987)

35.28 G. Schöner: A dynamic theory of coordination of discrete movement, Biol. Cybern. **63**, 257–270 (1990)

35.29 J.A.S. Kelso: *Dynamic Patterns: The Self-Organization of Brain and Behavior* (MIT Press, Cambridge 1995)

35.30 P. Morasso, V. Sanguineti, G. Spada: A computational theory of targeting movements based on force fields and topology representing networks, Neurocomputing **15**, 411–434 (1997)

35.31 J.P. Scholz, G. Schöner: The uncontrolled manifold concept: Identifying control variables for a functional task, Exp. Brain Res. **126**, 289–306 (1999)

35.32 M.A. Riley, M.T. Turvey: Variability and determinism in motor behavior, J. Mot. Behav. **34**, 99–125 (2002)

35.33 M.A. Riley, N. Kuznetsov, S. Bonnette: State-, parameter-, and graph-dynamics: Constraints and the distillation of postural control systems, Sci. Mot. **74**, 5–18 (2011)

35.34 E.C. Goldfield: *Emergent Forms: Origins and Early Development of Human Action and Perception* (Oxford Univ. Press, Oxford 1995)

35.35 J.A.S. Kelso, G.C. de Guzman, C. Reveley, E. Tognoli: Virtual partner interaction (VPI): Exploring novel behaviors via coordination dynamics, PLoS ONE **4**, e5749 (2009)

35.36 V.C. Ramenzoni, M.A. Riley, K. Shockley, A.A. Baker: Interpersonal and intrapersonal coordinative modes for joint and individual task performance, Human Mov. Sci. **31**, 1253–1267 (2012)

35.37 M.A. Riley, M.C. Richardson, K. Shockley, V.C. Ramenzoni: Interpersonal synergies, Front. Psychol. **2**(38), DOI 10.3389/fpsyg.2011.00038. (2011)

35.38 A.G. Feldman, M.F. Levin: The equilibrium-point hypothesis–past, present and future, Adv. Exp. Med. Biol. **629**, 699–726 (2009)

35.39 M.L. Latash: Motor synergies and the equilibrium-point hypothesis, Mot. Control **14**, 294–322 (2010)

35.40 C.E. Sherrington: *Integrative Actions of the Nervous System* (Yale Univ. Press, New Haven 1906)

35.41 C.E. Sherrington: Remarks on the reflex mechanism of the step, Brain **33**, 1–25 (1910)

35.42 C.E. Sherrington: Flexor-reflex of the limb, crossed extension reflex, and reflex stepping and standing (cat and dog), J. Physiol. **40**, 28–121 (1910)

35.43 S. Grillner, T. Deliagina, O. Ekeberg, A. El Manira, R.H. Hill, A. Lansner, G.N. Orlovsky, P. Wallén: Neural networks that co-ordinate locomotion and body orientation in lamprey, Trends Neurosci. **18**, 270–279 (1995)

35.44 P.J. Whelan: Control of locomotion in the decerebrate cat, Prog. Neurobiol. **49**, 481–515 (1996)

35.45 S. Grillner: The motor infrastructure: From ion channels to neuronal networks, Nat. Rev. Neurosci. **4**, 673–686 (2003)

35.46 S. Grillner: Biological pattern generation: The cellular and computational logic of networks in motion, Neuron **52**, 751–766 (2006)

35.47 A.J. Ijspeert, A. Crespi, D. Ryczko, J.M. Cabelguen: From swimming to walking with a salamander robot driven by a spinal cord model, Science **315**, 1416–1420 (2007)

35.48 A.P. Georgopoulos, J.F. Kalaska, R. Caminiti, J.T. Massey: On the relations between the direction of two-dimensional arm movements and cell discharge in primate motor cortex, J. Neurosci. **2**, 1527–1537 (1982)

35.49 A.P. Georgopoulos, R. Caminiti, J.F. Kalaska, J.T. Massey: Spatial coding of movement: A hypothesis concerning the coding of movement direction by motor cortical populations, Exp. Brain Res. Suppl. **7**, 327–336 (1983)

35.50 A.P. Georgopoulos, R. Caminiti, J.F. Kalaska: Static spatial effects in motor cortex and area 5: Quantitative relations in a two-dimensional space, Exp. Brain Res. **54**, 446–454 (1984)

35.51 A.P. Georgopoulos, R.E. Kettner, A.B. Schwartz: Primate motor cortex and free arm movements to visual targets in three-dimensional space. II: Coding of the direction of movement by a neuronal population, J. Neurosci. **8**, 2928–2937 (1988)

35.52   A.P. Georgopoulos, J. Ash, N. Smyrnis, M. Taira: The motor cortex and the coding of force, Science **256**, 1692–1695 (1992)

35.53   J. Ashe, A.P. Georgopoulos: Movement parameters and neural activity in motor cortex and area, Cereb. Cortex **5**(6), 590–600 (1994)

35.54   A.B. Schwartz, R.E. Kettner, A.P. Georgopoulos: Primate motor cortex and free arm movements to visual targets in 3-D space. I. Relations between singlecell discharge and direction of movement, J. Neurosci. **8**, 2913–2927 (1988)

35.55   D. Bullock, S. Grossberg: Neural dynamics of planned arm movements: emergent invariants and speed–accuracy properties during trajectory formation, Psychol. Rev. **95**, 49–90 (1988)

35.56   D. Bullock, S. Grossberg, F.H. Guenther: A self-organizing neural model of motor equivalent reaching and tool use by a multijoint arm, J. Cogn. Neurosci. **5**, 408–435 (1993)

35.57   D. Bullock, P. Cisek, S. Grossberg: Cortical networks for control of voluntary arm movements under variable force conditions, Cereb. Cortex **8**, 48–62 (1998)

35.58   S.H. Scott, J.F. Kalaska: Changes in motor cortex activity during reaching movements with similar hand paths but different arm postures, J. Neurophysiol. **73**, 2563–2567 (1995)

35.59   S.H. Scott, J.F. Kalaska: Reaching movements with similar hand paths but different arm orientations. I. Activity of individual cells in motor cortex, J. Neurophysiol. **77**, 826–852 (1997)

35.60   D.W. Moran, A.B. Schwartz: Motor cortical representation of speed and direction during reaching, J. Neurophysiol. **82**, 2676–2692 (1999)

35.61   R. Shadmehr, S.P. Wise: *The Computational Neurobiology of Reaching and Pointing: A Foundation for Motor Learning* (MIT Press, Cambridge 2005)

35.62   A. d'Avella, A. Portone, L. Fernandez, F. Lacquaniti: Control of fast-reaching movements by muscle synergy combinations, J. Neurosci. **26**, 7791–7810 (2006)

35.63   E. Bizzi, V.C. Cheung, A. d'Avella, P. Saltiel, M. Tresch: Combining odules for movement, Brain Res. Rev. **7**, 125–133 (2008)

35.64   S. Muceli, A.T. Boye, A. d'Avella, D. Farina: Identifying representative synergy matrices for describing muscular activation patterns during multidirectional reaching in the horizontal plane, J. Neurophysiol. **103**, 1532–1542 (2010)

35.65   S.F. Giszter, F.A. Mussa-Ivaldi, E. Bizzi: Convergent force fields organized in the frog's spinal cord, J. Neurosci. **13**, 467–491 (1993)

35.66   F.A. Mussa-Ivaldi, S.F. Giszter: Vector field approximation: A computational paradigm for motor control and learning, Biol. Cybern. **67**, 491–500 (1992)

35.67   M.C. Tresch, P. Saltiel, E. Bizzi: The construction of movement by the spinal cord, Nat. Neurosci. **2**, 162–167 (1999)

35.68   W.J. Kargo, S.F. Giszter: Rapid correction of aimed movements by summation of force-field primitives, J. Neurosci. **20**, 409–426 (2000)

35.69   A. d'Avella, P. Saltiel, E. Bizzi: Combinations of muscle synergies in the construction of a natural motor behavior, Nat. Neurosci. **6**, 300–308 (2003)

35.70   A. d'Avella, E. Bizzi: Shared and specific muscle synergies in natural motor behaviors, Proc. Natl. Acad. Sci. USA **102**, 3076–3081 (2005)

35.71   L.H. Ting, J.M. Macpherson: A limited set of muscle synergies for force control during a postural task, J. Neurophysiol. **93**, 609–613 (2005)

35.72   G. Torres-Oviedo, J.M. Macpherson, L.H. Ting: Muscle synergy organization is robust across a variety of postural perturbations, J. Neurophysiol. **96**, 1530–1546 (2006)

35.73   L.E. Sergio, J.F. Kalaska: Systematic changes in motor cortex cell activity with arm posture during directional isometric force generation, J. Neurophysiol. **89**, 212–228 (2003)

35.74   R. Ajemian, A. Green, D. Bullock, L. Sergio, J. Kalaska, S. Grossberg: Assessing the function of motor cortex: Single-neuron models of how neural response is modulated by limb biomechanics, Neuron **58**, 414–428 (2008)

35.75   B. Cesqui, A. d'Avella, A. Portone, F. Lacquaniti: Catching a ball at the right time and place: Individual factors matter, PLoS ONE **7**, e31770 (2012)

35.76   P. Morasso, F.A. Mussa-Ivaldi: Trajectory formation and handwriting: A computational model, Biol. Cybern. **45**, 131–142 (1982)

35.77   A.B. Schwartz: Motor cortical activity during drawing movements: Single unit activity during sinusoid tracing, J. Neurophysiol. **68**, 528–541 (1992)

35.78   A.B. Schwartz: Motor cortical activity during drawing movements: Population representation during sinusoid tracing, J. Neurophysiol. **70**, 28–36 (1993)

35.79   A.B. Schwartz: Direct cortical representation of drawing, Science **265**, 540–542 (1994)

35.80   D.W. Moran, A.B. Schwartz: Motor cortical activity during drawing movements: Population representation during spiral tracing, J. Neurophysiol. **82**, 2693–2704 (1999)

35.81   R.W. Paine, S. Grossberg, A.W.A. Van Gemmert: A quantitative evaluation of the AVITEWRITE model of handwriting learning, Human Mov. Sci. **23**, 837–860 (2004)

35.82   G. Torres-Oviedo, L.H. Ting: Muscle synergies characterizing human postural responses, J. Neurophysiol. **98**, 2144–2156 (2007)

35.83   L.H. Ting, J.L. McKay: Neuromechanics of muscle synergies for posture and movement, Curr. Opin. Neurobiol. **17**, 622–628 (2007)

35.84   N. Bernstein: *The Coordination and Regulation of Movements* (Pergamon, Oxford 1967)

35.85   P.D. Neilson, M.D. Neilson: On theory of motor synergies, Human Mov. Sci. **29**, 655–683 (2010)

35.86   W. Penfield, E. Boldrey: Somatic motor and sensory representation in the motor cortex of man as studied by electrical stimulation, Brain **60**, 389–443 (1937)

35.87   T.D. Sanger: Theoretical considerations for the analysis of population coding in motor cortex, Neural Comput. **6**, 29–37 (1994)

35.88   J.K. Chapin, R.A. Markowitz, K.A. Moxo, M.A.L. Nicolelis: Direct real-time control of a robot arm using signals derived from neuronal population recordings in motor cortex, Nat. Neurosci. **2**, 664–670 (1999)

35.89   A.A. Lebedev, M.A.L. Nicolelis: Brain-machine interfaces: Past, present, and future, Trends Neurosci. **29**, 536–546 (2006)

35.90   E. Salinas, L. Abbott: Transfer of coded information from sensory to motor networks, J. Neurosci. **15**, 6461–6474 (1995)

35.91   E. Salinas, L. Abbott: A model of multiplicative neural responses in parietal cortex, Proc. Natl. Acad. Sci. USA **93**, 11956–11961 (1996)

35.92   A. Pouget, T. Sejnowski: A neural model of the cortical representation of egocentric distance, Cereb. Cortex **4**, 314–329 (1994)

35.93   A. Pouget, T. Sejnowski: Spatial transformations in the parietal cortex using basis functions, J. Cogn. Neurosci. **9**, 222–237 (1997)

35.94   A. Pouget, L.H. Snyder: Computational approaches to sensorimotor transformations, Nat. Neurosci. Supp. **3**, 1192–1198 (2000)

35.95   A. Pouget, P. Dayan, R.S. Zemel: Information processing with population codes, Nat. Rev. Neurosci. **1**, 125–132 (2000)

35.96   A. Pouget, P. Dayan, R.S. Zemel: Inference and computation with population codes, Annu. Rev. Neurosci. **26**, 381–410 (2003)

35.97   W.J. Ma, J.M. Beck, P.E. Latham, A. Pouget: Bayesian inference with probabilistic population codes, Nat. Neurosci. **9**, 1432–1438 (2006)

35.98   J.M. Beck, W.J. Ma, R. Kiani, T. Hanks, A.K. Churchland, J. Roitman, M.N. Shadlen, P.E. Latham, A. Pouget: Probabilistic population codes for bayesian decision making, Neuron **60**, 1142–1152 (2008)

35.99   R. Ajemian, D. Bullock, S. Grossberg: Kinematic coordinates in which motor cortical cells encode movement direction, Neurophys. **84**, 2191–2203 (2000)

35.100   R. Ajemian, D. Bullock, S. Grossberg: A model of movement coordinates in the motor cortex: Posture-dependent changes in the gain and direction of single cell tuning curves, Cereb. Cortex **11**, 1124–1135 (2001)

35.101   B.B. Averbeck, M.V. Chafee, D.A. Crowe, A.P. Georgopoulos: Parallel processing of serial movements in prefrontal cortex, PNAS **99**, 13172–13177 (2002)

35.102   R. Caminiti, P.B. Johnson, A. Urbano: Making arm movements within different parts of space: Dynamic aspects in the primate motor cortex, J. Neurosci. **10**, 2039–2058 (1990)

35.103   K.M. Graham, K.D. Moore, D.W. Cabel, P.L. Gribble, P. Cisek, S.H. Scott: Kinematics and kinetics of multijoint reaching in nonhuman primates, J. Neurophysiol. **89**, 2667–2677 (2003)

35.104   A. Shah, A.H. Fagg, A.G. Barto: Cortical involvement in the recruitment of wrist muscles, J. Neurophysiol. **91**, 2445–2456 (2004)

35.105   M.S.A. Graziano, T. Aflalo, D.F. Cooke: Arm movements evoked by electrical stimulation in the motor cortex of monkeys, J. Neurophysiol. **94**, 4209–4223 (2005)

35.106   M.S.A. Graziano: The organization of behavioral repertoire in motor cortex, Annu. Rev. Neurosci. **29**, 105–134 (2006)

35.107   M.S.A. Graziano: *The Intelligent Movement Machine* (Oxford Univ. Press, Oxford 2008)

35.108   T.N. Aflalo, M.S.A. Graziano: Possible origins of the complex topographic organization of motor cortex: Reduction of a multidimensional space onto a two-dimensional array, J. Neurosci. **26**, 6288–6297 (2006)

35.109   K. Shima, J. Tanji: Both supplementary and pre-supplementary motor areas are crucial for the temporal organization of multiple movements, J. Neurophysiol. **80**, 3247–3260 (1998)

35.110   J.-W. Sohn, D. Lee: Order-dependent modulation of directional signals in the supplementary and presupplementary motor areas, J. Neurosci. **27**, 13655–13666 (2007)

35.111   H. Mushiake, M. Inase, J. Tanji: Neuronal Activity in the primate premotor, supplementary, and precentral motor cortex during visually guided and internally determined sequential movements, J. Neurophysiol. **66**, 705–718 (1991)

35.112   H. Mushiake, P.L. Strick: Pallidal neuron activity during sequential arm movements, J. Neurophysiol. **74**, 2754–2758 (1995)

35.113   H. Mushiake, P.L. Strick: Preferential activity of dentate neurons during limb movements guided by vision, J. Neurophysiol. **70**, 2660–2664 (1993)

35.114   D. Baldauf, H. Cui, R.A. Andersen: The posterior parietal cortex encodes in parallel both goals for double-reach sequences, J. Neurosci. **28**, 10081–10089 (2008)

35.115   T. Flash, B. Hochner: Motor primitives in vertebrates and invertebrates, Curr. Opin. Neurobiol. **15**, 660–666 (2005)

35.116   J.A.S. Kelso: Synergies: Atoms of brain and behavior. In: *Progress in Motor Control*, ed. by D. Sternad (Springer, Berlin, Heidelberg 2009) pp. 83–91

35.117   F.A. Mussa-Ivaldi: Do neurons in the motor cortex encode movement direction? An alternate hypothesis, Neurosci. Lett. **91**, 106–111 (1988)

35.118 F.A. Mussa-Ivaldi: From basis functions to basis fields: vector field approximation from sparse data, Biol. Cybern. **67**, 479489 (1992)

35.119 A. d'Avella, D.K. Pai: Modularity for sensorimotor control: Evidence and a new prediction, J. Mot. Behav. **42**, 361–369 (2010)

35.120 A. d'Avella, L. Fernandez, A. Portone, F. Lacquaniti: Modulation of phasic and tonic muscle synergies with reaching direction and speed, J. Neurophysiol. **100**, 1433–1454 (2008)

35.121 G. Torres-Oviedo, L.H. Ting: Subject-specific muscle synergies in human balance control are consistent across different biomechanical contexts, J. Neurophysiol. **103**, 3084–3098 (2010)

35.122 C.B. Hart: A neural basis for motor primitives in the spinal cord, J. Neurosci. **30**, 1322–1336 (2010)

35.123 T. Drew, J. Kalaska, N. Krouchev: Muscle synergies during locomotion in the cat: A model for motor cortex control, J. Physiol. **586**(5), 1239–1245 (2008)

35.124 P.D. Neilson, M.D. Neilson: Motor maps and synergies, Human Mov. Sci. **24**, 774–797 (2005)

35.125 J.J. Kutch, A.D. Kuo, A.M. Bloch, W.Z. Rymer: Endpoint force fluctuations reveal flexible rather than synergistic patterns of muscle cooperation, J. Neurophysiol. **100**, 2455–2471 (2008)

35.126 M.C. Tresch, A. Jarc: The case for and against muscle synergies, Curr. Opin. Neurobiol. **19**, 601–607 (2009)

35.127 A. Ijspeert, J. Nakanishi, S. Schaal: Learning rhythmic movements by demonstration using nonlinear oscillators, IEEE Int. Conf. Intell. Rob. Syst. (IROS 2002), Lausanne (2002) pp. 958–963

35.128 A. Ijspeert, J. Nakanishi, S. Schaal: Movement imitation with nonlinear dynamical systems in humanoid robots, Int. Conf. Robotics Autom. (ICRA 2002), Washington (2002) pp. 1398–1403

35.129 A. Ijspeert, J. Nakanishi, S. Schaal: Trajectory formation for imitation with nonlinear dynamical systems, IEEE Int. Conf. Intell. Rob. Syst. (IROS 2001), Maui (2001) pp. 752–757

35.130 A. Ijspeert, J. Nakanishi, S. Schaal: Learning attractor landscapes for learning motor primitives. In: *Advances in Neural Information Processing Systems 15*, ed. by S. Becker, S. Thrun, K. Obermayer (MIT Press, Cambridge 2003) pp. 1547–1554

35.131 S. Schaal, J. Peters, J. Nakanishi, A. Ijspeert: Control, planning, learning, and imitation with dynamic movement primitives, Proc. Workshop Bilater. Paradig. Humans Humanoids. IEEE Int. Conf. Intell. Rob. Syst. (IROS 2003), Las Vegas (2003)

35.132 S. Schaal, P. Mohajerian, A. Ijspeert: Dynamics systems vs. optimal control – a unifying view. In: *Computational Neuroscience: Theoretical Insights into Brain Function, Progress in Brain Research*, Vol. 165, ed. by P. Cisek, T. Drew, J.F. Kalaska (Elsevier, Amsterdam 2007) pp. 425–445

35.133 K.V. Byadarhaly, M. Perdoor, A.A. Minai: A neural model of motor synergies, Proc. Int. Conf. Neural Netw., San Jose (2011) pp. 2961–2968

35.134 K.V. Byadarhaly, M.C. Perdoor, A.A. Minai: A modular neural model of motor synergies, Neural Netw. **32**, 96–108 (2012)

35.135 G.E. Alexander, M.R. DeLong, P.L. Strick: Parallel organization of functionally segregated circuits linking basal ganglia and cortex, Annu. Rev. Neurosci. **9**, 357–381 (1986)

35.136 A.W. Flaherty, A.M. Graybiel: Input–output organization of the sensorimotor striatum in the squirrel monkey, J. Neurosci. **14**, 599–610 (1994)

35.137 S. Grillner, J. Hellgren, A. Ménard, K. Saitoh, M.A. Wikström: Mechanisms for selection of basic motor programs – roles for the striatum and pallidum, Trends Neurosci. **28**, 364–370 (2005)

35.138 W. Schultz, P. Dayan, P.R. Montague: A neural substrate of prediction and reward, Science **275**, 1593–1599 (1997)

35.139 W. Schultz, A. Dickinson: Neuronal coding of prediction errors, Annu. Rev. Neurosci. **23**, 473–500 (2000)

35.140 W. Schultz: Multiple reward signals in the brain, Nat. Rev. Neurosci. **1**, 199–207 (2000)

35.141 P.R. Montague, S.E. Hyman, J.D. Cohen: Computational roles for dopamine in behavioural control, Nature **431**, 760–767 (2004)

35.142 D.M. Wolpert, M. Kawato: Multiple paired forward and inverse models for motor control, Neural Netw. **11**, 1317–1329 (1998)

35.143 M. Kawato: Internal models for motor control and trajectory planning, Curr. Opin. Neurobiol. **9**, 718–727 (1999)

35.144 D.M. Wolpert, Z. Ghahramani: Computational principles of movement neuroscience, Nat. Neurosci. Supp. **3**, 1212–1217 (2000)

35.145 M. Kawato, K. Furukawa, R. Suzuki: A hierarchical neural network model for control and learning of voluntary movement, Biol. Cybern. **57**, 169–185 (1987)

35.146 R. Shadmehr, F.A. Mussa-Ivaldi: Adaptive representation of dynamics during learning of a motor task, J. Neurosci. **74**, 3208–3224 (1994)

35.147 D.M. Wolpert, Z. Ghahramani, M.I. Jordan: An internal model for sensorimotor integration, Science **269**, 1880–1882 (1995)

35.148 A. Karniel, G.F. Inbar: A model for learning human reaching movements, Biol. Cybern. **77**, 173–183 (1997)

35.149 A. Karniel, G.F. Inbar: Human motor control: Learning to control a time-varying, nonlinear, many-to-one system, IEEE Trans. Syst. Man Cybern. Part C **30**, 1–11 (2000)

35.150 Y. Burnod, P. Baraduc, A. Battaglia-Mayer, E. Guigon, E. Koechlin, S. Ferraina, F. Lacquaniti, R. Caminiti: Parieto-frontal coding of reaching: An integrated framework, Exp. Brain Res. **129**, 325–346 (1999)

35.151 M. Haruno, D.M. Wolpert, M. Kawato: MOSAIC model for sensorimotor learning and control, Neural Comput. **13**, 2201–2220 (2001)

35.152   A.G. Barto, A.H. Fagg, N. Sitkoff, J.C. Houk: A cerebellar model of timing and prediction in the control of reaching, Neural Comput. **11**, 565–594 (1999)

35.153   H. Ritter, T. Martinetz, K. Schulten: Topology-conserving maps for learning visuo-motor-coordination, Neural Netw. **2**, 159–168 (1989)

35.154   T. Martinetz, H. Ritter, K. Schulten: Three-dimensional neural net for learning visuo-motor coordination of a robot arm, IEEE Trans. Neural Netw. **1**, 131–136 (1990)

35.155   T. Kohonen: Self-organized formation of topologically correct feature maps, Biol. Cybern. **43**, 59–69 (1982)

35.156   P. Baraduc, E. Guignon, Y. Burnod: Recoding arm position to learn visuomotor transformations, Cereb. Cortex **11**, 906–917 (2001)

35.157   M.V. Butz, O. Herbort, J. Hoffmann: Exploiting redundancy for flexible behavior: Unsupervised learning in a modular sensorimotor control architecture, Psychol. Rev. **114**, 1015–1046 (2007)

35.158   A.F. Morse, J. de Greeff, T. Belpeame, A. Cangelosi: Epigenetic Robotics Architecture (ERA), IEEE Trans. Auton. Ment. Develop. 2 (2002) pp. 325–339

35.159   T. Matsubara, S.-H. Hyon, J. Morimoto: Learning parametric dynamic movement primitives from multiple demonstrations, Neural Netw. **24**, 493–500 (2011)

35.160   K.V. Byadarhaly, A.A. Minai: A Hierarchical Model of Synergistic Motor Control, Proc. Int. Joint Conf. Neural Netw., Dallas (2013)

35.161   T. Flash, N. Hogan: The coordination of arm movements: An experimentally confirmed mathematical model, J. Neurosci. **5**, 1688–1703 (1985)

35.162   Y. Uno, M. Kawato, R. Suzuki: Formation and control of optimal trajectories in human multi-joint arm movements: Minimum torque-change model, Biol. Cybern. **61**, 89–101 (1989)

35.163   S. Ben-Itzhak, A. Karniel: Minimum acceleration criterion with constraints implies bang-bang control as an underlying principle for optimal trajectories of arm reaching movements, Neural Comput. **20**, 779–812 (2008)

35.164   C.M. Harris, D.M. Wolpert: Signal-dependent noise determines motor planning, Nature **394**, 780–784 (1998)

35.165   E. Todorov, M.I. Jordan: Optimal feedback control as a theory of motor coordination, Nat. Neurosci. **5**, 1226–1235 (2002)

35.166   E. Todorov: Optimality principles in sensorimotor control, Nat. Neurosci. **7**, 907–915 (2004)

35.167   F.J. Valero-Cuevas, M. Venkadesan, E. Todorov: Structured variability of muscle activations supports the minimal intervention principle of motor control, J. Neurophysiol. **102**, 59–68 (2009)

35.168   E. Trainin, R. Meir, A. Karniel: Explaining patterns of neural activity in the primary motor cortex using spinal cord and limb biomechanics models, J. Neurophysiol. **97**, 3736–3750 (2007)

35.169   A. Biess, D.G. Libermann, T. Flash: A computational model for redundant arm three-dimensional pointing movements: Integration of independent spatial and temporal motor plans simplifies movement dynamics, J. Neurosci. **27**, 13045–13064 (2007)

35.170   J.J. Gibson: The Theory of Affordances. In: *Perceiving, Acting, and Knowing: Toward an Ecological Psychology*, ed. by R. Shaw, J. Bransford (Lawrence Erlbaum, Hillsdale 1977) pp. 67–82

35.171   M.L. Latash, J.P. Scholz, G. Schöner: Toward a new theory of motor synergies, Mot. Control **11**, 276–308 (2007)

35.172   S. Dehaene, M. Kerszberg, J.-P. Changeux: A neuronal model of a global workspace in effortful cognitive tasks, Proc. Natl. Acad. Sci. USA **95**, 14529–14534 (1998)

35.173   M. Botvinick, D.C. Plaut: Doing without schema hierarchies: A recurrent connectionist approach to normal and impaired routine sequential action, Psychol. Rev. **111**, 395–429 (2004)

35.174   F.A. Middleton, P.L. Strick: Basal ganglia output and cognition: Evidence from anatomical, behavioral, and clinical studies, Brain Cogn. **42**, 183–200 (2000)

35.175   F.A. Middleton, P.L. Strick: Basal ganglia 'projections' to the prefrontal cortex of the primate, Cereb. Cortex **12**, 926–945 (2002)

35.176   R.S. Sutton, A.G. Barto: *Reinforcement Learning* (MIT Press, Cambridge 1998)

35.177   R.S. Sutton: Learning to predict by the methods of temporal difference, Mach. Learn. **3**, 9–44 (1988)

35.178   N.D. Daw, Y. Niv, P. Dayan: Uncertainty-based competition between prefrontal and dorsolateral striatal systems for behavioral control, Nat. Neurosci. **8**, 1704–1711 (2005)

35.179   M.J. Frank, R.C. O'Reilly: A mechanistic account of striatal dopamine function in human cognition: Psychopharmacological studies with cabergoline and haloperidol, Behav. Neurosci. **120**, 497–517 (2006)

35.180   A.J. Gruber, P. Dayan, B.S. Gutkin, S.A. Solla: Dopamine modulation in the basal ganglia locks the gate to working memory, J. Comput. Neurosci. **20**, 153–166 (2006)

35.181   A. Baddeley: *Human Memory* (Lawrence Erlbaum, Hove, UK 1990)

35.182   A. Baddeley: The episodic buffer: A new component of working memory?, Trends Cogn. Sci. **4**, 417–423 (2000)

35.183   P.S. Goldman-Rakic: Cellular basis of working memory, Neuron **14**, 477–485 (1995)

35.184   P.S. Goldman-Rakic, A.R. Cools, K. Srivastava: The prefrontal landscape: implications of functional architecture for understanding human mentation and the central executive, Philos. Trans.: Biol. Sci. **351**, 1445–1453 (1996)

35.185   J. Duncan: An adaptive coding model of neural function in prefrontal cortex, Nat. Rev. Neurosci. **2**, 820–829 (2001)

35.186   R. Ratcliff, G. McKoon: The diffusion decision model: Theory and data for two-choice decision tasks, Neural Comput. **20**, 873–922 (2008)

35.187   G. Miller: The magical number seven, plus or minus two: Some limits of our capacity for processing information, Psychol. Rev. **63**, 81–97 (1956)

35.188   J.E. Lisman, A.P. Idiart: Storage of 7 ± 2 short-term memories in oscillatory subcycles, Science **267**, 1512–1515 (1995)

35.189   K. Ericsson, W. Kintsch: Long-term working memory, Psychol. Rev. **102**, 211–245 (1995)

35.190   C.J. Wilson: The contribution of cortical neurons to the firing pattern of striatal spiny neurons. In: *Models of Information Processing in the Basal Ganglia*, ed. by J.C. Houk, J.L. Davis, D.G. Beiser (MIT Press, Cambridge 1995) pp. 29–50

35.191   A.M. Graybiel, T. Aosaki, A.W. Flaherty, M. Kimura: The basal ganglia and adaptive motor control, Science **265**, 1826–1831 (1994)

35.192   A.M. Graybiel: The basal ganglia and chunking of action repertoires, Neurobiol. Learn. Mem. **70**, 119–136 (1998)

35.193   M.D. Humphries, K. Gurney: The role of intra-thalamic and thalamocortical circuits in action selection, Network **13**, 131–156 (2002)

35.194   R.C. O'Reilly, Y. Munakata: *Computational explorations in cognitive neuroscience: Understanding the mind by simulating the brain* (MIT Press, Cambridge 2000)

35.195   M.J. Frank, B. Loughry, R.C. O'Reilly: Interactions between frontal cortex and basal ganglia in working memory: A computational model, Cogn. Affect Behav. Neurosci. **1**, 137–160 (2001)

35.196   T.E. Hazy, M.J. Frank, R.C. O'Reilly: Banishing the homunculus: Making working memory work, Neuroscience **139**, 105–118 (2006)

35.197   R.C. O'Reilly, M.J. Frank: Making working memory work: A computational model of learning in the prefrontal cortex and basal ganglia, Neural Comput. **18**, 283–328 (2006)

35.198   M.J. Frank, E.D. Claus: Anatomy of a decision: Striato-orbitofrontal interactions in reinforcement learning, decision making and reversal, Psychol. Rev. **113**, 300–326 (2006)

35.199   R.C. O'Reilly: Biologically based computational models of high level cognition, Science **314**, 91–94 (2006)

35.200   R.C. O'Reilly, S.A. Herd, W.M. Pauli: Computational models of cognitive control, Curr. Opin. Neurobiol. **20**, 257–261 (2010)

35.201   M.E. Hasselmo: A model of prefrontal cortical mechanisms for goal-directed behavior, J. Cogn. Neurosci. **17**, 1–14 (2005)

35.202   M.E. Hasselmo, C.E. Stern: Mechanisms underlying working memory for novel information, Trends Cogn. Sci. **10**, 487–493 (2006)

35.203   J.R. Reynolds, R.C. O'Reilly: Developing PFC representations using reinforcement learning, Cognition **113**, 281–292 (2009)

35.204   T.S. Braver, D.M. Barch, J.D. Cohen: Cognition and control in schizophrenia: A computational model of dopamine and prefrontal function, Biol. Psychiatry **46**, 312–328 (1999)

35.205   S. Monsell: Task switching, Trends Cog. Sci. **7**, 134–140 (2003)

35.206   T.S. Braver, J.R. Reynolds, D.I. Donaldson: Neural mechanisms of transient and sustained cognitive control during task switching, Neuron **39**, 713–726 (2003)

35.207   H. Imamizu, T. Kuroda, T. Yoshioka, M. Kawato: Functional magnetic resonance imaging examination of two modular architectures for switching multiple internal models, J. Neurosci. **24**, 1173–1181 (2004)

35.208   R.P. Cooper, T. Shallice: Contention scheduling and the control of routine activities, Cogn. Neuropsychol. **17**, 297–338 (2000)

35.209   R.P. Cooper, T. Shallice: Hierarchical schemas and goals in the control of sequential behavior, Psychol. Rev. **113**, 887–916 (2006)

35.210   P. Dayan: Images, frames, and connectionist hierarchies, Neural Comput. **18**, 2293–2319 (2006)

35.211   P. Dayan: Simple substrates for complex cognition, Front. Neurosci. **2**, 255–263 (2008)

35.212   S. Helie, R. Sun: Incubation, insight, and creative problem solving: A unified theory and a connectionist model, Psychol. Rev. **117**, 994–1024 (2010)

35.213   S. Grossberg, L.R. Pearson: Laminar cortical dynamics of cognitive and motor working memory, sequence learning and performance: Toward a unified theory of how the cerebral cortex works, Psychol. Rev. **115**, 677–732 (2008)

35.214   B.J. Rhodes, D. Bullock, W.B. Verwey, B.B. Averbeck, M.P.A. Page: Learning and production of movement sequences: Behavioral, neurophysiological, and modeling perspectives, Human Mov. Sci. **23**, 683–730 (2004)

35.215   B. Ans, Y. Coiton, J.-C. Gilhodes, J.-L. Velay: A neural network model for temporal sequence learning and motor programming, Neural Netw. **7**, 1461–1476 (1994)

35.216   R.S. Bapi, D.S. Levine: Modeling the role of frontal lobes in sequential task performance. I: Basic Strucure and primacy effects, Neural Netw. **7**, 1167–1180 (1994)

35.217   J.G. Taylor, N.R. Taylor: Analysis of recurrent cortico-basal ganglia-thalamic loops for working memory, Biol. Cybern. **82**, 415–432 (2000)

35.218   R.P. Cooper: Mechanisms for the generation and regulation of sequential behaviour, Philos. Psychol. **16**, 389–416 (2003)

35.219   R. Nishimoto, J. Tani: Learning to generate combinatorial action sequences utilizing the initial sensitivity of deterministic dynamical systems, Neural Netw. **17**, 925–933 (2004)

35.220 P.F. Dominey: From sensorimotor sequence to grammatical construction: evidence from simulation and neurophysiology, Adapt. Behav. **13**, 347–361 (2005)

35.221 E. Salinas: Rank-order-selective neurons form a temporal basis set for the generation of motor sequences, J. Neurosci. **29**, 4369–4380 (2009)

35.222 S. Vasa, T. Ma, K.V. Byadarhaly, M. Perdoor, A.A. Minai: A Spiking Neural Model for the Spatial Coding of Cognitive Response Sequences, Proc. IEEE Int. Conf. Develop. Learn., Ann Arbor (2010) pp. 140–146

35.223 F. Chersi, P.F. Ferrari, L. Fogassi: Neuronal chains for actions in the parietal lobe: A computational model, PloS ONE **6**, e27652 (2011)

35.224 M.R. Silver, S. Grossberg, D. Bullock, M.H. Histed, E.K. Miller: A neural model of sequential movement planning and control of eye movements: Item-order-rank working memory and saccade selection by the supplementary eye fields, Neural Netw. **26**, 29–58 (2011)

35.225 B.J. Baars: *A Cognitive Theory of Consciousness* (Cambridge Univ. Press, Cambridge 1988)

35.226 B.J. Baars, S. Franklin: How conscious experience and working memory interact, Trends Cog. Sci. **7**, 166–172 (2003)

35.227 S. Franklin, F.G.J. Patterson: The LIDA Architecture: Adding New Modes of Learning to an Intelligent, Autonomous, Software Agent, IDPT-2006 Proc. (Integrated Design and Process Technology) (Society for Design and Process Science, San Diego 2006)

35.228 S. Dehaene, J.-P. Changeux: The Wisconsin card sorting test: Theoretical analysis and modeling in a neuronal network, Cereb. Cortex **1**, 62–79 (1991)

35.229 S. Dehaene, L. Naccache: Towards a cognitive neuroscience of consciousness: Basic evidence and a workspace framework, Cognition **79**, 1–37 (2001)

35.230 L.R. Iyer, S. Doboli, A.A. Minai, V.R. Brown, D.S. Levine, P.B. Paulus: Neural dynamics of idea generation and the effects of priming, Neural Netw. **22**, 674–686 (2009)

35.231 L.R. Iyer, V. Venkatesan, A.A. Minai: Neurocognitive spotlights: Configuring domains for ideation, Proc. Int. Conf. Neural Netw. (2011) pp. 2961–2968

35.232 S. Doboli, A.A. Minai, P.J. Best: Latent attractors: A model for context-dependent place representations in the hippocampus, Neural Comput. **12**, 1003–1037 (2000)

35.233 R. Ratcliff: A theory of memory retrieval, Psychol. Rev. **85**, 59–108 (1978)

35.234 F.G. Ashby: A biased random-walk model for two choice reaction times, J. Math. Psychol. **27**, 277–297 (1983)

35.235 J.R. Busemeyer, J.T. Townsend: Decision field theory, Psychol. Rev. **100**, 432–459 (1993)

35.236 J.L. McClelland, D.E. Rumelhart: An interactive activation model of context effects in letter perception. Part 1: An account of basic findings, Psychol. Rev. **88**, 375–407 (1981)

35.237 D.E. Rumelhart, J.L. McClelland: An interactive activation model of context effects in letter perception: Part 2. The contextual enhancement effect and some tests and extensions of the model, Psychol. Rev. **89**, 60–94 (1982)

35.238 M. Usher, J.L. McClelland: The time course of perceptual choice: The leaky, competing accumulator model, Psychol. Rev. **108**, 550–592 (2001)

35.239 J.I. Gold, M.N. Shadlen: The neural basis of decision making, Annu. Rev. Neurosci. **30**, 535–574 (2007)

35.240 M. Khamassi, L. Lachèze, B. Girard, A. Berthoz, A. Guillot: Actor–critic models of reinforcement learning in the basal ganglia: From natural to artificial rats, Adapt. Behav. **13**, 131–148 (2005)

35.241 N.D. Daw, K. Doya: The computational neurobiology of learning and reward, Curr. Opin. Neurobiol. **16**, 199–204 (2006)

35.242 P. Dayan, Y. Niv: Reinforcement learning: The Good, The Bad and The Ugly, Curr. Opin. Neurobiol. **18**, 185–196 (2008)

35.243 K. Doya: Modulators of decision making, Nat. Neurosci. **11**, 410–416 (2008)

35.244 E.M. Izhikevich: Solving the distal reward problem through linkage of STDP and dopamine signaling, Cereb. Cortex **17**, 2443–2452 (2007)

35.245 R. Urbanczik, W. Senn: Reinforcement learning in populations of spiking neurons, Nat. Neurosci. **12**, 250–252 (2009)

35.246 D. Durstewitz, J.K. Seamans, T.J. Sejnowski: Dopamine mediated stabilization of delay-period activity in a network model of prefrontal cortex, J. Neurophysiol. **83**, 1733–1750 (2000)

35.247 D. Durstewitz, J.K. Seamans: The computational role of dopamine D1 receptors in working memory, Neural Netw. **15**, 561–572 (2002)

35.248 J.J. Hopfield: Neural networks and physical systems with emergent collective computational abilities, Proc. Natl. Acad. Sci. USA **79**, 2554–2558 (1982)

35.249 D.J. Amit, N. Brunel: Learning internal representations in an attractor neural network with analogue neurons, Netw. Comput. Neural Syst. **6**, 359–388 (1995)

35.250 D.J. Amit, N. Brunel: Model of global spontaneous activity and local structured activity during delay periods in the cerebral cortex, Cereb. Cortex **7**, 237–252 (1997)

35.251 X.J. Wang: Synaptic basis of cortical persistent activity: The importance of NMDA receptors to working memory, J. Neurosci. **19**, 9587–9603 (1999)

35.252 D. Durstewitz, M. Kelc, O. Gunturkun: A neurocomputational theory of the dopaminergic modulation of working memory functions, J. Neurosci. **19**, 2807–2822 (1999)

35.253 N. Brunel, X.-J. Wang: Effects of neuromodulation in a cortical network model of object working memory dominated by recurrent inhibition, J. Comput. Neurosci. **11**, 63–85 (2001)

35.254   J.D. Cohen, T.S. Braver, J.W. Brown: Computational perspectives on dopamine function in prefrontal cortex, Curr. Opin. Neurobiol. **12**, 223–229 (2002)

35.255   D. Servan-Schreiber, H. Printz, J.D. Cohen: A network model of catecholamine effects: Gain, signal-to-noise ratio, and behavior, Science **249**, 892–895 (1990)

35.256   S. Hochreiter, J. Schmidhuber: Long short-term memory, Neural Comput. **9**, 1735–1780 (1997)

35.257   S.L. Moody, S.P. Wise, G. di Pellegrino, D. Zipser: A model that accounts for activity in primate frontal cortex during a delayedmatch-to-sample task, J. Neurosci. **18**, 399–410 (1998)

35.258   R. Hahnloser, R.J. Douglas, M. Mahowald, K. Hepp: Feedback interactions between neuronal pointers and maps for attentional processing, Nat. Neurosci. **2**, 746–752 (1999)

35.259   A. Compte, N. Brunel, P.S. Goldman-Rakic, X.-J. Wang: Synaptic mechanisms and network dynamics underlying spatial working memory in a cortical network model, Cereb. Cortex **10**, 910–923 (2000)

35.260   G. Mongillo, O. Barak, M. Tsodyks: Synaptic theory of working memory, Science **319**, 1543–1546 (2008)

36. Cognitive Architectures and Agents

Sebastien Hélie, Ron Sun

A cognitive architecture is the essential structures and processes of a domain-generic computational cognitive model used for a broad, multiple-level, multiple domain analysis of cognition and behavior. This chapter reviews some of the most popular psychologically-oriented cognitive architectures, namely adaptive control of thought-rational (ACT-R), Soar, and CLARION. For each cognitive architecture, an overview of the model, some key equations, and a detailed simulation example are presented. The example simulation with ACT-R is the initial learning of the past tense of irregular verbs in English (developmental psychology), the example simulation with Soar is the well-known missionaries and cannibals problem (problem solving), and the example simulation with CLARION is a complex mine field navigation task (autonomous learning). This presentation is followed by a discussion of how cognitive architectures can be used in multi-agent social simulations. A detailed cognitive social simulation with CLARION is presented to reproduce results from organizational decision-making. The chapter concludes with a discussion of the impact of neural network modeling on cognitive architectures and a comparison of the different models.

36.1 Background

Cognitive theories are often underdetermined by data [36.1]. As such, different theories, with very little in common, can sometimes be used to account for the very same phenomena [36.2]. This problem can be resolved by adding constraints to cognitive theories. The most intuitive approach to adding constraints to any scientific theory is to collect more data. While experimental psychologists have come a long way toward this goal in over a century of psychology research, the gap between empirical and theoretical progress is still significant.

Another tactic that can be adopted toward constraining psychological theories is *unification* [36.1]. Newell argued that more data could be used to constraint a theory if the theory was designed to explain a wider range of phenomena. In particular, these *unified* (i.e., integrative) cognitive theories could be put to the test against well-known (stable) regularities that have been observed in psychology. So far, these integrative theories have taken the form of *cognitive architectures*, and some of them have been very successful in explaining a wide range of data.

A cognitive architecture is the essential structures and processes of a domain-generic computational cognitive model used for a broad, multiple-level, multiple-domain analysis of cognition and behavior [36.3]. Specifically, cognitive architectures deal with componential processes of cognition in a structurally and mechanistically well-defined way. Their function is to provide an essential framework to facilitate more detailed exploration and understanding of various components and processes of the mind. In this way, a cognitive architecture serves as an initial set of assumptions to be used for further development. These assumptions may be based on available empirical data (e.g., psychological or biological), philosophical thoughts and arguments, or computationally inspired hypotheses concerning psychological processes. A cognitive architecture is useful and important precisely because it provides a comprehensive initial framework for further modeling and simulation in many task domains.

While there are all kinds of cognitive architectures in existence, in this chapter we are concerned specifically with psychologically-oriented cognitive architectures (as opposed to software engineering-oriented cognitive architectures, e.g., LIDA [36.4], or neurally-oriented cognitive architectures, e.g., ART [36.5]). Psychologically-oriented cognitive architectures are particularly important because they shed new light on human cognition and, therefore, they are useful tools for advancing the understanding of cognition. In understanding cognitive phenomena, the use of computational simulation on the basis of cognitive architectures forces one to think in terms of processes, and in terms of details. Instead of using vague, purely conceptual theories, cognitive architectures force theoreticians to think clearly. They are, therefore, critical tools in the study of the mind. Cognitive psychologists who use cognitive architectures must specify a cognitive mechanism in sufficient detail to allow the resulting models to be implemented on computers and run as simulations. This approach requires that important elements of the models be spelled out explicitly, thus aiding in developing better, conceptually clearer theories. It is certainly true that more specialized, narrowly-scoped models may also serve this purpose, but they are not

as generic and as comprehensive and thus may not be as useful to the goal of producing general intelligence.

It is also worth noting that psychologically-oriented cognitive architectures are the antithesis of expert systems. Instead of focusing on capturing performance in narrow domains, they are aimed to provide broad coverage of a wide variety of domains in a way that mimics human performance [36.6]. While they may not always perform as well as expert systems, business and industrial applications of intelligent systems increasingly require broadly-scoped systems that are capable of a wide range of intelligent behaviors, not just isolated systems of narrow functionalities. For example, one application may require the inclusion of capabilities for raw image processing, pattern recognition, categorization, reasoning, decision-making, and natural language communications. It may even require planning, control of robotic devices, and interactions with other systems and devices. Such requirements accentuate the importance of research on broadly-scoped cognitive architectures that perform a wide range of cognitive functionalities across a variety of task domains (as opposed to more specialized systems).

In order to achieve general computational intelligence in a psychologically-realistic way, cognitive architectures should include only minimal initial structures and independently learn from their own experiences. Autonomous learning is an important way of developing additional structure, bootstrapping all the way to a full-fledged cognitive model. In so doing, it is important to be careful to devise only minimal initial learning capabilities that are capable of *bootstrapping*, in accordance with whatever phenomenon is modeled [36.7]. This can be accomplished through environmental cues, structures, and regularities. The avoidance of overly complicated initial structures, and thus the inevitable use of autonomous learning, may often help to avoid overly representational models that are designed specifically for the task to be achieved [36.3]. Autonomous learning is thus essential in achieving generality in a psychologically-realistic way.

36.1.1 Outline

The remainder of this chapter is organized as follows. The next three sections review some of the most popular cognitive architectures that are used in psychology and cognitive science. Specifically, Sect. 36.2 reviews ACT-R, Sect. 36.3 reviews Soar, and Sect. 36.4 reviews CLARION. Each of these sections includes an overview of the architecture, some key equations, and a detailed simulation example. Following this presentation, Sect. 36.5 discusses how cognitive architectures can be used in multi-agent cognitive social simulations and presents a detailed example with CLARION. Finally, Sect. 36.6 presents a general discussion and compares the models reviewed.

36.2 Adaptive Control of Thought–Rational (ACT–R)

ACT-R is one of the oldest and most successful cognitive architectures. It has been used to simulate and implement many cognitive tasks and applications, such as the Tower of Hanoi, game playing, aircraft control, and human–computer interactions [36.8]. ACT-R is based on three key ideas [36.9]:

a) Rational analysis
b) The distinction between procedural and declarative memories
c) A modular structure linked with communication buffers (Fig. 36.1).

According to the rational analysis of cognition (first key idea; [36.10]), the cognitive architecture is optimally tuned to its environment (within its computational limits). Hence, the functioning of the architecture can be understood by investigating how optimal behavior in a particular environment would be implemented.

According to Anderson, such optimal adaptation is achieved through evolution [36.10].

The second key idea, the distinction between declarative and procedural memories, is implemented by having different modules in ACT-R, each with its own representational format and learning rule. Briefly, procedural memory is represented by production rules (similar to a production system) that can act on the environment through the initiation of motor actions. In contrast, declarative memory is passive and uses chunks to represent world knowledge that can be accessed by the procedural memory but does not interact directly with the environment through motor actions.

The last key idea in ACT-R is modularity. As seen in Fig. 36.1, procedural memory (i.e., the production system) cannot directly access information from the other modules: the information has to go through dedicated buffers. Each buffer can hold a single chunk at any time. Hence, buffers serve as information processing

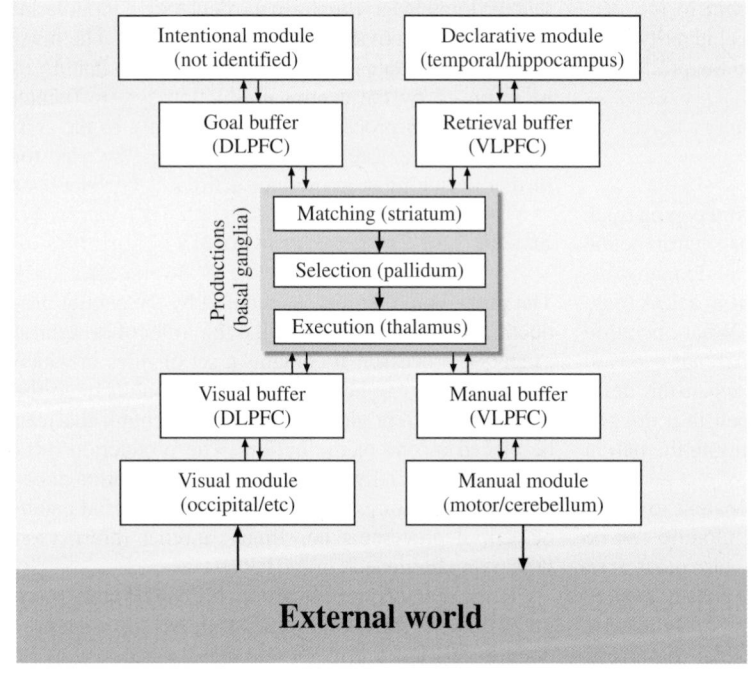

Fig. 36.1 General architecture of ACT-R. DLPFC = dorsolateral prefrontal cortex; VLPFC = ventrolateral prefrontal cortex (after [36.11], by courtesy of the American Psychological Association)

bottlenecks in ACT-R. This restricts the amount of information available to the production system, which in turn limits the processing that can be done by this module at any given time. Processing within each module is encapsulated. Hence, all the modules can operate in parallel without much interference. The following subsections describe the different ACT-R modules in more details.

36.2.1 The Perceptual–Motor Modules

The perceptual-motor modules in ACT-R include a detailed representation of the output of perceptual systems, and the input of motor systems [36.8]. The visual module is divided into the well-established ventral (what) and dorsal (where) visual streams in the primate brain. The main function of the dorsal stream is to find the location of features in a display (e.g., red-colored items, curved-shaped objects) without identifying the objects. The output from this module is a location chunk, which can be sent back to the central production system. The function of the ventral stream is to identify the object at a particular location. For instance, the central production system could send a request to the dorsal stream to find a red object in a display. The dorsal stream would search the display and return a chunk representing the location of a red object. If the central production system needs to know the identity of that object, the location chunk would be sent to the ventral stream. A chunk containing the object identity (e.g., a fire engine) would then be returned to the production system.

36.2.2 The Goal Module

The goal module serves as the context for keeping track of cognitive operations and supplement environmental stimulations [36.8]. For instance, one can do many different operations with a pen picked up from a desk (e.g., write a note, store it in a drawer, etc.). What operation is selected depends primarily on the goal that needs to be achieved. If the current goal is to clean the desk, the appropriate action is to store the pen in a drawer. If the current goal is to write a note, putting the pen in a drawer is not a useful action.

In addition to providing a mental context to select appropriate production rules, the goal module can be used in more complex problem-solving tasks that need subgoaling [36.8]. For instance, if the goal is to play a game of tennis, one first needs to find an opponent. The goal module must create this subgoal that needs

to be achieved before moving back to the original goal (i.e., playing a game of tennis). Note that goals are centralized in a unique module in ACT-R and that production rules only have access to the goal buffer. The current goal to be achieved is the one in the buffer, while later goals stored in the goal module are not accessible to the production. Hence, the *play a game of tennis* goal is not accessible to the production rules while the *find an opponent* subgoal is being pursued.

36.2.3 The Declarative Module

The declarative memory module contains knowledge about the world in the form of chunks [36.8]. Each chunk represents a piece of knowledge or a concept (e.g., fireman, bank, etc.). Chunks can be accessed effortlessly by the central production system, and the probability of retrieving a chunk depends on the chunk activation

$$P_i = \left[1 + e^{-\frac{(B_i + \sum_j W_j S_{ji} - \tau)}{\varepsilon}} \right]^{-1} , \qquad (36.1)$$

where P_i is the probability of retrieving chunk i, B_i is the base-level activation of chunk i, S_{ji} is the association strength between chunks j and i, W_j is the amount of attention devoted to chunk j, τ is the activation threshold, and ε is a noise parameter. It is important to note that the knowledge chunks in the declarative module are passive and do not do anything on their own. The function of this module is to store information so that it can be retrieved by the central production system (which corresponds to procedural memory). Only in the central production system can the knowledge be used for further reasoning or to produce actions.

36.2.4 The Procedural Memory

The procedural memory is captured by the central production rule module and fills the role of a central executive processor. It contains a set of rules in which the conditions can be matched by the chunks in all the peripheral buffers and the output is a chunk that can be placed in one of the buffers. The production rules are chained serially, and each rule application takes a fixed amount of psychological time. The serial nature of central processing constitutes another information processing bottleneck in ACT-R.

Because only one production rule can be fired at any given time, its selection is crucial. In ACT-R, each production rule has a utility value that depends on: a) its probability of achieving the current goal, b) the value (importance) of the goal, and c) the cost of using the production rule [36.8]. Specifically,

$$U_i = P_i G - C_i \, , \qquad (36.2)$$

where U_i is the utility of production rule i, P_i is the (estimated) probability that selecting rule i will achieve the goal, G is the value of the current goal, and C_i is the (estimated) cost of rule i. The most useful rule is always selected in every processing cycle, but the utility values can be noisy, which can result in the selection of suboptimal rules [36.12]. Rule utilities are learned online by counting the number of times that applying a rule has achieved the goal. *Anderson* [36.10] has shown that the selection according to these counts is optimal in the Bayesian sense. Also, production rules can be made more efficient by using a process called *production compilation* [36.12, 13]. Briefly, if two production rules are often fired in succession and the result is positive, a new production rule is created which directly links the conditions from the first production rule to the action following the application of the second production rule. Hence, the processing time is cut in half by applying only one rule instead of two.

36.2.5 Simulation Example: Learning Past Tenses of Verbs

General computational intelligence requires the abstraction of regularities to form new rules (e.g., rule induction). A well-known example in psychology is children learning English past tenses of verbs [36.13]. This classical result shows that children's accuracy in producing the past tense of irregular verbs follows a U-shaped curve [36.14]. Early in learning, children have a separate memory representation for the past tense of each verb (and no conjugation rule). Hence, the past tenses of irregular verbs are used mostly correctly (Phase 1). After moderate training, children notice that most verbs can be converted to their past tense by adding the suffix -ed. This leads to the formulation of a default rule (e.g., to find the past tense of a verb, add the suffix -ed to the verb stem). This rule is a useful heuristic and works for all regular verbs in English. However, children tend to overgeneralize and (incorrectly) apply the rule to irregular verbs. This leads to errors and the low point of the U-shaped curve (Phase 2). Finally, children learn that there are exceptions to the default rule and memorize the past tense of irregular verbs. Performance improves again (Phase 3).

In ACT-R, the early phase of training uses instance-based retrieval (i. e., retrieving the chunk representing each verb's past tense using (36.1)). The focus of the presentation is on the induction of the default rule, which is overgeneralized in Phase 2 and correctly applied in Phase 3. This is accomplished by joining two production rules. First, consider the following memory retrieval rule used in Phase 1 [36.12]:

1. *Retrieve-past-tense*:
   IF   the goal is to find the past tense of a word w:
   THEN issue a request to declarative memory for the past tense of w.

If a perfect match is retrieved from declarative memory, a second rule is used to produce the (probably correct) response. However, if Rule 1 fails to retrieve a perfect match, the verb past tense is unknown and an analogy rule is used instead [36.12]:

2. *Analogy-find-pattern*:
   IF   the goal is to find the past tense of word $w1$;
   AND the retrieval buffer contains past tense $w2$-suffix of $w2$:
   THEN set the answer to $w1$-($w2$-suffix).

This rule produces a form of generalization using an analogy. Because Rule 2 always follows Rule 1, they can be combined using production compilation [36.12]. Also, $w2$ is likely to be a regular verb, so $w2$-suffix is likely to be -ed. Hence, combining Rules 1 and 2 yields [36.12]:

3. *Learned-rule*:
   IF   the goal is to find the past tense of word w:
   THEN set the answer to w-ed

which is the default rule that can be used to accurately find the past tense of regular verbs. The U-shaped curve representing the performance of children learning irregular verbs can thus be explained with ACT-R as follows [36.13]: In Phase 1, Rule 3 does not exist and Rule 1 is applied to correctly conjugate irregular verbs. In Phase 2, Rule 3 is learned and has proven useful with regular verbs (thus increasing P_i in (36.2)). Hence, it is often selected to incorrectly conjugate irregular verbs. In phase 3, the irregular verbs become more familiar as more instances have been encountered. This increases their base-level activation in declarative memory (B_i in (36.1)), which facilitates retrieval and increases the likelihood that Rule 1 is selected to correctly conjugate the irregular verbs. More details about this simulation can be found in [36.13].

36.3 Soar

Soar was the original unified theory of cognition proposed by *Newell* [36.1]. Soar has been used successfully in many problem solving tasks such as Eight Puzzle, the Tower of Hanoi, Fifteen Puzzle, Think-a-Dot, and Rubik's Cube. In addition, Soar has been used for many military applications such as training models for human pilots and mission rehearsal exercises. According to the Soar theory of intelligence [36.15], human intelligence is an approximation of a knowledge system [36.9]. Hence, the most important aspect of intelligence (natural or artificial) is the use of all available knowledge [36.16], and failures of intelligence are failures of knowledge [36.17].

All intelligent behaviors can be understood in terms of problem solving in Soar [36.9]. As such, Soar is implemented as a set of *problem-space computational models* (PSCM) that partition the knowledge into goal relevant ways [36.16]. Each PSCM implicitly contains the representation of a problem space defined by a set of states and a set of operators that can be visualized using a decision tree [36.17]. In a decision tree representation, the nodes represent the states, and one moves around from state to state using operators (the branches/connections in the decision tree). The objective of a Soar agent is to move from an initial state to one of the goal states, and the best operator is always selected at every time step [36.16]. If the knowledge in the model is insufficient to select a single best operator at a particular time step, an impasse is reached, and a new goal is created to resolve the impasse. This new goal defines its own problem space and set of operators.

36.3.1 Architectural Representation

The general architecture of Soar is shown in Fig. 36.2. The main structures are a working memory and a long-term memory. Working memory is a blackboard where all the relevant information for the current decision cycle is stored [36.17]. It contains a goal representation, perceptual information, and relevant knowledge that can be used as conditions to fire rules. The outcome of rule firing can also be added to the working memory to cause more rules to fire. The long-term memory contains associative rules representing the knowledge in the system (in the form of *IF → THEN* rules). The rules in long-term memory can be grouped/organized to form operators.

36.3.2 The Soar Decision Cycle

In every time step, Soar goes through a six-step decision cycle [36.17]. The first step in Soar is to receive an input from the environment. This input is inserted into working memory. The second step is called the *elaboration phase*. During this phase, all the rules matching the content of working memory fire in parallel, and the result is put into working memory. This, in turn, can create a new round of parallel rule firing. The elaboration phase ends when the content of working memory is stable, and no new knowledge can be added in working memory by firing rules.

The third step is the proposal of operators that are applicable to the content in working memory. If no operator is applicable to the content of working memory, an impasse is reached. Otherwise, the potential operators are evaluated and ordered according to a symbolic preference metric. The fourth step is the selection of a single operator. If the knowledge does not allow for the selection of a single operator, an impasse is reached. The fifth step is to apply the operator. If the operator does not result is a change of state, an impasse is reached. Finally, the sixth step is the output of the model, which can be an external (e.g., motor) or an internal (e.g., more reasoning) action.

36.3.3 Impasses

When the immediate knowledge is insufficient to reach a goal, an impasse is reached and a new goal is created to resolve the impasse. Note that this subgoal produces its own problem space with its own set of states and

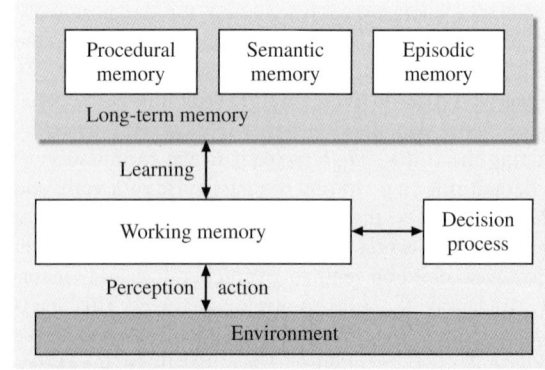

Fig. 36.2 The general architecture of Soar. The subdivision of long-term memory is a new addition in Soar 9

operators. If the subgoal reaches an impasse, another subgoal is recursively created to resolve the second impasse, and so on. There are four main types of impasses in Soar [36.17]:

1. No operator is available in the current state.
   New goal: Find an operator that is applicable in the current state.
2. An operator is applicable in the current state, but its application does not change the current state.
   New goal: Modify the operator so that its application changes the current state. Alternatively, the operator can be modified so that it is no longer deemed applicable in the current state.
3. Two or more operators are applicable in the current state but neither one of them is preferred according to the symbolic metric.
   New goal: Further evaluate the options and make one of the operators preferred to the others.
4. More than one operator is applicable, and there is knowledge in working memory favoring two or more operators in the current state.
   New goal: Resolve the conflict by removing from working memory one of the contradictory preferences.

Regardless of which type of impasse is reached, resolving an impasse is an opportunity for learning in Soar [36.17]. Each time a new result is produced while achieving a subgoal, a new rule associating the current state with the new result is added in the long-term memory to ensure that the same impasse will not be reached again in the future. This new rule is called a *chunk* to distinguish it from rules that were precoded by the modeler (and learning is called *chunking*).

36.3.4 Extensions

Unlike ACT-R (and CLARION, as described next), Soar was originally designed as an artificial intelligence model [36.16]. Hence, initially, more attention was paid to functionality and performance than to psychological realism. However, Soar has been used in psychology and Soar 9 has been extended to increase its psychological realism and functionality [36.18]. This version of the architecture is illustrated in Fig. 36.2. First, the long-term memory has been further subdivided in correspondence with psychology theory [36.19]. The associative rules are now part of the procedural memory. In addition to the procedural memory, the long-term memory now also includes a semantic and an episodic

memory. The semantic memory contains knowledge structures representing factual knowledge about the world (e.g., the earth is round), while the episodic memory contains a snapshot of the working memory representing an *episode* (e.g., Fido the dog is now sitting in front of me).

At the subsymbolic level, Soar 9 includes activations in working memory to capture recency/usefulness (as in ACT-R). In addition, Soar 9 uses non-symbolic (numerical) values to model operator preferences. These are akin to utility functions and are used when symbolic operators as insufficient to select a single operator [36.20]. When numerical preferences are used, an operator is selected using a Boltzmann distribution

$$P(O_i) = \frac{e^{S(O_i)/\tau}}{\sum_j e^{S(O_j)/\tau}} , \qquad (36.3)$$

where $P(O_i)$ is the probability of selecting operator i, $S(O_i)$ is the summed support (preference) for operator i, and τ is a randomness parameter. Numerical operator preferences can be learned using reinforcement learning.

Finally, recent work has been initiated to add a clustering algorithm that would allow for the creation of new symbolic structures and a visual imagery module to facilitate symbolic spatial reasoning (not shown in Fig. 36.2). Also, the inclusion of emotions is now being considered (via appraisal theory [36.21]).

36.3.5 Simulation Example: Learning in Problem Solving

Nason and *Laird* [36.20] proposed a variation of Soar that includes a reinforcement learning algorithm (following the precedents of CLARION and ACT-R) to learn numerical preference values for the operators. In this implementation (called Soar-RL), the preferences are replaced by Q-values [36.22] that are learned using environmental feedback. After the Q-value of each relevant operator has been calculated (i. e., all the operators available in working memory), an operator is stochastically selected (as in (36.3)).

Soar-RL has been used to simulate the missionaries and cannibals problem. The goal in this problem is to transport three missionaries and three cannibals across a river using a boat that can carry at most two persons at a time. Several trips are required, but the cannibals must never outnumber the missionaries on either riverbank. This problem has been used as a benchmark in problem-solving research because, if the desirability of a move is evaluated in terms of the number of people that have crossed the river (which is a common assumption), a step backward must be taken midway in solving the problem (i.e., a move that reduces the number of peoples that crossed the river must be selected).

In the Soar-RL simulation, the states were defined by the number of missionaries and cannibals on each side of the river and the location of the boat. The operators were boat trips transporting people, and the Q-values of the operators were randomly initialized. Also, to emphasize the role of reinforcement learning in solving this problem, chunking was disengaged. Hence, the only form of adaptation was the adjustment

of the operator Q-values. Success states (i.e., all people crossed the river) were rewarded, failure states (i.e., cannibals outnumbering missionaries on a riverbank) were punished, and all other states received neutral reinforcement.

Using this simulation methodology, Soar-RL generally learned to solve the missionaries and cannibals problem. Most errors resulted from the stochastic decision process [36.20]. *Nason* and *Laird* also showed that the model performance can be improved fivefold by adding a symbolic preference preventing an operator at time t from undoing the result of the application of an operator at time $t-1$. More details on this simulation can be found in [36.20].

36.4 CLARION

CLARION is an integrative cognitive architecture consisting of a number of distinct subsystems with a dual representational structure in each subsystem (implicit versus explicit representations). CLARION is the

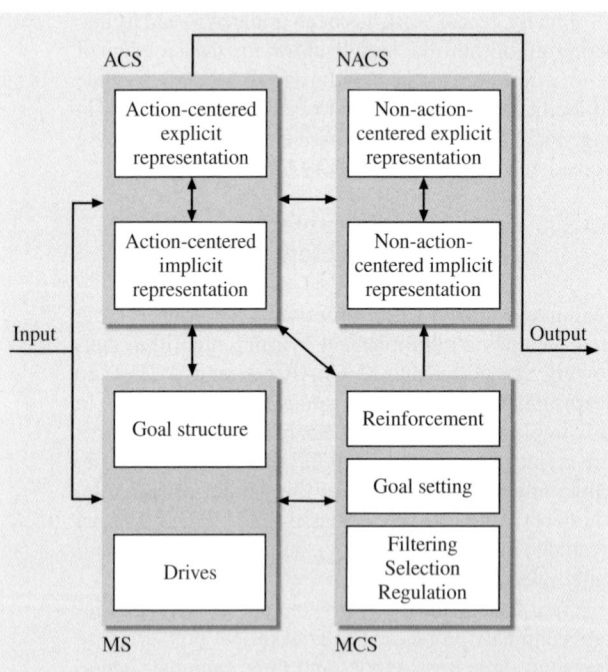

Fig. 36.3 The CLARION architecture. ACS stands for the action-centered subsystem, NACS the non-action-centered subsystem, MS the motivational subsystem, and MCS the meta-cognitive subsystem (after [36.23])

newest of the reviewed architectures, but it has already been successfully applied to several tasks such as navigation in mazes and mine fields, human reasoning, creative problem solving, and cognitive social simulations. CLARION is based on the following basic principles [36.24]. First, humans can learn with or without much a priori specific knowledge to begin with, and humans learn continuously from on-going experience in the world. Second, there are different types of knowledge involved in human learning (e.g., procedural vs. declarative, implicit vs. explicit; [36.24]), and different types of learning processes are involved in acquiring different types of knowledge. Third, motivational processes as well as meta-cognitive processes are important and should be incorporated in a psychologically realistic cognitive architecture. According to CLARION, all three principles are required to achieve general computational intelligence. An overview of the architecture is shown in Fig. 36.3.

The CLARION subsystems include the action-centered subsystem (ACS), the non-action-centered subsystem (NACS), the motivational subsystem (MS), and the meta-cognitive subsystem (MCS). The role of the ACS is to control actions, regardless of whether the actions are for external physical movements or for internal mental operations. The role of the NACS is to maintain general knowledge. The role of the MS is to provide underlying motivations for perception, action, and cognition, in terms of providing impetus and feedback (e.g., indicating whether or not outcomes are satisfactory). The role of the MCS is to monitor, direct, and modify dynamically the operations of the other subsystems.

Each of these interacting subsystems consists of two *levels* of representations. In each subsystem, the top level encodes explicit (e.g., verbalizable) knowledge and the bottom level encodes implicit (e.g., non-verbalizable) knowledge. The two levels interact, for example, by cooperating through a combination of the action recommendations from the two levels, respectively, as well as by cooperating in learning through bottom-up and top-down processes. Essentially, it is a dual-process theory of mind [36.24].

36.4.1 The Action-Centered Subsystem

The ACS is composed of a top and a bottom level. The bottom level of the ACS is modular. A number of neural networks co-exist, each of which is adapted to a specific modality, task, or group of input stimuli. These modules can be developed in interacting with the world (computationally, through various decomposition methods [36.25]). However, some of them are formed evolutionarily, reflecting hardwired instincts and propensities [36.26]. Because of these networks, CLARION is able to handle very complex situations that are not amenable to simple rules.

In the top level of the ACS, explicit symbolic conceptual knowledge is captured in the form of explicit symbolic rules (for details, see [36.24]). There are many ways in which explicit knowledge may be learned, including independent hypothesis testing and bottom-up learning. The basic process of bottom-up learning is as follows: if an action implicitly decided by the bottom level is successful, then the model extracts an explicit rule that corresponds to the action selected by the bottom level and adds the rule to the top level. Then, in subsequent interactions with the world, the model verifies the extracted rule by considering the outcome of applying the rule. If the outcome is not successful, then the rule should be made more specific; if the outcome is successful, the agent may try to generalize the rule to make it more universal [36.27]. After explicit rules have been learned, a variety of explicit reasoning methods may be used. Learning explicit conceptual representations at the top level can also be useful in enhancing learning of implicit reactive routines at the bottom level [36.7]. The action-decision cycle in the ACS can be described by the following steps:

1. Observe the current state of the environment
2. Compute the value of each possible action in the current state in the bottom level
3. Compute the value of each possible action in the current state in the top level
4. Choose an appropriate action by stochastically selecting or combining the values in the top and bottom levels
5. Perform the selected action
6. Update the top and bottom levels according to the received feedback (if any)
7. Go back to Step 1.

36.4.2 The Non-Action-Centered Subsystem

The NACS may be used for representing general knowledge about the world (i.e., the semantic memory and the episodic memory), and for performing various kinds of memory retrievals and inferences. The NACS is also composed of two levels (a top and a bottom level) and is under the control of the ACS (through its actions).

At the bottom level, *associative memory* networks encode non-action-centered implicit knowledge. Associations are formed by mapping an input to an output (such as mapping $2 + 3$ to 5). Backpropagation [36.7, 28] or Hebbian [36.29] learning algorithms can be used to establish such associations between pairs of inputs and outputs. At the top level of the NACS, a general knowledge store encodes explicit non-action-centered knowledge [36.29, 30]. In this network, chunks (passive knowledge structures, similar to ACT-R) are specified through dimensional values (features). A node is set up in the top level to represent a chunk. The chunk node connects to its corresponding features represented as individual nodes in the bottom level of the NACS [36.29, 30]. Additionally, links between chunk nodes encode explicit associations between pairs of chunks, known as *associative rules*. Explicit associative rules may be learned in a variety of ways [36.24].

During reasoning, in addition to applying associative rules, similarity-based reasoning may be employed in the NACS. Specifically, a known (given or inferred) chunk may be automatically compared with another chunk. If the similarity between them is sufficiently high, then the latter chunk is inferred. The similarity between chunks i and j is computed by using

$$s_{c_i \sim c_j} = \frac{n_{c_i \cap c_j}}{f\left(n_{c_j}\right)}, \tag{36.4}$$

where $s_{c_i \sim c_j}$ is the similarity from i to chunk j, $n_{c_i \cap c_j}$ counts the number of features shared by chunks i and j (i.e., the feature overlap), n_{c_j} counts the total number of features in chunk j, and $f(x)$ is a slightly super-linear, monotonically increasing, positive function (by default, $f(x) = x^{1.1}$). Thus, similarity-based reasoning in CLARION is naturally accomplished using (1) top-down activation by chunk nodes of their corresponding bottom-level feature-based representations, (2) calculation of feature overlap between any two chunks (as in (36.4)), and (3) bottom-up activation of the top-level chunk nodes. This kind of similarity calculation is naturally accomplished in a multi-level cognitive architecture and represents a form of synergy between the explicit and implicit modules. Each round of reasoning in the NACS can be described by the following steps:

1. Propagate the activation of the activated features in the bottom level
2. Concurrently, fire all applicable associative rules in the top level
3. Integrate the outcomes of top and bottom-level processing
4. Update the activations in the top and bottom levels (e.g., similarity-based reasoning)
5. Go back to Step 1 (if another round of reasoning is requested by the ACS).

36.4.3 The Motivational and Meta-Cognitive Subsystems

The motivational subsystem (MS) is concerned with drives and their interactions [36.31], which lead to actions. It is concerned with why an agent does what it does. Simply saying that an agent chooses actions to maximize gains, rewards, reinforcements, or payoffs leaves open the question of what determines these things. The relevance of the MS to the ACS lies primarily in the fact that it provides the context in which the goal and the reinforcement of the ACS are set. It thereby influences the working of the ACS, and by extension, the working of the NACS.

Dual motivational representations are in place in CLARION. The explicit goals (such as *finding food*) of an agent may be generated based on internal drives (for example, *being hungry*; see [36.32] for details). Beyond low-level drives (concerning physiological needs), there are also higher-level drives. Some of them are primary, in the sense of being *hard wired*, while others are secondary (*derived*) drives acquired mostly in the process of satisfying primary drives.

The meta-cognitive subsystem (MCS) is closely tied to the MS. The MCS monitors, controls, and regulates cognitive processes for the sake of improving cognitive performance [36.33, 34]. Control and regulation may be in the forms of setting goals for the ACS, setting essential parameters of the ACS and the NACS, interrupting and changing on-going processes in the ACS and the NACS, and so on. Control and regulation can also be carried out through setting reinforcement functions for the ACS. All of the above can be done on the basis of drive activations in the MS. The MCS is also made up of two levels: the top level (explicit) and the bottom level (implicit).

36.4.4 Simulation Example: Minefield Navigation

Sun et al. [36.7] empirically tested and simulated a complex minefield navigation task. In the empirical task, the subjects were seated in front of a computer monitor that displayed an instrument panel containing several gauges that provided current information on the status/location of a vehicle. The subjects used a joystick to control the direction and speed of the vehicle. In each trial, a random mine layout was generated, and the subjects had limited time to reach a target location without hitting a mine. Control subjects were trained for several consecutive days in this task. Sun and colleagues also tested three experimental conditions with the same amount of training but emphasized verbalization, over-verbalization, and dual-tasking (respectively). The human results show that learning was slower in the dual-task condition than in the single-task condition, and that a moderate amount of verbalization speeds up learning. However, the effect of verbalization is reversed in the over-verbalization condition; over-verbalization interfered with (slowed down) learning.

In the CLARION simulation, simplified (explicit) rules were represented in the form *state* → *action* in the top level of the ACS. In the bottom level of the ACS, a backpropagation network was used to (implicitly) learn the input–output function using reinforcement learning. Reinforcement was received at the end of every trial. The bottom-level information was used to create and refine top-level rules (with bottom-up learning). The model started out with no specific a priori knowledge about the task (the same as a typical subject). The bottom level contained randomly initialized weights. The top level started empty and contained no a priori knowledge about the task (either in the form of instructions or rules). The interaction of the two levels was not determined a priori either: there was no fixed weight in combining outcomes from the two levels. The weights were automatically set based on relative performance of the two levels on a periodic basis. The effects of the dual task and the various verbalization conditions were modeled using rule-learning thresholds so that more/less activities could occur at the top level. The CLARION simulation results closely matched the human results [36.7]. In addition, the human and simulated data were input into a common ANOVA and no statistically significant difference between human and simulated data was found in any of the conditions. Hence, CLARION did a good job of simulating detailed human data in the minefield navigation task. More details about this simulation can be found in [36.7].

36.5 Cognitive Architectures as Models of Multi-Agent Interaction

Most of the work in social simulation assumes rudimentary cognition on the part of agents. Agent models have frequently been custom-tailored to the task at hand, often with a restricted set of highly domain-specific rules. Although this approach may be adequate for achieving the limited objectives of some social simulations, it is overall unsatisfactory. For instance, it limits the realism, and hence applicability of social simulation, and more importantly it also precludes any possibility of resolving the theoretical question of the micro–macro link.

Cognitive models, especially cognitive architectures, may provide better grounding for understanding multi-agent interaction. This can be achieved by incorporating realistic constraints, capabilities, and tendencies of individual agents in terms of their psychological processes (and maybe even in terms of their physical embodiment) and their interactions with their environments (which include both physical and social environments). Cognitive architectures make it possible to investigate the interaction of cognition/motivation on the one hand and social institutions and processes on the other, through psychologically realistic agents. The results of the simulation may demonstrate significant interactions between cognitive-motivational factors and social-environmental factors. Thus, when trying to understand social processes and phenomena, it may be important to take the psychology of individuals into consideration given that detailed computational models of cognitive agents that incorporate a wide range of psychological functionalities have been developed in cognitive science.

For example, *Sun* and *Naveh* simulated an organizational classification decision-making task using the CLARION cognitive architecture [36.35]. In a classification decision-making task, agents gather information about problems, classify them, and then make further decisions based on the classification. In this case, the task is to determine whether a blip on a screen is a hostile aircraft, a flock of geese, or a civilian aircraft. In each case, there is a single object in the airspace. The object has nine different attributes, each of which can take on one of three possible values (e.g., its speed can be low, medium, or high). An organization must determine the status of an observed object: whether it is friendly, neutral, or hostile. There are a total of 19 683 possible objects, and 100 problems are chosen randomly from this set.

Critically, no one single agent has access to all the information necessary to make a choice. Decisions are made by integrating separate decisions made by different agents, each of which is based on a different subset of information. In terms of organizational structures, there are two major types of interest: teams and hierarchies. In teams, decision-makers act autonomously, individual decisions are treated as votes, and the organization decision is the majority decision. In hierarchies, agents are organized in a chain of command, such that information is passed from subordinates to superiors, and the decision of a superior is based solely on the recommendations of his/her subordinates. In this task, only a two-level hierarchy with nine subordinates and one supervisor is considered.

In addition, organizations are distinguished by the structure of information accessible to each agent. There are two types of information access: distributed access, in which each agent sees a different subset of three attributes (no two agents see the same subset of three attributes), and blocked access, in which three agents see exactly the same subset of attributes. In both cases, each attribute is accessible to three agents.

The human experiments by *Carley* et al. [36.36] were done in a 2×2 fashion (organization $\times$ information access). The data showed that humans generally performed better in team situations, especially when distributed information access was in place. Moreover, distributed information access was generally better than blocked information access. The worst performance occurred when hierarchical organizational structure and blocked information access were used in conjunction.

The results of the CLARION simulations closely matched the patterns of the human data, with teams outperforming hierarchal structures, and distributed access being superior to blocked access. As in the human data, the effects of organization and information access were present, but more importantly the interaction of these two factors with length of training was reproduced. These interactions reflected the following trends: (1) the superiority of team and distributed information access at the start of the learning process and, (2) either the disappearance or reversal of these trends towards the end. These trends persisted robustly across a wide variety of settings of cognitive parameters, and did not critically depend on any one setting of these parameters. Also, as in humans, performance was not grossly skewed towards one condition or the other.

36.5.1 Extention

One advantage of using a more *cognitive* agent in social simulations is that we can address the question of what happens when cognitive parameters are varied. Because CLARION captures a wide range of cognitive processes, its parameters are generic (rather than task specific). Thus, one has the opportunity of studying social and organizational issues in the context of a general theory of cognition. Below we present some of the results observed (details can be found in [36.35]).

Varying the parameter controlling the probability of selecting implicit versus explicit processing in CLARION interacted with the length of training. Explicit rule learning was far more useful at the early stages of learning, when increased reliance on rules tended to boost performance (compared with performance toward the end of the learning process). This is because explicit rules are crisp guidelines that are based on past success, and as such, they provide a useful anchor at the uncertain early stages of learning. However, by the end of the learning process, they become no more reliable than highly trained networks. This corresponds to findings in human cognition, where there are indications that rule-based learning is more widely used in the early stages of learning, but is later increasingly supplanted by similarity-based processes and skilled performance [36.37, 38]. Such trends may partially explain why hierarchies did not perform well initially; because a hierarchy's supervisor was burdened with a higher input dimensionality, it took a longer time to encode rules (which were, nevertheless, essential at the early stages of learning).

Another interesting result was the effect of varying the generalization threshold. The generalization threshold determines how readily an agent generalizes a successful rule. It was better to have a higher rule generalization threshold than a lower one (up to a point). That is, if one restricts the generalization of rules to those rules that have proven to be relatively successful, the result is a higher-quality rule set, which leads to better performance in the long run.

This CLARION simulation showed that some cognitive parameters (e.g., learning rate) had a monolithic, across-the-board effect under all conditions, while in other cases, complex interactions of factors were at work (see [36.35] for full details of the analysis). This illustrates the importance of limiting one's social simulation conclusions to the specific cognitive context in which human data were obtained (in contrast to the practice of some existing social simulations). By using CLARION, *Sun* and *Naveh* [36.35] were able to accurately capture organizational performance data and, moreover, to formulate deeper explanations for the results observed. In cognitive architectures, one can vary parameters and options that correspond to cognitive processes and test their effects on collective performance. In this way, cognitive architectures may be used to predict human performance in social/organizational settings and, furthermore, to help to improve collective performance by prescribing optimal or near-optimal cognitive abilities for individuals for specific collective tasks and/or organizational structures.

36.6 General Discussion

This chapter reviewed the most popular psychologically-oriented cognitive architectures with some example applications in human developmental learning, problem-solving, navigation, and cognitive social simulations. ACT* (ACT-R's early version; [36.39]) and Soar [36.15] were some of the first cognitive architectures available and have been around since the early 1980s, while CLARION was first proposed in the mid-1990s [36.30]. This chronology is crucial when exploring their learning capacity. ACT* and Soar were developed before the connectionist revolution [36.40], and were, therefore, implemented using knowledge-rich production systems [36.41]. In contrast, CLARION was proposed after the connectionist revolution and was implemented using neural networks. While some attempts have been made to implement ACT-R [36.42] and Soar [36.43] with neural networks, these architectures remain mostly knowledge-rich production systems grounded in the artificial intelligence tradition. One of the most important impacts of the connectionist revolution has been data-driven learning rules (e.g., backpropagation) that allows for autonomous learning. CLARION was created within this tradition, and every component in CLARION has been implemented using neural networks. For instance, explicit knowledge may be implemented using linear, two-layer neural networks [36.7, 23, 28, 29], while implicit knowledge has been implemented using nonlinear multilayer backpropagation networks in the ACS [36.7, 29] and recurrent associative memory networks in the NACS [36.23, 29]. This general philosophy has also been applied to modeling the MS and the MCS using linear (explicit) and nonlinear (implicit) neural networks [36.44]. As such, CLARION requires less pre-coded knowledge to achieve its goals, and can be considered more autonomous.

While the different cognitive architectures were motivated by different problems and took different implementation approaches, they share some theoretical similarities. For instance, Soar is somewhat similar to the top levels of CLARION. It contains production rules that fire in parallel and cycles until a goal is reached. In CLARION, top-level rules in the NACS fire in parallel in cycles (under the control of the ACS). However, CLARION includes a distinction between action-centered and non-action-centered knowledge. While this distinction has been added in Soar 9 [36.18], the additional distinction between explicit and implicit knowledge (one of the main assumptions in CLARION) was not. The inclusion of implicit knowledge in the bottom level of CLARION allows for an *automatic* representation of similarity-based reasoning, which is absent in Soar. While Soar can certainly account for similarity-based reasoning, adding an explicit (and ad hoc) representation of similarity can become cumbersome when a large number of items are involved.

ACT-R initially took a different approach. Work on the ACT-R cognitive architecture has clearly focused on psychological modeling from the very beginning and, as such, it includes more than one long-term memory store, distinguishing between procedural and declarative memories (similar to CLARION). In addition, ACT-R has a rudimentary representation of explicit and implicit memories: explicit memory is represented by symbolic structures (i. e., chunks and production rules), while implicit memory is represented by the activation of the structures. In contrast, the distinction between explicit and implicit memories in CLARION is one of the main focuses of the architecture, and a more detailed representation of implicit knowledge has allowed for a natural representation of similarity-based reasoning as well as natural simulations of many psychological data sets [36.7, 28, 29]. Yet, ACT-R memory structures have been adequate for simulating many data sets with over 30 years of research. Future work should be devoted to a detailed comparison of ACT-R, Soar, and CLARION using a common set of tasks to more accurately compare their modeling paradigms, capacities, and limits.

References

36.1   A. Newell: *Unified Theories of Cognition* (Harvard Univ. Press, Cambridge 1990)

36.2   S. Roberts, H. Pashler: How persuasive is a good fit? A comment on theory testing, Psychol. Rev. **107**, 358–367 (2000)

36.3   R. Sun: Desiderata for cognitive architectures, Philos. Psychol. **17**, 341–373 (2004)

36.4   S. Franklin, F.G. Patterson Jr.: The Lida architecture: Adding new modes of learning to an intelligent, autonomous, software agent, Integr. Design Process Technol. IDPT-2006, San Diego (Society for Design and Process Science, San Diego 2006) p. 8

36.5   G.A. Carpenter, S. Grossberg: A massively parallel architecture for a self-organizing neural pattern recognition machine, Comput. Vis. Graph. Image Process. **37**, 54–115 (1987)

36.6   P. Langley, J.E. Laird, S. Rogers: Cognitive architectures: Research issues and challenges, Cogn. Syst. Res. **10**, 141–160 (2009)

36.7   R. Sun, E. Merrill, T. Peterson: From implicit skills to explicit knowledge: A bottom-up model of skill learning, Cogn. Sci. **25**, 203–244 (2001)

36.8   J.R. Anderson, D. Bothell, M.D. Byrne, S. Douglass, C. Lebiere, Y. Qin: An integrated theory of the mind, Psychol. Rev. **111**, 1036–1060 (2004)

36.9   N.A. Taatgen, J.R. Anderson: Constraints in cognitive architectures. In: *The Cambridge Handbook of Computational Psychology*, ed. by R. Sun (Cambridge Univ. Press, New York 2008) pp. 170–185

36.10   J.R. Anderson: *The Adaptive Character of Thought* (Erlbaum, Hillsdale 1990)

36.11 J.R. Anderson, D. Bothell, M.D. Byrne, S. Douglass, C. Lebiere, Y. Qin: An integrated theory of the mind, Psychol. Rev. **111**, 1037 (2004)

36.12 N.A. Taatgen, C. Lebiere, J.R. Anderson: Modeling paradigms in ACT-R. In: *Cognition and Multi-Agent Interaction: From Cognitive Modeling to Social Simulation*, ed. by R. Sun (Cambridge Univ. Press, New York 2006) pp. 29–52

36.13 N.A. Taatgen, J.R. Anderson: Why do children learn to say "broke"? A model of learning the past tense without feedback, Cognition **86**, 123–155 (2002)

36.14 G.F. Marcus, S. Pinker, M. Ullman, M. Hollander, T.J. Rosen, F. Xu: Overregularization in language acquisition, Monogr. Soc. Res. Child Dev. **57**, 1–182 (1992)

36.15 J.E. Laird, A. Newell, P.S. Rosenbloom: Soar: An architecture for general intelligence, Artif. Intell. **33**, 1–64 (1987)

36.16 J.F. Lehman, J. Laird, P. Rosenbloom: *A Gentle Introduction to Soar, an Architecture for Human Cognition* (University of Michigan, Ann Arbor 2006)

36.17 R.E. Wray, R.M. Jones: Considering Soar as an agent architecture. In: *Cognition and Multi-Agent Interaction: From Cognitive Modeling to Social Simulation*, ed. by R. Sun (Cambridge Univ. Press, New York 2006) pp. 53–78

36.18 J.E. Laird: Extending the Soar cognitive architecture, Proc. 1st Conf. Artif. General Intell. (IOS Press, Amsterdam 2008) pp. 224–235

36.19 D.L. Schacter, A.D. Wagner, R.L. Buckner: Memory systems of 1999. In: *The Oxford Handbook of Memory*, ed. by E. Tulving, F.I.M. Craik (Oxford Univ. Press, New York 2000) pp. 627–643

36.20 S. Nason, J.E. Laird: Soar-RL: Integrating reinforcement learning with Soar, Cogn. Syst. Res. **6**, 51–59 (2005)

36.21 K.R. Scherer: Appraisal considered as a process of multi-level sequential checking. In: *Appraisal Processes in Emotion: Theory, Methods, Research*, ed. by K.R. Scherer, A. Schor, T. Johnstone (Oxford Univ. Press, New York 2001) pp. 92–120

36.22 C. Watkins: *Learning from Delayed Rewards* (Cambridge Univ., Cambridge 1990)

36.23 R. Sun, S. Hélie: Psychologically realistic cognitive agents: Taking human cognition seriously, J. Exp. Theor. Artif. Intell. **25**(1), 65–92 (2013)

36.24 R. Sun: *Duality of the Mind: A Bottom-up Approach Toward Cognition* (Lawrence Erlbaum Associates, Mahwah 2002)

36.25 R. Sun, T. Peterson: Multi-agent reinforcement learning: Weighting and partitioning, Neural Netw. **12**, 127–153 (1999)

36.26 L. Hirschfield, S. Gelman (Eds.): *Mapping the Mind: Domain Specificity in Cognition and Culture* (Cambridge Univ. Press, Cambridge 1994)

36.27 R. Michalski: A theory and methodology of inductive learning, Artif. Intell. **20**, 111–161 (1983)

36.28 R. Sun, P. Slusarz, C. Terry: The interaction of the explicit and the implicit in skill learning: A dual-process approach, Psychol. Rev. **112**, 159–192 (2005)

36.29 S. Hélie, R. Sun: Incubation, insight, and creative problem solving: A unified theory and a connectionist model, Psychol. Rev. **117**, 994–1024 (2010)

36.30 R. Sun: *Integrating Rules and Connectionism for Robust Commonsense Reasoning* (Wiley, New York 1994)

36.31 F. Toates: *Motivational Systems* (Cambridge Univ. Press, Cambridge 1986)

36.32 R. Sun: Motivational representations within a computational cognitive architecture, Cogn. Comput. **1**, 91–103 (2009)

36.33 T. Nelson (Ed.): *Metacognition: Core Readings* (Allyn and Bacon, Boston 1993)

36.34 J.D. Smith, W.E. Shields, D.A. Washburn: The comparative psychology of uncertainty monitoring and metacognition, Behav. Brain Sci. **26**, 317–373 (2003)

36.35 R. Sun, I. Naveh: Simulating organizational decision-making using a cognitively realistic agent model, J. Artif. Soc. Soc. Simul. **7**(3) (2004)

36.36 K.M. Carley, M.J. Prietula, Z. Lin: Design versus cognition: The interaction of agent cognition and organizational design on organizational performance, J. Artif. Soc. Soc. Simul. **1** (1998)

36.37 S. Hélie, J.G. Waldschmidt, F.G. Ashby: Automaticity in rule-based and information-integration categorization, Atten. Percept. Psychophys. **72**, 1013–1031 (2010)

36.38 S. Hélie, J.L. Roeder, F.G. Ashby: Evidence for cortical automaticity in rule-based categorization, J. Neurosci. **30**, 14225–14234 (2010)

36.39 J.R. Anderson: *The Architecture of Cognition* (Harvard Univ. Press, Cambridge 1983)

36.40 D. Rumelhart, J. McClelland, The PDP Research Group (Eds.): *Parallel Distributed Processing: Explorations in the Microstructure of Cognition. Vol. 1: Foundations* (MIT Pres, Cambridge 1986)

36.41 S. Russell, P. Norvig: *Artificial Intelligence: A Modern Approach* (Prentice Hall, Upper Saddle River 1995)

36.42 C. Lebiere, J.R. Anderson: A connectionist implementation of the ACT-R production system, Proc. 15th Annu. Conf. Cogn. Sci. Soc. (Lawrence Erlbaum Associates, Hillsdale 1993) pp. 635–640

36.43 B. Cho, P.S. Rosenbloom, C.P. Dolan: Neuro-Soar: A neural-network architecture for goal-oriented behavior, Proc. 13th Annu. Conf. Cogn. Sci. Soc. (Lawrence Erlbaum Associates, Hillsdale 1991) pp. 673–677

36.44 N. Wilson, R. Sun, R. Mathews: A motivationally-based simulation of performance degradation under pressure, Neural Netw. **22**, 502–508 (2009)

37. Embodied Intelligence

Angelo Cangelosi, Josh Bongard, Martin H. Fischer, Stefano Nolfi

Embodied intelligence is the computational approach to the design and understanding of intelligent behavior in embodied and situated agents through the consideration of the strict coupling between the agent and its environment (situatedness), mediated by the constraints of the agent's own body, perceptual and motor system, and brain (embodiment). The emergence of the field of embodied intelligence is closely linked to parallel developments in computational intelligence and robotics, where the focus is on morphological computation and sensory–motor coordination in evolutionary robotics models, and in neuroscience and cognitive sciences where the focus is on embodied cognition and developmental robotics models of embodied symbol learning. This chapter provides a theoretical and technical overview of some principles of embodied intelligence, namely morphological computation, sensory–motor coordination, and developmental embodied cognition. It will also discuss some tutorial examples on the modeling of body/brain/environment adaptation for the evolution of morphological computational agents, evolutionary robotics model of navigation and object discrimination, and developmental robotics

models of language and numerical cognition in humanoid robots.

Part D | 37.1

37.1 Introduction to Embodied Intelligence

Organisms are not isolated entities which develop their sensory–motor and cognitive skills in isolation from their social and physical environment, and independently from their motor and sensory systems. On the contrary, behavioral and cognitive skills are dynamical properties that unfold in time and arise from a large number of interactions between the agents' nervous system, body, and environment [37.1–7]. Embodied intelligence is the computational approach to the design and understanding of intelligent behavior in embodied and situated agents through the consideration of

the strict coupling between the agent and its environment (situatedness), mediated by the constraints of the agent's own body, perceptual and motor system, and brain (embodiment).

Historically, the field of embodied intelligence has its origin from the development and use of bio-inspired computational intelligence methodologies in computer science and robotics, and the overcoming of the limitations of symbolic approaches typical of classical artificial intelligence methods. As argued in *Brooks'* [37.2] seminal paper on *Elephants don't play chess*, the study of apparently simple behaviors, such as locomotion and motor control, permits an understanding of the embodied nature of intelligence, without the requirement to start from higher order abstract skills as those involved in chess playing algorithms. Moreover, the emergence of the field of embodied intelligence is closely linked to parallel developments in robotics, with the focus on morphological computation and sensory–motor coordination in evolutionary and developmental robotics models, and in neuroscience and cognitive sciences with the focus on embodied cognition (EC).

The phenomenon of *morphological computation* concerns the observation that a robot's (or animal's) *body plan* may perform computations: A body plan that allows the robot (or animal) to passively exploit interactions with its environment may perform computations that lead to successful behavior; in another body plan less well suited to the task at hand, those computations would have to be performed by the control policy [37.8–10]. If both the body plans and control policies of robots are evolved, evolutionary search may find robots that exhibit more morphological computation than an equally successful robot designed by hand (see more details in Sect. 37.2).

The principle of *sensory–motor coordination*, which concerns the relation between the characteristics of the agents' control policy and the behaviors emerging from agent/environmental interactions, has been demonstrated in numerous evolutionary robotics models [37.6]. Experiments have shown how adaptive agents can acquire an ability to coordinate their sensory and motor activity so as to self-select their forthcoming sensory experiences. This sensory–motor coordination can play several key functions such as enabling the agent to access the information necessary to make the appropriate behavioral decision, elaborating sensory information, and reducing the complexity of the agents' task to a manageable level. These two themes will be exemplified through the illustration of evolutionary robotics experiments in Sect. 37.3 in which the fine-grained characteristics of the agents' neural control system and body are subjected to variations (e.g. gene mutation) and in which variations are retained or discarded on the basis of their effects at the level of the overall behavior exhibited by the agent in interaction with the environment.

In cognitive and neural sciences, the term *embodied cognition* (EC) [37.11, 12] is used to refer to systematic relationships between an organism's cognitive processes and its perceptual and response repertoire. Notwithstanding the many interpretations of this term [37.13], the broadest consensus of the proponents of EC is that our knowledge representations encompass the bodily activations that were present when we initially acquired this knowledge (for differentiations, [37.14]). This view helps us to understand the many findings of modality-specific biases induced by cognitive computations. Examples of EC in psychology and cognitive science can be sensory–motor (e.g., a systematic increase in comparison time with angular disparity between two views of the same object [37.15]), or conceptual (e.g., better recall of events that were experienced in the currently adopted body posture [37.16]), or emotional in nature (e.g., interpersonal warmth induced by a warm handheld object [37.17]). Such findings were hard to accommodate under the more traditional views where knowledge was presumed symbolic, amodal and abstract and thus dissociated from sensory input and motor output processes.

Embodied cognition experiments in psychology have inspired the design of developmental robotics models [37.18] which exploit the ontogenetic interaction between the developing (baby) robot and its social and physical environment to acquire both simple sensory–motor control strategies and higher order capabilities such as language and number learning (Sect. 37.4).

To provide the reader with both a theoretical and technical understanding of the principles of morphological computation, sensory–motor coordination and developmental EC the following three sections will review the progress in these fields, and analyze in detail some key studies as examples. The presentation of studies on the modeling of both sensory–motor tasks (such as locomotion, navigation, and object discrimination) and of higher order cognitive capabilities (such as linguistic and numerical cognition) demonstrates the impact of embodied intelligence in the design of a variety of perceptual, motor, and cognitive skills.

37.2 Morphological Computation for Body–Behavior Coadaptation

Embodied intelligence dictates that there are certain body plans and control policies that, when combined, will produce some desired behavior. For example, imagine that the desired task is active categorical perception (ACP) [37.19, 20]. ACP requires a learner to actively interact with objects in its environment to classify those objects. This stands in contrast to passive categorization whereby an agent observes objects from a distance – perhaps it is fed images of objects or views them through a camera – and labels the objects according to their perceived class. In order for an animal or robot to perform ACP, it must not only possess a control policy that produces as output the correct class for the object being manipulated, but also some manipulator with which to physically affect (and be affected by) the object.

One consequence of embodied intelligence is that certain pairings of *body* and *brain* produce the desired behavior, and others do not. Returning to the example of ACP, if a robot's arm is too short to reach the objects then it obviously will not be able to categorize them. Imagine now a second robot that possesses an arm of the requisite length but can only bring the back of its hand into contact with the objects. Even if this robot's control policy distinguishes between round and edged objects based on the patterned firing of touch sensors embedded in its palm, this robot will also not be able to perform ACP.

A further consequence of embodied intelligence is that some body plans may require a complex control policy to produce successful behavior, while another body plan may require a simpler control policy. This has been referred to as the morphology and control tradeoff in the literature [37.7]. Continuing the ACP example, consider a third robot that can bring its palm and fingers into contact with the objects, but only possesses a single binary touch sensor in its palm. In order to distinguish between round and edged objects, this robot will require a control policy that performs some complex signal processing on the time series data produced by this single sensor during manipulation. A fourth robot however, equipped with multiple tactile sensors embedded in its palm and fingers, may be able to categorize objects immediately after grasping them: Perhaps round objects produce characteristic static patterns of tactile signals that are markedly different from those patterns produced when grasping edged objects.

The morphology and control tradeoff however raises the question as to what is being traded. It has been argued that what is being traded is computation [37.7, 8]. If two robots succeed at a given task, and each robot is equipped with the simplest control policy that will allow that robot to succeed, but one control policy performs fewer computations than the other control policy, then the body plan of the robot

equipped with the simpler control policy must perform the *missing* computations required to succeed at the task.

This phenomenon of a robot's (or animal's) body plan performing computation has been termed *morphological computation* [37.8–10]. *Paul* [37.8] outlined a theoretical robot that uses its body to compute the XOR function. In another study [37.9] it was shown how the body of a vacuum cleaning robot could literally replace a portion of its artificial neural network controller, thus subsuming the computation normally performed by that part of the control policy into the robot's body. *Pfeifer* and *Gomez* [37.21] describe a number of other robots that exhibit the phenomenon of morphological computation.

37.2.1 The Counterintuitive Nature of Morphological Computation

All of the robots outlined by *Pfeifer* and *Gomez* [37.21] were designed manually; in some cases the control policies were automatically optimized. If for each task there are a spectrum of robot body plan/control policy pairings that achieve the task, one might ask where along this spectrum the human-designed robots fall. That is, what mixtures of morphological computation and control computation do human designers tend to favor? The bulk of the artificial intelligence literature, since the field's beginnings in the 1950s, seems to indicate that humans exhibit a cognitive chauvinism: we tend to favor control complexity over morphological complexity. Classical artificial intelligence dispensed with the body altogether: it was not until the 1980s that the role of morphology in intelligent behavior was explicitly stated [37.2]. As a more specific example, object manipulation was first addressed by creating rigid, articulated robot arms that required complex control policies to succeed [37.22]. Later, it was realized that soft manipulators could simplify the amount of control required for successful manipulation (e.g., [37.23]). Most recently, a class of robot manipulators known as *jamming grippers'* was introduced [37.24]. In a jamming gripper, a robot arm is tipped with a bag of granular material such that when air is removed from the bag the grains undergo a phase transition into a *jammed*, solid-like state. The control policies for jamming grippers are much simpler than those required for rigid or even soft multifingered dexterous manipulators: at the limit, the controller must switch the manipulator between just two states (*grip* or *release*), regardless of the object.

Despite the fact that the technology for creating jamming grippers has existed for decades, it took a long time for this class of manipulators to be discovered. In other branches of robotics, one can discern a similar historical pattern: new classes of robot body plan were successively proposed that required less and less explicit control. In the field of legged locomotion for example, robots with *whegs* (wheel-leg hybrids) were shown to require less explicit control than robots with legs to enable travel over rough terrain [37.25].

These observations suggest that robots with more morphological computation are less intuitive for humans to formulate and then design than robots with less morphological computation. However, there may be a benefit to creating robots that exhibit significant amounts of morphological computation. For example, hybrid dynamic walkers require very little control and are much more energy efficient compared to fully actuated legged robots [37.26]. It has been argued that tensegrity robots also require relatively little control compared to robots composed of serially linked rigid components, and this class of robot has several desirable properties such as the ability to absorb and recover from external perturbations [37.9].

So, if robots that exhibit morphological computation are desirable, yet it is difficult for humans to navigate in this part of the space of possible robots, can an automated search method be used to discover such robots?

37.2.2 Evolution and Morphological Computation

One of the advantages of using evolutionary algorithms to design robots, compared to machine learning methods, is that both the body plan and the control policy can be placed under evolutionary control [37.27]. Typically, machine learning methods optimize some of the parameters of a control policy with a fixed topology. However, if the body plans and control policies of robots are evolved, and there is sufficient variation within the population of evolving robots, search may discover multiple successful robots that exhibit varying degrees of morphological computation. Or, alternatively, if morphological computation confers a survival advantage within certain contexts, a phylogeny of robots may evolve that exhibit increasing amounts of morphological computation.

A recent pair of experiments illustrates how morphological computation may be explored. An evolutionary algorithm was employed to evolve the body plans

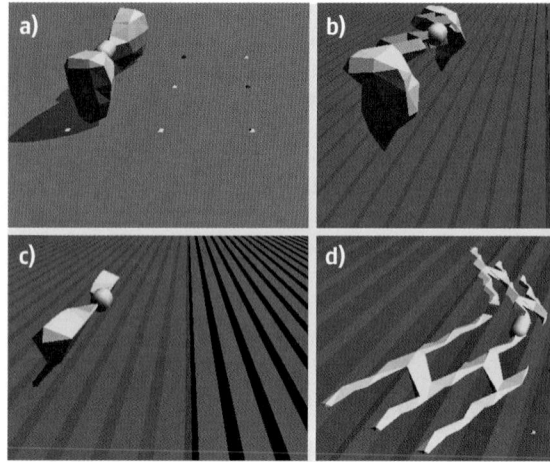

Fig. 37.1a–d A sample of four evolved robots with differing amounts of morphological complexity. (**a**) A simple-shaped robot that evolved to locomote over flat ground. (**b–d**) Three sample robots, more morphologically complex than the robot in (**a**), that evolved in icy environments (after *Auerbach* and *Bongard* [37.28]). To view videos of these robots see [37.29]

and control policies of robots that must move in one of two environments. The first environment included nothing else other than a flat, high-friction ground plane (Fig. 37.1a). The second environment was composed of a number of low-friction bars that sit atop the high-friction ground plane (Fig. 37.1b–d). These bars can be thought of as ice distributed across a flat landscape. In order for robots to move across the icy terrain, they must evolve appendages that are able to reach down between the icy blocks, come into contact with the high-friction ground, and push or pull themselves forward.

It was found that robots evolved to travel over the ice had more complex shapes than those evolved to travel over flat ground (compare the robot in Fig. 37.1a to those in Fig. 37.1b–d) [37.28]. However, it was also found that the robots that travel over ice had fewer mechanical degrees of freedom (DOFs) than the robots evolved to travel over flat ground [37.30]. If a robot possesses fewer mechanical DOFs, one can conclude that it has a simpler control policy, because there are fewer motors to control. It seems that the robots evolved to travel over ice do so in the following manner: the complex shapes of their appendages cause the appendages to *reach* down into the crevices between the ice without explicit control; the simple control policy then simply sweeps the appendages back and forth, horizontally, to in effect *skate* along the tops of the ice. In contrast, robots evolved to travel over flat ground must somehow push back, reach up, and pull forward – using several mechanical DOFs – to move forward.

One could conclude from these experiments that the robots evolved to travel over ice perform more morphological computation than those evolved to travel over flat ground: the former robots have more complex bodies but simpler control policies than the latter robots, yet both successfully move in their environments. Much more work is required to generalize this result to different robots, behaviors, and environments, but this initial work suggests that evolutionary robotics may be a unique tool for studying the phenomenon of morphological computation.

37.3 Sensory–Motor Coordination in Evolving Robots

The actions performed by embodied and situated agents inevitably modify the agent–environmental relation and/or the environment. The type of stimuli that an agent will sense at the next time step at t_{+1} crucially depends, for example, on whether the agent turns left or right at the current time t. Similarly, the stimuli that an agent will experience next at time t_{+1} when standing next to an object depend on the effort with which it will push the object at time t. This implies that actions might play direct and indirect adaptive roles. Actions playing a direct role are, for example, foraging or predator escaping behaviors that directly impact on the agent's own survival chances. Action playing indirect roles consists, for example, in wandering through the environment to spot interesting sensory information (e.g., the perception of a food area that might eventually afford foraging actions) or playing a fighting game with a conspecific that might enable the agent to acquire capacities that might later be exploited to deal with aggressive individuals. The possibility to self-select useful sensory stimuli through action is referred with the term sensory–motor coordination.

Together with morphological computation, sensory–motor coordination constitutes a fundamental property of embodied and situated agents and one of most important characteristic that can be used to differentiate these systems from alternative forms of intelligence. In the following sections, we illustrate three of the key roles that can be played by sensory–motor coordination:

i)   The discovery of parsimonious behavioral strategies
ii) The access and generation of useful sensory information through action and active perception
iii) The constraining and channeling of the learning process during evolution and development.

37.3.1 Enabling the Discovery of Simple Solutions

Sensory–motor coordination can be exploited to find solutions relying on more parsimonious control policies than alternative solutions not relying, or relying less, on this principle. An example is constituted by a set of experiments in which a Khepera robot [37.31] endowed with infrared and speed sensors, has been evolved for the ability to remain close to large cylindrical objects (food) while avoiding small cylindrical objects (dangers). From a passive perspective, that does not take into account the possibility to exploit sensory–motor coordination, the ability to discriminate between sensory stimuli experienced near small and large cylindrical objects requires a relatively complex control policy since the two classes of stimuli strongly overlap in the robot's perceptual space [37.32]. On the other hand, robots evolved for the ability to perform this task tend to converge on a solution relying on a rather simple control policy: the robots begin to turn around objects as soon as they approach them and then discriminate the size of the object on the basis of the sensed differential speed of the left and right wheels during the execution of the object-circling behavior [37.33]. In other words, the execution of the object-circling behavior allows the robots to experience sensory stimuli on the wheel sensors that are well differentiated for small and large objects. This, in turn, allows them to solve the object discrimination problem with a rather simple but reliable control policy.

Another related experiment in which a Khepera robot provided solely with infrared sensors was adapted for finding and remaining close to a cylindrical object, while avoiding walls, demonstrates how sensory–motor coordination can be exploited to solve tasks that require the display of differentiated behavior in different

Part D | 37.3

environmental circumstances, without discriminating the contexts requiring different responses [37.32, 34]. Indeed, evolved robots manage to avoid walls, find a cylindrical object, and remain near it simply by moving backward or forward when their frontal infrared sensors are activated or not, respectively, and by turning left or right when their right and left infrared sensors are activated, respectively (providing that the turning speed and the move forward speed is appropriately regulate on the basis of the sensors activation). Indeed, the execution of this simple control rule combined with the effects of the robot's actions lead to the exhibition of a move-forward behavior far from obstacles, an obstacle avoidance behavior near walls, and an oscillatory behavior near cylindrical objects (in which the robot remains near the object by alternating forward and backward and/or turn-left and turn-right movements). The differentiation of the behavior observed during the robot/wall and robot/cylinder interactions can be explained by considering that the execution of the same action produces different sensory effects in interaction with different objects. In particular, the execution of a turn-left action at time t elicited by the fact that the right infrared sensors are more activated than the left sensors near an object leads to the perception of: (i) a similar sensory stimulus eliciting a similar action at time t_{+1}, ultimately producing an object avoidance behavior near a wall object, (ii) a different sensory stimulus (in which left infrared sensors can become more activated than the left infrared sensors) eliciting a turn-right action at time t_{+1} ultimately producing an oscillatory behavior near the cylinder.

Examples of clever use of sensory–motor coordination abound in natural and artificial evolution. A paradigmatic example of the use of sensory–motor coordination in natural organisms are the navigation capabilities of flying insects that are based on the optic flow, i.e., the apparent motion of contrasting objects in the visual field caused by the relative motion of the agent [37.35]. Houseflies, for example, use this solution to navigate up to 700 body lengths per second in unknown 3D environment while using quite modest processing resources, i.e., about 0.001% of the number of neurons present in the human brain [37.36]. Examples in the evolutionary robotics literature include wheeled robots performing navigation tasks ([37.32], see below), artificial fingers and humanoid robotic arms evolved for the ability to discriminate between object varying in shapes [37.20, 37], and wheeled robots able to navigate visually by using a pan-tilt camera [37.38].

37.3.2 Accessing and Generating Information Through Action

A second fundamental role of sensory–motor coordination consists in accessing and/or generating useful sensory information though action. Differently from experimental settings in which stimuli are brought to the passive agent by the experimenter, in ecological conditions agents need to access relevant information through action. For example, infants access the visual information necessary to recognize the 3D structure of an object by rotating it in the hand and by keeping it at close distance so to minimize visual occlusions [37.39]. The use of sensory–motor coordination for this purpose is usually named *active perception* [37.37, 40, 41].

Interestingly, action can be exploited not only to access sensory information but also to generate it. To understand this aspect, we should consider that through their action agents can elaborate the information they access through their sensory system over time and store the result of the elaboration in their body state and/or in their posture or location. A well-known example of this phenomenon is constituted by depth perception as a result of convergence, i.e., the simultaneous inward movement of both eyes toward each other, to maintain a single binocular percept of a selected object. The execution of this behavior produces a kinesthetic sensation in the eye muscles that reliably correlates with the object's depth.

The careful reader might have recognized that the robot's behavioral discrimination strategies to perceive larger and smaller cylindrical objects, described in the previous section, exploit the same active perception mechanism. For a robot provided with infrared and wheel-speed sensors, the perception of object size necessarily requires a capacity to integrate the information provided by several stimuli. The elaboration of this information however is not realized internally, within the robot's nervous system, but rather externally through the exhibition of the object-circling behavior. It is this behavior that generates the corresponding kinesthetic sensation on the wheel sensors that is then used by the robot to decide to remain or leave, depending on the circumstances.

Examples of clever strategies able to elaborate the required information through action and active perception abound in evolutionary robotics experiments. By carrying out an experiment in which a robot needed to reach two foraging areas located in the northeast and southwest side of a rectangular environment surrounded by walls, *Nolfi* [37.34] observed that the evolved robots developed a clever strategy that allows them to compute the relative length of the two sides of the environment and to navigate toward the two right corners on the basis of a simple control policy. The strategy consists in leaving the first encountered corner with an angle of about $45°$ with respect to the two sides, moving straight, and then eventually following the left side of the next encountered wall ([37.34] for details). Another clever exploitation of sensory–motor coordination was observed in an experiment involving two cooperating robots that helped each other to navigate toward circular target areas [37.42]. Evolved robots discovered and displayed a behavior solution that allowed them to inform each other on the relative location of the center of their target navigation area despite their sensory system being unable to detect their relative position within the area [37.42].

37.3.3 Channeling the Course of the Learning Process

A third fundamental role of sensory–motor coordination consists in channeling the course of the forthcoming adaptive process.

The sensory states experienced during learning crucially determine the course and the outcome of the learning process [37.43]. This implies that the actions displayed by an agent, that co-determine the agent's forthcoming sensory states, ultimately affect how the agent changes ontogenetically. In other words, the behavior exhibited by an agent at a certain stage of its development constraints and channels the course of the agent's developmental process.

Indeed, evolutionary robotics experiments indicate how the evolution of plastic agents (agents that vary their characteristics while they interact with the environment [37.44]) lead to qualitatively different results with respect to the evolution of nonplastic individuals. The traits evolved in the case of nonplastic individuals are selected directly for enabling the agent to display the required capabilities. The traits evolved in the case of plastic individuals, instead, are selected primarily for enabling the agents to acquire the required capabilities through an ontogenetic adaptation process. This implies that, in this case, the selected traits do not enable the agent to master their adaptive task (agents tend to display rather poor performance at the beginning of their lifetime) but rather to acquire the required capacities through ontogenetic adaptation.

More generally, the behavioral strategies adopted by agents at a certain stage of their developmental process can crucially constrain the course of the adaptive process. For example, agents learning to reach and grasp objects might temporarily reduce the complexity of the task to be mastered by freezing (i.e., locking) selected DOFs and by then unfreezing them when their capacity reaches a level that allows them to master the task in its full complexity [37.45, 46]. This type of process can enable exploratory learning by encompassing variation and selection of either the general strategy displayed by the agent or the specific way in which the currently selected strategy is realized.

37.4 Developmental Robotics for Higher Order Embodied Cognitive Capabilities

37.4.1 Embodied Cognition and Developmental Robots

The previous sections have demonstrated the fundamental role of embodiment and of the agent–environment coupling in the design of adaptive agents and robots capable to perform sensory–motor tasks such as navigation and object discrimination. However, embodiment also plays an important role in higher order cognitive capabilities [37.12], such as object categorization and representation, language learning, and processing, and even the acquisition of abstract concepts such as numbers. In this section, we will consider some of the key psychological and neuroscience evidence of EC and its contribution in the design of linguistic and numerical skills in cognitive robots.

Intelligent behavior has traditionally been modeled as a result of activation patterns across distributed knowledge representations, such as hierarchical networks of interrelated propositional (symbolic) nodes that represent objects in the world and their attributes as abstract, amodal (nonembodied) entities [37.47]. For example, the response *bird* to a flying object with feathers and wings would result from perceiving its features and retrieving its name from memory on the basis of a matching process. Such traditional views were attractive for a number of reasons: They followed the predominant philosophical tradition of logical conceptual knowledge organization, according to which all objects are members of categories and category membership can be determined in an all-or-none fashion via defining features. Also, such hierarchical knowledge networks were consistent with cognitive performance in simple tasks such as speeded property verification, which were thought to tap into the retrieval of knowledge. For example, verifying the statement *a bird has feathers* was thought to be easier than verifying the statement *a bird is alive* because the feature *feathers* was presumably stored in memory as defining the category *bird*, while the feature *alive* applies to all animals and was therefore represented at a superordinate level of knowledge, hence requiring more time to retrieve after having just processed *bird* [37.47]. Finally, it was convenient to computationally model such networks by liking the human mind to an information processing device with systematic input, storage, retrieval, and output mechanisms. Thus, knowledge was considered as an abstract commodity independent of the physical device within which it was implemented.

More recent work called into question several of these assumptions about the workings of the human mind. For example, graded category memberships and prototypicality effects in categorization tasks pointed to disparities between the normative logical knowledge organization and the psychological reality of knowledge retrieval [37.48]. Computational modeling of cognitive processes has revealed alternative, distributed representational networks for computing intelligent responses in perceptual, conceptual, and motor tasks that avoid the neurophysiologically implausible assumption of localized storage of specific knowledge [37.49]. Most importantly, though, traditional propositional knowledge networks were limited to explaining the meaning of any given concept in terms of an activation pattern across other conceptual nodes, thus effectively defining the meaning of one symbol in terms of arbitrary other symbols. This process never referred to a concrete experience or event and essentially made the process of connecting internal and external referents arbitrary. In other words, traditional knowledge representations never make contact with specific sensory and motor modalities that is essential to imbue meaning to the activation pattern in a network. This limitation is known as the grounding problem [37.50] and points to a fundamental flaw in traditional attempts to model human knowledge representations.

A second reason for abandoning traditional amodal models of knowledge representation is the fact that these models cannot account for patterns of sensory and motor excitation that occur whenever we activate our knowledge. Already at the time when symbol manipulation approaches to intelligent behavior had their heyday there was powerful evidence for a mandatory link between intelligent thought and sensory–motor experience: When matching two images of the same object, the time we need to recognize that it is the same object is linearly related to the angular disparity between the two views [37.15]. This result suggests that the mental comparison process simulates the physical object rotation we would perform if the two images were manipulable in our hands. In recent years, there has been both more behavioral and also neuroscientific evidence of an involvement of sensory–motor processes in intelligent thought, leading to the influential notion of action simulation as an obligatory component of intelligent thought (for review, [37.51]).

To summarize, the idea that sensory and motor processes are an integral part of our knowledge is driven by both theoretical and empirical considerations. On the theoretical side, the EC stance addresses the grounding problem, a fundamental limitation of classical views of knowledge representation. Empirically, it is tough for traditional amodal conceptualizations of knowledge to address systematic patterns of sensory and motor biases that accompany knowledge activation.

Amongst the latest development in robotics and computational intelligence, the field of developmental robotics has specifically focused on the essential role of EC in the ontogenetic development of cognitive capabilities. Developmental robotics (also know as epigenetic robotics and as the field of autonomous mental development) is the interdisciplinary approach to the autonomous design of behavioral and cognitive capabilities in artificial agents (robots) that takes direct inspiration from the developmental principles and mechanisms observed in natural cognitive systems (children) [37.18, 52–54]. In particular, the key principle of developmental robotics is that the robot, using a set of intrinsic developmental principles regulating the real-time interaction between its body, brain, and environment, can autonomously acquire an increasingly complex set of sensorimotor and mental capabilities. Existing models in developmental robotics have covered the full range of sensory–motor and cognitive capabilities, from intrinsic motivation and motor control to social learning, language and reasoning with abstract knowledge ([37.18] for a full overview).

To demonstrate the benefits of combining EC with developmental robotics in the modeling of embodied intelligence, the two domains of the action bases of language and of the relationship between space and numerical cognition have been chosen. In Sect. 37.4.2, we will look at seminal examples of the embodied bases of language in psycholinguistics, neuroscience, and developmental psychology, and the corresponding developmental robotics models. Section 37.4.3 will consider EC evidence on the link between spatial and numerical cognition, and a developmental robotics model of embodied language learning.

37.4.2 Embodied Language Learning

In experimental psychology and psycholinguistics, an influential demonstration of action simulation as part of language comprehension was first carried out by *Glenberg* and *Kaschak* [37.55]. They asked healthy adults to move their right index finger from a button in their midsagittal plane either away from or toward their body to indicate whether a visually presented statement was meaningful or not. Sentences like *Open the drawer* led to faster initiation of movements toward than away from the body, while sentences like *Close the drawer* led to faster initiation of movements away from than toward the body. Thus, there was a congruency effect between the implied spatial direction of the linguistic description and the movement direction of the reader's motor response. This motor congruency effect in language comprehension has been replicated and extended (for review, [37.56]). It suggests that higher level cognitive feats (such as language comprehension) are ultimately making use of lower level (sensory–motor) capacities of the agent, as predicted by an embodied account of intelligence.

In parallel, growing cognitive neuroscience evidence has shown that the cortical areas of the brain specialized for motor processing are also involved in language processing tasks; thus supporting the EC view that action and language are strictly integrated [37.57, 58]. For example, *Hauk* et al. [37.59] carried out brain imaging experiments where participants read words referring to face, arm, or leg actions (e.g., *lick, pick, kick*). Results support the embodied view of language, as the linguistic task of reading a word differentially activated parts of the premotor area that were directly adjacent, or overlapped, with region activated by actual movement of the tongue, the fingers, or the feet, respectively.

The embodied nature of language has also been shown in developmental psychology studies, as in *Tomasello*'s [37.60] constructivist theory of language acquisition and in *Smith* and *Samuelson*'s [37.61] study on embodiment biases in early word learning. For example, *Smith* and *Samuelson* [37.61] investigated the role of embodiment factors such as posture and spatial representations during the learning of first words. They demonstrated the importance of the changes in postures involved in the interaction with objects located in different parts (left and right) of the child's peripersonal space. Experimental data with 18-month old children show that infants can learn new names also in the absence of the referent objects, when the new label is said whilst the child looks at the same left/right location where the object has previously appeared. This specific study was the inspiration of a developmental robotics study on the role of posture in the acquisition of object names with the iCub baby robot [37.62].

The iCub is an open source robotic platform developed as a benchmark experimental tool for cognitive and developmental robotics research [37.63]. It has a total of 53 DOF, with a high number of DOF (32) in the arms and hands to study object manipulation and the role of fine motor skills in cognitive development. This facilitates the replication of the experimental setup of *Smith* and *Samuelson*'s study [37.61]. In the iCub experiments, a human tutor shows two novel objects respectively in the left and right location of a table put in front of the robot. Initially the robot moves to look at each object and learns to categorize it according to its visual features, such as shape and color. Subsequently the tutor hides both objects, directs the robot's attention toward the right side where the first object was shown and says a new word: *Modi*. In the test phase both objects are presented simultaneously in the centre of the table, and the robot is asked *Find the modi*. The robot must then look and point at the object that was presented in the right location. Four different experiments were carried out, as in Smith and Samuelson's child study. Two experiments differ with regards to the frequency of the left/right locations used to show each objects: the Default Condition when each object always appears in the same location, and the Switch Condition when the position of the two objects is varied to weaken the object/location spatial association. In the other two experimental conditions, the object is named whilst in sight, so to compare the relative weighting of the embodiment spatial constraints and the time constraint.

The robot's behavior is controlled by a modular neural network consisting of a series of pretrained Kohonen self-organizing maps (SOMs), connected through Hebbian learning weights that are trained online during the experiment [37.64]. The first SOM is a *color map* as it is used to categorize objects according to their color (average RGB (red-green-blue) color of the foveal area). The second map, the *auditory map*, is used to represent the words heard by the robot, as the result of the automatic speech recognition system. The other SOM is the *body-hub map*, and this is the key component of the robot's neural system that implements the role of embodiment. The body-hub SOM has four inputs, each being the angle of a single joint. In the experiments detailed here only 2 degrees from the head (up/down and left/right motors), and 2 degrees from the eyes (up/down and left/right motors) are used. Embodiment is operationalized here as the posture of eye and head position when the robot has to look to the left and to the right of the scene.

During each experiment, the connection weight linking the color map and the auditory map to the body-hub map are adjusted in real time using a Hebbian learning rule. These Hebbian associative connections are only modified from the current active body posture node. As the maps are linked together in real time, strong connections between objects typically encountered in particular spatial locations, and hence in similar body postures, build up.

To replicate the four experimental conditions of *Smith* and *Samuelson* [37.61], 20 different robots were used in each condition, with new random weights for the SOM and Hebbian connections. Results from the four conditions show a very high match between the robot's data and the child experiment results, closely replicating the variations in the four conditions. For example, in the Default Condition 83% of the trials resulted in the robots selecting the spatially linked objects, whilst in the Switch condition, where the space/object association was weakened, the robots' choices were practically due to chance at 55%. *Smith* and *Samuelson* [37.61] reported 71% of children selected the spatially linked object, versus 45% in the Switch condition.

This model demonstrates that it is possible to build an embodied cognitive system that develops linguistic and sensorimotor capabilities through interaction with the world, closely resembling the embodiment strategies observed in children's acquisition early word learning. Other cognitive robotics models have also been developed which exploit the principle of embodiment in robots' language learning, as in models of compositionality in action and language [37.65–68], in models of the cultural evolution of construction grammar [37.69, 70], and the modeling of the grounding of abstract words [37.71].

37.4.3 Number and Space

Number concepts have long been considered as prototypical examples of abstract and amodal concepts because their acquisition would require generalizing across a large range of instances to discover the invariant cardinality meaning of words such as *two* and *four* [37.72]. Mental arithmetic would therefore appropriately be modeled as abstract symbol manipulation such as incrementing a counter or retrieving factual knowledge [37.73]. But evidence for an inescapable reference back from abstract number concepts to the sensori-motor experiences during concept acquisition has been present for a long time. Specifically, *Moyer* and *Landauer* [37.74] showed that the speed of deciding which of two visually presented digits represents the larger number depends on their numerical distance, with faster decisions for larger distances. Thus, even in the presence of abstract symbols we seem to refer to analog representations, as if comparing sensory impressions of small and large object compilations.

More recent studies provided further evidence that sensory–motor experiences have a strong impact on the availability of number knowledge. This embodiment signature can be documented by measuring the speed of classifying single digits as odd or even with lateralized response buttons. The typical finding is that small numbers (1, 2) are classified faster with left responses and large numbers (8, 9) are classified faster with right responses [37.76]. This spatial–numerical association response codes, or SNARCs effect, has been replicated across several tasks and extended to other effectors (for review [37.77]), including even attention shifts to the left or right side induced by small or large numbers, respectively [37.78].

Importantly, SNARC depends on one's sensory-motor experiences, such as directional scanning and finger counting habits, as well as current task demands. For example, the initial acquisition of number concepts in childhood occurs almost universally through finger counting and this learning process leaves a residue in the number knowledge of adults. Those who start counting on their left hand, thereby associating small numbers with left space, have a stronger SNARC than those who start counting on their right hand [37.79]. Similarly, reading direction modulates the strength of SNARC. In the original report by *Dehaene* et al. [37.76], it was noticed that adults from a right-to-left reading culture presented with weaker or even reversed SNARC. The notion of a spill-over of directional reading habits into the domain of num-

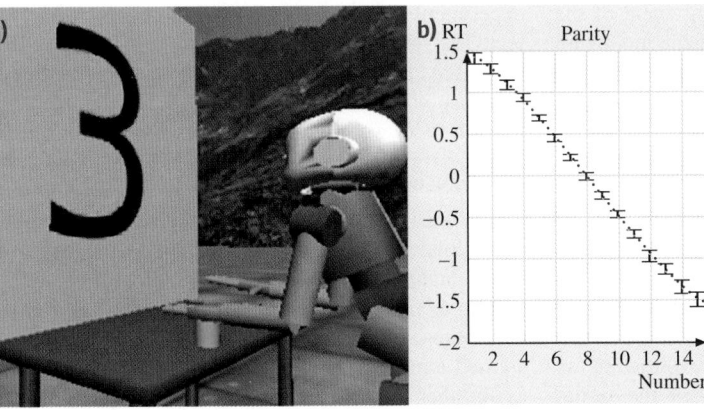

Fig. 37.2a,b (a) iCub simulation model of the SNARC (spatial–numerical association response code) effect; (b) SNARC effect results, with the difference in reaction times (right minus left hand) is plotted against number magnitude (after [37.75])

 per knowledge was further supported by developmental studies showing that it takes around 3 years of schooling before the SNARC emerges [37.80]. However, more recent work has found SNARC even in preschoolers (for review [37.81], thus lending credibility to the role of embodied practices such as finger counting in the formation of SNARC.

In a recent series of experiments with Russian–Hebrew bilinguals, *Shaki* et al. [37.82–84] (for review [37.85]) documented that both one's habitual reading direction and the most recent, task-specific scanning direction determine the strength of one's SNARC. These findings make clear that SNARC is a compound effect where embodied and situated (task-specific) factors add different weights to the overall SNARC.

SNARC and other biases extend into more complex numerical tasks such as mental arithmetic. For example, the association of larger numbers with right space is also present during addition (the operational momentum or OM effect). Regardless of whether symbolic digits or nonsymbolic dot patterns are added together, participants tend to over-estimate the sum, and this bias also influences spatial behavior [37.86]. More generally, intelligent behavior such as mental arithmetic seems to reflect component processes (distance effect, SNARC effect, OM effect) that are grounded in sensorimotor experiences.

The strong link between spatial cognition and number knowledge permits the modeling of the embodiment processes in the acquisition of number in robots. This has been the case with the recent developmental model developed by *Rucinski* et al. [37.75, 87] to model the SNARC effect and the contribution of pointing gestures in number acquisition. In the first study [37.75], a simulation model of the iCub is used. The robot is first trained to develop a body schema of the upper

body through motor babbling of its arms. The iCub is subsequently trained to learn to recognize numbers by associating quantities of objects with numerical symbols as *1* and *2*. In the SNARC test case, the robot has to perform a psychological-like experiment and press a left or right button to make judgments on number comparison and parity judgment (Fig. 37.2b).

The robot's cognitive architecture is based on a modular neural network controller with two main components, following inspiration from a connectionist model of numerical cognition [37.88] and the TRoPI-CALS cognitive architecture of *Caligiore* et al. [37.89, 90]. The two main components of the neural control system are: (i) *ventral* pathway network, responsible for processing of the identity of objects as well as task-dependent decision making and language processing; and (ii) *dorsal* pathway network, involved in processing of spatial information about locations and shapes of objects and processing for the robot's action.

The *ventral* pathway is modeled, following *Chen* and *Verguts* [37.88], with a symbolic input which encodes the alphanumerical number symbols of numbers from 1 to 15, a mental number line encoding the number meaning (quantity), a decision layer for the number comparison and parity judgment tasks, and a response layer, with two neurons for left/right hand response selection. The *dorsal* pathway is composed of a number of SOMs which code for spatial locations of objects in the robot peripersonal space. One map is associated with gaze direction, and two maps respectively for each of the robot's left and right arms. The input to the gaze map arrives from the 3-dimensional proprioceptive vector representing the robot gaze direction (azimuth, elevation and vergence). The input to each arm position map consists of a 7-dimensional proprioceptive vector representing the position of the relevant arm joints. This dorsal pathway constitutes the core component of the model where the embodied properties of the model are directly implemented as the robot's own sensorimotor maps.

To model the developmental learning processes involved in number knowledge acquisition, a series of training phases are implemented. For the embodiment part, the robot is first trained to perform a process equivalent to motor babbling, to develop the gaze and arm space maps. With motor babbling the robot builds its internal visual and motor space representations (SOMs) by performing random reaching movements to touch a toy in its peripersonal space, whilst following its hand's position. Transformations between the visual spatial map for gaze and the maps of reachable left and right spaces are implemented as connections between the maps, which are learned using the classical Hebbian rule. At each trial of motor babbling, gaze and appropriate arm are directed toward the same point and resulting co-activations in already developed spatial maps is used to establish links between them.

The next developmental training establishes the links between number words (modeled as activations in the ventral input layer) and the number meaning (activations in the mental number line hidden layer). Subsequently the robot is taught to count. This stage models the cultural biases that result in the internal association of *small* numbers with the left side of space and *large* numbers with the right side. As an example of these biases, we considered a tendency of children to count objects from left to right, which is related to the fact that European culture is characterized by left-to-right reading direction [37.91]. In order to model the process of learning to count, the robot was exposed to an appropriate sequence of number words (fed to the ventral input layer of the model network), while at the same time the robot's gaze was directed toward a specific location in space (via the input to the gaze visual map). These spatial locations were generated in such a way that their horizontal coordinates correlated with number magnitude (small numbers presented on the left, large numbers on the right) with a certain amount of Gaussian noise. During this stage, Hebbian learning established links between number word and stimuli location in the visual field.

Finally, the model is trained to perform number reasoning tasks, such as number comparison and parity judgment, which corresponds to establishing appropriate links between the mental number line hidden layer and neurons in the decision layer. Specifically, one

experiment focuses on the modeling of the SNARC effect. The robot's reaction time (i. e., amount of activity needed to exceed a response threshold in one of the two response nodes) in parity judgment and number comparison tasks were recorded to calculate the difference between right hand and left hand RTs for the same number. When difference values are plotted against number magnitudes the SNARC effect manifests itself in a negative slope as in Fig. 37.2. As the connections between visual and motor maps form a gradient from left to right, the links to the left arm map become weaker, while those to the right become stronger. Thus, when a small number is presented, internal connections lead to stronger automatic activation of the representations linked with the left arm than that of the right arm, thus causing the SNARC effect.

This model of space and number knowledge was also extended to include a more active interaction with the environment during the number learning process. This is linked to the fact that gestures such as pointing at the object being counted, or the actual touching of the objects enumerated, has been show to improve the overall counting performance in children [37.92]. In the subsequent model by *Rucinski* et al. [37.87], a simpler neural control architecture was used based on the Elman recurrent network to allow sequential number counting and the representation of gestures as proprioceptive states for the pointing gestures. The robot has to learn to produce a sequence of number words (from 1 to 10) with the length of the sequence equivalent to the number of objects present in the scene. Visual input to the model is a one-dimensional saliency map, which can be considered a simple model of a retina. In input, the additional proprioceptive signal was obtained from a pointing gesture performed by the iCub humanoid robot and is used to implement the gestural input to the model in the pointing condition. The output nodes encode the phonetic representation of the 10 numbers.

During the experiment, the robot is first trained to recite a sequence of number words. Then, in order to assess the impact of the proprioceptive information connected with the pointing gesture, the training is divided into two conditions: (i) training to count the number of objects shown to the visual input in the absence of the proprioceptive gesture signal, and (ii) counting though pointing, via the activation of the gesture proprioceptive input. Results show that such a simple recurrent architecture benefits from the input of the proprioceptive gesturing signal, with improved counting accuracy. In particular, the model reproduces the quantitative effects of gestures on the counted set size, replicating child psychology data reported in [37.92].

Overall, such a developmental robotics model clearly shows that the modeling of embodiment phenomena, such as the use of spatial representation in number judgments, and of the pointing gestures for number learning, can allow us to understand the acquisition of abstract concepts in humans as well as artificial agents and robots. This further demonstrates the benefit of the embodied intelligence approach to model a range of behavioral and cognitive phenomena from simple sensory–motor tasks to higher order linguistic and abstract cognition tasks.

37.5 Conclusion

This chapter has provided an overview of the three key principles of embodied intelligence, namely morphological computation, sensory–motor coordination, and EC, and of the experimental approaches and models from evolutionary robotics and developmental robotics. The wide range of behavioral and cognitive capabilities modeled through evolutionary and developmental experiments (e.g., locomotion in different environments, navigation and object discrimination, posture in early word learning and space and number integration) demonstrates the impact of embodied intelligence in the design of a variety of perceptual, motor and cognitive skills, including the potential to model the embodied basis of abstract knowledge as in numerical cognition.

The current progress of both evolutionary and developmental models of embodied intelligence, although showing significant scientific and technological advances in the design of embodied and situated agents, still has a series of open challenges and issues. These issues are informing ongoing work in the various fields of embodied intelligence.

One open challenge in morphological computation concerns how best to automatically design the body plans of robots so that they can best exploit this phenomenon. In parallel to this, much work remains to be done to understand what advantages morphological computation confers on a robot. For one, it is likely that a robot with a simpler control policy will be more robust to unanticipated situations: for example the jamming gripper is able to grasp multiple objects with the same control strategy; a rigid hand requires different control strategies for different objects. Secondly, a robot that performs more morphological computation may be more easily transferred from the simulation in which it was evolved to a physical machine: with a simpler control policy there is less that can go wrong when experiencing the different sensor signals and motor feedback generated by operation in the physical world.

Evolving robots provides a unique opportunity for developing rigorous methods for measuring whether and how much morphological computation a robot performs. For instance, if evolutionary algorithms can be designed that produce robots with similar abilities yet different levels of control and morphological complexity, and it is found that in most cases reduced control complexity implies greater morphological complexity, this would provide evidence for the evolution of morphological computation.

The emerging field of soft robotics [37.93] provides much opportunity for exploring the various aspects of morphological computation because the space of all possible soft robot body plans – with large variations in shape and continuous admixtures of hard and soft materials – is much larger than the space of rigid linkages traditionally employed in *classical* robots.

The design issue, i.e., the question of how systems able to exploit coordinated action and perception processes can be designed, represents an open challenge for sensory–motor coordination as well. As illustrated above, adaptive techniques in which the fine-grained characteristics that determine how agents react to current and previous sensory states are varied randomly and in which variations are retained or discarded on the basis of their effects at the level of the overall behavior exhibited by the agent/s interacting with their environment constitutes an effective method. However, this method might not scale up well with the number of parameters to be adapted. The question of how sensory–motor coordination capabilities can be acquired through the use of other learning techniques that relays on shorter term feedbacks represents an open issue. An interesting research direction, in that respect, consists in the hypothesis that the development of sensory–motor coordination can be induced through the use of task independent criteria such as information theoretic measures [37.94, 95].

Other important research directions concerns the theoretical elaboration of the different roles that morphological computation and sensory–motor coordination can play and the clarification of the relationship between processes occurring as a result of the agent/environmental interactions and processes occurring inside the agents' nervous systems

In developmental robotics models of EC the issues of open-ended, cumulative learning and of the scaling up of the sensory–motor and cognitive repertoires still requires significant efforts and novel methodological and theoretical approaches. Another issue, which combines both evolutionary and developmental approaches, is the interaction of phylogenetic and ontogenetic phenomena in the body/environment/brain adaptation.

Human development is characterized by cumulative, open-ended learning. This refers to the fact that learning and development do not start and stop at specific stages, but rather this is a life-long learning experience. Moreover, the skills acquired in various developmental stages are accumulated and integrated to support the further acquisition of more complex capabilities. One consequence of cumulative, open-ended learning is cognitive bootstrapping. For example in language development, the phenomenon of the vocabulary spurt exist, in which the knowledge and experience from the slow learning of the first 50−100 words causes a redefinition of the word learning strategy, and to syntactic bootstrapping, where children rely on syntactic cues and word context in verb learning to determine the meaning of new verbs [37.96]. Although some computational intelligence models of the vocabulary spurt exist [37.97], robotic experiments on language learning have been restricted to smaller lexicons, not reaching the critical threshold to allow extensive modeling of the bootstrapping of the agent's lexicon and grammar knowledge. These current limitations are also linked to the general issue of the scaling up of the robot's motor and cognitive capabilities and of cross-modal learning. Most of the current cognitive robotics models typically focus on the separate acquisition of only one task or modality (perception, or phonetics, or semantics etc.), often with limited repertoires rarely reaching 10 or slightly more learned actions or words. Thus a truly online, cross-modal, cumulative, open-ended developmental robotics model remains a fundamental challenge to the field.

Another key challenge for future research is the modeling of the interaction of the different timescales of adaptation in embodied intelligence, that is between phylogenetic (evolutionary) factors and ontogenetic (development, learning, maturation) phenomena. For example, maturation refers to changes in the anatomy and physiology of both the child's brain and the body, especially during the first years of life. Maturational phenomena related to the brain include the decrease of brain plasticity during early development, whilst maturation in the body is more evident due to the significant morphological growth changes a child goes through from birth to adulthood (see *Thelen* and *Smith*'s analysis of crawling and walking [37.98]). The ontogenetic changes due to maturation and learning have important implications for the interaction of development with phylogenetic changes due to evolution. Body morphology and brain plasticity variations can be in fact explained as evolutionary adaptations of the species to changing environmental context as with heterochronic changes [37.99]. For example, *Elman* et al. [37.43] discuss how genetic and heterochronic mechanisms provide an alternative explanation of the nature/nurture debate, where genetic phenomena produce architectural constraints of the organism's brain and body, which subsequently control and affects the results of learning interaction. Following this, *Cangelosi* [37.100] has tested the effects of heterochronic changes in the evolution of neural network architectures for simulated robotic agents.

The interaction between ontogenetic and phylogenetic factors has been investigated through evolutionary robotics models. For example, *Hinton* and *Nolan* [37.101] and *Nolfi* et al. [37.102] have developed evolutionary computational models explaining the effects of learning in evolution. The modeling of the evolution of varying body and brain morphologies in response to phylogenetic and ontogenetic requirements is also the goal of the *evo-devo* field of computational intelligence [37.7, 103–105]. These evolutionary/ontogenetic interaction models have, however, mostly focused on simple sensory–motor tasks such as navigation and foraging. Future work combining evolutionary and developmental robotics models can better provide theoretical and technological understanding of the contribution of different adaptation time scales and mechanisms in embodied intelligence.

References

37.1   R.D. Beer: A dynamical systems perspective on agent-environment interaction, Artif. Intell. **72**, 173–215 (1995)

37.2   R.A. Brooks: Elephants don't play chess, Robot. Auton. Syst. **6**(1), 3–15 (1990)

37.3   A. Cangelosi: Grounding language in action and perception: From cognitive agents to humanoid robots, Phys. Life Rev. **7**(2), 139–151 (2010)

37.4   H.J. Chiel, R.D. Beer: The brain has a body: Adaptive behavior emerges from interactions of nervous system, body and environment, Trends Neurosci. **20**, 553–557 (1997)

37.5   F. Keijzer: *Representation and Behavior* (MIT Press, London 2001)

37.6   S. Nolfi, D. Floreano: *Evolutionary Robotics: The Biology, Intelligence, and Technology of Self-Organizing Machines* (MIT/Bradford Books, Cambridge 2000)

37.7   R. Pfeifer, J.C. Bongard: *How the Body Shapes the Way We Think: A New View of Intelligence* (MIT Press, Cambridge 2006)

37.8   C. Paul: Morphology and computation, Proc. Int. Conf. Simul. Adapt. Behav. (2004) pp. 33–38

37.9   C. Paul: Morphological computation: A basis for the analysis of morphology and control requirements, Robot. Auton. Syst. **54**(8), 619–630 (2006)

37.10   R. Pfeifer, F. Iida: Morphological computation: Connecting body, brain and environment, Jpn. Sci. Mon. **58**(2), 48–54 (2005)

37.11   G. Pezzulo, L.W. Barsalou, A. Cangelosi, M.H. Fischer, K. McRae, M.J. Spivey: The mechanics of embodiment: A dialog on embodiment and computational modeling, Front. Psychol. **2**(5), 1–21 (2011)

37.12   D. Pecher, R.A. Zwaan (Eds.): *Grounding Cognition: The Role of Perception and Action in Memory, Language, and Thinking* (Cambridge Univ. Press, Cambridge 2005)

37.13   M. Wilson: Six views of embodied cognition, Psychon. Bull. Rev. **9**, 625–636 (2002)

37.14   L. Meteyard, S.R. Cuadrado, B. Bahrami, G. Vigliocco: Coming of age: A review of embodiment and the neuroscience of semantics, Cortex **48**(7), 788–804 (2012)

37.15   R. Shepard, J. Metzler: Mental rotation of three dimensional objects, Science **171**(972), 701–703 (1972)

37.16   K. Dijkstra, M.P. Kaschak, R.A. Zwaan: Body posture facilitiates the retrieval of autobiographical memories, Cognition **102**, 139–149 (2007)

37.17   L.E. Williams, J.A. Bargh: Keeping one's distance: The influence of spatial distance cues on affect and evaluation, Psychol. Sci. **19**, 302–308 (2008)

37.18   A. Cangelosi, M. Schlesinger: *Developmental Robotics: From Babies to Robots* (MIT Press, Cambridge 2012)

37.19   J. Bongard: The utility of evolving simulated robot morphology increases with task complexity for object manipulation, Artif. Life **16**(3), 201–223 (2010)

37.20   E. Tuci, G. Massera, S. Nolfi: Active categorical perception of object shapes in a simulated anthropomorphic robotic arm, IEEE Trans. Evol. Comput. **14**(6), 885–899 (2010)

37.21   R. Pfeifer, G. Gomez: Morphological computation – Connecting brain, body, and environment, Lect. Notes Comput. Sci. **5436**, 66–83 (2009)

37.22   V. Pavlov, A. Timofeyev: Construction and stabilization of programmed movements of a mobile robot-manipulator, Eng. Cybern. **14**(6), 70–79 (1976)

37.23   S. Hirose, Y. Umetani: The development of soft gripper for the versatile robot hand, Mech. Mach. Theor. **13**(3), 351–359 (1978)

37.24   E. Brown, N. Rodenberg, J. Amend, A. Mozeika, E. Steltz, M.R. Zakin, H. Lipson, H.M. Jaeger: Universal robotic gripper based on the jamming of granular material, Proc. Natl. Acad. Sci. USA **107**(44), 18809–18814 (2010)

37.25   T.J. Allen, R.D. Quinn, R.J. Bachmann, R.E. Ritzmann: Abstracted biological principles applied with reduced actuation improve mobility of legged vehicles, Proc. IEEE/RSJ Int. Conf. Intell. Robot. Syst. 2 (2003) pp. 1370–1375

37.26   M. Wisse, G. Feliksdal, J. Van Frankkenhuyzen, B. Moyer: Passive-based walking robot, IEEE Robot. Autom. Mag. **14**(2), 52–62 (2007)

37.27   K. Sims: Evolving 3d morphology and behavior by competition, Artif. Life **1**(4), 353–372 (1994)

37.28   J.E. Auerbach, J.C. Bongard: On the relationship between environmental and morphological complexity in evolved robots, Proc. 14th Int. Conf. Genet. Evol. Comput. (2012) pp. 521–528

37.29   GECCO 2012 Robot Videos: https://www.youtube.com/playlist?list=PLD5943A95ABC2C0B3

37.30   J.E. Auerbach, J.C. Bongard: On the relationship between environmental and mechanical complexity in evolved robots, Proc. 13th Int. Conf. Simul. Synth. Living Syst. (2012) pp. 309–316

37.31   F. Mondada, E. Franzi, P. Ienne: Mobile robot miniaturisation: A tool for investigation in control algorithms, Proc. 3rd Int. Symp. Exp. Robot. (Kyoto, Japan 1993)

37.32   S. Nolfi: Power and limits of reactive agents, Neurocomputing **49**, 119–145 (2002)

37.33   C. Scheier, R. Pfeifer, Y. Kunyioshi: Embedded neural networks: Exploiting constraints, Neural Netw. **11**, 1551–1596 (1998)

Part D | 37

37.34   S. Nolfi: Categories formation in self-organizing embodied agents. In: *Handbook of Categorization in Cognitive Science*, ed. by H. Cohen, C. Lefebvre (Elsevier, Amsterdam 2005) pp. 869–889

37.35   J.J. Gibson: *The Perception of the Visual World* (Houghton Mifflin, Boston 1950)

37.36   N. Franceschini, F. Ruffier, J. Serres, S. Viollet: Optic flow based visual guidance: From flying insects to miniature aerial vehicles. In: *Aerial Vehicles*, ed. by T.M. Lam (InTech, Rijeka 2009)

37.37   S. Nolfi, D. Marocco: Active perception: A sensorimotor account of object categorization. In: *From Animals to Animats 7*, (MIT Press, Cambridge 2002) pp. 266–271

37.38   D. Floreano, T. Kato, D. Marocco, S. Sauser: Coevolution of active vision and feature selection, Biol. Cybern. **90**(3), 218–228 (2004)

37.39   H.A. Ruff: Infants' manipulative exploration of objects: Effect of age and object characteristics, Dev. Psychol. **20**, 9–20 (1984)

37.40   R. Bajcsy: Active perception, Proc. IEEE **76**(8), 996–1005 (1988)

37.41   D.H. Ballard: Animate vision, Artif. Intell. **48**, 57–86 (1991)

37.42   J. De Greef, S. Nolfi: Evolution of implicit and explicit communication in a group of mobile robots. In: *Evolution of Communication and Language in Embodied Agents*, ed. by S. Nolfi, M. Mirolli (Springer, Berlin 2010)

37.43   J.L. Elman, E.A. Bates, M. Johnson, A. Karmiloff-Smith, D. Parisi, K. Plunkett: *Rethinking Innateness: A Connectionist Perspective on Development* (MIT Press, Cambridge 1996)

37.44   S. Nolfi, D. Floreano: Learning and evolution, auton, Robots **7**(1), 89–113 (1999)

37.45   N. Bernstein: *The Coordination and Regulation of Movements* (Pergamon, Oxford 1967)

37.46   P. Savastano, S. Nolfi: Incremental learning in a 14 DOF simulated iCub robot: Modelling infant reach/grasp development, Lect. Notes Comput. Sci. **7375**, 369–370 (2012)

37.47   A.M. Collins, M.R. Quillian: Retrieval time from semantic memory, J. Verb. Learn. Verb. Behav. **8**, 240–247 (1969)

37.48   E. Rosch: Cognitive representations of semantic categories, J. Exp. Psychol. Gen. **104**, 192–233 (1975)

37.49   D.E. Rumelhart, J.L. McClelland, P.D.P. Group: *Parallel Distributed Processing: Explorations in the microstructure of Cognition* (MIT Press, Cambridge 1986)

37.50   S. Harnad: The symbol grounding problem, Physica D **42**, 335–346 (1990)

37.51   L.W. Barsalou: Grounded cognition, Annu. Rev. Psychol. **59**, 617–645 (2008)

37.52   M. Asada, K. Hosoda, Y. Kuniyoshi, H. Ishiguro, T. Inui, Y. Yoshikawa, M. Ogino, C. Yoshida: Cognitive developmental robotics: A survey, IEEE Trans. Auton. Mental Dev. **1**, 12–34 (2009)

37.53   M. Lungarella, G. Metta, R. Pfeifer, G. Sandini: Developmental robotics: A survey, Connect. Sci. **15**(4), 151–190 (2003)

37.54   P.Y. Oudeyer: Developmental robotics. In: *Encyclopedia of the Sciences of Learning*, Springe References Series, ed. by N.M. Seel (Springer, New York 2012) p. 329

37.55   A. Glenberg, K. Kaschak: Grounding language i action, Psychon. Bull. Rev. **9**(3), 558–565 (2002)

37.56   M.H. Fischer, R.A. Zwaan: Embodied language A review of the role of the motor system in language comprehension, Q. J. Exp. Psychol. **61**(6) 825–850 (2008)

37.57   F. Pulvermüller: *The Neuroscience of Languag* (Cambridge Univ. Press, Cambridge 2003)

37.58   S.F. Cappa, D. Perani: The neural correlate of noun and verb processing, J. Neurolinguist **16**(2/3), 183–189 (2003)

37.59   O. Hauk, I. Johnsrude, F. Pulvermüller: Somatotopic representation of action words in huma motor and premotor cortex, Neuron **41**(2), 301–33 (2004)

37.60   M. Tomasello: *Constructing a Language* (Harvar Univ. Press, Cambridge 2003)

37.61   L.B. Smith, L. Samuelson: Objects in space an mind: From reaching to words. In: *Thinkin Through Space: Spatial Foundations of Language and Cognition*, ed. by K. Mix, L.B. Smith M. Gasser (Oxford Univ. Press, Oxford 2010)

37.62   A.F. Morse, T. Belpaeme, A. Cangelosi, L.B. Smith Thinking with your body: Modelling spatial biase in categorization using a real humanoid robot 2010 Annu. Meet. Cogn. Sci. Soc. (2010) pp. 33-38

37.63   G. Metta, L. Natale, F. Nori, G. Sandini, D. Vernon, L. Fadiga, C. von Hofsten, J. Santos-Victor A. Bernardino, L. Montesano: The iCub humanoi robot: An open-systems platform for research in cognitive development, Neural Netw. **23**, 1125–113 (2010)

37.64   A.F. Morse, J. de Greeff, T. Belpaeme, A. Cangelosi Epigenetic robotics architecture (ERA), IEEE Trans Auton. Mental Dev. **2**(4), 325–339 (2010)

37.65   Y. Sugita, J. Tani: Learning semantic combinatoriality from the interaction between linguistic and behavioral processes, Adapt. Behav. **13**(1), 33–5 (2005)

37.66   V. Tikhanoff, A. Cangelosi, G. Metta: Language understanding in humanoid robots: iCub simulatio experiments, IEEE Trans. Auton. Mental Dev. **3**(1) 17–29 (2011)

37.67   E. Tuci, T. Ferrauto, A. Zeschel, G. Massera, S. Nolfi An experiment on behavior generalization and the emergence of linguistic compositionality in evolving robots, IEEE Trans. Auton. Mental Dev **3**(2), 176–189 (2011)

Part D | 37

37.68   Y. Yamashita, J. Tani: Emergence of functional hierarchy in a multiple timescale neural network model: A humanoid robot experiment, PLoS Comput. Biol. **4**(11), e1000220 (2008)

37.69   L. Steels: Modeling the cultural evolution of language, Phys. Life Rev. **8**(4), 339–356 (2011)

37.70   L. Steels: *Experiments in Cultural Language Evolution, Advances in Interaction Studies*, Vol. 3 (John Benjamins, Amsterdam 2012)

37.71   F. Stramandinoli, D. Marocco, A. Cangelosi: The grounding of higher order concepts in action and language: A cognitive robotics model, Neural Netw. **32**, 165–173 (2012)

37.72   J. Piaget: *The Origins of Intelligence in Children* (International Univ. Press, New York 1952)

37.73   G.J. Groen, J.M. Parkman: A chronometric analysis of simple addition, Psychol. Rev. **79**(4), 329–343 (1972)

37.74   R.S. Moyer, T.K. Landauer: Time required for judgements of numerical inequality, Nature **215**, 1519–1520 (1967)

37.75   M. Rucinski, A. Cangelosi, T. Belpaeme: An embodied developmental robotic model of interactions between numbers and space, Expanding the Space of Cognitive Science, 23rd Annu. Meet. Cogn. Sci. Soc., ed. by L. Carlson, C. Hoelscher, T.F. Shipley (Cognitive Science Society, Austin 2011) pp. 237–242

37.76   S. Dehaene, S. Bossini, P. Giraux: The mental representation of parity and number magnitude, J. Exp. Psychol. Gen. **122**, 371–396 (1993)

37.77   G. Wood, H.C. Nuerk, K. Willmes, M.H. Fischer: On the cognitive link between space and number: A meta-analysis of the SNARC effect, Psychol. Sci. Q. **50**(4), 489–525 (2008)

37.78   M.H. Fischer, A.D. Castel, M.D. Dodd, J. Pratt: Perceiving numbers causes spatial shifts of attention, Nat. Neurosci. **6**(6), 555–556 (2003)

37.79   D.B. Berch, E.J. Foley, R. Hill, R.P. McDonough: Extracting parity and magnitude from Arabic numerals: Developmental changes in number processing and mental representation, J. Exp. Child Psychol. **74**, 286–308 (1999)

37.80   M.H. Fischer: Finger counting habits modulate spatial-numerical associations, Cortex **44**, 386–392 (2008)

37.81   S.M. Göbel, S. Shaki, M.H. Fischer: The cultural number line: A review of cultural and linguistic influences on the development of number processing, J. Cross-Cult. Psychol. **42**, 543–565 (2011)

37.82   S. Shaki, M.H. Fischer: Reading space into numbers – A cross-linguistic comparison of the SNARC effect, Cognition **108**, 590–599 (2008)

37.83   S. Shaki, M.H. Fischer, W.M. Petrusic: Reading habits for both words and numbers contribute to the SNARC effect, Psychon. Bull. Rev. **16**(2), 328–331 (2009)

37.84   M.H. Fischer, R. Mills, S. Shaki: How to cook a SNARC: Number placement in text rapidly changes spatial-numerical associations, Brain Cogn. **72**, 333–336 (2010)

37.85   M.H. Fischer, P. Brugger: When digits help digits: Spatial-numerical associations point to finger counting as prime example of embodied cognition, Front. Psychol. **2**, 260 (2011)

37.86   M. Pinhas, M.H. Fischer: Mental movements without magnitude? A study of spatial biases in symbolic arithmetic, Cognition **109**, 408–415 (2008)

37.87   M. Rucinski, A. Cangelosi, T. Belpaeme: Robotic model of the contribution of gesture to learning to count, Proc. IEEE ICDL-EpiRob Conf. Dev. (2012)

37.88   Q. Chen, T. Verguts: Beyond the mental number line: A neural network model of number-space interactions, Cogn. Psychol. **60**(3), 218–240 (2010)

37.89   D. Caligiore, A.M. Borghi, D. Parisi, G. Baldassarre: TRoPICALS: A computational embodied neuroscience model of compatibility effects, Psychol. Rev. **117**, 1188–1228 (2010)

37.90   D. Caligiore, A.M. Borghi, R. Ellis, A. Cangelosi, G. Baldassarre: How affordances associated with a distractor object can cause compatibility effects: A study with the computational model TRoPICALS, Psychol. Res. **77**(1), 7–19 (2013)

37.91   O. Lindemann, A. Alipour, M.H. Fischer: Finger counting habits in Middle-Eastern and Western individuals: An online survey, J. Cross-Cult. Psychol. **42**, 566–578 (2011)

37.92   M.W. Alibali, A.A. DiRusso: The function of gesture in learning to count: More than keeping track, Cogn. Dev. **14**(1), 37–56 (1999)

37.93   R. Pfeifer, M. Lungarella, F. Iida: The challenges ahead for bio-inspired 'soft' robotics, Commun. ACM **55**(11), 76–87 (2012)

37.94   M. Lungarella, O. Sporns: Information self-structuring: Key principle for learning and development, Proc. 4th Int. Conf. Dev. Learn. (2005)

37.95   P. Capdepuy, D. Polani, C. Nehaniv: Maximization of potential information flow as a universal utility for collective behaviour, Proc. 2007 IEEE Symp. Artif. Life (CI-ALife 2007) (2007) pp. 207–213

37.96   L. Gleitman: The structural sources of verb meanings, Lang. Acquis. **1**, 135–176 (1990)

37.97   J. Mayor, K. Plunkett: Vocabulary explosion: Are infants full of Zipf?, Proc. 32nd Annu. Meet. Cogn. Sci. Soc., ed. by S. Ohlsson, R. Catrambone (Cognitive Science Society, Austin 2010)

37.98   E. Thelen, L.B. Smith: *A Dynamic Systems Approach to the Development of Cognition and Action* (MIT Press, Cambridge 1994)

37.99   M.L. McKinney, K.J. McNamara: *Heterochrony, the Evolution of Ontogeny* (Plenum, New York 1991)

37.100 A. Cangelosi: Heterochrony and adaptation in developing neural networks, Proc. GECCO99 Genet.

Evol. Comput. Conf., ed. by W. Banzhaf (Morgan Kaufmann, San Francisco 1999) pp. 1241–1248

37.101   G.C. Hinton, S.J. Nowlan: How learning can guide evolution, Complex Syst. **1**, 495–502 (1987)

37.102   S. Nolfi, J.L. Elman, D. Parisi: Learning and evolution in neural networks, Adapt. Behav. **3**(1), 5–28 (1994)

37.103   J. Bongard: Morphological change in machines accelerates the evolution of robust behavior, Proc. Natl. Acad. Sci. USA **108**(4), 1234–1239 (2011)

37.104   S. Kumar, P. Bentley (Eds.): *On Growth, Form, and Computers* (Academic, London 2003)

37.105   K.O. Stanley, R. Miikkulainen: A taxonomy for artifcial embryogeny, Artif. Life **9**, 93–130 (2003)

38. Neuromorphic Engineering

Giacomo Indiveri

Neuromorphic engineering is a relatively young field that attempts to build physical realizations of biologically realistic models of neural systems using electronic circuits implemented in very large scale integration technology. While originally focusing on models of the sensory periphery implemented using mainly analog circuits, the field has grown and expanded to include the modeling of neural processing systems that incorporate the computational role of the body, that model learning and cognitive processes, and that implement large distributed spiking neural networks using a variety of design techniques and technologies. This emerging field is characterized by its multidisciplinary nature and its focus on the physics of computation, driving innovations in theoretical neuroscience, device physics, electrical engineering, and computer science.

Part D | 38.1

38.1 The Origins

Models of neural information processing systems that link the type of information processing that takes place in the brain with theories of computation and computer science date back to the origins of computer science itself [38.1, 2]. The theory of computation based on abstract neural networks models was developed already in the 1950s [38.3, 4], and the development of artificial neural networks implemented on digital computers was very popular throughout the 1980s and the early 1990s [38.5–8]. Similarly, the history of implementing electronic models of neural circuits extends back to the construction of perceptrons in the late 1950s [38.3] and retinas in the early 1970s [38.9]. However, the modern wave of research utilizing very large scale integration technology and emphasizing the nonlinear current characteristics of the transistor to study and implement neural computation began only in the mid-1980s, with the collaboration that sprung up be-

tween scientists such as Max Delbrück, John Hopfield, Carver Mead, and Richard Feynman [38.10]. Inspired by graded synaptic transmission in the retina, Mead sought to use the graded (analog) properties of transistors, rather than simply operating them as on–off (digital) switches, to build circuits that emulate biological neural systems. He developed *neuromorphic* circuits that shared many common physical properties with proteic channels in neurons, and that consequently required far fewer transistors than digital approaches to emulating neural systems [38.11]. Neuromorphic engineering is the research field that was born out of this activity and which carries on that legacy: it takes inspiration from biology, physics, mathematics, computer science, and engineering to design artificial neural systems for carrying out robust and efficient computation using low power, massively parallel analog very large scale integration (VLSI) circuits, that operate with the same *physics of computation* present in the brain [38.12]. Indeed, this young research field was born both out of the *Physics of Computation* course taught at Caltech by Carver Mead, John Hopfield, and Richard Feynman and with *Mead*'s textbook *Analog Very Large Scale Integration and Neural Systems* [38.11]. Prominent in the early expansion of the field were scientists and engineers such as Christof Koch, Terry Sejnowski, Rodney Douglas, Andreas Andreou, Paul Mueller, Jan van der Spiegel, and Eric Vittoz, training a generation of cross-disciplinary students. Examples of successes in neuromorphic engineering range from the first biologically realistic silicon neuron [38.13], or realistic silicon models of the mammalian retina [38.14], to more recent silicon cochlea devices potentially useful for cochlear implants [38.15], or complex distributed multichip architectures for implementing event-driven autonomous behaving systems [38.16].

It is now a well-established field [38.17], with two flagship workshops (the Telluride Neuromorphic Engineering [38.18] and Capo Caccia Cognitive Neuromorphic Engineering [38.19] workshops) that are currently still held every year. Neuromorphic circuits are now being investigated by many academic and industrial research groups worldwide to develop a new generation of computing technologies that use the same organizing principles of the biological nervous system [38.15, 20, 21]. Research in this field represents *frontier* research as it opens new technological and scientific horizons: in addition to basic science questions on the fundamental principles of computation used by the cortical circuits, neuromorphic engineering addresses issues in computer-science, and electrical engineering which go well beyond established frontiers of knowledge. A major effort is now being invested for understanding how these neuromorphic computational principles can be implemented using massively parallel arrays of basic computing elements (or *cores*) and how they can be exploited to create a new generation of computing technologies that takes advantage of future (nano)technologies and scaled VLSI processes, while coping with the problems of low-power dissipation, device unreliability, inhomogeneity, fault tolerance, etc.

38.2 Neural and Neuromorphic Computing

Neural computing (or neurocomputing) is concerned with the implementation of artificial neural networks for solving practical problems. Similarly, hardware implementations of artificial neural networks (neurocomputers) adopt mainly statistics and signal processing methods to solve the problem they are designed to tackle. These algorithms and systems are not necessarily tied to detailed models of neural or cortical processing. Neuromorphic computing on the other hand aims to reproduce the principles of neural computation by emulating as faithfully as possible the detailed biophysics of the nervous system in hardware. In this respect, one major characteristic of these systems is their use of *spikes* for representing and processing signals. This is not an end in itself: spiking neural networks represent a promising computational paradigm for solving complex pattern recognition and sensory processing tasks that are difficult to tackle using standard machine vision and machine learning techniques [38.22, 23]. Much research has been dedicated to software simulations of spiking neural networks [38.24], and a wide range of solutions have been proposed for solving real-world and engineering problems [38.25, 26]. Similarly, there are projects that focus on software simulations of large-scale spiking neural networks for exploring the computational properties of models of cortical circuits [38.27, 28]. Recently, several research projects have been established worldwide to develop large-scale hardware implementations of spiking neural systems using VLSI technologies, mainly for allowing neuroscientists to carry out simulations and virtual experiments in real time or even faster than real-time scales [38.29–31]. Although dealing with hardware implementations of neural systems, either with custom VLSI devices or with dedicated computer architectures, these projects represent the conventional neurocomputing approaches, rather than neuromorphic-computing ones. Indeed, these systems are mainly concerned with fast and large simulations of spiking neural networks. They are optimized for speed and precision, at the cost of size and power consumption (which ranges from megawatts to kilowatts, depending on which approach is followed). An example of an alternative large-scale spiking neural network implementation that follows the original neuromorphic engineering principles (i. e., that exploits the characteristics of VLSI technology to directly emulate the biophysics and the connectivity of cortical circuits) is represented by the *Neurogrid* system em [38.32]. This system comprises an array of 16 VLSI chips, each integrating mixed analog neuromorphic neuron and synapse circuits with digital asynchronous event routing logic. The chips are assembled on a $16.5 \times 9 \, \text{cm}^2$ printed circuit board, and the whole system can model over one million neurons connected by billions of synapses in real time, and using only about $\approx 3\text{W}$ of power [38.32].

Irrespective of the approach followed, these projects have two common goals: On one hand they aim to advance our understanding of neural processing in the brain by developing models and physically building them using electronic circuits, and on the other they aim to exploit this understanding for developing a new generation of radically different non-von Neumann computing technologies that are inspired by neural and cortical circuits. In this interdisciplinary journey neuroscience findings will influence theoretical developments, and these will determine specifications and constraints for developing new neuromorphic circuits and systems that can implement them optimally.

38.3 The Importance of Fundamental Neuroscience

The neocortex is a remarkable computational device [38.33]. It is the neuronal structure in the brain that most expresses biology's ability to implement perception and cognition. Anatomical and neurophysiological studies have shown that the mammalian cortex with its laminar organization and regular microscopic structure has a surprisingly uniform architecture [38.34]. Since the original work of *Gilbert* and *Wiesel* [38.35] on the neural circuits of visual cortex it has been argued that this basic architecture, and its underlying computational principles computational principles can be understood in terms of the laminar distribution of relatively few classes of excitatory and inhibitory neurons [38.34]. Based on these slow, unreliable and inhomogeneous computing elements, the cortex easily outperform today's most powerful computers in a wide variety of computational tasks such as vision, audition, or motor control. Indeed, despite the remarkable progress in information and communication technology and the vast amount of resources dedicated to information and communication technology research and development, today's most fastest and largest computers are still not able to match neural systems, when it comes to carrying out robust computations in real-world tasks. The reasons for this performance gap are not yet fully understood, but it is clear that one fundamental difference between the two types of computing systems lies in the style of computation. Rather than using Boolean logic, precise digital representations, and clocked operations, nervous systems carry out robust and reliable computation using hybrid analog/digital unreliable components; they emphasize distributed, event driven, collective, and massively parallel mechanisms, and make extensive use of adaptation, self-organization and learning. Specifically, the patchy organization of the neurons in the cortex suggests a computational machine where populations of neurons perform collective computation in individual clusters, transmit the results of this computation to neighboring clusters, and set the local context of the cluster by means of feedback connections from/to other relevant cortical areas. This overall graphical architecture resembles graphical processing models that perform Bayesian inference [38.36, 37]. However, the theoretical knowledge for designing and analyzing these models is limited mainly to graphs without loops, while the cortex is characterized by massive recurrent (loopy) connectivity schemes. Recent studies exploring loopy graphical models related to cortical architectures started to emerge [38.33, 38], but issues of convergence and accuracy remain unresolved, hardware implementations in a cortical architectures composed of spiking neurons have not been addressed yet.

Understanding the fundamental computational principles used by the cortex, how they are exploited for processing, and how to implement them in hardware, will allow us to develop radically novel computing paradigms and to construct a new generation of information and communication technology that combine the strengths of silicon technology with the performance of brains. Indeed fundamental research in neuroscience has already made substantial progress in uncovering these principles, and information and communication technologies have advanced to a point where it is possible to integrate almost as many transistors in a VLSI system as neurons in a brain. From the theoretical standpoint of view, it has been demonstrated that any Turing machine, and hence any conceivable digital computation, can be implemented by a noise-free network of spiking neurons [38.39]. It has also been

Part D | 38.3

shown that networks of spiking neurons can carry out a wide variety of complex state-dependent computations, even in the presence of noise [38.40–44]. However, apart from isolated results, a general insight into which computations can be carried out in a robust manner by networks of unreliable spiking elements is still missing. Current proposals in state-of-the-art computational and theoretical neuroscience research represent mainly approximate functional models and are implemented as abstract artificial neural networks [38.45, 46]. It is less clear how these functions are realized by the actual networks of neocortex [38.34], how these networks are interconnected locally, and how percep-

tual and cognitive computations can be supported by them.

Both additional neurophysiological studies on neuron types and quantitative descriptions of local and inter-areal connectivity patterns are required to determine the specifications for developing the neuromorphic VLSI analogs of the cortical circuits studied, and additional computational neuroscience and neuromorphic engineering studies are required to understand what level of detail to use in implementing spiking neural networks, and what formal methodology to use for synthesizing and programming these non-von Neumann computational architectures.

38.4 Temporal Dynamics in Neuromorphic Architectures

Neuromorphic spiking neural network architectures typically comprise massively parallel arrays of simple processing elements with memory and computation co-localized (Fig. 38.1). Given their architectural constraints, these neural processing systems cannot process signals using the same strategies used by the conventional von Neumann computing architectures, such as digital signal processor or central processing unit, that time-domain multiplex small numbers of highly complex processors at high clock rates and operate by transferring the partial results of the computation from and to external memory banks. The synapses and neurons in these architectures have to process input spikes and produce output responses as the input signals arrive, in real time, at the rate of the incoming data. It is not possible to virtualize time and transfer partial results in memory banks outside the architecture core, at higher rates. Rather it is necessary to employ resources that compute with time constants that are well matched to those of the signals they are designed to process. Therefore, to interact with the environment and process signals with biological timescales efficiently, hardware neuromorphic systems need to be able to compute using biologically realistic time constants. In this way, they are well matched to the signals they process, and are inherently synchronized with the real world events.

This constraint is not easy to satisfy using analog VLSI technology. Standard analog circuit design techniques either lead to bulky and silicon-area expensive solutions [38.47] or fail to meet this condition, resorting to modeling neural dynamics at *accelerated* unrealistic timescales [38.48–50].

One elegant solution to this problem is to use current-mode design techniques [38.51] and log-domain circuits operated in the weak-inversion regime [38.52]. When metal oxide semiconductor field effect transistors are operated in this regime, the main mechanism of carrier transport is that of diffusion, as it is for ions flowing through proteic channels across neuron membranes. In general, neuromorphic VLSI circuits operate in this domain (also known as the *subthreshold* regime and this is why they share many common physical properties with proteic channels in neurons [38.52]. For example, metal oxide semiconductor field effect transistor have an exponential relationship between gate-to-source voltage and drain current, and produce currents that range from femto- to nanoampere resolution. In this domain, it is therefore possible to integrate relatively small capacitors in VLSI circuits, to implement temporal filters that are both compact and have biologically realistic time constants, ranging from tens to hundreds of milliseconds.

A very compact subthreshold log-domain circuit that can reproduce biologically plausible temporal dynamics is the differential pair integrator circuit [38.53] shown in Fig. 38.2. It can be shown, by log-domain circuit analysis techniques [38.54, 55] that the response of this circuit is governed by the following first-order differential equation

$$\tau \left(1 + \frac{I_{\text{th}}}{I_{\text{out}}}\right) \frac{\text{d}}{\text{d}t} I_{\text{out}} + I_{\text{out}} = \frac{I_{\text{th}} I_{\text{in}}}{I_\tau} - I_{\text{th}} \,, \qquad (38.1)$$

where the time constant $\tau \triangleq CU_{\text{T}}/\kappa I_\tau$, the term U_{T} represents the thermal voltage and κ the subthreshold slope factor [38.52].

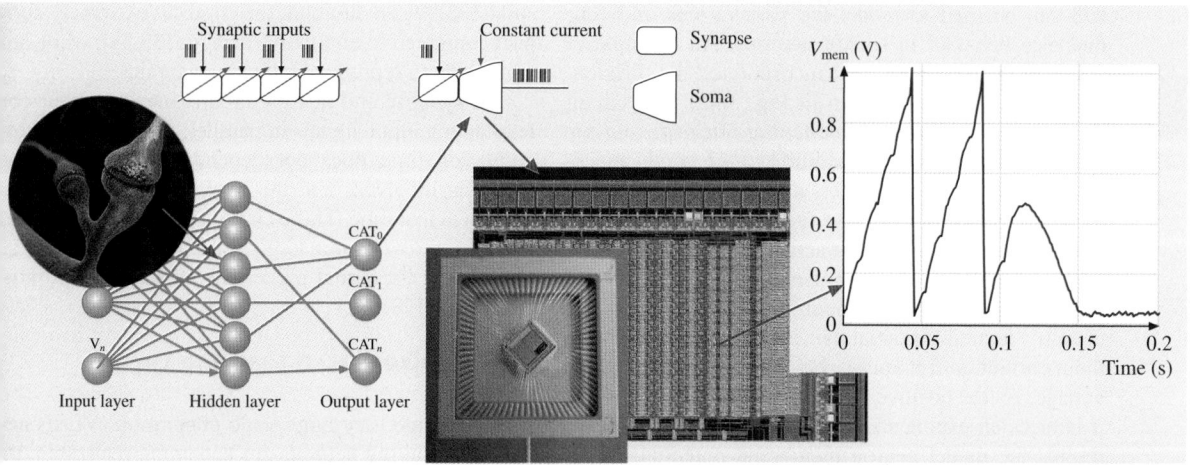

Fig. 38.1 Neuromorphic spiking neural network architectures: detailed biophysical models of cortical circuits are derived from neuroscience experiments; neural networks models are designed, with realistic spiking neurons and dynamic synapses; these are mapped into analog circuits, and integrated in large numbers on VLSI chips. Input spikes are integrated by synaptic circuits, which drive their target postsynaptic neurons, which in turn integrate all synaptic inputs and generate action potentials. Spikes of multiple neurons are transmitted off chip using asynchronous digital circuits, to eventually control in real-time autonomous behaving systems

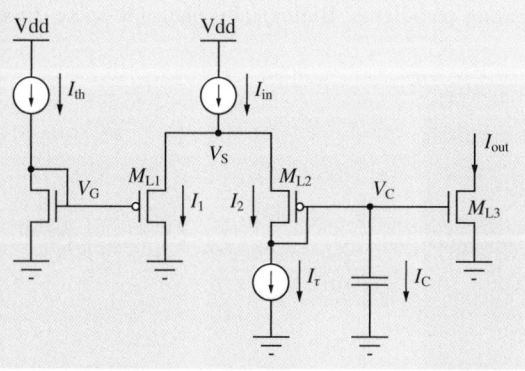

Fig. 38.2 Schematic diagram of neuromorphic integrator circuit. Input currents are integrated in time to produce output currents with large time constants, and with a tunable gain factor

Although this first-order *nonlinear* differential equation cannot be solved analytically, for sufficiently large input currents ($I_{in} \gg I_\tau$) the term $-I_{th}$ on the right-hand side of (38.1) becomes negligible, and eventually when the condition $I_{out} \gg I_{th}$ is met, the equation can be well approximated by

$$\tau \frac{\mathrm{d}}{\mathrm{d}t} I_{out} + I_{out} = \frac{I_{in} I_{th}}{I_\tau} \, . \tag{38.2}$$

Under the reasonable assumptions of nonnegligible input currents, this circuit implements therefore a compact linear integrator with time constants that can be set to range from microseconds to hundreds of milliseconds. It is a circuit that can be used to build neuromorphic sensory systems that interact with the environment [38.56], and most importantly, is is a circuit that reproduces faithfully the dynamics of synaptic transmission observed in biological synapses [38.57].

38.5 Synapse and Neuron Circuits

Synapses are fundamental elements for computation and information transfer in both real and artificial neural systems. They play a crucial role in neural coding and learning algorithms, as well as in neuromorphic neural network architectures. While modeling the nonlinear properties and the dynamics of real synapses can be extremely onerous for software simulations in terms of computational power, memory requirements, and simulation time, neuromorphic synapse circuits can faithfully reproduce synaptic dynamics using integrators such as the differential pair integrator shown in Fig. 38.2. The same differential pair integrator circuit can be used to model the passive leak and conductance behavior in silicon neurons. An example of a silicon neuron circuit that incorporated the differential pair integrator is shown in Fig. 38.3. This circuit implements an *adaptive exponential integrate-and-fire neuron* model [38.58]. In addition to the conductance-based behavior, it implements a spike-frequency adaptation mechanisms, a positive feedback mechanism that models the effect of sodium activation and inactivation channels, and a reset mechanism with a free parameter that can be used to set the neuron's reset potential. The neuron's input differential pair integrator integrates the input current until it approaches the neuron's threshold voltage. As the positive feedback circuit gets activated, it induces an exponential rise in the variable that represents the model neuron membrane potential, which in the circuit of Fig. 38.3 is the current I_{mem}. This quickly causes the neuron to produce an action potential and make a request for transmitting a spike (i. e., the REQ signal of Fig. 38.3 is activated). Once the digital request signal is acknowledged, the membrane capacitance C_{mem} is reset to the neuron's tunable reset potential V_{rst}. These types of neuron circuits have been shown to be extremely low power, consuming about 7 pJ per

spike [38.59]. In addition, the circuit is extremely compact compared to alternative designs [38.58], while still being able to reproduce realistic dynamics.

As synapse and neuron circuits integrate their corresponding input signals in parallel, the neural network emulation time does not depend on the number of elements involved, and the network response always happen in real time. These circuits can be therefore used to develop low-power large-scale hardware neural architectures, for signal processing and general purpose computing [38.58]

38.5.1 Spike-Based Learning Circuits

As large-scale very large scale integration (VLSI) networks of spiking neurons are becoming realizable, the development of robust spike-based learning methods, algorithms, and circuits has become crucial. Spike-based learning mechanisms enable the hardware neural systems they are embedded in to adapt to the statistics of their input signals, to learn and classify complex sequences of spatiotemporal patterns, and eventually to implement general purpose state-dependent computing paradigms. Biologically plausible spike-driven

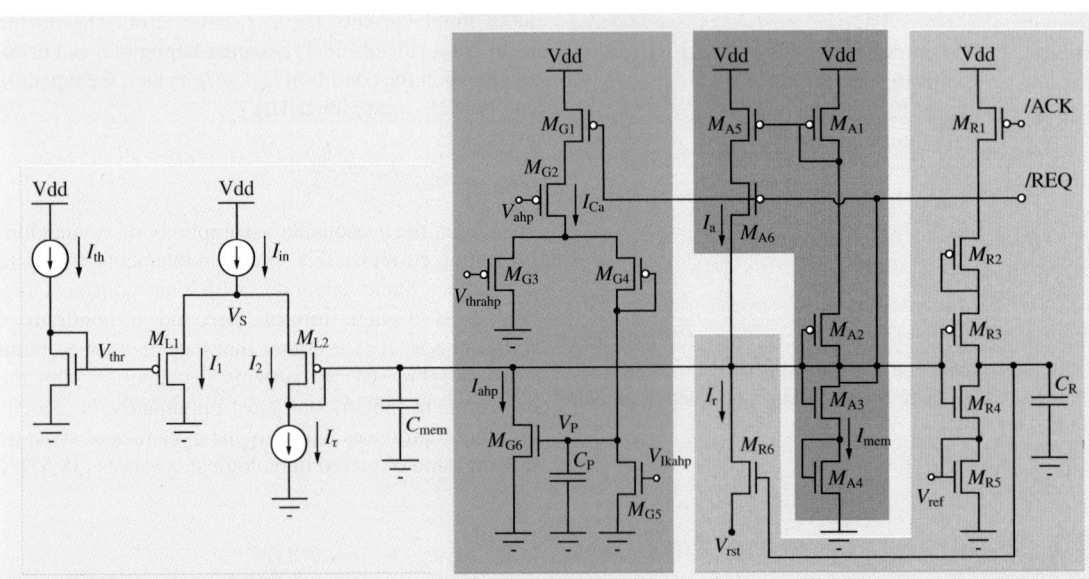

Fig. 38.3 Schematic diagrams of a conductance-based integrate-and-fire neuron. An input differential pair integrator low-pass filter (M_{L1-3}) implements the neuron leak conductance. A noninverting amplifier with current-mode positive feedback (M_{A1-6}) produces address events at extremely low-power operation. A reset block (M_{R1-6}) resets the neuron to the reset voltage V_{rst} and keeps it reset for a refractory period, set by the V_{ref} bias voltage. An additional differential pair integrator low-pass filter (M_{G1-6}) integrates the output events in a negative feedback loop, to implement a *spike frequency adaptation* mechanism

ynaptic plasticity mechanisms have been thoroughly nvestigated in recent years. It has been shown, for ex- ample, how spike-timing dependent plasticity (STDP) can be used to learn to encode temporal patterns of spikes [38.42, 60, 61]. In spike-timing dependent plasticity the relative timing of pre- and postsynap- ic spikes determine how to update the efficacy of a synapse. Plasticity mechanisms based on the timing of the spikes map very effectively onto silicon neuro- morphic devices, and so a wide range of spike-timing lependent plasticity models have been implemented n VLSI [38.62–67]. It is therefore possible to build

large-scale neural systems that can carry out signal processing and neural computation, and include adap- tation and learning. These types of systems are, by their very own nature, modular and scalable. It is pos- sible to develop very large scale systems by designing basic neural processing *cores*, and by interconnecting them together [38.68]. However, to interconnect mul- tiple neural network chips among each other, or to provide sensory inputs to them, or to interface them to conventional computers or robotic platforms, it is nec- essary to develop efficient spike-based communication protocols and interfaces.

38.6 Spike–Based Multichip Neuromorphic Systems

n addition to using spikes for signal efficient process- ng and computations, neuromorphic systems can use spiking representations also for efficient communica- ion. The use of asynchronous spike- or event-based epresentations in electronic systems can be energy effi- cient and fault tolerant, making them ideal for building modular systems and creating complex hierarchies of computation. In recent years, a new class of neuromor- phic multichip systems started to emerge [38.69–71]. These systems typically comprise one or more neuro- morphic sensors, interfaced to general-purpose neural network chips comprising spiking silicon neurons and dynamic synapses. The strategy used to transmit sig- nals across chip boundaries in these types of systems s based on asynchronous *address-events*: output events are represented by the addresses of the neurons that spiked, and transmitted in real time on a digital bus

(Fig. 38.4). The communication protocol used by these systems is commonly referred to as address event rep- resentation [38.72, 73]. The analog nature of the AER (address event representation) signals being transmitted is encoded in the mean frequency of the neurons spikes (spike rates) and in their precise timing. Both types of representations are still an active topic of research in neuroscience, and can be investigated in real time with these hardware systems. Once on a digital bus, the ad- dress events can be translated, converted or remapped to multiple destinations using the conventional logic and memory elements. Digital address event representa- tion infrastructures allow us to construct large multichip networks with arbitrary connectivity, and to seamlessly reconfigure the network topology. Although digital, the asynchronous real-time nature of the AER protocol poses significant technological challenges that are still

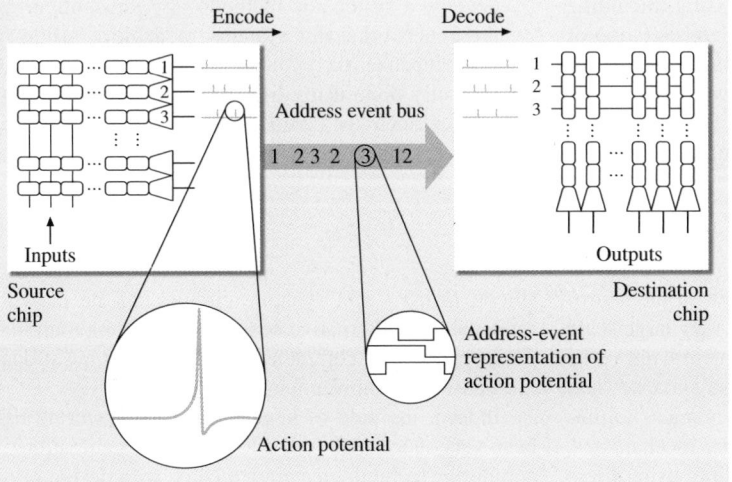

Fig. 38.4 Asynchronous communi- cation scheme between two chips: when a neuron on the source chip generates an action potential, its ad- dress is placed on a common digital bus. The receiving chip decodes the address events and routes them to the appropriate synapses

being actively investigated by the electrical engineering community [38.74]. But by using analog processing in the neuromorphic *cores* and asynchronous digital communication outside them, neuromorphic systems can exploit the best of both worlds, and implement compact low-power brain inspired neural processing systems that can interact with the environment in real time, and represent an alternative (complementary) computing technology to the more common and the conventional VLSI computing architectures.

38.7 State-Dependent Computation in Neuromorphic Systems

General-purpose cortical-like computing architectures can be interfaced to real-time autonomous behaving systems to process sensory signals and carry out event-driven state-dependent computation in real time. However, while the circuit design techniques and technologies for implementing these neuromorphic systems are becoming well established, formal methodologies for *programming* them, to execute specific procedures and solve user defined tasks, do no exist yet. A first step toward this goal is the definition of methods and procedures for implementing state-dependent computation in networks of spiking neurons. In general, state-dependent computation in autonomous behaving systems has been a challenging research field since the advent of digital computers. Recent theoretical findings and technological developments show promising results in this domain [38.16, 43, 44, 75, 76]. But the computational tasks that these systems are currently able to perform remain rather simple, compared to what can be achieved by humans, mammals, and many other animal species. We know, for instance, that nervous systems can exhibit context-dependent behavior, can execute *programs* consisting of series of flexible iterations, and can conditionally branch to alternative behaviors. A general understanding of how to configure artificial neural systems to achieve this sophistication of processing, including also adaptation, autonomous learning, interpretation of ambiguous input signals, symbolic manipulation, inference, and other characteristics that we could regard as effective cognition is still missing. But progress is being made in this direction by studying the computa-

tional properties of spiking neural networks configured as *attractors* or winner-take-all networks [38.33, 44, 77]. When properly configured, these architectures produce persistent activities, which can be regarded as computational states. Both software and VLSI event-driven soft-winner-take-all architectures are being developed to couple spike-based computational models among each other, using the asynchronous communication infrastructure, and use them to investigate their computational properties as neural finite-state machines in autonomous behaving robotic platforms [38.44, 78].

The theoretical, modeling, and VLSI design interdisciplinary activities is carried out with tight interactions, in an effort to understand:

1. How to use the analog, unreliable, and low-precision silicon neurons and synapse circuits operated in the weak-inversion regime [38.52] to carry out reliable and robust signal processing and pattern recognition tasks;
2. How to compose networks of such elements and how to embody them in real-time behaving systems for implementing sets of prespecified desired functionalities and behaviors; and
3. How to formalize these theories and techniques to develop a systematic methodology for configuring these networks and systems to achieve arbitrary state-dependent computations, similar to what is currently done using high-level programming languages such as Java or C++ for conventional digital architectures.

38.8 Conclusions

In this chapter, we presented an overview of the neuromorphic engineering field, focusing on very large scale integration implementations of spiking neural networks and on multineuron chips that comprise synapses and neurons with biophysically realistic dynamics, nonlinear properties, and spike-based plasticity mechanisms. We argued that the multineuron chips built using these

silicon neurons and synaptic circuits can be used to implement an alternative brain inspired computational paradigm that is complementary to the conventional ones based on von Neumann architectures.

Indeed, the field of neuromorphic engineering has been very successful in developing a new generation of computing technologies implemented with design prin-

:iples based on those of the nervous systems, and which :xploit the physics of computation used in biological ieural systems. It is now possible to design and im->lement complex large-scale artificial neural systems vith elaborate computational properties, such as spike->ased plasticity and soft winner-take-all behavior, or :ven complete artificial sensory-motor systems, able to obustly process signals in real time using neuromor->hic VLSI technology.

Within this context, neuromorphic VLSI technology :an be extremely useful for exploring neural processing strategies in real time. While there are clear advantages of this technology, for example, in terms of power bud-get and size requirements, there are also restrictions and limitations imposed by the hardware implemen-ations that limit their possible range of applications. [hese constraints include for example limited resolu-ion in the state variables or bounded parameters (e.g.,)ounded synaptic weights that cannot grow indefinitely or become negative). Also the presence of noise and nhomogeneities in all circuit components, place se-vere limitations on the precision and reliability of the :omputations performed. However, most, if not all, the imitations that neuromorphic hardware implementa-ions face, (e.g., in maintaining stability, in achieving obust computation using unreliable components, etc.)

are often the same one faced by real neural systems. So these limitations are useful for reducing the space of possible artificial neural models that explain or re-produce the properties of real cortical circuits. While in principle these *features* could be simulated also in soft-ware (e.g., by adding a noise term to each state variable, or restricting the resolution of variables to 3, 4, or 6 bits instead of using the floating point representations), they are seldom taken into account. So in addition to representing a technology useful for implementing hardware neural processing systems and solving practi-cal applications, neuromorphic circuits can be used as an additional tool for studying and understanding basic neuroscience.

As VLSI technology is widespread and readily ac-cessible, it is possible to easily learn (and train new generations of students) to design neuromorphic VLSI neural networks for building hardware models of neu-ral systems and sensory-motor systems. Understanding how to build real-time behaving neuromorphic systems that can work in real-world scenarios, will allow us to both gain a better understanding of the principles of neural coding in the nervous system, and develop a new generation of computing technologies that extend and that complement current digital computing devices, cir-cuits, and architectures.

Part D | 38

References

38.1 W.S. McCulloch, W. Pitts: A logical calculus of the ideas immanent in nervous activity, Bull. Math. Biophys. **5**, 115–133 (1943)
38.2 J. von Neumann: *The Computer and the Brain* (Yale Univ. Press, New Haven 1958)
38.3 F. Rosenblatt: The perceptron: A probabilistic model for information storage and organization in the brain, Psychol. Rev. **65**(6), 386–408 (1958)
38.4 M.L. Minsky: *Computation: Finite and Infinite Machines* (Prentice-Hall, Upper Saddle River 1967)
38.5 J.J. Hopfield: Neural networks and physical systems with emergent collective computational abilities, Proc. Natl. Acad. Sci. USA **79**(8), 2554–2558 (1982)
38.6 D.E. Rumelhart, J.L. McClelland: Foundations, par-allel distributed processing. In: *Explorations in the Microstructure of Cognition*, ed. by D.E. Rumelhart, J.L. McClelland (MIT, Cambridge 1986)
38.7 T. Kohonen: *Self-Organization and Associative Memory*, Springer Series in Information Sciences, 2nd edn. (Springer, Berlin Heidelberg 1988)
38.8 J. Hertz, A. Krogh, R.G. Palmer: *Introduction to the Theory of Neural Computation* (Addison-Wesley, Reading 1991)
38.9 K. Fukushima, Y. Yamaguchi, M. Yasuda, S. Nagata: An electronic model of the retina, Proc. IEEE **58**(12), 1950–1951 (1970)
38.10 T. Hey: Richard Feynman and computation, Con-temp. Phys. **40**(4), 257–265 (1999)
38.11 C.A. Mead: *Analog VLSI and Neural Systems* (Addison-Wesley, Reading 1989)
38.12 C. Mead: Neuromorphic electronic systems, Proc. IEEE **78**(10), 1629–1636 (1990)
38.13 M. Mahowald, R.J. Douglas: A silicon neuron, Na-ture **354**, 515–518 (1991)
38.14 M. Mahowald: The silicon retina, Sci. Am. **264**, 76–82 (1991)
38.15 R. Sarpeshkar: Brain power – borrowing from biol-ogy makes for low power computing – bionic ear, IEEE Spectrum **43**(5), 24–29 (2006)
38.16 R. Serrano-Gotarredona, T. Serrano-Gotarredona, A. Acosta-Jimenez, A. Linares-Barranco, G. Ji-ménez-Moreno, A. Civit-Balcells, B. Linares-Barranco: Spike events processing for vision sys-tems, Int. Symp. Circuits Syst. (ISCAS, Piscataway) (2007)
38.17 G. Indiveri, T.K. Horiuchi: Frontiers in neuromor-phic engineering, Front. Neurosci. **5**(118), 1–2 (2011)
38.18 Telluride neuromorphic cognition engineer-ing workshop, http://ine-web.org/workshops/workshops-overview
38.19 The Capo Caccia Workshops toward Cognitive Neu-romorphic Engineering. http://capocaccia.ethz.ch.

38.20 K.A. Boahen: Neuromorphic microchips, Sci. Am. **292**(5), 56–63 (2005)

38.21 R.J. Douglas, M.A. Mahowald, C. Mead: Neuromorphic analogue VLSI, Annu. Rev. Neurosci. **18**, 255–281 (1995)

38.22 W. Maass, E.D. Sontag: Neural systems as nonlinear filters, Neural Comput. **12**(8), 1743–1772 (2000)

38.23 A. Belatreche, L.P. Maguire, M. McGinnity: Advances in design and application of spiking neural networks, Soft Comput. **11**(3), 239–248 (2006)

38.24 R. Brette, M. Rudolph, T. Carnevale, M. Hines, D. Beeman, J.M. Bower, M. Diesmann, A. Morrison, P.H. Harris Jr., F.C. Goodman, M. Zirpe, T. Natschläger, D. Pecevski, B. Ermentrout, M. Djurfeldt, A. Lansner, O. Rochel, T. Vieville, E. Muller, A.P. Davison, S. El Boustani, A. Destexhe: Simulation of networks of spiking neurons: A review of tools and strategies, J. Comput. Neurosci. **23**(3), 349–398 (2007)

38.25 J. Brader, W. Senn, S. Fusi: Learning real world stimuli in a neural network with spike-driven synaptic dynamics, Neural Comput. **19**, 2881–2912 (2007)

38.26 P. Rowcliffe, J. Feng: Training spiking neuronal networks with applications in engineering tasks, IEEE Trans. Neural Netw. **19**(9), 1626–1640 (2008)

38.27 *The Blue Brain Project*. EPFL website. (2005) http://bluebrain.epfl.ch/

38.28 E. Izhikevich, G. Edelman: Large-scale model of mammalian thalamocortical systems, Proc. Natl. Acad. Sci. USA **105**, 3593–3598 (2008)

38.29 *Brain-Inspired Multiscale Computation in Neuromorphic Hybrid Systems (BrainScaleS)*. FP7 269921 EU Grant 2011–2015

38.30 Systems of Neuromorphic Adaptive Plastic Scalable Electronics (SyNAPSE). US Darpa Initiative (http://www.darpa.mil/dso/solicitations/baa08-28.html) (2009)

38.31 R. Freidman: Reverse engineering the brain, Biomed. Comput. Rev. **5**(2), 10–17 (2009)

38.32 B.V. Benjamin, P. Gao, E. McQuinn, S. Choudhary, A.R. Chandrasekaran, J.M. Bussat, R. Alvarez-Icaza, J.V. Arthur, P.A. Merolla, K. Boahen: Neurogrid: A mixed-analog-digital multichip system for large-scale neural simulations, Proc. IEEE **102**(5), 699–716 (2014)

38.33 R.J. Douglas, K. Martin: Recurrent neuronal circuits in the neocortex, Curr. Biol. **17**(13), R496–R500 (2007)

38.34 R.J. Douglas, K.A.C. Martin: Neural circuits of the neocortex, Annu. Rev. Neurosci. **27**, 419–451 (2004)

38.35 C.D. Gilbert, T.N. Wiesel: Clustered intrinsic connections in cat visual cortex, J. Neurosci. **3**, 1116–1133 (1983)

38.36 G.F. Cooper: The computational complexity of probabilistic inference using bayesian belief networks, Artif. Intell. **42**(2/3), 393–405 (1990)

38.37 D.J.C. MacKay: *Information Theory, Inference and Learning Algorithms* (Cambridge Univ. Press, Cambridge 2003)

38.38 A. Steimer, W. Maass, R. Douglas: Belief propagation in networks of spiking neurons, Neural Comput. **21**, 2502–2523 (2009)

38.39 W. Maass: On the computational power of winner-take-all, Neural Comput. **12**(11), 2519–2535 (2000)

38.40 W. Maass, P. Joshi, E.D. Sontag: Computational aspects of feedback in neural circuits, PLOS Comput. Biol. **3**(1), 1–20 (2007)

38.41 L.F. Abbott, W.G. Regehr: Synaptic computation, Nature **431**, 796–803 (2004)

38.42 R. Gütig, H. Sompolinsky: The tempotron: A neuron that learns spike timing–based decisions, Nat. Neurosci. **9**, 420–428 (2006)

38.43 T. Wennekers, N. Ay: Finite state automata resulting from temporal information maximization and a temporal learning rule, Neural Comput. **10**(17), 2258–2290 (2005)

38.44 U. Rutishauser, R. Douglas: State-dependent computation using coupled recurrent networks, Neural Comput. **21**, 478–509 (2009)

38.45 P. Dayan, L.F. Abbott: *Theoretical Neuroscience: Computational and Mathematical Modeling of Neural Systems* (MIT, Cambridge 2001)

38.46 M. Arbib (Ed.): *The Handbook of Brain Theory and Neural Networks*, 2nd edn. (MIT, Cambridge 2002)

38.47 G. Rachmuth, H.Z. Shouval, M.F. Bear, C.-S. Poon: A biophysically-based neuromorphic model of spike rate- and timing-dependent plasticity, Proc. Natl. Acad. Sci. USA **108**(49), E1266–E1274 (2011)

38.48 J. Schemmel, D. Brüderle, K. Meier, B. Ostendorf: Modeling synaptic plasticity within networks of highly accelerated I & F neurons, Int. Symp. Circuits Syst. (ISCAS, Piscataway) (2007) pp. 3367–3370

38.49 J.H.B. Wijekoon, P. Dudek: Compact silicon neuron circuit with spiking and bursting behaviour, Neural Netw. **21**(2/3), 524–534 (2008)

38.50 D. Brüderle, M.A. Petrovici, B. Vogginger, M. Ehrlich, T. Pfeil, S. Millner, A. Grübl, K. Wendt, E. Müller, M.-O. Schwartz, D.H. de Oliveira, S. Jeltsch, J. Fieres, M. Schilling, P. Müller, O. Breitwieser, V. Petkov, L. Muller, A.P. Davison, P. Krishnamurthy, J. Kremkow, M. Lundqvist, E. Muller, J. Partzsch, S. Scholze, L. Zühl, C. Mayr, A. Destexhe, M. Diesmann, T.C. Potjans, A. Lansner, R. Schüffny, J. Schemmel, K. Meier: A comprehensive workflow for general-purpose neural modeling with highly configurable neuromorphic hardware systems, Biol. Cybern. **104**(4), 263–296 (2011)

38.51 C. Tomazou, F.J. Lidgey, D.G. Haigh (Eds.): *Analogue IC Design: The Current-Mode Approach* (Peregrinus, Stevenage, Herts., UK 1990)

38.52 S.-C. Liu, J. Kramer, G. Indiveri, T. Delbruck, R.J. Douglas: *Analog VLSI: Circuits and Principles* (MIT Press, Cambridge 2002)

38.53 C. Bartolozzi, G. Indiveri: Synaptic dynamics in analog VLSI, Neural Comput. **19**(10), 2581–2603 (2007)

38.54 E.M. Drakakis, A.J. Payne, C. Toumazou: Log-domain state-space: A systematic transistor-level approach for log-domain filtering, IEEE Trans. Circuits Syst. II **46**(3), 290–305 (1999)

38.55 D.R. Frey: Log-domain filtering: An approach to current-mode filtering, IEE Proc G **140**(6), 406–416 (1993)

38.56 S.-C. Liu, T. Delbruck: Neuromorphic sensory systems, Curr. Opin. Neurobiol. **20**(3), 288–295 (2010)

38.57 A. Destexhe, Z.F. Mainen, T.J. Sejnowski: Kinetic models of synaptic transmission. In: *Methods in Neuronal Modelling, from Ions to Networks*, ed. by C. Koch, I. Segev (MIT Press, Cambridge 1998) pp. 1–25

38.58 G. Indiveri, B. Linares-Barranco, T.J. Hamilton, A. van Schaik, R. Etienne-Cummings, T. Delbruck, S.-C. Liu, P. Dudek, P. Häfliger, S. Renaud, J. Schemmel, G. Cauwenberghs, J. Arthur, K. Hynna, F. Folowosele, S. Saighi, T. Serrano-Gotarredona, J. Wijekoon, Y. Wang, K. Boahen: Neuromorphic silicon neuron circuits, Front. Neurosci. **5**, 1–23 (2011)

38.59 P. Livi, G. Indiveri: A current-mode conductance-based silicon neuron for address-event neuromorphic systems, Int. Symp. Circuits Syst. (ISCAS) (2009) pp. 2898–2901

38.60 L.F. Abbott, S.B. Nelson: Synaptic plasticity: Taming the beast, Nat. Neurosci. **3**, 1178–1183 (2000)

38.61 R.A. Legenstein, C. Näger, W. Maass: What can a neuron learn with spike-timing-dependent plasticity?, Neural Comput. **17**(11), 2337–2382 (2005)

38.62 S.A. Bamford, A.F. Murray, D.J. Willshaw: Spike-timing-dependent plasticity with weight dependence evoked from physical constraints, IEEE Trans, Biomed. Circuits Syst. **6**(4), 385–398 (2012)

38.63 S. Mitra, S. Fusi, G. Indiveri: Real-time classification of complex patterns using spike-based learning in neuromorphic VLSI, IEEE Trans. Biomed. Circuits Syst. **3**(1), 32–42 (2009)

38.64 G. Indiveri, E. Chicca, R.J. Douglas: A VLSI array of low-power spiking neurons and bistable synapses with spike-timing dependent plasticity, IEEE Trans. Neural Netw. **17**(1), 211–221 (2006)

38.65 A. Bofill, I. Petit, A.F. Murray: Synchrony detection and amplification by silicon neurons with STDP synapses, IEEE Trans. Neural Netw. **15**(5), 1296–1304 (2004)

38.66 S. Fusi, M. Annunziato, D. Badoni, A. Salamon, D.J. Amit: Spike–driven synaptic plasticity: Theory, simulation, VLSI implementation, Neural Comput. **12**, 2227–2258 (2000)

38.67 P. Häfliger, M. Mahowald: Weight vector normalization in an analog VLSI artificial neuron using a backpropagating action potential. In: *Neuro-morphic Systems: Engineering Silicon from Neurobiology*, ed. by L.S. Smith, A. Hamilton (World Scientific, London 1998) pp. 191–196

38.68 P.A. Merolla, J.V. Arthur, R. Alvarez-Icaza, A. Cassidy, J. Sawada, F. Akopyan, B.L. Jackson, N. Imam, A. Chandra, C. Guo, Y. Nakamura, B. Brezzo, I. Vo, S.K. Esser, R. Appuswamy, B. Taba, A. Amir, M.D. Flickner, W.P. Risk, R. Manohar, D.S. Modha: A million spiking-neuron integrated circuit with a scalable communication network and interface, Science **345**(6197), 668–673 (2014)

38.69 R. Serrano-Gotarredona, M. Oster, P. Lichtsteiner, A. Linares-Barranco, R. Paz-Vicente, F. Gómez-Rodriguez, L. Camunas-Mesa, R. Berner, M. Rivas-Perez, T. Delbruck, S.-C. Liu, R. Douglas, P. Häfliger, G. Jimenez-Moreno, A. Civit-Ballcels, T. Serrano-Gotarredona, A.J. Acosta-Jiménez, B. Linares-Barranco: CAVIAR: A 45k neuron, 5M synapse, 12G connects/s aer hardware sensory–processing–learning–actuating system for high-speed visual object recognition and tracking, IEEE Trans. Neural Netw. **20**(9), 1417–1438 (2009)

38.70 E. Chicca, A.M. Whatley, P. Lichtsteiner, V. Dante, T. Delbruck, P. Del Giudice, R.J. Douglas, G. Indiveri: A multi-chip pulse-based neuromorphic infrastructure and its application to a model of orientation selectivity, IEEE Trans. Circuits Syst. I **5**(54), 981–993 (2007)

38.71 T.Y.W. Choi, P.A. Merolla, J.V. Arthur, K.A. Boahen, B.E. Shi: Neuromorphic implementation of orientation hypercolumns, IEEE Trans. Circuits Syst. I **52**(6), 1049–1060 (2005)

38.72 M. Mahowald: *An Analog VLSI System for Stereoscopic Vision* (Kluwer, Boston 1994)

38.73 K.A. Boahen: Point-to-point connectivity between neuromorphic chips using address-events, IEEE Trans. Circuits Syst. II **47**(5), 416–434 (2000)

38.74 A.J. Martin, M. Nystrom: Asynchronous techniques for system-on-chip design, Proc. IEEE **94**, 1089–1120 (2006)

38.75 G. Schoner: Dynamical systems approaches to cognition. In: *Cambridge Handbook of Computational Cognitive Modeling*, ed. by R. Sun (Cambridge Univ. Press, Cambridge 2008) pp. 101–126

38.76 G. Indiveri, E. Chicca, R.J. Douglas: Artificial cognitive systems: From VLSI networks of spiking neurons to neuromorphic cognition, Cogn. Comput. **1**, 119–127 (2009)

38.77 M. Giulioni, P. Camilleri, M. Mattia, V. Dante, J. Braun, P. Del Giudice: Robust working memory in an asynchronously spiking neural network realized in neuromorphic VLSI, Front. Neurosci. **5**, 1–16 (2011)

38.78 E. Neftci, J. Binas, U. Rutishauser, E. Chicca, G. Indiveri, R. Douglas: Synthesizing Cognition in neuromorphic electronic Systems, Proc. Natl. Acad. Sci. USA **110**(37), E3468–E3476 (2013)

39. Neuroengineering

Damien Coyle, Ronen Sosnik

Neuroengineering of sensorimotor rhythm-based brain–computer interface (BCI) systems is the process of using engineering techniques to understand, repair, replace, enhance, or otherwise exploit the properties of neural systems, engaged in the representation, planning, and execution of volitional movements, for the restoration and augmentation of human function via direct interactions between the nervous system and devices.

This chapter reviews information that is fundamental for the complete and comprehensive understanding of this complex interdisciplinary research field, namely an overview of the motor system, an overview of recent findings in neuroimaging and electrophysiology studies of the motor cortical anatomy and networks, and the engineering approaches used to analyze motor cortical signals and translate them into control signals that computer programs and devices can interpret.

Specifically, the anatomy and physiology of the human motor system, focusing on the brain areas and spinal elements involved in the generation of volitional movements is reviewed. The stage is then set for introducing human prototypical motion attributes, sensorimotor learning, and several computational models suggested to explain psychophysical motor phenomena based on the current knowledge in the field of neurophysiology.

An introduction to invasive and non-invasive neural recording techniques, including functional and structural magnetic resonance imaging (fMRI and sMRI), electrocorticography (ECoG), electroencephalography (EEG), intracortical single unit activity (SU) and multiple unit extracellular recordings, and magnetoencephalography (MEG) is integrated with coverage aimed at elucidating what is known about sensory motor oscillations and brain anatomy, which are used to generate control signals for brain actuated devices and al-

ternative communication in BCI. Emphasis is on latest findings in these topics and on highlighting what information is accessible at each of the different scales and the levels of activity that are discernible or utilizable for the effective control of devices using intentional activation sensorimotor neurons and/or modulation of sensorimotor rhythms and oscillations.

The nature, advantages, and drawbacks of various approaches and their suggested functions as the neural correlates of various spatiotemporal motion attributes are reviewed. Sections dealing with signal analysis techniques, translation algorithms, and adaption to the brain's non-stationary dynamics present the reader with a wide-ranging review of the mathematical and statistical techniques commonly used to extract and classify the bulk of neural information recorded by the various recording techniques and the challenges that are posed for deploying BCI systems for their intended uses, be it alternative communication and control, assistive technologies, neurorehabilitation, neurorestoration or replacement, or recreation and entertainment, among other applications. Lastly, a discussion is presented on the future of the field, highlighting newly emerging research directions and their potential ability to enhance our understanding of the human brain and specifically the human motor system and ultimately how that knowledge may lead to more advanced and intelligent computational systems.

Part D | 39

39.1 Overview – Neuroengineering in General

Neuroengineering is defined as the interdisciplinary field of engineering and computational approaches to problems in basic and clinical neurosciences. Thus, education and research in neuroengineering encompasses the fields of engineering, mathematics, and computer science on the one hand, and molecular, cellular, and systems neurosciences on the other. Prominent goals in the field include restoration and augmentation of human functions via direct interactions between the nervous system and artificial devices. Much current research is focused on understanding the coding and processing of information in the sensory and motor systems, quantifying how this processing is altered in the pathological state, and how it can be manipulated through interactions with artificial devices, including brain–computer interfaces (BCIs) and neuroprostheses.

Although there are many topics that can be covered under neuroengineering umbrella, this chapter does not aim to cover them all. Focus is on providing a com-

prehensive overview of state-of-the-art in technologies and knowledge surrounding the human motor system. The motor system is extremely complex in terms of the functions it performs and the structure underlying the control it provides; however, there is an ever-increasing body of knowledge on how it works and is controlled. This has been facilitated by studies of animal models, computational models, electrophysiology, and neuroimaging of humans. Another important development that is extending the boundaries of our knowledge about the motor system is the development of brain–computer interface (BCI) technologies that involve intentional modulation of sensorimotor activity through executed as well as imagined movement (motor imagery). BCI research not only opens up a framework for non-muscular communication between humans and computers/machines but offers experimental paradigms for understanding the neuroscience of motor control, testing hypotheses, and gaining detailed insight into motor control from the activity of a single neuron, a small population of neurons, networks of neurons, and the spatial and spectral relationship across multiple brain regions and networks. This knowledge will undoubtedly lead to better diagnostics for motor-related pathologies, better BCIs for assistance and alternative non-muscular communication applications for the physically impaired, better rehabilitation for those capable of regaining lost motor function, and a better understanding of the brain as a whole.

Relevantly, for the context and scope of this handbook, BCI research will contribute to better gaining a better insight of information processing in the brain, resulting in better, more intelligent computational approaches to developing, truly intelligent systems – systems that perceive, reason, and act autonomously.

The motor system is often considered to be at the heart of human intelligence. From the *motor chauvinist's point of view the entire purpose of the brain is to produce movement* [39.1]. This assertion is based on the following observations about movement:

1. Interaction with the world is only achieved through movement.
2. All communication is mediated via the motor system including speech, sign language, gesture, and writing.
3. All sensory and cognitive processes can be considered inputs that determine future motor outputs.

Neuroscientists and researchers focusing on other areas and functions of the brain may refute this sugges-

tion given the fact that many regions related to general intelligence are located throughout the brain and that a single *intelligence center* is unlikely. No single neuroanatomical structure determines general intelligence, and different types of brain designs can produce equivalent intellectual performance [39.3]. Nevertheless, there is no doubt that the motor system is critical to the advancement of human level intelligence and, therefore, in the context of computational intelligence, this chapter focuses on reviewing studies and methodologies that elucidate some of the aspects that we know about sensorimotor systems and how these can be studied. Although the aim of the chapter is not to provide an exhaustive review of the available extensive literature, it does aim to provide insights into key findings using some of the state-of-the-art experimental and methodological approaches deployed in neuroscience and neuroengineering, whilst at the same time reviewing methodology that may lead to the development of practical BCIs. BCIs have revealed new ways of studying how the brain learns and adapts, which in turn have helped improve BCI designs and better computational intelligence for adapting the signal processing to the adaptation regime of the brain. One of the key findings in BCI research is that it can trigger plastic changes in different brain areas, suggesting that the brain has even greater flexibility than previously thought [39.4]. These findings can only serve to improve our understanding of how the brain, the most sophisticated and complex organism in the known universe, functions, undoubtedly leading to better computational systems.

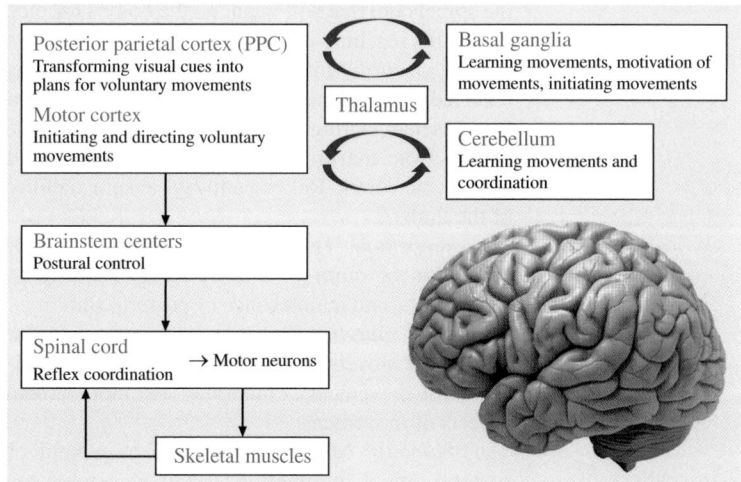

Fig. 39.1 Major components of the motor system (after [39.2], courtesy of *Shadmehr*)

Posterior parietal cortex (PPC)
Transforming visual cues into plans for voluntary movements

Motor cortex
Initiating and directing voluntary movements

Thalamus

Basal ganglia
Learning movements, motivation of movements, initiating movements

Cerebellum
Learning movements and coordination

Brainstem centers
Postural control

Spinal cord
Reflex coordination   → Motor neurons

Skeletal muscles

39.1.1 The Human Motor System

The human motor system produces action. It controls goal-directed movement by selecting the targets of action, generating a motor plan and coordinating the generation of forces needed to achieve those objectives. Genes encode a great deal of the information required by the motor system – especially for actions in-

volving locomotion, orientation, exploration, ingestion, defence, aggression, and reproduction – but every individual must learn and remember a great deal of motor information during his or her lifetime. Some of that information rises to conscious awareness, but much of it does not. Here we will focus on the motor system of humans, drawing on information from primates and other mammals, as necessary.

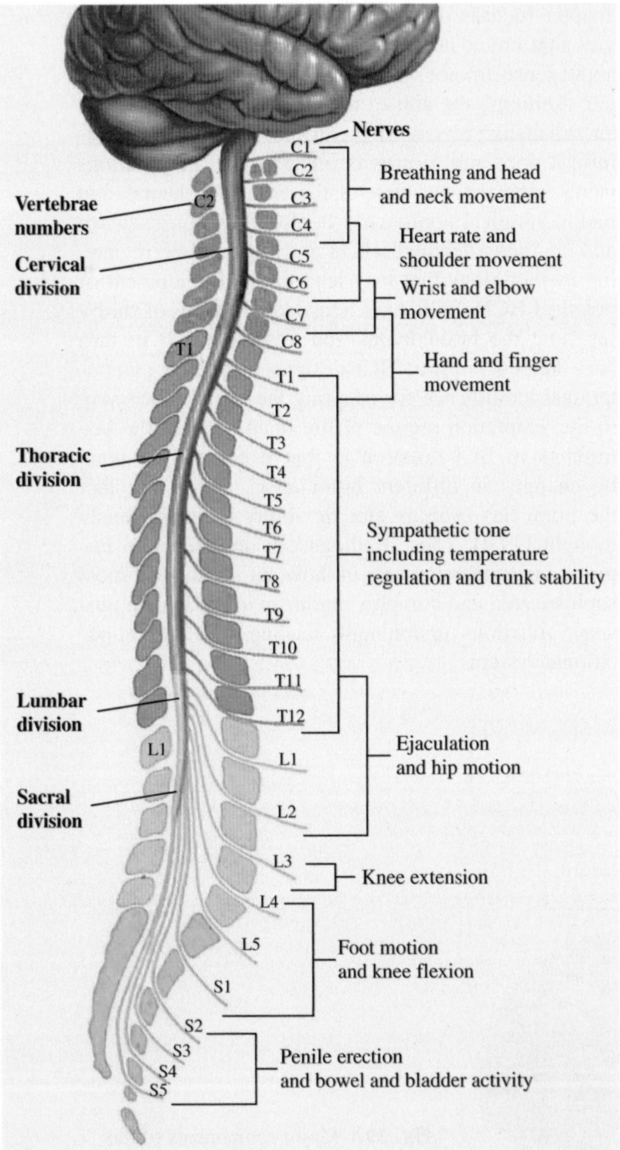

Major Components of the Motor System

The central nervous system that vertebrates have evolved comprises six major components: the spinal cord, medulla, pons, midbrain, diencephalon and telencephalon, the last five of which compose the brain. In a different grouping, the hindbrain (medulla and pons), the midbrain, and the forebrain (telencephalon plus diencephalon) constitute the brain. Taken together, the midbrain and hindbrain make up the brainstem. All levels of the central nervous system participate in motor control. However, let us take the simple act of reaching to pick up a cup of coffee to illustrate the function of the various components of the motor system (Fig. 39.1):

- The *parietal cortex*: Processes visual information and proprioceptive information to compute location of the cup with respect to the hand. Sends this information to the motor cortex.
- The *motor cortex*: Using the information regarding the location of the cup with respect to the hand, it computes forces that are necessary to move the arm. This computation results in commands that are sent to the brainstem and the spinal cord.
- The *brainstem motor center*: Sends commands to the spinal cord that will maintain the body's balance during the reaching movement.
- The *spinal cord*: Motor neurons send the commands received from the motor cortex and the brainstem to the muscles. During the movement, sensory information from the limb is acquired and transmitted back to the cortex. Reflex pathways ensure stability of the limb.
- The *cerebellum*: This center is important for coordination of multi-joint movements, learning of movements, and maintenance of postural stability.
- The *basal ganglia*: This center is important for learning of movements, stability of movements, initiation of movements, emotional, and motivational aspects of movements.
- The *thalamus*: May be thought of as a kind of switchboard of information. It acts as a relay between a variety of subcortical areas and the cerebral

Fig. 39.2 Divisions of the spinal cord (after [39.5], courtesy of *Shadmehr* and *McDonald*)

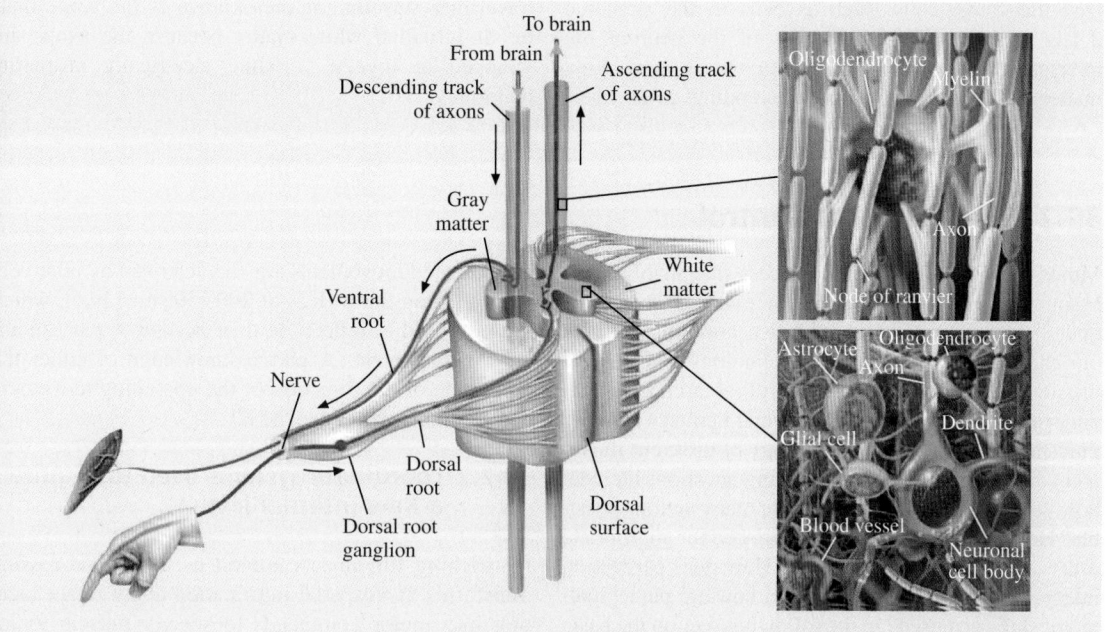

To brain

From brain

Ascending track of axons

Descending track of axons

Oligodendrocyte

Myelin

Gray matter

White matter

Axon

Ventral root

Node of ranvier

Nerve

Astrocyte

Oligodendrocyte

Axon

Glial cell

Dendrite

Dorsal root

Dorsal surface

Dorsal root ganglion

Blood vessel

Neuronal cell body

Fig. 39.3 A spinal segment (after [39.5], courtesy of *Shadmehr* and *McDonald*)

cortex, although recent studies suggest that thalamic function is more selective. The neuronal information processes necessary for motor control are proposed as a network involving the thalamus as a subcortical motor center. The nature of the interconnected tissues of the cerebellum to the multiple motor cortices suggests that the thalamus fulfills a key function in providing the specific channels from the basal ganglia and cerebellum to the cortical motor areas.

The spinal cord comprises four major divisions. From rostral to caudal, these are called: cervical, thoracic, lumbar, and sacral (Fig. 39.2). Cervix is the Latin word for neck. The cervical spinal segments intervene between the pectoral (or shoulder) girdle and the skull. Thorax means chest (or breast plate). Lumbar refers to the loins. Sacral, the most intriguing name of all refers to some sort of sacred bone.

In mammals, the cervical spinal cord has 8 segments; the thoracic spinal cord has 12, and the lumbar and sacral cords both have 5. The parts of the spinal cord that receive inputs from and control the muscles of the arms (more generally, forelimbs) and legs (more generally, hind limbs) show enlargements associated with an increasing number and size of neurons and fibers: the cervical enlargement for the arms and the lumbar enlargement for the legs. Each segment is labeled and numbered according to its order, from rostral to caudal, within each general region of spinal cord. Thus, the first cervical segment is abbreviated C1 and together the eight cervical segments can be designated as C1–C8.

In each spinal segment, one finds a ring of white matter (WM) surrounding a central core of gray matter (GM) (Fig. 39.3). White matter is so called because the high concentration of myelin in the fiber pathways gives it a lighter, shiny appearance relative to regions with many cell bodies. The spinal gray matter bulges at the dorsal and ventral surfaces to form the dorsal horn and ventral horn, respectively.

The cord has two major systems of neurons: descending and ascending. In the descending group, the neurons control both smooth muscles of the internal organs and the striated muscles attached to our bones. The descending pathway begins in the brain, which sends electrical signals to specific segments in the cord. Motor neurons in those segments then convey the impulses towards their destinations outside the cord.

The ascending group is the sensory pathways, sending sensory signals received from the limbs and muscles and our organs to specific segments of the cord and to the brain. These signals originate with special-

Part D | 39.1

ized transducer cells, such as cells in the skin that detect pressure. The cell bodies of the neurons are in a gray, butterfly-shaped region of the cord (gray matter). The ascending and descending axon fibers travel in a surrounding area known as the white matter. It is called white matter because the axons are wrapped in myelin, a white electrically insulating material.

39.2 Human Motor Control

Motor control is a complex process that involves the brain, muscles, limbs, and often external objects. It underlies motion, balance, stability, coordination, and our interaction with others and technology. The general mission of the human motor control research field is to understand the physiology of normal human voluntary movement and the pathophysiology of different movement disorders. Some of the opening questions include: how do we select our actions of the many actions possible? How are these behaviors sequenced for appropriate order and timing between them? How does perception integrate with motor control? And how are perceptual-motor skills acquired? In the following section the basic aspects of motor control – motor planning and motor execution are presented.

39.2.1 Motion Planning and Execution in Humans

Human goal-directed arm movements are fast, accurate, and can compensate for various dynamic loads exerted by the environment. These movements exhibit remarkable invariant properties, although a motor goal can be achieved using different combinations of elementary movements.

Models of goal-directed human arm movements can be divided into two major groups: feedback and feed-forward. Feedback schemes for motion planning assume that the motion is generated through a feedback control law, whereas feed-forward schemes of trajectory formation propose that the movement is planned in advance and then executed. While a comprehensive model of human arm movements should include feed-forward as well as feedback control mechanisms, pure feedback control mechanisms cannot account for the fast and smooth movements performed by adult humans. Although none of the existing models are able to account for all the characteristics of human motion, there is compelling evidence that mechanisms for feed-forward motion planning exist within the central nervous system (CNS). A further supporting argument for the existence of a pre-planned trajectory is that visu-

ally directed movements are characterized by relatively long reaction times (RT) of 200–500 ms [39.6], which are supposed to reflect the time needed to plan an adequate movement. A partial knowledge of either the amplitude or the direction of the upcoming movement can significantly reduce the RT.

39.2.2 Coordinate Systems Used to Acquire a New Internal Model

In reaching for objects around us, neural processing transforms visuospatial information about target location into motor commands to specify muscle forces and joint motions that are involved in moving the hand to the desired location [39.7]. In planar reaching movements, extent and direction have different variable errors, suggesting that the CNS plans the movement amplitude and direction independently and that the hand paths are initially planned in vectorial coordinates without taking into account joint motions. In this framework, the movement vector is specified as an extent and direction from the initial hand position. Kinematic accuracy depends on learning a scaling factor from errors in extent and reference axes from errors in direction, and the learning of new reference axes shows limited generalization [39.8]. Altogether these findings suggest that motor planning takes place in extrinsic, hand-centered, visually perceived coordinates. Finally however, vectorial information need to be converted into muscle forces for the desired movement to be produced. This transformation need to take into account the biomechanical properties of the moving arm, notably the interaction torques produced at all the joints by the motion of all limb segments. For multi-joint arms, there are significant inertial dynamic interactions between the moving skeletal segments, and several muscles pull across more than one joint. Clearly, these complexities raise complicated control problems since one needs to overcome or solve the inverse dynamics problem. The capacity to anticipate the dynamic effects is understood to depend on learning an internal models of musculoskeletal dynamics and other forces acting on the limb.

The equilibrium trajectory hypothesis for multi-joint arm motions [39.9] circumvented the complex dynamic problem mentioned above by using the spring-like properties of muscles and stating that multi-joint arm movements are generated by gradually shifting the hand equilibrium positions defined by the neuromuscular activity. The magnitude of the force exerted on the arm, at any time, depends on the difference between the actual and equilibrium hand positions and the stiffness and viscosity about the equilibrium position.

Neuropsychological studies indicate that for M1 (primary motor) region, the representations that mediate motor behavior are distributed, often in a graded manner, across extensive, overlapping cortical regions, so that different memory systems can underlie different coordinate systems, which are used at different hierarchical levels.

39.2.3 Spatial Accuracy and Reproducibility

Our ability to generate accurate and appropriate motor behavior relies on tailoring our motor commands to the prevailing movement context. This context embodies parameters of both our own motor system, such as the level of muscle fatigue, and the outside world, such as the weight of a bottle to be lifted. As the consequence of a given motor command depends on the current context, the CNS has to estimate this context so that the motor commands can be appropriately adjusted to attain accurate control. A current context can be estimated by integrating two sources of information: sensory feedback and knowledge about how the context

is likely to have changed from the previous estimate. In the absence of sensory feedback about the context, the CNS is able to extrapolate the likely evolution of the context without requiring awareness that the context is changing [39.10].

Although the CNS tries to maximize our motion accuracy, systematic directional errors are still found. These errors may result from a number of sources. One cause for not being accurate is a visual distortion, which could be the outcome of fatigue of the eyes or inherent optical distortion. A second cause could be imperfect control processes due to the noise in the neuromuscular system or blood flow pulsations, which cause our movements to be jerky. A third cause could be that each movement we utilize, consciously or unconsciously, may involve different motor plans, which result in slightly different trajectories and end-point accuracies.

In a simple aiming movement, the task is to minimize the final error, as measured by the variance about the target. The endpoint variability has an ellipsoid shape with two main axes perpendicular one to another. This finding led to the vectorial planning hypotheses stating that planning of visually guided reaches is accomplished by independent specification of extent and direction [39.8]. It was later suggested that the aim of the optimal control strategy is to minimize the volume of the ellipsoid, thereby being as accurate as possible. Non-smooth movements require increased motor commands, which generate increased noise; smoothness thereby leads to increased end-point accuracy but is not a goal in its own. Although the end-point-error cost function specifies the optimal movement, how one approaches this optimum for novel, unrehearsed movements is an open question.

39.3 Modeling the Motor System – Internal Motor Models

An internal model is a postulated neural process that simulates the response of the motor system in order to estimate the outcome of a motor command. The internal model theory of motor control argues that the motor system is controlled by the constant interactions of the *plant* and the *controller*. The plant is the body part being controlled, while the internal model itself is considered part of the controller. Information from the controller, such as information from the CNS, feedback information, and the efference copy, is sent to the plant which moves accordingly.

Internal models can be controlled through either feed-forward or feedback control. Feed-forward control computes its input into a system using only the current state and its model of the system. It does not use feedback, so it cannot correct for errors in its control. In feedback control, some of the output of the system can be fed back into the system's input, and the system is then able to make adjustments or compensate for errors from its desired output. Two primary types of internal models have been proposed: forward models and inverse models. In simulations, models can be combined together to solve more complex movement tasks.

The following section elaborates on the two internal models, introduces the concept of optimization principles and its use in modeling human motor behavior, presenting a well-established motor control model for 2-D volitional hand movement.

39.3.1 Forward Models, Inverse Models, and Combined Models

In their simplest form, forward models take the input of a motor command to the *plant* and output a predicted position of the body. The motor command input to the forward model can be an efference copy. The output from that forward model, the predicted position of the body, is then compared with the actual position of the body. The actual and predicted position of the body may differ due to noise introduced into the system by either internal (e.g., body sensors are not perfect, sensory noise) or external (e.g., unpredictable forces from outside the body) sources. If the actual and predicted body positions differ, the difference can be fed back as an input into the entire system again so that an adjusted set of motor commands can be formed to create a more accurate movement.

Inverse models use the desired and actual position of the body as inputs to estimate the necessary motor commands that would transform the current position into the desired one. For example, in an arm reaching task, the desired position (or a trajectory of consecutive positions) of the arm is input into the postulated inverse model, and the inverse model generates the motor commands needed to control the arm and bring it into this desired configuration.

Theoretical work has shown that in models of motor control, when inverse models are used in combination with a forward model, the efference copy of the motor command output from the inverse model can be used as an input to a forward model for further predictions. For example if, in addition to reaching with the arm, the hand must be controlled to grab an object, an efference copy of the arm motor command can be input into a forward model to estimate the arm's predicted trajectory. With this information, the controller can then generate the appropriate motor command telling the hand to grab the object. It has been proposed that if they exist, this combination of inverse and forward models would allow the CNS to take a desired action (reach with the arm), accurately control the reach, and then accurately control the hand to grip an object.

39.3.2 Adaptive Control Theory

With the assumption that new models can be acquired and pre-existing models can be updated, the efference copy is important for the adaptive control of a movement task. Throughout the duration of a motor task, an efference copy is fed into a forward model known as a dynamics predictor whose output allows prediction of the motor output. When applying adaptive control theory techniques to motor control, the efference copy is used in indirect control schemes as the input to the reference model.

39.3.3 Optimization Principles

Optimization theory is a valuable integrative and predictive tool for studying the interaction between the many complex factors, which result in the generation of goal-directed motor behavior. It provides a convenient way to formulate a model of the underlying neural computations without requiring specific details on the way those computations are carried out. The components of optimization problems are: a task goal (defined mathematically by a performance criterion or a cost function), a system to be controlled (a set of system variables that are available for modulation), and an algorithm capable of finding an analytical or a numerical solution. By rephrasing the learning problem within the framework of an optimization problem, one is forced to make explicit, quantitative hypotheses about the goals of motor actions and to articulate how these goals relate to observable behavior.

As indicated in the last section, goal-directed arm movements exhibit remarkable invariant properties despite the fact that a given point in space can be reached through an infinite number of spatial, articular, and muscle combinations. In order to account for this observation it is necessary to postulate the existence of a *regulator*, i. e., a functional constraint, to reduce the number of degrees of freedom available to perform the task. Most of the regulators proposed during the last decade refer to a general hypothesis that the nervous system *tries* to minimize some cost related to the movement performance. *Nelson* [39.11] first formulated this idea in an operative way by proposing to use mathematical cost functions to estimate the energy or other costs consumed during a movement. This approach was further developed by several investigators who proposed different criteria such as, for instance, minimum muscular energy, minimum effort, minimum torque, minimum work, or minimum variance. A model that is indisputably one of the most mentioned in the literature and that has proven to be very powerful in describing multi-joint movements is the minimum jerk model described in the next section.

39.3.4 Kinematic Features of Human Hand Movements and the Minimum Jerk Hypothesis

Human point-to-point arm movements that are re-stricted in the horizontal plane tend to be straight, smooth, with single-peaked, bell-shaped velocity pro-files and are invariant with respect to rotation, transla-tion, and spatial or temporal scaling. Motor adaptation studies in which unexpected static loads or velocity-dependent force fields were applied during horizontal reaching movements further supported the hypothesis that arm trajectories follow a kinematic plan formu-lated in an extrinsic Cartesian task-space. The mor-phological invariance of the movement in Cartesian space supported the hypothesis that the hand trajec-tory in task-space is the primary variable computed during movement planning. It is assumed that follow-ing the planning process, the CNS performs non-linear inverse kinematics computations, which convert time sequences of hand position into time sequences of joint positions.

The kinematic features of one-joint goal directed movements were successfully modeled by the mini-mum jerk hypothesis [39.13] and were later extended for planar hand motion [39.12]. The minimum jerk model states that the time integral of the squared mag-nitude of the first derivative of the Cartesian hand acceleration (jerk) is minimized,

$$C = \int_0^T \left(\frac{d^3 r}{dt^3} \right)^2 dt , \tag{39.1}$$

where $r(t) = (x(t), y(t))$ are the Cartesian coordinates of the hand and T is the movement duration. The solu-tion of this variational problem, assuming zero velocity and zero acceleration at the initial and final hand loca-tions r_i, r_f, is given by

$$r(t) = ri + \left(10 \left(\frac{t}{T} \right)^3 - 15 \left(\frac{t}{T} \right)^4 + 6 \left(\frac{t}{T} \right)^5 \right) \times (r_i - r_f) . \tag{39.2}$$

The experimental setup and the comparison between experimental data and the minimum-jerk model predic-tion for hand paths, tangential velocities, and accelera-tion components between different targets are depicted in Fig. 39.4.

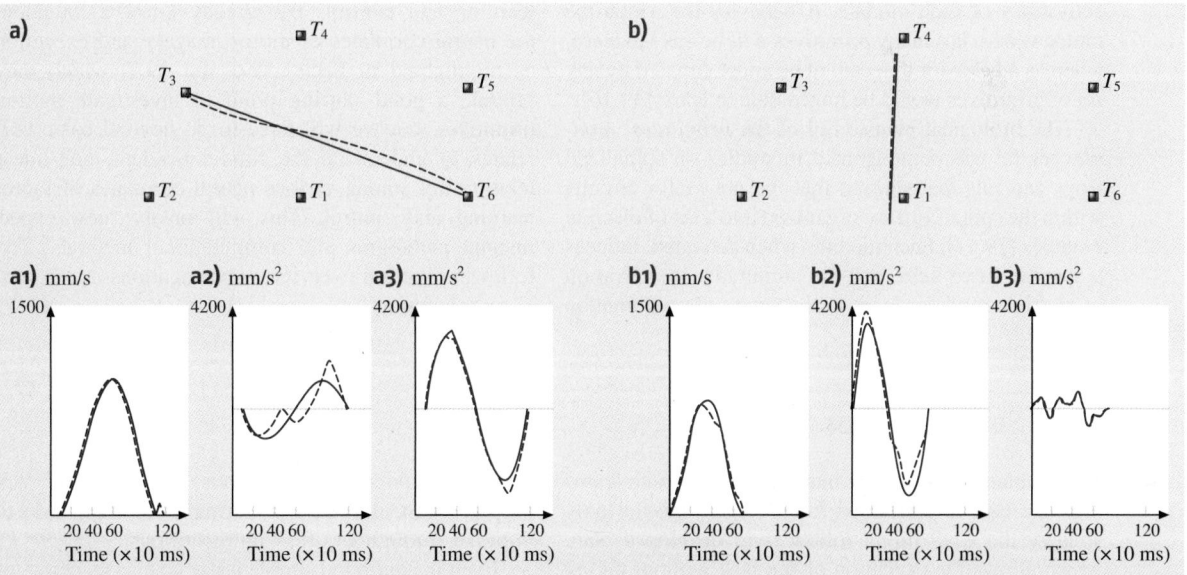

Fig. 39.4a,b Overlapped predicted (*solid lines*) and measured (*dashed lines*) hand paths (a_1, b_1), speeds (a_2, b_2), and acceleration components along the y-axis (a_3, b_3) and along the x-axis (d) for two unconstrained point-to-point movements. (**a**) A movement between targets 3 and 6. (**b**) A movement between targets 1 and 4 (after [39.12], courtesy of *Flash*)

Part D | 39.3

39.3.5 The Minimum Jerk Model, The Target Switching Paradigm, and Writing-like Sequence Movements

The stereotyped kinematic patterns of planar reaching movements are not the expression of a pre-wired or inborn motor pattern, but the result of learning during ontogenesis. When infants start to reach, their reaching is characterized by multiple accelerations and decelerations of the hand, while experienced infants reach with much straighter hand paths and with a single smooth acceleration and deceleration of the hand. It is possible to decompose a large proportion of infant reaches into an underlying sequence of sub-movements that resemble simple movements of adults. It is now believed that the CNS uses small, smooth sub-movements, commonly known as *motion primitives*, which are smoothly concatenated in time and space, in order to construct more complicated trajectories. Motor primitives can be considered neural control modules that can be flexibly combined to generate a large repertoire of behaviors. A primitive may represent the temporal profile of a particular muscle activity (*low level*, dynamic intrinsic primitive) or a geometrical shape in visually perceived Cartesian coordinates (*high level*, kinematic extrinsic primitive [39.14, 15]). The overall motor output will be the sum of all primitives, weighted by the level of activations of each module. A behavior for which the motor system has many primitives will be easy to learn, whereas a behavior that cannot be approximated by any set of primitives would be impossible to learn [39.16].

The biological plausibility of the *primitives' modules* model was demonstrated in studies on spinalized frogs and rats that showed that the pre-motor circuits within the spinal cord are organized into a set of discrete modules [39.17]. Each module, when activated, induces a specific force field, and the simultaneous activation of multiple modules leads to the vectorial combination of the corresponding fields. Other evidence for the existence of primitive sub-movements came from works on hemiplegic stroke patients, which showed that the patients' first movements were clearly segmented and exhibited a remarkably invariant speed vs. time profile.

The concept of superposition was further elaborated and modeled for target switching experiments [39.18]. It was found that arm trajectory modification in a double target displacement paradigm involves the vectorial summation of two independent plans, each coding for a maximally smooth point-to-point trajectory. The first plan is the initial unmodified plan for moving between the initial hand position and the first target location. The second plan is a time-shifted trajectory plan that starts and ends at rest and has the same amplitude and kinematic form as a simple point-to-point movement between the first and second target locations.

The minimum jerk model is also a powerful model for predicting the generated trajectory when subjects are instructed to generate continuous movements from one target to another through an intermediate target. It was also shown that, using the minimum jerk model, human handwriting properties can be faithfully reconstructed while specifying the velocities and the positions at via-points, taken at maximum curvature locations.

Understanding primitives may only be achieved by investigating the neural correlates of sensorimotor learning and control. We already know a lot about the neural correlates of motor imagery and execution as highlighted in Sects. 39.6 and 39.7, which may provide a good starting point to investigate motion primitives, but we will have to go beyond basic correlates to understand the time-dependent, non-linear relationship among various neural correlates of motor learning and control. This will involve new experimental paradigms and computational methods. The following section overviews investigations into sensorimotor learning.

39.4 Sensorimotor Learning

Motion planning strategies may also change with learning. If a task is performed for the first time the only strategy the CNS might follow is to develop a plan, which allows the execution of the task without taking into account the *computational cost*. A repetitive performance might result in a change in the coding of the movement and produce a more optimal behavior – at a lower computational cost. Thus, practice – the track for perfection, allows the performance of many tasks to improve, throughout life, with repetitions.

Even in adulthood simple tasks such as reaching towards a target or rapidly and accurately tapping a short sequence of finger movements, which appear when mastered to be effortlessly performed, often require extensive training before skilled performance develops. A performance gains asymptotes after a long training period and is usually kept intact for years to come. Many studies have focused on different aspects of motor learning: time scale in motor learning and development, task and effector specificity, effect of attention, and intention and explicit vs. implicit motor leaning. These topics are discussed in the next section in the context of motor sequence learning.

39.4.1 Explicit Versus Implicit Motor Learning

When considering sequence learning one needs to distinguish between explicit and implicit learning. Explicit learning is frequently assumed to be similar to the processes which operate during conscious problem solving, and includes: conscious attempts to construct a representation of the task; directed search of memory for similar or analogous task relevant information; and conscious attempts to derive and test hypotheses related to the structure of the task. This type of learning has been distinguished from alternative models of learning, termed *implicit* learning. The term *implicit* learning denotes learning phenomena in which more or less complex structures are reflected in behavior, although the learners are unable to verbally describe these structures. Numerous studies have examined implicit learning of serial-order information using the serial reaction time (SRT) task. In this task, learning is revealed as a decrease in reaction times for stimuli presented when needed to repeat a sequence versus those presented in a random order.

There is a vast literature debating what is really learned in the SRT task. The description of a given sequence structure is from a theoretical point of view not trivial because a given structure typically has several different structural components. Implicit learning may depend on each of these structural components. In sequence learning tasks these components may pertain to: frequency-based, statistical structures (i. e., redundancy), relational structures, and temporal and spatial patterns. A literature review shows that all of these components influence on the rate in which a sequence is learned.

Neuropsychological research suggests that implicit sequence learning in the SRT task is spared in patients with organic amnesia, so implicit SRT learning does not appear to depend on the medial temporal and diencephalic brain regions that are critical for explicit memory. Conversely, patients with Huntington or Parkinson diseases have consistently shown SRT im-

pairments, so the basal ganglia seem to be critically involved in SRT learning. Recent studies indicate that the anterior striatum affects learning of new sequences while the posterior striatum is engaged in recalling a well-learned sequence. In the following section the discussion is restricted to explicit motor learning.

39.4.2 Time Phases in Motor Learning

It is reasonable to assume that a gain in a motor performance reflects a change in brain processing which is triggered by practice. The verity that many skills, when acquired, are retained over long time intervals suggests that training can induce long-lasting neural changes. Previous results from neuroimaging studies in which performance was modified over time have shown that different learning stages can be defined by altered brain activations patterns. As an effect of repetition or practice, several studies report that specific brain areas showed an increase in the magnitude or extent of activation. Motor skill learning (e.g., sequential finger opposition tasks) requires prolonged training times and has two distinct phases, analogous to those subserving perceptual skill learning: an initial, fast improvement phase (*fast learning*) in which the extent of activation in the M1 area decreases (habituation-like response) and a slowly evolving, post training incremental performance gain (*slow learning*), in which the activation in M1 increases compared to control conditions [39.19].

39.4.3 Effector Dependency

Another fascinating enigma in the realm of motor learning is whether the representation of procedural memory in the brain changes throughout training and whether different neural correlates underlie the different learning stages. A study conducted on monkeys, in which a sequence of ten button presses is learned by trial and error, has shown that the time course of improvement of two performance measures: key press errors and reaction-time (RT), was different [39.20]. The key press errors reached an asymptote within a shorter period of training compared to the RTs, which continued to decrease throughout a longer time period. This finding suggested that the acquisition of sequence knowledge (as measured by key press errors) may take place quickly but long-term motor sequence learning (as measured by RT) may take longer to be established, thus different aspects of the task are learned in different time scales. Further studies on monkeys and humans demonstrated that although effector-dependent and independent learning occur simultaneously, effector-dependent representation might take longer to establish than effector-independent representation.

39.4.4 Coarticulation

After a motor sequence is extensively trained, most of the subjects undergo implicit or explicit anticipation, which results in a coarticulation – the spatial and temporal overlap of adjacent articulatory activities. It is well known that as we learn to speak, our speech becomes smooth and fluent. Coarticulation in speech production is a phenomenon in which the articulator movements for a given speech sound vary systematically with the surrounding sounds and their associated movements. Several models have tried to predict the movements of fleshy tissue points on the tongue, lips, and jaw during speech production. Coarticulation has also been studied in the hand motor sequence. It was shown that pianists could anticipate a couple of notes before playing, which resulted in hand and finger kinematic divergence (assuming an anticipatory position) prior to the depression of the last common note. Such a divergence implies an anticipatory modification of sequential movements of the hand, akin to the phenomenon of coarticulation in speech. Moreover, studies on fluent finger spelling has shown that rather than simply an interaction whereby a preceding movement affects the one following, the anticipated movement in a sequence could systematically affect the one preceding it.

39.4.5 Movement Cuing

Another important aspect of motor sequence performance is the type of movement cuing, external or internal. As internally cued movements are initiated at subject's will, they have, by definition, predictable timing. Externally triggered movements are performed in response to *go* signals; hence, they have unpredictable timing (unless the timing of the *go* signal is not random and follows some temporal pattern that can be learned implicitly or explicitly). Studies on movement cuing in animals and patients with movement disorders have showed that the basal ganglia are presumably internal-cue generators and that they are preferentially connected with the supplementary motor area (SMA), an area that is concerned more with internal than with external motor initiation. In normal subjects the type of movement cuing influences movement execution and performance. It has also been shown that teaching Parkinson disease (PD) patients, who are impaired with respect to tasks involving the spontaneous generation of appropriate strategies, to initiate movements concurrently with external cue improved their motor performance.

The preceding sections have provided a brief overview of the extensive literature available on understanding the motor system from an experimental psychophysics and model-based perspective. A focus on general high level modeling is critical to understanding motor control; however, the problem is being tackled from other perspectives, namely understanding the details of neuro and electrophysiology of brain regions and neural pathways involved in controlling motor function. In the context of developing brain–computer interfaces there have been significant efforts focused on understanding small network populations and structural, functional, and electrophysiological correlates of motor functions using epidural and subdural recordings, as well as non-invasively recorded electroencephalography (EEG), magnetoencephalography (MEG), and magnetic resonance imaging-based (MRI) technologies. Understanding the differences between imagined movement and motor execution, as well as the effects of movement feedback and no feedback have shed light on motor functioning. The following sections provide a snapshot of some recent findings.

39.5 MRI and the Motor System – Structure and Function

A new key phase of research is beginning to investigate how functional networks relate to structural networks, with emphasis on how distributed brain areas communicate with each other [39.21]. Structural methods have been powerful in indicating when and where changes occur in both gray and white matter with learning and recovery [39.22] and disease [39.23]. Here we review some of the findings in sensorimotor systems with an emphasis on elucidating regions engaged in motor execution and motor imagery (imagined movement), and motor sequence learning.

Even with identical practice, no two individuals are able to reach the same level of performance on a motor skill – nor do they follow the same trajectory of improvement as they learn [39.24]. These differences are related to brain structure and function, but individual differences in structure have rarely been explored. Studies have shown individual differences in white matter (WM) supporting visuospatial attention, motor cortical connectivity through the Corpus callosum, and connectivity between the motor regions of the cerebellum and motor cortex. *Steele* et al. [39.24] studied the structural characteristics of the brain regions that are functionally engaged in motor sequence performance along with the fiber pathways involved. Using diffusion tensor imaging (DTI), probabilistic tractography, and voxel-based morphometry they aimed to determine the structural correlates of skilled motor performance. DTI is used to asses white matter integrity and perform probabilistic tractography.

Fractional anisotropy (FA) is affected by WM properties, including axon myelination, diameter, and packing density. Differences in these properties may lead to individual differences in performance through pre-existing differences or training-induced changes in axon conduction velocity and synaptic synchronization, or density of innervation [39.24, 25]. Greater fiber integrity along the superior longitudinal fasciculus (SLF) would be consistent with the idea that greater myelination observed in relation to performance may underlie enhancements in synchronized activity between task-relevant regions.

Voxel-based morphometry is used to assess gray matter (GM) volume. Individual differences in GM volume may be influenced by multiple factors such as neuronal and glial cell density, synaptic density, vascular architecture, and cortical thickness [39.26].

The majority of structural studies of individual differences find that better performance is associated with higher FA or greater GM volume. Individual differences in structural measures reflect differences in the microstructural organization of tissue related to task performance. A greater FA, an index of fiber integrity, may represent a greater ability for neurons in connected regions to communicate. *Steele* et al. [39.24] found enhanced synchronization performance on a temporal motor sequence task related to greater fiber integrity of the SLF, where the rate of improvement on synchronization was positively correlated with GM volume in cerebellar lobules HVI and V-regions that showed training-related decreases in activity in the same sample. The synchronization performance on the task was negatively correlated with FA in WM underlying the bilateral sensorimotor cortex, in particular within the bilateral corticospinal tract (CST), such that participants with greater final synchronization performance on the tasks had lower FA in these clusters.

The results provide clear evidence of the importance of structure in learning skilled tasks and that a larger corticospinal tract does not necessarily mean better performance. Enhanced fiber integrity in the SLF may result in reduced FA in regions where it crosses the CST and, therefore, there is a trade-off between the two in the region of the CST-SLF fiber crossing, which enables better performance for some motor imagery and BCI participants – and is consistent with the idea of enhanced communication/synchronization between regions that are functionally important for this task. The causes of inter-individual variability in brain structure are not fully understood, but are likely to include pre-existing genetic contributions and contributions from learning and the environment [39.24]. *Ullén* et al. [39.27] attempted to address this by investigating whether millisecond variability in a simple, automatic timing task, isochronous finger tapping, correlates with intellectual performance and, using voxel-based morphometry, whether these two tasks share neuroanatomical substrates. Volumes of overlapping right prefrontal WM regions were found to be correlated with both stability of tapping and intelligence. These results suggest a bottom-up explanation where extensive pre-frontal connectivity underlies individual differences in both variables as opposed to top-down mechanisms such as attentional control and cognitive strategies.

Sensorimotor rhythm modulation is the most popular BCI control strategy, yet little is known about the structural and functional differences that separate motor areas related to motor output from higher-order motor control areas or about the functional neural correlates of high-order control areas during voluntary motor control. EEG and fMRI studies have shown the extent of motor regions that are active along with the temporal sequence of activations across different motor areas during a motor task and across different subjects [39.28, 29]. *Ball* et al. [39.28] have shown that all subjects in an EEG/fMRI study involving finger flexion had highly activated primary motor cortex areas along with activation of the frontal medial wall motor areas. They also showed that some subjects had anterior type activations as opposed to posterior activation for others, with some showing activity starting in the anterior cingulate motor area (CMA) and then shifting to the intermediate supplementary motor areas. The time sequence of these activations was noted where it was shown that $\approx 120\,\mathrm{ms}$ before movement onset there was a drop in source strength in conjunction with an immediate increase of source strength in the M1 area. Those who showed more posterior activations

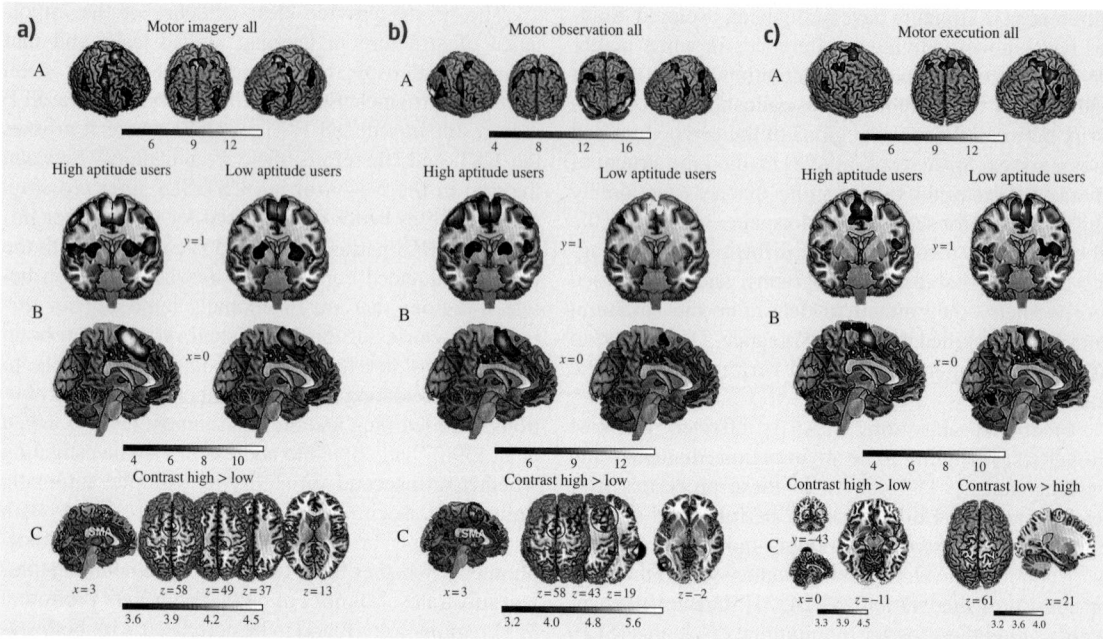

Fig. 39.5a–c Brain activation motor imagery (**a**) motor observation task (**b**) and motor execution task (**c**) showing mean activation of all participants (A), high aptitude users and low aptitude users individually (B), and the contrast of high aptitude users low aptitude users (C). The figure illustrates the maximum contrast between low aptitude and high aptitude BCI users (after [39.29], courtesy of *Halder*)

were restricted to the posterior SMA. Some subjects showed activation of the inferior parietal lobe (IPL) during early movement onset. In all subjects showing activation of higher-order motor areas (anterior CMA, intermediate SMA, IPL), these areas became active before the executive motor areas (M1 and posterior SMA). A number of these areas are related to attentional processing, others to triggering and others to executing. Understanding the sequence of these events for each individual in the context of rehabilitation and more advanced brain and neural computer interfacing will be important.

The neural mechanisms of brain–computer interface control were investigated by [39.29] in an fMRI study. It was shown that up to 30 different motor sites are significantly activated during motor execution, motor observation, and motor imagery and that the number of activated voxels during motor observation was significantly correlated with accuracy in an EEG sensorimotor rhythm-based (SMR) BCI task (see Sect. 39.7.1 for further details on SMR). Significantly higher activations of the supplementary motor areas for motor imagery and motor observation tasks were observed for high aptitude BCI users (see Fig. 39.5 for

an illustration [39.29]). The results demonstrate that acquisition of the sensorimotor program reflected in SMR-BCI control (Sect. 39.7.1) is tightly related to the recall of such sensorimotor programs during observation of movements and unrelated to the actual execution of these movement sequences.

Using such knowledge about sensorimotor control will be critical in understanding and developing successful learning and control models for robotic devices and BCIs, and fully closing the sensorimotor learning loop to enable finer manipulation abilities using BCIs and for retraining or enabling better relearning of motor actions after cortical damage. Understanding the neuroanatomy involved in motor execution/imagery/observation may also provide a means of enhancing our knowledge of motion primitives and their neural correlates as discussed in Sect. 39.4. MRI and fMRI, however, only provide part of the picture, at the level of large networks of neurons, and on relatively large time scales. Invasive electrophysiology, however, can target specific neuronal networks at millisecond time resolution. The following section highlights some of the most recent findings from motor cortical surface potentials investigations.

39.6 Electrocorticographic Motor Cortical Surface Potentials

The electroencephalogram (EEG) is derived from the action potentials of millions of neurons firing electrical pulses simultaneously. The human brain has more than $100\,000\,000\,000$ (10^{11}) neurons and thousands of spikes (electrical pulses) are emitted each millisecond. EEG reflects the aggregate activity of millions of individual neurons recorded with electrodes positioned in a standardized pattern on the scalp. Brainwaves are categorized into a number of different frequency bands including delta ($1-4$ Hz), theta ($5-8$ Hz), alpha ($8-12$ Hz), mu ($8-12$ Hz), beta ($13-30$ Hz), and gamma (> 30 Hz). Each of these brain rhythms can be associated with various brain processes and behavioral states, however, knowledge of exactly where brainwaves are generated in the brain, and if/how they communicate information, is very limited. By studying brain rhythms and oscillations we attempt to answer these questions and have realized that brain rhythms underpin almost every aspect of information processing in the brain, including memory, attention, and even our intelligence. We also observe that abnormal brain oscillations may underlie the problems experienced in diseases such as epilepsy or Alzheimer's disease and we know that certain changes in brain rhythms and oscillations are good indicators of brain pathology associated with these diseases. If we know more about the function of brainwaves we may be able to develop better diagnosis and treatments of these diseases. It may also lead to better computational tools and better bio-inspired processing tools to develop artificial cognitive systems.

Brain rhythmic activity can be recorded noninvasively from the scalp as EEG or intracranially from the surface of the cortex as cortical EEG or the electrocorticogram (EEG is described in Sect. 39.7).

Electrocorticography (ECoG), involving the clinical placement of electrode arrays on the brain surface (usually above the dura) enables the recording of, similar to EEG, large-scale field potentials that are primarily derived from the aggregate synaptic potential from large neuronal populations, whereby synaptic current produces a change in the local electric field. ECoG can characterize local cortical potentials with high spatiotemporal precision (0.5 cm^2 in ECoG compared to 1 cm^2 in EEG) and high amplitudes ($10-200\,\mu$V in ECoG compared to $10-100\,\mu$V in EEG). Furthermore, the ECoG spectral content can reach 300 Hz (compared to 60 Hz in EEG) due to the closer vicinity of the electrodes to the electric source (the non-homogenous,

anisotropic brain volume and tissues act as a low-pass filter). Independent individual finger movement dynamics can be resolved at the 20 ms time scale, which has been shown not to be possible with EEG (but has recently been demonstrated using MEG [39.30] as described in Sect. 39.7). Here we review some of the latest findings of ECoG studies involving human sensorimotor systems.

The power spectral density (PSD) of the cortical potential can reveal properties within neuronal populations. Peaks in the PSD indicate activity that is synchronized across a neuronal population, for example, movement decreases the lateral frontoparietal alpha ($8-12$ Hz) and beta rhythm ($12-30$ Hz) amplitudes with limited spatial specificity whereas high gamma changes, which are spatially more focused, are also observable during motor control. *Miller* et al. [39.31], however, observed through a range of studies investigating local gamma band-specific cortical processing, a lack of distinct peaks in the cortical potential PSD beyond 60 Hz and hypothesized the existence of broadband changes across all frequencies that were obscured at low frequencies by covariate fluctuations in θ ($4-7$ Hz), α ($8-12$ Hz), and β ($13-30$ Hz) band oscillations. They demonstrated that there is a phenomenon that obeys a broadband, power-law form extending across the entire frequency range. Even with local brain activity in the gamma band there is an increase in power across all frequencies, and the power law shape is conserved. This suggests that there are phenomena with no special timescale where the neuronal population beneath does not oscillate synchronously but may simply reflect a change in the population mean firing rate. *Miller* et al. [39.31] postulated that the power-law scaling during high γ activity is a reflection of changes in asynchronous activity and not necessarily synchronous, rhythmic, action potential activity changes, as is often hypothesized.

These findings suggest a fundamentally different approach to the way we consider the cortical potential spectrum: power-law scaling reflects asynchronous, averaged input to the local neural population, whereas changes in characteristic brain rhythms reflect synchronized populations that coherently oscillate across large cortical regions. *Miller* et al. [39.31] also augment the findings by demonstrating power-law scaling in simulated cortical potentials using small-scale, simplified integrate and fire neuronal models, an example of which is shown in Fig. 39.6.

Part D | 39.6

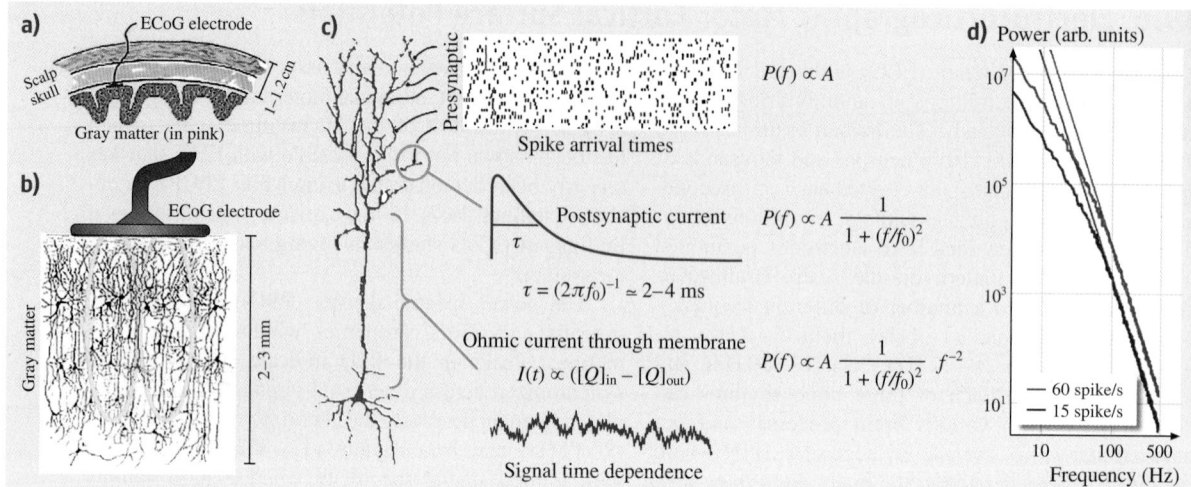

Fig. 39.6a–d An illustration of how the power-law phenomena in the cortical potential might be generated based on a simulation study (see [39.31] for details). Panel (*d*) shows the PSD of this signal and has a knee 70 Hz, with a power law of $P \approx 1/f^4$, which would normally be observed in ECoG PSD. The change in the spectra with increasing mean spike rate of synaptic input strongly resembles the change observed experimentally over motor cortex during activity, as demonstrated in [39.31] (after [39.31] with permission from *Miller*)

Knowledge of this power-law scaling in the brain surface electric potential was subsequently exploited in a number of further studies investigating differences in motor cortical processing during imagined and executed movements [39.32] and the role of rhythms and oscillations in sensorimotor activations [39.33].

As outlined, motor imagery to produce volitional neural signals to control external devices and for rehabilitation is one of the most popular approaches employed in brain–computer interfaces. As highlighted in the previous section neuroimaging using hemodynamic markers (positron emission tomography (PET) and fMRI) and extra cerebral magnetic and electric field studies (MEG and EEG) have shown that motor imagery activates many of the same neocortical areas as those involved in planning and execution of movements. *Miller* et al. [39.32] studied the execution-imagery similarities with electrocorticographic cortical surface potentials in eight human subjects during overt action and kinaesthetic imagery of the same movement to determine what and where are the neuronal substrates that underlie motor imagery-based learning and the congruence of cortical electrophysiologic change associated with motor movement and motor imagery. The results show that the spatial distribution of activation significantly overlaps between hand and tongue movement in the lower frequency bands (LFB) but not in the higher frequency bands (HFB), whereas during

Fig. 39.7a–e An illustration of modes of neural activity with cortical beta rhythm states. (**a**) Modulation of broadband amplitude by underlying rhythm can be thought of as population-averaged spike-field interaction. (**b**) *Released cortex* demonstrates a small amount of broadband power coupling to underlying rhythm phase, and the underlying spiking from pyramidal neurons is high in rate and only weakly coupled to the underlying rhythm phase. (**c**) *Suppressed cortex* demonstrates less broadband power but with higher modulation by the underlying rhythm, while underlying single unit spiking is low in rate but tightly coupled to the rhythm phase. (**d**) A simplified heuristic for how rhythms might influence cortical computation: During active computation, pyramidal neurons (PN) engage in asynchronous activity, where mutual excitation has a sophisticated spatio-temporal pattern. Averaged across the population, the ECoG signal shows broadband increase, with negligible beta. (**e**) During resting state, cortical neurons, via synchronized interneuron (IN) input, are entrained with the beta rhythm, which also involves extracortical circuits symbolized by the input froma synchronizing neuron in the thalamus (TN). The modulation of local activity with rhythms is revealed in the ECoG by significant broadband modulation with the phase of low frequency rhythms (after [39.33], courtesy of *Miller*; see [39.33] for further details) ▶

kinaesthetic imagination of the same movement task the magnitude of spectral changes were smaller (26% less

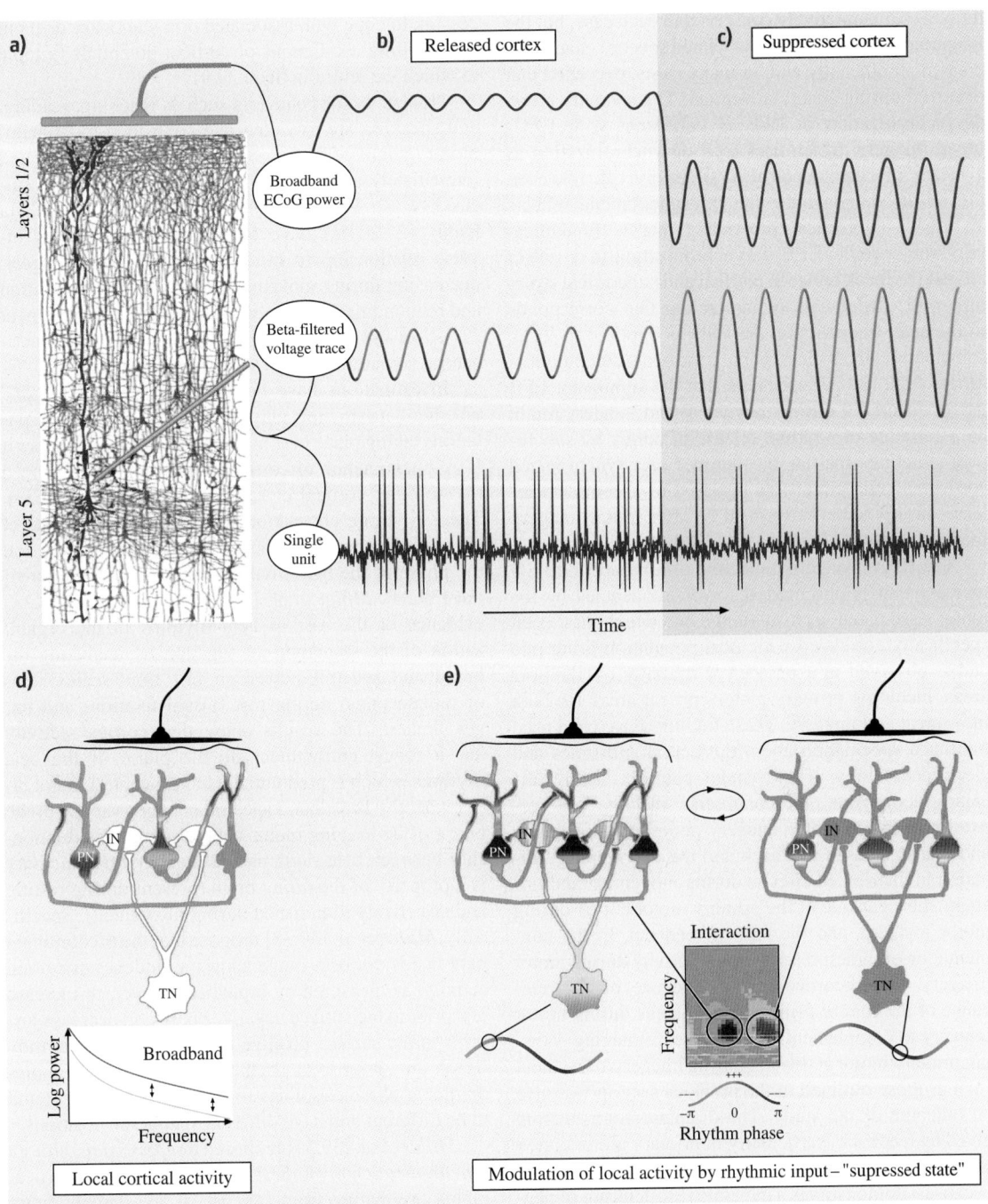

calculated across the electrode array) even though the spatially broad decrease in power in the LFB and the focal increase in HFB power were similar for movement and imagery.

During an imagery-based learning task involving real-time feedback of the magnitude of cortical activation of a particular electrode, in the form of cursor movement on screen, the spatial distribution of HFB activity was quantitatively conserved in each case, but the magnitude of the imagery associated spectral change increased significantly and, in many cases, exceeded that observed during actual movement. The spatially broad desynchronization in LFB is consistent with EEG-based imagery, which uses α/β desynchronization as a means of cursor control in BCIs [39.19]; however, the results demonstrate that this phenomena reflects an aspect of cortical processing that is fundamentally non-specific. LFB desynchronization may reflect altered feedback between cortical and subcortical structure with a timescale of interaction that corresponds to the peak frequency in the PSD as opposed to local, somatotopically distinct, population-scale computation. *Miller* et al. [39.32] speculate that the significant LFB power difference during movement and imagery might be a correlate of a partial release of cortex by subcortical structures (partial decoherence of a synchronized corticothalamic circuit) as opposed to a complete release during actual movement or after motor imagery feedback.

The HFB change is reflective of a broadband PSD increase that is obscured at lower frequencies by the motor associated α/β rhythms but which has been specifically correlated with local population firing rate and is observed in a number of spatially overlapping areas, including primary motor cortical areas for both movement and imagery. These findings have been used for much speculation about the neural substrates and electrophysiology of movement control. The results clearly demonstrate the congruence in large-scale activation between motor imagery and overt movement, and imagery-based feedback and the overlapping activation in distributed circuits during movement and imagery, the clear role of the primary motor cortex during motor imagery, and the role of feedback in the augmentation of widespread neuronal activity during motor imagery. Electrocorticographic evidence of the relevance of the role of primary motor areas during motor imagery to complement EEG and neuroimaging showing primary motor activation during imagery/movement such as those outlined in the previous section was also an outcome of the study. The dramatic augmentation given by feedback, particularly in primary motor cortex is significant, particularly in the context of BCI training, because it demonstrates a dynamic restructuring of neuronal dynamics across a whole population in the motor cortex on very short time scales (< 10 min) [39.32]. This augmentation and restructuring can, indeed, result in improved motor imager performance over time but leads to the necessity to co-adapt the BCI signal pro-

cessing to cope with associated non-stationary drifts in the resulting oscillations of cortical potentials (a topic to which we return in Sect. 39.9).

Human motor behaviors such as reaching, reading, and speaking are executed and controlled by somatomotor regions of the cerebral cortex, which are located immediately anterior and posterior to the central sulcus [39.33]. Electrical oscillations in the lower beta band (12−20 Hz) have been shown to have an inverse relationship to motor production and imagery, decreasing during movement initiation and production and rebounding (synchronization) following movement cessation and during imagery continuation in the pericentral somatomotor and somatosensory cortex.

Investigations have been conducted to determine whether beta rhythms play an active role in the computations taking place in somatomotor cortex or whether it is epiphenomenon of cortical state changes influenced by the other cortical or subcortical processes [39.33]. There is strong correlation between the firing time of individual neurons in the primary motor cortex and the phase of the beta rhythms in the local field potential [39.34]. *Miller* et al. [39.33] have acquired ECoG evidence of the role of beta rhythms in the organization of the somatomotor function by analyzing the broadband spectral power on fast time scales (tens of milliseconds) during rest (visual fixation) and finger flexion. The results show that cortical activity has a robust entrainment on the phase of the beta rhythms, which is predominant in peri-central motor areas whereby broadband spectral changes vary with the phase of underlying motor beta rhythm. This relationship between beta rhythms and local neuronal activity is a property of the *idling* brain (present during resting and selectively diminished during movement). Specifically, *Miller* et al. [39.33] propose that the predominant pattern for the beta range shows a tendency for brain activity, as measured by broadband power, to increase just prior to the surface negative phase and decrease just prior to the surface positive phase of the beta rhythm, which they refer to as rhythmic entrainment. The predominant phase couplings for $\theta/\alpha/\beta$ ranges are found to be different and have different spatial localizations.

Miller et al. [39.33] proposed a *suppression through synchronization hypothesis*, whereby diffuse cortical inputs originating from subcortical areas might functionally suppress large regions of the cortex, the advantage of which is to enable selective engagement of task-relevant and task-irrelevant brain circuits and for dynamical reallocation of metabolic resources. This shifting entrainment suggests that the β-rhythm is not imply a background process that is suppressed during movement, but rather that the beta rhythm plays an active and important role in motor processing. In recent years, there has been a growing focus on coupling between neuronal firing and rhythmic brain activity, and this study provides substantial evidence and methodology to support the important role of brain rhythms in neuronal functioning.

39.7 MEG and EEG — Extra Cerebral Magnetic and Electric Fields of the Motor System

The previous section highlighted a number of the most recent examples of ECoG-based studies that are shedding more light on the way in which the motor system processes information and is activated during imagery and movement. As electrocorticography is a highly invasive procedure involving surgery, a key question that has been addressed is whether the spectral findings and spatial specificity of ECoG will ever be possible using non-invasive extracerebrally acquired EEG, or whether ECoG findings can be used to develop better EEG-based processing methodologies to extract ECoG information. At the International BCI meeting in 2010, a workshop addressed the critical questions around the state-of-the-art BCI signal processing, in particular, *should future BCI research emphasize a shift from scalp-recorded EEG to ECoG, and how are the signals from the two modalities related?* [39.35].

There is still much debate around the future of EEG for BCI due to its limited spatial resolution and various noise-related issues, whereas ECoG shows much promise in addressing both of these issues. However, ECoG requires surgical implantation and the long-term effectiveness remains to be verified in humans. A step toward answering this question is to better understand the relationship between EEG and ECoG. In a workshop summary the question was addressed by comparing and contrasting the contribution of population synchronized (rhythmic) and asynchronous changes in the EEG and ECoG potential measurements [39.35]. The beta rhythm is robust in extracerebral EEG recordings, spatially synchronous across the pre and postcentral gyri, so this coherent rhythm is augmented with respect to background spatial averaging. The different states of the surface rhythms may represent switching between the stable modes observed in on-going surface oscillations.

In contrast, the broadband spectral change that accompanies movement is asynchronous at the local level and unrelated across cortical regions, so it is distorted or diminished by spatial averaging. *Krusienski et al.* [39.35] compared the contribution of population synchronized (rhythmic) and asynchronous (broadband, $1/f$) changes in the EEG and ECoG potential measurements using a number of simplifications and approximations. These approximations suggest that synchronized cortical oscillations may be differently reflected at the EEG scale than the ECoG scale. *Krusienski* et al. [39.35] show that to have the same contribution to EEG that a single cortical column has on ECoG, the spatial extent of cortical activity would have to span nearly the full width of a gyrus, and nearly a centimeter longitudinally. Based upon ECoG measurements of the $1/f$ change in the visual cortex, the findings confirm the possibility of detecting $1/f$ change in EEG during visual input directly over the occipital pole as an event-related potential. In the pre-central motor cortex, the movement of several digits in concert can produce a widespread change, which is dramatic enough to be measured in the EEG; however, based on these findings, the detection of single finger digit movement in EEG is not possible. This finding has been supported in other recent studies, a number of which involved magnetoencephalography (MEG). As with EEG, the magnetoencephalogram (MEG) is recorded non-invasively; however MEG is a record of magnetic fields, measured outside the head, produced by electrical activity within the brain, whereas EEG is a measure of the electrical potentials. Synchronized neuronal currents, produced primarily by the intracellular electrical currents within the dendrites of pyramidal cells in the cortex, induce weak magnetic fields. Neuronal networks of 50 000 or more active neurons are needed to produce a detectable signal using MEG. MEG has a number of advantages over EEG, most notably its spatial specificity as the magnetic flux detected at the surface of the head with MEG penetrates the skull and tissues without significant distortion, unlike the secondary volume currents detected outside the head with EEG [39.36]. MEG, however, is also less practical, requiring significant shielding from environmental electric magnetic interference and is not a wearable or mobile technology like EEG and, therefore, cannot be used for bedside recordings in a clinical setting or mobile BCI applications. MEG has been used in a range of clinical applications (cf. [39.36] for a review) and for research. Below we describe a number of studies with focus on motor cortical investigations in the context of developing brain–computer interfaces.

Quandt et al. [39.30] have investigated single trial brain activity in MEG and EEG recordings elicited by finger movement on one hand. The muscle mass involved in finger movement is smaller than in limb or hand movement, and neuronal discharges of motor cortex neurons are correspondingly smaller in finger movement than in arm or wrist movements. This makes detection of finger movement more difficult from non-invasive recordings. Using MEG *Kauhanen* et al. [39.37, 38] showed that left and right-hand index finger movement can be discriminated and that single trial brain activity recorded non-invasively can be used to decode finger movement; however, there are significant obstacles in non-invasive recordings in terms of the substantial overlapping activations in M1 when decoding individual finger movements on the same hand. *Miller* et al. [39.39] and *Wang* et al. [39.40] have shown that real-time representation of individual finger movements is possible using ECoG; however, finger movement discrimination from extracerebral neural recordings has only recently been shown to be possible. *Quandt* et al. [39.30] found using simultaneously recorded EEG and MEG that finger discrimination on the same hand is possible with MEG but EEG is not sufficient for robust classification. The lower spatial resolution of scalp signal EEG is due to the spatial blurring at the interface of tissues with different conductance. The issue cannot be overcome by increasing the density of EEG electrodes. It is speculated that the strong curvature of the cortical sheet in the finger knob (an omega-shaped knob of the central sulcus) contributes to the high decoding accuracy of MEG, whereby orientation change in the active tissue may change spatial patterns of magnetic flux measured in sensor space, but potentials caused by the same processes are not detectable at the scalp. Using different approaches four fingers on the same hand could be decoded with circa 57% accuracy using MEG and across all cases MEG performs better than EEG ($p < 0.005$), whilst EEG often only produced accuracies slightly above the upper confidence interval for guessing.

Analysis of the oscillations from MEG correspond to ECoG studies where the power of the lower oscillations (< 60 Hz) decreases around the movement, whereas power in the high gamma band increases

and that the effects in the high gamma band are more spatially focused than in the lower frequency bands [39.30]. Interestingly, the discrimination accuracy from the band power of the most informative frequency band between 6 and 11 Hz was clearly inferior to the accuracy derived from the time-series data indicating that slow movement-related neural activation modulations are most informative about which finger of a hand moves, and the inferior accuracy given by the band power is likely to be due to the lack of phase information contained in band-power features [39.30] (time embedding, temporal sequence information and exploiting phase information in discriminating motor signals are revisited in Sect. 39.9).

The above are just a few examples of what has been shown not to be possible to characterize sufficiently in EEG. The following section provides an overview of the known sensorimotor phenomena detectable from EEG and some of the most recent advances in decoding hand/arm movements non-invasively.

39.7.1 Sensorimotor Rhythms and Other Surrounding Oscillations

There are a number of rhythms and potentials that have been strongly linked with motor control, many of which have been exploited in EEG-based non-invasive BCI devices. As outlined previously, the sensorimotor area (SMA) generates a variety of rhythms that have specific functional and topographic properties. To reiterate, distinct rhythms are generated by hand movement over the post central somatosensory cortex. The μ (8–12 Hz) and β (13–30 Hz) bands are altered during sensorimotor processing [39.41–43]. Attenuation of the spectral power in these bands indicates an event related desynchronization (ERD), whilst an increase in power indicates event-related synchronization (ERS). ERD of the μ and β bands are commonly associated with activated sensorimotor areas and ERS in the μ band is associated with idle or resting sensorimotor areas. ERD/ERS has been studied widely for many cognitive studies and provides very distinctive lateralized EEG pattern differences, which form the basis of left hand vs. right hand or foot MI-based BCIs [39.44, 45]. However, as outlined above, later studies have shown the actual rhythmic activity generated by the sensorimotor system can be much more detailed. The α or μ component of the SMR also has a phase-coupled second peak in the beta band. Both the alpha and beta peaks can become independent at the offset of a movement, after which the beta band rebounds faster and with higher amplitude than the alpha band. Desynchronization of the beta band during a motor task can occur in different frequency bands than the subsequent resynchronization (rebound) after the motor task [39.41]. As previously outlined, many studies have shown that neural networks similar to those of executed movement are activated during imagery and observation of movement and thus similar sensorimotor rhythmic activity can be observed during motor imagery and execution.

Gamma oscillations of the electromagnetic field of the brain are known to be involved in a variety of cognitive processes and are believed to be fundamental for information processing within the brain. Gamma oscillations have been shown to be correlated with other brain rhythms at different frequencies and a recent study has shown the causal influences of gamma oscillation on sensorimotor rhythms (SMR) in healthy subjects using magnetoencephalography [39.46]. It has been shown that the modulation of sensorimotor rhythms is positively correlated with the power of frontal and occipital gamma oscillations, negatively correlated with the power of centro-parietal gamma oscillations and that simple causal structure can be attributed to a causal relationship or influence of gamma oscillations on the SMR. The behavioral correlate of the topographic alterations of gamma power, a shift of gamma power from centro-parietal to frontal and occipital regions, remains elusive, although increased gamma power over frontal areas has been associated with selective attention in auditory paradigms. *Grosse-Wentrup* et al. [39.46] postulated that neurofeedback of gamma activity may be used to enhance BCI performance to help low aptitude BCI users, i. e., those who appear incapable of BCI control using SMR.

39.7.2 Movement-Related Potentials

Signals observed during and before the onset of movement signify motor planning and preparation. For example, the bereitschafts potential or BP (from German, *readiness potential*), also called the pre-motor potential or readiness potential (RP), is a measure of activity in the motor cortex of the brain leading up to voluntary muscle movement [39.47, 48]. The BP is a manifestation of cortical contribution to the pre-motor planning of volitional movement. *Krauledat* et al. [39.49, 50] report on experiments carried out using the lateralized readiness potential (LRP) (i. e., Bereitschafts potential) for brain–computer interfaces. Before accomplishing motor tasks a negative readiness potential which reflects the preparation can be observed. They showed

it is possible to distinguish the pre-movement potentials from finger tapping experiments, even before the movement occurs or the onset of the movement, thus potentially improving accuracy and reducing latency in the BCI system. The BP is ten to a hundred times smaller than the α-rhythm of the EEG and it can only be identified by averaging across trials and has two components: an early component referred to as BP1 (sometime NS1) lasting from about -1.2 to -0.5 s before movement onset (negative slope (NS) of early BP) and a late component (BP2 or NS2) from -0.5 to shortly before 0 s (steeper negative slope of late BP) [39.48, 51, 52]. A pre-movement positivity can be observed along with a motor-potential which starts about 50 to 60 ms before the onset of movement and has its maximum over the contralateral precentral hand area.

39.7.3 Decoding Hand Movements from EEG

Movement-related cortical potentials (MRCP) have been used as control signals for BCIs [39.53]. MRCP and SMR have distinct changes during execution or imagination of voluntary movements. MRCP is considered a slow cortical potential where the surface negativity which develops 2 s before the movement onset is the Beireitschaftspotential referred to above. *Gu* et al. [39.53] studied MRCP and SMRs in the context of discriminating the movement type and speed from the same limb based on the hypothesis that if the imagined movements are related to the same limb, the control could be more natural than associating commands to movements of different limbs for BCIs. They focused on fast slow wrist rotation and extension and they found that average MRCPs rebounded more strongly when fast-speed movements were imagined compared with slow-speed movements; however, the rebound rate of MRCP was not substantially different between movement types. The peak negativity was more pronounced in the frontal (Fz) and central region (C1) than in the occipital region (Pz). The rebound rate of MRCP was greater in the central region (C1) when compared to the occipital region (Pz). MRCP and SMR are independent of each other as they originate from different brain sources and they occupy different frequency bands [39.52–54]. This renders them useful for multi-dimensional control in BCIs.

In accordance with the analysis of averaged MRCPs, the single-trial classification rate between two movements performed at the same speed was lower than when combining movements at different speeds. *Gu* et al. [39.53] suggest that selecting different speeds rather than different movements when these are executed at the same joint may be best for BCI applications. However, the task pair that was optimal in terms of classification accuracy is subject-dependent, and thus a subject-specific evaluation of the task pair should be conducted. The study by *Gu* et al. [39.53] is important as it is one of a limited number of studies that focus on discriminating different movements of the same limb as opposed to moving different limbs from EEG, which is much more common practice in BCI designs. However, *Lakany* and *Conway* [39.55] investigated the difference between imagined and executed wrist movements in 20 different directions using machine learning and found that the accuracy of discriminating wrist movement imagination is much less than for actual movement; however, they later found [39.56] time-frequency EEG features modulated by force direction in arm isometric exertions to four different directions in the horizontal plane can give better directional discrimination information and that $t–f$ features from the planning and execution phase may be most appropriate.

Although a limited number of works demonstrating EEG-based 2-D and 3-D continuous control of a cursor through biofeedback have been reported [39.57, 58] along with a few studies of classification of the direction/speed of 2-D hand/wrist movements outlined above, there are very few studies that have demonstrated continuous decoding of hand kinematics from EEG. Classification of different motor imagery tasks on single trial basis is more commonly reported. The signal-to-noise ratio, the bandwidth, and the information content of electroencephalography are generally thought to be insufficient to extract detailed information about natural, multi-joint movements of the upper limb. However, evidence from a study by *Bradberry* et al. [39.59] investigating whether the kinematics of natural hand movements are decodable from EEG challenges this assumption. They continuously extract hand velocity from signals collected during a three-dimensional (3-D) center-out reaching task and found that a linear EEG decoding model could reconstruct 3-D hand-velocity profiles reasonably well and that sensor CP3, which lies roughly above the primary sensori-

motor cortex contralateral to the reaching hand, made the greatest contribution. Using a time-lagged approach they found that EEG data from 60 ms in the past supplied the most information with 16.0% of the total contribution suggesting a linear decoding method such as the one used [39.59] rely on a sub-seconds history of neural data to reconstruct hand kinematics. Using a source localization technique they found that the primary sensorimotor cortex (pre-central gyrus and post-central gyrus) was indeed a major contributor along with the inferior parietal lobule (IPL), all of which have been found to be activated during motor execution and imagery in other investigations [39.12, 13, 16]. *Bradberry* et al. [39.59] also found that the movement variability is negatively correlated with decoding accuracy, suggesting two reasons; 1) increased movement variability could degrade decoding accuracy due to less similar pairs of EEG-kinematic exemplars, i. e., less movement variability results in reduced intra-class variability for training, and 2) subjects differ in their ability to perform the task without practice (motor learning is important for improving predictions of movement). Hence, the strengths of a priori neural representations of the required movements vary until learned or practiced, and these differences could directly relate to the accuracy with which the representations can be extracted. This study provides important evidence that decodable information about detailed, complex hand movements can be derived from the brain non-invasively using EEG; however, it remains to be determined whether these findings are consistent when using the same methodology in an imagined 3-D center-out task.

Although we know a lot about brain structure associated with sensorimotor activity, as well as the rhythms and potentials surrounding this activity, we have not yet systematically linked the neural correlates of these to specific motion primitives or motor control models. Modeling, using biological plausible neural models, the findings in relation to motor cortical structure, function, and dynamics along with linkage to the underlying motor psychophysics and advanced signal processing in BCI may help advance our knowledge on motion primitives, sensorimotor learning, and control.

39.8 Extracellular Recording – Decoding Hand Movements from Spikes and Local Field Potential

Although fMRI, MEG, and EEG offer low risk, non-surgical recording procedures they have inherent limita-

tions which many expect can be overcome with invasive approaches such as ECoG (described in Sect. 39.6) and y implanting electrodes to record the electrical activity f single neurons extracellularly (single unit recordngs). Here we focus on some recent studies aimed t testing this scale for use in sensorimotor-related 3CIs.

Extracellular recording has many advantageous, inluding high signal amplitude (up to $500\,\mu\mathrm{V}$), low usceptibility to external noise and artefact (eye movenents, cardiac activity, muscle activity) leading to high ignal-to-noise ratio, high spatial resolution ($50\,\mu\mathrm{m}^2$), igh temporal resolution ($\approx 1\,\mathrm{ms}$), and high spectral ontent (up to $2\,\mathrm{kHz}$) due to the close vicinity to he electric source. As a consequence, there is a high orrelation between the neural signals recorded and he generated/imagined hand movements, resulting in . short learning duration when employed in a mor BCI system. The disadvantages of the invasive ecording technique include a complex and expensive nedical procedure, susceptibility to infections (possily leading to meningitis, epilepsy), pain, prolonged ospitalization, direct damage to the neural tissue (e.g., flat $15\,\mu\mathrm{m} \times 60\,\mu\mathrm{m}$ electrode penetrating $2\,\mathrm{mm}$ deep its, on average, 5 neurons and 40 000 synapses), indiect damage to the neural tissue (small blood vessels re hit by the electrode causing ischemia for distant eurons and synapses and the evolution of an inflamnatory response), and evolvement of a scarred tissue vhich electrically isolates the electrodes from the suroundings and render the system non-responsive after eing implanted for extended durations. Furthermore, he electrode material itself, however biocompatible, ooner or later causes an inflammatory reaction and the volvement of scarred tissue.

Theoretically, however, extracellular recordings ofer more accurate information that may enable us to levise realistic BCI systems that allow for additional legrees of freedom and natural control of prosthetic levices, such as a hand and arm prostheses. To this nd, substantial efforts have been put into devising ovel biocompatible electrodes (e.g., platinum, iridum oxide, carbonic polymers) that will delay immune ystem stimulation, devising multi-functional microlectrodes that allow for recording/stimulating while njecting anti-inflammatory agents to suppress inflamnatory response, devising hybrid microelectrodes that llow for the inclusion of pre-amplifier and multiplexer on the electrode chip to allow wireless transmission of the data, thus avoiding the necessity for scalp drill hole used for taking out the flat cable arrying the neural data, which is prone to causing nfections.

Extracellular recording, being the most invasive recording technique (compared to non-invasive EEG and MEG recording and partially invasive ECoG recording) allows recording both the high-frequency content neural *output* activity, i.e., spikes, and the low frequency content neural *input* activity, denoted as local field potential (LFP), which is the voltage caused by electrical current flowing from all nearby dendritic synaptic activity across the resistance of the local extracellular space. In the following section, the neural coding schemes, in general, and the cortical correlates of kinematic and dynamic motion attributes, in specific, will be presented along with their suggested use for current and future BCI systems.

39.8.1 Neural Coding Schemes

A sequence, or *train*, of spikes may contain information based on different coding schemes. In motor neurons, for example, the strength at which an innervated muscle is flexed depends solely on the *firing rate*, the average number of spikes per unit time (a *rate code*). At the other end, a complex *temporal code* is based on the precise timing of single spikes. They may be locked to an external stimulus such as in the auditory system or be generated intrinsically by the neural circuitry. Whether neurons use rate coding or temporal coding is a topic of intense debate within the neuroscience community, even though there is no clear definition of what these terms mean. Neural schemes include rate coding, spike count rate, time-dependent firing rate, temporal coding, and population coding.

Rate coding

Rate coding is a traditional coding scheme, assuming that most, if not all, information about the stimulus is contained in the firing rate of the neuron. The concept of firing rates has been successfully applied during the last 80 years. It dates back to the pioneering work of *Adrian* and *Zotterman* who showed that the firing rate of stretch receptor neurons in the muscles is related to the force applied to the muscle [39.60]. In the following decades, measurement of firing rates became a standard tool for describing the properties of all types of sensory or cortical neurons, partly due to the relative ease of measuring rates experimentally.

Because the sequence of action potentials generated by a given stimulus varies from trial to trial, neuronal responses are typically treated statistically or probabilistically. They may be characterized by firing rates, rather than as specific spike sequences. In most sensory systems, the firing rate increases, generally non-linearly, with increasing stimulus intensity. Any information possibly encoded in the temporal structure of the spike train is ignored. Consequently, rate coding is inefficient but highly robust with respect to the inter-spike interval (ISI) *noise*. During recent years, more and more experimental evidences have suggested that a straightforward firing rate concept based on temporal averaging may be too simplistic to describe brain activity [39.61]. In rate coding, learning is based on activity-dependent synaptic weight modifications.

Spike-count rate

Spike-count rate also referred to as temporal average, is obtained by counting the number of spikes that appear during a trial and dividing by the duration of the trial. The length T of the time window is set by the experimenter and depends on the type of neuron recorded from and the stimulus. In practice, to obtain sensible averages, several spikes should occur within the time window. Typical values are $T = 100$ ms or $T = 500$ ms, but the duration may also be longer or shorter.

The spike-count rate can be determined from a single trial, but at the expense of losing all temporal resolution about variations in neural response during the course of the trial. Temporal averaging can work well in cases where the stimulus is constant or slowly varying and does not require a fast reaction of the organism – and this is the situation usually encountered in experimental protocols. Real-world input, however, is hardly stationary, but often changing on a fast time scale. For example, even when viewing a static image, humans perform saccades, rapid changes of the direction of gaze. The image projected onto the retinal photoreceptors changes, therefore, every few hundred milliseconds. Despite its shortcomings, the concept of a spike-count rate code is widely used not only in experiments, but also in models of neural networks. It has led to the idea that a neuron transforms information about a single input variable (the stimulus strength) into a single continuous output variable (the firing rate).

Time-dependent firing rate

Time-dependent firing rate is defined as the average number of spikes (averaged over trials) appearing during a short interval between times t and $t + \Delta t$, divided by the duration of the interval. It works for stationary as well as for time-dependent stimuli. To experimentally measure the time-dependent firing rate, the experimenter records from a neuron while stimulating with some input sequence. The same stimulation sequence is repeated several times and the neuronal response is re ported in a peri-stimulus-time histogram (PSTH). Th time t is measured with respect to the start of the stimu lation sequence. The Δt must be large enough (typicall in the range of 1 or a few milliseconds) so there i a sufficient number of spikes within the interval to ob tain a reliable estimate of the average. The number o occurrences of spikes $n_K(t; t + \Delta t)$ summed over all rep etitions of the experiment divided by the number K o repetitions is a measure of the typical activity of th neuron between time t and $t + \Delta t$. A further divisio by the interval length Δt yields the time-dependent fir ing rate $r(t)$ of the neuron, which is equivalent to th spike density of PSTH.

For sufficiently small Δt, $r(t)\Delta t$ is the average num ber of spikes occurring between times t and $t + \Delta$ over multiple trials. If Δt is small, there will neve be more than one spike within the interval between and $t + \Delta t$ on any given trial. This means that $r(t)\Delta$ is also the fraction of trials on which a spike occurre between those times. Equivalently, $r(t)\Delta t$ is the proba bility that a spike occurs during this time interval. A an experimental procedure, the time-dependent firin rate measure is a useful method to evaluate neurona activity, in particular in the case of time-dependen stimuli. The obvious problem with this approach is tha it cannot be the coding scheme used by neurons in th brain. Neurons cannot wait for the stimuli to repeat edly present in exactly the same manner as observe before generating the response. Nevertheless, the ex perimental time-dependent firing rate measure make sense, if there are large populations of independent neu rons that receive the same stimulus. Instead of recordin from a population of N neurons in a single run, it is ex perimentally easier to record from a single neuron an average over N repeated runs. Thus, the time-dependen firing rate coding relies on the implicit assumption tha there are always populations of neurons.

Temporal coding

When precise spike timing or high-frequency firing-rat fluctuations are found to carry information, the neura code is often identified as a temporal code. A numbe of studies have found that the temporal resolution of th neural code is on a millisecond time scale, indicating that precise spike timing is a significant element in neu ral coding [39.62]. Temporal codes employ those fea tures of the spiking activity that cannot be described b the firing rate. For example, the time to first spike afte the stimulus onset, characteristics based on the secon and higher statistical moments of the ISI probability dis ribution, spike randomness, or precisely timed groups of spikes (temporal patterns) are candidates for temporal codes. As there is no absolute time reference in the nervous system, the information is carried either in terms of the relative timing of spikes in a population of neurons or with respect to an ongoing brain oscillation.

The temporal structure of a spike train or firing rate evoked by a stimulus is determined both by the dynamics of the stimulus and by the nature of the neural encoding process. Stimuli that change rapidly tend to generate precisely timed spikes and rapidly changing firing rates no matter what neural coding strategy is being used. Temporal coding refers to temporal precision in the response that does not arise solely from the dynamics of the stimulus, but that nevertheless relates to properties of the stimulus. The interplay between stimulus and encoding dynamics makes the identification of a temporal code difficult. The issue of temporal coding is distinct and independent from the issue of independent-spike coding. If each spike is independent of all the other spikes in the train, the temporal character of the neural code is determined by the behavior of the time-dependent firing rate $r(t)$. If $r(t)$ varies slowly with time, the code is typically called a rate code, and if it varies rapidly, the code is called temporal.

Population coding

Population coding is a method to represent stimuli by using the joint activities of a number of neurons. In population coding, each neuron has a distribution of responses over some set of inputs, and the responses of many neurons may be combined to determine some value about the inputs.

Currently, BCI and BMI (brain machine interface) systems rely mostly on population coding. The description of one of the most famous population codes – the motor population vector along with its use in current and future BCI and BMI systems is presented in Sect. 39.8.2.

39.8.2 Single Unit Activity Correlates of Hand Motion Attributes

In 1982, *Georgopoulos* et al. [39.63] found that the activity of single cells in the motor cortex of monkeys, who were making arm movements in eight directions (at $45°$ intervals) in a two-dimensional apparatus, varied in an orderly fashion with the direction of the movement. Discharge was most intense with movements in a preferred direction and was reduced gradually when movements were made in directions farther and farther

away from the preferred movement. This resulted in a bell-shaped directional tuning curve. These relations were observed for cell discharge during the reaction time, the movement time, and the period that preceded the earliest changes in the electromyographic activity (approximately 80 ms before movement onset) (electromyography (EMG) is a technique for evaluating and recording the electrical activity produced by skeletal muscles). In about 75% of the 241 directionally tuned cells, the frequency of discharge D was a sinusoidal function of the direction of movement θ

$$D = b_0 + b_1 \sin\Theta + b_2 \cos\Theta \,, \tag{39.3}$$

or, in terms of the preferred direction Θ_0

$$D = b_0 + c_1 \cos(\Theta - \Theta_0) \,, \tag{39.4}$$

where b_0, b_1, b_2, and c_1 are regression coefficients. Preferred directions differed for different cells so that the tuning curves partially overlapped. The orderly variation of cell discharge with the direction of movement and the fact that cells related to only one of the eight directions of movement tested were rarely observed, indicated that movements in a particular direction are not subserved by motor cortical cells uniquely related to that movement. It was suggested, instead, that a movement trajectory in a desired direction might be generated by the cooperation of cells with overlapping tuning curves. The orderly variation in the frequency of discharge of a motor cortical cell with the direction of movement is shown in Fig. 39.8.

Later on, *Amirikian* et al. systematically examined the variation in the shape of the directional tuning profiles among a population of cells recorded from the arm area of the motor cortex of monkeys using movements in 20 directions, every $18°$ [39.64]. This allowed the investigation of tuning functions with extra parameters to capture additional features of the tuning curve (i.e., tuning breadth, symmetry, and modality) and determine an *optimal* tuning function. It was concluded that motor cortical cells are more sharply tuned than previously thought.

Paninski et al. [39.65] using a pursuit-tracking task (PTT) in which a monkey had to continuously track a randomly moving visual stimulus (thus providing a broad sample of velocity and position space) with invasive recordings from the M1 region showed that there is heterogeneity of position and velocity coding in that region, with markedly different temporal dynamics for each – velocity-tuned neurons were approximately sinusoidally tuned for direction, with linear speed scaling; other cells showed sinusoidal tuning for position,

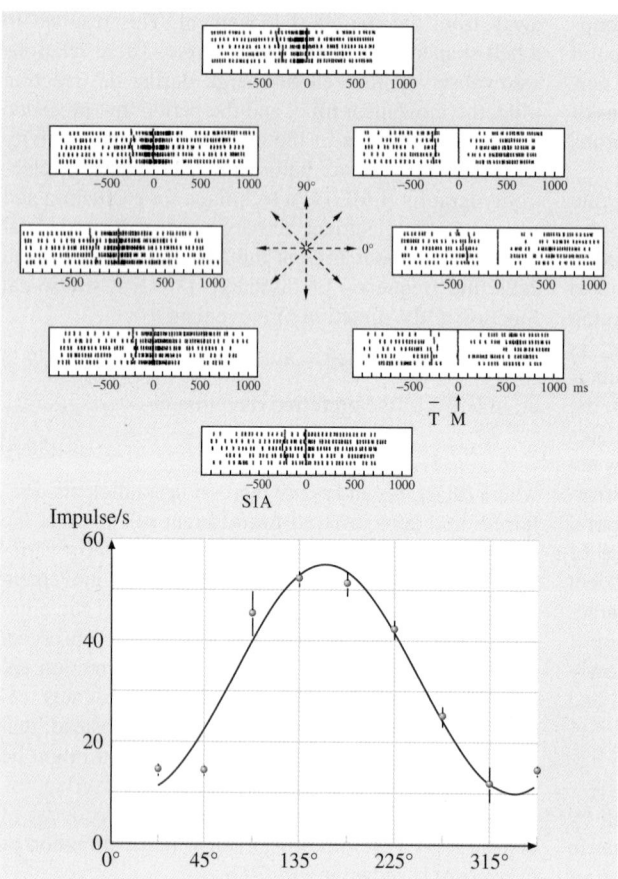

Fig. 39.8 Orderly variation in the frequency of discharge of a motor cortical cell with the direction of movement. *Upper half*: rasters are oriented to the movement onset M and show impulse activity during five repetitions of movements made in each of the eight directions indicated by the *center diagram*. Notice the orderly variation in cell's activity during the RT (reaction time), MOT (movement time) and TET (total experiment time; TET = RT + MOT). *Lower half*: directional tuning curve of the same cell. The discharge frequency is for TET. The data points are mean ± SEM. The regression equation for the fitted sinusoidal curve is $D = 32.37 + 7.281 \sin \Theta - 21.343 \cos \Theta$, where D is the frequency of discharge and Θ is the direction of movement, or, equivalently, $D = 32.37 + 22.5 \cos (\Theta - \Theta_0)$, where Θ_0 is the preferred direction ($\Theta_0 = 161°$) (after [39.63], courtesy of *A.P. Georgopoulos*)

with linear scaling by distance. Velocity encoding led behavior by about 100 ms for most cells, whereas position tuning was more broadly distributed, with leads and lags suggestive of both feed-forward and feedback coding. Linear regression methods confirmed that random, 2-D hand trajectories can be reconstructed from the firing of small ensembles of randomly selected neurons (3−19 cells) within the M1 arm area. These findings demonstrate that M1 carries information about evolving hand trajectory during visually guided pursuit tracking, including information about arm position both during and after its specification.

Georgopoulos et al. formulated a population vector hypothesis to explain how populations of motor cortex neurons encode movement direction [39.66]. In the population vector model, individual neurons *vote* for their preferred directions using their firing rate. The final vote is calculated by vectorial summation of individual preferred directions weighted by neuronal rates. This model proved to be successful in description of

motor-cortex encoding of 2-D and 3-D reach directions, and was also capable of predicting new effects, e.g., accurately describing mental rotations made by the monkeys that were trained to translate locations of visual stimuli into spatially shifted locations of reach targets [39.67, 68].

The population vector study actually divided the field of motor physiologists between *Evarts' upper motor neuron* group, which followed the hypothesis that motor cortex neurons contributed to control of single muscles [39.69] and the Georgopoulos group studying the representation of movement directions in the cortex. From the theoretical point of view, population coding is one of a few mathematically well-formulated problems in neuroscience. It grasps the essential features of neural coding and, yet, is simple enough for theoretic analysis. Experimental studies have revealed that this coding paradigm is widely used in the sensor and motor areas of the brain. For example, in the visual area medial temporal (MT) neurons are tuned to the movement direction. In response to an object moving in a particular direction, many neurons in MT fire, with a noise-corrupted and bell-shaped activity pattern across the population. The moving direction of the object is retrieved from the population activity, to be immune from the fluctuation existing in a single neuron's signal.

Population coding has a number of advantages, including reduction of uncertainty due to neuronal variability and the ability to represent a number of different stimulus attributes simultaneously. Population coding is also much faster than rate coding and can reflect changes in the stimulus conditions nearly instantaneously. Individual neurons in such a population typically have different but overlapping selectivities, so that many neurons, but not necessarily all, respond to a given stimulus. The Georgopoulos vector coding is an example of simple averaging. A more sophisticated mathematical technique for performing such a reconstruction is the method of maximum likelihood based on a multi-variate distribution of the neuronal responses. These models can assume independence, second-order correlations [39.70], or even more detailed dependencies such as higher-order maximum entropy models [39.71]

The finding that arm movement is well represented in populations of neurons recorded from the motor cortex has resulted in a rapid advancement in extracellular recording-based BCI in non-human primates and in a limited number of human studies. Several groups have been able to capture complex brain motor cortex signals by recording from neural ensembles (groups of neurons) and using these to control external devices. First, cortical activity patterns have been used in BCIs to show how cursors on computer displays can be moved in two and three-dimensional space. It was later realized that the ability to move a cursor can be useful in its own right and that this technology could be applied to restore arm and hand function for amputees and the physically impaired.

Miguel Nicolelis has been a prominent proponent of using multiple electrodes spread over a greater area of the brain to obtain neuronal signals to drive a BCI. Such neural ensembles are said to reduce the variability in output produced by single electrodes, which could make it difficult to operate a BCI. After conducting initial studies in rats during the 1990s, Nicolelis et al. succeeded in building a BCI that reproduced owl monkey movements while the monkey operated a joystick or reached for food [39.72]. The BCI operated in real time and could also control a separate robot remotely over internet protocol. However, the monkeys could not see the arm moving and did not receive any feedback, a so-called open-loop BCI.

Other laboratories which have developed BCIs and algorithms that decode neuron signals include those run by John Donoghue, Andrew Schwartz, and Richard Andersen. These researchers have been able to produce working BCIs, even using recorded signals from far fewer neurons than Nicolelis used (15–30 neurons versus 50–200 neurons). *Donoghue* et al. reported training rhesus monkeys to use a BCI to track visual targets on a computer screen (closed-loop BCI) with or without the assistance of a joystick [39.73].

Later experiments by Nicolelis using rhesus monkeys succeeded in closing the feedback loop and reproduced monkey reaching and grasping movements in a robot arm. With their deeply cleft and furrowed brains, rhesus monkeys are considered to be better models for human neurophysiology than owl monkeys. The monkeys were trained to reach and grasp objects on a computer screen by manipulating a joystick, while corresponding movements by a robot arm were hidden [39.74, 75]. The monkeys were later shown the robot directly and learned to control it by viewing its movements. The BCI used velocity predictions to control reaching movements and simultaneously predicted hand gripping force.

The use of cortical signals to control a multi-jointed prosthetic device for direct real-time interaction with the physical environment (*embodiment*) was first demonstrated by *Schwartz* et al. [39.76]. Schwartz et al. implanted 96 intracortical microelectrodes in the proximal arm region of the primary motor cortex of monkeys (*Macaca mulatta*) and used their motor cortical activity to control a mechanized arm replica and control a gripper on the end of the arm. The monkey could feed itself pieces of fruit and marshmallows using a robotic arm controlled by the animal's own brain signals. Owing to the physical interaction between the monkey, the robotic arm, and the objects in the workspace, this new task presented a higher level of difficulty than previous virtual (cursor control) experiments.

In 2012 *Schwartz* et al. [39.68] showed that a 52-year-old individual with tetraplegia who was implanted with two 96-channel intracortical microelectrodes in the motor cortex could rapidly achieve neurological control of an anthropomorphic prosthetic limb with seven degrees of freedom (three-dimensional translation, three-dimensional orientation, one-dimensional grasping). The participant was able to move the prosthetic limb freely in the three-dimensional workspace on the second day of training. After 13 weeks, robust seven-dimensional movements were performed routinely. The participant was also able to use the prosthetic limb to do skillful and coordinated reach and grasp movements that resulted in clinically significant gains in tests of upper limb function. No adverse events were reported.

In addition to predicting kinematic and kinetic parameters of limb movements, BCIs that predict electromyographic or electrical activity of the muscles of primates are being developed [39.77]. Such BCIs may be used to restore mobility in paralyzed limbs by electrically stimulating muscles. Miguel Nicolelis and colleagues demonstrated that the activity of large neural ensembles can predict arm position. This work made possible the creation of BCIs that read arm movement intentions and translate them into movements of artificial actuators. *Carmena* et al. [39.74] programmed the neural coding in a BCI that allowed a monkey to control reaching and grasping movements by a robotic arm. *Lebedev* et al. [39.75] argued that brain networks reorganize to create a new representation of the robotic appendage in addition to the representation of the animal's own limbs.

The biggest impediment to BCI technology at present is the lack of a sensor modality that provides safe, accurate, and robust access to brain signals. It is conceivable or even likely, however, that such a sensor will be developed within the next 20 years. The use of such a sensor should greatly expand the range of communication functions that can be provided using a BCI.

To conclude, this demonstration of multi-degree-of-freedom embodied prosthetic control paves the way towards the development of dexterous prosthetic devices that could ultimately achieve arm and hand function at a near-natural level.

39.8.3 Local Field Potential Correlates of Hand Motion Attributes

Local field potentials can be recorded with extracellular recordings, and a number of studies have shown their application; however, as ECoG and EEG (covered in Sects. 39.5 and 39.6) are indirect measures of LFPs we do not cover LFPs here again for brevity.

39.9 Translating Brainwaves into Control Signals − BCIs

Heretofore the chapter has focused on the characteristics of the neural correlates of motor control and how these might be deployed in SMR-based EEG, ECoG, and MEG BCI designs, providing evidence of activations at various scales of the brain and a brief outline of individual methodologies for attaining this evidence.

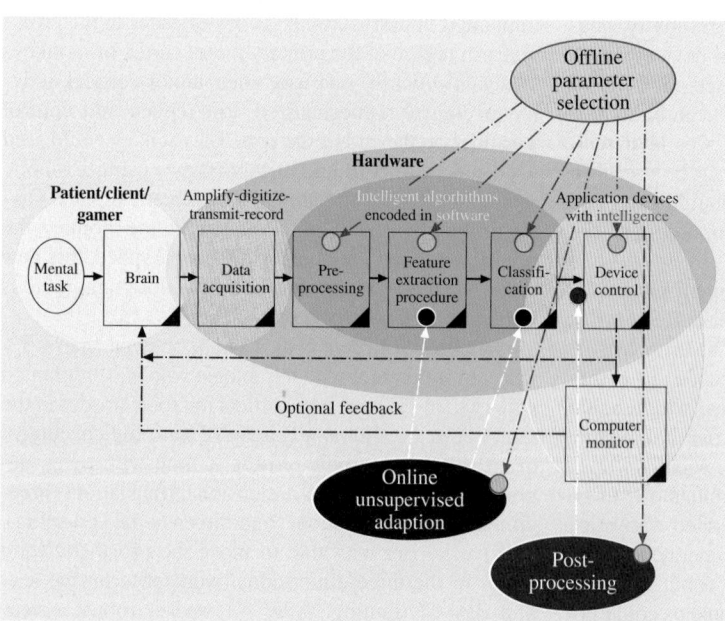

Fig. 39.9 Illustration of the various components of a BCI involving a closed-loop learning system as well as offline and online parameter optimization and system adaptation

Brain-computer interfaces, however, require a number of stages of signal processing and components to be effective and robust. Figure 39.9 shows common components of a complete BCI system. Although not all components shown are deployed together in every system there is increasing evidence that combining the best approaches deployed for each component and process in a multi-stage framework as well as ensemble methods or multi-classifier approaches can lead to significant performance gains when discriminating sensorimotor rhythms and translating brain oscillations into stable and accurate control signals. Performance here can be considered from various perspectives, including system accuracy in producing the correct response, the speed at which a response is detected (or the number of correct detections possible in a given period), the adaptability to each individual and the inherent non-stationary dynamics of the mutual interaction between the brain and the translating algorithm, the length of training required to reach an acceptable performance, the number of sensors required to derive a useful control signal, and the amount of engagement needed by the participant, to name but a few. The following sections highlight some of the methods which have been tried and tested in sensorimotor rhythm BCIs; however the coverage is by no means exhaustive. Also, the main emphasis is on EEG-based BCI designs as EEG-based BCI has been the driving force behind much of the novel signal processing research conducted in the field over the last 20 years, with some of the more invasive approaches considered less usable in the short term, high risk for experimentation and deployment in humans, with less funding to develop invasive strategies and less data availability.

EEG being the least informative, spectrally and spatially, about the underlying brain processes and subject to deterioration and spatial diffusion by the physical properties of the cerebrospinal fluid, skull, and skin, as well as the ominous susceptibility to contamination from other sources such as muscle and eye movements, poses the most challenges for engineers, mathematicians, and computer scientists. Researchers in these, among many other disciplines, are eager to solve a problem which has dogged the field for long namely, creating an EEG-based BCI which is accurate and robust across time for individual subjects and can be deployed across multiple subjects easily to offer a communication channel which matches or surpasses, at least, other basic, tried and tested computer peripheral input devices and/or basic assistive communication technologies. Signal processing, as shown in Fig. 39.10,

is only one piece of the puzzle with a range of other components being equally as important, including electrode technologies and hardware being critical to data quality, usability, and acceptability of the system. Additionally, the technologies and devices under the control of the BCI are another aspect, which is not dealt with here but is a topic which requires investigation to determine how applications can be adapted to cope with the, as yet, inevitable inconsistencies in the communication and control signals derived from the BCI. Here our intention is not to deal with these elements of brain computer interface but only to provide the reader with an indicative overview of key signal processing and discrimination topics under consideration in the area, perhaps not topics that have received the attention deserved, but show promise. Interested readers are referred to [39.78–87] for comprehensive surveys of BCI-control strategies and signal-processing strategies.

39.9.1 Pre-Processing and Feature Extraction/Selection

Oscillatory and rhythmic activity in various frequency bands are a predominant feature in sensorimotor rhythm-based BCIs, as outlined in Sect. 39.7.1. Whilst amplitude of power in subject-specific sub-bands has proven to be a reliable feature to enable discrimination of the lateralized brain activity associated with

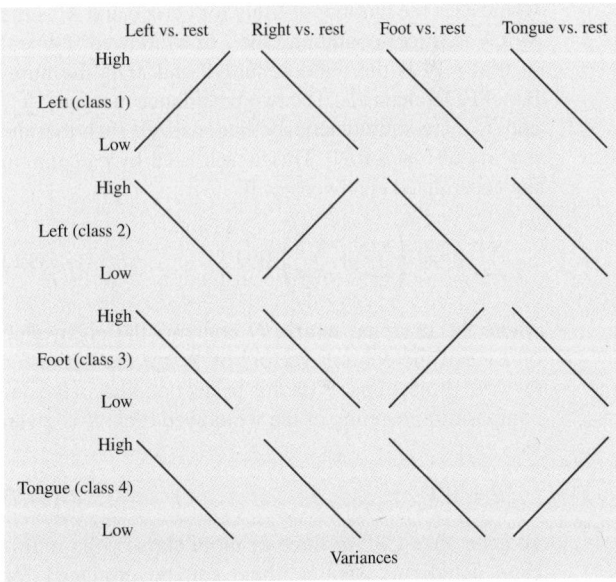

Fig. 39.10 Hypothetical relative variance levels of the CSP transformed surrogate data

Part D | 39.9

gross arm movement imagination from EEG, there is a general consensus that there is a necessity to extract much more information about spatial and temporal relationships by correlating the synchronicity, amplitude, phase, and coherence of oscillatory activity across distributed brain regions. To that end, spectral filtering is often accompanied with spatial pattern estimation techniques, channel selection techniques, along with other preprocessing techniques to detect signal sources and for noise removal. These include principle component analysis (PCA) and independent component analysis (ICA), among others, whilst the most commonly used is the common spatial patterns (CSP) approach [39.88–91].

Many of these methods involve linear transformations where a set of possibly correlated observations are transformed into a set of uncorrelated variables and can be used for feature dimensionality reduction, artifact removal, channel selection, and dimensionality reduction. CSP is by far the most commonly deployed of all these filters in sensorimotor rhythm-based BCIs.

CSP maximizes the ratio of class-conditional variances of EEG sources [39.88, 89]. To utilize CSP, $\sum_1$ and $\sum_2$ are the pooled estimates of the covariance matrices for two classes, as follows

$$\sum_c = \frac{1}{I_c} \sum_{i=1}^{I_c} X_i X_i^t \quad c \in \{1, 2\}, \tag{39.5}$$

where I_c is the number of trials for class c and X_i is the $M \times N$ matrices containing the i-th windowed segment of trial i; N is the window length and M is the number of EEG channels. The two covariance matrices, $\sum_1$ and $\sum_2$, are simultaneously diagonalized such that the eigenvalues sum to 1. This is achieved by calculating the generalized eigenvectors W

$$\sum_1 W = \left(\sum_1 + \sum_2 \right) W D, \tag{39.6}$$

where the diagonal matrix D contains the eigenvalue of Σ_1 and the column vectors of W are the filters for the CSP projections. With this projection matrix the decomposition mapping of the windowed trials X is given as

$$E = WX. \tag{39.7}$$

To generalize CSP to three or more classes (the multiclass paradigm), spatial filters can be produced for each class vs. the remaining classes (one vs. rest approach). If q is the number of filters used then there are $q \times C$ surrogate channels from which to extract features. To illustrate how CSP enhances separability among four classes the hypothetical relative variance level of the data in each of the four classes are shown in Fig. 39.10.

CSP has been modified and improved substantially using numerous techniques and deployed and tested in BCIs [39.88–92]. CSP is commonly applied with spectral filters. One of the more successful approaches to spectral filtering combined with CSP is the filter bank CSP approach [39.93, 94]. Another promising technique for the analysis of multi-modal, multi-channel, multi-tasks, multi-subject, and multi-dimensional data is multi-way (array) tensor factorization/decomposition [39.95]. The technique has been shown to have the ability to discriminate between different conditions, such as right hand motor imagery, left hand motor imagery, or both hands motor imagery, based on the spatiotemporal features of the different EEG tensor factorization components observed.

Due to the short sequences of events during motor control it is likely that assessment of activity at a fine granularity such as the optimal embedding parameters for prediction as well as the predictability of EEG over short and long time spans and across channels will also provide clues about the temporal sequences of motor planning and activations and the motion primitives involved in different hand movement trajectories. Work has shown that subject, channel, and class-specific optimal time embedding parameter selection using partial mutual information improves the performance of a predictive framework for EEG classification in SMR-based BCIs [39.92, 96–102]. Many other time series modeling, embedding, and prediction through traditional and computational intelligence techniques such as fuzzy and recurrent neural networks (FNN and RNN) have been promoted for EEG preprocessing and feature extraction to maximize signal separability [39.19, 66–69, 103].

The above preprocessing or filtering frameworks have been used extensively, yet rarely independently, but in conjunction with a stream of other signal processing methodologies to extract reliable information from neural data. It is well known that the amplitude and the phase of neural oscillations are spatially and temporally modulated during sensorimotor processing (see Sects. 39.6 and 39.7 for further details). Spectral information and band power extraction have been commonly used as features ([39.82, 83] for reviews); however, phase and cross frequency coupling less so, even though a number of non-invasive and intracortical studies have emphasized the importance of phase information [39.33, 35, 104, 105]. Furthermore, amplitude-phase cross-frequency coupling has been suggested to play an important role in neural coding [39.106]. While neural representations of movement kinematics and movement imagination by amplitude information in sensorimotor cortex have been extensively reported using different oscillatory signals (LFP, ECoG, MEG, EEG) [39.33, 35, 39, 107–111] and used extensively in non-invasive motor imagery-based BCI designs phase information has not been given as much attention as possibly deserved [39.35]. As reported in [39.35] there have been some recent developments describing synchronized activity between M1 and hand speed [39.81, 82], corticomuscular coupling [39.112], and the LFP beta oscillations phase locked to target cue onset in an instructed-delay reaching task [39.113], in addition to the studies covered in Sects. 39.6 and 39.7, among others. The role of phase coding in the sensorimotor cortex should be further explored to fully exploit the complementary information encoded by amplitude and phase [39.35].

Parameter optimization can be made more proficient through global searches of the parameter space using evolutionary computation-based approaches such as particle swarm optimization (PSO) and genetic algorithms (GAs). The importance of features can be assessed and ranked for different tasks using various feature selection techniques using information theoretic approaches such as partial mutual information-based (PMI) input variable selection [39.98, 114]. Parameter optimization and feature selection approaches such as these enable coverage of a large parameter space when additional features are identified to enhance performance. Heuristic-based approaches can be used to determine the relative increase in classification associated with each variable along with other more advanced methods for feature selection such as Fisher's criterion and partial mutual information to estimate the level of redundancy among features. Verifying the feature landscape using global heuristic searches is important initially and automated intelligent approaches enable efficient and automated system optimization during application at a later time and easy application to a large sample of participant data, i.e., removing the necessity to conduct global parameter searches.

39.9.2 Classification

Various classifier techniques can be applied to the sampled data to determine classification/prediction performance, including standard linear methods such as linear discriminant analysis (LDA), support vector machines (SVM), and probabilistic-based approaches [39.115], as well as non-linear approaches such as backpropagation neural networks (NN) and self-organizing fuzzy neural networks (SOFNN) [39.116]. Other adaptive methods and approaches to classifier combination have been investigated [39.87, 88, 117, 118] along with Type-2 fuzzy logic approaches to deal with uncertainty [39.119, 120]. Recent evidence has shown that probabilistic classifier vector machines (PCVM) have significant potential to outperform other tried and tested classifiers [39.121, 122]. These are just a few of the available approaches (see [39.82] for a more detailed review). Here we focus on one of the latest trends in BCI translation algorithms, i.e., automated adaptation to non-stationary changes in the EEG dynamics over time.

39.9.3 Unsupervised Adaptation in Sensorimotor Rhythms BCIs

EEG signals deployed in BCI are inherently non-stationary resulting in substantial change over time, both within a single session and between sessions, resulting in significant challenges in maintaining BCI system robustness. There are various sources of non-stationarities: short-term changes related to modification to the strategy that users apply to motor imagery to enhance performance, drifts in attention, attention to different stimuli or processing other thoughts or stimuli/feedback, slow cortical potential drifts and less specific long-term changes related to fatigue, small day to day differences in the placement of electrodes, among others. However, one which is considered a potential source of change over time is user adaption through motor learning to improve BCI performance over time, sometimes referred to as the effects of feedback training [39.123, 124] and sensorimotor learning.

The effects of feedback on the user's ability to produce consistent EEG, as he/she begins to become more confident and learns to develop more specific communication and control signals, can have a negative effect on the BCI's feature extraction procedure and classifier. During sensorimotor learning the temporal and spatial activity of the brain continually adapts and the features which were initially suited to maximizing the discrimination accuracy may not remain stable as time evolves, thus degradation in communication occurs. For this reason, the BCI must have the ability to adapt and interact with the adaptations that the brain makes in response to the feedback. According to *Wol-*

paw et al. [39.125] the BCI operation depends on the interaction of two adaptive controllers, the user's brain, which produces the signals measured by the BCI, and the BCI itself, which translates these signals into specific commands [39.125].

With feedback, even though classification accuracy is expected to improve with an increasing number of experiments, the performance has been shown to decrease with time if the classifier is not updated [39.43]. This has been referred to as the *man–machine learning dilemma* (MMLD), meaning that the two systems involved (man and machine) are strongly interdependent, but cannot be controlled or adapted in parallel [39.43]. The experiments shown in many studies show that feedback results in changing EEG patterns, and thus adaptation of the pattern recognition methods is required. It is, therefore, paramount to adapt a BCI periodically or continuously if possible. Autonomous adaptive system design is required but a challenge. The recognition and productive engagement of adaptation will be important for successful BCI operation. According to *Wolpaw* et al. [39.125] there are three levels of adaptation which are not always accounted for but have great importance for future adoption of BCI systems:

1. When a new user first accesses the BCI, the algorithm adapts to the user's signal features.
   - No two people are the same physiologically or psychologically, therefore brain topography differs among individuals, and the electrophysiological signals that are produced from different individuals are unique to each individual, even though they may be measured from the same location on the scalp whilst performing the same mental tasks at the same time. For each new user the BCI has to adapt specifically to the characteristics of each particular person's EEG. This adaptation may be to find subject-specific frequency bands which contain frequency components that enable maximal discrimination accuracy between two mental tasks or train a static classifier on a set of extracted features.
2. The second level of adaptation requires that the BCI system components be periodically adjusted or adapted online to reduce the impact of spontaneous variations in the EEG.
   - Any BCI system which only possesses the first level of adaptation will continue to be effective only if the user's performance is very stable. Most electrophysiological signals display short and long-term variations due to the complexity

of the physiological functioning of the underlying processes in the brain, among other sources of change as outlined above. The BCI system should have the ability to accommodate these variations by adapting to the signal feature values which maximally express the user's intended communication.
3. The third level of adaptation accommodates and engages the adaptive capacities of the brain.
   - The BCI depends on the interaction of two adaptive controllers, the BCI and the user's brain [39.125]:

   When an electrophysiological signal feature that is normally merely a reflection of brain function becomes the end product of that function, that is, when it becomes an output that carries the user intent to the outside world, it engages the adaptive capacities of the brain.

   This means that, as the user develops the skill of controlling their EEG, the brain has learned a new function and, hopefully, the brain's newly learned function will modify the EEG so as to improve BCI operation. The third level of adaptation should accommodate and encourage the user to develop and maintain the highest possible level of correlation between the intended communication and the extracted signal features that the BCI employs to decipher the intended communication. Due to the nature of this adaptation (the continuous interaction between the user and the BCI) it can only be assessed online and its design is among the most difficult problems confronting BCI research.

McFarland et al. [39.126] further categorize adaptation into system adaptation, user adaptation, and system and user co-adaption, asking the question: is it necessary to continuously adapt the parameters of the BCI translation algorithm? Their findings show that for sensorimotor rhythms BCI it is, whereas perhaps it is not for other stimulus-based BCIs.

A review of adaptation methods is included by *Hasan* [39.127] focusing on questions: what, how, and when to adapt and how to evaluate adaptation success. A range of studies has been aimed at addressing the adaption requirements [39.123, 124, 128–137]. *Krusienski* et al. [39.35] define the various types of possible adaptation as follows:

- *Covariate shift adaptation/minimization*: *Covariate shift* refers to when the distribution of the training features and test features follow different distributions, while the conditional distribution of the output values (of the classifier) and the features is unchanged [39.138]. The shift in feature distribution from session to session can be significant and can result in substantive biasing effects. Without some form of adaption to the features and/or classifier, the classifier trained on a past session would perform poorly in a more recent session. *Satti* et al. [39.139] proposed a method for covariate shift minimization (CSM), where features can be adapted so that the feature distribution is always consistent with the distribution of the features that were used to train the classifier in the first session. This can be achieved in an unsupervised manner by estimating the shift in distribution using a least squares fitting polynomial for each feature and removing the shift by adding the common mean of the training feature distribution so that the feature space distribution remains constant over time as described in [39.139]. *Mohammadi* et al. [39.140] applied CSM in self-paced BCI updated features to account for short terms (within trial) drifts in signal dynamics. In [39.138] an importance-weighted cross-validation for accommodating covariate shift under a number of assumptions is described but is not adaptively updated online in an unsupervised manner whereas other offline approaches have been investigated to enable feature extraction methods to accommodate non-stationarity and covariate shifts [39.90, 91].

- *Feature adaptation/regression*: Involves adapting the parameters of the feature extraction methods to account for subject learning, e.g., modifying the subject-specific frequency bands can be easily achieved in a supervised manner but this is not necessarily easily achieved online, unsupervised. An approach to adaptively weight features based on mu and beta rhythm amplitudes and their interactions using regression [39.4] resulted in significant performance improvements and may be adapted for unsupervised feature adaptation.

Covariate shift adaptation/minimization can be considered an *anti-biasing* method because it prevents the classifier biasing, whereas *feature adaption/regression* is likely to result in the need to adapt the classifier to suit the new feature distributions. Both methods help to improve the performance over time, but it is uncertain if feature adaption followed by *covariate shift minimization* (to shift features towards earlier distribution) would limit the need for

classifier adaptation and/or provide stable performance or negate the benefits of feature regressions. An interesting discussion on the interplay between feature regression and adapting bias and gain terms in the classifier is presented in [39.4].

- *Classifier adaptation*: Unsupervised classifier adaptation has received more attention than feature adaptation with a number of methods having been proposed [39.124, 128]. Classifier adaptation is required when significant learning (or relearning)-induced plasticity in the brain significantly alters the brain dynamics, resulting in a shift in feature distribution, as well as significant changes in the conditional distribution between features and classifier output as opposed to cases where only covariate shift has occurred. In such cases, classifier adaptation can neither be referred to as anti-biasing or de-biasing.
- *Post-processing adaptation*: De-biasing the classifier output, in its simplest form, can be performed in an unsupervised manner by removing the mean calculated from a window of recent classifier outputs from the instantaneous value of the classifier [39.141], also referred to as normalization in [39.142], where the data from recent trials are used to predict the mean and standard deviation of the next trial and the data of the next trial is then normalized by these estimates to produce a control signal which is assumed to be stationary. De-biasing is suitable when covariate shift has not been accounted for and can improve the online feedback response but may only provide a slight performance improvement.
- *EEG data space adaptation* (EEG-DSA): acts on the raw data space and is a new approach to linearly transform the EEG data from the target space (evaluation/testing session), such that the distribution difference to the source space (training session) is minimized [39.143]. The Kullback–Leibler (KL) divergence criterion is the main method deployed in this approach and it can be applied in a supervised or unsupervised manner either periodically or continuously. Other adaptations (feature space or classifier) can be applied in tandem but accurate minimization of feature space adaptation should negate the need for further anti-biasing and or de-biasing adaptations.

Classifier adaptation (anti-biasing) negates the need for post processing (de-biasing) if the classifier is updated continuously, which is a challenging task to undertake in an unsupervised manner (with no class labels) and may result in maladaptation, whereas de-biasing can be conducted easily, unsupervised, regardless of the classifier used. Because post-processing-based de-biasing only results in removal of bias (shifts it to mean zero) in the feedback signal and not necessarily a change in the dynamics of the feedback signal, feature adaptation or classifier adaptation is necessary during subject learning and adaptation as the conditional distribution between features and classifier output evolves as outlined above.

All of the above methods are heavily dependent upon the context in which the BCI is used. For example, for a BCI applied in alternative communication the objective is to maximize the probability of interpreting the user's intent correctly; therefore the adaptation is performed with that objective, whereas, if the BCI is aimed at inducing neuroplastic changes in specific cortical areas, e.g., a BCI which is aimed at supporting stroke survivors perform motor imagery as means of enhancing the speed or level of rehabilitation post stroke, the objective is to not only provide accurate feedback but to encourage the user to activate regions of cortex which do not necessarily provide optimal control signals [39.35]. The latter may require electrode/channel adaptation strategies but not necessarily in a fast online unsupervised manner. Abrupt changes

to classifier performance may also lead to negative learning where the user cannot cope with the rate at which the feedback dynamics change, in such cases consistent feedback, even though less accurate, may be appropriate. As outlined in [39.126], there is still debate around whether mutual adaptation of a system and user is a necessary feature of successful BCI operation or if fast adaptation of parameters during training is not necessary. A recent study in animal models suggests that there is no negative correlation between decoding performance and the time between model generation and model testing, which suggests that the neural representations that encode kinematic parameters of reaching movements are stable across the months of study [39.147, 148], which further suggests little adaptation is needed for ECoG decoding in animal models, but this may not necessarily translate to humans and non-invasive BCIs involving motor imagery. Much more research on the issue of what type of adaption methods to apply and at what rate adaptation is necessary. Another important factor is to consider a person's level of ability to control a BCI and those persons close to chance levels may actually benefit from an incorrect belief on their performance level [39.149]. This would imply adapting the classifier output based on knowledge of the targets in a supervised manner such that the user thinks they are performing better – a method which may help in the initial training phase to improve BCI performance [39.149]. Most of the techniques outlined above have been tested offline and therefore, there is need to assess how the techniques improve performance as the user and BCI are mutually adapted. Table 39.1 provides a summary of the categories of adaptation and their interrelationships and requirements.

39.9.4 BCI Outlook

Translating brain signals into control signals is a complex task. The communication bandwidth given by BCI is still lagging behind most other communication methods rates between humans and the external world where the maximum BCI communication rate is $\approx 0.41 \, \text{bit/s}$ ($\approx 25 \, \text{bit/min}$) [39.144] (see Fig. 39.11 for an illustration that nicely illustrates the gap in communication bandwidth between BCI and other communication methods, as well as the relatively low communication bandwidth across all human–human and human–computer interaction methods).

Nevertheless with the many developments and studies highlighted throughout this chapter (a selected few

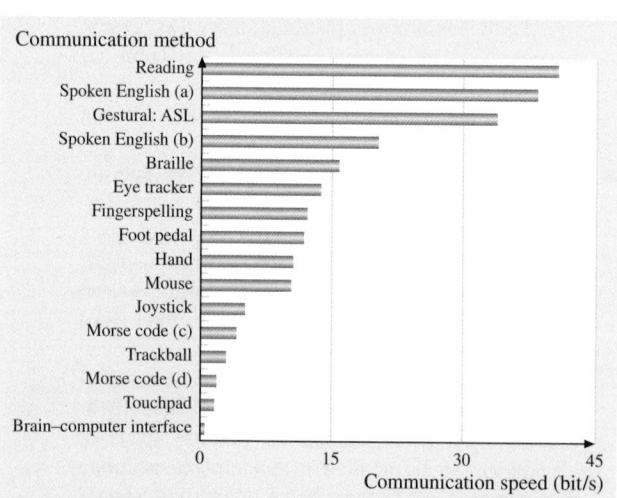

Fig. 39.11 Comparison of communication rates between humans and the external world: (**a**) speech received auditorily; (**b**) speech received visually using lip reading and supplemented by cues; (**c**) Morse code received auditorily; (**d**) Morse code received through vibrotactile stimulation (figure adapted from [39.144] with permission and other sources [39.145, 146])

Table 39.1 BCI components that can be adapted and the way in which they can be adapted. Interrelationship between components, i. e., indicating when one is adapted which other component or stages of the signal processing pipeline it might be necessary to adapt (whether a calibration session is needed for offline setup or there is certain number of trials needed before adaption begins is not specified in the criteria but is another consideration)

Adaptation type	Anti-biasing	De-biasing	Subject relearning	Feature updates	Classifier adaptation	Online	Super-vised	Unsuper-vised	Performance improvement likely
Channel adaptation			Y	Y	Y		Y		Y
Data space adaptation	Y					Y		Y	Y
Feature regression	Y		Y	Y	Y	Y		Y	Y
Covariate shift minimization	Y			Y		Y		Y	Y
Covariate shift adaptation			Y		Y		Y		Y
Classifier adaptation	Y		Y		Y		Y	Y	Y
Gain/bias		Y						Y	Slight

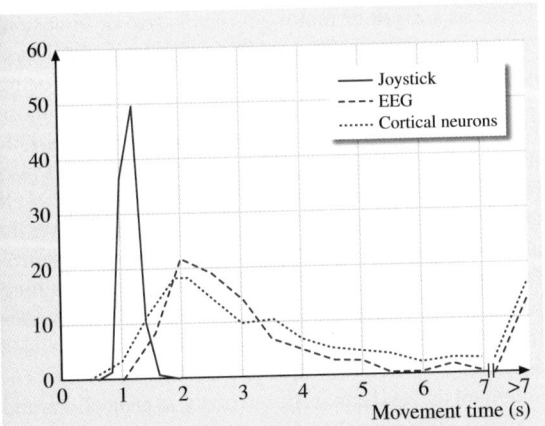

Fig. 39.12 Distributions of target-acquisition times (i.e., time from target appearance to target hit) on a 2-D center-out cursor-movement task for joystick control, EEG-based BCI control, and cortical neuron-based BCI control. The EEG-based and neuron-based BCIs perform similarly and both are slower and much less consistent than the joystick. For both BCIs in a substantial number of trials, the target is not reached even in the 7 s allowed. Such inconsistent performance is typical of movement control by present-day BCIs, regardless of what brain signals they use. (Joystick data and neuron-based BCI data from *Hochberg* et al. [39.150]; EEG-based BCI data from *Wolpaw* and *McFarland* [39.57]; figure after [39.58], courtesy of *McFarland* et al.)

among many) there has been progress, yet there is still debate around whether invasive recordings are more appropriate for BCI, with findings showing that performance to date is not necessarily better or communications faster with invasive or extracellular recordings compared to EEG (see Fig. 39.12 for an illustration [39.58]). As shown, performance is far less consistent than a joystick for 2-D center-out tasks using both methods, however the performance is remarkably similar even though the extracellular recordings are high resolution and EEG is low resolution. Training rates/durations with invasive BCI are probably less onerous on the BCI user compared to EEG-based approaches, which often require longer durations, however only a select few are willing to undergo surgery for BCI implants due to the high risk associated with the surgery required, at least with the currently available technology. This is likely to change in the future and information transfer between humans and machines is likely to increase to overcome the communication bottleneck human–human and human–computer interaction by directly interfacing brain and machine [39.144]. There is one limitation that dogs many movement or motor-related BCI studies and that is that in a large part control relies only a signal from single cortical area [39.58]. Exploiting multiple cortical areas may offer much more and this may be achieved more easily and successfully by exploiting information acquired at different scales using both invasive and non-invasive technologies (many of the studies reported in this chapter have shown advantages that are unique at the various scales of recording). *Carmena* [39.151] recommends that non-invasive BCIs should not be pitted against invasive ones as both have pros and cons and have gone beyond pitching resolution as an argument to use one type or the other. In the future, BCI systems may very well become a hybrid of different kinds of neural signals, be able to benefit from local, high-resolution information (for generating motor commands) and more global information (arousal, level of attention, and other cognitive states) [39.151].

In summary, BCI technology is developing through a better understanding of the motor system and sensorimotor control, better recording technologies, better signal processing, more extensive trials with users, long-term studies, more multi-disciplinary interactions, among many other reasons. According to report conducted by *Berger* et al. [39.152] the magnitude of BCI research throughout the world will grow substantially, if not dramatically, in future years with multiple driving forces:

- Continued advances in underlying science and technology
- Increasing demand for solutions to repair the nervous system
- Increase in the aging population worldwide; a need for solutions to age-related, neurodegenerative disorders, and for *assistive* BCI technologies
- Commercial demand for non-medical BCIs.

BCI has the potential to meet many of these challenges in healthcare and is already growing in popularity for non-medical applications. BCI is considered by many as a revolutionary technology.

An analysis of the history of technology shows that technological change is exponential, and accord-

ing to the law of accelerating returns as the technology performance increases, more and more users group begin to adopt the technology and prices begin to fall [39.153]. In terms of BCI there has been significant progress over recent years, and these trends are being observed with increasing technology diffusion [39.144]. In terms of research there has been an exponential growth in the number of peer reviewed publications since 2000 [39.84].

Many studies over the past two decades have demonstrated that non-muscular communication, based on brain–computer interfaces (BCIs), is possible and, despite the nascent nature of BCIs there is already a range of products, including alternative communication and control for the disabled stroke rehabilitation, electrophysiologically interactive computer systems, neurofeedback therapy, and BCI-controlled robotics/wheelchairs. A range of case studies have also shown that head trauma victims diagnosed as being in a persistent vegetative state (PVS) or a minimally conscious state and patients suffering *locked-in syndrome* as a result of motor neuron disease or brainstem stroke can specifically benefit from current BCI systems, although, as BCIs improve and surpass existing assistive technologies, they will be beneficial to those with less severe disabilities. In addition, the possibility of enriching computer game play through BCI also has immense potential, and computer games as well as other forms of interactive multi-media are currently an engaging interface techniques for therapeutic neurofeedback and improving BCI performance and training paradigms. Brain–computer games interaction provides motivation and challenge during training, which is used as a stepping stone towards applications that offer enablement and assistance. Based on these projections and the ever-increasing knowledge of the brain the future looks bright for BCIs.

39.10 Conclusion

The scientific approaches described throughout this chapter often overlook the underpinning processes and rely on correlations between a minimal number of factors only. As a result, current sensorimotor rhythms BCIs are of limited functionality and allow basic motor functions (a two degrees-of-freedom (DOF) limited control of a wheelchair/mouse cursor/robotic arm) and limited communication abilities (word dictation). It is assumed that BCI systems could greatly benefit

from the inclusion of multi-modal data and multi-dimensional signal processing techniques, which would allow the introduction of additional data sources and data from multiple brain scales, and enable detection of more subtle features embedded in the signal. Furthermore, using knowledge about sensorimotor control will be critical in understanding and developing successful learning and control models for robotic devices and BCI, fully closing the sensorimotor learning loop to enable finer manipulation abilities using BCIs and for retraining or enabling better relearning of motor actions after cortical damage. As demonstrated throughout the chapter, many remarkable studies have been conducted with truly inspirational engineering and scientific methodologies resulting in many very useful and interesting findings.

There are many potential advantages of understanding motor circuitry, not to mention the many clinical and quality of life benefits that a greater understanding of the motor systems may provide. Such knowledge may offer better insights into treating motor pathologies that occur as a result of injury or diseases such as spinal cord injury, stroke, Parkinson's disease, Guillain Barre syndrome, motor diseases, and Alzheimer's disease, to mention just a few. Understanding sensorimotor systems can provide significant gains in developing more intelligent systems that can provide multiple benefits for humanity in general. However, there are still lacunae in our biological account of how the motor system works.

Animals have superb innate abilities to choose and execute simple and extended courses of action and the ability to adapt their actions to a changing environment. We are still a long way from understanding how that is achieved and are exploiting this to tackle the issues outlined above comprehensively. There are number of key questions that need to be addressed [39.154]:

- What are the roles of the cortex, the basal ganglia, and the cerebellum – the three major neural control structures involved in movement planning and generation?
- How do these structures in the brain interact to deliver seamless adaptive control?
- How do we specify how hierarchical control structures can be learned?
- What is the relationship between reflexes, habits, and goal-directed actions?
- Is there anything to be gained for robotic control by thinking about how interactions are organized in sensorimotor regions?
- Is it essential to replicate this lateralized structure in sensorimotor areas to produce better motor control in an artificial cognitive system?
- How can we create more accurate models of how the motor cortex works? Can such models be implemented to provide human-like motor control in an artificial system?
- How can we decode motor activity to undertake tasks that require accurate and robust three dimensional control under multiple different scenarios?

Wolpert et al. [39.16] elaborate on some of these questions, in particular, one which has not been addressed in this chapter, namely modeling sensorimotor systems. Although substantial progress has been made in computational sensorimotor control, the field has been less successful in linking computational models to neurobiological models of control. Sensorimotor control has traditionally been considered from a control theory perspective, without relation to neurobiology [39.155]. Although neglected in this chapter, computational motor cortical circuit modeling will be a critical aspect of research into understanding sensorimotor control and learning, and is likely to fill parts of the lacunae in our understanding that are not accessible with current imaging, electrophysiology, and experimental methodology. Likewise, understanding the computations undertaken in many of sensorimotor areas will depend heavily on computational modeling. *Doya* [39.156] suggested the classical notion that the cerebellum and the basal ganglia are dedicated solely to motor control. This is now under dispute given increasing evidence of their involvement in non-motor functions. However, there is enough anatomical, physiological and theoretical evidence to support the hypotheses that the cerebellum is a specialized organism that may support supervised learning, the basal ganglia may perform reinforcement learning role, and the cerebral cortex may perform unsupervised learning. Alternative theories that enable us to comprehend the way the cortex, cerebellum, and the basal ganglia participate in motor, sensory or cognitive tasks are required [39.156].

Additionally, as has been illustrated throughout this work, investigating brain oscillations is key to understanding brain coordination. Understanding the coordination of multiple parts of an extremely complex system such as the brain is a significant challenge. Models of cortical coordination dynamics can show how brain areas may cooperate (integration) and at the same time retain their functional specificity (segregation). Such models can exhibit properties that the brain is known to exhibit, including self-organization, multi-functionality, meta-stability, and switching. Cortical coordination can be assessed by investigating the collective phase relationships among brain oscillations and rhythms in neurophysiological data. Imaging and electrophysiology can be used to tackle the challenge of understanding how different brain areas interact and cooperate.

Ultimately better knowledge of the motor system through neuroengineering sensorimotor–computer interfaces may lead to better methods of understanding brain dysfunction and pathology, better brain–computer interfaces, biological plausible neural circuit models, and inevitably more intelligent systems and machines that can perceive, reason, and act autonomously. It is too early to know the overarching control mech-

anisms and exact neural processes involved in the motor system but through the many innovations of scientists around the world, as highlighted in this chapter, pieces of the puzzle are being understood and slowly assembled to reach this target and go beyond it.

References

39.1   D.M. Wolpert, Z. Ghahramani, J.R. Flanagan: Perspectives and problems in motor learning, Trends Cogn. Sci. **5**, 487–494 (2001)

39.2   Laboratory for Computational Motor Control, John Hopkins University, Baltimore, USA: http://www.shadmehrlab.org/Courses/medschoollectures.html

39.3   R.E. Jung, R.J. Haier: The Parieto-Frontal Integration Theory (P–FIT) of intelligence: Converging neuroimaging evidence, Behav. Brain Sci. **30**, 135–154 (2007)

39.4   D.J. McFarland, J.R. Wolpaw: Sensorimotor rhythm-based brain-computer interface (BCI): Feature selection by regression improves performance, IEEE Trans. Neural Syst. Rehabil. Eng. **13**, 372–379 (2005)

39.5   J.W. McDonald: Repairing the damaged spinal cord, Sci. Am. **281**, 64–73 (1999)

39.6   A.P. Georgopoulos, J. Ashe, N. Smyrnis, M. Taira: The motor cortex and the coding of force, Science **256**, 1692–1695 (1992)

39.7   M. Desmurget, D. Pélisson, Y. Rossetti, C. Prablanc: From eye to hand: Planning goal-directed movements, Neurosci. Biobehav. Rev. **22**, 761–788 (1998)

39.8   C. Ghez, J.W. Krakauer, R.L. Sainburg, M.-F. Ghilasdi: Spatial representations and internal models of limb dynamics in motor learning. In: *The New Cognitive Neurosciences*, ed. by M.S. Gazzaniga (MIT Press, Cambridge 2000) pp. 501–514

39.9   N. Hogan: The mechanics of multi-joint posture and movement control, Biol. Cybern. **52**, 315–331 (1985)

39.10   P. Vetter, D.M. Wolpert: Context estimation for sensorimotor control, J. Neurophysiol. **84**, 1026–1034 (2000)

39.11   W.L. Nelson: Physical principles for economies of skilled movements, Biol. Cybern. **46**, 135–147 (1983)

39.12   T. Flash, N. Hogan: The coordination of arm movements: An experimentally confirmed mathematical model, J. Neurosci. **5**, 1688–1703 (1985)

39.13   N. Hogan: An organizing principle for a class of voluntary movements, J. Neurosci. **4**, 2745–2754 (1984)

39.14   R. Sosnik, T. Flash, B. Hauptmann, A. Karni: The acquisition and implementation of the smooth-

ness maximization motion strategy is dependent on spatial accuracy demands, Exp. Brain Res. **176**, 311–331 (2007)

39.15   R. Sosnik, M. Shemesh, M. Abeles: The point of no return in planar hand movements: An indication of the existence of high level motion primitives, Cogn. Neurodynam. **1**, 341–358 (2007)

39.16   D.M. Wolpert, J. Diedrichsen, J.R. Flanagan: Principles of sensorimotor learning, Nat. Rev. Neurosci. **12**, 739–751 (2011)

39.17   F.A. Mussa-Ivaldi, E. Bizzi: Motor learning through the combination of primitives, Philos. Trans. R. Soc. B **355**, 1755–1769 (2000)

39.18   E.A. Henis, T. Flash: Mechanisms underlying the generation of averaged modified trajectories, Biol. Cybern. **72**, 407–419 (1995)

39.19   A. Karni, G. Meyer, C. Rey-Hipolito, P. Jezzard, M.M. Adams, R. Turner, L.G. Ungerleider: The acquisition of skilled motor performance: Fast and slow experience-driven changes in primary motor cortex, Proc. Natl. Acad. Sci. USA **95**, 861–868 (1998)

39.20   O. Hikosaka, M.K. Rand, S. Miyachi, K. Miyashita: Learning of sequential movements in the monkey: Process of learning and retention of memory, J. Neurophysiol. **74**, 1652–1661 (1995)

39.21   R. Colom, S. Karama, R.E. Jung, R.J. Haier: Human intelligence and brain networks, Dialogues Clin. Neurosci. **12**, 489–501 (2010)

39.22   H. Johansen-Berg: The future of functionally-related structural change assessment, NeuroImage **62**, 1293–1298 (2012)

39.23   X. Li, D. Coyle, L. Maguire, D.R. Watson, T.M. McGinnity: Gray matter concentration and effective connectivity changes in Alzheimer's disease: A longitudinal structural MRI study, Neuroradiology **53**, 733–748 (2011)

39.24   C.J. Steele, J. Scholz, G. Douaud, H. Johansen-Berg, V.B. Penhune: Structural correlates of skilled performance on a motor sequence task, Front. Human Neurosci. **6**, 289 (2012)

39.25   R.D. Fields: Imaging learning: the search for a memory trace, Neuroscientist **17**, 185–196 (2011)

39.26   S.A. Huettel, A.W. Song, G. McCarthy: *Functional Magnetic Resonance Imaging*, 2nd edn. (Sinauer Associates, Sunderland 2009)

9.27 F. Ullén, L. Forsman, O. Blom, A. Karabanov, G. Madison: Intelligence and variability in a simple timing task share neural substrates in the prefrontal white matter, J. Neurosci. **28**, 4238–4243 (2008)

9.28 T. Ball, A. Schreiber, B. Feige, M. Wagner, C.H. Lücking, R. Kristeva-Feige: The role of higher-order motor areas in voluntary movement as revealed by high-resolution EEG and fMRI, NeuroImage **10**, 682–694 (1999)

9.29 S. Halder, D. Agorastos, R. Veit, E.M. Hammer, S. Lee, B. Varkuti, M. Bogdan, W. Rosenstiel, N. Birbaumer, A. Kübler: Neural mechanisms of brain-computer interface control, NeuroImage **55**, 1779–1790 (2011)

9.30 F. Quandt, C. Reichert, H. Hinrichs, H.J. Heinze, R.T. Knight, J.W. Rieger: Single trial discrimination of individual finger movements on one hand: A combined MEG and EEG study, NeuroImage **59**, 3316–3324 (2012)

9.31 K.J. Miller, L.B. Sorensen, J.G. Ojemann, M. den Nijs: Power-law scaling in the brain surface electric potential, PLoS Comput. Biol. **5**, e1000609 (2009)

9.32 K.J. Miller, G. Schalk, E.E. Fetz, M. den Nijs, J.G. Ojemann, R.P.N. Rao: Cortical activity during motor execution, motor imagery, and imagery-based online feedback, Proc. Natl Acad. Sci. USA **107**, 4430–4435 (2010)

9.33 K.J. Miller, D. Hermes, C.J. Honey, A.O. Hebb, N.F. Ramsey, R.T. Knight, J.G. Ojemann, E.E. Fetz: Human motor cortical activity is selectively phase-entrained on underlying rhythms, PLoS Comput. Biol. **8**, e1002655 (2012)

9.34 V.N. Murthy, E.E. Fetz: Synchronization of neurons during local field potential oscillations in sensorimotor cortex of awake monkeys, J. Neurophysiol. **76**, 3968–3982 (1996)

9.35 D.J. Krusienski, M. Grosse-Wentrup, F. Galán, D. Coyle, K.J. Miller, E. Forney, C.W. Anderson: Critical issues in state-of-the-art brain-computer interface signal processing, J. Neural Eng. **8**, 025002 (2011)

9.36 A.C. Papanicolaou, E.M. Castillo, R. Billingsley-Marshall, E. Pataraia, P.G. Simos: A review of clinical applications of magnetoencephalography, Int. Rev. Neurobiol. **68**, 223–247 (2005)

9.37 L. Kauhanen, T. Nykopp, J. Lehtonen, P. Jylänki, J. Heikkonen, P. Rantanen, H. Alaranta, M. Sams: EEG and MEG brain-computer interface for tetraplegic patients, IEEE Trans. Neural Syst. Rehabil. Eng. **14**, 190–193 (2006)

9.38 L. Kauhanen, T. Nykopp, M. Sams: Classification of single MEG trials related to left and right index finger movements, Clin. Neurophysiol. **117**, 430–439 (2006)

9.39 K.J. Miller, S. Zanos, E.E. Fetz, M. den Nijs, J.G. Ojemann: Decoupling the cortical power spectrum reveals real-time representation of individual finger movements in humans, J. Neurosci. **29**, 3132–3137 (2009)

39.40 Z. Wang, Q. Ji, K.J. Miller, G. Schalk: Prior knowledge improves decoding of finger flexion from electrocorticographic signals, Front. Neurosci. **5**, 127 (2011)

39.41 G. Pfurtscheller, C. Neuper, C. Brunner, F. da Lopes Silva: Beta rebound after different types of motor imagery in man, Neurosci. Lett. **378**, 156–159 (2005)

39.42 G. Pfurtscheller, C. Brunner, A. Schlögl, F.H. da Lopes Silva: Mu rhythm (de)synchronization and EEG single-trial classification of different motor imagery tasks, NeuroImage **31**, 153–159 (2006)

39.43 G. Pfurtscheller, C. Neuper, A. Schlögl, K. Lugger: Separability of EEG signals recorded during right and left motor imagery using adaptive autoregressive parameters, IEEE Trans. Rehabil. Eng. **6**, 316–325 (1998)

39.44 D. Coyle, G. Prasad, T.M. McGinnity: A time-frequency approach to feature extraction for a brain-computer interface with a comparative analysis of performance measures, EURASIP J. Appl, Signal Process. **2005**, 3141–3151 (2005)

39.45 G. Pfurtscheller, C. Guger, G. Müller, G. Krausz, C. Neuper: Brain oscillations control hand orthosis in a tetraplegic, Neurosci. Lett. **292**, 211–214 (2000)

39.46 M. Grosse-Wentrup, B. Schölkopf, J. Hill: Causal influence of gamma oscillations on the sensorimotor rhythm, NeuroImage **56**, 837–842 (2011)

39.47 H.H. Kornhuber, L. Deecke: Hirnpotentialänderungen bei Willkürbewegungen und passiven Bewegungen des Menschen: Bereitschaftspotential und reafferente Potentiale, Pflüg. Arch. **284**, 1–17 (1965)

39.48 L. Deecke, B. Grozinger, H.H. Kornhuber: Voluntary finger movement in man: Cerebral potentials and theory, Biol. Cybern. **23**, 99–119 (1976)

39.49 M. Krauledat, G. Dornhege, B. Blankertz, G. Curio, K.-R. Müller: The Berlin brain-computer interface for rapid response, Proc. 2nd Int. Brain-Comp. Interface Workshop Train. Course Biomed. Tech. (2004) pp. 61–62

39.50 M. Krauledat, G. Dornhege, B. Blankertz, G. Curio, F. Losch, K.-R. Müller: Improving speed and accuracy of brain-computer interfaces using readiness potential, Proc. 26th Int. IEEE Eng. Med. Biol. Conf. (2004) pp. 4512–4515

39.51 J.P.R. Dick, J.C. Rothwell, B.L. Day, R. Cantello, O. Buruma, M. Gioux, R. Benecke, A. Berardelli, P.D. Thompson, C.D. Marsden: The Bereitschaftspotential is abnormal in Parkinson's disease, Brain **112**(1), 233–244 (1989)

39.52 H. Shibasaki, M. Hallett: What is the Bereitschaftspotential?, Clin. Neurophysiol. **117**, 2341–2356 (2006)

39.53 Y. Gu, K. Dremstrup, D. Farina: Single-trial discrimination of type and speed of wrist move-

Part D | 39

ments from EEG recordings, Clin. Neurophysiol. **120**, 1596–1600 (2009)

39.54 D.J. McFarland, L.A. Miner, T.M. Vaughan, J.R. Wolpaw: Mu and beta rhythm topographies during motor imagery and actual movements, Brain Topogr. **12**, 177–186 (2000)

39.55 H. Lakany, B.A. Conway: Comparing EEG patterns of actual and imaginary wrist movements – A machine learning approach, Proc. 1st Int. Conf. Artif. Intell. Mach. Learn. AIML (2005) pp. 124–127

39.56 B. Nasseroleslami, H. Lakany, B.A. Conway: Identification of time-frequency EEG features modulated by force direction in arm isometric exertions, Proc. 5th Int. IEEE EMBS Conf. Neural Eng. (2011) pp. 422–425

39.57 J.R. Wolpaw, D.J. McFarland: Control of a two-dimensional movement signal by a noninvasive brain-computer interface in humans, Proc. Natl. Acad. Sci. USA **101**, 17849–17854 (2004)

39.58 D.J. McFarland, W.A. Sarnacki, J.R. Wolpaw: Electroencephalographic (EEG) control of three-dimensional movement, J. Neural Eng. **7**, 036007 (2010)

39.59 T.J. Bradberry, R.J. Gentili, J.L. Contreras-Vidal: Reconstructing three-dimensional hand movements from noninvasive electroencephalographic signals, J. Neurosci. **30**, 3432–3437 (2010)

39.60 E.D. Adrian, Y. Zotterman: The impulses produced by sensory nerve-endings: Part II. The response of a Single End-Organ, J. Physiol. **61**, 151–171 (1926)

39.61 R.B. Stein, E.R. Gossen, K.E. Jones: Neuronal variability: Noise or part of the signal?, Nat. Rev. Neurosci. **6**, 389–397 (2005)

39.62 D.A. Butts, C. Weng, J. Jin, C.-I. Yeh, N.A. Lesica, J.-M. Alonso, G.B. Stanley: Temporal precision in the neural code and the timescales of natural vision, Nature **449**, 92–95 (2007)

39.63 A.P. Georgopoulos, J.F. Kalaska, R. Caminiti, J.T. Massey: On the relations between the direction of two-dimensional arm movements and cell discharge in primate motor cortex, J. Neurosci. **2**, 1527–1537 (1982)

39.64 B. Amirikian, A.P. Georgopoulos, A.P. Georgopulos: Directional tuning profiles of motor cortical cells, Neurosci. Res. **36**, 73–79 (2000)

39.65 L. Paninski, M.R. Fellows, N.G. Hatsopoulos, J.P. Donoghue: Spatiotemporal tuning of motor cortical neurons for hand position and velocity, J. Neurophysiol. **91**, 515–532 (2004)

39.66 A.P. Georgopoulos, A.B. Schwartz, R.E. Kettner: Neuronal population coding of movement direction, Science **233**, 1416–1419 (1986)

39.67 H. Tanaka, T.J. Sejnowski, J.W. Krakauer: Adaptation to visuomotor rotation through interaction between posterior parietal and motor cortical areas, J. Neurophysiol. **102**, 2921–2932 (2009)

39.68 S.M. Chase, R.E. Kass, A.B. Schwartz: Behavioral and neural correlates of visuomotor adaptation observed through a brain-computer interface in

39.69 primary motor cortex, J. Neurophysiol. **108**, 624–644 (2012)

39.69 E.V. Evarts: Relation of pyramidal tract activity to force exerted during voluntary movement, J. Neurophysiol. **31**, 14–27 (1968)

39.70 E. Schneidman, M.J. Berry, R. Segev, W. Bialek: Weak pairwise correlations imply strongly correlated network states in a neural population, Nature **440**, 1007–1012 (2006)

39.71 S.-I. Amari: Information geometry on hierarchy of probability distributions, IEEE Trans. Inform. Theor. **47**, 1701–1711 (2001)

39.72 J. Wessberg, C.R. Stambaugh, J.D. Kralik, P.D. Beck, M. Laubach, J.K. Chapin, J. Kim, S.J. Biggs, M.A. Srinivasan, M.A. Nicolelis: Real-time prediction of hand trajectory by ensembles of cortical neurons in primates, Nature **408**, 361–365 (2000)

39.73 M.D. Serruya, N.G. Hatsopoulos, L. Paninski, M.R. Fellows, J.P. Donoghue: Instant neural control of a movement signal, Nature **416**, 141–142 (2002)

39.74 J.M. Carmena, M.A. Lebedev, R.E. Crist, J.E. O'Doherty, D.M. Santucci, D.F. Dimitrov, P.G. Patil, C.S. Henriquez, M.A.L. Nicolelis: Learning to control a brain-machine interface for reaching and grasping by primates, PLoS Biol. **1**, E42 (2003)

39.75 M.A. Lebedev, J.M. Carmena, J.E. O'Doherty, M. Zacksenhouse, C.S. Henriquez, J.C. Principe, M.A.L. Nicolelis: Cortical ensemble adaptation to represent velocity of an artificial actuator controlled by a brain-machine interface, J. Neurosci. **25**, 4681–4693 (2005)

39.76 M. Velliste, S. Perel, M.C. Spalding, A.S. Whitford, A.B. Schwartz: Cortical control of a prosthetic arm for self-feeding, Nature **453**, 1098–1101 (2008)

39.77 D.M. Santucci, J.D. Kralik, M.A. Lebedev, M.A.L. Nicolelis: Frontal and parietal cortical ensembles predict single-trial muscle activity during reaching movements in primates, Eur. J. Neurosci. **22**, 1529–1540 (2005)

39.78 J.R. Wolpaw, N. Birbaumer, D.J. McFarland, G. Pfurtscheller, T.M. Vaughan: Brain-computer interfaces for communication and control, Clin. Neurophysiol. **113**, 767–791 (2002)

39.79 A. Bashashati, S.G. Mason, J.F. Borisoff, R.K. Ward, G.E. Birch: A comparative study on generating training-data for self-paced brain interfaces, IEEE Trans. Neural Syst. Rehabil. Eng. **15**, 59–66 (2007)

39.80 S.G. Mason, A. Bashashati, M. Fatourechi, K.F. Navarro, G.E. Birch: A comprehensive survey of brain interface technology designs, Ann. Biomed. Eng. **35**(2), 137–169 (2007)

39.81 N. Brodu, F. Lotte, A. Lécuyer: Comparative study of band-power extraction techniques for Motor Imagery classification, IEEE Symp. Comput. Intell. Cogn. Algorithm, Mind, Brain (2011) pp. 1–6

9.82   F. Lotte, M. Congedo, A. Lécuyer, F. Lamarche, B. Arnaldi: A review of classification algorithms for EEG-based brain–computer interfaces, J. Neural Eng. **4**, R1–R13 (2007)

9.83   P. Herman, G. Prasad, T.M. McGinnity, D. Coyle: Comparative analysis of spectral approaches to feature extraction for EEG-based motor imagery classification, IEEE Trans. Neural Syst. Rehabil. Eng. **16**, 317–326 (2008)

9.84   J. Wolpaw, E.W. Wolpaw: *Brain–Computer Interfaces: Principles and Practice* (Oxford Univ. Press, Oxford 2012)

9.85   S. Sun: Ensemble learning methods for classifying EEG sign, Lect. Notes Comput. Sci. **4472**, 113–120 (2007)

9.86   L.F. Nicolas-Alonso, J. Gomez-Gil: Brain computer interfaces – A review., Sensors **12**, 1211–1279 (2012)

9.87   A. Soria-Frisch: A critical review on the usage of ensembles for BCI. In: *Towards Practical Brain-Computer Interfaces*, ed. by B.Z. Allison, S. Dunne, R. Leeb, J.R. Del Millán, A. Nijholt (Springer, Berlin, Heidelberg 2013) pp. 41–65

39.88   H. Ramoser, J. Muller-Gerking, G. Pfurtscheller: Optimal spatial filtering of single trial EEG during imagined hand movement, IEEE Trans. Rehabil. Eng. **8**, 441–446 (2000)

39.89   B. Blankertz, R. Tomioka, S. Lemm, M. Kawanabe, K.-R. Muller: Optimizing Spatial filters for Robust EEG Single-Trial Analysis, IEEE Signal Process. Mag. **25**, 41–56 (2008)

39.90   B. Blankertz, M. Kawanabe, R. Tomioka, F.U. Hohlefeld, V. Nikulin, K.-R. Müller: Invariant Common Spatial Patterns: Alleviating Nonstationarities in Brain–Computer Interfacing, Adv. Neural Inf. Process. **20**, 1–8 (2008)

39.91   F. Lotte, C. Guan: Regularizing common spatial patterns to improve BCI designs: Unified theory and new algorithms, IEEE Trans. Neural Syst. Rehabil. Eng. **58**, 355–362 (2011)

39.92   D. Coyle: Neural network based auto association and time-series prediction for biosignal processing in brain–computer interfaces, IEEE Comput. Intell. Mag. **4**(4), 47–59 (2009)

39.93   H. Zhang, Z.Y. Chin, K.K. Ang, C. Guan, C. Wang: Optimum spatio-spectral filtering network for brain–computer interface, IEEE Trans. Neural Netw. **22**, 52–63 (2011)

39.94   K.K. Ang, Z.Y. Chin, C. Wang, C. Guan, H. Zhang: Filter bank common spatial pattern algorithm on BCI competition IV datasets 2a and 2b, Front. Neurosci. **6**, 39 (2012)

39.95   A. Cichocki, Y. Washizawa, T. Rutkowski, H. Bakardjian, A.H. Phan, S. Choi, H. Lee, Q. Zhao, L. Zhang, Y. Li: Noninvasive BCIs: Multiway signal-processing array decompositions, Computer **41**, 34–42 (2008)

39.96   D.H. Coyle, G. Prasad, T.M. McGinnity: Improving information transfer rates of BCI by self-organising fuzzy neural network-based multi-step-ahead time series prediction, Proc. 3rd IEEE Syst. Man Cybern. Conf. (2004)

39.97   D. Coyle, G. Prasad, T.M. McGinnity: A time-series prediction approach for feature extraction in a brain–computer interface, IEEE Trans. Neural Syst. Rehabil. Eng. **13**, 461–467 (2005)

39.98   D. Coyle: Channel and class dependent time-series embedding using partial mutual information improves sensorimotor rhythm based brain–computer interfaces. In: *Time Series Analysis, Modeling and Applications – A Computational Intelligence Perspective*, ed. by W. Pedrycz (Springer, Berlin, Heidelberg 2013) pp. 249–278

39.99   A. Schlögl, D. Flotzinger, G. Pfurtscheller: Adaptive autoregressive modeling used for single-trial EEG classification – Verwendung eines Adaptiven Autoregressiven Modells für die Klassifikation von Einzeltrial-EEG-Daten, Biomed. Tech./Biomed. Eng. **42**, 162–167 (1997)

39.100   E. Haselsteiner, G. Pfurtscheller: Using time-dependent neural networks for EEG classification, IEEE Trans. Rehabil. Eng. **8**, 457–463 (2000)

39.101   E.M. Forney, C.W. Anderson: Classification of EEG during imagined mental tasks by forecasting with Elman Recurrent Neural Networks, Int. Joint Conf. Neural Netw. (2011) pp. 2749–2755

39.102   C. Anderson, E. Forney, D. Hains, A. Natarajan: Reliable identification of mental tasks using time-embedded EEG and sequential evidence accumulation, J. Neural Eng. **8**, 025023 (2011)

39.103   H.K. Kimelberg: Functions of mature mammalian astrocytes: A current view, Neuroscientist **16**, 79–106 (2010)

39.104   N.A. Busch, J. Dubois, R. VanRullen: The phase of ongoing EEG oscillations predicts visual perception, J. Neurosci. **29**, 7869–7876 (2009)

39.105   W.J. Freeman: Origin, structure, and role of background EEG activity. Part 1. Analytic amplitude, Clin. Neurophysiol. **115**, 2077–2088 (2004)

39.106   R.T. Canolty, E. Edwards, S.S. Dalal, M. Soltani, S.S. Nagarajan, H.E. Kirsch, M.S. Berger, N.M. Barbaro, R.T. Knight: High gamma power is phase-locked to theta oscillations in human neocortex, Science **313**, 1626–1628 (2006)

39.107   C. Mehring, J. Rickert, E. Vaadia, S. De Cardosa Oliveira, A. Aertsen, S. Rotter: Inference of hand movements from local field potentials in monkey motor cortex, Nat. Neurosci. **6**, 1253–1254 (2003)

39.108   K.J. Miller, E.C. Leuthardt, G. Schalk, R.P.N. Rao, N.R. Anderson, D.W. Moran, J.W. Miller, J.G. Ojemann: Spectral changes in cortical surface potentials during motor movement, J. Neurosci. **27**, 2424–2432 (2007)

39.109   S. Waldert, H. Preissl, E. Demandt, C. Braun, N. Birbaumer, A. Aertsen, C. Mehring: Hand movement direction decoded from MEG and EEG, J. Neurosci. **28**, 1000–1008 (2008)

Part D | 39

39.110 K. Jerbi, J.-P. Lachaux, K. N'Diaye, D. Pantazis, R.M. Leahy, L. Garnero, S. Baillet: Coherent neural representation of hand speed in humans revealed by MEG imaging, Proc. Natl. Acad. Sci. USA **104**, 7676–7681 (2007)

39.111 K. Jerbi, O. Bertrand: Cross-frequency coupling in parieto-frontal oscillatory networks during motor imagery revealed by magnetoencephalography, Front. Neurosci. **3**, 3–4 (2009)

39.112 S.N. Baker: Oscillatory interactions between sensorimotor cortex and the periphery, Curr. Opin. Neurobiol. **17**, 649–655 (2007)

39.113 D. Rubino, K.A. Robbins, N.G. Hatsopoulos: Propagating waves mediate information transfer in the motor cortex, Nat. Neurosci. **9**, 1549–1557 (2006)

39.114 R.J. May, H.R. Maier, G.C. Dandy, T.M.K. Gayani Fernando: Non-linear variable selection for artificial neural networks using partial mutual information, Environ. Model. Softw. **23**, 1312–1326 (2008)

39.115 S. Lemm, C. Schäfer, G. Curio: BCI competition 2003-data set III: Probabilistic modeling of sensorimotor μ rhythms for classification of imaginary hand movements, IEEE Trans. Biomed. Eng. **51**, 1077–1080 (2004)

39.116 D. Coyle, G. Prasad, T.M. McGinnity: Faster self-organizing fuzzy neural network training and a hyperparameter analysis for a brain-computer interface, IEEE Trans. Syst. Man Cybern. **39**, 1458–1471 (2009)

39.117 C. Sannelli, C. Vidaurre, K.-R. Müller, B. Blankertz: CSP patches: An ensemble of optimized spatial filters. An evaluation study, J. Neural Eng. **8**, 025012 (2011)

39.118 C. Vidaurre, M. Kawanabe, P. von Bünau, B. Blankertz, K.R. Müller: Toward unsupervised adaptation of LDA for brain-computer interfaces, IEEE Trans. Biomed. Eng. **58**, 587–597 (2011)

39.119 P. Herman, G. Prasad, T.M. McGinnity: Computational intelligence approaches to brain signal pattern recognition. In: *Pattern Recognition Techniques, Technology and Applications*, ed. by B. Verma (InTech, Rijeka 2008) pp. 91–120

39.120 J.M. Mendel: Type-2 fuzzy sets and systems: An overview, IEEE Comput. Intell. Mag. **2**(1), 20–29 (2007)

39.121 R. Mohammadi, A. Mahloojifar, H. Chen, D. Coyle: EEG based foot movement onset detection with the probabilistic classification vector machine, Lect. Notes Comput. Sci. **7666**, 356–363 (2012)

39.122 H. Chen, P. Tino, X. Yao: Probabilistic classification vector machines, IEEE Trans. Neural Netw. **20**, 901–914 (2009)

39.123 C. Vidaurre, B. Blankertz: Towards a cure for BCI illiteracy, Brain Topogr. **23**, 194–198 (2010)

39.124 A. Schlögl, C. Vidaurre, K.-R. Müller: Adaptive methods in BCI research: An introductory tutorial.

In: *Brain Computer Interfaces – Revolutionizing Human-Computer Interfaces*, ed. by B. Graimann, B. Allison, G. Pfurtscheller (Springer, Berlin, Heidelberg 2010) pp. 331–355

39.125 J.R. Wolpaw, N. Birbaumer, W.J. Heetderks, D.J. McFarland, P.H. Peckham, G. Schalk, E. Donchin, L.A. Quatrano, C.J. Robinson, T.M. Vaughan: Brain-computer interface technology: A review of the first international meeting, IEEE Trans. Rehabil. Eng. **8**, 164–173 (2000)

39.126 D.J. McFarland, W.A. Sarnacki, J.R. Wolpaw: Should the parameters of a BCI translation algorithm be continually adapted?, J. Neurosci. Methods **199**, 103–107 (2011)

39.127 B.A.S. Hasan: Adaptive Methods Exploiting the Time Structure in EEG for Interfaces (University of Essex, Colchester 2010)

39.128 C. Vidaurre, C. Sannelli, K.-R. Müller, B. Blankertz: Co-adaptive calibration to improve BCI efficiency, J. Neural Eng. **8**, 025009 (2011)

39.129 J.W. Yoon, S.J. Roberts, M. Dyson, J.Q. Gan: Adaptive classification for brain computer interface systems using sequential Monte Carlo sampling, Neural Netw. **22**, 1286–1294 (2009)

39.130 J.Q. Gan: Self-adapting BCI based on unsupervised learning, 3rd Int. Workshop Brain Comput. Interfaces (2006) pp. 50–51

39.131 S. Lu, C. Guan, H. Zhang: Unsupervised brain computer interface based on intersubject information and online adaptation, IEEE Trans. Neural Syst. Rehabil. Eng. **17**, 135–145 (2009)

39.132 S.E. Eren, M. Grosse-Wentrup, M. Buss: Unsupervised classification for non-invasive brain-computer-interfaces, Proc. Autom. Workshop (VDI Verlag, 2007) pp. 65–66

39.133 B.A.S. Hasan, J.Q. Gan: Unsupervised adaptive GMM for BCI, 4th Int. IEEE EMBS Conf. Neural Eng. (2009) pp. 295–298

39.134 J. Blumberg, J. Rickert, S. Waldert, A. Schulze-Bonhage, A. Aertsen, C. Mehring: Adaptive classification for brain computer interfaces, Annu. Int. Conf. IEEE Eng. Med. Biol. Soc. (2007) pp. 2536–2539

39.135 G. Liu, G. Huang, J. Meng, D. Zhang, X. Zhu: Unsupervised adaptation based on fuzzy c-means for brain-computer interface, 1st Int. Conf. Inform. Sci. Eng. (2009) pp. 4122–4125

39.136 P. Shenoy, M. Krauledat, B. Blankertz, R.P.N. Rao, K.-R. Müller: Towards adaptive classification for BCI, J. Neural Eng. **3**, R13–23 (2006)

39.137 T. Gürel, C. Mehring: Unsupervised adaptation of brain-machine interface decoders, Front. Neurosci. **6**, 164 (2012)

39.138 M. Sugiyama, M. Krauledat, K.-R. Müller: Covariate shift adaptation by importance weighted cross validation, J. Mach. Learn. Res. **8**, 985–1005 (2007)

39.139 A. Satti, C. Guan, D. Coyle, G. Prasad: A covariate shift minimisation method to alleviate nonstationarity effects for an adaptive brain–computer interface, 20th Int. Conf. Pattern Recogn. (2010) pp. 105–108

39.140 R. Mohammadi, A. Mahloojifar, D. Coyle: Unsupervised short-term covariate shift minimization for self-paced BCI, IEEE Symp. Comput. Intell. Cogn. Algorithm. Mind, Brain (2013) pp. 101–106

39.141 A. Satti, D. Coyle, G. Prasad: Continuous EEG classification for a self-paced BCI, 4th Int. IEEE/EMBS Conf. Neural Eng. (2009) pp. 315–318

39.142 G.E. Fabiani, D.J. McFarland, J.R. Wolpaw, G. Pfurtscheller: Conversion of EEG activity into cursor movement by a brain-computer interface (BCI), IEEE Trans. Neural Syst. Rehabil. Eng. **12**, 331–338 (2004)

39.143 M. Arvaneh, C. Guan, K.K. Ang, C. Quek: EEG data space adaptation to reduce intersession non-stationarity in brain-computer interface, Neural Comput. **25**(8), 2146–2171 (2013)

39.144 G. Schalk: Brain-computer symbiosis, J. Neural Eng. **5**, P1–P15 (2008)

39.145 C.M. Reed, N.I. Durlach: Note on information transfer rates in human communication, Presence: Teleoper. Virtual Environ. **7**, 509–518 (1998)

39.146 I.S. MacKenzie: Fitts' Law as a research and design tool in human-computer interaction, Human-Comput. Interact. **7**, 91–139 (1992)

39.147 Z.C. Chao, Y. Nagasaka, N. Fujii: Long-term asynchronous decoding of arm motion using electrocorticographic signals in monkeys, Front. Neuroeng. **3**, 3–13 (2010)

39.148 G. Schalk: Can electrocorticography (ECoG) support robust and powerful brain-computer interfaces?, Front. Neuroeng. **3**, 9 (2010)

39.149 A. Barbero, M. Grosse-Wentrup: Biased feedback in brain-computer interfaces, J. Neuroeng. Rehabil. **7**, 34 (2010)

39.150 L.R. Hochberg, M.D. Serruya, G.M. Friehs, J.A. Mukand, M. Saleh, A.H. Caplan, A. Branner, D. Chen, R.D. Penn, J.P. Donoghue: Neuronal ensemble control of prosthetic devices by a human with tetraplegia, Nature **442**, 164–171 (2006)

39.151 J.M. Carmena: Becoming bionic, IEEE Spect. **49**, 24–29 (2012)

39.152 T. W. Berger, J. K. Chapin, G. A. Gerhardt, D. J. McFarland, D. M. Taylor, P. A. Tresco: WTEC Panel Report on International Assessment of Research and Development in Brain-Computer Interfaces (2007)

39.153 R. Kurzweil: *The Age of Spiritual Machines: When Computers Exceed Human Intelligence* (Penguin, New York 2000)

39.154 P. Barnard, P. Dayan, P. Redgrave: Action. In: *Cognitive Systems: Information Processing Meets Brain Science*, ed. by R. Morris, L. Tarassenki, M. Kenward (Elsevier Academic, London 2006)

39.155 G.L. Chadderdon, S.A. Neymotin, C.C. Kerr, W.W. Lytton: Reinforcement learning of targeted movement in a spiking neuronal model of motor cortex, PloS ONE **7**, e47251 (2012)

39.156 K. Doya: What are the computations of the cerebellum, the basal ganglia and the cerebral cortex?, Neural Netw. **12**, 961–974 (1999)

40. Evolving Connectionist Systems: From Neuro-Fuzzy-, to Spiking- and Neuro-Genetic

Nikola Kasabov

This chapter follows the development of a class of neural networks (NN) called evolving connectionist systems (ECOS). The term *evolving* is used here in its meaning of *unfolding, developing, changing, revealing* (according to the Oxford dictionary) rather than *evolutionary*. The latter represents processes related to populations and generations of them. An ECOS is a neural network-based model that evolves its structure and functionality through incremental, adaptive learning and self-organization during its *lifetime*. In principle, it could be a simple NN or a *hybrid* connectionist system. The latter is a system based on neural networks that also integrate other computational principles, such as linguistically meaningful explanation features of fuzzy rules, optimization techniques for structure and parameter optimization, quantum-inspired methods, and gene regulatory networks. The chapter includes definitions and examples of ECOS such as: evolving neuro-fuzzy and hybrid systems; evolving spiking neural networks, neurogenetic systems, quantum-inspired systems, which are all discussed from the point of view of the structural and functional development of a connectionist-based model and the knowledge that it represents. Applications for knowledge engineering across domain areas, such as in bioinformatics, brain study, and intelligent machines are presented.

40.1 Principles of Evolving Connectionist Systems (ECOS)

Everything in Nature evolves, develops, unfolds, reveals, and changes in time. The brain is probably the ultimate evolving system, which develops during a lifetime, based on genetic information (Nature) and learning from the environment (nurture). Inspired by information principles of the developing brain, ECOS are adaptive, incremental learning and knowledge representation systems that evolve their structure and functionality from incoming data through interaction with the environment, where in the core of a system is a connectionist architecture that consists of neurons (information processing units) and connections between them [40.1]. An ECOS is a system based on neural networks and the use of also other techniques of computational intelligence (CI), which operates continuously in time and adapts its structure and functionality through continuous interaction with the environment and with other systems. The adaptation is defined through:

1. A set of evolving rules.
2. A set of parameters (*genes*) that are subject to change during the system operation.
3. An incoming continuous flow of information, possibly with unknown distribution.
4. Goal (rationale) criteria (also subject to modification) that are applied to optimize the performance of the system over time.

ECOS learning algorithms are inspired by brain-like information processing principles, e.g.:

1. They evolve in an open space, where the dimensions of the space can change.
2. They learn via incremental learning, possibly in an on-line mode.

3. They learn continuously in a lifelong learning mode.
4. They learn both as individual systems and as an evolutionary population of such systems.
5. They use constructive learning and have evolving structures.
6. They learn and partition the problem space locally thus allowing for a fast adaptation and tracing the evolving processes over time.
7. They evolve different types of knowledge representation from data, mostly a combination of memory-based and symbolic knowledge.

Many methods, algorithms, and computational intelligence systems have been developed since the conception of ECOS and many applications across disciplines. This chapter will review only the fundamental aspects of some of these methods and will highlight some principal applications.

40.2 Hybrid Systems and Evolving Neuro-Fuzzy Systems

40.2.1 Hybrid Systems

A hybrid computational intelligent system integrates several principles of computational intelligence to enhance different aspects of the performance of the system. Here we will discuss only hybrid connectionist systems that integrate artificial neural networks (NN) with other techniques utilizing the adaptive learning features of the NN.

Early hybrid connectionist systems combined NN with rule-based systems such as production rules [40.3]

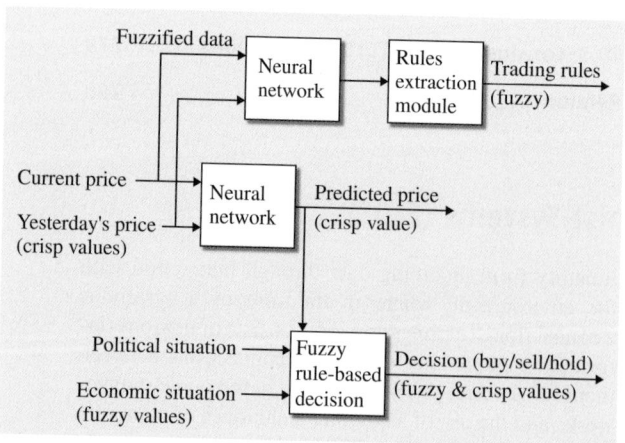

Fig. 40.1 A hybrid NN-fuzzy rule-based expert system for financial decision support (after [40.2])

or predicate logic [40.4]. They utilized NN modules for a lower level of information processing and rule-based systems for reasoning and explanation at a higher level.

The above principle is applied when fuzzy rules are used for higher-level information processing and for approximate reasoning [40.5–7]. These are expert systems that combine the learning ability of NN with the explanation power of linguistically plausible fuzzy rules [40.8–11]. A block diagram of an exemplar system is shown in Fig. 40.1, where at a lower level a neural network (NN) module predicts the level of a stock index and at a higher level a fuzzy reasoning module combines the predicted values with some macro-economic variables representing the political and the economic situations using the following types of fuzzy rules [40.2]

> IF <the predicted by the NN module stock value
>   in the future is high> AND
>   <the economic situation is good> AND
>   <the political situation is stable>
> THEN <buy stock> .   (40.1)

Along with the integration of NN and fuzzy rules for a better decision support, the system from Fig. 40.1 includes an NN module for extracting recent rules form data that can be used by experts to analyze the dy-

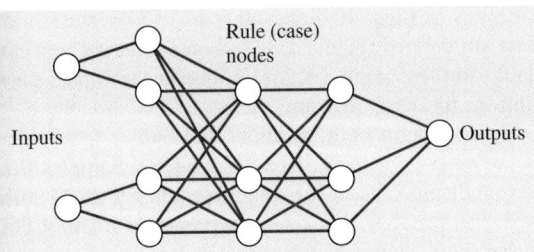

Fig. 40.2 A simple, feedforward EFuNN structure. The rule nodes evolve from data to capture cluster centers in the input space, while the output nodes evolve local models to learn and approximate the data in each of these clusters

namics of the stock and to possibly update the trading fuzzy rules in the fuzzy rule-based module. This NN module uses a fuzzy neural (FNN) network for the rule extraction.

Fuzzy neural networks (FNN) integrate NN and fuzzy rules into a single neuronal model tightly coupling learning and fuzzy reasoning rules into a connectionist structure. One of the first FNN models was initiated by *Yamakawa* and other Japanese scientists and promoted at a series of IIZUKA conferences in Japan [40.12, 13]. Many models of FNNs were developed based on these principles [40.2, 14, 15].

40.2.2 Evolving Neuro–Fuzzy Systems

The evolving neuro-fuzzy systems further extended the principles of hybrid neuro-fuzzy systems and the FNN, where instead of training a fixed connectionist structure,

the structure and its functionality evolve from incoming data, often in an on-line, one-pass learning mode. This is the case with evolving connectionist systems (ECOS) [40.1, 16–19].

ECOS are modular connectionist-based systems that evolve their structure and functionality in a continuous, self-organized, on-line, adaptive, and interactive way from incoming information [40.17]. They can process both data and knowledge in a supervised and/or unsupervised way. ECOS learn *local models* from data through clustering of the data and associating a local output function for each cluster represented in a connectionist structure. They can learn incrementally single data items or chunks of data and also incrementally change their input features [40.18].

Elements of ECOS have been proposed as part of the early, classical NN models, such as *Kohonen's* self organising maps (SOM) [40.20], redical basis function(RBF) [40.21], FuzyARTMap [40.22] by *Carpenter* et al. and *Fritzke's* growing neural gas [40.23], *Platt's* resource allocation networks (RAN) [40.24].

Some principles of ECOS are:

● Neurons are created (evolved) and allocated as centers of (fuzzy) data clusters. Fuzzy clustering, as a means to create local knowledge-based systems, was stimulated by the pioneering work of *Bezdek*, *Yager* and *Filev* [40.27–30].
● Local models are evolved and updated in these clusters.

Here we will briefly illustrate the concepts of ECOS on two implementations: evolving fuzzy neu-

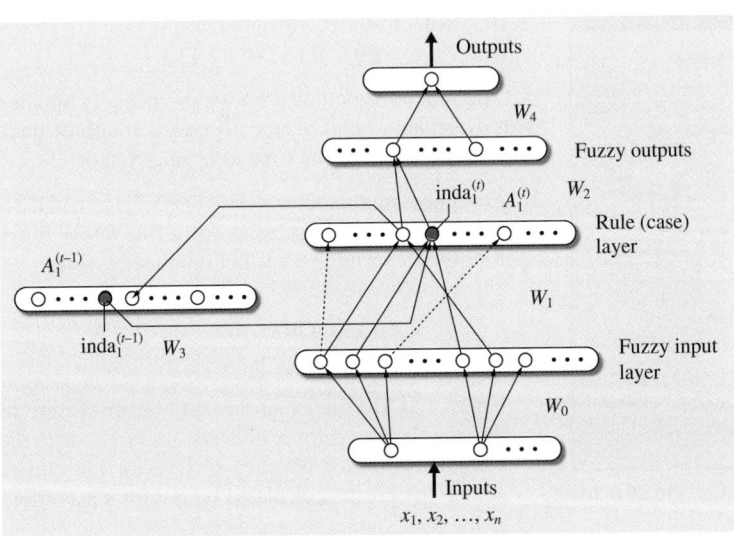

Fig. 40.3 An EFuNN structure with feedback connections (after [40.16])

tral networks (EFuNN) [40.16] and dynamic neuro-fuzzy inference systems (DENFIS) [40.25]. Examples of EFuNN are shown in Figs. 40.2 and 40.3 and of

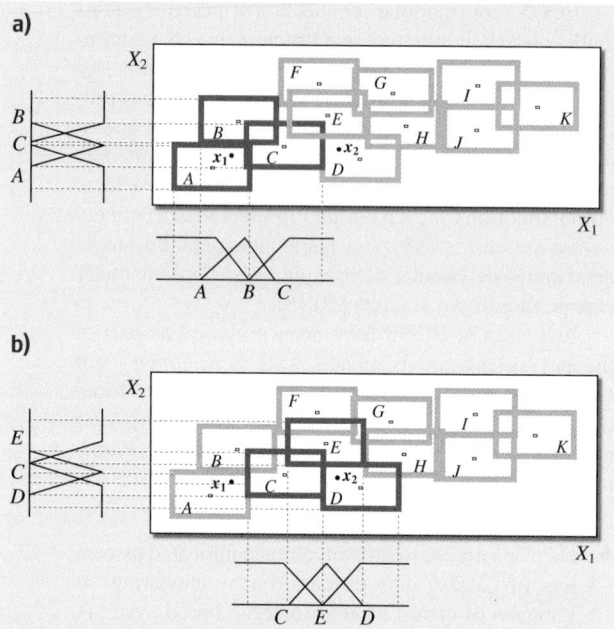

Fig. 40.4a,b Learning in DENFIS uses the evolving clustering method illustrated on a simple example of 2 inputs and 1 output and 11 data clusters evolved. The recall of the DENFIS for two new input vectors x_1 and x_2 is illustrated with the use of the 3 closets clusters to the new input vector (after [40.25]). (**a**) Fuzzy role group 1 for a DENFIS. (**b**) Fuzzy role group 2 for a DENFIS

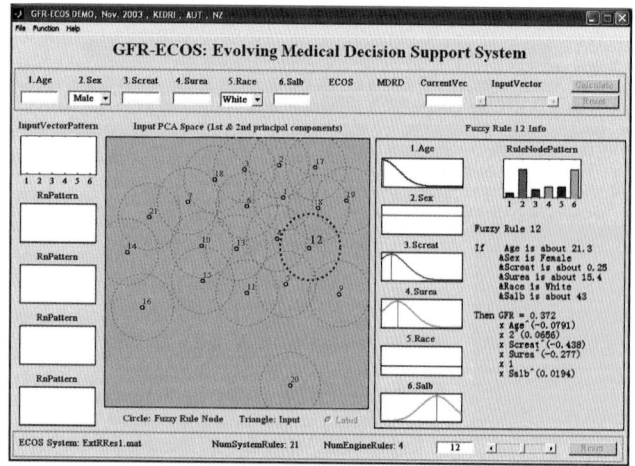

Fig. 40.5 An example of the DENFIS model (after [40.26]) for medical renal function evaluation

DENFIS in Figs. 40.4 and 40.5. In ECOS, clusters of data are created (evolved) based on similarity between data samples (input vectors) either in the input space (this is the case in some of the ECOS models, e.g. DENFIS), or in both the input and output space (this is the case, e.g., in the EFuNN models). Samples that have a distance to an existing node (cluster center, rule node, neuron) less than a certain threshold are allocated to the same cluster. Samples that do not fit into existing clusters, form (generate, evolve) new clusters. Cluster centers are continuously adjusted according to new data samples, others are created incrementally. ECOS learn from data and automatically create or update a *local* (fuzzy) model/function in each cluster, e.g.,

> *IF < data is in a (fuzzy) cluster Ci >*
>
> *THEN < the model is Fi>,*   (40.2)

where *Fi* can be a fuzzy value, a linear or logistic regression function (Fig. 40.5), or an NN model [40.25].

ECOS utilize evolving clustering methods. There is no fixed number of clusters specified a priori, but clusters are created and updated incrementally. Other ECOS that use this principle are: evolving self-organized maps (ESOM) [40.17], evolving classification function [40.18, 26], evolving spiking neural networks (Sect. [40.4]).

As an example, the following are the major steps for the training and recall of a DENFIS model:

Training:

1. Create or update a cluster from incoming data.
2. Create or update a Takagi–Sugeno fuzzy rule for each cluster:
   IF x is in cluster Cj THEN yj = fj (x),
   where: yi = $\beta 0 + \beta 1$ x1 + $\beta 2$ x2 + $\cdots + \beta$q.

The function coefficients are incrementally updated with every new input vector or after a chunk of data. Recall – fuzzy inference for a new input vector:

1. For a new input vector x = [x1, x2, . . . , xq] DENFIS chooses m fuzzy rules from the whole fuzzy rule set for forming a current inference system.
2. The inference result is

$$ y = \frac{\Sigma_{i=1,m} [\omega i \, fi(x1, x2, \ldots, xq)]}{\Sigma_{i=1,m} \omega i} , \quad (40.3)$$

where i is the index of one of the m closets to the new input vector **x** clusters, $\omega i = 1 - di$ is the weighted distance between this vector the cluster center, fi(**x**) is the calculated output for **x** according to the local model fi for cluster i.

0.2.3 From Local to Transductive (Individualized) Learning and Modeling

A special direction of ECOS is transductive reasoning and personalized modeling. Instead of building a set of local models fi (e.g., prototypes) to cover the whole problem space and then using these models to classify/predict any new input vector, in transductive modeling for every new input vector x a new model x is created based on selected nearest neighbor vectors from the available data. Such ECOS models are neuro-fuzzy inference systems (NFI) [40.31] and the transductive weighted neuro-fuzzy inference system (TWNFI) [40.32]. In TWNFI for every new input vector the neighborhood of the closest data vectors is optimized using both the distance between the new vector and the neighboring ones and the weighted importance of the input variables, so that the error of the model is minimized in the neighborhood area [40.33]. TWNFI is a further development of the weighted-weighted nearest neighbor method (WWKNN) proposed in [40.34]. The output for a new input vector is calculated based on the outputs of the k-nearest neighbors, where the weighting is based on both distance and a priori calculated importance for each variable using a ranking method such as signal-to-noise ratio or the t-test.

Other ECOS were been developed as improvements of EFuNN, DENFIS, or other early ECOS models by Ozawa et al. and Watts [40.35–37], including ensembles of ECOS [40.38]. A similar approach to ECOS was used by *Angelov* in the development of the (ETS) models [40.39].

40.2.4 Applications

ECOS have been applied to problems across domain areas. It is demonstrated that local incremental learning or transductive learning are superior when compared to global learning models and when compared in terms of accuracy and new knowledge obtained. A review of ECOS applications can be found in [40.26]. The applications include:

- Medical decision support systems (Fig. 40.5)
- Bioinformatics, e.g., [40.40]
- Neuroinformatics and brain study, e.g., [40.41]
- Evolvable robots, e.g., [40.42]
- Financial and economic decision support systems, e.g., [40.43]
- Environmental and ecological modeling, e.g., [40.44]
- Signal processing, speech, image, and multimodal systems, e.g., [40.45]
- Cybersecurity, e.g., [40.46]
- Multiple time series prediction, e.g., [40.47].

While classical ECOS use a simple McCulloch and Pitts model of a neuron and the Hebbian learning rule [40.48], evolving spiking neural network (eSNN) architectures use a spiking neuron model, applying the same or similar ECOS principles.

40.3 Evolving Spiking Neural Networks (eSNN)

40.3.1 Spiking Neuron Models

A single biological neuron and the associated synapses is a complex information processing machine that involves short-term information processing, long-term information storage, and evolutionary information stored as genes in the nucleus of the neuron. A spiking neuron model assumes input information represented as trains of spikes over time. When sufficient input spikes are accumulated in the membrane of the neuron, the neuron's post-synaptic potential exceeds a threshold and the neuron emits a spike at its axon (Fig. 40.6a,b). Some of the-state-of-the-art models of spiking neurons include: early models by *Hodgkin* and *Huxley* [40.49], and *Hopfield* [40.50]; and more recent models by *Maass*, *Gerstner*, *Kistler*, *Izhikevich*, *Thorpe* and *van Ruller* [40.51–54]. Such models are spike response models (SRMs),

the leaky integrate-and-fire model (LIFM) (Fig. 40.6), Izhikevich models, adaptive LIFM, and probabilistic IFM [40.55].

40.3.2 Evolving Spiking Neural Networks (eSNN)

Based on the ECOS principles, an evolving spiking neural network architecture (eSNN) was proposed in [40.26], which was initially designed as a visual pattern recognition system. The first eSNNs were based on Thorpe's neural model [40.54], in which the importance of early spikes (after the onset of a certain stimulus) is boosted, called rank-order coding and learning. Synaptic plasticity is employed by a fast supervised one-pass learning algorithm. An exemplar eSNN for multimodal auditory-visual information processing on

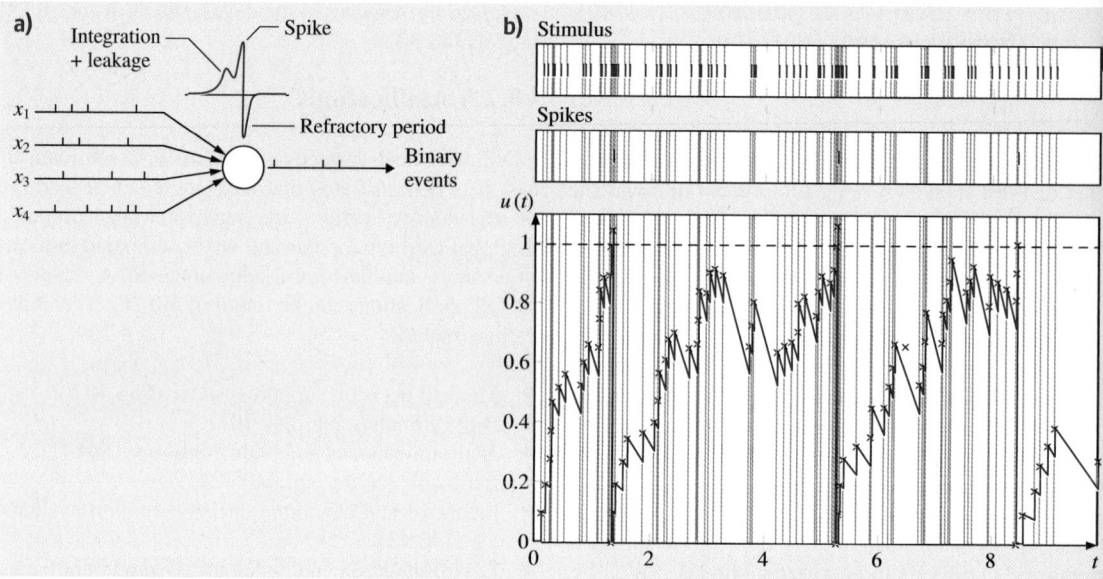

Fig. 40.6 (a) LIFM of a spiking neuron. **(b)** The LIFM increases its membrane potential $u(t)$ with every incoming spike at time t until the potential reaches a threshold, after which the neuron emits an output spike and its potential is reset to an initial value

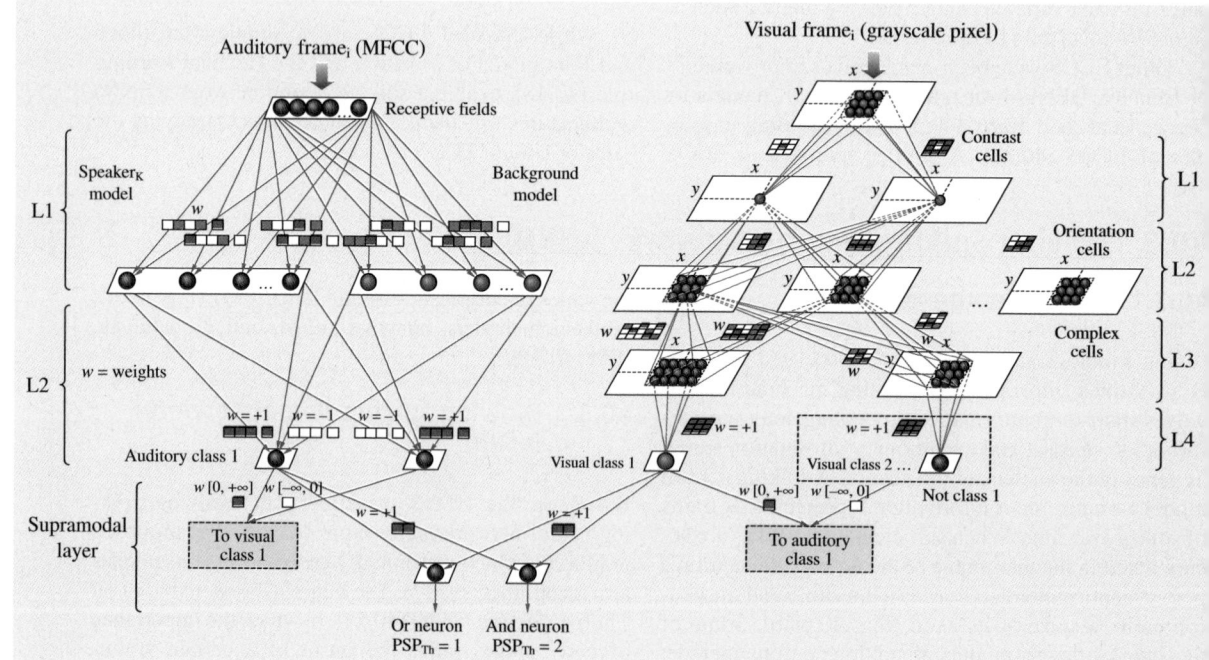

Fig. 40.7 An exemplar eSNN for multimodal auditory-visual information processing in the case study problem of speaker authentication (after [40.56])

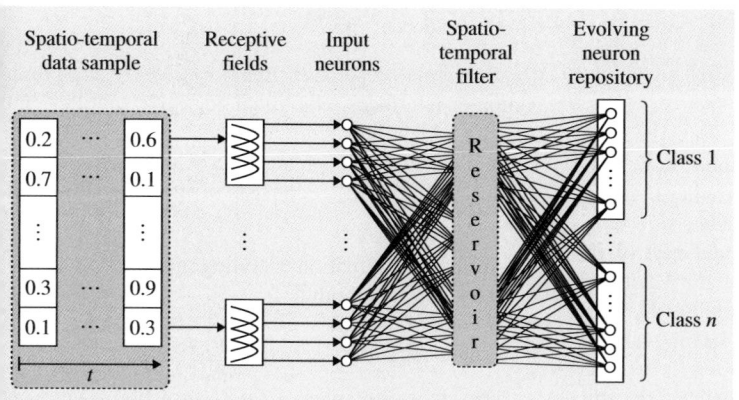

Fig. 40.8 A reservoir-based eSNN for spatio-temporal pattern classification (after [40.55])

The case study problem of speaker authentication is shown in Fig. 40.7.

Different eSNN models use different architectures. Figure 40.8 shows a reservoir-based eSNN for spatio-temporal pattern recognition where the reservoir [40.57] uses the spike-time-dependent plasticity (STDP) learning rule [40.58], and the output classifier that classifies spatio-temporal activities of the reservoir uses rank-order learning rule [40.54].

40.3.3 Extracting Fuzzy Rules from eSNN

Extracting fuzzy rules from an eSNN would make eSNN not only efficient learning models, but also knowledge-based models. A method was proposed in [40.59] and illustrated in Fig. 40.9a,b. Based on the connection weights w between the receptive field layer L1 and the class output neuron layer L2 fuzzy rules are extracted.

40.3.4 eSNN Applications

Different eSNN models and systems have been developed for different applications, such as:

- eSNN for spatio- and spectro-temporal pattern recognition – http://ncs.ethz.ch/projects/evospike
- Dynamic eSNN (deSNN) for moving object recognition – [40.60]
- Spike pattern association neuron(SPAN) for generation of precise time spike sequences as a response to recognized input spiking patterns – [40.61]
- Environmental and ecological modeling – [40.44]
- EEG data modeling – [40.62]
- Neuromorphic SNN hardware – [40.63, 64]
- Neurogenetic models (Sect. 40.4).

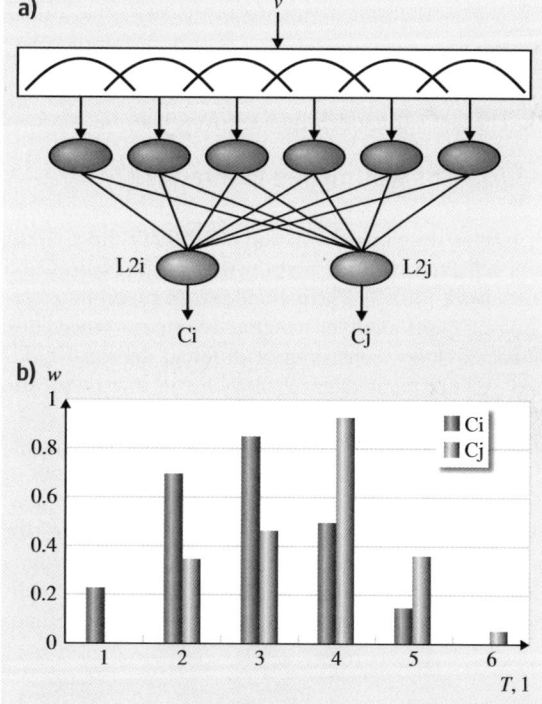

Fig. 40.9 (a) A simple structure of an eSNN for 2-class classification based on one input variable using six receptive fields to convert the input values into spike trains. **(b)** The connection weights of the connections to class Ci and Cj output neurons, respectively, are interpreted as fuzzy rules: IF(input variable v is SMALL) THEN class Ci; IF(v is LARGE)THEN class Cj

A review of eSNN methods, systems and their applications can be found in [40.65].

40.4 Computational Neuro-Genetic Modeling (CNGM)

40.4.1 Principles

A neuro-genetic model of a neuron was proposed in [40.41, 66]. It utilizes information about how some proteins and genes affect the spiking activities of a neuron such as *fast excitation, fast inhibition, slow excitation, and slow inhibition*. An important part of the model is a dynamic gene/protein regulatory network (GRN) model of the dynamic interactions between genes/proteins over time that affect the spiking activity of the neuron – Fig. 40.10.

A CNGM is a dynamical model that has two dynamical sub-models:

- GRN, which models dynamical interaction between genes/proteins over time scale T1
- eSNN, which models dynamical interaction between spiking neurons at a time scale T2.

The two sub-models interact over time.

40.4.2 The NeuroCube Framework

A further development of the eSNN and the CNGM was achieved with the introduction of the NeuroCube framework [40.67]. The main idea is to support the creation of multi-modular integrated systems, where different modules, consisting of different neuronal types and genetic parameters *correspond* in a way to different parts of the brain and different functions (e.g., vision, sensory information processing, sound recognition, motor-control) and the whole system works in an integrated mode for brain signal pattern recognition. A concrete model built with the use of the NeuroCube would have a specific structure and a set of algorithms depending on the problem and the application conditions, e.g., classification of EEG, recognition of functional magneto-resonance imaging (fMRI) data, brain computer interfaces, emotional cognitive robotics, and modeling Alzheimer's disease.

A block diagram of the NeuroCube framework is shown in Fig. 40.11. It consists of the following modules:

- An input information encoding module
- A NeuroCube module
- An output module
- A gene regulatory network (GRN) module.

The main principles of the NeuroCube framework are:

1. NeuroCube is a framework to model brain data (and not a brain model or a brain map).
2. NeuroCube is a selective, approximate map of relevant to the brain data brain regions, along with relevant genetic information, into a 3-D spiking neuronal structure.
3. An initial NeuroCube structure can include known connections between different areas of the brain.
4. There are two types of data used for both training a particular NeuroCube and to recall it on new data: (a) data, measuring the activity of the brain when certain stimuli are presented, e.g., (EEG, fMRI); (b) direct stimuli data, e.g., sound, spoken language video data, tactile data, odor data, etc.
5. A NeuroCube architecture, consisting of a NeuroCube module, (GRN)s at the lowest level, and a higher-level evaluation (classification) module.
6. Different types of neurons and learning rules can be used in different areas of the architecture.
7. Memory of the system is represented as a combination of: (a) short-term memory, represented as changes of the neuronal membranes and temporary changes of synaptic efficacy; (b) long-term memory represented as a stable establishment of synaptic efficacy; (c) genetic memory, represented as a change in the genetic code and the gene/protein expression level as a result of the above short-term and long-term memory changes and evolutionary processes.
8. Parameters in the NeuroCube are defined by genes/proteins that form dynamic GRN models.
9. NeuroCube can potentially capture in its internal representation both spatial and temporal characteristics from multimodal brain data.
10. The structure and the functionality of a NeuroCube architecture evolve in time from incoming data.

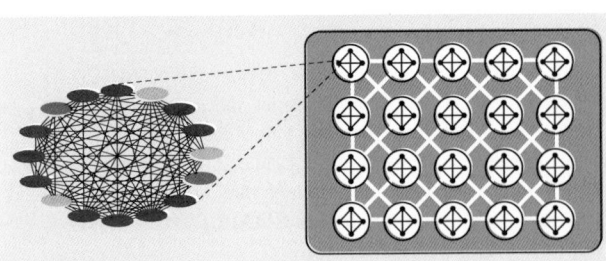

Fig. 40.10 A schematic diagram of a computational neuro-genetic modeling (CNGM) framework consisting of a gene/protein regulatory network (GRN) as part of an eSNN (after [40.41])

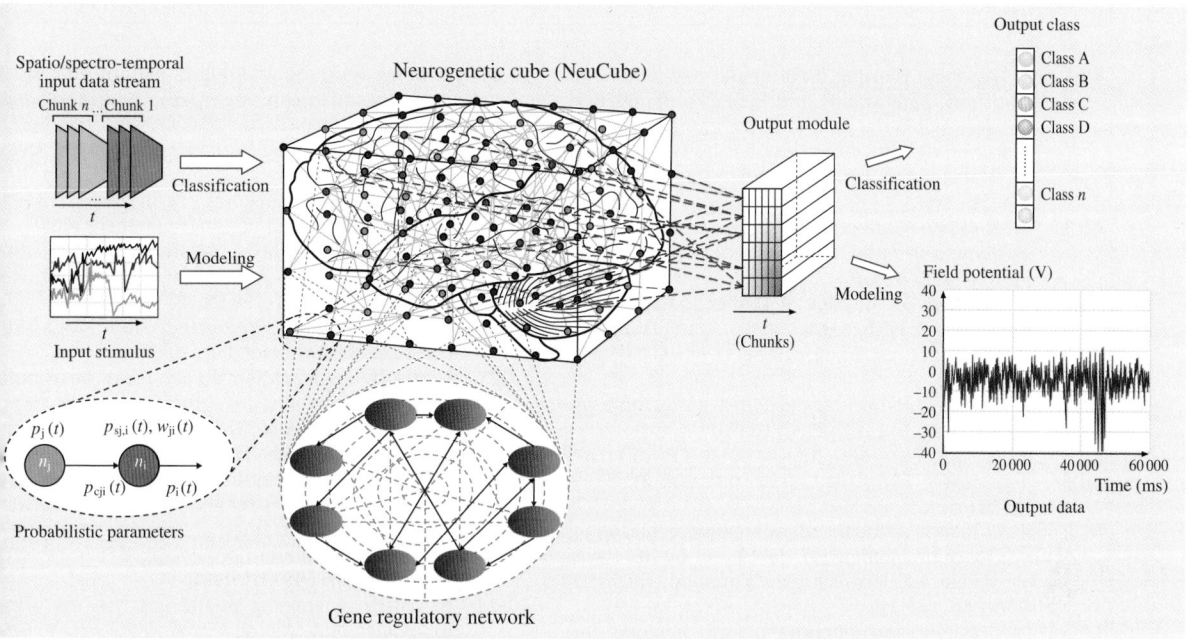

Fig. 40.11 The NeuroCube framework (after [40.67])

40.4.3 Quantum-Inspired Optimization of eSNN and CNGM

CNGM has a large number of parameters that need to be optimized for an efficient performance. Quantum-inspired optimization methods are suitable for this purpose as they can deal with a large number of variables and will converge in much faster time that any other optimization algorithms [40.68]. Quantum-inspired eSNN (QeSNN) use the principle of superposition of states to represent and optimize features (input variables) and parameters of the eSNN including genes in a GRN [40.44]. They are optimized through a quantum-inspired genetic algorithm [40.44]

or a quantum-inspired particle swarm optimization algorithm [40.69]. Features are represented as *qubits* in a superposition of 1 (selected), with a probability α, and 0 (not selected) with a probability β. When the model has to be calculated, the quantum bits *collapse* in 1 or 0.

40.4.4 Applications of CNGM

Various applications of CNGM have been developed such as:

- Modeling brain diseases [40.41, 70]
- EEG and fMRI spatio-temporal pattern recognition [40.67].

40.5 Conclusions and Further Directions

This chapter presented a brief overview of the main principles of a class of neural networks called evolving connectionist systems (ECOS) along with their applications for computational intelligence. ECOS facilitate fast and accurate learning from data and new knowledge discovery across application areas. They

integrate principles from neural networks, fuzzy systems, evolutionary computation, and quantum computing. The future directions and applications of ECOS are foreseen as a further integration of principles from information science-, bio-informatics, and neuro-informatics [40.71].

References

40.1   N. Kasabov: Evolving fuzzy neural networks – Algorithms, applications and biological motivation. In: *Methodologies for the Conception, Design Application of Soft Computing*, ed. by T. Yamakawa, G. Matsumoto (World Scientific, Singapore 1998) pp. 271–274

40.2   N. Kasabov: *Foundations of Neural Networks, Fuzzy Systems and Knowledge Engineering* (MIT, Cambridge 1996) p. 550

40.3   N. Kasabov, S. Shishkov: A connectionist production system with partial match and its use for approximate reasoning, Connect. Sci. **5**(3/4), 275–305 (1993)

40.4   N. Kasabov: Hybrid connectionist production system, J. Syst. Eng. **3**(1), 15–21 (1993)

40.5   L.A. Zadeh: Fuzzy sets, Inf. Control **8**, 338–353 (1965)

40.6   L.A. Zadeh: Fuzzy logic, IEEE Computer **21**, 83–93 (1988)

40.7   L.A. Zadeh: A theory of approximate reasoning. In: *Machine Intelligence*, Vol. 9, ed. by J.E. Hayes, D. Michie, L.J. Mikulich (Ellis Horwood, Chichester 1979) pp. 149–194

40.8   N. Kasabov: Incorporating neural networks into production systems and a practical approach towards realisation of fuzzy expert systems, Comput. Sci. Inf. **21**(2), 26–34 (1991)

40.9   N. Kasabov: Hybrid connectionist fuzzy production systems – Towards building comprehensive AI, Intell. Autom. Soft Comput. **1**(4), 351–360 (1995)

40.10   N. Kasabov: Connectionist fuzzy production systems, Lect. Notes Artif. Intell. **847**, 114–128 (1994)

40.11   N. Kasabov: Hybrid connectionist fuzzy systems for speech recognition and the use of connectionist production systems, Lect. Notes Artif. Intell. **1011**, 19–33 (1995)

40.12   T. Yamakawa, E. Uchino, T. Miki, H. Kusanagi: A neo fuzzy neuron and its application to system identification and prediction of the system behaviour, Proc. 2nd Int. Conf. Fuzzy Log. Neural Netw. (Iizuka, Japan 1992) pp. 477–483

40.13   T. Yamakawa, S. Tomoda: A fuzzy neuron and its application to pattern recognition, Proc. 3rd IFSA Congr., ed. by J. Bezdek (Seattle, Washington 1989) pp. 1–9

40.14   T. Furuhashi, T. Hasegawa, S. Horikawa, Y. Uchikawa: An adaptive fuzzy controller using fuzzy neural networks, Proc. 5th IFSA World Congr. Seoul (1993) pp. 769–772

40.15   N. Kasabov, J.S. Kim, M. Watts, A. Gray: FuNN/2 – A fuzzy neural network architecture for adaptive learning and knowledge acquisition, Inf. Sci. **101**(3/4), 155–175 (1997)

40.16   N. Kasabov: Evolving fuzzy neural networks for supervised/unsupervised online knowledge-based learning, IEEE Trans. Syst. Man Cybern. B **31**(6), 902–918 (2001)

40.17   D. Deng, N. Kasabov: On-line pattern analysis by evolving self-organising maps, Neurocomputing **51**, 87–103 (2003)

40.18   N. Kasabov: *Evolving Connectionist Systems: Methods and Applications in Bioinformatics, Brain Study and Intelligent Machines*, Perpective in Neural Computing (Springer, Berlin, Heidelberg 2003)

40.19   M. Watts: A decade of Kasabov's evolving connectionist systems: A review, IEEE Trans. Syst. Man Cybern. C **39**(3), 253–269 (2009)

40.20   N. Kohonen: *Self-Organizing Maps*, 2nd edn (Springer, Berlin, Heidelberg 1997)

40.21   F. Girosi: Regularization theory, radial basis functions and networks. In: *From Statistics to Neural Networks*, ed. by V. Cherkassky, J.H. Friedman, H. Wechsler (Springer, Heidelberg 1994) pp. 166–18

40.22   G.A. Carpenter, S. Grossberg, N. Markuzon, J.H. Reynolds, D.B. Rosen: Fuzzy ARTMAP: A neural network architecture for incremental supervised learning of analogue multidimensional maps, IEEE Trans. Neural Netw. **3**(5), 698–713 (1991)

40.23   B. Fritzke: A growing neural gas network learns topologies, Adv. Neural Inf. Process. Syst. **7**, 625–632 (1995)

40.24   J. Platt: A resource allocating network for function interpolation, Neural Comput. **3**, 213–225 (1991)

40.25   N. Kasabov, Q. Song: DENFIS: Dynamic, evolving neural-fuzzy inference Systems and its application for time-series prediction, IEEE Trans. Fuzzy Syst. **10**, 144–154 (2002)

40.26   N. Kasabov: *Evolving Connectionist Systems: The Knowledge Engineering Approach* (Springer, Berlin, Heidelberg 2007)

40.27   J. Bezdek: A review of probabilistic, fuzzy, and neural models for pattern recognition, J. Intell. Fuzzy Syst. **1**, 1–25 (1993)

40.28   J. Bezdek (Ed.): *Analysis of Fuzzy Information* (CRC, Boca Raton 1987)

40.29   J. Bezdek: *Pattern Recognition with Fuzzy Objective Function Algorithms* (Plenum, New York 1981)

40.30   R.R. Yager, D. Filev: Generation of fuzzy rules by mountain clustering, J. Intell. Fuzzy Syst. **2**, 209–219 (1994)

40.31   Q. Song, N. Kasabov: NFI: A neuro-fuzzy inference method for transductive reasoning, IEEE Trans. Fuzzy Syst. **13**(6), 799–808 (2005)

40.32   Q. Song, N. Kasabov: TWNFI – A transductive neuro-fuzzy inference system with weighted data normalisation for personalised modelling, Neural Netw. **19**(10), 1591–1596 (2006)

40.33   N. Kasabov, Y. Hu: Integrated optimisation method for personalised modelling and case studies for medical decision support, Int. J. Funct. Inf. Per Med. **3**(3), 236–256 (2010)

40.34   N. Kasabov: Global, local and personalised modelling and profile discovery in bioinformatics: A integrated approach, Pattern Recognit. Lett. **28**(6), 673–685 (2007)

40.35   S. Ozawa, S. Pang, N. Kasabov: On-line feature selection for adaptive evolving connectionist systems, Int. J. Innov. Comput. Inf. Control **2**(1), 181–192 (2006)

40.36   S. Ozawa, S. Pang, N. Kasabov: Incremental learning of feature space and classifier for online pattern recognition, Int. J. Knowl. Intell. Eng. Syst. **10**, 57–65 (2006)

40.37   M. Watts: Evolving Connectionist Systems: Characterisation, Simplification, Formalisation, Explanation and Optimisation, Ph.D. Thesis (University of Otago, Dunedin 2004)

40.38   N.L. Mineu, A.J. da Silva, T.B. Ludermir: Evolving neural networks using differential evolution with neighborhood-based mutation and simple sub-population scheme, Proc. Braz. Symp. Neural Netw. SBRN (2012) pp. 190–195

40.39   P. Angelov: *Evolving Rule-Based Models: A Tool for Design of Flexible Adaptive Systems* (Springer, Berlin, Heidelberg 2002)

40.40   N. Kasabov: Adaptive modelling and discovery in bioinformatics: The evolving connectionist approach, Int. J. Intell. Syst. **23**, 545–555 (2008)

40.41   L. Benuskova, N. Kasabov: *Computational Neuro-Genetic Modelling* (Springer, Berlin, Heidelberg 2007)

40.42   L. Huang, Q. Song, N. Kasabov: Evolving connectionist system based role allocation for robotic soccer, Int. J. Adv. Robot. Syst. **5**(1), 59–62 (2008)

40.43   N. Kasabov: Adaptation and interaction in dynamical systems: Modelling and rule discovery through evolving connectionist systems, Appl. Soft Comput. **6**(3), 307–322 (2006)

40.44   S. Schliebs, M. Defoin-Platel, S.P. Worner, N. Kasabov: Integrated feature and parameter optimization for evolving spiking neural networks: Exploring heterogeneous probabilistic models, Neural Netw. **22**, 623–632 (2009)

40.45   N. Kasabov, E. Postma, J. van den Herik: AVIS: A connectionist-based framework for integrated auditory and visual information processing, Inf. Sci. **123**, 127–148 (2000)

40.46   S. Pang, T. Ban, Y. Kadobayashi, K. Kasabov: LDA merging and splitting with applications to multi-agent cooperative learning and system alteration, IEEE Trans. Syst. Man Cybern. B **42**(2), 552–564 (2012)

40.47   H. Widiputra, R. Pears, N. Kasabov: Multiple time-series prediction through multiple time-series relationships profiling and clustered recurring trends, Lect. Notes Artif. Intell. **6635**, 161–172 (2011)

40.48   D. Hebb: *The Organization of Behavior* (Wiley, New York 1949)

40.49   A.L. Hodgkin, A.F. Huxley: A quantitative description of membrane current and its application to conduction and excitation in nerve, J. Physiol. **117**, 500–544 (1952)

40.50   J. Hopfield: Pattern recognition computation using action potential timing for stimulus representation, Nature **376**, 33–36 (1995)

40.51   W. Maass: Computing with spiking neurons. In: *Pulsed Neural Networks*, ed. by W. Maass, C.M. Bishop (MIT, Cambridge 1998) pp. 55–81

40.52   W. Gerstner: Time structure of the activity of neural network models, Phys. Rev. E **51**, 738–758 (1995)

40.53   E.M. Izhikevich: Which model to use for cortical spiking neurons?, IEEE Trans. Neural Netw. **15**(5), 1063–1070 (2004)

40.54   S. Thorpe, A. Delorme, R. van Ruller: Spike-based strategies for rapid processing, Neural Netw. **14**(6/7), 715–725 (2001)

40.55   N. Kasabov: To spike or not to spike: A probabilistic spiking neuron model, Neural Netw. **23**(1), 16–19 (2010)

40.56   S. Wysoski, L. Benuskova, N. Kasabov: Evolving spiking neural networks for audiovisual information processing, Neural Netw. **23**(7), 819–836 (2010)

40.57   D. Verstraeten, B. Schrauwen, M. d'Haene, D. Stroobandt: An experimental unification of reservoir computing methods, Neural Netw. **20**(3), 391–403 (2007)

40.58   S. Song, K. Miller, L. Abbott: Competitive Hebbian learning through spike-timing-dependent synaptic plasticity, Nat. Neurosci. **3**, 919–926 (2000)

40.59   S. Soltic, N. Kasabov: Knowledge extraction from evolving spiking neural networks with rank order population coding, Int. J. Neural Syst. **20**(6), 437–445 (2010)

40.60   N. Kasabov, K. Dhoble, N. Nuntalid, G. Indiveri: Dynamic evolving spiking neural networks for on-line spatio- and spectro-temporal pattern recognition, Neural Netw. **41**, 188–201 (2013)

40.61   A. Mohemmed, S. Schliebs, S. Matsuda, N. Kasabov: SPAN: Spike pattern association neuron for learning spatio-temporal spike patterns, Int. J. Neural Syst. **22**(4), 1250012 (2012)

40.62   N. Nuntalid, K. Dhoble, N. Kasabov: EEG classification with BSA spike encoding algorithm and evolving probabilistic spiking neural network, Lect. Notes Comput. Sci. **7062**, 451–460 (2011)

40.63   G. Indiveri, B. Linares-Barranco, T.J. Hamilton, A. van Schaik, R. Etienne-Cummings, T. Delbruck, S.-C. Liu, P. Dudek, P. Häfliger, S. Renaud, J. Schemmel, G. Cauwenberghs, J. Arthur, K. Hynna, F. Folowosele, S. Saighi, T. Serrano-Gotarredona, J. Wijekoon, Y. Wang, K. Boahen: Neuromorphic silicon neuron circuits, Front. Neurosci. **5**, 5 (2011)

40.64   G. Indiveri, E. Chicca, R.J. Douglas: Artificial cognitive systems: From VLSI networks of spiking neurons to neuromorphic cognition, Cogn. Comput. **1**(2), 119–127 (2009)

40.65   S. Schliebs, N. Kasabov: Evolving spiking neural networks – a survey, Evol. Syst. **4**(2), 87–98 (2013)

40.66   N. Kasabov, L. Benuskova, S. Wysoski: A computational neurogenetic model of a spiking neuron, Neural Netw. IJCNN'05. Proc. (2005) pp. 446–451

40.67 N. Kasabov: NeuCube EvoSpike architecture for spatio-temporal modelling and pattern recognition of brain signals, Lect. Notes Comput. Sci. **7477**, 225–243 (2012)

40.68 M. Defoin-Platel, S. Schliebs, N. Kasabov: Quantum-inspired evolutionary algorithm: A multimodel EDA, IEEE Trans. Evol. Comput. **13**(6), 1218–1232 (2009)

40.69 H. Nuzly, A. Hamed, S.M. Shamsuddin: Probabilistic evolving spiking neural network optimization using dynamic quantum inspired particle swarm optimization, Aust. J. Intell. Inf. Process. Syst. **11**(1), 5–15 (2010)

40.70 N. Kasabov, R. Schliebs, H. Kojima: Probabilistic computational neurogenetic framework: From modelling cognitive systems to Alzheimer's disease, IEEE Trans. Auton. Ment. Dev. **3**(4), 300–311 (2011)

40.71 N. Kasabov (Ed.): Springer Handbook of Bio/Neuroinformatics (Springer, Berlin, Heidelberg 2014)

41. Machine Learning Applications

Piero P. Bonissone

We describe the process of building computational intelligence (CI) models for machine learning (ML) applications. We use offline metaheuristics to design the models' run-time architectures and online metaheuristics to control/aggregate the object-level models (base models) in these architectures. CI techniques complement more traditional statistical techniques, which are the core of ML for unsupervised and supervised learning. We analyze CI/ML industrial applications in the area of prognostics and health management (PHM) for industrial assets, and describe two PHM case studies. In the first case, we address anomaly detection for aircraft engines; in the second one, we rank locomotives in a fleet according to their expected remaining useful life. Then, we illustrate similar CI-enabled capabilities as they are applied to risk management for commercial and financial assets. In this context, we describe three case studies in insurance underwriting, mortgage collateral valuation, and portfolio optimization. We explain the current trend favoring the use of model ensemble and fusion over individual models, and emphasize the need for injecting diversity during the model generation phase. We present a model−agnostic fusion mechanism, which can be used with commoditized models obtained from crowdsourcing, cloud−based evolution, and other sources. Finally, we explore research trends, and future challenges/opportunities for ML techniques in the emerging context of big data and cloud computing.

Part D | 41

41.1 Motivation

Based on the definition provided by the IEEE Computational Intelligence Society, computational intelligence (CI) covers *biologically and linguistically motivated computational paradigms*. Its scope broadly overlaps with that of soft computing (SC), a similar concept also conceived in the 1990s. The original definition of soft computing was [41.1]:

> An association of computing methodologies that includes as its principal members fuzzy logic (FL), neuro-computing (NC), evolutionary computing (EC) and probabilistic computing (PC).

Thus, in its original scope, CI excluded probabilistic reasoning systems, while including other nature-inspired methodologies, such as swarm computing, ant colony optimization, etc. More recently, however, CI has extended its scope to include statistically inspired machine-learning techniques. Throughout this review, we will adopt this less restrictive definition of CI techniques, increasing its overlapping with SC even more [41.2, 3]. Readers interested in the historical origins of the CI concept should consult [41.4–6].

In addressing real-world problems, we usually deal with physical systems that are difficult to model and possess large solution spaces. In these situations, we leverage two types of resources: domain knowledge of the process or product and field data that characterize the system's behavior. The relevant engineering knowledge tends to be a combination of first principles and empirical knowledge. Usually, it is captured in physics-based models, which tend to be more precise than data-driven models, but more difficult to construct and maintain. The available data are typically a collection of input–output measurements, representing instances of the system's behavior. Usually, data tend to be incomplete and noisy. Therefore, we often augment knowledge-driven models by integrating them with approximate solutions derived from CI methodologies, which are robust to this type of imperfect data. CI is a flexible framework that offers a broad spectrum of design choices to perform such integration.

Domain knowledge can be integrated within CI models in a variety of ways. Arguably, the simplest integration is the use of physics-based models (derived from domain knowledge) to predict expected values of variables of interest. By contrasting the expected values with the actual measured values, we compute the residuals for the same variables and use CI based models to explain the differences. Domain knowledge can also be used to design CI-based models: it can influence the selection of the features (functions of raw data) that are the inputs to the CI models; it can suggest certain topologies for graphical models (e.g., NN architectures) to approximate known functional dependences; it can be represented by linguistics fuzzy terms and relationships to provide coarse approximations; it can be used to define data structures of individuals in the population of an evolutionary algorithm (EA); it can be used explicitly in metaheuristics (MH's) that leverage such knowledge to focus its search in a more efficient way. For a more detailed discussion of the use of domain knowledge in EAs, see [41.7].

Computational intelligence started in the 1990s with three pillars: *Neural networks* (NNs), to create functional approximations from input–outputs training sets; *fuzzy systems*, to represent imprecise knowledge and perform approximate deductions with it; and *Evolutionary systems*, to create efficient global search methods based on optimization through adaptation. Over the last decade, the individual developments of these pillars have become intertwined, leading to successful hybridizations.

41.1.1 Building Computational Intelligence Object- and Meta-Models

Recently, as described in [41.21], this hybridization has been structured as a three-layer approach, in which each layer has a specific purpose:

- Layer 1: *Offline MHs*. They are used in batch mode during the model creation phase, to design, tune and optimize run-time model architectures for deployment. Then they are used to adapt them and maintain them over time. Examples of offline MHs are global search methods, such as EAs, scatter search, tabu search, swarm optimization, etc.
- Layer 2: *Online MHs*. They are part of the run-time model architecture, and they are designed by offline MHs. The *online MHs* are used to integrate/interpolate among multiple (local) object models, manage their complexity, and improve their overall performance. Examples of online MHs are fuzzy supervisory systems, fusion modules, etc.
- Layer 3: *Object-level Models*. They are also part of the run-time architecture, and they are designed by offline MHs to solve object-level prob-

lems. For simpler cases, we use *single object-level models* that provide an individual SC functionality (functional approximation, optimization, or reasoning with imperfect data). For complex cases, we use *multiple object-level models* in parallel configuration (ensemble) or sequential configuration (cascade, loop), to integrate functional approximation with optimization and reasoning with imperfect data (imprecise and uncertain).

The underlying idea is to reduce or eliminate manual intervention in any of these layers, while leveraging CI capabilities at every level. We can manage complexity by finding the best model architecture to support problem decomposition, create high-performance local models with limited competence regions, allow for smooth interpolations among them, and promote robustness to imperfect data by aggregating diverse

models. Let us examine some case studies that further illustrate this concept.

Examples of Offline MH, Online MH, and Object Models

In Table 41.1, we observe a variety of CI applications in which we followed the separation between object- and meta-level described earlier. In most of these applications, the object-level models were based on different technologies such as machine learning (support vector machines, random forest), statistics (multivariate adaptive regression splines, MARS), Hotelling's T^2), neural networks (feedforward, self-organizing maps), fuzzy systems, EAs, Case based. The online metaheuristics were mostly based on fuzzy aggregation (interpolation) of complementary local models or fusion of competing models. The offline MHs were mostly implemented by evolutionary search in the model design space. Descriptions of these applications can be found in the

Table 41.1 Examples of CI applications at meta-level and object-level

Case study	Problem instance	Problem type	Model design (offline MHs)	Model controller (online MHs)	Object-level models	References
	Anomaly detection (system)	Classification	Model T-norm tuning	Fuzzy aggregation	*Multiple Models*: SVM, NN, Case-Based, MARS	[41.8]
	Anomaly detection (system)	Classification	Manual design	Fusion	*Multiple Models*: Kolmogorov complexity, SOM. random forest, Hotteling T2, AANN	[41.9]
#1	Anomaly detection (model)	Classification and regression	EA-base tuning of fuzzy supervisory termset	Fuzzy supervisory	*Multiple Models*: Ensemble of AANN's	[41.10]
#2	Best units selection	Ranking	EA-base tuning of similarity function	None	*Single Model*: Fuzzy instance based models (Lazy Learning)	[41.11, 12]
#3	Insurance underwriting: Risk management	Classification	EA	Fusion	*Multiple Models*: NN, Fuzzy, MARS	[41.13, 14]
#4	Mortgage collateral valuation	Regression	Manual design	Fusion	*Multiple Models*: ANFIS, Fuzzy CBR, RBF	[41.15]
#5	Portfolio rebalancing	Multiobjective optimization	Seq. LP	None	*Single Model*: MOEA (SPEA)	[41.16]
	Load, HR, NO_x forecast	Regression	Multiple CART trees	Fusion	*Multiple Models*: Ensemble of NN's	[41.17]
	Aircraft engine fault recovery	Control/Fault accommodation	EA tuning of linear control gains	Crisp supervisory	*Multiple Models (Loop)*: SVM + linear control	[41.18]
	Power plant optimization	Optimization	Manual design	Fusion	*Multiple Models (Loop)*: MOEA + NN's	[41.19]
	Flexible manufacturing optimization	Optimization	Manual design	Fuzzy supervisory	*Single Model*: Genetic Algorithms	[41.20]

references listed in the last column of Table 41.1. The five case studies covered in this review are indicated in the first column of Table 41.1.

41.1.2 Model Lifecycle

In real-world applications, before using a model in a production environment we must address the model's complete life cycle, from its design and implementation to its validation, tuning, production testing, use, monitoring, and maintenance. By maintenance, we refer to all the steps required to keep the model vital (e.g., nonobsolete) and to adapt it to changes in the environment in which it is deployed. Many reasons justify this focus on model maintenance. Over the model's life cycle, maintenance costs are the by far most expensive ones (as software maintenance costs are the most expensive ones in the life of a software system). Furthermore, when dealing with mission-critical software we need to guarantee continuous operation or at least fast recovery from system failures or model obsolescence to avoid lost revenues and other business costs.

The use of MHs in the design stage allows us to create a process for automating the model building phase and subsequent model updates. This is a critical step to quickly deploy and maintain CI models in the field, and it will be further described in the case studies. Additional information on this topic can be found in [41.22].

41.2 Machine Learning (ML) Functions

Machine learning techniques can be roughly subdivided into *supervised*, *semisupervised reinforcement*, and *unsupervised learning*. The distinction among these categories depends on the complete, partial, or lack of available ground truth (i. e., correct outputs for each input vector) during the training phase.

Unsupervised learning techniques are used when no ground truth is available. Their goal is to identify structures in the input space that could be used to decompose the problem and facilitate local model building. Typical examples of unsupervised learning are cluster analysis, self-organizing maps (SOMs) [41.23], and dimension reduction techniques, such as principal components analysis (PCA), independent components analysis (ICA), multidimensional scaling (MDS), etc.

Reinforcement learning (RL) does not rely on ground truth. It assumes that an agent operates in an environment and after performing one or more actions it receives a reward that is a consequence of its actions, rather than an explicit expression of ground truth. *Sutton* and *Barto* [41.24] were among the first proponents of this technique, which is quite promising to model adversarial situations, but it has not generated many industrial or commercial applications. A succinct description of RL can be found in [41.25].

Semisupervised and *supervised learning* techniques are used when partial or complete ground truth is available, such as labels for *classification* problems and real-values for *regression* problems. There are many traditional linear models for classification and regression. For instance, we have linear discriminant analysis (LDA) and logistic regression (LR) for classification, and least-squares techniques – combined with feature subset selection or shrinkage methods (e.g., Ridge, least absolute shrinkage and selection operator (LASSO)) – for regressions. CI techniques usually generate nonlinear solutions to these problems. We can group the most of the commonly used nonlinear techniques, as

- *Directed graphical-based models*, such as neural networks, neural fuzzy systems, Bayesian belief networks, Bayesian neural networks, etc.
- *Tree based*: Classification analysis and regression trees (CARTs) [41.26], ID3/C4.5 [41.27], etc.
- *Grammar based*: Genetic programming, evolutionary programming, etc.
- *Similarity and metric learning:* Lazy learning (or instance-based learning) [41.28, 29], case-based reasoning, (fuzzy) *k*-means, etc.
- *Undirected graphical models:* Markov graphs, restricted Boltzmann machines [41.30], etc.

The reader can find a comprehensive treatment of these techniques in [41.31]. Some of these models are used as part of on ensemble, rather than individually. Such is the case of random forest [41.32], which is based on a collection of CART trees [41.26]. Similarly, a fuzzy extension of random forest using fuzzy decision trees [41.33] can be found in [41.34]. This trend toward the use of ensembles is covered in Sect. 41.5. We will now focus on a subset of the CI/ML applications. Specifically, we will analyze two case studies in industrial applications (Sect. 41.3) and three in financial applications (Sect. 41.4).

41.3 CI/ML Applications in Industrial Domains: Prognostics and Health Management (PHM)

To provide a coherent theme for the ML industrial application, we will focus on prognostics and health management (PHM). The main goal of PHM for assets such as locomotives, medical scanners, and aircraft engines is to maintain these assets' operational performance over time, improving their utilization while minimizing their maintenance cost. This tradeoff is critical for the proper execution of contractual service agreements (CSAs) offered by original equipment manufacturer's (OEM) to their valued customers.

PHM is a multidiscipline field, as it includes aspects of electrical engineering (reliability, design, service), computer and decision sciences (artificial intelligence, CI, MI, statistics, operations research (OR)), mechanical engineering (geometric models for fault propaga-

tion), material sciences, etc. Within this paper, we will focus on the role that CI plays in PHM functionalities.

PHM can be divided into two main components:

- *Health assessment*: the evaluation and interpretation of the asset's current and future health state.
- *Health management*: the control, operation, and logistic plans to be implemented in response to such assessment.

PHM functional architecture is illustrated in Fig. 41.1, adapted from [41.35].

The first two tasks:

(1) Remote monitoring, and
(2) Input data preprocessing, are platform dependent, as they need domain knowledge to identify and

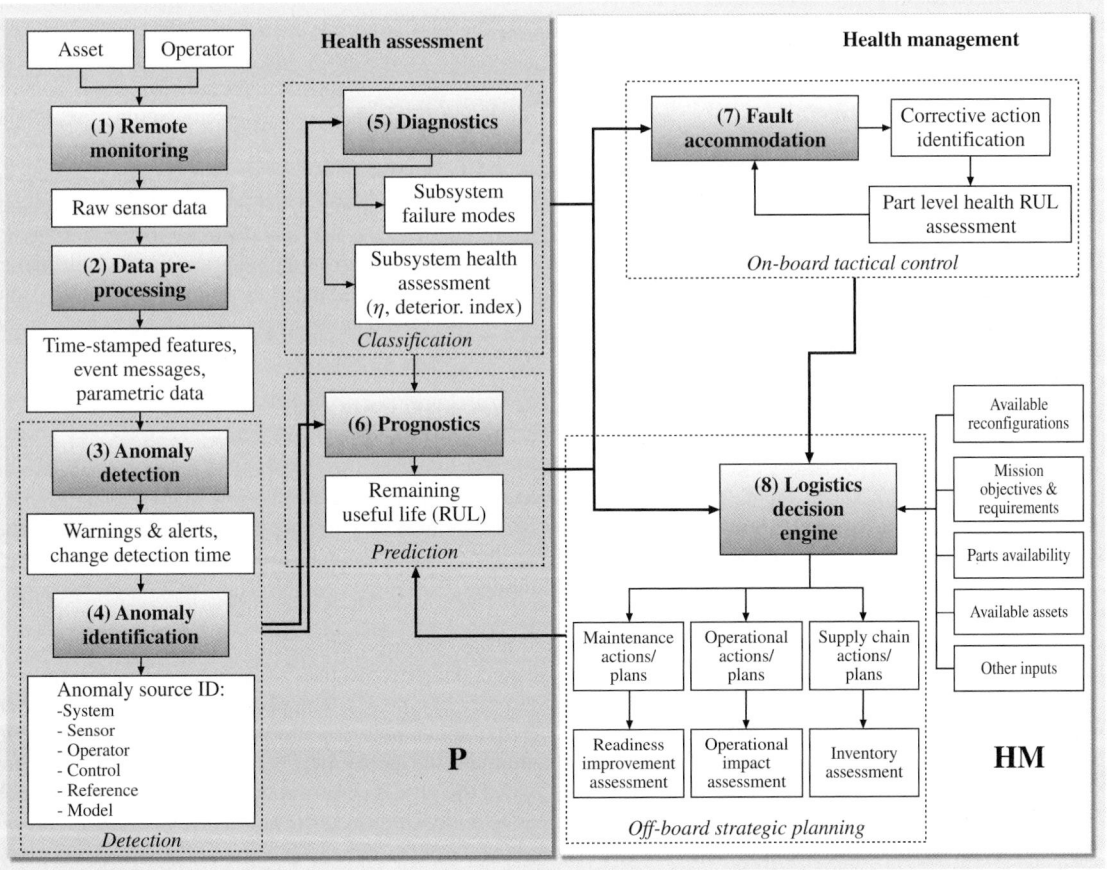

Fig. 41.1 PHM functional architecture

select the most informative inputs, perform data curation (de-noising, imputation, and normalization), aggregate them, and prepare them to be suitable inputs for the models.

The remaining decisional tasks could be considered platform independent (to the extent that their functions could be accomplished by data-driven models alone.) These tasks are:

(3) Anomaly detection and identification
(4) Anomaly resolution
(5) Diagnostics
(6) Prognostics
(7) Fault accommodation
(8) Logistics decisions.

Health assessment, which is based on *descriptive* and *predictive* analytics, is contained in the left block of Fig. 41.1 (annotated with P). Health management (HM), which is based on *prescriptive* analytics, is contained in the right block of Fig. 41.1 (annotated with HM). In the remaining of this section, we will cover two case studies related to anomaly identification and prognostics.

41.3.1 Health Assessment and Anomaly Detection: An Unsupervised Learning Problem

Anomaly Detection (AD)

Using platform-deployed sensors, we collect data remotely. We preprocess it, via segmentation, filtering, imputation, validation, and we summarize it by extracting feature subsets that provide a more succinct, robust representation of its information content. These features, which could contain a combination of categorical and numerical values, are analyzed by an anomaly detection model to assess the degree of abnormal behavior of each asset in the fleet. If the degree of abnormality exceeds a given threshold, the model will identify the asset, determine the time when the anomaly was first noticed and suggest possible causes of the anomaly (usually a coarse identification at the systems/subsystem level). Anomaly detection usually leverages unsupervised learning techniques, such as clustering. Its goal is to extract the underlying structural information from the data, define normal structures and regions, and identify departures from such regions.

Anomaly Identification (AI)

After detecting an abnormal change, e.g., a departure from a normal region of the state space, we need to identify its cause. There are many factors that could cause such change:

(a) A *system fault*, which could eventually lead to a failure.
(b) A *sensor fault*, which is creating incorrect measurements.
(c) An *inadequate* anomaly detection *model*, which is falsely reporting anomalies due to poor design, inadequate model update, execution outside its region of competence, etc.
(d) A sudden, unexpected *operational transient*, which is stressing the system by creating an abrupt load change. This transient could be originated by an operator error, who is requesting such sudden change; by an incorrect reference vector (in case of operation automation), which is also requesting such abrupt change; or by a poorly designed controller, which is either over- or under-compensating for a perceived state change.

The first factor (system fault) represents a correct anomaly classification and should trigger the rest of the workflow (diagnostics, prognostics, fault accommodation, and maintenance optimization), while the other three factors generate false alarms (false positives.) In the next case study, we will focus on how to improve the accuracy of an anomaly detection model, (third factor in the list) and decrease the probability of causing false positives. This increase in model fidelity will also create a sharper distinction between system faults and sensor faults (first and second factors in the list).

Anomaly Detection for Aircraft Engines

Problem Definition. As noted in Sect. 41.1, one of the best way to leverage domain knowledge is to create expected values using highly tuned physics based simulators, compare them with actual values and analyze the differences (residuals) using data-driven models.

Physics–Based Simulator. In this case study, we focused on the detection of anomalies in a simulated aircraft engine. A component level model (CLM), a thermodynamic model that has been widely used to simulate the performance of an aircraft engine, provided the physics-based model. Flight conditions, such as *altitude, Mach number, ambient temperature*, and *engine fan speed*, and a large variety of model parameters, such as module efficiency and flow capacity are inputs the CLM. The outputs of the CLM are the values for pressures, core speed and temperatures at various locations of engine, which simulate sensor measurements. Realistic values of sensor noise can be added after the CLM calculation. In this study, we used a steady state CLM model for a commercial, high-bypass, twin spool, turbofan engine.

Actual Values. We used engine data collected under cruise conditions to monitor engine health changes.

Data-Driven Model. We realized that a single, global model – regardless of the technology used to implement it – would be inadequate for large operating spaces of the simulated engine. Global models are designed to achieve a compromise among completeness (for coverage), high fidelity (for accuracy), and transparency (for maintainability). As a result, we usually end up with models that in order to maintain small biases exhibit large variability. This variability might be too large to distinguish between model error and anomalous system behavior and can be a significant factor in the generation of false alarms.

CI-Based Approach. To solve the model fidelity problem, we decomposed the engine's operating space into several, partially overlapping regions and developed a set of local models, trained on each region. This schema required a supervisory model (or meta model) to determine the competence region of each local model and select the appropriate one. In control problems, this supervisory module typically selects one controller (out of a collection of low-level controllers) to close the loop with the dynamic system. In many fuzzy controllers application [41.36, 37], a fuzzy supervisory module determines the applicability degree of the low-level controller and interpolates their outputs. Usually, this is done with a weighted, convex sum of the controllers' outputs. The weights used in the convex sum are the applicability degrees of the low-level controllers in the part of the state space that contains the input. The transition from *mode selection* to *mode melting* [41.38] generates a smoother response surface by avoiding discontinuities.

We applied the same concept to the problem of improving the fidelity of data-driven models for anomaly detection. First, we decided to use auto-associative NNs (AANN) to implement the local models. Then, we developed a fuzzy supervisory controller, defining the applicability of each AANN as a fuzzy region in the engine's three-dimensional operating space, defined by *altitude, Mach number*, and *Ambient temperature*. Finally, we used and evolutionary algorithm to tune the term set of the fuzzy supervisory and find the best fuzzy boundaries to interpolate between AANNs with overlapping applicability.

Local Models. Auto-Associative Neural Networks (AANN's) are feedforward neural networks with structure satisfying requirements for performing restricted auto-association. The inputs to the AANN go through a dimensionality reduction, as their information is combined and compressed in intermediate layers. For example, in Fig. 41.2 the nine nodes in the input layer are reduced to five and then three, in the second layer (encoding) and third layer (bottleneck), respectively. Then, the nodes in the 3rd layer are used to recreate the original inputs, by going through a dimensionality expansion (fourth layer, decoding, and fifth layer, outputs). In the ideal case, the AANN outputs should be identical to the inputs. Their difference (residuals) and their gradient information are used to train the AANN to minimize such difference.

This network computes the largest nonlinear principal components (NLPCA's) – the nodes in the intermediate layer – to identify and remove correlations among variables. Besides the generation of residuals this type of network can also be used in dimensionality reduction, visualization, and exploratory data analysis. As noted in [41.39]:

While (principal component analysis) PCA identifies only linear correlations between variables,

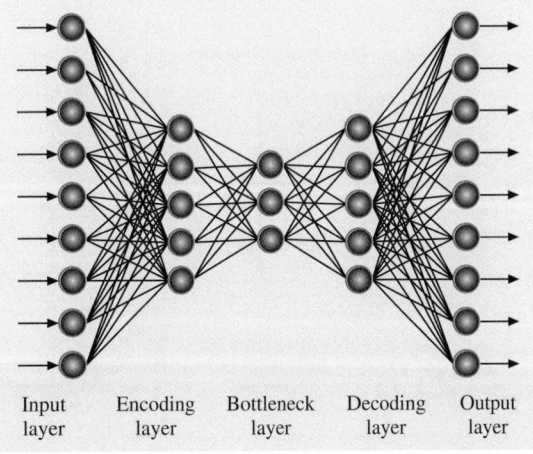

| Input layer | Encoding layer | Bottleneck layer | Decoding layer | Output layer |

Fig. 41.2 Architecture of a 9-5-3-5-9 auto associative neural network

NLPCA uncover both linear and nonlinear correlations, without restriction on the character of the nonlinearities present in the data.

NLPCA operates by training a feedforward neural network to perform the identity mapping, where the network inputs are reproduced at the output layer. The network contains an internal *bottleneck* layer (containing fewer nodes than input or output layers), which forces the network to develop a compact representation of the input data, and two additional hidden layers. Additional information about AANNs can be found in [41.39–41].

The complete CI approach is illustrated in Fig. 41.3, adapted from [41.10]. The left part of the figure shows the run-time anomaly detection (AD) model. The center part of Fig. 41.3 shows an instance of the term set used by the fuzzy supervisory system (the scale of the operational state variables was normalized as a percentage of the range of values to preserve proprietary information). In the right part of Fig. 41.3, we can see the evolutionary algorithm (EA) in a wrapper configuration, used to tune the membership functions (term sets). Each individual in the EA population is a set of parameters that represents an implementable term set configuration. For *Altitude*, we varied two pa-

rameters. For each of *ambient temperature* and *Mac number*, we varied four parameters, with a total of te search parameters. Each individual was a set of te parameters that created a corresponding set of men bership functions that controlled residuals behavic of the fuzzy supervisory model. The fitness of eac individual was computed based on the aggregate c the nine sensor residuals, with a goal toward max imizing fitness or minimizing overall residuals. Th EA used was based on the genetic algorithm opti mization toolbox (GAOT) toolkit. The population siz was set at 500, and the generation count was set a 1000. The EA execution was very efficient taking onl about 2 h of execution time on a standard deskto machine.

Results. As a result of this experiment, we wer able to drastically reduce the residuals generated un der steady state, no-fault assumption, and we improve the fidelity of the local model ensemble by more tha a factor of four with respect to a reference globa data-driven model. This fidelity allowed us to cre ate a sharper baseline used to identify true engin anomalies, distinguishing them from sensor anoma lies. For a more detailed description of these resul see [41.10].

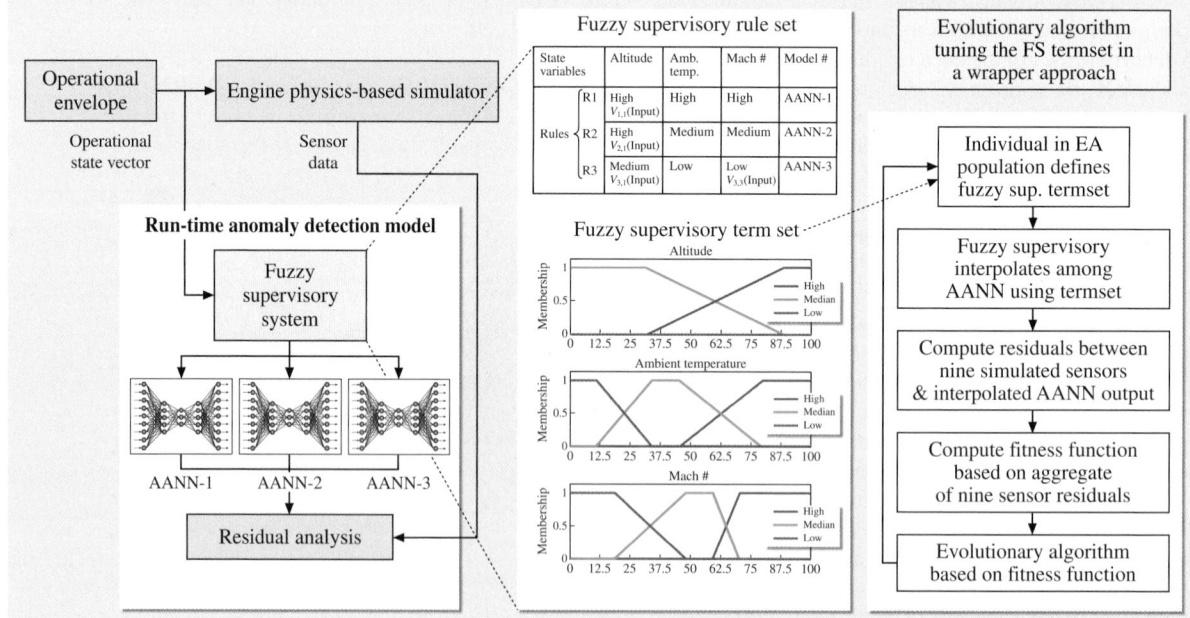

Fig. 41.3 Evolutionary algorithms tune the term sets of the fuzzy supervisory system to interpolate the outputs of an ensemble of local auto associative NNs

41.3.2 Health Assessment – Diagnostics: A Semisupervised and Supervised Learning Problem

The information generated by the anomaly identification model allows a diagnostic module to focus on a given unit's subsystem, analyzing key variables associated with the subsystem, and trying to match their patterns with a library of signatures associated with faults or incipient failure modes. The result is a ranked list of possible faults. Therefore diagnostics is a classification problem, mapping a feature space into a labeled fault space.

Usually, data-driven diagnostics leverages supervised learning techniques to extract potential signatures from the historical data and use them to recognize different failure modes automatically. A large variety of statistical and AI-based techniques can be used for automatic fault diagnostics, including neural networks, decision tree, random forest, Bayesian belief network, case-based reasoning, hidden Markov model, support vector machine, fuzzy logic etc. Those data-driven diagnostics methods are able to learn the faulty signatures or patterns from the training data and associate them with different failure modes when new data arrives.

Data-driven approaches have many benefits. First, they can be designed to be independent of domain knowledge related to a particular system. We could use this approach with data recorded for almost any component/system, as long as the recorded data is relevant to the health condition of the interested component. This reduces the effort involved with eliciting and incorporating domain specific knowledge. A second benefit is the use of fusion techniques to take advantage of diverse information from multiple data sources/models to boost diagnostics performance. The third benefit is the robustness to noise exhibited by most of data-driven methods, such as fuzzy logic and neural networks. However, all data-driven techniques require the availability of labeled historical data so these data collection step must precede the application of these methods.

Domain knowledge, when available, can still be leveraged to initialize the structures of the data-driven models (feature selection, network topology, etc.) and provide better initial conditions for optimization and tuning techniques applied to the data-driven diagnostics models.

Supervised learning is a very mature topic in ML. As a result, there are many diagnostics applications of CI techniques to medical [41.42–45] industrial [41.46, 47] automotive [41.48], and other domains. Given its widespread use, we will not provide additional case studies for diagnostics.

41.3.3 Health Assessment – Prognostics: A Regression Problem

Prognostics is the prediction of *remaining useful life* (RUL), when the anomaly detection and diagnostics modules can identify and isolate an incipient failure through its preceding faults. This incipient failure changes the graph of RUL versus time from a linear, normal-wear trajectory to an exponentially decaying one. The fault time and incipient failure mode determine the inflection point in such curve and the deterioration steepness, respectively. These estimates are usually in units of time or utilization cycles, and have an associated uncertainty, e.g., a probability density curve around the actual estimate. Typically, this uncertainty (e.g., RUL confidence interval) increases as the prediction horizon is extended. Operators can choose a confidence level that allows them to incorporate a risk level into their decision making. They can change operational characteristics, such as load, which may prolong the life of components at risk. They can also account for upcoming maintenance and set in motion a logistics process to support a smooth transition from faulted equipment to fully functioning.

Predicting RUL is not trivial, because RUL depends on current deterioration state and future usage, such as unit load and speed, among others. Prognostics is closely linked with diagnostics. In the absence of any evidence of damage or faulted condition, prognostics reverts to statistical estimation of fleet-wide life. It is common to employ prognostics in the presence of an indication of abnormal wear, faults, or other abnormal situation. Therefore, it is critical to provide accurate and quick diagnostics to allow prognostics to operate. At the heart of prognostics is the ability to properly model the accumulation and propagation of damage. A common approach to prognostics is to employ a model of damage propagation contingent on future use. Such models are often times based on detailed materials knowledge and makes use of finite element modeling. This requires an in depth understanding of the local conditions the particular component is exposed to.

For example, for spall propagation in bearings, we need to know the local load, speed, and temperature conditions at the site of the damage, e.g., at the outer race (or ball or cage). In addition, we need to know the geometry and local material properties at the suspected damage site. This information is used to derive the stresses that components are expected to experience, typically using a finite element approach. The potential benefit of this process is the promise of accurate prediction of when the bearing will fail. For a different fault mode, the process needs to be repeated. Because of the cost and effort involved, this method is reserved for a set of components that, if left undetected and without remaining life information, might experience catastrophic failure that transcends the entire system and causes system failure. However, there is a large set of components that will not benefit from this approach, either because a physics-based damage model is not achievable or is too costly to develop. Therefore, it is desirable to increase coverage of prognostics for a range of fault modes. To this end, the techniques would ideally utilize existing models and sensor data.

ML provides us with an alternative approach, which is based on analyzing time series data where the equipment behavior has been monitored via sensor measurements during the normal operation until equipment failure. When a reasonable set of these observations exists, ML algorithms can be employed to recognize these trends and predict remaining life (albeit, often times under the assumption of near-constant future load conditions.) Usually, specific faults have preferred directions in the health related feature space. By extrapolating the propagation in this parameter space and by mapping the extrapolation into the time domain, we can derive RUL information.

A prerequisite to leverage RUL estimation is to have a narrow confidence interval, so that this information is *actionable* and can be used in the asset health management part of PHM as a time horizon to optimize the logistics/maintenance scheduling plan. In most cases, however, we do not have run-to-failure data in the time series. Usually, when a failure is identified it is corrected promptly, causing the time series to be statistically censored on the right. The lack of run-to-failure data further compounds the technical difficulty of predicting RUL with a small variance.

We consider two options to address this problem. The first option is to use of an ensemble of *diverse* predictive models (Sect. 41.5.3 for a definition of diversity) such that the fusion of the ensemble will reduce the variance and make the output more actionable – Sect. 41.5 is devoted to this topic. The second option is to relax the problem formulation, by increasing the granularity of the models output. This granularity is determined by the actions that we will perform with such information. For example, we could formulate prognostics as:

(1) *A partial ordering* over RUL. This formulation could be used to estimate the risk of claims in term life policies. Insurance underwriters estimate the applicants' expected mortality at a coarse level by classifying each applicant into a given rate-class from a set of sorted classes that define decreasing RUL. Applicants inside each class are indistinguishable in terms of risk and are charged the same premium (gender and age being equal). This will be further described in the case study of Sect. 41.4.1. In a PHM context, this formulation could be used to price the contractual service agreement renewals for different units within a fleet, in a fashion similar to the risk-based pricing of insurance underwriting.

(2) *An ordinal ordering* over RUL (ranking). This formulation could be used to select the most reliable units of a fleet for mission-critical assignments. This will be further described in the case study of Sect. 41.3.3, where we illustrate how a train dispatcher could select the best locomotives to create a *hot train*, e.g., a freight train with a guaranteed arrival time.

(3) *A cardinal ordering* over RUL (rating). This formulation could be used to understand the relative level of readiness of units in a fleet, to prioritize the need for instruments calibration, power management assessment/verification, etc.

(4) *A binary classification* of whether a given event (causing the end of RUL) will happen within a given time window. This formulation could be used to generate a time-dependent risk assessment to optimize fleet scheduling and unit allocation.

(5) *A regression* on RUL, including the confidence interval of the prediction. This formulation provides the finest granularity. If we were able to reduce the confidence intervals of these predictions, we could refine and optimize the condition-based maintenance of the assets in the fleet.

The following case study will illustrate the second problem reformulation (ranking). In this case, genetic algorithms are used to evolve fuzzy instance-based models that will generate a ranking of the most reliable locomotives within a fleet.

Case Study 2: RUL–Driven Ranking of Locomotives in a Fleet

Problem Definition. The problem of selecting the best units from a fleet of equipment occurs in many military and commercial applications. Given a specific mission profile, a commander may have to decide which five armored vehicles to deploy in order to minimize the chance of a breakdown. In the commercial world, rail operators need to make decisions on which locomotives to use in a train traveling from coast to coast with time sensitive shipments.

The behavior of these complex electromechanical assets varies considerably across different phases of their life cycle. Assets that are identical at the time of manufacture will evolve into somewhat individual systems with unique characteristics based on their usage and maintenance history. Utilizing these assets efficiently requires a) being able to create a model characterizing their expected performance, and b) keeping this model updated as the behavior of the underlying asset changes.

In this problem formulation, RUL prediction for each individual unit is computed by aggregating its own track record with that of a number of *peer* units – units with similarities along three key dimensions: *system design, patterns of utilization, and maintenance history*. The notion of a *peer* is close to that of a neighbor in CBR, except the states of the peers are constantly changing. Odometer-type variables like mileage and age increase, and discrete events like major maintenance or upgrades occur. Thus, it is reasonable to assume that after every significant mission, the peers of a target unit may change based upon changes in both the unit itself, and the fleet at large. Our results suggest that estimating unit performance from peers is a practical, robust, and promising approach. We conducted two experiments – one for *retrospective* estimation and one for *prospective* estimation. In the first experiment, we explored how well the median RUL of any unit could be estimated from the medians of its peers. In the second experiment, for a given instant in time, we predicted the time to the next failure for each unit using the history of the peers. In these experiments, the retrospective (or prospective) RUL estimates were used to induce a ranking over the units. The selection of the best N units was based on this ranking. The precision of the selection was the percentage of the correctly selected units among the N units (based on ground truth).

CI-Based Approach. Our approach was based on fuzzy instance-based model (FIM), which can be found in [41.11]. We addressed the definition of similarity among peers by evolving the design of a similarity function in conjunction with the design of the attribute space in which the similarity was evaluated. Specifically, we used the following four steps:

(1) *Retrieval* of similar instances from the database (DB).
(2) *Evaluation of similarity measures* between the probe and the retrieved instances.
(3) *Creation of local models* based on the most similar instances.
(4) *Aggregation of local models outputs* (weighted by their similarity measures).

(1) Retrieval. We looked for all units in the fleet DB whose behavior was similar to the probe. These instances are the potential peers of the probe. The peers and probe can be seen as points in an n-dimensional feature space. For instance, let us assume that a probe Q is characterized by an n-dimensional vector of feature $\bar{X}_Q$, and $O(Q) = [D_{1,Q}, D_{2,Q}, \ldots, D_{k(Q),Q}]$ the history of its operational availability durations

$$
\begin{aligned}
Q &= [\bar{X}_Q; O(Q)] \\
&= [x_{1,Q}, \ldots, x_{n,Q}; D_{1,Q}, \ldots, D_{k(Q),Q}] . \quad (41.1)
\end{aligned}
$$

Any other unit u_i in the fleet has a similar characterization

$$
\begin{aligned}
u_j &= [\bar{X}_j; O(u_j)] \\
&= [x_{1,j}, x_{2,j}, \ldots, x_{n,j}; D_{1,j}, D_{2,j}, \ldots, D_{k(j),j}] . \\
&\quad (41.2)
\end{aligned}
$$

For each dimension i we defined a *truncated generalized Bell function*, $\text{TGBF}_i(x_i; a_i, b_i, c_i)$, centered at the value of the probe c_i, which represents the degree of similarity along that dimension. Specifically

$$
\begin{aligned}
&\text{TGBF}_i(x_i; a_i, b_i, c_i) \\
&= \begin{cases} \left[1 + \left| \dfrac{x_i - c_i}{a_i} \right|^{2b_i} \right]^{-1} \\ \quad \text{if } \left[1 + \left| \dfrac{x_i - c_i}{a_i} \right|^{2b_i} \right]^{-1} > \varepsilon \\ 0 \quad \text{otherwise} \end{cases} ,
\end{aligned}
$$

$$(41.3)$$

where e is the truncation parameter, e.g., $e = 10^{-5}$.

Since the parameters c_i in each TGBF_i are determined by the values of the probe, each TGBF_i has only two free parameters a_i and b_i to control its spread and curvature. In a coarse retrieval step, we extracted an instance in the DB if all of its features are within the *support* of the TGBF's. Then we formalized the retrieval step. *P(Q)*, the set of potential peers of *Q*, is composed of all units within a range from the value of $Q : P(Q) = \{u_j, j = 1, \ldots, m \mid u_j \in N(\bar{X}_Q)\}$ where $N(\bar{X}_Q)$ is the neighborhood of *Q* in the state space $\bar{X}$, defined by the constraint $\|x_{i,Q} - x_{i,j}\| < R_i$ for all potential attributes *i* for which the corresponding weight is nonzero. R_i is half of the support of the TGBF_i, centered on the probe's coordinate $x_{i,Q}$.

(2) Similarity Evaluation. Each TGBF_i is a membership function representing the partial degree of satisfaction of constraint $A_i(x_i)$. Thus, it represents the *closeness* of the instance around the probe value for that particular attribute. For a given peer P_j, we evaluated the function $S_{i,j} = \text{TGBF}_i(x_{i,j}; a_i, b_i, x_{i,Q})$ along each potential attribute *i*. The values (a_i, b_i) are design choices manually initialized, and later refined by the EAs. Since we wanted the most similar instances to be the closest to the probe along *all n* attributes, we used a similarity measure defined as the intersection of the constraint-satisfaction values. Furthermore, to represent the different relevance that each criterion should have in the evaluation of similarity, we attached a weight w_i to each attribute A_i. Therefore, we extended the notion of a similarity measure between P_j and the probe *Q* as a weighted minimum operator

$$S_j = \min_{i=1}^{n}\{\max\lfloor(1 - w_i), S_{j,i}\rfloor\}$$
$$= \min_{i=1}^{n}\{\max\lfloor(1 - w_i),$$
$$\text{TGBF}_i(x_{i,j}; a_i, b_i, x_{i,Q})\rfloor\}, \qquad (41.4)$$

where $w_i \in [0, 1]$. The set of values for the weights $\{w_i\}$ and parameters $\{(a_i, b_i)\}$ are critical design choices that impact the proper selection of peers.

(3) Local Models. The idea of creating a local model on demand can be traced back to memory-based approaches [41.28, 29] and lazy learning [41.49]. Within this case study, we focused on the creation of local predictive models used to forecast each unit's remaining life. First, we used each local model to generate an estimated value of the predicted variable. Then, we used an aggregation mechanism based on the similarities of the peers to determine the final output.

The generation of local models can vary in complexity, depending on the task difficulty. In the first experiment, we used the *Median* operator as the local model, hence we did not need to define any parameter

$$y_j = \text{Median}[D_{1,j}, D_{2,j}, \ldots, D_{k(j),j}]. \qquad (41.5)$$

In the second experiment we used an exponential average, requiring the definition of a forgetting factor α

$$y_i = D_{k(j)+1,j} = \bar{D}_{k(j),j} = \alpha \times D_{k(j),j}$$
$$+ (1 - \alpha) \times \bar{D}_{k(j)-1,j} \quad [\text{where } \bar{D}_{i,j} = D_{1,j}].$$
$$(41.6)$$

(4) Aggregation. We needed to combine the individual outputs y_j of the peers $P_i(Q)$ to generate the estimated output y_Q for the probe *Q*. median (for experiment I) or the prediction of the *next* availability duration, $D_{\text{Next}, Q}$ (for experiment II) for the probe *Q*. To this end, we computed the weighted average of the peers' individual outputs using their normalized similarity to the probe as a weight, namely

$$y_Q = \text{Median}_Q = \frac{\sum_{j=1}^{m} S_j \times y_j}{\sum_{j=1}^{m} S_j}$$

where $y_j = \text{Median}[D_{1,j}, D_{2,j}, \ldots, D_{k(j),j}]$
for Exp. I

$$y_Q = D_{\text{Next}, Q} = \frac{\sum_{j=1}^{m} S_j \times y_j}{\sum_{j=1}^{m} S_j}$$

where $y_j = D_{k(j)+1,j}$ for Exp. II . $\qquad (41.7)$

The entire process is summarized in Fig. 41.4, adapted from [41.11].

Structural and Parametric Tuning

Given the critical design roles of the weights $\{w_i\}$, the parameters $\{(a_i, b_i)\}$, and the forgetting factor α, it was necessary to create a methodology to generate their best values according to our metric, i. e., classification precision. After testing several manually created peer-based models, we decided to use evolutionary search to develop and maintain the fuzzy instance-based classifier, following a wrapper methodology detailed in [41.13]. In this application, however, we extended evolutionary to include structural search, via attribute selection and weighting [41.50], besides the parametric tuning.

The EAs were composed of a population of individuals, each of which containing a vector of elements that represented distinct tunable parameters within the FIM configuration. Examples of tunable parameters included the range of each parameter used to retrieve neighbor instances and the relative parameter weights used for similarity calculation. The EAs used two types of mutation operators (Gaussian and uniform), and no crossover. Its population (with 100 individuals) was evolved over 200 generations.

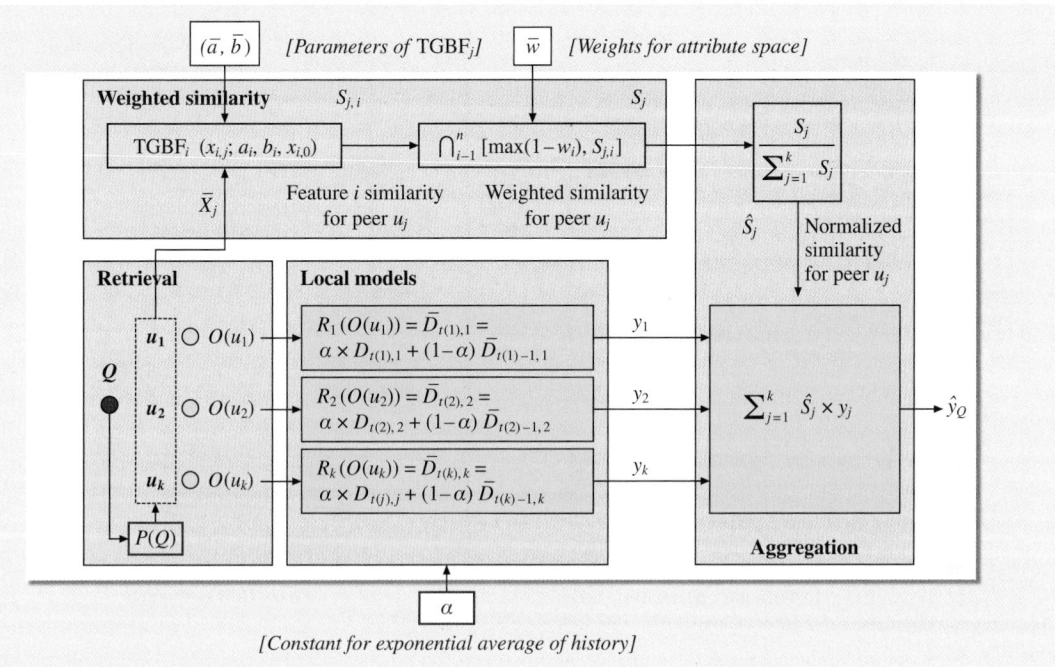

Fig. 41.4 Description of Fuzzy instance-based models (FIM) aggregated by convex sum

Each chromosome defined an instance of the attribute space used by the associated classifier by specifying a vector of weights $[w_1, w_2, \ldots, w_n]$. If $w_i \in \{0, 1\}$, we perform *attribute selection*, i.e., we select a crisp subset from the universe of potential attributes. If $w_i \in \{0, 1\}$, we perform *attribute weighting*, i.e., we define a fuzzy subset from the universe of potential attributes

$$[w_1, w_2, \ldots, w_n][(a_1, b_1), (a_2, b_2), \ldots, (a_n, b_n)][\alpha] \,, \tag{41.8}$$

where

- $w_i \in [0, 1]$ *for* attribute *weighting* and
- $w_i \in \{0, 1\}$ *for* attribute *selection*
- $n = $ Cardinarlity of universe of U, $|U| = n$
- $d = \sum_i^n w_i$ (fuzzy) cardinality of selected features
- $(a_i, b_i) = $ Parameters for GBF$_i$
- $\alpha = $ Parameter for exponential average.

The first part of the chromosome, containing the weights vector $[w_1, w_2, \ldots, w_n]$, defines the attribute space (the FIM structure) and the relevance of each attribute in evaluating similarity. The second part of the chromosome, containing the vector of pairs $[(a_1, b_1), \ldots (a_i, b_i), \ldots (a_n, b_n)]$ defines the parameter

for retrieval and similarity evaluation. The last part of the chromosome, containing the parameter α, defines the forgetting factor for the local models. The fitness function is computed using a *wrapper* approach [41.50]. For each chromosome, represented by (41.8) we instantiated its corresponding FIM. Following a *leave-one-out* approach, we used FIM to predict the expected life of the probe unit following the four steps described in the previous subsection. We repeated this process for all units in the fleet and ranked them in decreasing order, using their predicted duration $D_{\text{Next}, Q}$. We then selected the top 20%. The fitness function of the chromosome was the precision of the classification, $\text{TP}/(\text{TP} + \text{FP})$, where TP is the count of *True Positives* and FP is the count of *False Positives*. This is illustrated in Fig. 41.5.

Results. We used 18 months worth of data and performed the experiments at three different times, after 6, 12, and 18 months, respectively. We wanted to test the adaptability of the learning techniques to environmental, operational, or maintenance changes. We also wanted to determine if their performance would improve over time with incremental data acquisition. For each start-up time, we used *EAs* to generate an opti-

Part D | 41.3

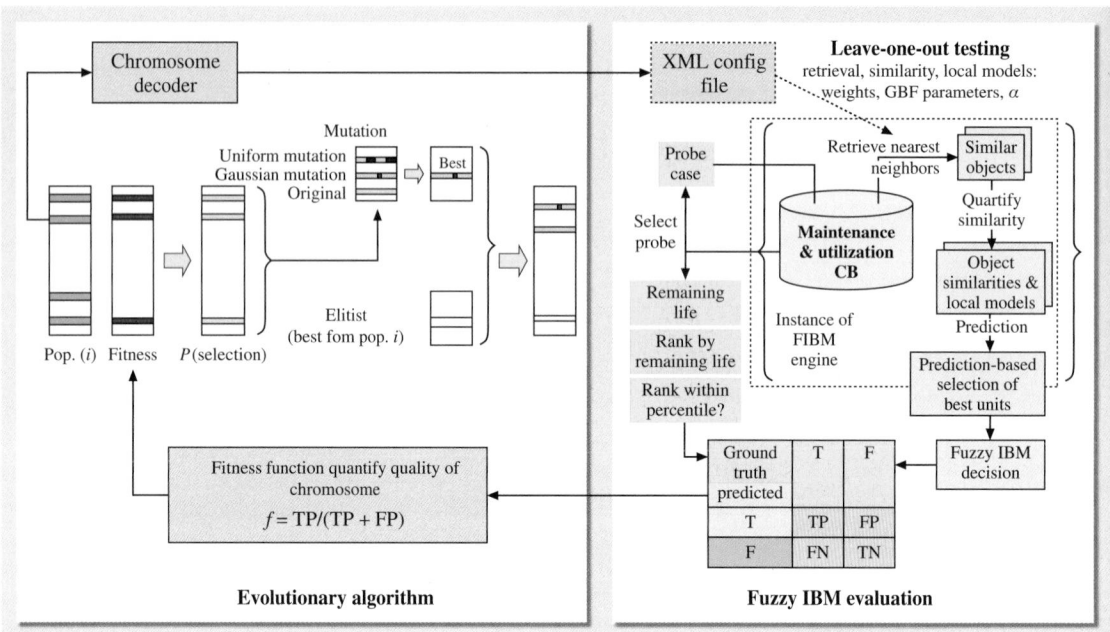

Fig. 41.5 FRC optimization using EA

mized weighted subset of attributes to define the *peers* of each unit.

Experiment 1: Retrospective Selection. The goal was to select the current best 20% units of the fleet based on their peers past performance. In this case, a random selection, which could be used as a baseline, would yield 20%. However, the size of the fleet at each start-up time was different, ranging from 262 (after 6 months) to 634 (after 12 months), to 845 (after 18 months.) We decided to keep the number of selected units constant (i. e.,

52 units) over the three start-up times. Thus the baseline random selection for each start-up time was [20%–8%–6%], i. e., $52/262 = 20\%$; $52/634 = 8\%$; $52/845 = 6\%$.

Experiment 2: Prospective Selection. We wanted to select the future best 20% units for the next-pulse duration. In this case, a random selection would yield 20%.

The peers designed by the EAs provided the best accuracy overall:

- Experiment 1 (*Retrospective selection*):
  Precision $= 63.5\%$, which was more than $10\times$ better than random selection, and $1.7\times$ better than existing heuristics.
- Experiment 2 (*Prospective selection*):
  Precision $= 55.0\%$, which was more than $2.5\times$ better than random selection, and $1.5\times$ better than existing heuristics.

Figures 41.6 and 41.7 illustrate the results of these two experiments.

As mentioned in Sect. 41.1.2, successfully deployed intelligent systems must remain valid and accurate over time, while compensating for drifts and accounting for contextual changes that might otherwise render their design stale or obsolete. In this case study, we repeated the last set of experiments using dynamic and static

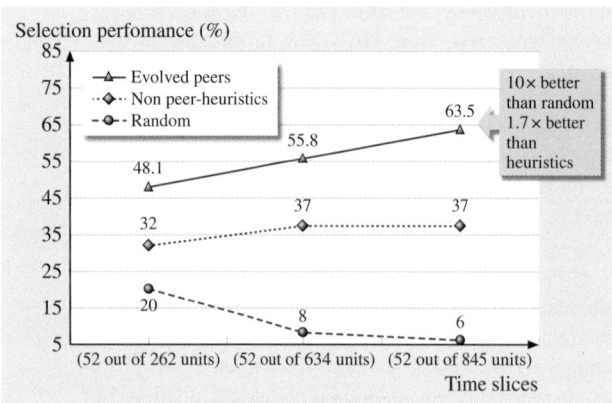

Fig. 41.6 Dynamic models first experiment

models. The dynamic models were *fresh* models, redeveloped at each time slice by using the methodology described. The static models were developed at time slice 1 and applied, unchanged, at time slices 2 and 3. In this experiment, the original models showed significant deterioration over time: $43\% \rightarrow 29\% \rightarrow 25\%$. In contrast, the dynamic models exhibited robust, improved precision: $43\% \rightarrow 43\% \rightarrow 55\%$. This is illustrated in Fig. 41.8.

This comparison shows the benefit of automated model updating. By using an offline metaheuristics such as EAs, we can automate model development and model re-tuning. This allows us to maintain model performance over time, through frequent updates, and avoid the obsolescence-driven model deterioration, which in this example occurred 1 year after the first deployment. A more detailed description of this case study can be found in [41.11, 12].

41.3.4 Health Management – Fault Accommodation and Optimization

All the functions described in Sects. 41.3.1–41.3.3 could be described as *descriptive* and *predictive* analytics, as they provide assessments and projections of the system's health state. These assessments lead to *prescriptive* analytics, as they determine the on-board control action and an off-board logistics, repair and planning actions. On-board control actions are usually focused on maintaining performance or safety margins, and are performed in real-time. Off-board maintenance/repair actions cover more offline decisions. They require a decision support system (DSS) performing multiobjective optimizations, exploring Pareto frontiers of corrective actions, and combining them with preference aggregations to generate the best decision tradeoffs. The underline techniques are intelligent control for fault accommodation [41.18] and multiobjective optimization

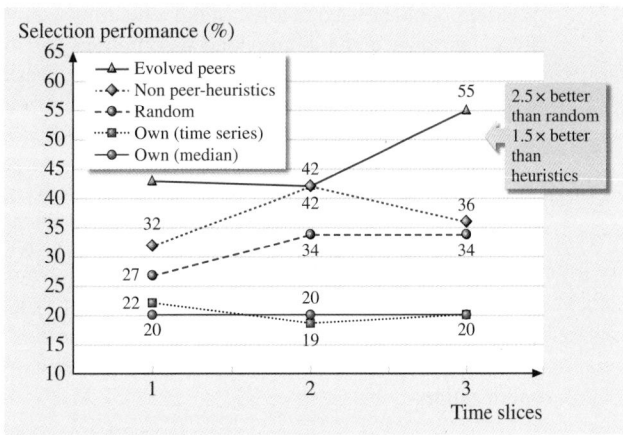

Fig. 41.7 Dynamic models second experiment

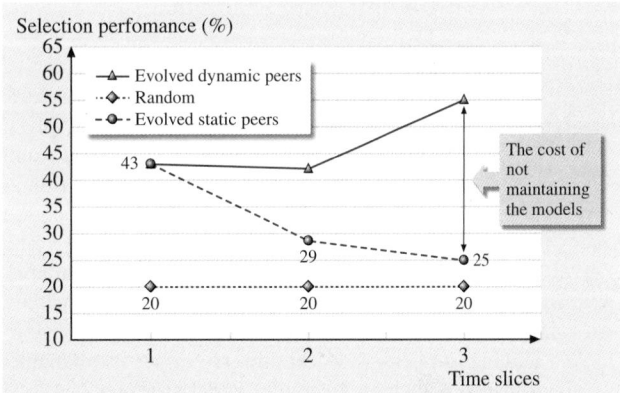

Fig. 41.8 Dynamic models versus static models in second experiment

techniques [41.51–54], aimed at minimizing the impact that maintenance and repairs event could cause to the profitable operation of the assets. For the sake of brevity, we will not provide a case study of optimization in the PHM domain, but we will present one in the financial domain (Sect. 41.4.3).

41.4 CI/ML Applications in Financial Domains: Risk Management

Prognostics and health management of industrial assets bears a strong analogy with risk management of financial and commercial assets. We have shown how *unsupervised learning* can be used to identify abnormal behaviors, i. e., deviations from normal states/structures. In PHM, units in a fleet that stray away from normal performance baselines are usually anomalies leading to incipient failure modes. In financial domains, nonconforming user behaviors could be precursors to fraudulent transactions and could be identified using similar techniques. Similarly, *supervised learning* could be used to classify the root cause of an anomaly (diagnostics) or to classify the risk class of an applicant for a financial/insurance product (risk classification). *Re-*

gressions could be used to forecast the remaining useful life of an asset under future load assumptions, or to forecast the residual value of assets after their lease period (or to create an instant valuation for an asset, such as in mortgage collateral valuation). *Multiobjective* optimization techniques could be used to balance production values with life erosion cost (or combustion efficiency with emissions) or to balance an investment portfolio using multiple metrics of returns and risk. We will illustrate this analogy with the following three case studies, in which we will describe the use of CI techniques in risk classification for insurance underwriting, residential property valuation, and portfolio rebalancing optimization.

41.4.1 Automation of Insurance Underwriting: A Classification Problem

Problem Definition

In many transaction-oriented processes, human decision makers evaluate new applications for a given service (mortgages, loans, credits, insurances, etc.) and assess their associated risk and price. The automation of these business processes is likely to increase throughput and reliability while reducing risk. The success of these ventures is depends on the availability of generalized decision-making systems that are not just able to reliably replicate the human decision-making process, but can do so in an explainable, transparent fashion. Insurance underwriting is one such high-volume application domain where intelligent automation can be highly beneficial, and reliability and transparency of decision-making are critical. Traditionally, highly trained individuals perform insurance underwriting. A given insurance application is compared against several standards put forward by the insurance company and classified into one of the risk categories (rate classes) available for the type of insurance requested. The risk categories then affect the premium paid by the applicant – the higher the risk category, the higher the premium. The *accept/reject* decision is also part of this risk classification, since risks above a certain tolerance level set by the company will simply be rejected.

There can be a large amount of variability in the underwriting process when performed by human underwriters. Typically the underwriting standards cannot cover all possible cases, and sometimes they might be ambiguous. The subjective judgment of the underwriter will almost always play a role in the process. Variation

in factors such as underwriter training and experience, and a multitude of other effects can cause different underwriters to issue inconsistent decisions. Sometimes these decisions fall in a *gray area* not explicitly covered by the standards. In these cases, the underwriter uses his/her own experience to determine whether the standards should be adjusted. Different underwriters could apply different assumption regarding the applicability of the adjustments, as they might use stricter or more liberal interpretations of the standards.

CI-Based Approach

To address these problems, we developed a system to automate the application placement process for cases of low or medium complexity. For more complex cases, the system provided the underwriter with an *assist* based on partial analysis and conclusions.

We used a fuzzy-rule-based classifier (FRC) to capture the underwriting standards derived from the actuarial guidance. Then we tuned the FRC with an evolutionarily algorithms to determine the best FRC parameters to maximize precision and recall, wile minimizing the cost of misclassification. The remaining of this section will summarize this solution.

Fuzzy Rule-Based Classifier (FRC). The fuzzy-rule based classifier (FRC), which is briefly described in [41.13, 14], uses rule sets to encode underwriting standards. Each rule set represents a set of fuzzy constraints defining the boundaries between rate classes. These constraints were first determined from the underwriting guidelines. They were then refined using knowledge engineering sessions with expert underwriters to identify factors such as blood pressure levels and cholesterol levels, which are critical in defining the applicant's risk and corresponding premium. The goal of the classifier is to assign an applicant to the most competitive rate class, providing that the applicant's vital data meet all of the constraints of that particular rate class to a minimum degree of satisfaction. The constraints for each rate class r are represented by n fuzzy sets: $A_i^r(x_i)$, $i = 1, \ldots, n$. Each constraint $A_i^r(x_i)$ can be interpreted as the *degree of preference* induced by value x_i for satisfying constraint $A_i^r(x_i)$. After evaluating all constraints, we compute two measures for each rate class r. The first one is the degree of intersection of all the constraints and measures the *weakest* constraint satisfaction

$$I(r) = \bigcap_{i=1}^{n} A_i^r(x_i) = \min_{i=1}^{r} A_i^r(x_i) . \tag{41.9}$$

This expression implies that each criterion has equal weight. If we want to attach a weight w_i to each criterion A_i we could use the weighted minimum operator:

$$I'(r) = \bigcap_{i=1}^{n} W_i A_i^r(x_i)$$

$$= \min_{i=1}^{n}(\max((1-w_i), A_i^r(x_i)))\,,$$

where $w_i \in [0, 1]$. The second one is a cumulative measure of missing points (the complement of the average satisfaction of all constraints), and measures the *overall tolerance* allowed to each applicant, i. e.,

$$MP(r) = \sum_{i=1}^{n}(1 - A_i^r(x_i)) = n\left(1 - 1/n \sum_{i=1}^{n} A_i^r(x_i)\right)$$

$$= n(1 - \bar{A}^r)\,. \tag{41.10}$$

The final classification is obtained by comparing the two measures, $I(r)$ and $MP(r)$ against two lower bounds

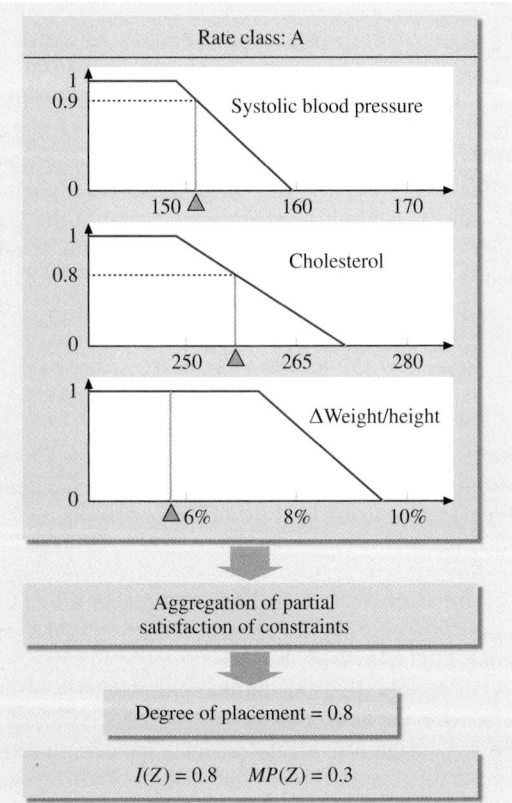

Fig. 41.9 Example of three fuzzy constraints for rate class Z

defined by thresholds τ_1 and τ_2. The parametric definition of each fuzzy constraint $A_i^r(x_i)$ and the values of τ_1 and τ_2 are design parameters that were initialized with knowledge engineering sessions.

Figure 41.9 – adapted from [41.13] – illustrates an example of three constraints (trapezoidal membership functions) associated with rate class Z, the input data corresponding to an application, and the evaluation of the first measure, indicating the weakest degree of satisfaction of all constraints.

Optimization of Design Parameters of the FRC Classifier. The FRC design parameters were tuned, monitored, and maintained to assure the classifier's optimal performance. To this end, we used EAs, composed of a population of *chromosomes*. Each chromosome contained a vector of elements that represent distinct tunable parameters to configure the FRC classifier, i. e., the parametric definition of the fuzzy constraints $A_i^r(x_i)$ and thresholds τ_1 and τ_2.

A chromosome, the genotypic representation of a model, defines the complete parametric configuration of the classifier. Thus, an instance of such classifier can be created for each chromosome, as shown in Fig. 41.10. Each chromosome c_i, of population $P(t)$ (left-hand side of Fig. 41.10), goes through a decoding process to allow them to create the classifier on the right. Each classifier is then tested on all the cases in the case base, assigning a rate class to each case. We can determine the quality of the configuration encoded by the chromosome, i. e., the *fitness* of the chromosome, by analyzing the results of the test. Our EA uses two types of mutations (uniform and Gaussian) to produce new individuals in the population by randomly varying parameters of a single chromosome. The more fit chromosomes in generation t will be more likely to be selected for this and pass their genetic material to the next generation $t + 1$. Analogously, the less fit solutions will be culled from the population. At the conclusion of the EAs execution the *best* chromosome of the *last* generation determines the classifier's configuration. Note the similarity between Figs. 41.10 and 41.5, which underlies the similar role that EAs play as the offline MHs to design the best fuzzy classifier (in this case study), or the best FIM (in the case of the second case study).

Standard Reference Dataset (SRD). To test and tune the classifiers, we needed to establish a benchmark. Therefore, we generated a *standard reference dataset* (SRD) of approximately 3000 cases taken from a stratified random sample of the historical case population.

Part D | 41.4

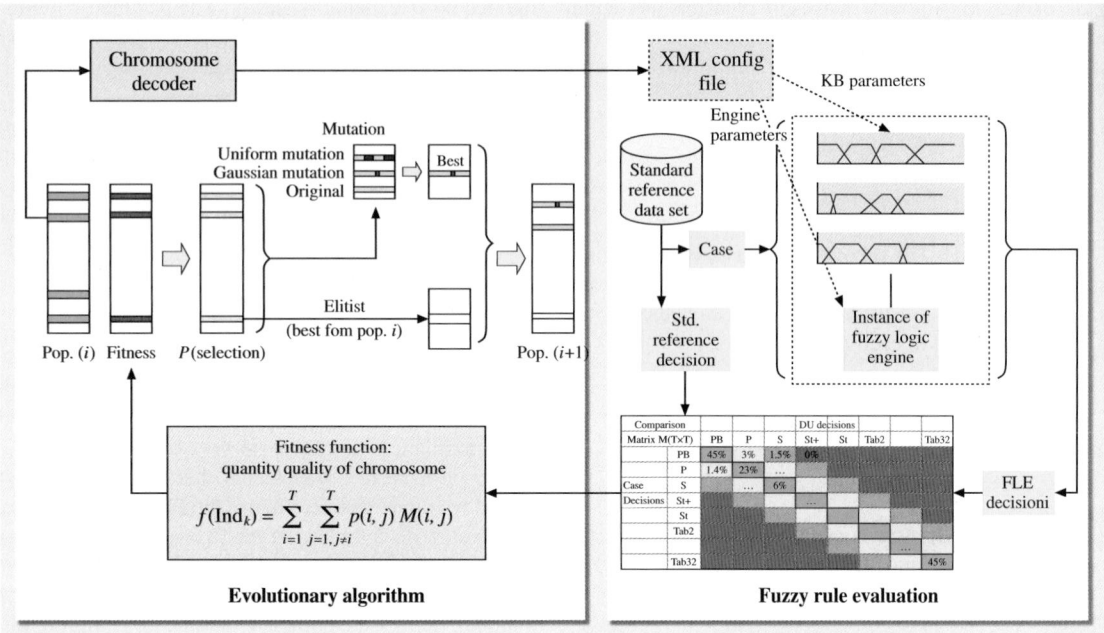

Fig. 41.10 FRC optimization using EA

Each of these cases received a rate class decision when it was originally underwritten. To reduce variability in these decisions, a team of experienced underwriters performed a *blind* review of selected cases to determine the *standard reference* decisions. These cases were then used to create and optimize the FRC model.

Fitness Function. In classification problems such as this one, we can use two matrices to construct the fitness function that we want to optimize. The first matrix is a *TxT confusion matrix M* that contains frequencies of correct and incorrect classifications for all possible combinations of the standard reference decisions (SRDs), which represent ground truth rate class decisions as reached by consensus among senior expert underwriters for a set of insurance applications and classifier decisions. The frequencies of correct classifications can be found on the main diagonal of matrix M. The first $(T-1)$ columns represent the rate classes available to the classifier. Column T represents the classifier's choice of not assigning any rate class, sending the case to a human underwriter. The same ordering is used to sort the rows for the SRD. The second matrix is a $T \times T$ *penalty matrix* P that contains the *value loss* due to misclassification. The entries in the *penalty matrix* P are zero or negative values. They were computed from actuarial data showing the net present value (NPV) for

each entry (j,k). The penalty value $P(j,k)$ was the difference between the NPV of the entry (j,kj) and the highest NPV – corresponding to the correct entry (j,j), located on the main diagonal. The fitness function f combined the values of M, resulted from a test run of the classifier configured with chromosome c_i, with the penalty matrix P to produce a single value

$$f(c_i) = \sum_{j=1}^{T} \sum_{k=1}^{T} M(j,k) * P(j,j) . \qquad (41.11)$$

Function f represents the expected value loss for that chromosome computed over the SRD and is the fitness function used to drive the evolutionary search.

Results
Testing and Validation of FRC. We defined *Coverage* as the percentage of cases as a fraction of the total number of input cases; Relative accuracy as the percentage of correct decisions on those cases that were not referred to the human underwriter; *Global accuracy* as the percentage of correct decisions, including making correct rate class decisions and making a correct decision to refer cases to human underwriters as a fraction of total input cases. Then we performed a comparison against the SRD. The results, reported in [41.13], show

Table 41.2 Typical performance of the un-tuned and tuned rule-based decision system (FRC)

Metrics	Initial parameters based on written guidelines (%)	Best knowledge engineered parameters	Optimized parameters
Coverage	94.01	90.38	91.71
Relative accuracy	75.92	92.99	95.52
Global accuracy	74.75	90.07	93.63

Table 41.3 Average FRC performance over 5 *tuning* case sets compared to five *disjoint test* sets

Metrics	Average performance on training sets	Average performance on disjoint test sets
Coverage	91.81	91.80
Relative accuracy	94.52	93.60
Global accuracy	92.74	91.60

a remarkable improvement in all measures. Specifically, we obtained the following results:

Using the initial parameters (first column of Table 41.2) we can observe a large moderate *Coverage* ($\approx$94%) associated with a low *relative accuracy* ($\approx$76%) and a lower *global accuracy* ($\approx$75%). These performance values are the result of applying a strict interpretation of the underwriter (UW) guidelines, without allowing for any tolerance. Had we implemented such crisp rules with a traditional rule-based system, we would have obtained these results. This strictness would prevent the insurer from being price competitive, and would not represent the typical *modus operandi* of human underwriters. However, by allowing each underwriter to use his/her own interpretation of such guidelines, we could introduce large underwriters' variability. One of our main goals was to provide a *uniform* interpretation, while still *allowing for some tolerance*. This goal is addressed in the second column of Table 41.2, which shows the results of performing *knowledge engineering* and encoding the desired tradeoff between risk and price competitiveness as fuzzy constraints with *preference* semantics. This intermediate stage shows a different tradeoff since both *Global* and *relative accuracy* have improved. *Coverage* slightly decreases ($\approx$90%) for a considerable gain in *relative accuracy* ($\approx$93%). Although we obtained this initial parameter set by interviewing the experts, we had no guarantee that such parameters were optimal. Therefore, we used EAs to tune them. We allowed the parameters to move within a predefined range centered on their initial values and, using the SRD and the fitness function described above, we obtained an optimized parameter set, whose results are described in the third column of Table 41.2. The results of the optimization show the point corresponding to the final parameter set dominates the second set point (in a Pareto sense), since both *cover-*

age and *relative accuracy* were improved. Finally, we can observe that the final metric, *global accuracy* (last row in Table 41.2), improves monotonically as we move from using the strict interpretation of the guidelines ($\approx$75%), through the knowledge-engineered parameters ($\approx$90%), to the optimized parameters ($\approx$94%).

While the reported performance of the optimized parameters, shown in Table 41.2, is typical of the performance achieved through the optimization, a five-fold cross-validation on the optimization was also performed to identify stable parameters in the design space and stable metrics in the performance space. This is shown in Table 41.3.

With this kind of automation, variability of the risk category decision was greatly reduced. This also eliminated a source of risk exposure for the company, allowing it to operate more competitively and profitably. The intelligent automation process, capable of determining its applicability for each new case, increased underwriting capacity by enabling them to handle larger volume of applications. Additional information on this approach can be found in [41.13, 14].

41.4.2 Mortgage Collateral Valuation: A Regression Problem

Problem Definition

Residential property valuation is the process of determining a dollar estimate of the property value for given market conditions. Within this case study, we will restrict ourselves to a single-family residence designed or intended for owner occupancy. The value of a property changes with market conditions, so any estimate of its value must be periodically updated to reflect those market changes. Any valuation must also be supported by current evidence of market conditions, e.g., recent real estate transactions.

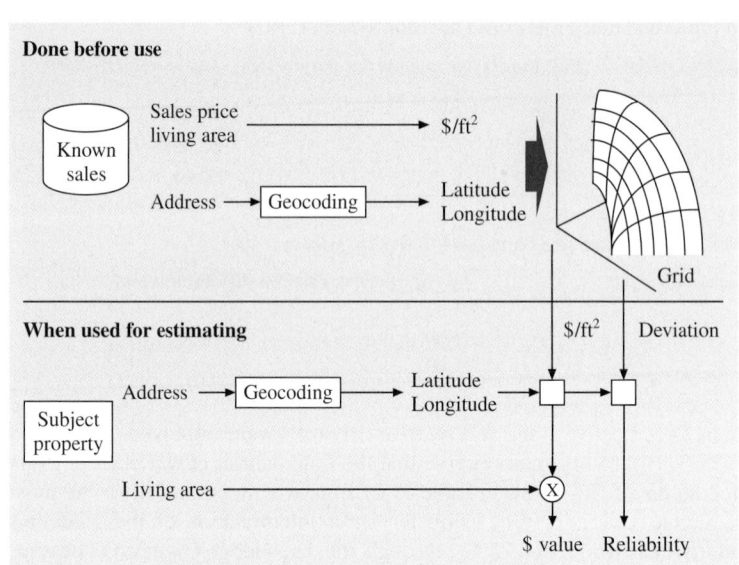

Fig. 41.11 Locational Value method (LOCVAL)

Current manual process for estimating the value of properties usually requires an on-site visit by a human appraiser. This process is slow and expensive for batch applications such as those used by banks for updating their loan and insurance portfolios, verifying risk profiles of servicing rights, or evaluating default risks for securitized packages of mortgages. The appraisal process for these batch applications is currently estimated, to a lesser degree of accuracy, by sampling techniques. Secondary buyers and mortgage insurers may also require verification of property value on individual transactions. Some of the applications also require that the output be qualified by a reliability measure and some justification, so that questionable data and unusual circumstances can be flagged for the human who uses the output of the system. Thus, the automation of residential property was motivated by a broad spectrum of application areas.

The most common and credible method used by appraisers is the *sales comparables* approach. This method consists of finding comparable cases, i. e., recent sales that are comparable to the subject property (using sales records); contrasting the subject property with the comparables; adjusting the comparables' sales price to reflect their differences from the subject property (using heuristics and personal experience); and reconciling the comparables adjusted sales prices to derive an estimate for the subject property (using any reasonable averaging method). This process assumes that the item's market value can be derived by the prices demanded by similar items in the same market.

CI–Based Approach: LOCVAL, AIGEN, AICOMP, Fusion

To automate the valuation process, we developed a program that combined the result of three independent estimators. The first one, locational value (LOCVAL), was a coarse estimator based on the locational value of the property. The second one, generative AI model (AIGEN), was a generative estimator based on neuro-fuzzy networks that only used five features from our training set. The third one, comparable based AI model (AICOMP), was a fuzzy case-based reasoned that followed the comparable-based approach of the appraisers. Finally, we fused the output of the estimators into a single estimate and reliability value.

Locational value model (LOCVAL). The first model was based solely on two features of the property: its location, expressed by a valid, geocoded address, and its living area, as shown in Fig. 41.11 (adapted from [41.15].)

A dollar per square foot measure was constructed for each point in the county, by suitably averaging the observed, filtered historical market values in the vicinity of that point. This locational value estimator (LOCVAL) produced two output values: *Locational_Value* (a $/ft^2 estimate) and *Deviation_from_prevailing_value*. The local averaging was done by an exponentially decreasing radial basis function with a *space constant* of 0.15–0.2 miles. It could be described as the weighted sum of radial basis functions (all of the same width), each situated at the site of a sale within the past 1-year and having amplitude equal to the sales price. Deviation from prevailing value was the standard deviation for houses within the area covered and was derived using a similar approach. The output of LOCVAL was a coarse estimate of the property value, which was used as an input for the generative approach (AIGEN).

AIGEN: Fuzzy–Neural Network. The generative AI model (AIGEN) relied on a fuzzy-neural net that, after a training phase, provided an estimate of the subject's value. The specific model was an extension of AN-FIS [41.55], which implemented a fuzzy system as a five-layer neural network so that the structure of the net could be interpreted in terms of high-level rules. The extension developed allowed the output to be linear functions of variables that did not necessarily occur in the input. In this fashion, we achieved more fidelity with the local models (linear functions) without incurring in the computational complexity caused by a large number of inputs. AIGEN inputs were five property features (*total_rooms, num_bedrooms, num_baths, living_area*, and *lot_size*) and the output of LOCVAL (*locational_value*).

AICOMP: Fuzzy Case–Based Reasoner (CBR). AICOMP is a fuzzy CBR system that used fuzzy predicates and fuzzy-logic-based similarity measures to estimate the value of residential property. This process consisted of selecting relevant cases (which would be nearby house sales), adapting them, and aggregating those adapted cases into a single estimate of the property value. AICOMP followed a process similar to the sales comparison used by certified appraisers to estimate a residential property's value. This approach, which is further described in [41.56], consisted of:

(1) *Retrieving recent sales from a case base.* Upon entering the subject property attributes, AICOMP retrieves potentially similar comparables from the case-base. This initial selection uses six attributes: address, date of sale, living area, lot area, number of bathrooms, and bedrooms.
(2) *Comparing the subject property with the retrieved cases.* The comparables are rated and ranked on a similarity scale to identify the most similar ones to the subject property. This rating is obtained from a weighted aggregation of the decision maker preferences, expressed as fuzzy membership distributions and relations.
(3) *Adjusting the sales price of the retrieved cases.* Each property's sales price is adjusted to reflect

their differences from the subject property. These adjustments are performed by a rule set that uses additional property attributes, such as construction quality, conditions, pools, fireplaces, etc.
(4) *Aggregating the adjusted sales prices of the retrieved cases.* The best four to eight comparables are selected. The adjusted sales price and similarity of the selected properties are combined to produce an estimate of the subject value with an associated reliability value.

Fusion. Each model produced a property value and an associated reliability value. The latter was a function of the *typicality* of the subject property based on its physical characteristics (such as lot size, living area, and total room). These typical values were represented by possibilistic distributions (fuzzy sets). We computed the degree to which each property satisfied each criterion. The overall property value reliability was obtained by considering the conjunction of these constraint satisfactions (i. e., the minimum of the individual reliability values).

The computation times, required inputs, errors, and reliability values for these three methods are shown in Fig. 41.12. The locational value (LOCVAL) model took the least time and information, but produced the largest error. The CBR approach (AICOMP) took the largest time and number of inputs, but produced the lowest error.

The fusion of the three estimators exhibited several advantages:

- The fusion process provided an indication of the reliability in the final estimate:
  - If reliability was high, the fused estimate was more accurate than any of the individual ones
  - If reliability was limited, the system generated an explanation in human terms
- The fused estimates were more robust.

These characteristics allowed the user to determine the suitability of the estimate within the given business application context. Knowledge-based rules were used for constructing this fusion at a supervisory level, and the few parameters were determined manually, by inspection and experimentation. A more detailed description of this process can be found in [41.15]. This case study was the oldest case study to use model ensemble and fusion. However, the use of metaheuristics to guide the design phase at that time was still not well understood as in the more recent case studies.

Part D | 41.4

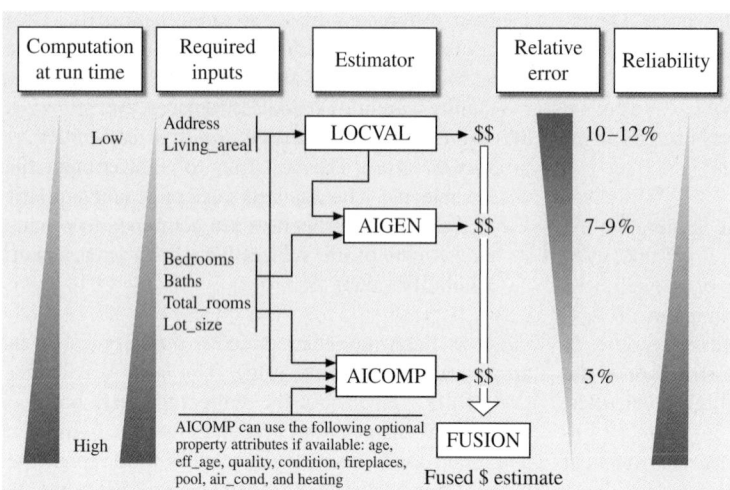

Fig. 41.12 Data comparison of multiple approaches

Results

The reliability values generated by the fusion were divided into three classes, labeled *good*, *fair*, and *poor*. From a test sample of 7293 properties, 63% were classified as *good*, with a median absolute error of 5.4% (an error that was satisfactory for the intended application.) Of the remaining subjects, 24% were classified as *fair*, and 13% as *poor*. The fair set had a medium error of 7.7%, and the poor set had a median error of 11.8%.

The reliability computation and the fusion increased the robustness and usefulness of the system, which achieved good accuracy and was scalable for thousands of automated transactions. This approach made it a transparent, interpretable, fast, and inexpensive choice for bulk estimates of residential property value for a variety of financial applications.

41.4.3 Portfolio Rebalancing: An Optimization Problem

Problem Definition

The goal of portfolio optimization is to manage risk through diversification and obtain an optimal risk-return tradeoff. In this case study, we address portfolio optimization within the context of an asset-liability management (ALM) application. The goal was to find the optimal allocation of available financial resources to a diversified portfolio of 1500+ long and short-term financial assets, in accordance with risk, liability, and regulatory constraints.

To characterize the investor's risk objectives and capture the potential risk-return tradeoffs, we used various measures to quantify different aspects of portfolio risk. For ALM applications, a typical measure of risk is *surplus variance*. We computed portfolio variance using an analytical method based on a multifactor risk framework. In this framework, the value of a security can be characterized as a function of multiple underlying risk factors. The change in the value of a security can be approximated by the changes in the risk factor values and risk sensitivities to these risk factors. The portfolio variance equation can be derived analytically from the underlying value change function.

In ALM applications, the portfolios have assets and liabilities that are affected by the changes in common risk factors. Since a majority of the assets are fixed-income securities, the dominant risk factors are *interest rates*. In ALM applications, in addition to maximizing return or minimizing risk, portfolio managers are constrained to match the characteristics of asset portfolios with those of the corresponding liabilities to preserve portfolio surplus due to interest rate changes. Therefore, the ALM portfolio optimization problem formulation has additional linear constraints that match the asset–liability characteristics when compared with the traditional Markowitz model. We use the following ALM portfolio optimization formulation

Maximize	Portfolio expected return	
Minimize	Surplus variance	
Minimize	Portfolio value at risk	(41.12)

Subject to:

    Duration mismatch $\leq$ target$_1$

    Convexity mismatch $\leq$ target$_2$; and

    Linear portfolio investment constraints.

To measure the three objectives in (41.12), namely portfolio expected return, surplus variance, and portfolio value at risk, we used *book yield, portfolio variance*, and *simplified value at risk* (SVaR), respectively. These metrics are defined as follows:

- *Portfolio book yield* represents its accounting yield to maturity and is defined as

$$\text{BookYield}_P = \frac{\sum_i \text{BookValue}_i \times \text{BookYield}}{\sum_i \text{BookValue}_i}$$

(41.13)

- *Portfolio variance* is a measure of its variability and is defined as the second moment of its value change ΔV

$$\sigma^2 = E\lfloor (\Delta V)^2 \rfloor - E[(\Delta V)]^2$$

(41.14)

- *Portfolio simplified value at risk* is a complex measure of the portfolio's catastrophic risk and is described in details in [41.57].

These metrics define the 3D optimization space. Now, let us analyze its constraints. The change in the value ΔV of a security can be approximated by a second order Taylor series expansion given by

$$\Delta V \approx \sum_{i=1}^{m} \left(\frac{\partial V}{\partial F_i} \right) \Delta F_i$$

$$+ \frac{1}{2} \sum_{i=1}^{m} \sum_{i=1}^{m} \left(\frac{\partial^2 V}{\partial F_i \partial F_j} \right) \Delta F_i \Delta F_j \, .$$

(41.15)

The first- and second-order partial derivatives in (41.15) are the risk sensitivities, i. e., the change in the security value with respect to the change in the risk factors F_i. These two terms are typically called *delta* and *gamma*, respectively [41.58]. For fixed-income securities, these measures are *duration* and *convexity*. The duration and convexity mismatches, which constrain our optimization space in (41.12), are the absolute

values of the differences between the effective durations and convexities of the assets and liabilities in the portfolio, respectively. Though they are nonlinear (because of the absolute value function), the constraints can easily be made linear by replacing each of them with two new constraints that each ensure that the actual value of the mismatch is less than the target mismatch and greater than the negative of the target mismatch, respectively. The other portfolio investment constraints include asset-sourcing constraints that impose a maximum limit on each asset class or security, overall portfolio credit quality, and other linear constraints.

CI–Based Approach

Given the explicit need for customization and hybridization in methods for portfolio optimization, we could not find an existing multiobjective optimization algorithms could be applied without extensive modifications. Specifically, the requirement to optimize while satisfying a large number of linear constraints excluded the ready application of prior evolutionary multiobjective optimization approaches. This aspect was a principal motivation to develop a novel hybrid techniques.

Figure 41.13 illustrates the process used to drive the search for the efficient frontier. The process consisted of three steps, corresponding to the three boxes in Fig. 41.13. The first step (box 1 in Fig. 41.13) was the *generation of the Pareto front*. It consisted of:

(a) Initializing the population of candidate portfolios using a randomized linear programming (RLP)
(b) Generating an interim Pareto front with a Pareto sorting evolutionary algorithm (PSEA)
(c) Completing gaps in the Pareto front with a target objective genetic algorithm (TOGA) and
(d) Storing the results in a repository. After many runs, we filtered the repository with an efficient dominance filter and generated the first efficient frontier.

The second step (box 2 in Fig. 41.13) was the *interactive densification of the Pareto front*. This richly sampled Pareto front was analyzed for possible gaps, and augmented with a last run of TOGA, leading to the generation of the second efficient frontier. Each point in this front represented a nondominated solution, i. e., a viable portfolio. The third step (box 3 in Fig. 41.13) was the *portfolio selection*. We needed to incorporate the decision maker's *preferences* in the return-risk tradeoff. Our goal is to reduce the large number of viable so-

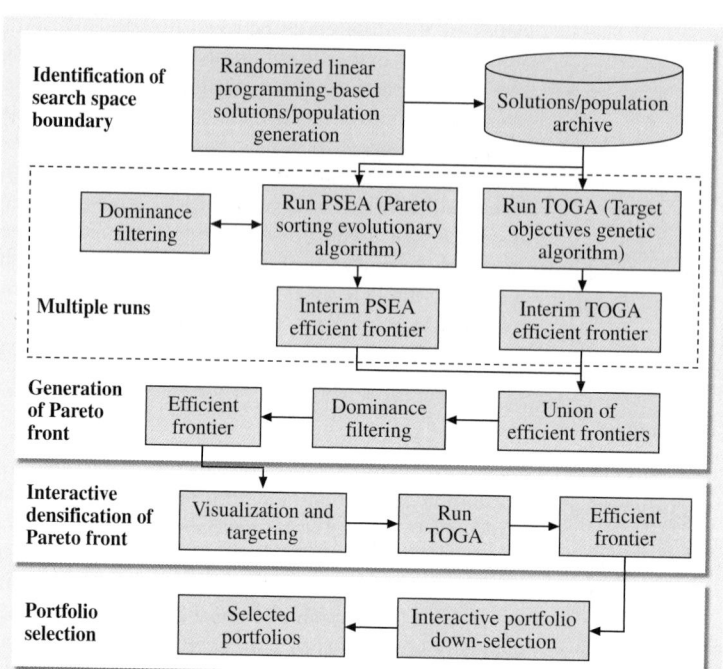

Fig. 41.13 Portfolio optimization process/workflow

lutions into a much smaller subset that could then be further analyzed for a final portfolio selection.

We will briefly describe the major components of this process.

Randomized Linear Programming (RLP). The key challenge in solving the portfolio optimization problem was presented by the large number of linear allocation constraints. The feasible space defined by these constraints is a high dimensional real-valued space (1500+ dimensions), and is a highly compact convex polytope, making for an enormously challenging constraint satisfaction problem. We leveraged our knowledge on the geometrical nature of the feasible space by designing a randomized linear programming algorithm that robustly sampled the boundary vertices of the convex feasible space. These extremity samples were seeded in the initial population of the PSEA and were exclusively used by the evolutionary multiobjective algorithm to generate interior points (via interpolative convex crossover) that were always geometrically feasible. This was similar in principle to the *preprocess phase*, proposed by *Kubalik* and *Lazansky* [41.59].

Pareto Sorting Evolutionary Algorithm (PSEA). We developed a Pareto sorting evolutionary algorithm (PSEA) that was able to robustly identify the Pareto front of optimal portfolios defined over a space of returns and risks. The algorithm used a secondary storage and maintains the diversity of the population by using a convex crossover operator, incorporating new random solutions in each generation of the search; and using a noncrowding filter. Given the reliance of the PSEA on the continuous identification of nondominated points, we developed a fast dominance filter to implement this function very efficiently.

Target Objectives Genetic Algorithm (TOGA). We further enhanced the quality of the Pareto front by using a target objectives genetic algorithm (TOGA), a non-Pareto nonaggregating function approach to multiobjective optimization. Unlike the PSEA, which was driven by the concept of dominance, the TOGA found solutions that were *as close as possible* to a predefined target for one or more criterion [41.60]. We used this to fill potential *gaps* in the Pareto front.

Decision Maker Preferences. We incorporated the decision-maker's *preferences* in the return-risk tradeoff to perform our selection. The goal was to reduce thousands of nondominated solutions into a much smaller subset (of ≈ 10 points), which could be further analyzed for a final portfolio selection. After obtaining a 3D Pareto front, we augmented this space with three additional metrics, to reflect additional constraints for use in the tradeoff process. This augmented 6D space was used for the down-selection problem. To incorporate progressive ordinal preferences, we used a graphical tool to visualize 2D projections of the Pareto front. After applying a set of constraints to further refine the best region, we used an *ordinal* preference, defined by the order in which we visited and executed *limited, local tradeoffs* in each of the available 2D projections of the Pareto front. In this approach, the decision maker could understand the available space of options and the costs/benefits of the available tradeoffs. The use of progressive preference elicitation provided a natural mechanism to identify a small number of the good solutions.

Results. The optimization process was successfully tested on large portfolios of fixed-income base securities – each portfolio involving over fifteen hundred financial assets, and investment decisions of several billion dollars. For a more complete description of this application refer to [41.16].

41.5 Model Ensembles and Fusion

Over the last decade, we have witnessed an emerging trend favoring the use of model ensembles over individual models. The elements of these ensembles are object-level models, the fusion mechanism is an *online MHs*, and their overall design is guided by *offline MHs*, as discussed in Sect. 41.1.

41.5.1 Motivations for Model Ensembles

This trend is driven by the improved performance obtained by ensembles. By fusing the outputs of an ensemble of *diverse* predictive models, we boost the overall prediction accuracy while reducing the variance. *Fumera* and *Roli* et al. [41.61] confirmed theoretically the claims of *Dietterich* [41.62]. They proved that averaging of classifiers outputs guarantees a better test set performance than the worst classifier of the ensemble. Moreover, under specific hypotheses, such as linear combiners of individual classifiers with unbiased and uncorrelated errors, the fusion of multiple classifiers can *improve the performance of the best individual classifiers*. Under ideal circumstances (e.g., with an infinite number of classifiers) the fusion can provide the optimal Bayes classifier [41.63]. All this is possible if individual classifiers make *different* errors (diversity), as we will discuss in Sect. 41.5.3.

There is also a computational motivation for using model ensembles. Many learning algorithms are based on local search and suffer from the problem of local minima, which is usually resolved by multiple independent initializations. In other cases, generating the optimal training might be computationally hard even with enough training data. The fusion of multiple classifiers trained from different starting points or training sets can better approximate the optimal classifier at a fraction of the computational cost.

41.5.2 Construction of Model Ensembles

The ensemble construction requires the creation of base models, an ensemble topology, and a fusion mechanism. Let us briefly review these concepts.

Base Models. Base models are the elements to be fused – they are the object-level models discussed in Sect. 41.1.1. They need to be *diverse*, e.g., they need to have low error correlations. They could differ in their parameters, and/or in their structure, and/or in the ML techniques used to create them. The process for injecting diversity in their design is described in Sect. 41.5.3.

Topology. The ensemble can be constructed by following a *parallel* or *serial* topology (or in some cases, a hybrid one). The most common topology is the parallel one, in which multiple models are fed the same inputs and their outputs are merged by the fusion mechanism. In the serial topology, the models are applied sequentially (as in the case when we first use a primary model, and in case of it failing to accept a pattern, a secondary model is used to attempt a classification).

Fusion Mechanism. We divide the fusion mechanisms based on two criteria: (a) the type of aggregation that they perform; (b) the dependency on their inputs. The former is concerned with the regions of competence of the base models to be aggregated and divides the fusion mechanisms into: *selection, interpolation*, and *integration*. The latter is concerned with the dependency of the meta-model (fusion) on the inputs to the ensemble and divides the fusion mechanisms into *static* and *dynamic* ones.

Based on the first criterion, we have the following types of fusion mechanisms:

- *Selection* – used to fuse *disjoint, complementary* models. In this case, the base models were trained on disjoint regions of the feature space and, for each pattern, just one model is responsible for the final decision. Selection determines a *binary relevance weight* of each complementary model (where all but one of the weights are zero). This mechanism is typically used in hierarchical control systems, in which a supervisory controller (meta controller) selects the most appropriate low-level controller for any given state. Another example of this mechanism is the use of decision trees [41.26], in which the leaf node reached by the input/state determines the selected model.
- *Interpolation* – used to fuse *overlapping complementary* models. In this case, the base models were trained on different but overlapping regions of the feature space and, for each state, a subset of models is responsible for the final decision. Interpolation determines a *fuzzy relevance weight* of each complementary model (where the weights are in the [0, 1] interval and they are usually normalized to add up to 1). By interpolating rather than switching between models, we introduce smoothness in the response surface induced by the ensemble. This interpolation mechanism is typical of hierarchical fuzzy systems, usually found in fuzzy control applications [41.36, 55].
- *Integration,* – used to fuse *competitive* models. In this case, all base models were trained on the same feature space and, for each input, all models contribute to the final decision according to their relevance weight. Integration determines the *relevance weight* of each competitive model.

Based on the second criterion (input dependency of the meta model), we have the following fusion mechanisms:

- *Static fusion.* The relevance weights are determined in a *batch mode* by a static fusion meta-model (online MHs). The mechanism is applied uniformly to all inputs. This is the typical case of algebraic expressions used to compute the relevance weights [41.64].

- *Dynamic fusion.* The relevance weights are determined at *run time*, by a dynamic fusion meta-model (online MHs). The weights vary according to the inputs. This is the typical case of dynamic systems used to compute the relevance weights [41.36–38].

As noted by *Roli* et al. [41.65], the design of a successful fusion system consists of three parts: design of the individual object-level models, selection of a set of diverse models, and design of the fusion mechanism. The operating word is *diverse*, where model diversity is defined by low correlation among the object-level model errors. In other words, these models should be as accurate as possible while avoiding coincident errors. This concept is described by *Kuncheva* and *Whitaker* [41.66], where the authors propose four *pairwise* and six *nonpairwise* diversity measures to determine the models difference. A more complete treatment of this topic can be found in [41.67].

41.5.3 Creating Diversity in the Model Ensembles

Let us consider a model as a mapping from an n-dimensional feature space $\mathbf{F}$ to a k-dimensional output space $\mathbf{Y}$. The model training dataset could be represented as a flat file, in which each row is a point in the cross-product $\mathbf{F} \times \mathbf{Y}$ and each column represent a coordinate dimension for such points (either in the feature space $\mathbf{F}$ or in the output space $\mathbf{Y}$.)

Among the many approaches for injecting diversity in the creation of an ensemble of models, we find *bagging, boosting, random subspace, randomization, and random forest*. Some of these approaches subsample the rows of the training set (points or examples), some other ones subsample the columns (features), and a few do both. Let us review some of these approaches in chronological order.

Bootstrap [41.68–70] or *bagging* [41.71] is arguably the oldest techniques for creating an ensemble of models. In this approach, diversity is obtained by building each model with a different set of examples, which are obtained from the original training dataset by *resampling the rows with replacement* (using a uniform probability distribution over the rows). *Bagging* combines the decisions of the classifiers using uniform-weighted voting. For each new training dataset, we must maintain the same number of rows as in the original training dataset, by sampling it that many times. Sampling with replacement leads $\approx 63.2\%$ unique rows. Sampling with replacement creates a series of indeendent Bernoulli trials, so the number of times a row is sampled from k trials out of N rows is $B(k, 1/N)$. For large values of N, the Bernoulli series can be approximated by a Poisson distribution with mean (k/N). Therefore, the proportion of rows not sampled will be approximately $e^{-k/N}$. In bootstrap, the number of samples is equal to the number of rows, i. e., $k = N$, so the Poisson approximation has a mean 1 and the proportion of sampled data is $(1 - e^{-k/N}) = (1 - e^{-1}) = 63.2\%$. As a result, we can achieve the same storage reduction by not duplicating the same rows and instead attach a count at the end of each sampled record to indicate the number of time it was selected. An interesting variation of this concept is the *Bag of Little Bootstraps* (BLBs) [41.72], which modifies the bootstrap approach to be usable with much larger data sets (where 63.2% of the original data would still be prohibitively large). Their proposed BLB approach performs a more drastic subsampling while maintaining the unbiased estimation and convergence rate of the original bootstrap method.

An alternative to *bagging* is *boosting*, which is rooted in the probably approximately correct (PAC) learning model [41.73–75]. Instead of training all classifiers in parallel (as in the case of bagging), we construct the ensemble in a serial fashion, by adding one model at a time. The model added to the ensemble at step j is trained on a dataset sampled selectively from the original dataset. The sampling distribution starts from a uniform distribution (as in bagging) and progresses toward increasing likelihood of misclassified examples in the new dataset. Thus, the distribution is modified at each step, increasing the likelihood of the examples misclassified by the classifier at step $(j-1)$ being in the training dataset at step j. Like *Bagging*, *Boosting* combines the decisions of the classifiers using uniform-weighted voting.

Adaboost (or *adapting boosting*) [41.74] extends *boosting* from binary to multiclass classification problems and regression models. It adapts the probability distribution over the rows in the training set to increase the difficulty of the training points by including more instances misclassified or wrongly predicted by previous models. *Adaboost* combines the decisions of the classifiers using a weighted voting. For regressions, it aggregates all the normalized confidences for the output. For multiclass classification it selects the class with the highest votes, calculated from the normalized classification errors of each class.

A different approach to inject diversity is to limit the number of columns (features), rather than the number of rows (points). *Ho's random subspaces* technique [41.76] selects random subsets of the available features to be used in training the individual classifiers in the ensemble.

Dietterich [41.77] introduced an approach called *randomization*. In this approach, at each node of each tree of the ensemble, the 20 best attributes to split the node are determined and one of them is randomly selected for use at that node.

Breiman [41.32] presented *random forest* ensembles, where bagging is used in combination with random feature subspace selection. At each node of each tree of the forest, a subset of m attributes (out of n available ones) is randomly selected, and the best split available based on the m attributes is selected for that node. Clearly, if m were too small, the tree performance would be severely affected, while if m were too close to the value of n, each tree performance would be higher, but diversity would suffer. In the case of *random forest*, a tradeoff between individual performance and overall diversity is achieved by using a value of m around the $\lfloor\sqrt{n}\rfloor$ for classification problems, and around $\lfloor n/3 \rfloor$ for regression problems.

Other approaches to increase diversity rely on the use of a high-level model to combine object-level models derived from different machine-learning techniques, e.g., *stacked generalization* [41.78]. Alternatively, we can inject structural diversity in the design of the object models by using different topologies/architectures in graphical models (e.g., neural networks) or different function sets/grammars in genetic programming algorithms to construct models [41.79].

The above approaches allow us to extract different types of information from the data, which should lead to lower error correlations among the models. With these approaches we can generate a space of diverse models, which can then be searched by *offline MHs* to tune and optimize the model ensemble, according to tradeoffs of performance and diversity.

41.5.4 Lazy Meta-Learning: A Model-Agnostic Fusion Mechanism

A second trend in the development of analytics models is the inevitable commoditization of object-level models. Multiple sources for model creation are now available, ranging from Crowdsourcing analytics by competition (e.g., [41.80]) or by collaboration (e.g., [41.81]) to cloud-based model automation tools, such as evolving model populations using genetic programming [41.79]. This situation creates different requirements for the fusion mechanism, which now should be *agnostic* with respect to the genesis of the object-level models in the ensemble.

We should note that all the previous approaches to fusion described in Sect. 41.5.3 use static fusion mechanisms, as they focus primarily on the creation of diverse base models (or object-level models). If we want to be agnostics with respect to these models, we need to have a *smarter* fusion mechanism, i. e., a meta model that can reason about the performance and applicability of the available object level models.

This issue is partially addressed by *Lazy Meta-Learning*, proposed in [41.82]. In this approach, for each query we instantiate a customized fusion mechanism. Such mechanism is a meta model, i. e., a model that operates on the object-level models whose predictions we want to fuse. Specifically, for a given query we dynamically (i. e., based on the query) create a model ensemble, followed by a customized fusion. The dynamic model ensemble consists of:

(1) Finding the most relevant object-level models from a DB of models, by matching their meta-information with the query.
(2) Identifying the relevant models with higher performance.
(3) Selecting a subset of models with highly uncorrelated errors to create the ensemble.

The customized fusion uses the meta-information of the models in the ensemble for dynamic bias compensation and relevance weighting. The output is a weighted interpolation or extrapolation of the outputs of the model ensemble.

More specifically the *Lazy Meta-Learning* process is divided into three stages:

- *Model creation*, an offline stage in which we create the initial building blocks for the assembly (or we collect them/acquire them from other sources) and we compile their meta-information
- *Dynamic model assembly*, an online stage in which, for a given query we select the best subset of models
- *Dynamic model fusion*, an online stage in which we evaluate the selected models and dynamically fuse them to solve the query.

Model Creation: The Building Blocks

We assume the availability of an initial training set that samples an underlying mapping from a feature space X to an output y. In the case of supervised learning, we also know the ground truth-value t for each record in the training set. We create a database DB of m diverse local or global models developed by any source. If we have control on the model creation, we can increase model diversity by any of the techniques described in Sect. 41.5.3. Every time, we add a model to the DB we need to capture its associated meta-information, i. e., information about the model itself, its training set, and its local/global performance. Such meta-information is used to create indices in the DB that will make its search more efficient. For each model M_i, we use a compiled summary of its performance, represented by a CART tree T_i, of depth d_i, and trained on the model error vector obtained during the validation of the model. To avoid overfitting, each tree is pruned to allow at least 25 points in each leaf node.

Dynamic Model Ensemble: Query–Driven Model Selection and Ensemble

This stage is divided into three steps:

- *Model Filtering*, in which we retrieve from the DB the applicable models for the given query. For a query q, the process starts with a set of constraints to define its related feasibility set. In this case the constraints are:

  (a) Model *soundness* and *competency* in its region of applicability (i. e., there must be sufficient points in the training set to develop a reliable model.

  (b) Model *vitality* (i. e., the model is up-to-date, not obsolete).

  (c) Model *applicability* to the query (i. e., the query is in the model's competence region).

  The intersection of these constraints satisfaction gives us a set of retrieved models for the query q. Let us denote the cardinality of this set as r.

- *Model Preselection*, in which we reduce the number of models, based on their local performance characteristics, such as bias, variability, and distance from the query. For a query q, having retrieved its feasible r models from the previous step, we classify the query using the same CART tree T_i, associated to each model, and reach leaf node $L_i(q)$. Each leaf node will be defined by its path to the root of the tree and will contain d_i constraints over (at most) d_i features. Leaf $L_i(q)$ provides performance estimates of model M_i in the region of the query. These estimates are used to retrieve the set of Pareto-best models to be used in the next step. Let us denote the cardinality of this set as p.

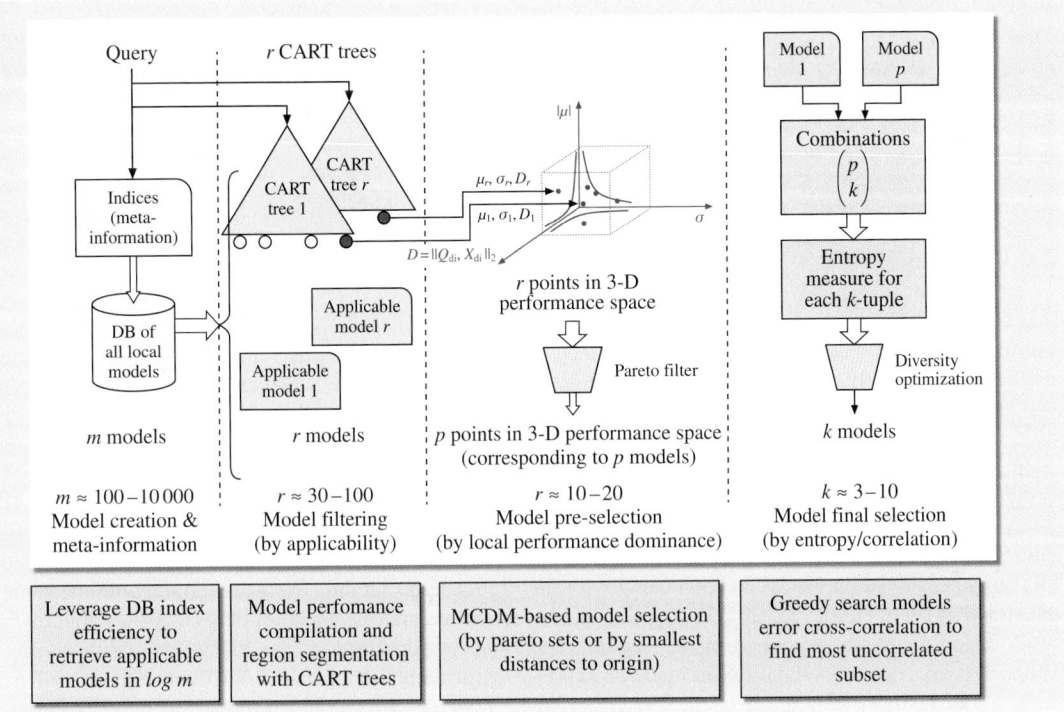

Fig. 41.14 Dynamic model ensemble on demand (filtering, selection)

- *Model Final Selection*, in which we define the final model subset. We need to use an ensemble whose elements have the most uncorrelated errors. We use the *Entropy Measure E*, proposed by Kuncheva and Whitaker, as the way to find the k most diverse models to form the ensemble. To avoid the intrinsic combinatorial complexity, we approximate our search via a greedy algorithm (further described in [41.82]).

The process is described in Fig. 41.14.

Dynamic Model Fusion: Generating the Answer
Finally, we evaluate the selected k models, compensate for their biases, and aggregate their outputs using a relevance weight that is a function of their proximity to the query to generate the solution to the query. This is illustrated in Fig. 41.15.

This approach was successfully tested against a regression problem for a coal-fired power plant optimization. The optimization problem, described in [41.83], required to adjust 20+ set-point values for the power

plant multivariable controller to match the generated power with the required *Load*, while minimizing emis-

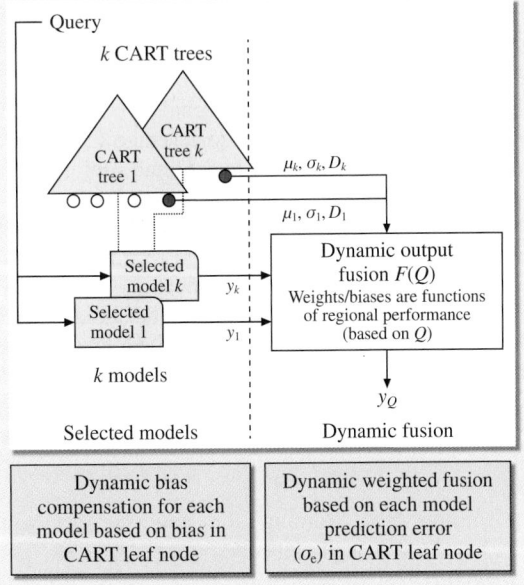

Fig. 41.15 Dynamic model fusion on demand ▶

Table 41.4 Static and dynamic fusion for: (a) 30 NNs (b) 45 SRs models; (c) 75 combined models

	NO$_x$			Heat rate			Load		
	(a) NN	(b) SR	(c) NN + SR	(a) NN	(b) SR	(c) NN + SR	(a) NN	(b) SR	(c) NN + SR
Baseline (average)	0.02279	0.03267	0.02877	91.79	109.15	101.14	1.0598	1.5275	1.3149
Dynamic fusion	0.01627	0.01651	0.01541	70.95	73.90	70.90	0.8474	0.8243	0.8209
Percentage gain	28.6%	49.5%	46.4%	22.7%	32.3%	29.9%	20.0%	46.0%	37.6%

sions (NO$_x$) and *heat rate* (inverse of efficiency). The optimization was predicated on having a reliable fitness function, i. e., a high fidelity mapping between the 20+ input vector and the three outputs (*load*, NO$_x$, *heat rate*).

In [41.82], we focused on generating this mapping via dynamic fusion. We used a data base of approximately 75 models: 30 neural networks (NNs) trained using bootstrapping and about 45 symbolic regression (SR) models evolved on the MIT Cloud using with the same training set of 5000+ records. We applied the dynamic fusion approach, described in this section, and evaluated it on a disjoint validation set made of 2200+ records. The results of the mean of the absolute error (MAE) computed over this validation set are summarized in Table 41.4.

The first conclusion is that dynamic fusion consistently outperformed static fusion, as shown by the percentage gain (last row in Table 41.4), which was computed as the difference between baseline and dynamic fusion, as a percentage of the baseline. In the cases of NO$_x$ and *load*, the baseline (average) for the 45 SR models was ≈50% worse than that of the 30 NNs. In creating the SR, we sacrificed individual performance to boost diversity, as described in Sect. 41.5.3. On the other hand, the NNs were trained for performance, while diversity was only partially addressed by bootstrap (but the NNs were trained with the same feature set and the same topology). After dynamic fusion the performance of the SR was roughly comparable with the NNs (within 2.7%), and the overall performance of the combined models was 3–5% better than that of the NNs alone. This experiment verified the importance of diversity during model creation. A more complete treatment of *Lazy Meta Learning*, including results of these experiments can be found in [41.82].

41.6 Summary and Future Research Challenges

We illustrated the use of CI techniques in ML applications. We explained how to leverage CI to build meta models (for offline design and online control/aggregation) and object-level models (for solving the problem at hand.) We described the most typical ML functions: unsupervised learning (clustering), supervised learning (classification and regressions), and optimization. To structure the cases studies described in this review, we presented two similar paradigms: PHM for industrial assets, and risk management for financial and commercial assets. We analyzed five case studies to show the use of CI models in:

(1) Unsupervised learning for anomaly detection (based on neural networks, fuzzy systems, and EAs).

(2) Supervised learning (classification) for assessing and pricing risk in insurance products (based on fuzzy systems and EAs).
(3) Supervised learning (regression) for valuating mortgage collaterals (based on radial basis functions, fuzzy systems, neural fuzzy systems, and fusion).
(4) Supervised learning (regression-induced ranking) for selecting the best units in a fleet (based on fuzzy systems and EAs).
(5) Multiobjective optimization for rebalancing a portfolio of investments (based on multiobjective EAs).

In the last section we covered model ensembles and fusion, and emphasize the need for injecting diversity during the model creation stage. We proposed a model–agnostics fusion mechanism that could be used to fuse commoditized models (such as the ones obtained y crowdsourcing). We will conclude this review with prospective view of research challenges for ML.

41.6.1 Future Research Challenges

All applications described in the five case studies were developed before the advent of cloud computing and big data. Since then, we have encountered situations in which we need to analyze very large data sets, enabled by the Internet of things (IoT) [41.84], machine-to-machine (M2M) connectivity, and social media. In this new environment, we need to scale up the CI/ML capabilities and address the underlying *three v's* in big data: *volume, velocity,* and *variability.* Large data volumes pose new challenges to data storage, organization, and query; data feed velocity requires novel streaming capabilities; data variability requires the collection and analysis of structured, unstructured, and semistructural (e.g., locational) data and the ability to learn across multiple modalities. The use of cloud computing will also result in the commoditization of analytics. In this context, ML applications will also include the delivery of *Analytics-as-a-Service* (AaaS) [41.85].

We will conclude this section with a view of five research challenges entailed by this new environment, as illustrated in Fig. 41.16:

1) *Data-driven Model Automation and Scalability*
   a. *Computation at the edge (velocity).* When the cost of moving large data set becomes significant, we need to perform analysis while the data are still in memory, via in-situ analytics and in-transit data transformation [41.86]. A variety of

solutions have been proposed, ranging from SW frameworks [41.87] to active Flash [41.88].
   b. *Technology stack for Big Data (volume).* We need to address scalability issues along many dimensions, such as data size, number of computational nodes over which to distribute the algorithms, number of models to be trained/deployed, etc. Among the research groups addressing this issue, we found the UC Berkeley AMP Lab [41.89] to be among the leaders in this area. The AMP researchers have developed the *Berkeley data analytics stack (BDAS)* [41.90], a technology stack composed of *Shark* (to run structured query language (SQL) and complex analytics on large clusters) [41.91, 92], *Spark* (to reuse working set of data across multiple parallel operations, typical of ML algorithms) [41.93], and *Mesos* (to share commodity clusters between multiple diverse cluster computing frameworks, such as *Hadoop* and message passing interface (*MPI*)) [41.94].
   c. *Parallelization of ML Algorithms (volume).* We need to design ML algorithms so that their computation can be distributed. Some algorithms are easy to parallelize, like population-based EAs (using an island [41.95, 96] or a diffusion grid models to distribute the subpopulations to many computational nodes), or *Random Forest* (growing subsets of trees on different computational nodes) [41.97]. Other algorithms will need to be redesigned for parallelization.
   d. *Multimodal learning (variability).* As the size of the information grows, its content will become

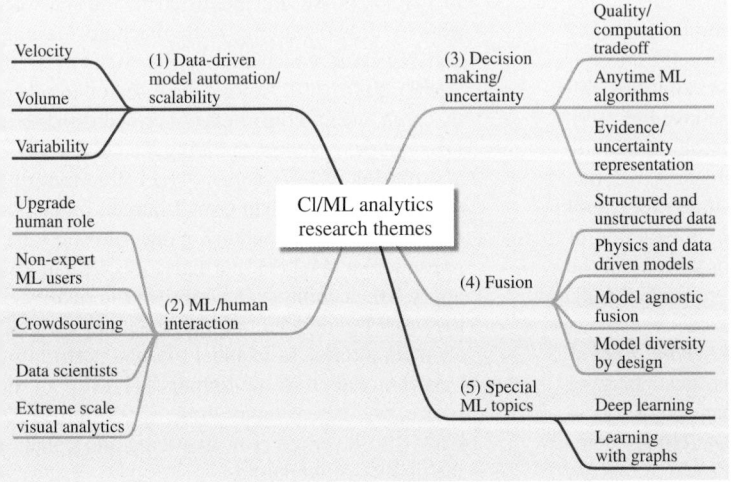

Fig. 41.16 Research challenges for CI/ML analytics

more heterogeneous. For instance, by navigating through web pages, we encounter information represented as text, images, audio, tables, video, applets, etc. There are preliminary efforts for representing and learning across multiple modalities using graphical approaches [41.98], and kernels [41.99, 100]. However, a comprehensive approach to this issue remains an open problem.

(2) *ML/Human Interactions*

a. *Upgrade the human role.* We need to remove the human modeler from the most time-consuming, iterative tasks, by automating data scrubbing (outliers removal, de-noising, imputation, or elimination of missing data) and data preparation (multiple sources integration, feature selection, feature generation). As noted in [41.86], while:

All high performance computing (HPC) components – power, memory, storage, bandwidth, concurrence, etc. – will improve performance by a factor of 3 to 4444 by 2018 . . . human cognitive capability will certainly remain constant.

We are obviously the bottleneck in any kind of automation process and we need to *upgrade* our role, by interacting with the process at higher levels of model design. For example, in active learning we maximize the information value of each additional question to be answered by a human expert [41.101, 102]. In interactive multicriteria decision making we can use *progressive preference articulation*. This allows the expert to guide the automated search in the design space, by interactively simplifying the problem, e.g., by transforming an objective into a constraint once the values of most solutions fall within certain ranges for that objective [41.103].

b. *Non-expert ML users.* For routine modeling task, we need to enable nonexpert users to define analytics in a declarative rather than procedural fashion. In [41.104] we can see an example of this concept, based on the analogy between the *MLbase* language for ML and traditional SQL languages for DBs.

c. *Integration of crowdsourcing with analytics engines.* Crowdsourcing is an emerging trend that is increasing human capacity in a manner similar to the way cloud computing is increasing computational capacity.

In this analogy, we could think of *Amazon Mechanical Turk* [41.105] as the dual of *Amazon Elastic Cloud Computing* [41.106]. Originated from the concept of *Wisdom of Crowds* [41.107], crowdsourcing has shown a tremendous growth [41.108]. According to *Malone* et al. [41.109], the crowd's contribution to the solution of a problem/task can be done via *collection*, *competition*, or *collaboration*.

i. *Collection* is used when the task can be decomposed into independent micro-tasks that are then executed by a large crowd to generate, edit, or augment information. An example of this case is the labeling of videos or images to create a training set for supervised learning [41.110]. The annotation of galaxy morphologies or lunar craters done at *Zooniverse* [41.111] is another example of this case.

ii. *Competition* is used when a single individual in the crowd can provide the complete solution to the task. An example of this case is the creation of data-driven models via contest hosted by sites such as *Kaggle* [41.80] or *CrowdANALYTICS* [41.81].

iii. *Collaboration* is used when single individuals in the crowd cannot provide a complete solution and the task cannot be decomposed into independent subtasks. This situation requires individuals to collaborate toward the generation of the solution. Examples of this case are usually large projects, such as the development of *Linux* or *Wikipedia*.

Additional crowdsourcing systems are surveyed in [41.112]. We are interested in the intersection of crowdsourcing with machine learning and big data, which is also the focus of UCB AMP Lab [41.89]. Among the research trends in this area, we find the impact of crowdsourcing in DBs queries, such as changing the closed-world assumption of DB queries [41.113], monitoring queries progress when crowdsourced inputs are expected [41.114], and using the crowd to interpret queries [41.115]. Furthermore, we can also apply ML techniques to improve the quality of the crowd-generated outputs, reducing variability in annotation tasks [41.116], and performing bias removal from the outputs [41.117]. In the future, we expect many more opportunities for CI/ML to leverage crowdsourcing and enhance the quality of its outputs.

d. *New generations of Data Scientists.* There is a severe skill gap, which we will need to overcome if we want to accelerate the applicability of ML to a broader set of problems. We need to train a new generation of data scientists, with skills in data flow (collection, storage, access, and mobility), data curation (preservation, publication, security, description, and cleanings), and basic analytics skills (applied statistics, MI/CI). Several universities are creating a customized curriculum to address this need. The National Consortium for Data Science is an illustrative example of this emerging trend [41.118].

e. *Extreme scale visual analytics.* We can cope with increases in data volume/velocity, and analysis complexity because we have benefited from similar increases in computing capacity. Unfortunately, as aptly noted in [41.86], *. . . there is no Moore's law for human cognitive abilities.* So, we face many challenges, described in the same reference, when we want to present/visualize the results of the complex analytics to the user (as in the case of multilevel hierarchies, time-dependent data, etc.) There are situations in which the user can select data with certain characteristics to be used in the analysis and steer data summarization and triage performed by the ML algorithms [41.86]. This will also overlap with the previous category of *upgrading the human role.*

(3) *Decision Making and Uncertainty*

a. *Quality/Computation tradeoff.* When faced with massively large data sets, we need to distribute data and models over multiple nodes. Often we also need to subsample the data sets while training the models. This might introduce biases and increase variances in the models results. The use of the *Bag of Little Bootstrap* [41.72] allows us to extend bootstrap (Sect. 41.5.3) to large data sets. However, not all queries or functions will work with bootstrap. *BlinkDB* is a useful tool to address this problem, as it allows us to understand if additional resources will actually improve the quality of the answer. *BlinkDB* is a [41.119]:

. . . massively parallel, sampling-based approximate query engine for running ad-hoc, interactive SQL queries on large volumes of data. It allows to trade-off query accuracy for response time, enabling interactive queries over massive data by running queries on data samples and presenting results annotated with meaningful error bars. . . .

b. *Anytime ML Algorithms.* Anytime algorithms are especially needed for online ML applications in which models need to produce results within a given real-time constraints [41.120]. In [41.121] the authors propose a method for determining when to terminate the learning phase of an algorithm (for a sequence of iid tasks) to optimize the expected average reward per unit time. In simpler situations, we want to be able to interrupt the algorithm and use its most recent cached answer. This idea is related to the quality/computation tradeoff of point (3.a), as we expect the quality of the answer to increase if we allow more computational resources. For example, in EAs, under convergence assumptions, we expect further generations to have a better fitness than the current one. At any time we can stop the EA and fetch the answer for the current population.

c. *Evidence and Uncertainty Representation.* This is one of the extreme-scale visual analytics challenges covered in [41.86]. In two previous points, we noted that as the data size increases we would need to perform data subsampling to meet real-time constraints. This subsampling will introduce even greater uncertainty in the process. We need better ways to quantify and visualize such uncertainty and provide the end-user with intuitive views of the information and its underlying risk.

(4) *Model Ensemble/Fusion*

a. *Integration of structured and unstructured data.* The simplest case is the integration of time dependent text (e.g., news, reports, logs) with time series data, e.g., text as a sensor. This topic is related to the multimodal learning discussion illustrated in (1.d). Alternatively, instead of learning across multiple data formats, we can use an ensemble of modality-specific learners and fuse their outputs. An example of this approach can be found in [41.122]

b. *Integration of physics-based and data-driven models.* There are at least two ways to perform such integration: using the models in parallel or serially. The simplest integration is based on a *parallel* architecture, in which both models are used simultaneously, yet separately. This case covers the use of data-driven models applied to the residuals between expected values (generated by physics-based models) and actual values (measured by sensors). This was illustrated in the first case study (Sect. 41.3.1). Another example of parallel integration is the use of an ensemble of physics-based and data-driven models, followed by an agnostic fusion mechanism (Sect. 41.5.4). A different type is the *serial integration*, in which one model is used to initialize the other one. An example of such integration is the use of data-driven models to generate estimates of parameters and initial conditions of physics based models. For instance, we could use data-driven RUL predictive models to estimate the current degree of degradation of key components in an electro-mechanical system. Then, we could run a physics-based model of the system, using these estimates as initial conditions, to determine the impact of future load scenarios to the RUL predictions. Another example of serial integration is the use of physics-based models to generate (offline) a large data set, usually following a Design of Experiment methodology. This data set becomes the training set for a data-driven model that will functionally approximate the physics-based model. Typically, a second data-driven model, frequently retrained with real-time data feeds, will be used to correct the outputs of the static approximation [41.123].

c. *Model-agnostics fusion.* In Sect. 41.5.4 we covered the concept of model-agnostic fusion to be deployed when predictive models are created by a variety of sources (such as crowd-sourcing via competition or cloud-based genetic programming). We showed that this type of fusion is a meta model that leverages each predictive model's meta-data, defining its region of applicability and relative level of performance [41.82]. Additional research is needed to prevent over-fitting of the meta models and to extend this concept to classification problems.

d. *Model diversity by design.* By leveraging the almost infinite computational capacity of the cloud, we should be able to construct model ensembles that are diverse by design. There are many techniques used to inject diversity in the models design, as described in Sect. 41.5.3. One of the most promising techniques con-

sists in evolving a large population of symbolic regression models by distributing genetic programming algorithms using an island approach [41.95, 96]. Random feature subsets can be assigned to each island, which also differ from the other islands through the use of distinct grammars, fitness functions, and functions sets. Additional information on this approach can be found in [41.79, 124, 125].

(5) *Special Topics in Machine Learning*

a. *Deep Learning.* Originally proposed by *Fukushima* [41.126], deep learning (DL) gained acceptance when Hinton's showed that DL training was decomposable [41.127, 128]. Hinton showed that each of the layers in the neural network could be pretrained one at a time, as an unsupervised *Restricted Boltzmann Machine*, and then fine-tuned using supervised backpropagation. This discovery allows us to use large (mostly unlabeled) data sets available from Big Data applications to train DL networks.

b. *Learning with Graphs.* There are many applications of Recommender Systems in social networks and targeted advertising. Typically these systems select information using collaborative filtering (CF), a technique based on collaboration among multiple agents, viewpoints, and data sources. Researchers have proposed various solutions to overcome some of the intrinsic challenges caused by data sparsity and network scalability. Among the most notable approaches we have:

(1) *Pregel* [41.129]: a synchronous message passing abstraction in which all vertex-programs run simultaneously in a sequence of super steps.

(2) *GraphLab* [41.130]: an asynchronous distributed, shared-memory abstraction, designed to leverage attributes typical of ML algorithms, such as sparse data with local dependences, iterative algorithms, and potentially asynchronous execution.

(3) *PowerGraph* [41.131] and its Spark implementation *GraphX* [41.132]: an abstraction combining the best features of *Pregel* and *GraphLab*, better suited for natural graphs applications with large neighborhoods and high degree vertices.

The above examples are just a sample of specialized ML algorithms and architectures for niche opportunities, exploiting the characteristic of their respective problems.

In conclusion, we need to shape CI research to address these new challenges. To remain a vital discipline during the next decade, we need to leverage the almost infinite capacity provided by cloud computing and crowdsourcing, understand the tradeoffs between solution quality and computational resource allocation, design fusion mechanisms for model ensembles derived from heterogeneous data, and create specialized architectures and algorithms to exploit problems characteristics.

References

41.1 L. Zadeh: Fuzzy logic and soft computing: Issues, contentions and perspectives, Proc. IIZUKA'94: 3rd Int. Conf. Fuzzy Logic Neural Nets Soft Comput. (1994) pp. 1–2

41.2 P. Bonissone: Soft computing: The convergence of emerging reasoning technologies, J. Res. Soft Comput. **1**(1), 6–18 (1997)

41.3 J.L. Verdegay, R. Yager, P. Bonissone: On heuristics as a fundamental constituent of soft computing, Fuzzy Sets Syst. **159**(7), 846–855 (2008)

41.4 J. Bezdek: On the relationship between neural networks, pattern recognition, and intelligence, Int. J. Approx. Reason. **6**, 85–107 (1992)

41.5 J. Bezdek: What is computational intelligence? In: *Computational Intelligence Imitating Life*, ed. by J. Zurada, R.I.I. Mark, C. Robinson (IEEE, New York 1994)

41.6 R.I.I. Mark: Computational versus artificial intelligence, IEEE Trans. Neural Netw. **4**(5), 737–739 (1993)

41.7 P. Bonissone, R. Subbu, N. Eklund, T. Kiehl: Evolutionary algorithms + domain knowledge = real-world evolutionary computation, IEEE Trans. Evol. Comput. **10**(3), 256–280 (2006)

41.8 P. Bonissone, K. Goebel, W. Yan: Classifier fusion using triangular norms, Proc. Multiple Classif. Syst. (MCS) (Cagliari, Italy 2004) pp. 154–163

41.9 P. Bonissone, N. Iyer: Soft computing applications to prognostics and health management (PHM): Leveraging field data and domain knowledge, 9th Int. Work-Conf. Artif. Neural Netw. (IWANN) (2007) pp. 928–939

41.10 P. Bonissone, X. Hu, R. Subbu: A systematic PHM approach for anomaly resolution: A hybrid neural fuzzy system for model construction, Proc. PHM 2009, San Diego (2009)

41.11 P. Bonissone, A. Varma: Predicting the best units within a fleet: Prognostic capabilities enabled by peer learning, fuzzy similarity, and evolutionary design process, Proc. FUZZ-IEEE 2005, Reno (2005)

41.12 P. Bonissone, A. Varma, K. Aggour, F. Xue: Design of local fuzzy models using evolutionary algorithms, Comput. Stat. Data Anal. **51**, 398–416 (2006)

41.13 P. Bonissone, R. Subbu, K. Aggour: Evolutionary optimization of fuzzy decision systems for automated insurance underwriting, Proc. FUZZ-IEEE 2002, Honolulu (2002) pp. 1003–1008

41.14 K. Aggour, P. Bonissone, W. Cheetham, R. Messmer: Automating the underwriting of insurance applications, AI Mag. **27**(3), 36–50 (2006)

41.15 P. Bonissone, W. Cheetham, D. Golibersuch, P. Khedkar: Automated residential property valuation: An accurate and reliable based on soft computing. In: *Soft Computing in Financial Engineering*, ed. by R. Ribeiro, H. Zimmermann, R.R. Yager, J. Kacprzyk (Springer, Heidelberg 1998)

41.16 R. Subbu, P. Bonissone, N. Eklund, S. Bollapragada, K. Chalermkraivuth: Multiobjective financial portfolio design: A hybrid evolutionary approach, Proc. IEEE Congr. Evol. Comput. (CEC 2005), Edinburgh (2005) pp. 1722–1729

41.17 P. Bonissone, F. Xue, R. Subbu: Fast meta-models for local fusion of multiple predictive models, Appl. Soft Comput. J. **11**(2), 1529–1539 (2011)

41.18 K. Goebel, R. Subbu, P. Bonissone: *Controller Adaptation to Compensate Deterioration Effects*, General Electric Global Research Technical Report, 2006GRC298 (2006)

41.19 R. Subbu, P. Bonissone, N. Eklund, W. Yan, N. Iyer, F. Xue, R. Shah: Management of complex dynamic systems based on model-predictive multi-objective optimization, Proc. IEEE CIMSA (La Coruña, Spain 2006) pp. 64–69

41.20 R. Subbu, P. Bonissone: A retrospective view of fuzzy control of evolutionary algorithm resources, Proc. FUZZ-IEEE 2003, St. Louis (2003) pp. 143–148

41.21 P. Bonissone: Soft computing: A continuously evolving concept, Int. J. Comput. Intell. Syst. **3**(2), 237–248 (2010)

41.22 A. Patterson, P. Bonissone, M. Pavese: Six sigma quality applied throughout the lifecycle of and automated decision system, J. Qual. Reliab. Eng. Int. **21**(3), 275–292 (2005)

41.23 T. Kohonen: Self-organized formation of topologically correct feature maps, Biol. Cybern. **43**, 59–69 (1982)

41.24 R.S. Sutton, A.G. Barto: *Reinforcement Learning* (MIT, Cambridge 1998)

41.25 F. Woergoetter, B. Porr: Reinforcement learning Scholarpedia **B**(3), 1448 (2008)

41.26   L. Breiman, J. Friedman, R.A. Olshen, C.J. Stone: *Classification and Regression Trees* (Wadsworth, Belmont 1984)

41.27   J.R. Quinlan: *C4.5: Programs for Machine Learning* (Morgan Kaufmann, San Francisco 1993)

41.28   C.G. Atkeson: Memory-based approaches to approximating continuous functions. In: *Nonlinear Modeling and Forecasting*, ed. by M. Casdagli, S. Eubank (Addison-Wesley, Redwood City 1992) pp. 503–521

41.29   C.G. Atkeson, A. Moore, S. Schaal: Locally weighted learning, Artif. Intell. Rev. **11**(1–5), 11–73 (1997)

41.30   B.D. Ripley: *Pattern Recognition and Neural Networks* (Cambridge Univ. Press, Cambridge 1996)

41.31   T. Hastie, R. Tibshirani, J. Friedman: *The Elements of Statistical Learning: Data Mining, Inference, and Prediction* (Springer, Berlin, Heidelberg 2010)

41.32   L. Breiman: Random forests, Mach. Learn. **45**(1), 5–32 (2001)

41.33   C.Z. Janikow: Fuzzy decision trees: Issues and methods, IEEE Trans. Syst. Man Cybern. – Part B **28**(1), 1–15 (1998)

41.34   P. Bonissone, J.M. Cadenas, M.C. Garrido, R.A. Diaz: A fuzzy random forest, Int. J. Approx. Reason. **51**(7), 729–747 (2010)

41.35   P. Bonissone: Soft computing applications in PHM. In: *Computational Intelligence in Decision and Control*, ed. by D. Ruan, J. Montero, J. Lu, L. Martinez, P.D. D'hondt, E. Kerre (World Scientific, Singapore 2008) pp. 751–756

41.36   P. Bonissone, V. Badami, K.H. Chiang, P.S. Khedkar, K.W. Marcelle, M.J. Schutten: Industrial applications of fuzzy logic at general electric, Proc. IEEE **83**(3), 450–465 (1995)

41.37   D. Filev: Gain scheduling based control of a class of TSK systems. In: *Fuzzy Control Synthesis and Analysis*, ed. by S. Farinwata, D. Filev, R. Langari (Wiley, New York 2000) pp. 321–334

41.38   P. Bonissone, K. Chiang: Fuzzy logic hierarchical controller for a recuperative turboshaft engine: From mode selection to mode melding, Industrial Applications of Fuzzy Control and Intelligent Systems, ed. by J. Yen, R. Langari, L. Zadeh (1995) pp. 131–156

41.39   M.A. Kramer: Autoassociative neural networks, Comput. Chem. Eng. **16**(4), 313–328 (1992)

41.40   J.W. Hines, I.E. Uhrig: Use of autoassociative neural networks for signal validation, J. Intell. Robot. Syst. **21**(2), 143–154 (1998)

41.41   B. Lerner, H. Guterman, M. Aladjem, I. Dinstein: A comparative study of neural network based feature extraction paradigms, Pattern Recognit. Lett. **20**, 7–14 (1999)

41.42   R.C. Eberhart, S. Yuhui: *Computational Intelligence: Concepts to Implementations* (Morgan Kaufmann, San Francisco 2007)

41.43   D. Poole, A. Mackworth, R. Goebel: *Computational Intelligence* (Oxford Univ. Press, Oxford 1998)

41.44   R. Tadeusiewicz: *Modern Computational Intelligence Methods for the Interpretation of Medical Images*, Vol. 84 (Springer, Berlin, Heidelberg 2008)

41.45   F. Steinman, K.-P. Adlassnig: Fuzzy medical diagnosis. In: *Handbook of Fuzzy Computation*, ed. by E.H. Ruspini, P.P. Bonissone, W. Pedrycz (IOP Publ., Bristol 1998) pp. 1–14

41.46   W. Yan, F. Xue: Jet engine gas path fault diagnosis using dynamic fusion of multiple classifiers, Neural Netw. 2008, IJCNN, Hong Kong (2008) pp. 1585–1591

41.47   D. Cayrac, D. Dubois, H. Prade: Possibilistic handling of uncertainty in fault diagnosis. In: *Handbook of Fuzzy Computation*, ed. by E.H. Ruspini, P.P. Bonissone, W. Pedrycz (Institute of Physics Publishing, Bristol 1998) pp. 1–7

41.48   D. Prokhorov (Ed.): *Computational Intelligence in Automotive Applications* (Springer, Berlin 2008)

41.49   H. Bersini, G. Bontempi, M. Birattari: Is readability compatible with accuracy? From neuro-fuzzy to lazy learning, Proc. Artif. Intell., Berlin, Vol. 7, ed. by C. Freksa (1998) pp. 10–25

41.50   A. Freitas: *Data Mining and Knowledge Discovery with Evolutionary Algorithms* (Springer, Berlin 2002)

41.51   K. Deb: *Multi-Objective Optimization Using Evolutionary Algorithms* (Wiley, Chichester 2001)

41.52   C.A.C. Coello, G.B. Lamont, D.A. Van Veldhuizen: *Evolutionary Algorithms for Solving Multi-Objective Problems* (Kluwer, New York 2002)

41.53   P. Bonissone, Y.-T. Chen, K. Goebel, P. Khedkar: Hybrid soft computing systems: Industrial and commercial applications, Proc. IEEE **87**(9), 1641–1667 (1999)

41.54   P. Bonissone, R. Subbu, J. Lizzi: Multi-criteria decision making (MCDM): A framework for research and applications, IEEE Comput. Intell. Mag. **4**(3), 48–61 (2009)

41.55   J.S.R. Jang: ANFIS: Adaptive-network-based-fuzzy-inference-system, IEEE Trans. Syst. Man Cybern. **23**(3), 665–685 (1993)

41.56   P. Bonissone, W. Cheetham: Fuzzy case-based reasoning for residential property valuation. In: *Handbook of Fuzzy Computing*, ed. by E.H. Ruspini, P.P. Bonissone, W. Pedrycz (Institute of Physics, Bristol 1998), Section G14.1

41.57   K.C. Chalermkraivuth: *Analytical Approaches for Multifactor Risk Measurements* GE Tech. Inform. 2004GRC184 (GE Global Research, Niskayuna 2004)

41.58   J.C. Hull: *Options, Futures and Other Derivatives* (Prentice Hall, Upper Saddle River 2000)

41.59   J. Kubalik, J. Lazansky: Genetic algorithms and their tuning. In: *Computing Anticipatory Systems*, ed. by D.M. Dubois (American Institute Physics, Liege 1999) pp. 217–229

41.60   N. Eklund: Multiobjective Visible Spectrum Optimization: A Genetic Algorithm Approach (Rensselaer, Troy 2002)

1.61   G. Fumera, F. Roli: A theoretical and experimental analysis of linear combiners for multiple classifier systems, IEEE Trans. Pattern Anal. Mach. Intell. **27**(6), 942–956 (2005)

1.62   T.G. Dieterich: Ensemble methods in machine learning, 1st Int. Workshop Mult. Classi. Syst. (2000) pp. 1–15

1.63   K. Tume, J. Ghosh: Error correlation and error reduction in ensemble classifiers, Connect. Sci. **8**, 385–404 (1996)

1.64   R. Polikar: Ensemble learning, Scholarpedia **4**(1), 2776 (2009)

1.65   F. Roli, G. Giacinto, G. Vernazza: Methods for designing multiple classifier systems, Lect. Notes Comput. Sci. **2096**, 78–87 (2001)

1.66   L. Kuncheva, C. Whitaker: Ten measures of diversity in classifier ensembles: Limits for two classifiers, Proc. IEE Workshop Intell. Sens. Process, Birmingham (2001), p. 10/1–6

1.67   L.I. Kuncheva: *Combining Pattern Classifiers: Methods and Algorithms* (Wiley, New York 2004)

1.68   B. Efron: Bootstrap methods: Another look at the jackknife, Ann. Stat. **7**(1), 1–26 (1979)

1.69   B. Efron: More efficient bootstrap computations, J. Amer. Stat. Assoc. **85**(409), 79–89 (1988)

1.70   B. Efron, R. Tibshirani: *An Introduction to the Bootstrap* (CRC, Boca Raton 1993)

1.71   L. Breiman: Bagging predictors, Mach. Learn. **24**(2), 123–140 (1996)

1.72   A. Kleiner, A. Talwalkar, P. Sarkar, M.I. Jordan: A scalable bootstrap for massive data, J. Roy. Stat. Soc. **76**(4), 795–816 (2014)

1.73   M.J. Kearns, U.V. Vazirani: *An Introduction to Computational Learning Theory* (MIT, Cambridge 1994)

1.74   Y. Freud, R.E. Schapire: A decision-theoretic generalization of on-line learning and an application to boosting, J. Comput. Syst. Sci. **55**(1), 119–139 (1997)

1.75   R.E. Schapire: Theoretical views of boosting, Proc. 4th Eur. Conf. Comput. Learn. Theory (1999) pp. 1–10

1.76   T.K. Ho: The random subspace method for constructing decision forests, IEEE Trans. Pattern Anal. Mach. Intell. **20**(8), 832–844 (1998)

1.77   T.G. Dieterich: An experimental comparison of three methods for constructing ensembles of decision trees: Bagging, boosting, and randomization, Mach. Learn. **40**(2), 139–157 (2000)

1.78   D.H. Wolpert: Stacked generalization, Neural Netw. **5**, 241–251 (1992)

1.79   D. Sherry, K. Veeramachaneni, J. McDermott, U.M. O'Reilly: FlexGP: Genetic programming on the cloud, Lect. Notes Comput. Sci. **7248**, 477–486 (2012)

1.80   A. Vance: Kaggle Contests: Crunching number for fame and glory, Businessweek, January 4, 2012

1.81   G. Stolovitsky, S. Friend: Dream Challenges: http://dreamchallanges.org

41.82   P. Bonissone: Lazy meta-learning: Creating customized model ensembles on demand, Lect. Notes Comput. Sci. **7311**, 1–23 (2012)

41.83   F. Xue, R. Subbu, P. Bonissone: Locally weighted fusion of multiple predictive models, IEEE Int. Jt. Conf. Neural Netw. (IJCNN'06), Vancouver (2006) pp. 2137–2143

41.84   M.A. Feki, F. Kawsar, M. Boussard, L. Trappeniers: The Internet of things: The next technological revolution, Computer **46**(2), 24–25 (2013)

41.85   Q. Chen, M. Hsu, H. Zeller: Experience in continuous analytics as a service (CaaaS), Proc. 14th Int. Conf. Extending Database Technol., EDBT/ICDT '11 (2011) pp. 509–514

41.86   P.C. Wong, H.-W. Shen, C.R. Johnson, C. Chen, R.B. Ross: The Top 10 challenges in extreme-scale visual analytics, IEEE Comput. Graph. Appl. **32**(4), 64–67 (2012)

41.87   M. Parashar: Addressing the petascale data challenge using in-situ analytics, Proc. 2nd Int. Workshop on Petascale Data Anal., PDAC'11 (2011) pp. 35–36

41.88   D. Tiwari, S. Boboila, S. Vazhkudai, Y. Kim, X. Ma, P. Desnoyers, Y. Solihin: Active flash: Towards energy-efficient, in-situ data analytics on extreme-scale machines, file and storage technologies (FAST), 11th USENIX Conf. File Storage Technol., FAST'13, San Jose (2013)

41.89   U. C. Berkeley: AMP Lab, https://amplab.cs.berkeley.edu/about/

41.90   Berkeley Data Analytics Stack (BDAS), https://amplab.cs.berkeley.edu/software/

41.91   C. Engle, A. Lupher, R. Xin, M. Zaharia, M. Franklin, S. Shenker, I. Stoica: Shark: Fast data analysis using coarse-grained distributed memory, ACM SIGMOD Conf. (2012)

41.92   R. Xin, J. Rosen, M. Zaharia, M. Franklin, S. Shenker, I. Stoica: Shark: SQL and rich analytics at scale, ACM SIGMOD Conf. (2013)

41.93   M. Zaharia, M. Chowdhury, M. Franklin, S. Shenker, I. Stoica: *Spark: Cluster Computing with Working Sets* (HotCloud, Boston 2010)

41.94   B. Hindman, A. Konwinski, M. Zaharia, A. Ghodsi, A. Joseph, R. Katz, S. Shenker, I. Stoica: Mesos: A platform for fine-grained resource sharing in the data center, Proc. Netw. Syst. Des. Implement., NSDI (2011)

41.95   D. Whitley, S. Rana, R.B. Heckendorn: Island model genetic algorithms and linearly separable problems, Lect. Notes Comput. Sci. **1305**, 109–125 (1997)

41.96   D. Whitley, S. Rana, R.B. Heckendorn: The Island model genetic algorithm: On separability, population size and convergence, J. Comput. Inf. Technol. **7**, 33–48 (1999)

41.97   L. Mitchell, T.M. Sloan, M. Mewissen, P. Ghazal, T. Forster, M. Piotrowski, A.S. Trew: A parallel random forest classifier for R, Proc. 2nd Int. Workshop

Emerg. Comput. Methods Life Sci., ECMLS '11, San Jose (2011) pp. 1–6

41.98   H. Tong, J. He, M. Li, C. Zhang, W.-Y. Ma: Graph based multi-modality learning, Proc. 13th ACM Int. Conf. Multimed., MULTIMEDIA '05 (2005) pp. 862–871

41.99   L. Barrington, D. Turnbull, G.R.C. Lanckriet: Game-powered machine learning, Proc. Natl. Acad. Sci. (2012) pp. 6411–6416

41.100   B. McFee, G.R.G. Lanckriet: Learning content similarity for music recommendation, IEEE Trans. Audio Speech Lang. Process. **20**, 2207–2218 (2012)

41.101   B. Settles: *Active Learning Literature Survey*, Comput. Sci. Tech. Rep. Vol. 1648 (University of Wisconsin, Madison 2009)

41.102   N. Rubens, D. Kaplan, M. Sugiyama: Active learning in recommender systems. In: *Recommender Systems Handbook*, ed. by F. Ricci, P.B. Kantor, L. Rokach, B. Shapira (Springer, Berlin, Heidelberg 2011) pp. 735–767

41.103   S. Adra, I. Griffin, P.J. Fleming: A comparative study of progressive preference articulation techniques for multiobjective optimisation, Lect. Notes Comput. Sci. **4403**, 908–921 (2007)

41.104   T. Kraska, A. Talwalkar, J. Duchi, R. Griffith, M. Franklin, M. Jordan: MLbase: A distributed machine-learning system, Conf. Innov. Data Syst. Res., CIDR, Asilomar (2013)

41.105   Amazon Mechanical Turk: https://www.mturk.com/mturk/welcome

41.106   Amazon Elastic Cloud Computing: http://aws.amazon.com/ec2/

41.107   J. Surowiecki: *The Wisdom of Crowds* (Random House, New York 2004)

41.108   J. Howe: *Crowdsourcing* (Random House, New York 2008)

41.109   T. Malone, R. Laubacher, C. Dellarocas: The collective intelligence genome, MIT Sloan Manag. Rev. **51**(3), 21–31 (2010)

41.110   L. Zhao, G. Sukthankar, R. Sukthankar: Robust active learning using crowdsourced annotations for activity recognition, AAAI Workshop Hum. Comput. (2011) pp. 74–79

41.111   Zooniverse: https://www.zooniverse.org, 2007

41.112   M.-C. Yuen, I. King, K.-S. Leung: A survey of crowdsourcing systems, 2011 IEEE Int. Conf. Priv. Secur. Risk Trust IEEE Int. Conf. Soc. Comput. (2011) pp. 766–773

41.113   A. Feng, M. Franklin, D. Kossmann, T. Kraska, S. Madden, S. Ramesh, A. Wang, R. Xin: CrowdDB: Query processing with the VLDB crowd, Proc. VLDB (2011)

41.114   B. Trushkowsky, T. Kraska, M. Franklin, P. Sarkar: Crowdsourced enumeration queries, Int. Conf. Data Eng., ICDE (2013)

41.115   G. Demartini, B. Trushkowsky, T. Kraska, M. Franklin: CrowdQ: Crowdsourced query understanding, Int. Conf. Data Eng. (ICDE) (2013)

41.116   Q. Liu, J. Peng, A. Ihler: Variational inference for crowdsourcing, Adv. Neural Inf. Process. Syst., NIPS 25, La Jolla, ed. by F. Pereira, C.J.C. Burges, L. Bottou, K.Q. Weinberger (2012) pp. 701–709

41.117   F. Wauthier, M. Jordan: Bayesian bias mitigation for crowdsourcing, Adv. Neural Inf. Process. Syst., NIPS 25, La Jolla, ed. by F. Pereira, C.J.C. Burges, L. Bottou, K.Q. Weinberger (2012) pp. 1800–1808

41.118   The National Consortium for Data Science: http://data2discovery.org

41.119   S. Agarwal, A. Panda, B. Mozafari, S. Madden, I. Stoica: BlinkDB: Queries with bounded errors and bounded response times on very large data, Proc. 8th ACM Eur. Conf. Comput. Syst. (2013) pp. 29–42

41.120   G.I. Webb: Anytime learning and classification for online applications, Proc. 2006 Conf. Adv. Intel. IT, Active Media Technol. 2006, ed. by Y. Li, M. Looi, N. Zhong (IOS, Amsterdam 2006)

41.121   B. Pöczos, Y. Abbasi-Yadkori, C. Szepesvári, R. Greiner, N. Sturtevant: Learning when to stop thinking and do something!, Proc. 26th Int. Conf. Mach. Learn., ICML'09 (2009) pp. 825–832

41.122   D.D. Palmer, M.B. Reichman, N. White: Multimedia information extraction in a live multilingual news monitoring system. In: *Multimedia Information Extraction: Advances in Video, Audio and Imagery Analysis for Search, Data Mining, Surveillance, and Authoring*, ed. by M.T. Maybury (Wiley, New York 2012) pp. 145–157

41.123   R.V. Subbu, L.M. Fujita, W. Yan, N.D. Ouellette, R.J. Mitchell, P.P. Bonissone, R.F. Hoskin: Method and System to Predict Power Plant Performance, US2 L3 A1 201 2008 3933 (2012)

41.124   D. Wilson, K. Veeramachaneni, U.M. O'Reilly: *Large Scale Island Model CMA-ES for High Dimensional Problems*, EVOPAR In EvoApplications 2013, Vienna (Springer, Berlin, Heidelberg 2013)

41.125   O. Derby, K. Veeramachaneni, E. Hemberg, U.M. O'Reilly: *Cloud Driven Island Model Genetic Programming*, EVOPAR In EvoApplications 2013, Vienna (Springer, Berlin, Heidelberg 2013)

41.126   K. Fukushima: Neocognitron: A self-organizing neural network model for a mechanism of pattern recognition unaffected by shift in position, Biol. Cybern. **36**(4), 193–202 (1980)

41.127   G. Hinton: A fast learning algorithm for deep belief nets, Neural Comput. **18**(7), 1527–1554 (2006)

41.128   G. Hinton: Learning multiple layers of representation, Trends Cogn. Sci. **11**, 10 (2007)

41.129   G. Malewicz, M.H. Austern, A.J.C. Bik, J.C. Dehnert, I. Horn, N. Leiser, G. Czajkowski: Pregel: A system for large-scale graph processing, SIGMOD'10, Proc. 2010 ACM SIGMOD Int. Conf. Manag. Data (2010) pp. 135–146

41.130   Y. Low, J. Gonzalez, A. Kyrola, D. Bickson, C. Guestrin, J.M. Hellerstein: Distributed GraphLab: A framework for parallel machine learning and data mining in the cloud, Proc. VLDB Endow. (2012) pp. 716–727

41.131   J. Gonzalez, Y. Low, H. Gu, D. Bickson, C. Guestrin: PowerGraph: Distributed graph-parallel compu-

tation on natural graphs, OSDI'12 (2012), Vol. 12, No. 1, p. 2

41.132   R. Xin, J. Gonzalez, M. Franklin, I. Stoica: GraphX: A Resilient Distributed Graph System on Spark, Proc. 1st Int. Workshop Graph Data Manag. Exp. Syst., GRADES 2013 (2013)

Part E

Evolution

Part E Evolutionary Computation

Ed. by Frank Neumann, Carsten Witt, Peter Merz, Carlos A. Coello Coello,
Thomas Bartz-Beielstein, Oliver Schütze, Jörn Mehnen, Günther Raidl

42. Genetic Algorithms

Jonathan E. Rowe

While much has been discovered about the properties of simple evolutionary algorithms (EAs), based on evolving a single individual, there is very little work touching on the properties of population-based systems. We highlight some of the work that does exist in this chapter.

Genetic algorithms (GA) are a particular class of evolutionary computation method often used for optimization problems. They were originally introduced by *Holland* [42.1] at around the same time when other evolutionary methods were being developed, and popularized by *Goldberg's* much-cited book [42.2]. They are characterized by the maintenance of a *population* of search points, rather than a single point, and the evolution of the system involves comparisons and interaction between the points in the population. They are usually used for combinatorial optimization problems, that is where the search space is a finite set (typically with some structure). The most common examples use fixed-length binary strings to represent possible solutions, though this is by no means always the case – the search space representation should be chosen to suit the particular problem class of interest. Members of the search space are then evaluated via a *fitness function*, which determines how well they solve the particular problem instance. This is, of course, by analogy with natural selection in which the fittest survives and evolves to produce better and better solutions. Notice that the efficiency of a genetic algorithm is therefore measured in terms of the number of evaluations of the fitness function required to solve the problem (rather than a more direct measure of the computational complexity). For any well-defined problem class, the maximum number of function evaluations required by an algorithm to solve a problem instance is called the *black box* complexity of the algorithm for that problem class. The black box complexity of a problem class is defined to be the maximum number of function evaluations required by the best possible black box algorithm. This is a research topic in its own right [42.3].

As an example, consider the subset sum problem, in which a set of n integers is given, along with a target integer T, and we have to find a subset whose sum is as close to T as possible. We can represent subsets as binary strings of length n, in which a 1 indicates that an element is in the subset and a 0 that it is excluded. The binary string forms the analog of the DNA (deoxyribonucleic acid) of the corresponding individual solution. Its fitness would then be given by the corresponding sum – which we are trying to minimize.

For a different example, consider the traveling salesman problem, in which there are a number of cities to be visited. We have to plan a route to visit each city once and return home, while minimizing the distance traveled. A potential solution here could be expressed as a permutation of the list of cities (acting as the DNA), with the fitness given by the corresponding distance.

A genetic algorithm maintains a population of such solutions and their corresponding fitnesses. By focusing on the better members of the population and introduc-

ing small variations (or mutations), we hope that the population will evolve good, or even optimal, solution in a reasonable time.

A popular general introduction to the field can be found in *Mitchell*'s book [42.4]. Much of what is known about the theory of genetic algorithms was developed by *Vose* and colleagues [42.5], which centers around a description of the changing population as a dynamic system. A gentle overview of this theory can be found in [42.6]. More recently, there has been a stronger emphasis on understanding the algorithmic aspects, with a particular focus on run-time analysis for optimization problems [42.7–9].

42.1 Algorithmic Framework

The genetic algorithm works by updating the population in discrete iterations, called *generations*. We begin with an initial, randomly generated, population. This acts as a set of *parents* from which a number of *offspring* are produced, from which the next generation is created. There are two basic schemes for doing this: the generational method and the steady-state method.

The generational approach is to repeatedly produce offspring from the parent population, until there are enough to fill up a whole new population. One generation, in this case, corresponds to the creation of all of these offspring. The steady-state approach, by contrast, produces a single offspring from the current parents, and then inserting it into the population, replacing some individual. Here, a generation consists of creating one new individual solution.

In either case, the population size stays fixed at its initial size, which is a parameter of the algorithm. There exists some theoretical work investigating a good choice of population size in different situations, but there are few general principles [42.10]. The correct size will depend on the problem to be optimized, and the particular details of the rest of the algorithm. It should be noted, however, that in a number of cases it can be shown that smaller population sizes are preferable and, indeed, in some cases a *population* of size one is sufficient.

The overall structure of the generational genetic algorithm is as follows:

1. Initialize population of size μ randomly with points from the search space.
2. Repeat until stopping criterion is satisfied:
   a) Repeat μ times:

   i. Choose a point from the population.
   ii. Modify the point with *mutation* and *crossover*.
   iii. Place resulting offspring in the new population.
3. Stop.

The critical points are therefore the selection method, used to choose points from the current population, and the mutation and crossover methods used to modify the chosen points. The idea is that selection will favor better solutions (in the sense that they provide better solutions to the optimization problem at hand), that mutation will introduce slight variations in the currently chosen solutions, and that crossover will combine together parts of different good solutions to, hopefully, form a better combination. We will look at different schemes for selection, mutation, and crossover in detail later, in Sects. 42.2, 42.4, and 42.6.

The overall structure of the steady-state genetic algorithm is as follows:

1. Initialize population of size μ randomly with points from the search space.
2. Repeat until stopping criterion is satisfied:
   a) Choose a point from the population.
   b) Modify the point with *mutation* and *crossover*.
   c) Choose an existing member of the population.
   d) Replace that member with the new offspring.
3. Stop.

It can be seen that, in addition to selection, mutation, and crossover, the steady-state genetic algorithm also requires us to specify a means for choosing an individual to be replaced. Suitable replacement strategies will be discussed later in Sect. 42.3, but for now we make a few general observations.

It is clear that in the generational genetic algorithm, progress is driven by the selection method. In this step of the algorithm, we choose to keep those solutions which we prefer, by dint of the degree to which they optimize the problem we are trying to solve. For the steady-state genetic algorithm, progress can also be maintained by the replacement strategy, if this is designed so as to affect the replacement of poorly performing individuals. In fact, it is possible to put the whole burden of evolution on the replacement step, for example, by always replacing the worst member of the population, and allowing the selection step to choose any individual uniformly at random. Conversely, one may use a stronger selection method and replace individuals randomly (or, of course, some combination of the two approaches). The steady-state algorithm therefore allows the user finer tuning of the strength of selective pressure.

In addition, the steady-state approach allows the user to guarantee that good individuals are never lost, by choosing a replacement strategy that protects such individuals. For example, replacing the worst individual each time ensures that copies of the best individual always remain. Any EA that has the property of preserving the best individual is called *elitist*. This would seem to be a desirable property as otherwise progress toward the optimum can be lost due to mutation and selection (Sect. 42.5). The generational framework, as it stands, offers no such guarantee. Indeed, depending on the method chose, the best individual may not even be selected, let alone preserved! It is quite a common strategy, therefore, to add elitism to the generational framework, for example, by making sure at least one copy of the best individual is copied across to the next generation each time.

The generational approach, without elitism explicitly added, is referred to as a *comma* strategy. In particular, it is a (μ, μ) EA. This means that the population has size μ, and a further μ offspring are created, which becomes the next generation population. More generally, we could have (μ, λ) algorithms, where $\lambda \geq \mu$, in which λ offspring are created and the best μ are taken to be the next generation. If λ is sufficiently large with respect to μ, then the probability of not selecting the best individual gets rather small. If, in addition, there is a reasonable chance that mutation and crossover do not make any changes, we get an approximation of elitism.

The steady-state algorithm, in which one replaces the worst individual, is referred to as a *plus* strategy.

In particular, it is a $(\mu + 1)$ EA. This means that one offspring is created from a population of size μ, and then the best μ are kept from the pool of parents plus offspring. More generally, we could have $(\mu + \lambda)$ algorithms, in which λ offspring are created and the best μ from the collection of parents and offspring are kept to be the next generation. *Plus* strategies are, of course, elitist.

Another advantage of the steady-state genetic algorithm is that progress can be immediately exploited. As soon as an improving individual is found, it is inserted into the population, and may be selected for further evolution. The generational algorithm, in contrast, has to produce a full set of offspring before any good discoveries can be built on. This overhead can be minimized by taking the population to be as small as possible, although one then risks losing good individuals in the selection process, unless some form of elitism is explicitly implemented.

A further extension often made to either scheme is to implement a mechanism for maintaining a level of diversity in the population. Clearly, a potential advantage of having a population is that it can cover a broad area of the search space and allow for more effective searching than if a single individual were used. Any such advantage would be lost, however, if the members of the population end up identical or very similar. In particular, the effectiveness of crossover is reduced, or even eliminated, if the population members are too similar. Some of the methods employed for maintaining diversity will be discussed in detail in Sect. 42.7.

Before we move on to details, it is worth asking whether or not maintaining a population is an effective approach to optimization. Indeed, such a question should be asked seriously whenever one attempts to use EAs for such problems. For certain classes of problem, it may well be the case that a simple local search strategy (that is, a $(1 + 1)$ evolutionary algorithm (EA)) may be as effective or better. For example, for some problems (such as OneMax, which simply totals the bit values in a string), the $(1 + 1)$-EA is provably optimal amongst evolutionary algorithms that only use standard bitwise mutation [42.11]. There is a small, but limited, amount of theoretical work on this issue to guide us [42.10]. In the first place, if the search space comprises islands of good solutions with small gaps containing poor solutions, then a population might provide an effective way to jump across the gaps. This is because the poorer offspring are not so readily rejected (especially with a generational approach), and may persist long enough for a lucky mutation to move into a neighboring good region. One might therefore also expect a population-based approach to be effective generally on highly multimodal problems but, unfortunately, there is very little analysis on this situation. If crossover is thought to be effective for your problem (Sect. 42.6), then a population is required, as we need to choose pairs of parents – although again, it might be the case that very small populations are sufficient. It certainly seems to be the case that for any such advantage, it is necessary to maintain a reasonable level of diversity, or the point of the population is lost [42.12].

One situation in particular, where a genetic algorithm may be helpful, is if the potential solutions to a problem are represented by the population as a whole. That is, one tries to find an optimal set of things, and we can use the population to represent that set. This is the case, for example, in *multiobjective* optimization, where one tries to determine the Pareto set of dominant solutions [42.13]. For single objective problems, it may be possible to represent solutions as a set of objects, each of which can be evaluated according to its contribution to the overall solution. There has been some recent progress on problems of this type [42.14] and will consider this case in Sect. 42.9.

42.2 Selection Methods

The selection method is the primary means a genetic algorithm has of directing the search process toward better solutions. It is usually defined in terms of a *fitness function* which assigns a positive score to each point in the search space, with the optimal solution having the maximum (or minimum) fitness. Often the fitness function is, in fact, just the objective function of the problem to be optimized, however, there are times when this can be modified. This typically happens when the problem involves some constraints, and so account has to be taken as to what extent the constraints are satisfied. There are a number of different ways to approach this situation:

1. Simply discard any illegal solution and try again. That is, if mutating (say) a solution produces an illegal solution, discard it and try mutating the original again, until a legal solution is obtained.
2. Repair the solution. This involves creating a special purpose heuristic which, given any illegal solution, modifies it until it becomes legal.
3. More generally, one can construct modified operators which are guaranteed to produce legal solutions. The above two methods are specific ways one might achieve this.
4. Adapt the fitness function by adding penalty terms.

It is this fourth approach which concerns us here, as it allows illegal solutions to be tolerated, but puts the onus on the selection method to drive the population away from illegal and toward legal solutions. The idea is to create a fitness function which is a weighted sum of the original objective function, and a measure of the extend to which constraints have been broken. That is,

if $h : X \rightarrow \mathbb{R}$ is the objective function, we might have a fitness function

$$f(x) = w_0 h(x) - \sum_{j=1}^{k} w_j c_j(x) \, ,$$

where c_j is a measure as to how far constraint j has been broken. A difficulty with this approach is in specifying the weights, since one would like to allow illegal solutions to be tolerated (at least in the early stages of the search), but do not want good, but illegal solutions to wipe out any legal ones. It is common, then, to try to fix the weights at least so that legal solutions are preferred to illegal ones (however good). This, then, suggests a fourth approach which works particularly well with tournament selection (see below), in which it is only necessary to say which of two solutions is to be preferred. We stipulate that legal solutions are to be preferred to illegal ones that two legal solutions should be compared with the objective function and that two illegal solutions should be compared by the extent of constraint violation.

The degree to which poor, or illegal, solutions are tolerated by a genetic algorithm is determined by the strength of the selection method chosen. A weak selection method will allow poor (that is, low fitness) solutions to be selected with high probability compared to a strong scheme, which will typically select better solutions. It is usual to insist that a selection scheme should have the property that a better solution should have a higher probability of being selected than a weaker one. A number of selection methods have been proposed, ranging from very weak to very strong, and we will consider them in this order.

2.2.1 Random Selection

The weakest selection method is simply to pick a member of the population uniformly at random. Of course, this has no selection strength at all and will not, by itself, guide the search process. It must therefore be combined with some other mechanism to achieve this. Typically, this would be used in a steady-state genetic algorithm, in which the replacement strategy imposes the selection pressure (Sect. 42.3). Another possibility is in a parallel genetic algorithm where offspring may replace parents if they are better, but the selection of partners for crossover is random (Sect. 42.8).

2.2.2 Proportional Selection

The fitness proportional selection method comes from taking the analogy of the role of fitness in natural evolution seriously. In biological evolution, fitness is literally a measure of how many offspring an individual expects to have. Within the fixed-sized population of a genetic algorithm, then, we model this by saying that the probability of an individual being selected should be proportional to its fitness within the population. That is, the probability of selecting item x is given by

$$\frac{f(x)}{\sum_y f(y)} \, ,$$

where the sum ranges over all members of the population. This selection method is often implemented by the *roulette wheel* algorithm as follows:

1. Let T be the total fitness of the population.
2. Let R be a random number in the range $0 \le R < T$.
3. Let $c = 0$. Let $i = 0$.
4. While $c < R$ do
   a) Let $c = c + f(i)$
   b) Let $i = i + 1$.
5. Return i.

where $f(i)$ is the fitness of the item with index i in the population.

2.2.3 Stochastic Universal Sampling

In a generational genetic algorithm, one needs to select μ individuals from the population in order to complete one generation. Using proportional selection to

do this therefore requires $O(\mu^2)$ time, which can be a significant burden on the running time. An alternative selection algorithm, which still ensures that the expected number of times an individual is selected is proportional to its fitness, is the stochastic universal sampling algorithm [42.15]. If T is again the total fitness of the population, then let

$$E[i] = \frac{f(i)}{T} \mu \, ,$$

which is the expected number of copies of item i. The selection algorithm guarantees that either $\lfloor E[i] \rfloor$ or $\lceil E[i] \rceil$ copies of i are selected, for each item i in the population:

1. Let r be a random number in the range $0 \le r < 1$.
2. Let $c = 0$.
3. For $i = 0$ to $\mu - 1$ do
   a) Let $c = c + E[i]$.
   b) While $r < c$ do: $r = r + 1$; Select(i).

By the time the algorithm terminates, μ items will have been selected, in $O(\mu)$ time (for a good introduction to asymptotic notation [42.16]).

42.2.4 Scaling Methods

In the early stages of a run of a genetic algorithm, there is usually considerable diversity in the population, and the fitness of the best individuals may be considerably greater than the others. When using fitness proportional selection, this can lead to strong selection of the better individuals. Later on when the algorithm is nearing the optimum, the population is less diverse, and fitnesses may be more or less constant. In this situation, proportional selection is very weak, and does not discriminate.

One idea to combat this problem is to scale the fitness function somehow, so as to adjust the selection strength during the run. Two proposals along these lines are *sigma scaling* and *Boltzmann selection*. Sigma scaling (invented by Stephanie Forrest and described in [42.2] and [42.4]) explicitly takes the diversity of the population into account via the standard deviation σ of the fitness in the population. Given the fitness function $f : X \to \mathbb{R}$, the new scaled fitness is

$$h(x) = 1 + \frac{f(x) - \overline{f}}{2\sigma} \, ,$$

where $\overline{f}$ is the average fitness of the population. A negative value of $h(x)$ might be clamped at zero or some small value. The idea behind sigma scaling is that now when there is a lot of diversity, σ is large, and so the best individuals will not dominate the selection process so much. However, when diversity is low, so is σ and the scaled fitness function can still discriminate effectively.

The second proposal, Boltzmann selection, makes use of the idea that the diversity (and therefore strength of selection) is lost over time [42.17]. We therefore seek to scale the fitness using the time (or generation number) as a parameter. This is usually done in the same way as simulated annealing, by controlling a *temperature* parameter T, which is initially large, but decreases over time. We have a scaled fitness of

$$h(x) = \exp\left(\frac{f(i)}{T}\right).$$

Of course, a difficulty with this approach is to select the appropriate *cooling schedule* by which T should decrease over time.

42.2.5 Rank Selection

We have seen that one of the major drawbacks of the proportional selection method, and stochastic universal sampling, is that the probability of choosing individuals is very sensitive to the relative scale of the fitness function. For example, an item with fitness 2 is twice as likely to be chosen as an item of fitness 1. But an item of fitness 101 is almost as likely to be chosen as one with fitness 100. It is therefore suggested that selection should depend only on the relative strength of the individual within the population. One way to achieve this is to choose an individual depending on its *rank*. That is, we sort the population using the fitness function, but then ascribe a rank to each member, with the best individual getting score μ down to 1, for the worst [42.18].

The simplest thing to do then is to choose individuals proportional to rank, using the roulette wheel or stochastic universal sampling algorithms. However, this is then sensitive to the population size. So a common alternative is to linearly scale the rank to achieve a score between two numbers a and b. We thus get a function

$$h(i) = \frac{(b-a)r(i) + \mu a - b}{\mu - 1},$$

where $r(i)$ is the rank of item i in the population. We then seek to select items in proportion to their h-value. This can be done with the roulette wheel or stochastic universal sampling method. Notice that since the sum

of ranks is known, so is the sum of h-values, and so the probability of selecting item i is

$$\frac{2((b-a)r(i) + \mu a - b)}{\mu(\mu - 1)(b + a)}.$$

42.2.6 Tournament Selection

A much simpler way to achieve a similar end as rank selection is to use tournament selection. We fix a parameter $k \leq \mu$. To select an item from the population we simply do the following:

1. Choose k items from the population, uniformly at random.
2. Return the *best* item from those chosen;

where *best* refers, of course, to assessment by the fitness function. Perhaps the most common version is binary tournament selection, where $k = 2$. In this case, it is not strictly necessary to have a fitness function assign a numerical value to points in the search space. All that is required is a means to compare two points and return the preferred one.

It is straightforward to show that the probability of choosing item i from the population is

$$\frac{2r(i) - 1}{\mu^2},$$

where $r(i)$ is the rank from 1 (the worst) to μ (the best). If one chooses the two items to be compared without replacement, then the probability of choosing item i becomes

$$\frac{2r(i) - 2}{\mu(\mu - 1)},$$

which is equivalent to rank selection, linearly scaled with $a = 0$ and $b = 1$.

At the other extreme, if one were to pick the tournament size to be very large (close to μ) then it becomes more and more likely that only the best individuals will be selected. Increasing the tournament size in this way is a good method for controlling the selection strength.

42.2.7 Truncation Selection

The strongest selection method of all would be to only select the best individual. Slightly more forgiving is truncation selection, where only individuals within a given fraction of top performers are selected. In a generational genetic algorithm, these must be repeatedly elected at random. In a steady-state algorithm one simply picks of them at random. Truncation selection, therefore, introduces a new parameter, which is the fraction of the population available for selection. This may vary from, say, a half (rather weak) to a tenth (rather strong). Usually, this must be done experimentally, as there is little theoretical analysis of this form of selection for combinatorial problems.

42.3 Replacement Methods

The steady-state genetic algorithm requires a method by which a new offspring solution can be placed into the population, replacing one of the existing members. As with selection, there are different approaches, which have different *strengths* in terms of the extent to which it drives the population to retain better solutions. Indeed, most of the methods for replacement are based on those already described for selection (Sect. 42.2).

42.3.1 Random Replacement

A simple method commonly found in steady-state genetic algorithms is for the new solution to replace an existing member chosen uniformly at random. If this is done, then the replacement phase does not push the search process in any particular direction and the onus for evolving toward better solutions is on the selection method chosen. Because of this, it is commonly supposed that a steady-state genetic algorithm with random replacement is more or less equivalent to a generational genetic algorithm, using the same selection method. This is not quite true; however, it can be shown theoretically that the long-term behavior of both algorithms will be the same [42.19]. What will not necessarily be the same is the short term, transient behavior and, in particular, the speed with which the algorithms will arrive at their long-term *equilibrium* may well be different.

42.3.2 Inverse Selection

Several replacement methods are based directly on selection methods, but changed so that the poorer performing solutions are more likely to be replaced than the better ones. For example, one can construct an *inverse fitness proportional* replacement method, where the probability of being replaced is determined by the fitness. In order to ensure that lower fitness means a higher chance of replacement, the reciprocal of the fitness might be used to determine the probability of replacement. Alternatively, the fitness can be subtracted from that of the global optimum (if known). This would have the advantage that the optimum, if found, would never be replaced. In this case, the probability that item i in the population is selected for replacement will be

$$\frac{f^* - f(i)}{\mu f^* - \sum_j f(j)} \, ,$$

where f^* is the optimum fitness value.

Replacement determined by fitness has the same drawbacks as selection done in this way. For example, toward the end of the search all the population members will have similar fitness values. Fitness proportional replacement will then be almost as likely to replace the best individual as the worst. Consequently, an alternative method using scaling or rank might be preferred. The simplest of these ideas is to use a tournament, but this time pick the worst of the sample:

1. Choose k items from the population, uniformly at random without replacement.
2. Return the worst item from those chosen.

This has the advantage that the best item in the population cannot be replaced.

42.3.3 Replace Worst

Perhaps the most common choice of replacement strategy is to simply replace the worst member of the population with the new offspring. This is a relatively strong approach, as it preserves all the better members of the population. Indeed, this strategy can well be combined with the random selection method, and using only replacement as the means of driving the evolution toward better individuals.

An even stronger variant would be to replace the worst member of the population only if the new offspring is better or equal in value. This has the property that the minimum fitness of the population can never decrease, and so we are guaranteed progress throughout the evolution. In some cases, this may lead to much faster evolution. However, it may also happen that for a long time, no new individuals are added to the population. This will be increasingly the case toward the end of a run, when it is increasingly hard to create better individuals. Replacing the worst member, regardless of how good or bad the new offspring, at least keeps adding new search points and creates new possibilities for further exploration, while retaining copies of the better solutions in the population.

42.3.4 Replace Parents

A different idea for choosing which element to replace is for the offspring to replace the parent (or parents). This would have an advantage in maintaining some level of diversity (Sect. 42.7) since the offspring is likely to be similar to the parent. The simplest way to do this, in the case when mutation but not crossover is used is for the offspring to replace the parent if it is better. If the selection process is purely random, then this amounts to running a number of local search algorithms in parallel, as there is no interaction between the individuals in the population.

A variation on this is when there is crossover, which requires the selection of a second parent (Sect. 42.6). In this case, it makes sense for the offspring to replace the worst of the two parents, guaranteeing that the best individual is never replaced. This is the idea behind the so-called *microbial* genetic algorithm [42.20], which is a steady-state genetic algorithm, with random selection, standard crossover, and mutation (Sects. 42.4 and 42.6), with the offspring replacing the worst parent:

1. Generate random population.
2. Repeat until stopping criterion satisfied:

a) Select two individuals from the population uniformly at random.
b) Perform crossover and mutation.
c) Let the new offspring replace the worst of the two parents.
3. Stop.

A further variation possible if crossover is used is to create two offspring and for them to replace both parents under some suitable conditions. Possibilities include:

● If at least one offspring is better than both parents.
● If both offspring are no worse than the worst parent.
● If one offspring is better than both parents and the other is better than one of them.

This is the idea behind the *gene invariant genetic algorithm* [42.21]. It is designed for use on search spaces given by fixed length binary strings. We arrange for the initial population to be constructed such that for every random string generated, we also include its bitwise complement. This ensures an equal number of ones and zeros at each bit position in the population. When crossover takes place between two parents, if we keep both possible offspring, then we maintain the number of ones and zeros. This arrangement naturally maintains a lot of diversity in the population, without even the necessity to include mutation. Early empirical studies suggested that this approach would be very good at avoiding certain kinds of *traps* in the search space. There seems to have been very little work following up this suggestion, however (although [42.22] for the analysis of a $(1+1)$ version of the algorithm).

42.4 Mutation Methods

Whereas selection and replacement focus the genetic algorithm on a subset of its population, the mutation and crossover operators enable it to sample new points in the search space. The idea behind mutation is that, having selected a good member of the population, we should try to create a variant with the hope that it is even better. To do this, it is often possible to make use of some natural or well-established local search operators for the problem class concerned. For example, for the traveling salesman problem, the well-known 2-opt operator works by reversing a random segment of the selected tour [42.23, 24]. This immediately provides us with a way of mutating solutions for this problem class.

Similarly, when solving the Knapsack problem, it is often helpful to exchange items, and this too gives us a good idea how to mutate.

Generally speaking, mutations are defined by choosing a representation for the points of the search space, and then defining a set of operators to act on that representation. For example, many combinatorial optimization problems concern choosing an optimal subset of some set. We can represent the search space, the collection of all subsets, using binary strings with length equal to the size of the set. Each position corresponds to a set element, and we use 0 and 1 to distinguish whether or not an element is included in a particular subset. We hen define a collection of operators with which to act upon the representation. For example, the action of removing or adding an item to the subset is very natural, and is given by simply flipping the bit at the appropriate position. If there are n bits in the representation, then this would give us n corresponding operators. In order to mutate a bitstring, one chooses an operator at random and applies it.

In order to exchange an item in a subset with one that is not, we must simultaneously flip a 1 to a 0 and a 0 to a 1. If the subset has k elements, then there are $k(n-k)$ ways to do this, giving us another possible set of operators.

Perhaps the most common choice of mutation for binary strings is to randomly flip each bit independently with a fixed probability u called the *mutation rate*. This corresponds to flipping a subset of bits of size k with probability $u^k(1-u)^{n-k}$. Very often, the mutation rate is set to $u = 1/n$. This is popular because, while it favors single bit mutations, there is a significant probability of flipping two or more bits, enabling exchanges to take place. Notice, however, that there is a probability of $(1-1/n)^n \approx 1/e$ that nothing happens at all. While this clearly slows down evolution by an almost constant factor, it is not necessarily a bad thing to have a significant probability of doing nothing – it can sometimes prevent evolution rushing off down the wrong path (Sect. 42.5).

While the mutation rate $u = 1/n$ is the most commonly recommended, the best value to choose depends on the details of the problem class, and the rest of the genetic algorithm being used [42.25]. For example, the simple $(1+1)$ EA maintains a single point of the search space which it repeatedly mutates – replacing it only if a better offspring occurs. For linear functions, the choice of $u = 1/n$ is provably optimal for the class of linear functions [42.26]. However, it can be shown that for the so-called leading ones problem, in which the fitness of a string is simply the position of the first zero, the optimal mutation rate for the $(1+1)$ algorithm is in fact close to $1.59/n$ [42.27].

For a more general approach to defining a mutation operator, one could assign a probability to each subset of bits, and flip such a subset with the given probability. This corresponds to defining a probability distribution over the set of binary strings π. To mutate a string x, we choose another string y with probability $\pi(y)$ and return the result $x \oplus y$, where the $\oplus$ symbol represents bitwise exclusive-or [42.28]. This general method has the property that it is invariant with respect to the labels 0 and 1 to represent whether or not an item is in a given subset. One might also wish mutation to be invariant with respect to the ordering of the n elements of the underlying set, so that the performance of the genetic algorithm is not sensitive to how this order is chosen. To do this requires a much more restricted mutation operator. One must specify a probability distribution over the numbers $0, 1, \ldots, n$ and then, having chosen a number according to this distribution, flip a random subset of bits of this size. Such a mutation is said to be *unbiased* with respect to bit labels (that is, the choice of 0 or 1) and bit ordering [42.29, 30].

Mutating each bit with a fixed rate is an example of an unbiased mutation operator, with the probability of flipping k bits being

$$\binom{n}{k} u^k (1-u)^{n-k} .$$

The following is an efficient algorithm to choose k according to this binomial distribution [42.31]:

1. Let $x = y = 0$.
2. Let $c = \log(1-u)$.
3. Let r be a random number from 0 to 1.
4. Let $y = y + 1 + \lfloor \log(r)/c \rfloor$.
5. If $y < n$ then let $x = x + 1$ and go to 3. Else return x.

We can use this algorithm to perform mutation by first selecting the number of bits to be flipped and then choosing which particular subset of that size will be mutated. The above algorithm has expected running time of $O(un)$. If u is relatively small, one can then sample the bit positions from $\{1, \ldots, n\}$ repeating if the same index is selected twice (use a hash table to detect the unlikely event of a repeated sample). For the choice $u = 1/n$ the random selection algorithm runs in constant time, and, for large n, the probability of having to do a repeat sample tends to zero. Consequently, performing mutation in this way is extremely efficient. An alternative method is to randomly choose the position of the next bit to be flipped [42.32].

42.5 Selection–Mutation Balance

We are now in a position to put together a simple genetic algorithm involving selection, replacement, and mutation. Selection and replacement will focus the search on good solutions, whereas mutation will explore the search space by generating new individuals. We will find that there is a balance between these two forces, but the exact nature of the balance depends on the details of the algorithm, as well as the problem to be solved. To simplify things a little, we will consider the well-known toy problem, OneMax, in which the fitness of a binary string is the sum of its bits. For this problem, there are a number of theoretical results concerning the balance between selection and mutation, for example [42.33, 34]. For a thorough analysis of the selection–mutation balance on a LeadingOnes type problem, [42.35]. Here, we will illustrate the effects with some empirical data.

The first algorithm we will look at is the steady state genetic algorithm, using binary tournament selection,

and random replacement. As described above, the long term behavior will be the same as for a generational genetic algorithm, with the same selection and mutation methods. Specifically, let us consider a population of size 10, fix the string length to 100 bit, and consider the effect of varying the mutation rate. In Fig. 42.1, we show four typical runs at mutation rates 0.03, 0.02, 0.0 and 0.001 respectively. We plot the fitness of the best member of the population at each generation.

Recall that this algorithm is not elitist. This means that it is possible to lose the best individual, by replacing it with a mutant of the selected individual. It is chosen for replacement with probability of 0.1 Whether it is replaced by something better or worse depends on the mutation rate. A higher mutation rate will tend to be more destructive. We see from the plotted trajectories, that for higher mutation rates, the algorithm converges to a steady state more quickly (around generation 200 in the case of mutation rate of 0.03) but t

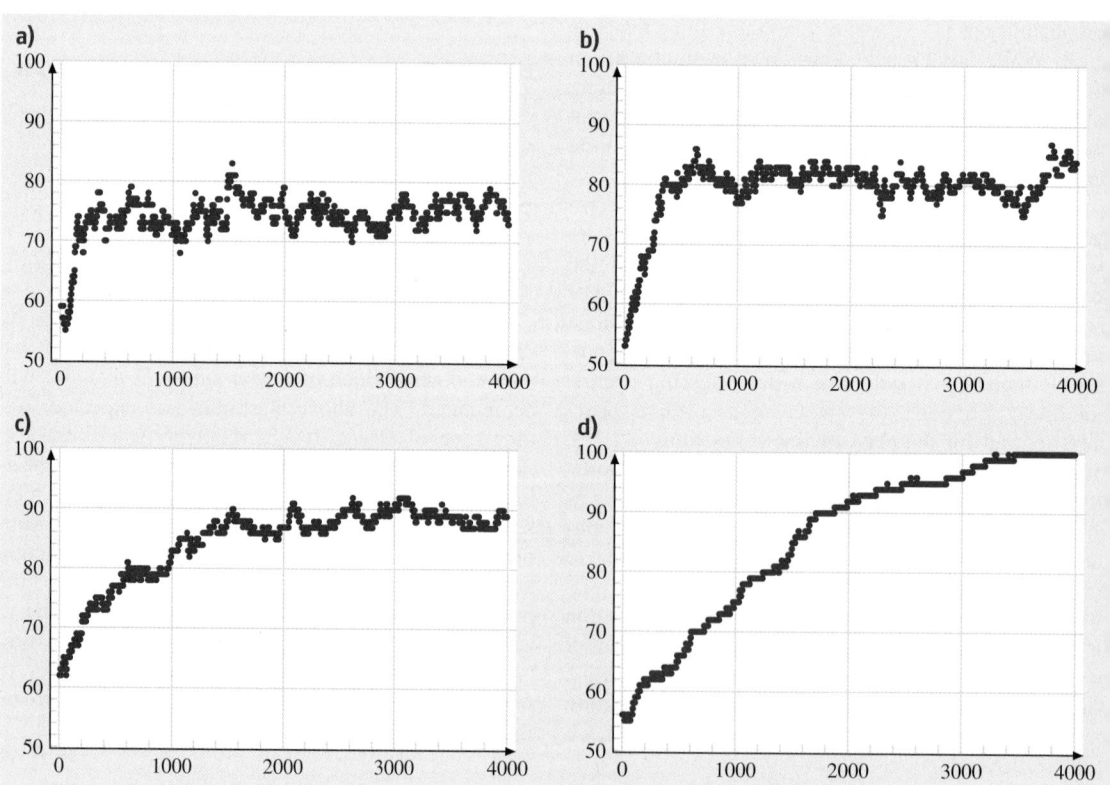

Fig. 42.1a–d Trajectories of best of population for steady-state GA with random replacement on OneMax problem (100 bit). (**a**) Mutation rate = 0.03; (**b**) mutation rate = 0.02, (**c**) mutation rate = 0.01, (**d**) mutation rate = 0.001

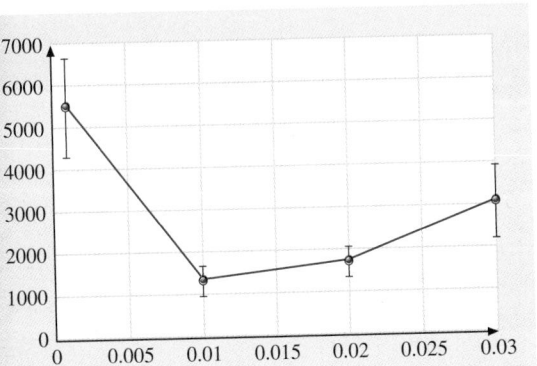

Fig. 42.2 Average time taken to optimize OneMax (100 bit) for steady-state GA with worst replacement, with varying mutation rates

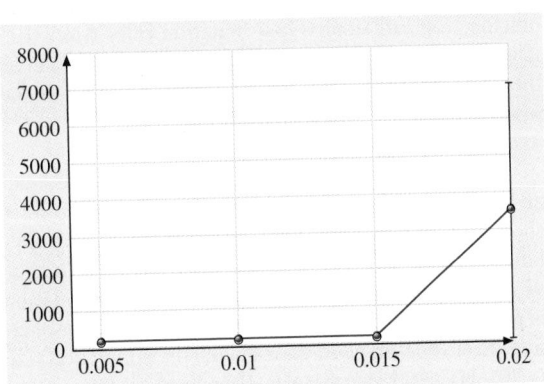

Fig. 42.3 Average time taken (generations) to optimize OneMax (100 bit) for generational GA with best selection, population size 10, and various mutation rates

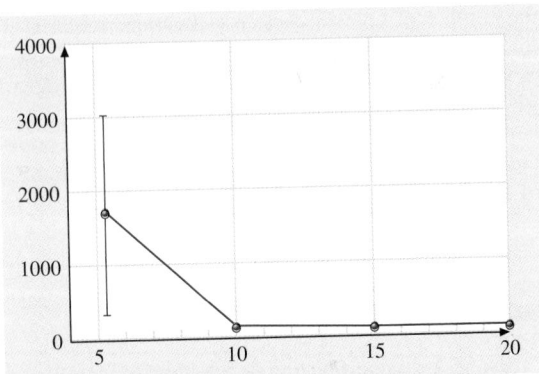

Fig. 42.4 Average time taken (generations) to optimize OneMax (100 bit) for generational GA with best selection, mutation rate 0.01, and various population sizes

a poor quality solution (average fitness around 74). As the mutation rate is reduced, the algorithm takes longer to find the steady state, but it is of higher quality. For the smallest mutation rate shown, 0.001, the run illustrated takes 3500 generations to converge, but this includes the optimal solution.

Since the OneMax problem is simply a matter of hill-climbing, we can improve matters by using an elitist algorithm. So we now consider the steady-state genetic algorithm, again with binary tournament selection, but this time replacing the worst individual in the population. In this setup, we cannot lose the current best solution in the population. We quickly find experimentally that for reasonable mutation rates, we can always find the optimum solution in reasonable time. However, again there is a balance to be struck between selection strength and mutation. If mutation is too high, it is again destructive, which slows down progress. If it is too low, we wait for a long time for progress to be made. There is now an optimal mutation rate to be sought. Figure 42.2 shows the average time to find the optimum for the same four different mutation rates (0.03, 0.02, 0.01, 0.001). The average is taken over 20 runs. We can clearly see that the best tradeoff is obtained with the mutation rate of 0.01 in this case, with an average of around 1400 generations required to find the optimum. This rate equals one divided by the string length, and is a common choice for EAs.

It is not strictly necessary to implement an elitist strategy to obtain reasonable optimisation performance. Consider instead a generational genetic algorithm in which our selection is the strongest possible – we always pick the best in the population. This is not technically elitist, as we apply muta-

tion to the selected individual, which means there is a chance it is lost. However, if there is a reasonable population size, and the mutation rate is not too large, then there is a good chance that a copy of the best individual will be placed in the next generation. This in effect simulates elitism [42.36]. Yet again there is a balance between selection and mutation, this time depending on the population size. If the mutation rate is high, then we will need a large population to have a good chance of preserving the best individual. Smaller mutation rates will allow smaller populations, but will slow down the evolution.

Consider first a population of size 10, and a range of mutation rates: 0.005, 0.01, 0.015 and 0.02. Figure 42.3 shows the average time to find the optimum for our generational genetic algorithm. We see that it is very efficient for sufficiently low mutation rates (only 125 generation when the mutation rate is 0.01). However, there is a transition to much longer run times when the mutation rate gets higher (2500 generations when the mutation rate is 0.02).

Conversely, we can consider fixing the mutation rate at 0.01 and varying the population size. Figure 42.4 shows the optimisation time in generations for populations of size 5, 10, 15 and 20. Again we see evidence of a transition from long times (when the population is too small) to very efficient times with larger populations (as low as 80 generations when the population size is 20). However, we have to bear in mind that a generational genetic algorithm has more fitness function evaluations per generation than a steady-state genetic algorithm. Thus we cannot keep increasing the population size indefinitely without cost. Figure 42.5 shows the number of fitness function evaluations required to find the optimum. We see that, of the examples shown, a population of size 10 is best (requiring an average of 1374 evaluations).

The exact tradeoff between mutation rate and population size can be calculated theoretically. If the number of bits is n, and the mutation rate is $1/n$, it can be

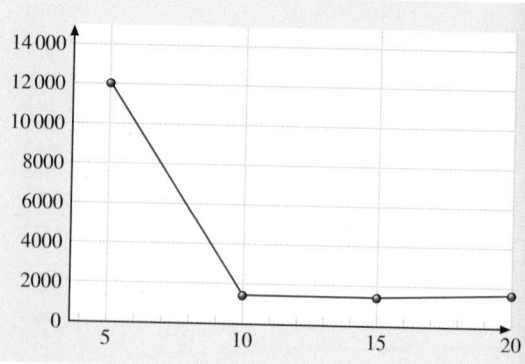

Fig. 42.5 Average time taken (fitness function evaluations) to optimise OneMax (100 bit) for generational GA with best selection, mutation rate 0.01, and various population sizes

shown that there is a transition between exponential and polynomial run time for the OneMax problem when the population size is approximately $5 \log_{10} n$. Indeed it can be shown that this is a lower bound on the required population size to efficiently optimize any fitness function with unique global optimum [42.37].

42.6 Crossover Methods

Crossover (or recombination) is a method for combining together parts of two different solutions to make a third. The hope is that good parts of each parent solution will combine to make an even better offspring. Of course, we might also be recombining the bad parts of each parent and come up with a worse solution – but then selection and replacement methods will filter these out.

Several methods exist for performing crossover, depending on the representation used. For binary strings, there are three common choices. One-point crossover chooses a bit position at random and combines all the bit values below this position from one parent, with all the remaining bit values from the other. Thus, given the parents

01001101,

11100111

if we choose the fifth position for our cut point, we obtain the offspring solution

01000111.

Similarly, for two-point crossover, we choose two bit positions at random. The bit values between these two cut points come from one parent, and the remaining values come from the other. Thus, with the same two parents, choosing cut positions 2 and 6 produces the offspring

01100101.

Both one- and two-point crossovers have the property of being biased with respect to the ordering of bits. If the problem representation has been chosen so that this order matters, then such a crossover may confer an advantage as they tend to preserve values that are next to each other. For example, if the problem relates to finding an optimal subset, and the elements of the set have been preordered according to some heuristic (e.g. a greedy algorithm), then one- or two-point crossover may be appropriate. If, however, the order of the bits is arbitrary, then one should choose a method which is unbiased with respect to ordering. The common choice is called uniform crossover, and involves choosing bit values from either parent at random. One way of implementing this would be to generate a random bit string and let the values in this string (or mask) determine which parents the bit values should come from. For example, using again the two parents above, if we generate the random mask 0 1 0 1 0 1 0 1, we get the offspring

01001101.

This leads to a more general view of crossover: We specify a probability distribution π over the set of binary strings and, to perform crossover, we select a string according to this distribution and use it as a mask [42.5]. Uniform crossover corresponds to a uniform distribution. One-point crossover corresponds to selecting only masks of the form 0 ...01 ...1. It can be shown that crossover by masks in general is always unbiased with respect to changes in labels of bit values (that is, exchanging ones for zeros). If we also require crossover to be unbiased with respect to bit ordering, then it is necessary and sufficient that masks containing the same number of ones are selected with the equal probability.

Any crossover by masks also has the nice property that if both parents agree on a bit position, then the offspring will also share the same value at that position. Such crossovers are called *respectful*, and emphasize the idea that one is trying to preserve structure found in the parents [42.38, 39]. It can be shown that such properties can be understood geometrically [42.40].

Our understanding on when crossover can be helpful is rather limited and there are many open questions. For example, continuing to look at the OneMax problem, we can examine experimentally whether or not crossover helps. Let us keep to the steady-state algorithm with tournament selection and replacement of the worst individual. We modify the algorithm by allowing, with a given probability, crossover to take place between the selected individual and a randomly chosen one. Our algorithm therefore is as follows:

1. Initialize population of size μ randomly with points from the search space.
2. Repeat until stopping criterion is satisfied:
   a) Choose a member of the population using binary tournament selection.
   b) With probability p, crossover selected individual with one chosen randomly from the population.
   c) Modify the result with *mutation* using rate $1/n$.
   d) Replace worst member of population with the new offspring.
3. Stop.

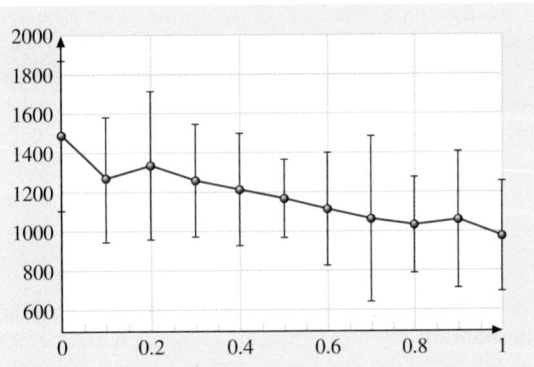

Fig. 42.6 Average time taken to optimize OneMax (100 bits) for generational GA with tournament selection, mutation rate 0.01, population size 10, and varying crossover probabilities

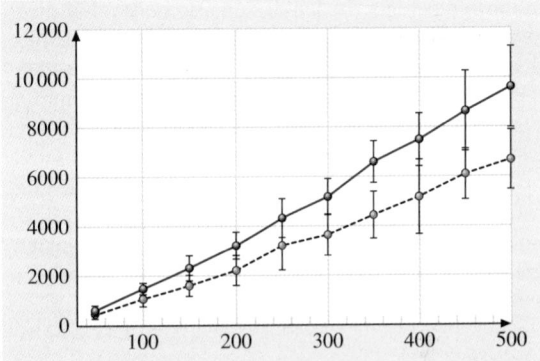

Fig. 42.7 Average time taken to optimize OneMax for various string lengths for generational GA with tournament selection, mutation rate $1/n$, population size 10. Solid line is with no crossover. Dotted line is with uniform crossover with probability 1

We first consider an experiment on OneMax with $n = 100$. We use a population of size 10, and vary the crossover probability between 0 and 1. The results are shown in Fig. 42.6, which shows the average time to find the optimum (averages over 20 runs). It can be seen that there is some improvement as the crossover rate increases, although the results are rather noisy (error bars represent one standard deviation). Examining this case further, we compare the steady-state genetic algorithm with no crossover ($p = 0$) with uniform crossover ($p = 1$) for different string lengths. The results are shown in Fig. 42.7. Here it is clear that there in significant improvement, which appears to be increasing as the string length grows. To date, there is no theoretical analysis of why this should be the case.

The first example of a problem class for which crossover can provably help is the following [42.12, 41]. We take the OneMax function, and then create a trap just before the optimum

$$jump(x) = \begin{cases} 0 & \text{if } \theta \le \|x\| < n, \\ \|x\| & \text{otherwise}, \end{cases}$$

where $\|x\|$ is the number of ones in x (Fig. 42.8 for an illustration). The idea is that the population first climbs the hill to the threshold θ just before the trap. It is then rather unlikely that a single mutation event will create a string that crosses the gap and finds the global optimum. However, crossing over two strings with just a few zeros in each may have a better chance of jumping the gap, especially if the zeros occur in different places in the two parents. To achieve this, some level of diversity must be maintained in the population – a subject discussed further in Sect. 42.7.

For problem classes in which solutions are not naturally represented by binary strings, one has to think carefully about the best way to design a crossover operator. For example, take the case of the traveling salesman problem, in which a solution is given by a permutation of the cities, indicating the order in which they are to be visited. Over the years, a number of different

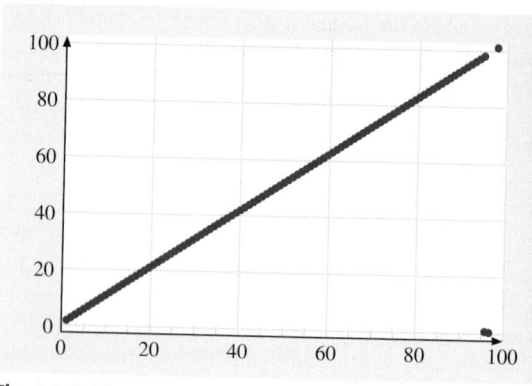

Fig. 42.8 The $jump(x)$ function for string length $n = 100$, with threshold $\theta = 98$

crossover methods have been proposed. For the most part, these were designed so as to be respectful of the positions of the cities along the route. That is, if a particular city was visited first by both parents, then it would also be visited first by the offspring. However, a much more effective approach is to try to preserve the *edges* between adjacent cities in a route. That is, if city A is followed immediately by city B by both parents, irrespective of where this takes place in the route, then we should try to ensure that this also happens in the case of offspring. This kind of crossover is called an edge recombination operator [42.42].

42.7 Population Diversity

For most forms of crossover, if one crosses an individual with itself, the resulting offspring will again be the same. Crossovers with this property are said to be pure [42.39]. This is certainly a property of one-point, two-point and uniform crossovers for binary strings. There is therefore no point in performing crossover if the population largely comprises copies of the same item. In fact, the whole idea of the population is rather wasted if this is the case. Rather, the hope is to gain advantage by having different members of the population search different parts of the search space. It seems important, therefore, to maintain a level of diversity in the population.

The importance of this can be seen in solving the $jump(x)$ problem described earlier in Sect. 42.6. Once the population has arrived at the local optimum, the individuals will typically contain $\theta - 1$ ones, and the rest zeros. If the zeros tend to fall in the same bit po-

sitions for each member of the population, then it will be impossible for crossover to jump across the gap. For example, crossing

1 1 1 1 1 1 1 0 0 0

and

1 1 1 1 1 0 1 0 0

cannot produce the optimum. If, however, we can ensure that diversity in the population, then two members at the local optimum might be

1 1 1 1 1 1 1 0 0 0

and

0 1 1 0 1 1 0 1 1 1 ,

which has a reasonable chance to jump the gap.

Several different mechanisms have been proposed ensure some diversity is maintained in a population. The simplest way is to enforce it directly by not allowing duplicate individuals in the population [42.43]. So an offspring is produced which is the same as something already in the population, we just discard it.

A second method is to adjust the replacement method in a steady-state genetic algorithm, by making sure the new offspring replaces something similar to itself. For example, one could have a replacement rule that makes the offspring replace the population member most similar to it (as measured by the Hamming distance). This method, called *crowding* will, of course, destroy the elitism property [42.44]. This can be salvaged by only doing the replacement if the offspring is at least as good as what it replaces.

A third approach to diversity is to explicitly modify the fitness function in such a way that individuals receive a penalty for being too similar to other population members [42.45]. This idea is called fitness sharing. We think of the fitness function as specifying a resource available to individuals. Similar individuals are competing for the resource, which has to be shared out between them.

A fourth approach is to limit the choice of the partner for crossover. So far, our algorithms have chosen the crossover partner uniformly at random. One could instead, try to explicitly choose the most different individual in the population [42.46].

We test these various methods using the *jump(x)* problem of the previous section. Recall that for the population to jump across the gap requires a crossover between two diverse individuals. We work with a string length $n = 100$ and a threshold $\theta = 98$. As a baseline, consider the steady-state genetic algorithm, with binary

Table 42.1 Success of various diversity methods in solving the *jump(x)* problem. Each is tested for 20 trials, to a maximum of 10 000 fitness function evaluations. Mean and standard deviation refer just to successful runs

Method	Percentage successful runs (%)	Mean evaluations	Standard deviation
None	5	8860	–
No duplicates	80	3808	2728
Crowding	100	1320	389
Sharing	65	3452	2652
Partner choice	10	688	113

tournament selection, replacement of the worst, uniform crossover (probability 1) and mutation rate 0.01. In 20 trials on the *jump(x)* function, only once did this algorithm succeed in finding the optimum within 10 000 generations. On that one successful run, it required 8860 evaluations to complete. We compare this result with the same algorithm, modified in each of the four diversity-preserving methods above. In the case of fitness sharing, we simply penalize any individual which has multiple copies in the population by subtracting the number of copies from the fitness (It should be noted that there are different ways of doing this, and the original method is far more complicated.). The results of the experiments are summarized in Table 42.1. The best result is given by the crowding mechanism, but notice that it is essential to preserve elitism here, otherwise the algorithm never solves the problem within 10 000 generations.

A different approach altogether is to structure the population in some way, so as to prevent certain individuals interaction. We take up this idea, in a more general context, in the following section.

42.8 Parallel Genetic Algorithms

There have been a number of studies of different ways to parallelize genetic algorithms. There are two basic methods. The first is the *island* model, in which we have several populations evolving in parallel, which occasionally exchange members [42.47]. To specify such an algorithm, one needs to decide on the topology of the network of populations (that is, which populations can exchange members) and the frequency with which migrations can take place. We also need a method to decide on which individuals should be passed, and how they should be incorporated into the new population.

That is, we need a form of selection and replacement for the migration stage.

As an example, consider having several steady-state genetic algorithms operating in parallel. After a certain number of generation, we choose a member of each population to migrate – for example, the best one in each population. We copy this individual to the neighboring populations, according to the chosen topology. To keep things simple, we choose the complete topology, in which every pair of populations is connected. Thus, each population receives a copy of the best from all the other populations. These now have to be incorporated into the home population somehow. An easy method is to take the best of all the incoming individuals, and use it to replace the worst in the current population. Such an algorithm will look like this:

1. Create m populations, each of size μ.
2. Update each population for c generations.
3. For each population, replace the worst individual by the best of the remaining populations.
4. Go to 2.

To take an extreme case, if the population size is $\mu = 1$ and we migrate every generation ($c = 1$), then this is rather similar to a $(1, m)$ EA.

There are two possible advantages of the island model. Firstly, it is straightforward to distribute it on a genuinely parallel processing architecture, leading to performance gains. Secondly, there may be some problem classes for which the use of different populations can help. The idea is that different populations may explore different parts of the search space, or develop different partial solutions. The parameter c is chosen large enough to allow some progress to be made. The migration stage then allows efforts in different directions to be shared, and workable partial solutions to be combined. Some recent theoretical progress has been made in analyzing this situation for certain problem classes [42.48].

A particular case of such a model is found in *co-evolutionary* algorithms. These come in two flavors: competitive and co-operative. In a competitive algorithm, we typically have two parallel populations. One represents solutions to a problem, and the other represents problem instances. The idea is that, as the former population is finding better solutions, so the latter is finding harder instances to test these solutions. An early example evolved sorting networks for sorting lists of integers [42.49]. While one population contained different networks, the other contained different lists to be sorted. The fitness of a network was judged by its ability to sort the problem instances. The fitness of a problem instance was its ability to cause trouble for the networks. As the instances get harder, the sorting networks become more sophisticated.

In a co-operative co-evolutionary algorithm, the different populations work together to solve a single problem [42.50]. This is done by dividing the problem into pieces, and letting each population evolve a solution for each piece. The fitness of a piece is judged by combining it with pieces from other populations and evaluating the success of the whole. Theoretical anal-

ysis shows that certain types of problems can benefit from this approach by allowing greater levels of exploration than in a single population [42.51].

The second parallel model for genetic algorithms is the *fine grained* model, in which there is a single population, but with the members of the population distributed spatially, typically on a rectangular grid [42.52]. At each time step, each individual is crossed over with a neighbor. The resulting offspring replaces the original if it is better, according to the fitness function. The algorithm looks like this:

1. Create an initial random population of size μ.
2. In parallel, for each individual x:
   a) Choose a random neighbor y of x.
   b) Cross over x and y to form z.
   c) Mutate z to form the offspring.
   d) Replace x with the offspring if it is better.
3. Go to 2.

Notice that such an algorithm is generational, but also elitist. There has been very little analysis of this kind of architecture, although there is some empirical evidence that it can be effective, especially for problems with multiple objectives, in which different tradeoffs can emerge in different parts of the population [42.53].

To illustrate the fine-grained parallel genetic algorithm, consider a ring topology, in which the kth member of the population has as neighbors the $(k-1)$th and $(k+1)$th member (wrapping round at the ends). We try it on OneMax, with a population of size 10, using uniform crossover and, as usual, a mutation rate of $1/n$. The results, for a variety of string lengths, are shown

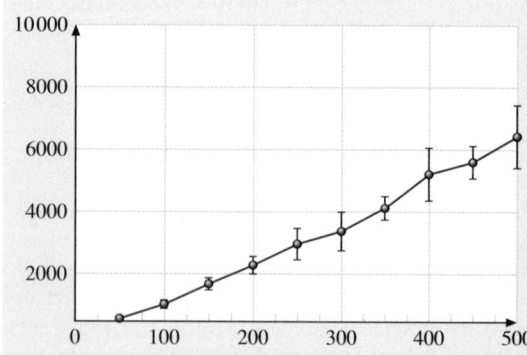

Fig. 42.9 The average time (in fitness function evaluations) for the ring-topology parallel genetic algorithm to find the optimum for OneMax for a variety of string lengths. Population size 10, mutation rate $= 1/n$, uniform crossover

n Fig. 42.9. We can see that the parallel algorithm
s competitive with the steady-state genetic algorithm.
Note that fitness function evaluations are plotted, and
not generations.

The distributed nature of the population should
help to maintain a level of diversity. Hence, we would
expect a parallel genetic algorithm (with crossover)
to perform reasonably well on the *jump*(*x*) function

of Sect. 42.6. Over 20 trials of our ring-based fine-
grained algorithm, on the *jump*(*x*) function with $n =
100$ and $\theta = 98$, we find that it solves the problem
on all trials, requiring an average of 2924 function
evaluations (standard deviation is 1483). It is therefore competitive with the simple diversity enforcement
method on this problem (compare with results in Table 42.1).

42.9 Populations as Solutions

We finish this chapter by considering a genetic algorithm for which the population as a whole represents the
solution to the problem, rather than it being a collection
of individual solutions. It is therefore an example of cooperative co-evolution taking place within a single population. Moreover, it is one of the very few examples
involving a population-based genetic algorithm using
crossover, for which a serious theoretical analysis exists. It is one of the highlights of the theory of genetic
algorithms to date [42.14, 54, 55].

The problem we are addressing is the classical All-
Pairs Shortest Path problem. We are given a graph with
vertex set V (containing n vertices) and edge set E with
positive weighted edges. The goal is to determine the
shortest path between every pair of vertices in the graph,
where length is given by summing the weights along the
path.

To clarify what is meant exactly by a path, first
consider a sequence of vertices, $v_1, \ldots, v_m$ such that
$(v_k, v_{k+1}) \in E$ for all $k = 1, \ldots, m-1$. Such a sequence
is called a *walk*. A *path* is then a walk with no repeated
vertices. Since for any walk between two vertices there
is a shorter path (by omitting any loops), we can equivalently consider the problem of finding the shortest walks
between any two vertices.

The population will represent a solution to the
problem for a given graph, by having each individual
representing a walk between two vertices. The problem is solved when the population contains exactly the
shortest paths for all of the $n(n-1)$ pairs of vertices.

The algorithm will be a steady-state genetic algorithm, with random selection and a replacement method
which enforce diversity. For any pair of vertices we will
allow at most one walk between them to exist in the
population. The outline of the algorithm is as follows:

1. Initialize the population to be E.
2. Select a population member uniformly at random.
3. With probability p do crossover, else do mutation.
4. If a walk with the same start and end is in population, replace it with offspring, if offspring length is
   no worse.
5. If a walk with same start and end is not in population, add offspring walk to population.
6. Go to 2.

We can see from line 5 that another unusual feature
of the algorithm is that the population size can grow.
Indeed, it starts with just the edge set from the graph
and has to grow to get the paths between all pairs of
vertices.

In line 3, we see there is a choice between mutation and crossover, governed by a parameter p. Given
that our individuals are walks in a graph (rather than binary strings) it is clear that we need to specify some
special purpose operators. We define mutation to be
a random lengthening or shrinking of a walk as follows.
Suppose that the selected walk is $v_1, v_2, \ldots, v_{m-1}, v_m$.
We randomly select a vertex from the neighbors of
v_1 and v_m. If this is neither v_2 nor v_{m-1} then we append it to the walk. If it is one of v_2 or v_{m-1} then we
truncate the path at that end. This process is repeated
a number of times, given by choosing an integer s according to a Poisson distribution with parameter $\lambda = 1$.
We then perform $s + 1$ mutations to generate the offspring walk.

As an example, consider the graph illustrated in
Fig. 42.10 (notice that the edge weights are not shown).
Suppose we have selected the walk $(3, 4, 5, 6)$ from the
population to mutate. We choose our random Poisson
variable s – let us say it is one. So we have two mutations to apply. We gather the set of vertices connected to
the two end points. That is, $\{1, 4, 8\}$ and $\{5, 7, 10, 11\}$,
and we pick one of these at random. Let us suppose
that 8 is selected. We therefore extend our walk to become $(8, 3, 4, 5, 6)$. For the second mutation, we again

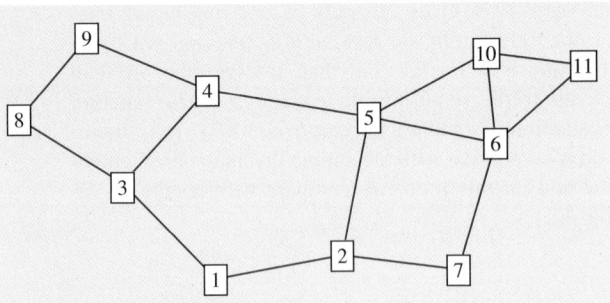

Fig. 42.10 An example graph for the All Pairs Shortest Path problem (edge weights not shown)

collect together the vertices attached to the end points: {3, 9} and {5, 7, 10, 11}. Choosing one at random we select, say, 5. Since this is a vertex in the walk prior to an end point, we truncate to produce the final offspring (8, 3, 4, 5). If there is already a walk from 8 to 5 in the population, we replace it with the new one, only if the new one is in fact shorter (that is, the sum of the weights is less). If there is no such walk, then we add the new one to the population.

To perform crossover, we have to be careful to ensure that we end up with a valid walk. Suppose that the individual we have selected is a walk from u to v. We consider all the members of the population that start from v, but exclude the one that goes back to u. We then choose a member of this set, uniformly at random, and concatenate it to our original walk. This guarantees that the offspring is again a valid walk.

For example, consider again the graph in Fig. 42.10 and the selected walk (3, 4, 5, 6). We need to first collect together all the walks in the population that start from vertex 6, excluding the one (if it exists) that goes from

6 to 3. Imagine that we find the following:

$$(6, 10, 5, 2)$$
$$(6, 5, 2, 7)$$
$$(6, 11, 10)$$

and we pick one at random – say, the second one. Concatenating this to the original walk produces the offspring (3, 4, 5, 6, 5, 2, 7). Notice that this is a walk rather than a path, since vertex 5 is repeated. If this is better than any existing walk from vertex 3 to vertex 7 then it replaces it. If there is no such walk in the population, then the new one gets added.

A considerable amount of theoretical analysis have been done for this genetic algorithm on the class of All Pairs Shortest Path problems. If we run the algorithm with no crossover (that is, we set $p = 0$) then it can be shown that it requires $\Theta(n^4)$ generations for the population to converge to the optimal set of paths. Adding crossover by choosing $0 < p < 1$ improves the performance to $O(n^3 \log n)$. Note that the classical approach to solving this problem (the Floyd–Warshall algorithm) requires $O(n^3)$ time, which is faster and includes all computations that have to be performed (i.e. this is not just a count of the *black box* function evaluations). Of course, the classical algorithm has full details of the problem instance on which it is working, whereas the genetic algorithm is operating *blind*. Despite this great disadvantage, the genetic algorithm only pays a cost of a factor of $\log n$ over the classical approach (in addition to any implementation overhead).

This example is one of the only cases where we have a proof that crossover helps for a naturally defined problem class. It is an important open problem to find others.

42.10 Conclusions

We have seen that the defining feature of a genetic algorithm is its maintenance of a population, which is used to search for an optimal (or at least sufficiently good) solution to the problem class at hand. For problems where solutions are naturally represented as binary strings, there is an obvious analogy with an evolving population of individuals in nature, with the strings providing the DNA. Analogs of mutation and crossover (recombination) are then readily definable and can be used as search operators. The theory that describes the

trajectory of a population under such operators, in general terms, is well developed [42.5].

What is much less clear is the question of when all this is worth doing? That is, if our primary interest is in solving problems efficiently, in what circumstances is a genetic algorithm a good choice? To begin to answer this question requires an in-depth theoretical analysis of algorithms and problem classes. The work on this area has only just begun – most known results relate to the so-called $(1 + 1)$ EA (that is, a *population* of size 1),

Part E | 42

nd very little work exists on the role of crossover. The ll Pairs Shortest Path example represents the current tate of the art in this respect.

Having said that, it is clear (at least empirically nd anecdotally) that genetic algorithms can be very effective for complex problems, where problem instance information is limited. Indeed, there are many successful applications of genetic algorithms presented every year at conferences, and they are a well-known tool for optimization in industry. There is, therefore, a desperate need for further theoretical work and understanding in this area.

References

2.1   J.H. Holland: *Adaptation in Natural and Artificial Systems* (MIT, Cambridge 1992)

2.2   D.E. Goldberg: *Genetic Algorithms in Search, Optimization and Machine Learning* (Addison Wesley, Indianapolis 1989)

2.3   S. Droste, T. Jansen, K. Tinnefeld, I. Wegener: A new framework for the valuation of algorithms for black-box optimization. In: *Foundations of Genetic Algorithms*, ed. by K.A. De Jong, R. Poli, J.E. Rowe (Morgan Kaufmann, Torremolinos 2002) pp. 253–270

2.4   M. Mitchell: *An Introduction to Genetic Algorithms* (MIT, Cambridge 1998)

2.5   M.D. Vose: *The Simple Genetic Algorithm* (MIT, Cambridge 1999)

2.6   C.R. Reeves, J.E. Rowe: *Genetic Algorithms: Principles and Perspectives* (Kluwer, Dordrecht 2003)

2.7   I. Wegener: Theoretical aspects of evolutionary algorithms, Lect. Notes Comput. Sci. **2076**, 64–78 (2001)

2.8   F. Neumann, C. Witt: *Bioinspired Computation in Combinatorial Optimization – Algorithms and Their Computational Complexity* (Springer, Berlin, Heidelberg 2010)

2.9   A. Auger, B. Doerr: *Theory of Randomized Search Heuristics* (World Scientific, Singapore 2011)

2.10   T. Jansen, I. Wegener: On the utility of populations in evolutionary algorithms, Proc. Genet. Evol. Comput. Conf. (GECCO) 2001 (Morgan Kaufmann, San Francisco 2001) pp. 1034–1041

2.11   D. Sudholt: General lower bounds for the running time of evolutionary algorithms, Lect. Notes Comput. Sci. **6238**, 124–133 (2010)

2.12   T. Jansen, I. Wegener: On the analysis of evolutionary algorithms – A proof that crossover really can help, Algorithmica **34**(1), 47–66 (2002)

2.13   K. Deb: *Multi-Objective Optimization Using Evolutionary Algorithms* (Wiley, New York 2009)

2.14   B. Doerr, E. Happ, C. Klein: Crossover can provably be useful in evolutionary computation, Proc. Genet. Evol. Comput. Conf. (GECCO) 2008 (Morgan Kaufmann, Atlanta 2008) pp. 539–546

2.15   J.E. Baker: Reducing bias and inefficiency in the selection algorithm, Proc. 2nd Int. Conf. Genet. Algorithms (ICGA) 1987 (Lawrence Erlbaum, Hillsdale 1987) pp. 14–21

42.16   G. Rawlins: *Compared to What?* (Freeman, New York 1991)

42.17   A. Prügel-Bennett, J.L. Shapiro: An analysis of genetic algorithms using statistical mechanics, Phys. Rev. Lett. **72**(9), 1305–1309 (1994)

42.18   L.D. Whitley: The GENITOR algorithm and selective pressure: Why rank-based allocation of reproductive trials is best, Proc. 3rd Int. Conf. Genet. Algorithms (ICGA) 1989 (Morgan Kaufmann, Atlante 1989) pp. 116–121

42.19   A.H. Wright, J.E. Rowe: *Continuous dynamical system models of steady-state genetic algorithms*, Foundations of Genetic Algorithms (Morgan Kaufmann, Charlottesville 2002) pp. 209–225

42.20   I. Harvey: The microbial genetic algorithm, Proc. 10th Eur. Conf. Adv. Artif. Life (Springer, Berlin, Heidelberg 2011) pp. 126–133

42.21   J. Culberson: *Genetic Invariance: A New Paradigm for Genetic Algorithm Design*, Univ. Alberta Tech. Rep. R92-02 (1992)

42.22   M. Dietzfelbinger, B. Naudts, C. Van Hoyweghen, I. Wegener: The analysis of a recombinative hill-climber on H-IFF, IEEE Trans. Evol. Comput. **7**(5), 417–423 (2003)

42.23   S. Lin: Computer Solutions to the travelling salesman problem, Bell Syst. Tech. J. **44**(10), 2245–2269 (1965)

42.24   C.H. Papadimitriou, K. Steiglitz: *Combinatorial Optimization* (Dover Publications, New York 1998)

42.25   T. Jansen, I. Wegener: On the choice of the mutation probability for the (1+1)EA, Lect. Notes Comput. Sci. **1917**, 89–98 (2000)

42.26   C. Witt: Optimizing linear functions with randomized search heuristics, 29th Int. Symp. Theor. Asp. Comp. Sci. (STACS 2012), Leibniz-Zentrum fuer Informatik (2012) pp. 420–431

42.27   S. Böttcher, B. Doerr, F. Neumann: Optimal fixed and adaptive mutation rates for the LeadingOnes problem, Lect. Notes Comput. Sci. **6238**, 1–10 (2010)

42.28   J.E. Rowe, M.D. Vose, A.H. Wright: Representation invariant genetic operators, Evol. Comput. **18**(4), 635–660 (2010)

42.29   J.E. Rowe, M.D. Vose: Unbiased black box algorithms, Genet. Evol. Comput. Conf. (GECCO 2011) (ACM, Dublin 2011) pp. 2035–2042

42.30   P.K. Lehre, C. Witt: Black box search by unbiased variation, Genet. Evol. Comput. Conf. (GECCO 2010) (ACM, Portland 2010) pp. 1441–1449

42.31   V. Kachitvichyanukul, B.W. Schmeiser: Binomial random variate generation, Communications ACM **31**(2), 216–222 (1988)

42.32   T. Jansen, C. Zarges: Analysis of evolutionary algorithms: From computational complexity analysis to algorithm engineering, Found. Genet. Algorithms (FOGA) (2011) pp. 1–14

42.33   J.E. Rowe: Population fixed-points for functions of unitation, Foundations of Genetetic Algorithms, Vol. 5 (Morgan Kaufmann, San Francisco 1998) pp. 69–84

42.34   F. Neumann, P. Oliveto, C. Witt: Theoretical analysis of fitness-proportional selection: Landscapes and efficiency, Genet. Evol. Comput. Conf. (GECCO 2009) (ACM, Portland 2009) pp. 835–842

42.35   P.K. Lehre, X. Yao: On the impact of the mutation-selection balance on the runtime of evolutionary algorithms, Found. Genet. Algorithms (FOGA 2009) (ACM, Portland 2009)

42.36   J. Jägersküpper, T. Storch: When the plus strategy outperforms the comma strategy and when not, Proc. IEEE Symp. Found. Comput. Intell. (FOCI 2007) (IEEE, Bellingham 2007) pp. 25–32

42.37   J.E. Rowe, D. Sudholt: The choice of the offspring population size in the $(1, \lambda)$ EA, Genet. Evol. Comput. Conf. (GECCO 2012) (ACM, Philadelphia 2012)

42.38   N.J. Radcliffe: Forma analysis and random respectful recombination, Fourth Int. Conf. Genet. Algorithms (Morgan Kaufmann, San Francisco 1991) pp. 31–38

42.39   J.E. Rowe, M.D. Vose, A.H. Wright: Group properties of crossover and mutation, Evol. Comput. **10**(2), 151–184 (2002)

42.40   A. Moraglio: Towards a Geometric Unification of Evolutionary Algorithms, Ph.D. Thesis (University of Essex, Colchester 2007)

42.41   T. Kötzing, D. Sudholt, M. Theile: How crossover helps in pseudo-Boolean optimization, Genet. Evol. Comput. Conf. (GECCO 2011) (ACM, Dublin 2011) pp. 989–996

42.42   D. Whitley, T. Starkweather, D. Shaner: The traveling salesman and sequence scheduling: Quality solutions using genetic edge recombination. In: *The Handbook of Genetic Algorithms*, ed. by L. Davis (Van Nostrand Reinhold, Amsterdam 1991) pp. 350–372

42.43   S. Ronald: Duplicate genotypes in a genetic algorithm, Proc. 1998 IEEE World Congr. Comput. Intell (1998) pp. 793–798

42.44   K. De Jong: An Analysis of the Behaviour of a Class of Genetic Adaptive System, Ph.D. Thesis (University of Michigan, Ann Arbor 1975)

42.45   D.E. Goldberg, J. Richardson: Genetic algorithm with sharing for multimodal function optimization, Proc. 2nd Int. Conf. Genet. Algorithms (ICGA) 1987 (Lawrence Erlbaum Associates, Hillsdale 1987) pp. 41–49

42.46   L.J. Eshelman, J.D. Shaffer: Preventing premature convergence in genetic algorithms by preventing incest, Proc. 4th Int. Conf. Genet. Algorithms (ICGA) 1991 (Morgan Kaufmann, San Diego 1991) pp. 115–122

42.47   M. Tomassini: *Spatially Structured Evolutionary Algorithms: Articial Evolution in Space and Time* (Springer, Berlin, Heidelberg 2005)

42.48   J. Lässig, D. Sudholt: Analysis of speedups in parallel evolutionary algorithms for combinatorial optimization, Proc. 22nd Int. Symp. Algorithms Comput. (ISAAC 2011, Yokohama 2011)

42.49   W.D. Hillis: Co-evolving parasites improve simulated evolution as an optimization procedure, Physica D **42**, 228–234 (1990)

42.50   M.A. Potter, K.A. De Jong: Cooperative coevolution: An architecture for evolving coadapted subcomponents, Evol. Comput. **8**(1), 1–29 (2000)

42.51   T. Jansen, R.P. Wiegand: The cooperative coevolutionary $(1 + 1)$ EA, Evol. Comput. **12**(4), 405–434 (2004)

42.52   H. Mühlenbein: Parallel genetic algorithms population genetics and combinatorial optimization, Proc. 3rd Int. Conf. Genet. Algorithms (ICGA) 1989 (Morgan Kaufmann, Fairfax 1989) pp. 416–421

42.53   J.E. Rowe, K. Vinsen, N. Marvin: Parallel GAs for multiobjective functions, Proc. 2nd Nordic Workshop Genet. Algorithms, ed. by J. Alander (Univ. Vaasa Press, Vaasa 1996) pp. 61–70

42.54   B. Doerr, M. Theile: Improved analysis methods for crossover-based algorithms, Genet. Evol. Comput. Conf. (GECCO 2009) (ACM, Portland 2009) pp. 247–254

42.55   B. Doerr, T. Kötzing, F. Neumann, M. Theile: More effective crossover operators for the all-pairs shortest path problem, Lect. Notes Comput. Sc. **6238**, 184–193 (2010)

43. Genetic Programming

ames McDermott, Una-May O'Reilly

Genetic programming (GP) is the subset of evolutionary computation in which the aim is to create
executable programs. It is an exciting field with
many applications, some immediate and practical,
others long-term and visionary. In this chapter,
we provide a brief history of the ideas of genetic
programming. We give a taxonomy of approaches
and place genetic programming in a broader taxonomy of artificial intelligence. We outline some
current research topics and point to successful use
cases. We conclude with some practical GP-related
resources including software packages and venues
for GP publications.

3.1 Evolutionary Search for Executable Programs

here have been many attempts to artificially emulate
uman intelligence, from symbolic artificial intelli-
ence (AI) [43.1] to connectionism [43.2, 3], to subcog-
tive approaches like behavioral AI [43.4] and statisti-
al machine learning (ML) [43.5], and domain-specific
chievements like web search [43.6] and self-driving
rs [43.7]. Darwinian evolution [43.8] has a type of
stributed intelligence distinct from all of these. It has
eated lifeforms and ecosystems of amazing diversity,
mplexity, beauty, facility, and efficiency. It has even
eated forms of intelligence very different from itself,
cluding our own.

The principles of evolution – fitness biased selec-
on and inheritance with variation – serve as inspiration
r the field of evolutionary computation (EC) [43.9],
 adaptive learning and search approach which is

general-purpose, applicable even with black-box performance feedback, and highly parallel. EC is a trial-
and-error method: individual solutions are evaluated
for fitness, good ones are selected as parents, and
new ones are created by inheritance with variation
(Fig. 43.1).

GP is the subset of EC in which the aim is to create
executable programs. The search space is a set of programs, such as the space of all possible Lisp programs
within a subset of built-in functions and functions composed by a programmer or the space of numerical C
functions. The program representation is an encoding of
such a search space, for example an abstract syntax tree
or a list of instructions. A program's fitness is evaluated
by executing it to see what it does. New programs are
created by inheritance and variation of material from

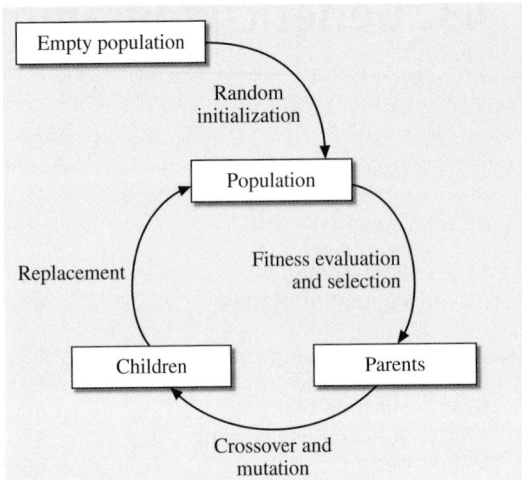

Fig. 43.1 The fundamental loop of EC

parent programs, with constraints to ensure syntactic correctness.

We define a program as a data structure capable of being executed directly by a computer, or of being compiled to a directly executable form by a compiler, or of interpretation, leading to execution of low-level code, by an interpreter. A key feature of some programming languages, such as Lisp, is *homoiconicity*: program code can be viewed as data. This is essential in GP, since when the algorithm operates on existing programs to make new ones, it is regarding them as data; but when they are being executed in order to determine what they do, they are being regarded as the program code. This double meaning echoes that of DNA (deoxyribonucleic acid), which is both data and code in the same sense.

GP exists in many different forms which differ (among other ways) in their executable representation. As in programming *by hand*, GP usually considers and composes programs of varying length. Programs are also generally hierarchical in some sense, with nesting of statements or control. These representation properties (variable length and hierarchical structure) raise a very different set of technical challenges for GP compared to typical EC.

GP is very promising, because programs are so general. A program can define and operate on any data structure, including numbers, strings, lists, dictionaries, sets, permutations, trees, and graphs [43.10–12]. Via Turing completeness, a program can emulate any mode of computation, including Turing machines, cellular automata, neural networks, grammars, and finite-state machines [43.13–18].

A program can be a data regression model [43.19] or a probability distribution. It can express the growth process of a plant [43.20], the gait of a horse [43.21], or the attack strategy of a group of lions [43.22]; it can model behavior in the Prisoner's Dilemma [43.23] or play chess [43.24], Pacman [43.25], or a car-racing game [43.26]. A program can generate designs for physical objects, like a space-going antenna [43.27], or plans for the organization of objects, like the layout of a manufacturing facility [43.28]. A program can implement a rule-based expert system for medicine [43.29], a scheduling strategy for a factory [43.30], or an exam timetable for a university [43.31]. A program can recognize speech [43.32], filter a digital signal [43.33], or process the raw output of a brain computer interface [43.34]. It can generate a piece of abstract art [43.35], a 3-D (three-dimensional) architectural model [43.36], or a piece of piano music [43.37].

A program can interface with natural or man-made sensors and actuators in the real world, so it can both act and react [43.38]. It can interact with a user or with remote sites over the network [43.39]. It can also introspect and copy or modify itself [43.40]. A program can be nondeterministic [43.41]. If true AI is possible, then a program can be intelligent [43.42].

43.2 History

GP has a surprisingly long history, dating back to very shortly after *von Neumann*'s 1945 description of the stored-program architecture [43.43] and the 1946 creation of ENIAC [43.44], sometimes regarded as the first general-purpose computer. In 1948, *Turing* stated the aim of machine intelligence and recognized that evolution might have something to teach us in this regard [43.45]:

Further research into intelligence of machinery will probably be very greatly concerned with searches [...] There is the genetical or evolutionary search by which a combination of genes is looked for, the criterion being survival value. The remarkable success of this search confirms to some extent the idea that intellectual activity consists mainly of various kinds of search.

However, Turing also went a step further. In 1950, he more explicitly stated the aim of automatic programming (AP) and a mapping between biological evolution and program search [43.46]:

> We have [...] divided our problem [automatic programming] into two parts. The child-program [Turing machine] and the education process. These two remain very closely connected. We cannot expect to find a good child-machine at the first attempt. One must experiment with teaching one such machine and see how well it learns. One can then try another and see if it is better or worse. There is an obvious connection between this process and evolution, by the identifications:
>
> Structure of the child machine = Hereditary material
> Changes = Mutations
> Natural selection = Judgment of the experimenter.

This is an unmistakeable, if abstract, description of GP though a computational fitness function is not envisaged).

Several other authors expanded on the aims and vision of AP and machine intelligence. In 1959 *Samuel* wrote that the aim was to be able to *Tell the computer what to do, not how to do it* [43.47]. An important early attempt at implementation of AP was the 1958 *learning machine* of *Friedberg* [43.48].

In 1963, *McCarthy* summarized [43.1] several representations with which machine intelligence might be attempted: neural networks, Turing machines, and calculator programs. With the latter, McCarthy was referring to Friedberg's work. McCarthy was prescient in identifying important issues such as representations,

operator behavior, density of good programs in the search space, sufficiency of the search space, appropriate fitness evaluation, and self-organized modularity. Many of these remain open issues in GP [43.49].

Fogel et al.'s 1960s *evolutionary programming* may be the first successful implementation of GP [43.50]. It used a finite-state machine representation for programs, with specialized operators to ensure syntactic correctness of offspring. A detailed history is available in *Fogel*'s 2006 book [43.51].

In the 1980s, inspired by the success of genetic algorithms (GAs) and learning classifier systems (LCSs), several authors experimented with hierarchically structured and program-like representations. *Smith* [43.52] proposed a representation of a variable-length list of rules which could be used for program-like behavior such as maze navigation and poker. *Cramer* [43.53] was the first to use a tree-structured representation and appropriate operators. With a simple proof of concept, it successfully evolved a multiplication function in a simple custom language. *Schmidhuber* [43.54] describes a GP system with the possibility of Turing completeness, though the focus is on meta-learning aspects. *Fujiki* and *Dickinson* [43.55] generated Lisp code for the prisoner's dilemma, *Bickel* and *Bickel* [43.56] used a GA to create variable-length lists of rules, each of which had a tree structure. An artificial life approach using machine-code genomes was used by *Ray* [43.57]. All of these would likely be regarded as on-topic in a modern GP conference.

However, the founding of the modern field of GP, and the invention of what is now called *standard* GP, are credited to *Koza* [43.19]. In addition to the abstract syntax tree notation (Sect. 43.3.3), the key innovations were subtree crossover (Sect. 43.3.3) and the description and set-up of many test problems. In this and

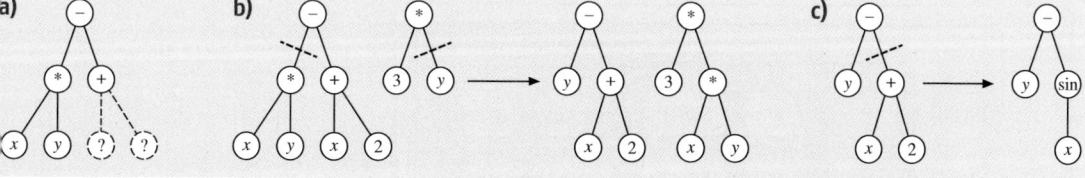

Fig. 43.2a–c The StdGP representation is an abstract syntax tree. The expression that will be evaluated in the *second tree from left* is, in inorder notation, $(x * y) - (x + 2)$. In preorder, or the notation of Lisp-style S-expressions, it is $- (* x y) (+ x 2)$. GP presumes that the variables x and y will be already bound to some value in the execution environment when the expression is evaluated. It also presumes that the operations $*$ and $-$, etc. are also defined. Note that, all interior tree nodes are effectively operators in some computational language. In standard GP parlance, these operators are called *functions* and the leaf tree nodes which accept no arguments and typically represent variables bound to data values from the problem domain are referred to as *terminals*

later research [43.10, 58, 59] symbolic regression of synthetic data and real-world time series, Boolean problems, and simple robot control problems such as the lawnmower problem and the artificial ant with Santa Fe trail were introduced as benchmarks and solved successfully for the first time, demonstrating that GP was a potentially powerful and general-purpose method capable of solving machine learning-style problems albeit conventional academic versions of them. Mutation was minimized in order to make it clear that GP was different from random search. GP took on its modern form in the years following Koza's 1992 book: many researchers took up work in the field, new types of GP were developed (Sect. 43.3), successful

applications appeared (Sect. 43.4), key research topics were identified (Sect. 43.5), further books were written, and conferences and journals were established (Sect. 43.6).

Another important milestone in the history of GP was the 2004 establishment of the *Humies*, the awards for human-competitive results produced by EC methods. The entries are judged for matching or exceeding human-produced solutions to the same or similar problems, and for criteria such as patentability and publishability. The impressive list of human-competitive results [43.60] again helps to demonstrate to researchers and clients outside the field of GP that it is powerful and general purpose.

43.3 Taxonomy of AI and GP

In this section, we present a taxonomy which firstly places GP in the context of the broader fields of EC, ML, and artificial intelligence (AI). It then classifies GP techniques according to their representations and their population models (Fig. 43.3).

43.3.1 Placing GP in an AI Context

GP is a type of EC, which is a type of ML, which is itself a subset of the broader field of AI (Fig. 43.3). *Carbonell* et al. [43.61] classify ML techniques according to the underlying learning strategy, which may be rote learning, learning from instruction, learning by analogy, learning from examples, and learning from observation and discovery. In this classification, EC and GP fit in the *learning from examples* category, in that an (individual, fitness) pair is an ex-

ample drawn from the search space together with its evaluation.

It is also useful to see GP as a subset of another field, AP. The term *automatic programming* seems to have had different meanings at different times, from automated card punching, to compilation, to template driven source generation, then generation techniques such as universal modeling language (UML), to the ambitious aim of creating software directly from a natural language English specification [43.62]. We interpret AP to mean creating software by specifying *what to do* rather than *how to do it* [43.47]. GP clearly fits into this category. Other nonevolutionary techniques also do so, for example inductive programming (IP). The main difference between GP and IP is that typically IP works only with programs which are known to be correct, achieving this using inductive methods over the spec

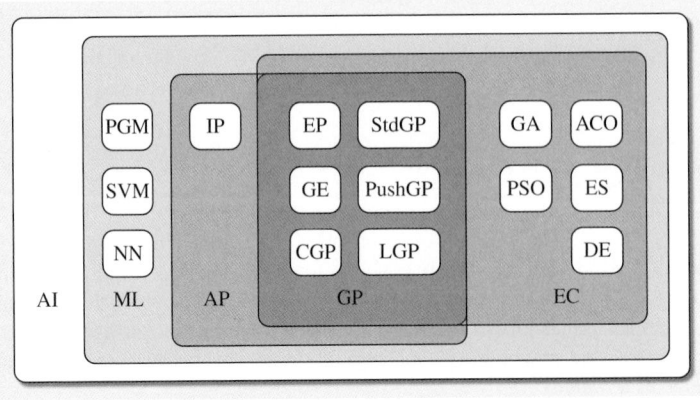

Fig. 43.3 A taxonomy of AI, EC, and GP

ifications, [43.63]. In contrast, GP is concerned mostly with programs which are syntactically correct, but behaviorally suboptimal.

43.3.2 Taxonomy of GP

It is traditional to divide EC into four main subfields: evolution strategies (ES) [43.64, 65], evolutionary programming (EP) [43.50], GAs [43.66], and GP. In this view, ES is chiefly characterized by real-valued optimization and self-adaptation of algorithm parameters; EP by a finite-state machine representation (later generalized) and the absence of crossover; GA by the bitstring representation; and GP by the abstract syntax tree representation. While historically useful, this classification is not exhaustive: in particular it does not provide a home for the many alternative GP representations which now exist. It also separates EP and GP, though they are both concerned with evolving programs. We prefer to use the term GP in a general sense to refer to all types of EC which evolve programs. We use the term *standard GP* (StdGP) to mean Koza-style GP with a tree representation. With this view, StdGP and EP are types of GP, as are several others discussed below. In the following, we classify GP algorithms according to their *representation* and according to their *population model*.

43.3.3 Representations

Throughout EC, it is useful to contrast *direct* and *indirect* representations. Standard GP is direct, in that the genome (the object created and modified by the genetic operators) serves directly as an executable program. Some other GP representations are indirect, meaning that the genome must be decoded or translated in some way to give an executable program. An example is grammatical evolution (GE, see below), where the genome is an integer array which is used to generate a program. Indirect representations have the advantage that they may allow an easier definition of the genetic operators, since they may allow the genome to exist in a rather simpler space than that of executable programs. Indirect representations also imitate somewhat more closely the mechanism found in nature, a mapping from DNA (deoxyribonucleic acid) to RNA (ribonucleic acid) to mRNA (messenger RNA) to codons to proteins and finally to cells. The choice between direct and indirect representations also affects the structure of the fitness landscape (Sect. 43.5.2). In the following, we present a nonexhaustive selection of the main representations used in GP, in each case describing initialization and the two key operators: mutation, and crossover.

Standard GP

In Standard GP (StdGP), the representation is an abstract syntax tree, or can be seen as a Lisp-style S-expression. All nodes are functions and all arguments are the same type. A function accepts zero or more arguments and returns a single value. Trees can be initialized by recursive random growth starting from a null node. StdGP uses parameterized initialization methods that diversify the size and structure of initial trees. Figure 43.2a shows a tree in the process of initialization.

Trees can be crossed over by cutting and swapping the subtrees rooted at randomly chosen nodes, as shown in Fig. 43.2b. They can be mutated by cutting and regrowing from the subtrees of randomly chosen nodes, as shown in Fig. 43.2c. Another mutation operator, HVL-Prime, is shown later in Fig. 43.11. Note that crossover or mutation creates an offspring of potentially different size and structure, but the offspring remains syntactically valid for evaluation. With these variations, a tree could theoretically grow to infinite size or height. To circumvent this, as a practicality, a hard parameterized threshold for size or height or some other threshold is used. Violations to the threshold are typically rejected. Bias may also be applied in the randomized selection of crossed-over subtree roots. A common variation of StdGP is *strongly typed GP* (STGP) [43.67, 68], which supports functions accepting arguments and returning values of specific types by means of specialized mutation and crossover operations that respect these types.

Executable Graph Representations

A natural generalization of the executable tree representation of StdGP is the executable graph. Neural networks can be seen as executable graphs in which each node calculates a weighted sum of its inputs and outputs the result after a fixed shaping function such as tanh(). *Parallel and distributed GP* (PDGP) [43.69] is more closely akin to StdGP in that nodes calculate different functions, depending on their labels, and do not perform a weighted sum. It also allows the topology of the graph to vary, unlike the typical neural network. *Cartesian GP* (CGP) [43.70] uses an integer-array genome and a mapping process to produce the graph. Each block of three integer genes codes for a single node in the graph, specifying the indices of its inputs and the function to be executed by the node (Fig. 43.4).

Neuro-evolution of augmenting topologies (NEAT) [43.71] again allows the topology to vary, and allows nodes to be labelled by the functions they perform, but in this case each node does perform a weighted sum of its inputs. Each of these representations uses different operators. For example, CGP uses simple array-oriented (GA-style) initialization, crossover, and mutation operators (subject to some customizations).

Finite-State Machine Representations

Some GP representations use graphs in a different way: the model of computation is the finite-state machine rather than the executable functional graph (Fig. 43.5). The original incarnation of *evolutionary programming* (EP) [43.72] is an example. In a typical implementation [43.72], five types of mutation are used: adding and deleting states, changing the initial state, changing the output symbol attached to edges, and changing the edges themselves. In this implementation, crossover is not used.

Grammatical GP

In *grammatical GP* [43.73], the context-free grammar (CFG) is the defining component of the representation. In the most common approach, search takes place in the space defined by a fixed nondeterministic CFG. The aim is to find a good program in that space. Often the CFG defines a useful subset of a programming language such as Lisp, C, or Python. Programs derived from the CFG can then be compiled or interpreted using either standard or special-purpose software. There are several advantages to using a CFG. It allows convenient definition of multiple data-types which are automatically respected by the crossover and mutation operators. It can introduce domain knowledge into the problem representation. For example, if it is known that good programs will consist of a conditional statement inside a loop, it is easy to express this knowledge using a grammar. The grammar can restrict the ways in which program expressions are combined, for example making the system aware of physical units in *dimensionally aware* GP [43.74, 75]. A grammatical GP system can conveniently be applied to new domains, or can incorporate new domain knowledge, through updates to the grammar rather than large-scale reprogramming.

In one early system [43.76], the derivation tree is used as the genome: initial individuals' genomes are randomly generated according to the rules of the grammar. Mutation works by randomly generating a new subtree starting from a randomly chosen internal node in the derivation tree. Crossover is constrained to exchange subtrees whose roots are identical. In this way, new individuals are guaranteed to be valid derivation trees. The executable program is then created from the genome by reading the leaves left to right. A later system, *grammatical evolution* (GEs) [43.77] instead uses an integer-array genome. Initialization, mutation and crossover are defined as simple GA-style array operations. The genome is mapped to an output program by using the successive integers of the genome to choose

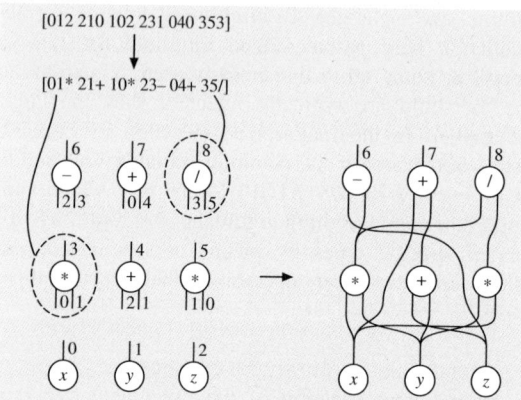

Fig. 43.4 Cartesian GP. An integer-array genome is divided into blocks: in each block the last integer specifies a function (*top-left*). Then one node is created for each input variable (x, y, z) and for each genome block. Nodes are arranged in a grid and outputs are indexed sequentially (*bottom-left*). The *first elements in each block* specify the indices of the incoming links. The *final graph* is created by connecting each node input to the node output with the same integer label (*right*). Dataflow in the graph is *bottom to top*. Multiple outputs can be read from the topmost layer of nodes. In this example node 6 outputs $xy - z + y$, node 7 outputs $x + z + y$, and node 8 outputs xy/xy

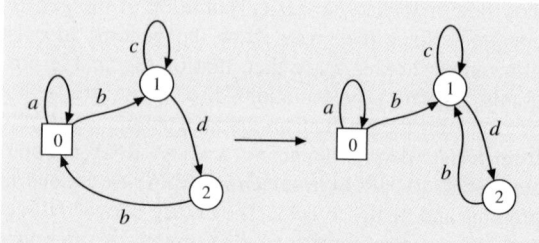

Fig. 43.5 EP representation: finite-state machine. In this example, a mutation changes a state transition

among the applicable production choices at each step of the derivation process. Figure 43.6 shows a simple grammar, integer genome, derivation process, and derivation tree. At each step of the derivation process, the left-most nonterminal in the derivation is rewritten. The next integer gene is used to determine, using the *mod rule*, which of the possible productions is chosen. The output program is the final step of the derivation tree.

Although successful and widely used, GE has also been criticized for the disruptive effects of its operators with respect to preserving the modular functionality of parents. Another system, *tree adjoining grammar-guided genetic programming* (TAG3P) has also been used successfully [43.78]. Instead of a string-rewriting CFG, TAG3P uses the tree-rewriting *tree adjoining grammars*. The representation has the advantage, relative to GE, that individuals are valid programs at every step of the derivation process. TAGs also have some context-sensitive properties [43.78]. However, it is a more complex representation.

Another common alternative approach, surveyed by *Shan* et al. [43.79], uses probabilistic models over grammar-defined spaces, rather than direct evolutionary search.

Linear GP

In *Linear GP* (LGP), the program is a list of instructions to be interpreted sequentially. In order to achieve complex functionality, a set of registers acting as state or memory are used. Instructions can read from or write to the registers. Several registers, which may be read-only, are initialized with the values of the input variables. One register is designated as the output: its value at the end of the program is taken as the result of the program. Since a register can be read multiple times after writing, an LGP program can be seen as having a graph structure. A typical implementation is that of [43.80]. It uses instructions of three registers each, which typically calculate a new value as an arithmetic function of some registers and/or constants, and assign it to a register (Fig. 43.7).

It also allows conditional statements and looping. It explicitly recognizes the possibility of nonfunctioning code, or *introns*. Since there are no syntactic constraints on how multiple instructions may be composed together, initialization can be as simple as the random generation of a list of valid instructions. Mutation can change a single instruction to a newly generated instruction, or change just a single element of an instruction. Crossover can be performed over the two parents' list structures, respecting instruction boundaries.

Stack–Based GP

A variant of linear GP avoids the need for registers by adding a stack. The program is again a list of instructions, each now represented by a single label. In a simple arithmetic implementation, the label may be one of the input variables (x_i), a numerical constant, or a function ($*$, $+$, etc.). If it is a variable or constant, the instruction is executed by pushing the value onto the stack. If a function, it is executed by popping the required number of operands from the stack, executing the function on them, and pushing the result back on. The result of the program is the value at the top of

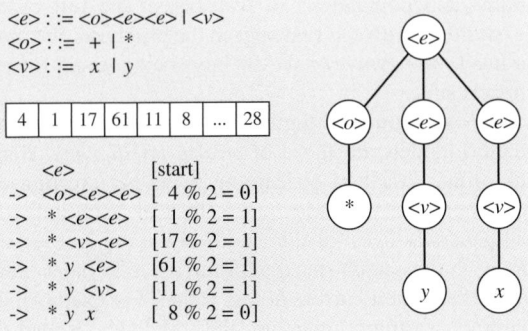

Fig. 43.6 GE representation. The grammar (*top-left*) consists of several rules. The genome (*center-left*) is a variable-length list of integers. At each step of the derivation process (*bottom-left*), the left-most nonterminal is rewritten as specified by a gene. The resulting derivation tree is shown on the *right*: reading just the leaves gives the derived program

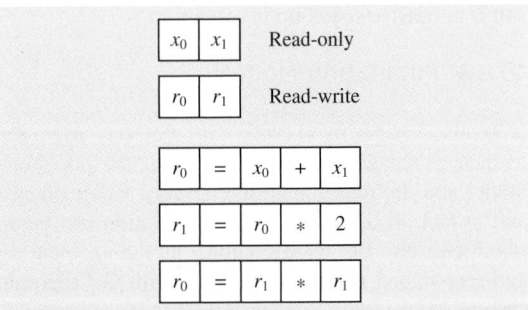

Fig. 43.7 Linear GP representation. This implementation has four registers in total (*top*). The representation is a list of register-oriented instructions (*bottom*). In this example program of three instructions, r_0 is the output register, and the formula $4(x_0 + x_1)^2$ is calculated

the stack after all instructions have been executed. With the stipulation that stack-popping instructions become no-ops when the stack is empty, one can again implement initialization, mutation, and crossover as simple list-based operations [43.81]. One can also constrain the operations to work on what are effectively subtrees, so that stack-based GP becomes effectively equivalent to a reverse Polish notation implementation of standard GP [43.82]. A more sophisticated type of stack-based GP is *PushGP* [43.83], in which multiple stacks are used. Each stack is used for values of a different type, such as integer, boolean, and float. When a function requires multiple operands of different types, they are taken as required from the appropriate stacks. With the addition of an *exec* stack which stores the program code itself, and the *code* stack which stores items of code, both of which may be both read and written, PushGP gains the ability to evolve programs with self-modification, modularity, control structures, and even self-reproduction.

Low-Level Programming

Finally, several authors have evolved programs directly in real-world low-level programming languages. *Schulte* et al. [43.84] automatically repaired programs written in Java byte code and in x86 assembly. *Orlov* and *Sipper* [43.85] evolved programs such as trail navigation and image classification de novo in Java byte code. This work made use of a specialized crossover operator which performed automated checks for compatibility of the parent programs' stack and control flow state. *Nordin* [43.86] proposed a machine-code representation for GP. Programs consist of lists of low-level register-oriented instructions which execute directly, rather than in a virtual machine or interpreter. The result is a massive speed-up in execution.

43.3.4 Population Models

It is also useful to classify GP methods according to their population models. In general the population model and the representation can vary independently, and in fact all of the following population can be applied with any EC representation including bitstrings and real-valued vectors, as well as with GP representations.

The simplest possible model, *hill-climbing*, uses just one individual at a time [43.87]. At each iteration, offspring are created until one of them is more highly fit than the current individual, which it then replaces. If at any iteration it becomes impossible to find an improve-

ment, the algorithm has *climbed the hill*, i.e. reached a local optimum, and stops. It is common to use a random restart in this case. The hill-climbing model can be used in combination with any representation. Note that it does not use crossover. Variants include ES-style (μ, λ) or $(\mu + \lambda)$ schemes, in which multiple parents each give rise to multiple offspring by mutation.

The most common model is an *evolving population*. Here a large number of individuals (from tens to many thousands) exist in parallel, with new generations being created by crossover and mutation among selected individuals. Variants include the steady-state and the generational models. They differ only in that the steady-state model generates one or a few new individuals at a time, adds them to the existing population and removes some old or weak individuals; whereas the generational model generates an entirely new population all at once and discards the old one.

The *island model* is a further addition, in which multiple populations all evolve in parallel, with infrequent migration between them [43.88].

In *coevolutionary* models, the fitness of an individual cannot be calculated in an endogenous way. Instead it depends on the individual's relationship to other individuals in the population. A typical example is in game-playing applications such as checkers, where the best way to evaluate an individual is to allow it to play against other individuals. Coevolution can also use fitness defined in terms of an individual's relationship to individuals in a population of a different type. A good example is the work of [43.89], which uses a type of *predator–prey* relationship between populations of programs and populations of test cases. The test cases (*predators*) evolve to find bugs in the programs; the programs (*prey*) evolve to fix the bugs being tested for by the test suites.

Another group of highly biologically inspired population models are those of *swarm intelligence*. Here the primary method of learning is not the creation of new individuals by inheritance. Instead, each individual generally lives for the length of the run, but *moves about* in the search space with reference to other individuals and their current fitness values. For example, in particle swarm optimization (PSO) individuals tend to move toward the global best and toward the best point in their own history, but tend to avoid moving too close to other individuals. Although PSO and related methods such as differential evolution (DE) are best applied in real-valued optimization, their population models and operators can be abstracted and applied in GP methods also [43.90, 91].

Finally, we come to *estimation of distribution algorithms* (EDAs). Here the idea is to create a population, select a subsample of the best individuals, model that subsample using a distribution, and then create a new population by sampling the distribution. This approach is particularly common in grammar-based GP [43.73], though it is also used with other representations [43.92–94]. The modeling-sampling process could be regarded as a type of whole-population crossover. Alternatively one can view EDAs as being quite far from the biological inspiration of most EC, and in a sense they bridge the gap between EC and statistical ML.

43.4 Uses of GP

Our introduction (Sect. 43.1) has touched on a wide array of domains in which GP has been applied. In this section, we give more detail on just a few of these.

43.4.1 Symbolic Regression

Symbolic regression is one of the most common tasks for which GP is used [43.19, 95, 96]. It is used as a component in techniques like data modeling, clustering, and classification, for example in the modeling application outlined in Sect. 43.4.2. It is named after techniques such as linear or quadratic regression, and can be seen as a generalization of them. Unlike those techniques it does not require a priori specification of the model. The goal is to find a function in symbolic form which models a data set. A typical symbolic regression is implemented as follows.

It begins with a dataset which is to be regressed, in the form of a numerical matrix (Fig. 43.8, left). Each row i is a data-point consisting of some input (explanatory) variables x_i and an output (response) variable y_i to be modeled. The goal is to produce a function $f(x)$ which models the relationship between x and y as closely as possible. Figure 43.8 (right) plots the existing data and one possible function f.

Typically StdGP is used, with a numerical *language* which includes arithmetic operators, functions like sinusoids and exponentials, numerical constants, and the input variables of the dataset. The internal nodes of each StdGP abstract syntax tree will be operators and functions, and the leaf nodes will be constants and variables.

To calculate the fitness of each model, the explanatory variables of the model are bound to their values at each of the training points x_i in turn. The model is executed, and the output $f(x_i)$ is the model's predicted response. This value $\hat{y}_i$ is then compared to the response of the training point y_i. The error can be visualized as the dotted lines in Fig. 43.8 (right). Fitness is usually defined as the root-mean-square error of the model's outputs versus the training data. In this formulation,

therefore, fitness is to be minimized

$$\text{fitness}(f) = \sqrt{\frac{\sum_{i=1}^{n}(f(x_i) - y_i)^2}{n}} \ .$$

Over the course of evolution, the population moves toward better and better models f of the training data. After the run, a testing data set is used to confirm that the model is capable of generalization to unseen data.

43.4.2 Machine Learning

Like other ML methods, GP is successful in quantitative domains where data is available for learning and both approximate solutions and incremental improvements are valued. In modeling or supervised learning, GP is preferable to other ML methods in circumstances where the form of the solution model is unknown a priori because it is capable of searching among possible forms for the model. Symbolic regression can be used as an approach to classification, regression modeling, and clustering. It can also be used to automatically extract influential features, since it is able to pare down the feature set it is given at initialization. GP-derived classifiers have been integrated into ensemble

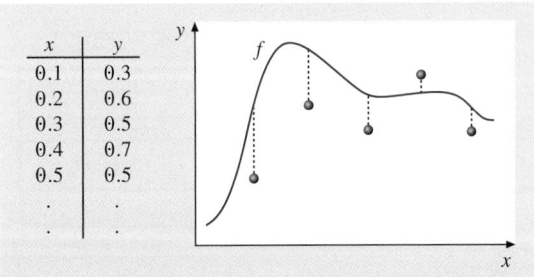

x	y
0.1	0.3
0.2	0.6
0.3	0.5
0.4	0.7
0.5	0.5
.	.
.	.

Fig. 43.8 Symbolic regression: a matrix of data (*left*) is to be modeled by a function. It is plotted as dots in the figure on the right. A candidate function f (*solid line*) can be plotted, and its errors (*dotted lines*) can be visualized

learning approaches and GP has been used in reinforcement learning (RL) contexts. Figure 43.9 shows GP as a means of ML which allows it to address problems such as planning, forecasting, pattern recognition, and modeling.

For the sensory evaluation problem described in [43.97], the authors use GP as the anchor of a ML framework (Fig. 43.10). A panel of assessors provides *liking scores* for many different flavors. Each flavor consists of a mixture of ingredients in different proportions. The goals are to discover the dependency of a liking score on the concentration levels of flavors' ingredients, identifying ingredients that drive liking, segmenting the panel into groups with similar liking preferences and optimizing flavors to maximize liking per group. The framework uses symbolic regression and ensemble methods to generate multiple diverse explanations of liking scores, with confidence information. It uses statistical techniques to extrapolate from the genetically evolved model ensembles to unobserved regions of the flavor space. It also segments the assessors into groups which either have the same propensity to like flavors, or whose liking is driven by the same ingredients.

Sensory evaluation data is very sparse and there is large variation among the responses of different assessors. A Pareto-GP algorithm (which uses multiobjective techniques to maximise model accuracy and minimise model complexity; [43.98]) was therefore used to evolve an ensemble of models for each assessor and to use this ensemble as a source of robust variable importance estimation. The frequency of variable occurrences in the models of the ensemble was interpreted as information about the ingredients that drive the liking of an assessor. Model ensembles with the same dominance of variable occurrences, and which demonstrate similar effects when the important variables are varied, were grouped together to identify assessors who are driven by the same ingredient set and in the same direction. Varying the input values of the important variables, while using the model ensembles of these panel segments, provided a means of conducting focused sensitivity analysis. Subsequently, the same model ensembles when clustered constitute the *black box* which is used by an evolutionary algorithm in its optimization of flavors that are well liked by assessors who are driven by the same ingredient.

43.4.3 Software Engineering

At least three areas of software engineering have been tackled with remarkable success by GP: bugfixing [43.99], parallelization [43.100, 101], and optimization [43.102–104]. These three areas are very different in their aims, scope, and methods; however, they all need to deal with two key problems in this domain: the very large and unconstrained search space, and the problem of program correctness. They therefore do not aim to evolve new functionality from scratch, but instead use existing code as material to be transformed in some way; and they either guarantee correctness of the evolved programs as a result of their representations, or take advantage of existing test suites in order to provide strong evidence of correctness.

Le Goues et al. [43.99] show that automatically fixing software bugs is a problem within the reach of GP. They describe a system called *GenProg*. It operates on C source code taken from open-source projects. It works by forming an abstract syntax tree from the original source code. The initial population is seeded with variations of the original. Mutations and crossover are constrained to copy or delete complete lines of code, rather than editing subexpressions, and they are constrained to alter only lines which are exercised by the failing test cases. This helps to reduce the search space size. The original test suites are used to give confidence that the program variations have not lost their original functionality. Fixes for several real-world bugs are produced, quickly and with high certainty of success, including bugs in HTTP servers, Unix utilities, and a media player. The fixes can be automatically processed to produce minimal patches. Best of all, the fixes

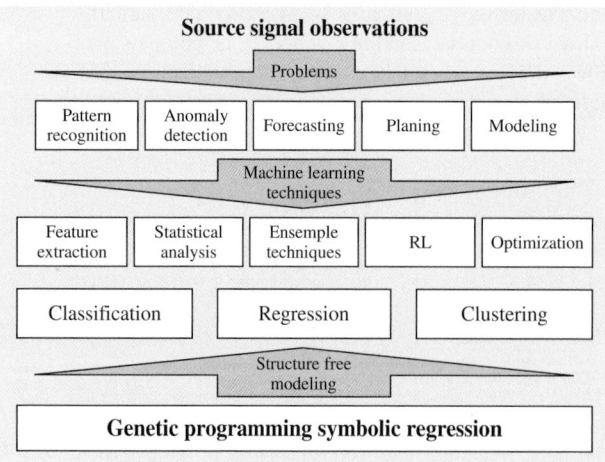

Fig. 43.9 GP as a component in ML. Symbolic regression can be used as an approach to many ML tasks, and integrated with other ML techniques

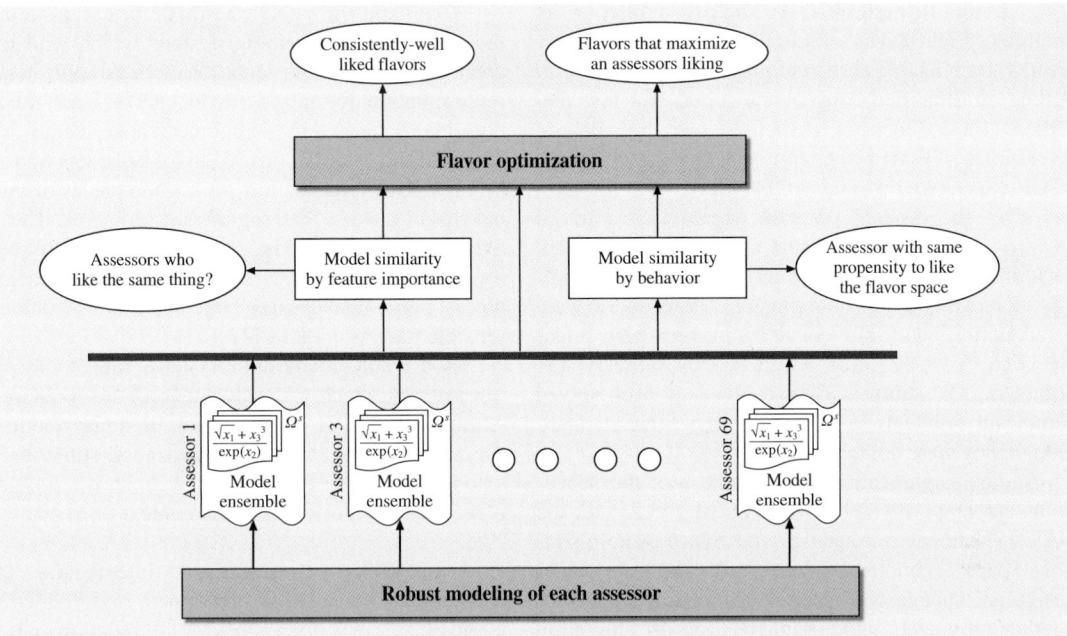

Part E | 43.4

Fig. 43.10 GP symbolic regression is unique and useful as an ML technique because it obviates the need to define the structure of a model prior to training. Here, it is used to form a personalized ensemble model for each assessor in a flavor evaluation panel

are demonstrated to be rather robust: some even generalize to fixing related bugs which were not explicitly encoded in the test suite.

Ryan [43.100] describes a system, *Paragen*, which automatically rewrites serial Fortran programs to parallel versions. In *Paragen I*, the programs are directly varied by the genetic operators, and automated tests are used to reward the preservation of the program's original semantics. The work of *Williams* [43.101] was in some ways similar to Paragen I. In *Paragen II*, correctness of the new programs is instead guaranteed, using a different approach. The programs to be evolved are sequences of transformations defined over the original serial code. Each transformation is known to preserve semantics. Some transformations however directly transform serial operations to parallel, while other transformations merely enable the first type.

A third goal of software engineering is optimization of existing code. *White* et al. [43.104] tackle this task using a multiobjective optimization method. Again, an existing program is used as a starting point, and the aim is to evolve a semantically equivalent one with improved characteristics, such as reduced memory usage, execution time, or power consumption. The system is capable of finding *nonobvious* optimizations, i. e. ones which cannot be found by optimizing compilers. A population of test cases is coevolved with the population of programs. *Stephenson* et al. [43.102, 103] in the Meta Optimization project improve program execution speed by using GP to refine priority functions within the compiler. The compiler generates better code which executes faster across the input range of one program and across the program range of a benchmark set.

A survey of the broader field of *search-based software engineering* is given by *Harman* [43.105].

43.4.4 Design

GP has been successfully used in several areas of design. This includes both engineering design, where the aim is to design some hardware or software system to carry out a well-defined task, and aesthetic design, where the aim is to produce art objects with subjective qualities.

Engineering Design
One of the first examples of GP design was the synthesis of analog electrical circuits by *Koza* et al. [43.106]. This work addressed the problem of automatically cre-

ating circuits to perform tasks such as a filter or an amplifier. Eight types of circuit were automatically created, each having certain requirements, such as outputting an amplified copy of the input, and low distortion. These functions were used to define fitness. A complex GP representation was used, with both STGP (Sect. 43.3.3) and ADFs (Sect. 43.5.3). Execution of the evolved program began with a trivial *embryonic circuit*. GP program nodes, when executed, performed actions such as altering the circuit topology or creating a new component. These nodes were parameterized with numerical parameters, also under GP control, which could be created by more typical arithmetic GP subtrees. The evolved circuits solved significant problems to a human-competitive standard though they were not fabricated.

Another significant success story was the space-going antenna evolved by *Hornby* et al. [43.27] for the NASA (National Aeronautics and Space Administration) Space Technology 5 spacecraft. The task was to design an antenna with certain beamwidth and bandwidth requirements, which could be tested in simulation (thus providing a natural fitness function). GP was used to reduce reliance on human labor and limitations on complexity, and to explore areas of the search space which would be rejected as not worthy of exploration by human designers. Both a GA and a GP representation were used, producing quite similar results. The GP representation was in some ways similar to a 3-D turtle graphics system. Commands included *forward* which moved the turtle forward, creating a wire component, and *rotate-x* which changed orientation. Branching of the antenna arms was allowed with special markers similar to those used in turtle graphics programs. The program composed of these primitives, when run, created a wire structure, which was rotated and copied four times to produce a symmetric result for simulation and evaluation.

Aesthetic Design

There have also been successes in the fields of graphical art, 3-D aesthetic design, and music. Given the aesthetic nature of these fields, GP fitness is often replaced by an interactive approach where the user performs *direct selection* on the population. This approach dates back to *Dawkins'* seminal *Biomorphs* [43.107] and has been used in other forms of EC also [43.108]. Early successes were those of *Todd* and *Latham* [43.109], who created pseudo-organic forms, and *Sims* [43.35] who created abstract art. An overview of evolutionary art is provided by *Lewis* [43.110].

A key aim throughout aesthetic design is to avoid the many random-seeming designs which tend to be created by typical representations. For example, a naive representation for music might encode each quarter note as an integer in a genome whose length is the length of the eventual piece. Such a representation will be capable of representing some good pieces of music but it will have several significant problems. The vast majority of pieces will be very poor and random sounding. Small mutations will tend to gradually degrade pieces, rather than causing large-scale and semantically sensible transformations [43.111].

As a result, many authors have tried to use representations which take advantage of forms of *reuse*. Although reuse is also an aim in nonaesthetic GP (Sect. 43.5.3), the hypothesis that good solutions will tend to involve reuse, even on new, unknown problems, is more easily motivated in the context of aesthetic design.

In one strand of research, the time or space to be occupied by the work is predefined, and divided into a grid of 1, 2, or 3 dimensions. A GP function of 1, 2 or 3 arguments is then evolved, and applied to each point in the grid with the coordinates of the point passed as arguments to the function. The result is that the function is reused many times, and all parts of the work are felt to be coherent. The earliest example of such work was that of *Sims* [43.35], who created fascinating graphical art (a 2-D grid) and some animations (a 3-D grid of two spatial dimensions and 1 time dimension). The paradigm was later brought to a high degree of artistry by *Hart* [43.112]. The same generative idea, now with a 1-D grid representing time, was used by *Hoover* et al. [43.113], *Shao* et al. [43.114] and *McDermott* and *O'Reilly* [43.115] to produce music as a function of time, and with a 3-D grid by *Clune* and *Lipson* [43.116] to produce 3-D sculptures.

Other successful work has used different approaches to reuse. *L-systems* are grammars in which symbols are recursively expanded in parallel: after several expansions (a *growth process*), the string will be highly patterned, with multiple copies of some substrings. Interpreting this string as a program can then yield highly patterned graphics [43.117], artificial creatures [43.118], and music [43.119]. Grammars have also been used in 3-D and architectural design, both in a modified L-system form [43.36] and in the standard GE form [43.120]. The *Ossia* system of *Dahlstedt* [43.37] uses GP trees with recursive pointers to impose reuse and a natural, *gestural* quality on short pieces of art music.

43.5 Research Topics

Many research topics of interest to GP practitioners are also of broader interest. For example, the self-adaptation of algorithm parameters is a topic of interest throughout EC. We have chosen to focus on four research topics of specific interest in GP: bloat, GP theory, modularity, and open-ended evolution.

43.5.1 Bloat

Most GP-type problems naturally require variable-length representations. It might be expected that selection pressure would effectively guide the population toward program sizes appropriate to the problem, and indeed this is sometimes the case. However, it has been observed that for many different representations [43.121] and problems, programs grow over time *without* apparent fitness improvements. This phenomenon is called *bloat*. Since the time complexity for the evaluation of a GP program is generally proportional to its size, this greatly slows the GP run down. There are also other drawbacks. The eventual solution may be so large and complex that is unreadable, negating a key advantage of symbolic methods like GP. Overly large programs tend to generalize less well than parsimonious ones. Bloat may negatively impact the rate of fitness improvement. Since bloat is a significant obstacle to successful GP, it is an important topic of research, with differing viewpoints both on the causes of bloat and the best solutions.

The competing theories of the causes of bloat are summarized by *Luke* and *Panait* [43.122] and *Silva* et al. [43.123]. A fundamental idea is that adding material rather than removing material from a GP tree is more likely to lead to a fitness improvement. The *hitchhiking* theory is that noneffective code is carried along by virtue of being attached to useful code. *Defense against crossover* suggests that large amounts of noneffective code give a selection advantage later in GP runs when crossover is likely to highly destructive of good, fragile programs. *Removal bias* is the idea that it is harder for GP operators to remove exactly the right (i.e., noneffective) code than it is to add more. The *fitness causes bloat* theory suggests that fitness-neutral changes tend to increase program size just because there are many more programs with the same functionality at larger sizes than at smaller [43.124]. The *modification point depth* theory suggests that children formed by tree crossover at deep crossover points are likely to have fitness similar to their parents and thus

more likely to survive than the more radically different children formed at shallow crossover points. Because larger trees have more very deep potential crossover points, there is a selection pressure toward growth. Finally, the *crossover bias* theory [43.125] suggests that after many crossovers, a population will tend toward a limiting distribution of tree sizes [43.126] such that small trees are more common than large ones – note that this is the opposite of the effect that might be expected as the basis of a theory of bloat. However, when selection is considered, the majority of the small programs cannot compete with the larger ones, and so the distribution is now skewed in favour of larger programs.

Many different solutions to the problem of bloat have been proposed, many with some success. One simple method is *depth limiting*, imposing a fixed limit on the tree depth that can be produced by the variation operators [43.19].

Another simple but effective method is *Tarpeian bloat control* [43.127]. Individuals which are larger than average receive, with a certain probability, a constant punitively bad fitness. The advantage is that these individuals are not evaluated, and so a huge amount of time can be saved and devoted to running more generations (as in [43.122]). The Tarpeian method does allow the population to grow beyond its initial size, since the punishment is only applied to a proportion of individuals – typically around 1 in 3. This value can also be set adaptively [43.127].

The *parsimony pressure* method evaluates all individuals, but imposes a fitness penalty on overly large individuals. This assumes that fitness is commensurable with size: the magnitude of the punishment establishes a *de facto exchange rate* between the two. *Luke* and *Panait* [43.122] found that parsimony pressure was effective across problems and across a wide range of exchange rates.

The choice of an exchange rate can be avoided using multiobjective methods, such as Pareto-GP [43.128], where one of the objectives is fitness and the other program length or complexity. The correct definition for complexity in this context is itself an interesting research topic [43.96, 129]. Alternatively, the pressure against bloat can be moved from the fitness evaluation phase to the the selection phase of the algorithm, using the *double tournament* method [43.122]. Here individuals must compete in one fitness-based tournament and one size-based one. Another approach is to incorporate tree size directly into fitness evaluation using a minimum description length principle [43.130].

Another technique is called *operator length equalization*. A histogram of program sizes is maintained throughout the run and is used to set the population's capacity for programs of different sizes. A newly created program which would cause the population's capacity to be exceeded is rejected, unless exceptionally fit. A *mutation-based* variation of the method instead mutates the overly large individuals using directed mutation to become smaller or larger as needed.

Some authors have argued that the choice of GP representation can avoid the issue of bloat [43.131]. Some aim to avoid the problem of bloat by speeding up fitness evaluation [43.82, 132] or avoiding wasted effort in evaluation [43.133, 134]. Sometimes GP techniques are introduced with other motivations but have the side-effect of reducing bloat [43.135].

In summary, researchers including *Luke* and *Panait* [43.122], *Poli* et al. [43.127], *Miller* [43.131], and *Silva* et al. [43.123] have effectively *declared victory* in the fight against bloat. However, their techniques have not yet become standard for new GP research and benchmark experiments.

43.5.2 GP Theory

Theoretical research in GP seeks to answer a variety of questions, for example: What are the drivers of population fitness convergence? How does the behavior of an operator influence the progress of the algorithm? How does the combination of different algorithmic mechanisms steer GP toward fitter solutions? What mechanisms cause bloat to arise? What problems are difficult for GP? How diverse is a GP population? Theoretical methodologies are based in mathematics and exploit formalisms, theorems, and proofs for rigor. While GP may appear simple, beyond its stochastic nature which it shares with all other evolutionary algorithms, its variety of representations each impose specific requirements for theoretical treatment. All GP representations share two common traits which greatly contribute to the difficulty it poses for theoretical analysis. First, the representations have no fixed size, implying a complex search space. Second, GP representations do not imply that parents will be equal in size and shape. While crossover accommodates this lack of synchronization, it generally allows the exchange of content from *anywhere* in one parent to *anywhere* in the other parent's

tree. This implies combinatorial outcomes and *likes not switching with likes*. This functionality contributes to complicated algorithmic behavior which is challenging to analyze.

Here, we select several influential methods of theoretical analysis and very briefly describe them and their results: schema-based analysis, Markov chain modeling, runtime complexity, and problem difficulty. We also introduce the No Free Lunch Theorem and describe its implications for GP.

Schema-Based Analysis

In schema-based analysis, the search space is conceptually partitioned into hyperplanes (also known as schemas) which represent sets of partial solutions. There are numerous ways to do this and, as a consequence, multiple schema definitions have been proposed [43.136–139]. The fitness of a schema is estimated as the average fitness of all programs in the sample of its hyperplane, given a population. The processes of fitness-based selection and crossover are formalized in a recurrence equation which describes the expected number of programs sampling a schema from the current population to the next. Exact formulations have been derived for most types of crossover [43.140, 141]. These alternatively depend on making explicit the effects and the mechanisms of schema creation. This leads to insight; however, tracking schema equations in actual GP population dynamics is infeasible. Also, while schema theorems predict changes from one generation to the next, they cannot predict further into the future to predict the long-term dynamics that GP practitioners care about.

Markov Chain Analysis

Markov chain models are one means of describing such long-term GP dynamics. They take advantage of the Markovian property observed in a GP algorithm: the composition of one generation's population relies only upon that of the previous generation. Markov chains describe the probabilistic movement of a particular population (state) to others using a probabilistic transition matrix. In evolutionary algorithms, the transition matrix must express the effects of any selection and variation operators. The transition matrix, when multiplied by itself k times, indicates which new populations can be reached in k generations. This, in principle, allows a calculation of the probability that a population with a solution can be reached. To date a Markov chain for a simplified GP crossover operator has been derived, see [43.142]. Another interesting Markov chain-based

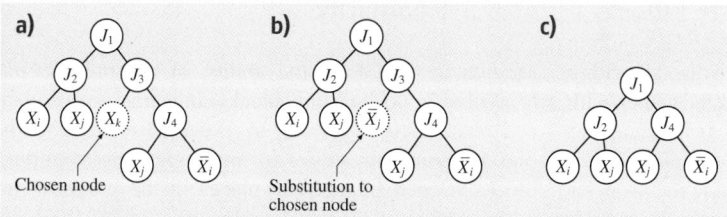

Fig. 43.11a–c HVL-prime mutation: substitution and deletion (**a**) Original parse tree, (**b**) Result of substitution (**c**) Result of deletion

result has revealed that the *distribution of functionality of non-Turing complete programs approaches a limit as length increases*. Markov chain analysis has also been the means of describing what happens with GP semantics rather than syntax. The influence of sub-tree crossover is studied in a semantic building block analysis by [43.143]. Markov chains, unfortunately, combinatorially explode with even simple extensions of algorithm dynamics or, in GP's case, its theoretically infinite search space. Thus, while they can support further analysis, ultimately this complexity is unwieldy to work with.

Runtime Complexity

Due to stochasticity, it is arguably impossible in most cases to make formal guarantees about the number of fitness evaluations needed for a GP algorithm to find an optimal solution. However, initial steps in the runtime complexity analysis of genetic programming have been made in [43.144]. The authors study the runtime of hill climbing GP algorithms which use a mutation operator called HVL-Prime (Figs. 43.11 and 43.12). Several of these simplified GP algorithms were analyzed on two separable model problems, Order and Majority introduced in [43.145]. Order and Majority each have an independent, additive fitness structure. They each admit multiple solutions based on their objective function, so they exhibit a key property of all real GP problems. They each capture a different relevant facet of typical GP problems. Order represents problems, such as classification problems, where the operators include conditional functions such as an IF-THEN-ELSE. These functions give rise to conditional execution paths which have implications for evolvability and the effectiveness of crossover. Majority is a GP equivalent of the GA OneMax problem [43.146]. It reflects a general (and thus weak) property required of GP solutions: a solution must have correct functionality (by evolving an aggregation of subsolutions) and no incorrect functionality. The analyses highlighted, in particular, the impact of accepting or rejecting neutral moves and the impor-

tance of a local mutation operator. A similar finding, [43.147], regarding mutation arose from the analysis of the Max problem [43.148] and hillclimbing. For a search process bounded by a maximally sized tree of n nodes, the time complexity of the simple GP mutation-based hillclimbing algorithms using HVL-Prime for the entire range of MAX variants are $O(n \log^2 n)$ when one mutation operation precedes each fitness evaluation. When multiple mutations are successively applied before each fitness evaluation, the time complexity is $O(n^2)$. This complexity can be reduced to $O(n \log n)$ if the mutations are biased to replace a random leaf with distance d from the root with probability 2^{-d}.

Runtime analyses have also considered parsimony pressure and multiobjective GP algorithms for generalizations of Order and Majority [43.149].

GP algorithms have also been studied in the PAC learning framework [43.150].

Problem Difficulty

Problem difficulty is the study of the differences between algorithms and problems which lead to differences in performance. Stated simply, the goal is to understand why some problems are easy and some are hard, and why some algorithms perform well on certain problems and others do not. Problem difficulty work in the field of GP has much in common with similar work in the broader field of EC. Problem difficulty is naturally related to the size of the search space; smaller spaces are easier to search, as are spaces in which

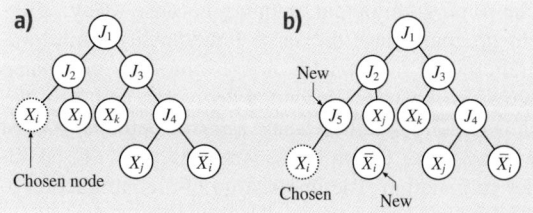

Fig. 43.12a,b HVL-prime mutation: insertion (**a**) Original parse tree, (**b**) Result of insertion

the solution is *over-represented* [43.151]. Difficulty is also related to the *fitness landscape* [43.152], which in turn depends on both the problem and the algorithm and representation chosen to solve it. Landscapes with few local optima (visualized in the fitness landscape as peaks which are not as high as that of the global optimum) are easier to search. *Locality*, that is the property that small changes to a program lead to small changes in fitness, implies a smooth, easily searchable landscape [43.151, 153].

However, more precise statements concerning problem difficulty are usually desired. One important line of research was carried out by *Vanneschi* et al. [43.154–156]. This involved calculating various measures of the *correlation* of the fitness landscape, that is the relationship between distance in the landscape and fitness difference. The measures include the *fitness distance correlation* and the *negative slope coefficient*. These measures require the definition of a distance measure on the search space, which in the case of standard GP means a distance between pairs of trees. Various tree distance measures have been proposed and used for this purpose [43.157–160]. However, the reliable prediction of performance based purely on landscape analysis remains a distant goal in GP as it does in the broader field of EC.

No Free Lunch
In a nutshell, the No Free Lunch Theorem [43.161] proves that, averaged over all problem instances, no algorithm outperforms another. Follow-up NFL analysis [43.162, 163] yields a similar result for problems where the set of fitness functions are closed under permutation. One question is whether the NFL theorem applies to GP algorithms: for some problem class, is it worth developing a better GP algorithm, or will this effort offer no extra value when all instances of the problem are considered? Research has revealed two conditions under which the NFL breaks down for GP because the set of fitness functions is not closed under permutation. First, GP has a many-to-one syntax tree to program output mapping because many different programs have the same functionality while program output functionality is not uniformly distributed across syntax trees. Second, a geometric argument has shown [43.164], that many *realistic* situations exist where a set of GP problems is provably not closed under permutation. The implication of a contradiction to the No Free Lunch theorem is that it is worthwhile investing effort in improving a GP algorithm for a class of problems.

43.5.3 Modularity

Modularity in GP is the ability of a representation to evolve good building blocks and then encapsulate and reuse them. This can be expected to make complex programs far easier to find, since good building blocks needed in multiple places in the program not be laboriously re-evolved each time. One of the best-known approaches to modularity is *automatically defined functions* (ADFs), where the building blocks are implemented as functions which are defined in one part of the evolving program and then invoked from another part [43.58]. This work was followed by automatically defined macros which are more powerful than ADFs and allow control of program flow [43.165]; automatically defined iteration, recursion, and memory stores [43.10]; modularity in other representations [43.166]; and demonstrations of the power of reuse, [43.167].

43.5.4 Open–Ended Evolution and GP

Biological evolution is a long-running exploration of the enormously varied and indefinitely sized DNA search space. There is no hint that a limit on new areas of the space to be explored will ever be reached. In contrast, EC algorithms often operate in search spaces which are finite and highly simplified in comparison to biology. Although GP itself can be used for a wide variety of tasks (Sect. 43.1), each specific instance of the GP algorithm is capable of solving only a very narrow problem. In contrast, some researchers see biological evolution as pointing the way to a more ambitious vision of the possibilities for GP [43.168]. In this vision, an evolutionary run would continue for an indefinite length of time, always exploring new areas of an indefinitely sized search space; always responding to changes in the environment; and always reshaping the search space itself. This vision is particularly well suited to GP, as opposed to GAs and similar algorithms, because GP already works in search spaces which are infinite in theory, if not in practice.

To make this type of GP possible, it is necessary to prevent convergence of the population on a narrow area of the search space. Diversity preservation [43.169], periodic injection of new random material [43.170], and island-structured population models [43.88] can help in this regard.

Open-ended evolution would also be facilitated by complexity and nonstationarity in the algorithm's evolutionary *ecosystem*. If fitness criteria are dynamic or coevolutionary [43.171–173], there may be no natural end-point to evolution, and so continued exploration under different criteria can lead to unlimited new results.

43.6 Practicalities

43.6.1 Conferences and Journals

Several conferences provide venues for the publication of new GP research results. The ACM *Genetic and Evolutionary Computation Conference* (GECCO) alternates annually between North America and the rest of the world and includes a GP track. *EuroGP* is held annually in Europe as the main event of *Evo**, and focuses only on GP. The IEEE *Congress on Evolutionary Computation* is a larger event with broad coverage of EC in general. *Genetic Programming Theory and Practice* is held annually in Ann Arbor, MI, USA and provides a focused forum for GP discussion. *Parallel Problem Solving from Nature* is one of the older, general EC conferences, held biennially in Europe. It alternates with the *Evolution Artificielle* conference. Finally, *Foundations of Genetic Algorithms* is a smaller, theory-focused conference.

The journal most specialized to the field is probably *Genetic Programming and Evolvable Machines* (published by Springer). The September 2010, 10-year anniversary issue included several review articles on GP. *Evolutionary Computation* (MIT Press) and the IEEE *Transactions on Evolutionary Computation* also publish important GP material. Other on-topic journals with a broader focus include *Applied Soft Computing* and *Natural Computing*.

43.6.2 Software

A great variety of GP software is available. We will mention only a few packages – further options can be found online.

One of the well-known Java systems is *ECJ* [43.174, 175]. It is a general-purpose system with support for many representations, problems, and methods, both within GP and in the wider field of EC. It has a helpful mailing list. Watchmaker [43.176] is another general-purpose system with excellent out-of-the-box examples. *GEVA* [43.177, 178] is another Java-based package, this time with support only for GE.

For users of C++ there are also several options. Some popular packages include *Evolutionary* *Objects* [43.179], *μGP* [43.180–182], and *OpenBeagle* [43.183, 184]. Matlab users may be interested in GPLab [43.185], which implements standard GP, while DEAP [43.186] provides implementations of several algorithms in Python. PushGP [43.187] is available in many languages.

Two more systems are worth mentioning for their deliberate focus on simplicity and understandability. *TinyGP* [43.188] and *PonyGE* [43.189] implement standard GP and GE respectively, each in a single, readable source file.

Moving on from open source, Michael Schmidt and Hod Lipson's Eureqa [43.190] is a free-to-use tool with a focus on symbolic regression of numerical data and the built-in ability to use cloud resources.

Finally, the authors are aware of two commercially available GP tools, each fast and industrial-strength. They have more automation and *it just works* functionality, relative to most free and open-source tools. Free trials are available. *DataModeler* (Evolved Analytics LLC) [43.191] is a notebook in Mathematica. It employs the ParetoGP method [43.128] which gives the ability to trade program fitness off against complexity, and to form ensembles of programs. It also exploits complex population archiving and archive-based selection. It offers means of dealing with ill-conditioned data and extracting information on variable importance from evolved models. *Discipulus* (Register Machine Learning Technologies, Inc.) [43.192] evolves machine code based on the ideas of *Nordin* et al. [43.193]. It runs on Windows only. The machine code representation allows very fast fitness evaluation and low memory usage, hence large populations. In addition to typical GP features, it can: use an ES to optimise numerical constants; automatically construct ensembles; preprocess data; extract variable importance after runs; automatically simplify results; and save them to high-level languages.

43.6.3 Resources and Further Reading

Another useful resource for GP research is the *GP Bibliography* [43.194]. In addition to its huge, regularly updated collection of BibTeX-formatted citations, it has lists of researchers' homepages [43.195] and co-authorship graphs. The *GP mailing list* [43.196] is one well-known forum for discussion.

Many of the traditional GP benchmark problems have been criticized for being unrealistic in various ways. The lack of standardization of benchmark problems also allows the possibility of cherry-picking of benchmarks. Effort is underway to bring some standardization to the choice of GP benchmarks [43.197, 198].

Those wishing to read further have many good options. The *Field Guide to GP* is a good introduction, walking the reader through simple examples,

scanning large amounts of the literature, and offering practical advice [43.199]. *Luke's Essentials of Meta-heuristics* [43.200] also has an introductory style, but is broader in scope. Both are free to download. Other broad and introductory books include those by *Fo-gel* [43.51] and *Banzhaf* et al. [43.201]. More specialized books include those by *Langdon* and *Poli* [43.202] (coverage of theoretical topics), *Langdon* [43.11] (narrower coverage of GP with data structures), *O'Neill* and *Ryan* [43.77] (GE), *Iba* et al. [43.203] (GP-style ML), and *Sipper* [43.204] (games). *Advances in Genetic Programming*, a series of four volumes, contains important foundational work from the 1990s.

References

43.1   J. McCarthy: Programs with Common Sense, Technical Report (Stanford University, Department of Computer Science, Stanford 1963)

43.2   F. Rosenblatt: The perceptron: A probabilistic model for information storage and organization in the brain, Psychol. Rev. **65**(6), 386 (1958)

43.3   D.E. Rumelhart, J.L. McClelland: *Parallel Distributed Processing: Explorations in the Microstructure of Cognition, Volume 1: Foundations* (MIT, Cambridge 1986)

43.4   R.A. Brooks: Intelligence without representation, Artif. Intell. **47**(1), 139–159 (1991)

43.5   C. Cortes, V. Vapnik: Support-vector networks, Mach. Learn. **20**(3), 273–297 (1995)

43.6   L. Page, S. Brin, R. Motwani, T. Winograd: *The Pagerank Citation Ranking: Bringing Order to the Web*, Technical Report 1999-66 (Stanford InfoLab, Stanford 1999), available online at http://ilpubs.stanford.edu:8090/422/. Previous number = SIDL-WP-1999-0120.

43.7   J. Levinson, J. Askeland, J. Becker, J. Dolson, D. Held, S. Kammel, J.Z. Kolter, D. Langer, O. Pink, V. Pratt, M. Sokolsky, G. Stanek, D. Stavens, A. Teichma, M. Werling, S. Thrun: Towards fully autonomous driving: Systems and algorithms, Intell. Veh. Symp. (IV) IEEE (2011) pp. 163–168

43.8   C. Darwin: *The Origin of Species by Means of Natural Selection: Or, the Preservation of Favored Races in the Struggle for Life* (John Murray, London 1859)

43.9   T. Bäck, D.B. Fogel, Z. Michalewicz (Eds.): *Handbook of Evolutionary Computation* (IOP Publ., Bristol 1997)

43.10 J.R. Koza, D. Andre, F.H. Bennett III, M. Keane: *Genetic Programming 3: Darwinian Invention and Problem Solving* (Morgan Kaufman, San Francisco 1999), available online at http://www.genetic-programming.org/gpbook3toc.html

43.11 W.B. Langdon: *Genetic Programming and Data Structures: Genetic Programming + Data Struc-*

tures = Automatic Programming! Genetic Programming, Vol. 1 (Kluwer, Boston 1998), available online at http://www.cs.ucl.ac.uk/staff/W.Langdon/gpdata

43.12 M. Suchorzewski, J. Clune: A novel generative encoding for evolving modular, regular and scalable networks, Proc. 13th Annu. Conf. Genet. Evol. Comput. (2011) pp. 1523–1530

43.13 J. Woodward: Evolving Turing complete representations, Proc. 2003 Congr. Evol. Comput. CEC2003, ed. by R. Sarker, R. Reynolds, H. Abbass, K.C. Tan, B. McKay, D. Essam, T. Gedeon (IEEE, Canberra 2003) pp. 830–837, available online at http://www.cs.bham.ac.uk/~jrw/publications/2003/EvolvingTuringCompleteRepresentations/cec032e.pdf

43.14 J. Tanomaru: Evolving Turing machines from examples, Lect. Notes Comput. Sci. **1363**, 167–180 (1993)

43.15 D. Andre, F.H. Bennett III, J.R. Koza: Discovery by genetic programming of a cellular automata rule that is better than any known rule for the majority classification problem, Proc. 1st Annu. Conf. Genet. Progr., ed. by J.R. Koza, D.E. Goldberg, D.B. Fogel, R.L. Riolo (MIT Press, Cambridge 1996) pp. 3–11, available online at http://www.genetic-programming.com/jkpdf/gp1996gkl.pdf

43.16 F. Gruau: Neural Network Synthesis Using Cellular Encoding and the Genetic Algorithm, Ph.D. Thesis (Laboratoire de l'Informatique du Parallilisme, Ecole Normale Supirieure de Lyon, France 1994), available online at ftp://ftp.ens-lyon.fr/pub/LIP/Rapports/PhD/PhD1994/PhD1994-01-E.ps.Z

43.17 A. Teller: Turing completeness in the language of genetic programming with indexed memory, Proc. 1994 IEEE World Congr. Comput. Intell., Orlando, Vol. 1 (1994) pp. 136–141, available online at http://www.cs.cmu.edu/afs/cs/usr/astro/public/papers/Turing.ps

43.18 S. Mabu, K. Hirasawa, J. Hu: A graph-based evolutionary algorithm: Genetic network programming (GNP) and its extension using reinforcement learning, Evol. Comput. **15**(3), 369–398 (2007)

43.19 J.R. Koza: *Genetic Programming: On the Programming of Computers by Means of Natural Selection* (MIT, Cambridge 1992)

43.20 P. Prusinkiewicz, A. Lindenmayer: *The Algorithmic Beauty of Plants (The Virtual Laboratory)* (Springer, Berlin, Heidelberg 1991)

43.21 J. Murphy, M. O'Neill, H. Carr: Exploring grammatical evolution for horse gait optimisation, Lect. Notes Comput. Sci. **5481**, 183–194 (2009)

43.22 T. Haynes, S. Sen: Evolving behavioral strategies in predators and prey, Lect. Notes Comput. Sci. **1042**, 113–126 (1995)

43.23 R. De Caux: Using Genetic Programming to Evolve Strategies for the Iterated Prisoner's Dilemma, Master's Thesis (University College, London 2001), available online at http://www.cs.ucl.ac.uk/staff/ W.Langdon/ftp/papers/decaux.masters.zip

43.24 A. Hauptman, M. Sipper: GP-endchess: Using genetic programming to evolve chess endgame players, Lect. Notes Comput. Sci. **3447**, 120–131 (2005), available online at http://www.cs.bgu.ac. il/~sipper/papabs/eurogpchess-final.pdf

43.25 E. Galván-López, J.M. Swafford, M. O'Neill, A. Brabazon: Evolving a Ms. PacMan controller using grammatical evolution, Lect. Notes Comput. Sci. **6024**, 161–170 (2010)

43.26 J. Togelius, S. Lucas, H.D. Thang, J.M. Garibaldi, T. Nakashima, C.H. Tan, I. Elhanany, S. Berant, P. Hingston, R.M. MacCallum, T. Haferlach, A. Gowrisankar, P. Burrow: The 2007 IEEE CEC simulated car racing competition, Genet. Program. Evol. Mach. **9**(4), 295–329 (2008)

43.27 G.S. Hornby, J.D. Lohn, D.S. Linden: Computer-automated evolution of an X-band antenna for NASA's space technology 5 mission, Evol. Comput. **19**(1), 1–23 (2011)

43.28 M. Furuholmen, K.H. Glette, M.E. Hovin, J. Torresen: Scalability, generalization and coevolution – experimental comparisons applied to automated facility layout planning, GECCO '09: Proc. 11th Annu. Conf. Genet. Evol. Comput., Montreal, ed. by F. Rothlauf, G. Raidl (2009) pp. 691–698, available online at http://doi.acm.org/10.1145/1569901. 1569997

43.29 C.C. Bojarczuk, H.S. Lopes, A.A. Freitas: Genetic programming for knowledge discovery in chest-pain diagnosis, IEEE Eng. Med. Biol. Mag. **19**(4), 38–44 (2000), available online at http:// ieeexplore.ieee.org/iel5/51/18543/00853480.pdf

43.30 T. Hildebrandt, J. Heger, B. Scholz-Reiter, M. Pelikan, J. Branke: Towards improved dispatching rules for complex shop floor scenarios: A genetic programming approach, GECCO '10: Proc. 12th Annu. Conf. Genet. Evol. Comput., Portland, ed. by J. Branke (2010) pp. 257–264

43.31 M.B. Bader-El-Den, R. Poli, S. Fatima: Evolving timetabling heuristics using a grammar-based genetic programming hyper-heuristic framework, Memet. Comput. **1**(3), 205–219 (2009), 10.1007/s12293-009-0022-y

43.32 M. Conrads, P. Nordin, W. Banzhaf: Speech sound discrimination with genetic programming, Lect. Notes Comput. Sci. **1391**, 113–129 (1998)

43.33 A. Esparcia-Alcazar, K. Sharman: Genetic programming for channel equalisation, Lect. Notes Comput. Sci. **1596**, 126–137 (1999), available online at http://www.iti.upv.es/~anna/papers/ evoiasp99.ps

43.34 R. Poli, M. Salvaris, C. Cinel: Evolution of a brain-computer interface mouse via genetic programming, Lect. Notes Comput. Sci. **6621**, 203–214 (2011)

43.35 K. Sims: Artificial evolution for computer graphics, ACM Comput. Gr. **25**(4), 319–328 (1991), available online at http://delivery.acm.org/10.1145/130000/ 122752/p319-sims.pdf SIGGRAPH '91 Proceedings

43.36 U.-M. O'Reilly, M. Hemberg: Integrating generative growth and evolutionary computation for form exploration, Genet. Program. Evol. Mach. **8**(2), 163–186 (2007), Special issue on developmental systems

43.37 P. Dahlstedt: Autonomous evolution of complete piano pieces and performances, Proc. Music AL Workshop (2007)

43.38 H. Iba: Multiple-agent learning for a robot navigation task by genetic programming, Genet. Program. Proc. 2nd Annu. Conf., Standord, ed. by J.R. Koza, K. Deb, M. Dorigo, D.B. Fogel, M. Garzon, H. Iba, R.L. Riolo (1997) pp. 195–200

43.39 T. Weise, K. Tang: Evolving distributed algorithms with genetic programming, IEEE Trans. Evol. Comput. **16**(2), 242–265 (2012)

43.40 L. Spector: Autoconstructive evolution: Push, pushGP, and pushpop, Proc. Genet. Evol. Comput. Conf. (GECCO-2001), ed. by L. Spector, E. Goodman (Morgan Kaufmann, San Francisco 2001) pp. 137–146, available online at http://hampshire.edu/ lspector/pubs/ace.pdf

43.41 J. Tavares, F. Pereira: Automatic design of ant algorithms with grammatical evolution. In: *Gnetic Programming. 15th European Conference, EuroGP*, ed. by A. Moraglio, S. Silva, K. Krawiec, P. Machado, C. Cotta (Springer, Berlin, Heidelberg 2012) pp. 206–217

43.42 M. Hutter: *A Gentle Introduction To The Universal Algorithmic Agent* AIXI. Technical Report IDSIA-01-03 (IDSIA, Manno-Lugano 2003)

43.43 J. Von Neumann, M.D. Godfrey: First draft of a report on the EDVAC, IEEE Ann. Hist. Comput. **15**(4), 27–75 (1993)

43.44 H.H. Goldstine, A. Goldstine: The electronic numerical integrator and computer (ENIAC), Math. Tables Other Aids Comput. **2**(15), 97–110 (1946)

43.45 A.M. Turing: Intelligent machinery. In: *Cybernetics: Key Papers*, ed. by C.R. Evans, A.D.J. Robertson (Univ. Park Press, Baltimore 1968), Written 1948

43.46   A.M. Turing: Computing machinery and intelligence, Mind **59**(236), 433–460 (1950)

43.47   A.L. Samuel: Some studies in machine learning using the game of checkers, IBM J. Res. Dev. **3**(3), 210 (1959)

43.48   R.M. Friedberg: A learning machine: Part I, IBM J. Res. Dev. **2**(1), 2–13 (1958)

43.49   M. O'Neill, L. Vanneschi, S. Gustafson, W. Banzhaf: Open issues in genetic programming, Genet. Program. Evol. Mach. **11**(3/4), 339–363 (2010), 10th Anniversary Issue: Progress in Genetic Programming and Evolvable Machines

43.50   L.J. Fogel, A.J. Owens, M.J. Walsh: *Artificial Intelligence Through Simulated Evolution* (Wiley, Hoboken 1966)

43.51   D.B. Fogel: *Evolutionary Computation: Toward a New Philosophy of Machine Intelligence*, Vol. 1 (Wiley, Hoboken 2006)

43.52   S.F. Smith: A Learning System Based on Genetic Adaptive Algorithms, Ph.D. Thesis (University of Pittsburgh, Pittsburgh 1980)

43.53   N.L. Cramer: A representation for the adaptive generation of simple sequential programs, Proc. Int. Conf. Genet. Algorithms Appl., Pittsburgh, ed. by J.J. Grefenstette (1985) pp. 183–187, available online at http://www.sover.net/~nichael/nlc-publications/icga85/index.html

43.54   J. Schmidhuber: Evolutionary Principles in Self-Referential Learning. On Learning Now to Learn: The Meta-Meta-Meta...-Hook, Diploma Thesis (Technische Universität, München 1987), available online at http://www.idsia.ch/~juergen/diploma.html

43.55   C. Fujiki, J. Dickinson: Using the genetic algorithm to generate lisp source code to solve the prisoner's dilemma, Proc. 2nd Int. Conf. Genet. Algorithms Appl., Cambridge, ed. by J.J. Grefenstette (1987) pp. 236–240

43.56   A.S. Bickel, R.W. Bickel: Tree structured rules in genetic algorithms, Proc. 2nd Int. Conf. Genet. Algorithms Appl., Cambridge, ed. by J.J. Grefenstette (1987) pp. 77–81

43.57   T.S. Ray: Evolution, Ecology and Optimization of Digital Organisms. Technical Report Working Paper 92-08-042 (Santa Fe Institute, Santa Fe 1992) available online at http://www.santafe.edu/media/workingpapers/92-08-042.pdf

43.58   J.R. Koza: *Genetic Programming II: Automatic Discovery of Reusable Programs* (MIT, Cambridge 1994)

43.59   J.R. Koza, M.A. Keane, M.J. Streeter, W. Mydlowec, J. Yu, G. Lanza: *Genetic Programming IV: Routine Human-Competitive Machine Intelligence* (Springer, Berlin, Heidelberg 2003), available online at http://www.genetic-programming.org/gpbook4toc.html

43.60   J. Koza: http://www.genetic-programming.org/hc2011/combined.html

43.61   J.G. Carbonell, R.S. Michalski, T.M. Mitchell: An overview of machine learning. In: *Machine Learning: An Artificial Intelligence Approach*, ed. by R.S. Michalski, J.G. Carbonell, T.M. Mitchell (Tioga, Palo Alto 1983)

43.62   C. Rich, R.C. Waters: Automatic programming: Myths and prospects, Computer **21**(8), 40–51 (1988)

43.63   S. Gulwani: Dimensions in program synthesis, Proc. 12th Int. SIGPLAN Symp. Princ. Pract. Declar. Program. (2010) pp. 13–24

43.64   I. Rechenberg: *Evolutionsstrategie – Optimierung Technischer Systeme nach Prinzipien der Biologischen Evolution* (Frommann-Holzboog, Stuttgart 1973)

43.65   H.-P. Schwefel: *Numerische Optimierung von Computer-Modellen* (Birkhäuser, Basel 1977)

43.66   J.H. Holland: *Adaptation in Natural and Artificial Systems* (University of Michigan, Ann Arbor 1975)

43.67   D.J. Montana: Strongly typed genetic programming, Evol. Comput. **3**(2), 199–230 (1995), available online at http://vishnu.bbn.com/papers/stgp.pdf

43.68   T. Yu: Hierachical processing for evolving recursive and modular programs using higher order functions and lambda abstractions, Genet. Program. Evol. Mach. **2**(4), 345–380 (2001)

43.69   R. Poli: Parallel distributed genetic programming. In: *New Ideas in Optimization*, Advanced Topics in Computer Science, ed. by D. Corne, M. Dorigo, F. Glover (McGraw-Hill, London 1999) pp. 403–431, Chapter 27, available online at http://citeseer.ist.psu.edu/328504.html

43.70   J.F. Miller, P. Thomson: Cartesian genetic programming, Lect. Notes Comput. Sci. **1802**, 121–132 (2000), available online at http://www.elec.york.ac.uk/intsys/users/jfm7/cgp-eurogp2000.pdf

43.71   K.O. Stanley: Compositional pattern producing networks: A novel abstraction of development, Genet. Program. Evol. Mach. **8**(2), 131–162 (2007)

43.72   L.J. Fogel, P.J. Angeline, D.B. Fogel: An evolutionary programming approach to self-adaptation on finite state machines, Proc. 4th Int. Conf. Evol. Program. (1995) pp. 355–365

43.73   R.I. McKay, N.X. Hoai, P.A. Whigham, Y. Shan, M. O'Neill: Grammar-based genetic programming: A survey, Genet. Program. Evol. Mach. **11**(3/4), 365–396 (2010), September Tenth Anniversary Issue: Progress in Genetic Programming and Evolvable Machines

43.74   M. Keijzer, V. Babovic: Dimensionally aware genetic programming, Proc. Genet. Evol. Comput. Conf., Orlando, Vol. 2, ed. by W. Banzhaf, J. Daida, A.E. Eiben, M.H. Garzon, V. Honavar, M. Jakiela, R.E. Smith (1999) pp. 1069–1076, available online at http://www.cs.bham.ac.uk/~wbl/biblio/gecco1999/GP-420.ps

43.75   A. Ratle, M. Sebag: Grammar-guided genetic programming and dimensional consistency: Application to non-parametric identification in mechanics, Appl. Soft Comput. **1**(1), 105–118 (2001), available online at http://www.sciencedirect.com/science/article/B6W86-43S6W98-B/1/38e0fa6ac503a5ef310e2287be01eff8

43.76   P.A. Whigham: Grammatically-based genetic programming, Proc. Workshop Genet. Program.: From Theory Real-World Appl., Tahoe City, ed. by J.P. Rosca (1995) pp. 33–41, available online at http://divcom.otago.ac.nz/sirc/Peterw/Publications/ml95.zip

43.77   M. O'Neill, C. Ryan: *Grammatical Evolution: Evolutionary Automatic Programming in a Arbitrary Language*, Genetic Programming, Vol. 4 (Kluwer, Boston 2003), available online at http://www.wkap.nl/prod/b/1-4020-7444-1

43.78   N. Xuan Hoai, R.I. McKay, D. Essam: Representation and structural difficulty in genetic programming, IEEE Trans. Evol. Comput. **10**(2), 157–166 (2006), available online at http://sc.snu.ac.kr/courses/2006/fall/pg/aai/GP/nguyen/Structdiff.pdf

43.79   Y. Shan, R.I. McKay, D. Essam, H.A. Abbass: A survey of probabilistic model building genetic programming. In: *Scalable Optimization via Probabilistic Modeling: From Algorithms to Applications*, Studies in Computational Intelligence, Vol. 33, ed. by M. Pelikan, K. Sastry, E. Cantu-Paz (Springer, Berlin, Heidelberg 2006) pp. 121–160, Chapter 6

43.80   M. Brameier, W. Banzhaf: *Linear Genetic Programming*, Genetic and Evolutionary Computation, Vol. 16 (Springer, Berlin, Heidelberg 2007), available online at http://www.springer.com/west/home/default?SGWID=4-40356-22-173660820-0

43.81   T. Perkis: Stack-based genetic programming, Proc. 1994 IEEE World Congr. Comput. Intell., Orlando, Vol. 1 (1994) pp. 148–153, available online at http://citeseer.ist.psu.edu/432690.html

43.82   W.B. Langdon: Large scale bioinformatics data mining with parallel genetic programming on graphics processing units. In: *Parallel and Distributed Computational Intelligence*, Studies in Computational Intelligence, Vol. 269, ed. by F. de Fernandez Vega, E. Cantu-Paz (Springer, Berlin, Heidelberg 2010) pp. 113–141, Chapter 5, available online at http://www.springer.com/engineering/book/978-3-642-10674-3

43.83   L. Spector, A. Robinson: Genetic programming and autoconstructive evolution with the push programming language, Genet. Program. Evol. Mach. **3**(1), 7–40 (2002), available online at http://hampshire.edu/lspector/pubs/push-gpem-final.pdf

43.84   E. Schulte, S. Forrest, W. Weimer: Automated program repair through the evolution of assembly code, Proc. IEEE/ACM Int. Conf. Autom. Softw. Eng. (2010) pp. 313–316

43.85   M. Orlov, M. Sipper: Flight of the FINCH through the Java wilderness, IEEE Trans. Evol. Comput. **15**(2), 166–182 (2011)

43.86   P. Nordin: A compiling genetic programming system that directly manipulates the machine code. In: *Advances in Genetic Programming*, ed. by K.E. Kinnear Jr. (MIT Press, Cambridge 1994) pp. 311–331, Chapter 14, available online at http://cognet.mit.edu/library/books/view?isbn=0262111888

43.87   U.-M. O'Reilly, F. Oppacher: Program search with a hierarchical variable length representation: Genetic programming, simulated annealing and hill climbing, Lect. Notes Comput. Sci. **866**, 397–406 (1994), available online at http://www.cs.ucl.ac.uk/staff/W.Langdon/ftp/papers/ppsn-94.ps.gz

43.88   M. Tomassini: *Spatially Structured Evolutionary Algorithms* (Springer, Berlin, Heidelberg 2005)

43.89   A. Arcuri, X. Yao: A novel co-evolutionary approach to automatic software bug fixing, IEEE World Congr. Comput. Intell., Hong Kong, ed. by J. Wang (2008)

43.90   A. Moraglio, C. Di Chio, R. Poli: Geometric particle swarm optimization, Lect. Notes Comput. Sci. **4445**, 125–136 (2007)

43.91   M. O'Neill, A. Brabazon: Grammatical differential evolution, Proc. Int. Conf. Artif. Intell. ICAI 2006, Las Vegas, Vol. 1, ed. by H.R. Arabnia (2006) pp. 231–236, available online at http://citeseerx.ist.psu.edu/viewdoc/summary?doi=10.1.1.91.3012

43.92   R. Poli, N.F. McPhee: A linear estimation-of-distribution GP system, Lect. Notes Comput. Sci. **4971**, 206–217 (2008)

43.93   M. Looks, B. Goertzel, C. Pennachin: Learning computer programs with the Bayesian optimization algorithm, GECCO 2005: Proc. Conf. Genet. Evol. Comput., Washington, Vol. 1, ed. by U.-M. O'Reilly, H.-G. Beyer (2005) pp. 747–748, available online at http://www.cs.bham.ac.uk/~wbl/biblio/gecco2005/docs/p747.pdf

43.94   E. Hemberg, K. Veeramachaneni, J. McDermott, C. Berzan, U.-M. O'Reilly: An investigation of local patterns for estimation of distribution genetic programming, Philadelphia, Proc. GECCO 2012 (2012)

43.95   M. Schmidt, H. Lipson: Distilling free-form natural laws from experimental data, Science **324**(5923), 81–85 (2009), available online at http://ccsl.mae.cornell.edu/sites/default/files/Science09_Schmidt.pdf

43.96   E.J. Vladislavleva, G.F. Smits, D. den Hertog: Order of nonlinearity as a complexity measure for models generated by symbolic regression via Pareto genetic programming, IEEE Trans. Evol. Comput. **13**(2), 333–349 (2009)

43.97   K. Veeramachaneni, E. Vladislavleva, U.-M. O'Reilly: Knowledge mining sensory evaluation data: Genetic programming, statistical techniques, and swarm optimization, Genet. Program. Evolvable Mach. **13**(1), 103–133 (2012)

43.98   M. Kotanchek, G. Smits, E. Vladislavleva: Pursuing the Pareto paradigm tournaments, algorithm variations & ordinal optimization. In: *Genetic Programming Theory and Practice IV*, Genetic and Evolutionary Computation, Vol. 5, ed. by R.L. Riolo, T. Soule, B. Worzel (Springer, Berlin, Heidelberg 2006) pp. 167–186, Chapter 12

43.99   C. Le Goues, T. Nguyen, S. Forrest, W. Weimer: GenProg: A generic method for automated software repair, IEEE Trans. Softw. Eng. **38**(1), 54–72 (2011)

43.100   C. Ryan: *Automatic Re-Engineering of Software Using Genetic Programming*, Genetic Programming, Vol. 2 (Kluwer, Boston 2000), available online at http://www.wkap.nl/book.htm/0-7923-8653-1

43.101   K.P. Williams: Evolutionary Algorithms for Automatic Parallelization, Ph.D. Thesis (University of Reading, Reading 1998)

43.102   M. Stephenson, S. Amarasinghe, M. Martin, U.-M. O'Reilly: Meta optimization: Improving compiler heuristics with machine learning, Proc. ACM SIGPLAN Conf. Program. Lang. Des. Implement. (PLDI '03), San Diego (2003) pp. 77–90

43.103   M. Stephenson, U.-M. O'Reilly, M.C. Martin, S. Amarasinghe: Genetic programming applied to compiler heuristic optimization, Lect. Notes Comput. Sci. **2610**, 238–253 (2003)

43.104   D.R. White, A. Arcuri, J.A. Clark: Evolutionary improvement of programs, IEEE Trans. Evol. Comput. **15**(4), 515–538 (2011)

43.105   M. Harman: The current state and future of search based software engineering, Proc. Future of Software Engineering FOSE '07, Washington, ed. by L. Briand, A. Wolf (2007) pp. 342–357

43.106   J.R. Koza, F.H. Bennett III, D. Andre, M.A. Keane, F. Dunlap: Automated synthesis of analog electrical circuits by means of genetic programming, IEEE Trans. Evol. Comput. **1**(2), 109–128 (1997), available online at http://www.genetic-programming.com/jkpdf/ieeetecjournal1997.pdf

43.107   R. Dawkins: *The Blind Watchmaker* (Norton, New York 1986)

43.108   H. Takagi: Interactive evolutionary computation: Fusion of the capabilities of EC optimization and human evaluation, Proc. IEEE **89**(9), 1275–1296 (2001)

43.109   S. Todd, W. Latham: *Evolutionary Art and Computers* (Academic, Waltham 1994)

43.110   M. Lewis: Evolutionary visual art and design. In: *The Art of Artificial Evolution: A Handbook on Evolutionary Art and Music*, ed. by J. Romero, P. Machado (Springer, Berlin, Heidelberg 2008) pp. 3–37

43.111   J. McDermott, J. Byrne, J.M. Swafford, M. O'Neill, A. Brabazon: Higher-order functions in aesthetic EC encodings, 2010 IEEE World Congr. Comput. Intell., Barcelona (2010), pp. 2816–2823, 18-23 July

43.112   D.A. Hart: Toward greater artistic control for interactive evolution of images and animation, Lect. Notes Comput. Sci. **4448**, 527–536 (2007)

43.113   A.K. Hoover, M.P. Rosario, K.O. Stanley: Scaffolding for interactively evolving novel drum tracks for existing songs, Lect. Notes Comput. Sci. **4974**, 412 (2008)

43.114   J. Shao, J. McDermott, M. O'Neill, A. Brabazon: Jive: A generative, interactive, virtual, evolutionary music system, Lect. Notes Comput. Sci. **6025**, 341–350 (2010)

43.115   J. McDermott, U.-M. O'Reilly: An executable graph representation for evolutionary generative music, Proc. GECCO 2011 (2011) pp. 403–410

43.116   J. Clune, H. Lipson: Evolving three-dimensional objects with a generative encoding inspired by developmental biology, Proc. Eur. Conf. Artif. Life (2011), available online at http://endlessforms.com

43.117   J. McCormack: Evolutionary L-systems. In: *Design by Evolution: Advances in Evolutionary Design*, ed. by P.F. Hingston, L.C. Barone, Z. Michalewicz, D.B. Fogel (Springer, Berlin, Heidelberg 2008) pp. 169–196

43.118   G.S. Hornby, J.B. Pollack: Evolving L-systems to generate virtual creatures, Comput. Graph. **25**(6), 1041–1048 (2001)

43.119   P. Worth, S. Stepney: Growing music: Musical interpretations of L-systems, Lect. Notes Comput. Sci. **3449**, 545–550 (2005)

43.120   J. McDermott, J. Byrne, J.M. Swafford, M. Hemberg, C. McNally, E. Shotton, E. Hemberg, M. Fenton, M. O'Neill: String-rewriting grammars for evolutionary architectural design, Environ. Plan. B **39**(4), 713–731 (2012), available online at http://www.envplan.com/abstract.cgi?id=b38037

43.121   W. Banzhaf, W.B. Langdon: Some considerations on the reason for bloat, Genet. Program. Evol. Mach. **3**(1), 81–91 (2002), available online at http://web.cs.mun.ca/~banzhaf/papers/genp_bloat.pdf

43.122   S. Luke, L. Panait: A comparison of bloat control methods for genetic programming, Evol. Comput. **14**(3), 309–344 (2006)

43.123   S. Silva, S. Dignum, L. Vanneschi: Operator equalisation for bloat free genetic programming and a survey of bloat control methods, Genet. Program. Evol. Mach. **3**(2), 197–238 (2011)

43.124   W.B. Langdon, R. Poli: Fitness causes bloat. In: *Soft Computing in Engineering Design and Manufacturing*, ed. by P.K. Chawdhry, R. Roy, R.K. Pant (Springer, London 1997) pp. 13–22, available online at http://www.cs.bham.ac.uk/~wbl/ftp/papers/WBL.bloat_wsc2.ps.gz

43.125 S. Dignum, R. Poli: Generalisation of the limiting distribution of program sizes in tree-based genetic programming and analysis of its effects on bloat, GECCO '07 Proc. 9th Annu. Conf. Genet. Evol. Comput., London, Vol. 2, ed. by H. Lipson, D. Thierens (2007) pp. 1588–1595, available online at http://www.cs.bham.ac.uk/~wbl/biblio/gecco2007/docs/p1588.pdf

43.126 W.B. Langdon: How many good programs are there? How long are they?, Found. Genet. Algorithms VII, San Francisco, ed. by K.A. De Jong, R. Poli, J.E. Rowe (2002), pp. 183–202, available online at http://www.cs.ucl.ac.uk/staff/W. Langdon/ftp/papers/wbl_foga2002.pdf

43.127 R. Poli, M. Salvaris, C. Cinel: Evolution of an effective brain-computer interface mouse via genetic programming with adaptive Tarpeian bloat control. In: Genetic Programming Theory and Practice IX, ed. by R. Riolo, K. Vladislavleva, J. Moore (Springer, Berlin, Heidelberg 2011) pp. 77–95

43.128 G. Smits, E. Vladislavleva: Ordinal pareto genetic programming, Proc. 2006 IEEE Congr. Evol. Comput., Vancouver, ed. by G.G. Yen, S.M. Lucas, G. Fogel, G. Kendall, R. Salomon, B.-T. Zhang, C.A. Coello Coello, T.P. Runarsson (2006) pp. 3114–3120, available online at http://ieeexplore.ieee.org/servlet/opac?punumber=11108

43.129 L. Vanneschi, M. Castelli, S. Silva: Measuring bloat, overfitting and functional complexity in genetic programming, GECCO '10: Proc. 12th Annu. Conf. Genet. Evol. Comput., Portland (2010) pp. 877–884

43.130 H. Iba, H. de Garis, T. Sato: Genetic programming using a minimum description length principle. In: Advances in Genetic Programming, ed. by K.E. Kinnear Jr. (MIT Press, Cambridge 1994) pp. 265–284, available online at http://citeseer.ist.psu.edu/327857.html, Chapter 12

43.131 J. Miller: What bloat? Cartesian genetic programming on boolean problems, 2001 Genet. Evol. Comput. Conf. Late Break. Pap., ed. by E.D. Goodman (2001) pp. 295–302, available online at http://www.elec.york.ac.uk/intsys/users/jfm7/gecco2001Late.pdf

43.132 R. Poli, J. Page, W.B. Langdon: Smooth uniform crossover, sub-machine code GP and demes: A recipe for solving high-order Boolean parity problems, Proc. Genet. Evol. Comput. Conf., Orlando, Vol. 2, ed. by W. Banzhaf, J. Daida, A.E. Eiben, M.H. Garzon, V. Honavar, M. Jakiela, R.E. Smith (1999) pp. 1162–1169, available online at http://www.cs.bham.ac.uk/~wbl/biblio/gecco1999/GP-466.pdf

43.133 M. Keijzer: Alternatives in subtree caching for genetic programming, Lect. Notes Comput. Sci. 3003, 328–337 (2004), available online at http://www.springerlink.com/openurl.asp?genre=article&issn=0302-9743&volume=3003&spage=328

43.134 R. Poli, W.B. Langdon: Running genetic programming backward. In: Genetic Programming Theory and Practice III, Genetic Programming, Vol. 9, ed. by T. Yu, R.L. Riolo, B. Worzel (Springer, Berlin, Heidelberg 2005) pp. 125–140, Chapter 9, available online at http://www.cs.essex.ac.uk/staff/poli/papers/GPTP2005.pdf

43.135 Q.U. Nguyen, X.H. Nguyen, M. O'Neill, R.I. McKay, E. Galván-López: Semantically-based crossover in genetic programming: Application to real-valued symbolic regression, Genet. Program. Evol. Mach. 12, 91–119 (2011)

43.136 L. Altenberg: Emergent phenomena in genetic programming, Evol. Progr. – Proc. 3rd Annu. Conf., San Diego, ed. by A.V. Sebald, L.J. Fogel (1994) pp. 233–241, available online at http://dynamics.org/~altenber/PAPERS/EPIGP/

43.137 U.-M. O'Reilly, F. Oppacher: The troubling aspects of a building block hypothesis for genetic programming, Working Paper 94-02-001 (Santa Fe Institute, Santa Fe 1992)

43.138 R. Poli, W.B. Langdon: A new schema theory for genetic programming with one-point crossover and point mutation, Proc. Second Annu. Conf. Genet. Progr. 1997, Stanford, ed. by J.R. Koza, K. Deb, M. Dorigo, D.B. Fogel, M. Garzon, H. Iba, R.L. Riolo (1997) pp. 278–285, available online at http://citeseer.ist.psu.edu/327495.html

43.139 J.P. Rosca: Analysis of complexity drift in genetic programming, Proc. 2nd Annu. Conf. Genet. Program. 1997, Stanford, ed. by J.R. Koza, K. Deb, M. Dorigo, D.B. Fogel, M. Garzon, H. Iba, R.L. Riolo (1997), pp. 286–294, available online at ftp://ftp.cs.rochester.edu/pub/u/rosca/gp/97.gp.ps.gz

43.140 R. Poli, N.F. McPhee: General schema theory for genetic programming with subtree-swapping crossover: Part I, Evol. Comput. 11(1), 53–66 (2003), available online at http://cswww.essex.ac.uk/staff/rpoli/papers/ecj2003partI.pdf

43.141 R. Poli, N.F. McPhee: General schema theory for genetic programming with subtree-swapping crossover: Part II, Evol. Comput. 11(2), 169–206 (2003), available online at http://cswww.essex.ac.uk/staff/rpoli/papers/ecj2003partII.pdf

43.142 R. Poli, N.F. McPhee, J.E. Rowe: Exact schema theory and Markov chain models for genetic programming and variable-length genetic algorithms with homologous crossover, Genet. Program. Evol. Mach. 5(1), 31–70 (2004), available online at http://cswww.essex.ac.uk/staff/rpoli/papers/GPEM2004.pdf

43.143 N.F. McPhee, B. Ohs, T. Hutchison: Semantic building blocks in genetic programming, Lect. Notes Comput. Sci. 4971, 134–145 (2008)

43.144 G. Durrett, F. Neumann, U.-M. O'Reilly: Computational complexity analysis of simple genetic programming on two problems modeling isolated program semantics, Proc. 11th Workshop Found. Genet. Algorithm. (ACM, New York 2011) pp. 69–

80, available online at http://arxiv.org/pdf/1007.4636v1 arXiv:1007.4636v1

43.145   D.E. Goldberg, U.-M. O'Reilly: Where does the good stuff go, and why? How contextual semantics influence program structure in simple genetic programming, Lect. Notes Comput. Sci. **1391**, 16–36 (1998), available online at http://citeseer.ist.psu.edu/96596.html

43.146   D.E. Goldberg: *Genetic Algorithms in Search, Optimization, and Machine Learning* (Addison-Wesley, Reading 1989)

43.147   T. Kötzing, F. Neumann, A. Sutton, U.-M. O'Reilly: The max problem revisited: The importance of mutation in genetic programming, GECCO '12 Proc. 14th Annu. Conf. Genet. Evolut. Comput. (ACM, New York 2012) pp. 1333–1340

43.148   C. Gathercole, P. Ross: *The Max Problem for Genetic Programming – Highlighting an Adverse Interaction Between the Crossover Operator and a Restriction on Tree Depth*, Technical Report (Department of Artificial Intelligence, University of Edinburgh, Edinburgh 1995) available online at http://citeseer.ist.psu.edu/gathercole95max.html

43.149   F. Neumann: Computational complexity analysis of multi-objective genetic programming, GECCO '12 Proc. 14th Annu. Conf. Genet. Evolut. Comput. (ACM, New York 2012) pp. 799–806

43.150   T. Kötzing, F. Neumann, R. Spöhel: PAC learning and genetic programming, Proc. 13th Annu. Conf. Genet. Evol. Comput. (ACM, New York 2011) pp. 2091–2096

43.151   F. Rothlauf: *Representations for Genetic and Evolutionary Algorithms*, 2nd edn. (Physica, Heidelberg 2006)

43.152   T. Jones: Evolutionary Algorithms, Fitness Landscapes and Search, Ph.D. Thesis (University of New Mexico, Albuquerque 1995)

43.153   J. McDermott, E. Galván-Lopéz, M. O'Neill: A fine-grained view of phenotypes and locality in genetic programming. In: *Genetic Programming Theory and Practice*, Vol. 9, ed. by R. Riolo, K. Vladislavleva, J. Moore (Springer, Berlin, Heidelberg 2011)

43.154   M. Tomassini, L. Vanneschi, P. Collard, M. Clergue: A study of fitness distance correlation as a difficulty measure in genetic programming, Evol. Comput. **13**(2), 213–239 (2005)

43.155   L. Vanneschi: Theory and Practice for Efficient Genetic Programming, Ph.D. Thesis (Université de Lausanne, Lausanne 2004)

43.156   L. Vanneschi, M. Tomassini, P. Collard, S. Verel, Y. Pirola, G. Mauri: A comprehensive view of fitness landscapes with neutrality and fitness clouds, Lect. Notes Comput. Sci. **4445**, 241–250 (2007)

43.157   A. Ekárt, S.Z. Németh: A metric for genetic programs and fitness sharing, Lect. Notes Comput. Sci. **1802**, 259–270 (2000)

43.158   S. Gustafson, L. Vanneschi: Crossover-based tree distance in genetic programming, IEEE Trans. Evol. Comput. **12**(4), 506–524 (2008)

43.159   J. McDermott, U.-M. O'Reilly, L. Vanneschi, K. Veeramachaneni: How far is it from here to there? A distance that is coherent with GP operators, Lect. Notes Comput. Sci. **6621**, 190–202 (2011)

43.160   U.-M. O'Reilly: Using a distance metric on genetic programs to understand genetic operators, Int. Conf. Syst. Man Cybern. Comput. Cybern. Simul. (1997) pp. 233–241

43.161   D.H. Wolpert, W.G. Macready: No free lunch theorems for optimization, Evol. Comput. IEEE Trans. **1**(1), 67–82 (1997)

43.162   C. Schumacher, M.D. Vose, L.D. Whitley: The no free lunch and problem description length, Proc. Genet. Evol. Comput. Conf. GECCO-2001 (2001) pp. 565–570

43.163   J.R. Woodward, J.R. Neil: No free lunch, program induction and combinatorial problems, Lect. Notes Comput. Sci. **2610**, 475–484 (2003)

43.164   R. Poli, M. Graff, N.F. McPhee: Free lunches for function and program induction, FOGA '09: Proc. 10th ACM SIGEVO Workshop Found. Genet. Algorithms, Orlando (2009) pp. 183–194

43.165   L. Spector: Simultaneous evolution of programs and their control structures. In: *Advances in Genetic Programming*, Vol. 2, ed. by P.J. Angeline, K.E. Kinnear Jr. (MIT, Cambridge 1996) pp. 137–154, Chapter 7, available online at http://helios.hampshire.edu/lspector/pubs/AiGP2-post-final-e.pdf

43.166   L. Spector, B. Martin, K. Harrington, T. Helmuth: Tag-based modules in genetic programming, Proc. Genet. Evol. Comput. Conf. GECCO-2011 (2011)

43.167   G.S. Hornby: Measuring, nabling and comparing modularity, regularity and hierarchy in evolutionary design, GECCO 2005: Proc. 2005 Conf. Genet. Evol. Comput., Washington, Vol. 2, ed. by H.-G. Beyer, U.-M. O'Reilly, D.V. Arnold, W. Banzhaf, C. Blum, E.W. Bonabeau, E. Cantu-Paz, D. Dasgupta, K. Deb, J.A. Foster, E.D. de Jong, H. Lipson, X. Llora, S. Mancoridis, M. Pelikan, G.R. Raidl, T. Soule, A.M. Tyrrell, J.-P. Watson, E. Zitzler (2005) pp. 1729–1736, available online at http://www.cs.bham.ac.uk/~wbl/biblio/gecco2005/docs/p1729.pdf

43.168   J.H. Moore, C.S. Greene, P.C. Andrews, B.C. White: Does complexity matter? Artificial evolution, computational evolution and the genetic analysis of epistasis in common human diseases. In: *Genetic Programming Theory and Practice Vol. VI*, ed. by R.L. Riolo, T. Soule, B. Worzel (Springer, Berlin, Heidelberg 2008) pp. 125–145, Chap. 9

43.169   S. Gustafson: An Analysis of Diversity in Genetic Programming, Ph.D. Thesis (School of Computer

Science and Information Technology, University of Nottingham, Nottingham 2004), available online at http://www.cs.nott.ac.uk/~smg/research/publications/phdthesis-gustafson.pdf

43.170 G.S. Hornby: A steady-state version of the age-layered population structure EA. In: *Genetic Programming Theory and Practice*, Vol. VII, Genetic and Evolutionary Computation, ed. by R.L. Riolo, U.-M. O'Reilly, T. McConaghy (Springer, Ann Arbor 2009) pp. 87–102, Chap. 6

43.171 J.C. Bongard: Coevolutionary dynamics of a multi-population genetic programming system, Lect. Notes Comput. Sci. **1674**, 154 (1999), available online at http://www.cs.uvm.edu/~jbongard/papers/s067.ps.gz

43.172 I. Dempsey, M. O'Neill, A. Brabazon: *Foundations in Grammatical Evolution for Dynamic Environments*, Studies in Computational Intelligence, Vol. 194 (Springer, Berlin, Heidelberg 2009), available online at http://www.springer.com/engineering/book/978-3-642-00313-4

43.173 J. Doucette, P. Lichodzijewski, M. Heywood: Evolving coevolutionary classifiers under large attribute spaces. In: *Genetic Programming Theory and Practice Vol. VII*, ed. by R.L. Riolo, U.-M. O'Reilly, T. McConaghy (Springer, Berlin, Heidelberg 2009) pp. 37–54, Chap. 3

43.174 S. Luke: http://cs.gmu.edu/~eclab/projects/ecj/

43.175 S. Luke: *The ECJ Owner's Manual – A User Manual for the ECJ Evolutionary Computation Library*, 0th edn. online version 0.2 edition, available online at http://www.cs.gmu.edu/~eclab/projects/ecj/docs/manual/manual.pdf

43.176 D.W. Dyer: https://github.com/dwdyer/watchmaker

43.177 E. Hemberg, M. O'Neill: http://ncra.ucd.ie/Site/GEVA.html

43.178 M. O'Neill, E. Hemberg, C. Gilligan, E. Bartley, J. McDermott, A. Brabazon: GEVA: Grammatical evolution in Java, SIGEVOlution **3**(2), 17–22 (2008), available online at http://www.sigevolution.org/issues/pdf/SIGEVOlution200802.pdf

43.179 J. Dréo: http://eodev.sourceforge.net/

43.180 G. Squillero: http://www.cad.polito.it/research/Evolutionary_Computation/MicroGP/index.html

43.181 M. Schillaci, E.E. Sanchez Sanchez: A brief survey of μGP, SIGEvolution **1**(2), 17–21 (2006)

43.182 G. Squillero: MicroGP - an evolutionary assembly program generator, Genet. Program. Evol. Mach. **6**(3), 247–263 (2005), Published online: 17 August 2005.

43.183 C. Gagné, M. Parizeau: http://beagle.sourceforge.net/

43.184 C. Gagné, M. Parizeau: Open BEAGLE A C++ framework for your favorite evolutionary algorithm, SIGEvolution **1**(1), 12–15 (2006), available online at http://www.sigevolution.org/2006/01/issue.pdf

43.185 S. Silva: http://gplab.sourceforge.net/

43.186 F.M. De Rainville, F.-A. Fortin: http://code.google.com/p/deap/

43.187 L. Spector: http://hampshire.edu/lspector/push.html

43.188 R. Poli: http://cswww.essex.ac.uk/staff/rpoli/TinyGP/

43.189 E. Hemberg, J. McDermott: http://code.google.com/p/ponyge/

43.190 H. Lipson: http://creativemachines.cornell.edu/eureqa

43.191 M.E. Kotanchek, E. Vladislavleva, G.F. Smits: http://www.evolved-analytics.com/

43.192 P. Nordin: http://www.rmltech.com/

43.193 P. Nordin, W. Banzhaf, F.D. Francone: Efficient evolution of machine code for CISC architectures using instruction blocks and homologous crossover. In: *Advances in Genetic Programming*, Vol. 3, ed. by L. Spector, W.B. Langdon, U.-M. O'Reilly, P.J. Angeline (MIT, Cambridge 1999) pp. 275–299, Chap. 12, available online at http://www.aimlearning.com/aigp31.pdf

43.194 W.B. Langdon: http://www.cs.bham.ac.uk/~wbl/biblio/

43.195 W.B. Langdon: http://www.cs.ucl.ac.uk/staff/W.Langdon/homepages.html

43.196 Genetic Programming Yahoo Group: http://groups.yahoo.com/group/genetic_programming/

43.197 J. McDermott, D. White: http://gpbenchmarks.org

43.198 J. McDermott, D.R. White, S. Luke, L. Manzoni, M. Castelli, L. Vanneschi, W. Jaśkowski, K. Krawiec, R. Harper, K. De Jong, U.-M. O'Reilly: Genetic programming needs better benchmarks, Proc. GECCO 2012, Philadelphia (2012)

43.199 R. Poli, W.B. Langdon, N.F. McPhee: *A Field Guide to Genetic Programming* (Lulu, Raleigh 2008), Published via http://lulu.com and available at http://www.gp-field-guide.org.uk (With contributions by J. R. Koza)

43.200 S. Luke: *Essentials of Metaheuristics*, 1st edn. (Lulu, Raleigh 2009), available online at http://cs.gmu.edu/~sean/books/metaheuristics/

43.201 W. Banzhaf, P. Nordin, R.E. Keller, F.D. Francone: *Genetic Programming – An Introduction; On the Automatic Evolution of Computer Programs and Its Applications* (Morgan Kaufmann, San Francisco 1998), available online at http://www.elsevier.com/wps/find/bookdescription.cws_home/677869/description#description

43.202 W.B. Langdon, R. Poli: *Foundations of Genetic Programming* (Springer, Berlin, Heidelberg 2002), available online at http://www.cs.ucl.ac.uk/staff/W.Langdon/FOGP/

43.203 H. Iba, Y. Hasegawa, T. Kumar Paul: *Applied Genetic Programming and Machine Learning*, CRC Complex and Enterprise Systems Engineering (CRC, Boca Raton 2009)

43.204 M. Sipper: *Evolved to Win* (Lulu, Raleigh 2011), available at http://www.lulu.com/

Part E | 43

44. Evolution Strategies

Nikolaus Hansen, Dirk V. Arnold, Anne Auger

Evolution strategies (ES) are evolutionary algorithms that date back to the 1960s and that are most commonly applied to black–box optimization problems in continuous search spaces. Inspired by biological evolution, their original formulation is based on the application of mutation, recombination and selection in populations of candidate solutions. From the algorithmic viewpoint, ES are optimization methods that sample new candidate solutions stochastically, most commonly from a multivariate normal probability distribution. Their two most prominent design principles are unbiasedness and adaptive control of parameters of the sample distribution. In this overview, the important concepts of success based step–size control, self–adaptation, and derandomization are covered, as well as more recent developments such as covariance matrix adaptation and natural ES. The latter give new insights into the fundamental mathematical rationale behind ES. A broad discussion of theoretical results includes progress rate results on various function classes and convergence proofs for evolution strategies.

Part E | 44.1

44.1 Overview

Evolution strategies [44.1–4], sometimes also referred to as evolutionary strategies, and *evolutionary programming* [44.5] are search paradigms inspired by the principles of biological evolution. They belong to the family of evolutionary algorithms that address optimization problems by implementing a repeated process of (small) stochastic variations followed by selection. In each generation (or iteration), new offspring (or candidate solutions) are generated from their parents (candidate solutions already visited), their fitness is

evaluated, and the better offspring are selected to become the parents for the next generation.

ES most commonly address the problem of *continuous black-box optimization*. The search space is the continuous domain, $\mathbb{R}^n$, and solutions in search space are n-dimensional vectors, denoted as $\boldsymbol{x}$. We consider an objective or fitness function $f : \mathbb{R}^n \to \mathbb{R}, \boldsymbol{x} \mapsto f(\boldsymbol{x})$ to be minimized. We make no specific assumptions on f, other than that f can be evaluated for each $\boldsymbol{x}$, and refer to this search problem as *black-box* optimization.

The objective is, loosely speaking, to generate solutions (x-vectors) with small f-values while using a small number of f-evaluations. Formally, we like to *converge* to an essential global optimum of f, in the sense that the best $f(x)$ value gets arbitrarily close to the *essential infimum* of f (i. e., the smallest f-value for which all larger, i. e., worse f-values have sublevel sets with positive volume).

In this context, we present an overview of methods that sample new offspring, or candidate solutions, from normal distributions. Naturally, such an overview is biased by the authors' viewpoints, and our emphasis will be on important design principles and on contemporary ES that we consider as most relevant in practice or future research. More comprehensive historical overviews can be found elsewhere [44.6, 7].

In the next section, the main principles are introduced and two *algorithm templates* for an evolution strategy are presented. Section 44.3 presents six ES that mark important conceptual and algorithmic developments. Section 44.4 summarizes important theoretical results.

44.1.1 Symbols and Abbreviations

Throughout this chapter, vectors like $z \in \mathbb{R}^n$ are column vectors, their transpose is denoted as z^T, and transformations like $\exp(z)$, z^2, or $|z|$ are applied component-wise. Further symbols are:

- $|z| = (|z_1|, |z_2|, \dots)^T$ absolute value taken component wise
- $\|z\| = \sqrt{\sum_i z_i^2}$ Euclidean length of a vector
- $\sim$ equality in distribution
- $\propto$ in the limit proportional to
- $\circ$ binary operator giving the component-wise product of two vectors or matrices (Hadamard product), such that for $a, b \in \mathbb{R}^n$ we have $a \circ b \in \mathbb{R}^n$ and $(a \circ b)_i = a_i b_i$.
- $\mathbb{1}$ the indicator function, $\mathbb{1}_\alpha = 0$ if α is false or 0 or empty, and $\mathbb{1}_\alpha = 1$ otherwise.
- $\lambda \in \mathbb{N}$ number of offspring, offspring population size
- $\mu \in \mathbb{N}$ number of parents, parental population size
- $\mu_w = \left(\sum_{k=1}^\mu |w_k|\right)^2 / \sum_{k=1}^\mu w_k^2$, the variance effective selection mass or *effective* number of parents, where always $\mu_w \le \mu$ and $\mu_w = \mu$ if all recombination weights w_k are equal in absolute value
- $(1+1)$ elitist selection scheme with one parent and one offspring, see Sect. 44.2.5

- $(\mu \overset{+}{,} \lambda)$, e.g., $(1+1)$ or $(1, \lambda)$, selection schemes, see Sect. 44.2.5
- $(\mu/\rho, \lambda)$ selection scheme with recombination (if $\rho > 1$), see Sect. 44.2.5
- $\rho \in \mathbb{N}$ number of parents for recombination
- $\sigma > 0$ a step-size and/or standard deviation
- $\sigma \in \mathbb{R}_+^n$ a vector of step-sizes and/or standard deviations
- $\varphi \in \mathbb{R}$ a progress measure, see Definition 44.2 and Sect. 44.4.2
- $c_{\mu/\mu, \lambda}$ the progress coefficient for the $(\mu/\mu, \lambda)$-ES [44.8] equals the expected value of the average of the largest μ order statistics of λ independent standard normally distributed random numbers and is of the order of $\sqrt{2 \log(\lambda/\mu)}$.
- $C \in \mathbb{R}^{n \times n}$ a (symmetric and positive definite) covariance matrix
- $C^{\frac{1}{2}} \in \mathbb{R}^{n \times n}$ a matrix that satisfies $C^{\frac{1}{2}} C^{\frac{1}{2}}{}^T = C$ and is symmetric if not stated otherwise. If $C^{\frac{1}{2}}$ is symmetric, the eigendecomposition $C^{\frac{1}{2}} = B\Lambda B^T$ with $BB^T = I$ and the diagonal matrix Λ exists and we find $C = C^{\frac{1}{2}} C^{\frac{1}{2}} = B\Lambda^2 B^T$ as eigendecomposition of C.
- e_i the i-th canonical basis vector
- $f : \mathbb{R}^n \to \mathbb{R}$ fitness or objective function to be minimized
- $I \in \mathbb{R}^{n \times n}$ the identity matrix (identity transformation)
- i.i.d. independent and identically distributed
- $\mathcal{N}(x, C)$ a multivariate normal distribution with expectation and modal value x and covariance matrix C, see Sect. 44.2.8.
- $n \in \mathbb{N}$ search space dimension
- $\mathcal{P}$ a multiset of individuals, a population
- $s, s_\sigma, s_c < \in \mathbb{R}^n$ a search path or evolution path
- s, s_k endogenous strategy parameters (also known as control parameters) of a single parent or the k-th offspring; they typically parametrize the mutation, for example with a step-size σ or a covariance matrix C
- $t \in \mathbb{N}$ time or iteration index
- $w_k \in \mathbb{R}$ recombination weights
- $x, x^{(t)}, x_k \in \mathbb{R}^n$ solution or object parameter vector of a single parent (at iteration t) or of the k-th offspring; an element of the search space $\mathbb{R}^n$ that serves as argument to the fitness function $f : \mathbb{R}^n \to \mathbb{R}$.
- $\text{diag} : \mathbb{R}^n \to \mathbb{R}^{n \times n}$ the diagonal matrix from a vector
- $\exp^\alpha : \mathbb{R}^{n \times n} \to \mathbb{R}^{n \times n}, A \mapsto \sum_{i=0}^\infty (\alpha A)^i / i!$ is the matrix exponential for $n > 1$, otherwise

the exponential function. If $\mathbf{A}$ is symmetric and $\mathbf{B}\mathbf{\Lambda}\mathbf{B}^{\mathrm{T}} = \mathbf{A}$ is the eigendecomposition of $\mathbf{A}$ with $\mathbf{B}\mathbf{B}^{\mathrm{T}} = \mathbf{I}$ and $\mathbf{\Lambda}$ diagonal, we have $\exp(\mathbf{A}) = \mathbf{B}\exp(\mathbf{\Lambda})\mathbf{B}^{\mathrm{T}} = \mathbf{B}\left(\sum_{i=0}^{\infty}\mathbf{\Lambda}^i/i!\right)\mathbf{B}^{\mathrm{T}} = \mathbf{I} + \mathbf{B}\mathbf{\Lambda}\mathbf{B}^{\mathrm{T}} + \mathbf{B}\mathbf{\Lambda}^2\mathbf{B}^{\mathrm{T}}/2 + \cdots$. Furthermore, we have $\exp^\alpha(\mathbf{A}) = \exp(\mathbf{A})^\alpha = \exp(\alpha\mathbf{A})$ and $\exp^\alpha(x) = (e^\alpha)^x = e^{\alpha x}$.

44.2 Main Principles

ES derive inspiration from principles of biological evolution. We assume a *population*, $\mathcal{P}$, of so-called *individuals*. Each individual consists of a solution or object parameter vector $\mathbf{x} \in \mathbb{R}^n$ (the visible traits) and further endogenous parameters, s (the hidden traits), and an associated fitness value, $f(\mathbf{x})$. In some cases, the population contains only one individual. Individuals are also denoted as *parents* or *offspring*, depending on the context. In a generational procedure:

1. One or several parents are picked from the population (mating selection) and new offspring are generated by duplication and recombination of these parents.
2. The new offspring undergo mutation and become new members of the population.
3. Environmental selection reduces the population to its original size.

Within this procedure, ES employ the following main principles that are specified and applied in the operators and algorithms further below.

44.2.1 Environmental Selection

Environmental selection is applied as so-called *truncation selection*. Based on the individuals' fitnesses, $f(\mathbf{x})$, only the μ best individuals from the population survive. In contrast to roulette wheel selection in *genetic algorithms* [44.9], only fitness *ranks* are used. In evolution strategies, environmental selection is deterministic. In evolutionary programming, like in many other evolutionary algorithms, environmental selection has a stochastic component. Environmental selection can also remove *overaged* individuals first.

44.2.2 Mating Selection and Recombination

Mating selection picks individuals from the population to become new parents. Recombination generates a single new offspring from these parents. Specifically, we differentiate two common scenarios for mating selection and recombination:

- *Fitness-independent* mating selection and recombination do not depend on the fitness values of the individuals and can be either deterministic or stochastic. *Environmental* selection is then essential to drive the evolution toward better solutions.
- *Fitness-based* mating selection and recombination, where the recombination operator utilizes the fitness ranking of the parents (in a deterministic way). *Environmental* selection can potentially be omitted in this case.

44.2.3 Mutation and Parameter Control

Mutation introduces small, random, and unbiased changes to an individual. These changes typically affect all variables. The average size of these changes depends on endogenous parameters that change over time. These parameters are also called control parameters, or *endogenous strategy parameters*, and define the notion of *small*, for example, via the *step-size* σ. In contrast, *exogenous strategy parameters* are fixed once and for all, for example, parent number μ. Parameter control is not always directly inspired by biological evolution, but is an indispensable and central feature of evolution strategies.

44.2.4 Unbiasedness

Unbiasedness is a generic design principle of evolution strategies. Variation resulting from mutation or recombination is designed to introduce new, unbiased *information*. Selection, on the other hand biases this information toward solutions with better fitness. Under neutral selection (i.e., fitness independent mating and environmental selection), all variation operators are desired to be unbiased. Maximum exploration and unbiasedness are in accord. ES are unbiased in the following respects:

- The type of mutation distribution, the Gaussian or normal distribution, is chosen in order to have rotational symmetry and maximum entropy (maximum exploration) under the given variances. Decreasing the entropy would introduce prior information and therefore a bias.
- Object parameters and endogenous strategy parameters are unbiased under recombination and unbiased under mutation. Typically, mutation has expectation zero.
- Invariance properties avoid a bias toward a specific representation of the fitness function, e.g., representation in a specific coordinate system or using specific fitness values (invariance to strictly monotonic transformations of the fitness values can be achieved). Parameter control in evolution strategies strives for invariance properties [44.10].

44.2.5 $(\mu/\rho \overset{+}{,} \lambda)$ Notation for Selection and Recombination

An evolution strategy is an iterative (generational) procedure. In each generation new individuals (offspring) are created from existing individuals (parents). A mnemonic notation is commonly used to describe some aspects of this iteration. The $(\mu/\rho \overset{+}{,} \lambda)$-ES, where μ, ρ and λ are positive integers, also frequently denoted as $(\mu \overset{+}{,} \lambda)$-ES (where ρ remains unspecified) describes the following:

- The parent population contains μ individuals.
- For recombination, ρ (out of μ) parent individuals are used. We have therefore $\rho \le \mu$.
- λ denotes the number of offspring generated in each iteration.
- $\overset{+}{,}$ describes whether or not selection is additionally based on the individuals' age. An evolution strategy applies *either plus- or comma*-selection. In *plus*-selection, age is not taken into account and the μ best of $\mu + \lambda$ individuals are chosen. Selection is elitist and, in effect, the parents are the μ all-time best individuals. In *comma*-selection, individuals die out after one iteration step and only the offspring (the youngest individuals) survive to the next generation. In that case, environmental selection chooses μ parents from λ offspring.

In a (μ, λ)-ES, $\lambda \ge \mu$ must hold and the case $\lambda = \mu$ requires *fitness-based* mating selection or recombination. In a $(\mu + \lambda)$-ES, $\lambda = 1$ is possible and known as *steady-state* scenario.

Occasionally, a subscript to ρ is used in order to denote the type of recombination, e.g., ρ_I or ρ_W for intermediate or weighted recombination, respectively. Without a subscript, we tacitly assume intermediate recombination, if not stated otherwise. The notation has also been expanded to include the maximum age, κ, of individuals as (μ, κ, λ)-ES [44.11], where *plus*-selection corresponds to $\kappa = \infty$ and *comma*-selection corresponds to $\kappa = 1$.

44.2.6 Two Algorithm Templates

Algorithm 44.1 gives pseudocode for the evolution strategy.

Algorithm 44.1 The $(\mu/\rho \overset{+}{,} \lambda)$-ES

1: **given** $n, \rho, \mu, \lambda \in \mathbb{N}_+$
2: **initialize** $\mathcal{P} = \{(\boldsymbol{x}_k, s_k, f(\boldsymbol{x}_k)) \mid 1 \le k \le \mu\}$
3: **while** not happy
4:    **for** $k \in \{1, \dots, \lambda\}$
5:       $(\boldsymbol{x}_k, s_k) = \text{recombine}(\text{select\_mates}(\rho, \mathcal{P}))$
6:       $s_k \leftarrow \text{mutate\_s}(s_k)$
7:       $\boldsymbol{x}_k \leftarrow \text{mutate\_x}(s_k, \boldsymbol{x}_k) \in \mathbb{R}^n$
8:    $\mathcal{P} \leftarrow \mathcal{P} \cup \{(\boldsymbol{x}_k, s_k, f(\boldsymbol{x}_k)) \mid 1 \le k \le \lambda\}$
9:    $\mathcal{P} \leftarrow \text{select\_by\_age}(\mathcal{P})$ // identity for '+'
10:   $\mathcal{P} \leftarrow \text{select\_}\mu\text{\_best}(\mu, \mathcal{P})$ // by f-ranking

---

Given is a population, $\mathcal{P}$, of at least μ individuals $(\boldsymbol{x}_k, s_k, f(\boldsymbol{x}_k))$, $k = 1, \dots, \mu$. Vector $\boldsymbol{x}_k \in \mathbb{R}^n$ is a solution vector and s_k contains the control or endogenous strategy parameters, for example, a success counter or a step-size that primarily serves to control the mutation of $\boldsymbol{x}$ (in Line 7). The values of s_k may be identical for all k. In each generation, first λ offspring are generated (Lines 4–7), each by recombination of $\rho \le \mu$ individuals from $\mathcal{P}$ (Line 2), followed by mutation of s (Line 6) and of $\boldsymbol{x}$ (Line 7). The new offspring are added to $\mathcal{P}$ (Line 8). Overaged individuals are removed from $\mathcal{P}$ (Line 9), where individuals from the same generation have, by definition, the same age. Finally, the best μ individuals are retained in $\mathcal{P}$ (Line 10).

The mutation of the $\boldsymbol{x}$-vector in Line 7 always involves a stochastic component. Lines 5 and 6 may have stochastic components as well.

When select_mates in Line 5 selects $\rho = \mu$ individuals from $\mathcal{P}$, it reduces to the identity. If $\rho = \mu$ and recombination is deterministic, as is commonly the case, the result of recombine is the same *parental centroid* for all offspring. The computation of the parental centroid can be done once before the **for** loop or as the

ast step of the **while** loop, simplifying the initialization
f the algorithm. Algorithm 44.2 shows the pseudocode
n this case.

---

lgorithm 44.2 The $(\mu/\mu \overset{+}{,} \lambda)$-ES

1: **given** $n, \lambda \in \mathbb{N}_+$
2: **initialize** $x \in \mathbb{R}^n$, s, $\mathcal{P} = \{\}$
3:   **while** not happy
4:     **for** $k \in \{1, \ldots, \lambda\}$
5:       $s_k = \mathsf{mutate\_s}(s)$
6:       $x_k = \mathsf{mutate\_x}(s_k, x)$
7:       $\mathcal{P} \leftarrow \mathcal{P} \cup \{(x_k, s_k, f(x_k))\}$
8:     $\mathcal{P} \leftarrow \mathsf{select\_by\_age}(\mathcal{P})$ // identity for '+'
9:     $(x, s) \leftarrow \mathsf{recombine}(\mathcal{P}, x, s)$

In Algorithm 44.2, only a single parental centroid
$x, s)$ is initialized. Mutation takes this parental cen-
roid as input (notice that s_k and x_k in Lines 5 and 6 are
now *assigned* rather than *updated*) and *recombination*
s postponed to the end of the loop, computing in Line 9
he new parental centroid. While (x_k, s_k) can contain all
necessary information for this computation, it is often
more transparent to use x and s as additional arguments
n Line 9. Selection based on f-values is now limited to
mating selection in procedure $\mathsf{recombine}$ (that is, procedure $\mathsf{select\_\mu\_best}$ is omitted and μ is the number
of individuals in $\mathcal{P}$ that are actually used by recom-
bine).

Using a single parental centroid has become the
most popular approach, because such algorithms are
simpler to formalize, easier to analyze, and even perform better in various circumstances as they allow for
maximum genetic repair (see in the following). All
nstances of ES given in Sect. 44.3 are based on Algorithm 44.2.

44.2.7 Recombination Operators

n ES, *recombination* combines information from several parents to generate a *single* new offspring. Often,
multirecombination is used, where more than two parents are recombined ($\rho > 2$). In contrast, in *genetic
algorithms* often *two* offspring are generated from the
recombination of two parents. In evolutionary programming, recombination is generally not used. The most
important recombination operators used in evolution
strategies are the following:

- *Discrete* or dominant recombination, denoted by
$(\mu/\rho_D \overset{+}{,} \lambda)$, is also known as *uniform crossover* in
genetic algorithms. For each variable (component

of the x-vector), a single parent is drawn uniformly
from all ρ parents to inherit the variable value. For
ρ parents that all differ in each variable value, the
result is uniformly distributed across ρ^n different
x-values. The result of discrete recombination depends on the given coordinate system.
- *Intermediate* recombination, denoted by $(\mu/\rho_I \overset{+}{,} \lambda)$, takes the average value of all ρ parents (computes the center of mass, the centroid).
- *Weighted* multirecombination [44.10, 12, 13], de-
noted by $(\mu/\rho_W \overset{+}{,} \lambda)$, is a generalization of intermediate recombination, usually with $\rho = \mu$. It takes
a weighted average of all ρ parents. The weight
values depend on the fitness ranking, in that better parents never get smaller weights than inferior
ones. With equal weights, intermediate recombination is recovered. By using comma selection and
$\rho = \mu = \lambda$, where some of the weights may be zero,
weighted recombination can take over the role of
fitness-based environmental selection and *negative*
weights become a feasible option [44.12, 13]. The
sum of weights must be either one or zero, or recombination must be applied to the vectors $x_k - x$
and the result added to x.

In principle, recombination operators from genetic
algorithms, like one-point and two-point crossover or
line recombination [44.14] can alternatively be used.
However, they have been rarely applied in ES.

In ES, the result of selection and recombination
is often deterministic (namely, if $\rho = \mu$ and recombination is intermediate or weighted). This means that
eventually all offspring are generated by mutation from
the same single solution vector (the parental centroid)
as in Algorithm 44.2. This leads, for given variances,
to maximum entropy because all offspring are independently drawn from the same normal distribution.
With discrete recombination, the offspring distribution
is generated from a mixture of normal distributions
with different mean values. The resulting distribution has lower entropy unless it has a larger overall
variance.

The role of recombination, in general, is to keep the
variation in a population high. Discrete recombination
directly introduces variation by generating different
solutions. Their distance resembles the distance between the parents. However, discrete recombination,
as it depends on the given coordinate system, relies
on separability: it can introduce variation *successfully*
only if values of disrupted variables do not strongly depend on each other. Solutions resulting from discrete

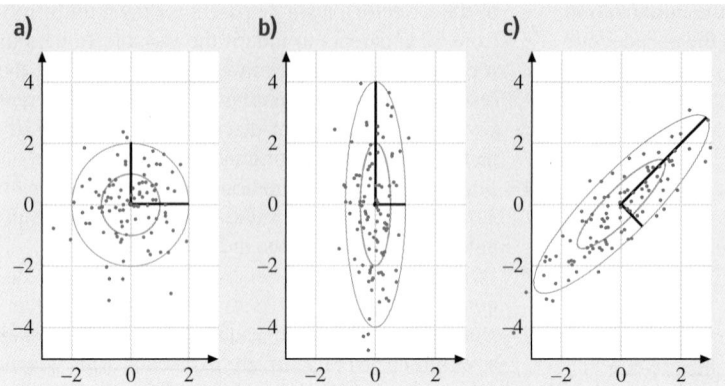

Fig. 44.1a–c Three two-dimensional multivariate normal distributions $\mathcal{N}(0, \mathbf{C}) \sim \mathbf{C}^{\frac{1}{2}}\mathcal{N}(0, \mathbf{I})$. The covariance matrix $\mathbf{C}$ of the distribution is, from left to right, the identity $\mathbf{I}$ (isotropic distribution), the diagonal matrix $\left(\begin{smallmatrix} 1/4 & 0 \\ 0 & 4 \end{smallmatrix}\right)$ (axis-parallel distribution) and $\left(\begin{smallmatrix} 2.125 & 1.875 \\ 1.875 & 2.125 \end{smallmatrix}\right)$ with the same eigenvalues $(1/4, 4)$ as the diagonal matrix. Shown are in each subfigure the mean at $\mathbf{0}$ as *small black dot* (a different mean solely changes the axis annotations), two eigenvectors of $\mathbf{C}$ along the principal axes of the ellipsoids (*thin black lines*), two ellipsoids reflecting the set of points $\{x : (x - 0)^{\mathrm{T}}\mathbf{C}^{-1}(x - 0) \in \{1, 4\}\}$ that represent the 1-σ and 2-σ lines of equal density, and 100 sampled points (however, a few of them are likely to be outside of the area shown)

recombination lie on the vertices of an *axis-parallel box*.

Intermediate and weighted multirecombination do not lead to variation within the new population as they result in the same single point for all offspring. However, they do allow the mutation operator to introduce *additional* variation by means of genetic repair [44.15]. Recombinative averaging reduces the effective step length taken in unfavorable directions by a factor of $\sqrt{\mu}$ (or $\sqrt{\mu_w}$ in the case of weighted recombination), but leaves the step length in favorable directions essentially unchanged, see also Sect. 44.4.2. This may allow increased variation by enlarging mutations by a factor of about μ (or μ_w) as revealed in (44.16), to achieve maximal progress.

44.2.8 Mutation Operators

The mutation operator introduces (*small*) variations by adding a point symmetric perturbation to the result of recombination, say a solution vector $x \in \mathbb{R}^n$. This perturbation is drawn from a multivariate normal distribution, $\mathcal{N}(\mathbf{0}, \mathbf{C})$, with zero mean (expected value) and covariance matrix $\mathbf{C} \in \mathbb{R}^{n \times n}$. Besides normally distributed mutations, Cauchy mutations [44.16–18] have also been proposed in the context of ES and evolutionary programming. We have $x + \mathcal{N}(\mathbf{0}, \mathbf{C}) \sim \mathcal{N}(x, \mathbf{C})$, meaning that x determines the expected value of the new offspring individual. We also have $x + \mathcal{N}(\mathbf{0}, \mathbf{C}) \sim$

$x + \mathbf{C}^{\frac{1}{2}}\mathcal{N}(\mathbf{0}, \mathbf{I})$, meaning that the linear transformation $\mathbf{C}^{\frac{1}{2}}$ generates the desired distribution from the vector $\mathcal{N}(\mathbf{0}, \mathbf{I})$ that has i.i.d. $\mathcal{N}(0, 1)$ components. (Using the normal distribution has several advantages. The $\mathcal{N}(\mathbf{0}, \mathbf{I})$ distribution is the most convenient way to implement an *isotropic* perturbation. The normal distribution is stable: sums of independent normally distributed random variables are again normally distributed. This facilitates the design and analysis of algorithms remarkably. Furthermore, the normal distribution has maximum entropy under the given variances.)

Figure 44.1 shows different normal distributions in dimension $n = 2$. Their lines of equal density are ellipsoids. Any straight section through the two-dimensional density recovers a two-dimensional Gaussian bell. Based on multivariate normal distributions, three different mutation operators can be distinguished:

- *Spherical/isotropic (Fig. 44.1a)* where the covariance matrix is proportional to the identity, i.e., the mutation distribution follows $\sigma\mathcal{N}(\mathbf{0}, \mathbf{I})$ with step-size $\sigma > 0$. The distribution is spherical and invariant under rotations about its mean. In the following, Algorithm 44.3 uses this kind of mutation.
- *Axis-parallel (Fig. 44.1b)* where the covariance matrix is a diagonal matrix, i.e., the mutation distribution follows $\mathcal{N}(\mathbf{0}, \mathrm{diag}(\sigma)^2)$, where σ is a vector of coordinate-wise standard deviations and the di-

agonal matrix $\mathrm{diag}(\boldsymbol{\sigma})^2$ has eigenvalues σ_i^2 with eigenvectors $\boldsymbol{e}_i$. The principal axes of the ellipsoid are parallel to the coordinate axes. This case includes the previous isotropic case. Below, Algorithms 44.4–44.6 implement this kind of mutation distribution.

• *General (Fig. 44.1c)* where the covariance matrix is symmetric and positive definite (i.e., $\boldsymbol{x}^\mathrm{T}\mathbf{C}\boldsymbol{x} > 0$ for all $\boldsymbol{x} \neq \boldsymbol{0}$), generally nondiagonal and has $(n^2 + n)/2$ degrees of freedom (control parameters). The general case includes the previous axis-parallel and spherical cases. Below, Algorithms 44.7 and 44.8 implement general multivariate normally distributed mutations.

In the first and the second cases, the variations of variables are independent of each other, they are uncorrelated. This limits the usefulness of the operator in practice. The third case is *incompatible* with discrete recombination: for a narrow, diagonally oriented ellipsoid (not to be confused with a diagonal covariance matrix), a point resulting from selection and *discrete* recombination lies within this ellipsoid only if each coordinate is taken from the same parent (which happens with probability $1/\rho^{n-1}$) or from a parent with a very similar value in this coordinate. The narrower the ellipsoid the more similar (i.e., correlated) the value needs to be. As another illustration consider sampling, neutral selection and discrete recombination based on Fig. 44.1c: after discrete recombination the points $(-2, 2)$ and $(2, -2)$ outside the ellipsoid have the same probability as the points $(2, 2)$ and $(-2, -2)$ inside the ellipsoid.

The mutation operators introduced are *unbiased* in several ways. They are all point symmetrical and have expectation zero. Therefore, mutation alone will almost certainly not lead to better fitness values *in expectation*. The isotropic mutation operator features the same distribution along any direction. The general mutation operator is, as long as $\mathbf{C}$ remains unspecified, unbiased toward the choice of a Cartesian coordinate system, i.e. unbiased toward the representation of solutions $\boldsymbol{x}$, which has also been referred to as invariance to affine coordinate system transformations [44.10]. This however depends on the way how $\mathbf{C}$ is adapted (see the following).

44.3 Parameter Control

Controlling the parameters of the mutation operator is key to the design of ES. Consider the isotropic operator (Fig. 44.1a), where the step-size σ is a scaling factor for the random vector perturbation. The step-size controls to a large extent the convergence speed. In situations where larger step-sizes lead to larger expected improvements, a step-size control technique should aim at increasing the step-size (and decreasing it in the opposite scenario).

The importance of step-size control is illustrated with a simple experiment. Consider a spherical function $f(\boldsymbol{x}) = \|\boldsymbol{x}\|^\alpha$, $\alpha > 0$, and a $(1+1)$-ES with constant step-size equal to $\sigma = 10^{-2}$, i.e., with mutations drawn from $10^{-2}\mathcal{N}(\boldsymbol{0}, \mathbf{I})$. The convergence of the algorithm is depicted in Fig 44.2 (constant σ graphs).

We observe, roughly speaking, three stages: up to 600 function evaluations, progress toward the optimum is slow. At this stage, the fixed step-size is too small. Between 700 and 800 evaluations, fast progress toward the optimum is observed. At this stage, the step-size is close to optimal. Afterward, the progress decreases and approaches the rate of the pure random search algorithm, well illustrated on the bottom subfigure. At this stage the fixed step-size is too large and the probability to sample better offspring becomes very small.

The figure also shows runs of the $(1+1)$-ES with $1/5$th success rule step-size control (as described in Sect. 44.3.1) and the step-size evolution associated to one of these runs. The initial step-size is far too small and we observe that the adaptation technique increases the step-size in the first iterations. Afterward, step-size is kept roughly proportional to the distance to the optimum, which is in fact optimal and leads to linear convergence on the top subfigure.

Generally, the goal of parameter control is to drive the endogenous strategy parameters close to their optimal values. These optimal values, as we have seen for the step-size in Fig. 44.2, can significantly change over time or depending on the position in search space. In the most general case, the mutation operator has $(n^2 + n)/2$ degrees of freedom (Sect. 44.2.8). The conjecture is that in the desired scenario lines of equal density of the mutation operator resemble locally the lines of equal fitness [44.4, pp. 242f.]. In the case of convex-quadratic fitness functions this resemblance can be perfect and, apart from the step-size, optimal parameters do not change over time (as illustrated in Fig. 44.3 below).

Control parameters like the step-size can be stored on different *levels*. Each individual can have its own step-size value (like in Algorithms 44.4 and 44.5), or a single step-size is stored and applied to all individuals in the population. In the latter case, sometimes different populations with different parameter values are run in parallel [44.19].

In the following, six specific ES are outlined, each of them representing an important achievement in parameter control.

44.3.1 The 1/5th Success Rule

The 1/5th success rule for step-size control is based on an important discovery made very early in the research of evolution strategies [44.1]. A similar rule had also been found independently before in [44.20]. As a control mechanism in practice, the 1/5th success rule has been mostly superseded by more sophisticated methods. However, its conceptual insight remains remarkably valuable.

Consider a linear fitness function, for example, $f : x \mapsto x_1$ or $f : x \mapsto \sum_i x_i$. In this case, any point symmetrical mutation operator has a success probability of $1/2$: in one-half of the cases, the perturbation will improve the original solution, in one half of the cases the solution will deteriorate. Following the Taylor's formula, we know that smooth functions with decreasing neighborhood size become more and more linear. Therefore, the success probability becomes $1/2$ for step-size $\sigma \to 0$. On most nonlinear functions, the success rate is indeed a monotonously decreasing function in σ and goes to zero for $\sigma \to \infty$. This suggests to control the step-size by increasing it for large success rates and decreasing it for small ones. This mechanism can drive the step-size close to the optimal value.

Rechenberg [44.1] investigated two simple but quite different functions, the corridor function

$$f : x \mapsto \begin{cases} x_1 & \text{if } |x_i| \le 1 \text{ for } i = 2, \dots, n \\ \infty & \text{otherwise}, \end{cases}$$

and the sphere function

$$f : x \mapsto \sum x_i^2 .$$

He found optimal success rates for the $(1+1)$-ES with isotropic mutation to be $\approx 0.184 > 1/6$ and $\approx 0.270 < 1/3$, respectively (for $n \to \infty$) [44.1]. Optimality here means to achieve the largest expected approach of the optimum in a single generation. This leads to approximately $1/5$ as being the success value where to switch between decreasing and increasing the step-size.

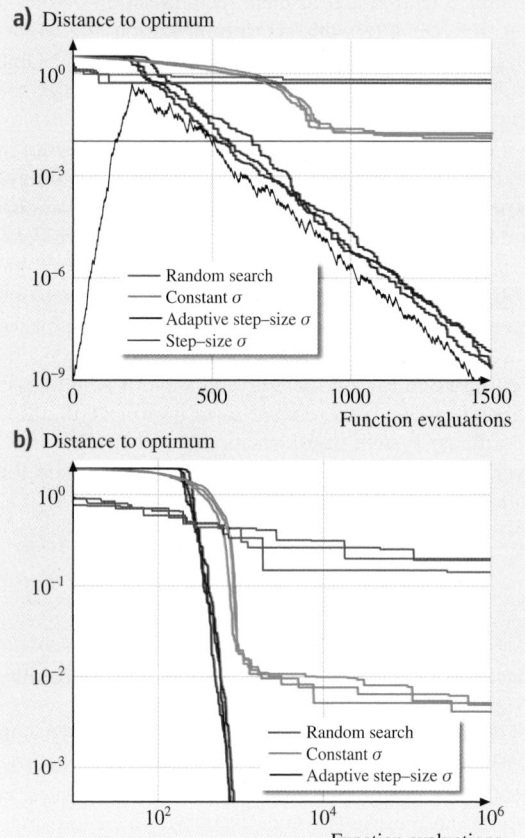

Fig. 44.2a,b Runs of the $(1+1)$-ES with constant step-size, of pure random search (uniform in $[-0.2, 1]^{10}$), and of the $(1+1)$-ES with 1/5th success rule (Algorithm 44.3) on a spherical function $f(x) = \|x\|^{\alpha}, \alpha > 0$ (because of invariance to monotonic f-transformation the same graph is observed for any $\alpha > 0$). For each algorithm, there are three runs in (**a**) and (**b**). The x-axis is linear in (**a**) and in log-scale in (**b**). For the $(1+1)$-ES with constant step-size, σ equals 10^{-2}. For the $(1+1)$-ES with 1/5th success rule, the initial step-size is chosen very small to 10^{-9} and the parameter d equals $1 + 10/3$. In (**a**) also the evolution of the step-size of one of the runs of the $(1+1)$-ES with 1/5th success rule is shown. All algorithms are initialized at **1**. Eventually, the $(1+1)$-ES with 1/5th success rule reveals linear behavior (**a**), while the other two algorithms reveal eventually linear behavior in (**b**)

Algorithm 44.3 The (1+1)-ES with 1/5th Rule

1: **given** $n \in \mathbb{N}_+$, $d \approx \sqrt{n+1}$
2: **initialize** $x \in \mathbb{R}^n$, $\sigma > 0$
3: **while** not happy
4:   $x_1 = x + \sigma \times \mathcal{N}(\mathbf{0}, \mathbf{I})$ // mutation
5:   $\sigma \leftarrow \sigma \times \exp^{1/d}(\mathbb{1}_{f(x_1) \leq f(x)} - 1/5)$
6:   **if** $f(x_1) \leq f(x)$ // select if better
7:     $x = x_1$ // x-value of new parent

Algorithm 44.3 implements the (1+1)-ES with 1/5th success rule in a simple and effective way [44.21]. Lines 5–7 implement Line 9 from Algorithm 44.2, including selection in Line 8. Line 5 in Algorithm 44.3 updates the step-size σ of the single parent. The step-size does not change if and only if the argument of exp is zero. While this cannot happen in a single generation, we still can find a stationary point for σ: $\log \sigma$ is unbiased if and only if the expected value of the argument of exp is zero. This is the case if $\mathbb{E}\mathbb{1}_{f(x_1) \leq f(x)} = 1/5$, in other words, if the probability of an improvement with $f(x_1) \leq f(x)$ is 20%. Otherwise, $\log \sigma$ increases in expectation if the success probability is larger than $1/5$ and decreases if the success probability is smaller than $1/5$. Hence, Algorithm 44.3 indeed implements the 1/5th success rule.

44.3.2 Self-Adaptation

A seminal idea in the domain of ES is parameter control via *self-adaptation* [44.3]. In self-adaptation, new control parameter settings are generated similar to new x-vectors by recombination and mutation. Algorithm 44.4 presents an example with adaptation of n coordinate-wise standard deviations (individual step-sizes).

Algorithm 44.4 The $(\mu/\mu, \lambda)$-σ SA-ES

1: **given** $n \in \mathbb{N}_+$, $\lambda \geq 5n$, $\mu \approx \lambda/4 \in \mathbb{N}$, $\tau \approx 1/\sqrt{n}$, $\tau_i \approx 1/n^{1/4}$
2: **initialize** $x \in \mathbb{R}^n$, $\sigma \in \mathbb{R}_+^n$
3: **while** not happy
4:   **for** $k \in \{1, \dots, \lambda\}$
     // random numbers i.i.d. for all k
5:     $\xi_k = \tau \, \mathcal{N}(0, 1)$ // global step-size
6:     $\boldsymbol{\xi}_k = \tau_i \, \mathcal{N}(\mathbf{0}, \mathbf{I})$ // coordinate-wise σ
7:     $z_k = \mathcal{N}(\mathbf{0}, \mathbf{I})$ // x-vector change
     // mutation
8:     $\sigma_k = \sigma \circ \exp(\boldsymbol{\xi}_k) \times \exp(\xi_k)$
9:     $x_k = x + \sigma_k \circ z_k$
10: $\mathcal{P} = \mathsf{sel\_}\mu\mathsf{\_best}(\{(x_k, \sigma_k, f(x_k)) \mid 1 \leq k \leq \lambda\})$
     // recombination

11: $\sigma = \dfrac{1}{\mu} \sum_{\sigma_k \in \mathcal{P}} \sigma_k$

12: $x = \dfrac{1}{\mu} \sum_{x_k \in \mathcal{P}} x_k$

First, for conducting the mutation, random events are drawn in Lines 5–7. In Line 8, the step-size vector for each individual undergoes (i) a mutation common for all components, $\exp(\xi_k)$, and (ii) a component-wise mutation with $\exp(\boldsymbol{\xi}_k)$. These mutations are unbiased, in that $\mathbb{E} \log \sigma_k = \log \sigma$. The mutation of x in Line 9 uses the mutated vector σ_k. After selection in Line 10, intermediate recombination is applied to compute x and σ for the next generation. By taking the average over σ_k we have $\mathbb{E}\sigma = \mathbb{E}\sigma_k$ in Line 11. However, the application of mutation *and* recombination on σ introduces a moderate bias such that σ tends to increase under neutral selection [44.22].

In order to achieve stable behavior of σ, the number of parents μ must be large enough, which is reflected in the setting of λ. A setting of $\tau \approx 1/4$ has been proposed in combination with ξ_k being uniformly distributed across the two values in $\{-1, 1\}$ [44.2].

44.3.3 Derandomized Self-Adaptation

Derandomized self-adaptation [44.23] addresses the problem of selection noise that occurs with self-adaptation of σ as outlined in Algorithm 44.4. Selection noise refers to the possibility that very good offspring may be generated with poor strategy parameter settings and vice versa. The problem occurs frequently and has two origins:

- A small/large component in $|\sigma_k \circ z_k|$ (Line 9 in Algorithm 44.4) does not necessarily imply that the respective component of σ_k is small/large. Selection of σ is disturbed by the respective realizations of z.
- Selection of a small/large component of $|\sigma_k \circ z_k|$ does not imply that this is necessarily a favorable setting: more often than not, the sign of a component is more important than its size and all other components influence the selection as well.

Due to selection noise, poor values are frequently inherited and we observe stochastic fluctuations of σ. Such fluctuations can in particular lead to very small values (very large values are removed by selection more quickly). The overall magnitude of these fluctuations can be implicitly controlled via the parent number μ,

Part E | 44.3

because intermediate recombination (Line 11 in Algorithm 44.4) effectively reduces the magnitude of σ-changes and biases $\log \sigma$ to larger values.

For $\mu \ll n$, the stochastic fluctuations become prohibitive and therefore $\mu \approx \lambda/4 \geq 1.25n$ is chosen to make σ-self-adaptation reliable.

Derandomization addresses the problem of selection noise on σ directly without resorting to a large parent number. The derandomized $(1, \lambda)$-σSA-ES is outlined in Algorithm 44.5 and addresses selection noise twofold.

Algorithm 44.5 Derandomized $(1, \lambda)$-σSA-ES

1: **given** $n \in \mathbb{N}_+, \lambda \approx 10, \tau \approx 1/3, d \approx \sqrt{n}, d_i \approx n$
2: **initialize** $x \in \mathbb{R}^n, \sigma \in \mathbb{R}^n_+$
3: **while** not happy
4:    **for** $k \in \{1, \ldots, \lambda\}$
         // random numbers i.i.d. for all k
5:       $\xi_k = \tau \, \mathcal{N}(0, 1)$
6:       $z_k = \mathcal{N}(\mathbf{0}, \mathbf{I})$
         // mutation, re-using random events
7:       $x_k = x + \exp(\xi_k) \times \sigma \circ z_k$
8:       $\sigma_k = \sigma \circ \exp^{1/d_i}\left(\dfrac{|z_k|}{\mathbb{E}|\mathcal{N}(0, 1)|} - \mathbf{1} \right)$
            $\times \exp^{1/d}(\xi_k)$
9:    $(x_1, \sigma_1, f(x_1)) \leftarrow$ select_single_best(
            $\{(x_k, \sigma_k, f(x_k)) \mid 1 \leq k \leq \lambda\})$
      // assign new parent
10:   $\sigma = \sigma_1$
11:   $x = x_1$

Instead of introducing new variations in σ by means of $\exp(\xi_k)$, the variations from z_k are directly used for the mutation of σ in Line 8. The variations are dampened compared to their use in the mutation of x (Line 7) via d and d_i, thereby mimicking the effect of intermediate recombination on σ [44.23, 24]. The order of the two mutation equations becomes irrelevant.

For Algorithm 44.5 also a $(\mu/\mu, \lambda)$ variant with recombination is feasible. However, in particular in the $(\mu/\mu_I, \lambda)$-ES, σ-self-adaptation tends to generate too small step-sizes. A remedy for this problem is to use nonlocal information for step-size control.

44.3.4 Nonlocal Derandomized Step-Size Control (CSA)

When using self-adaptation, step-sizes are associated with individuals and selected based on the fitness of each individual. However, step-sizes that serve individuals well by giving them a high likelihood to be selected are generally not step-sizes that maximize the progress of the entire population or the parental centroid x. We will see later that, for example, the optimal step-size may increase linearly with μ (Sect. 44.4.2 and (44.16)). With self-adaptation on the other hand, the step-size of the μ-th best offspring is typically even smaller than the step-size of the best offspring. Consequently, Algorithm 44.5 assumes often too small step-sizes and can be considerably improved by using nonlocal information about the evolution of the population. Instead of single (local) mutation steps z, an exponentially fading record, s_σ, of mutation steps is taken. This record, referred to as *search path* or *evolution path*, can be pictured as a sequence or sum of consecutive successful z-steps that is nonlocal in time and space. A search path carries information about the interrelation between single steps. This information can improve the adaptation and search procedure remarkably. Algorithm 44.6 outlines the $(\mu/\mu_I, \lambda)$-ES with *cumulative path length control*, also denoted as *cumulative step-size adaptation* (CSA), and additionally with nonlocal *individual* step-size adaptation [44.25, 26].

Algorithm 44.6 The $(\mu/\mu, \lambda)$-ES with Search Path

1: **given** $n \in \mathbb{N}_+,$ $\lambda \in \mathbb{N},$ $\mu \approx \lambda/4 \in \mathbb{N}$
      $c_\sigma \approx \sqrt{\mu/(n+\mu)},$ $d \approx 1 + \sqrt{\mu/n},$ $d_i \approx 3n$
2: **initialize** $x \in \mathbb{R}^n, \sigma \in \mathbb{R}^n_+, s_\sigma = \mathbf{0}$
3: **while** not happy
4:    **for** $k \in \{1, \ldots, \lambda\}$
5:       $z_k = \mathcal{N}(\mathbf{0}, \mathbf{I})$ // i.i.d. for each k
6:       $x_k = x + \sigma \circ z_k$
7:    $\mathcal{P} \leftarrow$ sel_μ_best($\{(x_k, z_k, f(x_k)) \mid 1 \leq k \leq \lambda\})$
      // recombination and parent update
8:    $s_\sigma \leftarrow (1 - c_\sigma) s_\sigma +$
            $\sqrt{c_\sigma (2 - c_\sigma)} \dfrac{\sqrt{\mu}}{\mu} \displaystyle\sum_{z_k \in \mathcal{P}} z_k$
9:    $\sigma \leftarrow \sigma \circ \exp^{1/d_i}\left(\dfrac{|s_\sigma|}{\mathbb{E}|\mathcal{N}(0, 1)|} - \mathbf{1} \right)$
10:            $\times \exp^{c_\sigma/d}\left(\dfrac{\|s_\sigma\|}{\mathbb{E}\|\mathcal{N}(\mathbf{0}, \mathbf{I})\|} - 1 \right)$
11:   $x = \dfrac{1}{\mu} \displaystyle\sum_{x_k \in \mathcal{P}} x_k$

In the $(\mu/\mu, \lambda)$-ES with search path, Algorithm 44.6, the factor ξ_k for changing the overall step-size has disappeared (compared to Algorithm 44.5) and the update of σ is postponed until after the *for* loop.

instead of the additional random variate ξ_k, the length of the search path $\|s_\sigma\|$ determines the global step-size change in Line 9. For the individual step-size change, $|z_k|$ is replaced by $|s_\sigma|$.

Using a search path is justified in two ways. First, it implements a low-pass filter for selected z-steps, removing high-frequency (most likely noisy) information. Second, and more importantly, it utilizes information that is otherwise lost: even if all single steps have the same length, the length of s_σ can vary, because it depends on the correlation between the directions of z-steps. If single steps point into similar directions, the path will be up to almost $\sqrt{2/c_\sigma}$ times longer than a single step and the step-size will increase. If they oppose each other the path will be up to almost $\sqrt{c_\sigma/2}$ times shorter and the step-size will decrease. The same is true for single components of s_σ.

The factors $\sqrt{c_\sigma(2-c_\sigma)}$ and $\sqrt{\mu}$ in Line 8 guaranty unbiasedness of s_σ under neutral selection, as usual.

All ES described so far are of somewhat limited value, because they feature only isotropic or axis-parallel mutation operators. In the remainder we consider methods that entertain not only an n-dimensional step-size vector σ, but also correlations between variables for the mutation of x.

44.3.5 Addressing Dependences Between Variables

The ES presented so far sample the mutation distribution independently in each component of *the given* coordinate system. The lines of equal density are either spherical or axis-parallel ellipsoids (compare Fig. 44.1). This is a major drawback, because it allows to solve problems with a long or elongated valley efficiently only if the valley is aligned with the coordinate system. In this section, we discuss ES that allow us to traverse nonaxis-parallel valleys efficiently by sampling distributions with correlations.

Full Covariance Matrix
Algorithms that adapt the complete covariance matrix of the mutation distribution (compare Sect. 44.2.8) are *correlated mutations* [44.3], the *generating set adaptation* [44.26], the *covariance matrix adaptation* (CMA) [44.27], a mutative invariant adaptation [44.28], and some instances of *natural evolution strategies* (NES) [44.29–31]. Correlated mutations and some natural ES are however not invariant under changes of the

coordinate system [44.10, 31, 32]. In the next sections, we outline two ES that adapt the full covariance matrix reliably and are invariant under coordinate system changes: the covariance matrix adaptation evolution strategy (CMA-ES) and the exponential natural evolution strategy (xNES).

Restricted Covariance Matrix
Algorithms that adapt nondiagonal covariance matrices, but are restricted to certain matrices, are the *momentum adaptation* [44.33], *direction adaptation* [44.26], *main vector adaptation* [44.34], and *limited memory CMA-ES* [44.35]. These variants are limited in their capability to shape the mutation distribution, but they might be advantageous for larger dimensional problems, say larger than a 100.

44.3.6 Covariance Matrix Adaptation (CMA)

The CMA-ES [44.10, 27, 36] is a de facto standard in continuous domain evolutionary computation. The CMA-ES is a natural generalization of Algorithm 44.6 in that the mutation ellipsoids are not constrained to be axis-parallel, but can take on a general orientation. The CMA-ES is also a direct successor of the *generating set adaptation* [44.26], replacing self-adaptation to control the overall step-size with cumulative step-size adaptation [44.37].

The $(\mu/\mu_W, \lambda)$-CMA-ES is outlined in Algorithm 44.7.

Algorithm 44.7 The $(\mu/\mu_W, \lambda)$-CMA-ES

1: **given** $n \in \mathbb{N}_+$, $\lambda \geq 5$, $\mu \approx \lambda/2$,
$w_k = w'(k)/\sum_k^\mu w'(k)$,
$w'(k) = \log(\lambda/2 + 1/2) - \log \text{rank}(f(x_k))$,
$\mu_w = 1/\sum_k^\mu w_k^2$, $c_\sigma \approx \mu_w/(n + \mu_w)$,
$d \approx 1 + \sqrt{\mu_w/n}$,
$c_c \approx (4 + \mu_w/n)/(n + 4 + 2\mu_w/n)$,
$c_1 \approx 2/(n^2 + \mu_w)$, $c_\mu \approx \mu_w/(n^2 + \mu_w)$, $c_m = 1$

2: **initialize** $s_\sigma = 0$, $s_c = 0$, $C = I$, $\sigma \in \mathbb{R}^n_+$, $x \in \mathbb{R}^n$

3: **while** not happy

4:     **for** $k \in \{1, \dots, \lambda\}$

5:         $z_k = \mathcal{N}(0, I)$ // i.i.d. for all k

6:         $x_k = x + \sigma C^{\frac{1}{2}} \times z_k$

7:     $\mathcal{P} = \text{sel\_}\mu\text{\_best}(\{(z_k, f(x_k)) \mid 1 \leq k \leq \lambda\})$

8:     $s_\sigma \leftarrow (1 - c_\sigma) s_\sigma +$ //search path for σ
$$\sqrt{c_\sigma(2 - c_\sigma)} \sqrt{\mu_w} \sum_{z_k \in \mathcal{P}} w_k z_k$$

Part E | 44.3

9: $s_c \leftarrow (1 - c_c) s_c +$ // search path for $\mathbf{C}$
$$h_\sigma \sqrt{c_c(2 - c_c)} \sqrt{\mu_w} \sum_{z_k \in P} w_k \mathbf{C}^{\frac{1}{2}} z_k$$

10: $x \leftarrow x + c_m \sigma \, \mathbf{C}^{\frac{1}{2}} \sum_{z_k \in P} w_k z_k$

11: $\sigma \leftarrow \sigma \, \exp^{c_\sigma / d} \left(\dfrac{\|s_\sigma\|}{\mathbb{E}\|\mathcal{N}(\mathbf{0}, \mathbf{I})\|} - 1 \right)$

12: $\mathbf{C} \leftarrow (1 - c_1 + c_h - c_\mu)\, \mathbf{C} +$
$$c_1 s_c s_c^{\mathsf{T}} + c_\mu \sum_{z_k \in P} w_k \mathbf{C}^{\frac{1}{2}} z_k (\mathbf{C}^{\frac{1}{2}} z_k)^{\mathsf{T}}$$

where $h_\sigma = \mathbb{1}_{\|s_\sigma\|^2 / n < 2 + 4/(n+1)}$, $c_h = c_1(1 - h_\sigma{}^2)c_c(2 - c_c)$, and $\mathbf{C}^{\frac{1}{2}}$ is the unique symmetric positive definite matrix obeying $\mathbf{C}^{\frac{1}{2}} \times \mathbf{C}^{\frac{1}{2}} = \mathbf{C}$. All c-coefficients are ≤ 1.

Two search paths are maintained, s_σ and s_c. The first path, s_σ, accumulates steps in the coordinate system where the mutation distribution is isotropic and which can be derived by scaling in the principal axes of the mutation ellipsoid only. The path generalizes s_σ from Algorithm 44.6 to nondiagonal covariance matrices and is used to implement cumulative step-size adaptation, CSA, in Line 10 (resembling Line 9 in Algorithm 44.6). Under neutral selection, $s_\sigma \sim \mathcal{N}(\mathbf{0}, \mathbf{I})$ and $\log \sigma$ is unbiased.

The second path, s_c, accumulates steps, disregarding σ, in the given coordinate system. Whenever s_σ is large and therefore σ is increasing fast, the coefficient h_σ prevents s_c from getting large and quickly changing the distribution shape via $\mathbf{C}$. Given $h_\sigma \equiv 1$, under neutral selection $s_c \sim \mathcal{N}(\mathbf{0}, \mathbf{C})$. The coefficient c_h in Line 12 corrects for the bias on s_c introduced by events $h_\sigma = 0$. The covariance matrix update consists of a rank-1 update, based on the search path s_c, and a rank-μ update with μ nonzero recombination weights w_k. Under neutral selection, the expected covariance matrix equals the covariance matrix before the update.

The updates of x and $\mathbf{C}$ follow a common principle. The mean x is updated such that the likelihood of successful offspring to be sampled again is maximized (or increased if $c_m < 1$). The covariance matrix $\mathbf{C}$ is updated such that the likelihood of successful steps $(x_k - x)/\sigma$ to appear again, or the likelihood to sample (in the direction of) the path s_c, is increased. A more fundamental principle for the equations is given in the next section.

Using not only the μ best but all λ offspring can be particularly useful for the *rank-μ* update of $\mathbf{C}$ in Line 12, where negative weights w_k for inferior offspring are advisable. Such an update has been introduced as *active CMA* [44.38].

The factor c_m in Line 10 can be equally written as a mutation scaling factor $\kappa = 1/c_m$ in Line 6, compare [44.39]. This means that the actual mutation steps are larger than the inherited ones, resembling the derandomization technique of damping step-size changes to address selection noise as described in Sect. 44.3.3.

An elegant way to replace Line 10 is

$$\sigma \leftarrow \sigma \, \exp^{(c_\sigma / d)/2} \left(\frac{\|s_\sigma\|^2}{n} - 1 \right) , \qquad (44.1)$$

and often used in theoretical investigations of this update as those presented in Sect. 44.4.2.

A single run of the $(5/5_w, 10)$-CMA-ES on a convex-quadratic function is shown in Fig. 44.3. For the sake of demonstration, the initial step-size is chosen far too small (a situation that should be avoided in practice) and increases quickly for the first 400 f-evaluations. After no more than 5500 f-evaluations the adaptation of $\mathbf{C}$ is accomplished. Then the eigenvalues of $\mathbf{C}$ (square roots of which are shown in the lower left) reflect the underlying convex-quadratic function and the convergence speed is the same as on the sphere function and about 60% of the speed of the $(1+1)$-ES as observed in Fig. 44.2. The resulting convergence speed is about 10 000 times faster than without adaptation of $\mathbf{C}$ and at least 1000 times faster compared to any of the algorithms from the previous sections.

44.3.7 Natural Evolution Strategies

The idea of using natural gradient learning [44.40] in ES has been proposed in [44.29] and further pursued in [44.31, 41]. *Natural evolution strategies* (NES) put forward the idea that the update of all distribution parameters can be based on the same fundamental principle. NES have been proposed as a more principled alternative to CMA-ES and characterized by operating on Cholesky factors of a covariance matrix. Only later was it discovered that also CMA-ES implements the underlying NES principle of natural gradient learning [44.31, 42].

For simplicity, let the vector θ represents all parameters of the distribution to sample new offspring. In the case of a multivariate normal distribution as above, we have a bijective transformation between θ and mean and covariance matrix of the distribution, $\theta \leftrightarrow (x, \sigma^2 \mathbf{C})$.

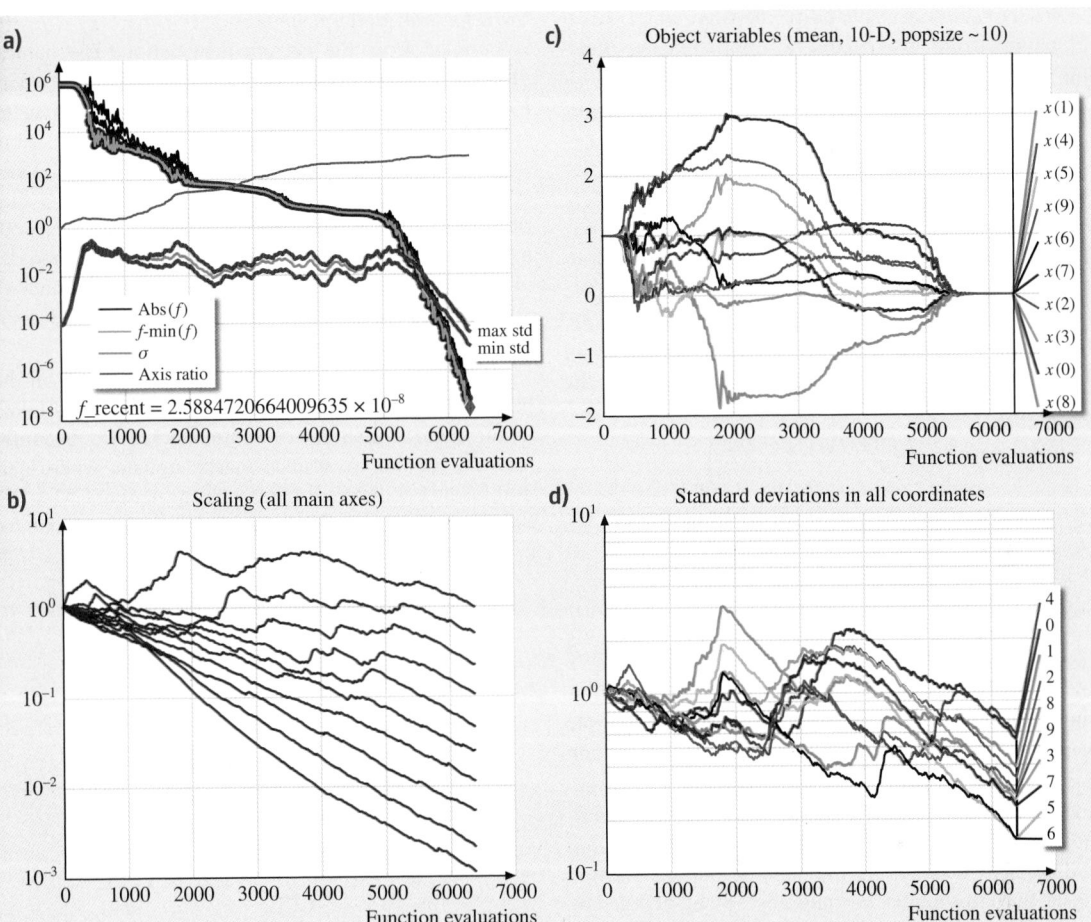

a)

Abs(f)
f-min(f)
σ
Axis ratio
max std
min std
f_recent $= 2.5884720664009635 \times 10^{-8}$

Function evaluations

b) Scaling (all main axes)

Function evaluations

c) Object variables (mean, 10-D, popsize ~10)

$x(1)$
$x(4)$
$x(5)$
$x(9)$
$x(6)$
$x(7)$
$x(2)$
$x(3)$
$x(0)$
$x(8)$

Function evaluations

d) Standard deviations in all coordinates

4
0
1
2
8
9
3
7
5
6

Function evaluations

Fig. 44.3a–d A single run of the $(5/5_w, 10)$-CMA-ES on the rotated ellipsoid function $x \mapsto \sum_{i=1}^{n} \alpha_i^2 y_i^2$ with $\alpha_i = 10^{3(i-1)/(n-1)}$, $y = \mathbf{R}x$, where $\mathbf{R}$ is a random matrix with $\mathbf{R}^T\mathbf{R} = \mathbf{I}$, for $n = 10$. Shown is the evolution of various parameters against the number of function evaluations. (**a**) best (*gray*), median and worst fitness value that reveal the final convergence phase after about 5500 function evaluations where the ellipsoid function has been reduced to the simple sphere; minimal and maximal coordinate-wise standard deviation of the mutation distribution and in between (mostly hidden) the step-size σ that is initialized far too small and increases quickly in the beginning, that increases afterward several times again by up to one order of magnitude and decreases with maximal rate during the last 1000 f-evaluations; axis ratio of the mutation ellipsoid (square root of the condition number of $\mathbf{C}$) that increases from 1 to 1000 where the latter corresponds to α_n/α_1. (**b**) sorted principal axis lengths of the mutation ellipsoid disregarding σ (square roots of the sorted eigenvalues of $\mathbf{C}$, see also Fig. 44.1) that adapt to the (local) structure of the underlying optimization problem; they finally reflect almost perfectly the factors α_i^{-1} up to a constant factor. (**c**) x (distribution mean) that is initialized with all ones and converges to the global optimum in zero while correlated movements of the variables can be observed. (**d**) standard deviations in the coordinates disregarding σ (square roots of diagonal elements of $\mathbf{C}$) showing the $\mathbf{R}$-dependent projections of the principal axis lengths into the given coordinate system. The straight lines to the right of the vertical line at about 6300 only annotate the coordinates and do not reflect measured data

We consider a probability density $p(.|\theta)$ over $\mathbb{R}^n$ parametrized by θ and a nonincreasing function $W_\theta^f : \mathbb{R} \to \mathbb{R}$. More specifically, $W_\theta^f : y \mapsto w(\mathrm{Pr}_{z \sim p(.|\theta)}(f(z) \leq y))$ computes the p_θ-quantile, or cumulative distribution function, of $f(z)$ with $z \sim p(.|\theta)$ at point y, composed with a nonincreasing predefined weight function $w : [0, 1] \to \mathbb{R}$ (where $w(0) > w(1/2) = 0$ is advisable). The value of $W_\theta^f(f(x))$ is invariant under strictly monotonous transformations of f. For $x \sim p(.|\theta)$ the distribution of $W_\theta^f(f(x)) \sim w(\mathcal{U}[0, 1])$ depends only on the predefined w; it is independent of θ and f and therefore also (time-)invariant under θ-updates. Given λ samples x_k, we have the rank-based consistent estimator

$$W_\theta^f(f(x_k)) \approx w\left(\frac{\mathrm{rank}(f(x_k)) - 1/2}{\lambda}\right).$$

We consider the expected $W_\theta^f p$-transformed fitness [44.43]

$$J(\theta) = \mathbb{E}(W_\theta^f p(f(x))) \qquad x \sim p(.|\theta)$$
$$= \int_{\mathbb{R}^n} W_\theta^f p(f(x)) \, p(x|\theta) \mathrm{d}x, \qquad (44.2)$$

where the expectation is taken under the given sample distribution. The maximizer of J w.r.t. $p(.|\theta)$ is, for any fixed $W_\theta^f p$, a Dirac distribution concentrated on the minimizer of f. A natural way to update θ is therefore a gradient ascent step in the $\nabla_\theta J$ direction. However, the *vanilla* gradient $\nabla_\theta J$ depends on the specific parametrization chosen in θ. In contrast, the natural gradient, denoted by $\widetilde{\nabla}_\theta$, is associated to the Fisher metric that is intrinsic to p and independent of the chosen θ-parametrization. Developing $\widetilde{\nabla}_\theta J(\theta)$ under mild assumptions on f and $p(.|\theta)$ by exchanging differentiation and integration, recognizing that the gradient $\widetilde{\nabla}_\theta$ does not act on $W_\theta^f p$, using the log-likelihood trick $\widetilde{\nabla}_\theta p(.|\theta) = p(.|\theta) \widetilde{\nabla}_\theta \ln p(.|\theta)$ and finally setting $\theta' = \theta$ yields

$$\widetilde{\nabla}_\theta J(\theta) = \mathbb{E}\left(W_\theta^f(f(x)) \widetilde{\nabla}_\theta \ln p(x|\theta)\right). \qquad (44.3)$$

We set $\theta' = \theta$ because we will estimate $W_{\theta'}$ using the current samples that are distributed according to $p(.|\theta)$. A Monte Carlo approximation of the expected value by the average finally yields the comparatively simple expression

$$\widetilde{\nabla}_\theta J(\theta) \approx \frac{1}{\lambda} \sum_{k=1}^{\lambda} \overbrace{W_\theta^f(f(x_k))}^{\text{preference weight}} \underbrace{\widetilde{\nabla}_\theta \ln p(x_k|\theta)}_{\text{intrinsic candidate direction}}$$
$$(44.4)$$

for a natural gradient update of θ, where $x_k \sim p(.|\theta)$ is sampled from the current distribution. The natural gradient can be computed as $\widetilde{\nabla}_\theta = \mathbf{F}_\theta^{-1} \nabla_\theta$, where $\mathbf{F}_\theta$ is the Fisher information matrix expressed in θ coordinates. For the multivariate Gaussian distribution, $\widetilde{\nabla}_\theta \ln p(x_k|\theta)$ can indeed be easily expressed and computed efficiently. We find that in CMA-ES (Algorithm 44.7), the rank-μ update (Line 12 with $c_1 = 0$) and the update in Line 10 are natural gradient updates of $\mathbf{C}$ and x, respectively [44.31, 42], where the k-th largest w_k is a consistent estimator for the k-th largest $W_\theta^f(f(x_k))$ [44.43].

While the natural gradient does not depend on the parametrization of the distribution, a finite step taken in the natural gradient direction does. This becomes relevant for the covariance matrix update, where natural ES take a different parametrization than CMA-ES. Starting from Line 12 in Algorithm 44.7, we find for $c_1 = c_h = 0$

$$\mathbf{C} \leftarrow (1 - c_\mu) \mathbf{C} + c_\mu \sum_{z_k \in P} w_k \mathbf{C}^{\frac{1}{2}} z_k (\mathbf{C}^{\frac{1}{2}} z_k)^{\mathrm{T}}$$

$$= \mathbf{C}^{\frac{1}{2}} \left((1 - c_\mu) \mathbf{I} + c_\mu \sum_{z_k \in P} w_k z_k z_k^{\mathrm{T}}\right) \mathbf{C}^{\frac{1}{2}}$$

$$\overset{\sum w_k = 1}{=} \mathbf{C}^{\frac{1}{2}} \left(\mathbf{I} + c_\mu \sum_{z_k \in P} w_k \left(z_k z_k^{\mathrm{T}} - \mathbf{I}\right)\right) \mathbf{C}^{\frac{1}{2}}$$

$$\overset{c_\mu \ll 1}{\approx} \mathbf{C}^{\frac{1}{2}} \exp^{c_\mu} \left(\sum_{z_k \in P} w_k \left(z_k z_k^{\mathrm{T}} - \mathbf{I}\right)\right) \mathbf{C}^{\frac{1}{2}}.$$

$$(44.5)$$

The term bracketed between the matrices $\mathbf{C}^{\frac{1}{2}}$ in the lower three lines is a multiplicative covariance matrix update expressed in the natural coordinates, where the covariance matrix is the identity and $\mathbf{C}^{\frac{1}{2}}$ serves as coordinate system transformation into the given coordinate system. Only the lower two lines of (44.5) do not rely on the constraint $\sum_k w_k = 1$ in order to satisfy a stationarity condition on $\mathbf{C}$. For a given $\mathbf{C}$ on the right-hand side of (44.5), we have under neutral selection the stationarity condition $\mathbb{E}(\mathbf{C}_{\mathrm{new}}) = \mathbf{C}$ for the first three lines and $\mathbb{E}(\log(\mathbf{C}_{\mathrm{new}})) = \log(\mathbf{C})$ for the last line, where $\log$ is the inverse of the matrix exponential $\exp$. The last line of (44.5) is used in the *exponential natural evolution strategy*, xNES [44.31] and guarantees positive definiteness of $\mathbf{C}$ even with negative weights, independent of c_μ and of the data z_k. The xNES is depicted in Algorithm 44.8.

Algorithm 44.8 The Exponential NES (xNES)

1: **given** $n \in \mathbb{N}_+, \quad \lambda \geq 5, \quad w_k = w'(k)/\sum_k |w'(k)|,$
   $w'(k) \approx \log(\lambda/2 + 1/2) - \log \text{rank}(f(\boldsymbol{x}_k)),$
   $\eta_c \approx (5+\lambda)/(5\,n^{1.5}) \leq 1, \, \eta_\sigma \approx \eta_c, \, \eta_x \approx 1$

2: **initialize** $\mathbf{C}^{\frac{1}{2}} = \mathbf{I}, \sigma \in \mathbb{R}_+, \boldsymbol{x} \in \mathbb{R}^n$

3: **while** not happy

4:     **for** $k \in \{1, \ldots, \lambda\}$

5:        $\boldsymbol{z}_k = \mathcal{N}(\mathbf{0}, \mathbf{I})$ // i.i.d. for all k

6:        $\boldsymbol{x}_k = \boldsymbol{x} + \sigma \mathbf{C}^{\frac{1}{2}} \times \boldsymbol{z}_k$

7:     $\mathcal{P} = \{(\boldsymbol{z}_k, f(\boldsymbol{x}_k)) \mid 1 \leq k \leq \lambda\}$

8:     $\boldsymbol{x} \leftarrow \boldsymbol{x} + \eta_x \sigma \, \mathbf{C}^{\frac{1}{2}} \sum_{\boldsymbol{z}_k \in \mathcal{P}} w_k \boldsymbol{z}_k$

9:     $\sigma \leftarrow \sigma \, \exp^{\eta_\sigma/2} \left(\sum_{\boldsymbol{z}_k \in \mathcal{P}} w_k \left(\frac{\|\boldsymbol{z}_k\|^2}{n} - 1 \right) \right)$

10:    $\mathbf{C}^{\frac{1}{2}} \leftarrow \mathbf{C}^{\frac{1}{2}} \times \exp^{\eta_c/2} \left(\sum_{\boldsymbol{z}_k \in \mathcal{P}} w_k \left(\boldsymbol{z}_k \boldsymbol{z}_k^{\mathrm{T}} - \frac{\|\boldsymbol{z}_k\|^2}{n} \mathbf{I} \right) \right)$

In xNES, sampling is identical to CMA-ES and environmental selection is omitted entirely. Line 9 resembles the step-size update in (44.1). Comparing the updates more closely, with $c_\sigma = 1$ (44.1) uses

$$\frac{\mu_w \| \sum_k w_k \boldsymbol{z}_k \|^2}{n} - 1$$

whereas xNES uses

$$\sum_k w_k \left(\frac{\|\boldsymbol{z}_k\|^2}{n} - 1 \right)$$

for updating σ. For $\mu = 1$ the updates are the same. For $\mu > 1$, the latter only depends on the lengths of the $\boldsymbol{z}_k$, while the former depends on their lengths and directions. Finally, xNES expresses the update (44.5) in Line 10 on the Cholesky factor $\mathbf{C}^{\frac{1}{2}}$, which does not remain symmetric in this case ($\mathbf{C} = \mathbf{C}^{\frac{1}{2}} \times \mathbf{C}^{\frac{1}{2}\mathrm{T}}$ still holds). The term $-\|\boldsymbol{z}_k\|^2/n$ keeps the determinant of $\mathbf{C}^{\frac{1}{2}}$ (and thus the trace of $\log \mathbf{C}^{\frac{1}{2}}$) constant and is of rather cosmetic nature. Omitting the term is equivalent to using $\eta_\sigma + \eta_c$ instead of η_σ in Line 9.

The exponential natural evolution strategy is a very elegant algorithm. Like CMA-ES it can be interpreted as an incremental estimation of distribution algorithm [44.44]. However, it performs generally inferior compared to CMA-ES because it does not use search paths for updating σ and $\mathbf{C}$.

44.3.8 Further Aspects

Internal Parameters

Adaptation and self-adaptation address the control of the most important internal parameters in ES. Yet, all algorithms presented have hidden and exposed parameters in their implementation. Many of them can be set to reasonable and robust default values. The population size parameters μ and λ however change the search characteristics of an evolution strategy significantly. Larger values, in particular for parent number μ, often help address highly multimodal or noisy problems successfully.

In practice, several experiments or restarts are advisable, where different initial conditions for $\boldsymbol{x}$ and σ can be employed. For exploring different population sizes, a schedule with increasing population size (IPOP) is advantageous [44.45–47], because runs with larger populations take typically more function evaluations. Preceding long runs (large μ and λ) with short runs (small μ and λ) leads to a smaller (relative) impairment of the later runs than vice versa.

Internal Computational Complexity

Algorithms presented in Sects. 44.3.1–44.3.4 that sample isotropic or axis-parallel mutation distributions have an internal computational complexity linear in the dimension. The internal computational complexity of CMA-ES and xNES is, for constant population size, cubic in the dimension due to the update of $\mathbf{C}^{\frac{1}{2}}$. Typical implementations of the CMA-ES however have quadratic complexity, as they implement a lazy update scheme for $\mathbf{C}^{\frac{1}{2}}$, where $\mathbf{C}$ is decomposed into $\mathbf{C}^{\frac{1}{2}}\mathbf{C}^{\frac{1}{2}}$ only after about n/λ iterations. An exact quadratic update for CMA-ES has also been proposed [44.48]. While never considered in the literature, a lazy update for xNES to achieve quadratic complexity seems feasible as well.

Invariance

Selection and recombination in ES are based solely on the ranks of offspring and parent individuals. As a consequence, the behavior of ES is invariant under order-preserving (strictly monotonous) transformations of the fitness function value. In particular, all spherical unimodal functions belong to the same function class, which the convex-quadratic sphere function is the most pronounced member of. This function is more thoroughly investigated in Sect. 44.4.

All algorithms presented are invariant under translations and Algorithms 44.3, 44.7, and 44.8 are invariant under rotations of the coordinate system, provided that the initial x is translated and rotated accordingly.

Parameter control can introduce yet further invariances. All algorithms presented are scale invariant due to step-size adaptation. Furthermore, ellipsoidal functions that are in the reach of the mutation operator of the ES presented in Sects. 44.3.2–44.3.7 are eventually transformed, effectively, into spherical functions. These

ES are invariant under the respective affine transformations of the search space, given the initial conditions are chosen respectively.

Variants

Evolution strategies have been extended and combined with other approaches in various ways. We mention here constraint handling [44.49, 50], fitness surrogates [44.51], multiobjective variants [44.52, 53] and exploitation of fitness values [44.54].

44.4 Theory

There is ample *empirical* evidence, that on many unimodal functions ES with step-size control, as those outlined in the previous section, converge fast and with probability one to the global optimum. Convergence proofs supporting this evidence are discussed in Sect. 44.4.3. On multimodal functions on the other hand, the probability to converge to the global optimum (in a single run of the same strategy) is generally smaller than one (but larger than zero), as suggested by observations and theoretical results [44.55]. Without parameter control on the other hand, elitist strategies always converge to the essential global optimum, however at a much slower rate (compare random search in Fig. 44.2). On a bounded domain and with mutation variances bounded away from zero, *nonelitist* strategies generate a *subsequence* of x-values converging to the essential global optimum.

In this section, we use a time index t to denote iteration and assume, for notational convenience and without loss of generality (due to translation invariance), that the optimum of f is in $x^* = \mathbf{0}$. This simplifies writing $x^{(t)} - x^*$ to simply $x^{(t)}$ and then $\|x^{(t)}\|$ measures the distance to the optimum of the parental centroid in time step t.

Linear convergence plays a central role for ES. For a *deterministic* sequence $x^{(t)}$ linear convergence (toward zero) takes place if there exists a $c > 0$ such that

$$\lim_{t \to \infty} \frac{\|x^{(t+1)}\|}{\|x^{(t)}\|} = \exp(-c) , \qquad (44.6)$$

which means, loosely speaking, that for t large enough, the distance to the optimum decreases in every step by the constant factor $\exp(-c)$. Taking the logarithm of (44.6), then exchanging the logarithm and the limit

and taking the Cesàro mean yields

$$\lim_{T \to \infty} \underbrace{\frac{1}{T} \sum_{t=0}^{T-1} \log \frac{\|x^{(t+1)}\|}{\|x^{(t)}\|}}_{= \frac{1}{T} \log \|x^{(T)}\| / \|x^{(0)}\|} = -c . \qquad (44.7)$$

For a *sequence of random vectors*, we define linear convergence based on (44.7) as follows.

Definition 44.1 Linear Convergence

The sequence of random vectors $x^{(t)}$ converges almost surely linearly to $\mathbf{0}$ if there exists a $c > 0$ such that

$$-c = \lim_{T \to \infty} \frac{1}{T} \log \frac{\|x^{(T)}\|}{\|x^{(0)}\|} \quad a.s.$$

$$= \lim_{T \to \infty} \frac{1}{T} \sum_{t=0}^{T-1} \log \frac{\|x^{(t+1)}\|}{\|x^{(t)}\|} \quad a.s. \qquad (44.8)$$

The sequence converges *in expectation* linearly to $\mathbf{0}$ if there exists a $c > 0$ such that

$$-c = \lim_{t \to \infty} \mathbb{E} \log \frac{\|x^{(t+1)}\|}{\|x^{(t)}\|} . \qquad (44.9)$$

The constant c is the convergence rate of the algorithm.

Linear convergence, hence, means that asymptotically in t, the logarithm of the distance to the optimum decreases linearly in t like $-ct$. This behavior has been observed in Fig. 44.2 for the (1+1)-ES with 1/5th success rule on a unimodal spherical function.

Note that λ function evaluations are performed per iteration and it is then often useful to consider a convergence rate *per function evaluation*, i.e., to normalize the convergence rate by λ.

The *progress rate* measures the reduction of the distance to optimum within a single generation [44.1].

Definition 44.2 Progress Rate
The normalized progress rate is defined as the expected relative reduction of $\|x^{(t)}\|$

$$\varphi^* = n \, \mathbb{E} \left(\frac{\|x^{(t)}\| - \|x^{(t+1)}\|}{\|x^{(t)}\|} \, \bigg| \, x^{(t)}, s^{(t)} \right)$$

$$= n \left(1 - \mathbb{E} \left(\frac{\|x^{(t+1)}\|}{\|x^{(t)}\|} \, \bigg| \, x^{(t)}, s^{(t)} \right) \right), \quad (44.10)$$

where the expectation is taken over $x^{(t+1)}$, given $(x^{(t)}, s^{(t)})$. In situations commonly considered in theoretical analyses, φ^* does not depend on $x^{(t)}$ and is expressed as a function of strategy parameters $s^{(t)}$.

Definitions 44.1 and 44.2 are related, in that for a given $x^{(t)}$

$$\varphi^* \leq -n \log \mathbb{E} \frac{\|x^{(t+1)}\|}{\|x^{(t)}\|} \quad (44.11)$$

$$\leq -n \, \mathbb{E} \log \frac{\|x^{(t+1)}\|}{\|x^{(t)}\|} = nc \,. \quad (44.12)$$

Therefore, progress rate φ^* and convergence rate nc do not agree and we might observe convergence ($c > 0$) while $\varphi^* < 0$. However for $n \to \infty$, we typically have $\varphi^* = nc$ [44.56].

The normalized progress rate φ^* for ES has been extensively studied in various situations, see Sect. 44.4.2. Scale-invariance and (sometimes artificial) assumptions on the step-size typically ensure that the progress rates do not depend on t.

Another way to describe how fast an algorithm approaches the optimum is to count the number of function evaluations needed to reduce the distance to the optimum by a given factor $1/\epsilon$ or, similarly, the runtime to hit a ball of radius ϵ around the optimum, starting, e.g., from the distance one.

Definition 44.3 Runtime
The runtime is the first hitting time of a ball around the optimum. Specifically, the runtime in number of func-

tion evaluations as a function of ϵ reads

$$\lambda \times \min \left\{ t : \|x^{(t)}\| \leq \epsilon \times \|x^{(0)}\| \right\}$$

$$= \lambda \times \min \left\{ t : \frac{\|x^{(t)}\|}{\|x^{(0)}\|} \leq \epsilon \right\}. \quad (44.13)$$

Linear convergence with rate c as given in (44.9) implies that, for $\epsilon \to 0$, the expected runtime divided by $\log(1/\epsilon)$ goes to the constant λ/c.

44.4.1 Lower Runtime Bounds

Evolution strategies with a fixed number of parent and offspring individuals cannot converge faster than linearly and with a convergence rate of $\mathcal{O}(1/n)$. This means that their runtime is lower bounded by a constant times $\log(1/\epsilon^n) = n \log(1/\epsilon)$ [44.57–61]. This result can be obtained by analyzing the branching factor of the tree of possible paths the algorithm can take. It therefore holds for any optimization algorithm taking decisions based solely on a bounded number of comparisons between fitness values [44.57–59].

More specifically, the runtime of any $(1 \overset{+}{,} \lambda)$-ES with isotropic mutations cannot be asymptotically faster than $\propto n \log(1/\epsilon) \, \lambda/\log(\lambda)$ [44.62]. Considering more restrictive classes of algorithms can provide more precise nonasymptotic bounds [44.60, 61]. Different approaches address in particular the $(1+1)$- and $(1, \lambda)$-ES and precisely characterize the fastest convergence rate that can be obtained with isotropic normal distributions on any objective function with any step-size adaptation mechanism [44.56, 63–65].

Considering the sphere function, the optimal convergence rate is attained with *distance proportional step-size*, that is, a step-size proportional to the distance of the parental centroid to the optimum, $\sigma = \text{const} \times \|x\| = \sigma^* \|x\|/n$. Optimal step-size and optimal convergence rate according to (44.8) and (44.9) can be expressed in terms of expectation of some random variables that are easily simulated numerically. The convergence rate of the $(1+1)$-ES with distance proportional step-size is shown in Fig. 44.4 as a function of the normalized step-size $\sigma^* = n\sigma/\|x\|$. The peak of each curve is the upper bound for the convergence rate that can be achieved on any function with any form of step-size adaptation. As for the general bound, the evolution strategy converges linearly and the convergence rate c decreases to zero like $1/n$ for $n \to \infty$ [44.56, 65, 66], which is equivalent to linear scaling of the runtime in the dimension. The asymptotic limit for the conver-

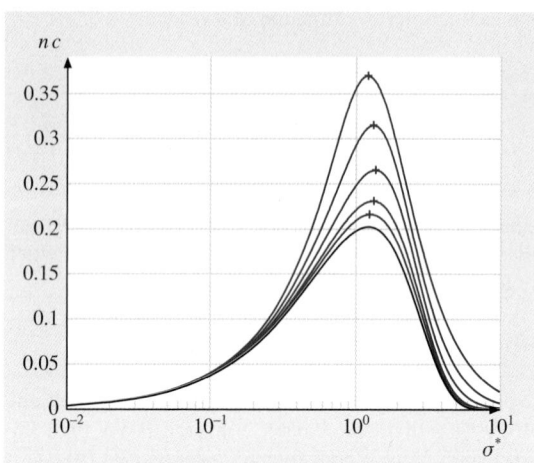

Fig. 44.4 Normalized convergence rate nc versus normalized step-size $n\sigma/\|x\|$ of the (1+1)-ES with distance proportional step-size for $n = 2, 3, 5, 10, 20, \infty$ (*top to bottom*). The *peaks* of the graphs represent the upper bound for the convergence rate of the (1+1)-ES with isotropic mutation (corresponding to the lower runtime bound). The limit curve for n to infinity (*lowest curve*) reveals the optimal normalized progress rate of $\varphi^* \approx 0.202$ of the (1+1)-ES on sphere functions for $n \to \infty$

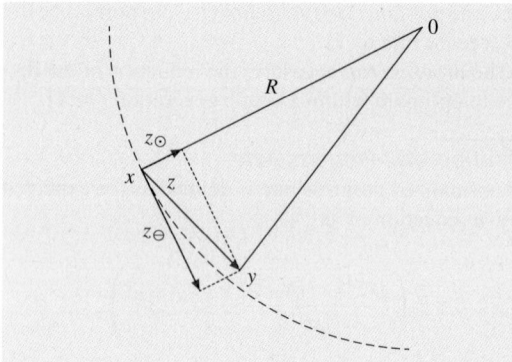

Fig. 44.5 Decomposition of mutation vector z into a component $z_\odot$ in the direction of the negative of the gradient vector of the objective function and a perpendicular component $z_\ominus$

gence rate of the (1+1)-ES, as shown in the lowest curve in Fig. 44.4, coincides with the progress rate expression given in the next section.

44.4.2 Progress Rates

This section presents analytical approximations to progress rates of ES for sphere, ridge, and cigar functions in the limit $n \to \infty$. Both one-generation results and those that consider multiple time steps and cumulative step-size adaptation are considered.

The first analytical progress rate results date back to the early work of *Rechenberg* [44.1] and *Schwefel* [44.3], who considered the sphere and corridor models and very simple strategy variants. Further results have since been derived for various ridge functions, several classes of convex quadratic functions, and more general constrained linear problems. The strategies that results are available for have increased in complexity as well and today include multiparent strategies employing recombination as well as several step-size adaptation mechanisms. Only strategy variants with isotropic mutation distributions have been considered up to this point. However, parameter control strategies that successfully adapt the shape of the mutation

distribution (such as CMA-ES) effectively transform ellipsoidal functions into (almost) spherical ones; thus lending extra relevance to the analysis of sphere and sphere-like functions.

The simplest convex quadratic functions to be optimized are variants of the sphere function (see also the discussion of invariance in Sect. 44.3.8)

$$f(x) = \|x\|^2 = \sum_{i=1}^{n} x_i^2 = R^2 ,$$

where R denotes the distance from the optimal solution. Expressions for the progress rate of ES on sphere functions can be computed by decomposing mutation vectors into two components $z_\odot$ and $z_\ominus$ as illustrated in Fig. 44.5. Component $z_\odot$ is the projection of z onto the negative of the gradient vector ∇f of the objective function. It contributes positively to the fitness of offspring candidate solution

$$y = x + z$$

if and only if

$$-\nabla f(x) \cdot z > 0 .$$

Component $z_\ominus = z - z_\odot$ is perpendicular to the gradient direction and contributes negatively to the offspring fitness. Its expected squared length exceeds that of $z_\odot$ by a factor of $n - 1$. Considering normalized quantities $\sigma^* = \sigma n/R$ and $\varphi^* = \varphi n/R$ allows giving concise mathematical representations of the scaling properties of various ES on spherical functions as shown below. Constant σ^* corresponds to the distance proportional step-size from Sect. 44.4.1.

(1+1)-ES on Sphere Functions

The normalized progress rate of the (1+1)-ES on sphere functions is

$$\varphi^* = \frac{\sigma^*}{\sqrt{2\pi}} e^{-\frac{1}{8}\sigma^{*2}} - \frac{\sigma^{*2}}{4}\left[1 - \mathrm{erf}\left(\frac{\sigma^*}{\sqrt{8}}\right)\right]$$

(44.14)

in the limit of $n \to \infty$ [44.1]. The expression in square brackets is the success probability (i.e., the probability that the offspring candidate solution is superior to its parent and thus replaces it). The first term in (44.14) is the contribution to the normalized progress rate from the component $z_\odot$ of the mutation vector that is parallel to the gradient vector. The second term results from the component $z_\ominus$ that is perpendicular to the gradient direction.

The black curve in Fig. 44.4 illustrates how the normalized progress rate of the (1+1)-ES on sphere functions in the limit $n \to \infty$ depends on the normalized mutation strength. For small normalized mutation strengths, the normalized progress rate is small as the short steps that are made do not yield significant progress. The success probability is nearly one-half. For large normalized mutation strengths, progress is near zero as the overwhelming majority of steps result in poor offspring that are rejected. The normalized progress rate assumes a maximum value of $\varphi^* = 0.202$ at normalized mutation strength $\sigma^* = 1.224$. The range of step-sizes for which close to optimal progress is achieved is referred to as the evolution window [44.1]. In the runs of the (1+1)-ES with constant step-size shown in Fig. 44.2, the normalized step-size initially is to the left of the evolution window (large relative distance to the optimal solution) and in the end to its right (small relative distance to the optimal solution), achieving maximal progress at a point in between.

(μ/μ, λ)-ES on Sphere Functions

The normalized progress rate of the $(\mu/\mu, \lambda)$-ES on sphere functions is described by

$$\varphi^* = \sigma^* c_{\mu/\mu,\lambda} - \frac{\sigma^{*2}}{2\mu}$$

(44.15)

in the limit $n \to \infty$ [44.2]. The term $c_{\mu/\mu,\lambda}$ is the expected value of the average of the μ largest order statistics of λ independent standard normally distributed random numbers. For λ fixed, $c_{\mu/\mu,\lambda}$ decreases with increasing μ. For the fixed truncation ratio μ/λ, $c_{\mu/\mu,\lambda}$ approaches a finite limit value as λ and μ increase [44.8, 15].

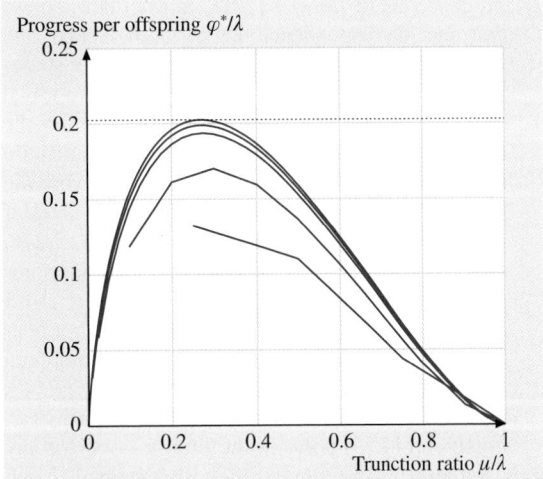

Fig. 44.6 Maximal normalized progress per offspring of the $(\mu/\mu, \lambda)$-ES on sphere functions for $n \to \infty$ plotted against the truncation ratio. The curves correspond to, *from bottom to top*, $\lambda = 4, 10, 40, 100, \infty$. The *dotted line* represents the maximal progress rate of the (1+1)-ES

It is easily seen from (44.15) that the normalized progress rate of the $(\mu/\mu, \lambda)$-ES is maximized by normalized mutation strength

$$\sigma^* = \mu c_{\mu/\mu,\lambda} .$$

(44.16)

The normalized progress rate achieved with that setting is

$$\varphi^* = \frac{\mu c_{\mu/\mu,\lambda}^2}{2} .$$

(44.17)

The progress rate is negative if $\sigma^* > 2\mu c_{\mu/\mu,\lambda}$. Figure 44.6 illustrates how the optimal normalized progress rate per offspring depends on the population size parameters μ and λ. Two interesting observations can be made from the figure:

● For all but the smallest values of λ, the $(\mu/\mu, \lambda)$-ES with $\mu > 1$ is capable of significantly more rapid progress per offspring than the $(1, \lambda)$-ES. This contrasts with findings for the $(\mu/1, \lambda)$-ES, the performance of which on sphere functions for $n \to \infty$ monotonically deteriorates with increasing μ [44.8].
● For large λ, the optimal truncation ratio is $\mu/\lambda = 0.27$, and the corresponding progress per offspring is 0.202. Those values are identical to the optimal success probability and resulting normalized

progress rate of the (1+1)-ES. *Beyer* [44.8] shows that the correspondence is no coincidence and indeed exact. The step-sizes that the two strategies employ differ widely, however. The optimal step-size of the (1+1)-ES is 1.224; that of the $(\mu/\mu,\lambda)$-ES is $\mu c_{\mu/\mu,\lambda}$ and for fixed truncation ratio μ/λ increases (slightly superlinearly) with the population size. For example, optimal step-sizes of $(\mu/\mu,4\mu)$-ES for $\mu \in \{1,2,3\}$ are 1.029, 2.276, and 3.538, respectively. If offspring candidate solutions can be evaluated in parallel, the $(\mu/\mu,\lambda)$-ES is preferable to the (1+1)-ES, which does not benefit from the availability of parallel computational resources.

Equation (44.15) holds in the limit $n \to \infty$ for any finite value of λ. In finite but high dimensional search spaces, it can serve as an approximation to the normalized progress rate of the $(\mu/\mu,\lambda)$-ES on sphere functions in the vicinity of the optimal step-size provided that λ is not too large. A better approximation for finite n is derived in [44.8, 15] (however compare also [44.56]).

The improved performance of the $(\mu/\mu,\lambda)$-ES for $\mu > 1$ compared to the strategy that uses $\mu = 1$ is a consequence of the factor μ in the denominator of the term in (44.15) that contributes negatively to the normalized progress rate. The components $z_\ominus$ of mutation vectors selected for survival are correlated and likely to point in the direction opposite to the gradient vector. The perpendicular components $z_\ominus$ in the limit $n \to \infty$ have no influence on whether a candidate solution is selected for survival and are thus uncorrelated. The recombinative averaging of mutation vectors results in a length of the $z_\ominus$-component similar to those of individual mutation vectors. However, the squared length of the components perpendicular to the gradient direction is reduced by a factor of μ, resulting in the reduction of the negative term in (44.15) by a factor of μ. *Beyer* [44.15] has coined the term *genetic repair* for this phenomenon.

Weighted recombination (compare Algorithms 44.7 and 44.8) can significantly increase the progress rate of $(\mu/\mu,\lambda)$-ES on sphere functions. If n is large, the k-th best candidate solution is optimally associated with a weight proportional to the expected value of the k-th largest order statistic of a sample of λ independent standard normally distributed random numbers. The resulting optimal normalized progress rate per offspring candidate solution for large values of λ then approaches a value of 0.5, exceeding that of optimal unweighted recombination by a factor of almost two and

a half [44.13]. The weights are symmetric about zero. If only positive weights are employed and $\mu = \lfloor \lambda/2 \rfloor$, the optimal normalized progress rate per offspring with increasing λ approaches a value of 0.25. The weights in Algorithms 44.7 and 44.8 closely resemble those positive weights.

$(\mu/\mu,\lambda)$-ES on Noisy Sphere Functions

Noise in the objective function is most commonly modeled as being Gaussian. If evaluation of a candidate solution $\boldsymbol{x}$ yields a noisy objective function value $f(\boldsymbol{x}) + \sigma_\epsilon \mathcal{N}(0,1)$, then inferior candidate solutions will sometimes be selected for survival and superior ones discarded. As a result, progress rates decrease with increasing noise strength σ_ϵ. Introducing normalized noise strength $\sigma_\epsilon^* = \sigma_\epsilon n/(2R^2)$, in the limit $n \to \infty$, the normalized progress rate of the $(\mu/\mu,\lambda)$-ES on noisy sphere functions is

$$\varphi^* = \frac{\sigma^* c_{\mu/\mu,\lambda}}{\sqrt{1+\vartheta^2}} - \frac{\sigma^{*2}}{2\mu}, \tag{44.18}$$

where $\vartheta = \sigma_\epsilon^*/\sigma^*$ is the noise-to-signal ratio that the strategy operates under [44.67]. Noise does not impact the term that contributes negatively to the strategy's progress. However, it acts to reduce the magnitude of the positive term stemming from the contributions of mutation vectors parallel to the gradient direction. Note that unless the noise scales such that σ_ϵ^* is independent of the location in search space (i.e., the standard deviation of the noise term increases in direct proportion to $f(\boldsymbol{x})$, such as in a multiplicative noise model with constant noise strength), (44.18) describes progress in single time steps only rather than a rate of convergence.

Figure 44.7 illustrates for different offspring population sizes λ how the optimal progress rate per offspring depends on the noise strength. The curves have been obtained from (44.18) for optimal values of σ^* and μ. As the averaging of mutation vectors results in a vector of reduced length, increasing λ (and μ along with it) allows the strategy to operate using larger and larger step-sizes. Increasing the step-size reduces the noise-to-signal ratio ϑ that the strategy operates under and thereby reduces the impact of noise on selection for survival. Through genetic repair, the $(\mu/\mu,\lambda)$-ES thus implicitly implements the rescaling of mutation vectors proposed in [44.2] for the $(1,\lambda)$-ES in the presence of noise. Compare c_m and η_x in Algorithms 44.7 and 44.8 that, for values smaller than one, implement the explicit rescaling. It needs to be emphasized though that in finite-dimensional search spaces, the ability to inrease λ without violating the assumptions made in the derivation of (44.18) is severely limited. Nonetheless, the benefits resulting from genetic repair are significant, and the performance of the $(\mu/\mu, \lambda)$-ES is much more robust in the presence of noise than that of the (1+1)-ES.

Cumulative Step–Size Adaptation

All progress rate results discussed up to this point consider single time steps of the respective ES only. Analyses of the behavior of ES that include some form of step-size adaptation are considerably more difficult. Even for objective functions as simple as sphere functions, the state of the strategy is described by several variables with nonlinear, stochastic dynamics, and simplifying assumptions need to be made in order to arrive at quantitative results.

In the following, we consider the $(\mu/\mu, \lambda)$-ES with cumulative step-size adaptation (Algorithm 44.6 with (44.1) in place of Line 9 for mathematical convenience) and parameters set such that $c_\sigma \to 0$ as $n \to \infty$ and $d = \Theta(1)$. The state of the strategy on noisy sphere functions with $\sigma_\epsilon^* = $ const (i.e., noise that decreases in strength as the optimal solution is approached) is described by the distance R of the parental centroid from the optimal solution, normalized step-size σ^*, the

length of the search path s parallel to the direction of the gradient vector of the objective function, and that path's overall squared length. After initialization effects have faded, the distribution of the latter three quantities is time invariant. Mean values of the time invariant distribution can be approximated by computing expected values of the variables after a single iteration of the strategy in the limit $n \to \infty$ and imposing the condition that those be equal to the respective values before that iteration. Solving the resulting system of equations for $\sigma_\epsilon^* \leq \sqrt{2}\mu c_{\mu/\mu,\lambda}$ yields

$$\sigma^* = \mu c_{\mu/\mu,\lambda} \sqrt{2 - \left(\frac{\sigma_\epsilon^*}{\mu c_{\mu/\mu,\lambda}}\right)^2} \tag{44.19}$$

for the average normalized mutation strength assumed by the strategy [44.69, 70]. The corresponding normalized progress rate

$$\varphi^* = \frac{\sqrt{2}-1}{2}\mu c_{\mu/\mu,\lambda}{}^2 \left[2 - \left(\frac{\sigma_\epsilon^*}{\mu c_{\mu/\mu,\lambda}}\right)^2\right] \tag{44.20}$$

is obtained from (44.18). Both the average mutation strength and the resulting progress rate are plotted against the noise strength in Fig. 44.8. For small noise strengths, cumulative step-size adaptation generates mutation strengths that are larger than optimal. The evolution window continually shifts toward smaller values of the step-size, and adaptation remains behind its target. However, the resulting mutation strengths achieve progress rates within 20% of optimal ones. For large noise strengths, the situation is reversed and the mutation strengths generated by cumulative step-size adaptation are smaller than optimal. However, increasing the population size parameters μ and λ allows shifting the operating regime of the strategy toward the left-hand side of the graphs in Fig. 44.8, where step-sizes are near optimal. As above, it is important to keep in mind the limitations of the results derived in the limit $n \to \infty$. In finite-dimensional search spaces the ability to compensate for large amounts of noise by increasing the population size is more limited than (44.19) and (44.20) suggest.

Parabolic Ridge Functions

A class of test functions that poses difficulties very different from those encountered in connection with

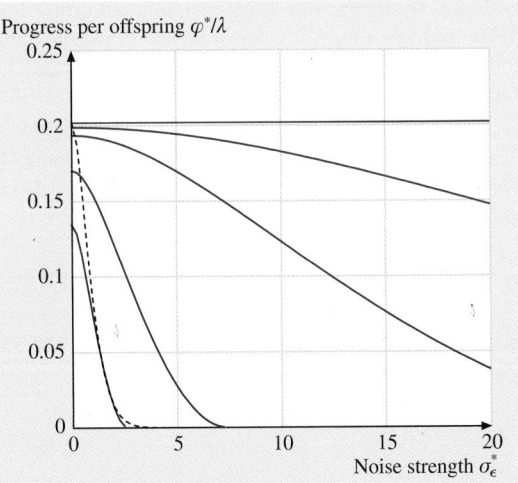

Fig. 44.7 Optimal normalized progress rate per offspring of the $(\mu/\mu, \lambda)$-ES on noisy sphere functions for $n \to \infty$ plotted against the normalized noise strength. The *solid lines* depict results for, from bottom to top, $\lambda = 4, 10, 40, 100, \infty$ and optimally chosen μ. The *dashed line* represents the optimal progress rate of the (1+1)-ES (after [44.68])

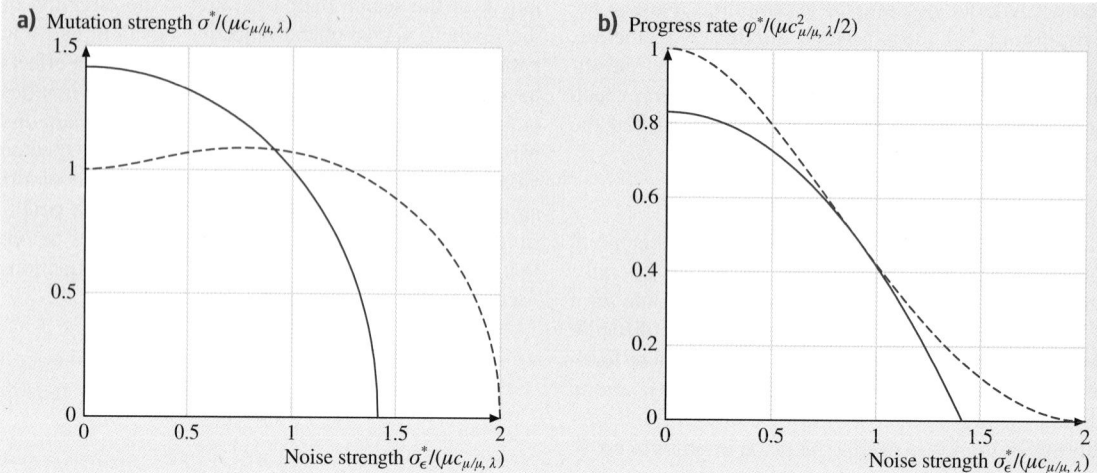

Fig. 44.8a,b Normalized mutation strength and normalized progress rate of the $(\mu/\mu, \lambda)$-ES with cumulative step size adaptation on noisy sphere functions for $n \to \infty$ plotted against the normalized noise strength. The *dashed lines* depic optimal values

sphere functions are ridge functions,

$$f(x) = x_1 + \xi \left(\sum_{i=2}^{n} x_i^2 \right)^{\alpha/2} = x_1 + \xi R^\alpha \,,$$

which include the parabolic ridge for $\alpha = 2$. The x_1-axis is referred to as the ridge axis, and R denotes the distance from that axis. Progress can be made by minimizing the distance from the ridge axis or by proceeding along it. The former requires decreasing step-sizes and is limited in its effect as $R \geq 0$. The latter allows indefinite progress and requires that the step-size does not decrease to zero. Short- and long-term goals may thus be conflicting, and inappropriate step-size adaptation may lead to stagnation.

As an optimal solution to the ridge problem does not exist, the progress rate φ of the $(\mu/\mu, \lambda)$-ES on ridge functions is defined as the expectation of the step made in the direction of the negative ridge axis. For constant step-size, the distance R of the parental centroid from the ridge axis assumes a time-invariant limit distribution. An approximation to the mean value of that distribution can be obtained by identifying that value of R for which the expected change is zero. Using this value yields

$$\varphi = \frac{2\mu c_{\mu/\mu,\lambda}^2}{n\xi(1 + \sqrt{1 + (2\mu c_{\mu/\mu,\lambda}/(n\xi\sigma))^2})} \qquad (44.21)$$

for the progress rate of the $(\mu/\mu, \lambda)$-ES on parabolic ridge functions [44.71]. The strictly monotonic behav-

ior of the progress rate, increasing from a value o zero for $\sigma = 0$ to $\varphi = \mu c_{\mu/\mu,\lambda}^2/(n\xi)$ for $\sigma \to \infty$, i fundamentally different from that observed on sphere functions. However, the derivative of the progress rate with regard to the step-size for large values of σ tends to zero. The limited time horizon of any search as well as the intent of using ridge functions as local rather than global models of practically relevant objective functions both suggest that it may be unwise to increase the step-size without bounds.

The performance of cumulative step-size adaptation on parabolic ridge functions can be studied using the same approach as described above for sphere functions yielding

$$\sigma = \frac{\mu c_{\mu/\mu,\lambda}}{\sqrt{2n\xi}} \qquad (44.22)$$

for the (finite) average mutation strength [44.72]. From (44.21), the corresponding progress rate

$$\varphi = \frac{\mu c_{\mu/\mu,\lambda}^2}{2n\xi} \qquad (44.23)$$

is greater than half of the progress rate attained with any finite step size.

Cigar Functions

While parabolic ridge functions provide an environment for evaluating whether step-size adaptation mechanisms are able to avoid stagnation, the ability to make continual meaningful positive progress with some constant nonzero step-size is, of course, atypical for practical optimization problems. A class of ridge-like functions that requires continual adaptation of the mutation strength and is thus a more realistic model of problems requiring ridge following are cigar functions:

$$f(\boldsymbol{x}) = x_1^2 + \xi \sum_{i=2}^{n} x_i^2 = x_1^2 + \xi R^2 ,$$

with parameter $\xi \geq 1$ being the condition number of the Hessian matrix. Small values of ξ result in sphere-like characteristics, large values in ridge-like ones. As above, R measures the distance from the x_1-axis.

Assuming successful adaptation of the step-size, ES exhibit linear convergence on cigar functions. The expected relative per iteration change in the objective function value of the population centroid is referred to as the quality gain Δ and determines the rate of convergence. In the limit $n \to \infty$ it is described by

$$\Delta^* = \begin{cases} \dfrac{\sigma^{*2}}{2\mu(\xi-1)} & \text{if } \sigma^* < 2\mu c_{\mu/\mu,\lambda} \dfrac{\xi-1}{\xi} \\[3ex] c_{\mu/\mu,\lambda}\sigma^* - \dfrac{\sigma^{*2}}{2\mu} & \text{otherwise} , \end{cases}$$

where $\sigma^* = \sigma n/R$ and $\Delta^* = \Delta n/2$ [44.73]. That relationship is illustrated in Fig. 44.9 for several values of the conditioning parameter. The parabola for $\xi = 1$ reflects the simple quadratic relationship for sphere functions seen in (44.15). (For the case of sphere functions, normalized progress rate and normalized quality gain are the same.) For cigar functions with large values of ξ, two separate regimes can be identified. For small step-sizes, the quality gain of the strategy is limited by the size of the steps that can be made in the direction of the x_1-axis. The x_1-component of the population centroid virtually never changes sign. The search process resembles one of ridge following, and we refer to that regime as the ridge regime. In the other regime, the step-size is such that the quality gain of the strategy is effectively limited by the ability to approach the optimal solution in the subspace spanned by the $x_2, \ldots, x_n$-axes. The x_1-component of the population centroid changes sign much more frequently than in the ridge regime, as is the case on sphere functions. We thus refer to the regime as the sphere regime.

The approach to the analysis of the behavior of cumulative step-size adaptation explained above for

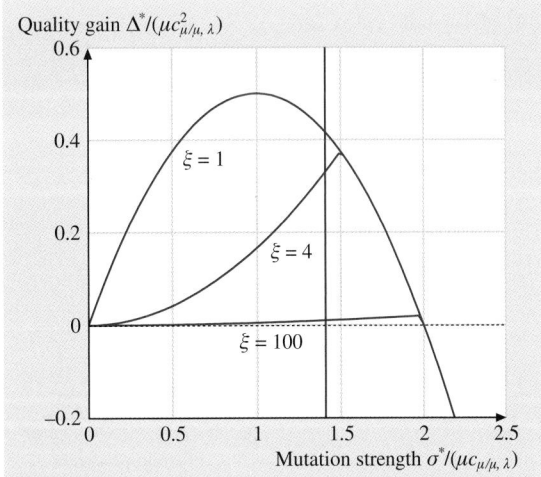

Fig. 44.9 Normalized quality gain of $(\mu/\mu,\lambda)$-ES on cigar functions for $n \to \infty$ plotted against the normalized mutation strength for $\xi \in \{1, 4, 100\}$. The *vertical line* represents the average normalized mutation strength generated by cumulative step-size adaptation

sphere and parabolic ridge functions can be applied to cigar functions as well, yielding

$$\sigma^* = \sqrt{2}\mu c_{\mu/\mu,\lambda}$$

for the average normalized mutation strength generated by cumulative step-size adaptation [44.73]. The corresponding normalized quality gain is

$$\Delta^* = \begin{cases} (\sqrt{2}-1)\mu c_{\mu/\mu,\lambda}^2 & \text{if } \xi < \dfrac{\sqrt{2}}{\sqrt{2}-1} \\[3ex] \dfrac{\mu c_{\mu/\mu,\lambda}^2}{\xi-1} & \text{otherwise} . \end{cases}$$

Both are compared with optimal values in Fig. 44.10. For small condition numbers, $(\mu/\mu,\lambda)$-ES operate in the sphere regime and are within 20% of the optimal quality gain as seen earlier. For large condition numbers, the strategy operates in the ridge regime and achieves a quality gain within a factor of 2 of the optimal one, in accordance with the findings for parabolic ridge above.

Further Work

Further research regarding the progress rate of ES in different test environments includes work analyzing the behavior of mutative self-adaptation for linear [44.22],

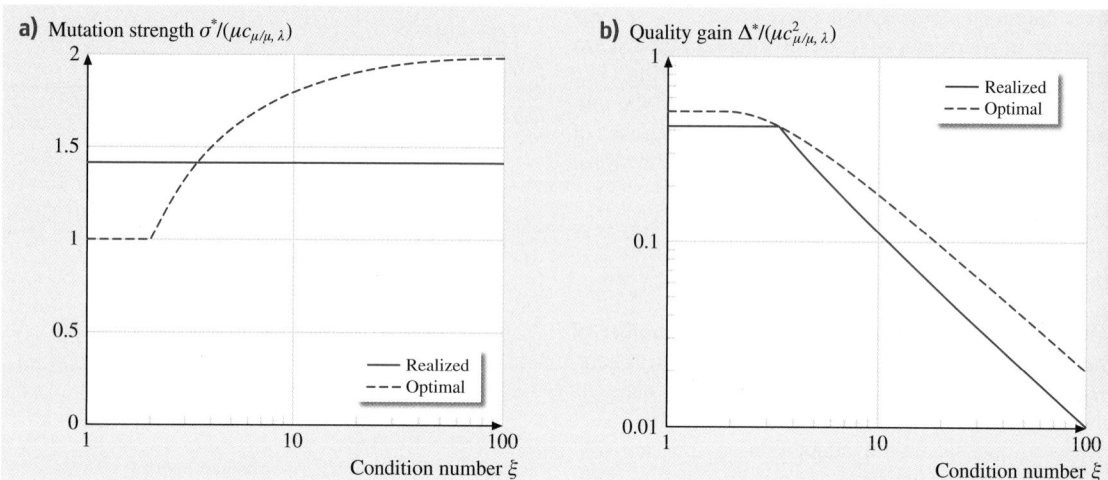

a) Mutation strength $\sigma^*/(\mu c_{\mu/\mu,\lambda})$

b) Quality gain $\Delta^*/(\mu c_{\mu/\mu,\lambda}^2)$

Condition number ξ

Condition number ξ

Realized
--- Optimal

Fig. 44.10a,b Normalized mutation strength and normalized quality gain of the $(\mu/\mu, \lambda)$-ES with cumulative step-size adaptation on cigar functions for $n \to \infty$ plotted against the condition number of the cigar. The *dashed curves* represent optimal values

spherical [44.74], and ridge functions [44.75]. Hierarchically organized ES have been studied when applied to both parabolic ridge and sphere functions [44.76, 77]. Several step-size adaptation techniques have been compared for ridge functions, including, but not limited to, parabolic ones [44.78]. A further class of convex quadratic functions for which quality gain results have been derived is characterized by the occurrence of only two distinct eigenvalues of the Hessian, both of which occur with high multiplicity [44.79, 80].

An analytical investigation of the behavior of the (1+1)-ES on noisy sphere functions finds that failure to re-evaluate the parental candidate solution results in the systematic overvaluation of the parent and thus in potentially long periods of stagnation [44.68]. Contrary to what might be expected, the increased difficulty of replacing parental candidate solutions can have a positive effect on progress rates as it tends to prevent the selection for survival of offspring candidate solutions solely due to favorable noise values. The convergence behavior of the (1+1)-ES on finite-dimensional sphere functions is studied by *Jebalia* et al. [44.81] who show that the additive noise model is inappropriate in finite dimensions unless the parental candidate solution is re-evaluated, and who suggest a multiplicative noise model instead. An analysis of the behavior of (μ, λ)-ES (without recombination) for noisy sphere functions finds that in contrast to the situation in the absence of noise, strategies with $\mu > 1$ can outperform $(1, \lambda)$-ES if there is noise present [44.82]. The use of nonsingle-

ton populations increases the signal-to-noise ratio and thus allows for more effective selection of good candidate solutions. The effects of non-Gaussian forms of noise on the performance of $(\mu/\mu, \lambda)$-ES applied to the optimization of sphere functions have also been investigated [44.83].

Finally, there are some results regarding the optimization of time-varying objectives [44.84] as well as analyses of simple constraint handling techniques [44.85–87].

44.4.3 Convergence Proofs

In the previous section, we have described theoretical results that involve approximations in their derivation and consider the limit for $n \to \infty$. In this section, exact results are discussed.

Convergence proofs with only mild assumptions on the objective function are easy to obtain for ES with a step-size that is effectively bounded from below and above (and, for nonelitist strategies, when additionally the search space is bounded) [44.12, 64]. In this case, the expected runtime to reach an ϵ-ball around the global optimum (see also Definition 44.3) cannot be faster than $\propto 1/\epsilon^n$, as obtained with pure random search for $\epsilon \to 0$ or $n \to \infty$. If the mutation distribution is not normal and exhibits a singularity in zero, convergence can be much faster than with random search even when the step-size is bounded away from zero [44.88]. Similarly, convergence proofs can be obtained for adapive strategies that include provisions for using a fixed step-size and covariance matrix with some constant probability.

Convergence proofs for strategy variants that do not explicitly ensure that long steps are sampled for a sufficiently long time typically require much stronger restrictions on the set of objective functions that they hold for. Such proofs, however, have the potential to reveal much faster, namely linear convergence. Evolution strategies with the *artificial* distance proportional step-size, $\sigma = \text{const} \times \|x\|$, exhibit, as shown above, linear convergence on the sphere function with an associated runtime proportional to $\log(1/\epsilon)$ [44.63, 65, 81, 89]. This result can be easily proved by using a law of large numbers, because $\|x^{(t+1)}\|/\|x^{(t)}\|$ are independent and identically distributed for all t.

Without the artificial choice of step-size, $\sigma/\|x\|$ becomes a random variable. If this random variable is a homogeneous Markov chain and stable enough to satisfy the law of large numbers, linear convergence is maintained [44.64, 89]. The stability of the Markov chain associated with the self-adaptive $(1, \lambda)$-ES on the sphere function has been shown in dimension

$n = 1$ [44.90] providing thus a proof of linear convergence of this algorithm. The extension of this proof to higher dimensions is straightforward.

Proofs that are formalized by upper bounds on the time to reduce the distance to the optimum by a given factor can also associate the linear dependency of the convergence rate in the dimension n. The $(1 + \lambda)$- and the $(1, \lambda)$-ES with common variants of the $1/5$th success rule converge linearly on the sphere function with a runtime of $\mathcal{O}(n \log(1/\epsilon) \lambda / \sqrt{\log \lambda})$ [44.62, 91]. When λ is smaller than $\mathcal{O}(n)$, the $(1 + \lambda)$-ES with a modified success rule is even $\sqrt{\log \lambda}$ times faster and therefore matches the general lower runtime bound $\Omega(n \log(1/\epsilon) \lambda / \log(\lambda))$ [44.62, Theorem 5]. On convex-quadratic functions, the asymptotic runtime of the $(1+1)$-ES is the same as on the sphere function and, at least in some cases, proportional to the condition number of the problem [44.92].

Convergence proofs of modern ES with recombination, of CSA-ES, CMA-ES, or xNES are not yet available; however, we believe that some of them are likely to be achieved in the coming decade.

Part E | 44

References

44.1   I. Rechenberg: *Evolutionstrategie: Optimierung technischer Systeme nach Prinzipien der biologischen Evolution* (Frommann-Holzboog, Stuttgart 1973)

44.2   I. Rechenberg: *Evolutionsstrategie '94* (Frommann-Holzboog, Stuttgart 1994)

44.3   H.-P. Schwefel: *Numerische Optimierung von Computer-Modellen mittels der Evolutionsstrategie* (Birkhäuser, Basel 1977)

44.4   H.-P. Schwefel: *Evolution and Optimum Seeking* (Wiley, New York 1995)

44.5   L.J. Fogel, A.J. Owens, M.J. Walsh: *Artificial Intelligence through Simulated Evolution* (Wiley, New York 1966)

44.6   H.-G. Beyer, H.-P. Schwefel: Evolution strategies – A comprehensive introduction, Nat. Comp. **1**(1), 3–52 (2002)

44.7   D.B. Fogel: *The Fossil Record* (Wiley, New York 1998)

44.8   H.-G. Beyer: *The Theory of Evolution Strategies* (Springer, Berlin, Heidelberg 2001)

44.9   D.E. Goldberg: *Genetic Algorithms in Search, Optimization and Machine Learning* (Addison Wesley, Reading 1989)

44.10   N. Hansen, A. Ostermeier: Completely derandomized self-adaptation in evolution strategies, Evol. Comp. **9**(2), 159–195 (2001)

44.11   H.-P. Schwefel, G. Rudolph: Contemporary evolution strategies. In: *Advances Artificial Life*, ed. by F. Morán, A. Moreno, J.J. Merelo, P. Chacón (Springer, Berlin, Heidelberg 1995) pp. 891–907

44.12   G. Rudolph: *Convergence Properties of Evolutionary Algorithms* (Dr. Kovač, Hamburg 1997)

44.13   D.V. Arnold: Weighted multirecombination evolution strategies, Theor. Comp. Sci. **361**(1), 18–37 (2006)

44.14   H. Mühlenbein, D. Schlierkamp-Voosen: Predictive models for the breeder genetic algorithm I. Continuous parameter optimization, Evol. Comp. **1**(1), 25–49 (1993)

44.15   H.-G. Beyer: Toward a theory of evolution strategies: On the benefits of sex – The $(\mu/\mu, \lambda)$ theory, Evol. Comp. **3**(1), 81–111 (1995)

44.16   C. Kappler: Are evolutionary algorithms improved by large mutations?, Lect. Notes Comput. Sci. **1141**, 346–355 (1996)

44.17   G. Rudolph: Local convergence rates of simple evolutionary algorithms with Cauchy mutations, IEEE Trans. Evol. Comp. **1**(4), 249–258 (1997)

44.18   X. Yao, Y. Liu, G. Lin: Evolutionary programming made faster, IEEE Trans. Evol. Comp. **3**(2), 82–102 (1999)

44.19 M. Herdy: The number of offspring as strategy parameter in hierarchically organized evolution strategies, ACM SIGBIO Newsl. **13**(2), 2–9 (1993)

44.20 M. Schumer, K. Steiglitz: Adaptive step size random search, IEEE Trans. Autom. Control **13**(3), 270–276 (1968)

44.21 S. Kern, S.D. Müller, N. Hansen, D. Büche, J. Ocenasek, P. Koumoutsakos: Learning probability distributions in continuous evolutionary algorithms – A comparative review, Nat. Comput. **3**(1), 77–112 (2004)

44.22 N. Hansen: An analysis of mutative σ-self-adaptation on linear fitness functions, Evol. Comp. **14**(3), 255–275 (2006)

44.23 A. Ostermeier, A. Gawelczyk, N. Hansen: A derandomized approach to self-adaptation of evolution strategies, Evol. Comp. **2**(4), 369–380 (1994)

44.24 T. Runarsson: Reducing random fluctuations in mutative self-adaptation, Lect. Notes Comput. Sci. **2439**, 194–203 (2002)

44.25 A. Ostermeier, A. Gawelczyk, N. Hansen: Step-size adaptation based on non-local use of selection information, Lect. Notes Comput. Sci. **866**, 189–198 (1994)

44.26 N. Hansen, A. Ostermeier, A. Gawelczyk: On the adaptation of arbitrary normal mutation distributions in evolution strategies: The generating set adaptation, Int. Conf. Gen. Algorith., ed. by L.J. Eshelman (Morgan Kaufmann, San Francisco 1995) pp. 57–64

44.27 N. Hansen, A. Ostermeier: Adapting arbitrary normal mutation distributions in evolution strategies: The covariance matrix adaptation, IEEE Int. Conf. Evol. Comp. (1996) pp. 312–317

44.28 A. Ostermeier, N. Hansen: An evolution strategy with coordinate system invariant adaptation of arbitrary normal mutation distributions within the concept of mutative strategy parameter control, Proc. Genet. Evol. Comput. Conf. (1999) pp. 902–909

44.29 D. Wierstra, T. Schaul, J. Peters, J. Schmidhuber: Natural evolution strategies, IEEE Cong. Evol. Comp. (CEC 2008) (2008) pp. 3381–3387

44.30 Y. Sun, D. Wierstra, T. Schaul, J. Schmidhuber: Efficient natural evolution strategies, Proc. Genet. Evol. Comput. Conf. (2009) pp. 539–546

44.31 T. Glasmachers, T. Schaul, Y. Sun, D. Wierstra, J. Schmidhuber: Exponential natural evolution strategies, Proc. Genet. Evol. Comput. Conf. (2010) pp. 393–400

44.32 N. Hansen: Invariance, self-adaptation and correlated mutations and evolution strategies, Lect. Notes Comput. Sci. **1917**, 355–364 (2000)

44.33 A. Ostermeier: An evolution strategy with momentum adaptation of the random number distribution. In: *Proc. 2nd Conf. Parallel Probl. Solving Nat.*, ed. by R. Männer, B. Manderick (North-Holland, Amsterdam 1992) pp. 199–208

44.34 J. Poland, A. Zell: Main vector adaptation: A CMA variant with linear time and space complexity, Proc. Genet. Evol. Comput. Conf. (2001) pp. 1050–1055

44.35 J.N. Knight, M. Lunacek: Reducing the space-time complexity of the CMA-ES, Proc. Genet. Evol. Comput. Conf. (2007) pp. 658–665

44.36 N. Hansen, S. Kern: Evaluating the CMA evolution strategy on multimodal test functions, Lect. Notes Comput. Sci. **3242**, 282–291 (2004)

44.37 H.G. Beyer, B. Sendhoff: Covariance matrix adaptation revisited – The CMSA evolution strategy, Lect. Notes Comput. Sci. **3199**, 123–132 (2008)

44.38 G.A. Jastrebski, D.V. Arnold: Improving evolution strategies through active covariance matrix adaptation, IEEE Cong. Evol. Comp. (CEC 2006) (2006) pp. 2814–2821

44.39 H.-G. Beyer: Mutate large, but inherit small! On the analysis of rescaled mutations in $(\tilde{1}, \tilde{\lambda})$-ES with noisy fitness data, Lect. Notes Comput. Sci. **1498**, 109–118 (1998)

44.40 S.I. Amari: Natural gradient works efficiently in learning, Neural Comput. **10**(2), 251–276 (1998)

44.41 Y. Sun, D. Wierstra, T. Schaul, J. Schmidhuber: Stochastic search using the natural gradient, Int. Conf. Mach. Learn., ed. by A.P. Danyluk, L. Bottou, M.L. Littman (2009) pp. 1161–1168

44.42 Y. Akimoto, Y. Nagata, I. Ono, S. Kobayashi: Bidirectional relation between CMA evolution strategies and natural evolution strategies, Lect. Notes Comput. Sci. **6238**, 154–163 (2010)

44.43 L. Arnold, A. Auger, N. Hansen, Y. Ollivier: Information-geometric optimization algorithms: A unifying picture via invariance principles, ArXiv e-prints (2011), DOI arXiv:1106.3708

44.44 M. Pelikan, M.W. Hausschild, F.G. Lobo: *Introduction to estimation of distribution algorithms* MEDAL Rep. No. 2012003 (University of Missouri, St Louis 2012)

44.45 G.R. Harik, F.G. Lobo: A parameter-less genetic algorithm, Proc. Genet. Evol. Comput. Conf. (1999) pp. 258–265

44.46 F.G. Lobo, D.E. Goldberg: The parameter-less genetic algorithm in practice, Inf. Sci. **167**(1), 217–232 (2004)

44.47 A. Auger, N. Hansen: A restart CMA evolution strategy with increasing population size, IEEE Cong. Evol. Comp. (CEC 2005) (2005) pp. 1769–1776

44.48 T. Suttorp, N. Hansen, C. Igel: Efficient covariance matrix update for variable metric evolution strategies, Mach. Learn. **75**(2), 167–197 (2009)

44.49 Z. Michalewicz, M. Schoenauer: Evolutionary algorithms for constrained parameter optimization problems, Evol. Comp. **4**(1), 1–32 (1996)

44.50 E. Mezura-Montes, C.A. Coello Coello: Constraint-handling in nature-inspired numerical optimization: Past, present, and future, Swarm Evol. Comp. **1**(4), 173–194 (2011)

44.51 M. Emmerich, A. Giotis, M. Özdemir, T. Bäck, K. Giannakoglou: Metamodel-assisted evolution strategies, Lect. Notes Comput. Sci. **2439**, 361–370 (2002)

44.52 C. Igel, N. Hansen, S. Roth: Covariance matrix adaptation for multi-objective optimization, Evol. Comp. **15**(1), 1–28 (2007)

44.53 N. Hansen, T. Voß, C. Igel: Improved step size adaptation for the MO-CMA-ES, Proc. Genet. Evol. Comput. Conf. (2010) pp. 487–494

44.54 R. Salomon: Evolutionary algorithms and gradient search: Similarities and differences, IEEE Trans. Evol. Comp. **2**(2), 45–55 (1998)

44.55 G. Rudolph: Self-adaptive mutations may lead to premature convergence, IEEE Trans. Evol. Comp. **5**(4), 410–414 (2001)

44.56 A. Auger, N. Hansen: Reconsidering the progress rate theory for evolution strategies in finite dimensions, Proc. Genet. Evol. Comput. Conf. (2006) pp. 445–452

44.57 O. Teytaud, S. Gelly: General lower bounds for evolutionary algorithms, Lect. Notes Comput. Sci. **4193**, 21–31 (2006)

44.58 H. Fournier, O. Teytaud: Lower bounds for comparison based evolution strategies using VC-dimension and sign patterns, Algorithmica **59**(3), 387–408 (2011)

44.59 O. Teytaud: Lower bounds for evolution strategies. In: *Theory of Randomized Search Heuristics: Foundations and Recent Developments*, ed. by A. Auger, B. Doerr (World Scientific Publ., Singapore 2011) pp. 327–354

44.60 J. Jägersküpper: Lower bounds for hit-and-run direct search, Lect. Notes Comput. Sci. **4665**, 118–129 (2007)

44.61 J. Jägersküpper: Lower bounds for randomized direct search with isotropic sampling, Oper. Res. Lett. **36**(3), 327–332 (2008)

44.62 J. Jägersküpper: Probabilistic runtime analysis of $(1 + \lambda)$ evolution strategies using isotropic mutations, Proc. Genet. Evol. Comput. Conf. (2006) pp. 461–468

44.63 M. Jebalia, A. Auger, P. Liardet: Log-linear convergence and optimal bounds for the (1+1)-ES, Lect. Notes Comput. Sci. **4926**, 207–218 (2008)

44.64 A. Auger, N. Hansen: Theory of evolution strategies: A new perspective. In: *Theory of Randomized Search Heuristics: Foundations and Recent Developments*, ed. by A. Auger, B. Doerr (World Scientific Publ., Singapore 2011) pp. 289–325

44.65 A. Auger, D. Brockhoff, N. Hansen: Analyzing the impact of mirrored sampling and sequential selection in elitist evolution strategies, Proc. 11th Workshop Found. Gen. Algorith. (2011) pp. 127–138

44.66 A. Auger, D. Brockhoff, N. Hansen: Mirrored sampling in evolution strategies with weighted recombination, Proc. Genet. Evol. Comput. Conf. (2011) pp. 861–868

44.67 D.V. Arnold, H.-G. Beyer: Local performance of the $(\mu/\mu_I, \lambda)$-ES in a noisy environment, Proc. 11th Workshop Found. Gen. Algorith. (2001) pp. 127–141

44.68 D.V. Arnold, H.-G. Beyer: Local performance of the $(1 + 1)$-ES in a noisy environment, IEEE Trans. Evol. Comp. **6**(1), 30–41 (2002)

44.69 D.V. Arnold: *Noisy Optimization with Evolution Strategies* (Kluwer Academic, Boston 2002)

44.70 D.V. Arnold, H.-G. Beyer: Performance analysis of evolutionary optimization with cumulative step length adaptation, IEEE Trans. Autom. Control **49**(4), 617–622 (2004)

44.71 A.I. Oyman, H.-G. Beyer: Analysis of the $(\mu/\mu, \lambda)$-ES on the parabolic ridge, Evol. Comp. **8**(3), 267–289 (2000)

44.72 D.V. Arnold, H.-G. Beyer: Evolution strategies with cumulative step length adaptation on the noisy parabolic ridge, Nat. Comput. **7**(4), 555–587 (2008)

44.73 D.V. Arnold, H.-G. Beyer: On the behaviour of evolution strategies optimising cigar functions, Evol. Comp. **18**(4), 661–682 (2010)

44.74 H.-G. Beyer: Towards a theory of evolution strategies: Self-adaptation, Evol. Comp. **3**, 3 (1995)

44.75 S. Meyer-Nieberg, H.-G. Beyer: Mutative self-adaptation on the sharp and parabolic ridge, Lect. Notes Comput. Sci. **4436**, 70–96 (2007)

44.76 D.V. Arnold, A. MacLeod: Hierarchically organised evolution strategies on the parabolic ridge, Proc. Genet. Evol. Comput. Conf. (2006) pp. 437–444

44.77 H.-G. Beyer, M. Dobler, C. Hämmerle, P. Masser: On strategy parameter control by meta-ES, Proc. Genet. Evol. Comput. Conf. (2009) pp. 499–506

44.78 D.V. Arnold, A. MacLeod: Step length adaptation on ridge functions, Evol. Comp. **16**(2), 151–184 (2008)

44.79 D.V. Arnold: On the use of evolution strategies for optimising certain positive definite quadratic forms, Proc. Genet. Evol. Comput. Conf. (2007) pp. 634–641

44.80 H.-G. Beyer, S. Finck: Performance of the $(\mu/\mu_I, \lambda)$-σSA-ES on a class of PDQFs, IEEE Trans. Evol. Comp. **14**(3), 400–418 (2010)

44.81 M. Jebalia, A. Auger, N. Hansen: Log-linear convergence and divergence of the scale-invariant (1+1)-ES in noisy environments, Algorithmica **59**(3), 425–460 (2011)

44.82 D.V. Arnold, H.-G. Beyer: On the benefits of populations for noisy optimization, Evol. Comp. **11**(2), 111–127 (2003)

44.83 D.V. Arnold, H.-G. Beyer: A general noise model and its effects on evolution strategy performance, IEEE Trans. Evol. Comp. **10**(4), 380–391 (2006)

44.84 D.V. Arnold, H.-G. Beyer: Optimum tracking with evolution strategies, Evol. Comp. **14**(3), 291–308 (2006)

44.85 D.V. Arnold, D. Brauer: On the behaviour of the (1 + 1)-ES for a simple constrained problem, Lect. Notes Comput. Sci. **5199**, 1–10 (2008)

Part E | 44

44.86   D.V. Arnold: On the behaviour of the $(1, \lambda)$-ES for a simple constrained problem, Lect. Notes Comput. Sci. **5199**, 15–24 (2011)

44.87   D.V. Arnold: Analysis of a repair mechanism for the $(1, \lambda)$-ES applied to a simple constrained problem, Proc. Genet. Evol. Comput. Conf. (2011) pp. 853–860

44.88   A. Anatoly Zhigljavsky: *Theory of Global Random Search* (Kluwer Academic, Boston 1991)

44.89   A. Bienvenüe, O. François: Global convergence for evolution strategies in spherical problems: Some simple proofs and difficulties, Theor. Comp. Sci. **306**(1-3), 269–289 (2003)

44.90   A. Auger: Convergence results for $(1, \lambda)$-SA-ES using the theory of φ-irreducible Markov chains, Theor. Comp. Sci. **334**(1-3), 35–69 (2005)

44.91   J. Jägersküpper: Algorithmic analysis of a basic evolutionary algorithm for continuous optimization, Theor. Comp. Sci. **379**(3), 329–347 (2007)

44.92   J. Jägersküpper: How the (1+1) ES using isotropic mutations minimizes positive definite quadratic forms, Theor. Comp. Sci. **361**(1), 38–56 (2006)

45. Estimation of Distribution Algorithms

Martin Pelikan, Mark W. Hauschild, Fernando G. Lobo

Estimation of distribution algorithms (EDAs) guide the search for the optimum by building and sampling explicit probabilistic models of promising candidate solutions. However, EDAs are not only optimization techniques; besides the optimum or its approximation, EDAs provide practitioners with a series of probabilistic models that reveal a lot of information about the problem being solved. This information can in turn be used to design problem-specific neighborhood operators for local search, to bias future runs of EDAs on similar problems, or to create an efficient computational model of the problem. This chapter provides an introduction to EDAs as well as a number of pointers for obtaining more information about this class of algorithms.

Part E | 45

Estimation of distribution algorithms (EDAs) [45.1–8], also called probabilistic model-building genetic algorithms (PMBGAs) and iterated density estimation evolutionary algorithms (IDEAs), view optimization as a series of incremental updates of a probabilistic model, starting with the model encoding the uniform distribution over admissible solutions and ending with the model that generates only the global optima. In the past decade and a half, EDAs have been applied to many challenging optimization problems [45.9–21]. In many of these studies, EDAs were shown to solve problems that were intractable with other techniques or no other technique could achieve comparable results. However, the motive for the use of EDAs in practice is not only that these algorithms can solve difficult optimization problems, but that in addition to the optimum or its approximation EDAs provide practitioners with a compact computational model of the problem represented by a series of probabilistic models [45.22–24]. These probabilistic models reveal a lot of information about the problem domain, which can in turn be used to bias optimization of similar problems, create problem-specific neighborhood operators, and many other tasks. While many metaheuristics exist that essentially sample implicit probability distributions by using a combination of stochastic search operators, the insight into the problem represented by a series of explicit probabilistic models of promising candidate solutions gives EDAs an edge over most of other metaheuristics.

This chapter provides an introduction to EDAs. Additionally, the chapter presents numerous pointers for obtaining additional information about this class of algorithms.

The chapter is organized as follows. Section 45.1 outlines the basic procedure of an EDA. Section 45.2 presents a taxonomy of EDAs based on the type of decomposition encoded by the model and the type of local distributions used in the model. Section 45.3 reviews some of the most popular EDAs. Section 45.4 discusses major research directions and the past results in theoretical modeling of EDAs. Section 45.5 focuses on efficiency enhancement techniques for EDAs. Section 45.6 gives pointers for obtaining additional information about EDAs. Section 45.7 summarizes and concludes the chapter.

45.1 Basic EDA Procedure

45.1.1 Problem Definition

An optimization problem may be defined by specifying (1) a set of potential solutions to the problem and (2) a procedure for evaluating the quality of these solutions. The set of potential solutions is often defined using a general representation of admissible solutions and a set of constraints. The procedure for evaluating the quality of candidate solutions can either be defined as a function that is to be minimized or maximized (often referred to as an objective function or fitness function) or as a partial ordering operator. The task is to find a solution from the set of potential solutions that maximizes quality as defined by the evaluation procedure.

As an example, let us consider the quadratic assignment problem (QAP), which is one of the fundamental NP-hard combinatorial problems [45.25]. In QAP, the input consists of distances between n locations and flows between n facilities. The task is to find a one-to-one assignment of facilities to locations so that the overall cost is minimized. The cost for a pair of locations is defined as the product of the distance between these locations and the flow between the facilities assigned to these locations; the overall cost is the sum of the individual costs for all pairs of locations. Therefore, in QAP, potential solutions are defined as permutations that define assignments of facilities to locations and the solution quality is evaluated using the cost function discussed above. The task is to minimize the cost. As another example, consider the maximum satisfiability problem for propositional logic formulas defined in the conjunctive normal form with 3 literals per clause (MAX3SAT). In MAX3SAT, each potential solution

defines one interpretation of propositions (making each proposition either true or false), and the quality of a solution is measured by the number of clauses that are satisfied by the specific interpretation. The task is to find an interpretation that maximizes the number of satisfied clauses.

Without additional assumptions about the problem, one way to find the optimum is to repeat three main steps:

- *Generate* candidate solutions.
- *Evaluate* the generated solutions.
- *Update* the procedure for generating new candidate solutions according to the results of the evaluation.

Ideally, the quality of generated solutions would improve over time and after a reasonable number of iterations, the execution of these three steps would generate the global optimum or its accurate approximation. Different algorithms implement the above three steps in different ways, but the key idea remains the same – iteratively update the procedure for generating candidate solutions so that generated candidate solutions continually improve in quality.

45.1.2 EDA Procedure

In EDAs, the central idea is to maintain an *explicit probabilistic model* that represents a probability distribution over candidate solutions. In each iteration, the model is adjusted based on the results of the evaluation of candidate solutions so that it will generate better candidate solutions in the subsequent iterations. Note that using an explicit probabilistic model makes EDAs quite diferent from many other metaheuristics, such as genetic algorithms [45.26, 27] or simulated annealing [45.28, 29], in which the probability distribution used to generate new candidate solutions is often defined *implicitly* by a search operator or a combination of several search operators. Researchers often distinguish two main types of EDAs:

● *Population-based EDAs*. Population-based EDAs maintain a population (multiset) of candidate solutions, starting with a population generated at random according to the uniform distribution over all admissible solutions. Each iteration starts by creating a population of promising candidate solutions using the selection operator, which gives preference to solutions of higher quality. Any popular selection method for evolutionary algorithms can be used, such as truncation or tournament selection [45.30, 31]. For example, truncation selection selects the top $\tau\%$ members of the population. A probabilistic model is then built for the selected solutions. New solutions are created by sampling the distribution encoded by the built model. The new solutions are then incorporated into the original population using a replacement operator. In full replacement, for example, the entire original population of candidate solutions is replaced by the new ones. A pseudocode of a population-based EDA is shown in Algorithm 45.1.

● *Incremental EDAs*. In incremental EDAs, the population of candidate solutions is fully replaced by a probabilistic model. The model is initialized so that it encodes the uniform distribution over all admissible solutions. The model is then updated incrementally by repeating the process of (1) sampling several candidate solutions from the current model and (2) adjusting the model based on the evaluation of these candidate solutions and their comparison so that the model becomes more likely to generate high-quality solutions in subsequent iterations. A pseudocode of an incremental EDA is shown in Algorithm 45.2.

Algorithm 45.1 Population-based estimation of distribution algorithm

1: $t \leftarrow 0$
2: generate population $P(0)$ of random solutions
3: **while** termination criteria not satisfied, repeat **do**
4:    evaluate all candidate solutions in $P(t)$
5:    select promising solutions $S(t)$ from $P(t)$
6:    build a probabilistic model $M(t)$ for $S(t)$
7:    generate new solutions $O(t)$ by sampling $M(t)$
8:    create $P(t+1)$ by combining $O(t)$ and $P(t)$
9:    $t \leftarrow t+1$
10: **end while**

Algorithm 45.2 Incremental estimation of distribution algorithm

1: $t \leftarrow 0$
2: initialize model $M(0)$ to represent the uniform distribution over admissible solutions
3: **while** termination criteria not satisfied, repeat **do**
4:    generate population $P(t)$ of candidate solutions by sampling $M(t)$
5:    evaluate all candidate solutions in $P(t)$
6:    create new model $M(t+1)$ by adjusting $M(t)$ according to evaluated $P(t)$
7:    $t \leftarrow t+1$
8: **end while**

Incremental EDAs often generate only a few candidate solutions at a time, whereas population-based EDAs often work with a large population of candidate solutions, building each model from scratch. Nonetheless, it is easy to see that the two approaches are essentially the same because even the population-based EDAs can be reformulated in an incremental-based manner.

The main components of a population-based EDA thus include:

(1) A selection operator for selecting promising solutions.
(2) An assumed class of probabilistic models to use for modeling and sampling.
(3) A procedure for learning a probabilistic model for the selected solutions.
(4) A procedure for sampling the built probabilistic model.
(5) A replacement operator for combining the populations of old and new candidate solutions.

The main components of an incremental EDA include:

(1) An assumed class of probabilistic models.
(2) A procedure for adjusting the probabilistic model based on new candidate solutions and their evaluations.
(3) A procedure for sampling the probabilistic model.

Part E | 45.1

The procedure for learning a probabilistic model usually requires two subcomponents: a metric for evaluating the probabilistic models from the assumed class, and a search procedure for choosing a particular model based on the metric used. EDAs differ mainly in the class of probabilistic models and the procedures used for evaluating candidate models and searching for a good model.

The general outline of a population-based EDA is quite similar to that of a traditional evolutionary algorithm (EA) [45.32]; both guide the search toward promising solutions by iteratively performing selection and variation, the two key ingredients of any EA. In particular, components (1) and (5) are precisely the same as those used in other EAs. Components (2), (3), and (4), however, are unique to EDAs, and constitute their way of producing variation, as opposed to using recombination and mutation operators as is often done with other EAs.

As we shall see, this alternative perspective opens a way for designing search procedures from principled grounds by bringing to the evolutionary computation domain a vast body of knowledge from the machine learning literature, and in particular from probabilistic graphical models. The key idea of EDAs is to look at a population of previously visited good solutions as data, learn a model (or theory) of that data, and use the resulting model to infer where other good solutions might be. This approach is powerful, allowing a search algorithm to learn and adapt itself with respect to the optimization problem being solved, while it is being solved.

45.1.3 Simulation of an EDA by Hand

To better understand the EDA procedure, this section presents a simple EDA simulation by hand. The purpose of presenting the simulation is to clarify the components of the basic EDA procedure and to build intuition about the dynamics of an EDA run.

The simulation assumes that candidate solutions are represented by binary strings of fixed length $n > 0$. The objective function to maximize is onemax, which is defined as the sum of the bits in the input binary string $(X_1, X_2, \ldots, X_n)$

$$f_{\text{onemax}}(X_1, X_2, \ldots, X_n) = \sum_{i=1}^{n} X_i . \tag{45.1}$$

The quality of a candidate solution improves with the number of 1s in the input string, and the optimum is the string of all 1s.

To model and sample candidate solutions, the simulation uses a *probability vector* [45.1, 6, 33]. A probability vector p for n-bit binary strings has n components $p = (p_1, p_2, \ldots, p_n)$. The component p_i represents the probability of observing a 1 in position i of a solution string. To learn the probability vector, p_i is set to the proportion of 1s in position i observed in the selected set of solutions. To sample a new candidate solution $(X_1, X_2, \ldots, X_n)$, the components of the probability vector are polled and each X_i is set to 1 with probability p_i and to 0 with probability $1 - p_i$.

The expected outcome of the learning and sampling of the probability vector is that the population of selected solutions and the population of new candidate solutions have the same proportion of 1s in each position. However, since the sampling considers each new candidate solution independently of others, the actual proportions may vary a little from their expected values. The probability-vector EDA described above is typically referred to as the *univariate marginal distribution algorithm* (UMDA) [45.6]; other EDAs based on the probability vector model [45.1, 33, 34] will be discussed in Sect. 45.3.1.

To keep the simulation simple, we consider a 5-bit onemax, a population of size $N = 6$, and truncation selection with threshold $\tau = 50\%$. Recall that the truncation selection with $\tau = 50\%$ selects the top half of the current population.

Figure 45.1 shows the first two iterations of the EDA simulation. The initial population of candidate solutions is generated at random. Truncation selection then selects the best 50% of candidate solutions based on their evaluation using onemax to form the set of promising solutions. Next, the probability vector is created based on the selected solutions and the distribution encoded by the probability vector is sampled to generate new candidate solutions. The resulting population replaces the original population and the procedure repeats.

In both iterations of the simulation, the average objective-function value in the new population is greater than the average value in the population before selection. The increase in the average quality of the population is good news for us because we want to maximize the objective function, but why does this happen? Since for onemax the solutions with more 1s are better than those with fewer 1s, selection should increase the number of 1s in the population. The learning and sampling of the probability vector is not expected to create or destroy any bits and that is why the new population of candidate solutions should contain more 1s than the

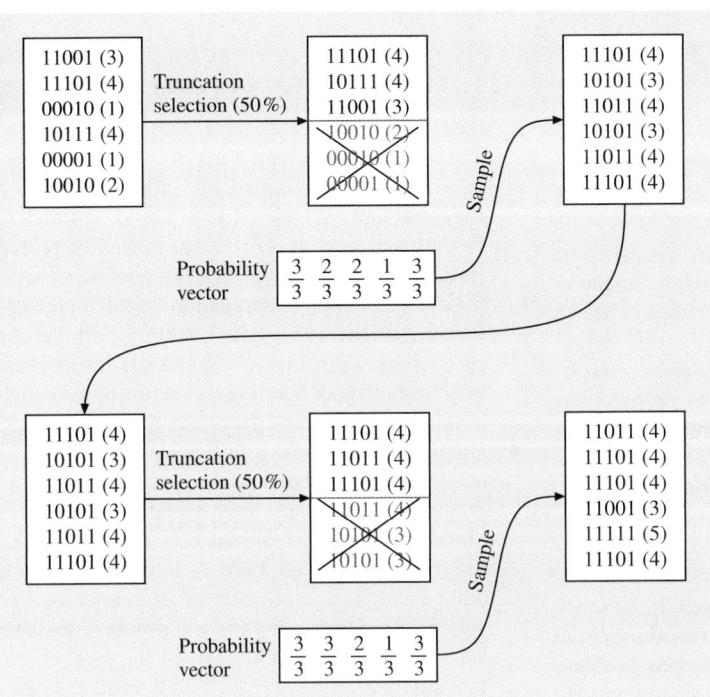

Fig. 45.1 Simple simulation of an EDA based on the probability-vector model for onemax. The fitness values of candidate solutions are shown inside parentheses

original population (both in the proportion and in the actual number). Since onemax value increases with the number of 1s, we can expect the overall quality of the population to increase over time. Ideally, every iteration should increase the objective-function values in the population unless no improvement is possible.

Nonetheless, the increase of the average objective-function value tells only half the story. A similar increase in the quality of the population in the first iteration would be achieved by just repeating selection alone without the use of the probabilistic model. However, by applying selection alone, no new solutions are ever created and the resulting algorithm produces no variation at all (i.e., there is no exploration of new candidate solutions). Since the initial population is generated at random, the EDA with selection alone would be just a poor algorithm for obtaining the best solution from the initial population. The learning and sampling of the probabilistic model provides a mechanism for

both (1) improving quality of new candidate solutions (under certain assumptions), and (2) facilitating exploration of the set of admissible solutions.

What we have seen in this simulation was an example of the simplest kind of EDAs. The assumed class of probabilistic models, the probability vector, has a fixed structure. Under these circumstances, the procedure for learning it becomes trivial because there are really no alternative models to choose from. This class of EDAs is quite limited in what it can do. As we shall see in a moment, there are other classes of EDAs that allow richer probabilistic models capable of capturing interactions among the variables of a given problem. More importantly, these interactions can be learned automatically on a problem by problem basis. This results of course in a more complex model building procedure, but the extra effort has been shown to be well worth it, especially when solving more difficult optimization problems [45.4, 5, 8, 22, 35–37].

45.2 Taxonomy of EDA Models

This section provides a high-level overview of the distinguishing characteristics of probabilistic models. The characteristics are discussed with respect to (1) the

types of interactions covered by the model and (2) the types of local distributions. This section only focuses on the key characteristics of the probabilistic models;

a more detailed overview of EDAs for various representations of candidate solutions will be covered by the following sections.

45.2.1 Classification Based on Problem Decomposition

To make the estimation and sampling tractable with reasonable sample sizes, most EDAs use probabilistic models that *decompose* the problem using unconditional or conditional independence. The way in which a model decomposes the problem provides one important characteristic that distinguishes different classes of probabilistic models. Classification of probabilistic models based on the way they decompose a problem is relevant regardless of the types of the underlying distributions or the representation of problem variables.

Most EDAs assume that candidate solutions are represented by fixed-length vectors of variables and they use graphical models to represent the underlying problem structure. Graphical models allow practitioners to represent both direct dependencies between problem variables as well as independence assumptions. One way to classify graphical models is to consider a hierarchy of model types based on the complexity of a model (see Fig. 45.2 for illustrative examples) [45.3, 4, 7]:

- *No dependencies.* In models that assume full independence, every variable is assumed to be independent of any other variable. That is, the probability distribution $P(X_1, X_2, \ldots, X_n)$ of the vector $(X_1, X_2, \ldots, X_n)$ of n variables is assumed to consist of a product of the distributions of individual variables

$$P(X_1, X_2, \ldots, X_n) = \prod_{i=1}^{n} P(X_i) . \tag{45.2}$$

The simulation presented in Sect. 45.1.3 was based on a model that assumed full independence of binary problem variables. EDAs based on univariate models that assume full independence of problem variables include the equilibrium genetic algorithm (EGA) [45.33], the population-based incremental learning (PBIL) [45.1], the UMDA [45.6], the compact genetic algorithm (cGA) [45.34], the stochastic hill climbing with learning by vectors of normal distributions [45.38], and the continuous PBIL [45.39].

- *Pairwise dependencies.* In this class of models, dependencies between variables form a tree or forest graph. In a tree graph, each variable except for the root of the tree is conditioned on its parent in a tree that contains all variables. A forest graph, on the other hand, is a collection of disconnected trees. Again, the forest contains all problem variables. Denoting by R the set of roots of the trees in a forest, and by $X = (X_1, X_2, \ldots, X_n)$ the entire vector of variables, the distribution from this class can be expressed as

$$P(X_1, X_2, \ldots, X_n)$$
$$= \prod_{X_i \in R} P(X_i)$$
$$\times \prod_{X_i \in X \setminus R} P(X_i | \text{parent}(X_i)) . \tag{45.3}$$

A special type of a tree model is sometimes distinguished, in which the variables form a sequence

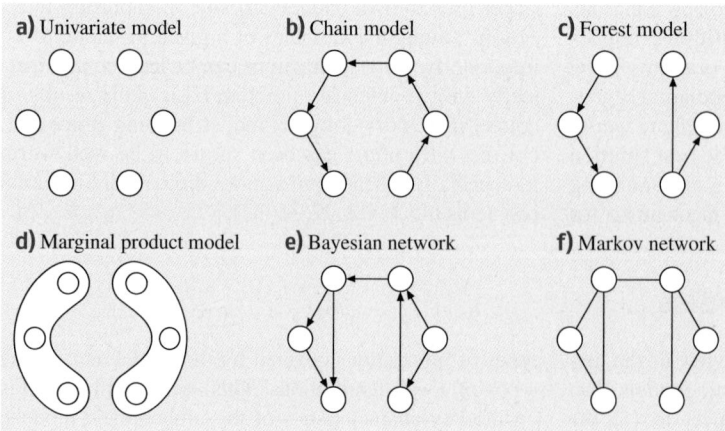

a) Univariate model   **b)** Chain model   **c)** Forest model

d) Marginal product model   **e)** Bayesian network   **f)** Markov network

Fig. 45.2a–f Illustrative examples of graphical models. Problem variables are displayed as circles and dependencies are shown as edges between variables or clusters of variables. (**a**) Univariate model. (**b**) Chain model. (**c**) Forest model. (**d**) Marginal product model. (**e**) Bayesian network. (**f**) Markov network

(or a chain), and each variable except for the first one depends directly on its predecessor. Denoting by $\pi(i)$ the index of the ith variable in the sequence, the distribution is given by

$$P(X_1, X_2, \ldots, X_n) = P(X_{\pi(1)}) \prod_{i=2}^{n} \times P(X_{\pi(i)} | X_{\pi(i-1)}) .$$

$$(45.4)$$

EDAs based on models with pairwise dependencies include the mutual information maximizing input clustering (MIMIC) [45.36], EDA based on dependency trees [45.35], and the bivariate marginal distribution algorithm (BMDA) [45.40].

• *Multivariate dependencies.* Multivariate models represent dependencies using either directed acyclic graphs or undirected graphs. Two representative models are popular in EDAs: (1) Bayesian networks and (2) Markov networks. A Bayesian network is represented by a directed acyclic graph where each node corresponds to a variable and each edge defines a direct conditional dependence. The probability distribution encoded by a Bayesian network can be written as

$$P(X_1, X_2, \ldots, X_n) = \prod_{i=1}^{n} P(X_i | \mathrm{parents}(X_i)) .$$

$$(45.5)$$

A Bayesian network represents problem decomposition by conditional independence assumptions; each variable is assumed to be independent of any of its antecedents in the ancestral ordering of the variables, given the values of the variable's parents. Note that all models discussed thus far were special cases of Bayesian networks. In fact, a Bayesian network can represent an arbitrary multivariate distribution; however, for such a model to be practical, it is often desirable to consider Bayesian networks of limited complexity.

In Markov networks (Markov random field models), two variables are assumed to be independent of each other given a subset of variables defining the condition if every path between these variables is separated by one or more variables in the condition. A special subclass of multivariate models is sometimes considered in which the variables are divided into disjoint clusters, which are independent of each

other. These models are called marginal product models (MPM). Polytrees also represent a subclass of multivariate models in which a directed acyclic graph is used as the basic dependency structure but the graph is restricted so that at most one undirected path exists between any two vertices.

EDAs based on models with multivariate dependencies include the factorized distribution algorithm (FDA) [45.37], the learning FDA (LFDA) [45.37], the estimation of Bayesian network algorithm (EBNA) [45.41], the Bayesian optimization algorithm (BOA) [45.42, 43] and its hierarchical version (hBOA) [45.44], the extended compact genetic algorithm (ECGA) [45.45], the polytree EDA [45.46], the continuous iterated density estimation algorithm [45.47], the estimation of multivariate normal algorithm (EMNA) [45.48], and the real-coded BOA (rBOA) [45.49].

• *Full dependence.* Models may be used that do not make *any* independence assumptions. However, such models must typically impose a number of other restrictions on the distribution to ensure that the models remain tractable for a moderate-to-large number of variables.

There are two additional types of probabilistic models that have been used in EDAs and that provide a somewhat different mechanism for decomposing the problem:

• *Grammar models.* Some EDAs use stochastic or deterministic grammars to represent the probability distribution over candidate solutions. The advantage of grammars is that they allow modeling of variable-length structures. Because of this, grammar distributions are mostly used as the basis for implementing genetic programming using EDAs [45.50], which represents candidate solutions using labeled trees of variable size. Grammar models are used, for example, in the probabilistic-grammar based EDA for genetic programming [45.51], the program distribution estimation with grammar model (PRODIGY) [45.52], or the EDA based on probabilistic grammars with latent annotations [45.53].

• *Feature-based models.* Feature-based models encode the distribution of the neighborhood of a candidate solution using position-independent substructures, which can be found in a variety of positions in fixed-length or variable-length solutions. This approach is used in the feature-based

BOA [45.54]. Other features may be discovered, encoded, and used for guiding the exploration of the space of candidate solutions. Model-directed neighborhood structures are also used in other EDA variants, as will be discussed in Sect. 45.5.2.

45.2.2 Classification Based on Local Distributions in Graphical Models

Regardless of how a graphical model decomposes the problem, each model must also assume one or more classes of distributions to encode local conditional and marginal distributions. Some of the most common classes of local distributions are discussed below:

- *Probability tables.* For discrete representations, conditional and marginal probabilities can be encoded using *probability tables*, which define a probability for each relevant combination of values in each conditional or marginal probability term. This was the case, for example, in the simulation in Sect. 45.1.3, in which the probability distribution for each string position i was represented by the probability p_i of a 1; the probability of a 0 in the same position was simply $1 - p_i$. As another example, in Bayesian networks, for each variable a probability table can be used to define conditional probabilities of any value of the variable given any combination of values of the variable's parents. While probability tables cannot directly represent continuous probability distributions, they can be used even for real-valued representations in combination with a discretization method that maps real-valued variables into discrete categories; each of the discrete categories can then be represented using a single probability entry. Probability tables are used, for example, in UMDA [45.6], BOA [45.43] and ECGA [45.45]. An example conditional probability table is shown in Fig. 45.3.

- *Decision trees or graphs, default tables.* To avoid excessively large probability tables when many probabilities are either similar or negligible, more advanced local structures such as decision trees, decision graphs, or default tables may be used. In decision trees, for example, probabilities are stored in leaves of a decision tree in which each internal node represents a test on a variable and the children of the node correspond to the different outcomes of the test. Decision trees and decision graphs can also be used in combination with real-valued variables in which the leaves store a continuous distribution in some way. More advanced structures such as decision trees and decision graphs are used, for example, in the decision-graph BOA (dBOA) [45.55], the hierarchical BOA (hBOA) [45.44], and the mixed BOA (mBOA) [45.56]. An example decision tree for representing conditional probabilities is shown in Fig. 45.3.

- *Multivariate, continuous distributions.* The normal distribution is by far the most popular distribution used in EDAs to represent univariate or multivariate distributions of real-valued variables. A multivariate normal distribution can encode a linear correlation between the variables using the covariance matrix, but it is often inefficient in representing many other types of interactions [45.56, 57]. Normal distributions were used in many EDAs for real-valued vectors [45.38, 39, 47, 48], although in many real-valued EDAs more advanced distributions were used as well. Examples of multivariate normal distributions are shown in Fig. 45.4a–c.

- *Mixtures of distributions.* A mixture distribution consists of multiple components. Each component is represented by a specific local probabilistic model, such as a normal distribution, and each component is assigned a probability. Mixture distributions were used in EDAs especially to enable EDAs for real-valued representations to deal with real-valued distributions with multiple basins of attraction, in which a single-peak distribution does not suffice. Mixture distributions were used for example, in the real-valued iterated density estimation algorithms [45.47] or the real-coded BOA [45.49]. The use of mixture distributions is more popular in EDAs for real-valued representations, although mixture distributions were also

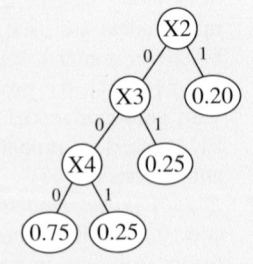

X2	X3	X4	p(X1 | X2, X3, X4)
0	0	0	0.75
0	0	0	0.25
0	0	0	0.25
0	0	0	0.25
1	1	1	0.20
1	1	1	0.20
1	1	1	0.20
1	1	1	0.20

Fig. 45.3 A conditional probability table for $p(X_1|X_2, X_3, X_4)$ and a corresponding decision tree that reduces the number of parameters (probabilities) from 8 to 4

a) Multivariate normal distribution with equal standard deviations and no covariance

b) Multivariate normal distribution with arbitrary standard deviations for each variable (diagonal covariance matrix)

c) Multivariate normal distribution with an arbitrary (nondiagonal) covariance matrix

d) Joint normal kernels distribution

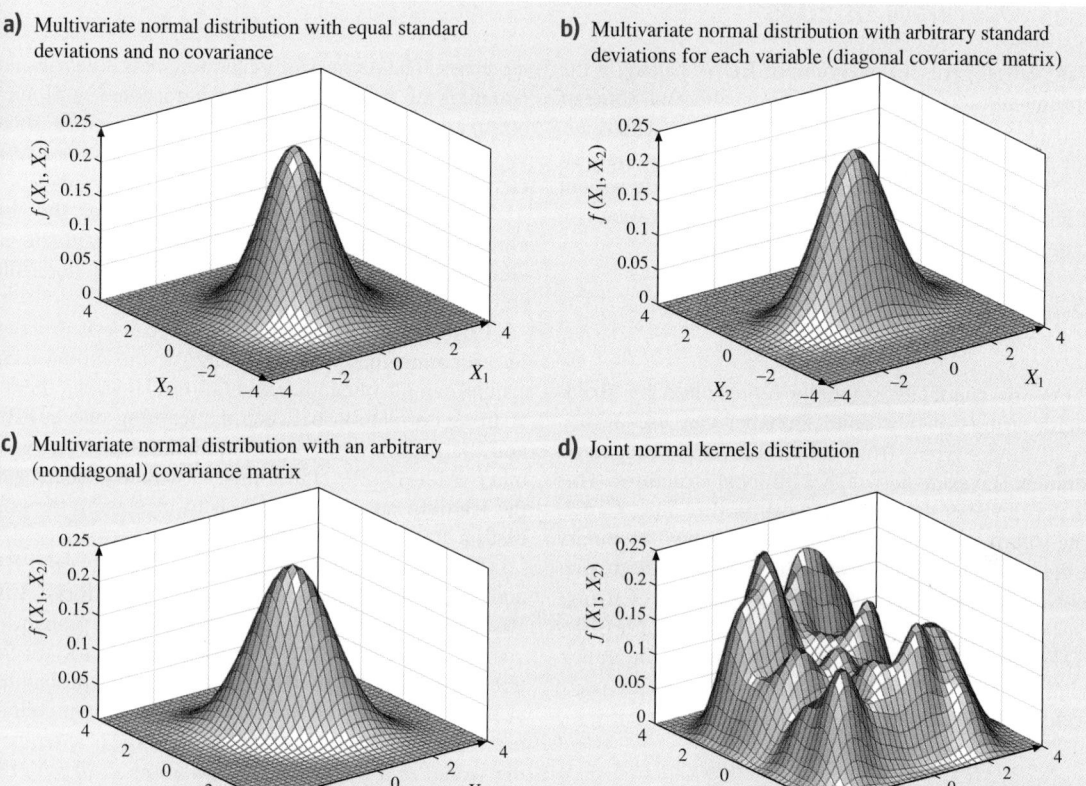

Fig. 45.4a–d Local models for continuous distributions over real-valued variables. **(a)** Multivariate normal distribution with equal standard deviations and no covariance, **(b)** Multivariate normal distribution with arbitrary standard deviations for each variable (diagonal covariance matrix), **(c)** Multivariate normal distribution with an arbitrary (nondiagonal) covariance matrix, **(d)** Joint normal kernels distribution

used to represent distributions over discrete representations in which the population consists of multiple dissimilar clusters [45.58] and in multiobjective EDAs [45.59, 60]. An example of a mixture of normal kernel distributions is shown in Fig. 45.4d.

- *Histograms.* In a number of EDAs for real-valued representations, to encode local distributions, real-valued variables or sets of such variables are divided into rectangular regions using a histogram-like model, and a separate probabilistic model is

used to represent the distribution in each region. Histogram models can be seen as a special subclass of the decision-tree models for real-valued variables. In real-valued EDAs, histograms were used, for example, in the histogram-based continuous EDA [45.61]. Histogram models can also be used for other representations; for example, when optimizing permutations, histograms can be used to represent different relative ordering constraints and their importance with respect to solution quality [45.62, 63].

45.3 Overview of EDAs

This section gives an overview of EDAs based on the representation of candidate solutions; although some of the EDAs can be used across several representations. Due to the large volume of work in EDAs in the past two decades, we do not aim to list every single variant of an EDA discussed in the past; instead, we focus on some of the most important representatives.

45.3.1 EDAs for Fixed–Length Strings over Finite Alphabets

EDAs for candidate solutions represented by fixed-length strings over a finite alphabet can use a variety of model types, from simple univariate models to complex Bayesian networks with local structures. This section reviews some of the work in this area. Candidate solutions are assumed to be represented by binary strings of fixed length n, although most methods presented here can be extended to optimization of strings over an arbitrary finite alphabet. The section classifies EDAs based on the order of interactions in the underlying dependency model along the lines discussed in Sect. 45.2.1 [45.3, 4, 7].

No Interactions

The EGA [45.33] and the population-based incremental learning (PBIL) [45.1] replace the population of candidate solutions represented as fixed-length binary strings by a probability vector $(p_1, p_2, \ldots, p_n)$, where n is the number of bits in a string and p_i denotes the probability of a 1 in the ith position of solution strings. Each p_i is initially set to 0.5, which corresponds to a uniform distribution over the set of all solutions. In each iteration, PBIL generates s candidate solutions according to the current probability vector where $s \geq 2$ denotes the selection pressure. Each value is generated independently of its context (remaining bits) and thus no interactions are considered (Fig. 45.2a). The best solution from the generated set of s solutions is then used to update the probability-vector entries using

$$p_i = p_i + \lambda(x_i - p_i),$$

where $\lambda \in (0, 1)$ is the learning rate (say, 0.02), and x_i is the ith bit of the best solution. Using the above update rule, the probability p_i of a 1 in the ith position increases if the best solution contains a 1 in that position and decreases otherwise. In other words, probability-vector entries move *toward* the best solution and, consequently, the probability of generating this solution

increases. The process of generating new solutions and updating the probability vector is repeated until some termination criteria are met; for instance, the run can be terminated if all probability-vector entries are sufficiently close to either 0 or 1.

Prior work refers to PBIL also as the hill climbing with learning (HCwL) [45.64] and the incremental univariate marginal distribution algorithm (IUMDA) [45.65].

PBIL is an incremental EDA, because it proceeds by executing incremental updates of the model using a small sample of candidate solutions. However, there is a strong correlation between the learning rate in PBIL and the population size in population-based EDAs or other evolutionary algorithms; essentially, decreasing the learning rate λ corresponds to increasing the population size.

The cGA [45.34, 66] reduces the gap between PBIL and traditional steady-state genetic algorithms. Like PBIL, cGA replaces the population by a probability vector and all entries in the probability vector are initialized to 0.5. Each iteration updates the probability vector by mimicking the effect of a single competition between two sampled solutions, where the best replaces the worst, in a hypothetical population of size N. Denoting the bit in the ith position of the best and worst of the two sampled solutions by x_i and y_i, respectively, the probability-vector entries are updated as follows:

$$p_i = \begin{cases} p_i + \dfrac{1}{N} & \text{if } x_i = 1 \text{ and } y_i = 0 \\ p_i - \dfrac{1}{N} & \text{if } x_i = 0 \text{ and } y_i = 1 \\ p_i & \text{otherwise .} \end{cases}$$

Although cGA uses the probability vector instead of a population, updates of the probability vector correspond to replacing one candidate solution by another one using a population of size N and shuffling the resulting population using a univariate model that assumes full independence of problem variables.

The UMDA [45.6] maintains a population of solutions. Each iteration of UMDA starts by selecting a population of promising solutions using an arbitrary selection method of evolutionary algorithms. A probability vector is then computed using the selected population of promising solutions and new solutions are generated by sampling the probability vector. Th new solutions replace the old ones and the process is repeated until termination criteria are met. Although UMDA uses a probabilistic model as an intermediate step between the original and new populations unlike PBIL and cGA, the performance, dynamics and limitations of PBIL, cGA, and UMDA are similar.

PBIL, cGA, and UMDA can solve problems decomposable into subproblems of order 1 in a linear or quadratic number of fitness evaluations. However, if decomposition into single-bit subproblems misleads the decision making away from the optimum, these algorithms scale up poorly with problem size [45.60, 67, 68].

Pairwise Interactions

EDAs based on pairwise probabilistic models, such as a chain, a tree or a forest, represent the first step toward EDAs being capable of learning variable interactions and therefore solving decomposable problems of bounded order (difficulty) in a scalable manner.

The MIMIC algorithm [45.36] uses a chain distribution (Fig. 45.2b) specified by

(1) an ordering of string positions (variables),
(2) a probability of a 1 in the first position of the chain, and
(3) conditional probabilities of every other position given the value in the previous position in the chain.

A chain probabilistic model encodes the probability distribution where all positions except the first are conditionally dependent on the previous position in the chain. After selecting promising solutions and computing marginal and conditional probabilities, MIMIC uses a greedy algorithm to maximize mutual information between the adjacent positions in the chain. In this fashion, the Kullback–Leibler divergence [45.69] between the chain and actual distributions is minimized. Nonetheless, the greedy algorithm does not guarantee global optimality of the constructed model (with respect to Kullback–Leibler divergence). The greedy algorithm starts in the position with the minimum unconditional entropy. The chain is expanded by adding a new position that minimizes the conditional entropy of the new variable given the last variable in the chain. Once the full chain is constructed for the selected population of promising solutions, new solutions are generated by sampling the distribution encoded by the chain. The use of pairwise interactions was one of the most important steps in the development of EDAs capable of solving decomposable problems of bounded difficulty scalably.

MIMIC was the first discrete EDA to not only learn and use a fixed set of st atistics, but it was also capable of identifying the statistics that should be considered to solve the problem efficiently.

Baluja and *Davies* [45.35] use dependency trees (Fig. 45.2c) to model promising solutions. Like in PBIL, the population is replaced by a probability vector but in this case the probability vector contains all *pairwise* probabilities. The probabilities are initialized to 0.25. Each iteration adjusts the probability vector according to new promising solutions acquired on the fly. A dependency tree encodes the probability distribution where every variable except for the root is conditioned on the variable's parent in the tree. A variant of *Prim*'s algorithm for finding the minimum spanning tree [45.70] can be used to construct an optimal tree distribution. Here the task is to find a tree that maximizes mutual information between parents (nodes with successors) and their children (successors). This can be done by first randomly choosing a variable to form the root of the tree, and *hanging* new variables to the existing tree so that the mutual information between the parent of the new variable and the variable itself is maximized. In this way, the Kullback–Leibler divergence between the tree and actual distributions is minimized as shown in [45.71]. Once a full tree is constructed, new solutions are generated according to the distribution encoded by the constructed dependency tree and the conditional probabilities computed from the probability vector.

The BMDA [45.40] uses a forest distribution (a set of mutually independent dependency trees, see Fig. 45.2c). This class of models is even more general than the class of dependency trees, because any forest that contains two or more disjoint trees cannot be generally represented by a tree. As a measure to determine whether to connect two variables, BMDA uses a Pearson's chi-square test [45.72]. This measure is also used to discriminate the remaining dependencies in order to construct the final model. To learn a model, BMDA uses a variant of *Prim*'s algorithm [45.70].

Pairwise models capture some interactions in a problem with reasonable computational overhead. EDAs with pairwise probabilistic models can identify, propagate, and juxtapose partial solutions of order 2, and therefore they work well on problems decomposable into subproblems of order at most two [45.35, 36, 40, 65, 73]. Nonetheless, capturing only some pairwise interactions has still been shown to be insufficient for solving all decomposable problems of bounded difficulty scalably [45.40, 73].

Multivariate Interactions

Using general multivariate models allows powerful EDAs capable of solving problems of bounded difficulty quickly, accurately, and reliably [45.4, 5, 8, 22, 37]. On the other hand, learning distributions with multivariate interactions necessitates more complex model-learning algorithms that require significant computational time and still do not guarantee global optimality of the resulting model. Nonetheless, many difficult problems are intractable using simple models and the use of complex models and algorithms is necessary.

The FDA [45.74] uses a fixed factorized distribution throughout the whole run. The model is allowed to contain multivariate marginal and conditional probabilities, but FDA learns only the probabilities, not the structure (dependencies and independencies). To solve a problem using FDA, we must first decompose the problem and then factorize the decomposition. While it is useful to incorporate prior information about the regularities in the search space, FDA necessitates that the practitioner is able to decompose the problem using a probabilistic model ahead of time. FDA does not learn *what statistics* are important to process within the EDA framework, it must be given that information in advance. A variant of FDA where probabilistic models are restricted to polytrees was also proposed [45.46].

The ECGA [45.45] uses an MPM that partitions the variables into disjoint subsets (Fig. 45.2d). Each partition (subset) is treated as a single variable and different partitions are considered to be mutually independent. To decide between alternative MPMs, ECGA uses a variant of the minimum description length (MDL) metric [45.75–77], which favors models that allow higher compression of data (in this case, the selected set of promising solutions). More specifically, the Bayesian information criterion (BIC) [45.78] is used. To find a good model, ECGA uses a greedy algorithm that starts with each variable forming one partition (like in UMDA). Each iteration of the greedy algorithm merges two partitions that maximize the improvement of the model with respect to BIC. If no more improvement is possible, the current model is used. ECGA provides robust and scalable solution for problems that can be decomposed into independent subproblems of bounded order (separable problems) [45.79–81]. However, many real-world problems contain overlapping dependencies, which cannot be accurately modeled by dividing the variables into disjoint partitions; this can result in poor performance of ECGA.

The dependency-structure matrix genetic algorithm (DSMGA) [45.82–84] uses a similar class of models as ECGA that splits the variables into independent clusters or linkage groups. However, DSMGA builds models via dependency structure matrix clustering techniques.

The BOA [45.42] builds a Bayesian network for the population of promising solutions (Fig. 45.2e) and samples the built network to generate new candidate solutions. BOA uses the Bayesian–Dirichlet metric subject to a maximum model-complexity constraint [45.85–87] to discriminate competing models, but other metrics (such as BIC) have been analyzed in BOA as well [45.88]. In all variants of BOA, the model is constructed by a greedy algorithm that iteratively adds a new dependency in the model that maximizes the model quality. Other elementary graph operators – such as edge removals and reversals – can be incorporated but edge additions are most important. The construction is terminated when no more improvement is possible. The greedy algorithm used to learn a model in BOA is similar to the one used in ECGA. However, Bayesian networks can encode more complex dependencies and independencies than models used in ECGA. Therefore BOA is also applicable to problems with overlapping dependencies. BOA uses an equivalent class of models as FDA; however, BOA learns both the structure and the probabilities of the model. Although BOA does not require problem-specific knowledge in advance, prior information about the problem can be incorporated using Bayesian statistics, and the relative influence of prior information and the population of promising solutions can be tuned by the user [45.89, 90].

A discussion of the use of Bayesian networks as an extension to tree models can also be found in *Baluja*'s and *Davies*' work [45.91]. An EDA that uses Bayesian networks to model promising solutions was independently developed by *Etxeberria* and *Larrañaga* [45.41], who called it the EBNA. *Mühlenbein* and *Mahnig* [45.37] improved the original FDA by using Bayesian networks together with the greedy algorithm for learning the networks described above; this modification of FDA was named the (LFDA). An incremental version of BOA, the incremental BOA (iBOA) was proposed by *Pelikan* et al. [45.92].

The hierarchical BOA (hBOA) [45.44] extends BOA by employing local structures to represent local distributions instead of using standard conditional probability tables. This enables hBOA to more efficiently represent distributions with high-order interactions. Furthermore, hBOA incorporates a niching technique called restricted tournament selection [45.93] to ensure effective diversity preservation. The two extensions enable hBOA to solve problems decomposable into subproblems of bounded order over a number of levels of difficulty of a hierarchy [45.44, 94].

Markov networks are yet another class of models that can be used to identify and use multivariate interactions in EDAs. Markov networks are undirected graphical models (Fig. 45.2f). Compared to Bayesian networks, Markov networks may sometimes cover the same distribution using fewer edges in the dependency model, but the sampling of these models becomes more complicated than the sampling of Bayesian networks. Markov networks are used, for example, in the Markov network EDA (MN-EDA) [45.95] and the density estimation using Markov random fields algorithm (DEUM) [45.96, 97].

Helmholtz machines used in the Bayesian evolutionary algorithm proposed by *Zhang* and *Shin* [45.98] can also encode multivariate interactions. Helmholtz machines encode interactions by introducing new, hidden variables, which are connected to every variable.

EDAs that use models capable of covering multivariate interactions can solve a wide range of problems in a scalable manner; promising results were reported on a broad range of problems, including several classes of spin–glass systems [45.22, 99–101], graph partitioning [45.90, 102, 103], telecommunication network optimization [45.104], silicon cluster optimization [45.80], scheduling [45.105], forest management [45.13], ground water remediation system design [45.106, 107], multiobjective knapsack [45.20], and others.

45.3.2 EDAs for Real-Valued Vectors

There are two basic approaches to extending EDAs for discrete, fixed-length strings to other domains such as real-valued vectors:

- Map the other representation to the domain of fixed-length discrete strings, solve the discrete problem, and map the solution back to the problem's original representation.
- Extend or modify the class of probabilistic models to other domains.

A number of studies have been published about the mapping of real-valued representations into a discrete one in evolutionary computation [45.26, 108–111]; this section focuses on EDAs from the second category. The approaches are classified along the lines presented in Sect. 45.2 [45.7, 22].

Single-Peak Normal Distributions

The stochastic hill climbing with learning by vectors of normal distributions (SHCLVND) [45.38] is a straightforward extension of PBIL to vectors of real-valued variables using a normal distribution to model each variable. SHCLVND replaces the population of real-valued solutions by a vector of means $\mu = (\mu_1, \ldots, \mu_n)$, where μ_i denotes a mean of the distribution for the ith variable. The same standard deviation σ is used for all variables. See Fig. 45.4a for an example model. In each generation (iteration), a random set of solutions is first generated according to μ and σ. The best solution out of this subset is then used to update the entries in μ by shifting each μ_i toward the value of the ith variable in the best solution using an update rule similar to the one used in PBIL. Additionally, each generation reduces the standard deviation to make the future exploration of the search space narrower. A similar algorithm was independently developed by *Sebag* and *Ducoulombier* [45.39], who also discussed several approaches to evolving a standard deviation for each variable.

Mixtures of Normal Distributions

The probability density function of a normal distribution is centered around its mean and decreases exponentially with square distance from the mean. If there are multiple *clouds* of values, a normal distribution must either focus on only one of these clouds, or it can embrace multiple clouds at the expense of including the low-density area between them. In both cases, the resulting distribution cannot model the data accurately. One way of extending standard single-peak normal-distribution models to enable coverage of multiple groups of similar points is to use a *mixture* of normal distributions. Each component of the mixture of normal distributions is a normal distribution by itself. A coefficient is specified for each component of the mixture to denote the probability that a random point belongs to this component. The probability density function of a mixture is thus computed by multiplying the density function of each mixture component by the probability that a random point belongs to the component, and adding these weighted densities together.

Gallagher et al. [45.112, 113] extended EDAs based on single-peak normal distributions by using an adaptive mixture of normal distributions to model each variable. The parameters of the mixture (including the number of components) evolve based on the discovered promising solutions. Using mixture distributions is a significant improvement compared to single-peak normal distributions, because mixtures allow simultaneous exploration of multiple basins of attraction for each variable.

Within the IDEA framework, *Bosman* and *Thierens* [45.47] proposed IDEAs using the joint normal kernels distribution, where a single normal distribution is placed around each selected solution (Fig. 45.4d). A joint normal kernels distribution can be therefore seen as an extreme use of mixture distributions with one mixture component per point in the training sample. The variance of each normal distribution can be fixed to a relatively small value, but it should be preferable to adapt variances according to the current state of search. Using kernel distributions corresponds to using a fixed zero-mean normally distributed mutation for each promising solution as is often done in evolution strategies [45.114]. That is why it is possible to directly take up strategies for adapting the variance of each kernel from evolution strategies [45.114–117].

Joint Normal Distributions and Their Mixtures

What changes when instead of fitting each variable with a separate normal distribution or a mixture of normal distributions, groups of variables are considered together? Let us first consider using a single-peak normal distribution. In multivariate domains, a joint normal distribution can be defined by a vector of n means (one mean per variable) and a covariance matrix of size $n \times n$. Diagonal elements of the covariance matrix specify the variances for all variables, whereas nondiagonal elements specify linear dependencies between pairs of variables. Considering each variable separately corresponds to setting all nondiagonal elements in a covariance matrix to 0. Using different deviations for different variables allows for *squeezing* or *stretching* the distribution along the axes. On the other hand, using nondiagonal entries in the covariance matrix allows rotating the distribution around its mean. Figure 45.4b and c illustrates the difference between a joint normal distribution using only diagonal elements of the covariance matrix and a distribution using the full covariance matrix. Therefore, using a covariance matrix introduces another degree of freedom and improves the expressiveness of a distribution. Again, one can use a number of joint normal distributions in a mixture, where each component consists of its mean, covariance matrix, and weight.

A joint normal distribution including a full or partial covariance matrix was used within the IDEA framework [45.47] and in the estimation of Gaussian networks algorithm (EGNA) [45.48]. Both these algorithms can be seen as extensions of EDAs that model each variable by a single normal distribution, which allow also the use of nondiagonal elements of the covariance matrix.

Bosman and *Thierens* [45.118] proposed *mixed IDEAs* as an extension of EDAs that use a mixture of normal distributions to model each variable. Mixed IDEAs allow multiple variables to be modeled by a separate mixture of joint normal distributions. At one extreme, each variable can have a separate mixture at another extreme, one mixture of joint distribution covering all the variables is used. Despite that learning such a general class of distributions is quite difficult and a large number of samples is necessary for reasonable accuracy, good results were reported on single-objective [45.118] as well as multiobjective problems [45.59, 119, 120]. Using mixture models for all variables was also proposed as a technique for reducing model complexity in discrete EDAs [45.58].

Real-valued EDAs presented so far are applicable to real-valued optimization problems without requiring differentiability or continuity of the underlying problem. However, if it is possible to at least partially differentiate the problem, gradient information can be used to incorporate some form of gradient-based local search and the performance of real-valued EDAs can be significantly improved. A study on combining real-valued EDAs within the IDEA framework with gradient-based local search can be found, for example in [45.121].

One of the crucial limitations of using estimation of real-valued distributions is that real-valued EDAs have a tendency to lose diversity too fast even when the problem is relatively easy to solve [45.122]; for example, maximum likelihood estimation and sampling of a normal distribution will lead to diversity loss even while climbing a simple linear slope. That is why several EDAs were proposed that aim to control variance of the probabilistic model so that the loss of variance is avoided and yet the effective exploration is not hampered by an overly large variance of the model. For example, the adapted maximum-likelihood Gaussian model iterated density-estimation evolutionary algorithm (AMaLGaM) scales up the covariance matrix to prevent premature convergence on slopes [45.123, 124].

Other Real-Valued EDAs

Using normal distributions is not the only approach to modeling real-valued distributions. Other density func tions are frequently used to model real-valued probability distributions, including histogram distributions, interval distributions, and others. A brief review of real-valued EDAs that use other than normal distributions or their mixtures follows.

In the algorithm proposed by *Servet* et al. [45.125], an interval (a_i, b_i) and a number $z_i \in (0, 1)$ are stored for each variable. By z_i, the probability that the ith variable is in the lower half of (a_i, b_i) is denoted. Each z_i is initialized to 0.5. To generate a new candidate solution, the value of each variable is selected randomly from the corresponding interval. The best solution is then used to update the value of each z_i. If the value of the ith variable of the best solution is in a lower half of (a_i, b_i), z_i is shifted toward 0; otherwise, z_i is shifted toward 1. When z_i gets close to 0, interval (a_i, b_i) is reduced to its lower half; if z_i gets close to 1, interval (a_i, b_i) is reduced to its upper half.

EDAs proposed in [45.47, 126] use empirical histograms to model each variable as opposed to using a single normal distribution or a mixture of normal distributions. In these approaches, a histogram for each single variable is constructed. New points are then generated according to the distribution encoded by the histograms for all variables. The sampling of a histogram proceeds by first selecting a particular bin based on its relative frequency, and then generating a random point from the interval corresponding to the bin. It is straightforward to replace the histograms in the above methods by various classification and discretization methods of statistics and machine learning (such as k-means clustering) [45.108].

Pelikan et al. [45.111, 127] use an adaptive mapping from the continuous domain to the discrete one in combination with discrete EDAs. The population of promising solutions is first discretized using equal-width histograms, equal-height histograms, k-means clustering, or other classification techniques. A population of promising discrete solutions is then selected. New points are created by applying a discrete recombination operator to the selected population of promising discrete solutions. For example, new solutions can be generated by building and sampling a Bayesian network like in BOA. The resulting discrete solutions are then mapped back into the continuous domain by sampling each class (a bin or a cluster) using the original values of the variables in the selected population of continuous solutions (before discretization). The resulting solutions are perturbed using one of the adaptive mutation operators of evolution strategies [45.114–117]. In this way, competent discrete EDAs can be combined with advanced methods based on adaptive local search in the continuous domain. A related approach was proposed by *Chen* and *Chen* [45.109], who propose a split-on-demand adaptive discretization method to use in combination with ECGA and report promising results on several benchmarks and one real-world problem.

The mixed Bayesian optimization algorithm (mBOA) developed by *Ocenasek* and *Schwarz* [45.56] models vectors of real-valued variables using an extension of Bayesian networks with local structures. A model used in mBOA consists of a decision tree for each variable. Each internal node in the decision tree for a variable is a test on the value of another variable. Each test on a variable is specified by a particular value, which is also included in the node. The test considers two cases: the value of the variable is greater or equal than the value in the node or it is smaller. Each internal node has two children, each child corresponding to one of the two results of the test specified in this node. Leaves in a decision tree thus correspond to rectangular regions in the search space. For each leaf, the decision tree for the variable specifies a single-variable mixture of normal distributions centered around the values of this variable in the solutions consistent with the path to the leaf. Thus, for each variable, the model in mBOA divides the space reduced to other variables into rectangular regions, and it uses a single-variable normal kernels distribution to model the variable in each region. The adaptive variant of mBOA (amBOA) [45.128] extends mBOA by employing variance adaptation with the goal of maximizing effectiveness of the search for the optimum on real-valued problems.

45.3.3 EDAs for Genetic Programming

In genetic programming [45.129], the task is to solve optimization problems with candidate solutions represented by labeled trees that encode computer programs or symbolic expressions. Internal nodes of a tree represent functions or commands; leaves represent functions with no arguments, variables, and constants. There are two key challenges that one must deal with when applying EDAs to genetic programming. Firstly, the length of programs is expected to vary and it is difficult to estimate how large the solution will be without solving the problem first. Secondly, small changes in parent–child relationships often lead to large changes in the performance of a candidate solution, and often the relationship between nodes in the program trees is more important than their actual position. Despite these challenges, even in this problem domain, EDAs have been quite successful. In this section, we briefly outline some EDAs for genetic programming.

The probabilistic incremental program evolution (PIPE) algorithm [45.130, 131] uses a probabilistic model in the form of a tree of a specified maximum allowable size. Nodes in the model specify probabilities of functions and terminals. PIPE does not capture any interactions between the nodes in the model. The model is updated by adjusting the probabilities based on the population of selected solutions using an update rule similar to the one in PBIL [45.1]. New program trees are generated in a top-down fashion starting in the root and continuing to lower levels of the tree. More specifically, if the model generates a function in a node and that function requires additional arguments, the successors (children) of the node are generated to form the arguments of the function. If a terminal is generated, the generation along this path terminates. An extension of PIPE named hierarchical probabilistic incremental program evolution (H-PIPE) was later proposed [45.132]. In H-PIPE, nodes of a model are allowed to contain subroutines, and both the subroutines as well as the overall program are evolved.

Handley [45.133] used tree probabilistic models to represent populations of programs (trees) in genetic programming. Although the goal of this work was to compress the population of computer programs in genetic programming, Handley's approach can be used within the EDA framework to model and sample candidate solutions represented by computer programs or symbolic expressions. A similar model was used in estimation of distribution programming (EDP) [45.134], which extended PIPE by employing parent–child dependencies in candidate labeled trees. Specifically, in EDP the content of each node is conditioned on the node's parent.

The extended compact genetic programming (ECGP) [45.135] assumes a maximum tree of maximum branching like PIPE. Nonetheless, ECGP uses an MPM which partitions nodes into clusters of strongly correlated nodes. This allows ECGP to capture and exploit interactions between nodes in program trees, and solve problems that are difficult for conventional genetic programming and PIPE. There are four main characteristics that distinguish ECGP and EDP. ECGP is able to capture dependencies between more than two nodes, it learns the dependency structure based on the promising candidate trees, and it is not restricted to the dependencies between parents and

their children. On the other hand, ECGP is somewhat limited in its ability to efficiently encode long-range interactions compared to probabilistic models that do not assume that groups of variables must be fully independent of each other.

Looks et al. [45.136] proposed to use Bayesian networks to model and sample program trees. Combinatory logic is used to represent program trees in a unified manner. Program trees translated with combinatory logic are then modeled with Bayesian networks of BOA, EBNA, and LFDA. Contrary to most other EDAs for genetic programming presented in this section, in the approach of Looks et al. the size of computer programs is not limited, but solutions are allowed to grow over time. *Looks* later developed a more powerful framework for competent program evolution using EDAs, which was named meta-optimizing semantic evolutionary search (MOSES) [45.54, 137, 138]. The key facets of MOSES include the division of the population into demes, the reduction of the problem of evolving computer programs to the one of building a representation with tunable features (knobs), and the use of hierarchical BOA [45.44] or another competent evolutionary algorithm to model demes and sample new candidate program solutions.

Several EDAs for genetic programming used probabilistic models based on grammar rules [45.51, 52, 139, 140]. Most grammar-based EDAs for genetic programming use a context-free grammar. The stochastic grammar-based genetic programming (SG-GP) [45.140, 140] started with a fixed context-free grammar with a default probability for each rule; the probabilities attached to the different rules were gradually adjusted based on the best candidate programs. The program evolution with explicit learning (PEEL) [45.139] used a probabilistic L-system with rules applicable at specific depths and locations; the probabilities of the rules were adapted using a variant of ant colony optimization (ACO) [45.141]. Another grammar-based EDA for genetic programming was proposed by *Bosman* and *de Jong* [45.51], who used a context-free grammar that was initialized to a minimum stochastic context-free grammar and adjusted to better fit promising candidate solutions by expanding rules and incorporating depth information into the rules. Grammar model-based program evolution (GMPE) [45.52, 142] also uses a probabilistic context-free grammar. In GMPE, new rules are allowed to be created and old rules may be eliminated from the model. A variant of the minimum-message-length metric is used in GMPE to compare grammars according to their quality. *Tanev* [45.143] incorporated tochastic context-sensitive grammars into the grammar-guided genetic programming [45.144–146].

45.3.4 EDAs for Permutation Problems

In many problems, candidate solutions are most naturally represented by permutations. This is the case, for example, in many scheduling or facility location problems. These types of problems often contain two specific types of features or constraints that EDAs need to capture. The first is the *absolute* position of a symbol in a string and the second is the *relative* ordering of specific symbols. In some problems, such as the traveling-salesman problem, relative ordering constraints matter the most. In others, such as the QAP, both the relative ordering and the absolute positions matter.

One approach to permutation problems is to apply an EDA for problems not involving permutations in combination with a mapping function between the EDA representation and the admissible permutations. For example, one may use the random key encoding [45.147] to transfer the problem of finding a good permutation into the problem of finding a high-quality real-valued vector, allowing the use of EDAs for optimization of real-valued vectors in solving permutation-based problems [45.148, 149]. Random key encoding represents a permutation as a vector of real numbers. The permutation is defined by the reordering of the values in the vector that sorts the values in ascending order. The main advantage of using random keys is that any real-valued vector defines a valid permutation and any EDA capable of solving problems defined on vectors of real numbers can thus be used to solve permutation problems. However, since EDAs do not process the aforementioned types of regularities in permutation problems directly their performance can often be poor [45.148, 150]. That is why several EDAs were developed that aim to encode either type of constraints for permutation problems explicitly.

To solve problems where candidate solutions are permutations of a string, *Bengoetxea* et al. [45.151] start with a Bayesian network model built using the same approach as in EBNA [45.41]. However, the sampling method is changed to ensure that only valid permutations are generated. This approach was shown to have promise in solving the inexact graph matching problem. In much the same way, the dependency-tree EDA (dtEDA) of *Pelikan* et al. [45.152] starts with a dependency-tree model [45.35, 71] and modifies the sampling to ensure that only valid permutations are

generated. dtEDA for permutation problems was used to solve structured QAPs with great success [45.152]. Bayesian networks and tree models are capable of encoding both the absolute position and the relative ordering constraints, although for some problem types, such models may turn out to be rather inefficient.

Bosman and *Thierens* [45.148] extended the real-valued EDA to the permutation domain by storing the dependencies between different positions in a permutation in the induced chromosome element exchanger (ICE). ICE works by first using a real-valued EDA, which encodes permutations as real-valued vectors using the random keys encoding. ICE extends the real-valued EDA by using a specialized crossover operator. By applying the crossover directly to permutations instead of simply sampling the model, relative ordering is taken into account. The resulting algorithm was shown to outperform many real-valued EDAs that use the random key encoding alone [45.148].

The edge-histogram-based sampling algorithm (EHBSA) [45.63, 153] works by creating an edge histogram matrix (EHM). For each pair of symbols, EHM stores the probabilities that one of these symbols will follow the other one in a permutation. To generate new solutions, EHBSA starts with a randomly chosen symbol. EHM is then sampled repeatedly to generate new symbols in the solution, normalizing the probabilities based on what values have already been generated. EHM does not take into account absolute positions at all; in order to address problems in which absolute positions are important, EHBSA was extended to use templates [45.153]. To generate new solutions, first a random string from the population was picked as a template. New solutions were then generated by removing random parts of the template string and generating the missing parts by sampling from EHM. The resulting algorithm was shown to be better than most other EDAs on the traveling salesman problem. In another study, the node-histogram based sampling algorithm (NHBSA) was proposed by *Tsutsui* et al. [45.63], which used a model capable of storing node frequencies in each position (thereby encoding absolute position constraints) and also used a template.

Zhang et al. [45.154–156] proposed to use guided mutation to optimize both permutation problems [45.154] as well as graph problems [45.156]. In guided mutation, the parts of the solution that are to be modified using a stochastic neighborhood operator are identified by analyzing a probabilistic model of the population of promising candidate solutions.

Part E | 45.3

45.4 EDA Theory

Along with the design and application of EDAs, the theoretical understanding of these algorithms has improved significantly since the first EDAs were proposed. One way to classify key areas of theoretical study of EDAs follows [45.3]:

● *Convergence proofs.* Some of the most important results in EDA theory focus on the number of iterations of an EDA on a particular class of problems or the conditions that allow EDAs to provably converge to a global optimum. The convergence time (number of iterations until convergence) of UMDA on onemax for selection methods with fixed selection intensity was derived by *Mühlenbein* and *Schlierkamp-Voosen* [45.157]. The convergence of FDA on separable additively decomposable functions (ADFs) was explored by *Mühlenbein* and *Mahnig* [45.158], who developed an exact formula for convergence time when using fitness-proportionate selection. Since, in practice, fitness-proportionate selection is rarely used because of its sensitivity to some linear and many other transformations of the objective function, truncation selection was also examined and an equation was derived giving the approximate time to convergence from the analysis of the onemax function. Later, *Mühlenbein* and *Mahnig* [45.37] adapted the theoretical model to the class of general ADFs where subproblems were allowed to interact. Under the assumption of Boltzmann selection, theory of graphical models was used to derive sufficient conditions for an FDA model so that FDA with a large enough population is guaranteed to converge to a model that generates only the global optima. *Zhang* [45.159] analyzed stability of fixed points of limit models of UMDA and FDA, and showed that at least for some problems the chance of converging to the global optimum is indeed increased when using higher order models of FDA rather than only the probability vector of UMDA. Convergence properties of PBIL were studied, for example, in [45.64, 160, 161].

● *Population sizing.* The convergence proofs mentioned above assumed infinite populations in order to simplify calculations. However, in practice using an infinite population is not possible and the choice of an adequate population size is crucial, similarly as for other population-based evolutionary algorithms [45.162–165]. Using a population that is too small can lead to convergence to solutions of low quality and inability to reliably find the global optimum. On the other hand, using a population that is too large can lead to an increased complexity of building and sampling probabilistic models, evaluating populations, and executing other EDA components. Similar to genetic algorithms, EDAs must have a population size sufficiently large to provide an adequate initial supply of partial solutions in an adequate problem decomposition [45.163, 166] and to ensure that good decisions are made between competing partial solutions [45.165]. However, the population must also be large enough for EDAs to make good decisions about the presence or the absence of statistically significant variable interactions. To examine this topic, *Pelikan* et al. [45.166] analyzed the population size required for BOA to solve decomposable problems of bounded difficulty with uniformly and nonuniformly scaled subproblems. The results showed that the population sizes required grew nearly linearly with the number of subproblems (or problem size). The results also showed that the approximate number of evaluations grew subquadratically for uniformly scaled subproblems but was quadratic on some nonuniformly scaled subproblems. *Yu* et al. [45.167] refined the model of *Pelikan* et al. [45.166] to provide a more accurate bound for the adequate population size in multivariate entropy-based EDAs such as ECGA and BOA, and also examined the effects of the selection pressure on the population size. Population sizing was also empirically analyzed in FDA by *Mühlenbein* [45.168].

● *Diversity loss.* Stochastic errors in sampling can lead to a loss of diversity that may sometimes hamper EDA performance. *Shapiro* [45.169] examined the susceptibility of UMDA to diversity loss and discussed how it is necessary to set the learning parameters in such a way that this does not happen. *Bosman* et al. [45.170] examined diversity loss in EDAs for solving real-valued problems and the approaches to alleviating this difficulty. The results showed that due to diversity loss some of the state-of-the-art EDAs for real-valued problems could still fail on slope-like regions in the search space. The authors proposed using anticipated mean shift (AMS) to shift the mean of new solutions each generation in order to effectively maintain diversity.

● *Memory complexity.* Another factor of importance in EDA problem solving is the memory required to solve the problem. *Gao* and *Culberson* [45.171] examined the space complexity of the FDA and BOA on additively decomposable functions where overlap was allowed between subfunctions. *Gao* and *Culberson* [45.171] proved that the space complexity of FDA and BOA is exponential in the problem size even with very sparse interaction between variables. While these results are somewhat negative, the authors point out that this only shows that EDAs have limitations and work best when the interaction structure is of bounded size. Note that one way to reduce the memory complexity of EDAs is to use incremental EDAs, such as PBIL [45.1], cGA [45.34] or the incremental Bayesian optimization algorithm (iBOA) [45.92].

• *Model accuracy.* A number of studies examined the accuracy of models in EDAs. *Hauschild* et al. [45.172] analyzed the models generated by hBOA when solving concatenated traps, random additively decomposable problems, hierarchical traps and two-dimensional Ising spin glasses. The models generated were then compared to the underlying problem structure by analyzing the number of spurious and correct dependencies. The results showed that the models corresponded closely to the structure of the underlying problems and that the models did not change significantly between consequent iterations of hBOA. The relationship between the probabilistic models learned by BOA and the underlying problem structure was also explored by *Lima* et al. [45.173]. One of the most important contributions of this study was to demonstrate the dramatic effect that selection has on spurious dependencies. The results showed that model accuracy was significantly improved when using truncation selection compared to tournament selection. Motivated by these results, the authors modified the complexity penalty of BOA model building to take into account tournament sizes when using binary tournament selection. *Echegoyen* et al. [45.174] also analyzed the structural accuracy of the models using EBNA on concatenated traps, two variants of Ising spin glass and MAXSAT. In this work, two variations of EBNA were compared, one that was given the complete model structure based on the underlying problem and another that learned the approximate structure. The authors then examined the probability at any generation that the models would generate the optimal solution. The results showed that it was not strictly necessary to have all the interactions that were in the complete model in order to solve the problems. Finally, the effects of spurious linkages on EDA performance were examined by *Radetic* and *Pelikan* [45.175]. The authors started by proposing a theoretical model to describe the effects of spurious (unnecessary) dependencies on the population sizing of EDAs. This model was then tested empirically on onemax and the results showed that while it would be expected that spurious dependencies would have little effect on population size, when niching was included the effects were substantial.

45.5 Efficiency Enhancement Techniques for EDAs

EDAs can solve many classes of important problems in a robust and scalable manner, oftentimes requiring only a low-order polynomial growth of the number of function evaluations with respect to the number of decision variables [45.4, 5, 8, 22, 74, 166, 176]. However, even a low-order polynomial complexity is sometimes insufficient for practical application of EDAs especially when the number of decision variables is extremely large, when evaluation of candidate solutions is computationally expensive, or when there are many conflicting objectives to optimize. The good news is that a number of approaches exist that can be used to further enhance efficiency of EDAs. Some of these techniques can be adopted from genetic and evolutionary algorithms with little or no change. However, some techniques are directly targeted at EDAs because these techniques exploit some of the unique advantages of EDAs over most other metaheuristics. Specifically, some efficiency enhancements capitalize on the facts that the use of probabilistic models in EDAs provides a rigorous and flexible framework for incorporating prior knowledge about the problem into optimization, and that EDAs provide practitioners with a series of probabilistic models that reveal a lot of information about the problem. This section reviews some of the most important efficiency enhancement techniques for EDAs with main focus on techniques developed specifically for EDAs.

45.5.1 Parallelization

One of the most straightforward approaches to speeding up any algorithm is to distribute the computation over a number of computational nodes so that several computational tasks can be executed in parallel. There are two main bottlenecks of EDAs that are typically addressed by parallelization: (1) fitness evaluation, and (2) model building and sampling. If fitness evaluation is computationally expensive, a master–slave architecture can be used for distributing fitness evaluations and collecting the results [45.177]. If most computational time is spent in model building and sampling, model building and sampling should be parallelized [45.4, 178, 179].

Many parallelization techniques and much of the theory can be adopted from research on parallelization in genetic and evolutionary algorithms [45.177]. In the context of EDAs, parallelization of model building was discussed, for example, by *Ocenasek* et al. [45.178–181] who proposed the parallel BOA, mBOA and hBOA, and by *Larrañaga* and *Lozano* [45.4] who parallelized model building in EBNA. One of the most impressive results in parallelization of EDAs was published by *Sastry* et al. [45.14, 182] who proposed a highly efficient, fully parallelized implementation of cGA to solve large-scale problems with millions to billions of variables even with a substantial amount of external noise in the objective function.

45.5.2 Hybridization

An optimization hybrid combines two or more optimizers in a single procedure [45.183–185]. Typically, a *global* procedure and a *local* procedure are combined; the global procedure is expected to find promising regions and the local procedure is expected to find local optima quickly within reasonable basins of attraction. Global and local search are used in concert to find good solutions faster and more reliably than would be possible using either procedure alone.

Numerous studies have proposed to combine EDAs with variants of local search both in the discrete domain [45.22, 99, 186] and in the real-valued domain [45.187]. The main reason for combining EDAs with local search is that by reducing the search space to the local optima, the structure of the problem can be identified more easily and the population-sizing requirements can be significantly decreased [45.22, 99]. Furthermore, the search reduces to the space of basins of attraction around each local optimum as opposed to the space of all admissible solutions.

However, hybridization of EDAs is not restricted to the combination of an EDA with simple local search. As was already pointed out, probabilistic models often contain a lot of information about the problem. By min-

ing these models for information about the structure and other properties of the problem landscape, decisions can be made about the nature and likely effectiveness of particular local search procedures and appropriate neighborhood structures for those procedures [45.188–192]. In turn, subsequent local search as well as the coordination of the global and local search in a hybrid can be managed so that excellent solutions are found quickly, reliably and accurately.

There are two main approaches to the design of EDA-based (model-directed) hybrids with advanced neighborhoods: (1) Belief propagation, which uses the probabilistic model to generate the maximum likely instance [45.189–191] and (2) local search with an advanced neighborhood structure derived from an EDA model [45.188, 192]. However, it is important to note that the use of EDA models is not limited to advanced neighborhood structures or belief propagation, and one may envision the use of probabilistic models to control the division of time resources between the global and local searcher and in a number of other tasks.

Local search based on advanced neighborhood structures in a hill-climbing like procedure [45.193, 194] is strongly related to model-directed hybridization using EDAs, although in this approach no estimation of distributions takes place. The basic idea is to use a linkage learning approach to detect important interactions between problem variables, and then run a local search based on a neighborhood defined by the underlying problem decomposition.

45.5.3 Time Continuation

To achieve the same solution quality, one may run an EDA or another population-based metaheuristic with a large population for one convergence epoch, or run the algorithm with a small population for a large number of convergence epochs with controlled restarts between these epochs [45.195]. Similar tradeoffs are involved in the design of efficient and reliable hybrid procedures where an appropriate division of computational resources between the component algorithms is critical. The term *time continuation* is used to refer to the tradeoffs involved [45.196].

Two important studies related to time continuation in EDAs were published by *Sastry* and *Goldberg* [45.197, 198]. Based on a theoretical model of an ECGA-based hybrid, Sastry showed that under certain assumptions, the neighborhoods created from EDA-built models provide sufficient information for local search to succeed on its own even on classes of problems for which local search with standard neighbor-hoods performs poorly. However, in many other cases, EDA-driven search in a hybrid with local search based on the adaptive neighborhood should perform better, especially if the structure of the problem is complex and the problem is affected by external noise.

One of the promising research directions related to time continuation in EDAs is to mine probabilistic models discovered by EDAs to find an optimal way to exploit time continuation tradeoffs, be it in an EDA alone or in an EDA-based hybrid.

45.5.4 Using Prior Knowledge and Learning from Experience

The use of prior knowledge has had longstanding study and use in optimization. For example, promising partial solutions may be used to bias the initial population of candidate solutions, specialized search operators can be designed to solve a particular class of problems, or representations can be biased in order to make the search for the optimum an easier task. However, one of the limitations of most of these approaches is that the prior knowledge must be incorporated by hand and the approaches are limited to one specific problem domain.

The use of probabilistic models provides EDAs with a unique framework for incorporating prior knowledge into optimization because of the possibility of using Bayesian statistics to combine prior knowledge with data in the learning of probabilistic models [45.23, 90, 199–201]. Furthermore, the use of probabilistic models in EDAs provides a basis for learning from previous runs in order to solve new problem instances of similar type with increased speed, accuracy, and reliability [45.22–24]. Practitioners can thus incorporate two sources of bias into EDAs: (1) prior knowledge and (2) information obtained from models from prior EDA runs on similar problems (or runs of some other algorithm); these two sources can of course be combined using Bayesian statistics or in another way [45.23, 24, 90, 200]. Then, the bias can be incorporated into EDAs either by restricting the class of allowable models [45.199] or by increasing scores of models that appear to be more likely than others [45.23, 24, 90, 200].

For example, *Hauschild* et al. [45.24, 89] proposed to use a probability coincidence matrix to store probabilities of Bayesian-network dependencies between pairs of problem variables in prior hBOA runs and to bias the model building in hBOA on future problem instances of similar type using the matrix. Other related approaches were proposed [45.24, 200] that were based on combining a distance metric on problem variables and the pool of models obtained in previous runs on problems of similar type. The use of a distance metric in combination with prior EDA runs is somewhat more broadly applicable and promises to be more useful for practitioners. One of the main reasons for this is that this approach allows the use of bias derived from prior runs on problems of smaller size to bias optimization of larger problems. Furthermore, the approach is applicable even in cases where the meaning of a variable and its context change significantly from one problem instance to another.

45.5.5 Fitness Evaluation Relaxation

To reduce the number of the objective function evaluations, a model of the objective function can be built [45.202–204]. While models of the objective function can be created for any optimization method, EDAs enable the use of probabilistic models for creating relatively complex computational models of the problem in a fully automated manner. Specifically, if an advanced EDA is used that contains a complex probabilistic model, the model can be mined to provide a set of statistics that can be estimated for an accurate, efficient computational model of the objective function. The model can then used to replace some of the evaluations, possibly most of them. It was shown that the use of adequate models of the objective function can yield multiplicative speedups of several tens [45.202–204].

45.5.6 Incremental and Sporadic Model Building

With sporadic model-building, the probabilities (parameters) of the model are updated in every iteration, but the structure of the probabilistic model is rebuilt only once in every few iterations (generations) [45.205]. Sporadic model building was shown to yield significant speedups that increased with problem size, mainly because building model structure is the most computationally expensive part of model building but model structure often changes only little between consequent iterations of an EDA. With incremental model building, the model is built incrementally starting from the structure discovered in the previous iteration [45.41]. This can often reduce computational resources required to learn an accurate model.

Part E | 45.5

45.6 Starting Points for Obtaining Additional Information

This section provides pointers for obtaining additional information about EDAs.

45.6.1 Introductory Books and Tutorials

Numerous books and other publications exist that provide introduction to EDAs and additional starting points. The following list of references includes some of them: [45.2–5, 7, 8, 22, 206].

45.6.2 Software

The following list includes some of the popular EDA implementations available online. These implementations should provide a good starting point for the interested reader. Entries in the list are ordered alphabetically. Note that the list is *not* exhaustive:

- Adapted maximum-likelihood Gaussian model iterated density estimation evolutionary algorithm (AMaLGaM) [45.124]: http://homepages.cwi.nl/~bosman/source_code.php
- Bayesian optimization algorithm (BOA) [45.43]; BOA with decision graphs [45.55]; dependency-tree EDA [45.35]: http://medal-lab.org/
- Demos of aggregation pheromone system (APS [45.207] and histogram-based EDAs for permutation-based problems (EHBSA) [45.63]: http://www.hannan-u.ac.jp/~tsutsui/research-e.html
- Distribution estimation using Markov random fields (DEUM) [45.96, 97]: http://sidshakya.com/Downloads/Main.html
- Extended compact genetic algorithm [45.45], ξ-ary ECGA, BOA [45.43], BOA with decision trees/graphs [45.55], and others: http://illigal.org/
- Mixed BOA (mBOA) [45.56], adaptive mBOA (am-BOA) [45.128]: http://jiri.ocenasek.com/
- Probabilistic incremental program evolution (PIPE) [45.131]: ftp://ftp.idsia.ch/pub/rafal/
- Real-coded BOA (rBOA) [45.49], multiobjective rBOA [45.208]: http://www.evolution.re.kr/
- Regularity model based multiobjective EDA (RM-MEDA) [45.209]; hybrid of differential evolution and EDA [45.210]; model-based multiobjective evolutionary algorithm (MMEA) [45.155],

and others: http://cswww.essex.ac.uk/staff/qzhang/mypublication.htm

45.6.3 Journals

The following journals are key venues for papers on EDAs and evolutionary computation, although papers on EDAs can be found in many other journals focusing on optimization, artificial intelligence, machine learning, and applications:

- *Evolutionary Computation* (MIT Press): http://www.mitpressjournals.org/loi/evco
- *Evolutionary Intelligence* (Springer): http://www.springer.com/engineering/journal/12065
- *Genetic Programming and Evolvable Machines* (Springer): http://www.springer.com/computer/ai/journal/10710
- *IEEE Transactions on Evolutionary Computation* (IEEE Press): http://ieeexplore.ieee.org/servlet/opac?punumber=4235
- *Natural Computing* (Springer): http://www.springer.com/computer/theoretical+computer+science/journal/11047
- *Swarm and Evolutionary Computation* (Elsevier): http://www.journals.elsevier.com/swarm-and-evolutionary-computation/

45.6.4 Conferences

The following conferences provide the most important venues for publishing papers on EDAs and evolutionary computation, although similarly as for journals, papers on EDAs are often published in other venues:

- *ACM SIGEVO Genetic and Evolutionary Computation Conference* (GECCO)
- *European Workshops on Applications of Evolutionary Computation* (EvoWorkshops)
- *IEEE Congress on Evolutionary Computation* (CEC)
- *Main European Events on Evolutionary Computation* (EvoStar)
- *Parallel Problem Solving in Nature* (PPSN)
- *Simulated Evolution and Learning* (SEAL)

45.7 Summary and Conclusions

EDAs are a class of stochastic optimization algorithms that have been gaining popularity due to their ability to solve a broad array of complex problems with excellent performance and scalability. Moreover, while many of these algorithms have been shown to perform well with little or no problem-specific information, such information can be used advantageously if available.

EDAs have their roots in the fields of evolutionary computation and machine learning. From evolutionary computation, EDAs borrow the idea of using a population of solutions that evolves through iterations of selection and variation. From machine learning, EDAs borrow the idea of learning models from data, and they use the resulting models to guide the search for better solutions. This approach is powerful especially because it allows the search algorithm to adapt to the problem being solved, giving EDAs the possibility of being an effective *black-box* search algorithm. Since most real-world problems have some sort of inherent structure (as opposed to being completely random), there is a hope that EDAs can learn such a structure, or at least parts of it, and put that knowledge to good use in searching for optima.

Another key characteristic of EDAs, and one that sets them apart from other metaheuristics, lies in the fact that the sequence of probabilistic models learned along a particular run (or a sequence or runs) yields important information that can be exploited for other means. For example, such information can be used for building surrogate models of the objective function leading to significant performance speedups, for designing effective neighborhoods for local search when conventional neighborhoods fail, and even for learning about characteristics of an entire class of problems that can in turn be used to solve other instances of the same problem class.

This chapter gave an introduction and reviewed both the history and the state-of-the-art in EDA research. The basic concepts of these algorithms were presented and a taxonomy was outlined from the views based on the model decomposition and the type of local distributions. The most popular EDAs proposed in the literature were then surveyed according to the most common representations for candidate solutions. Finally, the major theoretical research areas and efficiency enhancement techniques for EDAs were highlighted. This chapter should be valuable both for those who want to grasp the basic ideas of EDAs as well as for those who want to have a coherent view of EDA research.

References

45.1   S. Baluja: Population-based incremental learning: A method for integrating genetic search based function optimization and competitive learning, Tech. Rep. No. CMU-CS-94-163 (Carnegie Mellon, Pittsburgh 1994)

45.2   J. Grahl, S. Minner, P. Bosman: Learning structure illuminates black boxes: An introduction into estimation of distribution algorithms. In: *Advances in Metaheuristics for Hard Optimization*, ed. by Z. Michalewicz, P. Siarry (Springer, Berlin, Heidelberg 2008) pp. 365–396

45.3   M.W. Hauschild, M. Pelikan: An introduction and survey of estimation of distribution algorithms, Swarm Evol. Comput. **1**(3), 111–128 (2011)

45.4   P. Larrañaga, J.A. Lozano (Eds.): *Estimation of Distribution Algorithms: A New Tool for Evolutionary Computation* (Kluwer Academic, Boston 2002)

45.5   J.A. Lozano, P. Larrañaga, I. Inza, E. Bengoetxea (Eds.): *Towards a New Evolutionary Computation: Advances on Estimation of Distribution Algorithms* (Springer, Berlin, Heidelberg 2006)

45.6   H. Mühlenbein, G. Paaß: From recombination of genes to the estimation of distributions I. Binary parameters, Lect. Notes Comput. Sci. **1141**, 178–187 (1996)

45.7   M. Pelikan, D.E. Goldberg, F. Lobo: A survey of optimization by building and using probabilistic models, Comput. Optim. Appl. **21**(1), 5–20 (2002)

45.8   M. Pelikan, K. Sastry, E. Cantú-Paz (Eds.): *Scalable Optimization via Probabilistic Modeling: From Algorithms to Applications* (Springer, Berlin, Heidelberg 2006)

45.9   R. Armañanzas, Y. Saeys, I. Inza, M. García-Torres, C. Bielza, Y.V. de Peer, P. Larrañaga: Peakbin selection in mass spectrometry data using a consensus approach with estimation of distribution algorithms, IEEE/ACM Trans. Comput. Biol. Bioinform. **8**(3), 760–774 (2011)

45.10   J. Bacardit, M. Stout, J.D. Hirst, K. Sastry, X. Llorà, N. Krasnogor: Automated alphabet reduction method with evolutionary algorithms for protein structure prediction, Genet. Evol. Comput. Conf. (2007) pp. 346–353

45.11   I. Belda, S. Madurga, X. Llorà, M. Martinell, T. Tarragó, M.G. Piqueras, E. Nicolás, E. Giralt: ENPDA: An evolutionary structure-based de novo peptide design algorithm, J. Comput. Aided Mol. Des. **19**(8), 585–601 (2005)

45.12   Y. Chen, T.L. Yu, K. Sastry, D.E. Goldberg: A survey of genetic linkage learning techniques. IlliGAL Rep. No. 2007014 (University of Illinois, Urbana 2007)

45.13   E. Ducheyne, B. De Baets, R. De Wulf: Probabilistic models for linkage learning in forest management. In: *Knowledge Incorporation in Evolutionary Computation*, ed. by Y. Jin (Springer, Berlin, Heidelberg 2004) pp. 177–194

45.14   D.E. Goldberg, K. Sastry, X. Llorà: Toward routine billion-variable optimization using genetic algorithms, Complexity **12**(3), 27–29 (2007)

45.15   P. Lipinski: ECGA vs. BOA in discovering stock market trading experts, Genet. Evol. Comput. Conf. (2007) pp. 531–538

45.16   J.B. Kollat, P.M. Reed, J.R. Kasprzyk: A new epsilon-dominance hierarchical Bayesian optimization algorithm for large multi-objective monitoring network design problems, Adv. Water Resour. **31**(5), 828–845 (2008)

45.17   P.M. Reed, R. Shah, J.B. Kollat: Assessing the value of environmental observations in a changing world: Nonstationarity, complexity, and hierarchical dependencies, 5th Bienn. Meet. Int. Congr. Environ. Model. Soft. Model. Environ. Sake (2010)

45.18   R. Santana, P. Larrañaga, J.A. Lozano: Protein folding in simplified models with estimation of distribution algorithms, IEEE Trans. Evol. Comput. **12**(4), 418–438 (2008)

45.19   S. Santarelli, T.L. Yu, D.E. Goldberg, E.E. Altshuler, T. O'Donnell, H. Southall, R. Mailloux: Military antenna design using simple and competent genetic algorithms, Math. Comput. Model. **43**(9-10), 990–1022 (2006)

45.20   R. Shah, P. Reed: Comparative analysis of multiobjective evolutionary algorithms for random and correlated instances of multiobjective d-dimensional knapsack problems, Eur. J. Oper. Res. **211**(3), 466–479 (2011)

45.21   J. Sun, Q. Zhang, J. Li, X. Yao: A hybrid EDA for CDMA cellular system design, Int. J. Comput. Intell. Appl. **7**(2), 187–200 (2007)

45.22   M. Pelikan: *Hierarchical Bayesian Optimization Algorithm: Toward a New Generation of Evolutionary Algorithms* (Springer, Berlin, Heidelberg 2005)

45.23   M.W. Hauschild, M. Pelikan: Enhancing efficiency of hierarchical BOA via distance-based model restrictions, Lect. Notes Comput. Sci. **5199**, 417–427 (2008)

45.24   M.W. Hauschild, M. Pelikan, K. Sastry, D.E. Goldberg: Using previous models to bias structural learning in the hierarchical BOA, Evol. Comput. **20**(1), 135–160 (2012)

45.25   E.L. Lawler: The quadratic assignment problem, Manag. Sci. **9**(4), 586–599 (1963)

45.26   D.E. Goldberg: *Genetic Algorithms in Search, Optimization, and Machine Learning* (Addison-Wesley, Reading 1989)

45.27   J.H. Holland: *Adaptation in Natural and Artificial Systems* (University of Michigan, Ann Arbor 1975)

45.28   V. Černý: Thermodynamical approach to the traveling salesman problem: An efficient simulation algorithm, J. Optim. Theory Appl. **45**, 41–51 (1985), 10.1007/BF00940812

45.29   S. Kirkpatrick, C.D. Gelatt, M.P. Vecchi: Optimization by simulated annealing, Science **220**, 671–680 (1983)

45.30   E. Cantú-Paz: Comparing selection methods of evolutionary algorithms using the distribution of fitness, Tech. Rep. UCRL-JC-138582 (University of California, San Francisco 2000)

45.31   D.E. Goldberg, K. Deb: A comparative analysis of selection schemes used in genetic algorithms, Found. Genet. Algorithms **1**, 69–93 (1991)

45.32   A.E. Eiben, J.E. Smith: *Introduction to Evolutionary Computing* (Springer, Berlin, Heidelberg 2010)

45.33   A. Juels, S. Baluja, A. Sinclair: The equilibrium genetic algorithm and the role of crossover, Unpublished manuscript (1993)

45.34   G.R. Harik, F.G. Lobo, D.E. Goldberg: The compact genetic algorithm, Int. Conf. Evol. Comput. (1998) pp. 523–528

45.35   S. Baluja, S. Davies: Using optimal dependency-trees for combinatorial optimization: Learning the structure of the search space, Proc. Int. Conf. Mach. Learn. (1997) pp. 30–38

45.36   J.S. De Bonet, C.L. Isbell, P. Viola: MIMIC: Finding optima by estimating probability densities, Adv. Neural Inf. Proc. Syst. **9**, 424–431 (1997)

45.37   H. Mühlenbein, T. Mahnig: FDA – A scalable evolutionary algorithm for the optimization of additively decomposed functions, Evol. Comput. **7**(4), 353–376 (1999)

45.38   S. Rudlof, M. Köppen: Stochastic hill climbing with learning by vectors of normal distributions, 1st On-line Workshop Soft Comput. (Nagoya, Japan 1996)

45.39   M. Sebag, A. Ducoulombier: Extending population-based incremental learning to continuous search spaces, Lect. Notes Comput. Sci. **1498**, 418–427 (1998)

45.40   M. Pelikan, H. Mühlenbein: The bivariate marginal distribution algorithm. In: *Advances in Soft Computing—Engineering Design and Manufacturing*, ed. by R. Roy, T. Furuhashi, P.K. Chawdhry (Springer, Berlin, Heidelberg 1999) pp. 521–535

45.41   R. Etxeberria, P. Larrañaga: Global optimization using Bayesian networks, 2nd Symp. Artif. Intell. (1999) pp. 332–339

45.42 M. Pelikan, D.E. Goldberg, E. Cantú-Paz: Linkage problem, distribution estimation, and Bayesian networks. IlliGAL Rep. No. 98013 (University of Illinois, Urbana 1998)

45.43 M. Pelikan, D.E. Goldberg, E. Cantú-Paz: BOA: The Bayesian optimization algorithm, Genet. Evol. Comput. Conf. (1999) pp. 525–532

45.44 M. Pelikan, D.E. Goldberg: Escaping hierarchical traps with competent genetic algorithms, Genet. Evol. Comput. Conf. (2001) pp. 511–518

45.45 G. Harik: Linkage learning via probabilistic modeling in the ECGA. IlliGAL Rep. No. 99010 (University of Illinois, Urbana 1999)

45.46 M. Soto, A. Ochoa: A factorized distribution algorithm based on polytrees, IEEE Congr. Evol. Comput. (2000) pp. 232–237

45.47 P.A.N. Bosman, D. Thierens: Continuous iterated density estimation evolutionary algorithms within the IDEA framework, Workshop Proc. Genet. Evol. Comput. Conf. (2000) pp. 197–200

45.48 P. Larrañaga, R. Etxeberria, J.A. Lozano, J.M. Pena: Optimization in continuous domains by learning and simulation of Gaussian networks, Workshop Proc. Genet. Evol. Comput. Conf. (2000) pp. 201–204

45.49 C.W. Ahn, R.S. Ramakrishna, D.E. Goldberg: Real-coded Bayesian optimization algorithm: Bringing the strength of BOA into the continuous world, Genet. Evol. Comput. Conf. (2004) pp. 840–851

45.50 R.I. McKay, N.X. Hoai, P.A. Whigham, Y. Shan, M. O'Neill: Grammar-based genetic programming: A survey, Genet. Progr. Evol. Mach. 11(3-4), 365–396 (2010)

45.51 P.A.N. Bosman, E.D. de Jong: Learning probabilistic tree grammars for genetic programming, Lect. Notes Comput. Sci. 3242, 192–201 (2004)

45.52 Y. Shan: Program Distribution Estimation with Grammar Models, Ph.D. Thesis (Wuhan Cehui Technical University, China 2005)

45.53 Y. Hasegawa, H. Iba: Estimation of distribution algorithm based on probabilistic grammar with latent annotations, IEEE Congr. Evol. Comput. (2007) pp. 1043–1050

45.54 M. Looks: Levels of abstraction in modeling and sampling: The feature-based Bayesian optimization algorithm, Genet. Evol. Comput. Conf. (2006) pp. 429–430

45.55 M. Pelikan, D.E. Goldberg, K. Sastry: Bayesian optimization algorithm, decision graphs, and Occam's razor, Genet. Evol. Comput. Conf. (2001) pp. 519–526

45.56 J. Ocenasek, J. Schwarz: Estimation of distribution algorithm for mixed continuous-discrete optimization problems, 2nd Euro-Int. Symp. Comput. Intell. (2002) pp. 227–232

45.57 P.A.N. Bosman: On empirical memory design, faster selection of Bayesian factorizations and parameter-free Gaussian EDAs, Genet. Evol. Comput. Conf. (2009) pp. 389–396

45.58 M. Pelikan, D.E. Goldberg: Genetic algorithms, clustering, and the breaking of symmetry, Lect. Notes Comput. Sci. 1517, 385–394 (2000)

45.59 D. Thierens, P.A.N. Bosman: Multi-objective mixture-based iterated density estimation evolutionary algorithms, Genet. Evol. Comput. Conf. (2001) pp. 663–670

45.60 M. Pelikan, K. Sastry, D.E. Goldberg: Multiobjective hBOA, clustering, and scalability, Genet. Evol. Comput. Conf. (2005) pp. 663–670

45.61 S. Tsutsui, M. Pelikan, D.E. Goldberg: Evolutionary algorithm using marginal histogram models in continuous domain, Workshop Proc. Genet. Evol. Comput. Conf. (2001) pp. 230–233

45.62 S. Tsutsui, M. Pelikan, D.E. Goldberg: Probabilistic model-building genetic algorithms using histogram models in continuous domain, J. Inf. Process. Soc. Jpn. 43, 24–34 (2002)

45.63 S. Tsutsui, M. Pelikan, D.E. Goldberg: Node histogram vs. edge histogram: A comparison of pmbgas in permutation domains. MEDAL Rep. No. 2006009 (University of Missouri, St. Louis 2006)

45.64 V. Kvasnicka, M. Pelikan, J. Pospichal: Hill climbing with learning (An abstraction of genetic algorithm), Neural Netw. World 6, 773–796 (1996)

45.65 H. Mühlenbein: The equation for response to selection and its use for prediction, Evol. Comput. 5(3), 303–346 (1997)

45.66 G.R. Harik, F.G. Lobo, D.E. Goldberg: The compact genetic algorithm, IEEE Trans. Evol. Comput. 3(4), 287–297 (1999)

45.67 D. Thierens: Analysis and design of genetic algorithms, Ph.D. Thesis (Katholieke Universiteit Leuven, Leuven 1995)

45.68 D. Thierens: Scalability problems of simple genetic algorithms, Evol. Comput. 7(4), 331–352 (1999)

45.69 S. Kullback, R.A. Leibler: On information and sufficiency, Ann. Math. Stats. 22, 79–86 (1951)

45.70 R. Prim: Shortest connection networks and some generalizations, Bell Syst. Tech. J. 36, 1389–1401 (1957)

45.71 C. Chow, C. Liu: Approximating discrete probability distributions with dependence trees, IEEE Trans. Inf. Theory 14, 462–467 (1968)

45.72 L.A. Marascuilo, M. McSweeney: Nonparametric and Distribution. Free Methods for the Social Sciences (Brooks/Cole, Monterey 1977)

45.73 P.A.N. Bosman, D. Thierens: Linkage information processing in distribution estimation algorithms, Genet. Evol. Comput. Conf. (1999) pp. 60–67

45.74 H. Mühlenbein, T. Mahnig, A.O. Rodriguez: Schemata, distributions and graphical models in evolutionary optimization, J. Heuristics 5, 215–247 (1999)

45.75 J.J. Rissanen: Modelling by shortest data description, Automatica 14, 465–471 (1978)

45.76 J.J. Rissanen: Stochastic Complexity in Statistical Inquiry (World Scientific, Singapore 1989)

Part E | 45

45.77   J.J. Rissanen: Fisher information and stochastic complexity, IEEE Trans. Inf. Theory **42**(1), 40–47 (1996)

45.78   G. Schwarz: Estimating the dimension of a model, Ann. Stat. **6**, 461–464 (1978)

45.79   K. Sastry, D.E. Goldberg: On extended compact genetic algorithm. IlliGAL Rep. No. 2000026 (University of Illinois, Urbana 2000)

45.80   K. Sastry: Efficient atomic cluster optimization using a hybrid extended compact genetic algorithm with seeded population. IlliGAL Rep. No. 2001018 (University of Illinois, Urbana 2001)

45.81   K. Sastry, D.E. Goldberg, D.D. Johnson: Scalability of a hybrid extended compact genetic algorithm for ground state optimization of clusters, Mater. Manuf. Process. **22**(5), 570–576 (2007)

45.82   T.L. Yu, D.E. Goldberg, Y.P. Chen: A genetic algorithm design inspired by organizational theory: A pilot study of a dependency structure matrix driven genetic algorithm. IlliGAL Rep. No. 2003007 (University of Illinois, Urbana 2003)

45.83   T.L. Yu, D.E. Goldberg, K. Sastry, C.F. Lima, M. Pelikan: Dependency structure matrix, genetic algorithms, and effective recombination, Evol. Comput. **17**(4), 595–626 (2009)

45.84   T.L. Yu: A matrix approach for finding extrema: Problems with Modularity, Hierarchy, and Overlap, Ph.D. Thesis (University of Illinois at Urbana-Champaign, Urbana 2006)

45.85   G.F. Cooper, E.H. Herskovits: A Bayesian method for the induction of probabilistic networks from data, Mach. Learn. **9**, 309–347 (1992)

45.86   D. Heckerman, D. Geiger, D. M. Chickering: LearningBayesian networks: The combination of knowledge and statistical data, Tech. Rep. MSR-TR-94-09 (Microsoft Research, Redmond 1994)

45.87   D. Heckerman, D. Geiger, D.M. Chickering: Learning bayesian networks: The combination of knowledge and statistical data, Mach. Learn. **20**(3), 197–243 (1995)

45.88   M. Pelikan, D. E. Goldberg: A comparative study of scoring metrics in the Bayesian optimization algorithm: Minimum description length and Bayesian-Dirichlet. Unpublished Tech. Rep. (2000)

45.89   M.W. Hauschild, M. Pelikan: Intelligent bias of network structures in the hierarchical BOA, Genet. Evol. Comput. Conf. (2009) pp. 413–420

45.90   J. Schwarz, J. Ocenasek: A problem-knowledge based evolutionary algorithm KBOA for hypergraph partitioning, Proc. 4th Jt. Conf. Knowl.-Based Softw. Eng. (2000) pp. 51–58

45.91   S. Baluja, S. Davies: Fast probabilistic modeling for combinatorial optimization, Proc. 15th Natl. Conf. Artif. Intell. (1998) pp. 469–476

45.92   M. Pelikan, K. Sastry, D.E. Goldberg: iBOA: The incremental Bayesian optimization algorithm, Genet. Evol. Comput. Conf. (2008) pp. 455–462

45.93   G.R. Harik: Finding multimodal solutions using restricted tournament selection, Int. Conf. Genet. Algorith. (1995) pp. 24–31

45.94   R.A. Watson, G.S. Hornby, J.B. Pollack: Modeling building-block interdependency, Lect. Notes Comput. Sci. **1498**, 97–106 (1998)

45.95   R. Santana: Estimation of distribution algorithms with Kikuchi approximations, Evol. Comput. **13**(1), 67–97 (2005)

45.96   S. Shakya, A.E.I. Brownlee, J.A.W. McCall, F.A. Fournier, G. Owusu: A fully multivariate DEUM algorithm, IEEE Congr. Evol. Comput. (2009) pp. 479–486

45.97   S.K. Shakya: DEUM: A Framework for an Estimation of Distribution Algorithm based on Markov Random Fields, Ph.D. Thesis (Robert Gordon University, Aberdeen 2006)

45.98   B.T. Zhang, S.Y. Shin: Bayesian evolutionary optimization using Helmholtz machines, Lect. Notes Comput. Sci. **1917**, 827–836 (2000)

45.99   M. Pelikan, D.E. Goldberg: Hierarchical BOA solves Ising spin glasses and maxsat, Gene. Evol. Comput. Conf. (2003) pp. 1275–1286

45.100   M. Pelikan, A.K. Hartmann: Searching for ground states of Ising spin glasses with hierarchical BOA and cluster exact approximation. In: *Scalable Optimization via Probabilistic Modeling: From Algorithms to Applications*, ed. by E. Cantú-Paz, M. Pelikan, K. Sastry (Springer, Berlin, Heidelberg 2006)

45.101   S.K. Shakya, J.A. McCall, D.F. Brown: Solving the Ising spin glass problem using a bivariate EDA based on Markov random fields, IEEE Congr. Evol. Comput. (2006) pp. 908–915

45.102   H. Mühlenbein, T. Mahnig: Evolutionary optimization and the estimation of search distributions with applications to graph bipartitioning, Int. J. Approx. Reason. **31**(3), 157–192 (2002)

45.103   J. Schwarz, J. Ocenasek: Experimental study: Hypergraph partitioning based on the simple and advanced algorithms BMDA and BOA, Int. Conf. Soft Comput. (1999) pp. 124–130

45.104   F. Rothlauf, D.E. Goldberg, A. Heinzl: Bad codings and the utility of well-designed genetic algorithms. IlliGAL Rep. No. 200007 (University of Illinois, Urbana 2000)

45.105   J. Li, U. Aickelin: A Bayesian optimization algorithm for the nurse scheduling problem, IEEE Congr. Evol. Comput. (2003) pp. 2149–2156

45.106   R. Arst, B.S. Minsker, D.E. Goldberg: Comparing advanced genetic algorithms and simple genetic algorithms for groundwater management, Proc. Water Resour. Plan. Manag. Conf. (2002)

45.107   M.S. Hayes, B.S. Minsker: Evaluation of advanced genetic algorithms applied to groundwater remediation design, Proc. World Water Environ. Resour. Congr. 2005 (2005)

45.108   E. Cantú-Paz: Supervised and unsupervised discretization methods for evolutionary algorithms,

Workshop Proc. Genet. Evol. Comput. Conf. (2001) pp. 213–216

45.109   Y.P. Chen, C.H. Chen: Enabling the extended compact genetic algorithm for real-parameter optimization by using adaptive discretization, Evol. Comput. **18**(2), 199–228 (2010)

45.110   D.E. Goldberg: Real-coded genetic algorithms, virtual alphabets, and blocking, Complex Syst. **5**(2), 139–167 (1991)

45.111   M. Pelikan, K. Sastry, S. Tsutsui: Getting the best of both worlds: Discrete and continuous genetic and evolutionary algorithms in concert, Inf. Sci. **156**(3–4), 147–171 (2003)

45.112   M. Gallagher, M. Frean: Population-based continuous optimization, probabilistic modelling and mean shift, Evol. Comput. **13**(1), 29–42 (2005)

45.113   M. Gallagher, M. Frean, T. Downs: Real-valued evolutionary optimization using a flexible probability density estimator, Genet. Evol. Comput. Conf. (1999), pp. 840–846 13–17

45.114   I. Rechenberg: *Evolutionsstrategie: Optimierung technischer Systeme nach Prinzipien der biologischen Evolution* (Frommann-Holzboog, Stuttgart 1973)

45.115   N. Hansen, A. Ostermeier, A. Gawelczyk: On the adaptation of arbitrary normal mutation distributions in evolution strategies: The generating set adaptation, Int. Conf. Genet. Algorithms (1995) pp. 57–64

45.116   I. Rechenberg: *Evolutionsstrategie '94* (Frommann-Holzboog, Stuttgart 1994)

45.117   H.P. Schwefel: *Numerische Optimierung von Computer-Modellen mittels der Evolutionsstrategie* (Birkhäuser, Basel, Switzerland 1977)

45.118   P.A.N. Bosman, D. Thierens: Mixed IDEAs, Tech. Rep. UU-CS-2000-45 (Utrecht University, Utrecht 2000)

45.119   N. Khan, D.E. Goldberg, M. Pelikan: Multiobjective Bayesian optimization algorithm. IlliGAL Rep. No. 2002009 (University of Illinois, Urbana 2002)

45.120   M. Laumanns, J. Ocenasek: Bayesian optimization algorithms for multi-objective optimization, Lect. Notes Comput. Sci. **2433**, 298–307 (2002)

45.121   P.A.N. Bosman, D. Thierens: Exploiting gradient information in continuous iterated density estimation evolutionary algorithms, Proc. Belg.-Neth. Conf. Artif. Intell. (2001) pp. 69–76

45.122   P.A.N. Bosman, J. Grahl, F. Rothlauf: SDR: A better trigger for adaptive variance scaling in normal EDAs, Genet. Evol. Comput. Conf. (2007) pp. 492–499

45.123   P. A. N. Bosman, J. Grahl, D. Thierens: AMaLGaM IDEAs in noiseless black-box optimization benchmarking. Black Box Optim. Benchmarking BBOB Workshop Genet. Evol. Comput. Conf., GECCO-2009 (2009) pp. 2247–2254

45.124   P.A.N. Bosman, J. Grahl, D. Thierens: AMaLGaM IDEAs in noisy black-box optimization bench-

45.125   I. Servet, L. Trave-Massuyes, D. Stern: Telephone network traffic overloading diagnosis and evolutionary computation techniques, Proc. Eur. Conf. Artif. Evol. (1997) pp. 137–144

45.126   S. Tsutsui, M. Pelikan, D.E. Goldberg: Probabilistic model-building genetic algorithm using marginal histogram models in continuous domain, Knowl.-Based Intell. Inf. Eng. Syst. Allied Thech. (2001) pp. 112–121

45.127   M. Pelikan, D.E. Goldberg, S. Tsutsui: Combining the strengths of the Bayesian optimization algorithm and adaptive evolution strategies, Genet. Evol. Comput. Conf. (2002) pp. 512–519

45.128   J. Ocenasek, S. Kern, N. Hansen, P. Koumoutsakos: A mixed Bayesian optimization algorithm with variance adaptation, Lect. Notes Comput. Sci. **3242**, 352–361 (2004)

45.129   J.R. Koza: *Genetic programming: On the Programming of Computers by Means of Natural Selection* (MIT, Cambridge 1992)

45.130   R.P. Salustowicz, J. Schmidhuber: Probabilistic incremental program evolution, Evol. Comput. **5**(2), 123–141 (1997)

45.131   R.P. Salustowicz, J. Schmidhuber: Probabilistic incremental program evolution: Stochastic search through program space, Proc. Eur. Conf. Mach. Learn. (1997) pp. 213–220

45.132   R. Salustowicz, J. Schmidhuber: H-PIPE: Facilitating hierarchical program evolution through skip nodes, Tech. Rep. IDSIA-08-98 (IDSIA, Lugano 1998)

45.133   S. Handley: On the use of a directed acyclic graph to represent a population of computer programs, Int. Conf. Evol. Comput. (1994) pp. 154–159

45.134   K. Yanai, H. Iba: Estimation of distribution programming based on Bayesian network, IEEE Congr. Evol. Comput. (2003) pp. 1618–1625

45.135   K. Sastry, D.E. Goldberg: Probabilistic model building and competent genetic programming. In: *Genetic Programming Theory and Practise*, ed. by R.L. Riolo, B. Worzel (Kluwer Acadamic, Boston 2003) pp. 205–220

45.136   M. Looks, B. Goertzel, C. Pennachin: Learning computer programs with the Bayesian optimization algorithm, Genet. Evol. Comput. Conf. (2005) pp. 747–748

45.137   M. Looks: Competent Program Evolution, Ph.D. Thesis (Washington University, St. Louis 2006)

45.138   M. Looks: Scalable estimation-of-distribution program evolution, Genet. Evol. Comput. Conf. (2007) pp. 539–546

45.139   Y. Shan, R. McKay, H.A. Abbass, D. Essam: Program evolution with explicit learning: A new framework for program automatic synthesis, IEEE Congr. Evol. Comput. (2003) pp. 1639–1646

45.140   A. Ratle, M. Sebag: Avoiding the bloat with probabilistic grammar-guided genetic program-

Part E | 45

ming, 5th Int. Conf. Evol. Artif. (2001) pp. 255–266

45.141   M. Dorigo, G.D. Caro, L.M. Gambardella: Ant algorithms for discrete optimization, Artif. Life **5**(2), 137–172 (1999)

45.142   Y. Shan, R.I. McKay, R. Baxter: Grammar model-based program evolution, IEEE Congr. Evol. Comput. (2004) pp. 478–485

45.143   I. Tanev: Incorporating learning probabilistic context-sensitive grammar in genetic programming for efficient evolution and adaptation of snakebot, Proc. 8th Eur. Conf. Genet. Progr. (2005) pp. 155–166

45.144   F. Gruau: On using syntactic constraints with genetic programming. In: *Advances in Genetic Programming*, Vol. 2, ed. by P.J. Angeline, K.E. Kinnear Jr. (MIT, Cambridge 1996) pp. 377–394

45.145   P. Whigham: Grammatically-based genetic programming, Proc. Workshop Genet. Progr. Theory Real-World Appl. (1995) pp. 33–41

45.146   M.L. Wong, K.S. Leung: Genetic logic programming and applications, IEEE Expert **10**(5), 68–76 (1995)

45.147   J.C. Bean: Genetic algorithms and random keys for sequencing and optimization, ORSA J. Comput. **6**(2), 154–160 (1994)

45.148   P.A.N. Bosman, D. Thierens: New IDEAs and more ICE by learning and using unconditional permutation factorizations, Late-Breaking Pap. Genet. Evol. Comput. Conf. (2001) pp. 13–23

45.149   V. Robles, P. de Miguel, P. Larrañaga: Solving the traveling salesman problem with edas. In: *Estimation of Distribution Algorithms. A New Tool for Evolutionary Computation*, ed. by P. Larrañaga, J.A. Lozano (Kluwer Academic, Boston 2002) pp. 227–238

45.150   P.A.N. Bosman, D. Thierens: Crossing the road to efficient IDEAs for permutation problems, Genet. Evol. Comput. Conf. (2001) pp. 219–226

45.151   E. Bengoetxea, P. Larrañaga, I. Bloch, A. Perchant, C. Boeres: Inexact graph matching using learning and simulation of Bayesian networks, Proc. CaNew Workshop Conf. (2000)

45.152   M. Pelikan, S. Tsutsui, R. Kalapala: Dependency trees, permutations, and quadratic assignment problem. MEDAL Rep. No. 2007003 (University of Missouri, St. Louis 2007)

45.153   S. Tsutsui, D.E. Goldberg, M. Pelikan: Solving sequence problems by building and sampling edge histograms. IlliGAL Rep. No. 2002024 (University of Illinois, Urbana 2002)

45.154   A. Salhi, J.A.V. Rodríguez, Q. Zhang: An estimation of distribution algorithm with guided mutation for a complex flow shop scheduling problem, Genet. Evol. Comput. Conf. (2007) pp. 570–576

45.155   Q. Zhang, H. Li: MOEA/D: A multiobjective evolutionary algorithm based on decomposition, IEEE Trans. Evol. Comput. **11**(6), 712–731 (2007)

45.156   Q. Zhang, J. Sun, E.P.K. Tsang: An evolutionary algorithm with guided mutation for the maximum clique problem, IEEE Trans. Evol. Comput. **9**(2), 192–200 (2005)

45.157   H. Mühlenbein, D. Schlierkamp-Voosen: Predictive models for the breeder genetic algorithm I. Continuous parameter optimization, Evol. Comput. **1**(1), 25–49 (1993)

45.158   H. Mühlenbein, T. Mahnig: Convergence theory and applications of the factorized distribution algorithm, J. Comput. Inf. Tech. **7**(1), 19–32 (1998)

45.159   Q. Zhang: On stability of fixed points of limit models of univariate marginal distribution algorithm and factorized distribution algorithm, IEEE Trans. Evol. Comput. **8**(1), 80–93 (2004)

45.160   C. Gonzalez, J. Lozano, P. Larrañaga: Analyzing the PBIL algorithm by means of discrete dynamical systems, Complex Syst. **4**(12), 465–479 (2001)

45.161   M. Höhfeld, G. Rudolph: Towards a theory of population-based incremental learning, Int. Conf. Evol. Comput. (1997) pp. 1–6

45.162   D.E. Goldberg, K. Deb, J.H. Clark: Genetic algorithms, noise, and the sizing of populations, Complex Syst. **6**, 333–362 (1992)

45.163   D.E. Goldberg, K. Sastry, T. Latoza: On the supply of building blocks, Genet. Evol. Comput. Conf. (2001) pp. 336–342

45.164   G.R. Harik, E. Cantú-Paz, D.E. Goldberg, B.L. Miller: The gambler's ruin problem, genetic algorithms, and the sizing of populations, Int. Conf. Evol. Comput. (1997) pp. 7–12

45.165   G. Harik, E. Cantú-Paz, D.E. Goldberg, B.L. Miller: The gambler's ruin problem, genetic algorithms, and the sizing of populations, Evol. Comput. **7**(3), 231–253 (1999)

45.166   M. Pelikan, K. Sastry, D.E. Goldberg: Scalability of the Bayesian optimization algorithm, Int. J. Approx. Reason. **31**(3), 221–258 (2002)

45.167   T.L. Yu, K. Sastry, D.E. Goldberg, M. Pelikan: Population sizing for entropy-based model building in estimation of distribution algorithms, Genet. Evol. Comput. Conf. (2007) pp. 601–608

45.168   H. Mühlenbein: Convergence of estimation of distribution algorithms for finite samples. Tech. Rep. (Fraunhofer Institut, Sankt Augustin 2008)

45.169   J.L. Shapiro: Drift and scaling in estimation of distribution algorithms, Evol. Comput. **13**, 99–123 (2005)

45.170   P.A.N. Bosman, J. Grahl, D. Thierens: Enhancing the performance of maximum-likelihood Gaussian EDAs using anticipated mean shift, Lect. Notes Comput. Sci. **5199**, 133–143 (2008)

45.171   Y. Gao, J. Culberson: Space complexity of estimation of distribution algorithms, Evol. Comput. **13**, 125–143 (2005)

45.172   M.W. Hauschild, M. Pelikan, K. Sastry, C.F. Lima: Analyzing probabilistic models in hierarchical BOA, IEEE Trans. Evol. Comput. **13**(6), 1199–1217 (2009)

45.173   C. Lima, F. Lobo, M. Pelikan, D.E. Goldberg: Model accuracy in the Bayesian optimization algorithm, Soft Comput. **15**, 1351–1371 (2011)

45.174   C. Echegoyen, A. Mendiburu, R. Santana, J.A. Lozano: Toward understanding EDAs based on Bayesian networks through a quantitative analysis, IEEE Trans. Evol. Comput. **99**, 1–17 (2011)

45.175   E. Radetic, M. Pelikan: Spurious dependencies and EDA scalability, Genet. Evol. Comput. Conf. (2010) pp. 303–310

45.176   D.E. Goldberg: *The Design of Innovation: Lessons from and for Competent Genetic Algorithms* (Kluwer Academic, Boston 2002)

45.177   E. Cantú-Paz: *Efficient and Accurate Parallel Genetic Algorithms* (Kluwer Academic, Boston 2000)

45.178   J. Ocenasek: *Parallel Estimation of Distribution Algorithms: Principles and Enhancements* (Lambert Academic, Saarbrüchen 2010)

45.179   J. Ocenasek, J. Schwarz: The parallel Bayesian optimization algorithm, Proc. Eur. Symp. Comput. Intell. (2000) pp. 61–67

45.180   J. Ocenasek: Parallel Estimation of Distribution Algorithms, Ph.D. Thesis (Brno University of Technology, Brno 2002)

45.181   J. Ocenasek, E. Cantú-Paz, M. Pelikan, J. Schwarz: Design of parallel estimation of distribution algorithms. In: *Scalable Optimization via Probabilistic Modeling: From Algorithms to Applications*, ed. by M. Pelikan, K. Sastry, E. Cantú-Paz (Springer, Berlin, Heidelberg 2006)

45.182   K. Sastry, D.E. Goldberg, X. Llorà: Towards billion-bit optimization via a parallel estimation of distribution algorithm, Genet. Evol. Comput. Conf. (GECCO-2007) (2007) pp. 577–584

45.183   G.E. Hinton, S.J. Nowlan: How learning can guide evolution, Complex Syst. **1**, 495–502 (1987)

45.184   A. Sinha, D.E. Goldberg: A survey of hybrid genetic and evolutionary algorithms. IlliGAL Rep. No. 2003004 (University of Illinois, Urbana 2003)

45.185   C. Grosan, A. Abraham, H. Ishibuchi (Eds.): *Hybrid Evolutionary Algorithms. Studies in Computational Intelligence* (Springer, Berlin, Heidelberg 2007)

45.186   E. Radetic, M. Pelikan, D.E. Goldberg: Effects of a deterministic hill climber on hBOA, Genet. Evol. Comput. Conf. (2009) pp. 437–444

45.187   P.A.N. Bosman: On gradients and hybrid evolutionary algorithms for real-valued multi-objective optimization, IEEE Trans. Evol. Comput. **16**(1), 51–69 (2012)

45.188   C.F. Lima, M. Pelikan, K. Sastry, M.V. Butz, D.E. Goldberg, F.G. Lobo: Substructural neighborhoods for local search in the Bayesian optimization algorithm, Lect. Notes Comput. Sci. **4193**, 232–241 (2006)

45.189   C.F. Lima, M. Pelikan, F.G. Lobo, D.E. Goldberg: Loopy substructural local search for the Bayesian optimization algorithm. In: *Engineering Stochas-tic Local Search Algorithms. Designing, Implementing and Analyzing Effective Heuristics*, ed. by T. Stützle, M. Birattari, H.H. Hoos (Springer, Berlin, Heidelberg 2009) pp. 61–75

45.190   A. Mendiburu, R. Santana, J.A. Lozano: Introducing belief propagation in estimation of distribution algorithms: A parallel approach, Tech. Rep. EHU-KAT-IK-11-07 (University of the Basque Country, San Sebastián 2007)

45.191   A. Ochoa, R. Hüns, M. Soto, H. Mühlenbein: A maximum entropy approach to sampling in EDA, Lect. Notes Comput. Sci. **2905**, 683–690 (2003)

45.192   K. Sastry, D.E. Goldberg: Designing competent mutation operators via probabilistic model building of neighborhoods, Genet. Evol. Comput. Conf. (GECCO) (2004) pp. 114–125, Also IlliGAL Rep. No. 2004006

45.193   D. Iclanzan, D. Dumitrescu: Overcoming hierarchical difficulty by hill-climbing the building block structure, Genet. Evol. Comput. Conf. (2007) pp. 1256–1263

45.194   P. Posík, S. Vanícek: Parameter-less local optimizer with linkage identification for deterministic order-k decomposable problems, Genet. Evol. Comput. Conf. (2011) pp. 577–584

45.195   D.E. Goldberg: Using time efficiently: Genetic-evolutionary algorithms and the continuation problem, Genet. Evol. Comput. Conf. (1999) pp. 212–219

45.196   D.E. Goldberg, S. Voessner: Optimizing global-local search hybrids, Genet. Evol. Comput. Conf. (1999) pp. 220–228

45.197   K. Sastry, D.E. Goldberg: Let's get ready to rumble: Crossover versus mutation head to head, Genet. Evol. Comput. Conf. (GECCO) (2004) pp. 126–137

45.198   K. Sastry, D.E. Goldberg: Let's get ready to rumble redux: Crossover versus mutation head to head on exponentially scaled problems, Genet. Evol. Comput. Conf. (GECCO) (2007) pp. 114–125, Also IlliGAL Report No. 2004006

45.199   S. Baluja: Incorporating a priori knowledge in probabilistic-model based optimization. In: *Scalable Optimization via Probabilistic Modeling: From Algorithms to Applications*, ed. by E. Cantú-Paz, M. Pelikan, K. Sastry (Springer, Berlin, Heidelberg 2006) pp. 205–219

45.200   M. Pelikan, M. Hauschild: Distance-based bias in model-directed optimization of additively decomposable problems. MEDAL Rep. No. 2012001 (University of Missouri, St. Louis 2012)

45.201   M. Pelikan, M. Hauschild, P.L. Lanzi: Transfer learning, soft distance-based bias, and the hierarchical BOA, Lect. Notes Comput. Sci. **7491**, 173–183 (2012)

45.202   M. Pelikan, K. Sastry: Fitness inheritance in the Bayesian optimization algorithm, Genet. Evol. Comput. Conf. (2004) pp. 48–59

Part E | 45

45.203   K. Sastry, M. Pelikan, D.E. Goldberg: Efficiency enhancement of genetic algorithms via building-block-wise fitness estimation, IEEE Congr. Evol. Comput. (2004) pp. 720–727

45.204   K. Sastry, M. Pelikan, D.E. Goldberg: Efficiency enhancement of estimation of distribution algorithms. In: *Scalable Optimization via Probabilistic Modeling: From Algorithms to Applications*, ed. by E. Cantú-Paz, M. Pelikan, K. Sastry (Springer, Berlin, Heidelberg 2006) pp. 161–185

45.205   M. Pelikan, K. Sastry, D.E. Goldberg: Sporadic model building for efficiency enhancement of the hierarchical BOA, Genet. Progr. Evol. Mach. **9**(1), 53–84 (2008)

45.206   M. Pelikan: Probabilistic model-building genetic algorithms, Proc. 13th Annu. Conf. Companion Genet. Evol. Comput. (2011) pp. 913–940

45.207   S. Tsutsui, M. Pelikan, A. Ghosh: Performance of aggregation pheromone system on unimodal and multimodal problems, IEEE Congr. Evol. Comput. (2005) pp. 880–887

45.208   C.W. Ahn, R.S. Ramakrishna: Multiobjective real-coded Bayesian optimization algorithm revisited: Diversity preservation, Genet. Evol. Comput. Conf. (2007) pp. 593–600

45.209   Q. Zhang, A. Zhou, Y. Jin: RM-MEDA: A regularity model-based multiobjective estimation of distribution algorithm, IEEE Trans. Evol. Comput. **12**(1), 41–63 (2008)

45.210   H. Li, Q. Zhang: A multiobjective differential evolution based on decomposition for multiobjective optimization with variable linkages, Lect. Notes Comput. Sci. **4193**, 583–592 (2006)

46. Parallel Evolutionary Algorithms

Dirk Sudholt

Evolutionary algorithms (EAs) have given rise to many parallel variants, fuelled by the rapidly increasing number of CPU cores and the ready availability of computation power through GPUs and cloud computing. A very popular approach is to parallelize evolution in island models, or coarse-grained EAs, by evolving different populations on different processors. These populations run independently most of the time, but they periodically communicate genetic information to coordinate search. Many applications have shown that island models can speed up computation significantly, and that parallel populations can further increase solution diversity.

The aim of this book chapter is to give a gentle introduction into the design and analysis of parallel evolutionary algorithms, in order to understand how parallel EAs work, and to explain when and how speedups over sequential EAs can be obtained.

Understanding how parallel EAs work is a challenging goal as they represent interacting stochastic processes, whose dynamics are determined by several parameters and design choices. This chapter uses a theory-guided perspective to explain how key parameters affect performance, based on recent advances on the theory of parallel EAs. The presented results give insight into the fundamental working principles of parallel EAs, assess the impact of parameters and design choices on performance, and contribute to an informed design of effective parallel EAs.

Part E | 46

Recent years have witnessed the emergence of a huge number of parallel computer architectures. Almost every desktop or notebook PC, and even mobile phones, come with several CPU cores built in. Also GPUs have been discovered as a source of massive computation power at no extra cost. Commercial IT solutions often use clusters with hundreds and thousands of CPU cores and cloud computing has become an affordable and convenient way of gaining CPU power.

With these resources readily available, it has become more important than ever to design algorithms that can be implemented effectively in a parallel architecture. Evolutionary algorithms (EA) are popular general-purpose metaheuristics inspired by the natural evolution of species. By using operators like mutation, recombination, and selection, a multi-set of solutions – the *population* – is evolved over time. The hope is that this artificial evolution will explore vast regions of the search space and yet use the principle of *survival of the fittest* to generate good solutions for the problem at hand. Countless applications as well as theoretical results have demonstrated that these algorithms are effective on many hard optimization problems.

One of many advantages of EAs is that they are easy to parallelize. The process of artificial evolution can be implemented on parallel hardware in various ways. It is possible to parallelize specific operations, or to parallelize the evolutionary process itself. The latter approach has led to a variety of search algorithms called island models or cellular evolutionary algorithms. They differ from a sequential implementation in that evolution happens in a spatially structured network. Subpopulations evolve on different processors and good solutions are communicated between processors. The spread of information can be tuned easily via key parameters of the algorithm. A slow spread of information can lead to a larger diversity in the system, hence increasing exploration.

Many applications have shown that parallel EAs can speed up computation and find better solutions, compared to a sequential EA. This book chapter reviews the most common forms of parallel EAs. We highlight what distinguishes parallel EAs from sequential EAs. We also we make an effort to understand the search dynamics of parallel EA. This addresses a very hot topic since, as of today, even the impact of the most basic parameters of a parallel evolutionary algorithms are not well understood.

The chapter has a particular emphasis on theoretical results. This includes runtime analysis, or computa-

tional complexity analysis. The goal is to estimate the expected time until an EA finds a satisfactory solution for a particular problem, or problem class, by rigorous mathematical studies. This area has led to very fruitful results for general EAs in the last decade [46.1, 2]. Only recently have researchers turned to investigating parallel evolutionary algorithms from this perspective [46.3–7]. The results help to get insight into the search behavior of parallel EAs and how parameters and design choices affect performance. The presentation of these results is kept informal in order to make it accessible to a broad audience. Instead of presenting theorems and complete formal proofs, we focus on key ideas and insights that can be drawn from these analyses.

The outline of this chapter is as follows. In Sect. 46.1 we first introduce parallel models of evolutionary algorithms, along with a discussion of key design choices and parameters. Section 46.2 considers performance measures for parallel EAs, particularly notions for *speedup* of a parallel EA when compared to sequential EAs.

Section 46.3 deals with the spread of information in parallel EAs. We review various models used to describe how the number of *good* solutions increases in a parallel EA. This also gives insight into the time until the whole system is taken over by good solutions, the so-called *takeover time*.

In Sect. 46.4 we present selected examples where parallel EAs were shown to outperform sequential evolutionary algorithms. Drastic speedups were shown on illustrative example functions. This holds for various forms of parallelization, from independent runs to offspring populations and island models.

Section 46.5 finally reviews a general method for estimating the expected running time of parallel EAs. This method can be used to transfer bounds for a sequential EA to a corresponding parallel EA, in an automated fashion. We go into a bit more detail here, in order to enable the reader to apply this method by her-/himself. Illustrative example applications are given that also include problems from combinatorial optimization.

The chapter finishes with conclusions in Sect. 46.6 and pointers to further literature on parallel evolutionary algorithms.

46.1 Parallel Models

46.1.1 Master–Slave Models

There are many ways how to use parallel machines. A simple way of using parallelization is to execute operations on separate processors. This can concern variation operators like mutation and recombination as well as function evaluations. In fact, it makes most sense for function evaluations as these operations can be performed independently and they are often among the most expensive operations. This kind of architecture is known as *master–slave model*. One machine represents the master and it distributes the workload for executing operations to several other machines called slaves. It is well suited for the creation of offspring populations as offspring can be created and evaluated independently, after suitable parents have been selected.

The system is typically synchronized: the master waits until all slaves have completed their operations before moving on. However, it is possible to use asynchronous systems where the master does not wait for slaves that take too long.

The behavior of synchronized master–slave models is not different from their sequential counterparts. The implementation is different, but the algorithm – and therefore search behavior – is the same.

46.1.2 Independent Runs

Parallel machines can also be used to simulate different, independent runs of the same algorithm in parallel. Such a system is very easy to set up as no communication during the runtime is required. Only after all runs have been stopped, do the results need to be collected and the best solution (or a selection of different high-quality solutions) is output.

Alternatively, all machines can periodically communicate their current best solutions so that the system can be stopped as soon as a satisfactory solution has been found. As for master–slave models, this prevents us from having to wait until the longest run has finished.

Despite its simplicity, independent runs can be quite effective. Consider a setting where a single run of an algorithm has a particular *success probability*, i. e., a probability of finding a satisfactory solution within a given time frame. Let this probability be denoted p. By using several independent runs, this success probability can be increased significantly. This approach is commonly known as *probability amplification*.

The probability that in λ independent runs no run is successful is $(1-p)^{\lambda}$. The probability that there is at least one successful run among these is, therefore,

$$1 - (1-p)^{\lambda} . \tag{46.1}$$

Figure 46.1 illustrates this amplified success probability for various choices of λ and p.

We can see that for a small number of processors the success probability increases almost linearly. If the number of processors is large, a saturation effect occurs. The benefit of using ever more processors decreases with the number of processors used. The point where saturation happens depends crucially on p; for smaller success probabilities saturation happens only with a fairly large number of processors.

Furthermore, independent runs can be set up with different initial conditions or different parameters. This is useful to effectively explore the parameter space and to find good parameter settings in a short time.

46.1.3 Island Models

Independent runs suffer from obvious drawbacks: once a run reaches a situation where its population has become stuck in a difficult local optimum, it will most likely remain stuck forever. This is unfortunate since other runs might reach more promising regions of the search space at the same time. It makes more sense to

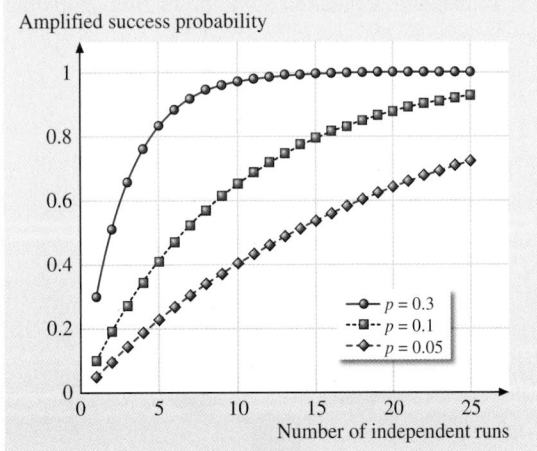

Fig. 46.1 Plots of the amplified success probability $1 - (1-p)^{\lambda}$ of a parallel system with λ independent runs, each having success probability p

establish some form of communication between the different runs to coordinate search, so that runs that have reached low-quality solutions can join in on the search in more promising regions.

In *island models*, also called *distributed EAs*, the *coarse-grained model*, or the *multi-deme model*, the population of each run is regarded an island. One often speaks of islands as *subpopulations* that together form the population of the whole island model. Islands evolve independently as in the independent run model, for most of the time. However, periodically solutions are exchanged between islands in a process called *migration*.

The idea is to have a *migration topology*, a directed graph with islands as its nodes and directed edges connecting two islands. At certain points of time selected individuals from each island are sent off to neighbored islands, i. e., islands that can be reached by a directed edge in the topology. These individuals are called migrants and they are included in the target island after a further selection process. This way, islands can communicate and compete with one another. Islands that get stuck in low-fitness regions of the search space can be taken over by individuals from more successful islands. This helps to coordinate search, focus on the most promising regions of the search space, and use the available resources effectively. An example of an island model is given in Fig. 46.2. Algorithm 46.1 shows the general scheme of a basic island model.

Algorithm 46.1 Scheme of an island model with migration interval τ
1: Initialize a population made up of subpopulations or islands, $P^{(0)} = \{P_1^{(0)}, \ldots, P_m^{(0)}\}$.
2: Let $t := 1$.
3: **loop**
4:    **for** each island i **do in parallel**
5:      **if** $t \bmod \tau = 0$ **then**

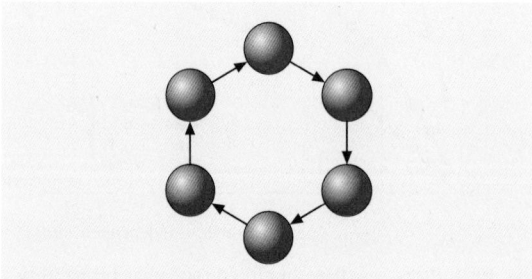

Fig. 46.2 Sketch of an island model with six islands and an example topology

6:      Send selected individuals from island $P_i^{(t)}$ to selected neighbored islands.
7:      Receive immigrants $I_i^{(t)}$ from islands for which island $P_i^{(t)}$ is a neighbor.
8:      Replace $P_i^{(t)}$ by a subpopulation resulting from a selection among $P_i^{(t)}$ and $I_i^{(t)}$.
9:    **end if**
10:    Produce $P_i^{(t+1)}$ by applying reproduction operators and selection to $P_i^{(t)}$.
11: **end for**
12: Let $t := t + 1$.
13: **end loop**

There are many design choices that affect the behavior of such an island model:

- *Emigration policy.* When migrants are sent, they can be removed from the sending island. Alternatively, copies of selected individuals can be emigrated. The latter is often called *pollination*. Also the selection of migrants is important. One might select the best, worst, or random individuals.
- *Immigration policy.* Immigrants can replace the worst individuals in the target population, random individuals, or be subjected to the same kind of selection used within the islands for parent selection or selection for replacement. Crowding mechanisms can be used, such as replacing the most similar individuals. In addition, immigrants can be recombined with individuals present on the island before selection.
- *Migration interval.* The time interval between migrations determines the speed at which information is spread throughout an island model. Its reciprocal is often called *migration frequency*. Frequent migrations imply a rapid spread of information, while rare migrations allow for more exploration. Note that a migration interval of ∞ yields independent runs as a special case.
- *Number of migrants.* The number of migrants, also called *migration size*, is another parameter that determines how quickly an island can be taken over by immigrants.
- *Migration topology.* Also the choice of the migration topology impacts search behavior. The topology can be a directed or undirected graph – after all, undirected graphs can be seen as special cases of directed graphs. Common topologies include unidirectional rings (a ring with directed edges

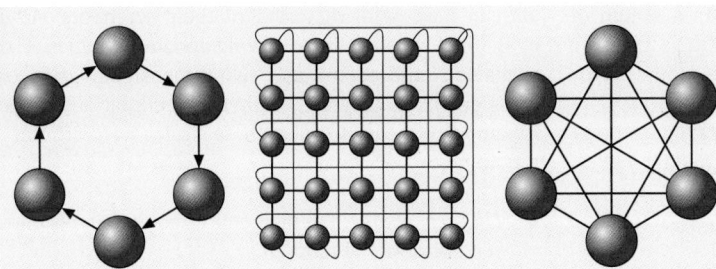

Part E | 46.1

Fig. 46.3 Sketches of common topologies: a unidirectional ring, a torus, and a complete graph. Other common topologies include bidirectional rings where all edges are undirected and grid graphs where the edges wrapping around the torus are removed

only in one direction), bidirectional rings, torus or grid graphs, hypercubes, scale-free graphs [46.8], random graphs [46.9], and complete graphs. Figure 46.3 sketches some of these topologies. An important characteristic of a topology $T = (V, E)$ is its *diameter*: the maximum number of edges on any shortest path between two vertices. Formally, $\text{diam}(T) = \max_{u,v \in V} \text{dist}(u, v)$, where $\text{dist}(u, v)$ is the graph distance, the number of edges on a shortest path from u to v. The diameter gives a good indication of the time needed to propagate information throughout the topology. Rings and torus graphs have large diameters, while hypercubes, complete graphs, and many scale-free graphs have small diameters.

Island models with non-complete topologies are also called *stepping stone models*. The impact of these design choices will be discussed in more detail in Sect. 46.3.

If all islands run the same algorithm under identical conditions, we speak of a *homogeneous island model*. *Heterogeneous island models* contain islands with different characteristics. Different algorithms might be used, different representations, objective functions, or parameters. Using heterogeneous islands might be useful if one is not sure what the best algorithm is for a particular problem. It also makes sense in the context of multiobjective optimization or when a diverse set of solutions is sought, as the islands can reflect different objective functions, or variations of the same objective functions, with an emphasis on different criteria.

Skolicki [46.10] proposed a two-level view of search dynamics in island models. The term *intra-island evolution* describes the evolutionary process that takes place within each island. On a higher level, *inter-island evolution* describes the interaction between different islands. He argues that islands can be regarded as individuals in a higher-level evolution. Islands compete with one another and islands can take over other islands, just like

individuals can replace other individuals in a regular population. One conclusion is that with this perspective an island models looks more like a compact entity.

The two levels of evolution obviously interact with one another. Which level is more important is determined by the migration interval and the other parameters of the system that affect the spread of information.

46.1.4 Cellular EAs

Cellular EAs represent a special case of island models with a more fine-grained form of parallelization. Like in the island model we have islands connected by a fixed topology. Rings and two-dimensional torus graphs are the most common choice. The most striking characteristic is that each island only contains a single individual. Islands are often called *cells* in this context, which explains the term *cellular EA*. Each individual is only allowed to mate with its neighbors in the topology. This kind of interaction happens in every generation. This corresponds to a migration interval of 1 in the con-

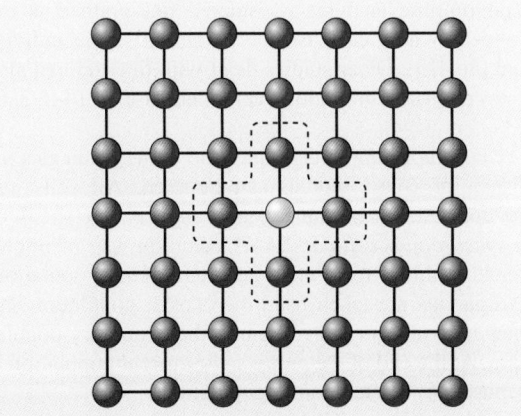

Fig. 46.4 Sketch of a cellular EA on a 7×7 grid graph. The *dashed line* indicates the neighborhood of the highlighted cell

text of island models. Figure 46.4 shows a sketch of a cellular EA. A scheme of a cellular EA is given in Algorithm 46.2.

Algorithm 46.2 Scheme of a cellular EA

1: Initialize all cells to form a population $P^{(0)} = \{P_1^{(0)}, \ldots, P_m^{(0)}\}$. Let $t := 0$.
2: **loop**
3:     **for** each cell i **do in parallel**
4:         Select a set S_i of individuals from $P_i^{(t)}$ out of all cells neighbored to cell i.
5:         Create a set R_i by applying reproduction operators to S_i.
6:         Create $P_i^{(t+1)}$ by selecting an individual from $\{P_i^{(t)}\} \cup R_i$.
7:     **end for**
8:     Let $t := t + 1$.
9: **end loop**

Cellular EAs yield a much more fine-grained system; they have therefore been called *fine-grained models*, *neighborhood models*, or *diffusion models*. The difference to island models is that no evolution takes place on the cell itself, i.e., there is no intra-island evolution. Improvements can only be obtained by cells interacting with one another. It is, however, possible that an island can interact with itself.

In terms of the two-level view on island models, in cellular EAs the intra-island dynamics have effectively been removed. After all, each island only contains a single individual. Fine-grained models are well suited for investigations of inter-island dynamics. In fact, the first runtime analyses considered fine-grained island models, where each island contains a single individual [46.4, 5]. Other studies dealt with fine-grained systems that use a migration interval larger than 1 [46.3, 6, 7].

For replacing individuals the same strategies as listed for island models can be used. All cells can be updated synchronously, in which case we speak of a *synchronous cellular EA*. A common way of implementing this is to create a new, temporary population. All parents are taken from the current population and new individuals are written into the temporary population. At the end of the process, the current population is replaced by the temporary population.

Alternatively, cells can be updated sequentially, resulting in an *asynchronous cellular EA*. This is likely to result in a different search behavior as individu-

als can mate with offspring of their neighbors. *Alba* et al. [46.11] define the following update strategies. The terms are tailored towards two-dimensional grids or torus graphs as they are inspired by cellular automata. It is, however, easy to adapt these strategies to arbitrary topologies:

- *Uniform choice*: the next cell to be updated is chosen uniformly at random.
- *Fixed line sweep*: the cells are updated sequentially, line by line in a grid/torus topology.
- *Fixed random sweep*: the cells are updated sequentially, according to some fixed order. This order is determined by a permutation of all cells. This permutation is created uniformly at random during initialization and kept throughout the whole run.
- *New random sweep*: this strategy is like fixed random sweep, but after each sweep is completed a new permutation is created uniformly at random.

A time step or generation is defined as the time needed to update m cells, m being the number of cells in the grid. The last three strategies ensure that within each time step each cell is updated exactly once. This yields a much more balanced treatment for all cells. With the uniform choice model is it likely that some cells must wait for a long time before being updated. In the limit, the waiting time for updates follows a Poisson distribution. Consider the random number of updates until the last cell has been updated at least once. This random process is known as the *coupon collector problem* [46.12, page 32], as it resembles the process of collecting coupons, which are drawn uniformly at random. A simple analysis shows that the expected number of updates until the last cell has been updated in the uniform choice model (or all coupons have been collected) equals

$$m \cdot \sum_{i=1}^{m} 1/i \approx m \cdot \ln(m) \, .$$

This is equivalent to

$$\sum_{i=1}^{m} 1/i \approx \ln m$$

time steps, which can be significantly larger than 1, the time for completing a sweep in any given order.

Cellular EAs are often compared to cellular automata. In the context of the latter, it is common practice to consider a two-dimensional grid and different neighborhoods. The neighborhood in Fig. 46.2 is called the *von Neumann neighborhood* or *Linear 5*. It includes the cell itself and its four neighbors along the directions north, south, west, and east. The *Moore neighborhood* or *Compact 9* in addition also contains the four cells to the north west, north east, south west, and south east. Also larger neighborhoods are common, containing cells that are further away from the center cell.

Note that using a large neighborhood on a two-dimensional grid is equivalent to considering a graph where, starting with a torus graph, for each vertex edges to nearby vertices have been added. We will, therefore, in the remainder of this chapter stick to the common notion of neighbors in a graph (i. e., vertices connected by an edge), unless there is a good reason not to.

46.1.5 A Unified Hypergraph Model for Population Structures

Sprave [46.13] proposed a unified model for population structures. It is based on hypergraphs; an extension of graphs where edges can connect more than two vertices. We present an informal definition to focus on the ideas; for formal definitions we refer to [46.13]. A hypergraph contains a set of vertices and a collection of hyperedges. Each hyperedge is a non-empty set of vertices. Two vertices are neighbored in the hypergraph if there is a hyperedge that contains both vertices. Note that the special case where each hyperedge contains two different vertices results in an undirected graph.

In Sprave's model each vertex represents an individual. Hyperedges represent the set of possible parents for each individual. The model unifies various common population models:

- *Panmictic populations*: for panmictic populations we have a set of vertices V and there is a single hyperedge that equals the whole vertex set. This reflects the fact that in a panmictic population each individual has all individuals as potential parents.
- *Island models with migration*: if migration is understood in the sense that individuals are removed, the set of potential parents for an individual contains all potential immigrants as well as all individuals from its own island, except for those that are being emigrated.
- *Island models with pollination*: if pollination is used, the set of potential parents contains all immigrants and all individuals on its own island.
- *Cellular EAs*: For each individual, the potential parents are its neighbors in the topology.

In the case of coarse-grained models, the hypergraph may depend on time. More precisely, we have different sets of potential parents when migration is used, compared to generations without migration. Sprave considers this by defining a dynamic population structure: instead of considering a single, fixed hypergraph, we consider a sequence of hypergraphs over time.

46.1.6 Hybrid Models

It is also possible to combine several of the above approaches. For instance, one can imagine an island model where each island runs a cellular EA to further promote diversity. Or one can think of hierarchical island models where islands are island models themselves. In such a system it makes sense that the inner-layer island models use more frequent migrations than the outer-layer island model. Island models and cellular EAs can also be implemented as master–slave models to achieve a better speedup.

46.2 Effects of Parallelization

An obvious effect of parallelization is that the computation time can be reduced by using multiple processors. This section describes performance measured that can be used to define this speedup. We also consider beneficial effects of using parallel EAs that can lead to superlinear speedups.

46.2.1 Performance Measures for Parallel EAs

The computation time of a parallel EA can be defined in various ways. It makes sense to use wall-clock time as the performance measure as this accounts for the

Part E | 46.2

overhead by parallelization. Under certain conditions, it is also possible to use the number of generations or function evaluations. This is feasible if these measures reflect the real running time in an adequate way, for instance if the execution of a generation (or a function evaluation) dominates the computational effort, including the effort for coordinating different machines. It is also feasible if one can estimate the overhead or the communication costs separately.

We consider settings where an EA is run until a certain goal is fulfilled. Goals can be reaching a global or local optimum or reaching a certain minimum fitness. In such a setting the goal is fixed and the running time of the EA can vary. This is in contrast to setups where the running time is fixed to a predetermined number of generations and then the quality or accuracy of the obtained solutions is compared. As *Alba* pointed out [46.14], performance comparisons of parallel and sequential EAs only make sense if they reach the same accuracy. In the following, we focus on the former setting where the same goal is used.

Still, defining speedup formally is far from trivial. It is not at all clear against what algorithm a parallel algorithm should be compared. However, this decision is essential to clarify the meaning of speedup. Not clarifying it, or using the wrong comparison, can easily yield misleading results and false claims. We present a taxonomy inspired by *Alba* [46.14], restricted to cases where a fixed goal is given:

● *Strong speedup*: the parallel run time of a parallel algorithm is compared against the sequential run time of the *best known sequential algorithm*. It was called *absolute speedup* by *Barr* and *Hickman* [46.15]. This measure captures in how far parallelization can improve upon the best known algorithms. However, it is often difficult to determine the best sequential algorithm. Most researchers, therefore, do not use strong speedup [46.14].
● *Weak speedup*: the parallel run time of an algorithm is compared against *its own sequential run time*. This gives rise to two subcases where the notion of its own sequential run time is made precise:
  – *Single machine/panmixia*: the parallel EA is compared against a canonical, panmictic version of it, running on a single machine. For instance, we might compare an island model with m islands against an EA running a single island. Thereby, the EA run on all islands is the same in both cases.

  – *Orthodox*: the parallel EA running on m machines is compared against the same parallel EA running on a single machine. This kind of speedup was called *relative speedup* by *Barr* and *Hickman* [46.15].

In the light of these essential differences, it is essential for researchers to clarify their notion of speedup.

Having clarified the comparison, we can now define the speedup and other measures. Let T_m denote the time for m machines to reach the goal. Let T_1 denote the time for a single machine, where the algorithm is chosen according to one of the definitions of speedup defined above.

The idea is to consider the ratio of T_m and the time for a single machine, T_1, as speedup. However, as we are dealing with randomized algorithms, T_1 and T_m are random variables and so the ratio of both is a random variable as well. It makes more sense to consider the ratio of expected times for both the parallel and the sequential algorithm as speedup

$$s_m = \frac{E(T_1)}{E(T_m)} .$$

Note that T_1 and T_m might have very dissimilar probability distributions. Even when both are re-scaled appropriately to obtain the best possible match between the two, they might still have different shapes and different variances. In some cases it might make sense to consider the median or other statistics instead of the expectation.

According to the speedup s_m we distinguish the following cases:

● *Sublinear speedup*: if $s_m < m$ we speak of a *sublinear speedup*. This implies that the total computation time across all machines is larger than the total computation time of the single machine (assuming no idle times in the parallel algorithm).
● *Linear speedup*: the case $s_m = m$ is known as *linear speedup*. There, the parallel and the sequential algorithm have the same total time. This outcome is very desirable as it means that parallelization does not come at a cost. There is no noticeable overhead in the parallel algorithm.
● *Superlinear speedup*: if $s_m > m$ we have a *superlinear speedup*. The total computation time of the parallel algorithm is even smaller than that of the single machine. This case is considered in more detail in the following section.

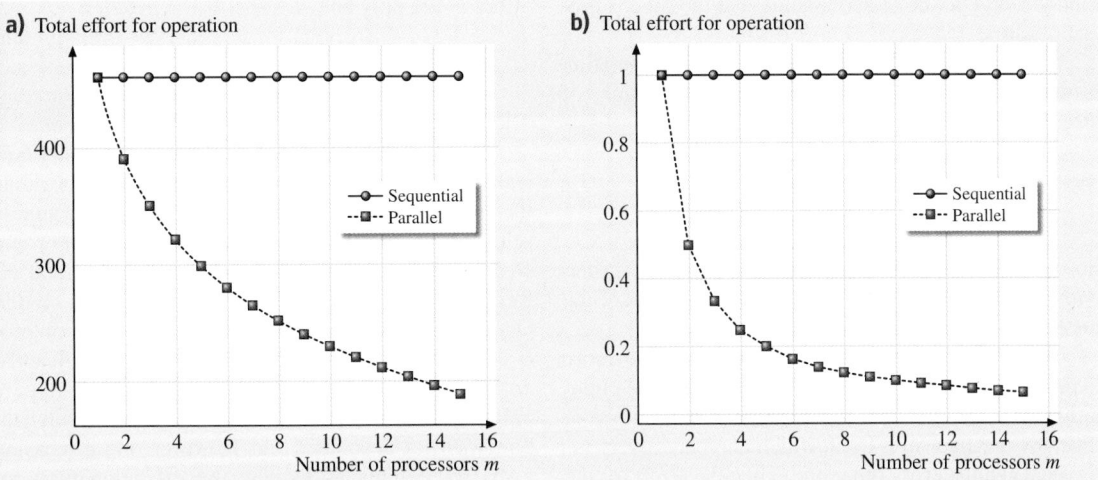

Fig. 46.5a,b Total effort for executing an operation on a single, panmictic population of size $\mu = 100$ (sequential algorithm) and a parallel algorithm with m processors and m subpopulations of size $\mu/m = 100/m$ each. The effort on a population of size n is assumed to be $n \ln n$ (**a**) and n^2 (**b**). Note that no overhead is considered for the parallel algorithm

Speedup is the best known measure, but not the only one used regularly. For the sake of completeness, we mention other measures. The *efficiency* is a normalization of the speedup

$$e_m = \frac{s_m}{m} .$$

Obviously, $e_m = 1$ is equivalent to a linear speedup. Lower efficiencies correspond to sublinear speedups, higher ones to superlinear speedups.

Another measure is called *incremental efficiency* and it measures the speedup when moving from $m-1$ processors to m processors

$$\mathrm{ie}_m = \frac{(m-1) \cdot \mathrm{E}(T_{m-1})}{m \cdot \mathrm{E}(T_m)} .$$

There is also a generalized form where $m-1$ is replaced by $m' < m$ in the above formula. This reflects the speedup when going from m' processors to m processors.

46.2.2 Superlinear Speedups

At first glance superlinear speedups seem astonishing. How can a parallel algorithm have a smaller total computation time than a sequential counterpart? After all, parallelization usually comes with significant overhead that slows down the algorithm. The existence of superlinear speedups has been discussed controversially in the literature. However, there are convincing reasons why a superlinear speedup might occur.

Alba [46.14] mentions physical sources as one possible reason. A parallel machine might have more resources in terms of memory or caches. When moving from a single machine to a parallel one, the algorithm might – purposely or not – make use of these additional resources. Also, each machine might only have to deal with smaller data packages. It might be that the smaller data fits into the cache while this was not the case for the single machine. This can make a significant performance difference.

When comparing a single panmictic population against smaller subpopulations, it might be easier to deal with the subpopulations. This holds even when the total population sizes of both systems are the same. In particular, a parallel system has an advantage if operations need time which grows faster than linearly with the size of the (sub)population.

We give two illustrative examples. Compare a single panmictic population of size μ with m subpopulations of size μ/m each. Some selection mechanisms, like ranking selection, might have to sort the individuals in the population according to their fitness. In a straightforward implementation one might use well-known sorting algorithms such as (randomized) *QuickSort*, *MergeSort*, or *HeapSort*. All of these are known to take time $\Theta(n \ln n)$ for sorting n elements, on average. Let us disregard the hidden constant and the randomness of randomized *QuickSort* and assume that the time is precisely $n \ln n$.

Now the effort of sorting the panmictic population is $\mu \ln \mu$. The total effort for sorting m populations of size μ/m each is

$$m \cdot \mu/m \cdot \ln(\mu/m) = \mu \cdot \ln(\mu/m)$$
$$= \mu \ln(\mu) - \mu \cdot \ln(m) .$$

So, the parallel system executes this operation faster, with a difference of $\mu \cdot \ln(m)$ time steps in terms of the total computation time.

This effect becomes more pronounced the more expensive operations are used (with respect to the population size). Assume that some selection mechanism or diversity mechanism is used, which compares every individual against every other one. Then the effort for the panmictic population is roughly μ^2 time steps. However, for the parallel EA and its subpopulations the total

effort would only be

$$m \cdot (\mu/m)^2 = \mu^2/m .$$

This is faster than the panmictic EA by a factor of m.

The above two growth curves are actually very typical running times for operations that take more than linear time. A table with time bounds for common selection mechanisms can be found in *Goldberg* and *Deb* [46.16]. Figure 46.5 shows plots for the total effort in both scenarios for a population size of $\mu = 100$. One can see that even with a small number of processors the total effort decreases quite significantly. To put this into perspective, most operations require only linear time. Also the overhead by parallelization was not accounted for. However, the discussion gives some hints as to why the execution time for smaller subpopulations can decrease significantly in practice.

46.3 On the Spread of Information in Parallel EAs

In order to understand how parallel EAs work, it is vital to get an idea on how quickly information is propagated. The spread of information is the most distinguishing aspect of parallel EAs, particularly distributed EAs. This includes island models and cellular EAs. Many design choices can tune the speed at which information is transmitted: the topology, the migration interval, the number of migrants, and the policies for emigration and immigration.

46.3.1 Logistic Models for Growth Curves

Many researchers have turned to investigating the selection pressure in distributed EAs in a simplified model. Assume that in the whole system we only have two types of solutions: current best individuals and worse solutions. No variation is used, i.e., we consider EAs using neither mutation nor crossover. The question is the following. Using only selection and migration, how long does it take for the best solutions to take over the whole system? This time, starting from a single best solution, is referred to as *takeover time*.

It is strongly related to the study of *growth curves*: how the number of best solutions increases over time. The takeover time is the first point of time at which the number of best solutions has grown to the whole population.

Growth curves are determined by both inter-island dynamics and intra-island dynamics: how quickly current best solutions spread in one island's population, and how quickly they populate neighbored islands, until the whole topology is taken over. Both dynamics are linked: intra-island dynamics can have a direct impact on inter-island dynamics as the fraction of best individuals can decide how many (if any) best individuals emigrate.

For intra-island dynamics one can consider results on panmictic EAs. Logistic curves have been proposed and found to fit simulations of takeover times very well for common selection schemes [46.16]. These curves are defined by the following equation. If $P(t)$ is the proportion of best individuals in the population at time t, then

$$P(t) = \frac{1}{1 + \left(\frac{1}{P(0)} - 1\right) e^{-at}} ,$$

where a is called the growth coefficient. One can see that the proportion of best individuals increases exponentially, but then the curve saturates as the proportion approaches 1.

Sarma and *De Jong* [46.17] considered growth curves in cellular EAs. They presented a detailed empirical study of the effects of the neighborhood size and the shape of the neighborhood for different selection schemes. They showed that logistic curves as defined above can model the growth curves in cellular EAs reasonably well.

Alba and *Luque* [46.18] proposed a logistic model called LOG tailored towards distributed EAs with periodic migration. If τ denotes the migration interval and m is the number of islands, then

$$P_{\text{LOG}}(t) = \sum_{i=0}^{m-1} \frac{1/m}{1 + a \cdot e^{-b(t-\tau \cdot i)}} .$$

In this model a and b are adjustable parameters. The model counts subsequent increases of the proportion of best individuals during migrations. However, it does not include any information about the topology and the authors admit that it only works appropriately on the ring topology [46.19, Section 4.2]. They, therefore, present an even more detailed model called TOP, which includes the diameter $\text{diam}(T)$ of the topology T.

$$P_{\text{TOP}}(t) = \sum_{i=0}^{\text{diam}(T)-1} \frac{1/m}{1 + a \cdot e^{-b(t-\tau \cdot i)}}$$
$$+ \frac{m - \text{diam}(T)/m}{1 + a \cdot e^{-b(t-\tau \cdot \text{diam}(T))}} .$$

Simulations show that this model yields very accurate fits for ring, star, and complete topologies [46.19, Section 4.3].

Luque and *Alba* [46.19, Section 4.3] proceed by analyzing the effect of the migration interval and the number of migrants. With a large migration interval, the growth curves tend to make jumps during migration and flatten out quickly to form plateaus during periods without migration. The resulting curves look like step functions, and the size of these steps varies with the migration interval.

Varying the number of migrants changes the slope of these steps. A large number of migrants has a better chance of transmitting best individuals than a small number of migrants. However, the influence of the number of migrants was found to be less drastic than the impact of the migration interval. When a medium or large migration frequency is used, the impact of the number of migrants is negligible [46.19, Section 4.5]. The same conclusion was made earlier by *Skolicki* and *De Jong* [46.20].

Luque and Alba also presented experiments with a model based on the *Sprave's* hypergraph formulation of distributed EAs [46.13]. This model gave a better fit

than the simple logistic model LOG, but it was less accurate than the model TOP that included the diameter.

For the sake of completeness, we also mention that *Giacobini* et al. [46.21] proposed an improved model for asynchronous cellular EAs, which is not based on logistic curves.

46.3.2 Rigorous Takeover Times

Rudolph [46.22, 23] rigorously analyzed takeover times in panmictic populations, for various selection schemes. In [46.22] he also dealt with the probability that the best solution takes over the whole population; this is not evident for non-elitistic algorithms. In [46.23] *Rudolph* considered selection schemes made elitistic by undoing the last selection in case the best solution would become extinct otherwise. Under this scheme the expected takeover time in a population of size μ is $O(\mu \log \mu)$.

In [46.24] *Rudolph* considered spatially structured populations in a fine-grained model. Each population has size 1, therefore vertices in the migration topology can be identified with individuals. Migration happens in every generation. Assume that initially only one vertex i in the topology is a best individual. If in every generation each non-best vertex is taken over by the best individual in its neighborhood, then the takeover time from vertex i equals

$$\max_{j \in V} \text{dist}(i, j) ,$$

where V is the set of vertices and $\text{dist}(i, j)$ denotes the graph distance, the number of edges on a shortest path from i to j.

Rudolph defines the takeover time in a setting where the initial best solution has the same chance of evolving at every vertex. Then

$$\frac{1}{|V|} \sum_{i \in V} \max_{j \in V} \text{dist}(i, j)$$

is the expected takeover time if, as above, best solutions are always propagated to their neighbors with probability 1. If this probability is lower, the expected takeover time might be higher. The above formula still represents a lower bound. Note that in non-elitist EAs it is possible that all best solutions might get lost, leading to a positive extinction probability [46.24].

Note that $\max_{j \in V} \text{dist}(i, j)$ is bounded by the *diameter* of the topology. The diameter is hence a trivial lower bound on the takeover times. *Rudolph* [46.24] conjectures that the diameter is more important than the selection mechanism used in the distributed EA.

In [46.25] the author generalizes the above arguments to coarse-grained models. Islands can contain larger populations and migration happens with a fixed frequency. In his model the author assumes that in each island new best individuals can only be generated by immigration. Migration always communicates best individuals. Hence, the takeover time boils down to a deterministic time until the last island has been reached, plus a random component for the time until all islands have been taken over completely.

Rudolph [46.25] gives tight bounds for unidirectional rings, based on the fact that each island with a best individual will send one such individual to each neighbored island. Hence, on the latter island the number of best individuals increases by 1, unless the island has been taken over completely. For more dense topologies he gives a general upper bound, which may not be tight for all graphs. If there is an island that receives best individuals from $k > 1$ other islands, the number of best individuals increases by k. (The number k could even increase over time.) It was left as an open problem to derive more tight bounds for interesting topologies other than unidirectional rings.

Other researchers followed up on Rudolph's seminal work. *Giacobini* et al. [46.26] presented theoretical and empirical results for the selection pressure on ring topologies, or linear cellular EAs. *Giacobini* et al. [46.27] did the same for toroidal cellular EAs. In particular, they considered takeover times for asynchronous cellular EAs, under various common update schemes. Finally, *Giacobini* et al. investigated growth curves for small-world graphs [46.9].

The assumption from Rudolph's model that only immigration can create new best individuals is not always realistic. If standard mutation operators are used, there is a constant probability of creating a clone of a selected parent simply by not flipping any bits. This can lead to a rapid increase in the number of high-fitness individuals.

This argument on the takeover of good solutions in panmictic populations has been studied as part of rigorous runtime analyses of population-based EAs. *Witt* [46.28] considered a simple $(\mu + 1)$ EA with uniform parent selection, standard bit mutations, no crossover, and cut selection at the end of the generation. From his work it follows that good solutions take over the population in expected time $O(\mu \log \mu)$. More precisely, if currently there is at least one individual with current best fitness i, then after $O(\mu \log \mu)$ generations all individuals in the population will have fitness i at least.

Sudholt [46.29, Lemma 2] extended these arguments to a $(\mu + \lambda)$ EA and proved an upper bound of $O(\mu/\lambda \cdot \log \mu + \log \mu)$. Note that, in contrast to other studies of takeover times, both results apply *real* EAs that actually use mutation. Extending these arguments to distributed EAs is an interesting topic for future work.

46.3.3 Maximum Growth Curves

Now, we consider inter-island dynamics in more detail. Assume for simplicity that intra-island takeover happens quickly: after each migration transmitting at least one best solution, the target island is completely taken over by best solutions before the next migra-

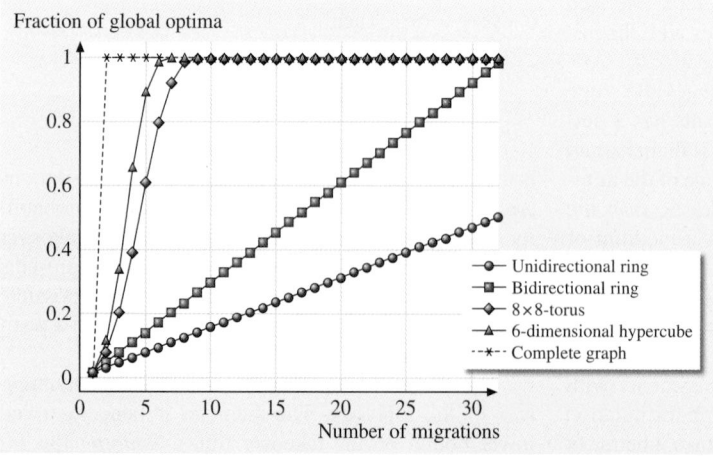

Fraction of global optima

Fig. 46.6 Plots of growth curves in an island model with 64 islands. We assume that in between two migrations all islands containing a current best solution completely take over all neighbored islands in the topology. Initially, one island contains a current best solution and all other islands are worse. The *curves* show the fraction of current best solutions in the system for different topologies: a unidirectional ring, a bidirectional ring, a square torus, a hypercube, and a complete graph

Legend:
- Unidirectional ring
- Bidirectional ring
- 8×8-torus
- 6-dimensional hypercube
- Complete graph

Number of migrations

tion. We start with only one island containing a best solution, assuming that all individuals on this island are best solutions. We call such an island an optimal island. If migrants are not subject to variation while emigrating or immigrating, we will always select best solutions for migration and, hence, successfully transmit best solutions.

These assumptions give rise to a deterministic spread of best solutions: after each migration, each optimal island will turn all neighbored islands into optimal islands. This is very similar to *Rudolph's* model [46.25], but it also accounts for a rapid intra-island takeover in between migrations.

We consider growth curves on various graph classes: unidirectional and bidirectional rings, square torus graphs, hypercubes, and complete graphs. Figure 46.6 shows these curves for all these graphs on 64 vertices. The torus graph has side lengths 8×8. The hypercube has dimension 6. Each vertex has a label made of 6 bits. All possible values for this bit string are present in the graph. Two vertices are neighbored if their labels differ in exactly one bit.

For the unidirectional ring, after $i-1$ migrations we have exactly i optimal islands, if $i \leq m$. The growth curve is, therefore, linear. For the bidirectional ring information spreads twice as fast as it can spread in two directions. After $i-1$ migrations we have $2i-1$ optimal islands if $2i-1 \leq m$.

The torus allows communication in two dimensions. After one migration there are $1+4=5$ optimal islands. After two migrations this number is $1+4+8$, and after three migrations it is $1+4+8+12$. In general, after $i-1$ migrations we have

$$1 + \sum_{j=1}^{i-1} 4j = 1 + 2i(i-1) = 1 + 2i^2 - 2i$$

optimal islands, as long as the optimal islands can freely spread out in all four directions, north, south, west, and east. At some point the ends of the region of optimal islands will meet, i.e., the northern tip meets the southern one and the same goes for west and east. Afterwards, we observe regions of non-optimal islands that constantly shrink, until all islands are optimal. The growth curve for the torus is hence quadratic at first and then it starts to saturate. The deterministic growth on torus graphs was also considered in [46.30].

For the hypercube, we can without loss of generality assume that the initial optimal island has a label

containing only zeros. After one migration all islands whose label contains a single one become optimal. After two migrations the same holds for all islands with two ones, and so on. The number of optimal islands after i migrations in a d-dimensional hypercube (i.e., $m = 2^d$) is hence $\sum_{j=0}^{i} \binom{d}{j}$. This number is close to d^j during the first migrations and then at some point starts to saturate. The complete graph is the simplest one to analyze here as it will be completely optimal after one migration.

These arguments and Fig. 46.6 show that the growth curves can depend tremendously on the migration topology. For sparse topologies like rings or torus graphs, in the beginning the growth is linear or quadratic, respectively. This is much slower than the exponential growth observed in logistic curves. Furthermore, for the ring there is no saturation; linear curves are quite dissimilar to logistic curves.

This suggests that logistic curves might not be the best models for growth curves across all topologies. The plots by *Luque* and *Alba* [46.19, Section 4.3] show a remarkably good overall fit for their TOP model. However, this might be due to the optimal choice of the parameters a and b and the fact that logistic curves are easily adaptable to various curves of roughly similar shape. We believe that it is possible to derive even more accurate models for common topologies, based on results by *Giacobini* et al. [46.9, 26, 27]. This is an interesting challenge for future work.

46.3.4 Propagation

So far, we have only considered models where migration always successfully transmits best individuals. For non-trivial selection of emigrants, this is not always given. Also if crossover is used during migration, due to disruptive effects migration is not always successful. If we consider randomized migration processes, things become more interesting.

Rowe et al. [46.31] considered a model of *propagation* in networks. Consider a network where vertices are either informed or not. In each round, each informed vertex tries to inform each of its neighbors. Every such trial is successful with a given probability p, and then the target island becomes informed. These decisions are made independently. Note that an uninformed island might obtain a probability larger than p of becoming informed, in case several informed islands try to inform it. The model is inspired by models from epidemiology; it can be used to model the spread of a disease.

The model of propagation of information directly applies to our previous setting where the network is the migration topology and p describes the probability of successfully migrating a current best solution. Note that when looking for estimations of growth curves and upper bounds on the takeover time, we can assume that p is a lower bound on the actual probability of a successful transmission. Then the model becomes applicable to a broader range of settings, where islands can have different transmission probabilities.

On some graphs like unidirectional rings, we can just multiply our growth curves by p to reflect the expected number of optimal islands after a certain time. It then follows that the time for taking over all m islands is by a factor of $1/p$ larger than in the previous, deterministic model.

However, this reasoning does not hold in general. Multiplying the takeover time in the deterministic setting by $1/p$ does not always give the expected takeover time in the random model. Consider a star graph (or *hub*), where initially only the center vertex is informed. In the deterministic case $p = 1$, the takeover time is clearly 1. However, if $0 < p < 1$, the time until the last vertex is informed is given by the maximum of $n - 1$ independent geometric distributions with parameter p. For constant p, this time is of order $\Theta(\log n)$, i.e., the time until the last vertex is informed is much larger than the expected time for any specific island to be informed.

Rowe et al. [46.31] presented a detailed analysis of hubs. They also show how to obtain a general upper bound that holds for all graphs. For every graph G with n vertices and diameter $\mathrm{diam}(G)$ the expected takeover time is bounded by

$$O\left(\frac{\mathrm{diam}(G) + \log n}{p}\right) .$$

Both terms $\mathrm{diam}(G)$ and $\log n$ make sense. The diameter describes what distance needs to be overcome in order to inform all vertices in the network. The factor $1/p$ gives the expected time until a next vertex is informed, assuming that it has only one informed neighbor. We also obtain $\mathrm{diam}(G)$ (without a factor $1/p$) as a lower bound on the takeover time. The additive term $+ \log n$ is necessary to account for a potentially large variance, as seen in the example for star graphs.

If the diameter of the graph is at least $\Omega(\log n)$, we can drop the $+ \log n$-term in the asymptotic bound, leading to an upper bound of $O(\mathrm{diam}(G)/p)$.

Interestingly, the concept of propagation also appears in other contexts. When solving shortest paths problems in graphs, metaheuristics like evolutionary algorithms [46.32–34] and ant colony optimization (ACO) [46.35, 36] tend to propagate shortest paths through the graph. In the single-source shortest paths problem (SSSP) one is looking for shortest paths from a source vertex to all other vertices of the graph. The EAs and ACO algorithms tend to find shortest paths first for vertices that are *close* to the source, in a sense that their shortest paths only contain few edges. If these shortest paths are found, it enables the algorithm to find shortest paths for vertices that are further away.

When a shortest paths to vertex u is found and there is an edge $\{u, v\}$ in the graph, it is easy to find a shortest path for v. In the case of evolutionary algorithms, an EA only needs to assign u as a predecessor of v on the shortest path in a lucky mutation in order to find a shortest path to v. In the case of ACO, pheromones enable an ant to follow pheromones between the source and u, and so it only has to decide to travel between u and v to find a shortest path to v, with good probability.

Doerr et al. [46.34, Lemma 3] used tail bounds to prove that the time for propagating shortest paths with an EA is highly concentrated. If the graph has diameter $\mathrm{diam}(G) \geq \log n$, the EA with high probability finds all shortest paths in time $O(\mathrm{diam}(G)/p)$, where $p = \Theta(n^{-2})$ in this case. This result is similar to the one obtained by *Rowe* et al. [46.31]; asymptotically, both bounds are equal. However, the result by *Doerr* et al. [46.33] also allows for conclusions about growth curves.

Lässig and *Sudholt* [46.6, Theorem 3] introduced yet another argument for the analysis of propagation times. They considered *layers* of vertices. The i-th layer contains all vertices that have shortest paths of at most i edges, and that are not on any smaller layer. They bound the time until information is propagated throughout all vertices of a layer. This is feasible since all vertices in layer i are informed with probability at least p if all vertices in layers $1, \ldots, i - 1$ are informed. If n_i is the number of vertices in layer i, the time until the last vertex in this layer is informed is $O(n_i \cdot \log n_i)$. This gives a bound for the total takeover time of $O(\mathrm{diam}(G) \cdot \ln(en/\mathrm{diam}(G)))$. For small $(\mathrm{diam}(G) = O(1))$ or large $(\mathrm{diam}(G) = \Omega(n))$ diameters, we get the same asymptotic bound as before. For other values it is slightly worse.

However, the layering of vertices allows for inclusion of intra-island effects. Assume that the transmission probability p only applies once islands have been taken over (to a significantly large degree) by best individuals. This is a realistic setting as with only a single best individual the probability of selecting it for emigration (or pollination, to be precise) might be very small. If all islands need time T_{intra} in order to reach this stage

after the first best individual has reached the island, we obtain an upper bound of

$$O(\text{diam}(G) \cdot \ln(en/\text{diam}(G))) + \text{diam}(G) \cdot T_{\text{intra}}$$

for the takeover time.

46.4 Examples Where Parallel EAs Excel

Parallel EAs have been applied to a very broad range of problems, including many NP-hard problems from combinatorial optimization. The present literature is immense; already early surveys like the one by *Alba* and *Troya* [46.37] present long lists of applications of parallel EAs. Further applications can be found in [46.38–40]. Research on and applications of parallel metaheuristics has increased in recent years, due to the emergence of parallel computer architectures.

Crainic and Hail [46.41] review applications of parallel metaheuristics, with a focus on graph coloring, partitioning problems, covering problems, Steiner tree problems, satisfiability problems, location and network design, as well as the quadratic assignment problems with its famous special cases: the traveling salesman problem and vehicle routing problems. *Luque* and *Alba* [46.19] present selected applications for natural language tagging, the design of combinatorial logic circuits, the workforce planning problem, and the bioinformatics problem of assembling DNA fragments.

The literature is too vast to be reviewed in this section. Also, for many hard practical problems it is often hard to determine the effect that parallelization has on search dynamics. The reasons behind the success of parallel models often remain elusive. We follow a different route and describe theoretical studies of evolutionary algorithms where parallelization was proven to be helpful. This concerns illustrative toy functions as well as problems from combinatorial optimization. All following settings are well understood and allow us to gain insights into the effect of parallelization. We consider parallel variants of the most simple evolutionary algorithm called $(1+1)$ evolutionary algorithm, shortly $(1+1)$ EA. It is described in Algorithm 46.3 and it only uses mutation and selection in a population containing just one current search point. We are interested in the *optimization time*, defined as the number of generations until the algorithm first finds a global optimum. Unless noted otherwise, we consider pseudo-

Boolean optimization: the search space contains all bit strings of length n and the task is to maximize a function $f: \{0,1\}^n \to \mathbb{R}$. We use the common notation $x = x_1 \ldots x_n$ for bit strings.

Algorithm 46.3 $(1+1)$ EA for maximizing $f: \{0,1\}^n \to \mathbb{R}$

1: Initialize $x \in \{0,1\}^n$ uniformly at random.
2: **loop**
3:   Create x' by copying x and flipping each bit independently with probability $1/n$.
4:   **if** $f(x') \geq f(x)$ **then** $x := x'$.
5: **end loop**

The presentation in this section is kept informal. For theorems with precise results, including all preconditions, we refer to the respective papers.

46.4.1 Independent Runs

Independent runs prove useful if the running time has a large variance. The reason is that the optimization time equals the time until *the fastest* run has found a global optimum.

The variance can be particularly large in the case when the objective function yields local optima that are very hard to overcome. Bimodal functions contain two local optima, and typically only one is a global optimum. One such example was already analyzed theoretically in the seminal runtime analysis paper by *Droste* et al. [46.42].

We review the analysis of a similar function that leads to a simpler analysis. The function *TwoMax* was considered by *Friedrich* et al. [46.43] in the context of diversity mechanisms. It is a function of unitation: the fitness only depends on the number of bits set to 1. The function contains two symmetric slopes that increase linearly with the distance to $n/2$. Only one of these slopes leads to a global optimum. Formally, the function is defined as the maximum of $OneMax := \sum_{i=1}^{n} x_i$ and its symmetric cousin $ZeroMax := \sum_{i=1}^{n}(1-x_i)$, with an additional fitness bonus for the all-ones bit string

$$TwoMax(x) := \max\left\{\sum_{i=1}^{n} x_i, \sum_{i=1}^{n}(1-x_i)\right\} + \prod_{i=1}^{n} x_i .$$

See Fig. 46.7 for a sketch.

The $(1+1)$ EA reaches either a local optimum or a global optimum in expected time $O(n \log n)$. Due to the perfect symmetry of the function on the remainder of the search space, the probability that this is the global optimum is exactly $1/2$. If a local optimum is reached, the $(1+1)$ EA has to flip all bits in one mutation in order to reach the global optimum. The probability for this event is exactly n^{-n}.

The authors consider deterministic crowding [46.43] in a population of size μ as a diversity mechanism. It has the same search behavior as μ independent runs of the $(1+1)$ EA, except that the running time is counted in a different way. Their result directly transfers to this parallel model. The only assumption is that the number of independent runs is polynomially bounded in n.

The probability of finding a global optimum after $O(n \log n)$ generations of the parallel system is amplified to $1 - 2^{-\mu}$. This means that only with probability $2^{-\mu}$ we arrive at a situation where the parallel EA needs to escape from a local optimum. When all m islands are in this situation, the probability that at least

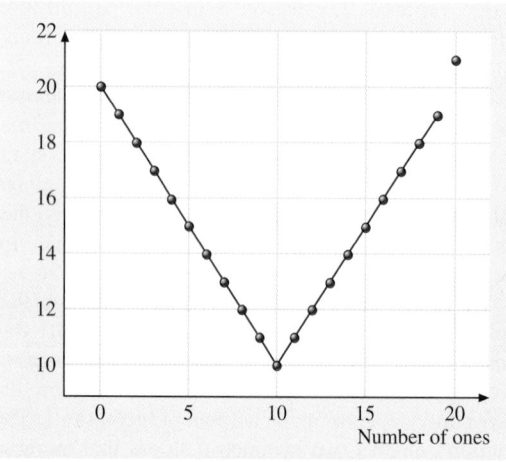

Fig. 46.7 Plots of the bimodal function *TwoMax* as defined in [46.43]

one island makes this jump in one generation is at most

$$1 - (1 - n^{-n})^m = \Theta(m \cdot n^{-n}) ,$$

where the last equality holds since m is asymptotically smaller than n^n.

This implies that the expected number of generations of a parallel system with m independent runs is

$$O(n \log n) + 2^{-m} \cdot \Theta\left(\frac{n^n}{m}\right) .$$

We can see from this formula that the number of runs m has an immense impact on the expected running time. Increasing the number of runs by 1 decreases the second summand by more than a factor of 2. The speedup is, therefore, exponential, up to a point where the running time is dominated by the first term $O(n \log n)$. Note in particular that $\log(n^n) = n \log n$ processors are sufficient to decrease the expected running time to $O(n \log n)$.

This is a very simple example of a superlinear speedup, with regard to the optimization time.

The observed effects also occur in combinatorial optimization. *Witt* [46.44] analyzed the $(1+1)$ EA on the NP-hard *PARTITION* problem. The task can be regarded as scheduling on two machines: given a sequence of jobs, each with a specific effort, the goal is to distribute the jobs on two machines to that the largest execution time (the makespan) is minimized.

On worst-case instances the $(1+1)$ EA has a constant probability of getting stuck in a bad local optimum. The expected time to find a solution with a makespan of less than $(4/3-\varepsilon) \cdot \text{OPT}$ is $n^{\Omega(n)}$, where $\varepsilon > 0$ is an arbitrary constant and OPT is the value of the optimal solution.

However, if the $(1+1)$ EA is lucky, it can, indeed, achieve a good approximation of the global optimum. Assume we are aiming at a solution with a makespan of at most $(1+\varepsilon) \cdot \text{OPT}$, for some $\varepsilon > 0$ we can choose. Witt's analysis shows that then $2^{(e \log e + e) \cdot \lceil 2/\varepsilon \rceil \ln(4/\varepsilon) + O(1/\varepsilon)}$ parallel runs output a solution of this quality with probability at least $3/4$. (This probability can be further amplified quite easily by using more runs.) Each run takes time $O(n \ln(1/\varepsilon))$. The parallel model represents what is known as a *polynomial-time randomized approximation scheme* (PRAS). The desired approximation quality $(1+\varepsilon)$ can be specified, and if ε is fixed, the total computation time is bounded by a polynomial in n. This was the first example that parallel runs of a randomized search heuristics constitute a PRAS for an NP-hard problem.

46.4.2 Offspring Populations

Using offspring populations in a master–slave architecture can decrease the parallel running time and lead to a speedup. We will discuss this issue further in Sect. 46.5 as offspring populations are very similar to island models on complete topologies. For now, we present one example where offspring populations decrease the optimization time very drastically.

Jansen et al. [46.45] compared the $(1+1)$ EA against a variant $(1+\lambda)$ EA that creates λ offspring in parallel and compares the current search point against the best offspring. They constructed a function SufSamp where offspring populations have a significant advantage. We refrain from giving a formal definition, but instead describe the main ideas. The vast majority of all search points tend to lead an EA towards the start of a path through the search space. The points on this path have increasing fitness, thus encouraging an EA to follow it. All points outside the path are worse, so the EA will stay on the path.

The path leads to a local optimum at the end. However, the function also includes a number of smaller paths that branch off the main path, see Fig. 46.8. All these paths lead to global optima, but they are difficult to discover. This makes a difference between the $(1+1)$ EA and the $(1+\lambda)$ EA for sufficiently large λ. The $(1+1)$ EA typically follows the main path without discovering the smaller paths branching off. At the end of the main path it thus becomes stuck in a local optimum. The analysis in [46.45] shows that the $(1+1)$ EA needs superpolynomial time, with high probability.

Contrarily, the $(1+\lambda)$ EA performs a more thorough search as it progresses on the main path. The many offspring tend to discover at least one of the smaller branches. The fitness on the smaller branches is larger than the fitness of the main path, so the EA will move away from the main path and follow a smaller path. It then finds a global optimum in polynomial time, with high probability.

Interestingly, this construction can be easily adapted to show an opposite result. We replace the local optimum at the end of the main path by a global optimum

and replace all global optima at the end of the smaller branches by local optima. This yields another function SufSamp', also shown in Fig. 46.8. By the same reasoning as above, the $(1+\lambda)$ EA will become stuck and the $(1+1)$ EA will find a global optimum in polynomial time, with high probability.

While the example is clearly constructed and artificial, it can be seen as a cautionary tale. The reader might be tempted to think that using offspring populations instead of creating a single offspring can never increase the number of *generations* needed to find the optimum. After all, evolutionary search with offspring population is more intense and improvements can be found more easily. As we focus on the number of generations (and do not count the effort for creating λ offspring), it is tempting to claim that offspring populations are never disadvantageous.

The second example shows that this claim – however obvious it may seem – does not hold for general problem classes. Note that this statement is also implied by the well-known *no free lunch theorems* [46.46], but the above results are much stronger and more concrete.

46.4.3 Island Models

The examples so far have shown that a more thorough search – by independent runs or increased sampling of offspring – can lead to more efficient running times. *Lässig* and *Sudholt* [46.3] presented a first example where communication makes the difference between exponential and polynomial running times, in a typical run. They constructed a family of problems called $\text{LOLZ}_{n,z,b,\ell}$ where a simple island model finds the optimum in polynomial time, with high probability. This holds for a proper choice of the migration interval and any migration topology that is not too sparse. The islands run $(1+1)$ EAs, hence the island model resembles a fine-grained model.

Contrarily, both a panmictic population as well as independent islands need exponential time, with high probability. This shows that the weak speedup versus panmixia is superlinear, even exponential (when considering speedups with respect to the typical running

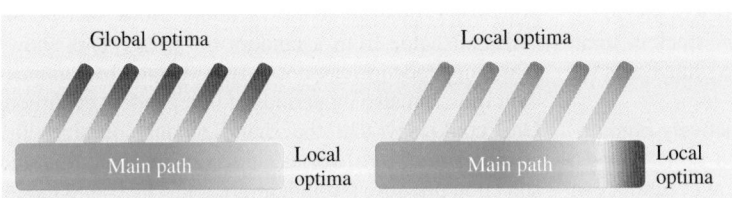

Fig. 46.8 Sketches of the functions SufSamp (*left*) and SufSamp' (*right*). The fitness is indicated by the *color*

Table 46.1 Examples of solutions for the function LOLZ with four blocks and $z = 3$, along with their fitness values. All blocks have to be optimized from left to right. The sketch shows in bold all bits that are counted in the fitness evaluation. Note how in x_3 in the third block only the first $z = 3$ zeros are counted. Further 0-bits are ignored. The only way to escape from this local optimum is to flip all z 0-bits in this block simultaneously

x_1	**11110**011	11010100	11010110	01011110	LOLZ$(x_1) = 4$
x_2	**11111111**	**110**10100	11010110	01011110	LOLZ$(x_2) = 10$
x_3	**11111111**	**11111111**	**00000**110	01011110	LOLZ$(x_3) = 19$

time instead of the expected running time). Unlike previous examples, it also shows that more sophisticated means of parallelization can be better than independent runs.

The basic idea of this construction is as follows. An EA can increase the fitness of its current solutions by gathering a prefix of bits with the same value. Generally, a prefix of i leading ones yields the same fitness as a prefix of i leading zeros. The EA has to make a decision whether to collect leading ones (LOs) or leading zeros (LZs). This not only holds for the $(1 + 1)$ EA but also for a (not too large) panmictic population as genetic drift will lead the whole population to either leading ones or leading zeros.

In the beginning, both decisions are symmetric. However, after a significant prefix has been gathered, symmetry is broken: after the prefix has reached a length of z, z being a parameter of the function, only leading ones lead to a further fitness increase. If the EA has gone for leading zeros, it becomes stuck in a local optimum. The parameter z determines the difficulty of escaping from this local optimum.

This construction is repeated on several blocks of the bit string that need to be optimized one-by-one. Each block has length ℓ. Only if the right decision towards the leading ones is made on the first block, can the block be filled with further leading ones. Once the first block contains only leading ones, the fitness depends on the prefix in the second block, and a further decision between leading ones and leading zeros needs to be made. Figure 46.1 illustrates the problem definition.

So, the problem requires an EA to make several decisions in succession. The number of blocks, b, is another parameter that determines how many decisions need to be made. Panmictic populations will sooner or later make a wrong decision and become stuck in some local optimum. If b is not too small, the same holds for independent runs.

However, an island model can effectively communicate the right decisions on blocks to other islands. Islands that have become stuck in a local optimum can be taken over by other islands that have made the correct decision. These dynamics make up the success of the island model as it can be shown to find global optima with high probability. A requirement is, though, that the migration interval is carefully tuned so that migration only transmits the right information. If migration happens before the symmetry between leading ones and leading zeros is broken, it might be that islands with leading zeros take over islands with leading ones. *Lässig* and *Sudholt* [46.3] give sufficient conditions under which this does not happen, with high probability.

An interesting finding is also how islands can regain independence. During migration, genetic information about future blocks is transmitted. Hence, after migration all islands contain the same genotype on future blocks. This is a real threat as this dependence might imply that all islands make the same decision after moving on to the next block. Then all diversity would be lost.

However, under the conditions given in [46.3] there is a period of independent evolution following migration, before any island moves on to a new block. During this period of independence, the genotypes of future blocks are subjected to random mutations, independently for each island. The reader might think of moving particles in some space. Initially, all bits are in the same position. However, then particles start moving around randomly. Naturally, they will spread out and separate from one another. After some time the distribution of particles will resemble a uniform distribution. In particular, an observer would not be able to distinguish whether the positions of particles were obtained by this random process or by simply drawing them from a uniform distribution.

The same effect occurs with bits of future blocks; after some time all bits of a future block will be indistinguishable from a random bit string. This shows that independence can not only be gained by independent runs, but also by periods of independent evolution. One could say that the island model combines the advantages of two worlds: independent evolution and selection pressure through migration. The island model

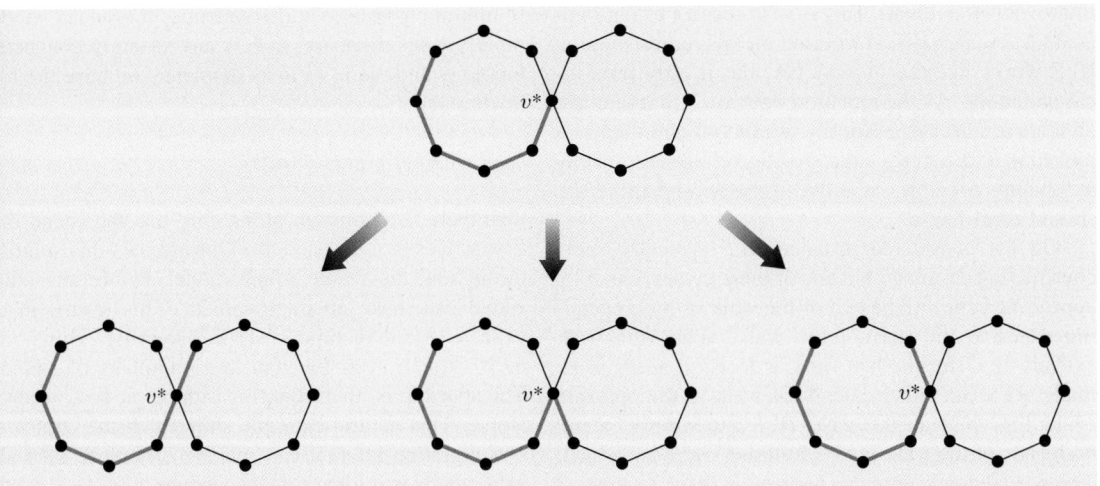

Fig. 46.9 Sketch of the graph G'. The *top* shows a configuration where a decision at v^* has to be made. The three configurations *below* show the possible outcomes. All these transitions occur with equal probability, but only the one *on the bottom right* leads to a solution where rotations are necessary

is only successful because it can use both migration and periods of independent evolution.

The theoretical results [46.3] were complemented by experiments in [46.47]. The aim was to look at what impact the choice of the migration topology and the choice of the migration interval have on performance, regarding the function LOLZ. The theoretical results made a statement about a broad class of dense topologies, but required a very precise migration interval. The experiments showed that the island model is far more robust with respect to the migration interval than suggested by theory.

Depending on the migration interval, some topologies were better than others. The topologies involved were a bidirectional ring, a torus with edges wrapping around, a hypercube graph, and the complete graph. We considered the success rate of the island model, stopping it as soon as all islands had reached local or global optima. We then performed statistical tests comparing these success rates. For small migration intervals, i.e., frequent migrations, sparse topologies were better than dense ones. For large migration intervals, i.e., rare migrations, the effect was the opposite. This effect was expected; however, we also found that the torus was generally better than the hypercube. This is surprising, as both have a similar density. Table 46.2 shows the ranking obtained for commonly used topologies.

Superlinear speedups with island models also occur in simpler settings. *Lässig* and *Sudholt* [46.6] also considered island models for the Eulerian cycle prob-

lem. Given an undirected Eulerian graph, the task is to find a Eulerian cycle, i.e., a traversal of the graph on which each edge is traversed exactly once. This problem can be solved efficiently by tailored algorithms, but it served as an excellent test bed for studying the performance of evolutionary algorithms [46.48–51].

Instead of bit strings, the problem representation by *Neumann* [46.48] is based on permutations of the edges of the graph. Each such permutation gives rise to a *walk*: starting with the first edge, a walk is the longest sequence of edges such that two subsequent edges in the permutation share a common vertex. The walk encoded by the permutation ends when the next edge does not share a vertex with the current one. A walk that contains all edges represents a Eulerian cycle. The length of the walk gives the fitness of the current solution.

Neumann [46.48] considered a simple instance that consists of two cycles of equal size, connected by one common vertex v^* (Fig. 46.9). The instance is interesting as it represents a worst case for the time until an

Table 46.2 Performance comparison according to success rates for commonly used migration topologies. The notion $A \prec B$ means that topology A has a significantly smaller success rate than topology B

Migration interval	Ranking
Small migration intervals	$K_\mu \prec$ hypercube $\prec$ torus $\prec$ ring
Medium migration intervals	hypercube $\prec K_\mu \prec$ ring $\prec$ torus
High migration intervals	ring $\prec$ torus $\prec$ hypercube $\prec K_\mu$

improvement is found. This is with respect to randomized local search (RLS) working on this representation. RLS works like the $(1 + 1)$ EA, but it only uses local mutations. As the mutation operator it uses jumps: an edge is selected uniformly at random and then it is moved to a (different) target position chosen uniformly at random. All edges in between the two positions are shifted accordingly.

On the considered instance RLS typically starts constructing a walk within one of these cycles, either by appending edges to the end of the walk or by prepending edges to the start of the walk. When the walk extends to v^* for the first time, a decision needs to be made. RLS can either extend the walk to the opposite cycle, Fig. 46.9. In this case, RLS can simply extend both ends of the walk until a Eulerian cycle is formed. The expected time until this happens is $\Theta(m^3)$, where m denotes the number of edges.

However, if another edge in the same cycle is added at v^*, the walk will evolve into one of the two cycles that make up the instance. It is not possible to add further edges to the current walk, unless the current walk starts and ends in v^*. However, the walk can be rotated so that the start and end vertex of the walk is moved to a neighbored vertex. Such an operation takes expected time $\Theta(m^2)$. Note that the fitness after a rotation is the same as before. Rotations that take the start and end closer to v^* are as likely as rotations that move it away from v^*. The start and end of the walk hence performs a fair random walk, and $\Theta(m^2)$ rotations are needed on average in order to reach v^*. The total expected time for rotating the cycle is hence $\Theta(m^4)$.

Summarizing, if RLS makes the right decision then expected time $\Theta(m^3)$ suffices in total. However, if rotations become necessary the expected time increases to $\Theta(m^4)$. Now consider an island model with m islands running RLS. If islands evolve independently for at least $\tau \geq m^3$ generations, all mentioned decisions are made independently, with high probability. The probability of making a wrong decision is $1/3$, hence with m islands the probability that all islands make the wrong decision is 3^{-m}. The expected time can be shown to be

$$\Theta(m^3 + 3^{-m} \cdot m^4) \, .$$

The choice $m := \log_3 m$ yields an expectation of $\Theta(m^3)$, and every value up to $\log_3 m$ leads to a superlinear speedup, asymptotically speaking. Technically, the speedup is even exponential.

Interestingly, this good performance only holds if migration is used rarely, or if independent runs are used.

If migration is used too frequently, the island model rapidly loses diversity. If T is any strongly connected topology and $\mathrm{diam}(T)$ is its diameter, we have the following. If

$$\tau \cdot \mathrm{diam}(T) \cdot m = O(m^2) \, ,$$

then there is a constant probability that the island that first arrives at a decision at v^* propagates this solution throughout the whole island model, before any other island can make an improvement. This results in an expected running time of $\Omega(m^4/\log(m))$. This is almost $\Theta(m^4)$, even for very large numbers of islands. The speedup is, therefore, logarithmic at best, or even worse. This natural example shows that the choice of the migration interval can make a difference between exponential and logarithmic speedups.

46.4.4 Crossover Between Islands

It has long been known that island models can also be useful in the context of crossover. Crossover usually requires a good diversity in the population to work properly. Due to the higher diversity between different islands, compared to panmixia, recombining individuals from different islands is promising.

Watson and *Jansen* [46.52] presented and analyzed a royal road function for crossover: a function where crossover drastically outperforms mutation-based evolutionary algorithms. In contrast to previous theoretically studied examples [46.53–57], their goal was to construct a function with a clear building-block structure. In order to prove that a GA was able to assemble all building blocks, they resorted to an island model with a very particular migration topology. In their *single-receiver model* all islands except one evolve independently. Each island sends its migrants to a designated island called the *receiver* (Fig 46.10). This way, all sending islands are able to evolve the right building blocks, and the receiver is used to assemble all these building blocks to obtain the optimum.

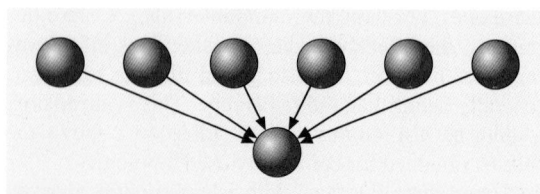

Fig. 46.10 The topology for *Watson* and *Jansen's* single-receiver model (after [46.52])

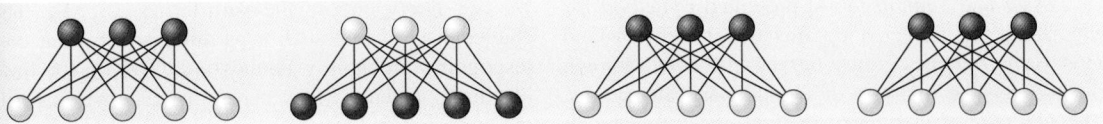

Fig. 46.11 Vertex cover instance with bipartite graphs. The *brown vertices* denote selected vertices. In this configuration the *second component* shows a locally optimal configuration while all other components are globally optimal

This idea was picked up later on by *Neumann* et al. [46.7] in a more detailed study of crossover in island models. We describe parts of their results, as their problem is more illustrative than the one by Watson and Jansen. The former authors considered instances of the NP-hard *Vertex cover* problem. Given an undirected graph, the goal is to select a subset of vertices such that each vertex is either selected or neighbored to a selected vertex. We say that vertices are *covered* if this property holds for them. The objective is to minimize the number of selected vertices. The problem has a simple and natural binary representation where each bit indicates whether a corresponding vertex is selected or not.

Prior work by *Oliveto* et al. [46.58] showed that evolutionary algorithms with panmictic populations even fail on simply structured instance classes like copies of bipartite graphs. An example is shown in Fig. 46.11. Consider a single bipartite graph, i. e., two sets of vertices such that each vertex in one set is connected to every vertex in the other set. If both sets have different sizes, the smaller set is an optimal Vertex cover. The larger set is another Vertex cover. It is, in fact, a non-optimal local optimum which is hard to overcome: the majority of bits has to flip in order to escape. If the instance consists of several independent copies of bipartite graphs, it is very likely that a panmictic EA will evolve a locally optimal configuration on at least one of the bipartite graphs. Then the algorithm fails to find a global optimum.

Island models perform better. Assume the topology is the single-receiver model. In each migration a 2-point crossover is performed between migrants and the individual on the target island. All islands have population size 1 for simplicity. We also assume that the bipartite subgraphs are encoded in such a way that each subgraph forms one block in the bit string. This is a natural assumption as all subgraphs can be clearly identified as building blocks. In addition, *Jansen* et al. [46.59] presented an automated way of encoding graphs in a crossover-friendly way, based on the degrees of vertices.

The analysis in [46.7] shows the following. Assume that the migration interval is at least $\tau \geq n^{1+\varepsilon}$ for some positive constant $\varepsilon > 0$. This choice implies that all islands will evolve to configurations where all bipartite graphs are either locally optimal or globally optimal. With probability $1 - e^{-\Omega(m)}$ we have that for each bipartite graph at least a constant fraction of all sender islands will have the globally optimal configuration.

All that is left to do for the receiver island is to rely on crossover combining all present good building blocks. As two-point crossover can select one block from an immigrant and the remainder from the current solution on the receiver island, all good building blocks have a good chance to be obtained. The island model finds a global optimum within a polynomial number of generations, with probability $1 - e^{-\Omega(\min\{n^{\varepsilon/2}, m\})}$.

46.5 Speedups by Parallelization

46.5.1 A General Method for Analyzing Parallel EAs

We now finally discuss a method for estimating the speedup by parallelization. Assume that, instead of running a single EA, we run an island model where each island runs the same EA. The question is by how much the expected optimization time (i. e., the number of generations until a global optimum is found) decreases,

compared to the single, panmictic EA. Recall that this speedup is called weak orthodox speedup [46.14].

In the following we sometimes speak of the expected *parallel* optimization time to emphasize that we are dealing with a parallel system. If the number of islands and the population size on each island is fixed, we can simply multiply this time by a fixed factor to obtain the expected number of function evaluations.

Lässig and *Sudholt* [46.4] presented a method for estimating the expected optimization time of island models. It combines growth curves with a well-known method for the analysis of evolutionary algorithms. The *fitness-level method* or *method of f-based partitions* [46.60] is a simple, yet powerful technique. The idea is to partition the search space into non-empty sets $A_1, A_2, \ldots, A_m$ such that the following holds:

- for each $1 \leq i < m$ each search point in A_i has a strictly worse fitness than each search point in A_{i+1} and
- A_m contains all global optima.

The described ordering with respect to the fitness f is often denoted

$$A_1 <_f A_2 <_f \cdots <_f A_m \, .$$

Note that A_m can also be redefined towards containing all search points of some desired quality if the goal is not global optimization.

We say that a population-based algorithm $\mathcal{A}$ (including populations of size 1) is in A_i or *on fitness level i* if the best search point in the population is in A_i. Now, assume that we know that s_i is a lower bound on the probability that the algorithm finds a solution in $A_{i+1} \cup \cdots \cup A_m$ if it is currently in A_i. Then the reciprocal $1/s_i$ is an upper bound on the expected time until this event happens. If the algorithm is *elitist* (i. e., it never loses the current best solution), then it will never decrease its current fitness level. A sufficient condition for finding an optimal solution is that all sets $A_1, A_2, \ldots, A_{m-1}$ are left in the described manner at least once. This implies the following bound on the expected optimization time.

Theorem 46.1 Wegener [46.60]

Consider an elitist EA and assume a fitness-level partition $A_1 <_f \cdots <_f A_m$ where A_m is the set of global optima. Let s_i be a lower bound for the probability that in one generation the EA finds a search point in $A_{i+1} \cup \cdots \cup A_m$ if the best individual in the parent population is in A_i. Then the expected optimization time is bounded by

$$\sum_{i=1}^{m-1} \frac{1}{s_i} \, .$$

The above bound applies to all elitist algorithms. It is generally applicable and often quite versatile, as

we can freely choose the partition $A_1, \ldots, A_m$. The challenge is to find such a partition and to find corresponding probability bounds $s_1, \ldots, s_{m-1}$ for finding improvements. Many papers have shown that this method – applied explicitly or implicitly – yields tight bounds on the expected optimization time of EAs for various problems [46.32, 42, 48]. It can also be used as part of a more general analysis [46.61, 62].

We are being pessimistic in assuming that every fitness level has to be left. In reality, several fitness levels might be skipped. The fitness-level method often yields good bounds if not too many levels are skipped, and if the probability bounds s_i are good estimates for the real probabilities of finding a better fitness-level set. Note that the lower bound s_i must apply regardless of the precise search point(s) in A_i present in the population, hence we need to consider the worst-case probability of escaping from A_i.

Nevertheless, the fitness-level method often yields tight bounds. *Sudholt* [46.63] recently developed a lower-bound method based on fitness levels, which in each case shows that the upper bound is tight. Also, *Lehre* [46.64] recently presented an extension of the method to non-elitist algorithms. Asymptotically, the same bound as in Theorem 46.1 applies, if some additional conditions on the selection pressure and the population size are fulfilled. For the sake of simplicity, we focus on elitist algorithms in the following.

If s_i denotes the probability of a single offspring finding an improvement, this probability can be increased by using λ offspring in parallel. We have already seen in Sect. 46.1 how λ independent trials can increase or amplify a success probability p to $1 - (1 - p)^\lambda$. The same reasoning applies to the probability s_i for finding an improvement on the current best level. Figure 46.1 has shown how this probability increases with the number of trials. Figure 46.12 shows how the expected time for having a success decreases with the number of offspring. In fact, the curves in Fig. 46.12 are just reciprocals of those in the previous Fig. 46.1.

Figure 46.12 shows that the speedup can be close to linear (in a strict, non-asymptotic sense), especially for low success probabilities. As the probability of increasing the current fitness level i is at least $1 - (1 - s_i)^\lambda$, we obtain the following.

Theorem 46.2

Consider an elitist EA creating λ offspring independently in each generation. Assume a fitness-level partition $A_1 <_f \cdots <_f A_m$, where A_m is the set of global optima. Let s_i be a lower bound for the probability that in one generation a single offspring finds a search point in $A_{i+1} \cup \cdots \cup A_m$ if the best individual in the parent population is in A_i. Then the expected optimization time is bounded by

$$\sum_{i=1}^{m-1} \frac{1}{1-(1-s_i)^\lambda} \le m-1 + \frac{1}{\lambda}\sum_{i=1}^{m-1} \frac{1}{s_i} \ .$$

Note that the first bound for $\lambda = 1$ reproduces the previous upper bound from Theorem 46.1. For the second bound we used

$$\frac{1}{1-(1-s_i)^\mu} \le 1 + \frac{1}{\mu} \cdot \frac{1}{s_i} \ , \tag{46.2}$$

where the inequality was proposed by Jon Rowe (personal communication, 2011); it can be proven by a simple induction.

Our estimate of the probability for an improvement increases with the number of islands on the current best fitness level. In a spatially structured EA these growth curves are non-trivial. Especially with a sparse migration topology, information about the current best fitness level is typically propagated quite slowly. The increased exploration slows down exploitation. Still, even sparse topologies lead to drastically improved upper bounds, when compared to the simple bound for a sequential EA from Theorem 46.1. The precise bounds crucially depend on the particular topology.

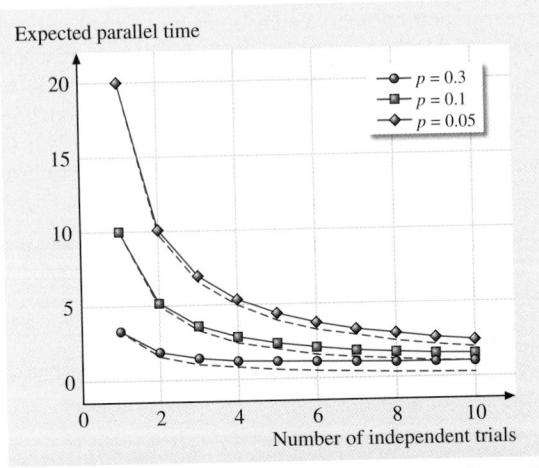

Fig. 46.12 Plots of the expected parallel time until an offspring population of size λ has a success, if each offspring independently has a success probability of p. The *dashed lines* indicate a perfect linear speedup

We first consider a setting where migration always transmits the current best fitness level and migration occurs in every generation. It is possible to adapt the results to account for larger migration intervals. One way of doing this is to redefine s_i to represent a lower bound of finding an improvement in a time period between migrations. Then we obtain an upper bound on the expected number of migrations. For the sake of simplicity, we only consider the case $\tau = 1$ in the following.

The following theorem was presented in *Lässig* and *Sudholt* [46.6]; it is a refined special case of previous results [46.4]. The main proof idea is to combine the investigation of growth curves with the consideration of amplified success probabilities.

Theorem 46.3 Lässig and Sudholt [46.6]
Consider an island model with μ islands where each island runs an elitist EA. In every iteration each island sends copies of its best individual to all neighbored islands (i. e., $\tau = 1$). Each island incorporates the best out of its own individuals and its immigrants.

For every partition $A_1 <_f \cdots <_f A_m$ if s_i is a lower bound for the probability that in one generation an island in A_i finds a search point in $A_{i+1} \cup \cdots \cup A_m$ then the expected parallel optimization time is bounded by:

1. $2\sum_{i=1}^{m-1} \frac{1}{s_i^{1/2}} + \frac{1}{\mu}\sum_{i=1}^{m-1} \frac{1}{s_i}$ for every unidirectional ring (a ring with edges in one direction) or any other strongly connected topology,
2. $3\sum_{i=1}^{m-1} \frac{1}{s_i^{1/3}} + \frac{1}{\mu}\sum_{i=1}^{m-1} \frac{1}{s_i}$ for every undirected grid or torus graph with side lengths at least $\sqrt{\mu} \times \sqrt{\mu}$,
3. $m-1 + \frac{1}{\mu}\sum_{i=1}^{m-1} \frac{1}{s_i}$ for the complete topology K_μ.

Note that the bound for the complete topology K_μ is equal to the upper bound for offspring populations, Theorem 46.2. This makes sense as an island model with a complete topology propagates the current best fitness level like an offspring population.

All bounds in Theorem 46.3 consist of two additive terms. The second term

$$\frac{1}{\mu}\sum_{i=1}^{m-1} \frac{1}{s_i}$$

represents a perfect linear speedup, compared to the upper bound from Theorem 46.1. The larger we choose the number of islands μ, the smaller this term becomes. The first additive term is related to the growth curves of the current best fitness level in the island model. The denser the topology, the faster information is spread, and the smaller this term becomes. Note that it is independent of μ. It can be regarded as the term limiting the degree of parallelizability. We can increase the number of islands in order to decrease the second term

$$\frac{1}{\mu} \sum_{i=1}^{m-1} \frac{1}{s_i},$$

but we cannot decrease the first term by changing μ.

This allows for immediate conclusions about cases where we obtain an asymptotic linear speedup over a single-island EA. For all choices of μ where the second term is asymptotically no smaller than the first term, the upper bound is smaller than the upper bound from Theorem 46.1 by a factor of order μ. This is an asymptotic linear speedup if the upper bound from Theorem 46.1 is asymptotically tight. (If it is not, we can only compare upper bounds for a sequential and a parallel EA.)

We illustrate this with a simple and well-known test function from pseudo-Boolean optimization. The algorithm considered is an island model where each island runs a $(1+1)$ EA; the island model is also called parallel $(1+1)$ EA. The function

$$\text{LO}(x) := \sum_{i=1}^{n} \prod_{j=1}^{i} x_j \quad (LeadingOnes)$$

counts the number of leading ones in the bit string. We choose the canonic partition where A_i contains all search points with fitness i, i.e., i leading ones. For any set A_i, $0 \le i \le n-1$ we use the following lower bound on the probability for an improvement.

An improvement occurs if the first 0-bit is flipped from 0 to 1 and no other bit flips. The probability of flipping the mentioned 0-bit is $1/n$ as each bit is flipped independently with probability $1/n$. The probability of not flipping any other bit is $(1-1/n)^{n-1}$. We use the common estimate $(1-1/n)^{n-1} \ge 1/e$, where $e = \exp(1) = 2.718\ldots$, so the probability of an improvement is at least $s_i \ge 1/(en)$ for all $0 \le i \le n-1$. Plugging this into Theorem 46.3, the second term is $\frac{1}{\mu} \cdot en^2$ for all bounds. The first terms are

$$2n \cdot (en)^{1/2} = 2e^{1/2}n^{3/2}$$

for the ring,

$$3n \cdot (en)^{1/3} = 3e^{1/3}n^{4/3}$$

for the torus, and n for the complete graph, respectively.

For the ring, choosing $\mu = O(n^{1/2})$ islands results in an expected parallel time of $O(\frac{1}{\mu} \cdot n^2)$ as the second term is asymptotically not smaller than the first one. This is asymptotically smaller by a factor of $1/\mu$ than the expected optimization time of a single $(1+1)$ EA, $\Theta(n^2)$ [46.42]. Hence, each choice of μ up to $\mu = O(n^{1/2})$ gives a linear speedup. For the torus we obtain a linear speedup for $\mu = O(n^{2/3})$ in the same fashion. For the complete graph this even holds for $\mu = O(n)$. One can see here that the island model can decrease the expected parallel running time by significant polynomial factors.

Table 46.3 lists expected parallel optimization time bounds for several well-known pseudo-Boolean functions. The above analysis for LO generalizes to all unimodal functions. A function is called unimodal here if every non-optimal search point has a better Hamming neighbor, i.e., a better search point can be reached by flipping exactly one specific bit. ONEMAX$(x) = \sum_{i=1}^{n} x_i$ counts the number of ones, hence modeling a simple hill climbing task. Finally, Jump$_k$ [46.42] is a multimodal function of tunable difficulty. An EA

Table 46.3 Upper bounds for expected parallel optimization times (number of generations) for the $(1+1)$ EA and the corresponding island model with μ islands in pseudo-Boolean optimization. The last but one column is for any unimodal function with d function values. The number of function evaluations in the island model is larger than the number of generations by a factor of μ

Algorithm	ONEMAX	LO	Unimodal, d values	Jump$_k$, $k \ge 3$
$(1+1)$ EA	$O(n \log n)$ [46.42]	$O(n^2)$ [46.42]	$O(nd)$	$O(n^k)$ [46.42]
Island model on ring	$O\left(n + \frac{n \log n}{\mu}\right)$	$O\left(n^{3/2} + \frac{n^2}{\mu}\right)$	$O\left(dn^{1/2} + \frac{dn}{\mu}\right)$	$O\left(n^{k/2} + \frac{n^k}{\mu}\right)$
Island model on torus	$O\left(n + \frac{n \log n}{\mu}\right)$	$O\left(n^{4/3} + \frac{n^2}{\mu}\right)$	$O\left(dn^{1/3} + \frac{dn}{\mu}\right)$	$O\left(n^{k/3} + \frac{n^k}{\mu}\right)$
Island model on $K_\mu/(1+\mu)$ EA	$O\left(n + \frac{n \log n}{\mu}\right)$	$O\left(n + \frac{n^2}{\mu}\right)$	$O\left(d + \frac{dn}{\mu}\right)$	$O\left(n + \frac{n^k}{\mu}\right)$

ypically has to make a *jump* by flipping k bits simultaneously, where $2 \le k \le n$. The $(1+1)$ EA has an expected optimization time of $\Theta(n^k)$, hence growing rapidly with increasing k.

One can see that the island model leads to drastically reduced parallel optimization times. This particularly holds for problems where improvements are hard to find.

We remark that *Lässig* and *Sudholt* [46.4] also considered parallel EAs where migration is not always successful in transmitting information about the current best fitness level. This includes the case where crossover is used during migration and crossover has a certain probability of being disruptive. We do obtain upper bounds on the expected optimization time if we know a lower bound p^+ on the probability of a successful transmission. The bounds depend on p^+; the degree of this dependence is determined by the topology. For simplicity we only focus on the deterministic case here.

46.5.2 Speedups in Combinatorial Optimization

The techniques are also applicable in combinatorial optimization. We review two examples here, presented in [46.6]. *Scharnow* et al. [46.32] considered the classical sorting problem as an optimization problem: given a sequence of n distinct elements from a totally ordered set, sorting is the problem of maximizing sortedness. Without loss of generality the elements are $1, \dots, n$; then the aim is to find the permutation π_{opt} such that $(\pi_{opt}(1), \dots, \pi_{opt}(n))$ is the sorted sequence.

The search space is the set of all permutations π on $1, \dots, n$. Two different operators are used for mutation. An exchange chooses two indices $i \ne j$ uniformly at random from $\{1, \dots, n\}$ and exchanges the entries at positions i and j. A jump chooses two indices in the same fashion. The entry at i is put at position j and all entries in between are shifted accordingly. For instance,

a jump with $i = 2$ and $j = 5$ would turn $(1, 2, 3, 4, 5, 6)$ into $(1, 3, 4, 5, 2, 6)$.

The $(1+1)$ EA draws S according to a Poisson distribution with parameter $\lambda = 1$ and then performs $S + 1$ elementary operations. Each operation is either an exchange or a jump, where the decision is made independently and uniformly for each elementary operation. The resulting offspring replaces its parent if its fitness is not worse. The fitness function $f_{\pi_{opt}}(\pi)$ describes the sortedness of $(\pi(1), \dots, \pi(n))$. As in [46.32], we consider the following measures of sortedness:

- INV(π) measures the number of pairs (i, j), $1 \le i < j \le n$, such that $\pi(i) < \pi(j)$ (pairs in correct order),
- HAM(π) measures the number of indices i such that $\pi(i) = i$ (elements at the correct position),
- LAS(π) equals the largest k such that $\pi(i_1) < \cdots < \pi(i_k)$ for some $i_i < \cdots < i_k$ (length of the longest ascending subsequence),
- EXC(π) equals the minimal number of exchanges (of pairs $\pi(i)$ and $\pi(j)$) to sort the sequence, leading to a minimization problem.

The expected optimization time of the $(1+1)$ EA is $\Omega(n^2)$ and $O(n^2 \log n)$ for all fitness functions. The upper bound is tight for LAS, and it is believed to be tight for INV, HAM, and EXC as well [46.32]. Theorem 46.3 yields the following. For INV, all topologies guarantee a linear speedup only in case $\mu = O(\log n)$ and the bound $O(n^2 \log n)$ for the $(1+1)$ EA is tight. The other functions allow for linear speedups up to $\mu = O(n^{1/2} \log n)$ (ring), $\mu = O(n^{2/3} \log n)$ (torus), and $\mu = O(n \log n)$ (K_μ), respectively (again assuming tightness, otherwise up to a factor of $\log n$). Note how the results improve with the density of the topology. HAM, LAS, and EXC yield much better guarantees for the island model than INV. This is surprising as there is no visible performance difference for a single $(1+1)$ EA. Theorem 46.3 yields the following results also shown in Tab. 46.4

Table 46.4 Upper bounds for expected parallel optimization times for the $(1+1)$ EA and the corresponding island model with μ islands for sorting n objects

Algorithm	INV	HAM, LAS, EXC
$(1+1)$ EA	$O(n^2 \log n)$ [46.32]	$O(n^2 \log n)$ [46.32]
Island model on ring	$O\left(n^2 + \frac{n^2 \log n}{\mu}\right)$	$O\left(n^{3/2} + \frac{n^2 \log n}{\mu}\right)$
Island model on torus	$O\left(n^2 + \frac{n^2 \log n}{\mu}\right)$	$O\left(n^{4/3} + \frac{n^2 \log n}{\mu}\right)$
Island model on $K_\mu/(1+\mu)$ EA	$O\left(n^2 + \frac{n^2 \log n}{\mu}\right)$	$O\left(n + \frac{n^2 \log n}{\mu}\right)$

Table 46.5 Worst-case expected parallel optimization times for the $(1+1)$ EA and the corresponding island model with μ islands for the SSSP on graphs with n vertices and m edges. The value ℓ is the maximum number of edges on any shortest path from the source to any vertex and $\ell^* := \max\{\ell, \ln n\}$. The second lines show a range of μ-values yielding a linear speedup, apart from a factor $\ln(en/\ell)$

Algorithm	Vertex-based mutation [46.32]	Edge-based mutation [46.65]
$(1+1)$ EA	$\Theta(n^2\ell^*)$ [46.34]	$\Theta(m\ell^*)$ [46.65]
Island model on ring	$O\left(n^{3/2}\ell^{1/2} + \frac{n^2\ell\ln(en/\ell)}{\mu}\right)$	$O\left(m^{1/2}n^{1/2}\ell^{1/2} + \frac{m\ell\ln(en/\ell)}{\mu}\right)$
	$\longrightarrow \mu = O\left((n\ell)^{1/2}\right)$	$\longrightarrow \mu = O\left((m/n\cdot\ell)^{1/2}\right)$
Island model on torus	$O\left(n^{4/3}\ell^{1/3} + \frac{n^2\ell\ln(en/\ell)}{\mu}\right)$	$O\left(m^{1/3}n^{2/3}\ell^{1/3} + \frac{m\ell\ln(en/\ell)}{\mu}\right)$
	$\longrightarrow \mu = O\left((n\ell)^{2/3}\right)$	$\longrightarrow \mu = O\left((m/n\cdot\ell)^{2/3}\right)$
Island model on $K_\mu/(1+\mu)$ EA	$O\left(n + \frac{n^2\ell\ln(en/\ell)}{\mu}\right)$	$O\left(n + \frac{m\ell\ln(en/\ell)}{\mu}\right)$
	$\longrightarrow \mu = O(n\ell)$	$\longrightarrow \mu = O(m/n\cdot\ell)$

An explanation is that INV leads to $\binom{n}{2}$ non-optimal fitness levels that are quite easy to overcome. HAM, LAS, and EXC have only n non-optimal fitness levels that are more difficult. For a single EA both settings are equally difficult, leading to asymptotically equal expected times (assuming all upper bounds are tight). However, the latter setting is easier to parallelize than the former as it is easier to amplify small success probabilities.

We also consider parallel variants of the $(1+1)$ EA for the single source shortest path problem (SSSP) [46.32]. An SSSP instance is given by an undirected connected graph with vertices $\{1,\ldots,n\}$ and a distance matrix $D = (d_{ij})_{1\le i,j\le n}$, where $d_{ij} \in \mathbb{R}_0^+ \cup \{\infty\}$ defines the length value for given edges from node i to node j. We are searching for shortest paths from a node s (without loss of generality $s = n$) to each other node $1 \le i \le n-1$.

A candidate solution is represented as a *shortest paths tree*, a tree rooted at s with directed shortest paths to all other vertices. We define a search point x as vector of length $n-1$, where position i describes the predecessor node x_i of node i in the shortest path tree. Note that infeasible solutions are possible if the predecessors do not encode a tree. An elementary mutation chooses a vertex i uniformly at random and replaces its predecessor x_i by a vertex chosen uniformly at random from $\{1,\ldots,n\}\setminus\{i,x_i\}$. We call this a vertex-based mutation. Doerr et al. [46.65] proposed an edge-based mutation operator. An edge is chosen uniformly at random, and the edge is made a predecessor edge for its end node.

The $(1+1)$ EA uses either vertex-based mutations or edge-based ones. It creates an offspring using S elementary mutations, where S is chosen according to

a Poisson distribution with $\lambda = 1$. The result of an offspring is accepted in case no distance to any vertex has gotten worse.

Applying Theorem 46.3 along with a layering argument as described at the end of Sect. 46.3.4 yields the bounds on the expected parallel optimization time shown in Table 46.5.

The upper bounds for the island models with constant μ match the expected time of the $(1+1)$ EA if $\ell = O(1)$ or $\ell = \Omega(n)$ as then $\ell\ln(en/\ell) = \Theta(\ell^*)$. In other cases, the upper bounds are off by a factor of $\ln(en/\ell)$. Table 46.5 also shows a range of μ-values for which the speedup is linear (if $\ell = O(1)$ or $\ell = \Omega(n)$) or almost linear, that is, when disregarding the $\ln(en/\ell)$ term.

Note how the possible speedups significantly increase with the density of the topology. The speedups also depend on the graph instance and the maximum number of edges ℓ on any shortest path. For a single $(1+1)$ EA edge-based mutations are more effective than vertex-based mutations [46.65]. Island models with edge-based mutations cannot be paral-

Table 46.6 Asymptotic bounds for expected parallel running times and expected sequential running times for the parallel $(1+1)$ EA with adaptive population models

	Scheme	Sequential	Parallel
ONEMAX	A	$\Theta(n\log n)$	$O(n\log n)$
	B	$\Theta(n\log n)$	$O(n)$
LO	A	$\Theta(n^2)$	$\Theta(n\log n)$
	B	$\Theta(n^2)$	$O(n)$
Unimodal f	A	$O(dn)$	$O(d\log n)$
with d f-values	B	$O(dn)$	$O(d+\log n)$
Jump$_k$	A	$O(n^k)$	$O(n\log n)$
with $k \ge 2$	B	$O(n^k)$	$O(n+k\log n)$

elized as effectively for sparse graphs as those with vertex-based mutations if the graph is sparse, i.e., $n = o(n^2)$. Then the number of islands that guarantees a linear speedup is smaller for edge-based mutations than for vertex-based mutations. The reason is that with a more efficient mutation operator there is less potential for further speedups with a parallel EA.

46.5.3 Adaptive Numbers of Islands

Theorem 46.3 presents a powerful tool for determining the number of islands that give an asymptotic linear speedup. However, it would be even more desirable to have an adaptive system that automatically finds the ideal number of islands throughout the run.

In [46.5] *Lässig* and *Sudholt* proposed and analyzed two simple adaptive schemes for choosing the number of islands. Both schemes check whether in the current generation some island has found an improvement over the current best fitness in the system. If no island has found an improvement, the number of islands is doubled. This can be implemented, for instance, by copying each island. New processors can be allocated to host these islands in large clusters or by using cloud computing.

If some island has found an improvement, the number of islands is reduced by removing selected islands from the system and de-allocating resources. Both schemes differ in the way they decrease the number of islands. The first scheme, simply called Scheme A, only keeps one island containing a current best solution. Scheme B halves the number of islands. Both schemes use complete topologies, so all remaining islands will contain current best individuals afterwards.

Both mechanisms lead to optimal speedups in many cases. Doubling the number of islands may seem aggressive, but the analysis shows that the probability of allocating far more islands than necessary is very very small. The authors considered the expected sequential optimization time, defined as the number of function evaluations, to measure the total effort over time. With both schemes it is guaranteed that the expected sequential time does not exceed the simple bound for a sequential EA from Theorem 46.1, asymptotically. The expected parallel times on each fitness level can, roughly speaking, be replaced by their logarithms.

The following is a slight simplification of results in [46.5].

Theorem 46.4 Lässig and Sudholt [46.5]
Given an f-based partition $A_1, \ldots, A_m$ and lower bounds $s_1, \ldots, s_{m-1}$ on the probability of a single island finding an improvement, the expected sequential times for island models using a complete topology and either Scheme A or Scheme B are bounded by

$$3 \sum_{i=1}^{m-1} \frac{1}{s_i} .$$

If each set A_i contains only a single fitness value then also the expected parallel time is bounded by

$$4 \sum_{i=1}^{m-1} \log \left(\frac{2}{s_i} \right) .$$

Actually, for Scheme A we can obtain slightly better constants than the ones stated in Theorem 46.4. However, with a more detailed analysis one can show that Scheme B can perform much better than Scheme A. *Lässig*'s and *Sundholt*'s work [46.5] contains a more refined upper bound for Scheme B. We only show a special case where the fitness levels become increasingly harder. Then it makes sense to only halve the number of islands when an improvement is found, instead of resetting the number of islands to 1.

Theorem 46.5 Lässig and Sudholt [46.5]
Given an f-based partition $A_1, \ldots, A_m$, where each set A_i contains only a single fitness value and for the probability bounds it holds $s_1 \geq s_2 \geq \cdots \geq s_{m-1}$. Then the expected parallel running time for an island model using a complete topology and Scheme B is bounded by

$$3(m - 2) + \log \left(\frac{1}{s_{m-1}} \right) .$$

Example applications for a parallel $(1 + 1)$ EA in Table 46.6 show that Scheme B can automatically lead to the same speedups as when using an optimal number of islands. This holds for ONEMAX, LO, and the general bound for unimodal functions. For Jump$_k$ it also holds in the most relevant cases, when $k = O(n/\log n)$, as then the expected parallel time is $O(n)$.

We conclude that simply doubling or halving the number of islands represents a simple and effective mechanism for finding optimal parameters adaptively.

46.6 Conclusions

Parallel evolutionary algorithm can effectively reduce computation time and at the same time lead to an increased exploration and better diversity, compared to sequential evolutionary algorithms.

We have surveyed various forms of parallel EAs, from independent runs to island models and cellular EAs. Different lines of research have been discussed that give insight into the working principles behind parallel EAs. This includes the spread of information, growth curves for current best solutions, and takeover times.

A recurring theme was the possible speedup that can be achieved with parallel EAs. We have elaborated on the reasons why superlinear speedups are possible in practice. Rigorous runtime analysis has given examples where parallel EAs excel over sequential algorithms, with regard to the number of generations or the number of function evaluations until a global optimum is found. The final section has covered a method for estimating the expected parallel optimization time of island models. The method is easy to apply as we can automatically transfer existing analyses for sequential EAs to a parallel version thereof. Examples have been given for pseudo-Boolean optimization and combinatorial optimization. The results have also led to the discovery of a simple, yet surprisingly powerful adaptive scheme for choosing the number of islands.

There are many possible avenues for future work. In the light of the development in computer architecture, it is important to develop parallel EAs that can run effectively on many cores. It also remains a crucial issue to increase our understanding of how design choices and parameters affect the performance of parallel EAs. Rigorous runtime analysis has emerged recently as a new line of research that can give novel insights in this respect and opens new roads. The present results should be extended towards further algorithms, further problems, and more detailed cost models that reflect the costs for communication in parallel architectures. It would also be interesting to derive further rigorous results on takeover times in settings where propagation through migration is probabilistic. Finally, it is important to bring theory and practice together in order to create synergetic effects between the two areas.

46.6.1 Further Reading

This book chapter does not claim to be comprehensive. In fact, parallel evolutionary algorithms represent a vast research area with a long history. Early variants of parallel evolutionary algorithms were developed, studied, and applied more than 20 years ago. We, therefore, point the reader to references that may complement this chapter. *Paz* [46.66] presented a review of early literature and the history of parallel EAs. The survey by *Alba* and *Troya* [46.37] contains detailed overviews of parallel EAs and their characteristics.

This chapter does not cover implementation details of parallel evolutionary algorithms. We refer to the excellent survey by *Alba* and *Tomassini* [46.38]. This survey also includes an overview of the theory of parallel EAs. The emphasis is different from this chapter and it can be used to complement this chapter.

Tomassini's text book [46.67] describes various forms of parallel EAs like island models, cellular EAs, and coevolution. It also presents many mathematical and experimental results that help understand how parallel EAs work. Furthermore, it contains an appendix dealing with the implementation of parallel EAs.

The book edited by *Alba* et al. [46.39] takes a broader scope on parallel models that also include parallel evolutionary multiobjective optimization and parallel variants of swarm intelligence algorithms like particle swarm optimization and ant colony optimization. The book contains a part on parallel hardware as well as a number of applications of parallel metaheuristics.

Alba's edited book on parallel metaheuristics [46.40] has an even broader scope. It covers parallel variants of many common metaheuristics such as genetic algorithms, genetic programming, evolution strategies, ant colony optimization, estimation-of-distribution algorithms, scatter search, variable-neighborhood search, simulated annealing, tabu search, greedy randomized adaptive search procedures (GRASPs), hybrid metaheuristics, multiobjective optimization, and heterogeneous metaheuristics.

The most recent text book was written by *Luque* and *Alba* [46.19]. It provides an excellent introduction into the field, with hands-on advice on how to present results for parallel EAs. Theoretical models of selection pressure in distributed GAs are presented. A large part of the book then reviews selected applications of parallel GAs.

References

46.1   P.S. Oliveto, J. He, X. Yao: Time complexity of evolutionary algorithms for combinatorial optimization: A decade of results, Int. J. Autom. Comput. **4**(3), 281–293 (2007)

46.2   F. Neumann, C. Witt: *Bioinspired Computation in Combinatorial Optimization – Algorithms and Their Computational Complexity* (Springer, Berlin, Heidelberg 2010)

46.3   J. Lässig, D. Sudholt: The benefit of migration in parallel evolutionary algorithms, Proc. Genet. Evol. Comput. Conf. (GECCO 2010) (ACM, New York 2010) pp. 1105–1112

46.4   J. Lässig, D. Sudholt: General scheme for analyzing running times of parallel evolutionary algorithms, 11th Int. Conf. Parallel Probl. Solving Nat. (PPSN 2010) (Springer, Berlin, Heidelberg 2010) pp. 234–243

46.5   J. Lässig, D. Sudholt: Adaptive population models for offspring populations and parallel evolutionary algorithms, Proc. 11th Workshop Found. Genet. Algorithms (FOGA 2011) (ACM, Berlin, Heidelberg 2011) pp. 181–192

46.6   J. Lässig, D. Sudholt: Analysis of speedups in parallel evolutionary algorithms for combinatorial optimization, 22nd Int. Symp. Algorithms Comput. (ISAAC '11) (Springer, Berlin, Heidelberg 2011) pp. 405–414

46.7   F. Neumann, P.S. Oliveto, G. Rudolph, D. Sudholt: On the effectiveness of crossover for migration in parallel evolutionary algorithms, Proc. Genet. Evol. Comput. Conf. (GECCO 2011) (ACM, New York 2011) pp. 1587–1594

46.8   M. De Felice, S. Meloni, S. Panzieri: Effect of topology on diversity of spatially-structured evolutionary algorithms, Proc. 13th Annu. Genet. Evol. Comput. Conf. (GECCO '11) (2011) pp. 1579–1586

46.9   M. Giacobini, M. Tomassini, A. Tettamanzi: Takeover time curves in random and small-world structured populations, Proc. Genet. Evol. Comput. Conf. (GECCO '05) (ACM, New York 2005) pp. 1333–1340

46.10   Z. Skolicki: An Analysis of Island Models in Evolutionary Computation, Ph.D. Thesis (George Mason University, Fairfax 2000)

46.11   E. Alba, M. Giacobini, M. Tomassini, S. Romero: *Comparing Synchronous and Asynchronous Cellular Genetic Algorithms, Parallel Problem Solving from Nature VII* (Springer, Berlin, Heidelberg 2002) pp. 601–610

46.12   M. Mitzenmacher, E. Upfal: *Probability and Computing* (Cambridge Univ. Press, Cambridge 2005)

46.13   J. Sprave: A unified model of non-panmictic population structures in evolutionary algorithms, Proc. 1999 Congr. Evol. Comput. (IEEE, Bellingham 1999) pp. 1384–1391

46.14   E. Alba: Parallel evolutionary algorithms can achieve super-linear performance, Inf. Process. Lett. **82**(1), 7–13 (2002)

46.15   R.S. Barr, B.L. Hickman: Reporting computational experiments with parallel algorithms: Issues, measures, and experts' opinion, ORSA J. Comput. **5**(1), 2–18 (1993)

46.16   D.E. Goldberg, K. Deb: A comparatative analysis of selection schemes used in genetic algorithms. In: *Foundations of Genetic Algorithms*, ed. by G.J.E. Rawlins (Morgan Kaufmann, Burlington 1991) pp. 69–93

46.17   J. Sarma, K. De Jong: An analysis of local selection algorithms in a spatially structured evolutionary algorithm, Proc. 7th Int. Conf. Genet. Algorithms (Morgan Kaufmann, Burlington 1997) pp. 181–186

46.18   E. Alba, G. Luque: Growth curves and takeover time in distributed evolutionary algorithms, Proc. Genet. Evol. Comput. Conf. (Springer, Berlin, Heidelberg 2004) pp. 864–876

46.19   G. Luque, E. Alba: *Parallel Genetic Algorithms – Theory and Real World Applications*, Studies in Computational Intelligence, Vol. 367 (Springer, Berlin, Heidelberg 2011)

46.20   Z. Skolicki, K.A. De Jong: The influence of migration sizes and intervals on island models, Proc. Genet. Evol. Comput. Conf. (GECCO '05) (ACM, New York 2005) pp. 1295–1302

46.21   M. Giacobini, E. Alba, M. Tomassini: Selection intensity in asynchronous cellular evolutionary algorithms, Proc. Genet. Evol. Comput. Conf. (GECCO '03) (Springer, Berlin, Heidelberg 2003) pp. 955–966

46.22   G. Rudolph: Takeover times and probabilities of non-generational selection rules, Proc. Genet. Evol. Comput. Conf. (GECCO '00) (Morgan Kaufmann, Burlington 2000) pp. 903–910

46.23   G. Rudolph: Takeover times of noisy non-generational selection rules that undo extinction, Proc. 5th Int. Conf. Artif. Neural Nets Genet. Algorithms (ICANNGA 2001) (Springer, Berlin, Heidelberg 2001) pp. 268–271

46.24   G. Rudolph: On takeover times in spatially structured populations: Array and ring, Proc. 2nd Asia-Pac. Conf. Genet. Algorithms Appl. (Global-Link Publishing, Hong Kong 2000) pp. 144–151

46.25   G. Rudolph: Takeover time in parallel populations with migration, Proc. 2nd Int. Conf. Bioinspired Optim. Methods Appl. (BIOMA 2006), ed. by B. Filipic, J. Silc (2006) pp. 63–72

46.26   M. Giacobini, M. Tomassini, A. Tettamanzi: Modelling selection intensity for linear cellular evolutionary algorithms, Proc. 6th Int. Conf. Artif. Evol., Evol. Artif. (Springer, Berlin, Heidelberg 2003) pp. 345–356

46.27   M. Giacobini, E. Alba, A. Tettamanzi, M. Tomassini: Selection intensity in cellular evolutionary algorithms for regular lattices, IEEE Trans. Evol. Comput. **9**, 489–505 (2005)

46.28   C. Witt: Runtime analysis of the $(\mu + 1)$EA on simple pseudo-Boolean functions, Evol. Comput. **14**(1), 65–86 (2006)

46.29   D. Sudholt: The impact of parametrization in memetic evolutionary algorithms, Theor. Comput. Sci. **410**(26), 2511–2528 (2009)

46.30   M. Giacobini, E. Alba, A. Tettamanzi, M. Tomassini: Modeling selection intensity for toroidal cellular evolutionary algorithms, Proc. Genet. Evol. Comput. Conf. (GECCO '04) (Springer, Berlin, Heidelberg 2004) pp. 1138–1149

46.31   J. Rowe, B. Mitavskiy, C. Cannings: Propagation time in stochastic communication networks, 2nd IEEE Int. Conf. Digit. Ecosyst. Technol. (2008) pp. 426–431

46.32   J. Scharnow, K. Tinnefeld, I. Wegener: The analysis of evolutionary algorithms on sorting and shortest paths problems, J. Math. Model, Algorithms **3**(4), 349–366 (2004)

46.33   B. Doerr, E. Happ, C. Klein: Crossover can provably be useful in evolutionary computation, Theor. Comput. Sci. **425**, 17–33 (2012)

46.34   B. Doerr, E. Happ, C. Klein: A tight analysis of the $(1+1)$-EA for the single source shortest path problem, Proc. IEEE Congr. Evol. Comput. (CEC '07) (IEEE, Bellingham 2007) pp. 1890–1895

46.35   C. Horoba, D. Sudholt: Ant colony optimization for stochastic shortest path problems, Proc. Genet. Evol. Comput. Conf. (GECCO 2010) (ACM, New York 2010) pp. 1465–1472

46.36   D. Sudholt, C. Thyssen: Running time analysis of ant colony optimization for shortest path problems, J. Discret. Algorithms **10**, 165–180 (2012)

46.37   E. Alba, J.M. Troya: A survey of parallel distributed genetic algorithms, Complexity **4**, 31–52 (1999)

46.38   E. Alba, M. Tomassini: Parallelism and evolutionary algorithms, IEEE Trans. Evol. Comput. **6**, 443–462 (2002)

46.39   E. Alba, N. Nedjah, L. de Macedo Mourelle: *Parallel Evolutionary Computations* (Springer, Berlin, Heidelberg 2006)

46.40   E. Alba: *Parallel Metaheuristics: A New Class of Algorithms* (Wiley-Interscience, New York 2005)

46.41   T.G. Crainic, N. Hail: Parallel metaheuristics applications. In: *Parallel Metaheuristics: A New Class of Algorithms*, (Wiley-Interscience, New York 2005)

46.42   S. Droste, T. Jansen, I. Wegener: On the analysis of the $(1+1)$ evolutionary algorithm, Theor. Comput. Sci. **276**, 51–81 (2002)

46.43   T. Friedrich, P.S. Oliveto, D. Sudholt, C. Witt: Analysis of diversity-preserving mechanisms for global exploration, Evol. Comput. **17**(4), 455–476 (2009)

46.44   C. Witt: Worst-case and average-case approximations by simple randomized search heuristics, Proc. 22nd Symp. Theor. Asp. Comput. Sci. (STACS '05) (Springer, Berlin, Heidelberg 2005) pp. 44–56

46.45   T. Jansen, K.A. De Jong, I. Wegener: On the choice of the offspring population size in evolutionary algorithms, Evol. Comput. **13**, 413–440 (2005)

46.46   C. Igel, M. Toussaint: A no-free-lunch theorem for non-uniform distributions of target functions, J. Math. Model, Algorithms **3**(4), 313–322 (2004)

46.47   J. Lässig, D. Sudholt: Experimental supplements to the theoretical analysis of migration in the island model, 11th Int. Conf. Parallel Probl. Solving Nat. (PPSN 2010) (Springer, Berlin, Heidelberg 2010) pp. 224–233

46.48   F. Neumann: Expected runtimes of evolutionary algorithms for the Eulerian cycle problem, Comput. Oper. Res. **35**(9), 2750–2759 (2008)

46.49   B. Doerr, N. Hebbinghaus, F. Neumann: Speeding up evolutionary algorithms through asymmetric mutation operators, Evol. Comput. **15**, 401–410 (2007)

46.50   B. Doerr, D. Johannsen: Adjacency list matchings – An ideal genotype for cycle covers, Proc. Genet. Evol. Comput. Conf. (GECCO '07) (ACM, New York 2007) pp. 1203–1210

46.51   B. Doerr, C. Klein, T. Storch: Faster evolutionary algorithms by superior graph representation, 1st IEEE Symp. Found. Comput. Intell. (FOCI '07) (2007) pp. 245–250

46.52   R.A. Watson, T. Jansen: A building-block royal road where crossover is provably essential, Proc. Genet. Evol. Comput. Conf. (GECCO '07) (ACM, New York 2007) pp. 1452–1459

46.53   T. Jansen, I. Wegener: On the analysis of evolutionary algorithms – A proof that crossover really can help, Algorithmica **34**(1), 47–66 (2002)

46.54   T. Jansen, I. Wegener: Real royal road functions – Where crossover provably is essential, Discret. Appl. Math. **149**, 111–125 (2005)

46.55   T. Storch, I. Wegener: Real royal road functions for constant population size, Theor. Comput. Sci. **320**, 123–134 (2004)

46.56   S. Fischer, I. Wegener: The one-dimensional ising model: Mutation versus recombination, Theor. Comput. Sci. **344**(2/3), 208–225 (2005)

46.57   D. Sudholt: Crossover is provably essential for the ising model on trees, Proc. Genet. Evol. Comput. Conf. (GECCO '05) (ACM, New York 2005) pp. 1161–1167

46.58   P.S. Oliveto, J. He, X. Yao: Analysis of the $(1+1)$-EA for finding approximate solutions to vertex cover problems, IEEE Trans. Evol. Comput. **13**(5), 1006–1029 (2009)

46.59   T. Jansen, P.S. Oliveto, C. Zarges: On the analysis of the immune-inspired B-cell algorithm for the vertex cover problem, Proc. 10th Int. Conf. Artif. Immune Syst. (ICARIS 2011) (Springer, Berlin, Heidelberg 2011) pp. 117–131

46.60   I. Wegener: Methods for the analysis of evolutionary algorithms on pseudo-Boolean functions. In: *Evolutionary Optimization*, ed. by R. Sarker,

X. Yao, M. Mohammadian (Kluwer, Dordrecht 2002) pp. 349–369

6.61 F. Neumann, I. Wegener: Randomized local search, evolutionary algorithms, and the minimum spanning tree problem, Theor. Comput. Sci. **378**(1), 32–40 (2007)

6.62 D. Sudholt, C. Zarges: Analysis of an iterated local search algorithm for vertex coloring, 21st Int. Symp. Algorithms Comput. (ISAAC 2010) (Springer, Berlin, Heidelberg 2010) pp. 340–352

6.63 D. Sudholt: General lower bounds for the running time of evolutionary algorithms, 11th Int. Conf. Parallel Probl. Solving Nat. (PPSN 2010) (Springer, Berlin, Heidelberg 2010) pp. 124–133

46.64 P.K. Lehre: Fitness-levels for non-elitist populations, Proc. 13th Annu. Genet. Evol. Comput. Conf. (GECCO '11) (ACM, New York 2011) pp. 2075–2082

46.65 B. Doerr, D. Johannsen, C. Winzen: Drift analysis and linear functions revisited, IEEE Congr. Evol. Comput. (CEC '10) (2010) pp. 1967–1974

46.66 E. Cantú Paz: A survey of parallel genetic algorithms, Tech. Rep., Illinois Genetic Algorithms Laboratory (University of Illinois at Urbana Champaign, Urbana 1997)

46.67 M. Tomassini: *Spatially Structured Evolutionary Algorithms: Artificial Evolution in Space and Time* (Springer, Berlin, Heidelberg 2005)

47. Learning Classifier Systems

Martin V. Butz

Learning Classifier Systems (LCSs) essentially combine fast approximation techniques with evolutionary optimization techniques. Despite their somewhat misleading name, LCSs are not only systems suitable for classification problems, but may be rather viewed as a very general, distributed optimization technique. Essentially, LCSs have very high potential to be applied in any problem domain that is best solved or approximated by means of a distributed set of local approximations, or predictions. The evolutionary component is designed to optimize a partitioning of the problem domain for generating maximally useful predictions within each subspace of the partitioning. The predictions are generated and adapted by the approximation technique. Generally any form of spatial partitioning and prediction are possible – such as a Gaussian-based partitioning combined with linear approximations, yielding a Gaussian mixture of linear predictions. In fact, such a solution is developed and optimized by XCSF (XCS for function approximation). The LCSs XCS (X classifier system) and the function approximation version XCSF, indeed, are probably the most well-known LCS architectures to date. Their optimization technique is very-well balanced with the approximation technique: as long as the approximation technique yields reasonably good solutions and evaluations of these solutions fast, the evolutionary component will pick-up on the evaluation signal and optimize the partitioning. This chapter provides historical background on LCSs. Then XCS and XCSF are introduced in detail providing enough information to be able to implement, understand, and apply these systems. Further LCS architectures are surveyed and their potential for future research and for applications is discussed. The conclusions provide an outlook on the many possible future LCS applications and developments.

Part E | 47

Learning classifier systems (LCSs) are machine learning algorithms that combine gradient-based approximation with evolutionary optimization. Due to this flexibility, LCSs have been successfully applied to classification and data mining problems, reinforcement learning (RL) problems, regression problems, cognitive map learning, and even robot control problems.

The main feature of LCSs is their innovative combination of two learning principles; whereas gradient-based approximation adapts local, predictive approximations of target function values, evolutionary optimization structures individual classifiers to enable the formation of effectively distributed and accurate approximations. The two learning methods interact bidirectionally in that the gradient-based approximations yield local fitness quality estimates of the generated approximations, which the evolutionary optimization technique uses for optimizing classifier structures. Concurrently, the evolutionary optimization technique is generating new classifier structures, which again need to be evaluated by the gradient-based approach in competition with the other, locally overlapping, interacting classifiers.

Due to the innovative combination of two learning and optimization techniques, LCSs are often perceived as being hard to understand. Facet-wise analyses of the individual LCS components and their interactions, however, give both mathematical scalability bounds for learning and an intuitive understanding of the systems in general. Moreover, the currently most common LCS, which is the XCS classifier system (note that the *X* in XCS does not really encode any particular acronym according to the system creator Wilson), is comparatively easy to understand, to tune, and to apply. Thus, the core of this chapter focuses on XCS, gives a facet-wise overview of its functionality, details several enhancements, and highlights various successful application domains. However, XCS is also compared with other LCS architectures and LCSs in general are compared with other machine learning techniques.

This chapter starts with a historical perspective providing information on the beginnings of LCSs and establishing some terminology background. We then introduce the XCS classifier system providing a detailed system overview as well as theoretical and facet-wise conceptual insights on its performance. Also tricks and tweaks are discussed to tune the system to the problem at hand. Next, the XCS counterpart for regression problems, XCSF, is introduced. Focusing then on the application-side, LCS applications to data mining tasks and to behavioral learning and cognitive modeling tasks are surveyed. We cover various LCS architectures that have been successfully applied in the data mining realm. With respect to behavioral learning, we point out the relation of LCSs to reinforcement learning. Moreover, we cover anticipatory learning classifier systems (ALCSs) – which learn predictive schema models of the environment rather than reward prediction maps – and we introduce the modified XCSF version that can effectively learn a redundant forward-inverse kinematics model of a robot arm. A summary and conclusions wrap up the chapter.

47.1 Background

Learning classifier systems (LCS) were proposed over 30 years ago by *Holland* [47.1–3]. Originally, *Holland* and *Reitman* actually called LCSs *cognitive systems* [47.4], focusing on problems related to reinforcement learning (RL) [47.5, 6]. His cognitive system developed a *memory* of *classifiers*, where each classifier consisted of a condition part (*taxon*), an action part (originally consisting of a *message*, and an *effector bit*), a payoff prediction part, and several other parameters that stored the age, the application frequency, and the *attenuation* of the classifier.

Concurrently with the development of temporal difference learning techniques in RL – such as the now well-known state-action-reward-state-action (SARSA) algorithm [47.6] – *Holland* and *Reitman* introduced the *bucket brigade* algorithm [47.4, 7], which also distributes reward backwards in time with a discounting mechanism. In addition, the *attenuation* parameter in a classifier realized something similar to an *eligibility trace* in RL – distributing a currently encountered reward also to classifiers that were active several time steps ago and that thus indirectly led to gaining the currently experienced reward. Meanwhile, Holland's cognitive system applied a genetic algorithm (GA) [47.1, 8] as its second learning mechanism. The GA modified the *taxa* in Holland and Reitman's cognitive system.

In sum, the first actual LCS implementation, i. e., the *cognitive system* by *Holland* and *Reitman* [47.4], was ahead of its time. It implemented various reward-related ideas that were later established in the reinforcement learning community – and can now partially be regarded as standard RL techniques. However, the combination with GAs yielded a highly interactive and very complex system that was and still is hard to analyze. Thus, while proposing a highly innovative cognitive learning approach, the applicability of the system remained limited at the time.

47.1.1 Early Applications

Nonetheless, early applications of LCSs were published in the 1980s. *Smith* developed a poker decision making system [47.9] based on *De Jong*'s approach to LCSs [47.10]. *Booker* worked on animal-like automation based on the cognitive systems architecture [47.11]. *Wilson* proposed and worked on the *animat problem* with LCS architectures derived from Holland and Reitman's cognitive systems approach [47.12, 13]. *Goldberg* solved a gas pipeline control task with a simplified version of the cognitive system architecture [47.8, 14]. Despite these successful early applications, a decade passed until a growing research community developed that worked on learning classifier systems.

47.1.2 The Pitt and Michigan Approach

Two fundamentally different LCS approaches were pursued from early on. The Pitt approach was fostered by the work of *De Jong* et al. [47.10, 15, 16]. On the other hand, the Michigan approach developed in the further years at Michigan under the supervision of *John H. Holland* [47.11, 14, 17, 18]. Diverse perspectives on the Michigan approach can be found in [47.19].

The essential difference between the two approaches is that in the Pitt approach rule sets are evolved where each particular rule set constitutes an individual for the GA. In contrast, in the Michigan approach one set of rules is evolved and each rule is an individual for

the GA. As a consequence, the Pitt-style LCSs are much closer to general GAs because each individual constitutes an overall problem solution. In the Michigan-style LCSs, on the other hand, each individual only applies in a subspace of the overall problem and only the whole set of rules that evolves constitutes the overall problem solution. Figure 47.1 illustrates this fundamental contrast between the two approaches.

As a consequence of this contrast, Pitt-style LCSs usually apply rather standard GA approaches. The whole population of rule sets is evolved. For fitness evaluation purposes, each set of rules needs to be evaluated in the problem environment addressed. On the other hand, Michigan-style LCSs need to continuously interact with an environment to sufficiently evaluate all the rules in the rule set – essentially exploring all the environmental subspaces to make sure all rules can develop a sufficiently useful fitness estimate. This continuous interaction and the typical interacting components of Michigan-style LCSs are illustrated in further detail in Fig. 47.2. Due to the continuously developing fitness estimates, often a more steady-state, niched GA is applied online in Michigan-style LCSs. The undertaken updates then depend directly on the current interaction and thus on the current subset of rules relevant in the experienced interaction. The steady-state, niched GA optimizes the internal knowledge base iteratively depending on the incoming learning samples.

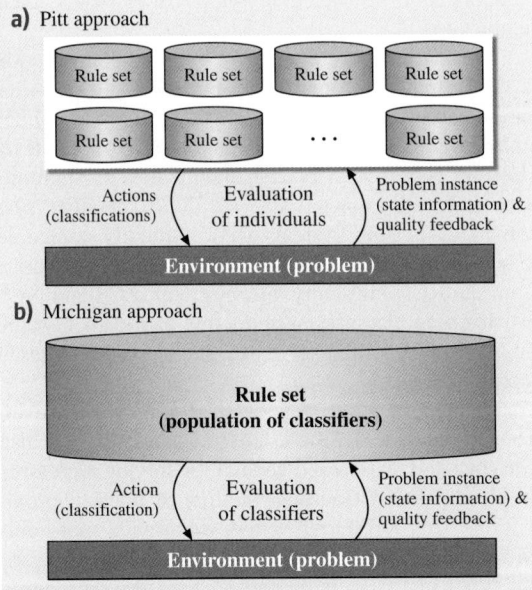

Fig. 47.1a,b While the Pitt approach to LCSs evolves a population of sets of rules, in the Michigan approach there is only one set of rules (i.e., the population) that is evolved

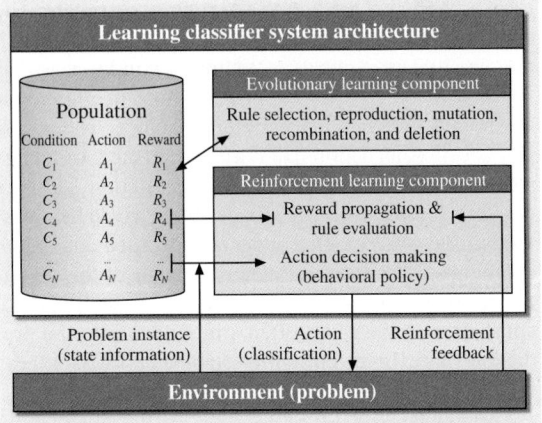

Fig. 47.2 LCSs consist of a knowledge base (population of classifiers), a genetic algorithm for rule structure evolution, and a reinforcement learning component for rule evaluation, reward propagation, and decision making. The system interacts with its environment or problem iteratively learning online

In summary, Pitt-style LCSs evaluate and optimize their rule sets globally based on sets of problem instances. They usually learn *offline*. Michigan-style LCSs evaluate and optimize their set of rules online while interacting with the problem, iteratively perceiving problem instances. The major qualities of Pitt-style LCSs are that they evolve competing global problem solutions in the form of sets of rules. Evolutionary rule structure optimization is used – typically evolving small sets of rules (10 s). Michigan-style LCSs, on the other hand, are designed to develop one distributed, locally optimized problem solution by combining local gradient-based approximation techniques with steady-state, niched GAs. In consequence, typically larger, more distributed sets of rules develop yielding problem solutions with potentially 1000 s of rules.

47.1.3 Basic Knowledge Representation

Because an exemplary knowledge representation was already discussed for the early *cognitive system* implementation of [47.4], we now provide a general sketch of the knowledge representation typically found in Michigan-style LCSs.

The knowledge representation of an LCS consists of a finite *population of classifiers* (that is, a finite set of rules). This population of classifiers essentially represents the current knowledge of the LCS about the problem the system is applied to. Each rule – or *classifier* – usually consists of a *condition* and an *action part*, as well as a *prediction* and a *fitness* estimate. The condition part specifies the problem subspace in which the classifier is applicable. When the condition part is satisfied given a particular problem instance, a classifier is said to *match* that problem instance. The action part specifies an action that may be executed, or a classification that may be tested. The prediction specifies the expected reward, or feedback value, given the specified action was executed under the specified contextual conditions. The fitness estimates the value of this classifier relative to other, competing classifiers. In the early approaches, fitness was often simply equal to the prediction value. In the currently established LCSs, fitness typically estimates the accuracy of the prediction.

Michigan-style LCSs usually learn online about a problem, iteratively perceiving or actively generating problem instances. Given a particular problem instance, first, the system forms a *match set* of those classifiers in the population whose conditions match. Next, the system decides on an action or classification and executes it. Classifiers in the match set that specify the executed action constitute the current *action set*. After feedback is received, the predictions of the classifiers in the action set are adjusted. From the classifier prediction estimates, a fitness estimate is derived for each classifier. Finally, the steady-state GA is applied to the match set or the population as a whole. The GA modifies classifier structures by reproducing, mutating, and recombining well-performing classifiers and by deleting ill-performing ones. In contrast to the Michigan approach, Pitt-style LCSs evaluate their sets of rules typically independently of each other in the provided problem. The GA exchanges rules and rule-structures within and across the sets of rules.

A Michigan-style LCS consequently is an interactive, online learning system. It maintains a population of classifiers as its knowledge base. It applies a niched, steady-state genetic algorithm for gradual rule structure evolution; it applies a gradient-based learning component for rule evaluation – yielding prediction and fitness estimates. Michigan-style LCSs are often applied in RL scenarios in which reward estimates need to be propagated and action decisions are made based on the learned reward prediction estimates. In this case, typically techniques similar to SARSA learning or Q-learning are applied. Figure 47.2 shows the basic components of a Michigan-style LCS as well as their interactions.

The earliest Michigan-style LCS implementation is the introduced cognitive system CS1 [47.4]. After various early applications of LCSs, *Wilson* set a milestone in LCS research by introducing the zeroth level classifier system ZCS [47.20] and the now most prominent and well-known LCS: the XCS classifier system [47.21]. Both systems were explicitly compared to the very well-known Q-learning [47.22] technique from the RL community, offering with ZCS and XCS two learning classifier systems that can learn Q-value functions with a compact highly generalized rule-based representation.

In the following, we now first give a precise introduction to the XCS classifier system. We then also introduce the real-valued version for solving regression problems, with a Gaussian mixture of linear approximations, i. e., XCSF. After that, we provide spot-lights on various current application domains where various types of LCSs, including XCS(F), have produced highly competitive problem solutions, when compared to other machine learning techniques and regression algorithms.

47.2 XCS

Wilson introduced the XCS classifier system in 1995 [47.21]. The two main novel features of XCS in comparison to earlier Michigan-style LCSs are its accuracy-based fitness estimation and its niche-based application of the evolutionary component. The introduction of accuracy-based fitness essentially decoupled the classifier fitness estimate from the reward prediction, enforcing that XCS learned *complete* payoff landscapes rather than only estimates for those subspaces where high reward is encountered. In addition, *Wilson* related XCS directly to Q-Learning [47.21, 22]. Much later, even a relation to Kalman filtering and general regression tasks was made mathematically explicit [47.23, 24]. The niche-based GA reproduction combined with population-wide deletion enabled a much more focused GA-based optimization of classifier structures as well as the generalization of classifier structures based on the sampling distribution [47.25]. In consequence, XCS is an LCS that is designed to evolve not only the best solution to a problem, but it evolves all alternative solutions with associated Q-value estimations and variance estimations of the respective Q-value estimates. Due to its GA design and fitness definition, XCS strives to approximate the full Q-table of a problem with a maximally accurate and maximally compact classifier-based representation.

Despite its original strong relation to Q-learning and RL in general, XCS has also been applied successfully to classification problems and regression problems. In the former case, XCS identifies locally relevant features for the generation of maximally accurate classification estimates. In the latter case, XCS optimizes the distribution and structure of local, typically linear estimators for a maximally accurate approximation of the function surface. Thus, despite its original strong relation to RL, XCS is a much more generally applicable learning system that can solve single-step classification or regression problems as well as multi-step RL problems, which are typically defined as Markov decision processes.

47.2.1 System Overview

XCS evolves one population of classifiers. Classifier structures are optimized by means of a steady-state GA. A classifier consists of a condition part C, an action part A, reward prediction r, reward prediction error ε, and fitness f estimates. While the condition and action structures are iteratively optimized by the steady-state GA, the estimates are adjusted using the Widrow–Hoff delta rule [47.26] based on an approximation of the Q-value signal.

While condition and action parts can be generally represented in any way desired [47.25], in this overview we focus on binary problems and a ternary representation of the condition part. Conventionally, the condition part C is coded by $C \in \{0, 1, \#\}^L$, where the # symbol matches zero and one. Condition C essentially specifies a hypercube within which the classifier *matches* and can be said to cover a certain *volume* of the complete problem space. Action part $A \in \mathcal{A}$ defines an action or classification from a provided finite set of possible actions $\mathcal{A}$. Reward prediction $r \in \mathbb{R}$ estimates the moving average of the received reward in the recent activations of the classifier. Reward prediction error ε estimates the moving average of the absolute error of the reward prediction. Finally, fitness $f \in [0, 1]$ estimates the moving average of the relative accuracy of the classifier compared to the competing classifiers in the activated match sets (or action sets). The larger the fitness estimate, the on average larger the accuracy of a classifier in comparison to all classifiers that encode the same action and whose condition parts define overlapping subspaces.

Each classifier also maintains several additional parameters. The *action set size estimate as* estimates the moving average of the action sets the classifier was part of. It is updated similarly to the reward prediction r. A *time stamp ts* specifies the last time the classifier was part of a GA competition. An *experience* counter *exp* specifies the number of applied parameter updates. The numerosity *num* specifies the number of (micro-) classifiers, this macro-classifier actually represents – mainly for saving computation time.

Learning usually starts with an empty population. The problem faced is sampled iteratively, encountering particular problem instances $s \in S$. The set of all matching classifiers in the classifier population $[P]$ is termed the *match set* $[M]$. If some action in $\mathcal{A}$ is not represented in $[M]$, a *covering mechanism* is applied. Covering creates classifiers that match s (inserting #-symbols in the new C with a probability $P_\#$ at each position) and that specify the unrepresented actions. $[M]$ essentially contains all the knowledge of XCS about the current problem instance. Given $[M]$, XCS estimates the payoff for each possible action forming a *prediction ar-*

ray $P(\mathcal{A})$,

$$P(A) = \frac{\sum_{\text{cl}.A=A \wedge \text{cl} \in [M]} \text{cl}.r \cdot \text{cl}.f}{\sum_{\text{cl}.A=A \wedge \text{cl} \in [M]} \text{cl}.f}, \qquad (47.1)$$

where classifier parameters are addressed using the dot notation. $P(A)$ computes the fitness-averaged Q-value estimates for each action in the current state s. Thus, $P(A)$ can be used to decide on the currently most promising action.

Any action selection policy may be applied, such as choosing the action with the largest Q-value expectation. Because XCS relies on exploring the complete problem spaces, however, it is important that all actions are applied sufficiently frequently. Alternatively, also the prediction error estimates may be considered for action selection – choosing, for example, that action with the highest fitness-averaged ε value with the aim of maximizing information gain (see also more elaborate techniques surveyed recently in the computational intelligence literature [47.27]).

After the choice of an action A, an *action set* $[A]$ is formed, which contains all classifiers in $[M]$ that specify the chosen action. Moreover, the chosen action is executed, feedback is received in the form of scalar reward $R \in \mathbb{R}$, and the next problem instance may be perceived. In conjunction with the maximum $P(A)$ derived from the resulting match set, the $[A]$ formed is updated according to the estimated Q-value signal, which is $R + \gamma \max_{A \in \mathcal{A}} P(A)$. Moreover, the steady-state GA may be applied, reproducing two classifiers in $[A]$, but choosing classifiers from $[P]$ for deletion. In classification problems – often also termed *single–step problems* – the Q-learning update only considers the immediate reward R. Figure 47.3 illustrates the iterative learning process applied in XCS.

Rule Evaluation

To evaluate the classifiers, it is crucial to update their parameter estimates and derive a relative fitness estimate. Parameter updates are applied iteratively in respective action sets. Usually, the prediction error is updated before the prediction and the fitness. Other parameters may be updated in any order.

In particular, the reward prediction error ε of each classifier in $[A]$ is updated by

$$\varepsilon \leftarrow \varepsilon + \beta(|\rho - R| - \varepsilon), \qquad (47.2)$$

where $\rho = R$ in classification problems and

$$\rho = R + \gamma \max_{A \in \mathcal{A}} P(A)$$

in multi-step reinforcement learning problems. Parameter $\beta \in [0, 1]$ specifies a learning rate, which is typically set to values between 0.05 and 0.2. The higher the value of β is, the more the ε value depends on the most recent problem interactions. Next, the reward prediction r of each classifier in $[A]$ is updated by

$$r \leftarrow r + \beta(\rho - r). \qquad (47.3)$$

Note that XCS essentially applies Q-learning updates, where Q-values are not approximated by a tabular entry but by a collection of rules expressed in the prediction array $P(\mathcal{A})$ [47.21].

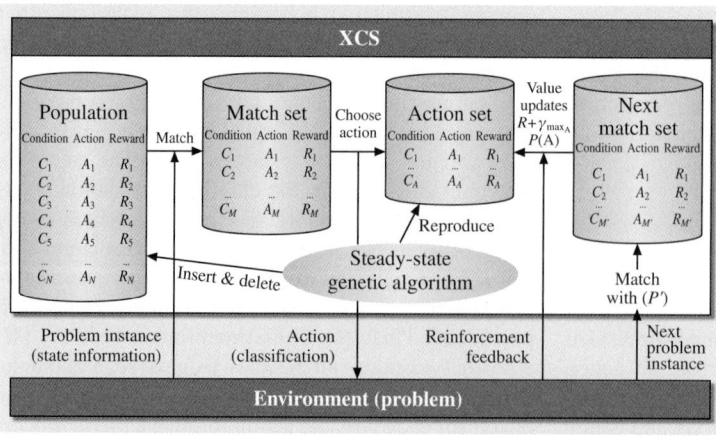

Fig. 47.3 The XCS classifier system learns iteratively online. With each iteration it forms a match set given the current problem instance. Next, it chooses an action or classification and applies it. After the perception of feedback, the classifiers in the corresponding action set $[A]$ are updated and the steady-state GA is applied. After that, the next problem iteration proceeds

To update the fitness estimate of each classifier in [A], a current scaled relative accuracy κ' is determined.

$$\kappa = \begin{cases} 1 & \text{if } \varepsilon < \varepsilon_0 \\ \alpha \left(\frac{\varepsilon_0}{\varepsilon} \right)^{\nu} & \text{otherwise} \end{cases}, \qquad (47.4)$$

$$\kappa' = \frac{\kappa \cdot \text{num}}{\sum\limits_{\text{cl} \in [A]} \text{cl}.\kappa \cdot \text{cl.num}}. \qquad (47.5)$$

κ essentially measures the current inverse error of a classifier. ε_0 specifies the targeted error below which a classifier is considered maximally accurate. κ' then determines the current relative accuracy with respect to all other classifiers in the current action set [A]. Thus, each classifier in [A] competes for a limited fitness resource, which is distributed relative to the current accuracy estimates κ. Finally, the fitness estimate f is updated given the current κ' by

$$f \leftarrow f + \beta(\kappa' - f). \qquad (47.6)$$

In effect, fitness reflects the moving average, set-relative accuracy of a classifier. As before, β controls the sensitivity of the fitness estimates to changes in the population.

The action set size estimate as is updated similarly to the reward prediction R but with respect to the current action set size $|[A]|$

$$as \leftarrow as + \beta(|[A]| - as), \qquad (47.7)$$

resulting in an action set size adaptation to changes $|[A]|$ in an order similar to the fitness changes. Parameters r, ε, and as are updated using the *moyenne adaptive modifiée* technique [47.28]. This technique sets parameter values directly to the average of the so far encountered cases until the resulting update is smaller than β (which is the case after $1/\beta$ updates). Finally, the experience counter *exp* is increased by one. If the GA is applied, the time stamps *ts* of all classifiers in [A] are set to the current iteration time t.

Rule Evolution

XCS applies a steady-state genetic algorithm (GA) for rule evolution. Given a current action set [A], the GA is invoked if the average time since the last GA application (stored in parameter *ts*) in [A] is larger than threshold θ_{GA}. This mechanism is applied to ensure sufficient evaluation of classifiers, as well as to control

unbalanced sampling. The higher the threshold θ_{GA} is, the slower evolution proceeds, but also the less prone XCS is to unbalanced problem sampling [47.29].

The steady-state GA first selects two parental classifiers for reproduction in [A]. While this selection process was done by proportionate selection based on fitness in the original XCS, more recently it was shown that tournament selection can improve the robustness of the system highly significantly [47.30]. Tournament selection in XCS chooses the classifier with the highest fitness from a tournament of randomly chosen classifiers from [A]. The tournament size is usually set relative to the current action set size $|[A]|$ to $\tau \cdot |[A]|$. Two classifiers are selected in two independent tournaments. The selected classifiers are reproduced generating the offspring. Crossover and mutation are applied to the offspring. The parents stay in the population. Mutation usually changes each condition and action symbol randomly with a certain probability μ. Crossover exchanges condition and action symbols. Often, simple uniform crossover is applied (exchanging each symbol with a probability of 0.5). However, also more sophisticated estimation of distribution (EDAs) algorithms have been applied for more effective building block processing [47.31].

The offspring parameters are initialized by setting prediction R, ε, f, and *as* to the parental values. Fitness f is often decreased to 10% of the parental fitness. Experience counter exp and numerosity num are set to one.

The resulting offspring classifiers are finally added to the population. In this case, *GA subsumption* may be applied [47.32] to stress generalization. *GA subsumption* searches for another classifier in [A] that may subsume an offspring classifier. This classifier must have a more general condition than the offspring classifier, its error estimate must be below ε_0, and its experience counter must be sufficiently high ($\exp > \theta_{\text{sub}}$). If such a classifier is found, the offspring is *subsumed*, increasing the numerosity of the more general classifier by one and discarding the offspring.

The population of classifiers [P] is maximally of finite size N. When this size is exceeded after offspring insertion, classifiers are deleted from [P]. Fitness proportionate selection is applied depending on the action set size estimates *as*. Note that tournament selection is not suitable in this case because a balance in the action set sizes is most desirable. The likelihood of deletion of a classifier is further increased by a factor $\bar{f}/f$ if this classifier is *experienced* $\exp > \theta_{\text{del}}$ and additionally if its fitness f is below a fraction δ of the average fitness $\bar{f}$ in the population.

47.2.2 When and How XCS Works

From the description above it may seem hard to understand why XCS learns successfully. This section provides intuition about when and how XCS works and points to relevant literature that quantifies the sketched-out intuition.

The two interacting learning components, which are gradient-based rule evaluation and evolutionary-based rule evolution, are strongly interactive. From an evolutionary point of view, several evolutionary pressures yield particular learning biases. Since reproduction is designed to maximize fitness, XCS strives to develop maximally accurate classifiers applying a *fitness pressure* [47.33]. Meanwhile, rules are selected in [A] for reproduction but they are selected in [P] for deletion. Since the classifier conditions in [A] will on average cover a larger subspace, i.e., they have a larger *volume* than the average condition volumes of classifiers in [P], more general classifiers will be reproduced on average (when ignoring the fitness pressure for the moment), yielding a sampling-dependent *generalization pressure* [47.33]. In consequence, it has been put forward that XCS strives to evolve a *complete problem solution* that is represented by *maximally general* classifiers that are meanwhile *maximally accurate* (error below the threshold ε_0). The resulting problem solution representation was previously termed the *optimal solution representation* [O] [47.34].

While these evolutionary pressures generally describe how the GA in XCS works, successful rule evolution still relies on sufficiently accurate fitness signals. Thus, rule evaluation needs to have enough time to estimate rule fitness before expected rule deletion. This leads to a *covering bound*, which quantifies the need for a sufficiently large population size given a particular initial condition volume. Moreover, each particular problem can be assumed to have a certain complexity in terms of subspace sizes that need to be separated for learning to take place, that is, for decreasing the error below the average deviation of the payoff signal to perceive an initial fitness signal towards higher accuracy. In consequence, the subspace size requires the generation of classifiers with condition volumes of maximally that size, consequently yielding a *schema bound* on the population size to be able to cover the full problem space with such condition volumes. Finally, better classifiers with a certain condition volume need to be able to grow, that is, have reproductive opportunities before deletion can be expected, consequently yielding a *reproductive opportunity bound*.

Together these bounds give estimates on the necessary initial condition volumes and the resulting maximal population size necessary to cover a problem space. For example, given the need for an initial classifier volume of 0.01 of the encountered problem space, the population size N should be set to about $10/0.01 = 1000$ to assure proper rule evolution. Given that these factors are satisfied, better classifiers are assured to be identified and to grow in the population with high probability. For binary and for real-valued problem domains, these considerations have been quantified, showing that XCS is an approximate polynomial-time learning algorithm in problem domains with bounded complexity [47.25, 35].

The considerations above ensure the theoretic growth of better classifiers. However, the evolutionary component may still destroy relevant classifier structures due to mutation and crossover. Thus, neither mutation nor crossover may be overly disruptive. In extreme cases, where highly unstructured subspaces may need to be identified and recombined, estimation of distribution algorithms can help to identify these subspaces [47.31, 36]. In most cases, though, a sufficiently low mutation rate and uniform crossover suffice to learn successfully. However, clearly mutation is mandatory to detect more accurate classifier structures over time. Thus, a good compromise is necessary to ensure that offspring is usually mutated but its structure is not fully destructed. In the binary domain, for example, the mutation probability is consequently often set to $1/l$, where l is the number of bits of a problem instance. This is a typical choice for the mutation strength used in genetic algorithms – essentially setting the expected number of attributes that will be mutated to one.

47.2.3 When and How to Apply XCS

From the reflections above it becomes clear that XCS is designed to learn the target function of a problem by a population of locally accurate predictors, that is, classifiers. This target function may be the Q-value function in RL problems, a *correctness* function in classification problems, or also any other type of function. XCS is best suited to be applied in problem domains that can be partitioned into subspaces within which simple predictions yield accurate values. Moreover, XCS is even better suited to be applied to problems where regularities in the target function can be well-represented in classifier conditions, that is, subspaces in which the unction values are approximately equal should be compactly representable with few classifiers. Overall, XCS thus strives to develop distributed problem solutions in the form of a set of locally partially overlapping classifier structures, which cover the whole sampled problem space in a generalized way.

As long as a condition representation can be chosen that identifies expectable regularities in a data set or also in a reinforcement learning problem well, XCS is a good candidate to optimize these local condition structures iteratively online. However, also in offline, data mining-based classification problems XCS was applied successfully and it was shown that the generalization and accuracy performance XCS yield is comparable to other state-of-the art machine learning algorithms [47.25, 37], such as decision tree learners, instance-based classifiers, or support vector machines. Thus, XCS may be applied to multi-step Q-learning problems but also to single-step classification problems and general regression problems. Online generalization and optimal condition structuring for accurate predictions are the major features of XCS. From a regression perspective, XCS is a non-parametric regression algorithm that strives to minimize the expected absolute function approximation error, or also the expected squared function approximation error as put forward elsewhere [47.24].

The two components, (a) gradient-based rule prediction approximation and evaluation and (b) evolutionary rule structure evolution, are the key to successful XCS applications. With respect to rule structure evolution, also the XCS system strongly depends on distance representations, which can be compared with general kernel representations as used in support vector machines and elsewhere [47.38, 39]. As long as the represented kernel-based condition structures can be meaningfully modified by genetic operators, evolution and thus also XCS can be applied. Meanwhile, also sensible value predictions need to be generated. Gradient-based methods work best to approximate these predictions, whether the prediction is a single value, is computed linearly or polynomially from input, or its structured otherwise depends on the problem at hand and the gradient-based approximation approach available. The more the prediction structure fits with the regularities in the target function, the faster and more robust learning can be expected. While such structural considerations can improve system performance, the successful applications of XCS to various problem domains show that successful learning is usually not precluded by suboptimal structural choices.

47.2.4 Parameter Tuning in XCS

While XCS does, indeed, specify many parameters, only few parameters are really crucial. All other parameter values can typically be set to standard values. Here we discuss some rules of thumb for tuning the critical parameter settings and also provide standard settings. While the following recommendations have not been published elsewhere so far, they can be derived from observations and other recommendations found in the literature [47.25, 35, 40].

The two most important parameters are the maximal population size N and the strived-for error threshold ε_0. The larger the population size N is, the more capacity XCS has for learning and thus the more complex problems XCS can learn. On the other hand, the larger N is, the slower XCS learns, because it reproduces and deletes only two classifiers in a typical learning iteration. Parameter ε_0 specifies the targeted approximation error. In continuous function approximation problems, smaller ε_0 values demand finer problem space partitionings and thus larger population sizes to cover the whole problem space and to enable reproductive opportunities (see above). Moreover, ε_0 can partially determine the fitness signal available to XCS: if ε_0 is chosen very small, (47.4) will yield values very close to zero for all highly inaccurate classifiers. Thus, overly small ε_0 values should be avoided. In noisy problems, ε_0 should thus also not be chosen much smaller than the standard deviation of the noise expected in the function value signal.

Without much knowledge of a problem, one may start with a rather small population size N – say 1000 – and evaluate learning progress in this setting with a desired ε_0. If the generated approximation error over time does not decrease, then ε_0 should be set to about $1/10$ of the encountered error. Next, the population size N should be progressively increased, for example, to $N = 5000$ or more. If still no error decrease is observed, further analysis is necessary. If the population is filled with classifiers but the match set sizes are very small (below 5), better classifiers probably do not receive enough reproductive opportunities. In this case, first the initial condition volume should be increased – for example, in the binary domain the probability $P_\#$ would need to be increased (up to close to 1). If the match set still decreases to sizes below 5, the problem is rather hard, requiring a further population size increase. On the other hand, if the match set sizes are very large (above 100), then over-generalization takes place and XCS apparently does not pick up the fitness signal. In this case, the initial condition volume should be decreased. If this does not help, then the GA application rate should be decreased to enforce a more accurate classifier evaluation before evolution applies. This can be accomplished by increasing the threshold θ_{GA} to say 100, 500, or even higher. An increase in θ_{GA} can also be crucial in problems where the problem domain is sampled highly unevenly, as is studied in detail elsewhere [47.29].

Several other parameter settings may be checked as well; the mutation rate should not be set overly high. As stated above, in the binary problem domain,

for example, a mutation rate of $\mu = 1/l$, where l denotes the condition size, is a good rule of thumb. Crossover can mostly be applied without restriction ($\chi = 1.0$) – especially when tournament selection for reproduction is chosen because in this case disruption is often prevented by choosing two equal classifiers. Other parameters can be safely set to somewhat standardized values. A typical initial parameter setting for XCS is: $N = 1000$, $\varepsilon_0 = 0.1$, $\mu = 1/l$, $\chi = 1$, $\alpha = 1$, $\beta = 0.2$, $\nu = 5$, $\theta_{GA} = 25$, $\gamma = 0.9$, $\theta_{del} = 20$, $\delta = 0.1$, $\theta_{sub} = 20$, $P_\# = 0.5$, and $\tau = 0.4$.

47.3 XCSF

The XCS classifier system for real-valued inputs was introduced by *Wilson* in 1999, introducing Michigan-style LCSs to the real-valued problem domain [47.41, 42]. It was further enhanced to approximate continuous real-valued function surfaces in 2001/2002 [47.43], yielding an iterative online learning non-parametric regression system. XCS for function approximation (XCSF) essentially enhances and modifies XCS by

changing its classifier condition structure to accept real valued input. Moreover, the prediction part no longer predicts single values, but it computes its prediction from the input using linear approximation techniques such as recursive least squares (RLS) [47.44]. Finally, the action part of the system is removed, applying the parts of the algorithm that were previously applied to [A] to the match set [M] in XCSF. Figure 47.4 illustrates the iterative learning process in XCSF.

XCSF is thus a regression system that solves function approximation problems by developing partially overlapping locally weighted projections in the form of a population of classifiers. In this form, XCS develops problem solutions that are similar to those developed by the locally-weighted projection regression algorithm (LWPR), which is rather well-known in the robotic community [47.45]. A comparative study has shown that XCSF can outperform LWPR in various problem domains [47.46], often yielding better problem space partitionings, as well as more accurate function value approximations with a comparable number of individual locally linear approximators (i.e., classifiers). In XCSF, each classifier specifies in its condition the subspace within which it is applicable. Thus, the condition may be compared with a receptive field determining the neural activity of the classifier. Moreover, each classifier specifies a linear approximator weighted within its subspace. In effect, the function approximation problem is approximated by locally-weighted, overlapping linear approximations. While typically the weighting

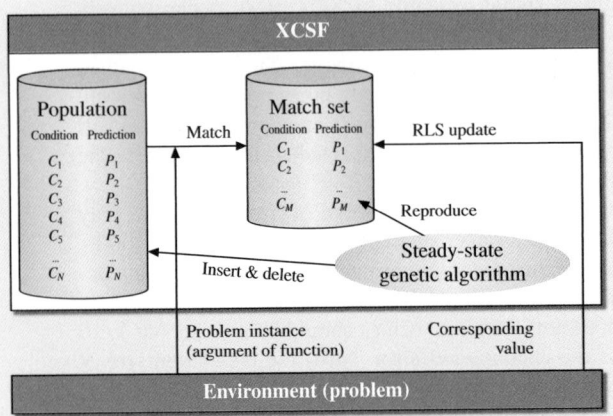

Fig. 47.4 The XCSF classifier system learns linear value predictions and usually does not specify actions. The feedback is the actual function value, which is used to update the linear approximators of the matching classifiers. The consequent error and fitness estimation updates are then considered in the evolutionary component for further optimization of the condition structures

Fig. 47.5a,b Screenshots of the XCSF program learning to approximate the crossed ridge function. Current performance values are plotted on the *top left*. The current approximation surface is approximated on the *bottom left*. On the *right hand side* the classifier condition structures are plotted. For visualization purposes, the receptive field sizes are plotted smaller than their actual size. *Darker* classifier conditions have higher fitness values ▶

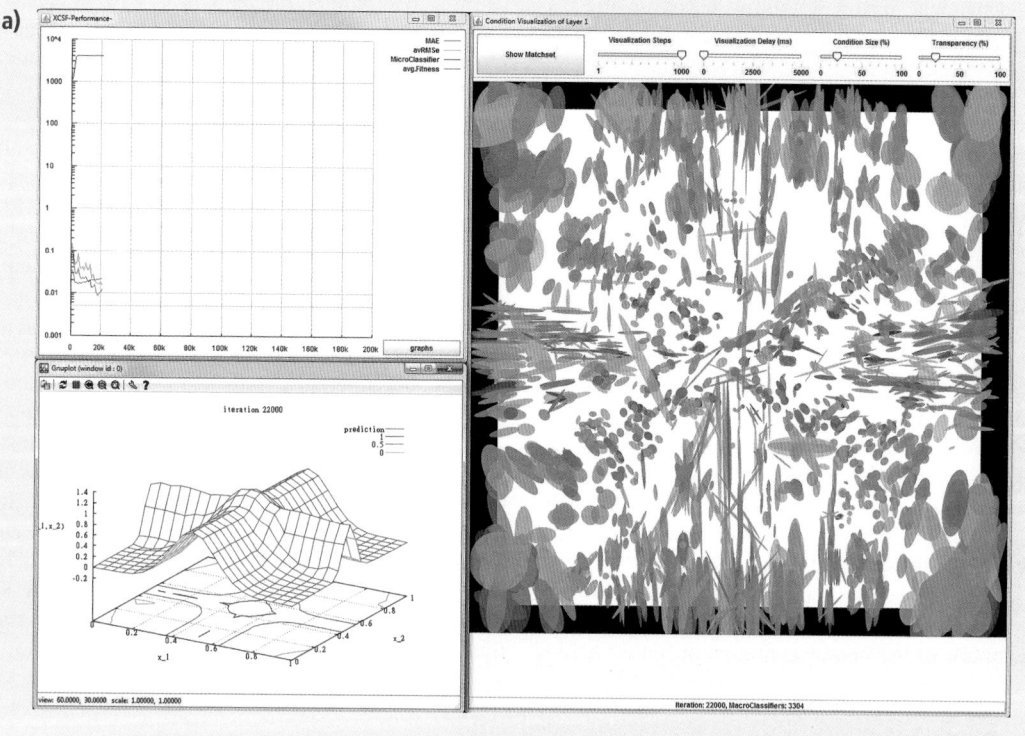

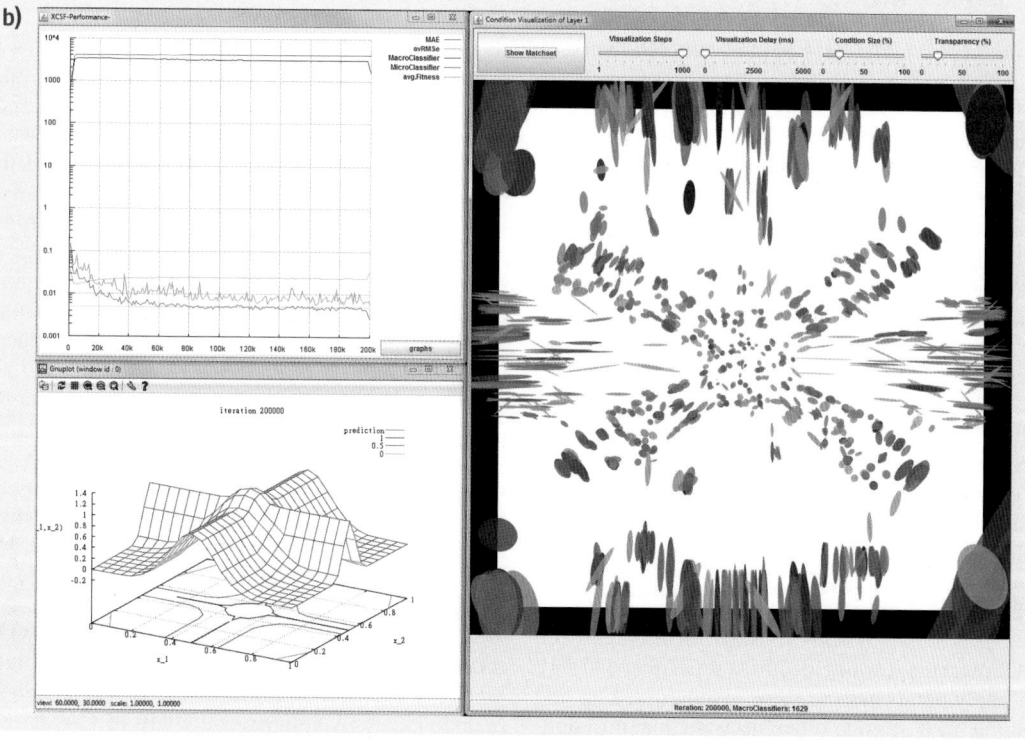

fitness-dependent, also a weighting based on the distance to the center of the classifier condition can be applied.

With this structure it has been shown that XCSF is very well suited for developing any type of kernel structure [47.47]. In effect, various condition structures have been applied, including rectangular structures with and without rotation and with various forms of representation [47.35, 48, 49]. Moreover, the linear approximations may be enhanced to polynomial approximations and others [47.50]. Finally, it is also possible to cluster a contextual space with conditions, while approximating linear (or other) predictions given totally different inputs. For example, the velocity kinematics of an arm can be predicted locally dependent on the angular arm constellation for redundancy resolution [47.51] (see further details below). Thus, XCSF is a highly flexible system with which other modifications in the condition and prediction parts of the classifiers may still yield highly vital system applications.

As an example, we applied XCSF to the *crossed ridge* function – a function that has been used as a benchmark in the neural computation and machine learning community for many years [47.45, 52]. The function contains a mix of linear and non-linear subspaces. It is specified in two dimensions as follows

$$f_1(x_1, x_2) = \max \{\exp(-10x_1^2), \exp(-50x_2^2),$$
$$1.25 \exp(-5(x_1^2 + x_2^2))\} . \quad (47.8)$$

We ran XCSF with a maximum population size $N = 4000$ and a target error $\epsilon_0 = 0.005$ on this function, applying a condensation mechanism late in the run. Figure 47.5 shows that XCSF is able to yield a good function approximation very early in the run. The evolving classifier structures learn to suitably partition the problem space into local subspaces. In consequence, a smooth overall approximation surface is generated. Note how the inverse exponential hill in the center is approximated with nearly circular receptive fields, while the fields are selectively elongated in the x_1 or x_2 dimension due to the non-linearities caused by the ridges extending to the four sides. Towards the corners of the input space, the function flattens out so that the receptive fields become increasingly wider.

47.4 Data Mining

Data mining is a rather large field of research that generally addresses the challenge of extracting knowledge from data. In the LCS realm, the addressed data usually consists of a set of data instances, where each instance specifies a set of features and a corresponding class the data instance belongs to. LCSs then typically learn to mine the data by predicting the class likelihoods of unseen data instances, as well as by identifying the most relevant features and feature interactions for classification. Particularly Pitt-style LCSs have proven to be highly valuable in data mining applications. However, also the XCS classifier system was successfully applied in this domain.

The XCS system was also converted to an offline learning system; the sUpervised classifier system (UCS) algorithm [47.53] determines classifier predictions and resulting fitness values in a supervised manner. Meanwhile, the other learning aspects of UCS were derived from XCS. Both, XCS and UCS have shown effective if not even superior prediction accuracies in various data mining tasks – most of them taken from the UCI machine learning database repository [47.54]. When applying always the same standard setting and comparing with various other decision making algorithms, such as support-vector machines, decision tree learning, naive Bayes classifiers, and others implemented in the WEKA machine learning tool [47.55], XCSF outperformed these competing techniques in many cases – often depending on the problem at hand [47.25]. A similar performance was achieved with UCS, outperforming XCS in some cases due to its more accurate classifier prediction estimates. XCS was also further enhanced to be able to deal with highly unbalanced datasets in data mining domains by automatically adjusting the threshold that controls the frequency of GA applications θ_{GA} [47.29].

Pitt-style LCSs have been evaluated and applied to data mining problems even more extensively. The typical offline-learning scenario faced in data mining particularly suits the Pitt approach. However, also the fact that often very compact rule sets are strived for is advantageous for the Pitt approach. More than 10 years ago, the GALE architecture [47.56, 57] yielded very good performance results on a collection of datasets from the UCI repository. GALE distributes its evolutionary process adding additional niching biases due to a grid-based spatial distribution of individuals. A comparative study of GALE, XCS

nd other machine learning algorithms can be found
n [47.58].

The GAassist architecture [47.59, 60] develops
priority list of classification rules. The advantage of
;Aassist is its developing compactness. A compara-
ve analysis with XCS is provided in [47.61]. Later,
ne architecture was enhanced with ensemble learn-
ng techniques [47.62] and memetic algorithms [47.63],
roving high scalability and fast learning of very com-
act rule sets.

Recently, many efficiency enhancement techniques
om the GA literature (cf. [47.64]) and from other
elds, including bioinformatics and systems biol-
gy [47.65] were applied to various LCSs. These tech-
iques can help tremendously to improve the learning
peed of LCSs, particularly in data mining realms. For

example, *windowing techniques* select subsets of data
instances to speed up the classifier evaluation process.
Fitness surrogates were used to make the fitness estimation even cheaper [47.66]. *Hybrid methods* were
already mentioned above; they combine traditional GA
operators with informed ones, as is done when applying memetic algorithms, which locally improve the
developing classifier structures when applied to LCSs.
In combination, such techniques can yield LCSs that
not only produce highly accurate classification performance and good generalizations, but they also offer
solution interpretability allowing mining of the knowledge developed in the LCS rules, and they generate
these results without requiring much computational
time – which is often comparable to the time needed
by much simpler machine learning techniques.

7.5 Behavioral Learning

/hile the application of LCSs to data mining problems
ill certainly still produce many further impressive re-
ults and promises to yield novel, deep insights into data
ructures, LCSs were originally designed as cognitive
ystems. Thus, in the following we will focus on LCSs
cognitive systems, their structures, and their poten-
al as neural cognitive models. As had been sketched
ut above, the XCS classifier system in particular was
ompared with Q-learning in RL. We start from this
erspective and detail various successful applications of
CS in reinforcement learning problems. Next, ALCSs
e surveyed. ALCSs learn generalized cognitive maps
at are suitable to apply Sutton's Dyna algorithm and
alue iteration techniques in general. A strong relation
factored RL techniques was pointed out recently in
is respect [47.67]. Finally, robotics applications of
CSs are discussed and their potential is revealed.

7.5.1 Reward–Based Learning with LCSs

om the beginning [47.2] a big appeal to LCSs lay in
e fact that they are designed for reward-based learn-
g. Once the original bucket-brigade algorithm was
placed by Q-learning techniques, a theory developed
the RL community also applied to LCSs to a certain
tent.

In XCS, in particular, it was shown that the sys-
m approximates the Q-value function by a collection
classifiers. The prediction array (47.1) calculation
sentially approximates the current Q-value estimates
r the current state in the environment. The fitness

weighting based on the relative accuracies, which are
normalized to one, assures that these Q-value estimates on average do not over or underestimate the
expected Q-value. Moreover, since Q-learning is an
off-policy learning technique, XCS is well-suited to
be combined with it because also XCS benefits from
exploring all possible state–action combinations in
the long run – striving to develop an approximation of the complete Q-value function in the problem
space.

As a result, XCS has been successfully applied to
learning optimal paths in various maze environments.
Starting from the Woods1 and Woods2 environments
proposed by *Wilson* [47.20, 21], XCS's performance
and generalization capabilities have been investigated
in various mazes [47.68]. For illustrative purposes such
mazes are shown in Fig. 47.6. These maze environments provide information about the surrounding grid
cells, indicating whether they are either free or occupied by an obstacle or by food. Reaching the latter cell
usually results in a reward trigger. Movements are typically possible to the eight surrounding cells, yielding
a rather large action space. The point of providing sensory state information rather that cell IDs or coordinates
is that XCS is then able to exhibit its generalization
capabilities. It essentially manages to generalize over
the sensory state space ignoring irrelevant bits and generalizing over the states with respect to state–action
combinations that yield the same reward.

Performance in many of these environments has
yielded extreme generalization capabilities. For exam-

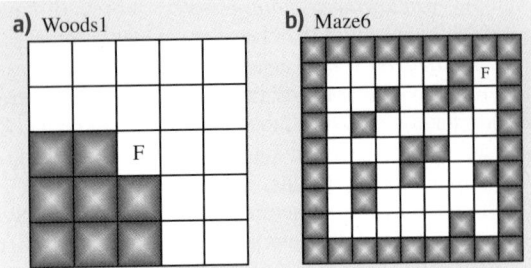

Fig. 47.6a,b Two highly typical maze environments used as benchmarks in the LCS literature for generalized reinforcement learning. Woods1 is a toroidal maze. In Maze6 the food location is much harder to find. In both cases, the LCS-controlled agent perceives information about the eight neighboring cells encoding free, blocked, and food cells by means of two bits. The agent can execute movements to each of these cells. Movements to blocked cells yield no reward. A movement to the food cell triggers reward and a reset of the agent

ple, in the Maze6 environment (Fig. 47.6) up to 90 irrelevant bits were introduced, which changed randomly while interacting with the environment. While learning was slightly delayed and a larger population size was needed for successful learning, the optimal Q-value function was still extracted from iterative interactions [47.25]. Thus, XCS learned the optimal Q-value function in a problem space that contained more than 10^{30} potential sensory state encodings. Also rather noisy action outcomes did not preclude learning success. Later, it was shown that highly effective generalizations are even possible when each bit in the sensory encoding is relevant. In [47.36] the encoding for each bit was changed to a nested Boolean function, such as the parity function. XCS was still able to learn the optimal Q-value function, while Q-learning without generalization failed miserably due to the large state space. Thus, XCS is able to identify those aspects of the available sensory information that are relevant for accurate reward predictions.

To successfully apply XCS in these scenarios, one crucial modification was necessary to stabilize the Q-values and thus the derived fitness values: the update of the classifier predictions had to be further modified by the error gradient factor, converting (47.3) to

$$r \leftarrow r + \beta(\rho - r) \frac{f}{\sum_{\text{cl}\in[A_{-1}]} \text{cl}.f} \, . \tag{47.9}$$

The exact derivation of this equation can be found in the literature [47.69]. The gradient term essentially re-

sults in much more stable performance and successful learning and generalization in problems that require the establishment of long reward chains. It stabilizes the reward learning by down-scaling updates of inaccurate and unreliable classifiers. Consequently, these rules do not tend to over-estimate reward, and thus learning progress is stabilized. As a further consequence, XCS with gradient-based reward predictions updates was also successfully applied to blocks world problems, in which even more generalizations are possible [47.25].

The generalization capabilities of LCSs reached even as far as being successfully applied to control simple light following behavior on a real robot platform [47.70, 71]. In this case, however, reward learning was maximized and no complete Q-value function approximation developed. Nonetheless, this work constituted one of the first successful application in the robotics domain.

Besides condition–action Michigan-style LCSs such as the XCS, other Michigan-style LCS techniques have been applied for behavioral learning and also for learning cognitive maps. Such anticipatory learning classifier systems are surveyed in the following.

47.5.2 Anticipatory Learning Classifier Systems

Anticipatory learning classifier systems (ALCSs) are learning systems that learn a generalized predictive model or *cognitive map* [47.72] of the encountered environment online. ALCSs are typical Michigan-style LCSs. However, in contrast to the usual classifier structure, classifiers in ALCSs have a state prediction or *anticipatory* part that predicts the environmental change in the environment caused when executing the specified action in the specified context. As in XCS, ALCSs derive classifier fitness estimates from the accuracy of their predictions. However, the accuracy of the anticipatory state predictions are considered, rather than the accuracy of the reward prediction. Figure 47.7 illustrates the typical structures and learning processes that apply in an ALCS architecture.

Rick Riolo originally proposed an ALCS that generated its cognitive map mediated by a *message list* storage system, which was also used in Holland's original classifier system architecture [47.73]. However, this approach appeared to not be sufficiently elegant to enable any serious learning. Starting with *Stolzmann*'s anticipatory classifier system [47.74], various ALCS architectures were developed. Particularly in maze problems, optimal behavior was achieved with

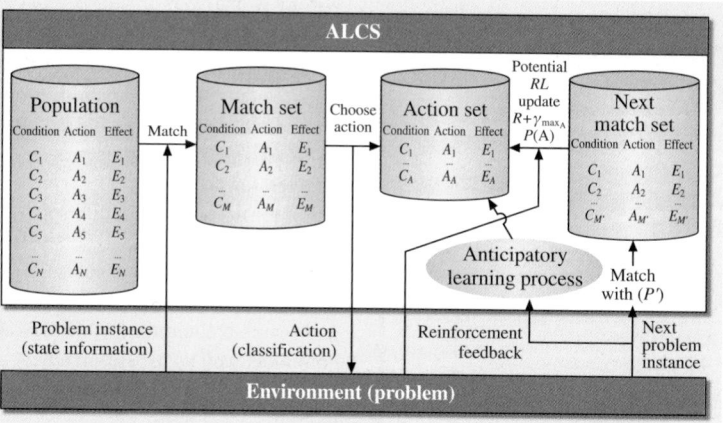

Fig. 47.7 Instead of the condition–action–reward prediction rules in typical LCSs, ALCSs encode and develop condition–action–effect rules. Typically, the structural optimization of these rules is done by a combination of evolutionary and heuristic techniques

various ALCSs [47.75–79]. To prevent the development of overgeneral models for concurrent reward learning, the reward learning process was often decoupled, yielding a system that learns a cognitive map based on LCS principles, and, additionally, a state value estimation system. In combination, DYNA-based learning techniques [47.80] were applied to improve the state value estimations also offline. These techniques allowed the simulation of animal-like behavioral patterns, such as reward adaptations based on knowledge about the behavioral consequences in rats in a T-maze environment [47.81], as well as in controlled devaluation or satiation experiments [47.82]. In these studies it was also pointed out that ALCSs do not only allow DYNA-based reward learning updates, but also enable the application of search and planning techniques for improving behavioral performance of the system. Even curiosity mechanisms have been added [47.83] to speed up the learning progress. Most approaches, however, never generalized the list of states with associated rewards.

The combination of the ACS2 system with the XCS system for state-value estimations, terming the resulting system XACS (x-anticipatory classifier system), may be the one with the most current potential for future research [47.84]. XACS essentially applies two LCS learning mechanisms: one being an ALCS architecture in the form of ACS2, which learns a cognitive model of the encountered environment, and the other one being the XCS system, which learns state-value estimations in this case. Figure 47.8 illustrates the components in the XACS architecture and their interactions.

XACS has been shown to develop optimal behavior in blocks world problems in which other approaches failed to yield proper generalizations and resulting

optimal behavior control. Moreover, the reward-based generalization mechanism in XACS is directly based on the XCS classifier system, thus enabling the incorporation of any tools and representations developed for XCS so far. The generalizations that were developed confirmed the identification of task-relevant perceptual attributes. In the XCS components, reward-distinguishing attributes were identified. In the ACS2 component, on the other hand, state prediction-relevant components were detected. In consequence, generalized detectors for prediction with respect to reward and state could be distinguished. The implementation of other anticipatory mechanisms in XACS, such as task-dependent attentional mechanisms, further interactions of the learning components, and multiple behavioral modules for the representation of multiple motivations (or needs) [47.84] are still open issues in the LCS realm. Further research with ALCSs is expected to yield highly promising, cognitive learning architectures.

47.5.3 Controlling a Robot Arm with an LCS

We end this section of behavioral learning with the XCSF system. Over the last decade or so it has become increasingly clear that XCS is extremely well suited to partition a contextual space for the generation of accurate predictions. Predictions, however, do not necessarily need to be reward predictions. Behavioral consequences serve just as well as a target for predictions. The forward kinematics mapping in the robotics domain [47.85] offers even another potential target for learning.

Consequently, XCSF was modified to learn the forward velocity kinematics of a robotic arm in simu-

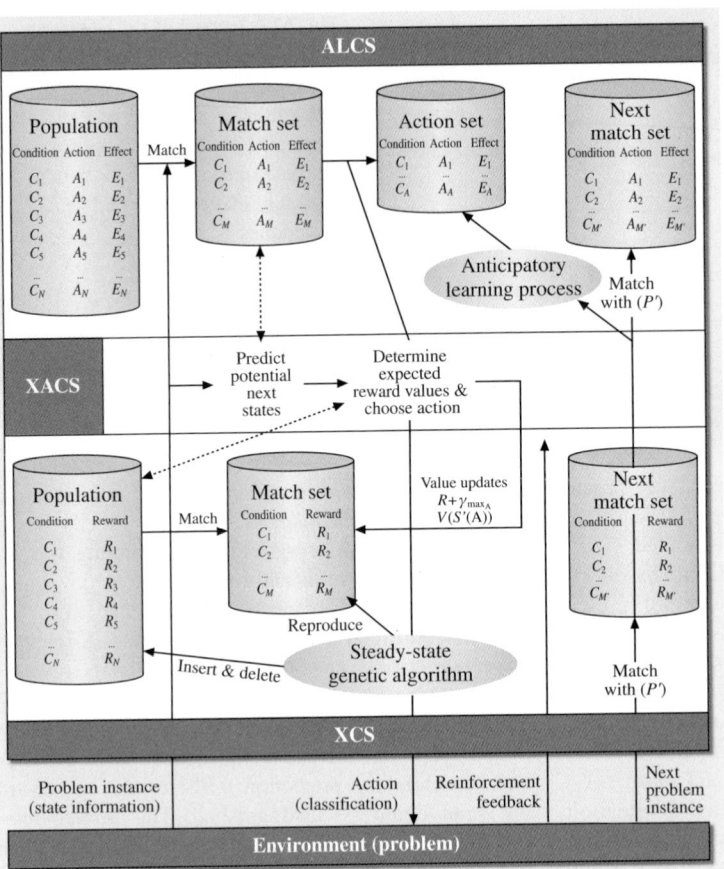

Fig. 47.8 The XACS system combines the model learning capabilities of ALCSs with the generalizing reinforcement learning capabilities of XCS. Consequently, generalizations in the two system components are targeted towards a compact representation for accurate predictions and for reward predictions, respectively. The combined system enables the application of lookahead planning and search techniques for behavioral control as well as of reinforcement learning techniques and combinations thereof

lation [47.86]. To do so, XCSF projects its condition parts into the joint angle space of the robotic arm. However, its locally linear predictions receive as input small joint movements, that is, changes in joint space and predict the consequent change in task space, that is, changes of the end-effector location. This mapping has the great advantage that it is locally linear so that given a current joint angle constellation of the arm not only location changes of joint angle movements can be predicted but also directional motion of the end-effector can be invoked by inverting the locally linear forward velocity mappings. Seeing that those are linear, the inversion can be rather easily done using linear algebra techniques. Given a redundant arm system – one that has more degrees of freedom (i.e., joint angles to manipulate) than actual locations to move to – it is possible to add additional constraints to the arm motion. For example, the arm can be driven to maintain a *relaxed* arm posture while pursuing a certain goal or it may be forced to prevent moving a certain joint

angle at all [47.87]. Recent advancements in the exploration strategy, which can be self-induced by the XCSF controller during learning, have shown that XCSF is able to learn to control all seven degrees of freedom of a humanoid arm highly effectively – flexibly adhering to different constraints while pursuing motion to certain goal locations. Moreover, mappings could be learned in different reference frame representations. For example, end-effector locations were either represented in a Cartesian coordinate system or in a distance plus angles encoding. XCSF learned different classifier structures due to the differences in the linearities encountered. Nonetheless, XCSF yielded equally good arm control in both cases [47.51]. Figure 47.9 illustrates the XCSF setup for arm control.

These results confirmed that XCSF may very well be further developed into a cognitive system architecture for behavioral control. While this type of architecture was probably not the one envisioned by Holland originally, it may still prove highly valuable. Various

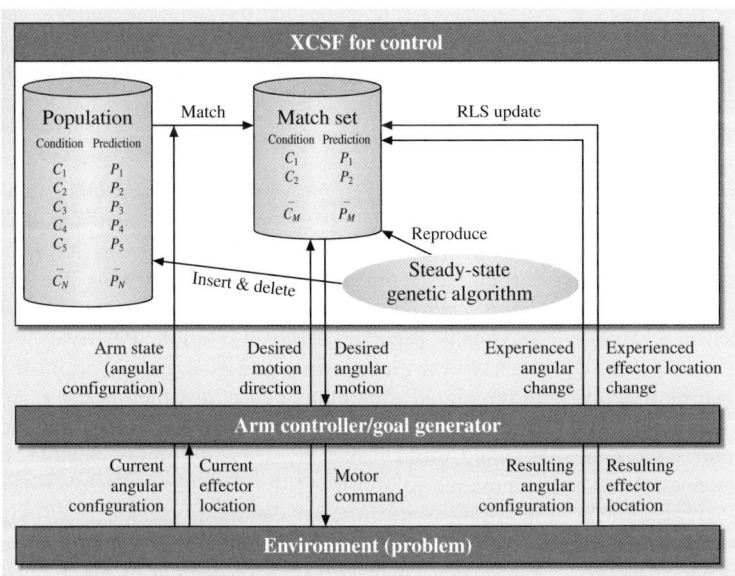

Fig. 47.9 In the published robot arm control applications, XCSF clusters the contextual configuration state of the arm and learns linear approximations of the average Jacobian in the respective subspaces. In consequence, the system can generate both forward predictions of movement consequences as well as inverse control commands when directional movements of the arm are desired

neuroscientific evidence points out that similar forward-inverse predictive-control structures may be found in the cerebellum [47.88, 89]. Only more detailed knowledge on cortical and cerebellar structures may allow the direct comparison of the shapes and orientations of the receptive fields developed by the XCSF system and potential cortical and neural structures found in the brain.

While the brain may not implement actual evolutionary techniques literally, as XCSF does, it appears plausible that local competitions take place [47.90]. Moreover, it is known that neurons populate novel information sources once available – as XCSF does. Further research in neural computation with LCSs may prove highly valuable.

47.6 Conclusions

While LCSs have been applied to a wide variety of problems, still there are many potential developments that have not been further evaluated. In the following, potential future research directions are summarized.

At the moment nearly all LCSs are flat in that they develop one population of classifiers (or competing sets of classifiers in the Pitt-style system). All of the classifiers, however, apply to the same problem granularity. Ever since the introduction of LCSs by Holland, the development of *default hierarchies* was envisioned. However, so far it was never convincingly or rigorously accomplished [47.19]. Default hierarchies refer to classifier systems in which general rules predict one thing but more specialized rules predict exceptions of the general rule. The emergent development of default hierarchies in LCSs remains an open challenge.

With the most recent understanding of LCSs and the XCS system in particular, it seems that at least

the development of a hierarchically-structured LCS architecture is within our grasp. We expect such a hierarchical LCS to progressively refine its predictions in a hierarchical way. Default rules may gain a certain level of accuracy, but more specialized rules may identify exceptions of the default prediction. Alternatively, the more specialized rules may also simply add further accuracy to the default predictions where and when necessary. In the latter case, a hierarchical predictive system may develop that allows the progressive refinement of activated predictions until the finest prediction granularity in the hierarchical representation is reached.

When developing hierarchical LCSs, also network LCSs seem to be of vital importance. For example, when developing classifier structures in spatial domains that are intricately structured, a network structure may provide additional hints on the connectivity of the space. Especially the case where XCSF learns velocity kinematics, or generally, contextually-dependent sensory-motor contingencies – as sketched out in the section on controlling a robot arm with XCSF above – a network structure can give additional hints on how the sensorimotor space is structured and may be traversed. Networks of LCS classifiers may allow the application of lookahead planning and goal-oriented control – as was pursued in early work in [47.91].

A network structure may also enable the speed-up of the XCS matching process. For example, when a problem space is sampled by means of a random walk process, overlapping classifiers may be directly identified within a classifier structure instead of applying a global matching process in each iteration. Also, when XCSF is used for goal-directed control – as mentioned above with respect to velocity kinematics – this may improve the efficiency of the system tremendously. Furthermore, given a hierarchically network structured LCS system matching may proceed from coarse-to-fine-grained levels. All these processes may speed up the matching, which is often considered a bottleneck in LCS research and has been improved by means of numerous approaches over the recent years [47.92, 93].

Besides these additions, also ALCSs may be pursued further, as sketched out above. From a cognitive modeling perspective, ALCSs essentially learn generalized schemata or production rules [47.94–96], which specify the expected state changes perceived after the execution of the specified action. Such rules may be applied by the cognitive science community for learning, for example, ACT-R structures [47.97]. The lookahead planning capabilities, the sensorimotor generalization capabilities, as well as the abstraction capabilities of these systems still ask for further development. The recent point that ALCSs can be very effectively applied to factored RL problems [47.98] should be further pursued. Also, the combination of ALCS-based cognitive map or concept learning and XCS-based reward learning promises further research advancements.

Even without the addition of hierarchies, network structures, or anticipations, however, LCSs can be successfully applied to various domains including reinforcement learning problems, classification and data mining problems, and regression problems. XCS, in particular, learns iteratively online, striving for the development of a compact, maximally general, and maximally accurate problem solution. Pitt-style systems typically learn offline and are thus most promising in large-scale data mining tasks in which rather small compact sets of rules are searched for. Seeing that the learning mechanisms of LCSs are highly flexible, it is possible to substitute the condition of a classifier with any other form or condition structure, as long as this structure can be mutated and recombined in a way that small structural changes also yield small changes in the defined subspace within which the condition matches. Similarly, the prediction structure can be replaced with any other prediction structure that can be quickly and accurately adapted by suitable learning techniques. Thus, the available LCS techniques – such as GALE and GAassist on the Pitt side and XCS, XCSF, or XACS on the Michigan side – can be further exploited and combined with novel structures and forms of representations. Learning promises to be robust due the combination of a flexible evolutionary component, which searches for optimal rule structures, and the gradient-based fitness estimation, which quickly yields useful prediction and fitness estimations. It seems only a matter of time until LCSs gain even more recognition and be successfully applied to even more diverse problem domains and challenging research tasks.

47.7 Books and Source Code

Further information about learning classifier systems can be found in the biannually published IWLCS (International Workshop on Learning Classifier Systems) workshop proceedings and yearly workshops on the topic. A book on LCSs and the XCS classifier system in particular covers XCS from a theoretical and application-oriented point of view and also provides a detailed algorithmic description of the system [47.25]. A more theoretical coverage of the approximation approach in XCS can be found in [47.23]. Several books also give further details on theoretical considerations [47.23, 99] as well as on successful applications of LCSs [47.100, 101]. The source code can be found online, for example, for XCS in C++ [47.102] as well as for XCSF in Java [47.103].

References

47.1   J.H. Holland: *Adaptation in Natural and Artificial Systems* (Univ. of Michigan, Ann Arbor 1975)

47.2   J.H. Holland: Adaptation. In: *Progress in Theoretical Biology*, Vol. 4, ed. by R. Rosen, F.M. Snell (Academic, New York 1976) pp. 263–293

47.3   L.B. Booker, D.E. Goldberg, J.H. Holland: Classifier systems and genetic algorithms, Artif. Intell. **40**, 235–282 (1989)

47.4   J.H. Holland, J.S. Reitman: Cognitive systems based on adaptive algorithms. In: *Pattern Directed Inference Systems*, ed. by D.A. Waterman, F. Hayes-Roth (Academic, New York 1978) pp. 313–329

47.5   L.P. Kaelbling, M.L. Littman, A.W. Moore: Reinforcement learning: A survey, J. Artif. Intell. Res. **4**, 237–285 (1996)

47.6   R.S. Sutton, A.G. Barto: *Reinforcement Learning: An Introduction* (MIT Press, Cambridge 1998)

47.7   J.H. Holland: Properties of the bucket brigade algorithm, Proc. Int. Conf. Genet. Algorithms Appl. (1985) pp. 1–7

47.8   D.E. Goldberg: *Genetic Algorithms in Search, Optimization and Machine Learning* (Addison-Wesley, Reading 1989)

47.9   S.F. Smith: A learning system based on genetic adaptive algorithms, Ph.D. Thesis (Univ. of Pittsburgh, Pittsburgh 1980)

47.10   K.A. De Jong: An analysis of the behavior of a class of genetic adaptive systems, Ph.D. Thesis (Univ. of Michigan, Ann Arbor 1975)

47.11   L.B. Booker: Intelligent behavior as an adaptation to the task environment, Ph.D. Thesis (The Univ. of Michigan, Ann Arbor 1982)

47.12   S.W. Wilson: Knowledge growth in an artificial animal, Proc. Int. Conf. Genet. Algorit. Appl. (1985) pp. 16–23

47.13   S.W. Wilson: Classifier systems and the animat problem, Mach. Learn. **2**, 199–228 (1987)

47.14   D.E. Goldberg: Computer-aided gas pipeline operation using genetic algorithms and rule learning, Diss. Abstr. Int. **44**, 3174B (1983)

47.15   K.A. De Jong: Learning with genetic algorithms: An overview, Mach. Learn. **3**, 121–138 (1988)

47.16   K.A. De Jong, W.M. Spears, D.F. Gordon: Using genetic algorithms for concept learning, Mach. Learn. **13**, 161–188 (1993)

47.17   R.L. Riolo: Bucket brigade performance: I. Long sequences of classifiers, Proc. 2nd Int. Conf. Genet. Algorithms (ICGA87), ed. by J.J. Grefenstette (Lawrence Erlbaum Associates, Cambridge 1987) pp. 184–195

47.18   R.E. Smith, H. Brown Cribbs: Is a learning classifier system a type of neural network?, Evol. Comput. **2**, 19–36 (1994)

47.19   J.H. Holland, L.B. Booker, M. Colombetti, M. Dorigo, D.E. Goldberg, S. Forrest, R.L. Riolo, R.E. Smith, P.L. Lanzi, W. Stolzmann, S.W. Wilson: What is a learning classifier system?, Lect. Notes Comput. Sci. **1813**, 3–6 (2000)

47.20   S.W. Wilson: ZCS: A zeroth level classifier system, Evol. Comput. **2**, 1–18 (1994)

47.21   S.W. Wilson: Classifier fitness based on accuracy, Evol. Comput. **3**, 149–175 (1995)

47.22   C.J.C.H. Watkins: Learning from delayed rewards, Ph.D. Thesis (King's College, Cambridge 1989)

47.23   J. Drugowitsch: Design and Analysis of Learning Classifier Systems: A Probabilistic Approach, Studies in Computational Intelligence (Springer, Berlin, Heidelberg 2008)

47.24   J. Drugowitsch, A. Barry: A formal framework and extensions for function approximation in learning classifier systems, Mach. Learn. **70**, 45–88 (2008)

47.25   M.V. Butz: *Rule-Based Evolutionary Online Learning Systems: A Principled Approach to LCS Analysis and Design* (Springer, Berlin, Heidelberg 2006)

47.26   B. Widrow, M. Hoff: Adaptive switching circuits, West. Electron. Show Conv. **4**, 96–104 (1960)

47.27   P.-Y. Oudeyer, F. Kaplan, V.V. Hafner: Intrinsic motivation systems for autonomous mental development, IEEE Trans. Evol. Comput. **11**, 265–286 (2007)

47.28   G. Venturini: Adaptation in dynamic environments through a minimal probability of exploration, from animals to animats 3, Proc. 3rd Int. Conf. Simul. Adapt. Behav. (1994) pp. 371–381

47.29   A. Orriols-Puig, E. Bernadó-Mansilla, D.E. Goldberg, K. Sastry, P.L. Lanzi: Facetwise analysis of XCS for problems with class imbalances, IEEE Trans. Evol. Comput. **13**, 1093–1119 (2009)

47.30   M.V. Butz, K. Sastry, D.E. Goldberg: Strong, stable, and reliable fitness pressure in XCS due to tournament selection, Genet. Program. Evol. Mach. **6**, 53–77 (2005)

47.31   M.V. Butz, M. Pelikan, X. Llorà, D.E. Goldberg: Automated global structure extraction for effective local building block processing in XCS, Evol. Comput. **14**, 345–380 (2006)

47.32   S.W. Wilson: Generalization in the XCS classifier system, genetic programming 1998, Proc. 3rd Ann. Conf. (1998) pp. 665–674

47.33   M.V. Butz, T. Kovacs, P.L. Lanzi, S.W. Wilson: Toward a theory of generalization and learning in XCS, IEEE Trans. Evol. Comput. **8**, 28–46 (2004)

47.34   T. Kovacs: XCS classifier system reliably evolves accurate, complete, and minimal representations for Boolean functions. In: *Soft Computing in Engineering Design and Manufacturing*, ed. by R. Roy, P.K. Chawdhry, R.K. Pant (Springer, Berlin, Heidelberg 1997) pp. 59–68

47.35   P.O. Stalph, X. Llorà, D.E. Goldberg, M.V. Butz: Resource management and scalability of the XCSF learning classifier system, Theor. Comput. Sci. **425**, 126–141 (2012)

47.36   M.V. Butz, P.L. Lanzi: Sequential problems that test generalization in learning classifier systems, Evol. Comput. **2**, 141–147 (2009)

47.37   L. Bull, E. Bernadó-Mansilla, J. Holmes (Eds.): *Learning Classifier Systems in Data Mining*, Studies in Computational Intelligence, Vol. 125 (Springer, Berlin, Heidelberg 2008)

47.38   B. Schökopf, A.J. Smola: *Learning with Kernels: Support Vector Machines, Regularization, Optimization, and Beyond* (MIT Press, Cambridge 2001)

47.39   W. Liu, J.C. Principe, S. Haykin: *Kernel Adaptive Filtering: A Comprehensive Introduction*, 1st edn. (Wiley, Hoboken 2010)

47.40   M.V. Butz, S.W. Wilson: An algorithmic description of XCS, Soft Comput. **6**, 144–153 (2002)

47.41   S.W. Wilson: Get real! XCS with continuous-valued inputs. In: *Festschrift in honor of John H. Holland*, ed. by L. Booker, S. Forrest, M. Mitchell, R.L. Riolo (Center for the Study of Complex Systems, Ann Arbor 1999) pp. 111–121

47.42   S.W. Wilson: Get real! XCS with continuous-valued inputs, Lect. Notes Comput. Sci. **1813**, 209–219 (2000)

47.43   S.W. Wilson: Classifiers that approximate functions, Nat. Comput. **1**, 211–234 (2002)

47.44   S. Haykin: *Adaptive Filter Theory*, 4th edn. (Prentice Hall, Upper Saddle River 2002)

47.45   S. Vijayakumar, A. D'Souza, S. Schaal: Incremental online learning in high dimensions, Neural Comput. **17**, 2602–2634 (2005)

47.46   P. Stalph, J. Rubinsztajn, O. Sigaud, M.V. Butz: Function approximation with LWPR and XCSF: A comparative study, Evol. Comput. **5**, 103–116 (2012)

47.47   M.V. Butz: Kernel-based, ellipsoidal conditions in the real-valued XCS classifier system, Proc. Genet. Evol. Comput. Conf. (GECCO 2005) (2005) pp. 1835–1842

47.48   C. Stone, L. Bull: For real! XCS with continuous-valued inputs, Evol. Comput. **11**, 299–336 (2003)

47.49   M.V. Butz, P.L. Lanzi, S.W. Wilson: Function Approximation With XCS: Hyperellipsoidal Conditions, Recursive Least Squares, and Compaction, IEEE Trans. Evol. Comput. **12**, 355–376 (2008)

47.50   D. Loiacono, P.L. Lanzi: Recursive least squares and quadratic prediction in continuous multistep problems, Lect. Notes Comput. Sci. **6471**, 70–86 (2010)

47.51   P.O. Stalph, M.V. Butz: Learning local linear Jacobians for flexible and adaptive robot arm control, Genet. Program. Evol. Mach. **13**, 137–157 (2012)

47.52   S. Schaal, C.G. Atkeson: Constructive incremental learning from only local information, Neural Comput. **10**, 2047–2084 (1998)

47.53   E. Bernadó-Mansilla, J.M. Garrell-Guiu: Accuracy-based learning classifier systems: Models, analysis, and applications to classification tasks, Evol. Comput. **11**, 209–238 (2003)

47.54   K. Bache, M. Lichman: *UCI Machine Learning Repository* (Univ. of California, School of Information and Computer Sciences 2013) http://archive.ics.uci.edu/ml

47.55   I.H. Witten, E. Frank: *Data Mining. Practical Machine Learning Tools and Techniques with Java Implementations* (Morgan Kaufmann, San Francisco 2000)

47.56   X. Llorà, J.M. Garrell: Knowledge independent data mining with fine-grained parallel evolutionary algorithms, Proc. Genet. Evol. Comput. Conf. (GECCO 2001) (2001) pp. 461–468

47.57   X. Llorà, J.M. Garrell: Inducing partially-defined instances with evolutionary algorithms, Proc. 18th Int. Conf. Mach. Learn. (ICML 2001) (2001)

47.58   E. Bernadó, X. Llorà, J.M. Garrell: XCS and GALE: A comparative study of two learning classifier systems and six other learning algorithms on classification tasks, Lect. Notes Comput. Sci. **2321**, 115–132 (2002)

47.59   J. Bacardit, J.M. Garrell: Evolving multiple discretizations with adaptive intervals for a Pittsburgh rule-based learning classifier system, Lect. Notes Comput. Sci. **2724**, 1818–1831 (2003)

47.60   J. Bacardit, M.V. Butz: Data mining in learning classifier systems: Comparing XCS with GAssist, Lect. Notes Comput. Sci. **4399**, 282–290 (2007)

47.61   J. Bacardit, M.V. Butz: *Data mining in learning classifier systems: Comparing XCS with GAssist* (Il-liGAL, Univ. of Illinois at Urbana-Champign 2004)

47.62   J. Bacardit, N. Krasnogor: Empirical evaluation of ensemble techniques for a Pittsburgh learning classifier system, Lect. Notes Comput. Sci. **4998**, 255–268 (2008)

47.63   J. Bacardit, N. Krasnogor: Performance and efficiency of memetic Pittsburgh learning classifier systems, Evol. Comput. **17**, 307–342 (2009)

47.64   K. Sastry, D.E. Goldberg, X. Llorá: Towards billion-bit optimization via a parallel estimation of distribution algorithm, Proc. Genet. Evol. Comput. Conf. (GECCO 2007) (2007) pp. 577–584

47.65   J. Bacardit, E. Burke, N. Krasnogor: Improving the scalability of rule-based evolutionary learning, Memet. Comput. **1**, 55–67 (2009)

47.66   X. Llorà, K. Sastry, T.-L. Yu, D.E. Goldberg: Do not match, inherit: Fitness surrogates for genetics-based machine learning techniques, Proc. Genet. Evol. Comput. Conf. (GECCO 2007) (2007) pp. 1798–1805

47.67   O. Sigaud, M.V. Butz, O. Kozlova, C. Meyer: *Anticipatory Learning Classifier Systems and Factored Reinforcement Learning* (Springer, Berlin, Heidelberg 2009) pp. 321–333

47.68   P.L. Lanzi: An analysis of generalization in the XCS classifier system, Evol. Comput. **7**, 125–149 (1999)

47.69   M.V. Butz, D.E. Goldberg, P.L. Lanzi: Gradient descent methods in learning classifier systems: Improving XCS performance in multistep problems, IEEE Trans. Evol. Comput. **9**, 452–473 (2005)

47.70 J. Hurst, L. Bull: Self-adaptation in classifier system controllers, Artif. Life Robot. **5**, 109–119 (2001)

47.71 J. Hurst, L. Bull: A neural learning classifier system with self-adaptive constructivism for mobile robot learning, Artif. Life **12**, 1–28 (2006)

47.72 E.C. Tolman: Cognitive maps in rats and men, Psychol. Rev. **55**, 189–208 (1948)

47.73 R.L. Riolo: Lookahead planning and latent learning in a classifier system, from animals to animats, Proc. 1st Int. Conf. Simul. Adapt. Behav. (1991) pp. 316–326

47.74 W. Stolzmann: Anticipatory classifier systems, Genetic Programming 1998, Proc. 3rd Ann. Conf. (1998) pp. 658–664

47.75 M.V. Butz: *Anticipatory Learning Classifier Systems* (Kluwer, Boston 2002)

47.76 M.V. Butz, D.E. Goldberg, W. Stolzmann: The anticipatory classifier system and genetic generalization, Nat. Comput. **1**, 427–467 (2002)

47.77 P. Gérard, O. Sigaud: YACS: Combining dynamic programming with generalization in classifier systems, Lect. Notes Comput. Sci. **1996**, 52–69 (2001)

47.78 P. Gérard, J.-A. Meyer, O. Sigaud: Combining latent learning and dynamic programming in MACS, Eur. J. Oper. Res. **160**, 614–637 (2005)

47.79 W. Stolzmann, M.V. Butz: Latent learning and action planning in robots with anticipatory classifier systems, Lect. Notes Comput. Sci. **1813**, 301–317 (2000)

47.80 R.S. Sutton: DYNA: an integrated architecture for learning, planning, and reacting, ACM SIGART Bull. **2**(4), 160–163 (1991)

47.81 W. Stolzmann, M.V. Butz, J. Hoffmann, D.E. Goldberg: First cognitive capabilities in the anticipatory classifier system, from animals to animats 6, Proc. 6th Int. Conf. Simul. Adapt. Behav. (2000) pp. 287–296

47.82 M.V. Butz, J. Hoffmann: Anticipations control behavior: Animal behavior in an anticipatory learning classifier system, Adapt. Behav. **10**, 75–96 (2002)

47.83 M.V. Butz: Biasing exploration in an anticipatory learning classifier system, Lect. Notes Comput. Sci. **2321**, 3–22 (2002)

47.84 M.V. Butz, D.E. Goldberg: Generalized state values in an anticipatory learning classifier system. In: *Anticipatory Behavior in Adaptive Learning Systems: Foundations, Theories, and Systems*, ed. by M.V. Butz, O. Sigaud, P. Gérard (Springer, Berlin, Heidelberg 2003) pp. 282–301

47.85 B. Siciliano, O. Khatib: *Springer Handbook of Robotics* (Springer, Berlin, Heidelberg 2007)

47.86 M.V. Butz, O. Herbort: Context-dependent predictions and cognitive arm control with XCSF, Proc. Genet. Evol. Comput. Conf. (GECCO 2008) (2008) pp. 1357–1364

47.87 M.V. Butz, G.K.M. Pedersen, P.O. Stalph: Learning sensorimotor control structures with XCSF: Redundancy exploitation and dynamic control, Proc. Genet. Evol. Comput. Conf. (GECCO 2009) (2009) pp. 1171–1178

47.88 D.M. Wolpert, R.C. Miall, M. Kawato: Internal models in the cerebellum, Trends Cogn. Sci. **2**, 338–347 (1998)

47.89 J.G. Fleischer: Neural correlates of anticipation in cerebellum, basal ganglia, and hippocampus, Lect. Notes Comput. Sci. **4520**, 19–34 (2007)

47.90 C.T. Fernando, E. Szathmary, P. Husbands: Selectionist and evolutionary approaches to brain function: A critical appraisal, Front. Comput. Neurosci. **6**, doi: 10.3389/fncom.2012.00024 (2012)

47.91 A. Tomlinson, L. Bull: A corporate XCS, Lect. Notes Comput. Sci. **1813**, 195–208 (2000)

47.92 X. Llorà, K. Sastry: Fast rule matching for learning classifier systems via vector instructions, Proc. Genet. Evol. Comput. Conf. (GECCO 2006) (2006) pp. 1513–1520

47.93 M.V. Butz, P.L. Lanzi, X. Llorà, D. Loiacono: An analysis of matching in learning classifier systems, Proc. Genet. Evol. Comput. Conf. (GECCO 2008) (2008) pp. 1349–1356

47.94 J.R. Anderson: *Rules of the Mind* (Lawrence Erlbaum Associates, Hillsdale 1993)

47.95 G.L. Drescher: *Made-Up Minds: A Constructivist Approach to Artificial Intelligence* (MIT Press, Cambridge 1991)

47.96 A. Newell: Physical symbol systems, Cogn. Sci. **4**, 135–183 (1980)

47.97 J.R. Anderson, D. Bothell, M.D. Byrne, S. Douglass, C. Lebiere, Y. Qin: An integrated theory of the mind, Psychol. Rev. **111**, 1036–1060 (2004)

47.98 O. Sigaud, S. Wilson: Learning classifier systems: A survey, soft computing – a fusion of foundations, Methodol. Appl. **11**, 1065–1078 (2007)

47.99 L. Bull, T. Kovacs (Eds.): *Foundations of Learning Classifier Systems*, Stud. Fuzziness and Soft Comput., Vol. 183 (Springer, Berlin, Heidelberg 2005)

47.100 L. Bull (Ed.): *Applications of Learning Classifier Systems* (Springer, Berlin, Heidelberg 2004)

47.101 L. Bull: On lookahead and latent learning in Simple LCS, Learn. Classif. Syst. Int. Workshops, IWLCS 2006-2007, ed. by J. Bacardit, E. Bernad-Mansilla, M.V. Butz (Springer, Berlin, Heidelberg 2008) pp. 154–168

47.102 P:L. Lanzi: xcslib – The XCS Library. http://xcslib.sourceforge.net/

47.103 P. O. Stalph, M. V. Butz: Documentation of JavaXCSF (COBOSLAB, University of Würzburg, Germany, Y2009N001 2009)

48. Indicator-Based Selection

Lothar Thiele

The goal of multiobjective evolutionary optimization is to determine a set of solutions that satisfies certain optimality properties. Recently, there is a growing number of very competitive search algorithms that are based on an explicit formulation of the optimization goal as a set property, i. e., they build on the concept of set indicators. These indicators are used to guide the selection process which is usually denoted as indicator-based selection. This major breakthrough leads to several advantages in terms of analysis and algorithm design: Algorithms are conceptually simpler and more robust as they are largely based on a single indicator; certain convergence properties can be proven; the optimization criterion is made explicit; by changing the set indicator, it is possible to explicitly consider preferences of a user. The chapter introduces step-by-step the concept of

set indicators and their use in indicator-based selection.

48.1 Motivation

Variation and selection are the main ingredients of evolutionary optimization algorithms. Despite of many variations that have been developed in the past, their basic iterative structure can simply be described by the following three steps:

1. From the current set of solutions (parent set), a subset is determined (mating pool) by mating selection.
2. From the solutions in the mating pool new solutions are generated (offspring set) through variation operators such as mutation and recombination.
3. Environmental selection determines the new parent set as a subset of the joined parent set and offspring set.

As can be seen in the above template, selection denotes the process of forming a subset of a set of solutions. Mating selection determines the set of candidate solutions that will be further explored by constructing new solutions, i. e., the offspring set. To this end, promising solutions need to be selected whose offsprings are expected to advance the optimization process most. In contrast, environmental selection combines parent and offspring sets toward the new parent set and thereby, it reduces the number of solutions that are considered in the next iteration. Loosely speaking, mating selection is involved in the exploration phase of the evolutionary optimization whereas environmental selection is central to the decision phase.

Set indicators map sets of solutions to scalar values. They characterize to which degree the set satisfies some desirable property. Therefore, they can be used to guide the selection process which is usually denoted as indicator-based selection. The following chapter concentrates on environmental selection as indicator-based methods have been applied in this context mainly.

Part E | 48.1

The goal of multiobjective evolutionary optimization is to determine a *set* of solutions that satisfies certain optimality properties. The corresponding notion of optimality is partially defined by *solution preference*, i. e., when we consider one single solution to be preferable to another single solution. One common choice of such a solution-oriented preference relation is Pareto–dominance. But there is still a large degree of freedom left in defining what an *optimal set* of solutions is, as there may be many more Pareto-optimal solutions than can be reasonably processed, stored, or presented to the user. Therefore, we need additional information that describes the preference of the user, i. e., what subset of Pareto-optimal solutions he/she is interested in. For example, the user may be interested in a diverse set of solutions or in solutions which cover a certain subspace of interest. Set indicators can now define such a preference relation and influence:

a) The result of the population-based optimization and
b) The characteristics of the sets of solutions during an optimization run and
c) The search efficiency.

Traditionally, multiobjective evolutionary optimization algorithms such as NSGA-II (nondominated sorting genetic algorithm II) [48.1] or SPEA2 (strength Pareto evolutionary algorithm 2) [48.2] start from solution preference, i. e., the Pareto–dominance relation between the solutions, and then attempt to consider set preferences such as diversity using heuristics. As a downside of this approach, deterioration and cyclic behavior have been reported [48.3], formal convergence results can not be obtained and unsatisfiable optimization results have been shown for high-dimensional objective spaces [48.4].

On the other hand, indicator-based selection treats multiobjective evolutionary optimization as a set-optimization with a *single optimization criterion*, namely the set–preference relation or its defining set indicator. In other words, instead focussing on individual solutions with multiple criteria, set-based methods consider sets of solutions as the object of optimization and a single set criterion, i. e., the set indicator. This is a radical change from the traditional approach. The set indicator directly represents the user preference and the optimization algorithm determines a set of solutions that optimizes this single set indicator. The advantages of this approach are obvious: Formal and unambiguous definition of the optimization goal, possibility to show strong convergence results, and a clear approach to consider user preferences in the search method.

48.2 Basic Concepts

Before discussing the role of indicators, selection, and archiving in multiobjective evolutionary algorithms, we will define the notation used in the forthcoming sections. In particular, we will define the underlying class of multiobjective optimization problems.

48.2.1 Notation

We will consider the minimization of a vector-valued objective function $f = (f_1, \ldots, f_n) : X \rightarrow \mathbb{R}$ which maps each point in the decision space X to an n-dimensional vector. The decision space X denotes the feasible set of alternatives for the optimization problem and n denotes the dimension of the minimization problem, i. e., the number of objectives. For simplicity of notation, we suppose in the following that X is finite. Often, we call an element of the decision space a *solution* and the corresponding objective value $z = f(x)$ is denoted as *objective vector*. The image of the decision space X under the objective function f is called objective space $Z \subseteq \mathbb{R}^n$ with $Z = \{f(x) | x \in X\}$, i. e., it contains all objective vectors corresponding to solutions in X.

In the above formulation, it is not yet clear what we understand as the *minimization* of a vector-valued function. In this chapter, we follow the usual concept of Pareto dominance which defines an order relation between all solutions based on a preference relation, i. e., it defines when we call a solution better than another one.

Definition 48.1

A solution $a \in X$ weakly Pareto dominates a solution $b \in X$, denoted as $a \preceq b$, if $f_i(a) \leq f_i(b)$ for all $1 \leq i \leq n$. Solution a strongly Pareto dominates b, denoted as $a \prec b$ if $(a \preceq b) \wedge (b \npreceq a)$.

We can rewrite the strong domination criterion as $(a \prec b) \Leftrightarrow (a \preceq b) \wedge (f(a) \neq f(b))$. We also say that solution a is *better than* or *weakly preferable to* b if

$a \prec b$ or $a \preceq b$, respectively. Note that the weak Pareto-dominance relation is suitable for optimization as it defines a preorder on the set of solutions $\mathcal{X}$. A preorder $\preceq$ on a given set $\mathcal{X}$ is reflexive and transitive: $a \preceq a$ and $(a \preceq b) \wedge (b \preceq c) \Rightarrow (a \preceq c)$ hold for all $a, b, c \in \mathcal{X}$.

In terms of optimization, we say that a solution $a \in \mathcal{X}$ is Pareto-optimal if there is no better solution in $\mathcal{X}$, i.e., if $b \preceq a$ for some $b \in \mathcal{X}$ then $a \preceq b$. The set of all Pareto-optimal solutions is denoted as the Pareto-optimal set and its image in the objective space as the Pareto-optimal front.

Ideally, a multiobjective optimizer determines the Pareto-optimal set for a given objective function f and the corresponding decision space $\mathcal{X}$. Traditionally, evolutionary multiobjective algorithms attempt to solve this problem by generating a suitable approximation of the Pareto-optimal set. To this end, they maintain and improve sets of solutions, denoted as populations. In this context, the following questions arise:

- If the set of Pareto-optimal solutions is too large to be determined efficiently, how do we select those which will be the result of the optimization process?
- How do we valuate a set of solutions, i.e., an approximation of the set of Pareto-optimal solutions, in terms of its degree of optimality in order to guide the optimization process?

The chapter describes how indicator-based selection can be used to answer the above questions. Therefore, it touches two core issues for multiobjective evolutionary algorithms: (a) how to formalize the optimization goal in the sense of specifying what type of set is sought; (b) how to efficiently determine a suitable subset to achieve the formalized optimization goal.

The following section introduces the concept of set indicators that can be used to valuate a set of solutions, i.e., associate a quality indicator which describes its degree of optimality.

48.2.2 Set Indicators

Preference relations between sets of solutions are the basis of set-based multiobjective optimization. They provide the information on the basis of which the search is carried out, i.e., for any two Pareto set approximations, they say whether one set is considered to be equal, better, or worse.

A set indicator can now be used to define such a preference relation and therefore to indicate whether one set of solutions is preferable to another one. In addi-tion, it also contains quantitative information about the degree of preference.

Depending on the particular definition of the preference relation, a set can be considered to be better than another one or even the other way round. With different definitions of such a preference relation, we can expect that the optimal result of a search process will be different as well. Therefore, the definition of an indicator is essential for formally defining the goal of the set-based optimization. In addition, it allows us to adjust the optimization goal according to the preferences of the user, i.e., to provide flexibility with respect to the subset of Pareto-optimal solutions searched for.

But the set indicator and the resulting preference relation can not be chosen arbitrarily as they need to conform to the concept of Pareto dominance. Otherwise, the search process may end up with a set which is weakly preferable to all other sets but does not contain any Pareto-optimal solution.

In order to derive the requirements for a *well-behaved* set indicator, let us first generalize the concept of Pareto dominance of solutions to Pareto dominance of sets.

Definition 48.2

A set of solutions $\mathcal{A} \subseteq \mathcal{X}$ weakly Pareto dominates a set of solutions $\mathcal{B} \subseteq \mathcal{X}$, denoted as $\mathcal{A} \preccurlyeq \mathcal{B}$, if $(\forall b \in \mathcal{B} : (\exists a \in \mathcal{A} : a \preceq b))$. Set $\mathcal{A}$ strongly Pareto-dominates set $\mathcal{B}$, denoted as $\mathcal{A} \prec \mathcal{B}$ if

$$(\mathcal{A} \preccurlyeq \mathcal{B}) \wedge (\mathcal{B} \npreccurlyeq \mathcal{A}) .$$

In other words, a set of solutions $\mathcal{A}$ weakly dominates a set of solutions $\mathcal{B}$ if every solution in $\mathcal{B}$ is weakly dominated by at least one solution in $\mathcal{A}$. Moreover, it can be shown that the set-based dominance relation defines a preorder, i.e., it is suited for optimization purposes.

Now, let us define the concept of a set indicator and its induced preference relation. In the first part of this section, we restrict ourselves to unary indicators. A more detailed discussion on the various aspects of indicators is provided in [48.5].

Definition 48.3

A unary indicator maps each set $\mathcal{A} \subseteq \mathcal{X}$ of the decision space to a real number $I(\mathcal{A}) \in \mathbb{R}$. Given an indicator, we can determine the corresponding preference relation $\preccurlyeq_I$ as

$$\mathcal{A} \preccurlyeq_I \mathcal{B} := (I(\mathcal{A}) \geq I(\mathcal{B})) .$$

In other words, the larger the set indicator of a set of solutions the better we consider the set. It can be shown that the preference relation induced by the indicator defines a total order on the set of solutions X.

As discussed above, not all preference relations can be used inside search methods as they at least need to *comply* to the definition of Pareto dominance in Def. 48.2. To this end, the following definition describes the notion of preference refinement:

Definition 48.4

A preference relation $\preceq_{ref}$ is denoted as a refinement of $\preceq$ if

$$\mathcal{A} \prec \mathcal{B} \Rightarrow \mathcal{A} \prec_{ref} \mathcal{B} .$$

What we need to guarantee can be formulated as follows: If a solution $\mathcal{A} \subseteq X$ is strictly better than a solution $\mathcal{B} \subseteq X$ in the sense of Pareto dominance, i.e., $\mathcal{A} \prec \mathcal{B}$, then the preference relation used for optimization should say so as well, i.e., $\mathcal{A} \prec_{ref} \mathcal{B}$, see also [48.5].

If we use this result for the unary indicator according to Def. 48.3, then we directly get the following condition for a *compliant* indicator, i.e., whose corresponding preference relation is a refinement of $\preceq$

$$\mathcal{A} \prec \mathcal{B} \Rightarrow (I(\mathcal{A}) > I(\mathcal{B})) . \tag{48.1}$$

In other words, if a solution $\mathcal{A}$ is strictly better than a solution $\mathcal{B}$, i.e., $\mathcal{A} \prec \mathcal{B}$, then the indicator should say so as well, i.e., $I(\mathcal{A}) > I(\mathcal{B})$. It has been shown in [48.5] that the Pareto-compliance guarantees that a set with the maximal indicator value is minimal with respect to the Pareto–dominance relation according to Def. 48.2.

Indicators have been introduced to the area of multiobjective evolutionary optimization first as a mean to compare different optimization runs [48.6–9]. The use of indicators to guide multiobjective search methods in general appeared in the year 2003, notably in [48.10] and later in [48.11–13]. In a more restricted setting, indicators have been used for archiving, i.e., maintaining a set of Pareto-approximate solutions [48.3, 14].

In several studies, the properties of set indicators have been investigated in terms of their compliance to the Pareto dominance [48.7, 15]. Whereas many well known and widely applied indicators do not fall into this class, there are various indicators that at least satisfy a weak refinement ($\mathcal{A} \prec \mathcal{B} \Rightarrow \mathcal{A} \preceq_{ref} \mathcal{B}$), e.g., the unary R_2 and R_3 indicators [48.16] and the multiplicative as well as the unary additive and multiplicative

epsilon indicators [48.3, 15]. The latter two indicators are related to additive or multiplicative approximation [48.17, 18].

Before discussing binary indicators, let us introduce an example of a set indicator that is compliant to Pareto dominance. The hypervolume indicator has been introduced to the field of multiobjective evolutionary optimization in [48.19] for the purpose of performance assessment. It can be defined as

$$I_H(\mathcal{A}, \mathcal{R}) = \int\limits_{z \in H(\mathcal{A}, \mathcal{R})} dz , \tag{48.2}$$

where $H(\mathcal{A}, \mathcal{R})$ denotes the objective space dominated by $\mathcal{A}$ and dominating $\mathcal{R}$

$$H(\mathcal{A}, \mathcal{R}) = \{z \in \mathbb{R}^n | \exists a \in \mathcal{A} : \exists r \in \mathcal{R} : f(a) \leq z \leq r\} .$$

In other words, we determine the volume covered by all points $z \in \mathbb{R}^n$ that are enclosed between the image of the solutions in objective space $f(\mathcal{A})$ and the reference set $\mathcal{R}$, where *enclosed* is interpreted in terms of weak Pareto dominance. Due to its compliance to Pareto dominance it has been used in most of the indicator-based selection schemes to date.

One of the major drawbacks of the hypervolume indicator is the associated computational overhead. *Bringmann* and *Friedrich* [48.20] have proven that the problem of computing the hypervolume is #P-complete, i.e., there exists no polynomial algorithm unless $NP = P$. Several algorithms have been proposed in the past to determine the hypervolume indicator, starting from the hypervolume by slicing objective approach independently proposed by several authors (*Knowles* and *Zitzler*) with complexity $\mathcal{O}(N^{n-1})$ where N is the number of solutions in the population and n is the number of objectives. Later on, improved version appeared with complexity $\mathcal{O}(N^{n-2} \log N)$ [48.21] and finally $\mathcal{O}(N \log N + N^{n/2} \log N)$ [48.22]. An approximation algorithm with proven bounds is presented in [48.23] which gives an ϵ-approximation of the hypervolume with probability $(1 - \delta)$ in time $\mathcal{O}(\log(1/\delta) nN/\epsilon^2)$.

Binary indicators $I(\mathcal{A}, \mathcal{B})$ can be used to compare two sets $\mathcal{A}$ and $\mathcal{B}$ as described in the following Def. 48.5.

Definition 48.5

A binary indicator maps an ordered pair of sets $\mathcal{A}, \mathcal{B} \subseteq X$ of the decision space to a real number $I(\mathcal{A}, \mathcal{B}) \in \mathbb{R}$.

Given a binary indicator, we can determine the corresponding preference relation $\preccurlyeq_I$ as

$$\mathcal{A} \preccurlyeq_I \mathcal{B} := (I(\mathcal{A}, \mathcal{B}) \geq I(\mathcal{B}, \mathcal{A})) \,.$$

In a similar way to (48.1), one can derive the condition for an indicator whose corresponding preference relation is a refinement of $\preccurlyeq$,

$$\mathcal{A} \prec \mathcal{B} \Rightarrow (I(\mathcal{A}, \mathcal{B}) > I(\mathcal{B}, \mathcal{A})) \,.$$

Two popular examples of a binary indicators that have been successfully used in indicator-based selection [48.11, 24, 25], archiving [48.3] and approximation schemes [48.18] are the additive and multiplicative epsilon indicators. They can be defined as

$$I_\epsilon^+ (\mathcal{A}, \mathcal{B}) = \min_{b \in \mathcal{B}} \max_{a \in \mathcal{A}} F_\epsilon^+ (a, b) \,,$$

$$F_\epsilon^+ (a, b) = \min_{1 \leq i \leq n} (f_i(b) - f_i(a)) \,, \qquad (48.3)$$

for the additive version and

$$I_\epsilon^* (\mathcal{A}, \mathcal{B}) = \min_{b \in \mathcal{B}} \max_{a \in \mathcal{A}} F_\epsilon^* (a, b) \,,$$

$$F_\epsilon^* (a, b) = \min_{1 \leq i \leq n} \frac{f_i(b)}{f_i(a)} \,, \qquad (48.4)$$

for the multiplicative one. Formally speaking, $I_\epsilon^+ (\mathcal{A}, \mathcal{B})$ (or $I_\epsilon^* (\mathcal{A}, \mathcal{B})$) denotes the maximum amount one can to add to (or multiply to) every objective value $f_i(a)$ of every solution $a \in \mathcal{A}$ such that the resulting set still weakly dominates $\mathcal{B}$.

Unfortunately, the above binary indicators do not induce a preorder as the resulting preference relation is not transitive in general [48.5]. This negative result needs to be considered when deciding to use it (and similar generalizations of unary indicators) in optimization algorithms.

Next, indicator-based selection and its integration into multiobjective search algorithms will be discussed.

48.3 Selection Schemes

48.3.1 Basic Search Algorithm

Let us start the discussion of indicator-based selection with a simple template of a multiobjective evolutionary algorithm (SPAM – set preference algorithm for multiobjective optimization [48.26]) as shown in Alg. 48.1.

Algorithm 48.1 Simple SPAM

1: generate initial set of solutions $\mathcal{P}$ of size μ
2: **while** termination criterion not fulfilled **do**
3:    generate λ offspring solutions $\mathcal{O} \in X$
4:    $S = select(\mathcal{P} \cup \mathcal{O}, \mu)$
5:    **if** $S \preccurlyeq_{\mathrm{ref}} \mathcal{P}$ **then** $\mathcal{P} \leftarrow S$
6: **return** $\mathcal{P}$

Obviously, the template is still very simplistic but it will help us to understand the integration of the concept of indicators and selection in multiobjective evolutionary algorithms. Line 3 in Alg. 48.1 refers to the variation of solutions that are in population $\mathcal{P}$, i.e., starting with mating selection and then applying variation operators such as mutation and recombination. This essential part of any evolutionary algorithm will not be discussed further here. Line 4 is denoted as environ-mental selection and reduces the union of parent set $\mathcal{P}$ and offspring set $\mathcal{O}$ from size $\mu + \lambda$ to μ again, i.e., $S \subseteq \mathcal{P} \cup \mathcal{O}$ and $|S| = \mu$. Finally, line 5 is responsible for selecting either the old population $\mathcal{P}$ or the new one S depending on the chosen preference relation $\preccurlyeq_{\mathrm{ref}}$.

In the following, we will stepwise refine the selection operator in line 4 and thereby, relate the above template to existing indicator-based selection schemes.

48.3.2 Exhaustive Selection

Let us first suppose that the selection operator in line 4 is exhaustive in the following sense: If there exists a subset $S \subseteq \mathcal{P} \cup \mathcal{O}$ with $|S| = \mu$ that satisfies $S \preccurlyeq_{\mathrm{ref}} \mathcal{P}$, then it will generate it with nonzero probability. Under this condition, one can proof an important convergence property of the algorithm that ensures that there is no deterioration behavior as reported for algorithms such as NSGA-II and SPEA2 [48.3]. The line of arguments is just sketched here as it closely follows the investigations in [48.27].

In most general terms, the goal of the optimization is to generate as large as possible subset of the Pareto-optimal solutions. Therefore, what we at least require is that Alg. 48.1 generates such a set provided that it runs long enough. Indeed, one can show that this is the case if:

a) The offspring generation in line 3 is exhaustive, i. e., all solutions are generated with nonzero probability,
b) The selection operator is exhaustive and
c) $\preccurlyeq_{ref}$ is a refinement of the Pareto dominance.

Let us suppose for simplicity of arguments, that there are more than μ Pareto-optimal solutions in $\mathcal{X}$. Moreover, suppose that the population at some point in time (still) contains a dominated solution. Then there is a nonzero-probability that in the set $\mathcal{O}$ of offsprings there is a Pareto-optimal solution not yet in $\mathcal{P}$. Replacing the dominated solution with the additional Pareto-optimal solution leads to a preferred set according to $\preccurlyeq_{ref}$ as it refines the Pareto dominance. Note, however, that the above convergence property does not mean that Alg. 48.1 determines an optimal set w.r.t. to $\preccurlyeq_{ref}$, i. e., that the resulting subset of Pareto-optimal solutions actually is minimal in terms of $\preccurlyeq_{ref}$.

Exhaustive selection can usually not be implemented efficiently, as all possible subsets must be tested, i. e., $\binom{\lambda+\mu}{\mu}$ possible preference relations. The following refinements of the basic algorithm lead to more efficient schemes.

48.3.3 Steady State and Greedy Selection

A first possibility has been proposed in the indicator-based selection and archiving schemes described in [48.12, 14]. The size of the offspring set is $\lambda = 1$ and therefore, at most $\mu + 1$ preference relations need to be constructed in each iteration. In particular, the hypervolume indicator I_H [48.9] (S measure) has been used to define the preference relation, i. e., $\preccurlyeq_{ref}:=\preccurlyeq_H$. In this case, the selection in line 4 just removes the solution that leads to the least loss in I_H.

Still, the convergence to a Pareto-optimal subset can be guaranteed if the offspring generation is exhaustive. On the other hand, it cannot be guaranteed that the algorithm determines an optimal set w.r.t. $\preccurlyeq_{ref}$, i. e., a set that is not strictly dominated by any other subset of size μ w.r.t. $\preccurlyeq_{ref}$. First counterexamples for various set indicators that show this property for $\lambda < \mu$ and especially for $\lambda = 1$ appeared in [48.5]. A more indepth discussion on this issue for the hypervolume indicator can be found in [48.28].

A second approach allows for general sizes λ of the offspring population $\mathcal{O}$ and employs a simple greedy strategy, i. e., solutions are removed one-by-one from the set $\mathcal{P} \cup \mathcal{O}$ until a set with size μ is obtained. The following template in Alg. 48.2 sketches the approach.

Algorithm 48.2 Greedy Selection

1: **procedure** $select(\mathcal{P} \cup \mathcal{O}, \mu)$
2:    $S \leftarrow \mathcal{P} \cup \mathcal{O}$
3:    **while** $|S| > \mu$ **do**
4:       **for all** $a \in S$ **do**
5:          $\delta_a \leftarrow loss(S, a)$
6:       choose $p \in S$ with $\delta_p = \min_{a \in S} \delta_a$
7:       $S \leftarrow S \setminus \{p\}$
8:    **return** S

If $\lambda = 1$, then this template covers the steady-state selection scheme in [48.12, 14]. The function $loss(S, a)$ quantifies the loss in set quality, if solution a is removed from it. In line 6, the solution with the smallest loss is chosen and removed from the population in line 7. If the preference relation is based on a unary indicator as shown in Def. 48.3, then the loss function can simply be determined as

$$loss(S, a) = I(S) - I(S \setminus \{a\}) .$$

For the more general case of preference relations that are not based on indicators, see also [48.5]. Note that convergence to the set with the maximal indicator value can now not be guaranteed anymore as the greedy selection is a heuristic. For the hypervolume indicator this is shown in [48.29]. On the other hand, we still can guarantee that SPAM with greedy selection generates an as large as possible subset of the Pareto-optimal solutions if the offspring generation in line 3 is exhaustive, i. e. all solutions are generated with nonzero probability, and $\preccurlyeq_{ref}$ is a refinement of the Pareto dominance.

As shown in Alg. 48.2, we do not need to determine the value of the set indicator but only the least contributor, i. e. the solutions that leads to the minimal loss. Depending on the choice of the indicator, this information may be easier to compute than evaluating the indicator for $\lambda + \mu$ different sets and comparing the values. In the context of the hypervolume indicator, a more detailed discussion on this issue is provided in [48.30].

As has been mentioned already, the indicator-based selection schemes described here have been applied to the problem of archiving as well. Archiving algorithms attempt to maintain a bounded set of solutions given a sequence of solutions [48.3, 14]. In analogy to the template in Alg. 48.1, the sequence of solutions would be generated by the offspring generation and the selection process would determine the new population $\mathcal{P}$. For archiving purposes, one is usually only interested in maintaining a subset of all nondominated solutions received so far. Dominated solutions in $\mathcal{P}$ are usually not considered in the underlying set preference relations.

The Pareto–dominance relation on sets is by definition insensitive to dominated solutions. The same holds for set preference relations based on popular indicators such as the hypervolume indicator which reflects the volume dominated by a set of solutions. On the other hand, preferences among dominated solutions may be of importance to guide the search. In particular, all solutions in $\mathcal{P}$ (dominated and nondominated ones) are candidates for the mating selection and therefore, may be chosen for variation in the generation of offspring solutions. Therefore, useful set preference relations that are refinements of Pareto dominance need to be constructed that allow us to consider preferences on dominated solutions as well.

48.3.4 Hierarchical Set Preferences

Most of the hierarchical indicator-based selection schemes that have been described so far combine set indicators with constructing a sequence of subsets of solutions [48.5, 12, 31]. For example, nondominated sorting [48.32, 33] starts with the whole set as the first element of the sequence, and then removes the nondominated solutions from the previous subset to construct the next subset in the sequence. The following Alg. 48.3 provides a template for a hierarchical selection scheme involving nondominated sorting. Many variants of the above basic scheme could be thought of such as other subset-constructions like dominance ranking [48.34].

Algorithm 48.3 Hierarchical Selection

1: **procedure** *select* $\mathcal{P} \cup \mathcal{O}, \mu$
2:     $S \leftarrow \mathcal{P} \cup \mathcal{O}$
3:     $S' \leftarrow \emptyset$
4:     $S'' \leftarrow \emptyset$
5:     **repeat**
6:         $S' \leftarrow S' \cup S''$
7:         $S'' \leftarrow \{a \in S|\ \not\exists b \in S : b \prec a\}$
8:         $S \leftarrow S \setminus S''$
9:     **until** $|S' \cup S''| \geq \mu$
10:     **while** $|S' \cup S''| > \mu$ **do**
11:         **for all** $a \in S''$ **do**
12:             $\delta_a \leftarrow loss(S'', a)$
13:         choose $p \in S''$ with $\delta_p = \min_{a \in S''} \delta_a$
14:         $S'' \leftarrow S'' \setminus \{p\}$

15:     **return** $S' \cup S''$

The set S in Alg. 48.3 before the execution of the iteration in lines 5−9 contains the last subset in the sequence of dominating subsets. When leaving the iteration with line 9, we have $\mathcal{P} \cup \mathcal{O} = S \cup S' \cup S''$ where S'' is the last dominating set that has been peeled off. The iteration in lines 10−14 removes solutions from S'' one-by-one as in Alg. 48.2 until the set $S' \cup S''$ contains μ solutions. In [48.5], a detailed analysis of the optimization and convergence properties of such constructions is provided. In particular, conditions are derived under which the corresponding preference relation is a refinement of Pareto dominance, i. e., can safely be used in an indicator-based selection.

48.3.5 Using Binary Indicators

The previous discussion concentrated on the use of unary indicators in multiobjective evolutionary algorithms. On the other hand, one of the first indicator-based selection schemes used for optimization was based on the concept of binary indicators following Def. 48.5 [48.11]. In particular, the use of a binary variant of the hypervolume indicator, (48.2), as well as the use of the binary additive epsilon indicator, (48.3), have been described.

The structure of IBEA (indicator-based evolutionary algorithm) follows directly the greedy selection scheme as described in Alg. 48.2. In the basic scheme of IBEA, i. e., without parameter adaptation, the *loss*-function is computed as follows,

$$loss(S, a) = - \sum_{b \in S \setminus \{a\}} e^{I(\{b\},\{a\})/\kappa} . \tag{48.5}$$

Let us interpret the above loss function by means of the binary additive epsilon indicator, i. e., $I(\mathcal{A}, \mathcal{B}) := I_\epsilon^+(\mathcal{A}, \mathcal{B})$. The solution with the smallest *loss*-function will be selected for removal from the set. This actually is the solution with the *largest* sum $\sum_{b \in S \setminus \{a\}} e^{I(\{b\},\{a\})/\kappa}$. If considering a large scaling factor κ, the sum of exponentials actually acts similar to sorting the indicators and just considers the largest one. If κ is smaller, then not only the largest indicator is taken into account but also smaller ones. As a result, the sum is dominated by the solution b which leads to the largest value of $I(\{b\}, \{a\})$.

Remember that $I(\{b\}, \{a\}) = \min_{1 \leq i \leq n}(f_i(a) - f_i(b))$ denotes the maximal amount one can add to every objective value $f_i(b)$ of solution b such that it still weakly dominates a. As a result for large values of κ we can summarize the selection as follows: For each solution a, we determine the solution b to which we can *add* the largest amount such that it still weakly dominates a. The solution a is removed for which this amount is largest. In some sense, the first step determines the *closest* solution (or strongest dominator) to a in the epsilon indicator and the second step then removes the solution which has the closest neighbor (or strongest dominator). If κ is smaller, also the closeness

(or domination) of other solutions b is taken into account.

Recently, an indicator-based algorithm has been developed [48.25], which uses a similar principle. It uses the binary epsilon indicator as defined in (48.3) and (48.4), but instead of comparing successive populations as in IBEA, it uses a possibly growing archive of the best Pareto-approximations found so far as the reference set. The solution to be removed according to the template in Alg. 48.2 is determined by sorting.

48.4 Preference-Based Selection

Recently, there has been increasing interest in constructing evolutionary optimization methods that allow to consider search preferences of the user. In other words, the resulting set of solutions should not contain an arbitrary subset of Pareto-optimal solutions but one that satisfies secondary criteria. For example, it may be desirable to preferably determine solutions that are close to some reference point, that are along a direction in the objective space, or have some other predefined distribution.

The choice of the *right* subset of solutions as a result of an evolutionary multiobjective optimization was of major concern since the beginning. In particular, it was a major goal to design algorithms that lead to well-distributed solutions that are close to the Pareto-optimal front. Various heuristics have been implemented in standard algorithms like SPEA2 [48.2] and NSGA-II [48.1] to achieve such an implicitly defined objective.

The concept of indicator-based selection changed this approach fundamentally. It not only allows to formalize the objective of population-based multiobjective optimization in general but also to design algorithms that optimize a set of solutions toward it. As a result of this achievement, the focus of research moved toward the following questions:

- What kind of user preference is useful in the context of preference-based search?
- How can these preferences be mathematically formulated and incorporated in set indicators?
- How can a preference-based algorithm be integrated into an interactive optimization approach that involves the decision maker?

Including preference information in multiobjective evolutionary methods has been investigated since the beginning [48.37] for a survey. In a very early attempt,

Fonseca and *Fleming* [48.34] suggested to assign ranks to the members of a population. Much later it was proposed in [48.38] to include preferences through the use of reference points, guided dominance schemes and a biased crowding scheme. Preference-based multiobjective evolutionary methods can be used within a hybrid approach that combine ideas from both, evolutionary and interactive multiobjective optimization [48.24]: In an iterative approach, several consecutive runs of the evolutionary algorithm are performed. The user is asked to give preference information in terms of his reference point consisting of desirable aspiration levels for objective functions. This information is used in a preference-based evolutionary algorithm that generates a new population by combining the fitness function and a so-called achievement scalarizing function containing the reference point.

In the meantime, many other possibilities to formalize user preferences have been investigated [48.35, 39], for example weight functions in the objective space which change the desired (nonuniform) density of solutions, stressing objectives, guiding the search toward preference points, transforming objective functions, weighted Tchebycheff approaches using ideal points, epsilon-constraint methods, and desirability functions, just to name a few.

In the following, two examples for considering preference information in selection schemes will be described in some more detail. In [48.24], the greedy selection scheme according to Alg. 48.2 with a binary indicator according to (48.5) has been used. In particular, a normalization according to

$$I_p(\{b\}, \{a\}) := I(\{b\}, \{a\})/s^*(g, f(a))$$

is proposed. The normalization function $s(g, f(x))$ is closely related to the concept of achievement scalariz-

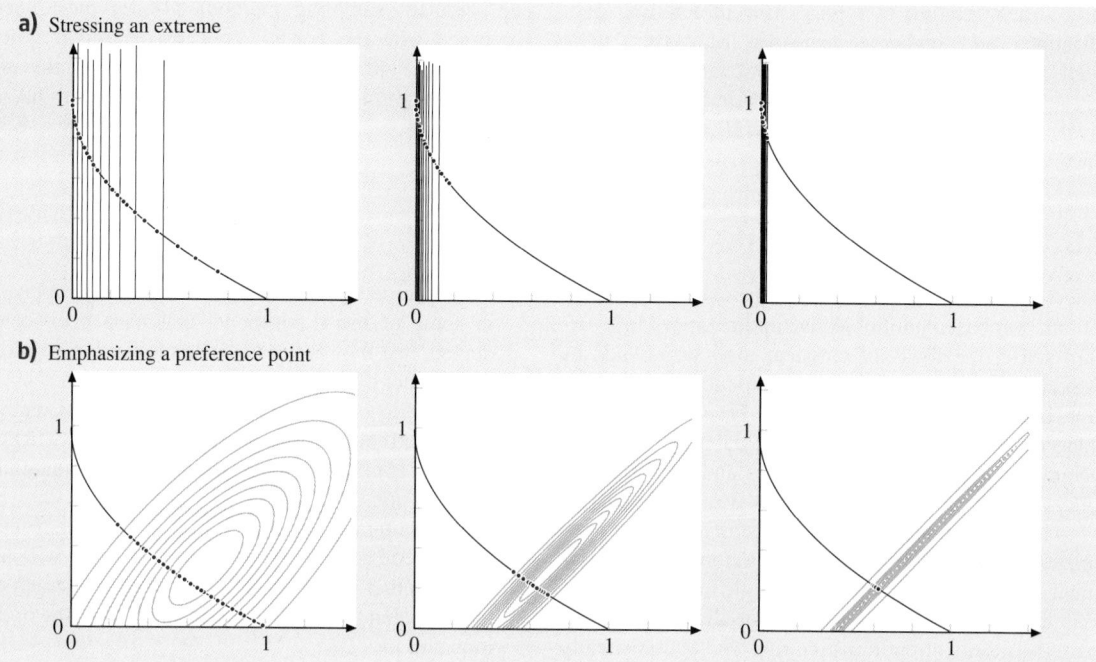

a) Stressing an extreme

b) Emphasizing a preference point

Fig. 48.1a,b The figures show the Pareto front approximations (*dots*) found by HypE (after [48.31]) using different weight distribution functions, shown as *contour lines* at intervals of 10% of the maximum weight value. For both rows one parameter of the sampled distribution was modified, i.e., on *top* the rate parameter of the exponential distribution, on the *bottom* the spread of a multivariate normal distribution (after [48.35]). The test problem is ZDT1 (after [48.36]) where the Pareto front is shown as a *solid line*. The graphics appeared in [48.35]

ing functions, first proposed by *Wierzbicki* [48.40]

$$s(g, f(a)) = \max_{1 \leq i \leq n} (f_i(a) - g_i),$$

where g denotes the reference point whose components represent the desired values of the objective functions. The function s^* is obtained from s by normalization such that only positive values are obtained, i.e., $s^*(g, f(a)) > 0$ [48.24].

A second approach uses the the concept of a weighted hypervolume indicator as proposed in [48.41]. In extension to (48.2), we now determine the weighted volume that is covered by all points $z \in \mathbb{R}^n$ that are enclosed between the image of the solutions in objective space $f(\mathcal{A})$ and the reference set $\mathcal{R}$, where enclosed is interpreted in terms of weak Pareto dominance:

Definition 48.6

Given a set of solutions $\mathcal{A} \subseteq X$, a set of reference points $\mathcal{R} \subset \mathbb{R}^n$ and a positive weight function $w: \mathbb{R} \to \mathbb{R}_{\geq 0}$. Then the weighted hypervolume indica-

tor $I_H^w(\mathcal{A}, \mathcal{R})$ of $\mathcal{A}$ with respect to $\mathcal{R}$ is defined as

$$I_H^w(\mathcal{A}, \mathcal{R}) = \int_{z \in H(\mathcal{A}, \mathcal{R})} w(z) \cdot \mathrm{d}z, \tag{48.6}$$

where $H(\mathcal{A}, \mathcal{R})$ denotes the objective space dominated by $\mathcal{A}$ and dominating $\mathcal{R}$

$$H(\mathcal{A}, \mathcal{R}) = \{z \in \mathbb{R}^n | \exists a \in \mathcal{A} : \exists r \in \mathcal{R} : f(a) \leq z \leq r\}.$$

The weight function is supposed to be integrable on any bounded set, i.e., $\int_{B(0,\gamma)} w(z)\mathrm{d}z < \infty$ for any $\gamma > 0$, where $B(0, \gamma)$ is the open ball centered in 0 and of radius γ.

In a similar way to (48.2), the weighted hypervolume indicator is compliant to Pareto dominance and can safely be used in the previously described indicator-based selection schemes.

In later work [48.35, 39], the approach [48.41] has been extended toward more general weight functions, their relation to typical user preference specifications and higher dimensions. Moreover, it is well known that the exact computation of the hypervolume is expensive in the number of objectives, i.e., it is exponential unless $P = NP$. To this end, efficient sampling methods [48.31] have been combined with the general concept of weight functions. Figure 48.1 shows some examples of the effect of weighting the hypervolume indicator, taken from [48.35].

48.5 Concluding Remarks

There has been a major shift in our understanding of population-based multiobjective optimization. In a certain sense, the focus of classical algorithms such as NSGA-II or SPEA2 was the Pareto–dominance relation between individual solutions. Properties such as a large diversity of solutions in the final population was achieved through (clever) heuristics and tuning of the selection mechanisms.

The role of set indicators in multiobjective optimization was first limited to the performance assessment. The possibility to assign a single measure to a set was used in elaborated methods that allow us to compare the results of optimization runs and to statistically verify whether one algorithms is preferable to another one. In this context, indicators have been compared in terms of their suitability for performance assessment, e.g., whether they comply to the underlying preference relation between solutions.

Recently, there is a growing number of very competitive search algorithms that are based on an explicit formulation of the optimization goal as a set property, i.e., they build on the concept of set indicators. In simplified terms, they can be regarded as optimization methods that deal with sets of solutions as their optimization object. In contrast, single-objective optimization traditionally works with single solutions. One can simply draw the correspondence between traditional single-objective optimization and population-based multiobjective optimization as follows: single solution versus set of solutions, and single objective function versus single set indicator.

This major breakthrough leads to several advantages in terms of analysis and algorithm design:

- Algorithms are conceptually simpler as they are based on a single indicator and do not rely on heuristics to a large extent. As a result, it can be expected that they are more robust and less parameter tuning is necessary.
- Certain convergence properties can be derived. As a result, the new class of algorithms does not show deterioration and/or cyclic behavior. It also appears in some of the experiments that have been conducted, that they are more robust toward increasing the number of objectives.
- The optimization criterion is made explicit, i.e., the discussion about convergence versus diversity in the research community can now be based on quantitative measures.
- By changing the set indicator, it is possible to explicitly consider preferences of a user. It will be seen whether this possibility will lead to interactive methods that involve the decision maker in the optimization process.

The purpose of the chapter was to introduce the concept of set indicators and to discuss several ways to use them as essential parts of a set-based multiobjective optimization algorithm. Many other important aspects have not been discussed in detail. In particular, due to its superior properties in terms of:

a) Compliance to Pareto dominance, and
b) Sensitivity to changes of single solutions and
c) Simple interpretation.

The hypervolume indicator has been very popular as a component of set-based optimization methods. Unfortunately, it has some detrimental properties, such as its computation complexity with respect to growing number of objectives. In addition, the complexity to determine a subset of solutions that has the least influence on the hypervolume increases exponentially with the size of the subset.

Finally, it has been shown that removing solutions one-by-one may lead to a major loss in optimization quality. Therefore, its use in indicator-based selection needs to be done with care. As described in this chapter, recent methods overcome some of these difficulties by using advanced methods such as sampling. A more detailed investigation and review of the hypervolume indicator can be found e.g., in [48.31, 35].

References

48.1   K. Deb, S. Agrawal, A. Pratap, T. Meyarivan: A fast elitist non-dominated sorting genetic algorithm for multi-objective optimization: NSGA-II, Lect. Notes Comput. Sci. **1917**, 849–858 (2000)

48.2   E. Zitzler, M. Laumanns, L. Thiele: SPEA2: Improving the strength Pareto evolutionary algorithm for multiobjective optimization, Evol. Methods Des. Optim. Control Appl. Ind. Probl. (2002) pp. 95–100

48.3   M. Laumanns, L. Thiele, K. Deb, E. Zitzler: Combining convergence and diversity in evolutionary multiobjective optimization, Evol. Comput. **10**(3), 263–282 (2002)

48.4   T. Wagner, N. Beume, B. Naujoks: Pareto-, aggregation-, and indicator-based methods in many-objective optimization, Lect. Notes Comput. Sci. **4403**, 742–756 (2007)

48.5   E. Zitzler, L. Thiele, J. Bader: On set-based multiobjective optimization, IEEE Trans. Evol. Comput. **14**(1), 58–79 (2010)

48.6   V. da Grunert Fonseca, C.M. Fonseca, A.O. Hall: Inferential performance assessment of stochastic optimisers and the attainment function, Conf. Evol. Multi-Criterion Optim. (EMO 2001), ed. by E. Zitzler, K. Deb, L. Thiele, C.A. Coello Coelle, D. Corne (Springer, Berlin, Zurich 2001) pp. 213–225

48.7   J. Knowles, D. Corne: On metrics for comparing non-dominated sets, Conf. Evol. Comput. (2002) pp. 711–716

48.8   D.A. Van Veldhuizen, G.B. Lamont: On measuring multiobjective evolutionary algorithm performance, Congr. Evol. Comput. (2000) pp. 204–211

48.9   E. Zitzler, L. Thiele: Multiobjective evolutionary algorithms: A comparative case study and the strength Pareto approach, IEEE Trans. Evol. Comput. **3**(4), 257–271 (1999)

48.10   M. Fleischer: The measure of Pareto optima. Applications to multi-objective metaheuristics, Conf. Evol. Multi-Criterion Optim. (2003) pp. 519–533

48.11   E. Zitzler, S. Künzli: Indicator-based selection in multiobjective search, Lect. Notes Comput. Sci. **3242**, 832–842 (2004)

48.12   M. Emmerich, N. Beume, B. Naujoks: An EMO algorithm using the hypervolume measure as selection criterion, Evol. Multi-Criterion Optim. 3rd Int. Conf. (2005) pp. 62–76

48.13   C. Igel, N. Hansen, S. Roth: Covariance matrix adaptation for multi-objective optimization, Evol. Comput. **15**(1), 1–28 (2007)

48.14   J.D. Knowles, D. Corne: Properties of an adaptive archiving algorithm for storing nondominated vectors, IEEE Trans. Evol. Comput. **7**(2), 100–116 (2003)

48.15   E. Zitzler, L. Thiele, M. Laumanns, C.M. Fonseca, V. da Grunert Fonseca: Performance assessment of multiobjective optimizers: An analysis and review, IEEE Trans. Evol. Comput. **7**(2), 117–132 (2003)

48.16   M.P. Hansen, A. Jaszkiewicz: *Evaluating the Quality of Approximations to the Non-dominated Set, Tech. Rep. IMM-REP-1998-7* (Technical Univ. of Denmark, Lyngby 2010) pp. 1–31

48.17   C.H. Papadimitriou, M. Yannakakis: On the approximability of trade-offs and optimal access of web sources, 41st Annu. Symp. Found. Comput. Sci. (2000) pp. 86–92

48.18   K. Bringmann, T. Friedrich: The maximum hypervolume set yields near-optimal approximation, Genet. Evol. Comput. Conf. (2010) pp. 511–518

48.19   E. Zitzler, L. Thiele: Multiobjective optimization using evolutionary algorithms – A comparative case study, Lect. Notes Comput. Sci. **1498**, 292–304 (1998)

48.20   K. Bringmann, T. Friedrich: Approximating the volume of unions and intersections of high-dimensional geometric objects, Lect. Notes Comput. Sci. **5369**, 436–447 (2008)

48.21   C.M. Fonseca, L. Paquete, M. López-Ibáñez: An improved dimension-sweep algorithm for the hypervolume indicator, Congr. Evol. Comput. (2006) pp. 1157–1163

48.22   N. Beume: S-Metric calculation by considering dominated hypervolume as Klee's measure problem, Evol. Comput. **17**(4), 477–492 (2009)

48.23   K. Bringmann, T. Friedrich: S-Metric calculation by considering dominated hypervolume as Klee's measure problem, Comput. Geom. **43**(6/7), 601–610 (2010)

48.24   L. Thiele, K. Miettinen, P.J. Korhonen, J. Molina: A preference-based evolutionary algorithm for multi-objective optimization, Evol. Comput. **17**(3), 411–436 (2009)

48.25   K. Bringmann, T. Friedrich, F. Neumann, M. Wagner: Approximation-guided evolutionary multiobjective optimization, Proc. 22nd Int. Jt. Conf. Artif. Intell. (2011) pp. 1198–1203

48.26   E. Zitzler, L. Thiele, J. Bader: SPAM: Set preference algorithm for multiobjective optimization, Lect. Notes Comput. Sci. **5199**, 847–858 (2008)

48.27   G. Rudolph, A. Agapie: Convergence properties of some multi-objective evolutionary algorithms, Congr. Evol. Comput. (2000) pp. 1010–1016

48.28   K. Bringmann, T. Friedrich: Convergence of hypervolume-based archiving algorithms I: Effectiveness, 13th Annu. Genet. Evol. Comput. Conf. (2011) pp. 745–752

48.29   K. Bringmann, T. Friedrich: An efficient algorithm for computing hypervolume contributions, Evol. Comput. **18**(3), 383–402 (2010)

48.30   K. Bringmann, T. Friedrich: Approximating the least hypervolume contributor: NP-hard in general, but fast in practice, evolutionary multi-criterion optimization, Lect. Notes Comput. Sci. **5467**, 6–20 (2009)

Part E | 48

48.31   J. Bader, E. Zitzler: HypE: An algorithm for fast hypervolume-based many-objective optimization, Evol. Comput. **19**(1), 45–76 (2011)

48.32   N. Srinivas, K. Deb: Multiobjective optimization using nondominated sorting in genetic algorithms, Evol. Comput. **2**(3), 221–248 (1994)

48.33   D.E. Goldberg: Multiobjective optimization. In: *Genetic Algorithms in Search, Optimization, and Machine Learning* (Addison-Wesley, Reading 1989) pp. 197–201

48.34   C.M. Fonseca, P.J. Fleming: Genetic algorithms for multiobjective optimization: Formulation, discussion and generalization, Proc. 5th Conf. Genet. Algorithms (1993) pp. 416–423

48.35   J. Bader: Hypervolume-Based Search for Multiobjective Optimization: Theory and Methods, Ph.D. Thesis (CreateSpace, ETH Zurich 2010)

48.36   E. Zitzler, K. Deb, L. Thiele: Comparison of multiobjective evolutionary algorithms: Empirical results, Evol. Comput. **8**(2), 173–195 (2000)

48.37   C.A. Coello Coello: Handling preferences in evolutionary multiobjective optimization: A survey, Congr. Evol. Comput. (2000) pp. 30–37

48.38   K. Deb, J. Sundar: Reference point based multiobjective optimization using evolutionary algorithms, Genet. Evol. Comput. Conf. (2006) pp. 635–642

48.39   A. Auger, J. Bader, D. Brockhoff, E. Zitzler: Articulating user preferences in many-objective problems by sampling the weighted hypervolume, Genet. Evol. Comput. Conf. (2009) pp. 555–562

48.40   A. Wierzbicki: The use of reference objectives in multiobjective optimization, Lect. Notes Econ. Math. Syst. **177**, 468–486 (1980)

48.41   E. Zitzler, D. Brockhoff, L. Thiele: The hypervolume indicator revisited: On the design of Pareto-compliant indicators via weighted integration, Lect. Notes Comput. Sci. **4403**, 862–876 (2007)

49. Multi-Objective Evolutionary Algorithms

Kalyanmoy Deb

Evolutionary algorithms (EAs) have amply shown their promise in solving various search and optimization problems for the past three decades. One of the hallmarks and niches of EAs is their ability to handle multi-objective optimization problems in their totality, which their classical counterparts lack. Suggested in the beginning of the 1990s, evolutionary multi-objective optimization (EMO) algorithms are now routinely used in solving problems with multiple conflicting objectives in various branches of engineering, science, and commerce. In this chapter, we provide an overview of EMO methodologies by first presenting principles of EMO through an illustration of one specific algorithm and its application to an interesting real-world bi-objective optimization problem. Thereafter, we provide a list of recent research and application developments of EMO to provide a picture of some salient advancements in EMO research. The development and application of EMO to multi-objective optimization problems and their continued extensions to solve other related problems has elevated EMO research to a level which may now undoubtedly be termed as an active field of research with a wide range of theoretical and practical research and application opportunities.

49.1 Preamble

Search and optimization problems, particularly involving nonlinear, non-convex and non-differentiable objective and constraint functions, provide a stiff challenge even today. No known mathematical algorithm exists to solve such problems to optimality. In such cases, the use of meta-heuristic optimization methods such as evolutionary algorithms [49.1–3], simulated annealing [49.4], tabu search [49.5, 6], and other methods motivated by another natural or physical phenomenon have been popularly applied.

EAs were traditionally used for solving problems having a single goal or objective. However, most real-world problems have multiple conflicting goals and theoretically they give rise to a set of trade-off solutions. The classical literature to solve multi-objective optimization problems has been mostly indirect, mainly due to the fact there did not exist any search and optimization methods which could find multiple optimal solutions in a single simulation. While the scientific community was waiting for a suitable algorithm for handling such problems, evolutionary algorithms with their population approach caught the eyes of a number of researchers. This spurred the development of a series of first generation evolutionary multi-objective optimization (EMO) algorithms around 1993–1995. A set of three different algorithms (but all motivated by a single idea portrayed by legendary EA researcher, Prof. *Goldberg* [49.1]) showed the world that EMOs are viable candidates for multi-objective optimization, and that there are meta-heuristic-based approaches for finding multiple trade-off solutions in a single simulation. EMO researchers, and in that spirit the whole EA research community, realized the niche of EAs in such problem-solving tasks and promoted the developmental and application studies using EMO. Subsequently, EMO methodologies were made to be better, faster, and more accessible. The algorithms were commercialized by various software companies and they made the field

of EMO more popular and applicable to many different problems, which academic researchers alone could not have done.

In this chapter, we provide a brief overview of the EMO principle, present one EMO algorithm in detail, and emphasize the importance of using EMO in practice. Besides this specific algorithm, there exist a number of other equally efficient EMO algorithms, which we do not describe here for brevity. Instead, in this chapter, we discuss a number of recent advancements of EMO research and applications that are driving researchers and practitioners ahead. Fortunately, researchers have utilized the EMO principle of solving multi-objective optimization problems in handling various other problem-solving tasks. The diversity of EMO's research is bringing together researchers and practitioners with different backgrounds, including computer scientists, mathematicians, economists, and engineers. The topics that we discuss here amply demonstrate why and how EMO researchers from different backgrounds must and should collaborate on complex problem-solving tasks, which have become the need of the hour in most branches of science, engineering, and commerce today.

49.2 Evolutionary Multi-Objective Optimization (EMO)

Before we discuss an evolutionary algorithm for multi-objective optimization, we present a generic problem that involves multiple conflicting objectives. A multi-objective optimization problem involves a number of objective functions that are to be either minimized or maximized, subject to a number of constraints and variable bounds

$$
\left.
\begin{aligned}
\text{Minimize/} \quad & f_m(\mathbf{x}), & m = 1, 2, \ldots, M \, ; \\
\text{Maximize} \\
\text{subject to} \quad & g_j(\mathbf{x}) \geq 0, & j = 1, 2, \ldots, J \, ; \\
& h_k(\mathbf{x}) = 0, & k = 1, 2, \ldots, K \, ; \\
& x_i^{(L)} \leq x_i \leq x_i^{(U)}, & i = 1, 2, \ldots, n \, .
\end{aligned}
\right\}
$$

$$(49.1)$$

A solution $\mathbf{x} \in \mathbf{R}^n$ is a vector of n decision variables: $\mathbf{x} = (x_1, x_2, \ldots, x_n)^T$. The solutions satisfying the constraints and variable bounds constitute a *fea-*

sible set S in the decision variable space $\mathbf{R}^n$. One of the striking differences between single-objective and multi-objective optimization is that in multi-objective optimization the objective function vectors belong to a multi-dimensional objective space $\mathbf{R}^M$. The objective function vectors constitute a feasible set Z in the objective space. For each solution $\mathbf{x}$ in S, there exists a point $\mathbf{z} \in Z$, denoted by $\mathbf{f}(\mathbf{x}) = \mathbf{z} = (z_1, z_2, \ldots, z_M)^T$. To make the descriptions clear, we refer to a decision variable vector as a solution and the corresponding objective vector as a point.

The optimal solutions in multi-objective optimization can be defined from the mathematical concept of *partial ordering* [49.7]. In the parlance of multi-objective optimization, the term *domination* is used for this purpose. In this section, we restrict ourselves to discussing unconstrained (without any equality, inequality or bound constraints) optimization problems. The domination between two solutions is defined as follows [49.8, 9]:

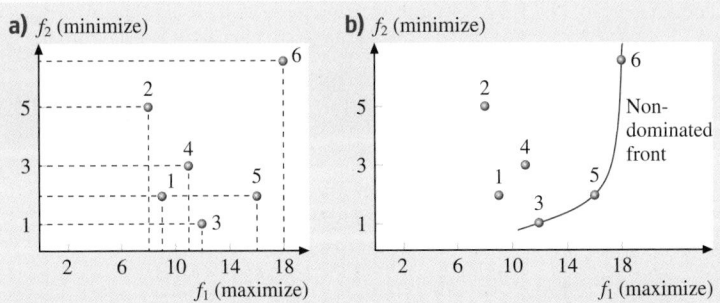

Fig. 49.1a,b A set of points and the first non-dominated front are shown

Definition 49.1
A solution $\mathbf{x}^{(1)}$ is said to dominate another solution $\mathbf{x}^{(2)}$, if both the following conditions are true.

1. The solution $\mathbf{x}^{(1)}$ is no worse than $\mathbf{x}^{(2)}$ in all objectives. Thus, the solutions are compared based on their objective function values (or location of the corresponding points ($\mathbf{z}^{(1)}$ and $\mathbf{z}^{(2)}$) in the objective function set Z).
2. The solution $\mathbf{x}^{(1)}$ is strictly better than $\mathbf{x}^{(2)}$ in at least one objective.

For a given set of solutions (or corresponding points in the objective function set Z, for example, those shown in Fig. 49.1a), a pair-wise comparison can be made using the above definition and whether one point dominates another point can be established. All points that are not dominated by any other member of the set are called non-dominated points of class one, or simply non-dominated points. For the set of six points shown in the figure, they are points 3, 5, and 6. One property of any two such points is that a gain in an objective from one point to the other happens only due to a sacrifice in at least one other objective. This *trade-off* property between non-dominated points makes practitioners interested in finding a wide variety of them before making a final choice. These points make up a front when viewed together on the objective space; hence non-dominated points are often visualized to represent a *non-dominated front*. The theoretical computational effort needed to select the points of the non-dominated front from a set of N points is $O(N \log N)$ for 2 and 3 objectives, and $O(N \log^{M-2} N)$ for $M > 3$ objectives [49.10], but for a moderate number of objectives, the procedure need not be particularly computationally effective in practice.

With the above concept, now it is easier to define the *Pareto-optimal solutions* in a multi-objective optimization problem. If the given set of points for the above task contain *all* feasible points in the objective space, the points lying on the first non-domination front, by definition, do not become dominated by any other point in the objective space; hence they are Pareto-optimal points (together they constitute the Pareto-optimal front), and the corresponding pre-images (decision variable vectors) are called Pareto-optimal solutions. However, more mathematically elegant definitions of Pareto-optimality (including the ones for continuous search space problems) exist in the multi-objective optimization literature [49.9, 11]. Some convergence analyses of EMO under certain assumptions can also be found elsewhere [49.12–15].

49.2.1 EMO Principles

In the context of multi-objective optimization, the extremist principle of finding the optimum solution cannot be applied to one objective alone, when the rest of the objectives are also important. This clearly suggests two ideal goals of multi-objective optimization.

- *Convergence*: find a (finite) set of solutions which lies on the Pareto-optimal front.
- *Diversity*: find a set of solutions which is diverse enough to represent the entire range of the Pareto-optimal front.

EMO algorithms attempt to follow both the above principles, similar to the a posteriori multiple criteria decision-making (MCDM) method. Figure 49.2 schematically shows the principles followed in an EMO procedure. Since EMO procedures are heuristic based, they may not guarantee finding exact Pareto-optimal points, as a theoretically provable optimization method

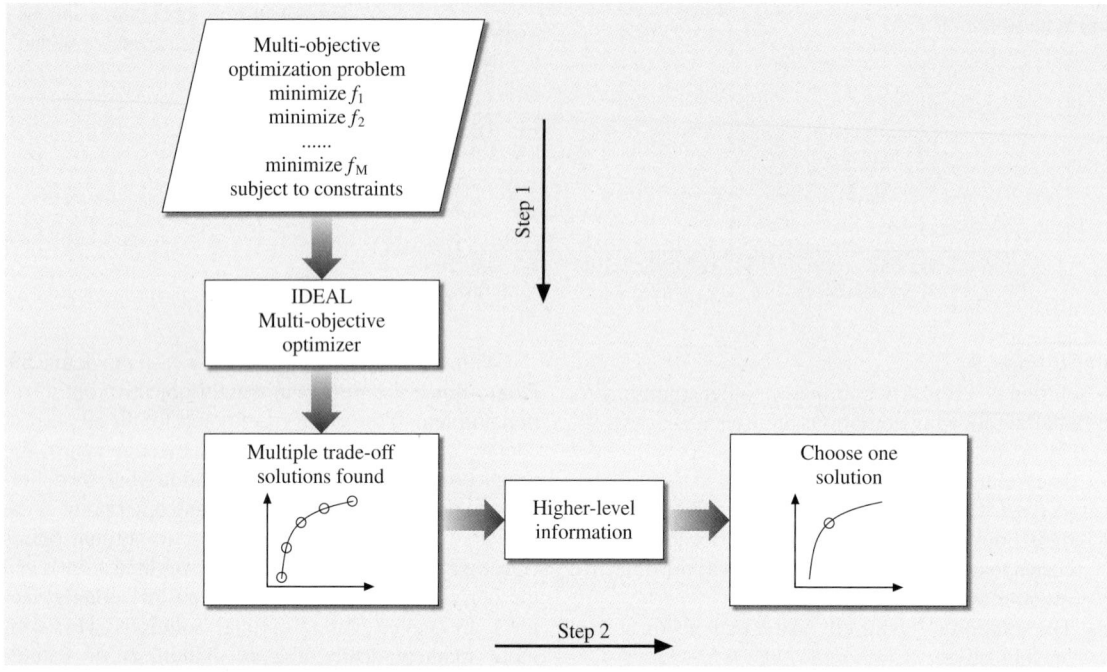

Fig. 49.2 Schematic of a two-step multi-criteria optimization and decision-making procedure

would do for tractable (for example, linear or convex) problems. However, EMO procedures have essential operators to constantly improve the evolving non-dominated points (from the point of view of convergence and diversity mentioned above) similar to how most natural and artificial evolving systems continuously improve their solutions. To this effect, a recent study [49.16] demonstrated that a particular EMO procedure, starting from random non-optimal solutions, can progress towards theoretical Karush–Kuhn–Tucker (KKT) points with iterations in real-valued multi-objective optimization problems. The main difference and advantage of using EMO compared to a posteriori MCDM procedures is that multiple trade-off solutions can be found in a single run of an EMO algorithm, whereas most a posteriori MCDM methodologies would require multiple independent runs.

In Step 1 of the EMO-based multi-objective optimization and decision-making procedure (the task shown vertically downwards in Fig. 49.2), multiple trade-off, non-dominated points are found. Thereafter, in Step 2 (the task shown horizontally, towards the right), higher-level information is used to choose one of the trade-off points obtained.

49.2.2 A Posteriori MCDM Methods and EMO

In the a posteriori MCDM approaches (also known as *generating MCDM methods*), the task of finding multiple Pareto-optimal solutions is achieved by executing many independent single-objective optimizations, each time finding a single Pareto-optimal solution [49.9]. A parametric scalarizing approach (such as the weighted-sum approach, ϵ-constraint approach, and others) can be used to convert multiple objectives into a parametric single-objective objective function. By simply varying the parameters (the weight vector or the ϵ-vector) and optimizing the scalarized function, different Pareto-optimal solutions can be found. In contrast, in an EMO, multiple Pareto-optimal solutions are attempted to be found in a single run of the algorithm by emphasizing multiple non-dominated and isolated solutions in each iteration of the algorithm and without the use of any scalarization of objectives.

Consider Fig. 49.3, in which we sketch how multiple independent parametric single-objective optimizations (through a posteriori MCDM method) may find different Pareto-optimal solutions. It is worth highlighting here that the Pareto-optimal front corresponds to the global optimal solutions of several problems,

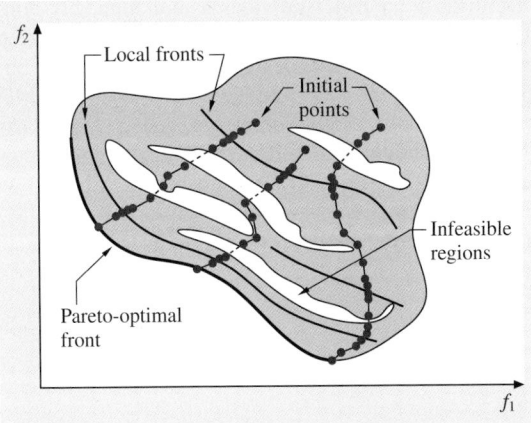

f_2
Local fronts
Initial points
Infeasible regions
Pareto-optimal front
f_1

Fig. 49.3 A posteriori MCDM methodology employing independent single-objective optimizations

each formed with a different scalarization of objectives. During the course of an optimization task, algorithms must overcome a number of difficulties, such as infeasible regions, local optimal solutions, flat or non-improving regions of objective landscapes, isolation of optimum, etc., to finally converge to the global optimal solution. Moreover, due to practical limitations, an optimization task must also be completed in a reasonable computational time. All these difficulties in a problem require that an optimization algo-

rithm strikes a good balance between exploring new search directions and exploiting the extent of search in currently-best search direction. When multiple runs of an algorithm need to be performed independently to find a set of Pareto-optimal solutions, the above balancing act must be performed in every single run. Since runs are performed independently from one another, no information about the success or failure of previous runs is utilized to speed up the overall process. In difficult multi-objective optimization problems, such memory-less, a posteriori methods may demand a large overall computational overhead to find a set of Pareto-optimal solutions [49.17]. Moreover, despite the issue of global convergence, independent runs may not guarantee achieving a good distribution among obtained points by an easy variation of scalarization parameters.

EMO, as was mentioned earlier, constitutes an inherent parallel search. When a particular population member overcomes certain difficulties and makes a progress towards the Pareto-optimal front, its variable values and their combination must reflect this fact. When a recombination takes place between this solution and another population member, such valuable information of variable value combinations is shared through variable exchanges and blending, thereby making the overall task of finding multiple trade-off solutions a parallelly processed task.

49.3 A Brief Timeline for the Development of EMO Methodologies

During the seventies and eighties, EA researchers realized the need for solving multi-objective optimization problems in practice and mainly resorted to using weighted-sum approaches to convert multiple objectives into a single goal [49.18, 19].

However, the first implementation of a real multi-objective evolutionary algorithm (vector-evaluated GA (genetic algorithm) or VEGA) was suggested by *Schaffer* in 1984 [49.20]. Schaffer modified the simple three-operator genetic algorithm [49.2] (with selection, crossover, and mutation) by performing independent selection cycles according to each objective. The selection method is repeated for each individual objective to fill up a portion of the mating pool. Then the entire population is thoroughly shuffled to apply crossover and mutation operators. This is performed to achieve the mating of individuals of different subpopulation groups. The algorithm worked efficiently for some generations but in some cases suffered from its bias towards

some individuals or regions (mostly individual objective champions). This does not fulfill the second goal of EMO, discussed earlier.

Ironically, no significant study was performed for almost a decade after the pioneering work of Schaffer, until a revolutionary 10-line sketch of a new non-dominated sorting procedure suggested by *Goldberg* in his seminal book on GAs [49.1]. Since an EA needs a fitness function for reproduction, the trick was to find a single metric from a number of objective functions. Goldberg's suggestion was to use the concept of *domination* to assign more copies to non-dominated individuals in a population. Since diversity is the other concern, he also suggested the use of a *niching* strategy [49.21] among solutions of a non-dominated class. To get this clue, at least three independent groups of researchers developed different versions of multi-objective evolutionary algorithms during 1993–1994 [49.22–24]. These algorithms differ in the way a fitness assignment scheme is introduced to each individual. Independently, *Poloni* [49.25] suggested a domination-based EMO approach (he called it multi-objective genetic algorithm (MOGA)) in which instead of niching, a toroidal grid-based local selection method was used to find multiple trade-off solutions.

These early EMO methodologies gave a good head-start to the research and application of EMO, but suffered from the fact that they did not use an elite-preservation mechanism in their procedures. Inclusion of elitists in an EMO provides a monotonically non-degrading performance [49.26]. The second generation EMO algorithms implemented an elite-preserving operator in different ways and gave birth to elitist EMO procedures, such as non-dominated sorting GA NSGA-II [49.27], strength Pareto EA (SPEA) [49.28], Pareto-archived ES (PAES) [49.29], and others. Since these EMO algorithms are state-of-the-art and commonly-used procedures, we describe one of these algorithms in detail.

49.4 Elitist EMO: NSGA-II

The NSGA-II procedure [49.27] is one of the popularly used EMO procedures which attempt to find multiple Pareto-optimal solutions in a multi-objective optimization problem and has the following three features:

1. It uses an elitist principle.
2. It uses an explicit diversity preserving mechanism.
3. It emphasizes non-dominated solutions.

At any generation t, the offspring population (say, Q_t) is first created by using the parent population (say, P_t) and the usual genetic operators. Thereafter, the two populations are combined to form a new population (say, R_t) of size $2N$. Then, the population R_t is classified into different non-dominated classes. Thereafter, the new population is filled by points of different non-dominated fronts, one at a time. The filling starts with the first non-dominated front (of class 1) and continues with points of the second non-dominated front, and so on. Since the overall population size of R_t is $2N$, not all fronts can be accommodated in the N slots available for the new population. All fronts that could not be accommodated are deleted. When the last allowed front is being considered, there may exist more points in the front than the slots remaining in the new population. This scenario is illustrated in Fig. 49.4. Instead of arbitrarily discarding some members from the last front, the points that will make the diversity of the selected points the highest are chosen.

The crowded-sorting of the points of the last front which could not be accommodated fully is achieved in the descending order of their *crowding distance* values, and points from the top of the ordered list are chosen. The crowding distance d_i of point i is a measure of the objective space around i which is not occupied by any other solution in the population. Here, we simply calcu-

late this quantity d_i by estimating the perimeter of the cuboid (Fig. 49.5) formed by using the nearest neighbors in the objective space as the vertices (we call this the *crowding distance*).

49.4.1 Sample Results

Here, we show results from several runs of the NSGA-II algorithm on two test problems. The first problem (ZDT2 – Zitzler–Deb–Thiele) is a two-objective, 30-

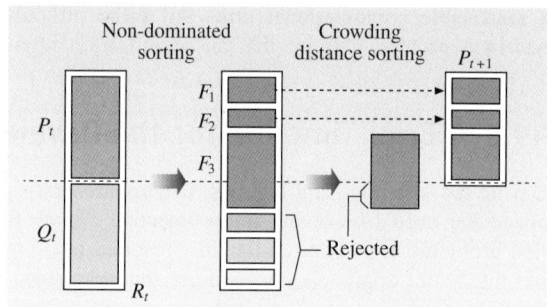

Fig. 49.4 Schematic of the NSGA-II procedure

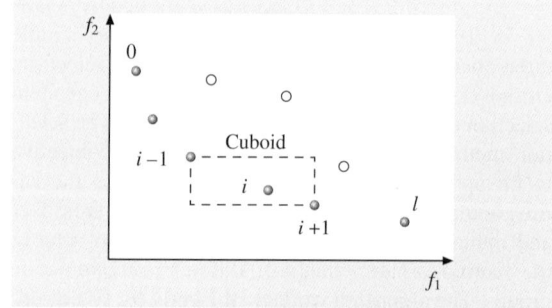

Fig. 49.5 The crowding distance calculation

variable problem with a concave Pareto-optimal front

$$\text{ZDT2:} \begin{cases} \text{minimize} & f_1(\mathbf{x}) = x_1 \,, \\ \text{minimize} & f_2(\mathbf{x}) = s(\mathbf{x})[1 - (f_1(\mathbf{x})/s(\mathbf{x}))^2] \,, \\ \text{where} & s(\mathbf{x}) = 1 + \frac{9}{29} \sum_{i=2}^{30} x_i \,, \\ & 0 \le x_1 \le 1 \,, \\ & -1 \le x_i \le 1 \,, \quad i = 2, 3, \dots, 30 \,. \end{cases}$$

$$(49.2)$$

The second problem (KUR – Kurswae), with three variables, has a disconnected Pareto-optimal front

$$\text{KUR:} \begin{cases} \text{minimize} & f_1(\mathbf{x}) = \sum_{i=1}^{2} \\ & \left[-10 \exp\left(-0.2 \sqrt{x_i^2 + x_{i+1}^2} \right) \right] \,, \\ \text{minimize} & f_2(\mathbf{x}) = \sum_{i=1}^{3} [|x_i|^{0.8} + 5 \sin(x_i^3)] \,, \\ & -5 \le x_i \le 5 \,, \quad i = 1, 2, 3 \,. \end{cases}$$

$$(49.3)$$

NSGA-II is run with a population size of 100 and for 250 generations. The variables are used as real numbers and a simulated binary crossover (SBX) recombination operator [49.30] with $p_c = 0.9$, a distribution index of $\eta_c = 10$, and a polynomial mutation operator [49.8] with $p_m = 1/n$ (n is the number of variables) and a distribution index of $\eta_m = 20$ are used. Figures 49.6 and 49.7 show that NSGA-II converges to the Pareto-optimal front and maintains a good spread of solutions in both test problems.

There also exist other competent EMOs, such as the strength Pareto evolutionary algorithm (SPEA) and its improved version SPEA2 [49.31], the Pareto-archived evolution strategy (PAES) and its improved versions pareto-envelope based selection algorithm (PESA) and PESA2 [49.32], multi-objective messy GA (MOMGA) [49.33], multi-objective micro-GA [49.34], neighborhood constraint GA [49.35], adaptive range MOGA (ARMOGA) [49.36], and others. Moreover, there exist other EA-based methodologies, such as particle swarm-based EMO [49.37, 38], ant-based EMO [49.39, 40], and differential evolution-based EMO [49.41]. Simulated annealing method is used to find multiple Pareto-optimal solutions for multi-objective optimization problems [49.42]. The tabu search method is also used for multi-objective optimization [49.43].

49.4.2 Constraint Handling in EMO

The constraint handling method modifies the binary tournament selection, where two solutions are picked from the population, and the better solution is chosen. In the presence of constraints, each solution can be either feasible or infeasible. Thus, there may be at most three situations:

i) Both solutions are feasible.
ii) One is feasible and other is not.
iii) Both are infeasible.

We consider each case by simply redefining the domination principle as follows (we call it the

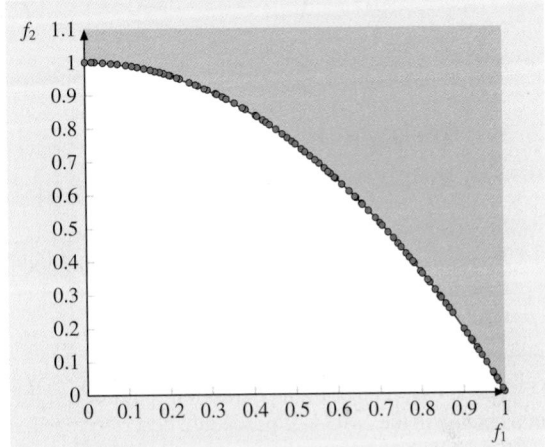

Fig. 49.6 NSGA-II on ZDT2

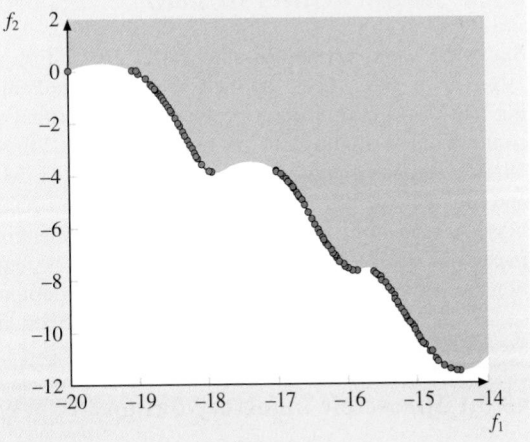

Fig. 49.7 NSGA-II on KUR

constrained-domination condition for any two solutions $\mathbf{x}^{(i)}$ and $\mathbf{x}^{(j)}$):

Definition 49.2

A solution $\mathbf{x}^{(i)}$ is said to be a *constrained-dominated* solution $\mathbf{x}^{(j)}$ (or $\mathbf{x}^{(i)} \preceq_c \mathbf{x}^{(j)}$), if any of the following conditions are true:

1. Solution $\mathbf{x}^{(i)}$ is feasible and solution $\mathbf{x}^{(j)}$ is not.
2. Solutions $\mathbf{x}^{(i)}$ and $\mathbf{x}^{(j)}$ are both infeasible, but solution $\mathbf{x}^{(i)}$ has a smaller constraint violation, which can be computed by adding the normalized violation of all constraints

$$CV(\mathbf{x}) = \sum_{j=1}^{J} \max\left(0, -\bar{g}_j(\mathbf{x})\right) + \sum_{k=1}^{K} \mathrm{abs}(\bar{h}_k(\mathbf{x})) \ .$$

The normalization is achieved with the population minimum ($\langle g_j \rangle_{\min}$) and maximum ($\langle g_j \rangle_{\max}$) constraint violations

$$\bar{g}_j(\mathbf{x}) = (\langle g_j(\mathbf{x}) \rangle - \langle g_j \rangle_{\min}) / (\langle g_j \rangle_{\max} - \langle g_j \rangle_{\min}) \ .$$

3. Solutions $\mathbf{x}^{(i)}$ and $\mathbf{x}^{(j)}$ are feasible and solution $\mathbf{x}^{(i)}$ dominates solution $\mathbf{x}^{(j)}$ in the usual sense (Definition 49.1).

The above change in the definition requires a minimal change in the NSGA-II procedure described earlier.

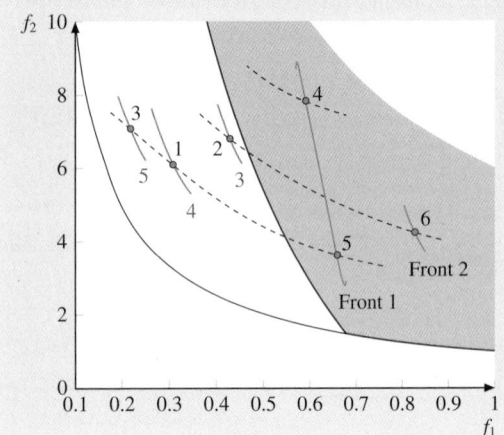

Fig. 49.8 Non-constrained-domination fronts

Figure 49.8 shows the non-dominated fronts on a six-member population due to the introduction of two constraints (the minimization problem is described as CONSTR elsewhere [49.8]). In the absence of the constraints, the non-dominated fronts (shown by dashed lines) would have been ((1,3,5), (2,6), (4)), but in their presence, the new fronts are ((4,5), (6), (2), (1), (3)). The first non-dominated front consists of the *best* (that is, non-dominated and feasible) points from the population and any feasible point lies on a better non-dominated front than an infeasible point.

49.5 Applications of EMO

Since the early development of EMO algorithms in 1993, they have been applied to many challenging real-world optimization problems. Descriptions of some of these studies can be found in books [49.8, 44–47], dedicated conference proceedings [49.48–53], and domain-specific books, journals, and proceedings. A repository of most research and application papers of EMO is available [49.54]. In this section, we describe one case study that clearly demonstrates the EMO philosophy which we described in Sect. 49.2.1.

49.5.1 Spacecraft Trajectory Design

Coverstone-Carroll et al. [49.55] proposed a multi-objective optimization technique using the original non-

dominated sorting algorithm (NSGA) [49.24] to find multiple trade-off solutions in a spacecraft trajectory optimization problem. To evaluate a solution (trajectory), the SEPTOP (solar electric propulsion trajectory optimization) software [49.56] is called, and the delivered payload mass and the total time of flight are calculated. The multi-objective optimization problem has eight decision variables controlling the trajectory and three objective functions:

i) Maximize the delivered payload at destination.
ii) Maximize the negative of the time of flight.
iii) Maximize the total number of heliocentric revolutions in the trajectory, and three constraints limiting the SEPTOP convergence error and minimum and maximum bounds on heliocentric revolutions.

On the Earth–Mars rendezvous mission, the study found interesting trade-off solutions [49.55]. Using a population of size 150, the NSGA was run for 30 generations. The non-dominated solutions obtained are shown in Fig. 49.9 for two of the three objectives, and some selected solutions are shown in Fig. 49.10. It is clear that there exist short-time flights with smaller delivered payloads (solution marked 44 with 1.12 years of flight and delivering 685.28 kg load) and long-time flights with larger delivered payloads (solution marked 36 with close to 3.5 years of flight and delivering about 900 kg load). While solution 44 can deliver a mass of 685.28 kg and requires about 1.12 years, solution 72 can deliver almost 862 kg with a travel time of about 3 years. In these figures, each continuous part of a trajectory represents a *thrusting* arc and each dashed part of a trajectory represents a *coasting* arc. It is interesting to note that only a small improvement in delivered mass occurs in the solutions between 73 and 72 with a sacrifice in flight time of about 1 year.

The multiplicity in trade-off solutions, as depicted in Fig. 49.10, is what we envisaged in discovering in a multi-objective optimization problem by using a posteriori procedure, such as a generating method or using an EMO procedure vis-a-vis an a priori approach in which a single scalarized problem is solved with a single preferred parameter setting to find a single Pareto-optimal solution. This aspect is also shown in Fig. 49.2. Once a set of solutions with a good trade-off among objectives is obtained, one can analyze them to choose a particular solution. For example, in this problem context, it makes sense not to choose a solution between points 73 and 72 due to poor trade-off between the objectives in this range, a matter which is only revealed after a representative set of trade-off solutions are found. On the other hand, choosing a solution within points 44 and 73 is worthwhile, but which particular solution to choose depends on other mission-related issues. However, by first finding a wide range of possible solutions thereby revealing the shape of front in a computationally quicker manner, EMO can help a decision-maker in narrowing down the choices

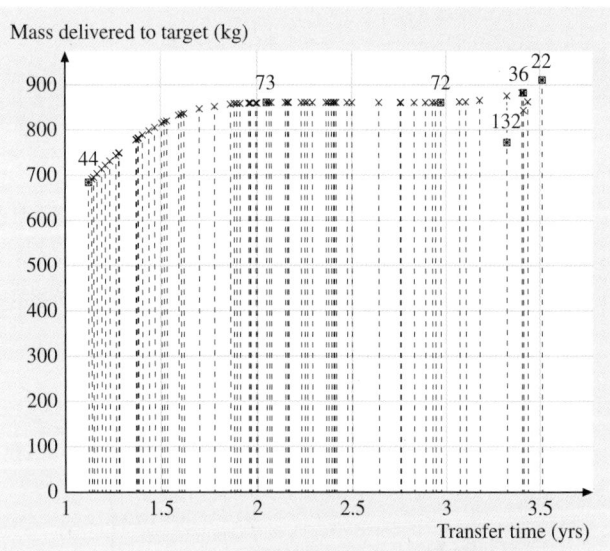

Fig. 49.9 Non-dominated solutions obtained using NSGA

and in allowing a better decision to be made. Without the knowledge of such a wide variety of trade-off solutions, proper decision-making may be a difficult task. With the use of an a priori approach to find a single solution using, for example, the ϵ-constraint method with a particular ϵ vector, the decision-maker will always wonder what solution would have been derived if a different ϵ vector had been chosen. For example, if $\epsilon_1 = 2.5$ years is chosen and the mass delivered to the target is maximized, a solution in between points 73 and 72 will be found. As discussed earlier, this part of the Pareto-optimal front does not provide the best trade-offs between the objectives that this problem can offer. A lack of knowledge of good trade-off regions before a decision is made may allow the decision-maker to settle for a solution which, although optimal, may not be a good compromise solution. The EMO procedure allows a flexible and a pragmatic procedure for finding a well-diversified set of solutions simultaneously so as to enable picking a particular region for further analysis or a particular solution for implementation.

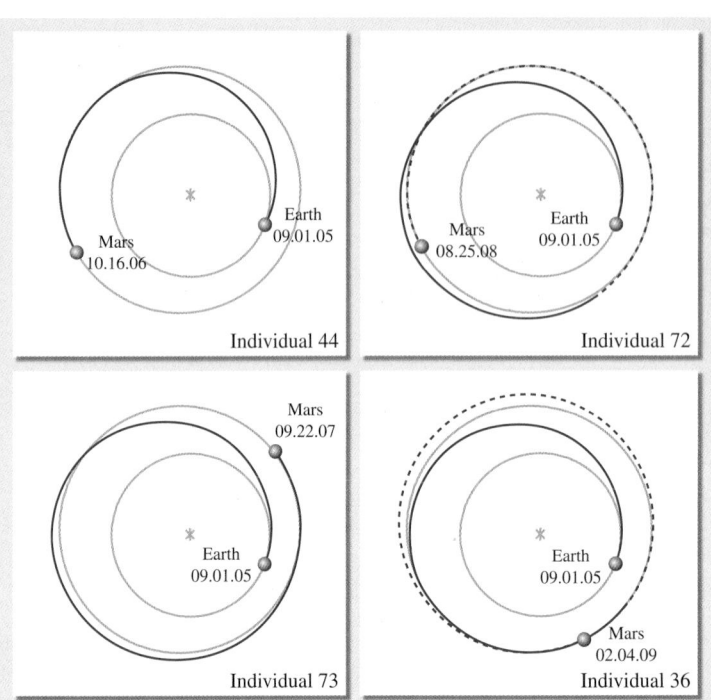

Fig. 49.10 Four trade-off trajectories (after [49.55])

49.6 Recent Developments in EMO

An interesting aspect regarding research and application of EMO is that soon after a number of efficient EMO methodologies had been suggested and applied in various interesting problem areas, researchers did not waste any time to look for opportunities to make the field broader and more useful by diversifying EMO applications to various other problem-solving tasks. In this section, we describe a number of such salient recent developments of EMO.

49.6.1 Hybrid EMO Algorithms

The search operators used in EMO are heuristic based. Thus, these methodologies are not guaranteed to find Pareto-optimal solutions with a finite number of solution evaluations in an arbitrary problem. In single-objective EA research, hybridization of EAs is common for ensuring convergence to an optimal solution; it is not surprising that studies on developing hybrid EMOs are now being pursued to ensure that true Pareto-optimal solutions are found by hybridizing them with mathematically convergent ideas.

EMO methodologies provide adequate emphasis on currently non-dominated and isolated solutions so that population members progress towards the Pareto-optimal front iteratively. To make the overall procedure faster and to perform the task with a more theoretical emphasis, EMO methodologies are combined with mathematical optimization techniques having local convergence properties. A simple-minded approach would be to start the process with an EMO and the solutions obtained from EMO could be improved by optimizing a composite objective derived from multiple objectives to ensure a good spread by using a local search technique [49.57]. Another approach would be to use a local search technique as a mutation-like operator in an EMO, so that all population members are at least guaranteed to be local optimal solutions [49.57, 58]. To save computational time, instead of performing the local search for every solution in a generation, a mutation can be performed only after a few generations. Some recent studies [49.58–60] have demonstrated the usefulness of such hybrid EMOs for a guaranteed convergence.

Although these studies concentrated on ensuring convergence to the Pareto-optimal front, some emphasis should now be placed on providing adequate diversity among the solutions obtained, particularly when a continuous Pareto-optimal front is represented by a finite set of points. Some ideas of maximizing the hypervolume measure [49.61] or the maintenance of a uniform distance between points are proposed for this purpose, but how such diversity-maintenance techniques would be integrated with convergence-ensuring principles in a synergistic way would be interesting and useful future research. Some relevant studies in this direction exist already [49.59, 62–65].

49.6.2 Multi-Objectivization

Interestingly, the act of finding multiple trade-off solutions using an EMO procedure has found its application outside the realm of solving multi-objective optimization problems. The concept of finding near-optimal trade-off solutions is applied to solve other kinds of optimization problems as well. For example, the EMO concept is used to solve constrained single-objective optimization problems by converting the task into a two-objective optimization task of additionally minimizing an aggregate constraint violation [49.66]. This eliminates the need to specify a penalty parameter while using a penalty-based constraint handling procedure. If viewed this way, the usual penalty function approach used in classical optimization studies is a special weighted-sum approach to the bi-objective optimization problem of minimizing the objective function and minimizing the constraint violation, for which the weight vector is a function of the penalty parameter. A well-known difficulty in genetic programming studies, called *bloating*, arises due to the continual increase in the size of *genetic programs* evolved with iteration. The reduction of bloating by minimizing the size of a program as an additional objective has helped find high-performing solutions with a smaller size of the code [49.67, 68]. In clustering algorithms, minimizing the intra-cluster distance and maximizing inter-cluster distance simultaneously in a bi-objective formulation of a clustering problem is found to yield better solutions than the usual single-objective minimization of the ratio of the intra-cluster distance to the inter-cluster distance [49.69]. An EMO is found to solve a minimum spanning tree problem better than a single-objective EA [49.70]. A recently edited book [49.71] describes many interesting applications in which EMO methodologies have helped to solve problems that are otherwise (or traditionally) not treated as multi-objective optimization problems.

49.6.3 Uncertainty-Based EMO

A major surge in EMO research has taken place in handling uncertainties among decision variables and problem parameters in multi-objective optimization. Practice is full of uncertainties and almost no parameter, dimension, or property can be guaranteed to be fixed at the value it is aimed at. In such scenarios, evaluation of a solution is not precise, and the resulting objective and constraint function values become probabilistic quantities. Optimization algorithms are usually designed to handle such stochastiticies by using crude methods, such as Monte Carlo simulation of stochasticities in uncertain variables and parameters and by sophisticated stochastic programming methods involving nested optimization techniques [49.72]. When these effects are taken care of during the optimization process, the resulting solution is usually different from the optimum solution of the problem and is known as a *robust* solution. Such an optimization procedure will then find a solution which may not be the true global optimum solution, but one which is less sensitive to uncertainties in decision variables and problem parameters. In the context of multi-objective optimization, a consideration of uncertainties for multiple objective functions will result in a robust frontier which may be different from the globally Pareto-optimal front. Each and every point on the robust frontier is then guaranteed to be less sensitive to uncertainties in decision variables and problem parameters. Some such studies in EMO are [49.73, 74].

When the evaluation of constraints under uncertainties in decision variables and problem parameters is considered, deterministic constraints become stochastic (they are also known as *chance constraints*) and involve a *reliability index* (R) to handle the constraints. A constraint $g(\mathbf{x}) \geq 0$ then becomes $\text{Prob}(g(\mathbf{x}) \geq 0) \geq R$. In order to find the left-hand side of the above chance constraint, a separate optimization methodology [49.75] is needed, thereby making the overall algorithm a bi-level optimization procedure. Approximate single-loop algorithms exist [49.76] and recently one such methodology was integrated with an EMO [49.72] and shown to find a *reliable* frontier corresponding a specified reliability index, instead of the Pareto-optimal frontier, in problems having uncertainty in decision variables and problem parameters. More such methodologies are needed, as uncertainties are an integral part of practical problem-solving, and multi-objective optimization researchers must look for better and faster algorithms to handle them.

49.6.4 EMO and Decision-Making

Searching for a set of Pareto-optimal solutions by using an EMO fulfills only one aspect of multi-objective optimization, as choosing a particular solution for an implementation is the remaining decision-making task, which is equally important. For many years, EMO researchers have postponed the decision-making aspect and concentrated on developing efficient algorithms for finding multiple trade-off solutions. Having pursued that part somewhat, now for the past couple of years or so, EMO researchers are putting efforts to design combined algorithms for optimization and decision-making. In the view of the author, the decision-making task can be considered from two main considerations in an EMO framework:

1. *Generic consideration*: there are some aspects that most practical users would like to use in narrowing down their choice. Above we discussed the importance of finding robust and reliable solutions in the presence of uncertainties in decision variables and/or problem parameters. In such scenarios, an EMO methodology can straightaway find a robust or a reliable frontier [49.72, 73] and no subjective preference from any decision maker may be necessary. Similarly, if a problem resorts to a Pareto-optimal front having *knee* points, such points are often the choice of decision-makers. Knee points demands a large sacrifice in at least one objective to achieve a small gain in another, thereby making it discouraging to move out from a knee point [49.77]. Other such generic choices are related to Pareto-optimal points depicting a certain pre-specified relationship between objectives, Pareto-optimal points having multiplicity (say, at least two or more solutions in the decision variable space mapping to identical objective values), Pareto-optimal solutions which do not lie close to variable boundaries, Pareto-optimal points having certain mathematical properties, such as all Lagrange multipliers with more or less identical magnitudes – a condition often desired to make an equal importance to all constraints, and others. These considerations are motivated from the fundamental and practical aspects of optimization and may be applied to most multi-objective problem-solving tasks, without any

consent of a decision-maker. These consideration may narrow down the set of non-dominated points A further subjective consideration (which is discussed below) may then be used to pick a preferred solution.

2. *Subjective consideration*: in this category, any problem-specific information can be used to narrow down the choices, and the process may even lead to a single preferred solution at the end. Most decision-making procedures use some preference information (utility functions, reference points [49.78], reference directions [49.79] marginal rate of return, and a host of other considerations [49.9]) to select a subset of Pareto-optimal solutions. A recent book [49.80] is dedicated to the discussion of many such multi-criteria decision analysis (MCDA) tools and collaborative suggestions of using EMO with such MCDA tools. Some hybrid EMO and MCDA algorithms have been suggested in the recent past [49.81–85].

Many other generic and subjective considerations are needed, and it is interesting that EMO and MCDM researchers are collaborating on developing such complete algorithms for multi-objective optimization [49.80].

49.6.5 EMO for Handling a Large Number of Objectives: Multi-Objective EMO

Initial studies of EMO amply showed that EMO algorithms can be used to find a wide spread of trade-off solutions on two and three-objective optimization problems. However, their performance on four or more objective problems have not been studied enough. Recently, such studies have become important and are known as many-objective optimization studies in the EMO literature.

A detailed study [49.86] made on eight-objective problems revealed somewhat negative results about the existing EMO methodologies. However, in his book [49.8] and recent other studies [49.87–90] the author has clearly explained the reasons for this behavior of EMO algorithms. EMO methodologies work by emphasizing non-dominated solutions in a population. Unfortunately, as the number of objectives increases most population members in a randomly created population tend to become non-dominated to each other. For example, in a three-objective scenario, about 10% of the members in a population of the size 200 are non-dominated, whereas in a 10-objective problem sce nario, as much as 90% of the members in a population of size of 200 are non-dominated. Thus, in a large-objective problem, an EMO algorithm runs out of room to introduce new population members into a generation, thereby causing a stagnation in the performance of an EMO algorithm. It has been argued that to make EMO procedures efficient, an exponentially large population size (with respect to the number of objectives) is needed. This makes the EMO procedure slow and computationally less attractive.

However, recent techniques use a fixed set of reference points [49.91–93] or reference directions [49.94] and are promising, as they are shown to find a widely distributed set of solutions in 3 to 15-objective test and real-world problems.

However, practically speaking, even if an algorithm can find tens of thousands of Pareto-optimal solutions for a multi-objective optimization problem, besides simply getting an idea of the nature and shape of the front, they are simply too many to be conceivable for any decision-making purposes. Keeping these views in mind, EMO researchers have taken two different approaches in dealing with many-objective problems.

Finding a Partial Set

Instead of finding the complete Pareto-optimal front in a problem having many objectives, EMO procedures can be used to find only a preferred part of the Pareto-optimal front. This can be achieved by indicating preference information by various means. Ideas such as reference point-based EMO [49.81, 85], *light beam search* [49.82], biased sharing approaches [49.95], cone-dominance [49.96], etc. have been suggested for this purpose. Each of these studies has shown that for up to 10 and 20-objective problems, although finding the complete frontier is a difficulty, finding a partial frontier corresponding to certain preference information is not that difficult a proposition.

The use of a parallel or a distributed computing platform can be used with the above idea, and the complete Pareto-optimal front can be obtained by a distributed computing procedure [49.96]. In the study, each processor in a distributed computing environment receives a unique cone defining domination. The cones are designed carefully so that at the end of such a distributed computing EMO procedure, solutions are found to exist in various parts of the complete Pareto-optimal front. A collection of these solutions is then able to provide a good representation of the entire original Pareto-optimal front.

Identifying and Eliminating Redundant Objectives

Many practical optimization problems can easily list a large of number of objectives (often more than 10), as many different criteria or goals are often of interest to practitioners. In most instances, it is not entirely sure whether or not the chosen objectives are all in conflict with each other. For example, the minimization of weight and the minimization of cost of a component or a system are often mistaken to have an identical optimal solution, but may lead to a range of trade-off optimal solutions. Practitioners do not take any chances and tend to include all (or as many as possible) objectives into the optimization problem formulation. There is another fact which is more worrisome. Two apparently conflicting objectives may show a good trade-off when evaluated with respect to some randomly created solutions. However, if these two objectives are evaluated for solutions close to their optima, they tend to show a good correlation. That is, although objectives can exhibit conflicting behavior for random solutions, near their Pareto-optimal front, the conflict vanishes and the optimum of one becomes close to the optimum of the other.

Thinking of the existence of such problems in practice, certain researchers [49.90, 97, 98] performed linear and non-linear principal component analysis (PCA) to a set of EMO-produced solutions. Objectives causing a positively correlated relationship between the the obtained NSGA-II solutions were identified and declared as redundant. The EMO procedure is then restarted with non-redundant objectives. This combined EMO-PCA procedure is continued until no further reduction in the number of objectives is possible. The procedure has handled practical problems involving five and more objectives and has shown to reduce the choice of real conflicting objectives to a few. On test problems, the proposed approach has been shown to reduce an initial 50-objective problem to the correct three-objective Pareto-optimal front by eliminating 47 redundant objectives. Another study [49.99] used an exact and a heuristic-based conflict identification approach on a given set of Pareto-optimal solutions. For a given error measure, an effort is made to identify a minimal subset of objectives that does not alter the original dominance structure on a set of Pareto-optimal solutions. This idea was recently introduced within an EMO [49.100], but a continual reduction of objectives through a successive application of the above procedure would be interesting.

This is a promising area of EMO research and more computationally faster objective-reduction techniques are definitely needed for the purpose. A recent approach uses previously-fixed multiple directional searches to find a widely distributed set of Pareto-optimal points [49.94]. In this direction, the use of alternative definitions of domination may be beneficial. One such idea redefined the definition of domination: a solution is said to dominate another solution, if the former solution is better than the latter one in more objectives. This certainly excludes finding the entire Pareto-optimal front and helps an EMO to converge near the intermediate and central part of the Pareto-optimal front. Another EMO study used a fuzzy dominance [49.101] relation (instead of Pareto-dominance), in which superiority of one solution over another in any objective is defined in a fuzzy manner. Many other such definitions are possible and can be implemented based on the problem context.

49.6.6 Knowledge Extraction Through EMO

One striking difference between single-objective optimization and multi-objective optimization is the cardinality of the solution set. In the latter, multiple solutions are the outcome and each solution is theoretically an optimal solution corresponding to a particular trade-off among the objectives. Thus, if an EMO procedure can find solutions close to the true Pareto-optimal set, what we have in our hands is a number of high-performing solutions trading-off the conflicting objectives considered in the study. Since these solutions are all near optimal, they can be analyzed for finding properties which are common to them. Such a procedure can then become a systematic approach in deciphering the important and hidden properties that optimal and high-performing solutions must have for that problem. In a number of practical problem-solving tasks, the so-called *innovation* procedure is shown to find important knowledge about high-performing solutions [49.102]. Such useful properties are expected to exist in practical problems, as they follow certain scientific and engineering principles at the core, but in the past not much attention had been paid to finding them through a systematic scientific procedure. The principle of first searching for multiple trade-off and high-performing solutions using a multi-objective optimization procedure and then analyzing them to discover useful knowledge certainly remains a viable way forward. The current efforts [49.103, 104] to automate the knowledge extraction procedure through a sophisticated data-mining task should make the overall approach more appealing and useful in practice.

49.6.7 Dynamic EMO

Dynamic optimization involves objectives, constraints or problem parameters that change over time. This means that as an algorithm approaches the optimum of the current problem, the problem definition changes and now the algorithm must solve a new problem. This is not equivalent to another optimization task in which a new and different optimization problem must be solved afresh. Often, in such dynamic optimization problems, an algorithm is usually not expected to find the optimum, instead it is best expected to track the optimum changing with the iteration. The performance of a dynamic optimizer then depends on how close it is able to track the true optimum (which changes with iteration or time). Thus, practically speaking, it may be hoped that optimization algorithms can handle problems that do not change significantly with time. With respect to the algorithm, since here the problem is not expected to change too much from one time instance to another and some good solutions to the current problem are already at hand in a population, researchers fancied solving such dynamic optimization problems using evolutionary algorithms [49.105].

A recent study [49.106] proposed the following procedure for dynamic optimization involving single or multiple objectives. Let $\mathcal{P}(t)$ be a problem that changes with time t (from $t = 0$ to $t = T$). Despite the continual change in the problem, we assume that the problem is fixed for a time period τ, which is not known a priori, and the aim of the (offline) dynamic optimization study is to identify a suitable value of τ for an accurate as well as a computationally faster approach. For this purpose, an optimization algorithm with τ as a fixed time period is run from $t = 0$ to $t = T$ with the problem assumed fixed for every τ time period. A measure $\Gamma(\tau)$ determines the performance of the algorithm and is compared with a pre-specified and expected value Γ_L. If $\Gamma(\tau) \geq \Gamma_L$, for the entire time domain of the execution of the procedure, we declare τ to be a permissible length of stasis. Then, we try with a reduced value of τ and check if a smaller length of statis is also acceptable. If not, we increase τ to allow the optimization problem to remain static for a longer time so that the chosen algorithm can now have more iterations (time) to perform better. Such a procedure will eventually come up with a time period τ^*, which would be the smallest time of statis allowed for the optimiza ion algorithm to work based on chosen performance requirement. Based on this study, a number of test problems and a hydro-thermal power dispatch problem were tackled recently [49.106].

In the case of dynamic multi-objective problem-solving tasks, there is an additional difficulty which is worth mentioning here. Not only does an EMO algorithm need to find or track the changing Pareto-optimal fronts, in a real-world implementation, it must also make an immediate decision about which solution to implement from the current front before the problem changes to a new one. Decision-making analysis is considered to be time-consuming, involving execution of analysis tools, higher-level considerations, and sometimes group discussions. If dynamic EMO is to be applied in practice, *automated* procedures for making decisions must be developed. Although it is not clear how to generalize such an automated decision-making procedure in different problems, problem-specific tools are certainly possible and a worthwhile and fertile area for research.

49.6.8 Quality Estimates for EMO

When algorithms are developed and test problems with known Pareto-optimal fronts are available [49.107–110], an important task is to have performance measures with which the EMO algorithms can be evaluated. Thus, a major focus of EMO research has been used to develop different performance measures. Since the focus in an EMO task is multi-faceted – convergence to the Pareto-optimal front and diversity of solutions along the entire front, it is also expected that one performance measure to evaluate EMO algorithms will be unsatisfactory. In the early years of EMO research, three different sets of performance measures were used:

1. Metrics evaluating convergence to the known Pareto-optimal front (such as error ratio, distance from reference set, etc.)
2. Metrics evaluating spread of solutions on the known Pareto-optimal front (such as spread, spacing, etc.).
3. Metrics evaluating certain combinations of convergence and spread of solutions (such as hypervolume, coverage, R-metric, etc.).

Some of these metrics are described in texts [49.8, 44]. A detailed study [49.111] comparing most existing performance metrics based on out-performance relations recommended the use of the S-metric (or the hypervolume metric) and the R-metric suggested

by [49.112]. A recent study argued that a single unary performance measure or any finite combination of them (for example, any of the first two metrics described above in the enumerated list or both together) cannot adequately determine whether one set is better than another [49.113]. That study also concluded that binary performance metrics (indicating usually two different values when a set of solutions A is compared against B and B is compared against A), such as an epsilon-indicator, a binary hypervolume indicator, utility indicators R1 to R3, etc., are better measures for multi-objective optimization. The flip side is that the chosen binary metric must be computed $K(K-1)$ times when comparing K different sets to make a fair comparison, thereby making the use of binary metrics computationally expensive in practice. Importantly, these performance measures have allowed researchers to use them directly as fitness measures within indicator-based EAs (IBEAs) [49.114]. In addition, the attainment indicators of [49.115, 116] provide further information about location and inter-dependencies among the solutions obtained.

The hypervolume metric is a popular metric used in EMO studies. However, the computation of the hypervolume metric for more than three-objective problems becomes a computationally challenging task. Recent studies on computationally fast estimation methods of the hypervolume metric have gained popularity among theoretical minds [49.62, 63, 117, 118]. These methods compute the proportion of randomly generated objective points that are dominated by the current set of non-dominated points to estimate the hypervolume metric. A reliable computation method of these studies will facilitate the use of the hypervolume metric in designing efficient EMO algorithms.

49.6.9 Exact EMO with Run-Time Analysis

Since they were first suggested, efficient EMO algorithms have been increasingly applied in a wide variety of problem domains to obtain trade-off frontiers. Simultaneously, some researchers have also devoted their efforts to developing exact EMO algorithms with a theoretical complexity estimate in solving certain discrete multi-objective optimization problems. The first such study [49.119] suggested a pseudo-Boolean multi-objective optimization problem – a two-objective LOTZ (leading ones trailing zeroes) – and a couple of EMO methodologies – a simple evolutionary multi-objective optimizer (SEMO) and an improved version fair evolutionary multi-objective optimizer (FEMO).

The study then estimated the worst-case computational effort needed to find all Pareto-optimal solutions of the LOTZ problem. This study spurred a number of improved EMO algorithms with run-time estimates and resulted in many other interesting test problems [49.120–123]. Although these test problems may not resemble common practical problems, the working principles of suggested EMO algorithms to handle specific problem structures bring in a plethora of insights about the working of multi-objective optimization, particularly in comprehensively finding all (not just one or a few) Pareto-optimal solutions.

49.6.10 EMO with Meta-Models

The practice of optimization algorithms is often limited by the computational overheads associated with evaluating solutions. Certain problems involve expensive computations, such as numerical solution of partial differential equations describing the physics of the problem, finite difference computations involving an analysis of a solution, computational fluid dynamics simulation to study the performance of a solution over a changing environment, etc. In some such problems, evaluation of each solution to compute constraints and objective functions may take a few hours to a complete day or two. In such scenarios, even if an optimization algorithm needs one hundred solutions to get anywhere close to a good and feasible solution, the application needs an easy 3 to 6 months of continuous computational time. In most practical purposes, this is considered a *luxury* in an industrial set-up. Optimization researchers are constantly on their toes in coming up with approximate, yet faster algorithms.

A little thought brings out an interesting fact about how optimization algorithms work. The initial iterations deal with solutions which may not be close to optimal solutions. Therefore, these solutions need not be evaluated with high precision. Meta-models for objective functions and constraints have been developed for this purpose. Mostly two different approaches are followed. In one approach, a sample of solutions is used to generate a meta-model (an approximate model of the original objectives and constraints), and then efforts are made to find the optimum of the meta-model, assuming that the optimal solutions of both the meta-model and the original problem are similar to each other [49.124, 125]. In another method, a successive meta-modeling approach is used in which the algorithm starts to solve the first meta-model obtained from a sample of the entire search space [49.126–128]. As the solutions start to focus near the optimum region of the meta-model, a new and more accurate meta-model is generated in the region dictated by the solutions of the previous optimization. A coarse-to-fine-grained meta-modeling technique based on artificial neural networks is shown to reduce the computational effort by about 30 to 80% on different problems [49.126]. Other successful meta-modeling implementations for multi-objective optimization are based on Kriging and response surface methodologies exist [49.128, 129].

49.7 Conclusions

The research and application in evolutionary multi-objective optimization (EMO) over the past 15 years have resulted in a number of efficient algorithms for finding a set of well-diversified, near Pareto-optimal solutions. EMO algorithms are now regularly being applied to different problems in most areas of science, engineering, and commerce. This chapter has discussed the principles of EMO and illustrated the principle by depicting one efficient and popularly used EMO algorithm. Results from an inter-planetary spacecraft trajectory optimization problem reveal the importance of the principles followed in EMO algorithms. Thereafter, a specific constraint handling procedure used in EMO studies was briefly described.

The main highlight of this chapter has been the description of some of the current research and application activities in EMO. One critical area of current research lies in collaborative EMO-MCDM algorithms for achieving a complete multi-objective optimization task of finding a set of trade-off solutions and finally arriving at a single preferred solution. Another direction taken by researchers is to address guaranteed convergence and diversity of EMO algorithms through hybridizing them with mathematical and numerical optimization techniques as local search algorithms. Interestingly, EMO researchers have discovered its potential in solving traditionally hard optimization problems, but not necessarily multi-objective ones in nature, in a convenient manner using EMO algorithms. So-called multi-objectivization studies are attracting researchers from various fields to develop and apply EMO algorithms in many innovative ways. Considerable interest n research and application has also been shown in addressing practical aspects in existing EMO algorithms. In this direction, handling uncertainty in decision variables and parameters, meeting an overall desired system reliability in obtained solutions, handling dynamically changing problems (on-line optimization), and handling a large number of objectives have been discussed in this chapter. Besides the practical aspects, EMO has also attracted mathematically-oriented theoreticians to develop EMO algorithms and design suitable problems for coming up with a computational complexity analysis. There are many other research directions which could not even mention due to space restrictions.

In the short span of about 15 years, it has become clear that the field of EMO research and application now has efficient algorithms and numerous interesting

and useful applications, and has been able to attract theoretically and practically-oriented researchers to come together and collaborate. The practical importance of EMO's working principle, the flexibility of evolutionary optimization, which lies at the core of EMO algorithms, and the demonstrated diversification of EMO's principle to a wide variety of different problem-solving tasks are the main cornerstones for their success so far. The scope of research and application in EMO and using EMO are enormous and open-ended. This chapter remains an open invitation to everyone who is interested in any type of problem-solving tasks to take a look at what has been done in EMO and to explore how one can contribute to collaborating with EMO to address problem-solving tasks that are still in need of a better solution procedure.

References

49.1   D.E. Goldberg: *Genetic Algorithms for Search, Optimization, and Machine Learning* (Addison-Wesley, Reading 1989)

49.2   J.H. Holland: *Adaptation in Natural and Artificial Systems* (MIT, Ann Arbor 1975)

49.3   K.A. De Jong: *Evolutionary Computation: A Unified Approach* (MIT, Cambridge 2006)

49.4   P.J.M. Laarhoven, E.H.L. Aarts: *Simulated Annealing: Theory and Applications* (Springer, Heidelberg 1987)

49.5   F. Glover: Tabu search – Part 1, ORSA J. Comput. **1**(2), 190–206 (1989)

49.6   F. Glover: Tabu search – Part 2, ORSA J. Comput. **2**(1), 4–32 (1990)

49.7   B.S.W. Schröder: *Ordered Sets: An Introduction* (Birkhäuser, Boston 2003)

49.8   K. Deb: *Multi-Objective Optimization Using Evolutionary Algorithms* (Wiley, Chichester 2001)

49.9   K. Miettinen: *Nonlinear Multiobjective Optimization* (Kluwer, Boston 1999)

49.10   H.T. Kung, F. Luccio, F.P. Preparata: On finding the maxima of a set of vectors, J. Assoc. Comput. Mach. **22**(4), 469–476 (1975)

49.11   J. Jahn: *Vector Optimization* (Springer, Berlin 2004)

49.12   G. Rudolph: On a multi-objective evolutionary algorithm and its convergence to the Pareto set, Proc. 5th IEEE Conf. Evol. Comput. (1998) pp. 511–516

49.13   G. Rudolph, A. Agapie: Convergence properties of some multi-objective evolutionary algorithms, Proc. 2000 Congr. Evol. Comput. (CEC2000) (2000) pp. 1010–1016

49.14   O. Schütze, M. Laumanns, C.A.C. Coello, M. Dellnitz, E.-G. Talbi: Convergence of stochastic search algorithms to finite size Pareto set approximations, J. Glob. Optim. **41**(4), 559–577 (2008)

49.15   O. Schütze, M. Laumanns, E. Tantar, C.A.C. Coello, E.-G. Talbi: Computing gap-free Pareto front approximations with stochastic search algorithms, Evol. Comput. J. **18**(1), 65–96 (2010)

49.16   K. Deb, R. Tiwari, M. Dixit, J. Dutta: Finding trade-off solutions close to KKT points using evolutionary multi-objective optimization, Proc. Congr. Evol. Comput. (CEC-2007) (2007) pp. 2109–2116

49.17   P. Shukla, K. Deb: On finding multiple Pareto-optimal solutions using classical and evolutionary generating methods, Eur. J. Oper. Res. (EJOR) **181**(3), 1630–1652 (2007)

49.18   R.S. Rosenberg: Simulation of Genetic Populations with Biochemical Properties, Ph.D. Thesis (University of Michigan, Ann Arbor 1967)

49.19   L.J. Fogel, A.J. Owens, M.J. Walsh: *Artificial Intelligence Through Simulated Evolution* (Wiley, New York 1966)

49.20   J.D. Schaffer: Some Experiments in Machine Learning Using Vector Evaluated Genetic Algorithms, Ph.D. Thesis (Vanderbilt University, Nashville 1984)

49.21   D.E. Goldberg, J. Richardson: Genetic algorithms with sharing for multimodal function optimization, Proc. First Int. Conf. Genet. Algorithms Their Appl. (1987) pp. 41–49

49.22   C.M. Fonseca, P.J. Fleming: Genetic algorithms for multiobjective optimization: Formulation, discussion, and generalization, Proc. Fifth Int. Conf. Genet. Algorithms (1993) pp. 416–423

49.23   J. Horn, N. Nafploitis, D.E. Goldberg: A niched Pareto genetic algorithm for multi-objective optimization, Proc. First IEEE Conf. Evol. Comput. (1994) pp. 82–87

49.24   N. Srinivas, K. Deb: Multi-objective function optimization using non-dominated sorting genetic algorithms, Evol. Comput. J. **2**(3), 221–248 (1994)

49.25   C. Poloni: Hybrid GA for multi-objective aerodynamic shape optimization. In: *Genetic Algorithms in Engineering and Computer Science*, ed. by G. Winter, J. Periaux, M. Galan, P. Cuesta (Wiley, Chichester 1997) pp. 397–414

49.26   G. Rudolph: Convergence analysis of canonical genetic algorithms, IEEE Trans. Neural Netw. **5**(1), 96–101 (1994)

49.27   K. Deb, S. Agrawal, A. Pratap, T. Meyarivan: A fast and elitist multi-objective genetic algorithm: NSGA-II, IEEE Trans. Evol. Comput. **6**(2), 182–197 (2002)

49.28   E. Zitzler, L. Thiele: Multiobjective evolutionary algorithms: A comparative case study and the strength Pareto approach, IEEE Trans. Evol. Comput. **3**(4), 257–271 (1999)

49.29   J.D. Knowles, D.W. Corne: Approximating the non-dominated front using the Pareto archived evolution strategy, Evol. Comput. J. **8**(2), 149–172 (2000)

49.30   K. Deb, R.B. Agrawal: Simulated binary crossover for continuous search space, Complex Syst. **9**(2), 115–148 (1995)

49.31   E. Zitzler, M. Laumanns, L. Thiele: SPEA2: Improving the strength Pareto evolutionary algorithm for multiobjective optimization. In: *Evolutionary Methods for Design Optimization and Control with Applications to Industrial Problems*, ed. by K.C. Giannakoglou, D.T. Tsahalis, J. Périaux, K.D. Papailiou, T. Fogarty (CIMNE, Athens 2001) pp. 95–100

49.32   D.W. Corne, J.D. Knowles, M. Oates: The Pareto envelope-based selection algorithm for multiobjective optimization, Proc. Sixth Int. Conf. Parallel Probl. Solving Nat. VI (PPSN-VI) (2000) pp. 839–848

49.33   D. Van Veldhuizen, G.B. Lamont: Multiobjective evolutionary algorithms: Analyzing the state-of-the-art, Evol. Comput. J. **8**(2), 125–148 (2000)

49.34   C.A.C. Coello, G. Toscano: A Micro-Genetic Algorithm for Multi-Objective Optimization, Technical Report Lania-RI-2000-06 (Laboratoria Nacional de Informatica Avanzada, Xalapa 2000)

49.35   D.H. Loughlin, S. Ranjithan: The neighborhood constraint method: A multiobjective optimization technique, Proc. Seventh Int. Conf. Genet. Algorithms (1997) pp. 666–673

49.36   D. Sasaki, M. Morikawa, S. Obayashi, K. Nakahashi: Aerodynamic shape optimization of supersonic wings by adaptive range multiobjective genetic algorithms, Proc. First Int. Conf. Evol. Multi-Criterion Optim. (EMO 2001) (2001) pp. 639–652

49.37   C.A.C. Coello, M.S. Lechuga: MOPSO: A proposal for multiple objective particle swarm optimization, Congr. Evol. Comput. (CEC'2002), Vol. 2 (IEEE Service Center, Piscataway 2002) pp. 1051–1056

49.38   S. Mostaghim, J. Teich: Strategies for finding good local guides in multi-objective particle swarm optimization (MOPSO), 2003 IEEE Swarm Intell. Symp. Proc. (IEEE Service Center, Indianapolis 2003) pp. 26–33

49.39   P.R. McMullen: An ant colony optimization approach to addessing a JIT sequencing problem with multiple objectives, Artifi. Intell. Eng. **15**, 309–317 (2001)

49.40   M. Gravel, W.L. Price, C. Gagné: Scheduling continuous casting of aluminum using a multiple objective ant colony optimization metaheuristic, Eur. J. Oper. Res. **143**(1), 218–229 (2002)

49.41   B.V. Babu, M.L. Jehan: Differential evolution for multi-objective optimization, Proc. 2003 Congr. Evol. Comput. (CEC'2003), Vol. 4 (IEEE, Canberra 2003) pp. 2696–2703

49.42   S. Bandyopadhyay, S. Saha, U. Maulik, K. Deb: A simulated annealing-based multiobjective optimization algorithm: Amosa, IEEE Trans. Evol. Comput. **12**(3), 269–283 (2008)

49.43   M.P. Hansen: Tabu search in multiobjective optimization: MOTS, Thirteenth Int. Conf. Multi-Criterion Decis. Mak. (MCDM'97) (University of Cape Town, Cape Town 1997)

49.44   C.A.C. Coello, D.A. VanVeldhuizen, G. Lamont: *Evolutionary Algorithms for Solving Multi-Objective Problems* (Kluwer, Boston 2002)

49.45   C.A.C. Coello, G.B. Lamont: *Applications of Multi-Objective Evolutionary Algorithms* (World Scientific, Singapore 2004)

49.46   A. Osyczka: *Evolutionary Algorithms for Single and Multicriteria Design Optimization* (Physica, Heidelberg 2002)

49.47   K.C. Tan, E.F. Khor, T.H. Lee: *Multiobjective Evolutionary Algorithms and Applications* (Springer, London 2005)

49.48   E. Zitzler, K. Deb, L. Thiele, C.A.C. Coello, D.W. Corne: *Evolutionary Multi-Criterion Optimization, 1st International Conference (EMO-2001)*, Lecture Notes in Computer Science, Vol. 1993 (Springer, Heidelberg 2001)

49.49   C.M. Fonseca, P. Fleming, E. Zitzler, K. Deb, L. Thiele: *Evolutionary Multi-Criterion Optimization, 2nd International Conference, (EMO-2003)*, Lecture Notes in Computer Science, Vol. 2632 (Springer, Heidelberg 2003)

49.50   C.A.C. Coello, A.H. Aguirre, E. Zitzler: *Evolutionary Multi-Criterion Optimization, 3rd International Conference (EMO-2005)*, Lecture Notes in Computer Science, Vol. 3410 (Springer, Heidelberg 2005)

49.51   S. Obayashi, K. Deb, C. Poloni, T. Hiroyasu, T. Murata: *Evolutionary Multi-Criterion Optimization, 4th International Conference (EMO-2007)*, Lecture Notes in Computer Science, Vol. 4403 (Springer, Heidelberg 2007)

49.52   M. Ehrgott, C.M. Fonseca, X. Gandibleux, J.-K. Hao, M. Sevaux: *Evolutionary Multi-Criterion Optimization, 5th International Conference (EMO-2009)*, Lecture Notes in Computer Science, Vol. 5467 (Springer, Heidelberg 2009)

49.53   R.H.C. Takahashi, K. Deb, E.F. Wanner, S. Greco: *Evolutionary Multi-Criterion Optimization, 6th International Conference (EMO-2011)*, Lecture Notes in Computer Science, Vol. 6576 (Springer, Heidelberg 2011)

49.54   C. A. Coello: List of references on evolutionary multiobjective optimization (emo), http://www.lania.mx/~ccoello/EMOO/EMOObib.html

49.55   V. Coverstone-Carroll, J.W. Hartmann, W.J. Mason: Optimal multi-objective low-thurst space-craft trajectories, Comput. Meth. Appl. Mech. Eng. **186**(2–4), 387–402 (2000)

49.56   C.G. Sauer: Optimization of multiple target electric propulsion trajectories, AIAA 11th Aerosp. Sci. Meet. (1973), Paper Number 73–205

49.57   K. Deb, T. Goel: A hybrid multi-objective evolutionary approach to engineering shape design, Proc. First Int. Conf. Evol. Multi-Criterion Optim. (EMO-01) (2001) pp. 385–399

49.58   K. Sindhya, K. Deb, K. Miettinen: A local search based evolutionary multi-objective optimization technique for fast and accurate convergence, Proc. Parallel Probl. Solving Nat. (PPSN-2008) (Springer, Berlin 2008)

49.59   H. Jin, M.-L. Wong: Adaptive diversity maintenance and convergence guarantee in multiobjective evolutionary algorithms, Proc. Congr. Evol. Comput. (CEC-2003) (2003) pp. 2498–2505

49.60   Z.M. Saul, C.A.C. Coello: A proposal to hybridize multi-objective evolutionary algorithms with non-gradient mathematical programming techniques, Proc. Parallel Probl. Solving Nat. (PPSN-2008) (2008) pp. 837–846

49.61   M. Fleischer: The measure of Pareto optima: Applications to multi-objective optimization, Proc. Second Int. Conf. Evol. Multi-Criterion Optim. (EMO-2003) (Springer, Berlin 2003) pp. 519–533

49.62   L. While, P. Hingston, L. Barone, S. Huband: A faster algorithm for calculating hypervolume, IEEE Trans. Evol. Comput. **10**(1), 29–38 (2006)

49.63   L. Bradstreet, L. While, L. Barone: A fast incremental hypervolume algorithm, IEEE Trans. Evol. Comput. **12**(6), 714–723 (2008)

49.64   M. Laumanns, L. Thiele, K. Deb, E. Zitzler: Combining convergence and diversity in evolutionary multi-objective optimization, Evol. Comput. **10**(3), 263–282 (2002)

49.65   P.A.N. Bosman, D. Thierens: The balance between proximity and diversity in multiobjective evolutionary algorithms, IEEE Trans. Evol. Comput. **7**(2), 174–188 (2003)

49.66   C.A.C. Coello: Treating objectives as constraints for single objective optimization, Eng. Optim. **32**(3), 275–308 (2000)

49.67   S. Bleuler, M. Brack, E. Zitzler: Multiobjective genetic programming: Reducing bloat using SPEA2, Proc. 2001 Congr. Evol. Comput. (2001) pp. 536–543

49.68   E.D. De Jong, R.A. Watson, J.B. Pollack: Reducing bloat and promoting diversity using multi-objective methods, Proc. Genet. Evol. Comput. Conf. (GECCO-2001) (2001) pp. 11–18

49.69   J. Handl, J.D. Knowles: An evolutionary approach to multiobjective clustering, IEEE Trans. Evol. Comput. **11**(1), 56–76 (2007)

49.70   F. Neumann, I. Wegener: Minimum spanning trees made easier via multi-objective optimization, GECCO'05: Proc. 2005 Conf. Genetic Evol. Comput. (ACM, New York 2005) pp. 763–769

49.71   J.D. Knowles, D.W. Corne, K. Deb: *Multiobjective Problem Solving from Nature*, Springer Natural Computing Series (Springer, Heidelberg 2008)

49.72   K. Deb, S. Gupta, D. Daum, J. Branke, A. Mall, D. Padmanabhan: Reliability-based optimization using evolutionary algorithms, IEEE Trans. Evol. Comput. **13**(5), 1054–1074 (2009)

49.73   K. Deb, H. Gupta: Introducing robustness in multi-objective optimization, Evol. Comput. J. **14**(4), 463–494 (2006)

49.74   M. Basseur, E. Zitzler: Handling uncertainty in indicator-based multiobjective optimization, Int. J. Comput. Intell. Res. **2**(3), 255–272 (2006)

49.75   T.R. Cruse: *Reliability-Based Mechanical Design* (Marcel Dekker, New York 1997)

49.76   X. Du, W. Chen: Sequential optimization and reliability assessment method for efficient probabilistic design, ASME Trans. J. Mech. Des. **126**(2), 225–233 (2004)

49.77   J. Branke, K. Deb, H. Dierolf, M. Osswald: Finding knees in multi-objective optimization, Lect. Notes Comput. Sci. **3242**, 722–731 (2004)

49.78   A.P. Wierzbicki: The use of reference objectives in multiobjective optimization. In: *Multiple Criteria Decision Making Theory and Applications*, ed. by G. Fandel, T. Gal (Springer, Berlin 1980) pp. 468–486

49.79   P. Korhonen, J. Laakso: A visual interactive method for solving the multiple criteria problem, Eur. J. Oper. Res. **24**, 277–287 (1986)

49.80   J. Branke, K. Deb, K. Miettinen, R. Slowinski: *Multiobjective Optimization: Interactive and Evolutionary Approaches* (Springer, Berlin 2008)

49.81   K. Deb, J. Sundar, N. Uday, S. Chaudhuri: Reference point based multi-objective optimization using evolutionary algorithms, Int. J. Comput. Intell. Res. (IJCIR) **2**(6), 273–286 (2006)

49.82   K. Deb, A. Kumar: Light beam search based multi-objective optimization using evolutionary algorithms, Proc. Congr. Evol. Comput. (CEC-07) (2007) pp. 2125–2132

49.83   K. Deb, A. Kumar: Interactive evolutionary multi-objective optimization and decision-making using reference direction method, Proc. Genet. Evol.

49.84 L. Thiele, K. Miettinen, P. Korhonen, J. Molina: A Preference-Based Interactive Evolutionary Algorithm for Multiobjective Optimization, Technical Report Working Paper W-412 (Helsingin School of Economics, Helsingin Kauppakorkeakoulu 2007)

49.85 M. Luque, K. Miettinen, P. Eskelinen, F. Ruiz: Incorporating preference information in interactive reference point based methods for multiobjective optimization, Omega **37**(2), 450–462 (2009)

49.86 V. Khare, X. Yao, K. Deb: Performance scaling of multi-objective evolutionary algorithms, Lect. Notes Comput. Sci. **2632**, 376–390 (2003)

49.87 J. Knowles, D. Corne: Quantifying the effects of objective space dimension in evolutionary multiobjective optimization, Lect. Notes Comput. Sci. **4403**, 757–771 (2007)

49.88 J.A. López, C.A.C. Coello: Some techniques to deal with many-objective problems, Proc. 11th Annu. Conf. Companion Genet. Evol. Comput. Conf. (ACM, New York 2009) pp. 2693–2696

49.89 E.J. Hughes: Evolutionary many-objective optimisation: Many once or one many?, IEEE Congr. Evol. Comput. (CEC-2005) (2005) pp. 222–227

49.90 D.K. Saxena, J.A. Duro, A. Tiwari, K. Deb, Q. Zhang: Objective reduction in many-objective optimization: Linear and nonlinear algorithms, IEEE Trans. Evol. Comput. **17**(1), 77–99 (2013)

49.91 K. Deb, H. Jain: *An Improved NSGA-II Procedure for Many-Objective Optimization, Part I: Problems with Box Constraints*, Tech. Rep. KanGAL Report, Vol. 2012009 (Indian Institute of Technology, Kanpur 2012)

49.92 K. Deb, H. Jain: *An Improved NSGA-II Procedure for Many-Objective Optimization, Part II: Handling Constraints and Extending to an Adaptive Approach*, Tech. Rep. KanGAL Report, Vol. 2012010 (Indian Institute of Technology, Kanpur 2012)

49.93 K. Deb, H. Jain: Handling many-objective problems using an improved NSGA-II procedure, Proc. World Congr. Comput. Intell. (WCCI-2012) (2012)

49.94 Q. Zhang, H. Li: MOEA/D: A multiobjective evolutionary algorithm based on decomposition, Evol. Comput. IEEE Trans. **11**(6), 712–731 (2007)

49.95 J. Branke, K. Deb: Integrating user preferences into evolutionary multi-objective optimization. In: *Knowledge Incorporation in Evolutionary Computation*, ed. by Y. Jin (Springer, Heidelberg 2004) pp. 461–477

49.96 K. Deb, P. Zope, A. Jain: Distributed computing of pareto-optimal solutions using multi-objective evolutionary algorithms, Lect. Notes Comput. Sci. **2632**, 535–549 (2003)

49.97 K. Deb, D. Saxena: Searching for Pareto-optimal solutions through dimensionality reduction for certain large-dimensional multi-objective optimization problems, Proc. World Congr.

49.84 Comput. Conf. (GECCO-2007) (ACM, New York 2007) pp. 781–788

49.98 D.K. Saxena, K. Deb: Non-linear dimensionality reduction procedures for certain large-dimensional multi-objective optimization problems Employing correntropy and a novel maximum variance unfolding, Proc. Fourth Int. Conf. Evol. Multi-Criterion Optim. (EMO-2007) (2007) pp. 772–787

49.99 D. Brockhoff, E. Zitzler: Dimensionality reduction in multiobjective optimization: The minimum objective subset problem. In: *Operations Research Proceedings 2006*, ed. by K.H. Waldmann, U.M. Stocker (Springer, Heidelberg 2007) pp. 423–429

49.100 D. Brockhoff, E. Zitzler: *Offline and Online Objective Reduction in Evolutionary Multiobjective Optimization Based on Objective Conflicts*, TIK Report, Vol. 269 (Institut für Technische Informatik und Kommunikationsnetze, ETH Zürich 2007)

49.101 M. Farina, P. Amato: A fuzzy definition of optimality for many criteria optimization problems, IEEE Trans. Syst., Man Cybern. Part A: Syst, Hum. **34**(3), 315–326 (2004)

49.102 K. Deb, A. Srinivasan: Innovization: Innovating design principles through optimization, Proc. Genet. Evol. Comput. Conf. (GECCO-2006) (ACM, New York 2006) pp. 1629–1636

49.103 S. Bandaru, K. Deb: Towards automating the discovery of certain innovative design principles through a clustering based optimization technique, Eng. Optim. **43**(9), 1–941 (2011)

49.104 S. Bandaru, K. Deb: Automated innovization for simultaneous discovery of multiple rules in bi-objective problems, Proc. Sixth Int. Conf. Evol. Multi-Criterion Optim. (EMO-2011) (Springer, Heidelberg 2011) pp. 1–15

49.105 J. Branke: *Evolutionary Optimization in Dynamic Environments* (Springer, Heidelberg 2001)

49.106 K. Deb, U.B. Rao, S. Karthik: Dynamic multi-objective optimization and decision-making using modified NSGA-II: A case study on hydro-thermal power scheduling bi-objective optimization problems, Proc. Fourth Int. Conf. Evol. Multi-Criterion Optim. (EMO-2007) (2007)

49.107 K. Deb: Multi-objective genetic algorithms: Problem difficulties and construction of test problems, Evol. Comput. J. **7**(3), 205–230 (1999)

49.108 K. Deb, L. Thiele, M. Laumanns, E. Zitzler: Scalable test problems for evolutionary multi-objective optimization. In: *Evolutionary Multiobjective Optimization*, ed. by A. Abraham, L. Jain, R. Goldberg (Springer, London 2005) pp. 105–145

49.109 S. Huband, L. Barone, L. While, P. Hingston: A scalable multi-objective test problem toolkit, Proc. Evol. Multi-Criterion Optim. (EMO-2005) (Springer, Berlin 2005)

49.110 T. Okabe, Y. Jin, M. Olhofer, B. Sendhoff: On test functions for evolutionary multi-objective optimization, Parallel Problem Solving from Nature (PPSN VIII) (2004) pp. 792–802

9.111   J.D. Knowles, D.W. Corne: On metrics for comparing nondominated sets, Congr. Evol. Comput. (CEC-2002) (IEEE, Piscataway 2002) pp. 711–716

9.112   M. P. Hansen, A. Jaskiewicz: Evaluating the Quality of Aapproximations to the Non-Dominated Set, Technical Report IMM-REP-1998-7 (Institute of Mathematical Modelling, Technical University of Denmark, Lyngby 1998)

9.113   E. Zitzler, L. Thiele, M. Laumanns, C.M. Fonseca, V.G. Fonseca: Performance assessment of multiobjective optimizers: An analysis and review, IEEE Trans. Evol. Comput. 7(2), 117–132 (2003)

9.114   E. Zitzler, S. Künzli: Indicator-based selection in multiobjective search, Lect. Notes Comput. Sci. 3242, 832–842 (2004)

9.115   C.M. Fonseca, P.J. Fleming: On the performance assessment and comparison of stochastic multiobjective optimizers. In: *Parallel Problem Solving from Nature (PPSN IV)*, ed. by H.-M. Voigt, W. Ebeling, I. Rechenberg, H.-P. Schwefel (Springer, Berlin 1996), pp. 584–593, Also available as Lecture Notes in Computer Science Vol. 1141

9.116   C.M. Fonseca, V. da Grunert Fonseca, L. Paquete: Exploring the performance of stochastic multiobjective optimisers with the second-order attainment function, Third Int. Conf. Evol. Multi-Criterion Optim. (EMO-2005) (Springer, Berlin 2005) pp. 250–264

9.117   A. Auger, J. Bader, D. Brockhoff: Theoretically investigating optimal μ-distributions for the hypervolume indicator: First results for three objectives, Lect. Notes Comput. Sci. 6238, 586–596 (2010)

9.118   J. Bader, K. Deb, E. Zitzler: Faster hypervolume-based search using monte carlo sampling, Lect. Notes Econ. Math. Syst. 634, 313–326 (2010)

9.119   M. Laumanns, L. Thiele, E. Zitzler, E. Welzl, K. Deb: Running time analysis of multi-objective evolutionary algorithms on a simple discrete optimization problem, Proc. Seventh Conf. Parallel Probl. Solving Nat. (PPSN-VII) (2002) pp. 44–53

49.120   O. Giel: Expected runtimes of a simple multi-objective evolutionary algorithm, Proc. Congr. Evol. Comput. (CEC-2003) (IEEE, Piscatway 2003) pp. 1918–1925

49.121   M. Laumanns, L. Thiele, E. Zitzler: Running time analysis of multiobjective evolutionary algorithms on pseudo-Boolean functions, IEEE Trans. Evol. Comput. 8(2), 170–182 (2004)

49.122   O. Giel, P.K. Lehre: On the effect of populations in evolutionary multi-objective optimization, Proc. 8th Annu. Genet. Evol. Comput. Conf. (GECCO 2006) (ACM, New York 2006) pp. 651–658

49.123   R. Kumar, N. Banerjee: Analysis of a multiobjective evolutionary algorithm on the 0-1 knapsack problem, Theor. Comput. Sci. 358(1), 104–120 (2006)

49.124   M.A. El-Beltagy, P.B. Nair, A.J. Keane: Metamodelling techniques for evolutionary optimization of computationally expensive problems: Promises and limitations, Proc. Genet. Evol. Comput. Conf. (GECCO-1999) (Morgan Kaufman, San Mateo 1999) pp. 196–203

49.125   K.C. Giannakoglou: Design of optimal aerodynamic shapes using stochastic optimization methods and computational intelligence, Prog. Aerosp. Sci. 38(1), 43–76 (2002)

49.126   P.K.S. Nain, K. Deb: Computationally effective search and optimization procedure using coarse to fine approximations, Proc. Congr. Evol. Comput. (CEC-2003) (2003) pp. 2081–2088

49.127   K. Deb, P.K.S. Nain: *An Evolutionary Multi-Objective Adaptive Meta-Modeling Procedure Using Artificial Neural Networks* (Springer, Berlin 2007) pp. 297–322

49.128   M. Emmerich, K.C. Giannakoglou, B. Naujoks: Single and multiobjective evolutionary optimization assisted by gaussian random field metamodels, IEEE Trans. Evol. Comput. 10(4), 421–439 (2006)

49.129   M. Emmerich, B. Naujoks: *Metamodel-assisted multiobjective optimisation strategies and their application in airfoil design*, Adaptive Computing in Design and Manufacture VI (Springer, London 2004) pp. 249–260

50. Parallel Multiobjective Evolutionary Algorithms

Francisco Luna, Enrique Alba

The use of evolutionary algorithms (EAs) for solving multiobjective optimization problems has been very active in the last few years. The main reasons for this popularity are their ease of use with respect to classical mathematical programming techniques, their scalability, and their suitability for finding trade-off solutions in a single run. However, these algorithms may be computationally expensive because (1) many real-world optimization problems typically involve tasks demanding high computational resources and (2) they are aimed at finding a whole front of optimal solutions instead of searching for a single optimum. Parallelizing EAs emerges as a possible way of reducing the CPU time down to affordable values, but it also allows researchers to use an advanced search engine – the parallel model – that provides the algorithms with an improved population diversity and enable them to cooperate with other (eventually nonevolutionary) techniques. The goal of this chapter is to provide the reader with an up-

to-date review of the recent literature on parallel EAs for multiobjective optimization.

50.1 Multiobjective Optimization and Parallelism

Multiobjective optimization arises in many real-world applications, especially in engineering, in which several performance criteria conflict with each other. These conflicting objectives make the optimization results in that no single solution can usually optimize them all simultaneously. Indeed, the aim of multiobjective optimization is to find a set of compromise solutions with different tradeoffs among criteria, also known as the *Pareto optimal set*. When this set is plotted in the objective space it is called the *Pareto front* [50.1, 2].

Many different techniques have been proposed in the multiobjective research community to address multiobjective optimization problems (MOPs). Unlike classical mathematical programming approaches, meta-

heuristics in general, and EAs (multiobjective evolutionary algorithms or MOEAs) in particular, have attracted growing attention over the last decade because of two main facts. On the one hand, EAs have the ability to generate several members of the Pareto optimal set in one single run, as opposed to classical multicriteria decision-making techniques. They are also less sensitive to the shape of the Pareto front so therefore can deal with a large variety of MOPs. On the other hand, as randomized black-box algorithms, EAs can address optimization problems with nonlinear, nondifferentiable, or noisy objective functions.

In spite of these advantages, these algorithms might be computationally expensive because, on the one hand, they need to explore larger portions of the search space since they seek the entire Pareto front, which usually results in more function evaluations being performed; on the other hand, and even more importantly, many real-world multiobjective problems typically use computationally expensive methods for computing the objective functions and constraints.

These issues are usually addressed in two different ways. First, one can use surrogate models of the fitness functions instead of true fitness function evaluations [50.3–5]. The second more important line lies in using parallel computing platforms to speed up the EA search [50.6]. This is the mainstream of this chapter.

Due to their population-based approach, EAs are very suitable for parallelization because their main operations (i.e., crossover, mutation, and in particular function evaluation) can be carried out independently on different individuals. There is a vast amount of literature on how to parallelize EAs; the reader is referred to [50.7–10] for surveys on this topic. However, parallelism here is not only a way for solving problems more rapidly, but also for developing new and more efficient search models: a parallel EA can be more effective than a sequential one, even when executed on a single processor. The advantages that parallelism offer to single-objective optimization also hold in multiobjective optimization. Of particular interest for parallel MOEAs is the improvement in population diversity that shall help to fully approximate the entire Pareto front of the given optimization problems.

The contribution of this chapter is to provide the reader with a recent review of publications related to parallel MOEAs, showing the latest advances in the field. Given the shear volume of papers, we have been forced to restrict ourselves to only those works which have been published since 2008/09, the years to which the two best known surveys date back [50.11, 12]. The structure of this chapter distinguishes between the numerical model of the parallel MOEA and its physical parallelization. In seminal papers in the fields [50.13], it was assumed that the model maps directly onto the parallel computing platform, but this is no longer true and any (MO)EA can be deployed in parallel, but not always resulting in a high performance. The next section is therefore devoted to presenting the classical models for parallel EAs and a recent proposal that not only considers EAs, but metaheuristics and exact algorithms in general. Section 50.3 dives into the details of more than 80 publications, analyzing particular features of MOEAs (fitness assignment, diversity preservation) as well as on their parallelization (model, topology, parallel platform). Finally, the last section presents the main conclusions and the trends for future research on parallel MOEAs.

50.2 Parallel Models for Evolutionary Multi-Objective Algorithms

Parallelism arises naturally when dealing with populations of individuals, since each individual is an independent unit. As a consequence, the performance of population-based algorithms is particularly improved when run in parallel. The main models for parallel MOEAs have been proposed within two clear scopes: especially EA-targeted models coming from the EA community [50.7, 14, 15] and those proposed for parallel metaheuristics in general (of which EAs are a subclass) [50.12, 16]. They are briefly presented in the following subsections.

50.2.1 Specialized Models for Parallel EAs

The most well-known models for parallel MOEAs have been inherited directly from the single-objective parallel EA community, in which two parallelizing strategies are defined for population-based algorithms: (1) parallelization of computation, in which the operations commonly applied to each individual are performed in parallel, and (2) parallelization of population, in which the population is split into different parts, each one evolving in semi-isolation (individuals can be exchanged between subpopulations).

The simplest parallelization scheme of EAs is the well-known *master–slave* or *global parallelization* method (Fig. 50.1a). In this scheme, a central processor performs the selection operations while the associated slave processors perform the recombination, mutation, and/or the evaluation of the fitness function. This algorithm is the same as the traditional (one population, panmictic), although it is faster, especially for time-consuming objective functions. Its simplicity has made it the most popular among practitioners.

However, other models for parallel EAs utilize some kind of spatial disposition of the individuals (it is said that the population is then *structured*), and afterward parallelize the resulting chunks in a pool of proces-

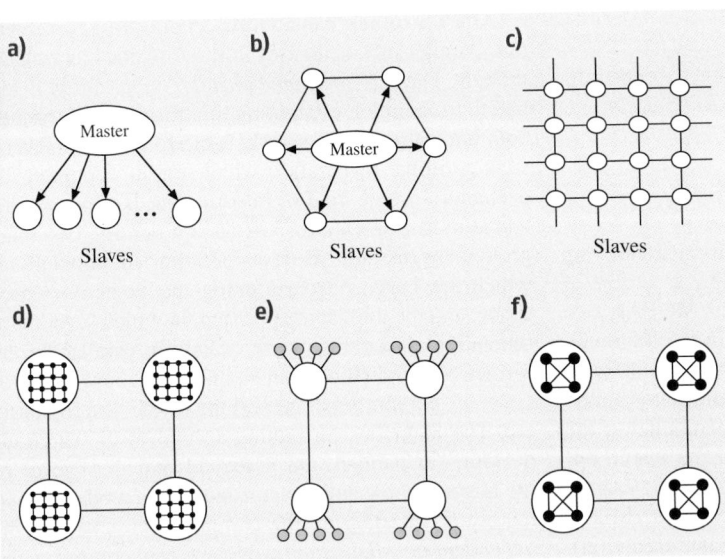

Fig. 50.1a-f Different models of parallel EAs: (**a**) global parallelization, (**b**) coarse grain, and (**c**) fine grain. Many hybrids have been defined by combining parallel EAs at two levels: (**d**) coarse and fine grain, (**e**) coarse grain and global parallelization, and (**f**) coarse grain at the two levels

sors. Among the most widely known types of structured EAs, the *distributed* (dEA) (or coarse-grain) and *cellular* (CEA) (fine-grain or diffusion) algorithms are very popular optimization procedures [50.7]. In the case of distributed EAs (Fig. 50.1b), the population is partitioned into a set of islands in which isolated EAs run in parallel. Sparse individual exchanges are performed among these islands, with the goal of inserting some diversity into the subpopulations, thus avoiding them getting stuck in local optima. Islands may apply the same (homogeneous) or different (heterogeneous) EAs [50.17]. In the case of a cellular EA (Fig. 50.1c), subpopulations are typically composed of one individual, which may only interact with its nearest neighbors in the breeding loop, i. e., the concept of *neighborhood* is introduced. These neighborhoods are overlapped, which implicitly defines a migration mechanism and allows a smooth diffusion of the best solutions throughout the population. This parallel scheme was targeted to massively parallel computers but nowadays it can be used sequentially on a regular computer or in parallel on graphic processing units (GPUs). Also, hybrid models have been proposed (Fig. 50.1d–f) in which a two-level approach of parallelization is undertaken. In these models, the higher level for parallelization uses to be a coarse-grain implementation and the basic island performs a CEA, a master–slave method, or even another distributed one.

This taxonomy holds as well for parallel MOEAs [50.15], so we can consider master–slave MOEAs (*msMOEAs*), distributed MOEAs (*dMOEAs*),

and cellular MOEAs (*cMOEAs*). Nevertheless, these two decentralized population approaches need a further particularization for MOPs [50.14]. As we stated before, the main goal of any multiobjective optimization algorithm is to find the optimal Pareto front for a given MOP. It is clear that in msMOEAs the management of this Pareto front is carried out by the master processor. But, when the search process is distributed among different subalgorithms, as happens in dMOEAs and cMOEAs, the management of the nondominated set of solutions during the optimization procedure becomes a capital issue. Hence, it can be distinguished when the Pareto front is distributed and locally managed by each sub-EA during the computation, or it is a centralized element of the algorithm. They have been called *centralized Pareto front* (CPF) structured MOEAs and *Distributed Pareto Front* (DPF) structured MOEAs, respectively [50.16].

For distributed MOEAs, very specialized models have been proposed in the literature which are aimed at capturing the different approaches for partitioning the search of each island so as to avoid them overlapping their exploration [50.18]. On the one hand, each island may consider a different subset of the objectives and then either aggregate them into a single-objective problem [50.19] or use a coevolutionary approach [50.20]. On the other hand, the search space (either the decision space or the objective space) can be explicitly partitioned and assigned to different islands. As stated in [50.11], in a general multiobjective problem it is difficult to design an a priori distribution so that it:

1. Covers the entire search space,
2. Assigns regions of equal size, and
3. Aggregates a minimum complexity to constraint demes to their assigned region.

50.2.2 General Models for Parallel Metaheuristics

Several models have been proposed for parallelizing metaheuristics [50.21, 22] in which EAs, as a type of metaheuristic, perfectly fit. For parallel MOEAs, two main approaches have been proposed in the literature. In [50.16], the authors distinguish between single-walk and multiple-walk parallelizations. The former is aimed at speeding up the computations by parallelizing the evaluation of the objective functions or the search operators. In the latter, several search threads (EAs or any other search method) cooperate to better explore the search space (not only accelerating the execution). The same issue with the Pareto front as in the parallel MOEA models emerges here, so the authors also subdivide into centralized and distributed Pareto front models (CPF and DPF, respectively).

On the other hand, *Talbi* et al. [50.12] categorize parallel metaheuristics in three major hierarchical models. The *self-contained parallel cooperation* is targeted to parallel computing platforms with limited communication. The search is performed by several subalgorithms in parallel, which might cooperate by exchanging some kind of information. It embraces the island model or dMOEAs explained before. Two main groups are distinguished: cooperating subpopulations, which are based on partitioning the objective/search space; and the multistart approach, in which several optimization algorithms run separately in parallel. In the former, subpopulations can be homogeneous or heterogeneous, explore separate regions of the search space, etc. The latter lies in running several local search algorithms in parallel. On a second and third level of the hierarchy, the authors consider those models aimed merely at speeding up the computations: *problem independent parallelization*, which mainly comprises the master–slave approach of parallel fitness evaluation, and *problem dependent parallelization*, which focuses on subdividing single evaluations into parallel tasks that speed up the evaluation step.

50.3 An Updated Review of the Literature

This section is devoted to presenting and analyzing the most recent contributions in the literature to the parallel evolutionary multiobjective optimization field. We have structured the published material according to the classical parallel EA models, i. e., master/slave, distributed, cellular, and hybrid models (Sect. 50.2.1) because this chapter is targeted precisely to EAs and, as a consequence, this classification better captures the design principles of the different contributions. Table 50.1 includes, ordered by the year of publication, an updated review of the field. Also, in order to help the reader with the terminology of this table, Table 50.2 displays the symbols used and their definitions. Then, for each row of Table 50.1, the following information is shown:

● FA-DP (Fitness assignment and diversity preservation): As two of the most important design issues in EMO algorithms, the fitness assignment and diversity preservation mechanisms allow, respectively, to better guide the search toward Pareto optimal solutions and to spread out these Pareto optimal solutions along the entire Pareto front. They are frequently merged into one single measure that translates the vector of objective functions value

of a multiobjective problem into one single scalar value which is used to rank solutions properly (from a Pareto optimality point of view, nondominated solutions are noncomparable).

● PM (Parallel model): It can take the values MS (Master/slave), Dis (distributed model), Cell (cellular model) or Hyb (hybrid), according to the classical parallel EA categorization.

● PFC (Pareto front computation): This column distinguishes between the CPF and DPF strategies defined before.

● PP (parallel platform): When applicable, this column indicates the kind of parallel computing platform in which the given algorithm is executed (GPUs, multicore, cluster, grid, etc.).

● Topology: Communication topology of the parallel MOEA (Star, Hybrid, all-to-all [A2A], etc.).

● Programming: When publicly reported, the programming language used to implement the parallel MOEA is included in this column.

● Description: The main features of the parallel MOEA in a few words.

● Application domain: The area in which the parallel MOEA has been applied.

Table 50.1 Parallel MOEAs since 2008

References	Year	FA-DP	PM	PFC	PP	Topology	Programming	Description	Application domain
[50.23–31]	2008	PT	MS	CPF	Cluster	Hyb	C/PVM	Hierarchical multiresolution parallel solver	Aerodynamic shape optimization
[50.32,33]	2008	RC	MS	CFP	Cluster	Star	C++/MPI	MS NSGA-II and IBEA	Molecular docking problem
[50.34]	2008	RC	Dis	DPF	Seq	Star	MatGrid	Grid-based NSGA-II with isolated islands	Allocation of chlorination stations
[50.35]	2008	SRF	Cell	CFP	Seq	Grid	Java	Hybridization of MOCell with DE	DTLZ, WFG
[50.36–38]	2008	Ib	Dis	DPF	Cluster	A2A	MPI	Parallel Hypervolume-based hyperheuristic	ZDT, WFG
[50.39]	2008	Ib	Dis	DPF	Cluster	Rand	MPI	Parallel Hypervolume-based hyperheuristic	Broadcasting protocol optimization
[50.40]	2008	RC	Dis	DPF	Grid	Ring	MPICH-G2	Island-based MOEA to be deployed on Grids	Protein design
[50.41,42]	2008	ε	MS	CPF	Cluster	Star	N/A	MS ε-NSGA-II for heterogeneous clusters	Operational satellite constellations
[50.43,44]	2008	RC	Dis	DPF	Multicore	Ring	DEVS/SOA	Parallel MOEA combining NSGA-II and SPEA2	Multimedia-embedded system design
[50.45]	2008	RC	MS	CPF	Cluster	Star	MPI	Master/slave NSGA-II with local search	Trajectory tracing
[50.5]	2008	RC	MS	CPF	N/A	Star	N/A	Surrogate-assisted steady-state MS MOEA	ZDT, production planning
[50.46]	2009	SRF	Cell	CPF	Cluster	Mesh	N/A	Cellular-like MOEA with poles and agents	ZDT, aerodynamic optimization
[50.47]	2009	RC	MS	CPF	Cluster	Star	MPI	Master/slave NSGA-II	Morphing-wing design
[50.48,49]	2009	RC	Dis	DPF	Multicore	Star	C++/MPI	Parallel MOEA for dynamic optimization	FDA benchmark
[50.50]	2009	WS	Dis	CPF	N/A	Isol	N/A	Independent WS island and a global Pareto front	Scheduling and knapsack
[50.51]	2009	Ib	Dis	DFP	Multicore	N/A	N/A	Set-based MOEA	DTLZ, WSN deployment
[50.52]	2009	ε	MS	CPF	Cluster	Star	Matlab	Master/slave ε-MOEA	Sonic crystal attenuation opt.
[50.53,54]	2009	WS	Cell	DPF	Seq	Eucl	C	Cellular version of MOEA/D	Multiobjective knapsack
[50.55]	2009	RS	Dis	DPF	Multicore	Torus	MPI	Two archives for migration on different neighbors	VRP with route rebalancing
[50.56]	2009	R	Dis	DPF	Multicore	Star	Python	Parallel graph-based MOEA	Drug design
[50.57,58]	2009	P	Dis	DPF	Cluster	Rand	MPI	Parallel skeleton for island-based MOEAs	Knapsack
[50.59]	2009	SRF	Dis	DPF	Multicore	A2A	MPI	Heterogeneous distributed SPEA2	Mesh partitioning
[50.60]	2009	RC	Dis	DPF	Cluster	Rand	C++	Parallel cooperative strategy of MOEAs	Broadcast optimization in MANETs
[50.61]	2009	SRF	Dis	DPF	N/A	Hier	N/A	Multiple resolution subpopulation pMOEA	ZDT
[50.62]	2009	R	MS	CPF	Cluster	Star	OpenMP	Parallel MOEA with Cluster OpenMP	Air-bearing design
[50.63,64]	2009	RC	Dis	DPF	N/A	A2A	N/A	Bi-population MOEA with adaptive migration	ZDT
[50.65]	2009	RC	MS	CPF	Cluster	Star	C	MS NSGA-II parallelization on a Beowulf cluster	Agriculture research
[50.66,67]	2009	RC	MS	CPF	GPU	Star	C/CUDA	Parallel EMO algorithm on GPU	ZDT, DTLZ, data mining
[50.68]	2010	Tcheb	Dis	DFP	N/A	Rand	N/A	Parallel MO DE-based algorithm	ZDT, DTLZ

Table 50.1 (continued)

References	Year	FA-DP	PM	PfC	PP	Topology	Programming	Description	Application domain
[50.69, 70]	2010	RC	MS	CPF	Cluster	Star	C++/MPI	Hybrid MS parallelization of a MOEA	Phylogenetic inference
[50.71, 72]	2010	RC	Dis	DFP	Multicore	Ring	JavaSpaces	Island based NSGA-II using JavaSpaces	Protein structure prediction
[50.73]	2010	Ib	Dis	DPF	Cluster	Star	C++/MPI	Parallel multiple reference point MO solver	Flow-shop scheduling problem
[50.74]	2010	RC	MS	CPF	Cluster	Star	N/A	Parallel meta-model assisted NSGA-II	Mechanical forging industry
[50.75]	2010	RC	Cell	CPF	Seq	N/A	N/A	Peer-to-peer population structure in a CMOGA	ZTD
[50.76]	2010	R	Cell	CPF	Seq	Grid	N/A	cGA with adaptive fuzzy fitness granulation	Benchmark functions
[50.77]	2010	RC	MS	CPF	N/A	N/A	N/A	Master/slave NSGA-II	Project scheduling
[50.77]	2010	RC	Dis	DPF	N/A	N/A	N/A	Multideme NSGA-II	Project scheduling
[50.78]	2010	RC	Dis	DPF	Grid	N/A	OpenMOLE	Island-based MOEA	Industrial cheese ripening process
[50.79]	2010	RC	Dis	DPF	Multicore	N/A	N/A	dNSGA-II with several constraint handling techs.	Software deployment
[50.80, 81]	2010	Tcheb	Dis	CPF	Multicore	Star	Java	Parallel MOEA/D for multicore processors	LZ
[50.82]	2010	R	Dis	DPF	N/A	A2A	N/A	Combination of MO and SO islands	ZDT, DTLZ
[50.19]	2010	WS	Dis	DFP	N/A	Isol	N/A	DGA with different weights in each island	Flow-shop scheduling
[50.83]	2011	RC	MS	CPF	Cluster	Star	N/A	Parallel master/slave NSGA-II and PAES	Protein structure prediction
[50.20]	2011	RC	Dis	DPF	Multicore	A2A	Java	Parallel cooperative coevolutionary MOEA	ZDT, DTLZ, Task scheduling
[50.84]	2011	WS	Dis	DPF	Seq	Ring	N/A	Parallel MOEA for many-objective problems	DTLZ
[50.85]	2011	R	MS	CPF	Cluster	Star	MATLAB	Parallel MOGA	Combustion of hybrid vehicles
[50.86]	2011	PT	MS	CPF	Cluster	Star	N/A	Parallel computational intelligence framework	Multilayered composite structures
[50.87]	2011	RC	MS	CPF	Multicore	Star	C#	Parallel MOEAs using Parallel.FX	Academic timetabling
[50.88]	2011	RC	Dis	CPF	Multicore	Star	C/MPI	Parallel NSGA-II with Cone Separation	Associative classification rule mining
[50.89]	2011	RC	Dis	DPF	Multicore	A2A	C++	Parallel island-based MOEA	Antenna positioning problem
[50.90]	2011	RC	Hyb	DPF	Multicore	Ring	C/MPI	Parallel hybrid MS/Island NSGA-II	0/1 Knapsack Problem
[50.91]	2011	RC	MS	CPF	Cluster	Star	N/A	Async. MOEA with het. evaluation costs	ZDT, Combustion opt. in engines
[50.92]	2011	ϵ	MS	CPF	Multicore	Star	Visual C	Parallel ϵ dominance EA	Voltage/reactive power control
[50.93, 94]	2011	SRF	MS	CPF	GPU	Star	C/CUDA	DE Markov chain Monte Carlo MO algorithm	ZDT, DTLZ

Table 50.2 includes a list of the symbols used here and their definitions.

50.3.1 Analysis by Year

The first point of analysis of the published material is done with respect to the number of publications over the years considered in this chapter, i.e., the period between 2008 and 2011. Figure 50.2 displays this information not only for the period analyzed in this chapter,

Table 50.2 List of symbols used in Table 50.1

Column	Symbol	Definition
FA-DP	R	Ranking
	RS	Ranking and sharing
	RC	Ranking and crowding
	SRF	Strength raw fitness
	WS	Weighted sum (aggregation)
	Ib	Indicator-based
	PT	Pareto tournaments
	ϵ	Epsilon dominance
	Tcheb	Tchebycheff aggregation
PM	MS	Master/slave model
	Dis	Distributed model
	Cell	Cellular model
	Hyb	Hybrid model
PFC	CPF	Centralized Pareto front
	DPF	Distributed Pareto front
PP	Seq	Sequential algorithm
	GPU	Graphics processing unit
Topology	A2A	All-to-All
	Rand	Random
	Isol	Isolated
	Eucl	Euclidean
	Hier	Hierarchical
	Hyb	Hybrid topology
Programming	PVM	Parallel virtual machine
	MPI	Message Passing Interface
	Mpich-G2	An MPI implementation for grid computing
	DEVS	Discrete event system
	SOA	Service-oriented architecture
	OpenMP	Open multiprocessing
	CUDA	Compute unified device architecture
	OpenMOLE	Open MOdeL Experiment
Description	MS	Master/slave
	MO	Multiobjective
	SO	Single-objective
	DE	Differential evolution
	dNSGA-II	Distributed NSGA-II
	Tech.	techniques
	Async.	Asynchronous
	Het.	Heterogeneous
–	N/A	Not available

but also for the period 1993–2005 presented in [50.14]. The trend is fairly clear: this research topic has been active during the last few years. Indeed, if one compares this evolution with that presented in [50.14] by 2006, where the highest number of works per year was 10, it can be seen that the published material is doubled (more than 20 publications/year in 2009, 2010, and 2011). Despite the relative lack of novel, attractive approaches in the field, parallelism remains as a powerful tool in the EMO community because of one major factor: the optimization problems addressed require to reduce the execution times to affordable values. This is emphasized with the current availability of cheap parallel computing platforms such as multicore processors and, lately, GPUs. Indeed, the keyword *multicore* in column *PP* in Table 50.1 is the second that appears the most.

50.3.2 Analysis of the Parallel Models

In this section, the different contributions are analyzed from the point of view of the characteristics of the parallel model. We will pay particular attention to the columns *FA-DP*, *PM*, *PFC*, and *Topology*.

The fitness assignment and diversity preservation is a major issue in parallel MOEAs because, in many cases, the Pareto front is spread between different subalgorithms (especially in the distributed models). The management of optimal solutions (via fitness assignment) and how they are distributed along the Pareto front (diversity) deserves a brief review. The FA-DP column shows that the Ranking and Crowding mechanisms inherited from the most widely used algorithm in the area, namely NSGA-II [50.95], are also the most present in the literature as long as NSGA-II is the base algorithm for many of these parallel MOEAs. In the case of the distributed and cellular models, it is worth mentioning that Crowding is applied locally, i.e., diversity if kept within the same subalgorithm. If no advanced mechanism is devised to partition the search space (such as in [50.96, 97]), the algorithm will be accepting/discarding solutions that should probably be in the same region as those computed by the other subalgorithm components. The same happens with classical FA-DP methods such as the strength raw fitness (SRF) or the indicator-based (IB) in column *FA-DP*, respectively. As a final note, we strongly believe that these algorithms based on decomposition such as MOEA/D [50.98] are especially well suited to profit from parallel platforms. Indeed, they are based on decomposing the multiobjective problems into a num-

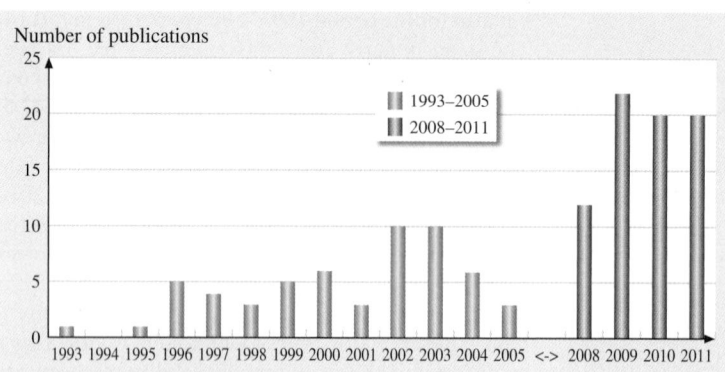

Number of publications

Fig. 50.2 Number of publications on parallel MOEAs grouped by the year of publication in the periods 1993–2005 (after [50.14]) and 2008–2011 (this chapter)

ber of scalar subproblems that are distributed along the Pareto front. Therefore, the partition of the search space is implicitly done and they take full advantage of multicore processors. The texts [50.80, 81] follow this line of research.

If one analyzes the usage of the different parallel models in the revised publications (Fig. 50.3), i.e., MS, distributed (Dis), and cellular (Cell), several clear conclusions can be drawn. First, despite the simplicity of the MS model, it appears in almost half of the related literature (45%). Multiobjective optimization problems are becoming more and more complex and demand high-end computational resources, what makes this approach very suitable in this context. Indeed, the underlying search model remains unchanged (entirely located at the master process), because authors are usually only interested at speeding up the computations. Second, the distributed models are still receiving much attention from the multiobjective community (47% of the analyzed publications use this model) in the quest for engineering an improved algorithm that reaches the Pareto fronts in a more effective way (not only reducing execution times). The promising results published with

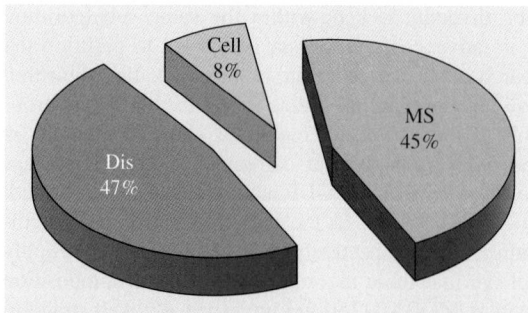

Fig. 50.3 Percentage of use of the different parallel MOEA models in the revised literature

their single objective counterparts [50.7] have pushed forward the research on this area. The major issue with distributed models arises with the difficult management of the Pareto front, as stated in Sect. 50.2.1 and as will be discussed below. Finally, a special comment about the cellular model: even though it is to be exploited in future literature, its percentage of use has doubled since the previous literature review in 2006 [50.14]. The point is that these algorithms are usually executed sequentially, with no parallelism at all, because they were originally targeted to massively parallel machines, and these kind of machines fell into disuse. There are, of course, some exceptions such as [50.46], where a cellular-like MOEA for aerodynamic optimization is deployed on a cluster of computers.

The PFC column is a hot topic in parallel MOEAs. Handling the nondominated solutions found during the search, when the search is distributed in probably separate processors, has promoted and is still promoting fundamental research within the community. The most widely used strategy, however, is to keep a central pool of nondominated solutions (CPF), that is, there is one single front. This approach appears in 55% of the analyzed publications and totally matches both the master/slave and the cellular parallel models. This design option is straightforward and makes common sense.

Almost the same happens with the DPF strategy (45% of the papers), which is mostly used with the distributed model of parallel MOEAs: the Pareto front is approximated separately for each of the subalgorithmic components during the search, only merged into one single front at the end of the exploration. In general, all DPF strategies are complemented with eventual CPF phases, which allows the search of the different subalgorithms to get overlapped [50.11]. A couple of exceptions are to be found among the revised literature, in which a distributed computation of the Pareto front is endowed with a fully centralized Pareto front. In [50.50], a distributed MOEA has one single Pareto front that is computed with several isolated islands. Each island uses a weighted sum approach (with different weights) and is targeted to one region of the search space. The second exception appears in [50.80, 81], in which a multithreaded parallelization of MOEA/D is presented. The Pareto front is stored in global memory and the different threads are in charge of separate groups of weight combinations.

The final aspect under analysis in this section is the interconnection topology of the different components of the parallel algorithms. The Star topologies are widely used in two scenarios: (i) master/slave models and (ii) in distributed models with periodic gathering operations required to generate a single Pareto front. A star topology is able to capture the idea of topologies with a central master that delivers tasks to a set of worker nodes and this is why it is so popular in these two previous cases. The column *Topology* in Table 50.1 also reveals that All-to-All (A2A) and Random topologies also exist in the literature. The former enables the different components of the parallel algorithm to be tightly coupled, thus quickly spreading the nondominated solutions found for a faster convergence toward the optimal Pareto front. The later implies that the genetic material may take longer to reach all the algorithmic components, thus promoting diversity.

50.3.3 Review of the Software Implementations

This section is mainly targeted to summarize the contents of the column *Programming* in Table 50.1, in which a note on the implementation of the algorithms is given. A quick look at the items of the column clearly states that the combination of C/C++ as the programming language and MPI (Message Passing Interface) [50.99] as the technology for enabling the parallel communication between the different components of the parallel algorithms are the preferred options. This can be explained by the strong engineering background of most of the MOPs addressed (and researchers), a field in which C/C++ has had a dominant position for many years. Indeed, C/C++ allows researchers to include very low level routines (even assembler code) that enable full control of all parts of their applications. MPI, in turn, is a standard (not just a library) for which many implementations exist (MPICH, LAM, etc.), so its use always guarantee correctness and efficiency.

Despite this clear fact, only two novels, relevant trends on this topic that are worth mentioning in detail can be found. On the one hand, even though clusters of computers are able to provide researchers with a large computational power, there are MOPs that require still more additional resources. These resources can only be supplied by grid computing platforms [50.100]. This has promoted the parallelization of EMO algorithms with grid-enabled technologies such as MPICH-G2 [50.40] or MatGrid [50.34]. On the other hand, there already are several seminal works on the parallelization of multiobjective optimizers in GPUs, as stated in Sect. 50.3.1. To the best of our knowledge, only implementations with C and CUDA (compute unified device architecture) [50.101] have been proposed in [50.66, 93], but nowadays other opportunities have also emerged such as, for example, OpenCL [50.102].

50.3.4 Main Application Domains

One of the main reasons, if not the main one for the popularity of MOEAs, is their success in solving real-world problems. Parallel EMO algorithms are no exception. As a consequence, the variability in the application domains is very large, which makes the task of classification rather difficult. By partially following the categorization proposed in [50.18], three main areas of application are distinguished: *engineering*, *industrial*, and *scientific*. Figure 50.4 summarizes the percentage of revised publications that fall into each area. Besides these three categories, we have also displayed in this figure a fourth item devoted to bench-

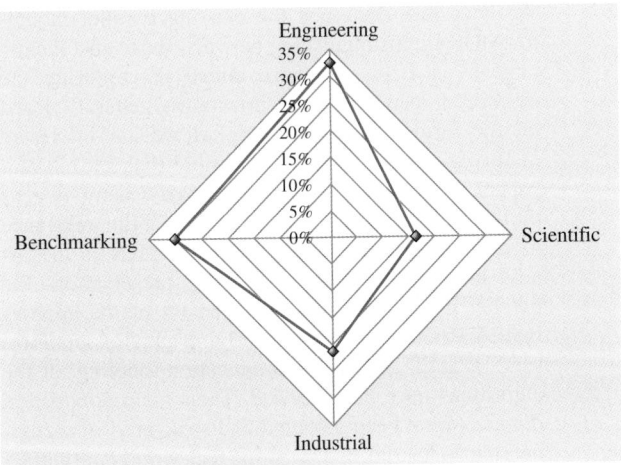

Fig. 50.4 Application domains

marking. This latter is not an *application* but it appears a lot in the revised literature. There are well-established testbeds such as Zitzler–Deb–Thiele (ZDT) [50.103], Deb–Thiele–Laumanns–Zitzler (DTLZ) [50.104], or the walking fish group (WFG) [50.105], that have been widely used as a comparison basis for introducing new algorithmic proposals.

Among the real-world applications, engineering applications are, by far, the most popular domain within the parallel EMO literature (also in the entire EMO field), principally because they usually have suitable mathematical models for this kind of algorithms. Indeed, Fig. 50.4 shows that almost 35% of the papers analyzed address an optimization problem from the engineering domain. Several relevant works among those analyzed are devoted to aerodynamic shape optimization [50.27, 30, 46], reconfiguration of operational satellite constellations [50.41], or the combustion control for different types of vehicles [50.85, 91].

The second place in terms of popularity is occupied by the industrial applications, that appear in 21% of the papers reviewed. These applications are related to the fields of manufacturing, scheduling, and management.

Very interesting problems have been addressed in this industrial domain, such as the optimization of sonic crystal attenuation properties [50.52], the Camembert cheese ripening process [50.78], and evaluation of the input tax to regulate pollution from agricultural production [50.65].

Scientific applications, the third category of real-world applications analyzed (16% of the papers), is intended to group optimization problems in bioinformatics, chemistry, and computer science. The most successful applications here are devoted to the bioinformatic and chemistry fields, in problems on molecular docking [50.33], protein design [50.40], drug design [50.56], and phylogenetic inference [50.69, 70]. In our view, the reason these applications (either engineering, industrial or scientific) have been tackled with parallel MOEAs is precisely the tasks involved to compute their objective functions. When a new enhanced parallel search model is sought, then authors usually rely on benchmarking functions. Indeed, as stated above, 30% of the revised papers (Fig. 50.4) use these testbeds either exclusively [50.61, 68, 82, 94] or prior to solving a real-world application [50.5, 20, 51].

50.4 Conclusions and Future Works

50.4.1 Summary

In this chapter, we have carried out a comprehensive survey of the literature concerning parallel MOEAs since 2008/2009, the year when two of the most well-known comprehensive surveys were published. We have first described the existing parallel models for MOEAs, distinguishing between those specifically targeted at EAs and those aimed at capturing the essence of parallel metaheuristics in general. Based on the former model (as long as we are interested in surveying parallel MOEAs), more than 80 relevant papers have been carefully analyzed (many dozens more studied but left out because of little relevance to this survey). Fundamental aspects such as the fitness assignment and the diversity preservation, the parallel model used, the management of the approximated Pareto front, the underlying parallel platform used (if any), and the communication topology of the algorithms have been revised. Their main application domains have been gathered and structured into engineering, industrial, and scientific real-world multiobjective problems.

Despite the lack of new, attractive ideas in the area, this survey has revealed that the research on parallel MOEAs is still moving forward for two main reasons. On the one hand, researchers think of parallelism as a way to not only speed up computations, but also as a strategy to enhance the search engines of the algorithms. On the other hand, the computational demands of many multiobjective problems (dimensionality, uncertainty, simulations, etc.) means that the parallelism is the only suitable option to address them.

In the following section, some topics for future research for engineers interested in parallel MOEAs are outlined. In the authors' opinion, these topics merit particular attention so that the current state-of-the-art can be improved.

50.4.2 Future Trends

We have structured the future trends section as a bottom-up approach, proposing research lines for parallel MOEAs that range from low-level algorithmic details to more complex enhanced strategies at a higher level. We will also suggest that different studies that are missing in the literature need to be carried out and, in our opinion, once completed, might have a great impact within the community.

Let us start with one of the main issues of parallel MOEAs which is such that, in the end, the algorithms have to compute one single approximation to the Pareto front, namely, they have an inherent centralized structure. When several subalgorithms cooperate to approximate such a Pareto front (i. e., the distributed models), they may coordinate themselves somehow in order to effectively sample different regions of the front in order to avoid overlaps. These overlaps are the main reason for the distributed algorithms being outperformed by centralized approaches (such as the master/slave). This issue has been partially addressed in the literature [50.96, 97] with some degree of success. However, these cited results have not been widely used yet and new advances are needed.

We propose here two lines of research that might help the distributed models of parallel MOEAs to overcome this issue. The first one is based on designing a fully distributed diversity preservation method (e.g., a distributed crowding). The research question here is whether it is possible to devise a density estimator that considers both local and global information from the other components of the distributed MOEA. We think so. Instead of trying to allocate a given portion of the Pareto front to each island, these islands should periodically broadcast a list with the objective values of their local solutions (but not the decision variables). Then, when checking whether a solution is to be stored in its local Pareto front, the island has to consider both the local information and the global information received from the other islands. To the best of our knowledge, such a mechanism does not exist in the literature. The second proposal is to use rough set theory [50.106] to effectively partition the search space and allocate different portions to different components of the parallel MOEA. There does exist a preliminary work [50.107] that uses a multiobjective simulated annealing, and

these ideas will have a great impact on the design of parallel MOEAs.

On a higher algorithmic level, this survey has also revealed that, despite their advanced search model and accurate results on many benchmarking functions [50.108], cellular models of parallel MOEAs have been marginally used in most of the application domains. We strongly believe that these algorithms might improve the state-of-the-art in many of these unexplored research areas. On a related matter, designing heterogeneous algorithms [50.17] is also a line of research with a high potential. Indeed, few contributions in the literature have considered parallel heterogeneous algorithms for multiobjective optimization [50.59]. To the best of our knowledge, there is no published analysis on the impact of heterogeneity (i. e., several different subalgorithms cooperating, namely, NSGA-II, SPEA2, PAES, MOEA/D, etc.) while approximating Pareto fronts. Our experience with these totally different algorithms is that each one is usually better suited to explore given regions of the search space, so their collaboration may result in a newly improved algorithm.

Finally, the last group of open research lines are devoted to well-grounded studies on the behavior of parallel MOEAs concerning their scalability with respect to the number of decision variables or their convergence speed toward the optimal Pareto front, as done for sequential MOEAs in [50.109, 110], respectively. Again, the influence of heterogeneity on these two search capabilities of parallel MOEAs is of interest.

Last but not least, there is a distinct lack of theoretical work in this area. For example, to the best of our knowledge, there is no analysis of the takeover time [50.111] in the multiobjective context, and we believe that characterizing parallel MOEAs based on this metric may help researchers in future developments. The difficulty of the landscapes in MOPs and the relative theoretical advantages for different types of problems are open research lines.

References

50.1   C.A. Coello Coello, D.A. Van Veldhuizen, G.B. Lamont: *Evolutionary Algorithms for Solving Multi-Objective Problems* (Kluwer, Boston 2002)

50.2   K. Deb: *Multi-Objective Optimization Using Evolutionary Algorithms* (Wiley, New York 2001)

50.3   R.R. Coelho, P. Bouillard: Multi-objective reliability-based optimization with stochastic metamodels, Evol. Comput. **19**(4), 525–560 (2011)

50.4   T. Goel, R. Vaidyanathan, R. Haftka, W. Shyy: Response surface approximation of Pareto optimization front in multi-objective optimization, 10th AIAA/ISSMO Multidiscip. Anal. Optim. Conf. (2004)

50.5   A. Syberfeldt, H. Grimm, A. Ng, R.I. John: A parallel surrogate-assisted multi-objective evolutionary algorithm for computationally expensive

optimization problems, IEEE Congr. Evol. Comput.
(2008) pp. 3177–3184

50.6   E. Alba: *Parallel Metaheuristics: A New Class of Algorithms* (Wiley, New York 2005)

50.7   E. Alba, M. Tomassini: Parallelism and evolutionary algorithms, IEEE Trans. Evol. Comput. **6**(5),
443–462 (2002)

50.8   E. Alba, J.M. Troya: A Survey of parallel distributed genetic algorithms, Complexity **4**(4), 31–
52 (1999)

50.9   E. Cantú-Paz: *Efficient and Accurate Parallel Genetic Algorithms* (Kluwer, New York 2000)

50.10   G. Luque, E. Alba: *Parallel Genetic Algorithms:
Theory and Real World Applications* (Springer,
Berlin, Heidelberg 2011)

50.11   A. Lopez-Jaimes, C.A. Coello Coello: Applications
of parallel platforms and models in evolutionary
multi-objective optimization. In: *Biologically-
Inspired Optimisation Methods*, ed. by A. Lewis,
S. Mostaghim, M. Randall (Springer, Berlin, Heidelberg 2009) pp. 23–29

50.12   E.-G. Talbi, S. Mostaghim, T. Okabe, H. Ishibuchi,
G. Rudolph, C.A. Coello Coello: Parallel approaches
for multiobjective optimization, Lect. Notes Comput. Sci. **5252**, 349–372 (2008)

50.13   A.J. Chipperfield, P.J. Fleming: Parallel genetic al-
gorithms. In: *Parallel and Distributed Computing
Handbook*, ed. by A.Y. Zomaya (McGraw Hill, New
York 1996) pp. 1118–1143

50.14   F. Luna, A.J. Nebro, E. Alba: Parallel evolutionary
multiobjective optimization. In: *Parallel Evolutionary Computations*, ed. by N. Nedjah, E. Alba,
L. de Macedo (Springer, Berlin, Heidelberg 2006)
pp. 33–56, Chapter 2

50.15   D.A. Van Veldhuizen, J.B. Zydallis, G.B. Lamont:
Considerations in engineering parallel multiob-
jective evolutionary algorithms, IEEE Trans. Evol.
Comput. **87**(2), 144–173 (2003)

50.16   A.J. Nebro, F. Luna, E.-G. Talbi, E. Alba: Parallel
multiobjective optimization. In: *Parallel Metaheuristics*, ed. by E. Alba (Wiley, New York 2005)
pp. 371–394

50.17   F. Luna, E. Alba, A.J. Nebro: Parallel heterogeneous metaheuristics. In: *Parallel Metaheuristics*,
ed. by E. Alba (Wiley, New York 2005) pp. 395–
422

50.18   C.A. Coello Coello, G.B. Lamont, D.A. Van Veldhuizen: *Evolutionary Algorithms for Solving
Multi-Objective Problems*, Genetic and Evolutionary Computation (Springer, Berlin, Heidelberg
2007)

50.19   E. Rashidi, M. Jahandar, M. Zandieh: An improved
hybrid multi-objective parallel genetic algorithm
for hybrid flow shop scheduling with unrelated
parallel machines, Int. J. Adv. Manuf. Technol. **49**,
1129–1139 (2010)

50.20   B. Dorronsoro, G. Danoy, P. Bouvry, A.J. Nebro:
Multi-objective cooperative coevolutionary evolutionary algorithms for continuous and com-

binatorial optimization. In: *Intelligent Decision
Systems in Large-Scale Distributed Environments*,
Studies in Computational Intelligence, Vol. 362,
(Springer, Berlin, Heidelberg 2011) pp. 49–74

50.21   T.G. Crainic, M. Toulouse: Parallel strategies for
metaheuristics. In: *Handbook of Metaheuristics*,
ed. by F.W. Glover, G.A. Kochenberger (Kluwer,
Boston 2003)

50.22   V.-D. Cung, S.L. Martins, C.C. Ribeiro, C. Rou-
cairol: Strategies for the parallel implementation
of metaheuristics. In: *Essays and Surveys in Metaheuristics*, ed. by C.C. Ribeiro, P. Hansen (Kluwer,
Boston 2003) pp. 263–308

50.23   L.F. Gonzalez: Robust Evolutionary Methods for
Multi-objective and Multidisciplinary Design in
Aeronautics, Ph.D. Thesis (University of Sydney,
Sydney 2005)

50.24   D.S. Lee, L.F. Gonzalez, J. Periaux, G. Bugeda:
Double-shock control bump design optimization
using hybridized evolutionary algorithms, Proc.
Inst. Mech. Eng. G: J. Aerosp. Eng. (2011) pp. 1175–
1192

50.25   D.S. Lee, L.F. Gonzalez, J. Periaux, K. Srinivas: Evolutionary optimisation methods with uncertainty
for modern multidisciplinary design in aeronautical engineering, Notes Numer. Fluid Mech. Multidiscip. Des. **100**, 271–284 (2009)

50.26   D.S. Lee, L.F. Gonzalez, J. Periaux, K. Srinivas:
Efficient hybrid-game strategies coupled to evolutionary algorithms for robust multidisciplinary
design optimization in aerospace engineering,
IEEE Trans. Evol. Comput. **15**(2), 133–150 (2011)

50.27   D.S. Lee, L.F. Gonzalez, J. Periaux, K. Srinivas, E. Onate: Hybrid-game strategies for multi-
objective design optimization in engineering,
Comput. Fluids **47**, 189–204 (2011)

50.28   D.S. Lee, L.F. Gonzalez, K. Srinivas, J. Periaux: Robust design optimisation using multi-objective
evolutionary algorithms, Comput. Fluids **37**(5),
565–583 (2008)

50.29   D.S. Lee, L.F. Gonzalez, K. Srinivas, J. Periaux: Robust evolutionary algorithms for UAV/UCAV aerodynamic and RCS design optimisation, Comput.
Fluids **37**(5), 547–564 (2008)

50.30   D.S. Lee, J. Periaux, L.F. Gonzalez, K. Srinivas,
E. Onate: Robust multidisciplinary UAS design optimisation, Struct. Multidiscip. Optim. **45**(3), 433–
450 (2012)

50.31   D.S. Lee, J. Periaux, E. Onate, L.F. Gonzalez,
N. Qin: Active transonic aerofoil design optimization using robust multiobjective evolutionary algorithms, J. Aircr. **48**(3), 1084–1094 (2011)

50.32   J.-C. Boisson, L. Jourdan, E.-G. Talbi, D. Horvath: Parallel multi-objective algorithms for the
molecular docking problem, IEEE Symp. Comput.
Intell. Bioinform. Comput. Biol. (2008) pp. 187–
194

50.33   J.-C. Boisson, L. Jourdan, E.-G. Talbi, D. Horvath:
Single- and multi-objective cooperation for the

flexible docking problem, J. Math. Model. Algorith. **9**, 195–208 (2010)

50.34   G. Ewald, W. Kurek, M.A. Brdys: Grid implementation of a parallel multiobjective genetic algorithm for optimized allocation of chlorination stations in drinking water distribution systems: Chojnice case study, IEEE Trans. Syst. Man Cybern. C: Appl. Rev. **38**(4), 497–509 (2008)

50.35   J.J. Durillo, A.J. Nebro, F. Luna, E. Alba: Solving three-objective optimization problems using a new hybrid cellular genetic algorithm, Lect. Notes Comput. Sci. **5199**, 661–670 (2008)

50.36   C. Leon, G. Miranda, E. Segredo, C. Segura: Parallel hypervolume-guided hyperheuristic for adapting the multi-objective evolutionary island model, Nat. Inspir. Coop. Strat. Optim. (2009) pp. 261–272

50.37   C. Leon, G. Miranda, C. Segura: A self-adaptive island-based model for multi-objective optimization, Genet. Evol. Comput. Conf. (2008) pp. 757–758

50.38   C. Leon, G. Miranda, C. Segura: Hyperheuristics for a dynamic-mapped multi-objective island-based model, Lect. Notes Comput. Sci. **5518**, 41–49 (2009)

50.39   C. Leon, G. Miranda, C. Segura: Optimizing the configuration of a broadcast protocol through parallel cooperation of multi-objective evolutionary algorithms, Int. Conf. Adv. Eng. Comput. Appl. Sci. (2008) pp. 135–140

50.40   P. Liu, S. Dong: Parallel multi-objective GA based rotamer optimization on grid, Int. Coll. Comput. Comm. Control. Manag. (CCCM) (2008) pp. 238–241

50.41   M.P. Ferringer, D.B. Spencer, P. Reed: Many-objective reconfiguration of operational satellite constellations with the large-cluster epsilon non-dominated sorting genetic algorithm II, IEEE Congr. Evol. Comput. (2009) pp. 340–349

50.42   P.M. Reed, J.B. Kollat, M.P. Ferringer, T.G. Thompson: Parallel evolutionary multi-objective optimization on large, heterogeneous clusters: An applications perspective, J. Aerosp. Comput. Inf. Commun. **5**, 460–478 (2008)

50.43   J.L. Risco-Martin, D. Atienza, J.I. Hidalgo, J. Lanchares: A parallel evolutionary algorithm to optimize dynamic data types in embedded systems, Soft Comput. **12**, 1157–1167 (2008)

50.44   J.L. Risco-Martin, D. Atienza, J.I. Hidalgo, J. Lanchares: Parallel and distributed optimization of dynamic data structures for multimedia embedded systems. In: *Parallel and Distributed Computational Intelligence*, ed. by F.F. Vega, E. Cantú-Paz (Springer, Berlin, Heidelberg 2010) pp. 263–290

50.45   D. Sharma, K. Deb, N.N. Kishore: Towards generating diverse topologies of path tracing compliant mechanisms using a local search based multi-objective genetic algorithm procedure, IEEE Congr. Evol. Comput. (2008) pp. 2004–2011

50.46   V.G. Asouti, K.C. Giannakoglou: Aerodynamic optimization using a parallel asynchronous evolutionary algorithm controlled by strongly interacting demes, Eng. Optim. **41**(3), 241–257 (2009)

50.47   S. Bharti, M. Frecker, G. Lesieutre: Optimal morphing-wing design using parallel nondominated sorting genetic algorithm II, AIAA J. **47**(7), 1627–1634 (2009)

50.48   M. Camara, J. Ortega, F. de Toro: A single front genetic algorithm for parallel multi-objective optimization in dynamic environments, Neurocomputing **72**, 3570–3579 (2009)

50.49   M. Camara, J. Ortega, F. de Toro: Approaching dynamic multi-objective optimization problems by using parallel evolutionary algorithms. In: *Advances in Multi-Objective Nature Inspired Computing*, ed. by C.A. Coello Coello, C. Dhaenes, L. Jourdan (Springer, Berlin, Heidelberg 2010) pp. 63–86

50.50   P.-C.S.-H. Chand Chen: The development of a sub-population genetic algorithm ii (SPGA II) for multi-objective combinatorial problems, Appl. Soft Comput. **9**, 173–181 (2009)

50.51   J. Bader, D. Brockhoff, S. Welten, E. Zitzler: On using populations of sets in multiobjective optimization, Lect. Notes Comput. Sci. **5467**, 140–154 (2009)

50.52   J.M. Herrero, S. Garcia-Nieto, X. Blasco, V. Romero-Garcia, J.V. Sanchez-Perez, L.M. Garcia-Raffi: Optimization of sonic crystal attenuation properties by ev-MOGA multiobjective evolutionary algorithm, Struct. Multidiscip. Optim. **39**, 203–215 (2009)

50.53   H. Ishibuchi, Y. Sakane, N. Tsukamoto, Y. Nojima: Effects of using two neighborhood structures on the performance of cellular evolutionary algorithms for many-objective optimization, IEEE Congr. Evol. Comput. (2009) pp. 2508–2515

50.54   H. Ishibuchi, Y. Sakane, N. Tsukamoto, Y. Nojima: Implementation of cellular genetic algorithms with two neighborhood structures for single-objective and multi-objective optimization, Soft Comput. **15**, 1749–1767 (2011)

50.55   N. Jozefowiez, F. Semet, E.-G. Talbi: An evolutionary algorithm for the vehicle routing problem with route balancing, Eur. J. Oper. Res. **195**, 761–769 (2009)

50.56   C.C. Kannas, C.A. Nicolaou, C.S. Pattichis: A parallel implementation of a multi-objective evolutionary algorithm, 9th Int. Conf. Inform. Technol. Appl. Biomed. (2009) pp. 1–6

50.57   C. Leon, G. Miranda, E. Segredo, C. Segura: Parallel library of multi-objective evolutionary algorithms, 17th Euromicro Int. Conf. IEEE (2009) pp. 28–35

50.58   C. Leon, G. Miranda, C. Segura: METCO: A parallel plugin-based framework for multi-objective optimization, Int. J. Artif. Intell. Tools **18**(4), 569–588 (2009)

50.59   A. Rama Mohan Rao: Distributed evolutionary multi-objective mesh-partitioning algorithm for parallel finite element computations. Comput, Struct. **87**(3), 1469–1473 (2009)

50.60   C. Segura, A. Cervantes, A.J. Nebro, M.D. Jaraíz-Simón, E. Segredo, S. García, F. Luna, J.A. Gómez-Pulido, G. Miranda, C. Luque, E. Alba, M.Á. Vega-Rodríguez, C. León, I.M. Galván: Optimizing the DFCN broadcast protocol with a parallel cooperative strategy of multi-objective evolutionary algorithms, Lect. Notes Comput. Sci. **5467**, 305–319 (2009)

50.61   E. Szlachcic, W. Zubik: Parallel distributed genetic algorithm for expensive multi-objective optimization problems, Lect. Notes Comput. Sci. **5717**, 938–946 (2009)

50.62   N. Wang, C.-M. Tsai, K.-C. Cha: Optimum design of externally pressurized air bearing using cluster OpenMP, Tribol. Int. **42**, 1180–1186 (2009)

50.63   T. Qiu, G. Ju: A selective migration parallel multi-objective genetic algorithm, Chin. Control Decis. Conf. (2010) pp. 463–467

50.64   Z.X. Wang, G. Ju: A parallel genetic algorithm in multi-objective optimization, Chin. Control Decis. Conf. (2009) pp. 3497–3501

50.65   G. Whittaker, R. Confesor Jr., S.M. Griffith, R. Fare, S. Grosskopf, J.J. Steiner, G.W. Mueller-Warrant, G.M. Banow: A hybrid genetic algorithm for multiobjective problems with activity analysis-based local search, Eur. J. Oper. Res. **193**, 195–203 (2009)

50.66   M.L. Wong: Parallel multi-objective evolutionary algorithms on graphics processing units, Genet. Evolut. Comput. Conf. (2009) pp. 2515–2522

50.67   M.L. Wong: Data mining using parallel multi-objective evolutionary algorithms on graphics hardware, IEEE Congr. Evol. Comput. (2010) pp. 1–8

50.68   A.A. Montaño, C.A. Coello Coello, E. Mezura-Montes: pMODE-LD+SS: An effective and efficient parallel differential evolution algorithm for multi-objective optimization, Lect. Notes Comput. Sci. **6239**, 21–30 (2010)

50.69   W. Cancino, L. Jourdan, E.-G. Talbi, A.C.B. Delbem: Parallel multi-objective approaches for inferring phylogenies, Lect. Notes Comput. Sci. **6023**, 26–37 (2010)

50.70   W. Cancino, L. Jourdan, E.-G. Talbi, A.C.B. Delbem: Parallel multi-objective evolutionary algorithm for phylogenetic inference, Lect. Notes Comput. Sci. **6073**, 196–199 (2010)

50.71   D. Becerra, A. Sandoval, D. Restrepo-Montoya, L.F. Nino: A parallel multi-objective ab initio approach for protein structure prediction, IEEE Int. Conf. Bioinform. Biomed. (2010) pp. 137–141

50.72   D. Dasgupta, D. Becerra, A. Banceanu, F. Nino, J. Simien: A parallel framework for multi-objective evolutionary optimization, IEEE Congr. Evol. Comput. (2010) pp. 1–8

50.73   J.R. Figueira, A. Liefooghe, E.-G. Talbi, A.P. Wierzbicki: A parallel multiple reference point approach for multi-objective optimization, Eur. J. Op. Res. **205**, 390–400 (2010)

50.74   L. Fourment, R. Ducloux, S. Marie, M. Ejday, D. Monnereau, T. Masse, P. Montmitonnet: Mono and multi-objective optimization techniques applied to a large range of industrial test cases using metamodel assisted evolutionary algorithms, 10th Int. Conf. Numer. Methods Ind. Form. (2010) pp. 833–840

50.75   T. Hiroyasu, T. Noda, M. Yoshimi, M. Miki, H. Yokouchi: Examination of multi-objective genetic algorithm using the concept of a peer-to-peer network, 2nd World Congr. Nat. Biol. Inspir. Comput. (2010) pp. 508–512

50.76   I. Kamkar, M.-R. Akbarzadeh-T: Multiobjective cellular genetic algorithm with adaptive fuzzy fitness granulation, IEEE Int. Conf. Syst. Man Cybern. (2010) pp. 4147–4153

50.77   A. Kandil, K. El-Rayes, O. El-Anwar: Optimization research: Enhancing the robustness of large-scale multiobjective optimization in construction, J. Constr. Eng. Manag. **136**(1), 17–25 (2009)

50.78   S. Mesmoudi, N. Perrot, R. Reuillon, P. Bourgine, E. Lutton: Optimal viable path search for a cheese ripening process using a multi-objective EA, Int. Conf. Evol. Comput. (2010)

50.79   J. Montgomery, I. Moser: Parallel constraint handling in a multiobjective evolutionary algorithm for the automotive deployment problem, 6th IEEE Int. Conf. e-Sci. Workshops (2010) pp. 104–109

50.80   J.J. Durillo, Q. Zhang, A.J. Nebro, E. Alba: Distribution of computational effort in parallel MOEA/D, Learn. Intell. Optim. (2011) pp. 488–502

50.81   A.J. Nebro, J.J. Durillo: A study of the parallelization of the multi-objective metaheuristic MOEA/D, Lect. Notes Comput. Sci. **6073**, 303–317 (2010)

50.82   M. Pilat, R. Neruda: Combining multiobjective and single-objective genetic algorithms in heterogeneous island model, IEEE Congr. Evol. Comput. (2010) pp. 1–8

50.83   J.C. Calvo, J. Ortega, M. Anguita: Comparison of parallel multi-objective approaches to protein structure prediction, J. Supercomput. **58**, 253–260 (2011)

50.84   M. Garza-Fabre, G. Toscano-Pulido, C.A. Coello Coello, E. Rodriguez-Tello: Effective ranking + speciation = many-objective optimization, IEEE Congr. Evol. Comput. (2011) pp. 2115–2122

50.85   D. Gladwin, P. Stewart, J. Stewart: Internal combustion engine control for series hybrid electric vehicles by parallel and distributed genetic programming/multiobjective genetic algorithms, Int. J. Syst. Sci. **42**(2), 249–261 (2011)

50.86   D.S. Lee, C. Morillo, G. Bugeda, S. Oller, E. Onate: Multilayered composite structure design optimisation using distributed/parallel multi-objective evolutionary algorithms, Compos. Struct. **94**(3), 1087–1096 (2012)

50.87 A.L. Márquez, C. Gil, R. Baños, J. Gómez: Parallelism on multicore processors using Parallel.FX, Adv. Eng. Softw. **42**, 259–265 (2011)

50.88 B.S.P. Mishra, A.K. Addy, R. Roy, S. Dehuri: Parallel multi-objective genetic algorithms for associative classification rule mining, Int. Conf. Commun. Comput. Secur. (2011) pp. 409–414

50.89 E. Segredo, C. Segura, C. Leon: On the comparison of parallel island-based models for the multiobjectivised antenna positioning problem, 15th Int. Conf. Knowl. Intell. Inf. Eng. Syst. (2011) pp. 32–41

50.90 G.N. Shinde, S.B. Jagtap, S.K. Pani: Parallelizing multi-objective evolutionary genetic algorithms, Proc. World Congr. Eng. (2011) pp. 1534–1537

50.91 M. Yagoubi, L. Thobois, M. Schoenauer: Asynchronous evolutionary multi-objective algorithms with heterogeneous evaluation costs, IEEE Congr. Evol. Comput. (2011) pp. 21–28

50.92 A. Zhang, H. Li, C. Xiao: Parallel computing model for time-varied coordinated voltage/reactive power control, J. Electr. Syst. **7**(1), 1–11 (2011)

50.93 W. Zhu, Y. Li: GPU-accelerated differential evolutionary Markov chain Monte Carlo method for multi-objective optimization over continuous space, 2nd Workshop Bio-Inspir. Algorithms Distrib. Syst. (2010) pp. 1–8

50.94 W. Zhu, A. Yaseen, Y. Li: DEMCMC-GPU: An efficient multi-objective optimization method with GPU acceleration on the fermi architecture, New Gener. Comput. **29**, 163–184 (2011)

50.95 K. Deb, A. Pratap, S. Agarwal, T. Meyarivan: A fast and elitist multiobjective genetic algorithm: NSGA-II, IEEE Trans. Evol. Comput. **6**(2), 182–197 (2002)

50.96 J. Branke, H. Schmeck, K. Deb, M.S. Reddy: Parallelizing multi-objective evolutionary algorithms: Cone separation, Congr. Evol. Comput. (2004) pp. 1952–1957

50.97 F. Streichert, H. Ulmer, A. Zell: Parallelization of multi-objective evolutionary algorithms using clustering algorithms, Lect. Notes Comput. Sci. **3410**, 92–107 (2005)

50.98 Q. Zhang, H. Li: MOEA/D: A multi-objective evolutionary algorithm based on decomposition, IEEE Trans. Evol. Comput. **11**(6), 712–731 (2007)

50.99 W. Gropp, E. Lusk, A. Skjellum: *Using MPI: Portable Parallel Programming with the Message-Passing Interface* (MIT, London 2000)

50.100 F. Berman, G.C. Fox, A.J.G. Hey: *Grid Comptuing Making the Global Infrastructure A Reality*, Communications Networking and Distributed Systems (Wiley, New York 2003)

50.101 NVIDIA Corporation: *NVIDIA CUDA Compute Unified Device Architecture Programming Guide* (NVIDIA Corporation, Santa Clara 2007)

50.102 R. Tsuchiyama, T. Nakamura, T. Iizuka, A. Asahara, S. Miki: *The OpenCL Programming Book* (Fixstars Corporation, Synnyvale 2010)

50.103 E. Zitzler, K. Deb, L. Thiele: Comparison of Multiobjective Evolutionary Algorithms: Empirical Results Evol, Comput. **8**(2), 173–195 (2000)

50.104 K. Deb, L. Thiele, M. Laumanns, E. Zitzler: Scalable test problems for evolutionary multiobjective optimization. In: *Evolutionary Multiobjective Optimization. Theoretical Advances and Applications*, ed. by A. Abraham, L. Jain, R. Goldberg (Springer, Berlin, Heidelberg 2005) pp. 105–145

50.105 S. Huband, P. Hingston, L. Barone, L. While: A review of multiobjective test problems and a scalable test problem toolkit, IEEE Trans. Evol. Comput. **10**(5), 477–506 (2006)

50.106 Z. Pawlak: Rough sets, Int. J. Parallel Program. **11**, 341–356 (1982)

50.107 U. Maulik, A. Sarkar: Evolutionary rough parallel multi-objective optimization algorithm, Fundam. Inform. **99**(1), 13–27 (2010)

50.108 A.J. Nebro, J.J. Durillo, F. Luna, B. Dorronsoro, E. Alba: A cellular genetic algorithm for multiobjective optimization, Int. J. Intell. Syst. **24**(7), 723–725 (2009)

50.109 J.J. Durillo, A.J. Nebro, C.A. Coello, J. Garcia-Nieto, F. Luna, E. Alba: A study of multiobjective metaheuristics when solving parameter scalable problems, IEEE Trans. Evol. Comput. **14**(4), 618–635 (2010)

50.110 J.J. Durillo, A.J. Nebro, F. Luna, C.A. Coello Coello, E. Alba: Convergence speed in multi-objective metaheuristics: Efficiency criteria and empirical study, Int. J. Numer. Methods Eng. **84**(11), 1344–1375 (2010)

50.111 D.E. Goldber, K. Deb: A comparative analysis of selection schemes used in genetic algorithms. In: *Foundations of Genetic Algorithms*, ed. by G.J.E. Rawlins (Morgan Kaufmann, San Mateo 1991) pp. 69–93

51. Many–Objective Problems: Challenges and Methods

Antonio López Jaimes, Carlos A. Coello Coello

This chapter presents a short review of the state-of-the-art efforts for understanding and solving problems with a large number of objectives (usually known as many-objective optimization problems, $\mathbb{M}$OP s). The first part of the chapter presents the current studies aimed at discovering the sources that make a multiobjective optimization problem (MOP) harder when more objectives are added, degrading in this way, the performance of a multiobjective evolutionary algorithm (MOEA). Next, some of the most relevant techniques designed to deal with $\mathbb{M}$OPs are presented and categorized.

51.1 Background

Since the first implementation of an MOEA in the mid-1980s [51.1], a wide variety of new MOEAs have been proposed, gradually improving in both their effectiveness and efficiency to solve MOPs [51.2]. However, most of these algorithms have been evaluated and applied to problems with only two or three objectives, in spite of the fact that many real-world problems have more than three objectives [51.3–6].

Recent experimental [51.7–9] and analytical [51.10, 11] studies have shown that MOEAs based on Pareto optimality [51.12] scale poorly in MOPs with a high number of objectives (4 or more). These MOPs are usually known in the community as $\mathbb{M}$OPs. Although those scalability issues seem mainly to affect Pareto-based MOEAs, as we will see later in this chapter, optimization problems with a large number of objec-

tives introduce some difficulties common to any other multiobjective optimizer.

The goal of this chapter is to present a general view of the difficulties posed by many-objective problems for Pareto-based MOEAs. Specifically, we present a review of the potential sources of difficulty currently found in the specialized literature. Likewise, we present a brief review of the current proposals to deal with these sources of difficulty. These proposals are classified into five classes. Among the most common approaches to deal with $\mathbb{M}$OPs, we can find the use of preference relations to further rank nondominated solutions, the removal of redundant objectives during or after the search, and the incorporation of preference information. Finally, at the end of the chapter some future research paths are outlined.

51.2 Basic Concepts and Notation

In this section, we will introduce the concepts and notation that will be used throughout the rest of the paper. Since some of these proposals are based on conflict information among the objectives, some definitions of conflict are also provided.

51.2.1 Multiobjective Optimization Problems

Definition 51.1 Multiobjective optimization problem

An MOP is defined as

$$\text{Minimize } f(x) = [f_1(x), f_2(x), \ldots, f_k(x)]^\mathsf{T},$$
$$\text{subject to } x \in X. \tag{51.1}$$

The vector $x \in \mathbb{R}^n$ is formed by n *decision variables* representing the quantities for which values are to be chosen in the optimization problem. The *feasible set* $X \subseteq \mathbb{R}^n$ is implicitly determined by a set of equality and inequality constraints. The vector function $f : X \to \mathbb{R}^k$ is composed of $k \geq 2$ scalar *objective functions* $f_i : X \to \mathbb{R}$ $(i = 1, \ldots, k)$. In multiobjective optimization, the sets $\mathbb{R}^n$ and $\mathbb{R}^k$ are known as the *decision variable space* and *objective function space*, respectively. The image of X under the function f is a subset of the objective function space denoted by $Z = f(X)$ and referred to as the *feasible set in the objective function space*.

In order to define precisely the multiobjective optimization problem stated in Definition 51.1, we have to establish the meaning of minimization in $\mathbb{R}^k$. That is to say, we need to define how vectors $z = f(x) \in \mathbb{R}^k$ have to be compared for different solutions $x \in \mathbb{R}^n$. In single-objective optimization the relation *less than or equal* ($\leq$) is used to compare the scalar objective values. By using this relation there may be many different optimal solutions $x \in X$, but only one optimal value $f^{\min} = \min\{f(x) \mid x \in X\}$ since the relation $\leq$ induces a total order in $\mathbb{R}$ (i. e., every pair of solutions is comparable, and thus, we can sort solutions from the best to the worst one). In contrast, in multiobjective optimization problems, there is no canonical order of $\mathbb{R}^k$, and thus, we need weaker definitions of order to compare vectors in $\mathbb{R}^k$.

In multiobjective optimization, the *Pareto dominance relation* is usually adopted. This relation was originally proposed by *Edgeworth* in 1881 [51.13], but generalized by the French-Italian economist *Pareto* in 1896 [51.12].

Definition 51.2 Pareto dominance relation

We say that a vector z^1 dominates vector z^2, denoted by $z^1 \prec z^2$, if and only if

$$\forall i \in \{1, \ldots, k\} : z_i^1 \leq z_i^2 \tag{51.2}$$

and

$$\exists i \in \{1, \ldots, k\} : z_i^1 < z_i^2. \tag{51.3}$$

If $z^1 = z^2$ or $z_i^1 > z_i^2$ for some i, then we say that z^1 does not dominate z^2 (denoted by $z^1 \nprec z^2$). Thus, to solve an MOP, we have to find those solutions $x \in X$ whose images, $z = f(x)$, are not dominated by any other vector in the feasible space. It is said that two vectors, z^1 and z^2, are *mutually nondominated vectors* if $z^1 \nprec z^2$ and $z^2 \nprec z^1$.

Definition 51.3 Pareto optimality

A solution $x^* \in X$ is Pareto optimal if there does not exist another solution $x \in X$ such that $f(x) \prec f(x^*)$.

Definition 51.4 Pareto optimal set

The Pareto optimal set, P_{opt}, is defined as

$$P_{\text{opt}} = \{x \in X \mid \nexists y \in X : f(y) \prec f(x)\}. \tag{51.4}$$

Definition 51.5 Pareto front

For a Pareto optimal set P_{opt}, the Pareto front PF_{opt} is defined as

$$PF_{\text{opt}} = \{z = (f_1(x), \ldots, f_k(x)) \mid x \in P_{\text{opt}}\}. \tag{51.5}$$

In decision variable space, these vectors are referred to as decision vectors of the Pareto optimal set, while in objective space, they are called objective vectors of the Pareto optimal set. In practice, the goal of MOEAs is to find the *best* approximation set of the Pareto optimal front. An approximation set is a finite subset of Z composed of mutually nondominated vectors and is denoted by PF_{approx}. Currently, it is well accepted that the best approximation set is determined by the closeness to the Pareto optimal front, and the spread over the entire Pareto optimal front [51.2, 14, 15].

A common approach to deal with multiobjective optimization problems is formulating it as a single optimization problem by means of a kind of function called scalarizing function.

Definition 51.6 Scalarizing function

A scalarizing function is a parameterized function $s : \mathbb{R}^k \to \mathbb{R}$. Thus, the multiobjective problem is transformed into the following scalar problem:

$$\text{Minimize } s(z),$$
$$\text{subject to } z \in Z. \tag{51.6}$$

It is worth noting, however, that scalarizing functions generate one point at a time (instead of several, as happens when using the definition of Pareto optimality). A common scalarizing function is based on the Chebyshev distance (L_∞ metric) [51.16, 17].

Definition 51.7 Weighted Chebyshev scalarizing function

The weighted Chebyshev scalarizing function (or Chebyshev function for short) is defined by

$$s_\infty(z, z^{\text{ref}}) = \max_{i=1,\dots,k} \{\lambda_i(z_i - z_i^{\text{ref}})\}, \tag{51.7}$$

where z^{ref} is a reference point, $\lambda = [\lambda_1, \dots, \lambda_k]$ is a vector of weights such that $\forall i \ \lambda_i \geq 0$ and, for at least one i, $\lambda_i > 0$.

51.2.2 Notions of Conflict Among Objectives

One important condition of a multiobjective problem is the conflict among their objectives. If the objectives have no conflict among them, then we could solve the problem optimizing each objective function independently. Nonetheless, it has been found that in some problems, although a conflict exists elsewhere, some objectives behave in a nonconflicting manner. Although different authors have proposed definitions for conflict (nonconflict) among objectives [51.18–21], in this chapter we only present conflict (nonconflict) definitions relevant to this document.

Definition 51.8

Let S_X be a subset of X, then, according to Carlsson and Fullér, two objectives can be related in the following ways (assuming minimization):

1. f_i is in conflict with f_j on S_X if $f_i(x^1) \leq f_i(x^2)$ implies $f_j(x^1) \geq f_j(x^2)$ for all $x^1, x^2 \in S_X$.

2. f_i supports f_j on S_X if $f_i(x^1) \geq f_i(x^2)$ implies $f_j(x^1) \geq f_j(x^2)$ for all $x^1, x^2 \in S_X$.
3. f_i and f_j are independent on S_X, otherwise.

In the cases 2 and 3, those objectives are also called nonconflicting objectives. When $S_X = X$, it is said that f_i is in conflict with (or supports) f_j globally. However, in many MOPs the relation among the objectives changes when comparing different subsets of X. Figure 51.1 shows an example in which two functions are in conflict in some subsets of X, while in others, they support each other.

Nonconflicting objectives are also known as nonessential or redundant objectives because, as pointed out by *Gal* and *Hanne* [51.22], when a nonconflicting objective is removed from the original set of objectives, the resulting Pareto front does not change. Based on the notion of nonessential objectives, *Brockhoff* and *Zitzler* [51.21] proposed a conflict definition that verifies whether the Pareto dominance relation changes when some objectives are removed, or not. The Pareto dominance relation induced by a given set of objectives, $F \subseteq \{f_1, f_2, \dots, f_k\}$, is defined as

$$\preceq_F = \{(x, y) \mid x, y \in X \text{ and } \forall f_i \in F : f_i(x) \leq f_i(y)\}.$$

Definition 51.9

Let $F_1, F_2 \subseteq \Phi$ be two subsets of objectives, where Φ is the entire set of objectives $\Phi = \{f_1, f_2, \dots, f_k\}$. Then, we call F_1 nonconflicting with F_2 iff $(\preceq_{F_1} \subseteq \preceq_{F_2}) \wedge (\preceq_{F_2} \subseteq \preceq_{F_1})$.

In other words, F_1 and F_2 are called nonconflicting if and only if the corresponding relations $\preceq_{F_1}$ and $\preceq_{F_2}$

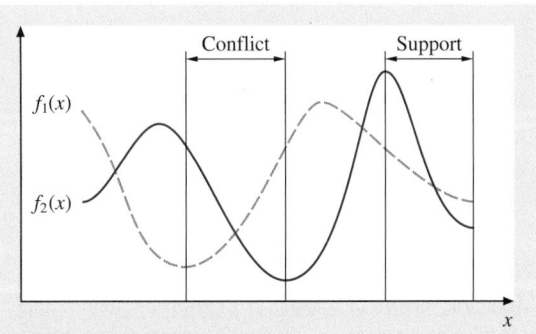

Fig. 51.1 Two objective functions can be in conflict in some subsets of the feasible space, and can be supportive in other subsets

are identical, but not necessarily $F_1 = F_2$. The noncon-flicting definition is useful since if F and $F' \subset F$ are nonconflicting, then we can replace F with F' and obtain the same Pareto optimal front. The objectives in F' are then called essential objectives, whereas the objectives in $F \setminus F'$ are known as nonessential or redundant objectives.

In practice, however, it is useful to allow a certain extent of change on the Pareto front when an objective is omitted in order to define degrees of nonconflict among objectives. In this direction, Brockhoff and Zitzler proposed to use the additive ϵ-dominance indicator to measure the change between two dominance rela-tions. The ϵ-dominance relation induced by a set F is defined by $\preceq_F^\epsilon = \{(x, y) \mid x, y \in X \text{ and } \forall f_i \in F : f_i(x) - \epsilon \leq f_i(y)\}$.

Definition 51.10
Let $F_1, F_2 \subseteq F$ be two subsets of objectives, where F is the entire set of objectives. Then, we call F_1 δ-nonconflicting with F_2 iff $(\preceq_{F_1} \subseteq \preceq_{F_2}^\delta) \wedge (\preceq_{F_2} \subseteq \preceq_{F_1}^\delta)$.

In this case, if an objective subset $F' \subset F$ is δ-nonconflicting with F, then we can omit all objectives in $F \setminus F'$ without causing a larger error than δ in the omitted objectives.

51.3 Sources of Difficulty to Solve Many-Objective Optimization Problems

51.3.1 Deterioration of the Search Ability

A widespread explanation for this problem is based on the fact that the proportion of nondominated solu-tions (i. e., equally good solutions according to Pareto dominance) in a population increases rapidly with the number of objectives [51.23, 24]. In order to illustrate this condition, Fig. 51.2 shows the nondominated re-gions with respect to a given solution z.

In general, as presented by *Farina* and *Am-ato* [51.23], the expression to compute the proportion, e, of mutually nondominated regions and the whole search space is given by $e = (2^k - 2)/2^k$, where k is the num-ber of objectives. This proportion goes to infinity when the number of objectives approaches infinity.

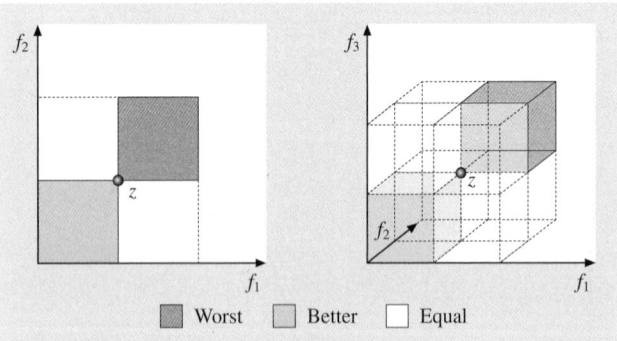

Fig. 51.2 Example of the increasing proportion of nondominated solutions: for 2 objectives $1/2$ of the search space is composed of nondominated regions, whereas for 3 objectives $3/4$ of the search space consists of nondominated regions. In general, for k objectives, $(2^k - 2)/2^k$ of the objective space comprises nondominated regions

Therefore, since in MOPs with a high number of objectives almost all solutions are equivalent, many re-searchers have suggested [51.11, 23, 25–28] that in such problems, the selection of the appropriate individuals for steering the population toward the Pareto optimal set gets more difficult. As a result, an MOP gets harder to solve as more objectives are added.

However, as pointed out by *Schütze* et al. [51.29], the increase of the number of nondominated individ-uals is not a sufficient condition for an increase of the hardness of a problem. Specifically, they conclude that in a class of uni-modal problems, their diffi-culty is marginally increased when more objectives are added despite the exponential growth of the propor-tion of nondominated solutions with k. Nonetheless, they suggest that the hardness increase observed in ex-perimental studies might be the result of the addition of local optima to the problem as more objectives are aggregated.

Therefore, although the rise of the proportion of incomparable solutions does not significantly deter-mine the difficulty of an MOP per se, it seems that the addition of objectives aggravates some particular difficulties observed in the context of 2 or 3 objec-tives. This is the case of the so-called dominance resistant solutions (DRSs) or outliers [51.14, 30–32]. DRSs are solutions with a poor value in at least one of the objectives, but with near optimal values in the others. In other words, those are nondomi-nated solutions, but far from the Pareto optimal front. Figure 51.3 shows an example of DRSs in the well-known test problem DTLZ2 [51.14]. These kinds of solutions represent potential difficulty since, as many researchers have pointed out [51.14, 30–32], the number of DRSs grows as the number of objectives is increased.

51.3.2 Effectiveness of Crossover Operators

In a combinatorial class of MOPs, *Sato* et al. [51.33] performed a series of experiments that revealed that solutions in the variable space become more distant (in terms of the Hamming distance between binary encoded solutions.) from each other as more objectives are added to the problem. In this scenario, the recombination of two parents close to the Pareto front might generate an offspring far from the Pareto front since a conventional crossover operator might be too disruptive.

51.3.3 Dimensionality of the Pareto Front

Due to the *curse of dimensionality*, the number of points required to represent accurately a Pareto front increases exponentially with the number of objectives. Formally, the number of points necessary to represent a Pareto front with k objectives and resolution r is bounded by $O(kr^{k-1})$ [51.34]. This expression is derived assuming that each solution is contained inside a hypercube to preserve an even distribution. As can be seen in Fig. 51.4, the number of hypercubes determines the resolution of the Pareto front, i. e., r is the number of hypercubes per dimension. An example of the shortest connected and nondegenerated 2-objective Pareto front (a straight line) is shown on the left side of Fig. 51.4. The figure also shows a bound for the largest Pareto front for 2 and 3 objectives. In general, the bounding Pareto front is formed by k hyperplanes containing r^{k-1} hypercubes each (see, for example, the 3-objective case shown on the right side of Fig. 51.4). This way, the maximum number of points of a 2-objective Pareto front with resolution $r = 6$ is $2 \times 6^{2-1} = 12$, whereas for 3 objectives and $r = 5$ is $3 \times 5^{3-1} = 75$. Table 51.1 shows the maximum number of points required to represent a Pareto front for different numbers of objectives using a resolution of $r = 25$, which is a conservative number considering that a resolution of $r = 50$ is usually used in several studies to obtain 100 solutions in 2-objective problems. Notwithstanding, for 5 objectives, we would require approximately 2 million points to represent a Pareto front with resolution $r = 25$. There are other formulations leading to a similar exponential expression with respect to k. For example, using the concept of ϵ-dominance, *Laumanns* et al. [51.35]

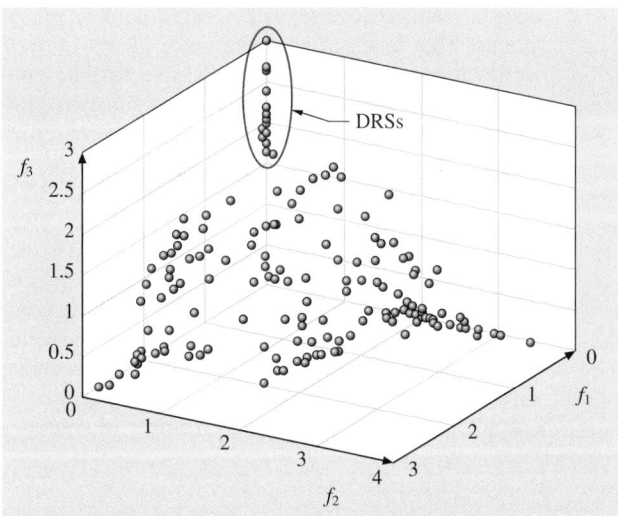

Fig. 51.3 Illustration of some DRSs in problem DTLZ2: although solutions marked as DRSs seem to be dominated by some solution in the lower part of the circled solutions, they achieve marginal improvements in objectives f_1 or f_2, and therefore, they are nondominated solutions, but having poor values in objective f_3, though

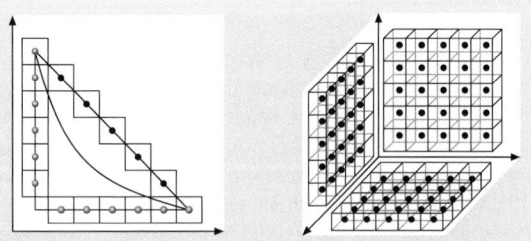

Fig. 51.4 Number of points required to represent a Pareto front with a resolution r, i. e., the number of hypercubes per dimension

Table 51.1 Bound for the number of points required to represent a Pareto front with resolution $r = 25$

k	Points
2	50
4	62 500
5	1 953 125
7	1 708 984 375

and *Schütze* et al. [51.36] give a similar exponential bound for the size of an approximation of a Pareto front.

This poses some difficulties to solve MOPs. The most important one is the number of function evaluations required to deal with a large number of solutions. This is a serious issue since plenty of real-world problems (e.g., [51.37–43]), due to time constraint reasons, have a small budget of function evaluations. In fact, there is an important research effort toward designing MOEAs that generate good approximations of the Pareto front using less than 1000 function evaluations (e.g., [51.44–47]). Other challenges are related to the design of both data structures to efficiently manage that number of points, and density estimators to achieve an even distribution of the solutions along the Pareto front. Unfortunately, even if we could efficiently obtain an accurate approximation of the Pareto front, the selection of one solution among such a huge number of solutions would be a very difficult task for a decision maker (DM).

51.3.4 Visualization of the Pareto Front

Clearly, with more than three objectives it is not possible to plot the Pareto front as usual. This is a serious problem since visualization plays a key role for a proper decision-making process. Parallel coordinates [51.48] and self-organizing maps [51.49] are some of the methods proposed to ease decision making in high dimensional problems. The reader is referred to Chapters 8 and 9 of [51.50] for a good review of various visualization techniques. Nevertheless, more research in the many-objective optimization context is still required.

51.4 Current Approaches to Deal with Many-Objective Problems

Besides studies about the scalability of Pareto-based MOEAs, in the current literature we can find several proposals to overcome those scalability issues. The most common approaches can be categorized as follows:

1. Adopt or propose a preference relation that yields a finer solution ordering than the one yielded by Pareto optimality. In other words, these relations are able to further rank nondominated solutions. In addition, most of these preference relations share the property that their optimal set of solutions is a subset of the Pareto optimal set. Therefore, these techniques can also be used as a remedy to cope with the dimensionality of Pareto fronts in MOPs.
2. Reduce the number of objectives of the problem during the search process or, a posteriori, once an approximation of the Pareto front has been found [51.21, 26, 51]. The main goal of these kinds of reduction techniques is to identify the nonconflicting objectives (at least to a certain extent) in order to discard them.
3. Scalarizing decomposition of an MOP. As described in the previous section, the degradation observed on MOEAs when dealing with many-objective problems is mainly attributed to the inefficiency of the Pareto relation in high-dimensional spaces. Therefore, methods that do not rely on Pareto dominance, like scalarizing decomposition methods, have been suggested as an alternative to

deal with many-objective problems. The underlying idea of these types of methods is to perform a number of single-objective searches along different search vectors evenly distributed over the objective space. Each single-objective search is formulated by means of a scalarizing function. This way, the approximation of the Pareto front is composed of the optima found by every single-objective search.

4. Incorporation of preference information interactively throughout the course of the optimization process. By incorporating preferences we can cope with MOPs in two aspects. First, the search can be focused on the decision maker's region of interest, avoiding this way, the evaluation of a huge number of solutions. Second, the preference relations usually used in interactive methods help to deal with a large number of objectives since they are able to rank incomparable nondominated solutions.
5. Use of specialized recombination operators or strategies to control the mating among parents. The first approach tries to diminish the disruptive effect of recombination operators by regulating the proportion in which the traits of each parent contribute to create the offspring. The second approach restricts which individuals can be paired for recombination, for instance, using the similarity as mating criteria or the location in the objective space.

In the remainder of this section, some of the most relevant approaches to deal with many-objective problems are presented.

51.4.1 Preference Relations to Deal with Many-Objective Problems

Bentley and *Wakefield* [51.52] proposed the average ranking (AR) and the maximum ranking (MR) preference relations. The AR relation computes, for each solution, a different rank considering each objective independently. The final rank is obtained by summing up the ranks on each objective. In turn, the MR relation takes the best rank as the global rank. Clearly, this method favors extreme solutions, i.e., solutions with high performance in some of the objectives, although with poor overall performance. Although it is less evident, the average ranking relation also favors extreme solutions.

In the *favor relation*, proposed by *Drechsler* et al. [51.53], a vector z^1 is preferred to vector z^2 with respect to the favor relation ($z^1 \prec_{\text{favor}} z^2$), if and only if

$$\sharp\{i : z_i^1 < z_i^2, 1 \leq i \leq k\}$$
$$> \sharp\{j : z_j^1 > z_j^2, 1 \leq j \leq k\} \, .$$

In other words, the favored vector is that which outperforms the other one in more objectives. Unfortunately, this relation emphasizes extreme solutions.

The preference order relation (POR), developed by *di Pierro* [51.54], is based on the concept of *efficiency of order* proposed by *Das* [51.55], which states that: A vector z^* is efficient of the order q if it is not dominated by any other vector in all the $\binom{k}{q}$ objective subsets of size q.

Based on that definition, it is said that the vector z^1 is preferred to the vector z^2 ($z^1 \prec_{\text{POR}} z^2$), if and only if, for some integer q and $\forall I \subseteq \{1, 2, \ldots, k\}$ such that $|I| = q$

$$z_i^1 \leq z_i^2 \quad \forall i \in I, \quad \text{and} \quad \exists i \in I : z_i^1 < z_i^2 \, .$$

In other words, if z^1 and z^2 do not dominate each other, then the solutions are compared in a lower dimensional space in order to break the tie.

Sato et al. [51.56] proposed a preference relation to control the dominance area of solutions. This method controls the degree of expansion or contraction of the dominance area by modifying each objective vector z with the expression

$$z_i' = \frac{r \cdot \sin(\omega_i + s_i \cdot \pi)}{\sin(s_i \cdot \pi)} \quad \forall i = 1, 2, \ldots, k \, ,$$

where $s \in \mathbb{R}^k$ is a user-defined vector, $r = ||z||$, and ω_i is the declination angle between z and the axis of f_i.

If the user adopts values $s_i < 0.5$ ($\forall i = 1, 2, \ldots, k$), the dominance area is expanded and produces a more fine-grained ranking of solutions which would strengthen the selection process. Thus, we can say that vector z is preferred to vector y with respect to the *expansion relation* ($z \prec_{\text{expansion}} y$), if and only if $z' \prec y'$.

Farina and *Amato* [51.57] proposed an alternative relation which takes into account the number of improved objectives between two solutions. This relation employs three quantities, $n_b(\mathbf{x}_1, \mathbf{x}_2)$, $n_e(\mathbf{x}_1, \mathbf{x}_2)$ and $n_w(\mathbf{x}_1, \mathbf{x}_2)$, which denote the objectives where $\mathbf{x}_1$ is better, equal or worse than $\mathbf{x}_2$, respectively. Using these quantities, the concepts of $(1-k)$-*dominance* and k-*optimality* are defined. A solution $\mathbf{x}_1$ $(1-k)$ dominates $\mathbf{x}_2$ if and only if

$$\begin{cases} n_e(\mathbf{x}_1, \mathbf{x}_2) < M \\ n_b(\mathbf{x}_1, \mathbf{x}_2) \geq \dfrac{M - n_e}{k + 1} \end{cases} .$$

In a similar way to Pareto optimality, a solution $\mathbf{x}^*$ is a k-*optimum* if and only if there is no $\mathbf{x}$ in the decision variable space such that $\mathbf{x}$ k-dominates $\mathbf{x}^*$.

An important remark that we have to keep in mind with respect to a new preference relation is that in spite of the fact that some preference relations contribute to converge faster to the Pareto front than the Pareto dominance relation, they also stress the generation of solutions far from the knee region (usually the middle region of the Pareto front). This condition limits the applicability of these relations since, in the general case, it is commonly assumed that the DM prefers solutions from the knee region [51.58–61].

51.4.2 Objective Reduction Approaches

Deb and *Saxena* [51.26] proposed a method for reducing the number of objectives based on principal component analysis. The main assumption is that if two objectives are negatively correlated (taking the generated Pareto front as the data set), then these objectives are in conflict with each other. To determine the most conflicting objectives (i. e., the most essential), the authors analyze in turn the eigenvectors (i. e., the principal components) of the correlation matrix. That is, by picking the most negative and the most positive elements from the first eigenvector, we can identify the two most important conflicting objectives. To aggregate more objectives to the set of essential objectives, the remainder of the eigenvectors are analyzed in a similar way

Part E | 51.4

until the cumulative contribution of the eigenvalues exceeds a threshold cut (*TC*). This method is incorporated into an iterative scheme which uses a multiobjective optimizer (the actual implementation uses the nondominated sorting genetic algorithm II (NSGA-II) [51.62]) to obtain a reduced objective set containing only the nonredundant objectives according to the analysis of the eigenvectors. In this scheme, the evolutionary multiobjective optimizer is first run and then, the correlation analysis is carried out to obtain a reduced set of objectives. This process is repeated using the new reduced set of objectives. The process stops when the current subset is equal to the subset generated in the previous iteration.

Brockhoff and *Zitzler* [51.21] defined two kinds of objective reduction problems and two corresponding algorithms to solve them. The problems proposed are the following:

1. *The δ-MOSS problem.* Given an MOP, the δ-minimum objective subset problem is defined as follows.
   - *Input*: A Pareto front approximation of the MOP and a $\delta \in \mathbb{R}$.
   - *Task*: Compute the minimum objective subset $F' \subseteq F$ such that F' is δ-nonconflicting with F.
2. *The K-EMOSS problem.* Given an MOP, the problem of finding the minimum objective subset of size K with minimum error is defined as follows.
   - *Input*: A Pareto front approximation of the MOP and a $\mathcal{K} \in \mathbb{N}$.
   - *Task*: Compute an objective subset $F' \subseteq F$ with size $|F'| \leq \mathcal{K}$, such that F' is δ-nonconflicting with F with the minimum possible δ.

Since both problems are $\mathcal{NP}$-hard, the authors proposed both an exact and a greedy algorithm for each of them. The exact algorithms for both problems have time complexity $O(m^2 k \cdot 2^k)$, where m is the size of the given nondominated set and k is the number of objectives. On the other hand, the greedy algorithm for the δ-MOSS problem has time complexity $O(\min\{m^2 k^3, m^4 k^2\})$, while the greedy algorithm for the K-EMOSS problem has time complexity $O(m^2 k^3)$.

A similar approach was proposed by *López Jaimes* et al. [51.51]. They proposed two different objective reduction algorithms:

1. An algorithm that finds a minimum subset of nonredundant objectives with the minimum error possible.

2. An algorithm that finds a K-size subset of nonredundant objectives, yielding the minimum error possible.

Both algorithms are based on an unsupervised feature selection technique proposed by *Mitra* et al. [51.63], in which the correlation coefficient is used to estimate the conflict among objectives. Specifically, a negative correlation between a pair of objectives means that one objective increases, while the other decreases and vice versa (see, for example, the functions in Fig. 51.1). On the other hand, if the correlation is positive, then both objectives increase or decrease at the same time. This way, we could interpret that the more negative the correlation between two objectives, the more the conflict between them.

These two algorithms were designed to be used after an approximation of the Pareto front has been found. From a general point of view, the removal of the nonconflicting objectives can help to the problem designer or the decision maker to gain knowledge about the relation and importance of the objectives according to the conflict. With regard to the decision-making process, the removal of the nonconflicting objectives eases the visualization of the approximation of the Pareto front. In cases with a moderate number of objectives (i. e., 4–7), the reduced objective set might be visualized using the traditional 3D plots.

However, an objective reduction technique can also be used in the course of the search. In [51.64], for instance, the authors proposed the incorporation of an objective reduction technique into a Pareto-based MOEA in order to cope with many-objective problems during the search. One possible approach is gradually reducing the number of objectives throughout different stages of the search until a target objective subset size has been reached. In each reduction stage, an objective reduction method is applied on the current Pareto front approximation. Toward the end of the search, the original objective set is used again to approximate the entire Pareto front. This kind of approach can be advantageous for solving real-world problems with expensive objective functions since only a small subset of the objective functions is evaluated. Additionally, the use of a small set of objectives throughout the course of the search makes possible the adoption of expensive ranking schemes (e.g., those based on the hypervolume indicator) in problems with a high number of objectives [51.65].

A further approach, presented in [51.66], consists in partitioning the objective set into several subsets so that a different portion of the population focuses the search on a different subspace. The partitioning of the set of objectives is based on the analysis of the conflict information obtained from the current Pareto front approximation.

51.4.3 Preference Incorporation Approaches

Like the alternative preference relations reviewed in Sect. 51.4.1, the integration of DM's preferences provides a finer rank of the solutions. However, unlike preference relation approaches, in an interactive approach, the region of interest can be changed during the search according to the requirements of the decision maker.

Among the earliest attempts to incorporate preferences in an MOEA, we can find *Fonseca* and *Fleming*'s proposal [51.67, 68]. This proposal consisted of extending the ranking mechanism of multiobjective genetic algorithm (MOGA) [51.69] using the so-called *preferability relation*. This relation accommodates goal information (equivalent to a reference point in other methods) and priorities in a single preference relation. The DM should define goal values and group objectives according to its priority. Using the preferability relation, two solutions are first compared in terms of the group of objectives with the highest priority. If the objectives of both solutions meet all their goal values or, contrarily, violate some or all of their goal values in a similar way, the next priority objective group is considered. This process continues until reaching the lowest priority group, where solutions are compared using the Pareto dominance relation. By setting particular goals and priorities the authors derived the following special cases: the usual Pareto relation, lexicographic relation, constrained optimization, and goal programming. One disadvantage of this relation is that it is affected by the feasibility of the goal provided by the decision maker. If the given goal is far away from the feasible region, then the solutions will be mainly compared in terms of the objective priorities, reducing the relation to the lexicographic relation. In addition, if two solutions either do or do not meet their goals, the relation does not take into account the degree of under- or over-attainment.

Deb [51.70] proposed a technique to transform goal programming problems into multiobjective optimization problems which are then solved using an MOEA. In goal programming, the DM has to assign goals that wishes to achieve for each objective, and these values are incorporated into the problem as additional constraints. The objective function then attempts to minimize the absolute deviations from the goals to the objectives. Unfortunately, as the previous method, this approach is sensitive to the feasibility of the goal values. If the goal is contained in the feasible space, it could prevent the generation of a better solution. On the other hand, if the goal is located far away from the feasible space, the effect of the method is practically nonexistent.

More recently, *Deb* and *Sundar* [51.71] incorporated a reference point approach into the NSGA-II [51.72]. They introduced a modification in the crowding distance operator in order to select from the last nondominated front the solutions that would take part of the new population. They used the Euclidean distance to sort and rank the population accordingly (the solution closest to the reference point receives the best rank). This method was designed to take into account a set of reference points. The drawback of this scheme is that it only guarantees weak Pareto optimality. That is to say, besides Pareto optimal solutions, the method might generate some weakly Pareto optimal solutions, particularly in MOPs with disconnected Pareto fronts. A similar approach was also proposed by *Deb* and *Kumar* [51.73], in which the light beam search procedure [51.74] was incorporated into the NSGA-II. Similar to the previous approach, they modified the crowding operator to incorporate DM's preferences. They used a weighted achievement function to assign a crowding distance to each solution in each front. Thus, the solution with the least distance will have the best crowding rank. Like in the previous approach, this algorithm finds a subset of solutions around the optimum of the achievement function adopting the usual outranking relation. A vector z^1 outranks vector z^2 if z^1 is considered to be at least as good as z^2. In [51.74], three kinds of thresholds are defined to determine if one solution outranks another one, namely, indifference, preference, and veto threshold. However, in [51.73] the veto threshold is the only one used. This relation depends on the crowding comparison operator. In contrast, the new preference relation presented in this work does not depend on external methods, and, therefore, it can be used in every Pareto-based MOEA.

Recently, *Thiele* et al. [51.75] proposed a variant of the indicator-based evolutionary algorithm (IBEA) [51.76], in which preference information is incorporated by means of an achievement scalarization function. The basic idea is to divide the original indicator value (which is to be maximized) by the

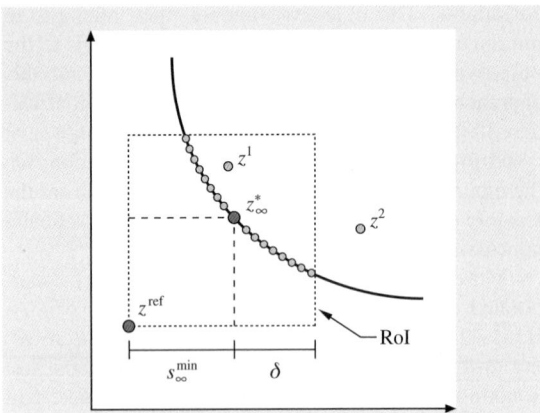

Fig. 51.5 Nondominated solutions with respect to the Chebyshev relation

achievement value (which is to be minimized). Thus, solutions with a smaller achievement value will be preferred since the modified indicator value is larger. In a further paper, the new IBEA of Thiele et al. was used in [51.77] in order to approximate the entire Pareto front by defining several reference points.

A recent interactive optimization method was proposed by *López Jaimes* et al. [51.78] to deal with

MOPs. This method is based on a Chebyshev achievement function. The basic idea of the Chebyshev preference relation is to combine the Pareto dominance relation and the achievement function to compare solutions in objective function space. The Chebyshev preference relation is defined as follows.

Definition 51.11
A solution z^1 is preferred to solution z^2 with respect to the Chebyshev relation ($z^1 \prec_{\text{cheby}} z^2$), if and only if:

1. $s_\infty(z^1, z^{\text{ref}}) < s_\infty(z^2, z^{\text{ref}}) \wedge \{z^1 \notin R(z^{\text{ref}}, \delta) \vee z^2 \notin R(z^{\text{ref}}, \delta)\}$, or,

2. $z^1 \prec z^2 \wedge \{z^1, z^2 \in R(z^{\text{ref}}, \delta)\}$,

where

$$R(z^{\text{ref}}, \delta) = \{z \mid s_\infty(z, z^{\text{ref}}) \leq s^{\text{min}} + \delta\}$$

is the region of interest (ROI) with respect to the vector of aspiration levels z^{ref}.

As an illustration of the preference relation, consider solutions z^1 and z^2 presented in Fig. 51.5. Since $z^2 \notin R(z^{\text{ref}}, \delta)$ and $s_\infty(z^1, z^{\text{ref}}) < s_\infty(z^2, z^{\text{ref}})$, then $z^1 \prec_{\text{cheby}} z^2$.

51.5 Recombination Operators and Mating Restrictions

The idea of restricted mating is not new in the field of evolutionary optimization. For instance, in 1989 *Deb* and *Goldberg* [51.79] suggested the use of restrictive mating with respect to the phenotypic (i. e., using the decoded values of the variables) distance using some metric. A different approach consisted in distributing solutions on a logical topology. For example, *Baita* et al. [51.80] placed solutions on a grid and restricted the area within which each solution could mate. For more examples of restricted mating, the reader is referred to [51.2].

Recently, specific mating techniques to deal with many-objective problems have been proposed. *Sato* et al. [51.81] described a local recombination scheme that recombines individuals if they have similar search directions in the objective space. The search direction is defined by the polar coordinates of each solution, i. e., its norm and declination angles to the axis associated with the first $k-1$ objectives.

In order to control the disruptive effect of recombination, in [51.33], a crossover operator for binary representation was proposed, namely the controlling crossed genes (CCG) operator. This technique was applied into the two-point and uniform-crossover operators. In two-point crossover, from the three binary segments in which two parents are divided, the middle segment is exchanged between the parents to produce two children. Thus, in the CCG operator for two-point crossover, the length of the middle segment is regulated by a user parameter. This way, as the middle segment gets shorter, the generated children become more similar to each parent.

Regarding uniform crossover, the number of exchanged bits between parents is regulated with the probability of writing a 1 or a 0 in the bit mask string that determines which parent bit will be copied into the produced offspring.

51.6 Scalarization Methods

Most of the scalarization methods have in common the following mechanisms (although they differ in the way in which they are implemented):

- A class of scalarizing function to evaluate solutions.
- A mechanism to generate a uniform distribution of search direction vectors.
- A mechanism to obtain an overall ranking of the solutions derived from the evaluation of each scalarizing function.

Hughes [51.82] proposed a method in which the weighted Chebyshev function and the vector angle distance scaling are used as scalarizing functions. The method to generate the search direction is formulated as the problem of maximizing the angle between each pair of neighboring search vectors. The fitness of each solution in the current population is based on the best result obtained over all the scalarizing function, i. e., the search direction in which the solution performs better.

Another algorithm that has been recently tested in many-objective problems is the multiobjective evolutionary algorithm based on decomposition (MOEA/D) [51.83]. In [51.84], the performance of MOEA/D using either a weighted sum function or a Chebyshev function was studied using several instances of a knapsack problem. The results showed that the weighted sum function provided better results than the Chebyshev function, while in nonconvex problems, the Chebyshev function helped to achieve a better performance of MOEA/D.

51.7 Conclusions and Research Paths

This chapter presented a short review of the current advances to cope with optimization problems with a high number of objectives MOPs using MOEA. We covered results aimed at discovering and studying the causes that make an MOP more difficult as more objectives are aggregated. We also described and classified some of the current techniques to deal with MOPs.

Regarding the sources of difficulty of many-objective optimization problems, we can realize that most of the initial works are based on experimental analysis, and only a few studies are focused on investigating the nature of the problem using theoretical considerations. When the interest on many-objective optimization problems begun, some hypotheses about the causes of the poor performance of MOEA on MOPs were suggested. Although some of them were considered highly probable and may turn out to be true, further investigation is still needed to confirm or refute these hypotheses. This was the case of the proportion of nondominated solutions, which was often taken as a sufficient condition to increase the difficulty of an MOP. However, recent studies have shown that there exists some problems, in which this proportion rises exponentially, while the hardness of the problem only increases marginally. In this sense, future research paths must be channeled to investigate other sources of difficulty. Some promising areas of future research are, for example, the following:

- Since DRS are not present in every MOP, a characterization of the problems that promote the creation of DRSs is required.
- Investigate if recombination operators in continuous spaces also represent an issue as observed in discrete spaces.

Regarding the methods to solve MOPs, many proposals have been designed to improve the search ability of MOEAs in high-dimensional scenarios. However, a few efforts are perceived for developing visualization methods specialized for MOPs. Similarly, more proposals for coping with the dimensionality of the Pareto front are needed. For instance, diversity mechanisms that are effective in large spaces or data structures to efficiently manage a large number of solutions. With respect to the assessment of a new MOEA in many-objective scenarios, our recommendation is adopting a diverse set of MOPs, taking instances from different families of test suites.

References

51.1   J.D. Schaffer: Multiple objective optimization with vector evaluated genetic algorithms, Proc. 1st Int. Conf. Genet. Algorithms (1985) pp. 93–100

51.2   C.A. Coello Coello, G.B. Lamont, D.A. Van Veldhuizen: *Evolutionary Algorithms for Solving Multi-Objective Problems*, 2nd edn. (Springer, New York 2007)

51.3   C.M. Fonseca, P.J. Fleming: Multiobjective optimization and multiple constraint handling with evolutionary algorithms – Part II: Application Example, IEEE Trans. Syst. Man Cybern. Part A **28**(1), 38–47 (1998)

51.4   E.J. Hughes: Radar waveform optimisation as a many-objective application benchmark, Lect. Notes Comput. Sci. **4403**, 700–714 (2007)

51.5   T. Stewart, O. Bandte, H. Braun, N. Chakraborti, M. Ehrgott, M. Göbelt, Y. Jin, H. Nakayama, S. Poles, D. Di Stefano: Real-world applications of multiobjective optimization, Lect. Notes Comput. Sci. **5252**, 285–327 (2009)

51.6   R.A. Shah, P.M. Reed, T.W. Simpson: Many-objective evolutionary optimisation and visual analytics for product family design. In: *Multiobjective Evolutionary Optimisation for Product Design and Manufacturing*, ed. by L. Wang, A.H.C. Ng, K. Deb (Springer, London 2011) pp. 137–159

51.7   E.J. Hughes: Evolutionary many-objective optimisation: Many once or one many?, IEEE Congr. Evol. Comput. (CEC'2005), Vol. 1 (IEEE Service Center, Edinburgh 2005) pp. 222–227

51.8   T. Wagner, N. Beume, B. Naujoks: Pareto-, aggregation-, and indicator-based methods in many-objective optimization, Lect. Notes Comput. Sci. **4403**, 742–756 (2007)

51.9   K. Praditwong, X. Yao: How well do multi-objective evolutionary algorithms scale to large problems, IEEE Congr. Evol. Comput. (CEC'2007) (IEEE, Singapore 2007) pp. 3959–3966

51.10   O. Teytaud: On the hardness of offline multi-objective optimization, Evol. Comput. **15**, 475–491 (2007)

51.11   J. Knowles, D. Corne: Quantifying the effects of objective space dimension in evolutionary multiobjective optimization, Lect. Notes Comput. Sci. **4403**, 757–771 (2007)

51.12   V. Pareto: *Cours D'Economie Politique* (F. Rouge, Paris 1896)

51.13   F.Y. Edgeworth: *Mathematical Psychics* (P. Keagan, London 1881)

51.14   K. Deb, L. Thiele, M. Laumanns, E. Zitzler: Scalable multi-objective optimization test problems, Congr. Evol. Comput. (CEC'2002), Vol. 1 (IEEE Service Center, New Jersey 2002) pp. 825–830

51.15   E. Zitzler, L. Thiele, M. Laumanns, C.M. Fonseca, V.G. da Fonseca: Performance assessment of multiobjective optimizers: An analysis and review, IEEE Trans. Evol. Comput. **7**(2), 117–132 (2003)

51.16   K.M. Miettinen: *Nonlinear Multiobjective Optimization* (Kluwer, Boston 1998)

51.17   M. Ehrgott: *Multicriteria Optimization*, 2nd edn. (Springer, Berlin 2005)

51.18   C. Carlsson, R. Fullér: Multiple criteria decision making: The case for interdependence, Comput. Oper. Res. **22**(3), 251–260 (1995)

51.19   R.C. Purshouse, P.J. Fleming: Conflict, harmony, and independence: Relationships in evolutionary multi-criterion optimisation, Lect. Notes Comput. Sci. **2632**, 16–30 (2003)

51.20   K.C. Tan, E.F. Khor, T.H. Lee: *Multiobjective Evolutionary Algorithms and Applications* (Springer, London 2005)

51.21   D. Brockhoff, E. Zitzler: Are all objectives necessary? On dimensionality reduction in evolutionary multiobjective optimization, Lect. Notes Comput. Sci. **4193**, 533–542 (2006)

51.22   T. Gal, T. Hanne: Consequences of dropping nonessential objectives for the application of MCDM methods, Eur. J. Oper. Res. **119**(2), 373–378 (1999)

51.23   M. Farina, P. Amato: On the optimal solution definition for many-criteria optimization problems, Proc. NAFIPS-FLINT Int. Conf. 2002 (IEEE Service Center, New Jersey 2002) pp. 233–238

51.24   P. Winkler: Random orders, Order **1**(4), 317–331 (1985)

51.25   R.C. Purshouse, P.J. Fleming: Evolutionary multi-objective optimisation: An exploratory analysis, Proc. Congr. Evol. Comput. (CEC'2003), Vol. 3 (IEEE, Canberra 2003) pp. 2066–2073

51.26   K. Deb, D.K. Saxena: Searching for Pareto-optimal solutions through dimensionality reduction for certain large-dimensional multi-objective optimization problems, IEEE Congr. Evol. Comput. (CEC'2006) (IEEE, Vancouver 2006) pp. 3353–3360

51.27   M. Köppen, K. Yoshida: Substitute distance assignments in NSGA-II for handling many-objective optimization problems, Lect. Notes Comput. Sci. **4403**, 727–741 (2007)

51.28   H. Ishibuchi, N. Tsukamoto, Y. Nojima: Evolutionary many-objective optimization: A short review, Congr. Evol. Comput. (CEC'2008) (IEEE Service Center, Hong Kong 2008) pp. 2424–2431

51.29   O. Schütze, A. Lara, C.A. Coello Coello: On the influence of the number of objectives on the hardness of a multiobjective optimization problem, IEEE Trans. Evol. Comput. **15**(4), 444–455 (2011)

51.30   K. Ikeda, H. Kita, S. Kobayashi: Failure of Pareto-based MOEAs: Does non-dominated really mean near to optimal?, Proc. IEEE Congr. Evol. Comput. (CEC'2001), Vol. 2 (IEEE Service Center, Piscataway 2001) pp. 957–962

51.31   T. Hanne: Global Multiobjective optimization with evolutionary algorithms: Selection mechanisms and mutation control, Lect. Notes Comput. Sci. **1993**, 197–212 (2001)

51.32   S. Huband, L. Barone, L. While, P. Hingston: A scalable multi-objective test problem toolkit, Lect. Notes Comput. Sci. **3410**, 280–295 (2005)

51.33   H. Sato, H. Aguirre, K. Tanaka: Genetic diversity and effective crossover in evolutionary many-objective optimization, Lect. Notes Comput. Sci. **6683**, 91–105 (2011)

51.34   P. Sen, J.-B. Yang: *Multiple Criteria Decision Support in Engineering Design* (Springer, London 1998)

51.35   M. Laumanns, L. Thiele, K. Deb, E. Zitzler: Combining convergence and diversity in evolutionary multiobjective optimization, Evol. Comput. **10**, 263–282 (2002)

51.36   O. Schütze, M. Laumanns, E. Tantar, C.A. Coello Coello, E.-G. Talbi: Computing gap free Pareto front approximations with stochastic search algorithms, Evol. Comput. **18**(1), 65–96 (2010)

51.37   A. Arias Montaño, C.A. Coello Coello, E. Mezura-Montes: Multi-objective evolutionary algorithms in aeronautical and aerospace engineering, IEEE Trans. Evol. Comput. **16**(5), 662–694 (2012)

51.38   J. Braun, J. Krettek, F. Hoffmann, T. Bertram: Multiobjective optimization with controlled model assisted evolution strategies, Evol. Comput. **17**(4), 577–593 (2009)

51.39   K. Chiba, A. Oyama, S. Obayashi, K. Nakahashi, H. Morino: Multidisciplinary design optimization and data mining for transonic regional-jet wing, AIAA J. Aircr. **44**(4), 1100–1112 (2007)

51.40   T. Kipouros, G.T. Parks, A.M. Savill, D.M. Jaeggi: Multi-objective aerodynamic design optimisation, ERCOF-TAC design optimization: Methods Appl. Conf. Proc., ed. by K.C. Giannakoglou, W. Haase (2004), on CD-ROM

51.41   P.M. Kruse, J. Wegener, S. Wappler: A highly configurable test system for evolutionary black-box testing of embedded systems, Proc. 11th Annu. Conf. Genet. Evol. Comput., GECCO '09 (ACM, New York 2009) pp. 1545–1552

51.42   P. Stewart, D.A. Stone, P.J. Fleming: Design of robust fuzzy-logic control systems by multi-objective evolutionary methods with hardware in the loop, Eng. Appl. Artif. Intell. **17**(3), 275–284 (2004)

51.43   P. Wozniak: Preferences in multi-objective evolutionary optimisation of electric motor speed control with hardware in the loop, Appl. Soft Comput. **11**(1), 49–55 (2011)

51.44   M. Emmerich, B. Naujoks: Metamodel assisted multiobjective optimisation strategies and their application in airfoil design. In: *Adaptive Computing in Design and Manufacture VI*, ed. by I.C. Parmee (Springer, London 2004) pp. 249–260

51.45   J. Knowles: ParEGO: A hybrid algorithm with online landscape approximation for expensive multiobjective optimization problems, IEEE Trans. Evol. Comput. **10**(1), 50–66 (2006)

51.46   C.A. Georgopoulou, K.C. Giannakoglou: A multiobjective metamodel-assisted memetic algorithm with strengthbased local refinement, Eng. Optim. **41**(10), 909–923 (2009)

51.47   S. Zapotecas Martínez, C.A. Coello Coello: A memetic algorithm with non gradient-based local search assisted by a meta-model, Lect. Notes Comput. Sci. **6238**, 576–585 (2010)

51.48   E.J. Wegman: Hyperdimensional data analysis using parallel coordinates, J. Am. Stat. Assoc. **85**, 664–675 (1990)

51.49   S. Obayashi, D. Sasaki: Visualization and data mining of pareto solutions using self-organizing map, Lect. Notes Comput. Sci. **2632**, 796–809 (2003)

51.50   J. Branke, K. Deb, K. Miettinen, R. Slowiński (Eds.): *Multiobjective Optimization: Interactive and Evolutionary Approaches*, Lecture Notes in Computer Science, Vol. 5252 (Springer, New York 2008)

51.51   A. López Jaimes, C.A. Coello Coello, D. Chakraborty: Objective reduction using a feature selection technique, Genet. Evol. Comput. Conf. (GECCO'2008) (ACM Press, Atlanta 2008) pp. 674–680

51.52   P.J. Bentley, J.P. Wakefield: Finding acceptable solutions in the Pareto-optimal range using multiobjective genetic algorithms. In: *Soft Computing in Engineering Design and Manufacturing Part 5*, ed. by P.K. Chawdhry, R. Roy, R.K. Pant (Springer, London 1997) pp. 231–240

51.53   N. Drechsler, R. Drechsler, B. Becker: Multi-objected optimization in evolutionary algorithms using satisfyability classes, Lect. Notes Comput. Sci. **1625**, 108–117 (1999)

51.54   F. di Pierro, S.-T. Khu, D.A. Savić: An investigation on preference-order ranking scheme for multiobjective evolutionary optimization, IEEE Trans. Evol. Comput. **11**(1), 17–45 (2007)

51.55   I. Das: A preference ordering among various Pareto optimal alternatives, Struct. Multidiscip. Optim. **18**(1), 30–35 (1999)

51.56   H. Sato, H.E. Aguirre, K. Tanaka: Controlling dominance area of solutions and its impact on the performance of MOEAs, Lect. Notes Comput. Sci. **4403**, 5–20 (2007)

51.57   M. Farina, P. Amato: A fuzzy definition of "optimality" for many-criteria optimization problems, IEEE Trans. Syst. Man Cybern. Part A **34**(3), 315–326 (2004)

51.58   I. Das: On characterizing the "knee" of the Pareto curve based on normal-boundary intersection, Struct. Optim. **18**(2/3), 107–115 (1999)

51.59   C.A. Mattson, A.A. Mullur, A. Messac: Smart Pareto filter: Obtaining a minimal representation of multiobjective design space, Eng. Optim. **36**(6), 721–740 (2004)

51.60   J. Branke, K. Deb, H. Dierolf, M. Osswald: Finding knees in multi-objective optimization, Lect. Notes Comput. Sci. **3242**, 722–731 (2004)

Part E | 51

51.61   O. Schütze, M. Laumanns, C.A. Coello Coello: Approximating the knee of an MOP with stochastic search algorithms, Lect. Notes Comput. Sci. **5199**, 795–804 (2008)

51.62   K. Deb, S. Agrawal, A. Pratap, T. Meyarivan: A fast elitist non-dominated sorting genetic algorithm for multi-objective optimization: NSGA-II, Lect. Notes Comput. Sci. **1917**, 849–858 (2000)

51.63   P. Mitra, C.A. Murthy, S.K. Pal: Unsupervised feature selection using feature similarity, IEEE Trans. Pattern Anal. Mach. Intell. **24**(3), 301–312 (2002)

51.64   A. López Jaimes, C.A. Coello Coello, J.E. Urías Barrientos: Online objective reduction to deal with many-objective problems, Lect. Notes Comput. Sci. **5467**, 423–437 (2009)

51.65   D. Brockhoff, E. Zitzler: Improving hypervolume-based multiobjective evolutionary algorithms by using objective reduction methods, 2007 IEEE Congr. Evol. Comput. (CEC'2007) (IEEE, Singapore 2007) pp. 2086–2093

51.66   A. López Jaimes, H. Aguirre, K. Tanaka, C.A. Coello Coello: Objective space partitioning using conflict information for many-objective optimization, Lect. Notes Comput. Sci. **6238**, 657–666 (2010)

51.67   C.M. Fonseca, P.J. Fleming: Genetic algorithms for multiobjective optimization: Formulation, discussion and generalization, Proc. 5th Int. Conf. Genet. Algorithms, ed. by S. Forrest (Morgan Kauffman, San Mateo 1993) pp. 416–423

51.68   C.M. Fonseca, P.J. Fleming: Multiobjective optimization and multiple constraint handling with evolutionary algorithms – Part I: A unified formulation, IEEE Trans. Syst. Man Cybern. Part A **28**(1), 26–37 (1998)

51.69   C.M. Fonseca, P.J. Fleming: An overview of evolutionary algorithms in multiobjective optimization, Evol. Comput. **3**(1), 1–16 (1995)

51.70   K. Deb: Solving goal programming problems using multi-objective genetic algorithms, Congr. Evol. Comput. 1999 (IEEE Service Center, Washington 1999) pp. 77–84

51.71   K. Deb, J. Sundar: Reference Point Based Multi-Objective Optimization Using Evolutionary Algorithms, 2006 Genetic Evol. Comput. Conf. (GECCO'2006), ed. by M. Keijzer (ACM, Seattle 2006) pp. 635–642

51.72   K. Deb, A. Pratap, S. Agarwal, T. Meyarivan: A fast and elitist multiobjective genetic algorithm:

NSGA-II, IEEE Trans. Evol. Comput. **6**(2), 182–197 (2002)

51.73   K. Deb, A. Kumar: Light beam search based multiobjective optimization using evolutionary algorithms, IEEE Congr. Evol. Comput. (CEC'2007) (IEEE, Singapore 2007) pp. 2125–2132

51.74   A. Jaszkiewicz, R. Słowiński: The light beam search approach – An overview of methodology and applications, Eur. J. Oper. Res. **113**(2), 300–314 (1999)

51.75   L. Thiele, K. Miettinen, P.J. Korhonen, J. Molina: A preference-based evolutionary algorithm for multi-objective optimization, Evol. Comput. **17**, 411–436 (2009)

51.76   E. Zitzler, S. Künzli: Indicator-based selection in multiobjective search, Lect. Notes Comput. Sci. **3242**, 832–842 (2004)

51.77   J.R. Figueira, A. Liefooghe, E.-G. Talbi, A.P. Wierzbicki: A parallel multiple reference point approach for multi-objective optimization, Eur. J. Oper. Res. **205**(2), 390–400 (2010)

51.78   A. López-Jaimes, A. Arias-Montaño, C.A. Coello Coello: Preference incorporation to solve many-objective airfoil design problems, IEEE Congr. Evol. Comput. (CEC'2011) (2011) pp. 1605–1612

51.79   K. Deb, D.E. Goldberg: An investigation of niche and species formation in genetic function optimization, Proc. 3rd Int. Conf. Genet. Algorithms (Morgan Kaufmann, San Francisco 1989) pp. 42–50

51.80   F. Baita, F. Mason, C. Poloni, W. Ukovich: Genetic algorithm with redundancies for the vehicle scheduling problem. In: *Evolutionary Algorithms in Management Applications*, ed. by J. Biethahn, V. Nissen (Springer, Berlin 1995) pp. 341–353

51.81   H. Sato, H.E. Aguirre, K. Tanaka: Local dominance and local recombination in MOEAs on 0/1 multiobjective knapsack problems, Eur. J. Oper. Res. **181**(3), 1708–1723 (2007)

51.82   E.J. Hughes: MSOPS-II: A general-purpose many-objective optimiser, IEEE Congr. Evol. Comput. (CEC'2007) (IEEE, Singapore 2007) pp. 3944–3951

51.83   Q. Zhang, H. Li: MOEA/D: A multiobjective evolutionary algorithm based on decomposition, IEEE Trans. Evol. Comput. **11**(6), 712–731 (2007)

51.84   H. Ishibuchi, Y. Sakane, N. Tsukamoto, Y. Nojima: Adaptation of scalarizing funtions in MOEA/D: An adaptive scalarizing funtion-based multiobjective evolutionary algorithm, Lect. Notes Comput. Sci. **5467**, 438–452 (2009)

52. Memetic and Hybrid Evolutionary Algorithms

Jhon Edgar Amaya, Carlos Cotta Porras, Antonio J. Fernández Leiva

This chapter presents an overview of hybridization mechanisms in evolutionary algorithms. Such mechanisms are aimed to introducing problem knowledge in the optimization technique by means of the synergistic combination of general–purpose methods and problemspecific add-ons. This combination is presented in this work from two wide perspectives: memetic algorithms and cooperative optimization models. Memetic algorithms are based on the smart orchestration of global (population-based) and local (trajectorybased) techniques, using an algorithmic scheme in which the latter are often subordinated to the former. As to cooperative models, they are based on the collaboration of different optimization techniques that exchange information in order to boost their respective performances. Both approaches, memetic algorithms and cooperative models, provide a framework to achieve synergistic algorithmic combinations for the resolution of large-scale combinatorial problems.

52.1 Overview

Heuristic methods are aimed to efficiently produce near-optimal solutions for hard problems (optimization problems in particular). We are here specifically concerned with those methods used to solve an optimization problem by means of an intelligent exploration of the search space and the fruitful exploitation of knowledge about the problem structure. This is admittedly a very broad class of methods that comprise – among others – classical artificial intelligence tools such a the A^* algorithm as well as modern optimization techniques such as metaheuristics [52.1]. The latter are general-purpose techniques for optimization that guide some underlying basic heuristics for intelligently exploring the search space of the problem under consideration.

There exists a plethora of metaheuristic methods, each of them with its own distinctive features and governing parameters, typically (yet not always) based on some analogy of a real-world phenomenon (be it in the area of biology, zoology, physics, etc.) Indeed, there have been several attempts in the literature to classify these techniques according to different criteria, e.g. whether they are inspired by nature or not, use of memory, neighborhood structure, use of single solutions or populations thereof, etc. *Blum* and *Roli* [52.1] proposed a classification in which a distinction was firstly made between trajectory-based (or single-point search) and population-based techniques (see also Fig. 52.1). The former can be depicted as following a particular trajectory (sequence of points) in the search space by the smart exploration of the neighborhood of a single solution (this is to some extent an oversimplification, since trajectory-based techniques are often endowed with intensification/diversification mechanisms that may turn this trajectory into complex branching paths; nevertheless, it serves as an initial analogy). The latter are, however, better imagined as a cloud of points moving through the search space, expanding and contracting according to some internal dynamics.

Despite what the above depiction may suggest in terms of the superiority or adequateness of methods falling within some particular class, it is not possible to state that any method is better than any other one, at least not in a general sense. This is a somewhat counterintuitive result that was formally derived by *Wolpert* and *Macready* [52.2] in the so-called "no free lunch theorem" (NFL). This theorem can be formulated as

$$\sum_f P(x_m \,|\, f, A, e) = \sum_f P(x_m \,|\, f, B, e) \,, \qquad (52.1)$$

where $P(x_m \,|\, f, A, e)$ is the probability that algorithm A detects the optimal solution for a generic objective function f using computational effort e (i.e., generating e different solutions) and $P(x_m \,|\, f, B, e)$ is the analogous probability for algorithm B. In other words, the average performance of any pair of algorithms across all possible problems defined on particular domains and co-domains is identical. Hence, whenever an algorithm performs well on a certain problem or class of problems, it follows that it will exhibit degraded performance on the set of all remaining problems. While the initial assumptions from which the NFL theorem is derived are questionable (most importantly, the consideration of all possible problems include many functions that are random or incompressible in an algorithmic information-complexity sense, and hence cannot be efficiently calculated, thus rendering them irrelevant from an optimization point of view), the concept that there is no universal optimizer had a significant impact on the scientific community and provides a safe ground onto which particular optimization procedures can be built. To be precise, the NFL theorem highlights the limitations of black-box optimization procedures, i.e., techniques whose search strategy is independent or unaware of the internal working of the objective function that is being optimized, and emphasizes the need for trying to exploit domain knowledge within the search algorithm in order to tailor the optimization process to the problem under consideration.

The argument above is commonly used to support the development and utilization of hybrid metaheuristics, where the term *hybrid* is used to denote in a broad sense the exploitation of problem-dependent knowledge (typically attained via the sensible combination of general-purpose and problem-specific mechanisms). Indeed, these hybrid methods can be shown to provide an efficient behavior and notable flexibility for dealing with real-world problems. The general idea here is achieving a synergetic combination of complementary techniques in order to enhance their strengths and having their weaknesses alleviated. Roughly speaking, such hybrid approaches can be attained via two different (and complementary) approaches: cooperation (the techniques involved exchange information in order to boost their respective performances) and integration (one of the techniques is subordinated to the other one, which uses the former as a tool to achieve some internal goal) [52.3].

Arguably, one of the advantages (if not from the performance point of view at least from the design point of view) of population-based methods over trajectory-based methods is their greater flexibility when it comes to integrating different metaheuristics. For example, cooperative methods can often be defined as a population of (possibly heterogenous) search agents exchanging information according to some underlying connection topology. Different architectures for such cooperative methods have been defined, e.g. MAGMA [52.4] or COSEARCH [52.5], depending on the communication strategy and the intervening methods. We can also cite hyper-heuristics [52.6, 7] in this regard, i.e., the use of a high-level heuristic to control the application of a population of low-level heuristics.

The above ideas fit nicely with the notion of memetic algorithm (MA). MAs are a family of metaheuristics that try to blend several concepts from population-based and trajectory-based techniques. The term *memetic* comes from *meme*, a word coined by *Dawkins* [52.8] as an analogy to the *gene* in the context of cultural evolution. In this sense, there is, indeed, a connection between cultural evolution and memetic

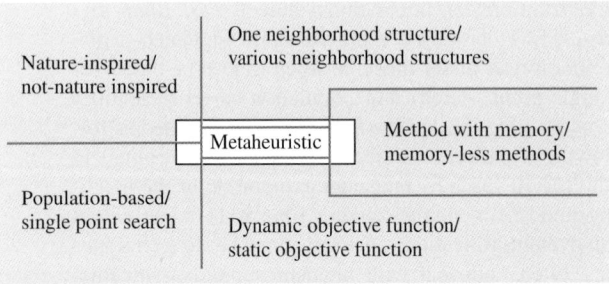

Fig. 52.1 Classification of metaheuristics according to *Blum* and *Roli* (after [52.1])

algorithms, in the sense that memes are much more plastic and flexible than genes – and hence evolve faster – and can be subject to lifetime learning, thus leading to the transmission of acquired traits (much unlike biological evolution). Due to the way in which this can be implemented, MAs are often termed hybrid EAs or Lamarckian EAs, among other fancy terms. From a general perspective, we can say that an MA is a search strategy in which a population of optimizing agents – explicitly concerned with using knowledge from the problem being solved – synergistically cooperate and compete [52.9].

Focusing on combinatorial optimization problems, that is, problems whose solution space is composed of combinatorial structures such as graphs, trees, sets, lists, permutations, etc., built on a discrete collection of variables, MAs and hybrid metaheuristics are very well suited to their resolution. On the one hand, the solutions to these problems are information-rich structures that the algorithmic designer can analyze in order to extract problem information to be later used in the optimizer (some attempts have been made to automatically extract this information in combina-

torial contexts as well [52.10]). This contrasts with most continuous optimization problems in which the high-dimensionality and highly non-linear coupling of variables makes them much more opaque in general (not to mention that black-box scenarios are more frequent in this continuous domain, e.g., optimization of physical or industrial processes via simulations of the system). On the other hand, it is very often the case that the objective function for combinatorial problems is decomposable or at least incrementally computable, meaning that after a small perturbation has been introduced in a solution the latter does not need to be fully evaluated from scratch (only an incremental term dependent on the modification done must be computed). This makes the use of local search strategies much more computationally amenable than in most continuous domains. Before describing MAs in more detail, let us first overview some generic ideas about evolutionary algorithms (EAs) and hybrid metaheuristics. Throughout the discussion we will focus on combinatorial problems and provide illustrative examples on the instantiation of these techniques for discrete optimization.

52.2 A Bird's View of Evolutionary Algorithms

An EA is a stochastic iterative procedure for generating candidate solutions for a certain problem. The algorithm manipulates a pool *pop* of individuals (the so-called population), each of them carrying one or more chromosomes. Chromosomes are, in turn, composed of smaller pieces called genes, each of them taking a value from a certain domain (the allele set). Chromosomes represent a solution for the problem at hand via an encoding/decoding process. More precisely, EAs assume the existence of a phenotype space comprising the solutions for the problem under consideration and a genotype space, comprising all possible chromosomes. It is between these two sets that the growth (or expression) function is defined so as to have the mapping between chromosomes and solutions. While in some cases these two spaces may be identical, this does not generally happen. In this general situation, the growth function is merely required to be surjective.

The pool of solutions is initialized either at random or by means of some heuristic seeding procedure. Each individual then receives a fitness value quantifying how good the solution it carries is. This value will be used by the EA for guiding the search. The ini-

tial population is actually the playground on which the EA will subsequently work, iteratively applying some evolutionary operators to modify its contents. More precisely, the process comprises two major stages: selection (promising solutions are selected for breeding and survival), and reproduction (new solutions are created by modifying selected solutions using some reproductive operators). Selection is further decomposed into two sub-stages: the first one is selection for reproduction (often simply called *selection*) in which solutions from the population are picked and fed to the reproduction stage; the second one is selection for survival (commonly called *replacement*) in which the new solutions obtained in the reproduction stage are inserted in the population at the expense of removing some older solutions. Both selection sub-stages are present in EAs (although in some cases one of these sub-stages may take a very simplistic form; e.g., random selection for reproduction is sometimes used in evolution strategies). This selection–production cycle is repeated until a certain termination criterion (usually reaching a maximum number of fitness computations; some more complex criteria based on stagnation detection are also possible [52.11]) is fulfilled. Each iteration of this process is commonly termed a generation. The whole process is illustrated in Algorithm 52.1. Every possible instantiation of this algorithmic template leads to a different EA.

Algorithm 52.1 A Basic Evolutionary Algorithm
1: **function** BasicEA (**in** P: Problem, **in** par: Parameters): Solution;
2: **begin**
3:    $pop \leftarrow$ INITIALIZE(par, P) ;
4:    **repeat**
5:        $newpop_1 \leftarrow$ SELECT(pop, par, P) ;
6:        $newpop_2 \leftarrow$ REPRODUCE$(newpop_1, par, P)$;
7:        $pop \leftarrow$ REPLACE $(pop, newpop_2)$;
8:    **until** TERMINATIONCRITERION(par);
9:    **return** GETBEST(pop) ;
10: **end**

52.3 From Hybrid Metaheuristics to Hybrid EAs

As was mentioned before, hybrid metaheuristics (and in particular hybrid EAs) are developed aiming to attain a synergistic combination of several techniques, exploiting their strengths and mitigating their weaknesses [52.12]. Besides the theoretical justifications for hybrid metaheuristics (arising, for example, from the NFL theorem sketched before), these techniques have been repeatedly vindicated by their practical success. Before getting to hybrid EAs, let us first focus on how hybridization can be approached.

52.3.1 Hybridization Mechanisms

As was already mentioned in Sect. 52.1, attempts to classify hybrid metaheuristics are manifold [52.13–17]. We will focus in the following on two of these, namely the classification of *Talbi* [52.13] and that of *Raidl* [52.16].

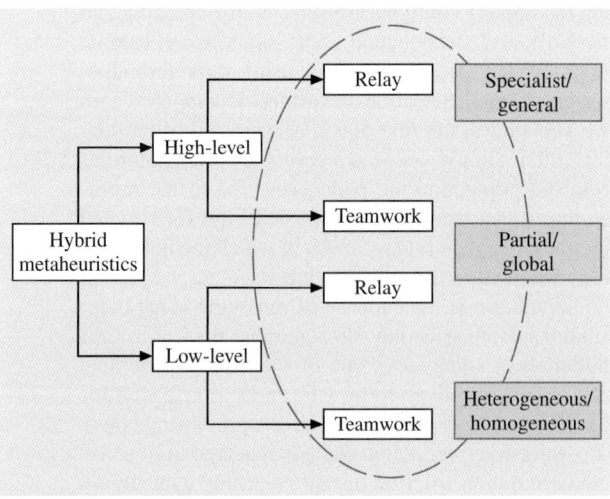

Fig. 52.2 Classification of hybrid metaheuristics by *Talbi* (after [52.13])

Talbi proposed a hierarchical taxonomy based on two design issues: functionality and algorithmic architecture. According to this, we can distinguish between high/low-level hybrids and relay/team-work hybrids. Low-level hybridization addresses the functional composition of a single optimization method in which a certain function of a metaheuristic is replaced by another metaheuristic. On the contrary, in high-level hybrids, the internals of different metaheuristics are nonintersecting. As for relay hybridization, it comprises models in which a set of metaheuristics are sequentially applied, each using the output of the previous as its input. On the other hand, teamwork hybridization represents cooperative optimization models. These two distinctions (low versus high, relay versus teamwork) are orthogonal, and hence lead to four different combinations. These four classes can, in turn, be refined using three additional dichotomies, namely homogeneous versus heterogeneous (referring to the type of metaheuristics involved in the hybrid), global versus partial (referring to whether or not each technique explores the whole search space) and specialist versus general (referring to whether or not all algorithms solve the same optimization problem). Figure 52.2 shows this taxonomy.

Raidl [52.16], in turn, proposed a hybrid metaheuristic classification centered around four elements: type of hybridization, level/strength of hybridization, control strategy, and execution order. Regarding the type of hybridization, we can distinguish between:

1) Combinations of different metaheuristics
2) Combination of metaheuristics and problem-specific algorithms
3) Combinations of metaheuristics with general operational research (OR), artificial intelligence (AI), or constraint programming (CP) techniques.

Regarding the hybridization strength, we can distinguish high-level/weakly-coupled hybrids and low-level/strongly-coupled hybrids. As to the control strategy, there are two possibilities: integrative (a technique takes a subordinate role) and collaborative (exchange of information without subordination). Finally, the order of execution captures the temporal aspect of the interaction among techniques. Thus, we can have sequential execution (a technique takes as input the output of another technique), intertwined execution (both techniques alternate parts of their execution at a computational or algorithmic level), and parallel execution (the techniques run in parallel). Figure 52.3 shows this classification.

52.3.2 Hybrid EAs

One of the most classical hybridization approaches for EAs is defined in the context of knowledge-augmented representations, particularly in the case that the solutions sought have an extremely complex structure for which a direct search does not seem adequate, or with problems that exhibit constraints. In the latter case, these can be handled in three ways:

i) By using penalty functions that guide the search to feasible solutions
ii) By using repairing mechanisms that turn infeasible solutions into feasible ones

iii) By defining reproductive operators that always remain in the feasible region.

While the complexity of the representation and the operators can be kept low in the first two cases (i. e., the complexity is moved to the fitness function and the repairing function, respectively), the third case requires either a careful representation safeguarding feasibility, or complex operators intelligently handling the constraints of the problem. Focusing on representations, decoders [52.18] are commonly used. These provide a complex genotype-to-phenotype mapping that may not just produce feasible solutions, but can also provide better quality solutions. Consider, for example, the knapsack problem: solutions are sets of objects in this case, but clearly a random set may be infeasible due to the knapsack capacity constraint. This could be handled with a penalty term to account for this capacity violation or by adding/removing some objects to turn the solution into a feasible one [52.19]. A decoder approach could, however, encode solutions as permutations, indicating the order in which objects are to be considered for inclusion in the knapsack. Since any object violating the capacity constraint would be skipped, a feasible solution would be always obtained. Problem-space search [52.20] – the use of a construction heuristic that is guided through problem-space – also falls within this class of low-level/strong hybrids. Following with the knapsack problem, solutions could in this case be represented as

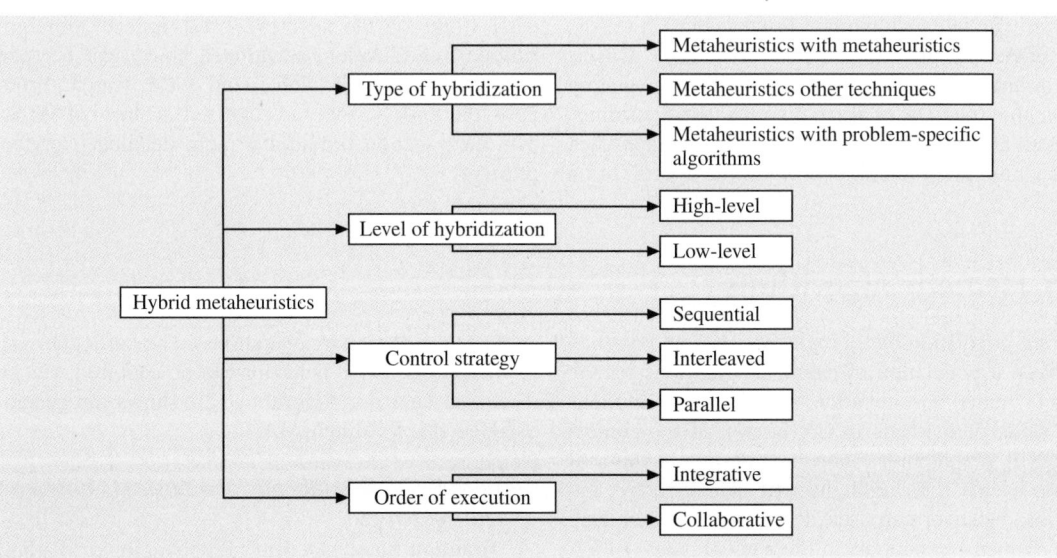

Fig. 52.3 Classification of hybrid metaheuristics by *Raidl* (after [52.16])

perturbations of the value of objects. Each of these solutions would be evaluated by constructing the so-modified problem instance, solving it with a greedy heuristic and using the original instance to evaluate the quality of the solution obtained. This strategy is very competitive for this problem, as is shown in [52.21].

On the other hand, high-level/weak hybrid evolutionary algorithms are most typically obtained either by integrating within the EA a local-search (single-point or trajectory-based) method, or other techniques from the realm of OR/AI/CP/. . . , etc. Regarding the former approach, the underlying idea is to boost the intensification capabilities of the algorithm by improving the solutions generated by the population-based search engine. This kind of combination dates back to the late 1980s, when it used to take the form of a genetic algorithm hybridized with simulated annealing (SA), [52.22]. A particularly interesting hybrid EA along this line is the parallel recombinative simulated annealing algorithm [52.23], in which a pool of SA algorithms cooperate/compete in a genetic algorithm framework. Tabu search (TS) is another popular local search metaheuristic to be hybridized with EAs (see [52.24–29], to mention just a few).

Other EA techniques such as estimation of distribution algorithms (EDAs) have also been hybridized with local search approaches, e.g., *Campelo* et al. [52.30] for the design of electromagnetic devices and *Laguna* et al. [52.31] for maximum cut. It is also worth mentioning the work by *Santana* et al. [52.32] on the combination of variable neighborhood search [52.33] (VNS) with EDAs for protein structure prediction. This is done in different ways, most notably either integrating VNS within the EDA or alternating the two algorithms. *Zhang* et al. [52.34] propose an analogous approach for quadratic programming based on the hybridization

of EDAs and 2-opt hill climbing. A very interesting approach was also proposed by *Peña* et al. [52.35], who hybridized a steady-state genetic algorithm and an EDA; each of these algorithms is responsible for generating a part of the population. On the other hand, *Zhou* et al. [52.36] and *Ahn* et al. [52.37] propose the hybridization of an EDA with particle swarm optimization (PSO) where the latter is used for intensification purposes.

As for hybridization with techniques from the realms of AI/OR or constraint programming, examples date back to the mid 1990s. Particularly interesting is the combination of EAs with exact techniques and derivatives thereof. For example, branch and bound (BnB) can be integrated within an EA as a recombination operator [52.38, 39] or in the decoding process [52.40]. Conversely, an EA can be used for the strategic guidance of BnB [52.41]. As for collaborative combinations, intertwined approaches were considered in [52.42] by combining EAs and BnB within a parallel multiagent system, and in [52.38, 43] by defining a model in which the exact technique provided partial promising solutions, and the EA returned improved bounds. A related multilevel approach involving beam search and an EA hybridized with local-search algorithm can be found in [52.44–46]. For further details on this kind of exact/metaheuristic hybridization we refer the reader to [52.3].

Most of the above hybrid EAs can be safely described as memetic algorithms, if only under the broad interpretation of MAs emanating from seminal and early works on the topic [52.9, 47]. Indeed, the algorithmic hybridizations mentioned above can be seen as combinations of global and local search, probably the most widely recognized feature of MAs. The next section provides a more detailed overview of MAs.

52.4 Memetic Algorithms

MAs are population-based metaheuristics and as such they keep a population of candidate solutions for the problem under consideration. While these solutions were called individuals in EA jargon, in the context of MAs it is sometimes more appropriate to think of them as agents, thus highlighting their more active nature (i. e., behavior purposefully directed at optimizing some problem) in contrast to the passive nature of EA individuals (which are mere information placeholders

subject to evolutionary operations). The particular way in which this active behavior can be captured will be discussed later on. Algorithm 52.2 shows the general pseudocode of a simple MA.

Algorithm 52.2 Pseudocode of a basic MA based on a local search LS

1: **function** Basic MA (**in** *P*: Problem, **in** *par*: Parameters): Solution;

2: **for** $i \in \mathbb{N}_\mu$ **do**
3:      $pop[i] \leftarrow$ GENERATE-SOLUTION(P);
4:      $pop[i] \leftarrow$ LOCAL-IMPROVEMENT   ($pop[i]$, P,
        par);
5: **end for**
6: $i \leftarrow 0$;
7: **while** $i < MaxEvals$ **do**
8:      $auxpop[0] \leftarrow$ SELECT (pop);
9:      **for** $j \leftarrow 1$ **to** $\#op$ **do**
10:         $auxpop[j] \leftarrow$ APPLY($op[j]$, $auxpop[j-1]$, P,
            par);
11:     **end for**
12:     $newpop \leftarrow$ LOCAL-IMPROVEMENT
        ($auxpop[\#op]$, P, par);
13:     $pop \leftarrow$ REPLACE(pop, $newpop$);
14:     **if** DEGENERATED(pop) **then**
15:         RESTART (pop, P);
16:     **end if**
17: **end while**
18: **return** GetBest (pop);

First of all, the population must be initialized. Problem knowledge can be introduced in this stage by using constructive heuristics. For example, greedy strategies based in the nearest neighbor heuristic [52.48] could be used to generate solutions for the traveling salesman problem (TSP) – see also [52.49–51] for other examples in the context of scheduling and timetabling. Then, the population of agents is subject to processes of competition and mutual cooperation much like in EAs. Competition (i. e., selection and replacement) can be done in general using any of the well-known strategies used in EAs, e.g., tournament, ranking, or fitness-proportionate selection, and/or comma replacement, etc. As for cooperation, it is accomplished by using a number of reproductive operators. Many different such operators can be used in an MA, as illustrated in the general pseudocode shown in Algorithm 52.2: an array op of operators is sequentially applied to the population in a pipeline fashion. Note also how these operators receive as input not just the solutions they act on but also problem data, thus emphasizing the usage of problem knowledge. While it is possible to consider local improvement as one of these operators, it plays such a distinctive role in most MAs that it is independently depicted in the pseudocode.

Recombination is the algorithmic component that best captures cooperation among two (or more [52.52]) agents in MAs. By using this operation, the relevant information contained in the parents is combined to produce new solutions. Relevance here amounts to be significant when it comes to evaluating the quality of solutions. As an example, consider again the TSP. While solutions can be encoded as permutations, a standard permutational recombination operator will not perform adequately in general. The reason is that permutations are information-rich structures carrying positional, precedence, and adjacency information [52.53]. Clearly, the latter is the really relevant piece of information when the TSP is involved. Hence, an edge-manipulation operator such as edge recombination [52.54] (ER) will perform better than position-based operators such as partially-mapped crossover [52.55] (PMX) or uniform cycle crossover [52.56] (UCX). There are several principled approaches to define measures capturing the goodness of different representations (that is, the way a particular encoding is interpreted) among which we can cite epistasis (nonadditive influence on the fitness function of combining several information units) [52.57, 58], fitness variance of formae (variance of the fitness values of a representative subset of solutions carrying a particular information unit) [52.59], and fitness correlation (correlation in the fitness values of parents and offspring) [52.60, 61].

Mutation is the other classical reproductive operator. Its role is that of injecting new material in the population (at a low rate to prevent the search degrading to a random walk in the solution space). This view of mutation as an important operator but it is, nevertheless, secondary to recombination and departs from the interpretation of the search process made in, e.g. evolutionary programming [52.62]. In either case, mutation plays an important role in EAs since it favors the effectiveness of recombination (particularly in some unstructured landscapes). Furthermore, if the problem exhibits constraints it is commonly much easier to handle these in a local way and maintain/achieve feasibility by introducing small perturbations in a solution than via recombination (e.g., consider a university timetabling problem [52.63]: given a feasible solution, it is easier to exchange a couple of slots, and keep them feasible, than to produce a new feasible solution that comes from the combination of two feasible assignments). However, it must be noted that unlike classical EAs, in which recombination is a mere random shuffler of information (and hence can be arguably cast as a macro-mutational process), MAs usually utilize intelligent problem-aware mechanisms for recombination, and thus play a crucial role in the search. Broadly speaking, this inclusion of problem knowledge during recombination can be projected on two aspects of the process, namely the selection of the pieces of information from the parents that will be transmitted to the offspring, and the selection of the external information that will be added to it. Regarding the former issue, it is commonly assumed that transmission of common features is beneficial for some problems [52.54, 64]. Further completion of the descendant can be done in several ways. *Radcliffe* and *Surry* [52.59] proposed the use of local improvers or implicit enumeration schemes. *Cotta* and *Troya* suggested the use of exact techniques to find the best way of combining the information present in the parents [52.39]. *Ibaraki* [52.65] and *Gallardo* et al. [52.66] used dynamic programming for this purpose.

The use of local search (LS) is one of the most distinctive components of MAs, to the extent that MAs are often equated to EAs endowed with LS. While this is certainly a very popular implementation of MAs, several authors [52.47, 67, 68] advocate a broader interpretation of the paradigm in which an explicit local search algorithm need not be present (e.g., local improvement can take place during recombination as in the edge assembly crossover defined in [52.69] for the TSP). In its simplest incarnation, these local improvers can be hill climbers, exploring the neighborhood of the current solution and performing uphill moves in the corresponding fitness landscape [52.70] until a local optimum is found or the computational budget assigned to this operator is exhausted. Obviously, much more complex mechanisms can be defined for this purpose, such as the use of fully-fledged metaheuristics, such as, for example, TS, SA, or VNS, just to mention a few. It must be also noted that it is mainly because of the use of this mechanism for improving solutions on a local (and even autonomous) basis that the term *agent* is deserved. Under this interpretation, the MA can be viewed as a collection of agents that autonomously explore the search space, cooperate via recombination, and compete for computational resources via selection and replacement. This also provides an interesting link to cooperative models for optimization and to memetic computing in general [52.71, 72].

One of the crucial elements governing the successful application of local search within an MA is achieving a good balance between global and local search. This amounts to determining when to apply local search (how often and on which solutions) and how intense this local search has to be. This parameterization problem is very hard and constitutes an active area of research [52.73]. An additional issue is the selection of a particular local search scheme within the MA. This has actually led to a very fruitful line of research in

so-called multimemetic algorithms (MMAs). Therein, a meme is interpreted as a lifetime learning procedure capable of improving individual solutions [52.74–79]. Each solution in a MMA carries a gene indicating the particular LS operator that has to be applied on it (a pointer to an existing operator, or the parameterization of a generic local search template). Thus, they constitute a generalization of meta-lamarckian EAs [52.80] (in which the selection of the LS operator – from a pre-fixed set – is made using some rules that are hard-wired into the MA) and an intermediate step in the direction of co-evolving MAs [52.75] (in which a population of LS operators co-evolve along with a population of solutions). Finally, it is essential from a purely computational perspective to be able to apply LS in an efficient way. As was mentioned in Sect. 52.1 this is normally attained in combinatorial problems by incrementally evaluating solutions belonging to the neighborhood area. For example, consider the 2-opt neighborhood in the TSP [52.48]: each neighbor of a given solution is obtained by a 2-opt move that removes two edges and adds two new edges; the fitness of such a neighbor can thus be computed by taking the fitness of the initial solution and adding a term accounting for the difference between added edges and removed edges.

Another interesting element of MAs is the restarting process invoked whenever the population is deemed degenerate due to a lack of diversity or any other factor impairing the subsequent performance of the algorithm. This restarting process can be done in numerous ways (for example, triggering hypermutation [52.81] or introducing random solutions in the population [52.82]) and can be often found in plain EAs as well (indeed the use of restarting procedures in EAs can be traced back to the CHC algorithm [52.83] in the early 1990s). This said, it constitutes a generic element to be routinely included in MAs. Indeed, scatter search [52.84] (a technique that can reasonably be termed memetic, despite having an independent origin from MAs and its fair share of distinctive features such as the emphasis on using deterministic strategies) has a restart as a crucial element in its algorithmic cycle. Note also that it is not unusual to have MAs without mutation, given the fact that new information can also be injected in the population via local search, and the availability of restarting mechanisms. Indeed, in some applications, it may be better to converge quickly and then restart, rather than continuously diversifying the search. This is not the general norm at any rate. As a matter of fact, one can easily find MAs that use several mutation operators, either by considering different basic neighborhoods [52.27, 85] or by defining *light* and *heavy* mutations that introduce different amounts of new information [52.86, 87] – cf. hypermutation. Needless to say, the use of restarting strategies is a corrective measure that is taken once a diversity problem is encountered and can be complemented with preventive measures aimed to hinder (or even avoid) this problem taking place for the first time. For example, structured populations [52.88] could be used to cause a slowdown in the propagation of information across the population, hence hindering the apparition of *super agents* that might quickly take the population over and destroy diversity. Also, population management strategies based on the use of distance measures have been utilized with notable success in combinatorial problems [52.89]. More traditional strategies for maintaining diversity during selection and replacement, such as crowding [52.90] or sharing [52.91], can be used as well.

52.5 Cooperative Optimization Models

As stated in previous section, the interpretation of MAs as a collection of interacting agents that autonomously explore the search space while cooperating/competing with each other seamlessly integrates with the more general notion of memetic computing and cooperative optimization models. According to the definition in [52.92], memetic computing is:

> *a paradigm that uses the notion of meme(s) as units of information encoded in computational representations for the purpose of problem solving,*

where *meme* should be interpreted as local-search operator as mentioned before. This orchestration of different LS operators naturally links with cooperative models dating back to the late 1990s [52.93]. These attempts to attain an effective mechanism for exploring the search space try to escape from local optima by combining search agents that have diverse intensification/diversification characteristics and that start from different points in the search space [52.94]. According to [52.95], the distinctive features of this kind of models are (1) a collection of autonomous algorithms (agents), each of them supporting a different optimization method, and (2) a cooperative scheme for combining these autonomous elements into an unified problem-solving strategy.

Early cooperative models involve an algorithmically homogenous collection of algorithms exchanging information. For example, *Toulouse* et al. [52.93] considered a collection of TS algorithms exchanging tabu attributes (notice the relation of this model with the parallel recombinative SA algorithm mentioned in Sect. 52.3.2) and later proposed a hierarchical decomposition approach [52.96]. A related model was also proposed by *Crainic* and *Gendreau* [52.97]. *Crainic*

et al. [52.98] put forward an asynchronous cooperative search procedure on the basis of VNS. A different approach based on the used of a central manager was proposed by *Pelta* et al. [52.99]. This central manager gathers information about the performance of the different agents and acts on them, altering their behavior – see also [52.100]. Other centralized approaches were defined by *LeBouthillier* and *Crainic* [52.101] by means of maintaining a *solution warehouse* upon which individual heuristics act. More recently, *Barbucha* [52.102] explored synchronous and asynchronous versions of an analogous memory-centralized approach in the context of vehicle routing problems. *Leung* et al. [52.103] proposed, in turn, a cooperative/competitive scheme in which the problem space is partitioned and a pool of agents is structured into several subgroups which repel each other, thus contributing to keeping diversity.

Multi-level models have also received a lot of attention in the last years. These models consist of layered algorithmic approaches and are not to be confused with multilevel partitioning strategies proposed for combinatorial optimization [52.104, 105], in which the resolution of the problem is attained via its incremental reduction and further reconstruction, using some solver at each level and the solution obtained therein as seeds for solving the next higher level. *Hulianytskyi* and *Sirenko* [52.106] presented a two-level cooperative approach: the lower level corresponds to basic algorithms, whereas the upper level combines the information found by these and broadcasts a refined version back to the basic algorithms. *Milano* and *Roli* [52.4] developed a multiagent system called MAGMA (multiagent metaheuristic architecture) allowing the use of metaheuristics at different levels (generating solutions, improving them, defining search strategies, and coordinating lower-level agents). Each level (or layer) provides a different abstraction level and can contain several agents loaded with a particular search algorithm. The lowest layer (level 0) generates solutions to be fed to level 1. The latter provides local improvement of these solutions. Level 2 has a global view of the search space and provides the means for escaping from local optima. The upmost level (level 3) coordinates the functioning of the underlying agents, rewarding those which perform well or adapting and improving their functioning. They specifically adapt this framework for deployment of MAs within it. Finally, *Amaya* et al. [52.107] defined a multilevel model in which heterogenous *simple* MAs (i. e., MAs obtained from the hybridization of an EA and a local search method) are combined in a cooperative model, and exchange information following an underlying arbitrary topology.

52.6 Conclusions

Memetic algorithms in particular, and memetic computing in general, constitute a flexible and powerful optimization approach. Rather than being competitors for existing methods and/or paradigms, they are a very suitable framework for integrating such existing techniques in order to attain synergistic combinations or being able to deal with the curse of dimensionality in large-scale optimization settings. They are also a very active research area in which, in addition to a steadily growing number of application works, new fundamental issues are attracting the interest of the research community. Among these we can cite the theoretical study of their self-adaptation capabilities and their deployment on the emerging computational platforms that are available nowadays. We refer to [52.72, 108, 109] for recent reviews of the field. For an overview of the literature dealing with the application of these techniques to combinatorial optimization problems we refer the reader to [52.67, 108] for a general perspective and to [52.63] for a review of scheduling and timetabling applications. Finally, we refer the reader to [52.110, 111] for further information on the deployment of MAs on combinatorial optimization problems.

References

52.1   C. Blum, A. Roli: Metaheuristics in combinatorial optimization: Overview and conceptual comparison, ACM Comput. Surv. **35**(3), 268–308 (2003)

52.2   D.H. Wolpert, W.G. Macready: No free lunch theorems for optimization, IEEE Trans. Evol. Comput. **1**(1), 67–82 (1997)

52.3   J. Puchinger, G.R. Raidl: Combining metaheuristics and exact algorithms in combinatorial optimization: A survey and classification, Lect. Notes Comput. Sci. **3562**, 113–124 (2005)

52.4   M. Milano, A. Roli: MAGMA: A multiagent architecture for metaheuristics, IEEE Trans. Syst. Man Cybern. Part B **34**(2), 925–941 (2004)

52.5   E.-G. Talbi, V. Bachelet: COSEARCH: A parallel cooperative metaheuristic, J. Math. Model, Algorithms **5**(1), 5–22 (2006)

52.6   P. Cowling, G. Kendall, E. Soubeiga: A hyperheuristic approach to scheduling a sales summit, Lect. Notes Comput. Sci. **2079**, 176–190 (2001)

52.7   K. Chakhlevitch, P.I. Cowling: Hyperheuristics: Recent developments. In: *Adaptive and Multilevel Metaheuristics*, Studies in Computational Intelligence, Vol. 136, ed. by C. Cotta, M. Sevaux, K. Sörensen (Springer, Berlin 2008) pp. 3–29

52.8   R. Dawkins: *The Selfish Gene* (Clarendon, Oxford 1976)

52.9   P. Moscato: *On Evolution, Search, Optimization, Genetic Algorithms and Martial Arts: Towards Memetic Algorithms. Technical Report Caltech Concurrent Computation Program, Report. 826* (California Institute of Technology, Pasadena 1989)

52.10   R. Santana, C. Bielza, P. Larranaga: *Network Measures for Re-using Problem Information in EDAs*. Technical Report UPM-FI/DIA/2010-3 (Department of Artificial Intelligence, Faculty of Informatics, Technical University of Madrid 2010)

52.11   C. Cotta, E. Alba, J.M. Troya: Stochastic reverse hillclimbing and iterated local search, Proc. 1999 Congr. Evol. Comput. (IEEE Neural Network Council – Evolutionary Programming Society – Institution of Electrical Engineers, Washington 1999) pp. 1558–1565

52.12   C. Blum, J. Puchinger, G. Raidl, A. Roli: A brief survey on hybrid metaheuristics, 4th Int. Conf. Bioinspired Optim. Methods Appl. (BIOMA 2010), ed. by B. Filipic, J. Silc (Ljubljana, Slovenia 2010) pp. 3–16

52.13 E.-G. Talbi: A taxonomy of hybrid metaheuristics, J. Heuristics **8**, 541–564 (2002)

52.14 C. Cotta, E.G. Talbi, E. Alba: Parallel hybrid metaheuristics. In: *Parallel Metaheuristics*, ed. by E. Alba (Wiley-Interscience, Hoboken 2005) pp. 347–370

52.15 M. El-Abd, M. Kamel: A taxonomy of cooperative search algorithms, Lect. Notes Comput. Sci. **3636**, 32–41 (2005)

52.16 G. Raidl: A unified view on hybrid metaheuristics, Lect. Notes Comput. Sci. **4030**, 1–12 (2006)

52.17 L. Jourdan, M. Basseur, E.-G. Talbi: Hybridizing exact methods and metaheuristics: A taxonomy, Eur. J. Oper. Res. **199**(3), 620–629 (2009)

52.18 Z. Michalewicz: Decoders. In: *Handbook of Evolutionary Computation*, ed. by T. Bäck, D.B. Fogel, Z. Michalewicz (Institute of Physics Publishing and Oxford Univ. Press, Bristol 1997)

52.19 P.C. Chu, J.E. Beasley: A genetic algorithm for the multidimensional knapsack problem, J. Heuristics **4**, 63–86 (1998)

52.20 R.H. Storer, S.D. Wu, R. Vaccari: New search spaces for sequencing problems with application to job-shop scheduling, Manag. Sci. **38**, 1495–1509 (1992)

52.21 C. Cotta, J.M. Troya: A hybrid genetic algorithm for the 0-1 multiple knapsack problem. In: *Artificial Neural Nets and Genetic Algorithms 3*, ed. by G.D. Smith, N.C. Steele, R.F. Albrecht (Springer, Wien 1998) pp. 251–255

52.22 M.G. Norman, P. Moscato: A competitive and cooperative approach to complex combinatorial search, Proc. 20th Inf. Oper. Res. Meet., Buenos Aires (1989), pp. 3.15–3.29

52.23 S.W. Mahfoud, D.E. Goldberg: Parallel recombinative simulated annealing: A genetic algorithm, Parallel Comput. **21**(1), 1–28 (1995)

52.24 C. Fleurant, J.A. Ferland: Genetic and hybrid algorithms for graph coloring, Ann. Oper. Res. **63**, 437–461 (1996)

52.25 H. Kim, Y. Hayashi, K. Nara: The performance of hybridized algorithm of genetic algorithm simulated annealing and Tabu search for thermal unit maintenance scheduling, 2nd IEEE Conf. Evol. Comput. ICEC'95 (Perth, Australia 1995) pp. 114–119

52.26 J. Thiel, S. Voss: Some experiences on solving multiconstraint zero-one knapsack problems with genetic algorithms, INFOR **32**(4), 226–242 (1994)

52.27 C.-F. Liaw: A hybrid genetic algorithm for the open shop scheduling problem, Eur. J. Oper. Res. **124**, 28–42 (2000)

52.28 E.K. Burke, A.J. Smith: A memetic algorithm to schedule planned maintenance for the national grid, J. Exp. Algorithmics **4**, 1–13 (1999)

52.29 J.E. Gallardo, C. Cotta, A.J. Fernández: Finding low autocorrelation binary sequences with memetic algorithms, Appl. Soft Comput. **9**(4), 1252–1262 (2009)

52.30 F. Campelo, F.G. Guimaraes, J.A. Ramirez, H. Igarashi: Hybrid estimation of distribution algorithm using local function approximations, IEEE Trans. Magn. **45**(3), 1558–1561 (2009)

52.31 M. Laguna, A. Duarte, R. Mart: Hybridizing the cross-entropy method: An application to the max-cut problem, Comput. Oper. Res. **36**(2), 487–498 (2009)

52.32 R. Santana, P. Larrañaga, J.A. Lozano: Combining variable neighborhood search and estimation of distribution algorithms in the protein side chain placement problem, J. Heuristics **14**, 519–547 (2008)

52.33 P. Hansen, N. Mladenović: Variable neighborhood search: Principles and applications, Eur. J. Oper. Res. **130**(3), 449–467 (2001)

52.34 Q. Zhang, J. Sun, E. Tsang, J. Ford: Estimation of distribution algorithm with 2-opt local search for the quadratic assignment problem. In: *Towards a New Evolutionary Computation*, Studies in Fuzziness and Soft Computing, Vol. 192, ed. by J. Lozano, P. Larrañaga, I. Inza, E. Bengoetxea (Springer, Berlin, Heidelberg 2006) pp. 281–292

52.35 J.M. Peña, V. Robles, P. Larrañaga, V. Herves, F. Rosales, M.S. Prez: GA-EDA: Hybrid evolutionary algorithm using genetic and estimation of distribution algorithms, Lect. Notes Comput. Sci. **3029**, 361–371 (2004)

52.36 Y. Zhou, J. Wang, J. Yin: A discrete estimation of distribution particle swarm optimization for combinatorial optimization problems, 3rd Int. Conf. Nat. Comput. (ICNC 2007) (2007) pp. 80–84

52.37 C.W. Ahn, J. An, J.-C. Yoo: Estimation of particle swarm distribution algorithms: Combining the benefits of PSO and EDAs, Inf. Sci. **192**, 109–119 (2012)

52.38 C. Cotta, J.F. Aldana, A.J. Nebro, J.M. Troya: Hybridizing genetic algorithms with branch and bound techniques for the resolution of the TSP. In: *Artificial Neural Nets and Genetic Algorithms 2*, ed. by D.W. Pearson, N.C. Steele, R.F. Albrecht (Springer, Wien 1995) pp. 277–280

52.39 C. Cotta, J.M. Troya: Embedding branch and bound within evolutionary algorithms, Appl. Intell. **18**(2), 137–153 (2003)

52.40 J. Puchinger, G.R. Raidl, G. Koller: Solving a real-world glass cutting problem, Lect. Notes Comput. Sci. **3004**, 165–176 (2004)

52.41 K. Kostikas, C. Fragakis: Genetic programming applied to mixed integer programming, Lect. Notes Comput. Sci. **3003**, 113–124 (2004)

52.42 J. Denzinger, T. Offermann: On cooperation between evolutionary algorithms and other search paradigms, 6th Int. Conf. Evol. Comput. IEEE (1999) pp. 2317–2324

52.43 J.E. Gallardo, C. Cotta, A.J. Fernández: Solving the multidimensional knapsack problem using an evolutionary algorithm hybridized with branch and bound, Lect. Notes Comput. Sci. **3562**, 21–30 (2005)

52.44   J.E. Gallardo, C. Cotta, A.J. Fernández: A multi-level memetic/exact hybrid algorithm for the still life problem, Lect. Notes Comput. Sci. **4193**, 212–221 (2006)

52.45   J.E. Gallardo, C. Cotta, A.J. Fernández: On the hybridization of memetic algorithms with branch-and-bound techniques, IEEE Trans. Syst. Man Cybern. Part B **37**(1), 77–83 (2007)

52.46   J.E. Gallardo, C. Cotta, A.J. Fernández: Reconstructing phylogenies with memetic algorithms and branch-and-bound. In: *Analysis of Biological Data: A Soft Computing Approach*, ed. by S. Bandyopadhyay, U. Maulik, J.T.-L. Wang (World Scientific, Singapore 2007) pp. 59–84

52.47   P. Moscato: Memetic algorithms: A short introduction. In: *New Ideas in Optimization*, ed. by D. Corne, M. Dorigo, F. Glover (McGraw-Hill, Maidenhead 1999) pp. 219–234

52.48   G. Reinelt: *The Traveling Salesman. Computational Solutions for TSP Applications* (Springer, Berlin, Heidelberg 1994)

52.49   W.-C. Yeh: A memetic algorithm of the $n/2/\text{Flowshop}/\alpha F + \beta C_{max}$ scheduling problem, Int. J. Adv. Manuf. Technol. **20**, 464–473 (2002)

52.50   R. Varela, J. Puente, C.R. Vela, A. Gómez: A knowledge-based evolutionary strategy for scheduling problems with bottlenecks, Eur. J. Oper. Res. **145**(1), 57–71 (2003)

52.51   O. Rossi-Doria, B. Paechter: A memetic algorithm for university course timetabling. In: *Combinatorial Optimisation 2004 Book of Abstracts*, Lancaster 2004, p. 56, ed. by Lancaster University

52.52   A.E. Eiben, P.-E. Raue, Z. Ruttkay: Genetic algorithms with multi-parent recombination, Lect. Notes Comput. Sci. **866**, 78–87 (1994)

52.53   B.R. Fox, M.B. McMahon: Genetic operators for sequencing problems. In: *Foundations of Genetic Algorithms I*, ed. by G.J.E. Rawlins (Morgan Kaufmann, San Mateo 1991) pp. 284–300

52.54   K. Mathias, L.D. Whitley: Genetic operators, the fitness landscape and the traveling salesman problem. In: *Parallel Problem Solving From Nature II*, ed. by R. Männer, B. Manderick (Elsevier Science B.V., Amsterdam 1992) pp. 221–230

52.55   D.E. Goldberg, R. Lingle Jr.: Alleles, loci and the traveling salesman problem, Proc. 1st Int. Conf. Genet. Algorithms, ed. by J.J. Grefenstette (Lawrence Erlbaum Associates, Hillsdale 1985) pp. 154–159

52.56   C. Cotta, J.M. Troya: Genetic forma recombination in permutation flowshop problems, Evol. Comput. **6**(1), 25–44 (1998)

52.57   Y. Davidor: Epistasis variance: Suitability of a representation to genetic algorithms, Complex Syst. **4**(4), 369–383 (1990)

52.58   Y. Davidor: Epistasis variance: A viewpoint on GA-hardness. In: *Foundations of Genetic Algorithms I*, ed. by G.J.E. Rawlins (Morgan Kaufmann, San Mateo 1991) pp. 23–35

52.59   N.J. Radcliffe, P.D. Surry: Fitness variance of formae and performance prediction. In: *Foundations of Genetic Algorithms III*, ed. by L.D. Whitley, M.D. Vose (Morgan Kaufmann, San Francisco 1994) pp. 51–72

52.60   B. Manderick, M. de Weger, P. Spiessens: The genetic algorithm and the structure of the fitness landscape, Proc. 4th Int. Conf. Genet. Algorithms, ed. by R.K. Belew, L.B. Booker (Morgan Kaufmann, San Mateo 1991) pp. 143–150

52.61   J. Dzubera, L.D. Whitley: Advanced correlation analysis of operators for the traveling salesman problem, Lect. Notes Comput. Sci. **866**, 68–77 (1994)

52.62   L.J. Fogel, A.J. Owens, M.J. Walsh: *Artificial Intelligence Through Simulated Evolution* (Wiley, New York 1966)

52.63   C. Cotta, A.J. Fernández: Memetic algorithms in planning, scheduling, and timetabling. In: *Evolutionary Scheduling*, Studies in Computational Intelligence, Vol. 49, ed. by K. Dahal, K.C. Tan, P.I. Cowling (Springer, Berlin, Heidelberg 2007) pp. 1–30

52.64   C. Oğuz, M.F. Ercan: A genetic algorithm for hybrid flow-shop scheduling with multiprocessor tasks, J. Sched. **8**, 323–351 (2005)

52.65   T. Ibaraki: Combination with dynamic programming. In: *Handbook of Evolutionary Computation*, ed. by T. Bäck, D. Fogel, Z. Michalewicz (Oxford Univ. Press, New York 1997), pp. D3.4:1–2

52.66   J.E. Gallardo, C. Cotta, A.J. Fernández: A memetic algorithm with bucket elimination for the still life problem, Lect. Notes Comput. Sci. **3906**, 73–85 (2006)

52.67   P. Moscato, C. Cotta: A gentle introduction to memetic algorithms. In: *Handbook of Metaheuristics*, ed. by F. Glover, G. Kochenberger (Kluwer, Boston 2003) pp. 105–144

52.68   P. Moscato, C. Cotta, A.S. Mendes: Memetic algorithms. In: *New Optimization Techniques in Engineering*, ed. by G.C. Onwubolu, B.V. Babu (Springer, Berlin, Heidelberg 2004) pp. 53–85

52.69   Y. Nagata, S. Kobayashi: Edge assembly crossover: A high-power genetic algorithm for the traveling salesman problem, Proc. 17th Int. Conf. Genet. Algorithms (ICGA), ed. by T. Bäck (Morgan Kaufmann, San Mateo 1997) pp. 450–457

52.70   T.C. Jones: Evolutionary Algorithms, Fitness Landscapes and Search, Ph.D. Thesis (University of New Mexico, Albuquerque 1995)

52.71   F. Neri, C. Cotta: A primer on memetic algorithms. In: *Handbook of Memetic Algorithms*, Studies in Computational Intelligence, Vol. 379, ed. by F. Neri, C. Cotta, P. Moscato (Springer, Berlin, Heidelberg 2012) pp. 43–52

52.72   F. Neri, C. Cotta: Memetic algorithms and memetic computing optimization: A literature review, Swarm Evol. Comput. **2**, 1–14 (2012)

52.73   D. Sudholt: Parametrization and balancing local and global search. In: *Handbook of Memetic Algorithms*, Studies in Computational Intelligence, Vol. 379, ed. by F. Neri, C. Cotta, P. Moscato (Springer, Berlin, Heidelberg 2012) pp. 55–72

52.74   N. Krasnogor, B.P. Blackburne, E.K. Burke, J.D. Hirst: Multimeme algorithms for protein structure prediction, Lect. Notes Comput. Sci. **2439**, 769–778 (2002)

52.75   J.E. Smith: Co-evolution of memetic algorithms: Initial investigations, Lect. Notes Comput. Sci. **2439**, 537–548 (2002)

52.76   N. Krasnogor: Self generating metaheuristics in bioinformatics: The proteins structure comparison case, Genet. Program. Evol. Mach. **5**(2), 181–201 (2004)

52.77   N. Krasnogor, S.M. Gustafson: A study on the use of "self-generation" in memetic algorithms, Nat. Comput. **3**(1), 53–76 (2004)

52.78   J.E. Smith: Coevolving memetic algorithms: A review and progress report, IEEE Trans. Syst. Man Cybern. Part B **37**(1), 6–17 (2007)

52.79   J.E. Smith: Credit assignment in adaptive memetic algorithms, GECCO '07: Proc. 9th Annu. Conf. Genet. Evol. Comput. Conf., ed. by H. Lipson (2007) pp. 1412–1419

52.80   Y.-S. Ong, A.J. Keane: Meta-Lamarckian learning in memetic algorithms, IEEE Trans. Evol. Comput. **8**(2), 99–110 (2004)

52.81   H.G. Cobb: *An Investigation into the Use of Hypermutation as an Adaptive Operator in Genetic Algorithms Having Continuous, Time-Dependent Nonstationary Environments*. Technical Report AIC-90-001 (Naval Research Laboratory, Washington, DC 1990)

52.82   J.J. Grefenstette: Genetic algorithms for changing environments. In: *Parallel Problem Solving from Nature II*, ed. by R. Männer, B. Manderick (Elsevier, Amsterdam 1992) pp. 137–144

52.83   L.J. Eshelman: The CHC adaptive search algorithm: How to have safe search when engaging in nontraditional genetic recombination. In: *Foundations of Genetic Algorithms I*, ed. by G.J.E. Rawlins (Morgan Kaufmann, San Mateo 1991) pp. 265–283

52.84   M. Laguna, R. Marti: Scatter search. In: *Methodology and Implementations in C*, Operations Research/Computer Science Interfaces, Vol. 24, ed. by R. Sharda, S. Voß (Kluwer, Boston 2003)

52.85   M. Sevaux, S. Dauzère-Pérès: Genetic algorithms to minimize the weighted number of late jobs on a single machine, Eur. J. Oper. Res. **151**, 296–306 (2003)

52.86   E.K. Burke, J. Newall, R. Weare: A memetic algorithm for university exam timetabling, Lect. Notes Comput. Sci. **1153**, 241–250 (1996)

52.87   P.M. França, J.N.D. Gupta, A.S. Mendes, P. Moscato, K.J. Veltnik: Evolutionary algorithms for scheduling a flowshop manufacturing cell with sequence dependent family setups, Comput. Ind. Eng. **48**, 491–506 (2005)

52.88   M. Tomassini: *Spatially Structured Evolutionary Algorithms: Artificial Evolution in Space and Time* (Springer, New York 2005)

52.89   K. Sörensen, M. Sevaux: MA|PM: Memetic algorithms with population management, Comput. Oper. Res. **33**(5), 1214–1225 (2006)

52.90   O.J. Mengshoel, D.E. Goldberg: The crowding approach to niching in genetic algorithms, Evol. Comput. **16**(3), 315–354 (2008)

52.91   D.E. Goldberg, J. Richardson: Genetic algorithms with sharing for multimodal function optimization, Proc. 2nd Int. Conf. Genet. Algorithms Genet. Algorithms Appl. (L. Erlbaum Associates, Hillsdale 1987) pp. 41–49

52.92   Y.-S. Ong, M.-H. Lim, X. Chen: Memetic computation – Past, present and future, IEEE Comput. Intell. Mag. **5**(2), 24–31 (2010)

52.93   M. Toulouse, T.G. Crainic, B. Sanso, K. Thulasiraman: Self-organization in cooperative Tabu search algorithms, IEEE Int. Conf. Syst. Man Cybern., Vol. 3 (1998) pp. 2379–2384

52.94   M. Toulouse, T.G. Crainic, B. Sans: Systemic behavior of cooperative search algorithms, Parallel Comput. **30**(1), 57–79 (2004)

52.95   T.G. Crainic, M. Toulouse: Explicit and emergent cooperation schemes for search algorithms, Lect. Notes Comput. Sci. **5313**, 95–109 (2008)

52.96   M. Toulouse, K. Thulasiraman, F. Glover: Multilevel cooperative search: A new paradigm for combinatorial optimization and an application to graph partitioning, Lect. Notes Comput. Sci. **1685**, 533–542 (1999)

52.97   T.G. Crainic, M. Gendreau: Cooperative parallel tabu search for capacitated network design, J. Heuristics **8**(6), 601–627 (2002)

52.98   T.G. Crainic, M. Gendreau, P. Hansen, N. Mladenović: Cooperative parallel variable neighborhood search for the *p*-median, J. Heuristics **10**, 293–314 (2004)

52.99   D. Pelta, C. Cruz, A. Sancho-Royo, J. Verdegay: Using memory and fuzzy rules in a co-operative multi-thread strategy for optimization, Inf. Sci. **176**, 1849–1868 (2006)

52.100   C. Cruz, D. Pelta: Soft computing and cooperative strategies for optimization, Appl. Soft Comput. **9**(1), 30–38 (2009)

52.101   A. LeBouthillier, T.G. Crainic: A cooperative parallel meta-heuristic for the vehicle routing problem with time windows, Comput. Oper. Res. **32**(7), 1685–1708 (2005)

52.102   D. Barbucha: Synchronous vs. asynchronous cooperative approach to solving the vehicle routing problem, Lect. Notes Comput. Sci. **6421**, 403–412 (2010)

Part E | 52

52.103   K.S. Leung, I. King, Y.B. Wong: A probabilistic cooperative-competitive hierarchical model for global optimization, Appl. Math. Comput. **175**(2), 1092–1124 (2006)

52.104   S.T. Barnard, H.D. Simon: Fast multilevel implementation of recursive spectral bisection for partitioning unstructured problems, Concurr. Pract. Exp. **6**(2), 101–117 (1994)

52.105   C. Walshaw: A multilevel approach to the travelling salesman problem, Oper. Res. **50**(5), 862–877 (2002)

52.106   L. Hulianytskyi, S. Sirenko: Cooperative model-based metaheuristics, Electron. Notes Discret. Math. **36**, 33–40 (2010)

52.107   J. Amaya, C. Cotta, A.J. Fernández-Leiva: Memetic cooperative models for the tool switching problem, Memetic Comput. **3**, 199–216 (2011)

52.108   P. Moscato, C. Cotta: A modern introduction to memetic algorithms. In: *Handbook of Meta-heuristics*, International Series in Operations Research and Management Science, Vol. 146, ed. by M. Gendreau, J.Y. Potvin (Springer, New York, Dordrecht, Heidelberg, London 2010) pp. 141–183

52.109   F. Neri, C. Cotta, P. Moscato: *Handbook of Memetic Algorithms*, Studies in Computational Intelligence, Vol. 379 (Springer, Berlin, Heidelberg 2012)

52.110   J.-K. Hao: Memetic algorithms in discrete optimization. In: *Handbook of Memetic Algorithms*, Studies in Computational Intelligence, Vol. 379, ed. by F. Neri, C. Cotta, P. Moscato (Springer, Berlin, Heidelberg 2012) pp. 73–95

52.111   P. Merz: Memetic algorithms and fitness landscapes in combinatorial optimization. In: *Handbook of Memetic Algorithms*, Studies in Computational Intelligence, Vol. 379, ed. by F. Neri, C. Cotta, P. Moscato (Springer, Berlin, Heidelberg 2012) pp. 96–122

53. Design of Representations and Search Operators

Franz Rothlauf

Successful and efficient use of evolutionary algorithms depends on the choice of genotypes and the representation – that is, the mapping from genotype to phenotype – and on the choice of search operators that are applied to the genotypes. These choices cannot be made independently of each other. This chapter gives recommendations on the design of representations and corresponding search operators and discusses how to consider problem-specific knowledge. For most problems in the real world, similar solutions have similar fitness values. This fact can be exploited by evolutionary algorithms if they ensure that the representations and search operators used are defined in such a way that similarities between phenotypes correspond to similarities between genotypes. Furthermore, the performance of evolutionary algorithms can be increased by problem-specific knowledge. We discuss how properties of high-quality solutions can be exploited by biasing representations and search operators.

53.1 Representations

Successful and efficient use of evolutionary algorithms (EA) and other types of modern heuristics [53.1, 2] depends on the choice of genotypes and the representation – that is, the mapping from genotype to phenotype – and on the choice of search operators that are applied to the genotypes. These choices cannot be made independently of each other [53.2]. The question whether a certain representation leads to a better performing EA than an alternative representation can only be answered when the operators applied are taken into account. The reverse is also true: deciding between alternative operators is only meaningful for a given representation.

In practice, one can distinguish two complementary approaches to the design of representations and search operators [53.3]. The first approach defines *representations* (also known as *decoders* or *indirect representations*) where a solution is encoded in a standard data structure, such as strings or vectors, and applies standard off-the-shelf search operators to these genotypes. To evaluate a solution, the genotype needs to be mapped to the phenotype space. The proper choice of this genotype–phenotype mapping is important for the performance of the search process. The second approach encodes solutions to the problem in its most *natural* problem space and designs search operators to operate on this search space. In this case, often no additional mapping between genotypes and phenotypes is necessary, but domain-specific search operators need to be defined. The resulting combination of representation and operator is often called *direct representation*.

This section focuses on representations. It introduces genotypes and phenotypes (Sect. 53.1.1) and discusses properties of the resulting genotype and phenotype space (Sect. 53.1.2). Section 53.1.3 lists the benefits of using (indirect) representations. Finally, Sect. 53.1.4 gives an overview of standard genotypes.

53.1.1 Genotypes and Phenotypes

In 1866, *Mendel* recognized that nature stores the complete genetic information for an individual in pairwise alleles [53.4]. The genetic information that determines the properties, appearance, and shape of an individual is stored by a number of strings. Later, it was discovered that the genetic information is formed by a double string of four nucleotides, called DNA (deoxyribonucleic acid).

Mendel found that nature distinguishes between the genetic code of an individual and its outward appearance. The genotype represents all the information stored in the chromosomes and allows us to describe an individual on the level of genes. The phenotype describes the outward appearance of an individual. A transformation exists – a genotype–phenotype mapping or a representation – that uses the genotype information to construct the phenotype. To represent the large number of possible phenotypes with only four nucleotides, the genotype information is not stored in the allele itself, but in the sequence of alleles. By interpreting the sequence of alleles, nature can encode a large number of different phenotypes using only a few different types of alleles.

Figure 53.1 illustrates the differences between *chromosome*, *gene*, and *allele*. A chromosome is a string of some length l where all the genetic information of an individual is stored. Although nature often uses more than one chromosome, many EAs use only one chromosome for encoding all phenotype information. Each chromosome consists of many alleles. Alleles are the smallest information units in a chromosome. In nature, alleles exist pairwise, whereas in most EA implementations an allele is represented by only one symbol. For example, binary genotypes only have alleles with value zero or one. If a phenotypic property of an individual (solution), like its hair color or eye size is determined by one or more alleles, then these alleles together are called a gene. A gene is a region on a chromosome that must be interpreted together and which is responsible for a specific property of a phenotype.

We must carefully distinguish between genotypes and phenotypes. The phenotypic appearance of a solution determines its objective value. Therefore, when comparing the quality of different solutions, we must judge them on the phenotype level. However, when it comes to the application of variation operators we must view solutions on the genotype level. New solutions that are created using variation operators do not *inherit* the phenotypic properties of its parents, but only the genotype information regarding the phenotypic properties. Therefore, search operators work on the genotype level, whereas the evaluation of the solutions is performed on the phenotype level.

Formally, we define Φ_g as the genotype space where the variation operators are applied. An optimization problem on Φ_g could be formulated as $f(x) : \Phi_g \to \mathbb{R}$, where f assigns an element (fitness value) in $\mathbb{R}$ to every element in the genotype space Φ_g. A maximization problem is defined as finding the optimal solution $x^* = \{x \in \Phi_g | \forall y \in \Phi_g : f(y) \leq f(x)\}$, where x is usually a vector or string of decision variables (alleles) and $f(x)$ is the objective or fitness function. x^* is the global maximum. To be able to apply EAs to a problem, the inverse function f^{-1} does not need to exist.

53.1.2 Genotype and Phenotype Search Spaces

When using a representation, we have to define – in analogy to nature – genotypes and a genotype–phenotype mapping [53.5, 6]. Therefore, the fitness function f can be decomposed into two parts. f_g maps the genotype space Φ_g to the phenotype space Φ_p, and f_p maps Φ_p to the fitness space $\mathbb{R}$

$$f_g(x^g) : \Phi_g \to \Phi_p \,,$$
$$f_p(x^p) : \Phi_p \to \mathbb{R} \,, \tag{53.1}$$

where $f = f_p \circ f_g = f_p(f_g(x^g))$. The genotype–phenotype mapping f_g is determined by the type of genotype used. f_p represents the fitness function and assigns a fitness value $f_p(x^p)$ to each solution $x^p \in \Phi_p$. The search operators are applied to the genotypes [53.7, 8].

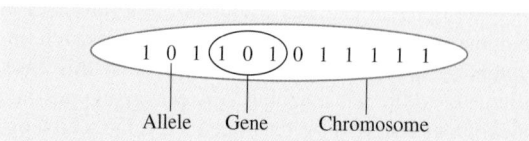

Fig. 53.1 Alleles, genes, and chromosomes

The *search space* describes the set of feasible solutions of an optimization problem and defines relationships (for example, distances) between solutions. A *metric* defined on a search space can be used for measuring similarities between solutions [53.2]. Usually, defining a search space Φ also defines a metric. Using a metric, the distance $d(x, y)$ between two solutions $x, y \in \Phi$ measures how different the two solutions are. The larger the distance, the more different two individuals are with respect to the metric used. In principle, different metrics can be used for the same search space. Different metrics result in different distances and different measurements for the similarity of solutions.

In *metric search spaces*, the similarities between solutions are measured by a distance. Therefore, we have a set X of solutions and a real-valued distance function (also called a metric) $d : X \times X \to \mathbb{R}$ that assigns a real-valued distance to any combination of two elements $x, y \in X$.

An example of a metric space is the set of real numbers $\mathbb{R}$. Here, a metric can be defined by $d(x, y) := |x - y|$. Therefore, the distance between any solutions $x, y \in \mathbb{R}$ is the absolute value of their differences. Extending this definition to two-dimensional spaces $\mathbb{R}^2$, we obtain the *city-block metric* (also known as the taxicab metric or the Manhattan distance). It is defined for two-dimensional spaces as $d(x, y) := |x_1 - y_1| + |x_2 - y_2|$, where $x = (x_1, x_2)$ and $y = (y_1, y_2)$. This metric is named the city-block metric as it describes the distance between two points on a two-dimensional plane in a city like Manhattan or Mannheim with a rectangular ground plan. On n-dimensional search spaces $\mathbb{R}^n$, the city-block metric becomes $d(x, y) := \sum_{i=1}^{n} |x_i - y_i|$, where $x, y \in \mathbb{R}^n$.

Another example of a metric that can be defined on $\mathbb{R}^n$ is the *Euclidean metric*. In Euclidean spaces, a solution $x = (x_1, \ldots, x_n)$ is a vector of continuous values ($x_i \in \mathbb{R}$). The Euclidean distance between two solutions x and y is defined as $d(x, y) := \sqrt{\sum_{i=1}^{n} (x_i - y_i)^2}$. For $n = 1$, the Euclidean metric coincides with the city-block metric. For $n = 2$, we have a standard two-dimensional search space and the distance between two elements $x, y \in \mathbb{R}^2$ is just a direct line between two points on a two-dimensional plane.

If we assume that we have a binary space ($x \in \{0, 1\}^n$), a commonly used metric is the binary *Hamming metric* [53.9] $d(x, y) = \sum_{i=1}^{n} |x_i - y_i|$, where $d(x, y) \in \{0, \ldots, n\}$. The binary Hamming distance between two binary vectors x and y of length n is just the number of binary decision variables on which x and y

differ. For continuous and discrete decision variables, it becomes $d(x, y) = \sum_{i=1}^{n} z_i$, where

$$z_i = \begin{cases} 0, & \text{for } x_i = y_i, \\ 1, & \text{for } x_i \neq y_i. \end{cases} \tag{53.2}$$

In general, the Hamming distance measures the number of decision variables on which x and y differ. Two individuals are neighbors if the distance between them is minimal. For the binary Hamming metric, the minimal distance between two individuals is $d_{min} = 1$. Therefore, two individuals x and y are neighbors if their distance $d(x, y) = 1$.

Using the Euclidean or the Hamming metric only makes sense for measuring distances between solutions of the same length n. If solutions have different lengths, the *Levenshtein* distance (or edit distance) [53.10] can be used. This distance counts the minimum number of insertion, deletion, or substitution operations that transform one solution into the other. The Levenshtein distance between two solutions can be calculated with polynomial effort using dynamic programming [53.11]. For fixed-length solutions, the Levenshtein distance is equivalent to the Hamming distance.

Using a representation f_g, we obtain two different search spaces, Φ_g and Φ_p. Therefore, different metrics can be defined for the phenotype and the genotype space. The metric used on the phenotype search space Φ_p is usually determined by the specific problem to be solved and describes which problem solutions are similar to each other. Examples of common phenotypes and corresponding metrics are given in Sect. 53.2.5. In contrast, the metric defined on Φ_g is not defined by the specific problem but can be defined by the search operators selected for the optimization method. As we can define different types of genotypes to represent the phenotypes, we are able to define different metrics on Φ_g. However, if the metrics on Φ_p and Φ_g are different, different neighborhoods can exist on Φ_g and Φ_p. For example, when encoding phenotype integers using genotype bitstrings, the phenotype $x^p = 5$ has two neighbors, $y^p = 6$ and $z^p = 4$. When using the Hamming metric and binary genotypes, the corresponding binary string $x^g = 101$ has three different neighbors, $y^g = 001$, $z^g = 111$, and $w^g = 100$ [53.12].

Therefore, the metric on the genotype space should be chosen such that it fits the metric on the phenotype space well. A representation introduces an additional genotype–phenotype mapping and thus modifies the fit.

Part E | 53.1

When designing optimization methods, we have to ensure that the metric on the genotype search space fits the original problem metric. We should choose the genotype metric in such a way that phenotypic neighbors remain neighbors in the genotype search space. Representations that ensure that neighboring phenotypes are also neighboring genotypes are called high-locality representations (Sect. 53.3.1).

53.1.3 Benefits of Representations

In principle, a representation is not necessary for the application of EAs as search operators may also be directly applied to phenotypes. However, the use of an additional genotype–phenotype mapping has some benefits:

- The use of representations is necessary for problems where a phenotype cannot be depicted as a string or in another way that is accessible to variation operators. A representative example is the shape of an object, for example the wing of an airplane. EAs that are used to find the optimal shape usually require a representation as the direct application of search operators to the *shape* of a wing is difficult. Therefore, additional genotype–phenotype mappings are used and variation operators are applied to genotypes that indirectly determine the shape.
- The introduction of a representation can be useful if there are constraints or restrictions on the phenotype space that can be advantageously modeled by a specific encoding. An example is a tree problem where the optimal solution is a star. Instead of applying search operators directly to trees, we can introduce genotypes that only encode stars resulting in a much smaller search space.
- The use of the same genotypes for different types of problems, and only interpreting them differently by using a different genotype–phenotype mapping, allows us to use standard search operators (Sect. 53.2.5) with known properties. In this case, we do not need to develop new operators with unknown properties and behavior.
- Finally, using an additional genotype–phenotype mapping can change the difficulty of a problem. A representation can reduce the difficulty of the problem and make it easier to solve for a particular optimization method. However, usually the definition of a proper representation is difficult and problem specific.

53.1.4 Standard Genotypes

We characterize some of the most important and widely used genotypes. For a more detailed overview of different types of genotypes, we refer to [53.13, Sect. C1].

Binary Genotypes
Binary genotypes are commonly used in genetic algorithms [53.14, 15]. Such EA types use recombination as the main search operator and mutation only serves as background noise. A typical search space is $\Phi_g = \{0, 1\}^l$, where l is the length of a binary vector $x^g = (x_1^g, \ldots, x_l^g)$. The genotype–phenotype mapping f_g depends on the specific optimization problem to be solved. For many combinatorial optimization problems using binary genotypes allows a direct and very natural encoding.

When using binary genotypes for encoding integer phenotypes, specific genotype–phenotype mappings are necessary. Different types of binary representations for integers assign the integers $x^p \in \Phi_p$ (phenotypes) in different ways to the binary vectors $x^g \in \Phi_g$ (genotypes). The most common binary genotype–phenotype mappings are binary, Gray, and unary encoding [53.3, 16, Chap. 5].

When using binary genotypes to encode continuous phenotypes, the accuracy (precision) depends on the number of bits that represent one phenotype variable. By increasing the number of bits that are used to represent one continuous variable the accuracy of the representation can be increased.

Integer Genotypes
Instead of using binary strings with cardinality $\chi = 2$, higher χ-ary alphabets, where $\{\chi \in \mathbb{N} | \chi > 2\}$, can also be used for the genotypes. Then, instead of a binary alphabet a χ-ary alphabet is used for a string of length l. Instead of encoding 2^l different individuals with a binary alphabet, we are able to encode χ^l different possibilities. The size of the search space increases from $|\Phi_g| = 2^l$ to $|\Phi_g| = \chi^l$.

For integer problems, users sometimes prefer to use binary instead of integer genotypes because schema processing is maximally efficient with binary alphabets when using standard recombination operators in genetic algorithms [53.17]. *Goldberg* [53.17] qualified this recommendation and emphasized that the alphabet used in the encoding should be as small as possible while still allowing a natural representation of solutions. To give general recommendations is difficult, as users often do not know a priori whether binary genotypes allow a natural encoding of integer phenotypes [53.18, 19]. We recommend that users use binary genotypes for encoding binary decision variables and integer genotypes for integer decision variables.

Continuous Genotypes

When using continuous genotypes, the search space is $\Phi_g = \mathbb{R}^l$, where l is the size of a real-valued string or vector. Continuous genotypes are often used in local search methods like evolution strategies or evolutionary programming. These types of optimization methods mainly rely on local search and search through the search space by adding a multivariate zero-mean Gaussian random variable to each continuous variable. In contrast, when using recombination-based genetic algorithms, continuous decision variables are often represented by using binary genotypes.

Continuous genotypes cannot only be used for encoding continuous problems, but also for permutation and combinatorial problems. Trees, schedules, tours, or other combinatorial problems can easily be represented by using continuous genotypes and special genotype–phenotype mappings (for an example, see weighted encodings for trees [53.20, 21]).

Messy Representations

In all previously discussed genotypes, the position of each allele is fixed along the chromosome and only the corresponding value is specified. A first gene-independent genotype was proposed by [53.22], where an inversion operator changes the relative order of the alleles in the string. The position of an allele and the corresponding value are coded together as a tuple in a string. This concept can be used for all types of genotypes such as binary, integer, and real-valued alleles, and allows an encoding which is independent of the position of the alleles in the chromosome. Later, *Goldberg* et al. [53.23] used this position-independent representation for the messy genetic algorithm.

53.2 Search Operators

This section distinguishes between standard search operators, which are applied to genotypes, and problem-specific search operators that can also be applied to phenotypes (often called *direct representations*). We start with an overview of general design guidelines. Sections 53.2.2 and 53.2.3 discuss local and recombination-based search operators. In Sect. 53.2.4, we focus on direct representations, where search operators are directly applied to phenotypes and no explicit genotype–phenotype mapping exists. Finally, Sect. 53.2.5 gives an overview of standard search operators.

53.2.1 General Design Guidelines

During the 1990s, Radcliffe developed guidelines for the design of search operators. It is important for search operators that the representation used is taken into account as search operators are based on the metric that is defined on the genotype space. *Radcliffe* introduced the principle of *formae*, which are subsets of the search space [53.24–29]. Formae are defined as *equivalence classes* that are induced by a set of equivalence relations. Any possible solution of an optimization problem can be identified by specifying the equivalence class to which it belongs for each of the equivalence re-

lations. For example, if we have a search space of faces [53.30], basic equivalence relations might be *same hair color* or *same eye color*, which would induce the formae *red hair*, *dark hair*, *green eyes*, etc. Formae of higher order like *red hair and green eyes* are then constructed by composing simple formae. The search space, which includes all possible faces, can be constructed with strings of alleles that represent the different formae. For the definition of formae, the structure of the phenotypes is relevant. For example, for binary problems, possible formae would be *bit i is equal to one/zero*.

It is an unsolved problem to find appropriate equivalences for a particular problem. From the equivalences, the genotype search space Φ_g and the genotype–phenotype mapping f_g can be constructed. Usually, a solution is encoded as a string of alleles. The value of an allele indicates whether the solution satisfies a particular equivalence. *Radcliffe* [53.25] proposed several design guidelines for creating appropriate equivalences for a given problem. The most important one is that the generated formae should group together solutions of related fitness [53.28], in order to create a fitness landscape or structure of the search space that can be exploited by search operators.

Radcliffe recognized that the genotype search space, the genotype–phenotype mapping, and the search operators belong together, and their design cannot be separated from each other [53.26]. He assumed that search operators create offspring solutions from a set of parent solutions. For the development of appropriate search operators that are based on predefined formae, he formulated the following four design principles [53.25, 29]:

- *Respect*: offspring produced by recombination should be members of all formae to which both their parents belong. For the *face example* this means that offspring should have red hair and green eyes if both parents have red hair and green eyes.
- *Transmission*: an offspring should be equivalent to at least one of its parents under each of the basic equivalence relations. This means that every gene should be set to an allele which is taken from one of the parents. If one parent has dark hair and the other red hair, then the offspring has either dark or red hair.
- *Assortment*: an offspring can be formed with any compatible characteristics taken from the parents. Assortment is necessary as some combinations of equivalence relations may be infeasible. This means, for example, that the offspring inherits dark hair from the first parent and blue eyes from the second parent only if dark hair and blue eyes are compatible. Otherwise, the alleles are set to feasible values taken from a random parent.
- *Ergodicity*: an iterative use of search operators allows us to reach any point in the search space from all possible starting solutions.

Radcliffe developed a consistent concept of how to design efficient EAs once appropriate equivalence classes (formae) are defined. However, the finding of appropriate equivalence classes, which is equivalent to either defining the genotype search space and the genotype–phenotype mapping or appropriate direct search operators on the phenotypes, is often difficult and remains an unsolved problem.

As long as the genotypes are either binary, integer, or real-valued strings, standard recombination and mutation operators can be used. The situation is different if direct representations (Sect. 53.2.4) are used for problems whose phenotypes are not binary, integer, or real-valued. Specialized operators are necessary that allow offspring to inherit important properties from their parents [53.24, 25, 27, 31]. In general, these operators

are problem-specific and must be developed separately for every optimization problem.

53.2.2 Local Search Operators

Local search and the use of local search operators are at the core of EAs. The goal of local search is to find fitter individuals by performing neighborhood search [53.32]. Usually, a local search operator creates offspring that have a small or sometimes even minimal distance to their parents. Therefore, local search operators and the metric on the corresponding search space cannot be decided independently of each other but determine each other. A metric defines possible local search operators and a local search operator determines the metric. As search operators are applied to the genotypes, the metric on Φ_g is relevant for the definition of local search operators.

The basic idea behind using local search operators is that the structure of a fitness landscape should guide a search heuristic to high-quality solutions [53.33], and that good solutions can be found by performing small iterated changes. We assume that in most real-world problems high-quality solutions are not isolated in the search space but grouped together [53.34, 35]. Therefore, better solutions can be found by searching in the neighborhood of already found good solutions. The search steps must be small because too large search steps would result in randomization of the search, and guided search around good solutions would become impossible. In contrast, when using search operators that perform large steps in the search space it would not be possible to find better solutions by searching around already found good solutions but the search algorithm would jump randomly around the search space (Sect. 53.3.1).

The following paragraphs review some common local search operators for binary, integer, and continuous genotypes and illustrate how they are designed based on the underlying metric. The local search operators (and underlying metrics) are commonly used and are usually a good choice. However, in principle, we are free to choose other metrics and to define corresponding search operators. Then, the metric should be chosen such that high-quality solutions are neighboring solutions and the resulting fitness landscape leads guided search methods to an optimal solution. The choice of a proper metric and corresponding search operators are always problem specific and the ultimate goal is to choose a metric such that the problem becomes easy for EAs. However, we want to emphasize that for most practical applications the illustrated search operators are a good choice and allow us to design efficient and effective EAs.

Binary Genotypes

When using binary genotypes, the distance between two solutions $x, y \in \{0, 1\}^l$ is often measured using the Hamming distance. Local search operators based on this metric generate new solutions with the Hamming distance $d(x, y) = 1$. This type of search operator is also known as a *standard mutation operator* for binary strings or a *bit-flipping* operator. As each binary solution of length l has l neighbors, this search operator can create l different offspring. For example, applying the bit-flipping operator to $(0, 0, 0, 0)$ can result in four different offspring $(1, 0, 0, 0), (0, 1, 0, 0), (0, 0, 1, 0)$, and $(0, 0, 0, 1)$.

Reeves [53.36] proposed another local search operator for binary strings based on a different neighborhood definition: for a randomly chosen $k \in \{0, \ldots, l\}$, it complements the bits $x_k, \ldots, x_l$. Again, each solution has l neighbors. For example, applying this search operator to $(0, 0, 0, 0)$ can result in $(1, 1, 1, 1)$, $(0, 1, 1, 1), (0, 0, 1, 1)$, or $(0, 0, 0, 1)$. Although the operator is of minor practical importance, it has some interesting theoretical properties. First, it is closely related to the one-point recombination crossover (see below) as it chooses a random point and inverts all x_i with $i \geq k$. Therefore, it has also been called the *complementary crossover operator*. Second, if all genotypes are encoded using Gray code [53.37, 38], the neighbors of a solution in the Gray-coded search space using Hamming distance are identical to the neighbors in the original binary-coded search space using the complementary crossover operator. Therefore, Hamming distances between Gray encoded solutions are equivalent to the distances between the original binary encoded solutions using the metric induced by the complementary crossover operator (neighboring solutions have distance one). For more information regarding the equivalence of different neighborhood definitions and search operators we refer to the literature [53.36, 39, 40].

Integer Genotypes

For integer genotypes, different metrics are common, leading to different local search operators. When using the binary Hamming metric, two individuals are neighbors if they differ in one decision variable. Search operators based on this metric can assign a random value to a randomly chosen allele. Therefore, each solution $x \in \{0, \ldots, k\}^l$ has lk neighbors. For example,

$x = (0, 0)$ with $x_i \in \{0, 1, 2\}$ has four different neighbors $((1, 0), (2, 0), (0, 1)$, and $(0, 2))$.

The situation changes when defining local search operators based on the city-block metric. Then, a local search operator can create new solutions by slightly increasing or decreasing one randomly chosen decision variable. For example, new solutions are generated by adding +/-1 to a randomly chosen variable x_i. Each solution of length l has a maximum of $2l$ different neighbors. For example, $x = (0, 0)$ with $x_i \in \{0, 1, 2\}$ has only two different neighbors $(0, 1)$ and $(1, 0)$.

Finally, we can define search operators such that they do not modify values of decision variables but exchange values of two decision variables x_i and x_j. Therefore, using the Hamming distance, two neighbors have distance $d = 2$ and each solution has a maximum of $\binom{l}{2}$ different neighbors. For example, $x = (3, 5, 2)$ has three different neighbors $((5, 3, 2), (2, 5, 3)$, and $(3, 2, 5))$.

Continuous Genotypes

For continuous genotypes, we can define local search operators analogously to integer genotypes. Based on the binary Hamming metric, the application of a local search operator can assign a random value $x_i \in [x_{i,\min}, x_{i,\max}]$ to the i-th decision variable. Furthermore, we can define a local search operator such that it exchanges the values of two decision variables x_i and x_j. The binary Hamming distance between old and new solutions is $d = 2$.

The situation is a little more complex in comparison to integer genotypes when designing a local search operator based on the city-block metric. We must define a search operator such that its iterative application allows us to reach all solutions in reasonable time. Therefore, a search step should be not too small (we want to have some progress in search) and not too large (the offspring should be similar to the parent solution). A commonly used concept for such search operators is to add a random variable with zero mean to the decision variables. This results in $x'_i = x_i + m$, where m is a random variable and x' is the offspring generated from x. Sometimes m is uniformly distributed in $[-a, a]$, where $a < (x_{i,\max} - x_{i,\min})$. More common is the use of a normal distribution $\mathcal{N}(0, \sigma)$ with zero mean and standard deviation σ. The addition of zero-mean Gaussian random variables generates offspring that have, on average, the same statistical properties as their parents. For more information on local search operators for continuous variables, we refer to [53.41].

53.2.3 Recombination Operators

To be able to use recombination operators, a set of solutions (*population*) must exist as the goal of recombination is to recombine meaningful properties of parent solutions. Thus, for the application of recombination operators at least two parent solutions are necessary; otherwise local search operators are the only option. Recombination operators should be designed according to Radcliffe's recommendations (Sect. 53.2.1).

Analogously to local search operators, recombination operators should be designed based on the metric used [53.6, 30]. Given two parent solutions x^{p1} and x^{p2} and one offspring solution x^o, recombination operators should be designed such that

$$\max(d(x^{p1}, x^o), d(x^{p2}, x^o)) \leq d(x^{p1}, x^{p2}) . \quad (53.3)$$

Therefore, the application of recombination operators should result in offspring where the distances between offspring and its parents are equal to or smaller than the distance between the parents. When viewing the distance between two solutions as a measurement of dissimilarity, this design principle ensures that offspring solutions are similar to parents. Consequently, applying a recombination operator to the same parent solutions $x^{p1} = x^{p2}$ should also result in the same offspring ($x^o = x^{p1} = x^{p2}$).

In the last few years, this basic concept of the design of recombination operators has been interpreted as *geometric crossover* [53.42–44]. This work builds upon previous work [53.6, 30, 45] and defines crossover and mutation representation-independently using the notion of distance associated with the search space.

Why should we use recombination operators in EAs? The motivation is that we assume that many real-world problems are decomposable. Therefore, problems can be solved by decomposing them into smaller subproblems, solving these smaller subproblems, and combining the optimal solutions of the subproblems to obtain overall problem solutions. The purpose of recombination operators is to form new overall solutions by recombining solutions of smaller subproblems that exist in different parent solutions. If this juxtaposition of smaller, highly fit, partial solutions (often denoted as building blocks) does not result in good solutions, search strategies that are based on recombination operators will show low performance. However, as many problems of practical relevance can be decomposed into smaller problems (they are decomposable), the use of recombination operators often results in good performance of EAs.

Common recombination operators for standard genotypes are *one-point crossover* [53.22] and *uniform crossover* [53.46–48]. We assume a vector or string x of decision variables of length l. When using one-point crossover, a *crossover point* $c = \{1, \ldots, l-1\}$ is initially chosen randomly. Usually, two offspring solutions are created from two parent solutions by swapping the partial strings. As a result, we obtain for the parents $x^{p1} = [x_1^{p1}, x_2^{p1}, \ldots, x_l^{p1}]$ and $x^{p2} = [x_1^{p2}, x_2^{p2}, \ldots, x_l^{p2}]$ the offspring $x^{o1} = [x_1^{p1}, x_2^{p1}, \ldots, x_c^{p1}, x_{c+1}^{p2}, \ldots, x_l^{p2}]$ and $x^{o2} = [x_1^{p2}, x_2^{p2}, \ldots, x_c^{p2}, x_{c+1}^{p1}, \ldots, x_l^{p1}]$. A generalized version of one-point crossover is *n-point crossover*. For this type of crossover operator, we choose n different crossover points and create an offspring by alternately selecting alleles from parent solutions. For uniform crossover, we decide independently for every single allele of the offspring from which parent solution it inherits the value of the allele. In most implementations, no parent is preferred and the probability of an offspring inheriting the value of an allele from a specific parent is $p = 1/m$, where m denotes the number of parents that are considered for recombination. For example, when two possible offspring are considered with the same probability ($p = 1/2$), we could obtain as offspring $x^{o1} = [x_1^{p1}, x_2^{p1}, x_3^{p2}, \ldots, x_{l-1}^{p1}, x_l^{p2}]$ and $x^{o2} = [x_1^{p2}, x_2^{p2}, x_3^{p1}, \ldots, x_{l-1}^{p2}, x_l^{p1}]$.

Figure 53.2 presents examples for the three crossover variants. All three recombination operators are based on the binary Hamming distance and follow (53.3) as $d(x^{p1}, x^{p2}) \geq \max(d(x^{p1}, x^o), d(x^{p2}, x^o))$. Therefore, the similarity between offspring and parent is not lower than between the parents.

Uniform and n-point crossover can be used independently of the type of decision variables (binary, integer, continuous, etc.), since these operators only exchange alleles between parents. In contrast, *intermediate recombination* operators attempt to average or blend components across multiple parents and are designed for continuous and integer problems. Given two parents x^{p1} and x^{p2}, a crossover operator known as *arithmetic crossover* [53.49] creates an offspring x^o as $x_i^o = \alpha x_i^{p1} + (1-\alpha)x_i^{p2}$, where $\alpha \in [0, 1]$. If $\alpha = 0.5$, the crossover just takes the average of both parent solutions. In general, for m parents, this operator becomes $x_i^o = \sum_{i=1}^{m} \alpha_i x_i^{pi}$, where $\sum_{i=1}^{m} \alpha_i = 1$. Arithmetic crossover is based on the city-block metric. With respect to this metric, the distance between offspring and parent is smaller than the distance between the parents. Another type of crossover operator that is based on the Euclidean metric is *geometrical crossover* [53.49].

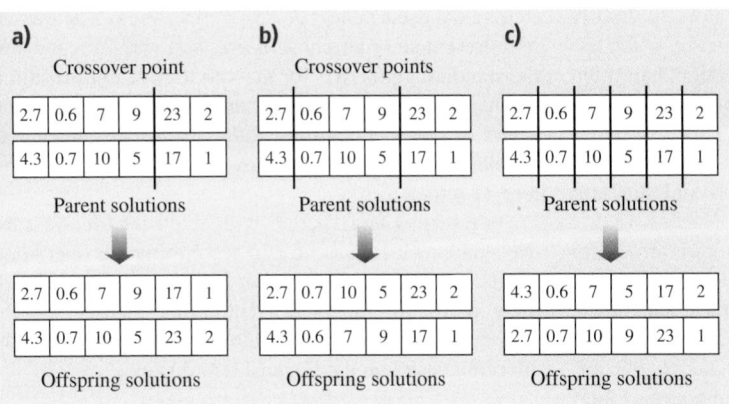

Fig. 53.2a–c Different crossover variants. (**a**) One-point crossover. (**b**) Two-point crossover. (**c**) Uniform crossover

Part E | 53.2

Given two parents, an offspring is created as $x_i^o = \sqrt{x_i^{p1} x_i^{p2}}$. For further information on crossover operators we refer to [53.50] (binary crossover) and [53.41] (continuous crossover).

53.2.4 Direct Representations

If we apply search operators directly to phenotypes, it is not necessary to specify a representation and a genotype space. Then, phenotypes are the same as genotypes

$$f(x^g) : \Phi_g \to \mathbb{R} . \qquad (53.4)$$

f_g does not exist and we have a *direct representation*. Because there is no longer an additional mapping between Φ_g and Φ_p, a direct representation does not change any aspect of the phenotype problem such as difficulty or metric. However, when using direct representations, we often cannot use standard search operators, but have to define problem-specific operators. Therefore, important for the success of EAs using a direct representation is not finding a *good* representation for a specific problem, but developing proper search operators defined on phenotypes.

The definition of the variation operators are relevant for different implementations of direct representations. Since we assume that local search operators always generate neighboring solutions, the definition of a local search operator induces a metric on the genotypes. Therefore, the metric that we use on the genotype space should be chosen in such a way that new solutions that are generated by local search operators have a small distance to the old solutions and the solutions are neighbors with respect to the metric used. Furthermore, the distance between two solutions $x \in \Phi_g$ and $y \in \Phi_g$ should be proportional to the minimal number of local search steps that are necessary to move from x to y. Analogously, the definition of a recombination operator also induces a metric on the search space. The metric used should guarantee that the application of a recombination operator to two solutions $x^p \in \Phi_g$ and $y^p \in \Phi_g$ creates a new solution $x^o \in \Phi_g$ whose distances to the parents are not larger than the distance between the parents (53.3).

For the definition of variation operators, we should also consider that for many problems we have a natural notion of similarity between phenotypes. When we create a problem model, we often *know* which solutions are similar to each other. Such a notion of similarity should be considered for the definition of variation operators. We should design local search operators in such a way that their application creates solutions which we *view* as similar. Such a definition of local search operators ensures that neighboring phenotypes are also neighbors with respect to the metric that is induced by the search operators.

At a first glance, it seems that the use of direct representations makes life easier as direct representations release us from the challenge to design efficient representations. However, we are confronted with some problems:

- There are many phenotypes to which no standard variation operators can be applied.
- The design of high-quality problem-specific search operators is difficult.
- We cannot use EAs that only work on standard genotypes.

For indirect representations with standard genotypes, the definition of search operators is straightforward as these are usually based on the metric of the genotype space (Sects. 53.2.2 and 53.2.3). The behavior of EAs using standard search operators is usually well studied and well understood. However, when using direct representations, standard operators can often no longer be used. Instead, problem-specific operators must be developed for each phenotype. This is difficult, as we cannot use most of our knowledge about the behavior of EAs using standard genotypes and standard operators.

The design of proper search operators is often demanding as phenotypes are usually not string-like but are more complicated structures like trees, schedules, time tables, or other structures (Sect. 53.2.5). In this case, phenotypes cannot be depicted as a string or in another way that is accessible to variation operators. Other representative examples are the form or shape of an object. Search operators that can be directly applied to the shape of an object are often difficult to design.

Finally, using specific variants of EAs like estimation of distribution algorithms (EDAs) becomes very difficult. These types of EAs do not use standard search operators that are applied to genotypes but build new solutions according to a probabilistic model of previously generated solutions [53.51–56]. These search methods were developed for a few standard genotypes (usually binary and continuous) and result in better performance than, for example, traditional simple genetic algorithms for decomposable problems [53.57, 58]. However, because direct representations with non-standard phenotypes and problem-specific search operators can hardly be implemented in EDAs, direct representations cannot benefit from these optimization methods.

53.2.5 Standard Search Operators

We provide an overview of standard search spaces and the corresponding search operators. The search spaces can either represent genotypes (indirect representation) or phenotypes (direct representation). We order the search spaces by increasing complexity. With increasing complexity, the design of search operators becomes more demanding. An alternative to designing complex search operators for complex search spaces is to introduce additional mappings that map complex search spaces to simpler ones. Then, the design of the corresponding search operators becomes easier, however, a proper design of the additional mapping (representation) becomes more important.

Strings and Vectors

Strings and vectors of either fixed or variable length are the most elementary search spaces. They are the most frequently used genotype structures. Vectors allow us to represent an ordered list of decision variables and are the standard genotypes for the majority of optimization problems. Strings are appropriate for sequences of characters or patterns. Consequently, strings are suited for problems where the objects modeled are *text, characters*, or *patterns*.

For strings and vectors with fixed length, we can use standard local search and recombination operators (Sects. 53.2.2 and 53.2.3) that are based on the Hamming metric or the binary Hamming metric. Search operators for strings and vectors with variable length are often based on the Levenshtein distance.

Coordinates/Points

To represent locations in a geometric space, coordinates can be used. Coordinates can be either integer or continuous. Common examples are locations of cities or other spots on two-dimensional grids. Coordinates are appropriate for problems that work on *sites, positions*, or *locations*.

We can use standard local and recombination operators for continuous variables and integers, respectively. For coordinates, the Euclidean metric is often used to measure the similarity of solutions.

Graphs

Graphs allow us to represent relationships between arbitrary objects. Usually, the structure of a graph is described by listing its edges. An edge represents a relationship between a pair of objects. Given a graph with n nodes (objects), there are $n(n-1)/2$ possible edges. Using graphs is appropriate for problems that seek a *network, circuit*, or *relationship*.

Common genotypes for graphs are lists of edges indicating which edges are used. For example, the characteristic vector representation encodes graphs of fixed size using a binary vector of length $n(n-1)/2$ [53.3, Sect. 6.3]. Standard search operators for the characteristic vector representation are based on the Hamming metric as the distance between two graphs can be calculated as the number of different edges. Standard search operators can be used if there are no additional constraints.

Subsets

Subsets represent selections from a set of objects. Given n different objects, the number of subsets having exactly k elements is equal to $\binom{n}{k}$. Thus, the number of possible subsets is $\sum_{k=0}^{n} \binom{n}{k} = 2^n$. For subsets, the order of the objects does not matter. The two example subsets $\{1, 3, 5\}$ and $\{3, 5, 1\}$ represent the same phenotype solution. Local search operators that can be applied directly to subsets often either modify the objects in the subset, or increase/reduce the number of objects in one subset. Recombination operators that are directly applied to subsets are more sophisticated as no standard operators can be used. We refer to [53.59] for detailed information on their design. Subsets are often used for problems that seek a *cluster, collection, partition, group, packaging*, or *selection*.

Given n different objects, a subset of fixed size k can be represented using an integer vector x of length k, where the x_i indicate the selected objects and $x_i \neq x_j$, for $i \neq j$ and $i, j \in [1, k]$. Then, standard local search operators can be applied if we assume that each of the k selected objects is unique. The application of recombination operators is more demanding as each subset is represented by $k!$ different genotypes (integer vectors) and the distances between the $k!$ different genotypes that represent the same subset are large [53.60]. Recombination operators must be designed such that the distances between offspring and parents are smaller than the distances between parents (53.3) and the recombination of two genotypes that represent the same subset always results in the same subset. For guidelines on the design of appropriate recombination operators and examples, we refer to [53.61] and [53.60].

Permutations

A large variety of EAs have been developed for permutation problems as many such problems are of practical relevance but NP-hard (NP: non-deterministic polynomial-time). Permutations are orderings of items. The order of the objects is relevant for permutations. The number of permutations on a set of n elements is $n!$. 1-2-3 and 1-3-2 are two examples of permutations of three integer numbers $x \in \{1, 2, 3\}$. The traveling salesperson problem (TSP) is a prominent example of a permutation problem. Permutations are commonly used for problems that seek an *arrangement, tour, ordering*, or *sequence*.

The design of appropriate search operators for permutations is demanding. In many approaches, permutations are encoded using an integer genotype vector of length n, where each decision variable x_i indicates an ob-

ject and has a unique value ($x_i \neq x_j$ for $i \neq j$ and $i, j \in \{1, \ldots, l\}$). Standard recombination and mutation operators applied to such genotypes fail since the resulting solutions usually represent no permutations. Therefore, in the literature a variety of different permutation-specific variation operators have been developed. They are either based on the absolute or relative ordering of the objects in a permutation. When using the absolute ordering of objects in a permutation as the distance metric, two solutions are similar to each other if the objects have the same position in the two solutions ($x_i^1 = x_i^2$). For example, 1-2-3-4 and 2-3-4-1 have a maximum absolute distance of $d = 4$, as the two solutions have no common absolute positions. In contrast, when using relative ordering, two solutions are similar if the sequence of objects is similar for the two solutions. For example, 1-2-3-4 and 2-3-4-1 have distance $d = 1$ as the two permutations are shifted by one position. Based on the metric used (relative versus absolute ordering), a large variety of different recombination and local search operators have been developed. Examples are the *order crossover* [53.62], *partially mapped crossover* [53.63], *cycle crossover* [53.64], *generalized order crossover* [53.65], or *precedence preservative crossover* [53.66]. For more information on the design of such permutation-specific variation operators, we refer to [53.67, 68], and [53.60].

Trees

Trees are used to describe hierarchical relationships between objects. Trees are a specialized variant of graphs where only one path exists between each pair of nodes. As standard search operators cannot be applied to tree structures, we either need to define problem-specific search operators that are directly applied to trees or additional genotype–phenotype mappings that map each tree to simpler genotypes where standard variation operators can be applied.

We can distinguish between trees of fixed and variable size. For trees of fixed size, we refer to [53.2]. Search operators for tree structures of variable size are at the core of *genetic programming*. They are often based on the Levenshtein distance. Further information about appropriate search operators for trees of variable size can be found in [53.69] and [53.70].

53.3 Problem–Specific Design of Representations and Search Operators

Jones and *Forrest* [53.71] assumed that the difficulty of an optimization problem is determined by how the ob-

jective values are assigned to the solutions $x \in X$ and what metric is defined on X. They classified fitness landscapes into three classes, straightforward, difficult, and misleading.

1. Straightforward. For such problems, the fitness of a solution is correlated with the distance to the optimal solution. With lower distance, the fitness difference to the optimal solution decreases. As the structure of the search space guides search methods towards the optimal solution such problems are usually easy for guided search methods.
2. Difficult. There is no correlation between the fitness difference and the distance to the optimal solution. The fitness values of neighboring solutions are uncorrelated and the structure of the search space provides no information about which solutions should be sampled next by the search method.
3. Misleading. The fitness difference is negatively correlated to the distance to the optimal solution. Therefore, the structure of the search space misleads a guided search method to sub-optimal solutions.

The general idea is to measure how well the metric defined on the search space fits the structure of the objective function. A high fit between metric and structure of the fitness function makes a problem easy for guided search methods.

A fundamental assumption about the application of EAs is that the vast majority of optimization problems that we can observe in the real world are:

● Neither misleading nor difficult,
● Have high locality (distance between solutions is correlated with their fitness difference).

We assume that misleading problems have no importance in the real world as usually optimal solutions are not isolated in the search space surrounded by only low-quality solutions. Furthermore, we assume that the metric of a search space is meaningful and, on average, the fitness differences between neighboring solutions are smaller than between randomly chosen solutions. It is only because most real-world problems are neither difficult nor misleading that guided search methods which use information about previously sampled solutions can outperform random search for real-world problems [53.2, 34].

Since we assume that high locality is a general property of real-world problems, EAs must ensure that their design does not destroy the high locality of a problem. If the high locality of a problem is destroyed, straightforward problems turn into difficult problems and cannot be solved better than by random search [53.2]. Therefore, EAs must ensure that the search operators

used fit the metric on the search space and representations have high locality; this means phenotype distances must correspond to genotype distances.

The second aspect of this section is how we can consider knowledge about problem-specific properties for the design of EAs. For example, we have knowledge about the character and properties of high-quality (or low-quality) solutions. Such problem-specific knowledge can be exploited by introducing a *bias* into EAs. The bias should consider this knowledge and, for example, concentrate search on solutions that are expected to be of high quality or avoid solutions expected to be of low quality. A bias can be considered in the representation as well as the search operator. However, EAs should only be biased if we have obtained some particular knowledge about an optimization problem or problem instance. If we have no knowledge about properties of a problem, we should not bias EAs as this will mislead the search heuristics.

Section 53.3.1 discusses how the design of EAs can modify the locality of a problem. To ensure guided search, the search operators must fit the problem metric. Local search operators must generate neighboring solutions and recombination operators must generate offspring where the distances between offspring and parents do not exceed the distances between parents. Section 53.3.2 focuses on the possibility of biasing EAs. We discuss how problem-specific construction heuristics can be used as genotype–phenotype mappings (Sect. 53.3.2) and how redundant representations affect heuristic search (Sect. 53.3.2).

53.3.1 High Locality

The *locality* of a problem measures how well the distances $d(x, y)$ between any two solutions $x, y \in X$ correspond to the difference in their fitness values $|f(x) - f(y)|$. Locality is high if neighboring solutions have similar fitness values and fitness differences correlate positively with distances. In contrast, the locality of a problem is low if low distances do not correspond to low differences in the fitness values. Important for the locality of a problem is the metric defined on the search space.

The performance of guided search methods is high if the locality of a problem is relatively high; this means that the structure of the fitness landscape leads search algorithms to high quality solutions [53.2]. Local search methods show especially good performance if either high-quality or low-quality solutions are grouped together in the solution space. Optimization problems with high locality can usually be solved well using EAs, as all EAs have some kind of local search elements.

The following paragraphs provide design guidelines for search operators and representations. Search operators must fit the metric of a search space, because otherwise EAs show low performance as they behave like random search. A representation introduces an additional genotype–phenotype mapping. The locality of a representation describes how well the metric on the phenotype space fits to the metric on the genotype space. Low locality, which means there is a poor fit, randomizes the search and also leads to low performance of EAs.

Search Operator

EAs rely on the concept of local search. Local search iteratively generates new solutions similar to existing ones. Local search is a reasonable and successful search approach for real-world problems, as most real-world problems have high locality and are neither misleading nor difficult. In addition, to avoid being trapped in local optima, EAs use diversification steps. Diversification steps randomize search and allow EAs to jump through the search space.

Different types of EAs use different concepts for controlling intensification and diversification [53.2, Chap. 5]. Local search intensifies the search as it allows incremental improvements of already found solutions. Diversification steps must be relatively rare as they usually lead to inferior solutions. When designing search operators, we must have in mind that EAs use local search operators as well as recombination operators for *intensifying* the search. Solutions that are generated should be similar to the existing ones. Therefore, we must ensure that search operators (local search operators as well as recombination operators) generate similar solutions and do not jump around in the search space. This can be done by ensuring that local search operators generate neighboring solutions and recombination operators generate solutions where the distances between parent and offspring are smaller or equal to the distances between parents (Sect. 53.2.3 and 53.3).

Applying search operators to solutions defines a metric on the corresponding search space. With respect to the search operators, solutions are similar to each other if only a few local search steps suffice to transform one solution into another. Therefore, when designing search operators, it is important that the metric induced by the search operators fits the problem metric. If both metrics are similar (this means a local search operator creates neighboring solutions with

respect to the problem metric), guided search will perform well as it can systematically explore promising areas of the search space.

Therefore, we should make sure that local search operators generate neighboring solutions. The fit between the problem metric and the metric induced by the search operators should be high. Then, most real-world problems, where neighboring solutions have, on average, similar fitness values, are easy to solve for EAs.

For real-world problems, the design or choice of proper search operators can be difficult if it is unclear what a *natural* problem metric is. We want to illustrate this issue for a scheduling problem. Given a number of tasks, we want to find an optimal schedule. There are different metrics that can be relevant for such a permutation problem. We have the choice between metrics based either on the relative or absolute ordering of the tasks (Sect. 53.2.5). The choice of the *right* problem metric depends on the properties of the scheduling problem. For example, if we want to find an optimal class schedule, usually it is more natural to use a metric based on the absolute ordering of the tasks (classes). The relative ordering of the tasks is much less important as we have fixed time slots. The situation is reversed if we want to find an optimal schedule for a paint shop. For example, when painting different cars consecutively, color changes are time-consuming as paint tools have to be cleaned before a new color can be used. Therefore, the relative ordering of the tasks (paint jobs) is important, as the tasks should be grouped together such that tasks that require the same color are painted consecutively and ordered such that the paint shop starts with the brightest colors and ends with the darkest ones.

This example makes it clear that the most *natural* problem metric does not depend on the set of possible solutions but on the character of the underlying optimization problem and fitness function. The goal is to choose a metric such that the locality of the problem is high. The same is true for the design of operators. A high-quality local search operator should generate solutions with similar fitness. Then, problems become easy to solve for EAs.

Representation

Representations introduce an additional genotype–phenotype mapping and thus modify the fit between the metric on the genotype space (which is induced by the search operators used), and the original problem metric on the phenotype space. High-quality representations ensure that the metric on the genotype space fits the original problem metric. The *locality of a rep-*

resentation describes how well neighboring genotypes correspond to neighboring phenotypes [53.72–76]. In contrast to the locality of a problem, which measures the fit between fitness differences and phenotype distances, the locality of a representation measures the fit between phenotype distances and genotype distances.

The use of a representation can change the difficulty of problems (Sect. 53.1.2) [53.6]. The ability of representations to change the difficulty of a problem is closely related to their locality. The locality of a representation is high if all neighboring genotypes correspond to neighboring phenotypes. In contrast, it is low if neighboring genotypes do not correspond to neighboring phenotypes. Therefore, the locality d_m of a representation can be defined as [53.3, Sect.3.3]

$$d_m = \sum_{d_{x,y}^g = d_{\min}^g} |d_{x,y}^p - d_{\min}^p| , \qquad (53.5)$$

where $d_{x,y}^p$ is the phenotype distance between the phenotypes x^p and y^p, $d_{x,y}^g$ is the genotypic distance between the corresponding genotypes, and $d_{\min}^p$ and $d_{\min}^g$ are the minimum distances between two (neighboring) phenotypes and genotypes, respectively. Without loss of generality, we assume that $d_{\min}^g = d_{\min}^p$. For $d_m = 0$, all genotypic neighbors correspond to phenotypic neighbors and the encoding has perfect (high) locality.

We want to emphasize that the locality of a representation depends on the representation f_g and the metrics that are defined on Φ_g and Φ_p. f_g alone only determines which phenotypes are represented by which genotypes and cannot be used for measuring similarities between solutions. To describe or measure the locality of a representation, a metric must be defined on Φ_g and Φ_p.

Figure 53.3 illustrates the difference between high-locality and low-locality representations. We assume 12 different phenotypes (a–l) and measure distances between solutions using the Euclidean metric. Each phenotype (lower case symbol) corresponds to one genotype (upper case symbol). The representation f_g has perfect (high) locality if neighboring genotypes correspond to neighboring phenotypes. Then, local search steps have the same effect in the phenotype and genotype search space.

If we assume that f_g is a one-to-one mapping, every phenotype is represented by exactly one genotype and there are $|\Phi_g|! = |\Phi_p|!$ different representations. Each of these many different representations assigns the genotypes to the phenotypes in a different way.

We want to ask how the locality of a representation influences the performance of EAs. Often, there is a *natural* problem metric describing which phenotypes are similar to each other. A representation introduces a new genotype metric based on the genotypes and search operators used. This metric can be different from the problem (phenotype) metric. Therefore, the character of search operators can be different for genotypes versus phenotypes. If the locality of a representation is high, then a search operator has the same effect on the phenotypes as on the genotypes. As a result, the original problem difficulty remains unchanged by a representation f_g. Easy (straightforward) problems remain easy and misleading problems remain misleading. Figure 53.4 (left) illustrates the effect of local search operators for high-locality representations. A local search step has the same effect on the phenotypes as on the genotypes.

For low-locality representations, the situation is different and the influence of a representation on the

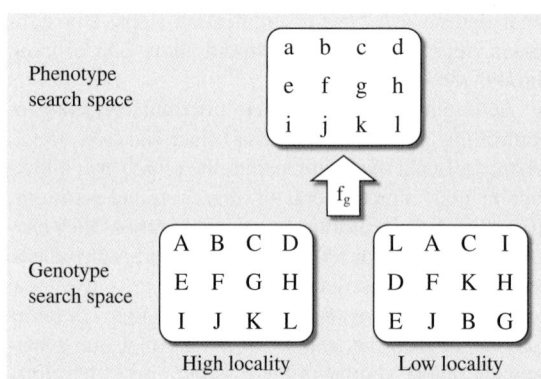

Fig. 53.3 High versus low-locality representations

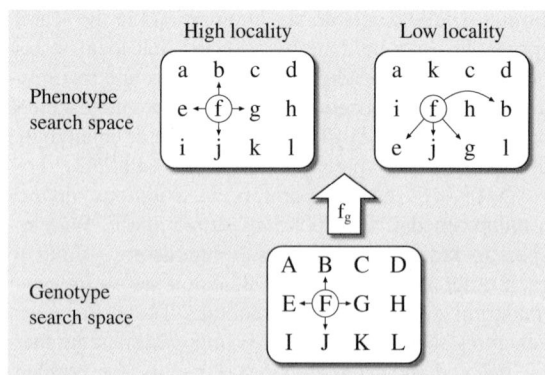

Fig. 53.4 The effect of local search operators for high versus low-locality representations

difficulty of a problem depends on its character. If a problem f_p is straightforward, a low-locality representation f_g randomizes the problem by destroying the correlation between distance and fitness and making the problem $f = f_p(f_g(x^g))$ more difficult. When using low-locality representations, a small change in a genotype does not correspond to a small change in the phenotype, but larger changes in the phenotype are possible (Fig. 53.4, right). Therefore, when using low-locality representations, straightforward problems become, on average, difficult as low-locality representations lead to a more uncorrelated fitness landscape and heuristics can no longer extract meaningful information about the structure of the problem. Guided search becomes more difficult as many genotypic search steps do not result in a similar solution but in a random one.

Summarizing the results, low-locality representations have the same effect as using random search. Therefore, on average, straightforward problems become more difficult for guided search methods. As most real-world problems are straightforward, the use of low-locality representations makes these problems more difficult. Therefore, we strongly encourage users of EAs to use high-locality representations for problems of practical relevance. Of course, low-locality representations make misleading problems easier for guided search [53.3]; however, these are problems which we do not expect to meet in reality and we do not really want to solve.

For more information on the influence of the locality of representations on the performance of EAs, we refer the interested reader to [53.3, Sect. 3.3] and [53.2].

53.3.2 Biasing Search

This section discusses how to bias EAs. If we have some knowledge about the properties of either high-quality or low-quality solutions, we can make use of this knowledge for the design of EAs. For representations, we can incorporate heuristics or introduce redundant encodings and assign a larger number of genotypes to high-quality phenotypes. Search operators can be designed in such a way that they distinguish between high-quality and low-quality solution features (building blocks) and prefer the high-quality ones [53.2].

A representation or search operator is *biased* if the application of a variation operator generates some solutions in the search space with higher probability [53.12]. We can bias representations by incorporating heuristics into the genotype–phenotype mapping. Furthermore, representations can be biased if the num-

ber of genotypes exceeds the number of phenotypes. Then, representations are called *redundant* [53.77–80]. Redundant representations are biased if some phenotypes are represented by a larger number of genotypes. Analogously, search operators are biased if some solutions are generated with higher probability.

When biasing EAs, we must make sure that we have a priori knowledge about the problem and the bias exploits this knowledge in an appropriate way. Introducing an inappropriate or wrong bias into EAs would mislead search and result in low solution quality. Furthermore, we must make sure that a bias is not too strong. Using a bias can focus the search on specific areas of the search space and exclude solutions from consideration. If the bias is too strong, EAs can easily fail.

The following paragraphs discuss biasing representations and search operators. The next one gives an overview of how problem-specific construction heuristics can be used as genotype–phenotype mappings. Then, heuristic search varies either the input (problem space search) or the parameters (heuristic space search) of the construction heuristic. The following paragraph discusses redundant representations. Redundant representations with low locality randomize guided search and thus should not be used. Redundant representations with high locality can be biased by overrepresenting solutions similar to optimal solutions.

Incorporating Construction Heuristics in Representations

We focus on combining problem-specific construction heuristics with genotype–phenotype mappings. The possibility to design problem-specific representations and to incorporate relevant knowledge about the problem into the genotype–phenotype mapping by using construction heuristics is a promising line of research and is continuously discussed in the operations research and evolutionary computation communities [53.6, 22, 81–84].

Genotype–phenotype mappings map genotypes to phenotypes and can incorporate problem-specific construction heuristics. An early example of a problem-specific representation is the *ordinal representation* of [53.85], who studied the performance of genetic algorithms for the TSP. The ordinal representation encodes a tour (permutation of n integers) by a genotype x^g of length n, where $x_i^g \in \{1, \ldots, n-i\}$ and $i \in \{0, \ldots, n-1\}$. For constructing a phenotype, a predefined permutation x^s of n integers representing the n different cities is used. x^s can be problem-specific and, for example, consider edge weights. A phenotype (tour) is constructed from x^g by subsequently adding (starting with $i = 0$) the x_i^g-th element of x^s to the phenotype (which initially contains no elements) and removing the x_i^g-th element of x^s. Problem-specific knowledge can be considered by choosing an appropriate x^s as genotypes define perturbations of x^s and using small integers for the x_i^g results in a bias of the resulting phenotypes towards x^s. For example, for $x_i^g = 1$ ($i \in \{0, \dots, n-1\}$), the resulting phenotype is x^s.

Other early examples of problem-specific representations can be found in [53.86], where a problem-specific schedule builder was incorporated into representations for job shop scheduling problems, in [53.87] where representations that use a greedy adding heuristic for partitioning problems, and the more general *adaptive representation genetic optimization technique* (ARGOT) strategy was studied in [53.88, 89]. ARGOT dynamically changes either the structure of the genotypes or the genotype–phenotype mapping according to the progress made during search.

In parallel to, and independently from representations, *Storer* et al. [53.90] proposed *problem space search* (PSS) and *heuristic space search* (HSS) as approaches that also combine problem-specific heuristics with problem-independent EAs. PSS and HSS apply in each search iteration of a modern heuristic a problem-specific base heuristic H that exploits known properties of the problem. The base heuristic H should be fast and creates a phenotype from a genotype. Results presented for different applications show that this approach can lead to improved performance of EAs [53.82, 91, 92]. For PSS, H is applied to perturbed versions of the genotype. The perturbations of the genotypes are usually small and based on a definition of neighborhood in the genotype space. For HSS, in each search step of the modern heuristic, the genotypes remain unchanged, but (slightly) different variants of the base heuristic H are used. For scheduling problems, linear combinations of priority dispatching rules with different weights, or the application of different base heuristics to different parts of the genotype, have been proposed [53.91].

PSS and HSS use the same underlying concepts as problem-specific representations. The base heuristic H is equivalent to a (usually problem-specific) genotype–phenotype mapping and assigns phenotypes to genotypes. PSS performs heuristic search by modifying (perturbing) the genotypes, which is equivalent to the concept of representations originally proposed by [53.22] (for an early example, see [53.85]). HSS perturbs the base heuristic (genotype–phenotype mapping), which is similar to the concept of adaptive representations (for early examples, see [53.88] or [53.93]).

Redundant Representation

We assume a combinatorial optimization problem with a finite number of phenotypes. If the size of the genotype search space is equal to the size of the phenotype search space ($|\Phi_g| = |\Phi_p|$) and the representation maps all genotypes to all phenotypes (bijection), a representation cannot be biased. All solutions are represented with the same probability and a bias can only be a result of the search operator used.

The situation is different for representations where the number of genotypes exceeds the number of phenotypes. We still assume that all phenotypes are encoded by at least one genotype. Such representations are usually called (e.g., in [53.78, 94, 95], or [53.80]). *Radcliffe* and *Surry* [53.28] introduced a different notion of redundancy and distinguished between *degenerated representations*, where more than one genotype encodes one phenotype, and *redundant representations* where parts of the genotypes are not used for the construction of a phenotype. However, this distinction has not generally been accepted in the EA community. Therefore, we follow the majority of the literature and define encodings to be redundant if the number of genotypes exceeds the number of phenotypes (which is equivalent to the notion of degeneracy of [53.28]).

Rothlauf and *Goldberg* [53.80] distinguished between different types of redundant representations based on the similarity of the genotypes that are assigned to the same phenotype. A representation is defined to be *synonymously redundant* if the genotypes that are assigned to the same phenotype are similar to each other. Consequently, representations are *non-synonymously redundant* if the genotypes that are assigned to the same phenotype are not similar to each other. Therefore, the synonymity of a representation depends on the genotype and phenotype metric. Figure 53.5 illustrates the differences between synonymous and non-synonymous redundant encodings. Distances between solutions are measured using a Euclidean metric. The symbols indicate different genotypes and their corresponding phenotypes. When using synonymously redundant representations (left), genotypes that represent the same phenotype are similar to each other. When using non-synonymously redundant representations (right), genotypes that represent the same phenotype are not similar to each other but distributed over the whole search space.

Formally, a redundant representation f_g assigns a phenotype x^p to a set of different genotypes $x^g \in \Phi_g^{x^p}$, where $\forall x^g \in \Phi_g^{x^p} : f_g(x^g) = x^p$. All genotypes x^g in the genotype set $\Phi_g^{x^p}$ represent the same phenotype x^p. A representation is synonymously redundant if the genotype distances between all $x^g \in \Phi_g^{x^p}$ are small for all different x^p. Therefore, if for all phenotypes the sum over the distances between all genotypes that correspond to the same phenotype

$$\sum_{x^p} \frac{1}{2} \left(\sum_{x^g \in \Phi_g^{x^p}} \sum_{y^g \in \Phi_g^{x^p}} d(x^g, y^g) \right), \qquad (53.6)$$

where $x^g \neq y^g$, is reasonably small, a representation is called synonymously redundant. $d(x^g, y^g)$ depends on the metric used and measures the distance between two genotypes $x^g \in \Phi_g^{x^p}$ and $y^g \in \Phi_g^{x^p}$, which both represent the same phenotype x^p.

Non–Synonymously Redundant Representations. The synonymity of a representation can have a large influence on the performance of EAs. When using non-synonymously redundant representations, a local search operator can result in solutions that are phenotypically completely different from their parents. For recombination operators, the distances between offspring and parents are not necessarily smaller than the distances between parents.

Local search methods outperform random search if solutions with similar fitness are grouped together in the search space and are not scattered over the whole search space [53.25, 33–35, 96, 97]. Furthermore, problems are easy for guided search methods if distances between solutions are related to corresponding fitness differences. However, non-synonymously redundant representations destroy existing correlations between solutions and their corresponding fitness values. Thus, search heuristics cannot use any information learned during the search for determining future search steps. As a result, it makes no sense for guided search approaches to search around already found high-quality genotypes and guided search algorithms become random search. A local search step does not result in a solution with similar properties but in a random solution. Analogously, recombination is not able to create new solutions with similar properties to their parents, but creates new, random solutions. Therefore, non-synonymously redundant representations have the same effect on EAs as low-locality representations (Sect. 53.3.1).

The use of non-synonymously redundant representations allows us to reach many different phenotypes in a single local search step (Fig. 53.6). However, increasing the *connectivity* between phenotypes results in random search and decreases the efficiency of EAs. As for low-locality representations, a genotype search step does not result in a similar phenotype but creates a random solution. Therefore, guided search is no longer possible but becomes random search. As a result, we obtain reduced performance of EAs on straightforward problems when using non-synonymously redundant representations.

On average, non-synonymously redundant representations transform straightforward (as well as misleading) problems into difficult problems where the fitness differences between two solutions are not cor-

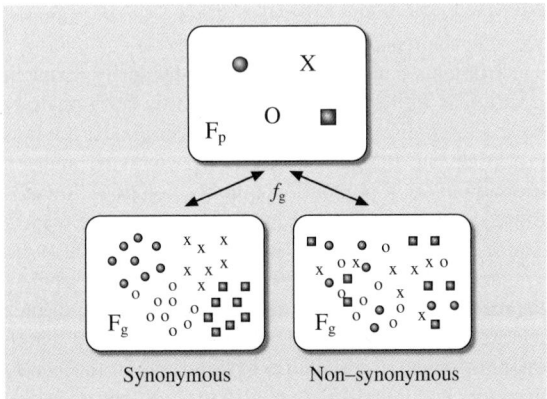

Fig. 53.5 Synonymous versus non-synonymous redundancy

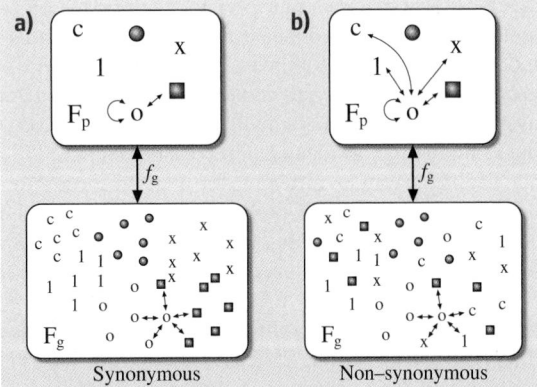

Fig. 53.6a,b The effects of local search steps for (**a**) synonymously versus (**b**) non-synonymously redundant representations. The *arrows* indicate search steps

related to their distances. Easy problems become more difficult. Therefore, we do not recommend using non-synonymously redundant encodings. A more detailed discussion of non-synonymously redundant representations can be found in [53.3, Sect. 3.1] and [53.60].

Bias of Synonymously Redundant Representations. The use of synonymously redundant representations allows local search to generate neighboring solutions. Small variations of genotypes cannot result in large phenotypic changes but result either in the same or a similar phenotype (Fig. 53.6, left).

To describe relevant properties of synonymously redundant representations, we can use the *order of redundancy*, k_r, which is defined as $k_r = \log(|\Phi_g|)/\log(|\Phi_p|)$ [53.3, Sect. 3.1]. k_r measures the amount of redundant information in the encoding. Furthermore, we are especially interested in biases of synonymously redundant representations. r measures a bias and denotes the number of genotypes that represent the optimal solution. When using non-redundant representations, every phenotype is assigned to exactly one genotype ($r = 1$). In general, $1 \leq r \leq |\Phi_g| - |\Phi_p| + 1$.

Synonymously redundant representations are unbiased (*uniformly redundant*) if each phenotype is, on average, encoded by the same number of genotypes. In contrast, encodings are biased (*non-uniformly redundant*) if some phenotypes are represented by a different number of genotypes. *Rothlauf* and *Goldberg* [53.80] studied how the bias of synonymously redundant representations influence the performance of EAs. If representations are uniformly redundant, unbiased search operators generate each phenotype with the same probability as for a non-redundant representation. Furthermore, variation operators have the same effect on the genotypes and phenotypes and the performance of EAs using a uniformly and synonymously redundant encoding is similar to non-redundant representations.

The situation is different for non-uniformly redundant encodings. The probability P of finding the correct solution depends on $P \propto 1 - \exp(-r/2^{k_r})$ [53.3]. Therefore, uniformly redundant representations do not change the behavior of EAs. Only by increasing r, which means overrepresenting optimal solutions, does the performance of EAs increase. In contrast, the performance of EAs decreases if the optimal solution is underrepresented. Therefore, non-uniformly redundant representations can only be used advantageously if a-priori information exists regarding optimal solutions.

For more information on redundant representations, we refer the reader to [53.80].

Search Operators

Search operators are applied either to genotypes or phenotypes and subsequently create new solutions. Search operators can be either *biased* or *unbiased*. In the unbiased case, each solution in the search space has the same probability of being created. If some phenotypes have higher probabilities to be created by applying a search operator to a randomly chosen solution, we call this a *bias* towards those phenotypes [53.3, 98].

Using biased search operators can be helpful if some knowledge about the structure of high-quality solutions exists and the search operators are biased such that high-quality solutions are preferred. Then, the average fitness of solutions that are generated by the biased search operator is higher than randomly generated solutions or search operators without a bias. Identifying high-quality solutions is often difficult, as exact optimization methods usually need exponential effort to find optimal solutions for relevant (often NP-hard) problems, and heuristic optimization methods usually do not provide any guarantee on solution quality. Possible approaches to overcome these problems are to exactly solve small problem instances and to deduce the structure of high-quality solutions from the solutions obtained. However, this assumes that relevant properties of optimal solutions are independent from the problem size. Second, solutions of higher quality can be identified by heuristic optimization methods if we increase the time spent on the heuristic search. Although heuristic optimization methods usually do not provide any guarantee of finding optimal solutions, the probability of finding high-quality solutions increases with the time spent on the heuristics search. For an example, we refer the reader to [53.99].

Problems can occur if a bias induced by a search operator is either too strong or towards solutions that have a large distance to an optimal solution. If the bias is too strong, the of a population is reduced and the individuals in a population quickly converge towards those solutions to which the search operators are biased. Then, after a few search steps, heuristic search is no longer possible. Furthermore, the performance of EAs decreases if a bias exists towards solutions that have a large distance to optimal solutions. The biased search operators push a population of solutions in the *wrong direction* and it is difficult for EAs to find optimal solutions. Therefore, biased search operators should be used with care.

Standard search operators for binary, integer, or continuous genotypes (Sect. 53.2.5) are unbiased. The situation is slightly different for standard recombination operators. Recombination operators never introduce new solution features but only recombine existing properties. Thus, once some solution features are lost in a population of solutions they can never be regained later by using recombination operators alone. Given a randomly generated and unbiased initial population of solutions, the iterative application of recombination operators can result in a random fixation of some decision variables, which reduces the of solutions that exist in a population. This effect is known as *genetic drift*. The existence of genetic drift is widely known and has been addressed in the field of population genetics [53.100–104] and also in the field of evolutionary algorithms [53.105–108].

When using more sophisticated phenotypes and direct search operators, identifying bias of search operators may be difficult. For example, in standard approaches, programs and syntactical expressions are encoded as trees of variable size. *Daida* et al. [53.109] showed that the two standard search operators in ge-netic programming (sub-tree swapping crossover and sub-tree mutation) are biased as they do not effectively search all tree shapes. In particular, very full or very narrow tree solutions are extraordinarily difficult to find, even when the fitness function provides good guidance to the optimum solutions. Therefore, genetic programming approaches will perform poorly on problems where optimal solutions require full or narrow trees. Furthermore, since the search operators do not find solutions that are at both ends of this *fullness* spectrum, problems may arise if we use those search operators to solve problems whose solutions are restricted to a particular shape, of whatever degree of fullness. *Hoai* et al. [53.110] studied approaches to overcome these problems and introduced a new tree-based representation and local insertion and deletion search operators with a lower bias.

In general, the bias of search operators can be measured by comparing the properties of randomly generated solutions with solutions that are created by subsequent applications of search operators. Examples of how to analyze the bias of search operators can be found in [53.2].

53.4 Summary and Conclusions

This chapter discusses the design of representations and search operators. Representations and search operators cannot be designed independently of each other, as together they define the structure of the search space. Section 53.1 summarized the benefits of representations and gave an overview of standard genotypes. Section 53.2 reviewed design guidelines for local, as well as recombination operators and gave an overview of standard search operators. Section 53.3 discussed possibilities for a problem-specific design of representations and search operators.

Representations and search operators should have high locality; this means that applying a local search operator to a genotype should result in a neighboring phenotype. If we have some knowledge about the properties of high-quality solutions, we can bias evolutionary algorithms by either incorporating problem-specific heuristics in the representation, using biased representations, or biased search operators.

For problems of practical relevance, we assume that the metric of a search space is meaningful and, on average, the fitness differences between neighboring solutions are smaller than between randomly chosen solutions. Search operators and representations should be designed in such a way that they fit the metric of the search space. If local as well as recombination-based search operators are not able to generate similar solutions, intensification of search is not possible, and EAs behave like random search. For a representation which introduces an additional genotype–phenotype mapping, we must make sure that it does not alter the character of the search operators. Therefore, its locality must be high; this means that the phenotype metric must fit the genotype metric. Low locality of a representation randomizes the search and leads to low performance of evolutionary algorithms.

There is a general trade-off between the effectiveness and application range of optimization methods. Usually, the more problems that can be solved with one particular optimization method, the lower its resulting average performance. Therefore, standard EAs that are not problem-specific often only work for small or toy problems. As the problem becomes larger and more realistic, performance degrades. To improve the performance for selected optimization problems, we must design them in a more problem-specific way. By assuming that most problems in the real world have high locality, EAs already exploit a specific property of problems, namely their high locality. We can further increase the performance of EAs if we have some idea about properties of high-quality solutions. Such problem-specific knowledge can be exploited by introducing a bias into EAs. The bias should consider this knowledge and, for example, concentrate search on solutions that are expected to be of high quality or avoid solutions expected to be of low quality.

References

53.1   Z. Michalewicz, D.B. Fogel: *How to Solve It: Modern Heuristics*, 2nd edn. (Springer, Berlin 2004)

53.2   F. Rothlauf: *Design of Modern Heuristics* (Springer, Berlin, Heidelberg 2011)

53.3   F. Rothlauf: *Representations for Genetic and Evolutionary Algorithms*, 2nd edn. (Springer, Berlin, Heidelberg 2006)

53.4   G. Mendel: Versuche über Pflanzen-Hybriden, Verhandlungen Des Naturforschenden Vereins (Naturforschender Verein zu Brünn) **4**(1), 3–47 (1866)

53.5   R.C. Lewontin: *The Genetic Basis of Evolutionary Change*, Columbia Biological Series, Vol. 25 (Columbia Univ. Press, New York 1974)

53.6   G.E. Liepins, M.D. Vose: Representational issues in genetic optimization, J. Exp. Theor. Artif. Intell. **2**, 101–115 (1990)

53.7   J.D. Bagley: The Behavior of Adaptive Systems Which Employ Genetic and Correlation Algorithms, Ph.D. Thesis (University of Michigan, College of Literature, Science, and the Arts. Department of Communication Sciences. Ann Arbor, Michigan 1967)

53.8   M.D. Vose: Modeling simple genetic algorithms. In: *Foundations of Genetic Algorithms*, Vol. 2, ed. by L.D. Whitley (Morgan Kaufmann, San Mateo 1993) pp. 63–73

53.9   R. Hamming: *Coding and Information Theory* (Prentice Hall, Englewood Cliffs, New Jersey 1980)

53.10   V.I. Levenshtein: Binary codes capable of correcting deletions, insertions and reversals, Sov. Phys. Dokl. **10**(8), 707–710 (1966), Doklady Akademii Nauk SSSR, V163 No4 845–848 1965

53.11   R.A. Wagner, M.J. Fisher: The string-to-string correction problem, J. ACM **21**, 168–174 (1974)

53.12   R.A. Caruana, J.D. Schaffer: Representation and hidden bias: Gray vs. binary coding for genetic algorithms, Proc. 5th Int. Workshop Mach. Learn., ed. by L. Laird (Morgan Kaufmann, San Francisco 1988) pp. 153–161

53.13   T. Bäck, D.B. Fogel, Z. Michalewicz: *Handbook of Evolutionary Computation* (Institute of Physics Publishing/Oxford Univ. Press, Bristol, New York 1997)

53.14   D.E. Goldberg: *The Design of Innovation*, Series on Genetic Algorithms and Evolutionary Computation (Kluwer, Boston 2002)

53.15   D.E. Goldberg: *Genetic Algorithms in Search, Optimization, and Machine Learning* (Addison-Wesley, Reading 1989)

53.16   J.E. Rowe, L.D. Whitley, L. Barbulescu, J.-P. Watson: Properties of Gray and binary representations, Evol. Comput. **12**(1), 46–76 (2004)

53.17   D.E. Goldberg: Real-coded genetic algorithms, virtual alphabets, and blocking, Complex Syst. **5**(2), 139–167 (1991)

53.18   N.J. Radcliffe: Theoretical foundations and properties of evolutionary computations: Schema processing. In: *Handbook of Evolutionary Computation*, ed. by T. Bäck, D.B. Fogel, Z. Michalewicz (Institute of Physics Publishing/Oxford Univ. Press, Bristol, New York 1997), pp. B2.5:1–B2.5:10

53.19   D.B. Fogel, L.C. Stayton: On the effectiveness of crossover in simulated evolutionary optimization, BioSystems **32**, 171–182 (1994)

53.20   G.R. Raidl, B.A. Julstrom: Edge-sets: An effective evolutionary coding of spanning trees, IEEE Trans. Evol. Comput. **7**(3), 225–239 (2003)

53.21   F. Rothlauf: A problem-specific and effective encoding for metaheuristics for the minimum communication spanning tree problem, INFORMS J. Comput. **21**(4), 575–584 (2009)

53.22   J.H. Holland: *Adaptation in Natural And Artificial Systems* (University of Michigan, Ann Arbor 1975)

53.23   D.E. Goldberg, B. Korb, K. Deb: Messy genetic algorithms: Motivation, analysis, and first results, Complex Syst. **3**(5), 493–530 (1989)

53.24   N.J. Radcliffe: Forma analysis and random respectful recombination. In: *Foundations of Genetic Algorithms*, ed. by G.J.E. Rawlins (Morgan Kaufmann, San Mateo 1991) pp. 222–229

53.25   N.J. Radcliffe: Equivalence class analysis of genetic algorithms, Complex Syst. **5**(2), 183–205 (1991)

53.26   N.J. Radcliffe: Non-linear genetic representations, Parallel Problem Solving from Nature – PPSN II, ed. by R. Männer, B. Manderick (Springer, Berlin, Heidelberg 1992) pp. 259–268

53.27   N.J. Radcliffe: Genetic set recombination. In: *Foundations of Genetic Algorithms*, Vol. 2, ed. by L.D. Whitley (Morgan Kaufmann, San Mateo 1993) pp. 203–219

53.28   N.J. Radcliffe, P.D. Surry: Fitness variance of formae and performance prediction. In: *Foundations of Genetic Algorithms 3*, ed. by L.D. Whitley, M.D. Vose (Morgan Kaufmann, San Mateo 1994) pp. 51–72

53.29   N.J. Radcliffe: The algebra of genetic algorithms, Ann. Maths. Artif. Intell. **10**, 339–384 (1994)

53.30   N.J. Radcliffe, P.D. Surry: Formal algorithms + formal representations = search strategies, Parallel Problem Solving from Nature – PPSN IV, ed. by H.-M. Voigt, W. Ebeling, I. Rechenberg, H.-P. Schwefel (Springer, Berlin 1996) pp. 366–375

53.31 H. Kargupta, K. Deb, D.E. Goldberg: Ordering genetic algorithms and deception, Parallel Problem Solving from Nature – PPSN IV, ed. by H.-M. Voigt, W. Ebeling, I. Rechenberg, H.-P. Schwefel (Springer, Berlin 1996) pp. 47–56

53.32 J. Doran, D. Michie: Experiments with the graph traverser program, Proc. R. Soc. Lond. (A) 294, 235–259 (1966)

53.33 B. Manderick, M. de Weger, P. Spiessens: The genetic algorithm and the structure of the fitness landscape, Proc. 4th Int. Conf. Genet. Algorithm., ed. by R.K. Belew, L.B. Booker (Morgan Kaufmann, Burlington 1991) pp. 143–150

53.34 S. Christensen, F. Oppacher: What can we learn from no free lunch?, Proc. Genet. Evol. Comput. Conf. (GECCO 2001), ed. by L. Spector, E. Goodman, A. Wu, W.B. Langdon, H.-M. Voigt, M. Gen, S. Sen, M. Dorigo, S. Pezeshk, M. Garzon, E. Burke (Morgan Kaufmann, San Francisco 2001) pp. 1219–1226

53.35 D. Whitley, K. Mathias, S. Rana, J. Dzubera: Evaluating evolutionary algorithms, Artif. Intell. 85, 245–276 (1996)

53.36 C.R. Reeves: Landscapes, operators and heuristic search, Ann. Oper. Res. 86, 473–490 (1999)

53.37 F. Gray: Pulse code communications, U.S. Patent 263 2058 (1953)

53.38 R.A. Caruana, J.D. Schaffer, L.J. Eshelman: Using multiple representations to improve inductive bias: Gray and binary coding for genetic algorithms, Proc. 6th Int. Workshop Mach. Learn., ed. by A.M. Segre (Morgan Kaufmann, San Francisco 1989) pp. 375–378

53.39 C. Höhn, C.R. Reeves: Are long path problems hard for genetic algorithms?, Parallel Problem Solving from Nature – PPSN IV, ed. by H.-M. Voigt, W. Ebeling, I. Rechenberg, H.-P. Schwefel (Springer, Berlin 1996) pp. 134–143

53.40 C. Höhn, C.R. Reeves: The crossover landscape for the onemax problem, Proc. 2nd Nordic Workshop Genet. Algorithm. Appl. (2NWGA), ed. by J.T. Alander (Vaasa, Finland 1996), pp. 27–43, Department of Information Technology and Production Economics, University of Vaasa

53.41 D.B. Fogel: Real-valued vectors. In: Handbook of Evolutionary Computation, ed. by T. Bäck, D.B. Fogel, Z. Michalewicz (Institute of Physics Publishing/Oxford Univ. Press, Bristol, New York 1997), pp. C3.2:2–C3.2:5

53.42 A. Moraglio, R. Poli: Topological interpretation of crossover. In: Proc. Genet. Evol. Comput. Conf. (GECCO 2004), ed. by K. Deb, R. Poli, W. Banzhaf, H.-G. Beyer, E. Burke, P. Darwen, D. Dasgupta, D. Floreano, J. Foster, M. Harman, O. Holland, P.L. Lanzi, L. Spector, A. Tettamanzi, D. Thierens, A. Tyrrell (Springer, Berlin, Heidelberg 2004) pp. 1377–1388

53.43 A. Moraglio, Y.-H. Kim, Y. Yoon, B.R. Moon: Geometric crossovers for multiway graph partitioning, Evol. Comput. 15(4), 445–474 (2007)

53.44 A. Moraglio: Towards a Geometric Unification of Evolutionary Algorithms, Ph.D. Thesis (Department of Computer Science, University of Essex, Colchester 2007)

53.45 F. Rothlauf: Representations for Genetic and Evolutionary Algorithms, Studies on Fuzziness and Soft Computing, Vol. 104, 1st edn. (Springer, Berlin, Heidelberg 2002)

53.46 J. Reed, R. Toombs, N.A. Barricelli: Simulation of biological evolution and machine learning: I. Selection of self-reproducing numeric patterns by data processing machines, effects of hereditary control, mutation type and crossing, J. Theor. Biol. 17, 319–342 (1967)

53.47 D.H. Ackley: A Connectionist Machine for Genetic Hill Climbing (Kluwer Academic, Boston 1987)

53.48 G. Syswerda: Uniform crossover in genetic algorithms, Proc. 3rd Int. Conf. Genet. Algorithm. (ICGA), ed. by J.D. Schaffer (Morgan Kaufmann, Burlington 1989) pp. 2–9

53.49 Z. Michalewicz: Genetic Algorithms + Data Structures = Evolution Programs (Springer, New York 1996)

53.50 L.B. Booker: Binary strings. In: Handbook of Evolutionary Computation, ed. by T. Bäck, D.B. Fogel, Z. Michalewicz (Institute of Physics Publishing/Oxford Univ. Press, Bristol, New York 1997), pp. C3.3:1–C3.3:10

53.51 H. Mühlenbein, G. Paaß: From recombination of genes to the estimation of distributions I, binary parameters, Parallel Problem Solving from Nature – PPSN IV, ed. by H.-M. Voigt, W. Ebeling, I. Rechenberg, H.-P. Schwefel (Springer, Berlin 1996) pp. 178–187

53.52 H. Mühlenbein, T. Mahnig: FDA – a scalable evolutionary algorithm for the optimization of additively decomposed functions, Evol. Comput. 7(4), 353–376 (1999)

53.53 M. Pelikan, D.E. Goldberg, E. Cantú-Paz: BOA: The Bayesian optimization algorithm, IlliGAL Report No. 99003 (University of Illinois, Urbana 1999)

53.54 M. Pelikan, D.E. Goldberg, F. Lobo: A survey of optimization by building and using probabilistic models, IlliGAL Report No. 99018 (University of Illinois, Urbana 1999)

53.55 P. Larrañaga, R. Etxeberria, J.A. Lozano, J.M. Pea: Optimization by learning and simulation of Bayesian and Gaussian networks. Technical Report EHU-KZAA-IK-4/99 (University of the Basque Country, San Sebastián 1999)

53.56 P.A.N. Bosman: Design and Application of Iterated Density-Estimation Evolutionary Algorithms, Ph.D. Thesis (Universiteit Utrecht, Utrecht 2003)

53.57 P. Larrañaga, J.A. Lozano: Estimation of Distribution Algorithms: A New Tool for Evolutionary Computation (Springer, Berlin 2001)

Part E | 53

53.58   M. Pelikan: *Hierarchical Bayesian Optimization Algorithm: Toward a New Generation of Evolutionary Algorithms*, Studies in Fuzziness and Soft Computing (Springer, New York 2006)

53.59   E. Falkenauer: *Genetic Algorithms and Grouping Problems* (Wiley, Chichester 1998)

53.60   S.-S. Choi, B.-R. Moon: Normalization for genetic algorithms with non-synonymously redundant encodings, IEEE Trans. Evol. Comput. **12**(5), 604–616 (2008)

53.61   S.-S. Choi, B.-R. Moon: Normalization in genetic algorithms, Proc. Genet. Evol. Comput. Conf. (GECCO 2003), ed. by E. Cantú-Paz, J.A. Foster, K. Deb, D. Davis, R. Roy, U.-M. O'Reilly, H.-G. Beyer, R. Standish, G. Kendall, S. Wilson, M. Harman, J. Wegener, D. Dasgupta, M.A. Potter, A.C. Schultz, K. Dowsland, N. Jonoska, J. Miller (Springer, Berlin, Heidelberg 2003) pp. 862–873

53.62   L. Davis: Applying adaptive algorithms to epistatic domains, Proc. 9th Int. Joint Conf. Artif. Intell., ed. by A. Joshi (Morgan Kaufmann, San Francisco 1985) pp. 162–164

53.63   D.E. Goldberg, R. Lingle Jr.: Alleles, loci, and the traveling salesman problem, Proc. Int. Conf. Genet. Algorithm. Their Appl., ed. by J.J. Grefenstette (Lawrence Erlbaum, Hillsdale 1985) pp. 154–159

53.64   I.M. Oliver, D.J. Smith, J.R.C. Holland: A study of permutation crossover operators on the traveling salesman problem, Proc. 2nd Int. Conf. Genet. Algorithm. (ICGA), ed. by J.J. Grefenstette (Lawrence Erlbaum, Hillsdale 1987)

53.65   C. Bierwirth: A generalized permutation approach to job shop scheduling with genetic algorithms, OR Spektrum **17**, 87–92 (1995)

53.66   C. Bierwirth, D.C. Mattfeld, H. Kopfer: On permutation representations for scheduling problems, Parallel Problem Solving from Nature – PPSN IV, ed. by H.-M. Voigt, W. Ebeling, I. Rechenberg, H.-P. Schwefel (Springer, Berlin 1996) pp. 310–318

53.67   L.D. Whitley: Permutations. In: *Handbook of Evolutionary Computation*, ed. by T. Bäck, D.B. Fogel, Z. Michalewicz (Institute of Physics Publishing/Oxford Univ. Press, Bristol, New York 1997), pp. C3.3:114–C3.3:20

53.68   D.C. Mattfeld: *Evolutionary Search and the Job Shop: Investigations on Genetic Algorithms for Production Scheduling* (Physica, Berlin, Heidelberg 1996)

53.69   J.R. Koza: *Genetic Programming: On the Programming of Computers by Natural Selection* (MIT, Cambridge 1992)

53.70   W. Banzhaf, P. Nordin, R.E. Keller, F.D. Francone: *Genetic Programming, An Introduction* (Morgan Kaufmann, Burlington 1997)

53.71   T. Jones, S. Forrest: Fitness distance correlation as a measure of problem difficulty for genetic algorithms, Proc. 6th Int. Conf. Genet. Algorithms, ed. by L. Eschelman (Morgan Kaufmann, San Francisco 1995) pp. 184–192

53.72   F. Rothlauf, D.E. Goldberg: Tree network design with genetic algorithms – an investigation in the locality of the prüfernumber encoding. In: *Late Breaking Papers at the Genetic and Evolutionary Computation Conference 1999*, ed. by S. Brave, A.S. Wu (Omni, Orlando 1999) pp. 238–244

53.73   J. Gottlieb, G.R. Raidl: Characterizing locality in decoder-based EAs for the multidimensional knapsack problem, Lect. Notes Comput. Sci. **1829**, 38–52 (1999)

53.74   J. Gottlieb, G.R. Raidl: The effects of locality on the dynamics of decoder-based evolutionary search, Proc. Genet. Evolu. Comput. Conf. (GECCO 2000), ed. by L.D. Whitley, D.E. Goldberg, E. Cantú-Paz, L. Spector, L. Parmee, H.-G. Beyer (Morgan Kaufmann, San Francisco 2000) pp. 283–290

53.75   J. Gottlieb, B.A. Julstrom, G.R. Raidl, F. Rothlauf: Prüfer numbers: A poor representation of spanning trees for evolutionary search, Proc. Genet. Evol. Comput. Conf. (GECCO 2001), ed. by L. Spector, E. Goodman, A. Wu, W.B. Langdon, H.-M. Voigt, M. Gen, S. Sen, M. Dorigo, S. Pezeshk, M. Garzon, E. Burke (Morgan Kaufmann, San Francisco 2001)

53.76   F. Rothlauf, D.E. Goldberg: Prüfernumbers and genetic algorithms: A lesson on how the low locality of an encoding can harm the performance of GAs. In: *Parallel Problem Solving from Nature – PPSN VI*, ed. by M. Schoenauer, K. Deb, G. Rudolph, X. Yao, E. Lutton, J.J. Merelo, H.-P. Schwefel (Springer, Berlin 2000) pp. 395–404

53.77   M. Gerrits, P. Hogeweg: Redundant coding of an NP-complete problem allows effective genetic algorithm search, Lect. Notes Comput. Sci. **496**, 70–74 (1991)

53.78   S. Ronald, J. Asenstorfer, M. Vincent: Representational redundancy in evolutionary algorithms, Proc. 1995 IEEE Int. Conf. Evol. Comput. 2, Piscataway, ed. by D.B. Fogel, Y. Attikiouzel (1995) pp. 631–636

53.79   R. Shipman: Genetic redundancy: Desirable or problematic for evolutionary adaptation?, Proc. 4th Int. Conf. Artif. Neural Netw. Genet. Algorithm. (ICANNGA), ed. by A. Dobnikar, N.C. Steele, D.W. Pearson, R.F. Albrecht (Springer, Berlin 1999) pp. 1–11

53.80   F. Rothlauf, D.E. Goldberg: Redundant representations in evolutionary computation, Evol. Comput. **11**(4), 381–415 (2003)

53.81   Z. Michalewicz, C.Z. Janikow: Handling constraints in genetic algorithm, Proc. 3rd Int. Conf. Genet. Algorithm. (ICGA), ed. by J.D. Schaffer (Morgan Kaufmann, Burlington 1989) pp. 151–157

53.82   R.H. Storer, S.D. Wu, R. Vaccari: Problem and heuristic space search strategies for job shop scheduling, ORSA J. Comput. **7**(4), 453–467 (1995)

53.83   Z. Michalewicz, M. Schoenauer: Evolutionary computation for constrained parameter optimization problems, Evol. Comput. **4**(1), 1–32 (1996)

53.84   P.D. Surry: A Prescriptive Formalism for Constructing Domain-Specific Evolutionary Algorithms, Ph.D. Thesis (University of Edinburgh, Edinburgh 1998)

53.85   J.J. Grefenstette, R. Gopal, B.J. Rosmaita, D. Van Gucht: Genetic algorithms for the traveling salesman problem, Proc. Int. Conf. Genet. Algorithm. Their Appl., ed. by J.J. Grefenstette (Lawrence Erlbaum, Hillsdale 1985) pp. 160–168

53.86   S. Bagchi, S. Uckun, Y. Miyabe, K. Kawamura: Exploring problem-specific recombination operators for job shop scheduling, Proc. 4th Int. Conf. Genet. Algorithm., ed. by R.K. Belew, L.B. Booker (Morgan Kaufmann, Burlington 1991) pp. 10–17

53.87   D.R. Jones, M.A. Beltramo: Solving partitioning problems with genetic algorithms, Proc. 4th Int. Conf. Genet. Algorithm., ed. by R.K. Belew, L.B. Booker (Morgan Kaufmann, Burlington 1991) pp. 442–449

53.88   C.G. Shaefer: The ARGOT strategy: Adaptive representation genetic optimizer technique, Proc. 2nd Int. Conf. Genet. Algorithm. (ICGA), ed. by J.J. Grefenstette (Lawrence Erlbaum, Hillsdale 1987) pp. 50–58

53.89   C. G. Shaefer, J. S. Smith: The ARGOT strategy II: Combinatorial optimizations, Technical Report RL90-1 (Thinking Machines Inc., Cambridge, 1990)

53.90   R.H. Storer, S.D. Wu, R. Vaccari: New search spaces for sequencing problems with application to job shop scheduling, Manag. Sci. **38**(10), 1495–1509 (1992)

53.91   K.S. Naphade, S.D. Wu, R.H. Storer: Problem space search algorithms for resource-constrained project scheduling, Ann. Oper. Res. **70**, 307–326 (1997)

53.92   A.T. Ernst, M. Krishnamoorthy, R.H. Storer: Heuristic and exact algorithms for scheduling aircraft landings, Netw.: Int. J. **34**, 229–241 (1999)

53.93   N.N. Schraudolph, R.K. Belew: Dynamic parameter encoding for genetic algorithms, Mach. Learn. **9**, 9–21 (1992)

53.94   J.P. Cohoon, S.U. Hegde, W.N. Martin, D. Richards: Distributed genetic algorithms for the floorplan design problem, IEEE Trans. Comput.-Aided Des. Integr. Circuits Syst. **10**(4), 483–492 (1991)

53.95   B.A. Julstrom: Redundant genetic encodings may not be harmful, Proc. Genet. Evol. Comput. Conf. (GECCO '99), ed. by W. Banzhaf, J. Daida, A.E. Eiben, M.H. Garzon, V. Honavar, M. Jakiela, R.E. Smith (Morgan Kaufmann, Burlington 1999) p. 791

53.96   J. Horn: *Genetic algorithms, problem difficulty, and the modality of fitness landscapes*, *IlliGAL Report No. 95004* (University of Illinois, Urbana 1995)

53.97   K. Deb, L. Altenberg, B. Manderick, T. Bäck, Z. Michalewicz, M. Mitchell, S. Forrest: Theoretical foundations and properties of evolutionary computations: Fitness landscapes. In: *Handbook of Evolutionary Computation*, ed. by T. Bäck, D.B. Fogel, Z. Michalewicz (Institute of Physics Publishing/Oxford Univ. Press, Bristol, New York 1997), pp. B2.7:1–B2.7:25

53.98   G.R. Raidl, J. Gottlieb: Empirical analysis of locality, heritability and heuristic bias in evolutionary algorithms: A case study for the multidimensional knapsack problem, Evol. Comput. **13**(4), 441–475 (2005)

53.99   F. Rothlauf: On optimal solutions for the optimal communication spanning tree problem, Oper. Res. **57**(2), 413–425 (2009)

53.100   M. Kimura: On the probability of fixation of mutant genes in a population, Genetics **47**, 713–719 (1962)

53.101   M. Kimura: Diffusion models in population genetics, J. Appl. Prob. **1**, 177–232 (1964)

53.102   J.S. Gale: *Theoretical Population Genetics* (Unwin Hyman, London 1990)

53.103   T. Nagylaki: *Introduction to Theoretical Population. Genetics* (Springer, Berlin, Heidelberg 1992)

53.104   D.L. Hartl, A.G. Clark: *Principles of Population Genetics*, 3rd edn. (Sinauer, Sunderland 1997)

53.105   D.E. Goldberg, P. Segrest: Finite Markov chain analysis of genetic algorithms, Proc. 2nd Int. Conf. Genet. Algorithm. (ICGA), ed. by J.J. Grefenstette (Lawrence Erlbaum, Hillsdale 1987) pp. 1–8

53.106   H. Asoh, H. Mühlenbein: On the mean convergence time of evolutionary algorithms without selection and mutation. In: *Parallel Problem Solving from Nature – PPSN III*, Lecture Notes in Computer Science, Vol. 866, ed. by Y. Davidor, H.-P. Schwefel, R. Männer (Springer, Berlin 1994) pp. 88–97

53.107   D. Thierens, D.E. Goldberg, Â.G. Pereira: Domino convergence, drift, and the temporal-salience structure of problems, Proc. 1998 IEEE Int. Conf. Evol. Comput., Piscataway, ed. by D.B. Fogel (1998) pp. 535–540

53.108   F.G. Lobo, D.E. Goldberg, M. Pelikan: Time complexity of genetic algorithms on exponentially scaled problems, Proc. Genet. Evolu. Comput. Conf. (GECCO 2000), ed. by L.D. Whitley, D.E. Goldberg, E. Cantú-Paz, L. Spector, L. Parmee, H.-G. Beyer (Morgan Kaufmann, San Francisco 2000) pp. 151–158

53.109   J.M. Daida, R. Bertram, S. Stanhope, J. Khoo, S. Chaudhary, O. Chaudhri, J. Polito: What makes a problem GP-hard?, Analysis of a tunably difficult problem in genetic programming, Genet. Program. Evol. Mach. **2**(2), 165–191 (2001)

53.110   N.X. Hoai, R.I. McKay, D.L. Essam: Representation and structural difficulty in genetic programming, IEEE Trans. Evol. Comput. **10**(2), 157–166 (2006)

54. Stochastic Local Search Algorithms: An Overview

Holger H. Hoos, Thomas Stützle

In this chapter, we give an overview of the main concepts underlying the stochastic local search (SLS) framework and outline some of the most relevant SLS techniques. We also discuss some major recent research directions in the area of stochastic local search. The remainder of this chapter is structured as follows. In Sect. 54.1, we situate the notion of SLS within the broader context of fundamental search paradigms and briefly review the definition of an SLS algorithm. In Sect. 54.2, we summarize the main issues and trends in the design of greedy constructive and iterative improvement algorithms, while in Sects. 54.3–54.5, we provide a concise overview of some of the most widely used simple, hybrid, and population-based SLS methods. Finally, in Sect. 54.6, we discuss some recent topics of interest, such as the systematic design of SLS algorithms and methods for the automatic configuration of SLS algorithms.

Part E | 54

Stochastic local search (SLS) algorithms are the method of choice for solving computationally hard decision and optimization problems from a wide range of areas, including computing science, operations research, engineering, chemistry, biology and physics. SLS comprises a spectrum of techniques ranging from simple constructive and iterative improvement procedures to more complex methods, such as simulated annealing (SA), iterated local search or evolutionary algorithms (EAs). As evident from the term *stochastic* local search, randomization can, and often does, play a prominent role in these methods. Randomized choices may be used in the generation of initial solutions or in the decision which of several possible

search steps to perform next – sometimes merely to break ties between equivalent alternatives, and sometimes to heuristically and probabilistically select from large and diverse sets of possible candidates. Judicious use of randomization can arguably simplify algorithm design and help achieve robust algorithm behavior.

The concept of an *SLS algorithm* has been defined formally [54.1] and not only provides a unifying framework for many different types of algorithms, including the previously mentioned constructive and iterative improvement procedures, but also provides a wide range of more complex search methods commonly known as metaheuristics.

Greedy constructive and iterative improvement procedures are important SLS algorithms, since they typically serve as building blocks for more complex SLS algorithms, whose performance critically depends on the design choices and fine tuning of these underlying components. Greedy constructive algorithms and iterative improvement procedures terminate naturally when a complete solution has been generated or a local optimum of a given evaluation function is reached, respectively. One possible way to obtain better solutions is to restart these basic SLS procedures from randomly chosen initial search positions. However, this approach has shown to be relatively ineffective in practice for reasonably sized problem instances (and it breaks down for large instances [54.2]).

To overcome these limitations, over the last decades, a large number of more sophisticated, general-purpose SLS methods [54.1] have been introduced; these are often called metaheuristics [54.3], since they are based on higher level schemes for controlling one or more subsidiary heuristic search procedures. We divide these general-purpose SLS methods into three broad classes: simple, hybrid and population-based SLS methods. Simple SLS methods typically use one neighborhood relation during the search and either modify the acceptance criterion for search steps, allowing to occasionally accept worsening steps, or modify the evaluation function that is used during the local search process. Examples of simple SLS methods include SA [54.4, 5] and (simple) tabu search [54.6–9]. A number of SLS methods combine different types of search steps – for example, construction steps and perturbative local search steps – or introduce occasional larger modifications into current candidate solutions, to provide appropriate starting points for subsequent iterative improvement search. Examples of such hybrid SLS methods include greedy randomized adaptive search procedures (GRASPs) [54.10] and iterated local search [54.11]. Finally, several SLS methods maintain and manipulate at each iteration a set, or population, of candidate solutions, which provides a natural way of increasing search diversification. Examples of such population-based SLS methods include EAs [54.12–15], scatter search [54.16, 17] and ant colony optimization [54.18–20].

Our classification into simple, hybrid and population-based SLS methods is not the only possible one, and certain SLS algorithms could be seen as belonging to more than one category. For example, many population-based SLS methods are also hybrid, as they use different search operators or combine the manipulation of the population of candidate solutions with iterative improvement on members of the population to achieve increased performance. In fact, there is an increasing trend to design and apply SLS algorithms that are not merely based on a single, well-established general-purpose SLS method, but rather combine flexibly elements of different SLS methods or incorporate mechanisms taken from systematic search algorithms, such as branch and bound or dynamic programming. The conceptual framework of SLS naturally accommodates this development, and the composition of more complex SLS algorithms from conceptually simpler components is explicitly supported, for example, by the concept of generalized local search machines [54.1]. In this context, methodological issues concerning the engineering of SLS algorithms [54.21, 22] are increasingly gaining importance. Similarly, the exploitation of automatic algorithm configuration techniques and, more generally, the programming by optimization paradigm [54.23] enable the systematic development of high-performance SLS algorithms.

54.1 The Nature and Concept of SLS

Computational approaches for the solution of hard, combinatorial problems can all be viewed as performing some form of search. Essentially, search algorithms generate and evaluate candidate solutions for the problem instance at hand. For combinatorial decision problems, the evaluation of a candidate solution requires to check whether the candidate solution is a feasible solution satisfying all given constraints; for combinatorial optimization problems, it involves computing the value of the given objective function. For NP-complete decision problems and NP-equivalent optimization problems, even the most efficient algorithms known to date require running time exponential in the instance size in the worst case, while candidate solutions can be evaluated in polynomial time.

A candidate solution for an instance of a combinatorial problem is generally composed of *solution components*. Consider, for example, the well-known traveling salesperson problem (TSP). In the TSP, one is given a weighted, fully connected graph $G = (V, E, w)$, where $V = \{v_1, v_2, \ldots, v_n\}$ is the set of $|V| = n$ vertices, $E \subset V \times V$ is the set of edges that fully connects the graph, and $w : E \mapsto \mathbb{R}_0^+$ is a function that assigns to each edge $e \in E$ a nonnegative weight $w(e)$. The objective is to find a minimum-weight Hamiltonian cycle in G. A candidate solution for a TSP instance can be represented by a permutation $\pi = (\pi(1), \pi(2), \ldots, \pi(n))$ of the vertex indices, and the objective function w is given as

$$w(\pi) := w(v_{\pi(n)}, v_{\pi(1)})$$
$$+ \sum_{i=1}^{n-1} w(v_{\pi(i)}, v_{\pi(i+1)}) . \qquad (54.1)$$

In the TSP, a (complete) candidate solution, commonly also called a tour, can be seen as consisting of n out of the $n \cdot (n-1)$ possible edges, and each edge represents a solution component.

Any given tour can be modified by removing two edges and introducing two unique new edges such that another valid tour is obtained. This modification is an example of a *perturbation* of a complete candidate solution, and we refer to search algorithms that make systematic use of such solution modifications as *perturbative search methods*. In practice, such perturbative search methods iteratively modify a current candidate solution according to some rule, and this process ends when a given termination criterion is met.

Perturbative search methods start from some complete candidate solution. The task of generating such candidate solutions is commonly accomplished by *constructive search methods* or *construction heuristics*. Constructive search methods iteratively extend an initially *empty* candidate solution by one or several solution components until a complete candidate solution is obtained. Constructive search methods can thus be seen as operating in a search space of *partial candidate solutions*. An example of a constructive search method is the nearest neighbor heuristic for the TSP. An initial vertex is chosen randomly, and at each construction step, the nearest neighbor heuristic follows a minimal weight edge to one of the vertices that have not yet been visited. These steps are iterated until all vertices have been visited, and the tour is completed by returning to the initial vertex.

Generally speaking, local search algorithms start at some initial search position and iteratively move, based on local information, from the current position to neighboring positions in the search space. Both perturbative and constructive search methods match this general description. While in the literature, the term local search is mostly used for perturbative search methods, it also

applies to constructive search methods: A partial solution corresponds to a position in the search space of partial candidate solutions, and the neighbors of a partial solutions are obtained by extending it with one or more solution components. In fact, there are a number of well-known generic SLS methods, such as GRASP, iterated greedy and ant colony optimization, that are based on constructive local search.

Many local search algorithms use randomized decisions, for example, for generating initial solutions or when determining search steps. We therefore refer to such methods as stochastic local search (SLS) *algorithms*. The following components need to be specified to define an SLS algorithm (for a formal definition, we refer to Chap. 1 of [54.1]).

- *Search space* – comprises the set of *candidate solutions* (or *search positions*) for the given problem instance.
- *Solution set* – consists of the search positions that are considered to be solutions of the given problem instance. In the case of decision problems, the solution set comprises all feasible candidate solutions; in the case of optimization problems, the solution set typically comprises all optimal feasible candidate solutions.
- *Neighborhood relation* – specifies the direct neighbors of each candidate solution s, i.e., the search positions that can be reached from s in a single search step of the SLS algorithm.
- *Memory states* – hold additional information about the search beyond the search position. If an algorithm is memoryless, the memory may consist of a single, constant state.
- *Initialization function* – specifies the search initialization in the form of a probability distribution over initial search positions and memory states.
- *Step function* – determines the computation of search steps by mapping each search position and memory state to a probability distribution over its neighboring search positions and memory states.
- *Termination predicate* – used to decide search termination based on the current search position and memory state.

The formal definition of an SLS algorithm specifies the initialization function, the step function, and the termination predicate as probability distributions, which the algorithm samples at each step during any given run. In practice, however, the initialization function, the step function, and the termination predicate will be specified by procedures, and the corresponding probability distributions are only implicitly defined. Note that the definition of an SLS algorithm is general enough to include deterministic local search algorithms. In fact, formally we can describe deterministic local search algorithms as special cases of SLS algorithms – deterministic decisions can be modeled using degenerate probability distributions (Dirac delta).

The working principle of an SLS algorithm is then as follows. The search process starts from some initial search state that is generated by the initialization function. While some termination criterion is not satisfied, search steps are performed according to the step function. In the case of optimization problems, the SLS algorithm keeps track of the best solution found so far, which is then returned upon termination of the algorithm. In the case of decision problems, the SLS algorithm typically stops as soon as a (feasible) solution is found or another termination criterion is satisfied.

In all but the simplest cases, the search process is guided by an *evaluation function*, which measures the quality of candidate solutions. The efficacy of this guidance depends on the properties of the evaluation function and the way in which it is integrated into the search process. Evaluation functions are generally problem specific. For many optimization problems, the objective function given by the problem definition is used; however, different evaluation functions can sometimes provide better guidance, for example, in the sense of approximation guarantees [54.24]. In decision problems, an appropriate evaluation function has to be defined by the algorithm designer. Often, the objective function used for optimization variants of the decision problem can provide useful guidance. For example, for the satisfiability problem in propositional logic (SAT), the objective function of MAX-SAT, which, in a nutshell, counts the number of constraint violations, provides effective guidance. Some SLS methods, such as dynamic local search (briefly discussed in Sect. 54.3), modify the evaluation function during the search process.

The general concept of SLS algorithms, as introduced above and discussed in depth by *Hoos* and *Stützle* [54.1], provides a unified view of constructive and perturbative local search techniques that range from rather simplistic greedy constructive heuristics and iterative improvement algorithms to rather complex hybrid and population-based SLS methods. Population-based algorithms, which manipulate sets of candidate solutions at each iteration, fall under the definition of an SLS algorithm by considering search positions consisting of sets of candidate solutions. In this case, the step function also operates on sets of candidate solutions for the given problem instance. For example, in the case of typical EAs, recombination, mutation, and selection can all be modeled as operations on sets of candidate solutions, which are formally parts of a single-step function used for mapping one generation to the next.

It is instructive to contrast the concept of an SLS algorithm with that of a metaheuristic. Metaheuristics have been described as heuristics that are *superimposed on another heuristic* [54.6], a [54.25]:

> *master strategy that guides and modifies other heuristics to produce solutions beyond those that are normally generated in a quest for local optimality,*

as [54.20]:

> *a set of algorithmic concepts that can be used to define heuristic methods applicable to a wide set of different problems,*

and as [54.26]:

> *a high-level problem-independent algorithmic framework that provides a set of guidelines or strategies to develop heuristic optimization algorithms.*

However, the term metaheuristic [54.26]:

> *is also used to refer to a problem-specific implementation of a heuristic optimization algorithm according to the guidelines expressed in such a framework.*

As is evident from these characterizations, there is no formal definition of the term metaheuristic, and its precise meaning has evolved over time. The term metaheuristic is commonly used to refer to the high-level guidance strategies that in many occasions are used to extend underlying greedy constructive or perturbative search procedures. Hence, the scope of the term metaheuristic differs from that of an SLS algorithm; it comprises what can be similarly loosely characterized as general-purpose SLS methods, but extends naturally to higher-level search strategies involving paradigms other than SLS, such as systematic search methods based on backtracking.

Conversely, the term metaheuristic is usually not applied to simple SLS procedures (such as random sampling, random walk and iterative improvement), nor to problem-specific SLS algorithms with provable properties. Therefore, there are SLS algorithms based on metaheuristics (such as ant colony optimization, iterated local search or EAs for various problems), SLS algorithms that are not metaheuristics (such as 2-opt for the TSP or conflict-directed random walk for SAT) and metaheuristics that are not based on SLS (such as various branch and bound

methods and hybrids between systematic and local search).

Because the notion of an SLS algorithm explicitly refers to aspects that are not related to the high-level guidance of the search process, such as the choice of a neighborhood relation, evaluation function and termination predicate, research on SLS also covers the design, implementation and analysis of these more problem-specific components.

54.2 Greedy Construction Heuristics and Iterative Improvement

The main SLS techniques underlying more complex SLS methods (or metaheuristics) comprise (greedy) constructive search and iterative improvement algorithms. In the following, we discuss the main principles and choices underlying these methods.

Constructive search procedures (or construction heuristics) typically evaluate at each construction step the quality of the available solution components based on a heuristic function. *Greedy construction heuristics* choose to add at each step a solution component with best heuristic value, breaking ties either randomly or by means of a secondary heuristic function. For several polynomially solvable problems, such as the minimum spanning tree problem, greedy construction heuristics (for example, Kruskal's algorithm) are guaranteed to produce optimal solutions [54.27]; unfortunately, for NP-hard problems, this is generally not the case, due to the myopic decisions taken during solution construction.

A useful distinction can be made between static and adaptive construction heuristics. In *static* construction heuristics, the heuristic values associated with solution components are precomputed before the actual construction process is executed and remain unchanged throughout. In *adaptive* construction heuristics, the heuristic values are recomputed at each construction step to take into account the impact of the current partial solution. Adaptive construction heuristics tend to be more accurate and result in better quality candidate solutions than static heuristics, but they are also computationally more expensive.

Construction heuristics are often used to provide good initial candidate solutions for perturbative local search algorithms. One of the most basic SLS methods is to iteratively improve a candidate solution for a given problem instance. Such an *iterative improvement* algorithm starts from some initial search position and iteratively replaces the current candidate solution s by an improving neighboring candidate solution s'. The local search is terminated once no improving neighbor is available, that is, $\forall s' \in N(s) : g(s) \leq g(s')$, where $g(\cdot)$ is the evaluation function to be minimized, and $N(s)$ denotes the set of neighbors of s. In the literature, iterated improvement algorithms are also referred to as *iterated descent* or (in the case of maximization problems) *hill-climbing procedures*.

Neighborhoods are problem specific, and it is generally difficult to predict a priori which of several possible neighborhoods results in best performance. However, for most problems, standard neighborhoods exist. Under the k-exchange neighborhood, two candidate solutions are neighbors if they differ by at most k solution components. An example is the 2-exchange neighborhood for the TSP, where two tours are neighbors if they differ by a pair of edges. Figure 54.1 illustrates a move in this neighborhood. In a k-exchange neighborhood, each candidate solution has $\mathcal{O}(n^k)$ direct neighbors, where n is the number of solution components in each candidate solution. Thus, the neighborhood size is exponential in k, as is the time to identify improving neighbors. While using larger neighborhoods typically makes it possible to reach better solutions, finding those solutions also takes more time. In other words, there is a tradeoff between the quality of the local optima

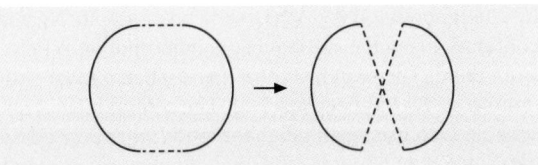

Fig. 54.1 A 2-exchange move for the symmetric TSP. Note that the pair of edges to be introduced is uniquely determined to ensure that the neighbor is again a tour

reachable by an iterative improvement algorithm and its run time. In practice, neighborhoods that involve a quadratic or cubic time-complexity may already result in prohibitive computation times for large problem instances.

The overall time-complexity of searching a given neighborhood is determined by its size and the cost of evaluating each neighbor. The power of local search crucially relies on the fact that *caching and incremental updating techniques* can significantly reduce the cost of evaluating neighbors compared to computing the respective evaluation function values from scratch. For example, the quality of a 2-exchange neighbor of a tour for a TSP instance with n vertices can be computed from the quality of the current tour by subtracting and adding two edge weights (that is, two numbers) each; computing the weight of such a tour from scratch, on the other hand, requires n additions. Sometimes, to render the computation of the incremental updates as efficient as possible, additional data structures need to be implemented, but the net effect is often a very large reduction in computational effort.

A second important technique for reducing the time-complexity of evaluating a given neighborhood is based on the idea of excluding from consideration neighbors that are unlikely or provably unable to lead to improvements. These *neighborhoods pruning techniques* play a crucial role in many high-performance SLS algorithms. Examples of such pruning techniques are the fixed radius searches and nearest neighbors lists used for the TSP [54.28–30], the use of so-called *don't look bits* [54.28], as well as reduced neighborhoods for the job-shop scheduling problem [54.31] and pre-tests for search steps, as done for the single machine total weighted tardiness problem [54.32].

The speed and performance of iterative improvement algorithms also depends on the mechanism used to determine search steps, the so-called *pivoting rule* [54.33]. *Iterative best improvement* chooses at each step a neighboring candidate solution that mostly improves the evaluation function value. Any ties that occur can be broken either randomly, according to the order in which the neighborhood is searched, or based on a secondary criterion (as in [54.34]). In order to find a most improving neighbors, iterative best improvement needs to examine the entire neighborhood in each step. *Iterative first improvement*, in contrast, examines the neighborhood in some given order and moves to the first improving neighboring candidate solution found during this neighborhood scan. Iterative first improvement applies improving search steps earlier than iterative best

improvement, but the amount of improvement achieved in each step tends to be smaller; therefore, it usually requires more improvement steps to reach a local optimum. If a candidate solution is a local optimum, first- and best-improvement algorithms detect this only by inspecting the entire neighborhoods of that solution; don't look bits [54.28, 29] offer a particularly useful mechanism for reducing the time required by this final check, the so-called check-out time.

Interestingly, the local optimum found by iterative first improvement depends on the order in which the neighborhood is examined. This property can be exploited by using a random order for scanning the neighborhood, and repeated runs of random-order first improvement algorithms can identify very different local optima, even if each run is started from the same initial position [54.1, Sect. 2.1]. Thus, the search process in random-order first improvement is more diversified than in the first improvement algorithms that scan local neighborhoods in fixed order.

The notion of local optimality is defined with respect to a specific neighborhood. Thus, changing the neighborhood during the local search process may provide an effective means for escaping from poor quality local optima, and offers the opportunity to benefit from the advantages of large neighborhoods without incurring the computational burden associated with using them exclusively. In the context of iterative improvement algorithms, this idea forms the basis of *variable neighborhood descent* (VND), a variant of a general-purpose SLS method known as variable neighborhood search (VNS) [54.35, 36]. VND uses a sequence of neighborhoods $N_1, N_2, \ldots, N_k$; this sequence is typically ordered according to increasing neighborhood size or increasing time complexity of searching the neighborhoods. VND starts by using the first neighborhood, N_1, until a local optimum is reached. Every time the exploration of a neighborhood N_i does not identify an improving local search step, that is, a local optimum w.r.t. neighborhood N_i is found, VND switches to the next neighborhood, N_{i+1} in the given sequence. Whenever an improving move has been made in a neighborhood N_i, VND switches back to N_1 and continues using the subsequent neighborhoods, N_2 etc., from there. The search is terminated when a local optimum w.r.t. N_k has been reached. The central idea of this scheme is to use small neighborhoods whenever possible, since they allow for the most efficient local search process. The VND scheme typically results in a significant reduction of computation time when compared to an iterative improvement algorithm that uses the largest neighborhood only. VND typically finds high-quality local optima, because upon termination, the resulting candidate solution is locally optimal with respect to all k neighborhoods examined.

Finally, recent years have seen an explosion in the development of iterative improvement methods that exploit *very large scale neighborhood*, whose size is typically exponential in the size of the given problem instance [54.37]. In fact, there are two main approaches to searching these neighborhoods. The first is to perform a heuristic search in the neighborhood, since a exact search would be computationally too demanding. This idea forms the basis of variable-depth search algorithms, where the number of solution components that are modified in each step is not determined a priori. Interestingly, the two best-known variable-depth search algorithms, the *Kernighan–Lin* algorithm for graph partitioning [54.38] and the *Lin–*

Kernighan algorithm for the TSP [54.39], have been devised about in the early 1970s, a fact that illustrates the lasting interest in these types of methods. The more recent concept of ejection chains [54.40] is related to variable-depth search. Another interesting approach is to devise neighborhoods with a special structure that allows them to be searched either in polynomial time or at least very efficiently in practice [54.37, 41–43]. This is the central idea behind many recent developments in very large scale neighborhoods, which include techniques such as Dynasearch [54.32, 44] and cyclic exchange neighborhoods [54.45, 46]. As a result of these research efforts, current state-of-the-art methods for a variety of combinatorial problems such as the TSP [54.47] or the single machine total weighted tardiness problem [54.48] rely on iterative improvement algorithms based on very large scale neighborhoods.

54.3 *Simple* SLS Methods

Iterative improvement algorithms accept only improving neighbors as new current candidate solutions, and they terminate when encountering a local optimum. To allow the search process to progress beyond local optima, many SLS methods permit moves to worsening neighbors. We refer to the methods discussed in the following as *simple* SLS methods, because they essentially only use one type of search steps, in a single, fixed neighborhood relation.

54.3.1 Randomized Iterative Improvement

The key idea behind randomized iterative improvement (RII) is to occasionally perform moves to random neighboring candidate solutions irrespective of their evaluation function value. The simplest way of implementing this idea is to apply, with a given probability w_p, a so-called *uninformed random walk* step, which chooses a neighbor of the current candidate solution uniformly at random, while with probability $1 - w_p$, an improvement step is performed. Often, the improvement step will correspond to one iteration of a best improvement procedure. The parameter w_p is referred to as *walk probability* or, simply, *noise parameter*. RII algorithms have the property that they can perform arbitrarily long sequences of random walk steps; the length of these sequences (i.e., the number of consecutive random walk steps) follows a geometric distribution

with parameter w_p. This allows effective escapes from local optima and renders RII probabilistically approximately complete [54.1, Sect. 4.1]. A main advantage of RII is ease of implementation – often, only a few additional lines of code are required to extend an iterative improvement procedure to an RII procedure – and its behavior is effectively controlled by a single parameter.

RII algorithms have been shown to perform quite well in a number of applications. For example, in the 1990s, minor variations of RII, in which random walk steps are determined based on the status of constraint violations rather than chosen uniformly at random, have been *state of the art* for solving the SAT [54.49, 50] and other constraint satisfaction problems [54.51]. Due to their simplicity, RII algorithms also facilitate theoretical analyses, including characterization of performance in dependence of parameter settings [54.52].

54.3.2 Probabilistic Iterative Improvement

Instead of accepting worsening search steps regardless of the amount of deterioration in evaluation function value they caused (as is the case for random walk steps), it may be preferable to have the probability of acceptance depend on the change of the evaluation function value incurred. This is the key idea underlying probabilistic iterative improvement (PII). Unlike RII, each step of PII involves two phases: first, a neighboring candidate solution $s' \in N(s)$ is selected uniformly at random (proposal mechanism); then, a probabilistic decision is made whether to accept s' as the new search position (acceptance test). For minimization problems, the acceptance probability is often based on the *Metropolis condition* and defined as

$$
p_{\text{accept}}(T, s, s')
$$
$$
:= \begin{cases} 1 & \text{if } g(s') < g(s) \\ \exp\left(\dfrac{g(s) - g(s')}{T}\right) & \text{otherwise ,} \end{cases}
$$
(54.2)

where $p_{\text{accept}}(T, s, s')$ is the acceptance probability, g is the evaluation function to be minimized, and T is a parameter that influences the probability of accepting a worsening search step. PII is closely related to simulated annealing (SA), discussed next; in fact, when using the acceptance mechanism given above, PII is equivalent to constant-temperature SA. In light of this connection, parameter T is also called *temperature*. For various applications, such PII procedures have been shown to perform quite well, provided that T is chosen carefully [54.53, 54]. It is worth noting that in the limit for $T = 0$, PII effectively turns into an iterative improvement procedure (i.e., never accepts worsening steps), while for $T = \infty$, it performs a uniform random walk.

54.3.3 Simulated Annealing

Simulated annealing (SA) [54.4, 5] is similar to PII, except that the parameter T is modified at run time. Following the analogy of the physical annealing of solid materials (e.g., metals and glass), which inspired SA, the temperature T is initially set to some high value and then gradually decreased. At the beginning of the search process, high temperature values result in relatively high probabilities of accepting worsening candidate solutions. As the temperature is decreased, the search process becomes increasingly greedy; for very low settings of the temperature, almost only improving neighbors or neighbors with evaluation function value equal to the current candidate solution are accepted.

Standard SA algorithms iterate over the same two stage process as PII, typically using uniform sampling (with or without replacement) from the neighborhood as a proposal mechanism and a parameter-

ized acceptance test based on the Metropolis condition (54.2) [54.4, 5]. The modification of temperature T is managed by a so-called *annealing (or cooling) schedule*, which is a function that determines the temperature value at each search step. One of the most common choices is a geometric cooling schedule, defined by an initial temperature, T_0, a parameter α between 0 and 1, and a value k, called the temperature length, which defines the number of candidate solutions that are proposed at each fixed value of the temperature; every k steps, the temperature is updated as $T := \alpha \cdot T$. Important parameters of SA are often determined based on characteristics of the problem instance to be solved. For example, the initial temperature may be based on statistics derived from an initial, short random walk; the temperature length may be set to a multiple of the neighborhood size, and the search process may be terminated when the frequency with which proposed search steps are accepted falls below a given threshold.

SA is one of the oldest and most studied SLS methods. It has been applied to a very broad range of computational problems, and many types of annealing schedules, proposal mechanisms, and acceptance tests have been investigated. SA has also been subject to a substantial amount of theoretical analysis, which has yielded various convergence results. For more details on SA, we refer to [54.55, 56].

54.3.4 Tabu Search

Tabu search (TS) differs significantly from the previously discussed SLS methods, in that it makes a direct and systematic use of memory to direct the search process [54.25]. In its most basic form, which is also called *simple tabu search*, TS expands an iterative improvement procedure with a short-term memory to prevent the local search process from returning to recently visited search positions. Instead of memorizing complete candidate solutions and forbidding these explicitly, TS usually associates a tabu status with specific solution components. In the latter case, TS stores for each solution component the time (i.e., the iteration number) at which it was last modified. Each solution component is then considered as potentially tabu if the difference between the stored iteration number and the current iteration number is larger than the value of a parameter called *tabu tenure* (or tabu list length). The tabu status of a local search step is then determined based on specific tabu criteria, which are a function of the tabu status of solution components that are affected by it. One efect is that once a search step has been performed, it is tabu in that it cannot be reversed for a certain number of iterations.

Seen from a neighborhood perspective, TS dynamically restricts the set of neighbors permissible at each local search step by excluding neighbors that are currently tabu. Since the tabu mechanism through prohibition of solution components is quite restrictive, many simple TS algorithms use an *aspiration criterion*, which overrides the tabu status of neighbors if specific conditions are satisfied; for example, if a local search step leads to a new best solution, aspiration allows it to be accepted regardless of its tabu status.

As an example, consider a simple TS algorithm for the TSP, based on the 2-exchange neighborhood. Edges that are removed (or introduced) by a 2-exchange step may then not be reintroduced into (or removed from) the current tour for tt search steps, where tt is the tabu tenure.

For several problems, even simple TS algorithms have been shown to perform quite well. However, the performance of TS strongly depends on the tabu tenure setting. To avoid the difficulty of finding fixed settings suitable for a given problem, mechanisms such as reactive tabu search [54.57] have been devised to adapt the tabu tenure at run time. Simple TS algorithms can be improved in many different ways. In particular, various mechanisms have been developed that make use of intermediate-term and long-term memory to further enhance the performance of simple TS. For a detailed description of such techniques, which aim either at intensifying the search in specific areas of the search space or at diversifying the search to explore unvisited search space regions, we refer to the book by *Glover* and *Laguna* [54.25].

54.3.5 Dynamic Local Search

In contrast to the *simple* SLS methods discussed so far, dynamic local search (DLS) does not accept worsening search steps, but rather modifies the evaluation function during the search in order to escape from local optima. These modifications of the evaluation function g are commonly triggered whenever the underlying local search algorithm, typically an iterative improvement procedure, has reached a locally optimal solution with respect to g', the current evaluation function. Next, the evaluation function is modified and the subsidiary local search algorithm is run until a local optimum (with respect to the new g') is encountered. These local search phases and evaluation function updates are

iterated until some termination criterion is met (see Algorithm 54.1).

Algorithm 54.1 High-level outline of dynamic local search

Dynamic local search (DLS):
determine initial candidate solution s
initialize penalties
while termination criterion is not satisfied **do**
    compute modified evaluation function g'
        from g and penalties
    perform subsidiary local search on s using g'
    update penalties based on s
end while

The modified evaluation function g' is typically computed as the sum of the original evaluation function and penalties associated with each solution component, that is

$$g'(s) := g(s) + \sum_{i \in SC(s)} \text{penalty}(i), \qquad (54.3)$$

where g is the original evaluation function, $SC(s)$ is the set of solution components of candidate solution s, and penalty(i) is the penalty of solution component i. Initially, all penalties are set to zero. Variants of DLS differ in the details of their penalty update mechanism (e.g., additive vs. multiplicative updates, occasional reduction of penalties) and the choice of the solution components whose penalties are adjusted. For example, *guided local search* [54.58, 59] uses the following mechanism for choosing the solution components whose penalties are increased: First, a utility value $u(i) := g_i(s)/(1 + \text{penalty}(i))$ is computed for each solution component i, where $g_i(s)$ measures the impact of i on the evaluation function; then, the penalties of solution components with maximal utility are increased.

DLS algorithms are sometimes referred to as a soft form of tabu search, since solution components are not strictly forbidden, but the effect of the penalties resembles a soft prohibition. There are also conceptual links to Lagrangian methods [54.60, 61]. DLS algorithms have been shown to reach state-of-the-art performance for SAT [54.62] and for the maximum clique problem [54.63].

54.4 Hybrid SLS Methods

The performance of basic SLS techniques can often be improved by combining them with each other. In fact, even RII can be seen as a combination of iterative improvement and random walk, using the same neighborhood. Several other SLS methods combine different types of search steps, and in the following, we briefly discuss some prominent examples.

54.4.1 Greedy Randomized Adaptive Search Procedures

As mentioned previously, construction heuristic can be easily and effectively combined with perturbative local search procedures. While greedy construction heuristics generally generate only one or very few different candidate solutions, randomization of the construction process makes it possible to generate many different high-quality solutions. The idea underlying GRASP [54.10, 64] is to combine randomized greedy construction with a subsequent perturbative local search phase, whose goal is to improve the candidate solutions produced by the construction heuristic. The two phases of solution construction and perturbative local search are repeated until a termination criterion, e.g., maximum computation time, is met. The term *adaptive* in GRASP refers to the fact that the hybrid search process typically uses an adaptive construction heuristic. Randomization in GRASP is realized based on the concept of a *restricted candidate list*, which contains the best-scoring solution components according to the given heuristic function. In the simplest and most common GRASP variants, elements are chosen uniformly at random from this restricted candidate list during the construction process. For a detailed description, various extensions, and an overview of applications of GRASP, we refer to [54.64].

54.4.2 Iterated Greedy Algorithms

A disadvantage of GRASP is that new candidate solutions are constructed from scratch and independently of previously found solutions. Iterated greedy (IG) algorithms iteratively apply greedy construction heuristics to generate a chain of high-quality candidate solutions. The central idea is to alternate between solution construction and destruction phases, and thus to combine at least two different types of search steps. IG algorithms first build an initial, complete candidate solution s. Then, they iterate over the following phases, until a termination criterion is met:

1. Starting from the current candidate solution, s, a destruction phase is executed, during which some solution components are removed from s, resulting in a partial candidate solution s'. The solution components that are removed in this phase may be chosen at random or, for example, based on their impact on the evaluation function.
2. Starting from s', a construction heuristic is used to generate another candidate solution, s''. This construction heuristic may differ from the one used to generate the initial candidate solution.
3. Based on an acceptance criterion, a decision is made whether to continue the search from s or s''. Additionally, it is often useful to further improve complete candidate solutions by means of a subsidiary perturbative local search procedure (see Algorithm 54.2 for a high-level outline of IG).

Algorithm 54.2 High-level outline of an iterated greedy (IG) algorithm

Iterated greedy (IG):
construct initial candidate solution s
perform subsidiary local search on s
while termination criterion is not satisfied **do**
    apply destruction to s, resulting in s'
    apply constructive heuristic starting from s', resulting in s''
    perform subsidiary local search on s'' *(optional)*
    based on acceptance criterion, keep s or accept $s := s''$
end while

The principle underlying IG methods has been rediscovered several times, and consequently, can be found under various names, including *ruin-and-recreate* [54.65], *iterative flattening* [54.66], and *iterative construction heuristic* [54.67]; it has also been used in the context of SA [54.68]. IG algorithms, especially when combined with perturbative local search methods, have reached state-of-the-art performance for a number of problems, including several variants of flowshop scheduling [54.69, 70].

54.4.3 Iterated Local Search

Iterated local search (ILS) generates a sequence of solutions by alternating applications of a perturbation mechanism and of a subsidiary local search algorithm. Consequently, ILS can be seen as a hybrid between the search methods underlying the local search and perturbation phases.

An ILS algorithm is specified by four main components. The first is the mechanism used for generating an initial solution, for example, a greedy constructive heuristic. The second is a subsidiary (perturbative) local search procedure; typically, this is an iterative improvement algorithm, but often, other simple SLS methods are used. The third component is a perturbation procedure that introduces a modification to a given candidate solution. These perturbations should be complementary to the modifications introduced by the subsidiary local search procedure; in particular, the effect of the perturbation procedure should not be easily reversible by the local search procedure. The fourth component is an acceptance criterion, which is used to decide whether to accept the outcome of the latest perturbation and local search phase.

ILS starts by generating an initial candidate solution, to which then subsidiary local search is applied. It then iterates over the following phases, until a termination criterion is met:

1. Perturbation is applied to the current candidate solution s, to obtain an intermediate candidate solution s'.
2. Subsidiary local search is applied to s'.
3. Based on the acceptance criterion, a decision is made whether to continue the search from s or s' (see Algorithm 54.3 for a high-level outline of ILS).

Often, the subsidiary search is based on iterative improvement and ends in a local optimum; ILS can therefore be seen as performing a biased random walk in the space of local optima produced by the given subsidiary local search procedure. The acceptance criterion (together with the strength of the perturbation mechanism) then determines the degree of search intensification: if only improving candidate solutions are accepted, ILS performs a randomized first-improvement search in the space of local optima; if any new local optimum is accepted, ILS performs a random walk in the space of local optima.

Algorithm 54.3 High–level outline of iterated local search

Iterated local search (ILS):
generate initial candidate solution s
perform subsidiary local search on s
while termination criterion is not satisfied **do**
　　apply perturbation to s, resulting in s'
　　perform subsidiary local search on s'
　　based on acceptance criterion, keep s
　　　　or accept $s := s'$
end while

An attractive feature of ILS is that basic versions can be quickly and easily implemented, especially if a simple SLS algorithm or an iterative improvement procedure is already available. Using some additional refinements, ILS methods define the current state of the art for solving many combinatorial problems, including the TSP [54.71]. Similar to IG, ILS is based on an idea that has been rediscovered several times and is known under various names, including *large-step Markov chains* [54.29] and *chained local optimization* [54.72]. There is also a close conceptual connection with several variants of variable neighborhood search (VNS) [54.35]; in fact, the so-called basic VNS and skewed VNS algorithms can be seen as variants of ILS that adapt the perturbation strength at run time. For more details on iterated local search, we refer to [54.73].

54.5 Population–Based SLS Methods

The use of a population of candidate solutions offers a convenient way to increase diversification in SLS. For example, population-based extensions of ILS algorithms have been proposed with this aim in mind [54.74, 75]. A further potential benefit comes from the inherent parallelizability of the most population-based SLS methods, although the parallelization thus achieved is not necessarily more effective than the simple and generic approach of performing multiple independent runs of an SLS algorithm in parallel (see also [54.1], Sect. 4.4). As previously remarked, population-based methods can be cast into the SLS framework described in Sect. 54.1 by defining search positions to consist of sets of candidate solutions and by using neighborhood relations, initialization, and step functions that operate on such populations.

Unfortunately, the benefits derived from the use of populations come at the cost of increased complexity, in terms of implementation effort, and parameters that need to be set appropriately. In what follows, we describe two of the most prominent population-based methods, one based on a constructive search paradigm (ant colony optimization), and the other based on a perturbative search paradigm (evolutionary algorithms).

54.5.1 Ant Colony Optimization

Ant colony optimization (ACO) algorithms have originally been inspired by the trail-following behavior of real ant species, which allows them to find shortest paths [54.76, 77]. This biological phenomenon gave rise to a surprisingly effective algorithm for combinatorial optimization [54.18, 19]. In ACO, the artificial ants perform a randomized constructive search that is biased by (artificial) pheromone trails and heuristic information derived from the given problem instance. The pheromone trails are numerical values associated with solution components that are adapted at run time to reflect experience gleaned from the search process so far.

During solution construction, at each step every ant chooses a solution component, probabilistically preferring those with high-pheromone trail and heuristic information values. For illustration, consider the TSP – the first problem to which ACO has been applied [54.18]. Each edge (i,j) has an associated pheromone value τ_{ij} and a heuristic value η_{ij}, which for the TSP is typically defined as $1/w(i,j)$, that is, the inverse of the edge weight. In ant system [54.19], the first ACO algorithm for the TSP, an ant located at vertex i would add vertex j to its current partial tour s' with probability

$$p_{ij} = \frac{\tau_{ij}^{\alpha} \cdot \eta_{ij}^{\beta}}{\sum_{l \in N(i)} \tau_{il}^{\alpha} \cdot \eta_{il}^{\beta}}, \qquad (54.4)$$

where $N(i)$ is the feasible neighborhood of vertex i, i.e., the set of all vertices that have not yet been visited in s', and α and β are parameters that control the relative importance of pheromone trails and heuristic information, respectively. Note that the tour construction procedure

implemented by the artificial ants is a randomized version of the nearest neighbor construction heuristic. In fact, randomizing a greedy construction heuristic based on pheromone trails associated with the decisions to be made would generally be a good initial step toward an effective ACO algorithm for a combinatorial problem.

Once every ant has constructed a complete candidate solution, it is typically highly advantageous to apply an iterative improvement procedure or a simple SLS algorithm [54.20, 78]. Next, the pheromone trail values are updated by means of two counteracting mechanisms. The first models pheromone evaporation and decreases some or all pheromone trail values by a constant factor. The second models pheromone deposit and increases the pheromone trail levels of solution components that have been used by one or more ants. The amount of pheromone deposited typically depends on the quality of the respective solutions. In the best performing ACO algorithms, only some of the ants with the highest quality solutions are allowed to deposit pheromone. The overall result of the pheromone update is an increased probability of choosing solution components in subsequent solution constructions that have previously been found to occur in high-quality solutions. ACO algorithms then cycle through these phases of solution construction, application of local search, and pheromone update until some termination criterion is met (see Algorithm 54.4 for a high-level outline of ACO).

Algorithm 54.4 High–level outline of ant colony optimization

Ant colony optimization (ACO):
initialize pheromone trails
while termination criterion is not satisfied **do**
    generate population *sp* of candidate solutions
        using subsidiary randomized
        constructive search
    perform subsidiary local search on *sp*
    update pheromone trails
end while

Many different variants of ACO algorithms have been studied. Along with many additional details on ACO, these are described in the book by *Dorigo* and *Stützle* [54.20]; for more recent surveys, we refer the reader to [54.79, 80]. The ACO metaheuristic [54.81, 82] provides a general framework for these variants and a generic view of how to apply ACO algorithms. ACO is also one of the most successful algorithmic techniques within the broader field of swarm intelligence [54.83].

54.5.2 Evolutionary Algorithms

Evolutionary algorithms (EAs) are a prominent class of population-based SLS methods that are loosely inspired by concepts from biological evolution. Unlike ACO algorithms, EAs work with a population of complete candidate solutions. The initial set of candidate solutions is typically created randomly, but greedy construction heuristics may also be used to seed the population. This population then undergoes an artificial evolution, where at each iteration, the population of candidate solutions is modified by means of mutation, recombination and selection.

Mutation operators typically introduce small, random perturbations into individual candidate solutions. The strength of these perturbations is usually controlled by a parameter called *mutation rate*; alternatively, a specific, fixed perturbation, akin to a random walk step in RII, may be performed. *Recombination operators* generate one or more new candidate solutions by combining information from two or more parent candidate solutions. The most common type of recombination is *crossover*, inspired by the homonymous mechanism in biological evolution; it generates offspring by assembling partial candidate solutions from linear representations of two parents. In addition to mutation and recombination, *selection mechanisms* are used to determine the candidate solutions that will undergo mutation and recombination, as well as those that will form the population used in the next iteration of the evolutionary process. Selection is based on the *fitness*, i. e., evaluation function values, of the candidate solutions, such that better candidate solutions have a higher probability to be selected.

Details of the mutation, recombination and selection mechanisms all have a strong impact on the performance of an EA. Generally, the use of problem specific knowledge within these mechanisms leads to better performance. In fact, much research in EAs has been devoted to the design of effective mutation and

recombination operators; a good example for this is the TSP [54.84, 85]. To achieve cutting-edge performance in an EA, it is often useful to improve at least the best candidate solutions in a given population by means of a perturbative local search method, such as iterative improvement. The resulting class of hybrid algorithms, which are also known as *memetic algorithms* (MA) [54.86], are enjoying increasing popularity as a broadly applicable method for solving solving combinatorial problems (see Algorithm 54.5 for a high-level outline of an MA).

Algorithm 54.5 High-level outline of a memetic algorithm

Memetic algorithm (MA):
initialize population p
perform subsidiary local search on each
    candidate solution in p
while termination criterion is not satisfied **do**
    generate set pr of candidate solutions
        through recombination
    perform subsidiary local search on each
        candidate solution of pr
    generate set pm of candidate solutions
        from $p \cup pr$ through mutation
    perform subsidiary local search on each
        candidate solution of pm
    select new population p from candidate
        solutions in $p \cup pr \cup pm$
end while

Several other techniques are conceptually related to evolutionary algorithms but have different roots. *Scatter search* and *path relinking* are SLS methods whose origins can be traced back to the mid-1970s [54.16]. Scatter search can be seen as a memetic algorithm that uses special types of recombination and selection mechanisms. Path relinking corresponds to a specific form of interpolation between two (or possibly more) candidate solutions and is thus conceptually related to recombination operators. Both methods have recently become increasingly popular; details can be found in [54.17, 87].

54.6 Recent Research Directions

In this section, we concisely discuss three research directions that we regard as particularly timely and promising: combinations of SLS and systematic search

techniques, SLS algorithm engineering, and automated configuration and design of SLS algorithms. For other topics of interests, such as SLS algorithms for multiobjective [54.88–90], stochastic [54.91] or dynamic problems [54.92, 93], we refer to the literature for more details.

54.6.1 Combination of SLS Algorithms with Systematic Search Techniques

Systematic search and SLS are traditionally seen as two distinct approaches for solving challenging combinatorial problems. Interestingly, the particular advantages and disadvantages of each of these approaches render them rather complementary. Therefore, it is hardly surprising that over the last few years, there has been increased interest in the exploration and development of hybrid algorithms that combine ideas from both paradigms. For example, related to the area of mathematical programming, the term *Matheuristics* has recently been coined to refer to methods that combine elements from mathematical programming techniques (which are primarily based on systematic search) and (meta)heuristic search algorithms [54.94].

Hybrids between SLS and systematic search fall into two main classes. The first of these consists of approaches where the systematic search algorithm plays the role of the master process, and an SLS procedure is used to solve subproblems that arise during the systematic search process. Probably, the simplest, yet potentially effective method is to use an SLS algorithm to provide an initial high-quality (primal) bound on the optimal solution of the problem, which is then used by the systematic search algorithm for pruning parts of the search tree. Several more elaborate schemes have been devised, e.g., in the context of column generation and separation routines in integer programming [54.95]. Other approaches introduce the spirit of local search into integer programming solvers; examples of these include *local branching* [54.96] and *relaxation-induced neighborhood search* [54.97]. We refer to [54.95] for a recent overview of such combinations.

The second class of hybrid approaches is based on the idea of using systematic search procedures to deal with specific tasks arising while running an SLS algorithm. Very-large neighborhood search [54.37], as discussed in Sect. 54.2, is probably one of the best-known examples. Elements of tree search methods can also be exploited within constructive search algorithms, as exemplified by the use of branch and bound techniques in ACO algorithms [54.98, 99]. Other examples include *tour merging* [54.100] and the usage of information derived from integer programming formulations

of optimization problems in heuristic methods [54.101]. We refer to [54.102] for a survey of this general approach. A taxonomy of the possible combinations of exact and local search algorithms has been introduced by *Jourdan* et al. [54.103].

Despite an increasing number of efforts on combining systematic search methods and SLS methods, as reviewed in [54.94], much work remains to be done in this direction, especially considering that the two underlying fundamental search paradigms are developed primarily in rather disjoint communities. We believe that much can be gained by overcoming the traditional view of these two approaches as being competing with each other in favour of focusing on synergies due to their complementarity.

54.6.2 SLS Algorithm Engineering

Despite the impressive successes in SLS research and applications – SLS algorithms are now firmly established as the method of choice for tackling a broad range of combinatorial problems – there are still significant shortcomings. Perhaps most prominently, there is a lack of guidelines and best practices regarding the design and development of effective SLS algorithms. Current practice is to implement one specific SLS method, based on one or more construction heuristics or iterative improvement procedures. However, general-purpose SLS methods are not fully defined recipes: they leave many design choices open, and typically only specific combinations of these choices will result in an effective algorithms for a given problem. Even worse, the underlying basic construction and iterative improvement procedures have a tremendous influence on the final performance of the SLS algorithms built on them, and this influence is frequently neglected.

We firmly believe that a more methodological approach needs to be taken toward the design and implementation of SLS algorithms. The research direction dedicated to developing such an approach is called *stochastic local search algorithm engineering* or, for short, *SLS engineering*; it is conceptually related to algorithm engineering [54.104] and software engineering [54.105], where similar methodological issues are tackled in a different context. Algorithm engineering is rather closely related to SLS engineering; it has been conceived as an extension to the traditionally more theoretically oriented research on algorithms. Algorithm engineering, according to [54.104], deals with the iterative process of designing, analyzing, implementing, tuning and experimentally evaluating algorithms. SLS engineering shares this motivation; however, the algorithms that are dealt with in the context of SLS approaches have substantially more complex and unpredictable behavior than those typically considered in algorithm engineering. There are several reasons for this: SLS algorithms are usually used for solving NP-hard problems, they allow for many more degrees of freedom in the choice of algorithm components, and their stochasticity makes analysis more complex.

From a high-level perspective, an initial approach to a successful *SLS engineering process* would proceed in a bottom-up fashion. Starting from knowledge about the problem, it would build SLS algorithms by iteratively adding complexity to simple, basic algorithms. More concretely, a tentative first attempt at such a process could be as follows:

1. Study existing knowledge on the problem to be solved and its characteristics;
2. Implement basic and advanced constructive and iterative improvement procedures;
3. Starting from these, add complexity (for example, by moving to simple SLS methods);
4. Improve performance by gradually adding concepts from more complex SLS techniques (for example, perturbations, prohibition mechanisms, populations);
5. Further configure and fine-tune parameters and design choices;
6. If found to be useful: iterate over steps 4–5.

Obviously, such a process would not necessarily strictly follow this outline, but insights gained at later stages could prompt revisiting earlier design decisions. Several high-performance SLS algorithms have already been developed following roughly the process outlined above (see [54.106] for an explicit example).

The SLS engineering process can be supported in various ways. Algorithm development, implementation and testing is facilitated by the use of programming frameworks like Paradiseo [54.107, 108] and EasyLocal++ [54.109, 110], dedicated languages and systems like COMET [54.111], libraries of data types (such as LEDA [54.112]), and statistical tools, such as the comprehensive, open-source R environment [54.113]. We expect that software environments specifically designed for the automated empirical analysis and design of algorithms, such as HAL [54.114, 115], will be especially useful in this context. Tools for the automatic configuration and tuning of algorithms, discussed

further in the next section are also of considerable importance.

Furthermore, we see an improved understanding of the relationship between problem and instance features on the one side, and the properties and the behavior of SLS methods on the other side as key enabling factors for advanced SLS engineering approaches. The potential insights to be gained are not only of practical value to SLS engineering but also of considerable scientific interest. Progress in this direction is facilitated by advanced search space analysis techniques, statistical methods and machine learning approaches (see, e.g., *Merz* and *Freisleben* [54.116], *Xu* et al. [54.117] and *Watson* et al. [54.118]). Another promising avenue for future research involves the integration of theoretical insights into the design process, for example, by restricting design alternatives or parameter choices.

It is important to note that research toward SLS engineering adopts a *component-wise view of SLS methods*. For example, iterated local search (ILS) uses perturbations to diversify the search as well as acceptance tests (components: perturbations, acceptance tests), while evolutionary algorithms prominently involve the use of a population of solutions (component: population of solutions). Each of these components can be instantiated in different ways, and various combinations are possible. An effective SLS engineering process should provide guidance to the algorithm designer regarding the choice and configuration of these components. It would naturally and incrementally lead to combinations of algorithmic components taken from different SLS methods (or other paradigms, such as mathematical programming – [54.94]), if these contribute to desirable performance characteristics of the algorithm under design. Such an engineering process would therefore rather naturally produce hybrid algorithms that are effective for solving the given computational problem.

Finally, SLS engineering highlights more the importance of decisions concerning the underlying basic SLS techniques (such as construction heuristics, neighborhoods, efficient data structures, etc.) than the general-purpose SLS methods (or metaheuristics) used in a given algorithm design scenario. In fact, in our experience, such fundamental choices together with: (i) the level of expertise of the SLS algorithm developer and implementer, (ii) the time invested in designing and configuring the SLS algorithm, (iii) the creative use of insights into algorithm behavior and interaction with problem characteristics play a considerably more important role in the design of effective SLS algorithms than the focus on specific features prescribed by so-called metaheuristics.

54.6.3 Automatic Configuration of SLS Algorithms

The performance of algorithms for virtually any computationally challenging problem (and in particular, for any NP-hard problem) depends strongly on appropriate settings of algorithm parameters. In many cases, there are tens of such parameter; for example, the well-known commercial CPLEX solver for integer programming problems has more than 130 user-specifiable parameters that influence its search behavior. Likewise, the behavior of most SLS algorithms is controlled by parameters, and many design choices can be exposed in the form of parameters. This gives rise to algorithms with many categorical and numerical parameters. Categorical parameters are used to make choices from a discrete set of design variants, such as search strategies, neighborhoods or perturbation mechanisms. Numerical parameters often arise as subordinate parameters that directly control the behavior of a search strategy (e.g., temperature in SA and tabu tenure in simple tabu search). The goal in automated algorithm configuration is to find settings of these parameters that achieve optimized performance w.r.t. a performance metric of interest (for example, solution quality or computation time).

Automated algorithm configuration methods are an active area of research and have been demonstrated to achieve very substantial performance gains on many widely studied and challenging problems [54.119]. So-called *offline configuration methods*, which determine performance-optimizing parameter settings on a representative set of benchmark instances during a training phase before algorithm deployment, have arguably been studied most intensely been studied. These in-clude procedures that are limited to tuning numerical parameters, such as CALIBRA [54.120], experimental design-based approaches [54.69, 121], SPO [54.122] and SPO$^+$ [54.123]. Methods that can handle categorical as well as numerical parameters are considerably more versatile; these include racing procedures [54.124, 125], model-free configuration procedures [54.126–128], and recent sequential model-based techniques [54.129, 130].

In *online configuration*, algorithm parameters are modified while attempting to solve a given problem instance. There are some inherent advantages of online configuration methods w.r.t. offline methods, especially when targeting very heterogeneous instances, where appropriate algorithm parameters may depend strongly on the problem instance to be solved. Some of these methods fall into the realm of reactive search methods [54.131]; others have been studied in the area of evolutionary computation (for an overview, we refer to [54.132]). Unfortunately, most online configuration methods presently available deal with very few parameters primarily responsible for algorithm performance (often only one) and rely on specific insight into the working principles of the given algorithm.

There are also various approaches for determining configurations of a given algorithm that result in good performance on a given problem instance. These *per-instance configuration methods* typically make use of computationally cheap instance features, which provide the basis for selecting the configuration to be used for solving a given instance [54.133–135].

Finally, we believe that there is significant promise in approaches for automating large parts of the design process of performance-optimized SLS algorithms as, for example, outlined in recent work on computer-aided algorithm design for generalized local search machines [54.136] and the *programming by optimization* (PbO) software design paradigm [54.23].

References

54.1   H.H. Hoos, T. Stützle: Stochastic Local Search—Foundations and Applications (Morgan Kaufmann, San Francisco 2004)

54.2   G.R. Schreiber, O.C. Martin: Cut size statistics of graph bisection heuristics, SIAM J. Optim. **10**(1), 231–251 (1999)

54.3   M. Gendreau, J.-Y. Potvin (Eds.): *Handbook of Metaheuristics*, International Series in Opera-tions Research & Management Science, Vol. 146 (Springer, New York 2010)

54.4   S. Kirkpatrick, C.D. Gelatt Jr., M.P. Vecchi: Optimization by simulated annealing, Science **220**, 671–680 (1983)

54.5   V. Cerný: A thermodynamical approach to the traveling salesman problem, J. Optim. Theory Appl. **45**(1), 41–51 (1985)

54.6   F. Glover: Future paths for integer programming and links to artificial intelligence, Comput. Oper. Res. **13**(5), 533–549 (1986)

54.7   F. Glover: Tabu search – Part I, ORSA J. Comput. **1**(3), 190–206 (1989)

54.8   F. Glover: Tabu search – Part II, ORSA J. Comput. **2**(1), 4–32 (1990)

54.9   P. Hansen, B. Jaumard: Algorithms for the maximum satisfiability problem, Computing **44**, 279–303 (1990)

54.10   T.A. Feo, M.G.C. Resende: A probabilistic heuristic for a computationally difficult set covering problem, Oper. Res. Lett. **8**(2), 67–71 (1989)

54.11   H.R. Lourenço, O. Martin, T. Stützle: Iterated local search. In: *Handbook of Metaheuristics*, ed. by F. Glover, G. Kochenberger (Kluwer, Norwell 2002) pp. 321–353

54.12   J.H. Holland: *Adaption in Natural and Artificial Systems* (The University of Michigan, Ann Arbor 1975)

54.13   D.E. Goldberg: *Genetic Algorithms in Search, Optimization, and Machine Learning* (Addison-Wesley, Reading 1989)

54.14   I. Rechenberg: *Evolutionsstrategie – Optimierung technischer Systeme nach Prinzipien der biologischen Information* (Fromman, Freiburg, Germany 1973)

54.15   H.-P. Schwefel: *Numerical Optimization of Computer Models* (Wiley, Chichester 1981)

54.16   F. Glover: Heuristics for integer programming using surrogate constraints, Decis. Sci. **8**, 156–164 (1977)

54.17   F. Glover, M. Laguna, R. Martí: Scatter search and path relinking: Advances and applications. In: *Handbook of Metaheuristics*, ed. by F. Glover, G. Kochenberger (Kluwer, Norwell 2002) pp. 1–35

54.18   M. Dorigo, V. Maniezzo, A. Colorni: Positive feedback as a search strategy. Techn. Rep. 91-016, Dipartimento di Elettronica, Politecnico di Milano, Italy, 1991

54.19   M. Dorigo, V. Maniezzo, A. Colorni: Ant System: Optimization by a colony of cooperating agents, IEEE Trans. Syst. Man. Cybern. B **26**(1), 29–41 (1996)

54.20   M. Dorigo, T. Stützle: *Ant Colony Optimization* (MIT, Cambridge 2004)

54.21   T. Stützle, M. Birattari, H.H. Hoos: Engineering stochastic local search algorithms – designing, implementing and analyzing effective heuristics, Lect. Notes Comput. Sci. **4638**, 1–221 (2007)

54.22   T. Stützle, M. Birattari, H.H. Hoos: Engineering stochastic local search algorithms – designing, implementing and analyzing effective heuristics, Lect. Notes Comput. Sci. **5217**, 1–155 (2009)

54.23   H.H. Hoos: Programming by optimization, Commun. ACM **55**, 70–80 (2012)

54.24   S. Khanna, R. Motwani, M. Sudan, U. Vazirani: On syntactic versus computational views of approximability, Proc. 35th Annu. IEEE Symp. Found.

54.25   Comput. Sci. (IEEE Computer Society, Los Alamitos 1994) pp. 819–830

54.25   F. Glover, M. Laguna: *Tabu Search* (Kluwer, Boston 1997)

54.26   K. Sörensen, F. Glover: Metaheuristics. In: *Encyclopedia of Operations Research and Management Science*, ed. by S.I. Gass, M.C. Fu (Springer, Berlin 2013) pp. 960–970

54.27   C.H. Papadimitriou, K. Steiglitz: *Combinatorial Optimization – Algorithms and Complexity* (Prentice Hall, Englewood Cliffs 1982)

54.28   J.L. Bentley: Fast algorithms for geometric traveling salesman problems, ORSA J. Comput. **4**(4), 387–411 (1992)

54.29   O.C. Martin, S.W. Otto, E.W. Felten: Large-step Markov chains for the traveling salesman problem, Complex Syst. **5**(3), 299–326 (1991)

54.30   D.S. Johnson, L.A. McGeoch: The traveling salesman problem: A case study in local optimization. In: *Local Search in Combinatorial Optimization*, ed. by E.H.L. Aarts, J.K. Lenstra (Wiley, Chichester 1997) pp. 215–310

54.31   A.S. Jain, B. Rangaswamy, S. Meeran: New and "stronger" job-shop neighbourhoods: A focus on the method of Nowicki and Smutnicki, J. Heuristics **6**(4), 457–480 (2000)

54.32   R.K. Congram, C.N. Potts, S. van de Velde: An iterated dynasearch algorithm for the single-machine total weighted tardiness scheduling problem, INFORMS J. Comput. **14**(1), 52–67 (2002)

54.33   M. Yannakakis: The analysis of local search problems and their heuristics, Lect. Notes Comput. Sci. **415**, 298–310 (1990)

54.34   R. Battiti, M. Protasi: Reactive search, a history-based heuristic for MAX-SAT, ACM J. Exp. Algorithmics **2**, 2 (1997)

54.35   P. Hansen, N. Mladenović: Variable neighborhood search: Principles and applications, Eur. J. Oper. Res. **130**(3), 449–467 (2001)

54.36   P. Hansen, N. Mladenović: Variable neighborhood search. In: *Handbook of Metaheuristics*, ed. by F. Glover, G. Kochenberger (Kluwer, Norwell 2002) pp. 145–184

54.37   R.K. Ahuja, O. Ergun, J.B. Orlin, A.P. Punnen: A survey of very large-scale neighborhood search techniques, Discrete Appl. Math. **123**(1–3), 75–102 (2002)

54.38   B.W. Kernighan, S. Lin: An efficient heuristic procedure for partitioning graphs, Bell Syst. Technol. J. **49**, 213–219 (1970)

54.39   S. Lin, B.W. Kernighan: An effective heuristic algorithm for the traveling salesman problem, Oper. Res. **21**(2), 498–516 (1973)

54.40   F. Glover: Ejection chain, reference structures and alternating path methods for traveling salesman problems, Discrete. Appl. Math. **65**(1–3), 223–253 (1996)

54.41   R.K. Ahuja, O. Ergun, J.B. Orlin, A.P. Punnen: Very large-scale neighborhood search. In: *Handbook*

Part E | 54

54.42   *of Approximation Algorithms and Metaheuristics*, Computer and Information Science Series, ed. by T.F. Gonzalez (Chapman Hall/CRC, Boca Raton 2007) pp. 1–12

54.42   I. Dumitrescu: Constrained Path and Cycle Problems, Ph.D. Thesis (University of Melbourne, Department of Mathematics and Statistics 2002)

54.43   M. Chiarandini, I. Dumitrescu, T. Stützle: Very large-scale neighborhood search: Overview and case studies on coloring problems. In: *Hybrid Metaheuristics – An Emergent Approach to Optimization*, Studies in Computational Intelligence, Vol. 117, ed. by C. Blum, M.J. Blesa Aguilera, A. Roli, M. Sampels (Springer, Berlin 2008) pp. 117–150

54.44   C.N. Potts, S. van de Velde: Dynasearch: Iterative local improvement by dynamic programming; Part I, the traveling salesman problem. Techn. Rep. LPOM–9511, Faculty of Mechanical Engineering, University of Twente, Enschede, The Netherlands, 1995

54.45   P.M. Thompson, J.B. Orlin: The theory of cycle transfers, Working Paper OR 200-89, Operations Research Center, MIT, Cambridge 1989

54.46   P.M. Thompson, H.N. Psaraftis: Cyclic transfer algorithm for multivehicle routing and scheduling problems, Oper. Res. **41**, 935–946 (1993)

54.47   K. Helsgaun: An effective implementation of the Lin-Kernighan traveling salesman heuristic, Eur. J. Oper. Res. **126**(1), 106–130 (2000)

54.48   A. Grosso, F. Della Croce, R. Tadei: An enhanced dynasearch neighborhood for the single-machine total weighted tardiness scheduling problem, Oper. Res. Lett. **32**(1), 68–72 (2004)

54.49   B. Selman, H. Kautz: Domain-independent extensions to GSAT: Solving large structured satisfiability problems, Proc. 13th Int. Jt. Conf. Artif. Intell., ed. by R. Bajcsy (Morgan Kaufmann, San Francisco 1993) pp. 290–295

54.50   B. Selman, H. Kautz, B. Cohen: Noise strategies for improving local search, Proc. 12th Natl. Conf. Artif. Intell., AAAI/The MIT (1994) pp. 337–343

54.51   O. Steinmann, A. Strohmaier, T. Stützle: Tabu search vs. random walk, Lect. Notes Artif. Intell. **1303**, 337–348 (1997)

54.52   O.J. Mengshoel: Understanding the role of noise in stochastic local search: Analysis and experiments, Artif. Intell. **172**(8/9), 955–990 (2008)

54.53   D.T. Connolly: An improved annealing scheme for the QAP, Eur. J. Oper. Res. **46**(1), 93–100 (1990)

54.54   M. Fielding: Simulated annealing with an optimal fixed temperature, SIAM J. Optim. **11**(2), 289–307 (2000)

54.55   E.H.L. Aarts, J.H.M. Korst, P.J.M. van Laarhoven: Simulated annealing. In: *Local Search in Combinatorial Optimization*, ed. by E.H.L. Aarts, J.K. Lenstra (Wiley, Chichester 1997) pp. 91–120

54.56   A.G. Nikolaev, S.H. Jacobsen: Simulated annealing. In: *Handbook of Metaheuristics*, International Series in Operations Research & Management Science, Vol. 146, ed. by M. Gendreau, J.-Y. Potvin (Springer, New York 2010) pp. 1–40 2 edition, chapter 8

54.57   R. Battiti, G. Tecchiolli: Simulated annealing and tabu search in the long run: A comparison on QAP tasks, Comput. Math. Appl. **28**(6), 1–8 (1994)

54.58   C. Voudouris: Guided Local Search for Combinatorial Optimization Problems, Ph.D. Thesis (University of Essex, Department of Computer Science, Colchester 1997)

54.59   C. Voudouris, E. Tsang: Guided local search and its application to the travelling salesman problem, Eur. J. Oper. Res. **113**(2), 469–499 (1999)

54.60   Y. Shang, B.W. Wah: A discrete Lagrangian-based global-search method for solving satisfiability problems, J. Glob. Optim. **12**(1), 61–100 (1998)

54.61   D. Schuurmans, F. Southey, R.C. Holte: The exponentiated subgradient algorithm for heuristic boolean programming, Proc. 17th Int. Jt. Conf. Artif. Intell., ed. by B. Nebel (Morgan Kaufmann, San Francisco 2001) pp. 334–341

54.62   F. Hutter, D.A.D. Tompkins, H.H. Hoos: Scaling and probabilistic smoothing: Efficient dynamic local search for SAT, Lect. Notes Comput. Sci. **2470**, 233–248 (2002)

54.63   W.J. Pullan, H.H. Hoos: Dynamic local search for the maximum clique problem, J. Artif. Intell. Res. **25**, 159–185 (2006)

54.64   M.G.C. Resende, C.C. Ribeiro: Greedy randomized adaptive search procedures: Advances and applications. In: *Handbook of Metaheuristics*, International Series in Operations Research & Management Science, Vol. 146, ed. by M. Gendreau, J.-Y. Potvin (Springer, New York 2010) pp. 281–317

54.65   G. Schrimpf, J. Schneider, H. Stamm-Wilbrandt, G. Dueck: Record breaking optimization results using the ruin and recreate principle, J. Comput. Phys. **159**(2), 139–171 (2000)

54.66   A. Cesta, A. Oddi, S.F. Smith: Iterative flattening: A scalable method for solving multi-capacity scheduling problems, Proc. 17th Natl. Conf. Artif. Intell., AAAI/The MIT (2000) pp. 742–747

54.67   A.J. Richmond, J.E. Beasley: An iterative construction heuristic for the ore selection problem, J. Heuristics **10**, 153–167 (2004)

54.68   L.W. Jacobs, M.J. Brusco: A local search heuristic for large set-covering problems, Nav. Res. Logist. **42**(7), 1129–1140 (1995)

54.69   R. Ruiz, T. Stützle: A simple and effective iterated greedy algorithm for the permutation flow-shop scheduling problem, Eur. J. Oper. Res. **177**(3), 2033–2049 (2007)

54.70   R. Ruiz, T. Stützle: An iterated greedy heuristic for the sequence dependent setup times flowshop problem with makespan and weighted tardiness objectives, Eur. J. Oper. Res. **187**(3), 1143–1159 (2008)

54.71   D.S. Johnson, L.A. McGeoch: Experimental analysis of heuristics for the STSP. In: *The Travel-*

54.72 D. Applegate, W. Cook, A. Rohe: Chained Lin-Kernighan for large traveling salesman problems, INFORMS J. Comput. **15**(1), 82–92 (2003)

54.73 H.R. Lourenço, O. Martin, T. Stützle: Iterated local search: Framework and applications. In: *Handbook of Metaheuristics*, International Series in Operations Research & Management Science, Vol. 146, ed. by M. Gendreau, J.-Y. Potvin (Springer, New York 2010) pp. 363–397

54.74 T. Stützle: Iterated local search for the quadratic assignment problem, Eur. J. Oper. Res. **174**(3), 1519–1539 (2006)

54.75 I. Hong, A.B. Kahng, B.R. Moon: Improved large-step Markov chain variants for the symmetric TSP, J. Heuristics **3**(1), 63–81 (1997)

54.76 S. Goss, S. Aron, J.L. Deneubourg, J.M. Pasteels: Self-organized shortcuts in the Argentine ant, Naturwissenschaften **76**, 579–581 (1989)

54.77 J.-L. Deneubourg, S. Aron, S. Goss, J.-M. Pasteels: The self-organizing exploratory pattern of the Argentine ant, J. Insect Behav. **3**, 159–168 (1990)

54.78 T. Stützle, H.H. Hoos: *MAX–MIN* ant system, Future Gener. Comput. Syst. **16**(8), 889–914 (2000)

54.79 M. Dorigo, M. Birattari, T. Stützle: Ant colony optimization: Artificial ants as a computational intelligence technique, IEEE Comput. Intell. Mag. **1**(4), 28–39 (2006)

54.80 M. Dorigo, T. Stützle: Ant colony optimization: Overview and recent advances. In: *Handbook of Metaheuristics*, International Series in Operations Research & Management Science, Vol. 146, ed. by M. Gendreau, J.-Y. Potvin (Springer, New York 2010) pp. 227–263

54.81 M. Dorigo, G. Di Caro: The ant colony optimization meta-heuristic. In: *New Ideas in Optimization*, ed. by D. Corne, M. Dorigo, F. Glover (McGraw Hill, London 1999) pp. 11–32

54.82 M. Dorigo, G. Di Caro, L.M. Gambardella: Ant algorithms for discrete optimization, Artif. Life **5**(2), 137–172 (1999)

54.83 E. Bonabeau, M. Dorigo, G. Theraulaz: *Swarm Intelligence: From Natural to Artificial Systems* (Oxford Univ. Press, New York 1999)

54.84 J.-Y. Potvin: Genetic algorithms for the traveling salesman problem, Ann. Oper. Res. **63**, 339–370 (1996)

54.85 P. Merz, B. Freisleben: Memetic algorithms for the traveling salesman problem, Complex Syst. **13**(4), 297–345 (2001)

54.86 P. Moscato: Memetic algorithms: A short introduction. In: *New Ideas in Optimization*, ed. by D. Corne, M. Dorigo, F. Glover (McGraw Hill, London 1999) pp. 219–234

54.87 M. Laguna, R. Martí: *Scatter Search: Methodology and Implementations in C*, Vol. 24 (Kluwer, Boston 2003)

54.88 M. Ehrgott, X. Gandibleux: Approximative solution methods for combinatorial multicriteria optimization, TOP **12**(1), 1–88 (2004)

54.89 M. Ehrgott, X. Gandibleux: Hybrid metaheuristics for multi-objective combinatorial optimization. In: *Hybrid Metaheuristics: An emergent approach for optimization*, ed. by C. Blum, M.J. Blesa, A. Roli, M. Sampels (Springer, Berlin, Germany 2008) pp. 221–259

54.90 L. Paquete, T. Stützle: Stochastic local search algorithms for multiobjective combinatorial optimization: A review. In: *Handbook of Approximation Algorithms and Metaheuristics*, Computer and Information Science Series, ed. by T.F. Gonzalez (Chapman Hall/CRC, Boca Raton 2007) pp. 1–15

54.91 L. Bianchi, M. Dorigo, L.M. Gambardella, W.J. Gutjahr: A survey on metaheuristics for stochastic combinatorial optimization, Nat. Comput. **8**(2), 239–287 (2009)

54.92 D. Ouelhadj, S. Petrovic: A survey of dynamic scheduling in manufacturing systems, J. Sched. **12**(4), 417–431 (2009)

54.93 V. Pillac, M. Gendreau, C. Guéret, A. L. Medaglia: A review of dynamic vehicle routing problems. Techn. Rep. CIRRELT-2011-62, Interuniversity Research Centre on Enterprise Networks, Logistics and Transportation, Montréal, Canada, October 2011

54.94 V. Maniezzo, T. Stützle, S. Voß (Eds.): *Matheuristics – Hybridizing Metaheuristics and Mathematical Programming*, Annals of Information Systems, Vol. 10 (Springer, New York 2010)

54.95 J. Puchinger, G.R. Raidl, S. Pirkwieser: MetaBoosting: Enhancing integer programming techniques by metaheuristics. In: *Matheuristics – Hybridizing Metaheuristics and Mathematical Programming*, Annals of Information Systems, Vol. 10, ed. by V. Maniezzo, T. Stützle, S. Voß (Springer, New York 2010) pp. 71–102

54.96 M. Fischetti, A. Lodi: Local branching, Math. Program. **98**(1/3), 23–47 (2003)

54.97 E. Danna, E. Rothberg, C. Le Pape: Exploring relaxation induced neighborhoods to improve mip solutions, Math Program. **102**(1), 71–90 (2005)

54.98 V. Maniezzo: Exact and approximate nondeterministic tree-search procedures for the quadratic assignment problem, INFORMS J. Comput. **11**(4), 358–369 (1999)

54.99 C. Blum: Beam-ACO for simple assembly line balancing, INFORMS J. Comput. **20**(4), 618–627 (2008)

54.100 W. Cook, P. Seymour: Tour merging via branch-decomposition, INFORMS J. Comput. **15**(3), 233–248 (2003)

54.101 M.A. Boschetti, V. Maniezzo: Benders decomposition, Lagrangean relaxation and metaheuristic design, J. Heuristics **15**(3), 283–312 (2009)

54.102 I. Dumitrescu, T. Stützle: Usage of exact algorithms to enhance stochastic local search algorithms. In: *Matheuristics – Hybridizing Meta-*

heuristics and Mathematical Programming, Annals of Information Systems, Vol. 10, ed. by V. Maniezzo, T. Stützle, S. Voß (Springer, New York 2010) pp. 103–134

54.103   L. Jourdan, M. Basseur, E.-G. Talbi: Hybridizing exact methods and metaheuristics: A taxonomy, Eur. J. Oper. Res. **199**(3), 620–629 (2009)

54.104   C. Demetrescu, I. Finocchi, G.F. Italiano: Algorithm engineering, Bulletin EATCS **79**, 48–63 (2003)

54.105   I. Sommerville (Ed.): *Software Engineering*, 7th edn. (Addison Wesley, Boston 2004)

54.106   P. Balaprakash, M. Birattari, T. Stützle: Engineering stochastic local search algorithms: A case study in estimation-based local search for the probabilistic traveling salesman problem. In: *Recent Advances in Evolutionary Computation for Combinatorial Optimization*, Studies in Computational Intelligence, Vol. 153, ed. by C. Cotta, J. van Hemert (Springer, Berlin 2008) pp. 55–69

54.107   S. Cahon, N. Melab, E.-G. Talbi: ParadisEO: A framework for the reusable design of parallel and distributed metaheuristics, J. Heuristics **10**(3), 357–380 (2004)

54.108   Paradiseo: A Software Framework for Metaheuristics, http://paradiseo.gforge.inria.fr

54.109   L. Di Gaspero, A. Schaerf: Writing local search algorithms using EASYLOCAL++. In: *Optimization Software Class Libraries*, ed. by S. Voß, D.L. Woodruff (Kluwer, Boston, 2002) pp. 155–175

54.110   Atlassian Bitbucket: https://bitbucket.org/satt/easylocal-3

54.111   P. Van Hentenryck, L. Michel: *Constraint-Based Local Search* (MIT, Cambridge 2005)

54.112   K. Mehlhorn, S. Näher: *LEDA: A Platform for Combinatorial and Geometric Computing* (Cambridge Univ. Press, Cambridge 1999)

54.113   The R Project for Statistical Computing, http://www.r-project.org

54.114   C.W. Nell, C. Fawcett, H.H. Hoos, K. Leyton-Brown: HAL: A framework for the automated design and analysis of high-performance algorithms, Lect. Notes Comput. Sci. **6683**, 600–615 (2011)

54.115   HAL: The High-performance Algorithm Laboratory, http://hal.cs.ubc.ca/

54.116   P. Merz, B. Freisleben: Fitness landscapes and memetic algorithm design. In: *New Ideas in Optimization*, ed. by D. Corne, M. Dorigo, F. Glover (McGraw Hill, London 1999) pp. 244–260

54.117   L. Xu, H. Hoos, K. Leyton-Brown: Hierarchical hardness models for SAT, Lect. Notes Comput. Sci. **4741**, 696–711 (2007)

54.118   J.-P. Watson, L.D. Whitley, A.E. Howe: Linking search space structure, run-time dynamics, and problem difficulty: A step towards demystifying tabu search, J. Artif. Intell. Res. **24**, 221–261 (2005)

54.119   H.H. Hoos: Automated algorithm configuration and parameter tuning. In: *Autonomous Search*, ed. by Y. Hamadi, E. Monfroy, F. Saubion (Springer, Berlin 2012) pp. 37–71

54.120   B. Adenso-Díaz, M. Laguna: Fine-tuning of algorithms using fractional experimental designs and local search, Oper. Res. **54**(1), 99–114 (2006)

54.121   S.P. Coy, B.L. Golden, G.C. Runger, E.A. Wasil: Using experimental design to find effective parameter settings for heuristics, J. Heuristics **7**(1), 77–97 (2001)

54.122   T. Bartz-Beielstein: *Experimental Research in Evolutionary Computation – The New Experimentalism* (Springer, Berlin 2006)

54.123   F. Hutter, H.H. Hoos, K. Leyton-Brown, K.P. Murphy: An experimental investigation of model-based parameter optimisation: SPO and beyond, Genet. Evol. Comput. Conf., GECCO 2009, ed. by F. Rothlauf (ACM, New York 2009) pp. 271–278

54.124   M. Birattari, T. Stützle, L. Paquete, K. Varrentrapp: A racing algorithm for configuring metaheuristics, Proc. Genet. Evol. Comput. Conf. (GECCO-2002), ed. by W.B. Langdon, E. Cantú-Paz, K.E. Mathias, R. Roy, D. Davis, R. Poli, K. Balakrishnan, V. Honavar, G. Rudolph, J. Wegener, L. Bull, M.A. Potter, A.C. Schultz, J.F. Miller, E.K. Burke, N. Jonoska (Morgan Kaufmann, San Francisco 2002) pp. 11–18

54.125   M. Birattari, Z. Yuan, P. Balaprakash, T. Stützle: F-Race and iterated F-Race: An overview. In: *Experimental Methods for the Analysis of Optimization Algorithms*, ed. by T. Bartz-Beielstein, M. Chiarandini, L. Paquete, M. Preuss (Springer, Berlin, Germany 2010) pp. 311–336

54.126   F. Hutter, H.H. Hoos, T. Stützle: Automatic algorithm configuration based on local search, Proc. 22nd Conf. Artif. Intell. (AAAI), ed. by R.C. Holte, A. Howe (AAAI / The MIT, Menlo Park 2007) pp. 1152–1157

54.127   C. Ansótegui, M. Sellmann, K. Tierney: A gender-based genetic algorithm for the automatic configuration of algorithms, Proc. 15th Int. Conf. Princ. Pract. Constraint Program. (CP 2009) (2009) pp. 142–157

54.128   F. Hutter, H.H. Hoos, K. Leyton-Brown, T. Stützle: Param ILS: An automatic algorithm configuration framework, J. Artif. Intell. Res. **36**, 267–306 (2009)

54.129   F. Hutter, H.H. Hoos, K. Leyton-Brown: Sequential model-based optimization for general algorithm configuration, Lect. Notes Comput. Sci. **6683**, 507–523 (2011)

54.130   F. Hutter, H.H. Hoos, K. Leyton-Brown: Parallel algorithm configuration, Lect. Notes Comput. Sci. **7219**, 55–70 (2011)

54.131   R. Battiti, M. Brunato, F. Mascia: *Reactive Search and Intelligent Optimization*, Operations Research/Computer Science Interfaces Series, Vol. 45 (Springer, New York 2008)

54.132   A.E. Eiben, Z. Michalewicz, M. Schoenauer, J.E. Smith: Parameter control in evolutionary algorithms. In: *Parameter Setting in Evolutionary Algorithms*, ed. by F. Lobo, C.F. Lima, Z. Michalewicz (Springer, Berlin, Germany 2007) pp. 19–46

54.133 F. Hutter, Y. Hamadi, H.H. Hoos, K. Leyton-Brown: Performance prediction and automated tuning of randomized and parametric algorithms, Lect. Notes Comput. Sci. **4204**, 213–228 (2006)

54.134 L. Xu, H.H. Hoos, K. Leyton-Brown: Hydra: Automatically configuring algorithms for portfolio-based selection, Proc. 24th AAAI Conf. Artif. Intell. (AAAI-10) (2010) pp. 210–216

54.135 S. Kadioglu, Y. Malitsky, M. Sellmann, K. Tierney: ISAC – Instance-specific algorithm configuration, Proc. 19th Eur. Conf. Artif. Intell. (ECAI 2010) (2010) pp. 751–756

54.136 H.H. Hoos: Computer-aided algorithm design using generalised local search machines and related design patterns. Techn. Rep. TR-2009-26, University of British Columbia, Department of Computer Science, 2009

55. Parallel Evolutionary Combinatorial Optimization

El-Ghazali Talbi

In this chapter, a clear difference is made between the parallel design aspect and the parallel implementation aspect of evolutionary algorithms (EAs). From the algorithmic design point of view, the main parallel models for EAs are presented. A unifying view of parallel models for EAs is outlined. This chapter is organized as follows. In Sect. 55.2, the main parallel models for designing EAs are presented. Section 55.3 deals with the implementation issues of parallel EAs. In this section, the main concepts of parallel architectures and parallel programming paradigms, which interfere with the design and implementation of parallel EAs, are outlined. The main performance indicators that can be used to evaluate a parallel EAs in terms of efficiency are detailed. Finally, Sect. 55.4 deals with the design and implementation of different parallel models for EAs based on the software framework ParadisEO.

55.1 Motivation

On one hand, optimization problems are more and more complex and their resource requirements to solve them are ever increasing. Real-life optimization problems are often NP-hard, and CPU time, and/or memory consuming. Although the use of evolutionary algorithms (EAs) allows us to significantly reduce the computational complexity of the solving algorithm, the latter remains time-consuming for many problems in diverse domains of application, where the objective function and the constraints associated with the problem are resource (e.g., CPU, memory) intensive and the size of the search space is huge. Moreover, more and more complex and resource intensive EAs are developed (e.g., hybrid EAs, multiobjective EAs) [55.1].

On the other hand, the rapid development of technology in designing processors (e.g. multicore processors, dedicated architectures), networks (local networks (LAN) such as Myrinet and Infiniband or wide area networks (WAN) such as optical networks), and data storage make the use of parallel computing more and more popular. Such architectures represent an effective opportunity for the design and implementation of parallel EAs. Indeed, sequential architectures are reaching physical limitations (speed of light, thermodynamics). Nowadays, even laptops and workstations are equipped with multicore processors, which represent one class of parallel architecture. Moreover, the ratio cost/performance is constantly decreasing. The proliferation of powerful workstations and fast communication networks have shown the emergence of dedicated architectures (e.g., GPUs), clusters of processors (COPs), networks of workstations (NOWs), and large-scale networks of machines (Grids) as platforms for high-performance computing.

Parallel and distributed computing can be used in the design and implementation of EAs for the following reasons:

- *Speedup the search*: One of the main goals in parallelizing an EA is to reduce the search time. This helps designing real time and interactive optimization methods. This is a very important aspect for some class of problems where there are hard requirements on search time such as in dynamic optimization problems and time-critical control problems such as *real-time* planning.
- *Improve the quality of the obtained solutions*: Some parallel models for EAs allow us to improve the quality of solutions. Indeed, exchanging information between algorithms will alter their behavior in terms of searching in the landscape associated with the problem. The main goal in the cooperation between algorithms is to improve the quality of solutions. Both convergence to better solutions and reduced search time may happen. Let us note that a parallel model for EAs may be more effective than a sequential algorithm even on a single processor [55.2].
- *Improve the robustness*: A parallel EA may be more robust in terms of solving in an effective manner different optimization problems and different instances of a given problem. Robustness may be measured in terms of the sensitivity of the algorithm to its parameters.

- *Solve large-scale problems*: Parallel EAs allow to solve large-scale instances of complex optimization problems. A challenge here is to solve very large instances that cannot be solved on a sequential machine. Another similar challenge is to solve more accurate mathematical models associated with different optimization problems. Improving the accuracy of mathematical models increases in general the size of the associated problems to be solved. Moreover, some optimization problems need the manipulation of huge databases such as data mining problems.

The implementation point of view deals with the efficiency of a parallel EAs on a target parallel architecture using a given parallel language, programming environment, or middleware. The focus is on the parallelization of EAs on general-purpose parallel and distributed architectures, since this is the most widespread computational platform. This chapter also deals with the implementation of EAs on dedicated architectures such as reconfigurable architectures and GPUs (graphical processing units). Different architectural criteria, which affect the efficiency of the implementation, will be considered: shared memory versus distributed memory, homogeneous versus heterogeneous, shared versus nonshared by multiple users, local network versus large network. Indeed, those criteria have a strong impact on the deployment technique employed such as load balancing and fault tolerance. Depending on the type of parallel architecture used, different parallel and distributed languages, programming environments, and middlewares may be used such as message passing (e.g., MPI), shared memory (e.g., multithreading, OpenMP, CUDA), remote procedural call (e.g., Java RMI, RPC), high-throughput computing (e.g., Condor), and grid computing (e.g., Globus).

55.2 Parallel Design of EAs

In terms of designing parallel EAs, three major parallel models are identified. They follow the following three hierarchical levels (Table 55.1):

- *Algorithmic level*: In this model, independent or cooperating self-contained EAs are used. It is a problem-independent interalgorithm parallelization. If the different EAs are independent, the search will be equivalent to the sequential execution of the

algorithms in terms of the quality of solutions. However, the cooperative model will alter the behavior of the EAs and enable the improvement in terms of the quality of solutions.

- *Iteration level*: In this model, each iteration of an EA is parallelized. It is a problem-independent intra-algorithm parallelization. The behavior of the EA is not altered. The main objective is to speedup the algorithm by reducing the search time. Indeed,

Table 55.1 Parallel models of EAs

Parallel model	Problem dependency	Behavior	Granularity	Goal
Algorithmic level	Independent	Altered	EA	Effectiveness
Iteration level	Independent	Nonaltered	Iteration	Efficiency
Solution level	Dependent	Nonaltered	Solution	Efficiency

the iteration cycle of EAs on large populations, especially for real-world problems, requires a large amount of computational resources.

- *Solution level*: In this model, the parallelization process handles a single solution of the search space. It is a problem-dependent intra-algorithm parallelization. In general, evaluating the objective function(s) or constraints for a generated solution is frequently the most costly operation in EAs. In this model, the behavior of the EA is not altered. The objective is mainly the speedup of the search.

In the following sections, different parallel models are detailed and analyzed in terms of algorithmic design.

55.2.1 Algorithmic–Level Parallel Model

In this model, many EAs are launched in parallel. They may cooperate or not to solve the target optimization problem.

Independent Algorithmic–Level Parallel Model
In the *independent* algorithmic-level parallel model, different EAs are executed without any cooperation. The different EAs may be initialized with different populations. Different parameter settings may be used for the EAs such as the mutation and crossover probabilities, etc. Moreover, each search component of an EA may be designed differently: encoding, search operators (e.g., variation operators), objective function, constraints, stopping criteria, etc.

This parallel model is straightforward to design and implement. The master/worker paradigm is well suited to this model. A worker implements an EA. The master defines different parameters to use by the workers and determines the best found solution from those obtained by different workers. In addition to speeding up the algorithm, this parallel model enables us to improve its robustness [55.3].

This model raises particularly the following question: Is it equivalent to execute k EAs during a time t and to execute a single EA during $k * t$? The answer depends on the landscape properties of the problem (e.g., the presence of multiple basins of attraction,

distribution of the local optima, and fitness distance correlation) [55.4].

Cooperative Algorithmic–Level Parallel Model
In the *cooperative* model for parallel EAs, different algorithms are exchanging informations related to the search with the intent to compute better and more robust solutions.

In designing a parallel cooperative model for any EA, the same design questions need to be answered:

- *The exchange decision criterion (When?)*: The exchange of information between the EAs can be decided either in a *blind* (periodic or probabilistic) way or according to an *intelligent* adaptive criterion. Periodic exchange occurs in each algorithm after a fixed number of iterations; this type of communication is synchronous. Probabilistic exchange consists in performing a communication operation after each iteration with a given probability. Conversely, adaptive exchanges are guided by some run-time characteristics of the search. For instance, it may depend on the evolution of the quality of the solutions or the search memory. A classical criterion is related to the improvement of the best found local solution.
- *The exchange topology (Where?)*: The communication exchange topology indicates for each EA its neighbor(s) regarding the exchange of information, i. e., the source/destination algorithm(s) of the information. Several works have been dedicated to the study of the impact of the topology on the quality of the provided results, and show that cyclic graphs are better [55.5, 6]. The ring, mesh, and hypercube regular topologies are often used.
- *The information exchanged (What?)*: This parameter specifies the information to be exchanged between the EAs. In general, this information can be composed of:
  - *Solutions*: This information deals with a selection of the solutions found during the search. In general, it contains elite solutions that have been found such as the best solution at the current iteration, local best solutions, global best solution, neighborhood best solution, best diversi-

a) Parallel insular model for EAs   **b)** Parallel cellular model for EAs

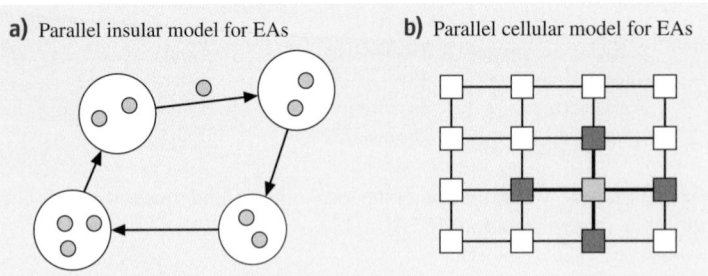

Fig. 55.1a,b The traditional parallel (**a**) island and (**b**) cellular models for evolutionary algorithms

fied solutions, and randomly selected solutions. The number of solutions to exchange may be an absolute value or a given percentage of the population. Any selection mechanism can be used to select the solutions.

- *Search memory*: This information deals with any element of the search memory that is associated with the involved EA.

● *The integration policy* (*How?*): Analogously to the information exchange policy, the integration policy deals with the usage of the received information. In general, there is a local copy of the received information. The local variables are updated using the received ones. For instance, the best found solution is simply updated by the best between the local best solution and the neighboring best solution. Any replacement strategy may be applied on the local population by the set of received solutions.

Traditional Parallel Models for EAs. Historically, the cooperative parallel model has been largely used in EAs [55.7]. In sequential genetic algorithms (the sequential model is known as the panmictic genetic algorithm), the selection takes place globally. Any individual can potentially reproduce with any other individual of the population. Among the best-known parallel algorithmic-level models for evolutionary algorithms are the island model and the cellular model. In the island model (also known as the migration model, distributed model, multideme EA, or coarse-grained EA) for genetic algorithms, the population is decomposed into several subpopulations distributed among different nodes (Fig. 55.1). Each node is responsible of the evolution of one subpopulation. It executes all the steps of a classical EA from the selection to the replacement on the subpopulation. Each island may use different parameter values and different strategies for any search component such as selection, replacement, variation operators (mutation, crossover), and encodings. After a given number of generations (synchronous exchange),

or when a condition holds (asynchronous exchange), the migration process is activated. Then, exchanges of some selected individuals between subpopulations are realized, and received individuals are integrated into the local subpopulation. The selection policy of emigrants indicates for each island in a *deterministic* or *stochastic* way the individuals to be migrated. The stochastic or random policy does not guarantee that the best individuals will be selected, but its associated computation cost is relatively lower. The deterministic strategy allows the selection of the best individuals. The number of emigrants can be expressed as a fixed or variable number of individuals, or through a percentage of individuals from the population. The choice of the value of such parameter is crucial. Indeed, if the number of emigrants is low, the migration process will be less efficient as the islands will have the tendency to evolve in an independent way. Conversely, if the number of emigrants is high, the EAs will likely converge to the same solutions [55.8]. In EAs, the replacement/integration policy of immigrants indicates in a stochastic or deterministic way the local individuals to be replaced by the newcomers. The objective of the model is to delay the global convergence and encourage diversity [55.9, 10].

The other well-known parallel model for EAs, the cellular model (also known as the diffusion or fine-grained model), may be seen as a special case of the island model where an island is composed of a single individual. Traditionally, an individual is assigned to a cell of a grid. The selection occurs in the neighborhood of the individual [55.11–13]. Hence, the selection pressure is less important than in sequential EAs. The overlapped small neighborhood in cellular EAs helps exploring the search space because a slow diffusion of solutions through the population provides a kind of exploration, while exploitation takes place inside each neighborhood. Cellular models applied to complex problems can have a higher convergence probability to better solutions than panmictic EAs [55.14, 15].

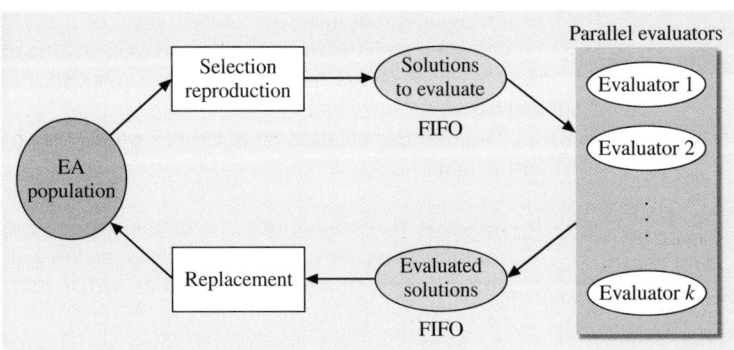

Parallel evaluators

Fig. 55.2 Parallel asynchronous evaluation of a population

55.2.2 Iteration–Level Parallel Model

In this parallel model, a focus is made on the parallelization of each iteration of EAs. The iteration-level parallel model is generally based on the distribution of the handled solutions. Indeed, the most resource-consuming part in an EA is the evaluation of the generated solutions. Our concerns in this model are only search mechanisms that are problem-independent operations such as the generation of successive populations. Any *search operator* of an EA which is not specific to the tackled optimization problem is involved in the iteration-level parallel model. This model keeps the sequentiality of the original algorithm, and, hence, the behavior of the EA is not altered.

Parallel iteration level models arise naturally when dealing with EAs, since each element belonging to the population is an independent unit. The iteration-level parallel model involves the distribution of the population. The operations commonly applied to each of the population elements are performed in parallel.

The population of individuals can be decomposed and handled in parallel. In the beginning of the parallelization of EAs the well-known *master-worker* (also known as *global parallelization*) method was used. In this scheme, a master performs the selection operations and the replacement. The selection and replacement are generally sequential procedures, as they require a global management of the population. The associated workers perform the recombination, mutation and the evaluation of the objective function. The master sends the partitions (subpopulations) to the workers. The workers return back newly evaluated solutions to the master.

According to the order in which the evaluation phase is performed in comparison with the other parts of the EA, two modes can be distinguished:

- *Synchronous*: In the synchronous mode, the worker manages the evolution process and performs in a serial way the different steps of selection and replacement. At each iteration, the master distributes the set of new generated solutions among the workers and waits for the results to be returned back. After the results are collected, the evolution process is restarted. The model does not change the behavior of the EA compared to a sequential model. The synchronous execution of the model is always synchronized with the return back of the last evaluated solution.
- *Asynchronous*: In the asynchronous mode, the evaluation phase is not synchronized with the other parts of the EA. The worker does not wait for the return of all evaluations to perform the selection, reproduction, and replacement steps. The *steady-state* EA is a good example illustrating the asynchronous model and its advantages. In the asynchronous model applied to a steady-state EA, the recombination and the evaluation steps may be done concurrently. The master manages the evolution engine and two queues of individuals of a given fixed size: individuals to be evaluated, and solutions being evaluated. The individuals of the first queue wait for a free evaluating node. When the queue is full the process blocks. The individuals of the second queue are assimilated into the population as soon as possible (Fig. 55.2). The reproduced individuals are stored in a FIFO data structure, which represents the individuals to be evaluated. The EA continues its execution in an asynchronous manner, without waiting for the results of the evaluation phase. The selection and reproduction phase are carried out until the queue of nonevaluated individuals is full. Each evaluator agent picks an individual from the data structure, evaluates it, and stores the results into another data structure storing the evaluated in-

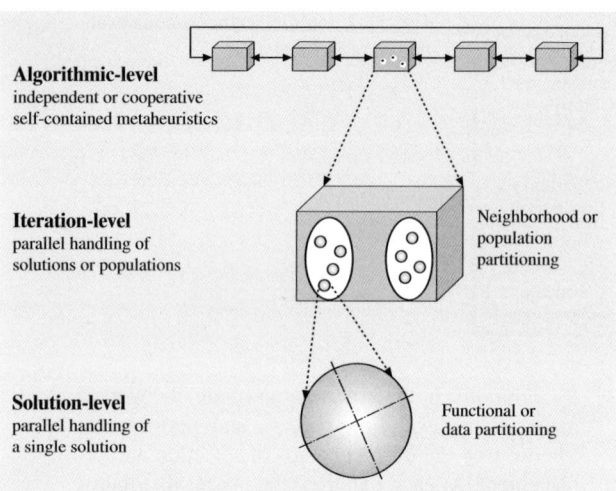

Algorithmic-level
independent or cooperative
self-contained metaheuristics

Iteration-level
parallel handling of
solutions or populations

Neighborhood or
population
partitioning

Solution-level
parallel handling of
a single solution

Functional or
data partitioning

Fig. 55.3 Combination of the three parallel hierarchical models of EAs

dividuals. The order of evaluation defined by the selection phase may not be the same as in the replacement phase. The replacement phase consists in receiving, in a synchronous manner, the results of the evaluated individuals, and applying a given replacement strategy of the current population.

In some EAs (e.g., blackboard-based ones) some information must be shared. For instance, in ant colony optimization (ACO), the pheromone matrix must be shared by all ants. The master has to broadcast the pheromone trails to each worker. Each worker handles an ant process. It receives the pheromone trails, constructs a complete solution, and evaluates it. Finally, each worker sends back to the master the constructed and evaluated solution. When the master receives all the constructed solutions, it updates the pheromone trails [55.16–19].

55.2.3 Solution–Level Parallel Model

In this model, problem-dependent operations performed on solutions are parallelized. In general, the interest here is the parallelization of the evaluation of a single solution (also called acceleration move parallel model; objective and/or constraints). This model is particularly interesting when the objective function or the constraints are time and/or memory consuming,

and/or input/output intensive. Indeed, most of real-life optimization problems need the intensive calculation of the objectives and/or the access to large input files or databases.

Two different solution-level parallel models may be carried out:

- *Functional decomposition*: In functional oriented parallelization, the objective function(s) and/or constraints are partitioned into different partial functions. The objective function(s) or the constraints are viewed as the aggregation of some partial functions. Each partial function is evaluated in parallel. Then, a reduction operation is performed on the results returned back by the computed partial functions. By definition, this model is synchronous, so one has to wait the termination of all workers calculating the partial functions.
- *Data partitioning*: For some problems, the objective function may require the access to a huge database that could not be managed on a single machine. Due to a memory requirement constraint, the database is distributed among different sites, and data parallelism is exploited in the evaluation of the objective function. In data-oriented parallelization, the same identical function is computed on different partitions of the input data of the problem. The data is then partitioned or duplicated over different workers.

In the solution-level parallel model, the maximum number of parallel operations will be equal to the number of partial functions or the number of data partitions. A hybrid model can also be used in which a functional decomposition and a data partitioning are combined.

55.2.4 Hierarchical Combination of the Parallel Models

The three presented models for parallel EAs may be used in conjunction within a hierarchical structure [55.20, 21] (Fig. 55.3). The parallelism degree associated with this hybrid model is very important. Indeed, this hybrid model is very scalable; the degree of concurrency is $k * m * n$, where k is the number of EAs used, m is the size of the population, and n is the number of partitions or tasks associated with the evaluation of a single solution.

55.3 Parallel Implementation of EAs

Parallel implementation of EAs deals with the efficient mapping of a parallel model of EAs on a given parallel architecture.

55.3.1 Parallel and Distributed Architectures

Parallel architectures are evolving quickly. The main criteria of parallel architectures, which will have an impact on the implementation of parallel EAs, are: memory sharing, homogeneity of resources, resource sharing by multiple users, scalability, and volatility (Fig. 55.4). Those criteria will be used to analyze different parallel models and their efficient implementation. A guideline is given for the efficient implementation of each parallel model of EAs according to each class of parallel architectures.

Shared Memory/Distributed Memory Architectures. In shared memory parallel architectures, the processors are connected by a shared memory. There are different interconnection schemes for the network (e.g., bus, crossbar, multistage crossbar). This architecture is easy to program. Conventional operating systems and programming paradigms of sequential programming can be used. There is only one address space for data exchange but the programmer must take care of synchronization in memory access, such as the mutual exclusion in critical sections. This type of architecture has a poor scalability (from 2 to 128 processors in current technologies) and a higher cost. An example of such shared memory architectures are symmetric multiprocessors (SMPs) machines and multicore processors.

In distributed memory architectures, each processor has its own memory. The processors are connected by a given interconnection network using different topologies (e.g., hypercube, 2D or 3D torus, fat-tree, and multistage crossbars). This architecture is harder to program; data and/or tasks have to be explicitly distributed to processors. Exchanging information is also explicitly handled using message passing between nodes (synchronous or asynchronous communications). The cost of communication is not negligible and must be minimized to design an efficient parallel EA. However, this architecture has a good scalability in terms of the number of processors. In recent years, clusters of processors (COWs) became one of the most popular parallel distributed memory architectures. A good ratio between cost and performance is obtained with this class of architectures.

Homogeneous/Heterogenous Parallel Architectures. Parallel architectures may be characterized by the homogeneity of the used processors, communication networks, operating systems, etc. For instance, COWs are in general homogeneous parallel architectures. The proliferation of powerful workstations and fast communication networks have shown the emergence of heterogeneous networks of workstations (NOWs) as platforms for high-performance computing. This type of architecture is present in any laboratory, company, campus, institution, etc. These parallel platforms are generally composed of an important number of owned heterogeneous workstations shared by many users.

Shared/Nonshared Parallel Architectures. Most massively parallel machines (MPP) and clusters of workstations (COWs) are generally nonshared by the applications. Indeed, at a given time, the processors composing those architectures are dedicated to the

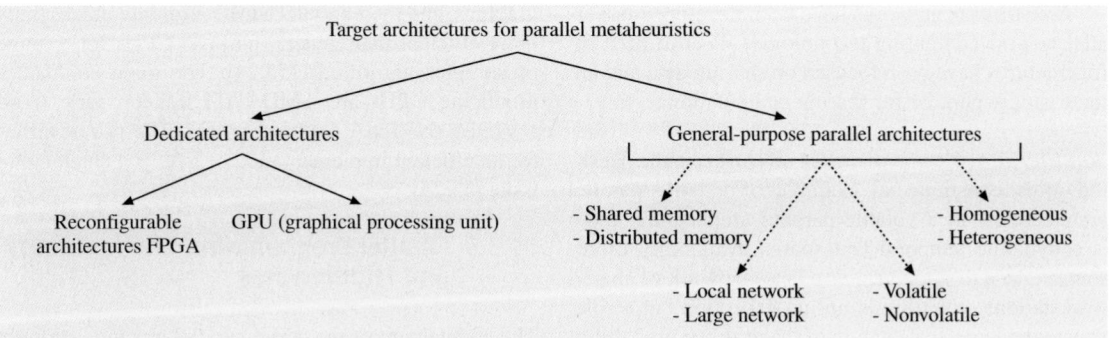

Fig. 55.4 Hierarchical and flat classification of target parallel architectures for EAs

Part E | 55.3

Table 55.2 Characteristics of the main parallel architectures. Hom: Homogeneous, Het: Heterogeneous

Criteria	Memory	Homogeneity	Sharing	Network	Volatility
SMP Multicore	Shared	Hom	Yes or No	Local	No
COW	Distributed	Hom or Het	No	Local	No
NOW	Distributed	Het	Yes	Local	Yes
HPC Grid	Distributed	Het	No	Large	No
Desktop grid	Distributed	Het	Yes	Large	Yes

execution of a single application. NOWs constitute a low-cost hardware alternative to run parallel algorithms but are in general shared by multiple users and applications.

Local Network (LAN)/Wide–Area Network (WAN). Massively parallel machines, clusters, and local networks of workstations may be considered as tightly coupled architectures. Large networks of workstations and grid computing platforms are loosely coupled and are affected by a higher cost of communication. During the last decade, *grid computing* systems have been largely deployed to provide high-performance computing platforms. A computational grid is a scalable pool of heterogeneous and dynamic resources geographically distributed across multiple administrative domains and owned by different organizations [55.22]. Two types of grids may be distinguished:

● *High-Performance Computing Grid* (HPC grid): This grid interconnect supercomputers or clusters via a dedicated high-speed network. In general, this type of grid is nonshared by multiple users (at the level of processors).
● *Desktop Grid*: This class of grids is composed of numerous owned workstations connected via nondedicated network such as the internet. This grid is volatile and shared by multiple users and applications.

Peer-to-peer networks have been developed in parallel to grid computing technologies. Peer-to-peer infrastructures have been focused on sharing data and are increasingly popular for sharing computation.

Volatile/Nonvolatile Parallel Architectures. Desktop grids constitute an example of volatile parallel architectures. In a volatile parallel architecture, there is a dynamic temporal and spatial availability of resources. In a desktop grid or a large network of shared workstations, volatility is not an exception but a rule. Due to the large-scale nature of the grid, the probability of resource failure is high. For instance, desktop grids

have a faulty nature (e.g., reboot, shutdown, and failure).

Table 55.2 recapitulates the characteristics of the main parallel architectures according to the presented criteria. Those criteria will be used to analyze the efficient implementation of the different parallel models of EAs.

55.3.2 Dedicated Architectures

Dedicated hardware represents programmable hardware or specific architectures that can be designed or reused to execute a parallel EA. The best-known dedicated hardware is represented by field programmable gate arrays (FPGA) and GPU (Fig. 55.4).

FPGAs are hardware devices that can be used to implement digital circuits by means of a programming process (do not confuse with evolvable hardware where the architecture is reconfigured using EAs) [55.23]. The use of the Xilinx's FPGAs to implement different EAs is more and more popular. The design and the prototyping of a FPGA-based hardware board to execute parallel EAs may restrict the design of some search components. However, for some specific challenging optimization problems with a high use rate such as in bioinformatics, dedicated hardware may be a good alternative.

GPU is a dedicated graphics rendering device for a workstation, personal computer, or game console. Recent GPUs are very efficient at manipulating computer graphics, and their parallel SIMD structure makes them more efficient than general-purpose CPUs for a range of complex algorithms [55.24]. The main companies producing GPUs are AMD (ATI Radeon series) and NVIDIA (NVIDIA Geforce series). The use of GPUs for an efficient implementation of EAs is a challenging issue [55.25].

55.3.3 Parallel Programming Environments and Middlewares

The architecture of the target parallel machine strongly influences the choice of the parallel programming

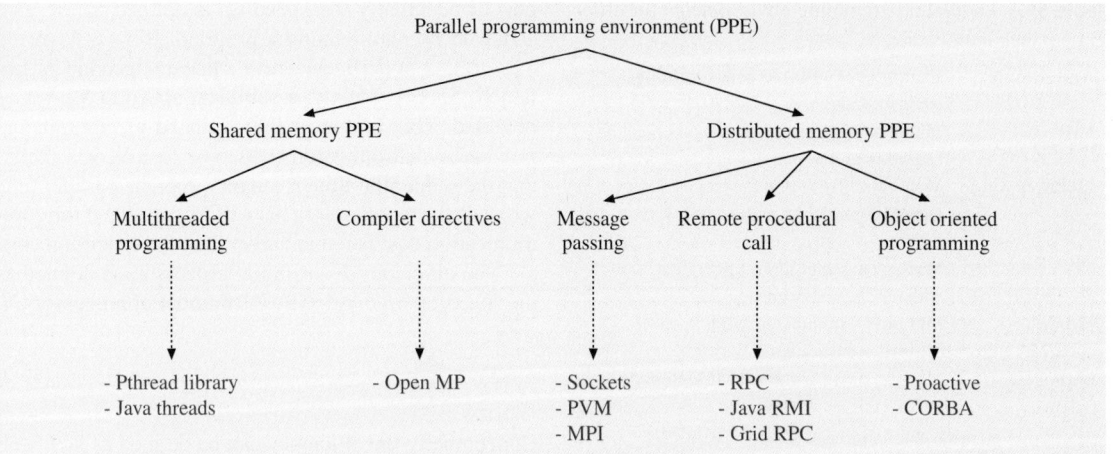

Fig. 55.5 Main parallel programming languages, programming environments and middlewares

model to use. There are two main parallel programming paradigms: shared memory and message passing (Fig. 55.5).

Two main alternatives exist to program shared memory architectures:

- *Multithreading*: A thread may be viewed as a lightweight process. Different threads of the same process share some resources and the same address space. The main advantages of multithreading are the fast context switch, the low resource usage, and the possible recovery between communication and computation. Each thread can be executed on a different processor or core. Multithreaded programming may be used within libraries such as the standard Pthreads library [55.26] or programming languages such as Java threads [55.27].
- *Compiler directives*: One of the standard shared memory paradigms is OpenMP and CUDA. It represents a set of compiler directives interfaced with the languages Fortran, C and C++ [55.28]. Those directives are integrated in a program to specify which sections of the program to be parallelized by the compiler.

Distributed memory parallel programming environments are based mainly on the following three paradigms:

- *Message passing*: Message passing is probably the most widely used paradigm to program parallel architectures. In the message passing paradigm, processes of a given parallel program communicate by exchanging messages in a synchronous or asynchronous way. The well-known programming environments based on message passing are sockets and message passing interface (MPI).
- *Remote Procedure Call*: Remote procedure call (RPC) represents a traditional way of programming parallel and distributed architectures. It allows a program to cause a procedure to execute on another processor.
- *Object-oriented models*: As in sequential programming, parallel object oriented programming is a natural evolution of RPC. A classical example of such a model is Java RMI (Remote Method Invocation).

In the last decade, great work has been carried out on the development of grid middlewares. The Globus toolkit represents the *de facto* standard grid middleware. It supports the development of distributed service-oriented computing applications [55.29].

It is not easy to propose a guideline on which environment to use in programming a parallel EA. It will depend on the target architecture, the parallel model of EAs, and the user preferences. Some languages are more system oriented such as C and C++. More portability is obtained with Java but the price is less efficiency. This tradeoff represents the classical efficiency/portability compromise. A Fortran programmer will be more comfortable with OpenMP. RPC models are more adapted to implement services. Condor represents an efficient and easy way to implement parallel programs on shared and volatile distributed architectures such as large networks of heterogeneous workstations and desktop grids, where fault tolerance is

Table 55.3 Parallel programming environments for different parallel architectures

Architecture	Examples of suitable programming environment
SMP	Multithreading library within an operating system (e.g., Pthreads)
Multicore	Multithreading within languages: Java OpenMP interfaced with C, C++ or Fortran
COW	Message passing library: MPI interfaced with C, C++, Fortran
Hybrid ccNUMA	MPI or Hybrid models: MPI/OpenMP, MPI/Multithreading
NOW	Message passing library: MPI interfaced with C, C++, Fortran Condor or object models (JavaRMI)
HPC grid	MPICH-G (Globus) or GridRPC models (Netsolve, Diet)
Desktop grid	Condor-G or object models (Proactive)

ensured by a checkpoint/recovery mechanism. The use of MPI within Globus is more or less adapted to high-performance computing (HPC) grids. However, the user has to deal with complex mechanisms such as dynamic load balancing and fault tolerance. Table 55.3 presents a guideline depending on the target parallel architecture.

55.3.4 Performance Evaluation

For sequential algorithms, the main performance measure is the execution time as a function of the input size. In parallel algorithms, this measure also depends on the number of processors and the characteristics of the parallel architecture. Hence, some classical performance indicators such as speedup and efficiency have been introduced to evaluate the scalability of parallel algorithms [55.30]. The scalability of a parallel algorithm measures its ability to achieve performance proportional to the number of processors.

The speed-up S_N is defined as the time T_1 it takes to complete a program with one processor divided by the time T_N it takes to complete the same program with N processors

$$S_N = \frac{T_1}{T_N} \,. \tag{55.1}$$

One can use *wall-clock time* instead of *CPU time*. The CPU time is the time a processor spends in the execution of the program, and the wall-clock time is the time of the whole program including the input and out-

put. Conceptually the speed-up is defined as the gain achieved by parallelizing a program. If $S_N > N$ (resp., $S_N = N$), a superlinear (resp., linear) speedup is obtained [55.14]. Mostly, a sublinear speedup $S_N < N$ is obtained. This is due to the overhead of communication and synchronization costs. The case $S_N < 1$ means that the sequential time is smaller than the parallel time which is the worst case. This will be possible if the communication cost is much higher than the execution cost.

The efficiency E_N using N processors is defined as the speed-up S_N divided by the number of processors N.

$$E_N = \frac{S_N}{N} \tag{55.2}$$

Conceptually the efficiency can be defined as how well N processors are used when the program is computed in parallel. An efficiency of 100% means that all of the processors are fully used all the time. For some large real-life applications, it is impossible to have the sequential time as the sequential execution of the algorithm cannot be performed. Then, the incremental efficiency E_{NM} may be used to evaluate the efficiency extending the number of processors from N to M processors

$$E_{NM} = \frac{N \times E_N}{M \times E_M} \,. \tag{55.3}$$

Different definitions of speedup may be used depending on the definition of the sequential time reference T_1. Asking what is the best measure is useless; there is no global dominance between the different measures. The choice of a given definition depends on the objective of the performance evaluation analysis. Then, it is important to specify clearly the choice and the objective of the analysis.

The *absolute speedup* is used when the sequential time T_1 corresponds to the best-known sequential time to solve the problem. Unlike other scientific domains such as numerical algebra where for some operations the best sequential algorithm is known, in EA search, it is difficult to identify the best sequential algorithm. So, the absolute speedup is rarely used. The *relative speedup* is used when the sequential time T_1 corresponds to the parallel program executed on a single processor.

Moreover, different stopping conditions may be used:

• *Fixed number of iterations*: This condition is the most used to evaluate the efficiency of a parallel EA.

Using this definition, a superlinear speedup is possible $S_N > N$. This is due to the characteristics of the parallel architecture where there is more resources (e.g. size of main memory and cache) than in a single processor (Fig. 55.6a). For instance, the search memory of an EA executed on a single processor may be larger than the main memory of a single processor and then some swapping will be carried out, which represents an overhead in the sequential time. When using a parallel architecture, the whole memory of the EA may fit in the main memory of its processors, and then the memory swapping overhead will not occur.

- *Convergence to a solution with a given quality*: This measure is interesting to evaluate the effectiveness of a parallel EA. It is only valid for parallel models of EAs based on the algorithmic level, which alters the behavior of the sequential EA. A superlinear speedup is possible and is due to the characteristics of the parallel search (Fig. 55.6b). Indeed, the order of searching different regions of the search space may be different from sequential search. The sequences of visited solutions in parallel and sequential search are different. This is similar to the superlinear speedups obtained in exact search algorithms such as branch and bound (this phenomenon is called speedup anomaly) [55.31].

Most of evolutionary algorithms are stochastic algorithms (scatter search, if considered as an evolutionary algorithm, is a deterministic algorithm). When the stopping condition is based on the quality of the solution, one cannot use the speedup metric as defined previously. The original definition may be extended to the average speedup

$$S_N = \frac{E(T_1)}{E(T_N)} . \tag{55.4}$$

The same *seed* for the generation of random numbers must be used for a more fair experimental performance evaluation.

The speedup metrics have to be reformulated for heterogeneous architectures. The efficiency metric may be used for this class of architectures. Moreover, it can be used for shared parallel machines with multiple users.

55.3.5 Main Properties of Parallel EAs

The performance of a parallel EA on a given parallel architecture depends mainly on its *granularity*.

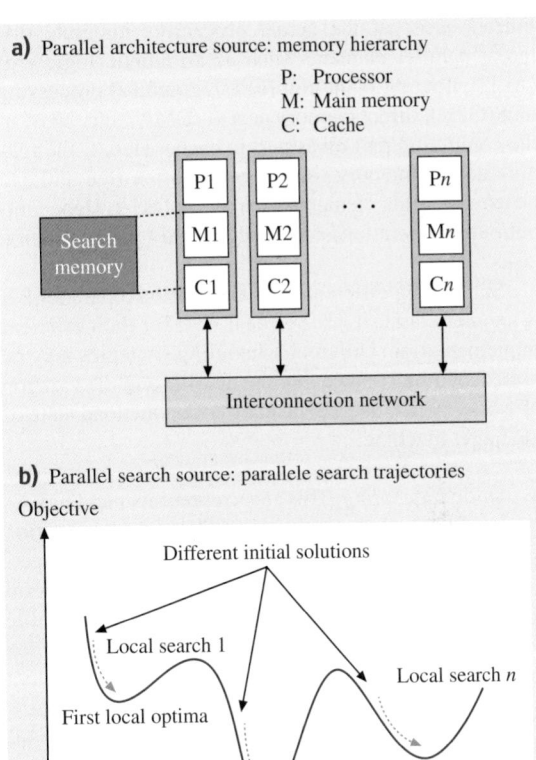

a) Parallel architecture source: memory hierarchy

P: Processor
M: Main memory
C: Cache

b) Parallel search source: parallele search trajectories

Fig. 55.6a,b Superlinear speedups for a parallel EA. **(a)** Parallel architecture source. **(b)** Parallel search source

The granularity of a parallel program is the amount of computation performed between two communications. It computes the ratio between the computation time and the communication time. The three parallel models (algorithmic level, iteration level, and solution level) have a decreasing granularity from coarse-grained to fine-grained. The granularity indicator has an important impact on the speedup. The larger is the granularity the better is the obtained speedup.

The *degree of concurrency* of a parallel EA is represented by the maximum number of parallel processes at any time. This measure is independent from the target parallel architecture. It is an indication of the number of processors that can employed usefully by the parallel EA.

Asynchronous communications and the recovery between computation and communication is also an important issue for a parallel efficient implementation.

Indeed, most of the actual processors integrate different parallel elements such as arithmetic logic unit (ALU), floating point unit (FPU), graphical processing unit (GPU), direct memory access (DMA), etc. Most of the computing part takes part in cache. Hence, the random access memory (RAM) bus is often free and can be used by other elements such as the DMA. Hence, input/output operations can be recovered by computation tasks.

Scheduling different tasks composing a parallel EA is another classical issue to deal with for their efficient implementation. Different scheduling strategies may be used depending on whether the number and the location of works (tasks, data) depend or not on the load state of the target machine:

- *Static scheduling*: This class represents parallel EAs in which both the number of tasks of the application and the location of work (tasks, data) are generated at compile time. Static scheduling is useful for homogeneous, and nonshared and nonvolatile heterogeneous parallel architectures. Indeed, when there are noticeable load or power differences between processors, the search time of an iteration is derived by the maximum execution time over all processors, presumably on the most highly loaded processor or the least powerful processor. A significant number of tasks are often idle waiting for other tasks to complete their work.
- *Dynamic scheduling*: This class represents parallel EAs for which the number of tasks is fixed at compile time, but the location of work is determined and/or changed at run time. The tasks are dynamically scheduled on different processors of the parallel architecture. Dynamic load balancing is important for shared (multiuser) architectures, where the load of a given processor cannot be determined at compile time. Dynamic scheduling is also important for *irregular* parallel EAs in which the execution time cannot be predicted at compile time and varies during the search. For instance, this happens when the evaluation cost of the objective function depends on the solution.

  Many dynamic load-balancing strategies may be applied. For instance, during the search, each time a processor finishes its work, it proceeds to a work-demand. The degree of parallelism of this class of scheduling algorithms is not related to load variations in the target machine. When the number of tasks exceeds the number of idle nodes, multiple tasks are assigned to the same node. Moreover,

when there are more idle nodes than tasks, some of them will not be used.
- *Adaptive scheduling*: Parallel adaptive algorithms are parallel computations with a dynamically changing set of tasks. Tasks may be created or killed as a function of the load state of the parallel machine. A task is created automatically when a node becomes idle. When a node becomes busy, the task is killed. Adaptive load balancing is important for volatile architectures such as desktop grids.

For some parallel and distributed architectures such as shared networks of workstations and grids, fault tolerance is an important issue. Indeed, in volatile shared architectures and large-scale parallel architectures, the fault probability is relatively important. Checkpointing and recovery techniques constitute one answer to this problem. Application-level checkpointing is much more efficient than system-level checkpointing. Indeed, in system-level checkpointing, a checkpoint of the global state of a distributed application composed of a set of processes is carried out. In application-level checkpointing, only minimal information will be checkpointed (e.g., population of individuals, generation number). Compared to system-level checkpointing, a reduced cost is then obtained in terms of memory and time.

Finally, security issues may be important for large-scale distributed architectures such as grids and peer-to-peer systems (multidomain administration, firewall, etc.) and some specific applications such as medical and bioinformatics research applications of industrial concern.

55.3.6 Algorithmic–Level Parallel Model

Granularity

The algorithmic-level parallel model has the largest granularity. Indeed, the time for exchanging the information is in general much less than the computation time of an EA. There are relatively low communication requirements for this model. The more important is the frequency of exchange and the size of exchanged information, the smaller is the granularity. This parallel model is the most suited to large-scale distributed architectures over internet such as grids. Moreover, the trivial model with independent algorithms is convenient for low-speed networks of workstations over intranet. As there is no essential dependency and communication between the algorithms, the speedup is generally linear for this parallel model.

For an efficient implementation, the frequency of exchange (resp., the size of the exchanged data) must be correlated to the latency (resp., bandwidth) of the communication network of the parallel architecture.

To optimize the communication between processors, the exchange topology can be specified according to the interconnection network of the parallel architecture. The specification of the different parameters associated with the blind or intelligent migration decision criterion (migration frequency/probability and improvement threshold) is particularly crucial on a computational grid. Indeed, due to the heterogeneous nature of computational grids these parameters must be specified for each EA in accordance with the machine it is hosted on.

Scalability

The degree of concurrency of the algorithmic-level parallel model is limited by the number of EAs involved in solving the problem. In theory, there is no limit. However, in practice, it is limited by the owned resources of the target parallel architectures, and also by the effectiveness aspect of using a large number of EAs.

Synchronous Versus Asynchronous Communications

The implementation of the algorithmic level model is either *asynchronous* or *synchronous*. The asynchronous mode associates with each EA an exchange decision criterion, which is evaluated at each iteration of the EA from the state of its memory. If the criterion is satisfied, the EA communicates with its neighbors. The exchange requests are managed by the destination EAs within an undetermined delay. The reception and integration of the received information is thus performed during the next iterations. However, in a computational grid context, due to the material and/or software heterogeneity issue, the EAs could be at different evolution stages leading to the *noneffect* and/or *supersolution* problem. For instance, the arrival of poor solutions at a very advanced stage will not bring any contribution as these solutions will likely not be integrated. In the opposite situation, the cooperation will lead to premature convergence.

From another point of view, as it is nonblocking, the model is more efficient and fault tolerant to such a degree a threshold of wasted exchanges is not exceeded. In the synchronous mode, the EAs perform a synchronization operation at a predefined iteration by exchanging some data. Such operation guarantees that the EAs are at the same evolution stage, and so prevents

the noneffect and supersolution problem quoted before. However, in heterogeneous parallel architectures, the synchronous mode is less efficient in term of consumed CPU time. Indeed, the evolution process is often hanging on powerful machines waiting the less powerful ones to complete their computation. The synchronous model is also not fault tolerant as a fault of a single EA implies the blocking of the whole model in a volatile environment. Then, the synchronous mode is globally less efficient on a computational grid.

Asynchronous communication is more efficient than synchronous communication for shared architectures such as NOWs and desktop grids (e.g., multiple users, multiple applications). Indeed, as the load of networks and processors is not homogeneous, the use of synchronous communication will degrade the performances of the whole system. The least powerful machine will determine the performance.

On a volatile computational grid, it is difficult to efficiently maintain topologies such as rings and torus. Indeed, the disappearance of a given node (i. e., EA) requires a dynamic reconfiguration of the topology. Such reconfiguration is costly and makes the migration process inefficient. Designing a cooperation between a set of EAs without any topology may be considered. For instance, a communication scheme in which the target EA is selected randomly is more efficient for volatile architecture such as desktop grids. Many experimental results show that such topology allows a significant improvement of the robustness and quality of solutions. The random topology is therefore thinkable and even commendable in a computational grid context.

Scheduling

Concerning the scheduling aspect, in the algorithmic-level parallel model the tasks correspond to EAs. Hence, the different scheduling strategies will differ as follows:

- *Static scheduling*: The number of EAs is constant and correlated to the number of processors of the parallel machine. A static mapping between the EAs and the processors is realized. The localization of EAs will not change during the search.
- *Dynamic scheduling*: EAs are dynamically scheduled on different processors of the parallel architecture. Hence, the migration of EAs during the search between different machines may happen.
- *Adaptive scheduling*: The number of EAs involved into the search will vary dynamically. For example, when a machine becomes idle, a new EA is

launched to perform a new search. When a machine becomes busy or faulty, the associated EA is stopped.

Fault Tolerance

The memory state of the algorithmic-level parallel model required for the checkpointing mechanism is composed of the memory of each EA and the information being migrated (i.e., population, generation number).

55.3.7 Iteration-Level Parallel Model

Granularity

A medium granularity is associated with the iteration-level parallel model. The ratio between the evaluation of a partition and the communication cost of a partition determines the granularity. This parallel model is then efficient if the evaluation of a solution is time-consuming and/or there are a large number of candidate solutions to evaluate. The granularity will depend on the number of solutions in each subpopulation.

Scalability

The degree of concurrency of this model is limited by the size of the population. The use of large populations will increase the scalability of this parallel model.

Synchronous Versus Asynchronous Communications

Introducing asynchronism in the iteration-level parallel model will increase the efficiency of parallel EAs. In the iteration-level parallel model, asynchronous communications are related to the asynchronous evaluation of partitions and construction of solutions. Unfortunately, this model is more or less synchronous. Asynchronous evaluation is more efficient for heterogeneous or shared or volatile parallel architectures. Moreover, asynchronism is necessary for optimization problems where the computation cost of the objective function (and constraints) depends on the solution and different solutions may have different evaluation cost.

Asynchronism may be introduced by relaxing the synchronization constraints. For instance, steady-state algorithms may be used in the reproduction phase.

The two main advantages of the asynchronous model over the synchronous model are fault tolerance and robustness if the fitness computation takes very different computations time. Whereas some time-out detection can be used to address the former issue, the latter one can be partially overcome if the grain is set

to very small values, as individuals will be sent out for evaluations upon request of the workers. Therefore, the model is blocking and, thus, less efficient on a heterogeneous computational grid. Moreover, as the model is not fault tolerant, the disappearance of an evaluating agent requires the redistribution of its individuals to other agents. As a consequence, it is essential to store all the solutions not yet evaluated. The scalability of the model is limited to the size of the population.

Scheduling

In the iteration-level parallel model, tasks correspond to the construction/evaluation of a set of solutions. Hence, the different scheduling strategies will differ as follows:

- *Static scheduling*: Here, a static partitioning of the population is applied. For instance, the population is decomposed into equal size partitions depending on the number of processors of the parallel homogeneous nonshared machine. A static mapping between the partitions and the processors is realized. For a heterogeneous nonshared machine, the size of each partition must be initialized according to the performance of the processors. The static scheduling strategy is not efficient for variable computational costs of equal partitions. This happens for optimization problems where different costs are associated with the evaluation of solutions. For instance, in genetic programming individuals may widely vary in size and complexity. This makes a static scheduling of the parallel evaluation of the individuals not efficient [55.32, 33].

- *Dynamic scheduling*: A static partitioning is applied but a dynamic migration of tasks can be carried out depending on the varying load of processors. The number of tasks generated may be equal to the size of the population. Many tasks may be mapped on the same processor. Hence, more flexibility is obtained for the scheduling algorithm. For instance, the approach based on the master-workers cycle stealing may be applied. To each worker is first allocated a small number of solutions. Once it has performed its iterations, the worker requests from the master additional solutions. All the workers are stopped once the final result is returned. Faster and less loaded processors handle more solutions than the others. This approach allows us to reduce the execution time compared to the static one.

- *Adaptive scheduling*: The objective in this model is to adapt the number of partitions generated to the load of the target architecture. More efficient scheduling strategies are obtained for shared, volatile, and heterogeneous parallel architectures such as desktop grids.

Fault Tolerance
The memory of the iteration-level parallel model required for the checkpointing mechanism is composed of different partitions. The partitions are composed of a set of (partial) solutions and their associated objective values.

55.3.8 Solution-Level Parallel Model

Granularity
This parallel model has a fine granularity. There is a relatively high communication requirements for this model. In the functional decomposition parallel model, the granularity will depend on the ratio between the evaluation cost of the subfunctions and the communication cost of a solution. In the data decomposition parallel model, it depends on the ratio between the evaluation of a data partition and its communication cost.

The fine granularity of this model makes it less suitable for large-scale distributed architectures where the communication cost (in terms of latency and/or bandwidth) is relatively important, such as in grid computing systems. Indeed, its implementation is often restricted to clusters or network of workstations or shared memory machines.

Scalability
The degree of concurrency of this parallel model is limited by the number of subfunctions or data partitions. Although its scalability is limited, the use of the solution-level parallel model in conjunction with the two other parallel models enables to extend the scalability of a parallel EA.

Synchronous Versus Asynchronous Communications
The implementation of the solution-level parallel model is always synchronous following a master-workers paradigm. Indeed, the master must wait for all partial results to compute the global value of the objective function. The execution time T will be bounded by the maximum time T_i of the different tasks. An exception occurs for hard-constrained optimization problems, where feasibility of the solution is first tested. The master terminates the computations as soon as a given task detects that the solution does not satisfy a given hard constraint. Due to its heavy synchronization steps, this

parallel model is worth applying to problems in which the calculations required at each iteration are time consuming. The relative speedup may be approximated as follows:

$$S_n = \frac{T}{\alpha + T/n} , \qquad (55.5)$$

where α is the communication cost.

Scheduling
In the solution-level parallel model, tasks correspond to subfunctions in the functional decomposition and to data partitions in the data decomposition model. Hence, different scheduling strategies will differ as follows:

- *Static scheduling*: Usually, the subfunctions or data are decomposed into equal size partitions depending on the number of processors of the parallel machine. A static mapping between the subfunctions (or data partitions) and the processors is applied. As for the other parallel models, this static scheme is efficient for parallel homogeneous nonshared machines. For a heterogeneous nonshared machine, the size of each partition in terms of subfunctions or data must be initialized according to the performance of the processors.
- *Dynamic scheduling*: Dynamic load balancing will be necessary for shared parallel architectures or variable costs for the associated subfunctions or data partitions. Dynamic load balancing may be easily achieved by evenly distributing at run time the subfunctions or the data among the processors. In optimization problems, where the computing cost of the subfunctions is unpredictable, dynamic load balancing is necessary. Indeed, a static scheduling cannot be efficient because there is no appropriate estimation of the task costs (i. e., unpredictable costs).
- *Adaptive scheduling*: In adaptive scheduling, the number of subfunctions or data partitions generated is adapted to the load of the target architecture. More efficient scheduling strategies are obtained for shared, volatile and heterogeneous parallel architectures such as desktop grids.

Fault Tolerance
The memory of the solution-level parallel model required for the checkpointing mechanism is straightforward. It is composed of the solution and its partial objective value calculations.

Depending on the target parallel architecture, Table 55.4 presents a general guideline for the efficient

Table 55.4 Efficient implementation of parallel EAs according to some performance metrics and used strategies

Property	Algorithmic level	Iteration level	Solution level
Granularity	Coarse (frequency of exchange, size of information)	Medium (nb. of solutions per partition)	Fine (eval. subfunctions, eval. data partitions)
Scalability	Number of EAs	Neighborhood size, populations size	Nb. of subfunctions, nb. data partitions
Asynchronism	High (information exchange)	Moderate (eval. of solutions)	Exceptional (feasibility test)
Scheduling and fault tolerance	EA	Solution(s)	Partial solution(s)

implementation of the different parallel models of EAs. For each parallel model (algorithmic level, iteration level, and solution level), the table shows its characteristics according to the outlined criteria (granularity, scalability, asynchronism, scheduling and fault tolerance).

55.4 Parallel EAs Under ParadisEO

Designing generic software frameworks to deal with the design and efficient *transparent* implementation of parallel and distributed EAs is an important challenge. Indeed, efficient implementation of parallel EAs is a complex task, which depends on the type of the parallel architecture used. In designing a software framework for parallel EAs, one has to keep in mind the following important properties: portability, efficiency, easiness of use, and flexibility in terms of parallel architectures and models.

Several white-box frameworks for the reusable design of parallel EAs have been proposed and are available from the Web. The most important of them are: DREAM (distributed resource evolutionary algorithm machine) [55.34], ECJ (Java evolutionary computation) [55.35], JDEAL (Java distributed evolutionary algorithms library) and Distributed BEAGLE (distributed Beagle engine advanced genetic learning environment) [55.36]. These frameworks are reusable as they are based on a clear object-oriented conceptual separation. They are also portable as they are developed in Java, an exception is the last system, which is programmed in C++. However, they are limited regarding the parallel distributed models. Indeed, in DREAM and ECJ only the island model is implemented using Java threads and TCP/IP sockets. DREAM is particularly deployable on peer-to-peer platforms. Furthermore, JDEAL provides only the master-worker model (iteration-level parallel model) using TCP/IP sockets. The latter also designs the synchronous migration-based island model, but implemented on a single processor.

Few frameworks available on the Web are devoted to EAs, and their hybridization. MALLBA [55.37],

MAFRA (Java MimeticAlgorithms Framework) [55.38] and ParadisEO are good examples of such frameworks. MAFRA is developed in Java using design patterns [55.39]. It is strongly hybridization-oriented, but it is very limited regarding parallelism and distribution. MALLBA and ParadisEO have numerous similarities. They are C++/MPI open source frameworks. They provide all the previously presented distributed models, and different hybridization mechanisms. However, they are quite different as ParadisEO is more flexible thanks to the finer granularity of its classes. Moreover, ParadisEO also provides the MPI-based communication layer and Pthreads-based multithreading. MALLBA is deployable on wide area networks using *NetStream*, a message passing service upon MPI [55.37]. ParadisEO is deployable on grid computing platforms using the Globus toolkit [55.21].

ParadisEO-PEO offers transparent implementation of the different parallel models on different architectures using suitable programming environments. ParadisEO-PEO offers an easy implementation of the three main parallel models. The algorithmic-level parallel model allows several optimization algorithms to cooperate and exchange any kind of data. The iteration-level parallel model proposes to parallelize and distribute a set of identical operations. In the solution-level parallel model, any calculation block specific to the optimization problem can be divided into smaller units to speed-up the treatment and gain efficiency.

ParadisEO contains three interconnected modules (Fig. 55.7): EO for evolutionary algorithms (population-based metaheuristics), MO for single solution-based metaheuristics (e.g., local search, tabu search simulated annealing), and MOEO for multi-

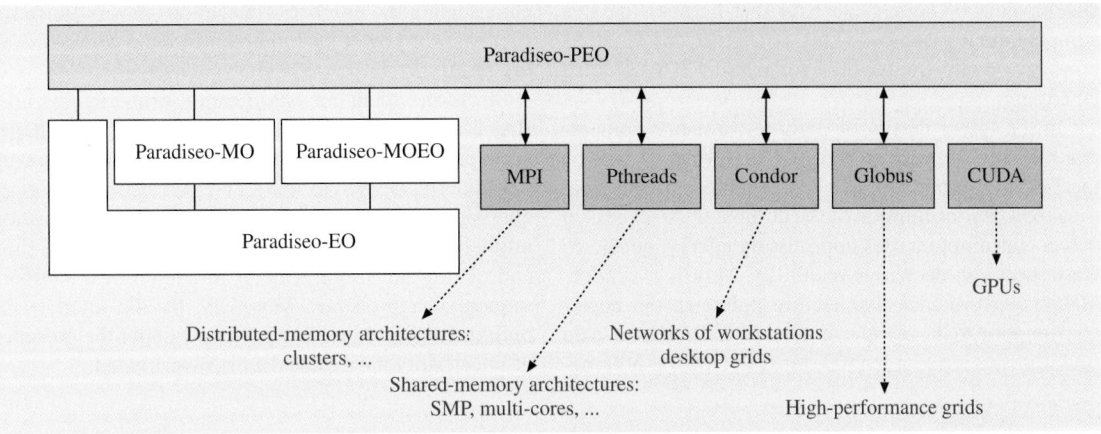

Fig. 55.7 ParadisEO-PEO implementation under different parallel programming environments and middlewares

objective evolutionary algorithms. ParadisEO offers transparency in the sense that the user has not to deal explicitly with parallel programming. One has just to instantiate the needed ParadisEO components. The implementation is portable on distributed-memory machines as well as on shared-memory multiprocessors. The user has not to manage the communications and threads-based concurrency. Moreover, the same parallel design (i. e., the same program) is portable over different architectures. Hence, ParadisEO-PEO has been implemented on different parallel programming environments and middlewares (MPI, Pthreads, Condor, Globus, CUDA) which are adapted to different target architectures (shared and distributed memory, cluster and network of workstations, Desktop and high-performance grid computing platforms, GPUs) (Fig. 55.7). The deployment of the presented parallel and distributed models is transparent for the user.

55.5 Conclusions and Perspectives

Parallel and distributed computing can be used in the design and implementation of EAs to speedup the search, to improve the quality of the obtained solutions, to improve the robustness, and to solve large-scale problems. The clear separation between parallel design and parallel implementation aspects of EAs is important to analyze parallel EAs. The most important lessons of this chapter can be summarized as follows:

- In terms of parallel design, the different parallel models for mono-objective EAs have been unified. Three hierarchical parallel models have been extracted: algorithmic level, iteration level, and solution level parallel models.
- In terms of parallel implementation, the question of an efficient mapping of a parallel model of EAs on a given parallel architecture and programming environment (i. e., language, library, and middleware) is handled. The focus was made on the key criteria of parallel architectures that influence the efficiency of an implementation of parallel EAs.
- The use of the ParadisEO-PEO software framework allows the parallel design of the different parallel models of EAs. It also allows their transparent and efficient implementation on different parallel and distributed architectures (e.g., clusters and networks of workstations, multicores, GPUs, high-performance computing and desktop grids) using suitable programming environments (e.g., MPI, Threads, Globus, Condor, CUDA).

One of the perspectives in the coming years is to achieve Petascale performance. The emergence of heterogeneous platforms composed of multicore chips and many-core chips technologies will speedup the achievement of this goal. In terms of programming models, cloud computing will become an important alternative to traditional high-performance computing for the development of large-scale EAs that harness massive computational resources. This is a great challenge as nowadays cloud frameworks for parallel EAs are just emerging.

In the future design of high-performance computers, the ratio between power and performance will be increasingly important. The power represents the electrical power consumption of the computer. An excess in power consumption uses unnecessary energy, generates waste heat and decreases reliability. Very few vendors of high-performance architecture publicize the power consumption data compared to the performance data (the web site www.green500.org ranks the top 500 ma-

chines using the number of megaflops they produce for each watt of power and complements the www.top500. org site).

In terms of target optimization problems, parallel EAs constitute unavoidable approaches to solve large-scale real-life challenging problems (e.g., engineering design, drug design) [55.23]. They are also an important alternative to solve dynamic and robust optimization problems, in which the complexities in terms of time and quality are more difficult to handle by traditional sequential approaches. Moreover, parallel models for optimization and learning problems under the presence of uncertainty have to be deeply investigated.

References

55.1   E.-G. Talbi: *Metaheuristics: From Design to Implementation* (Wiley, Hoboken 2009)

55.2   H. Mühlenbein: Parallel genetic algorithms, population genetics and combinatorial optimization, 3rd Int. Conf. Genet. Algorithms (1989) pp. 416–421

55.3   E. Alba, M. Tomassini: Parallelism and evolutionary algorithms, IEEE Trans. Evol. Comput. **6**(5), 443–462 (2002)

55.4   E. Alba, E.-G. Talbi, G. Luque, N. Melab: Metaheuristics and parallelism. In: *Parallel Metaheuristics*, ed. by E. Alba (Wiley, Hoboken 2005)

55.5   J. Cohoon, S. Hedge, W. Martin, D. Richards: Punctuated equilibria: A parallel genetic algorithm, Second Int. Conf. Genet. Algorithms (1987) pp. 148–154

55.6   T. Belding: The distributed genetic algorithm revisited, 6th Int. Conf. Genet. Algorithms (1995)

55.7   E. Cantú-Paz: *Efficient and Accurate Parallel Genetic Algorithms* (Kluwer, Boston 2000)

55.8   E. Alba, J.M. Troya: Influence of the migration policy in parallel distributed GAs with structured and panmictic populations, Appl. Intell. **12**(3), 163–181 (2000)

55.9   T. Hiroyasu, M. Miki, M. Negami: Distributed genetic algorithms with randomized migration rate, Proc. IEEE Conf. Systems, Man Cybern. 1 (1999) pp. 689–694

55.10   S.-L. Lin, W.F. Punch, E.D. Goodman: Coarse-grain parallel genetic algorithms: Categorization and new approach, 6th IEEE Symp. Parallel Distrib. Proces. (1994) pp. 28–37

55.11   P. Spiessens, B. Manderick: A massively parallel genetic algorithm, Proc. 4th Int. Conf. Genet. Algorithms (1991) pp. 279–286

55.12   G. von Laszewski, H. Mühlenbein: Partitioning a graph with parallel genetic algorithm, Lect. Notes Comput. Sci. **496**, 165–169 (1990)

55.13   E.G. Talbi, P. Bessière: A parallel genetic algorithm for the graph partitioning problem, Proc. 5th Int. Conf. Supercomput. (1991) pp. 312–320

55.14   E.G. Talbi, P. Bessière: Superlinear speedup of a parallel genetic algorithm on the supernode, SIAM News **24**(4), 12–27 (1991)

55.15   J.M. Ahuactzin, E.G. Talbi, P. Bessière, E. Mazer: Using genetic algorithms for robot motion planning, Lect. Notes Comput. Sci. **708**, 84–93 (1993)

55.16   K.F. Doerner, R.F. Hartl, G. Kiechle, M. Lucka, M. Reimann: Parallel ant systems for the capacited vehicle routing problem, Lect. Notes Comput. Sci. **3004**, 72–83 (2004)

55.17   M. Rahoual, R. Hadji, V. Bachelet: Parallel ant system for the set covering problem, Lect. Notes Comput. Sci. **2463**, 262–267 (2002)

55.18   M. Randall, A. Lewis: A parallel implementation of ant colony optimization, J. Parallel Distrib. Comput. **62**(9), 1421–1432 (2002)

55.19   E.-G. Talbi, O. Roux, C. Fonlupt, D. Robillard: Parallel ant colonies for combinatorial optimization problems, Lect. Notes Comput. Sci. **1586**, 239–247 (1999)

55.20   E.G. Talbi, S. Cahon, N. Melab: Designing cellular networks using a parallel hybrid metaheuristic on the computational grid, Comput. Commun. **30**(4), 698–713 (2007)

55.21   N. Melab, S. Cahon, E.-G. Talbi: Grid computing for parallel bioinspired algorithms, J. Parallel Distrib. Comput. **66**(8), 1052–1061 (2006)

55.22   I. Foster, C. Kesselman (Eds.): *The Grid: Blueprint for a New Computing Infrastructure* (Morgan Kaufmann, San Mateo 1999)

55.23   R. Zeidman: *Designing with FPGAs and CPLDs* (CMP, Lawrence 2002)

55.24   M. Pharr, R. Fernando: *GPU Gems 2: Programming Techniques for High-Performance Graphics and General-Purpose Computation* (Addison-Wesley, Upper Saddle River 2005)

55.25   T.-V. Luong, N. Melab, E.-G. Talbi: Parallel hybrid evolutionary algorithms on GPU, IEEE Congr. Evol. Comput. (2010) pp. 1–8

55.26 D.R. Butenhof: *Programming with POSIX Threads* (Addison-Wesley, Upper Saddle River 1997)

55.27 P. Hyde: *Java Thread Programming* (Sams, Indianapolis 1999)

55.28 B. Chapman, G. Jost, R. VanderPas, D.J. Kuck: *Using OpenMP: Portable Shared Memory Parallel Programming* (MIT, Cambridge 2007)

55.29 B. Sotomayor, L. Childers: *Globus Toolkit 4: Programming Java Services* (Morgan Kaufmann, San Mateo 2005)

55.30 V. Kumar, A. Grama, A. Gupta, G. Karypis: *Introduction to Parallel Computing: Design and Analysis of Algorithms* (Addison-Wesley, Upper Saddle River 1994)

55.31 E.-G. Talbi: *Parallel Combinatorial Optimization* (Wiley, Hoboken 2006)

55.32 H. Juille, J.B. Pollack: Massively parallel genetic programming. In: *Advances in Genetic Programming 2*, ed. by P.J. Angeline, K.E. Kinnear Jr. (MIT, Cambridge 1996) pp. 339–358

55.33 G. Folino, C. Pizzuti, G. Spezzano: CAGE: A tool for parallel genetic programming applications, Lect. Notes Comput. Sci. **2038**, 64–73 (2001)

55.34 M.G. Arenas, P. Collet, A.E. Eiben, M. Jelasity, J.J. Merelo, B. Paechter, M. Preuss, M. Schoenauer: A framework for distributed evolutionary algorithms, Lect. Notes Comput. Sci. **2439**, 665–675 (2002)

55.35 G.C. Wilson, A. McIntyre, M.I. Heywood: Resource review: Three open source systems for evolving programs-Lilgp, ECJ and grammatical evolution, Genet. Program. Evol. Mach. **5**(19), 103–105 (2004)

55.36 C. Gagné, M. Parizeau, M. Dubreuil: Distributed Beagle: An environment for parallel and distributed evolutionary computations, Proc. 17th Ann. Int. Symp. High Perform. Comput. Syst. Appl. (2003) pp. 201–208

55.37 E. Alba, F. Almeida, M. Blesa, C. Cotta, M. Díaz, I. Dorta, J. Gabarró, J. González, C. León, L. Moreno, J. Petit, J. Roda, A. Rojas, F. Xhafa: MALLBA: A library of skeletons for combinatorial optimisation, Lect. Notes Comput. Sci. **2400**, 927–932 (2002)

55.38 N. Krasnogor, J. Smith: MAFRA: A Java memetic algorithms framework, Workshop Proc. GECCO (2002)

55.39 E. Gamma, R. Helm, R. Johnson, J. Vlissides: *Design Patterns, Elements of Reusable Object-Oriented Software* (Addison-Wesley, Upper Saddle River 1994)

56. How to Create Generalizable Results

Thomas Bartz-Beielstein

Basically, this chapter tries to find answers for the following fundamental questions in experimental research.

(Q-1) How can problem instances be generated?
(Q-2) How can experimental results be generalized?

The chapter is structured as follows. Section 56.2 introduces real-world and artificial optimization problems. Algorithms are described in Sect. 56.3. Objective functions and statistical models are introduced in Sect. 56.4; these models take problem and algorithm features into consideration. Section 56.5 presents case studies that illustrate our methodology. The chapter closes with a summary and an outlook.

56.1 Test Problems in Computational Intelligence

Computational intelligence (CI) methods have gained importance in several real-world domains such as process optimization, system identification, data mining, or statistical quality control. Tools to determine the applicability of CI methods in these application domains in an objective manner are missing. Statistics provide methods for comparing algorithms on certain data sets. In the past, several test suites were presented and considered as state of the art. However, these test suites have several drawbacks, namely:

- Problem instances are mostly artificial and have no direct link to real-world settings.
- Since there is a fixed number of test instances, algorithms can be fitted or tuned to this specific and very limited set of test functions. As a consequence, studies (benchmarks) provide insight how these algorithms perform on this specific set of test instances, but no insight on how they perform in general.
- Statistical tools for comparisons of several algorithms on several test problem instances are relatively complex and not easy to analyze.

We propose a methodology to overcome these difficulties. This methodology, which generates problem classes rather than uses one instance, is constructed as follows:

1. First, we pre-process the underlying real-world data.
2. In a second step, features from these data are extracted. This extraction relies on the assumption that mathematical variables can be used to represent real-world features. For example, decomposition techniques can be applied to model the underlying data structures, if we are using time-series data. The

original time series is deconstructed into a number of component series, where each of these reflects a certain type of behavior, e.g., a trend or seasonality [56.1]. We obtain an analytical model of the data.

3. Then, we parameterize this model. Based on this parametrization and randomization, we can generate infinitely many new problem instances.

4. If no real-world data are available, problem instances can be generated using test-problem generators. The generation of test problems, which are well-founded and have practical relevance, has been an on-going field of research for several decades.

5. From this infinite set, we can draw a limited number of problem instances which will be used for the comparison.

6. Since problem instances are selected randomly, we apply random and mixed models for the analysis [56.2]. Mixed models include fixed and random effects. A fixed effect is an unknown constant. Its estimation from the data is a common practice in analysis of variance (ANOVA) or regression. A random effect is a random variable. We estimate the parameters that describe its distribution, because – in contrast to fixed effects –

it makes no sense to estimate the random effect itself.

This chapter combines ideas from two approaches: problem generation and statistical analysis of computer experiments. The work presented by *Chiarandini* and *Goegebeur* [56.3] provides the basis of our statistical analysis. They present a systematic and well-developed framework for mixed models. Related modeling approaches were suggested by *McGeoch* [56.4] and *Birattari* [56.5]. *Gallagher* and *Yuan* [56.6] present a problem instance (landscape) generator that is parameterized by a small number of parameters, and the values of these parameters have a direct and intuitive interpretation in terms of the geometric features of the landscapes that they produce. *Castiñeiras* et al. [56.7] present a parameterizable benchmark generator for bin packing instances based on the well-known Weibull distribution. Using the shape and scale parameters of the Weibull distribution, the authors generate benchmarks that contain a variety of item size distributions. They report that for all bin capacities, the number of bins required in an optimal solution increases as the Weibull shape parameter increases. Using this feature, scalability is enabled.

56.2 Features of Optimization Problems

56.2.1 Problem Classes and Instances

Nowadays, it is common practice in optimization to choose a *fixed* set of problem instances in advance and to apply classical ANOVA or regression analysis. In many experimental studies a few problem instances π_i ($i = 1, 2, \ldots, q$) are used and the results of some runs of the algorithms α_j ($j = 1, 2, \ldots, h$) on these instances are collected. The instances can be treated as *blocks* and all algorithms are run on each single instance. Results are grouped per instance π_i. Analyses of these experiments shed some light on the performance of the algorithms on those specific instances. However, the interest of the researcher should not be just the performance of the algorithms on those specific instances chosen, but rather on the generalization of the results to the entire class Π. Generalizations about the algorithm's performance on new problem instances are difficult or impossible in this setting.

Based on ideas from *Chiarandini* and *Goegebeur* [56.3], to overcome this difficulty, we propose the following approach: a small set of problem instances $\{\pi_i \in \Pi \mid i = 1, 2, \ldots, q\}$ is chosen at random from a large set, or class Π, of possible instances of the problem. Problem instances are considered as factor levels. However, this factor is of a different nature from the fixed algorithmic factors in the classical ANOVA setting. Indeed, the levels are chosen at random and the interest is not in these specific levels but in the problem class Π from which they are sampled. Therefore, the levels and the factor are *random*. Consequently, our results are not based on a limited, fixed number of problem instances. They are randomly drawn from an infinite set, which enables generalization.

56.2.2 Feature Extraction and Instance Generation

A problem class Π can be generated in different manners. We will consider *artificial* and *natural* problem class generators. Artificially generated problems allow feature generation based on some predefined characteristics. They are basically theory driven, i.e., the researcher defines certain features such as linearity or multi modality. Based on these features, a model (formula) is constructed. By integrating parameters into this formula, many problem instances can be generated by parameter variation. We will exemplify this approach in the following paragraph. The second way, which will generate natural problem classes, uses a three-stage approach. First, the real-word system and its components are described. Then, features are extracted from a real-world system. Based on this feature set, a model is defined. Adding parameters to this model, new problem instances can be generated. There is also a third way to *generate* test instances: if we are lucky, many data are available. In this case, we can sample a limited number of problem instances from the larger set of real-world data. The statistical analysis is similar for these three cases.

Artificial Test Functions

Several problem instance generators have been proposed over the last years. For example, *Gallagher* and *Yuan* present a landscape test generator, which can be used to set up problem instances for continuous, bound-constrained optimization problems [56.6]. The *Max-set of Gaussian landscape generator* (MSG) uses the maximum of m weighted Gaussian functions

$$G(x) = \max_{i \in 1,2,\ldots,m} (w_i g_i(x)),$$

where $g : \mathbb{R}^n \to \mathbb{R}$ denotes an n-dimensional Gaussian function

$$g(x) = \left(\frac{\exp\left(-\frac{1}{2}(x-\mu)\Sigma^{-1}(x-\mu)^T\right)}{(2\pi)^{n/2}|\Sigma|^{1/2}} \right)^{1/n},$$

μ is an n-dimensional vector of means, and Σ is an $(n \times n)$ covariance matrix. The mean of each Gaussian corresponds to an optimum on the landscape and the location of all optima is known. The global optimum is the one with the largest value. We will use the MSG problem instance generator in Sect. 56.5 to demonstrate our approach.

Natural Problem Classes

This section exemplifies the three fundamental steps for generating real-world problem instances, namely:

1. Describing the real-world system and its data
2. Feature extraction and model construction
3. Instance generation.

We will illustrate this procedure by using the classic Box and Jenkins airline data [56.8]. These data contain the monthly totals of international airline passengers from 1949 to 1961. The feature extraction is based on methods from time-series analysis. Because of its simplicity the Holt–Winters method is popular in many application domains. It is able to adapt to changes in trends and seasonal patterns. The Holt–Winters prediction function requires the estimation of three parameters, i.e., α, β and γ, which can be estimated from original time-series data. Their optimal values are determined by minimizing the squared one-step prediction error. To generate new problem instances, these parameters can be slightly modified. Based on these modified values, the model is re-fitted. Finally, we can extract the new time series. One typical result from this instance generation is shown in Fig. 56.1. *Bartz-Beielstein* [56.9] describes this procedure in detail.

To illustrate the wide applicability of this approach, we will list further real-work problem domains, which are subject of our current research:

- *Smart metering*: The development of accurate forecasting methods for electrical energy consumption profiles is an important task. We consider time series collected from a manufacturing process. Each time series contains quarter-hourly samples of the energy consumption of a bakery. A detailed data description can be found in [56.10].
- *Water industry*: Canary is a software developed by the *United States Environmental Protection Agency* (US EPA) and Sandia National Laboratories. Its purpose is to detect events in the context of wa-

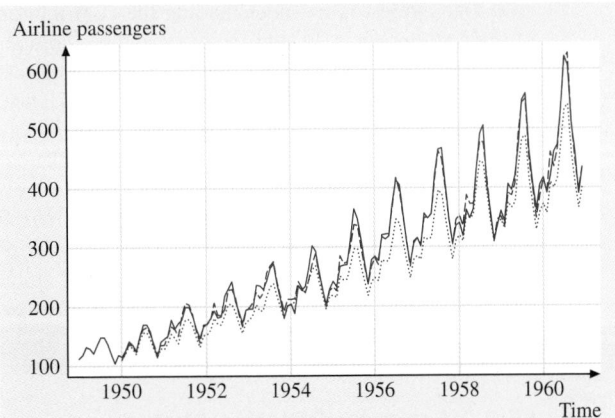

Fig. 56.1 Holt–Winters problem instance generator. The *solid line* represents the real data, the *dotted line* predictions from the Holt–Winters model and the *fine dotted line* modified predictions, respectively

ter contamination. An event is in this context defined as a certain time period where a contaminant significantly deteriorates the water quality. Distinguishing events from (i) background changes, (ii) maintenance and modification due to operation, and (iii) outliers is an essential task, which was implemented in the Canary software. Therefore, deviations are compared to regular patterns and short term changes. The corresponding data contains multi-variate time-series data. It is a selection from a larger dataset shipped with the open source event-detection software CANARY developed by US EPA and Sandia National Laboratories [56.11].

- *Finance*: The data are real-world data from intraday *foreign exchange* (FX) trading. The FX market is a financial market for trading currencies to enable international trade and investment. It is the largest and most liquid financial market in the world. Currencies can be traded via a wide variety of different financial instruments, ranging from simple spot trades over to highly complex derivatives.

We use three foreign exchange (currency rate) time series collected from Bloomberg. Each time series contains hourly samples of the change in currency exchange rate [56.12].

One typical goal in forecasting is the minimization of the forecast errors or the differences between real (observed) values, say y_i, and predicted values, say $\hat{y}_i$. This goal can be considered as an optimization problem.

As stated in Sect. 56.2.2, the statistical analysis is similar for artificial and natural problem classes. Our goal can be stated as follows: For a given problem class Π, which can be artificial or natural, we try to determine if an optimization algorithm α or several algorithm instances α_i show similar behavior on randomly selected problem instances $\pi_i \in \Pi$. This question will be formulated as a statistical hypothesis. Based on the related statistical framework, we can determine confidence intervals for the performance of the algorithm on unseen problem instances.

56.3 Algorithm Features

56.3.1 Factors and Levels

Evolutionary algorithms (EA) belong to the large class of bio-inspired search heuristics. They combine specific components, which may be *qualitative*, like the recombination operator or *quantitative*, like the population size. Our interest is in understanding the contribution of these components. In statistical terms, these components are called *factors*. The interest is in the effects of the specific *levels* chosen for these factors. Hence, we say that the levels and, consequently, the factors are *fixed*. Although modern search techniques like sequential parameter optimization or Pareto genetic programming [56.13] allow multi-objective performance measures (solution quality versus variability or description length), we restrict ourselves to analyzing the effect of these factors on a univariate measure of performance. We will use the quality of the solutions returned by the algorithm at termination as the performance measure.

56.3.2 Example: Evolution Strategy

Evolution strategies (ES) are prominent representatives of evolutionary algorithms, which includes genetic algorithms and genetic programming as well [56.15].

They can be classified as generic population-based metaheuristic optimization algorithms for global optimization that in some sense mimics the natural evolution. Evolution strategies are applied to hard real-valued optimization problems. Mutation is performed by adding a normally distributed random value to each vector component. The standard deviation of these random values is modified by self-adaptation. Evolution

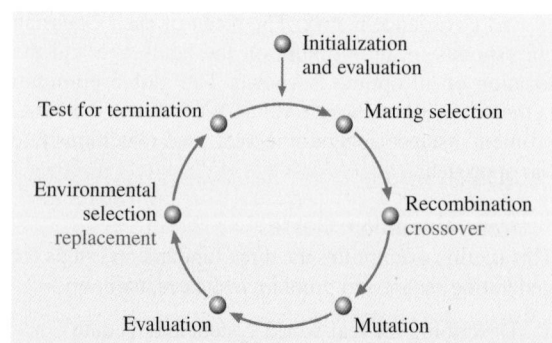

Fig. 56.2 The evolutionary cycle, basic working scheme of all ES and EA. Terms common for describing evolution strategies are used, alternative terms are added below in *brown*

Table 56.1 Settings of exogenous parameters of an ES. Recombination operators are labeled as follows: $1 =$ no, $2 =$ dominant, $3 =$ intermediate, $4 =$ intermediate as in [56.14]. Mutation uses the following encoding: $1 =$ no mutation, $2 =$ self adaptive mutation

Parameter	Symbol	Name	Range	Value
mue	μ	Number of parent individuals	$\mathbb{N}$	5
nu	$\nu = \lambda/\mu$	Offspring–parent ratio	$\mathbb{R}_+$	2.0
sigmaInit	$\sigma_i^{(0)}$	Initial standard deviations	$\mathbb{R}_+$	1.0
nSigma	n_σ	Number of standard deviations. d denotes the problem dimension	$\{1, d\}$	1
	c_τ	Multiplier for individual and global mutation parameters	$\mathbb{R}_+$	1.0
tau0			$\mathbb{R}_+$	0.0
tau			$\mathbb{R}_+$	1.0
rho	ρ	Mixing number	$\{1, \mu\}$	2
sel	κ	Maximum age	$\mathbb{R}_+$	1.0
mutation		Mutation	$\{1, 2\}$	2
sreco	r_σ	Recombination operator for strategy variables	$\{1, 2, 3, 4\}$	3
oreco	r_x	Recombination operator for object variables	$\{1, 2, 3, 4\}$	2

strategies can use a population of several solutions. Each solution is considered as individual and consists of object and strategy variables. Object variables represent the position in the search space, whereas strategy variables store the step sizes, i.e., the standard deviations for the mutation. We analyze the ES basic variant, which was proposed in [56.14].

Mutation means neighborhood-based movement in search space, which includes the exploration of the *outer space* currently not covered by a population, whereas recombination rearranges existing information and so focuses on the *inner space*. Selection is meant to introduce a bias towards better fitness values. A concrete ES may contain specific mutation, recombination, or selection operators, or call them only with a certain probability, but the control flow is usually left unchanged. Each of the consecutive cycles is termed a *generation*. The control flow is shown in Fig. 56.2.

Concerning the representation, it should be noted that most empiric studies are based on canonical forms as binary strings or real-valued vectors, whereas many real-world applications require specialized, problem-dependent ones. Table 56.1 summarizes important ES parameters. This chapter presents two case studies. The first case study is based on a fixed ES parameter setting, whereas the second case study modifies the recombination operator for object variables. We are convinced that the applicability of the methods presented in this chapter goes far beyond the simplified case studies. Our main contribution is a framework, which allows conclusions that are not limited to a small number of problem instances but to problem classes.

56.4 Objective Functions

We will use the following optimization framework: an ES is applied as a minimizer on the test function $f(x)$. Formally speaking, let S denote some set, e.g., $S \subseteq \mathbb{R}^n$. We are seeking for values f^* and x^*, such that $\min_{x \in S} f(x)$ with $f^* = \min_{x \in S} f(x)$ and $x^* = \arg \min f(x)$. This approach can be extended in many ways. For example, if S denotes times-series data, then an optimization algorithm can be applied to minimize the empirical mean squared prediction error.

Test problem instances will be drawn from Gallagher's and Yuan's MSG test function generator. The following parameters can be used to specify the MSG generator:

- The number of Gaussian components m.
- The mean vector μ of each component.
- The covariance matrix Σ of each component.
- The weight of each component w_i.
- A maximum threshold $t \in [0; 1]$ can be specified for local optima and the fitness value of the global optimum G^*. Local optima are randomly generated within $[0; t \times G^*]$.

The following tuple can be used to specify an MSG generator

$$\Pi := \left([-c, c]^n, n, m, D_\mu, \{D_\Sigma\}, \{t, G^*\}\right), \quad (56.1)$$

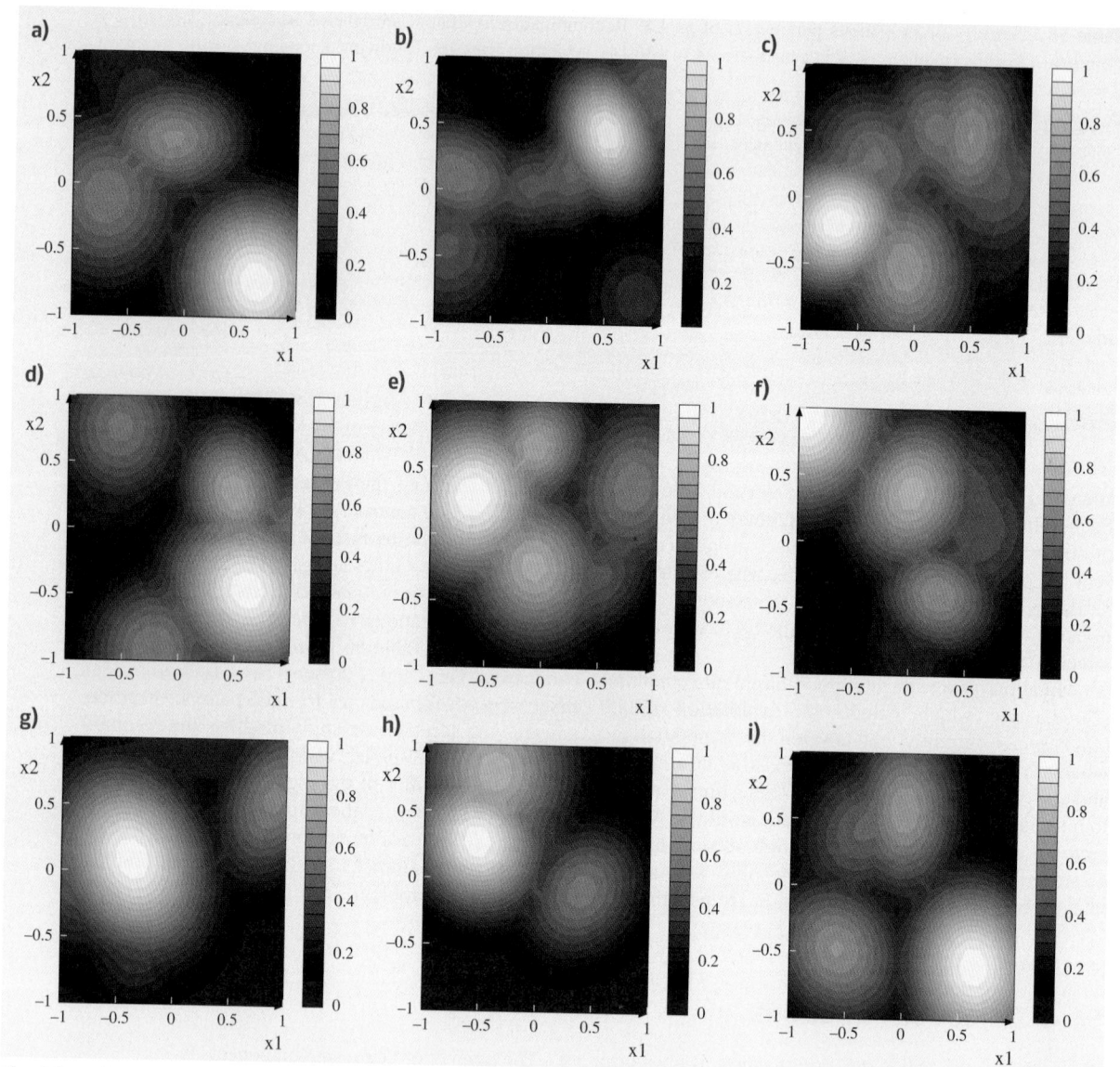

Fig. 56.3a–i Nine test problem instances from Π_{MSG}, generated with the MSG landscape generator as specified in (56.2). These figures exemplify how numbers and locations of the randomly generated optima can vary. Usually, the optima are evenly distributed in the search space. In some settings, there are a few dominating optima as can be seen in part (**g**)

where $c \in \mathbb{R}$ defines the boundary constraints of the search space, n the search space dimensionality, m the number of Gaussian components, D_μ the distribution used to generate the mean vectors of components, D_Σ the distribution or procedures used to generate covariances of components, $t \in [0; 1]$ the threshold for local optima, and G^* the function value of the global optimum.

Based on (56.1), we have specified the following MSG landscape generator for our experiments

$$\Pi_{MSG} := \big([-1; 1]^2, 2, 10, \mathcal{U}[-1; 1],$$
$$\{\mathcal{U}[0.05; 0.15], \mathcal{U}[-\pi/4, \pi/4]\}, \{0.8, 1\}\big).$$

$$(56.2)$$

With this setting, the mean vector of each component is generated randomly within $[-1, 1]^2$. The covariance matrix of each component is generated with the procedure D_Σ in three steps:

1. A diagonal matrix $\mathbf{S}$ with eigenvalues is generated.
2. An orthogonal matrix $\mathbf{T}$ is generated through $n(n-1)/2$ rotations with random angles between $[-\pi/4, \pi/4]$.
3. The covariance matrix is generated as $\mathbf{T}^\mathsf{T}\mathbf{S}\mathbf{T}$.

The weight w_i of the component corresponding to the global optimum is set to 1 while other weights are randomly generated within $[0; 0.8]$. The nine problem instances, $\pi_i \in \Pi_{\mathrm{MSG}}$, $(i = 1, \ldots, 9)$

from Fig. 56.3 were generated with this parametrization.

Note that we are using the distance to the optimum as an objective function in our experiments. Our objective function reads $G^* - f(x)$, because we are considering minimization problems. Other measures of interest might be the *gap percent of optimality*

$$\frac{(G^* - f(x))}{G^*} \times 100 ,$$

or computation time, etc., see, e.g., [56.16].

56.5 Case Studies

Bartz-Beielstein [56.9] introduced the acronyms:

- SASP: one single algorithm and one single problem instance
- SAMP: one single algorithm and multiple problems instances
- MASP: multiple algorithms and one single problem instance
- MAMS: multiple algorithms and multiple problem instances

for classifying optimization designs [56.17].

56.5.1 Single Problem Designs: SASP and MASP

In SASP we analyze the performance of an *optimization algorithm* α on a single problem instance π. An optimization problem has a set of input data which instantiate the problem. This might be a function in continuous optimization or the location and distances between cities in a traveling salesman problem. In the following, we will use Y to denote the random performance measure obtained by r runs of algorithm α on problem instance π. Because many optimization algorithms such as evolutionary algorithms are randomized, their performance Y on one instance is a random variable. It might be described by a probability density/mass function $p(y|\pi)$. Running the algorithm with different random seeds on one problem instance, we collect sample data $y_1, \ldots, y_r$, which are *independent and identically distributed* (i.i.d.).

There are situations, in which SASP is the method of first choice. Real-world problems, which have to be solved only once in a very limited time, are good ex-

amples for using SASP optimizations. MASP shares several characteristics with SASP. Because of their limited capacities for generalization, SASP and MASP will not be investigated further in this study.

56.5.2 SAMP: Single Algorithm, Multiple Problems

Fixed–Effects Models
This setup is commonly used for testing an algorithm on a given (fixed) set of problem instances. Standard assumptions from *analysis of variance* (ANOVA) lead us to propose the following *fixed-effects model* [56.2]

$$Y_{ij} = \mu + \tau_i + \varepsilon_{ij} , \tag{56.3}$$

where μ is an overall mean, τ_i is a parameter unique to the i-th treatment (problem instance factor), and ε_{ij} is a random error term for replication j on problem instance i. Usually, the model errors ε_{ij} are assumed to be normally and independently distributed with mean zero and variance σ^2. If problem instance factors are considered fixed, i.e., non random, the stochastic behavior of the response variable originates from the algorithm. This implies the experimental results

$$Y_{ij} \approx N(\mu + \tau_i, \sigma^2) , \quad i = 1, \ldots, q, \ j = 1, \ldots, r , \tag{56.4}$$

and that the Y_{ij} are mutually independent. Results from statistical analyses remain valid only on the specific instances. Furthermore, SAMP with a fixed set of problem instances is subject to criticism, e.g., that algorithms are trained for this specific set up test instances (over fitting).

In order to make the results of the analysis independent of the specific instances and dependent instead

Part E | 56.5

on the class of instances from which the specific instances are drawn, *Chiarandini* and *Goegebeur* propose randomized and mixed models for the experimental analysis of optimization algorithms as an extension of (56.3) [56.3]. In contrast to model (56.3), these models allow generalizations of results to the whole class of instances.

Randomized Models

In the following, we consider a population or *class* of instances Π. The class Π consists of a large, possibly an infinite, number of problem instances $\pi_i, i = 1, 2, 3, \ldots$ Let $p(\pi)$ denote the probability of sampling instance π. The performance Y of the algorithm α on the class Π is described by the probability function

$$p(y) = \sum_{\pi \in \Pi} p(y|\pi)p(\pi) \,. \tag{56.5}$$

If we run an algorithm α r times on instance π, then we receive r replicates of α's performance, denoted by $Y_1, \ldots, Y_r$. These r observations are i.i.d., i.e.,

$$p(y_1, \ldots, y_r|\pi) = \prod_{j=1}^{r} p(y_j|\pi) \,. \tag{56.6}$$

So far, we have considered r replicates of the performance measure Y on *one* problem instance π. Now we consider *several*, randomly sampled problem instances. Over all the instances the joint probability distribution of the observed performance measures is obtained by marginalizing over all instances

$$p(y_1, \ldots, y_r) = \sum_{\pi \in \Pi} p(y_1, \ldots, y_r|\pi)p(\pi) \,. \tag{56.7}$$

Extending the model (56.7) to the case where one algorithm with several parameter settings or several algorithms are analyzed leads to mixed models, which will be discussed in Sect. 56.5.3.

Example SAMP: ES on Π_1 (Random-Effects Design)

The simplest random-effects experiment is performed as follows. For $i = 1, \ldots, q$ a problem instance π_i is drawn randomly from the class of problem instances Π. On each of the sampled π_i, the algorithm α is run r times using different seeds for α. Due to α's stochastic nature, we obtain, *conditionally on the sampled instance*, r replications of the performance measure that are i.i.d.

Let Y_{ij} $(i = 1, \ldots, q; j = 1, \ldots, r)$ denote the random performance measure obtained in the j-th replication of α on π_i. We are interested in drawing conclusions about α's performance on a larger set of problem instances from Π and not just on those q problem instances included in the experiment. A systematic approach to accomplish this task comprehends the following steps:

- SAMP-1 algorithm and problem instances
- SAMP-2 ANOVA and restricted maximum likelihood estimator (REML) model building
- SAMP-3 validation of the model assumptions
- SAMP-4 hypothesis testing
- SAMP-5 Confidence intervals and prediction.

SAMP-1 Algorithm and Problem Instances. The goal of this case study is to analyze if one algorithm shows a similar performance on a class of problem instances, say Π_{MSG}. A random-effects design will be used to model the results. We illustrate the decomposition of the variance of the response values in (i) the variance due to problem instance and (ii) the variance due to the algorithm and derive results, which are based on hypotheses testing as introduced in (56.12).

We consider one algorithm, an ES, which is run $r = 10$ times on a set of randomly generated problem instances. The ES is parameterized with the default setting from Table 56.1. These parameters are kept constant during the experiment. Nine instances are drawn from the set of problem instances Π_{MSG}. Problem instances were generated with the MSG landscape generator as specified in (56.2). The corresponding problem instances are shown in Fig. 56.3.

The null hypothesis reads *There is no instance effect*. Since we are considering the SAMP case, our experiment is based on one ES instance only. There are 90 observations, because 10 repeats were performed on 9 problem instances. Figure 56.4 shows the performance of the ES on these nine instances. The variable fSeed is used to denote the problem instance number π_i.

SAMP-2 ANOVA and REML Model Building.

ANOVA Model Building. The following analysis is based on the linear statistical model

$$Y_{ij} = \mu + \tau_i + \varepsilon_{ij} \quad \begin{cases} i = 1, \ldots, q \,, \\ j = 1, \ldots, r \,, \end{cases} \tag{56.8}$$

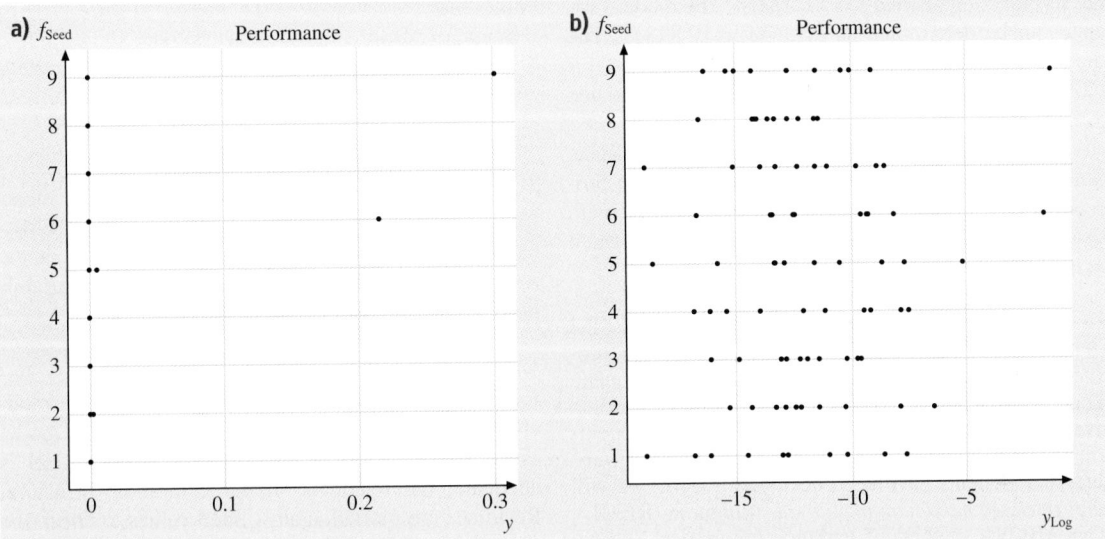

Fig. 56.4a,b Performance of the ES on nine test problem instances. (**a**) Problem instances plotted versus algorithm performance. (**b**) Problem instances plotted against logarithmic performance. Smaller values are better

where μ is an overall mean and ε_{ij} is a random error term for replication j on instance i. Note that in contrast to the fixed-effects model from (56.3), τ_i is a random variable representing the effect of instance i. The stochastic behavior of the response variable originates from both the instance and the algorithm. This is reflected in (56.8), where both τ_i and ϵ_{ij} are random variables. The model (56.8) is the so-called *random-effects model*, cf. [56.2] or [56.3].

We assume that $\tau_1, \ldots, \tau_q$ are i.i.d. $\mathcal{N}(0, \sigma_\tau^2)$ and ε_{ij}, $i = 1, \ldots, q, j = 1, \ldots, r$, are i.i.d. $\mathcal{N}(0, \sigma^2)$. If τ_i is independent of ϵ_{ij} and has variance $V(\tau_i) = \sigma_\tau^2$, the variance of any observation is $V(Y_{ij}) = \sigma^2 + \sigma_\tau^2$. Similar to the partition in classical ANOVA, the variability in the observations can be partitioned into a component that measures the variation between treatments and a component that measures the variation within treatments. Based on the fundamental ANOVA identity $\mathrm{SS}_{\text{total}} = \mathrm{SS}_{\text{treat}} + \mathrm{SS}_{\text{err}}$, we define

$$\mathrm{MS}_{\text{treat}} = \frac{\mathrm{SS}_{\text{treat}}}{q-1} = \frac{r\sum_{i=1}^{q}(\bar{Y}_{i.} - \bar{Y}_{..})^2}{q-1},$$

and

$$\mathrm{MS}_{\text{err}} = \frac{\mathrm{SS}_{\text{err}}}{q(r-1)} = \frac{\sum_{i=1}^{q}\sum_{j=1}^{r}(Y_{ij} - \bar{Y}_{i.})^2}{q(r-1)}.$$

It can be shown that

$$E(\mathrm{MS}_{\text{treat}}) = \sigma^2 + r\sigma_\tau^2 \quad \text{and} \quad E(\mathrm{MS}_{\text{err}}) = \sigma^2,$$

$$(56.9)$$

cf. [56.2]. Therefore, the estimators of the variance components are

$$\hat{\sigma}^2 = \mathrm{MS}_{\text{err}}, \tag{56.10}$$

$$\hat{\sigma}_\tau^2 = \frac{\mathrm{MS}_{\text{treat}} - \mathrm{MS}_{\text{err}}}{r}. \tag{56.11}$$

The corresponding ANOVA table is shown in Table 56.2. Based on ANOVA calculations, with (56.10) we obtain an estimator of the first variance component $\hat{\sigma}^2 = -0.4848257$, and from (56.11), we obtain

Table 56.2 ANOVA table for a one-factor fixed and random effects models

Source of variation	Sum of squares	Degrees of freedom	Mean square	EMS fixed	EMS random
Treatment	$\mathrm{SS}_{\text{treat}}$	$q-1$	$\mathrm{MS}_{\text{treat}}$	$\sigma^2 + r\frac{\sum_{i=1}^{q}\tau_i^2}{q-1}$	$\sigma^2 + r\sigma_\tau^2$
Error	SS_{err}	$q(r-1)$	MS_{err}	σ^2	σ^2
Total	$\mathrm{SS}_{\text{total}}$	$qr-1$			

Part E | 56.5

the second component $\hat\sigma_\tau^2 = 11.32854$. The model variance can be determined as $\hat\sigma^2 + \hat\sigma_\tau^2 = 10.84372$. The mean $\mu = -12.05554$ from (56.8) can be extracted. Finally, the p value in the ANOVA table is calculated as 0.7979083.

Note that we have obtained a negative variance. Since negative variances are not feasible, we can proceed by setting their values to zero and proceed with these modified values. A more elegant way is presented in the following.

Restricted Maximum Likelihood. In some cases, the standard ANOVA, which was used in our example, produces a negative estimate of a variance component. This can be seen in (56.11): if $MS_{err} > MS_{treat}$, negative values occur. By definition, variance components are positive. Methods that always yield positive variance components have been developed. Here, we will use *restricted maximum likelihood estimators* (REML). The ANOVA method of variance component estimation, which is a method of moments procedure, and REML estimation may lead to different results. Output from an R-based analysis with the function lme from the package lme4 reads as follows (fSeed denotes the problem instance) [56.18]:

```
Linear mixed model fit by REML
Formula: yLog ~ 1 + (1 | fSeed)
   Data: samp.df
   AIC   BIC logLik deviance REMLdev
 475.6 483.1 -234.8    469.3   469.6
Random effects:
 Groups    Name        Variance Std.Dev.
 fSeed     (Intercept) 0.000    0.0000
```

```
 Residual               10.893    3.3004
Number of obs: 90, groups: fSeed, 9
```

```
Fixed effects:
            Estimate Std. Error t value
(Intercept) -12.0555     0.3479  -34.65
```

Compared to the ANOVA setting, different values for $\hat\sigma^2$, $\hat\sigma_\tau^2$, and μ were obtained. However, the REML-based analysis also shows that the variability in the response observations can be attributed to the variability of the algorithm.

SAMP-3 Validation of the Model Assumptions. Before performing hypothesis testing based on the models introduced in SAMP-2, the validity of the model assumptions has to be investigated. If the model is adequate, the residuals should exhibit no structure. Residuals are plotted against fitted values to check the assumption of homoscedasticity and quantile–quantile (Q–Q) plots are used to check if residuals meet the normality assumption. Quantile–quantile plots of the residuals are shown in Fig. 56.5 for the raw and the log-transformed responses. These plots provide a good way to compare the distribution of a sample with a distribution. Large deviations from the line indicate non-normality of the sample data. These Q–Q plots indicate that a log transformation of the response might be useful in our setting.

SAMP-4 Hypothesis Testing. Testing hypotheses about individual treatments (instances) is useless be-

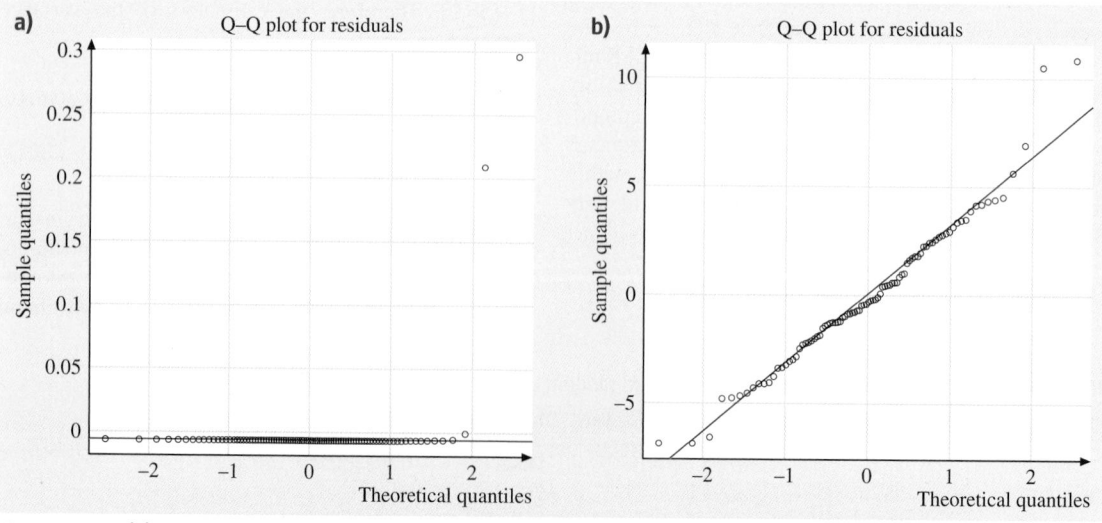

Fig. 56.5a,b (a) Q–Q plot of the residuals for raw data. (b) Q–Q plot for the log-transformed responses

cause the problem instances π_i are here considered as samples from some larger population of instances Π. We test hypotheses about the variance component σ_τ^2, i.e., the null hypothesis

$$H_0 : \sigma_\tau^2 = 0 \text{ versus } H_1 : \sigma_\tau^2 > 0 . \qquad (56.12)$$

Under H_0, the algorithm performance is identical on all problem instances (*all treatments are identical*), i.e., $r\sigma_\tau^2$ is very small. Based on (56.9), we conclude that $E(\text{MS}_{\text{treat}}) = \sigma^2 + r\sigma_\tau^2$ and $E(\text{MS}_{\text{err}}) = \sigma^2$ are similar. Under the alternative, variability exists between treatments. Standard analysis shows that $\text{SS}_{\text{err}}/\sigma^2$ is distributed as chi-square with $q(r-1)$ degrees of freedom. Let $F_{u,v}$ denote the F distribution with u numerator and v denominator degrees of freedom. Under H_0, the ratio

$$F_0 = \frac{\text{SS}_{\text{treat}}/q - 1}{\text{SS}_{\text{err}}/q(r-1)} = \frac{\text{MS}_{\text{treat}}}{\text{MS}_{\text{err}}}$$

is distributed as $F_{q-1,q(r-1)}$. To test hypotheses in (56.8), we require that $\tau_1, \ldots, \tau_q$ are i.i.d. $\mathcal{N}(0, \sigma_\tau^2)$, $\varepsilon_{ij}, i = 1, \ldots, q, j = 1, \ldots, r$, are i.i.d. $\mathcal{N}(0, \sigma^2)$, and all τ_i and ε_{ij} are independent of each other. These considerations lead to the decision rule to reject H_0 at the significance level α if

$$f_0 > F(1 - \alpha; q - 1, q(r - 1)) , \qquad (56.13)$$

where f_0 is the realization of F_0 from the data observed. An intuitive motivation for the form of statistic F_0 can be obtained from the expected mean squares. Under H_0 both MS_{treat} and MS_{err} estimate σ^2 in an unbiased way, and F_0 can be expected to be close to one. On the other hand, large values of F_0 give evidence against H_0.

Regarding the SAMP case, we obtain the following values: Based on (56.9) and (56.13), we can determine the F statistic and the p value. We get $\text{MS}_{\text{treat}} = \text{MS}_{\text{err}} = 10.89275$ and $f_0 = 1$, which results in a large p value: 0.4426363. The null hypothesis $H_0 : \sigma_\tau^2 = 0$ from (56.12) cannot be rejected, i.e., we conclude that there is no instance effect. A similar conclusion was obtained from the ANOVA method of variance component estimation as introduced in Table 56.2.

SAMP-5 Confidence Intervals and Prediction. An unbiased estimator of the overall mean μ is

$$\hat{\mu} = \bar{y}_{..} = \sum_{i=1}^q \sum_{j=1}^r \frac{y_{ij}}{(qr)} .$$

Its variance is given by

$$V(\bar{y}_{..}) = V\left(\sum_{i=1}^q \sum_{j=1}^r \frac{y_{ij}}{(qr)} \right) = \frac{r\sigma_\tau^2 + \sigma^2}{qr} .$$

With (56.9) and (56.10), we obtain an estimator of the variance of the overall mean μ as

$$\hat{V}(\bar{y}_{..}) = \frac{\text{MS}_{\text{treat}}}{qr} .$$

Since

$$\frac{\bar{Y}_{..} - \mu}{\sqrt{\frac{\text{MS}_{\text{treat}}}{qr}}} \approx t_{q(r-1)} ,$$

the confidence limits for μ can be derived as

$$\bar{y}_{..} \pm t_{1-\alpha/2;q(r-1)} \sqrt{\frac{\text{MS}_{\text{treat}}}{qr}} . \qquad (56.14)$$

We conclude the SAMP case study with prediction of the algorithm's performance on a new instance from the same class. Based on (56.14), we obtain the following 95% confidence interval: $[2.6773 \times 10^{-6}; 1.262 \times 10^{-5}]$. Again, confidence intervals from the REML and ANOVA methods are very similar. Summarizing, we can conclude that the ES performs similarly on instances from Π_{MSG}, which were generated with (56.2).

56.5.3 MAMP: Multiple Algorithms, Multiple Problems

In the MAMP case study, fixed effects are included in the conditional structure of (56.6), which leads to a mixed model. Instead of one fixed algorithm as in the SAMP case, we consider either several algorithms or algorithms with several parameters. Both situations can be treated while considering algorithms as levels of a fixed factor, whereas problem instances are drawn randomly from the population of instances Π_{MSG}:

- MAMP-1 algorithm and problem instances
- MAMP-2 ANOVA and REML model building
- MAMP-3 validation of the model assumptions
- MAMP-4 hypothesis testing:
  1. Random effects
  2. Fixed effects
- MAMP-5 confidence intervals and prediction.

MAMP-1 Algorithm and Problem Instances

We aim at comparing the performance of the ES with different recombination operators over an instance class. More precisely, we have four ES instances using recombination operators $\{1, 2, 3, 4\}$ and nine instances randomly sampled from the class Π_{MSG} as illustrated in Fig. 56.3. Each run is repeated ten times. In this study $4 \times 9 \times 10 = 360$ data were used. We are interested in the following questions:

- Is there an instance effect?
- Do the mean performances of the ES with different recombination operators differ?
- Do the instance–algorithm interactions contribute to the variability of the response?

A first visual inspection, which plots the performance of the algorithm within each problem instance, is shown in Fig. 56.6. In eight of the nine instances the linear regression line has a negative slope and the intercepts do not differ very much. This indicates that there is no significant interaction between the fixed and the random factors.

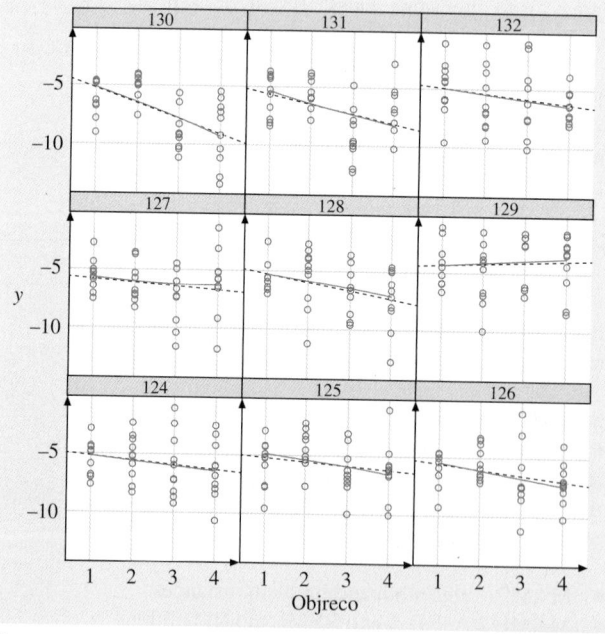

Fig. 56.6 Four algorithms (ES with modified recombination operators) on nine test problem instances. Each *panel* represents one problem instance and problem instances are labeled from 124 to 130. Performance is plotted against the level of the recombination operator

MAMP-2 ANOVA and REML Model Building

The variability in the performance measure can be decomposed according to the following *mixed-effects ANOVA model*

$$Y_{ijk} = \mu + \alpha_j + \tau_i + \gamma_{ij} + \varepsilon_{ijk}, \qquad (56.15)$$

where μ is an overall performance level common to all observations, α_j is a fixed effect due to the algorithm j, τ_i is a random effect associated with instance i, γ_{ij} is a random interaction between instance i and algorithm j, and ε_{ijk} is a random error for replication k of algorithm j on instance i. We assume that the α_j's are fixed effects such that $\sum_{j=1}^{h} \alpha_j = 0$ and that the random elements τ_i are i.i.d. $\mathcal{N}(0, \sigma_\tau^2)$, γ_{ij} are i.i.d. $\mathcal{N}(0, \sigma_\gamma^2)$, ε_{ijk} are i.i.d. $\mathcal{N}(0, \sigma^2)$ and τ_i, γ_{ij} and ε_{ijk} are mutually independent random variables. Similarly to (56.6) the conditional distribution of the performance measure given the instance and the instance–algorithm interaction is given by

$$Y_{ijk} | \tau_i, \gamma_{ij} \approx \mathcal{N}\left(\mu + \alpha_j + \tau_i + \gamma_{ij}, \sigma^2\right), \qquad (56.16)$$

with $i = 1, \ldots, q$, $j = 1, \ldots, h$, and $k = 1, \ldots, r$. The marginal model reads (after integrating out the random effects τ_i and γ_{ij}):

$$Y_{ijk} \approx \mathcal{N}\left(\mu + \alpha_j, \sigma^2 + \sigma_\tau^2 + \sigma_\gamma^2\right). \qquad (56.17)$$

Based on these statistical assumptions, hypothesis tests can be performed about fixed and random factor effects. Using the mixed model (56.16), we are interested in testing whether there is a difference between the factor level means $\mu + \alpha_j$ ($j = 1, \ldots, h$). The hypotheses for testing the fixed effects can be formulated as

$$H_0 : \alpha_i = 0 \ \forall i \quad \text{against} \quad H_1 : \exists \alpha_j \neq 0. \qquad (56.18)$$

Regarding random effects, tests about particular levels are useless. This is similar to the random-effects model (56.8). Again, we perform tests on the variance components σ_τ^2 and σ_γ^2 instead. These can be formulated as follows

$$
\begin{aligned}
H_0: \quad & \sigma_\tau^2 = 0, \quad \text{and} \quad H_0: \quad \sigma_\gamma^2 = 0, \\
H_1: \quad & \sigma_\tau^2 > 0, \quad \text{and} \quad H_1: \quad \sigma_\gamma^2 > 0,
\end{aligned} \qquad (56.19)
$$

respectively. If all treatment (problem instances) combinations have the same number of observations, i.e., if the design is balanced, the test statistics for these hypotheses are ratios of mean squares that are chosen such

Table 56.3 Expected mean squares and consequent appropriate test statistics for a mixed two-factor model with h fixed factors, q random factors, and r repeats (after [56.3])

Effects	Mean squares	Df	Expected mean squares	Test statistics
Fixed factor	MSA	$h-1$	$\sigma^2 + r\sigma_\gamma^2 + rq\frac{\sum_{j=1}^h \alpha_j^2}{h-1}$	MSA/MSAB
Random factor	MSB	$q-1$	$\sigma^2 + r\sigma_\gamma^2 + rh\sigma_\tau^2$	MSB/MSAB
Interaction	MSAB	$(h-1)(q-1)$	$\sigma^2 + r\sigma_\gamma^2$	MSAB/MSE
Error	MSE	$hq(r-1)$	σ^2	

Table 56.4 ANOVA for the MAMP case

Mean squares	Factors	Df	Sum Sq	Mean Sq	F value	Pr(>F)
MSA	objreco	3	154.59	51.53	11.05	0.0000
MSB	fSeed	8	251.79	31.47	6.75	0.0000
MSAB	objreco:fSeed	24	185.60	7.73	1.66	0.0288
MSE	Residuals	324	1511.27	4.66		

that the expected mean squares of the numerator differs from the expected mean squares of the denominator only by the variance components of the random factor under test. *Chiarandini* and *Goegebeur* [56.3] present the resulting analysis of variance, which is shown in Table 56.3.

ANOVA Model Building
The ANOVA table for the experiments from the MAMP case study is shown in Table 56.4. Equating the observed mean squares in the lines of the ANOVA table to their expected values and solving for the variance components leads to the following equations [56.2]

$$\hat{\sigma}_\tau^2 = \frac{MSB - MSAB}{hr} = 0.593502 \,,$$

$$\hat{\sigma}_\gamma^2 = \frac{MSAB - MSE}{r} = 0.306907 \,,$$

$$\hat{\sigma}^2 = MSE = 4.664423 \,.$$

Next, we will compare these results to the REML-based analysis of the mixed model.

REML Model Building
We have specified sum contrasts instead of the default treatment contrasts used in `lmer()`. Again, `fSeed` represents the problem instance, whereas the algorithm instance $\alpha_j, j = 1, \dots, 4$, is represented by `objreco`.

```
Linear mixed model fit by REML
Formula: yLog ~ objreco + (1 | fSeed)
                + (1 | fSeed:objreco)

Random effects:
 Groups          Name        Variance Std.Dev.
 fSeed:objreco (Intercept)  0.30691  0.55399
 fSeed         (Intercept)  0.59351  0.77039
```

```
 Residual                   4.66442  2.15973
Number of obs: 360,
groups: fSeed:objreco, 36; fSeed, 9

Fixed effects:
             Estimate Std. Error t value
(Intercept)  -6.0222     0.2956 -20.370
objreco1      0.6176     0.2539   2.433
objreco2      0.6918     0.2539   2.725
objreco3     -0.6671     0.2539  -2.628
```

As can be seen from the `Random effects` section of the REML model output, the estimated variances for the problem instance and the instance-interaction random effects are $\hat{\sigma}_\tau^2 = 0.59351$ and $\hat{\sigma}_\gamma^2 = 0.30691$, respectively. The `Random effects` section presents the estimates of the fixed effects model parameters, i.e., `objreco`.

MAMP-3 Validation of the Model Assumptions
Again, a check of the diagnostic plots (Fig. 56.7) reveals that a log transformation of the response improves the model adequacy.

MAMP-4a Hypothesis Testing: Random Effects
We will consider random effects first. Regarding problem instances, test about levels are meaningless. Hence, we perform tests about the variance components σ_τ^2 and σ_γ^2, which were presented in (56.19). First, we test the null hypothesis, which states that the components of the random effects are zero. Based on the ANOVA from Table 56.3, we obtain the values for the MAMP case that are shown in Table 56.4. The values reveal that there are main factor effects (fixed and random), but no significant interaction effects.

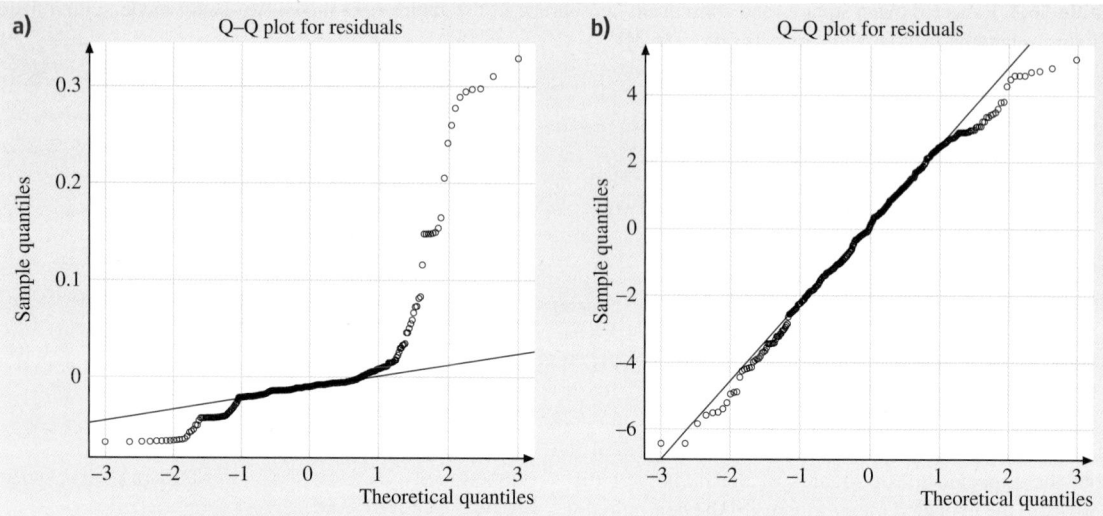

Fig. 56.7a,b (a) Q–Q plot of the residuals for raw data. (b) Q–Q plot for the log-transformed responses

Alternatively, we can compute the likelihood ratios of models with and without the factors under observation.

```
Data: mamp.df
Models:
mamp.lmer2: yLog ~ objreco + (1 | fSeed)
mamp.lmer3: yLog ~ objreco + (1 | fSeed)
                 + (1 | fSeed:objreco)
            Df    AIC     BIC   logLik
mamp.lmer2   6 1616.7 1640.0 -802.35
mamp.lmer3   7 1616.6 1643.8 -801.31
   Chisq Chi Df Pr(>Chisq)
   2.0929      1     0.148
```

These tests indicate that there are also no significant instance-algorithm interactions. Additional likelihood-ratio tests show that the fixed factor and random factor effects are significant.

MAMP–4b Hypothesis Testing: Fixed Factor Effects

Regarding fixed factors, we are interested in testing for differences in the factor level means $\mu + \alpha_i$. These tests were formulated in (56.18), i.e., we are testing H_0: all α_i are equal to 0 versus H_1: at least one $\alpha_j \neq 0$. Here, we use the test statistic from [56.2, p. 523] for testing that the means of the fixed factor effects are equal. The appropriate test statistic for testing that the means of the fixed factor effects are equal, i.e., H_0 is true, is

$$F_0 = \frac{\text{MSA}}{\text{MSAB}} = \frac{154.59/3}{185.6/24} = 6.663\,362 ,$$

with values taken from Table 56.4. The reference distribution is $F_{n-1,(n-1)(q-1)}$. We calculate the p value for the test on the fixed-effect term. The p value obtained is 0.002, hence the results collected indicate that the factor recombination (objreco) has a statistically significant impact on the performance of the algorithm. Using sum of contrasts implies that $\sum \alpha_j = 0$. The point estimates for the mean algorithm performance with the j-th fixed factor setting can be obtained by $\mu_{\cdot j} = \mu + \alpha_j$. The fixed factor effects can be estimated

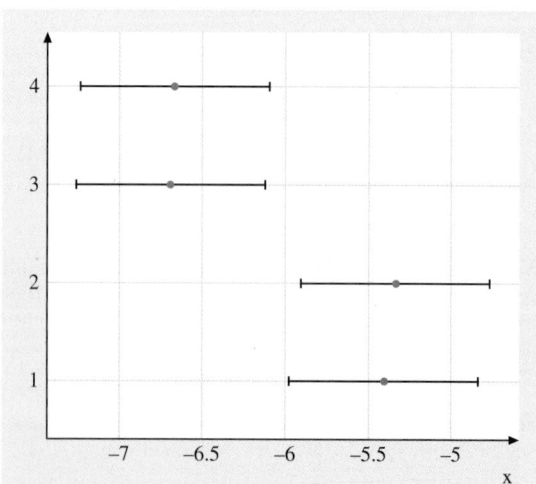

Fig. 56.8 Paired comparison plots. Results from four ES instances with different recombination operators are shown in this plot

in the mixed model as

$$\hat{\mu} = \overline{y}_{...}$$
$$\hat{\alpha}_j = \overline{y}_{j.} - \overline{y}_{...} ,$$

which results in the following estimates: $\hat{\alpha}_1 = 0.6175519$, $\hat{\alpha}_2 = 0.6918047$, $\hat{\alpha}_3 = -0.6671266$, and $\hat{\alpha}_4 = -0.6423659$.

The same estimates were obtained with the REML analysis as can be seen from the REML model output in Sect 56.5.3. The corresponding fixed effects are shown in the `Fixed effects` section of the REML output. For example, we obtain the following value: `objreco1` $= \hat{\alpha}_1 = 0.6176$.

MAMP-5 Confidence Intervals and Prediction

We generate paired comparisons plots, which are based on confidence intervals. The wrapper function `intervals()` from *Chiarandini* and *Goegebeur* [56.3] was used for visualizing these confidence intervals as shown in Fig. 56.8. When intervals overlap we conclude that there is no significant difference. Here, we can conclude that the recombination operators (1) and (2) show a similar performance, whereas performances between (3) and (2) are different.

Intermediate recombination of the object variables, i.e., (3) and (4), results in a significant improvement of the performance.

56.6 Summary and Outlook

In order to answer question (Q-1), we propose an approach to generate natural problem classes, which are based on real-world data. If no such data are available, artificial problem generators such as MSG can be used. Since our approach uses a model, say M, to generate new problem instances, one conceptual problem arises: this approach is not applicable, if the final goal is the determination of a model for the data, because M is per definition the best model in this case, and the search for good models will result in M. However, there is a simple solution to this problem. In this case, the feature extraction and model generation should be skipped and the original data should be modified by adding some noise or performing transformations on the data. Nevertheless, if applicable, the model-based approach is preferred, because it sheds some light on the underlying problem structure.

The model-based approach can be used to generate infinitely many test-problem instances. Instead of using a fixed number of problem instances, we propose:

1. Using randomly generated problem instances
2. Treating the problem instance as a random factor.

Algorithms with different parameterizations are tested on this set of randomly generated problem instances. This experimental setup requires modified statistics, so-called random-effects models or mixed models. This approach may lead to objective evaluations and comparisons. If normality assumptions are met, confidence intervals can be determined, which *forecast* the behavior of an algorithm on unseen problem instances. Furthermore, results can be generalized in real-world settings. This gives an answer to question (Q-2).

In order to demonstrate the applicability of our approach, the performance of an evolution strategy was analyzed. The first SAMP example illustrates that the selection of the problem instance from the problem class Π_{MSG} has no significant impact on the performance of the optimization algorithm. Furthermore, confidence intervals, which can be used to predict the performance of the algorithm on a problem class, were determined. The MAMP case exemplifies how to analyze the effect of different algorithm parameter settings on the performance. Four variants of the recombination operator and nine problem instances were used. The analysis reveals that the choice of the recombination operator has a significant effect on the algorithm's performance: the performance of the algorithm differs with different recombination operators. Intermediate recombination of the object variables results in an performance improvement. We demonstrated that the problem instances contribute significantly to the variability in the response and that there is no significant instance–algorithm interaction.

The software that was used in this study will be integrated into the R package SPOT (sequential parameter optimization toolbox) [56.19].

References

56.1   P.J. Brockwell, R.A. Davis: *Introduction to Time Series and Forecasting* (Springer, New York 2002)

56.2   D.C. Montgomery: *Design and Analysis of Experiments* (Wiley, New York 2001)

56.3   M. Chiarandini, Y. Goegebeur: Mixed models for the analysis of optimization algorithms. In: *Experimental Methods for the Analysis of Optimization Algorithms*, ed. by T. Bartz-Beielstein, M. Chiarandini, L. Paquete, M. Preuss (Springer, Berlin Heidelberg 2010) pp. 225–264, preliminary version available as Tech. Rep. DMF-2009-07-001 at the The Danish Mathematical Society

56.4   C.C. McGeoch: Toward an experimental method for algorithm simulation, INFORMS J. Comput. **8**(1), 1–15 (1996)

56.5   M. Birattari: *On the Estimation of the Expected Performance of a Metaheuristic on a Class of Instances 2004*, Tech. Rep. (IRIDIA, Bruxelles 2004)

56.6   M. Gallagher, B. Yuan: A general-purpose tunable landscape generator, IEEE Trans. Evol. Comput. **10**(5), 590–603 (2006)

56.7   I. Castiñeiras, M.D. Cauwer, B. O'Sullivan: Weibull-based benchmarks for bin packing, Lect. Notes Comput. Sci. **7514**, 207–222 (2012)

56.8   G.E.P. Box, G.M. Jenkins, G.C. Reinsel: *Time Series Analysis, Forecasting and Control* (Holden-Day, San Francisco 1976)

56.9   T. Bartz-Beielstein: *Beyond Particular Problem Instances: How to Create Meaningful and Generalizable Results*, Tech. Rep. TR 03/2012 (Cologne University of Applied Sciences, 2012)

56.10   T. Bartz-Beielstein, M. Friese, B. Naujoks, M. Zaefferer: SPOT applied to non-stochastic optimization problems – An experimental study, Genet.

56.11   M. Zaefferer: Optimization and Empirical Analysis of an Event Detection Software for Water Quality Monitoring, Master Thesis (University of Applied Sciences, Cologne 2012)

56.12   O. Flasch, T. Bartz-Beielstein, D. Bicker, W. Kantschik, C. von Strachwitz: *Results of the GECCO 2011 Industrial Challenge: Optimizing Foreign Exchange Trading Strategies*. CIOP Tech. Rep. 10/11, Res. Center CIOP (Cologne University of Applied Science, Cologne 2011)

56.13   E. Vladislavleva: Model-based Problem Solving through Symbolic Regression via Pareto Genetic Programming, Ph.D. Thesis (Tilburg University, Tilburg 2008)

56.14   H.-G. Beyer, H.-P. Schwefel: Evolution strategies – A comprehensive introduction, Nat. Comput. **1**, 3–52 (2002)

56.15   H.-P. Schwefel: *Numerical Optimization of Computer Models* (Wiley, Chichester 1981)

56.16   T. Bartz-Beielstein: *Experimental Research in Evolutionary Computation – The New Experimentalism*, Natural Computing Series (Springer, Berlin, Heidelberg, New York 2006)

56.17   T. Bartz-Beielstein, M. Preuss: Automatic and interactive tuning of algorithms, Proc. 13th Annu. Conf. Companion Genet. Evol. Comput. (GECCO) (ACM, New York 2011) pp. 1361–1380

56.18   J. Pinheiro, D. Bates: *Mixed-Effects Models in S and S-PLUS* (Springer, New York 2000)

56.19   T. Bartz-Beielstein, M. Zaefferer: A Gentle Introduction to Sequential Parameter Optimization, Tech. Rep. TR 01/2012 CIplus Bd. 1/2012 (Cologne University of Applied Sciences, Cologne 2012)

Evol. Comput. Conf. (GECCO 2012) (ACM, Philadelphia 2012)

57. Computational Intelligence in Industrial Applications

Ekaterina Vladislavleva, Guido Smits, Mark Kotanchek

In this chapter, we review the progress and the impact of computational intelligence for industrial applications sampled from the last 10 years of our personal careers and areas of research (all authors of this chapter do computational modeling for a living). This chapter is structured as follows. Section 57.2 introduces a classification of data-driven predictive analytics problems into three groups based on the goals and the information content of the data. Section 57.3 briefly covers most frequently used methods for predictive modeling and compares them in the context of available a priori knowledge and required execution time. Section 57.4 focuses on the importance of good workflows for successful predictive analytics projects. Section 57.5 provides several examples of such workflows. Section 57.6 concludes the chapter.

57.1 Intelligence and Computation

Developments in computational intelligence (CI) are driven by real-world applications. Over the years a lot of CI has become ubiquitous to the average user and is deeply interwoven into the way modern design, research and development is done.

In our view, CI is human intelligence assisted and (dramatically) enhanced by computational modeling. Intelligence is the capability to predict, and, in theory, there are two directions to get to prediction through computing – data-driven modeling and first principle modeling. In reality though, since even fundamental models and theories have to be validated by data, everything is data driven. For this reason, from now on we will focus on data-driven computational modeling, and say that it exists to enhance predictive capabilities

of the human or business. While prediction is the ultimate goal and computational modeling is the means to achieve this goal, we will use concepts of predictive analytics and (data-driven) computational modeling as if they were the same.

Computational modeling methods allow us to generate various hypotheses about a specific problem based on observations in an objective way. The mental models that the scientists develop during this process help them to filter through these hypotheses and come up with new experiments that either support or falsify some of the previous hypotheses or lead to new ones. This process supports the scientific method and significantly accelerates technological development and innovation.

There are many examples of new computational methods empowering problem solving in the areas of material science, energy management, plant optimization, sensory evaluation science, broadband technology, social science (economic modeling), infectious disease prevention, etc. And while success in many cases is undeniable, two main challenges still remain.

First, there is an education gap to bridge before modern CI techniques can reach their full potential, are widely accepted, and become as natural as performing experiments in the lab. While many engineering educational programs are embracing these techniques and help raise awareness of the useful methods in data-driven modeling and computational statistics, the majority of programs in pure sciences tend to ignore them for the most part. There is still a considerable (psychological, cognitive, educational) barrier for experimental scientists – biologists, chemists, physicists, computer scientists – to fully exploit the potential of CI. People will happily save an hour of computing time by spend-ing an additional week in the lab, while in some cases it makes much more sense to spend a week of computing time to spare one experiment in the lab (consider, e.g., an expensive car *crash-test*). We appeal to educational programs to nurture the interest in computation among graduates and facilitate the joint projects of academia with industry targeted at the use and further development of computational intelligence methods for real-world problems.

Second, there is a development gap in the production of scalable off-the-shelf CI algorithms. The parallelization bottleneck seems to affect most CI methods when they are executed on massively parallel architectures. The fact that computational advances in hardware (exa-scale computing) happen at a much faster pace than advances in the design of scalable CI algorithms raises the question: *Up to which moment can we get more intelligence, i. e., more predictive capability, with more computational power?*

57.2 Computational Modeling for Predictive Analytics

While many barriers remain in improving the incorporation of CI in classical education, in solving the new (previously unthinkable) challenges, and in further innovating CI technology, the current time is a perfect moment to make this happen.

First of all, the realization for the indispensability of CI across all industries and all sciences grows as does the number of required CI practitioners (computational statisticians, data scientists, modelers). The report of *Manyika* et al. on Big Data [57.1] predicts a potential gap of 50–60% (300 000 people) in demand relative to the supply of well-educated analytical talent in the USA by 2018. The *data science* and *big data* movement have grown in the last decade to become a buzz-word omnipresent in scientific magazines, technology reviews, and business offerings.

While we are happy that the attention of the average user is being focused on the importance and impact of computational modeling, we are also concerned with the fact that too many details are omitted and almost everything (business strategies, CI methods, targets for predictive analytics, etc.) gets thrown onto one pile. While Big Data is occupying the minds of future engineers, data scientists, and business majors as a *next big thing to watch* and a synonym of predictive analytics, we want to balance the story some more and provide a full picture of what we think constitutes predictive an-alytics by computational modeling. While business and industry is striving to become data driven these days, it seeks CI strategies to compete, innovate, and capture value. Success and impact of CI will be generated only if the right strategies are used in the right place.

Success of CI in industry will be awarded to methods that create impact measured in attaining the new level of understanding and knowledge, in units of dollars. In Fig. 57.1 we sketch a relation between the degree of intelligence and the level of competitive advantage from [57.2]. Further on, we will use the terms predictive analytics and computational modeling (for predictive analytics to sustain human intelligence) as if they were the same.

We distinguish three pillars of computational modeling for predictive analytics: business analytics applied to big data (millions to billions of records, dozens to hundreds of variables), process analytics applied to medium-sized data (tens of thousands of records, hundreds of variables), and research analytics applied to precious data (tens to thousands of records, dozens to hundreds to thousands of variables) (Fig. 57.2).

57.2.1 Business Analytics

Business analytics is the part of predictive analytics associated with big data. In recent years, other sciences also created big data problems, so the field could be called big data analytics. The distinguishing feature of business analytics is the fact that it is used to inspect big data streams to provide a quick and simple analysis with immediate value reliably and consistently. Because of the size, big data already offers tremendous challenges in stages preceding analytics – in storage, retrieval, and visualization. These imply that the predictive goals can only be modest, except when big computing facilities and specialized data bases are available (like it happens in environmental and biological research, Internet search, smart grids, etc.). Main goals here are:

- Visualization (e.g., dashboards).
- Recommendation (e.g., studying customer habits and preferences to recommend a new suitable product item). Recommendation uses network analysis to select relevant or similar items.
- Identification of (simple) trends to enhance customer experience and increase surplus. Trends are typically found using time series analysis.
- Binary classification to distinguish out-of-the-ordinary data points from the prototypes following the trends (credit risk analysis, fraud detection, spam identification).

Because of the memory limitations, the challenge in business analytics is to quickly give an answer to simple questions with the main focus on algorithms for in- and out-of-memory computation and visualization. Industries benefitting most from business analytics are retail, banking, insurance, health-care, telecommunications, and social networks.

For example, at large multinational manufacturing companies, business analytics predominantly revolves around the multivariate forecasting of supply and demand. The expected prices and volumes of feedstocks and raw materials as well as the expected demand for various products are important to minimize risk and optimize production as well as the supply chain. Classical statistical forecasting techniques are the main workhorse for this area and the main challenges consist of being able to gather the required data, dealing with possibly large numbers of candidate inputs and outputs for the models and properly dealing with the hierarchies that exist, e.g., products-markets-industry resulting in an explosion in the number of models that have to be built and maintained.

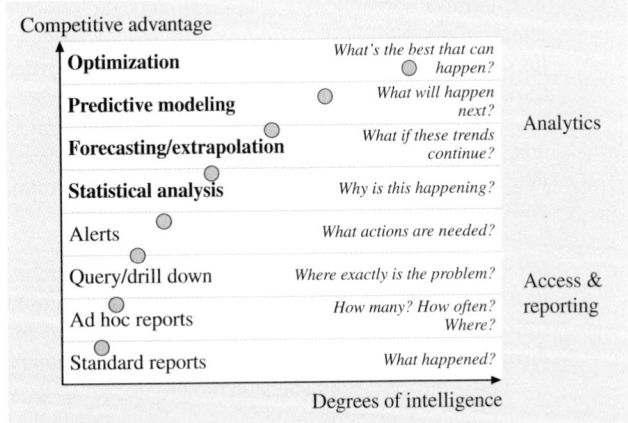

Fig. 57.1 *Davenport* and *Harris* [57.2] have wonderfully adapted the graphics from SAS software. The *graph above* eloquently explains why to use predictive modeling to excel, compete, and capture value

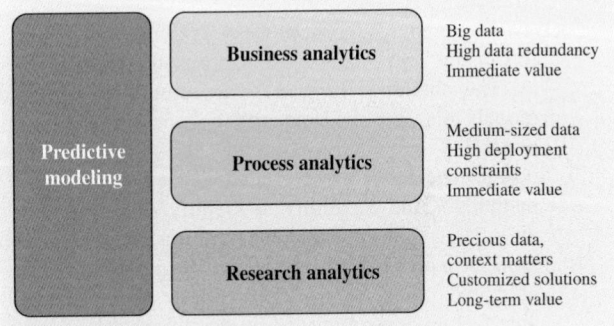

Fig. 57.2 Predictive modeling has three components: Business analytics, predictive analytics, and research analytics

57.2.2 Process Analytics

Process analytics exploits medium-size data to generate time-sensitive prediction of an industrial process (e.g., manufacturing, process monitoring, remote sensing, etc.) with immediate value.

Process analytics models must be very robust, simple (mostly linear), and concise to be deployed in real industrial processes.

This well understood and probably most conservative area of predictive analytics has experienced a big change in the last years. A couple of decades ago, process optimization and control groups had more people and less pressure. Nowadays, pressure for integrating production workflows has increased together with the need to meet tighter quality specifications, much tighter emission thresholds, requirements to reduce production, operation, and energy costs, and to maximize throughput. The sensor's side has changed – sensors have become much more sophisticated and much more abundant. The human interference has also decreased due to (sometimes exaggerated) drive to automation, and cost reduction.

All these factors have dramatically increased the demand for reliable optimization and control. In general, process analytics models must be very robust, simple (mostly linear) and concise to be deployed in real industrial processes. The main challenge for this industry is to integrate more sophisticated models and adopt new computational methods for process analytics to adapt to the changing world of new requirements while maintaining robustness over a wide process range. At the time when this chapter was written data-driven modeling was still considered exotic for the field of process analytics, and model deployment is still heavily constrained.

The main goals in process analytics are process forecasting and process optimization and control.

The challenge in process forecasting is to build models that hit the tradeoffs between model interpretability and their long-term (real-time) predictive power. The technological challenge of successful CI methods is the capability to identify driving features in a large set of correlated features. For example, think of a problem of predicting the quality of a manufactured plastic using the smallest subset of available factors controlling the production process – pressures, temperatures, flows. Robust feature selection is as important as good prediction accuracy – models that are too bulky will never be accepted by process engineers.

The main challenge in process control is the multiobjective nature of control specifications and subsequent optimization problems. Consider an example of manufacturing and wholesaling thin sheets of metal. The thickness of the sheet is an important quality characteristic that should not fall below a predefined minimum, or the product will be considered off-spec. If due to the processing condition the thickness variability is high (sheets are several meters wide and tens of meters long, rolled at high speeds, high temperatures), penalty for off-spec material is high, and costs for raw steel are also high – the manufacturer faces a delicate problem of making the sheet thicker than the allowed minimum to keep the clients happy but not too thick to keep the production costs down. These competing objectives usually require a multiobjective approach to process optimization.

Process analytics relies on a rich data set coming from the many sensors in a typical plant. Mature platforms exist that store this, often high-frequency, sensor data in databases and plant information systems. The primary intent for this data is to run the various plant control and quality control systems but archived data are often available for predictive modeling as well. The use of models that predict the aging and lifetime of catalysts and the associated changes in optimal settings for the plant are good examples.

Another example is the building of the so-called soft sensors that link difficult measurements, such as, e.g., grab samples that need to be brought to the lab for analysis with results only becoming available after some time to some of the easier high-frequency measurements, such as, e.g., temperatures, flows, and pressures. These models then serve as substitutes for the difficult measurements at a high frequency and can be calibrated if needed when the slow measurements become available. There are also many opportunities to use the demand and supply forecasting models from the business analytics side to optimize production and product mix that is most optimal for a given scenario. As an extreme example, it may be cheaper to shutdown a plant for a while vs continued production when demand is forecasted to be very weak. The level and amount of coupling that is possible between demand–supply forecasts and actual production can vary significantly and depends on many factors, but it is clear that much more is possible in this area.

Examples of industries employing process analytics are manufacturing, chemical engineering, energy, environmental science.

57.2.3 Research Analytics

Research analytics is used to accelerate the development of new products and systems. This task is fundamentally different from all the ones mentioned previously as it is usually applied to small, complex, and precious data, is heavily dependent on problem context and provides long-term value without immediate rewards. (By *small* we mean any data set where the number of record is comparable or even smaller than the number of features. In this way, gene expression data with thousands of variables taken over dozens of individuals is small.)

Research analytics provides very customized solutions and requires a close collaboration between analysts/modelers and subject matter experts.

Research analytics is by nature much less generic and becomes very dependent on the specific product that is being developed. In research, once you have predictive analytics, then there is only a small step to make from optimization of existing products to the design of new ones. One example of a research analytics success story is the development of an application to predict the exact color of a plastic part based on the composition of the colorants and the specific grade a plastic being used, see [57.3]. Robust color prediction models led to the capability of actually designing colorant compositions in silico directly from customer specifications. The models also provided the specifications that were necessary to even let the customer produce that part himself.

How far one is able to take this depends on the fidelity of the models as well as the quality of the data that is available. Another example of research analytics at work is the design of new coatings and catalysts based on high throughput experimentation where all the available data is being used to build models on the fly. These models are than used to design the next experiments such that the information gain is maximum. The requirements for the modeling process are quite high because everything is embedded in a high-throughput workflow but the benefits are also huge. Significant speedups in the total design time as well as the performance of new products can be achieved this way.

We stress that because research analytics is an enhancement to human intelligence for the development of new products and systems, the benefits of its application scale proportionally with the size of the problem and the impact of that particular product or system. For big enough problems the benefits quickly get into the hundreds of thousands to millions of dollars.

Research analytics can help drive innovation in all industry segments, particularly in materials science, formulation design, pharmaceuticals, engineering, simulation-based optimization in research, bio-engineering, healthcare, telecommunications, etc. In the coming 10 years, we will continue to see the trend of innovation enabled to a large extent by predictive modeling.

57.3 Methods

Over many years of exercising process and research analytics, we built up a practice of using predictive modeling as the integration technology for real-world problem solving. In the last 8–10 years, predictive modeling for computational intelligence has evolved from the solution of last resort to the main stream approach to industrial problem solving (prediction, control, and optimization). It is technology that glues together fundamental modeling and domain expertise, high-performance computing and computer science, empirical modeling and mathematics – a heaven for an inquiring mind and interdisciplinary enthusiast.

Predictive modeling is a bridge that connects theory and facts (data) to enable insight and system understanding. The theory for poorly understood problems is often based on simplifying assumptions, on which the fundamental models are built. The facts, or empirical evidence, are often affected by high uncertainty and a limited observability of the system's behavior.

Predictive modeling applied iteratively to a growing set of facts tests the theory against the data and *extrapolates* models build on the data to confirm or adjust the theory until the theory and facts start to agree.

The validation always lies in the hands of a subject matter expert who in the case of success accepts both the theory and the designed predictive models as plausible and interesting. While the real validation comes when models are deployed and keep generating value, without the first step of intriguing the subject matter expert the project does not have a chance to succeed.

To clear any obstacles toward the acceptance of models by the domain expert the models should be:

1. Interpretable
2. Parsimonious
3. Accurate
4. Extrapolative
5. Trustable, and
6. Robust.

In an industrial setting, the capability of having a trustable prediction of the output within and outside the training range is as important as interpretability and the possibility of integrating information from first principles, low maintenance and development costs with no (or negligible) operator interference, robustness with respect to the variability in data, and the ability to detect novelties in data to attune itself toward changes in system's behavior.

There is no single technique producing models that are guaranteed to fulfill all of the requirements above, but rather there is a continuum of methods (and hybrids) offering different tradeoffs in these competing objectives.

Commonly used predictive modeling techniques include linear regression, and nonlinear regression [57.4], boosting, regression random forests [57.5], radial-basis functions [57.6], neural networks [57.7], support vector machines (SVM) [57.8, 9], and symbolic regression [57.10, 11] (see more in [57.12]).

In Fig. 57.3, we place some of the most common methods for predictive modeling for process and research analytics in the objective space of development time versus the level of a priori knowledge about the problem. When identifying which methods to use other objectives (like interpretability and extrapolative capability) must also be taken into account. The time axis is depicted on a log scale, and the exact development time depends on implementation and a particular algorithm flavor.

Support vector machines and ensemble-based neural networks lose to linear, nonlinear, and regularized regression in interpretability, but have advantages for problems where little a priori information is known about the system, and no assumptions on model structures can be made (see Fig. 57.3).

Regression random forests, and symbolic regression [57.13–15] have further advantages for problems where not only model structures but also the variable drivers (significant factors) are unknown.

Random forests proved to be robust and very efficient for predicting the response within the training range and for identifying the most significant variables, but because they do not possess extrapolative properties they can only be used in problems where no extrapolation is necessary. Recent studies [57.16] indicate that variable selection information obtained by random forests can loose meaning if correlated variables are present in the data and affect the response differently.

In business analytics, when the speed of model development is the main goal, linear regression and regularized learning are the only remaining options. (Recent developments for predictive modeling for big data are also focusing on the feature generation problem, where the set of original data variables gets expanded to a much larger set of new features – transformations of the original variables, for which regularized linear regression is applied. Much like in support vector regression).

In process analytics when the driving input factors are known – ensemble-based neural networks, support vector regression, and ensemble-based symbolic regression are the modeling alternatives.

If very little is known about the process or system, experiments are demonstrating correlations among input variables, and concise interpretable models are required – symbolic regression is the only resort, which comes at a price of a higher development time (Fig. 57.3).

We stress the importance of using ensembles of predictive models irrespective of which modeling method is used. Ensemble disagreement used as a trustability measure defines the confidence of prediction and is crucial for reliable extrapolation. (It cannot be stressed enough that all prediction in a space of dimensionality above 3 is mostly extrapolation, even when evaluated inside the training range.)

We deal mostly with process and research analytics. In our experience, the aspect of trustablility via ensembles of global transparent models, coupled with the massive algorithmic efficiency gains and the ability to easily handle real-world data with spurious and correlated inputs has led to symbolic regression largely replacing neural networks and support vector machines in our industrial modeling. Our experience also is that symbolic regression models tend to extrapolate well as well as provide warning of that extrapolation.

The reason for symbolic regression being successful for process and research analytics is the fact that all real-world modeling problems we have seen up to now contained only a dozen of relevant inputs (never more

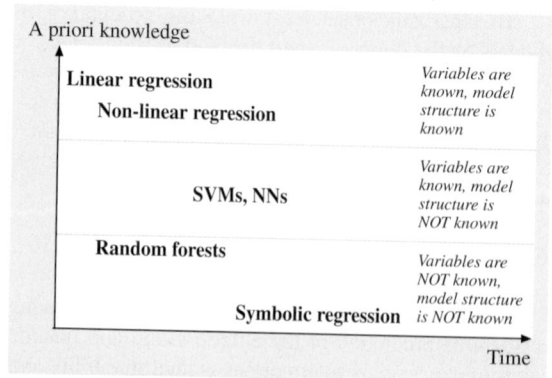

Fig. 57.3 Predictive modeling methods as competing tradeoffs in development time versus the level of a priori knowledge about the problem

than 25 variables, in most cases less than 10) which were truly significantly related to the response. Because symbolic regression searches for plausible models in a space of all possible structures from the given set of potential inputs, and allowed functional transforms, the computational complexity increases nonlinearly with the dimensionality of the true design space. For this reason, symbolic regression effortlessly identifies dozens of driving variables among tens to hundreds of candidates (albeit using hours of multicore computing time). But it should not be used for problems where hundreds of inputs are significantly related to the response and should be filtered out of thousands of candidates. We claim though that no methods are available to solve the latter kind of problems because the necessary amount of data capturing true input–output relationships will never be collected.

Although tremendous progress has been made over the past decade in terms of the efficiency and quality of symbolic regression model development, we also have made corresponding advances from a holistic perspective encompassing the overall modeling workflows from data collection through model deployment.

57.4 Workflows

Although there is no universal solution for predictive modeling and no size fits all, especially for research analytics, nothing is as important for a successful solution as a good modeling workflow.

We would like to make a case for the utmost importance of workflows and the need to nurture and actively proliferate them through all CI projects. In the next section, we give an example on how a successful workflow developed in a project from flavor science could be seamlessly applied to a project in video quality prediction. And because predictive modeling for CI will soon be used in nearly all industry segments and research domains, we believe that it is the responsibility of CI practitioners to facilitate innovation through proliferation and popularization of (interpretable) workflows allowing straightforward application in new domains.

The most general approach to practical predictive modeling is depicted in Fig. 57.3.

We view this generic framework as an iterative feedback loop between three stages of problem solving (just as it usually happens in real-life applications):

1. Data generation, analysis and adaptation
2. Model development, and
3. Problem analysis and reduction.

An important observation is made in the *Toward 2020 Science* report edited by *Emmott* and *Rison* [57.17]:

> *What is surprising is that science largely looks at data and models separately, and as a result, we miss the principal challenge – the articulation of modelling and experimentation. Put simply, models both consume experimental data, in the form of the con-*

> *text or parameters with which they are supplied, and yield data in the form of the interpretations that are the product of analysis or execution. Models themselves embed assumptions about phenomena that are subject of experimentation. The effectiveness of modeling as a future scientific tool and the value of data as a scientific resource are tied into precisely how modelling and experimentation will be brought together.*

This is exactly the challenge of predictive modeling workflows – a holistic approach to bring together data, models, and problem analysis into one generic framework. Ultimately, we want to automate this iterative feedback loop over data analysis and generation, model development, and problem reduction as much as possible, not in order to eliminate the expert, but in order to free as much thinking time for the expert as possible.

This philosophical shift away from human replacement in the modeling workflow toward human augmentation has been very important in the last decade. A successful workflow must offer suites which mine the developed models to identify driving factors, variable combinations, and key variable transforms that lead to insight as well as robust prediction.

57.4.1 Data Collection and Adaptation

Very often, especially in big companies, and especially for process analytics, CI practitioners do not have access to data creation and experiment planning. This gap is a typical example of a situation, where multivariate data is given and there is no possibility to gather better sampled data.

In other situations, there is a possibility to plan the experiments, and gather new observations of the response for desired combinations of input variables, but the assumption always is that these experiments are very expensive, i.e., require long computation, simulation, or experimentation time. Such a situation is most common in research analytics and meta modeling for the design and analysis of simulation experiments.

The questions to ask at the data collection and adaptation stage are: How to design experiments within the available timing and cost budget to optimally cover the design space (possibly containing spurious variables)? How can available data and developed models guide design-space exploration in the next iterations? Is available data well sampled? Is it balanced? What is the information content of performed experiments? Is there redundancy in the data and how to minimize it?

57.4.2 Model Development

In model development, the focus is on automatic creation of collections of diverse data-driven models that infer hidden dependencies on given data and provide insight into the problem, process, or system in question.

Irrespective, of which modeling engines are used at this stage, the questions on how to best generate, evaluate, select, and validate models given particular data features (size and dimensionality) are of great importance. Model quality, in general, i.e., generalization, interpretability, efficiency, trustworthiness, and robustness is the main focus for model analysis leading to the next stage.

57.4.3 Problem Analysis and Reduction

The stage of problem analysis and reduction supposes that developed models are carefully scrutinized, filtered, and validated, to infer preliminary conclusions on problem difficulty. The focus is on driving inputs, assessment of variable contribution, linkages among variables, dimensionality analysis, and construction of trustable model ensembles. The latter if defined well will contribute to intelligent data collection in the style of active learning.

With a goal to augment human intelligence by computation, we emphasize the critical need for a human, an inquiring mind who will test the theory, the facts (data) and their interpretations (models) against each other to iteratively develop a convincing story where all elements fit and agree.

57.5 Examples

57.5.1 Hybrid Intelligent Systems for Process Analytics

A good example of a unified workflow for process analytics is the hybrid intelligent systems framework popularized at the Core R&D department of the Dow Chemical Company in the late 1990s.

The methodology was developed to improve soft sensor performance (performance of predictive models), to shorten its development time, and minimize maintenance. It employed different intelligent system components – genetic programming, support vector machines, and analytic neural networks [57.18].

The process analytics in this framework consists of three steps following data collection:

1. Data preprocessing and compression. Support vector regression using the ϵ-insensitive margin is used to identify and remove data outliers and compress data to a representative set of prototypes (support vectors). The result is a clean and compressed data set.

2. Preliminary variable selection using ensemble-based stacked analytic neural networks [57.19]. The result of this step is a ranking of input variables and quantification of variable contribution based on iterative input elimination and re-training.

3. Convolution parameter estimation to identify relevant time-lags of significant inputs using appropriate convolution functions.

4. Development of transparent predictive models using symbolic regression via genetic programming and final variable selection using symbolic regression models.

5. Model selection and analytical function validation.

6. Online implementation.

7. Soft sensor maintenance to guarantee robustness of process prediction.

Examples of the use of this workflow for reactor modeling can be found in [57.18].

We all practiced the hybrid intelligent systems workflow in the past, but the massive algorithmic efficiency gains in ensemble-based symbolic regression via genetic programming of the last decade [57.14, 20] have led us to simplify the workflow and largely eliminate steps one and two to replace them by direct application of symbolic regression.

57.5.2 Symbolic-Regression Workflow for Process Analytics

The major modeling engine breakthrough was the incorporation of a multiobjective viewpoint; this introduced orders of magnitude improvements in model development speed while simultaneously allowing the analyst to choose the proper balance between complexity and accuracy post facto. In essence, the data could now define the appropriate model structure and driving inputs, which became the main reason for symbolic regression's success for predictive modeling.

Other conceptual advances ordinal genetic programming, interval arithmetic, Lamarckian evolution and secondary optimization objectives, such as age, model dimensionality, nonlinearity, etc., have brought us to the current situation where we can largely inject data into a (properly designed) symbolic regression engine and interesting and useful models will emerge.

The symbolic regression workflow has become as depicted in Fig. 57.4, but with model development done using Pareto-aware symbolic regression [57.14].

Distillation Tower Example

The dataset comes from an industrial problem on modeling gas chromatography measurements of the composition of a distillation tower and is available online at http://www.symbolicregression.com.

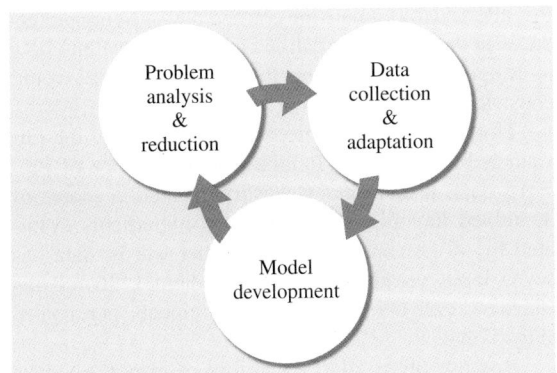

Fig. 57.4 Generic iterative model-based problem solving workflow (after [57.15])

A chemical reaction typically generates a variety of chemicals along with the one (or several) of interest. One method of isolating the mixture coming from the reactor into various purified components is to use a distillation column. The (hot gaseous) input stream is fed into the bottom and on the way to the top goes through a series of trays having successively cooler temperatures. The temperature at the top is the coolest. Along the way, different components will condense at different temperatures and be isolated (with some statistical distribution on the actual components). With vapors rising and liquids falling through the column, purified fractions (different chemical compounds) can be retrieved from the various trays. The distillation column is very important for the chemical industry because it allows continuous operation as opposed to a batch process and is relatively efficient.

This distillation column problem contains nearly 7000 records and 23 potential input variables – mixture of flows, pressures, and temperatures – in addition to the quality metric and material balance. The response variable is the concentration of a purified component at the top of the distillation tower. This quality variable needs to modeled as a function of relevant inputs only. The range of the measured quality metric is very broad and covers most of the expected operating conditions in the distillation column.

To design the test data, we sorted the samples by their response values and selected every third and seventh samples for the validation set and every fourth and eight samples for the test set. The remaining points formed the training set.

Many input variables in the data are heavily correlated. Because symbolic regression can deal with correlated variables, we used all 23 inputs in the first round of modeling to perform initial variable importance analysis.

The workflow that follows exploratory data analysis is described below:

1. *Initial modeling*: We allocated 2 hours of computing time on a quad-core machine to perform 24 20-minute independent runs of symbolic regression by genetic programming using Evolved-Analytics' DataModeler [57.14]. All symbolic regression runs used basic arithmetic operators augmented by a negation and a square as primitives. All models were stored on disk, and all other settings set to default settings of the symbolic regression function of [57.14]. In total, more than 3000 symbolic

regression models were generated during 24 independent runs.

2. *Variable importance analysis*: For all models presence-based importances were computed. Figure 57.5 demonstrates that only a handful of variables is identified as drivers ([57.14] suggests to use importance threshold of 20%).

3. *Variable combination analysis*: All developed models were analyzed for dimensionality and most frequent variable combinations. In Fig. 57.6, one can see model subsets niched according to constituting variable combinations. The bottom graph suggests that variables colTemp1, colTemp3, and colTemp5 might be sufficient for describing the response, since they cover the *knee* of the Pareto front in complexity vs. accuracy space.

4. *Variable contribution analysis*: Models were simplified by identifying and eliminating the least contributing variable. Variable combination analysis was repeated for simplified models and resulted in identifying colTemp1 and colTemp3 as new candidates for a sufficient subspace.

5. *New runs performed on a subset of input variables identified as drivers*: The new batch of independent symbolic regression runs was applied to the same data but only using colTemp1 and colTemp3 as the candidate input variables. As expected, models generated in this experiment demonstrated that the same complexity–accuracy tradeoffs can be achieved in only two-

6. *Ensemble generation using developed models and a validation set*: Final model ensemble was generated automatically using developed symbolic regression models and validation data set. It was augmented by quadratic and cubic models on two variable drivers.

7. *Ensemble prediction validation using test data*: Ensemble prediction and ensemble disagreement were finally evaluated on the test data. Initial requirements for prediction accuracy to not exceed 5–7% of standard deviation were met by all ensemble models. Ensemble prediction is graphed in Fig. 57.7.

This example demonstrates the use of a good model development workflow. An ensemble similar to the one described here has been deployed for controlling a gas chromatography measurement in a real distillation column.

57.5.3 Sensory Evaluation Workflow for Research Analytics

A flavor design case study is an example of a more specialized workflow [57.21]. In sensory evaluation, scientifically designed experiments are used to define a small set of mixtures that can be presented aromatically to evaluators to identify the ingredients that drive hedonic response (positively or negatively) of a target panel of consumers. Each panelist is asked how much they like the flavor, ranging from like extremely to dislike extremely with 9 distinctions. Details of the study can be found in [57.21]. Our focus here is the workflow that allowed to evaluate the consistency of liking preferences in the target population and gain insight into how to design or identify flavors that most consumers would consistently like.

The data for this project was provided by the Givaudan Flavors Corp. It falls into a category of precious data. It consists of sensory evaluation scores of 36 mixed flavors containing seven ingredients evaluated by 69 human panelists. In other words, data has seven input variables (flavor ingredients), 36 records (flavors), and 69 response measurements per record (Fig. 57.8).

Because of the high variability of response values per flavor, panelist responses were modeled individually. Because transparent and diverse input response models were required to approximate this challeng-

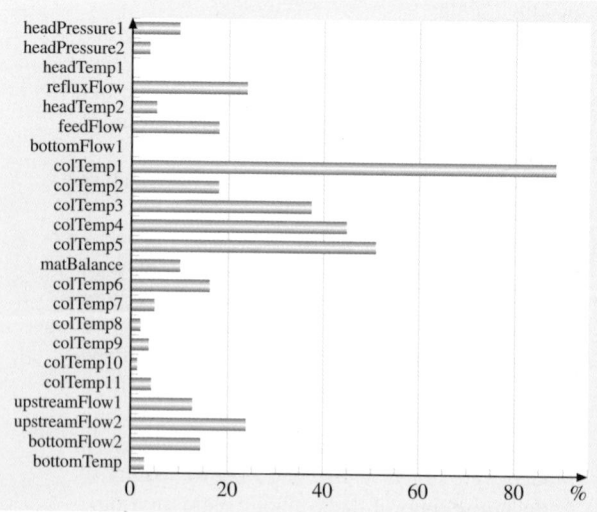

Fig. 57.5 Variable presence in developed symbolic regression models

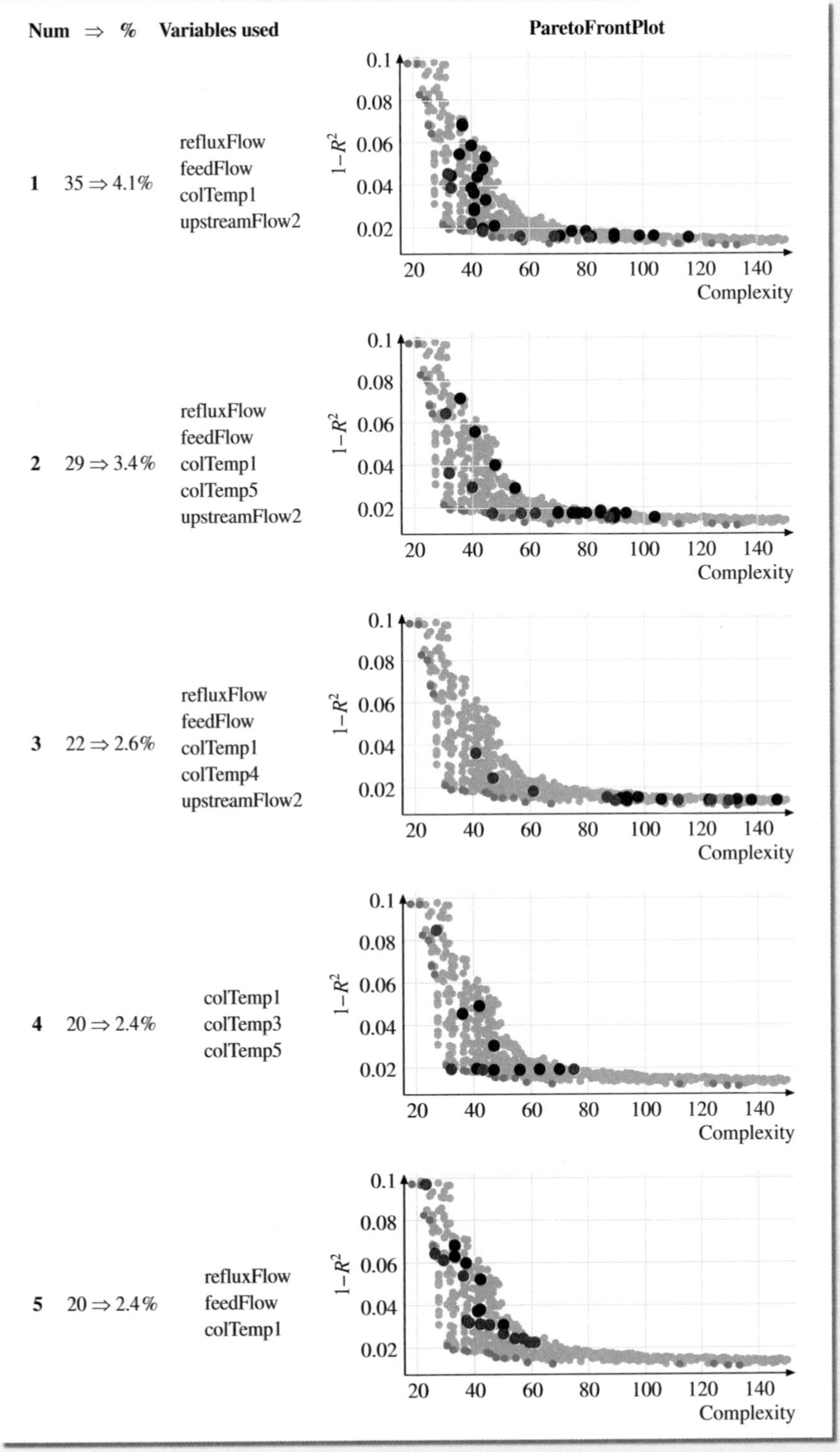

Num	⇒	%	Variables used	ParetoFrontPlot
1	35 ⇒	4.1%	refluxFlow feedFlow colTemp1 upstreamFlow2	
2	29 ⇒	3.4%	refluxFlow feedFlow colTemp1 colTemp5 upstreamFlow2	
3	22 ⇒	2.6%	refluxFlow feedFlow colTemp1 colTemp4 upstreamFlow2	
4	20 ⇒	2.4%	colTemp1 colTemp3 colTemp5	
5	20 ⇒	2.4%	refluxFlow feedFlow colTemp1	

Fig. 57.6 Complexity–accuracy tradeoffs for most frequent variable combinations in the distillation column example

Part E | 57.5

Predicted
2
1.5
1
0.5
0
0
0.5
1
1.5
2
Observed

Fig. 57.7 Prediction of the final ensemble of symbolic regression models on test data. All models seem to agree on unseen test data set. This should not be surprising, because the training, validation, and the test set were designed to cover the full range of operating conditions ▶

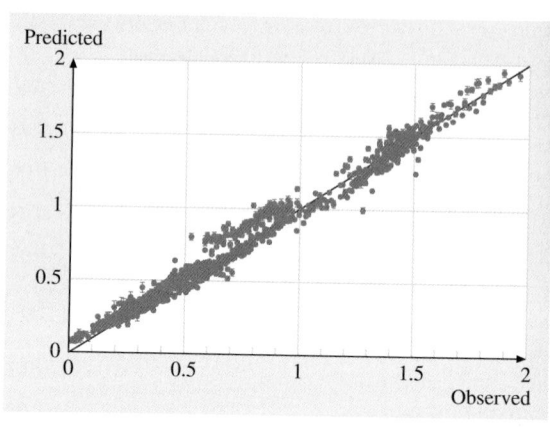

ing data set, modeling was done using ensemble-based symbolic regression.

For each panelist, a standard workflow was applied to identify driving ingredients which changes in panelist's liking [57.22].

When developed, model ensembles predicting individual responses could be bootstrapped to a richer set of virtual mixtures (tens of thousands of flavors instead of the available 36). The bootstrapped responses

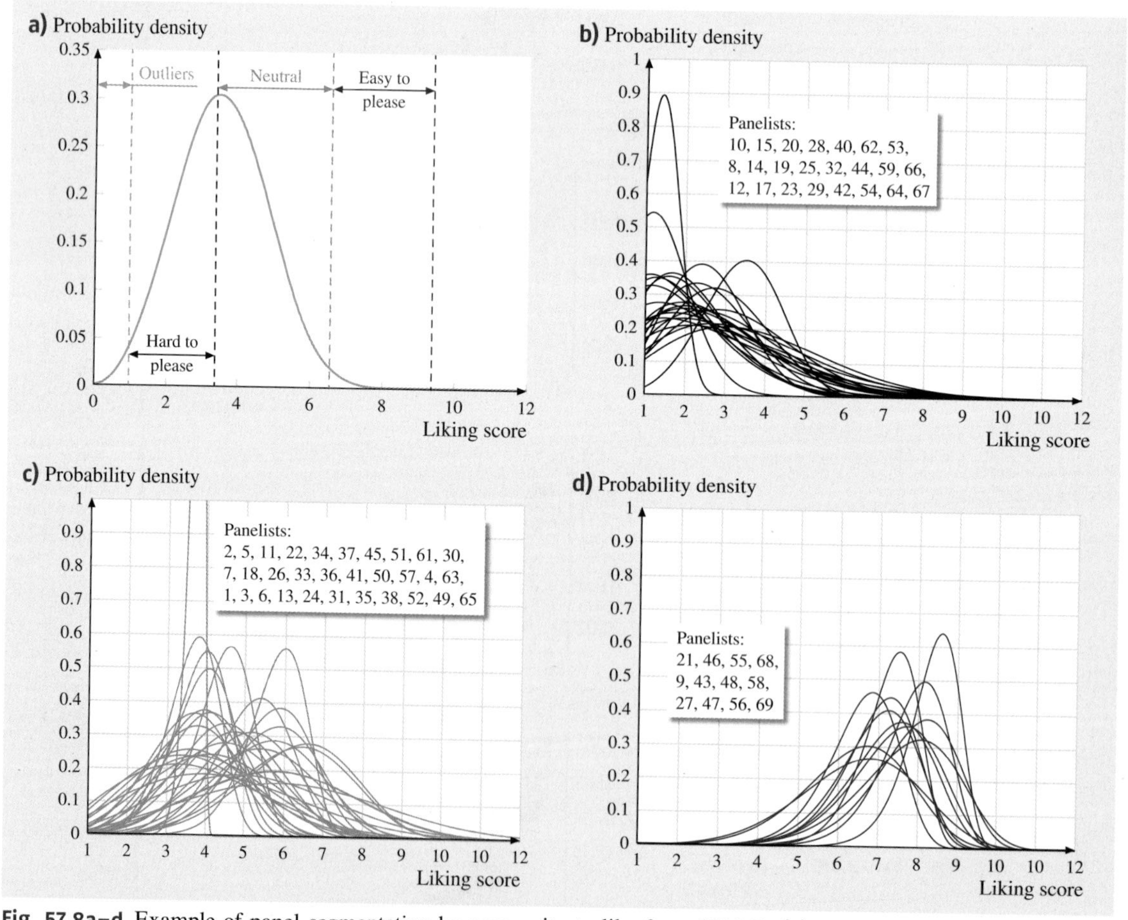

a) Probability density
Outliers
Neutral
Easy to please
Hard to please
Liking score
b) Probability density
Panelists:
10, 15, 20, 28, 40, 62, 53,
8, 14, 19, 25, 32, 44, 59, 66,
12, 17, 23, 29, 42, 54, 64, 67
Liking score
c) Probability density
Panelists:
2, 5, 11, 22, 34, 37, 45, 51, 61, 30,
7, 18, 26, 33, 36, 41, 50, 57, 4, 63,
1, 3, 6, 13, 24, 31, 35, 38, 52, 49, 65
Liking score
d) Probability density
Panelists:
21, 46, 55, 68,
9, 43, 48, 58,
27, 47, 56, 69
Liking score

Fig. 57.8a–d Example of panel segmentation by propensity to like from [57.22]: (**a**) Decision regions for evaluating cumulative distribution for liking score density model (**b**) hard to please panelist (**c**) neutral panelists, (**d**) easy to please panelists

were used to cluster the target population into three segments: easy to please – (cyber)individuals who consistently give high ratings to most flavors, hard to please – individuals that consistently use a low range of scores for all flavors, and neutral panelists whose preferential range is centered around the medium score – *neither like, nor dislike*. Such segmentation of the target population by people's propensity to like products turned out to be very useful in several other applications beyond flavor design. It focuses product development by giving insight into the fundamental variability in the preferences of the target audience.

The standard workflow for variable importance estimation applied to model ensembles forecasting the scores of individual panelists also allowed to segment the target population by ingredients that drive liking in the same direction. Such segmentation of the consumer market combined with the cost analysis for new product design offers visualization and analysis of beneficial tradeoffs for product specialization.

The third outcome of this study was the development of a model-guided optimization workflow for designing optimal virtual mixtures. Multi-objective optimization using swarm intelligence was used to find tradeoffs in the flavor design space that simultaneously maximize the average liking score and minimize variance in the liking across virtual panelists.

Such model-guided optimization workflow combined with the standard ensemble-based modeling workflow presents a strong motivation for the development of a targeted data collection system for designing new products.

We should point out that despite a very custom design and specialized domain of sensory evaluation in food science, the workflow could successfully be applied in the very different domain of video quality prediction. Ensemble-based symbolic regression was used to model the perceived quality of perturbed video frames and results were used to predict customer satisfaction and segment the representative population of video viewers by propensity to notice perturbations and sensitivity to particular perturbations [57.23].

57.6 Conclusions

In this chapter, we discussed how computational intelligence leads to predictive analytics to produce business impact. We identified three main areas of predictive analytics: business analytics that deals mainly with visualization and forecasting, process analytics which aims to improve optimization and control of manufacturing processes, and research analytics which aims at speeding up and improving product and process design. All three areas have the potential to save and earn many millions of dollars but deal with very different data sources, context, information content, amount of available domain knowledge, and time and prediction requirements for value generation. Driven by different motivations, the areas are subsequently employing different predictive modeling methods.

We presented several predictive modeling methods in the context of different prediction requirements, solution development, and deployment constraints. We emphasized that there is no single method that fits all problems, but rather there is a continuum of methods, and each problem dictates selection of a method by specific time requirements and the amount of available a priori subject-matter knowledge (Fig. 57.3).

We stressed the importance of good and stable predictive modeling workflows for success in CI projects and provided several examples of such workflows for process and research analytics, illustrating that research analytics projects require highly customized approaches.

We point out that successful CI projects are amplifiers, that necessarily keep the human in the loop and vastly enhance her/his capabilities. Because of this, integrating CI in the various process and business workflows is essential!

It is clear that our ability to generate data as well as our ability to analyze it and produce actionable knowledge are quickly expanding. The challenge remains on how to develop scalable CI algorithms that keep up with the ever rising tide of data, given that computational advances in hardware (massive parallelization, exa-scale computing) are developing at a much faster pace than the CI algorithms.

A question that still puzzles us is: *Can we get more intelligence with more computational power, and where (and whether) it stops?* Undoubtedly, the right answer lies in the development of new algorithms that can tackle the new challenges – advanced material design, problems in bio-informatics, complex-system modeling in social sciences, and social networks. We expect the largest impact of predictive modeling to happen in the areas of research and process analytics – in design of new products and new processes. Examples of design problems that can be assisted by data-driven CI methods for research analytics are the development of advanced materials – photovoltaic cells, alternative fuels, bio-degradable replacements for paints and plastics, composite materials, sustainable food sources. From the process analytics side, we would like to see CI methods used for optimization of water purification, emission control in combustion processes, simulation-based optimization of social events on a world scale (terror attacks, revolutions, pandemics spread), efficiency optimization of manufacturing cycles, garbage minimization, and recycling.

It cannot be stressed enough that the dynamics around CI is changing – instead of CI being an optional addition to the arsenal of problem solving tools and methods, CI is becoming indispensable to deal and make progress with this new breed of real-world problems. The only way for CI practitioners to bring CI to prime time is to develop scalable algorithms, proliferate good workflows, and implement them in great applications.

References

57.1   M. J. Manyika, M. Chui, B. Brown, J. Bughin, R. Dobbs, C. Roxburgh, A. H. Byers: Big data: The next frontier for innovation, competition, and productivity, available online at http://www.mckinsey.com/mgi (2011)

57.2   T.H. Davenport, J.G. Harris: *Competing on Analytics: The New Science of Winning*, 1st edn. (Harvard Business School, Boston 2007)

57.3   J.C. Torfs, G.J. Brands, E.G. Goethals, E.M. Dedeyne: Method for characterizing the appearance of a particular object, for predicting the appearance of an object, and for manufacturing an object having a predetermined appearance, which has optionally been determined on a basis of a reference object, WO Patent Ser 20 0204 2750 A1 (2004)

57.4   R.A. Johnson, D.W. Wichern: *Applied Multivariate Statistical Analysis* (Prentice Hall, Englewood Cliffs 1988)

57.5   L. Breiman: Random forests, Mach. Learn. **45**, 5–32 (2001)

57.6   M.D.J. Powell: Radial basis functions for multivariable interpolation: A review. In: *Algorithms for Approximation*, ed. by J. Mason, M.G. Cox (Clarendon, Oxford 1987) pp. 143–167

57.7   S. Haykin: *Neural Networks and Learning Machines*, 3rd edn. (Pearson Educ., Harlow 2008)

57.8   V. Vapnik: *Estimation of Dependences Based on Empirical Data* (Springer, Berlin, Heidelberg 1982)

57.9   V. Vapnik: The support vector method, Proc. 7th Int. Conf. Artif. Neural Netw. (1997) pp. 263–271

57.10   J.R. Koza: *Genetic Programming: On the Programming of Computers by Means of Natural Selection* (MIT, Cambridge 1992)

57.11   R. Poli, W.B. Langdon, N.F. McPhee: *A Field Guide to Genetic Programming* (Lulu, Raleigh 2008)

57.12   A.K. Kordon: *Applying Computational Intelligence: How to Create Value* (Springer, Berlin, Heidelberg 2010)

57.13   W. Banzhaf, P. Nordin, R.E. Keller, F.D. Francone: *Genetic Programming: An Introduction on the Automatic Evolution of Computer Programs and its Applications* (Morgan Kaufmann, San Francisco 1998)

57.14   M. Kotanchek: Evolved Analytics LLC: *DataModeler Release 8.0* (Evolved Analytics LLC, Midland 2010)

57.15   E. Vladislavleva: *Model-based Problem Solving through Symbolic Regression via Pareto Genetic Programming* (Tilburg Univ., Tilburg 2008)

57.16   S. Stijven, W. Minnebo, K. Vladislavleva: Separating the wheat from the chaff: On feature selection and feature importance in regression random forests and symbolic regression, Proc. 13th Annu. Conf. Companion Genet. Evol. Comput. (2011) pp. 623–630

57.17   S. Emmott, S.Rison: *Towards 2020 Science*, Microsoft, Cambridge (2006)

57.18   A.K. Kordon, G.F. Smits: Soft sensor development using genetic programming, Proc. Genet. Evolut. Comput. Conf. (2001) pp. 1346–1351

57.19   A.K. Kordon, G.F. Smits, A.N. Kalos, E.M. Jordaan: Robust soft sensor development using genetic programming. In: *Nature-Inspired Methods in Chemometrics: Genetic Algorithm and Artificial Neural Networks*, ed. by R. Leardi (Elsevier, Amsterdam 2003) pp. 69–108

57.20   M. Kotanchek: Real-world data modeling, Proc. 12th Annu. Conf. Companion Genet. Evol. Comput. (2010) pp. 2863–2896

57.21   K. Veeramachaneni, E. Vladislavleva, U.-M. O'Reilly: Knowledge mining sensory evaluation data: Genetic programming, statistical techniques, and swarm optimization, Genet. Progr. Evol. Mach. **13**(1), 103–133 (2012)

57.22   K. Vladislavleva, K. Veeramachaneni, U.-M. O'Reilly: Learning a lot from only a little: Genetic programming for panel segmentation on sparse sensory evaluation data, Proc. 13th Eur. Conf. Genet. Progr. (2010) pp. 244–255

57.23   N. Staelens, D. Deschrijver, E. Vladislavleva, B. Vermeulen, T. Dhaene, P. Demeester: Constructing a no-reference H.264/AVC bitstream-based video quality metric using genetic programming-based symbolic regression, IEEE Trans. Circuits Syst. Video Technol. **23**(8), 1322–1333 (2013)

58. Solving Phase Equilibrium Problems by Means of Avoidance-Based Multiobjectivization

Mike Preuss, Simon Wessing, Günter Rudolph, Gabriele Sadowski

Phase-equilibrium problems are good examples for real-world engineering optimization problems with a certain characteristic. Despite their low dimensionality, finding the desired optima is difficult as their basins of attraction are small and surrounded by the much larger basin of the global optimum, which unfortunately resembles a physically impossible and therefore unwanted solution. We tackle such problems by means of a multiobjectivization-assisted multimodal optimization algorithm which explicitly uses problem knowledge concerning where the sought solutions are not in order to find the desired ones. The method is successfully applied to three phase-equilibrium problems and shall be suitable also for tackling difficult multimodal optimization problems from other domains.

58.1 Coping with Real-World Optimization Problems

A multitude of methods from within and beyond *evolutionary computation* (EC) has been applied to real-valued multimodal optimization problems. These are generally considered the harder, the more basins of attraction they contain, and the less smooth the fitness landscape is. Additionally, a search space that extends over a large number of dimensions is said to complicate search for the desired global or good local optima [58.1].

However, in a real-world setting, even a low-dimensional problem may turn out to be quite difficult. This can stem from different factors, one of which would be a very small extent of the basins that contain the sought optima. Figure 58.1 visualizes the fitness landscape of an optimization problem that possesses this property. The application background will be detailed in Sect. 58.2, but for now it suffices to know

that there are only two variables a and b, and that the desired minima (function values do not depend on variable order and are thus symmetric to the main diagonal) are located near $(0.650, 0.001)$ and $(0.001, 0.650)$, respectively. It is easy to see that the appropriate basins are small; in the figure, they are hardly recognizable at all.

Another complicating factor would be uncertainty about the relative target function value of the sought optima. If it is not a priori known whether we are looking for a global or only a certain local optimum, there is no way around enumerating all existing optima and choosing the *right* solution out of these afterward. Such difficulties may occur in cases where it is not possible to integrate the whole available application specific knowledge into the established target function, i.e., if its value must be obtained by simulation and the exist-

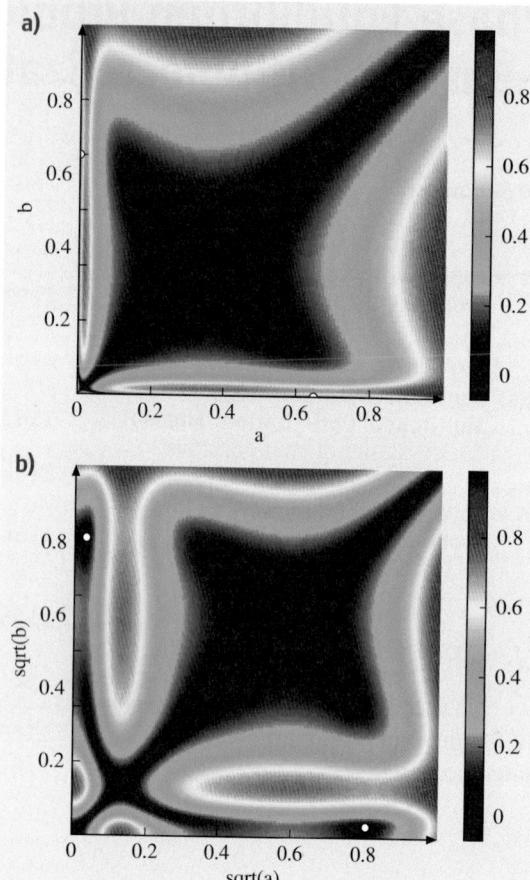

Fig. 58.1a,b Visualizations of the two-dimensional example problem. In the bottom panel, the search space is transformed by a square root. The desired optima are marked with *white dots*. Note that the diagonal consists of globally optimal but undesired (trivial) solutions

ing simulation tool is not able to represent all important features of the real system.

Obviously, there are several workarounds to overcome the difficulties imposed by this problem:

- Applying a transformation to the search space, so that the local optima at the lower boundaries occupy more space. This is shown in Fig. 58.1.
- Only initializing the optimization algorithm with solutions on the boundaries of the search space. In this case, we sometimes start from very near to the local optima, and thus have a higher chance to find them.
- Exploiting the symmetry of the landscape by a special representation. This can be done by enforcing

$a \geq b$ and would help, e.g., recombination operators of evolutionary algorithms (EAs).

However, all these approaches are dependent on the location of the desired optima. Any algorithm exploiting this expert knowledge would neccessarily show a worse performance on problems without these special features, as predicted by the *no free lunch* theorem [58.2]. Instead, a more general method, which uses information on where the desired optima is *not*, will be discussed and evaluated in this chapter.

Many different EAs may be used to tackle this global or multimodal optimization problem because they are able to detect several optima simultaneously or subsequently. The latter may be achieved by multistart approaches as sequential niching [58.3], whereas the former is established by means of diversity maintenance. That is, candidate solutions of the search populations are prevented from converging to the same region by implicitly or explicitly keeping them apart [58.4]. Prominent examples are crowding [58.5] and fitness sharing [58.6], and their successors. More recent approaches include, but are not limited to UEGO [58.7], clearing [58.8], species conservation [58.9], clustering-based niching [58.10], and cellular EA (CEA) [58.11]. Although there is no commonly accepted formal definition of what a niching method is [58.12], most of these algorithms may be subsumed under the term *niching EA*. They all use the distance between candidate solutions (diversity) as an implicit criterion which shall be maximized.

However, nothing prevents us from utilizing a diversity criterion directly. A step into this direction has been taken in the shifting balance GA [58.13]. But although it employs a separate diversity evaluation via subpopulation distance computation, it finally resorts to a single objective by weighting the distance and target function values.

In [58.14], we established a more radical approach and employ diversity in search space as an additional objective and treat the resulting combined problem by an *evolutionary multiobjective algorithm* (EMOA). The expected benefit is twofold:

- It enables placing solution candidates in basins that would otherwise go unnoticed due to their small size.
- We obtain a good overview of the available *interesting* search space regions in a single run.

As we presume that this approach is not only applicable to the thermodynamic problems treated in this work but also to real-valued engineering problems with similar properties, it is also followed and further extended here. Other related multiobjectivization approaches are discussed in Sect. 58.3 after introducing the problem context.

58.2 The Phase-Equilibrium Calculation Problem

The knowledge of phase equilibria is required for the design and optimization of separation processes which are essential parts of typical chemical plants. The aim of a *phase-equilibrium calculation* is to quantitatively relate the variables (in particular, temperature T, pressure p, and mole fraction x) which describe the state of equilibrium of two or more homogenous phases [58.15].

In any problem concerning the equilibrium distributions of k components between two phases, one must always begin with the equality of the chemical potential μ as

$$\forall i \in \{1, \ldots, k\} : \mu_i' = \mu_i'' \, . \tag{58.1}$$

To establish the relation of μ_i' (We use the domain-specific notation with upper index denoting different phases and lower index standing for separate substances.) to T, p, and x_i', it is convenient to introduce a certain auxiliary function such as the fugacity coefficient $\varphi_i'(T, p, x_i')$, which can be calculated by a thermodynamic model. Then, (58.1) can be rewritten as

$$\forall i \in \{1, \ldots, k\} : x_i' \cdot \varphi_i' = x_i'' \cdot \varphi_i'' \, . \tag{58.2}$$

Typically, the calculation is performed at constant temperature and pressure, and the remaining concentrations x_i' and x_i'', respectively, are to be found. The fugacity coefficient φ_i of component i in the mixture is calculated as

$$\ln \varphi_i = \frac{\mu_i^{\text{res}}}{RT} - \ln Z \, , \tag{58.3}$$

with Z being the compressibility factor, defined as

$$Z \equiv \frac{pv}{RT} \, , \tag{58.4}$$

where v is the molar volume, and R is the gas constant. The residual chemical potential μ_i^{res} is given by

$$\mu_i^{\text{res}} = a^{\text{res}} + RT(Z - 1) \\ + \frac{\partial a^{\text{res}}}{\partial x_i} - \sum x_\ell \left(\frac{\partial a^{\text{res}}}{\partial x_\ell} \right) \, , \tag{58.5}$$

where $(\partial a^{\text{res}} / \partial x_i)$ is a partial derivative of the residual Helmholtz energy with respect to the mole fraction stated in the denominator, while all other mole fractions are considered constant.

The residual Helmholtz energy according to the perturbed chain statistical associating fluid theory (PC-SAFT) is considered as the sum of different contributions resulting from repulsion (hard chain), van der Waals attraction (dispersion), and hydrogen bonding (association)

$$a^{\text{res}} = a^{\text{hc}} + a^{\text{disp}} + a^{\text{assoc}} \, . \tag{58.6}$$

The detailed equations for each contribution can be found in [58.16] and [58.17].

Solving phase-equilibrium problems according to (58.2) may lead to trivial solutions, i. e., $x_i' = x_i''$, which are mathematically correct but have no physical meaning (except at the so-called critical demixing point). To avoid obtaining them, the initial guesses for the minimization procedure may not be too far away from the correct solutions, provided that the correct solutions are known.

In the case of polymer solutions, initialization is very critical, because the concentration of the polymer in the solvent-rich phase can be in the magnitude of 10^{-20}, which is a numerical challenge for simulation programs [58.18]. Another difficulty arises as the number of components in the mixture increases. All these challenges point out the need for a robust algorithm to solve the phase-equilibrium calculation for an arbitrary number of components and phases, and which is also applicable to polymer solutions.

Figure 58.1 actually shows a phase-equilibrium problem, namely a simple two-component mixture of water and pentanol. This type of liquid–liquid equilibrium (LLE) data are necessary for the design and optimization of liquid–liquid extractors and decanters. The two variables correspond to the concentrations of water in the water-rich phase (for the larger of the two) and in the pentanol-rich phase (for the smaller one). Under the assumption that $a > b$, and that w stands for water and p for pentanol, we have $a = x_w'$, and $b = x_w''$.

The remaining mole fractions x'_p and x''_p can be obtained indirectly as $x'_p = 1 - x'_w$ and $x''_p = 1 - x''_w$, because for every phase, the following equality holds:

$$\sum_{i=1}^{k} x'_i = \sum_{i=1}^{k} x''_i = 1 .$$ (58.7)

For this two-component problem, two equations of type (58.2) have to be satisfied, resulting in two error values $e_w = |x'_w \varphi'_w - x''_w \varphi''_w|$ and $e_p = |x'_p \varphi'_p - x''_p \varphi''_p|$. A feasible solution to the problem shall exhibit errors below 10^{-10} due to practical requirements. In the following, e_w and e_p are aggregated into a single target function value by using the sum of squares, which is to be minimized (note the vector notation)

$$f_1(x', x'') = e_w^2 + e_p^2 .$$ (58.8)

Table 58.1 The sought optima at different temperatures

Mole fraction	40 °C	60 °C	90 °C
x'_w	0.74698	0.7097	0.65084
x''_w	0.00020913	0.00038142	0.00082809

In Fig. 58.1, (58.8) is modeled at a temperature of 90 °C, for which the sought optimum is located near the coordinates (0.650,0.001). As system properties change with temperature and pressure, the pursued optimum also moves through the search space. Table 58.1 depicts approximate solutions for different temperatures and constant pressure of 1.0132 bar. The trivial solutions are the only feature representative for all phase-equilibrium problems. Thus, this is the only information that shall be exploited in the following.

58.3 Multiobjectivization–Assisted Multimodal Optimization: MOAMO

As seen in Sect. 58.1, the optimization problem at hand is inherently multimodal. That is, local optimization schemes are only successful if started from a region near the desired nontrivial solution. To make things worse, the basin of attraction of the undesired trivial solutions may largely dominate the search space as found for the very simple LLE problem (two phases, two components: water/pentanol). Hitting the basin of attraction of the desired solution can be very difficult, and if failing on this, the final outcome of quasi-Newton or similar algorithms will be a trivial solution.

Stochastic optimization methods like EAs and other metaheuristics employ a more globally oriented optimization scheme. Several attempts using these methods have been tried on equilibrium detection problems in recent years, namely genetic algorithms (GA) and simulated annealing in [58.19] or differential evolution (DE) and tabu search (TS) in [58.20]. The algorithms have been mostly used in their canonical form with some parameters tuning and a concluding local optimization step by means of a quasi-Newton method. Alternative approaches applied artificial neural networks for learning and predicting phase equilibria as in [58.21], the authors of which evolve the neural networks by means of genetic programming (GP), and [58.22], where the authors employ a real-coded GA to optimize initial weights and biases of the neural network before it is further refined using a quasi-Newton method. Where

enough training data is available, the binodal curves of equilibria can be learned and predicted for the missing areas.

Some recent metaheuristic attempts concentrate on the global (multimodal) nature of the optimization problem to find equilibrium points for rather difficult systems where global optima are located in relatively small basins. [58.23] use tabu search, [58.24] a random tunneling method, and [58.25] a DE hybrid with TS components. While we agree that *looking elsewhere* for even better solutions is mandatory for a multimodal problem, it may be even more rewarding to obtain a good overview over large portions of the search space before climbing down into the individual optima. This has been attempted by using a refined version of the algorithm of [58.26] which has been applied to phase stability problems by [58.27]. The base algorithm GLOBAL has been developed further in [58.28]. As the latter methods start from a random sample, it may however happen that either the initial sample is too small so that important optima are missed, or it is relatively large and thus costly.

The optimization concept suggested in this work therefore relies on an evolutionary multiobjective algorithm (EMOA) approach in order to generate a spectrum of possible near-optimal solutions before applying a local search method on these. We term it *multiobjectivization-assisted multimodal optimization* (MOAMO). The method was successfully applied

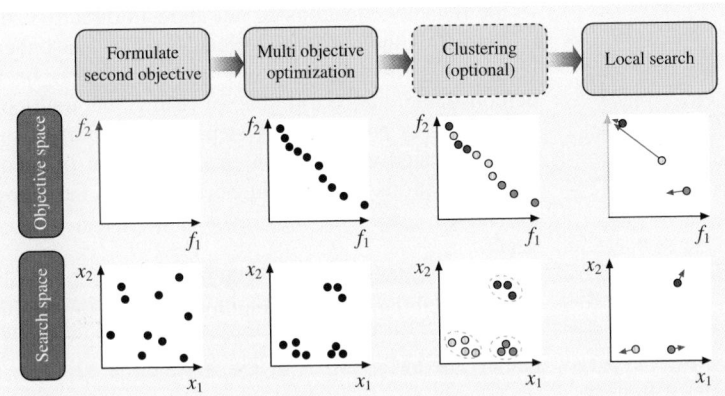

Fig. 58.2 The general concept of MOAMO and its influence on search and objective space

onto the two-phase 2-component water/pentanol system in [58.14]. Here, we demonstrate that it is viable for more complicated equilibrium problems with more phases and components. Although not yet tried on polymer problems, this ultimate goal seems to be in reach as very small basins of attraction can be attained reliably.

Figure 58.2 shows the main concept of the MOAMO approach. The key idea is to use a population-based multiobjective algorithm as a preprocessing step for generating search points in the different basins of attraction of the tackled problem, the basin of the nontrivial optimum being among them. To do this, the practitioner first has to formulate an additional objective function. This second objective is then employed to obtain good coverage of the search space despite the high attraction of certain areas. We label this type of multiobjectivization *avoidance-based* because application knowledge about where the sought optimum is *not* helps to transform the single-objective optimization problem into a multiobjective one that is easier to solve. More precisely, it enables detecting several different basins, among them many that would most likely have gone unnoticed with the single-objective approach alone.

For this specific application, the distance to the trivial solution (equal concentrations) is taken into account. From then on, the system can work autonomously. In the next step, the multiobjective optimization is carried out. The obtained search points then are fed one by one into a local optimization method, until a satisfying nontrivial solution is found. For this local search, only the original objective is relevant. We employed the algorithm of [58.29] and the covariance matrix adaptation evolution strategy (CMA-ES) of [58.30] for

this last step. The experimental results suggest that especially the latter seems well suited for the task. However, one may resort to another method here (e.g., quasi-Newton or similar standard optimization algorithms as described in [58.31]) if it is deemed more appropriate. To avoid superfluous local optimization steps on candidate solutions that are close to each other, this phase may be prepended with a clustering step so that one tries a representative of each *group* of solutions first and then proceeds in a round robin fashion.

The idea of simplifying a difficult single-objective problem by a multiobjective approach has some precursors in evolutionary computation and has been coined as multiobjectivization by [58.32]. The approach can be divided into two general categories, namely multiobjectivization by adding objectives and multiobjectivization by the decomposition of a scalar objective function.

For the latter one, it can be proven that the approach can only decrease the number of local optima [58.33]. It was for example successfully applied to protein structure prediction problems in [58.34, 35]. MOAMO belongs to the category of multiobjectivization by adding objectives. No theoretic guarantees of benefits can be given [58.36] for this approach, but nonetheless it has already been tried in several different ways [58.37–40]. However, these applications somewhat remain in the tradition of evolutionary multiobjective algorithms that already contain diversity preserving mechanisms. The second objectives suggested all refer to the current population or single individuals thereof and do not take characteristics of the actual problem into account. MOAMO strongly differs as instead of a *population-relative*, it employs an *absolute*

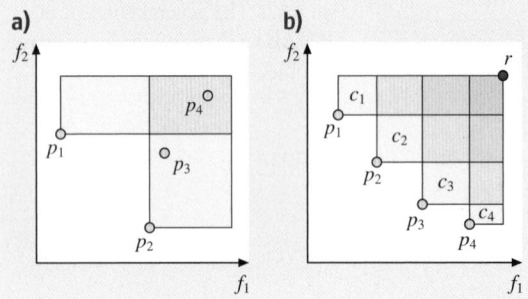

Fig. 58.3 (a) Pareto dominance for minimization: p_1 and p_2 are non-dominated, p_3 is dominated by p_2, and p_4 is dominated by p_1 and p_2. (b) A non-dominated front between objectives f_1 and f_2, consisting of points p_1 to p_4. c_1 to c_4 denote the hypervolume contribution of each point (the space not covered by any other point) against the reference point r

distance objective, namely the distance to the known trivial solutions. The MOAMO approach is therefore especially well suited to phase equilibrium problems,

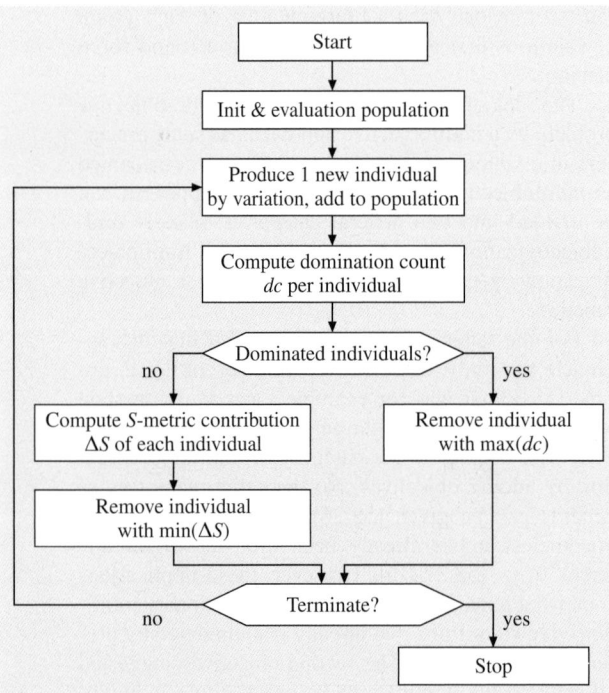

Fig. 58.4 Working scheme of the SMS-EMOA. Termination is done according to predefined conditions, e.g., a certain budget of fitness evaluations

as the fugacity equations do not allow to directly conclude where the sought solution is, but at least they provide information on where it is not. It has been demonstrated in [58.14] that by using the multiobjective EA as preprocessing step, the important basin can be located with a much smaller amount of function evaluations than would be needed by sampling the search space randomly, even if the basin is very small.

In the following, basic EMOA concepts are summarized and the particular multiobjective optimization algorithm employed in our experiments is introduced, namely the SMS-EMOA by [58.41] and [58.42].

58.3.1 Basics of Multiobjective Optimization

Multi-objective optimization fundamentally relies on Pareto dominance. A point in the objective space of two or more objective functions is dominated, if there is at least one other that is not worse in all objectives and better in at least one (Fig. 58.3a). As the optimization progresses, the population approaches the Pareto front which resembles the set of optimal compromises and consists of non-dominated points only.

Several criteria exist to judge the quality of whole populations within the algorithm run (as means to determine the next search steps) and thereafter to assess optimization success. One of the most popular is the hypervolume, the amount of objective space covererd by the population with regard to a reference point as documented in the right panel of Fig. 58.3.

The S-metric selection evolutionary multiobjective algorithm (SMS-EMOA) is a further development of the popular NSGA2 (nondominated sorting genetic algorithm 2) by [58.43]. Figure 58.4 displays its major steps. Starting from a usually randomly placed population, a loop begins with deriving one new individual (search point) and adding it to the population. The domination count of each individual is computed by counting how many other individuals dominate it. If such dominated individuals exist, the one with the largest domination count is deleted. Otherwise, the hypervolume contribution of each individual is determined (Fig. 58.3b), and the individual with the smallest contribution is deleted. If the current state does not fulfill the termination criterion (e.g., a predefined budget of function evaluations) the loop starts over. After terminating, the remaining population is the result set.

58.4 Solving General Phase-Equilibrium Problems

We present the results of phase-equilibrium calculations for the three-component system water/methanol/MMA as well as for the three-phase systems water/MMA and water/furfural. The corresponding optimization problems have four, three, and three decision variables.

PC-SAFT uses statistical mechanics for its simulation of thermodynamic systems and thus requires a calibration of some pure-component parameters and one binary parameter. The aim of this calibration is to achieve a consistency between the calculated phase equilibria and results of physical experiments. Carrying out this task manually for a single substance takes up several days of work for a chemical engineer, although the data of the physical experiments are already available in the literature [58.44]. These data contain series of measurements of temperature, density, and pressure for the vapor and the liquid phase of each substance. Among the several parameters that model the molecular properties, there are five per substance that have to be estimated. These are the number of sphere segments m, the segment diameter σ, the segment energy parameter ϵ/k, an association energy ϵ^{AiBi}/k, and the effective association volume κ^{AiBi}. Two different association sites are assigned to all the considered substances. If the substance is non-self-associating, then association energy as well as association volume are set to zero. Besides the five (three) parameters per substance, the

model requires one parameter k_{ij} that is characteristic for each binary mixture. The respective values for all these parameters were taken from [58.45, 46] and are summarized in Tables 58.2 and 58.3. The applicability of PC-SAFT to model the mentioned systems in good agreement with experimental data has been proved in [58.46].

The following experiments show that the MOAMO approach provides a reliable and fast tool for the detection of equilibrium points which are difficult to find with standard optimization tools as a gradient or quasi-Newton search.

58.4.1 Ternary Liquid–Liquid Equilibrium: Water/Methanol/MMA

In Fig. 58.5, the ternary phase diagram of water/methanol/MMA at 50 °C and 1.013 bar with two liquid phases is shown. The calculation of the tie-lines was performed for different fixed concentrations of MMA in one liquid phase (x'_{MMA}), see Table 58.4, at constant temperature and pressure.

Pre-Experimental Planning
The first objective (58.9) is generated from the error values output by PC-SAFT. These refer to the departure from the equilibrium state between every two phases of

Table 58.2 PC-SAFT pure-component parameters for considered components

Substance	m	σ	ϵ/k	ϵ^{AiBi}/k	κ^{AiBi}
Water	1.0656	3.0007	366.5121	2500.6706	0.0349
Methyl methacrylate (MMA)	3.0632	3.6238	265.6874	0	0.0349
Methanol	1.5255	3.2300	188.9046	2899.4906	0.0352
Furfural	4.1604	3.0204	270.0700	0	0.0349

Table 58.3 PC-SAFT binary parameters

Binary system	k_{ij}
Water/MMA	0
Water/methanol	−0.05
Water/furfural	−0.006
MMA/methanol	0

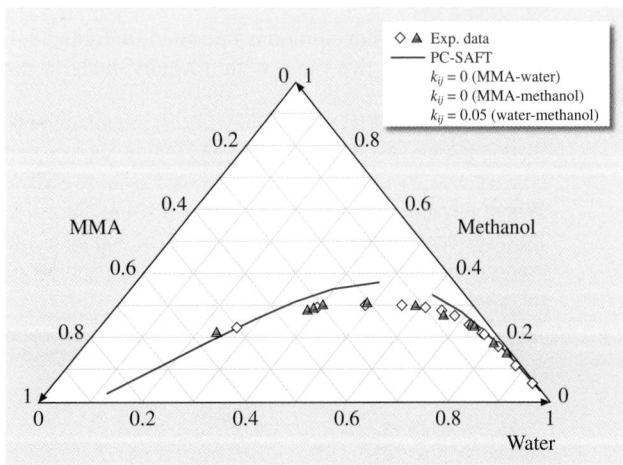

Fig. 58.5 Phase diagram of water/methanol/MMA system at 50 °C and 1.013 bar. The symbols are experimental data from [58.47] and [58.48]. The line is the calculation result of PC-SAFT with MOAMO-approach

one component as given by (58.1).

$$f_1(\mathbf{x}', \mathbf{x}'') = \sum_{i=1}^{3} |x_i' \varphi_i' - x_i'' \varphi_i''| \,. \qquad (58.9)$$

Different formulations of the second objective for the SMS-EMOA were tried and several of them work well. Therefore, a generalization of the distance criterion for the two-component two-phase case in Sect. 58.1 was chosen. It measures the Euclidean norm of a vector of concentration differences (slightly shifted to allow for minimization) and is easily extendable for more components

$$f_2(\mathbf{x}', \mathbf{x}'') = \sqrt{2} - \|x' - x''\|_2$$
$$= \sqrt{2} - \sqrt{\sum_{i=1}^{3} (x_i' - x_i'')^2} \,. \qquad (58.10)$$

Experimental Task
The task for MOAMO in this experiment is to reliably reach the sought optimum for all indicated MMA concentrations, that is the number of individuals converging to the optimum in the local search phase shall be considerably larger than 1 on average. Furthermore, the MOAMO-based approach shall find the optimum considerably faster than a naïve multistart local search procedure.

Setup
For each of the concentrations indicated in Table 58.4, MOAMO is run five times with 30 individuals in the

Table 58.4 MOAMO with 30 individuals, remaining population put into local optimization and rate of success and convergence to trivial solution, averaged over five runs. Where the sum of optimum and trivial is below 30, some local searches did not converge. The last column gives the empirical success probabilities for random start points of the local search

x'_{MMA}	Optimum	Trivial	Success rate (%)
0.05	25.2	3.8	45.0
0.15	0.0	30.0	2.7
0.25	5.8	24.0	3.6
0.35	19.6	10.2	3.9
0.45	25.8	4.0	3.5
0.55	29.4	0.6	2.9
0.65	28.8	0.0	3.1
0.75	29.0	0.6	3.1
0.85	23.2	0.6	2.3

multiobjective first step. Each search point contained in the last population is then optimized by a local search procedure (CMA-ES is employed for this second step). For each local search, it is recorded if either the undesired trivial solution or the sought optimum is obtained or if the search did not converge. Other than population size and run length (30 and 5000), the SMS-EMOA parameters are chosen as in [58.41].

In order to perform a comparison, the local search procedure (CMA-ES) is started 1000 times for each MMA concentration from a randomized start point and the rate of success for converging to the sought optimum is recorded. The CMA-ES terminates if progress or adapted stepsizes decrease below 10^{-12} as usual.

Observations
Table 58.4 comprises the results for the MOAMO approach and in comparison the success rates for the random start local search procedure. Run lengths of the CMA-ES are not given in detail, but mostly range between 2600 and 5000 evaluations.

For the MMA concentrations from 0.25 to 0.85, both methods are consistent: MOAMO obtains the sought optimum from at least 60% of the last population's search points, while the success rates of the random start local search vary between 2 and 4%. However, 0.05 and 0.15 are special cases: In the first case, the problem is obviously not that hard as the random start local search also detects the sought optimum often, and in the second case, the MOAMO approach completely fails.

Discussion
The most striking result of the experiment is that hardness of the problem for the two compared approaches seems uncorrelated. An MMA concentration of 0.05 is much more easily solved by the random start local search than any other, but the success rates for MOAMO do not reflect this. For 0.15, the opposite happens as the problem poses average difficulties for the random start local search procedure, but is very hard for MOAMO. We conjecture that this is an exception as we are almost at the critical point here, where concentrations in both phases differ less and less. Presumably, trivial solution and sought optimum are too equal to separate them in the SMS-EMOA phase via the distance objective. However, we can be satisfied with the results for the other concentrations, where the MOAMO approach reliably detects the sought optimum and is much faster than the random start local search procedure, even if the effort for the first (multiobjective)

phase is considered (which is on the order of one or two local searches).

58.4.2 Three Phase Equilibria: Water/MMA and Water/Furfural

We now turn to an application of the MOAMO approach on 2 component/3 phase systems in order to detect the heteroazeotrope point (a 3-phase equilibrium). The first objective is again obtained from the phase equilibrium equations and differs from the one chosen for the 3 component/2 phase system (58.9) in the number of relevant phase equations. Due to transitivity, four error values remain here. Additionally, a quadratic form is chosen here instead of the absolute value form used in the previous case, under the assumption that the quadratic form simplifies the local optimization task (Quasi-Newton as well as evolutionary optimization methods usually perform better in this case).

$$f_1(\mathbf{x}', \mathbf{x}'', \mathbf{x}''') = \sum_{i=1}^{2} \left[(x_i' \varphi_i' - x_i'' \varphi_i'')^2 \right.$$
$$\left. + (x_i' \varphi_i' - x_i''' \varphi_i''')^2 \right] . \quad (58.11)$$

As for the previous system, it is necessary to determine a suitable second (distance) criterion for the multiobjective first step. However, for three phases, the approach

taken in [58.14] has to be generalized in a different way than done for three components. Interestingly, our preliminary test showed that it is sufficient to consider only one component and its three phases to create a distance criterion. We may use mutual phase concentration differences of phases 1 and 2, 2 and 3, and 1 and 3 to aggregate an objective function. (Note that Euclidean distances have been employed in the previous section, however our tests show that for the multiobjective MOAMO step, the choice of the distance norm itself is not very important and Manhattan distances as used here are also sufficient.)

$$f_2(\mathbf{x}', \mathbf{x}'', \mathbf{x}''') = 2 - \sum_{i=1}^{2} \left(|x_i' - x_i''| \right.$$
$$\left. + |x_i'' - x_i'''| + |x_i' - x_i'''| \right) . \quad (58.12)$$

Alternatively, the phase concentration differences can also be stated as three separate criteria, resulting in a four-objective problem for the SMS-EMOA

$$f_2(\mathbf{x}', \mathbf{x}'', \mathbf{x}''') = 1 - \sum_{i=1}^{2} |x_i' - x_i''|$$

$$f_3(\mathbf{x}', \mathbf{x}'', \mathbf{x}''') = 1 - \sum_{i=1}^{2} |x_i'' - x_i'''|$$

$$f_4(\mathbf{x}', \mathbf{x}'', \mathbf{x}''') = 1 - \sum_{i=1}^{2} |x_i' - x_i'''| . \quad (58.13)$$

The following experiment will show whether the aggregated formulation or the separate criteria are more advisable.

The binary system water/MMA in Fig. 58.6 exhibits a heteroazeotrope behavior at 1 bar. According to the phase rule, only one variable can be fixed to determine the heteroazeotrope, as in this case the pressure. The temperature of the heteroazeotrope was found at 81.93 °C and the concentrations of MMA in the three phases were $x'_{\mathrm{MMA}} = 0.841826$, $x''_{\mathrm{MMA}} = 0.488033$, and $x'''_{\mathrm{MMA}} = 0.002577$.

The identification of the heteroazeotrope point for water/furfural at 1 bar was more complicated than the previous system due to the fact that two sought water concentrations are close to each other ($x'_{\mathrm{water}} = 0.911822$ and $x''_{\mathrm{water}} = 0.973374$), see Fig. 58.7. The third water concentration was found at $x'''_{\mathrm{water}} = 0.507017$ and the heteroazeotrope temperature was determined at 97.64 °C.

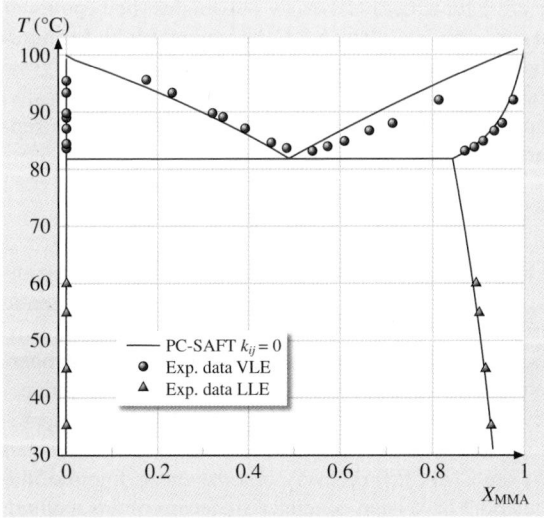

Fig. 58.6 Phase diagram of water/MMA system at 1 bar. The symbols are experimental data from [58.49] and [58.50]. Lines are calculation results of PC-SAFT with MOAMO-approach

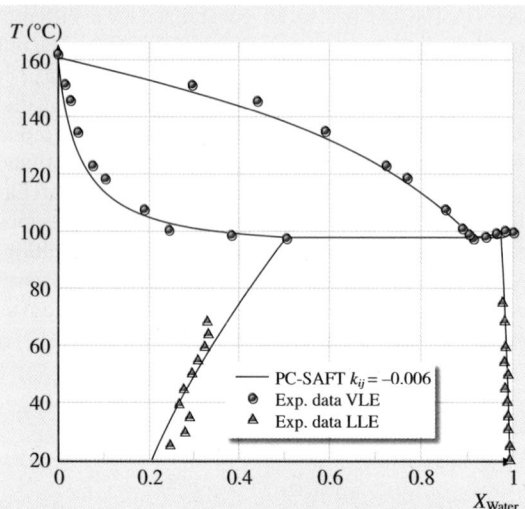

Fig. 58.7 Phase diagram of water/furfural system at 1 bar. The symbols are experimental data from [58.51]. Lines are calculation results of PC-SAFT with MOAMO-approach

Pre-Experimental Planning
Taking over the SMS-EMOA parameters (population size and run length) from the previous experiment led to an unreliable behavior for the two systems tested here. Seemingly, they are more difficult to solve than the given three-component/two-phase system. Therefore, population size is doubled to 60 individuals and run length is accordingly slightly increased to 6000 evaluations.

Experimental Task
As the last paragraph indicated that the problems in this section are even more difficult than the that of Sect. 58.4.1, there is no point in testing against random start local search again. Instead, it shall be determined if the aggregated (58.12) or the separate criteria approach (58.13) is more suitable for solving the problems with MOAMO. To enable a decision between the two, a significant difference in success rates is required.

Setup
For each of the two systems (water/MMA and water/furfural) and each of the problem formulations (aggregated/separate), 30 MOAMO runs are performed and the number of successes is recorded. A run is successful if at least one of the local search steps obtains the sought optimum the number of successful local searches is not recorded. As before, we employ the combination of SMS-EMOA and CMA-ES. The

Table 58.5 Success rates for detecting the heteroazeotrope point via MOAMO approach under different formulation of the distance criterion

System	Distance criterion	Success rate (%)
Water/MMA	Aggregated	100.0
	Separate	50.0
Water/furfural	Aggregated	93.3
	Separate	36.7

resulting values for the first objective function (computed from the error values output by PC-SAFT) shall be below 10^{-15} in this case, requiring to modify the CMA-ES internal stopping criteria accordingly. Its initial step size is set to 0.01. The SMS-EMOA parameters are set as in the previous experiment except population size and run length which are modified as documented above.

Observations
The number of successful runs is given in Table 58.5. The aggregated approach seems to consistently perform better than the one with separate criteria, and success rates hint to the fact that the second system poses more difficulty than the first one.

Discussion
Fortunately, the much simpler (aggregated) approach is also the more reliable for both systems. The much larger objective space in the first phase (four instead of two objective functions) obviously outweights the benefits of a *correct mapping* by far. Furthermore, for higher numbers of phases, the number of objectives would grow faster than linear, so that in conclusion, the aggregated approach is much more suitable than the one with separate objective functions.

58.4.3 Obtaining the Phase Diagrams

Once the heteroazeotrope point is detected, a phase diagram of the system may be obtained by systematic exploration of the two-phase equilibria at different temperatures. We simply increase or decrease the temperature (which is a free variable for two-phase systems) by 1 °C and take the solution for the last temperature step as initial point for a local search (executed by the CMA-ES) on every binodal curve. Figures 58.6 and 58.7 have been generated by means of this method. (Note that this is different from the common approach of detecting several two-phase equilibria by means of a quasi-Newton method first and then to conclude on the heteroazeotrope point from these.)

58.5 Conclusions and Outlook

In this chapter, a multistage method named MOAMO (multiobjectivization-assisted multimodal optimization) was presented. It is especially designed for difficult multimodal direct search problems as arising in phase equilibrium detection. However, the method is very well applicable whenever some problem knowledge is available concerning where the global optimum is *not*. The experimental analysis, performed on three different systems with either three components and two phases or two components and three phases, has shown that the approach is reliable and fast. It outperforms random multistart local search by a large margin under nearly all tested conditions. Two important properties of the approach need to be emphasized:

- Unlike many attempts to solve phase-equilibrium problems by means of evolutionary or related algorithms, MOAMO utilizes known features of the problem to direct the search and thereby avoids spending too much effort in repeatedly approaching

trivial solutions. However, it does not make any assumptions about the *location* of the sought optima and is thus still a generic approach.
- Unlike in some other multiobjectivization approaches, the second objective is population independent. Moving a single individual does not change the objective function values of any other. This prevents unwanted feedback loops. The optimization focuses on the problem and not on the current population.

Our results indicate the MOAMO approach as remarkably independent of the actual formulation of the second objective. Performance increases or decreases only gradually for alternative objectives, the overall concept remains intact. However, obtaining better guidelines for setting up a matching second objective is an area for future research, as is the comparison with more different algorithms and the adoption for other problems, not necessarily restricted to phase equilibrium detection.

References

58.1   A.A. Törn, A. Žilinskas (Eds.): *Global Optimization*, Lecture Notes in Computer Science, Vol. 350 (Springer, Berlin, Heidelberg 1989)

58.2   D.H. Wolpert, W.G. Macready: No free lunch theorems for optimization, IEEE Trans. Evol. Comput. **1**(1), 67–82 (1997)

58.3   D. Beasley, D.R. Bull, R.R. Martin: A sequential niche technique for multimodal function optimization, Evol. Comput. **1**(2), 101–125 (1993)

58.4   A.E. Eiben, J.E. Smith: *Introduction to Evolutionary Computing* (Springer, Berlin, Heidelberg 2003)

58.5   K.A. De Jong: An Analysis of the Behavior of a Class of Genetic Adaptive Systems, Ph.D. Thesis (University of Michigan, Ann Arbor 1975)

58.6   D.E. Goldberg, J. Richardson: Genetic algorithms with sharing for multimodal function optimization, Proc. Second Int. Conf. Genet. Algorithm. Their Appl. (1987) pp. 41–49

58.7   M. Jelasity: UEGO, an abstract niching technique for global optimization, Lect. Notes Comput. Sci. **1498**, 378–387 (1998)

58.8   A. Pétrowski: A clearing procedure as a niching method for genetic algorithms, Proc. 1996 IEEE Int. Conf. Evol. Comput. (1996) pp. 798–803

58.9   J.-P. Li, M.E. Balazs, G.T. Parks, P.J. Clarkson: A species conserving genetic algorithm for multimodal function optimization, Evol. Comput. **10**(3), 207–234 (2002)

58.10   F. Streichert, G. Stein, H. Ulmer, A. Zell: A clustering based niching method for evolutionary algorithms, Proc. Genet. Evol. Comput. (2003) pp. 644–645

58.11   M. Tomassini: *Spatially Structured Evolutionary Algorithms Artificial Evolution in Space and Time* (Springer, Berlin, Heidelberg 2005)

58.12   M. Preuss: Niching prospects, bioinspired optimization methods and their applications, BIOMA 2006 (2006) pp. 25–34

58.13   F. Oppacher, M. Wineberg: The shifting balance genetic algorithm: Improving the GA in a dynamic environment, Proc. Genet. Evol. Comput. Conf. (1999) pp. 504–510

58.14   M. Preuss, G. Rudolph, F. Tumakaka: Solving multimodal problems via multiobjective techniques with application to phase equilibrium detection, IEEE Cong. Evol. Comput. (CEC 2007) (2007) pp. 2703–2710

58.15   E.G. de Azevedo, J.M. Prausnitz, R.N. Lichtenthaler: *Molecular Thermodynamics of Fluid Phase Equilibria* (Prentice Hall, Englewood Cliffs 1986)

58.16   J. Gross, G. Sadowski: Perturbed-chain saft: An equation of state based on a perturbation theory for chain molecules, Ind. Eng. Chem. Res. **40**(4), 1244–1260 (2001)

58.17   M. Kleiner, F. Tumakaka, G. Sadowski, H. Latz, M. Buback: Phase equilibria in polydisperse and

58.18 associating copolymer solutions: Poly(ethene-co-(meth)acrylic acid)–monomer mixtures, Fluid Ph. Equilib. **241**(1/2), 113–123 (2006)

58.18 S. Behme: Thermodynamik von Polymersystemen bei hohen Drücken, Ph.D. Thesis (Technische Universität, Berlin 2000)

58.19 G.P. Rangaiah: Evaluation of genetic algorithms and simulated annealing for phase equilibrium and stability problems, Fluid Ph. Equilib. **187/188**, 83–109 (2001)

58.20 M. Srinivas, G.P. Rangaiah: A study of differential evolution and tabu search for benchmark, phase equilibrium and phase stability problems, Comput. Chem. Eng. **31**(7), 760–772 (2007)

58.21 L. Gao, N.W. Loney: New hybrid neural network model for prediction of phase equilibrium in a two-phase extraction system, Ind. Eng. Chem. Res. **41**(1), 112–119 (2002)

58.22 X. He, X. Zhanga, S. Zhanga, J. Liub, C. Lia: Prediction of phase equilibrium properties for complicated macromolecular systems by HGALM neural networks, Fluid Ph. Equilib. **238**(1), 52–57 (2005)

58.23 Y.S. Teh, G.P. Rangaiah: Tabu search for global optimization of continuous functions with application to phase equilibrium calculations, Comput. Chem. Eng. **27**(11), 1665–1679 (2003)

58.24 M. Srinivas, G.P. Rangaiah: Implementation and evaluation of random tunneling algorithm for chemical engineering applications, Comput. Chem. Eng. **30**(9), 1400–1415 (2006)

58.25 M. Srinivas, G.P. Rangaiah: Differential evolution with tabu list for global optimization and its application to phase equilibrium and parameter estimation problems, Ind. Eng. Chem. Res. **46**(10), 3410–3421 (2007)

58.26 C.G.E. Boender, A.H.G. Rinnooy Kan, G.T. Timmer, L. Stougie: A stochastic method for global optimization, Math. Program. **22**(1), 125–140 (1982)

58.27 J. Balogh, T. Csendes, R.P. Stateva: Application of a stochastic method to the solution of the phase stability problem: cubic equations of state, Fluid Ph. Equilib. **212**(1/2), 257–267 (2003)

58.28 T. Csendes, L. Pál, J.O.H. Sendín, J.R. Banga: The global optimization method revisited, Optim. Lett. **2**(4), 445–454 (2008)

58.29 R. Hooke, T.A. Jeeves: Direct search solution of numerical and statistical problems, J. ACM **8**, 212–229 (1961)

58.30 N. Hansen, A. Ostermeier: Completely derandomized self-adaptation in evolution strategies, Evol. Comput. **9**(2), 159–195 (2001)

58.31 J.C. Nash: *Compact Numerical Methods for Computers: Linear Algebra and Function Minimisation*, 2nd edn. (Adam Hilger, Bristol 1990)

58.32 J.D. Knowles, R.A. Watson, D.W. Corne: Reducing local optima in single-objective problems by multi-objectivization, Lect. Notes Comput. Sci. **1993**, 269–283 (2001)

58.33 J. Handl, S. Lovell, J. Knowles: Multiobjectivization by decomposition of scalar cost functions, Lect. Notes Comput. Sci. **5199**, 31–40 (2008)

58.34 J. Handl, S. Lovell, J. Knowles: Investigations into the effect of multiobjectivization in protein structure prediction, Lect. Notes Comput. Sci. **5199**, 702–711 (2008)

58.35 V. Cutello, G. Narzisi, G. Nicosia: Computational studies of peptide and protein structure prediction problems via multiobjective evolutionary algorithms. In: *Multiobjective Problem Solving from Nature. From Concepts to Applications*, ed. by J. Knowles, D. Corne, K. Deb (Springer, Berlin, Heidelberg 2008) pp. 93–114

58.36 D. Brockhoff, T. Friedrich, N. Hebbinghaus, C. Klein, F. Neumann, E. Zitzler: Do additional objectives make a problem harder?, Proc. 9th Annu. Conf. Genet. Evol. Comput. (2007) pp. 765–772

58.37 H.A. Abbass, K. Deb: Searching under multi-evolutionary pressures, Lect. Notes Comput. Sci. **2632**, 391–404 (2003)

58.38 A. Toffolo, E. Benini: Genetic diversity as an objective in multi-objective evolutionary algorithms, Evol. Comput. **11**(2), 151–167 (2003)

58.39 L.T. Bui, J. Branke, H.A. Abbass: Diversity as a selection pressure in dynamic environments, Proc. 2005 Conf. Genet. Evol. Comput. (2005) pp. 1557–1558

58.40 K. Deb, A. Saha: Multimodal optimization using a bi-objective evolutionary algorithm, Evol. Comput. **20**(1), 27–62 (2012)

58.41 M. Emmerich, N. Beume, B. Naujoks: An EMO algorithm using the hypervolume measure as selection criterion, Lect. Notes Comput. Sci. **3410**, 62–76 (2005)

58.42 N. Beume, B. Naujoks, M. Emmerich: SMS-EMOA: Multiobjective selection based on dominated hypervolume, Eur. J. Oper. Res. **181**(3), 1653–1669 (2007)

58.43 K. Deb: *Multi-Objective Optimization Using Evolutionary Algorithms* (Wiley, New York 2001)

58.44 T.E. Daubert, R.P. Danner: *Data Compilation Tables of Properties of Pure Compounds, Design Institute for Physical Property Data* (American Institute of Chemical Engineers, New York 1985)

58.45 J. Gross, G. Sadowski: Application of the perturbed-chain saft equation of state to associating systems, Ind. Eng. Chem. Res. **41**(22), 5510–5515 (2002)

58.46 M. Kleiner, G. Sadowski: Modeling of polar systems using PC-SAFT: An approach to account for induced-association interactions, J. Phys. Chem. C **111**(43), 15544–15553 (2007)

58.47 G.A. Chubarov, S.M. Danov, G.V. Brovkina, T.V. Kupriyanov: Equilibrium in system methanol methyl methacrylate water, J. Appl. Chem. USSR **51**(2), 434–437 (1978)

58.48 J. Kooi: The system methylmethacrylate – methanol – water, J. R. Neth. Chem. Soc. **68**(1), 34–42 (1949)

58.49   S.M. Danov, T.N. Obmelyukhina, G.A. Chubarov, A.L. Balashov, A.A. Dolgopolov: Investigation and calculations of liquid–vapor-equilibrium in binary methyl-methacrylate impurity systems, J. Appl. Chem. USSR **63**(3), 566–568 (1990)

58.50   J. Fu, K. Wang, Y. Hu: Studies on the vapor-liquid equilibrium and vapor-liquid-liquid equilibrium for a methanol-methyl methacrylate-water ternary system (II) Ternary system, J. Chem. Ind. Eng. (China) **4**(1), 14–25 (1988)

58.51   A.C.G. Marigliano, M.B.G. de Doz, H.N. Solimo: Influence of temperature on the liquid–liquid equilibria containing two pairs of partially miscible liquids – water + furfural + 1-butanol ternary system, Fluid Ph. Equilib. **153**(2), 279–292 (1998)

59. Modeling and Optimization of Machining Problems

Dirk Biermann, Petra Kersting, Tobias Wagner, Andreas Zabel

In this chapter, applications of computational intelligence methods in the field of production engineering are presented and discussed. Although a special focus is set to applications in machining, most of the approaches can be easily transferred to respective tasks in other fields of production engineering, e.g., forming and coating. The complete process chain of machining operations is considered: The design of the machine, the tool, and the workpiece, the computation of the tool paths, the model selection and parameter optimization of the empirical or simulation-based surrogate model, the actual optimization of the process parameters, the monitoring of important properties during the process, as well as the posterior multicriteria decision analysis. For all these steps, computational intelligence techniques provide established tools. Evolutionary and genetic algorithms are commonly utilized for the internal optimization tasks. Modeling problems can be solved using artificial neural networks. Fuzzy logic represents an intuitive way to formalize expert knowledge in automated decision systems.

In production engineering and particularly in the field of machining, improvements in materials, coatings, tools, and machines continuously provide potentials for improving the processes. In order to exploit these potentials, however, optimal setups of the changing processes have to be found. Since modern production processes involve many complex subsystems, as well as preceding and subsequent steps, all these systems and steps have to be adapted for achieving the optimal result.

In this chapter, it is shown that computational intelligence (CI) provides methods to assist in achieving this ambitious aim. A particular focus is on the applications of evolutionary computation (EC) in machining, but also artificial neural networks (NN) and fuzzy logic are considered. A comprehensive overview is presented by

considering several subsystems, as well as the preceding and subsequent steps in the operating sequence. In this aspect, the chapters contribute to common surveys in the literature [59.1–5], which are often only focused on the modeling and optimization of the actual process.

In order to assist interested engineers in choosing a suitable method for their problem, the solutions offered by CI are structured according to the specific subproblems to be solved in a machining problem. To keep the big picture still apparent, these subproblems are integrated into the complete operating sequence in the following section. They are then discussed according to their chronological order in the sequence. The chapter is concluded with summarizing remarks on CI applications in the field of production engineering.

59.1 Elements of a Machining Process

An overview of the elements and steps to be considered when optimizing a machining process is shown in Fig. 59.1. In the focus of the considerations is the actual process. The results of this process, however, significantly depend on its elements, in particular on the mechanical properties and the dynamic characteristics of the machine, geometry, and the properties of the tools, as well as the layout of the workpiece which determines the required machining operations. All these elements can be individually optimized to improve the results of the process. For the latter, often complex numerical control (NC) paths for the machines have to be generated using computer-assisted manufacturing (CAM) software. To accomplish this, a model of the final workpiece geometry is required. If no such model is available, e.g., after manual modifications of a prototype, CI-based methods can assist in computing an optimized workpiece model for the CAM software. However, even if a model is available, the NC paths computed by the CAM software can be far from optimal due to the complexity of the process, e.g., in five-axis milling operations. In this case, the subsequent optimization of the position-dependent parameters of the NC code, such as the inclination angles α and β, and the feed rate f [59.6], can significantly increase the efficiency of the process.

When all the components of the actual process are selected and fixed therewith, the optimization of the adjustable process parameters can begin. Thereby, CI-based techniques are usually based on a self-organizing process. In order to let the self-organization take effect, a high number of experiments is required. Since a real-world experiment involves high costs, it can become necessary to use a surrogate model on which the method is applied. In this case, however, additional problems have to be solved. It has to be selected which kind of model (empirical, analytical, physical, numerical) is applied and which type or realization of this kind of model is implemented, e.g., an empirical model can be computed using artificial neural networks, Gaussian processes, or regression techniques. As soon as a model is chosen, the parameters of this model (internal coefficients, material constants, etc.) have to be adapted with respect to the given application. This often represents an additional nonlinear optimization problem which can be solved using techniques of EC.

Moreover, the robustness of the process can be increased by a monitoring-based process control. To accomplish this, dynamic characteristics of the process, such as acoustic emission signals and force measurements, are analyzed online and control operations are initiated as soon as these characteristics show suspicious patterns. In this kind of application, however, it is necessary to automatically detect what indeed is a suspicious pattern. Fuzzy logic and NNs have proven to be capable of performing these tasks.

A lot of information can be obtained in order to analyze the process and its results. This information can either be achieved by measurements during and after the process or by performing simulation studies. They usually build the basis for the calculation of the actual objectives. In this context, machining processes have to be optimized with respect to several conflicting aims, e.g., a simultaneous minimization of tool wear and maximization of the material removal rate. Even if multiobjective optimization techniques are used, a lot of details can be lost in this formulization step. Often the first version of the objectives does not result in the desired results. Additional objectives have to be defined or preferences have to be integrated. In order to allow a deeper understanding of the process to be obtained and a refinement of the objectives to be made, an intuitive visualization and exploration of the detail information is required. For this task, again CI-based techniques can be used.

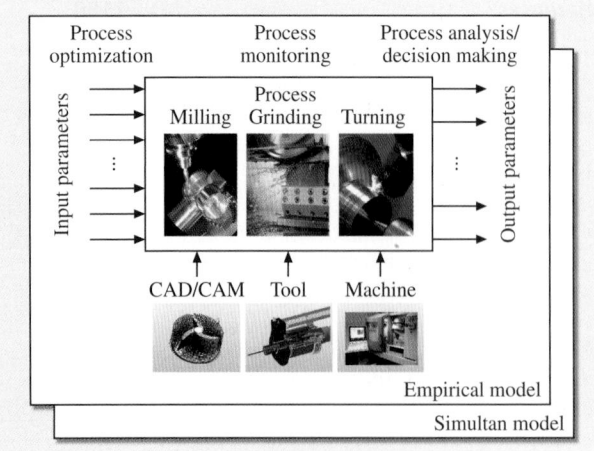

Fig. 59.1 Overview of the elements and steps of an arbitrary machining process

59.2 Design Optimization

The optimal design of a machine, tool, or workpiece is a great challenge in the field of production engineering. The optimization task is often conducted as an iterative manual process which is based on expert knowledge and which can be very cost and time consuming. *Roy* et al. [59.7] gave an extensive overview of the recent advances in automated and interactive design optimization. They presented a classification of the optimization problems and discussed the most important optimization approaches and techniques. In the following subsections, examples of successful applications of CI for the optimization of machine, tool, and workpiece designs are provided.

59.2.1 Optimal Design of Machines

Designing machines necessitates the consideration of multiple objectives, such as geometric accuracy and costs. *Liu* and *Liang* [59.8], for instance, presented an approach combining a modified Chebyshev programming method for the scalarization of these objectives and a particle swarm optimization algorithm for evolving the machine designs. They were dealing with reconfigurable machine tools, so not only the process accuracy and investment costs of the machine layouts, but also the configurability was considered. Significant changes in the shape of the product could thus be easily adapted. *Mekid* and *Khalid* [59.9] discussed an optimization method based on a multiobjective genetic algorithm for the design of three-axis micromilling machines. They took user requirements (for example the workspace volume), axis positions, and geometric errors of the machine into account. For the latter, they used a mathematical error model of the three-axis milling machines.

59.2.2 Tool Optimization

Designing machining tools is a very difficult optimization task since not only complex geometries, but also different machining criteria have to be taken into account [59.10]. *Abele* and *Fujara*, for example, presented a simulation approach for optimizing the drill geometry based on a genetic algorithm [59.11]. They considered not only the structural stiffness of the tool during their optimization run, but also took the coolant flow resistance and the chip evacuation capability into ac-

count. They also defined the machinability, especially the grindability of the chip flute, as constraint. In order to take all these criteria into account, different simulation approaches have to be used (Sect. 59.4). *Abele* and *Fujara* used, for example, the finite element method in order to analyze the structural stiffness. The cutting forces were computed using a semiempirical cutting force model. Additionally, a model of the grinding wheel had to be determined in order to evaluate the grindability of the optimized drill geometry. Another application was presented by *Jared* et al. [59.12] who integrated GA into the computer-aided design software CATIA. In one of their case studies, the volume and the tip deflection of a cutting tool were minimized by automatically parameterizing length and angles between segments of a 2-D (two-dimensional) profile which were then extruded to the actual tool.

59.2.3 Workpiece Layout Optimization

The layout of products can usually be described as multiobjective optimization problem. For example, the design of aerospace structures always faces a tradeoff between the stiffness and the weight of the products [59.13]. The layout of a cooling system, e.g., for a turbine blade [59.13] is a tradeoff between the machining quality, the cooling effect, and the production costs. *Weinert* et al. [59.14–17] developed a simulation system for optimizing the layout of mold temperature control systems in order to minimize the production cycle times and costs, and to maximize the product quality. They developed an efficient simulation system in order to evaluate the effect and homogeneity of the tempering of the design layout and to estimate the manufacturing costs [59.18]. Using fast but sufficiently accurate evaluation methods, a computer-aided optimization of the temperature control system based on multiobjective optimization methods, like NSGA-II [59.19] and SMS-EMOA [59.20], became possible [59.21–24]. Nevertheless, this optimization task is very complex and the engineer's experience is still necessary. Due to this, *Biermann* et al. [59.25] combined the computer-aided optimization system with the possibility of user interaction so that a visual real-time manipulation of target functions is possible. *Dürr* and *Jurklies* [59.26] presented a fuzzy expert system in order to use the expert knowledge in a computer-assisted way.

59.3 Computer-Aided Design and Manufacturing

In the modern construction process, computer-aided design (CAD) software is used for all design tasks – for example for the model of the workpiece. This model is the basis for the generation of the NC paths by CAM software. However, if only a physical prototype exists or manual modifications of the original model have been performed, methods to compute a respective model are required. To accomplish this, the original object is scanned and a point-based representation is obtained. From this point data, a new CAD model has to be calculated or the original model has to be adapted. This process is called *surface reconstruction* or *reverse engineering*.

When a model of the workpiece is available, NC paths can be generated based on CAM software for most machining processes. For complex five-axis milling processes, however, the results of standard CAM software are not always optimal with respect to the requirements of the specific machine and process. In this case, CI-based techniques can be used to improve the NC paths generated by the CAM software.

59.3.1 Surface Reconstruction

The optimization of the visual quality of triangulations with respect to different quality criteria was successfully performed using evolutionary algorithms by *Weinert* et al. [59.27]. Based on an initial triangulation, as provided by the software of the scanning system, edges were flipped in order to minimize the total length of all edges, the surface area, the sum of angles between normals, and the total absolute curvature. It was found that the latter is best suited for generating visually smooth surfaces.

Small tolerances in the representation of the original object, however, result in a huge number of required scan points. Current scanners are able to provide this dense and precise set of scan points, but the resulting triangulations become very large and difficult to handle. Approximating triangulations tackle this problem. The number of control points for the triangles is independent of the size of the point set and usually considerably smaller than the number of scan points. *Weinert* et al. [59.28] documented the capabilities of a standard evolution strategy to optimize the control point positions of approximating triangulations. In order to avoid an uncontrolled expansion of the triangulation, balancing strategies based on mass–spring systems were integrated.

Unfortunately, even approximating triangulations produce a nonsmooth surface and are therefore not convenient for the later computation of NC paths. *Nonuniform rational B-splines* (NURBS) [59.29] are another popular mathematical model for free-form surfaces in CAD software. The most important advantages of NURBS over triangulations are their smoothness, their compact definition, the possibility for an intuitive local manipulation, as well as the ability to combine NURBS patches to larger structures. *Mehnen* et al. [59.30, 31] applied an evolution strategy to the coordinates of the NURBS's control points in order to minimize the distance between the scan points and their projection to the NURBS. *Wagner* et al. [59.32] did the same using a real-valued genetic algorithm. They also proposed another distance indicator that is based on a sampling of the NURBS and that is much cheaper to evaluate. The use of the sampling-based distance measure in combination with a equation-solver-based hybrid real-valued genetic algorithm significantly reduced the runtime of the optimization. This approach was further enhanced [59.16] to a two-step approach, in which the single-objectively optimized solution is used as initial individual for a multiobjective optimization. As additional objective, the smoothness of the NURBS was considered. This objective was also considered by *Jared* et al. [59.12] in their GA-based optimization of NURBS in CATIA.

In addition, *Weinert* et al. [59.33] combined NURBS with constructive solid geometries [59.34] in a hybrid evolutionary algorithm/genetic programming approach. By these means, the constructional logic behind the workpiece could also be approximated.

59.3.2 Optimization of NC Paths

The five-axis milling process offers the possibilities to tilt the milling tool and, thus, to use shorter and therewith stiffer tools. This allows complex free-form surfaces to be machined in one workpiece clamping, and the engagement conditions to be adapted [59.35]. An improvement of the machining results and a reduction of the machining time can be achieved. However, in contrast to the three-axis process, the generation of the NC paths particularly for the machining of free-form surfaces is much more complex [59.6].

Weinert and *Stautner* [59.36] presented an algorithm for converting three- into five-axis milling paths in which the position of the tool tip is kept from the three-axis NC program. An optimization approach based on an evolutionary strategy was used to improve the tool movement [59.37]. To accomplish this, they developed a fast simulation system of the five-axis milling process based on a discrete dexel model of the workpiece (Sect. 59.4) [59.38].

The NC paths generated for a five-axis milling process are often not smooth enough since the kinematic behavior of the specific milling machine is not taken into account. *Zabel* et al. developed a simulation approach which is placed in the process chain between the CAM system and the real-milling process [59.39]. The five-axis tool movement is optimized taking the tool axis configuration and the dynamic behavior of the milling machine into account. For this purpose, methods of evolutionary computation and wavelet the-

ory were combined [59.35]. In 2007, *Mehnen* et al. integrated a multiobjective optimization algorithm into this simulation system which combined the variation of a modern single-objective approach with the selection mechanism of a classical multiobjective optimization algorithm in order to optimize the tool movement [59.40].

One challenging task during the optimization of the five-axis milling process is the avoidance of collisions between the milling tool and the workpiece. *Kersting* and *Zabel* [59.6] developed an efficient simulation approach, which maps the high-dimensional restriction area on a two-dimensional matrix structure. They showed that the use of a multipopulation multiobjective evolutionary algorithm in the restriction-free area improved the corresponding Pareto fronts [59.41].

59.4 Modeling and Simulation of the Machining Process

The optimization of real-world applications using CI-based or classical optimization approaches requires that a performance value or vector can be obtained for all possible settings of the input parameters, whereby the performance values are usually calculated based on measurements during or after the actual process. In order to achieve a near-optimal result, however, far more than 100 different parameter vectors have to be evaluated – even for low-dimensional problems. This amount of real experiments is often impossible due to the costs related to them. As a possible solution, the use of empirical or physical (simulation) models can significantly reduce the number of required experiments since most of the evaluations can be performed on the model. For both kinds of approaches, CI techniques have already been successfully used. Some examples are presented in the following subsections.

59.4.1 Empirical Modeling

For the use of empirical models, real or simulated experiments are still required in order to build up a data base for the training of the model. In contrast to the direct optimization of the process, however, these experiments are performed as a block of moderate size in the beginning of the optimization. Afterward, the model allows new parameter settings to be predicted based on the information obtained from training data. The determination of near-optimal solutions can be performed on the model.

The number of empirical models is exhaustive [59.42]. Nevertheless, NNs often showed their capability to empirically model responses from machining processes. For instance, the material removal rate of an abrasive jet drilling process was successfully predicted by using an NN with back error propagation [59.43]. As input parameters, varying gas pressure, nozzle inside diameter, abrasive flow rate, size of the medium particle, and standoff distance were considered. Accordingly, the ablating depth obtained for specific values of the peak power, pulsing frequency, and overlapping in a laser drilling process could be predicted using NN [59.44]. *Casalino* et al. [59.45] showed that NN achieve higher prediction accuracies than regression techniques in predicting surface roughness and resultant forces for varying cutting speed, feed rate, and radial depth in milling. In the same line, NN were used for the prediction of the specific cutting constants resulting from different cutting speeds, feeds, inclination angles α and β, cutting depths, and cutting widths [59.46]. With respect to tool wear, the wheel life of a cylindrical grinding wheel was modeled using a feedforward backpropagation NN. A direct prediction of the tool wear was also accomplished using NN [59.47, 48]. Moreover, the thermal expansion of the Y-axis ball screw was predicted based on temperature measurements at different points of the machine structure [59.49].

In addition, CI-based techniques can also indirectly be used for empirical modeling. As soon as complex empirical models, such as Gaussian processes, support vector or other kernel machines, are used, the determination of the optimal model parameters is an individual nonlinear optimization problem. Evolutionary algorithms, in particular the covariance matrix adaption evolution strategy (CMA-ES) [59.50], showed to be suitable for solving these problems [59.51, 52].

59.4.2 Physical Modeling for Simulation

In cases where sufficient knowledge about the physical laws of the process is available, simulation models based on equations representing these physical laws are likely to be superior to the very general formulations of the empirical models. Nevertheless, also these models have parameters that are related to the properties of the material, tool, and machine. Since these parameters can often not be measured, their values are usually set by minimizing the error between the predictions of the simulation and a training set of observations from real-world experiments. As consequence, EC is a valuable tool for calibrating simulation models which was shown to be superior to classical data fitting tools [59.53].

In an exemplary application, the dynamic behavior of manufacturing systems was characterized by its frequency response function. This function can be modeled by a superposition of decoupled damped harmonic oscillators, whereby each oscillator has three parameters (mass, natural frequency, and damping) [59.54]. In order to minimize the deviation between the measured frequency response function and one of the oscillators, an interactive approach based on evolutionary algorithms was successfully implemented [59.54].

An open issue in the simulation of machining processes is the modeling of the extremely high strain rates which can only rarely be covered by classical material models and tensile tests. As a possible solution, EC can be used as a submodule of a simulation in order to predict the deformation and flow characteristics for high strain rates. For instance, *Weinert* et al. used symbolic regression by means of a genetic programming system to evolve mathematical formulae that describe the trajectories of single particles of steel based on recordings of a high-speed camera during the turning process [59.55, 56]. *Teti* et al. [59.57] employed NN to reconstruct the stress–strain curve of the workpiece material from experimental data of tensile tests. They found out that the learned NN is capable of predicting workpiece material properties in a wide range of temperature and strain rate values. A hybrid simulation model based on physical equations and the empirical stress–strain prediction was finally proposed. Two recent overviews of hybrid models for simulation which also incorporate CI techniques were provided by *Jawahir* et al. [59.58, 59].

59.5 Optimization of the Process Parameters

In this section, possible applications of EC methods for the optimization of the actual process parameters are discussed. Since a recent survey book for the model-based optimization of process parameters exist [59.1], only a short summary of possible applications is provided. In contrast to this survey, the following presentation does not distinguish between different processes, as the aspects related to the use of EC are independent of the actual process, e.g., milling, turning, or grinding.

As already discussed in the previous section, it is mandatory to approximate the process quality indicators by means of analytical, empirical, or physical models. In the literature, no direct application of EC optimization techniques to machining processes was reported until now. Instead, polynomial or process-related equations were usually fitted to experimental data [59.60–78]. Neural networks [59.63, 79–83], other empirical models [59.51, 62, 84], and simulation models [59.85, 86] were also popular to accomplish this task.

For the actual optimization, two important decisions on the formulation of the problem have to be taken in order to choose the EC method. These decisions are concerned with the representation of the input parameters and the objectives. In most cases, continuously defined input parameters, such as feed and cutting speed, are to be optimized. This relates to techniques such as evolution strategies, particle swarm optimization, and real-valued genetic algorithms (GAs). If also discrete parameters, such as the cooling concept or tool material, are considered, special evolution strategies [59.65, 87] or binary GAs may better be suited. With respect to the objectives, it has to be decided whether a single optimal solution or a set of tradeoffs is desired. In the former case, almost all EC techniques can directly be used. Due to the complexity of production engineering problems, however, a suitable scalarization of the different objectives has to be found in order to achieve reasonable results. In the latter case of searching for an approximation of the trade-off structure, it is important that the algorithm is capable of coping with multiple objectives which have to be considered in parallel [59.51, 72, 74, 78, 79, 84, 86].

In the literature, the use of continuous input variables and single-objective formulations is established. The most popular EC methods are particle swarm optimization (PSO) [59.63, 68, 75, 76, 81–83, 85, 88] and standard GA or evolutionary algorithm (EA) [59.60, 62, 64, 67, 69, 77, 80]. The use of specifically designed heuristics [59.71, 75, 89] is rather uncommon. Nevertheless, the formulation of the problem and the design of the algorithm should aim at incorporating as much knowledge as possible into the optimization [59.16].

Unfortunately, the generality of CI-based techniques often results in problem formulations which are not completely sophisticated. An important factor often neglected when optimizing production engineering problems is the uncertainty about the external process variables, e.g., properties of the tool or material. Although modern algorithms are capable of incorporating them into the optimization [59.90], only a few applications actually take these uncertainties into account [59.70]. More specifically, two sources of uncertainty can be considered [59.91]: perturbations in the input variables, e.g., due to online control, and environmental uncertainties, such as outdoor temperature, humidity, and the already mentioned external process variables. A detailed overview of such factors can be found in the literature [59.92]. A comprehensive survey of possible problem formulations and respective optimization approaches was presented by *Beyer* and *Sendhoff* [59.91]. In production-engineering applications, however, classical statistical methods are usually used to cope with these problems. The potential of CI-based techniques has not yet been exploited.

59.6 Process Monitoring

The analysis of different process variables – like for example the cutting forces, acoustic emission, or temperatures – allows conclusions about the process-dependent state of the machining processes and its components (tools, machines, workpieces, etc.) to be drawn and provides the possibility for an adaptive process control [59.93]. The idea of process monitoring is to measure, visualize, and analyze the values of these variables during the machining process. *Teti* et al. [59.93] gave an extensive overview of *advanced monitoring of machining operations* describing sensor systems for machining, signal processing, monitoring scopes, and the decision-making support systems. In order to evaluate the measured values, cognitive computing methods – for example genetic algorithms, fuzzy logic, or NNs – can be used. In contrast to the rule-based fuzzy logic approach, NNs do not store the knowledge in an explicit form. A survey of the successful applications of these techniques for the advanced monitoring of machining operations was provided by *Teti* et al. [59.93]. It is thus omitted in this section.

59.7 Visualization

In the field of production engineering, the complex optimization problems are often characterized by multiple objectives and restrictions. Additionally, the decision space can be high dimensional – like for example in the case of optimizing NC paths (Sect. 59.3.2) [59.6]. In order to analyze the optimization problems and the applied optimization approach, an intuitive visualization of the data resulting from the evolutionary process is advisable [59.94]. For this purpose, *Pohlheim* [59.95] reviewed several visualization techniques in order to obtain a better understanding of the optimization process of real-world problems. He recommended the use of three diagrams in order to analyze the optimization algorithm: A convergence diagram, visualization of the change of the best individual during the optimization approach, and a diagram of the objective values of all individuals in the population of all generations.

Müller et al. discussed techniques for an *intuitive visualization and interactive analysis of Pareto sets ap-*

plied on production engineering systems [59.94]. They analyzed different visualization and analysis methods in order to gain insight into both the optimization problem and the optimization algorithm, and to support an intuitive decision-making process. For this purpose, they presented a simultaneous visualization of the decision and the objective space. An interactive navigation through the solution sets supports the user to detect specific process characteristics [59.94]. This also helps to redesign the objective formulation in cases where the optimization results are not in agreement with the actual preferences of the decision maker.

In order to support the trade-off analysis in multiple dimensions, *Obayashi* and *Sasaki* [59.96] presented a visualization approach based on self-organizing maps (SOMs). The idea is to map from the high-dimensional objective function space to two-dimensional map units. They showed the applicability of this approach analyzing two multiobjective aerodynamic design problems [59.96].

The *innovization* approach of *Deb* [59.97] provides an automated identification of design principles by searching for common features of the optimal trade-offs in a multiobjective optimization problem. These features are provided by means of analytical relations between the design variables. A successful application of innovization in machining was already reported [59.78]. Another possibility to learn about the structure of the objectives and the effect of the input parameters is provided by visualizations and analyses based on the surrogate models of the process (Sect. 59.4) [59.51, 98].

59.8 Summary and Outlook

This chapter focused on applications of CI in the optimization of machining problems. For this purpose, the whole process chain – from the design of a machine, tool, or workpiece, as well as the corresponding optimization of process parameters, to the process monitoring and subsequent analysis of the results – was taken into account. Different modeling and simulation techniques, which are necessary to optimize real-world problems, were discussed. Successful examples in the field of production engineering were compiled to present the applicability of the CI methods. In conclusion, evolutionary and genetic algorithms are general and powerful solvers for nonlinear optimization tasks,

artificial neural networks can be used for continuous modeling problems, and fuzzy logic provides an intuitive way to represent expert knowledge.

Unfortunately, the generality of CI-based techniques often results in problem formulations which are not completely sophisticated. For instance, possibilities of creating good initial solutions, uncertainty in the design variables, and specific aspects of the quality indicators resulting in undesirable scalarizations, are often neglected. EC provides the means to appropriately consider these aspects. A proper analysis of the results can assist in identifying such pitfalls and in improving the problem formulation.

References

59.1   E. Venkata Rao: *Advanced Modeling and Optimization of Manufacturing Processes* (Springer, Berlin, Heidelberg 2011)

59.2   D. Dasgupta, Z. Michalewicz: *Evolutionary Algorithms in Engineering Applications* (Springer, Berlin, Heidelberg 1997)

59.3   W. Banzhaf, M. Brameier, M. Stautner, K. Weinert: Genetic programming and its application in machining technology. In: *Advances in Computational Intelligence: Theory and Practice*, ed. by H.-P. Schwefel, I. Wegener, K. Weinert (Springer, Berlin, Heidelberg 2003) pp. 194–244, Chap. 7

59.4   I. Mukherjee, P. Ray: A review of optimization techniques in metal cutting processes, Comput. Ind. Eng. **50**(1-2), 15–34 (2006)

59.5   H. Aytug, M. Khouja, F.E. Vergara: Use of genetic algorithms to solve production and operations management problems: A review, Int. J. Prod. Res. **41**(17), 3955–4009 (2003)

59.6   P. Kersting, A. Zabel: Optimizing NC-tool paths for simultaneous five-axis milling based on multi-population multi-objective evolutionary algorithms, Adv. Eng. Softw. **40**(6), 452–463 (2009)

59.7   R. Roy, S. Hinduja, R. Teti: Recent advances in engineering design optimisation: Challenges and future trends, CIRP Ann. Manuf. Technol. **57**(2), 697–715 (2008)

59.8   W. Liu, M. Liang: A particle swarm optimization approach to a multi-objective reconfigurable machine tool design problem, IEEE Cong. Evol. Comput. (2006) pp. 2222–2229

59.9   S. Mekid, A. Khalid: Robust design with error optimization analysis of CNC micromilling machine, 5th CIRP Int. Semin. Intell. Comput. Manuf. Eng. (2006) pp. 583–587

59.10   H. Schulz, A.K. Emrich: Optimization of the chip flute of drilling tools using the principle of genetic algorithms, 2nd CIRP Int. Semin. Intell. Comput. Manuf. Eng. (2000) pp. 371–376

59.11   E. Abele, M. Fujara: Simulation-based twist drill design and geometry optimization, CIRP Ann. Manuf. Technol. **59**(1), 145–150 (2010)

59.12   G. Jared, R. Roy, J. Grau, T. Buchannan: Flexible optimization within the CAD/CAM environment, CIRP Int. Seminar Intell. Comput. Manuf. Eng. (1998) pp. 503–508

59.13   R. Roy, A. Tiwari, J. Corbett: Designing a turbine blade cooling system using a generalised regression genetic algorithm, CIRP Ann. Manuf. Technol. **52**(1), 415–418 (2003)

59.14   J. Mehnen, K. Weinert, H.-W. Meyer: Evolutionary optimization of deep drilling strategies for mold temperature control, 3rd CIRP Int. Semin. Intell. Comput. Manuf. Eng. (2002)

59.15   T. Michelitsch, J. Mehnen: Evolutionary optimization of cooling circuit layouts based on the electrolytic tank method, 4th CIRP Int. Semin. Intell. Comput. Manuf. Eng. (2004)

59.16   K. Weinert, A. Zabel, P. Kersting, T. Michelitsch, T. Wagner: On the use of problem-specific candidate generators for the hybrid optimization of multi-objective production engineering problems, Evol. Comput. **17**(4), 527–544 (2009)

59.17   J. Mehnen, T. Michelitsch, K. Weinert: Production engineering: Optimal structures of injection molding tools. In: *Emergence, Analysis and Optimization of Structures – Concepts and Strategies across Disciplines*, ed. by K. Lucas, P. Roosen (Springer, Berlin, Heidelberg 2010) pp. 75–90

59.18   T. Michelitsch, J. Mehnen: Optimization of production engineering problems with discontinuous cost-functions, 5th CIRP Int. Semin. Intell. Comput. Manuf. Eng. (2006) pp. 275–280

59.19   K. Deb, A. Pratap, S. Agarwal, T. Meyarivan: A fast and elitist multiobjective genetic algorithm: NSGA-II, IEEE Trans. Evol. Comput. **6**(2), 182–197 (2002)

59.20   N. Beume, B. Naujoks, M. Emmerich: SMS-EMOA: Multiobjective selection based on dominated hypervolume, Eur. J. Oper. Res. **181**(3), 1653–1669 (2007)

59.21   J. Mehnen, T. Michelitsch, T. Bartz-Beielstein, N. Henkenjohann: Systematic analyses of multiobjective evolutionary algorithms applied to real-world problems using statistical design of experiments, 4th CIRP Int. Semin. Intell. Comput. Manuf. Eng. (2004)

59.22   J. Mehnen, H. Trautmann: Integration of expert's preferences in pareto optimization by desirability function techniques, 5th CIRP Int. Semin. Intell. Comput. Manuf. Eng. (2006) pp. 293–298

59.23   T. Michelitsch, T. Wagner, D. Biermann, C. Hoffmann: Designing memetic algorithms for real-world applications using self-imposed constraints, Proc. 2007 IEEE Congr. Evol. Comput. (2007) pp. 3050–3057

59.24   D. Biermann, R. Joliet, T. Michelitsch, T. Wagner: Sequential parameter optimization of an evolution strategy for the design of mold temperature control systems, Proc. 2010 IEEE Congr. Evol. Comput. (2010) pp. 4071–4078

59.25   D. Biermann, R. Joliet, T. Michelitsch: Interactive manipulation of target functions for the optimization of mold temperature control systems, 2nd Int. Conf. Manuf. Eng., Qual. Prod. Syst. (2010) pp. 239–244

59.26   H. Dürr, I. Jurklies: A fuzzy expert system assist the CAD/CAM/CAE process chain in the tool and mould making industry, 3rd CIRP Int. Semin. Intell. Comput. Manuf. Eng. (2002)

59.27   K. Weinert, J. Mehnen, F. Albersmann, P. Drerup: New solutions for surface reconstruction from discrete point data by means of computational intelligence, CIRP Int. Seminar Intell. Comput. Manuf. Eng. (1998) pp. 431–438

59.28   K. Weinert, J. Mehnen, M. Schneider: Evolutionary optimization of approximating triangulations for surface reconstruction from unstructured 3D Data, Proc. 6th Jt. Conf. Inf. Sci. (2002) pp. 578–581

59.29   L. Piegl, W. Tiller: *The NURBS Book* (Springer, Berlin, Heidelberg 1997)

59.30   K. Weinert, J. Mehnen: NURBS-surface approximation of discrete 3D-point data by means of evolutionary algorithms, 2nd CIRP Int. Semin. Intell. Comput. Manuf. Eng. (2000) pp. 263–268

59.31   T. Beielstein, J. Mehnen, L. Schönemann, H.-P. Schwefel, T. Surmann, K. Weinert, D. Wiesmann: Design of evolutionary algorithms and applications in surface reconstruction. In: *Advances in Computational Intelligence: Theory and Practice*, ed. by H.-P. Schwefel, I. Wegener, K. Weinert (Springer, Berlin, Heidelberg 2003) pp. 164–193

59.32   T. Wagner, T. Michelitsch, A. Sacharow: On the design of optimisers for surface reconstruction, Proc. 9th Annu. Genet. Evol. Comput. Conf. (2007) pp. 2195–2202

59.33   K. Weinert, T. Surmann, J. Mehnen: Evolutionary surface reconstruction using CSG-NURBS-Hybrids, Proc. Genet. Evol. Comput. Conf. (2001) pp. 1456–1463

59.34   C.M. Hoffmann: *Geometric & Solid Modeling* (Kaufmann Publ., San Mateo 1989)

59.35   K. Weinert, A. Zabel, H. Müller, P. Kersting: Optimizing NC tool paths for five-axis milling using evolutionary algorithms on wavelets, 8th Annu. Genet. Evol. Comput. Conf. (2006) pp. 1809–1816

59.36   K. Weinert, M. Stautner: Generating multiaxis tool paths for die and mold making with evolutionary algorithms, Lect. Notes Comput. Sci. **3103**, 1287–1298 (2004)

Part E | 59

59.37 K. Weinert, M. Stautner: A new system optimizing tool paths for multi-axis die and mould making by using evolutionary algorithms, production engineering, Res. Dev. **12**(1), 15–20 (2005)

59.38 A. Zabel, M. Stautner: Optimizing the multi-axis milling process via evolutionary algorithms, Berichte aus dem IWU, 8th CIRP Int. Workshop Model. Mach. Oper. (2005) pp. 363–370

59.39 A. Zabel, H. Müller, M. Stautner, P. Kersting: Improvement of machine tool movements for simultaneous five-axes milling, 5th CIRP Inter. Semin. Intell. Comput. Manuf. Eng. (2006) pp. 159–164

59.40 J. Mehnen, R. Roy, P. Kersting, T. Wagner: ICS-PEA: Evolutionary five-axis milling path optimisation, 9th Annu. Genet. Evol. Comput. Conf. (2007) pp. 2122–2128

59.41 D. Biermann, A. Zabel, T. Michelitsch, P. Kersting: Intelligent process planning methods for the manufacturing of moulds, Inter. J. Comput. Appl. Technol. **40**(1/2), 64–70 (2011)

59.42 T. Hastie, R. Tibshirani, J. Friedman: *The Elements of Statistical Learning: Data Mining, Inference, and Prediction*, 2nd edn. (Springer, Berlin, Heidelberg 2009)

59.43 M. Gheorghe, C. Neagu, S. Antoniu, C. Ionita: Modeling of abrasive jet drilling by applying a neural network method, 2nd CIRP Int. Seminar Intell. Comput. Manuf. Eng. (2000) pp. 221–226

59.44 S.L. Campanelli, A.D. Ludovico, C. Bonserio, P. Cavalluzzi: Artificial neural network modelling of the laser milling process, 5th CIRP Int. Semin. Intell. Comput. Manuf. Eng. (2006) pp. 107–111

59.45 G. Casalino, A.D. Ludovico, F.M.C. Minutolo, A. Rotondo: On the numerical modelling of a milling operation: Data recoveringand interpolation, 5th CIRP Int. Semin. Intell. Comput. Manuf. Eng. (2006) pp. 193–197

59.46 P. Clayton, M.A. Elbestawi, T.I. El-Wardany, D. Viens: An innovative calibration technique using neural networks for a mechanical model of the 5-Axis milling process, 2nd CIRP Int. Semin. Intell. Comput. Manuf. Eng. (2000) pp. 391–396

59.47 G. Casalino, A.D. Ludovico: Tool life estimation in single point turning using artificial neural networks, 4th CIRP Int. Semin. Intell. Comput. Manuf. Eng. (2004)

59.48 G. Ambroglio, D. Umbrello, L. Filice: Diffusion wear modelling in machining using ANN, 5th CIRP Int. Semin. Intell. Comput. Manuf. Eng. (2006) pp. 69–73

59.49 C. Bruni, A. Forcellese, F. Gabrielli, M. Simoncini: Thermal error prediction in a machining center using statistical and neural network-based models, 4th CIRP Int. Semin. Intell. Comput. Manuf. Eng. (2004)

59.50 N. Hansen, A. Ostermeier: Completely derandomized self-adaptation in evolution strategies, Evol. Comput. **9**(2), 159–195 (2001)

59.51 D. Biermann, K. Weinert, T. Wagner: Model-based optimization revisited: Towards real-world processes, Proc. 2008 IEEE Congr. Evol. Comput. (2008) pp. 2980–2987

59.52 O. Kramer: Covariance matrix self-adaptation and kernel regression – perspectives of evolutionary optimization in kernel machines, J. Fundam. Inform. **98**(1), 87–106 (2010)

59.53 T. Özel, Y. Karpat: Identification of constitutive material model parameters for high-strain rate metal cutting conditions using evolutionary computational algorithms, Mater. Manuf. Process. **22**, 659–667 (2007)

59.54 D. Biermann, T. Surmann, G. Kehl: Oscillator model of machine tools for the simulation of self excited vibrations in machining processes, 1st Int. Conf. Process Mach. Interact. (2008) pp. 23–29

59.55 K. Weinert, M. Stautner: Reconstruction of particle flow mechanisms with symbolic regression via genetic programming, Proc. Genet. Evol. Comput. Conf. (2001) pp. 1439–1443

59.56 K. Weinert, M. Stautner, J. Mehnen: Automatic generation of mathematical descriptions of cutting processes from video data, production engineering, Res. Dev. **9**(2), 55–58 (2002)

59.57 R. Teti, G. Giorleo, U. Prisco, D. DAddona: Integration of neural network material modelling into the FEM simulation of metal cutting, 3rd CIRP Int. Semin. Intell. Comput. Manuf. Eng. (2002)

59.58 I.S. Jawahir, X. Wang: Development of hybrid predictive models and optimization techniques for machining operations, J. Mater. Process. Technol. **185**(1–3), 46–59 (2007)

59.59 A.D. Jayal, I.S. Jawahir: Analytical and computational challenges for developing predictive models and optimization strategies for sustainable machining, 7th CIRP Int. Conf. Intell. Comput. Manuf. Eng. (2010)

59.60 G. Celano, S. Fichera, E.L. Valvo: Optimization of cutting parameters in multi pass turning operations for continuous forms, 2nd CIRP Int. Semin. Intell. Comput. Manuf. Eng. (2000) pp. 417–422

59.61 C.W. Lee, Y.C. Shin: Evolutionary modelling and optimization of grinding processes, Intern. J. Prod. Res. **38**(12), 2787–2813 (2000)

59.62 X. Wang, Z.J. Da, A.K. Balaji, I.S. Jawahir: Performance-based optimal selection of cutting parameters and cutting tools in multi-pass turning operations using genetic algorithms, 2nd CIRP Int. Semin. Intell. Comput. Manuf. Eng. (2000) pp. 409–414

59.63 V. Tandon, H. El-Mounayri, H. Kishawy: NC end milling optimization using evolutionary computation, Int. J. Mach. Tools Manuf. **42**(5), 595–605 (2002)

59.64 X. Wang, I.S. Jawahir: Web-based optimization of milling operations for the selection of cutting conditions using genetic algorithms, 3rd CIRP Int. Semin. Intell. Comput. Manuf. Eng. (2002)

59.65 C.W. Lee, T. Choi, Y.C. Shin: Intelligent model-based optimization of the surface grinding process for heat-treated 4140 steel alloys with aluminum oxide grinding wheels, J. Manuf. Sci. Eng. **125**(1), 65–76 (2003)

59.66 K. Vijayakumar, G. Prabhaharan, P. Asokan, R. Saravanan: Optimization of multi-pass turning operations using ant colony system, Inter. J. Mach. Tools Manuf. **43**(15), 1633–1639 (2003)

59.67 X. Wang, A. Kardekar, I.S. Jawahir: Performance-based optimization of multi-pass face-milling operations using genetic algorithms, 4th CIRP Int. Semin. Intell. Comput. Manuf. Eng. (2004)

59.68 Y. Karpat, T. Özel: Hard turning optimization using neural network modeling and swarm intelligence, Trans. North Am. Manuf. Res. Inst. **33**, 179–186 (2005)

59.69 T.-H. Hou, C.-H. Su, W.-L. Liu: Parameters optimization of a nano-particle wet milling process using the Taguchi method, response surface method and genetic algorithm, Powder Technol. **173**(3), 153–162 (2007)

59.70 J.L. Vigouroux, L. Deshayes, S. Foufou, L.A. Welsh: An approach for optimization of machining parameters under uncertainties using intervals and evolutionary algorithms, CIRP J. Manuf. Syst. **5**(36), 395–399 (2007)

59.71 A.R. Yildiz: A novel hybrid immune algorithm for global optimization in design and manufacturing, Robot. Comput. Integr. Manuf. **25**(2), 261–270 (2009)

59.72 R. Roy, J. Mehnen: Dynamic multi-objective optimisation for machining gradient materials, CIRP Ann. Manuf. Technol. **57**(1), 429–432 (2008)

59.73 F. Cus, J. Balic, U. Zuperl: Hybrid ANFIS-ants system based optimisation of turning parameters, J. Achiev. Mater. Manuf. Eng. **36**(1), 79–86 (2009)

59.74 R. Datta, K. Deb: A classical-cum-evolutionary multi-objective optimization for optimal machining parameters, Nat. Biol. Inspir. Comput. (2009) pp. 607–612

59.75 A.R. Yildiz: A novel particle swarm optimization approach for product design and manufacturing, Inter. J. Adv. Manuf. Technol. **40**(5-6), 617–628 (2009)

59.76 A.N. Sait: Optimization of machining parameters of GFRP pipes using evolutionary techniques, Int. J. Precis. Eng. Manuf. **11**(6), 891–900 (2010)

59.77 A.M. Zain, H. Haron, S. Sharif: Application of GA to optimize cutting conditions for minimizing surface roughness in end milling machining process, Expert Syst. Appl. **37**(6), 4650–4659 (2010)

59.78 K. Deb, R. Datta: Hybrid evolutionary multi-objective optimization of machining parameters, Eng. Optim. **44**(6), 685–706 (2011)

59.79 A.A. Krimpenis, P.I.K. Liakopoulos, K.C. Giannakoglou, G.-C. Vosniakos: Multi-objective design of optimal sculptured surface rough machining through pareto and nash techniques, 6th Conf. Evol. Determ. Methods Des. Optim. Contr. Appl. Ind. Soc. Probl. (2005)

59.80 I.N. Tansel, B. Ozcelik, W.Y. Bao, P. Chen, D. Rincon, S.Y. Yang, A. Yenilmez: Selection of optimal cutting conditions by using GONNS, Inter. J. Mach. Toolls Manuf. **46**(1), 26–35 (2006)

59.81 F. Cus, U. Zuperl, V. Gecevska: High speed end-milling optimisation using particle swarm intelligence, J. Achiev. Mater. Manuf. Eng. **22**(2), 75–78 (2007)

59.82 U. Zuperl, F. Cus, V. Gecevska: Optimization of the characteristic parameters in milling using the PSO evolution technique, J. Mech. Eng. **6**, 354–368 (2007)

59.83 F. Cus, U. Zuperl: Particle swarm intelligence based optimisation of high speed end-milling, Arch. Comput. Mater. Sci. Surf. Eng. **1**(3), 148–154 (2009)

59.84 T. Wagner, H. Trautmann: Integration of preferences in hypervolume-based multiobjective evolutionary algorithms by means of desirability functions, IEEE Trans. Evol. Comput. **14**(5), 688–701 (2010)

59.85 E.L. Valvo, B. Martuscelli, M. Piacentini: NC end milling optimization within CAD/CAM system using particle swarm optimization, 4th CIRP Int. Semin. Intell. Comput. Manuf. Eng. (2004) pp. 357–362

59.86 E. Borsetto, N. Gramegna: Multi-objective optimization of machining process using advantedge FEM tool, 7th CIRP Int. Conf. Intell. Comput. Manuf. Eng. (2010)

59.87 R. Li, M.T.M. Emmerich, J. Eggermont, T. Bäck, M. Schütz, J. Dijkstra, J.H.C. Reiber: Mixed-integer evolution strategies for parameter optimization, Evol. Comput. **21**(1), 29–64 (2013)

59.88 R.V. Rao, P.J. Pawar: Parameter optimization of a multi-pass milling process using non-traditional optimization algorithms, Appl. Soft Comput. **10**(2), 445–456 (2010)

59.89 Z.G. Wang, M. Rahman, Y.S. Wong, J. Sun: Optimization of multi-pass milling using parallel genetic algorithm and parallel genetic simulated annealing, Int. J. Mach. Tool. Manuf. **45**(15), 1726–1734 (2005)

59.90 R. Roy, Y.T. Azene, D. Farrugia, C. Onisa, J. Mehnen: Evolutionary multi-objective design optimisation with real life uncertainty and constraints, CIRP Annu. Manuf. Tech. **58**(1), 169–172 (2009)

59.91 H.-G. Beyer, B. Sendhoff: Robust optimization – A comprehensive survey, Comput. Method. Appl. Mech. Eng. **196**(33/34), 3190–3218 (2007)

59.92 S. Mekid, T. Ogedengbe: A review of machine tool accuracy enhancement through error compensation in serial and parallel kinematic machines, Int. J. Precis. Technol. **1**(314), 251–286 (2010)

59.93 R. Teti, K. Jemielniak, G. O'Donnell, D. Dornfeld: Advanced monitoring of machining operations, CIRP Ann. Manuf. Technol. **59**, 717–739 (2010)

59.94 H. Müller, D. Biermann, P. Kersting, T. Michelitsch, C. Begau, C. Heuel, R. Joliet, J. Kolanski,

M. Kröller, C. Moritz, D. Niggemann, M. Stöber, T. Stönner, J. Varwig, D. Zhai: Intuitive visualization and interactive analysis of pareto sets applied on production engineering. In: *Success in Evolutionary Computation*, Studies Computational Intelligence, Vol. 92, ed. by A. Yang, Y. Shan, L.T. Bui (Springer, Berlin, Heidelberg 2008) pp. 189–214

59.95 H. Pohlheim: Understanding the course and state of evolutionary optimizations using visualization: Ten years of industry experience with evolutionary algorithms, Artif. Life **12**(2), 217–227 (2006)

59.96 S. Obayashi, D. Sasaki: Evolutionary multi-criterion optimization. In: *Visualization and Data Mining of Pareto Solutions Using Self-Organizing Map*, ed. by C. Fonseca, P. Fleming, E. Zitzler, L. Thiele, K. Deb (Springer, Berlin Heidelberg 2003) pp. 796–809

59.97 K. Deb: *Innovization: Discovering Innovative Solution Principles Through Optimization* (Springer, Berlin, Heidelberg 2011)

59.98 B. Sieben, T. Wagner, D. Biermann: Empirical modeling of hard turning of AISI 6150 steel using design and analysis of computer experiments, Prod. Eng. Res. Dev. **4**(2–3), 115–125 (2010)

60. Aerodynamic Design with Physics-Based Surrogates

Emiliano Iuliano, Domenico Quagliarella

Details, references and guidelines are given about the adoption of surrogate models and reduced-order models within the aerodynamic shape optimization context. The aerodynamic design problem and its approximated version are introduced and discussed and then, an overview of various surrogate models and surrogate-based optimization methods is given. Subsequently, the concept of model order reduction is recalled, and the performance analysis of reduced-order models based on proper orthogonal decomposition (POD) is discussed. Within this context, some techniques to adaptively and globally improve the accuracy of POD-based surrogates are illustrated. Finally, an aerodynamic shape design problem of a transonic airfoil is used to practically analyze and compare the performances of various surrogate-based optimization methods.

Modern air vehicle design has been increasingly driven by environmental as well as operational constraints. Environmental concerns, including emissions and noise, are gaining increasing importance in the design and operations of commercial aircraft. Taking into account the current prognoses for the growth in air traffic, the above-mentioned challenges become even more significant [60.1–4]. In this context, the development and assessment of new theoretical methodologies represents a cornerstone for reducing the experimental load, exploring trade-offs, and proposing alternatives along the design path. The fidelity of such methods is essential to reproduce *real-life* phenomena with a significant degree of accuracy and to take them into account from the very beginning of the design process. Due to the intrinsic complexity of aircraft design, the design space is often huge and difficult to explore fully, so that fast semi-empirical tools and rules [60.5–7], derived from classical configuration data, have been traditionally applied. However, they exhibit a severe lack of accuracy when designing novel and unconventional concepts. Therefore, highly accurate analysis methods have been continuously introduced both in geometric representation and physical modeling, but the main drawback is that they are computationally expensive. For example, the solution of the Navier–Stokes equations around complex aerodynamic configurations requires a huge amount of computational resources even on modern state-of-art computing platforms. This turns out to be an even bigger issue when hundreds or thousands of analysis evaluations, like in parametric or optimization studies, have to be performed. In order to speed up the computation while keeping a high level of fidelity, the scientific community is increasingly focusing on surrogate methodologies like meta-models, multi-fidelity models, or reduced-order models. These can provide a compact, accurate, and computationally efficient representation of aircraft design performance.

The present chapter will give details and references about the adoption of surrogate models and, in particular, reduced-order models within the aerodynamic shape optimization context. In Sect. 60.1, the aerodynamic design problem and its approximated version will be introduced. Then, an overview of various surrogate models and surrogate-based optimization methods will be given. In Sects. 60.3 and 60.4, the concept of model order reduction will be recalled and the performance analysis of reduced-order models based on proper orthogonal decomposition (POD) will be discussed. In the Sect. 60.5, some techniques to adaptively and globally improve the accuracy of POD-based surrogates will be presented. Finally, Sect. 60.6 will be devoted to the analysis and comparison of the performances of various surrogate-based optimization methods with respect to the aerodynamic shape design of a transonic airfoil.

60.1 The Aerodynamic Design Problem

A broad class of aircraft design applications can be numerically modeled with the minimization of a function f which depends on two sets of variables: the design variables w, which the designer can directly control it, and the state variables x, which provide the evolution of the system representing the underlying physics. The design problem can be formulated as the non-linear programming problem

$$
\begin{aligned}
&\min_{w,x} && f(w,x) \\
&\text{subject to} && r(w,x)=0, \quad h(w,x)=0, \\
& && g(w,x)\le 0, \\
& && w_L \le w \le w_U .
\end{aligned}
$$

(60.1)

f is the objective function which the designer wants to minimize to improve performance. In aircraft design, typical objective functions are weight, noise, drag, aerodynamic efficiency, or a combination of thereof. $r(w,x)$ is the state equations set, which links the design variables and the state variables and it usually represents the governing laws of physics. In aerodynamic design, the state equations are modeled through computational fluid dynamics, e.g., the Navier–Stokes equations, which relate scalar or vector field (state) variables, like pressure or velocity, to the design variable vector. In a shape optimization problem, the design vector is made dependent on the aircraft component shape

by means of a parameterization approach. The vectors $g(w,x)$ and $h(w,x)$ are filled, respectively, with inequality and equality constraint functions, which must be satisfied for a design candidate to be considered feasible. Typical constraint functions in aircraft design are related to the generation of a minimum lift level to balance the weight or a threshold pitching moment coefficient to allow for trim. w_L and w_U are the lower and upper bounds of the design variables and thus specify the range of allowable values for the design vector w.

60.1.1 Problem Approximation

The computational time required to solve this problem is basically affected by two parameters: the number of function evaluations required to minimize the objective function and the cost of a single evaluation. Given a vector w^*, the latter is dominated by the computational effort needed to solve the state equations

$$
r(w^*,x)=0 .
$$

The adoption of a surrogate model reduces the cost per objective function evaluation. A surrogate model consists in replacing the expensive objective f and constraint functions g, h with less expensive, lower-fidelity models $\hat{f}$ and $\hat{g}, \hat{h}$. Concerning reduced-order modeling, it can be observed that the dimensionality of the optimization problem is twofold: the state vector and design vector dimension. As the first one is usually much bigger than the second one, model reduction can be applied to make explicit the dependency of x on w and solve the state variables as functions of the design ones. In other words, unlike response surfaces and meta-models, an approximation $\hat{x}$ of the state variables is available thanks to the model order reduction. As a consequence, the reduced-order approximate form of the optimization problem (60.1) can be cast as

$$\min_{w} \quad \hat{f}(w)$$
$$\text{subject to} \quad \hat{h}(w) = 0, \qquad \hat{g}(w) \leq 0 , \qquad (60.2)$$
$$w_L \leq w \leq w_U ,$$

where the dependence on the state variables has been dropped and the state vector has an explicit approximate relation with the design vector: $\hat{x} = k(w)$.

60.2 Literature Review of Surrogate-Based Optimization

This section proposes a survey of the most relevant surrogate-based optimization concepts. The topics have been widely discussed in the recent past, thanks to their innovative character and broad application areas. The introduction of surrogate models as fitness approximation within an evolutionary optimization system mitigates the demand for large computational resources associated to such search algorithms, allowing us to find a proper balance between the complete exploration of huge design spaces and limited cost. To this aim, reduced-order modeling through POD is a step forward, as a modal decomposition of an ensemble of functions, derived from numerical simulations, is performed to extract the most relevant patterns in the data set. Hence, compared to standard, interpolating meta-models, which are usually trained on an integral function representing the objective, reduced-order models should assure a deeper insight into the phenomena modeled.

Surrogate-based optimization (SBO) has been introduced to tackle the number of function evaluations in many engineering optimization problems. In aircraft design common practice, they can be used as a quick evaluator in several tasks: parametric analyses over the design space, optimization and control, and uncertainty quantification. A special challenge is represented by their use in global optimization as state-of-the-art methods, which often requires more function evaluations than can be comfortably affordable. A well-established approach consists in fitting some kind of response functions to basic data obtained by evaluating the objectives and constraints at a few points. The resulting surfaces, affordable at low cost, can provide fast answers in terms of trade-off analysis and optimization, as well as just an intuitive sketch behavior by means of simple visualization. The basic process consists of the following steps:

sampling the design space – once the design variables have been chosen, a sampling plan is defined and some initial sample designs are analyzed with an accurate solver; surrogate model selection and construction – a surrogate model type is selected and used to build a meta-model of the underlying problem; model validation – the model is checked according to some statistical metrics and, if not accurate enough, a search is carried out using the model to identify new design points for analysis; model updating – the new results are added to those already available and a new meta-model is built (repeating the last three steps); optimization – the refined surrogate is used to provide objective/constraint functions.

As SBO covers so many topics, the literature on the subject is huge. Many ideas have been proposed in the last 20 years, which are classified for design space dimensions, surrogate methods, search algorithms, updating algorithms, application areas. Hence, an exhaustive survey of all the possible ideas for each topic and all their possible combinations would go beyond the scope of the present research. Generally speaking, surrogate models can be roughly divided into three classes: data fit surrogates, multi-fidelity models, and reduced-order models. Data fitting models rely on the approximation of data (response values, gradients, and Hessians) generated from the high-fidelity model. In order to give a global behavior to surrogate methods, the whole design space must be sampled in advance by using design of experiments. Global approximations, often referred to as *response surface methods*, can be obtained with polynomial regression [60.8], Gaussian processes, Kriging interpolation [60.9], radial basis function networks [60.10], multi-adaptive regression splines [60.11], and artificial neural networks [60.12].

A second class of surrogates is the hierarchical one (also called multi-fidelity or variable fidelity). Unlike the data fit surrogates, they do not need to be trained on a sampling dataset, but they rely on a lower fidelity approximation which, however, is still inspired by the physical behavior of the system. Multi-fidelity models are classified according to the way they operate the fidelity reduction: examples in aerodynamics are coarser mesh discretization, partially converged solution [60.13], and model fidelity reduction [60.14–16] (e.g., using the Euler model instead of the Navier–Stokes equations by neglecting the effects of fluid viscosity and heat transfer). The name *multi-fidelity* usually refers to the capability of mixing and exploiting both high-fidelity and lower-fidelity models in an efficient way so as to keep the fidelity of the former only when it is needed and to take advantage of the higher speed of the latter otherwise.

A third class is represented by reduced-order models. A reduced-order model (ROM) is mathematically derived from a high-fidelity model using a projection technique. It consists in computing a set of basis functions (e.g., eigenmodes, left singular vectors) upon which the available dataset (ensemble) is projected to compute the unknown model parameters. The model reduction is obtained by capturing the principal dynamics of the system and neglecting the less significant from a physical point of view. Hence, similarly to data fit surrogates, reduced-order models require the a-priori solution of the expensive high-fidelity model. The advantage of reduced-order models with respect to data fits is that the most significant features of the flow field can be derived by approximation, thus offering the potential to keep more physics within the surrogate. The proper orthogonal decomposition or principal component analysis (PCA) is an elegant and powerful data-reduction method for non-linear physical systems. Its application as a surrogate to the aerodynamic optimization of aircraft components is the core of the present chapter.

Hereinafter, a more in depth look is given at the various methods of constructing a surrogate model and, in particular, at optimization assisted with the surrogate. *Jones* et al. [60.17] was among the first to propose a response surface methodology based on modeling the objective and constraint functions with stochastic processes (Kriging). The so-called design and analysis of computer experiments (DACE) stochastic process model was built as a sum of regression terms and normally distributed error terms. The main conceptual assumption was that the lack of fit due only to the regression terms can be considered as entirely due to modeling error, not measurement error or noise, because the training data are derived from a deterministic simulation. Hence, by assuming that the errors at different points in the design space are not independent and the correlation between them is related to the distance between the computed points, the authors came up with an interpolating surrogate model that is able to provide not only the prediction of objectives/constraints at a desired sample point, but also an estimation of the approximation error. After the construction of such a surrogate model, the latter powerful property is exploited to perform an efficient global optimization (EGO), which can be considered as the progenitor of a long and still in development chain of SBO methods. Indeed, they found a proper balance between the need to exploit the approximation surface (by sampling where it is minimized) with the need to improve the approximation (by sampling where prediction error may be high). This was done by introducing the expected improvement (EI) concept, already proposed by *Schonlau* et al. [60.18], which is an auxiliary function to be maximized instead of the original objective. The EI function is designed in order to provide a proper balance between exploration and exploitation.

In a further work, *Jones* [60.19] proposed a taxonomy of global SBO methods. Seven methods were identified and classified according to whether they were interpolating (cubic splines, thin-plate splines, multiquadrics, Kriging) or not (quadratic polynomials), whether they provided statistical information (Kriging) or not (splines), and whether the method for selecting search points (updating the model by adding new sample points) was two stage (probability/expected improvement) or one stage (goal-seeking, credibility function).

Gutmann [60.10] reported excellent numerical results for a spline-based implementation of Method 7 and proved the convergence of the method. Compared to previous methods, Method 7 required a high number of true function evaluations to find the global optimum, but, as Jones wrote, *this is the price we pay for the additional robustness*. An overview of SBO techniques was also presented by *Queipo* et al. [60.20] and *Simpson* et al. [60.21]. They covered some of the most popular methods in design space sampling, surrogate model construction, model selection and validation, sensitivity analysis, and surrogate-based optimization. *Forrester* and *Keane* [60.22] recently proposed a review of some advances in surrogate-based optimization. An important lesson learned is that only calling the true function can confirm the results coming from the surrogate model. Indeed, the path towards the global optimum is made of iterative steps where, even exploiting some surrogate model, only the best results coming from the true function evaluations are taken as optimal or sub-optimal design. The true function evaluation also has to be invoked to improve the surrogate model. With the term *in-fill criteria* we usually mean some principles that allow us to intelligently place new points (infill points) at which the true function should be called. The selection of infill points, also referred to as adaptive sampling or model updating, represent the core of a surrogate-based optimization method and helps to improve the surrogate prediction in promising areas of the objective space.

The right choice of the number of points which the initial sampling plan would comprise and the ratio between initial/in-fill points has been the focus of several recent studies. However, it must be emphasized that no universal rules exist, as each choice should be carefully evaluated according to the design problem (e.g., the number of variables, computational budget, type of surrogate). *Forrester* and *Keane* [60.22] assumed that there is a maximum budget of function evaluations, so as to define the number of points as a fraction of this budget. They identified three main cases according to the aim of the surrogate construction: pure visualization and design space comprehension, model exploitation, and balanced exploration/exploitation. In the first case, the sampling plan should contain all of the budgeted points, as no further refinement of the model is foreseen. In the exploitation case, the surrogate can be used as the basis for an in-fill criterion, which means some computational budget must be saved for adding points to improve the model. They also proposed to reserve less than one half of the points for the exploitation phase, as a small amount of surrogate enhancement is possible during the in-fill process. In the third case, that is the two-stage balanced exploitation/exploration in-fill criterion, as also shown by *Sóbester* et al. [60.23], they suggested employing one third of the points in the initial sample while saving the remaining for the in-fill stage. Indeed, such balanced methods rely less on the initial prediction, and so fewer points are required. Concerning the choice of the surrogate, the authors observed that it should depend on the problem size, i.e., the dimensionality of the design space, the expected complexity, the cost of the true analyses, and the in-fill strategy to be adopted.

However, for a given problem, there is no general rule. The proper choice could come up after various

model selection and validation criteria. The accuracy of a number of surrogates could be compared by assessing their ability to predict a validation data set. Therefore, part of the true computed data should be used for validation purposes only and not for model training. This approach can be infeasible when the true evaluations are computationally expensive. To overcome this issue, *Goel* et al. [60.24] proposed a weighted average of an ensemble of surrogates. For example, a better model can be achieved by combining Kriging, which might accurately predict the non-linear aspects of a function, and polynomials to better capture the regression trends. Forrester also underlined that some in-fill criteria and certain surrogate models are somewhat intimately connected. For a surrogate model to be considered suitable for a give in-fill criterion, the mathematical machinery of the surrogate should exhibit the capability to adapt to unexpected, local non-linear behavior of the true function to be mimicked. From this point of view, polynomials can be immediately excluded since a very high order would be required to match this capability, implying a high number of sampling points. In principle, the convergence to a local optimum might be achieved by simply minimizing the surrogate, evaluating the true function at the minimum point and updating the model database with the new point. Conversely, a global search would require a surrogate model able to provide an estimate of the error it commits when predicting. Thus, the authors suggested the use of Gaussian process-based methods like Kriging, although citing the work of *Gutmann* [60.10] as an example of a one-stage goal seeking approach employing various radial basis functions. Finally, some interesting suitable convergence criterion to stop the surrogate in-fill process were proposed. In an exploitation case, i.e., when minimizing the surrogate prediction, one can rather obviously choose to stop when no further significant improvement is detected. On the other hand, when an exploration method is employed, one is interested in obtaining a satisfying prediction everywhere, so that one can decide to stop the in-filling when some generalization error metrics, e.g., cross-validation, fall below a certain threshold. When using the probability or expectation of improvement, a natural choice is to consider the algorithm converged when the probability is very low or the expected improvement drops below a percentage of the range of objective function values observed. However, the authors also observed that a discussion on convergence criteria may be interesting and fruitful, but *in many real engineering problems we actually stop when we run out of available time or resources, as dictated by design cycle scheduling or costs.* This is what typically happens in aerodynamic design, where the high-dimensionality of the design space and expensive computer simulations often do not allow us to reach the global optimum of the design problem but suggest that we consider even a premature, sub-optimal solution as a converged point.

60.3 POD–Based Surrogates

In this section a review of the mathematical core of POD is presented. POD is a mathematical procedure that allows us to perform a modal decomposition of a large set of multi-dimensional data so as to derive a dimensionality reduction and describe the original system with much fewer unknowns. The mathematical development of POD for fluid flow applications, in particular, is described in some detail in [60.25]. Here, the main aspects related to the construction of a reduced-order model through singular value decomposition are presented and mainly the use of this technique for steady-state problems is addressed.

60.3.1 Model Order Reduction

Physics-based approximation concepts require a deep understanding of the governing equations and the numerical methods employed for their solution. The substantial difference between a reduced-order model and data fit model consists in retaining an explicit dependency between state variables, related to the governing equations and design parameters. In other words, reduced-order models operate on the dimensionality of the discretization of the state equations rather than on the design space. Thus, such models are partially independent of a notable increase in the number of design variables. A reduced-order model, in fact, mimics the basic structure of the problem and not just a functional relationship between input and output parameters. Hence, the main advantage of using reduced-order models lies in their being mostly insensitive to the curse of dimensionality.

To illustrate the model order reduction concept, consider the discrete mathematical model (e.g., the Navier–Stokes equations) of a physical system written in the form

$$R(w, x(w)) = 0 \,, \qquad (60.3)$$

where $w \in R^t$ is the vector of design variables and $x \in R^q$ the discretized vector of state (or field) variables (velocity, energy, density). Note that x is an implicit vector

function of the design variable vector. Unlike classical data fit methods (e.g., Kriging, RBF) which work on local or integral values of the state variables, reduced-order methods, instead, provide an approximation of the state vector in the form

$$\hat{x} = c_1 \boldsymbol{\phi}_1 + \cdots + c_M \boldsymbol{\phi}_M = \boldsymbol{\Phi} \boldsymbol{c} \,,$$

where

$$\boldsymbol{\Phi} = \{\boldsymbol{\phi}_1, \ldots, \boldsymbol{\phi}_M\} \in R^{q \times M}$$

is a matrix of known basis vectors and

$$c = \{c_1, \ldots, c_M\} \in R^M$$

is a vector of unknown coefficients. The underlying approximation is that the state vector lies in the subspace spanned by a set of basis vectors. Obviously, this is not true for each state vector, but a proper choice of an orthonormal basis can lead to the minimization of the approximation error in a least squares sense. This is how a proper orthogonal decomposition is derived. Following this approach, the problem of representing a state vector with q unknowns can be recast into a problem with M unknowns and, as usually $q \gg M$, it is possible to obtain an approximation of x very efficiently. The estimation of the vector c can be obtained with different techniques, classified as intrusive and non-intrusive. The first introduce the approximation in the governing equations and find the coefficients by minimization of the residual norm; the second employ data fit techniques trained on a set of known coefficients.

The basis vectors can be computed starting from state solutions of the discrete governing equations which correspond to M different values of the parameters w. As a consequence, the matrix $\boldsymbol{\Phi}$ contains the basis vectors of the subspace

$$\boldsymbol{\Phi} = \text{span}\{x(w_1), x(w_2), \ldots, x(w_M)\} \in R^{q \times M} \,. \qquad (60.4)$$

For instance, the state solutions $x_i = x(w_i)$ are obtained by solving the Reynolds-averaged Navier–Stokes (RANS) equations on M different configurations generated by applying the parameterization method employed on M design vectors w_i. The definition of the M design sites where to compute the solutions is not a trivial issue; generally speaking, standard design of experiments techniques are used to sample the design space with good coverage properties, but, as will be discussed in next sections, this approach may lead to erroneous results when facing highly multi-modal, highly non-linear problems. Indeed, the quality of the approximation strongly depends on the location of training data in the design space.

60.3.2 POD Theory and Solution

The construction and training of POD-based surrogate models for aerodynamic applications are described in detail in [60.26, 27]. The singular value decomposition (SVD) solution of the POD basis vectors and coefficients is used for steady-state problems. This approach is normally preferred to the eigenvalue/eigenvector solution, as it is faster and easier to implement. POD modeling is specifically focused on compressible aerodynamic problems, hence the space domain will be represented by the discretized volume occupied by the flowing air, and the snapshot vectors will be defined from computed flow fields. The column vectors of the snapshot matrix (here also referred to as the ensemble matrix) contain the volume grid and flow variables as computed with a computational fluid dynamics (CFD) solver.

The SVD solution allows us to obtain an optimal basis in the sense of the maximization of the averaged projection of the ensemble onto it. Hence, each snapshot vector can be retrieved as linear combination of the POD basis. If a fluid dynamics problem is approxi-

mated with a suitable number of snapshots from which a rich set of basis vectors is available, the singular values become small rapidly and a small number of basis vectors are adequate to reconstruct and approximate the snapshots as they preserve the most significant ensemble energy contribution. In this way, POD provides an efficient mean of capturing the dominant features of a multi-degree of freedom system and representing it to the desired precision by using the relevant set of modes. The reduced-order model is derived by projecting the CFD model onto a reduced space spanned by only some of the proper orthogonal modes or POD eigenfunctions. This process realizes a kind of lossy data compression of the original ensemble.

The resulting POD-based reduced-order model can be used in an optimization process to predict state solutions that are not included in the original ensemble. This useful feature requires the transformation of the projection coefficients from the discrete sample space for which they have been computed to a continuous space. In other words, by itself the POD model does not have a predictive feature globally, i.e., over the whole design space. Among the possible options to accomplish this task, here a functional relation is established between the POD coefficients, which represent the projection of a generic CFD flow field onto the set of POD basis vectors, and the design variables. It is well known that regression techniques are particularly suitable to fit experimental data, as they filter the random noise out from the data. This behavior is less desirable when working with computer simulations based on determinism. In this case, one asks the data fit model to exactly reproduce the sample data used for training and to consistently catch the local data trends. A radial basis function (RBF) network answers to these criteria and, therefore, is used here to interpolate the POD coefficients over the whole design space.

60.4 Application Example of POD-Based Surrogates

This section proposes an application of POD-based reduced-order models to the transonic flow around an airfoil. Indeed, as the POD model should introduce more physics within a surrogate approximation, one is interested in comparing such a novel approach to standard methodologies in order to establish, though in a preliminary way, its advantages and drawbacks. We consider a typical case in aircraft aerodynamics which,

far from being considered of industrial interest, retains the main physics features of it. The aerodynamic case is represented by the steady, viscous air flow around a scaled RAE 2822 airfoil. This geometry was selected as it is a standard object in CFD numerical modeling and validation [60.28]. The POD snapshots are obtained by perturbing the RAE 2822 airfoil by means of the parameterization described later on. A mixed

POD/CFD approach (zonal POD) is proposed to increase the accuracy level of the surrogate model in transonic conditions.

60.4.1 Parameterization and Design Space Definition

In the present context, surrogate modeling is aimed at providing a fast and accurate tool to speed up the process in aerodynamic shape design. As a consequence, one of the most important issues is to show its suitability and applicability to the shape optimization problem. Indeed, the definition of the design space through shape modification parameters typically involves a complex, often highly non-linear relation between the flow field and the design variables. Moreover, modifying an aircraft component (e.g., a wing airfoil) requires several parameters, thus enlarging the dimensions of the design space. It is straightforward, then, that the complexity of the problem increases and approaches a real-world application level. The class-shape transformation (CST) method [60.29] provides an analytical form to represent various geometries of aeronautical interest and it shows the interesting properties of continuity, differentiability, and reproducibility of a huge number of test shapes. It allows us to specify the airfoil contour as a product of a class function, which in the proposed case defines the rounded leading edge/pointed trailing edge airfoil class, and a shape function obtained as a linear combination of n-th-order Bernstein polynomials. The design vector is

$$w = \left(A_0^u, A_1^u, \ldots, A_n^u, A_0^l, A_1^l, \ldots, A_n^l\right) ,$$

where the first and last parameters $A_0^{u,l}, A_n^{u,l}$ are related, respectively, to the upper and lower leading edge radius

$$R_{le}^{u,l} \left(A_0^{u,l} = \pm\sqrt{\frac{2R_{le}^{u,l}}{c}}\right)$$

and to the trailing edge closure angle β ($A_n^{u,l} = \tan \beta^{u,l}$), as is shown in detail in [60.29].

In the present context, seventh-order Bernstein polynomials are considered, hence each airfoil side is described by eight design variables. The design space DW is then a subset of $\mathbb{R}^{16}$. A scaled 14% thickness ratio RAE 2822 airfoil is selected as baseline airfoil. The airfoil geometry is shown in Fig. 60.1, where the x-coordinate is the abscissa along the airfoil chord and y is given by the aforementioned approach.

The values of the corresponding design parameters, which define the RAE 2822 profile according to the parameterization, and their range of variation, which defines the design space, are reported in Table 60.1.

60.4.2 Design of Experiments

The location of sample points is an important issue with respect to the cost and accuracy of any surrogate model. The design of experiment theory (DOE) [60.30] was developed to provide experimentalists with a tool to optimally choose the independent variable values for a limited number of experiments. The aim is generally to use the results of the experiments to study and investigate the response and sensitivity of some dependent quantity to the independent variables. Classical DOE methods were originally designed to alleviate the effects of noise, so they have been generally employed in conjunction with regression techniques. However, computer experiments are not subject to random errors, hence it is worthwhile using a different strategy to obtain as much information as possible about the input–output dependence. A variety of methods have been developed to fill the design space in an optimal sense. One of the most widely adopted is the Latin hypercube sampling (LHS) technique. It was first proposed by *McKay* et al. [60.31] as an alternative to Monte Carlo techniques for the design of computer experiments. The basic principle of LHS is to bound the randomness of the sample selection in a given region. In fact, given that t is the number of design variables, each design variable range is divided into m intervals or bins of equal probability. This generates a total of $m \times t$ bins in the whole space. Within each bin only one sample is allocated randomly. This ensures that a one-dimensional projection onto the parameter space will produce one sample in each bin, thus eliminating the correlations between the design variables. LHS is useful for the initialization of POD-based surrogate models, but, as will be detailed later on, it exhibits some major

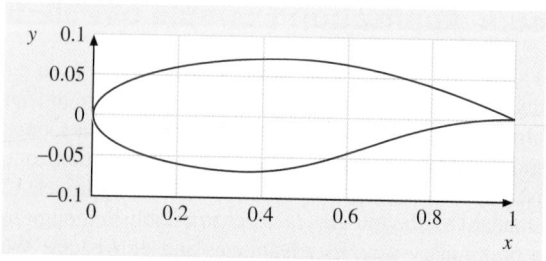

Fig. 60.1 Baseline geometry, RAE 2822 airfoil

Table 60.1 Design parameters, values, and ranges

Parameter	Baseline value	Range	Parameter	Baseline value	Range
A_0^u	0.1293	[0.1293, 0.2293]	A_0^l	−0.1280	[−0.2280, −0.1280]
A_1^u	0.1282	[0.1282, 0.2282]	A_1^l	−0.1483	[−0.2483, −0.1483]
A_2^u	0.1771	[0.1771, 0.2771]	A_2^l	−0.1080	[−0.2080, −0.1080]
A_3^u	0.1219	[0.1219, 0.2219]	A_3^l	−0.2580	[−0.3580, −0.2580]
A_4^u	0.2393	[0.2393, 0.3393]	A_4^l	−0.0918	[−0.1918, −0.0918]
A_5^u	0.1662	[0.1662, 0.2662]	A_5^l	−0.1079	[−0.2079, −0.1079]
A_6^u	0.1976	[0.1976, 0.2976]	A_6^l	−0.0561	[−0.1561, −0.0561]
A_7^u	0.2110	[0.2110, 0.3110]	A_7^l	0.0638	[−0.0362, 0.0638]

limits that prevent it from being used as a standard sampling technique for optimization purposes. Indeed, LHS is optimal in the sense of design space coverage, but it does not allow for refining the sampling distribution according to enrichment or improvement criteria, e.g., design space exploration or objective function minimization.

60.4.3 Zonal POD

The POD surrogate model is mainly designed as a ROM within a shape optimization process, where the geometry and, hence, the volume mesh vary with the design site. Moreover, the application is focused on transonic aerodynamics with potential flow separations and shock waves. Therefore, care must be taken with the definition of the snapshot domain and how to extract the integral quantities of interest (e.g., aerodynamic force coefficients) from the snapshot structure. Indeed, as the snapshots are expressed through a linear combination of POD modes, shock waves, flow separations, and other non-linearities present in the training ensemble would be captured and replied in the POD modes, so that any prediction of a new snapshot would be likely to bring the footprint of those flow features with it. This is desirable behavior on average for a physics-based approach, but when approaching the optima, which should be featured with a shockless and fully attached flow profile, a POD approximation of this type would hide the potential improvement behind the trace of the original snapshots. This issue is of paramount importance and can be tackled by introducing and combining two concepts: zonal POD and adaptive sampling. The first will allow us to reduce the inherent variability of the snapshots by means of a domain partitioning, thus avoiding the POD basis to capture all the physics within the field. The second technique will allow us to enrich the POD approximation by sampling at new points, which are *optimal* in the sense of exploration/model improvement balance. In this section, the discussion will be focused

on the zonal POD approach. The basic idea proposed in [60.26] is to use a mixed full-order/reduced-order model (FOM/ROM) by splitting the solution domain into two sub-domains: the FOM (i. e., the CFD RANS model) is used only in the vicinity of the surface to accurately solve the near wall boundary layer, non-linearities (e.g., shock waves), and flow separations where they occur; the ROM (i. e., the POD surrogate model) is exploited to reconstruct the flow field far from the solid wall, where a smoother and weakly varying solution is expected.

Figure 60.2 shows a sketch of the domain decomposition. The POD-based surrogate model is built on the spatial domain defined in the light gray region. Once the POD model has been trained, the surrogate response on the FOM/ROM boundary interface is extracted and used as boundary conditions to iterate the full-order CFD solver in the inner domain (dark gray). Details about the specific boundary condition formulation across the two domains can be found in [60.26].

A useful advantage of the zonal POD is that any aerodynamic coefficient or surface distribution of interest (e.g., pressure or skin friction distributions) can be

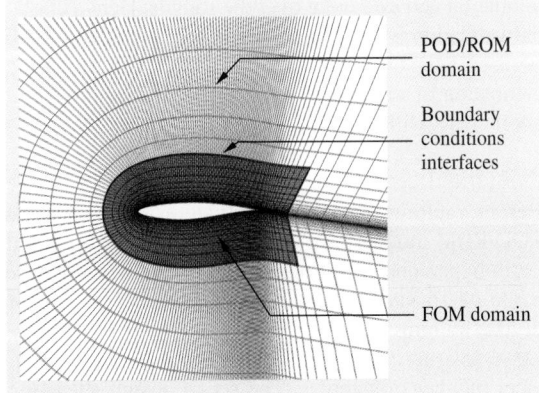

Fig. 60.2 Zonal approach, FOM/ROM domains and volume mesh

POD/ROM domain

Boundary conditions interfaces

FOM domain

directly extracted from the CFD solution in the inner domain. On the other hand, when the full POD model is used (i.e., trained on the whole domain), the surrogate model would provide a prediction of the mesh and state variables, so that a properly designed *condensation* procedure has to be applied to retrieve the integral coefficients like lift (C_l), drag (C_d), and pitching moment (C_m) coefficients.

60.4.4 Model Training, Validation, and Error Analysis

Training and validation are key phases that POD surrogates must undergo for assessment of their performance and potential. Validation means measuring the goodness of a surrogate model with respect to a so-called *truth* response (e.g., the CFD solution) and, therefore, drawing information to eventually optimize it. The goal is to evaluate the potential of the model to globally approximate the design space. Once a surrogate model has been trained, classical validation is carried out by sampling the design space once more, estimating the full and reduced-order models on the new sampling set, and computing a set of statistics from the data obtained. This approach requires computing new CFD solutions and, for this reason, it is computationally expensive. In order to reduce the number of full order computations, the validation points could be represented by the same set used for training, provided that a cross-validation technique is used (e.g., leave-one-out). Indeed, cross-validation implies the partitioning of a sample of data into complementary subsets: one subset is used for training, the other one for validation or testing. The variability due to the choice of the partitions is usually reduced by performing multiple rounds of cross-validation and averaging over the rounds. Here a classical validation is performed, while cross-validation will be used later on in auto-adaptive sampling, when the estimation of the quality of the POD model basis and coefficients will be required.

Model Training and Setup
Before getting into the validation process, the POD/ROM models have to be trained, so that an initial Latin hypercube sampling is done on the design space made of 16 variables (see Sect. 60.4.1). The size M of the training sampling is chosen to be very large ($M = 180$) to cover each design variable with a sufficient number of samples. The set of design sites $\{w_i\}$ is then transformed into the physical representation of the airfoil geometry due to the chosen parameteriza-

tion. The baseline geometry is a modified RAE 2822 airfoil, scaled to 14% thickness-to-chord ratio to amplify compressibility effects. 180 calls of the volume mesh generator and CFD solver are launched in parallel at fixed flow conditions to compute the flow field around each airfoil shape. Due to a proper selection of the baseline geometry and design weight ranges, a wide and varied distribution of shock wave locations and flow separations is obtained through the training dataset. This is a highly desirable feature to test the predictive capability of such a physics-based surrogate model. The Mach number is 0.729, the Reynolds number is 6 500 000, and the flow angle of attack is 2°. Fully turbulent flow is assumed. For each airfoil shape, a single-block structured volume mesh made of 25 186 points (12 288 cells) is computed by means of an automatic hyperbolic grid generator. Using fixed topology, mesh parameters, and sizes, standard quality grids are obtained for each geometry. The first cell at the wall is placed so as to have a unit y^+ at the specified flow conditions. A sketch of the volume mesh around the baseline airfoil is shown in Fig. 60.2.

With reference to Fig. 60.3, the mesh partitioning is applied to define the FOM/ROM domains, which are required to be non-overlapping and adjoining. This can be easily done when a structured mesh is available as the grid lines can be used as interfaces between domains. To this aim, the d parameter is introduced as the distance of the FOM/ROM interface from the airfoil leading edge. Indeed, different POD-based reduced-order models can be defined by varying this distance and, hence, reducing or increasing not only the size but also the inherent variability of the snapshot set. This mechanism has to be carefully considered beside the coexistence of eight heterogeneous variables (spatial coordinates, density, pressure and velocity) in the same snapshots, as it could introduce a bias in the correlation process. For example, the POD reduction could give more importance to the flow features related to the snapshot variables which exhibit the largest absolute values or the widest range of variation. To avoid this, a scaling operator is applied to the snapshot set prior to feeding the POD model. The scaling factors are designed so as to map each variable to the interval [0, 1] by normalizing as follows

$$x_h^* = \frac{x_h - \min(x_h)}{\max(x_h) - \min(x_h)},\qquad(60.5)$$

where x_h is the vector containing the h-th flow variable in the snapshot $s = (x_1, x_2, \ldots, x_8)^T$ and the minimum

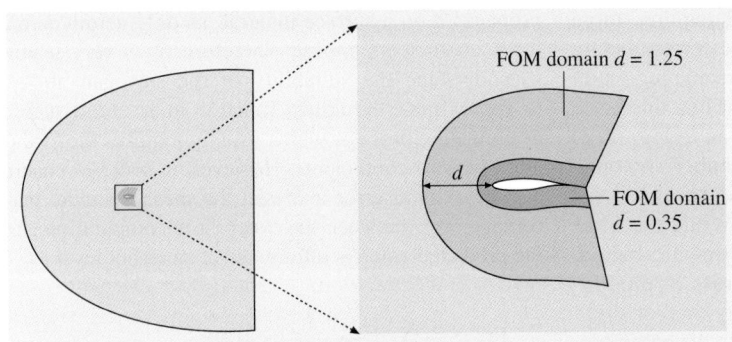

Fig. 60.3 FOM/ROM domains with varying interface

and maximum are taken over the vector x_h. In the present investigation the scaling factors are defined once for each variable and kept constant even when varying the FOM/ROM domains (and hence the snapshot size).

Dealing with a zonal approach, we do not know a priori the optimal distance of the domain interface from the airfoil surface. Hence, in order to assess the effects of changing the interface position, three domain partitions are considered, corresponding to three different settings of the domain interface. As a consequence, three POD-surrogate models (here referred to with the initials *SM*) are built exploiting the CFD data obtained in the training phase: *SM1* – a POD model with $d = 0$, i.e., the full-order model is not invoked in the validation as the POD approximation is used to get the flow field everywhere and no boundary condition is exchanged, the snapshot size N is 201 488; *SM2* – a POD model with $d = 0.35$, i.e., the full-order domain is the dark gray one in Fig. 60.3, while the POD approximation is used to get the flow field anywhere else, the snapshot size N is 91 792; *SM3* – a POD surrogate model with $d = 1.25$, similar to the previous one but now the full-order domain is the light gray one in Fig. 60.3, the snapshot size N is 75 232. Besides, two more surrogates are introduced and trained on the same dataset to act as standard meta-models: *SM4* – a Kriging interpolation model with Gaussian correlation using the aerodynamic efficiency C_l/C_d as response function (Dakota package implementation [60.32]); *SM5* – a quadratic polynomial regression model using the aerodynamic efficiency C_l/C_d as response function. Given t the number of design variables, at least $(t + 1) \times (t + 2)/2$ design sites should be evaluated to train this type of model. In the present case, $(t + 1) \times (t + 2)/2 = 153$, hence the size of the a-priori sampling is sufficient.

For each of the POD-based approximations, the ensemble energy content threshold ϵ is reported in

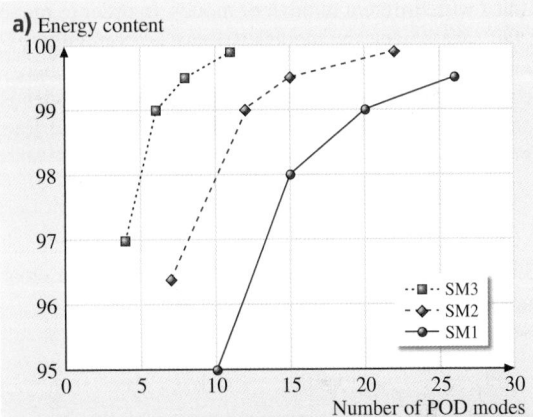

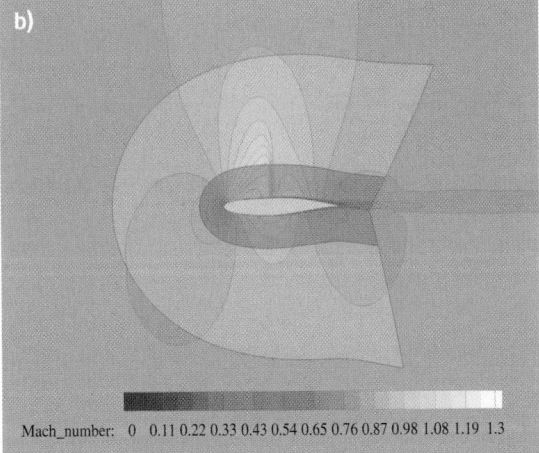

Fig. 60.4a,b Effect of zonal interface on the energy amount captured by POD

Fig. 60.4a as a function of the number of POD modes. It is clearly evident that SM1 requires a big number of modes even to reproduce a relatively low energy level (95%), while SM3 performs considerably better (97%)

with just four modes preserved. SM2 requires more modes with respect to SM3 because the corresponding ROM domain embeds part of the supersonic region on the airfoil suction side; Fig. 60.4b clarifies this issue as it reports the FOM/ROM domains (as in Fig. 60.3) superimposed with the local Mach number contours. The solution is here computed around an airfoil selected within the ensemble database. While the SM3 ROM domain is quite far off the supersonic region, the SM2 FOM/ROM interface lies across it, thus introducing a stronger source of variability (and of slight discontinuity due to the shock wave) into the ensemble. Therefore, for each model a given energy level is obtained with different number of modes. In order to make a fair assessment, the models will not be compared using a pre-defined number of modes, but at a fixed energy level (95%). Indeed, the number of preserved modes is ten for SM1 (95%), seven for SM2 (96.4%), and four for SM3 (97%).

Error Estimation

For each design candidate, the aerodynamic efficiency $E = C_l/C_d$ is computed and used to assess some error measures: the percentage error

$$PE_i = \left| \frac{E_i - \hat{E}_i}{E_i} \right| \times 100, \quad i = 1, \ldots, \bar{M} ;$$

the mean percentage error (MPE)

$$MPE = \frac{1}{\bar{M}} \sum_{i=1}^{\bar{M}} PE_i ;$$

the standard deviation of the percentage error (SDPE)

$$SDPE = \frac{1}{\bar{M}-1} \sum_{i=1}^{\bar{M}} [PE_i - MPE]^2 ;$$

the R-squared coefficient of determination

$$R^2 = \frac{1 - \sum_{i=1}^{\bar{M}} \left(E_i - \hat{E}_i \right)^2}{\sum_{i=1}^{\bar{M}} \left(E_i - \frac{1}{\bar{M}} \sum_{i=1}^{\bar{M}} E_i \right)^2} ;$$

where index i denotes the i-th sample of the DOE validation plan, the hat quantities refer to the surrogate predictions, while the hatless ones to the full-order predictions. This type of error measure provides a picture of how the POD model reconstruction error is propagated on a surface integral, as only aerodynamic force coefficients appear. Therefore, it is very useful to understand the suitability of the surrogate model to approximate the fitness function in an aerodynamic optimization process, which usually requires the evaluation of aero-coefficients. However, in order to ensure a more general error analysis, the mean absolute percentage error between the exact CFD computation and the predicted value is introduced at snapshot level as

$$Er_i = \frac{1}{N} \sum_{j=1}^{N} \left| \frac{s_{i,j} - \hat{s}_{i,j}}{s_{i,j}} \right| \times 100,$$

where N is the snapshot size, $s_{i,j}$ ($\hat{s}_{i,j}$) is the j-th element of the computed (predicted) snapshot vector at the i-th validation site.

Finally, monotonicity is one of the properties a good surrogate should have in an optimization process. Given two *true* data $f(w_i)$ and $f(w_j)$ and the corresponding surrogate predictions $\hat{f}(w_i)$ and $\hat{f}(w_j)$, a surrogate model is monotonic when

$$f(w_i) \leq f(w_j) \implies \hat{f}(w_i) \leq \hat{f}(w_j) . \tag{60.6}$$

This property can be global (i. e., valid for each $\mathbf{w}_i, \mathbf{w}_j \in DW \subset \mathbb{R}^t$) or local. In order to measure the monotonicity, the following metric is introduced

$$G = \sum_{i=1}^{\bar{M}} \sum_{j=1}^{i} - \min \left(0, \frac{\Delta \hat{E}_{ij}}{\Delta E_{ij}} \right) , \tag{60.7}$$

where $\Delta E_{ij} = E_i - E_j$ and $\Delta \hat{E}_{ij} = \hat{E}_i - \hat{E}_j$. The G index can assume any non-negative value, zero value indicates global monotonicity, and the higher the magnitude, the more significant the monotonicity loss.

Validation Results and Analysis

The validation plan is generated with a new Latin hypercube sampling of size $\bar{M} = 50$. The goodness-of-fit for each model is estimated; the results are summarized in Table 60.2. The surrogate models can be

Table 60.2 Surrogate goodness-of-fit estimation

Surrogate	R-squared	MPE	SDPE	G	Ranking
SM1	0.5876	10.33	48.14	1597.22	4
SM2	0.8899	4.55	12.85	647.83	2
SM3	0.9791	2.30	1.61	171.10	1
SM4	0.8657	4.56	26.61	853.98	3
SM5	0.06074	15.62	171.62	1761.64	5

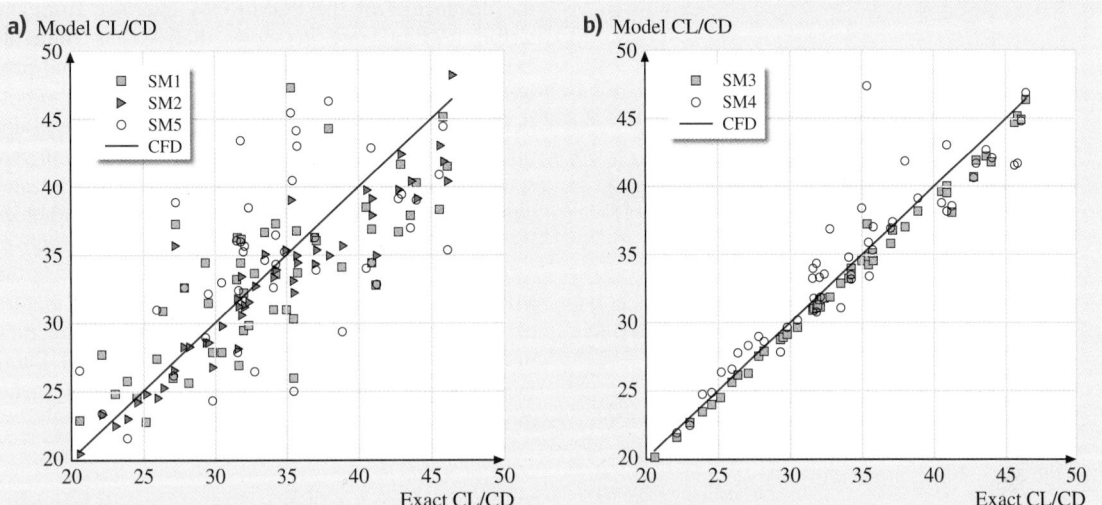

Fig. 60.5a,b Correlation plot of the prediction of surrogate models

ranked as reported in the rightmost column: SM3 exhibits superior performances for each quality index, while the quadratic polynomial surface is very poor in approximating the objective function. SM3 performs very well even on the SDPE estimate, which measures the variation of the percentage prediction error along the validation sampling. Hence, the prediction errors at any validation site are comparable and close to the mean value. This is a very desirable feature for a surrogate model designed for optimization. On the other hand, SM5 shows very poor performances because such a polynomial regression is unable to approximate a multi-modal, rapidly changing objective function.

Looking at the figures in Table 60.2, models SM2 and SM4 show similar results, even if they differ completely in methodology and construction. This is a useful indication when seeking the proper balance between the FOM and ROM domains (i.e., to determine the distance d): the POD surrogate accuracy increases by moving the FOM/ROM interface away from the airfoil

surface, and there exists a peculiar value of the distance d for which its predictive power is very close to standard and efficient interpolation techniques. As shown in Table 60.2, the monotonicity measure is coherent with the previously introduced indicators and, considering the big difference between SM3 and other models, it provides additional evidence of the quality of this model.

Figure 60.5 reports the correlation plot between the prediction of the model and the *true* CFD data. Again, SM2, SM3, and SM4 are globally closer to the linear trend, resulting in a better fit. The correlation plot highlights another significant feature of SM3 model, as it generally underestimates the aerodynamic efficiency. For further comparisons, Table 60.3 summarizes the validation set indices where each model predicts the highest and lowest efficiencies, the corresponding values of aerodynamic efficiency, and the percentage error with respect to the CFD datum. This is useful for evaluating the capability of the model identifying the global extrema of the objective function. It is observed that only SM4 leads to a wrong estimation of the position of the *optimal* airfoil while SM5 underestimates the performance of the worst profile.

The last two properties, i.e., the capability to preserve the monotonicity of the dataset and to correctly identify the best/worst candidates, are crucial aspects in SBO, so that models SM2 and SM3 seem to be more suitable for this purpose.

The accuracy of POD models is also evaluated and compared in terms of the point-to-point snapshot er-

Table 60.3 Surrogate estimations of aerodynamic efficiency for best and worst validation airfoils

Surrogate	ID of max	ID of min	max	min	Δ max (%)	Δ min (%)
CFD	12	22	46.43	20.61	0	0
SM1	12	22	52.19	22.84	12.40	10.81
SM2	12	22	48.27	20.45	3.95	−0.77
SM3	12	22	46.40	20.12	−0.07	−2.39
SM4	26	22	47.40	19.86	2.08	−3.65
SM5	12	39	54.62	16.78	17.63	−18.59

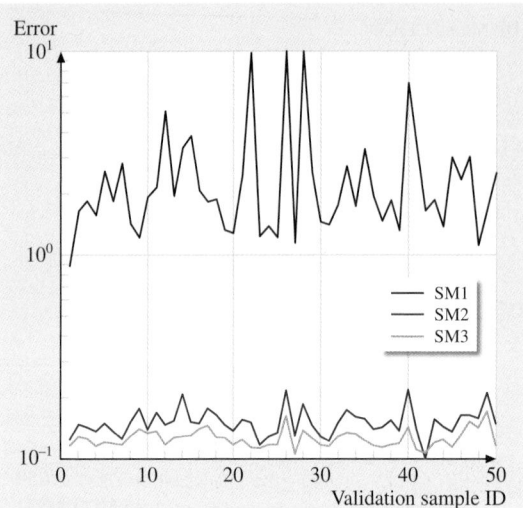

Fig. 60.6 Snapshot error prediction

ror. Figure 60.6 shows the results for each snapshot belonging to the validation plan (again ranging from 1 to 50). The error index is plotted in logarithmic scale. It turns out that, in strong transonic conditions, training a POD model on the full CFD domain (SM1) would lead to misleading results in the prediction phase, as the model would not be able to catch the highly non-linear trends that characterize this kind of flow. Indeed, the high number of POD modes required and the low goodness-of-fit performance suggest that further modeling is needed to optimize the computation of the basis vectors and modal coefficients in transonic aerodynamics. In the next sections, we will introduce some adaptive sampling concepts to globally improve the reduced-order models predictions.

A final comparison is possible in terms of POD model accuracy versus computational time and cell saving. In particular, the R-squared prediction error can be taken as a measure of the model accuracy, while the time saving index TS and the cell saved index CS are defined as

$$TS = \frac{T_{\text{FULL}} - T_{\text{SM}}}{T_{\text{FULL}}}, \quad CS = 1 - \frac{N_{\text{SM}}}{N_{\text{FULL}}}, \quad (60.8)$$

where T and N are, respectively, the computational time for 1000 CFD iterations and the number of solved computational cells. The subscripts FULL and SM refer to the full grid CFD computation and the CFD computation on the smaller FOM domain. In Fig. 60.7a the three indices (R-squared, TS, and CS) are plotted against

the distance d of the FOM/ROM interface from the airfoil leading edge. It shows that a clear trade-off exists between accuracy and time/cell saving and provides useful guidelines to tailor the choice of the best POD model to the basic requirements of the target application. For instance, if the target is to do a pre-screening of the objective space, one could use a faster and less accurate POD model that, however, guarantees the preservation of the physics. Figure 60.7 proposes a comparison between surrogate models in terms of the MPE and SDPE. The plot shows a graphical picture of the results in Table 60.2 concerning the accuracy of the POD models with moving FOM/ROM interface and the comparison with more classical meta-models.

Conclusions

Three POD/ROM models have been trained and compared: the first one consisted in feeding the POD ensemble with the full field, hence without any domain decomposition; in the second and third one, the zonal approach was applied by defining two different values of the distance of the interface from the airfoil leading edge. Results showed that the model accuracy is strongly dependent on the distance parameter, mainly because of the presence of the supersonic expansion lobe and the pressure jump across the shock wave on the airfoil suction side. In fact, the SM3 model showed superior performance with respect to both the other POD models and standard interpolation techniques like Kriging and regression methods like quadratic polynomial fitting. It also allows us to obtain a very accurate reconstruction of airfoil surface distributions and, hence, of aerodynamic coefficients, which are very often the actual target of aerodynamic design. Another important conclusion of the work is that it seems completely misleading to base the POD ensemble on the full flow field when transonic conditions and shape modifications act together. Indeed, as the POD reconstruction is a linear combination of POD modes, capturing the combined non-linear effects of boundary layer and compressibility is hardly possible when the position and intensity of the shock wave and its interaction with the boundary layer vary too much. Globally, the proposed POD surrogate model was shown to have many characteristics that make it suitable for aerodynamic design. However, a trade-off was found between POD model accuracy and resource saving as a function of the distance parameter: the smaller the full-order domain, the shorter the computational time required but also the less accurate the reconstruction.

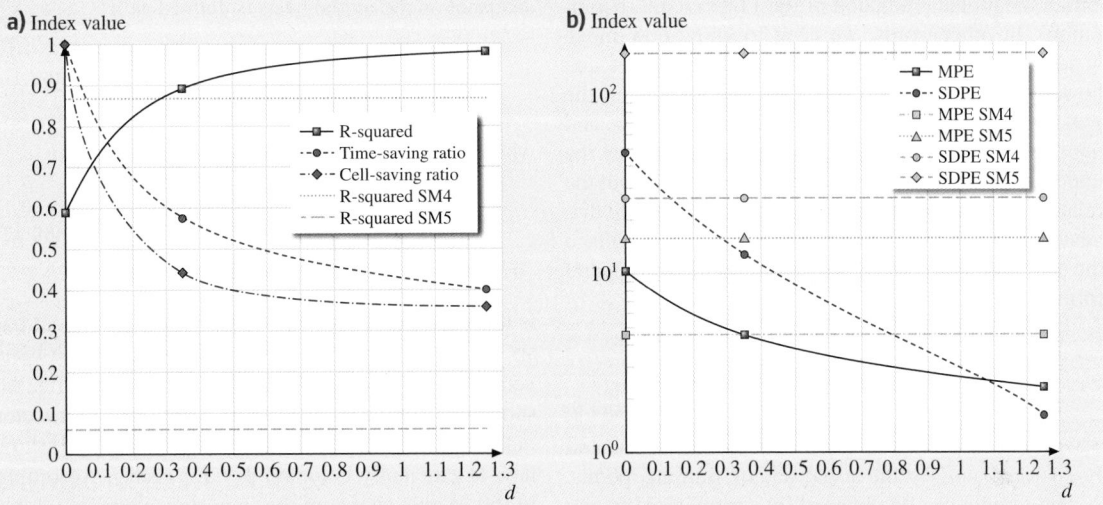

Fig. 60.7a,b Performance of surrogate models as a function of FOM/ROM interface positioning

Indeed, one of the key points of the research is to recover the accuracy issues by *optimally* selecting the training candidates. In the proposed example, we selected 180 sites to a-priori sample the 16-dimensional design space, but in principle we do not have any information about the appropriate size and locations of the sample points. Intuitively, we would like to have a sampling strategy that would fill the space in an efficient manner and would allocate more points in regions of the design space where the simulation response is strongly non-linear or is likely to find an optimum. In industrial practice, this would mean, given a computational budget, improving the quality of the POD surrogate by *intelligently* choosing the training samples. Conversely, given a POD model with a certain quality level, the rationale would be to reach the same accuracy with less high-fidelity computations.

60.5 Strategies for Improving POD Model Quality: Adaptive Sampling

In order to get rid of the issues raised in the previous section, a set of strategies has been proposed to update and enhance the surrogate/POD model through adaptive DOE techniques. Indeed, the selection of the design sites to be included in the POD ensemble, instead of being fully derived from an a-priori sampling strategy, can be tailored to match specific POD-related improvement requirements. Adaptive sampling strategies can be properly designed to account for these requirements by means of the so-called *in-fill* criteria. While a-priori sampling techniques do not use any information about the model prediction, adaptive techniques incrementally select new sampling points by exploiting the input/output relation observed at the previous stages. Hence, some adaptive DOE strategies for POD-based reduced-order models are proposed. The main references are the works by *Goblet* and *Lepot* [60.33] and *Sainvitu* et al. [60.34].

60.5.1 Rationale

Adaptive sampling is aimed at improving the modal basis or the modal coefficients set, which represents the core of POD modeling. Indeed, given a POD model built on a snapshot ensemble $\{s(w_1), s(w_2), \dots, s(w_M)\}$, the aim is to find a new point w_{new} in the design space so that the new POD model, built on the new set $\{s(w_1), s(w_2), \dots, s(w_M), s(w_{\text{new}})\}$, will provide for improved predictions and better exploration of the design space at the same time. The fundamental idea is to realize a proper trade-off between local accuracy and global design space exploration. On the one hand, we would like to sample *near* those design sites whose *importance* is higher (exploitation). On the other hand, knowing the relative distances between new sampling points and snapshot sites, we would like to sample far away from the known points in order to potentially enrich the global prediction of the POD model (exploration). In other words, we need to know how much a new potential sample is *near* the training set and how much the nearest training sample *weighs* on the POD model. Of course, the meaning of *distance* and *importance* needs to be mathematically defined, but the underlying idea is to combine the information about the relative *importance* of the snapshot and the *nearest* distance in order to satisfy both requirements. This leads to the definition of a *potential of enrichment* in the general form

$$V_\bullet(w_1, w_2, \ldots, w_M, y) = R(w_1, w_2, \ldots, w_M, y)$$
$$\times I_\bullet(w_1, w_2, \ldots, w_M, y),$$
$$(60.9)$$

where y is a generic point in the design space and $\{w_1, w_2, \ldots, w_M\}$ is the usual set of training points. A new sample can be obtained by simply maximizing the enrichment function. The function R gives a measure of the distance of $\mathbf{y}$ from the set $\{w_1, w_2, \ldots, w_M\}$, according to a certain norm, and helps to obtain good space-filling properties. The function $I_\bullet$ can be referred to as an *importance* function, as it has to properly provide the information about the quantity to be improved. A natural candidate could be a measure of the error at each new point if the surrogate model would be designed so as to provide for an error estimation at each y. Otherwise, in a surrogate-based optimization process, the function $I_\bullet$ could be directly linked to the approximation of the objective function to drive the search for a new optimized design sample. With reference to the POD modeling presented so far, the *importance* function should be closely tied to the quality of the modal basis $\{\boldsymbol{\phi}_1, \ldots, \boldsymbol{\phi}_M\}$ or the quality of the RBF models built on the modal coefficient $\{\alpha_1(w), \alpha_2(w), \ldots, \alpha_M(w)\}$. Hence, two different approaches can be followed, both based on the leave-one-out cross-validation technique. For the sake of clarity, the superscript $^{-j}$ will indicate that the referenced element (basis vector, coefficient model, SVD matrices) has been obtained by means of a leave-one-out process, i. e., by removing the j-th sample from the training set and re-computing the model.

60.5.2 Improvement of the Modal Basis

The first strategy consists in defining the *importance* function as a measure of the relative influence of each snapshot on the modal basis. This requires evaluating how much the modal basis changes when removing the snapshots one by one. The relative influence of the j-th

snapshot on the modal basis is defined as

$$I_b^r(w_j) = \frac{I_b(w_j)}{\sum_{k=1}^M I_b(w_k)},$$
$$(60.10)$$

where

$$I_b(w_j) = \sum_{i=1}^M \sigma_i \left(\frac{1}{\left| \left(\boldsymbol{\phi}_i, \boldsymbol{\phi}_i^{-j} \right) \right|} - 1 \right),$$
$$(60.11)$$

is the influence of the j-th snapshot on the modal basis, σ_i is the singular values associated to the i-th basis vector, and $\boldsymbol{\phi}_i^{-j}$ is the i-th basis vector obtained after the substitution of the j-th snapshot vector with a null vector in the ensemble matrix. *Goblte* and *Lepot* [60.33] show how to efficiently compute $\boldsymbol{\phi}_i^{-j}$ for each j. According to (60.9), this choice of the *importance* function drops the dependency on y, so that we need to condense it in the distance function.

As was mentioned above, a new optimal sample can be found by maximizing the potential of enrichment. However, in order to avoid solving a maximization problem, the design space is heavily sampled with a Latin hypercube technique, e.g., 100 times the dimension t of the design space. Then, the Euclidean distance of each new sampled point y_i, $i = 1, \ldots, l = 100t$ from each of the snapshot sites w_k, $k = 1, \ldots, M$ is computed and, for each y_i, the distance from the nearest snapshot $w_{\bar{k}}$ is stored as $\Delta(w_{\bar{k}}, y_i)$. This represents the distance function. Hence, for each new candidate y_i, the potential of enrichment can be written as $V_{\boldsymbol{\phi}}(y_i) = \Delta(w_{\bar{k}}, y_i) I_b^r(w_{\bar{k}})$. Finally, a new sample point is selected at $w_{\text{new}} = \text{argmax}_{y_i} V_{\boldsymbol{\phi}}(y_i)$.

60.5.3 Improvement of the Modal Coefficients

The second adaptive method is conceived to improve the quality of the RBF networks built on the POD modal coefficients. Two sub-strategies are proposed: the first aims at improving the worst modal coefficient, the second is designed to improve all coefficients simultaneously.

First Sub-Strategy
This strategy is applied when one of the coefficient model $\{\alpha_1(w), \ldots, \alpha_M(w)\}$ exhibits low quality with respect to the others. Therefore, we first need to select a modal coefficient. The quality of the i-th modal coefficient $\alpha_i(w)$ is estimated by using a leave-one-out

process and computing a weighted form of the Pearson correlation coefficient as

$$P_c(\alpha_i) = \frac{\sigma_i}{\sum_{k=1}^{M} \sigma_k}$$

$$\times \frac{\mu\left(\alpha_i \alpha_i^{-j}\right) - \mu(\alpha_i)\mu\left(\alpha_i^{-j}\right)}{\sqrt{\mu(\alpha_i \alpha_i) - [\mu(\alpha_i)]^2}\sqrt{\mu\left(\alpha_i^{-j}\alpha_i^{-j}\right) - \left[\mu\left(\alpha_i^{-j}\right)\right]^2}},$$

$$(60.12)$$

where μ is the arithmetic mean operator applied to a generic dataset $\kappa = \{\kappa_1, \ldots, \kappa_M\}$

$$\mu(\kappa) = \frac{1}{M}\sum_{j=1}^{M}\kappa_i .$$

Weighting the correlation coefficient is needed because the POD modal coefficients are not equally important but they can be ranked according to the magnitude of the corresponding singular values.

The modal coefficient with the lowest value of the weighted correlation coefficient is selected and tagged as $\bar{\imath}$. The *importance* function is then defined as

$$I_{\alpha_{\bar{\imath}}}(w_j) = \left|\alpha_{\bar{\imath}}(w_j) - \alpha_{\bar{\imath}}^{-j}(w_j)\right| ,$$

i.e., the absolute error of the $\bar{\imath}$-th model when leaving out the j-th snapshot. The choice of the distance function is the same as in the previous case. Hence, for each new candidate y_i, the potential of enrichment (with respect to the worst coefficient model) is defined as

$$V_{\alpha_{\bar{\imath}}}(y_i) = \Delta(w_{\bar{k}}, y_i)I_{\alpha_{\bar{\imath}}}(w_{\bar{k}}) .$$

Finally, a new sample point is selected at $w_{\text{new}} = $ argmax$_{y_i} V_{\alpha_{\bar{\imath}}}(y_i)$.

The evaluation of the quantities $\alpha_i^{-j}(w)$ is very expensive as, for each j, a new model has to be computed. However, when using RBF network interpolators for POD coefficient models, the leave-one-out procedure can be performed at no extra cost by using the efficient formula provided by *Rippa* [60.35].

Second Sub-Strategy
This sub-strategy is used when the quality of all coefficient models is comparable and it is very similar to the improvement of the modal basis. The *importance* function is defined as a measure of the relative influence of each snapshot on the whole set of coefficient models. This requires evaluating the absolute error of each modal coefficient model when removing the snapshots one by one. The relative influence of the j-th snapshot on the whole set of coefficient models is defined as

$$I_c^r(w_j) = \frac{I_c(w_j)}{\sum_{k=1}^{M} I_c(w_k)} ,$$

where

$$I_c(w_j) = \sum_{i=1}^{M} \sigma_i I_{\alpha_i}(w_j) = \sum_{i=1}^{M} \sigma_i \left|\alpha_i(w_j) - \alpha_i^{-j}(w_j)\right| .$$

$$(60.13)$$

As in the previous cases, the potential of enrichment V_α (with respect to all POD coefficients) is defined as

$$V_\alpha(y_i) = \Delta(w_{\bar{k}}, y_i)I_c^r(w_{\bar{k}}) .$$

Finally, a new sample point is selected at $w_{\text{new}} = $ argmax$_{y_i} V_\alpha(y_i)$.

60.6 Aerodynamic Shape Optimization by Surrogate Modeling and Evolutionary Computing

The POD surrogates as well as the adaptive sampling techniques described in Sects. 60.3 and 60.5 have been included within an evolutionary optimization loop. The aim is to assess POD-based surrogate models as fitness function evaluators in a shape optimization problem in transonic flow. Several approaches are proposed, differing for the key ingredients of the methodology: the construction of the POD model (full/zonal approach),

the strategy chosen to compute the training sample (a-priori, auto-adaptive), and the strategy to exploit the optimization results (single optimization, iterative optimization with real-time updating). The optimization approaches share the same target, i.e., to improve the aerodynamic performance of the scaled RAE 2822 airfoil. The surrogate-based shape optimization process consists of an a-priori design of the experiment module (e.g., LHS), the CST parameterization module, an in-house developed automatic mesh generator, the ZEN CFD flow solver [60.36], the POD/ROM module, which also encloses the adaptive sampling techniques, and the in-house adaptive genetic algorithm optimization library ADGLIB [60.15, 37–39].

60.6.1 Problem Definition

The geometry parameters, the design variable ranges, the parameterization technique, and the design point from Sect. 60.4.1 will be used. Here, we define the airfoil shape optimization problem in terms of objective/constraint function specifications as

$$
\begin{aligned}
&\underset{w \in DW \subset R^{16}}{\text{minimize}} \quad -\frac{C_1}{C_d} \\
&\text{subject to} \quad \left(\frac{t}{c}\right)_{\text{max}} = 0.14, \quad C_1 \geq 0.5, \\
&\qquad\qquad C_m \geq -0.05, \\
&\qquad\qquad C_m \leq 0.05.
\end{aligned}
\tag{60.14}
$$

In other words, the goal is to maximize the aerodynamic efficiency C_1/C_d while keeping a minimum level of lift generation ($C_1 \geq 0.5$) and pitching moment controllability ($|C_m| \leq 0.05$). Moreover, a geometric constraint is added in order to set the airfoil maximum thickness-to-chord ratio t/c at 14%: this constraint is implicitly treated within the parameterization. The constraint functions are actually treated as quadratic penalties, hence the constrained optimization is transformed into the following unconstrained problem

$$
\begin{aligned}
\underset{w \in DW \subset R^{16}}{\text{minimize}} -\frac{C_1}{C_d} &+ K[\min(C_1 - 0.5, 0)]^2 \\
&+ K[\min(C_m + 0.05, 0)]^2 \\
&+ K[\min(-C_m + 0.05, 0)]^2,
\end{aligned}
\tag{60.15}
$$

where K is a constant weight (equal to 10^4) which amplifies the relative importance of possible constraint violations. For instance, a unit penalty will be applied to the objective function in the case of an airfoil having a pitching moment of ± 0.06.

60.6.2 Optimization Strategies and Setup

In Sect. 60.4, three POD-based surrogate models were introduced, trained, and validated against an independent dataset. Here, we propose a set of numerical experiments to assess their potential as fitness evaluator and

their suitability for an evolutionary optimization problem. Several optimization approaches were set up and tested in order to possibly cover all the issues concerning surrogate/ROM training and prediction. Table 60.4 summarizes the characteristics of each optimization in terms of: fitness evaluator, optimization algorithm, POD energy threshold (when using POD as surrogate), high-fidelity computational budget, i. e., the total number of computations with the ZEN RANS solver during the optimization process, number of a-priori LHS samples M_{apr}, number of adaptively added samples M_{adp}, and number of surrogate-based optima M_{opt} which are iteratively added to the ensemble database. It must be noted that not all the optimization strategies use POD as a surrogate; in particular, optimizations tagged as Kriging-driven Genetic Algorithm (KGA) and EGO have been performed by using a Kriging method as the fitness evaluator and the EGO algorithm [60.17], based on Kriging and expected improvement evaluation, to compute new optimal samples. The EGO algorithm represents a modern state-of-art method in surrogate-based global optimization. In the following, with the term *truth* or *true* we will indicate the results obtained with the Zonal Euler–Navier–Stokes (ZEN) CFD solver as it is adopted as the reference high-fidelity simulation tool. Each optimization method is described in detail here:

- DGA (Direct Genetic Algorithm) – a plain, brute-force genetic optimization with the full high-fidelity solver ZEN called as fitness evaluator.
- FPGA1 (Full POD Genetic Algorithm 1) – a surrogate-based optimization where the aerodynamic analysis is carried out through a POD model built on the complete flow field of a set of 180 initial samples. This case corresponds to the POD-driven *standalone* mode and the surrogate POD evaluator is the one presented as SM1. No zonal approach is used. The POD energy content is 85%. The snapshot size N is 201 488.
- FPGA2 (Full POD Genetic Algorithm 2), FPGA3 (Full POD Genetic Algorithm 3) – same as FPGA1, but the POD models are defined by increasing the energy content (95 and 99%, respectively).
- MPGA1 (Mixed-flow POD Genetic Algorithm 1) – a surrogate-based optimization where the zonal CFD/POD model is trained on the initial design space sampling (180 snapshots) and adopted as the objective function evaluator throughout the optimization cycle. The FOM domain is defined at a distance $d = 1.25$ chord length from the airfoil's

Table 60.4 Optimization approaches

Opt Tag	Fitness evaluator	Optimizer	POD energy (%)	Budget hi-fi	M_{apr}	M_{adp}	M_{opt}
DGA	ZEN	ADGLIB	–	9600	0	0	0
FPGA1	standalone POD	ADGLIB	85	180	180	0	0
FPGA2	standalone POD	ADGLIB	95	180	180	0	0
FPGA3	standalone POD	ADGLIB	99	180	180 ·	0	0
MPGA1	zonal POD	ADGLIB	95	180	180	0	0
MPGA2	zonal POD	ADGLIB	99	180	180	0	0
KGA	Kriging	Dakota SOGA	–	190	180	0	10
EGO	Kriging	Dakota EGO	–	553	153	400	0
AFPGA1	standalone POD	ADGLIB	99	96	32	16	48
AFPGA2	standalone POD	ADGLIB	99	96	16	32	48
AFPGA3	standalone POD	ADGLIB	99	96	4	44	48
AMPGA1	zonal POD	ADGLIB	99	112	8	56	48
AMPGA2	zonal POD	ADGLIB	99	96	8	40	48

leading edge. The POD model used here has been already validated as SM3 in previous sections. The POD energy threshold is set at 95%. The snapshot size is 75 232.

- MPGA2 (Mixed-flow POD Genetic Algorithm 2) – same as MPGA1, but the POD energy content is increased up to 99%.

- KGA – a surrogate-based optimization where a Kriging meta-model, built on the objective function, is coupled to the genetic optimization. Here, the DAKOTA package [60.32] is used both for optimization process control and algorithm capabilities. The John Eddy Genetic Algorithm (JEGA) library [60.40] was used for optimization purposes. In particular, the single-objective genetic algorithm (SOGA) was used to perform optimization on a single objective function with general constraints. Kriging is initially trained on the 180 samples dataset. Then, a classical surrogate-based iterative optimization scheme is performed, consisting in building the surrogate, optimizing the surrogate objective, evaluating the minimizers with the *truth* (CFD) model, and rebuilding the surrogate. In the present optimization, 10 SBO iterations are performed.

- EGO – the key idea in EGO [60.17–19] is to exploit the Gaussian process capability to provide both the prediction at a new input location as well as the uncertainty associated with that prediction.

- AFPGA1 (Adaptive Full POD Genetic Algorithm 1), AFPGA2 (Adaptive Full POD Genetic Algorithm 2), AFPGA3 (Adaptive Full POD Genetic Algorithm 3) – the surrogate model employed is the same as FPGA3, but the training method is different and an adaptive sampling strategy is

added. In particular, it was decided to follow a different approach: the aim was to check whether, with a limited computational budget, better results can be obtained by adaptively training the POD model. Hence, the surrogate training phase was split into three contributions: an a-priori contribution, sampling the design space with the LHS technique and producing M_{apr} samples, an iterative, adaptive sampling aimed at improving the modal basis and enriching the ensemble dataset with M_{adp} samples, and a series of M_{opt} genetic optimizations, each producing an optimal candidate to update the ensemble and recompute the surrogate. The last phase is also called real-time updating. The three strategies differ for the relative amount of these three contributions as highlighted in Table 60.4, keeping fixed the total computational budget. The POD energy content is 99%. The snapshot size N is 201 488.

- AMPGA1 – the surrogate model employed is the SM2. The FOM/ROM interface is defined at $d = 0.35$ chord length from the airfoil's leading edge. However, the training method is different as it embeds a-priori, auto-adaptive, and optimal samples as described earlier. The POD energy content is 99%. The snapshot size N is 91 792;

- AMPGA2 – the surrogate model employed is again the SM3, but it differs from MPGA2 because the training method embeds a-priori, auto-adaptive, and optimal samples as described before. The POD energy content is 99%. The snapshot size N is 75 232.

The optimization setup is the same for all the approaches, except for AMPGA1 and AMPGA2. A population of 64 individuals is let to evolve for 150 generations with an 80% bit crossover rate and a 2% bit mutation rate. The genetic evolution is repeated every time a new optimal sample has to be added to the ensemble. Hence, a total number of 9600 evaluations are required for each optimization process. The setup of AMPGA1 and AMPGA2 differ slightly because the surrogate models adopted are more expensive (Fig. 60.7a). In order to increase the frequency of model updating stages, a population of 48 individuals is let to evolve for just 10 generations and the process is repeated 48 times to iteratively provide new optimal samples. The new feature is that each optimization step is a restart of the previous one with re-evaluation of the population candidates as the surrogate model has meanwhile been updated. In other words, the idea is to update the surrogate model more frequently (after just 10 genetic algorithm (GA) generations instead of 150) even if with smaller amounts of improvement (10 generations are not enough to converge the GA).

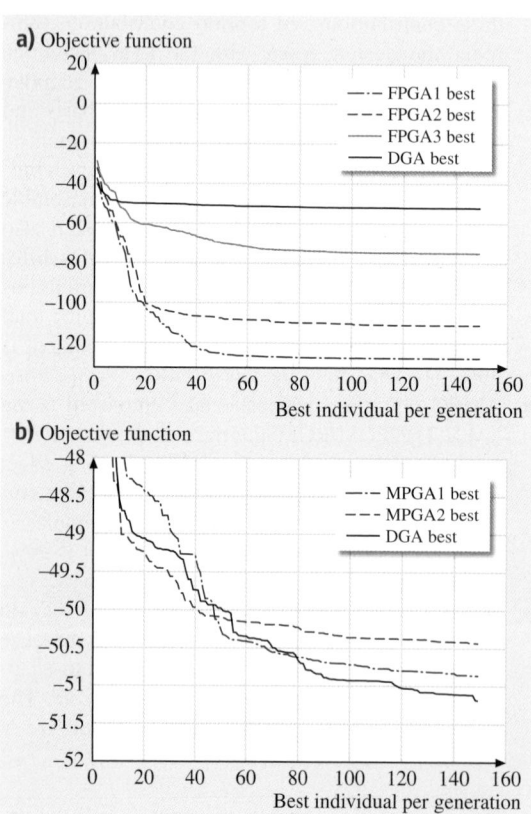

Fig. 60.8a,b Non-adaptive POD-driven optimization history

By looking at the details of the SBO approaches described so far, it seems quite natural to divide them into two main classes: non-adaptive (FPGAx, MPGAx), i.e., those without any adaptation/real-time updating, and adaptive optimizations (KGA, EGO, AFPGAx, AMPGAx). Consequently, the presentation of the results obtained will follow this logical sequence.

60.6.3 Non–Adaptive Optimization Results

Figure 60.8 shows the convergence history of the three FPGA optimizations compared to the plain DGA (solid black line) on the left and the two MPGA optimization histories on the right. The graphs show the sequence of the best candidates for each generation. It must be pointed out that while the DGA predictions (solid black lines) are obtained with the CFD solver, the POD-based predictions (dash, dotted, and dash-dotted lines) are the surrogate ones. For example, the dash-dotted line does not indicate that FPGA1 reached objective levels significantly lower than DGA, but simply that the predicted values of the airfoil performances were significantly overestimated. The plot clearly highlights that whatever the energy content, the full-POD approximation is not able to match the *true* data during the search process. Moreover, the general trend is towards an overestimation of the aerodynamic characteristics, which leads to lower values of the objective function. On the other hand, the MPGA model agreement with the CFD progress is very satisfying, both in terms of trends and accuracy.

60.6.4 Adaptive Optimization Results

Figure 60.9a shows the convergence history of the iterative SBSO (which stands for surrogate based shape optimization) KGA run. As was already mentioned, it is made of ten sequential surrogate optimizations; at the end of each of them, the optimal candidate is re-evaluated with the CFD solver and injected in the training set, so that an updated surrogate is available. The left-hand figure compares the surrogate and *true* prediction of the optimal candidate at each iteration. After about 6–7 SBO iterations, the Kriging model has been improved enough to predict very closely to the CFD solver. In Fig. 60.9b, the convergence history corresponding to the tenth SBO iteration is superimposed with the DGA run: a noticeable agreement is found, both in the initial drop in the fitness function and in the final plateau. Among the SBO minimizers, the ninth iteration shows

Fig. 60.9a–c Kriging-based optimization ▶

the lowest *true* objective function value, so that it will be considered as the actual KGA optimum.

Figure 60.9c reports the convergence history of the EGO optimization. Dark gray circles depict the initial DOE sampling (153 candidates), while light gray circles denote the subsequent 400 candidates found by minimizing the expected improvement. The graph also reports the expected improvement values in gray white-filled circles and logarithmic scale (right axis). It is clearly evident how the progressive decrease of the EIF produces a better quality of the Kriging model, which in turn results in a minimization of the *true* objective function.

The convergence histories of the AFPGA1, AFPGA2, and AFPGA3 optimizations are reported in Fig. 60.10a together with the objective function values computed on the training points. In the plot, each point represents a single high-fidelity evaluation, the squares depict the a-priori and adaptive training sites, while the circles connected with lines represent the sequence of optima from M_{opt} GA optimizations. It is fairly evident that adaptive sampling is often helpful as it allows us to find sub-optimal solutions even before optimization (see AFPGA1 and AFPGA2). On the other hand, this somewhat disappointing behavior in the optimization step is due to the fact that the surrogate underestimates the objective function, thus pushing the surrogate-based optimizer to explore uninteresting design space regions. In particular, results show that the more adapted the initial sampling, the smaller the underestimation. Hence, the ratio M_{apr}/M_{adp} should be kept low. However, another important feature is related to the AFPGA3 method; it shows that by lowering the ratio M_{apr}/M_{adp} too much (up to 0.09), the performance of the method deteriorates, as the final AFPGA3 optimum is worse than the previous ones. Indeed, leaving too much room for adaptive criteria seems to produce a sampling with very poor exploratory capabilities.

These considerations give a helpful hint about the right combination of a-priori and adaptive sampling: the ratio M_{apr}/M_{adp} should be kept between 0.1 and 0.5, which is in line with the value of the EGO and goal-seeking methods proposed by *Jones* [60.19]. This information is exploited in tuning the parameters for AMPGA1 and AMPGA2 optimization. Figure 60.10b shows the *true* objective functions of the training samples and of the sequence of optima candidates. Even if the AMPGA1 performs quite well, it exhibits simi-

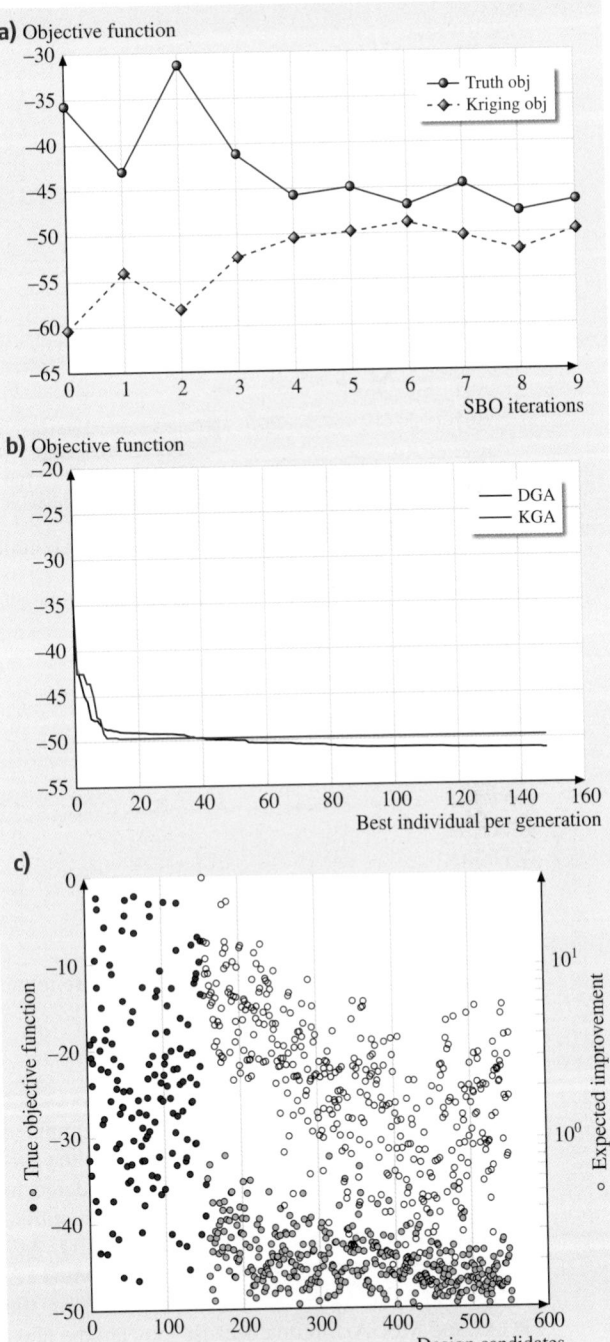

lar characteristics to the AFPGAx optimization. On the other hand, the AMPGA2 optimum outperforms the optima seen so far and, as it will be clear in the next

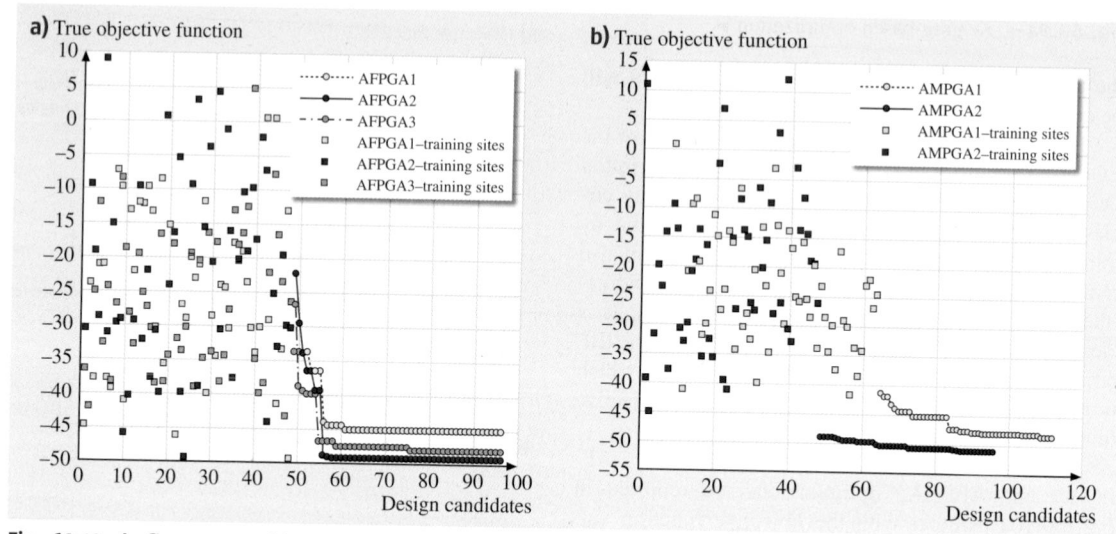

Fig. 60.10a,b Convergence history of POD-based optimization

Table 60.5 Optimal candidates, objective function breakdown

Opt. run ID	Truth obj.	Predicted obj.	Penalty	C_l	C_d	C_m
DGA	−51.18	−51.18	1.025	0.619	0.0118	−0.0602
MPGA1	−48.70	−50.86	1.13	0.578	0.0116	−0.0606
FPGA3	−38.33	−73.45	0.608	0.553	0.0142	−0.0578
KGA	−47.65	−51.94	0.585	0.612	0.0127	−0.0576
EGO	−49.71	−49.71	0.530	0.618	0.0123	−0.0573
AFPGA1	−49.24	−47.14	1.12	0.635	0.0126	−0.0606
AFPGA2	−49.20	−52.61	0.551	0.631	0.0127	−0.0574
AFPGA3	−48.13	−47.88	1.29	0.583	0.0118	−0.0614
AMPGA1	−48.58	−44.61	0.567	0.576	0.0117	−0.0575
AMPGA2	−51.13	−50.31	0.947	0.612	0.0117	−0.0597

section, it is the only candidate to get very close to the *truth* optimum, i.e., the DGA optimum.

60.6.5 Optima Analysis

This section gives details about the optima computed with each of the methodologies presented. In the following, ten optimal candidates will be considered to assess the optimization results, namely the optima from runs DGA, FPGA3, MPGA1, KGA, EGO, AFPGA1, AFPGA2, AFPGA3, AMPGA1, and AMPGA2. FPGA3 and MPGA1 have been selected among the FPGAx and MPGAx optima because they are the closest to the high-fidelity DGA optimum. The objective function breakdown for each optimal candidate is summarized in Table 60.5. The table reports both the *true* data, obtained by re-computing each design with the CFD solver, and the predicted objective function as cal-

culated by the surrogate model. Each optimum does not satisfy the pitching moment constraint because the quadratic penalty function and its weight, chosen in the problem definition, purposely do not enforce this constraint strictly to have a less stiff optimization problem. Indeed, getting precisely into the constraint boundaries would probably have penalized the aerodynamic efficiency to much, i.e., the actual objective function, while applying small penalties near a constraint boundary gives more flexibility to the search of the optimal design.

Non-Adaptive Optima

Among the non-adaptive methods (here KGA is considered as non-adaptive to set a comparison), optimal designs coming from MPGA1 and KGA are closer to the plain one in terms of global performance. The MPGA1 optimum catches the DGA constraint viola-

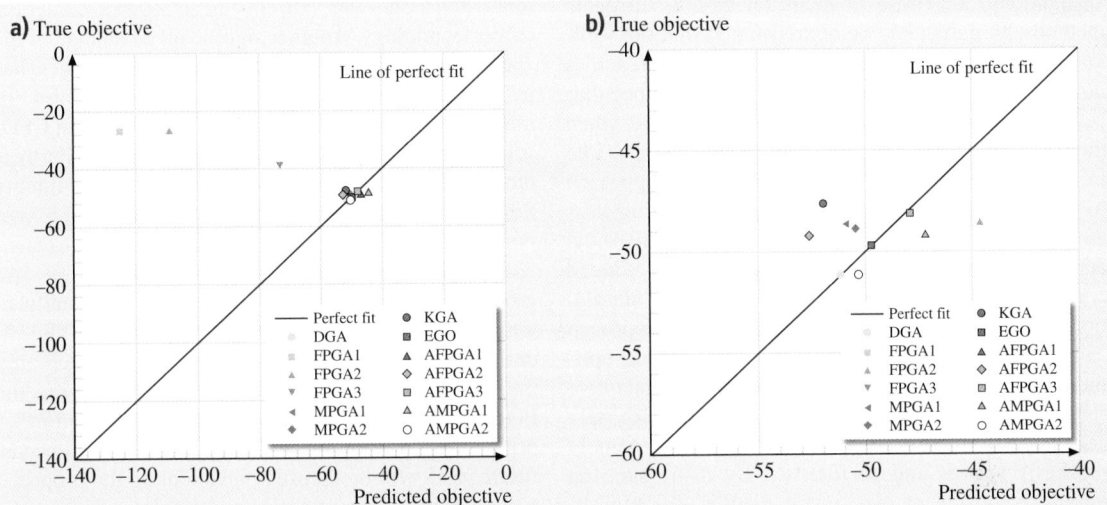

Fig. 60.11a,b Computed optima in the surrogate vs truth objective plane

tion almost perfectly, while KGA design performs even better on pitching moment but at the cost of a slightly lower aerodynamic efficiency. FPGA3 design, although using 75 POD modes, does not belong to an optimal sub-set but exhibits a small penalty. The best surrogate solution is the MPGA1, where a weak shock appears on the suction side but at a lower lift level. Indeed, the optimal leading edge radius is almost twice the DGA value, and this feature causes an over-expansion on the suction side, which in turn makes the shock wave occur more upstream and more strongly. The Kriging-based best candidate shows a reduced rear loading to limit nose-down pitching moment and trim drag associated with the rear location of the center of pressure. This beneficial feature is counterbalanced by the lift production increase in the fore airfoil part and, consequently, by a more pronounced pressure jump across the shock wave.

Adaptive Optima

In order to highlight the undertaken improvement path, Fig 60.11 reports a correlation plot where the whole set of optima is depicted in the surrogate objective – the true objective plane. Two different zooming levels

are set, as they reflect the non-adaptive and adaptive process: the FPGA1, FPGA2, and FPGA3 optima show very large discrepancies between true value and surrogate prediction, hence they are located very far from the line of perfect fit. However, a trend is observable as, increasing the POD energy content (i.e., passing from FPGA1 to FPGA3), the best candidate gets closer to the true optimum. By looking at the top part of the figure, a clustering of the remaining optima is observable, so that a closer look is offered in the bottom figure for better understanding. Among the adaptive optima, AMPGA2 and EGO produce the best results and demonstrate the benefits of opportunely coupling the zonal approach and an *intelligent* design space sampling. Indeed, these optimal candidates are the closest to the target point in the sense of the Euclidean distance in the objective plane. From a more strict aerodynamic point of view, turning on the adaptive criteria brought the quality of the optimization result to a level that is very close to the expectations of the designer, as a shockless profile featured with a gentle re-compression on the suction side represents the *golden* goal for the proposed optimization problem.

60.7 Conclusions

The aim of the present book chapter was to review and investigate ad-hoc computational techniques to ease the solution of complex aerodynamic shape optimiza-

tion problems such as those commonly encountered in aerospace design at industrial level. Among the various approaches that are the subject of research investigations, we chose to focus on ad-hoc surrogate methods. In particular, we demonstrated that the well-known proper orthogonal decomposition approach is not adequate to provide reliable predictions in peculiar aerodynamic conditions like transonic flow and when the boundary of the computational domain changes like in shape optimization. We proposed a zonal approach to de-couple the strong non-linearities occurring near the body-wall from the POD approximation. This zonal approach proved to give reliable results at a reduced computational cost compared to the full CFD simulation. Furthermore, we showed that the zonal approach can give an accurate approximation of the true optimum when trained with specifically designed adaptive sampling techniques. The latter have been purposely conceived to improve POD model machinery, namely the basis vectors and coefficients. By using such an *intelligent* design of experiment method, the high-fidelity computational budget can be further reduced and the overall performance of the design loop increases. The beneficial effects of this approach have been illustrated by a comparison of several surrogate-based optimization processes on the shape design of a two-dimensional airfoil. The extension of the methodology to complex three-dimensional problems is straightforward and under way. Indeed, one of the main advantages of the proposed methodology is its relative insensitivity to the curse of dimensionality of the design parameter space. On the other hand, the larger snapshot size required by three-dimensional CFD flow fields, where millions of unknowns may be handled, does not repre-

sent a big issue with current linear algebra numerical solver technology. Another significant advantage of the zonal approach with respect to other surrogates lies in its favorable scaling property when the third dimension is introduced because the ratio between CFD-solved and POD-predicted points decreases. Furthermore, zonal POD allows us to solve the high-fidelity flow field locally, i.e., only where it is required by geometry-driven considerations. This represents a tremendous benefit when the complexity of the design case grows. As an example, if the goal is to optimally fit a nacelle body in an already optimal wing, the high-fidelity computation zone can be restricted to catch only the wing–nacelle interaction phenomena, leaving the POD model to predict the flow field outside. To further bridge the gap with real-world applications and needs, future work will be focused on validating the proposed methodology in large-scale, multi-point aerodynamic problems involving huge design parameter spaces and aiming at predicting the aerodynamic characteristics in deep transonic flow. Finally, improvements towards a more efficient exploitation of approximate models in the search algorithm are under investigation. In particular, a bi-objective approach seems promising where the first objective is the true function evaluation and the second is the surrogate one. An asymmetric algorithm, i.e., an algorithm that invokes the surrogate many more times than the high-fidelity one, was already proposed in a multi-fidelity environment [60.16] and will be extended to adaptive POD-based surrogate models.

References

60.1   Advisory Council for Aviation Research and innovation in Europe: European aeronautics: A vision for 2020 (European Communities, Luxembourg 2001)

60.2   G. Schrauf: *KATnet. Key Aerodynamic Technologies for Aircraft Performance Improvement* (DGLR – Deutsche Gesellschaft für Luft- und Raumfahrt, Vienna 2006)

60.3   O. von Tronchin: The Airbus Global Market Forecast (Herndon 2010)

60.4   Boeing: Current Market Outlook 2009–2028 (Seattle 2009)

60.5   E. Torenbeek: *Synthesis of Subsonic Airplane Design: An Introduction to the Preliminary Design of Subsonic General Aviation and Transport Aircraft, with Emphasis on Layout, Aerodynamic Design, Propulsion and Performance* (Delft Univ. Press, Delft 1982) p. 620

60.6   D.P. Raymer: *Aircraft Design: A Conceptual Approach* (AIAA, Reston 1999)

60.7   D.P. Raymer: *RDS-Student, Software for Aircraft Design, Sizing, and Performance* (AIAA, Reston 2000)

60.8   R.H. Myers, D.C. Montgomery: *Response Surface Methodology: Process and Product Optimization Using Designed Experiments* (Wiley, New York 1995)

60.9   A. Giunta, L.T. Watson: A comparison of approximation modeling techniques: Polynomial versus interpolating models, Proc. 11th AIAA/ISSMO Multidiscip. Anal. Optim. Conf., St. Louis (1998) pp. 392–404

60.10   H.M. Gutmann: A radial basis function method for global optimization, J. Global Optim. **19**, 201–227 (2001)

60.11   J.H. Friedman: Multivariate adaptive regression splines, Ann. Stat. **19**(1), 1–141 (1991)

60.12 M.H. Hassoun: *Fundamentals of Artificial Neural Networks, A* (MIT Press, Cambridge 1995)

60.13 A.I.J. Forrester, N.W. Bressloff, A.J. Keane: Optimization using surrogate models and partially converged computational fluid dynamics simulations, Proc. R. Soc. A Math. Phys. Eng. Sci. **462**(2071), 2177–2204 (2006)

60.14 P.I.K. Liakopoulos, I.C. Kampolis, K.C. Giannakoglou: Grid enabled, hierarchical distributed metamodel-assisted evolutionary algorithms for aerodynamic shape optimization, Future Gener. Comput. Syst. **24**(7), 701–708 (2008)

60.15 N.M. Alexandrov, R.M. Lewis, C.R. Gumbert, L.L. Green, P.A. Newman: Optimization with variable-fidelity models applied to wing design, ICASE Re. No. 99-49 (Institute for Computer Applications in Science and Engineering, Hampton 1999)

60.16 D. Quagliarella: Airfoil design using Navier–Stokes equations and an asymmetric multi-objective genetic algorithm, Proc. EUROGEN 2003 Conf., Barcelona (2003)

60.17 D.R. Jones, M. Schonlau, W.J. Welch: Efficient global optimization of expensive black-box functions, J. Global Optim. **13**, 455–492 (1998)

60.18 M. Schonlau, W.J. Welch, D.R. Jones: Global versus local search in constrained optimization of computer models, IMS Lect. Notes **34**, 11–25 (1998)

60.19 D.R. Jones: A taxonomy of global optimization methods based on response surfaces, J. Global Optim. **21**, 345–383 (2001)

60.20 N. Queipo, R. Haftka, W. Shyy, T. Goel, R. Vaidyanathan, P. Kevintucker: Surrogate-based analysis and optimization, Prog. Aerosp. Sci. **41**(1), 1–28 (2005)

60.21 T.W. Simpson, V.V. Toropov, V. Balabanov, F.A.C. Viana: Design and analysis of computer experiments in multidisciplinary design optimization: A review of how far we have come – or not, Proc. 12th AIAA/ISSMO Multidiscip. Anal. Optim. Conf. (2008) pp. 1–22

60.22 A.I.J. Forrester, A.J. Keane: Recent advances in surrogate-based optimization, Prog. Aerosp. Sci. **45**(1–3), 50–79 (2009)

60.23 A. Sóbester, S.J. Leary, A.J. Keane: A parallel updating scheme for approximating and optimizing high fidelity computer simulations, Struct. Multidiscip. Optim. **27**, 371–383 (2004)

60.24 T. Goel, R.T. Haftka, W. Shyy, N.V. Queipo: Ensemble of surrogates, Struct. Multidiscip. Optim. **33**(3), 199–216 (2007)

60.25 J.L. Lumley: The structure of inhomogeneous turbulent flows, Atmos. Turbul. Radio Wave Propag.-Proc. Int. Colloq. (1965) pp. 166–178

60.26 E. Iuliano: Towards a pod-based surrogate model for CFD optimization, Proc. ECCOMAS CFD Optim. Conf. (2011)

60.27 E. Iuliano, D. Quagliarella: Surrogate-based aerodynamic optimization via a zonal pod model, Proc. EUROGEN 2011 Conf., Barcelona (2011)

60.28 P.H. Cook, M.C.P. Firmin, M.A. McDonald: *Aerofoil RAE 2822: Pressure Distributions, and Boundary Layer and Wake Measurements*, Technical memorandum (Royal Aircraft Establishment, Famborough 1977)

60.29 B.M. Kulfan: Universal parametric geometry representation method, J. Aircr. **45**(1), 142–158 (2008)

60.30 D.C. Montgomery: *Design and Analysis of Experiments* (Wiley, New York 2006)

60.31 M.D. McKay, W.J. Conover, R.J. Beckman: A comparison of three methods for selecting values of input variables in the analysis of output from a computer code, Technometrics **21**, 239–245 (1979)

60.32 M.S. Eldred, B.J. Bichon, B.M. Adams, S. Mahadevan: Overview of reliability analysis and design capabilities in DAKOTA with application to shape optimization of MEMS. In: *Structural Design Optimization Considering Uncertainties*, ed. by Y. Tsompanakis, N.D. Lagaros, M. Papadrakakis (Taylor Francis, New York 2008) pp. 401–432

60.33 J. Goblet, I. Lepot: *Two Adaptive DOE Strategies for POD-Based Surrogate Models*, Tech. Rep. (CENAERO 2010)

60.34 C. Sainvitu, M. Guenòt, I. Lepot, J. Goblet: Adaptive sampling strategies for POD-based surrogate models in an optimization framework, Proc. EUROGEN 2011 Conf. (2011)

60.35 S. Rippa: An algorithm for selecting a good value for the parameter c in radial basis function interpolation, Adv. Comput. Math. **11**, 193–210 (1999)

60.36 M. Amato, P. Catalano: Non linear $\kappa-\varepsilon$ turbulence modeling for industrial applications, Proc. ICAS 2000 Conf. (2000)

60.37 D. Quagliarella, P. Iannelli, P.L. Vitagliano, G. Chinnici: Aerodynamic shape design using hybrid evolutionary computation and fitness approximation, AAIA 1st Int. Syst. Tech. Conf. (2004)

60.38 D. Quagliarella, A. Vicini: GAs for aerodynamic shape design I: General issues, shape parametrization problems and hybridization techniques, Lect. Ser. van Kareman Inst. Fluid Dyn. (2000)

60.39 D. Quagliarella, A. Vicini: GAs for aerodynamic shape design II: Multiobjective optimization and multi-criteria design, Lect. Ser. van Kareman Inst. Genet. Algorithm. Optim. Aeronaut. Turbomach. (2000)

60.40 J.E. Eddy, K. Lewis: Effective generation of Pareto sets using genetic programming, Proc. DETC '01 ASME Comput. Inform. Eng. Conf. (2001)

Part E | 60

61. Knowledge Discovery in Bioinformatics

Julie Hamon, Julie Jacques, Laetitia Jourdan, Clarisse Dhaenens

Biomedical research progresses rapidly, in particular in the area of genomic and postgenomic research. Hence many challenges appear for biostatistics and bioinformatics to deal with the large amount of data generated. After presenting some of these challenges, this chapter aims at presenting evolutionary combinatorial optimization approaches proposed to deal with knowledge discovery in bioinformatics. Therefore, the chapter will focus on three main tasks of data mining (association rules, feature selection, and clustering) widely encountered in bioinformatics applications. For each of them, a description of the task will be given as well as information about their uses in bioinformatics. Then, some evolutionary approaches proposed to cope with such a task will be exposed and discussed.

61.1 Challenges in Bioinformatics

Biomedical research progresses rapidly, in particular in the area of genomic and postgenomic research. Hence many challenges appear for biostatistics and bioinformatics to deal with the large amount of data generated. This data, related to the sequencing of the genome, may deal, for example, with the identification of more than 1 million single nucleotide polymorphisms (SNPs) – corresponding to genetic variations – that can be used to carry out genome-wide association studies (GWAS). Analyzing such data requires advanced methods able to deal with such a large number of information and with their specificities. This is the reason why knowledge discovery approaches have

been proposed to either:

1. Extract interesting rules or
2. Reduce the dimensionality of the data or
3. Classify/cluster data.

All these knowledge discovery tasks have been addressed by several communities: statistics, machine learning, and combinatorial optimization. This latest is the subject of this chapter and a recent review reports synergies between operations research and data mining [61.1]. In this chapter, we focus on evolutionary combinatorial optimization and see how it may be used to extract knowledge for bioinformatics.

Many problems arise in bioinformatics. In order to illustrate this chapter, three types of applications will be mainly used. They are described hereafter:

- *Microarray – Gene expression data*: A typical bioinformatics application requiring knowledge discovery deals with DNA microarray data analysis. Indeed, DNA microarray experiments are of great interest and importance for biologists; thanks to their ability to simultaneously measure the expressions and interactions of thousands of genes. Such experiments are used to point out, for example, genes of predisposition to some diseases such as diabetes, cancer, etc. These experiments are generating huge amounts of data that need to be analyzed. Those data are mainly represented in gene expression matrix. Some experiments add the time parameter (to analyze the evolution of the expressions after a stress, for example) and report results at different time points. This special case is sometimes called 3-D-microarray (three-dimensional microarray). For microarray data analysis several data mining approaches have been proposed (association rule discovery, feature selection, clustering, and bi-clustering) [61.2] and benchmarks are available to compare efficiency of methods. Hence it will provide a good illustrative application along this chapter. As the number of genes to consider is huge many heuristics, and in particular evolutionary algorithms, have been proposed to deal with such data.
- *Genome-wide association studies*: Another interesting approach to find genetic susceptibility for disease is to track genetic variations. Indeed, as indicated by *Moore* et al. in their recent study of *Bioinformatics challenges for genome-wide association studies*, the sequencing of the human genome has made possible to identify more than 1 million SNPs (genetic variations) across the genome that can be used to carry out GWAS in order to reveal genetic basis of disease susceptibility [61.3]. First approaches used to deal with these massive amounts of GWAS data, mainly based on biostatistics have enabled the discovery of new associations. However, as such approaches consider only one SNP at a time and most of the time ignore the genomic and environmental contexts, more complex approaches, that consider genotype–phenotype relationships have to be proposed. Regarding the large number of SNPs to consider and the complexity of the relationships to discover, the knowledge discovery paradigm has been used to deal with such data and optimization approaches have been proposed.
- *Protein analysis*: There are now plenty of identified proteins that are not completely known. For example, their function may still be unknown (even if the sequence may be known). However, the knowledge of their functions is crucial for the development of new drugs. Hence, *automated function prediction* is an active research field and computational techniques that use high-throughput experimental data (protein and genome sequences, protein interaction networks, phylogenetic profiles, etc.), have been developed. Once again such experiments produce a large amount of data that need to be analyzed.

Considering this variety of applications, this chapter aims at presenting evolutionary combinatorial optimization approaches proposed to deal with knowledge discovery in bioinformatics. Therefore, the chapter will focus on three main tasks of data mining widely encountered in bioinformatics applications. These tasks are: association rules, feature selection, and clustering (unsupervised classification). For each of them, a description of the task will be given as well as information about their uses in bioinformatics. Then, some evolutionary approaches proposed to cope with such a task will be exposed and discussed. A table provides an overview of these approaches and serves as a guideline for the reader to know which type of approach to use in a specific context.

61.2 Association Rules by Evolutionary Algorithm in Bioinformatics

61.2.1 Association Rules Discovery

Task Description

The problem of discovering association rules was first formulated in [61.4] and was called the market-basket problem. The initial problem was the following: given a set of items and a large collection of sales records, which consist of a transaction date and the items purchased in that transaction, the task is to find significant relationships between the items contained in different transactions. Since this first application, many other problems, in particular in bioinformatics, have been studied with association rules that may be defined in a more general way. Let us consider a database composed of transactions (records or objects) described according to several – maybe many – attributes (features or columns). Association rules provide a very simple (but useful) way to present correlations or other relationships among attributes (features) expressed in the form $A \Rightarrow C$, where A is the antecedent part (condition) and C the consequent part (prediction). A and C are sets of attributes that are disjoint. The best-known algorithm to mine association rules is A-priori, proposed by *Agrawal* and *Srikant* [61.5]. This two-phase algorithm first finds all *frequent item sets* (sets of items – or attributes – that often occur together within transactions) that have at least a given minimum level of *confidence*. This is done via an efficient search exploiting the downward closure property of *support* (which measures the frequency of the rules). A lot of improvements upon the initial method, as well as efficient implementations (including parallel implementations) have been proposed to be able to deal with very large databases [61.6–8].

We note that a specific case of rule mining deals with classification rules where the consequent is the same for every rule. This may be seen as a straightforward classification task; however, the models and methods used for this are closed to those used more generally in rule mining; hence it will be considered in this section.

The task of discovering effective association rules may be seen as a combinatorial optimization problem, as rules are combinations of attributes. Each attribute may participate to the rule in the antecedent or the consequent part. Each attribute may have several values that have to be checked. As the number of attributes may be very large (up to several thousands), the number of possible rules (choice of the attributes that participate to the rule and their values) may be very large. Therefore, efficient methods (heuristic approaches and in particular evolutionary approaches) are direly needed.

Use in Bioinformatics
In their survey, *Atluri* et al. present different types of association patterns and discuss some of their applications in bioinformatics [61.9]. They indicate that association rules discovery has not been widely used yet in bioinformatics except to deal with microarray data and data on genetic variations (SNP data) for which several works exist. Their feeling is that association rules have been underutilized in bioinformatics and they propose, in their article, to give hints on how to exploit the

potential benefits of such an approach to deal with protein function prediction and in particular to address the noise and the incompleteness issues of currently protein interaction network data.

In addition, association rules discovery allows the integration of external biological information with gene expression data. For example, *Carmona-Saez* et al. propose an approach based on co-occurrence patterns, that integrates gene annotations and expression data to discover intrinsic associations among both data sources [61.10].

61.2.2 Evolutionary Approaches for Association Rules in Bioinformatics

Motivations
Association rules are a very general model and may overcome some drawbacks of other classical knowledge discovery tasks such as classification. For example, considering microarray data in which relationships between genes are searched for, using classification will impose a gene participating to several relations to be classified in a single group. Classification will also have difficulty to point out relations between genes belonging to a same group and finally, classification will be made according to the whole set of experiments which do not allow to exhibit relationships between genes in a subset of conditions. Association rules may overcome these drawbacks by providing relationships between genes that occur in certain conditions.

However, one of the drawback of classical association rules discovery approaches (algorithm A-priori for example), is the role played by the *support* measure. Indeed, allowing to identify low support rules (but still interesting information as rare rules may be very important in the context of bioinformatics) will generate a huge number of rules that will be difficult to interpret. In this sense other types of approaches, using different quality measures, have to be proposed.

In this sense, for example, *Khabzaoui* et al. proposed to analyze microarray data with an association rule-based technique. They modeled this problem as a multiobjective combinatorial optimization problem (which allows us to use other quality criteria than the *support*) and solved it using an evolutionary algorithm based on a genetic algorithm. Therefore, specific mechanisms (mutation and crossover operators, elitism, and so on) are designed for this task [61.11]. In order to improve the quality of the rules obtained, cooperative approaches are proposed [61.12].

Overview

Firstly, this section will introduce common approaches used in evolutionary rule mining, like learning classifier systems (LCSs), rough sets approach or genetic programming. Secondly, Pittsburgh and Michigan rule designs will be detailed. Finally, implementation details of genetic algorithms for rule mining with applications in bioinformatics will be presented, and a summary table will be provided.

Some Classical Approaches. *Learning classifier systems* (LCS) come from the machine learning community and are useful for classification tasks using classification rules. LCS evolve a population of classifiers – decision trees, rules, or rule sets – using a genetic algorithm and a credit assignment module that awards good classifiers. A more detailed introduction to LCS can be found in [61.13]. Some bioinformatics applications have been realized with GAssist algorithm [61.14] and its successor (BioHEL) (*bioinformatics-oriented hierarchical evolutionary Learning*) [61.15].

The rough set approach consists in finding approximation sets of features: a lower set, whose features allows us to identify objects that certainly belong to approximated set, and an upper set whose features describe objects that probably belong to approximated set. Rules can be generated from these resulting sets. More complete information about rough set theory and applications can be found in [61.16]. In [61.17, 18], *Rosetta* toolkit was used to solve bioinformatics problems; *Vinterbo* and *Øhrn* explained in [61.19] the implementation of a genetic algorithm for rough sets in Rosetta toolkit and show that this algorithm allows us to produce smaller rules with better predictability. They measured the predictability with the AUC score (area under ROC curve). The ROC curve (receiver operating characteristic) is often used in data mining to assess the performance of classification algorithms. It is plotted using true positive rate (known as *sensitivity*) and false positives rate (also called *1 – specificity*) as axes. More details can be found in [61.20].

ROC curve is often used in data mining to assess the performance of classification algorithms, especially ranking algorithms. It is plotted using true positive rate (known as *sensitivity*) and false positives rate (also called *1 – specificity*) as axes. More details can be found in [61.20].)).

Genetic programming has been used to extract rules from biological data. For example, *Pappa* and *Freitas* proposed an original approach to predict protein postsynaptic activity [61.22]. Since there are a lot of classification algorithms, and the impact of the choice of the algorithm is important, they chose to design an algorithm that searches for a good classification algorithm. Therefore, they use *grammar-based genetic programming* (GGP). The main difference with the genetic programming approach implemented by *Yang* et al. [61.23] is the use of a grammar.

Rule Design. When mining rules, two designs are available: in *Michigan design*, each solution is a rule, while in *Pittsburgh design* each solution is a rule set. Pittsburgh design has a larger search space; moreover, fitness and operators are harder to implement. However, with this design there is no need to use a covering algorithm to encourage rules from the same solution (rule set) to cover different objects. With the Michigan design, without any covering algorithm, solutions (rules) can cover the same objects. This point may cause problems when searching for classification rules. In [61.26], *Bacardit* and *Butz* compared Michigan and Pittsburgh LCS. They concluded that both are suitable for data mining. Michigan LCS tend to overfit the data – rules are too specific – while Pittsburgh LCS are sometimes too general and miss some search subspaces.

Genetic Algorithm Design. As many genetic algorithms have been proposed to deal with rule mining (Table 61.1), this paragraph presents the main components used:

- *Initial population*: Most of the time, the initial population is composed of random individuals. However, *Cho* et al. initialized their population with rules generated by a neural network [61.21]. In [61.14], the population is initialized by iteratively choosing one object and generating a rule covering it.
- *Fitness function*: It often contains an objective on the rule size to limit bloat effect and overfitting, both responsible of generating specific and complicated rules. This is an application of the *minimum description length* (MDL) principle. It is frequently associated to a performance measure: quality of hitting sets in rough sets, accuracy, coverage, confidence, or AUC. As many measures have been proposed, we will not detail all of them, but refer to the review of *Geng* and *Hamilton* on rule interestingness measures [61.27]. *Pappa* and *Freitas* recommend to use sensitivity × specificity as the fitness function when class distribution is unbalanced (in their protein data, only 6.04% of objects had the positive class) [61.22]. The majority of bioinformat-

Table 61.1 Overview of applications of evolutionary rule mining in bioinformatics

Application	EA	Approach	Design	Evaluation function	Encoding	Operators	Reference
Protein structure prediction	GA	LCS	Pittsburgh	Size accuracy	Hyper rectangle	Tournament selection, 1-point crossover	[61.15]
Protein binding	GA	NN (hybrid)	Michigan	Confidence, size	Fixed size value-vectors	RWS, 1-point crossover	[61.21]
Protein classification	GGP			Normalized accuracy	Grammar derivation tree		[61.22]
Protein binding	GA	Rough sets	Michigan	Size, coverage	Binary	SUS	[61.18]
Microarray	GP		Michigan	Size, coverage	Binary	Elite selection, cut & splice crossover	[61.23]
Microarray	MOEA		Michigan	Support, jmeasure, interest surprise, confidence		RWS	[61.11]
Microarray	GA	LCS	Pittsburgh	Accuracy, misclassification rate	Binary	Tournament selection, 1-point crossover	[61.24]
3-D-Microarray (time series)	GA	Rough sets	Michigan	Size, coverage	Binary	SUS	[61.25]

MOEA: multiobjective evolutionary algorithm, GA: genetic algorithm, GP: genetic programming, GGP: grammar-based genetic programming, SUS: stochastic universal sampling, LCS: learning classifier systems, NN: neural network, RWS: roulette wheel selection

ics applications use an aggregation to combine these multiple objectives. Weights are sometimes introduced to balance between objectives. For example, *Bacardit* uses an automatic weighting function that changes weights while the algorithm is running [61.28]. However, weights can be difficult to configure; therefore, multiobjective algorithms can overcome this problem [61.11, 29].

- *Encoding*: For the greater part, encoding is binary. In rough sets approach, it is fixed size and matches selected features of the approximated sets. Rules can be encoded as binary or list of values or features and values. In [61.15] a rule representation for real-values features is used: hyper-rectangle instead

of fuzzy rules approaches or data preparation with discretization methods. *Pappa* and *Freitas* encoding differentiates itself from others because of a grammar derivation tree encoding [61.22]. This is needed since they search for classification algorithms, and not for rules.

- *Operators*: Mostly used crossover operators are 1-point crossovers. *Casillas* et al. proposed adapted crossover and mutation operators to deal with rule overlapping when using rule sets [61.29]. Parents are mainly selected using fitness proportionate selection, as stochastic universal sampling (SUS) or roulette wheel. Less frequently, elite selection and tournament are used.

61.3 Feature Selection for Classification and Regression by Evolutionary Algorithm in Bioinformatics

61.3.1 Feature Selection

Feature selection is an active research domain in statistics (variable selection) and data mining communities. Feature selection can, jointly used with classification (or clustering), significantly improve the comprehensibility of the resulting classifier models and often build a model that generalizes better unseen points. The main

idea of feature selection is to choose a subset of input variables by eliminating features with little or no predictive information. Hence, finding the correct subset of predictive features is an important problem in itself.

Feature selection for classification can be classified in three classes depending on how the selection process is combined with the classifier: the wrapper approach, the filter approach, and the embedded approach.

The *wrapper approach* model uses learning algorithms during the feature selection process and assesses the selected features by the learning algorithm's performance using, for example, accuracy, sensitivity, or specificity. The *filter approach model* considers statistical characteristics of a dataset directly without involving any learning algorithm. In the *embedded approach* model, the learning algorithm uses its own embedded feature selection algorithm (either explicit or implicit). Let us remark that an *hybrid approach* model is sometimes used to, first adopt a filter approach that will reduce the number of features to consider, and then realize, with the remaining features, a wrapper approach that will select features in a more accurate way.

The general task of feature selection can be formulated as an optimization problem. Binary values of the variables x_i are used in order to indicate the presence ($x_i = 1$) or the absence ($x_i = 0$) of the feature i in the optimal feature set. Then, the problem is formulated as $\max_{x=(x_1,...,x_n) \in \{0;1\}^n} F(x)$ for a function F that has to be determined regarding the context (filter, wrapper, or embedded approach and application under study). In filter approaches, many different statistical feature selection measures, such as the correlation feature selection (CFS) measure, the minimal-redundancy-maximal-relevance (mRMR) measure, the discriminant function, or the Mahalanobis distance have been used to assess to each feature a score. In wrapper approaches, classification algorithms may be used to assign to a selection of features a score that represents the ability of the selection to lead to a correct classification. Such classical algorithms are KNNs (k nearest neighbors), SVMs (support vector machines), NN (neural networks), etc.

As reported by *Kim* et al., traditional approaches to feature selection with a single criterion have shown some limitations [61.30]. Therefore, they propose to consider this problem as a multiobjective one and present an adaptation of (ELSA) (evolutionary local selection algorithm) , inspired from artificial life models of adaptive agents to cope with this multiobjective problem. Another multiobjective approach may be found in *García-Nieto* et al. where a multiobjective genetic algorithm is used for cancer diagnosis from gene selection in microarray datasets [61.31].

Use in Bioinformatics

As indicated by *Saeys* et al., feature selection in bioinformatics is motivated by the high-dimensional nature of modeling tasks (sequence analysis over microarray analysis, spectral analyses, literature mining,

etc.) [61.32]. Let us remark, that in contrast with other dimensionality reduction techniques (based on projection, or compression, for example), feature selection techniques do not modify data. Thus they preserve the original semantics of the variables which helps the interpretability of results.

In their review of *Feature selection techniques in bioinformatics*, *Saeys* et al. identify three classes of problems where feature selection is involved [61.32]:

- Sequence analysis
- Microarray analysis
- Mass spectra analysis.

Sequence analysis deals with the study of either the content of the sequence or its signal. As far as the content is concerned, the prediction of subsequences that code for proteins requires a feature selection to cope with the large number of features that can be extracted from a sequence and the lack of samples available. Recently, feature selection approaches have also been used for other applications such as the recognition of promoter regions or the prediction of microRNA target. Regarding the signal analysis, the aim is mainly to identify more or less conserved signals in the sequence (motifs), representing binding sites for proteins. Therefore, regression approaches are proposed to relate motifs to gene expression levels and feature selection can be used to search for the best motif.

Microarray analysis, as already said, poses a great challenge for computational techniques because of their large dimensionality. *Saeys* et al. give in their survey an overview of the most influential techniques [61.32]. In particular, genetic algorithms can be used to deal with microarray data in wrapper type approaches.

Mass spectra analysis deals with the analysis of thousands of signal intensity measures. This context, even if data are different, is very similar to microarray analysis and feature selection is an important step to reduce the dimensionality of the problem. Genetic algorithms and nature inspired algorithms have been proposed to deal with such data, using wrapper approaches.

61.3.2 Evolutionary Approaches for Feature Selection for Classification and Regression in Bioinformatics

Motivations

With the development of technologies, many bioinformatics applications deal with large datasets, often with more features than objects (samples). However, among these features, some are irrelevant or redundant. That is why feature selection aims to select a subset of relevant features. Reducing the dimension of the problem, this method can reduce the computational time and improve prediction accuracy. Indeed, including nonsignificant features can induce a noise and may mask significant ones.

Traditional feature subset selection methods are sequential and based on greedy heuristics. For example, sequential forward selection (SFS) starts with an empty subset and iteratively adds some features, whereas the sequential backward selection (SBS) starts with the full feature set and iteratively removes features [61.47]. An important drawback of these methods is that they consider one feature at a time, ignoring possible interactions between features. Fairly recently, more advanced methods such as evolutionary algorithms have been proposed to explore the space of feature subsets [61.35, 48].

Overview

Evolutionary algorithms for feature selection in bioinformatics are rarely used in a filter approach as such approaches ignore the effects of the selected feature subset on the performance of the classifier and do not consider existing correlations between features. Hence,

Table 61.2 Overview of evolutionary feature selection applications in bioinformatics

Application	EA	Approach	Classifier	Evaluation function	Encoding	Operators	Reference
Microarray (Cancer)	GA	W	KNN	CA (LOOCV)	Binary		[61.33]
Mass spectra (Cancer)	GA	W	KNN	CA (LOOCV)	Discrete		[61.34]
Microarray (Cancer)	Hybrid GA	W	KNN	CA (LOOCV), # features	Binary	Rank-based RWS, m-point crossover	[61.35]
Microarray	GA	W	SVM	Sensitivity, specificity, geometric mean	Binary	RWS, 1-point/2-point crossover	[61.36]
Microarray (Cancer)	GA	W	AP/SVM	CA (LOOCV)		SUS and RWS, Uniform and 1-point crossover	[61.37]
Microarray (Cancer)	GA	W	SVM	CA (10-fold), # features, feature cost	Binary	RWS, 2-point crossover, elitist replacement	[61.38]
Microarray (Cancer)	GA	W	SVM	CA (10-fold)	Binary	Specific SSOCF crossover	[61.39]
Microarray (Cancer)	GA	E	SVM	CA (10-fold), # features	Binary + coefficient vector	SUS, Specific crossover, Specific mutation	[61.40]
Microarray (Cancer)	GA	H	SVM	CA (LOOCV)	Binary	RWS, random 1-point crossover, multiuniform mutation	[61.41]
Microarray (Cancer)	GP			CA (10-fold), # features	Binary	Reproduction, homo(hetero)geneous crossover	[61.42]
Microarray (Cancer)	GP			AUC-ROC		Generational, tournament selection	[61.43]
Microarray (Cancer)	MOEA	W	GS	CA (LOOCV), # features	Binary	Elitist + ranking selection	[61.44]
Microarray (Cancer)	MOEA (NSGAII)	W	SVM	Sensitivity, specificity (10-fold) # features	Binary	SSOCF crossover, bit-flip mutation (uniform, one reduction, zero reduction)	[61.31]
Microarray (Diabetes/obesity)	Parallel GA			EH-DIALL, CLUMP	Discrete	Uniform crossover, specific mutation	[61.45]
Mass spectra (Regression)	GA	W	MLR PLS	RMSEP (data-splitting)	Binary		[61.46]

EA: evolutionary algorithm, MOEA: multiobjective evolutionary algorithm, GA: genetic algorithm, GP: genetic programming, GPSO: geometric particle swarm optimization. Approach: wrapper (W), embedded (E), hybrid (H). Classifiers: KNN: *k* nearest neighbor, SVM: support vector machine, AP: all paired, MLR: multiple linear regression, PLS: partial least square. Evaluation functions: CA: classification accuracy, AUC: area under curve, LOOCV: leave-one-out cross-validation, RMSEP: root-mean-square error of prediction. Operators: SUS: stochastic universal sampling, RWS: roulette wheel selection, SSOCF: subset size-oriented common features considering jointly the feature subset selection and the classification (or regression) process is more promising. This can be performed by three different approaches: embedded [61.40], hybrid [61.41], or more frequently wrapper approaches.

Any evolutionary algorithm can be used for feature selection. For example, some methods use genetic programming [61.42], but the vast majority uses genetic algorithms (GAs). Table 61.2 reports some works about evolutionary feature selection applications in bioinformatics. These works are described according to the application field, the evolutionary algorithm used, the approach (embedded, hybrid, wrapper), the classifier used (when one is used), the evaluation function and the specificities about encoding and operators. This table helps to identify tendencies of the use of evolutionary algorithms for feature selection in bioinformatics:

- *Encoding*: Solutions are mainly encoded with binary vectors of size n (initial number of features), each bit indicating if a feature is selected or not. However, in their studies, *Jourdan* et al. [61.45] and *Li* et al. [61.34] propose to use a discrete vector encoding, where each solution is described by the list of the selected features that is particularly well adapted for large datasets as encountered in bioinformatics.
- *Evaluation functions*: As explained before, evolutionary algorithms are mainly used jointly with a classification algorithm such as KNN [61.35] or SVM [61.36]. Using such a classifier, allows us to evaluate the potential of the selection to lead to a good classification by the computation of the classification accuracy. This accuracy can be computed with various methodologies such as k-fold cross-validation (10-fold, for example), leave-one-out

cross-validation (LOOCV) or bootstrap methodology. The 0.632 bootstrap has been proven to be the best estimator in [61.49], but the drawback of this method is its computational cost in comparison to LOOCV. For this reason, most of the authors use LOOCV which is fast and almost unbiased [61.35]. When dealing with larger datasets, 10-fold cross-validation can also be used [61.38].

The evaluation function can also take into account other parameters such as the number of features [61.39]. In this context (feature subset size minimization and performance maximization), feature selection can be viewed as a multiobjective optimization problem [61.31, 44].

- *Operators*: In terms of operators, some works deal with specific ones [61.39], but classical operators are mainly used. For example, we may cite the SUS or the roulette wheel selection (RWS) [61.37] for the selection, the 1-point or 2-point crossovers [61.36] for the evolution of solutions, the bit-flip mutation etc.

Feature selection is often used for classification, in order to predict a discrete trait and to classify samples (disease or not, for example). However, to predict a quantitative trait (a value indicating the good disposition for a treatment, for example), regression is used instead of classification. The problem is the same, as if too many features are available, including nonsignificant ones, the regression method will have difficulties to give good results. Hence, feature selection may also be used in a regression context. For example, *Broadhurst* et al. combined a genetic algorithm with a multiple linear regression (MLR) or with a partial least square (PLS) regression on a mass spectrometry problem [61.46].

61.4 Clustering by Evolutionary Algorithm in Bioinformatics

61.4.1 Clustering

Task Description

Clustering or unsupervised classification aims at decomposing or partitioning a (usually multivariate) dataset into groups so that objects in a group are similar to each other, and are as different as possible from objects of other groups. A survey of clustering algorithms can be found in [61.50]; thus we will just introduce generalities below.

Clustering techniques can be broadly divided into three main types: partitional, hierarchical, and overlap-ping. Partitional and hierarchical clusterings produce a hard partition of data as an object must belong to one and a single group, whereas in overlapping clustering objects may belong to several groups. In clustering, the number of groups can be known and fixed before realizing the clustering or must be determined directly by the algorithm.

For partitional-based methods, the most common algorithm is k-means [61.51], which is often described as a local search. For hierarchical clustering, two distinct types of hierarchical methods are identifiable: The agglomerative ones and the divisive ones.

Use in Bioinformatics

Clustering is the most popular method currently used in the first step of gene expression matrix analysis. Clustering is appropriate when there is no a priori knowledge about the data. In such circumstances, the only possible approach is to study the similarity between different samples or experiments. There are two straightforward ways to study the gene expression matrix: comparing expression profiles of genes by comparing rows in the expression matrix and comparing expression profiles of samples by comparing columns in the matrix. By comparing rows, we may find similarities or differences between different genes and thus conclude about the correlation between the two genes. If we find that two rows are similar, we can hypothesize that the respective genes are co-regulated and possibly functionally related. By comparing samples, we can find which genes are differentially expressed in different situations.

61.4.2 Evolutionary Approaches for Clustering in Bioinformatics

Motivations

Evolutionary clustering has been particularly used in bioinformatics as datasets are particularly large and classical methods are inefficient as they often lead to suboptimal solutions. A good survey of the use of evolutionary algorithms for clustering can be found in [61.52]. The authors proposed a classification of algorithms taking into consideration different aspects of evolutionary data clustering:

- Fixed or variable number of clusters
- Cluster-oriented or nonoriented operators
- Context-sensitive or context-insensitive operators
- Binary, integer, or real encoding
- Centroid-based, medioid-based, label-based, tree-based, or graph-based representations.

Other surveys can be found on genetic based [61.53] and on multiobjective clustering [61.54, 55].

Overview

Evolutionary algorithms for clustering bioinformatics data are applied to both, fixed number of clusters and variable number of clusters. The majority of the applications concerns microarray data. In Table 61.3, some works are presented through important components (application field, evolutionary algorithm used, fixed or variable number of clusters (k is known or not?), etc.). Here below an attempt to describe tendencies of existing methods is proposed by separating the two cases:

fixed or variables number of clusters. At the end, information will be given about biclustering that is more and more used in bioinformatics.

Fixed Number of Clusters. The number of clusters can be fixed before finding a clustering model through evolutionary algorithms. Therefore, the number of clusters (often denoted by k) can be fixed by an expert of the domain (here a biologist for example) or by using some specific criteria like a naive criterion $k \approx \sqrt{\frac{n}{2}}$, where n is the number of objects, or more specific ones based on an information criterion approach such as the Akaike information criterion (AIC), Bayesian information criterion (BIC), or the deviance information criterion (DIC):

- *Encoding*: For a fixed number of clusters, there exist several possible encodings: binary, integer, and real. For binary encoding, both prototype or partition could be realized. For integer encoding, two usual representations are used: *label-based encoding* where each gene represents an object and the value indicates the label of the cluster it is assigned to [61.60]; the *medioid-based encoding* represents the prototype of each cluster (the object that depicts the cluster). *The real encoding* is used to represent the coordinates of the center of each cluster and corresponds to the *centroid-based representation* [61.57, 62, 63].

- *Fitness function*: One other specific component for evolutionary clustering of bioinformatics data is the fitness function. Many clustering validity criteria exist and can be adapted to measure the quality of a solution of an evolutionary algorithm. Some examples are: minimization of the sum of within-cluster distances, minimization of the sum of the squared Euclidian distance of the objects from their respective cluster means [61.57], minimization of the distortion of the cluster (intracluster diversity), sum of the within cluster distances, etc.

- *Operators*: Concerning operators, some authors use classical operators like the 1-point crossover but a lot of articles show the drawbacks of classical genetic operators [61.57] and prefer to use context-sensitive operators [61.56].

Variable Number of Clusters.

- *Encoding*: For a variable number of clusters, where evolutionary algorithms aim at optimizing both the number of clusters and the partition of objects, the previously mentioned representations can be used

Table 61.3 Overview of applications of evolutionary clustering in bioinformatics

Application	EA	K?	Evaluation function	Encoding	Operators	Reference
Microarrays	GA	Y	Interestingness measure	Set of clusters + label of objects	Specific crossover, specific mutation	[61.56]
Microarrays (HL-60, HD-4cl, Yeast)	Memetic (GA)	Y	Minimum sum of square	Centroid	Kmean (LS), Uniform + specific crossover, split mutation	[61.57]
Microarrays (AD400_10_10, Yeast, Human, Rats)	GA	N	XB (Cluster validity index)	Centroid	RWS, specific crossover, mutation value	[61.58]
Microarrays	MOEA	N	Overall deviation + connectivity	Set of clusters	Binary tournament, specific crossover, no mutation	[61.59]
Microarrays	GA	N	Silhouette + K	Label	Mutation: split + eliminate a cluster, specific crossover	[61.60]
Microarrays (Lung + Leukemia)	GA	N	Silhouette + VRC	Label + K + distance + centroid	Centroid-based crossover	[61.61]
Microarrays	GA	N	Bayesian validation	Centroid	RWS, 1-point crossover, mutation value	[61.62, 63]
Protein structure	Chaotic GA	N	Max clustering coefficient	Binary	Specific crossover, specific mutation	[61.64]
Protein–Protein functional interaction	MOEA	N	Cluster size + 3 problem related functions	Centroid	No crossover, mutation: split + delete, merge	[61.65]

K?: is the number of clusters known? (Y: yes, N: no), MOEA: multiobjective evolutionary algorithm, GA: genetic algorithm, K: number of clusters, VCR: variance ratio criterion, XB: Xie-Beni cluster validity index, LS: local search, RWS: roulette wheel selection

but there are also some new ones. For example, the number of clusters can be stored in the representation [61.61]. There are also some rule-based representations [61.66], graph-based representations...

- *Operators*: As the encodings can be more complex than in the case of a fixed number of clusters [61.61], operators are adapted to the representation and the context of clustering.
- *Fitness function*: Concerning the evaluation, the authors often use criteria of validity of clustering [61.67]. We can also observe that the silhouette coefficient is often used to evaluate the quality of a clustering [61.60]. The authors can also add some problem-related functions as in [61.65].

Biclustering. Biclustering, (also called co-clustering or two-mode clustering) has for objective to compute biclusters (or co-clusters) that are associations of (possibly overlapping) sets of objects with sets of features. A biclustering algorithm computes simultaneously linked partitions on both rows and columns. Many formulations of the biclustering problem have been proposed, such as hierarchical model, biclustering model, and pattern-based model. The term biclustering has been introduced by *Cheng* and *Church* in [61.68]. Up to now, in the context of bioinformatics, biclustering approaches have been proposed mainly to deal with microarray data [61.69]. As clusters may overlap in the two dimensions of the matrix and no constraint is given about their size, it may be possible to find a very large number of significant biclusters. Hence, to have a concise description of the data through biclusters, the size aspect is often considered as an additional objective. This leads to multiobjective models for which MOEAs have been proposed [61.70, 71].

61.5 Conclusion

Bioinformatics research generates a lot of data and knowledge extracted from this data is still basic. Much more knowledge could be discovered with the proposition of new data mining methods. Many of these data mining problems can be modelized as combinatorial optimization problems and efficient algorithms such as evolutionary algorithms can be used to explore the huge search space of these problems. Some research has been conducted in this sense and the aim of this chapter was to present the tendencies of these works. It shows that some promising results have been obtained for several applications in bioinformatics.

However, there is much room for future research since the problems in bioinformatics presented in this chapter require even more effective approaches to gain important knowledge from the biological and biomedical experiments. In particular, information about the domain under study is still underutilized within research methods. Biological aspects should be more present and this may be done thanks to a more accurate modeling or the incorporation of biological concepts within

evaluation functions, for example. This could lead to multiobjective modelizations of these problems, where classical data mining criteria are jointly used with biological ones. Some interesting works on multiobjective optimization in bioinformatics and computational biology are reported in [61.72].

Another interesting perspective is the cooperation between methods coming from different domains. Indeed, several communities are working of these problematics and each of them acquired experience that can be exploited by making them cooperating.

References

61.1   D. Corne, C. Dhaenens, L. Jourdan: Synergies between operations research and data mining: The emerging use of multi-objective approaches, Eur. J. Oper. Res. **221**(3), 469–479 (2012)

61.2   F. Valafar: Pattern recognition techniques in microarray data analysis, Ann. N. Y. Acad. Sci. **980**(1), 41–64 (2002)

61.3   J.H. Moore, F.W. Asselbergs, S.M. Williams: Bioinformatics challenges for genome-wide association studies, Bioinformatics **26**(4), 445–455 (2010)

61.4   R. Agrawal, T. Imielinski, A.N. Swami: Mining association rules between sets of items in large databases, Proc. 1993 ACM SIGMOD Int. Conf. Manag. Data (ACM, New York 1993) pp. 207–216

61.5   R. Agrawal, R. Srikant: Fast algorithms for mining association rules in large databases, VLDB '94: Proc. 20th Int. Conf. Very Large Data Bases (Morgan Kaufmann, 1994) pp. 487–499

61.6   C. Borgelt: Efficient implementations of a priori and eclat, Proc. 1st IEEE ICDM Workshop Freq. Item Set Min. Implement. (FIMI 2003) (2003), p. 90

61.7   Y. Ye, C.-C. Chiang: A parallel apriori algorithm for frequent itemsets mining, Proc. 4th Int. Conf. Softw. Eng. Res. Manag. Appl. (2006) pp. 87–94

61.8   M.J. Zaki: Parallel sequence mining on shared-memory machines, J. Parallel Distrib. Comput. **61**(3), 401–426 (2001)

61.9   G. Atluri, R. Gupta, G. Fang, G. Pandey, M. Steinbach, V. Kumar: Association analysis techniques for bioinformatics problems, Proc. 1st Int. Conf. Bioinform. Comput. Biol. (BICoB '09) (Springer, Berlin, Heidelberg 2009) pp. 1–13

61.10   P. Carmona-Saez, M. Chagoyen, A. Rodriguez, O. Trelles, J. Carazo, A. Pascual-Montano: Integrated analysis of gene expression by association rules discovery, BMC Bioinformatics **7**(1), 54 (2006)

61.11   M. Khabzaoui, C. Dhaenens, E.-G. Talbi: A multicriteria genetic algorithm to analyze microarray data, Evol. Comput., CEC2004. Congr., Vol. 2 (2004) pp. 1874–1881

61.12   L. Jourdan, M. Khabzaoui, C. Dhaenens, E.-G. Talbi: A hybrid evolutionary algorithm for knowledge discovery in microarray experiments. In: *Handbook of Bioinspired Algorithms and Applications*, ed. by S. Olariu, A.Y. Zomaya (CRC, London 2005) pp. 491–508

61.13   P. Lanzi: Learning classifier systems: Then and now, Evol. Intell. **1**, 63–82 (2008)

61.14   M. Stout, J. Bacardit, J.D. Hirst, R.E. Smith, N. Krasnogor: Prediction of topological contacts in proteins using learning classifier systems, Soft Comput. J. **13**(3), 245–258 (2009)

61.15   J. Bacardit, E.K. Burke, N. Krasnogor: Improving the scalability of rule-based evolutionary learning, Memet. Comput. **1**(1), 55–67 (2008)

61.16   R. Slowinski, S. Greco, B. Matarazzo: Rough sets in decision making. In: *Encyclopedia of Complexity and Systems Science*, ed. by R.A. Meyers (Springer, New York 2009) pp. 7753–7787

61.17   J. Komorowski, A. Øhrn, A. Skowron: *The ROSETTA Rough Set Software System* (Oxford Univ. Press, New York 2002), Chap. D.2.3.

61.18   H. Strömbergsson, P. Prusis, H. Midelfart, M. Lapinsh, J.E.S. Wikberg, J. Komorowski: Rough set-based proteochemometrics modeling of *G-protein-coupled* receptor-ligand interactions, Proteins: Struct. Funct. Bioinform. **63**(1), 24–34 (2006)

61.19   S. Vinterbo, A. Øhrn: Minimal approximate hitting sets and rule templates, Int. J. Approx. Reason. **25**(2), 123–143 (2000)

61.20   T. Fawcett: An introduction to ROC analysis, Pattern Recognit. Lett. **27**(8), 861–874 (2006)

61.21   Y.J. Cho, H. Kim, H.-B. Oh: Generating rules for predicting *MHC* class *I* binding peptide using *ANN* and knowledge-based *GA*, JDCTA: Int. J. Dig. Content Technol. Appl. **3**, 111–119 (2009)

61.22   G.L. Pappa, A.A. Freitas: Automatically evolving rule induction algorithms tailored to the prediction of postsynaptic activity in proteins, Intell. Data Anal. **13**, 243–259 (2009)

61.23 Z.R. Yang, G. Lertmemongkolchai, G. Tan, P.L. Fel-gner, R.W. Titball: A genetic programming approach for *Burkholderia pseudomallei* diagnostic pattern discovery, Bioinformatics **25**(17), 2256–2262 (2009)

61.24 X. Llorá, R. Reddy, B. Matesic, R. Bhargava: Towards better than human capability in diagnosing prostate cancer using infrared spectroscopic imaging, GECCO '07 Proc. 9th Annu. Conf. Genet. Evol. Comput. (2007)

61.25 A. Laegreid, T.R. Hvidsten, H. Midelfart, J. Komorowski, A.K. Sandvik: Predicting gene ontology biological process from temporal gene expression patterns, Genome Res. **13**(5), 965–979 (2003)

61.26 J. Bacardit, M.V. Butz: Data mining in learning classifier systems: Comparing *XCS* with *GAssist*. IWLCS 2003–2005, Lect. Notes Artif. Intell. **4399**, 282–290 (2007)

61.27 L. Geng, H.J. Hamilton: Interestingness measures for data mining: A survey, ACM Comput. Surv. (CSUR) **38**(3), 9 (2006)

61.28 J. Bacardit: Pittsburgh Genetic-Based Machine Learning in the Data Mining Era: Representations, Generalization, and Run-Time, Ph.D. Thesis (Universitat Ramon Llull, Barcelona 2004)

61.29 J. Casillas, P. Martínez, A. Benítez: Learning consistent, complete and compact sets of fuzzy rules in conjunctive normal form for regression problems, Soft Comput. Fus. Found. Methodol. Appl. **13**, 451–465 (2009)

61.30 Y.S. Kim, W.M. Street, F. Menczer: Feature selection in data mining. In: *Data Mining: Opportunities and Challenges*, ed. by J. Wang (Idea Group, Hershey 2002) pp. 80–105

61.31 J. García-Nieto, E. Alba, L. Jourdan, E.-G. Talbi: Sensitivity and specificity based multiobjective approach for feature selection: Application to cancer diagnosis, Inf. Process. Lett. **109**, 887–896 (2009)

61.32 Y. Saeys, I. Inza, P. Larraaga: A review of feature selection techniques in bioinformatics, Bioinformatics **23**(19), 2507–2517 (2007)

61.33 T.J. Umpai, S. Aitken: Feature selection and classification for microarray data analysis: Evolutionary methods for identifying predictive genes, BMC Bioinformatics **6**(1), 148 (2005)

61.34 L. Li, D.M. Umbach, P. Terry, J.A. Taylor: Application of the GA/KNN method to SELDI proteomics data, Bioinformatics **20**(10), 1638–1640 (2004)

61.35 I.-S. Oh, J.-S. Lee, B.-R. Moon: Hybrid genetic algorithms for feature selection, IEEE Trans. Pattern Anal. Mach. Intell. **26**(11), 1424–1437 (2004)

61.36 P. Xuan, M.Z. Guo, J. Wang, C.Y. Wang, X.Y. Liu, Y. Liu: Genetic algorithm-based efficient feature selection for classification of pre-miRNAs, Genet. Mol. Res. **10**(2), 588–603 (2011)

61.37 S. Peng: Molecular classification of cancer types from microarray data using the combination of genetic algorithms and support vector machines, FEBS Letters **555**(2), 358–362 (2003)

61.38 C.-L. Huang, C.-J. Wang: A GA-based feature selection and parameters optimization for support vector machines, Expert Syst. Appl. **31**(2), 231–240 (2006)

61.39 E.-G. Talbi, L. Jourdan, J. Garca-Nieto, E. Alba: Comparison of population based metaheuristics for feature selection: Application to microarray data classification, IEEE/ACS Int. Conf. Comput. Syst. Appl. (2008) pp. 45–52

61.40 J.C.H. Hernandez, B. Duval, J.-K. Hao: A genetic embedded approach for gene selection and classification of microarray data, Proc. 5th Eur. Conf. Evol. Comput. Mach. Learn. Data Min. Bioinform. (EvoBIO'07) (Springer, Berlin, Heidelberg 2007) pp. 90–101

61.41 E.B. Huerta, B. Duval, J.-K. Hao: A hybrid GA/SVM approach for gene selection and classification of microarray data, Lect. Notes Comput. Sci. **3907**, 34–44 (2006)

61.42 D.P. Muni, N.R. Pal, J. Das: Genetic programming for simultaneous feature selection and classifier design, IEEE Trans. Syst. Man Cybern. Part B **36**(1), 106–117 (2006)

61.43 J. Yu, J. Yu, A.A. Almal, S.M. Dhanasekaran, D. Ghosh, W.P. Worzel, A.M. Chinnaiyan: Feature selection and molecular classification of cancer using genetic programming, Neoplasia **9**(4), 292–303 (2007)

61.44 J. Liu, H. Iba, M. Ishizuka: Selecting informative genes with parallel genetic algorithms in tissue classification, Genome Inform. Ser. **9**, 14–23 (2001)

61.45 L. Jourdan, C. Dhaenens, E.-G. Talbi: Linkage disequilibrium study with a parallel adaptive GA, Int. J. Found. Comput. Sci. **16**(2), 241–260 (2004)

61.46 D. Broadhurst, R. Goodacre, A. Jones, J.-J. Rowland, D.B. Kelp: Genetic algorithms as a method for variable selection in multiple linear regression and partial least squares regression, with applications to pyrolysis mass spectrometry, Anal. Chim. Acta **348**, 71–86 (1997)

61.47 A.W. Whitney: A direct method of nonparametric measurement selection, IEEE Trans. Comput. **C-20**(9), 1100–1103 (1971)

61.48 M. Pei, E.D. Goodman, W.F. Punch: Feature extraction using genetic algorithms, Proc. 1st Int. Symp. Intell. Data Eng. Learn. (IDEAL), Vol. 98 (1998) pp. 371–384

61.49 U.M. Braga-Neto, E.R. Dougherty: Is cross-validation valid for small-sample microarray classification?, Bioinformatics **20**(3), 374–380 (2004)

61.50 R. Xu, D. Wunsch: Survey of clustering algorithms, IEEE Trans. Neural Netw. **16**, 645–678 (2005)

61.51 J.B. MacQueen: Some methods for classification and analysis of multivariate observations, Proc. 5th Berkeley Symp. Math. Stat. Probab. (1967) pp. 281–297

61.52 E.R. Hruschka, R.J. Campello, A.A. Freitas, A.C. de Carvalho: A survey of evolutionary algorithms for clustering, IEEE Trans. Syst. Man Cybern. Part C **39**(2), 133–155 (2009)

61.53   R.H. Sheikh, M.M. Raghuwanshi, A.N. Jaiswal: Genetic algorithm based clustering: A survey, 1st Int. Conf. Emerg. Trends Eng. Technol. ICETET '08. (2008) pp. 314–319

61.54   J. Handl, J. Knowles: An evolutionary approach to multiobjective clustering, IEEE Trans. Evol. Comput. **11**(1), 56–76 (2007)

61.55   J. Handl, J. Knowles: Evolutionary multiobjective clustering, Parallel Problem Solving Nat. **3242**, 1081–1091 (2004)

61.56   P.C. Ma, K.C. Chan, Y. Xin, D.K. Chiu: An evolutionary clustering algorithm for gene expression microarray data analysis, IEEE Trans. Evol. Comput. **10**(3), 296–314 (2006)

61.57   P. Merz, A. Zell: Clustering gene expression profiles with memetic algorithms, Proc. 7th Int. Conf. Parallel Problem Solving Nat. (PPSN VII) (Springer, London 2002) pp. 811–820

61.58   S. Bandyopadhyay, A. Mukhopadhyay, U. Maulik: An improved algorithm for clustering gene expression data, Bioinformatics **23**(21), 2859–2865 (2007)

61.59   K. Faceli, M. de Souto, D. de Araujo, A. de Carvalho: Multi-objective clustering ensemble for gene expression data analysis, Neurocomputing **72**(13–15), 2763–2774 (2009)

61.60   E. Hruschka, L. de Castro, R. Campello: Evolutionary algorithms for clustering gene-expression data, 4th IEEE Int. Conf. Data Min. (ICDM '04) (2004) pp. 403–406

61.61   M.C. Naldi, A. de Carvalho: Clustering using genetic algorithm combining validation criteria, Proc. 15th Eur. Symp. Artif. Neural Netw. (2007) pp. 139–147

61.62   H.S. Park, S.H. Yoo, S.B. Cho: Evolutionary fuzzy clustering algorithm with knowledge-based evaluation and applications for gene expression profiling, J. Comput. Theor. Nanosci. **2**(4), 524–533 (2005)

61.63   H.S. Park, S.B. Cho: Evolutionary fuzzy cluster analysis with bayesian validation of gene expression profiles, J. Intell. Fuzzy Syst. **18**(6), 543–559 (2007)

61.64   D. Hutchison, T. Kanade, J. Kittler, J.M. Kleinberg, F. Mattern, J.C. Mitchell, M. Naor, O. Nierstrasz, C. Pandu Rangan, B. Steffen, M. Sudan, D. Terzopoulos, D. Tygar, M.Y. Vardi, G. Weikum, H. Liu, J. Liu: Clustering protein interaction data through chaotic genetic algorithm. In: *Simulated Evolution and Learning*, Vol. 4247, ed. by T.-D. Wang, X. Li, S.-H. Chen, X. Wang, H. Abbass, H. Iba, G.-L. Chen, X. Yao (Springer, Berlin, Heidelberg 2006) pp. 858–864

61.65   J.J. Tapia, E.E. Vallejo, E. Morett: MOCEA: A multi-objective clustering evolutionary algorithm for inferring protein-protein functional interactions, Proc. 11th Annu. Conf. Genet. Evol. Comput. (2009) pp. 1793–1794

61.66   I.A. Sarafis, P.W. Trinder, A.M.S. Zalzala: NOCEA: A rule-based evolutionary algorithm for efficient and effective clustering on massive high-dimensional databases (invited paper), Int. J. Appl. Soft Comput. **7**(3), 668–710 (2007)

61.67   J.J. Tapia, E. Morett, E.E. Vallejo: A clustering genetic algorithm for genomic data mining. In: *Foundations of Computational Intelligence (4)*, Studies in Computational Intelligence, Vol. 204, ed. by A. Abraham, A.E. Hassanien, A.C.P.L. de Ferreira Carvalho (Springer, Berlin, Heidelberg 2009) pp. 249–275

61.68   Y. Cheng, G.M. Church: Biclustering of expression data, Proc. 8th Int. Conf. Intell. Syst. Mol. Biol. (ISMB 2000), San Diego (2000) pp. 93–103

61.69   F. Divina, J.S. Aguilar-Ruiz: Biclustering of expression data with evolutionary computation, IEEE Trans. Knowl. Data Eng. (2006) p. 18

61.70   S. Mitra, H. Banka: Multi-objective evolutionary biclustering of gene expression data, Pattern Recognit. **39**(12), 2464–2477 (2006)

61.71   K. Seridi, L. Jourdan, E.-G. Talbi: Multi-objective evolutionary algorithm for biclustering in microarrays data, IEEE Congr. Evol. Comput. (2011) pp. 2593–2599

61.72   J. Handl, D.B. Kell, J. Knowles: Multiobjective optimization in bioinformatics and computational biology, IEEE/ACM Trans. Comput. Biol. Bioinform. **4**(2), 279–292 (2007)

Part E | 61

62. Integration of Metaheuristics and Constraint Programming

Luca Di Gaspero

A promising research line in the optimization community regards the hybridization of exact and heuristics methods. In this chapter we survey the specific integration of two complementary optimization paradigms, namely Constraint Programming, for the exact part, and metaheuristics.

62.1 Constraint Programming and Metaheuristics

Constraint programming (CP) [62.1, 2] is an effective methodology for the solution of combinatorial problems that has been successfully applied in many domains. In a nutshell, CP is a declarative programming paradigm based on the idea of describing the relations (i. e., constraints) between variables that must hold in all solutions of the combinatorial problem at hand. For example, in the solution to a Sudoku puzzle, the numbers to be placed must be unique with respect to columns, rows, and blocks of the board.

CP has an interdisciplinary nature, since it relies on contributions and methods from the communities of logic programming (LP), artificial intelligence (AI), and operations research (OR). Indeed, the simple declarative modeling language of CP, consisting of variables and constraints, is very similar to those available in classical LP languages such as Prolog. The solution method features constraint propagation which, in its essence, is a reasoning or inference procedure typical of AI. Finally, especially for optimization problems, the solution process makes use of OR inspired branch and bound procedures and/or of dedicated OR solvers for specific types of variables/constraints (e.g., the simplex method for real variables and linear constraints).

A CP model is an encoding of the problem statement using the basic CP building blocks, i. e., variables and constraints. Once a CP model of the problem under consideration has been stated, a CP solver is used to systematically search the solution space by alternating deterministic phases (constraint propagation) and nondeterministic phases (variable assignment, tree search), thus exploring implicitly or explicitly the whole search space. To this respect, CP belongs to the family of *complete* (or *exact*) solution methods. In other words, CP guarantees finding the (optimal) solution of the problem or proving that the problem is not satisfiable.

A different approach is usually taken by metaheuristics [62.3], such as local search [62.4], evolutionary algorithms [62.5], and ant colony optimization [62.6], just to name a few. These methods are *incomplete*, since they rely on heuristic information to focus on *interesting areas* of the search space and, in general, do not explore it entirely but are stopped after a given time limit. As a consequence, these algorithms do not guarantee finding the (optimal) solution, trading completeness for a (possibly) greater (empirical) efficiency in the solution process.

Just looking at completeness, it seems that the clear choice for solving combinatorial problems would

Part E | 62.1

be to always prefer CP over metaheuristics as the solution method. However, in practice completeness is hindered by the high computational effort due to the worst case complexity of the problems considered (usually NP-complete or NP-hard). Therefore, for practical purposes, also the execution of CP solvers is terminated before the whole search space has been explored and a number of heuristics is used to focus the search in the regions where it is more likely to find the solutions of the problem. Consequently, CP and metaheuristics could be seen as complementary approaches.

Although these two kinds of methods are have been individually studied by separated scientific communities (for historical reasons), in recent years we have witnessed an increasing interest in the integration of the methods. In many cases, indeed, each approach has its own strengths and weaknesses, and the general aim of method integration is to create hybrid algorithms that enhance the strengths of both approaches and (possibly) overcome some of the weaknesses. To this respect, *Yunes* maintains a web page listing a number of success stories of hybrid solution methods [62.7], that is, papers describing integrated approaches that outperform single optimization methods.

A number of conferences and workshops specifically aiming at bringing together researchers working on the integration of solution techniques for combinatorial problems have also recently started. Notable examples are the series of CP-AI-OR conferences [62.8, 9], started in 1999, and the Hybrid Metaheuristics workshops [62.10–16], started in 2004. The scope of these conferences is not limited to the integration of CP

techniques with metaheuristics, but they also consider hybridization among other methods.

Additionally, a few surveys on the integration of complete methods with metaheuristics have appeared in the literature [62.17–19]. However, these surveys either deal with a particular class of metaheuristics (i. e., local search) [62.17, 19] and/or with a different class of complete methods (integer linear programming) [62.17, 18]. *Jourdan* et al. [62.20] also took CP methods into account, but they provide mostly a taxonomy of cooperation between optimization methods rather than surveying the specific integrations. *Wallace* and *Azevedo* et al. [62.21, 22] surveys hybrid algorithms, but from a constraint programming viewpoint and mainly in the settings of hybrid exact methods. In their recent review of hybrid metaheuristics *Blum* et al. [62.23] include a section on the integration of CP with local search and ant colony optimization (ACO). However, to the best of our knowledge, at present no specific survey on the integration of metaheuristics and constraint programming has been published in the literature. This work tries to overcome this lack and to review the different approaches specifically employed in the integration of CP methods within metaheuristics.

The chapter is organized as follows. In Sect. 62.2 the basic concepts of the constraint programming paradigm are introduced. They include modeling (Sect. 62.2.1), solution methods (Sect. 62.2.2), and CP systems (Sect. 62.2.3). The integration of CP with metaheuristics is presented in Sect. 62.3, which is organized on the basis of the metaheuristic type involved in the integration. Finally, in Sect. 62.4 some conclusions are drawn.

62.2 Constraint Programming Essentials

In this section, we will briefly describe the essential concepts of CP, which are needed to understand the following sections. The readers interested in a more detailed introduction to CP are referred to the book of *Apt* [62.1] and to the recent comprehensive *Handbook of Constraint Programming* [62.2].

In order to apply constraint programming to a combinatorial problem one first needs to model it through the specific formalism of constraint satisfaction or constrained optimization problems. Afterwards, the model can be solved by a CP solver, which alternates the analysis of constraints with tree search. Let us review these basic concepts.

62.2.1 Modeling

Constraint satisfaction problems (CSPs) are a useful formalism for modeling many real-world problems, either discrete or continuous. Remarkable examples are planning, scheduling, timetabling, just to name a few.

A CSP is generally defined as the problem of associating values (taken from a set of domains) to variables subject to a set of constraints. A solution of a CSP is an assignment of values to all the variables so that the constraints are satisfied. In some cases not all solutions are equally preferable and we can associate a cost function to the variable assignments. In these cases, we talk about constrained optimization problems (COPs), and we are looking for a solution that (without loss of generality) minimizes the cost value. These concepts are formally introduced in the following.

Constraint Satisfaction Problems

Given:

- $X = \{x_1, \ldots, x_k\}$ is a set of *variables*.
- $\mathcal{D} = \{D_1, \ldots, D_k\}$ is a set of *domains* associated to the variables. In other words, each variable x_i can assume value d_i if and only if $d_i \in D_i$.
- C is a set of *constraints*, i.e., mathematical relations over $\mathbf{Dom} = D_1 \times \cdots \times D_k$.

We say that a tuple $\langle d_1, \ldots, d_k \rangle \in \mathbf{Dom}$ *satisfies* a constraint $C \in C$ if and only if $\langle d_1, \ldots, d_k \rangle \in C$.

A *constraint satisfaction problem* (CSP) $\mathcal{P}$, described by the triple $\langle X, \mathcal{D}, C \rangle$, is the problem of finding the tuples $\bar{d} = \langle d_1, \ldots, d_k \rangle \in \mathbf{Dom}$ that satisfy every constraint $C \in C$. Such tuples are called *solutions* of the CSP, and the set of solutions of $\mathcal{P}$ is denoted by $\mathbf{sol}(\mathcal{P})$.

$\mathcal{P}$ is said to be *consistent* or *satisfiable* if and only if $\mathbf{sol}(\mathcal{P}) \neq \emptyset$.

Notice that, depending on the modeling of the combinatorial problem at hand, we could be interested in determining different properties of the CSP. In the extreme case, for example, one could just want to know whether the problem is satisfiable, regardless of the actual solutions. The most common case is to search and provide a single solution to the problem, whereas sometimes one could be interested in all the solutions.

Constrained Optimization Problems

A *constrained optimization problem* (COP) $\mathcal{O} = \langle X, \mathcal{D}, C, f \rangle$ is a CSP $\mathcal{P} = \langle X, \mathcal{D}, C \rangle$ with an associated *objective* function $f : \mathbf{sol}(\mathcal{P}) \to E$, where $\langle E, \leq \rangle$ is a well-ordered set (typically, E is one of the sets $\mathbb{N}, \mathbb{Z}, \mathbb{R}$).

Differently from the previous case, the tuples $\bar{d} \in \mathbf{sol}(\mathcal{O})$ that satisfy every constraint are called *feasible solutions*, and the set of these tuples is usually assumed to be non-empty. A *solution* of the COP $\mathcal{O}$ is a feasible solution $\bar{e} \in \mathbf{sol}(\mathcal{O})$ for which the value of the objective function f is minimized, i.e.,

$$\forall \bar{d} \in \mathbf{sol}(\mathcal{O}) f(\bar{e}) \leq f(\bar{d}) .$$

Observations

A few observations about this formalism are worth noting. First, notice that the general framework does not impose any restriction on either the type of domains and constraints or on the form of the objective function that can be used to express the problem. The basic type of domain is a finite set of integer values (also known as a *finite domain*), but there are other possibilities that enhance the expressive power of the modeling framework and capture some combinatorial substructures of the problem more naturally. For example, it is possible to deal with variables whose values are finite (multi)sets, (hyper)graphs, real valued intervals, or resources of a scheduling problem. Moreover, also the kind of constraints that can be employed is quite rich and includes arithmetic constraints, set constraints, permutation, counting and other types of combinatorial constraints, resource scheduling constraints, path constraints on graphs, and constraints expressible through regular expressions, just to name a few possibilities (see [62.24] for a comprehensive set of constraints and their implementation in actual CP systems).

These features clearly make the modeling phase easier and more precise with respect to other formalisms such as integer linear programming. Indeed, part of the combinatorial structure of the problem can be directly captured by the use of complex domains/constraints and, as for the objective function, there is no general limitation on its form, in particular, there is no assumption of linearity.

Another important point to be noticed regards the role of constraints. Differently from other modeling formalisms, which distinguish between constraints that *must* be satisfied (called *hard* constraints) and that *should preferably* be satisfied (*soft* constraints), in the original CSP/COP framework constraints are all hard and the solution methods, described in the following section, consider it mandatory to satisfy all of them. There have been several attempts in the CP literature to include soft constraints in the general framework (see, e.g., [62.25] for a review) but the most common way to handle them is to include a measure of their violation in the objective function of the problem.

62.2.2 Solution Methods

CP solution methods basically exploit a form of tree search that interleaves a branching phase with an analysis of constraints called constraint propagation. These two components are described in the following.

Branching and Tree Search

Once the combinatorial problem has been modeled as a CSP or a COP, CP solves it by constructing a solution by a process that exploits a non-deterministic *variable assignment*, where one value is selected together with one value in its current domain. This phase is also called *labeling* using (constraint) logic programming terminology, and a solution to the problem is a complete labeling. The process proceeds by recursively checking whether the current labeling can be extended to a consistent solution or, in the negative case, undoing the current assignment.

The pseudocode of the procedure, called (chronological) *backtracking*, is given in Algorithm 62.1. The procedure is at first called with the full set of variables and empty labeling as follows Backtracking($X, \emptyset, C, \mathbf{Dom}$). The procedure performs an implicit form of tree search, where a *branch* is identified by the selection of one variable (a node of the search tree) and all the possible values for that variable (the edges).

Note that, at each step of the recursive procedure, the choice of the variable and the value to branch on is non-deterministic. Therefore, these choices are susceptible to heuristics to enhance performances.

In addition, there are also other possibilities to define a branching rule. For example, instead of selecting a possible value for the variable selected (i. e., the assignment $x_i := v$), the branching rule could split the domain of a given variable x_i in two by selecting a value $v \in D_i$ and adding the constraint $x_i \leq v$ on one branch and $x_i > v$ on the other.

Consistency and Constraint Propagation

The check for solution *consistency* does not need all the variables to be instantiated, in particular for detecting the unsatisfiability of the CSP with respect to some constraint. For example, in Algorithm 62.2, the most straightforward implementation of the procedure Consistent($L, C, \mathbf{Dom}$) is reported. The procedure simply checks whether the satisfiability of a given constraint can be ascertained according to the current labeling (i. e., if all of the constraint variables are assigned). However, the reasoning about the current labeling with respect to the constraints of the problem and the domains of the unlabeled variables does not necessarily need all the variables appearing in a constraint to be instantiated. Moreover, the analysis can prune (as a side effect) the domains of the unlabeled variables while preserving the set of solutions $\mathbf{sol}(\mathcal{P})$, making the exploration of the subtree more effective. This phase is called *constraint propagation* and is interleaved with the variable assignment. In general, the analysis of each constraint is repeated until a fixed point for the current

situation is achieved. In the case that one of the domains becomes empty consistency cannot be achieved and, consequently, the procedure returns a fail.

Different notions of consistency can be employed. For example, one of the most common and most studied notions is hyper-arc consistency [62.26]. For a k-ary constraint C it checks the compatibility of a value v in the domain of one of the variables with the currently possible combinations of values for the remaining $k-1$ variables, pruning v from the domain if no supporting combination is found. The algorithms that maintain hyper-arc consistency have a complexity that is polynomial in the size of the problem (measured in terms of number of variables/constraints and size of domains). Other consistency notions have been introduced in the literature, each having different pruning capabilities and computational complexity, which are, usually, proportionally related to their effectiveness.

One of the major drawbacks of (practical) consistency notions is that they are *local* in nature, that is, they just look at the current situation (partial labeling and current domains). This means that it would be impossible to detect future inconsistencies due to the interaction of variables. A basic technique, called *forward checking*, can be used to mitigate this problem. This method exploits a one-step *look-ahead* with respect to the current assignment, i. e., it simulates the assignment of pair of variables, instead of a single one, thus evaluating the next level of the tree through a consistency notion. This technique can be generalized to several other problems.

Algorithm 62.1 Backtracking (U, L, C, Dom)
1: **if** $U = \emptyset$ **then**
2:    **return** L
3: **end if**
4: pick variable $x_i \in U$
      /*possibly x_i is selected non-deterministically*/
5: **for** $v \in D_i$ /*Try to label x_i with value v*/ **do**
6:    $\mathbf{Dom}' \leftarrow \mathbf{Dom}$
7:    **if** Consistent($L \cup \{x := v\}, C, \mathbf{Dom}'$)
         /*consistency notions can be different and have side effects on **Dom***/ **then**
8:       $r \leftarrow$ Backtracking($U \setminus \{x\}, L \cup \{x := v\}, C, \mathbf{Dom}'$)
9:       **if** $r \neq$ fail **then**
10:         **return** r /*a consistent assignment has been found for the variables in $U \setminus \{x_i\}$ with respect to $x_i := v$*/
11:      **end if**
12:   **end if**
13: **end for**

14: **return** fail /*backtrack to the previous variable (no consistent assignment has been found for x_i)*/

---

*Algorithm 62.2 Consistent (L, C, **Dom**)*

1: **for** $C \in C$ **do**
2:     **if** all variables in C are labeled in $L \wedge C$ is not satisfied by L **then**
3:         **return** fail
4:     **end if**
5: **end for**
6: **return** true

---

*Algorithm 62.3 BranchAndBound (U, L, C, **Dom**, f, b, L_b)*

1: **if** $U = \emptyset$ **then**
2:     **if** $f(L) < b$ **then**
3:         $b \leftarrow f(L)$ $L_b \leftarrow L$
4:     **end if**
5: **else**
6:     pick variable $x_i \in U$
        /*possibly x_i is selected non-deterministically*/
7:     **for** $v \in D_i$ /*Try to label x_i with value v*/ **do**
8:         **Dom'** $\leftarrow$ **Dom**
9:         **if**   Consistent($L \cup \{x := v\}, C, $**Dom'**$) \wedge$ bound($f, L \cup \{x := v\}, $**Dom'**$) < b$
        /*additionally verify whether the current solution is bounded*/ **then**
10:             BranchAndBound($U \setminus \{x\}, L \cup \{x := v\}, C,$ **Dom'**$, f, b, L_b$)
11:         **end if**
12:     **end for**
13: **end if**

Branch and Bound

In the case of a COP, the problem is solved by exploring the set **sol**$(\mathcal{O})$ in the way above, storing the best value for f found as sketched in Algorithm 62.3. However, a constraint analysis (bound($f, L \cup \{x := v\}, $**Dom'**)) based on a partial assignment and on the best value already computed, might allow to sensibly prune the search tree. This complete search heuristic is called (with a slight ambiguity with respect to the same concept in operations research) *branch and bound*.

62.2.3 Systems

A number of practical CP systems are available. They mostly differ with regards to the targeted pro-

gramming language and modeling features available. For historical reasons, the first constraint programming systems were built around a Prolog system. For example, SICStus Prolog [62.27], was one of the first logic programming systems supporting constraint programming which is still developed and released under a commercial license. Another Prolog-based system specifically intended for constraint programming is ECL^iPS^e [62.28], which differently from SICStus Prolog is open source. Thanks to their longevity, both systems cover many of the modeling features described in the previous sections (such as different type of domains, rich sets of constraints, etc.).

Another notable commercial system specifically designed for constraint programming is the ILOG CP optimizer, now developed by IBM [62.29]. This system offers modeling capabilities either by means of a dedicated modeling language (called OPL [62.30]) or by means of a callable library accessible from different imperative programming languages such as C/C++, Java, and C#. The modeling capabilities of the system are mostly targeted to scheduling problems, featuring a very rich set of constructs for this kind of problems. Interestingly, this system is currently available at no cost for researchers through the IBM Academic Initiative.

Open source alternatives that can be interfaced with the most common programming languages are the C++ libraries of Gecode [62.31], and the Java libraries of Choco [62.32]. Both systems are well documented and constantly developed.

A different approach has been taken by other authors, who developed a number of modeling languages for constraint satisfaction and optimization problems that can be interfaced to different type of general purpose CP solvers. A notable example is MiniZinc [62.33], which is an expressive modeling language for CP. MiniZinc models are translated into a lower level language, called FlatZinc, that can be compiled and executed, for example, by Gecode, ECL^iPS^e or SICStus prolog.

Finally, a mixed approach has been taken by the developers of Comet [62.34]. Comet is a hybrid CP system featuring a specific programming/modeling language and a dedicated solver. The system has been designed with hybridization in mind, and, among other features, it natively supports the integration of metaheuristics (especially in the family of local search methods) with CP.

62.3 Integration of Metaheuristics and CP

Differently from *Wallace* [62.21], we will review the integration of CP with metaheuristics from the perspective of metaheuristics, and we classify the approaches on the basis of the type of metaheuristic employed. Moreover, following the categorization of *Puchinger* and *Raidl* [62.18], we are mostly interested in reviewing the *integrative* combinations of metaheuristics and constraint programming, i. e., those in which constraint programming is embedded as a component of a metaheuristic to solve a subproblem or vice versa.

Indeed, the types of collaborative combinations are either straightforward (e.g., collaborative-sequential approaches using CP as a constructive algorithm for finding a feasible initial solution of a problem) or rather uninvestigated (e.g., parallel or intertwined hybrids of metaheuristics and CP).

62.3.1 Local Search and CP

Local search methods [62.4] are based on an iterative scheme in which the search moves from the current solution to an adjacent one on the basis of the exploration of a *neighborhood* obtained by perturbing the current solutions.

The hybridization of constraint programming with local search metaheuristics is the most studied one and there is an extensive literature on this subject.

CP Within Local Search
The integration of CP within local search methods is the most mature form of integration. It dates back to the mid 1990s [62.35], and two main streams are identifiable to this respect. The first one consists in defining the search of the candidate neighbor (e.g., the best one) as a constrained optimization problem. The neighborhoods induced by these definitions can be quite large, therefore, a variant of this technique is known by the name of *large neighborhood search* (LNS) [62.36]. The other kind of integration, lately named *constraint-based local search* (CBLS) [62.34], is based on the idea of expressing local search algorithms by exploiting constraint programming primitives in their control (e.g., for constraint checks during the exploration of the neighborhood) [62.37]. In fact, the two streams have a non-empty intersection, since the CP primitives employed in CBLS could be used to explore the neighborhood in a LNS fashion. In the following sections we review some of the work in these two areas.

A few surveys on the specific integration between local search and constraint programming exist, for example [62.38, 39].

Large Neighborhood Search. In LNS [62.36, 40] an existing solution is not modified just by applying small perturbations to solutions but a large part of the problem is perturbed and searched for improving solutions in a sort of *re-optimization* approach. This part can be represented by a set $\mathcal{F} \subseteq X$ of *released* variables, called *fragment*, which determines the neighborhood relation $\mathcal{N}$. Precisely, given a solution $\bar{s} = \langle d_1, \ldots, d_k \rangle$ and a set $\mathcal{F} \subseteq \{X_1, \ldots, X_k\}$ of free variables, then

$$\mathcal{N}(s, \mathcal{F}) = \{\langle e_1, \ldots, e_k \rangle \in \mathbf{sol}(O) \\ : (X_i \notin \mathcal{F}) \to (e_i = d_i)\} \,.$$

Given $\mathcal{F}$, the neighborhood exploration is performed through CP methods (i. e., propagation and tree search).

The pseudocode of the general LNS procedure is shown in Algorithm 62.4. Notice that in the procedure there are a few hotspots that can be customized. Namely, one of the key issues of this technique consists in the criterion for the selection of the set $\mathcal{F}$ given the current solution $\bar{s}$, which is denoted by SelectFragment($\bar{s}$) in the algorithm. The most straightforward way to select it is to randomly release a percentage of the problem variables. However, the variables in $\mathcal{F}$ could also be chosen in a *structured* way, i. e., by releasing related variables simultaneously. In [62.41], the authors compare the effectiveness of these two alternative choices in the solution of a jobshop scheduling problem.

Also the upper bounds employed for the branch and bound procedure can be subject to a few design alternatives. A possibility, for example, is to set the bound value to $f(\bar{s}_b)$, the best solution value found that far, so that the procedure is forced to search at each step only for improving solutions. This alternative can enhance the technique when the propagation on the cost functions is particularly effective in pruning the domains of the released variables. At the opposite extreme, instead, the upper bound could be set to an infinite value so that a solution is searched regardless whether or not it is improving the cost function with respect to the current incumbent.

Moreover, another design point is the solution acceptance criterion, which is implemented by the AcceptSolution function. In general, all the classical local search solution acceptance criteria are applicable, obviously in dependence on the neighborhood selection criterion employed. For example, in the case of randomly released variables a Metropolis acceptance criterion could be adequate to implement a sort of simulated annealing.

Finally, the TerminateSearch criterion is one of those usually adopted in non-systematic search methods, such as the expiration of a time/iteration budget, either absolute or relative, or the discovery of an optimal solution.

Algorithm 62.4 LargeNeighborhoodSearch (X, C, **Dom**, *f)*

1: create a (feasible) initial solution $\bar{s}_0 = \langle d_1^0, \ldots, d_k^0 \rangle$
    /*possibly random or finding the first feasible solution of the full CP model*/
2: $\bar{s}_b \leftarrow \bar{s}_0$
3: $i \leftarrow 0$
4: **while** not TerminateSearch($i, \bar{s}_i, f(\bar{s}_i), \bar{s}_b$) **do**
5:    $\mathcal{F} \leftarrow$ SelectFragment($\bar{s}_i$)
      /*strategy for selecting the released variables*/
6:    $L \leftarrow \{x_j := d_j^i : x_j \notin \mathcal{F}\}$
7:    $U \leftarrow \mathcal{F}$
8:    BranchAndBound($U, L, C,$ **Dom**, $f,$ ChooseBounds($\bar{s}_i, \bar{s}_b$))
      /*neighborhood exploration*/
9:    **if** AcceptSolution(L) **then**
10:     $\bar{s}_{i+1} \leftarrow L$
11:     **if** $f(\bar{s}_{i+1}) < f(\bar{s}_b)$ **then**
12:      $\bar{s}_b \leftarrow \bar{s}_{i+1}$
13:     **end if**
14:    **else**
15:     $\bar{s}_{i+1} \leftarrow \bar{s}_i$
16:    **end if**
17:    $i \leftarrow i + 1$
18: **end while**
19: **return** $\bar{s}_b$

LNS has been successfully applied to routing problems [62.36, 42–45], nurse rostering [62.46], university course timetabling [62.47], protein structure prediction [62.48, 49], and car sequencing [62.50].

Cipriano et al. propose GELATO, a modeling language and a hybrid solver specifically designed for LNS [62.51–53]. The system has been tested on a set of benchmark problems, such as the asymmetric traveling salesman problem, minimum energy broadcast, and university course timetabling.

The developments of the LNS technique in the wider perspective of very large neighborhood search

(VLNS) was recently reviewed by *Pisinger* and *Ropke* [62.54]. *Charchrae* and *Beck* [62.55] also propose a methodological contribution to this area with some design principles for LNS.

Constraint-Based Local Search. The idea of encoding a local search algorithm by means of constraint programming primitives was originally due to *Pesant* and *Gendreau* [62.35, 56], although in their papers they focus on a framework that allows neighborhoods to be expressed by means of CP primitives. The basic idea is to extend the original CP model of the problem with a sort of surrogate model comprising a set of variables and constraints that intentionally describe a *neighborhood* of the current solution.

A pseudocode of CBLS defined along these lines is reported in Algorithm 62.5. The core of the procedure is at line 5, which determines the neighborhood model on the basis of the current solution. The main components of the neighborhood model are the new set of variables Y and constraints $C_{X,Y}$ that act as an interface of the neighborhood variables Y with respect to those of the original problem X. For example, the classical *swap* neighborhood, which perturbs the value of two variables of the problem by exchanging their values, can be modeled by the set $Y = \{y_1, y_2\}$, consisting of the variables to exchange, and with the interface constraints

$$(y_1 = i \wedge y_2 = j) \iff (x_i = s_j \wedge x_j = s_i)$$
$$\forall i, j \in \{1, \ldots, n\} \ .$$

Moreover, an additional component of the neighborhood model is the evaluator of the move impact Δf, which can be usually computed incrementally on the basis of the single move.

It is worth noticing that the use of different modeling viewpoints is common practice in constraint programming. In classical CP modeling the different viewpoints usually offer a convenient way to express some constraint in a more concise or more efficient manner. The consistency between the viewpoints is maintained through the use of *channeling* constraints that link the different modelings. Similarly, although with a different purpose, in CBLS the linking between the full problem model and the neighborhood model is achieved through interface constraints.

Algorithm 62.5 ConstraintBasedLocalSearch (X, C_X, **Dom**_X, *f)*

1: create a (feasible) initial solution $\bar{s}_0 = \langle d_1^0, \ldots, d_k^0 \rangle$
    /*possibly random or finding the first feasible solution of the original CP model*/

2: $\bar{s}_b \leftarrow \bar{s}_0$
3: $i \leftarrow 0$
4: **while** not TerminateSearch$(i, \bar{s}_i, f(\bar{s}_i), \bar{s}_b)$ **do**
5:     $\langle Y, C_{X,Y}, \mathbf{Dom}_Y, \Delta f \rangle \leftarrow$
        NeighborhoodModel$(\bar{s}_i)$
6:     $L \leftarrow \emptyset$
7:     $U \leftarrow Y$
8:     BranchAndBound$(U, L, C_{X,Y}, \mathbf{Dom}_Y, \Delta f)$
       /*neighborhood exploration*/
9:     **if** AcceptSolution(L) **then**
10:        $\bar{s}_{i+1} \leftarrow$ Apply$(L, \bar{s}_i)$
11:        **if** $f(\bar{s}_{i+1}) < f(\bar{s}_b)$ **then**
12:            $\bar{s}_b \leftarrow \bar{s}_{i+1}$
13:        **end if**
14:    **else**
15:        $\bar{s}_{i+1} \leftarrow \bar{s}_i$
16:    **end if**
17:    $i \leftarrow i + 1$
18: **end while**
19: **return** $\bar{s}_b$

This stream of research has been revamped thanks to the design of the Comet language [62.34, 57], the aim of which is specifically to support declarative components inspired from CP primitives for expressing local search algorithms. An example of such primitives are *differentiable invariants* [62.58], which are declarative data structures that support incremental differentiation to effectively evaluate the effect of local moves (i. e., the Δf in Algorithm 62.5). Moreover, Comet support control abstractions [62.59, 60] specifically designed for local search such as the *neighbors* construct, which aims at expressing the unions of heterogeneous neighborhoods. Finally, Comet has been extended also to support distributed computing [62.61].

The embedding of local search within a constraint programming environment and the employment of a common programming language makes it possible to automatize the synthesis of CBLS algorithms from a high-level model expressed in Comet [62.62, 63]. The synthesizer analyzes the combinatorial structure of the problem, expressed through the variables and the constraints, and combines a set of basic *recommendations*, which are the basic constituents of the synthesized algorithm.

Other Integrations. The idea of exploring with local search a space of incomplete solutions (i. e., those where not all variables have been assigned a value) exploiting constraint propagation has been pursued, among others, by *Jussien* and *Lhomme* [62.64] for

an open-shop scheduling problem. Constraint propagation employed in the spirit of forward checking and, more in general, look-ahead has been effectively employed, among others, by *Schaerf* [62.65] and *Prestwich* [62.66], respectively, for scheduling and graph coloring problems.

Local Search Within CP

Moving to the integration of local search within constraint programming, the most common utilization of local search-like techniques consists in limiting the exploration of the tree search only to paths that are "close" to a reference one. An example of such a procedure is *limited discrepancy search* (LDS) [62.67], an incomplete method for tree search in which only *neighboring* paths of the search tree are explored, where the proximity is defined in terms of different decision points called *discrepancies*. Only the paths (i. e., complete solutions) with at most k discrepancies are considered, as outlined in Algorithm 62.6.

Algorithm 62.6 LimitedDiscrepancySearch $(X, C, \mathbf{Dom}, f, k)$

1: $\bar{s}^* \leftarrow$ FirstSolution$(X, C, \mathbf{Dom}, f)$
2: $\bar{s}_b \leftarrow \bar{s}^*$
3: **for** $i \in \{1, \ldots, k\}$ **do**
4:     **for** $\bar{t} \in \{\bar{s} : \bar{s}$ differs w.r.t. $\bar{s}^*$ for
                       exactly i variables$\}$ **do**
5:         **if** Consistent$(\bar{t}, \mathbf{Dom}) \wedge f(\bar{t}) < f(\bar{s}_b)$ **then**
6:             $\bar{s}_b \leftarrow \bar{t}$
7:         **end if**
8:     **end for**
9: **end for**
10: **return** $\bar{s}_b$

Another approach due to *Prestwich* [62.68] is called *incomplete dynamic backtracking*. Differently from LDS, in this approach proximity is defined among *partial* solutions, and when backtracking needs to take place it is executed by randomly unassigning (at most) b variables. This way, the method could be intended as a local search on partial solutions. In fact, the method also features other CP machinery, such as forward checking, which helps in boosting the search.

An alternative possibility is to employ local search in constraint propagation. Local probing [62.69, 70] is based on the partition of constraints into the set of *easy* and *hard* ones. At each choice point in the search tree the set of easy constraints is dealt with a local search metaheuristic (namely simulated annealing), while the hard constraints are considered by classical constraint propagation. This idea generalizes the approach of *Zhang* and *Zhang* [62.71], who first presented such a combination. Another similar approach was taken by *Sellmann* and *Harvey* [62.72], who used local search to propagate redundant constraints.

In [62.73] the authors discuss the incorporation of the tabu search machinery within CP tree search. In particular, they look at the memory mechanisms for limiting the size of the tree and the elite candidate list for keeping the most promising choices in order to be evaluated first.

62.3.2 Genetic Algorithms and CP

A genetic algorithm [62.5] is an iterative metaheuristic in which a population of strings, which represent candidate solutions, evolves toward better solutions in a process that mimics natural evolution. The main components of the evolution process are crossover and mutation operators, which, respectively, combine two parent solutions generating an offspring and mutate a given solution. Another important component is the strategy for the offspring selection, which determines the population at the next iteration of the process.

To the best of our knowledge, one of the first attempts to integrate constraint programming and genetic algorithms is due to *Barnier* and *Brisset* [62.74]. They employ the following genetic representation: given a CSP with variables $\{X_1, \ldots, X_k\}$, the i-th gene in the chromosomes is related to the variable X_i and it stores a subset of the domain D_i that is allowed to be searched. Each chromosome is then decoded by CP, which searches for the best solution of the sub-CSP induced by the restrictions in the domains. The genetic operators used are a mutation operator that changes values on the subdomain of randomly chosen genes and a crossover operator that is based on a recombination of the set-union of the subdomains of each pair of genes. The method was applied to a vehicle routing problem and outperformed both a CP and a GA solver.

A different approach, somewhat similar to local probing, was used in [62.75] for tackling a production scheduling problem. In this case, the problem variables are split into two sets, defining two coupled subproblems. The first set of variables is dealt with by the genetic algorithm, which determines a partial schedule. This partial solution is then passed to CP for completing (and optimizing) the assignment of the remaining variables.

Finally, CP has been used as a post-processing phase for optimizing the current population in the spirit of memetic algorithms. In [62.76] CP actually acts as an unfeasibility repairing method for a university course timetabling problem, whereas in [62.77] the optimization performed by CP on a flow-shop scheduling problem is an alternative to the classical local search applied in memetic algorithms. This approach is illustrated in Algorithm 62.7.

Algorithm 62.7 A Memetic Algorithm with CP for Flow-Shop scheduling (adapted from [62.77])

1: generate an initial population $P = \{p_1, \ldots, p_l\}$ of permutations of n jobs (each composed of k tasks τ_{ij} whose start time and end time are denoted by σ_{ij} and η_{ij} respectively)
2: $g \leftarrow 0$
3: **while** not TerminateSearch$(g, P, \min_{p \in P} f(p))$ **do**
4:     select p_1 and p_2 from P by binary tournament
5:     $c \leftarrow p_1 \otimes p_2$ /*apply crossover*/
6:     **if** $f(c) \geq \min_{p \in P} f(p)$ **then**
7:         mutate c under probability p_m
8:     **end if**
9:     decode $c = \langle c_1, \ldots, c_n \rangle$ to the set of precedence constraints $C = \{\eta_{kc_j} \leq \sigma_{1c_{j+1}} : j = 1, \ldots, n-1\}$
10:    $L \leftarrow \emptyset$
11:    $U \leftarrow \{\sigma_{ij}, \eta_{ij} : i = 1, \ldots, k, j = 1, \ldots, n\}$
12:    BranchAndBound$(U, L, C \cup \{\eta_{ij} \leq \sigma_{i+1j} : i = 1, \ldots, k\}, \textbf{Dom}, f)$
13:    **if** $f(c) \geq \max_{p \in P} f(p)$ **then**
14:       discard c
15:    **else**
16:       select r by reverse binary tournament
17:       c replaces r in P
18:    **end if**
19:    $g \leftarrow g + 1$
20: **end while**
21: **return** arg $\min_{p \in P} f(p)$

62.3.3 ACO and CP

Ant colony optimization [62.6] is an iterative constructive metaheuristic, inspired by ant foraging behavior. The ACO construction process is driven by a probabilistic model, based on pheromone trails, which are dynamically adjusted by a learning mechanism.

The first attempt to integrate ACO and CP is due to *Meyer* and *Ernst* [62.78], who apply the method for solving a job-shop scheduling problem. The proposed procedure employs ACO to learn the variable and value ordering used by CP for branching in the tree search. The solutions found by the CP procedure are fed back

Part E | 62.3

to the ACO in order to update its probabilistic model. In this approach, ACO can be conceived as a master online-learning branching heuristic aimed at enhancing the performance of a slave CP solver.

A slightly different approach was taken by *Khichane* et al. [62.79, 80]. Their hybrid algorithm works in two phases. At first, CP is employed to sample the space of feasible solutions, and the information collected is processed by the ACO procedure for updating the pheromone trails according to the solutions found by CP. In the second phase, the learned pheromone information is employed as the value ordering used for CP branching. This approach, differently from the previous one, uses the learning capabilities of ACO in an offline-learning fashion.

More standard approaches in which CP is used to keep track of the feasibility of the solution constructed by ACO and to reduce the domains through constraint propagation have been used by a number of authors. *Khichane* et al. apply this idea to job-shop scheduling [62.78] and car sequencing [62.79, 81]. Their general idea is outlined in Algorithm 62.8, where each ant maintains a partial assignment of values to variables. The choice to extend the partial assignment with a new variable/value pair is driven by the pheromone trails and the heuristic factors in lines 7–8 through a standard probabilistic selection rule. Propagation is employed at line 10 to prune the possible values for the variables not included in the current assignment.

Another work along this line is due to *Benedettini* et al. [62.82], who integrate a constraint propagation phase for Boolean constraints to boost a ACO approach for a bioinformatics problem (namely, haplotype inference). Finally, in the same spirit of the previous idea, *Crawford* et al. [62.83, 84] employ a look-ahead technique within ACO and apply the method to solve set covering and set partitioning problems.

Algorithm 62.8 Ant Constraint Programming (adapted from [62.79])

1: initialize all pheromone trails to $\tau_{\max}$
2: $g \leftarrow 0$
3: **repeat**
4:    **for** $k \in \{1, \ldots, n\}$ **do**
5:       $\mathcal{A}_k \leftarrow \emptyset$
6:       **repeat**
7:          select a variable $x_j \in X$ so that $x_j \notin \mathrm{var}(\mathcal{A}_k)$ according to the pheromone trail τ_j
8:          choose a value $v \in D_j$ according to the pheromone trail τ_{jv} and a heuristic factor η_{jv}
9:          add $\{x_j := v\}$ to $\mathcal{A}_k$
10:         Propagate($\mathcal{A}_k, C$)
11:      **until** $\mathrm{var}(\mathcal{A}_k) = X$ or Failure
12:      update pheromone trails using $\{\mathcal{A}_1, \ldots, \mathcal{A}_n\}$
13:   **end for**
14: **until** $\mathrm{var}(\mathcal{A}_i) = X$ for some $i \in \{1, (\ldots), n\}$ or TerminateSearch($g, \mathcal{A}_i$)

62.4 Conclusions

In this chapter we have reviewed the basic concepts of constraint programming and its integration with metaheuristics. Our main contribution is the attempt to give a comprehensive overview of such integrations from the viewpoint of metaheuristics.

We believe that the reason why these integrations are very promising resides in the complementary merits of the two approaches. Indeed, on the one hand, metaheuristics are, in general, more suitable to deal with optimization problems, but their treatment of constraints can be very awkward, especially in the case of tightly constrained problems. On the other hand, constraint programming is specifically designed for finding

feasible solutions, but it is not particularly effective for handling optimization. Consequently, a hybrid algorithm that uses CP for finding feasible solutions and metaheuristics to search among them has good chances to outperform its single components.

Despite the important steps made in this field during the last decade, there are still promising research opportunities, especially in order to investigate topics such as collaborative hybridization of CP and metaheuristics and validate existing integration approaches in the yet uninvestigated area of multiobjective optimization. We believe that further research should devote more attention to these aspects.

References

62.1   K.R. Apt: *Principles of Constraint Programming* (Cambridge Univ. Press, Cambridge 2003)

62.2   F. Rossi, P. van Beek, T. Walsh: *Handbook of Constraint Programming*, Foundations of Artificial Intelligence (Elsevier Science, Amsterdam 2006)

62.3   M. Dorigo, M. Birattari, T. Stützle: Metaheuristic. In: *Encyclopedia of Machine Learning*, ed. by C. Sammut, G.I. Webb (Springer, Berlin, Heidelberg 2010) p. 662

62.4   H.H. Hoos, T. Stützle: *Stochastic Local Search: Foundations & Applications* (Morgan Kaufmann, San Francisco 2004)

62.5   C. Sammut: Genetic and evolutionary algorithms. In: *Encyclopedia of Machine Learning*, ed. by C. Sammut, G.I. Webb (Springer, Berlin, Heidelberg 2010) pp. 456–457

62.6   M. Dorigo, M. Birattari: Ant colony optimization. In: *Encyclopedia of Machine Learning*, ed. by C. Sammut, G.I. Webb (Springer, Berlin, Heidelberg 2010) pp. 36–39

62.7   T. Yunes: Success stories in integrated optimization (2005) http://moya.bus.miami.edu/~tallys/integrated.php

62.8   W. J. van Hoeve: CPAIOR conference series (2010) available online from http://www.andrew.cmu.edu/user/vanhoeve/cpaior/

62.9   P. van Hentenryck, M. Milano (Eds.): *Hybrid Optimization: The Ten Years of CPAIOR*, Springer Optimization and Its Applications, Vol. 45 (Springer, Berlin 2011)

62.10   C. Blum, A. Roli, M. Sampels (Eds.): Hybrid Metaheuristics, First International Workshop (HM 2004), Valencia (2004)

62.11   M.J. Blesa, C. Blum, A. Roli, M. Sampels (Eds.): *Hybrid Metaheuristics: Second International Workshop (HM 2005)*, Lecture Notes in Computer Science, Vol. 3636 (Springer, Berlin, Heidelberg 2005)

62.12   F. Almeida, M.J. Blesa Aguilera, C. Blum, J.M. Moreno-Vega, M. Pérez, A. Roli, M. Sampels (Eds.): *Hybrid Metaheuristics: Third International Workshop*, Lecture Notes in Computer Science, Vol. 4030 (Springer, Berlin, Heidelberg 2006)

62.13   T. Bartz-Beielstein, M.J. Blesa Aguilera, C. Blum, B. Naujoks, A. Roli, G. Rudolph, M. Sampels (Eds.): *Hybrid Metaheuristics: 4th International Workshop (HM 2007)*, Lecture Notes in Computer Science, Vol. 4771 (Springer, Berlin, Heidelberg 2007)

62.14   M.J. Blesa, C. Blum, C. Cotta, A.J. Fernández, J.E. Gallardo, A. Roli, M. Sampels (Eds.): *Hybrid Metaheuristics: 5th International Workshop (HM 2008)*, Lecture Notes in Computer Science, Vol. 5296 (Springer, Berlin, Heidelberg 2008)

62.15   M.J. Blesa, C. Blum, L. Di Gaspero, A. Roli, M. Sampels, A. Schaerf (Eds.): *Hybrid Metaheuristics: 6th International Workshop (HM 2009)*, Lecture Notes in Computer Science, Vol. 5818 (Springer, Berlin, Heidelberg 2009)

62.16   M.J. Blesa, C. Blum, G.R. Raidl, A. Roli, M. Sampels (Eds.): *Hybrid Metaheuristics: 7th International Workshop (HM 2010)*, Lecture Notes in Computer Science, Vol. 6373 (Springer, Berlin, Heidelberg 2010)

62.17   I. Dumitrescu, T. Stützle: Combinations of local search and exact algorithms, Lect. Notes Comput. Sci. **2611**, 211–223 (2003)

62.18   J. Puchinger, G. Raidl: Combining metaheuristics and exact algorithms in combinatorial optimization: A survey and classification, Lect. Notes Comput. Sci. **3562**, 113–124 (2005)

62.19   S. Fernandes, H. Ramalhinho Dias Lourenço: Hybrids combining local search heuristics with exact algorithms, V Congr. Esp. Metaheurísticas, Algoritm. Evol. Bioinspirados (MAEB2007), Tenerife, ed. by F. Rodriguez, B. Mélian, J.A. Moreno, J.M. Moreno (2007) pp. 269–274

62.20   L. Jourdan, M. Basseur, E.-G. Talbi: Hybridizing exact methods and metaheuristics: A taxonomy, Eur. J. Oper. Res. **199**(3), 620–629 (2009)

62.21   M. Wallace: Hybrid algorithms in constraint programming, Lect. Notes Comput. Sci. **4651**, 1–32 (2007)

62.22   F. Azevedo, P. Barahona, F. Fages, F. Rossi (Eds.): *Recent Advances in Constraints: 11th Annual ERCIM International Workshop on Constraint Solving and Constraint Logic Programming (CSCLP 2006)*, Lecture Notes in Computer Science, Vol. 4651 (Springer, Berlin, Heidelberg 2007)

62.23   C. Blum, J. Puchinger, G.R. Raidl, A. Roli: Hybrid metaheuristics in combinatorial optimization: A survey, Appl. Soft Comput. **11**(6), 4135–4151 (2011)

62.24   N. Beldiceanu, H. Simonis: Global constraint catalog (2011), available online from http://www.emn.fr/z-info/sdemasse/gccat/

62.25   P. Meseguer, F. Rossi, T. Schiex: Soft constraints. In: *Handbook of Constraint Programming*, Foundations of Artificial Intelligence, ed. by F. Rossi, P. van Beek, T. Walsh (Elsevier, Amsterdam 2006)

62.26   A.K. Mackworth: Consistency in networks of relations, Artif. Intell. **8**(1), 99–118 (1977)

62.27   SICStus prolog homepage, available online from http://www.sics.se/isl/sicstuswww/site/index.html

62.28   K.R. Apt, M. Wallace: *Constraint Logic Programming Using Eclipse* (Cambridge Univ. Press, Cambridge 2007)

62.29   ILOG CP optimizer, available online from http://www-01.ibm.com/software/integration/optimization/cplex-cp-optimizer/

62.30   P. van Hentenryck: *The OPL Optimization Programming Language* (MIT Press, Cambridge 1999)

62.31   Gecode Team: Gecode: Generic constraint development environment (2006), available online from http://www.gecode.org

62.32   CHOCO Team: *Choco: An open source java constraint programming library*, Res. Rep. 10-02-INFO (Ecole des Mines de Nantes, Nantes 2010)

62.33   N. Nethercote, P.J. Stuckey, R. Becket, S. Brand, G.J. Duck, G. Tack: Minizinc: Towards a standard CP modelling language, Lect. Notes Comput. Sci. **4741**, 529–543 (2007)

62.34   P.V. Hentenryck, L. Michel: *Constraint-Based Local Search* (MIT Press, Cambridge 2005)

62.35   G. Pesant, M. Gendreau: A view of local search in constraint programming, Lect. Notes Comput. Sci. **1118**, 353–366 (1996)

62.36   P. Shaw: Using constraint programming and local search methods to solve vehicle routing problems, Lect. Notes Comput. Sci. **1520**, 417–431 (1998)

62.37   B.D. Backer, V. Furnon, P. Shaw, P. Kilby, P. Prosser: Solving vehicle routing problems using constraint programming and metaheuristics, J. Heuristics **6**(4), 501–523 (2000)

62.38   F. Focacci, F. Laburthe, A. Lodi: Local search and constraint programming. In: *Handbook of Metaheuristics*, ed. by F. Glover, G. Kochenberger (Kluwer, Boston 2003) pp. 369–403

62.39   P. Shaw: Constraint programming and local search hybrids. In: *Hybrid Optimization*, Springer Optimization and Its Applications, Vol. 45, ed. by P. van Hentenryck, M. Milano (Springer, Berlin, Heidelberg 2011) pp. 271–303

62.40   L. Perron, P. Shaw, V. Furnon: Propagation guided large neighborhood search, Lect. Notes Comput. Sci. **3258**, 468–481 (2004)

62.41   E. Danna, L. Perron: Structured vs. unstructured large neighborhood search: A case study on job-shop scheduling problems with earliness and tardiness costs, Lect. Notes Comput. Sci. **2833**, 817–821 (2003)

62.42   Y. Caseau, F. Laburthe, G. Silverstein: A meta-heuristic factory for vehicle routing problems, Lect. Notes Comput. Sci. **1713**, 144–158 (1999)

62.43   L.M. Rousseau, M. Gendreau, G. Pesant: Using constraint-based operators to solve the vehicle routing problem with time windows, J. Heuristics **8**(1), 43–58 (2002)

62.44   S. Jain, P. van Hentenryck: Large neighborhood search for dial-a-ride problems, Lect. Notes Comput. Sci. **6876**, 400–413 (2011)

62.45   J.H.-M. Lee (Ed.): *Principles and Practice of Constraint Programming – CP 2011 – 17th International Conference, CP 2011, Perugia, Italy, September 12-16, 2011, Proceedings*, Lecture Notes in Computer Science, Vol. 6876 (Springer, Berlin, Heidelberg 2011)

62.46   R. Cipriano, L. Di Gaspero, A. Dovier: Hybrid approaches for rostering: A case study in the integration of constraint programming and local search, Lect. Notes Comput. Sci. **4030**, 110–123 (2006)

62.47   H. Cambazard, E. Hebrard, B. O'Sullivan, A. Papadopoulos: Local search and constraint programming for the post enrolment-based course timetabling problem, Ann. Oper. Res. **194**(1), 111–135 (2012)

62.48   I. Dotu, M. Cebrián, P. van Hentenryck, P. Clote: Protein structure prediction with large neighborhood constraint programming search. In: *Principles and Practice of Constraint Programming*, ed. by I. Dotu, M. Cebrián, P. van Hentenryck, P. Clote (Springer, Berlin, Heidelberg 2008) pp. 82–96

62.49   R. Cipriano, A. Dal Palù, A. Dovier: A hybrid approach mixing local search and constraint programming applied to the protein structure prediction problem, Proc. Workshop Constraint Based Methods Bioinform. (WCB 2008), Paris (2008)

62.50   L. Perron, P. Shaw: Combining forces to solve the car sequencing problem, Lect. Notes Comput. Sci. **3011**, 225–239 (2004)

62.51   R. Cipriano, L. Di Gaspero, A. Dovier: A hybrid solver for Large Neighborhood Search: Mixing Gecode and EasyLocal++, Lect. Notes Comput. Sci. **5818**, 141–155 (2009)

62.52   R. Cipriano: On the hybridization of constraint programming and local search techniques: Models and software tools, Lect. Notes Comput. Sci. **5366**, 803–804 (2008)

62.53   R. Cipriano: On the Hybridization of Constraint Programming and Local Search Techniques: Models and Software Tools, Ph.D. Thesis (PhD School in Computer Science – University of Udine, Udine 2011)

62.54   D. Pisinger, S. Ropke: Large neighborhood search. In: *Handbook of Metaheuristics*, ed. by M. Gendreau, J.-Y. Potvin (Springer, Berlin, Heidelberg 2010) pp. 399–420, 2nd edn., Chap. 13

62.55   T. Carchrae, J.C. Beck: Principles for the design of large neighborhood search, J. Math. Model, Algorithms **8**(3), 245–270 (2009)

62.56   G. Pesant, M. Gendreau: A constraint programming framework for local search methods, J. Heuristics **5**(3), 255–279 (1999)

62.57   L. Michel, P. van Hentenryck: A constraint-based architecture for local search, Proc. 17th ACM SIGPLAN Object-oriented Program. Syst. Lang. Appl. (OOPSLA '02), New York (2002) pp. 83–100

62.58   P. van Hentenryck, L. Michel: Differentiable invariants, Lect. Notes Comput. Sci. **4204**, 604–619 (2006)

62.59   P. van Hentenryck, L. Michel: Control abstractions for local search, J. Constraints **10**(2), 137–157 (2005)

62.60   P. van Hentenryck, L. Michel: Nondeterministic control for hybrid search, Lect. Notes Comput. Sci. **3524**, 863–864 (2005)

62.61   L. Michel, A. See, P. van Hentenryck: Distributed constraint-based local search, Lect. Notes Comput. Sci. **4204**, 344–358 (2006)

62.62   P. van Hentenryck, L. Michel: Synthesis of constraint-based local search algorithms from high-level models, 22nd Natl. Conf. Artif. Intell. AAAI, Vol. 1 (2007) pp. 273–278

62.63 S.A. Mohamed Elsayed, L. Michel: Synthesis of search algorithms from high-level cp models, Lect. Notes Comput. Sci. **6876**, 256–270 (2011)

62.64 N. Jussien, O. Lhomme: Local search with constraint propagation and conflict-based heuristic, Artif. Intell. **139**(1), 21–45 (2002)

62.65 A. Schaerf: Combining local search and look-ahead for scheduling and constraint satisfaction problems, 15th Int. Joint Conf. Artif. Intell. (IJCAI-97), Nagoya (1997) pp. 1254–1259

62.66 S. Prestwich: Coloration neighbourhood search with forward checking, Ann. Math. Artif. Intell. **34**, 327–340 (2002)

62.67 W.D. Harvey, M.L. Ginsberg: Limited discrepancy search, 14th Int. Joint Conf. Artif. Intell., Montreal (1995) pp. 607–613

62.68 S. Prestwich: Combining the scalability of local search with the pruning techniques of systematic search, Ann. Oper. Res. **115**(1), 51–72 (2002)

62.69 O. Kamarainen, H. Sakkout: Local probing applied to scheduling, Lect. Notes Comput. Sci. **2470**, 81–103 (2006)

62.70 O. Kamarainen, H. El Sakkout: Local probing applied to network routing, Lect. Notes Comput. Sci. **3011**, 173–189 (2004)

62.71 J. Zhang, H. Zhang: Combining local search and backtracking techniques for constraint satisfaction, Proc. 13th Natl. Conf. Artif. Intell. (AAAI96) (1996) pp. 369–374

62.72 M. Sellmann, W. Harvey: Heuristic constraint propagation, Lect. Notes Comput. Sci. **2470**, 319–325 (2006)

62.73 M. Dell'Amico, A. Lodi: On the integration of metaheuristic stratgies in constraint programming. In: *Metaheuristic Optimization Via Memory and Evolution: Tabu Search and Scatter Search*, Operations Research/Computer Science Interfaces, Vol. 30, ed. by C. Rego, B. Alidaee (Kluwer, Boston 2005) pp. 357–371, Chap. 16

62.74 N. Barnier, P. Brisset: Combine & conquer: Genetic algorithm and CP for optimization, Lect. Notes Comput. Sci. **1520**, 463–463 (1998)

62.75 H. Hu, W.-T. Chan: A hybrid GA-CP approach for production scheduling, 5th Int. Conf. Nat. Comput. (2009) pp. 86–91

62.76 S. Deris, S. Omatu, H. Ohta, P. Saad: Incorporating constraint propagation in genetic algorithm for university timetable planning, Eng. Appl. Artif. Intell. **12**(3), 241–253 (1999)

62.77 A. Jouglet, C. Oguz, M. Sevaux: Hybrid flowshop: a memetic algorithm using constraint-based scheduling for efficient search, J. Math. Model Algorithms **8**(3), 271–292 (2009)

62.78 B. Meyer, A. Ernst: Integrating ACO and constraint propagation, Lect. Notes Comput. Sci. **3172**, 166–177 (2004)

62.79 M. Khichane, P. Albert, C. Solnon: CP with ACO. In: *Integration of AI and OR Techniques in Constraint Programming for Combinatorial Optimization Problems*, ed. by L. Perron, M.A. Trick (Springer, Berlin, Heidelberg 2008) pp. 328–332

62.80 M. Khichane, P. Albert, C. Solnon: Strong combination of ant colony optimization with constraint programming optimization, Lect. Notes Comput. Sci. **6140**, 232–245 (2010)

62.81 M. Khichane, P. Albert, C. Solnon: Integration of ACO in a constraint programming language, Lect. Notes Comput. Sci. **5217**, 84–95 (2008)

62.82 S. Benedettini, A. Roli, L. Di Gaspero: Two-level ACO for haplotype inference under pure parsimony, Lect. Notes Comput. Sci **5217**, 179–190 (2008)

62.83 B. Crawford, C. Castro: Integrating lookahead and post processing procedures with ACO for solving set partitioning and covering problems, Lect. Notes Comput. Sci. **4029**, 1082–1090 (2006)

62.84 B. Crawford, C. Castro, E. Monfroy: Constraint programming can help ants solving highly constrained combinatorial problems, ICSOFT 2008 – Proc. 3rd Int. Conf. Software Data Technol., INSTICC, Porto (2008) pp. 380–383

Part E | 62

63. Graph Coloring and Recombination

Rhyd Lewis

It is widely acknowledged that some of the most powerful algorithms for graph coloring involve the combination of evolutionary-based methods with exploitative local search-based techniques. This chapter conducts a review and discussion of such methods, principally focussing on the role that recombination plays in this process. In particular we observe that, while in some cases recombination seems to be usefully combining substructures inherited from parents, in other cases it is merely acting as a macro perturbation operator, helping to reinvigorate the search from time to time.

63.1 Graph Coloring

Graph coloring is a well-known NP-hard combinatorial optimization problem that involves using a minimal number of colors to paint all vertices in a graph such that all adjacent vertices are allocated different colors. The problem is more formally stated as follows: given an undirected simple graph $G = (V, E)$, with vertex set V and edge set E, our task is to assign each vertex $v \in V$ an integer $c(v) \in \{1, 2, \ldots, k\}$ so that:

- $c(v) \neq c(u) \forall \{v, u\} \in E$
- k is minimal.

Though essentially a theoretical problem, graph coloring is seen to underpin a wide variety of seemingly unrelated operational research problems, including satellite scheduling [63.1], educational timetabling [63.2, 3], sports league scheduling [63.4], frequency assignment problems [63.5, 6], map coloring [63.7], airline crew scheduling [63.8], and compiler register allocation [63.9]. The design of effective algorithms for graph coloring thus has positive implications for a large range of real-world problems.

Some common terms used with graph coloring are as follows:

- A coloring of a graph is called *complete* if all vertices $v \in V$ are assigned a color $c(v) \in \{1, \ldots, k\}$; else it is considered *partial*.
- A *clash* describes a situation where a pair of adjacent vertices $u, v \in V$ are assigned the same color (that is, $\{u, v\} \in E$ and $c(v) = c(u)$). If a coloring contains no clashes, then it is considered *proper*; else it is *improper*.
- A coloring is *feasible* if and only if it is both complete and proper.
- The *chromatic number* of a graph G, denoted $\chi(G)$, is the minimal number of colors required in a feasible coloring. If a feasible coloring uses $\chi(G)$ colors, it is considered *optimal*.
- An *independent set* is a subset of vertices $I \subseteq V$ that are mutually non-adjacent. That is, $\forall u, v \in I$, $\{u, v\} \notin E$. Similarly, a *clique* is a subset of vertices $C \subseteq V$ that are mutually adjacent: $\forall u, v \in C$, $\{u, v\} \in E$.

Given these definitions, we might also view graph coloring as a type of partitioning/grouping problem where the aim is to split the vertices into a set of subsets $\mathcal{U} = \{U_1, \ldots, U_k\}$ such that $U_i \cap U_j = \emptyset (1 \leq i < j \leq k)$.

Part E | 63.1

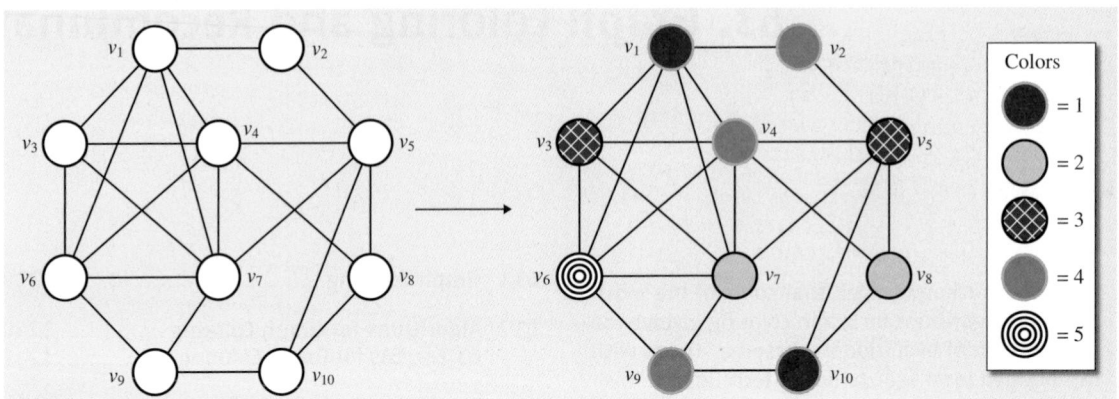

Fig. 63.1 A simple graph (*left*) and a feasible five-coloring (*right*)

If $\bigcup_{i=1}^{k} U_i = V$, then the partition represents a complete coloring. Moreover, if all subsets $U_1, \ldots, U_k$ are independent sets, the coloring is also feasible.

To exemplify these concepts, Fig. 63.1 shows an example graph with ten vertices, together with a corresponding coloring. In this case the presented coloring is both complete and proper, and therefore feasible. It is also optimal because it uses just five colors, which happens to be the chromatic number in this case. The graph also contains one clique of size 5 (vertices v_1, v_3, v_4, v_6, and v_7), and numerous independent sets, such as vertices v_2, v_3, v_8,

and v_9. As a partition, this coloring is represented $\mathcal{U} = \{\{v_1, v_{10}\}, \{v_7, v_8\}, \{v_3, v_5\}, \{v_2, v_4, v_9\}, \{v_6\}\}$.

It should be noted that various subsidiary problems related to the graph coloring problem are also known to be NP-hard. These include computing the chromatic number itself, identifying the size of the largest clique, and determining the size of the largest independent set in a graph [63.10, 11]. In addition, the decision variant of the graph coloring problem, which asks: *given a fixed positive integer k, is there a feasible k-coloring of the vertices?* is NP-complete.

63.2 Algorithms for Graph Coloring

Graph coloring has been studied as an algorithmic problem since the late 1960s and, as a result, an abundance of methods have been proposed. Loosely speaking, these methods might be grouped into two main classes: *constructive methods*, which build solutions step-by-step, perhaps using various heuristic and backtracking operators; and *stochastic search-based methods*, which attempt to navigate their way through a space of candidate solutions while optimizing a particular objective function.

The earliest proposed algorithms for graph coloring generally belong to the class of constructive methods. Perhaps the simplest of these is the *first-fit* (or *greedy*) algorithm. This operates by taking each vertex in turn in a specified order and assigning it to the lowest indexed color where no clash is induced, creating new colors when necessary [63.12]. A development on this method is the DSATUR algorithm [63.13, 14] in

which the ordering of the vertices is determined dynamically – specifically, by choosing at each step the uncolored vertex that currently has the largest number of different colors assigned to adjacent vertices, breaking ties by taking the vertex with the largest degree. Other constructive methods have included backtracking strategies, such as those of *Brown* [63.15] and *Korman* [63.16], which may ultimately perform complete enumerations of the solution space given excess time. A survey of backtracking approaches was presented by *Kubale* and *Jackowski* [63.17] in 1985.

Many of the more recent methods for graph coloring have followed the second approach mentioned above, which is to search a space of candidate solutions and attempt to identify members that optimize a specific objective function. Such methods can be further classified according to the composition of their search spaces, which can comprise:

(a) The set of all feasible solutions (using an undefined number of colors)
(b) The set of complete colorings (proper and improper) for a fixed number of colors k.
(c) The set of proper solutions (partial and complete), also for a fixed number of colors k.

Algorithms following scheme (a) have been considered by, among others, *Culberson* and *Luo* [63.18], *Mumford* [63.19], *Erben* [63.20], and *Lewis* [63.21]. Typically, these methods consider different permutations of the vertices, which are then fed into a constructive method (such as first-fit) to form feasible solutions. An intuitive cost function in such cases is simply the number of colors used in a solution, though other more fine-grained functions have been suggested, such as the following due to *Erben* [63.20]

$$f_1 = \frac{\sum_{U_i \in \mathcal{U}} \left[\sum_{v \in U_i} \deg(v) \right]^2}{|\mathcal{U}|} . \qquad (63.1)$$

Here, the term $\left(\sum_{v \in U_i} \deg(v) \right)$ gives the sum of the degrees of all vertices assigned to a color class U_i. The aim is to maximize f_1 by making increases to the numerator (by forming large color classes that contain high-degree vertices), and decreases to the denominator (by reducing the number of color classes).

On the other hand, algorithms following scheme (b) operate by first proposing a fixed number of colors k. At the start of a run, each vertex will be assigned to one of the k colors using heuristics, or randomly. However, this may involve the introduction of one or more clashes, resulting in a complete, improper k-coloring. The cost of such a solution might then be evaluated using the following cost function, which is simply a count on the number of clashes

$$f_2 = \sum_{\forall \{v,u\} \in E} g(v,u) \quad \text{where}$$

$$g(v,u) = \begin{cases} 1 & \text{if } c(v) = c(u) \\ 0 & \text{otherwise} . \end{cases} \qquad (63.2)$$

The strategy in such approaches is to make alterations to a solution such that the number of clashes is reduced to zero. If this is achieved k can be reduced; alternatively if all clashes cannot be eliminated, k can be increased. This strategy has been quite popular in the literature, involving the use of various stochastic search methodologies, including simulated annealing [63.22, 23], tabu search [63.24], greedy randomized adaptive search procedure (GRASP) methods [63.25],

iterated local search [63.26, 27], variable neighborhood search [63.28], ant colony optimization [63.29], and evolutionary algorithms (EA) [63.30–35].

Finally, scheme (c) also involves using a fixed number of colors k; however in this case, rather than allowing clashes to occur in a solution, vertices that cannot be feasibly assigned to a color are placed into a set of uncolored vertices S. The aim is, therefore, to make changes to a solution so that these vertices can eventually be feasibly colored, resulting in $S = \emptyset$. This approach has generally been less popular in the literature than scheme (b), though some prominent examples include the simulated annealing approach of *Morgenstern* [63.36], the tabu search method of *Blochliger* and *Zufferey* [63.37], and the EA of *Malaguti* et al. [63.38]. More recently, *Hertz* et al. [63.39] also suggested an algorithm that searches different solution spaces during different stages of a run. The idea is that when the search is deemed to have stagnated in one space, a procedure is used to alter the current solution so that it becomes a member of another space (e.g., clashing vertices are *uncolored* by transferring them to S). Once this has been done, the search can then be continued in this new space where further improvements might be made.

63.2.1 EAs for Graph Coloring

In this section we now examine the ways in which EAs have been applied to the graph coloring problem, particularly looking at issues surrounding the recombination of solutions.

Assignment-Based Operators

Perhaps the most intuitive way of applying EAs to the graph coloring problem is to view the task as one of assignment. In this case, a candidate solution can be viewed as a mapping of vertices to colors $c : V \rightarrow \{1, \ldots, k\}$, and a natural chromosome representation is a vector $(c(v_1), c(v_2), \ldots, c(v_{|V|}))$, where $c(v_i)$ gives the color of vertex v_i (the solution given in Fig. 63.1 would be represented by $(1, 4, 3, 4, 3, 5, 2, 2, 4, 1)$ under this scheme). However, it has long been argued that this sort of approach brings disadvantages, not least because it contradicts a fundamental design principle of EAs: the principle of minimum redundancy [63.40], which states that each member of the search space should be represented by as few distinct chromosomes as possible. To expand upon this point, we observe that under this *assignment-based* representation, if we are given a solution using $l \leq k$ colors, the number of different chromosomes representing this solution will be ^kP_l due to the arbitrary way in which colors are allocated labels. (For example, swapping the labels of colors 2 and 4 in Fig. 63.1's solution would give a new chromosome (1,2,3,2,3,5,4,4,2,1), but the same solution.) Of course, this implies a search space that is far larger than necessary.

Furthermore, authors such as *Falkenauer* [63.41] and *Coll* et al. [63.42] have also argued that traditional recombination schemes such as 1, 2, and n-point crossover with this representation have a tendency to recklessly break up building-blocks that we might want promoted in a population. As an example, consider a recombination of the two example chromosomes given in the previous paragraph using two-point crossover: (1,4,3,4,3,**5,2,2**,4,1) crossed with (**1,2,3,2,3**,5,4,4,**2,1**) would give (1,4,3,4,3,5,4,4,4,1) as one of the offspring. Here, despite the fact that the two parent chromosomes actually represent the same feasible solution, the resultant offspring seems to have little in common with its parents, having lost one of its colors, and seen a number of clashes having been introduced. Thus, it is concluded by these authors that such operations actually constitute more of a random perturbation operator, rather than a mechanism for combining meaningful substructures from existing solutions. Nevertheless, recent algorithms following this scheme are still reported in the literature [63.43].

In recognition of the proposed disadvantages of the assignment-based representation, *Coll* et al. [63.42] proposed a procedure for relabeling the colors of one of the parent chromosomes before applying crossover. Consider two (not necessarily feasible) parent solutions represented as partitions: $\mathcal{U}_1 = \{U_{1,1}, \ldots, U_{1,k}\}$, and $\mathcal{U}_2 = \{U_{2,1}, \ldots, U_{2,k}\}$. Now, using $\mathcal{U}_1$ and $\mathcal{U}_2$, a complete bipartite graph $K_{k,k}$ is formed. This bipartite graph has k vertices in each partition, and the weights between two vertices i, j from different partitions are defined as $w_{i,j} = |U_{1,i} \cap U_{2,j}|$. Given $K_{k,k}$, a maximum weighted matching can then be determined using any suitable algorithm (e.g., the Hungarian algorithm [63.44] or auc-

tion algorithm [63.45]), and this matching can be used to re-label the colors in one of the chromosomes.

Figure 63.2 gives an example of this procedure and shows how the second parent can be altered so that its color labelings maximally match those of parent 1. In this case, we note that the color classes $\{v_1, v_{10}\}$, $\{v_3, v_5\}$, and $\{v_6\}$ occur in both parents and will be preserved in any offspring produced via a traditional crossover operator. However, this will not always be the case and will depend very much on the best matching that is available in each case.

A further scheme for color relabeling that also addresses the issue of redundancy was proposed by *Tucker* et al. [63.46]. This method involves representing solutions using the assignment-based scheme, but under the following restriction

$$c(v_1) = 1 , \tag{63.3}$$
$$c(v_{i+1}) \leq \max\{c(v_1), \ldots, c(v_i)\} + 1 . \tag{63.4}$$

Chromosomes obeying these labeling criteria might, therefore, be considered as being in their canonical form such that, by definition, vertex v_1 is always colored with color 1, v_2 is always colored with color 1 or 2, and so on. (The solution given in Fig. 63.1 would be represented by (1,2,3,2,3,4,5,5,2,1) under this scheme.) However, although this ensures a one-to-one correspondence between the set of chromosomes and the set of vertex partitions (thereby removing any redundancy), research by *Lewis* and *Pullin* [63.47] demonstrated that this scheme is not particularly useful for graph coloring, not least because minor changes to a chromosome (such as the recoloring a single vertex) can lead to major changes to the way colors are labeled, making the propagation of useful solution substructures more difficult to achieve when applying traditional crossover operators.

Partition–Based Operators

Given the proposed issues with the assignment-based approach, the last 15 years or so have also seen a num-

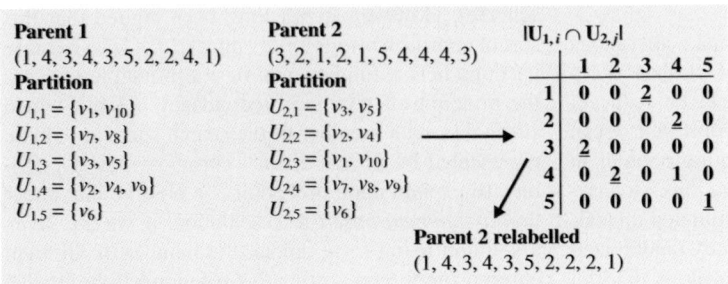

Parent 1	Parent 2	$|U_{1,i} \cap U_{2,j}|$					
(1, 4, 3, 4, 3, 5, 2, 2, 4, 1)	(3, 2, 1, 2, 1, 5, 4, 4, 4, 3)		1	2	3	4	5
Partition	**Partition**	1	0	0	2	0	0
$U_{1,1} = \{v_1, v_{10}\}$	$U_{2,1} = \{v_3, v_5\}$	2	0	0	0	2	0
$U_{1,2} = \{v_7, v_8\}$	$U_{2,2} = \{v_2, v_4\}$	3	2	0	0	0	0
$U_{1,3} = \{v_3, v_5\}$	$U_{2,3} = \{v_1, v_{10}\}$	4	0	2	0	1	0
$U_{1,4} = \{v_2, v_4, v_9\}$	$U_{2,4} = \{v_7, v_8, v_9\}$	5	0	0	0	0	1
$U_{1,5} = \{v_6\}$	$U_{2,5} = \{v_6\}$						

Parent 2 relabelled
(1, 4, 3, 4, 3, 5, 2, 2, 2, 1)

Fig. 63.2 Example of the relabeling procedure proposed by *Coll* et al. [63.42]. Here, parent 2 is relabeled as $1 \to 3, 2 \to 4, 3 \to 1, 4 \to 2$, and $5 \to 5$

ber of articles presenting recombination operators focussed on the partition (or grouping) interpretation of graph coloring. The philosophy behind this approach is that it is actually the color classes (and the vertices that are assigned to them) that represent the underlying building blocks of the graph coloring problem. In other words, it is not the color of individual vertices per se, but the *way in which vertices are grouped* that form the meaningful substructures. Consequently, the focus should be on the design of operators that are successfully able to combine and promote these within a population.

Perhaps the first major work in this area was due to *Falkenauer* [63.48] in 1994 (and later [63.41]) who argued in favor of the partition interpretation in the justification of his grouping genetic algorithm (GGA) – an EA methodology specifically designed for use with partitioning problems. *Falkenauer* applied this GGA to two important operational research problems: the bin-packing problem and bin-balancing problem, with strong results being reported. In subsequent work, *Erben* [63.20] also tailored the GGA for graph coloring. Erben's approach operates in the space of feasible colorings and allows the number of colors in a solution to vary. Solutions are then stored as partitions, and evaluated using (63.1). In this approach, recombination operates by taking two parent solutions and randomly selecting a subset of color classes from the second. These color classes are then copied into the first parent, and all color classes coming from the first parent containing duplicate vertices are deleted. This operation results in an offspring solution that is proper, but most likely partial. Thus uncolored vertices are then reinserted into the solution, in this case using the first-fit algorithm. A number of other recombination operators for use in the space of feasible solutions have also been suggested by *Mumford* [63.19]. These operate on permutations of vertices, which are again decoded into solutions using the first-fit algorithm.

Another recombination operator that focusses on the partition interpretation of graph coloring is due to *Galinier* and *Hao*, who in 1999 proposed an EA that, at the date of writing, is still understood to be one of the best performing algorithms for graph coloring [63.33, 38, 49, 50]. Using a fixed number of colors k, Galinier and Hao's method operates in the space of complete (proper and improper) k-colorings using cost function f_2 (63.2). A population of candidate solutions is then evolved using local search (based on tabu search) together with a specialized recombination operator called greedy partition crossover (GPX). The latter is used as a global operator and is intended to guide the search over the long term, gently directing it towards favorable regions of the search space (exploration), while the local search element is used to identify high quality solutions within these regions (exploitation).

The idea behind GPX is to construct offspring using large color classes inherited from the parent solutions. A demonstration of how this is done is given in Fig. 63.3. As is shown, the largest (not necessarily proper) color class in the parents is first selected and copied into the offspring. Then, in order to avoid dupli-

		Parent 1	Parent 2	Offspring	
a)	$U_1 =$	$\{v_1, v_2, v_3\}$	$\{v_3, v_4, v_5, v_7\}$	$\{\}$	Select the color with most vertices and copy to the
	$U_2 =$	$\{v_4, v_5, v_6, v_7\}$	$\{v_1, v_6, v_9\}$	$\{\}$	child (U_2 from parent 1 here).
	$U_3 =$	$\{v_8, v_9, v_{10}\}$	$\{v_2, v_8, v_{10}\}$	$\{\}$	Delete copied vertices from both parents.
b)	$U_1 =$	$\{v_1, v_2, v_3\}$	$\{v_3\}$	$\{v_4, v_5, v_6, v_7\}$	Select the color with most vertices in parent 2 and
	$U_2 =$	$\{\}$	$\{v_1, v_9\}$	$\{\}$	copy to child.
	$U_3 =$	$\{v_8, v_9, v_{10}\}$	$\{v_2, v_8, v_{10}\}$	$\{\}$	Delete copied vertices from both parents.
c)	$U_1 =$	$\{v_1, v_3\}$	$\{v_3\}$	$\{v_4, v_5, v_6, v_7\}$	Select the color with most vertices in parent 1 and
	$U_2 =$	$\{\}$	$\{v_1, v_9\}$	$\{v_2, v_8, v_{10}\}$	copy to the child.
	$U_3 =$	$\{v_9\}$	$\{\}$	$\{\}$	Delete copied vertices from both parents.
d)	$U_1 =$	$\{\}$	$\{\}$	$\{v_4, v_5, v_6, v_7\}$	Having formed k colors, assign any missing vertices
	$U_2 =$	$\{\}$	$\{v_9\}$	$\{v_2, v_8, v_{10}\}$	to random colors.
	$U_3 =$	$\{v_9\}$	$\{\}$	$\{v_1, v_3\}$	
e)	$U_1 =$	$\{\}$	$\{v_9\}$	$\{v_4, v_5, v_6, v_7\}$	A complete (though not necessarily proper) solution
	$U_2 =$	$\{v_9\}$	$\{\}$	$\{v_2, v_8, v_{10}, v_9\}$	results.
	$U_3 =$	$\{\}$	$\{\}$	$\{v_1, v_3\}$	

Fig. 63.3 Demonstration of the GPX operator using $k = 3$

cate vertices occurring in the offspring at a later stage, these copied vertices are removed from both parents. To form the next color, the other (modified) parent is then considered and, again, the largest color class is selected and copied into the offspring, before again removing these vertices from both parents. This process is continued by alternating between the parents until the offspring's k color classes have been formed. At this point, each color class in the offspring will be a subset of a color class existing in one or both of the parents. That is

$$\forall U_i \in \mathcal{U}_c \; \exists U_j \in (\mathcal{U}_1 \cup \mathcal{U}_2) : U_i \subseteq U_j , \qquad (63.5)$$

where $\mathcal{U}_c$, $\mathcal{U}_1$, and $\mathcal{U}_2$ represent the offspring, and parents 1 and 2, respectively.

One feature of the GPX operator is that on production of an offspring's k color classes, some vertices may be missing (this occurs with vertex v_9 in Fig. 63.3). *Galinier* and *Hao* [63.33] suggest assigning these uncolored vertices to random classes, which of course could introduce further clashes. This element of the procedure might, therefore, be viewed as a type of perturbation (mutation) operator in which the number of random assignments (the size of the perturbation) is determined by the construction stages of GPX. However, *Glass* and *Prugel-Bennett* [63.49] observe that GPX's strategy of inheriting the largest available color class at each step (as opposed to a random color class) generally reduces the number of uncolored vertices. This means that the amount of information inherited directly from the parents is increased, reducing the potential for disruption. Once a complete offspring is formed, it is then modified and improved via a local search procedure before being inserted into the population.

Since the proposal of GPX by *Galinier* and *Hao* [63.33], further recombination schemes based on this method have also been suggested, differing primarily in the criteria used for selecting the color classes that are inherited by the offspring. *Lü* and *Hao* [63.34], for example, extended the GPX operator to allow more

than two parents to play a part in producing a single offspring (Sect. 63.5). On the other hand, *Porumbel* et al. [63.35] suggest that instead of choosing the largest available color class at each stage of construction, classes with the least number of clashes should be prioritized, with class size (and information regarding the degrees of the vertices) then being used to break ties. *Malaguti* et al. [63.38] also use a modified version of GPX with an EA that navigates the space of partial, proper solutions. In all of these cases the authors combined their recombination operators with a local search procedure in the same manner as *Galinier* and *Hao* [63.33] and, with the problem instances considered, the reported results are generally claimed to be competitive with the state of the art.

Assessing the Effectiveness of EAs for Graph Coloring

In recent work carried out by the author of this chapter [63.50], a comparison of six different graph coloring algorithms was presented. This study was quite broad and used over 5000 different problem instances. Its conclusions were also rather complex, with each method outperforming all others on at least one class of problems. However, a salient observation was that the GPX-based EA of *Galinier* and *Hao* [63.33] was by far the most consistent and high-performing algorithm across the comparison.

In the remainder of this chapter we pursue this matter further, particularly focussing on the role that GPX plays in this performance. Under a common EA framework, described in Sect. 63.3, we first evaluate the performance of GPX by comparing it to two other recombination operators (Sect. 63.4). Using information gained from these experiments, Sect. 63.5 then looks at how the performance of the GPX-based EA might be enhanced, particularly by looking at ways in which population diversity might be prolonged during a run. Finally, conclusions and a further discussion surrounding the virtues of recombination in this problem domain are presented in Sect. 63.6.

63.3 Setup

The EA used in the following experiments operates in the same manner as *Galinier* and *Hao's* [63.33]. To form an initial population, a modified version of the DSATUR algorithm is used. Specifically, each individual is formed by taking the vertices in turn according to

the DSATUR heuristic and then assigning it to the lowest indexed color $i \in \{1, \ldots, k\}$ where no clash occurs. Vertices for which no clash-free color exists are assigned to random colors at the end of this process. Ties in the DSATUR heuristic are broken randomly, providing di-

Table 63.1 Details of the five problem instances used in our analysis

#: Name	$|V|$	Density	Vertex degree min; med; max	Mean	SD	Best known (colors)
1: Random	1000	0.499	450; 499; 555	499.4	16.1	83
2: Flat(10)	500	0.103	36; 52; 61	51.7	4.4	10
3: Flat(100)	500	0.841	393; 421; 445	420.7	7.6	100
4: TT(A)	682	0.128	0; 77; 472	87.4	62.0	27
5: TT(B)	2419	0.029	0; 47; 857	71.3	92.3	32

versity in the initial population. Each individual is then improved by the local search routine.

The EA evolves the population using recombination, local search, and replacement pressure. In each iteration two parent solutions are selected at random, and the selected recombination operator is used to produce one offspring. This offspring is then improved via local search and inserted into the population by replacing the weaker of its two parents.

The local search element of this EA makes use of tabu search – specifically the TABUCOL algorithm of *Hertz* and *de Werra* [63.24], run for a fixed number of iterations. In this method, moves in the search space are achieved by selecting a vertex v whose assignment to color i is currently causing a clash, and moving it to a new color $j \neq i$. The inverse of this move is then marked as tabu for the next t steps of the algorithm (meaning that v cannot be re-assigned to color i until at least t further moves have been performed). In each iteration, the complete neighborhood is considered, and the non-tabu move that is seen to invoke the largest decrease in cost (or failing that, the smallest increase) is performed. Ties are broken randomly, and tabu moves are also carried out if they are seen to improve on the best solution observed so far in the process. The tabu search routine terminates when the iteration limit is reached (at which case the best solution found during the process is taken), or when a zero cost solution is achieved. Further descriptions of this method, including implementation details, can be found in [63.51].

In terms of parameter settings, in all cases we use a population size of 20 (as in [63.34, 35]) and set the tabu search iteration limit to $16|V|$, which approximates the settings used in the best reported runs in [63.33]. As with other algorithms that use this local search technique [63.29, 33, 37], the tabu tenure t is made proportional to the current solution cost: specifically, $t = \lceil 0.6f_2 \rceil + r$, where r is an integer uniformly selected from the range 0–9 inclusive.

Finally, because this algorithm operates in the space of complete k-colorings (proper and improper), values for k must be specified. In our case, initial values are

determined by executing DSATUR on each instance and setting k to the number of colors used in the resultant solution. During runs, k is then decremented by 1 as soon as a feasible k-coloring is found, and the algorithm is restarted. Computational effort is measured by counting the number of constraint checks carried out by the algorithm, which occur when the algorithm requests information about a problem instance, including checking whether two vertices are adjacent (by accessing an adjacency list or matrix), and referencing the degree of a vertex. In all trials a cut-off point of 5×10^{11} checks is imposed, which is roughly double the length of the longest run performed in [63.33]. In our case, this led to run times of ≈ 1 h on our machines (algorithms were coded in C++ and executed on a PC under Windows XP using a 3.0 GHz processor with 3.18 GB of RAM).

63.3.1 Problem Instances

For our trials a set of five problem instances is considered. Though this set is quite small, its members should be considered as case studies that have been deliberately chosen to cover a wide range of graph structure – a factor that we have found to be very important in influencing the relative performance of graph coloring algorithms [63.50]. The first three graphs are generated using the publicly available software of *Culberson* [63.52], while the remaining two are taken from a collection of real-world timetabling problems compiled by *Carter* et al. [63.53]. Names and descriptions of these graphs now follow. Further details are also given in Table 63.1:

#1: *Random.* This graph features $|V| = 1000$ and is generated such that each of the $\binom{|V|}{2}$ pairs of vertices is linked by an edge with probability 0.5. Graphs of this nature are nearly always considered in comparisons of coloring algorithms.

#2: *Flat(10).* Flat graphs are generated by partitioning the vertices into K equi-sized groups and then adding edges between vertices in different groups with probability p. This is done such that the variance in vertex degrees is kept to a minimum. It is well known that feasible K-colored solutions to such graphs are generally easy to achieve except in cases where p is within a specific range of values, which results in problems that are notoriously difficult. Such ranges are commonly termed *phase transition regions* [63.54]. This particular instance is generated so that it features a relatively small number of large color classes (using $V = 500$ and $K = 10$, implying ≈ 50 vertices per color). A value of $p = 0.115$ is used, which has been observed to provide very difficult instances for a range of different graph coloring algorithms [63.50].

#3 *Flat(100)*. This graph is generated in the same manner as the previous one, using $|V| = 500, K = 100$, and $p = 0.85$. Solutions thus feature a relatively large number of small color classes (≈ 5 vertices per color).

#4: *TT(A)*. This graph is named *car_s_91* in the original dataset of *Carter* et al. [63.53]. It is chosen because it is quite large and, unlike the previous three graphs, the variance in vertex degrees is quite high. This problem's structure is also much less regular than the previous three graphs, which are generated in a fairly regimented manner.

#5: *TT(B)*. This graph, originally named *pur_s_93*, is the largest problem in Carter's dataset, with $|V| = 2419$. It is also quite sparse compared to the previous graph, though it still features a high variance in vertex degrees (Table 63.1).

The rightmost column of Table 63.1 also gives information on the best solutions known for each graph. These values were determined via extended runs of our algorithms, or due to information provided by the problem generator.

63.4 Experiment 1

Our first set of experiments looks at the performance of GPX by comparing it to two additional recombination operators. To gauge the advantages of using a global operator (recombination in this case), we also consider the performance of TABUCOL on its own, which iterates on a single solution until the run cut-off point is met.

Our first additional recombination operator follows the assignment-based scheme discussed in Sect. 63.2.1 and, in each application, utilizes the procedure of *Coll* et al. [63.42] (Fig. 63.2) to relabel the second parent. Offspring are then formed using the classical n-point crossover, with each gene being inherited from either parent with probability 0.5.

Our second recombination operator is based on the grouping genetic algorithm (GGA) methodology (Sect. 63.2.1), adapted for use in the space of k-colorings. An example is given in Fig. 63.4. Given

two parents, the color classes in the second parent are first relabeled using Coll et al.'s procedure. Using the partition-based representations of these solutions, a subset of colors in parent 2 is then chosen randomly, and these replace the corresponding colors in a copy of parent 1. Duplicate vertices are then removed from color classes originating from parent 1 and uncolored vertices are assigned to random color classes. Note that like GPX, before uncolored vertices are assigned, the property defined by (63.5) is satisfied by this operator; however, unlike GPX there is no requirement to inherit larger color classes or to inherit half of its color classes from each parent.

A summary of the results achieved by the three recombination operators (together with TABUCOL) is given in Table 63.2. For each instance the same set of 20 initial populations was used with the EAs, and entries in bold signify samples that are significantly different to the non-bold EA entries according to a Wilcoxon signed-rank test at the 0.01 significance level. For graph #1 we see that GPX has clearly produced the best results – indeed, even its worst result features two fewer colors than the next best solution. However, for graphs #2 and #5, no significant difference between the EAs is observed, while for #3 and #4, better results are produced by the GGA and the n-point crossover.

Figure 63.5 shows run profiles for two example graphs. We see that in both cases TABUCOL provides

Parent 1	Parent 2	Offspring
$U_1 = \{v_1, v_{10}\}$	$U_1 = \{v_1, v_9\}$	$U_1 = \{v_1, v_{10}\}$
$U_2 = \{v_7, v_8\}$	$U_2 = \{v_7\}$	$U_2 = \{v_7, \not v_8\}$
$U_3 = \{v_3, v_5\}$	⇨$U_3 = \{v_3, v_5\}$	$U_3 = \{v_3, v_5\}$
$U_4 = \{v_2, v_4, v_9\}$	⇨$U_4 = \{v_2, v_4, v_9\}$	$U_4 = \{v_2, v_4, v_9\}$
$U_5 = \{v_6\}$	$U_5 = \{v_6, v_{10}\}$	$U_5 = \{v_6\}$ **Uncolored** $= \{v_9\}$

Fig. 63.4 Demonstration of the GGA recombination operator. Here, color classes in parent 2 are labeled to maximally match those of parent 1

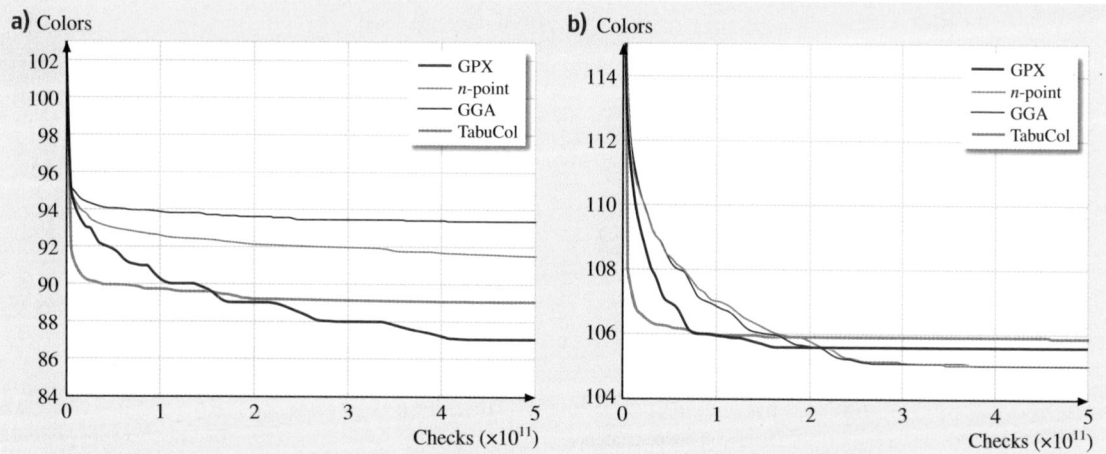

Fig. 63.5a,b Run profiles for the instances (mean of 20 runs): (a) #1 (random). (b) #3 (Flat 100)

the fastest rates of improvement, though it is eventually overtaken by at least one of the EAs. Table 63.2, however, also reveals that TABUCOL performs very poorly with graphs #4 and #5. This seems due to the high degree variance in these cases, which we observe makes the cost of neighboring solutions in the search space vary more widely. This suggests a more *spiky* cost landscape in which the use of local search in isolation exhibits a susceptibility for becoming trapped at local optima (see also [63.50]).

An important factor behind the differing performances of these EAs is the effect that recombination has on the population diversity. To examine this, we first define a metric for measuring the distance between two solutions: Given a solution $\mathcal{U}$, let $P_{\mathcal{U}} = \{\{u, v\} : c(u) = c(v)\}$, for $\forall u, v \in V$, $u \neq v$. The *distance* between two solutions $\mathcal{U}_1$ and $\mathcal{U}_2$ can then be defined,

$$D(\mathcal{U}_1, \mathcal{U}_2) = \frac{|P_{\mathcal{U}_1} \cup P_{\mathcal{U}_2}| - |P_{\mathcal{U}_1} \cap P_{\mathcal{U}_2}|}{|P_{\mathcal{U}_1} \cup P_{\mathcal{U}_2}|} . \quad (63.6)$$

This measure gives the proportion of vertex pairings (assigned to the same color) that exist in just one of the two solutions. Consequently, if $\mathcal{U}_1$

and $\mathcal{U}_2$ are identical, then $P_{\mathcal{U}_1} \cup P_{\mathcal{U}_2} = P_{\mathcal{U}_1} \cap P_{\mathcal{U}_2}$, giving $D(\mathcal{U}_1, \mathcal{U}_2) = 0$. Conversely, if no vertex pair is assigned the same color, $P_{\mathcal{U}_1} \cap P_{\mathcal{U}_2} = \emptyset$, implying $D(\mathcal{U}_1, \mathcal{U}_2) = 1$. Population diversity can also be defined as the mean distance between each pair of solutions in the population. That is, given a set of m individuals $\mathbf{U} = \{\mathcal{U}_1, \mathcal{U}_2, \dots, \mathcal{U}_m\}$

$$\text{Diversity}(\mathbf{U}) = \frac{1}{\binom{m}{2}} \sum_{\forall \mathcal{U}_i, \mathcal{U}_j \in \mathbf{U}: i < j} D(\mathcal{U}_i, \mathcal{U}_j) .$$

$$(63.7)$$

Considering our results, the two scatter plots of Fig. 63.6 demonstrate the positive correlation that exists between parental distance and the number of uncolored vertices that result in applications of the GPX and GGA operators. This data was derived from graph #4, though similar patterns were observed for the other instances. Note that the correlation is weaker for GGA due to two reasons. First, unlike GPX, which requires half of the color classes to be inherited from each parent, with GGA this proportion can vary. Thus if the majority of color classes are inherited from just one par-

Table 63.2 Number of colors in the best feasible solution achieved at the cut-off point (mean (min; median; max) of 20 runs)

	GPX		**n-point**		**GGA**		**TABUCOL**	
#1	87.00	(87; 87; 87)	93.35	(93; 93; 94)	91.55	(91; 92; 92)	89.10	(89; 89; 90)
#2	12.95	(12; 13; 13)	13.00	(13; 13; 13)	13.00	(13; 13; 13)	13.00	(13; 13; 13)
#3	105.60	(105; 106; 106)	**105.05**	**(105; 105; 106)**	105.05	**(105; 105; 106)**	105.90	(105; 106; 106)
#4	29.05	(28; 29; 30)	**28.00**	(28; 28; 28)	27.90	(27; 28; 29)	38.20	(32; 37.5; 46)
#5	33.30	(33; 33; 34)	33.15	(32; 33; 34)	33.10	(32; 33; 34)	52.05	(47; 52; 56)

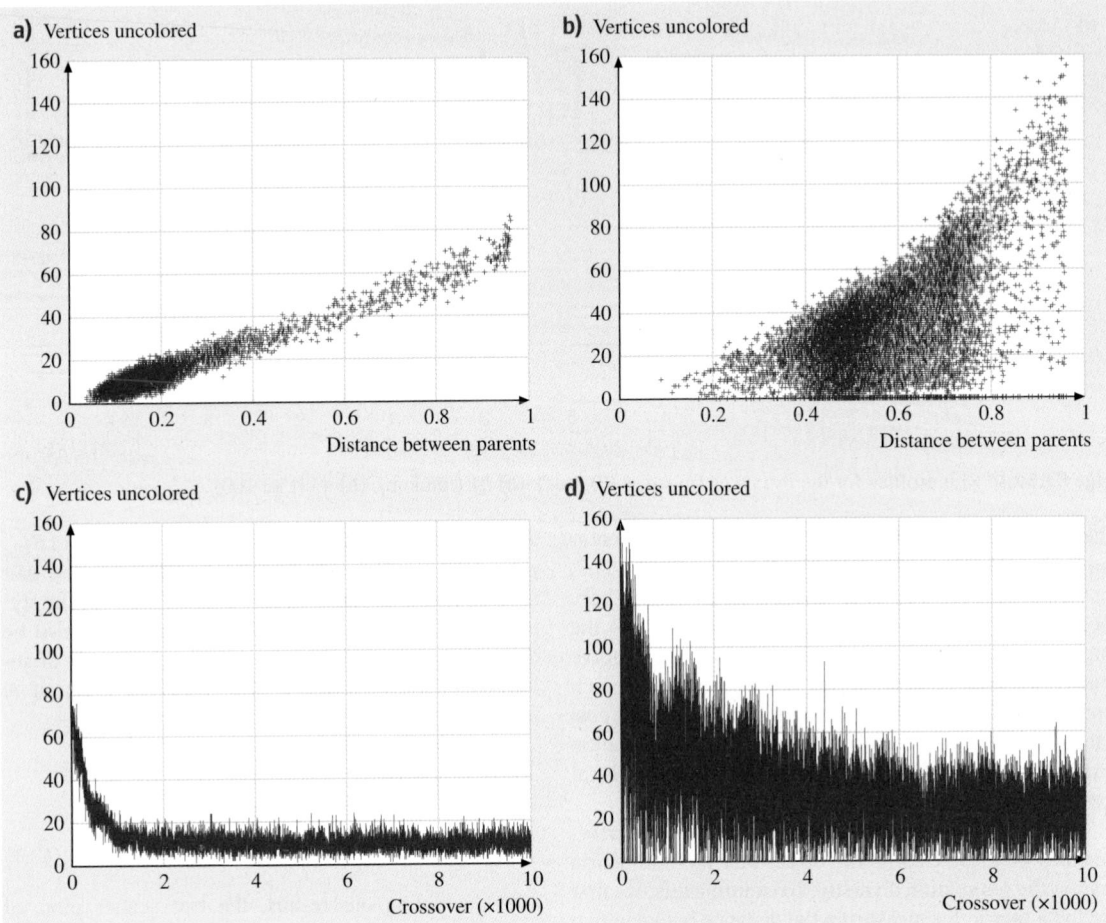

Fig. 63.6a–d Relationship between parental distance and number of uncolored vertices with the GPX (**a**) and GGA (**b**) operators. Also shown is the number of uncolored vertices in the first 10 000 applications of GPX (**c**) and GGA (**d**)

ent, it is possible to have two very different parents, but only a small number of uncolored vertices. Second, as mentioned earlier GGA shows no bias towards inheriting larger color classes, meaning that the number of uncolored vertices can also be higher than GPX, particularly when inheriting around half of the color classes from each parent. An effect of these patterns is shown in the lower graphs of Fig. 63.6, where throughout the evolutionary process, the number of uncolored vertices occurring during recombination is fewer and less varied with GPX. In comparison to GGA, this behavior leads to a more rapid loss of diversity, as is demonstrated in Fig. 63.7 for two example graphs.

Whether sustained diversity is a help or hindrance with these EAs thus seems to depend on the type of graph being tackled. As can be seen in Fig. 63.7, for

graph #1 GPX is the only recombination operator that leads to any sort of population convergence, and it is also the algorithm that produces the best solutions given sufficient time, suggesting that is suitably *homing in* on high-quality regions of the search space. On the other hand, for graphs #3 and #4, GGA's more sustained diversity (caused and perpetuated by the greater number of uncolored vertices that occur during recombination) causes the operator to be more disruptive. However, in these cases this factor also seems to provide a useful diversification mechanism, allowing the algorithm to sample wider areas of the search space, leading to better results. An extreme case of diversity loss occurs with graph #5, which we recall has a low density and high degree variance. In this case, when using GPX large color classes of low-degree vertices that are formed in

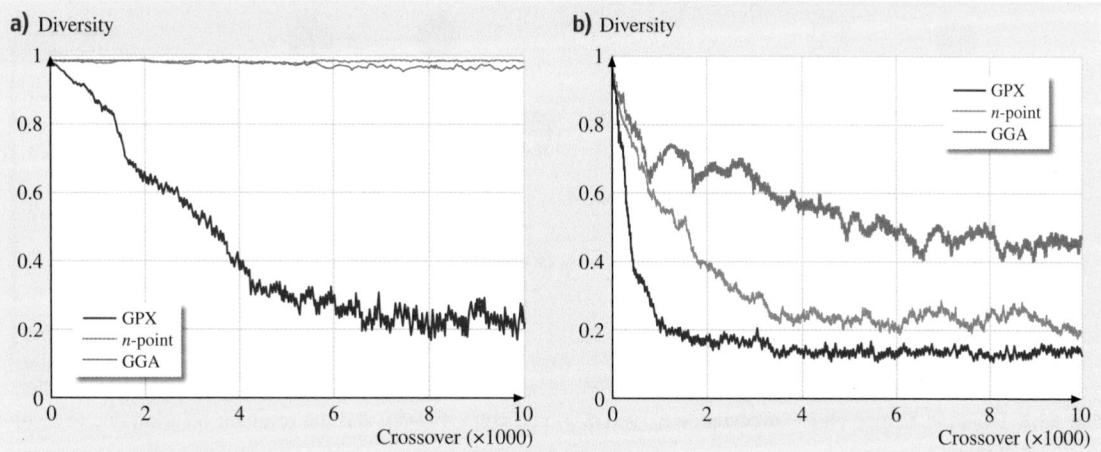

Fig. 63.7a,b Population diversity during the first 10 000 recombinations with **(a)** the random (#1) and **(b)** TT(A) (#4) instances

early stages of the algorithm quickly come to dominate the population limiting the exploration that then takes place – indeed, in many runs the algorithm was actually unable to improve on costs achieved in the initial population.

Figure 63.7 also shows that n-point crossover tends to maintain diversity for longer periods than GPX in this case, allowing it to produce superior results for graphs #3 and #4. However, the sustained diversity is not due to uncolored vertices (which do not occur with this operator); rather, it seems due to the naturally occurring disruption that results from the color labeling issues mentioned in Sect. 63.2.1.

Finally, we also mention that during our runs with these EA's, the local search element was observed to be by far the most expensive part of the algorithm, with none of the recombination operators consuming more than 1.8% of the available run time.

63.5 Experiment 2

In this section we now consider ways in which the results of the GPX operator might be improved, particularly looking at how we might encourage diversity to be sustained in the population.

As mentioned in Sect. 63.2.1, *Lü* and *Hao* [63.34] previously proposed extending the GPX operator to allow offspring to be produced using $m \geq 2$ parents. In this operator, which we call MULTIX, offspring are constructed in the same manner as GPX, except that at each stage the largest color class from *multiple parents* is chosen to be copied into the offspring. The intention behind this increased choice is that larger color classes will be identified, resulting in fewer uncolored vertices once the k color classes have been constructed. In order to prohibit too many colors being inherited from one particular parent, Lü and Hao also make use of a parameter q, specifying that if the i-th color class in an offspring is copied from a particular parent, then this parent should not be considered for further q colors. In our application of MULTIX we follow the recommendations of the Lü and Hao, choosing m randomly from the set $\{2, \dots, 6\}$ in each application and using $q = \lfloor m/2 \rfloor$. Note also that GPX is simply an application of MULTIX using $m = 2$ and $q = 1$.

Though having the potential to produce good results [63.34], an issue with MULTIX is that it could result in diversity being lost even more rapidly than with GPX, particularly if fewer vertices need to be randomly recolored at the end of each application. In [63.34], *Lü* and *Hao* attempt to deal with this using a mechanism whereby offspring are only inserted into the population if they are seen to be sufficiently different or better than existing members. However, in our case, we suggest two alternative methods.

The first of these involves altering the MULTIX operator so that it works exclusively with proper col-

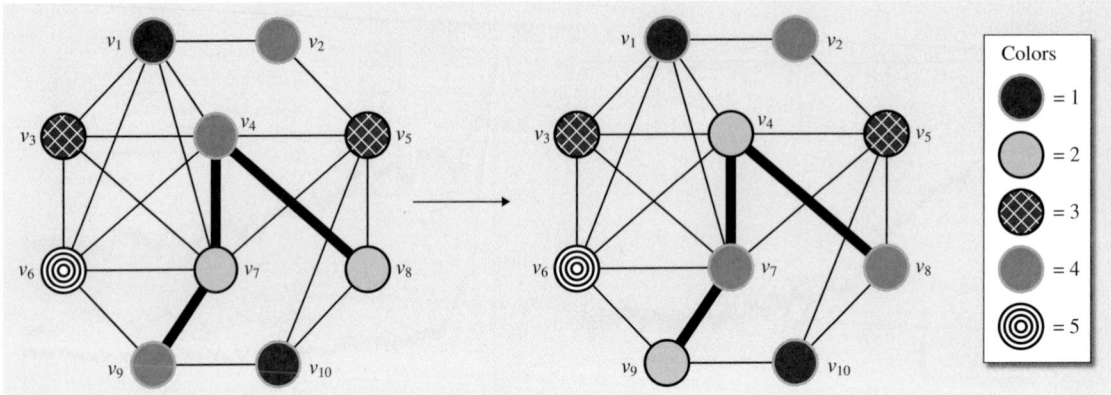

Fig. 63.8 Example Kempe chain involving, e.g., vertex v_7 and color 4 (*left*), and the resultant coloring due to a color interchange (*right*)

orings. As noted, GPX and MULTIX currently operate on colorings in which clashes are permitted; however, this could in theory result in large color classes that feature many clashes being unduly promoted in the population, when perhaps the real emphasis should be on the promotion of large color classes that are *independent sets*. The ISETS approach thus operates by first iteratively removing clashing vertices from each parent (in a random order, until proper colorings are achieved), and then using the MULTIX operator to produce an offspring as before. This implies that, before recoloring missing vertices, offspring will also be proper, since subsets of independent sets are themselves independent sets. A further effect is that a greater number of vertices might need to be recolored, since vertices originally removed from the parents could also be missing in the resultant offspring.

Our second proposal for prolonging diversity is to make changes directly to an offspring to try to increase its distance from its parents before reinsertion into the population. One way of doing this would be to increase the iteration limit of the local search procedure, as demonstrated by *Galinier* and *Hao* [63.33]. However, we find that such an approach can slow the algorithm unnecessarily, particularly because as the procedure progresses, movements in the search space (due to improving or sideways moves) become less frequent. An alternative in this case is to exploit the structure of the graph coloring problem via the use of a Kempe chain interchange operator. Kempe chains define connected sub-graphs that involve exactly two colors, and can be generated by taking an arbitrary vertex v and color i, such that $c(v) \neq i$. An example is given in Fig. 63.8. Note that when interchanging the colors of vertices in

a Kempe chain, if the original coloring is proper, then so is the new coloring. Thus we have the opportunity to quickly alter colorings without compromising their quality.

Our KEMPE approach operates in the same manner as ISETS, except that before reassigning uncolored vertices, a series of randomly selected Kempe chain interchanges are performed on the existing proper coloring. In our case, $2k$ such moves are applied.

The results achieved by our three modifications are summarized in Table 63.3, where bold entries signify samples that are significantly different to GPX at significance level 0.01. We see that improvements over GPX were only obtained on graph #1, where all three variants were successful, and graph #4 using the KEMPE variant. In practice, we found that MULTIX causes diversity to be lost more quickly than GPX with these graphs – however, the ISETS mechanism did not seem to alter this behavior a great deal, usually because the number of clashing vertices needing to be removed was quite small (less than 10).

Surprisingly, we also found that the KEMPE variant was only able to maintain higher levels of diversity with instances #4 and #5. For graphs #1, #2, and #3, it turns out that when using a suitably low number of colors k, the bipartite graphs induced by most pairs of color classes in a solution are connected. In these cases, all of the vertices belonging to the two color classes are included in the Kempe chain, meaning that a color interchange does not alter the structure of the solution, but merely produces a relabeling of the two color classes. (An example of such a Kempe chain would occur in Fig. 63.8 using vertex v_3 and color 2.) This is not the case for the less structured graphs #4 and #5, where we

Table 63.3 Number of colors in the best feasible coloring achieved at the cut-off point (mean (min; median; max) from 20 runs)

	GPX		MULTIX		ISETS		KEMPE	
#1	87.00	(87; 87; 87)	**85.00**	(85; 85; 85)	85.05	(85; 85; 86)	85.15	(85; 85; 86)
#2	12.95	(12; 13; 13)	13.00	(13; 13; 13)	13.00	(13; 13; 13)	12.90	(12; 13; 13)
#3	105.60	(105; 106; 106)	105.55	(105; 106; 106)	105.85	(105; 106; 106)	105.30	(105; 105; 106)
#4	29.05	(28; 29; 30)	29.10	(29; 29; 30)	29.00	(28; 29; 30)	**28.00**	(28; 28; 28)
#5	33.30	(33; 33; 34)	33.30	(33; 33; 34)	33.30	(33; 33; 34)	33.30	(33; 33; 34)

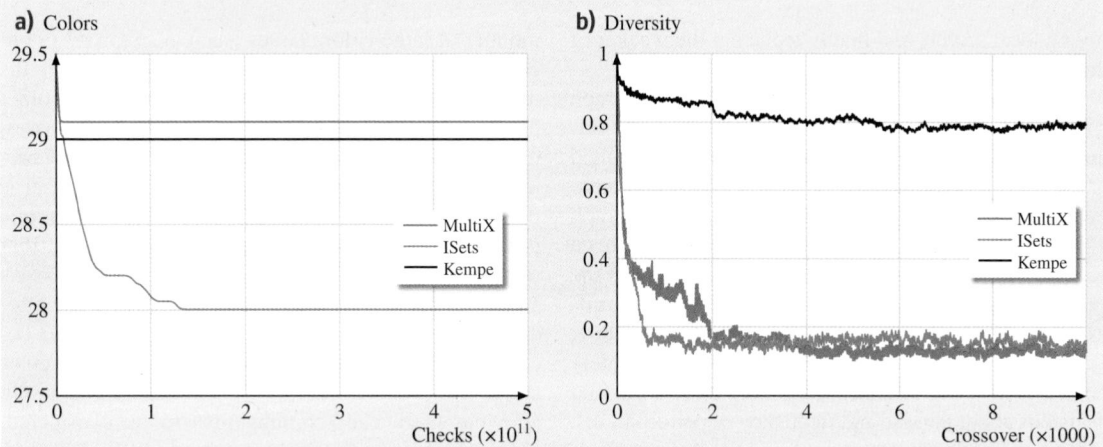

Fig. 63.9 (**a**) Run profile for TT(A) (graph #4, left), and (**b**) its diversity over the first 10 000 recombinations

found that diversity could be maintained for longer periods. However, this only led to significant improvements in the results for graph #4, whose run profiles are shown in Fig. 63.9. Also note that these enhanced results still fail to beat those of the GGA and n-point operators, as shown in Table 63.2.

63.6 Conclusions and Discussion

In this chapter we have examined the relative performance of a number of different graph coloring recombination operators. Using a common evolutionary framework, we have seen that this performance varies, particularly due to the underlying structures of the graphs being tackled.

A desirable property of recombination is that it should be able to combine meaningful substructures of existing candidate solutions (parents) in the production of new, hopefully fitter, offspring. However, does that process actually occur with any of these operators? Or, by involving the random reassignment of some vertices, do the operators simply provide a mechanism by which large random perturbations are periodically applied to a solution, helping to re-invigorate the search process?

Again, the answer to such a question seems to depend on the problem instance at hand. In Table 63.4 we compare the costs of solutions achieved by the best available recombination operator for each instance, together with those produced by a corresponding random perturbation operator. Specifically, for each graph we identified the best run from the EA's sample of 20 and recorded the number of uncolored vertices that resulted in each application of recombination. We then used these figures, together with the same k-value, to specify the number of vertices that would be randomly selected and reassigned in each corresponding application of our random perturbation operator. In each iteration this algorithm then operated by selecting two parents, making a copy of parent 1, randomly perturbing this copy, ap-

Table 63.4 Comparison of the best EA and corresponding random perturbation operator. (Cost of best solutions using f_2 (63.2); (mean, (min; median; max) from 20 runs), and proportion of runs where $f_2 = 0$ (feasibility) was achieved)

	k		EA				Random		
		Type	Cost		Feas.		Cost		Feas.
#1	85	MULTIX	**0.00**	**(0; 0; 0)**	**1.00**		16.80	(4; 17.5; 31)	0.00
#2	12	GPX	**2.40**	**(0; 2; 4)**	0.05		7.60	(5; 8; 10)	0.00
#3	105	GPX	0.90	(0; 1; 2)	0.40		1.75	(0; 2; 3)	0.15
#4	27	GGA	1.10	(0; 1; 2)	0.15		1.35	(0; 1; 2)	0.05
#5	32	GGA	1.75	(0; 2; 3)	0.05		1.50	(0; 1.5; 3)	0.15

plying local search, and finally replacing the weaker of the two parents.

The results in Table 63.4 indicate that, for graph #1, recombination is clearly doing more than just randomly perturbing solutions since all runs have resulted in feasible 85-colorings. However, although recombination has achieved significantly lower *costs* with graph #2, the proportion of runs where feasibility has been achieved shows no significant difference for any of the graphs #2 to #5 (according to McNemar's test at significance level 0.01). We find this observation compelling as it might suggest that better results might ultimately be achieved using schemes that make more informed decisions about the size and frequency of perturbations. Indeed, currently the size of random perturbations tends to fall as the run progresses (Fig. 63.6); however, it may be useful to allow this trend to be reversed, particularly if improvements are not achieved for a lengthy period of time. In addition, the *way* in which vertices are chosen for random reassignment might also influence performance – for example, we might target those belonging to a specific color, those that are causing clashes, those that have been assigned to a particular color for the longest, and so on. This requires further research.

An interesting point regarding the structure of solutions was raised previously by *Porumbel* et al. [63.35], who considered the sizes of the color classes. Specifically, they propose that when solutions involve a small number of large color classes (such as graph #2 in our case), good quality colorings tend to result through the identification of large independent sets. On the other hand, if a solution involves many small color classes, quality is determined more by the *productive interaction* between classes. In other words, the proposal is that small independent sets in isolation do not constitute good features in these cases; rather, quality results from appropriate combinations of these sets. Such an observation might provide evidence as to why the GGA recombination has outperformed GPX with graph #3 because, unlike GPX, it does not require half of the color classes to be inherited from each parent, thus potentially allowing more class-combinations to be considered. However, this argument is countered by the fact that, according to Table 63.4, GGA has not outperformed the random perturbation operator, suggesting that it is actually this mechanism that influences the search. Clearly, further research in this area is also required.

Given such observations, another important avenue of future research will be to increase our understanding of the links between a graph's structure and the best algorithms that can then be used to color it. This might, for example, be derived by increasing our understanding of the behavior, strengths, and weaknesses of the various algorithmic operators available for graph coloring, and also via more empirical means such as data mining, as discussed by *Smith-Miles* and *Lopes* [63.55].

References

63.1   N. Zufferey, P. Amstutz, O. Giaccari: Graph colouring approaches for a satellite range scheduling problem, J. Sched. **11**(4), 263–277 (2008)

63.2   M. Carter: A survey of practical applications of examination timetabling algorithms, Oper. Res. **34**(2), 193–202 (1986)

63.3   R. Lewis: A survey of metaheuristic-based techniques for university timetabling problems, OR Spectrum **30**(1), 167–190 (2008)

63.4   R. Lewis, J. Thompson: On the application of graph colouring techniques in round-robin sports scheduling, Comput. Oper. Res. **38**(1), 190–204 (2010)

63.5   K. Aardel, S. van Hoesel, A. Koster, C. Mannino, A. Sassano: Models and solution techniques for the frequency assignment problems, 4OR: Q. J. Belg. Fr. Ital. Oper. Res. Soc. **1**(4), 1–40 (2002)

63.6 C.M. Valenzuela: A study of permutation operators for minimum span frequency assignment using an order based representation, J. Heuristics **7**, 5–21 (2001)

63.7 K. Appel, W. Haken: Solution of the four color map problem, Sci. Am. **4**, 108121 (1977)

63.8 M. Gamache, A. Hertz, J. Ouellet: A graph coloring model for a feasibility problem in monthly crew scheduling with preferential bidding, Comput. Oper. Res. **34**, 2384–2395 (2007)

63.9 G. Chaitin: Register allocation and spilling via graph coloring, ACM SIGPLAN Notices **39**(4), 66–74 (2004)

63.10 M.R. Garey, D.D. Johnson: *Computers and Intractability – A guide to NP-completeness*, 1st edn. (W. H. Freeman, San Francisco 1979)

63.11 M. Karp: Reducibility among combinatorial problems. In: *Complexity of Computer Computations*, The IBM Research Symposia Series, Vol. 1972, ed. by R.E. Miller, J.W. Thatcher, J.D. Bohlinger (Plenum Press, New York 1972) pp. 85–103

63.12 D. Welsh, M. Powell: An upper bound for the chromatic number of a graph and its application to timetabling problems, Comput. J. **12**, 317–322 (1967)

63.13 D. Brélaz: New methods to color the vertices of a graph, Commun. ACM **22**(4), 251–256 (1979)

63.14 P. Spinrad, G. Vijayan: Worse case analysis of a graph colouring algorithm, Discrete Appl. Math. **12**, 89–92 (1984)

63.15 R. Brown: Chromatic scheduling and the chromatic number problem, Manag. Sci. **19**(4), 451–463 (1972)

63.16 S. Korman: The graph-colouring problem. In: *Combinatorial Optimization*, ed. by N. Christofides, A. Mingozzi, P. Toth, C. Sandi (Wiley, New York 1979) pp. 211–235

63.17 M. Kubale, B. Jackowski: A generalized implicit enumeration algorithm for graph colouring, Communications ACM **28**(28), 412–418 (1985)

63.18 J. Culberson, F. Luo: Exploring the k-colorable landscape with iterated greedy, Proc. 2nd DIMACS Implement. Chall. (1996), pp. 245–284

63.19 C. Mumford: New order-based crossovers for the graph coloring problem, Lect. Notes Comput. Sci. **4193**, 880–889 (2006)

63.20 E. Erben: A grouping genetic algorithm for graph colouring and exam timetabling, Lect. Notes Comput. Sci. **2079**, 132–158 (2001)

63.21 R. Lewis: A general-purpose hill-climbing method for order independent minimum grouping problems: A case study in graph colouring and bin packing, Comput. Oper. Res. **36**(7), 2295–2310 (2009)

63.22 M. Chams, A. Hertz, O. Dubuis: Some experiments with simulated annealing for coloring graphs, Eur. J. Oper. Res. **32**, 260–266 (1987)

63.23 D. Johnson, C. Aragon, L. McGeoch, C. Schevon: Optimization by simulated annealing: An experimental evaluation; part II, graph coloring and number partitioning, Oper. Res. **39**, 378–406 (1991)

63.24 A. Hertz, D. de Werra: Using tabu search techniques for graph coloring, Computing **39**(4), 345–351 (1987)

63.25 M. Laguna, R. Marti: A GRASP for coloring sparse graphs, Comput. Optim. Appl. **19**, 165–178 (2001)

63.26 M. Chiarandini, T. Stützle: An application of iterated local search to graph coloring, Proc. Comput. Symp. Graph Color. Gen. (2002) pp. 112–125

63.27 L. Paquete, T. Stützle: An experimental investigation of iterated local search for coloring graphs, applications of evolutionary computing, Lect. Notes Comput. Sci. **2279**, 121–130 (2002)

63.28 C. Avanthay, A. Hertz, N. Zufferey: A variable neighborhood search for graph coloring, Eur. J. Oper. Res. **151**, 379–388 (2003)

63.29 J. Thompson, K. Dowsland: An improved ant colony optimisation heuristic for graph colouring, Discrete Appl. Math. **156**, 313–324 (2008)

63.30 R. Dorne, J.-K. Hao: A new genetic local search algorithm for graph coloring, Lect. Notes Comput. Sci. **1498**, 745–754 (1998)

63.31 A.E. Eiben, J.K. van der Hauw, J.I. van Hemert: Graph coloring with adaptive evolutionary algorithms, J. Heuristics **4**(1), 25–46 (1998)

63.32 C. Fleurent, J. Ferland: Genetic and hybrid algorithms for graph colouring, Ann. Oper. Res. **63**, 437–461 (1996)

63.33 P. Galinier, J.-K. Hao: Hybrid evolutionary algorithms for graph coloring, J. Comb. Optim. **3**, 379–397 (1999)

63.34 Z. Lü, J.-K. Hao: A memetic algorithm for graph coloring, Eur. J. Oper. Res. **203**(1), 241–250 (2010)

63.35 D. Porumbel, J.-K. Hao, P. Kuntz: An evolutionary approach with diversity guarantee and well-informed grouping recombination for graph coloring, Comput. Oper. Res. **37**, 1822–1832 (2010)

63.36 C. Morgenstern: Distributed coloration neighborhood search, Discrete Math. Theor. Comput. Sci. **26**, 335–358 (1996)

63.37 I. Blochliger, N. Zufferey: A graph coloring heuristic using partial solutions and a reactive tabu scheme, Comput. Oper. Res. **35**, 960–975 (2008)

63.38 E. Malaguti, M. Monaci, P. Toth: A metaheuristic approach for the vertex coloring problem, INFORMS J. Comput. **20**(2), 302–316 (2008)

63.39 A. Hertz, M. Plumettaz, N. Zufferey: Variable space search for graph coloring, Discrete Appl. Math. **156**(13), 2551–2560 (2008)

63.40 N.J. Radcliffe: Forma analysis and random respectful recombination, Proc. 4th Int. Conf. Genet. Algorithms (1991) pp. 222–229

63.41 E. Falkenauer: *Genetic Algorithms and Grouping Problems*, 1st edn. (Wiley, New York 1998)

63.42 E. Coll, G. Duran, P. Moscato: A discussion on some design principles for efficient crossover operators for graph coloring problems, An. XXVII Simp. Bras. Pesqui. Oper. (1995)

63.43   R. Abbasian, M. Mouhoub, A. Jula: Solving graph coloring problems using cultural algorithms, Proc. 24th Florida Artif. Intell. Res. Soc. Conf. (2011)

63.44   J. Munkres: Algorithms for the assignment and transportation problems, J. Soc. Ind. Appl. Math. **5**(1), 32–38 (1957)

63.45   D. Bertsekas: Auction algorithms for network flow problems: A tutorial introduction, Comput. Optim. Appl. **1**, 7–66 (1992)

63.46   A. Tucker, J. Crampton, S. Swift: RGFGA: An efficient representation and crossover for grouping genetic algorithms, Evol. Comput. **13**(4), 477–499 (2005)

63.47   R. Lewis, E. Pullin: Revisiting the restricted growth function genetic algorithm for grouping problems, Evol. Comput. **19**(4), 693–704 (2011)

63.48   E. Falkenauer: A new representation and operators for genetic algorithms applied to grouping problems, Evol. Comput. **2**(2), 123–144 (1994)

63.49   C. Glass, A. Prugel-Bennett: Genetic algorithms for graph coloring: Exploration of Galnier and Hao's algorithm, J. Comb. Optim. **7**, 229–236 (2003)

63.50   R. Lewis, J. Thompson, C. Mumford, J. Gillard: A wide-ranging computational comparison of high-performance graph colouring algorithms, Comput. Oper. Res. **39**(9), 1933–1950 (2012)

63.51   P. Galinier, A. Hertz: A survey of local search algorithms for graph coloring, Comput. Oper. Res. **33**, 2547–2562 (2006)

63.52   J. Culberson: Graph coloring page, http://web.cs.ualberta.ca/~joe/Coloring/ (2010)

63.53   M. Carter, G. Laporte, S.Y. Lee: Examination timetabling: Algorithmic strategies and applications, J. Oper. Res. Soc. **47**, 373–383 (1996)

63.54   T. Hogg, B. Huberman, C. Williams: Refining the phase transition in combinatorial search, Artif. Intell. **81**(1/2), 127–154 (1996)

63.55   K. Smith-Miles, L. Lopes: Measuring instance difficulty for combinatorial optimization problems, Comput. Oper. Res. **39**(5), 875–889 (2012)

64. Metaheuristic Algorithms and Tree Decomposition

Thomas Hammerl, Nysret Musliu, Werner Schafhauser

This chapter deals with the application of evolutionary approaches and other metaheuristic techniques for generating tree decompositions. Tree decomposition is a concept introduced by *Robertson* and *Seymour* [64.1] and it is used to characterize the difficulty of constraint satisfaction and NP-hard problems that can be represented as a graph. Although, in general, no polynomial algorithms have been found for such problems, particular instances can be solved in polynomial time if the treewidth of their corresponding graph is bounded by a constant. The process of solving problems based on tree decomposition comprises two phases. First, a decomposition with small width is generated. Basically in this phase the problem is divided into several subproblems, each included in one of the nodes of the tree decomposition. The second phase includes solving a problem (based on the generated tree decomposition) with a particular algorithm such as dynamic programming. The main idea is that by decomposing a problem into subproblems of limited size, the whole problem can be solved more efficiently. The time for solving the problem based on its tree decomposition usually depends on the width of the tree decomposition. Thus, it is of high interest to generate tree decompositions having small widths.

Finding the treewidth of a graph is an NP-hard problem [64.2]. In order to solve this problem, different algorithms have been proposed in the literature. Exact methods such as branch and bound techniques can be used only for small graphs. Therefore, metaheuristic algorithms based on genetic algorithms [64.3], simulated annealing [64.4], tabu search [64.5], iterated local search [64.6], and ant colony optimization

(ACO) [64.7, 8] have been proposed in the literature to generate good upper bounds for larger graphs. Such techniques have been applied very successfully and they are able to find the best existing upper bounds for many benchmark problems in the literature.

In this chapter, we will first introduce the concept of tree decomposition, and then give a survey on metaheuristic techniques used to generate tree decompositions. Three approaches based on genetic algorithms, iterated local search, and ACO that were proposed in the literature will be described in detail. Finally, we will also mention briefly two recent approaches that exploit tree decompositions within metaheuristic search.

64.1 Tree Decompositions

We start with an informal description of tree decomposition. Suppose that we have to find solutions for the graph coloring problem (GCP), which is a well-known constraint satisfaction problem (CSP) in the literature. For this problem, we have to find a coloring of vertices of a given graph in such a way that no two vertices connected by an edge share the same color. An instance of the GCP is shown on the left-hand side of Fig. 64.1. The task is now to find a valid coloring just using the colors red, green, and blue.

A naive approach to solve this problem might be to try out all possible combinations of variable assignments and see which ones are valid. In general, there are d^n possible combinations, where d is the number of available colors and n is the number of vertices.

To solve this problem by tree decomposition, first we generate the tree decomposition of the corresponding problem graph. Informally, a tree decomposition is a tree containing a group of graph vertices where each tree node fulfils the following conditions: each vertex of the graph appears in one of the nodes of the tree; if two vertices are connected in the graph, they must appear together in some of the tree nodes; connectedness condition must be fulfilled, i. e., if a vertex appears in two different nodes of the tree, it must appear also in other nodes between these two nodes. The formal definition of tree decomposition is given in the next section.

The corresponding constraint graph of a coloring problem and a possible tree decomposition is shown in Fig. 64.1a,b. If we want to solve the graph coloring problem based on this tree decomposition, we can start out by solving the subproblems given by each node in the tree decomposition. Using a naive approach of trying out all possible combinations of variable assignments, one has to generate 3^3 (27) different solution candidates for the vertex containing A, B, and C. Because of the constraints $A \neq B, A \neq C$, and $B \neq C$ only six of them are valid. For the subproblem containing the

vertices C and D we generate 3^2 (9) solution candidates and rule out three of them because of the constraint $C \neq D$. We can now get all solutions to the whole problem by joining the subproblem solutions. Therefore, we will take a look at the variables that both subproblems have in common. In this case, that is the variable C. Each solution for the subproblem A, B, C is joined with the solutions for the subproblem C, D sharing the same color for the vertex C. By using the tree decomposition, we have to generate 36 combinations of variable assignments in order to determine all solutions compared to the 81 combinations we would have to generate without the tree decomposition. This difference increases very quickly with the size of the graph coloring problem and constraint satisfaction problems in general. The smaller the subproblems in the tree decomposition the more efficiently we can solve a particular problem. This motivates our interest in finding tree decompositions of small *width*.

Note that tree decompositions have been applied for several applications, like combinatorial optimization problems, expert systems, computational biology etc. The use of tree decomposition for inference problems in probabilistic networks is shown in [64.9]. *Koster* et al. [64.10] propose the application of tree decompositions for frequency assignment problem. Tree decomposition has also been applied for the vertex cover problem on planar graphs [64.11]. Furthermore, solving partial constraint satisfaction problems (e.g. MAX-SAT) with tree-decomposition-based method has been investigated in [64.12]. In computational biology tree decompositions has been used for protein structure prediction [64.13]. Recently, the application of tree decomposition in Answer-Set Programming has been investigated in [64.14].

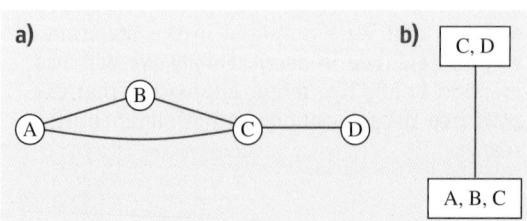

Fig. 64.1a,b Instance of the graph coloring problem and a possible tree decomposition

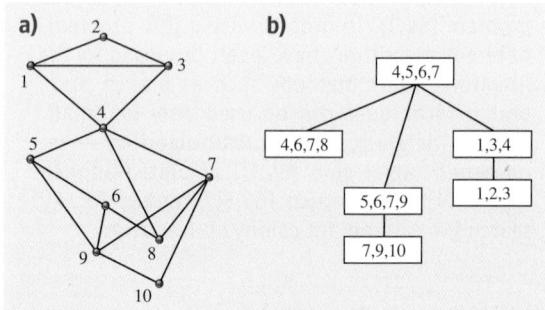

Fig. 64.2a,b A graph G (**a**) and a tree decomposition of G (**b**)

64.1.1 Formal Definitions

The concept of tree decompositions has been first introduced by *Robertson* and *Seymour* [64.1]. The formal definition of tree decomposition is given as follows [64.1, 15].

Definition 64.1

Let $G = (V, E)$ be a graph. A tree decomposition of G is a pair (T, χ), where $T = (I, F)$ is a tree with node set I and edge set F, and $\chi = \{\chi_i : i \in I\}$ is a family of subsets of V, one for each node of T, such that:

1. $\bigcup_{i \in I} \chi_i = V$,
2. for every edge $(v, w) \in E$, there is an $i \in I$ with $v \in \chi_i$ and $w \in \chi_i$, and
3. for all $i, j, k \in I$, if j is on the path from i to k in T, then $\chi_i \cap \chi_k \subseteq \chi_j$.

The width of a tree decomposition is $\max_{i \in I} |\chi_i| - 1$. The treewidth of a graph G, denoted by $tw(G)$, is the minimum width over all possible tree decompositions of G.

Figure 64.2 shows a graph G and a possible tree decomposition of G. The width of shown tree decomposition is 3.

For the given graph G, the treewidth can be found from its triangulation. In the following, we will give basic definitions, explain how the triangulation of graph can be constructed, and give lemmas which give relation between the treewidth and the triangulated graph.

Two vertices u and v of graph $G(V, E)$ are neighbors, if they are connected by an edge $e \in E$. The neighborhood of a vertex v is defined as $N(v) := \{w | w \in V, (v, w) \in E\}$. A set of vertices is clique if they are fully connected. An edge connecting two nonadjacent vertices in the cycle is called chord. The graph is triangulated if there exists a chord in every cycle of length larger than 3.

A vertex of a graph is simplicial if its neighbors form a clique. An ordering of nodes $\sigma(1, 2, \ldots, n)$ of V is called a perfect elimination ordering for G if for any $i \in \{1, 2, \ldots, n\}$, $\sigma(i)$ is a simplicial vertex in $G[\sigma(i), \ldots, \sigma(n)]$ [64.16]. In [64.17] it is proved that the graph G is triangulated if and only if it has a perfect elimination ordering. Given an elimination ordering of nodes the triangulation H of graph G can be constructed as following. Initially $H = G$, then in the process of elimination of vertices, the next vertex in order to be eliminated is made simplicial vertex by adding of new edges to connect all its neighbors in current G and H. The vertex is then eliminated from G. This process is repeated for all vertices in the ordering.

The treewidth of a triangulated graph can be calculated based on its cliques. For the given triangulated graph, the treewidth is equal to its largest clique minus 1 [64.18]. Moreover, the largest clique of a triangulated graph can be calculated in polynomial time. The complexity of calculating the largest clique for the triangulated graphs is $O(|V| + |E|)$ [64.18]. For every graph $G = (V, E)$, there exists a triangulation of G, $\overline{G} = (V, E \bigcup E_t)$, with $tw(\overline{G}) = tw(G)$. Thus, finding the treewidth of a graph G is equivalent to finding a triangulation $\overline{G}$ of G with the minimum clique size (for more information see [64.15]).

The process of elimination of nodes from the given graph G is illustrated in Fig. 64.3. Suppose that we have given the following elimination ordering: 10, 9, 8, 7, 2, 3, 6, 1, 5, 4. The vertex 10 is first eliminated from G. When this vertex is eliminated no new edges are added to the graph G and H (graph H is not shown in the figure), as all neighbors of node 10 are connected. From the remained graph G the vertex 9 is eliminated. To connect all neighbors of vertex 9, two new edges are added in G and H (edges $(5, 7)$ and $(6, 7)$). The process of elimination continues until the triangulation H is obtained. A more detailed description of the algorithm for constructing a graph's triangulation for a given elimination ordering is found in [64.15].

For generating the tree decomposition during the vertex elimination process, first the nodes of the tree decomposition are created. This is illustrated in Fig. 64.3. When vertex 10 is eliminated a new tree decomposition node is created. This node contains the vertex 10 and all other vertices which are connected with this vertex in the current graph G. Further, the next tree node with vertices $\{5, 6, 7, 9\}$ is created when the vertex 9 is eliminated. To the end of elimination process all tree decomposition nodes will be created. The created tree nodes should be connected, such that the connectedness condition for vertices is fulfilled. This is the third condition in the tree decomposition definition. To fulfil this condition, the tree decomposition nodes are connected as following. The tree decomposition node with vertices $\{7, 9, 10\}$ that is created when vertex 10 is eliminated, is connected with the tree decomposition node which will be created when the next vertex which appears in $\{7, 9, 10\}$ is eliminated. In this case, the node $\{7, 9, 10\}$ should be connected with the node created when vertex 9 is eliminated, because this is the next vertex in the ordering that is contained in

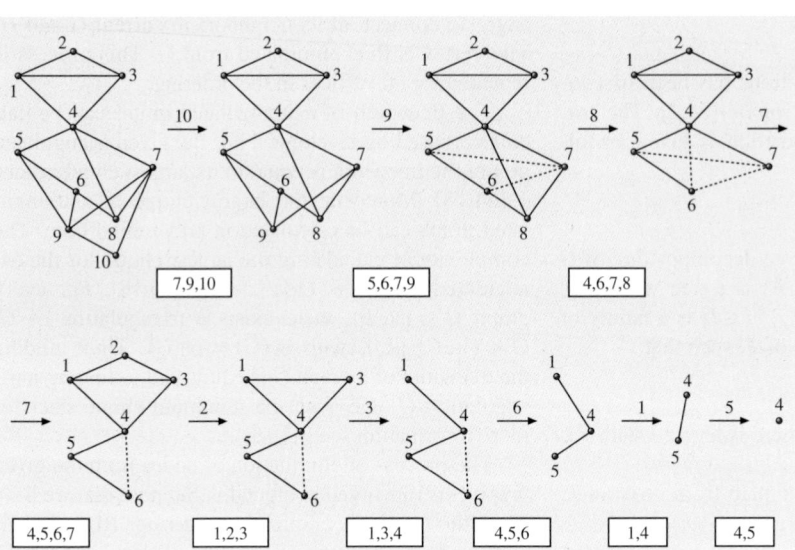

Fig. 64.3 Elimination of vertices 10, 9, 8, 7, 2, 3, 6, 1, 5, 4. When a vertex is eliminated a tree node containing eliminated vertex and its neighbors is created

$\{7, 9, 10\}$. This rule is further applied for connection of other tree decomposition nodes, and from the graph the tree decomposition in Fig. 64.2 will be constructed. Note that some of tree nodes that are created in the elimination process are not presented in the tree decomposition, because they are contained in larger tree nodes. For example, the node $\{4, 5, 6\}$ which is created by eliminating vertex 6 is already contained in the node $\{4, 5, 6, 7\}$ which is created by eliminating vertex 7. Moreover, tree nodes which are created by eliminating vertices 1, 5, 4 are also contained in other larger tree nodes.

64.2 Generating Tree Decompositions by Metaheuristic Techniques

As described in the previous section, the width of the tree decomposition depends on the elimination ordering of vertices. Therefore, the task of finding tree decomposition with minimal width consists of finding the best permutation of graph vertices. This problem is similar to the traveling salesman problem, but with a different objective function.

In the last two decades, researchers have been proposing different techniques to find tree decompositions for different benchmark examples. This includes the exact techniques based on tree search and branch and bound, the simple greedy techniques and metaheuristic techniques. In this chapter, we focus on metaheuristic techniques. At the end of this section, we will also shortly describe other approaches used for tree decompositions.

The metaheuristic techniques applied for tree decomposition can be divided in two groups: population based/nature inspired techniques, and local search techniques. Regarding nature inspired techniques the ap-

plication of genetic algorithms has been investigated in [64.19, 20], and ACO has been used in [64.21]. Examples of local search techniques for tree decompositions are [64.16, 22, 23].

64.2.1 Genetic Algorithms for Tree Decomposition

Application of genetic algorithm for tree decompositions has been first investigated in [64.19]. This algorithm tried to minimize a weight associated with the decompositions of Bayesian networks which is not exactly the same as the width of the tree decomposition. In [64.20], this algorithm has been extended for generating hypertree decompositions and with some changes in fitness function (the width of tree decompositions has been used as a objective function) has been tested on different problems from the literature. The following description of genetic algorithm for tree decomposition is based on our previous work in [64.20].

Genetic algorithms (GAs) were developed by [64.3]. They try to find a good solution for an optimization problem by imitating the principle of evolution. Genetic algorithms alter and select individuals from a population of solutions for the optimization problem. In the following, we describe frequently used terms within the field of genetic algorithms:

Population ... set of candidate solutions.
Individual ... a single candidate solution.
Chromosome ... set of parameters determining the properties of a solution.
Gene ... single parameter.

A genetic algorithm tends to optimize the value of an objective function of an optimization problem, in terms of genetic algorithms also called *fitness function*. At the beginning a genetic algorithm creates an initial population containing randomly or heuristically created individuals. These individuals are evaluated and assigned a *fitness* value, which is the value of the fitness function for the solution represented by the individual. The population is evolved over a number of generations until a halting criterion is satisfied. At each generation, the population undergoes *selection* and *recombination*, also called *crossover* and *mutation*.

During the selection process, the genetic algorithm decides which individuals from the current population are allowed to enter the next population. This decision is based on the fitness value of the individuals and individuals of better fitness should enter the next population with higher probability than individuals of lower fitness. Not selected individuals are discarded and will not be evolved further.

The recombination process or crossover combines different properties of several parent solutions within one or more children solutions, also denoted as *offsprings*. Crossover exchanges properties between the individuals with the aim of increasing the average quality of the population.

During the mutation process, individuals are slightly altered. Mutation is used to explore new regions of the search space and to avoid early convergence to local optima.

In practice, parameters are used in order to control the behavior of a genetic algorithm. Typical *control parameters* are mutation rate, crossover rate, population size, and parameters for selection techniques. The choice of the control parameters has a crucial effect on the quality of the best solution found by a genetic algorithm.

The genetic algorithm for tree decomposition presented below is named *GA-tw* and was implemented in [64.20]. Algorithm 64.1 presents algorithm *GA-tw* in pseudo code notation.

The algorithm takes as input a graph and several control parameters. Individual solutions are vertex orderings. Each individual is assigned the width of the tree decomposition returned from the corresponding vertex ordering as its fitness value.

Initially *GA-tw* generates a population consisting of randomly created individuals. Tournament selection was chosen as the selection technique. Tournament selection selects an individual by randomly choosing a group of several individuals from the former population. The individual of highest fitness (smallest width) within this group is selected to join the next population. This process is applied until enough individuals have entered the next population. Finally, after a certain number of generations, algorithm *GA-tw* will return the best fitness (smallest width) of an individual found during the search process.

Crossover and Mutation Operators

Within the genetic algorithms in [64.20] nearly all types of crossover operators and all mutation operators were implemented. The same operators were also applied in [64.19] for decomposing the moral graph of Bayesian networks.

Algorithm 64.1 Genetic Algorithm for Tree Decompositions – GA-tw

Input: a graph $G = (V, E)$
control parameters for the GA n, p_m, p_c, s
and *max_iterations*

Output: an upper bound on the treewidth
of the graph

$t = 0$
initialize $(population(t), n)$
evaluate $population(t)$
while $t <$ *max_iterations* **do**
$t = t + 1$
$population(t) =$ tournament_selection
$(population(t - 1), s)$
recombine $(population(t), p_c)$
mutate $(population(t), p_m)$
evaluate $population(t)$
end while
return the smallest width found during the search

Crossover Operators

- Partially mapped crossover (PMX)
- Cycle crossover (CX)

- Order crossover (OX1)
- Order-based crossover (OX2)
- Position-based crossover (POS)
- Alternating-position crossover (AP).

Mutation Operators

- Displacement mutation operator (DM)
- Exchange mutation operator (EM)
- Insertion mutation operator (ISM)
- Simple-inversion mutation operator (SIM)
- Inversion mutation operator (IVM)
- Scramble mutation operator (SM).

We will describe the crossover and mutation operators which returned the best results of algorithm *GA-tw* in more detail.

Order Crossover (OX1)

The order crossover operator determines a crossover area within the parents by randomly selecting two positions within the ordering. The elements in the crossover area of the first parent are copied to the offspring. Starting at the end of the crossover area all elements outside the area are inserted in the same order in which they occur in the second parent.

Order-Based Crossover (OX2)

The order-based crossover operator selects at random several positions in the parent orderings by tossing a coin for each position. The elements of the first parent at these positions are deleted in the second parent. Afterward they are reinserted in the order of the second parent.

Position-Based Crossover (POS)

The position-based crossover operator also starts with selecting a random set of positions in the parent strings by tossing a coin for each position. The elements at the selected positions are exchanged between the parents in order to create the offsprings. The elements missing after the exchange are reinserted in the order of the second parent.

Exchange Mutation Operator (EM)

The exchange mutation operator randomly selects two elements in the solution and exchanges them.

Insertion Mutation Operator (ISM)

The insertion mutation operator randomly chooses an element in the solution and moves it to a randomly selected position (Fig. 64.5).

The genetic algorithm implemented in [64.19] was applied to two artificial graphs. This genetic approach returned competitive results when compared to results obtained by simulated annealing [64.22]. The algorithm implemented in [64.20] was evaluated on 62 graphs of the second DIMACS graph coloring challenge [64.24]. Different experiments were performed to find the best parameter values for parameters of the genetic algorithm and it turned out that the position-based crossover operator(POS) and the insertion mutation operator(ISM) were best suited for finding tree decompositions of small width. Existing upper bounds for treewidth for several DIMACS instances could be improved.

64.2.2 Ant Colony Optimization for Tree Decomposition

Ant colony optimization (ACO has been applied for tree decompositions in [64.21, 25]. The current section is based on [64.21] and describes different ant colony optimization variants applied for tree decomposition.

ACO is a population-based metaheuristic introduced by *Dorigo* et al. [64.7, 8]. As the name suggests, the technique was inspired by the behavior of *real* ants. Ant colonies are able to find the shortest path between their nest and a food source just by depositing and reacting to pheromones while they are exploring their environment. The basic principles driving this system can also be applied to many combina-

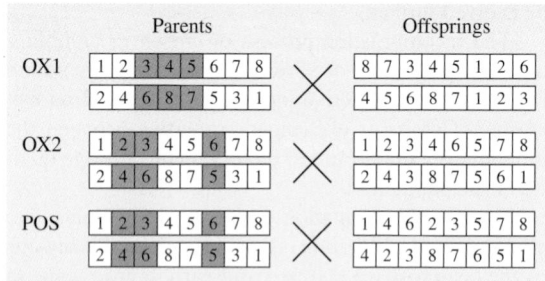

Fig. 64.4 Selected crossover operators for vertex orderings

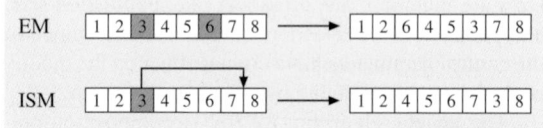

Fig. 64.5 Selected mutation operators for vertex orderings

torial optimization problems. For a detailed description of different ACO algorithms and their applications the reader is referred to the book *ant colony optimization* [64.26].

The following variants of ACO algorithms for finding good upper bounds for tree decompositions were investigated in [64.21, 25]: simple ant system [64.7, 8], elitist ant system [64.7, 8], rank-based ant system [64.27], max–min ant system [64.28, 29], and ant colony system [64.30]. Two different pheromone update strategies were proposed and two stagnation measures were implemented that indicate the degree of diversity of the solutions constructed by the ants. Furthermore, two constructive heuristics (Min-Degree, Min-Fill) were implemented and incorporated alternatively into every ACO variant as a guiding function, and the combination of ACO with two existing local search methods: Hill Climbing and Iterated Local Search [64.23] were investigated.

A simple constraint graph and the corresponding ACO construction tree are shown in Fig. 64.6. The construction tree can be obtained from the constraint graph as follows:

1. Create a root node s that will be the starting point of every ant in the colony.
2. For every vertex of the constraint graph append a child node to the root node s.
3. To every leaf node append a child node for every vertex of the constraint graph that is neither represented by the leaf node itself nor by an ancestor of this node.
4. Repeat step 3 until there are no nodes left to append.

All possible elimination orderings for the constraint graph can now be represented as a path from the root node s to one of the leaf nodes in the construction tree. Therefore, each of the ants finds such a path and at each node on its way the ant decides where to move next probabilistically based on the pheromone trails and a heuristic value both associated with the outgoing edges.

Pheromone Trails

A pheromone trail gives information how favorable it is to eliminate a certain vertex x after another vertex y. The more pheromone is located on a trail the more likely the corresponding vertex will be chosen by the ant. A way to represent the pheromone trails of the construction tree in Fig. 64.6 is the matrix as shown in the

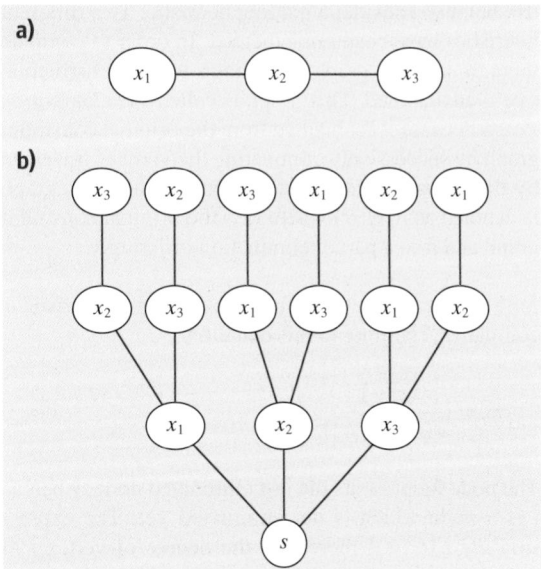

Fig. 64.6a,b Constraint graph G and the ACO construction tree

following,

$$\mathcal{T} = \begin{pmatrix} \tau_{x_1x_1} & \tau_{x_1x_2} & \tau_{x_1x_3} \\ \tau_{x_2x_1} & \tau_{x_2x_2} & \tau_{x_2x_3} \\ \tau_{x_3x_1} & \tau_{x_3x_2} & \tau_{x_3x_3} \\ \tau_{sx_1} & \tau_{sx_2} & \tau_{sx_3} \end{pmatrix}. \tag{64.1}$$

In this matrix, each row contains the amounts of pheromone located on the trails connecting a certain node with all the other nodes. For example, the first row contains the pheromone levels related to the node x_1 describing the desirability of eliminating $x_2(\tau_{x_1x_2})$, respectively, $x_3(\tau_{x_1x_3})$ immediately after x_1. The last row is related to the root node s that is the starting point for every ant.

All pheromone trails are initialized to the same value in the beginning of the algorithm that is computed according to the following equation,

$$\tau_{ij} = \frac{m}{W_\eta}, \qquad \forall \tau_{ij} \in \mathcal{T}. \tag{64.2}$$

where W_η is the width of the decomposition obtained using the guiding heuristic (min-degree or min-fill) while m is the size of the ant colony.

Heuristic Information

The ants make their decision about which vertex to eliminate next not solely based on the pheromone matrix but also consider a guiding heuristic. Two different heuristics have been implemented. In order to compute them, a separate graph in addition to the construction tree is maintained. This graph is called the *elimination graph* because it is obtained from the original constraint graph by successively eliminating the vertices traversed by the ant in the construction tree. Further, this graph is denoted as $E(G, \sigma)$ where G is the original constraint graph and σ is a partial elimination ordering.

Min-Degree. The value for the min-degree heuristic is computed according to this equation

$$\eta_{ij} = \frac{1}{d(j, E(G, \sigma)) + 1} . \tag{64.3}$$

The node i represents the last eliminated node, whereas j is a node which is not eliminated yet. The expression $d(j, E(G, \sigma))$ represents the degree of vertex j in the elimination graph $E(G, \sigma)$.

Min-Fill. The value for the min-fill heuristic is computed according to this equation

$$\eta_{ij} = \frac{1}{f(j, E(G, \sigma)) + 1} . \tag{64.4}$$

The expression $f(j, E(G, \sigma))$ represents the number of edges that would be added to the elimination graph due to the elimination of vertex j.

Probabilistic Vertex Elimination

In the following it is shown how exactly the ants move from node to node on the construction tree. All of the ACO variants with the exception of ant colony system use (64.5) alone to compute the probability p_{ij} of moving from a node i to another node j where α and β are parameters that can be passed to the algorithm in order to weight the pheromone trails and the heuristic values

$$p_{ij} = \frac{[\tau_{ij}]^\alpha [\eta_{ij}]^\beta}{\sum\limits_{l \in E(G, \sigma)} [\tau_{il}]^\alpha [\eta_{il}]^\beta} , \quad \text{if } j \in E(G, \sigma) . \tag{64.5}$$

This probability is computed for each vertex left in the elimination graph. According to these probabilities, the ant decides which vertex to eliminate next.

Ant colony system introduces an additional parameter q_0 that constitutes the probability that the ant

makes a greedy move instead of making a probabilistic decision

$$j = \begin{cases} \arg\max\limits_{l \in E(G, \sigma)} \{[\tau_{il}]^\alpha [\eta_{il}]^\beta\} , & \text{if } q \le q_0 ; \\ (64.5) , & \text{otherwise} . \end{cases}$$
$$\tag{64.6}$$

If a randomly generated number q in the interval of $[0, 1]$ is less or equal q_0 then the ant moves to the node that otherwise would have the highest probability to be chosen. Ties are broken randomly.

Ant colony system also introduces a so-called local pheromone update. After an ant has constructed its solution it removes pheromone from the trails belonging to its solution according to the following equation whereas ξ is a variant-specific parameter and τ_0 is the initial amount of pheromone

$$\tau_{ij} \leftarrow (1 - \xi)\tau_{ij} + \xi\tau_0 . \tag{64.7}$$

The motivation is to diversify the search so that subsequent ants will more likely choose other branches of the construction tree.

Pheromone Update

After each of the ants has constructed an elimination ordering (that optionally has been improved by a local search thereafter) the values in the pheromone matrix are updated reflecting the quality of the constructed solutions which will enable the subsequent ants in the following iteration to make decisions in a more informed manner. Moreover, pheromone is gradually removed from the pheromone trails so that solutions that might have been the best known solutions in earlier iterations of the algorithm can be forgotten.

Pheromone Deposition

In this step for an elimination ordering σ_k that was constructed by an ant k the amount of pheromone that will be deposited for each (i, j) in σ_k is determined. An edge-independent and an edge-specific pheromone update strategy were considered. The first adds the same amount of pheromone to all trails belonging to σ_k while the latter adds more or less pheromone to individual trails depending on the quality of a certain elimination.

The edge-independent pheromone update strategy adds the reciprocal value of the tree decomposition's width to all pheromone trails that are part of σ_k

$$\Delta \tau_{ij}^k = \begin{cases} \dfrac{1}{W(\sigma_k)}, & \text{if } (i,j) \text{ belongs to } \sigma_k ; \\ 0, & \text{otherwise} . \end{cases} \quad (64.8)$$

In contrast to the edge-independent update strategy the edge-specific update strategy deposits different amounts of pheromone onto the trails belonging to the same elimination ordering

$$\Delta \tau_{ij}^k = \begin{cases} \dfrac{1}{d(j, E(G, \sigma_{kj}))/|E(G, \sigma_{kj})|} \cdot \dfrac{1}{W(\sigma_k)}, \\ \qquad \text{if } (i,j) \text{ belongs to } \sigma_k ; \\ 0, \quad \text{otherwise} . \end{cases}$$

$$(64.9)$$

This amount depends on the ratio between the degree of the vertex j when it was eliminated $d(j, E(G, \sigma_{kj}))$ and the number of vertices left in the elimination graph $|E(G, \sigma_{kj})|$ at that time (σ_{kj} is the partial elimination ordering that is obtained from σ_k by omitting j and all vertices that are eliminated after j).

The selection of ants that deposit pheromone and the weighting of this pheromone varies between the different ACO variants. The reader is referred to [64.26] for description of these variants.

Pheromone Evaporation
After the pheromone has been added to the trails, a certain amount of pheromone is removed. This amount is determined based on the pheromone evaporation rate ρ

$$\tau_{ij} = (1 - \rho)\tau_{ij}, \quad \forall \tau_{ij} \in \mathcal{T} . \quad (64.10)$$

Ant colony system only removes pheromone from the trails belonging to the best known elimination ordering σ_{bs}

$$\tau_{ij} = (1 - \rho)\tau_{ij}, \quad \forall (i,j) \in \sigma_{bs} . \quad (64.11)$$

Hybridization with Local Search
All ACO variants were extended with two local search methods for tree decompositions. Both of these algorithms try to improve the quality of the solutions that were constructed by the ant colony by changing the positions of certain vertices in the elimination orderings. Two local search techniques were used: an hill climbing algorithm and an iterated local search similar to the algorithm proposed in [64.23].

Stagnation Measures
If the distribution of the pheromone on the trails becomes too unbalanced due to the pheromone depositions, the ants will generate very similar solutions causing the search to stagnate. In order to enable the algorithm to detect such situations two stagnation measures were implemented (variation coefficient and $\overline{\lambda}$ branching factor) proposed by *Dorigo* and *Stützle* [64.26] that indicate how explorative the search behavior of the ants is. A detailed description of stagnation measures is given in [64.25, page 67].

All described ACO variants in [64.21] were evaluated experimentally with DIMACS Graph Coloring Challenge instances. Max–Min ant system and ant colony system performed slightly better than the other variants. Although the ant colony optimization in general could not compete with iterated local search and genetic algorithms, it could improve the upper bound for one of problems.

64.2.3 Iterated Local Search for Tree Decomposition

The application of iterated local search for generating tree decompositions has been investigated in [64.23, 31]. In this section, we give the description of this algorithm based on these references.

The algorithm is based on the iterated local search framework and it includes a simple local search heuristic to generate good orderings, and an iterative process in which the algorithm calls a local search technique with the initial solution produced in the previous iteration. The algorithm also includes a mechanism for acceptance of a candidate solution for the next iteration. Although the constructing phase is very important, choosing the appropriate perturbation at each iteration as well as the mechanism for acceptance of solution are also crucial to obtain good results for an iterative local search algorithm. The iterated local search algorithm for tree decomposition is presented below.

Algorithm 64.2 Iterative Heuristic Algorithm – IHA
    Generate initial solution $S1$
    BestSolution $= S1$
    while Termination Criteria is not fulfilled **do**
        $S2 = ConstructionPhase(S1)$
        if Solution $S2$ fulfils the acceptance criteria **then**
            $S1 = S2$
        else
            $S1 = BestSolution$
        end if

Apply perturbation in solution *S1*
Update *BestSolution* if solution *S2* has better
(or equal) width than the current best solution
end while
return *BestSolution*

The algorithm starts with an initial solution which takes an order of nodes as they appear in the input. Better initial solutions can also be constructed by using other heuristics which run in polynomial time, such as maximum cardinality search, min-fill heuristic, etc. However, as the proposed method usually finds a solution produced by these heuristics in a very short time, the algorithm starts with an ordering of nodes given in the input.

After constructing the initial solution the iterative phase starts. In this phase, the local search method is called iteratively, and then the selected solution is perturbed. Two different local search techniques that can be used in the construction phase were proposed. The solution returned from the construction phase is accepted for the next iteration if it fulfils the specific criteria determined by the solution acceptance mechanism. Experiments with different possibilities for the acceptance of the solution returned from the construction phase were performed. If the solution does not fulfil the acceptance criteria this solution is discarded and the currently best solution is selected. In the selected solution, the perturbation mechanism is applied. Different possibilities are used for perturbation. The perturbed solution is given as an input solution in the next call of the construction phase. This process continues until the termination criterion is fulfilled.

Two local search methods were proposed for generating a good solution which is used as an initial solution with some perturbation in the next call of the same local search algorithm. Both techniques are based on the idea of moving only vertices in the ordering which cause the largest clique during the elimination process. The motivation for using this method is to reduce the number of solutions that should be evaluated. The first proposed technique named LS1 is presented below.

Algorithm 64.3 Local Search Algorithm 1 − LS1 (InputSolution)

BestLSSolution = InputSolution
NrNotImprovments = 0
while   *NrNotImprovments < MAXNotImprovments*
do
    In the current solution (*InputSolution*) select a vertex in the elimination ordering which causes the

largest clique when eliminated − ties are broken randomly if there are several vertices which cause the clique equal with the largest clique
Swap this vertex with another vertex located in a randomly chosen position
if the current solution is better than *BestLSSolution*
then
    BestLSSolution = InputSolution
    NrNotImprovments = 0
else
    NrNotImprovments = NrNotImprovements + 1
end if
end while
return *BestLSSolution*

The proposed algorithm applies a simple heuristic. In the current solution a vertex is chosen randomly among the vertices that produce the largest clique in the elimination process. Then the selected vertex is moved from its position. Two types of moves were used. In the first variant, the vertex is inserted in a random position in the elimination ordering, while in the second variant the vertex is swapped with another vertex located in a randomly selected position, i. e., the two chosen vertices change their position in the elimination ordering. The swap move was shown to give better results. The heuristic stops if the solution does not improve for a certain number of iterations. Experiments with different *MAXNotImprovments* were performed. LS1 alone is a simple heuristic and usually cannot produce good results for tree decompositions. However, by using this heuristic as a local search heuristic in the iterated local search algorithm good results for tree decompositions are obtained.

The second proposed heuristic (LS2) is similar to algorithm LS1. However, this technique differs from LS1 regarding the exploration of the neighborhood. In LS2 in some of iterations the neighborhood of solution consists of only one solution which is generated by swapping a vertex (that causes the largest clique) in the elimination ordering with another vertex located in the randomly chosen position. This neighborhood is used in a particular iteration with probability p. Experiments with different values for parameter p were performed. With probability $1 − p$, the other type of neighborhood will be explored. The neighborhood of current solution in this case consists of all solutions which can be obtained by swapping of a vertex (which causes the largest clique) in the elimination ordering with its neighbors. The best solution from the generated neighborhood is selected for the next iteration in the LS2. Note that in this technique the number of solutions that have to be evaluated is much larger than in LS1. In particular, in the first phase of search the node which causes the largest clique usually has many neighbors and therefore the number of solutions to be evaluated when the second type of neighborhood is used is equal to the size of the largest clique produced during the elimination process.

Perturbation

During the perturbation phase the solution obtained by local search procedure is perturbed and the newly obtained solution is used as an initial solution for the new call of the local search technique. The main idea is to avoid the random restart. Instead of random restart the solution is perturbed with a bigger move(s) as those applied in the local search technique. This enables some diversification that helps to escape from the local optimum, but avoids beginning from scratch (as in the case of random restart), which is very time consuming. Three perturbation mechanisms were proposed:

- *RandPert*: N vertices are chosen randomly and they are moved into new random positions in the ordering.
- *MaxCliquePer*: All nodes that produce the maximal clique in the elimination ordering are inserted in a new randomly chosen positions in the ordering.
- *DestroyPartPert*: All nodes between two positions (selected randomly) in the ordering are inserted in the new randomly chosen positions in the ordering.

The perturbation RandPert just perturbs the solution with a larger random move and would be kind or random restart if N is very large. Keeping N smaller avoids restarting from completely new solution, and the perturbed solution does not differ much from the previous solution. MaxCliquePer concentrates on moving only vertices which produce maximal clique in the elimination ordering. The basic idea for this perturbation is to apply a technique similar to min-conflict heuristic, by moving only the vertices that cause large treewidth. DestroyPartPert is similar to RandPert, except that the selected nodes to be moved are located near each other in the elimination ordering.

Determining the number of nodes N that will be moved is complex and may be dependent on the problem. To avoid this problem an adaptive perturbation mechanism was proposed that takes into consideration the feedback from the search process. The number of nodes N varies from 2 to some number y (determined experimentally), and the algorithm begins with small perturbation ($N = 2$). If during the iterative process (for a determined number of iterations) the local search technique produces solutions with same tree width for more than 20% of cases, the size of perturbation is increased by 1, otherwise the size of N will be decreased by 1. This enables an automatic change of perturbation size based on the repetition of solutions with the same width.

The combination of two perturbations was considered. The mixed perturbation applies two perturbations: RandPert and MaxCliquePer. The algorithm starts with RandPert, and switches alternatively between two perturbations if the solution is not improved for a determined number of iterations. Experiments with different sizes of perturbation sizes for each type of perturbation were performed.

Acceptance Criterion

Different techniques can be applied for accepting the solution obtained by the local search technique. Following variants for acceptance of solution for the next iteration were used:

- Solution returned from the construction phase is accepted only if it has a better width than the best current existing solution.
- Solution returned from the construction phase is always accepted.
- Solution is accepted if its treewidth is smaller than the treewidth of the best yet found solution plus x, where x is an integer.

The first variant for accepting a solution is very restrictive. In this variant, the solution from the construction phase is accepted only if it improves the best existing solution. Otherwise, the best existing solution is perturbed and it is used as input solution for next call of the construction phase. In the second variant, the iterated local search applies the perturbation in a solution returned from the construction phase, independently from the quality of produced solution. The third variant is between the first and the second variant, and in this case the solution which does not improve the best existing solution can be accepted for the next iteration, if its width is smaller than the best found width plus some bound.

64.2.4 Other Techniques for Tree Decomposition

This section gives a short overview on other approaches applied for tree decomposition. Examples of complete algorithms for tree decompositions are [64.32–34]. *Gogate* and *Dechter* [64.33] reported good results for tree decompositions by using branch and bound algorithms. They showed that their algorithm is superior compared to the algorithm proposed in [64.32]. The branch and bound algorithm proposed in [64.33] applies different pruning techniques, and provides anytime solutions, which are good upper bounds for tree decompositions. The algorithm proposed in [64.34] includes several other pruning and reduction rules and is successful on small graphs. The complete techniques described earlier have exponential running time in the worst case and can only be used to find the optimal width for not too large graphs.

To generate good upper bounds (which can be sufficient for many applications) for treewidth several greedy heuristic techniques that run in polynomial time have been proposed. These heuristics select the ordering of nodes step by step based on different criteria, such as the degree of the nodes, the number of edges to be added to make the node simplicial etc. Most popular techniques are maximum cardinality search (MCS), min-fill heuristic and minimum degree heuristic.

MCS [64.35] initially selects a random vertex of the graph to be the first vertex in the elimination ordering (the elimination ordering is constructed from right to left). The next vertex will be picked such that it has the highest connectivity with the vertices previously selected in the elimination ordering. Ties are broken randomly. MCS repeats this process iteratively until all vertices are selected.

The min-fill heuristic first picks the vertex which adds the smallest number of edges when eliminated (ties are broken randomly). The selected vertex is made simplicial (a vertex of a graph is simplicial if its neighbors form a clique) and it is eliminated from the graph. The next vertex in the ordering will be any vertex that adds the minimum number of edges when eliminated from the graph. This process is repeated iteratively until the whole elimination ordering is constructed.

The minimum degree heuristic picks first the vertex with the minimum degree. The selected vertex is made simplicial and it is removed from the graph. Further, the vertex that has the minimum number of unselected neighbors will be chosen as the next node in the elimination ordering. This process is repeated iteratively.

MCS, min-fill, and min-degree heuristics run in polynomial time and usually produce tree decompositions in a reasonable amount of time. According to [64.33], the min-fill heuristic performs better than MCS and min-degree heuristic. Although these heuristics sometimes give good upper bounds for tree decompositions, more advanced techniques usually provide better upper bounds for most problems. Min-degree heuristic has been improved by *Clautiaux* et al. [64.16] by adding a new criterion based on the lower bound of the treewidth for the graph obtained when the node is eliminated. Recently, *Kask* et al. [64.36] proposed an iterative greedy variable ordering algorithm to improve the greedy heuristics given earlier. We refer to [64.15, 37] for a survey of different upper bounds algorithms.

64.2.5 Comparison of Algorithms for Tree Decomposition

In this section, we compare results obtained with metaheuristic aproaches described in this chapter and other existing methods in the literature. The results of these methods for 62 DIMACS vertex coloring instances are given. These instances have been used for testing several methods for tree decompositions proposed in the literature. The compared methods have been executed in different computers and we give here only results regarding the width of the tree decomposition. The reader is referred to [64.15, 16, 20, 23, 25, 33], for the information about the computers used and the time needed to generate solutions.

In Tables 64.1 and 64.2, the results for DIMACS graph coloring instances are presented. First and second columns of the tables present the instances and the number of nodes and edges for each instance. In column KBH are shown the best results obtained by algorithms in [64.15]. The TabuS column presents the results reported in [64.16], and column BB shows the results obtained with the branch and bound algorithm proposed in [64.33]. Finally, columns GA, IHA, and ACO represent, respectively, results obtained with a genetic algorithm [64.20], iterated local search [64.23], and ant colony optimization [64.21, 25].

Based on the results given in Tables 64.1 and 64.2 we conclude that regarding the width of tree decomposition, the metaheuristic techniques described in this paper give very good results and for many instances the best existing upper bounds for the treewidth.

Table 64.1 Algorithms comparison regarding treewidth for DIMACS graph coloring instances

Instance	$|V|/|E|$	KBH	TabuS	BB	GA	IHA	ACO
anna	138/986	12	12	12	12	12	12
david	87/812	13	13	13	13	13	13
huck	74/602	10	10	10	10	10	10
homer	561/3258	31	31	31	9	31	30
jean	80/508	9	9	9	9	9	9
games120	120/638	37	33	–	32	32	37
queen5_5	25/160	18	18	18	18	18	18
queen6_6	36/290	26	25	25	26	25	25
queen7_7	49/476	35	35	35	35	35	35
queen8_8	64/728	46	46	46	45	45	46
queen9_9	81/1056	59	58	59	58	58	59
queen10_10	100/1470	73	72	72	72	72	73
queen11_11	121/1980	89	88	89	87	87	89
queen12_12	144/2596	106	104	110	104	103	109
queen13_13	169/3328	125	122	125	121	121	128
queen14_14	196/4186	145	141	143	141	140	150
queen15_15	225/5180	167	163	167	162	162	174
queen16_16	256/6320	191	186	205	186	186	201
fpsol2.i.1	269/11 654	66	66	66	66	66	66
fpsol2.i.2	363/8691	31	31	31	32	31	31
fpsol2.i.3	363/8688	31	31	31	31	31	31
inithx.i.1	519/18 707	56	56	56	56	56	56
inithx.i.2	558/13 979	35	35	31	35	35	31
inithx.i.3	559/13 969	35	35	31	35	35	31
miles1000	128/3216	49	49	49	50	49	50
miles1500	128/5198	77	77	77	77	77	77
miles250	125/387	9	9	9	10	9	9
miles500	128/1170	22	22	22	24	22	25
miles750	128/2113	37	36	37	37	36	38
mulsol.i.1	138/3925	50	50	50	50	50	50
mulsol.i.2	173/3885	32	32	32	32	32	32
mulsol.i.3	174/3916	32	32	32	32	32	32
mulsol.i.4	175/3946	32	32	32	32	32	32
mulsol.i.5	176/3973	31	31	31	31	31	31
myciel3	11/20	5	5	5	5	5	5
myciel4	23/71	11	10	10	10	10	10
myciel5	47/236	20	19	19	19	19	19
myciel6	95/755	35	35	35	35	35	35
myciel7	191/2360	74	66	54	66	66	66

64.2.6 Application of Tree Decomposition in Metaheuristic Techniques

Traditionally, tree decompositions have been used to solve constraint satisfaction problems exactly by dynamic programming algorithms. Recently, researchers have been investigating the incorporation of tree decomposition within metaheuristics techniques. The work in this direction is just in the starting phase and to the best of our knowledge only two papers investigated

yet the application of tree decomposition in metaheuristic search.

In [64.38] tree-decomposition-based heuristics have been developed for the two-dimensional bin packing problem with conflicts. The aim is to find a conflict-free packing of given items by using minimal number of bins. Tree decomposition is applied to decompose a problem instance into subproblems which can be solved independently. First a tree decomposition is obtained, and then each item is assigned to a specific

Table 64.2 Algorithms comparison regarding treewidth for DIMACS graph coloring instances

Instance	$|V|/|E|$	KBH	TabuS	BB	GA	IHA	ACO
school1	385/19 095	244	188	–	185	178	228
school1_nsh	352/14 612	192	162	–	157	152	185
zeroin.i.1	126/4100	50	50	–	50	50	50
zeroin.i.2	157/3541	33	32	–	32	32	33
zeroin.i.3	157/3540	33	32	–	32	32	33
le450_5a	450/5714	310	256	307	243	244	304
le450_5b	450/5734	313	254	309	248	246	308
le450_5c	450/9803	340	272	315	265	266	309
le450_5d	450/9757	326	278	303	265	265	290
le450_15a	450/8168	296	272	–	265	262	288
le450_15b	450/8169	296	270	289	265	258	292
le450_15c	450/16 680	376	359	372	351	350	368
le450_15d	450/16 750	375	360	371	353	355	371
le450_25a	450/8260	255	234	255	225	216	249
le450_25b	450/8263	251	233	251	227	219	245
le450_25c	450/17 343	355	327	349	320	322	346
le450_25d	450/17 425	356	336	349	327	328	355
dsjc125.1	125/736	67	65	64	61	60	63
dsjc125.5	125/3891	110	109	109	109	108	108
dsjc125.9	125/6961	119	119	119	119	119	119
dsjc250.1	250/3218	179	173	176	169	167	174
dsjc250.5	250/15 668	233	232	231	230	229	231
dsjc250.9	250/,897	243	243	243	243	243	243

cluster (this phase is called cluster separation). Then these clusters are considered as subproblems which are solved iteratively. Finally, the partial solutions from subproblems are merged to obtain solutions for the whole problem.

Another application of tree decomposition includes the approach introduced by *Fontaine* et al. [64.39] where tree decomposition is used to guide the explo-

ration for the search space. Authors propose a method called decomposition guided VNS (variable neighborhood search) that exploits the graph of clusters to build neighborhood structures. By using clusters better intensification and diversification is achieved. For example, the moves are favored in regions that are closely linked and the search is diversified by selecting new clusters and therefore exploring new regions of the search space.

64.3 Conclusion

Several metaheuristic approaches based on nature inspired strategies and local search have been used successfully in the literature for generating tree decompositions. Among these approaches, genetic algorithms and iterated local search-based algorithms provide best upper bounds for many benchmark instances.

Although metaheuristic techniques currently provide state-of-the-art upper bounds for most problems, the runtime of such algorithms for large graphs is still high. Greedy heuristic approaches generate slightly worse upper bounds, but are more efficient. Therefore, developing more efficient metaheuristics for tree de-

compositions is still a challenging task. Moreover, for many problems the treewidth is still not known, and the question is if the current metaheuristics can still be improved to find new upper bounds for such problems. To obtain better upper bounds, it would be interesting to investigate some other approaches such as memetic algorithms, large neighborhood search, and other hybrid techniques. Furthermore, the iterative improvement of the initial generated tree decomposition (based on vertex ordering) is an interesting question.

Finally, in some applications, the treewidth is not the only important parameter for solving problems based on tree decompositions efficiently. Therefore, the development of metaheuristics for generating tree decompositions which optimize other features of tree decomposition would be of interest in the future.

References

64.1   N. Robertson, P.D. Seymour: Graph minors II: Algorithmic aspects of tree-width, J. Algorithms **7**, 309–322 (1986)

64.2   S. Arnborg, D.G. Corneil, A. Proskurowski: Complexity of finding embeddings in a k-tree, SIAM J. Algebr. Discrete Methods **8**, 277–284 (1987)

64.3   J.H. Holland: *Adaptation in Natural and Artificial Systems* (Univ. of Michigan Press, Ann Arbor 1975)

64.4   S. Kirkpatrick, C.D. Gelaff, M.P. Vecchi: Optimization by simmulated annealing, Science **220**(4598), 671–680 (1983)

64.5   F. Glover: Future paths for integer programming and links to artificial intelligence, Comput. Oper. Res. **5**, 533–549 (1986)

64.6   H. Lourenço, O. Martin, T. Stützle: Iterated local search. In: *Handbook of Metaheuristics*, Vol. 57, ed. by F. Glover, G.A. Kochenberger (Springer, New York 2003) pp. 320–353

64.7   M. Dorigo: Optimization, Learning and Natural Algorithms, Ph.D. Thesis (Dipartimento di Elettronica, Politecnico di Milano, Italy 1992), in Italian

64.8   M. Dorigo, V. Maniezzo, A. Colorni: The ant system: Optimization by a colony of cooperating agents, IEEE Trans. Syst. Man Cybern. B **26**(1), 29–41 (1996)

64.9   S. Lauritzen, D. Spiegelhalter: Local computations with probabilities on graphical structures and their application to expert systems, J. R. Stat. Soc. Ser. B **50**, 157–224 (1988)

64.10   A.M. Koster, S.P. van Hoesel, A.W. Kolen: Optimal solutions for frequency assignment problems via tree decomposition, Lect. Notes Comput. Sci. **1665**, 338–350 (1999)

64.11   J. Alber, F. Dorn, R. Niedermeier: Experimental evaluation of a tree decomposition based algorithm for vertex cover on planar graphs, Discrete Appl. Math. **145**, 210–219 (2004)

64.12   A. Koster, S. van Hoesel, A. Kolen: Solving partial constraint satisfaction problems with tree-decomposition, Networks **40**(3), 170–180 (2002)

64.13   J. Xu, F. Jiao, B. Berger: A tree-decomposition approach to protein structure prediction, Proc. IEEE Comput. Syst. Bioinform. Conf. (2005) pp. 247–256

64.14   M. Morak, N. Musliu, R. Pichler, S. Rümmele, S. Woltran: Evaluating tree-decomposition based algorithms for answer set programming, Proc. Learn. Intell. Optim. Conf. (LION 6) (2012)

64.15   A. Koster, H. Bodlaender, S. van Hoesel: *Treewidth: Computational Experiments, Electronic Notes in Discrete Mathematics*, Vol. 8 (Elsevier Science, Amsterdam 2001)

64.16   F. Clautiaux, A. Moukrim, S. Négre, J. Carlier: Heuristic and meta-heurisistic methods for com-

puting graph treewidth, RAIRO Oper. Res. **38**, 13–26 (2004)

64.17   D.R. Fulkerson, O. Gross: Incidence matrices and interval graphs, Pac. J. Math. **15**, 835–855 (1965)

64.18   F. Gavril: Algorithms for minimum coloring, maximum clique, minimum coloring cliques and maximum independent set of a chordal graph, SIAM J. Comput. **1**, 180–187 (1972)

64.19   P. Larranaga, C. Kuijpers, M. Poza, R. Murga: Decomposing Bayesian networks: Triangulation of the moral graph with genetic algorithms, Stat. Comput. **7**(1), 19–34 (1997)

64.20   N. Musliu, W. Schafhauser: Genetic algorithms for generalized hypertree decompositions, Eur. J. Ind. Eng. **1**(3), 317–340 (2007)

64.21   T. Hammerl, N. Musliu: Ant colony optimization for tree decompositions. In: *EvoCOP*, ed. by P. Cowling, P. Merz (Springer, Berlin, Heidelberg 2010) pp. 95–106

64.22   U. Kjaerulff: Optimal decomposition of probabilistic networks by simulated annealing, Stat. Comput. **2**(1), 2–17 (1992)

64.23   N. Musliu: An iterative heuristic algorithm for tree decomposition. In: *Studies in Computational Intelligence, Recent Advances in Evolutionary Computation for Combinatorial Optimization*, Vol. 153, ed. by C. Cotta, J.I. van Hemert (Springer, Berlin, Heidelberg 2008) pp. 133–150

64.24   D.S. Johnson, M.A. Trick: *The Second Dimacs Implementation Challenge: NP-Hard Problems: Maximum Clique, Graph Coloring, and Satisfiability*, Series in Discrete Mathematics and Theoretical Computer Science (American Mathematical Society, Boston 1993)

64.25   T. Hammerl: Ant Colony Optimization for Tree and Hypertree Decompositions, M.S. Thesis (Vienna University of Technology, Vienna 2009)

64.26   M. Dorigo, T. Stützle: *Ant Colony Optimization*, A Bradford Book (MIT Press, Cambridge 2004)

64.27   B. Bullnheimer, R.F. Hartl, C. Strauss: A new rank based version of the ant system: A computational study, Cent. Eur. J. Oper. Res. Econ. **7**(1), 25–38 (1999)

64.28   T. Stützle, H. Hoos: Max-min ant system and local search for the traveling salesman problem, IEEE Int. Conf. Evol. Comput. (1997) pp. 309–314

64.29   T. Stützle, H. Hoos: Max-min ant system, Future Gener. Comput. Syst. **16**(9), 889–914 (2000)

64.30   M. Dorigo, L.M. Gambardella: Ant colony system: A cooperative learning approach to the traveling salesman problem, IEEE Trans. Evol. Comput. **1**(1), 53–66 (1997)

64.31   N. Musliu: Generation of tree decompositions by iterated local search. In: *EvoCOP*, ed. by C. Cotta, J. van Hemert (Springer, Berlin, Heidelberg 2007) pp. 130–141

64.32   K. Shoikhet, D. Geiger: A practical algorithm for finding optimal triangulations, Proc. Natl. Conf. Artif. Intell. (AAAI'97) (1997) pp. 185–190

64.33   V. Gogate, R. Dechter: A complete anytime algorithm for treewidth, Proc. 20th Annu. Conf. Uncertain. Artif. Intell. UAI-04 (2004) pp. 201–208

64.34   E. Bachoore, H. Bodlaender: A branch and bound algorithm for exact, upper, and lower bounds on treewidth, Lect. Notes Comput. Sci. **4041**, 255–266 (2006)

64.35   R. Tarjan, M. Yannakakis: Simple linear-time algorithm to test chordality of graphs, testacyclicity of hypergraphs, and selectively reduce acyclic hypergraphs, SIAM J. Comput. **13**, 566–579 (1984)

64.36   K. Kask, A. Gelfand, L. Otten, R. Dechter: Pushing the power of stochastic greedy ordering schemes for inference in graphical models, Proc. Natl. Conf. Artif. Intell. (AAAI) (2011) pp. 54–60

64.37   H.L. Bodlaender, A.M.C.A. Koster: Treewidth computations I. Upper bounds, Inf. Comput. **208**(3), 259–275 (2010)

64.38   A. Khanafer, F. Clautiaux, E.-G. Talbi: Tree-decomposition based heuristics for the two-dimensional bin packing problem with conflicts, Comput. Oper. Res. **39**(1), 54–63 (2012)

64.39   M. Fontaine, S. Loudni, P. Boizumault: Guiding VNS with tree decomposition, 23rd IEEE Int. Conf. Tools Artif. Intell. (ICTAI) (IEEE, Boca Raton 2011) pp. 505–512

65. Evolutionary Computation and Constraint Satisfaction

Jano I. van Hemert

In this chapter we will focus on the combination of *evolutionary computation* (EC) techniques and *constraint satisfaction problems* (CSPs). *Constraint programming* (CP) is another approach to deal with constraint satisfaction problems. In fact, it is an important prelude to the work covered here as it advocates itself as an alternative approach to programming [65.1]. The first step is to formulate a problem as a CSP such that techniques from CP, EC, combinations of the two, often referred to as hybrids [65.2, 3], or other approaches can be deployed to solve the problem. The formulation of a problem has an impact on its complexity in terms of effort required to either find a solution or that proof no solution exists. It is, therefore, vital to spend time on getting this right.

CP defines search as iterative steps over a search tree where nodes are partial solutions to the problem where not all variables are assigned values. The search then maintains a partial solution that satisfies all variables with assigned values. Instead, in EC algorithms sample a space of candidate solutions where for each sample point variables are all assigned values. None of these candidate solutions will satisfy all constraints in the problem until a solution is found. Such algorithms are often classified as Davis–Putnam–Logemann–Loveland (DPLL) algorithms, after the first backtracking algorithm for solving CSP [65.4].

Another major difference is that many constraint solvers from CP are sound, whereas EC solvers are not. A solver is sound if it always finds

a solution if it exists. Furthermore, most constraint solvers from CP can easily be made complete, although this is often not a desired property for a constraint solver. A constraint solver is complete if it can find every solution to a problem.

65.1 Informal Introduction to CSP

For a formal definition please skip to the next section. A constraint satisfaction problem consists of a set of variables and each variable must be assigned one value from its finite set of values, called its domain.

A set of constraints restricts certain simultaneous assignments. In most CSPs, the objective is to search for a simultaneous assignment of all the variables such that all constraints are satisfied, i.e., no forbidden simultaneous assignment from the set of constraints is used.

A famous example is the SEND MORE MONEY puzzle, where each letter must be replaced by a unique number such that the following sum holds [65.5]

$$
\begin{array}{ccccc}
 & S & E & N & D \\
+ & M & O & R & E \\
= M & O & N & E & Y.
\end{array}
$$

In this CSP, the variables are S, E, N, D, M, O, R, Y and the domains are $\{1, \ldots, 9\}$ for S, M and $\{0, \ldots, 9\}$ for E, N, D, O, R, Y. The constraint can be also written as $1000 \times S + 100 \times E + 10 \times N + D + 1000 \times M + 100 \times O + 10 \times R + E = 10\,000 \times M + 1000 \times O + 100 \times N + 10 \times E + Y$. Every CSP A can be rewritten into an another CSP B where a bijective mapping exists between the solutions of A and B, which follows

from the reducibility theorem from complexity theory [65.6]. The solution to this CSP is the assignment $S = 9, E = 5, N = 6, D = 7, M = 1, O = 0, R = 8, Y = 2$, which uniquely satisfies the constraint.

Other very well-known constraint satisfaction problems are map coloring, more commonly known as vertex coloring (Sect. 65.5.2), and the recreational game Sudoku, which is equivalent to completing a graph 9-coloring problem on a given specific graph with 81 vertices. A specific EC solution is provided by *Lewis* [65.7]. Quite a lot of constraint satisfaction problems exist; we will first look at CSP in general within the context of EC as problem solvers. Then we will discuss several specific constraint satisfaction problems and the particular EC approaches applied to these problems. Last, we will provide a brief overview on using EC for generating problem instances for CSP.

65.2 Formal Definitions

Slightly different, but equivalent, formal definitions of CSP exist. The most common definition is:

Definition 65.1 (Constraint Satisfaction Problem) is a triple $\langle V, D, C \rangle$:

- V is an n-tuple of *variables* $V = \langle v_1, v_2, \ldots, v_n \rangle$,
- Each $v \in V$ has a corresponding m-tuple of values called its *domains*, $D_v = \langle d_1, d_2, \ldots, d_m \rangle$ of which it can be assigned one and
- $C = \langle C_1, \ldots, C_t \rangle$ is a t-tuple of *constraints* where each $c \in C$ *restricts* certain simultaneous variable assignments to occur.

The definition of a constraint is often reversed in the literature, where generic CSP is discussed in that constraints are defined as the set of assignments that are *allowed* rather than restricted. Note, in generic CSP literature, variables are often denoted with X, whereas in graph-oriented problem domains such as graph coloring and maximum clique, V is adopted.

Definition 65.2 (Solution to a CSP) is an assignment of variables $(d_1, \ldots, d_n) \in D_1 \times \cdots \times D_n$ such that for every constraint $c \in C$ on $x_{i_1}, \ldots, x_{i_m}$: $(d_{i_1}, \ldots, d_{i_m}) \notin c$.

In the context of one constraint c, we say an assignment of variables *satisfies* the constraint c if the

assignment is in c or *violates* the constraint c if the assignment is *not* in c. A CSP can be *insoluble* – more commonly written as *insolvable*, which means every assignment of variables will violate at least one constraint.

A *constraint solver* is an algorithm that takes as input a CSP and produces as output either a solution or a proof that no solution exists or a notification of failure. The input is often referred to as a *problem instance*, as a CSP is often defined to cover a class of problems such as, 3-satisfiability. The output can be more than one solution, in fact it could be every solution. However, as EC techniques are based on sampling, in principle they cannot proof that every solution has been found, which is referred to as not complete. Moreover, they cannot proof no solution exists, which is referred to as not sound. Therefore, constraint solvers based on EC and other heuristic approaches often terminate after a certain criterion is met, e.g., a predefined elapsed time is reached in terms of the number of solutions evaluated, the computation time spent, or a certain convergence of the population reached.

We recommend the following books for further reading on constraint satisfaction. For the foundations of the problem and basic algorithms, *Tsang* [65.8]; for an introduction with comprehensive overview of constraint programming techniques, *Dechter* [65.9] and *Lecoutre* [65.10]; and for a more theoretical approach *Apt* [65.1] and *Chen* [65.11].

65.3 Solving CSP with Evolutionary Algorithms

In this chapter we will restrict ourself to covering the conceptual mapping required to solve a CSP with an evolutionary algorithm. This mapping will consist of choosing a representation for the problem and a corresponding fitness function to determine the quality of a solution. Once this mapping is complete, the evolutionary algorithm will require other components, such as appropriate variation operators, selection mechanisms, and a suitable initialization method for the population and termination criteria. All these, and other optional variants can be found elsewhere in the handbook.

We will explain the two most common mappings using the well-known n-queens on an $n \times n$-chessboard problem. These mappings are *direct encoding* and *indirect encoding*. First we introduce a conceptual definition of the problem.

The *n*-queens problem requires the placing of n queens on an $n \times n$ chessboard such that no queen attacks any of the other $n-1$ queens. Thus, a solution requires that no two queens share the same row, column, or diagonal. Several common formal definitions of the problem exist. The most common is to define n variables $\{q_1, \ldots, q_n\}$, where each variable q has a domain that consists of the row position the queen will be placed on in its corresponding unique column, i.e., $q \in \{1, \ldots, n\} \forall i = 1, \ldots, n$. The set of constraints consists of $q_i \neq q_j$ (i.e., not in the same row) and $|q_i - q_j| \neq |i - j| \forall i, j = 1, \ldots, n$ (i.e., not in the same diagonal).

The n-queens problem is no longer considered a challenging problem as it has a structure that can be exploited to solve very large problems of over 9 million queens by repeating a pattern [65.12]. It is, however, an excellent problem for explaining characteristics of constraint satisfaction problems and their solvers due to the simple 2-D spatial nature of the problem. For instance, to explain symmetry in CSP, the 8-queens problem can be used to show it has 12 unique solutions, as shown in Fig. 65.1 out of the 92 distinct solutions when removing variants due to rotational and reflection symmetry.

65.3.1 Direct Encoding

With a direct encoding the genotype consists of a vector $\vec{g}$ where each element corresponds uniquely to one variable of the CSP; an element g_i contains values directly from the domain of its corresponding variable D_i. A wide variety of genetic operators both for mutation and recombination are applicable to this encoding and

can be found in [65.13]. Most of these operators will be called discrete or mixed-integer operations.

The genotype is mapped to the phenotype by taking into consideration the constraints; it requires a measurement for determining the quality of candidate solutions. Thus, we need to introduce a fitness function. The most common fitness function takes the sum of all constraints violated by a candidate solution

$$\text{fitness}(\vec{g}) = \sum_{c \in C} \text{violated}(c) ,$$

where $\text{violated}(c) = \begin{cases} 1 & \text{if } c \text{ violated by } \vec{g} \\ 0 & \text{if } c \text{ satisfied by } \vec{g} \end{cases} .$

The fitness should be minimized and once it reaches zero, a solution has been found.

65.3.2 Indirect Encoding

With an indirect encoding the genotype first needs to be transformed into a full or partial assignment of the variables of the CSP. It is also referred to as local search depending on the level of sophistication; these transformations range from as simple as a greedy assignment all the way to sound search algorithms evaluating a small part of the CSP.

The most common approach for this representation takes as a genotype the permutation of variables of the CSP. Many genetic operators are designed to maintain a permutation and several are explained in the Handbook of Evolutionary Computation [65.13]. The permutation is the input to the local search and determines the order in which variables are processed; processing a variable involves trying to assign a value such that no constraint is violated and perhaps further steps if no value can be assigned without violating at least one constraint.

More advanced encodings may also include the ordering in which to consider values from each variable's domain. From constraint programming we know that the order in which variables and values are considered has a huge impact on the efficiency of search algorithms [65.14]; more often it is the search method that determines the order using a particular heuristics such as choosing the next vertex with the *maximum saturation degree*, as is used in DSatur [65.15]. The saturation degree for a vertex is defined as the total number of colors used for coloring its neighbors. The principle

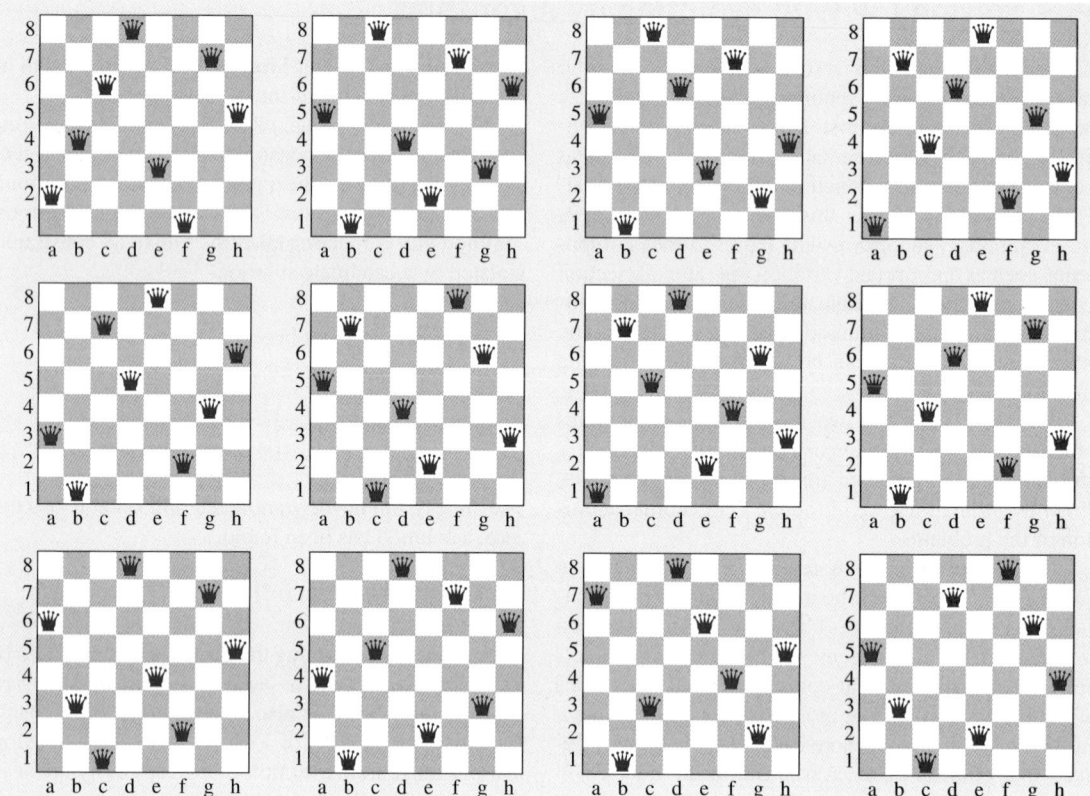

Fig. 65.1 The 12 unique solutions under symmetry via rotations and reflections for the 8-queens problem

has been used in many algorithms since its introduction in 1979.

The most common fitness function used with indirect encoding simply counts the number of unassigned variables after the local search terminates. Note that two different strategies will influence the resolution of this function. If the local search terminates after it first encounters a variable it cannot assign, then many candidate solutions will have the same fitness but can still be very different. On the other hand, terminating after all variables have been considered will give a richer landscape to consider but may incur more computational effort. See [65.16] for a comprehensive theoretical and empirical analysis of sampling in EC.

65.3.3 General Techniques to Improve Performance

Over the past two decades, many techniques were developed to improve the efficiency and/or the effectiveness of EC for solving constraint satisfaction problems.

Only a handful of these techniques were evaluated on more than one problem. Hence, we cannot draw any general conclusions about the success of these techniques. Even worse is that many studies will show improvement only compared to their previous results or compare their results with an algorithm that has already been superseded in terms of performance by many other techniques. Often the set of competitor algorithms is chosen to fall within EC, which severely limits the strength of the competition. Therefore, we will discuss techniques for improving performance in the context of the problems they were developed for. Section 65.5 reviews several popular CSPs used for developing more efficient and effective evolutionary algorithms.

One approach that has been applied to several CSPs with varying success is that of assigning weights to constraints to allow biasing the search towards satisfying certain constraints; in the first experiments this approach was referred to as penalty functions [65.17]. Moreover, the search can be influenced dynamically by adapting weights according to heuristics, such as increasing the weight of the constraint that has been satisfied the least number of times recently [65.18]. The origin of this idea can be found in the self-adaptation used in evolution strategies [65.19].

With penalty functions, the optimization objectives replacing the constraints are traditionally viewed as penalties for constraint violation, hence to be minimized [65.20]. There are two basic types of penalties:

1. Penalty for violated constraints
2. Penalty for wrongly instantiated variables.

Formally, let us assume that we have constraints c_i $(i = \{1, \ldots, m\})$ and variables v_j $(j = \{1, \ldots, n\})$. Let C^j be the set of constraints involving variable v_j. Then the penalties relative to the two options described above can be expressed as follows:

1. $f_1(s) = \sum_{i=1}^{m} w_i \times \chi(s, c_i)$, where

$$\chi(s, c_i) = \begin{cases} 1 & \text{if } s \text{ violates } c_i \\ 0 & \text{otherwise} \end{cases},$$

2. $f_2(s) = \sum_{j=1}^{n} w_j \times \chi(s, C^j)$, where

$$\chi(s, C^j) = \begin{cases} 1 & \text{if } s \text{ violates at least one } c \in C^j \\ 0 & \text{otherwise}, \end{cases},$$

where the w_i and w_j are weights that correspond to a constraint and a variable, respectively. These will be important later on, for now we assume all these weights equal to 1.

Obviously, for each of the above functions $f \in \{f_1, f_2\}$ and for each $s \in S$ we have that $\phi(s) = true$ if and only if $f(s) = 0$. For instance, in the graph 3-coloring problem the vertices of a given graph $G = (V, E)$, $E \subseteq V \times V$, have to be colored by three colors in such a way that no neighboring vertices, i.e., graph nodes connected by an edge, have the same color. This problem can be formalized by means of a CSP with $n = |V|$ variables, each with the same domain $D = \{1, 2, 3\}$. Furthermore, we have $m = |E|$ constraints, one for each edge $e = (k, l) \in E$, with $c_e(s) = true$ if and only if $s_k \neq s_l$. Then the corresponding CSP is $\langle S, \phi \rangle$, where $S = D^n$ and $\phi(s) = \bigwedge_{e \in E} c_e$. Using the constraint-oriented penalty function f_1 with $w_i = 1$ for all $i = \{1, \ldots, m\}$ we count the incorrect edges that connect two vertices with the same color. The variable-oriented penalty function f_2 with $w_i = 1$ for all $i = \{1, \ldots, m\}$ amounts to counting the incorrect vertices that have a neighbor with the same color.

Advantages of indirect encoding:

- Introduces in general, e.g., f_1, f_2 are problem-independent penalty functions
- Reduces problem to *simple* optimization
- Allows user preferences by weights.

Disadvantages of indirect encoding:

- Loss of information by packing everything in a single number
- In the case of constrained optimization (as opposed to CSP as we are handling here) f_1, f_2 are reported to be weak [65.21].

65.4 Performance Indicators

An understanding of the efficiency and effectiveness is vital when choosing which solver to use or when developing an algorithm to deal with a specific CSP. In this section we briefly explain measures for determining these properties in the context of solving CSP. However, these properties must be measured using a suite of benchmark instances and, as EAs are generally randomized algorithms, with multiple independent runs of the algorithm on each instance. Choosing an appropriate suite of benchmark instances is paramount to making decisions on which algorithm, parameter setting, or next algorithmic feature to add.

In a sense, the search for a good algorithm is in itself an optimization problem. The suite of benchmark instances represents only the problem, just like training data in a machine learning problem represents all data possibly encountered. Changing an algorithm and tuning its parameters on the same small suite of instances could lead to over-fitting [65.22, 23], which in turn means the algorithm will have a poorer performance in the general case. Therefore, the first step should be to characterize the problem well and have a good representation, e.g., spread, of the instances possibly encountered when deployed.

65.4.1 Efficiency

The time taken by an algorithm to provide a solution is an important factor. Even more so in situations where solutions are required in real time. Much research is devoted to speeding up algorithms, either by cleverly exploiting properties of the problem, by parallelization, or by balancing aspects of the quality of the solution.

The most common approach to measuring the efficiency of evolutionary algorithms is by counting the number of evaluations, i.e., the number of times the fitness function is executed. This approach has several drawbacks. First, the approach allows comparison only with algorithms that use the exact same fitness function and spend the most significant part of their time on computing that function. Second, the computational complexity of the evolutionary algorithm may not be dependent on the fitness function. For instance, with the indirect encoding described in Sect. 65.3.2, much computational effort will go into the local search, whereas the computation of the fitness is trivial.

Another common approach is to measure time spent as reported by the operating system. This has even more drawbacks as the reported numbers will depend on the computer programming language used for implementing the algorithm, the compiler and its setting for translating the implementation into machine code, the architecture of the computer for executing the machine code, and the operating system for hosting the execution environment. Variations of these will have an affect on the reported results and, moreover, as these environments themselves change over time, future studies will find it hard to reproduce results accurately or even create meaningful comparisons to reported results.

A more meaningful solution is to count all the atomic operations that are directly related to the problem. The operations that must be included should be those that in theory increase exponentially in numbers with larger problems, as CSP fall under the class of nonpolynomial deterministic problems. The most common operation will be a *conflict check*; this is also referred to as a constraint check, but in the strictest sense, a constraint check consists of multiple conflict checks [65.8]. For example, when solving the n-queens problem, every time the algorithm checks $q_i \neq q_j$ for any q_i and q_j, this should be recorded as one check. The same procedure should be followed for the constraint concerning diagonal attacks $|q_i - q_j| \neq |i - j|$. The sum of all checks when the algorithm terminates is the computational effort spent.

By reporting the number of conflict checks we assure future studies can compare with current results as this measurement will not be affected by future changes in hardware and software environments. We are measuring a property of the algorithm here as opposed to a property of one implementation of the algorithm running in one particular environment.

It is important to note that there are subtle differences in the reporting used in different studies. Some studies report the average number of operations over all independent runs, including runs that are unsuccessful, i.e., where no solution was found. Other studies report the average number of operations to a solution, where only the runs that yield a solution are taken into account. The former method will produce higher averages than the latter if the success rate is less than 1.

65.4.2 Effectiveness

Efficiency is only one aspect of which to measure the success of a constraint solver. The other most important aspect is that of effectiveness, which measures how successful an algorithm is in finding or approximating a solution. The easiest and most commonly used measurement is that of the *success rate*, which is defined for an experiment as the number of runs in which an algorithm finds a solution divided by the total of number of runs of the same algorithm in that experiment. As no prior knowledge is required about whether problem instances are insolvable, this measurement is straightforward to implement.

Another popular measurement in combinatorial optimization is *distance to the optimal solution*. This measurement poses two challenges in the context of constraint satisfaction. Unlike a combinatorial optimization problem, which has the function to optimize, a CSP has no such function. As an alternative we could use the fitness function, but that is not an inherent property of the problem. Also, we often do not know whether a CSP has a solution and when it does not, then we do not know the optimal fitness function. Distance to the optimal solution is rarely used when solving CSP due to these impracticalities.

65.5 Specific Constraint Satisfaction Problems

Many specific constraint satisfaction problems have been addressed in the literature. A full overview of these would not provide much benefit, as the most likely scenario is that one is looking for papers that provide descriptions of algorithms and results with those algorithms on a certain problem. The exceptions to this are several problems that in the literature are used to drive the development of algorithms in terms of efficiency and effectiveness. These *core* problems are used over and over to test whether new algorithms are better than existing algorithms.

Several reasons exist for the choice of these problems. Their compact definition means that the problem is easy to replicate by everyone and quick to introduce in papers. The most popular problems were used in the 1970s when the theory on non-polynomial deterministic problems was developed, which were consequently seen as important intelligent building blocks. Also, test sets and later problem generators were released in the public domain, thereby providing easy access to test suites.

We will use several of these *core* problems to describe the progress of development in evolutionary computation for constraint satisfaction problems. For each problem we will provide a quick introduction, a justification of its importance in terms of practical applications, and a set of pointers to problem suites before describing the approaches used.

65.5.1 Boolean Satisfiability Problem

Given a Boolean formula ϕ determine whether an assignment of the variables in ϕ exists that makes it TRUE. It is often referred to as satisfiability and abbreviated to SAT [65.24]. In SAT variables are often referred to as literals. Most often the problem is studied in conjunctive normal form (CNF) where ϕ is a conjunction of clauses where each clause is a disjunction of variables. Every SAT problem can be reduced to a 3-CNF-SAT (three variables/clause-conjunctive normal form-satisfiability) [65.25], where each clause has three literals.

3-CNF-SAT was the first problem to be shown to be NP-complete [65.26]. It serves as an important basis to proving that other problems are NP-complete, such as the maximal clique problem. Such a proof involves a polynomial-time reduction from 3-CNF-SAT to the other problem [65.6].

The following is an example of 3-CNF-SAT:

- $\phi = (x_1 \vee \neg x_3 \vee x_4) \wedge (\neg x_2 \vee x_1 \vee \neg x_6) \wedge (x_3 \vee x_2 \vee \neg x_5)$
- A solution: $x_1 = 1$, $x_2 = 0$, $x_3 = 1$, $x_4 = 0$, $x_5 = 0$, $x_6 = 0$.

Important practical applications of SAT are model checking [65.27], for example, in mathematical proof planning [65.28], generic planning problems, especially using the planning domain definition language (PDDL) [65.29], test pattern generation [65.30], and haplotyping in the scientific field of bioinformatics [65.31].

As far as the development of efficient and effective CSP solvers go, SAT is the most active field. It has an annual conference – The International Conference on Theory and Applications of Satisfiability Testing, which also hosts an annual competition to determine the current best solvers. The latter also ensures that new problem instances are continuously added, which prevents what is called *overfitting* [65.32] of the solvers to an existing set of problem instances.

The general approach to solve satisfiability with EC is to directly represent the variables in ϕ and assign these either TRUE or FALSE, i.e., these form the domain. The fitness function used is the number of clauses violated, which should be minimized.

The earliest evolutionary algorithm for SAT was reported in 1994 by [65.33] and was soon followed by the work of *Gottlieb* and *Voss* [65.34, 35], who were looking to improve its performance. Soon after, independent efforts led to parallelized algorithms [65.36, 37]. In 2000, the first adaptive evolutionary algorithms were applied [65.38], which was 3 years after they were applied to graph coloring (Sect. 65.5.2).

The introduction of hybrid evolutionary algorithms with local search created a real boost of research activity [65.39–43]. However, a major issue remains with research on solving satisfiability with EC, as all studies include only local search and evolutionary algorithms without comparing to the state-of-art DPLL and heuristic solvers from the annual satisfiability community. This holds true even for recent studies such as [65.44]. Due to this major gap between the two communities of EC and CP, we do not comment on the comparison in terms of effectiveness and efficiency.

New research [65.45] focusses on using EC to evolve parameter settings for existing sound SAT solvers, mostly ones based on the Davis–Putnam–Logemann–Loveland algorithm [65.46]. All modern SAT solvers have many parameters to tune how the search is organized. These parameters are often tuned manually, which allows for only a small exploration. Using EC, a much larger space can be explored in order to create fast SAT solvers for a given benchmark.

65.5.2 Graph Coloring

Graph coloring has several variants. The most commonly used definition is that of graph k-coloring, also known as the vertex coloring problem. Given a graph of vertices and edges $\langle V, E \rangle$ the goal is to find a coloring of the vertices V of the graph such that no two adjacent vertices have the same coloring. If $c(v)$ provides the color assigned to v, then $\forall v, w \in V : c(v) \neq c(w)$ iff $(v, w) \in E$. The objective is to make use of k or less colors. The problem is known to be NP-complete for $k \geq 3$ and to be decidable in linear time for $k \leq 2$.

Graph coloring is an abstract problem that lies at the core of many applications. Well-known applications are scheduling, most specifically timetabling [65.47], register allocation in compilers [65.48], and frequency assignment in wireless communication [65.49]. It is a well-studied problem as is shown by the number of entries in the best-kept bibliography source until April 2010 with over 450 publications contributing to vertex coloring [65.50].

The Second DIMACS Implementation Challenge in 1992–1993 focused on maximum clique, graph coloring, and satisfiability. The challenge provided not only a standard format for graph k-coloring problem instances, but also provided a set of problem instances that is still popular today. Soon after, in 1994, *Culberson* and *Luo* [65.51] created a problem instance generator, which can create problem instances with a known k and various other properties. Several other generators exist with specific goals, such as to hide cliques [65.52], to create register-interference graphs [65.53], and to create timetabling problems (Sect. 65.5.4).

The most straightforward approach to solving graph k-coloring with EC is to represent a genome as a vector of all variables of the problem. This vector can then undergo genetic operators suitable for integer representations. The fitness function is simply the number of violated constraints, which should be minimized until a solution is found when the fitness is equal to zero. Unfortunately, this approach leads to algorithms that are inefficient and ineffective [65.54].

To make EC more efficient and effective for solving graph k-coloring, new algorithms have been developed; these broadly fall into two categories. The first category consists of adding mechanisms that prevent the stagnation of search due to premature convergence. The second category consists of alternative representations that make use of decoders to map genotypes to phenotypes. The two categories are not mutually exclusive, and studies have included algorithms that combine mechanisms from both categories.

The earliest work on solving graph k-coloring with EC includes the following. *Fleurent* and *Ferland* successfully considered various hybrid evolutionary algorithms [65.55] with Tabu search and extended their work into a general implementation of heuristic search methods in [65.56]. *Von Laszewski* looked at structured operators and used adaption to improve the convergence rate of a genetic algorithm [65.57]. Davis designed an algorithm [65.58] to maximize the total of weights of nodes in a graph colored with a fixed number of colors. *Coll* et al. [65.59] discussed graph coloring and crossover operators in a more general context.

Juhos and *van Hemert* introduced several heuristics [65.60, 61] for guiding the search of an evolutionary algorithm. All these heuristics depend on their novel representation that collapses the graph by combining nodes assigned with the same color into one hypernode, which speeds up further constraint checking as edges are merged into hyperedges [65.62]. This representation benefits both complete and heuristic methods.

Moreover, as shown in the results in Fig. 65.2, the evolutionary algorithms developed by Juhos and van Hemert are able to outperform a complete method (Backtracking-DSatur) on very difficult problem instances where the chromatic number is 10 or 20. These algorithms are unable to compete with the complete method for smaller chromatic numbers of 3 and 5.

65.5.3 Binary Constraint Satisfaction Problems

A *binary constraint satisfaction problem* (BINCSP) is a CSP where every constraint $c \in C$ restricts at most two variables [65.63]. Often, network graphs are used to visualize (CSP) instances. In Fig. 65.3, we provide an example of a restricting hypergraph of a BINCSP. It consists of three variables $V = \{v_1, v_2, v_3\}$, all of which have domain $D = \{a, b\}$. In a hypergraph every vertex corresponds to a possible variable assignment, i.e., $\langle v, d \rangle$, where $v \in V$ and $d \in D_v$. Every edge indicates the variable assignments that are forbidden by the

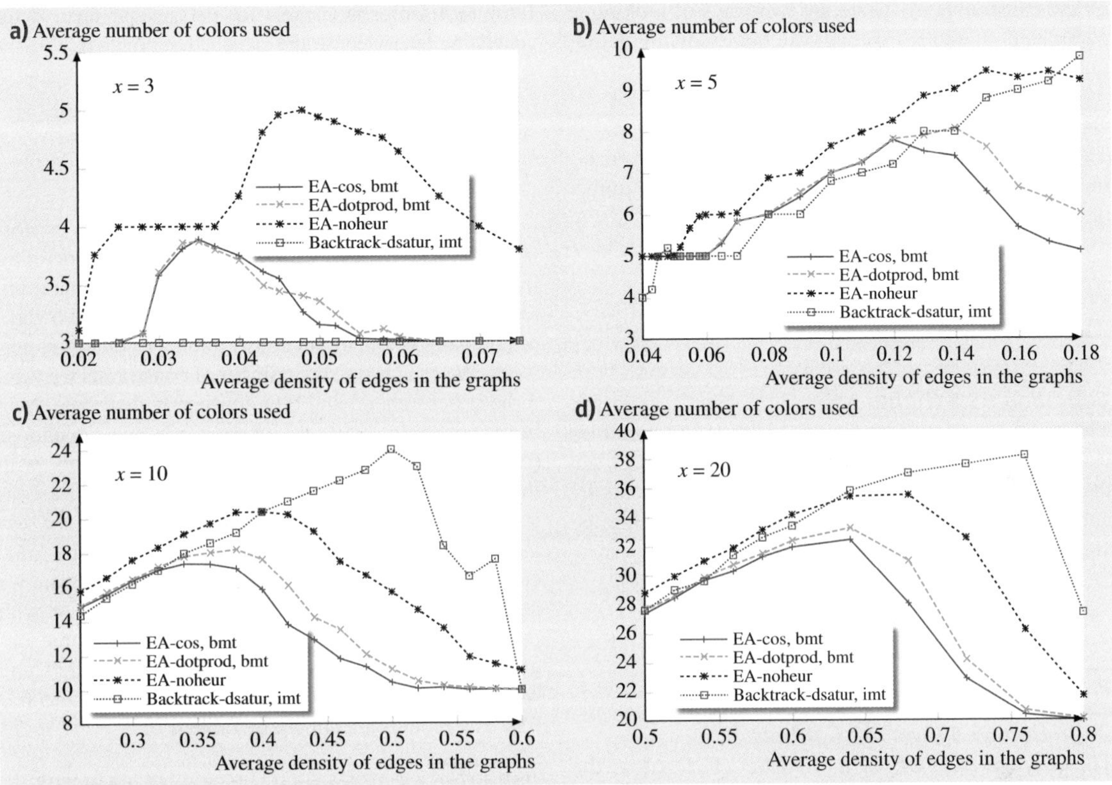

a) Average number of colors used

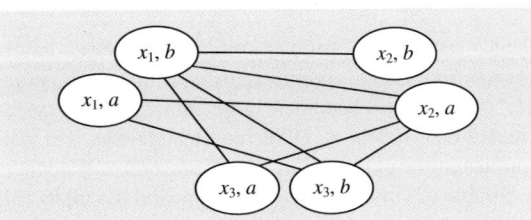

b) Average number of colors used

c) Average number of colors used

d) Average number of colors used

Fig. 65.2a–d Results of several evolutionary algorithms against the complete method Backtracking-DSatur; average minimum number of colors used through the phase transition

set of constraints C. In the example, we show all the edges that correspond to the following set of forbidden value pairs $C = \{ \{\langle v_1, a \rangle, \langle v_2, a \rangle\}, \{\langle v_1, a \rangle, \langle v_3, b \rangle\}, \{\langle v_1, b \rangle, \langle v_2, a \rangle\}, \{\langle v_1, b \rangle, \langle v_2, b \rangle\}, \{\langle v_1, b \rangle, \langle v_3, a \rangle\}, \{\langle v_1, b \rangle, \langle v_3, b \rangle\}, \{\langle v_2, a \rangle, \langle v_3, a \rangle\}, \{\langle v_2, a \rangle, \langle v_3, b \rangle\} \}.$

For problem instances, studies on BINCSP generally create large sets of instances using one of many problem instance generators. Several models to randomly create BINCSPs have been designed and analyzed [65.63–65]. All of these incorporate a set of

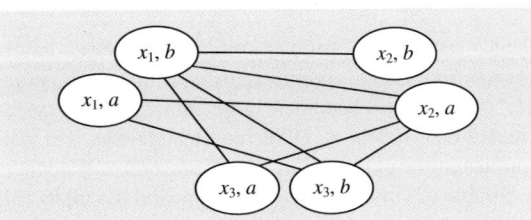

Fig. 65.3 Example of a $|V|$-partite hypergraph of a (BINCSP) with one solution: $\{\langle v_1, a \rangle, \langle v_2, b \rangle, \langle v_3, a \rangle\}$

parameters that may be used to control the size and difficulty of the problems. Often, these parameters can be used to create a set of problems that go through a *phase transition*. That is, we order the set on the parameters and observe how the algorithms behave when we move through the parameter space. In most constraint satisfaction problems we observe that the performance drops gradually until it reaches a minimum, after which it rises again. Most researchers test their algorithms in the region where the minimum is reached. Here the set of most difficult to solve problem instances is found. We will discuss these methods next.

The model most often used in empirical research on binary constraint satisfaction problems is one that uses four parameters to control, to some degree, the difficulty of an instance. By varying these global parameters one can characterize instances that are more likely to be either more or less difficult to solve. These parameters are: the number of variables $n = |V|$, the size of each variable's domain $m = |D_{v_1}| = |D_{v_2}| = \cdots = |D_{v_n}|$, the density of constraints p_1, and the average tightness of all the constraints p_2. There are two ways of looking at parameters p_1 and p_2. We will use the following definitions.

Definition 65.3 (Density)

The *density* of a BINCSP is the ratio between the maximum number of constraints $\binom{n}{2}$ and the actual number of constraints $|C|$,

$$p_1 = \frac{|C|}{\binom{n}{2}}.$$

Definition 65.4 (Tightness)

The *tightness* of a constraint $c \subset C$ over the variables $v, w \in V$ of a BINCSP $\langle V, D, C \rangle$ is the ratio between the total number of forbidden variable assignments $|c|$ and the total number of combinations of variable assignments possible $m = |D_v||D_w|$,

$$p_2(c) = \frac{|c|}{m^2}.$$

Definition 65.5 (Average Tightness)

The *average tightness* of a BINCSP $\langle V, D, C \rangle$ is the sum of the tightness over all constraints divided by the number of constraints,

$$\overline{p_2} = \frac{\sum_{c \in C} p_2(c)}{|C|}.$$

These definitions give the density and tightness in terms of a ratio, or in other words, as the percentages of the maximum. Another way of looking at these two properties uses probabilities [65.66]. We could define the density of a BINCSP as the probability that a constraint exists between two variables. The tightness can be alternatively defined in an analogous way, as the probability that a conflict exists between two instantiations of two variables. The differences in these viewpoints becomes apparent in the different implementations of algorithms that generate BINCSPs, as with uniform generation the ratio in an instance is determined beforehand, while with probability the ratio will vary according to a normal distribution. When comparing studies it is important to know when probabilities are used whether the results reported are against the probability set or the actual measured ratio in the whole instance.

Table 65.1 Different models for the general method for generating binary constraint satisfaction problems

Constraints		Nogoods	
		Probability	Uniform
	Probability	Model A	Model C
	Uniform	Model D	Model B

The simplest way to empirically test the performance of an algorithm on solving CSPs is by generating instances using different settings for the four main parameters, n, m, p_1, and $\overline{p_2}$. However, there are two ways of choosing where to put constraints in a constraint network. We can choose the number of constraints we want to have beforehand and then uniformly distribute them in the constraint network. Alternatively, we can choose for each *possible* edge in the constraint network with the probability p_1 if this edge is inserted, i.e., a constraint is added. We will call the first model the *uniform model* and the second the *probability model*. The same categorization holds for nogoods. Given a constraint we can either distribute $\overline{p_2}m^2$ nogoods uniformly or with probability $\overline{p_2}$ decide which value pairs become nogoods. Now we can define four different models and we will name them according to the models in [65.63, 65]. The models are shown in Table 65.1.

Definition 65.6 (Parameter Vector of a BINCSP)

A *parameter vector* of a binary constraint satisfaction problem (BINCSP) with n variables and m as each variable's domain size is a 4-tuple $\langle n, m, p_1, \overline{p_2} \rangle$ of four parameters: the number of variables n, the domain size of each variable m, the density p_1, and the average tightness $\overline{p_2}$.

We can also characterize a set of binary constraints satisfaction problems using the parameter vector as a set B of BINCSP instances where

$$\forall \langle n, m, p_1, \overline{p_2} \rangle, \langle n', m', p_1', \overline{p_2}' \rangle \in B$$
$$\Leftrightarrow n = n' \wedge m = m' \wedge p_1 = p_1' \wedge \overline{p_2} = \overline{p_2}'.$$

Such a set we call a *suite of problem instances*.

Achlioptas et al. proves in [65.64] that as the number of variables becomes large almost all instances created by Models A–D become unsolvable. The reason lies in the existence of *flawed variables*. Whenever a variable v is involved in a constraint and has all its values incompatible with a value of an adjacent variable w, this variable is called *flawed*. In terms of compound labels using the constraint c over variables v and w this is written as,

$$\forall v \in D_v : \nexists w \in D_w :$$
$$satisfies(((\langle v, v \rangle, \langle w, w \rangle), c) \wedge c \in C .$$

When the number of variables is increased without changing the other parameters, the number of flawed variables will increase, thus making it easy to prove instances have no solution. To overcome the problems a new model is proposed [65.64]:

Definition 65.7 (Model E)
The graph C^Π is a random n-partite graph with m nodes in each part that is constructed by uniformly, independently, and with repetitions selecting $p_e \binom{n}{2} m^2$ edges out of the $\binom{n}{2} m^2$ possible ones.

The idea behind this model is that the difficulty is controlled by the tightness and not influenced by the structure of the constraint network. The parameter p_e is responsible for the average tightness of the BINCSP. However, it is not the same parameter as the average tightness $\overline{p_2}$. Because we allow repetitions in the process we end up with an average tightness smaller than or at most equal to p_e.

Parameter p_e also influences the value of p_1. In [65.65] we find the proof that using Model E with fairly small values ($p_e < 0.05$) will result in a fully connected constraint network ($p_1 = 1$). This is seen as a flaw in Model E, as many problems do not require a fully connected constraint network. This has led to yet another model.

MacIntyre et al. propose a more generalized version of Model E called *Model F* [65.65]. This model starts out the same way as Model E by generating $p_1\overline{p_2}m\binom{n}{2}$ nogoods. Afterwards, a constraint network is generated with exactly $p_1\binom{n}{2}$ edges in the uniform way. All nogoods that are not in a constraint in the constraint network are removed from the problem instance. Model E is the special case of Model F where $p_1 = 1$. The benefit of Model F is the ability to generate problems where $p_1 < 1$, which is more realistic towards real-world problems.

Craenen et al. [65.67] present the largest comparison study of EC and CP approaches for the BINCSP. In this study they compare the success rate and average number of conflict checks to a solution of 11 evolutionary algorithms. The best four evolutionary algorithms are compared with forward checking with conflict-directed backjumping [65.68], and the authors concluded the latter has a superior performance on every problem instance in the benchmark.

The following heuristic approaches are included in the study. In [65.69, 70], *Eiben* et al. propose to incorporate existing CSP heuristics into genetic operators. A study on the performance of these heuristic-based operators when solving binary CSPs was published in [65.71]. Two heuristic-based genetic operators are specified: an asexual operator that transforms one individual into a new one and a multi-parent operator that generates one offspring using a number of parents. In [65.72–74], *Riff-Rojas* introduced an EA for solving CSPs that uses information about the constraint network in the fitness function and in the genetic operators (crossover and mutation). The fitness function is based on the notion of the *error evaluation* of a constraint. *Marchiori* et al. introduced and investigated EAs for solving CSPs based on pre-processing and post-processing techniques [65.75–77]. Included in the comparison is the variant form [65.75, 78] that transforms constraints into a canonical form in such a way that there is only one single (type of) primitive constraint; we call this algorithm glass-box. This approach is used in constraint programming, where CSPs are given in implicit form by means of formulas of a given specification language. In [65.79, 80] *Handa* et al. formulate a coevolutionary algorithm where a population of schemata are parasitic on the host population. Schemata in this algorithm are individuals where a portion of variables in the individual has values while all other variables have do-not-care symbols represented by asterisks.

The following approaches with emphasis on adaptive features are included in the comparison; a coevolutionary approach invented by *Paredis* and evaluated on different problems, such as neural net learning [65.81], constraint satisfaction [65.81, 82], and searching for cellular automata that solve the density classification task [65.83]. Furthermore, results on the performance of the co-evolutionary approach when facing the task of solving binary CSPs are reported in [65.84, 85]. In the co-evolutionary approach for CSPs two populations evolve according to a predator-prey model: a population of candidate solutions and a population of constraints. In the approach proposed by *Dozier* et al. in [65.86] and further refined and applied in [65.87–89], information about the constraints is incorporated both in the genetic operators and in the fitness function. In the microgenetic iterative descent algorithm the fitness function is adaptive and employs *Morris'* breakout creating mechanism [65.90] to escape from local optima. The stepwise adaptation of weights mechanism was introduced by *Eiben* and *van der Hauw* [65.91, 92] as an improved version of the weight adaptation mechanism of *Eiben* et al. [65.93, 94]. The approach has been studied in several comparisons and often proved to be a robust technique for solving several specific CSPs [65.95–97]. A comprehensive study of different parameters and genetic operators can be found in [65.98]. The basic idea is that constraints that are not satisfied or variables causing constraint violations after a certain number of steps must be hard, thus must be given a high weight (penalty) in the fitness function.

65.5.4 Examination Timetabling

Examination timetabling has been studied for many years as it is a common problem in many organizations. Already in 1986, *Carter* gave an extended survey of work on automated timetabling [65.99]. He is also responsible for providing problem instances, which are still available and popular today [65.100], although a more diverse benchmark is used in the annual timetabling competition [65.101]. *Burke* et al. provide the most extensive recent surveys of automated timetabling in [65.102, 103]. Examination timetabling is just one of many problems under the topic of timetabling [65.104].

Timetabling as a problem has many different definitions due to different kinds of constraints and objectives. The definition that is most relevant for constraint satisfaction is often referred to as examination timetabling. The most abstract definition simply consists of a matrix C where $C_{i,j} = 1$ if exam i conflicts with exam j by having common students that must take both exams, $C_{i,j} = 0$ otherwise. This definition is equivalent to a graph coloring problem if the objective is to minimize the number of exam slots required, where the number of slots equals the number of colors required for coloring the graph with incidence matrix C. Hence, an appropriate approach to performance testing is via graph coloring instances based on examination timetabling, such as the problem instances labeled SCH (school) in the graph coloring instances suite provided by *Lewandowski* [65.105].

Many problem instances and problem instance generators exist. Infrequently, an International Timetabling Competition is organized by The International Series of Conferences on the Practice and Theory of Automated Timetabling. At each event, another definition of timetabling problems is tackled. The differences between definitions are in the objectives and the soft and hard constraints used. Hard constraints are treated the same as in constraint satisfaction, whereas soft constraints may be violated but will either incur an additional penalty on the objective function or be used to prioritize solutions otherwise, for instance, using a Pareto front. *Corne* et al. [65.106] identified five categories of constraints, unary, binary, capacity, event spread, and agent preference.

Three approaches exist to solving timetabling problems. The first approach is called *one-stage optimization*. It aggregates all types of constraints of one problem, often by summation, into one objective function where each type is assigned a weight. The advantage is that, in principle, the approach can be applied to any set of constraints. In practice, it may prove difficult to optimize such a function. Representations of the problem fall into the two main categories direct encoding (Sect. 65.3.1) [65.107] and indirect encoding (Sect. 65.3.2) [65.106, 108].

The second approach is called *two-stage optimization*. It first solves the problem of finding a feasible solution where all the hard constraints are satisfied. In the second stage it searches within the space set with these hard constraints and optimizes only against the soft constraints. The benefits are that during search we do not have to distinguish between feasible and infeasible constraints and, therefore, are not in danger of the search wandering off into an infeasible part of the search space. *Thompson* and *Dowsland* [65.109] were the first to report on this approach using simulated annealing, closely followed by the first EA by *Yu* and *Sung* [65.110].

The third approach uses *relaxation of constraints*. Typically, relaxation in timetabling is achieved by not assigning events to slots or by adding additional time slots. An early example of an EA is by *Burke* et al. [65.111], where an indirect encoding is used and additional time slots are used to relax the problem.

65.6 Creating Rather than Solving Problems

So far we have covered evolutionary computation for solving CSP. A contrasting idea proposed first for constraint satisfaction in [65.112] is to use evolutionary computation to generate problem instances. Such an approach allows a search for problem instances that adhere to certain properties as long as these can be measured efficiently by a fitness function.

A straightforward use for such an approach is to evolve problem instances that are difficult to solve for a particular algorithm. By measuring the efficiency of an algorithm to solve instances of a certain problem we can then change the instances with the aim of decreasing the efficiency. Measurements for efficiency of EC for CSP are discussed in Sect. 65.4.1. It is important to note that the algorithm we are evolving problem instances for can be of any kind, as long as we can execute it on problem instances generated and we can measure its efficiency.

Such hard problem instances identify the weak spots in the algorithm that tries to solve it. Moreover, if we can characterize a set of problem instances where all members of the set are hard for an algorithm, then we can use that characterization to decide what algorithm is suitable for solving a new problem instance. That is, if the work required to obtain the characteristics of one instance takes less effort than solving the actual problem instance itself [65.113].

65.6.1 Evolving Binary Constraint Satisfaction Problem Instances

The first application to constrained problems was for the binary constraint satisfaction problem (Sect. 65.5.3), where problem instances are represented as a binary vector with each element corresponding to the element of a conflict matrix between two variables [65.114]. Even the small instances investigated in the study led to large vectors, i.e., with 15 variables each with a domain of size 15, the corresponding vector has $\binom{15}{2} \cdot 15^2 = 23\,625$ elements. Results with problem instances of this size show problem instances can be created that are far more difficult to solve than when creating a much larger set of randomly generated instances [65.112]. Furthermore, analysis of these instances provides an insight as to what structure is responsible for making instances difficult for the algorithm; two well-known algorithms from constraint programming were tested: chronological backtracking [65.115] and

forward checking with conflict-directed backjumping [65.116].

65.6.2 Evolving Boolean Satisfiability Problem Instances

In [65.114] an evolutionary algorithm is used to evolve solvable Boolean satisfiability problem instances that are in conjunctive normal form and have three variables per clause. A 3-SAT problem is represented by a list of natural numbers. A number in the list, i. e., a gene, corresponds to a unique clause with three different literals. The number of possible unique clauses depends on the number of variables and the size of the clause. Here, the number of variables is set to 100 and the size of the clause is 3, hence there are 1 313 400 unique clauses. This representation has strong advantages over a simple *one gene for every literal* approach. Most importantly, it prevents duplicate variables in clauses, which reduces the state space and could otherwise introduce trivial clauses, e.g., $(x \vee \neg x \vee y)$, or 2-SAT clauses, e.g., $(x \vee x \vee y)$. Also, the variation operators now simply become mutation and uniform crossover for lists of natural numbers over a fixed domain.

Two problem solvers are used from the annual SAT competition [65.117]; both are based on the *Davis–Putnam* procedure [65.4]. zChaff [65.118] is based on Chaff [65.119], a SAT solver that employs a particularly efficient implementation of Boolean constraint propagation and a novel low overhead decision strategy. Relsat [65.120] is explained in [65.121, 122]. In both solvers, the number of states of instantiations are enumerated to determine the search effort required.

The change of certain structural properties over the duration of evolution was analyzed. Two established properties were used: the number of solutions [65.123, 124] and the backbone size [65.125]. No clear relationship was identified with these properties.

However, a new relationship was identified: when problem instances are becoming more difficult to solve, the variance in the frequency in variable usage decreases. In other words, the distribution of variables throughout the instances is more uniform when problems are more difficult to solve.

65.6.3 Further Investigations

The application of evolutionary computation in problem generation is widespread. *Smith-Miles* and

Lopes [65.126] provide an extensive review in terms of measuring instance difficulty in combinatorial optimization problems, which also discusses studies that evolve problem instances for constrained optimization as well as for constraint satisfaction problems.

The maximization of the effort required to solve a problem instance highlights only one aspect of the problem difficulty. Another aspect that looks at the effectiveness is to maximize the distance a solver is able to reach to the optimal solution. To compute this distance, we require the fitness of the optimal solution a priori. Note, however, we do not need to know what the optimal solution is, only its fitness. Another approach is to directly compare solvers by maximizing the difference in some aspect, e.g., efficiency or effectiveness, between two solvers.

65.7 Conclusions and Future Directions

Research on solving constraint satisfaction problems with evolutionary computation has produced a rich set of research papers that contribute solvers, insights into solvers and their performance, and heuristic subroutines. One major flaw in this research has remained consistent over the past 20 years: most studies compare performance results only to other evolutionary or closely associated techniques. Even recent studies, such as [65.127–129], restrict themselves to comparing only results from other heuristic methods or have not included alternative techniques at all.

Many studies report on the promising performance of a particular evolutionary algorithm over another existing heuristic technique. The few systematic studies that do compare evolutionary and constraint programming techniques conclude that constraint programming is superior in terms of efficiency [65.60, 67]. Also, constraint programming techniques are generally sound and, therefore, given sufficient time, always find a solution or proof that none exists. Hence, these solvers are more effective unless they are bounded by time. Recent efforts have shown success in speeding up modern DPLL-based techniques using heuristics for guiding the search [65.130, 131].

In Sect. 65.5 we reviewed many techniques that were developed and studied for the purpose of improving EC in terms of efficiency and effectiveness. The vast majority of these techniques was applied to one problem only. A huge benefit would come from studies that show the success of a technique across several CSPs. Such studies would be especially opportune for the SAT problem, which is still the most actively used CSP for benchmarking algorithms [65.132].

References

65.1   K. Apt: *Principles of Constraint Programming* (Cambridge Univ. Press, Cambridge 2003)

65.2   B.G.W. Craenen, A.E. Eiben: Hybrid evolutionary algorithms for constraint satisfaction problems: Memetic overkill?, 2005 IEEE Congr. Evol. Comput., Vol. 3 (2005) pp. 1922–1928

65.3   R. Kibria, Y. Li: Optimizing the initialization of dynamic decision heuristics in DPLL SAT solvers using genetic programming, Lect. Notes Comput. Sci. **3905**, 331–340 (2006)

65.4   M. Davis, H. Putnam: A computing procedure for quantification theory, Journal ACM **7**, 201–215 (1960)

65.5   H.E. Dudeney: Cryptarithm, Strand Mag. **68**, 97 and 214 (1924)

65.6   M.R. Garey, D.S. Johnson: *Computers and Intractability: A Guide to the Theory of NP-Completeness* (W.H. Freeman, San Francisco 1979)

65.7   R. Lewis: Metaheuristics can solve Sudoku puzzles, J. Heuristics **13**, 387–401 (2007)

65.8   E. Tsang: *Foundations of Constraint Satisfaction* (Academic, London 1993)

65.9   R. Dechter: *Constraint Processing* (Morgan Kaufmann, San Francisco 2003) pp. 1–481

65.10 C. Lecoutre: *Constraint Networks: Techniques and Algorithms* (Wiley, Hoboken 2009)

65.11 H. Chen: A rendezvous of logic, complexity, and algebra, ACM Comput. Surv. **42**(1), 2 (2009)

65.12 B. Bernhardsson: Explicit solutions to the *n*-queens problem for all *n*, SIGART Bull. **2**, 7 (1991)

65.13 T. Bäck, D. Fogel, Z. Michalewicz (Eds.): *Handbook of Evolutionary Computation* (Oxford Univ. Press, New York 1997)

65.14 F. Rossi, P. Van Beek, T. Walsh: *Handbook of Constraint Programming* (Elsevier, Amsterdam 2006)

65.15 D. Brélaz: New methods to color the vertices of a graph, Communications ACM **22**, 251–256 (1979)

65.16   D.B. Fogel: *Evolutionary Computation: Towards a New Philosophy of Machine Intelligence*, 2nd edn. (Wiley, Hoboken 1999)

65.17   A.E. Eiben, Z. Ruttkay: Self-adaptivity for constraint satisfaction: Learning penalty functions, Int. Conf. Evol. Comput. (1996) pp. 258–261

65.18   R. Hinterding, Z. Michalewicz, A.E. Eiben: Adaptation in evolutionary computation: A survey, Proc. 4th IEEE Conf. Evol. Comput. (1997) pp. 65–69

65.19   T. Bäck: Introduction to the special issue: Self-adaptation, Evol. Comput. **9**(2), 3–4 (2001)

65.20   T. Runnarson, X. Yao: Constrained evolutionary optimization – The penalty function approach. In: *Evolutionary Optimization*, ed. by R. Sarker, M. Mohammadian, X. Yao (Kluwer, Boston 2002) pp. 87–113, Chap. 4

65.21   J.T. Richardson, M.R. Palmer, G. Liepins, M. Hilliard: Some guidelines for genetic algorithms with penalty functions, Proc. 3rd Int. Conf. Genet. Algoritms. (1989) pp. 191–197

65.22   M.L. Braun, J.M. Buhmann: The noisy Euclidean traveling salesman problem and learning, Proc. 2001 Neural Inf. Process. Syst. Conf. (2002)

65.23   D. Whitley, J.P. Watson, A. Howe, L. Barbulescu: Testing, evaluation and performance of optimization and learning systems. In: *Adaptive Computing in Design and Manufacturer*, ed. by I.C. Parmee (Springer, Berlin, Heidelberg 2002) pp. 27–39

65.24   A. Biere, M. Heule, H. van Maaren, T. Walsh: *Handbook of Satisfiability* (IOS, Amsterdan 2009)

65.25   V. Malek: *Introduction to Mathematics of Satisfiability* (Chapman Hall, Boca Raton 2009)

65.26   S.A. Cook: The complexity of theorem-proving procedures, Proc. 3rd Annu. ACM Symp. Theory Comput. (1971) pp. 151–158

65.27   M. Utting, B. Legeard: *Practical Model-Based Testing: A Tools Approach* (Morgan Kaufmann, San Francisco 2007)

65.28   A. Bundy: A science of reasoning: extended abstract, Proc. 10th Int. Conf. Autom. Deduc. (1990) pp. 633–640

65.29   D. McDermott, M. Ghallab, A. Howe, C. Knoblock, A. Ram, M. Veloso, D. Weld, D. Wilkins: PDDL – The planning domain definition language, Tech. Rep. TR-98-003, Yale Center for Computational Vision and Control (1998)

65.30   R. Drechsler, S. Eggersglüß, G. Fey, D. Tille: *Test Pattern Generation using Boolean Proof Engines* (Springer, Berlin, Heidelberg 2009) pp. 1–192

65.31   D. He, A. Choi, K. Pipatsrisawat, A. Darwiche, E. Eskin: Optimal algorithms for haplotype assembly from whole-genome sequence data, Bioinformatics **26**(12), i183–i190 (2010)

65.32   I.V. Tetko, D.J. Livingstone, A.I. Luik: Neural network studies. 1. Comparison of overfitting and overtraining, J. Chem Inf. Comput. Sci. **35**, 826–833 (1995)

65.33   J.-K. Hao, R. Dorne: An empirical comparison of two evolutionary methods for satisfiability problems, Int. Conf. Evol. Comput. (1994) pp. 451–455

65.34   J. Gottlieb, N. Voss: Fitness functions and genetic operators for the satisfiability problem, Lect. Notes Comput. Sci. **1363**, 55–68 (1997)

65.35   J. Gottlieb, N. Voss: Improving the performance of evolutionary algorithms for the satisfiability problem by refining functions, Lect. Notes Comput. Sci. **1498**, 755–764 (1998)

65.36   G. Folino, C. Pizzuti, G. Spezzano: Solving the satisfiability problem by a parallel cellular genetic algorithm, Proc. 24th Euromicro Conf. (1998) pp. 715–722

65.37   N. Nemer-Preece, R.W. Wilkerson: Parallel genetic algorithm to solve the satisfiability problem, Proc. 1998 ACM Symp. Appl. Comput. (1998) pp. 23–28

65.38   C. Rossi, E. Marchiori, J.N. Kok: An adaptive evolutionary algorithm for the satisfiability problem, Proc. 2000 ACM Symp. Appl. Comput. (2000) pp. 463–469

65.39   J.-K. Hao, F. Lardeux, F. Saubion: Evolutionary computing for the satisfiability problem, Lect. Notes Comput. Sci. **2611**, 258–267 (2003)

65.40   M.E. Bachir Menai: An evolutionary local search method for incremental satisfiability, Lect. Notes Comput. Sci. **3249**, 143–156 (2004)

65.41   L. Aksoy, E.O. Günes: An evolutionary local search algorithm for the satisfiability problem, Lect. Notes Comput. Sci. **3949**, 185–193 (2005)

65.42   M.E. Bachir Menai, M. Batouche: Solving the maximum satisfiability problem using an evolutionary local search algorithm, Int. Arab J. Inf. Technol. **2**(2), 154–161 (2005)

65.43   P. Guo, W. Luo, Z. Li, H. Liang, X. Wang: Hybridizing evolutionary negative selection algorithm and local search for large-scale satisfiability problems, Lect. Notes Comput. Sci. **5821**, 248–257 (2009)

65.44   Y. Kilani: Comparing the performance of the genetic and local search algorithms for solving the satisfiability problems, Appl. Soft. Comput. **10**(1), 198–207 (2010)

65.45   R.H. Kibria: Soft Computing Approaches to DPLL SAT Solver Optimization, Ph.D. Thesis (TU Darmstadt, Darmstadt 2011)

65.46   M. Davis, G. Logemann, D. Loveland: A machine program for theorem-proving, Communications ACM **5**(7), 394–397 (1962)

65.47   R. Lewis, J. Thompson: On the application of graph colouring techniques in round-robin sports scheduling, Comput. Oper. Res. **38**, 190–204 (2011)

65.48   S.S. Muchnick: *Advanced Compiler Design and Implementation* (Morgan Kaufmann, San Fransisco 1997)

65.49   W.K. Hale: Frequency assignment: Theory and applications, Proc. IEEE **68**(12), 1497–1514 (1980)

65.50   J. Culberson: Graph Coloring Page (2010), available online at http://webdocs.cs.ualberta.ca/~joe/Coloring/

65.51   J.C. Culberson, F. Luo: Exploring the k-colorable landscape with iterated greedy. In: *Cliques, Coloring, and Satisfiability: Second DIMACS Implementation Challenge*, DIMACS Series in Discrete Mathematics and Theoretical Computer Science, Vol. 26, ed. by D.S. Johnson, M.A. Trick (American Mathematical Society, Providence 1996) pp. 245–284

65.52   M. Brockington, J.C. Culberson: Camouflaging independent sets in quasi-random graphs. In: *Cliques, Coloring, and Satisfiability: Second DIMACS Implementation Challenge*, DIMACS Series in Discrete Mathematics and Theoretical Computer Science, Vol. 26, ed. by D.S. Johnson, M.A. Trick (American Mathematical Society, Providence 1996) pp. 75–88

65.53   G.J. Chaitin, M.A. Auslander, A.K. Chandra, J. Cocke, M.E. Hopkins, P.W. Markstein: Register allocation via coloring, Comput. Lang. **6**(1), 47–57 (1981)

65.54   D.J.A. Welsh, M.B. Powell: An upper bound for the chromatic number of a graph and its application to timetabling problems, Comput. J. **10**(1), 85–86 (1967)

65.55   C. Fleurent, J. Ferland: Genetic and hybrid algorithms for graph coloring, Ann. Oper. Res. **63**(3), 437–461 (1996)

65.56   C. Fleurent, J.A. Ferland: Object-oriented implementation of heuristic search methods for graph coloring, maximum clique, and satisfiability. In: *Cliques, Coloring, and Satisfiability: Second DIMACS Implementation Challenge*, DIMACS Series in Discrete Mathematics and Theoretical Computer Science, Vol. 26, ed. by D.S. Johnson, M.A. Trick (American Mathematical Society, Providence 1996) pp. 619–652

65.57   G. von Laszewski: Intelligent structural operators for the k-way graph partitioning problem, Proc. 4th Int. Conf. Genet. Algorithms (1991) pp. 45–52

65.58   L. Davis: Order-based genetic algorihms and the graph coloring problem. In: *Handbook of Genetic Algorithms*, ed. by L. Davis (Van Nostrand Reinhold, New York 1991) pp. 72–90

65.59   P.E. Coll, G.A. Durán, P. Moscato: A discussion on some design principles for efficient crossover operators for graph coloring problems, An. XXVII Simp. Brasil. Pesqui. Oper. (1995)

65.60   I. Juhos, J.I. van Hemert: Contraction-based heuristics to improve the efficiency of algorithms solving the graph colouring problem. In: *Recent Advances in Evolutionary Computation for Combinatorial Optimization*, ed. by C. Cotta, J.I. van Hemert (Springer, Berlin, Heidelberg 2008) pp. 167–184

65.61   I. Juhos, J.I. van Hemert: Graph colouring heuristics guided by higher order graph properties, Lect. Notes Comput. Sci. **4972**, 97–109 (2008)

65.62   I. Juhos, J.I. van Hemert: Increasing the efficiency of graph colouring algorithms with a representation based on vector operations, J. Softw. **1**(2), 24–33 (2006)

65.63   E.M. Palmer: *Graphical Evolution* (Wiley, New York 1985)

65.64   D. Achlioptas, L.M. Kirousis, E. Kranakis, D. Krizanc, M.S.O. Molloy, Y.C. Stamatiou: Random constraint satisfaction: A more accurate picture, Lect. Notes Comput. Sci. **1330**, 107–120 (1997)

65.65   E. MacIntyre, P. Prosser, B.M. Smith, T. Walsh: Random constraint satisfaction: Theory meets practice. In: *Principles and Practice of Constraint Programming – CP98*, ed. by M. Maher, J.-F. Puget (Springer, Berlin, Heidelberg 1998) pp. 325–339

65.66   E. Freuder, R.J. Wallace: Partial constraint satisfaction, Artif. Intell. **65**, 363–376 (1992)

65.67   B.G.W. Craenen, A.E. Eiben, J.I. van Hemert: Comparing evolutionary algorithms on binary constraint satisfaction problems, IEEE Trans. Evol. Comput. **7**(5), 424–444 (2003)

65.68   R. Haralick, G. Elliot: Increasing tree search efficiency for constraint-satisfaction problems, Artif. Intell. **14**(3), 263–313 (1980)

65.69   A.E. Eiben, P.-E. Raué, Z. Ruttkay: Heuristic Genetic Algorithms for Constrained Problems, Part I: Principles, Tech. Rep. IR-337 (Vrije Universiteit Amsterdam 1993)

65.70   A.E. Eiben, P.-E. Raué, Z. Ruttkay: Solving constraint satisfaction problems using genetic algorithms, Proc. 1st IEEE Conf. Evol. Comput. (1994) pp. 542–547

65.71   B.G.W. Craenen, A.E. Eiben, E. Marchiori: Solving constraint satisfaction problems with heuristic-based evolutionary algorithms, Congr. Evol. Comput. (2000)

65.72   M.C. Riff-Rojas: Using the knowledge of the constraint network to design an evolutionary algorithm that solves CSP, Proc. 3rd IEEE Conf. Evol. Comput. (1996) pp. 279–284

65.73   M.C. Riff-Rojas: Evolutionary search guided by the constraint network to solve CSP, Proc. 4th IEEE Conf. Evol. Comput. (1997) pp. 337–348

65.74   M.-C. Riff-Rojas: A network-based adaptive evolutionary algorithm for constraint satisfaction problems. In: *Meta-heuristics: Advances and Trends in Local Search Paradigms for Optimization*, ed. by S. Voss (Kluwer, Boston 1998) pp. 325–339

65.75   E. Marchiori: Combining constraint processing and genetic algorithms for constraint satisfaction problems, Proc. 7th Int. Conf. Genet. Algorithms (1997) pp. 330–337

65.76   E. Marchiori, A. Steenbeek: A genetic local search algorithm for random binary constraint satisfaction problems, Proc. ACM Symp. Appl. Comput. (2000) pp. 458–462

65.77   B.G.W. Craenen, A.E. Eiben, E. Marchiori, A. Steenbeek: Combining local search and fitness function adaptation in a GA for solving binary constraint satisfaction problems, Proc. Genet. Evol. Comput. Conf. (2000)

65.78   P. van Hentenryck, V. Saraswat, Y. Deville: Constraint processing in cc(FD). In: *Constraint Programming: Basics and Trends*, ed. by A. Podelski (Springer, Berlin, Heidelberg 1995)

65.79   H. Handa, C.O. Katai, N. Baba, T. Sawaragi: Solving constraint satisfaction problems by using coevolutionary genetic algorithms, Proc. 5th IEEE Conf. Evol. Comput. (1998) pp. 21–26

65.80   H. Handa, N. Baba, O. Katai, T. Sawaragi, T. Horiuchi: Genetic algorithm involving coevolution mechanism to search for effective genetic information, Proc. 4th IEEE Conf. Evol. Comput. (1997)

65.81   J. Paredis: Co-evolutionary computation, Artif. Life 2(4), 355–375 (1995)

65.82   J. Paredis: Coevolutionary constraint satisfaction, Lect. Notes Comput. Sci. **866**, 46–55 (1994)

65.83   J. Paredis: Coevolving cellular automata: Be aware of the red queen, Proc. 7th Int. Conf. Genet. Algorithms (1997)

65.84   A.E. Eiben, J.I. van Hemert, E. Marchiori, A.G. Steenbeek: Solving binary constraint satisfaction problems using evolutionary algorithms with an adaptive fitness function, Lect. Notes Comput. Sci. **1498**, 196–205 (1998)

65.85   J.I. van Hemert: Applying Adaptive Evolutionary Algorithms to Hard Problems, M.Sc. Thesis (Leiden University, Leiden 1998)

65.86   G. Dozier, J. Bowen, D. Bahler: Solving small and large constraint satisfaction problems using a heuristic-based microgenetic algorithm, Proc. 1st IEEE Conf. Evol. Comput. (1994) pp. 306–311

65.87   J. Bowen, G. Dozier: Solving constraint satisfaction problems using a genetic/systematic search hybride that realizes when to quit, Proc. 6th Int. Conf. Genet. Algorithms (Morgan Kaufmann, Burlington 1995) pp. 122–129

65.88   G. Dozier, J. Bowen, D. Bahler: Solving randomly generated constraint satisfaction problems using a micro-evolutionary hybrid that evolves a population of hill-climbers, Proc. 2nd IEEE Conf. Evol. Comput. (1995) pp. 614–619

65.89   P.J. Stuckey, V. Tam: Improving evolutionary algorithms for efficient constraint satisfaction, Int. J. Artif. Intell. Tools 8(4), 363–384 (1999)

65.90   P. Morris: The breakout method for escaping from local minima, Proc. 11th Natl. Conf. Artif. Intell. (1993) pp. 40–45

65.91   A.E. Eiben, J.K. van der Hauw: Adaptive penalties for evolutionary graph-coloring, Lect. Notes Comput. Sci. **1363**, 95–106 (1998)

65.92   J.K. van der Hauw: Evaluating and Improving Steady State Evolutionary Algorithms on Constraint Satisfaction Problems, M.Sc. Thesis (Leiden University, Leiden 1996)

65.93   A.E. Eiben, P.-E. Raué, Z. Ruttkay: Constrained problems. In: *Practical Handbook of Genetic Algorithms*, ed. by L. Chambers (Taylor Francis, Boca Raton 1995) pp. 307–365

65.94   A.E. Eiben, Z. Ruttkay: Self-adaptivity for constraint satisfaction: Learning penalty functions, Proc. 3rd IEEE Conf. Evol. Comput. (1996) pp. 258–261

65.95   T. Bäck, A.E. Eiben, M.E. Vink: A superior evolutionary algorithm for 3-SAT, Lect. Notes Comput. Sci. **1477**, 125–136 (1998)

65.96   A.E. Eiben, J.K. van der Hauw, J.I. van Hemert: Graph coloring with adaptive evolutionary algorithms, J. Heuristics 4(1), 25–46 (1998)

65.97   A.E. Eiben, J.I. van Hemert: SAW-ing EAs: Adapting the fitness function for solving constrained problems. In: *New Ideas in Optimization*, ed. by D. Corne, M. Dorigo, F. Glover (McGraw Hill, New York 1999) pp. 389–402

65.98   B.G.W. Craenen, A.E. Eiben: Stepwise adaption of weights with refinement and decay on constraint satisfaction problems, Proc. Genet. Evol. Comput. Conf. (2001) pp. 291–298

65.99   M.W. Carter: A survey of practical applications of examination timetabling algorithms, Oper. Res. **34**, 193–202 (1986)

65.100 M.W. Carter, G. Laporte, S.Y. Lee: Examination timetabling: Algorithmic strategies and application, J. Oper. Res. Soc. **47**(3), 373–383 (1996)

65.101 International Timetabling Competition 2011: available online at http://www.utwente.nl/ctit/itc2011/

65.102 E.K. Burke, S. Petrovic: Recent research directions in automated timetabling, Eur. J. Oper. Res. **140**(2), 266–280 (2002)

65.103 R. Qu, E.K. Burke, B. Mccollum, L.T. Merlot, S.Y. Lee: A survey of search methodologies and automated system development for examination timetabling, J. Sched. **12**, 55–89 (2009)

65.104 E.K. Burke, D. Corne, B. Paechter, P. Ross (Eds.): *Proc. 1st Int. Conf. Pract. Theory Autom. Timetabling* (Napier University, Edinburgh 1995)

65.105 G. Lewandowski: *Course scheduling: Metrics, Models, and Methods* (Xavier University, Cincinnati 1996)

65.106 D. Corne, P. Ross, H.-L. Fang: Evolving timetables. In: *Practical Handbook of Genetic Algorithms: Applications*, Vol. I, ed. by L. Chambers (Taylor Francis, Boca Raton 1995) pp. 219–276

65.107 A. Colorni, M. Dorigo, V. Maniezzo: Metaheuristics for high school timetabling, Comput. Optim. Appl. **9**(3), 275–298 (1998)

65.108 M.P. Carrasco, M.V. Pato: A multiobjective genetic algorithm for the class/teacher timetabling problem, Proc. 3rd Int. Conf. Pract. Theory Autom. Timetabling (2001) pp. 3–17

65.109   J.M. Thompson, K.A. Dowsland: A robust simulated annealing based examination timetabling system, Comput. Oper. Res. **25**, 637–648 (1998)

65.110   E. Yu, K.-S. Sung: A genetic algorithm for a university weekly courses timetabling problem, Int. Trans. Oper. Res. **9**(6), 703–717 (2002)

65.111   E.K. Burke, D. Elliman, R.F. Weare: A hybrid genetic algorithm for highly constrained timetabling problems, Proc. 6th Int. Conf. Genet. Algorithms (1995) pp. 605–610

65.112   J.I. van Hemert: Evolving binary constraint satisfaction problem instances that are difficult to solve, Proc. IEEE 2003 Congr. Evol. Comput. (New York) (2003) pp. 1267–1273

65.113   K. Smith-Miles, J.I. van Hemert: Discovering the suitability of optimisation algorithms by learning from evolved instances, Ann. Math. Artif. Intell. **61**(2), 87–104 (2011)

65.114   J.I. van Hemert: Evolving combinatorial problem instances that are difficult to solve, Evol. Comput. **14**(4), 433–462 (2006)

65.115   S.W. Golomb, L.D. Baumert: Backtrack programming, Journal ACM **12**(4), 516–524 (1965)

65.116   P. Prosser: Hybrid algorithms for the constraint satisfaction problem, Comput. Intell. **9**(3), 268–299 (1993)

65.117   D. Le Berre, L. Simon: *Sat Competitions* http://www.satcompetition.org. (2005)

65.118   Z. Fu: *zChaff* (Princeton University) Version 2004.11.15 http://www.princeton.edu/~chaff/zchaff.html (2004)

65.119   M. Moskewicz, C. Madigan, Y. Zhao, L. Zhang, S. Malik: Chaff: Engineering an efficient SAT solver, Proc. 38th Design Autom. Conf. (2001) pp. 530–535

65.120   R. Bayardo: *Relsat. Version 2.00* (IBM, San Jose 2005), available online at http://www.almaden.ibm.com/cs/people/bayardo/resources.html

65.121   R. Bayardo, R.C. Schrag: Using CSP look-back techniques to solve real world SAT instances, Proc. 14th Natl. Conf. Artif. Intell. (1997) pp. 203–208

65.122   R. Bayardo, J. Pehoushek: Counting models using connected components, Proc. 17th Natl. Conf. Artif. Intell. (2000)

65.123   D. Achlioptas, C.P. Gomes, H.A. Kautz, B. Selman: Generating satisfiable problem instances, Proc. 17th Natl. Conf. Artif. Intell. 12th Conf. Innov. Appl. Artif. Intell. (2000) pp. 256–261

65.124   D. Achlioptas, H. Jia, C. Moore: Hiding satisfying assignments: Two are better than one, J. Artif. Intell. Res. **24**, 623–639 (2005)

65.125   S. Boettcher, G. Istrate, A.G. Percus: Spines of random constraint satisfaction problems: Definition and impact on computational complexity, 8th Int. Symp. Artif. Intell. Math. (2005), extended version

65.126   K. Smith-Miles, L. Lopes: Review: Measuring instance difficulty for combinatorial optimization problems, Comput. Oper. Res. **39**, 875–889 (2012)

65.127   R. Abbasian, M. Mouhoub: An efficient hierarchical parallel genetic algorithm for graph coloring problem, Proc. 13th Annu. Conf. Genet. Evol. Comput. (2011) pp. 521–528

65.128   D.C. Porumbel, J.-K. Hao, P. Kuntz: An evolutionary approach with diversity guarantee and well-informed grouping recombination for graph coloring, Comput. Oper. Res. **37**(10), 1822–1832 (2010)

65.129   M. Mouhoub, B. Jafari: Heuristic techniques for variable and value ordering in CSPs, Proc. 13th Annu. Conf. Genet. Evol. Comput. (2011) pp. 457–464

65.130   J. Chen: Building a hybrid sat solver via conflict-driven, look-ahead and XOR reasoning techniques, Lect. Notes Comput. Sci. **5584**, 298–311 (2009)

65.131   A. Balint, M. Henn, O. Gableske: A novel approach to combine a SLS- and a DPLL-solver for the satisfiability problem, Lect. Notes Comput. Sci. **5584**, 284–297 (2009)

65.132   O. Kullmann (Ed.): *Theory and Applications of Satisfiability Testing – SAT 2009*, Lecture Notes in Computer Science, Vol. 558 (Springer, Berlin, Heidelberg 2009)

Part F Swarm Intelligence

Ed. by Christian Blum, Roderich Groß

66. Swarm Intelligence in Optimization and Robotics

Christian Blum, Roderich Groß

Swarm intelligence is an artificial intelligence discipline, which was created on the basis of the laws that govern the behavior of, for example, social insects, fish schools, and flocks of birds. The organization of these animal societies has always mesmerized humans. Therefore, it is surprising that it has only been in the second half of the last century that some of the most important principles of swarm intelligent behavior have been unraveled. A prime example is stigmergy, which refers to a self-organization of the animal society via changes applied to the environment.

In this chapter, we provide a concise introduction to swarm intelligence, with two main research lines in mind: optimization and robotics. Popular examples of optimization algorithms based on swarm intelligence principles are ant colony optimization and particle swarm optimization. On the other side, the field of robotics has adopted var-

ious swarm intelligent behaviors for problem solving and organizing groups of robots. This has resulted in a separate research field nowadays known as swarm robotics.

66.1 Overview

Swarm intelligence (SI) [66.1–3] is a subfield of the more general field of artificial intelligence [66.4]. The term *swarm intelligence* was introduced and used for the first time by *Beni* et al. [66.5–7] in the context of cellular robotic systems. Nowadays, SI research is generally concerned with the design of intelligent multiagent systems whose inspiration is taken from the collective behavior of social – or even eusocial – insects and other animal populations. Examples include ant colonies, bee hives, wasp colonies, frog populations, flocks of birds, and fish schools. Among these, social insects have always played a prominent role in the inspiration of SI techniques. Even though their intrinsic ways of functioning have fascinated researchers for many years, the mechanisms that govern their behavior remained unknown for a long time. In colonies of social insects, for example, single colony members are

unsophisticated individuals, yet they are able to achieve complex tasks in cooperation. Essential colony behaviors emerge from relatively simple interactions between the colony's individual members.

An important aspect of any SI system is *self-organization* [66.8]. Originally, the term *self-organization* was introduced by the German philosopher *Immanuel Kant* [66.9] in an attempt to characterize what makes organisms so different from other objects. Nowadays, the term self-organization refers to a process where some form of global order or coordination emerges from rather simple interactions between low-level components of an initially unordered system. Self-organizing processes are neither directed nor controlled by any agent or component, neither from inside nor from outside the system. They are often triggered by random fluctuations that are amplified by *positive feedback* and

Fig. 66.1 Ants cooperate for retrieving a heavy prey (photo courtesy of M. J. Blesa)

tallization, molecular self-assembly, and the way in which neural networks learn to recognize complex patterns.

During the last 50 years or so, biologists discovered that many aspects of the collective activities of social insects are self-organized as well, that is, they function without a central control. For example, the African weaver ant constructs nests by pulling leaves together. Where the gap between leaves exceeds the body length of an individual ant, multiple ants organize into pulling chains. Once the leaves are in contact, they are glued together using silk from larvae, which are carried to the site by other workers of the colony [66.10]. Other examples concern the recruitment of fellow colony members for prey retrieval (Fig. 66.1), the capabilities of termites and wasps to build sophisticated nests, or the ability of bees and ants to orient themselves in their environment. For more examples, we refer the interested reader to [66.1, 2].

In the meantime, some of the above mentioned behaviors have been used as inspiration for the resolution of technical problems, especially in the context of optimization and in robotics. This chapter is dedicated to reviewing some of the – in the opinion of the authors – most interesting algorithms/systems from these two fields.

possibly counterbalanced by *negative feedback*, which generally aids in stabilizing the system. The global properties exhibited by self-organizing systems are thus the result of this distributed interplay of their components. As such, self-organization is typically robust and able to survive and self-repair damage or perturbations. Historically, self-organization processes have been studied in physical, chemical, biological, social, and cognitive systems. Well known examples are crys-

66.2 SI in Optimization

The use of SI techniques for solving optimization problems has already a rather extensive history. SI techniques have been used for both solving combinatorial and continuous optimization problems in static and in distributed settings. Two of the most well-known SI techniques for solving optimization problems are ant colony optimization (ACO) and particle swarm optimization (PSO). More recently, other techniques such as the artificial bee colony algorithm have been developed. Apart from solving optimization problems, SI techniques are being used for management tasks, for example, in distributed settings or in online optimization. The following sections will give a brief overview of this application field of SI.

66.2.1 Ant Colony Optimization

ACO [66.11] is one of the earliest SI techniques for optimization. Dorigo and colleagues developed the first ACO algorithms in the early 1990s [66.12–14]. The

development of these algorithms was inspired by the observation of ant colonies. Ants are social insects. They live in colonies and their behavior is governed by the goal of colony survival rather than being focused on the survival of individuals. The behavior that provided the inspiration for ACO is the ants' foraging behavior, and in particular, how ants of many species can find shortest paths between food sources and their nest. In order to search for food, ants initially explore the area around their nest by means of random walks. While moving, ants leave tiny drops of a pheromone substance on the ground. Ants are also able to scent these pheromones. When choosing their way, they are attracted by paths marked by strong pheromone concentrations. When having identified a food source, ants evaluate the quantity and the quality of the food and carry some of it back to their nest. During the return trip, the quantity of pheromone that ants leave on the ground may depend on the quantity and quality of the food. The pheromone trails will guide other ants to the food source. It has been shown in [66.15] that the indirect communication between the ants via pheromone trails – known as *stigmergy* [66.16] – enables them to find the shortest paths between their nest and food sources. Initially, ACO algorithms were developed with the aim of solving discrete optimization problems. It should be mentioned, however, that nowadays the class of ACO algorithms also comprise methods for the application to problems arising in networks, such as routing and load balancing [66.17], and for the application to continuous optimization problems [66.18].

ACO algorithms may be regarded from different perspectives. First of all, as mentioned above, they are SI techniques. However, seen from an operations research perspective, ACO algorithms belong to the class of metaheuristics [66.19–21]. The term *metaheuristic*, first introduced in [66.22], has been derived from the composition of two Greek words. *Heuristic* derives from the verb *heuriskein* ($\epsilon \upsilon \rho \iota \sigma \kappa \epsilon \iota \nu$) which means *to find*, while the prefix *meta* means *beyond, in an upper level*. Before this term was widely adopted, metaheuristics were often called *modern heuristics* [66.23]. In addition to ACO, other algorithms such as evolutionary computation, iterated local search, simulated annealing, and tabu search, are often regarded as metaheuristics. For books and surveys on metaheuristics, we refer the reader to [66.19–21, 23].

Algorithm 66.1 Ant colony optimization (ACO)
1: **while** termination conditions not met **do**
2:    **ScheduleActivities**
3:       AntBasedSolutionConstruction()
4:       PheromoneUpdate()
5:       DaemonActions(){optional}
6:    **end ScheduleActivities**
7: **end while**

From a technical perspective, ACO algorithms work as follows. Given a combinatorial optimization problem to be solved, first a finite set C of the so-called solution components, used for assembling solutions to the problem, must be defined. Second, a set $\mathcal{T}$ of *pheromone values* must be defined. This set of values is commonly called the *pheromone model*, which is – from a mathematical point of view – a parameterized probabilistic model. The pheromone model is one of the central components of any ACO algorithm. The pheromone values $\tau_i \in \mathcal{T}$ are commonly associated with solution components. The pheromone model is used to probabilistically generate solutions to the problem under

consideration by assembling them from the set of solution components. In general, ACO algorithms attempt to solve an optimization problem by iterating the following two steps:

- Candidate solutions are constructed using a pheromone model, that is, a parameterized probability distribution over the search space.
- The candidate solutions are used to update the pheromone values in a way that is deemed to bias future sampling toward high-quality solutions.

The pheromone update aims to concentrate the search in regions of the search space containing high-quality solutions. In particular, the reinforcement of solution components depending on the solution quality is an important ingredient of ACO algorithms. It implicitly assumes that good solutions consist of good solution components. To learn which components contribute to good solutions can help assemble them into better solutions. The main steps of any ACO algorithm are shown in Algorithm 66.1. DaemonActions (see line 5 of Algorithm 66.1) may include, for example, the application of local search to solutions constructed in function AntBasedSolutionConstruction().

The class of ACO algorithms comprises several variants. Among the most popular ones are $\mathcal{MAX}$–$\mathcal{MIN}$ Ant System ($\mathcal{MMAS}$) [66.24] and ant colony system (ACS) [66.25]. For more comprehensive information, we refer the interested reader to [66.26].

66.2.2 Particle Swarm Optimization

PSO [66.2, 27] is an SI technique for optimization that is inspired by the collective behavior of flocks of birds and/or fish schools. The first PSO algorithm was introduced in 1995 by *Kennedy* and *Eberhart* [66.28] for the purpose of optimizing the weights of a neural network, that is, for continuous optimization. In the meantime, PSO has also been adapted for its application to discrete optimization problems [66.29].

In PSO, solutions to the problem under consideration are labeled *particles*. The algorithm works on a whole set of particles at the same time, the so-called *swarm*. Therefore, PSO can be seen as a population-based optimization technique. During the run time of the algorithm, particles move through the search space on the search for an optimal, or good enough, solution. Moreover, particles communicate their current positions to neighboring particles. The position of each particle is updated according to three terms: its so-called *velocity*, the difference between its current position and the best position it has found so far, and that from the best position found by its neighbors. This has the effect that, during the execution of the algorithm, the swarm increasingly focuses on areas of the search space containing high-quality solutions. The term *particle swarm* was chosen by Kennedy and Eberhart for the following reason. Their initial intention was to model the movements of flocks of birds and fish schools. As their model further evolved toward an algorithm for optimization, the visual plots produced from the results of the algorithm rather resembled swarms of mosquitoes. The term *particle* was used due to making use of the term velocity, and *particle* seemed to be the most appropriate term in this context.

PSO is closely related to *artificial life* models. Early works by *Reynolds* on the flocking model known as *boids* [66.30], and *Heppner* and *Grenander's* studies on rules governing large numbers of birds flocking synchronously [66.31], suggested that bird flocking is an emergent behavior resulting from local interactions between the birds. These studies laid the foundation for the development of PSO for solving optimization problems. PSO is – in some way – similar to *cellular automata* (CA), which are often used for generating astonishing self-replicating patterns based on simple local rules. CAs may be characterized by the following three main attributes:

1. Cells are updated in parallel.
2. The value of each new cell depends only on the old values of the cell and its neighbors.
3. There is no difference in rules for updating different cells [66.32].

These three attributes also hold for the particles in PSO.

Henceforth, v_i denotes the velocity of the ith particle in the swarm, x_i denotes its position, p_i denotes the *personal best* position, and p_g is the best position found by particles in its neighborhood. In the original PSO algorithm, v_i and x_i, for $i = 1, \ldots, n$, are updated according to the following two equations [66.28]:

$$v_i \leftarrow v_i + c_1 R_1 \otimes (p_i - x_i) + c_2 R_2 \otimes (p_g - x_i) , \tag{66.1}$$

$$x_i \leftarrow x_i + v_i , \tag{66.2}$$

where R_1 and R_2 are independent functions returning a vector of values, generated uniformly at random, from the range $[0, 1]$. Moreover, c_1 and c_2 are the so-called acceleration coefficients. The symbol $\otimes$ refers to

point-wise vector multiplication. As shown in (66.1), the velocity term v_i of a particle is composed of three components: the *momentum*, the *cognitive* and the *social* terms. The *momentum* term v_i carries the particle toward the previous direction; the *cognitive* term,

$$c_1 R_1 \otimes (p_i - x_i) ,$$

represents a force that pulls the particle toward its personal-best position; finally, the *social* part,

$$c_2 R_2 \otimes (p_g - x_i) ,$$

represents a force that influences the new direction toward the best position of neighboring particles. Various different neighborhood topologies may be used for this purpose. Examples include ring, star, and *von Neumann*. The use of rather small neighborhood topologies – such as the one induced by the *von Neumann* neighborhood – has generally been shown to lead to better results when rather complex problems are addressed, whereas larger neighborhoods generally lead to a better performance for simpler problems [66.33]. Algorithm 66.2 summarizes the basic PSO algorithm.

Algorithm 66.2 Particle swarm optimization (PSO)
  1: Randomly generate an initial swarm
  2: **while** termination conditions not met **do**
  3:    **for** each particle i **do**
  4:       **if** $f(x_i) < f(p_i)$ **then** $p_i \leftarrow x_i$
  5:       $p_g = \min(p_{\text{neighbors}})$
  6:       Update velocity (66.1)
  7:       Update position (66.2)
  8:    **end for**
  9: **end while**

The class of PSO algorithms is characterized by a multitude of different variants, rendering it impossible to mention all of them here. However, popular variants include the *Inertia Weight PSO* [66.34], *fully informed PSO* [66.33], and *adaptive hierarchical particle swarm optimizer* [66.35]. Moreover, *Frankenstein's PSO* [66.36] is a PSO variant that was created by analyzing the components of existing PSO variants and combining (some of) them in a beneficial way. For more information, the interested reader may consult [66.37].

66.2.3 Artificial Bee Colony Algorithm

The artificial bee colony (ABC) algorithm was first proposed by *Karaboga* and *Basturk* in 2005 [66.38,

39]. The inspiration for the ABC algorithm is to be found in the foraging behavior of honey bees, which essentially consists of three components: food source positions, amount of nectar and three types of honey bees, that is, *employed bees*, *onlookers*, and *scouts*. In short, the algorithm works as follows. Feasible solutions to the problem under consideration are modeled as *food source positions*. Moreover, the quality of a feasible solution is modeled as the *amount of nectar* present at the corresponding food source position. Each type of bee is responsible for one particular operation in the context of generating new candidate food source positions, that is, new candidate solutions. Specifically, employed bees will search in the vicinity of the food source position that is presently in their memory; meanwhile they pass information about good food source positions to onlooker bees. Onlooker bees tend to select good food source positions from those found by the employed bees, and then further search for better food source positions around the selected food source position. In case the employed bee and the onlookers associated with a food source position are not able to find a better food source position, their current food source position is abandoned and the employed bee associated with this food source becomes a scout bee that performs a search for discovering new food source positions. If a scout identifies a new food source position, it turns into an employed bee again.

Essentially, the difference between the ABC algorithm and other population-based optimization techniques is to be found in the specific way of managing the resources of the algorithm, as suggested by the foraging behavior of honey bees. Due to its simplicity and ease of implementation, the ABC algorithm has captured much attention recently. It should also be mentioned that, although the algorithm has initially been introduced for continuous optimization, in the meantime it has been adapted for its application to combinatorial optimization problems as well [66.40, 41]. For a recent survey, we refer to [66.42].

66.2.4 Other SI Techniques for Optimization and Management Tasks

In the following, we briefly mention other applications of SI techniques for optimization and management tasks, the latter especially for what concerns distributed environments. They are grouped with respect to their natural inspiration.

Division of Labor (Ants/Wasps)

In colonies of ants and wasps, for example, there are various tasks to be dealt with by the colony members. However, the urgency to engage in certain tasks may change over time. In 1984, *Wilson* [66.43] showed that the concept of *division of labor* in colonies of *Pheidole* genus ants allows the colony to adapt to these changing demands. Division of labor was later modeled in [66.44, 45] by means of response threshold models.

These models were later used in several technical applications. In the following, we mention a few of them. *Nouyan* et al. [66.46] consider static and dynamic task allocation problems in which trucks have to be painted in painting booths. Another application concerns media streaming in peer-to-peer networks [66.47]. A multiagent system for the scheduling of dynamic job shops with flexible routing and sequence-dependent setups is considered in [66.48]. *Merkle* et al. [66.49] made use of a response threshold model for self-organized task allocation in the context of computing systems with reconfigurable components. Finally, [66.50] present a system for task allocation in distributed environments.

Cemetery Formation (Ants)

The term *cemetery formation* refers to a behavior which has been observed in ant colonies of the species *Pheidole pallidula*, among others, which cluster the bodies of dead nest mates. This self-organized behavior has given rise to several applications, especially in the context of clustering and sorting. In 1991, a model for the clustering and sorting behavior of ants was published in [66.51]. Note that clustering refers in this context to the formation of piles, and sorting, on the other hand, refers to the spatial arrangement of objects according to their properties.

Mainly based on the model from [66.51], several algorithms for clustering and sorting were proposed in the literature. The first one was presented in [66.52], extending the original model to handle numerical data. More recent papers include [66.53] which deals with clustering and topographic mapping. Finally, the cemetery formation behavior of ants has also inspired an algorithm for dynamic load balancing [66.54].

Flashing in Fireflies

Fireflies are winged beetles that make use of bioluminescence to attract mates or prey. Moreover, tropical fireflies, in particular the ones from Southeast Asia, synchronize their light flashes in large groups of individuals. This is a self-organized phenomenon which is mathematically described by the so-called *phase-coupled oscillator* models [66.55]. The benefits of this self-synchronization are not yet fully understood. Current hypotheses consider diet, social interaction, and altitude.

The literature contains, at least, two types of technical applications that are inspired by different aspects of the flashing of fireflies. First, there are applications that require some type of self-synchronization. Examples include, but are not limited to, a synchronization protocol in sensor networks [66.56], the synchronization in overlay networks [66.57], and dynamic pricing in online markets [66.58]. Second, the literature offers the so-called firefly algorithm (FA) [66.59], which is inspired by the way in which fireflies attract mates or prey. This algorithm was initially introduced for continuous optimization. It has, however, been adapted for the application to combinatorial optimization as well [66.60].

Fish Schooling

A group of fish that have gathered are commonly called an *aggregation of fish*. Such a fish aggregation is called unstructured in the case in which the group consists of various species of fish having randomly gathered, for example, in the vicinity of a food source. If there is some social component to this gathering, the fish are said to be *shoaling*. Shoaling fish are aware of each other's presence, adjusting, for example, their swimming behavior to each other in order to stay together. However, their relation is rather loose. If, in contrast, an aggregation of fish is more tightly organized, for example, when all fish move at the same speed in the same direction, then the aggregation is said to be *schooling*. Schooling is a self-organized behavior that results from local interactions between the fish. This behavior comes with several advantages such as providing a means for social interactions, more successful foraging, and predator avoidance.

There are basically two different algorithms for optimization based on fish schooling to be found in the literature. The first algorithm is referred to as the *artificial fish swarm algorithm* (AFSA). It has, for example,

been applied to the training of feed-forward neural networks [66.61], multiuser detection [66.62], image segmentation [66.63], and generally to continuous optimization [66.64]. The second algorithm is known as *fish school search* [66.65].

Self-Desynchronized Croaking (Japanese Tree Frogs)

Different biological studies – for example, [66.66] – have dealt with the croaking of Japanese tree frogs. The male individuals make use of their croaks in order to attract females. Moreover, females of this family of frogs can recognize the source of such a croak and are able to determine the current location of the corresponding male. However, this is only possible if no two frogs (that are close enough to the female) croak at the same time. In such a case, the female is not able to detect where the croaks came from. This is why, over time, male frogs evolved a self-organized way of desynchronizing their croaks. *Aihara* et al. [66.67] introduced a first formal model based on a set of pulse-coupled oscillators for capturing this behavior. So far, this model has only been applied to distributed graph coloring [66.68, 69]. However, the algorithm proposed in [66.69] is currently the state of the art for this problem.

Nest Building (Termites/Wasps)

Both termites and wasps build highly complex nests in cooperation. The construction of such nests is well beyond the capabilities of an individual insect. The nests of both termites and wasps have a very complex internal structure. Moreover, termite nests are extremely large in comparison to individual insects. Scientists studying the nest-building behavior came up with probabilistic models for describing (parts of) the behavior [66.70]. It is nowadays generally accepted that stigmergy plays a central role in nest building.

Models for nest building based on stigmergy have been used mainly in software tools for simulating the automated building of certain structures. Examples can be found in [66.71–74].

66.3 SI in Robotics: Swarm Robotics

Swarm robotics refers to the study and use of SI techniques for the coordination of groups of robots. The following sections provide a brief overview of this field, with a focus on swarm robotic systems and the tasks they accomplish.

66.3.1 Systems

In the late 1940s, *Walter* [66.75] built two autonomous robots called *Machina speculatrix*, or simply tortoise, which exhibited behaviors resembling those of simple animals. The robots had a driving/steering mechanism, a head light, a photoreceptor, and a bump sensor. They were designed to search for and approach light sources of moderate intensity. If a robot observed such a source, its head light was turned off, otherwise it was turned on. In an experiment, the robots were set up in a dark environment, where they approached each other exhibiting complex motion patterns. Such *mutual recognition* allowed *a population of machines* to form *a sort of community*, which broke up once an external light source was introduced [66.75, p. 129]. This two-robot system may be the first self-organizing multirobot system. Interestingly, even a single robot was reported to exhibit complex interactions when facing its mirror image – such a behavior, if *observed in an animal, might be accepted as evidence of some degree of self-awareness* [66.75, pp. 128–129].

In the 1950s, inspired by *von Neumann's* kinematic model of machine replication [66.76], the first physical models of self-replication were built. *Penrose* and *Penrose* [66.77] studied a system in which passive mechanical parts move on a linear track when the latter is subjected to side-to-side agitation. In their default position, the parts do not link under the influence of shaking alone. If a seed object composed of two complementary parts, one hooked up to the other, is added, it replicates by interacting with the other parts on the track. *Jacobson* [66.78] implemented a system in which self-propelled electromechanical parts move on a circular track with several branches. A seed object composed of two parts could trigger other parts to assemble into identical objects without human intervention.

In the late 1980s, studies of *Fukuda* and *Nakagawa* [66.79–81], *Beni* [66.5], and *Wang* and *Beni* [66.82] provided an enormous impetus for the field that developed into *swarm robotics*. *Fukuda* and *Nakagawa* proposed a novel type of robotic system called *dynamically reconfigurable robotic system (DRRS)*, which can *dynamically reorganize its shape and structure [...] for a given task and strategic purpose*. DRRS is made of *several cells* with built-in intelligence and the ability to autonomously connect to and detach from one another [66.81, pp. 55–56]. The authors also presented a first prototype of this system, the CEBOT Mark I [66.80]. At the same time, *Beni* introduced the term *cellular robotic system*, referring to a system that can *encode information as patterns of its own structural units* [66.5, p. 59]; the units would be structural elements, each with built-in intelligence, able to move in space and act asynchronously under distributed control. *Beni* and *Wang* also used the terms

swarm and *swarm intelligence* in this context [66.83, 84].

Other early physical implementations of distributed robotic systems include the CEBOT Mark II [66.85], ACTRESS [66.86], and GOFER [66.87].

Hardware Architectures

Advances in technology, for example, in computers, manufacturing and mobile devices have made it affordable to study swarms of around 20–1000 physical robots [66.88] and up to around 1 000 000 robots in simulation [66.93–95]. At present, most swarm robotic systems consist of mobile robots that operate on the ground. An example is the Kilobot platform (Fig. 66.2a), which was designed to facilitate the fabrication and operation of thousands of robots – including their charging, programming and activation all at once [66.88]. Other state-of-the-art robotic systems include the r-one (Fig. 66.2b), which features, among others, a set of IR transmitters and receivers for communication and relative localization [66.89], and the Khepera I-IV [66.96] and e-puck [66.97], which feature a range of sensors including a camera. Increasingly, swarm robotic systems operate in spaces other than on the ground, such as underwater [66.90, 98] (Fig. 66.2c) or in the air [66.99, 100]. In some robotic systems, the swarms operate and collaborate across multiple spaces, such as on the ground and in the air [66.91, 101] (Fig. 66.2d,e).

According to their system architecture, most swarm robotic systems can be categorized into either *multirobot* systems or *modular reconfigurable robot* systems. Multirobot systems are composed of multiple distinct robots, which are typically mobile and able to perform (collectively) more than one task in parallel (Fig. 66.2a–c). Modular reconfigurable robot systems are composed of component modules that can be physically linked together to form a robot (Fig. 66.2f). A few *hybrid* systems exist, sharing properties of both multirobot and modular reconfigurable robot systems [66.91, 102–104] (Fig. 66.2d).

Of particular interest among systems of modular reconfigurable robots are those where the robots can build themselves [66.105, 106]. The term *self-reconfigurable* denotes the general ability of physical modules to reconfigure themselves, regardless of whether the process is centrally controlled, for example, by an external computer, or decentralized and autonomous. In the following, we use the term *self-assembly* to refer to processes by which pre-existing components (separate or distinct parts of a disordered structure) autonomously organize

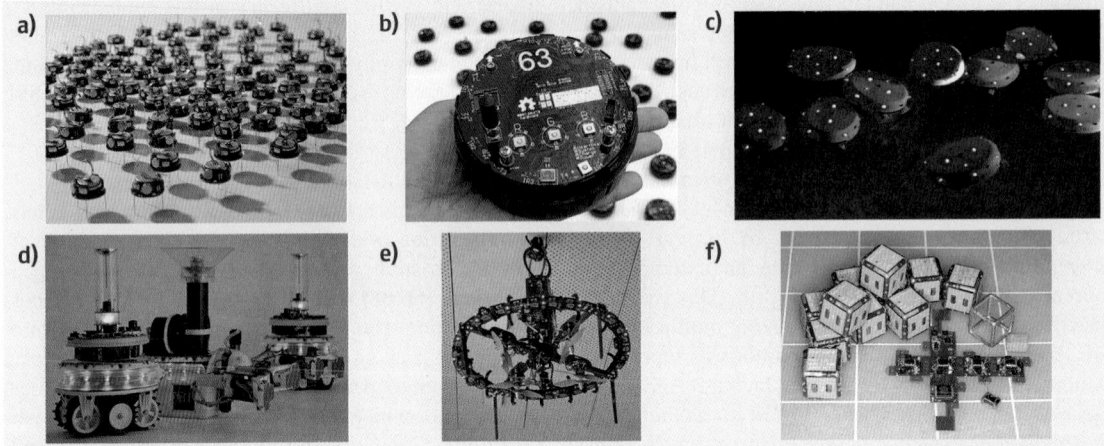

Fig. 66.2a–f Examples of swarm robotic systems: (**a**) Kilobots developed by Harvard University [66.88]; (**b**) r-one (after [66.89], photo courtesy of J. McLurkin, Rice University); (**c**) Lily developed in the CoCoRo project (after [66.90], photo courtesy of T. Schmickl, University of Graz); (**d,e**) a heterogeneous system studied in the Swarmanoid project (after [66.91], photo courtesy of M. Dorigo, Université Libre de Bruxelles); (**f**) Pebbles (after [66.92], photo courtesy of D. Rus, MIT)

into patterns or structures without external intervention. Self-assembly is responsible for the generation of much of the order in nature [66.107] and has widely been applied in the synthesis of products from molecular components. Increasingly, the potential of self-assembly processes involving larger components – up to the centimeter-scale – is being recognized [66.108]. In robotic systems, two distinct classes of self-assembling systems exist [66.109]: (i) systems in which the components that self-assemble are externally propelled, and (ii) systems in which the components that self-assemble are self-propelled.

Sensing and Communication

In most multirobot systems, robots interact with each other by using their sensors or some form of communication. *Dudek* et al. [66.110] presented a detailed taxonomy considering communication range, topology, and bandwidth. In the following, we adopt a simpler categorization proposed by *Cao* et al. [66.111]:

- *Interaction via environment* refers to the transfer of information that is mediated through the *memory* of the environment. In this case, robots leave persistent signs that stimulate the activity of other robots. This kind of indirect communication is also referred to as *stigmergy* [66.16]. Stigmergic communication is widely used in social insect societies, for example, during the construction of mounds by

termites of *Macrotermes bellicosus* [66.8], and has been implemented in several swarm robotic systems [66.112–116].

- *Interaction via sensing refers to local interactions that occur between agents as a result of agents sensing one another, but without explicit communication* [66.111, p. 12]. We include in this category interactions where agents sense each other indirectly, that is, where the current presence or motion of another agent can be inferred from changes in the environment. Note that the boundary to stigmergic communication is blurred; for example, consider the situation where multiple agents push an object simultaneously [66.117–119].

In some social animals, the members of a group observe a common leader individual. Their actions can be highly dependent on the observed behavior of the leader, as, for instance, during an attack of the group [66.120]. In other animals, no recognizable leader individual exists; instead, individuals observe nearby group members. The latter situation is typical for swarm systems. It is reported, for instance, for animal groups that exhibit herding, flocking, and schooling behavior [66.8]. Note that where the groups are not homogeneous, even a minority of individuals may be able to influence the rest of the group [66.121].

In principle, interaction via sensing can be considered an implicit form of communication, in particular, as an observed agent can change action and thereby influence the behavior of its observers. *Arkin* [66.122] referred to the interaction via sensing category as *cooperation without communication*, and showed that it is sufficient to accomplish tasks, that require the cooperation of multiple robots. Other examples of swarm robotic studies using interaction via sensing include [66.123–126].

- *Interaction via communication* refers to interactions involving explicit communication. Thereby, information is either broadcast or transferred to specific teammates. Information transfer can take place through direct physical interactions, such as touch. This latter form of communication can also be referred to as *direct interaction* [66.127]. Explicit communication can improve the performance of a multirobot system. This is typically the case where the system benefits from robots being recruited to certain areas of the environment. *Balch* and *Arkin* [66.128] studied such an environment and showed that it can be sufficient for each robot to signal its overall state. The transfer of more elaborate information however would not result in any significant increase in task performance. Explicit communication is commonly used in modular reconfigurable robot systems, for example, to exchange information between inter-connected modules or to support the docking process of separate modules [66.129].

Control and Coordination

Over the last two decades, a range of design methods have been proposed for the control of swarm robotic systems. They can be broadly classified into behavior-based design methods and automated design methods [66.130].

In behavior-based design methods, the user approaches the problem in a bottom-up manner [66.131]. A repertoire of behaviors for individual robots is defined and often refined through a trial-and-error process. A common approach is the use of finite state machines. Each state defines a basic behavior. Transitions between states can be triggered by probability, external events, time-outs, and combinations of these [66.132–134]. A prominent example is the use of response threshold functions, for example,

$$1 - \exp^{-s_i/\theta_i} \text{ or } \frac{s_i^2}{s_i^2 + \theta_i^2} \, ,$$

which define the probability for an individual to engage in task i based on the perceived task stimulus s_i and

threshold θ_i. The particular threshold value θ_i can either be fixed for each individual from the outset [66.135] or learned during its lifetime [66.136, 137]. In both cases, the mechanism can facilitate the emergent allocation of tasks in groups of otherwise identical individuals (see also Sect. 66.2.4). In addition, *intentional* approaches to task allocation have been considered [66.138, 139]. These require the agents to cooperate explicitly with each other. For example, the decentralized ALLIANCE algorithm [66.140, 141] can be used for groups of heterogenous robots to perform tasks and subtasks, which may have ordering dependencies, in a fault-tolerant way. It assumes that the robots detect with some probability the effect of their own actions as well as the actions of other team members.

Virtual potential fields [66.142, 143], and physicomimetics [66.144], is another widely used behavior-based design method. The robots mimic a physical particle under the influence of a potential field. The latter guides the particle toward a point of minimal potential energy. While the goal point, which the robot shall reach, would exert an attractive force on the particle, any obstacle would exert a repulsive force. Other robots can exert forces on the particle as well. Using this concept, a wide repertoire of behaviors can be realized, such as the collective movement of robots arranged in particular formations [66.145], or the tracking of multiple moving targets [66.146]. The properties of the resulting swarm systems, for example, the cohesion of the swarm, can also be formally analyzed [66.147].

Other design methods include the Growing Point Language [66.148], the Origami Shape Language [66.149], and Proto [66.150]. These languages were developed in the context of Amorphous Computing [66.151], which considers systems of massively distributed, disordered, asynchronous, and locally interacting computational devices. The Proto language has been extended for use on mobile devices. This extension was validated with a swarm of 40 iRobot robots [66.152]. Some amorphous computing approaches allow users to specify desired global system properties in the language. A compiler then produces the local rule set for the agents to achieve these properties [66.149].

Automated design methods can be grouped into reinforcement learning and evolutionary robotics. In reinforcement learning [66.153], an agent interacts with its environment by choosing actions and receiving rewards. *Matarić* [66.154, 155] applied reinforcement learning in a swarm robotic context. The robots had to learn how to collaborate in a foraging task. The robots were provided with a set of hand-coded behaviors (as in a behavior-based approach) and were required to learn how *to correlate appropriate conditions for each of these behaviors in order to optimize the higher-level behavior* [66.155]. The difficulties of using reinforcement learning in a swarm robotic context are discussed in [66.130]. A recent survey of reinforcement learning in robotics is reported in [66.156].

Evolutionary robotics is an approach to designing robots, or aspects of them (e.g., morphology, control) using evolutionary algorithms [66.157, 158]. This approach can also be applied to the design of swarm robotic systems. In principle, evolution can bypass both the problem of decomposing a given task and the problem of identifying basic behaviors that achieve the subtasks [66.159, 160]. Early studies in evolutionary robotics developed collective behavior such as herding or flocking in simplistic simulation environments [66.161–163]. Simulation environments with physically embodied agents were considered in [66.159], where neural network controllers for aggregation were first evolved using a group of five robots in a simple simulation environment; the best of these controllers were subsequently validated using a more detailed simulation model of the robots. *Quinn* et al. [66.164] evolved neural network controllers for collective motion using a group of three simulated robots and subsequently tested the best-rated network in 100 trials with a group of three physical robots. *Watson* et al. [66.165] went a step further in that controllers for a simple phototaxis task were directly evolved on a group of eight physical robots. Working toward a distributed evolution of robot morphologies in hardware, *Griffith* et al. [66.166] demonstrated a system of template-replicating polymers, which were made of reconfigurable modules that slid passively on an air table and executed a finite state machine to control their connectivity. Recent work on evolutionary swarm robotics considers cultural evolution, for example, where behaviors that can be imitated (memes) are subject to an evolutionary process. In these, the robots engage as both teachers and learners to exchange memes [66.167].

Several design methods were developed specifically for, or mainly adopted in, the context of modular reconfigurable robot systems. One class of algorithms addresses the problem of how to adjust the relative positions of modules without changing their connection topology. *Yim* [66.168] proposed the use of *gait control tables* to produce a range of animal-like locomotion patterns, such as the walking gaits of hexapods. Each gait control table specifies for each control cycle and module a basic action to be performed. The controller is executed either from a central place or in a distributed fashion. In the latter case, the modules synchronize their actions using internal timers. *Shen* et al. [66.169] proposed *hormone-inspired* communication and control, in which artificial hormones help modules to synchronize actions and discover changes in their topology. For example, a set of independent caterpillar-like robots could be connected into a single entity, which would adapt its gait to the new topology. In a similar experiment, a connected entity was manually split into smaller entities that continued to move as independent caterpillars. *Støy* [66.170] proposed a *role-based control* algorithm to let modular robots display periodic locomotion patterns. A module's role specifies its actions and how to synchronize them with neighbor modules. For communication, a parent–child architecture is used; thus, modules need to be arranged in acyclic graphs. An extended version of the control algorithm can also cope with cycles.

Another class of algorithms addresses the problem of how to adjust the relative positions of modules by changing the connection topology [66.106]. One approach is to formulate the problem as a search problem. For example, in order to reconfigure a lattice-based robot from one topology to another, a graph search is performed, where the start node of the graph corresponds to the initial topology of the robot and the end node corresponds to the desired topology of the robot [66.171]. Due to the combinatorial explosion of possibilities, an exhaustive search of such graphs is impractical whenever the number of modules is not small. State-of-the-art approaches are thus heuristic and consider ways of reducing the problem complexity. For example, *Yoshida* et al. [66.172] proposed a two-level *motion planner*. A global planner ensures that the robot as a whole follows a predefined 3D trajectory. To do so, it specifies several candidate paths that bring individual modules from the tail to the head of the robot. A motion scheme selector chooses a feasible path for each module based on a rule database. Another example is to merge logically a group of nearby modules into *meta-modules*, which, typically, have more advanced locomotion abilities than the individual modules. The problem is then reduced to developing controllers for both meta-modules and modular robots composed of meta-modules [66.173]. In principle, modular robots can solve the search problem on the fly [66.174]. Other than by search, the reconfiguration problem can also be attempted by local movement strategies, for example, random walks [66.175, 176], cellular automata rules [66.177], gradient rules [66.178, 179], or combinations of these [66.180]. These approaches naturally lead to decentralized implementations, as is desired in swarm robotics.

66.3.2 Tasks

A range of capabilities have already been demonstrated with swarm robotic systems. In the following, a brief overview is given. More detailed information is provided in Chaps. 71–74 of Part F of this handbook. *Garnier* et al. [66.189] demonstrated how a group of 20 Alice robots *aggregate* in a homogeneous environment. The robots mimic the aggregation behavior of cockroaches, which are reported to join and leave clusters with probabilities that depend on the sizes of clusters [66.190]. Such probabilistic algorithms have the advantage that, as long as the environment is bounded, it is not required that the robots initially form a connected graph in terms of their sensing and/or communication. A deterministic algorithm for aggregation is considered in [66.181]. It requires robots to have one binary sensor, which informs them whether or not there is another robot in their line of sight. The robots do not need memory and do not need to perform arithmetic computation. They rotate on the spot when they perceive another robot, and move backward along a circular trajectory otherwise. This algorithm was validated with groups of 40 e-puck robots (Fig. 66.3a).

Werfel et al. [66.116] developed a system of robots that can simultaneously *construct* and navigate structures from a supply of building blocks (Fig. 66.3b). The robots are inspired by termites, which use stigmergic rules to construct sophisticated structures, in particular, the mounds they inhabit. Given a desired target structure, it is possible to generate automatically a set of rules to be uploaded onto each robot. Using only local information, these rules allow the robots to coordinate their activities in a way that avoids conflict. A group of three robots constructed several structures, one resembling a castle.

Halloy et al. [66.182] showed that hybrid societies comprising both cockroaches and robots can collectively *decide* to aggregate under either of two shelters (Fig. 66.3c) and that it is possible for the robots to influence the decision-making process. In general, such interactive robots could be used to study and control animal groups [66.182, 191], including livestock [66.192, 193], and to inform ecological conservation policy.

Following the pioneering simulation works on *boids* [66.30], *Turgut* et al. [66.183] demonstrated how a group of robots can *flock* through a real environment using simple rules. To align with each other, the robots used *virtual heading sensors*, each comprising a digital compass and a wireless communication module. Flocking was demonstrated with 9 Kobot robots in a bounded environment (Fig. 66.3d).

Krieger et al. [66.184] studied algorithms that allow a group of robots to *forage* (Fig. 66.3e). The robots rested in a central place, the nest. A robot would leave the nest if the total energy of the colony dropped below a threshold. Each robot had its own threshold, which effectively enabled the division of labor within the group. In addition, a robot would leave the nest when being recruited by another robot that had found a cluster of food. The pair of robots would then perform a tandem run to reach the cluster. The algorithms were tested on groups of up to 12 Khepera robots. The groups were reported to perform more efficiently when employing the division of labor and recruitment mechanisms than without such mechanisms.

Groß et al. demonstrated how a group of 16 s-bot robots *self-assemble* into a single composite entity [66.185]. The process was seeded by one of the robots activating its light emitting diode (LED) ring in red. Other robots activated their LED rings in blue. Once a robot would connect to the seed structure, it became red too, thereby attracting other robots to the structure as it grows (Fig. 66.3f). The problem of self-assembling into arbitrary morphologies of s-bot robots was considered in [66.194].

Holland and *Melhuish* [66.186] studied algorithms that allow groups of robots to *sort* (and cluster) objects of different types (Fig. 66.3g). Six robots were programmed using simple rules, which regulated the conditions under which objects of different types were picked up and deposited.

Following the pioneering work of *Kube* et al. [66.195, 196], *Chen* et al. [66.187] proposed an algorithm for a group of robots to *transport* objects larger than themselves toward a goal location (Fig. 66.3h). The robots were programmed to only push the object across the portion of its surface where the direct line of sight to the goal is occluded by the object. The algorithm was proven to work for objects of arbitrary convex shape and it was tested with 20 e-puck robots.

Ijspeert et al. [66.188] studied an algorithm that allows a group of robots to *pull sticks* out of the ground collaboratively (Fig. 66.3i). Upon encountering a stick,

Fig. 66.3a–i Examples of capabilities demonstrated by swarm robotic systems: (**a**) aggregation (after [66.181]); (**b**) construction (after [66.116]; reprinted with permission from AAAS); (**c**) decision making (after [66.182]; photo courtesy of J. Halloy, Université Libre de Bruxelles); (**d**) flocking (after [66.183]; photo courtesy of E. Şahin, Middle East Technical University); (**e**) foraging (after [66.184]; photo courtesy of L. Keller, University of Lausanne); (**f**) self-assembly (after [66.185]); (**g**) sorting of objects (after [66.186]; photo courtesy of C. Melhuish, Bristol Robotics Laboratory); (**h**) transport of objects (after [66.187]); (**i**) pulling sticks out of the ground (after [66.188]; reprinted with permission from Springer)

a robot would only be able to pull it partially out of the ground. It would then wait for a second robot to arrive and pull the stick out completely. The optimal waiting time for the first robot was derived from an analytic model of the system. The algorithm was validated using a system of six Khepera robots.

66.4 Research Challenges

Research challenges concerning the use of swarm intelligence in optimization are mainly related to increasing their efficiency. More specifically, in addition to providing an innovative way of problem solving, swarm intelligence approaches must also be efficient concerning, for example, computation time in order to be able to compete with state-of-the-art optimization techniques. This may often be achieved by hybridizing swarm intelligence approaches with components taken from optimization algorithms in other fields such as, for example, operations research. The interested reader may find various references to such kind of techniques in [66.197].

With regard to swarm robotics, a major challenge is the transition from systems operating in structured indoor environments, as typically found in laboratories, to the more complex environments found in the real world. Over the next decades, swarms of robots are expected to have impact in a range of application scenarios, including cognitive factories, deep sea exploration, disaster management, precision farming, and space systems. Working toward more complex environments also concerns the ability of swarms of robots to interact safely with humans. Another challenge concerns the miniaturization of swarm robotic systems. Most of the current systems comprise of centimeter-sized robots. The swarm robotics approach, however, should be equally applicable to intelligent autonomous devices operating at scales from a millimeter down to a micrometer. This could have profound implications, for example, on advanced materials and healthcare technologies.

References

66.1   E. Bonabeau, M. Dorigo, G. Theraulaz: *Swarm Intelligence: From Natural to Artificial Systems* (Oxford Univ. Press, New York 1999)

66.2   J. Kennedy, R.C. Eberhart, Y. Shi: *Swarm Intelligence* (Morgan Kaufmann, San Francisco 2001)

66.3   C. Blum, D. Merkle (Eds.): *Swarm Intelligence: Introduction and Applications* (Springer, Berlin, Heidelberg 2008)

66.4   S.J. Russell, P. Norvig: *Artificial Intelligence. A Modern Approach* (Simon Schuster Co., Englewood Cliffs 1995)

66.5   G. Beni: The concept of cellular robotic systems, Proc. 3rd IEEE Int. Symp. Intell. Syst., Piscataway (1988) pp. 57–62

66.6   G. Beni, J. Wang: Swarm intelligence, Proc. 7th Annu. Meet. Robot. Soc. Japan, RSJ, Tokyo (1989) pp. 425–428

66.7   G. Beni, S. Hackwood: Stationary waves in cyclic swarms, Proc. 1992 IEEE Int. Symp. Intell. Control, Los Alamitos (1992) pp. 234–242

66.8   S. Camazine, J.-L. Deneubourg, N.R. Franks, J. Sneyd, G. Theraulaz, E. Bonabeau: *Self-Organization in Biological Systems* (Princeton Univ. Press, New Jersey 2001)

66.9   I. Kant: *Critique of Judgement* (Hackett, Indianapolis 1987), Translated by W. S. Pluhar

66.10   B. Hölldobler, E.O. Wilson (Eds.): *The Ants* (Springer, Berlin, Heidlberg 1990)

66.11   M. Dorigo, T. Stützle: *Ant Colony Optimization* (MIT, Cambridge 2004)

66.12   M. Dorigo: Optimization, Learning and Natural Algorithms, Ph.D. Thesis (Dipartimento di Elettronica, Politecnico di Milano, Italy 1992), in Italian

66.13   M. Dorigo, V. Maniezzo, A. Colorni: Positive feedback as a search strategy, Tech. Rep. 91-016, Dipartimento di Elettronica, Politecnico di Milano, Italy, 1991

66.14   M. Dorigo, V. Maniezzo, A. Colorni: Ant System: Optimization by a colony of cooperating agents, IEEE Trans. Syst. Man Cybern. Part B **26**(1), 29–41 (1996)

66.15   J.-L. Deneubourg, S. Aron, S. Goss, J.-M. Pasteels: The self-organizing exploratory pattern of the argentine ant, J. Insect Behav. **3**, 159–168 (1990)

66.16   P.-P. Grassé: La reconstruction du nid et les coordinations interindividuelles chez *Bellicositermes natalensis* et *Cubitermes* sp. La théorie de la stig-mergie: Essai d'interprétation du comportement des termites constructeurs, Insectes Soc. **6**(1), 41–80 (1959), in French

66.17   G. Di Caro, M. Dorigo: AntNet: Distributed stigmergetic control for communications networks, J. Artif. Intell. Res. **9**, 317–365 (1998)

66.18   K. Socha: ACO for continuous and mixed-variable optimization, Lect. Notes Comput. Sci. **3172**, 25–36 (2004)

66.19   C. Blum, A. Roli: Metaheuristics in combinatorial optimization: Overview and conceptual comparison, ACM Comput. Surv. **35**(3), 268–308 (2003)

66.20   F. Glover, G. Kochenberger (Eds.): *Handbook of Metaheuristics* (Kluwer, Boston 2002)

66.21   H.H. Hoos, T. Stützle: *Stochastic Local Search: Foundations and Applications* (Elsevier, Amsterdam 2004)

66.22   F. Glover: Future paths for integer programming and links to artificial intelligence, Comput. Oper. Res. **13**, 533–549 (1986)

66.23   C.R. Reeves (Ed.): *Modern Heuristic Techniques for Combinatorial Problems* (Wiley, New York 1993)

66.24   T. Stützle, H.H. Hoos: $\mathcal{MAX}$-$\mathcal{MIN}$ Ant System, Futur. Gener. Comput. Syst. **16**(8), 889–914 (2000)

66.25   M. Dorigo, L.M. Gambardella: Ant colony system: A cooperative learning approach to the traveling salesman problem, IEEE Trans. Evol. Comput. **1**(1), 53–66 (1997)

66.26   M. Dorigo, T. Stützle: Ant colony optimization: Overview and recent advances. In: *Handbook of Metaheuristics*, ed. by M. Gendreau, J.-Y. Potrin (Springer, Berlin, Heidelberg 2010) pp. 227–264

66.27   M. Clerc (Ed.): *Particle Swarm Optimization* (ISTE, Newport Beach 2006)

66.28   J. Kennedy, R.C. Eberhart: Particle swarm optimization, Proc. 1995 IEEE Int. Conf. Neural Netw., Piscataway, Vol. 4 (1995) pp. 1942–1948

66.29   Q.-K. Pan, M. Fatih Tasgetiren, Y.-C. Liang: A discrete particle swarm optimization algorithm for the no-wait flowshop scheduling problem, Comput. Oper. Res. **35**(9), 2807–2839 (2008)

66.30   C.W. Reynolds: Flocks, herds and schools: A distributed behavioral model, Comput. Graph. **21**(4), 25–34 (1987)

66.31   F. Heppner, U. Grenander: A stochastic nonlinear model for coordinated bird flocks. In: *The Ubiq-*

66.32   *uity of Chaos*, ed. by S. Krasner (AAAS, Washington DC 1990)

66.32   R. Rucker: *Seek!* (Four Walls Eight Windows, New York 1999)

66.33   R. Mendes, J. Kennedy, J. Neves: The fully informed particle swarm: Simpler, maybe better, IEEE Trans. Evol. Comput. **8**(3), 204–210 (2004)

66.34   Y. Shi, R. Eberhart: A modified particle swarm optimizer, Proc. 1998 IEEE World Congr. Comput. Intell. (1998) pp. 69–73

66.35   S. Janson, M. Middendorf: A hierarchical particle swarm optimizer and its adaptive variant, IEEE Trans. Syst. Man Cybern. Part B Cybern. **35**(6), 1272–1282 (2005)

66.36   M.A. de Montes Oca, T. Stützle, M. Birattari, M. Dorigo: Frankenstein's PSO: A composite particle swarm optimization algorithm, IEEE Trans. Evol. Comput. **13**(5), 1120–1132 (2009)

66.37   M. Dorigo, M.A. de Montes Oca, A. Engelbrecht: Particle swarm optimization, Scholarpedia **3**(11), 1486 (2008)

66.38   D. Karaboga, B. Basturk: A powerful and efficient algorithm for numerical function optimization: artificial bee colony (ABC) algorithm, J. Glob. Optim. **39**(3), 459–471 (2007)

66.39   D. Karaboga, B. Basturk: On the performance of artificial bee colony (ABC) algorithm, Appl. Soft Comput. **8**(1), 687–697 (2008)

66.40   Q.-K. Pan, M.F. Tasgetiren, P.N. Suganthan, T.J. Chua: A discrete artificial bee colony algorithm for the lot-streaming flow shop scheduling problem, Inf. Sci. **181**(12), 2455–2468 (2011)

66.41   F.J. Rodriguez, C. García-Martínez, C. Blum, M. Lozano: An artificial bee colony algorithm for the unrelated parallel machines scheduling problem, Lect. Notes Comput. Sci. **7492**, 143–152 (2012)

66.42   D. Karaboga, B. Gorkemli, C. Ozturk, N. Karaboga: A comprehensive survey: Artificial bee colony (ABC) algorithm and applications, Artif. Intell. Rev. **42**, 21–57 (2014)

66.43   E.O. Wilson: The relation between caste ratios and division of labour in the ant genus *phedoile*, Behav. Ecol. Sociobiol. **16**(1), 89–98 (1984)

66.44   G. Theraulaz, E. Bonabeau, J.-L. Deneubourg: Response threshold reinforcement and division of labour in insect societies, Proc. Biol. Sci. **265**(1393), 327–332 (1998)

66.45   E. Bonabeau, G. Theraulaz, J.-L. Deneubourg: Fixed response thresholds and the regulation of division of labor in social societies, Bull. Math. Biol. **60**, 753–807 (1998)

66.46   S. Nouyan, R. Ghizzioli, M. Birattari, M. Dorigo: An insect-based algorithm for the dynamic task allocation problem, Künstl. Intell. **4**, 25–31 (2005)

66.47   M. Sasabe, N. Wakamiya, M. Murata, H. Miyahara: Effective methods for scalable and continuous media streaming on peer-to-peer networks, Eur. Trans. Telecommun. **15**, 549–558 (2004)

66.48   X. Yu, B. Ram: Bio-inspired scheduling for dynamic job shops with flexible routing and sequence-dependent setups, Int. J. Prod. Res. **44**(22), 4793–4813 (2006)

66.49   D. Merkle, M. Middendorf, A. Scheidler: Self-organized task allocation for computing systems with reconfigurable components, Proc. 20th Int. Parallel Distrib. Proc. Symp., IPDPS 2006 (2006) p. 8

66.50   R. Klazar, A.P. Engelbrecht: Dynamic load balancing inspired by division of labour in ant colonies, Proc. 2011 IEEE Symp. Swarm Intell., SIS (2011) pp. 1–8

66.51   J.-L. Deneubourg, S. Goss, N. Franks, A. Sendova-Franks, C. Detrain, L. Chrétien: The dynamics of collective sorting: Robot-like ants and ant-like robots, Proc. 1st Int. Conf. Simul. Adapt. Behav.: From Animals to Animats 1, SAB 91 (MIT, Cambridge 1991) pp. 356–365

66.52   E.D. Lumer, B. Faieta: Diversity and adaptation in populations of clustering ants, Proc. 3rd Int. Conf. Simul. Adapt. Behav.: From Animals to Animats 3, SAB 94, MIT Diversity, ed. by D. Cliff, P. Husbands, J.-A. Meyer, S.W. Wilson (1994) pp. 501–508

66.53   J. Handl, J. Knowles, M. Dorigo: Ant-based clustering and topographic mapping, Artif. Life **12**(1), 35–62 (2006)

66.54   R. Klazar, A.P. Engelbrecht: Dynamic load balancing inspired by cemetery formation in ant colonies, Lect. Notes Comput. Sci. **7461**, 236–243 (2012)

66.55   R.E. Mirollo, S.H. Strogatz: Synchronization of pulse-coupled biological oscillators, SIAM J. Appl. Math. **50**(6), 1645–1662 (1990)

66.56   Y.-W. Hong, A. Scaglione: A scalable synchronization protocol for large scale sensor networks and its applications, IEEE J. Sel. Areas Commun. **23**(5), 1085–1099 (2005)

66.57   O. Babaoglu, T. Binci, M. Jelasity, A. Montresor: Firefly-inspired heartbeat synchronization in overlay networks, Proc. SASO 2007 – 1st Int. Conf. Self-Adapt. Self-Organ. Syst. (2007) pp. 77–86

66.58   J. Jumadinova, P. Dasgupta: Firefly-inspired synchronization for improved dynamic pricing in online markets, Proc. SASO 2008 – 2nd IEEE Int. Conf. Self-Adapt. Self-Organ. Syst. (2008) pp. 403–412

66.59   X.S. Yang: *Nature Inspired Metaheuristic Algorithms* (Luniver, UK 2010)

66.60   G.K. Jati, S. Suyanto: Evolutionary discrete firefly algorithm for travelling salesman problem, Lect. Notes Comput. Sci. **6943**, 393–403 (2011)

66.61   C.-R. Wang, C.-L. Zhou, J.-W. Ma: An improved artificial fish-swarm algorithm and its application in feed-forward neural networks, Proc. 2005 Int. Conf. Mach. Learn. Cybern., Vol. 5 (2005) pp. 2890–2894

66.62   M. Jiang, Y. Wang, S. Pfletschinger, M.A. Lagunas, D. Yuan: Optimal multiuser detection with artificial fish swarm algorithm. In: *Advanced Intelli-*

gent Computing Theories and Applications. With Aspects of Contemporary Intelligent Computing Techniques, Communications in Computer and Information Science, Vol. 2, ed. by D.-S. Huang, L. Heutte, M. Loog (Springer, Berlin, Heidelberg 2007) pp. 1084–1093

66.63   M. Jiang, N.E. Mastorakis, D. Yuan, M.A. Lagunas: Image segmentation with improved artificial fish swarm algorithm, Proc. Eur. Comput. Conf., Lect. Notes Electr. Eng., Vol. 28, ed. by N. Mastorakis, V. Mladenov, V.T. Kontargyri (2009) pp. 133–138

66.64   A.M.A.C. Rocha, T.F.M.C. Martins, E.M.G.P. Fernandes: An augmented Lagrangian fish swarm based method for global optimization, J. Comput. Appl. Math. **235**(16), 4611–4620 (2011)

66.65   C.J.A.B. Filho, F.B. de Lima Neto, A.J.C.C. Lins, A.I.S. Nascimento, M.P. Lima: A novel search algorithm based on fish school behavior, Proc. SMC 2008 – IEEE Int. Conf. Syst. Man Cybern. (2008) pp. 2646–2651

66.66   K.D. Wells: The social behaviour of anuran amphibians, Anim. Behav. **25**, 666–693 (1977)

66.67   I. Aihara, H. Kitahata, K. Yoshikawa, K. Aihara: Mathematical modeling of frogs' calling behavior and its possible application to artificial life and robotics, Artif. Life Robot. **12**(1), 29–32 (2008)

66.68   S.A. Lee, R. Lister: Experiments in the dynamics of phase coupled oscillators when applied to graph coloring, ACSC 2008 – Proc. 31st Australas. Conf. Comput. Sci., Darlinghurst (2008) pp. 83–89

66.69   H. Hernández, C. Blum: Distributed graph coloring: An approach based on the calling behavior of Japanese tree frogs, Swarm Intell. **6**, 117–150 (2012)

66.70   E. Boneabeau, G. Theraulaz, J.-L. Deneubourg, N.-R. Franks, O. Rafelsberger, J.L. Joly, S. Blanco: A model for the emergence of pillars, walls and royal chambers in termite nests, Philos. Trans. R. Soc. **353**(1375), 1561–1576 (1997)

66.71   Z. Mason: Programming with stigmergy: Using swarms for construction, Proc. Artif. Life VIII – 8th Int. Conf. Artif. Life (2003) pp. 371–374

66.72   R.L. Stewart, R.A. Russell: A distributed feedback mechanism to regulate wall construction by a robotic swarm, Adapt. Behav. **14**(1), 21–51 (2006)

66.73   A. Grushin, J.A. Reggia: Stigmergic self-assembly of prespecified artificial structures in a constrained and continuous environment, J. Integr. Comput.-Aided Eng. **13**(4), 289–312 (2006)

66.74   E. Bonabeau, S. Guerin, D. Snyers, P. Kuntz, G. Theraulaz: Three-dimensional architectures grown by simple *stigmergic* agents, Biosystems **56**(1), 13–32 (2000)

66.75   W.G. Walter: *The Living Brain* (W. W. Norton, New York 1953)

66.76   J. von Neumann: The general and logical theory of automata, Cerebral Mechanisms in Behavior: The Hixon Symposium, ed. by L.A. Jeffress (Wiley, New York 1951) pp. 1–41

66.77   L.S. Penrose, R. Penrose: A self-reproducing analogue, Nature **179**(4571), 1183 (1957)

66.78   H. Jacobson: On models of reproduction, Am. Sci. **46**, 255–284 (1958)

66.79   T. Fukuda, S. Nakagawa: A dynamically reconfigurable robotic system (concept of a system and optimal configurations), Proc. 1987 IEEE Int. Conf. Ind. Electron. Control Instrum., Piscataway (1987) pp. 588–595

66.80   T. Fukuda, S. Nakagawa: Dynamically reconfigurable robotic system, Proc. 1988 IEEE Int. Conf. Robot. Autom., Piscataway, Vol. 3 (1988) pp. 1581–1586

66.81   T. Fukuda, S. Nakagawa: Approach to the dynamically reconfigurable robotic system, J. Intell. Robot. Syst. **1**(1), 55–72 (1988)

66.82   J. Wang, G. Beni: Pattern generation in cellular robotic systems, Proc. 3rd IEEE Int. Symp. Intell. Control (IEEE, Piscataway 1988) pp. 63–69

66.83   G. Beni, J. Wang: Swarm intelligence in cellular robotic systems, Proc. NATO Adv. Workshop Robot. Biol. Syst., Il Ciocco (1989) pp. 703–712

66.84   G. Beni, J. Wang: Swarm intelligence, Proc. 7th Annu. Meet. Robot. Soc. Japan, RSJ, Tokyo (1989) pp. 425–428, in Japanese

66.85   T. Fukuda, S. Nakagawa, Y. Kawauchi, M. Buss: Self organizing robots based on cell structures – CEBOT, Proc. 1988 IEEE Int. Workshop Intell. Robot., Piscataway (1988) pp. 145–150

66.86   H. Asama, A. Matsumoto, Y. Ishida: Design of an autonomous and distributed robot system: ACTRESS, Proc. 1989 IEEE/RSJ Int. Workshop Intell. Robot. Syst., Piscataway (1989) pp. 283–290

66.87   P. Caloud, W. Choi, J.-C. Latombe, C. Le Pape, M. Yim: Indoor automation with many mobile robots, Proc. 1990 IEEE Int. Workshop Intell. Robot. Syst., Piscataway, Vol. 1 (1990) pp. 67–72

66.88   M. Rubenstein, A. Cornejo, R. Nagpal: Programmable self-assembly in a thousand-robot swarm, Science **345**, 795–799 (2014)

66.89   J. McLurkin, A.J. Lynch, S. Rixner, T.W. Barr, A. Chou, K. Foster, S. Bilstein: A low-cost multi-robot system for research, teaching, and outreach, Proc. 10th Int. Symp. Distrib. Auton. Robot. Syst. (DARS 2010), ed. by A. Martinoli, F. Mondada, N. Correll, G. Mermoud, M. Egerstedt, M. Ani Hsieh, L.E. Parkes, K. Støy (2010) pp. 597–609

66.90   T. Schmickl, R. Thenius, C. Möslinger, J. Timmis, A. Tyrrell, M. Read, J. Hilder, J. Halloy, A. Campo, C. Stefanini, L. Manfredi, S. Orofino, S. Kernbach, T. Dipper, D. Sutantyo: Cocoro: The self-aware underwater swarm, 2011 Firth IEEE Conf. Self-Adapt. Self-Organ. Syst. Workshop, SASOW (2011) pp. 120–126

66.91   M. Dorigo, D. Floreano, L.M. Gambardella, F. Mondada, S. Nolfi, T. Baaboura, M. Birattari, M. Bonani, M. Brambilla, A. Brutschy, D. Burnier, A. Campo, A.L. Christensen, A. Decugnière, G.A. Di Caro, F. Ducatelle, E. Ferrante, A. Förster, J. Guzzi,

V. Longchamp, S. Magnenat, J. Martinez Gonzalez, N. Mathews, M.A. de Montes Oca, R. O'Grady, C. Pinciroli, G. Pini, P. Rétornaz, J. Roberts, V. Sperati, T. Stirling, A. Stranieri, T. Stuetzle, V. Trianni, E. Tuci, A.E. Turgut, F. Vaussard: Swarmanoid: A novel concept for the study of heterogeneous robotic swarms, IEEE Robot. Autom. Mag. **20**(4), 60–71 (2013)

66.92   K. Gilpin, A. Knaian, D. Rus: Robot pebbles: One centimeter modules for programmable matter through self-disassembly, IEEE Int. Conf. Robot. Auton., ICRA (2010) pp. 2485–2492

66.93   R. Vaughan: Massively multi-robot simulation in stage, Swarm Intell. **2**(2–4), 189–208 (2008)

66.94   R. Fitch, Z. Butler: Million module march: Scalable locomotion for large self-reconfiguring robots, Int. J. Robot. Res. **27**(3–4), 331–343 (2008)

66.95   M.P. Ashley-Rollman, P. Pillai, M.L. Goodstein: Simulating multi-million-robot ensembles, Proc. 2011 IEEE Int. Conf. Robotic. Autom. (2011) pp. 1006–1013

66.96   F. Mondada, E. Franzi, A. Guignard: The Development of Khepera, Exp. Mini-Robot Khepera, Proc. 1st Int. Khepera Workshop (HNI-Verlagsschriftenreihe, Heinz Nixdorf Institut 1999) pp. 7–14

66.97   F. Mondada, M. Bonani, X. Raemy, J. Pugh, C. Cianci, A. Klaptocz, S. Magnenat, J.-C. Zufferey, D. Floreano, A. Martinoli: The e-puck, a robot designed for education in engineering, Proc. 9th Conf. Mobile Robot. Compet., ROBOTICA 2009, Castelo Branco (2009) pp. 59–65

66.98   N. Kottege, U.R. Zimmer: Underwater acoustic localization for small submersibles, J. Field Robot. **28**(1), 40–69 (2011)

66.99   J.F. Roberts, T. Stirling, J.-C. Zufferey, D. Floreano: 3-D relative positioning sensor for indoor flying robots, Auton. Robot. **33**(1–2), 5–20 (2012)

66.100   A. Kushleyev, D. Mellinger, C. Powers, V. Kumar: Towards a swarm of agile micro quadrotors, Auton. Robot. **35**(4), 287–300 (2013)

66.101   L. Chaimowicz, V. Kumar: Aerial shepherds: Coordination among UAVs and swarms of robots, 7th Int. Symp. Distrib. Auton. Robot. Syst. (2004) pp. 23–25

66.102   F. Mondada, L.M. Gambardella, D. Floreano, S. Nolfi, J.-L. Deneubourg, M. Dorigo: The cooperation of swarm-bots: Physical interactions in collective robotics, IEEE Robot. Autom. Mag. **12**(2), 21–28 (2005)

66.103   S. Kernbach, E. Meister, F. Schlachter, K. Jebens, M. Szymanski, J. Liedke, D. Laneri, L. Winkler, T. Schmickl, R. Thenius, P. Corradi, L. Ricotti: Symbiotic robot organisms: REPLICATOR and SYMBRION projects, Proc. 8th Workshop Perform. Metr. Intell. Syst. (2008) pp. 62–69

66.104   H. Wei, Y. Chen, J. Tan, T. Wang: Sambot: A self-assembly modular robot system, IEEE/ASME Trans. Mechatr. **16**(4), 745–757 (2011)

66.105   M. Yim, W.-M. Shen, B. Salemi, D. Rus, M. Moll, H. Lipson, E. Klavins, G.S. Chirikjian: Modular self-reconfigurable robot systems, IEEE Robot. Autom. Mag. **14**(1), 43–52 (2007)

66.106   K. Støy, D. Brandt, D.J. Christensen: *Self-Reconfigurable Robots: An Introduction* (MIT, Cambridge 2010)

66.107   G.M. Whitesides, B. Grzybowski: Self-assembly at all scales, Science **295**(5564), 2418–2421 (2002)

66.108   G.M. Whitesides, M. Boncheva: Beyond molecules: Self-assembly of mesoscopic and macroscopic components, Proc. Natl. Acad. Sci. USA **99**(8), 4769–4774 (2002)

66.109   R. Groß, M. Dorigo: Self-assembly at the macroscopic scale, Proc. IEEE **96**(9), 1490–1508 (2008)

66.110   G. Dudek, M. Jenkin, E. Milios: A taxonomy of multirobot systems. In: *Robot teams: From Diversity to Polymorphism*, ed. by T. Balch, L.E. Parker (A. K. Peters, Natick 2002) pp. 3–22

66.111   Y.U. Cao, A.S. Fukunaga, A.B. Kahng: Cooperative mobile robotics: Antecedents and directions, Auton. Robot. **4**(1), 7–27 (1997)

66.112   S. Goss, J.-L. Deneubourg: Harvesting by a group of robots, Proc. 1st Eur. Conf. Artif. Life (MIT, Cambridge 1992) pp. 195–204

66.113   R. Beckers, O. Holland, J.-L. Deneubourg: From local actions to global tasks: Stigmergy and collective robotics, Proc. 4th Int. Workshop Synth. Simul. Living Syst. (Artificial Life IV) (MIT, Cambridge 1994) pp. 181–189

66.114   A. Martinoli: Swarm Intelligence in Autonomous Collective Robotics: From Tools to the Analysis and Synthesis of Distributed Control Strategies, Ph.D. Thesis (EPFL, Lausanne, Switzerland 1999)

66.115   I.A. Wagner, Y. Altshuler, V. Yanovski, A.M. Bruckstein: Cooperative cleaners: A study in ant robotics, Int. J. Robot. Res. **27**(1), 127–151 (2008)

66.116   J. Werfel, K. Petersen, R. Nagpal: Designing collective behavior in a termite-inspired robot construction team, Science **343**(6172), 754–758 (2014)

66.117   D.J. Stilwell, J.S. Bay: Toward the development of a material transport system using swarms of ant-like robots, Proc. 1993 IEEE Int. Conf. Robot. Autom. 1 (1993) pp. 766–771

66.118   S. Sen, M. Sekaran, J. Hale: Learning to coordinate without sharing information, Proc. 12th Natl. Conf. Artif. Intell., AAAI'94, Menlo Park (1994) pp. 426–431

66.119   R. Groß, M. Dorigo: Evolution of solitary and group transport behaviors for autonomous robots capable of self-assembling, Adapt. Behav. **16**(5), 285–305 (2008)

66.120   M.R.A. Chance, J.J. Clifford: *Social Groups of Monkeys, Apes and Men* (E. P. Dutton, New York 1970)

66.121   I.D. Couzin, J. Krause, N.R. Franks, S.A. Levin: Effective leadership and decision-making in animal groups on the move, Nature **433**(7025), 513–516 (2005)

66.122   R.C. Arkin: Cooperation without communication: Multiagent schema-based robot navigation, J. Robot. Syst. **9**(3), 351–364 (1992)

66.123   M.J. Matarić: Designing emergent behaviors: From local interactions to collective intelligence, Proc. 2nd Int. Conf. Simul. Adapt. Behav. (MIT, Cambridge 1992) pp. 432–441

66.124   M.J. Matarić: Minimizing complexity in controlling a mobile robot population, Proc. 1992 IEEE Int. Conf. Robot. Autom., Piscataway (1992) pp. 830–835

66.125   Y. Kuniyoshi, N. Kita, S. Rougeaux, S. Sakane, M. Ishii, M. Kakikura: Cooperation by observation: The framework and basic task patterns, Proc. 1994 IEEE Int. Conf. Robot. Autom., Piscataway (1994) pp. 767–774

66.126   B.P. Gerkey, M.J. Matarić: Sold!: Auction methods for multirobot coordination, IEEE Trans. Robot. Autom. **18**(5), 758–768 (2002)

66.127   V. Trianni, M. Dorigo: Self-organisation and communication in groups of simulated and physical robots, Biol. Cybern. **95**(3), 213–231 (2006)

66.128   T. Balch, R.C. Arkin: Communication in reactive multiagent robotic systems, Auton. Robot. **1**(1), 27–52 (1994)

66.129   M. Rubenstein, K. Payne, P. Will, W.-M. Shen: Docking among independent and autonomous CONRO self-reconfigurable robots, Proc. 2004 IEEE Int. Conf. Robot. Autom. 3 (2004) pp. 2877–2882

66.130   M. Brambilla, E. Ferrante, M. Birattari, M. Dorigo: Swarm robotics: A review from the swarm engineering perspective, Swarm Intell. **7**(1), 1–41 (2013)

66.131   R.A. Brooks: Intelligence without representation, Artif. Intell. **47**(1–3), 139–159 (1991)

66.132   S. Berman, Á. Halász, M.A. Hsieh, V. Kumar: Optimized stochastic policies for task allocation in swarms of robots, IEEE Trans. Robot. **25**(4), 927–937 (2009)

66.133   S. Nouyan, R. Groß, M. Bonani, F. Mondada, M. Dorigo: Teamwork in self-organized robot colonies, IEEE Trans. Evol. Comput. **13**(4), 695–711 (2009)

66.134   W. Liu, A.F.T. Winfield: Modeling and optimization of adaptive foraging in swarm robotic systems, Int. J. Robot. Res. **29**(14), 1743–1760 (2010)

66.135   E. Bonabeau, G. Theraulaz, J.-L. Deneubourg: Quantitative study of the fixed threshold model for the regulation of division of labour in insect societies, Proc. R. Soc. B **263**(1376), 1565–1569 (1996)

66.136   G. Theraulaz, E. Bonabeau, J.-L. Deneubourg: Response threshold reinforcements and division of labour in insect societies, Proc. R. Soc. B **265**(1393), 327–332 (1998)

66.137   T.H. Labella, M. Dorigo, J.-L. Deneubourg: Division of labor in a group of robots inspired by ants' foraging behavior, ACM Trans. Auton. Adapt. Syst. **1**(1), 4–25 (2006)

66.138   L.E. Parker, F. Tang: Building multirobot coalitions through automated task solution synthesis, Proc. IEEE **94**(7), 1289–1305 (2006)

66.139   G.A. Korsah, A. Stentz, M. Bernardine Dias: A comprehensive taxonomy for multi-robot task allocation, Int. J. Robot. Res. **32**(12), 1495–1512 (2013)

66.140   L.E. Parker: ALLIANCE: An architecture for fault-tolerant multi-robot cooperation, IEEE Trans. Robot. Autom. **14**(2), 220–240 (1998)

66.141   L.E. Parker: Adaptive heterogeneous multi-robot teams, Neurocomputing **28**(1–3), 75–92 (1999)

66.142   O. Khatib: Real-time obstacle avoidance for manipulators and mobile robots, Int. J. Robot. Res. **5**(1), 90–98 (1986)

66.143   J.H. Reif, H. Wang: Social potential fields: A distributed behavioral control for autonomous robots, Robot. Auton. Syst. **27**(3), 171–194 (1999)

66.144   W.M. Spears, D.F. Spears (Eds.): *Physicomimetics: Physics-Based Swarm Intelligence* (Springer, Berlin, Heidelberg 2011)

66.145   W.M. Spears, D.F. Spears: Distributed, physics-based control of swarms of vehicles, Auton. Robot. **17**(2–3), 137–162 (2004)

66.146   A. Kolling, S. Carpin: Cooperative observation of multiple moving targets: An algorithm and its formalization, Int. J. Robot. Res. **26**(9), 935–953 (2007)

66.147   V. Gazi, K.M. Passino: Stability analysis of social foraging swarms, IEEE Trans. Syst. Man Cybern. Part B Cybern. **34**(1), 539–557 (2004)

66.148   D. Coore: Botanical Computing: A Developmental Approach to Generating Interconnect Topologies on an Amorphous Computer, Ph.D. Thesis (MIT, Cambridge 1999)

66.149   R. Nagpal: Programmable self-assembly using biologically-inspired multiagent control, Proc 1st Int. Joint Conf. Auton. Agents Multiagent Syst.: Part 1, AAMAS '02, New York (2002) pp. 418–425

66.150   J. Beal, J. Bachrach: Infrastructure for engineered emergence on sensor/actuator networks, IEEE Intell. Syst. **21**(2), 10–19 (2006)

66.151   H. Abelson, D. Allen, D. Coore, C. Hanson, G. Homsy, T.F. Knight Jr., R. Nagpal, E. Rauch, G.J. Sussman, R. Weiss: Amorphous computing, Commun. ACM **43**(5), 74–82 (2000)

66.152   J. Bachrach, J. Beal, J. McLurkin: Composable continuous-space programs for robotic swarms, Neural Comput. Appl. **19**(6), 825–847 (2010)

66.153   R.S. Sutton, A.G. Barto: *Reinforcement Learning: An Introduction* (MIT, Cambridge 1998)

66.154   M.J. Matarić: Reward functions for accelerated learning, Proc. 11th Int. Conf. Mach. Learn. (Morgan Kaufmann, San Francisco 1994) pp. 181–189

66.155   M.J. Matarić: Learning social behavior, Robot. Auton. Syst. **20**(2–4), 191–204 (1997)

66.156   J. Kober, J.A. Bagnell, J. Peters: Reinforcement learning in robotics: A survey, Int. J. Robot. Res. **32**(11), 1238–1274 (2013)

66.157   I. Harvey, P. Husbands, D. Cliff, A. Thompson, N. Jakobi: Evolutionary robotics: The Sussex approach, Robot. Auton. Syst. **20**(2–4), 205–224 (1997)

66.158   S. Nolfi, D. Floreano: *Evolutionary Robotics – The Biology, Intelligence, and Technology of Self-Organizing Machines* (MIT, Cambridge 2000)

66.159   M. Dorigo, V. Trianni, E. Şahin, R. Groß, T.H. Labella, G. Baldassarre, S. Nolfi, J.-L. Deneubourg, F. Mondada, D. Floreano, L.M. Gambardella: Evolving self-organizing behaviors for a *swarm-bot*, Auton. Robot. **17**(2–3), 223–245 (2004)

66.160   V. Trianni: Evolutionary Swarm Robotics: Evolving Self-Organising Behaviours in Groups of Autonomous Robots. In: *Studies in Computational Intelligence*, Vol. 108, (Springer, Berlin, Heidelberg 2008)

66.161   C.W. Reynolds: An evolved, vision-based behavioral model of coordinated group motion, From Animals to Animats 2. Proc. 2nd Int. Conf. Simul. Adapt. Behav. (SAB92) (MIT, Cambridge 1993) pp. 384–392

66.162   G.M. Werner, M.G. Dyer: Evolution of herding behavior in artificial animals, From Animals to Animats 2. Proc. 2nd Int. Conf. Simul. Adapt. Behav. (SAB92) (MIT, Cambridge 1993) pp. 393–399

66.163   L. Spector, J. Klein, C. Perry, M. Feinstein: Emergence of collective behavior in evolving populations of flying agents, Genet. Program. Evol. Mach. **6**(1), 111–125 (2005)

66.164   M. Quinn, L. Smith, G. Mayley, P. Husbands: Evolving controllers for a homogeneous system of physical robots: Structured cooperation with minimal sensors, Philos. Trans. R. Soc. A **361**(1811), 2321–2343 (2003)

66.165   R.A. Watson, S.G. Ficici, J.B. Pollack: Embodied evolution: Distributing an evolutionary algorithm in a population of robots, Robot. Auton. Syst. **39**(1), 1–18 (2002)

66.166   S. Griffith, D. Goldwater, J.M. Jacobson: Self-replication from random parts, Nature **437**(7059), 636 (2005)

66.167   A.F.T. Winfield, M. Dincer Erbas: On embodied memetic evolution and the emergence of behavioural traditions in robots, Memet. Comput. **3**(4), 261–270 (2011)

66.168   M. Yim: Locomotion with a Unit-Modular Reconfigurable Robot, Ph.D. Thesis (Dept. Mech. Eng., Stanford Univ., Stanford 1994)

66.169   W.-M. Shen, B. Salemi, P. Will: Hormone-inspired adaptive communication and distributed control for CONRO self-reconfigurable robots, IEEE Trans. Robot. Autom. **18**(5), 700–712 (2002)

66.170   K. Støy: Emergent Control of Self-Reconfigurable Robots, Ph.D. Thesis (The Maersk Mc-Kinney Moller Institute for Production Technology, Univ. Southern Denmark, Denmark 2004)

66.171   G. Chirikjian, A. Pamecha: Evaluating efficiency of self-reconfiguration in a class of modular robots, J. Robot. Syst. **13**(5), 317–338 (1996)

66.172   E. Yoshida, S. Murata, A. Kamimura, K. Tomita, H. Kurokawa, S. Kokaji: A motion planning method for a self-reconfigurable modular robot, Proc. 2001 IEEE/RSJ Int. Conf. Intell. Robot. Syst., Piscataway, Vol. 1 (2001) pp. 590–597

66.173   D. Brandt, D.J. Christensen: A new meta-module for controlling large sheets of ATRON modules, Proc. 2007 IEEE Int. Workshop Intell. Robot. Syst. (2007) pp. 2375–2380

66.174   Z. Butler, D. Rus: Distributed planning and control for modular robots with unit-compressible modules, Int. J. Robotic. Res. **22**(9), 699–715 (2003)

66.175   S. Murata, H. Kurokawa, S. Kokaji: Self-assembling machine, Proc. 1994 IEEE Int. Conf. Robot. Autom. 1 (1994) pp. 441–448

66.176   M.D. Rosa, S. Goldstein, P. Lee, J. Campbell, P. Pillai: Scalable shape sculpturing via hole motion: Motion planning in lattice-constrained modular robots, Proc. 2006 IEEE Int. Conf. Robot. Autom. (2006) pp. 1462–1468

66.177   Z. Butler, K. Kotay, D. Rus, K. Tomita: Generic decentralized control for a class of self-reconfigurable robots, Proc. 2002 IEEE Int. Conf. Robot. Autom., Piscataway, Vol. 1 (2002) pp. 809–816

66.178   K. Hosokawa, T. Tsujimori, T. Fujii, H. Kaetsu, H. Asama, Y. Kuroda, I. Endo: Self-organizing collective robots with morphogenesis in a vertical plane, Proc. 1998 IEEE Int. Conf. Robot. Autom., Piscataway, Vol. 4 (1998) pp. 2858–2863

66.179   M. Yim, Y. Zhang, J. Lamping, E. Mao: Distributed control for 3D metamorphosis, Auton. Robot. **10**(1), 41–56 (2001)

66.180   K. Støy: Using cellular automata and gradients to control self-reconfiguration, Robot. Auton. Syst. **54**(2), 135–141 (2006)

66.181   M. Gauci, J. Chen, W. Li, T.J. Dodd, R. Groß: Self-organised aggregation without computation, Int. J. Robot. Res. **33**, 1145–1161 (2014)

66.182   J. Halloy, G. Sempo, G. Caprari, C. Rivault, M. Asadpour, F. Tche, I. Sad, V. Durier, S. Canonge, J.M. Am, C. Detrain, N. Correll, A. Martinoli, F. Mondada, R. Siegwart, J.L. Deneubourg: Social integration of robots into groups of cockroaches to control self-organized choices, Science **318**(5853), 1155–1158 (2007)

66.183   A.E. Turgut, H. Çelikkanat, F. Gökçe, E. Şahin: Self-organized flocking in mobile robot swarms, Swarm Intell. **2**(2–4), 97–120 (2008)

66.184   M.J.B. Krieger, J.-B. Billeter, L. Keller: Ant-like task allocation and recruitment in cooperative robots, Nature **406**(6799), 992–995 (2000)

66.185   R. Groß, M. Bonani, F. Mondada, M. Dorigo: Autonomous self-assembly in swarm-bots, IEEE Trans. Robot. **22**(6), 1115–1130 (2006)

66.186 O. Holland, C. Melhuish: Stigmergy, self-organization, and sorting in collective robotics, Artif. Life **5**(2), 173–202 (1999)

66.187 J. Chen, M. Gauci, W. Li, A. Kolling, R. Groß: Occlusion-based cooperative transport with a swarm of miniature mobile robots, IEEE Trans. Robotic. (in press)

66.188 A.J. Ijspeert, A. Martinoli, A. Billard, L.M. Gambardella: Collaboration through the exploitation of local interactions in autonomous collective robotics: The stick pulling experiment, Auton. Robot. **11**(2), 149–171 (2001)

66.189 S. Garnier, C. Jost, J. Gautrais, M. Asadpour, G. Caprari, R. Jeanson, A. Grimal, G. Theraulaz: The embodiment of cockroach aggregation behavior in a group of micro-robots, Artif. Life **14**(4), 387–408 (2008)

66.190 R. Jeanson, C. Rivault, J.-L. Deneubourg, S. Blanco, R. Fournier, C. Jost, G. Theraulaz: Self-organized aggregation in cockroaches, Anim. Behav. **69**, 169–180 (2005)

66.191 J. Krause, A.F.T. Winfield, J.-L. Deneubourg: Interactive robots in experimental biology, Trends Ecol. Evol. **26**(7), 369–375 (2011)

66.192 R. Vaughan, N. Sumpter, J. Henderson, A. Frost, S. Cameron: Experiments in automatic flock control, Robot. Auton. Syst. **31**(1/2), 109–117 (2000)

66.193 A. Gribovskiy, J. Halloy, J.-L. Deneubourg, H. Bleuler, F. Mondada: Towards mixed societies of chickens and robots, Proc. 2010 IEEE Int. Workshop Intell. Robot. Syst. (2010) pp. 4722–4728

66.194 A.L. Christensen, R. O'Grady, M. Dorigo: Swarmorph-script: A language for arbitrary morphology generation in self-assembling robots, Swarm Intell. **2**(2–4), 143–165 (2008)

66.195 C.R. Kube, H. Zhang: Collective robotics: From social insects to robots, Adapt. Behav. **2**(2), 189–218 (1993)

66.196 C.R. Kube, E. Bonabeau: Cooperative transport by ants and robots, Robot. Auton. Syst. **30**(1–2), 85–101 (2000)

66.197 C. Blum, J. Puchinger, G. Raidl, A. Roli: Hybrid metaheuristics in combinatorial optimization: A survey, Appl. Soft Comput. **11**(6), 4135–4151 (2011)

67. Preference–Based Multiobjective Particle Swarm Optimization for Airfoil Design

Robert Carrese, Xiaodong Li

A significant challenge to the application of evolutionary multiobjective optimization (EMO) for transonic airfoil design is the often excessive number of computational fluid dynamic (CFD) simulations required to ensure convergence. In this study, a multiobjective particle swarm optimization (MOPSO) framework is introduced, which incorporates designer preferences to provide further guidance in the search. A reference point is projected onto the Pareto landscape by the designer to guide the swarm towards solutions of interest. The framework is applied to a typical transonic airfoil design scenario for robust aerodynamic performance. Time-adaptive Kriging models are constructed based on a high-fidelity Reynolds-averaged Navier–Stokes (RANS) solver to assess the performance of the solutions. The successful integration of these design tools is facilitated through the reference point, which ensures that the swarm does not deviate from the preferred search trajectory. A comprehensive discussion on the proposed optimization framework is provided, highlighting its viability for the intended design application.

67.1 Airfoil Design

Airfoil design originates from an understanding of the fundamental physics of flight, where the aim is to identify or conform to the best possible shape for the given operating requirements. It has evolved from the use of wind tunnel catalogs and traditional *cut-and-try* methods to automated computational frameworks. While automated frameworks effectively simplify the design process, success is still largely dependent on the fidelity of the computational methods, as well as the experience of the designer in formulating the problem [67.1]. This section is devoted to a discussion

of airfoil design optimization architecture. The concepts that are especially applicable to this study are introduced, laying the foundations for the proposed methodology.

67.1.1 Airfoil Design Architecture

The direct method of airfoil design, pioneered by the work of *Hicks* and *Henne* [67.2], refers to the philosophy of using mathematical optimization methods to identify the optimal shape that achieves the prescribed

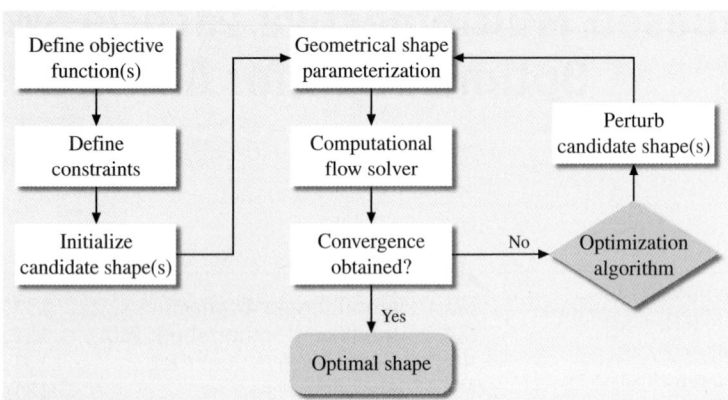

Fig. 67.1 Generalized process flowchart for direct airfoil shape optimization

design criteria. The generalized framework for an aerodynamic shape optimization process is demonstrated in Fig. 67.1. The success of the direct approach is essentially dependent on three main components within the design loop:

- *Shape parameterization.* All design strategies share the common requirement that the geometry is represented by a finite number of design variables. A method to mathematically parameterize shapes is, therefore, required so that modifications can be made via direct manipulation of the design variables. The number of design variables is directly proportional to the geometrical degrees of freedom and, therefore, governs the dimensionality of the problem.
- *Computational flow solver.* The objective function is obtained from the flow solver and it is, therefore, up to the discretion of the designer to appropriately formulate the objective and constraint functions, such that they reflect the design and operating requirements. The choice of the flow solver ultimately governs the overall fidelity and efficiency of the optimization process, since repeated evaluations of the objective function are required for each candidate shape.
- *Optimization algorithm.* The responsibility of the optimizer is to iteratively determine the shape modifications required to satisfy the objective, whilst adhering to any shape or performance constraints. The optimizer should be robust and applicable to a wide operational spectrum, yet efficient to guarantee convergence with the least computational expense.

The integration of high-fidelity flow solvers and flexible parameterization methods for numerical op-

timization is still a computationally challenging and intensive undertaking. The extension to multiple objectives leads to a more generalized problem formulation, yet significantly increases the computational cost of convergence. While all elements of the design loop influence the efficiency of the process, arguably the most important element is the optimizer itself. The following section introduces the optimization paradigm adopted in this study, derived from the field of computational swarm intelligence.

67.1.2 Intelligent Optimization: PSO

The formation of hierarchies within groups of animals is a naturally occurring phenomenon and is simple to comprehend. Even humans have the intuitive tendency to appoint leaders (e.g., political leaders, military generals, etc.). Another interesting phenomenon, which is more difficult to perceive, is the self-organized behavior of groups where a leader *cannot* be identified. This is known as swarming and is evident from the flocking behavior of birds or fish moving in unison. The increasingly cited field of computational swarm intelligence focuses on the artificial simulation of swarming behavior to model a wide range of applications, including optimization [67.3].

Particle swarm optimization (PSO) is the stochastic population-based technique described by *Kennedy* and *Eberhart* [67.4] in accordance with the principles of swarm intelligence. The PSO architecture was derived from a synthesis of the fields of social psychology and engineering optimization. As was eloquently stated by the authors in their original paper [67.4]:

Why is social behavior so ubiquitous in the animal kingdom? Because it optimizes. What is a good way to solve engineering optimization problems? Modeling social behavior.

The dynamics of the swarm are modeled on the social-psychological tendency of individuals to learn from previous experience and emulate the success of others. Similar to most evolutionary techniques, the swarm is initialized with a population of random individuals (particles) sampled over the design space. The particles navigate the multi-dimensional design space over a number of iterations or time steps. Each particle maintains knowledge of its current position in the design space. This is analogous to the *fitness* concept of conventional evolutionary algorithms (EAs). Each particle also records its personal best position, which is where the particle has experienced the greatest success. Aside from recording personal information, each particle also tracks the position of other members in the swarm. This level of social interaction between particles is coined the swarm *topology*. Particles may either be confined to share information only with their immediate neighbors, or they may be encouraged to share their experiences with the entire swarm. Utilizing this information, each particle adjusts its position in the design space by *accelerating* towards the successful areas of the design space. The absence of selection is compensated by this use of leaders to guide the swarm to converge to the most successful position. In this way, a solution which initially performs poorly may possibly be on the future road to success.

PSO has steadily gained popularity as a global optimization technique [67.3]. Its increasing use in the literature is due to its simple and straightforward implementation (despite its intricate origins) and its efficient and accurate convergence rates [67.5].

67.1.3 Multiobjective Optimization

Airfoil design problems are often characterized by several interacting or conflicting requirements, which must be satisfied simultaneously. In the case of an airfoil operating within the transonic regime, airfoil shape optimization is performed to limit shock and viscous drag (C_d) losses, and reduce shock-induced boundary layer instability at the design Mach number (M) and lift coefficient (C_l). This often occurs at the expense of excessive pitching moments (C_m) due to aft loading and performance degradation under off-design conditions. To facilitate adequate performance over a wide operational spectrum requires a search algorithm capable of handling multiple conflicting objectives.

Let $S \in \mathbb{R}^n$ denote the design space and let $x = \{x_1, x_2, \ldots, x_n\} \in S$ denote the decision vector with lower and upper bounds x_{min} and x_{max}, respectively. The generic unconstrained multiobjective problem (MOP) is thus expressed as,

$$\min f(x) = \{f_1(x), \ldots, f_m(x)\} , \qquad (67.1)$$

where $f_i(x) : \mathbb{R}^n \to \mathbb{R}$ is the i-th component of the objective vector and m is the number of objectives. The definition of the optimum must be redefined since in the presence of conflicting objectives, improvement in one objective may cause a deterioration in another. It is often necessary to identify a set of trade-off solutions, which can all be considered equally optimal. A solution is termed non-dominated or Pareto optimal (after the nineteenth century Italian economist Vilfredo Pareto) if the value of any objective cannot be improved without deteriorating at least one other objective. The candidate solutions are defined as a and $b \in S$. The candidate a *dominates* the candidate b (denoted by $a \prec b$) if,

$$\forall j = 1, \ldots, m \quad f_j(a) \leq f_j(b) \wedge \exists j : f_j(a) < f_j(b) . \qquad (67.2)$$

The concept of dominance is illustrated in Fig. 67.2. The shaded area denotes the area of objective vectors dominated by a. A decision vector a^* is, therefore, non-dominated or Pareto optimal if there is no other feasible decision vector $a \neq a^* \in S$ such that $f(a) \prec f(a^*)$. The Pareto front is the set of objective vectors which correspond to all non-dominated solutions. Multiobjective algorithms aim to identify the closest approximation to the true Pareto front, while ensuring a diverse Pareto optimal set.

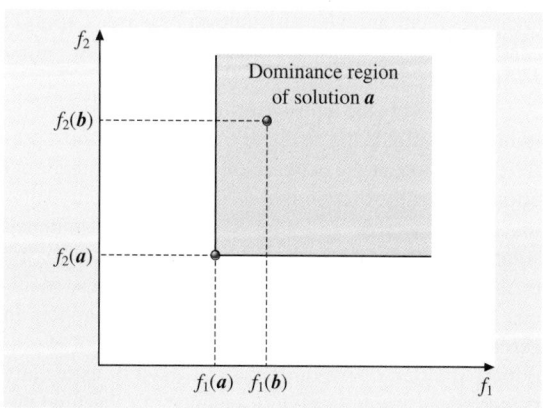

Fig. 67.2 Illustration of dominance on a two-objective landscape

Techniques for Solving MOP

From a design perspective, the primary aim of multiobjective optimization is to obtain Pareto optimal solutions which are in the preferred interests of the designer, or best suit the intended application. Methods for solving MOPs are, therefore, characterized by how the designer preferences are articulated. As suggested by *Fonseca* and *Fleming* [67.6], there are three generic classes of methods for solving multiobjective problems:

- A priori methods. The preferences of the designer are expressed by aggregating the objective functions into a single scalar through weights or bias, ultimately making the problem single objective.
- A posteriori methods. The algorithm first identifies a set of non-dominated solutions, subsequently providing the designer greater flexibility in selecting the most appropriate solution.
- *Interactive* methods. The decision making and optimization processes occur at interleaved steps, and the preferences of the designer are interactively refined.

The a priori strategy requires the designer to indicate the relative importance of each objective before performing the optimization. A notable method that falls into this category is the weighted aggregation method, which is a fairly popular choice for airfoil design applications due to its simplicity and capability of handling many flight conditions [67.7–9]. Despite its popularity, there are recognized deficiencies with this strategy [67.10]. The prior selection of weights does not necessarily guarantee that the final solution will faithfully reflect the preferred interests of the designer, and varying the weights continuously will not necessarily result in an even distribution of Pareto optimal solutions, nor a complete representation of the Pareto front [67.11].

Alternatively the a posteriori methods provide maximum flexibility to the designer to identify the most preferred solution, at the expense of greater computational effort. Generally, these methods involve explicitly solving each objective to obtain a set of non-dominated solutions, a concept which is ideal for population-based evolutionary algorithms [67.12–14]. While these methods are computationally more complex, researchers in aerodynamic design are realizing the benefits of evolutionary multiobjective optimization (EMO), especially if there is a certain ambiguity in selecting the final design [67.15–17]. However, it poses the challenge of identifying and exploiting the entire Pareto front, which

may be impractical for design applications due to the excessive number of function evaluations.

While conventional EMO techniques may be computationally demanding, *Fonseca* and *Fleming* [67.12] argue that their most attractive aspect is the intermediate information generated, which can be exploited by the designer to refine preferences and improve convergence. These *interactive* methods involve the progressive articulation of preferences, which originates from the multicriteria decision making literature [67.18]. The optimization and decision making processes are interleaved, exploiting the intermediate information provided by the optimizer to refine preferences [67.6].

Handling Multiple Objectives with PSO

PSO has been demonstrated to be an effective tool for single-objective optimization problems due to its fast convergence [67.5]. It has also gained rapid popularity in the field of multi-objective optimization (MOO) [67.19]. Since PSO is a population-based technique, it could ideally be tailored to identify a number of trade-off solutions to a MOP in one single run, similar to EMO techniques. Comprehensive surveys on extending PSO to handle multiple objectives have been provided by *Engelbrecht* [67.20], and more recently by *Sierra* and *Coello Coello* [67.19]. It was established that the primary ambiguity in specifically tailoring PSO to handle multiple objectives was the selection of guides for each particle to avoid convergence to a single solution. The selection process for particle leaders must, therefore, be restructured, to encourage search diversity and to ensure that non-dominated solutions found during the search are maintained.

Initial attempts to design a multiobjective particle swarm optimization (MOPSO) algorithm were motivated by the archive strategy by [67.21]. *Coello Coello* and *Lechuga* [67.22] incorporated the concept of Pareto dominance in PSO by maintaining two independent populations: the particle swarm and the elitist archive. Non-dominated solutions are stored in the archive and subsequently used as neighborhood leaders. The objective space is separated into hypercubes, which serve as a particle anti-clustering mechanism. Solutions in sparsely populated hypercubes have a higher selection pressure to be leaders, and solutions in densely populated hypercubes are removed if the archive limit is exceeded. This initial approach was later extended by *Mostaghim* and *Teich* [67.23], who studied the concept of ϵ-dominance and compared it to existing clustering techniques for fixing the archive size, with favorable results.

Fieldsend and *Singh* [67.24] addressed the computational complexity of maintaining a restricted archive, by incorporating the *dominated tree* method. This data structure allows for an unrestricted archive size, which interacts with the population to define global leaders. A *turbulence* operator (similar to the concept of mutation in EA) was also implemented, where swarm members were randomly displaced on the design space to reduce the probability of premature stagnation. In the non-dominated sorting particle swarm optimization (NSPSO) algorithm of *Li* [67.25], the non-dominated sorting mechanisms of non-dominated sorting genetic algorithm (NSGA-II) are incorporated. The population and the personal best position of each particle are consolidated to form one single population, and the non-dominated sorting scheme is utilized to rank each solution. Global guides are selected based on particle clustering, where a niching or crowding distance metric is used to further classify non-dominated solutions. *Li* later proposed the maximinPSO algorithm [67.26], which does not use any niching method to maintain diversity.

Sierra and *Coello Coello* [67.27] proposed an elitist archive incorporating the ϵ-dominance strategy to maintain global leaders for the swarm. A crowding distance operator is employed to classify non-dominated solutions and maintain uniformity. The crowding distance operator is also used to limit the number of candidate leaders after each population update, simplifying the mechanism to control the set of candidate leaders. A turbulence operator is implemented to encourage diversity, whereby particles are randomly mutated. A similar approach by [67.28] was developed in parallel (although this method does not implement ϵ-dominance), where the crowding distance was used to both define the global guides and truncate the size of the archive. The proposed algorithm is primarily influenced by the two latter studies.

Preference-Based Optimization

The concept of interactive optimization has led to an increasing interest in coupling classical interactive methods to EMO as an intuitive way of reflecting the designer preferences and identifying solutions of interest to the designer. This has led to the development of the preference-based optimization philosophy, which provides the motivation for the current study. Comprehensive surveys on preference-based optimization are provided by *Coello Coello* [67.29] and *Rachmawati* and *Srinivasan* [67.30]. The first recorded attempt at incorporating preferences within an evolutionary multiobjective optimization framework was made by *Fonseca* and *Fleming* [67.31] using the goal programming approach. Goal programming [67.11] is an ideal approach to indicate desired levels of performance for each objective, since they closely relate to the final solution of the problem. Goals may either represent target or ideal values. *Fonseca* and *Fleming* later extended the approach where an online decision making strategy was proposed based on goal and priority information [67.6]. A goal programming mechanism for identifying preferred solutions for MOP was also proposed by [67.32]. While the reported frameworks draw on the preferred interests of the designer to aid the optimization process, the goal programming approach is computationally complex, and there is no means of specifying any relation or trade-off between the objectives [67.30].

Thiele et al. [67.33] proposed another variant of interactive evolutionary multiobjective optimization. A coarse representation of the Pareto front is initially presented to the designer. The most interesting regions are subsequently isolated, on which the algorithm continues to focus on exclusively. This proposal effectively removes the necessity to predefine target values for each objective and provides the designer with a means of isolating the preferred trade-offs. However, it is a two-stage approach requiring an initial approximation to the Pareto front, which may be unnecessarily expensive. The integration of other classical preference articulation methods has also been proposed in the literature. A reference point-based evolutionary multiobjective optimization framework was proposed by [67.34]. The crowding distance operator of the NSGA-II algorithm was modified to include the reference point information and the extent of the preferred region was controlled by ϵ-dominance. *Deb* and *Kumar* also experimented with the use of other classical preference methods, such as the reference direction method [67.35] and the light beam search method [67.36].

Recently, the use of interactive methods has also been integrated within PSO frameworks. *Wickramasinghe* and *Li* [67.37] integrated the reference point method to both the NSPSO [67.25] and maximinPSO [67.26] algorithms. Significant improvement in convergence efficiency was highlighted, and it was demonstrated that final solutions are of higher relevance to the designer. *Wickramasinghe* and *Li* [67.38] later extended their approach to handle MOP, by replacing the dominance criteria entirely with the simpler distance metric. It was conclusively demonstrated that without the use of the reference point, obtaining a final set of preferred solutions solely through conventional dominance-based techniques is improbable.

67.1.4 Surrogate Modeling

The most prohibiting factor of design optimization is the cost of evaluating the objective and constraint functions. For high-fidelity airfoil design, function evaluations may very well be measured in hours. This computational burden ultimately questions the practicality of performing an optimization study, and is often alleviated by simply reducing the level of sophistication of the solver. This consequently reduces the fidelity of the final design, which is undesirable. Another mitigating strategy which has steadily gained popularity in design is the use of inexpensive surrogates or metamodels [67.39]. These models emulate the response of the expensive function at an unobserved location, based on observations at other locations. Surrogate models are not specifically optimization methods, but rather they may ideally be used in lieu of the expensive function to extract information from the design space during the optimization process.

The insightful texts by *Keane* and *Nair* [67.39] and *Forrester* et al. [67.40] provide a detailed account of the use of surrogates in design. The most common use is to construct a *curve fit* of an expensive function landscape, which can be used to predict results without recourse to the original function. This is supported by the assumption that the inexpensive surrogate will still be usefully accurate when predicting sufficiently far from observed data points [67.40]. Figure 67.3 illustrates the use of

a surrogate to fit the one-dimensional multi-modal function, based on four sample observations. It is important to note, however, that the original function landscape could potentially represent any deterministic quantity of the design space. Rather than exactly emulating the response of a high-fidelity flow solver, the surrogate may, in fact, be used to bridge the gap between flow solvers of varying fidelity [67.40]. Alternatively, a surrogate may be used to interpret or filter noisy landscapes, so as to eliminate the adverse effects of flow solver convergence or grid discretization. Surrogates may also be used for data mining and design space visualization. Such methodologies are applied to extract useful information about the relationship between the design space and the objective space, allowing informed decisions to be made, which could simplify a seemingly complex problem.

For the aforementioned uses of surrogate modeling, the common requirement is to replicate the function relationship between the variable inputs and the output quantity of interest. This is typically achieved by sampling the design space using the exact function to sufficiently model the underlying relationship within the allowable computational budget. Whether the aim is to locally model the design space surrounding an existing design or tune a surrogate to replicate the global design space is entirely dependent on the formation of the sampling plan [67.39]. The construction of a surrogate model in either case should ideally make use of a parallel computing structure. A suitable surrogate model $\hat{f}$ of the precise objective function f should then be constructed to fit the dataset.

There are a multitude of popular techniques for constructing surrogates in the literature. For a comprehensive review of different methods, the reader is referred to (among others) [67.39–42]. The selection of the surrogate model is dependent on the information that the designer is attempting to extract from the design space. Polynomial response surfaces and radial basis functions are fairly popular techniques for constructing local surrogates, especially if some level of regression is desirable. Techniques such as Kriging or support vector machines are more ideally suited to global optimization studies, since they offer greater flexibility in tuning model parameters and provide a confidence interval of the predicted output. Neural networks require extensive training and validation, yet have also been a popular technique for design applications, notably in aerodynamic modeling [67.43] and visualization techniques [67.44, 45].

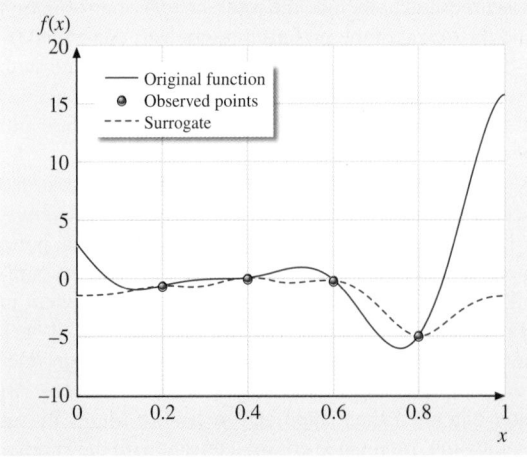

Fig. 67.3 Constructing an interpolation-based surrogate to fit a one-dimensional function

67.2 Shape Parameterization and Flow Solver

It was established in Sect. 67.1.1 that the shape parameterization method essentially governs the dimensionality of the problem and the attainable shapes, whereas the objective flow solver dictates the overall fidelity of the optimum design. In this section, we present a discussion on these elements of the design loop to be used in conjunction with the developed optimizer for the subsequent design process.

67.2.1 The PARSEC Parameterization Method

Geometry manipulation is of particular importance in aerodynamic design. The selection of the shape parameterization method is an important contributing factor, since it will effectively define the objective landscape and the topology of the design space [67.46]. If the aim of the optimization process is to improve on an established design, then perhaps local parameterization methods, which offer a greater number of geometrical degrees of freedom, are desirable. However, the large number of variables may cause the convergence rate for global design applications to deteriorate. The development of efficient parameterization models has, therefore, been given significant attention, to increase the flexibility of geometrical control with a minimum number of design variables.

For certain applications, it is possible to make use of fundamental aerodynamic theory to refine the parameterization method, such that the design variables relate to important aerodynamic or geometric quantities. A common method for airfoil shape parameterization is the PARSEC method [67.47]. It has the advantage of strict control over important aerodynamic features, and it allows independent control over the airfoil geometry for imposing shape constraints. The methodology

is characterized by 11 design variables (Fig. 67.4), including leading edge radius (r_{LE}), upper and lower thickness locations ($x_{UP}, z_{UP}, x_{LO}, z_{LO}$) and curvatures ($z_{xxUP}, z_{xxLO}$), trailing edge direction (α_{TE}) and wedge angle (β_{TE}), and trailing edge coordinate (z_{TE}) and thickness (Δz_{TE}). The shape function is modeled via a sixth-order polynomial function

$$z_k = \sum_{n=1}^{6} a_{n,k} \cdot x_k^{n-\frac{1}{2}} , \qquad (67.3)$$

where (x, z) are the shape coordinates and k denotes either the upper (suction) or lower (pressure) airfoil surface. The coefficients a_n are determined from the geometric parameters. A modification by *Jahangirian* and *Shahrokhi* [67.48] was introduced to provide additional control over the trailing edge curvature. For supercritical transonic airfoils, this is beneficial to reduce the probability of downstream boundary layer separation, which results in increased drag values. A new variable $\Delta\alpha_{TE}$ was introduced, which directly influences the additional curvature of the trailing edge. The modification decouples the trailing edge parameterization by first defining a smoother upper surface contour and then constraining the lower surface to intersect the trailing edge coordinate. Figure 67.5 illustrates the modification to the trailing edge curvature. The modification is applied to the upper and lower surfaces as follows

$$\delta_z = \frac{L \cdot \tan \Delta\alpha_{TE}}{2\mu\tau} \left[1 + \eta \cdot x^\tau - \left(1 - x^\tau\right)^\mu \right] , \quad (67.4)$$

where the constants η, μ, and τ are set to 0.8, 2, and 6, respectively. The modification is applied over the entire

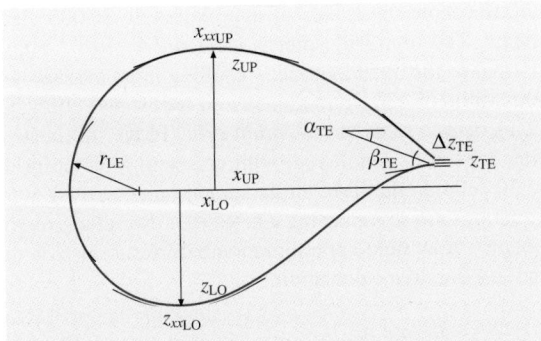

Fig. 67.4 Airfoil representation via the PARSEC method

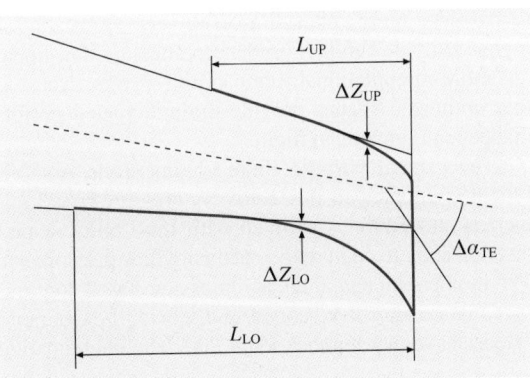

Fig. 67.5 Additional trailing edge curvature via the modified PARSEC method

Table 67.1 PARSEC parameter ranges for transonic optimization

Description	Variable	Lower bound	Upper bound
Leading edge radius	r_{LE}	0.0063	0.0151
Trailing edge direction	α_{TE}	0.2405(−)	0.0026(−)
Trailing edge wedge angle	β_{TE}	0.0655	0.2618
Upper-crest abscissa	x_{UP}	0.3170	0.5250
Upper-crest ordinate	z_{UP}	0.0497	0.0683
Upper-crest curvature	z_{xxUP}	0.5135(−)	0.2393(−)
Lower-crest abscissa	x_{LO}	0.2835	0.3418
Lower-crest ordinate	z_{LO}	0.0603(−)	0.0478(−)
Lower-crest curvature	z_{xxLO}	0.2535	0.8405
Trailing edge curvature	$\delta_{\alpha TE}$	0.0080(−)	0.3696

surface, such that $L_{UP} = L_{LO} = c$, where c is the airfoil chord length.

Table 67.1 presents the upper and lower boundaries for the subsequent optimization case study. These boundaries have been selected based on a thorough screening study involving a statistical sample of a number of benchmark airfoils.

67.2.2 Transonic Flow Solver

The optimization process is ultimately dependent on the selection of the flow solver, since it is the most computationally expensive component, and repeated evaluations of the objective and constraint functions are required for each candidate shape. However if the flow solver is not sufficiently accurate, the optimization process will converge to shapes that exploit the numerical errors or limitations, rather than the fundamental physics of the problem. For this reason, it is desirable to maintain the correct balance between solution accuracy and computational expense, which is dictated by the flow regime. For certain problems where the aerodynamic flow field is well behaved, it may be sufficient to consider more robust linear solvers. However for high-fidelity design it is prudent to consider non-linear and more computationally demanding solvers, to ensure that optimized shapes provide the anticipated performance requirements in flight.

The general purpose finite volume code ANSYS Fluent is adopted in this study. A pressure-based numerical procedure is adopted with third-order spatial discretization to capture the occurring flow phenomena. The momentum equations and pressure-based continuity equation are solved concurrently, with the Courant–Friedrichs–Lewy number set at 200. The one-equation

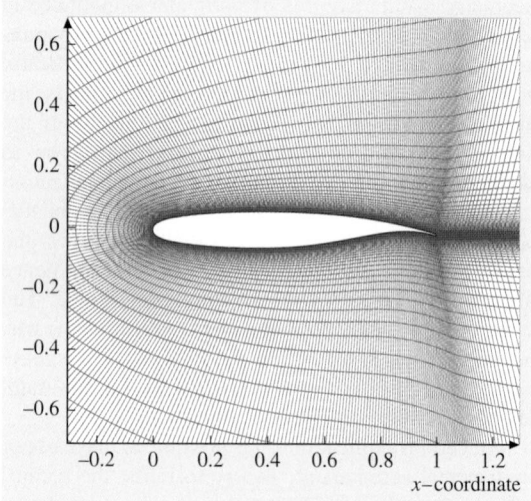

Fig. 67.6 C-type grid for transonic simulation

Spalart–Allmaras turbulence model [67.49] is selected, and turbulent flow is modeled over the entire airfoil surface. The C-type grid (as represented in Fig. 67.6) stretches 25 chord lengths aft and normal of the airfoil section. Resolution of the C-grid is 460×65, providing an affordable mesh size of approximately 30 000 elements. The first grid point is located 2.5×10^{-4} units normal to the airfoil surface, resulting in an average y-plus value of 120. In the interest of robust and efficient convergence rates, a full multi-grid (FMG) initialization scheme is employed, with coarsening of the grid to 30 cells. In the FMG initialization process, the Euler equations are solved using a first-order discretization to obtain a flow field approximation before submission to the full iterative calculation.

67.3 Optimization Algorithm

The proposed algorithm was primarily motivated by the studies of *Wickramasinghe* and *Li* [67.38]. The principal argument is that for most design applications, to explore the entire Pareto front is often unnecessary, and the computational burden can be alleviated by considering the immediate interests of the designer. In Sect. 67.1.3, a discussion on the benefits of preference-based optimization was provided. Drawing on these concepts, a preference-based algorithm is proposed, where a designer-driven distance metric is used to scalar quantify the success of a solution. The multiobjective search effort is coordinated via a MOPSO algorithm. The swarm is guided by a reference point, which is an intuitive means of articulating the preferences of the designer and can ideally be based on an existing or target design. This section provides a comprehensive discussion on the proposed algorithm, highlighting its viability for the intended domain of application.

67.3.1 The Reference Point Method

In this research, the swarm is guided by a reference point to confine its search focus exclusively on the preferred region of the Pareto front as dictated by the preferences of the designer. Introducing the preferred region provides the designer flexibility to explore other interesting alternatives. This hybrid methodology is advantageous for navigating high-dimensional and multimodal landscapes, which are typical of aerodynamic design problems. Furthermore, inherently considering the preferences of the designer provides a feasible means of quantifying the practicality of a design.

The Reference Point Distance Metric
The reference point method has been integrated into MOO algorithms, notably by *Deb* and *Sundar* [67.34] and *Wickramasinghe* and *Li* [67.37, 38]. These studies highlight the benefits of incorporating preference information via the reference point in terms of convergence. Guided by the information provided by the reference point, the swarm can simultaneously identify multiple solutions in the preferred region. This provides the designer flexibility to explore several *preferred* designs, while alleviating the computational burden of identifying the entire Pareto front. A reference point distance metric following the work of *Wickramasinghe* and *Li* [67.37] is proposed. This metric provides an intuitive criterion to select global leaders and assists the swarm to identify only solutions of interest to the de-

signer. The distance of a particle x to the reference point $\bar{z}$ is defined as

$$d_z(x) = \max_{i=1:m} \{(f_i(x) - \bar{z}_i)\} \ . \tag{67.5}$$

A solution a is, therefore, *preferred* to solution b if $d_z(a) < d_z(b)$. This condition is an extension of the condition $f(a) \prec f(b)$, therefore, the distance metric may, in fact, substitute the dominance criteria entirely [67.38]. Using this distance metric, the swarm is guided to preferred regions of the Pareto optimal front. Figure 67.7 illustrates the search directions of the algorithm when guided by a reference point, and the corresponding preferred design as a direct result of minimizing the distance metric d_z.

The distinguishing feature of the reference point distance metric over the mathematical Euclidean distance is that solutions do not converge to the reference point, but on the preferred region of the Pareto front as dictated by the search direction. This is illustrated in Fig. 67.8. All solutions are non-dominated and lie on the circular arc surrounding the reference point $\bar{z}$, and thus the Euclidean distance to the reference point is equal. However, since solution i has the smallest maximum translational distance to the reference point compared to any other solution, it is considered preferred. The definition of the reference point distance also suggests negative values. If the distance of the preferred solution $d_z(z') < 0$, then it can simply be considered that the reference point is dominated or $z' \prec \bar{z}$. Since the designer generally has no prior knowledge of the topology and location of the Pareto front,

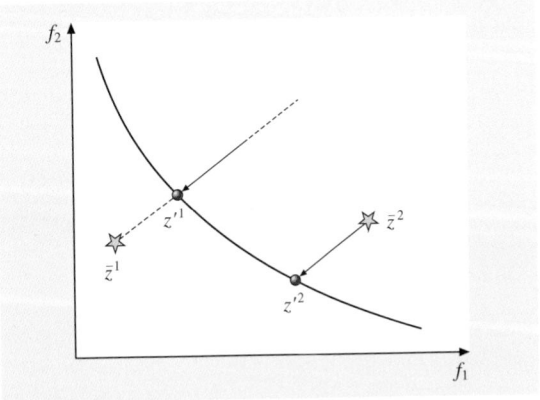

Fig. 67.7 Illustration of the search direction governed by the reference point

a reference point may be ideally placed in any feasible or infeasible region, as is shown in Fig. 67.7. It is, therefore, the consensus that the reference point draws on the experience of the designer to express the preferred compromise, rather than specific target values or goals. Similarly, the reference point distance metric ranks or assesses the success of a particle as one single scalar, instead of an array of objective values.

Defining the Preferred Region

As is demonstrated in Fig. 67.7, if there is no control over the solution spread the swarm will explore the preferred search direction and converge to the single solution z' as dictated by the reference point $\bar{z}$. The advantage of maintaining a population of particles provides the designer the possibility to explore a range of interesting alternatives within a preferred region of the Pareto front. The aim is, therefore, to identify a set of solutions surrounding the intersection point z'. A threshold parameter $\delta > 0$ is defined, such that a solution x is within the preferred region if the following conditional statement is true

$$d_z(x) \le d_z(z') + \delta . \tag{67.6}$$

Figure 67.9 illustrates the preferred region for a bi-objective problem. The extent of the solution spread is proportional to δ and evidently as $\delta \to 0$, the designer is interested in determining only the most preferred solution z'. Conversely, as $\delta \to \infty$, the designer is interested in determining all solutions along the Pareto front, and thus the influence of the reference point location diminishes.

67.3.2 User–Preference Multiobjective PSO: UPMOPSO

The proposed algorithm combines the searching proficiency of PSO and the guidance of the reference point method. The swarm is guided by the user-defined reference point to confine its search to focus exclusively on the identified preferred region of the Pareto front. While the concept of the reference point is fairly intuitive, ensuring that the swarm is guided by this information to identify preferred solutions is more ambiguous. The algorithm function is consolidated in Algorithm 67.1 and further described in the subsequent steps.

Algorithm 67.1 The UPMOPSO algorithm

1: **OBTAIN** user-defined preferences
2: **INITIALIZE** swarm
3: **EVALUATE** fitness and distance metric
4: **ASSIGN** personal best

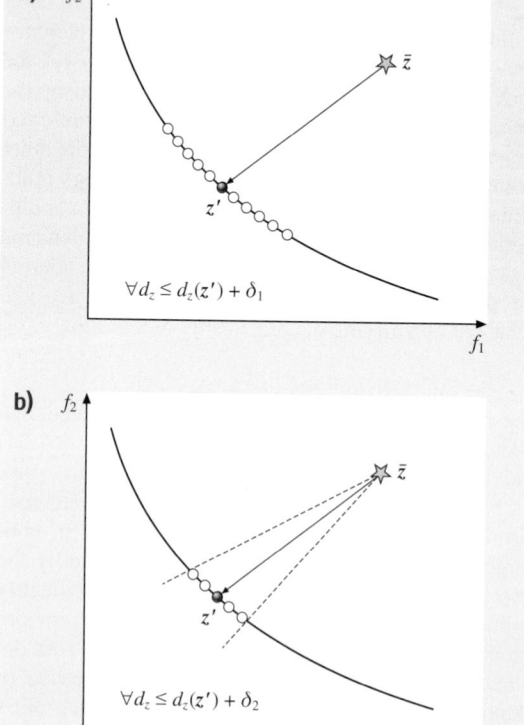

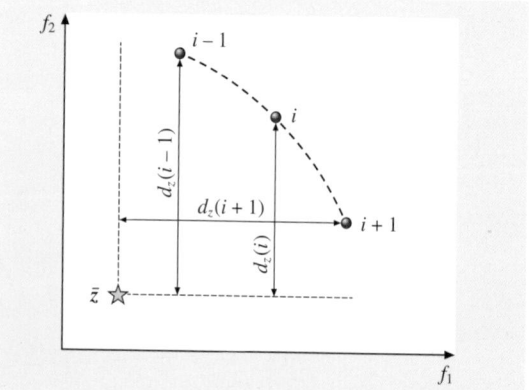

Fig. 67.8 Illustration of the reference point distance for solutions with equal Euclidean distance

Fig. 67.9a,b Definition of the preferred region via the parameter δ. **(a)** $\delta_1 = 0.01$, **(b)** $\delta_2 = 0.001$

5: CONSTRUCT archive
6: $t = 1$
7: **repeat**
8:      SELECT global leaders
9:      UPDATE particle velocity
10:     UPDATE particle position
11:     ADJUST boundary violation
12:     EVALUATE fitness and distance metric
13:     UPDATE personal best
14:     UPDATE archive
15:     $t = t + 1$
16: **until** $t = t_{max}$ OR f_{max}

OBTAIN User-Defined Preferences

The designer stipulates the reference point $\bar{z}$ and the corresponding solution spread δ to define the location and extent of the preferred region. For airfoil design applications, designers can exploit the existing domain knowledge to determine the most feasible performance compromise for the desired operating conditions.

INITIALIZE the Particles

A swarm of N particles is required to navigate the design space S bounded by x_{min} and x_{max}. To safeguard against magnitude and scaling issues, all variables are normalized into the unit cube, such that $S = [0, 1]^n$. The i-th particle in the swarm is characterized by the n-dimensional vectors x_i and v_i, which are the particle position and velocity, respectively. These vectors are randomly initialized within the unit cube at time $t = 0$. The particle personal best position is recorded as the particle position, such that $p_i = x_i$. The particles are then evaluated with the objective functions and fitness is assigned. The reference point distance metric is computed for each particle to measure the individual preference value.

UPDATE Archive and SELECT Global Leaders

A secondary population of non-dominated solutions in the form of an elitist archive is maintained at time, t. The non-dominated solutions identified by the particles are appended to the archive. A non-dominated sorting procedure is applied, where all members pertaining to local inferior fronts are omitted. The archive serves as a mutually accessible memory bank for the particles of the swarm. Each member is a potential candidate for global leadership of the particles during the subsequent velocity update.

Defining the global leaders ultimately governs the direction of the search. The swarm should efficiently navigate the design space such that the search effort is locally focused within the preferred region and provides a uniform spread of solutions. Since all members of the archive are mutually non-dominated, a ranking procedure is necessary to distinguish the most appropriate candidates for leadership from the remaining members. At each time step t, the most preferred solution $z'(t)$ is recorded. The subset of members $X_g(t)$ selected for global leadership satisfy the condition of (67.6), such that

$$X_g(t) \in d_z(Q(t)) \le d_z(z'(t)) + \delta \,. \tag{67.7}$$

Since not every member will initially satisfy this condition, the number of candidate leaders may fluctuate over time. This condition provides the necessary selection pressure for particles to locally focus the search effort within the preferred region, avoiding the unnecessary computational effort of exploring undesired regions of the design space. Each swarm particle is randomly assigned a leader to promote diversity in the search. In the case where all non-dominated solutions satisfy the condition of (67.7), additional guidance through a crowding distance metric (as described in [67.27]) is provided to promote a uniform spread.

As the particles are guided to converge to the preferred region, the number of identified non-dominated solutions will steadily increase. To avoid this number escalating unnecessarily and to maintain high competitiveness within the archive, there is a restriction (denoted by K_{max}) on the number of solutions permitted for entry. If the number of members $K > K_{max}$, the newest solution is permitted entry, and the existing least preferred member is removed. If all archive members exist within the preferred region, the most crowded solutions are removed. This ensures that solutions in densely populated regions are removed in favor of solutions which exploit sparsely populated regions, to further promote a uniform spread.

UPDATE Particle Position

The update equations of PSO adjust the position of the i-th particle from time t to $t+1$. In this algorithm, the constriction *type 1* framework of *Clerc* and *Kennedy* [67.50] is adopted. In their studies, the authors studied particle behavior from an eigenvalue analysis of swarm dynamics. The velocity update of the i-th particle is a function of acceleration components to both the personal best position, p_i, and the global best position, p_g. The updated velocity vector is given by the expression,

$$v_i(t+1) = \chi\{v_i(t) + R_1[0, \varphi_1] \otimes (p_i(t) - x_i(t)) + R_2[0, \varphi_2] \otimes (p_g(t) - x_i(t))\} . \quad (67.8)$$

The velocity update of (67.8) is quite complex and is composed of many quantities that affect certain search characteristics. The previous velocity $v_i(t)$ serves as a memory of the previous flight direction and prevents the particle from drastically changing direction and is referred to as the inertia component. The cognitive component of the update equation $(p_i(t) - x_i(t))$ quantifies the performance of the i-th particle relative to past performances. The effect of this term is that particles are drawn back to their own best positions, which resembles the tendency of individuals to return to situations where they experienced most success. The social component $(p_g(t) - x_i(t))$ quantifies the performance of the i-th particle relative to the global (or neighborhood) best position. This resembles the tendency of individuals to emulate the success of others.

The two functions $R_1[0, \varphi_1]$ and $R_2[0, \varphi_2]$ return a vector of uniform random numbers in the range $[0, \varphi_1]$ and $[0, \varphi_2]$, respectively. The constants φ_1 and φ_2 are equal to $\varphi/2$ where $\varphi = 4.1$. This randomly affects the magnitude of both the social and cognitive components. The constriction factor χ applies a dampening effect as to how far the particle explores within the search space and is given by

$$\chi = 2/|2 - \varphi - \sqrt{\varphi^2 - 4\varphi}| . \quad (67.9)$$

Once the particle velocity is calculated, the particle is displaced by adding the velocity vector (over the unit time step) to the current position,

$$x_i(t+1) = x_i(t) + v_i(t+1) . \quad (67.10)$$

Particle flight should ideally be confined to the feasible design space. However, it may occur during flight that a particle involuntarily violates the boundaries of the design space. While it is suggested that particles which leave the confines of the design space should simply be ignored [67.51], the violated dimension is restricted such that the particle remains within the feasible design space without affecting the flight trajectory.

UPDATE Personal Best
The ambiguity in updating the personal best using the dominance criteria lies in the treatment of the case

when the personal best solution $p_i(t)$ is mutually non-dominated with the solution $x_i(t + 1)$. The introduction of the reference point distance metric elegantly deals with this ambiguity. If the particle position $\hat{x}_i(t + 1)$ is preferred to the existing personal best $p_i(t)$, then the personal best is replaced. Otherwise the personal best is remained unchanged.

67.3.3 Kriging Modeling

Airfoil design optimization problems benefit from the construction of inexpensive surrogate models that emulate the response of exact functions. This section presents a novel development in the field of preference-based optimization. Adaptive Kriging models are incorporated within the swarm framework to efficiently navigate design spaces restricted by a computational budget. The successful integration of these design tools is facilitated through the reference point distance metric, which provides an intuitive criterion to update the Kriging models during the search.

In most engineering problems, to construct a globally accurate surrogate of the original objective landscape is improbable due to the weakly correlated design space. It is more common to locally update the prediction accuracy of the surrogate as the search progresses towards promising areas of the design space [67.40]. For this purpose, the Kriging method has received much interest, because it inherently considers confidence intervals of the predicted outputs. For a complete derivation of the Kriging method, readers are encouraged to follow the work of *Jones* [67.41] and *Forrester* et al. [67.40]. We provide a very brief introduction to the ordinary Kriging method, which expresses the unknown function $y(x)$ as,

$$y(x) = \beta + z(x) , \quad (67.11)$$

where $x = [x_1, \dots, x_n]$ is the data location, β is a constant global mean value, and $z(x)$ represents a local deviation at the data location x based on a stochastic process with zero-mean and variance σ^2 following the Gaussian distribution. The approximation $\hat{y}(x)$ is obtained from

$$\hat{y}(x) = \hat{\beta} + r^T R^{-1}(Y - 1\hat{\beta}) , \quad (67.12)$$

where $\hat{\beta}$ is the approximation of β, R is the correlation matrix, r is the correlation vector, Y is the training dataset of N observed samples at location X, and 1 is a column vector of N elements of 1. The correlation matrix is a modification of the Gaussian basis function,

$$R(\mathbf{x}^i, \mathbf{x}^j) = \exp\left(-\sum_{k=1}^{n} \theta_k |x_k^i - x_k^j|^2\right), \quad (67.13)$$

where $\theta_k > 0$ is the k-th element of the correlation parameter $\boldsymbol{\theta}$. Following the work of *Jones* [67.41], the correlation parameter $\boldsymbol{\theta}$ (and hence the approximations $\hat{\beta}$ and $\hat{\sigma}^2$) are estimated by maximizing the concentrated ln-likelihood of the dataset $\mathbf{Y}$, which is an n-variable single-objective optimization problem, solved using a pattern search method. The accuracy of the prediction $\hat{y}$ at the unobserved location $\mathbf{x}$ depends on the correlation distance with sample points $\mathbf{X}$. The closer the location of $\mathbf{x}$ to the sample points, the more confidence in the prediction $\hat{y}(\mathbf{x})$. The measure of uncertainty in the prediction is estimated as

$$\hat{s}^2(\mathbf{x}) = \hat{\sigma}^2 \left[1 - \mathbf{r}^T \mathbf{R}^{-1} \mathbf{r} + \frac{(1 - \mathbf{1}^T \mathbf{R}^{-1} \mathbf{r})^2}{\mathbf{1}^T \mathbf{R}^{-1} \mathbf{1}}\right] \quad (67.14)$$

if $\mathbf{x} \subset \mathbf{X}$, it is observed from (67.14) that $\hat{s}(\mathbf{x})$ reduces to zero.

67.3.4 Reference Point Screening Criterion

Training a Kriging model from a training dataset is time consuming and is of the order $O(N^3)$. Stratified sampling using a maximin Latin hypercube (LHS [67.52] is used to construct a global Kriging approximation $[\mathbf{X}, \mathbf{Y}]$. The non-dominated subset of $\mathbf{Y}$ is then stored within the elitist archive. This ensures that candidates for global leadership have been precisely evaluated (or with negligible prediction error) and, therefore, offer no false guidance to other particles. Adopting the concept of individual-based control [67.42], Kriging predictions are then used to pre-screen each candidate particle after the population update (or after mutation) and subsequently flag them for precise evaluation or rejection. The Kriging model estimates a lower-confidence bound for the objective array as

$$\{\hat{f}_1(\mathbf{x}), \ldots, \hat{f}_m(\mathbf{x})\}_{lb} = [\{\hat{y}_1(\mathbf{x}) - \omega\hat{s}_1(\mathbf{x})\}, \ldots, \{\hat{y}_m(\mathbf{x}) - \omega\hat{s}_m(\mathbf{x})\}], \quad (67.15)$$

where $\omega = 2$ provides a 97% probability for $\hat{f}_{lb}(\mathbf{x})$ to be the lower bound value of $\hat{f}(\mathbf{x})$. An approximation to the reference point distance, $\hat{d}_z(\mathbf{x})$, can thus be obtained using (67.5). This value, whilst providing a means of ranking each solution as a single scalar, also gives an estimate to the improvement that is expected from the solution. At time t, the archive member with the highest ranking according to (67.5) is recorded as d_{min}. The candidate $\mathbf{x}$ may then be accepted for precise evaluation, and subsequent admission into the archive if $\hat{d}_z(\mathbf{x}) < d_{min}$. Particles will thus be attracted towards the areas of the design space that provide the greatest resemblance to $\bar{z}$, and the direction of the search will remain consistent.

As the search begins in the explorative phase and the prediction accuracy of the surrogate model(s) is low, depending on the deceptivity of the objective landscape(s) there will initially be a large percentage of the swarm that is flagged for precise evaluation. Subsequently, as the particles begin to identify the preferred region and the prediction accuracy of the surrogate model(s) gradually increases, the screening criterion becomes increasingly difficult to satisfy, thereby reducing the number of flagged particles at each time step. To restrict saturation of the dataset used to train the Kriging models, a limit is imposed of $N = 200$ sample points where lowest ranked solutions according to (67.5) are removed.

67.4 Case Study: Airfoil Shape Optimization

The parameterization method and transonic flow solver described in the preceding section are now integrated within the Kriging-assisted UPMOPSO algorithm for an efficient airfoil design framework. The framework is applied to the re-design of the NASA-SC(2)0410 airfoil for robust aerodynamic performance. A three-objective constrained optimization problem is formulated, with $f_1 = C_d$ and $f_2 = -C_m$ for $M = 0.79$, $C_l = 0.4$, and $f_3 = \partial C_d / \partial M$ for the design range $M = [0.79, 0.82]$, $C_l = 0.4$. The lift constraint is satisfied internally within the solver, by allowing Fluent to determine the angle of incidence required. A constraint is imposed on the allowable thickness, which is defined through the parameter ranges (see Table 67.2) as approximately 9.75% chord. The reference point is logically selected as the NASA-SC(2)0410, in an attempt to improve on the performance characteristics of the airfoil, whilst still maintaining a similar level of compromise between the design objectives. The solution variance is controlled by $\delta = 5 \times 10^{-3}$.

The design application is segregated into three phases: pre-optimization and variable screening; optimization and; post-optimization and trade-off screening.

67.4.1 Pre-Optimization and Variable Screening

Global Kriging models are constructed for the aerodynamic coefficients from a stratified sample of $N = 100$ design points based on a Latin hypercube design. This sampling plan size is considered sufficient in order to obtain sufficient confidence in the results of the subsequent design variable screening analysis. Whilst a larger sampling plan is essential to obtain fairly accurate correlation, the interest here is to quantify the elementary effect of each variable to the objective landscapes. The global Kriging models are initially trained via cross-validation. The cross-validation curves for the Kriging models are illustrated in Fig. 67.10. The subscripts to the aerodynamic coefficients refer to the respective angle of incidence.

It is observed in Fig. 67.10 that the Kriging models constructed for the aerodynamic coefficients are able to reproduce the training samples with sufficient confidence, recording error margin values between 2 to 4%. It is hence concluded that the Kriging method is very adept at modeling complex landscapes represented by a limited number of precise observations.

To investigate the elementary effect of each design variable on the metamodeled objective landscapes, we present a quantitative design space visualization technique. A popular method for designing preliminary experiments for design space visualization is the screening method developed by *Morris* [67.53]. This algorithm calculates the elementary effect of a variable x_i and establishes its correlation with the objective space f as:

a) Negligible
b) Linear and additive
c) Nonlinear
d) Nonlinear and/or involved in interactions with x_j.

Table 67.2 NASA-SC(2)0410 airfoil results for the formulated objectives

Airfoil	Mach number, M	f_1	f_2	f_3
NASA-SC(2)0410	0.79	0.008708	0.1024	0.189625

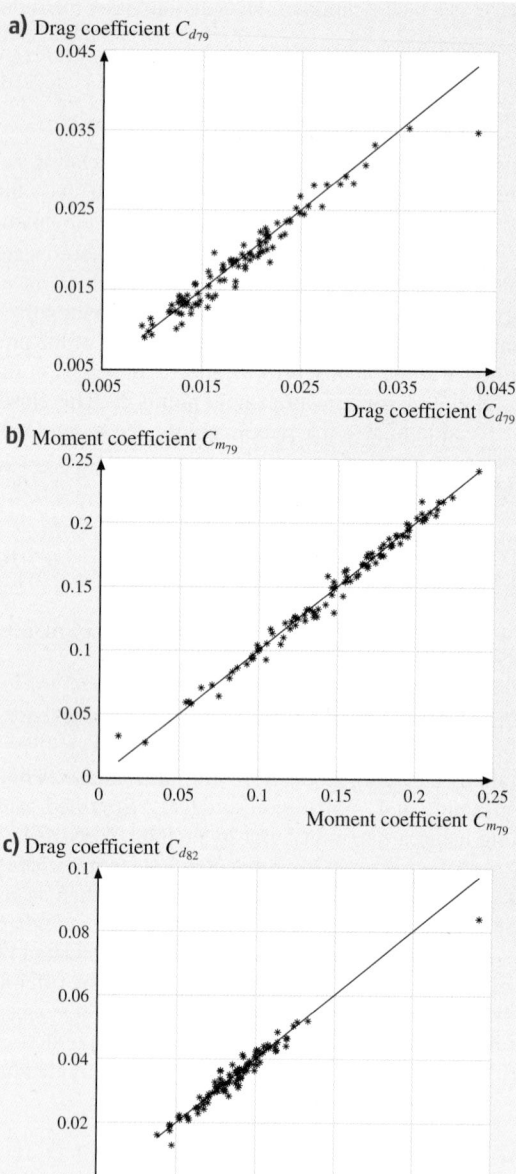

Fig. 67.10a–c Cross-validation curves for the constructed Kriging models. (**a**) Training sample for C_d at $M = 0.79$. (**b**) Training sample for C_m at $M = 0.79$. (**c**) Training sample for C_d at $M = 0.82$

In plain terminology, the Morris algorithm measures the sensitivity of the i-th variable to the objective landscape f. For a detailed discussion on the Morris al-

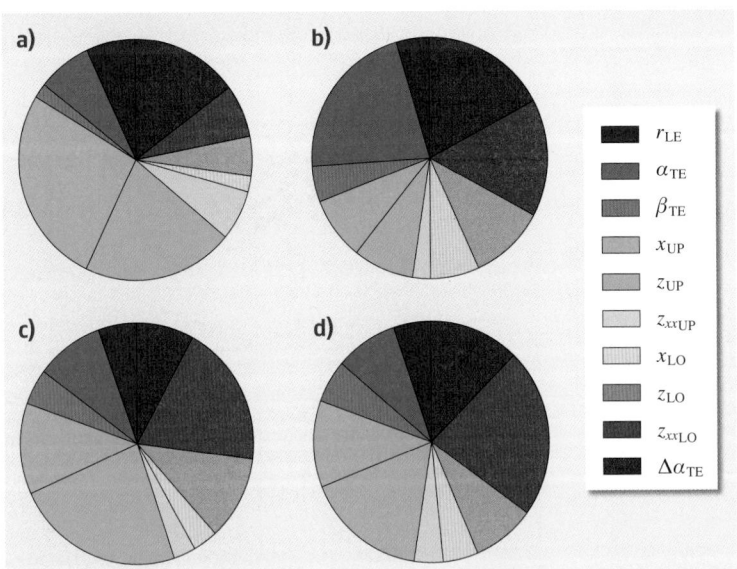

Fig. 67.11a–d Variable influence on aerodynamic coefficients (subscripts refer to the operating Mach number). (**a**) Drag $C_{d_{79}}$. (**b**) Moment $C_{m_{79}}$. (**c**) Drag $C_{d_{82}}$. (**d**) d_z

gorithm the reader is referred to *Forrester* et al. [67.40] and *Campolongo* et al. [67.54]. Presented here are the results of the variable screening analysis using the Morris algorithm for the proposed design application.

Figure 67.11 graphically shows the results obtained from the design variable screening study. It is immediately observed that the upper thickness coordinates have a relatively large influence on the drag coefficient for both design conditions. At higher Mach numbers the effect of the lower surface curvature $z_{xx_{LO}}$ is also significant. It is demonstrated, however, that the variables $z_{xx_{LO}}$ and α_{TE} have the largest effect on the moment coefficient – variables which directly influence the aft camber (and hence the aft camber) on the airfoil. These variables will no doubt shift the loading on the airfoil forward and aft, resulting in highly fluctuating moment values.

Similar deductions can be made by examining the variable influence on d_z shown in Fig. 67.11d. The variable influence on d_z is case specific and entirely dependent on the reference point chosen for the proposed optimization study. Since the value of d_z is a means of ranking the success of a multiobjective solution as one single scalar, variables may be ranked by influence, which is otherwise not possible when considering a multiobjective array. Preliminary conclusions to the priority weighting of the objectives to the reference point compromise can also be made. It is observed that the variable influence on d_z most closely resembles the plots of the drag coefficients $C_{d_{79}}$ and $C_{d_{82}}$, suggesting

that the moment coefficient is of least priority for the preferred compromise. It is interesting to see that the trailing edge modification variable $\Delta\alpha_{TE}$ is of particular importance for all design coefficients, which validates its inclusion in the subsequent optimization study.

67.4.2 Optimization Results

A swarm population of $N_s = 100$ particles is flown to solve the optimization problem. The objective space is normalized for the computation of the reference point distance by $f_{max} - f_{min}$. Instead of specifying a maximum number of time steps, a computational budget of 250 evaluations is imposed. A stratified sample of $N = 100$ design points using an LHS methodology was used to construct the initial global Kriging approximations for each objective. A further 150 precise updates were performed over $t \approx 100$ time steps until the computational budget was breached. As is shown in Fig. 67.12a, the largest number of update points was recorded during the initial explorative phase. As the preferred region becomes populated and $\hat{s} \to 0$, the algorithm triggers exploitation, and the number of update points steadily reduces.

The UPMOPSO algorithm has proven to be very capable for this specific problem. Figure 67.12b features the progress of the highest ranked solution (i.e., d_{min}) as the number of precise evaluations increase. The reference point criterion is shown to be proficient in filtering out poorer solutions during exploration, since it

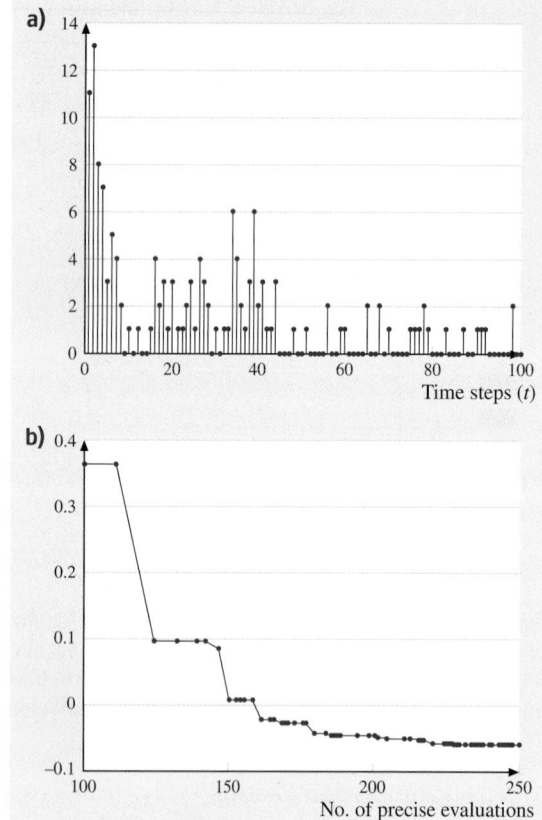

Fig. 67.12a,b UPMOPSO performance for transonic air-foil shape optimization. (**a**) History of precise updates. (**b**) Progress of most preferred solution

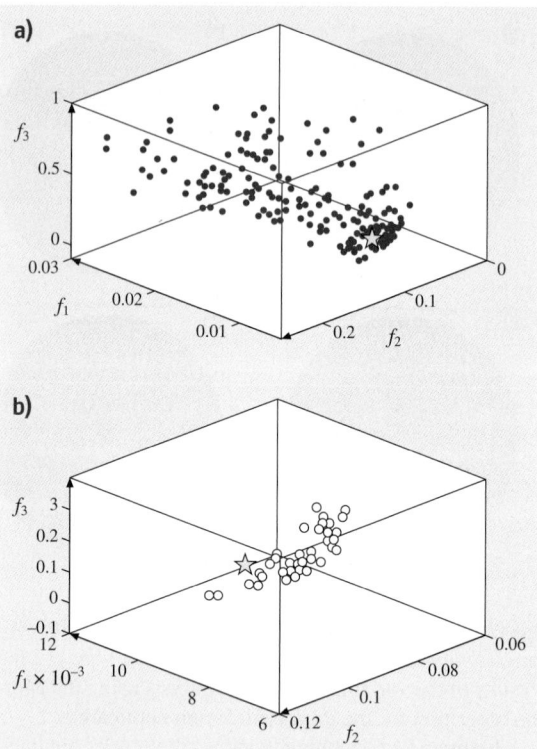

Fig. 67.13a,b Precise evaluations performed and the re-sulting non-dominated solutions. (**a**) Scatter plot of all precise evaluations. (**b**) Preferred region 250 evaluations

is only required to reach 50 update evaluations within 15% of d_{min}, and to reach a further 50 evaluations within 3%. Furthermore, no needless evaluations as a result of the lower-bound prediction are performed during the exploitation phase. This conclusion is further comple-mented by Fig. 67.13a, as a distinct attraction to the preferred region is clearly visible. A total of 30 non-dominated solutions were identified in the preferred region, which are shown in Fig. 67.13b.

67.4.3 Post–Optimization and Trade–Off Visualization

The reference point distance also provides a feasi-ble means of selecting the most appropriate solutions. For example, solutions may be ranked according to how well they represent the reference point compro-mise. To illustrate this concept, self-organizing maps

(SOMs) [67.44] are utilized to visualize the interac-tion of the objectives with the reference point com-promise. Clustering SOM techniques are based on a technique of unsupervised artificial neural networks that can classify, organize, and visualize large sets of data from a high to low-dimensional space [67.45]. A neuron used in this SOM analysis is associated with the weighted vector of m inputs. Each neuron is connected to its adjacent neurons by a neighborhood relation and forms a two-dimensional hexagonal topol-ogy [67.45]. The SOM learning algorithm will attempt to increase the correlation between neighboring neurons to provide a global representation of all solutions and their corresponding resemblance to the reference point compromise.

Each input objective acts as a neuron to the SOM. The corresponding output measures the reference point distance (i. e., the resemblance to the reference point compromise). A two-dimensional representation of the data is presented in Fig. 67.14, organized by six SOM-ward clusters. Solutions that yield negative d_z values

Table 67.3 Preferred airfoil objective values with measure of improvement

Airfoil	f_1	f_2	f_3
NASA-SC(2)0410	0.008708	0.1024	0.189625
Preferred design	0.008106	0.0933	0.168809
% Improvement	6.9	8.8	10.9

indicate success in the improvement over each aspiration value. Solutions with positive d_z values do not surpass each aspiration value but provide significant improvement in at least one other objective. Each of the node values represent one possible Pareto-optimal solution that the designer may select. The SOM chart colored by d_z is a measure of how far a solution deviates from the preferred compromise. However, the concept of the preferred region ensures that only solutions that slightly deviate from the compromise dictated by $\bar{z}$ are identified. Following the SOM charts, it is possible to

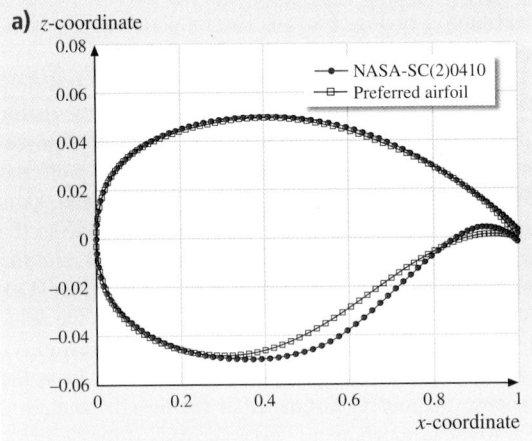

a) *z*-coordinate

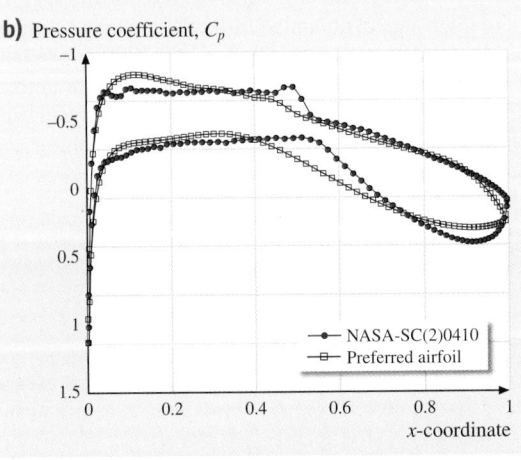

b) Pressure coefficient, C_p

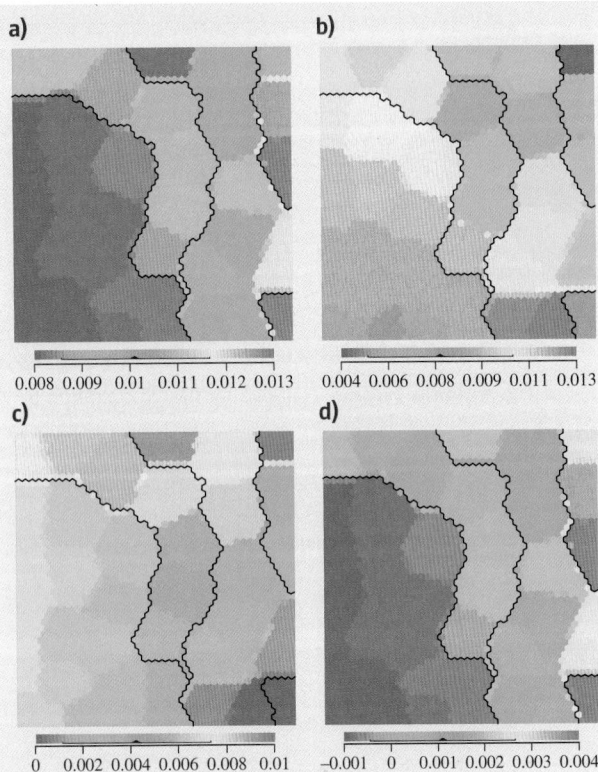

Fig. 67.14 SOM charts to visualize optimal trade-offs between the design objectives. **(a)** f_1, **(b)** f_2, **(c)** f_3, **(d)** d_z

visualize the preferred compromise between the design objectives that is obtained. The chart of d_z closely follows the f_1 chart, which suggests that this objective has the highest priority. If the designer were slightly more inclined towards another specific design objective, then solutions that perhaps place more emphasis on the other objectives should be considered. In this study, the most preferred solution is ideally selected as the highest ranked solution according to (67.5).

67.4.4 Final Designs

Table 67.3 shows the objective comparisons with the NASA-SC(2)0410. Of interest to note is that the most active objective is f_1, since the solution which provides the minimum d_z values also provides the minimum f_1 value. This implies that the reference point was sit-

Fig. 67.15a,b The most preferred solution observed by the UPMOPSO algorithm. **(a)** Preferred airfoil geometry. **(b)** C_p distributions for $M = 0.79$, $C_l = 0.4$ ◄

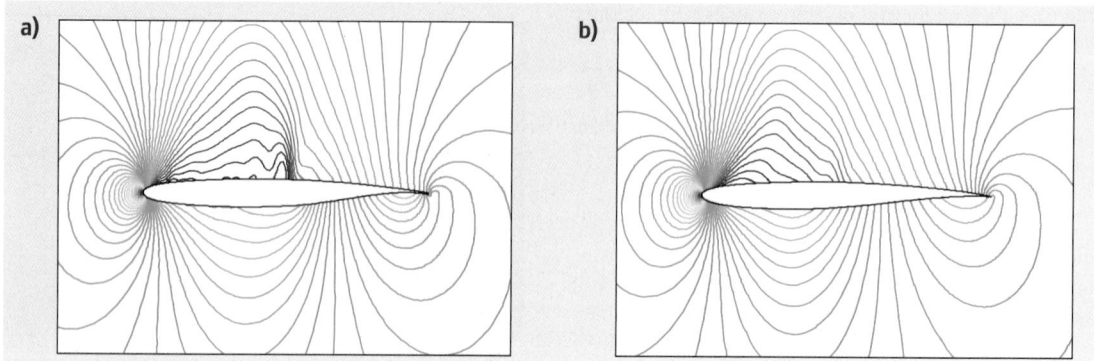

Fig. 67.16a,b Pressure contours for design condition of $M = 0.79$, $C_l = 0.4$. (**a**) NASA-SC(2)0410, (**b**) Preferred airfoil

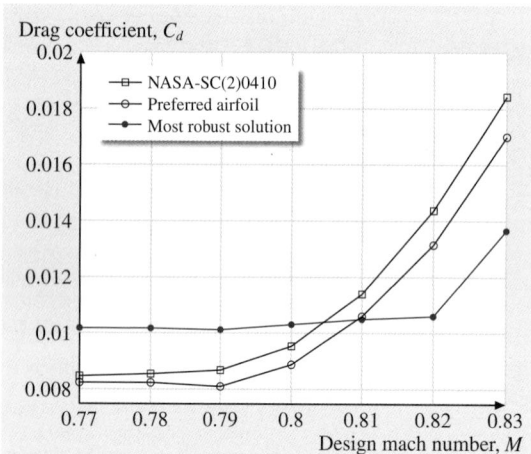

Fig. 67.17 Drag rise curves for $C_l = 0.4$

uated near the f_1 Pareto boundary. Of the identified set of Pareto-optimal solutions, the largest improvements obtained in objectives f_2 and f_3 are 36.4 and 91.6%, respectively, over the reference point. The preferred airfoil geometry is shown in Fig. 67.15a in comparison with the NASA-SC(2)0410. The preferred airfoil has a thickness of 9.76% chord and maintains a moderate curvature over the upper surface. A relatively small aft curvature is used to gener-

ate the required lift, whilst reducing the pitching moment.

Performance comparisons between the NASA-SC(2)0410 and the preferred airfoil at the design condition of $M = 0.79$ can be made from the static pressure contour output in Fig. 67.16, and the surface pressure distribution of Fig. 67.15b. The reduction in C_d is attributed to the significantly weaker shock that appears slightly upstream of the supercritical shock position. The reduction in the pitching moment is clearly visible from the reduced aft loading. Along with the improvement at the required design condition, the preferred airfoil exhibits a lower drag rise by comparison, as is shown in Fig. 67.17. There is a notable decline in the drag rise at the design condition of $M = 0.79$, and the drag is recorded as lower than the NASA-SC(2)0410, even beyond the design range. Also visible is the solution that provides the most robust design (i. e., min f_3). The most robust design is clearly not obtained at the expense of poor performance at the design condition, due to the compromising influence of $\bar{z}$. If the designer were interested in obtaining further alternative solutions which provide greater improvement in either objective, it would be sufficient to re-commence (at the current time step) the optimization process with a larger value of δ, or by relaxing one or more of the aspiration values, $\bar{z}_i$.

67.5 Conclusion

In this chapter, an optimization framework has been introduced and applied to the aerodynamic design of transonic airfoils. A surrogate-driven multiobjective particle swarm optimization algorithm is applied to navigate the design space to identify and exploit preferred regions of the Pareto frontier. The integration of all components of the optimization framework is entirely achieved through the use of a reference point distance metric which provides a scalar measure of the preferred interests of the designer. This effectively allows for the scale of the design space to be reduced, confining it to the interests reflected by the designer.

The developmental effort that is reported on here is to reduce the often prohibitive computational cost of multiobjective optimization to the level of practical affordability in computational aerodynamic design. A concise parameterization model was considered to perform the necessary shape modifications in conjunction with a Reynolds-averaged Navier–Stokes flow solver. Kriging models were constructed based on a stratified sample of the design space. A pre-optimization visualization tool was then applied to

screen variable elementary influence and quantify its relative influence to the preferred interests of the designer. Initial design drivers were easily identified and an insight to the optimization landscape was obtained. Optimization was achieved by driving a surrogate-assisted particle swarm towards a sector of special interest on the Pareto front, which is shown to be an effective and efficient mechanism. It is observed that there is a distinct attraction towards the preferred region dictated by the reference point, which implies that the reference point criterion is adept at filtering out solutions that will disrupt or deviate from the optimal search path.

Non-dominated solutions that provide significant improvement over the reference geometry were identified within the computational budget imposed and are clearly reflective of the preferred interest. A post-optimization data mining tool was finally applied to facilitate a qualitative trade-off visualization study. This analysis provides an insight into the relative priority of each objective and their influence on the preferred compromise.

References

67.1   T.E. Labrujère, J.W. Sloof: Computational methods for the aerodynamic design of aircraft components, Annu. Rev. Fluid Mech. **51**, 183–214 (1993)

67.2   R.M. Hicks, P.A. Henne: Wing design by numerical optimization, J. Aircr. **15**(7), 407–412 (1978)

67.3   J. Kennedy, R.C. Eberhart, Y. Shi: *Swarm Intelligence* (Morgan Kaufmann, San Francisco 2001)

67.4   J. Kennedy, R.C. Eberhart: Particle swarm optimization, Proc. IEEE Intl. Conf. Neural Netw. (1995) pp. 1942–1948

67.5   I.C. Trelea: The particle swarm optimization algorithm: Convergence analysis and parameter selection, Inform. Proces. Lett. **85**(6), 317–325 (2003)

67.6   C.M. Fonseca, P.J. Fleming: Multiobjective optimization and multiple constraint handling with evolutionary algorithms. A unified formulation, IEEE Trans. Syst. Man Cybern. A **28**(1), 26–37 (1998)

67.7   M. Drela: Pros and cons of airfoil optimization, Front. Comput. Fluid Dynam. **19**, 1–19 (1998)

67.8   W. Li, L. Huyse, S. Padula: Robust airfoil optimization to achieve drag reduction over a range of mach numbers, Struct. Multidiscip. Optim. **24**, 38–50 (2002)

67.9   M. Nemec, D.W. Zingg, T.H. Pulliam: Multipoint and multi-objective aerodynamic shape optimization, AIAA J. **42**(6), 1057–1065 (2004)

67.10   I. Das, J.E. Dennis: A closer look at drawbacks of minimizing weighted sums of objectives for Pareto set generation in multicriteria optimization problems, Struct. Optim. **14**, 63–69 (1997)

67.11   R.T. Marler, J.S. Arora: Survey of multi-objective optimization methods for engineering, Struct. Multidiscip. Optim. **26**, 369–395 (2004)

67.12   C.M. Fonseca, P.J. Fleming: An overview of evolutionary algorithms in multiobjective optimization, Evol. Comput. **3**, 1–16 (1995)

67.13   K. Deb: *Multi-Objective Optimization Using Evolutionary Algorithms* (Wiley, New york 2001)

67.14   C.A. Coello Coello: *Evolutionary Algorithms for Solving Multi-Objective Problems* (Springer, Berlin, Heidelberg 2007)

67.15   S. Obayashi, D. Sasaki, Y. Takeguchi, N. Hirose: Multiobjective evolutionary computation for supersonic wing-shape optimization, IEEE Trans. Evol. Comput. **4**(2), 182–187 (2000)

67.16   A. Vicini, D. Quagliarella: Airfoil and wing design through hybrid optimization strategies, AIAA J. **37**(5), 634–641 (1999)

67.17   A. Ray, H.M. Tsai: Swarm algorithm for single and multiobjective airfoil design optimization, AIAA J. **42**(2), 366–373 (2004)

67.18   M. Ehrgott, X. Gandibleux: *Multiple Criteria Optimization: State of the Art Annotated Bibliographic Surveys* (Kluwer, Boston 2002)

67.19   M.R. Sierra, C.A. Coello Coello: Multi-objective particle swarm optimizers: A survey of the state-of-the-art, Int. J. Comput. Intell. Res. **2**(3), 287–308 (2006)

67.20   A.P. Engelbrecht: *Fundamentals of Computational Swarm Intelligence* (Wiley, New York 2005)

67.21   J. Knowles, D. Corne: Approximating the non-dominated front using the Pareto archived evolution strategy, Evol. Comput. **8**(2), 149–172 (2000)

67.22   C.A. Coello Coello, M.S. Lechuga: Mopso: A proposal for multiple objective particle swarm optimization, IEEE Cong. Evol. Comput. (2002) pp. 1051–1056

67.23   S. Mostaghim, J. Teich: The role of ϵ-dominance in multi-objective particle swarm optimization methods, IEEE Cong. Evol. Comput. (2003) pp. 1764–1771

67.24   J.E. Fieldsend, S. Singh: A multi-objective algorithm based upon particle swarm optimisation, an efficient data structure and turbulence, U.K Workshop Comput. Intell. (2002) pp. 37–44

67.25   X. Li: A non-dominated sorting particle swarm optimizer for multiobjective optimization, Genet. Evol. Comput. Conf. (2003) pp. 37–48

67.26   X. Li: Better spread and convergence: Particle swarm multiobjective optimization using the maximin fitness function, Genet. Evol. Comput. Conf. (2004) pp. 117–128

67.27   M.R. Sierra, C.A. Coello Coello: Improving pso-based multi-objective optimization using crowding, mutation and ϵ-dominance, Lect. Notes Comput. Sci. **3410**, 505–519 (2005)

67.28   C.R. Raquel, P.C. Naval: An effective use of crowding distance in multiobjective particle swarm optimization, Genet. Evol. Comput. Conf. (2005) pp. 257–264

67.29   C.A. Coello Coello: Handling preferences in evolutionary multiobjective optimization: A survey, IEEE Cong. Evol. Comput. (2000) pp. 30–37

67.30   L. Rachmawati, D. Srinivasan: Preference incorporation in multi-objective evolutionary algorithms: A survey, IEEE Cong. Evol. Comput. (2006) pp. 962–968

67.31   C.M. Fonseca, P.J. Fleming: Handling preferences in evolutionary multiobjective optimization: A survey, Proc. IEEE 5th Int. Conf. Genet. Algorithm (1993) pp. 416–423

67.32   K. Deb: Solving goal programming problems using multi-objective genetic algorithms, IEEE Cong. Evol. Comput. (1999) pp. 77–84

67.33   L. Thiele, P. Miettinen, P.J. Korhonen, J. Molina: A preference-based evolutionary algorithm for multobjective optimization, Evol. Comput. **17**(3), 411–436 (2009)

67.34   K. Deb, J. Sundar: Reference point based multi-objective optimization using evolutionary algo-

67.35   K. Deb, A. Kumar: Interactive evolutionary multi-objective optimization and decision-making using reference direction method, Genet. Evol. Comput. Conf. (2007) pp. 781–788

67.36   K. Deb, A. Kumar: Light beam search based multi-objective optimization using evolutionary algorithms, Genet. Evol. Comput. Conf. (2007) pp. 2125–2132

67.37   U.K. Wickramasinghe, X. Li: Integrating user preferences with particle swarms for multi-objective optimization, Genet. Evol. Comput. Conf. (2008) pp. 745–752

67.38   U.K. Wickramasinghe, X. Li: Using a distance metric to guide pso algorithms for many-objective optimization, Genet. Evol. Comput. Conf. (2009) pp. 667–674

67.39   A.J. Keane, P.B. Nair: *Computational Approaches for Aerospace Design: The Pursuit of Excellence* (Wiley, New York 2005)

67.40   A. Forrester, A. Sóbester, A.J. Keane: *Engineering Design Via Surrogate Modelling: A Practical Guide* (Wiley, New York 2008)

67.41   D.R. Jones: A taxomony of global optimization methods based on response surfaces, J. Glob. Optim. **21**, 345–383 (2001)

67.42   Y. Jin: A comprehensive survey of fitness approximation in evolutionary computation, Soft Comput. **9**(1), 3–12 (2005)

67.43   R.M. Greenman, K.R. Roth: High-lift optimization design using neural networks on a multi-element airfoil, J. Fluids Eng. **121**(2), 434–440 (1999)

67.44   T. Kohonen: *Self-Organizing Maps* (Springer, Berlin, Heidelberg 1995)

67.45   S. Jeong, K. Chiba, S. Obayashi: Data mining for aerodynamic design space, J. Aerosp. Comput. Inform. Commun. **2**, 452–469 (2005)

67.46   W. Song, A.J. Keane: A study of shape parameterisation methods for airfoil optimisation, Proc. 10th AIAA/ISSMO Multidiscip. Anal. Optim. Conf. (2004)

67.47   H. Sobjieczky: Parametric airfoils and wings, Notes Numer. Fluid Mech. **68**, 71–88 (1998)

67.48   A. Jahangirian, A. Shahrokhi: Inverse design of transonic airfoils using genetic algorithms and a new parametric shape model, Invers. Probl. Sci. Eng. **17**(5), 681–699 (2009)

67.49   P.R. Spalart, S.R. Allmaras: A one-equation turbulence model for aerodynamic flows, Rech. Aerosp. **1**, 5–21 (1992)

67.50   M. Clerc, J. Kennedy: The particle swarm - explosion, stability, and convergence in a multidimensional complex space, IEEE Trans. Evol. Comput. **6**(1), 58–73 (2002)

67.51   D. Bratton, J. Kennedy: Defining a standard for particle swarm optimization, IEEE Swarm Intell. Symp. (2007) pp. 120–127

67.52 M.D. Mckay, R.J. Beckman, W.J. Conover: A comparison of three methods for selecting values of input variables in the analysis of output from a computer code, Invers. Prob. Sci. Eng. **21**(2), 239–245 (1979)

67.53 M.D. Morris: Factorial sampling plans for preliminary computational experiments, Technometrics **33**(2), 161–174 (1991)

67.54 F. Campolongo, A. Saltelli, S. Tarantola, M. Ratto: *Sensitivity Analysis in Practice* (Wiley, New York 2004)

68. Ant Colony Optimization for the Minimum–Weight Rooted Arborescence Problem

Christian Blum, Sergi Mateo Bellido

The minimum–weight rooted arborescence problem is an *NP*-hard combinatorial optimization problem which has important applications, for example, in computer vision. An example of such an application is the automated reconstruction of consistent tree structures from noisy images. In this chapter, we present an ant colony optimization approach to tackle this problem. Ant colony optimization is a metaheuristic which is inspired by the foraging behavior of ant colonies. By means of an extensive computational evaluation, we show that the proposed approach has advantages over an existing heuristic from the literature, especially for what concerns rather dense graphs.

Part F | 68.1

68.1 Introductiory Remarks

Solving combinatorial optimization problems with approaches from the swarm intelligence field has already a considerably long tradition. Examples of such approaches include particle swarm optimization (PSO) [68.1] and artificial bee colony (ABC) optimization [68.2]. The oldest – and most widely used – algorithm from this field, however, is ant colony optimization (ACO) [68.3]. In general, the ACO metaheuristic attempts to solve a combinatorial optimization problem by iterating the following steps: (1) Solutions to the problem at hand are constructed using a pheromone model, that is, a parameterized probability distribution over the space of all valid solutions, and (2) (some of) these solutions are used to change the pheromone values in a way being aimed at biasing subsequent sampling toward areas of the search space containing high quality solutions. In particular, the reinforcement of solution components depending on the quality of the solutions in which they appear is an important aspect of ACO algorithms. It is implicitly assumed that good solutions consist of good solution components. To learn

which components contribute to good solutions most often helps assembling them into better solutions.

In this chapter, ACO is applied to solve the minimum-weight rooted arborescence (MWRA) problem, which has applications in computer vision such as, for example, the automated reconstruction of consistent tree structures from noisy images [68.4]. The structure of this chapter is as follows. Section 68.2 provides a detailed description of the problem to be tackled. Then, in Sect. 68.3 a new heuristic for the MWRA problem is presented which is based on the deterministic construction of an arborescence of maximal size, and the subsequent application of dynamic programming (DP) for finding the best solution within this constructed arborescence. The second contribution is to be found in the application of ACO [68.3] to the MWRA problem. This algorithm is described in Sect. 68.4. Finally, in Sect. 68.5 an exhaustive experimental evaluation of both algorithms in comparison with an existing heuristic from the literature [68.5] is presented. The chapter is concluded in Sect. 68.6.

68.2 The Minimum-Weight Rooted Arborescence Problem

As mentioned before, in this work we consider the MWRA problem, which is a generalization of the problem proposed by *Venkata Rao* and *Sridharan* in [68.5, 6]. The MWRA problem can technically be described as follows. Given is a directed acyclic graph $G = (V, A)$ with integer weights on the arcs, that is, for each $a \in A$ exists a corresponding weight $w(a) \in \mathbb{Z}$. Moreover, a vertex $v_r \in V$ is designated as the *root vertex*. Let $\mathcal{A}$ be the set of all arborescences in G that are rooted in v_r. In this context, note that an arborescence is a directed, rooted tree in which all arcs point away from the root vertex (see also [68.7]). Moreover, note that $\mathcal{A}$ contains all arborescences, not only those with maximal size. The objective function value (that is, the weight) $f(T)$ of an arboresence $T \in \mathcal{A}$ is defined as follows:

$$f(T) := \sum_{a \in T} w(a) . \tag{68.1}$$

The goal of the MWRA problem is to find an arboresence $T^* \in \mathcal{A}$ such that the weight of T^* is smaller or equal to all other arborescences in $\mathcal{A}$. In other words, the goal is to minimize objective function $f(\cdot)$. An example of the MWRA problem is shown in Fig. 68.1.

The differences to the problem proposed in [68.5] are as follows. The authors of [68.5] require the root vertex v_r to have only one single outgoing arc. Moreover, numbering the vertices from 1 to $|V|$, the given acyclic graph G is restricted to contain only arcs $a_{i,j}$ such that $i < j$. These restrictions do not apply to the MWRA problem. Nevertheless, as a generalization of the problem proposed in [68.5], the MWRA problem is NP-hard. Concerning the existing work, the literature only offers the heuristic proposed in [68.5], which can also be applied to the more general MWRA problem.

The definition of the MWRA problem as previously outlined is inspired by a novel method which was recently proposed in [68.4] for the automated reconstruction of consistent tree structures from noisy images, which is an important problem, for example, in Neuroscience. Tree-like structures, such as dendritic, vascular, or bronchial networks, are pervasive in biological systems. Examples are 2D retinal fundus images and 3D optical micrographs of neurons. The approach proposed in [68.4] builds a set of candidate arborescences over many different subsets of points likely to belong to the optimal delineation and then chooses the best one according to a global objective function that combines image evidence with geometric priors (Fig. 68.2, for example). The solution of the MWRA problem (with additional hard and soft constraints) plays an important role in this process. Therefore, developing better algorithms for the MWRA problem may help in composing better techniques for the problem of the automated reconstruction of consistent tree structures from noisy images.

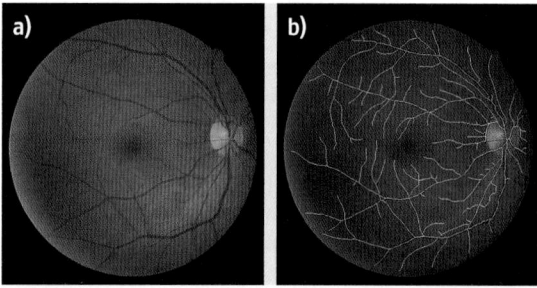

Fig. 68.1a,b (a) An input DAG with eight vertices and 14 arcs. The uppermost vertex is the root vertex v_r. (b) The optimal solution, that is, the arborescence rooted in v_r which has the minimum weight among all arborescence rooted in v_r that can be found in the input graph

Fig. 68.2a,b (a) A 2D image of the retina of a human eye. The problem consists in the automatic reconstruction (or delineation) of the vascular structure. (b) The reconstruction of the vascular structure as produced by the algorithm proposed in [68.4]

68.3 DP-HEUR: A Heuristic Approach to the MWRA Problem

In this section, we propose a new heuristic approach for solving the MWRA problem. First, starting from the root vertex v_r, a spanning arborescence T' in G is constructed as outlined in lines 2–9 of Algorithm 68.1. Second, a DP algorithm is applied to T' in order to obtain the minimum-weight arborescence T that is contained in T' and rooted in v_r. The DP algorithm from [68.8] is used for this purpose. Given an undirected tree $T = (V_T, E_T)$ with vertex and/or edge weights, and any integer number $k \in [0, |V_T| - 1]$, this DP algorithm provides – among all trees with exactly k edges in T – the minimum-weight tree T^*. The first step of the DP algorithm consists in artificially converting the input tree T into a rooted arborescence. Therefore, the DP algorithm can directly be applied to arborescences. Morever, as a side product, the DP algorithm also provides the minimum-weight arborescences for all l with $0 \leq l \leq k$, as well as the minimum-weight arborescences rooted in v_r for all l with $0 \leq l \leq k$. Therefore, given an arborescence of maximal size T', which has $|V| - 1$ arcs (where V is the vertex set of the input graph G), the DP algorithm is applied with $|V| - 1$. Then, among all the minimum-weight arborescences rooted in v_r for $l \leq |V| - 1$, the one with minimum weight is chosen as the output of the DP algorithm. In this way, the DP algorithm is able to generate the minimum-weight arborescence T (rooted in v_r) which can be found in arborescence T'. The heuristic described above is henceforth labeled DP-HEUR. As a final remark, let us mention that for the description of this heuristic, it was assumed that the input graph is connected. Appropriate changes have to be applied to the description of the heuristic if this is not the case.

Algorithm 68.1 Heuristic DP-HEUR for the MWRA problem

1: **input:** a DAG $G = (V, A)$, and a root node v_r
2: $T'_0 := (V'_0 = \{v_r\}, A'_0 = \emptyset)$
3: $A_{\text{pos}} := \{a = (v_q, v_l) \in A \mid v_q \in V'_0, v_l \notin V'_0\}$
4: **for** $i = 1, \ldots, |V| - 1$ **do**
5: $\quad a^* = (v_q, v_l) := \text{argmin}\{w(a) \mid a \in A_{\text{pos}}\}$
6: $\quad A'_i := A'_{i-1} \cup \{a^*\}$
7: $\quad V'_i := V'_{i-1} \cup \{v_l\}$
8: $\quad T'_i := (V'_i, A'_i)$
9: $\quad A_{\text{pos}} := \{a = (v_q, v_l) \in A \mid v_q \in V'_i, v_l \notin V'_i\}$
10: **end for**
11: $T := \text{Dynamic\_Programming}(T'_{|V|-1}, k = |V| - 1)$
12: **output:** arborescence T

68.4 Ant Colony Optimization for the MWRA Problem

The ACO approach for the MWRA problem which is described in the following is a $\mathcal{MAX}$-$\mathcal{MIN}$ Ant System ($\mathcal{MMAS}$) [68.9] implemented in the hyper-cube framework (HCF) [68.10]. The algorithm, whose pseudocode can be found in Algorithm 68.2, works roughly as follows. At each iteration, a number of n_a solutions to the problem is probabilistically constructed based on both pheromone and heuristic information. The second algorithmic component which is executed at each iteration is the pheromone update. Hereby, some of the constructed solutions – that is, the iteration-best solution T^{ib}, the restart-best solution T^{rb}, and the best-so-far solution T^{bs} – are used for a modification of the pheromone values. This is done with the goal of focusing the search over time on high-quality areas of the search space. Just like any other MMAS algorithm, our approach employs restarts consisting of a re-initialization of the pheromone values. Restarts are controlled by the so-called convergence factor (cf) and a Boolean control variable called bs_update. The main functions of our approach are outlined in detail in the following.

Algorithm 68.2 Ant Colony Optimization for the MWRA Problem

1: **input:** a DAG $G = (V, A)$, and a root node v_r
2: $T^{\text{bs}} := (\{v_r\}, \emptyset), \qquad T^{\text{rb}} := (\{v_r\}, \emptyset), \qquad cf := 0,$
$\quad$ bs_update $:=$ **false**
3: $\tau_a := 0.5$ for all $a \in A$
4: **while** termination conditions not met **do**
5: $\quad S := \emptyset$
6: $\quad$ **for** $i = 1, \ldots, n_a$ **do**
7: $\quad\quad T := \text{Construct\_Solution}(G, v_r)$
8: $\quad\quad S := S \cup \{T_i\}$
9: $\quad$ **end for**
10: $\quad T^{\text{ib}} := \text{argmin}\{f(T) \mid T \in S\}$
11: $\quad$ **if** $T^{\text{ib}} < T^{\text{rb}}$ **then** $T^{\text{rb}} := T^{\text{ib}}$
12: $\quad$ **if** $T^{\text{ib}} < T^{\text{bs}}$ **then** $T^{\text{bs}} := T^{\text{ib}}$

Part F | 68.4

13:    ApplyPheromoneUpdate
       $(cf,bs\_update,\mathcal{T},T^{\mathrm{ib}},T^{\mathrm{rb}},T^{\mathrm{bs}})$
14:    $cf := $ ComputeConvergenceFactor$(\mathcal{T})$
15:    **if** $cf > 0.99$ **then**
16:        **if** $bs_update = $ **true then**
17:            $\tau_a := 0.5$ for all $a \in A$
18:            $T^{\mathrm{rb}} := (\{v_r\}, \emptyset)$
19:            $bs_update := $ **false**
20:        **else**
21:            $bs_update := $ **true**
22:        **end if**
23:    **end if**
24: **end while**
25: **output:** T^{bs}, the best solution found by the algorithm

Construct_Solution(G, v_r): This function, first, constructs a spanning arborescence T' in the way which is shown in lines 2–9 of Algorithm 68.1. However, the choice of the next arc to be added to the current arborescence at each step (see line 5 of Algorithm 68.1) is done in a different way. Instead of deterministically choosing from A_{pos}, the arc which has the smallest weight value, the choice is done probabilistically, based on pheromone and heuristic information. The pheromone model $\mathcal{T}$ that is used for this purpose contains a pheromone value τ_a for each arc $a \in A$. The heuristic information $\eta(a)$ of an arc a is computed as follows. First, let

$$w_{\max} := \max\{w(a) \mid a \in A\} . \tag{68.2}$$

Based on this maximal weight of all arcs in G, the heuristic information is defined as follows:

$$\eta(a) := w_{\max} + 1 - w(a) . \tag{68.3}$$

In this way, the heuristic information of all arcs is a positive number. Moreover, the arc with minimal weight will have the highest value concerning the heuristic information. Given an arborescence T'_i (obtained after the ith construction step), and the nonempty set of arcs A_{pos} that may be used for extending T'_i, the probability for choosing arc $a \in A_{\mathrm{pos}}$ is defined as follows

$$\mathbf{p}(a \mid T'_i) := \frac{\tau_a \cdot \eta(a)}{\sum_{\hat{a} \in A_{\mathrm{pos}}} \tau_{\hat{a}} \cdot \eta(\hat{a})} . \tag{68.4}$$

However, instead of choosing an arc from A_{pos} always in a probabilistic way, the following scheme is applied at each construction step. First, a value $r \in [0, 1]$ is chosen uniformly at random. Second, r is compared to

a so-called *determinism rate* $\delta \in [0, 1]$, which is a fixed parameter of the algorithm. If $r \leq \delta$, arc $a^* \in A_{\mathrm{pos}}$ is chosen to be the one with the maximum probability, that is

$$a^* := \mathrm{argmax}\{\mathbf{p}(a \mid T'_i) \mid a \in A_{\mathrm{pos}}\} . \tag{68.5}$$

Otherwise, that is, when $r > \delta$, arc $a^* \in A_{\mathrm{pos}}$ is chosen probabilistically according to the probability values.

The output T of the function Construct_Solution(G, v_r) is chosen to be the minimum-weight arborescence which is encountered during the process of constructing T', that is,

$$T := \mathrm{argmin}\{f(T'_i) \mid i = 0, \ldots, |V| - 1\} . $$

ApplyPheromoneUpdate$(cf, bs\_update, \mathcal{T}, T^{\mathrm{ib}}, T^{\mathrm{rb}}, T^{\mathrm{bs}})$: The pheromone update is performed in the same way as in all $\mathcal{MMAS}$ algorithms implemented in the HCF. The three solutions T^{ib}, T^{rb}, and T^{bs} (as described at the beginning of this section) are used for the pheromone update. The influence of these three solutions on the pheromone update is determined by the current value of the convergence factor cf, which is defined later. Each pheromone value $\tau_a \in \mathcal{T}$ is updated as follows:

$$\tau_a := \tau_a + \rho \cdot (\xi_a - \tau_a) , \tag{68.6}$$

where

$$\xi_a := \kappa_{\mathrm{ib}} \cdot \Delta(T^{\mathrm{ib}}, a) + \kappa_{\mathrm{rb}} \cdot \Delta(T^{\mathrm{rb}}, a) + \kappa_{\mathrm{bs}} \cdot \Delta(T^{\mathrm{bs}}, a) , \tag{68.7}$$

where κ_{ib} is the weight of solution T^{ib}, κ_{rb} the one of solution T^{rb}, and κ_{bs} the one of solution T^{bs}. Moreover, $\Delta(T, a)$ evaluates to 1 if and only if arc a is a component of arborescence T. Otherwise, the function evaluates to 0. Note also that the three weights must be chosen such that $\kappa_{\mathrm{ib}} + \kappa_{\mathrm{rb}} + \kappa_{\mathrm{bs}} = 1$. After the application

Table 68.1 Setting of κ_{ib}, κ_{rb}, and κ_{bs} depending on the convergence factor cf and the Boolean control variable *bs_update*

	bs_update = FALSE			*bs_update*
	cf < 0.7	cf ∈ [0.7, 0.9)	cf ≥ 0.9	TRUE
κ_{ib}	2/3	1/3	0	0
κ_{rb}	1/3	2/3	1	0
κ_{bs}	0	0	0	1

of (68.6), pheromone values that exceed $\tau_{max} = 0.99$ are set back to τ_{max}, and pheromone values that have fallen below $\tau_{min} = 0.01$ are set back to τ_{min}. This prevents the algorithm from reaching a state of complete convergence. Finally, note that the exact values of the weights depend on the convergence factor cf and on the value of the Boolean control variable bs_update. The standard schedule as shown in Table 68.1 has been adopted for our algorithm.

ComputeConvergenceFactor($\mathcal{T}$): The convergence factor (cf) is computed on the basis of the pheromone values

$$cf := 2 \left(\left(\frac{\sum_{\tau_a \in \mathcal{T}} \max\{\tau_{max} - \tau_a, \tau_a - \tau_{min}\}}{|\mathcal{T}| \cdot (\tau_{max} - \tau_{min})} \right) - 0.5 \right).$$

This results in $cf = 0$ when all pheromone values are set to 0.5. On the other side, when all pheromone values have either value τ_{min} or τ_{max}, then $cf = 1$. In all other cases, cf has a value in $(0, 1)$. This completes the description of all components of the proposed algorithm, which is henceforth labeled ACO.

68.5 Experimental Evaluation

The algorithms proposed in this chapter – that is, DP-HEUR and ACO – were implemented in ANSI C++ using GCC 4.4 for compiling the software. Moreover, the heuristic proposed in [68.5] was reimplemented. As mentioned before, this heuristic – henceforth labeled VENSRI – is the only existing algorithm which can directly be applied to the MWRA problem. All three algorithms were experimentally evaluated on a cluster of PCs equipped with Intel Xeon X3350 processors with 2667 MHz and 8 Gb of memory. In the following, we first describe the set of benchmark instances that have been used to test the three algorithms. Afterward, the algorithm tuning and the experimental results are described in detail.

68.5.1 Benchmark Instances

A diverse set of benchmark instances was generated in the following way. Three parameters are necessary for the generation of a benchmark instance $G = (V, A)$. Hereby, n and m indicate, respectively, the number of vertices and the number of arcs of G, while $q \in [0, 1]$ indicates the probability for the weight of any arc to be positive (rather than negative). The process of the generation of an instance starts by constructing a random arborescence T with n vertices. The root vertex of T is called v_r. Each of the remaining $m - n + 1$ arcs was generated by randomly choosing two vertices v_i and v_j, and adding the corresponding arc $a = (v_i, v_j)$ to T. In this context, $a = (v_i, v_j)$ may be added to T, if and only if by its addition no directed cycle is produced, and neither (v_i, v_j) nor (v_j, v_i) form already part of the graph. The weight of each arc was chosen by, first, deciding with probability q if the weight is to be positive (or nonpositive). In the case of a positive weight, the weight value was chosen uniformly at random from $[1, 100]$, while in the case of a nonpositive weight, the weight value was chosen uniformly at random from $[-100, 0]$.

In order to generate a diverse set of benchmark instances, the following values for n, m, and q were considered:

- $n \in \{20, 50, 100, 500, 1000, 5000\}$;
- $m \in \{2n, 4n, 6n\}$;
- $q \in \{0.25, 0.5, 0.75\}$.

For each combination of n, m, and q, a total of 10 problem instances were generated. This resulted in a total of 540 problem instances, that is, 180 instances for each value of q.

68.5.2 Algorithm Tuning

The proposed ACO algorithm has several parameters that require appropriate values. The following parameters, which are crucial for the working of the algorithm, were chosen for tuning:

- $n_a \in \{3, 5, 10, 20\}$: the number of ants (solution constructions) per iteration;
- $\rho \in \{0.05, 0.1, 0.2\}$: the learning rate;
- $\delta \in \{0.0, 0.4, 0.7, 0.9\}$: the determinism rate.

We chose the first problem instance (out of 10 problem instances) for each combination of n, m, and q for tuning. A full factorial design was utilized. This means that ACO was applied (exactly once) to each of the problem instances chosen for tuning. The stopping criterion was fixed to 20 000 solution evaluations for each application of ACO. For analyzing the results, we used a rank-based analysis. However, as the set of problem instances is quite diverse, this rank-based analysis was performed separately for six subsets of instances. For defining these subsets, we refer to the instances with $n \in \{20, 50, 100\}$ as *small instances*, and the remaining ones as *large instances*. With this definition, each of the three subsets of instances concerning the three different values for q, was further separated into two subsets concerning the instance size. For each of these six subsets, we used the parameter setting with which ACO achieved the best average rank for the corresponding tuning instances. These parameter settings are given in Table 68.2.

68.5.3 Results

The three algorithms considered for the comparison were applied exactly once to each of the 540 problem instances of the benchmark set. Although ACO is a stochastic search algorithm, this is a valid choice, because results are averaged over groups of instances that were generated with the same parameters. As in the case of the tuning experiments, the stopping criterion for ACO was fixed to 20 000 solution evaluations. Tables 68.3–68.5 present the results averaged – for each algorithm – over the 10 instances for each combination of n and m (as indicated in the first two table columns). Four table columns are used for presenting the results of each algorithm. The column with heading *value* provides the average of the objective function values of the best solutions found by the respective algorithm for the 10 instances of each combination of n and m. The second column (with heading *std*) contains the corresponding standard deviation. The third column (with heading *size*) indicates the average size (in terms of the number or arcs) of the best solutions found by the respective algorithm (remember that solutions – that is, arborescences – may have any number of arcs between 0 and $|V| - 1$, where $|V|$ is the number of the input DAG $G = (V, A)$). Finally, the fourth column (with heading *time (s)*) contains the average computation time (in sec-

onds). For all three algorithms, the computation time indicates the time of the algorithm termination. In the case of ACO, an additional table column (with heading *evals*) indicates at which solution evaluation, on average, the best solution of a run was found. Finally, for each combination of n and m, the result of the best-performing algorithm is indicated in bold font.

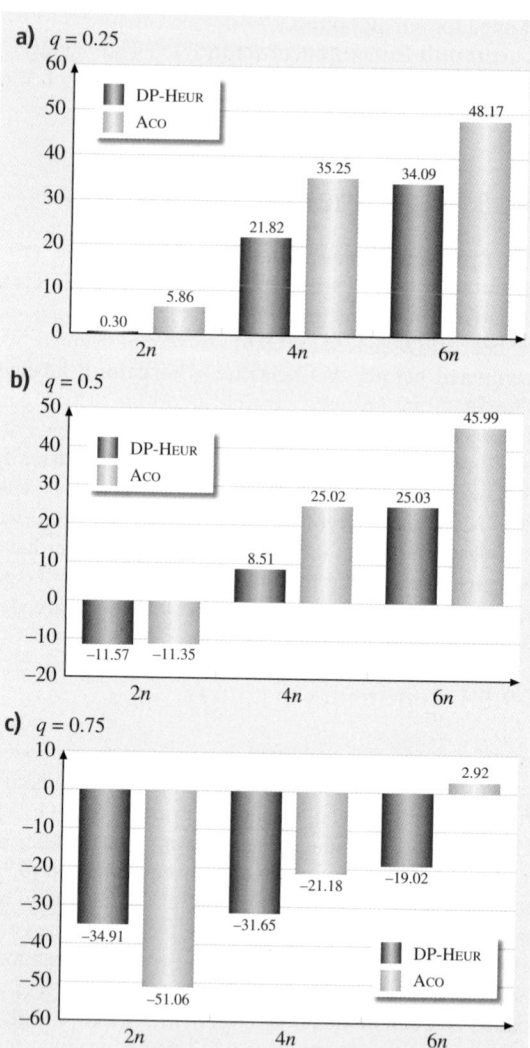

Fig. 68.3a–c Average improvement (in %) of ACO and DP-HEUR over VENSRI. Positive values correspond to an improvement, while negative values indicate that the respective algorithm is inferior to VENSRI. The improvement is shown for the three different arc-densities that are considered in the benchmark set, that is, $m = 2n$, $m = 4n$, and $m = 6n$

Table 68.2 Parameter setting (concerning ACO) used for the final experiments

	$q = 0.25$	$q = 0.5$	$q = 0.75$
Small instances	$n_a = 20$	$n_a = 20$	$n_a = 5$
	$\rho = 0.2$	$\rho = 0.2$	$\rho = 0.05$
	$\delta = 0.7$	$\delta = 0.7$	$\delta = 0.4$
Large instances	$n_a = 20$	$n_a = 20$	$n_a = 20$
	$\rho = 0.2$	$\rho = 0.2$	$\rho = 0.2$
	$\delta = 0.9$	$\delta = 0.9$	$\delta = 0.9$

Table 68.3 Experimental results for the 180 instances with $q = 0.25$. ACO is compared to the heuristic proposed in this work (DP-HEUR), and the algorithm from [68.5] (VENSRI)

n	m	DP-HEUR				VENSRI				ACO				
		Value	Std	Size	Time (s)	Value	Std	Size	Time (s)	Value	Std	Size	Evals	Time (s)
20	2n	−855.40	(133.68)	17.00	<0.01	−845.60	(150.48)	17.40	<0.01	−962.70	(147.81)	17.70	2543.70	0.62
	4n	−1093.10	(155.68)	18.10	<0.01	−999.20	(85.56)	18.30	<0.01	−1248.10	(113.66)	18.10	4435.00	0.74
	6n	−1138.20	(201.72)	17.70	<0.01	−1024.00	(131.08)	18.10	<0.01	−1321.00	(139.52)	18.30	5904.00	0.80
50	2n	−2089.30	(250.78)	42.10	<0.01	−2066.70	(284.37)	44.40	<0.01	−2379.10	(237.14)	44.00	6522.60	1.21
	4n	−3022.80	(258.93)	47.00	<0.01	−2560.10	(266.46)	47.60	<0.01	−3314.40	(188.43)	47.60	9299.60	1.41
	6n	−3281.40	(221.73)	48.20	<0.01	−2522.60	(215.18)	48.60	<0.01	−3620.30	(131.79)	48.50	13101.90	1.58
100	2n	−4260.00	(402.24)	87.70	<0.01	−4325.40	(326.54)	91.20	<0.01	−4944.00	(436.59)	90.20	10949.00	2.19
	4n	−5386.50	(364.51)	92.70	<0.01	−4598.90	(209.84)	95.70	0.01	−6161.10	(291.93)	94.80	16187.30	2.99
	6n	−6222.80	(315.85)	95.30	<0.01	−5015.80	(304.61)	96.40	0.01	−7048.60	(229.84)	96.20	17028.70	3.66
500	2n	−20899.10	(1169.37)	438.70	0.06	−21040.80	(625.26)	461.50	1.79	−23501.40	(796.48)	454.70	17159.70	30.52
	4n	−27856.50	(1275.43)	473.30	0.07	−22648.40	(696.45)	484.70	1.78	−30853.00	(908.66)	480.00	18153.90	48.36
	6n	−30372.50	(889.92)	482.00	0.06	−23487.20	(717.32)	490.00	1.83	−33892.20	(793.20)	485.10	16940.30	56.65
1000	2n	−42022.20	(2058.38)	876.60	0.24	−41817.70	(1178.70)	922.60	14.21	−45498.70	(1950.29)	916.20	17749.30	110.68
	4n	−53728.80	(1987.03)	946.90	0.26	−44435.50	(1431.45)	971.80	14.64	−60574.80	(960.59)	960.10	16568.40	172.62
	6n	−61829.30	(1621.11)	965.70	0.25	−45723.40	(884.88)	981.90	14.44	−67974.40	(1054.47)	973.30	16376.90	242.84
5000	2n	−206085.80	(2554.80)	4365.00	4.99	−205287.40	(2691.58)	4591.30	1939.72	−214243.30	(2850.45)	4576.90	15441.40	2715.48
	4n	−266109.50	(8892.83)	4728.30	6.51	−217967.60	(1604.04)	4857.60	2009.05	−294401.20	(4810.59)	4818.90	14802.70	4518.43
	6n	−297185.10	(7903.69)	4815.00	6.41	−220559.40	(3876.02)	4909.20	2015.12	−328173.90	(4492.29)	4874.30	16287.20	5752.23

Part F | 68.5

Table 68.4 Experimental results for the 180 instances with $q = 0.5$. ACO is compared to the heuristic proposed in this work (DP-HEUR), and the algorithm from [68.5] (VENSRI)

n	m	DP-HEUR Value	Std	Size	Time (s)	VENSRI Value	Std	Size	Time (s)	ACO Value	Std	Size	Evals	Time (s)
20	2n	−524.50	(134.16)	12.60	< 0.01	−569.10	(156.69)	14.90	< 0.01	−631.30	(171.28)	15.00	4766.50	0.62
	4n	−831.60	(230.68)	15.90	< 0.01	−806.30	(108.14)	17.40	< 0.01	−1009.90	(149.29)	17.30	4812.50	0.72
	6n	−1031.10	(197.50)	17.70	< 0.01	−947.10	(151.05)	17.80	< 0.01	−1210.70	(152.26)	17.90	3920.30	0.83
50	2n	−1246.30	(273.88)	33.60	< 0.01	−1476.70	(295.11)	38.50	< 0.01	−1584.70	(294.18)	39.20	9133.00	1.28
	4n	−1912.30	(432.79)	39.70	< 0.01	−1812.30	(208.43)	43.80	< 0.01	−2466.90	(278.53)	43.50	11 435.80	1.40
	6n	−2372.70	(368.03)	43.60	< 0.01	−2166.10	(307.75)	45.70	< 0.01	−2923.40	(265.90)	44.90	10 244.00	1.62
100	2n	−2523.10	(442.91)	67.10	< 0.01	−2828.70	(409.73)	76.20	0.01	−3187.60	(436.84)	74.90	12 870.70	2.33
	4n	−3903.00	(659.69)	82.30	< 0.01	−3871.70	(305.29)	89.90	0.02	−5031.30	(375.58)	89.20	14 445.10	3.08
	6n	−4819.40	(582.18)	87.30	< 0.01	−4059.70	(374.22)	93.10	0.02	−5867.70	(423.46)	91.20	16 744.20	3.64
500	2n	−12 404.50	(1308.74)	348.90	0.06	−14 085.50	(608.59)	398.70	2.12	−14 381.90	(750.38)	408.60	18 360.90	31.48
	4n	−18 321.80	(2222.19)	402.00	0.06	−17 256.00	(703.46)	449.20	2.28	−22 194.00	(1901.21)	447.80	18 880.40	40.71
	6n	−22 386.60	(2202.23)	434.90	0.06	−18 896.40	(739.65)	471.60	2.38	−27 130.80	(688.55)	458.80	16 207.20	53.55
1000	2n	−24 493.80	(1577.30)	671.60	0.23	−26 995.80	(995.40)	770.10	17.40	−26 046.20	(1336.03)	786.30	17 380.20	115.79
	4n	−37 715.40	(3030.59)	811.80	0.23	−34 317.50	(1461.89)	905.10	18.69	−44 061.00	(1602.97)	893.20	17 454.10	160.78
	6n	−45 280.10	(2376.76)	875.00	0.27	−36 790.50	(846.78)	941.40	19.41	−53 867.00	(1368.28)	920.20	16 219.20	212.90
5000	2n	−119 122.90	(4980.74)	3371.60	5.23	−135 333.80	(2296.56)	3921.10	2440.70	−114 887.00	(4657.51)	3733.50	15 223.20	2674.64
	4n	−177 605.60	(7388.53)	4045.10	6.42	−163 385.60	(2153.92)	4550.00	2585.65	−202 096.40	(7870.45)	4399.70	14 275.70	3982.99
	6n	−217 112.00	(12 667.37)	4325.60	7.29	−171 483.70	(2839.81)	4707.20	2679.99	−251 128.00	(4891.47)	4568.70	16 443.00	4905.29

Table 68.5 Experimental results for the 180 instances with $q = 0.75$. ACO is compared to the heuristic proposed in this work (DP-HEUR), and the algorithm from [68.5] (VENSRI)

n	m	DP-HEUR				VENSRI				ACO				
		Value	Std	Size	Time (s)	Value	Std	Size	Time (s)	Value	Std	Size	Evals	Time (s)
20	2n	−186.50	(134.06)	7.80	< 0.01	−229.70	(157.66)	10.00	< 0.01	**−242.20**	(164.32)	9.90	1431.00	1.27
	4n	−391.50	(133.13)	9.90	< 0.01	−479.30	(107.17)	12.80	< 0.01	**−560.70**	(106.44)	13.60	6451.30	1.44
	6n	−549.10	(212.91)	12.60	< 0.01	−698.10	(174.68)	15.70	< 0.01	**−838.30**	(147.90)	15.70	5398.70	1.51
50	2n	−488.00	(170.61)	19.70	< 0.01	−580.80	(167.41)	24.70	< 0.01	**−642.70**	(202.15)	23.80	7511.10	1.74
	4n	−861.10	(287.25)	26.60	< 0.01	−1125.50	(160.68)	35.50	< 0.01	**−1336.60**	(146.24)	34.00	10451.90	1.89
	6n	−1231.40	(344.87)	31.20	< 0.01	−1449.90	(110.20)	41.10	< 0.01	**−1817.80**	(239.20)	38.60	11572.90	2.08
100	2n	−907.90	(342.95)	34.10	< 0.01	−1182.90	(384.00)	45.50	< 0.01	**−1224.50**	(449.23)	45.30	12586.30	2.66
	4n	−1766.80	(226.75)	53.90	< 0.01	−2216.20	(272.32)	70.90	< 0.01	**−2605.40**	(311.13)	68.20	15318.80	3.30
	6n	−2787.80	(527.96)	67.00	< 0.01	−2938.20	(294.57)	80.50	< 0.01	**−3811.80**	(416.14)	80.80	16809.60	3.92
500	2n	−4647.40	(804.52)	198.00	0.06	**−6495.10**	(828.27)	263.40	0.02	−5183.40	(900.98)	268.40	17705.80	31.59
	4n	−7723.40	(1393.23)	263.30	0.06	**−10393.40**	(599.96)	367.50	0.02	−10386.10	(1215.50)	340.20	17027.00	41.39
	6n	−11642.30	(1255.86)	311.20	0.07	−12691.90	(608.40)	403.50	0.03	**−15203.50**	(849.56)	396.80	17841.20	53.84
1000	2n	−8323.40	(1555.82)	367.30	0.23	**−11908.10**	(1371.03)	483.60	2.85	−8009.90	(1106.18)	473.70	18558.30	116.37
	4n	−14434.90	(1298.71)	518.90	0.28	**−20508.10**	(1527.07)	725.20	3.39	−18301.00	(1484.42)	693.30	16070.90	158.25
	6n	−22564.10	(1500.62)	621.80	0.29	−25288.30	(629.85)	822.50	3.72	**−28023.90**	(1994.36)	755.10	16971.50	194.97
5000	2n	−36776.60	(2336.68)	1831.10	5.43	**−58465.20**	(1821.57)	2473.00	24.04	−23289.00	(1874.45)	1896.70	16259.30	2710.57
	4n	−68235.80	(9123.95)	2563.50	6.80	**−101955.80**	(1607.50)	3663.00	28.28	−74538.30	(5125.67)	3172.50	16524.40	3656.49
	6n	−96284.50	(7465.28)	2994.70	4.25	**−123714.50**	(1310.01)	4120.80	30.79	−121960.40	(4233.52)	3707.60	14190.90	4459.02

Part F | 68.5

Concerning the 180 instances with $q = 0.25$, the results allow us to make the following observations. First, ACO is for all combinations of n and m the best-performing algorithm. Averaged over all problem instances ACO obtains an improvement of 29.8% over VENSRI. Figure 68.3a shows the average improvement of ACO over VENSRI for three groups of input instances concerning the different arc densities. It is interesting to observe that the advantage of ACO over VENSRI seems to grow when the arc density increases. On the downside, these improvements are obtained at the cost of a significantly increased computation time. Concerning heuristic DP-HEUR, we can observe that it improves over VENSRI for all combinations of n and m, apart from $(n = 100, m = 2n)$ and $(n = 500, m = 2n)$. This seems to indicate that, also for DP-HEUR, the sparse instances pose more of a challenge than the dense instances. Averaged over all problem instances, DP-HEUR obtains an improvement of 18.6% over VENSRI. The average improvement of DP-HEUR over VEN-SRI is shown for the three groups of input instances concerning the different arc-densities in Fig. 68.3a. Concerning a comparison of the computation times, we can state that DP-HEUR has a clear advantage over VENSRI especially for large-size problem instances.

Concerning the remaining 360 instances ($q = 0.5$ and $q = 0.75$), we can make the following additional observations. First, both ACO and DP-HEUR seem to experience a downgrade in performance (in comparison to the performance of VENSRI) when q increases. This holds especially for rather large and rather sparse graphs. While both algorithms still obtain an average improvement over VENSRI in the case of $q = 0.5$ – that is, 19.9% improvement in the case of ACO and 7.3% in the case of DP-HEUR – both algorithms are on average inferior to VENSRI in the case of $q = 0.75$.

Finally, Fig. 68.4 presents the information which is contained in column *size* of Tables 68.3–68.5 in graphical form. It is interesting to observe that the solutions produced by DP-HEUR consistently seem to be the smallest ones, while the solutions produced by VENSRI seem generally to be the largest ones. The size of the solutions produced by ACO is generally in between these two extremes. Moreover, with growing q the difference in solution size as produced by the three algorithms seems to be more pronounced. We currently have no explanation for this aspect, which certainly deserves further examination.

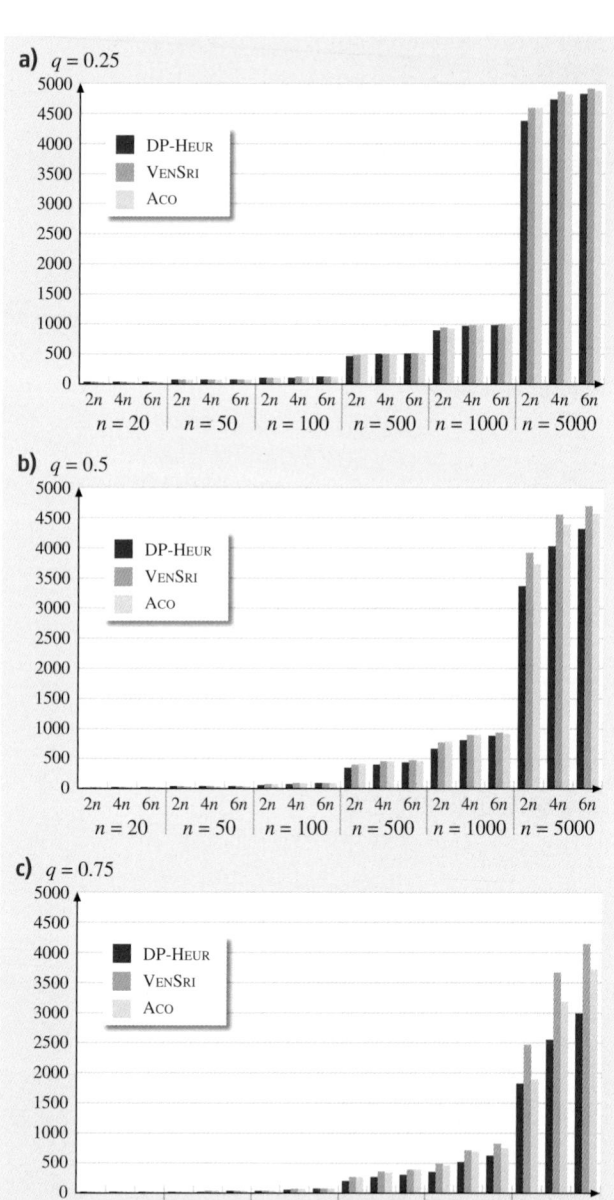

Fig. 68.4 These graphics show, for each combination of n and m, information about the average size – in terms of the number of arcs – of the solutions produced by DP-HEUR, ACO, and VENSRI

68.6 Conclusions and Future Work

In this work, we have proposed a heuristic and an ACO approach for the minimum-weight rooted arboresence problem. The heuristic makes use of dynamic programming as a subordinate procedure. Therefore, it may be regarded as a hybrid algorithm. In contrast, the proposed ACO algorithm is a pure metaheuristic approach. The experimental results show that both approaches are superior to an existing heuristic from the literature in those cases in which the number of arcs with positive weights is not too high and in the case of rather dense graphs. However, as far as sparse graphs with a rather large fraction of positive weights are concerned, the existing heuristic from the literature seems to have advantages over the algorithms proposed in this chapter.

Concerning future work, we plan to develop a hybrid ACO approach which makes use of dynamic programming as a subordinate procedure, in a way similar to the proposed heuristic. Moreover, we plan to implement an integer programming model for the tackled problem – in the line of the model proposed in [68.11] for a related problem – and to solve the model with an efficient integer programming solver.

References

68.1   R. Poli, J. Kennedy, T. Blackwell: Particle swarm optimization – an overview, Swarm Intell. **1**(1), 33–57 (2007)

68.2   M.F. Tasgetiren, Q.-K. Pan, P.N. Suganthan, A.H.-L. Chen: A discrete artificial bee colony algorithm for the total flowtime minimization in permutation flow shops, Inf. Sci. **181**(16), 3459–3475 (2011)

68.3   M. Dorigo, T. Stützle: *Ant Colony Optimization* (MIT, Cambridge 2004)

68.4   E. Türetken, G. González, C. Blum, P. Fua: Automated reconstruction of dendritic and axonal trees by global optimization with geometric priors, Neuroinformatics **9**(2/3), 279–302 (2011)

68.5   V. Venkata Rao, R. Sridharan: Minimum-weight rooted not-necessarily-spanning arborescence problem, Networks **39**(2), 77–87 (2002)

68.6   V. Venkata Rao, R. Sridharan: The minimum weight rooted arborescence problem: Weights on arcs case, IIMA Working Papers WP1992-05-01_01106 (Indian Institute of Management Ahmedabad, Research and Publication Department, Ahmedabad 1992)

68.7   W.T. Tutte: *Graph Theory* (Cambridge Univ. Press, Cambridge 2001)

68.8   C. Blum: Revisiting dynamic programming for finding optimal subtrees in trees, Eur. J. Oper. Res. **177**(1), 102–114 (2007)

68.9   T. Stützle, H.H. Hoos: $\mathcal{MAX}$-$\mathcal{MIN}$ ant system, Future Gener. Comput. Syst. **16**(8), 889–914 (2000)

68.10   C. Blum, M. Dorigo: The hyper-cube framework for ant colony optimization, IEEE Trans. Syst. Man Cybern. B **34**(2), 1161–1172 (2004)

68.11   C. Duhamel, L. Gouveia, P. Moura, M. Souza: Models and heuristics for a minimum arborescence problem, Networks **51**(1), 34–47 (2008)

69. An Intelligent Swarm of Markovian Agents

Dario Bruneo, Marco Scarpa, Andrea Bobbio, Davide Cerotti, Marco Gribaudo

We define a Markovian agent model (MAM) as an analytical model formed by a spatial collection of interacting Markovian agents (MAs), whose properties and behavior can be evaluated by numerical techniques. MAMs have been introduced with the aim of providing a flexible and scalable framework for distributed systems of interacting objects, where both the local properties and the interactions may depend on the geographical position. MAMs can be proposed to model biologically inspired systems since they are suited to cope with the four common principles that govern swarm intelligence: positive feedback, negative feedback, randomness, and multiple interactions. In the present work, we report some results of a MAM for a wireless sensor network (WSN) routing protocol based on swarm intelligence, and some preliminary results in utilizing MAs for very basic ant colony optimization (ACO) benchmarks.

69.1 Swarm Intelligence: A Modeling Perspective

Swarm intelligent (SI) algorithms are variously inspired from the way in which colonies of biological organisms self-organize to produce a wide diversity of functions [69.1, 2]. Individuals of the colony have a limited knowledge of the overall behavior of the system and follow a small set of rules that may be influenced by the interaction with other individuals or by modifications produced in the environment. The collective behavior of large groups of relatively simple individuals, interacting only locally with few neighboring elements, produces global patterns. Even if many approaches have been proposed that differentiate in many respects, four basic common principles have been isolated that govern SI:

- *Positive feedback*
- *Negative feedback*
- *Randomness*
- *Multiple interactions.*

The same four principles also govern a class of algorithms inspired by the expansion dynamics of slime molds in the search for food [69.3, 4], that have been utilized as the base for the generation of routing protocols in wireless sensor networks (WSNs).

Through the adoption of the above four principles, it is possible to design distributed, self-organizing, and fault tolerant algorithms able to self-adapt to the environmental changes, that present the following main properties [69.1]:

i) Single individuals are assumed to be simple with low computational intelligence and communication capabilities.

ii) Individuals communicate indirectly, through modification of the environment (this property is known as *stigmergy* [69.2]).
iii) The range of the interaction may be very short; nevertheless, a robust global behavior emerges from the interaction of the nodes.
iv) The global behavior adapts to topological and environmental changes.

The usual way to study such systems is through simulation, due to the large number of involved individuals that lead to the well-known state explosion problem. Analytical techniques are preferable if, starting from the peculiarities of SI systems, they allow to realize effective and scalable models. Along this line, new stochastic entities, called Markovian agents (MAs) [69.5, 6] have been introduced with the aim of providing a flexible, powerful, and scalable technique for modeling complex systems of distributed interacting objects, for which feasible analytical and numerical solution algorithms can be implemented. Each object has its own local behavior that can be modified by the mutual interdependences with the other objects. MAs are scattered over a geographical area and retain their spatial position so that the local behavior and the mutual interdependencies may be related to their geographical positions and other features like the transmittance characteristics of the interposed medium. MAs are modeled by a discrete-state continuous-time finite Markov chain (CTMC) whose infinitesimal generator is influenced by the interaction with other MAs. The interaction among agents is represented by a *message passing model* combined with a *perception function*. When residing in a state or during a transition, an MA is allowed to send messages that are perceived by the other MAs, according to a spatial-dependent *perception function*, modifying their behavior. Messages may model real physical messages (as in WSNs) or simply the mutual influences of an MA over the other ones.

The flexibility of the MA representation, the spatial dependency, and the mutual interaction through message passing and perception function, make MA models suited to cope with various biologically inspired mechanisms governed by the four aforementioned principles. In fact, the MAM, whose constituent elements are the MAs, was specifically studied to cope with the following needs [69.6]:

i) Provide analytical models that can be solved by numerical techniques, thus avoiding the need of long and expensive simulation runs.
ii) Provide a flexible and scalable modeling framework for distributed systems of interacting objects.
iii) Provide a framework in which local properties can be coupled with global properties.
iv) Local and global properties and interactions may depend on the position of the objects in the space (space-sensitive models).
v) The solution algorithm self-adapts to variations in the system topology and in the interaction mechanisms.

Interactive Markovian agents have been first introduced in [69.5, 7] for single class MAs and then extended to multiclass multimessage Markovian agent model in successive works [69.8–10]. In [69.9, 11, 12], MAs have been applied to routing algorithms in WSNs, adopting SI principles [69.13].

This work describes the structure of MAMs and the numerical solution algorithms in Sect. 69.2. Then, applications derived from biological models are presented: a swarm intelligent algorithm for routing protocols in WSNs (Sect. 69.3) and a simple ant colony optimization (ACO) example (Sect. 69.4).

69.2 Markovian Agent Models

The structure of a single MA is represented in Fig. 69.1. States $i, j, \ldots, k$ are the states of the CTMC representing the MA. The transitions among the states are of two possible types and are drawn in a different way:

- Solid lines (like the transition from i to j or the self-loops in i or in j) indicate the fixed component of the infinitesimal generator and represent the local or autonomous behavior of the object that is independent of the interaction with the other MAs (like, for instance, the time-to-failure distribution, or the reaction to an external stimulus). Note that we include in the representation also self-loop transitions that require a particular notation since they are not visible in the infinitesimal generator of the CTMC [69.14].
- Dashed lines (like the transition from i to k or the transitions entering into i or j from other states

not shown in the figure) represent the transitions induced by the interaction with the other MAs. The way in which the rates of the induced transitions are computed is explained in the following section.

During a local transition (or a self-loop) an MA can emit a message of any type with an assigned probability, as represented by the dotted arrows in Fig. 69.1 emerging from the solid transitions. The pair $\langle g_{ij}, m \rangle$ denotes both the message generation probability and the message type. Messages generated by an MA may be perceived by other MAs with a given probability, according to a suitable perception function, and the interaction mechanism between emitted messages and perceived messages generates the induced transitions (dashed lines). The pair $\langle m, a_{ik} \rangle$ denotes both the type of the perceived message and the corresponding acceptance probability.

An MAM is a collection of interacting MAs defined over a geographical space $\mathcal{V}$. Given a position $\mathbf{v}$ inside $\mathcal{V}$, $\rho(\mathbf{v})$ denotes the density of MAs in $\mathbf{v}$. According to the definition of the density $\rho(\mathbf{v})$, we can classify a MAM with the following taxonomy:

- An MAM is *static* if $\rho(\mathbf{v})$ does not depend on time, and *dynamic* if it does depend on time.
- An MAM is *discrete* if the geographical area on which the MAs are deployed is discretized and $\rho(\mathbf{v})$ is a discrete function of the space or it is *continuous* if $\rho(\mathbf{v})$ is a continuous function of the space.

Further, MAs may belong to a single class or to different classes with different local behaviors and interaction capabilities, and messages may belong to different types where each type induces a different effect on the interaction mechanism. The perception function describes how a message of a given type emitted by an MA of a given class in a given position in the space

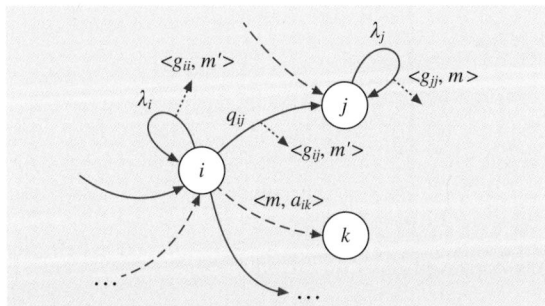

Fig. 69.1 Schematic structure of a Markovian agent

is perceived by an MA of a given class in a different position.

69.2.1 Mathematical Formulation

A *multiple agent class, multiple message type* MAM is defined by the tuple [69.12]

$$\text{MAM} = \{C, \mathcal{M}, \mathcal{V}, \mathcal{U}, \mathcal{R}\}, \tag{69.1}$$

where $C = \{1, \ldots, C\}$ is the set of agent classes. We denote with MA^c an agent of class $c \in C$. $\mathcal{M} = \{1, \ldots, M\}$ is the set of message types. Each agent (independently of its class) can send or receive messages of type $m \in \mathcal{M}$. $\mathcal{V}$ is the finite space over which Markovian agents are spread. $\mathcal{U} = \{u^1(\cdot), \ldots, u^M(\cdot)\}$ is a set of M perception functions (one for each message type). $\mathcal{R} = \{\rho^1(\cdot), \ldots, \rho^C(\cdot)\}$ is a set of C agent density functions (one for each agent class). Each agent MA^c of class c is characterized by a state space with n_c states, and it is defined by the tuple

$$\text{MA}^c = \{\mathbf{Q}^c(\mathbf{v}), \mathbf{\Lambda}^c(\mathbf{v}), \mathbf{G}^c(\mathbf{v}, m), \mathbf{A}^c(\mathbf{v}, m), \pi_0^c(\mathbf{v})\}, \tag{69.2}$$

where $\mathbf{Q}^c(\mathbf{v})$ is the local component of the infinitesimal generator; $\mathbf{\Lambda}^c(\mathbf{v})$ is the vector of the self-jump transition rates; $\mathbf{G}^c(\mathbf{v}, m)$ is the matrix containing the probabilities of generating a message of type m; $\mathbf{A}^c(\mathbf{v}, m)$ is the matrix containing the probabilities of accepting a message of type m; $\pi_0^c(\mathbf{v})$ is the initial probability vector.

Note that even if the structure of the CTMC associated to each MA of a given class is the same for all the objects, the values of the parameters may depend on position $\mathbf{v}$ and, therefore, may vary from MAs belonging to the same class.

An MAM can be analyzed solving a set of coupled differential equations. Let us call $\rho_i^c(t, \mathbf{v})$ the density of agents of class c, in state i, located in position $\mathbf{v}$ at time t. In the following, we will focus on static MAMs thus assuming that the total density of agents in position $\mathbf{v}$ remains constant over the time; we have that

$$\sum_{i=1}^{n_c} \rho_i^c(t, \mathbf{v}) = \rho^c(\mathbf{v}), \quad \forall \mathbf{v}, \forall t \geq 0. \tag{69.3}$$

We collect the state densities into a vector $\rho^c(t, \mathbf{v}) = [\rho_i^c(t, \mathbf{v})]$ and we are interested in computing the transient evolution of $\rho^c(t, \mathbf{v})$.

From the above definitions, we can compute the total rate $\beta_j^c(\mathbf{v}, m)$ at which messages of type m are generated by an agent of class c in state j in position $\mathbf{v}$

$$\beta_j^c(\mathbf{v}, m) = \lambda_j^c(\mathbf{v}) \, g_{jj}^c(\mathbf{v}, m)$$
$$+ \sum_{k \neq j} q_{jk}^c(\mathbf{v}) \, g_{jk}^c(\mathbf{v}, m) \,, \qquad (69.4)$$

where the first term on the right-hand isde is the contribution of the messages of type m emitted during a self-loop from j and the second term is the contribution of messages of type m emitted during a transition from j to any k $(\neq j)$.

The interdependences among MAs are ruled by a set of perception functions whose general form is

$$u^m(c, \mathbf{v}, i, c', \mathbf{v}', j) \,. \qquad (69.5)$$

The perception function $u^m(\cdot)$ in (69.5) represents how an MA of class c in position $\mathbf{v}$ in state i perceives the messages of type m emitted by an MA of class c' in position $\mathbf{v}'$ in state j. The functional form of $u^m(\cdot)$ identifies the perception mechanisms and must be specified for any given application since it determines how an MA is influenced by the messages emitted by the other MAs. The transition rates of the induced transitions are primarily determined by the structure of the perception function.

A pictorial and intuitive representation of how the perception function $u^m(c, \mathbf{v}, i, c', \mathbf{v}', j)$ acts, is given in Fig. 69.2. The MA in the top right portion of the figure in position $\mathbf{v}'$ broadcasts a message of type m from state j that propagates in the geographical area until reaches the MA in the bottom left portion of the figure in position $\mathbf{v}$ and in state i. Upon acceptance of the message according to the acceptance probability $a_{ik}(\mathbf{v}, m)$, an induced transition from state i to state k (represented by a dashed line) is triggered in the model.

With the above definitions we are now in the position to compute the components of the infinitesimal generator of an MA that depends on the interaction with the other MAs and that constitutes the original and innovative part of the approach.

We define $\gamma_{ii}^c(t, \mathbf{v}, m)$ the total rate at which messages of type m coming from the whole volume $\mathcal{V}$ are perceived by an MA of class c in state i in location $\mathbf{v}$.

$$\gamma_{ii}^c(t, \mathbf{v}, m) = \int_{\mathcal{V}} \sum_{c'=1}^{C} \sum_{j=1}^{n_{c'}} u^m(c, \mathbf{v}, i, c', \mathbf{v}', j)$$
$$\times \beta_j^{c'}(m) \rho_j^{c'}(t, \mathbf{v}') d\mathbf{v}' \,, \qquad (69.6)$$

where $\gamma_{ii}^c(t, \mathbf{v}, m)$ is computed by taking into account the total rate of messages of type m emitted by all the MAs in state j and in a given position $\mathbf{v}'$ (the term $\beta_j^c(\mathbf{v}, m)$) times the density of MAs in $\mathbf{v}'$ (the term $\rho_j(t, \mathbf{v}')$) times the perception function (the term $u^m(c, \mathbf{v}, i, c', \mathbf{v}', j)$) summed over all the possible states j and class c' of each MA and integrated over the whole space $\mathcal{V}$. From an MA of class c in position $\mathbf{v}$ and in state i an induced transition to state k (drawn in dashed line) is triggered with rate $\gamma_{ii}^c(t, \mathbf{v}, m) \, a_{ik}(\mathbf{v}, m)$ where $a_{ik}(\mathbf{v}, m)$ is the appropriate entry of the acceptance matrix $\mathbf{A}(\mathbf{v}, m)$.

We collect the rates (69.6) in a diagonal matrix $\boldsymbol{\Gamma}^c(t, \mathbf{v}, m) = \mathrm{diag}(\gamma_{ii}^c(t, \mathbf{v}, m))$. This matrix can be used to compute $\mathbf{K}^c(t, \mathbf{v})$, the infinitesimal generator of a class c agent at position $\mathbf{v}$ at time t

$$\mathbf{K}^c(t, \mathbf{v}) = \mathbf{Q}^c + \sum_m \boldsymbol{\Gamma}^c(t, \mathbf{v}, m)[\mathbf{A}^c(m) - \mathbf{I}] \,. \quad (69.7)$$

The first term on the right-hand side is the local transition rate matrix and the second term contains the rates induced by the interactions.

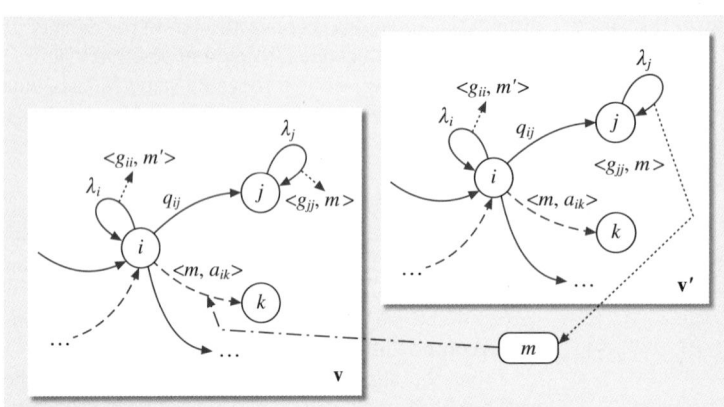

Fig. 69.2 Message passing mechanism ruled by a perception function

The evolution of the entire model can be studied by solving $\forall \mathbf{v}, c$ the following differential equations:

$$\boldsymbol{\rho}^c(0, \mathbf{v}) = \rho^c(\mathbf{v})\boldsymbol{\pi}_0^c, \tag{69.8}$$

$$\frac{\mathrm{d}\boldsymbol{\rho}^c(t, \mathbf{v})}{\mathrm{d}t} = \boldsymbol{\rho}^c(t, \mathbf{v})\mathbf{K}^c(t, \mathbf{v}). \tag{69.9}$$

From the density of agents in each state, we can compute the probability of finding a class c agent at time t in position $\mathbf{v}$ in state i as

$$\pi_i^c(t, \mathbf{v}) = \frac{\rho_i^c(t, \mathbf{v})}{\rho^c(\mathbf{v})}. \tag{69.10}$$

We collect all the terms in a vector $\boldsymbol{\pi}^c(t, \mathbf{v}) = [\pi_i^c(t, \mathbf{v})]$. Note that the definition of (69.10) together with (69.3) ensures that $\sum_i \pi_i^c(t, \mathbf{v}) = 1$, $\forall t$, $\forall \mathbf{v}$.

Note that each equation in (69.9) has the dimension n_c of the CTMC of a single MA of class c. In this way, a problem defined over the product state space of all the MAs is decomposed into several subproblems, one for each MA, having decoupled the interaction by means of (69.6). Equation (69.9) provides the basic time-dependent measures to evaluate more complex performance indices associated to the system. Equation (69.9) is discretized both in time and space and are solved by resorting to standard numerical techniques for differential equations.

69.3 A Consolidated Example: WSN Routing

In this section, we present our first attempt to model swarm intelligence inspired mechanisms through the MAM formalism. This application describes an MAM model for the analysis of a swarm intelligence routing protocol in WSNs and was first proposed in [69.9] and then enriched in [69.12]. In this work, we show new experiments to illustrate the self-adaptability of the MAM model to the changing of environmental conditions.

WSNs are large networks of tiny sensor nodes that are usually randomly distributed over a geographical region. The network topology may vary in time in an unpredictable manner due to many different causes. For example, in order to reduce power consumption, battery-operated sensors undergo cycles of sleeping – active periods; additionally, sensors may be located in hostile environments increasing their likelihood of failure; furthermore, data might also be collected from different sources at different times and directed to different sinks. For this reason, multihop routing algorithms used to route messages from a sensor node to a sink should be rapidly adaptable to the changing topology. Swarm intelligence has been successfully used to face these problems thanks to its ability in converging to a single global behavior starting from the interaction of many simple local agents.

69.3.1 A Swarm Intelligence Based Routing

In [69.15], a new routing algorithm, inspired by the biological process of *pheromone* emission (a chemical substance produced and layed down by ants and other biological entities), has been proposed. The routing table in each node stores the *pheromone level* owned by each neighbor, coded as a natural integer quantity [69.15]; when a data packet has to be sent it is forwarded to the neighbor with the highest pheromone level. This approach correctly works only if a sequence of increasing values of pheromone levels toward the sinks exists; in other words, the sinks must have the maximum pheromone level in the WSN and a decreasing pheromone gradient must be established around the sinks covering all the net.

To build the pheromone gradient, the initial setting of the WSN is as follows: the sinks are set to a fixed maximum pheromone level, whereas the sensor nodes' pheromone levels are set to 0. When the WSN is operating, each node periodically sends a signaling packet with its pheromone level and updates its value based on the level of its neighbors.

More specifically, the algorithm for establishing the pheromone gradient is based on two types of nodes in the WSN, called *sinks* and *sensors*, respectively, and the pheromone is assumed discretized into P different levels, ranging from 0 to $P-1$. In this way, routing paths toward the sink are established through the exchange of pheromone packets containing the pheromone level $p (0 \leq p \leq P-1)$ of each node.

Sink nodes, once activated, set their internal pheromone level to the highest value $p = P-1$. Then, they, at fixed time interval, broadcast a pheromone message to their neighbors with the value p. We assume $T1$ is the time interval incurring between two consecutive sending of pheromone message.

Instead, the pheromone level of a sensor node is initially set to 0 and then it is periodically updated according to two distinct actions – *excitation action* (the positive feedback) and *evaporation action* (the negative feedback):

- *Excitation action*: Sensor nodes periodically broadcast to the neighbors a pheromone message containing their internal pheromone level p. Like the sink node, sensor nodes perform the sending at regular time interval $T1$. When a sensor node receives a pheromone level p_n sent by a neighbor it compares p_n with its own level p and updates the latter if $p_n > p$. The new value is computed as a function of the current and the received pheromone level $update(p, p_n)$. In this context, we use the average of the sender and the receiver level as the new updating value, thus the function is assumed to be $update(p, p_n) = round((p + p_n)/2)$.
- *Evaporation action*: it is triggered at regular time interval $T2$ and it simply decreases the current value of p by one unit assuring it maintains a value greater or equal to 0.

We note that, despite all nodes perform their excitation action with the same mean time interval $T1$, no synchronization activity is required among the nodes; all of them act asynchronously in accordance with the principles of biological systems where each entity acts autonomously with respect to the others.

The excitation–evaporation process, like in biological systems, assures the stability of the system and the adaptability to possible changes in the environment or in some nodes. Any change in the network condition is captured by an update of the pheromone level of the involved nodes that modifies the pheromone gradient automatically driving the routing decisions toward the new optimal solution. In this way, the network can self-organize its topology and adapt to environmental changes. Moreover, when link failures occur, the network reorganization task is accomplished by those nodes near the broken links. This results in a robust and self-organized architecture.

The major drawback of this algorithm is the difficulty in appropriately setting the parameter $T1$ and $T2$: as shown in [69.12, 15], the stability of the system and the *quality* of the produced pheromone gradient is strictly dependent on the parameters ratio. When $T1$ decreases and $T2$ is fixed, pheromone messages are exchanged more rapidly among the nodes and their pheromone level tends to the maximum level because the sink node always sends the same maximum value. Without an appropriate balancing action, the pheromone level saturates all the nodes of the WSN.

At the opposite, let us suppose $T1$ is fixed and $T2$ decreases; in this case the pheromone level in each sensor node decreases more quickly than its updating according to the value of the neighbors. As a result all the levels will be close to zero. From this behavior, we note that: (1) both timers are necessary to ensure that the algorithm could properly work, and (2) a smart setting of both timers is necessary in order to have the best gradient shape all over the network. The MAM model we are going to describe in the next section helps us to determine the best parameter values.

69.3.2 The MAM Model

The MAM model used to study the gradient formation is based on two agent classes: the class *sink node* denoted by a superscript s and the class *sensor node* denoted by a superscript n. The message exchange is modeled by using M different message types. As we will explain later, since each message is used to send a pheromone level, we set $M = P$, where P is the number of different pheromone intensities considered in the model.

Geographical Space

The geographical space $\mathcal{V}$ where the N agents are located is modeled as a $n_h \times n_w$ rectangular grid, and each cell has a square shape with side d_s. Sensors can only be located in the center of each cell and we allow at most one node per cell: i. e., some cell might be empty, and $N \leq n_h \times n_w$. Moreover, sink nodes are very few with respect to the number of sensor nodes.

Agent's Structure and Behavior

Irrespective of the MA class considered, we model the pheromone level of a node with a state and this choice determines two different MA structures.

The *sink* class (Fig. 69.3a) is very simple and is characterized by a single state labeled $P - 1$ with a self-loop of rate $\lambda = \frac{1}{T1}$. The sink has always the same maximum pheromone level, and emits a single message of type $P - 1$ with rate λ.

Instead, the *sensor* class (Fig. 69.3b) has P states identifying the range of all the possible pheromone levels. Each state is labeled with the pheromone intensity i ($i = 0, \ldots, P - 1$) in the corresponding node and has a self-loop of rate $\lambda = \frac{1}{T1}$ that represents the firing of timer at regular intervals equal to $T1$. This event causes the sending of a message (Sect. 69.3.2). The evaporation phenomenon is modeled by the solid arcs (local transitions) connecting state i with state $i - 1$

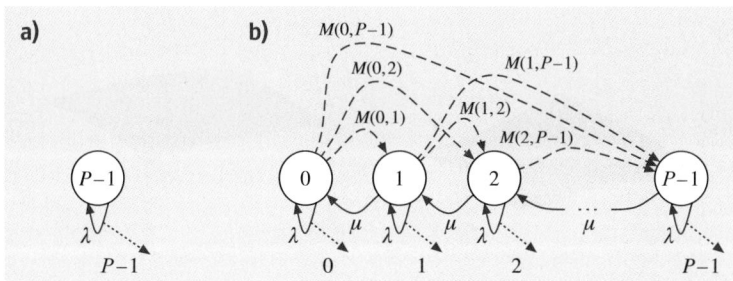

Fig. 69.3a,b Markovian agent models. (a) Agent class = sink, (b) Agent class = sensor

$(0 < i \leq P-1)$. The evaporation rate is set to $\mu = \frac{1}{T2}$; in such a way we represent the firing of timer $T2$.

Message Types

The types of messages in the model correspond to the different levels of pheromone a node can store, thus we define $\mathcal{M} = \{0, 1, \ldots, P-1\}$. Any self-loop transition in state i emits a message of the corresponding type i at a constant rate λ, both in sink and in sensor nodes. The sink message is always of type $P-1$, representing the maximum pheromone intensity, whereas the messages emitted by a sensor node corresponds to the state where it actually is.

When a message of type m is emitted, neighboring nodes are able to receive it changing their state accordingly. This behavior is implemented through the dashed arcs (whose labels are defined through (69.11)) that model the transitions induced by the reception of a message. In particular, when a node in state i receives a message of type m, it immediately jumps to state j if $m \in M(i,j)$, with

$$M(i,j) = \{m \in [0, \ldots, P-1] : \text{round}((m+i)/2) = j\}$$
$$\forall i, j \in [0, \ldots, P-1] : j > i. \qquad (69.11)$$

In other words, an MA in state i jumps to the state j that represents the pheromone level equal to the mean between the current level i and the level m encoded in the perceived message.

Perception Function

Messages of any type sent by a node are characterized by the same transmission range t_r that defines the radius of the area in which an MA can perceive a message produced by another MA. This property is reflected in the perception function $u^m(\cdot)$ that, $\forall m \in [1, \ldots, M]$, is defined as

$$u^m(\mathbf{v}, c, i, \mathbf{v}', c', i') = \begin{cases} 0 & \text{dist}(\mathbf{v}, \mathbf{v}') > t_r \\ 1 & \text{dist}(\mathbf{v}, \mathbf{v}') \leq t_r, \end{cases} \qquad (69.12)$$

where $\text{dist}(\mathbf{v}, \mathbf{v}')$ represents the Euclidean distance between the two nodes at position $\mathbf{v}$ and $\mathbf{v}'$.

As can be observed, the perception function in (69.12) is defined irrespective of the message type, because in this kind of application the reception of a message of any type i depends only on the distance between the emitting and the perceiving node. The transmission range t_r depends on the properties of the sensor and it influences the number η of neighbors perceiving the message. In the numerical experimentation, we consider $d_s \leq t_{r4} < \sqrt{2}\,d_s$ corresponding to $\eta = 4$.

Generation and Acceptance Probabilities

In this application, messages are only generated during self-loop transitions with probability 1, so that $\forall i, j$, $g_{ii}^c(m) = 1$ and $g_{ij}^c(m) = 0$, $(i \neq j)$. Similarly, we assume either $a_{ij}^c(m) = 0$ or $a_{ij}^c(m) = 1$, that is incoming messages are always accepted or always ignored.

69.3.3 Numerical Results

In order to analyze the behavior of the WSN model, the main measure of interest is the evolution of $\pi_i^n(t, \mathbf{v})$ i.e., the distribution of the pheromone intensity of a sensor node over the entire area $\mathcal{V}$ as a function of the time. The value of $\pi_i^n(t, \mathbf{v})$ can be computed from (69.10) and allows us to obtain several performance indices like average pheromone intensity $\phi(t, \mathbf{v})$ at time t for each cell $\mathbf{v} \in \mathcal{V}$

$$\phi(t, \mathbf{v}) = \sum_{i=0}^{P-1} i \cdot \pi_i^n(t, \mathbf{v}). \qquad (69.13)$$

The distribution of the pheromone intensity over the entire area $\mathcal{V}$ depends both on the pheromone emission rate λ and on the pheromone evaporation rate μ; furthermore, the excitation–evaporation process depends on the transmission range t_r that determines the number of neighboring cells η perceived by an MA in a given position. To take into account this physical mechanism,

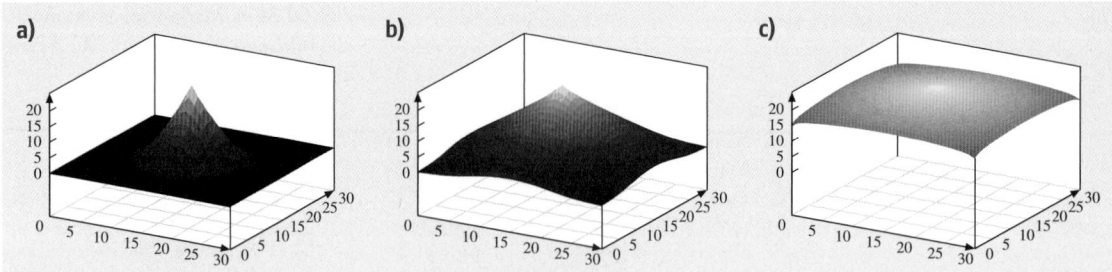

Fig. 69.4a–c Distribution of the pheromone intensity varying r. (**a**) $r = 1.2$, (**b**) $r = 1.8$, (**c**) $r = 2.4$

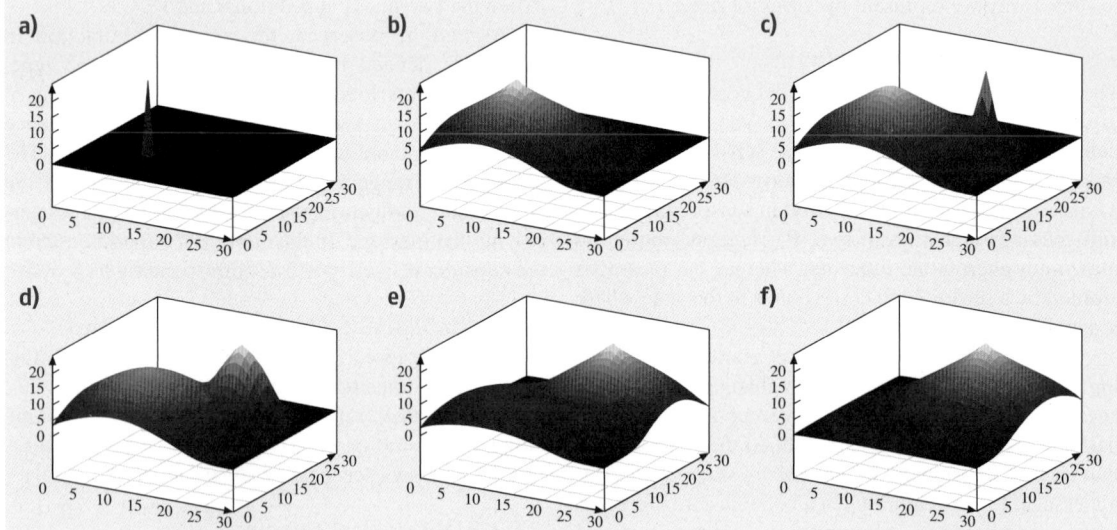

Fig. 69.5a–f Distribution of the pheromone intensity with respect to t when two sinks are alternately activated. The change is applied at time $t = 17.5$ s. (**a**) $t = 0$ s, (**b**) $t = 17$ s, (**c**) $t = 17.5$ s, (**d**) $t = 19$ s, (**e**) $t = 24$ s, (**f**) $t = 29$ s

we define the following quantity,

$$r = \frac{\lambda \cdot \eta}{\mu} , \tag{69.14}$$

which regulates the balance between the pheromone emission and evaporation in the SI routing algorithm. For a complete discussion about the performance indices that can be derived and analyzed using the described MAM, refer to [69.12].

The numerical results have been obtained with the following experimental setting. The geographical space is a square grid of sizes $n_h = n_w = 31$, where $N = 961$ sensors are uniformly distributed with a spatial density equal to 1 (one sensor per cell). Further, we set $\lambda = 4.0$, $P = 20$, and $\eta = 4$. The first experiment aims at investigating the formation of the pheromone gradient around

the sink as a function of the model parameters. To this end, a single sink node is placed at the center of the area and the pheromone intensity distribution is evaluated as a function of the parameter r, by varying μ being λ and η fixed.

Figure 69.4 shows the distribution of the pheromone intensity $\phi(t, \mathbf{v})$ measured in the stable state for three different values of r. If the value of r is small ($r = 1.2$) or high ($r = 2.4$), the quality of the gradient is poor. This is due to the prevalence of one of the two feedbacks: negative (with $r = 1.2$ evaporation prevails) or positive (with $r = 2.4$ excitation prevails and all sensors saturate). On the contrary, intermediate values ($r = 1.8$) generate well-formed pheromone gradients able to cover the whole area, thanks to the correct balance between the two feedbacks. Then, an opportune evaluation of the value of r has to be carried

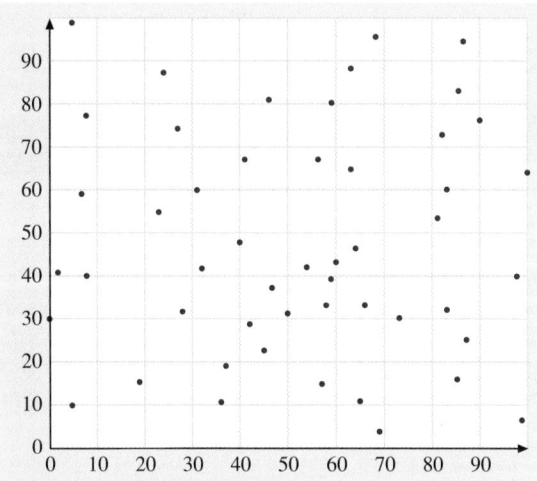

Fig. 69.6 The 100×100 grid with 10 000 cells and 50 randomly scattered sinks

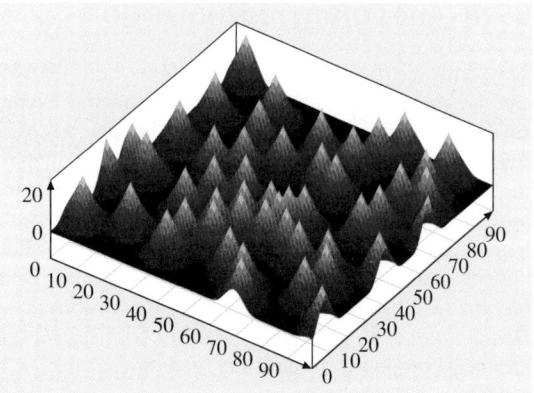

Fig. 69.7 Distribution of the pheromone intensity when the network is composed by a grid of 10 000 sensor nodes with 50 sinks

out in order to generate a pheromone gradient that fits with the topological specification of the WSN under study.

In order to understand the dynamic behavior of the SI algorithm, we carried out a transient analysis able to highlight different phases of the gradient construction process when the position of the sink changes in time. In particular, in the following experiment (Fig. 69.5) we analyzed how the algorithm self-adapts to topological modifications by recalculating the pheromone gradient when two different sinks are present in the network and they are alternately activated. Figure 69.5a,b show how the pheromone signal is spread on the space $\mathcal{V}$ until the stable state is reached. At this point ($t = 17.5\,\mathrm{s}$), we deactivated the old sink and we activated a new one in a different position (Fig. 69.5c). Figure 69.5d,e describe the evolution of the gradient modification. It is possible to observe that, thanks to the properties of the SI algorithm, the WSN is able to rapidly discover the new sink and to change the pheromone gradient by forgetting the old information until a new stable state is reached (Fig. 69.5f).

Finally, in order to test the scalability of the MAM in more complex scenarios, we have assumed a rectangular grid with $n_{\mathrm{h}} = n_{\mathrm{w}} = 100$ hence with $N = 100 \times 100 = 10\,000$ sensors, and we have randomly scattered 50 sinks in the grid. The grid is represented in Fig. 69.6, where the sinks are drawn as black spots. Since each sensor is represented by an MA with $P = 20$ states (Fig. 69.3b), the product state space of the overall system has $N = 20^{10\,000}$ states!

The steady pheromone intensity distribution for the geographical space represented in Fig. 69.6 is reported in Fig. 69.7. Through this experiment, we can assess that the pheromone gradient is also reached when no symmetries are present in the network and that the proposed model is able to capture the behavior of the protocol in generating a correct pheromone gradient also in the presence of different maximums. Using the same protocol configurations found for a simple scenario, the SI algorithm is able to create a well-formed pheromone gradient also in a completely different situation, making such routing technique suitable in nonpredictable scenarios. This scenario also demonstrates the scalability of the proposed analytical technique that can be easily adopted in the analysis of very large networks.

69.4 Ant Colony Optimization

The aim of this section is to show how MAMs can be adopted to represent one of the more classical swarm intelligence algorithm known as ACO [69.2], that was inspired by the foraging behavior of ant colonies which, during food search, exhibit the ability to solve simple shortest path problems. To this end, in this work, we simply show how to build a MAM that solves the famous *Double Bridge Experiment* which was first proposed by *Deneubourg* et al. in the early 90s [69.16, 17], and that has been proposed as an entry benchmark for ACO models.

In the experiment, a nest of Argentine ants is connected to a food source using a double bridge as shown in Fig. 69.8. Two scenarios are considered: in the first one the bridges have equal length (Fig. 69.8a), in the second one the lengths of the bridges are different (Fig. 69.8b). The collective behavior can be explained by the way in which ants communicate indirectly among them (*stigmergy*). During the journey from the nest to the food source and vice versa, ants release on the ground an amount of *pheromone*. Moreover ants can perceive pheromone and they choose with greater probability a path marked by a stronger concentration of pheromone. As a results, ants releasing pheromone on a branch, increase the probability that other ants choose it. This phenomenon is the realization of the positive feedback process described in Sect. 69.1 and it is the reason for the convergence of ants to the same branch in the equal length bridge case. When lengths are different, the ants choosing the shorter path reach the food source quicker than those choosing the longer path. Therefore, the pheromone trail grows faster on the shorter bridge and more ants choose it to reach food. As a result, eventually all ants converge to follow the shortest path.

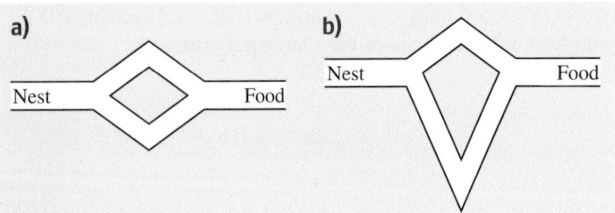

Fig. 69.8a,b Experiment scenarios. Modified from *Goss* et al. [69.17]. (**a**) Equal branches, (**b**) Different branches

69.4.1 The MAM Model

We represent the double bridge experiment through a multiple agent class and multiple message type MAM. We model ants by messages, and locations that ants traverse by MAs. Three different MA classes are introduced: the class *Nest* denoted by superscript n, the class *Terrain* denoted by superscript t, and the class *Food* denoted by superscript f. Two types of messages are used: ants walking from the nest to the food source correspond to messages of type fw (*forward*), whereas ants coming back to the nest correspond to messages of type bw (*backward*).

Geographical Space

Agents (either nest, terrain, or food source) are deployed on a discrete geographical space $\mathcal{V}$ represented as an undirected graph $G = (V, E)$, where the elements in the set V are the vertices and the elements in the set E are the edges of the graph. In Fig. 69.9a,b, we show the locations of agents for the equal and the different length bridge scenarios, respectively. The squares are the vertices of the graph and the labels inside them indicate the class of the agent residing on the vertex. In this model, we assume that only a single agent resides on each vertex. Message passing from a node to another is depicted as little arrows labeled by the message type. As shown in Fig. 69.9, different lengths of the branches are represented by a different number of hops needed by a message to reach the food source starting from the nest. Figure 69.9c represents a three branches bridge with branches of different length.

Agent's Structure and Behavior

The structure of the three MA classes is described in the following:

- *MA Nest*: The nest is represented by a single MA of class n, shown in Fig. 69.10a. The nest MA^n is composed by a single state that emits messages of type fw at a constant rate λ, modeling ants leaving the nest in search for food.
- *MA Terrain*: An MA of class t (Fig. 69.10c) represents a portion of terrain on which an ant walks and encodes in its state space the concentration of the pheromone trail on that portion of the ground. We assume that the intensity of the pheromone trail is discretized in P levels numbered $0, 1, \ldots P - 1$.

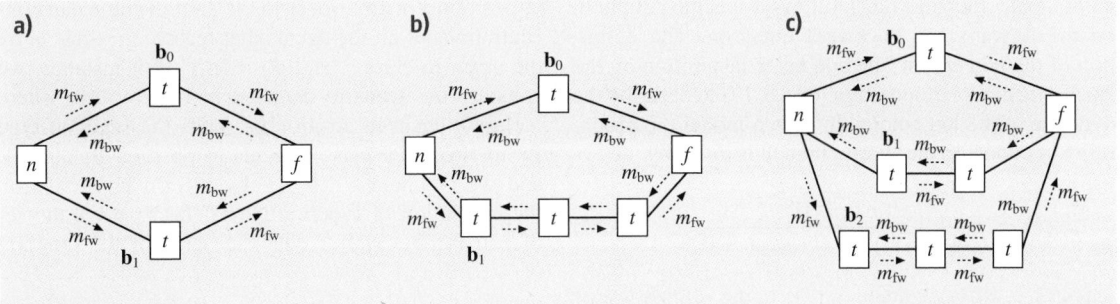

Fig. 69.9a–c Graph used to model the experiment scenarios. (**a**) Equal branches, (**b**) two different branches, (**c**) three different branches

With reference to Fig. 69.10c, the meaning of the states is the following:

- t_0 denotes no pheromone on the ground and no ant walking on it;
- t_i denotes a concentration of pheromone of level i and no ant on the ground;
- t_{if} denotes an ant of forward type residing on the terrain while the pheromone concentration is at level i;
- t_{ib} denotes an ant of backward type residing on the terrain while the pheromone concentration is at level i.

The behavior of the MA^t agent at the reception of the messages is the following:

- *fw – forward ant*: A message of type fw perceived by an MA^t in states t_i, induces a transition to state $t_{(i+1)f}$ meaning that the arrival of a forward ant increases the pheromone concentration of one level *(positive feedback)*.
- *bw – backward ant*: A message of type bw perceived by an MA^t in states t_i, induces a transition to state $t_{(i+1)b}$ meaning that the arrival of a backward ant increases the pheromone concentration of one level *(positive feedback)*.

Ants remain on a single terrain portion for a mean time of $1/\eta$ s, then they leave toward another destination. The local transitions from states t_{if} to states t_i and the generation of message fw model this behavior for forward ants. An analogous behavior is represented for backward ants by local transitions from states t_{ib} to states t_i. The local transitions at constant rate μ from states t_i to states t_{i-1} indicate the decreasing of one unit of the concentration of pheromone due to evaporation *(negative feedback)*:

- *MA Food source*: An MA of class f represents the food source (Fig. 69.10b). The reception of a message of type fw in state f_0 indicates that a forward ant has reached the food source. After a mean time of $1/\eta$ s, such an ant leaves the food and starts the way back to the nest becoming a backward ant (emission of message of type bw).

In order to keep model complexity low thus increasing the model readability, we have chosen to limit to 1 the number of ants that can reside at the same time on a portion of terrain (or in the food source). For this reason, message reception is not enabled in states t_{if} or t_{ib} for MAs of class t and in the state f_1 for MAs of class f. In future works, we will study effective techniques (e.g., intervening on MA density) in order to release such an assumption.

Perception Function

The perception function rules the interactions among agents and, in this particular example, defines the prob-

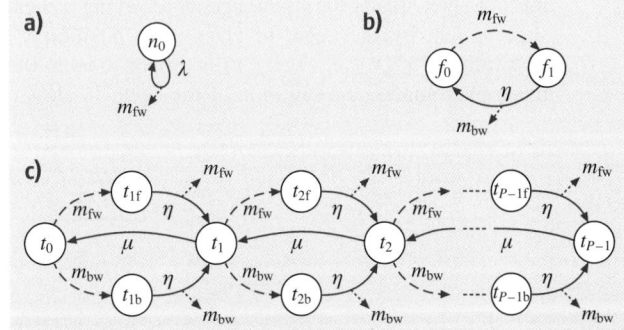

Fig. 69.10a–c Markovian agent models for the ACO experiment. (**a**) MA^n: Agent of class nest. (**b**) MA^f: Agent of class food. (**c**) MA^t: Agent of class terrain

ability that a message (ant) follows a specific path both on the forward and backward direction. The definition of the perception function takes inspiration on the stochastic model proposed in [69.16, 17] to describe the dynamic of the ant colony. In such a model the probability of choosing the shorter branch is given by

$$p_{is}(\tau) = \frac{(k + \varphi_{is}(\tau))^{\alpha}}{(k + \varphi_{is}(\tau))^{\alpha} + (k + \varphi_{il}(\tau))^{\alpha}}, \quad (69.15)$$

where $p_{is}(\tau)$ (respectively $p_{il}(\tau)$) is the probability of choosing the shorter (longer) branch, $\varphi_{is}(\tau)$ ($\varphi_{il}(\tau)$) is the total amount of pheromone on the shorter (longer) branch at a time τ. The parameter k is the degree of attraction attributed to an unmarked branch. It is needed to provide a non-null probability of choosing a path not yet marked by pheromone. The exponent α provides a nonlinear behavior.

In our MA model, the perception function $u^m(\cdot)$ is defined, $\forall m \in \{\text{fw}, \text{bw}\}$, as

$$u^m(\mathbf{v}, c, i, \mathbf{v}', c', j, \tau)$$
$$= \frac{(k + E[\boldsymbol{\pi}^c(\tau, \mathbf{v})])^{\alpha}}{\sum_{(c'', \mathbf{v}'') \in \text{Next}^m(\mathbf{v}', c')} (k + E[\boldsymbol{\pi}^{c''}(\tau, \mathbf{v}'')])^{\alpha}}, \quad (69.16)$$

where k and α have the same meaning as in (69.15), $E[\boldsymbol{\pi}^c(\tau, \mathbf{v})]$ gives the mean value of the concentration of pheromone at a time τ in position $\mathbf{v}$ on the ground, and corresponds to $\varphi(\tau)$. The computation of $E[\boldsymbol{\pi}^c(\tau, \mathbf{v})]$ will be addressed in (69.18). The function $\text{Next}^m(\mathbf{v}', c')$ gives the set of pairs $\{(c'', \mathbf{v}'')\}$ such that the agent of class c'' in position $\mathbf{v}''$ perceives a message of type m emitted by the agent of class c' in position $\mathbf{v}'$. Figure 69.11a helps to interpret (69.16). The multiple box stands for all the agents receiving a message m sent by the agent of class c' in position $\mathbf{v}'$. The value of $u^m(\mathbf{v}, c, i, \mathbf{v}', c', j, \tau)$ is proportional to the mean pheromone concentration of the agent in class c

at position $\mathbf{v}$ with respect to the sum of the mean concentrations of all the agents that receive message m by the agent in class c' and position $\mathbf{v}'$. For instance, we consider the scenario depicted in Fig. 69.11b, where a class n agent in position $\mathbf{b}_0$ sends messages of type fw to two other class t agents at position $\mathbf{b}_1$ and $\mathbf{b}_2$, and we compute $u^{\text{fw}}(\mathbf{b}_2, t, i, \mathbf{b}_0, n, j, \tau)$. In such case, the evaluation of function $\text{Next}^{\text{fw}}(\mathbf{b}_0, n)$ gives the set of pair $\{(t, \mathbf{b}_1), (t, \mathbf{b}_2)\}$ and the value of the function is

$$u^{\text{fw}}(\mathbf{b}_2, t, i, \mathbf{b}_0, n, j, \tau)$$
$$= \frac{(k + E[\boldsymbol{\pi}^t(\tau, \mathbf{b}_2)])^{\alpha}}{(k + E[\boldsymbol{\pi}^t(\tau, \mathbf{b}_1)])^{\alpha} + (k + E[\boldsymbol{\pi}^t(\tau, \mathbf{b}_2)])^{\alpha}}. \quad (69.17)$$

As a final remark, we highlight that $u^m(\cdot)$ does not depend on the state variables i and j of the sender and receiver agents even if these variables appear in the definition of $u^m(\cdot)$ ((69.16)). Instead, $u^m(\cdot)$ depends on the whole probability distribution $\boldsymbol{\pi}^c(\tau, \mathbf{v})$ needed to compute the mean value $E[\boldsymbol{\pi}^c(\tau, \mathbf{v})]$.

Generation and Acceptance Probabilities

As in Sect. 69.3, also in this ACO-MAM model we only allow $g^c_{i,j}(m) = 0$ or $g^c_{i,j}(m) = 1$ and $a^c_{i,j}(m) = 0$ or $a^c_{i,j}(m) = 1$ $\forall c, m$. In particular, for the terrain agent MA^t, messages of type fw are sent with probability $g^t_{\text{if},i}(\text{fw}) = 1$, and are accepted with probability $a^t_{i,(i+1)\text{f}}(\text{fw}) = 1$ only in a t_i state inducing a transition to a $t_{(i+1)\text{f}}$ state. An analogous behavior is followed during emission and reception of messages of type bw.

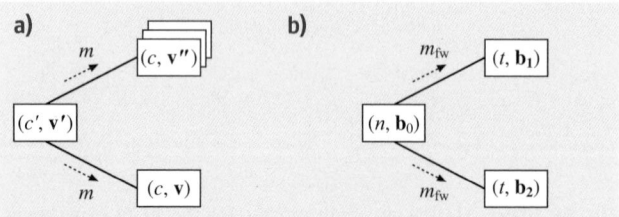

Fig. 69.11a,b Perception function description. (a) General case, (b) example of scenario in Fig. 69.9b

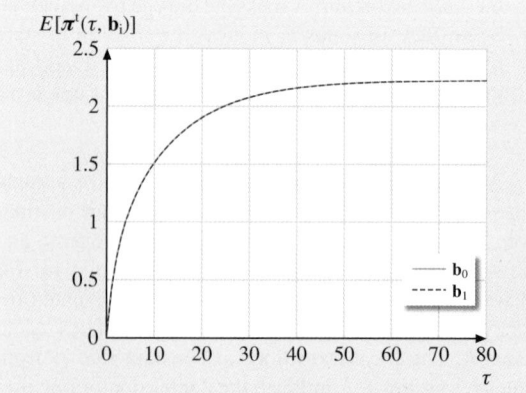

Fig. 69.12 Mean pheromone concentration with $\lambda = 1.0$, $\mu = 1$ and $\eta = 1$ for the equal branches experiment

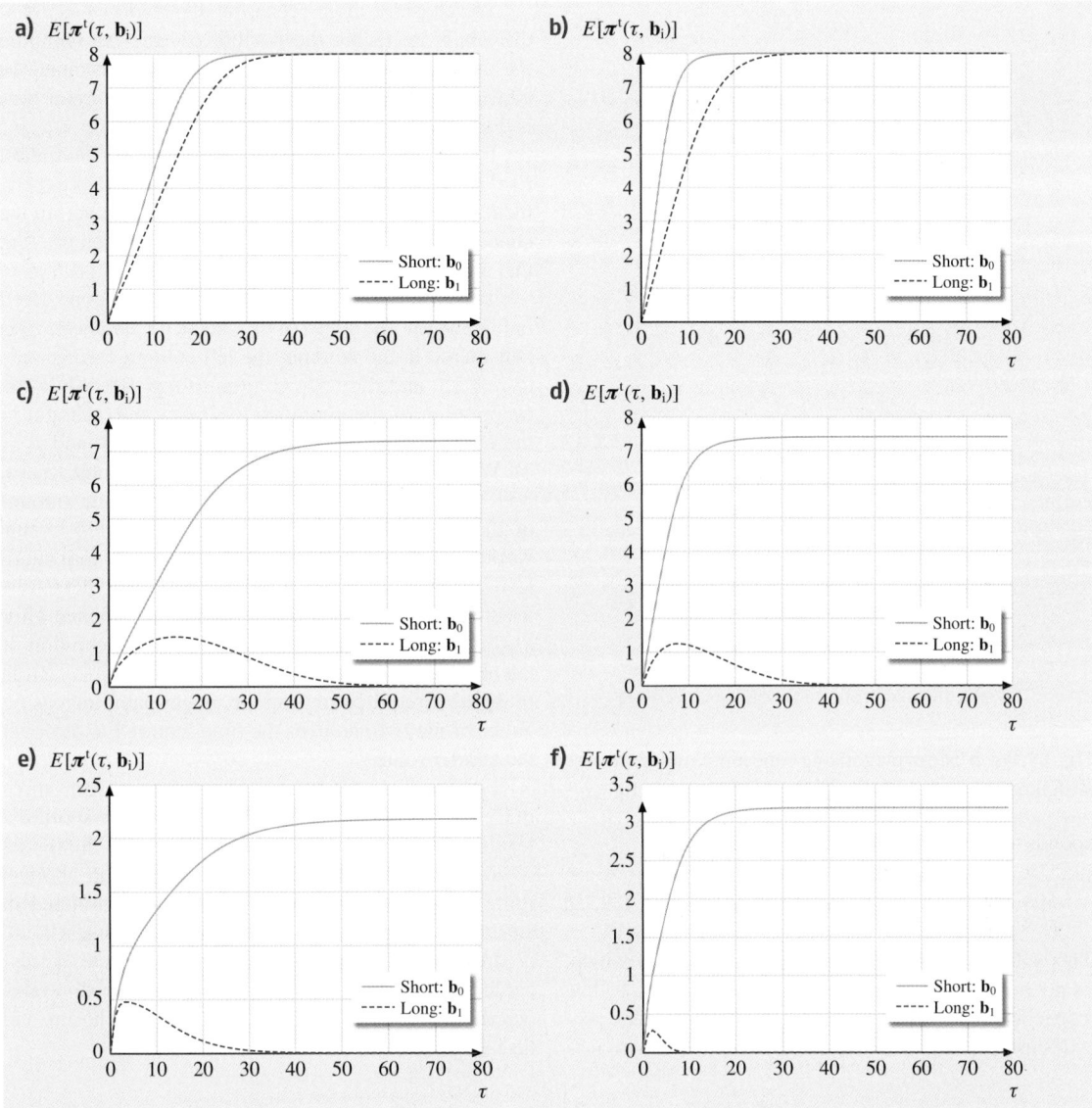

Fig. 69.13a–f Mean pheromone concentration for the case with two different branches. (**a**) Mean pheromone concentration $\lambda = 1.0$, $\mu = 0$ and $\eta = 1$, (**b**) mean pheromone concentration $\lambda = 1.0$, $\mu = 0$ and $\eta = 10$, (**c**) mean pheromone concentration $\lambda = 1.0$, $\mu = 0.5$ and $\eta = 1$, (**d**) mean pheromone concentration $\lambda = 1.0$, $\mu = 0.5$ and $\eta = 10$, (**e**) mean pheromone concentration $\lambda = 1.0$, $\mu = 2$ and $\eta = 1$, (**f**) mean pheromone concentration $\lambda = 1.0$, $\mu = 2$ and $\eta = 10$

69.4.2 Numerical Results for ACO Double Bridge Experiment

We have performed several experiments on the ACO model. In particular, we study *the mean value of the concentration of pheromone* at a time τ in position $\mathbf{v}$

for a class c agent, $E[\pi^c(\tau, \mathbf{v})]$, defined as

$$E[\pi^c(\tau, \mathbf{v})] = \sum_{s \in S^c} \pi_s(\mathbf{v}, c) I(s) ,$$ (69.18)

where S^c denotes the state space of a class c agent, $I(s)$ represents the pheromone level in state s, and it corre-

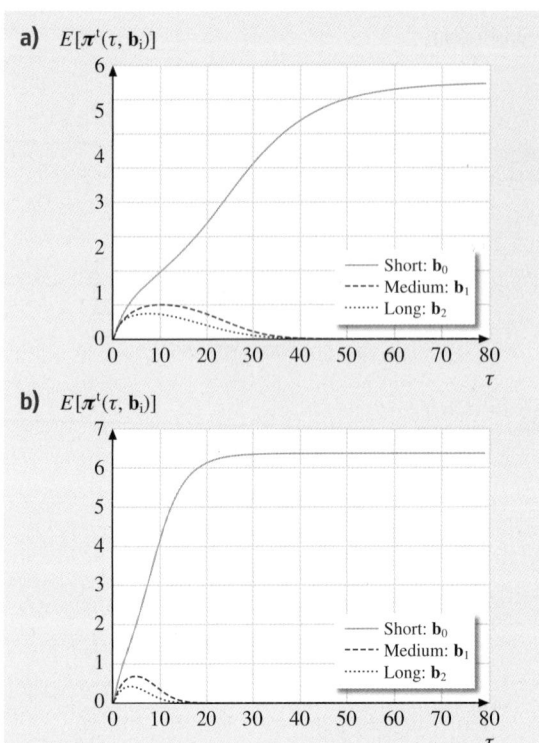

a) $E[\pi^t(\tau, \mathbf{b}_i)]$

— Short: $\mathbf{b}_0$
--- Medium: $\mathbf{b}_1$
···· Long: $\mathbf{b}_2$

b) $E[\pi^t(\tau, \mathbf{b}_i)]$

— Short: $\mathbf{b}_0$
--- Medium: $\mathbf{b}_1$
···· Long: $\mathbf{b}_2$

Fig. 69.14a,b Mean pheromone concentration for the case with three different branches. **(a)** $\eta = 10$, **(b)** $\eta = 10$

sponds to

$$I(s) = i, \quad \forall s \in \{t_i\} \cup \{t_{if}\} \cup \{t_{ib}\}. \tag{69.19}$$

This value is used in (69.16) to compute $u^m(\cdot)$ which, as previously said, rules the ant's probability to follow a specific path; therefore, such performance index provides useful insights of the modeled ant's behavior.

We consider three scenarios depicted in Fig. 69.9, the labels $\mathbf{b}_i$ denote the positions where we compute the mean value of the concentration of pheromone. In all the experiments, the intensity of the pheromone trail is discretized in $P = 8$ levels.

In Fig. 69.12, the mean pheromone concentration $E[\pi^c(\tau, \mathbf{b}_i)]$ over the time for the equal branches experiment is plotted. As it can be seen, both mean pheromone concentrations have exactly the same evolution proving that ants do not prefer one of the routes.

The case with two different branches is considered in Fig. 69.13. The speed of the ants (i. e., parameter η) is considered in the column (the left column corresponds to $\eta = 1.0$ and the right column to $\eta = 10$), while the evaporation of the pheromone is taken into account in the rows (respectively with $\mu = 0$, $\mu = 0.5$, and $\mu = 2$). When no evaporation is considered (Fig. 69.13a,b), both paths are equally chosen due to the finite amount of the maximum pheromone level considered in this work. However the shorter path reaches its maximum level earlier than the longer route. In all the other cases, it can be seen that the longer path is abandoned after a while in favor of the shorter one. The evaporation of the pheromone and the speed of the ants both play a role in the time required to drop the longer path. Increasing either of the two, reduces the time required to discover the shorter route.

Finally, Fig. 69.14 considers a case with three branches of different length and different evaporation levels ($\eta = 1$ and $\eta = 10$). Also in this case the model is able to predict that ants will choose the shortest route. It also shows that longer paths are dropped in an order proportional to their length: the longest route is dropped first, and the intermediate route is discarded second. Also in this case, the evaporation rates determine the speed at which paths are chosen and discarded.

69.5 Conclusions

In this work, we have presented how the Markovian agents performance evaluation formalism can be used to study swarm intelligent algorithms. Although the formalism was developed to study largely distributed systems like sensor networks, or physical propagation phenomena like fire or earthquakes, it has been proven to be very efficient in capturing the main features of swarm intelligence.

Beside the two cases presented in this chapter, routing in WSNs and ant colony optimiza-

tion, the formalism is capable of considering other cases like Slime Mold models. Future research lines will try to emphasize the relations between Markovian agents and swarm intelligence, trying to integrate both approaches: using Markovian agents to formally study new swarm intelligent algorithms, and use swarm intelligent techniques to study complex Markovian agents models in order to find optimal operation points and best connection strategies.

References

69.1 M.G. Hinchey, R. Sterritt, C. Rouff: Swarms and swarm intelligence, IEEE Comput. **40**(4), 111–113 (2007)

69.2 M. Dorigo, T. Stützle: *Ant Colony Optimization* (MIT, Cambridge 2004)

69.3 K. Li, K. Thomas, L.F. Rossi, C.-C. Shen: Slime mold inspired protocols for wireless sensor networks, 2nd IEEE Int. Conf. Self-Adapt. Self-Organ. Syst. (SASO), Venice (2008) pp. 319–328

69.4 K. Li, C.E. Torres, K. Thomas, L.F. Rossi, C.-C. Shen: Slime mold inspired routing protocols for wireless sensor networks, Swarm Intell. **5**(3/4), 183–223 (2011)

69.5 D. Cerotti, M. Gribaudo, A. Bobbio: Performability analysis of a sensor network by interacting Markovian agents, 4th IEEE Int. Workshop Sensor Netw. Syst. Pervasive Comput. (PerSens), Hong Kong (2008) pp. 300–305

69.6 A. Bobbio, D. Bruneo, D. Cerotti, M. Gribaudo: Markovian agents: A new quantitative analytical framework for large-scale distributed interacting systems, IIIS Int. Conf. Design Model. Sci. Educ. Technol. (DeMset), Orlando (2011) pp. 327–332

69.7 M. Gribaudo, A. Bobbio: Performability analysis of a sensor network by interacting markovian agents, 8th Int. Workshop Perform. Model. Comput. Commun. Syst. (PMCCS), Edinburgh (2007)

69.8 A. Bobbio, D. Cerotti, M. Gribaudo: Presenting dynamic Markovian agents with a road tunnel application, 17th IEEE/ACM Int. Symp. Model. Anal. Simul. Comput. Telecommun. Syst. (MASCOTS), London (2009)

69.9 D. Bruneo, M. Scarpa, A. Bobbio, D. Cerotti, M. Gribaudo: Analytical modeling of swarm intel-ligence in wireless sensor networks, 4th Int. Conf. Perform. Eval. Methodol. Tools (Valuetools), Pisa (2009)

69.10 D. Cerotti, M. Gribaudo, A. Bobbio, C.T. Calafate, P. Manzoni: A Markovian agent model for fire prop-agation in outdoor environments, 7th Eur. Perform. Eng. Workshop (EPEW), Bertinoro (2010) pp. 131–146

69.11 D. Bruneo, M. Scarpa, A. Bobbio, D. Cerotti, M. Grib-audo: Adaptive swarm intelligence routing algo-rithms for WSN in a changing environment, 9th Annu. IEEE Conf. Sensors (SENSORS), Waikoloa (2010) pp. 1813–1818

69.12 D. Bruneo, M. Scarpa, A. Bobbio, D. Cerotti, M. Grib-audo: Markovian agent modeling swarm intelli-gence algorithms in wireless sensor networks, Per-form. Eval. **69**(3/4), 135–149 (2011)

69.13 M. Saleem, G.A. Di Caro, M. Farooq: Swarm intel-ligence based routing protocol for wireless sensor networks: Survey and future directions, Inf. Sci. **181**(20), 4597–4624 (2011)

69.14 K. Trivedi: *Probability & Statistics with Reliabil-ity, Queueing & Computer Science Applications*, 2nd edn. (Wiley, New York 2002)

69.15 M. Paone, L. Paladina, D. Bruneo, A. Puliafito: A swarm-based routing protocol for wireless sensor networks, 6th IEEE Int. Symp. Netw. Comput. Appl. (NCA), Cambridge (2007) pp. 265–268

69.16 J. Deneubourg, S. Aron, S. Goss, J.M. Pasteels: The self-organizing exploratory pattern of the argen-tine ant, J. Insect Behav. **3**(2), 159–168 (1990)

69.17 S. Goss, S. Aron, J. Deneubourg, J. Pasteels: Self-organized shortcuts in the Argentine ant, Natur-wissenschaften **76**(12), 579–581 (1989)

70. Honey Bee Social Foraging Algorithm for Resource Allocation

Jairo Alonso Giraldo, Nicanor Quijano, Kevin M. Passino

Bioinspired mechanisms are an emerging area in the field of optimization, and various algorithms have been developed in the last decade. We introduce a novel bioinspired model based on the social behavior of honey bees during the foraging process, and we show how this algorithm solves a class of dynamic resource allocation problems. To illustrate the practical utility of the algorithm, we show how it can be used to solve a dynamic voltage allocation problem to achieve a maximum uniform temperature in a multizone temperature grid. Its behavior is compared with other evolutionary algorithms.

Over several decades researchers' interest in understanding the patterns and collective behaviors of some organisms has increased because of the possibility of generating mathematical models that can be used for solving problems [70.1]. These bioinspired models have been used to develop robust technological solutions in different research fields [70.2]. One of the first models based on natural behaviors is the genetic algorithm (GA), proposed by *Holland* in [70.3]. This method reproduces the concepts of evolution considering natural selection, reproduction, and mutation in organisms. Many variations have been developed since then [70.4], and a wide variety of applications have been implemented [70.5, 6]. A sub-field of bioinspired algorithms is the so-called *swarm intelligence* [70.1, 7], which is inspired by the collective behavior of social animals that are able to solve distributed and complex problems following individual simple rules and producing emerging behaviors. Swarm intelligence mainly refers to those techniques

inspired by the social behavior of insects, such as ants [70.8] and bees [70.9–11], or the social interaction of different animal societies (e.g., flocks of birds) [70.12]. Ant colony optimization (ACO), as introduced by *Dorigo* et al. [70.13], mimics the foraging behavior of a colony of ants, based on pheromone proliferation, and it has been used in the solution of optimization problems [70.1, 14], and in some engineering applications [70.15–17]. Another common approach is particle swarm optimization (PSO), which mimics the behavior of social organisms that move according to the knowledge of their neighbors' goodness, and it is able to solve continuous optimization problems [70.18]. This technique has been widely implemented in a variety of applications, such as economical dispatch [70.19, 20], feature selection [70.21], and some resource allocation problems [70.22, 23], to name just a few.

There are also several bioinspired techniques based on the collective behavior of foraging bees, and each one has different characteristics and applications. In [70.24], a decentralized honey bee algorithm is presented, which is based on the distribution of forager bees amongst flower patches, which occurs in such a way that the nectar intake is maximized. This technique has been applied to Internet servers hosting distribution. *Tereshko* in [70.25] also developed a model of the foraging behavior of a honey bee colony based only on the recruitment and abandonment process, taking into account just the local information of a food source. However, in [70.26], the algorithm was improved by considering either local and global information of food sources. Another approach of honey bee foraging algorithms was developed by *Karaboga* [70.27], and it is called the artificial bee colony (ABC), which can be used to solve unconstrained optimization problems. In [70.28], a comparison of the ABC algorithm performance was made with other common heuristic algorithms, such as genetic algorithms and particle swarm optimization. The authors conclude that ABC can be used for multivariable and multimodal function optimization. Several applications have been developed [70.29–31], and some improvements have been made in order to solve constrained optimization problems [70.32]. *Teodorović* and *Dell'Orco* in [70.33] introduced another algorithm based on honey bee foraging called bee colony optimization (BCO). This technique is very similar to ABC and follows almost the same steps of exploring, foraging, and recruitment based on the waggle dance. However, in ABC, the initial population is distributed in such a way that scouts and foragers are in equal proportion, while in BCO the initial population distribution is not fixed. Some applications have been developed using BCO to solve difficult optimization problems, such as combined heat and power dispatch [70.34], and job scheduling [70.35]. There are many other applications, and we refer the reader to [70.36] for an extensive literature review of the field.

In general, none of the previous optimization methods based on honey bee social foraging attempt to mimic the whole behavior of the foraging process. They mainly concentrate on the communication between the agents (bees), which is achieved through the waggle dance. One of the goals of this chapter is to show another swarm intelligence method (i.e., a honey bee social foraging) that mimics very closely the real behavior of a hive (or even multiple hives) of bees, in order to solve dynamic resource allocation problems. This method is based on the models obtained by *Seeley* and *Passino* in [70.37, 38], where each bee can be an explorer, an employed forager, an observer, or a rester. The foraging process consists of exploring a landscape with different profitability sites. Hence, if a site is good enough, the explorer will try to recruit other bees in the hive using the waggle dance, which varies its intensity according to the quality of the site and the nectar unloading time. The observers will tend to follow the bees with the higher dance intensity and they may become employed foragers. If a site is no longer good enough, the bees may tend to become observers and will try to follow another waggle dance. One of the advantages of this method is that each bee only considers local information about its position and the profitability of the forage site. Besides, the communication is only considered in the waggle dance process, which depends on the nectar unloading wait time. Hence, we do not need to have full information of each agent. However, with only the unloading wait time information, an emerging behavior is produced, and complex resource allocation problems can be solved. This method is based on experimental results and imitates almost the whole behavior of honey bees during the foraging process, which is not the case with the other approaches presented before, which only considered a few actions of the foraging activity. On the other hand, the utility of the theoretical concepts that are introduced in this chapter are illustrated in an engineering application, which consists of a multizone temperature control grid. These kinds of problems are very important in commercial and industrial applications, including the distributed control of thermal processes, semiconductor processing, and smart building temperature control [70.39–42]. Here, we use a multizone grid similar to the one in [70.43], with four zones, each one with a temperature sensor, and a lamp that varies its temperature. The complexity of these kinds of problems arises mainly due to the interzone effects (e.g., lamps affecting the temperature in neighboring zones), ambient temperature and external wind currents, zone component differences, and sensor noise. This is why common control strategies cannot be applied. For this reason, different experiments are implemented in order to observe the performance of the algorithm under different conditions. Besides, we compare its behavior with two common evolutionary algorithms, i.e., genetic algorithm and PSO, which have been selected because of their low computational cost and their high capability to solve optimization problems. These algorithms have been modified in order to solve dynamic resource allocation problems, and their behavior can be compared with the honey bee social foraging algorithm.

This chapter is organized as follows. First, in Sect. 70.1, we introduce the honey bee social foraging algorithm. Then, in Sect. 70.2 the multizone temperature problem is presented and the other two evolutionary strategies, genetic algorithm and PSO for resource allocation are introduced. The results and comparisons are presented in Sects. 70.3 and 70.4, and in Sect. 70.5 some conclusions are drawn.

70.1 Honey Bee Foraging Algorithm

The honey bee social foraging algorithm models the behavior of social honey bees during nectar foraging, based on experimental studies summarized in [70.37] and some ideas from other mathematical models. This algorithm models some activities such as exploration and foraging, nectar unload, dance strength decisions, explorer allocation, recruitment on the dance floor, and interactions with other hives. The theory and the experiments are based on the work developed by *Quijano* and *Passino* in [70.43].

70.1.1 Landscape of Foraging Profitability

The landscape is assumed as a spatial distribution of forage sites with encoded information of the foraging profitability that quantifies the distance from the hive, nectar sugar content, nectar abundance, and any other relevant site variables. There is a number of B bees that are represented by a two-dimensional position $\theta^i \in \Re^2$, for $i = 1, 2, \ldots, B$. During foraging, bees sample a *foraging profitability landscape* denoted by $J_f(\theta) \in [0, 1]$, which is proportional to the profitability of nectar at location θ. Hence, $J_f(\theta) = 1$ represents a location with the highest possible profitability and $J_f(\theta) = 0$ represents a location with no profitability.

As an example, assume the foraging landscape $J_f(\theta)$ is zero everywhere except at forage sites. We could have four forage sites, indexed by $j = 1, 2, 3, 4$, centered at various positions that are initially unknown to the bees. Each site can be represented as a cylinder with radius ϵ_f^j, and height $N_f^j \in [0, 1]$ that is proportional to nectar profitability. We may also assume that the profitability of a bee being at site j decreases as the number of bees visiting that site increases. This can be denoted by s_j, which in behavioral ecology theory is called the *suitability function* [70.44].

70.1.2 Roles and Expedition of Bees

There are several kinds of bees involved in the foraging process during an expedition, and each kind has a different function. An expedition can be considered as a time instant where each bee executes a function according to its role. There are $B_f(k)$ *employed foragers* that actively bring nectar back from some site, and some of them dance to recruit new bees if the site is good. $B_e(k)$ explorer foragers go to random positions in the environment, bring their nectar back if they find any, dance to recruit, and they can become foragers if they find a relative good site. There are $B_u(k) = B_o(k) + B_r(k)$ *unemployed foragers*, with $B_r(k)$ bees that rest (or are involved in some other activity), and $B_o(k)$ that observe the dances of employed and explorer foragers on the dance floor. Some of the observers will follow the dances.

We ignore the specific path used by the foragers on expeditions and we assume that a bee samples the foraging profitability landscape once on its expedition, and this value is held when the bee returns to the hive. Let the foraging profitability assessment by the employed forager or explorer i be

$$
F^i(k) = \begin{cases}
1 & \text{if } J_f(\theta^i(k)) \\
& \quad +\omega_f^i(k) \geq 1, \\
J_f(\theta^i(k)) + \omega_f^i(k) & \text{if } 1 > J_f(\theta^i(k)) \\
& \quad +\omega_f^i(k) > \epsilon_n, \\
0 & \text{if } J_f(\theta^i(k)) \\
& \quad +\omega_f^i(k) \leq \epsilon_n,
\end{cases}
$$

where $\theta^i(k)$ represents the position of the i-th bee at the k-th expedition, and $\omega_f^i(k)$ is the profitability assessment noise, which can be considered uniformly distributed between $(-0.1, 0.1)$. The value ϵ_n sets a lower threshold on site profitability, and here we use $\epsilon_n = 0.1$.

70.1.3 Dance Strength Determination

The number of waggle runs of bee i at the k-th expedition is called *dance strength* and is denoted by $L_f^i(k)$. The unemployed foragers have $L_f^i(k) = 0$, and the employed foragers that have $F^i(k) = 0$ will have $L_f^i(k) = 0$

since they do not find a location above the profitability threshold ϵ_n, and for this reason they will become unemployed foragers.

Unloading Waiting Time

Now, we will explain dance strength decisions for the employed foragers and explorers that find a site of sufficiently good profitability and have $F^i(k) > \epsilon_n$. Firstly, we have to model the unloading wait time in order to relate it with the dance strength. Let $F_t(k) = \sum_{i=0}^{B} F^i(k)$ be the total nectar profitability assessment at time k for the hive and $F_q^i(k)$ be the quantity of nectar gathered for a profitability assessment $F^i(k)$. We assume that $F_q^i(k) = \alpha F^i(k)$, where $\alpha > 0$ is a proportionality constant. We may choose $\alpha = 1$, such that the total hive nectar influx $F_{tq}(k)$ is equal to the total nectar profitability assessment. Suppose that the number of food-storer bees is sufficiently large so the wait time $W^i(k)$ that bee i experiences is given by

$$W^i(k) = \psi \max \left\{ F_{tq}(k) + \omega_\omega^i(k), 0 \right\} , \qquad (70.1)$$

where ψ is a scale factor and $\omega_\omega^i(k)$ is a random variable uniformly distributed in $(-\omega_\omega, \omega_\omega)$ that represents variations in the wait time a bee experiences. When the total nectar influx is maximum, the value of the wait time is approximately 30 s, based on the experiments in [70.45]. With this assumption we can obtain the values of ψ and ω_ω from the fact that the maximum wait time from (70.1) is given by $\psi(B + \omega_\omega) = 30$. Hence, it can be noted that $\psi \omega_\omega$ is the variation in the number of seconds in wait time due to the noise, and ω_ω has to be set adequately. If we let $\psi \omega_\omega = 5$ and we have assumed that $B = 200$, we obtain two equations and two unknowns, which gives $\psi = 52/200$ and $\omega_\omega = 40$.

Dance Decision Function

Now, we assume that each successful forager converts the wait time it experienced into a scaled version of an estimate of the total nectar influx that we define as

$$\hat{F}_{tq}(k) = \delta W^i(k) . \qquad (70.2)$$

The value $\hat{F}_{tq}(k)$ provides bee i a noisy estimate of the whole colony's foraging performance, since it provides an indication of how many successful foragers are waiting to be unloaded [70.37]. The proportionality constant is $\delta > 0$, and since $W^i(k) \in [0, \psi(B + \omega_\omega)] = [0, 30]$ s, it implies that $\hat{F}_{tq}^i(k) \in [0, 30\delta]$. In order to ensure that $\hat{F}_{tq}^i(k) \in [0, 1]$, we consider that $0 < \delta < \frac{1}{30}$.

With this estimation, each bee has to decide how long to dance according to some forage site variables that determine the energetic profitability (e.g., distance from hive, sugar content of nectar, nectar abundance), and some conditions that determine the threshold of the dance response (e.g., weather, time of day, colony's nectar influx). The *decision function* is

$$L_f^i = \max \left\{ \beta \left(F^i(k) - \hat{F}_{tq}(k) \right), 0 \right\} , \qquad (70.3)$$

which indicates the number of waggle runs of bee i at expedition k. The parameter $\beta > 0$ has the effect of a gain on the rate of recruitment for sites above the dance threshold, and experimentally [70.37] we can set $\beta = 100$.

When a bee has $L_f^i(k) > 0$, it may consider dancing for her forage site. The probability that bee i will choose to dance for the site it is dedicated to is given by

$$p_r(i, k) = \frac{\phi}{\beta} L_f^i(k) ,$$

where $\phi \in [0, 1]$; matching the behavior of what is found in experiments, we choose $\phi = 1$.

70.1.4 Explorer Allocation and Forager Recruitment

Bees that are not successful on an expedition, or those that do not consider dancing, become unemployed foragers. Some of these bees will start to rest or they will become observers and they will start seeking dancing bees in order to get recruited. The probability that an unemployed forager or current rester bee will become an observer is $p_o \in [0, 1]$. Based on the results in [70.46], we choose $p_o = 0.35$ in such a way that in times where there are no forage sites being harvested there can be about 35% of the bees performing as forage explorers.

If an observer bee does not find any dance to follow, it will go exploring. So we take the $B_o(k)$ observer bees and each one can become an explorer with probability $p_e(k)$ or can follow the dance and become an employed forager with probability $1 - p_o(k)$. We choose

$$p_e(k) = \exp \left(-\frac{1}{2} \frac{L_t^2(k)}{\sigma^2} \right) , \qquad (70.4)$$

where $L_t(k) = \sum_{i=1}^{B_f(k)} L_f^i(k)$ is the total number of waggle runs on the dance floor at step k. Notice that if $L_t(k) = 0$, there are no dancing bees on the dance floor, so $p_e(k) = 1$ and all the observers will explore (i. e., 35% of the unemployed foragers). Here, we choose $\sigma = 1000$ since it produces patterns of foraging behavior in simulations that correspond to experiments.

Now, we take the observer bees that did not go to explore and some of them will be recruited in order to follow the dance of bee i with probability

$$p_i(k) = \frac{L_f^i(k)}{\sum_{j=1}^{B_f(k)} L_f^j(k)} . \tag{70.5}$$

In this way, bees that dance more strongly will tend to recruit more foragers for their site.

Algorithm 70.1 summarizes the pseudo-code of the honey bee social foraging algorithm described above.

Algorithm 70.1 Honey Bee Social Foraging Algorithm

1: *Set the parameter values*
2: **while** *Stopping criterium is not reached* **do**
3:      Determine number of bees at each forage site, and compute the suitability of each site.
4:      **for** *Each employed forager and explorer* **do**
5:          Define a noisy profitability assessment $F^i(k)$ according to the location
6:          **if** $F^i(k) > \epsilon_n$ **then**
7:              **if** *Bee is an employed forager* **then**
8:                  Stays that way
9:              **else**
10:                 Bee becomes an employed forager
11:             **end if**
12:         **else**
13:             Bee becomes an observer or rester
14:         **end if**
15:     **end for**
16:     Compute the total nectar profitability, and total nectar influx
17:     **for** *All employed foragers* **do**
18:         Compute wait time W^i, and the noise for unload wait time w_w.
19:         Compute estimate of scaled total nectar influx $\hat{F}_{tq}$
20:         Compute dance decision function L_f^i
21:         **if** $L_f^i = 0$ **then**
22:             Bee becomes unemployed
23:         **end if**
24:         **if** *Employed forager should not recruit* **then**
25:             $L_f^i = 0$. Bee i is removed from those that dance
26:         **end if**
27:     **end for**
28:     Determine L_t. Employed foragers and successful forager explorers may dance based on sampling of profitability
29:     Send all employed foragers back the their previous site after recruitment for the next expedition
30:     **for** *Unemployed foragers* **do**
31:         We set $W^i = L_f^i = \hat{F}_{tq}^i = 0$
32:         Unemployed foragers become observers with probability p_o. The remaining unemployed foragers become resters
33:     **end for**
34:     Set p_e
35:     **for** *Unemployed foragers* **do**
36:         **if** $rand < p_e$ **then**
37:             Bee becomes explorer. Set explorer location for the next expedition
38:         **end if**
39:         **for** *Unemployed observers* **do**
40:             Unemployed observer will be recruited by bee i with probability p_i
41:         **end for**
42:     **end for**
43: **end while**

70.2 Application in a Multizone Temperature Control Grid

In order to apply the proposed algorithm in the context of a physical resource allocation problem, we implement the multi-zone temperature control grid introduced in [70.43], with four zones as shown in Fig. 70.1. The relations between our problem and the proposed algorithm can be summarized as follows:

1. We assume that there is a population of B individuals (i. e., bees, chromosomes, or particles) that contains the information of a position θ_i, where $i = 1, \ldots, B$.
2. The search space is composed of *allocation sites*, which are denoted with a position R_j and a width ϵ_f^j.

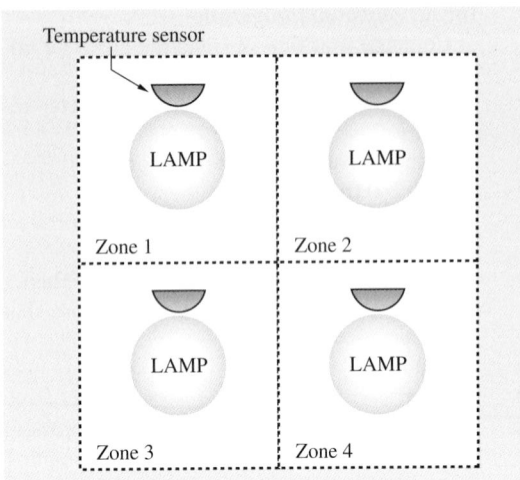

Fig. 70.1 Layout for the multizone temperature control grid

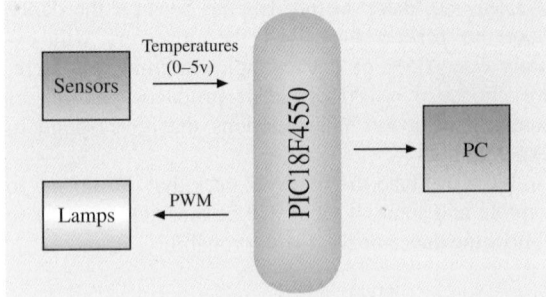

Fig. 70.2 Layout of data acquisition and temperature control

3. Each allocation site corresponds to the zone j in the temperature grid, for $j = 1, \ldots, 4$.
4. Let T_j^d and T_j be a temperature reference and the temperature for each zone j, respectively.
5. We consider the temperature error for zone j as $e_j = T_j^d - T_j$, and if an individual is located in a zone, its fitness is given by γe_j, for γ being a positive constant that sets the fitness value in the range of $[0, 1]$.

70.2.1 Hardware Description

A zone contains a temperature sensor LM35 and a lamp that varies its intensity in order to increase or decrease the temperature of the zone. The data acquisition and the lamps' intensity variations are performed using a microcontroller PIC18F4550, which receives the temperature values (voltages between 0 and 5 V) and transmits them through the USB port to a PC using the USB-bulk communication class. We cannot guarantee that the four sensors have the same characteristics (they have $\pm 0.2\,°C$ typical accuracy, and $\pm 0.5\,°C$ guaranteed). With these temperature values, a Matlab program executes an iteration of one of the algorithms and sends pulse width modulation (PWM) width information back to the peripheral interface controller (PIC) (Fig. 70.2). PWM signals are generated using four digital outputs of the PIC and a couple of transistors that drive the amount of current and voltage necessary to control the lamps. The width of the PWM signal depends on the number of individuals (i. e., bees,

chromosomes, or particles) allocated on a site. Each individual is equivalent to a portion of the PWM, and a 100% duty cycle corresponds to 12 V of direct current (DC).

We assume that there is a total amount of voltage that can be distributed between the four zones. The goal is to allocate that voltage in such a way that the reference temperature for each zone can be achieved. However, to achieve this goal is complicated due to external effects, such as ambient temperature and wind currents, interzone effects, differences between the components of a zone, and sensor noise. For this reason, the total voltage amount has to be dynamically allocated, despite the external and internal effects.

70.2.2 Other Algorithms for Resource Allocation

In order to compare the behavior of the honey bee social foraging algorithm to solve dynamic resource allocation problems, two evolutionary algorithms were selected, i. e., the genetic algorithm (GA) and particle swarm optimization (PSO). These methods have been implemented in a wide variety of applications because of their low computational cost and their huge capability for solving optimization problems. Some implementations for resource allocation can be found in [70.22, 23, 47]. We will show below that these algorithms can be adapted in such a way that these resource allocation problems can be solved.

Genetic Algorithms
A genetic algorithm (GA) is a random search algorithm based on the mechanics of natural selection, genetics, and evolution [70.48]. The basic structure of the population is the chromosome. During each generation, chromosomes are evaluated based on a fitness function

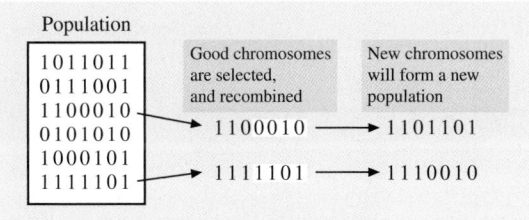

Population

```
1011011
0111001
1100010
0101010
1000101
1111101
```

Good chromosomes are selected, and recombined	New chromosomes will form a new population

1100010 → 1101101

1111101 → 1110010

Fig. 70.3 Selection and crossover in GA

and some of them are stochastically selected depending on their fitness values. The population evolves from generation to generation through the application of genetic operators [70.49].

The simplest GA form uses chromosome information, which is encoded into a binary string. The chromosomes are modified using three operators: selection, crossover, and mutation [70.3]. *Selection* is an artificial version of natural selection, and only the fittest chromosomes from the population are selected. With *crossover*, two parents are chosen for reproduction, and a crossover site (a bit position) is randomly selected. The subsequences after the crossover site

are exchanged with a probability p_c, producing two offspring with information from both parents. Then, *mutation* randomly inverts a bit on a string with a very low probability, and introduces new information into the population at the bit level. Figure 70.3 illustrates the GA selection and crossover process for the simplest GA algorithm.

To solve a resource allocation problem, we adjust this algorithm as follows: we set a population of B individuals, each one with binary encoded information about its position θ_i, for $i = 1, \ldots, B$. There is a landscape that contains 4 different resources sites, each one located in a position R_j with a width ϵ_f^j, for $j = 1, \ldots, 4$. Each resource site corresponds to a temperature zone in the multizone control grid. When an individual is located in a resource site j, the fitness of that individual is given by the error between the current temperature T_j and the reference temperature T_j^d, otherwise, the fitness is 0. As was pointed out before, each individual corresponds to a portion of the total amount of voltage. For that reason, if the fitness is good, the population evolves, most of the individuals will have the same genetic information (position), and they will be allocated

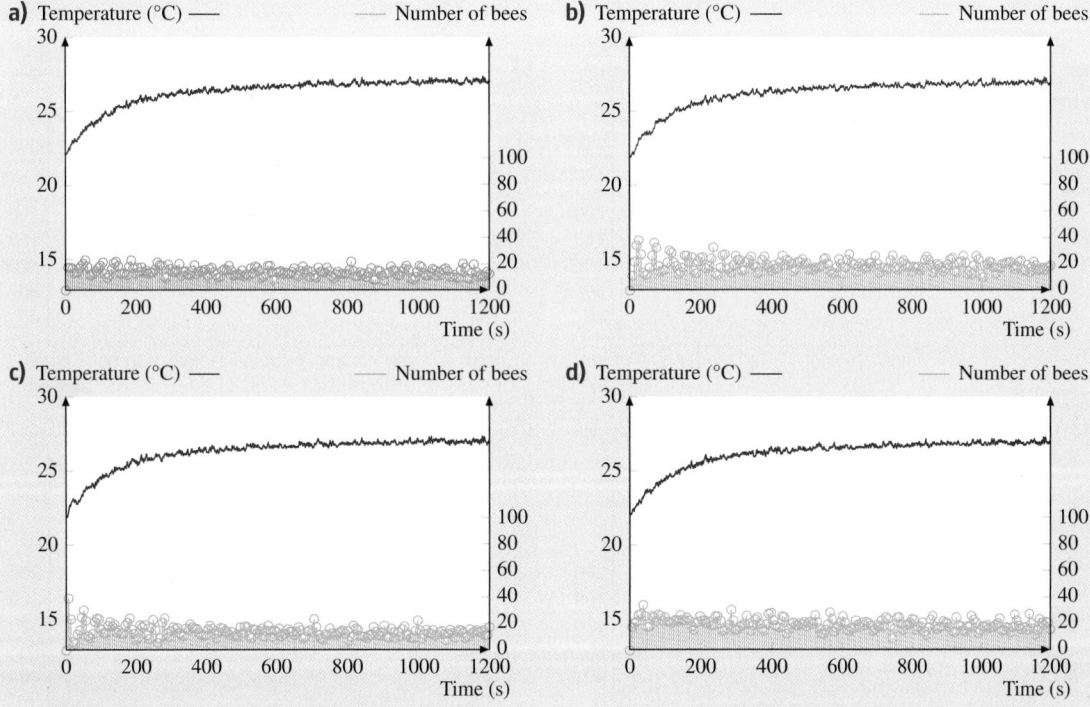

Fig. 70.4a-d Average temperature (*solid lines*) and number of bees per zone (*stem plots*) using the honey bee foraging algorithm

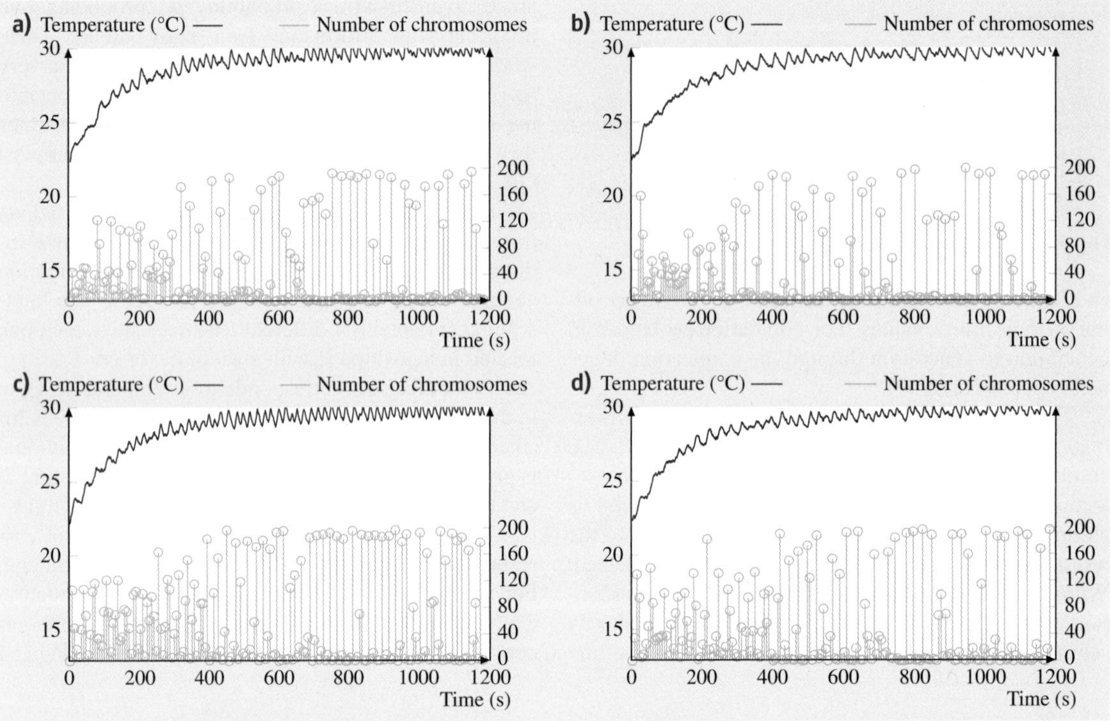
a) Temperature (°C)
Number of chromosomes
b) Temperature (°C)
Number of chromosomes
c) Temperature (°C)
Number of chromosomes
d) Temperature (°C)
Number of chromosomes
30
25
20
15
200
160
120
80
40
0
0 200 400 600 800 1000 1200
Time (s)

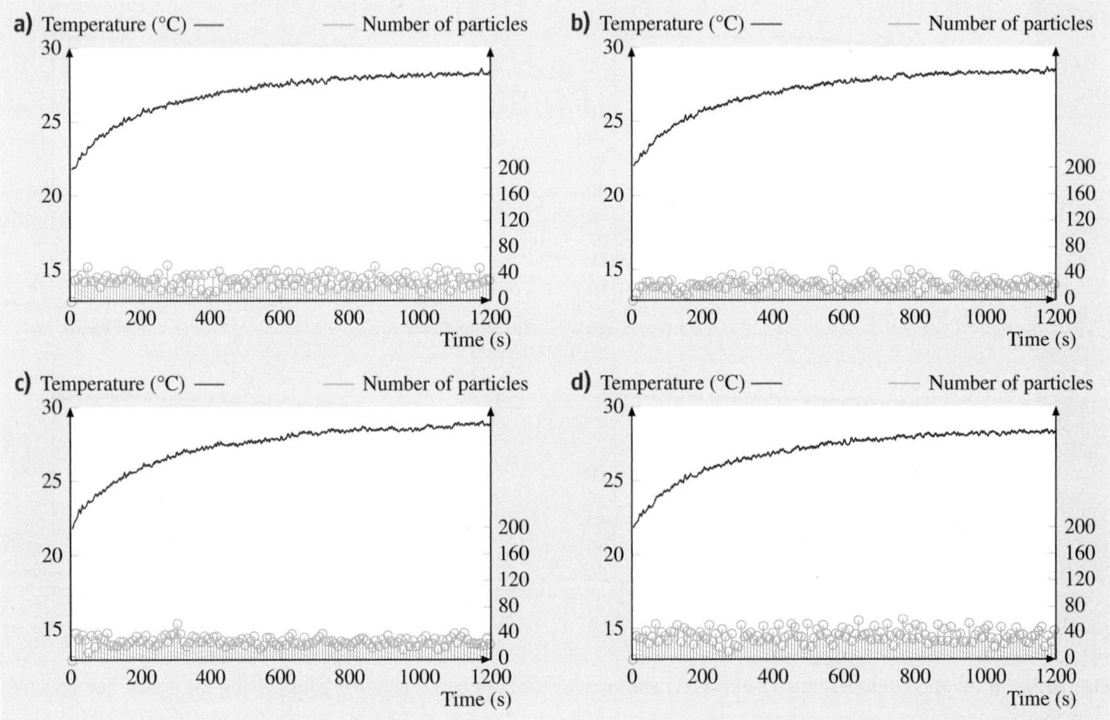
a) Temperature (°C)
Number of particles
b) Temperature (°C)
Number of particles
c) Temperature (°C)
Number of particles
d) Temperature (°C)
Number of particles
30
25
20
15
200
160
120
80
40
0
0 200 400 600 800 1000 1200
Time (s)

Fig. 70.5a–d Average temperature (*solid lines*) and number of chromosomes per zone (*stem plots*) using the genetic algorithm ◄

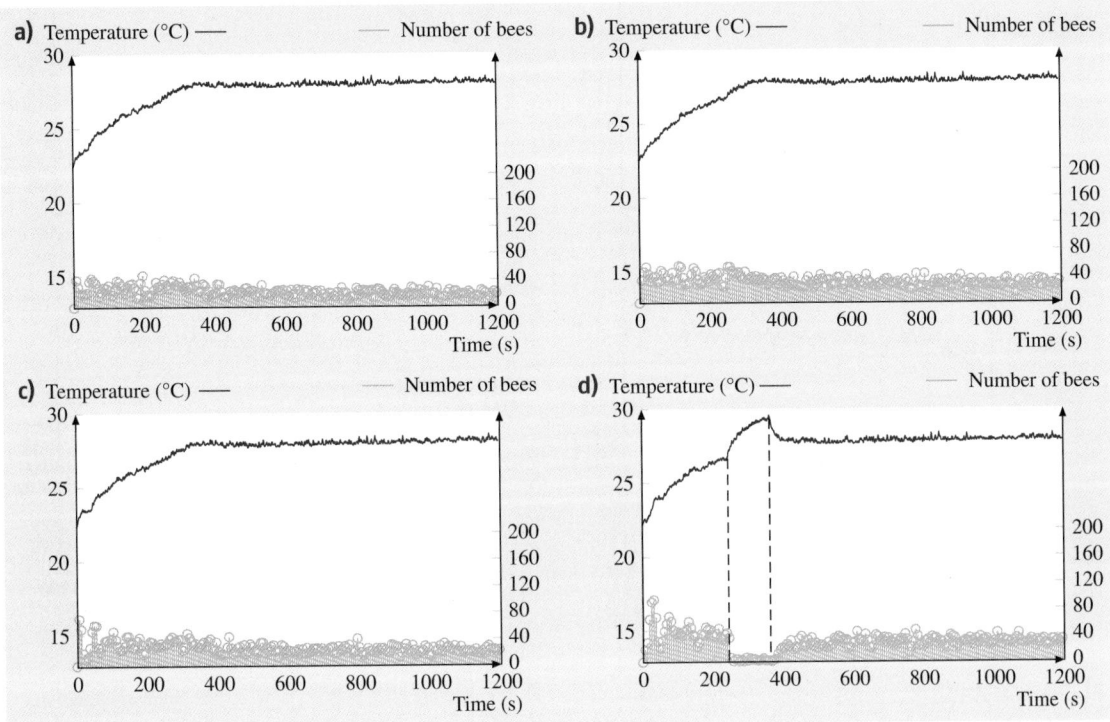

Fig. 70.7a–d Average temperature (*solid lines*) and number of bees per zone (*stem plots*) using the honey bee foraging algorithm. *Dashed lines* indicate the beginning and end of the disturbance

Fig. 70.6a–d Average temperature (*solid*) and number of particles per zone (*stem plots*) using the PSO algorithm ◄

at that site. Then, the amount of voltage applied to that zone increases as well as the temperature, which provokes that the fitness associated to that site decreases, and it will become less profitable. For the next generation, the individuals tend to be reallocated into another more profitable zone, and after some generations, the population is distributed in such a way that a uniform temperature is achieved for all sites.

Particle Swarm Optimization

Particle swarm optimization (PSO) is a population-based stochastic optimization technique, inspired by the social behavior of animals (e.g., bird flocks, fish schools, or even human groups) [70.7]. In PSO, the potential solutions, called *particles*, *fly* through the problem space by following the currently best particles. They have two essential reasoning capabilities: mem-

ory of their own best position and knowledge of the global or their neighborhood's best. Each particle in a population has the information about its current position that defines a potential solution, and its fitness value associated to that position. A change of position of a particle is defined by the velocity, which is a vector of numbers that are added to the position coordinates in order to move the particle from one time step to another. At each iteration, a particle's velocity is updated depending on the difference between the individual's previous best and current positions, and the difference between the neighborhood's best and the individual's current position. With these simple rules, individuals tend to follow the particles associated to the more profitable sites, and optimization problems can be solved. The details of this method are summarized in [70.7].

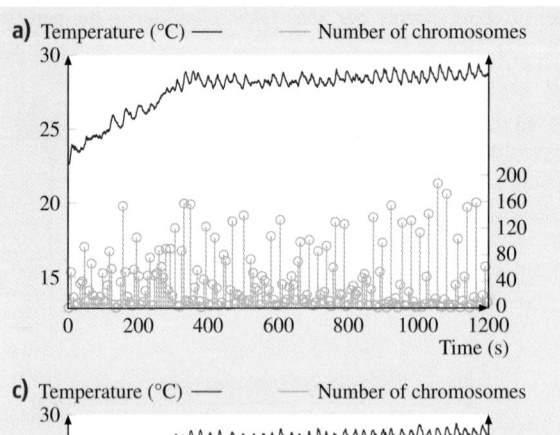

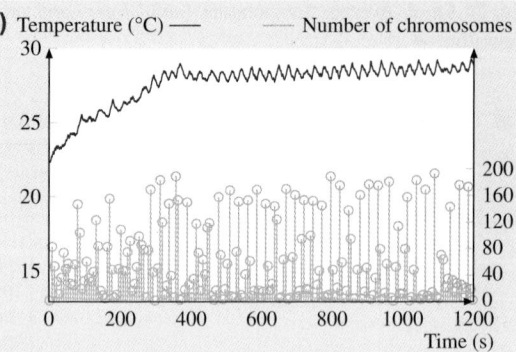

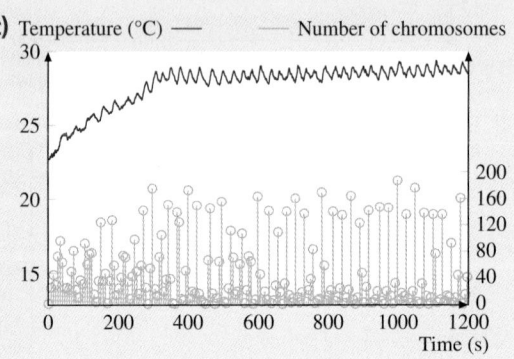

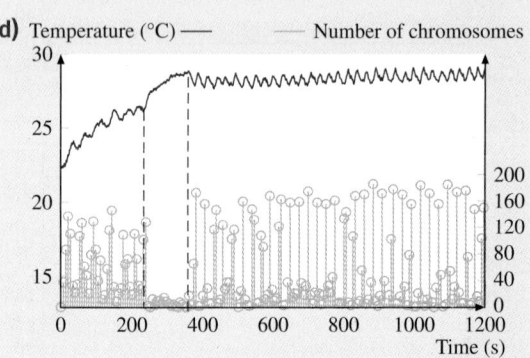

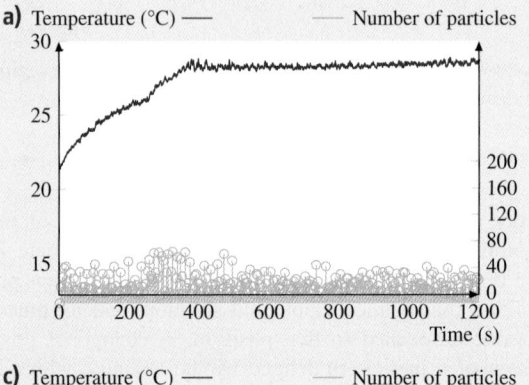

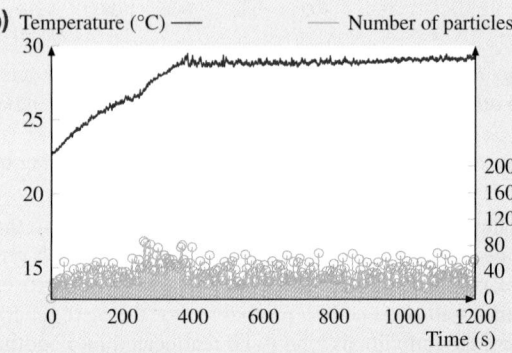

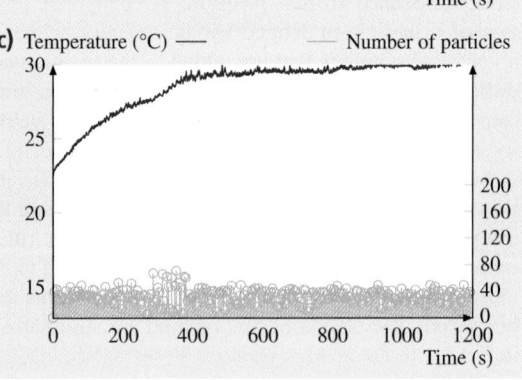

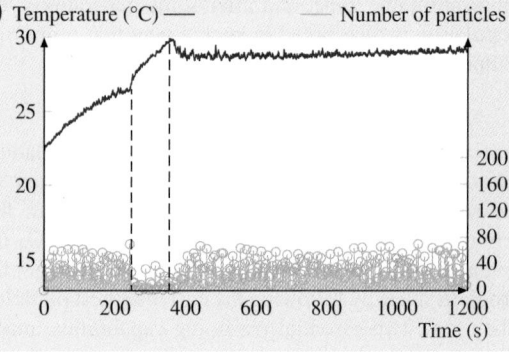

Fig. 70.8a–d Average temperature (*solid lines*) and number of chromosomes per zone (*stem plots*) using the genetic algorithm. *Dashed lines* indicate the beginning and end of the disturbance ◄

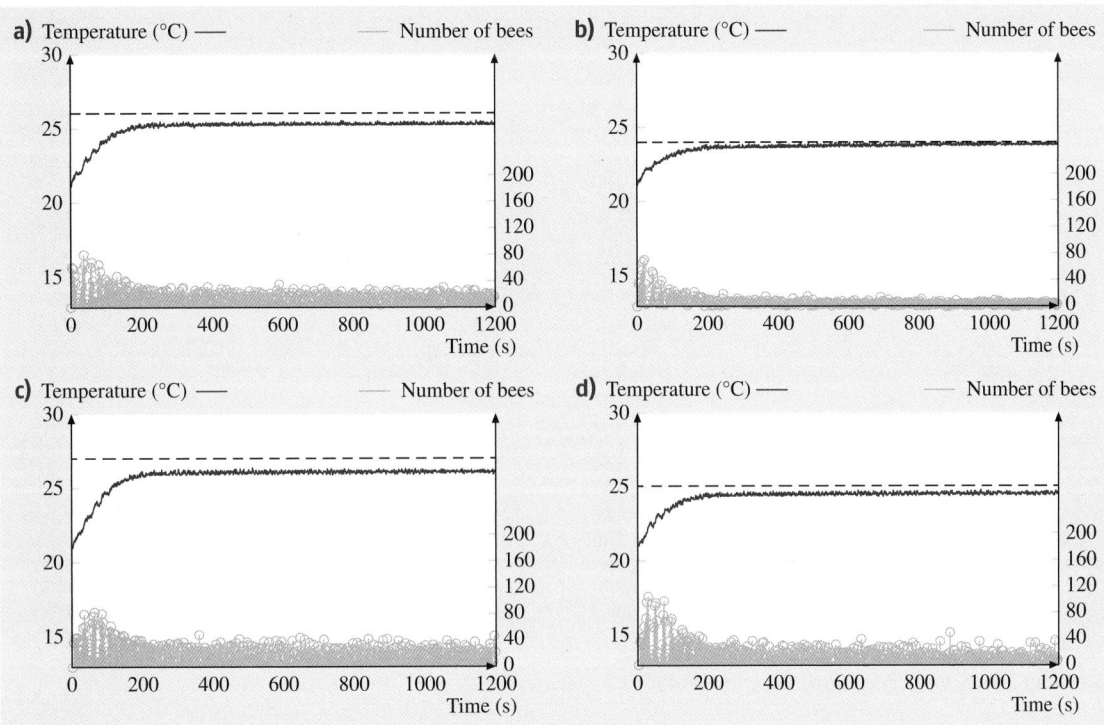

Fig. 70.10a–d Average temperature (*solid lines*) and number of bees per zone (*stem plots*) using the honey bee foraging algorithm. The *dashed lines* indicate the different set points for each zone

Fig. 70.9a–d Average temperature (*solid lines*) and number of particles per zone (*stem plots*) using the PSO algorithm. *Dashed lines* indicate the beginning and end of the disturbance ◄

To solve a dynamic resource allocation problem, an adjustment to this algorithm is made as follows. First, we assume a population of B particles, where each particle represents a one-dimensional position θ_i in the search space. The fitness of each individual is defined by the temperature error γe_j shown above. When a particle has a good fitness, most of the individuals will tend to fly to the same position, provoking a temperature increment and a decrease of the error. For that reason, the particles are reallocated to a more profitable place, and after some generations, the population is distributed in such a way that a uniform temperature is achieved for all sites.

70.3 Results

In order to compare the behavior of the proposed algorithms to solve the dynamic resource allocation problem, three experiments are performed using the multizone temperature control grid described in Sect. 70.2.1. In the first experiment, we seek the maximum uniform temperature with a single population of $B = 200$ individuals. The second experiment illustrates the response of the individuals when a disturbance is applied in the fourth zone. Finally, multiple temperature set points are assigned to each zone. Our results show the behavior

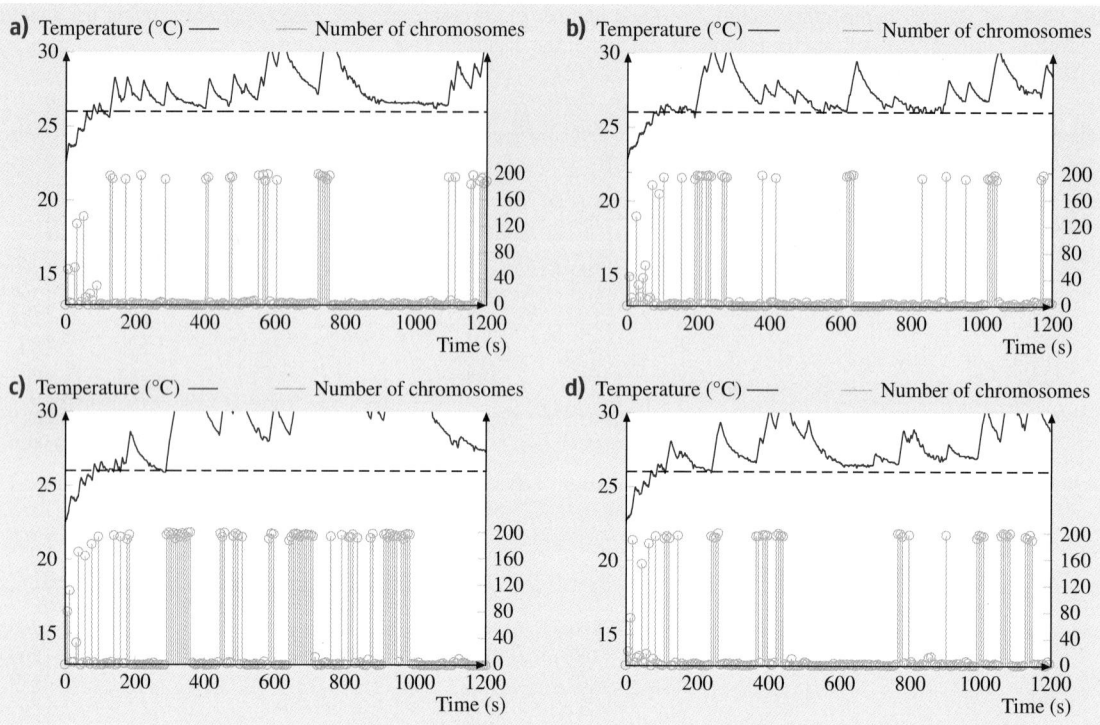

Fig. 70.11a–d Average temperature (*solid lines*) and number of chromosomes per zone (*stem plots*) using the genetic algorithm. The *dashed lines* show the set point for each zone

of the three algorithms applied in a resource allocation implementation under the same circumstances.

70.3.1 Experiment I: Maximum Uniform Temperature

In this experiment, we want to achieve a maximum uniform temperature for all zones. For this reason, we set reference temperatures $T_j^d = 30$, for all $j = 1, \ldots, 4$ (this value cannot be achieved by the system). We have a population of 200 individuals and a PWM frequency of 70 Hz, where each individual corresponds to a duty cycle of 0.5% (i.e., each individual corresponds to 0.06 V.) For example, 50 individuals in a zone are equal to 25% of the duty cycle and they correspond to 3 volts. Figures 70.4–70.6 show the temperature results and the number of individuals allocated in each zone.

70.3.2 Experiment II: Disturbance

This experiment is similar to the first one, but now we add a controlled disturbance. An extra lamp is placed next to zone 4; it is turned on after 4 min, and is turned

off 2 min later. When we apply this disturbance, the temperature in zone 4 increases drastically and site 4 becomes the least profitable. Then, the number of individuals in that site are reallocated, provoking a small increase in temperatures of the other three zones. Figures 70.7–70.9 illustrate the behavior of the system applying the three algorithms.

70.3.3 Experiment III: Multiple Set Points

In this experiment we want to achieve multiple set points (26, 24, 27, and 25 °C, respectively, for each of the four zones), which are lower than the ones achieved before. Figure 70.10 presents the results obtained using the honey bee foraging algorithm. We can observe that the set points are never achieved, but the temperatures get very close to it, and the steady states are reached quickly. This is because the algorithm requires an error for each zone $e_j > 0$, which implies that the temperatures are always below the set point.

On the other hand, with the GA and PSO, the behavior is very different. When one of the set points is achieved, the resources tend to reallocate to the other

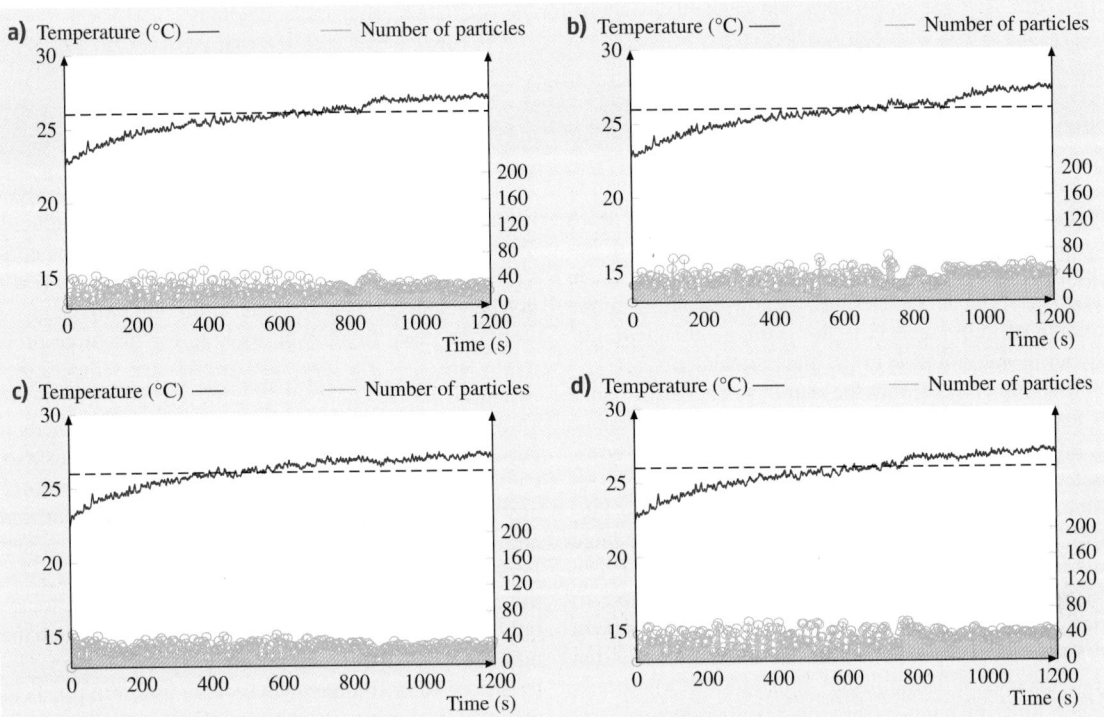

Fig. 70.12a–d Average temperature (*solid lines*) and number of chromosomes per zone (*stem plots*) using the PSO. The *dashed lines* show the set point for each zone

zones. Therefore, when all the zones reach their set points, all the fitnesses (which are proportional to the error values) are equal to 0, and the remaining resources should not be allocated. However, for these two methods, most of the resources need to be allocated to any site, and the remaining resources stay in the last site, provoking a drastic increase of the temperature in only that zone. When the temperature in any other zone de-

creases, agents tend to be reallocated to that new site very quickly, until the temperature increases again. Figures 70.11 and 70.12 show this behavior for the GA and PSO, respectively, with a fixed set point of 26 °C for all four zones. The drastic changes produced by the non-needed resources can be observed, and it can be seen that these methods are not feasible for these kinds of problems.

70.4 Discussion

The experiments that were performed using the bioinspired algorithms show behaviors common to each of the three techniques, and some advantages and disadvantages can be discussed. In Sect. 70.3.1, we saw how the maximum uniform temperature can be achieved by the three methods implemented. We obtained the average values and standard deviation of temperatures after ten experiments (Table 70.1). We observe that the genetic algorithm achieves the highest average temperatures, but the standard deviations are also high, which means that the GA behavior is unsteadier. This is be-

cause the GA is very susceptible to changes and, as soon as a zone becomes more profitable (the temperature decreases), most chromosomes tend to abandon their current positions and go to the more profitable one. That is why the number of chromosomes changes abruptly (Fig. 70.5), and the temperature variation is also high for each one of the ten experiments. On the other hand, the other algorithm's reactions are slower, and the number of individuals remains almost constant (very low variations). Table 70.1 illustrates that the honey bee social foraging algorithm has the lowest

Table 70.1 Average temperatures and standard deviations for the last 10 min, using each algorithm

Algorithms	Zone 1	Zone 2	Zone 3	Zone 4
Honey bee	27.28 ± 0.12	27.26 ± 0.12	27.22 ± 0.11	27.18 ± 0.11
GA	29.78 ± 0.38	29.8 ± 0.3	29.75 ± 0.34	29.76 ± 0.36
PSO	27.84 ± 0.17	27.86 ± 0.18	28.55 ± 0.14	27.81 ± 0.16

temperature variation, which means that for every experiment, the maximum temperature achieved for each zone is practically the same. This is an advantage, because repeatability is a very important characteristic in practical applications. Besides, the low variations in the amount of individuals (i. e., low voltage changes), imply a low deterioration of the electric elements.

Section 70.3.2 shows the results when a disturbance is applied to the fourth zone. When this disturbance is turned on, the temperature increases in that zone, and that site becomes less profitable, provoking a reallocation of the individuals into the other zones. Figures 70.7–70.9 show how the resources are reallocated and the temperatures in the other sites increase. It can be observed that during the disturbance, the number of individuals in the fourth zone tends to 0 and, as soon as the disturbance is turned off, the temperature in that zone decreases. These results illustrate the robustness of the three techniques for external disturbances, and we can also observe that regardless of the type of disturbance, the maximum uniform temperature is achieved.

Experiment 3 illustrates the behavior of the algorithms when low temperature set points are considered. We observed that most of the resources should be allocated to the sites, and low temperature references could not be achieved. This is because the individuals move from one place to another, looking for the more profitable site, i. e., the one with the lowest temperature. When all the set points are achieved, all profitability values are 0, and the individuals will remain in their current position. Hence, temperatures continue to rise, even if the reference has been achieved. Figures 70.11 and 70.12 illustrate this behavior for the GA algorithm and PSO when a reference temperature of $26\,°C$ is set. However, the honey bee foraging algorithm can solve this kind of problem because of its capability to allocate only the *necessary* bees into the foraging places, and the non-needed resources are simply not used. This characteristic of our technique may be very useful in practical applications, such as in smart building temperature control, where multiple temperature references for different rooms need to be achieved.

70.5 Conclusions

In this chapter, a novel bioinspired method for dynamic resource allocation based on the social behavior of honey bees during the foraging process was presented, and an application that illustrates the validity of the approach was studied. The application that we used is a multizone temperature control grid, where the objective is to achieve some reference temperatures for each one of the zones, taking into account the complexity induced by the interzone effects and external or internal noise. Some comparative analyses have been developed

with two evolutionary algorithms, i. e., GA and PSO. We can see that the proposed method has some advantages compared to other bioinspired methods due to its capability of allocating only the necessary resources, and the low variability of the number of individuals in each one of the four zones. Clearly, there are other applications for the social foraging method for resource allocation. For instance, in the area of task allocation of agents, formation control, economic dispatch, and smart building temperature control.

References

70.1   E. Bonabeau, M. Dorigo, G. Theraulaz: *Swarm Intelligence: From Natural to Artificial Systems* (Oxford Univ. Press, New York 1999)

70.2   K.M. Passino: *Biomimicry for Optimization, Control and Automation* (Springer, London 2005)

70.3   J.H. Holland: *Adaptation in Natural and Artificial Systems: An Introductory Analysis with Applications to Biology, Control, and Artificial Intelligence*, 1st edn. (Univ. Michigan Press, Ann Arbor 1975)

70.4   T.P. Hong, W.Y. Lin, S.M. Liu, J.H. Lin: Dynamically adjusting migration rates for multi-population genetic algorithms, J. Adv. Comput. Intell. Intell. Inform. **11**, 410–415 (2007)

70.5   M. Affenzeller, S. Winkler: *Genetic Algorithms and Genetic Programming: Modern Concepts and Practical Applications*, Vol. 6 (Chapman Hall/CRC, Boca Raton 2009)

70.6   M.D. Higgins, R.J. Green, M.S. Leeson: A genetic algorithm method for optical wireless channel control, J. Ligthwave Technol. **27**(6), 760–772 (2009)

70.7   J. Kennedy, R.C. Eberhart: *Swarm Intelligence* (Morgan Kaufmann, San Francisco 2001)

70.8   M. Dorigo, C. Blum: Ant colony optimization theory: A survey, Theor. Comput. Sci. **344**(2), 243–278 (2005)

70.9   D. Sumpter, D.S. Broomhead: Formalising the link between worker and society in honey bee colonies. In: *Multi-Agent Systems and Agent-Based Simulation*, ed. by J. Sichman, R. Conte, N. Gilbert (Springer, Berlin 1998) pp. 95–110

70.10   T.D. Pham, A. Ghanbarzadeh, E. Koc, S. Otri, S. Rahim, M. Zaidi: The bees algorithm – a novel tool for complex optimisation problems, Proc. 2nd Int. Virtual Conf. Intell. Prod. Mach. Syst. (Elsevier, Oxford 2006) pp. 454–461

70.11   D. Teodorovic: *Bee Colony Optimization (BCO)*, Studies in Computational Intelligence, Vol. 248 (Springer, Berlin/Heidelberg 2009)

70.12   R.C. Eberhart, Y. Shi, J. Kennedy: *Swarm Intelligence*, 1st edn. (Morgan Kaufmann, San Francisco 2001)

70.13   M. Dorigo, V. Maniezzo, A. Colorni: Ant system: Optimization by a colony of cooperating agents, IEEE Trans. Syst. Man Cybern. B **26**(1), 29–41 (1996)

70.14   M. Dorigo, L.M. Gambardella: Ant colony system: A cooperative learning approach to the traveling salesman problem, IEEE Trans. Evol. Comput. **1**(1), 53–66 (1997)

70.15   K.M. Sim, W.H. Sun: Ant colony optimization for routing and load-balancing: Survey and new directions, IEEE Trans. Syst. Man Cybern. A **33**(5), 560–572 (2003)

70.16   J. Zhang, H.S.-H. Chung, A.W.-L. Lo, T. Huang: Extended ant colony optimization algorithm for power electronic circuit design, IEEE Trans. Power Syst. **24**(1), 147–162 (2009)

70.17   D. Merkle, M. Middendorf, H. Schmeck: Ant colony optimization for resource-constrained project scheduling, IEEE Trans. Evol. Comput. **6**(4), 333–346 (2002)

70.18   R. Poli, J. Kennedy, T. Blackwell: Particle swarm optimization: An overview, Swarm Intell. **1**(1), 33–57 (2007)

70.19   A.I. Selvakumar, K. Thanushkodi: A new particle swarm optimization solution to nonconvex economic dispatch problems, IEEE Trans. Power Syst. **22**(1), 42–51 (2007)

70.20   Z.L. Gaing: Particle swarm optimization to solving the economic dispatch considering the generator constraints, IEEE Trans. Power Syst. **18**(3), 1187–1195 (2003)

70.21   Y. Liu, G. Wang, H. Chen, H. Dong, X. Zhu, S. Wang: An improved particle swarm optimization for feature selection, J. Bionic Eng. **8**(2), 191–200 (2011)

70.22   P.Y. Yin, J.Y. Wang: A particle swarm optimization approach to the nonlinear resource allocation problem, Appl. Math. Comput. **183**(1), 232–242 (2006)

70.23   S. Gheitanchi, F. Ali, E. Stipidis: Particle swarm optimization for adaptive resource allocation in communication networks, EURASIP J. Wirel. Commun. Netw. **2010**, 9–21 (2010)

70.24   S. Nakrani, C. Tovey: From honeybees to internet servers: Biomimicry for distributed management of internet hosting centers, Bioinspir. Biomim. **2**(4), S182–S197 (2007)

70.25   V. Tereshko: Reaction-diffusion model of a honeybee colony's foraging behaviour, Proc. 6th Int. Conf. Parallel Probl. Solving Nat. PPSN VI (Springer, London 2000) pp. 807–816

70.26   V. Tereshko, A. Loengarov: Collective decision making in honey-bee foraging dynamics, Comput. Inf. Syst. **9**(3), 1–7 (2005)

70.27   B. Basturk, D. Karaboga: An artificial bee colony (abc) algorithm for numeric function optimization, IEEE Swarm Intell. Symp. 2006 (2006) pp. 12–14

70.28   D. Karaboga, B. Basturk: A powerful and efficient algorithm for numerical function optimization: Artificial bee colony (ABC) algorithm, J. Global Optim. **39**(3), 459–471 (2007)

70.29   C. Ozturk, D. Karaboga, B. Gorkemli: Probabilistic dynamic deployment of wireless sensor networks by artificial bee colony algorithm, Sensors **11**(6), 6056–6065 (2011)

70.30   W.Y. Szeto, Y. Wu, S.C. Ho: An artificial bee colony algorithm for the capacitated vehicle routing problem, Eur. J. Oper. Res. **215**(1), 126–135 (2011)

70.31   C. Zhang, D. Ouyang, J. Ning: An artificial bee colony approach for clustering, Expert Syst. Appl. **37**(7), 4761–4767 (2010)

70.32   D. Karaboga, B. Akay: A modified artificial bee colony (ABC) algorithm for constrained optimization problems, Appl. Soft Comput. **11**(3), 3021–3031 (2011)

70.33   D. Teodorović, M. Dell'Orco: Bee colony optimization – A cooperative learning approach to complex transportation problems. In: *Advanced OR and AI Methods in Transportation*, ed. by A. Jaszkiewicz, M. Kaczmarek, J. Zak, M. Kubiak (Publishing House of Poznan University of Technology, Poznan 2005) pp. 51–60

70.34   M. Basu: Bee colony optimization for combined heat and power economic dispatch, Expert Syst. Appl. **38**(11), 13527–13531 (2011)

Part F | 70

70.35   C.S. Chong, A.I. Sivakumar, M.Y. Low, K.L. Gay: A bee colony optimization algorithm to job shop scheduling, Proc. 38th Conf. Winter Simul. (2006) pp. 1954–1961

70.36   A. Kaur, S. Goyal: A survey on the applications of bee colony optimization techniques, Int. J. Comput. Sci. Eng. **3**(8), 3037–3046 (2011)

70.37   T.D. Seeley: *The Wisdom of the Hive* (Harvard Univ. Press, Cambridge 1995)

70.38   K.M. Passino, T.D. Seeley: Modeling and analysis of nest-site selection by honeybee swarms: The speed and accuracy trade-off, Behav. Ecol. Sociobiol. **59**(3), 427–442 (2006)

70.39   G. Obando, A. Pantoja, N. Quijano: Evolutionary game theory applied to building temperature control, Proc. Nolcos (IFAC, Bologna 2010) pp. 1140–1145

70.40   A. Pantoja, N. Quijano, S. Leirens: A bioinspired approach for a multizone temperature control system, Bioinspir. Biomim. **6**(1), 16007–16020 (2011)

70.41   N. Quijano, K.M. Passino: The ideal free distribution: Theory and engineering application, IEEE Trans. Syst. Man Cybern. B **37**(1), 154–165 (2007)

70.42   N. Quijano, A.E. Gil, K.M. Passino: Experiments for dynamic resource allocation, scheduling, and control, IEEE Control Syst. Mag. **25**(1), 63–79 (2005)

70.43   N. Quijano, K.M. Passino: Honey bee social foraging algorithms for resource allocation: Theory and application, Eng. Appl. Artif. Intell. **23**(6), 845–861 (2010)

70.44   S.D. Fretwell, H.L. Lucas: On territorial behavior and other factors influencing habitat distribution in bird. I. Theoretical development, Acta Biotheor. **19**, 16–36 (1970)

70.45   T.D. Seeley, C.A. Tovey: Why search time to find a food-storer bee accurately indicates the relative rates of nectar collecting and nectar processing in honey bee colonies, Animal Behav. **47**(2), 311–316 (1994)

70.46   T.D. Seeley: Division of labor between scouts and recruits in honeybee foraging, Behav. Ecol. Sociobiol. **12**, 253–259 (1983)

70.47   J. Alcaraz, C. Maroto: A robust genetic algorithm for resource allocation in project scheduling, Ann. Oper. Res. **102**, 83–109 (2001)

70.48   S.N. Sivanandam, S.N. Deepa: *Introduction to Genetic Algorithms* (Springer, Berlin 2007)

70.49   M. Mitchel: *An Introduction to Genetic Algorithms* (MIT Press, Cambridge 1996)

71. Fundamental Collective Behaviors in Swarm Robotics

Vito Trianni, Alexandre Campo

In this chapter, we present and discuss a number of types of fundamental collective behaviors studied within the swarm robotics domain. Swarm robotics is a particular approach to the design and study of multi-robot systems, which emphasizes decentralized and self-organizing behavior that deals with limited individual abilities, local sensing, and local communication. The desired features for a swarm robotics system are flexibility to variable environmental conditions, robustness to failure, and scalability to large groups. These can be achieved thanks to well-designed collective behavior – often obtained via some sort of bio-inspired approach – that relies on cooperation among redundant components. In this chapter, we discuss the solutions proposed for a limited number of problems common to many swarm robotics systems – namely aggregation, synchronization, coordinated motion, collective exploration, and decision making. We believe that many real-word applications subsume one or more of these problems, and tailored solutions can be developed starting from the studies we review in this chapter. Finally, we propose possible directions for future research and discuss the relevant challenges to be addressed in order to push forward the study and the applications of swarm robotics systems.

Part F | 71

71.1 Designing Swarm Behaviours

Imagine the following scenario: in a large area there are multiple items that must be reached, and possibly moved elsewhere or processed in some particular way. There is no map of the area to be searched, and the area is rather unknown, unstructured, and possibly dangerous for the intervention of humans or any valuable asset. The items must be reached and processed as quickly as possible, as a timely intervention would correspond to a higher overall performance. This is the typical scenario to be tackled with swarm robotics. It contains all the properties and complexity issues that make a swarm robotics solution particularly appropriate. Parallelism, scalability, robustness, flexibility, and adaptability to unknown conditions are features that are required from a system confronted with such a scenario, and exactly those features are sought in swarm robotics research.

Put in other terms, swarm robotics promises the solution of complex problems through robotic systems made up of multiple cooperating robots. With respect to other approaches in which multiple robots are exploited at the same time, swarm robotics emphasizes aspects like decentralization of control, limited individual abilities, lack of global knowledge, and scalability to large groups.

One important aspect that characterizes a swarm robotics system concerns the robotic units, which are unable to solve the given problem individually. The limitation is given either by physical constraints that would prevent the single robot to individually tackle the problem (e.g., the robot has to move some items that are too heavy), or by time constraints that would make a solitary action very inefficient (e.g., there are too many items to be collected in a limited time). Another source of limitation for the individual robot comes from its inability to acquire a global picture of the problem, having only access to partial (local) information about the environment and about the collective activity. These limitations imply the need for cooperation to ensure task achievement and better efficiency. Groups of autonomous cooperating robots can be exploited to synergistically achieve a complex task, by joining forces and sharing information, and to distributedly undertake the given task and achieve higher efficiency through parallelism.

The second important aspect in swarm robotics is redundancy in the system, which is intimately connected with robustness and scalability. Swarm robotics systems are made by homogeneous robots (or by relatively few heterogeneous groups of homogeneous robots). This means that the failure of a single or a few robots is not a relevant fact for the system as a whole, because the failing robot can easily be replaced by another teammate. Differently from a centralized system, in a swarm robotics system there is no single point of failure, and every component is interchangeable with other components. Redundancy, distributed control, and local interactions also allow for scalability, enabling the robotic system to seamlessly adapt to varying group sizes. This is a significant advantage with respect to centralized systems, which would present an exponential increase in complexity for larger group sizes.

Because all the above features are desiderata, the problem remains as to how to design and implement such a robotic system. The common starting point in swarm robotics is the biological metaphor, for which the fundamental mechanisms that govern the organization of animal societies can be distilled in simple rules to be implemented in the robotic swarm. This approach allowed us to extract the basic working principles for many types of collective behavior, and several examples will be presented in this chapter. However, it is worth noting that swarm robotics systems are not constrained to mimicking nature. Indeed, in many cases there is no biological example to be taken as reference, or the mechanisms observed in the natural system are too difficult to be implemented in the robotic swarm (e.g., odor perception is an open problem in robotics, preventing easy exploitation of pheromone-based mechanisms by using real chemicals). Still, even in those systems that have no natural counterpart, the relevant property that should be present is self-organization, for which group behavior is the emergent result of the numerous interactions among different individuals. Thanks to self-organization, simple control rules repeatedly executed by the individual robots may result in complex group behavior.

If we consider the scenario presented at the beginning of this chapter, it is possible to recognize a number of problems common to many swarm robotics systems, which need to be addressed in order to develop suitable controllers. One first problem in swarm robotics is having robots get together in some place, especially when the robotic system is composed by potentially many individuals. Getting together (i.e., aggregation) is the precondition for many types of collective behavior, and needs to be addressed according to the particular characteristics of the robotic system and of the environment in which it must take place. The aggregation problem is discussed in Sect. 71.2. Once groups are formed, robots need some mechanism to stay together and to keep a coherent organization while performing their task. A typical problem is, therefore, how to maintain such coherence, which corresponds to ensuring the synchronization of the group activities (Sect. 71.3), and to keep the group in coordinated motion when the swarm must move across the environment (Sect. 71.4). Another common problem in swarm robotics corresponds to searching together and processing some items in the environment. To this aim, different strategies can be adopted to cover the available space, and to identify relevant navigation routes without resorting to maps

and global knowledge (Sect. 71.5). Finally, to maintain coherence and efficiency, the swarm robotics system is often confronted with the necessity to behave as a single whole. Therefore, it must be endowed with collective perception and collective decision mechanisms. Some examples are discussed in Sect. 71.6. For each of these problems, we describe some seminal work that produced solutions in a swarm robotics context. In each section, we describe the problem along with some possible variants, the biological inspiration and the theoretical background, the relevant studies in swarm robotics, and a number of other works that are relevant for some particular contribution given to the specific problem.

71.2 Getting Together: Aggregation

Aggregation is a task of fundamental importance in many biological systems. It is the basic behavior for the creation of functional groups of individuals, and therefore, supports the emergence of various forms of cooperation. Indeed, it can be considered a prerequisite for the accomplishment of many collective tasks. In swarm robotics too, aggregation has been widely studied, both as a standing-alone problem or within a broader context. Speaking in general terms, aggregation is a collective behavior that leads a group of agents to gather in some place. Therefore, from a (more or less) uniform distribution of agents in the available space, the system converges to a varied distribution, with the formation of well recognizable aggregates. In other words, during aggregation there is a transition from a homogeneous to a heterogeneous distribution of agents.

71.2.1 Variants of Aggregation Behavior

Aggregation can be achieved in many different ways. The main issue to be considered is whether or not the environment contains pre-existing heterogeneities that can be exploited for aggregation: light or humidity gradients (think of flies or sow bugs), corners, shelters, and so forth represent heterogeneities that can be easily exploited. Their presence can, therefore, be at the basis of a collective aggregation behavior, which, however, may not exploit interactions between different agents. Instead, whenever heterogeneities are not present (or cannot be exploited for the aggregation behavior), the problem is more complex. The agents must behave in order to create the heterogeneities that support the formation of aggregates. In this case, the basic mechanism

of aggregation relies on a self-organizing process based on a positive feedback mechanism. Agents are sources of some small heterogeneity in the environment (e.g., being the source of some signal that can be chemical, tactile or visual). The more aggregated agents, the higher the probability to be attracted by the signal. This mechanism leads to amplification of small heterogeneities, leading to the formation of large aggregates.

71.2.2 Self-Organized Aggregation in Biological Systems

Several biological systems present self-organized aggregation behavior. One of the best studied examples is given by the cellular slime mold *Dictyostelium discoideum*, in which aggregation is enabled by self-generated biochemical signals that support the migration of cells and the formation of a multi-cellular body [71.1, 2]. A similar aggregation process can be observed in many other unicellular organisms [71.3]. Social and pre-social insects also present multiple forms of aggregation [71.4, 5]. In all these systems, it is possible to recognize two main variants of the aggregation process. On the one hand, the agents can emit a signal that creates an intensity gradient in the surrounding space. This gradient enables the aggregation process: agents react by moving in the direction of higher intensity, therefore aggregating with their neighbors (Fig. 71.1). On the other hand, aggregation may result from agents modulating their stopping time in response to social cues. Agents have a certain probability to stop and remain still for some time. The vicinity to other agents increases the probability of stopping and of remaining

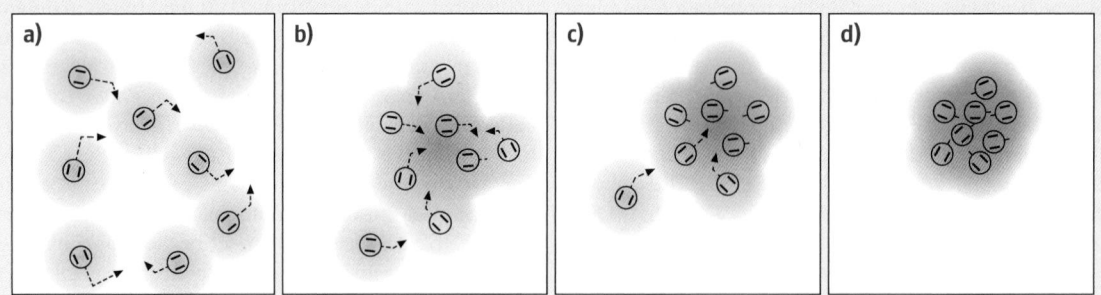

Fig. 71.1a–d Aggregation process based on a diffusing signal that creates an intensity gradient. (**a**) Agents individually emit a signal and move in the direction of higher concentration. (**b**) The individual signals sum up to form a stronger intensity gradient in correspondence with forming aggregates. (**c**) A positive feedback loop amplifies the aggregation process until all agents are in the same cluster (**d**)

within the aggregate, eventually producing an aggregation process mediated by social influences (Fig. 71.2). In both cases, the same general principle is at work. Aggregation is dependent on two main probabilities: the probability to enter an aggregate, which increases with the aggregate size, and the probability to leave an aggregate, which decreases accordingly. This creates a positive feedback loop that makes larger aggregates more and more attractive with respect to small ones. Some randomness in the system helps in breaking the symmetry and reaching a stable configuration.

71.2.3 Self-Organized Aggregation in Swarm Robotics

On the basis of the studies of aggregation in biological systems, various robotic implementations have been presented, based on either of the two behavioral mod-

els described above. Of particular interest is the work presented in [71.6], in which the robotic system was developed to accurately replicate the dynamics observed in the cockroach aggregation experiments presented in [71.5]. In this work, a group of Alice robots [71.7] was used and their controller was implemented by closely following the behavioral model derived from experiments with cockroaches. The behavioral model consists of four main conditions:

i)   Moving in the arena center
ii) Moving in the arena periphery
iii) Stopping in the center
iv) Stopping in the periphery.

When stopping, the mean waiting time is influenced by the number of perceived neighbors (for more details, see [71.6]). The group behavior resulting from the

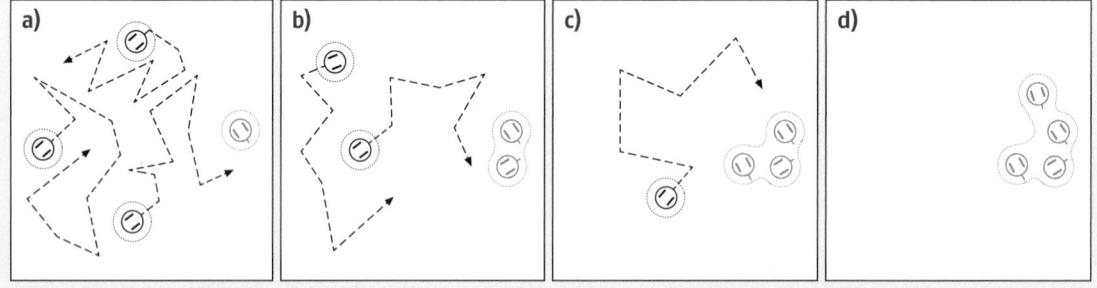

Fig. 71.2a–d Aggregation process based on variable probability of stopping within an aggregate. (**a**) Agents move randomly and may stop for some time (*gray agent*). (**b**) When encountering a stopped agent, other agents stop as well, therefore increasing the size of the aggregate. (**c**) The probability of meeting an aggregate increases with the aggregate size for geometric reasons. Social interactions modulate the probability of leaving the aggregate, which diminishes with the increasing number of individuals. (**d**) Eventually, all agents are in the same aggregate

interaction among Alice robots was analyzed with the same tools used for cockroaches [71.5, 6]. The comparison of the robotic system with the biological model shows a very good correspondence, demonstrating that the mechanisms identified by the behavioral model are sufficient to support aggregation in a group of robots, with dynamics that are comparable to that observed in the biological system. Additionally, the robotic model constitutes a constructive proof that the identified mechanisms really work as suggested.

This study demonstrates, in terms of simple rules, the approach of distilling the relevant mechanisms that produce a given self-organizing behavior. A different approach consists in exploiting artificial evolution to synthesize the controllers for the robotic swarm. This allows the user to simply define some performance metric for the group and let the evolutionary algorithm find the controllers capable of producing the desired behavior. This generic approach has been exploited to evolve various self-organizing behaviors, including aggregation [71.8]. In this case, robots were rewarded to minimize their distance from the geometric center of the group and to keep moving. The analysis of the evolved behavior revealed that in all cases robots are attracted by teammates and repelled by obstacles. When a small aggregate forms, robots keep on moving thanks to the delicate balance between attractive and repulsive forces. This makes the aggregate continuously expand and shrink, moving slightly across the arena. This slow motion of the aggregate makes it possible to attract other robots or other aggregates formed in the vicinity, and results in a very good scalability of the ag-

gregation behavior with respect to the group size. This experiment revealed a possible alternative mechanism for aggregation, which is not dependent on the probability of joining or leaving an aggregate. In fact, robots here never quit an aggregate to which they are attracted. Rather, the aggregates themselves are dynamic structures capable of moving within the environment, and in doing so they can be attracted by neighboring aggregates, until all robots belong to the same group.

71.2.4 Other Studies

The seminal papers described above are representative of other studies, which either exploit a probabilistic approach [71.9, 10], or rely on artificial evolution [71.11]. Approaches grounded on mathematical models and control theory are also worth mentioning [71.12, 13]. Other variants of the aggregation behavior can be considered. The aggregate may be characterized by an internal structure, that is, agents in the aggregate are distributed on a regular lattice or form a specific shape. In such cases, we talk about *pattern/shape formation* [71.14]. Another possibility is given by the admissibility of multiple aggregates. In the studies mentioned so far, multiple aggregates may form at the beginning of the aggregation process, but as time goes by smaller aggregates are disbanded in favor of larger ones, eventually leading to a single aggregate for the whole swarm. However, it could be desirable to obtain multiple aggregates forming functional groups of a specific size. In this case, it is necessary to devise mechanisms for controlling the group size [71.15].

71.3 Acting Together: Synchronization

Synchronization is a common phenomenon observed both in the animate and inanimate world. In a synchronous system, the various components present a strong time coherence between the individual types of behavior. In robotics, synchronization can be exploited for the coordination of actions, both within a single or a multi-robot domain. In the latter case, synchronization may be particularly useful to enhance the system efficiency and/or to reduce the interferences among robots.

71.3.1 Variants of Synchronization Behavior

Synchronization in a multi-agent system can be of mainly two forms: *loose* and *tight*. In the case of loose

synchronization, we observe a generic coordination in time of the activities brought forth by different agents. In this case, single individuals do not present a periodic behavior, but as a group it is possible to observe bursts of synchronized activities. Often in this case there are external cues that influence synchrony, such as the daylight rhythm. On the other hand, it is possible to observe tight synchronization when the individual actions are perfectly coherent. To ensure tight synchronization in a group, it is possible to rely on either a centralized or a distributed approach. In the former, one agent acts as a reference (e.g., a conductor for the orchestra or the music theme for a ballet) and drives the behavior of the other system components. In the latter, a self-organizing process is in place, and the system shows the ability to synchronize without an externally-imposed rhythm. It is worth noting that tight synchronization does not necessitate individual periodic behavior, neither in the centralized nor in the self-organized case. For instance, synchronization has also been studied between coupled chaotic systems [71.16]. In the following, we focus on self-organized synchronization of periodic behavior, which is the most studied phenomenon as it is commonly observed in many different systems.

71.3.2 Self-Organized Synchronization in Biology

Although synchronization has always been a well-known phenomenon [71.17], its study did not arouse much interest until the late 1960s, when *Winfree* began investigating the mechanisms underlying biological rhythms [71.18]. He observed that many systems in biology present periodic oscillations, which can get entrained when there is some coupling between the oscillators. A mathematical description of this phenomenon was first introduced by *Kuramoto* [71.19], who developed a very influential model that was afterwards refined and applied to various domains [71.17].

Similar mechanisms are at the base of the synchronous signaling behavior observed in various animal species [71.3]. *Chorusing* is a term commonly used to refer to the coordinated emission of acoustic communication signals by large groups of animals. To cite a few examples, chorusing has been observed in frogs, crickets, and spiders. However, probably the most fascinating synchronous display is the synchronous flashing of fireflies from South-East Asia. This phenomenon was

thoroughly studied until a self-organizing explanation was proposed to account for the emergence of synchrony [71.20].

A rather simple model describes the behavior of fireflies as the interactions between pulse-coupled oscillators [71.21]. In Fig. 71.3, the activity of two oscillators is represented as a function of time. Each oscillator is of the integrate-and-fire type, which well represents a biological oscillator such as the one of fireflies. The oscillator is characterized by a voltage-like variable that is integrated over time until a threshold is reached. At this point, a pulse is fired and the variable is reset to the base level (Fig. 71.3). Interactions between oscillators take the form of constant phase shifts induced by incoming pulses, which bring other oscillators close to the firing state, or make them directly fire. These simple interactions are sufficient for synchronization; in a group of similarly pulse-coupled oscillators, constant adjustments of the phase made by all the individuals lead to a global synchronization of pulses (for a detailed description of this model, see [71.21]).

71.3.3 Self-Organized Synchronization in Swarm Robotics

The main purpose of synchronization in swarm robotics is the coordination of the activities in a group. This can be achieved in different ways, and mechanisms inspired by the behavior of pulse-coupled oscillators have been developed. In [71.15], synchronization is exploited to regulate the size of traveling robotic aggregates. Robots can emit a short sound signal (a chirp), and enter a refractory state for a short time after signaling. Then, robots enter an active state in which they may signal

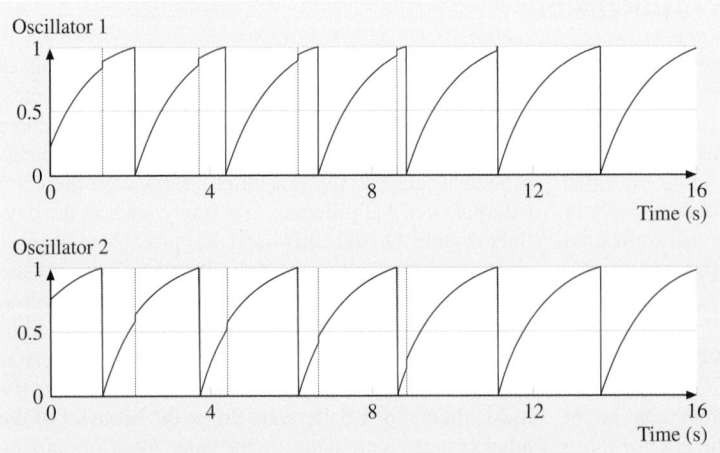

Fig. 71.3 Synchronization between pulse-coupled oscillators. The oscillator emits pulses each time its state variable reaches the threshold level (corresponding to 1 in the plot). When one oscillator emits a pulse, its state is reset while the state of the other oscillator is advanced by a constant amount, which corresponds to a phase shift, or to the oscillator firing if it overcomes the threshold

at any time, on the basis of a constant probability per time-step. Therefore, the chirping period is not constant and depends on the chirping probability. In this state, robots also listen to external signals and react by immediately emitting a chirp. This mechanism, similar to chorusing in frogs and crickets, leads to synchronized emission of signals. Thanks to this simple synchronization mechanism, the size of an aggregate can somehow be estimated. Given the probabilistic nature of chirping, a robot has a probability of independently initiating signaling that depends on the number of individuals in the group; estimating this probability by listening to own and others' chirps allows an approximate group size estimation. Synchronization, therefore, ensures a mechanism to keep coherence in the group, which is the precondition for group size estimation.

In [71.22], synchronization is instead necessary to reduce the interferences between robots, which periodically perform foraging and homing movements in a cluttered environment. Without coordination, the physical interferences between robots going toward and away from the home location lead to a reduced overall performance. Therefore, a synchronization mechanism based on the firefly behavior was devised. Robots emit a signal in correspondence to the switch from foraging to homing. This signal can be perceived by neighboring robots within a limited radius and induces a reset of the internal rhythm that corresponds to a behavioral shift to homing. Despite the limited range of communication among robots, a global synchronization is quickly achieved, which leads the group to reduce interferences and increase the system performance [71.22].

A different approach to the study of synchronization is described in [71.23]. Here, artificial evolution is exploited to synthesize the behavior of a group of robots, with the objective of obtaining minimal communication strategies for synchronization. Robots were rewarded to present an individual periodic movement and to signal in order to synchronize the individual oscillations. The results obtained through artificial evolution are then analyzed to understand the mechanisms that can support

synchronization, showing that two types of strategies are evolved: one is based on a modulation of the oscillation frequency, the other relies on a phase reset. These two strategies are also observed in biological oscillators: for instance, different species of fireflies present different synchronization mechanisms, based on delayed or advanced phase responses [71.20].

71.3.4 Other Studies

While self-organized synchronization is a well-known phenomenon, its application in collective and swarm robotics has not been largely exploited. The coupled-oscillator synchronization mechanism was applied to a cleaning task to be performed by a swarm of micro robots [71.24]. Another interesting implementation of the basic model can be found in [71.25]. Here, synchronization is exploited to detect and correct faults in a swarm robotics system. It is assumed that robots can synchronize a periodic flashing behavior while moving in the arena and accomplishing their task. If a robot incurs some fault, it will forcedly stop synchronizing. This fault can be detected and recovered by neighboring robots. Similar to the heartbeat in distributed computing, correct synchronization corresponds to a well-functioning system, while the lack of synchronization corresponds to a faulty condition.

Finally, synchronization behavior may emerge spontaneously in an evolutionary robotics setup, even if they are not explicitly rewarded. In [71.26], synchronization of group activities evolved spontaneously as a result of the need to limit the interferences among robots in a foraging task. In [71.27], robots were rewarded to maximize the mean mutual information between their motor actions. Mutual information is a statistical measure derived in information theory, and roughly corresponds to the correlation between the output of two stochastic processes. Evolution, therefore, produced synchronous movements among the robots, which could actually maximize the mutual information while maintaining a varied behavior.

71.4 Staying Together: Coordinated Motion

Another fundamental problem for a swarm is ensuring coherence in space. This means that the individuals in the swarm must display coordinated movement in order to maintain a consistent spatial structure. Coordinated motion is often observed in groups of animals. Flocks of

birds or schools of fish are fascinating examples of self-organized behavior producing a collective motion of the group. Similar problems need to be tackled in robotics, for instance for moving in formation or for distributedly deciding a common direction of motion.

71.4.1 Variants of the Coordinated Motion Behavior

The coordinated motion of a group of agents can be achieved in different ways. Also in this case, we can distinguish mainly between a centralized and a distributed approach. In a centralized approach, one agent can be considered the leader and the other agents follow (e.g., the mother duck with her ducklings). In the distributed approach, instead, there is no single leader and some coordination mechanism must be found to let the group move in a common direction. Of particular interest for swarm robotics are the coordinated motion models based on self-organization. Such models consider multi-agent systems that are normally homogeneous and characterized by a uniform distribution of information: no agent is more informed than the others, and there exists no a priori preference for any direction of motion (i. e., agents start being uniformly distributed in space). However, through self-organization and amplification of shared information, the system can break the symmetry and converge to a common direction of motion. A possible variant of the self-organized coordinated motion consists in having a non-uniform distribution of information, which corresponds to hav-

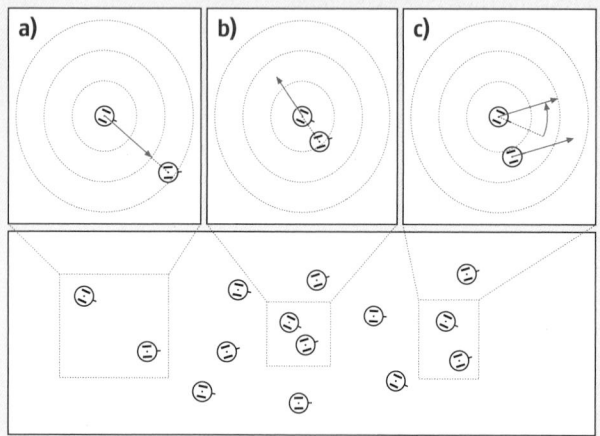

Fig. 71.4a–c Self-organized coordinated motion in a group of agents. In the *bottom part*, a group of agents is moving in roughly the same direction. According to the model presented in [71.28], agents react to the closest neighbor within their perception range and follow three main rules: (**a**) agents move toward a neighbor when it is too far; (**b**) agents move away from a neighbor when it is too close; (**c**) agents rotate and align with a neighbor situated at intermediate distances. The iterated application of these rules leads the group to move in a same direction

ing some agents that are more informed than the others on a preferred direction of motion. In this case, a few informed agents may influence the motion of the entire group.

71.4.2 Coordinated Motion in Biology

Many animal species present coordinated motion behavior, ranging from bacteria to fish and birds. Not all animal species employ the same mechanisms, but in general it is possible to recognize various types of interactions among individuals that have a bearing on the choice of the motion direction. Coordinated motion has mainly been studied in various species of fish, in birds, and in insect swarms [71.29, 30]. The most influential model was introduced by *Huth* and *Wissel* to describe the behavior of various species of fish observed [71.28]. In this model, it is assumed that each fish is influenced solely by its nearest neighbor. Also, the movement of each fish is based on the same behavioral model, which also includes some inherent random fluctuation. According to the proposed behavioral model, each fish follows essentially three rules:

i) Approach a far away individual
ii) Get away from individuals that are too close
iii) Align with the neighbor direction (Fig. 71.4).

When the nearest neighbor is within the closest region, the fish reacts by moving away. When the nearest neighbor is in the farthest region, the fish reacts by approaching. Otherwise, if the neighbor is within the intermediate region, the fish reacts by aligning. These simple rules are sufficient to produce collective group motion, and the final direction emerges from the interactions among the individuals.

Starting from the above model, a number of variants have been proposed, which take into account different parameters and different numbers of individuals. In [71.31], a model including all individuals in the perceptual range was introduced, and a broad analysis of the parameters was performed, showing how minor differences at the individual level correspond to large differences at the group level. In [71.32], an experimental study on bird flocks in the field was performed, and position and velocity data were obtained for each bird in a real flock through stereo-photography and 3-D mapping. The data obtained data were used to verify the assumption about the number of individuals that each bird monitors during flocking, showing that this number is constant (and corresponds to about 7 individuals) notwithstanding the varying density of the flock. Finally, in [71.33], a model was developed in which some of the group members have individual knowledge on a preferential direction. The model describes the outcome of a consensus decision in the flock as a result of the interaction between informed and uninformed individuals.

71.4.3 Coordinated Motion in Swarm Robotics

The models introduced for characterizing the self-organized behavior of fish schools or bird flocks have also inspired a number of interesting studies. The most influential work is definitely that of *Reynolds*, who developed virtual creatures called *boids* [71.34]. In this work, each creature executes three simple types of behavior:

i) Collision avoidance, to avoid crashing with nearby flockmates
ii) Velocity matching, to move in the same way of nearby flockmates
iii) Flock centering, to stay close to nearby flockmates.

Notice that the behavioral model corresponds to the models proposed in biological studies. The merit of this work is that it is the first implementation of the rules studied for real flocks in a virtual 3-D world, showing a close correspondence of the behavior of boids with that of flocks, herds, and schools. Reynolds' research has been taken as inspiration by many other studies on coordinated motion, mainly in simulation. In [71.35], an implementation of the flocking behavioral model was proposed and tested on real robots. Robots use infrared proximity sensors to recognize the presence of other robots and their distance, which is necessary for collision avoidance and flock centering behavior. Additionally, a dedicated sensor to perceive the heading of neighbors was developed to support aligning behavior. This system, called the virtual heading system (VHS), is based on a digital compass and wireless communication. Despite the fact that a digital compass cannot reliably work in an indoor environment, it is assumed that neighboring robots have similar perceptions. The heading perceived with respect to the local north is communicated over the wireless channel, and it is exploited for alignment behavior. This system al-

lowed testing the flocking behavior of small robotic groups in a physical setting and studying the dynamics of flocking with up to 1000 simulated robots. This work was later extended in [71.36], by having a subgroup of informed individuals which could steer the whole flock, following the model presented in [71.33]. The dynamics of steered flocking have been studied by varying the percentage of informed robots in simulation, and tests with real robots have been performed as well.

71.4.4 Other Studies

As mentioned above, there exist numerous studies that were inspired by the schooling/flocking models. All these studies adopt some variants of the behavioral rules described above, or analyze the group dynamics under some particular perspective. A different approach to coordinated motion can be found in [71.37]. In this work, robots have to transport a heavy object and have imperfect knowledge of the direction of motion. They can, however, negotiate the goal direction by displaying their own preferred direction of motion and by adjusting it on the basis of the direction displayed by others. On the whole, this mechanism implements similar dynamics to the alignment behavior of the classical flocking model. Here, however, robots are connected together to the object to be transported, adding a further constraint to the system that obliges a good negotiation to allow motion. A similar constraint characterizes the coordinated motion studies with physically assembled robots presented in [71.38, 39]. Here, robots form a physical structure of varying shape and can rotate their chassis in order to match the direction of motion of the other robots. In this case, there is no direct detection of the motion direction of neighbors. Instead, robots can sense the pulling and pushing forces that are exerted by the other connected robots through the physical connections. These pulling/pushing forces are naturally averaged by the force sensor, which returns their resultant. Artificial evolution was exploited to synthesize an artificial neural network that could transform the forces sensed to motor commands. The results obtained show the impressive capability of self-organized coordination between the robots, as well as scalability and generalization to different size and shapes [71.38], and the ability to cope with obstacles and to avoid falling outside the borders of the arena [71.39].

71.5 Searching Together: Collective Exploration

Exploring and searching the environment is an important behavior for robot swarms. In many tasks, the swarm must interact with the environment, sometimes only to monitor it, but sometimes also to process materials or other kinds of resources. Usually, the swarm cannot completely perceive the environment, and the environment may also change during the operation of the robots. Hence, robots need to explore and search the environment to monitor for changes or in order to detect new resources.

To cope with its partial perception of the environment, a swarm can move, for instance using flocking, in order to explore new places (some locations may be unavailable, though). Hence, most of the environment can be perceived, but not at the same time. As in many other artificial systems, a tradeoff between exploration and exploitation exists and requires careful design choices.

71.5.1 Variants of Collective Exploration Behavior

There is no perfect exploration and search strategy because the structure of the environment in which the swarm is placed can take many different shapes. Strategies only perform more or less well as a function of the situation with which they are faced [71.40]. For instance, the swarm could be in a maze, in a open environment with few obstacles, or in an environment with many obstacles.

We identified a restricted number of environmental characteristics that play an important role in the choice of searching behavior in swarm robotics. These characteristics are commonly found in swarm robotics scenarios, and are the presence of a central place, the size of the environment, the presence of obstacles.

The central place is a specific location where robots must come back regularly, for instance for maintenance or to deposit foraged items. A scenario that involves a central place requires a swarm able to either remember or keep track of that location.

If the environment is closed (finite area) and not too large, the swarm may use random motion to explore, with fair chances to rapidly locate resources (or even the central place). In an open environment, robots can get lost very quickly. In this type of environment, it is necessary to use a behavior that allows robots to stay together and maintain connectivity.

Obstacles are environmental elements that constrain the motion of the swarm. If the configuration of the obstacles is known in advance, the swarm can move in the environment following appropriate patterns. In most cases, however, obstacles are unexpected or might be dynamic and may prevent the swarm from exploring parts of the environment.

71.5.2 Collective Exploration in Biology

In nature, animals are constantly looking for resources such as food, sexual partners, or nesting sites. Animals living in groups may use several types of behavior to explore their environment and locate these resources.

For instance, fish can take advantage of the number of individuals in a shoal to improve their capabilities to find food [71.41–43]. To do so, they move and maintain large interdistances between individuals. In this way, fish increase their perceptual coverage as well as their chances to find new resources.

Animals also heavily rely on random motion to explore their environment [71.44–46]. Usually the exploratory pattern is not fully random (that is, isotropic), because animals use all possible environmental cues at hand to guide themselves. Random motion can be biased towards a given direction, or it can be constrained in a specific area, for instance around a previously memorized location [71.47]. Some desert ants achieve high localization performance with odometry (counting their footsteps) and relying on gravity and the polarization of natural light. They may move randomly to look for resources but they are able to quickly return to their nest and also to return to an interesting location previously identified.

71.5.3 Collective Exploration in Swarm Robotics

One of the most common exploration strategies used in robotics is random exploration. In a typical implementation, robots wander in the environment until they perceive a feature of interest [71.48–50]. By doing this, robots possibly lose contact with each other and, therefore, their ability to work together. Hence this strategy is not suited for large or open environments. Due to the stochastic nature of the strategy, its performance can only be evaluated statistically. On average, the time to locate a feature is proportional to the squared distance with robots [71.44].

Systematic exploration strategies are very different. Robots use some a priori knowledge about the structure

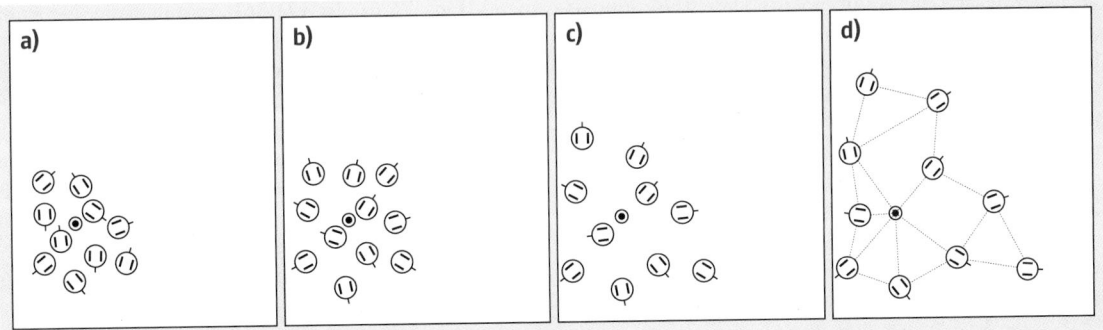

Fig. 71.5a–d Gas expansion behavior to monitor the surroundings of a central place. (**a**) The swarm starts aggregated around a central place (represented by a *black spot*). (**b,c**) Robots try to move as far as possible from their neighbors, while maintaining some visual or radio connection. (**d**) As a result, the whole swarm expands in the environment, like a gas, covering part of the environment

of the environment in order to methodically sweep it and find features. To ensure that robots do not repeatedly cover the same places, they may need to memorize which places have already been explored. This is often implemented with localization techniques and mapping of the environment [71.51]. The advantage of this technique is that an answer will be found with certainty, and the time of exploration has a lower and upper bound if the environment is not open. However, memory requirements may be excessive, and the strategy is not suited for open environments.

Between the two extreme strategies reported in the previous paragraph lie a number of more specialized strategies that present advantages and drawbacks depending on the structure of the environment and the distribution of the resources.

Collective motion (which has already been detailed in Sect. 71.4) allows swarms to maintain their cohesion while moving through the environment. Flocking behavior can be employed in an open environment with a limited risk of losing contact between robots. The swarm behaves like a sort of physical mesh that covers part of the environment; to maximize the area covered during exploration robots can increase their interdistance during motion as much as possible.

Gas expansion behavior (Fig. 71.5) allows robots to quickly and exhaustively explore the surroundings of a central place [71.52–55]. While one or several robots keep track of a central place, other robots try to move as far as possible from their neighbors, while still maintaining direct line of sight with at least one of them. The swarm behaves like a fluid or gas that penetrates

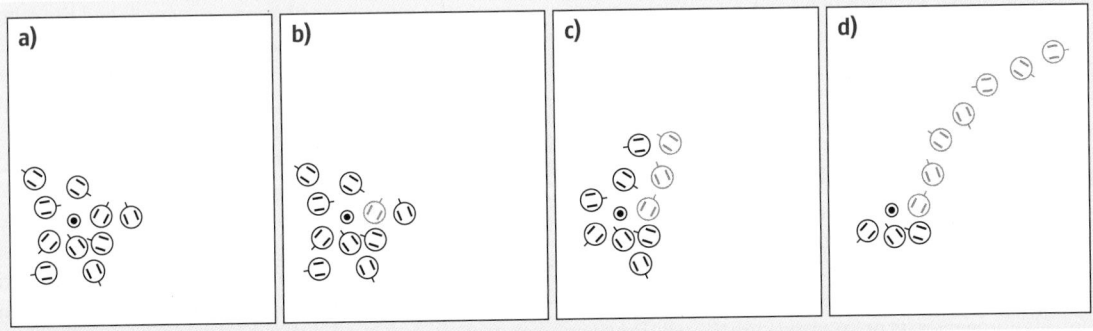

Fig. 71.6a–d Chaining behavior in action with a central place represented by a *large black dot, bottom left*. (**a**) Robots start aggregated around the central place. (**b**) While maintaining visual or radio contact with neighbors, some robots change role and become part of a chain (*grayed out*). (**c**) Other robots move around the central place and encounter the early chain of robots. With some probability, they also turn into new parts of the chain. (**d**) At the end of the iterative process, robots form a long chain that spans through the environment and maintains a physical link to the central place

the asperities of the environment. The exploration is very effective and any change or new resource within the perception range of the swarm is immediately perceived. However, since robots are bound to the central place, the area that they can explore is limited by the number of robots in the swarm. If robots do not stick to a central place, the resulting behavior shifts to a type of flocking or moving formation.

With chaining behavior, swarms can form a chain with one end that sticks to a central place and the other end that freely moves through the environ-ment (Fig. 71.6). In [71.56], minimalistic behavior produces a static chain, but different types of chain motions can be imagined. In [71.57], for instance, chains can build up, move, and disaggregate until a resource is found. Contrary to gas expansion behavior, a chaining swarm may not immediately perceive changes in the environment because it has to constantly sweep the space. Chaining allows robots to cover a more important area than gas expansion behavior, ideally a disc of radius proportional to the number of robots.

71.6 Deciding Together: Collective Decision Making

Decision making is a behavior used by any artificial system that must produce an adapted response when facing new or unexpected situations. Because the best action depends on the situation encountered, a swarm cannot rely on a pre-programmed and systematic re-action. Monolithic artificial systems make decisions all the time, by gathering information and then evaluating the different options at hand. However, when it comes to swarms, each group member might have its own opin-ion about the correct decision. If all individuals perceive the same information and process it in the same way, then they might independently make the same decision. However, in practice, the more common case is that in-dividuals perceive partial and noisy information about the situation. Thus, if no coordination among group members occurs, a segregation based on differing opin-ions might take place, thereby removing the advantages of being a swarm. Therefore, the challenge is to have the whole group collaborate to make a collective deci-sion and take action accordingly.

71.6.1 Variants of Collective Decision Making Behavior

There are mainly three mechanisms reported in the literature that allow swarms to make collective deci-sions. The first and most simple mechanism is based on opinion propagation. As soon as a group member has enough information about a situation to make up its mind, it propagates its opinion through the whole group.

The second mechanism is based on opinion averag-ing. All individuals constantly share their opinion with their neighbors and also adjust their own opinion in con-sequence. This iterative process leads to the emergence of a collective decision. The adjustment of the opinion is typically achieved with an average function, espe-cially if opinions are about quantitative values such as a location, a distance, or a weight, for instance.

The third and last mechanism relies on amplifi-cation to produce a collective decision. In a nutshell, all individuals start with an opinion, and may decide to change their opinion to another one. The switch to a new opinion happens with a probability calculated on the basis of the frequency of this opinion in the swarm. Practically, this means that if an opinion is represented often in the group, it has also more chances of being adopted by an individual, which is why the term ampli-fication is used.

Each of the three aforementioned mechanisms has some advantages over the others and may be preferred, depending on the situation faced. The factors that play an important role in collective decision processes in-clude the speed and the accuracy needed to make the choice, the robustness of communication, and the relia-bility of individual information.

In terms of speed, opinion propagation allows fast collective decisions, in contrast with the two other mechanisms, which require numerous interac-tions among individuals. However, this speed generally comes at the cost of robustness or accuracy [71.58–60]. If communication is not robust enough, messages can be corrupted. The mechanism of opinion propagation is particularly sensitive to such effects, and a wrong or random collective decision might be made by a swarm in that case.

The averaging mechanism would produce a more robust decision because wrong information from erro-neous messages is diluted in the larger amount of infor-mation present in the swarm [71.61]. However, opinion averaging works best if all individuals have roughly identical knowledge. If a small proportion of individuals have excellent knowledge to make the decision, while the remaining individuals have poor information, opinion propagation may produce better results than opinion averaging [71.33].

Lastly, the amplification mechanism is the main choice for a gradually emerging collective decision if opinions cannot be merged with some averaging function. Instead of adjusting opinions, individuals simply adopt new opinions with some probability. It is worth noting that this mechanism can produce good decisions even if individuals have poor information.

71.6.2 Collective Decision Making in Biology

The powerful possibilities of decision making in groups were already suggested by *Galton* back in 1905 [71.62]. In that paper, Galton reports the results of a weight-judging competition in which competitors had to estimate the weight of a fat ox. With slightly less than 800 independent estimates, Galton observed that the average estimate was accurate to 1% of the real weight of the ox. This early observation opened interesting perspectives about the accuracy of collective estimations, but it did not describe a collective decision mechanism, since Galton himself had to gather the estimates and apply some calculation to evaluate the estimate of the crowd.

More recent studies about group navigation have shown that groups of animals cohesively moving together towards a goal direction reach their objective faster than independent individuals [71.63, 64]. The mechanism of collective navigation not only allows the individuals to move and stay together, but it also acts as a distributed averaging function that locally fuses the opinions of individuals about the direction of motion, allowing them to improve their navigation performance.

In the last decades, the amplification mechanism has been identified as a source of collective decision in a broad range of animal species such as ants [71.65, 66], honeybees [71.67, 68], spiders [71.69], cockroaches [71.70], monkeys [71.71], and sheep [71.72].

Ants that choose one route to a resource probably constitute the most well-known example of the amplification mechanism. In [71.66], an ant colony is offered two paths to two identical resource sites. Initially, the two resources are exploited equally, but after a short time ants focus on a single resource. This collective choice happens because ants that have found the resource come back to the nest, marking the ground with a pheromone trail. The next ants that try to reach the

resource are sensitive to this odor and have higher chances of following the path with higher pheromone concentration. As a result of this amplified response, a collective decision rapidly emerges. In addition, it was shown in [71.73] that when ants are presented two paths of different lengths to the same resource, the same pheromone-based mechanism allows them to choose the shortest path. This can be explained by the fact that ants using the shortest path need less time to make round trips, making the pheromone concentration on this path grow faster.

Quorum sensing is a special case of the amplification mechanism which has been notably used to explain nest site selection in ants and bees [71.74–76]. The most basic example of quorum sensing uses a threshold to dictate if individuals should change their opinion. If an individual perceives enough neighbors (above the threshold) that already share the opposite opinion, then it will in turn adopt this opinion. It has been shown that this threshold makes quorum sensing more robust to the propagation of erroneous information during the decision process. In addition, the accuracy of collective decisions made with quorum sensing may improve with group size, and cognitive capabilities of groups may outperform the ones of single individuals [71.77, 78]. In the case of nest site selection, cohesion is mandatory for the group. A cross inhibition mechanism complementing amplification was identified as a key feature to ensure that groups do not split [71.79].

71.6.3 Collective Decision Making in Swarm Robotics

In swarm robotics, opinion averaging has been used to improve the localization capabilities of robots. In [71.50], a swarm of robots carries out a foraging task between a central place and a resource site. The robots have to navigate back and forth between the two places and use odometry to estimate their location. As odometry provides noisy estimates, robots using solely this technique may quickly get lost. Here, robots can share and merge their localization opinions when they meet, by means of local infrared communication. By doing so, robots manage better localization and improve their performance in the foraging task. Moreover, robots associate a confidence level to their estimates, which is used to decide how information is merged. If a robot advertises an opinion with very high confidence, then the mechanism produces opinion propagation. Hence the two mechanisms of averaging and propagation are

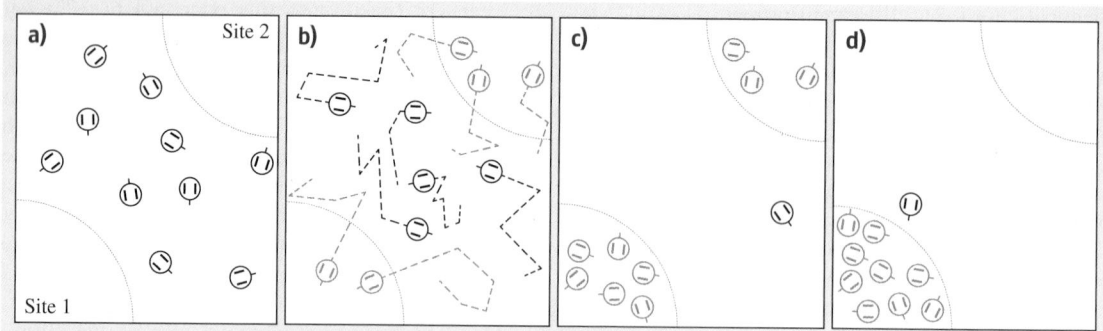

Fig. 71.7a–d A swarm of robots is presented with two resource sites in its environment and must collectively choose one. (**a**) Initially, robots are randomly scattered. (**b**) Using a random walk, they move until a resource site is found. On average, the swarm is split in the two sites. (**c**) The more neighbors they perceive, the longer the robots stay. A competition between the two sites takes place and any random event may change the situation. Here a robot just left the *top right site*, further reducing the chances that other robots stay there. (**d**) The swarm has made a choice in favor of the *bottom left site*. The choice is stable, although some robots may frequently leave the site for exploration

blended in a single behavior, and the balance between them is tuned by the user with a control parameter.

The aggregation behavior previously mentioned in Sect. 71.2.3 can be exploited to trigger collective decision making in situations where there are several environmental heterogeneities. In [71.80], the robots are presented two shelters and they choose one of them as a resting site by aggregating there. The behavior of the robots closely follows the one observed in cockroaches (Fig. 71.7). In [71.81], both robots and cockroaches are introduced in an arena with two shelters, demonstrating the influence of the two groups on each other when making the collective decision. The collective decision is the result of an amplification mechanism, implemented via the probability of a robot leaving an aggregate. This probability diminishes with the number of perceived neighbors, allowing larger aggregates to attract more robots.

71.6.4 Other Studies

The opinion averaging mechanism was deeply investigated with a general mathematical approach in [71.82, 83]. These studies demonstrate convergence of the mechanism and emphasize the importance of the topology of the communication network through which interactions take place.

Another amplification mechanism inspired from the behavior of honeybees was implemented in [71.84]. With this mechanism, it was shown that robots are able to make a collective decision and between two sites reliably choose the one offering the best illumination conditions.

The amplification mechanism based on pheromone trails, which is used by ants, has also inspired several swarm robotics studies. In [71.85, 86], the pheromone is replaced by light projected by a beamer. This implementation is limited to laboratory studies, but it allowed demonstrating path selection with robot swarms. In [71.87], the process is abstracted inside a network of robots that are deployed in the environment. Virtual ants hop from robots to robots and deposit pheromone inside them. Eventually, the shortest path to a resource is marked out by robots with high and sustained levels of pheromone.

71.7 Conclusions

In this chapter, we have presented a broad overview of the common problems faced in a swarm robotics context, and we pointed to possible approaches to obtain solutions based on a self-organizing process. We have discussed aggregation, synchronization, co-ordinated motion, collective exploration, and decision making, and we argued that many application scenarios could be solved by a mix of the above solutions. So, are we done with swarm robotics research? Definitely not.

First of all, the fact that possible solutions exist does not mean that they are the most suitable for any possible application scenario. Hardware constraints, miniaturization, environmental contingencies, and performance issues may require the design of different solutions, which may strongly depart from the examples given above. Still, the approaches we presented constitute a logical starting point, as well as a valid benchmark against which novel approaches can be compared.

Another important research direction consists in characterizing the self-organizing behavior we presented in terms of abstract properties, such as the time of convergence toward a stable state, sensitivity to parameter changes, robustness to failures, and so forth. From this perspective, the main problem is to ensure a certain functionality of the system with respect to the needs of the application and to predict the system features before actual development and testing. In many cases, a precise characterization of the system is not possible, and only a statistical description can be achieved. Still, such an enterprise would bring swarm robotics closer to an engineering practice, eventually allowing us to guarantee a certain performance of the developed system, as well as other properties that engineering commonly deals with.

The examples we presented all refer to homogeneous systems, in which all individuals are physically identical and follow exactly the same rules. This is, however, a strong simplification, which follows the tradition of biological modeling of self-organizing behavior. However, instead of being a limitation, heterogeneity is potentially a richness to be exploited in a swarm robotics system, which can lead to more complex group behavior. For instance, different individual reactions to features in the environment can be at the basis of optimal decision making at the group level [71.88]. Otherwise, heterogeneity between groups of individuals can be exploited for performing tasks that require specialized abilities, but maintaining an overall redundancy of the system that ensures robustness and scalability [71.89].

In conclusion, swarm robotics research still has many challenges to address, which range from the need for a more theoretical understanding of the relation between individual behavior and group dynamics, to the autonomy and adaptation to varied real-world conditions in order to face complex application scenarios (e.g., due to harsh environmental conditions such as planetary or underwater exploration, or to strong miniaturization down to the micro scale). Whatever the theoretical or practical driver is, we believe that the studies presented in this chapter constitute fundamental reference points that teach us how self-organization can be achieved in a swarm robotics system.

References

71.1   J.T. Bonner: Chemical signals of social amoebae, Sci. Am. **248**, 114–120 (1983)

71.2   C. van Oss, A.V. Panfilov, P. Hogeweg, F. Siegert, C.J. Weijer: Spatial pattern formation during aggregation of the slime mould *Dictyostelium discoideum*, J. Theor. Biol. **181**(3), 203–213 (1996)

71.3   S. Camazine, J.-L. Deneubourg, N. Franks, J. Sneyd, G. Theraulaz, E. Bonabeau: *Self-Organization in Biological Systems* (Princeton Univ. Press, Princeton 2001)

71.4   J.-L. Deneubourg, J.C. Grégoire, E. Le Fort: Kinetics of larval gregarious behavior in the bark beetle *Dendroctonus micans* (coleoptera, scolytidae), J. Insect Behav. **3**(2), 169–182 (1990)

71.5   R. Jeanson, C. Rivault, J.-L. Deneubourg, S. Blanco, R. Fournier, C. Jost, G. Theraulaz: Self-organized aggregation in cockroaches, Anim. Behav. **69**(1), 169–180 (2005)

71.6   S. Garnier, C. Jost, J. Gautrais, M. Asadpour, G. Caprari, R. Jeanson, A. Grimal, G. Theraulaz: The embodiment of cockroach aggregation behavior in a group of micro-robots, Artif. Life **14**(4), 387–408 (2008)

71.7   G. Caprari, R. Siegwart: Mobile micro-robots ready to use: Alice, Proc. 2005 IEEE/RSJ Int. Conf. Intell. Robot. Syst. (IROS 2005), Piscataway (2005) pp. 3295–3300

71.8   M. Dorigo, V. Trianni, E. Şahin, R. Groß, T.H. Labella, G. Baldassarre, S. Nolfi, J.-L. Deneubourg, F. Mondada, D. Floreano, L.M. Gambardella: Evolving self-organizing behaviors for a swarm-bot, Auton. Robot. **17**(2/3), 223–245 (2004)

71.9   S. Kernbach, R. Thenius, O. Kernbach, T. Schmickl: Re-embodiment of honeybee aggregation behavior in an artificial micro-robotic system, Adapt. Behav. **17**(3), 237–259 (2009)

71.10   O. Soysal, E. Şahin: A macroscopic model for self-organized aggregation in swarm robotic systems, Lect. Notes Comput. Sci. **4433**, 27–42 (2007)

71.11   E. Bahçeci, E. Şahin: Evolving aggregation behaviors for swarm robotic systems: A systematic case study, Proc. IEEE Swarm Intell. Symp. (SIS 2005), Piscataway (2005) pp. 333–340

71.12   H. Ando, Y. Oasa, I. Suzuki, M. Yamashita: Distributed memoryless point convergence algorithm for mobile robots with limited visibility, IEEE Trans. Robot. Autom. **15**(5), 818–828 (1999)

71.13   V. Gazi: Swarm aggregations using artificial potentials and sliding-mode control, IEEE Trans. Robot. **21**(6), 1208–1214 (2005)

71.14   W.M. Spears, D.F. Spears, J.C. Hamann, R. Heil: Distributed, physics-based control of swarms of vehicles, Auton. Robot. **17**(2), 137–162 (2004)

71.15   C. Melhuish, O. Holland, S. Hoddell: Convoying: Using chorusing to form travelling groups of minimal agents, Robot. Auton. Syst. **28**, 207–216 (1999)

71.16   A. Pikovsky, M. Rosenblum, J. Kurths: Phase synchronization in regular and chaotic systems, Int. J. Bifurc. Chaos **10**(10), 2291–2305 (2000)

71.17   S.H. Strogatz: *Sync: The Emerging Science of Spontaneous Order* (Hyperion, New York 2003)

71.18   A.T. Winfree: Biological rhythms and the behavior of populations of coupled oscillators, J. Theor. Biol. **16**(1), 15–42 (1967)

71.19   Y. Kuramoto: Phase dynamics of weakly unstable periodic structures, Prog. Theor. Phys. **71**(6), 1182–1196 (1984)

71.20   J. Buck: Synchronous rhythmic flashing of fireflies. II, Q. Rev. Biol. **63**(3), 256–289 (1988)

71.21   R.E. Mirollo, S.H. Strogatz: Synchronization of pulse-coupled biological oscillators, SIAM J. Appl. Math. **50**(6), 1645–1662 (1990)

71.22   S. Wischmann, M. Huelse, J.F. Knabe, F. Pasemann: Synchronization of internal neural rhythms in multi-robotic systems, Adapt. Behav. **14**(2), 117–127 (2006)

71.23   V. Trianni, S. Nolfi: Self-organising sync in a robotic swarm. A dynamical system view, IEEE Trans. Evol. Comput. **13**(4), 722–741 (2009)

71.24   M. Hartbauer, H. Roemer: A novel distributed swarm control strategy based on coupled signal oscillators, Bioinspiration Biomim. **2**(3), 42–56 (2007)

71.25   A.L. Christensen, R. O'Grady, M. Dorigo: From fireflies to fault-tolerant swarms of robots, IEEE Trans. Evol. Comput. **13**(4), 754–766 (2009)

71.26   S. Wischmann, F. Pasemann: The emergence of communication by evolving dynamical systems, Lect. Notes Artif. Intell. **4095**, 777–788 (2006)

71.27   V. Sperati, V. Trianni, S. Nolfi: Evolving coordinated group behaviours through maximization of mean mutual information, Swarm Intell. **2**(2–4), 73–95 (2008)

71.28   A. Huth, C. Wissel: The simulation of the movement of fish schools, J. Theor. Biol. **156**(3), 365–385 (1992)

71.29   I. Aoki: A simulation study on the schooling mechanism in fish, Bull. Jpn. Soc. Sci. Fish. **48**(8), 1081–1088 (1982)

71.30   A. Okubo: Dynamical aspects of animal grouping: Swarms, schools, flocks, and herds, Adv. Biophys. **22**, 1–94 (1986)

71.31   I.D. Couzin, J. Krause, R. James, G.D. Ruxton, N.R. Franks: Collective memory and spatial sorting in animal groups, J. Theor. Biol. **218**(1), 1–11 (2002)

71.32   M. Ballerini, N. Calbibbo, R. Candeleir, A. Cavagna, E. Cisbani, I. Giardina, V. Lecomte, A. Orlandi, G. Parisi, A. Procaccini, M. Viale, V. Zdravkovic: Interaction ruling animal collective behavior depends on topological rather than metric distance: Evidence from a field study, Proc. Natl. Acad. Sci. USA **105**(4), 1232–1237 (2008)

71.33   I.D. Couzin, J. Krause, N.R. Franks, S.A. Levin: Effective leadership and decision-making in animal groups on the move, Nature **433**(7025), 513–516 (2005)

71.34   C.W. Reynolds: Flocks, herds, and schools: A distributed behavioral model, Comput. Graph. **21**(4), 25–34 (1987)

71.35   A.E. Turgut, H. Çelikkanat, F. Gökçe, E. Şahin: Self-organized flocking in mobile robot swarms, Swarm Intell. **2**(2–4), 97–120 (2008)

71.36   H. Çelikkanat, E. Şahin: Steering self-organized robot flocks through externally guided individuals, Neural Comput. Appl. **19**(6), 849–865 (2010)

71.37   A. Campo, S. Nouyan, M. Birattari, R. Groß, M. Dorigo: Negotiation of goal direction for cooperative transport, Lect. Notes Comput. Sci. **4150**, 191–202 (2006)

71.38   G. Baldassarre, V. Trianni, M. Bonani, F. Mondada, M. Dorigo, S. Nolfi: Self-organised coordinated motion in groups of physically connected robots, IEEE Trans. Syst. Man Cybern. B **37**(1), 224–239 (2007)

71.39   V. Trianni, M. Dorigo: Self-organisation and communication in groups of simulated and physical robots, Biol. Cybern. **95**, 213–231 (2006)

71.40   D.H. Wolpert, W.G. Macready: No free lunch theorems for search. Technical Report SFI-TR-95-02-010 (Santa Fe Institute 1995)

71.41   T.J. Pitcher, A.E. Magurran, I.J. Winfield: Fish in larger shoals find food faster, Behav. Ecol. Sociobiol. **10**(2), 149–151 (1982)

71.42   T.J. Pitcher, J.K. Parrish: Functions of shoaling behaviour in teleosts, Behav. Teleost Fishes **2**, 369–439 (1993)

71.43   D.J. Hoare, I.D. Couzin, J.-G.J. Godin, J. Krause: Context-dependent group size choice in fish, Anim. Behav. **67**(1), 155–164 (2004)

71.44   E.A. Codling, M.J. Plank, S. Benhamou: Random walk models in biology, J. R. Soc. Interface **5**(25), 813–834 (2008)

71.45   P. Turchin: *Quantitative Analysis of Movement: Measuring and Modeling Population Redistribution in Animals and Plants* (Sinauer Associates Sunderland, Massachusetts 1998)

71.46   A. Ôkubo, S.A. Levin: *Diffusion and Ecological Problems: Modern Perspectives*, Vol. 14 (Springer, Berlin, Heidelberg 2001)

71.47   S. Benhamou: Spatial memory and searching efficiency, Animal Behav. **47**(6), 1423–1433 (1994)

71.48   J.-L. Deneubourg, S. Goss, N. Franks, A. Sendova-Franks, C. Detrain, L. Chrétien: The dynamics of collective sorting robot-like ants and ant-like robots, Proc. 1st Int. Conf. Simul. Adapt. Behav. Anim. Animat. (1991) pp. 356–363

71.49 T. Schmickl, K. Crailsheim: Trophallaxis within a robotic swarm: Bio-inspired communication among robots in a swarm, Auton. Robot. **25**(1), 171–188 (2008)

71.50 Á. Gutiérrez, A. Campo, F. Santos, F. Monasterio-Huelin Maciá, M. Dorigo: Social odometry: Imitation based odometry in collective robotics, Int. J. Adv. Robot. Syst. **6**(2), 129–136 (2009)

71.51 W. Burgard, M. Moors, D. Fox, R. Simmons, S. Thrun: Collaborative multi-robot exploration, Proc. IEEE Int. Conf. Robot. Autom. (ICRA '00), San Francisco (2000) pp. 476–481

71.52 D. Payton, M. Daily, R. Estowski, M. Howard, C. Lee: Pheromone robotics, Auton. Robot. **11**(3), 319–324 (2001)

71.53 D. Payton, R. Estkowski, M. Howard: Progress in pheromone robotics. In: *Intelligent Autonomous Systems 7*, ed. by M. Gini, W.-M. Shen, C. Torras, H. Yuasa (IOS, Amsterdam 2002) pp. 256–264

71.54 A. Howard, M.J. Matarić, G.S. Sukhatme: An incremental self-deployment algorithm for mobile sensor networks, Auton. Robot. **13**(2), 113–126 (2002)

71.55 M. Batalin, G. Sukhatme: Spreading out: A local approach to multi-robot coverage. In: *Distributed Autonomous Robotic Systems 5*, ed. by H. Asama, T. Arai, T. Fukuda, T. Hasegawa (Springer, Berlin, Heidelberg 2002) pp. 373–382

71.56 B.B. Werger, M.J. Matarić: Robotic *food* chains: Externalization of state and program for minimal-agent foraging, Proc. 4th Int. Conf. Simul. Adapt. Behav. Anim. Animat., ed. by P. Maes, M.J. Matarić, J. Meyer, J. Pollack, S. Wilson (MIT, Cambridge 1996) pp. 625–634

71.57 S. Nouyan, R. Groß, M. Bonani, F. Mondada, M. Dorigo: Teamwork in self-organized robot colonies, IEEE Trans. Evol. Comput. **13**(4), 695–711 (2009)

71.58 L. Chittka, P. Skorupski, N.E. Raine: Speed–accuracy tradeoffs in animal decision making, Trends Ecol. Evol. **24**(7), 400–407 (2009)

71.59 N.R. Franks, A. Dornhaus, J.P. Fitzsimmons, M. Stevens: Speed versus accuracy in collective decision making, Proc. R. Soc. B **270**(1532), 2457–2463 (2003)

71.60 J.A.R. Marshall, A. Dornhaus, N.R. Franks, T. Kovacs: Noise, cost and speed-accuracy trade-offs: Decision-making in a decentralized system, J. R. Soc. Interface **3**(7), 243–254 (2006)

71.61 A. Gutiérrez, A. Campo, F.C. Santos, F. Monasterio-Huelin, M. Dorigo: Social odometry: Imitation based odometry in collective robotics, Int. J. Adv. Robot. Syst. **6**(2), 1–8 (2009)

71.62 F. Galton: Vox populi, Nature **75**, 450–451 (1907)

71.63 A.M. Simons: Many wrongs: The advantage of group navigation, Trends Ecol. Evol. **19**(9), 453–455 (2004)

71.64 E.A. Codling, J.W. Pitchford, S.D. Simpson: Group navigation and the many-wrongs principle in models of animal movement, Ecology **88**(7), 1864–1870 (2007)

71.65 J.-L. Deneubourg, S. Goss: Collective patterns and decision-making, Ethol. Ecol. Evol. **1**, 295–311 (1989)

71.66 R. Beckers, J.-L. Deneubourg, S. Goss, J.M. Pasteels: Collective decision making through food recruitment, Insectes Soc. **37**(3), 258–267 (1990)

71.67 T.D. Seeley, S. Camazine, J. Sneyd: Collective decision-making in honey bees: How colonies choose among nectar sources, Behav. Ecol. Sociobiol. **28**(4), 277–290 (1991)

71.68 T.D. Seeley, S.C. Buhrman: Nest-site selection in honey bees: How well do swarms implement the *best-of-N* decision rule?, Behav. Ecol. Sociobiol. **49**(5), 416–427 (2001)

71.69 F. Saffre, R. Furey, B. Krafft, J.-L. Deneubourg: Collective decision-making in social spiders: Dragline-mediated amplification process acts as a recruitment mechanism, J. Theor. Biol. **198**(4), 507–517 (1999)

71.70 J.M. Amé, J. Halloy, C. Rivault, C. Detrain, J.-L. Deneubourg: Collegial decision making based on social amplification leads to optimal group formation, Proc. Natl. Acad. Sci. USA **103**(15), 5835–5840 (2006)

71.71 O. Petit, J. Gautrais, J.B. Leca, G. Theraulaz, J.-L. Deneubourg: Collective decision-making in white-faced capuchin monkeys, Proc. R. Soc. B **276**(1672), 3495 (2009)

71.72 P. Michelena, R. Jeanson, J.-L. Deneubourg, A.M. Sibbald: Personality and collective decision-making in foraging herbivores, Proc. R. Soc. B **277**(1684), 1093 (2010)

71.73 S. Goss, S. Aron, J.-L. Deneubourg, J.M. Pasteels: Self-organized shortcuts in the Argentine ant, Naturwissenschaften **76**(12), 579–581 (1989)

71.74 D.J.T. Sumpter, J. Krause, R. James, I.D. Couzin, A.J.W. Ward: Consensus decision making by fish, Curr. Biol. **18**(22), 1773–1777 (2008)

71.75 A.J.W. Ward, D.J.T. Sumpter, I.D. Couzin, P.J.B. Hart, J. Krause: Quorum decision-making facilitates information transfer in fish shoals, Proc. Natl. Acad. Sci. USA **105**(19), 6948 (2008)

71.76 S.C. Pratt, E.B. Mallon, D.J. Sumpter, N.R. Franks: Quorum sensing, recruitment, and collective decision-making during colony emigration by the ant *Leptothorax albipennis*, Behav. Ecol. Sociobiol. **52**(2), 117–127 (2002)

71.77 D.J.T. Sumpter, S.C. Pratt: Quorum responses and consensus decision making, Philos. Trans. R. Soc. B **364**(1518), 743–753 (2009)

71.78 S. Canonge, J.-L. Deneubourg, S. Sempo: Group living enhances individual resources discrimination: The use of public information by cockroaches to assess shelter quality, PLoS ONE **6**(6), e19748 (2011)

71.79 T.D. Seeley, P.K. Visscher, T. Schlegel, P.M. Hogan, N.R. Franks, J.A.R. Marshall: Stop signals provide cross inhibition in collective decision-making by honeybee swarms, Science **335**(6064), 108–111 (2012)

71.80   S. Garnier, J. Gautrais, M. Asadpour, C. Jost, G. Theraulaz: Self-organized aggregation triggers collective decision making in a group of cockroach-like robots, Adapt. Behav. **17**(2), 109–133 (2009)

71.81   J. Halloy, G. Sempo, G. Caprari, C. Rivault, M. Asadpour, F. Tâche, I. Said, V. Durier, S. Canonge, J.M. Amé, C. Detrain, N. Correll, A. Martinoli, F. Mondada, R. Siegwart, J.L. Deneubourg: Social integration of robots into groups of cockroaches to control self-organized choices, Science **318**(5853), 1155 (2007)

71.82   R. Olfati-Saber, R.M. Murray: Consensus problems in networks of agents with switching topology and time-delays, IEEE Trans. Autom. Control **49**(9), 1520–1533 (2004)

71.83   R. Olfati-Saber, J.A. Fax, R.M. Murray: Consensus and cooperation in networked multi-agent systems, Proc. IEEE **95**(1), 215–233 (2007)

71.84   T. Schmickl, R. Thenius, C. Moeslinger, G. Radspieler, S. Kernbach, M. Szymanski, K. Crailsheim: Get in touch: Cooperative decision making based on robot-to-robot collisions, Auton. Agents Multi-Agent Syst. **18**(1), 133–155 (2009)

71.85   K. Sugawara, T. Kazama, T. Watanabe: Foraging behavior of interacting robots with virtual pheromone, Proc. Int. Conf. Intell. Robot. Syst. (IROS 2004) (2004) pp. 3074–3079

71.86   S. Garnier, F. Tâche, M. Combe, A. Grimal, G. Theraulaz: Alice in pheromone land: An experimental setup for the study of ant-like robots, Proc. IEEE Swarm Intell. Symp. (SIS 2007), Piscataway (2007) pp. 37–44

71.87   A. Campo, Á. Gutiérrez, S. Nouyan, C. Pinciroli, V. Longchamp, S. Garnier, M. Dorigo: Artificial pheromone for path selection by a foraging swarm of robots, Biol. Cybern. **103**(5), 339–352 (2010)

71.88   E.J.H. Robinson, N.R. Franks, S. Ellis, S. Okuda, J.A.R. Marshall: A simple threshold rule is sufficient to explain sophisticated collective decision-making, PLoS ONE **6**(5), e19981 (2011)

71.89   M. Dorigo, D. Floreano, L.M. Gambardella, F. Mondada, S. Nolfi, T. Baaboura, M. Birattari, M. Bonani, M. Brambilla, A. Brutschy, D. Burnier, A. Campo, A.L. Christensen, A. Decugnire, G.A. Di Caro, F. Ducatelle, E. Ferrante, A. Fröster, J.M. Gonzales, J. Guzzi, V. Longchamp, S. Magnenat, N. Mathews, M.A. de Montes Oca, R. O'Grady, C. Pinciroli, G. Pini, P. Rétornaz, J. Roberts, V. Sperati, T. Stirling, A. Stranieri, T. Stützle, V. Trianni, E. Tuci, A.E. Turgut, F. Vaussard: Swarmanoid: A novel concept for the study of heterogeneous robotic swarms, IEEE Robot. Autom. Mag. **20**(4), 60–71 (2012)

72. Collective Manipulation and Construction

Lynne Parker

Many practical applications can make use of robot collectives that can manipulate objects and construct structures. Examples include applications in warehousing, truck loading and unloading, transporting large objects in industrial environments, and assembly of large-scale structures. Creating such systems, however, can be challenging. When collective robots work together to manipulate physical objects in the environment, their interactions necessarily become more tightly coupled. This need for tight coupling can lead to important control challenges, since actions by some robots can directly interfere with those of other robots. This chapter explores techniques that have been developed to enable robot swarms to effectively manipulate and construct objects in the environment. The focus in this chapter is on decentralized manipulation and construction techniques that would likely scale to large robot swarms (at least 10 robots), rather than approaches aimed primarily at smaller teams that attempt the same objectives. This chapter first discusses the swarm task of object transportation; in this domain, the objective is for

robots to collectively move objects through the environment to a goal destination. The chapter then discusses object clustering and sorting, which requires objects in the environment to be aggregated at one or more locations in the environment. The final task discussed is that of collective construction and wall building, in which robots work together to build a prespecified structure. While these different tasks vary in their specific objectives for collective manipulation, they also have several commonalities. This chapter explores the state of the art in this area.

72.1 Object Transportation

Some of the earliest work in swarm robotics was aimed at the object transportation task [72.1–6], which requires a swarm of robots to move an object from its current position in the environment to some goal destination. The primary benefit of using collective robots for this task is that the individual robots can combine forces to move objects that are too heavy for individual robots working alone or in small teams. However, the task is not without its challenges; it is nontrivial to design decentralized robot control algorithms that can effectively coordinate robot team members during object transportation. A further complication is that the interaction dynamics of the robots with the object can

be sensitive to certain object geometries [72.7, 8] and object rotations during transportation [72.8], thus exacerbating the control problem.

There are many ways to compare and contrast alternative distributed techniques to collective object transport. The most common distinctions are:

- Local knowledge only versus some required global knowledge (e.g., of team size, state, position).
- Homogeneous swarms versus heterogeneous swarms (e.g., teams with leaders and followers).
- Manual controller design versus autonomously learned control.

Part F | 72.1

- 2-D (two-dimensional) vs. 3-D (three-dimensional) environments.
- Obstacle-free environments versus cluttered environments.
- Static environments versus dynamic environments.
- Dependent on fully functioning robots versus systems robust to error.

Alternatively, we can compare transportation techniques by focusing on the specific manipulation technique employed. The manipulation techniques used for collective object transportation can be grouped into three primary methods [72.9]: pushing, grasping, and caging. The pushing approach requires contact between each robot and the object, in order to impart force in the goal direction; however, the robots are not physically connected with the object. In the grasping approach, each robot in the swarm is physically attached to the object being transported. Finally, the caging approach involves robots encircling the object so that the object moves in the desired direction, even without the constant contact of all the robots with the object.

This section outlines some of the key techniques developed to address this object transportation task, organized according to these three main techniques.

72.1.1 Transport by Pushing

A canonical task often used as a testbed in distributed robotics is the box pushing task. The number, size, or weight of the boxes can be varied to explore different types of multirobot cooperation. This task typically involves robots first locating a box, positioning themselves at the box, and then moving the box cooperatively toward a goal position. Typically, this task is explored in 2-D. The domain of box pushing is also popular because it has relevance to several real-world applications [72.10], including warehouse stocking, truck loading and unloading, transporting large objects in industrial environments, and assembling of large-scale structures.

The pushing technique was first demonstrated in the early work of *Kube* and *Zhang* [72.1], inspired by the cooperative transport behavior in ants [72.7]. They proposed a behavior-based approach that combined behaviors for seeking out the object (illuminated by a light), avoiding collisions, following other robots, and motion control. An additional behavior to detect stagnation was used to ensure that the collective did not work consistently against each other. In this approach, all robots acted similarly; there was no concept

of a leader and followers. While some of the robots in the swarm might not contribute to the pushing task due to poor alignment or positioning along the nondominant pushing direction, Kube and Zhang showed that careful design of these behaviors enabled the robot swarm to distribute along the boundary of the object and push it. Figure 72.1 shows five robots cooperatively pushing a lighted box.

Other researchers have explored different aspects of box pushing in multirobot systems. While much of this early work involved demonstrations of smaller robot teams, many of these techniques could theoretically scale to larger numbers of robots. Task allocation and action selection are often demonstrated using collective box pushing experiments; examples of this work include that of *Parker* [72.11, 12], who illustrated aspects of adaptive task allocation and learning; *Gerkey* and *Mataric* [72.13], who present a publish/subscribe dynamic task allocation method; and *Yamada* and *Saito* [72.14], who develop a behavior-based action selection technique that does not require any communication.

Other work using box pushing as an implementation domain for multirobot studies includes *Donald* et al. [72.15], who illustrates concepts of information invariance and the interchangeability of sensing, communication, and control; *Simmons* et al. [72.16], who demonstrate the feasibility of cooperative control for building planetary habitats, *Brown* and *Jennings* [72.17], and *Böhringer* et al. [72.18], who explored notions of strong cooperation without communication in pusher/steerer models, *Rus* et al. [72.19], who

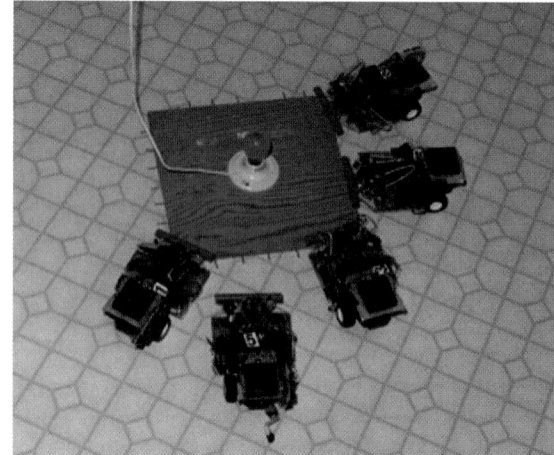

Fig. 72.1 Demonstration of five robots collectively pushing a lighted box (after [72.7])

studied different cooperative manipulation protocols in robot teams that make use of different combinations of state, sensing, and communication, and *Jones* and *Mataric* [72.20], who developed general methods for automatically synthesizing controllers for multirobot systems.

Most of this existing work in box pushing has focused, not on box pushing as the end objective, but rather on using box pushing for demonstrating various techniques for multirobot control. However, for studies whose primary objective is to generate robust cooperative transport techniques, work has more commonly focused on manipulation techniques involving grasping and caging, rather than pushing, since grasping and caging provide more controllability by the robot team.

72.1.2 Transport by Grasping

Grasping approaches for object transportation in swarm robotics typically make use of form closure and force closure properties [72.21]. In *form closure*, the object motion is constrained via frictionless contact constraints; in *force closure*, frictional contact forces exerted by the robots prevent unwanted motions of the manipulated object. The earliest work representing the grasping technique is that of *Wang* et al. [72.4]. This approach uses form closure, along with a behavior-based control approach that is similar to the early swarm robot pushing technique of *Kube* and *Zhang* [72.1]. The technique of Wang et al. called BeRoSH (for Behavior-based Multiple Robot System with Host for Object Manipulation), incorporates behaviors for pushing, maintaining contact, moving, and avoiding objects. In this approach, the goal pose of the object is provided directly to each robot from an external source (i. e., the Host); otherwise, the robots work independently according to their designed behaviors. As a collective, the swarm exhibits form closure. Wang et al. showed that this form closure technique can successfully transport an object to its desired goal pose from a variety of different starting locations.

Another early work using the force closure grasping technique is that of *Stilwell* and *Bay* [72.2] and *Johnson* and *Bay* [72.3]. They developed distributed leader–follower techniques that enable swarms of tank-like robots to transport pallets collectively while maintaining a level height of the pallet during transportation (Fig. 72.2). In their approaches, a pallet sits atop several tank-like robots; the weight of the pallet creates a coupling with the robots that could be viewed similar to a grasp. To transport the pallet, one vehicle is

designated as the leader. This leader then perturbs the dynamics of the system to move the swarm in the desired direction, and with the desired pallet height. The remaining robots in the swarm react to the perturbations to stabilize the forces in the system. The system is fully distributed, and requires robots to only use local force information to achieve the collective motion. The individual robots do not require knowledge of the pallet mass or inertia, the size of the collective, or the robot positions relative to the pallet's center of gravity. They showed the control stability of their approach for this application, even in the presence of inaccurate sensor data.

A related approach is that of *Kosuge* and *Oosumi* [72.5], who also used a decentralized leader–follower approach for multiple holonomic robots grasping and moving an object, in a manner similar to that of [72.2]. Their approach defines a compliant motion control algorithm for each velocity-controlled robot. The main difference of this work compared to [72.2] is that the control algorithm specifies the desired internal force as part of the coordination algorithm. This approach was validated in simulation for robots carrying an aluminum steel pipe.

Another related approach is that of *Miyata* et al. [72.6], who addressed the need for nonholonomic vehicles to regrasp the object during transport. Their approach includes a hybrid system that makes use of both centralized and decentralized planners. The centralized planner develops an approximate motion plan for the object, along with a regrasping plan at low resolution; the decentralized planner precisely estimates object motion and robot control at a much higher resolution.

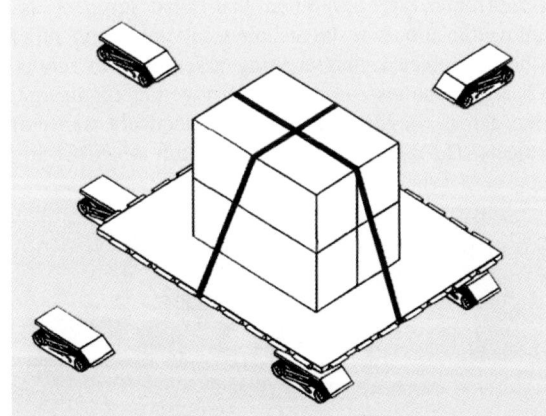

Fig. 72.2 Cooperative transport of a pallet using tank-like robots (after [72.2])

They demonstrated the effectiveness of this approach in simulation.

Sugar and *Kumar* [72.22] developed distributed control algorithms enabling robots with manipulators to grasp and cooperatively transport a box. In this work, a novel manipulator design enables the locomotion control to be decoupled from the manipulation control. Only a small number of the team members need to be equipped with actively controlled end effectors. This approach was shown to be robust to positioning errors related to the misalignment between the two platforms and errors in the measurement of the box size.

Cooperative stick pulling [72.23, 24] was explored by *Ijspeert* et al.; this task requires robots to pull sticks out of the ground (Fig. 72.3). The robot controllers are behavior-based, and include actions such as looking for sticks, detecting sticks, gripping sticks, obstacle avoidance, and stick release. Experiments show that the dynamics are dependent on the ratio between the number of robots and sticks; that collaboration can increase superlinearly with certain team sizes; that heterogeneity in the robots can increase the collaboration rate in certain circumstances; and that a simple signalling scheme can increase the effectiveness of the collaboration for certain team sizes. A main objective of this research was to explore the effectiveness of various modeling techniques for group behavior. These modeling techniques are discussed in more detail in a separate chapter.

The SWARM-BOTS project is a more recent example of the use of grasping for collective transport; it also makes use of self-assembly as a novel approach for achieving distributed transport. In this work [72.25], *s-bot* robots are developed that have grippers enabling the robots to create physical links with other s-bots or objects, thus creating assemblies of robots. These assemblies can then work together for navigation across rough terrain, or to collectively transport objects. The s-bots are cylindrical, with a flexible arm

and toothed gripper that can connect one s-bot to another (Fig. 72.4).

The decentralized control of the SWARM-BOT robots is learned using evolutionary techniques in simulation, then ported to the physical robots. The learned s-bot control [72.26] consists of an assembly module, which is a neural network that controls the robot prior to connection, and a transport module, which is a neural network that enables the s-bot to move the object toward the goal after a grasp connection is made. The self-assembly process involves the use of a red-colored seed object, to which other s-bots are attracted. S-bots initially light themselves with a blue ring, and then are attracted to the red color, while being repulsed by the blue color. Once robots make a connection, they color themselves red.

The interaction of these attractive and repulsive forces across the s-bots enables the robots to self-assemble into various connection patterns. Once the s-bots have self-assembled, they use the transport module to align toward a light source, which indicates the target position. The s-bots then apply pushing and pulling motions to transport the object to the destination. Similar to the approach of *Kube* and *Zhang* [72.1], the s-bots also check for stagnation and execute a recovery move when needed. The authors demonstrate [72.8] how the evolutionary learning approach allows the collective

Fig. 72.4 An s-bot, developed as part of the SWARM-BOTS project (after [72.25])

Fig. 72.3 Stick pulling experiment using robot collectives (after [72.23])

to successfully deal with different object geometries, adapt to changes in target location, and scale to larger team sizes.

This technique for collective transport using self-assembly was demonstrated [72.25] in an interesting application of object transport, in which 20 s-bots self-assembled into four chains in order to pull a child across the floor (Fig. 72.5). In this experiment, the user specifies the number of assembled chains, the distribution of the s-bots into the chains, the global localization of the child, and the global action timing. The s-bots then autonomously form the chains using self-assembly and execute the pull.

Several additional interesting phenomena regarding collective transport were discovered in related studies with the SWARM-BOTS. *Nouyan* et al. [72.27] showed that the different collective tasks of path formation, self-assembly, and group transport can be solved in a single system using a homogeneous robot team. They further introduce the notion of *chains with cycle directional patterns*, which facilitate swarm exploration in unknown environments, and assist in establishing paths between the object and goal. The paths established by the robot-generated chains mimic pheromone trails in ants. In [72.28], *Groß* and *Dorigo* determined that, while robots that behave as if they are solitary robots can still collectively move objects, robots that learn transport behaviors in a group can achieve a better performance. In [72.29], *Campo* et al. showed that the SWARM-BOTS robots could effectively transport objects even with only partial knowledge of the direction of the goal. They investigated four alternative control strategies, which vary in the degree to which the robots negotiate regarding the goal position during transport. Their results showed that negotiating throughout object transport can improve motion coordination. All of these works are based on inspiration from biological systems.

The work of *Berman* et al. [72.31] is not only bio-inspired, but also seeks to directly model the group retrieval behavior in ants. Their studies examined the ants' roles during transport in order to define rules that govern the ants' actions. They further explored measurements of individual forces used by the ants to guide food to their nest. They found that the distributed ant transport behavior exhibits an initial disordered phase, which then transitions to a more highly coordinated phase of increased load speed. From these studies, a computational dynamic model of the ant behavior was designed and implemented in simulations, showing that the derived model matches ant behavior. Ultimately, this approach could be adapted for use on physical robot teams.

Once a robot collective has begun transporting an object, the question arises as to how new robots can join the group to help with the transport task. *Esposito* [72.30] addresses this challenge by adapting a grasp quality function from the multifingered hand literature. This approach assumes that robots know the object geometry, the total number of robots in the swarm, and the actuator limitation. Individual robot contact configurations are defined relative to the object center and object boundary. The objective is to find an optimal position for a new robot by optimizing across the grasping wrench space. A numerical algorithm was developed to address this problem, which incorporates the force closure criteria. This ap-

Fig. 72.5 SWARM-BOTS experiment in which s-bots self-assemble to pull a child across the floor (after [72.25])

Fig. 72.6 Illustration of unmanned tugboats autonomously transporting a barge (after [72.30])

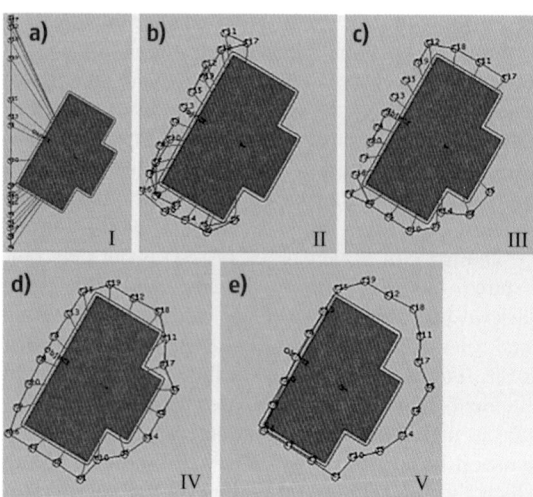

Fig. 72.7a–e Illustration in simulation of object closure by 20 robots (after [72.32])

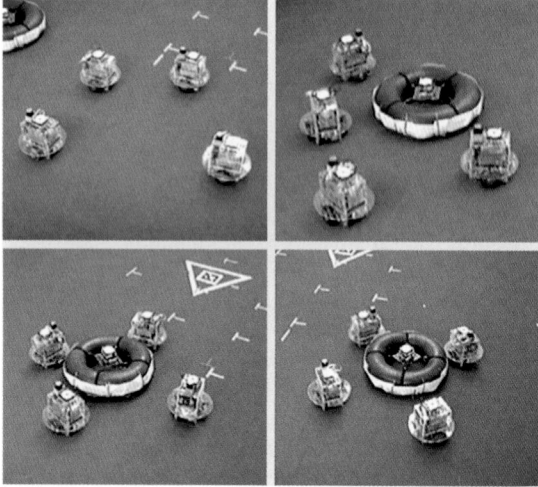

Fig. 72.8 Demonstration of the use of vector fields for collective transport via caging (after [72.33])

and median runtimes polynomial in the number of robots.

72.1.3 Transport by Caging

The caging approach simplifies the object manipulation task, compared to the grasping approach, by making use of the concept of *object closure* [72.34]. In object closure, a bounded movable area is defined for the object by the robots surrounding it. The benefit of this approach is that continuous contact between the object and the robots is not needed, thus making for simpler motion planning and control techniques, compared to grasping techniques based on the form or force closure. *Wang* and *Kumar* [72.32] developed this object-closure technique under the assumptions that the robots are circular and holonomic, the object is star-shaped, the robots know the number of robots in the collective, and can estimate the geometric properties of the object, along with the distance and orientation to other robots and the object. Their approach causes the robots to first approach the object independently, and then search for an *inescapable formation*, which is a configuration of the robots from which the object cannot escape. Finally, the robots execute a formation control strategy to guide the object to the goal destination. The object approach technique is based on potential fields, in which force vectors attract the robot toward the object and generally away from other robots. *Song* and *Kumar* [72.35] proved the stability of this potential field approach for collective transport. Robots search for proper configurations around the object by representing the problem as a path finding problem in configuration space. This work describes a necessary and sufficient condition for testing for object closure. Later work [72.36] presents a fast algorithm to test for object closure. Experiments with 20 robots validate the proposed approach (Fig. 72.7).

A further enhancement of this vector-based control strategy was developed in [72.33], which can account for inter-robot collisions. This latter strategy implements three primary behaviors – approach, surround, and transport. In this variant of the work, robots converge to a smooth boundary using control-theoretic techniques. This work was implemented on a collective of physical robots, as illustrated in Fig. 72.8.

proach was demonstrated on unmanned tugboats collectively moving a barge, as illustrated in Fig. 72.6. In this demonstration, the robots are equipped with articulated magnetic attachments that allow them to grasp the barge. This approach is scalable to larger numbers of robots, with constant best case runtime,

72.2 Object Sorting and Clustering

Collective object sorting and clustering requires robot teams to sort objects from multiple classes, typically into separate physical clusters. There are different types of related tasks in this domain [72.37], including clustering, segregation, patch sorting, and annular sorting. Early discussions of this task in robot swarms were given by *Deneubourg* et al. [72.38], with the ideas inspired by similar behaviors in ant colonies. The objective is to achieve clustering and sorting behaviors without any need for hierarchical decision making, inter-robot communication, or global representations of the environment. Deneubourg et al. showed that stigmergy could be used to cluster scattered objects of a single type, or to sort objects of two different types. To achieve the sorting behavior, the robots sensed the local densities of the objects, as well as the type of object they were carrying. Clustering resulted from a similar mechanism operating on a single type of object. *Beckers* et al. [72.39] achieved clustering from even simpler robots and behaviors, via stigmergic threshold mechanisms.

Holland and *Melhuish* [72.37] explored the effect of stigmergy and self-organization in swarms of homogeneous physical robots. The robots are programmed with simple rule sets with no ability for spatial orientation or memory. The experiments show the ability of the robots to achieve effective sorting and clustering, as illustrated in Fig. 72.9. In this work, a variety of influences were explored, including boundary effects and the distance between objects when deposited. The authors concluded that the effectiveness of the developed sorting behaviors is critically dependent on the exploitation of real-world physics. An implication of this finding is that simulators must be used with care when exploring these behaviors.

Wang and *Zhang* [72.40, 41] explored similar aims, but focused on discovering a general approach to the sorting problem. They conjecture that the outcome of the sorting task is dependent primarily on the capabilities of the robots, rather than the initial configuration. This conjecture is validated in simulation experiments, as illustrated in Fig. 72.10.

Other work on this topic includes that of *Yang* and *Kamel* [72.42], who present research using three colonies of ants having different speed models. The approach is a two-step process. The first step is for clusterings to be visually formed on the plane by agents walking, picking up, or setting down objects according

to a probabilistic model, which is based on *Deneubourg* et al. [72.38]. The second step is for clusters to be combined using a hypergraph model. Experiments were conducted in simulation to show the viability of the approach. The authors also discovered that having too many agents can lead to a deterioration in the swarm performance.

Martinoli and *Mondada* [72.43] implemented another bio-inspired approach to clustering, in which the robot behavior is similar to that of a Braitenberg vehicle. They also discovered that large numbers of robots can cause interference in this task, concluding that noncooperative task cannot always be improved with more robots.

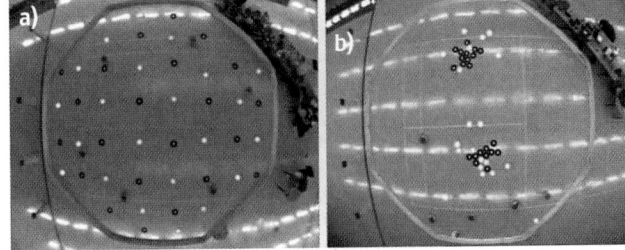

Fig. 72.9 Results of physical robot experiments in sorting. Panel **(a)** shows the starting configuration, while **(b)** shows the sorting results after 1.75 h (after [72.37])

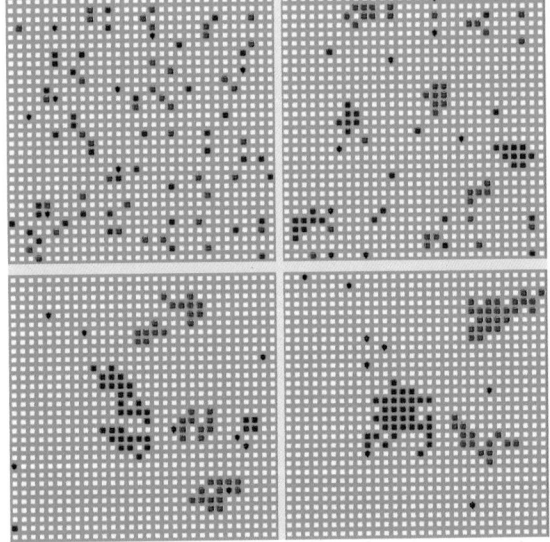

Fig. 72.10 Results of simulations of sorting tasks, with 8 robots and 40 objects of two types (after [72.40])

72.3 Collective Construction and Wall Building

The objective of the collective construction and wall building task is for robots to build structures of a specified form, in either 2-D or 3-D. This task is distinguished from self-reconfigurable robots, whose bodies themselves serve as the dynamic structure. This section is focused on the former situation, in which manipulation is required to create the desired structure. The argument in favor of this separation of mobility and structure is that, once formed, the structure does not need to move again, and thus the ability to move could serve as a liability [72.44]. Furthermore, robotic units that serve both as mobility and structure might not be effective as passive structural elements.

Werfel et al. have extensively explored this topic, developing distributed algorithms that enable simplified robots to build structures based on provided blueprints, both in 2-D [72.45–47] and in 3-D [72.44]. In their 3-D approach, the system consists of idealized mobile robots that perform the construction, and smart blocks that serve as the passive structure. The robots' job is to provide the mobility, while the blocks' role is to identify places in the growing structure at which an additional block can be placed that is on the path toward obtaining the desired final structure. The goal of their work is to be able to deploy some number of robots and free blocks into a construction zone, along with a single block that serves as a seed for the structure, and then have the construction to proceed autonomously according to the provided blueprint of the desired structure.

Several simplifying assumptions are made in this work [72.44], such as the environment being weightless and the robots being free to move in any direction in 3-D, including along the surface of the structure under construction. This work does not address physical robot navigation and locomotion challenges, grasping challenges, etc.

In this approach, blocks are smart cubes; they can communicate with attached neighbors, they share a global coordinate system, and they can communicate with passing robots regarding the validity of block attachments to exposed faces. Once robots have transported a free block to the structure, they locate attachment points in one of three ways: random movement, systematic search, or gradient following. A significant contribution of this work is the development of the block algorithm that enables the blocks to specify

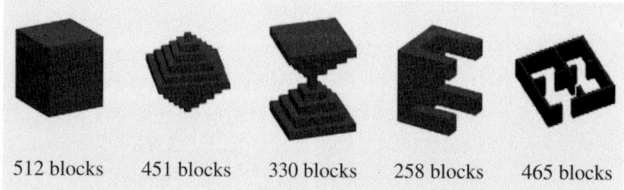

512 blocks   451 blocks   330 blocks   258 blocks   465 blocks

Fig. 72.11 Experiments for a variety of 3-D structures, built autonomously by a system of simple robots and blocks (after [72.44])

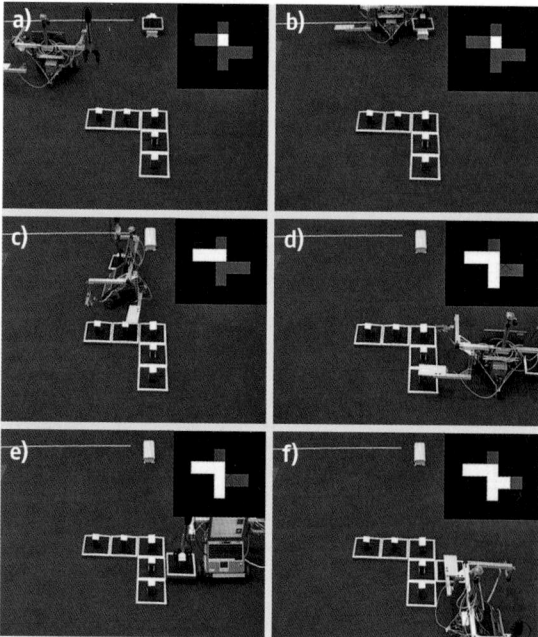

Fig. 72.12a–f Proof-of-principle experiments for 2-D construction, using a single robot (after [72.45])

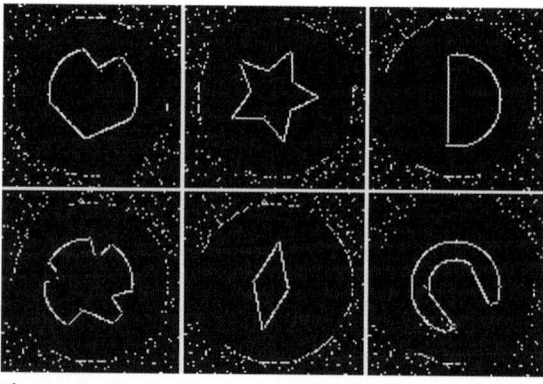

Fig. 72.13 Geometric structures built by a team of 30 robots, in simulation (after [72.48])

Fig. 72.14 Experiments with prototype hardware designed for multirobot construction tasks (after [72.49])

how to grow the developing structure with guarantees, and with only limited required communication. More specifically, the communication requirement between blocks scales linearly in the size of the structure, while no explicit communication between the mobile robots is needed.

Experiments using this approach have shown the ability of the system to build a variety of structures in simulation, as illustrated in Fig. 72.11. A proof-of-principle physical robot experiment using a single robot in the 2-D case [72.45] is illustrated in Fig. 72.12.

Werfel [72.48] also describes a system for arranging inert blocks into arbitrary shapes. The input to the robot system is a high-level geometric program, which is then translated by the robots into an appropriate arrangement of blocks using their programmed behaviors. The desired structure is communicated to the robots as a list of corners, the angles between corners, and whether the connection between corners is to be straight or curved. Robots are provided with behaviors such as *clearing*, *doneClearing*, *beCorner*, *collect*, *seal*, and *off*. Figure 72.13 shows some example structures built using this system in simulation.

Hardware challenges of collective robot construction are addressed by *Terada* and *Murata* [72.49]. In this work, a hardware design is proposed that defines passive building blocks, along with an assembler robot that constructs structures with the robots. Figure 72.14 shows the prototype hardware completing an assembly task. In principle, multiple assembler robots could work together to create larger construction teams more closely aligned with the concept of swarm construction.

Other related work on the topic of collective construction includes the work of *Wawerla* et al. [72.51], in which robots use a behavior-based approach to build a linear wall using blocks equipped with either positive or negative Velcro, distinguished by block color. Their results show that adding 1 bit of state information to communicate the color of the last attached block provides a significant improvement in the collective performance. The work by *Stewart* and *Russell* [72.50, 52] proposes a distributed approach to building a loose wall structure with a robot swarm. The approach makes use of a spatiotemporal varying light-field template, which is generated by an organizer robot to help direct the actions of the builder robots. Builder robots deposit objects in locations indicated by the template.

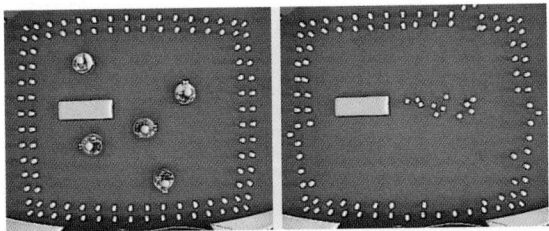

Fig. 72.15 Experimental trial demonstrating a swarm building a loose wall via a spatiotemporal varying template (after [72.50])

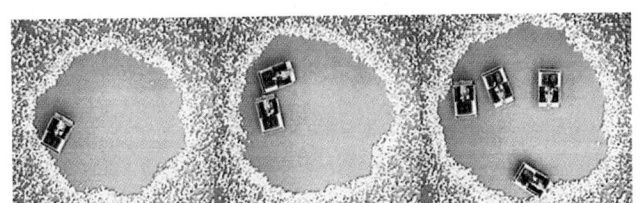

Fig. 72.16 Experiments of blind bulldozing for site clearing using physical robots (after [72.53])

Figure 72.15 shows the results from one of the experiments using this approach on physical robots.

Another type of construction is called *blind bulldozing*, which is inspired by a behavior observed in certain ant colonies. Rather than constructing by accumulating materials, this approach achieves construction by removing materials. This task has practical application in site clearing, such as would be needed for planetary exploration [72.54]. Early ideas of this concept were discussed by *Brooks* et al. [72.55], which argues for large numbers of small robots to be delivered to the lunar surface for site preparation. *Parker* et al. [72.53],

further developed this idea by proposing robots using force sensors to clear an area by pushing material to the edges of the work site. In this approach, the robot system collective behavior is modeled in terms of how the nest grows over time. Stigmergy is used to control the construction process, in that the work achieved by each robot affects the other robots' behaviors through the environment. Figure 72.16 shows some experiments using this approach on physical robots. The authors argue that blind bulldozing is appropriate in applications where the cost, complexity, and reliability of the robots is a concern.

72.4 Conclusions

This chapter has surveyed some of the important techniques that have been developed for collective object transport and manipulation. While many advances have been made, there are still many open challenges that remain. Some open problems include: How to deal with faults in the robot team members during task execution; how to address construction in dynamic and cluttered

environments; how to enable humans to interact with the robot swarms; how to extend more of the existing techniques to 3-D applications; how to design formal techniques for predicting and guaranteeing swarm behavior; how to realize larger scale systems on physical robots; and how to apply swarm techniques for manipulation and construction to practical applications.

References

72.1 C.R. Kube, H. Zhang: Collective robotics: From social insects to robots, Adapt. Behav. **2**(2), 189–218 (1993)
72.2 D. Stilwell, J.S. Bay: Toward the development of a material transport system using swarms of ant-like robots, IEEE Int. Conf. Robot. Autom. (1993) pp. 766–771
72.3 P.J. Johnson, J.S. Bay: Distributed control of simulated autonomous mobile robot collectives in payload transportation, Auton. Robot. **2**(1), 43–63 (1995)
72.4 Z. Wang, E. Nakano, T. Matsukawa: Realizing cooperative object manipulation using multiple behaviour-based robots, Proc. IEEE/RSJ Int. Conf. Intell. Robot. Syst. (1996) pp. 310–317
72.5 K. Kosuge, T. Oosumi: Decentralized control of multiple robots handling an object, Proc. IEEE/RSJ Int. Conf. Intell. Robot. Syst. (1996) pp. 318–323
72.6 N. Miyata, J. Ota, Y. Aiyama, J. Sasaki, T. Arai: Cooperative transport system with regrasping car-like mobile robots, Proc. IEEE/RSJ Int. Conf. Intell. Robot. Syst. (1997) pp. 1754–1761
72.7 C.R. Kube, E. Bonabeau: Cooperative transport by ants and robots, Robot. Auton. Syst. **30**(1), 85–101 (2000)
72.8 R. Groß, M. Dorigo: Towards group transport by swarms of robots, Int. J. Bio-Inspir. Comput. **1**(1), 1–13 (2009)

72.9 Y. Mohan, S.G. Ponnambalam: An extensive review of research in swarm robotics, World Congr. Nat. Biol. Inspir. Comput. 2009 (2009) pp. 140–145
72.10 D. Nardi, A. Farinelli, L. Iocchi: Multirobot systems: A classification focused on coordination, IEEE Trans. Syst. Man Cybern. B **34**(5), 2015–2028 (2004)
72.11 L.E. Parker: ALLIANCE: An architecture for fault tolerant, cooperative control of heterogeneous mobile robots, Proc. IEEE/RSJ/GI Int. Conf. Intell. Robot. Syst. (1994) pp. 776–783
72.12 L.E. Parker: Lifelong adaptation in heterogeneous teams: Response to continual variation in individual robot performance, Auton. Robot. **8**(3), 239–269 (2000)
72.13 B.P. Gerkey, M.J. Mataric: Sold! Auction methods for multi-robot coordination, IEEE Trans. Robot. Autom. **18**(5), 758–768 (2002)
72.14 S. Yamada, J. Saito: Adaptive action selection without explicit communication for multirobot box-pushing, IEEE Trans. Syst. Man Cybern. C **31**(3), 398–404 (2001)
72.15 B. Donald, J. Jennings, D. Rus: Analyzing teams of cooperating mobile robots, IEEE Int. Conf. Robot. Autom. (1994) pp. 1896–1903
72.16 R. Simmons, S. Singh, D. Hershberger, J. Ramos, T. Smith: First results in the coordination of hetero-

geneous robots for large-scale assembly, ISER 7th Int. Symp. Exp. Robot. (2000)

72.17 R.G. Brown, J.S. Jennings: A pusher/steerer model for strongly cooperative mobile robot manipulation, Proc. 1995 IEEE Int. Conf. Intell. Robot. Syst. (1995) pp. 562–568

72.18 K. Böhringer, R. Brown, B. Donald, J. Jennings, D. Rus: Distributed robotic manipulation: Experiments in minimalism, Lect. Notes Comput. Sci. **223**, 11–25 (1997)

72.19 D. Rus, B. Donald, J. Jennings: Moving furniture with teams of autonomous robots, Proc. IEEE/RSJ Int. Conf. Intell. Robot. Syst. (1995) pp. 235–242

72.20 C. Jones, M.J. Mataric: Automatic synthesis of communication-based coordinated multi-robot systems, Proc. IEEE/RJS Int. Conf. Intell. Robot. Syst. (2004) pp. 381–387

72.21 A. Bicchi: On the closure properties of robotic grasping, Int. J. Robot. Res. **14**(4), 319–334 (1995)

72.22 T.G. Sugar, V. Kumar: Control of cooperating mobile manipulators, IEEE Trans. Robot. Autom. **18**(1), 94–103 (2002)

72.23 A.J. Ijspeert, A. Martinoli, A. Billard, L.M. Gambardella: Collaboration through the exploitation of local interactions in autonomous collective robotics: The stick pulling experiment, Auton. Robot. **11**(2), 149–171 (2001)

72.24 A. Martinoli, K. Easton, W. Agassounon: Modeling swarm robotic systems: A case study in collaborative distributed manipulation, Int. J. Robot. Res. **23**(4/5), 415–436 (2004)

72.25 F. Mondada, L.M. Gambardella, D. Floreano, S. Nolfi, J.L. Deneuborg, M. Dorigo: The cooperation of swarm-bots: Physical interactions in collective robotics, IEEE Robot. Autom. Mag. **12**(2), 21–28 (2005)

72.26 R. Groß, E. Tuci, M. Dorigo, M. Bonani, F. Mondada: Object transport by modular robots that self-assemble, IEEE Int. Conf. Robot. Autom. (2006) pp. 2558–2564

72.27 S. Nouyan, R. Groß, M. Bonani, F. Mondada, M. Dorigo: Group transport along a robot chain in a self-organised robot colony, Proc. 9th Int. Conf. Intell. Auton. Syst. (2006) pp. 433–442

72.28 R. Groß, M. Dorigo: Evolution of solitary and group transport behaviors for autonomous robots capable of self-assembling, Adapt. Behav. **16**(5), 285–305 (2008)

72.29 A. Campo, S. Nouyan, M. Birattari, R. Groß, M. Dorigo: Negotiation of goal direction for cooperative transport, Lect. Notes Comput. Sci. **4150**, 191–202 (2006)

72.30 J.M. Esposito: Distributed grasp synthesis for swarm manipulation with applications to autonomous tugboats, IEEE Int. Conf. Robot. Autom. (2008) pp. 1489–1494

72.31 S. Berman, Q. Lindsey, M.S. Sakar, V. Kumar, S.C. Pratt: Experimental study and modeling of group retrieval in ants as an approach to collec-

tive transport in swarm robotic systems, Proc. IEEE **99**(9), 1470–1481 (2011)

72.32 Z. Wang, V. Kumar: Object closure and manipulation by multiple cooperating mobile robots, IEEE Int. Conf. Robot. Autom. (2002) pp. 394–399

72.33 J. Fink, N. Michael, V. Kumar: Composition of vector fields for multi-robot manipulation via caging, Robot. Sci. Syst. (2007) pp. 25–32

72.34 Z.D. Wang, V. Kumar: Object closure and manipulation by multiple cooperating mobile robots, IEEE Int. Conf. Robot. Autom. (2002) pp. 394–399

72.35 P. Song, V. Kumar: A potential field based approach to multi-robot manipulation, IEEE Int. Conf. Robot. Autom. (2002) pp. 1217–1222

72.36 Z. Wang, Y. Hirata, K. Kosuge: Control a rigid caging formation for cooperative object transportation by multiple mobile robots, Proc. IEEE Int. Conf. Robot. Autom. (2004) pp. 1580–1585

72.37 O. Holland, C. Melhuish: Stigmergy, self-organization, and sorting in collective robotics, Artif. Life **5**(2), 173–202 (1999)

72.38 J.L. Deneubourg, S. Goss, N. Franks, A. Sendova-Franks, C. Detrain, L. Chretien: The dynamics of collective sorting robot-like ants and ant-like robots, Proc. 1st Int. Conf. Simul. Adapt. Behav. Anim. Anim. (1990)

72.39 R. Beckers, O. Holland, J. Deneubourg: From local actions to global tasks: Stigmergy and collective robotics, Proc. 14th Int. Workshop Synth. Simul. Living Syst. (1994) pp. 181–189

72.40 T. Wang, H. Zhang: Multi-robot collective sorting with local sensing, IEEE Intell. Autom. Conf. (2003)

72.41 T. Wang, H. Zhang: Collective Sorting with Multiple Robots, IEEE Int. Conf. Robot. Biomim. (2004) pp. 716–720

72.42 Y. Yang, M. Kamel: Clustering ensemble using swarm intelligence, IEEE, Swarm Intell. Symp. (2003) pp. 65–71

72.43 A. Martinoli, F. Mondada: Collective and cooperative group behaviours: Biologically inspired experiments in robotics, Lect. Notes Comput. Sci. **223**, 1–10 (1997)

72.44 J. Werfel, R. Nagpal: Three-dimensional construction with mobile robots and modular blocks, Int. J. Robot. Res. **27**(3/4), 463–479 (2008)

72.45 J. Werfel, Y. Bar-Yam, D. Rus, R. Nagpal: Distributed construction by mobile robots with enhanced building blocks, IEEE Int. Conf. Robot. Autom. (2006) pp. 2787–2794

72.46 J. Werfel: Building patterned structures with robot swarms, Proc. 19th Int. Joint Conf. Artif. Intell. (2005) pp. 1495–1502

72.47 J. Werfel, R. Nagpal: Extended stigmergy in collective construction, IEEE Intell. Syst. **21**(2), 20–28 (2006)

72.48 J. Werfel: Building blocks for multi-robot construction, Distrib. Auton. Robot. Syst. **6**, 285–294 (2007)

72.49   Y. Terada, S. Murata: Automatic modular assembly system and its distributed control, Int. J. Robot. Res. **27**, 445–462 (2008)

72.50   R.L. Stewart, R.A. Russell: A distributed feedback mechanism to regulate wall construction by a robotic swarm, Adapt. Behav. **14**(1), 21–51 (2006)

72.51   J. Wawerla, G.S. Sukhatme, M.J. Mataric: Collective construction with multiple robots, IEEE/RSJ Int. Conf. Intell. Robot. Syst. (2002) pp. 2696–2701

72.52   R.L. Stewart, R.A. Russell: Building a loose wall structure with a robotic swarm using a spatio-temporal varying template, IEEE/RSJ Int. Conf. Intell. Robot. Syst. (2004) pp. 712–716

72.53   C.A.C. Parker, H. Zhang, C.R. Kube: Blind bulldozing: Multiple robot nest construction, IEEE/RSJ Int. Conf. Intell. Robot. Syst. (2003) pp. 2010–2015

72.54   T. Huntsberger, G. Rodriguez, P.S. Schenker: Robotics challenges for robotic and human mars exploration, Proc. Robot. (2000) pp. 340–346

72.55   R.A. Brooks, P. Maes, M.J. Mataric, G. More: Lunar based construction robots, Proc. Int. Conf. Intell. Robot. Syst. (1990)

73. Reconfigurable Robots

Kasper Støy

Reconfigurable robots are robots built from mechatronics modules that can be connected in different
ways to create task-specific robot morphologies.
In this chapter we introduce reconfigurable robots
and provide a brief taxonomy of this type of robot.
However, the main focus of this chapter is on the
four most important challenges in realizing reconfigurable robots. The first two are mechatronics
challenges, namely the challenge of connector design and energy. Connectors are the most important
design element of any reconfigurable robot because they provide it with much of its functionality,
but also many of its limitations. Supplying energy
to a connected, distributed multi-robot system
such as a reconfigurable robot is an important, but
often underestimated problem. The third challenge
is distributed control of reconfigurable robots. It is
examined both how reconfigurable robots can be
controlled in static configurations to produce locomotion and manipulation and how configurations
can be transformed through a self-reconfiguration
process. The fourth challenge that we will discuss
is programability and debugging of reconfigurable
robot systems. The chapter is concluded with a brief

perspective. Overall, the chapter provides a general
overview of the field of reconfigurable robots and
is a perfect starting point for anyone interested in
this exciting field.

Reconfigurable robots are a kind of robot built from modules that can be connected in different ways to form different morphologies for different purposes. The motivation for this is that conventional robots are limited by their morphology. E.g., the size of the wheels of a wheeled robot determines what terrain it can traverse. If it has small wheels it can operate in confined spaces, but not traverse rugged terrain and, vice versa, if it has large wheels. A robot arm may be limited in terms of reach or inability to move around in the environment. Reconfigurable robots aim to solve this problem by providing a robotic system that can be manually reconfigured to be physically suited for the specific task at hand. It is conceivable that a reconfigurable robot on-

site can be fitted with several types of wheels, different segments can be added to form the body or arms. In this way, the reconfigurable robot can be a perfect physical fit for its task despite the task not being known or defined in advance.

While this practically oriented motivation is sufficient for developing and working with reconfigurable robots, the vision for the field is much deeper. The underlying, long-term vision is to develop robots that are robust, self-healing, versatile, cheap, and autonomous. Hence most research in the field of reconfigurable robots has been on self-reconfigurable robots. These robots consist of modules just like reconfigurable robots, but in addition the modules can automatically

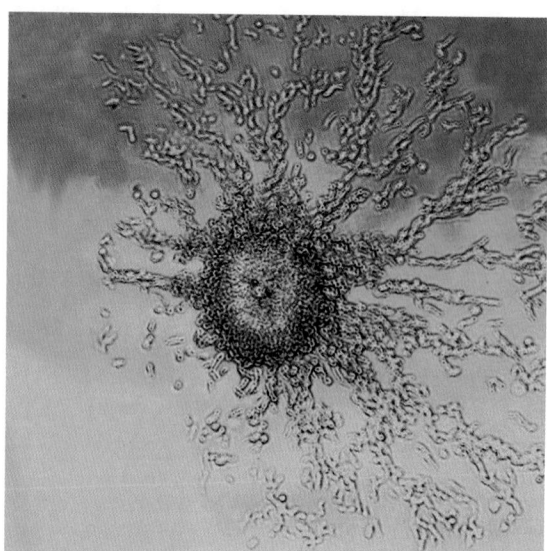

Fig. 73.1 A gathering of Dictyostelium discoideum amoebae cells can be seen migrating (some individually, and some in streams) toward a central point. Aggregation territories can be as much as a centimeter in diameter (after [73.2])

Fig. 73.2 The CKbot in a chain configuration with wheels attached (courtesy of Modlab at University of Pennsylvania)

connect, disconnect, and move with respect to neighboring modules, and thus the robot as a whole can change shape not unlike the robot systems envisioned in science fiction movies such as Terminator and Transformers or the children's cartoon Barbapapa. Another way to view self-reconfigurable robots is as multicellular robots [73.1]. In this case, each module is comparable to a cell in an organism. In fact, the small animal hydra is able to self-reconfigure in the sense that if it is cut in half, the two sections each form a smaller, but complete hydra. While not strictly a multi-cellular organism, the slime mold Dictyostelium has also been of significant inspiration to the self-reconfigurable robot community. In this slime mold individual cells search for food in a local area, but once the food sources are used up the cells aggregate, as is shown in Fig. 73.1, to form slugs that number in the hundreds of thousands of cells, however, is still able to function as one organism whose mission it is in fact to find a suitable place to disperse spores for the next generation of slime molds.

The modular design gives reconfigurable robots a number of useful features. First of all the robots are robust since if modules fail, the rest of the modules will still continue working and thus the robot can maintain a level of functionality, although it is slightly reduced. Self-reconfigurable robots extend on this by being able to eject failed modules from the robot and replace them with modules from other parts of their bodies and effectively achieve self-healing. A powerful demonstration of this is the CKBot shown in Fig. 73.2, which after a kick has broken into several clusters of modules. These clusters are able to locate each other again and recreate the original structure [73.3]. Another vision of reconfigurable robots is that they are versatile due to their modular structure. Reconfigurable robots are not limited by a fixed shape, but the number and capabilities of the modules are available. A final feature is that the individual modules can be produced relatively cheaply due to production at scale. The implication is that even though the individual module can be quite complex they can be mass produced and thus their cost can be relatively low compared to their complexity.

It is also important to point out that reconfigurable robots can never be a universal robot that can take on the functionality of any robot. It is, of course, the ultimate dream, but given a known task-environment a robot can be custom-designed for this and thus will be simpler and better performing than a comparable reconfigurable robot. Thus reconfigurable robots are best suited for situations where the task or environment is not known in advance or in locations where many different types of robots are needed, but it is not possible to bring them all. Optimal applications for reconfigurable robots are thus in extra-planetary missions, disaster areas, and so on. However, they may also find their use in more down to earth applications such as educational robotics [73.4] or as robot construction kits [73.5].

It is important to note that these are all visions that the reconfigurable robot community is striving towards, but has only realized in limited ways. However, it is the vision of these truly amazing reconfigurable robots that drives us forward.

73.1 Mechatronics System Integration

In mechatronics system integration the main challenge is how to make a trade-off between different potential features of a module and the need to fit everything in a mechatronics module, which typically has a radius on the order of centimeters. The different classes of reconfigurable robots reflect different trade-offs.

The oldest class is the *mobile* reconfigurable robot and can be traced back to the cellular robot (CEBOT) developed in the late 1980s [73.6], but new instantiations of this class, which are an order of magnitude smaller, have also recently been published [73.7–9]. This class of reconfigurable robots is characterized by the modules having a high-degree of self-mobility, typically obtained by providing each module with a set of wheels. Modules of other classes also have a limited degree of self-mobility, e.g., a module can perform inchworm like gaits, but it is so limited that we do not consider it mobile.

Chain-type reconfigurable robots (Fig. 73.3a) were the first of the modern reconfigurable robots in the sense that they successfully demonstrated versatility: given the same reconfigurable robot several locomotion gaits could be implemented, including inch-worming, rolling, and even walking, as was demonstrated using the PolyPod robot [73.10, 11] and later its descendent PolyBot [73.12] and CONRO [73.13]. The chain-type modules are characterized by having a high degree of internal actuation which typically allow a module to bend or twist. However, there are examples of modules which provide for both bending and twisting [73.14]. These modules are also typically elongated, which makes them suitable for forming chains of modules with many degrees of freedom, making them appropriate for making limbs.

Another class of reconfigurable robots is the *lattice-type* robots (Fig. 73.3b). This class of robot addresses one of the short-comings of chain-type robots. Namely, that it is difficult for chains to align and connect because this requires precision that they often do not have. This makes it difficult to achieve self-reconfiguration with chain-type modular robots. The solution lattice-type reconfigurable robots represents is to have a geometric design that allows them to fit in a lattice just

like atoms in a crystal. The movement of the modules is then limited to moving from lattice to lattice position, a task that only requires limited precision, sensing,

Fig. 73.3a–c Examples of the three main types of reconfigurable robots: (**a**) CONRO chain-type, (**b**) molecule lattice-type, (**c**) M-TRAN hybrid, courtesy of USC's Information Sciences Institute (**a**); Distributed Robotics Laboratory, MIT (**b**); AIST, Japan (**c**) ▶

and actuation. However, this often means that modules have limited functionality outside of the lattice. Early two-dimensional lattice-type robots include the Fracta and Metamorphic robots [73.15, 16]; the first three-dimensional lattice-type robots were the Molecule and the 3-D-Unit [73.17, 18].

Chain-type and lattice-type robots (Fig. 73.3c) have largely been superseded by the *hybrid* reconfigurable robots. These robots combine the characteristics of chain-type and lattice-type robots in one system. That is, they can both fit in a lattice structure, which allows for relatively easy self-reconfiguration, and out-of-lattice, which allows for efficient locomotion or manipulation. The recent generation of self-reconfigurable robots, including M-TRAN III, ATRON shown in Fig. 73.8, and SuperBot, are all of the hybrid type [73.14, 19, 20].

A final type of reconfigurable robot are the actuation-less robots. These robots depend on external forces to provide reconfiguration capabilities and are not able to move once they are in a lattice. There are stochastic versions of actuation-less modules that are suspended in a fluid [73.21] or float on an air-hockey table [73.22], and the random movements of the modules in the medium allow them by chance to get close enough to form connections. A slight twist of this approach is that modules only use the external forces to swing from lattice position to lattice position, but maintain control of when to disconnect and connect themselves [73.23]. There are also deterministic versions, which employ an assembly-by-disassembly process where modules start connected in a lattice and modules that are not needed in the specific configuration can then deterministically decide to disconnect from the structure (typically based on electro-magnetic forces) [73.24]. Finally, there are the manually reconfigurable robots that depend on the human user (or another robot) to perform the reconfiguration [73.25, 26].

An orthogonal classification of reconfigurable robots is according to whether they are homogeneous or heterogeneous. Homogeneous modular robots consist of identical modules and have been favored in the community because they lend themselves to self-reconfiguration. However, it is becoming clear that if we want to keep modules simple and provide a certain level of functionality, we need to focus more on heterogeneous systems. It is not cost effective to provide all modules with the same level of functionality, and more importantly, the modules become too complex, heavy, and large if they are to contain all the functionality needed, which in practice make them unsuited for practical applications [73.27].

Another emerging idea is that of soft modules. In fact, quite a number of rigid modules that have been built come out of projects that aimed to build soft reconfigurable robots. The motivation for soft modules is that they provide a certain level of compliance in the interaction with the environment and also within the robot. However, a good way to realize soft reconfigurable robots has not yet been discovered.

73.2 Connection Mechanisms

An element of the mechatronic design of reconfigurable modular robots that has turned out to be a significant challenge is the mechanism that connects modules to one another. This may appear puzzling at first, but individual modules are functionally limited and, hence, reconfigurable robots perform most tasks using groups of modules. This means that everything has to be passed across connectors from module to module, including forces, communication, and energy. For self-reconfigurable robots connector design is even more difficult because the connector also has to be able to actively connect and disconnect. The optimal connector would have the following features:

- Small size
- Fast
- Strong
- Robust to wear and tear
- High tolerance to alignment errors
- Energy use only in the transition phases
- Transferal of electrical and/or communication signals between modules
- Genderless
- Allows connection with different orientations
- Disconnects from both sides
- Dirt resistant.

While there are a few connectors that incorporate most of these features, none has implemented them all. It is easy to imagine that a solution is something along the lines of self-cleaning, conducting, with active velcro or gecko skin. Unfortunately, this does not exist (yet) and connector designs are, therefore, based on conventional electro-magnetic, mechanical, or electro-static principles of connection.

Magnetic connectors and combinations with electro-magnets for active connection and disconnection have been quite successful given that they meet most of the above requirements except for being gender-less and strong. The gender issue is normally solved by laying out magnets in a geometrical pattern on the surface of the connector that allow male connectors and female connectors to be connected at a discrete number of angles. However, the main shortcoming of the magnetic solution is the strength it provides. It is clear that if the magnetic force is too weak modules will fall apart easily. However, it is, in fact, also a problem if the magnetic force is too strong, because for active mechanisms the modules have to overcome the magnetic force to disconnect (or for manually re-configurable robots, the human user has to overcome the magnetic force). Therefore, a compromise has to be found, which is not optimal for any of the situations. A recent solution to this problem is the use of switchable magnets. Switchable magnets come in simple mechanical forms where magnets are physically turned to change the direction of the magnetic flux and thus the connection strength. The more advanced type use electro-magnets to change the magnetic polarization of a permanent magnet achieving the same but in a much smaller form factor. These developments have opened up the possibility of using magnetic connectors again, but it remains largely unexplored in the community except for the robot pebbles system (where the technology originated) [73.24].

The most recent generations of self-reconfigurable robots have all favored mechanical solutions. A mechanical solution is based on hooks coming out of one connector surface attaching to holes in the opposing connector surface. A mechanical solution immediately solves the problem of having strong connectors, but unfortunately introduces others. The most important problems are they are large and slow, e.g., in the ATRON self-reconfigurable robots the connector mechanism and associated actuators and electronics account for up to 60% of the modules' size and weight and it takes 2 s to connect. However, the *Terada* connector [73.28] used in M-TRAN III and the SINGO connector [73.29] used in SuperBot appear to have provided potential solutions to the problem of size, but the time issue still remains.

A last class of connectors are based on electro-static forces [73.30]. The idea is to charge two opposing metal surfaces causing the two surfaces to connect strongly. While being an interesting option, the realized systems are impractical, because they are large and sensitive to the distance between the connection surfaces. This approach makes most sense at smaller scales, but despite some effort this has not been realized. Also, in this category of non-standard connector unisex velcro connectors [73.31] should also be mentioned, but here, of course, the main problem is to obtain enough connection strength.

Overall, connector technology is fairly advanced, but there is certainly room for improvement. However, at this point for significant progress to be made, new results probably have to emerge from material science and not from the reconfigurable robotic community itself.

73.3 Energy

Reconfigurable robots are typically designed for autonomy and hence rely on on-board batteries for power. The challenges here are to enable the modules to share the available energy and to allow the robot to recharge once batteries are depleted.

It is important for the modules of a reconfigurable robot to be able to share energy because modules may have very different activity levels and hence very different levels of energy consumption. Therefore, the life of the robot can be extended significantly by allowing inactive modules to donate their energy to more active ones. The issue is largely unexplored but there has been attempts of passing energy across connectors [73.32] through physical connections, and it has

been discussed in the context of electro-static connectors [73.30]. For self-reconfigurable robots there is also likely to be an algorithmic solution where modules change roles over time and, hence, distribute the energy consumption equally among modules over time.

For recharging, a solution is to run an energy bus through the robot that both allows modules to recharge their batteries and run off an external power supply [73.33]. It has also been proposed for more stationary applications that modules do not have onboard batteries, but are powered by the external power supply. A way to achieve this while still giving the modules some autonomy of movement is to charge them through the floor; this has been investigated both mechatronically [73.15, 34] and algorithmically [73.35]. A more flexible, but challenging approach is to allow a subset of robot modules to return to the charger and then return to charge the remaining modules [73.36].

At the more explorative end of the spectrum there may be interesting possibilities in wireless energy transfer [73.37], solar energy, or other alternative forms of obtaining or harvesting energy.

73.4 Distributed Control

Reconfigurable robots were born out of the distributed autonomous robot systems community and there has, therefore, been a focus on distributed control algorithms since the early beginning. The reason why distributed control is such a good match for reconfigurable robots is that, if designed well, they have characteristics such as robustness and scalability. Robustness in this context refers to the ability of the control system and, hence, the robot to continue to function despite module failures and communication errors. This may seem as a relatively modest advantage, however, it turns out to be crucial.

Communication systems on reconfigurable robots tend to be unreliable because communication in the case of wired communication has to be passed across the physical connector between modules and the connection may not have perfectly connected and, hence, the electrical terminals for passing the communication signals have not made a completely stable physical connection. It may also be that dust has temporarily ruined the physical interface between the two modules, not allowing electrical communication signals to pass through. For these reasons, reconfigurable robots often rely on wireless communication in the form of either infrared communication or more global forms of wireless communication such as Bluetooth. However, this does not solve the problem, it just changes it. For infrared communication the transmitter and receiver on modules that are to communicate may not be aligned perfectly. There may be crosstalk caused by reflections of signals that cause interference between signals and even cause modules to receive messages that were not for them in the first place [73.38]. Communication relying on electro-magnetic waves do not have these problems, but then often the interference between modules and even background wireless signals can cause communication to fail. The point here is that communication errors cannot only be attributed to poor design or the immature nature of module prototypes, but are fundamental problems that the algorithms have to be able to handle robustly.

Scalability is the other advantage of distributed control algorithms. The motivation here is that while current reconfigurable robots consist of tens of modules, the ambition is that eventually we will have robots consisting of hundreds, thousands, or maybe even millions of modules. It is, therefore, important that the controller does not rely on a central module for control, since this module would be both a bottleneck for the responsiveness of the system and also be a single point of failure. Therefore, scalability of control algorithms is crucial, and distributed control algorithms have the potential to provide just that. Also, it is important here to understand that algorithms also have to be able to deal with module failure, since as the number of modules increases the chance of module failures and communication errors increase. In fact, given a high enough number of modules, modules will fail. Assume that the probability that a single module fails is p_1, the probability that one module out of n fails, p_n is given by

$$p_n = 1 - (1 - p_1)^n . \tag{73.1}$$

This is a very basic consideration, but it is important to understand that the probabilities are working against a controller that is not fault tolerant. For example, given ten modules with just a 1% probability of failing, the probably of one of them failing is 9.6%. Given this background we come to realize that distributed control algorithms are not just a luxury, but absolutely required if we are to realize reconfigurable robots.

73.4.1 Communication

A fundamental basis for all distributed control systems is the supporting communication system. In reconfigurable robots there are local, global, and hybrid communication systems, see Fig. 73.4a-c, respectively.

Local communication systems are based on module to module communication based on, for instance, infrared transceivers. While fairly primitive, this form of communication is essential in reconfigurable robots

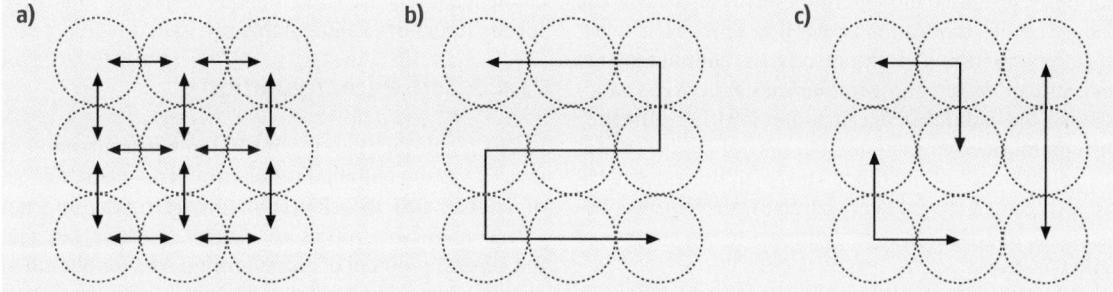

Fig. 73.4a–c The underlying communication models used in reconfigurable robots are (**a**) local, (**b**) global, or (**c**) hybrid (after [73.39])

because it allows two modules to determine their relative positions and orientations. This is not possible using global communication systems independently of whether they are wired or wireless. Local communication also scales since there is no common communication medium that becomes saturated.

Global communication systems using wired bus systems or wireless communication are also useful for real-time control of the reconfigurable robot. Because in a purely local communication system there may significant lag even if the communication volume is low, because messages may have to travel many links to arrive. Hence, an optimal design is one that has both a local communication system to support topology discovery and a global communication system to perform high-speed global coordination.

An alternative is hybrid communication [73.39]. The idea behind this is to make a bus system whose topology can be changed dynamically. Initially, modules can connect to neighbors to discover the local topology. As the need arises modules may connect or disconnect, i.e., reconfigure, their busses to match a given communication load and distribution across the system. While the idea seems to hold potential, it has not been thoroughly investigated.

73.4.2 Locomotion

One of the basic tasks of a reconfigurable robot is locomotion, hence let us take a look a some of the algorithms which have been proposed for controlling locomotion. One of the first distributed control algorithms was gait control tables [73.11, 40] (Fig 73.5). Despite being very simple, the algorithm is a powerful demonstration that often practical and robust algorithms are more useful than theoretically sound algorithms.

Each cell in a control table corresponds to the position of one actuator of one module of a specific time interval. The column identifies the actuator and the row identifies the time interval. The algorithm is based on the assumption that all modules are synchronized. When the algorithm is activated each module moves its actuators to the position identified in the first row of the gait control table. It then waits until the time interval has passed and then move actuators to the position identified by the second row and so on. When the end of the gait control table is reached the controller loops back to the first row.

Gait control tables are a simple form of distributed control since they only work with the specific number of modules for which they were designed and they make the relatively large assumption that all modules' clocks are synchronized. There is no way around the first limitation; however, the second one in practice often holds long enough to make successful experiments relying on modules being initialized at the same time. While be-

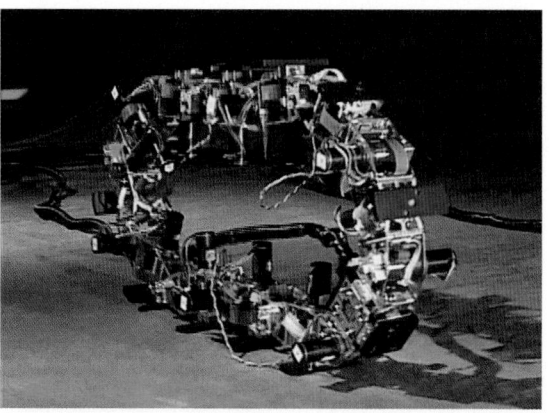

Fig. 73.5 The PolyBot robot in a loop configuration controlled by a gait control table (after [73.41])

ing fairly primitive the algorithm has been successful, and given our motivation above it is fairly clear why: it is very robust. It does not rely on communication and, in fact there is no communication between modules so the failure of one module does not influence the control of other modules, but may of course reduce the performance of the robot. However, it is often what we refer to as graceful degradation of performance, because the degradation is proportional to the number of failed modules.

Above we have described gait control tables in their most basic form, however, in reality each cell of a control table may refer to a general behavior instead of a specific actuator position. For example, it is possible that an actuator should implement a spring-like function, be turned off completely, or be completely stiff. In this way, the behavior of individual modules can be influenced by the behavior of other modules and the environment through which the robot is navigating.

Theoretically, the main problem of gait control tables is that modules have no mechanism to stay synchronized and, hence, over time modules will lose synchronization. Another more serious consequence of this is that the robot as a whole cannot react to the environment as a whole and, for instance, change locomotion pattern or shape. One solution to this problem is represented by hormone-based control algorithms [73.42]. Slightly simplified, the idea is that before each module executes a row of the gait control table a synchronizing hormone is passed through the robot. This ensures that modules stay synchronized and, in addition, it is also possible to pass different hormones to reflect different desired locomotion patterns. While theoretically well-developed hormone-based control slows down the robot due to the overhead of synchronization and even worse if synchronization hormones are lost the robot may stop for a while until a new hormone is generated. Role-based control [73.43] is a compromise between the two. The main idea of role-based control is to have a looser coupling between action and synchronization. The modules have autonomy like in gait control tables and achieve synchronization over time. However, the robot is not able to react globally as fast as hormone-based controllers because synchronization signals are slow compared to the movement of the robot.

These algorithms are mainly suited for open-looped control. However, an important challenge is to understand how to adapt locomotion patterns and configuration to the environment. This is a less explored challenge [73.44, 45] and is a very important challenge for the future of reconfigurable robots.

73.4.3 Self-Reconfiguration

A challenge that has received significant attention is that of self-reconfiguration control. This challenge is, of course, tied to self-reconfigurable robots and not to reconfigurable robots in general. It turns out that the general problem of reconfiguring one configuration into another is computationally intractable. In fact, it is currently believed that to find the optimal solution is NP-hard [73.46]. However, it is not entirely clear if there exists a subspace of the problem where it is computationally more tractable. There may be special cases where this is the case, but we currently do not know. The current status is that self-reconfiguration control remains difficult, in particular because we also aim for distributed, not just centralized, algorithms for solving this believed-to-be NP-hard problem. In Fig. 73.6 we shown an example of a short self-reconfiguration sequence.

Definition

One way to define the distributed self-reconfiguration problem is:

> Given a start configuration A and a final configuration B, distributedly find and execute a sequence of disconnections, moves, and connections that transforms A into B.

This formulation shies away from the optimality criteria because, in practice, good-enough solutions are what we are after. Of course, optimal solutions would be better, but given that the problem is NP-hard we cannot find them efficiently. A likely false assumption this formulation implies is that configurations A and B are known. While this is true for very simple cases where the self-reconfigurable robot is to transform itself into a pre-specified object like a chair, in general the robot should be able to distributedly discover suitable configurations. That is, the final configuration B is often not known beforehand. This leads to the flip-side of the self-reconfiguration problem, which has only been addressed to a limited degree:

> Given a start configuration A, a task T, and an environment E, distributedly find a configuration B better suited for task T in environment E.

It is likely that the split between configuration discovery and self-reconfiguration control is not entirely productive and thus maybe a formulation like this is better:

Given a start configuration A, a task T, and an environment E, distributedly find an action that makes configuration A better suited for task T in environment E.

A final comment on the problem formulation is that we may need all three variations of the problem. The last formulation is useful for incremental improvement, but occasionally the robot has to go through a paradigm shift, e.g., from a snake to a walking robot, and to achieve this we need the first two formulations of the problem.

Fig. 73.6 A self-reconfiguration sequence that transforms M-TRAN from a walker to a snake configuration (after [73.47])

Algorithms

A self-reconfiguration algorithm consists of a representation of the final configuration and a movement strategy.

The movement strategies that have been researched so far are random movement [73.15], local rules [73.48], coordinate attractors [73.49], gradients [73.50], and recruitment [73.51]. The most conceptually simple algorithm is one where modules know the global coordinates of the positions contained in the goal configuration. In this case, modules can move around randomly and stop when they are at a coordinate which is contained in the goal configuration. While this strategy is attractive due to its simplicity, it has a number of drawbacks. In a three-dimensional self-reconfigurable robot random movement is dangerous because modules may by accident disconnect from the structure and fall down. While this problem may be solved by building sturdy, soft modules that are able to reattach to the structure after a fall it is a difficult solution (at least nobody has attempted it so far). Another problem is that when modules settle randomly hollow subspaces or sealed off caves may be created where modules cannot enter. Again, in practice this may not be a problem since these subspaces are likely to be relatively small. Finally, and this is probably the least problematic, random walk is inefficient for a large number of modules and self-reconfiguration sequences consisting of a large number of actions. In other words, a movement strategy based on random walk has scalability issues. Coordinate attractors, gradients, and recruitments are all designed to improve scalability. The idea behind coordinate attractors is that modules that have reached a coordinate contained in the goal configuration that through local sensing discover that an adjacent position in the goal configuration is unfilled and can broadcast this coordinate to all the modules to attract free modules to this location.

Gradients are again an improvement over this strategy because coordinate attractors are prone to local minima. That is, there may be no direct path from a free module to the free goal position. A gradient-based strategy does not broadcast a coordinate, but communicates an integer to neighbor modules; these communicate this integer minus 1 to their neighbors, and so on. Modules listen for integers and pass the highest one they have received minus 1 on. Once this process is complete, free modules can climb the gradient to find the location of the available goal position and, importantly, they can do so by following the structure of the robot and avoid local minima (this is the strategy used in the self-reconfiguration sequence shown in Fig. 73.7).

Finally, the recruitment strategy is a more conservative version of gradients because here the module next to the goal position sends out a single message whose purpose it is to recruit a single free module for the unfilled goal position that it knows about. Whether it pays off to be conservative or not is a matter of priorities. However, it may often be a good strategy to attract more modules because where one module is needed probably more are needed later, and then accept the movement overhead when this is not the case. The final movement strategy is based on local rules. These rules only allow a module to move if the local configuration satisfies a rule in a rule set. In this case, the rule will fire and an action will be executed. By defining the local rules cleverly the resulting configurations can be constrained. The movement strategy of local rules is typically used to control cluster-flow or water-flow locomotion where modules from the back of the robot move towards the front, resulting in a forward locomotion of the robot [73.52].

All these movement strategies, except for local rules, rely on a representation of the goal configuration. A simple representation is to represent all the goal coordinates in the final configuration. However, given that all modules need a copy of this representation it is important that it is space efficient and, therefore, a direct representation like this is only suited for small configurations. Another representation used is one of overlapping cubes [73.50], but other representations could also be used. Typically, which representation to use depends on a trade-off between space and computational complexity. It is also possible to have indirect

representations that code for growth patterns instead, but these are explored to a lesser degree.

The standing challenge for self-reconfiguration is to make algorithms that are practical to use on physical self-reconfigurable robots. The range of algorithms covered here is sufficient for robots consisting of tens of modules and, thus the self-reconfiguration challenge is currently more of a mechatronics problem than an algorithmic problem [73.53]; however there is certainly room for algorithmic improvements as well.

73.4.4 Manipulation

Manipulation is another task that is suitable for reconfigurable robots. If modules are connected in a chain configuration, they can form a serial manipulator with properties not unlike those of traditional robot manipulators. However, two important problems have to be solved before they can work as a traditional robot manipulator.

One problem is how to calculate the inverse kinematics for a chain of modules. The other is how to increase the strength of a modular manipulator since it is relatively weak because of the limited actuator strength of the individual module. Inverse kinematics provide a way to calculate the position of the internal joint angles of the module needed for the outermost module to reach a given position and orientation in space [73.54]. Several answers to the question of how to calculate inverse kinematics have been explored. One option is to fit the serial chain of modules to a curve [73.55]. Another is to use constrained optimization techniques [73.56]. A third option is to use a fast method based on what is defined as dexterous workspaces of subchains [73.57]. An important aspect of this work is also the potential for the robot to dis-

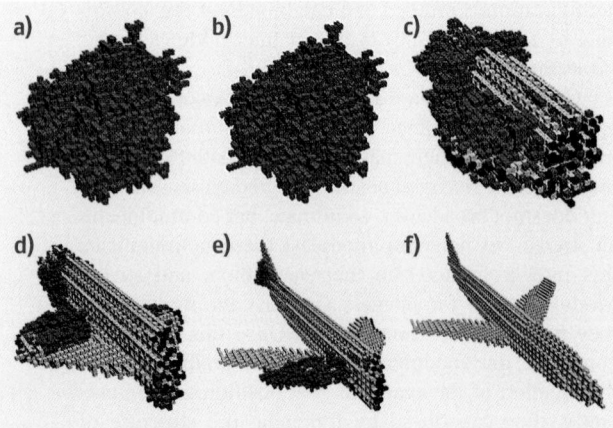

Fig. 73.7a–f Simulated, large-scale self-reconfiguration

Fig. 73.8 The ATRON hybrid self-reconfigurable robot

cover its own kinematics [73.58]. However, maintaining a correspondences between the kinematic model and the physical robot remains a significant challenge for these approaches given that a chain of modules is often not rigid enough in part due to the connectors.

The other problem that needs to be addressed is the relative weakness of the modules and, as a consequence, the modular manipulator as a whole. We need to find a way to make modules work together to produce cooperative actuation that allows them to produce larger forces than those that the modules can produce individually. This question is a little harder to answer, but one option is to exploit the large mechanical advantages formed near singularities of the joints [73.59]. The idea is that a closed loop of modules forms a manipulator. If specific sets of modules are alternately moved and locked, the chain of modules can continually maintain a large mechanical advantage and then generate a much larger force than the one an individual module can provide. However, prob-

lems remain with internal forces of the chain and the weight of the modules involved. Alternative, mostly theoretical approaches include a biologically inspired approach to mechanical design where modules form the equivalent of bones, muscles, and tendons [73.60] and use these construction elements to propagate forces to where they are needed in the configuration. It may also be possible to coordinate the movement of all modules so that the movements add up to produce a larger movement on the global level, i.e., perform collective actuation [73.61].

Besides the traditional arm-based approach to manipulation, it is also possible to use reconfigurable robots as a distributed actuator array [73.62]. In this approach, modules are spread over a surface. When an object is placed on the surface, many modules can work together to manipulate it. The array can handle heavy objects as long as their surface area is relatively large. In combination with self-reconfiguration, it is also possible to use this approach in three dimensions [73.63].

73.5 Programmability and Debugging

An area that is receiving more and more focus in the community is the challenge of how to efficiently program and debug reconfigurable robots. In the previous section, we discussed how to control reconfigurable robots distributedly; while these distributed control algorithms are desired for their robustness and scalability they are notoriously difficult to implement and debug in general, but even more so on modular robots that are resource constrained, embedded platforms that often only allow for debugging output in the form of blinking LEDs.

73.5.1 Iterative, Incremental Programming and Debugging

Conventionally, the challenge is met by developing applications for reconfigurable robots using an iterative, incremental approach to programming and debugging. This ensures that we can locate errors relatively quickly and correct them before we introduce additional functionality and hence complexity. A good programming practice is also to develop an application programming interface that hides the low-level hardware interface for the programmer more interested in the higher-level control algorithms. This conventional approach is suitable for small demonstration programs, but as the complex-

ity of the task and thus the program increases, this approach becomes intractable. The main reason is that it becomes increasingly difficult to obtain reliable debugging information due to the distributed and dynamic nature of reconfigurable robots and also more often than not due to the immaturity of the physical platforms.

73.5.2 Simulation

Given the shortcomings of iterative, incremental development, researchers use simulations, e.g., [73.64], to ensure that the distributed and dynamic aspects of the controller are thoroughly debugged before being deployed on the physical platform. However, building a reliable simulator is a feat in its own right. The simplest form of simulators are logic in nature where events are discrete and instantaneous in time, e.g., the transmission of a message or the movement of a module from one lattice position to another. These logic-based simulators are useful for experimentally convincing oneself that the logic of the distributed algorithm under development is correct. If this algorithm is then transferred to the physical, reconfigurable robot and allowed to run on a carefully debugged application programming interface, it is only real-world issues and hardware limitations that stand in the way of success.

However, these issues and limitations should not be underestimated and include, but are not limited to: unreliable communication, limited communication bandwidth, time, parallelism, hardware failures, differences between modules in terms of actuation, sensing, and communication performance. These issues and limitations can to some degree be handled by proper simulation, but typically are not considered and thus leave algorithms stranded in simulation because they are unable to deal with the real-world constraints of a physical reconfigurable robot. The gap between the simulated world and the real world is often referred to as the reality-gap [73.65]. This gap is widened even more by using simulations based on simplified physics engines, because precise modeling of the physics of a reconfigurable modular robot is almost impossible due to the complex interactions between the modules themselves and the modules and the environment. While the physics-based simulations increase the reality gap, they do allow for a wider area of study, e.g., study of locomotion algorithms. However, often at the cost of reduced transfer of results to the physical reconfigurable robot.

73.5.3 Emerging Solutions

Above we have presented current approaches and their advantages and in particular their disadvantages. It is clear that we are far from meeting the challenge of efficient programming and debugging of physical reconfigurable robots. However, there are currently two approaches under investigation that may provide some solutions.

One is the use of domain specific programming languages [73.66, 67]. The fundamental idea is to expose programming primitives at the level of abstraction preferred by the programmer and hide the implementation of the primitives. For instance, communication is not necessarily central to the programmer and can, therefore, be implicitly handled by the programming language. The advantage of this approach is that it frees the programmer from error-prone, repetitive programming and allows him to focus more energy on the programming challenges related to the task at hand. The language can also to some degree help deal with hardware limitations, allowing the programmer to build on reliable programming primitives. However, there may be a problem of leaky abstractions where it can be hard for a programmer to discover hardware problems because the software that interfaces with the hardware is hidden from the programmer.

Another development is that reconfigurable robots increasingly get more and more powerful processors and communication hardware, a development which is primarily driven and made possible by the cell-phone industry. This development opens an opportunity for efficient programming and debugging of reconfigurable robots. The reason is that up until now reconfigurable robots have been resource constrained to the degree that it was not possible to run programs targeted at debugging in parallel to the executing program because either there was simply not enough available memory and processing energy or the debugging tool would interfere with the executing program to the degree that its behavior would be completely altered or simply not work at all. However, with the increase in processing and communication energy this problem may be reduced, and this will open the door to new forms of programming middleware that has been instrumental to the success of other areas of robotics such as Player/Stage [73.68] and robot operating system ROS [73.69].

In the broader context breakthroughs in programming and debugging of reconfigurable robots will be a significant contribution that can help us develop solutions to complex tasks that are currently beyond our reach.

73.6 Perspective

This concludes the overview of reconfigurable robots. The question is where does this leave us? What does the future of reconfigurable robots and reconfigurable robot research look like?

First of all, the field of reconfigurable robots has matured significantly over the last decade, leading to the first applications of modular robot technology: Cubelets [73.4], a construction kit teaching children about emergent behavior in complex systems. Another example is LocoKit that has been developed to efficiently explore morphology related questions in the context of robot locomotion [73.5, 70]. The community, in general, is very engaged in understanding how reconfigurable robots can be adapted to this specific application, which in time is likely to lead to more applications.

From a research point of view, the vision of an autonomously distributed reconfigurable robot still remains to be realized. This requires advances in all areas covered in this chapter. Like other fields of robotics, reconfigurable robotics is nurtured by the progressive development of rapid prototyping technology and smartphone technology, including wireless charging, which may open the path to novel mechatronic designs. Also, the emerging field of soft robotics may hold potential for radical new designs of reconfigurable robots. Overall, there seems to be a growing opportunity to exploit these advances to design a new generation of reconfigurable robots.

In the area of distributed control, which is probably the best understood area of reconfigurable robots, there

is a clear understanding of how the modules of a reconfigurable robot can be coordinated internally. However, there is still a significant open challenge in making reconfigurable robots adapt to and interact with unknown environments through sensors. Also, a more engineering oriented area of programmability and debugging is crucial for the field if we are to handle the complexity of reconfigurable robots more efficiently.

The potential of autonomously distributed reconfigurable robots is as exciting as ever, and their realization today appears much more realistic than when the idea was conceived 25 years ago, thanks to the hard work of the community. However, there are still important discoveries to be made before the potential of autonomously distributed reconfigurable robots is realized.

73.7 Further Reading

This chapter has provided a high-level overview of the field of reconfigurable robots. Those readers interested in a complementary introduction should con-

sider reading [73.71] and those needing a more detailed introduction are referred to the books [73.72, 73].

References

73.1   T. Fukuda, T. Ueyama: *Cellular Robotics and Micro Robotics Systems* (World Scientific, Singapore 1994)
73.2   R.H. Kessin: Cell motility: Making streams, Nature **422**, 482 (2003)
73.3   M. Yim, B. Shirmohammadi, J. Sastra, M. Park, M. Dugan, C.J. Taylor: Towards robotic self-reassembly after explosion, Proc. IEEE/RSJ Int. Conf. Intell. Robot. Syst. (2007) pp. 2767–2772
73.4   E. Schweikardt et al.: Cubelets robot construction kit, https://modrobotics.com/ (2015)
73.5   J.C. Larsen, D. Brandt, K. Stoy: LocoKit: A construction kit for building functional morphologies for robots, Proc. 12th Int. Conf. Adapt. Behav. (2012) pp. 12–24
73.6   T. Fukuda, Y. Kawauchi, M. Buss: Self organizing robots based on cell structures – CEBOT, Proc. IEEE/RSJ Int. Workshop Intell. Robot. Syst. (1988) pp. 145–150
73.7   R. Groß, M. Bonani, F. Mondada, M. Dorigo: Autonomous self-assembly in swarm-bots, IEEE Trans. Robot. **22**(6), 1115–1130 (2006)
73.8   M.D.M. Kutzer, M.S. Moses, C.Y. Brown, D.H. Scheidt, G.S. Chirikjian, M. Armand: Design of a new independently-mobile reconfigurable modular robot, Proc. IEEE Int. Conf. Robot. Autom. (2010) pp. 2758–2764

73.9   G.G. Ryland, H.H. Cheng: Design of imobot, an intelligent reconfigurable mobile robot with novel locomotion, Proc. IEEE Int. Conf. Robot. Autom. (2010) pp. 60–65
73.10   M. Yim: A reconfigurable modular robot with many modes of locomotion, Proc. JSME Int. Conf. Adv. Mechatron. (1993) pp. 283–288
73.11   M. Yim: Locomotion with a Unit-Modular Reconfigurable Robot, Ph.D. Thesis (Department of Mechanical Engineering, Stanford University, Stanford 1994)
73.12   M. Yim, D.G. Duff, K. Roufas, Y. Zhang, C. Eldershaw: Evolution of PolyBot: A modular reconfigurable robot, Proc. Harmon. Drive Int. Symp. (2001)
73.13   A. Castano, R. Chokkalingam, P. Will: Autonomous and self-sufficient CONRO modules for reconfigurable robots, Proc. 5th Int. Symp. Distrib. Auton. Robot. Syst. (2000) pp. 155–164
73.14   B. Salemi, M. Moll, W.-M. Shen: SuperBot: A deployable, multi-functional, and modular self-reconfigurable robotic system, Proc. IEEE/RSJ Intl. Conf. Intell. Robot. Syst. (2006) pp. 3636–3641
73.15   S. Murata, H. Kurokawa, S. Kokaji: Self-assembling machine, Proc. IEEE Int. Conf. Robot. Autom. (1994) pp. 441–448

73.16 G.S. Chirikjian: Kinematics of a metamorphic robotic system, Proc. IEEE Int. Conf. Robot. Autom. (1994) pp. 449–455

73.17 K. Kotay, D. Rus, M. Vona, C. McGray: The self-reconfiguring robotic molecule, Proc. IEEE Int. Conf. Robot. Autom. (1998) pp. 424–431

73.18 S. Murata, H. Kurokawa, E. Yoshida, K. Tomita, S. Kokaji: A 3-d self-reconfigurable structure, Proc. IEEE Int. Conf. Robot. Autom. (1998) pp. 432–439

73.19 H. Kurokawa, K. Tomita, A. Kamimura, S. Kokaji, T. Hasuo, S. Murata: Distributed self-reconfiguration of M-TRAN III modular robotic system, Int. J. Robot. Res. 27(3/4), 373–386 (2008)

73.20 E.H. Østergaard, K. Kassow, R. Beck, H.H. Lund: Design of the ATRON lattice-based self-reconfigurable robot, Auton. Robot. 21(2), 165–183 (2006)

73.21 P. White, V. Zykov, J. Bongard, H. Lipson: Three dimensional stochastic reconfiguration of modular robots, Proc. Robot. Sci. Syst. (2005) pp. 161–168

73.22 J. Bishop, S. Burden, E. Klavins, R. Kreisberg, W. Malone, N. Napp, T. Nguyen: Self-organizing programmable parts, Proc. Int. Conf. Intell. Robot. Syst. (2005) pp. 3684–3691

73.23 P.J. White, M. Yim: Reliable external actuation for full reachability in robotic modular self-reconfiguration, Int. J. Robot. Res. 29(5), 598–612 (2010)

73.24 K. Gilpin, A. Knaian, D. Rus: Robot pebbles: One centimeter modules for programmable matter through self-disassembly, Proc. IEEE Int. Conf. Robot. Autom. (2010) pp. 2485–2492

73.25 V. Zykov, A. Chan, H. Lipson: Molecubes: An open-source modular robotics kit, Proc. IEEE/RSJ Int. Conf. Robot. Syst., Self-Reconfig. Robot. Workshop (2007)

73.26 A. Lyder, R.F.M. Garcia, K. Stoy: Mechanical design of ODIN, an extendable heterogeneous deformable modular robots, Proc. IEEE/RSJ Int. Conf. Int. Robot. Syst., Nice (2008) pp. 883–888

73.27 A. Lyder, K. Stoy, R.F.M. Garciá, J.C. Larsen, P. Hermansen: On sub-modularization and morphological heterogeneity in modular robotics, Proc. 12th Int. Conf. Intell. Auton. Syst. (2012) pp. 1–14

73.28 Y. Terada, S. Murata: Automatic modular assembly system and its distribution control, Int. J. Robot. Res. 27, 445–462 (2008)

73.29 W.-M. Shen, R. Kovac, M. Rubenstein: SINGO: A single-end-operative and genderless connector for self-reconfiguration, self-assembly and self-healing, Proc. IEEE/RSJ Int. Conf. Intell. Robot. Syst., Workshop Self-Reconfig. Robot., Syst. Appl. (2008) pp. 64–67

73.30 M.E. Karagozler, J.D. Campbell, G.K. Fedder, S.C. Goldstein, M.P. Weller, B.W. Yoon: Electrostatic latching for inter-module adhesion, power transfer, and communication in modular robots, Proc. IEEE/RSJ Int. Conf. Intell. Robot. Syst. (2007) pp. 2779–2786

73.31 A. Ishiguro, M. Shimizu, T. Kawakatsu: Don't try to control everything! An emergent morphology control of a modular robot, Proc. IEEE/RSJ Int. Conf. Intell. Robot. Syst. (2004) pp. 981–985

73.32 M.W. Jørgensen, E.H. Ostergaard, H.H. Lund: Modular ATRON: Modules for a self-reconfigurable robot, Proc. IEEE/RSJ Int. Conf. Robot. Syst. (2004) pp. 2068–2073

73.33 R.F.M. Garcia, A. Lyder, D.J. Christensen, K. Stoy: Reusable electronics and adaptable communication as implemented in the ODIN modular robot, Proc. IEEE Int. Conf. Robot. Autom. (2009)

73.34 B. Kirby, B. Aksak, J. Hoburg, T. Mowry, P. Pillai: A modular robotic system using magnetic force effectors, Proc. IEEE/RSJ Int. Conf. Intell. Robot. Syst. (2007) pp. 2787–2793

73.35 J. Campbell, P. Pillai, S.C. Goldstein: The robot is the tether: Active, adaptive power routing for modular robots with unary inter-robot connectors, Proc. IEEE/RSJ Int. Conf. Intell. Robot. Syst. (2005) pp. 4108–4115

73.36 S. Kernbach, O. Kernbach: Collective energy homeostasis in a large-scale micro-robotic swarm, Robot. Auton. Syst. 59, 1090–1101 (2011)

73.37 M.P.O. Cabrera, R.S. Trifonov, G.A. Castells, K. Stoy: Wireless communication and power transfer in modular robots, Proc. IROS Workshop Reconfig. Modul. Robot. (2011)

73.38 D.J. Christensen, U.P. Schultz, D. Brandt, K. Stoy: Neighbor detection and crosstalk elimination in self-reconfigurable robots, Proc. 1st Int. Conf. Robot Commun. Coord. (2007)

73.39 R.F.M. Garcia, D.J. Christensen, K. Stoy, A. Lyder: Hybrid approach: A self-reconfigurable communication network for modular robots, Proc. 1st Int. Conf. Robot Commun. Coord. (2007) pp. 23:1–23:8

73.40 M. Yim: New locomotion gaits, Proc. Int. Conf. Robot. Autom. (1994) pp. 2508–2514

73.41 M. Yim, Y. Zhang, K. Roufas, D.G. Duff, C. Eldershaw: Connecting and disconnecting for chain self-reconfiguration with PolyBot, IEEE/ASME Trans. Mechatron. 7(4), 442 (2002)

73.42 W.-M. Shen, B. Salemi, P. Will: Hormone-inspired adaptive communication and distributed control for conro self-reconfigurable robots, IEEE Trans. Robot. Autom. 18(5), 700–712 (2002)

73.43 K. Støy, W.-M. Shen, P. Will: Using role based control to produce locomotion in chain-type self-reconfigurable robot, IEEE Trans. Mechatron. 7(4), 410–417 (2002)

73.44 K. Støy, W.-M. Shen, P. Will: On the use of sensors in self-reconfigurable robots, Proc. 7th Int. Conf. Simul. Adapt. Behav. (2002) pp. 48–57

73.45 A. Kamimura, H. Kurokawa, E. Yoshida, S. Murata, K. Tomita, S. Kokaji: Automatic locomotion design and experiments for a modular robotic system, IEEE/ASME Trans. Mechatron. 10(3), 314–325 (2005)

73.46 F. Hou, W.-M. Shen: On the complexity of optimal reconfiguration planning for modular reconfigurable robots, Proc. IEEE Int. Conf. Robot. Autom. (2010) pp. 2791–2796

73.47 H. Kurokawa et al. (2010): M-TRAN (Modular Transformer), Research, https://unit.aist.go.jp/is/frrg/dsysd/mtran3/research.htm

73.48 Z. Butler, K. Kotay, D. Rus, K. Tomita: Generic de-centralized control for lattice-based self-reconfigurable robots, Int. J. Robot. Res. **23**(9), 919–937 (2004)

73.49 M. Yim, Y. Zhang, J. Lamping, E. Mao: Distributed control for 3-D metamorphosis, Auton. Robot. **10**(1), 41–56 (2001)

73.50 K. Støy, R. Nagpal: Self-reconfiguration using directed growth, Proc. Int. Conf. Distrib. Auton. Robot Syst. (2004) pp. 1–10

73.51 Z. Butler, R. Fitch, D. Rus: Experiments in distributed control of modular robots, Proc. Int. Symp. Exp. Robot. (2003) pp. 307–316

73.52 E.H. Østergaard, H.H. Lund: Distributed cluster walk for the ATRON self-reconfigurable robot, Proc. 8th Conf. Intell. Auton. Syst. (2004) pp. 291–298

73.53 K. Stoy, H. Kurokawa: Current topics in classic self-reconfigurable robot research, Proc. IROS/RSJ Workshop Reconfig. Modul. Robot. (2011)

73.54 J.J. Craig: *Introduction to Robotics: Mechanics and Control*, 3rd edn. (Prentice Hall, Reading 2003)

73.55 G.S. Chirikjian, J.W. Burdick: The kinematics of hyper-redundant robot locomotion, IEEE Trans. Robot. Autom. **11**(6), 781–793 (1995)

73.56 Y. Zhang, M. Fromherz, L. Crawford, Y. Shang: A general constraint-based control framework with examples in modular self-reconfigurable robots, Proc. IEEE/RSJ Int. Conf. Intell. Robot. Syst. (2002) pp. 2163–2168

73.57 S.K. Agrawal, L. Kissner, M. Yim: Joint solutions of many degrees-of-freedom systems using dextrous workspaces, Proc. IEEE Int. Conf. Robot. Autom. (2001) pp. 2480–2485

73.58 M. Bordignon, U.P. Schultz, K. Stoy: Model-based kinematics generation for modular mechatronic toolkits, Proc. 9th Int. Conf. Gener. Progr. Compon. Eng. (2010) pp. 157–166

73.59 M. Yim, D. Duff, Y. Zhang: Closed chain motion with large mechanical advantage, Proc. IEEE/RSJ Int. Conf. Intell. Robot. Syst. (2001) pp. 318–323

73.60 D.J. Christensen, J. Campbell, K. Stoy: Anatomy-based organization of morphology and control in self-reconfigurable modular robots, Neural Comput. Appl. **19**(6), 787–805 (2010)

73.61 J. Campbell, P. Pillai: Collective actuation, Int. J. Robot. Res. **27**(3/4), 299–314 (2007)

73.62 M. Yim, J. Reich, A. Berlin: Two approaches to distributed manipulation. In: *Distributed Manipulation*, ed. by H. Choset, K. Bohringer (Kluwer Academic, Boston 2000) pp. 237–260

73.63 J. Kubica, A. Casal, T. Hogg: Agent-based control for object manipulation with modular self-reconfigurable robots, Proc. Int. Jt. Conf. Artif. Intell. (2001) pp. 1344–1352

73.64 D. Christensen, U.P. Schultz, D. Brandt, K. Stoy: A unified simulator for self-reconfigurable robots, Proc. IEEE/RSJ Int. Conf. Intell. Robot. Syst. (2008) pp. 870–876

73.65 N. Jakobi, P. Husbands, I. Harvey: Noise and the reality gab: The use of simulation in evolutionary robotics, Adv. Artif. Life: Proc. Third Eur. Conf. Artif. Life (1995) pp. 704–720

73.66 U.P. Schultz, M. Bordignon, K. Stoy: Robust and reversible execution of self-reconfiguration sequences, Robotica **29**(1), 35–57 (2011)

73.67 M.P. Ashley-Rollman, P. Lee, S.C. Goldstein, P. Pillai, J.D. Campbell: A language for large ensembles of independently executing nodes, Proc. Int. Conf. Log. Progr. (2009) pp. 265–280

73.68 B.P. Gerkey, R.T. Vaughan, K. Støy, A. Howard, G.S. Sukhatme, M.J. Mataric: Most valuable player: A robot device server for distributed control, Proc. IEEE/RSJ Int. Conf. Intell. Robot. Syst. (2001) pp. 1226–1231

73.69 M. Quigley, K. Conley, B. Gerkey, J. Faust, T.B. Foote, J. Leibs, R. Wheeler, A.Y. Ng: ROS: An open-source robot operating system, Proc. ICRA Workshop Open Source Softw. (2009)

73.70 J. C. Larsen et al.: LocoKit: Robots that move, http://www.locokit.sdu.dk (2013)

73.71 M. Yim, W.-M. Shen, B. Salemi, D. Rus, M. Moll, H. Lipson, E. Klavins, G.S. Chirikjian: Modular self-reconfigurable robot systems, IEEE Robot. Autom. Mag. (2007) pp. 43–52

73.72 S. Murata, H. Kurokawa: *Self-Organizing Robots* (Springer, Berlin, Heidelberg 2012)

73.73 K. Støy, D.J. Christensen, D. Brandt: *Self-Reconfigurable Robots: An Introduction* (MIT, Cambridge 2010)

74. Probabilistic Modeling of Swarming Systems

Nikolaus Correll, Heiko Hamann

This chapter provides on overview on probabilistic modeling of swarming systems. We first show how population dynamics models can be derived from the master equation in physics. We then present models with increasing complexity and with varying degrees of spatial dynamics. We will first introduce a model for collaboration and show how macroscopic models can be used to derive optimal policies for the individual robot analytically. We then introduce two models for collective decisions; first modeling spatiality implicitly by tracking the number of robots at specific sites and then explicitly using a Fokker–Planck equation. The chapter is concluded with open challenges in

combining non-spatial with spatial probabilistic modeling techniques.

74.1 From Bioligical to Artificial Swarms

The swarming behavior of ants, wasps, and bees demonstrates the emergence of stupendously complex spatio–temporal patterns ranging from a swarm finding the shortest paths to the assembly of three-dimensional structures with intricate architecture and well-regulated thermodynamics [74.1, 2]. In the bigger scheme of things, these systems represent just the tip of the iceberg; their behavior is considerably less complex than that of the brain, cities, or galaxies, all of which are essentially swarming systems (and all of which can be reduced to first principles and interactions on atomic scale). Yet, social insects make the world of self-organization accessible to us as they are comparably easy to observe. Studying these systems is interesting from an engineering perspective as they demonstrate how collectives can transcend the abilities of the individual member and let the organism as a whole exhibit cognitive behavior.

Cognition is derived from the Latin word *cognescere* and means *to know*, *to recognize*, and also to *conceptualize*. In the human brain, cognition emerges – to the best of our knowledge – from the complex interactions of highly connected, large-scale distributed neural activity. We argue that *cognition* can manifest itself at multiple different levels of complexity, ranging from conceptualizing collective decisions such as assuming a certain shape or deciding between different abstract choices in social insects to reasoning on complex problems and expressing emotions in humans, the combination of the latter two often framed as the *Turing test* in artificial intelligence. This chapter aims at developing formal models to capture the characteristic properties of the most simple cognitive primitives in swarming systems. In particular, we wish to understand the relationship between the activities of the individual member of the swarm and the dynamics that arise at collective level. The resulting models can be matched to data recorded from physical systems, be used to predict the outcome of a robot's individual behavior on a larger swarm, and used in an optimization framework to determine the best parameters that help improve a certain metric [74.3].

This chapter reviews probabilistic models of three swarming primitives that are examples of conceptualizations that are exclusively represented at the collective level: collaboration, collective decision making, and collective optimization. Guided by examples from social insects, we present models that generalize to arbitrary agent systems and can serve as building blocks for more complex systems. The probabilistic component of the models arises from:

1. The agent's motion, which often has a random component
2. Explicit random decisions made by individual agents
3. Random encounters between agents.

Randomness in an agent's motion can be introduced, for example, by physical properties such as slip, by deficits in robot hardware, or by explicitly explorative behavior, e.g., based on random turns. It is, therefore, reasonable to model at least the single-agent behavior with probabilistic methods. Yet, it is possible to model the expected swarm-level behavior using deterministic models. In such a swarm-level model the underlying stochastic motion of agents is summarized in macroscopic properties, which are averages such as the expected swarm fraction in a certain state or at a certain position [74.4–7].

Such probabilistic models are in contrast with deterministic models of swarming systems, which explicitly model the positions of individual robots. Representative examples include controllers for flocking [74.8], consensus [74.9], and optimal sensor distribution for sampling a given probability density function [74.10]. While the robots' spatial distribution is explicitly modeled, those models have difficulties dealing with randomness or robot populations in which robots can be in different states at the same time.

After providing a brief background on phenomenological probabilistic models based on the master equation, this chapter will first review population dynamic models that ignore the spatial distribution of the individual robots and the swarm and then present models that explicitly model the spatial distribution of the robot swarm using time-dependent, spatial probability density functions.

74.2 The Master Equation

Let a robot be in a discrete set of states with probability $p_i \in P$, with P a vector maintaining the probabilities of all possible states and $\sum P = 1$. These states model internal states of the robot, determined by its program, or external states, determined by the state of the robot within its environment. Actions of the robot and environmental effects will change these probabilities. This is captured by a phenomenological set of first-order differential equations, also known as the master equation [74.11],

$$\frac{\mathrm{d}P}{\mathrm{d}t} = \mathbf{A}(t)P , \qquad (74.1)$$

where $\mathbf{A}(t)$ is the transition matrix consisting of entries $p_{ij}(t)$ that correspond to the probability of a transition from state i to state j at time t. Multiplying both sides with the total number of robots N_0 allows us to calculate the average number of robots in each state. For brevity, we write

$$N_i(t) = N_0 p_i(t) .$$

Similarly, when expanding the master equation for a continuous space variable, one finds the Fokker–Planck equation, also known as the Kolmogorov forward equation or the Smoluchowski equation [74.12, 13].

74.3 Non-Spatial Probabilistic Models

We will first consider two models that assume the spatial distribution of the agents in the environment to be uniform: collaboration and collective decision.

74.3.1 Collaboration

An important swarming primitive is *collaboration*, which requires a number of agents (n) to get together at a site. Collaboration is different from the more general *task allocation* problem, in which the number of agents is not explicitly specified. In swarm robotics, a *site* can have spatial meaning, but can also be understood in an abstract way as means to form teams. Although there are many different algorithms for team formation, we focus on a collaboration mechanism that was introduced in the *stick-pulling experiment* [74.14] and turned out to be a recurrent primitive in swarm robotic systems, e.g., in swarm robotic inspection, where robots can serve as temporary markers in the environment [74.15]. Here, collaboration happens when an inspecting robot encounters a marker, which informs it that this specific area has already been inspected. The collaboration model, therefore, finds application in studying trade-offs between serving as *memory* to the swarm and actively contributing to the swarming behavior's metric.

In the stick-pulling experiment N_0 robots are concerned with pulling M_0 sticks out of the ground in a bounded environment. This task requires exactly $n = 2$ robots. Physically, this can be understood as a stick that is too long to be extracted from the ground by a single robot. Rather, every robot that grabs the stick can pull it out a little further and keeps it there until the next robot arrives. In this work, we abstract the classical stick-pulling experiment to a generic collaboration model in which robots are simply required to meet, see also Fig. 74.1. Intuitively, the amount of time spent waiting for collaboration to happen is a trade-off between (1) waiting at a site to find a collaborator and (2) having the chance to find a collaborator oneself by actively browsing the environment. Finding the collaboration rate, and the individual parameters that lead to it, that is *optimal* for a given environment, i.e., the number of collaboration sites and the number of agents, illustrates how probabilistic models can be employed to design this process and find optimal collaboration policies.

The following model is loosely based on the development in [74.16], which applies discrete time difference equations. For simplicity, we assume that collaboration happens instantaneously and focus on a continuous-time representation and stochastic waiting times. The reader is referred to [74.16] for an extensive treatment of deterministic waiting times and [74.17] for an extension to $n > 2$ agents. Variables used in the equations that follow are summarized in Table 74.1.

Let $n_s(t)$ with $n_s(0) = N_0$ be the number of searching agents at time $t \in \mathbb{R}^+$ and N_0 the total number of agents. Let $n_w(t) = N_0 - n_s(t)$ be the number of waiting agents at time t. With p the probability to encounter or match a waiting agent and T_r the average time an agent will wait for collaboration, we can write

$$\dot{n}_s(t) = -p(M_0 - n_w(t))n_s(t) \tag{74.2}$$

$$+ \frac{1}{T_r} n_w(t)$$

$$+ p n_s(t) n_w(t) . \tag{74.3}$$

Thus, $n_s(t)$ decreases by the rate at which searching agents encounter empty collaboration sites (of which there exist $M_0 - n_w(t)$ at time t), and it increases by those agents that return either from unsuccessful (at rate $1/T_r$) or successful collaboration, i.e., find any of the $n_w(t)$ waiting agents.

In order to maximize the collaboration rate in the system we are interested in maximizing the rate at which robots return from successful collaboration, i.e., $c(t) = p n_s(t) n_w(t)$.

Solving for $\dot{n}_s(t) = 0$ and substituting $n_w(t) = N_0 - n_s(t)$ allows us to calculate the number of robots at steady-state n_s^*

$$n_s^* = \frac{\left[\begin{array}{c}(2N_0 - M_0)pT_r - 1 \\ + \sqrt{8N_0 pT_r(1 + (M_0 - 2N_0)pT_r)^2}\end{array}\right]}{4pT_r} . \tag{74.4}$$

As $n_w^* = N_0 - n_s^*$ by definition, we can write

$$c^* = p\left(n_s^* N_0 - n_s^{*2}\right) . \tag{74.5}$$

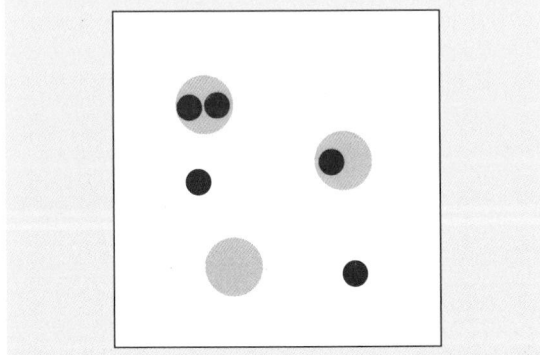

Fig. 74.1 A collaboration example. $N_0 = 5$ robots (*black*) in a bounded environment with $M_0 = 3$ collaboration sites, each requiring $n = 2$ robots to be present simultaneously for collaboration to happen

Table 74.1 Notation used in the collaboration model

$n_s(t)$	Average number of searching agents
$n_w(t)$	Average number of waiting agents
p	Probab. to encounter/match a waiting agent
$c(t)$	Average rate of collaboration matches
N_0	Total number of agents
M_0	Total number of collaboration sites
T_w	Waiting time

The collaboration rate as a function of T_r and N_0 is shown in Fig. 74.2. By solving $dc^*/dn_s^* = 0$, we can calculate $n_{s,\,opt}^* = \frac{1}{2}N_0$ that maximizes c^*. Substituting $n_{s,\,opt}^*$ into (74.4) and solving for T_r, we can calculate the optimal waiting time $T_{r,\,opt}$ as

$$T_{r,\,opt} = \frac{1}{(M_0 - N_0)p} \,. \qquad (74.6)$$

As $T_{r,\,opt}$ cannot be negative, an optimal waiting time can only exist if $M_0 > N_0$. This intuitively makes sense, because if there are less agents than collaboration sites, waiting too long might consume all agents in waiting states. We can also see that the more collaboration sites there are, the less an agent should wait. There are two interesting special cases: first, $N_0 = M_0$. In this case $T_{r,\,opt}$ is undefined. Considering that collaboration sites exceed agents by exactly one, $T_{r,\,opt}$ is fully defined by $1/p$. Thus, the higher the likelihood is that agents find a collaboration site, the lower the waiting time should be. In this case, it makes sense to release searching agents from wait states to find another agent to collaborate. If this likelihood is low, however, agents are better off waiting to serve as collaborators for few searching agents.

With $T_{r,\,opt}$ given by (74.6) we can derive the following guidelines for agent behavior. First, an optimal wait time exists only if there are less agents than collaboration sites. Otherwise, longer waits improve the chance of collaboration. Second, if the number of agents, the number of collaboration sites, and the likelihood to encounter a collaboration site are known to each agent at all times, e.g., due to global communication or shared memory, agents could cal-

Table 74.2 Notation used in the collective decision model

$n_s(t)$	Average number of searching agents
$n_i(t)$	Average number of agents committed to choice i
p_i	Unbiased probability to select choice i
T_i	Unbiased time to stay with choice i
N_0	Total number of agents
M_0	Total number of choices
T_w	Waiting time

culate $T_{r,\,opt}$ at all time. If these quantities are not known, however, agents can estimate these quantities based on their interactions in the environment by observing the rates at which they encounter collaboration and empty sites. Individual agent learning algorithms that accomplish this goal are discussed in detail in [74.18].

74.3.2 Collective Decisions

Another collective intelligent swarming primitive is *collective decisions*. These can be observed in the path selection of ants [74.19] or shelter selection of cockroaches [74.20] or robots [74.21], but can also have non-spatial meaning, for example when a consensus on M_0 different discrete values is needed. An example of such a situation is shown in Fig. 74.3. While the above references provide models that are specific to their application, this chapter provides a generalized model for collective decisions that rely on different ways of social amplification, i.e., a change of the behavior based on the activities of other swarm members, or the absence thereof.

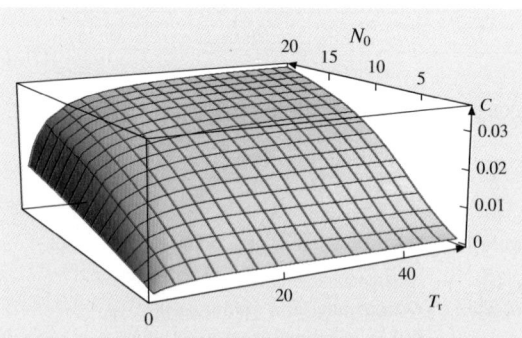

Fig. 74.2 The collaboration rate as a function of T_r and N_0 for $M_0 = 10$ collaboration sites. There exists an optimal T_r for $N_0 < M_0$, whereas the collaboration rate increases steadily otherwise for increasing values of T_r

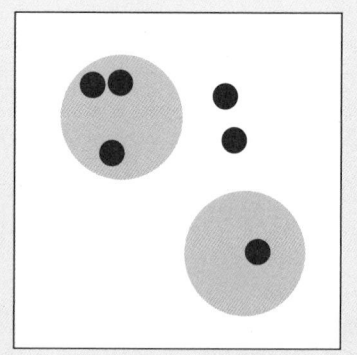

Fig. 74.3 Collective decision example. $N_0 = 6$ robots decide between $M_0 = 2$ choices. Three plus one robot have already made decisions, two robots remain undecided

Part F | 74.3

Model parameters for the collective decision model are summarized in Tab. 74.2. Let $n_s(t)$ with $n_s(0) = N_0$ be the number of searching/undecided agents at time $t \in \mathbb{R}_{\not\sim}^+$ and N_0 the total number of agents. Let p_i, $0 < i \leq M_0$, be the unbiased probability for an agent to select value i from M_0 different values. This probability is unbiased as it does not depend on social amplification. We can then write the following differential equations for the number of agents $n_i(t)$ that have the selected value i

$$\dot{n}_i(t) = n_s(t)p_iR_i(t) - \frac{1}{T_i}n_i(t)Q_i(t) , \quad n_i(0) = 0 ,$$

(74.7)

$$n_s(t) = N_0 - \sum_{i=1}^{M_0} n_i(t) ,$$

(74.8)

where T_i is the average time spent on solution i before resuming search, and $R_i(t), Q_i(t) : n_i(t), n_s(t) \to \mathbb{R}^+$ are functions that might or might not depend on the number of agents in other states, and therefore making the differential equation for $n_i(t)$ linear or non-linear, respectively. There are four interesting cases: both $R_i(t)$ and $Q_i(t)$ being constant, both being functions of one or more states of the system, e.g., $n_i(t)$ or $n_s(t)$, and combinations thereof.

If both $R_i(t)$ and $Q_i(t)$ are constants, one can show that the number of agents selecting choice i at steady-state n_i^* is given by

$$n_i^* = \frac{R_i}{Q_i}n_s^* ,$$

(74.9)

with n_s^* the number of agents that remain undecided at steady-state. (This results from agents discarding choices at rate $1/T_i$.) For example, for a two-choice system, using $n_s^* = N_0 - n_1^* - n_2^*$ yields the steady states

$$n_1^* = \frac{Q_2R_1}{Q_1Q_2 + Q_2R_1 + Q_1R_2} ,$$

(74.10)

$$n_2^* = \frac{Q_1R_2}{Q_1Q_2 + Q_2R_1 + Q_1R_2} .$$

(74.11)

A solution for $R_1 = 0.01$, $R_2 = 0.04$, and $Q_1 = Q_2 = 1/10$ is depicted in Fig. 74.4 and leads to ≈ 7 and $\approx 27\%$ of agents in states one and two, respectively, while most agents remain undecided. In this system, the speed at which the steady-state is reached depends on the values of R_i, with higher values of R_i leading

to faster decisions, whereas the steady-state of undecided agents is determined by Q_i, with lower values of Q_i corresponding to lower values of n_s^*. In particular, values for $Q_i = 1/100$ or $Q_i = 1/1000$ will drastically increase convergence, in this example to 67 and 78% for the majority choice, respectively.

If $Q_i(t)$ is constant, but $R_i(t)$ is a non-linear function of the form $R_i(t) = f[n_i(t)]_i^\alpha$ with $\alpha_i > 1$ a constant, we observe $n_i(t)$ to grow faster due to social amplification of attraction; the larger $n_i(t)$, the larger the positive influx into $\dot{n}_i(t)$. Systems with this property usually converge much faster than linear systems. For example, a system with

$$R_i(t) = \left(1 + \frac{n_i(t)}{N_0}\right)^{\alpha_i}$$

(74.12)

shows faster convergence than a linear system for $\alpha_i \geq 1$. Here, normalizing social attraction with N_0 provides independence of the dynamics of the number of agents. An example with $\alpha_i = 5$ is shown in Fig. 74.4b.

Similarly, if $R_i(t)$ is constant, but $Q_i(t)$ is a non-linear function of the form $Q_i(t) = f[n_i(t)]^\beta$ with $\beta < 0$

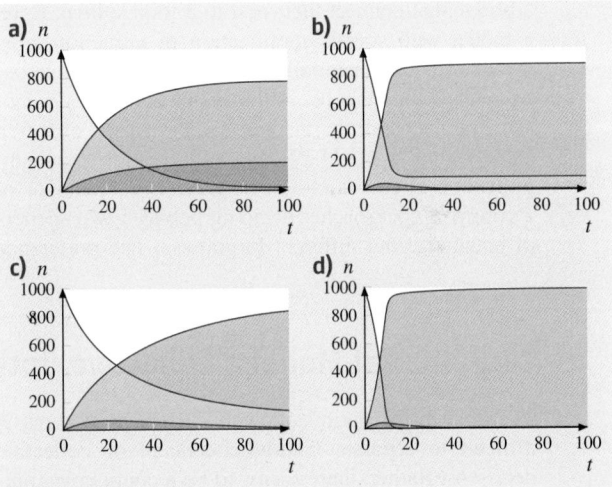

Fig. 74.4a-d Time evolution of a collective decision where solution two is picked four times as likely as solution one, and both solutions are re-evaluated after an average of 100 s, for different non-linear dynamics. *Graphs* show the fraction of agents picking solutions one and two. (a) Linear system achieving steady-states of $\approx 20\%$ and $\approx 80\%$, matching analytical results. (b) Time evolution of a system with social amplification of attraction using R_i given by (74.12) and $\alpha_i = 5$ for both choices. (c) Social amplification of rest using (74.13) and $\beta_i = 5$. (d) Social amplification of both attraction and rest with $\alpha_i = \beta_i = 5$

a constant, we observe $n_i(t)$ to grow faster due to social amplification of rest; the larger $n_i(t)$, the smaller the out-flux from $\dot{n}_i(t)$. For example, a system with

$$Q_i(t) = \left(1 + \frac{n_i(t)}{N_0}\right)^{\beta_i} \qquad (74.13)$$

also shows faster convergence than a linear system. Notice that we do not consider positive exponents for β_i, as this will drive agents away from decisions exponentially fast and will simply increase $n_s(t)$, i.e., the number of undecided agents. Results for a two-choice system with $\beta_i = 5$ are shown in Fig. 74.4c.

Finally, systems that rely both on social amplification of attraction *and* rest exhibit the best convergence, when compared with a purely linear system as well as systems that rely only on either social amplification mechanism. Results for a two-choice system with $\alpha_i = \beta_i = 5$ are shown in Fig. 74.4d.

Similar models, i.e., models that rely on non-linear amplification of either attraction, rest, or both have been proposed for a series of social insect experiments. For example, in [74.19] an ant colony is presented with a binary choice to select the shortest of two branches of a bridge that connect their nest to a food source. Here, a model with social amplification of attraction – by means of an exponentially higher likelihood to choose a branch with higher pheromone concentration – is chosen and successfully models the dynamics observed experimentally. In [74.20] a model that uses social amplification of rest is chosen to model the behavior of a swarm of cockroaches deciding between two shelters of equal size but different brightness. The preference

of cockroaches for darker shelters is expressed with a higher p_i for this shelter. Convergence to the dark shelter is then achieved by social amplification of rest, increasing the time cockroaches remain in a shelter exponentially with the number of individuals that are already in the shelter. Here, all cockroaches converge to a single shelter, even though the model proposed in [74.20] employs negative social amplification of attraction by introducing a notion of shelter capacity, which cancels the positive term in $\dot{n}_i(t)$ when the shelter reaches a constant carrying capacity. Finally, [74.22] presents a model for cockroach aggregation in which the likelihood to join an aggregate of cockroaches increases with the size of the aggregate, whereas the likelihood to leave a cluster exponentially decreases with its size.

The examples from the social insect domain are trade-offs between the expressiveness of the model and its complexity. As the true parameter values of α_i and β_i are unknown, the same experimental data can be accurately matched by models with different dynamics. For example, social amplification of attraction as observed on larvae of German cockroaches in [74.22] was deemed to have negligible influence on mature American cockroaches selecting shelters with limiting carrying capacity in [74.20].

With respect to artificial agent and robotic systems, the models presented can instead provide design guidelines for achieving a desired convergence rate. At the same time, the models are able to support decisions on sensing and communication sub-systems that are required to implement one or the other social amplification mechanism.

74.4 Spatial Models: Collective Optimization

The concept of optimization in collective systems is difficult to separate from the concept of collective decisions. Rather, there seems to be a continuous transition. Collective decisions are made between several distinct alternatives, implying a discrete world of options (e.g., left and right branches in path selection, two shelters, etc.). Typically, one refers to the term *optimization* in collective systems in the case of tasks that allow for a vast (possibly even infinite) number of alternatives implying a continuous world of options.

For this optimization scenario we apply the probabilistic model reported in [74.4, 5, 23–25]. It is based

on a stochastic differential equation (SDE, the Langevin equation) and a partial differential equation (PDE, the Fokker–Planck equation), which can be derived from the former. While the Langevin equation is a stochastic description of the trajectory in space over time of a single robot, the Fokker–Planck equation describes the temporal evolution of the probability density in space for these trajectories. Hence, it can be interpreted as the average over many samples of robot trajectories (i.e., ensembles of trajectories). Even a second, more daring interpretation arises. We can interpret this probability density directly as a swarm density, that is, the expected fraction of the robot swarm for a given area and time.

The deterministic PDE describes the mean swarm fraction in space and time. Interactions between robots can be modeled via dependence on the swarm density itself [74.4].

We introduce our formalism (see Table 74.3 for a summary of all variables used). The Langevin equation that gives the position of a robot $\mathbf{R}$ at time t is

$$\dot{\mathbf{R}}(t) = -\mathbf{A}(\mathbf{R}(t), t) + B(\mathbf{R}(t), t)\mathbf{F}(t) , \qquad (74.14)$$

where $\mathbf{A}$ defines directed motion via drift depending on the current position $\mathbf{R}$ and $B(\mathbf{R}(t), t)\mathbf{F}$ defines random motion based on $\mathbf{F}$, which is a stochastic process (e.g., white noise). Based on the Langevin equation the Fokker–Planck equation can be derived [74.4, 11, 12, 26]

$$\begin{aligned}\frac{\partial \rho(\mathbf{r}, t)}{\partial t} &= -\nabla(\mathbf{A}(\mathbf{r}, t)\rho(\mathbf{r}, t)) \\ &\quad + \frac{1}{2}Q\nabla^2(B^2(\mathbf{r}, t)\rho(\mathbf{r}, t)) , \end{aligned} \qquad (74.15)$$

for a swarm density $\rho(\mathbf{r}, t)$ (according to the above interpretation) at position $\mathbf{r}$ and time t, a drift term $(-\nabla(\mathbf{A}(\mathbf{r}, t)\rho(\mathbf{r}, t)))$ due to directed motion and a diffusion term $(\frac{1}{2}Q\nabla^2(B^2(\mathbf{r}, t)\rho(\mathbf{r}, t)))$ due to random motion, whereas typically we set $Q = 2$ for simplicity. According to our general approach [74.4] we introduce a Fokker–Planck equation for each robot state and manage the transitions between states by rates similar to the rate equation approach of the above sections.

The optimization scenario considered here was inspired by the behavior of young honeybees. The algorithm, which defines the robots' behavior, is derived from a behavioral model of honeybees [74.27, 28]. Honeybees of an age of less than 24 h stay in the hive, cannot yet fly, navigate towards spots of a preferred warmth of 36 °C, and stay mostly inactive. An interesting example of swarm intelligent behavior is how they

search and find the right temperature that their bodies need. It turns out that they do not seem to do a gradient ascent in the temperature field but rather a correlated random walk with inactive periods triggered by social interaction. Both the above-mentioned behavioral model and the robot controller – called BEECLUST – are defined by the following:

1. Each robot moves straight until it perceives an obstacle Ω within sensor range.
2. If Ω is a wall the robot turns away and continues with step 1.
3. If Ω is another robot, the robot measures the local temperature. The higher the temperature is the longer the robot stays stopped. When the waiting elapses, the robot turns away from the other robot and continues with step 1.

The temperature field that we investigate in the scenario here has one global optimum (36 °C) at the right end of the arena and one local optimum (32 °C) at the left end of the arena. In analogy to the behavior observed in young honeybees, the swarm is desired to aggregate fully at the global optimum but, at the same time, should also stay flexible within a possibly dynamic environment. The latter is implemented by robots (bees) that leave the cluster from time to time and explore the remaining arena. If a more preferable spot were to emerge elsewhere they would start to aggregate there, and the former cluster might shrink in size and finally vanish fully.

Now we apply the above modeling approach to this scenario. We have two states: *moving* and *stopped*. It turns out that in the *moving* state we do not have any directed motion, hence, we will turn off the bias in the Langevin equation (74.14, $\mathbf{A} = 0$). Without any directed motion in BEECLUST (no gradient ascent, actually movement fully independent from the temperature field) the Fokker–Planck equation can be reduced to a mere diffusion equation in order to model the moving robots

$$\frac{\partial \rho(\mathbf{r}, t)}{\partial t} = \nabla^2(B^2(\mathbf{r}, t)\rho(\mathbf{r}, t)) . \qquad (74.16)$$

This equation is our approach for state *moving* yet without addressing state transition rates.

The state *stopped* is even easier to model as it naturally lacks motion. That way it can be viewed as a reduction to a mere rate equation defined in each position $\mathbf{r}$. The state transition rates are defined by a stopping rate φ, which can be determined, for exam-

Table 74.3 Notation used in the optimization model

$\mathbf{R}$	Robot position
$\mathbf{A}$	Direction and intensity of robots' directed motion
B	Intensity of robots' random motion
$\mathbf{F}$	Stochastic process (fluctuating directions)
$\mathbf{r}$	Point in space
Q	Theoretic term describing intensity of collisions
ρ_s	Expected density of robots in state *stopped*
ρ_m	Expected density of robots in state *moving*
w	Waiting time
φ	Rate of stopping robots

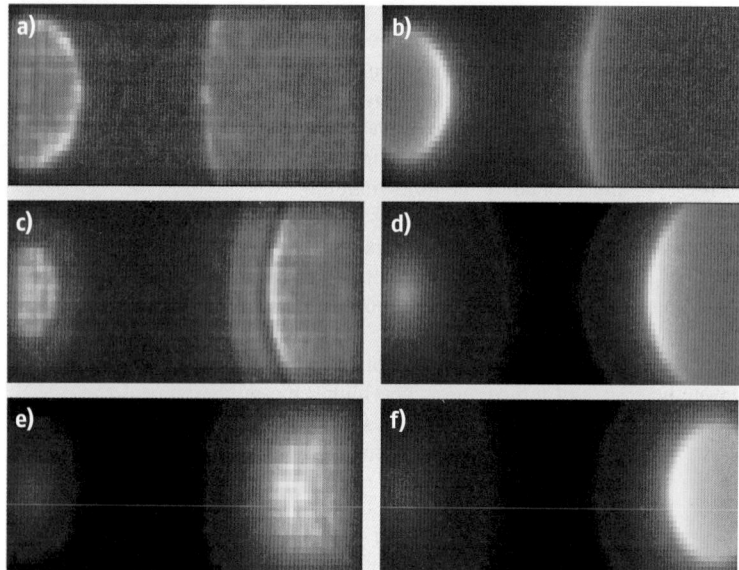

Fig. 74.5a-f Comparison of histograms of swarm density obtained by an agent-based simulation and the corresponding model based on (74.18) for different times and an initial state with equal distribution of robots. An optimal temperature peak of $36\,^\circ$C is at the *right end* of the arena, at the *left end* there is a suboptimal peak in temperature of $32\,^\circ$C, the *middle part* is cooler. We observe that on average at first clusters form at *both ends* of the arena, but later those on the *left* vanish. Swarm size is $N = 25$. The histograms obtained by simulation are based on 10^6 samples. (a) Simulation, $t = 30$; (b) model, $t = 30$; (c) simulation, $t = 130$; (d) model, $t = 130$ (e) simulation, $t = 200$; (f) model, $t = 200$

ple, empirically or by geometrical investigations (e.g., calculation of collision probabilities) [74.4]. For the *stopped* state we obtain

$$\frac{\partial \rho_s(\mathbf{r}, t)}{\partial t} = \rho_m(\mathbf{r}, t)\varphi - \rho_m(\mathbf{r}, t - w(\mathbf{r}))\varphi, \quad (74.17)$$

for a stopping swarm fraction $\rho_s(\mathbf{r}, t)\varphi$ at spot $\mathbf{r}$ and time t, and an awakening swarm fraction $\rho_s(\mathbf{r}, t - w(\mathbf{r}))\varphi$. The robots stop and wait for a time period $w(\mathbf{r})$, which depends on the temperature at spot $\mathbf{r}$.

Here, we choose to approximate the robots' correlated random walk as mere diffusion in a rough estimation. The function B in (74.16) is reduced to a diffusion constant D. We add the rates of stopping/awakening and obtain the equation for state *moving*

$$\frac{\partial \rho_m(\mathbf{r}, t)}{\partial t} = D\nabla^2 \rho_m(\mathbf{r}, t) - \rho_m(\mathbf{r}, t)\varphi$$
$$+ \rho_m(\mathbf{r}, t - w(\mathbf{r}))\varphi. \quad (74.18)$$

If we ignore diffusion and focus on one point in space we would have a mere rate equation similar to the above sections (except for the time delay)

$$\dot{\rho}_m(t) = -\rho_m(t)\varphi + \rho_m(t - w(\mathbf{r}))\varphi. \quad (74.19)$$

Using (74.18) ((74.17) is mathematically not necessary) we can model the BEECLUST behavior. For a provided initial distribution of the robots we end up with an initial value problem for a PDE that we can solve numerically.

The solution of this initial value problem is the temporal evolution of the swarm density. In Fig. 74.5 we compare the model to the results obtained by a simple agent-based simulation of BEECLUST. This comparison is meant to be qualitative only. The model catches most of the qualitative features that occur in simulation, although we approximate the robots' motion in a rough estimation by diffusion.

Our approach shows how borders between the fields of engineering and biology vanish in swarm robotics. The BEECLUST algorithm is at the same time a controller for robots but also a behavioral model of an animal. The same Fokker–Planck model is used to model the macroscopic behavior of honeybees and robot swarms.

The Fokker–Planck model gives good estimates for expected swarm densities in space, the transient/asymptotic behavior of the swarm, and density flows. Modeling space explicitly allows for specific investigations such as objective areas and obstacles of certain shapes. Other case studies included an emergent taxis task which relies on one group of robots that is *pushing* another group by collision avoidance [74.4], a collective perception task in which robots have to discriminate aggregation areas of different sizes [74.29], and a foraging task [74.30]. This model is mostly relevant to scenarios with spatially inhomogeneous swarm densities, that is, swarms forming particular spatial structures that cannot be averaged over several runs.

74.5 Conclusion

We presented mathematical models for three distributed swarming behaviors: collaboration, deciding between different choices, and optimization. Each of these processes are collective decisions of increasing complexity.

While the behaviors and trajectories of individual robots might be erratic and probabilistic, the average swarm behavior might be considered deterministic. This holds for both the models and the observed reality in robotic and biological experiments. An analogy is the distinction between the complex, microscopic dynamics of multi-particle systems and the much simpler properties of the corresponding ensembles of such systems in thermodynamics. This insight is important as it allows us to design the individual behavior so

that the expected value of collective performance is maximized.

Although we presented models with increasing level of spatiality – from collaboration sites in the environment to modeling the distribution of robots over continuous space – modeling swarming systems with heterogeneous spatial and state distributions using closed form expressions is still a major challenge. A better understanding of swarming systems with non-uniform spatial distributions will help us to better understand the impact of environmental patterns such as terrain, winds, or currents, thereby enabling swarm engineering for a series of real-world applications that swarming systems have yet to tackle.

References

74.1   E. Bonabeau, M. Dorigo, G. Theraulaz: *Swarm Intelligence: From Natural to Artificial Systems*, SFI Studies in the Science of Complexity (Oxford Univ. Press, New York 1999)

74.2   S. Camazine, J.-L. Deneubourg, N.R. Franks, J. Sneyd, G. Theraulaz, E. Bonabeau: *Self-Organization in Biological Systems*, Princeton Studies in Complexity (Princeton Univ. Press, Princeton 2001)

74.3   N. Correll, A. Martinoli: Towards optimal control of self-organized robotic inspection systems, 8th Int. IFAC Symp. Robot Control (SYROCO), Bologna (2006)

74.4   H. Hamann: *Space-Time Continuous Models of Swarm Robotics Systems: Supporting Global-to-Local Programming* (Springer, Berlin, Heidelberg 2010)

74.5   A. Prorok, N. Correll, A. Martinoli: Multi-level spatial modeling for stochastic distributed robotic systems, Int. J. Robot. Res. **30**(5), 574–589 (2011)

74.6   D. Milutinovic, P. Lima: *Cells and Robots: Modeling and Control of Large-Size Agent Populations* (Springer, Berlin, Heidelberg 2007)

74.7   A. Kettler, H. Wörn: A framework for Boltzmann-type models of robotic swarms, Proc. IEEE Swarm Intell. Symp. (SIS'11) (2011) pp. 131–138

74.8   A. Jadbabaie, J. Lin, A.S. Morse: Coordination of groups of mobile autonomous agents using nearest neighbor rules, IEEE Trans. Autom. Control **48**(6), 988–1001 (2003)

74.9   R. Olfati-Saber, R. Murray: Consensus problems for networks of dynamic agents with switching topology and time-delays, IEEE Trans. Autom. Control **49**, 1520–1533 (2004)

74.10 J. Cortés, S. Martínez, T. Karatas, F. Bullo: Coverage control for mobile sensing networks, IEEE Trans. Autom. Control **20**(2), 243–255 (2004)

74.11 N.G. van Kampen: *Stochastic Processes in Physics and Chemistry* (Elsevier, Amsterdam 1981)

74.12 H. Haken: *Synergetics – An Introduction* (Springer, Berlin, Heidelberg 1977)

74.13 F. Schweitzer: *Brownian Agents and Active Particles. On the Emergence of Complex Behavior in the Natural and Social Sciences* (Springer, Berlin, Heidelberg 2003)

74.14 A.J. Ijspeert, A. Martinoli, A. Billard, L. Gambardella: Collaboration through the exploitation of local interactions in autonomous collective robotics: The stick pulling experiment, Auton. Robot. **11**, 149–171 (2001)

74.15 N. Correll, A. Martinoli: Modeling and analysis of beaconless and beacon-based policies for a swarm-intelligent inspection system, Proc. 2005 IEEE Int. Conf. Robot. Autom. (ICRA 2005) (2005) pp. 2477–2482

74.16 A. Martinoli, K. Easton, W. Agassounon: Modeling of swarm robotic systems: A case study in collaborative distributed manipulation, Int. J. Robot. Res. **23**(4), 415–436 (2004)

74.17 K. Lerman, A. Galstyan, A. Martinoli, A.-J. Ijspeert: A macroscopic analytical model of collaboration in distributed robotic systems, Artif. Life **7**(4), 375–393 (2001)

74.18 L. Li, A. Martinoli, Y. Abu-Mostafa: Learning and Measuring Specialization in Collaborative Swarm Systems, Adapt. Behav. **12**(3/4), 199–212 (2004)

74.19 J.-L. Deneubourg, S. Aron, S. Goss, J.M. Pasteels: The self-organizing exploratory pattern of the argentine ant, J. Insect Behav. **3**, 159–168 (1990)

74.20 J. Halloy, J.-M. Amé, G.S.C. Detrain, G. Caprari, M. Asadpour, N. Correll, A. Martinoli, F. Mondada, R. Siegwart, J.-L. Deneubourg: Social integration of robots in groups of cockroaches to control self-organized choice, Science **318**(5853), 1155–1158 (2009)

74.21   S. Garnier, C. Jost, R. Jeanson, J. Gautrais, M. Asad-pour, G. Caprari, J.-L. Deneubourg, G. Ther-aulaz: Collective decision-making by a group of cockroach-like robots, 2nd IEEE Swarm Intell. Symp. (SIS) (2005)

74.22   R. Jeanson, C. Rivault, J.-L. Deneubourg, S. Blanco, R. Fournier, C. Jost, G. Theraulaz: Self-organized aggregation in cockroaches, Anim. Behav. **69**, 169–180 (2005)

74.23   H. Hamann, H. Wörn: A framework of space-time continuous models for algorithm design in swarm robotics, Swarm Intell. **2**(2–4), 209–239 (2008)

74.24   H. Hamann, H. Wörn, K. Crailsheim, T. Schmickl: Spatial macroscopic models of a bio-inspired robotic swarm algorithm, IEEE/RSJ 2008 Int. Conf. Intell. Robot. Syst. (IROS'08), Los Alamitos (2008) pp. 1415–1420

74.25   T. Schmickl, H. Hamann, H. Wörn, K. Crailsheim: Two different approaches to a macroscopic model

of a bio-inspired robotic swarm, Robot. Auton. Syst. **57**(9), 913–921 (2009)

74.26   J.L. Doob: *Stochastic Processes* (Wiley, New York 1953)

74.27   T. Schmickl, R. Thenius, C. Möslinger, G. Radspieler, S. Kernbach, K. Crailsheim: Get in touch: Coop-erative decision making based on robot-to-robot collisions, Auton. Agents Multi-Agent Syst. **18**(1), 133–155 (2008)

74.28   S. Kernbach, R. Thenius, O. Kornienko, T. Schmickl: Re-embodiment of honeybee aggregation behav-ior in an artificial micro-robotic swarm, Adapt. Behav. **17**, 237–259 (2009)

74.29   H. Hamann, H. Wörn: A space- and time-continuous model of self-organizing robot swarms for design support, 1st IEEE Int. Conf. Self-Adapt. Self-Organ. Syst. (SASO'07), Boston, Los Alamitos (2007) pp. 23–31

74.30   H. Hamann, H. Wörn: An analytical and spatial model of foraging in a swarm of robots, Lect. Notes Comput. Sci. **4433**, 43–55 (2007)

Hybrid Sy Part G

Part G Hybrid Systems

Ed. by Oscar Castillo, Patricia Melin

75. A Robust Evolving Cloud-Based Controller

Plamen P. Angelov, Igor Škrjanc, Sašo Blažič

In this chapter a novel online self-evolving cloud-based controller, called Robust Evolving Cloud-based Controller (RECCo) is introduced. This type of controller has a parameter-free antecedent (IF) part, a locally valid PID consequent part, and a center-of-gravity based defuzzification. A first-order learning method is applied to consequent parameters and reference model adaptive control is used locally in the ANYA type fuzzy rule-based system. An illustrative example is provided mainly for a proof of concept. The proposed controller can start with no pre-defined fuzzy rules and does not need to pre-define the range of the output, number of rules, membership functions, or connectives such as AND, OR. This RECCo controller learns autonomously from its own actions while controlling the plant. It does not use any off-line pre-training or explicit models (e.g. in the form of differential equations) of the plant. It has been demonstrated that it is possible to fully autonomously and in an unsupervised manner (based only on the data density and selecting representative prototypes/focal points from the control hypersurface acting as a data space) generate and self-tune/learn a non-linear controller structure and evolve it in online mode. Moreover,

the results demonstrate that this autonomous controller has no parameters in the antecedent part and surpasses both traditional PID controllers being a non-linear, fuzzy combination of locally valid PID controllers, as well as traditional fuzzy (Mamdani and Takagi–Sugeno) type controllers by their lean structure and higher performance, lack of membership functions, antecedent parameters, and because they do not need off-line tuning.

75.1 Overview of Some Adaptive and Evolving Control Approaches

Fuzzy logic controllers where proposed some four decades ago by *Mamdani* and *Assilian* [75.1]. Their main advantage is that they do not require the model of the plant to be known and their linguistic form is closer to the way human reasoning is expressed and formalized. It is difficult to identify all possible events or the frequency of their occurrences while modeling a system. The lack of this knowledge requires use of an approximate model of a system.

Due to the fact that a fuzzy logic algorithm has the characteristic of a universal approximator, it is possible to model systems containing unknown nonlinearities using a set of IF-THEN fuzzy rules.

The main challenges in designing conventional fuzzy controllers are that they are sometimes designed to work in certain modeling conditions [75.2]. Moreover, fuzzy controllers include at least two parameters per fuzzy set, which are usually predefined in advance and tuned off-line [75.3]. Many techniques have been presented for auto-tuning of the parameters of controllers in batch mode [75.4], mostly using genetic algorithms [75.5, 6] or neural networks [75.7, 8] offline. From a practical point of view, however, there is no guarantee that pre-training parameters have satisfactory performance in online applications when the environment or the object of the controller changes. To tackle this problem several approaches have been proposed for online adaptation of fuzzy parameters [75.9–15].

Nevertheless, only a few approaches have been introduced for online adaptation of fuzzy controller structures when no prior knowledge of the system is available. Evolving fuzzy rule-based controllers were introduced in 2001 by *Angelov* et al. [75.16]. They allow the controller structure (fuzzy rules, fuzzy sets, membership functions, etc.) to be created based on data collected online. This is based on a combination of inverse plant dynamic modeling [75.17] using self-evolving fuzzy rule-based systems [75.18]. The proposed approach is applied to autonomously learning controllers that are self-designed online.

The advantage of this method is that there is no need for pre-tuning of the control parameters. Moreover, the proposed method can start with an empty topology, and the structure of the controller is modified online based on the data obtained during the operation of the closed loop system. Two main phases were introduced for parameter learning of the controller's consequents and modifying the structure of the controller. The proposed approach was successfully applied to a nonlinear servo system consisting of a DC motor and showed satisfactory performance [75.19], as well as to control of mobile robots [75.20]. The drawback of this approach is that the addition of new membership functions increases the number of rules exponentially, and each membership requires at least two parameters in the antecedent part to be specified, plus connectives such as AND, OR, NOT. Many of these problems have been overcome with the latest version of the approach [75.21], which combines the Angelov–Yager (ANYA) type fuzzy rule-based system (FRB) [75.22] with the inverse plant dynamics model. ANYA can be seen as the next form of FRB system types after the two well-known Mamdani and Takagi–Sugeno type FRBs. It does not require the membership functions to be defined for the antecedent part, nor the connectives such as AND, OR, NOR. It still has a linguistic form and is non-linear. It is fuzzy in terms of the defuzzification. In order to clarify what that means let us

recall that all three types of FRB: Mamdani, Takagi–Sugeno, and ANYA can be represented as a set of fuzzy rules of the form IF (antecedent) and THEN (consequent). While in the Mamdani type FRB both antecedent and consequent parts are fuzzy, in the so-called Takagi–Sugeno type FRB the consequent part is a functional, $f(x)$ (most often linear) with the antecedent part being fuzzy. In both types of FRB the defuzzification can be either of so-called center-of-gravity (COG) or *winner takes all* (WTA) type. There are variations such as few winners take all, etc., but usually COG is applied unless a classification problem is considered, where WTA performs better. In ANYA type FRBs the antecedent part is defined using an alternative, density-based representation which is parameter-free and reduces the problems of definition and tuning of membership functions (one of the stumbling blocks in the application of the fuzzy set theory overall). The consequent part of the ANYA type FRB can still be same as in Takagi–Sugeno type FRBs. For more detail, the reader is referred to [75.21, 22]. The so-called SPARC self-evolving controller, however, has poorer performance in the first moments when applied *from scratch* (with no pre-trained model and no rules). In this chapter, model reference control and gradient-based learning of the consequents of the individual locally valid rules. In the proposed method the antecedent part is determined using focal points/prototypes (selected descriptive actual data points) instead of pre-defining the membership functions in an explicit manner. The fuzzy rules are formed around selected representative points from the control surface; thus, there is no need to define the membership functions per variable. It has a much simplified antecedent part which is formed using so-called data clouds. Fuzzy data clouds are fuzzy sets of data samples which have no specific shape, parameters, or boundaries. With ANYA type FRBs the relative density is used to define the relative membership to a particular cloud. It takes into account the distance to all previous data samples and can be calculated recursively.

In order to show the effective performance of the proposed controller, it is applied to a simulated problem of temperature control in a water bath [75.21].

The remainder of this chapter is organized as follows. In Sect. 75.2 the new simplified FRB system is introduced, including rule representation and the associated inference process. The evolving methodology used for the online learning of both the structure and the parameters of RECCo is described in Sect. 75.3. First, we present the mechanism for the online adaptation of the consequents in Sect. 75.3.1 and then, we illustrate the structure evolution process in Sect. 75.3.2. In Sect. 75.4, the simulation exam- ple is presented as a proof of concept for the proposed methodology. Finally, conclusions are drawn in Sect. 75.5.

75.2 Structure of the Cloud–Based Controller

ANYA [75.22] is a recently proposed type of FRB system characterized by the use of non-parametric antecedents. Unlike traditional Mamdani and Takagi–Sugeno FRB systems, ANYA does not require an explicit definition of fuzzy sets (and their corresponding membership functions) for each input variable. On the contrary, ANYA applies the concepts of fuzzy data clouds and relative data density to define antecedents that represent exactly the real data density and distribution and that can be obtained recursively from the streaming data online.

Data clouds are subsets of previous data samples with common properties (closeness in the data space). Contrary to traditional membership functions, they represent directly and exactly all the previous data samples. Some given data can belong to all the data clouds with a different degree $\gamma \in [0, 1]$, thus the fuzziness in the model is preserved and used in the defuzzification, as will be shown later. It is important to stress that clouds are different from traditional clusters in that they do not have specific shapes and, thereby, do not require the definition of boundaries.

First it was proposed to use ANYA to design fuzzy controllers for the situations in which the lack of knowledge about the plant makes it difficult to define the rule antecedents in [75.21]. SPARC autonomous controllers have a rule base with N rules of the following form

$$\mathcal{R}^i : \text{IF } \left(x \sim X^i \right) \text{ THEN } \left(u^i \right) , \qquad (75.1)$$

where $\sim$ denotes the fuzzy membership expressed linguistically as *is associated with*, $X^i \in \mathcal{R}^n$ is the i-th data cloud defined in the input space, $x = [x_1, x_2, \ldots, x_n]^T$ is the controller's input vector, and u^i is the control action defined by the i-th rule.

It is to be noted that no aggregation operator is required to combine premises of the form IF x_j is X_j^k, as in traditional fuzzy systems. All the remaining components of the FRB system (e.g., the consequents and the defuzzification method) can be selected as in any of the traditional fuzzy systems.

A rule base of the form (75.1) can describe complex, generally non-linear, non-stationary, non-deterministic systems that can be only observed through their inputs and outputs. Hence, autonomous controllers based on ANYA type FRB systems are suitable to describe dependence of the type *IF X THEN U* based on the history of pairs of data observations of the form $z_j = [x_j^T; u_j]^T$ (with $j = 1, \ldots, k-1$ and $z \in \mathbb{R}^{n+1}$), and the current k-th input x_k^T.

The degree of membership of the data sample x_k to the cloud X^i is measured by the normalized relative density as follows

$$\lambda_k^i = \frac{\gamma_k^i}{\sum_{j=1}^{N} \gamma_k^i} , \qquad i = 1, \ldots, N , \qquad (75.2)$$

where γ_k^i is the local density of the i-th cloud for that data sample.

This local density is defined by a suitable kernel over the distance between x_k and all the other samples in the cloud, i.e.,

$$\gamma_k^i = K \left(\sum_{j=1}^{M^i} d_{kj}^i \right) , \qquad i = 1, \ldots, N , \qquad (75.3)$$

where d_{kj}^i denotes the distance between the data samples x_k and x_j, and M^i is the number of input data samples associated with the cloud X^i.

In a similar manner, we consider that a sample is associated with the cloud with the highest local density. In addition, we use the Euclidean distance, i.e., $d_{kj}^i = \|x_k - x_j\|^2$. Nonetheless, any other type of distance could also be used [75.22].

In this study we used a Cauchy kernel. Thereby, (75.3) can be recursively determined as follows [75.23]

$$\gamma_k^i = \frac{1}{1 + \|x_k - \mu_k\|^2 + \Sigma_k - \|\mu_k\|^2} , \qquad (75.4)$$

where γ_k^i denotes the relative density to the i-th data cloud calculated in the k-th time instant, and Σ_k denotes the scalar product of the data x_k

$$\Sigma_k = \frac{k-1}{k} \Sigma_{k-1} + \frac{1}{k} \|x_k\|^2 , \qquad (75.5)$$

with starting condition $\Sigma_1 = ||x_1||^2$. The update of the mean value, μ is straightforward

$$\mu_k = \frac{k-1}{k}\mu_{k-1} + \frac{1}{k}x_k , \tag{75.6}$$

with the starting condition $\mu_1 = x_1$.

As for the defuzzification, it is to be noted that ANYA can work with both Mamdani and Takagi–Sugeno–Kang (TSK) consequents. In this case, we use the latter type, as is usual in control applications [75.19, 24–26]. Hence, if we consider the weighted average for the defuzzification, the output of the ANYA controller is

$$u_k = \sum_{i=1}^{N} \lambda_k^i u^i = \frac{\sum_{i=1}^{N} \gamma_k^i u^i}{\sum_{i=1}^{N} \gamma_k^i} , \tag{75.7}$$

where u^i denotes the i-th rule consequent.

From a *local* point of view, the goal of the controller is to bring the plant's output from its current value to the desired reference value as soon as possible, i.e., ideally $y_{k+1} = r_k$, where k and $k+1$ represent consecutive control steps. It is well known that it is not possible to do this immediately due to several limitations in the system (most notably – the actuator ones). A useful practice is to introduce a reference model that represents the desired closed-loop dynamics of the controlled systems [75.27]. The simplest choice is to use a linear reference model of the first order. Then the prediction of the reference output y^r can be obtained using the following equations

$$y_{k+1}^r = a_r y_k^r + (1 - a_r)r_k , \quad 0 < a_r < 1 , \tag{75.8}$$

where a_r is the pole of the first-order filter.

It can be tuned according to the desired speed of the closed-loop system. Comparing the output of the plant y_k to the output of the reference model y_k^r, the tracking error ε is obtained

$$\varepsilon_k = y_k - y_k^r . \tag{75.9}$$

The goal of the controller in terms of the tracking error, ε_k is to keep it as low as possible. Since the reference model output y_k^r is a filtered version of the reference

signal r_k, this means that the tracking error has no step changes due to reference signal changes. It also has to be noted that tracking error is used as a driving error during parameter adaptation, as we shall see in Sect. 75.3.1.

As noted above, the proposed approach is compatible with a wide spectrum of control laws in rule consequents. Here, the PID-based rule consequents are proposed

$$u_k^i = P_k^i \varepsilon_k + I_k^i \Sigma_k + D_k^i \Delta_k , \quad i = 1, \ldots, N , \tag{75.10}$$

where Σ_k and Δ_k denote the discrete-time integral and derivative of the tracking error, respectively,

$$\Sigma_k = \sum_{\kappa=0}^{k-1} \varepsilon_\kappa ,$$

$$\Delta_k = \varepsilon_k - \varepsilon_{k-1} , \tag{75.11}$$

while P_k^i, I_k^i, and D_k^i are parameters that will be tuned by means of adaptation of rule consequents.

The approach offers the possibility of implementing several subsets of PID-based controllers such as P, PI, PD, etc. For simplicity, only proportional controllers will be used in the rule consequent for the rest of this chapter

$$u_k^i = P_k^i \varepsilon_k , \quad i = 1, \ldots, N . \tag{75.12}$$

It has to be kept in mind that most real-life controllers are limited in their operation and can only provide control actions within a specific range, namely, the actuator's interval $[u_{min}, u_{max}]$. If the computed control signal u_k is outside the actuator's interval, it can simply be projected onto the interval if P or PD type controllers are used. If integral controllers are used as well, some classical approaches to avoid integral windup should be implemented. When the violations of the actuator's constraints are more drastic, because of the chosen dynamics of the reference model and a narrow actuator interval, then also the interruption of the control parameters adaptation can be employed to make the adaptation more robust. This modification will also be explained in detail later.

75.3 Evolving Methodology for RECCo

In this section, we present the methodology applied for evolving the structure and parameters of the consequents of RECCo online. Initially, the controller is empty, so it has to be initialized from the first data sample received. After this, the same steps are repeated for all incoming data. First, the consequents of the current rules are updated according to the error at the plant output. Then, a new control action is generated by applying the inference process described in Sect. 75.2. Finally, the structure of the controller is updated. If the appropriate conditions are satisfied, a new cloud (and hence, a rule) is created; otherwise, the new sample is used to update the information about the data density and the consequent parameters of the current configuration of the controller.

In the following sections, the entire process is described in more detail. Section 75.3.1 is devoted to the mechanism for online adaptation of the rule consequents. In Sect. 75.3.2 the process of adding new clouds is described.

75.3.1 Online Adaptation of the Rule Consequents

Assuming that the plant is monotonic with respect to the control signal, the partial derivative of the plant's output with respect to the control signal has a definite constant sign $G_{sign} = \pm 1$, which has to be known in advance. Therefore, the combination of the error at the plant's output and the sign of the monotonicity of the plant with respect to the control signal provides information about the right direction in which to move the rule consequents to achieve the local control objective [75.10].

As is already known, the parameters of the rule consequents are obtained by means of adaptation. In normal circumstances the parameter changes are calculated as follows

$$\Delta P_k^i = \gamma_P G_{sign} \lambda_k^i(\boldsymbol{x}) \varepsilon_k, \quad i = 1, \dots N, \quad (75.13)$$

where γ_P is an adaptive gain for the proportional controller gains.

Equation (75.13) is obtained by using gradient descent and having the square of the tracking error as a cost function. The controller gains are obtained by summing up the terms obtained in (75.13)

$$P_k^i = P_{k-1}^i + \Delta P_k^i, \quad i = 1, \dots N. \quad (75.14)$$

Note that parameters keep changing until the tracking error is driven towards 0. Note also that only parameters corresponding to the active clouds are adapted, while the others are kept constant.

Systems with parameter adaptation are subjected to parameter drift, which can lead to performance degradation and, eventually, to system instability [75.28]. There exist many known approaches to make adaptive laws more robust [75.29–31]. We will employ parameter projection, parameter leakage, introduce dead zone into adaptive laws, and employ the saturation of the adaptive parameters when the actuator is in saturation.

Dead Zone in the Adaptive Law

As already has already been said, adaptation of the parameters in the closed loop always presents potential danger to the system's stability. The adaptation is driven by an error signal that is always composed of the useful component and the harmful one. The latter is due to disturbances and parasitic dynamics, and is usually bounded. Large errors are usually mostly composed of useful components, while small tracking error is very often due to harmful signals. Having the adaptation active during the time that the error is small results in a false adaptation. The idea behind the dead zone in the adaptive law is that the adaptation is simply switched off if the absolute value of the error that governs the adaptation is small [75.32]

$$\Delta P_k^i = \begin{cases} \gamma_P G_{sign} \lambda_k^i(\boldsymbol{x}) \varepsilon_k & |\varepsilon_k| \geq d_{dead} \\ 0 & |\varepsilon_k| < d_{dead} \end{cases},$$
$$i = 1, \dots N. \quad (75.15)$$

Parameter Projection

Parameter projection is a natural way to prevent parameter drift. The idea is to project the parameters onto a compact set [75.27]. In our case, each individual parameter is projected on a certain interval or a ray. When projecting the parameters some prior knowledge must always be available. Since in our case proportional controller gains are adapted, their sign is always known and is equal to G_{sign}. In the case of positive plant gain all the consequent parameters should be bounded by 0 from below, while an upper bound may or may not (if not enough prior knowledge is available) be provided. The adaptive law given by (75.14) is generalized as follows if the controller gains are projected onto the interval $[\underline{P}, \overline{P}]$

$$P_k^i = \begin{cases} P_{k-1}^i + \Delta P_k^i & \underline{P} \leq P_{k-1}^i + \Delta P_k^i \leq \overline{P} \\ \underline{P} & P_{k-1}^i + \Delta P_k^i < \underline{P} \\ \overline{P} & P_{k-1}^i + \Delta P_k^i > \overline{P} \end{cases},$$

$$i = 1, \dots N,$$

$$\tag{75.16}$$

where $\underline{P}$ and $\overline{P}$ are two design parameters. In our approach $\underline{P} = 0$ and $\overline{P} = \infty$ will be used.

Leakage in the Adaptive Law

The idea of leakage is that the discrete integration in the adaptive law in (75.14) presents a potential danger to the adaptive system, and the pole due to the integrator should be pushed inside the unit disc [75.33], which results in the adaptive law

$$P_k^i = (1 - \sigma_P)P_{k-1}^i + \Delta P_k^i, \quad i = 1, \dots N, \quad (75.17)$$

where σ_P defines the extent of the leakage. The introduction of the leakage results in adaptive parameter boundedness. This is why leakage is sometimes referred to as soft projection.

Interruption of Adaptation

When the chosen dynamics and the actuator constraints are in conflict, such that tracking of the reference model cannot be achieved in a sufficiently small interval, then drift of the control parameters often occurs, because the adaptive law is driven with a tracking error ε_k, which cannot be reduced, due to the control signal constraints. The interruption of the adaptation results in the following modification

$$\Delta P_k^i = \begin{cases} \gamma_P G_{sign} \lambda_k^i(\boldsymbol{x}) \varepsilon_k, & u_{min} \leq u_k \leq u_{max} \\ 0, & \text{else} \end{cases},$$

$$i = 1, \dots N.$$

$$\tag{75.18}$$

75.3.2 Evolution of the Structure: Adding New Clouds

The adaptation of the parameters of the consequents is performed online in a closed loop manner (while the controller operates over the real plant). The control is applied from the first moment (no a-priori information or controller structure is needed). Adaptive systems

traditionally [75.34] concern tuning parameters of the controllers for which the structure has been pre-selected by the designer. Self-evolving controllers [75.16, 24] offer the possibility of evolving the structure of the controller as well as adapting parameters. This helps design *on the fly* controllers, which are non-linear and with no pre-defined structure or knowledge about the plant model. This requires us to define a mechanism for the online evolution of the controller's structure, i.e., for adding new antecedents and fuzzy rules.

We already defined the local density earlier. Now, the global density Γ, will be defined. Its definition is analogous to the one given for the local density, except that it takes into account the distance to all the previously observed samples z_j ($j = 1, \dots, k-1$). It has to be noted that the global density is computed for the points $z_k = [\boldsymbol{x}_k^T; u_k]^T$, whilst the local density is defined only for the input vectors $\boldsymbol{x}_k$. Using again the Cauchy kernel, this density can be defined as

$$\Gamma_k = \frac{1}{1 + \frac{\sum_{j=1}^{k-1}(d_{kj})^2}{k-1}}, \tag{75.19}$$

and can be computed recursively by [75.23]

$$\Gamma_k = \frac{1}{1 + ||z_k - \mu_k||^2 + \Sigma_k^G - ||\mu_k^G||^2}, \tag{75.20}$$

where Γ_k denotes the global density to all the data calculated in the k-th time instant, and Σ_k^G denotes the scalar product of the data z_k

$$\Sigma_k^G = \frac{k-1}{k}\Sigma_{k-1}^G + \frac{1}{k}||z_k||^2, \tag{75.21}$$

with starting condition $\Sigma_1^G = ||z_1||^2$. The update of the mean value μ^G is straightforward

$$\mu_k = \frac{k-1}{k}\mu_{k-1} + \frac{1}{k}z_k, \tag{75.22}$$

with starting condition $\mu_1^G = z_1$.

Since this measure considers all existing samples, it provides an indication of how representative a given point z_k is with respect to the entire data distribution.

Additionally, and only for learning purposes, a focal point X_t^i and a radius r_{jk}^i are defined for each cloud. The focal point is a real data sample that has highly representative qualities. The fact that the focal point is always a real sample is important, as it avoids problems that may appear when only a descriptive measure is used instead (as in the case of the average of the points in the cloud). In order to follow the philosophy of the proposed methodology (i. e., avoiding the need to pre-define the parameters of the rules), the focal point is updated online. Thus, for each new sample z_k, the following process is applied:

1. Find the associated cloud C^i, according to

$$C^i = \arg\max_i \left(\gamma_k^i \right) . \tag{75.23}$$

2. Check the representative qualities of the new data point using the following conditions,

$$\gamma_k^i > \gamma_f^i , \tag{75.24a}$$
$$\Gamma_k^i > \Gamma_f^i , \tag{75.24b}$$

where γ_f^i and Γ_f^i represent the local and global density of the current focal point, respectively.
3. If both conditions are satisfied, then replace the focal point by applying $X_f^C \leftarrow x_k$.

The radius provides an idea of the spread of the cloud. Since the cloud does not have a definite shape or boundary, the radius represents only an approximation of the spread of the data in the different dimensions. It is also recursively updated as follows [75.35]

$$r_{jk}^i = \rho \cdot r_{j(k-1)}^i + (1-\rho)\sigma_{jk}^i ; \quad r_{j1}^i = 1 , \tag{75.25}$$

where ρ is a constant that regulates the compatibility of the new information with the old one and is usually set to $\rho = 0.5$ [75.35]. The value σ_{jk}^i denotes the cloud's local scatter over the input data space and is given by

$$\sigma_{jk}^i = \sqrt{\frac{1}{M^i} \sum_{l=1}^{M^i} ||X_f^i - x_l||_j^2} ; \quad \sigma_{j0}^i = 1 . \tag{75.26}$$

It is important to note that the radii and focal points are only used to provide an idea of the location and distribution of the data in the clouds during the structure evolution process. However, they are not actually used to represent the clouds or the fuzzy rules and do not affect the inference process at any point.

The structure-learning mechanism applied for the proposed RECCo is based on the following principles [75.36, 37]:

a) Good generalization and summarization are achieved by forming new clouds from data samples with high global density Γ.
b) Excessive overlap between clouds is avoided by controlling the minimum distance between them.

Hence, the evolution of the structure is based on the addition of new clouds and the associated rules. First, RECCo is initialized by creating a cloud C^1 from the first data sample $z_1 = [x_1^T; u_1]^T$. The antecedent of the first rule is then defined by this cloud and its consequent equals the value u_1. Next, for all the further incoming data samples z_k, the following steps are applied:

1. The sample $z_k = [x_k^T; u_k]^T$ is considered to have good generalization and summarization capabilities if its global density is higher than the global density of all the existing clouds. Thus, the following condition is defined

$$\Gamma_k > \Gamma_f^i , \quad \forall i = 1, \ldots, N . \tag{75.27}$$

Note that this is a very restrictive condition that requires that the inequality is satisfied for all the existing clouds, which is not very often for real data.
2. Check if the existing clouds are sufficiently far from z_k with the following condition

$$d_{ki} > \frac{r_{jk}^i}{2} , \quad \forall i = 1, \ldots, N , \tag{75.28}$$

where d_{ki} represents the distance from the current sample to the focal point of the associated cloud, X_f^i.
3. According to the result of the previous steps, take one of the following actions:
a) If conditions (75.27) and (75.28) are both satisfied, then create a new cloud C^{N+1}. The focal point of the new cloud is $X_f^{N+1} = x_k$. Its local scatter is initialized based on the average of the local scatters of the existing clouds [75.35]

$$\sigma_{jk}^{N+1} = \frac{1}{N} \sum_{l=1}^{N} \sigma_{jk}^l ,$$
$$j = 1, \ldots, n . \tag{75.29}$$

Additionally, the corresponding rule has to be added to the rule base. The antecedent of the new rule is defined by the newly created cloud C^{N+1}. For the consequent, we provide an initial value that guarantees that the output of the new controller when the input vector is equal to the focal point (i. e., $x = X_f^{N+1}$) equals the controller's output under its previous configuration for that same input [75.21]. The rationale behind this initialization is to provide a smooth transition from the old configuration of the controller to the new one. This avoids sudden changes in the output surface that could damage the controller's performance in the first time instants after the rule has been added (and before the consequents are adapted) [75.25].

b) If the conditions are not satisfied, update the parameters of the cloud C^i associated with z_k, as previously explained.

It is important to stress that the methodology presented for the evolution of the controller's structure starts from an empty controller. However, if an initial set of rules is known beforehand (e.g., provided by an expert or obtained from any other training method), it can be used for the initial controller. In this case, the algorithm's initialization step can be omitted.

75.4 Simulation Study

A simulation study of the proposed self-organizing controller is presented in this section. The main attention is given to the study of different modifications of self-organizing controllers to make the adaptive laws more robust. In the study, we show the implementation of parameter projection, parameter leakage, and the introduction of dead zone into the adaptive laws. The study was carried out with the assumption of no prior knowledge about the plant dynamics. The mathematical model was only used to simulate the plant dynamics.

The plant for the simulation study is the thermal process of a water bath. The main goal is the control of the temperature in the water bath. The plant is described with the following mathematical model in a discrete form

$$y(k+1) = ay(k) + bu(k) + (1-a) y_o , \qquad (75.30)$$

where $a = e^{-\alpha T_s}$ and $b = \frac{\beta(1-e^{-\alpha T_s})}{\alpha}$. The parameters of the plant are estimated as $\alpha = 10^{-4}$, $\beta = 8.7 \times 10^{-3}$, $\gamma = 40$, and $y_o = 20\,^\circ\text{C}$. The sampling period, T_s, is equal to 25 s.

The open loop response of the plant is shown in Fig. 75.1. It is shown that the behavior of the plant exhibits a huge nonlinearity in the static gain of the process. The reference signal is chosen to show the ability of self-learning and dealing with nonlinearity, which is the main advantage of the proposed algorithm. The

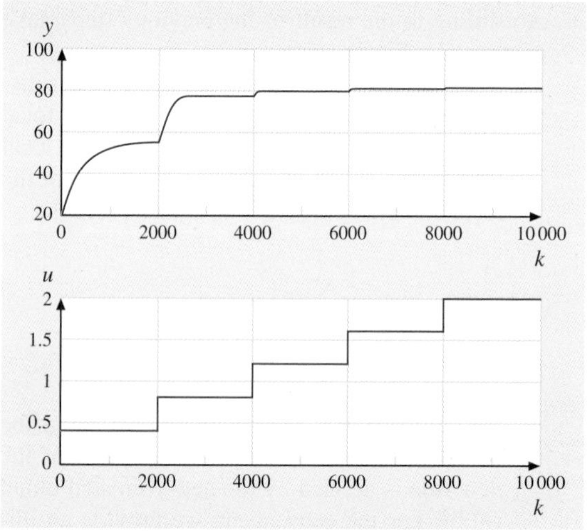

Fig. 75.1 Open-loop response of the plant: output temperature and input variable

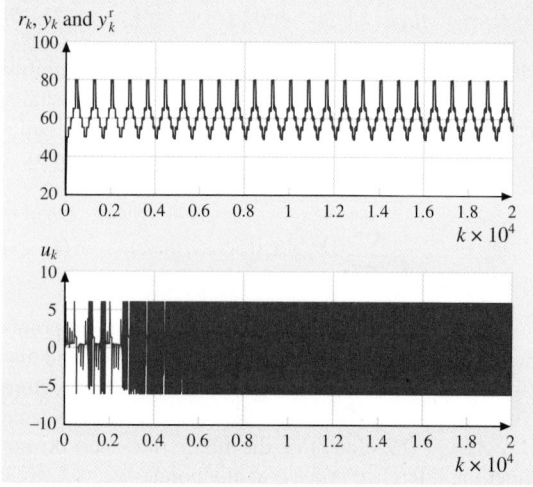

Fig. 75.2 Reference, model reference, output signal tracking, and control signal in the case of no robust adaptive laws

clouds are defined in a way so as to enable dealing with nonlinearity, and for that reason the input variables for the controller are the reference value r_k and the tracking error ε_k.

All the simulations started from zero fuzzy rules and membership functions, and new rules were generated during the process. The first simulation was done for the case without any robust adaptive laws. The parameters of the control law do not converge and drift, which leads to performance degradation and, after some time, also to instability. In this example, the actuator interval was given as $[-6, 6]$, the reference model parameter was defined as $a_r = 0.925$, and the adaptive gain $\gamma_P = 0.1$. The upper plot of Fig. 75.2 shows the reference r, the

model reference y_r, the output signal y, and the lower plot shows the control signal. The seven clouds generated during the self-learning procedure are given in Fig. 75.3.

The drifts of the adaptive parameters P_k^i are shown in Fig. 75.4, where the parameters for all clouds are shown, and in Fig. 75.5, where the adaptive parameter P_k is shown.

The second simulation is done with dead zone modification to make the adaptation robust. The dead zone was chosen as $d_{\mathrm{dead}} = 2$. All the other parameters of the algorithm were the same as in the first simulation:

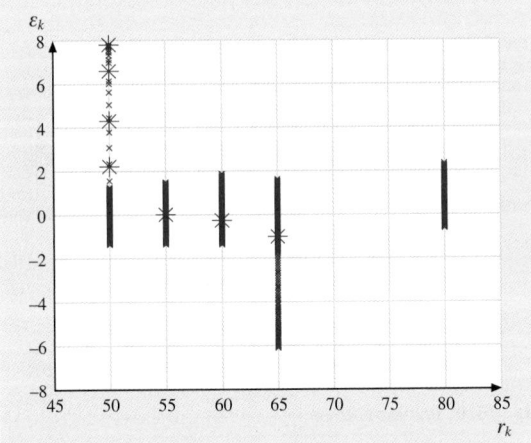

Fig. 75.3 Clouds in the case of no robust adaptive laws

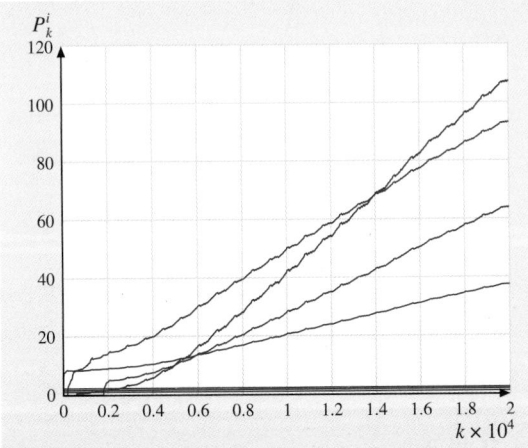

Fig. 75.4 Adaptive parameters P_i^k in the case of no robust laws

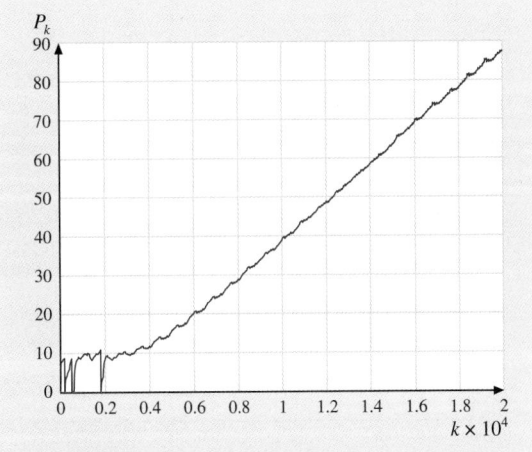

Fig. 75.5 Drift of the adaptive parameter P_k in the case of no robust laws

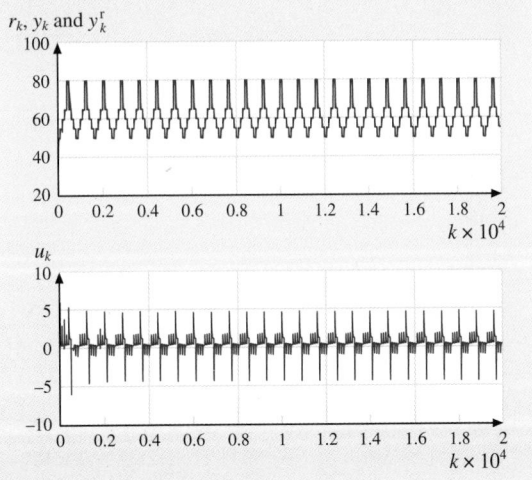

Fig. 75.6 Reference, model reference, output signal tracking, and control signal in the case of a dead zone

the actuator's interval was given as $[-6, 6]$, the reference model parameter was defined as $a_r = 0.925$, and the adaptive gain $\gamma_P = 0.1$. In the upper plot, Fig. 75.6 shows the reference, the model reference, and the output signal, and in the lower part the control signal. The model reference tracking is suitable and the parameters of the control law converge and enable a reasonable performance.

The clouds generated during the self-learning procedure are the same as those obtained in the first example in Fig. 75.3. During the procedure seven clouds were generated again. The adaptive parameters P_k^i are shown in Fig. 75.7, where the parameters for all clouds are shown, and in Fig. 75.8, where the adaptive parameter P_k is shown.

The tracking error e_k in the case of dead zone modification is shown in Fig. 75.9.

The results of the last 500 samples are shown in detail in Fig. 75.10. It can be seen that the tracking using the proposed modification of the adaptive laws has a very good control performance.

The relatively big dead zone stops the adaptation of the control parameters and results in a bigger tracking error. On the other hand, a smaller dead zone would result in longer settling of the adaptive parameters and

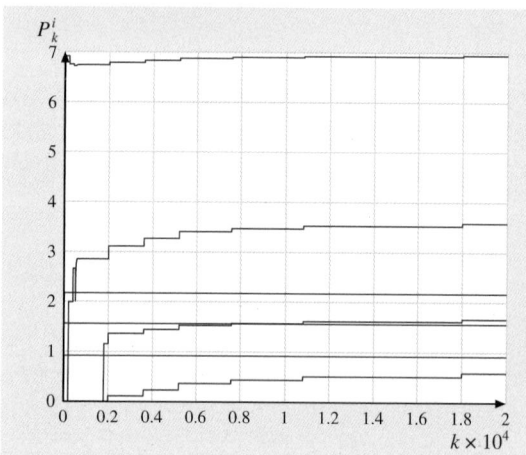

Fig. 75.7 Adaptive parameters P_i^k in the case of a dead zone

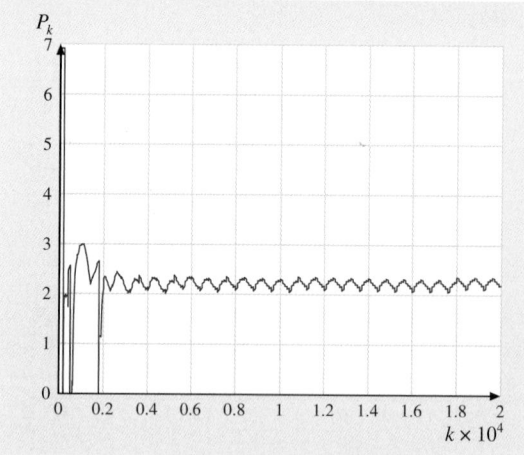

Fig. 75.8 Drift of the adaptive parameter P_k in the case of dead zone

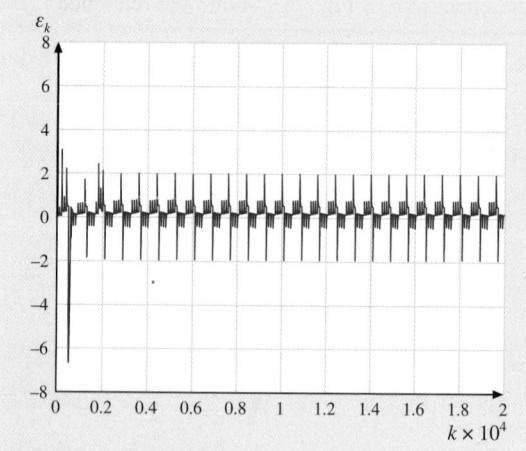

Fig. 75.9 Tracking error

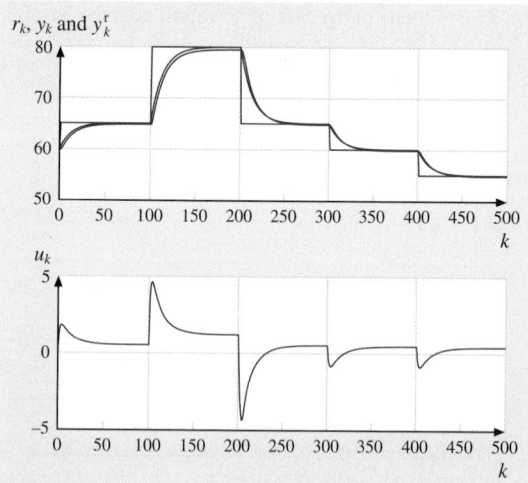

Fig. 75.10 Reference, model reference, output signal tracking, and control signal in the case of a dead zone, in detail

also in possible drifting. This way, the combination of a dead zone and a leakage adaptive law is proposed in the third simulation, where the rest of the parameters are the same as in the previous simulation, except the dead zone, which is now chosen to be $d_{\mathrm{dead}} = 0.25$, and the leakage term which is defined as $\sigma_P = 10^{-5}$.

Figure 75.11 shows the reference, the model reference, the output signal, and the control signal. The model reference tracking is satisfactory and the parameters of the control law converge and enable a reasonable performance.

The clouds generated during the self-learning procedure with leakage in adaptive law are the same as those obtained in the previous two examples, and are given in Fig. 75.3. The positions of the clouds remain the same in all three approaches using different robust modifications of the adaptive laws. The adaptive parameters P_k^i are shown in Fig. 75.12, where the parameters for all clouds are shown, and in Fig. 75.13 where the adaptive parameter P_k is shown.

The tracking error e_k in the case of leakage in the adaptive law is shown in Fig. 75.14. Due to the use of a smaller dead zone and leakage in the adaptive law, the tracking is better and also the parameter convergence is good.

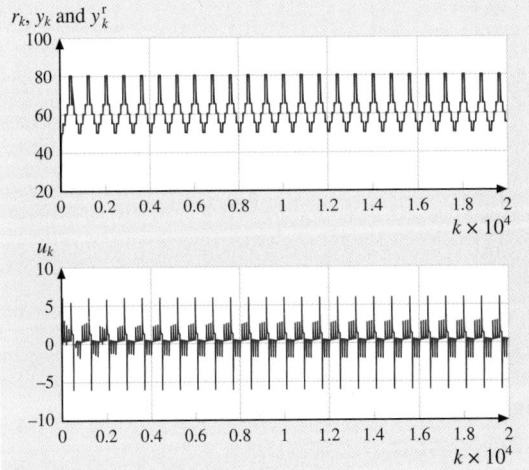

Fig. 75.11 Reference, model reference, output signal tracking, and control signal in the case of leakage in the adaptive law

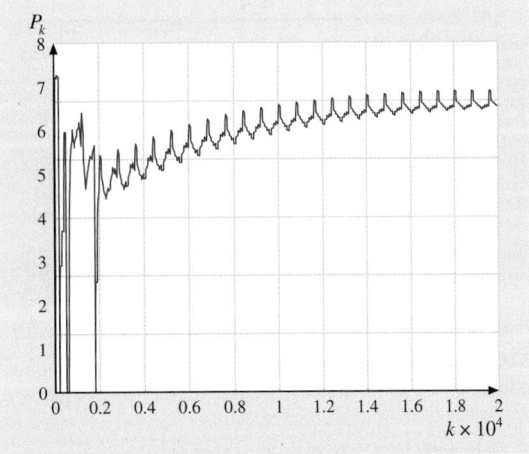

Fig. 75.13 Drift of the adaptive parameter P_k in the case of leakage in the adaptive law

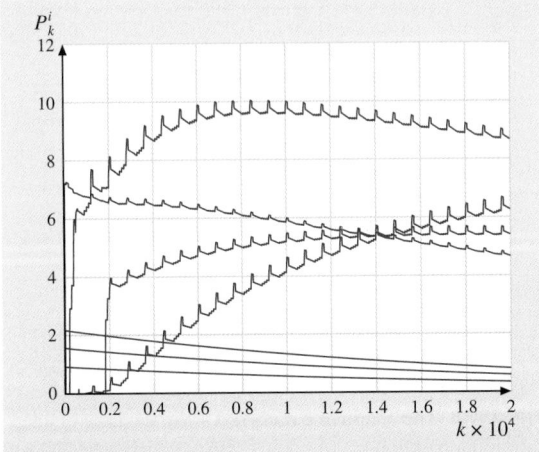

Fig. 75.12 Adaptive parameters P_i^k in the case of leakage in the adaptive law

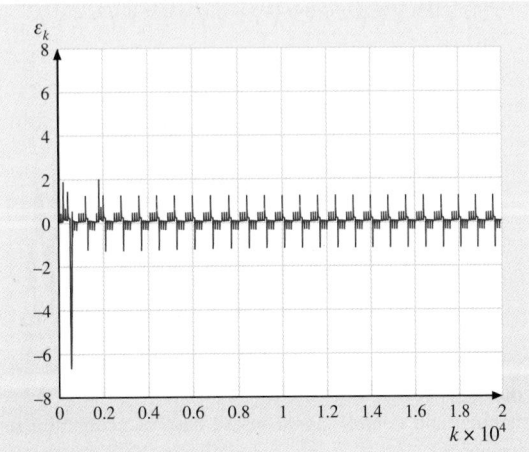

Fig. 75.14 Tracking error

The results of the last 500 samples are shown in detail in Fig. 75.15. It is shown that the tracking using the proposed leakage in the modification of the adaptive laws has a high control performance.

In the fourth simulation study we would like to show an example of drastic constraints in the process actuator. In this case, the actuator constraints are given by the interval $[-1, 2]$. The dead zone is now chosen

to be $d_{\text{dead}} = 0.5$. Figure 75.16 shows the reference, the model reference, the output signal, and the control signal in the case of adaptation interruption. The model reference tracking is satisfactory and the parameters of the control law converge and enable a reasonable performance.

The clouds generated during the self-learning procedure are shown in Fig. 75.17. The positions of the clouds is now different because of the constraints and

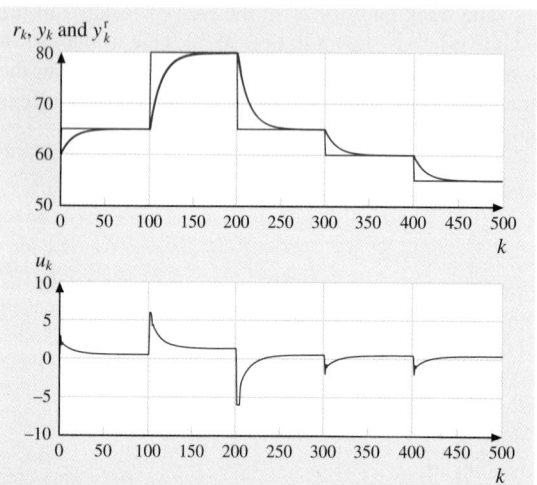

Fig. 75.15 Reference, model reference, output signal tracking, and control signal in the case of leakage in the adaptive law, in detail

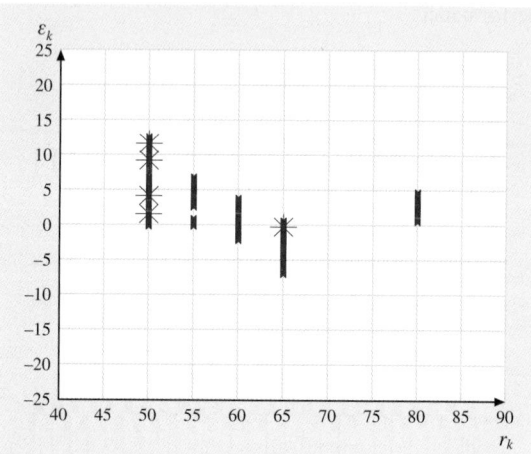

Fig. 75.17 Clouds in the case of adaptation interruption

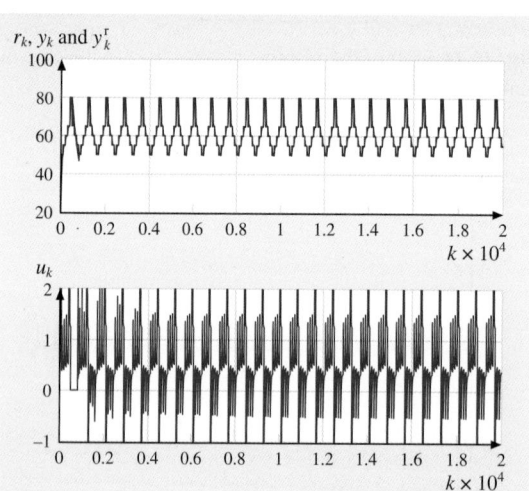

Fig. 75.16 Reference, model reference, output signal tracking, and control signal in the case of adaptation interruption

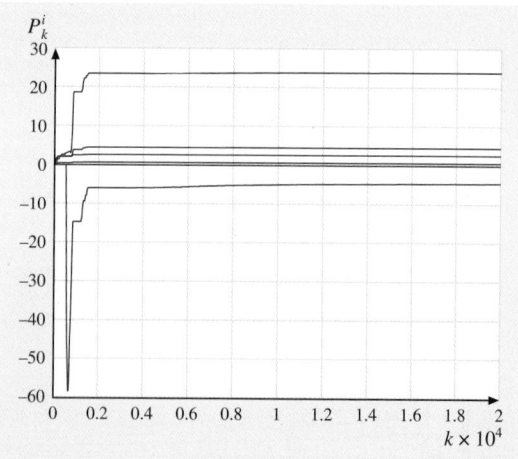

Fig. 75.18 The adaptive parameters P_i^k in the case of adaptation interruption

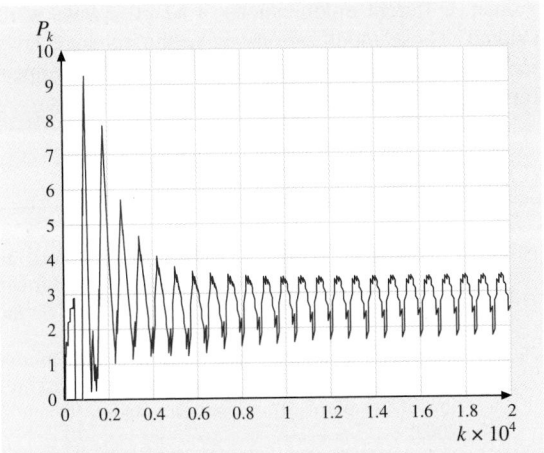

Fig. 75.19 The drift of the adaptive parameter P_k in the case adaptation interruption

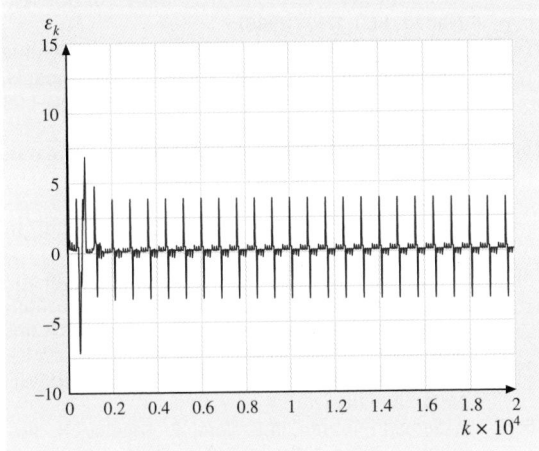

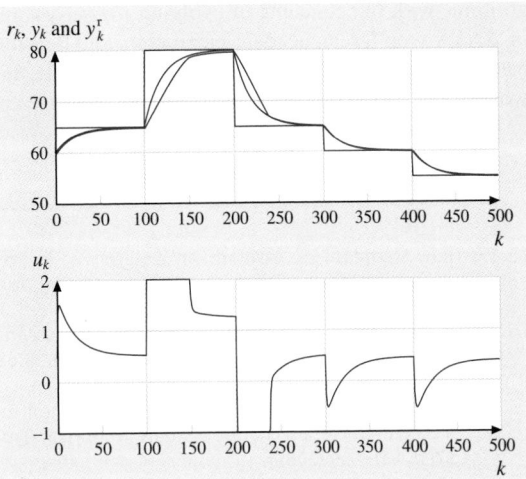

Fig. 75.21 Reference, model reference, output signal tracking, and control signal in the case of adaptation interruption

different tracking errors. The adaptive parameters P_k^i are shown in Fig. 75.18, where the parameters for all clouds are shown, and in Fig. 75.19, where the adaptive parameter P_k is shown.

The tracking error ε_k in the case of adaptation interruption is shown in Fig. 75.20.

The results of the last 500 samples are shown in detail in Fig. 75.21. It is shown that perfect tracking in the case of a highly constrained control signal can be achieved by using the proposed modification based on adaptation interruption.

Fig. 75.20 Tracking error ◄

75.5 Conclusions

In this chapter, a new approach for an online self-evolving cloud-based fuzzy rule-based controller (RECCo), which has no antecedent parameters, was proposed. One illustrative example was provided to support the concept. It has been shown that the proposed controller can start with no a-priori knowledge. All the fuzzy rules are defined during the self-evolving phase. The controller performs the self-evolving algorithm simultaneously with the control of the plant. The advantage of the proposed controller is the self-evolving procedure, which enables a working algorithm that starts from no a-priori knowledge; it can cope perfectly with nonlinearity because of the use of fuzzy data clouds, which actually divide the input space and enable the use of different control parameters in each cloud, and also enables adaptation to changes of the process parameters during the control. No explicit membership function is needed, no pre-training or any explicit model in any form. The proposed algorithm combines the well-known concept of model-reference adaptive control algorithms with the concepts of evolving fuzzy systems of ANYA type (no antecedent parameters and density-based fuzzy aggregation of the linguistic rules). In this work, we analyzed problems related to the adaptive ap-

proach. Different modifications of adaptive laws were studied. Those modifications make the adaptive laws more robust to parameter drift which often leads to performance degradation and instability.

References

75.1 E.H. Mamdani, S. Assilian: An experiment in linguistic synthesis with a fuzzy logic controller, Int. J. Man–Mach. Stud. **7**(1), 1–13 (1975)

75.2 S. Sugawara, T. Suzuki: Applications of fuzzy control to air conditioning environment, J. Therm. Biol. **18**(5), 456–472 (1993)

75.3 Y. Liu, Y. Zheng: Adaptive robust fuzzy control for a class of uncertain chaotic systems, Nonlinear Dyn. **57**(3), 431–439 (2009)

75.4 M. Mucientes, J. Casillas: Quick design of fuzzy controllers with good interpretability in mobile robotics, IEEE Trans. Fuzzy Syst. **15**(4), 636–651 (2007)

75.5 K. Shimojima, T. Fukuda, Y. Hashegawa: Self-tuning modelling with adaptive membership function, rules, and hierarchical structure based on genetic algorithm, J. Fuzzy Sets Syst. **71**(3), 295–309 (1995)

75.6 P. Angelov, R. Guthke: A genetic-algorithm-based approach to optimization of bioprocesses described by fuzzy rules, Bioprocess Eng. **16**, 299–303 (1996)

75.7 M. Huang, J. Wan, Y. Ma, Y. Wang, W. Li, X. Sun: Control rules of aeration in a submerged biofilm wastewater treatment process using fuzzy neural networks, Expert Syst. Appl. **36**(7), 10428–10437 (2009)

75.8 C. Li, C. Lee: Self-organizing neuro-fuzzy system for control of unknown plants, IEEE Trans. Fuzzy Syst. **11**(1), 135–150 (2003)

75.9 H. Pomares, I. Rojas, J. González, F. Rojas, M. Damas, F.J. Fernández: A two-stage approach to self-learning direct fuzzy controllers, Int. J. Approx. Reason. **29**(3), 267–289 (2002)

75.10 I. Rojas, H. Pomares, J. Gonzalez, L. Herrera, A. Guillen, F. Rojas, O. Valenzuela: Adaptive fuzzy controller: Application to the control of the temperature of a dynamic room in real time, Fuzzy Sets Syst. **157**(16), 2241–2258 (2006)

75.11 W. Wang, Y. Chien, I. Li: An on-line robust and adaptive T-S fuzzy-neural controller for more general unknown systems, Int. J. Fuzzy Syst. **10**(1), 33–43 (2008)

75.12 I. Škrjanc, K. Kavaek-Biasizzo, D. Matko: Real-time fuzzy adaptive control, Eng. Appl. Artif. Intell. **10**(1), 53–61 (1997)

75.13 I. Škrjanc, S. Blažič, D. Matko: Direct fuzzy model-reference adaptive control, Int. J. Intell. Syst. **17**(10), 943–963 (2002)

75.14 I. Škrjanc, S. Blažič, D. Matko: Model-reference fuzzy adaptive control as a framework for nonlinear system control, J. Intell. Robot. Syst. **36**(3), 331–347 (2003)

75.15 S. Blažič, I. Škrjanc, D. Matko: Globally stable model reference adaptive control based on fuzzy description of the plant, Int. J. Syst. Sci. **33**(12), 995–1012 (2002)

75.16 P. Angelov, R. Buswell, J.A. Wright, D. Loveday: Evolving rule-based control, Proc. EUNITE Symp. (2001) pp. 36–41

75.17 D. Pasaltis, A. Sideris, A. Yamamura: A multilayer neural network controller, IEEE Trans. Control Syst. Manag. **8**(2), 17–21 (1988)

75.18 P. Angelov, D.P. Filev: An approach to online identification of Takagi–Sugeno fuzzy models, IEEE Trans. Syst. Man Cybern. **34**(1), 484–498 (2004)

75.19 A.B. Cara, Z. Lendek, R. Babuska, H. Pomares, I. Rojas: Online self-organizing adaptive fuzzy controller: Application to a nonlinear servo system, IEEE Int. Conf. Fuzzy Syst. (2010) pp. 1–8

75.20 P. Angelov, P. Sadeghi-Tehran, R. Ramezani: An approach to automatic real-time novelty detection, object identification, and tracking in video streams based on recursive density estimation and evolving Takagi–Sugeno fuzzy systems, Int. J. Intell. Syst. **26**(3), 189–205 (2011)

75.21 P. Sadeghi-Tehran, A.B. Cara, P. Angelov, H. Pomares, I. Rojas, A. Prieto: Self-evolving parameter-free rule-based controller, IEEE Proc. World Congr. Comput. Intell. (2012) pp. 754–761

75.22 P. Angelov, R. Yager: Simplified fuzzy rule-based systems using non-parametric antecedents and relative data density, IEEE Workshop Evol. Adapt. Intell. Syst. (2011) pp. 62–69

75.23 P. Angelov: Anomalous system state identification, Patent GB120 8542.9 (2012)

75.24 P. Angelov: A fuzzy controller with evolving structure, Inf. Sci. **161**(1/2), 21–35 (2004)

75.25 A.B. Cara, H. Pomares, I. Rojas: A new methodology for the online adaptation of fuzzy self-structuring controllers, IEEE Trans. Fuzzy Syst. **19**(3), 449–464 (2011)

75.26 H. Pomares, I. Rojas, J. Gonzalez, M. Damas, B. Pino, A. Prieto: Online global learning in direct fuzzy controllers, IEEE Trans. Fuzzy Syst. **12**(2), 218–229 (2004)

75.27   G. Kreisselmeier, K.S. Narendra: Stable model refer-
        ence adaptive control in the presence of bounded
        disturbances, IEEE Trans. Autom. Control **27**(6),
        1169–1175 (1982)

75.28   C.E. Rohrs, L. Valavani, M. Athans, G. Stein: Ro-
        bustness of continuous-time adaptive control al-
        gorithms in the presence of unmodeled dynam-
        ics, IEEE Trans. Autom. Control **30**(9), 881–889
        (1985)

75.29   P.A. Ioannou, J. Sun: *Robust Adaptive Control*
        (Prentice Hall, Upper Saddle River 1996)

75.30   S. Blažič, I. Škrjanc, D. Matko: Globally stable direct
        fuzzy model reference adaptive control, Fuzzy Sets
        Syst. **139**(1), 3–33 (2003)

75.31   S. Blažič, I. Škrjanc, D. Matko: A new fuzzy adap-
        tive law with leakage, IEEE Conf. Evol. Adapt. Intell.
        Syst. (2012) pp. 47–50

75.32   B.B. Peterson, K.S. Narendra: Bounded error adap-
        tive control, IEEE Trans. Autom. Control **27**(6), 1161–
        1168 (1982)

75.33   P.A. Ioannou, P.V. Kokotovic: Instability analysis
        and improvement of robustness of adaptive con-
        trol, Automatica **20**(5), 583–594 (1984)

75.34   K. Åström, B. Wittenmark: *Adaptive Control* (Addi-
        son Wesley, Reading 1989)

75.35   P. Angelov, X. Zhou: Evolving fuzzy systems from
        data streams in real-time, IEEE Int. Symp. Evol.
        Fuzzy Syst. (2006) pp. 29–35

75.36   P. Angelov: On line learning fuzzy rule-based sys-
        tem structure from data streams, IEEE Int. Conf.
        Fuzzy Syst. (2008) pp. 915–922

75.37   P. Angelov, D.P. Filev, N.K. Kasabov: *Evolving In-
        telligent Systems: Methodology and Applications*
        (Wiley, Hoboken 2010)

76. Evolving Embedded Fuzzy Controllers

Oscar H. Montiel Ross, Roberto Sepúlveda Cruz

The interest in research and implementations of type-2 fuzzy controllers (T2FCs) is increasing. It has been demonstrated that these controllers provide more advantages in handling uncertainties than type-1 FCs (T1FCs). This characteristic is very appealing because real-world problems are full of inaccurate information from diverse sources. Nowadays, it is no problem to implement an intelligent controller (IC) for microcomputers since they offer powerful operating systems, high-level languages, microprocessors with several cores, and co-processing capacities on graphic processing units (GPUs), which are interesting characteristics for the implementation of fast type-2 ICs (T2ICs). However, the above benefits are not directly available for the design of embedded ICs for consumer electronics that need to be implemented in devices such as an application-specific integrated circuit (ASIC), a field-programmable gate array (FPGAs), etc. Fortunately, for T1FCs there are platforms that generate code in VHSIC hardware description language (VHDL; VHSIC: very high speed integrated circuit), C++, and Java. This is not true for the design of T2ICs, since there are no specialized tools to develop the inference system as well as to optimize it.

The aim of this chapter is to present different ways of achieving high-performance computing for evolving T1 and T2 ICs embedded into FPGAs. Therefore, we provide a compiled introduction to T1 and T2 FCs, with emphasis on the well-known bottle neck of the interval T2FC (IT2FC), and software and hardware proposals to minimize its effect regarding computational cost. An overview of learning systems and hosting technology for their implementation is given. We explain different ways to achieve such implementations: at the circuit level using a hardware description lan-

guage, using a multiprocessor system and a high-level language, and combining both methods. We explain how to use the IT2FC developed in VHDL as a standalone system, and as a coprocessor for the FPGA Fusion of Actel, Spartan 6, and Virtex 5. We present the methodology and two new proposals to achieve evolution of the IT2FC for FPGA, one for the static region of the FPGA, and the other one for the reconfigurable region using the dynamic partial reconfiguration methodology.

76.1 Overview

An intelligent system and evolution are intrinsically related since it is difficult to conceive intelligence without evolution because intelligence cannot be static. Human beings create, adapt, and replace their own rules throughout their whole lives. The idea to apply evolution to a fuzzy system is an attempt to construct a mathematical assembly that can approximate human-like reasoning and learning mechanisms [76.1]. A mathematical tool that has been successfully applied to better represent different forms of knowledge is fuzzy logic (FL); also if-then rules are a good way to express human knowledge, so the application of FL to a rule-based system leads to a Fuzzy Rule-Based System (FRBS). Unfortunately, an FRBS is not able to learn by itself, the knowledge needs to be derived from the expert or generated automatically with an evolutionary algorithm (EA) such as a genetic algorithm (GA) [76.2].

The use of GAs to design machine learning systems constitutes the soft computing paradigm known as the genetic fuzzy system where the goal is to incorporate learning to the system or tuning different components of the FRBS. Other proposals in the same line of work are: genetic fuzzy neural networks, genetic fuzzy clustering, and fuzzy decision trees. A system with the capacity to evolve can be defined as a self-developing, self-learning, fuzzy rule-based or neuro-fuzzy system with the ability to self-adapt its parameters and structure online [76.3].

Figure 76.1 shows the general structure of an evolutionary FRBS (EFRBS) that can be used for tuning or learning purposes. Although, it is difficult to make a clear distinction between tuning and learning, the particular aspect of each process can be summarized as follows. The tuning process is assumed to work on a predefined rule base having the target to find the optimal set of parameters for the membership functions and/or scaling functions. On the other hand, the learning process requires that a more elaborated search in the space of possible rule bases, or in the whole knowledge base be achieved, as well as for the scaling functions. Since the learning approach does not depend on a predefined set of rules and knowledge, the system can change its fundamental structure with the aim of improving its performance according to some criteria. The idea of using scaling functions for input and output variables is to normalize the universe of discourse in which membership functions were defined.

According to *De Jong* [76.4]:

the common denominator in most learning systems is their capability of making structural changes to themselves over time with the intent of improving performance on tasks defined by the environment, discovering and subsequently exploiting interesting concepts, or improving the consistency and generality of internal knowledge structures.

Hence, it is important to have a clear understanding of the strengths and limitations of a particular learning system, to achieve a precise characterization of all the

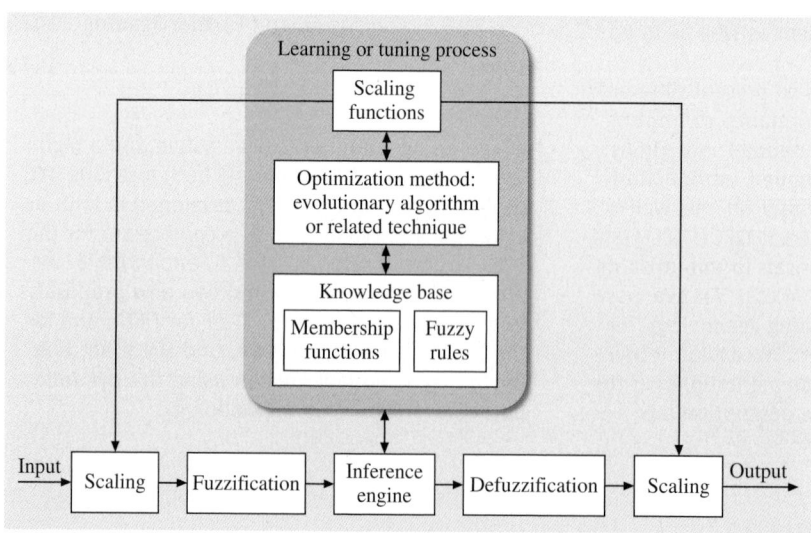

Fig. 76.1 General structure of an evolutionary fuzzy rule-based system

permitted structural changes and how they are going to be made.

De Jong sets three different levels of complexity where the GA can perform legal structural changes in following a goal, these are [76.4]:

1. By changing critical parameters' values
2. By changing key data structures
3. By changing the program itself with the idea of achieving effective behavioral changes in a task subsystem where a prominent representative of this branch is the *learning production-systems program*.

A good reason behind the success of production systems in machine learning is due to the fact that they have a representation of knowledge that can simultaneously support two kinds of activities: (1) the knowledge can be treated as data that can be manipulated according to some criteria; (2) for a particular task, the knowledge can be used as an executable entity.

The two classical approaches for working with evolutionary FRBS (EFRBS) for a learning system are the Pittsburgh and Michigan approaches. Historically, in 1975 *Holland* [76.5] affirmed that a natural way to represent an entire rule set is to use a string, i.e., an individual; so, the population is formed by candidate rule sets, and to achieve evolution it is necessary to use selection and genetic operators to produce new generations of rule sets. This was the approach taken by De Jong at the University of Pittsburgh, hence the name of *Pittsburgh approach*. During the same period, Holland developed a model of cognition in which the members of population are individual rules, and the entire population is conformed with the rule set; this quickly became the *Michigan approach* [76.6, 7].

There are extensive pioneering and recent work about tuning and learning using FRBS most of them fall in some way in the Michigan or in the Pittsburgh approaches, for example, the supervised inductive algorithm [76.8, 9], the iterative rule learning approach [76.10], coverage-based genetic induction (COGIN) [76.11, 12], the relational genetic algorithm learner (REGAL) system [76.13], the compact fuzzy classification system [76.14], with applications to fuzzy control [76.15, 16], and about tuning type-2 fuzzy controllers [76.17–20].

The focus of this chapter is on evolving embedded fuzzy controllers; this subclassification reduces the number of related works; however, they are still a big quantity, since by an embedding system (ES), we can understand a combination of computer hardware (HW)

and software (SW) devoted to a specific control function within a larger system. Typically, the HW of an ES can be a dedicated computer system, a microcontroller, a digital signal processor, or a FPGA-based system. If the SW of the ES is fixed, it is called firmware; because there are no strict boundaries between firmware and software, and the ES has the capability of being reprogrammed, the firmware can be low level and high level. Low-level firmware tells the hardware how to work and typically resides in a read only memory (ROM) or in a programmable logic array (PLA); high-level firmware can be updated, hence is usually set in a flash memory, and it is often considered software.

In the literature, there is extensive work on successful applications of type-1 and type-2 fuzzy systems; with regards to evolving embedded fuzzy systems, they were applied in a control mechanism for autonomous mobile robot navigation in real environments in [76.21]. For the sake of limiting more the content of this chapter, we have focused on EFRBSs to be implemented in an FPGA HW platform, with special emphasis on type-2 FRBSs. In this last category, with respect to type-1 FRBS took our attention to the following proposals: The development of an FPGA-based proportional-differential (PD) fuzzy look-up table controller [76.22], FPGA implementation of embedded fuzzy controllers for robotic applications [76.23], a non-fixed structure fuzzy logic controller is presented in [76.24], a flexible architecture to implement a fuzzy controller into an FPGA [76.25], a very simple method for tuning the input membership function (MF) for modifying the implemented FPGA controller response [76.26]; how to test and simulate the different stages of a FRBS for future implementation into an FPGA are explained in [76.27–29]. On type-1 EFRBS there are some works like: A reconfigurable hardware platform for evolving a fuzzy system by using a cooperative coevolutionary methodology [76.30], the tuning of input MFs for an incremental fuzzy PD controller using a GA [76.31]. In the type-2 FRBS category, the amount of reported work is less; representative work can be listed as follows: an architectural proposal of hardware-based interval type-2 fuzzy inference engine for FPGA is presented in [76.32], the use of parallel HW implementation using bespoke coprocessors handled by a soft-core processor of an interval type-2 fuzzy logic controller is explored in [76.33], a high-performance interval type-2 fuzzy inference system (IT2-FIS) that can achieve the four stages fuzzification, inference, KM-type reduction, and defuzzification in four clock cycles is shown in [76.34]; the same system is suitable for implementation in pipelines providing the complete IT2-FIS process in just one clock cycle.

This work deals with the development of evolving embedded type-1 and type-2 fuzzy controllers. In the chapter, a broad exploration of several ways to implement evolving embedded fuzzy controllers are presented. We choose to work with the Mamdani fuzzy controller proposal since it provides a highly flexible means to formulate knowledge.

The organization of this chapter is as follows. In Sect. 76.2 we present the basis of T1 and T2 FL to explain how to achieve the HW implementation of an FRBS. In Sect. 76.3 a brief description of the state of the art in hosting technology for high-performance embedded systems is given.

76.2 Type-1 and Type-2 Fuzzy Controllers

The type-2 fuzzy sets (T2FS) were developed with the aim of handling uncertainty in a better way than T1 FS does, since a T1FS has crisp grades of membership, whereas a T2FS has fuzzy grades of membership. An important point to note is that if all uncertainty disappears, a T2 FS can be reduced to a T1FS. A type-2 membership function (T2MF) is an FS that has primary and secondary membership values; the primary MF is a representation of an FS, and serves to create a linguistic representation of some concept with linguistic and random uncertainties with limited capabilities; the secondary MF allows capturing more about linguistic uncertainty than a T1MF.

There are two common ways to use a T2FS, the generalized T2FS (GT2), and the interval T2FS (IT2FS). The former has secondary membership grades of different values to represent more accurately the existing uncertainty; on the other hand, in an IT2FS the secondary membership value always takes the value of 1. Unfortunately, to date for GT2 no one knows yet how to choose their best secondary MFs; moreover, this method introduces a lot of computations, making it inappropriate for current application in real-time (RT) systems, even those with small time constraints; in contrast, the calculations are easy to perform in an IT2FS.

A T2MF can be represented using a 3-D figure that is not as easy to sketch as a T1MF. A more common way to visualize a T2MF is to sketch its footprint of uncertainty (FOU) on the 2-D domain of the T2FS. We illustrate this concept in Fig. 76.2, where we show a vertical slice sketch of the FOU at the primary MF value x'; in the case of a GT2, in the right upper part of the figure, the secondary MF shows different height values of the GT2; in the case of an IT2F2, just below is the secondary MF with uniform values for the IT2FS. Note that the secondary values sit on top of its FOU.

Figure 76.3 shows the main components of a fuzzy logic system showing the differences between the T1 and T2 FC. For T1 systems, there are three components: fuzzifier, inference engine, and the defuzzifier which is

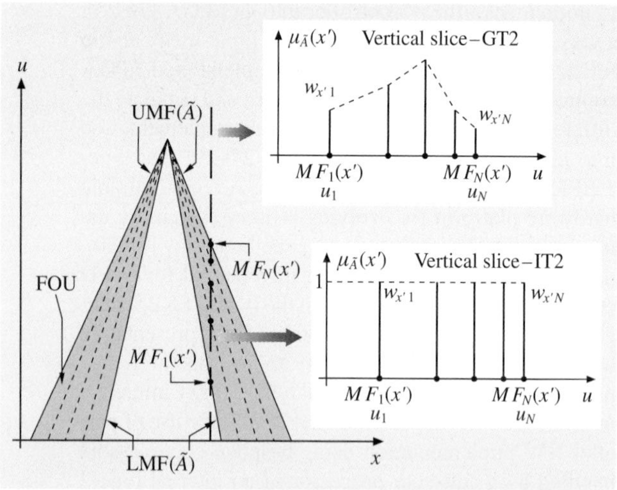

Fig. 76.2 Type-2 membership function. For the triangular MF the FOU is shown. The FOU is bounded by the upper part UMF($\widetilde{A}$) and the lower part LMF($\widetilde{A}$). A vertical slice at x' is illustrated. *Right, top*: secondary MF values for a generalized T2MF; *bottom*: secondary MF values of an IT2MF

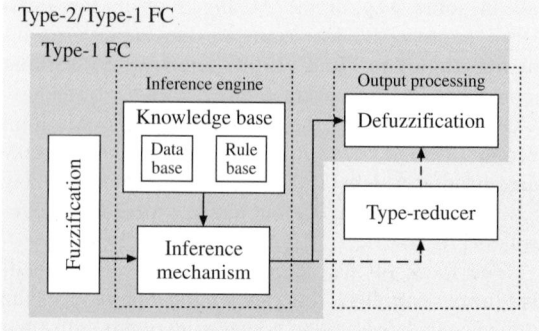

Fig. 76.3 Type-1 and type-2 FC. The T2FC at the output processing has the type reducer block

the only output processing unit; whereas for a T2 system there are four components, since the output processing has interconnected the type reducer (TR) block and the defuzzifier to form the output processing unit.

Ordinary fuzzy sets were developed by *Zadeh* in 1965 [76.35]; they are an extension of classical set theory where the concept of membership was extended to have various grades of membership on the real continuous interval $[0, 1]$. The original idea was to use a fuzzy set (FS); i.e., a linguistic term to model a word; however, after almost 10 years, *Zadeh* introduced the concept of type-n FS as an extension of an ordinary FS (T1FS) with the idea of blurring the degrees of membership values [76.36].

T1FSs have been demonstrated to work efficiently in many applications; most of them use the mathematics of fuzzy sets but lose the focus on words that are mainly used in the context to represent a function which is more mathematical than linguistic [76.37].

A T1FS is a set of ordered pairs represented by (76.1) [76.38],

$$A = \{(x, \mu_A(x)) | x \in X\} , \qquad (76.1)$$

where each element is mapped to $[0, 1]$ by its MF μ_A, where $[0, 1]$ means real numbers between 0 and 1, including the values 0 and 1,

$$\mu_A(x) : X \to [0, 1] . \qquad (76.2)$$

A pointwise definition of a T2FS is given as follows, $\widetilde{A}$ is characterized by a T2MF $\mu_{\widetilde{A}}(x, u)$, where $x \in X$ and $u \in J_x \subseteq [0, 1]$, i.e. [76.39],

$$\widetilde{A} = \{(x, u), \mu_{\widetilde{A}}(x, u) | \forall x \in X, \forall u \in J_x \subseteq [0, 1]\} , \qquad (76.3)$$

where $0 \leq \mu_{\widetilde{A}}(x, u) \leq 1$.

Another way to express $\widetilde{A}$ is

$$\widetilde{A} = \int_{x \in X} \int_{u \in J_x} \mu_{\widetilde{A}}(x, u)/(x, u) \quad J_x \subseteq [0, 1] , \qquad (76.4)$$

where $\int \int$ denote the union over all admissible input variables x' and u'. For discrete universes of discourse $\int$ is replaced by $\sum$ [76.39]. In fact, $J_x \subseteq [0, 1]$ represents the primary membership of $x \in X$ and $\mu_{\widetilde{A}}(x', u)$ is a T1FS known as the secondary set. Hence, a T2MF can be any subset in [0,1], the primary membership, and corresponding to each primary membership, there is a secondary membership (which can also be in [0,1]) that defines the uncertainty for the primary membership.

When $\mu_{\widetilde{A}}(x, u) = 1$, where $x \in X$ and $u \in J_x \subseteq [0, 1]$, we have the IT2MF shown in Fig. 76.2. The uniform shading for the FOU represents the entire IT2FS and it can be described in terms of an upper membership function and a lower membership function

$$\bar{\mu}_{\widetilde{A}}(x) = \overline{\text{FOU}(\widetilde{A})} \ \forall x \in X , \qquad (76.5)$$

$$\underline{\mu}_{\widetilde{A}}(x) = \underline{\text{FOU}(\widetilde{A})} \ \forall x \in X . \qquad (76.6)$$

Figure 76.2 shows an IT2MF, the shadow region is the FOU. At the points x_1 and x_2 are the primary MFs J_{x_1} and J_{x_2}, and the corresponding secondary MFs $\mu_{\widetilde{A}}(x_1)$ and $\mu_{\widetilde{A}}(x_2)$ are also shown.

The basics and principles of fuzzy logic do not change from T1FSs to T2FSs [76.37, 40, 41], they are independent of the nature of the membership functions, and in general, will not change for any type-*n*. When a FIS uses at least one type-2 fuzzy set, it is a type-2 FIS.

In this chapter we based our study on IT2FSs, so the IT2 FIS can be seen as a mapping from the inputs to the output and it can be interpreted quantitatively as $Y = f(X)$, where $X = \{x_1, x_2, \ldots, x_n\}$ are the inputs to the IT2 FIS f, and $Y = \{y_1, y_2, \ldots, y_n\}$ are the defuzzified outputs. These concepts can be represented by rules of the form

$$\text{If } x_1 \text{ is } \widetilde{F}_1 \text{ and } \ldots \text{ and } x_p \text{ is } \widetilde{F}_p, \text{ then } y \text{ is } \widetilde{G} . \qquad (76.7)$$

In a T1FC, where the output sets are T1FS, the defuzzification produces a number, which is in some sense a crisp representation of the combined output sets. In the T2 case, the output sets are T2, so the extended defuzzification operation is necessary to get T1FS at the output. Since this operation converts T2 output sets to a T1FS, it is called type reduction, and the T1FS is called a type-reduced set, which may then be defuzzified to obtain a single crisp number.

The TR stage is the most computationally expensive stage of the T2FC; therefore, several proposals to improve this stage have been developed. One of the first proposals was the iterative procedure known as the Karnik–Mendel (KM) algorithm.

In general, all the proposals can be classified into two big groups. Group I embraces all the algorithmic improvements and Group II all the hardware improvements, as follows [76.42]:

1. Improvements to software algorithms, where the dominant idea is to reduce computational cost of IT2-FIS based on algorithmic improvements. This group can be subdivided into three subgroups.

(a) Enhancements to the KM TR algorithm. As the classification's name claims, the aim is to improve the original KM TR algorithm directly, to speed it up. The best known algorithms in this classification are:

i. Enhanced KM (EKM) algorithms. They have three improvements over the original KM algorithm. First, a better initialization is used to reduce the number of iterations. Second, the termination condition of the iterations is changed to remove unnecessary iterations (one). Finally, a subtle computing technique is used to reduce the computational cost of each iteration.

ii. The enhanced Karnik–Mendel algorithm with new initialization (EKMANI) [76.43]. It computes the generalized centroid of general T2FS. It is based on the observation that for two alpha-planes close to each other, the centroids of the two resulting IT2FSs are also closed to each other. So, it may be advantageous to use the switch points obtained from the previous alpha-plane to initialize the switch points in the current alpha-plane. Although EKMANI was primarily intended for computing the generalized centroid, it may also be used in the TR of IT2-FIS, because usually the output of an IT2-FIS changes only a small amount at each step.

iii. The iterative algorithm with stop condition (IASC). This was proposed by *Melgarejo* et al. [76.44] and is based on the analysis of behavior of the firing strengths.

iv. The enhaced IASC [76.45] is an improvement of the IASC.

v. Enhanced opposite directions searching (EODS), which is a proposal to speed up KM algorithms. The aim is to search in both directions simultaneously, and in each iteration the points L and R are the switching points.

(b) Alternative TR algorithms. Unlike iterative KM algorithms, most alternative TR algorithms have a closed-form representation. Usually, they are faster than KM algorithms. Two representative examples are:

i. The Gorzalczany method. A polygon using the firing strengths $[\underline{f}^n, \overline{f}^n]$ and $[(y^1, y^n)$, which can be viewed as an IT2FS. It computes an approximate membership value for each point. Here, $\underline{y}^n = \overline{y}^n = y^n$,

for $n = 1, 2, 3 \ldots, N$.

$$\mu(y) = \frac{\underline{f} + \overline{f}}{2} \cdot [1 - (\overline{f} - \underline{f})], \qquad (76.8)$$

where $\overline{f} - \underline{f}$ is called the bandwidth. Then the defuzzified output can be computed as

$$y_G = \arg \max_y \mu(y). \qquad (76.9)$$

ii. The Wu–Tan (WT) method. It searches an equivalent T1FS. The centroid method is applied to obtain the defuzzification. This is the faster method in this category.

2. Hardware implementation. The main idea is to take advantage of the intrinsic parallelism of the hardware and/or combinations of hardware and parallel programming. Here, we divided this group into four main approaches that embrace the existing proposals of reducing the computational time of the type reduction stage by the use of parallelism at different levels.

(a) The use of multiprocessor systems, including multicore systems that enable the same benefits at a reduced cost. In this category are personal and industrial computers with processors such as the Intel Pentium Core Processor family, which includes the Intel Core i3, i5 and i7; the AMD Quad-Core Optetron, the AMD Phenom X4 Quad-Core processors, multicore microcontrollers such as the Propeller P8X32A from Parallax, or the F28M35Hx of the Concerto Microcontrollers family of Texas Instruments. Multicore processors also can be implemented into FPGAs.

(b) The use of a general-purpose GPU (GPGPU), and compute unified device architecture (CUDA). In general, GPU provides a new way to perform high performance computing on hardware. In particular IT2FCs can take the most advantage of this technology because their complexity. Traditionally, before the development of the CUDA technology, the programming was achieved by translating a computational procedure into a graphic format with the idea to execute it using the standard graphic pipeline; a process known as encoding data into a texture format. The CUDA technology of NVIDIA offers a parallel programming model for GPUs that does not require the use of a graphic application programming interface (API), such as OpenGL [76.46].

(c) The use of FPGAs. This approach offers the best processing speed and flexibility. One of the main advantages is that the developer can determine the desired parallelism grade by a trade-off analysis. Moreover, this technology allows us to use the strength of all platforms in tight integration to provide the large performance available at the present time. It is possible to have a standalone T1/IT2FC, or to integrate the same T1/T2FC as a coprocessor as part of a high performance computing system.

(d) The use of ASICs. The T1/T2FC is factory integrated using complementary metal-oxide-semiconductor (CMOS) technology. The main advantages are that they are cheaper than FPGAs. Differently to FPGA technology, ASIC solutions are not field reprogrammable.

A system based on an FPGA platform allows us to program all the Group I algorithms since modern FPGAs have embedded hard and/or soft processors; this kind of system can be programmed using high-level languages such as C/C++ and also they can incorporate operating systems such as Linux. On the other hand, T1/T2 FC hardware implementations have the advantage of providing competitive faster systems in comparison to ASIC systems and the in field reconfigurability.

76.3 Host Technology

Until the beginnings of this century, general-purpose computers with a single-core processor were the systems of choice for high-performance computing (HPC) for many applications; they replaced existing big and expensive computer architectures [76.47]. In 2001, IBM introduced a reduced intstruction set computer (RISC) microarchitecture named POEWER4 (performance optimization with enhanced RISC) [76.48]. This was the first dual core processor embedded into a single die, and subsequently other companies introduced different multicore microprocessor architectures to the market, such as the Arm Cortex A9 [76.49], Sparc64 [76.50], Intel and AMD Quad Core processors, Intel i7 processors, and others [76.51]. These developments, together with the rapid development of GPUs that offer massively parallel architectures to develop high-performance software, are an attractive choice for professionals, scientists, and researchers interested in speeding up applications. Undoubtedly, the use of a generic computer with GPU technology has many advantages for implementing an embedded learning fuzzy system [76.46], and disadvantages are mainly related to size and power consumption. A solution to the aforementioned problems is the use of application specific integrated circuits (ASICs) fuzzy processors [76.52–54], or reprogrammable hardware based on microcontrollers and/or FPGAs.

The orientation of this paper is towards tuning and learning using FRBS for embedded applications; for now, we are going to focus on FPGAs and ASIC technology [76.55], since they provide the best level of parallelization. Both families of devices provide characteristics for HPC that the other options cannot. Each technology has its own advantages and disadvantages, which are narrowing down due to recent developments. In general, ASICs are integrated circuits that are designed to implement a single application directly in fixed hardware; therefore, they are very specialized for solving a particular problem. The costs of ASIC implementations are reduced for high volumes; they are faster and consume less power; it is possible to implement analog circuitry, as well as mixed signal design, but the time to market can take a year or more. There are several design issues that need to be carried out that do not need to be achieved using FPGAs, the tools for development are very expensive. On the other hand, FPGAs can be introduced to the market very fast since the user only needs a personal computer and low-cost hardware to burn the HDL (HDL) code to the FPGA before it is ready to work. They can be remotely updated with new software since they are field reprogrammable. They have specific dedicated hardware such as blocks of random access memory (RAM); they also provide high-speed programmable I/O, hardware multipliers for digital signal processing (DSP), intellectual property (IP) cores, microprocessors in the form of hard cores (factory implemented) such as PowerPC and ARM for Xilinx, or Microblaze and Nios softcore (user implemented) for Xilinx and Altera, respectively. They can have built-in analog digital converters (ADCs). The synthesis process is easier. A significant point is that the HDL tested code developed for FPGAs may be used in the design process of an ASIC.

There are three main disadvantages of the FPGAs versus ASICs, they are: FPGA devices consume more power than ASICs, it is necessary to use the resources available in the FPGA which can limit the design, and they are good for low-quantity production. To overcome these disadvantages it is very important to achieve optimized designs, which can only be attained by coding efficient algorithms.

During the last decade, there has been an increasing interest in evolving hardware by the use of evolutionary computations applied to an embedded digital system. Although different custom chips have been proposed for this plan, the most popular device is the FPGA because its architecture is designed for general-purpose commercial applications. New FGAs allow modification of part of the programmed logic, or add new logic at the running time. This feature is known as dynamic or active reconfiguration, and because in an FPGA we can combine a multiprocessor system and coprocessors, FPGAs are very attractive for implementing evolvable hardware algorithms. Therefore, in the next sections, we shall put special emphasis on multiprocessor systems and FPGAs.

76.4 Hardware Implementation Approaches

In this section, an overview of the three main lines of attack to do a hardware implementation of an intelligent system is given.

76.4.1 Multiprocessor Systems

Multiprocessor systems consist of multiple processors residing within one system; they have been available for many years. Multicore processors have equivalent benefits to multiprocessors at a lower cost; they are integrated in the same electronic component. At the present time, most modern computer systems have many processors that can be single core or multicore processors; therefore, we can have three different layouts for multiprocessing; a multicore system, a multiprocessor system, and a multiprocessor/multicore system.

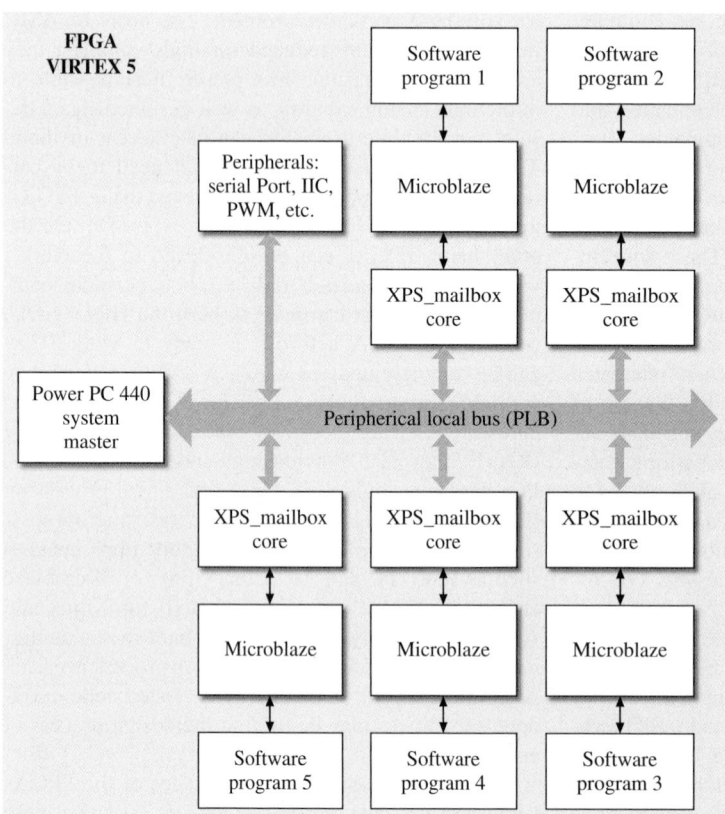

Fig. 76.4 Multicore system embedded into an FPGA. Embedded is a hard-processor PowerPC440 and five MicroBlaze soft-processors. In this system we can process an EA using the island model

Fig. 76.5 The whole embedded evolutionary IT2FC implemented in the program memory of the multiprocessor system, similarly as in a desktop computer ▶

Figure 76.4 shows a multicore system embedded into a Virtex 5 FPGA XC5VFX70; it has the capacity to integrate a distributed multicore system with a hard-processor PowerPC 440 as the master, five Microblaze 32-bit soft-processor slaves, coprocessors, and peripherals. The FPGA capacity to integrate devices is, of course, limited by the size of the FPGA. Figure 76.5 shows the full implementation in the program memory of the multiprocessor system.

76.4.2 Implementations into FPGAs

The architecture of FPGAs offers massive parallelism because they are composed of a large array of configurable logic blocks (CLBs), digital signal processing blocks (DSPs), block RAM, and input/output blocks (IOBs). Similarly, to a processor's arithmetic unit (ALU), CLBs and DSPs can be programmed to perform arithmetic and logic operations like compare, add/subtract, multiply, divide, etc. In a processor, ALU architectures are fixed because they have been designed in a general-purpose manner to execute various operations. CLBs can be programmed using just the operations that are needed by the application, which results in increased computation efficiency. Therefore, an FPGA consists of a set of programmable logic cells manufactured into the device according to a connection paradigm to build an array of computing resources; the resulting arrangement can be classified into four categories: symmetrical array, row-based, hierarchy-based, and sets of gates [76.56]. Figure 76.6 shows a symmetrical array-based FPGA that consists of a two-dimensional array of logic blocks immersed in a set of vertical and horizontal lines; examples of FPGAs in this category are Spartan and Virtex from Xilinx, and Atmel AT40K. In Fig. 76.6 three main parts can be identified: a set of programmable logic cells also called logic blocks (LBs) or configurable logic blocks (CLBs), a programmable interconnection network, and a set of input and output cells around the device.

Embedded programmable logic devices usually integrate one or several processor cores, programmable logic and memory on the same chip (an FPGA) [76.56]. Developments in the field of FPGA have been very amazing in the last two decades, and for this reason, FPGAs have moved from tiny devices with a few thousand gates that were used in small applications such as

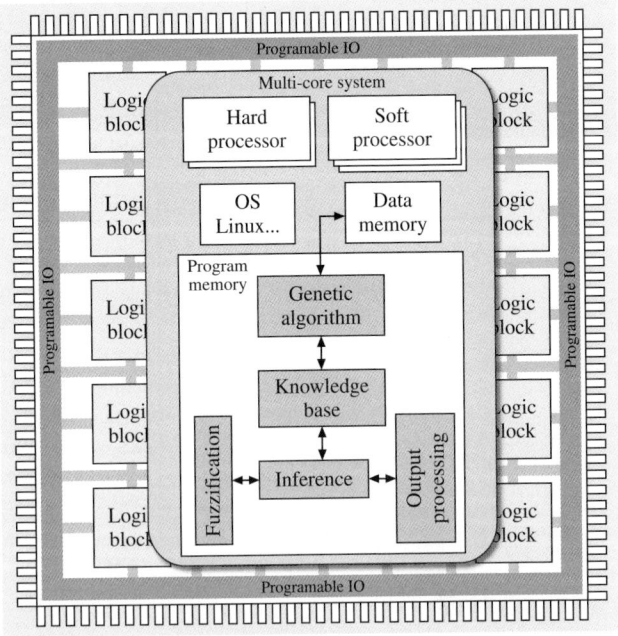

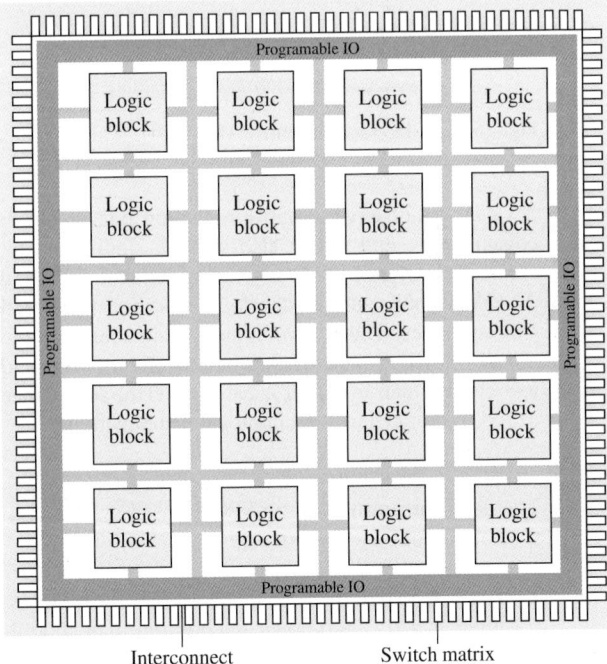

Fig. 76.6 Symmetric array-based FPGA architecture *island style*

finite state machines, glue-logic for complex devices, and very limited CPUs. In a 10-year period of time, a 200% growth rate in the capacity of Xilinx FPGAs

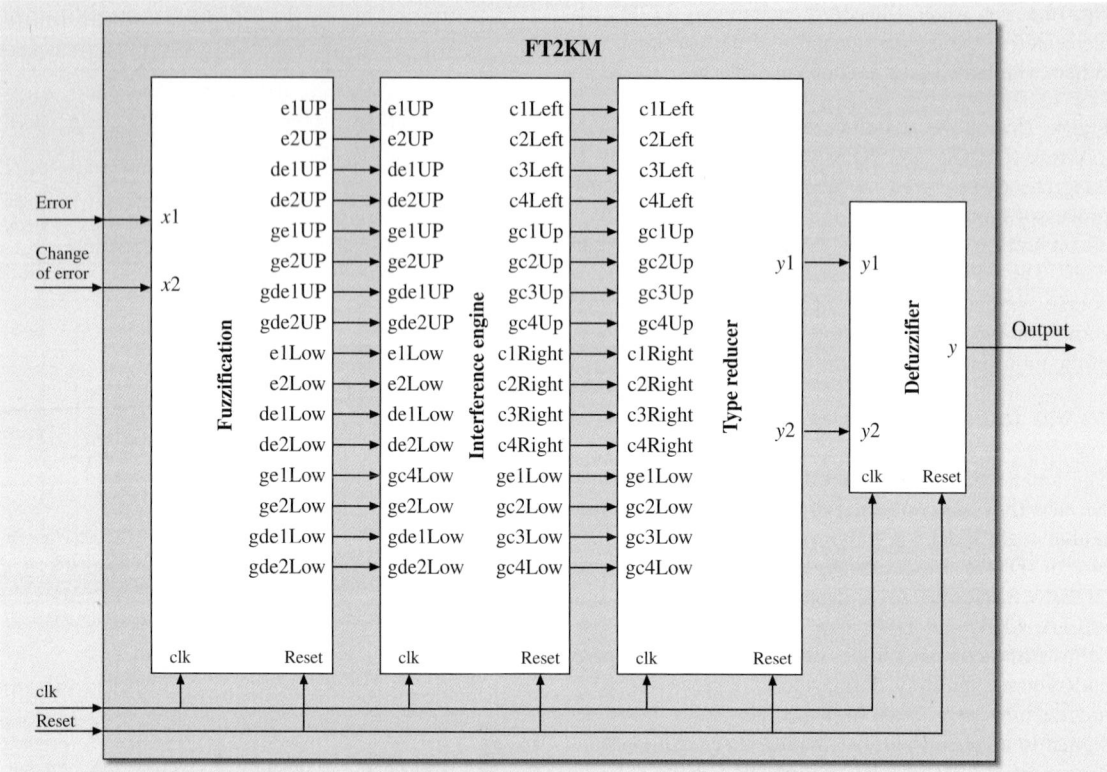

Fig. 76.7 IT2FC design entity (FT2KM). This top-level module contains instances of the four fuzzy controller submodules

devices was observed, a 50% reduction rate in power consumption, and prices also show a significant decrease rate. Other FPGA vendors, such as ACTEL, and ALTERA show similar developments, and this trend still continues. These developments, together with the progress in development tools that include software and low-cost evaluation boards, have boosted the acceptance of FPGAs for different technological applications.

Development Flow

The development flow of an FPGA-based system consists of the following major steps:

1. Write in VHDL the code that describes the systems' logic; usually a top-down and bottom-up methodology is used. For example, to design an IT2FC, we need to achieve the following procedure:
   (a) Describe the design entity where the designer defines the input and output of the top VHDL module. The idea is to present the complex object in different hierarchical levels of abstrac-

tion. For our example, the top design entity is *FT2KM*.
   (b) Once the design entity has been defined, it is required to define its architecture, where the description of the design entity is given; in this step, we define its behavior, its structure, or a mixture of both. For the case of the IT2 FLS, we define the system's internal behavior, so we determined the necessity to achieve a logic design formed by four interconnected modules: fuzzification, inference engine, type reduction, and defuzzification. The VHDL circuits (submodules) are described using a register transfer logic (RTL) sequence, since we can divide the functionality in a sequence of steps. At each step, the circuit achieves a task consisting in data transference between registers and evaluation of some conditions in order to go to the next step; in other words, each VHDL module (design entity) can be divided into two areas: data and control. Each of the four modules needs

to be conceptualized, so we need to define its own design entity and, therefore, its particular architecture as well interconnections with internal modules. This process is achieved when we have reached the last system component.

(c) Integrate the system. It is necessary to create a main design entity (top level) that integrates the submodules defining their interconnections. In Fig. 76.7 the integration of the four modules is shown.

2. Develop the test bench in VHDL and perform RTL simulations for each submodule of the main design entity. It is necessary to achieve timing and functional simulations to create reliable internal design entities.

3. Perform synthesis and implementation. In the synthesis process, the software transforms the VHDL constructs to generic gate-level components, such as simple logic gates and flip-flops. The implementation process is composed of three small subprocesses: translate, map, and place, and route. In the translate process the multiple design files of a project are merged into a single netlist. The map process maps the generic gates in the netlist to the FPGA's logic cells and IOBs, this process is also known as technology mapping. In the place and route process, using the physical layout inside the FPGA chip, the process places the cells in physical locations and determines the routes to connect diverse signals. In the Xilinx flow, the static timing analysis performed at the end of the implantation process determines various timing parameters such as maximal clock frequency and maximal propagation delay [76.57].

4. Generate the programming file and download it to the FPGA. According to the final netlist a configuration file is generated, which is downloaded to the FPGA serially.

5. Test the design entity using a simulation program such as Simulink of Matlab and the Xilinx system generator (XSG) for Xilinx devices. The idea here is first to plot the surface control in order to analyze the general behavior of the design (a controller in

our example), and second to integrate the design entity as a block of the desired system to be controlled. Although, this fifth step, is not in the current literature of logic design for FPGA implementation, it is the authors's recommendation since we have experienced good results following this practice.

Using the design entity FT2KM.vhd, which was created and tested using the aforementioned development flow, we can integrate it an FPGA in two ways:

1. As a standalone system. Here, we mean an independent system that does not require the support of any microprocessor to work, the system itself is a specialized circuit that can produce the desired output. The IT2FC is implemented using the FPGA flow design; therefore, it is programmed using the complete development flow for a specific application.

2. As a coprocessor. The coprocessor performs specialized functions in such a way that the main system processor cannot perform as well and faster. For IT2FCs, given an input, the time to produce an output is big enough to achieve an adequate control of many plants when the IT2FC is programmed using high-level language, even we have used a parallel programming paradigm. Since a coprocessor is a dedicated circuit designed to offload the main processor, and the FPGA can offer parallelism on the circuit level, the designer of the IT2FC coprocessor can have control of the controller performance. The coprocessor can be physically separated, i.e., in a different FPGA circuit (or module), or it can be part of the system, in the same FPGA circuit. In this work, we show two methods to develop a system with an IT2FC as a coprocessor. In both methods, we consider that we have a tested IT2FC design entity. In the first case, we shall use the *FT2KM* design entity to incorporate the fuzzy controller as a coprocessor of an ARM processor into an FPGA Fusion. In the second case, we are going to create the IT2FC IP core using the Xilinx Platform Studio; the core will serve as a coprocessor of the MicroBlaze processor embedded into a Spartan 6 FPGA.

76.5 Development of a Standalone IT2FC

Figure 76.7 shows the top-level design entity (FT2KM) of the IT2FC and its components (submodules) for FPGA implementation. The entity codification of the top-level entity and its components are given in

Sect. 76.5.1. All stages include the clock (clk) and reset (rst) signals. In the defuzzifier, we have included these two signals to illustrate that a full process takes only four clock cycles, one for each stage. In practice, we did not add these two signals, since when we used it as a coprocessor, in order to incorporate it to the system, one 8-bit data latch is added at the output. For a detailed description of the IT2FC stages consult [76.34].

The fuzzification stage has two input variables, x_1 and x_2. This module contains a fuzzifier for the upper MFs, and another for the lower MFs of the IT2FC. For the upper part, for the first input x_1, considering that a crisp value can be fuzzified by two MFs because it may have membership values in two contiguous T2MFs, the linguistic terms are assigned to the VHDL variables e_{1up} and e_{2up}, and their upper membership values are ge_{1up} and ge_{2up}. For the second input x_2, the linguistic terms are assigned to the VHDL variables de_{1up} and de_{2up}, and gde_{1up} and gde_{2up} are the upper membership values. The lower part of the fuzzifier is similar; for example, for the input variable x_1 the VHDL assigned variables are e_{1low} and e_{2low}, and their lower MF values are ge_{1low} and ge_{2low}, etc. The fuzzification stage entity *only needs one clock cycle* to perform the fuzzification. These eight variables are the inputs of the inference engine stage [76.58].

The inference engine is divided into two parallel inference engine entities IEEup is used to manage the upper bound of the IT2FC, and IEElow for the lower bound of the IT2FCs. Each entity has eight inputs from the corresponding fuzzifier stage, and eight outputs; four belong to the output linguistic terms, the rest correspond to their firing strengths. All the inputs enter into a parallel selection VHDL process, the circuits into the process are placed in parallel; the degree of parallelism can be tailored by an adequate codification style. In our case, all the rules are processed in parallel and the eight outputs of each inference engine section (upper bound and lower bound) are obtained at the same time because the *clk* signal synchronizes the process, hence *this stage needs only one clock cycle to perform a whole inference and provide the output to the next stage*. In the upper bound, the four antecedents are formed at the same time, for example, for the first rule, the antecedent is formed using the concatenation operator &, so it looks like *ante := e*1 & *de*1. Each antecedent can address up to four rules and depending on the combination, one of the four rules is chosen; the upper inference engine output provides the active consequents and its firing strengths. The lower bound of the inference engine is treated in the same way [76.59].

At the input of the TR, we have the equivalent values of the pre-computed y_l^i, i.e., the linguistic terms of the active consequents (C_{1left}, C_{2left}, C_{3left}, and C_{4left}), the upper firing strength (gc_{1up}, gc_{2up}, gc_{3up}, and gc_{4up}), in addition to the equivalent values of the pre-computed y_r^i(C_{1right}, C_{2right}, C_{3right}, and C_{4right}), the lower firing strength (gc_{1low}, gc_{2low}, gc_{3low}, and gc_{4low}) [76.60]. All the above-mentioned signals go to a parallel selection process to perform the KM algorithm [76.39]. There are parallel blocks to obtain the average of the upper and lower firing strength for the active consequents, required to obtain the average of the y_r and y_l; a block to obtain the different defuzzified values of y_r and y_l; parallel comparator blocks to obtain the final result of y_r and y_l [76.61].

The final result of the IT2FC is obtained using the defuzzification block, which computes the average of the y_r and y_l, and produces the only output y.

76.5.1 Development of the IT2 FT2KM Design Entity

Figure 76.8 shows the implementation of a static IT2FC that can work as a standalone system. By static, we mean that the only way to reconfigure (modify) the FC is to stop the application and uploading the whole

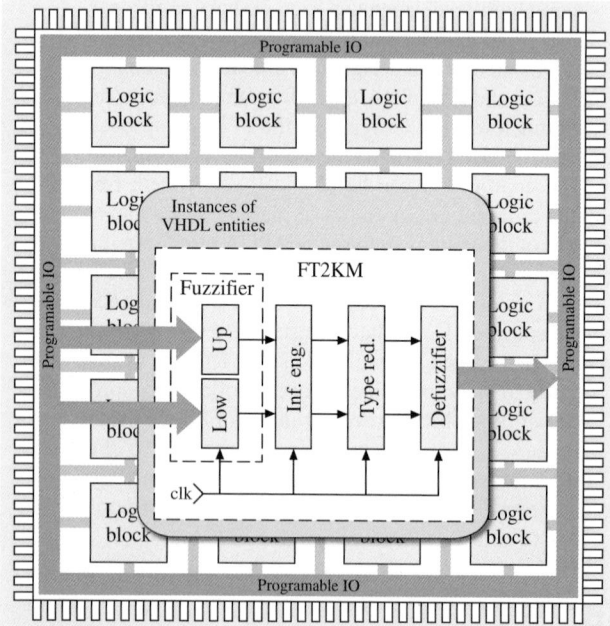

Fig. 76.8 A standalone IT2FC is embedded into an FPGA. The fuzzifier reads the inputs directly from the FPGA terminals. The defuzzifier sends the crisp output to the FPGA terminals. The system may be embedded in the static region or in the reprogrammable region

configuration bit file (bitstream). In this system, the inputs of the fuzzifier and the defuzzifier output are connected directly to the FPGA terminals. The assignment of the terminals is achieved in accordance with the internal architecture of the chosen FPGA. Hence, it is necessary to provide to the Xilinx Integrated Synthesis Environment (ISE) program, special instructions (constraints) to carry through the synthesis process. They are generally placed in the user constraint file (UCF), although they may exist in the HDL code. In general, constraints are instructions that are given to the FPGA implementation tools with the purpose of directing the mapping, placement, timing, or other guidelines for the implementation tools to follow while processing an FPGA design. In Fig. 76.7 the overall entity of design of the IT2FC (FTK2M) was defined as follows,

```
entity FT2KM is
  Port(clk, reset    : in std_logic;
        x1, x2       : in std_logic_vector(8 downto 1);
        y            : out std_logic_vector (8 downto 1)
  );
end FT2KM;
```

The architecture of FT2KM has four components, and all of them have two common input ports: clock (clk), and reset (rst). All ports in an entity are signals by default. This is important since a signal serves to pass values in and out of the circuit; a signal represents circuit interconnects (wires). A component is a simple piece of customized code formed by entities as corresponding architectures, as well as library declarations. To allow a hierarchical design, each component must be declared before been used by another circuit, and to use a component it is neccesary to instatiate it first. In this approach the components are:

1. The component labeled as *fuzzyUpLw*. It is the T2 fuzzifier that consists of one fuzzifier for the upper MF of the FOU and one for the lower MF. It has two input ports *x1* and *x2*; these are 16: *e1Up* to *de2Low*.

```
component fuzzyUpLw is
   port(clk, reset : in std_logic;
        x1, x2, ge1Up, ge2Up, gde1Up, gde2Up :
            in std_logic_vector(n downto 1);
        e1Up, e2Up, de1Up, de2Up, e1Low,
           e2Low, de1Low,
        de2Low : out std_logic_vector(3 downto 1);
        ge1Up, ge2Up, gde1Up, gde2Up, ge1Low,
           ge2Low, gde1Low,
        gde2Low : out std_logic_vector(n downto 1);
   );
   end component;
```

The instantiation of this component is achieved using nominal mapping and the name of this instance is *fuzt2*. Note that ports *clk*, *reset*, and *x1* and *x2* are mapped (connected) directly to the entity of design FT2KM, since as we explained before, all ports are signals by default, which represent wires. The piece of code that defines the instantiation of the *fuzzyUpLw* component is as follows,

```
fuzt2 : fuzzyUpLw port map(
        clk => clk, reset=> reset, x1 => x1, x2 => x2,
        e1Up => e1upsig, e2Up => e2upsig, de1Up => de1upsig,
        de2Up => de2upsig, ge1Up => ge1upsig, ge2Up => ge2upsig,
        gde1Up => gde1upsig, gde2Up => gde2upsig, e1Low   => e1lowsig,
        e2Low   => e2lowsig, de1Low => de1lowsig, de2Low => de2lowsig,
        ge1Low => ge1lowsig, ge2Low => ge2lowsig, gde1Low => gde1lowsig,
        gde2Low => gde2lowsig
    );
```

2. The component *Infer_type_2* corresponds to the T2 inference the controller. It has 16 inputs that match to the 16 outputs of the fuzzification stage. This component has 16 outputs to be connected to the type reduction stage. The piece of code to include this component is:

```
component Infer_type_2 is
   port(rst, clk : in std_logic;
        e1, e2, de1, de2, e1_2, e2_2, de1_2, de2_2 : in  STD_LOGIC_VECTOR (m downto 1);
        g_e1, g_e2, g_de1, g_de2, g_e1_2, g_e2_2,
        g_de1_2, g_de2_2 : in  STD_LOGIC_VECTOR (n downto 1);
        c1, c2, c3, c4, c1_2, c2_2, c3_2, c4_2 : out  STD_LOGIC_VECTOR (m downto 1);
        gc1_2, gc2_2, gc3_2, gc4_2, gc1, gc2, gc3, gc4 : out  STD_LOGIC_VECTOR (n downto 1);
   );
end component;
```

This component is instantiated with the name *Infer_type_2* as follows,

```
inft2: Infer_type_2 port map(
        rst => reset, clk => clk, e1 => e1upsig, e2 => e2upsig, de1 => de1upsig,
        de2 => de2upsig, g_e1 => ge1upsig, g_e2 => ge2upsig, g_de1 => gde1upsig,
        g_de2 => gde2upsig, e1_2 => e1lowsig, e2_2 => e2lowsig, de1_2 => de1lowsig,
        de2_2 => de2lowsig, g_e1_2 => ge1lowsig, g_e2_2 => ge2lowsig, g_de1_2 => gde1lowsig,
        g_de2_2 => gde2lowsig, c1 => c1sig, c2 => c2sig, c3   => c3sig, c4   => c4sig,
        gc1 => gc1sig, gc2 => gc2sig, gc3 => gc3sig, gc4 => gc4sig, c1_2 => c12sig,
        c2_2 => c22sig, c3_2 => c32sig, c4_2 => c42sig, gc1_2 => gc12sig,
        gc2_2 => gc22sig, gc3_2 => gc32sig, gc4_2 => gc42sig
    );
```

To connect the instances *fuzt2* and *Infer_type_2* it is necessary to define some signals (wires),

```
signal e1upsig, e2upsig, de1upsig, de2upsig : std_logic_vector (m-1 downto 0);
signal ge1upsig, ge2upsig, gde1upsig, gde2upsig :std_logic_vector (7 downto 0);
signal e1lowsig, e2lowsig, de1lowsig, de2lowsig :std_logic_vector (m-1 downto 0);
signal ge1lowsig, ge2lowsig, gde1lowsig, gde2lowsig : std_logic_vector (7 downto 0);
```

3. The component *TypeRed* corresponds to the type reduction stage of the T2FC. It has 16 inputs that should connect the inference engine's outputs and it has two outputs *yr* and *yl* that should be connected to the deffuzifier through signals, once both have been instantiated. The piece of code to include this component is:

```
component TypeRed is
    Port (clk, rst : in std_logic;
          c1,  c2,  c3, c4, c1_2, c2_2, c3_2,  c4_2 : in  STD_LOGIC_VECTOR (3 downto 1);
          gc1, gc2, gc3, gc4, gc1_2, gc2_2,  gc3_2, gc4_2 : in  STD_LOGIC_VECTOR (7 downto 0);
          yl, yr : out std_logic_vector (8 downto 1));
end component;
```

This component is instantiated with the name *trkm* as follows,

```
inft2: Infer_type_2 port map(
         rst => reset, clk => clk, e1 => e1upsig, e2 => e2upsig, de1 => de1upsig,
         de2 => de2upsig, g_e1 => ge1upsig, g_e2 => ge2upsig, g_de1 => gde1upsig,
         g_de2 => gde2upsig, e1_2 => e1lowsig, e2_2 => e2lowsig, de1_2 => de1lowsig,
         de2_2 => de2lowsig, g_e1_2 => ge1lowsig, g_e2_2 => ge2lowsig, g_de1_2 => gde1lowsig,
         g_de2_2 => gde2lowsig, c1 => c1sig, c2 => c2sig, c3 => c3sig, c4 => c4sig,
         gc1 => gc1sig, gc2 => gc2sig, gc3 => gc3sig, gc4 => gc4sig, c1_2 => c12sig,
         c2_2 => c22sig, c3_2 => c32sig, c4_2 => c42sig, gc1_2 => gc12sig,
         gc2_2 => gc22sig, gc3_2 => gc32sig, gc4_2 => gc42sig
      );
```

The signals that connect the instance *Infer_type_2* to the instance *trkm* are

```
signal c1sig, c2sig, c3sig, c4sig : std_logic_vector (m-1 downto 0);
signal gc1sig, gc2sig, gc3sig, gc4sig : std_logic_vector (7 downto 0);
signal c12sig, c22sig, c32sig, c42sig : std_logic_vector (m-1 downto 0);
signal gc12sig, gc22sig, gc32sig, gc42sig :std_logic_vector (7 downto 0);
```

4. The last component *defit2* corresponds to the de-fuzzifier stage of the T2FLC. It has two inputs and one output.

```
component defit2 is
    Port ( yl, yr : in  std_logic_vector (n-1 downto 0);
           y  : out std_logic_vector (n-1 downto 0));
end component;
```

This component is instantiated with the name *dfit2* as follows,

```
dfit2 : defit2 port map(yl => ylsig, yr => yrsig, y => y);
```

We did not define any signal for the port *y* since it can be connected directly to the entity of design *FT2KM*. The instances *trkm* and *dfit2* are connected using the following signals,

```
signal ylsig, yrsig : std_logic_vector (n-1 downto 0);
```

This approach of implementing an IT2FC provides the faster response. The whole process consisting of fuzzification, inference, type reduction, and defuzzification is achieved in four clock cycles, which for a Spartan family implementation using a 50 MHz clock represents 80×10^{-9} s, and for a Virtex 5 FPGA-based system is 40×10^{-9} s.

76.6 Developing of IT2FC Coprocessors

The use of IT2FC embedded into an FPGA can certainly be the option that offers the best performance and flexibility. As we shall see, the best performance can be obtained when the embedded FC is used as standalone system. Unfortunately, this gain in performance can present some drawbacks; for example, for people who were not involved in the design process of the controller or who are not familiar with VHDL codification, or the code owners simply want to keep the codification secret. All these obstacles can be overcome by the use of IP cores. Next, we shall explain two methods of implementing IT2FC as coprocessors.

76.6.1 Integrating the IT2FC Through Internal Ports

In Fig. 76.9, we show a control system that integrates the FT2KM design entity embedded into the Actel Fusion FPGA [76.62] as a coprocessor of an ARM processor. This FPGA allows incorporating the soft processor ARM Cortex, as well as other IP cores to make a custom configuration. The embedded system contains the ARM processor, two memory blocks, timers, interrupt controller (IC), a Universal Asynchronous Receiver/Transmitter (UART) serial port, IIC, pulse width modulator/tachometer block, and a general-purpose input/output interface (GPIO) interfacing the FT2KM block. All the factory embedded components are soft IP cores. The FT2KM is a VHDL module that together with the GPIO form the Ft2km_core soft coprocessor, handled as an IP core; however, in this case, it is necessary to have the VHDL code. In the system, the IT2 coprocessor is composed of the GPIO and the FT2KM modules, forming the Ft2km_core. In the system, moreover, are a DC motor with a high-resolution quadrature optical encoder, the system's power supply, an H-bridge for power direction, a personal computer, and a digital display.

The Ft2km_core has six inputs and two outputs. The inputs are *error*, *c.error*, *ce*, *rst*, *w*, and *clk*. The 8-bit inputs *error* and *c.errror* are the controller input for the error and change of error values. *ce* input is used to enable/disable the fuzzy controller, the input *rst* restores all the internal registers of the IT2FC, and the input *w* allows starting a fuzzy inference cycle. The outputs are *out*, and *IRQ/RDY*; the first one is the crisp output value, which is 8-bit wide. *IRQ/RDY* is produced when the output data corresponding to the respective input is ready to be read. *IRQ* is a pulse used to request an interrupt, whereas *RDY* is a signal that can be programmed to be active in high or low binary logic level, indicating that valid output was produced; this last signal can be used in a polling mode. In Fig. 76.9 we used only 1 bit for the *IRQ/RDY* signal, at the moment of designing the system the designer will have to decide on one method. It is possible to use both, modifying the logic or separating the signal and adding an extra 1-bit output.

The GPIO IP has two 32 bit wide ports, one for input (reading bus) and one for output (write bus). The output bus connects the GPIO IP to the ARM cortex using the 32 bit bus APB. The input bus connects the IT2FC IP to the GPIO IP. The ARM cortex uses the Ft2km_core as a coprocessor.

76.6.2 Development of IP Cores

In Sect. 76.6.1, we showed how to integrate the fuzzy coprocessor through an input/output port, i.e., the IP GPIO. We also commented on the existence of IP cores such as the UART and the timers that are connected directly to the system bus as in any microcontroller system with integrated peripherals. In this section, we shall show how to implement an IT2FC connected to the system bus to obtain an IT2FC IP core integrated to the system architecture. The procedure is basically the same for any FPGA of the Xilinx family. We worked with the Spartan 6 and Virtex 5, so the Xilinx ISE Design Suite was used.

The whole process to start an application that includes a microprocessor and a coprocessor can be broadly divided into three steps:

1. Design and implement the design entity that will be integrated as an IP core in further steps, then follows

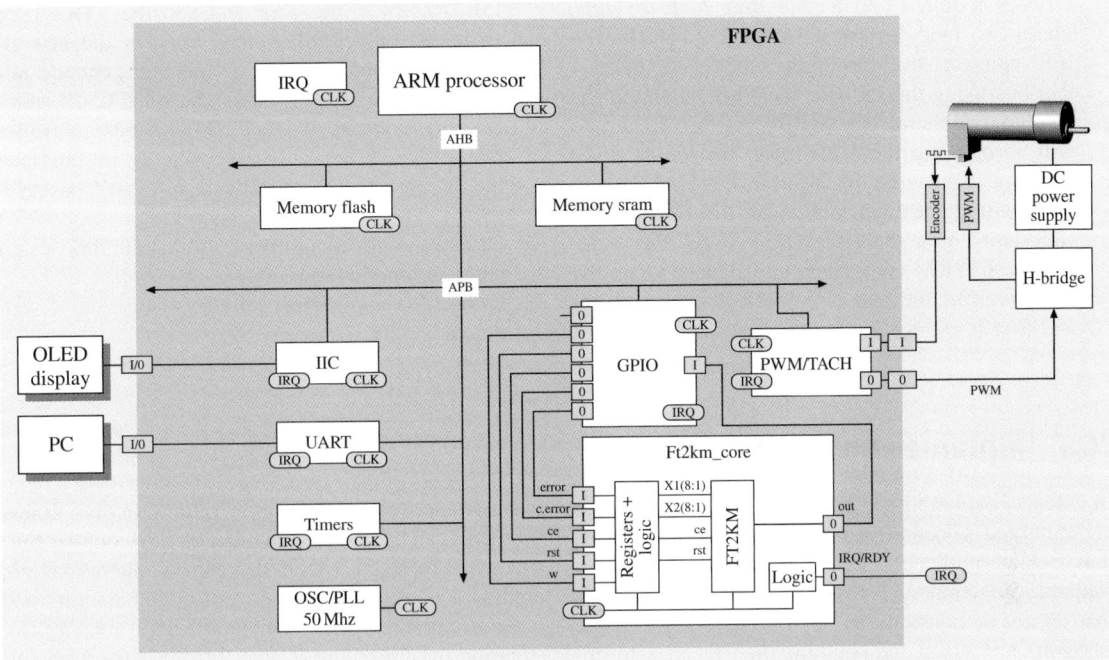

Fig. 76.9 A coprocessor implemented into the Actel Fusion FPGA. The system has an ARM processor, the IT2FC coprocessor implemented through the general-purpose input/output port, and some peripherals

the development flow explained in Sect. 76.4.2. In our case, the design entity is FT2KM.

2. Create the basic embedded microcontroller system tailored for our application. We already know the kind and amount of memory that we will need, as

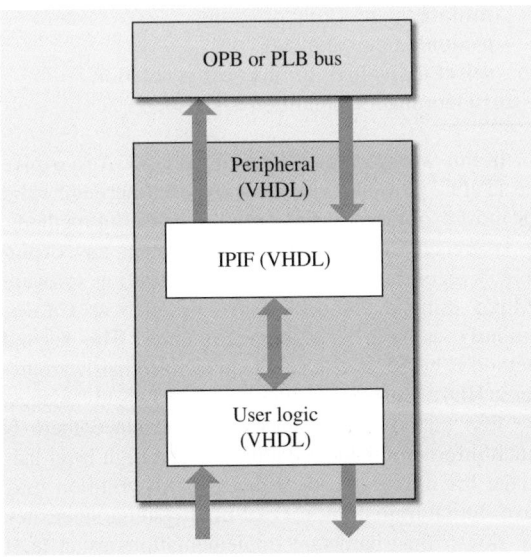

well as the peripherals. This step is achieved as follows: we create the microprocessor system using the base system builder (BSB) of the Xilinx Platform Studio (XPS) software. The system contains a Microblaze softcore, 16 KB of local memory, the data controller bus (dlmb_cntlr), and the instruction controller bus (ilmb_cntlr).

3. Create the IP core, which should contain the desired design entity, in our case the FT2KM. This step is achieved using the *Import Peripheral Wizard* found in the Hardware option in the XPS. The idea is to connect the FTKM design entity to the processor local bus (PLB V4.6) through three registers, one for each input (two registers) and one for the output. Upon the completion, this tool will create synthesizable HDL file (ft2km_core) that implements the intellectual property interface (IPIF)

Fig. 76.10 IP Core implementation of a user defined peripheral. The IT2FC coprocessor is implemented into the user logic module. This module achieves communication with the rest of the system through the PLB or the on-chip peripheral bus OPB. For a static coprocessor, use the PLB. For an implementation in the reconfigurable region, use the OPB ◄

services required and a stub *user_logic_module*. These two modules are shown in Fig. 76.10. The IPIF connects the user logic module to the system bus using the OPB or the PLB bus or to the on-chip peripheral bus (OPB). At this stage, we will need to use the *ISE Project Navigator* (ISE) software to integrate to the *user_logic_module* all the required files that implement the FT2KM design entity. Edit the *User_Logic_I.vhd* file to define the FT2KM component and signals. Open the *ftk2_core.vhd* file and create the *ftk2_core* entity and user logic. Synthesize the HDL code and exit

ISE. Return to the XSP and add the FTK2_core IP to the embedded system, connect the new IP core to the *mb_plb* bus system and generate address. Figure 76.10 shows the IT2FC IP core; the IPIF consists of the PLB V4.6 bus controller that provides the necessary signals to interface the IP core to the embedded soft core bus system.

4. Design the drivers (software) to handle this design entity as a peripheral.
5. Design the application software to use the design entity.

76.7 Implementing a GA in an FPGA

In essence, evolution is a two-step process of random variation and selection of a population of individuals that responds with a collection of behaviors to the environment. Selection tends to eliminate those individuals that do not demonstrate an appropriate behavior. The survivors reproduce and combine their features to obtain better offspring. In replication random mutation always occurs, which introduces novel behavioral characteristics. The evolution process optimizes behavior and this is a desirable characteristic for a learning system. Although the term *evolutionary computation* dates back to 1991, the field has decades of history, *genetic algorithms* being one avenue of investigation in simulated evolution [76.63]. GAs are family of computational models, which imitates the principles of natural evolution. For consistency they adopt biological terminology to describe operations. There are six main steps of a GA: population initialization, evaluation of candidates using a fitness function, selection, crossover, and termination judgment, as is shown in Algorithm 76.1. The first step is to decide how to code a solution to the problem that we want to optimize; hence, each individual is represented using a chromosome that contains the parameters. Common encoding of solutions are binary, integer, and real value. In binary encoding, every chromosome is a string of bits. In real-value encoding, every chromosome is a string than can contain one or several parameters encoded as real numbers. Algorithm 76.1 starts initializing a population with random solutions, and then each individual of the population is evaluated using a fitness function, which is selected according to the optimization goals. For example, for tuning a controller it may be enough to check if the actual output controller is minimizing errors between the target and

the reference. However, one or more complex fitness functions can be designed in order to carry out the control goal. In steps 3 to 5 the genetic operations are applied, i.e., selection, crossover (recombination), and mutation. In step 6, the termination criteria are checked, stopping the procedure if such criteria have been fulfilled.

Algorithm 76.1 General scheme of a GA

initialize population with random candidate solutions

evaluate each candidate

repeat

  select parents

  recombine pairs of parents

  mutate the resulting offspring

  evaluate new candidates

  select individuals for the next generation

until termination condition is satisfied

In this work, we have chosen work a GA to evolve the IT2FC. However, the ideas exposed here are valid for most evolutionary and natural computing methods. So, there are two methods to implement any evolutionary algorithm. One is based on executing software written using a computer language such as C/C++, similarly as with a desktop computer. The second method is based on designing specialized hardware using a HDL. Both have advantages and disadvantages; the first method is the easier method since there is much information about coding using a high level language for different EAs. However, this solution may have similar limitations for real-time systems since they are slower than hardware implementations by at least

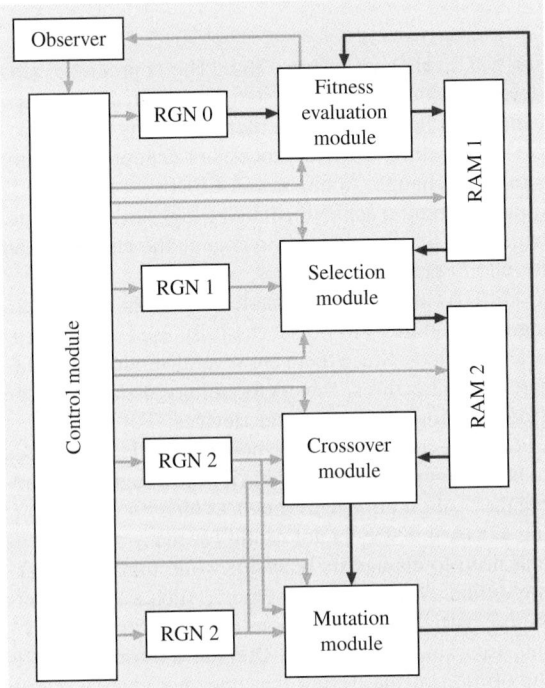

Fig. 76.11 High-level view of the structure of a GA for FPGA implementation

a factor of magnitude of five. On the other hand, state machine hardware-based designs are more complex to implement and use. In this section we shall present a small overview of both methods.

76.7.1 GA Software Based Implementations

It is well known that a GA can run in parallel, taking advantage of the two types of known parallelism: data and control parallelism. Data parallelism refers to executing one process over several instances of the EA, while control parallelism works with separate instances.

Coarse-grained parallelism and fine-grained parallelism are two methods often associated with the use of EA in parallel. The use of both methods is called a hybrid approach. Coarse-grained parallelism entails the EA cores to work in conjunction to solve a problem. The nodes swap individuals of their population with another node running the same problem. The cores

can exchange individuals with each other to improve diversity. The amount of information, frequency of exchange, direction, data pattern, etc., are factors that can affect the efficiency of this approach.

In fine-grained parallelism, the approach is to share mating partners instead of populations. The members of populations across the parallel cores select to mate their fittest members with the fittest found in a neighboring node's population. Then, the offspring of the selected individuals are distributed. The distribution of this next generation can go to one of the parents' populations, both parents' population, or all cores' populations, based on the means of distribution.

Figure 76.4 shows a six-core architecture design for the Virtex 5. Here, we can make fine or coarse-grained implementations of an EA. For example, for coarse-grained implementation, the island model with one processor per island can be used.

76.7.2 GA Hardware Implementations

Figure 76.11 shows a high-level view of the architecture of a GA for hardware implementation. The system has eight basic modules: selection module, crossover module, mutation module, fitness evaluation module, control module, observer module, four random generation number (RGN) modules, and two random access memory modules.

The control module is a Mealy state machine designed to feed all other modules with the necessary control signals to synchronize the algorithm execution. The selection module can have any existing method of selection, for example the Roulette Wheel Selection Algorithm. This method picks the genes of the parents of the current population, and the parents are processed to create new individuals. At the current generation, the crossover and genetic modules achieve the corresponding genetic operation on the selected parents. The fitness evaluation module computes the fitness of each offspring and applies elitism to the population. The observer module determines the stopping criterion and observes its fulfilment. RNGs are indispensable to provide the randomness that EAs require. Additionally, RAM 1 is necessary to store the current population and RAM 2 to store the selected parents of each generation.

76.8 Evolving Fuzzy Controllers

In Sect. 76.1 the general structure of an EFRBS was presented. It was mentioned that the common denominator in most learning systems is their capability of making structural changes to themselves over time to improve their performance for defined tasks. It also was mentioned that the two classical approaches for fuzzy learning systems are the Michigan and Pittsburgh approaches, and there exist newer proposals with the same target. Although to programm a learning system in a computer using high-level language, such as C/C++, requires some skill, system knowledge, and experimentation, there are no technical problems with achieving a system with such characteristics. This can be also true for hardware implementation, if the EFRBS was developed in C/C++ and executed by a hard or soft processor such as PowerPC or Microblaze, it is similarly as it is done in a computer. How to develop a coproces-

sor was explained in Sect. 76.6. The coprocessor was developed in the FPGA's static (base) region, which cannot be changed during a partial reconfiguration process. Therefore, such coprocessors cannot suffer any structural change. Achieving an EFRBS in hardware is quite different to achieving it using high-level language, because it is more difficult to change the circuitry than to modify programming lines.

FPGAs are reprogrammable devices that need a design methodology to be successfully used as reconfigurable devices. Since there are several vendors with different architectures, the methodology usually change from vendor to vendor and devices. For the Xilinx FPGAs the configuration memory is volatile, so, it needs to be configured every time that it is powered by uploading the configuration data known as *bitstream*. Configuring FPGA this way is not useful for many applications that need to change its behavior while they still working online. A solution to overcome such a limitation is to use partial reconfiguration, which splits the FPGA into two kinds of regions. The static (base) region is the portion of the design that does not change during partial reconfiguration, it may include logic that controls the partial reconfiguration process. In other words, partial reconfiguration (PR) is the ability to reconfigure select areas of an FPGA any time after its initial configuration [76.64]. It can be divided into two groups: dynamic partial reconfiguration (DPR) and static partial reconfiguration (SPR). DPR is also known as active partial reconfiguration. It allows changing a part of the device while the rest of the FPGA is still running. DPR is accomplished to allow the FPGA to adapt to changing algorithms and enhance performance, or for critical missions that cannot be disrupted while some subsystems are being defined. On the other hand, in SPR the static section of the FPGA needs to be stopped, so autoreconfiguration is impossible (Fig. 76.12).

For Xilinx FPGAs, there are basically three ways to achieve DPR for devices that support this feature. The two basic styles are difference-based partial reconfiguration and module-based partial reconfiguration. The first one can be used to achieve small changes to the design, the partial bitstream only contains information about differences between the current design structure that resides in the FPGA and the new content of the FPGA. Since the bitstream differences are usually small, the changes can be made very quickly. Module-based partial reconfiguration is useful for reconfiguring large blocks of logic using modular design

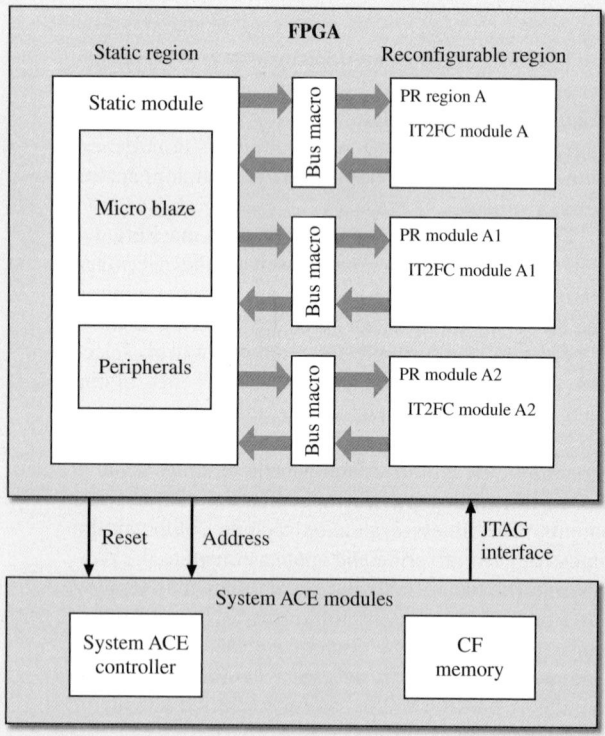

Fig. 76.12 The FPGA is divided into two regions: static and reconfigurable. The soft processor and peripherals are in the static region. Different fuzzy controller architectures are in the reconfigurable region. The bus macro are fixed data paths for signals going between a reconfigurable module and another module

concepts. The third style is also based on modular design but is more flexible and less restrictive. This new style was introduced by Xilinx in 2006 and it is known as early access partial reconfiguration (EAPR) [76.65, 66]. There are two key differences between the design flow EAPR and the module-based one. (1) In the EAPR flow the shape and size of partially reconfigurable regions (PRRs) can be defined by the user. Each PRR has at least one, and usually multiple, partially reconfigurable modules (PRMs) that can be loaded into the PRR. (2) For modules that communicate with each other, a special bus macro allows signals to cross over a partial reconfiguration boundary. This is an important consideration, since without this feature intermodule communication would not be feasible, as it is impossible to guarantee routing between modules. The bus macro provides a fixed *bus* of inter-design communication. Each time partial reconfiguration is performed, the bus macro is used to establish unchanging routing channels between modules, guaranteeing correct connections [76.65].

An important core that enables embedded microprocessors such as MicroBlaze and PowerPC to achieve reconfiguration at run time is HWICAP (hardware internal configuration access point) for the OPB. The HWICAP allows the processors to read and write the FPGA configuration memory through the ICAP (internal configuration access point). Basically it allows writing and reading the configurable logic block (CLB) look-up table (LUT) of the FPGA.

The process to achieve reconfigurable computing with application to IT2FC will be explained with more detail in Sect. 76.8.2. Moreover, how to evolve an IT2FC embedded into an FPGA, whether it resides in the static or in the reconfigurable region, will be also explained in therein.

76.8.1 EAPR Flow for Changing the Controller Structure

Figure 76.12 shows the basic idea of using EAPR flow for reconfigurable computing to change from one IT2FC structure to a different one. In this figure the Microblaze soft processor can evaluate each controller structure according to single or multiobjective criteria. The processor communicates with a PR region using the bus macro, which provides a means of locking the routing between the PRM and the base design. The system can achieve fast reconfiguration operations since partial bitstream are transferred between the FPGA and the compact flash memory (CF) where bitstreams are stored.

In general, the EAPR design flow is as follows [76.64, 67, 68]:

1. Hardware description language design and synthesis. The first steps in the EAPR design flow are very similar to the standard modular design flow. We can summarize this in three steps:
   (a) Top-level design. In this step, the design description must only contain black-box instantiations of lower-level modules. Top-level design must contain: I/O instantiations, clock primitives instantiations, static module instantiations, PR module instantiations, signal declarations, and bus macro instantiations, since all nonglobal signals between the static design and the PRMs must pass through a bus macro.
   (b) Base design. Here, the static modules of the system contain logic that will remain constant during reconfiguration. This step is very similar to the design flow explained in Sect. 76.4.2. However, the designer must consider input and output assignment rules for PR.
   (c) PRM design. Similarly to static modules, PR modules must not include global clock signals either, but may use those from top-level modules. When designing multiple PRMs to take advantage of the same reconfigurable area, for each module, the component name and port configuration must match the reconfigurable module instantiation of the top-level module.

2. Set design constraints. In this step, we need to place constraints in the design for place and route (PAR). The constraints included are: area group, reconfiguration mode, timing constraint, and location constraints. The area group constraint specifies which modules in the top-level module are static and which are reconfigurable. Each module instantiated by the top-level module is assigned to a group. The reconfiguration mode constraint is only applied to the reconfigurable group, which specifies that the group is reconfigurable. Location constraints must be set for all pins, clocking primitives, and bus macros in top-level design. Bus macros must be located so that they straddle the boundary between the PR region and the base design.

3. Implement base design. Before the implementation of the static modules, the top level is translated to ensure that the constraints file has been created properly. The information generated by implementing the base design is used for the PRM implemen-

tation step. Base design implementation follows three steps: translate, map, and PAR.

4. Implement PRMs. Each of the PRMs must be implemented separately within its own directory, and follows base design implementation steps: i. e., translate, map and PAR.

5. Merge. The final step in the partial reconfiguration flow is to merge the top level, base, and PRMs. During the merge step, a complete design is built from the base design and each PRM. In this step, many partial bitstreams for each PRM and initial full bitstreams are created to configure the FPGA.

Partial dynamic reconfigurable computing allows us to achieve online reconfiguration. By selecting a certain bitstream is possible to change the full controller structure, or any of the stages (fuzzification, inference engine, type reduction, and defuzzification), as well as any individual section of each stage, for example, different membership functions for the fuzzification stage, etc. However, we need to have all the reconfigurable modules previously synthesized because they are loaded using partial bitstreams. Therefore, to have the capability to evolve reconfigurable modules we need to provide them with a control register (CR) to change the desired parameters.

Next, a flexible coprocessor (FlexCo) prototype of an IT2FC (FlexCo IT2FC) that can be implemented either in the static region as well as in the PR is presented.

76.8.2 Flexible Coprocessor Prototype of an IT2FC

Figure 76.13 illustrates the FlexCo IT2FC, which contains the four stages (fuzzification, inference engine, type reduction, and defuzzification). They are connected depending on the target region, to the PLB or to the OPB through a 32 bits command register (CR), which is formed by four 8 bit registers named R1 to R4 (Fig. 76.14). The parameters of each stage can be changed by the programmer since they are not static as they were defined previously for the FT2KM (Sect. 76.5). Now, they are volatile registers connected through signals to save parameter values. The processor (MicroBlaze) can send through the PLB or the OPB, two kinds of commands to the CR: control words (CWs) and data words (DWs). The state machine of the FlexCo IT2FC interprets the command.

Figure 76.14 illustrates the CR coding for static and reconfigurable FC. This register is used to perform

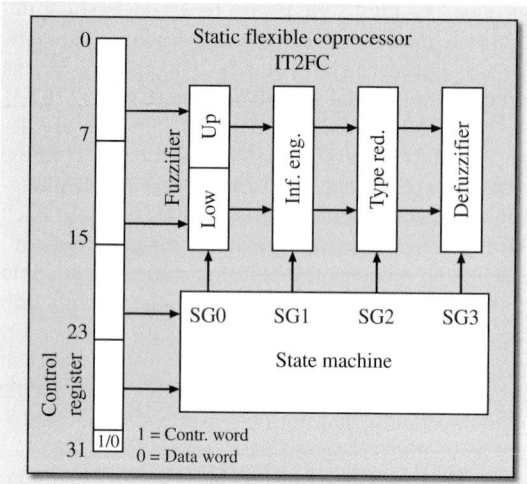

Fig. 76.13 Flexible coprocessor proposal of an IT2FC for the static region

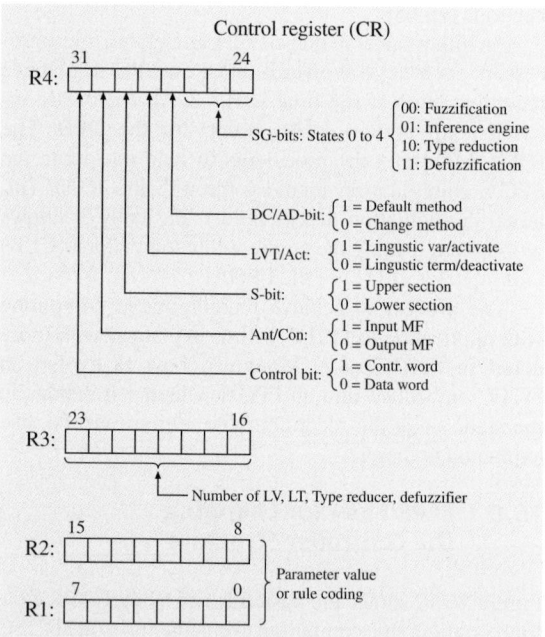

Fig. 76.14 The control register is used for both styles of implementation, in the static region or in the reconfigurable region

parameter modification in both modes, static and reconfigurable. In general, bit 7 of R4 is used to differentiate between a CW or a DW, *1* means a CW, whereas *0* means a DW. The StaGe bits (SG-bits) serves to identify the IT2FC stage that is to be modified.

Fig. 76.15 In the static region of the FPGA a multiprocessor system (MPS) with operating system. The GA resides in the program memory, it is executed by the MPS. The IT2FC may be implemented in the reconfigurable region, Fig. 76.16, or in the static region, Fig. 76.13 ►

- *SG-bits = 00*: The fuzzification stage has been chosen, then it is necessary to set the bit Ant/Con to *1* to indicate that the antecedent MFs are going to be modified. With the section-bit (S-bit) we indicate which part of the FOU (upper or lower) will be modified. The bit linguistic-variable-term/active (LVT/Active) is to indicate whether we want to modify a linguistic variable (LV) or the linguistic term (LT), the *Act* option is for the inference engine (IE). In accordance to the LV/LT bit value, in the register R3 we set the number of the LV or the LT that will be changed. Finally, with registers R1 and R2, the parameter value of the LV or the LT is given, R1 is the least significant byte.
- *SG-bits = 01*: With this setting, the state machine identifies that the IE will be modified. It works in conjunction with Ant/Con, S-bit, and the registers R1, R2, and R3. Set a *0* value in the Ant/Con bit to change the consequent parameters of a Mamdani inference system, in S-bit choose the upper or lower MF, using R3 indicate the number of MF, and with R1 and R2 set the corresponding value or static implementation. It is possible to activate and deactivate rules using the bit LVT/Active. With bit dynamic change/activate-deactivate (DC/AD), it is possible to change the combination of antecedents and consequents of a specific rule provided that we have made this part flexible by using registers. For an implementation in the reconfigurable region, it is possible to add or remove rules. These two features need to work in conjunction with registers R1, R2, and R3.
- *SG-bits = 10*: This selection is to modify the type reduction stage. It is possible to have more than one type reducer. By setting the DC/AD-bit to *1*, we indicate that we wish to change the method at running time without the necessity of achieving a reconfiguration process that implies uploading partial bitstreams. The methods can be selected using register R3. By using a DC/AD-bit equal to *0* and LVT/Act equal to *0*, in combination with registers R1 to R3 we can indicate that we wish to change the preloaded values that the KM-algorithm needs to achieve the TR.

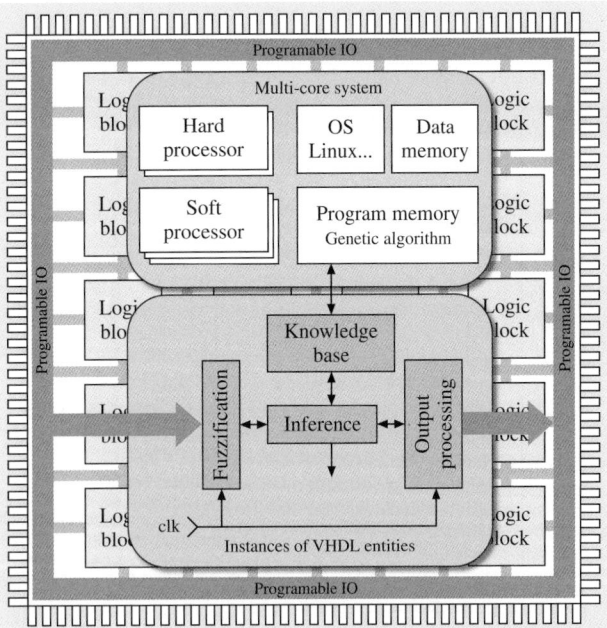

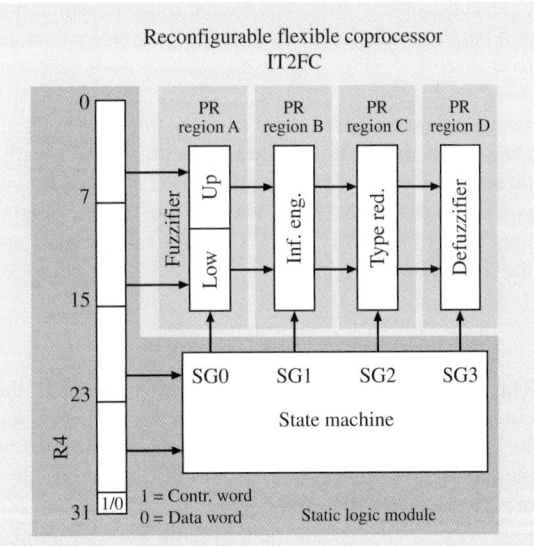

Fig. 76.16 Flexible coprocessor proposal of an IT2FC for the reconfigurable region

- *SG-bits = 11*: Similarly to the type reduction stage, we can change the defuzzifier at running time.

With respect to the type reducer and defuzzification stages, we give the option to have more than one module, which has the advantage of making the process

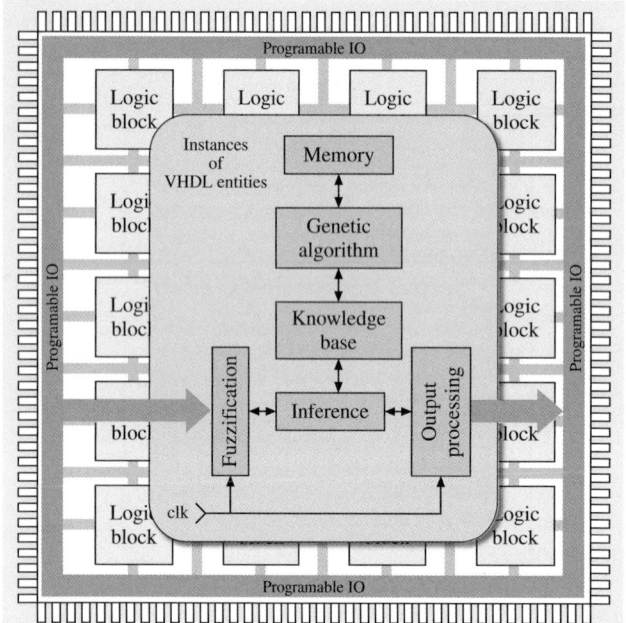

Fig. 76.17 This design may be implemented in both regions to have a dynamic reconfigurable system. For a static implementation, the system must have registers for all the variable parameters to make possible to change their values, Fig. 76.13

easier and possible for static designs, but the disadvantage is that the design will consume more macrocells, increasing the cost of the required FPGAs, boards, and power consumption. Next, we will explain the implementation of the FlexCo IT2FC for the static region and the reconfigurable region.

Implementing the FlexCo IT2FC on the Static Region

The IT2FC of is connected to the PLB. Although the controller structure is static, this system can be evolved for tuning and learning because it is possible to achieve parametric modifications to all the IT2FC stages. Figure 76.13 shows the architecture of this system and Fig. 76.15 a conceptual model of the possible implementation.

Implementing the FlexCo IT2FC on the PR

Figure 76.16 illustrates a more flexible architecture for FlexCo IT2FC. The IT2FC is implemented in the reconfigurable region, using a partially reconfigurable region (PRR) for each stage. This is convenient since each region can have multiple modules that can be swapped

in and out of the device on the fly. This is the most recommended method to achieve the evolving IT2FC since it is more flexible. One disadvantage is that at running time it is slower than the static implementation because more logic circuits are incorporated.

Figure 76.17 is an evolutive standalone system; as it was mentioned, the IT2FC and the GA can be in the static or in the reconfigurable region.

76.8.3 Conclusion and Further Reading

FPGAs combine the best parts of ASICs and processor-based systems, since they do not require high volumes to justify making a custom design. Moreover, they also provide the flexibility of software, running on a processor-based system, without being limited by the number of cores available. They are one of the best options to parallelize a system since they are parallel in nature. In an IT2FC, a typical whole T2-inference, computed using an industrial computer equipped with a quad-core processor, lasts about 18×10^{-3} s. A whole IT2FC (fuzzification, inference, KM-type reducer, and defuzzification) lasts only four clock cycles, which for a Spartan implementation using a 50 MHz clock represents 80×10^{-9} s, and for a Virtex 5 FPGA-based system represents 40×10^{-9} s. For the Spartan family the typical implementation speedup is 225 000, whereas for the Virtex 5 it is 450 000. Using a pipeline architecture, the speedup of the whole IT2 process can be obtained in just one clock cycle, so using the same criteria to compare, the speedup for Spartan is 90 000 and 2 400 000 for Virtex. Reported speedups of GAs implemented into an FPGA, are at least 5 times higher than in a computer system. For all these reasons, FPGAs are suitable devices for embedding evolving fuzzy logic controllers, especially the IT2FC, since they are computationally expensive. There are some drawbacks with the use of this technology, mostly with respect to the need to have a highly experienced development team because its implementation complexity. Achieving an evolving intelligent system using reconfigurable computing is not as direct as it is using a computer system. It requires the knowledge of FPGA architectures, VHDL coding, soft processor implementation, the development of coprocessors, high-level languages, and reconfigurable computing bases. Therefore, people interested in achieving such implementations require expertise in the above fields, and further reading must focus on these topics, FPGA vendor manuals and white papers, as well as papers and books on reconfigurable computing.

References

76.1   P.P. Angelov, X. Zhou: Evolving fuzzy-rule-based classifiers from data streams, IEEE Trans. Fuzzy Syst. **16**(6), 1462–1475 (2008)

76.2   O. Cordón, F. Herrera, F. Hoffman, L. Magdalena: *Genetic Fuzzy Systems: Evolutionary Tuning and Learning of Fuzzy Knowledge Bases* (World Scientific, Singapore 2001)

76.3   P. Angelov, R. Buswell: Evolving rule-based models: A tool for intelligent adaptation, IFSA World Congr. 20th NAFIPS Int. Conf. 2001. Jt. 9th, Vancouver, Vol. 2 (2001) pp. 1062–1067

76.4   K. De Jong: Learning with genetic algorithms: An overview, Mach. Learn. **3**(2), 121–138 (1988)

76.5   J.H. Holland: *Adaptation in Natural and Artificial Systems: An Introductory Analysis with Applications to Biology, Control, and Artificial Intelligence* (MIT Press/Bradford Books, Cambridge 1998)

76.6   K.A. De Jong: *Evolutionary Computation: A Unified Approach* (MIT Press, Cambridge 2006)

76.7   O. Cordón, F. Gomide, F. Herrera, F. Hoffmann, L. Magdalena: Ten years of genetic fuzzy systems: Current framework and new trends, Fuzzy Sets Syst. **141**(1), 5–31 (2004)

76.8   V. Gilles: SIA: A supervised inductive algorithm with genetic search for learning attributes based concepts, Lect. Notes Comput. Sci. **667**, 280–296 (1993)

76.9   J. Juan Liu, J. Tin-Yau Kwok: An extended genetic rule induction algorithm, Proc. 2000 Congr. Evol. Comput., Vol. 1 (2000) pp. 458–463

76.10   O. Cordón, M.J. del Jesus, F. Herrera, M. Lozano: MOGUL A methodology to obtain genetic fuzzy rule based systems under the iterative rule learning approach, Int. J. Intell. Syst. **14**(11), 1123–1153 (1999)

76.11   G.D. Perry, F.S. Stephen: Competition-based induction of decision models from examples, Mach. Learn. **13**, 229–257 (1993)

76.12   G.D. Perry, F.S. Stephen: Using coverage as a model building constraint in learning classifier systems, Evol. Comput. **2**, 67–91 (1994)

76.13   A. Giordana, F. Neri: Searc-intensive concept induction, Evol. Comput. **3**, 375–416 (1995)

76.14   H. Ishibuchi, K. Nozaki, N. Yamamoto, H. Tanaka: Selecting fuzzy if-then rules for classification problems using genetic algorithms, IEEE Trans. Fuzzy Syst. **3**(3), 260–270 (1995)

76.15   A. Homaifar, E. McCormick: Simultaneous design of membership functions and rule sets for fuzzy controllers using genetic algorithms, IEEE Trans. Fuzzy Syst. **3**(2), 129–139 (1995)

76.16   D. Park, A. Kandel, G. Langholz: Genetic-based new fuzzy reasoning models with application to fuzzy control, IEEE Trans. Syst. Man Cybern. **24**(1), 39–47 (1994)

76.17   O. Castillo, R. Sepúlveda, P. Melin, O. Montiel: Evolutionary optimization of interval type-2 member-

ship functions, Proc. 2006 Int. Conf. Artif. Intell. ICAI 2006, Las Vegas (2006) pp. 558–564

76.18   R. Sepúlveda, O. Castillo, P. Melin, O. Montiel, L.T. Aguilar: Evolutionary optimization of interval type-2 membership functions using the human evolutionary model, FUZZ-IEEE (2007) pp. 1–6

76.19   R. Sepúlveda, O. Montiel-Ross, O. Castillo, P. Melin: Optimizing the MFs in type-2 fuzzy logic controllers, using the human evolutionary model, Int. Rev. Autom. Control **3**(1), 1–10 (2010)

76.20   O. Castillo, P. Melin, A.A. Garza, O. Montiel, R. Sepúlveda: Optimization of interval type-2 fuzzy logic controllers using evolutionary algorithms, Soft Comput. **15**(6), 1145–1160 (2011)

76.21   C. Wagner, H. Hagras: A genetic algorithm based architecture for evolving type-2 fuzzy logic controllers for real world autonomous mobile robots, Fuzzy Syst. Conf, Proc. 2007. FUZZ-IEEE 2007, London (2007) pp. 1–6

76.22   J.E. Bonilla, V.H. Grisales, M.A. Melgarejo: Genetic tuned FPGA based PD fuzzy LUT controller, 10th IEEE Int. Conf. Fuzzy Syst. (2001) pp. 1084–1087

76.23   S. Sánchez-Solano, A.J. Cabrera, I. Baturone: FPGA implementation of embedded fuzzy controllers for robotic applications, IEEE Trans. Ind. Electron. **54**(4), 1937–1945 (2007)

76.24   J.L. González, O. Castillo, L.T. Aguilar: FPGA as a tool for implementing non-fixed structure fuzzy logic controllers, IEEE Symp. Found. Comput. Intell. 2007. FOCI 2007 (2007) pp. 523–530

76.25   O. Montiel, Y. Maldonado, R. Sepúlveda, O. Castillo: Simple tuned fuzzy controller embedded into an FPGA, Fuzzy Inf. Proc. Soc. 2008. NAFIPS 2008. Annu. Meet. North Am. (2008) pp. 1–6

76.26   O. Montiel, J. Olivas, R. Sepúlveda, O. Castillo: Development of an embedded simple tuned fuzzy controller, IEEE Int. Conf. Fuzzy Syst., FUZZ-IEEE 2008, IEEE World Congr. Comput. Intell. (2008) pp. 555–561

76.27   Y. Maldonado, O. Montiel, R. Sepúlveda, O. Castillo: Design and simulation of the fuzzification stage through the Xilinx system generator. In: *Soft Computing for Hybrid Intelligent Systems*, Studies in Computational Intelligence, Vol. 154, ed. by O. Castillo, P. Melin, J. Kacprzyk, W. Pedrycz (Springer, Berlin, Heidelberg 2008) pp. 297–305

76.28   J.A. Olivas, R. Sepúlveda, O. Montiel, O. Castillo: Methodology to test and validate a VHDL inference engine through the Xilinx system generator. In: *Soft Computing for Hybrid Intelligent Systems*, Studies in Computational Intelligence, Vol. 154, ed. by O. Castillo, P. Melin, J. Kacprzyk, W. Pedrycz (Springer, Berlin, Heidelberg 2008) pp. 325–331

76.29   G. Lizárraga, R. Sepúlveda, O. Montiel, O. Castillo: Modeling and simulation of the defuzzification stage using Xilinx system generator and simulink.

In: *Soft Computing for Hybrid Intelligent Systems*, Studies in Computational Intelligence, Vol. 154, ed. by O. Castillo, P. Melin, J. Kacprzyk, W. Pedrycz (Springer, Berlin, Heidelberg 2008) pp. 333–343

76.30   M. Grégory, U. Andres, P. Carlos-Andres, S. Eduardo: A dynamically-reconfigurable FPGA platform for evolving fuzzy systems, Lect. Notes Comput. Sci. **3512**, 296–359 (2005)

76.31   Y. Maldonado, O. Castillo, P. Melin: Optimization of membership functions for an incremental fuzzy PD control based on genetic algorithms. In: *Soft Computing for Intelligent Control and Mobile Robotics*, Studies in Computational Intelligence, ed. by O. Castillo, J. Kacprzyk, W. Pedrycz (Springer, Berlin Heidelberg 2011) pp. 195–211

76.32   R.M.A. Melgarejo, C.A. Peña-Reyes: Hardware architecture and FPGA implementation of a type-2 fuzzy system, Proc. 14th ACM Great Lakes Symp. VLSI (2004) pp. 458–461

76.33   C. Lynch, H. Hagras, V. Callaghan: Parallel type-2 fuzzy logic co-processors for engine management, IEEE Int. Conf. Fuzzy Syst., FUZZ-IEEE (2007) pp. 1–6

76.34   R. Sepúlveda, O. Montiel, O. Castillo, P. Melin: Embedding a high speed interval type-2 fuzzy controller for a real plant into an FPGA, Appl. Soft Comput. **12**(3), 988–998 (2012)

76.35   L.A. Zadeh: Fuzzy sets, Inf. Control **8**(3), 338–353 (1965)

76.36   L.A. Zadeh: The concept of a linguistic variable and its application to approximate reasoning –I, Inf. Sci. **8**(3), 199–249 (1975)

76.37   J.M. Mendel: Type-2 fuzzy sets: Some questions and answers, IEEE Connect. Newsl. IEEE Neural Netw. Soc. **1**, 10–13 (2003)

76.38   J.S.R. Jang, C.T. Sun, E. Mizutani: *Neuro-Fuzzy and Soft Computing: A Computational Approach to Learning and Machine Intelligence* (Prentice Hall, Upper Saddle River 1997)

76.39   J.M. Mendel: *Uncertainty Rule-Based Fuzzy Logic Systems: Introduction and New Directions* (Prentice Hall, Upper Saddle River 2001)

76.40   J.M. Mendel: Type-2 fuzzy sets and systems: An overview, IEEE Comput. Intell. Mag. **2**(2), 20–29 (2007)

76.41   J.M. Mendel, R.I.B. John: Type-2 fuzzy sets made simple, IEEE Trans. Fuzzy Syst. **10**(2), 117–127 (2002)

76.42   D. Wu: Approaches for reducing the computational cost of interval type-2 fuzzy logic controllers: overview and comparison, IEEE Trans. Fuzzy Syst. **21**(1), 80–99 (2013)

76.43   D. Wu: Approaches for Reducing the Computational Cost of Interval Type-2 Fuzzy Logic Systems: Overview and Comparisons, IEEE Trans. Fuzzy Syst. **21**(1), 80–99 (2013)

76.44   K. Duran, H. Bernal, M. Melgarejo: Improved iterative algorithm for computing the generalized centroid of an interval type-2 fuzzy set, Fuzzy Inf. Proc. Soc. 2008. NAFIPS 2008. Annu. Meet. North Am. (2008) pp. 1–6

76.45   D. Wu, M. Nie: Comparison and practical implementation of type reduction algorithms for type-2 fuzzy sets and systems, Proc. IEEE Int. Conf. Fuzzy Syst. (2008) pp. 2131–2138

76.46   L.T. Ngo, D.D. Nguyen, L.T. Pham, C.M. Luong: Speed up of interval type-2 fuzzy logic systems based on GPU for robot navigation, Adv. Fuzzy Syst. **2012**, 475894 (2012), doi: 10.1155/2012/698062

76.47   P. Sundararajan: High Performance Computing Using FPGAs, Xilinx. White Paper: FPGA. WP375 (v1.0), 1–15 (2010)

76.48   IBM Redbooks: *The Power4 Processor Introduction and Tuning Guide*, 1st edn. (IBM, Austin 2001)

76.49   J. Yiu: *The Definitive Guide To The ARM CORTEX-M0* (Newnes, Oxford 2011)

76.50   S.P. Dandamudi: *Fundamentals of Computer Organization and Design* (Springer, Berlin, Heidelberg 2003)

76.51   T. Tauber, G. Runger: *Parallel Programming: For Multicore and Cluster Systems* (Springer, Berlin, Heidelberg 2010)

76.52   K.P. Abdulla, M.F. Azeem: A novel programmable CMOS fuzzifiers using voltage-to-current converter circuit, Adv. Fuzzy Syst. **2012**, 419370 (2012), doi: 10.1155/2012/419370

76.53   D. Fikret, G.Z. Sezgin, P. Banu, C. Ugur: ASIC implementation of fuzzy controllers: A sampled-analog approach, 21st Eur. Solid-State Circuits Conf. 1995 ESSCIRC '95. (1995) pp. 450–453

76.54   L. Kourra, Y. Tanaka: Dedicated silicon solutions for fuzzy logic systems, IEE Colloquium on 2 Decades Fuzzy Contr. Part 1 **3**, 311–312 (1993)

76.55   M. Khosla, R.K. Sarin, M. Uddin: Design of an analog CMOS based interval type-2 fuzzy logic controller chip, Int. J. Artif. Intell. Expert Syst. **2**(4), 167–183 (2011)

76.56   C. Bobda: *Introduction to Reconfigurable Computing. Architectures, Algorithms, and Applications* (Springer, Berlin, Heidelberg 2007)

76.57   P.C. Pong: *FPGA Prototyping by VHDL Examples* (Wiley, Hoboken 2008)

76.58   M. Oscar, S. Roberto, M. Yazmin, C. Oscar: Design and simulation of the type-2 fuzzification stage: Using active membership functions. In: *Evolutionary Design of Intelligent Systems in Modeling, Simulation and Control*, Studies in Computational Intelligence, Vol. 257, ed. by O. Castillo, W. Pedrycz, J. Kacprzyk (Springer, Berlin, Heidelberg 2009) pp. 273–293

76.59   S. Roberto, M. Oscar, O. José, C. Oscar: Methodology to test and validate a VHDL inference engine of a type-2 FIS, through the Xilinx system generator. In: *Evolutionary Design of Intelligent Systems in Modeling, Simulation and Control*, Studies in Computational Intelligence, Vol. 257, ed. by O. Castillo, W. Pedrycz, J. Kacprzyk (Springer, Berlin/Heidelberg 2009) pp. 295–308

76.60   S. Roberto, M.-R. Oscar, C. Oscar, M. Patricia: Embedding a KM type reducer for high speed fuzzy controller into an FPGA. In: *Soft Computing in Industrial Applications*, Advances in Intelligent and Soft Computing, Vol. 75, ed. by X.-Z. Gao, A. Gaspar-Cunha, M. Köppen, G. Schaefer, J. Wang (Springer, Berlin/Heidelberg 2010) pp. 217–228

76.61 S. Roberto, M. Oscar, L. Gabriel, C. Oscar: Modeling and simulation of the defuzzification stage of a type-2 fuzzy controller using the Xilinx system generator and simulink. In: *Evolutionary Design of Intelligent Systems in Modeling, Simulation and Control*, Studies in Computational Intelligence, Vol. 257, ed. by O. Castillo, W. Pedrycz, J. Kacprzyk (Springer, Berlin/Heidelberg 2009) pp. 309–325

76.62 O. Montiel-Ross, J. Quiñones, R. Sepúlveda: Designing high-performance fuzzy controllers combining ip cores and soft processors, Adv. Fuzzy Syst. **2012**, 1–11 (2012)

76.63 D.B. Fogel, T. Back: An introduction to evolutionary computation. In: *Evolutionary Computation. The Fossile Record*, ed. by D.F. Fogel (IEEE, New York 1998)

76.64 W. Lie, W. Feng-yan: Dynamic partial reconfiguration in FPGAs, 3rd Int. Symp. Intell. Inf. Technol. Appl. (2009) pp. 445–448

76.65 D. Lim, M. Peattie: Two Flows for Partial Reconfiguration: Module Based or Small Bit Manipulations. XAPP290. May 17, 2007

76.66 Emil Eto: Difference-Based Partial Reconfiguration. XAPP290. December 3, 2007

76.67 Xilinx: Early Access Partial Reconfiguration User Guide For ISE 8.1.01i, UG208 (v1.1). May 6, 2012

76.68 C.-S. Choi, H. Lee: A self-reconfigurable adaptive FIR filter system on partial reconfiguration platform, IEICE Trans. **90-D**(12), 1932–1938 (2007)

77. Multiobjective Genetic Fuzzy Systems

Hisao Ishibuchi, Yusuke Nojima

This chapter explains evolutionary multiobjective
design of fuzzy rule-based systems in comparison
with single-objective design. Evolutionary algorithms have been used in many studies on fuzzy
system design for rule generation, rule selection,
input selection, fuzzy partition, and membership
function tuning. Those studies are referred to as
genetic fuzzy systems because genetic algorithms
have been mainly used as evolutionary algorithms.
In many studies on genetic fuzzy systems, the accuracy of fuzzy rule-based systems is maximized.
However, accuracy maximization often leads to the
deterioration in the interpretability of fuzzy rule-
based systems due to the increase in their complexity. Thus, multiobjective genetic algorithms
were used in some studies to maximize not only
the accuracy of fuzzy rule-based systems but also
their interpretability. Those studies, which can be
viewed as a subset of genetic fuzzy system studies, are referred to as multiobjective genetic fuzzy
systems (MoGFS). A number of fuzzy rule-based
systems with different complexities are obtained
along the interpretability–accuracy tradeoff curve.
One extreme of the tradeoff curve is a simple highly
interpretable fuzzy rule-based system with low
accuracy while the other extreme is a complicated
highly accurate one with low interpretability. In
MoGFS, multiple accuracy measures such as a true
positive rate and a true negative rate can be simultaneously used as separate objectives. Multiple
interpretability measures can also be simultaneously used in MoGFS.

77.1 Fuzzy System Design

A fuzzy rule-based system is a set of fuzzy rules, which
has been successfully used as a nonlinear controller in
various real-world applications. The basic structure of
fuzzy rules for multi-input and single-output fuzzy control can be written as follows [77.1–3]

$$\text{Rule } R_q : \text{If } x_1 \text{ is } A_{q1} \text{ and } \dots \text{ and } x_n \text{ is } A_{qn} \\ \text{then } y \text{ is } B_q , \quad (77.1)$$

where q is a rule index, R_q is the label of the qth rule, n is the number of input variables, x_i is the ith input variable ($i = 1, 2, \ldots, n$), A_{qi} is an antecedent fuzzy set for the ith input variable x_i, y is an output variable, and B_q is a consequent fuzzy set for the output variable y. The antecedent and consequent fuzzy sets A_{qi} and B_q are specified by their membership functions $\mu_{A_{qi}}(x_i)$ and $\mu_{B_q}(y)$, respectively. Examples of antecedent fuzzy sets are shown in Fig. 77.1 where the domain interval $[0, 1]$ of the input variable x_i is partitioned into three fuzzy sets *small*, *medium*, and *large* with triangular membership functions.

Fuzzy rules of the form in (77.1) are based on the concept of linguistic variables by *Zadeh* [77.4–6]. According to *Zadeh* [77.4–6], a fuzzy set with a linguistic meaning such as *small* and *large* is referred to as a linguistic value while a variable with linguistic values is called a linguistic variable. For example, in Fig. 77.1, the three fuzzy sets are linguistic values while x_i is a linguistic variable. In our daily life, we almost always use linguistic variables and linguistic values. When we say *your car is fast but my car is slow*, the speed of cars is a linguistic variable while *fast* and *slow* are linguistic values. When we say *it is hot today*, the temperature is a linguistic variable while *hot* is a linguistic value. Of course, we use those linguistic values without explicitly specifying their meanings by membership functions. However, we have our own vague definitions of those linguistic values, which may be approximately represented by membership functions.

The main advantage of fuzzy rule-based systems over other nonlinear models such as multilayer feedforward neural networks is their linguistic interpretability. In Fig. 77.2, we show a two-input and single-output fuzzy rule-based system with the following nine fuzzy rules

Rule R_1 : If x_1 is *small* and x_2 is *small*
then y is *medium* ,

Rule R_2 : If x_1 is *small* and x_2 is *medium*
then y is *small* ,

Rule R_3 : If x_1 is *small* and x_2 is *large*
then y is *medium* ,

Rule R_4 : If x_1 is *medium* and x_2 is *small*
then y is *small* ,

Rule R_5 : If x_1 is *medium* and x_2 is *medium*
then y is *medium* ,

Rule R_6 : If x_1 is *medium* and x_2 is *large*
then y is *large* ,

Rule R_7 : If x_1 is *large* and x_2 is *small*
then y is *medium* ,

Rule R_8 : If x_1 is *large* and x_2 is *medium*
then y is *large* ,

Rule R_9 : If x_1 is *large* and x_2 is *large*
then y is *medium* .

A linguistic value in each cell in Fig. 77.2 shows the consequent fuzzy set of the corresponding fuzzy rule.

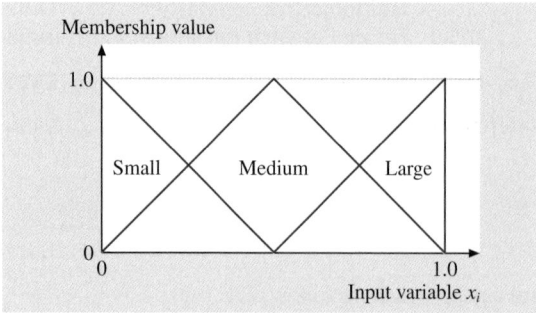

Fig. 77.1 Three antecedent fuzzy sets *small*, *medium*, and *large*

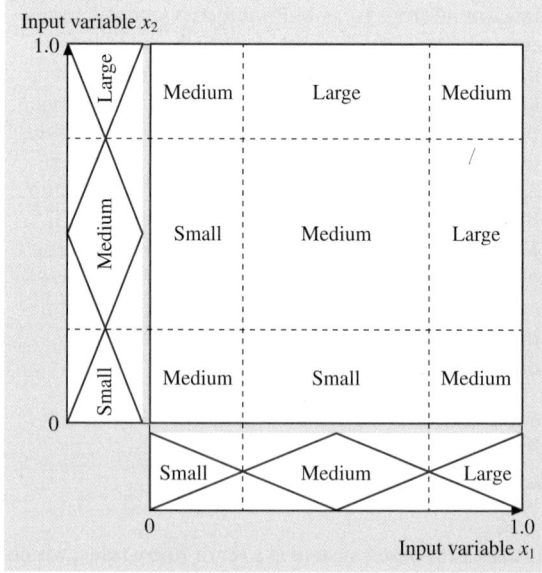

Fig. 77.2 A two-input and single-output fuzzy rule-based system with nine fuzzy rules

For example, *medium* in the bottom-right cell shows the consequent fuzzy set of the fuzzy rule R_7 with the antecedent fuzzy set *large* for x_1 and *small* for x_2. Let us assume that the consequent fuzzy sets in Fig. 77.2 are defined by the triangular membership functions in Fig. 77.1. Then, we can roughly understand the shape of the two-input and single-output nonlinear function represented by the fuzzy rule-based system in Fig. 77.2 (even when we do not know anything about fuzzy reasoning).

It is easy to linguistically understand the input–output relation of the fuzzy rule-based system in Fig. 77.2. That is, the fuzzy rule-based system in Fig. 77.2 has high interpretability. However, it is difficult to approximate a complicated highly nonlinear function by such a simple 3×3 fuzzy rule-based system. More membership functions for the input and output variables may be needed for improving the accuracy of fuzzy rule-based systems. The tuning of each membership function may be also needed. Theoretically, fuzzy rule-based systems are universal approximators of nonlinear functions. This property has been shown for fuzzy rule-based systems [77.7–9] and multilayer feedforward neural networks [77.10–12]. This means that fuzzy rule-based systems as well as neural networks have high approximation ability of nonlinear functions.

In Fig. 77.3, we show an example of a tuned 7×7 fuzzy partition of the two-dimensional input space $[0, 1] \times [0, 1]$. We can design a much more ac-

curate fuzzy rule-based system by using such a tuned fuzzy partition than the simple 3×3 fuzzy partition in Fig. 77.2. That is, we can say that Fig. 77.3 is a better fuzzy partition than Fig. 77.2 with respect to the accuracy of fuzzy rule-based systems. However, it is very difficult to linguistically interpret each antecedent fuzzy set in Fig. 77.3. In other words, it is very difficult to assign an appropriate linguistic value such as *small* and *large* to each antecedent fuzzy set in Fig. 77.3. Thus, we can say that the fuzzy partition in Fig. 77.3 does not have high linguistic interpretability. That is, Fig. 77.2 is a better fuzzy partition than Fig. 77.3 with respect to the linguistic interpretability of fuzzy rule-based systems. As shown by the comparison between the two fuzzy partitions in Figs. 77.2 and 77.3, accuracy maximization usually conflicts with interpretability maximization in the design of fuzzy rule-based systems.

Let us denote a fuzzy rule-based system by S. The fuzzy rule-based system S is a set of fuzzy rules. In fuzzy system design, the accuracy of S is maximized. The accuracy maximization of S is usually formulated as the following error minimization

$$\text{Minimize } f(S) = Error(S) , \qquad (77.2)$$

where $f(S)$ is an objective function to be minimized, and $Error(S)$ is an error measure.

As shown in Fig. 77.3, the accuracy maximization often leads to a complicated fuzzy rule-based system with low interpretability. Thus, a complexity measure is combined into the objective function in (77.2) as follows [77.13, 14]

$$\text{Minimize } f(S) = w_1 \, Complexity(S) + w_2 \, Error(S) , \qquad (77.3)$$

where w_1 and w_2 are nonnegative weights, and $Complexity(S)$ is a complexity measure.

In the late 1990s, the idea of multiobjective fuzzy system design [77.15] was proposed where the accuracy maximization and the complexity minimization were handled as separate objectives

$$\text{Minimize } f_1(S) = Complexity(S) \quad \text{and}$$
$$f_2(S) = Error(S) , \qquad (77.4)$$

where $f_1(S)$ and $f_2(S)$ are separate objectives to be minimized.

The two-objective optimization problem in (77.4) does not have a single optimal solution that simultaneously optimizes the two objectives $f_1(S)$ and $f_2(S)$.

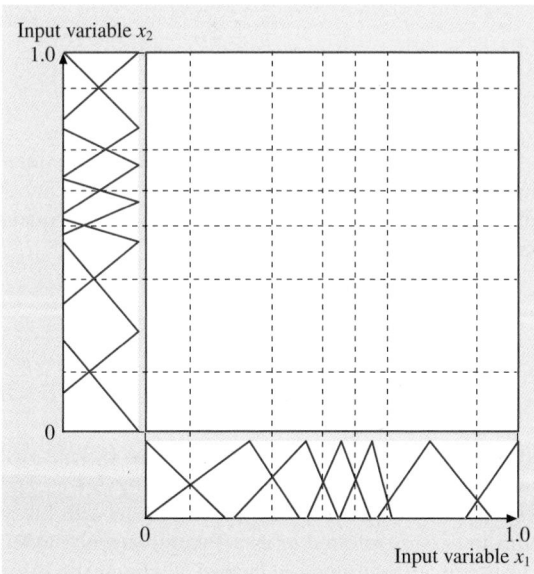

Fig. 77.3 A tuned 7×7 fuzzy partition

This is because the error minimization increases the complexity of fuzzy rule-based systems (i. e., the optimization of $f_2(S)$ deteriorates $f_1(S)$). That is, the two objectives $f_1(S)$ and $f_2(S)$ in (77.4) are conflicting with each other. In general, a multiobjective optimization problem has a number of nondominated solutions with different tradeoffs among the conflicting objectives. Those solutions are referred to as Pareto optimal solutions. The two-objective optimization problem in (77.4) has a number of nondominated fuzzy rule-based systems with different complexities (Figs. 77.2 and 77.3).

In Fig. 77.4, we illustrate the concept of complexity-accuracy tradeoff in the design of fuzzy rule-based systems. The horizontal axis of Fig. 77.4 shows the values of the complexity measure (i. e., $Complexity(S)$) while the vertical axis shows the values of the error measure (i. e., $Error(S)$). Around the top-left corner of Fig. 77.4, we have simple fuzzy rule-based systems with high interpretability and low accuracy (e.g., a simple 3×3 fuzzy rule-based system in Fig. 77.2). The improvement in their accuracy increases their complexity. By minimizing the error measure $Error(S)$, we have complicated fuzzy rule-based systems with high accuracy and low interpretability around the bottom-right corner of Fig. 77.4 (e.g., a tuned 7×7 fuzzy rule-based system in Fig. 77.3). In Fig. 77.4, we have many nondominated fuzzy rule-based systems along the complexity–accuracy tradeoff curve. It should be noted that there exist no fuzzy rule-based systems around the bottom-left corner (i. e., no ideal fuzzy rule-based sys-

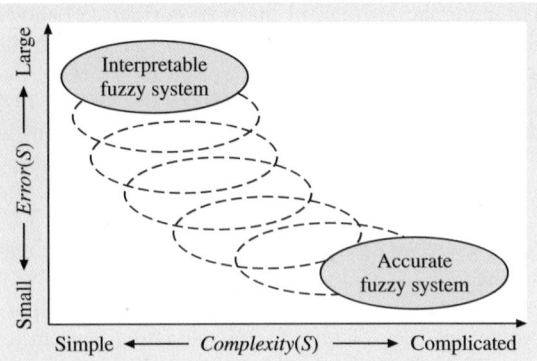

Fig. 77.4 Nondominated fuzzy rule-based systems with different complexity–accuracy tradeoffs

tems with high accuracy and high interpretability). This is because the two objectives in Fig. 77.4 are conflicting with each other.

Since the late 1990s, a number of multiobjective approaches have been proposed for fuzzy system design [77.16–19]. In this chapter, we explain the basic idea of multiobjective fuzzy system design using multiobjective evolutionary algorithms [77.20–22]. Whereas we started with fuzzy rules for fuzzy control in (77.1), our explanations in this chapter are mainly about multiobjective design of fuzzy rule-based systems for pattern classification. This is because early multiobjective approaches were mainly proposed for pattern classification problems.

77.2 Accuracy Maximization

In this section, we briefly explain various approaches proposed for improving the accuracy of fuzzy rule-based systems. Those approaches often deteriorate the interpretability.

77.2.1 Types of Fuzzy Rules

Fuzzy rules of the form in (77.1) have been successfully used in fuzzy controllers since *Mamdani*'s pioneering work in 1970s [77.23, 24]. Those fuzzy rules have often been called Mamdani-type fuzzy rules or Mamdani fuzzy rules. A heuristic rule generation method of such a fuzzy rule from numerical data was proposed by *Wang* and *Mendel* [77.25], which has been used for function approximation.

A well-known idea for improving the approximation ability of fuzzy rules in (77.1) is the use of a linear function instead of a linguistic value in the consequent part

$$\text{Rule } R_q : \text{If } x_1 \text{ is } A_{q1} \text{ and } \dots \text{ and } x_n \text{ is } A_{qn}$$
$$\text{then } y = b_{q0} + b_{q1}x_1 + b_{q2}x_2 + \dots + b_{qn}x_n , \tag{77.5}$$

where b_{qi} is a real number coefficient ($i = 0, 1, \dots, n$). Fuzzy rules of this type were proposed by *Takagi* and *Sugeno* [77.26]. A fuzzy rule-based system with fuzzy rules in (7.5) is referred to as a Takagi–Sugeno model. The use of a linear function instead of a linguistic value in the consequent part of fuzzy rules clearly increases the accuracy of fuzzy rule-based systems. However, it degrades their interpretability.

The following simplified version of fuzzy rules in Takagi–Sugeno models has been also used

$$\text{Rule } R_q : \text{If } x_1 \text{ is } A_{q1} \text{ and } \dots \text{ and } x_n \text{ is } A_{qn} \\ \text{then } y = b_q ,$$ (77.6)

where b_q is a consequent real number. It is easy to tune the consequent real number of each fuzzy rule. This is the main advantage of simplified fuzzy rules in (77.6). Thus, simplified fuzzy rules have often been used in trainable fuzzy rule-based systems called *neuro-fuzzy systems* [77.27–29]. In those studies, antecedent fuzzy sets as well as consequent real numbers are adjusted in the same manner as the learning of neural networks.

Due to their simple structure, simplified fuzzy rules in (77.6) may have higher interpretability than Takagi–Sugeno fuzzy rules in (77.5). However, it is usually difficult to linguistically interpret a consequent real number. Thus, the linguistic interpretability of simplified fuzzy rules in (77.6) is usually viewed as being limited if compared with Mamdani fuzzy rules with a linguistic value in their consequent part in (77.1).

For pattern classification problems, three types of fuzzy rules have been used in the literature [77.30]. The simplest structure of fuzzy rules for pattern classification problems is as follows

$$\text{Rule } R_q : \text{If } x_1 \text{ is } A_{q1} \text{ and } \dots \text{ and } x_n \text{ is } A_{qn} \\ \text{then Class } C_q ,$$ (77.7)

where C_q is a consequent class.

The compatibility grade of an input pattern $x_p = (x_{p1}, x_{p2}, \dots, x_{pn})$ with the antecedent part of the fuzzy rule R_q in (77.7) is usually calculated by the minimum or product operator. In this chapter, we use the following product operator

$$\mu_{A_q}(x_p) = \mu_{A_{q1}}(x_{p1}) \mu_{A_{q2}}(x_{p2}) \dots \mu_{A_{qn}}(x_{pn})$$ (77.8)

where $A_q = (A_{q1}, A_{q2}, \dots, A_{qn})$ is an antecedent fuzzy set vector, and $\mu_{A_q}(x_p)$ shows the compatibility of x_p with the antecedent fuzzy set vector A_p.

Let S be a set of fuzzy rules of the form in (77.7). The rule set S can be viewed as a fuzzy rule-based classifier. When an input pattern $x_p = (x_{p1}, x_{p2}, \dots, x_{pn})$ is presented to S, x_p is classified

by a single winner rule with the maximum compatibility. Such a single winner-based fuzzy reasoning method has been frequently used in fuzzy rule-based classifiers.

Let us assume that we have nine fuzzy rules in Fig. 77.5 for a pattern classification problem with the two-dimensional pattern space $[0, 1] \times [0, 1]$. A different consequent class is assigned to each rule in Fig. 77.5 for explanation purposes. The grid lines in the pattern space in Fig. 77.5 show the classification boundary between different classes when we use the single winner-based fuzzy reasoning method together with the product operator-based compatibility calculation. It should be noted that the classification boundary by the nine fuzzy rules in Fig. 77.5 can also be generated by nine non-fuzzy rules with interval antecedent conditions [77.31, 32].

The second type of fuzzy rules for pattern classification problems has a rule weight [77.30]

$$\text{Rule } R_q : \text{If } x_1 \text{ is } A_{q1} \text{ and } \dots \text{ and } x_n \text{ is } A_{qn} \\ \text{then Class } C_q \text{ with } CF_q ,$$ (77.9)

where CF_q is a real number in the unit interval $[0, 1]$, which is called a rule weight or a certainty factor. This type of fuzzy rules has been used in many studies on fuzzy rule-based classifiers since the early 1990s [77.33, 34].

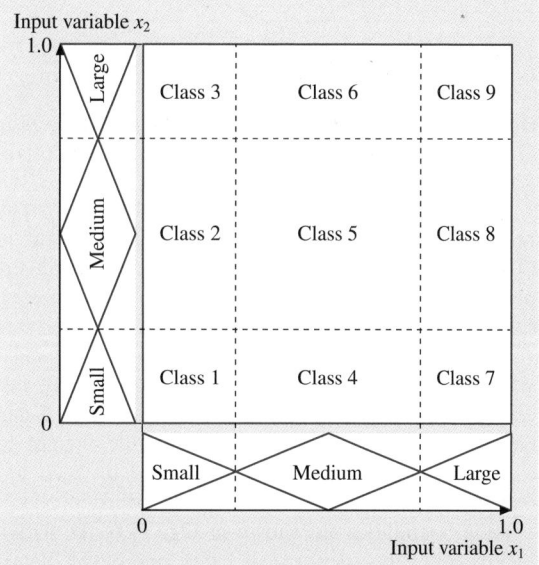

Fig. 77.5 A fuzzy rule-based classifier with nine fuzzy classification rules

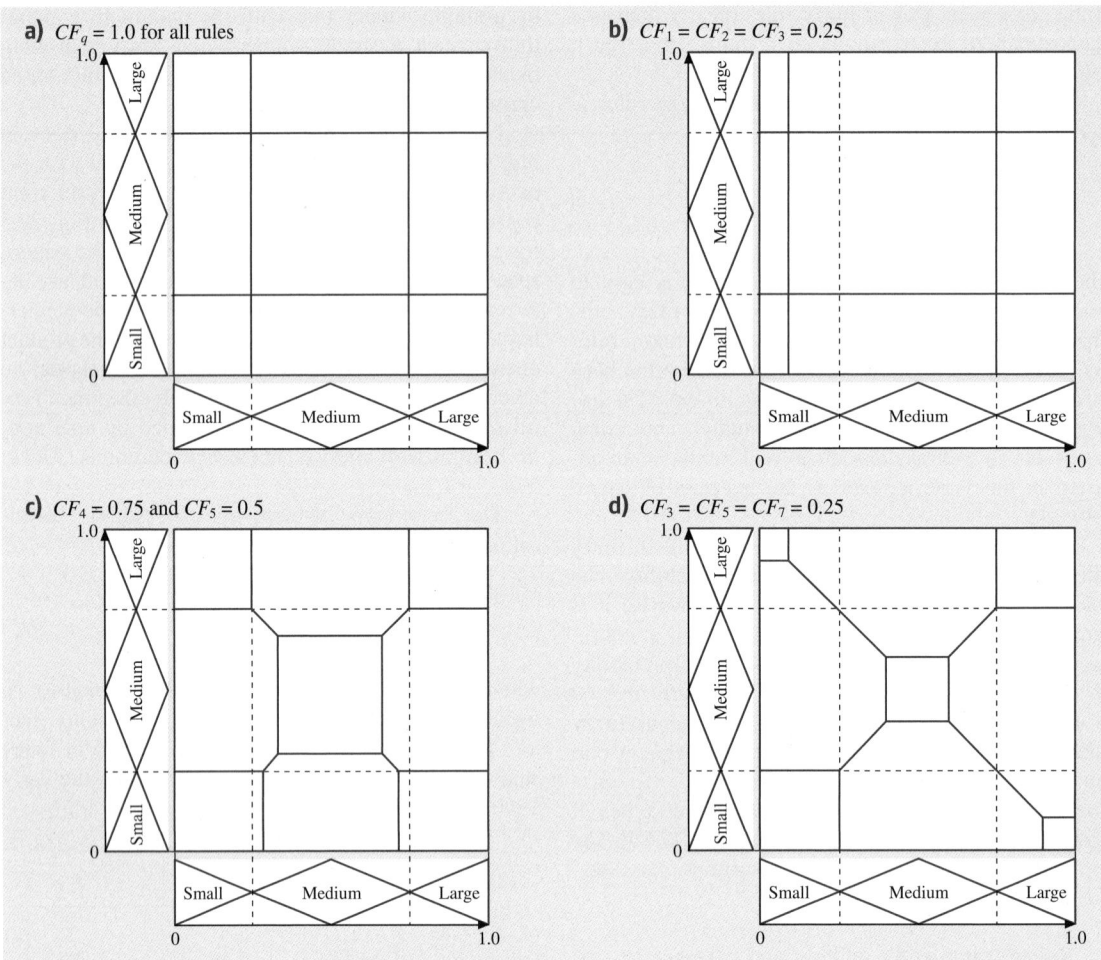

Fig. 77.6a–d Classification boundaries generated by assigning a different rule weight to each of the nine fuzzy rules in Fig. 77.5. In each plot, the default setting of CF_q is 1.0 (e.g., $CF_q = 1.0$ for $q = 4, 5, 6, 7, 8, 9$ in (**b**))

When an input pattern x_p is presented to a fuzzy rule-based classifier with fuzzy rules of the form in (77.9), a single winner rule is determined using the product of the compatibility $\mu_{A_q}(x_p)$ of x_p with each rule R_q and its rule weight CF_q: $\mu_{A_q}(x_p)CF_q$.

Fuzzy rules with a rule weight have higher classification ability than those with no rule weight. For example, the classification boundary in Fig. 77.5 can be adjusted by assigning a different rule weight to each rule (without changing the shape of each antecedent fuzzy set). Examples of the adjusted classification boundaries are shown in Fig. 77.6. As shown in Fig. 77.6, the accuracy of fuzzy rule-based classifiers can be improved by using fuzzy rules with a rule weight. However, the use of a rule weight degrades

the interpretability of fuzzy rule-based classifiers. It is a controversial issue to compare the interpretability of fuzzy rule-based classifiers between the following two approaches: One is the use of fuzzy rules with a rule weight and the other is the modification of antecedent fuzzy sets [77.35, 36].

The third type of fuzzy rules has multiple rule weights as follows [77.30]

> Rule R_q : If x_1 is A_{q1} and $\ldots$ and x_n is A_{qn}
>
> then Class C_1 with $CF_{q1}, \ldots,$ Class C_m (77.10)
>
> with CF_{qm} ,

where m is the number of classes and CF_{qj} is a real number in the unit interval $[0, 1]$, which can be viewed

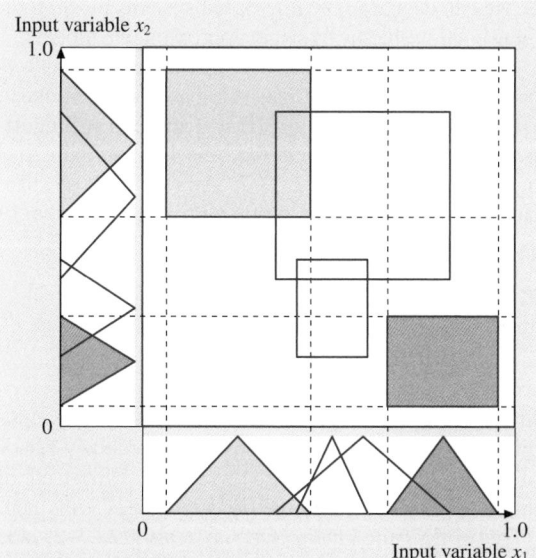

Fig. 77.7 Examples of fuzzy rules in an approximative fuzzy rule-based system

as a rule weight for the jth class C_j ($j = 1, 2, \ldots, m$). When we use the single winner-based fuzzy reasoning method, the classification result of each pattern depends only on the maximum rule weight CF_q of each rule (i.e., $CF_q = \max\{CF_{q1}, CF_{q2}, \ldots, CF_{qm}\}$). Thus, the use of multiple rule weights in (77.9) is meaningless under the single winner rule-based fuzzy reasoning method. However, they can improve the accuracy of fuzzy rule-based classifiers when we use a voting-based fuzzy reasoning method [77.30, 37]. Of course, the use of multiple rule weights further degrades the interpretability of fuzzy rule-based classifiers.

77.2.2 Types of Fuzzy Partitions

Since *Mamdani's* pioneering work in the 1970s [77.23, 24], grid-type fuzzy partitions have frequently been used in fuzzy control (e.g., the 3×3 fuzzy partition in Fig. 77.2). Such a grid-type fuzzy partition has high interpretability when it is used for two-dimensional problems (i.e., for the design of two-input single-output fuzzy rule-based systems). However, grid-type fuzzy partitions have the following two difficulties. One difficulty is the inflexibility of membership function tuning. Since each antecedent fuzzy set is used in multiple fuzzy rules, membership function tuning for improving the accuracy of one fuzzy rule may degrade the accuracy of some other fuzzy rules. The other difficulty is

the exponential increase in the number of fuzzy rules with respect to the number of input variables. Let L be the number of antecedent fuzzy sets for each of the n variables. In this case, the number of cells in the corresponding n-dimensional fuzzy grid is L^n (e.g., $5^{10} = 9\,765\,625$ when $L = 5$ and $n = 10$).

These two difficulties can be removed by assigning different antecedent fuzzy sets to each fuzzy rule as shown in Fig. 77.7. Each fuzzy rule has its own antecedent fuzzy sets. That is, no antecedent fuzzy set is shared by multiple fuzzy rules.

Fuzzy rule-based systems with this type of fuzzy rules are referred to as approximative models whereas grid-type fuzzy rule-based systems such as Fig. 77.2 are called descriptive models [77.38, 39]. If the accuracy of fuzzy rule-based systems is much more important than their interpretability, approximative models may be a better choice than descriptive models. Approximative models have been used as fuzzy rule-based classifiers since the early 1990s [77.40, 41].

One limitation of approximative models with respect to accuracy maximization is that every antecedent fuzzy set is defined on a single input variable. As a result, the shape of a fuzzy subspace covered by the antecedent part of each fuzzy rule is rectangular as shown in Fig. 77.7. This means that such a fuzzy subspace cannot handle any correlation among input variables. One approach to the handling of correlated subspaces is the use of a single high-dimensional antecedent fuzzy set in each fuzzy rule

$$\text{Rule } R_q : \text{If } \boldsymbol{x} \text{ is } A_q \text{then Class } C_q \,, \tag{77.11}$$

where $\boldsymbol{x}$ is an n-dimensional input vector (i.e., $\boldsymbol{x} = (x_1, x_2, \ldots, x_n)$) and A_q is a n-dimensional antecedent fuzzy set directly defined in the n-dimensional input space. This type of fuzzy rules has also been used for pattern classification problems since the 1990s [77.42]. Figure 77.8 illustrates an example of the n-dimensional antecedent fuzzy set A_q in the case of $n = 2$. As we can see from Fig. 77.8, antecedent fuzzy sets in fuzzy rules of the type in (77.11) can cover correlated fuzzy subspaces of the input space. This characteristic feature is an advantage over single-dimensional antecedent fuzzy sets with respect to the accuracy of fuzzy rule-based systems. However, as we can see from Fig. 77.8, it is almost impossible to linguistically interpret a high-dimensional antecedent fuzzy set. That is, the use of high-dimensional antecedent fuzzy sets may improve the accuracy of fuzzy rule-based systems but degrade their interpretability.

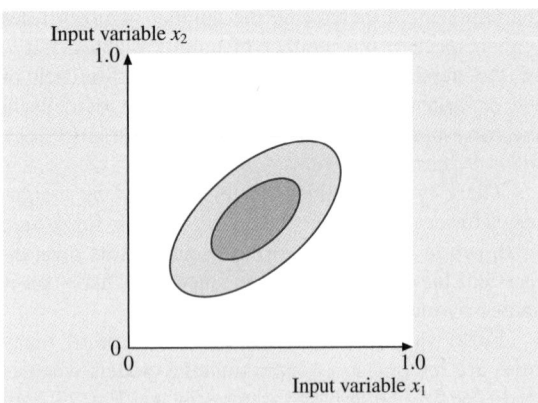

Fig. 77.8 Illustration of an *n*-dimensional antecedent fuzzy set A_q in the two-dimensional input space

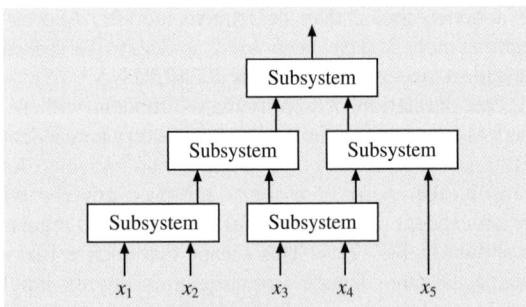

Fig. 77.9 A five-input single-output fuzzy rule-based system with a hierarchical structure

77.2.3 Handling of High-Dimensional Problems with Many Input Variables

As we have already explained, the number of fuzzy rules exponentially increases with the number of input variables when we use a descriptive model with a grid-based fuzzy partition. Thus, it looks impractical to design a descriptive model for high-dimensional problems.

Approximative models do not have such a difficulty of grid-based fuzzy partitions. This is because the number of fuzzy rules in an approximative model is independent from the number of input variables. That

is, we can design fuzzy rule-based systems for high-dimensional problems by using approximative models.

One difficulty in the use of approximative models is poor interpretability of fuzzy rule-based systems due to the following two reasons: (i) it is difficult to linguistically interpret antecedent fuzzy sets in approximative models as shown in Fig. 77.7, and (ii) it is also difficult to understand a fuzzy rule with a large number of antecedent conditions.

77.2.4 Hybrid Approaches with Neural Networks and Genetic Algorithms

In the 1990s, a large number of learning and optimization methods were proposed for accuracy maximization of fuzzy rule-based systems. Almost all of those approaches were hybrid approaches with neural networks called *neuro-fuzzy systems* [77.27–29, 43, 44] and with genetic algorithms called *genetic fuzzy systems* [77.45–48]. In neuro-fuzzy systems, learning algorithms of neural networks were utilized for parameter tuning (e.g., for membership function tuning). As shown in Fig. 77.3, parameter tuning in fuzzy rule-based systems usually leads to accuracy improvement and interpretability deterioration.

Genetic fuzzy systems can be used not only for parameter tuning but also for structure optimization such as rule selection, input selection and fuzzy partition. As we will explain in the next section, rule selection, and input selection can improve the interpretability of fuzzy rule-based systems by decreasing their complexity whereas parameter tuning almost always deteriorates their interpretability. Genetic fuzzy systems were also used for constructing a hierarchical structure of fuzzy rule-based systems [77.49]. Figure 77.9 shows an example of a fuzzy rule-based system with a hierarchical structure. The use of hierarchical structures can prevent the exponential increase in the number of fuzzy rules because each subsystem has only a few inputs (e.g., in Fig. 77.9, each subsystem has only two inputs). However, it significantly degrades the interpretability of fuzzy rule-based systems. This is because the interpretation of intermediate variables between subsystems is usually impossible.

77.3 Complexity Minimization

In this section, we briefly explain various approaches proposed for decreasing the complexity of fuzzy rule-based systems. Those approaches improve the interpretability of fuzzy rule-based systems but often degrade their accuracy.

77.3.1 Decreasing the Number of Fuzzy Rules

A simple idea for complexity reduction of fuzzy rule-based systems is to decrease the number of fuzzy rules. Let us consider a three-class pattern classification problem in Fig. 77.10. All patterns in Fig. 77.10 can be correctly classified by the following nine fuzzy rules with the 3×3 fuzzy grid in Fig. 77.10

Rule R_1 : If x_1 is *small* and x_2 is *small*

   then Class 2 ,

Rule R_2 : If x_1 is *small* and x_2 is *medium*

   then Class 2 ,

Rule R_3 : If x_1 is *small* and x_2 is *large*

   then Class 1 ,

Rule R_4 : If x_1 is *medium* and x_2 is *small*

   then Class 2 ,

Rule R_5 : If x_1 is *medium* and x_2 is *medium*

   then Class 2 ,

Rule R_6 : If x_1 is *medium* and x_2 is *large*

   then Class 1 ,

Rule R_7 : If x_1 is *large* and x_2 is *small*

   then Class 3 ,

Rule R_8 : If x_1 is *large* and x_2 is *medium*

   then Class 3 ,

Rule R_9 : If x_1 is *large* and x_2 is *large*

   then Class 3 .

That is, all patterns in Fig. 77.10 can be correctly classified by a fuzzy rule-based classifier with these nine fuzzy rules. It is also possible to correctly classify all patterns in Fig. 77.10 using a simple fuzzy rule-based classifier only with the four fuzzy rules around the top-right corner (i. e., fuzzy rules R_5, R_6, R_8, and R_9). This example illustrates the simplification of fuzzy rule-based systems through rule selection.

The use of genetic algorithms for fuzzy rule selection was proposed by *Ishibuchi* et al. [77.13, 14] in the

1990s. Let S_{All} be a set of all fuzzy rules. Since an arbitrary subset of S_{All} can be represented by a binary string of length $|S_{\text{All}}|$, standard genetic algorithms for binary strings can be directly applied to fuzzy rule selection [77.13, 14]. The number of fuzzy rules, which should be minimized, was used as a part of a fitness function in single-objective approaches [77.13, 14]. It was also used as a separate objective in multiobjective approaches [77.15].

77.3.2 Decreasing the Number of Antecedent Conditions

In Fig. 77.10, the rightmost three fuzzy rules (i. e., R_7, R_8 and R_9 with the same antecedent condition on x_1) can be combined into a single fuzzy rule: If x_1 is *large* then Class 3. This fuzzy rule has no condition on the second input variable x_2. In this manner, the 3×3 fuzzy rule-based classifier with the nine fuzzy rules can be simplified to a simpler classifier with the seven fuzzy rules.

The fuzzy rule If x_1 is *large* then Class 3 is viewed as having a *don't care* condition on the second input variable x_2: If x_1 is *large* and x_2 is *don't care* then Class 3. In this fuzzy rule, *don't care* is a special antecedent fuzzy set that is fully compatible with

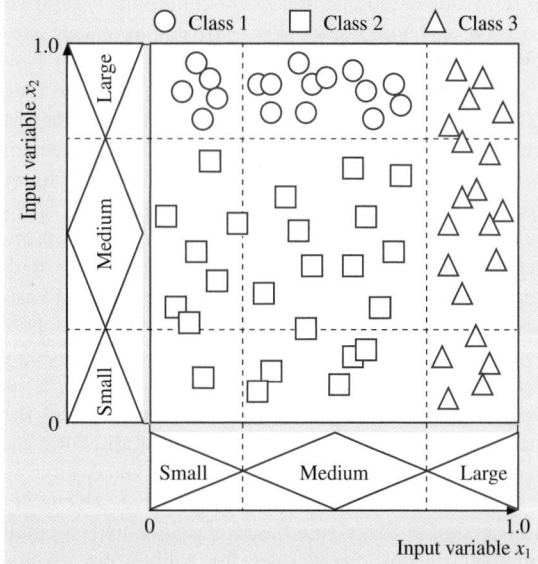

Fig. 77.10 A three-class pattern classification problem and a 3×3 fuzzy grid

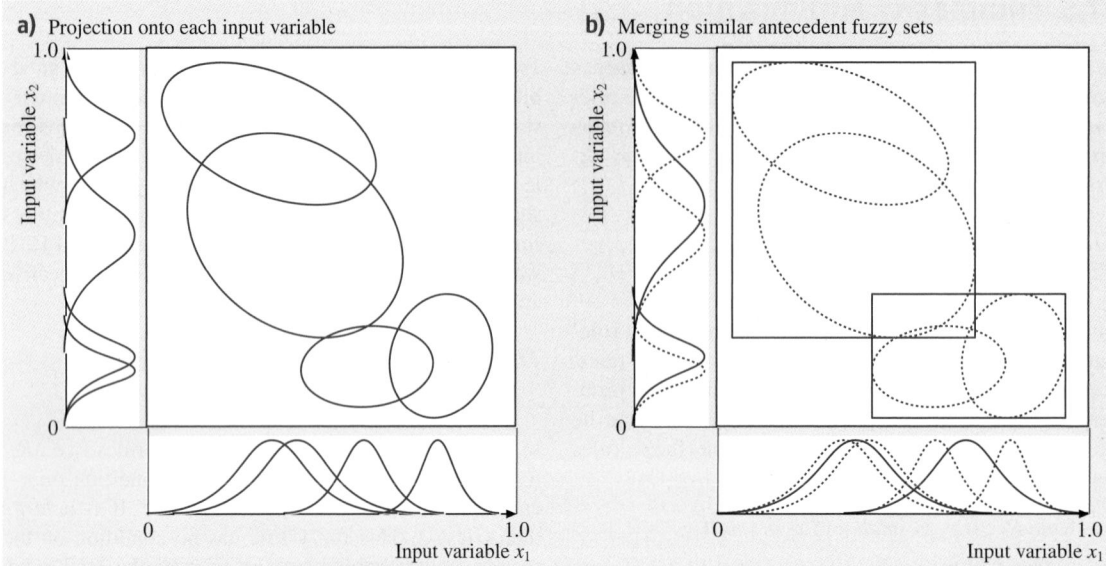

Fig. 77.11a,b Projection of two-dimensional fuzzy sets and the merge of similar fuzzy sets. (**a**) Projection onto each input variable, (**b**) merging similar antecedent fuzzy sets

any input values. The use of *don't care* enables us to perform rule-level input selection, which significantly improves the applicability of descriptive fuzzy rule-based systems to high-dimensional problems [77.50].

When we use *don't care* as a special antecedent fuzzy set, the number of antecedent conditions in a fuzzy rule excluding *don't care* conditions is referred to as the rule length since *don't care* conditions are usually omitted (e.g., If x_1 is *large* and x_2 is *don't care* then Class 3 is usually written as If x_1 is *large* then Class 3). A short fuzzy rule with a small number of antecedent conditions covers a large fuzzy subspace of a high-dimensional pattern space while a long fuzzy rule covers a small fuzzy subspace. For example, let us consider a 50-dimensional pattern classification problem with the pattern space $[0, 1]^{50}$. A fuzzy rule with the antecedent fuzzy set *small* on all the 50 input variables covers less than $1/10^1$ of the pattern space $[0, 1]^{50}$. However, a short fuzzy rule with the antecedent fuzzy set *small* on only two input variables (e.g., If x_1 is *small* and x_{49} is *small* then Class 3) covers 1/4 of the pattern space $[0, 1]^{50}$. As a result, almost all of the entire high-dimensional pattern space can be covered by a small number of short fuzzy rules. That is, we can design a simple fuzzy rule-based classifier with a small number of short fuzzy rules for a high-dimensional pattern classification problem. It should be noted that different fuzzy rules may have antecedent conditions

on different input variables. Moreover, the rule length of each fuzzy rule may be different (e.g., one fuzzy rule has an antecedent condition only on x_1 while another fuzzy rule has antecedent conditions on x_2, x_3 and x_4).

The total rule length (i. e., the total number of antecedent conditions), which should be minimized, was used as a part of a fitness function in single-objective approaches [77.51]. It was also used as a separate objective in multiobjective approaches [77.52, 53]. In multiobjective approaches, the total rule length instead of the average rule length has been used in the literature. This is because the minimization of the average rule length does not necessarily mean the complexity minimization of fuzzy rule-based systems. In many cases, the average rule length can be decreased by adding a new fuzzy rule with a single antecedent condition, which leads to the increase in the complexity of a fuzzy rule-based system.

77.3.3 Other Interpretability Improvement Approaches

For the design of accurate fuzzy rule-based systems for high-dimensional problems, clustering techniques such as fuzzy c-means [77.54–56] have often been used to generate fuzzy rules [77.42, 57–61]. Fuzzy rules with ellipsoidal high-dimensional antecedent fuzzy sets are often obtained from clustering-based fuzzy rule generation methods. Fuzzy rules of this type have high accuracy but low interpretability. Their interpretability is improved by projecting high-dimensional antecedent fuzzy sets onto each input variable. As a result, we have approximative fuzzy rule-based systems (Fig. 77.11a). The interpretability of the obtained fuzzy rule-based systems can be further improved by merging similar antecedent fuzzy sets on each input variable into a single one (Fig. 77.11b). Each of the generated antecedent fuzzy sets by a merging procedure is replaced with a linguistic value to further improve the interpretability of fuzzy rule-based systems.

It should be noted that each of the abovementioned interpretability improvement steps (i. e., projec-

tion of high-dimensional antecedent fuzzy sets, merging similar fuzzy sets, and replacement with linguistic values) deteriorates the accuracy of fuzzy rule-based systems. Thus, the design of fuzzy rule-based systems can be viewed as being the search for a good tradeoff solution between accuracy and interpretability. From this viewpoint, some sophisticated approaches were proposed [77.62–68] after a large number of accuracy improvement algorithms were proposed in 1990s. Some of those approaches tried to improve the accuracy of fuzzy rule-based systems without severely deteriorating their interpretability. Other approaches tried to improve the interpretability of fuzzy rule-based systems without severely deteriorating their accuracy.

77.4 Single-Objective Approaches

As we have already explained, the simplest multiobjective formulation of fuzzy system design has two objectives (i. e., error minimization and complexity minimization) as follows

$$\text{Minimize } \boldsymbol{f}(S) = (f_1(S), f_2(S))$$
$$= (Complexity(S), Error(S)) ,$$
$$(77.12)$$

where $\boldsymbol{f}(S)$ shows an objective vector. In this section, we explain how the two-objective problem in (77.12) can be handled by single-objective approaches. For more general and comprehensive explanations on the handling of multiobjective problems through single-objective optimization, see textbooks on multicriteria decision making such as *Miettinen* [77.69].

77.4.1 Use of Scalarizing Functions

One of the most frequently used approaches to multiobjective optimization is the use of scalarizing functions. Multiple objective functions are combined into a single scalarizing function. That is, a multiobjective problem is handled as a single-objective problem. Our two objectives in multiobjective fuzzy system design are combined as follows

$$\text{Minimize } f(S) = f(f_1(S), f_2(S))$$
$$= f(Complexity(S), Error(S)) ,$$
$$(77.13)$$

where $f(S)$ is a scalarizing function to be minimized. A simple but frequently used scalarizing function is the weighted sum

$$\text{Minimize } f(S) = w_1 f_1(S) + w_2 f_2(S)$$
$$= w_1 \, Complexity(S)$$
$$+ w_2 \, Error(S) , \qquad (77.14)$$

where w_1 and w_2 are nonnegative weights ($\boldsymbol{w}$ is a weight vector: $\boldsymbol{w} = (w_1, w_2)$).

Single-objective optimization algorithms such as genetic algorithms are used to search for the optimal solution (i. e., optimal fuzzy rule-based system) of the minimization problem in (77.13). In Fig. 77.12, we

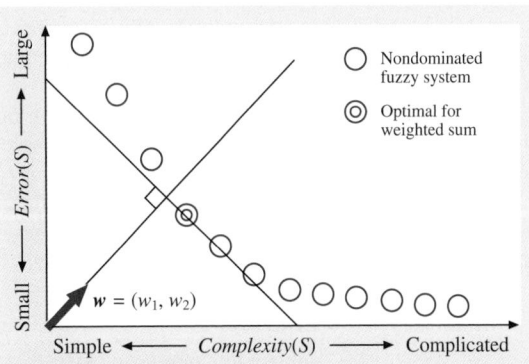

Fig. 77.12 The optimal fuzzy rule-based system of the weighted-sum minimization problem in (77.14) and the nondominated fuzzy rule-based systems of the original two-objective problem

illustrate the search for the optimal fuzzy rule-based system of the weighed-sum minimization problem in (77.14) together with the nondominated fuzzy rule-based systems of the original two-objective problem in (77.12).

As shown in Fig. 77.12, a single optimal fuzzy rule-based system is obtained from a scalarizing function-based approach. The main difficulty of this approach is the dependency of the obtained fuzzy rule-based system on the choice of a scalarizing function. A different fuzzy rule-based system is likely to be obtained from a different scalarizing function. For example, a different specification of the weight vector in Fig. 77.12 leads to a different fuzzy rule-based system. Moreover, an appropriate choice of a scalarizing function is not easy.

77.4.2 Handling of Objectives as Constraint Conditions

If we have a pre-specified requirement about the complexity or the accuracy, we can use it as a constraint condition. For example, let us assume that the error measure $Error(S)$ in our two-objective problem is the classification error rate. We also assume that the upper bound of the allowable error rate is given as $\alpha\%$. In this case, our two-objective problem can be reformulated as the following single-objective problem with a constraint condition

$$\text{Minimize } f_1(S) = Complexity(S)$$
$$\text{subject to } Error(S) \leq \alpha \ . \tag{77.15}$$

This single-objective problem is to find the simplest fuzzy rule-based system among those with a pre-specified accuracy (i. e., with error rates smaller than or equal to $\alpha\%$).

It is also possible to use a constraint condition on the complexity measure $Complexity(S)$. For example, let us assume that $Complexity(S)$ is the number of fuzzy rules. We also assume that the upper bound of the allowable number of fuzzy rules is given as β. In this case, the following single-objective problem is formulated

$$\text{Minimize } f_2(S) = Error(S)$$
$$\text{subject to } Complexity(S) \leq \beta \ . \tag{77.16}$$

This formulation is illustrated in Fig. 77.13 where the optimal solution is the most accurate fuzzy rule-based

system under the constraint condition $Complexity(S) \leq \beta$.

When we have more than two objectives, only a single objective is used as an objective function while all the others are used as constraint conditions in this approach. That is, an m-objective problem is reformulated as a single-objective problem with $(m-1)$ constraint conditions. The main difficulty in this constraint condition-based approach is an appropriate specification of the upper bound for each objective.

77.4.3 Minimization of the Distance to the Reference Point

In the abovementioned constraint condition-based approach, the right-hand side constant for each objective is the upper bound of the allowable error or complexity (e.g., the error rate should be at least smaller than or equal to $\alpha\%$). The right-hand side constant should be specified so that the formulated constrained optimization problem has feasible fuzzy rule-based systems.

A single-objective problem can be also formulated when an ideal fuzzy rule-based system is given as a reference point in the objective space. We assume that the given reference point is outside the feasible region of the original two-objective problem in (77.12). That is, the ideal fuzzy rule-based system does not exist as a feasible solution of the two-objective problem. Let the reference point in the two-dimensional objective space be $\boldsymbol{f}^* = (f_1^*, f_2^*)$. The following single-objective problem can be formulated to search for the fuzzy rule-based system closest to the reference

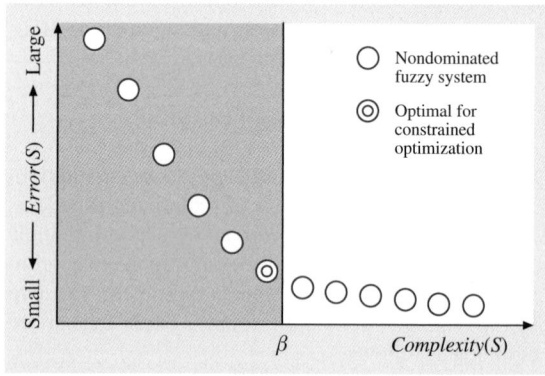

Fig. 77.13 The optimal fuzzy rule-based system of the constrained optimization problem in (77.15) and the nondominated fuzzy rule-based systems of the original two-objective problem

point

$$\text{Minimize distance } (\boldsymbol{f}(S), \boldsymbol{f}^*) \,, \qquad (77.17)$$

where $\boldsymbol{f}(S)$ is the objective vector (i.e., $\boldsymbol{f}(S) = (f_1(S), f_2(S))$, and *distance*(A, B) is a distance measure between the two points A and B in the objective space. Various distance measures can be used in (77.17). We illustrate the reference point-based approach in Fig. 77.14 where the Euclidean distance is used. As shown in Fig. 77.14, the fuzzy rule-based system closest to the given reference point (f_1^*, f_2^*) is the optimal solution of the single-objective problem in (77.17).

The main difficulty of the reference point-based approach is an appropriate specification of the reference point. When we have no information about the complexity and the accuracy of fuzzy rule-based systems, it is very difficult to appropriately specify the reference point in the reference point-based approach as well as the right-hand side constant for each objective in the constraint condition-based approach. However, if we know the shape of the complexity–accuracy trade-

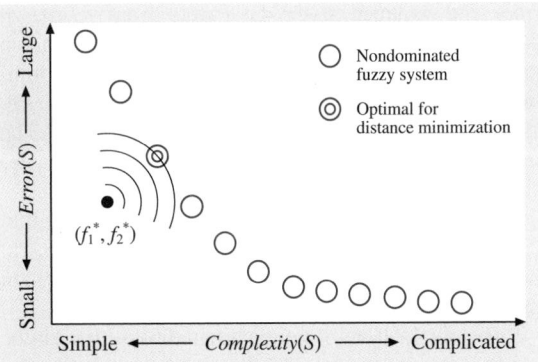

Fig. 77.14 The optimal fuzzy rule-based system of the distance minimization problem from the reference point in (77.17) and the nondominated fuzzy rule-based systems of the original two-objective problem

off surface in the objective space (i.e., if we know the nondominated fuzzy rule-based systems in the objective space), such a parameter specification becomes much easier.

77.5 Evolutionary Multiobjective Approaches

Since an early study in the 1990s [77.15], various multiobjective approaches have been proposed to search for a large number of nondominated solutions of multiobjective fuzzy system design problems. In this section, we explain the basic idea of those multiobjective approaches, recent studies on multiobjective fuzzy system design, and future research directions.

77.5.1 Basic Idea of Evolutionary Multiobjective Approaches

Multiobjective fuzzy system design was first formulated as a two-objective optimization problem to maximize the accuracy of fuzzy rule-based classifiers and to minimize the number of fuzzy rules in the 1990s [77.15]. Then this two-objective optimization problem was extended to a three-objective problem by including an additional objective to minimize the total rule length (i.e., the total number of antecedent conditions) in [77.52].

The main characteristic feature of evolutionary multiobjective approaches to fuzzy system design is that a number of nondominated fuzzy rule-based systems are obtained by a single run of an evolutionary multiobjective optimization (EMO) algorithm. This is clearly

different from the single-objective approaches where a single fuzzy rule-based system is obtained by a single run of a single-objective optimization algorithm. In Fig. 77.15, we illustrate the search for nondominated fuzzy rule-based systems in evolutionary multiobjective approaches. The population of solutions (i.e., fuzzy rule-based systems) is pushed toward the Pareto

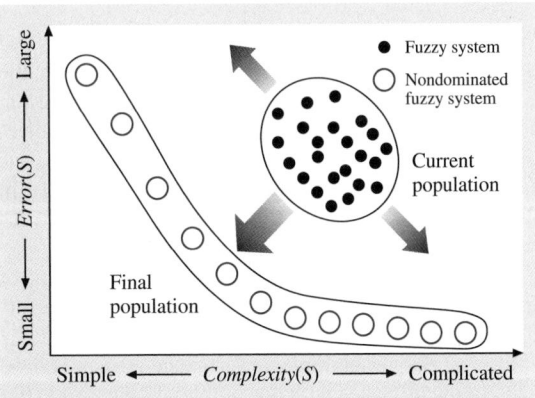

Fig. 77.15 Search for a variety of nondominated fuzzy rule-based systems along the Pareto front by evolutionary multiobjective approaches

front and widened along the Pareto front to search for a variety of nondominated solutions. Well-known and frequently used EMO algorithms such as nondominated sorting genetic algorithm II (NSGA-II) [77.70], strength Pareto evolutionary algorithm (SPEA) [77.71], multiobjective evolutionary algorithm based on decomposition (MOEA/D) [77.72], and S metric selection evolutionary multiobjective optimisation algorithm (SMS-EMOA) [77.73] have their own mechanisms to push the population toward the Pareto front and widen the population along the Pareto front.

The obtained set of nondominated solutions can be used to examine the complexity–accuracy tradeoff relation in the design of fuzzy rule-based systems [77.53]. A human decision maker is supposed to choose a final fuzzy rule-based system from the obtained nondominated ones according to his/her preference. It should be noted that the decision maker's preference is needed in the problem formulation phase in the single-objective approaches in the previous section (i. e., in the form of a scalarizing function, the upper bound of the allowable values for each objective, and the reference point in the objective space). However, the evolutionary multiobjective approaches do not need any information on the decision maker's preference in their search for nondominated fuzzy rule-based systems. That is, a number of nondominated fuzzy rule-based systems can be obtained with no information on the decision maker's preference. A human decision maker is needed only in the solution selection phase after a number of nondominated solutions are obtained.

77.5.2 Various Evolutionary Multiobjective Approaches

We have explained multiobjective fuzzy rule-based design using the two-objective formulation with the complexity minimization and the error minimization in Fig. 77.15. However, various evolutionary multiobjective approaches have been proposed for multiobjective fuzzy system design (for their review, see [77.19]). In this subsection, we briefly explain some of those evolutionary multiobjective approaches.

In some real-world applications, the design of fuzzy rule-based systems involves multiple performance measures. Especially in multiobjective fuzzy controller design, multiple performance measures have been frequently used with no complexity measures. For example, in *Stewart* et al. [77.74], multiobjective fuzzy controller design was formulated as a three-objective problem with three performance measures: a current

tracking error, a velocity tracking error, and a power consumption. In *Chen* and *Chiang* [77.75], fuzzy controller design was formulated using no complexity measure and three accuracy measures: the number of collisions, the distance between the target and lead points of the new path, and the number of explored actions. Whereas multiple performance measures have been frequently used in multiobjective fuzzy controller design, a single performance measure such as the error rate has been mainly used in multiobjective fuzzy classifier design. However, for the handling of classification problems with imbalanced and cost-sensitive data sets, multiple performance measures were used in some studies on multiobjective fuzzy classifier design. For example, a true positive rate and a false positive rate were used as separate performance measures together with a complexity measure in three-objective fuzzy classifier design in [77.76].

Multiple complexity measures have been frequently used in multiobjective fuzzy classifier design. In the first study on multiobjective fuzzy classifier design [77.15], the number of fuzzy rules was used as a complexity measure. Then the total rule length (i. e., the total number of antecedent conditions) was added as another complexity measure in three-objective fuzzy classifier design [77.52, 53]. The number of fuzzy rules and the total rule length have been used in many other studies on multiobjective fuzzy classifier design [77.77–79]. In some studies, the number of antecedent fuzzy sets was used instead of the total rule length [77.80, 81].

When membership function tuning is performed together with fuzzy rule generation in fuzzy classifier design, complexity measures such as the number of fuzzy rules and the total rule length are not always enough to evaluate the interpretability of fuzzy rule-based systems. Let us compare two fuzzy partitions in Fig. 77.16 with each other. The 5×5 fuzzy partition in Fig. 77.16a has 25 fuzzy rules while the 4×4 fuzzy partition in Fig. 77.16b has 16 fuzzy rules. Thus, the fuzzy partition in Fig. 77.16a is evaluated as being more complicated than that of Fig. 77.16b when the abovementioned simple complexity measures are used. However, we intuitively feel that the simple 5×5 fuzzy partition in Fig. 77.16a is more interpretable than the tuned 4×4 fuzzy partition in Fig. 77.16b. This is because the tuned antecedent fuzzy sets in Fig. 77.16b are not easy to interpret linguistically. These discussions on the comparison between the two fuzzy partitions show the necessity of interpretability measures in addition to the abovementioned simple complexity measures in

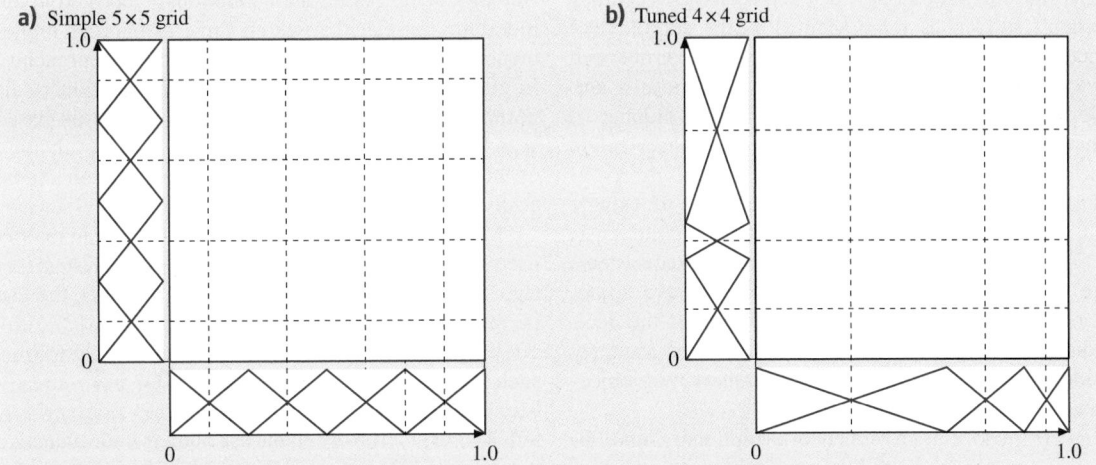

Fig. 77.16a,b Two fuzzy partitions: (**a**) simple 5×5 grid, (**b**) tuned 4×4 grid

fuzzy classifier design when membership function tuning is performed.

Interpretability of fuzzy rule-based systems has been a hot topic in the field of fuzzy systems [77.82]. Various aspects of fuzzy rule-based systems are related to their interpretability [77.83–88]. Some studies focus on the explanation ability of fuzzy rule-based classifiers to explain why each pattern is classified as a particular class in an understandable manner [77.89].

Whereas a number of studies have already addressed the interpretability of fizzy rule-based systems [77.82–89], it is still a very difficult open problem to quantitatively define all aspects of the interpretability of fuzzy rule-based systems. This is because the interpretability is totally subjective. That is, its definition totally depends on human users. Each human user may have a different idea about the interpretability of fuzzy rule-based systems.

A number of approaches have been proposed to incorporate the interpretability into evolutionary multiobjective fuzzy system design [77.90–94]. The basic idea is to significantly improve the accuracy of fuzzy rule-based systems by slightly deteriorating their interpretability (e.g., by slightly tuning antecedent fuzzy sets). Since the interpretability is totally subjective, it is not easy to compare those approaches. In this sense, experimental studies on the interpretability of fuzzy rule-based systems seem to be one of the promising research directions [77.83].

Whereas multiobjective genetic algorithms have been mainly used in evolutionary multiobjective fuzzy system design, the use of other algorithms was also ex-

amined. This is closely related to the increase in the popularity of not only multiobjective genetic algorithms but also other multiobjective algorithms. For example, multiobjective versions of particle swarm optimization (PSO) have been actively studies in the field of evolutionary computation [77.95–99]. In response to those active studies, multiobjective PSO algorithms were used for multiobjective fuzzy system design [77.100–104].

77.5.3 Future Research Directions

Formulation of interpretability is still an important issue to be further studied. As pointed out by many studies [77.83–88], various aspects are related to the interpretability of fuzzy rule-based systems. One problem is to quantitatively formulate those aspects so that they can be used as objectives in evolutionary multiobjective fuzzy system design. Another problem is how to use them. We may have several options: the use of all aspects as separate objectives, the choice of only a few aspects as separate objectives, and the integration of all or some aspects into a few interpretability measures. If we use all aspects as separate objectives, multiobjective fuzzy system design is formulated as a many-objective problem. It is well-known that many-objective problems are usually very difficult for evolutionary multiobjective optimization problems [77.105–107]. However, both the choice of only a few aspects and the integration into a few interpretability measures are also difficult.

The main advantage of multiobjective approaches to fuzzy system design over single-objective approaches is that a number of nondominated fuzzy rule-based systems are obtained along the interpretability–accuracy tradeoff surface as we explained in the complexity–accuracy objective space. One issue, which has not been discussed in many studies, is how to choose a single fuzzy rule-based system from a large number of obtained ones. It is implicitly assumed that a single fuzzy rule-based system is to be selected by a human decision maker. However, the selection of a single fuzzy rule-based system is an important issue especially when a large number of fuzzy rule-based systems are obtained in a high-dimensional objective space. A related research topic is the elicitation of the decision maker's preference about interpretability–accuracy tradeoffs and its utilization in evolutionary multiobjective fuzzy system design.

Performance improvement of evolutionary multiobjective approaches is still an important research topic. Since multiobjective fuzzy system design is often formulated as complicated multiobjective optimization problems with many discrete and continuous decision variables, it is very difficult to search for their true Pareto optimal solutions. Thus, it is likely that better fuzzy rule-based systems than reported results in the literature would be obtained by more efficient multiobjective algorithms and/or better problem formulations.

Actually better results are continuously reported in the literature. A related research topic is parallel implementation of evolutionary multiobjective approaches. In general, parallel implementation of evolutionary algorithms is not difficult due to their population-based search mechanisms (i. e., because the fitness evaluation of multiple individual in the current population can be easily performed in parallel).

Multiobjective genetic algorithms have been mainly used for evolutionary multiobjective fuzzy system design. As we have already mentioned, recently the use of multiobjective PSO has been examined [77.100–104]. Since other population-based search algorithms such as ant colony optimization (ACO) have already been used in single-objective approaches to fuzzy system design [77.108–112], the use of their multiobjective versions will be examined for multiobjective fuzzy system design.

A very important and promising research direction is multiobjective design of type-2 fuzzy systems [77.113]. A number of single-objective approaches have already been proposed for type-2 fuzzy system design [77.114–116]. However, multiobjective type-2 fuzzy system design has not been discussed in many studies.

77.6 Conclusion

We explained the basic idea of evolutionary multiobjective fuzzy system design using a simple two-objective formulation for complexity and error minimization in comparison with single-objective approaches. The main advantage of multiobjective approaches is that a large number of fuzzy rule-based systems with different complexity–accuracy tradeoffs are obtained from a single run of a multiobjective approach. A human user can choose a single fuzzy rule-based system based on

his/her preference and the requirement in each application field. Highly interpretable fuzzy systems may be needed in some application fields while highly accurate ones may be preferred in other application fields. See [77.19] for more comprehensive review on evolutionary multiobjective approaches to fuzzy rule-based system design, [77.88] for single-objective and multiobjective approaches, and [77.115, 116] for type-2 fuzzy system design.

References

77.1   C.C. Lee: Fuzzy logic in control systems: Fuzzy logic controller – Part I, IEEE Trans. Syst. Man Cybern. **20**(2), 404–418 (1990)

77.2   C.C. Lee: Fuzzy logic in control systems: Fuzzy logic controller – Part II, IEEE Trans. Syst. Man Cybern. **20**(2), 419–435 (1990)

77.3   J.M. Mendel: Fuzzy logic systems for engineering: A tutorial, Proc. IEEE **83**(3), 345–377 (1995)

77.4   L.A. Zadeh: The concept of a linguistic variable and its application to approximate reasoning – I, Inf. Sci. **8**(3), 199–249 (1975)

77.5   L.A. Zadeh: The concept of a linguistic variable and its application to approximate reasoning – II, Inf. Sci. **8**(4), 301–357 (1975)

77.6   L.A. Zadeh: The concept of a linguistic variable and its application to approximate reasoning – III, Inf. Sci. **9**(1), 43–80 (1975)

77.7   B. Kosko: Fuzzy systems as universal approxima-tors, Proc. 1992 IEEE Int. Conf. Fuzzy Syst. (IEEE, San Diego 1992) pp. 1153–1162

77.8   L.X. Wang: Fuzzy systems are universal approxi-mators, Proc. 1992 IEEE Int. Conf. Fuzzy Syst. (IEEE, San Diego 1992) pp. 1163–1170

77.9   L.X. Wang, J.M. Mendel: Fuzzy basis functions, universal approximation, and orthogonal least-squares learning, IEEE Trans. Neural Netw. **3**(5), 807–814 (1992)

77.10   K. Funahashi: On the approximate realization of continuous mappings by neural networks, Neural Netw. **2**(3), 183–192 (1989)

77.11   K. Hornik, M. Stinchcombe, H. White: Multilayer feedforward networks are universal approxima-tors, Neural Netw. **2**(5), 359–366 (1989)

77.12   J. Park, I.W. Sandberg: Universal approxima-tion using radial-basis-function networks, Neu-ral Comput. **3**(2), 246–257 (1991)

77.13   H. Ishibuchi, K. Nozaki, N. Yamamoto, H. Tanaka: Construction of fuzzy classification systems with rectangular fuzzy rules using genetic algorithms, Fuzzy Sets Syst. **65**(2/3), 237–253 (1994)

77.14   H. Ishibuchi, K. Nozaki, N. Yamamoto, H. Tanaka: Selecting fuzzy if-then rules for classification problems using genetic algorithms, IEEE Trans. Fuzzy Syst. **3**(3), 260–270 (1995)

77.15   H. Ishibuchi, T. Murata, I.B. Türkşen: Single-objective and two-objective genetic algorithms for selecting linguistic rules for pattern classifi-cation problems, Fuzzy Sets Syst. **89**(2), 135–150 (1997)

77.16   H. Ishibuchi: Multiobjective genetic fuzzy sys-tems: Review and future research directions, Proc. 2007 IEEE Int. Conf. Fuzzy Syst. (IEEE, London 2007) pp. 913–918

77.17   H. Ishibuchi, Y. Nojima, I. Kuwajima: Evolutionary multiobjective design of fuzzy rule-based clas-sifiers. In: *Computational Intelligence: A Com-pendium*, ed. by J. Fulcher, L.C. Jain (Springer, Berlin 2008) pp. 641–685

77.18   H. Ishibuchi, Y. Nojima: Multiobjective genetic Fuzzy Systems. In: *Computational Intelligence: Collaboration, Fusion and Emergence*, ed. by C.L. Mumford, L.C. Jain (Springer, Berlin 2009) pp. 131–173

77.19   M. Fazzolari, R. Alcalá, Y. Nojima, H. Ishibuchi, F. Herrera: A review of the application of multiob-jective evolutionary fuzzy systems: Current status and further directions, IEEE Trans. Fuzzy Syst. **21**(1), 45–65 (2013)

77.20   K. Deb: *Multi-Objective Optimization Using Evo-lutionary Algorithms* (Wiley, Chichester 2001)

77.21   K.C. Tan, E.F. Khor, T.H. Lee: *Multiobjective Evo-lutionary Algorithms and Applications* (Springer, Berlin 2005)

77.22   C.A.C. Coello, G.B. Lamont: *Applications of Multi-Objective Evolutionary Algorithms* (World Scien-tific, Singapore 2004)

77.23   E.H. Mamdani, S. Assilian: An experiment in lin-guistic synthesis with a fuzzy logic controller, Int. J. Man-Mach. Stud. **7**(1), 1–13 (1975)

77.24   E.H. Mamdani: Application of fuzzy logic to ap-proximate reasoning using linguistic synthesis, IEEE Trans. Comput. **C-26**(12), 1182–1191 (1977)

77.25   L.X. Wang, J.M. Mendel: Generating fuzzy rules by learning from examples, IEEE Trans. Syst. Man Cy-bern. **22**(6), 1414–1427 (1992)

77.26   T. Takagi, M. Sugeno: Fuzzy identification of sys-tems and its applications to modeling and con-trol, IEEE Trans. Syst. Man Cybern. **15**(1), 116–132 (1985)

77.27   C.T. Lin, C.S.G. Lee: Neural-network-based fuzzy logic control and decision system, IEEE Trans. Comput. **40**(12), 1320–1336 (1991)

77.28   S. Horikawa, T. Furuhashi, Y. Uchikawa: On fuzzy modeling using fuzzy neural networks with the back-propagation algorithm, IEEE Trans. Neural Netw. **3**(5), 801–806 (1992)

77.29   J.S.R. Jang: ANFIS: Adaptive-network-based fuzzy inference system, IEEE Trans. Syst. Man Cybern. **23**(3), 665–685 (1993)

77.30   O. Cordón, M.J. del Jesus, F. Herrera: A proposal on reasoning methods in fuzzy rule-based clas-sification systems, Int. J. Approx. Reason. **20**(1), 21–45 (1999)

77.31   L.I. Kuncheva: How good are fuzzy If-then clas-sifiers?, IEEE Trans. Syst. Man Cybern. B **30**(4), 501–509 (2000)

77.32   L.I. Kuncheva: *Fuzzy Classifier Design* (Physica, Heidelberg 2000)

77.33   H. Ishibuchi, K. Nozaki, H. Tanaka: Distributed representation of fuzzy rules and its application to pattern classification, Fuzzy Sets Syst. **52**(1), 21–32 (1992)

77.34   H. Ishibuchi, T. Nakashima, M. Nii: *Classifica-tion and Modeling with Linguistic Information Granules: Advanced Approaches to Linguistic Data Mining* (Springer, Berlin 2004)

77.35   D. Nauck, R. Kruse: How the learning of rule weights affects the interpretability of fuzzy sys-tems, Proc. IEEE Int. Conf. Fuzzy Syst. (IEEE, An-chorage 1998) pp. 1235–1240

77.36   H. Ishibuchi, T. Nakashima: Effect of rule weights in fuzzy rule-based classification systems, IEEE Trans. Fuzzy Syst. **9**(4), 506–515 (2001)

77.37   H. Ishibuchi, T. Nakashima, T. Morisawa: Voting in fuzzy rule-based systems for pattern classifi-cation problems, Fuzzy Sets Syst. **103**(2), 223–238 (1999)

77.38   O. Cordón, F. Herrera: A three-stage evolutionary process for learning descriptive and approximate fuzzy-logic-controller knowledge bases from ex-amples, Int. J. Approx. Reason. **17**(4), 369–407 (1997)

77.39   J.G. Marin-Blázquez, Q. Shen: From approxima-tive to descriptive fuzzy classifiers, IEEE Trans. Fuzzy Syst. **10**(4), 484–497 (2002)

Part G | 77

77.40   P.K. Simpson: Fuzzy min-max neural networks – Part 1: Classification, IEEE Trans. Neural Netw. **3**(5), 776–786 (1992)

77.41   S. Abe, M.S. Lan: A method for fuzzy rules extraction directly from numerical data and its application to pattern classification, IEEE Trans. Fuzzy Syst. **3**(1), 18–28 (1995)

77.42   S. Abe, R. Thawonmas: A fuzzy classifier with ellipsoidal regions, IEEE Trans. Fuzzy Syst. **5**(3), 358–368 (1997)

77.43   D. Nauck, F. Klawonn, R. Kruse: *Foundations of Neuro-Fuzzy Systems* (Wiley, New York 1997)

77.44   S. Abe: *Pattern Classification: Neuro-Fuzzy Methods and Their Comparison* (Springer, Berlin 2001)

77.45   O. Cordón, F. Herrera, F. Hoffmann, L. Magdalena: *Genetic Fuzzy Systems* (World Scientific, Singapore 2001)

77.46   O. Cordón, F. Gomide, F. Herrera, F. Hoffmann, L. Magdalena: Ten years of genetic fuzzy systems: Current framework and new trends, Fuzzy Sets Syst. **141**(1), 5–31 (2004)

77.47   F. Herrera: Genetic fuzzy systems: Status, critical considerations and future directions, Int. J. Comput. Intell. Res. **1**(1), 59–67 (2005)

77.48   F. Herrera: Genetic fuzzy systems: Taxonomy, current research trends and prospects, Evol. Intell. **1**(1), 27–46 (2008)

77.49   K. Shimojima, T. Fukuda, Y. Hasegawa: Self-tuning fuzzy modeling with adaptive membership function, rules, and hierarchical structure based on genetic algorithm, Fuzzy Sets Syst. **71**(3), 295–309 (1995)

77.50   H. Ishibuchi, T. Nakashima, T. Murata: Performance evaluation of fuzzy classifier systems for multidimensional pattern classification problems, IEEE Trans. Syst. Man Cybern. B **29**(5), 601–618 (1999)

77.51   H. Ishibuchi, T. Yamamoto, T. Nakashima: Hybridization of fuzzy GBML approaches for pattern classification problems, IEEE Trans. Syst. Man Cybern. B **35**(2), 359–365 (2005)

77.52   H. Ishibuchi, T. Nakashima, T. Murata: Three-objective genetics-based machine learning for linguistic rule extraction, Inf. Sci. **136**(1–4), 109–133 (2001)

77.53   H. Ishibuchi, Y. Nojima: Analysis of interpretability-accuracy tradeoff of fuzzy systems by multiobjective fuzzy genetics-based machine learning, Int. J. Approx. Reason. **44**(1), 4–31 (2007)

77.54   J.C. Dunn: A fuzzy relative of the ISODATA process and its use in detecting compact well-separated clusters, J. Cybern. **3**(3), 32–57 (1973)

77.55   J.C. Bezdek: *Pattern Recognition with Fuzzy Objective Function Algorithms* (Plenum Press, New York 1981)

77.56   J.C. Bezdek, R. Ehrlich, W. Full: FCM: The fuzzy c-means clustering algorithm, Comput. Geosci. **10**(2/3), 191–203 (1984)

77.57   S.L. Chiu: Fuzzy model identification based on cluster estimation, J. Intell. Fuzzy Syst. **2**(3), 267–278 (1994)

77.58   J.A. Dickerson, B. Kosko: Fuzzy function approximation with ellipsoidal rules, IEEE Trans. Syst. Man Cybern. **26**(4), 542–560 (1996)

77.59   C.J. Lin, C.T. Lin: An ART-based fuzzy adaptive learning control network, IEEE Trans. Fuzzy Syst. **5**(4), 477–496 (1997)

77.60   M. Delgado, A.F. Gomez-Skarmeta, F. Martin: A fuzzy clustering-based rapid prototyping for fuzzy rule-based modeling, IEEE Trans. Fuzzy Syst. **5**(2), 223–233 (1997)

77.61   M. Setnes: Supervised fuzzy clustering for rule extraction, IEEE Trans. Fuzzy Syst. **8**(4), 416–424 (2000)

77.62   M. Setnes, R. Babuska, H.B. Verbruggen: Rule-based modeling: Precision and transparency, IEEE Trans. Syst. Man Cybern. C **28**(1), 165–169 (1998)

77.63   M. Setnes, R. Babuska, U. Kaymak, H.R. van Nauta Lemke: Similarity measures in fuzzy rule base simplification, IEEE Trans. Syst. Man Cybern. B **28**(3), 376–386 (1998)

77.64   Y. Jin, W. von Seelen, B. Sendhoff: On generating FC3 fuzzy rule systems from data using evolution strategies, IEEE Trans. Syst. Man Cybern. B **29**(6), 829–845 (1999)

77.65   M. Setnes, H. Roubos: GA-fuzzy modeling and classification: Complexity and performance, IEEE Trans. Fuzzy Syst. **8**(5), 509–522 (2000)

77.66   H. Roubos, M. Setnes: Compact and transparent fuzzy models and classifiers through iterative complexity reduction, IEEE Trans. Fuzzy Syst. **9**(4), 516–524 (2001)

77.67   J. Abonyi, J.A. Roubos, F. Szeifert: Data-driven generation of compact, accurate, and linguistically sound fuzzy classifiers based on a decision-tree initialization, Int. J. Approx. Reason. **32**(1), 1–21 (2003)

77.68   R. Alcalá, J. Alcalá-Fdez, F. Herrera, J. Otero: Genetic learning of accurate and compact fuzzy rule based systems based on the 2-tuples linguistic representation, Int. J. Approx. Reason. **44**(1), 45–64 (2007)

77.69   K. Miettinen: *Nonlinear Multiobjective Optimization* (Kluwer, Boston 1999)

77.70   K. Deb, A. Pratap, S. Agarwal, T. Meyarivan: A fast and elitist multiobjective genetic algorithm: NSGA-II, IEEE Trans. Evol. Comput. **6**(2), 182–197 (2002)

77.71   E. Zitzler, L. Thiele: Multiobjective evolutionary algorithms: A comparative case study and the strength Pareto approach, IEEE Trans. Evol. Comput. **3**(4), 257–271 (1999)

77.72   Q. Zhang, H. Li: MOEA/D: A multiobjective evolutionary algorithm based on decomposition, IEEE Trans. Evol. Comput. **11**(6), 712–731 (2007)

77.73   N. Beume, B. Naujoks, M. Emmerich: SMS-EMOA: Multiobjective selection based on dominated hypervolume, Eur. J. Oper. Res. **181**(3), 1653–1669 (2007)

77.74   P. Stewart, D.A. Stone, P.J. Fleming: Design of robust fuzzy-logic control systems by multiobjective evolutionary methods with hardware in the loop, Eng. Appl. Artif. Intell. **17**(3), 275–284 (2004)

77.75   L.H. Chen, C.H. Chiang: An intelligent control system with a multi-objective self-exploration process, Fuzzy Sets Syst. **143**(2), 275–294 (2004)

77.76   P. Ducange, B. Lazzerini, F. Marcelloni: Multiobjective genetic fuzzy classifiers for imbalanced and cost-sensitive datasets, Soft Comput. **14**(7), 713–728 (2010)

77.77   C. Setzkorn, R.C. Paton: On the use of multiobjective evolutionary algorithms for the induction of fuzzy classification rule systems, BioSystems **81**(2), 101–112 (2005)

77.78   H. Wang, S. Kwong, Y. Jin, W. Wei, K.F. Man: Agent-based evolutionary approach for interpretable rule-based knowledge extraction, IEEE Trans. Syst. Man Cybern. C **35**(2), 143–155 (2005)

77.79   C.H. Tsang, S. Kwong, H.L. Wang: Genetic-fuzzy rule mining approach and evaluation of feature selection techniques for anomaly intrusion detection, Pattern Recognit. **40**(9), 2373–2391 (2007)

77.80   H. Wang, S. Kwong, Y. Jin, W. Wei, K.F. Man: Multiobjective hierarchical genetic algorithm for interpretable fuzzy rule-based knowledge extraction, Fuzzy Sets Syst. **149**(1), 149–186 (2005)

77.81   Z.Y. Xing, Y. Zhang, Y.L. Hou, L.M. Jia: On generating fuzzy systems based on Pareto multi-objective cooperative coevolutionary algorithm, Int. J. Control Autom. Syst. **5**(4), 444–455 (2007)

77.82   J. Casillas, O. Cordón, F. Herrera, L. Magdalena (Eds.): *Interpretability Issues in Fuzzy Modeling* (Springer, Berlin 2003)

77.83   J.M. Alonso, L. Magdalena, G. González-Rodríguez: Looking for a good fuzzy system interpretability index: An experimental approach, Int. J. Approx. Reason. **51**(1), 115–134 (2009)

77.84   H. Ishibuchi, Y. Kaisho, Y. Nojima: Design of linguistically interpretable fuzzy rule-based classifiers: A short review and open questions, J. Mult.-Valued Log. Soft Comput. **17**(2/3), 101–134 (2011)

77.85   J.M. Alonso, L. Magdalena: Editorial: Special issue on interpretable fuzzy systems, Inf. Sci. **181**(20), 4331–4339 (2011)

77.86   M.J. Gacto, R. Alcalá, F. Herrera: Interpretability of linguistic fuzzyrule-based systems: An overview of interpretability measures, Inf. Sci. **181**(20), 4340–4360 (2011)

77.87   C. Mencar, C. Castiello, R. Cannone, A.M. Fanelli: Interpretability assessment of fuzzy knowledge bases: A cointension based approach, Int. J. Approx. Reason. **52**(4), 501–518 (2011)

77.88   O. Cordón: A historical review of evolutionary learning methods for Mamdani-type fuzzy rule-based systems: Designing interpretable genetic fuzzy systems, Int. J. Approx. Reason. **52**(6), 894–913 (2011)

77.89   H. Ishibuchi, Y. Nojima: Toward quantitative definition of explanation ability of fuzzy rule-based classifiers, Proc. 2011 IEEE Int. Conf. Fuzzy Syst. (IEEE, Taipei 2011) pp. 549–556

77.90   M. Antonelli, P. Ducange, B. Lazzerini, F. Marcelloni: Learning concurrently partition granularities and rule bases of Mamdani fuzzy systems in a multi-objective evolutionary framework, Int. J. Approx. Reason. **50**(7), 1066–1080 (2009)

77.91   A. Botta, B. Lazzerini, F. Marcelloni, D.C. Stefanescu: Context adaptation of fuzzy systems through a multi-objective evolutionary approach based on a novel interpretability index, Soft Comput. **13**(5), 437–449 (2009)

77.92   M.J. Gacto, R. Alcalá, F. Herrera: Integration of an index to preserve the semantic interpretability in the multiobjective evolutionary rule selection and tuning of linguistic fuzzy systems, IEEE Trans. Fuzzy Syst. **18**(3), 515–531 (2010)

77.93   Y. Zhang, X.B. Wu, Z.Y. Xing, W.L. Hu: On generating interpretable and precise fuzzy systems based on Pareto multi-objective cooperative coevolutionary algorithm, Appl. Soft Comput. **11**(1), 1284–1294 (2011)

77.94   R. Alcalá, Y. Nojima, F. Herrera, H. Ishibuchi: Multiobjective genetic fuzzy rule selection of single granularity-based fuzzy classification rules and its interaction with the lateral tuning of membership functions, Soft Comput. **15**(12), 2303–2318 (2011)

77.95   C.A.C. Coello, G.T. Pulido, M.S. Lechuga: Handling multiple objectives with particle swarm optimization, IEEE Trans. Evol. Comput. **8**(3), 256–279 (2004)

77.96   D. Liu, K.C. Tan, C.K. Goh, W.K. Ho: A multiobjective memetic algorithm based on particle swarm optimization, IEEE Trans. Syst. Man Cybern. B **37**(1), 42–50 (2007)

77.97   Y. Wang, Y. Yang: Particle swarm optimization with preference order ranking for multi-objective optimization, Inf. Sci. **179**(12), 1944–1959 (2009)

77.98   A. Elhossini, S. Areibi, R. Dony: Strength Pareto particle swarm optimization and hybrid EA-PSO for multi-objective optimization, Evol. Comput. **18**(1), 127–156 (2010)

77.99   C.K. Goh, K.C. Tan, D.S. Liu, S.C. Chiam: A competitive and cooperative co-evolutionary approach to multi-objective particle swarm optimization algorithm design, Eur. J. Oper. Res. **202**(1), 42–54 (2010)

77.100   A.R.M. Rao, K. Sivasubramanian: Multi-objective optimal design of fuzzy logic controller using a self configurable swarm intelligence algorithm, Comput. Struct. **86**(23/24), 2141–2154 (2008)

77.101   M. Marinaki, Y. Marinakis, G.E. Stavroulakis: Fuzzy control optimized by a multi-objective particle swarm optimization algorithm for vibration suppression of smart structures, Struct. Multidiscip. Optim. **43**(1), 29–42 (2011)

77.102   C.N. Nyirenda, D.S. Dawoud, F. Dong, M. Negnevitsky, K. Hirota: A fuzzy multiobjective particle swarm optimized TS fuzzy logic congestion controller for wireless local area networks, J. Adv. Comput. Intell. Intell. Inf. **15**(1), 41–54 (2011)

77.103   Q. Zhang, M. Mahfouf: A hierarchical Mamdani-type fuzzy modelling approach with new training data selection and multi-objective optimisation mechanisms: A special application for the prediction of mechanical properties of alloy steels, Appl. Soft Comput. **11**(2), 2419–2443 (2011)

77.104   B.J. Park, J.N. Choi, W.D. Kim, S.K. Oh: Analytic design of information granulation-based fuzzy radial basis function neural networks with the aid of multiobjective particle swarm optimization, Int. J. Intell. Comput. Cybern. **5**(1), 4–35 (2012)

77.105   H. Ishibuchi, N. Tsukamoto, Y. Nojima: Evolutionary many-objective optimization: A short review, Proc. 2008 IEEE Congr. Evol. Comput. (IEEE, Hong Kong 2008) pp. 2424–2431

77.106   H. Ishibuchi, N. Akedo, H. Ohyanagi, Y. Nojima: Behavior of EMO algorithms on many-objective optimization problems with correlated objectives, Proc. 2011 IEEE Congr. Evol. Comput. (IEEE, New Orleans 2011), pp. 1465–1472

77.107   O. Schutze, A. Lara, C.A.C. Coello: On the influence of the number of objectives on the hardness of a multiobjective optimization problem, IEEE Trans. Evol. Comput. **15**(4), 444–455 (2011)

77.108   C.F. Juang, C.M. Lu, C. Lo, C.Y. Wang: Ant colony optimization algorithm for fuzzy controller design and its FPGA implementation, IEEE Trans. Ind. Electron. **55**(3), 1453–1462 (2008)

77.109   C.F. Juang, C.Y. Wang: A self-generating fuzzy system with ant and particle swarm cooperative optimization, Expert Syst. Appl. **36**(3), 5362–5370 (2009)

77.110   C.F. Juang, P.H. Chang: Designing fuzzy-rule-based systems using continuous ant-colony optimization, IEEE Trans. Fuzzy Syst. **18**(1), 138–149 (2010)

77.111   C.F. Juang, P.H. Chang: Recurrent fuzzy system design using elite-guided continuous ant colony optimization, Appl. Soft Comput. **11**(2), 2687–2697 (2011)

77.112   G.M. Fathi, A.M. Saniee: A fuzzy classification system based on ant colony optimization for diabetes disease diagnosis, Expert Syst. Appl. **38**(12), 14650–14659 (2011)

77.113   S. Wang, M. Mahfouf: Multi-objective optimisation for fuzzy modelling using interval type-2 fuzzy sets, Proc. 2012 IEEE Int. Conf. Fuzzy Syst. (IEEE, Brisbane 2012) pp. 722–729

77.114   O. Castillo, P. Melin, A.A. Garza, O. Montiel, R. Sepúlveda: Optimization of interval type-2 fuzzy logic controllers using evolutionary algorithms, Soft Comput. **15**(6), 1145–1160 (2011)

77.115   O. Castillo, P. Melin: A review on the design and optimization of interval type-2 fuzzy controllers, Appl. Soft Comput. **12**(4), 1267–1278 (2012)

77.116   O. Castillo, R. Martínez-Marroquín, P. Melin, F. Valdez, J. Soria: Comparative study of bio-inspired algorithms applied to the optimization of type-1 and type-2 fuzzy controllers for an autonomous mobile robot, Inf. Sci. **192**(1), 19–38 (2012)

78. Bio-Inspired Optimization of Type-2 Fuzzy Controllers

Oscar Castillo

A review of the bio-inspired optimization methods used in the design of type-2 fuzzy systems, which are relatively novel models of imprecision, is considered in this chapter. The fundamental focus of the work is based on the basic reasons for the need for optimization of type-2 fuzzy systems for different areas of application. Recently, bio-inspired methods have emerged as powerful optimization algorithms for solving complex problems. In the case of designing type-2 fuzzy systems for particular applications, the use of bio-inspired optimization methods has helped in the complex task of finding the appropriate parameter values and structure of fuzzy systems. In this chapter, we consider the application of genetic algorithms, particle swarm optimization, and ant colony optimization as three different paradigms that help in the design of optimal type-2 fuzzy systems. We also provide a comparison of the different optimization methods for the case of designing type-2 fuzzy systems.

78.1 Related Work in Type-2 Fuzzy Control

Uncertainty affects decision-making and appears in a number of different forms. The concept of information is fully connected with the concept of uncertainty [78.1]. The most fundamental aspect of this connection is that the uncertainty involved in any problem-solving situation is a result of some information deficiency, which may be incomplete, imprecise, fragmentary, not fully reliable, vague, contradictory, or deficient in some other way. Uncertainty is an attribute of information [78.2]. The general framework of fuzzy reasoning allows handling much of this uncertainty, and fuzzy systems that employ type-1 fuzzy sets represent uncertainty by numbers in the range [0, 1]. When something is uncertain, like a measurement, it is difficult to determine its exact value, and of course type-1 using fuzzy sets make more sense than using crisp sets [78.3].

However, it is not reasonable to use an accurate membership function for something uncertain, so in this case what we need are higher-order fuzzy sets, which are able to handle these uncertainties, like the so-called type-2 fuzzy sets [78.3]. So, the degree of uncertainty can be managed by using type-2 fuzzy logic because this offers better capabilities to handle linguistic uncertainties by modeling vagueness and unreliability of information [78.4–6].

Recently, we have seen the use of type-2 fuzzy sets in fuzzy logic systems (FLS) in different areas of application [78.7–11]. In this paper we deal with the application of interval type-2 fuzzy control to non-linear dynamic systems [78.4, 12–15]. It is a well-known fact that in the control of real systems, the instrumentation elements (instrumentation amplifier, sensors, digital to analog, analog to digital converters, etc.) introduce some sort of unpredictable values in the information that has been collected [78.16]. So, the controllers designed under idealized conditions tend to behave in an inappropriate manner [78.17].

78.2 Fuzzy Logic Systems

In this section, a brief overview of type-1 and type-2 fuzzy systems is presented. This overview is considered to be necessary to understand the basic concepts needed to develop the methods and algorithms presented later in the chapter.

78.2.1 Type-1 Fuzzy Logic Systems

Soft computing techniques have become an important research topic that can be applied in the design of intelligent controllers, which utilize human experience in a more natural form than the conventional mathematical approach [78.18, 19]. An FLS described completely in terms of type-1 fuzzy sets is called a type-1 fuzzy logic system (type-1 FLS). In this paper, the fuzzy controller has two input variables, which are the error $e(t)$ and the error variation $\Delta e(t)$,

$$e(t) = r(t) - y(t) , \tag{78.1}$$

$$\Delta e(t) = e(t) - e(t-1) , \tag{78.2}$$

so the control system can be represented as shown in Fig. 78.1.

78.2.2 Type-2 Fuzzy Logic Systems

If for a type-1 membership function, as in Fig. 78.2, we blur it to the left and to the right, as illustrated in Fig. 78.3, then a type-2 membership function is obtained. In this case, for a specific value x', the membership function (u') takes on different values, which are not all weighted the same, so we can assign an amplitude distribution to all of those points.

A type-2 fuzzy set $\tilde{A}$ is characterized by the membership function [78.1, 3]

$$\tilde{A} = \left\{ \left((x, u), \mu_{\tilde{A}}(x, u) \right) \mid \forall x \in X , \quad \forall u \in J_x \subseteq [0, 1] \right\} , \tag{78.3}$$

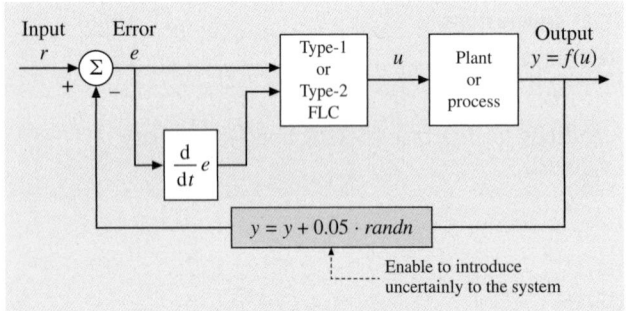

Fig. 78.1 System used to obtain the experimental results

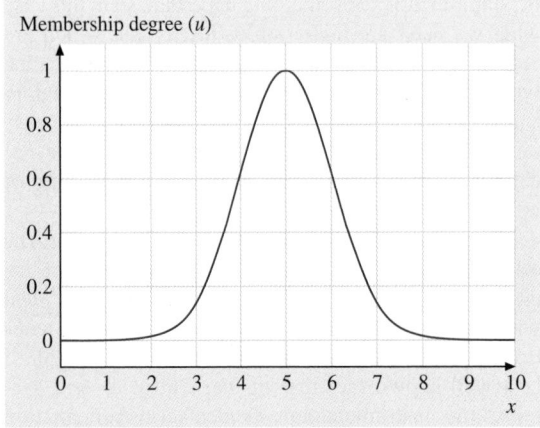

Fig. 78.2 Type-1 membership function

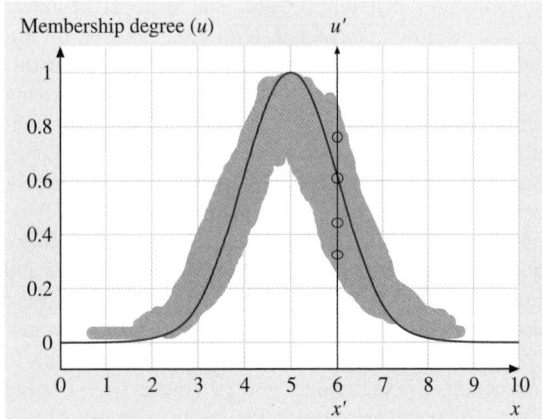

Fig. 78.3 Blurred type-1 membership function

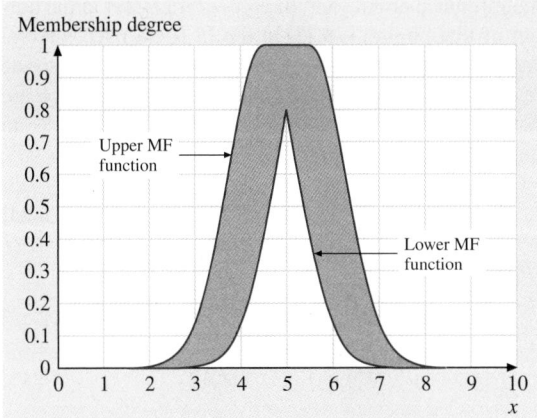

Fig. 78.4 Interval type-2 membership function

in which $0 \leq \mu_{\tilde{A}}(x, u) \leq 1$. Another expression for $\tilde{A}$ is

$$\tilde{A} = \int_{x \in X} \int_{u \in J_x} \frac{\mu_{\tilde{A}}(x, u)}{(x, u)}, \qquad J_x \subseteq [0, 1], \qquad (78.4)$$

where $\int \int$ denotes the union over all admissible input variables x and u. For discrete universes of discourse $\int$ is replaced by $\sum$. In fact $J_x \subseteq [0, 1]$ represents the primary membership of x, and $\mu_{\tilde{A}}(x, u)$ is a type-1 fuzzy set known as the secondary set. Hence, a type-2 membership grade can be any subset in $[0, 1]$, the primary membership, and corresponding to each primary membership, there is a secondary membership (which can also be in $[0, 1]$) that defines the possibilities for the primary membership. Uncertainty is represented by a region, which is called the footprint of uncertainty (FOU). When $\mu_{\tilde{A}}(x, u) = 1, \forall u \in J_x \subseteq [0, 1]$ we have an interval type-2 membership function, as shown in Fig. 78.4. The uniform shading for the FOU rep-

resents the entire interval type-2 fuzzy set and it can be described in terms of an upper membership function $\bar{\mu}_{\tilde{A}}(x)$ and a lower membership function $\underline{\mu}_{\tilde{A}}(x)$.

A FLS described using at least one type-2 fuzzy set is called a type-2 FLS. Type-1 FLSs are unable to directly handle rule uncertainties, because they use type-1 fuzzy sets that are certain [78.3]. On the other hand, type-2 FLSs, are very useful in circumstances where it is difficult to determine an exact membership function and there are measurement uncertainties [78.14, 20, 21].

A type-2 FLS is again characterized by IF-THEN rules, but its antecedent or consequent sets are now of type-2. Similar to a type-1 FLS, a type-2 FLS includes a fuzzifier, a rule base, fuzzy inference engine, and an output processor, as we can see in Fig. 78.5. The output processor includes a type-reducer and a defuzzifier; it generates a type-1 fuzzy set output (type-reducer) or a crisp number (defuzzifier).

Fuzzifier
The fuzzifier maps a crisp point $\mathbf{x} = (x_1, \ldots, x_p)^T \in X_1 \times X_2 \times \ldots \times X_p \equiv \mathbf{X}$ into a type-2 fuzzy set $\tilde{A}_x$ in $\mathbf{X}$ [78.1], interval type-2 fuzzy sets in this case. We will use type-2 singleton fuzzifier, in a singleton fuzzification, the input fuzzy set has only a single point on nonzero membership [78.3]. $\tilde{A}_x$ is a type-2 fuzzy singleton if $\mu_{\tilde{A}_x}(\mathbf{x}) = 1/1$ for $\mathbf{x} = \mathbf{x}'$ and $\mu_{\tilde{A}_x}(\mathbf{x}) = 1/0$ for all other $\mathbf{x} \neq \mathbf{x}'$ [78.1].

Rules
The structure of rules in a type-1 FLS and a type-2 FLS is the same, but in the latter the antecedents and the consequents will be represented by type-2 fuzzy sets. So for a type-2 FLS with p inputs $x_1 \in X_1, \ldots, x_p \in X_p$ and one output $y \in Y$, multiple input single output (MISO). If we assume that there are M rules, the l-th rule in the

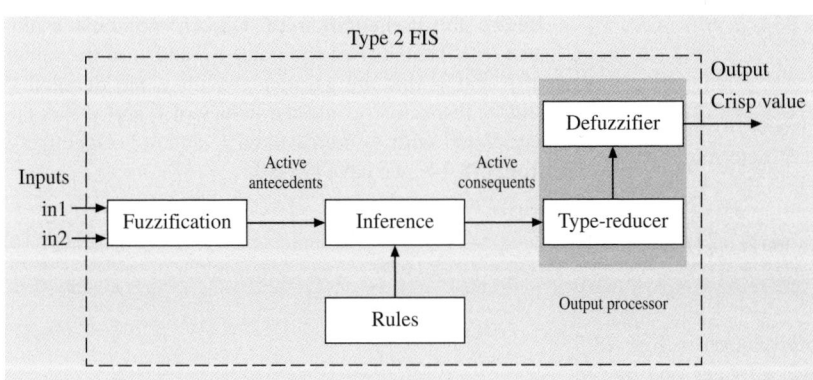

Fig. 78.5 Type-2 FLS

type-2 FLS can be written as follows [78.3]

$$R^l : \text{IF } x_1 \text{ is } \tilde{F}_1^l \text{ and } \dots \text{ and } x_p \text{ is } \tilde{F}_p^l ,$$

$$\text{THEN } y \text{ is } \tilde{G}^l ,$$

$$l = 1, \dots, M . \tag{78.5}$$

Inference

In the type-2 FLS, the inference engine combines rules and gives a mapping from input type-2 fuzzy sets to output type-2 fuzzy sets. It is necessary to compute the join ⊔, (unions) and the meet Π (intersections), as well as extended sup-star compositions of type-2 relations [78.3]. If

$$\tilde{F}_1^l \times \dots \times \tilde{F}_p^l = \tilde{A}^l ,$$

(78.5) can be re-written as

$$R^l : \tilde{F}_1^l \times \dots \times \tilde{F}_p^l \rightarrow \tilde{G}^l = \tilde{A}^l \rightarrow \tilde{G}^l , \quad l = 1, \dots, M . \tag{78.6}$$

R^l is described by the membership function $\mu_{R^l}(\mathbf{x}, y) = \mu_{R^l}(x_1, \dots, x_p, y)$, where

$$\mu_{R^l}(\mathbf{x}, y) = \mu_{\tilde{A}^l \rightarrow \tilde{G}^l}(\mathbf{x}, y) \tag{78.7}$$

can be written as [78.3]

$$\mu_{R^l}(\mathbf{x}, y) = \mu_{\tilde{A}^l \rightarrow \tilde{G}^l}(\mathbf{x}, y) = \mu_{\tilde{F}_1^l}(x_1)$$
$$\times \Pi \dots \Pi \mu_{\tilde{F}_p^l}(x_p) \Pi \mu_{\tilde{G}^l}(y)$$
$$= \left[\Pi_{i=1}^p \mu_{\tilde{F}_i^l}(x_i) \right] \Pi \mu_{\tilde{G}^l}(y) . \tag{78.8}$$

In general, the p-dimensional input to R^l is given by the type-2 fuzzy set $\tilde{A}_x$ whose membership function is

$$\mu_{\tilde{A}_x}(\mathbf{x}) = \mu_{\tilde{x}_1}(x_1) \Pi \dots \Pi \mu_{\tilde{x}_p}(x_p) = \Pi_{i=1}^p \mu_{\tilde{x}_i}(x_i) , \tag{78.9}$$

where $\tilde{X}_i (i = 1, \dots, p)$ are the labels of the fuzzy sets describing the inputs. Each rule R^l determines a type-2 fuzzy set $\tilde{B}^l = \tilde{A}_x \circ R^l$ such that [78.3]

$$\mu_{\tilde{B}^l}(y) = \mu_{\tilde{A}_x \circ R^l} = \sqcup_{x \in \mathbf{x}} \left[\mu_{\tilde{A}_x}(\mathbf{x}) \Pi \mu_{R^l}(\mathbf{x}, y) \right] ,$$
$$y \in Y, \quad l = 1, \dots, M . \tag{78.10}$$

This equation is the input/output relation in Fig. 78.5 between the type-2 fuzzy set that activates one rule in

the inference engine and the type-2 fuzzy set at the output of that engine [78.3]. In the FLS we used interval type-2 fuzzy sets and meet under product t-norm, so the result of the input and antecedent operations, which are contained in the firing set $\Pi_{i=1}^p \mu_{\tilde{F}_{l_i}}(x_i' \equiv F^l(\mathbf{x}'))$, is an interval type-1 set [78.3],

$$F^l(\mathbf{x}') = \left[\underline{f}^l(\mathbf{x}'), \bar{f}^l(\mathbf{x}') \right] \equiv \left[\underline{f}^l, \bar{f}^l \right] , \tag{78.11}$$

where

$$\underline{f}^l(\mathbf{x}') = \underline{\mu}_{\tilde{F}_1^l}(x_1') * \dots * \underline{\mu}_{\tilde{F}_p^l}(x_p') , \tag{78.12}$$

$$\bar{f}^l(\mathbf{x}') = \bar{\mu}^{\tilde{F}_1^l}(x_1') * \dots * \bar{\mu}^{\tilde{F}_p^l}(x_p') , \tag{78.13}$$

where $*$ is the product operation.

Type-Reducer

The type-reducer generates a type-1 fuzzy set output, which is then converted in a crisp output through the defuzzifier. This type-1 fuzzy set is also an interval set, for the case of our FLS we used center of sets (coss) type reduction, $Y_{\cos}$, which is expressed as [78.3]

$$Y_{\cos}(\mathbf{x}) = [y_l, y_r] = \int_{y^1 \in [y_l^1, y_r^1]} \dots \int_{y^M \in [y_l^M, y_r^M]}$$
$$\times \int_{f^1 \in [\underline{f}^1, \bar{f}^1]} \dots \int_{f^M \in [\underline{f}^M, \bar{f}^M]} 1 \bigg/ \frac{\sum_{i=1}^M f^i y^i}{\sum_{i=1}^M f^i} . \tag{78.14}$$

This interval set is determined by its two end points, y_l and y_r, which correspond to the centroid of the type-2 interval consequent set $\tilde{G}^i$ [78.3],

$$C_{\tilde{G}^i} = \int_{\theta_1 \in J_{y1}} \dots \int_{\theta_N \in J_{yN}} 1 \bigg/ \frac{\sum_{i=1}^N y_i \theta_i}{\sum_{i=1}^N \theta_i} = [y_l^i, y_r^i] . \tag{78.15}$$

Before the computation of $Y_{\cos}(\mathbf{x})$, we must evaluate (78.15) and its two end points, y_l and y_r. If the values of f_i and y_i that are associated with y_l are denoted f_l^i and y_l^i, respectively, and the values of f_i and y_i that are associated with y_r are denoted f_r^i and y_r^i, respectively, from (78.14), we have [78.3]

$$y_l = \frac{\sum_{i=1}^M f_l^i y_l^i}{\sum_{i=1}^M f_l^i} , \tag{78.16}$$

$$y_r = \frac{\sum_{i=1}^M f_r^i y_r^i}{\sum_{i=1}^M f_r^i} . \tag{78.17}$$

Defuzzifiers

From the type-reducer we obtain an interval set $Y_{\cos}$; to defuzzify it we use the average of y_l and y_r, so the defuzzified output of an interval singleton type-2 FLS is [78.3]

$$y(\mathbf{x}) = \frac{y_l + y_r}{2} .$$ (78.18)

78.3 Bio-Inspired Optimization Methods

In this section a brief overview of the basic concepts from bio-inspired optimization methods needed for this work is presented.

78.3.1 Particle Swarm Optimization

Particle swarm optimization is a population-based stochastic optimization technique, which was developed by Eberhart and Kennedy in 1995. It was inspired by the social behavior of bird flocking or fish schooling [78.7]. (PSO) shares many similarities with evolutionary computation techniques such as the genetic algorithm (GA) [78.22].

The system is initialized with a population of random solutions and searches for optima by updating generations. However, unlike the GA, PSO has no evolution operators such as crossover and mutation. In PSO, the potential solutions, called particles, fly through the problem space by following the current optimum particles [78.18]. Each particle keeps track of its coordinates in the problem space, which are associated with the best solution (fitness) it has achieved so far (the fitness value is also stored). This value is called *pbest*. Another *best* value that is tracked by the particle swarm optimizer is the best value, obtained so far by any particle in the neighbors of the particle. This location is called *lbest*. When a particle takes all the population as its topological neighbors, the best value is a global best and is called *gbest* [78.15].

The particle swarm optimization concept consists of, at each time step, changing the velocity of (accelerating) each particle toward its *pbest* and *lbest* locations (local version of PSO). Acceleration is weighted by a random term, with separate random numbers being generated for acceleration toward *pbest* and *lbest* locations [78.7]. In the past several years, PSO has been successfully applied in many research and application areas. It has been demonstrated that PSO obtains better results in a faster, cheaper way compared with other methods [78.15]. Another reason that PSO is attractive is that there are few parameters to

adjust. One version, with slight variations, works well in a wide variety of applications. Particle swarm optimization has been considered for approaches that can be used across a wide range of applications, as well as for specific applications focused on a specific requirement.

The basic algorithm of PSO has the following nomenclature:

- x_z^i: Particle position
- v_z^i: Particle velocity
- w_{ij}: Inertia weight
- p_z^i: Best *remembered* individual particle position
- p_z^g: Best *remembered* swarm position
- c_1, c_2: Cognitive and social parameters
- r_1, r_2: Random numbers between 0 and 1.

The equation to calculate the velocity is

$$v_{z+1}^i = w_{ij}v_z^i + c_1 r_1 \left(p_z^i - x_z^i\right) + c_2 r_2 \left(p_z^g - x_z^i\right) ,$$ (78.19)

and the position of the individual particles is updated as follows

$$x_{z+1}^i = x_z^i + v_{z+1}^i .$$ (78.20)

The basic PSO algorithm is defined as follows:

1) *Initialize*
   a) *Set constants $z_{\max}, c_1, c_2$*
   b) *Randomly initialize particle position $x_0^i \in D$ in R^n for $i = 1, \ldots, p$*
   c) *Randomly initialize particle velocities $0 \leq v_0^i \leq v_0^{\max}$ for $i = 1, \ldots, p$*
   d) *Set $Z = 1$*
2) *Optimize*
   a) *Evaluate function value f_k^i using design space coordinates x_k^i*
   b) *If $f_z^i \leq f_{\text{best}}^i$ then $f_{\text{best}}^i = f_z^i, p_z^i = x_z^i$.*
   c) *If $f_z^i \leq f_{\text{best}}^g$ then $f_{\text{best}}^g = f_z^i, p_z^g = x_z^i$.*
   d) *If stopping condition is satisfied then go to 3.*

e) *Update all particle velocities v_z^i for $i = 1, \ldots, p$*
f) *Update al particle positions x_z^i for $i = 1, \ldots, p$*
g) *Increment z.*
h) *Goto 2(a).*
3) *Terminate.*

78.3.2 Genetic Algorithms

Genetic algorithms (GAs) are adaptive heuristic search algorithms based on the evolutionary ideas of natural selection and genetic processes [78.21]. The basic principles of GAs were first proposed by Holland in 1975, inspired by the mechanism of natural selection, where stronger individuals are likely to be the winners in a competing environment [78.22]. GA assumes that the potential solution of any problem is an individual and can be represented by a set of parameters. These parameters are regarded as the genes of a chromosome and can be structured by a string of values in binary form. A positive value, generally known as a fitness value, is used to reflect the degree of *goodness* of the chromosome for the problem, which would be highly related with its objective value. The pseudocode of a GA is as follows:

1) *Start with a randomly generated population of n chromosomes (candidate solutions to a problem).*
1. *Calculate the fitness of each chromosome in the population.*
2. *Repeat the following steps until n offspring have been created:*
   a) *Select a pair of parent chromosomes from the current population, the probability of selection being an increasing function of fitness. Selection is done with replacement, meaning that the same chromosome can be selected more than once to become a parent.*
   b) *With probability (crossover rate), perform crossover to the pair at a randomly chosen point to a form two offspring.*
   c) *Mutate the two offspring at each locus with probability (mutation rate), and place the resulting chromosomes in the new population.*
2) *Replace the current population with the new population.*
3) *Go to step 2.*

The simple procedure just described above is the basis for most applications of GAs found in the literature [78.23, 24].

78.3.3 Ant Colony Optimization

Ant colony optimization (ACO) is a probabilistic technique that can be used for solving problems that can be reduced to finding good paths along graphs. This method was inspired from the behavior exhibited by ants in finding paths from the nest or colony to the food source.

Simple ant colony optimization (S-ACO) is an algorithmic implementation that adapts the behavior of real ants to solutions of minimum cost path problems on graphs [78.11]. A number of artificial ants build solutions for a certain optimization problem and exchange information about the quality of these solutions making allusion to the communication system of real ants [78.25].

Let us define the graph $G = (V, E)$, where V is the set of nodes and E is the matrix of the links between nodes. G has $n_G = |V|$ nodes. Let us define L^k as the number of hops in the path built by the ant k from the origin node to the destiny node. Therefore, it is necessary to find

$$Q = \{q_a, \ldots, q_f | q_1 \in C\} \, , \tag{78.21}$$

where Q is the set of nodes representing a continuous path with no obstacles; $q_a, \ldots, q_f$ are former nodes of the path, and C is the set of possible configurations of the free space. If $x^k(t)$ denotes a Q solution in time t, $f(x^k(t))$ expresses the quality of the solution. The S-ACO algorithm is based on (78.22)–(78.24)

$$p_{ij}^k(t) = \begin{cases} \dfrac{\tau_{ij}^k}{\sum_{j \in N_{ij}^k} \tau_{ij}^\alpha (t)} & \text{if } j \in N_i^k \\ 0 & \text{if } j \notin N_i^k \end{cases} , \tag{78.22}$$

$$\tau_{ij}(t) \leftarrow (1 - \rho)\tau_{ij}(t) \, , \tag{78.23}$$

$$\tau_{ij}(t+1) = \tau_{ij}(t) + \sum_{k=1}^{n_k} \tau_{ij}(t) \, . \tag{78.24}$$

Equation (78.22) represents the probability for an ant k located on a node i selects the next node denoted by j, where, N_i^k is the set of feasible nodes (in a neighborhood) connected to node i with respect to ant k, τ_{ij} is the total pheromone concentration of link ij, and α is a positive constant used as a gain for the pheromone influence.

Equation (78.23) represents the evaporation pheromone update, where $\rho \in [0, 1]$ is the evaporation rate value of the pheromone trail. The evaporation is added to the algorithm in order to force the exploration of the ants and avoid premature convergence to sub-optimal solutions. For $\rho = 1$ the search becomes completely random.

Equation (78.24), represents the concentration pheromone update, where $\Delta \tau_{ij}^k$ is the amount of pheromone that an ant k deposits in a link ij in a time t.

The general steps of S-ACO are the following:

1. *Set a pheromone concentration τ_{ij} to each link (i,j).*
2. *Place a number $k = 1, 2, \ldots, n_k$ in the nest.*
3. *Iteratively build a path to the food source (destiny node), using (78.22) for every ant.*
   - *Remove cycles and compute each route weight $f(x^k(t))$. A cycle could be generated when there are no feasible candidates nodes, that is, for any i and any k, $N_i^k = \emptyset$; then the predecessor of that node is included as a former node of the path.*
4. *Apply evaporation using (78.2).*
5. *Update of the pheromone concentration using (78.24)*
6. *Finally, finish the algorithm in any of the three different ways:*

- *When a maximum number of epochs has been reached.*
- *When it has found an acceptable solution, with $f(x_k(t)) < \varepsilon$.*
- *When all ants follow the same path.*

78.3.4 General Remarks About Optimization of Type-2 Fuzzy Systems

The problem of designing type-2 fuzzy systems can be solved with any of the above-mentioned optimization methods. The main issue in any of these methods is to decide on the appropriate representation of the type-2 fuzzy system in the corresponding optimization paradigm. For example, in the case of GAs, the type-2 fuzzy systems must be represented in the chromosomes. On the other hand, in PSO the fuzzy system is represented as a particle in the optimization process. In the ACO method, the fuzzy system can be represented as one of the paths that the ants can follow in a graph. Also, the evaluation of the fuzzy system must be represented as an objective function in any of the methods.

78.4 General Overview of the Area and Future Trends

In this section, a general overview of the area of type-2 fuzzy system optimization is presented. Also, possible future trends that we can envision based on the review of this area are presented. It has been well known for a long time that to design fuzzy systems is a difficult task, and this is especially true in the case of type-2 fuzzy systems [78.4]. The use of GAs, ACO, and PSO in designing type-1 fuzzy systems has become standard practice for automatically designing this sort of system [78.7, 8, 23, 25]. This trend has also continued to the type-2 fuzzy systems area, which has been accounted for with the review of papers presented in the previous sections. In the case of designing type-2 fuzzy systems the problem is more complicated due to the higher number of parameters to consider, making it of very important to use bio-inspired optimization techniques to achieve the optimal designs of this sort of system. In this section, a summary of the total number of papers published in the area of type-2 fuzzy system optimization is presented, so that the increasing trend occurring in this area can be better appreciated. Also, the distribution of papers according to the optimization technique used is presented, so that a general

idea of how these different techniques contribute to the automatic design of optimal type-2 fuzzy systems is obtained.

Figure 78.6 shows the distribution of the papers published on the optimization of type-2 fuzzy systems according to the different bio-inspired optimization techniques previously mentioned. From Fig. 78.6 it can

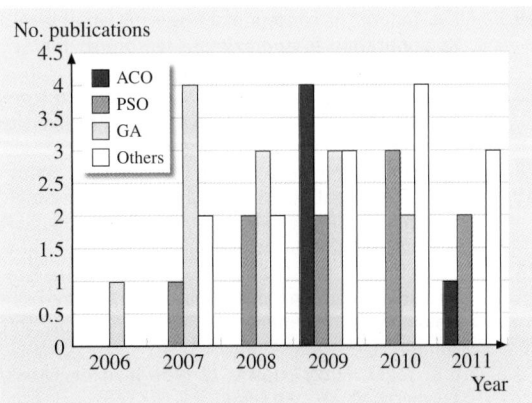

Fig. 78.6 Distribution of publications per area and year

be noted that the use of GAs has been decreasing recently. On the other hand, the use of PSO, ACO, and other methods have been increasing. The reason for the increase in use of PSO and ACO may be due to recent work in which either PSO or ACO have been able to outperform GAs for different applications. Regarding the question of which method would be the most appropriate for optimizing type-2 fuzzy systems, there is no easy answer. At the moment, what we can be sure of is that the techniques mentioned in this paper, and probably newer ones that may appear in the future, would certainly be tested in the optimization of type-2 fuzzy systems because the problem of automatically designing these types of systems is complex enough to require their use.

There are other bio-inspired or nature-inspired techniques that at the moment have not been applied to the optimization of type-2 fuzzy systems that may be worth mentioning. For example, membrane computing, harmony computing, electromagnetism-based computing, and other similar approaches have not been applied (to date) in the optimization of type-2 fuzzy systems. It is expected that these approaches and similar ones could be applied in the near future in the area of type-2 fuzzy system optimization. Of course, as new bio-inspired and nature-inspired optimization methods are continuously being proposed in this fruitful area of research, it is expected that newer optimization techniques will also be tried in the near future in the automatic design of optimal type-2 fuzzy systems.

78.5 Conclusions

In this chapter we have presented a representative account of the different optimization methods that have been applied in the optimal design of type-2 fuzzy systems. To date, genetic algorithms have been used more frequently to optimize type-2 fuzzy systems. However, more recently PSO and ACO have attracted more attention and have also been applied with some degree of success to the problem of the optimal design of type-2 fuzzy systems. There have been also other optimization methods applied

to the optimization of type-2 fuzzy systems, like artificial immune systems and the chemical optimization paradigm. At this time, it would be very difficult to declare one of these optimization techniques as the best for optimizing type-2 fuzzy systems, because different techniques have had success in different applications of type-2 fuzzy logic. In any case, the need for bio-inspired optimization methods is justified due to the complexity of designing type-2 fuzzy systems.

References

78.1 J.M. Mendel: Uncertainty, fuzzy logic, and signal processing, Signal Process. J. **80**, 913–933 (2000)

78.2 L.A. Zadeh: The concept of a linguistic variable and its application to approximate reasoning, Inf. Sci. **8**, 43–80 (1975)

78.3 N.N. Karnik, J.M. Mendel: *An Introduction to Type-2 Fuzzy Logic Systems*, Technical Report (University of Southern California, Los Angeles 1998)

78.4 O. Castillo, P. Melin, A. Alanis, O. Montiel, R. Sepulveda: Optimization of interval type-2 fuzzy logic controllers using evolutionary algorithms, J. Soft Comput. **15**(6), 1145–1160 (2011)

78.5 R. Sepulveda, O. Montiel, O. Castillo, P. Melin: Embedding a high speed interval type-2 fuzzy controller for a real plant into an FPGA, Appl. Soft Comput. **12**(3), 988–998 (2012)

78.6 R.R. Yager: Fuzzy subsets of type II in decisions, J. Cybern. **10**, 137–159 (1980)

78.7 Z. Bingül, O. Karahan: A fuzzy logic controller tuned with PSO for 2 DOF robot trajectory control, Expert Syst. Appl. **38**(1), 1017–1031 (2011)

78.8 J. Cao, P. Li, H. Liu, D. Brown: Adaptive fuzzy controller for vehicle active suspensions with particle swarm optimization, Proc. SPIE Int. Soc. Opt. Eng., Vol. 7129 (2008)

78.9 J.R. Castro, O. Castillo, P. Melin: An interval type-2 fuzzy logic toolbox for control applications, Proc. FUZZ-IEEE (2007) pp. 1–6

78.10 T. Dereli, A. Baykasoglu, K. Altun, A. Durmusoglu, I.B. Turksen: Industrial applications of type-2 fuzzy sets and systems: A concise review, Comput. Ind. **62**, 125–137 (2011)

78.11 C.-F. Juang, C.-H. Hsu: Reinforcement ant optimized fuzzy controller for mobile-robot wall-following control, IEEE Trans. Ind. Electron. **56**(10), 3931–3940 (2009)

78.12 O. Castillo, G. Huesca, F. Valdez: Evolutionary computing for topology optimization of type-2 fuzzy controllers, Stud. Fuzziness Soft Comput. **208**, 163–178 (2008)

78.13 O. Castillo, L.T. Aguilar, N.R. Cazarez-Castro, S. Cardenas: Systematic design of a stable type-2 fuzzy logic controller, Appl. Soft Comput. J. **8**, 1274–1279 (2008)

78.14 R. Martinez, O. Castillo, L.T. Aguilar: Optimization of interval type-2 fuzzy logic controllers for a perturbed autonomous wheeled mobile robot using genetic algorithms, Inf. Sci. **179**(13), 2158–2174 (2009)

78.15 S.-K. Oh, H.-J. Jang, W. Pedrycz: A comparative experimental study of type-1/type-2 fuzzy cascade controller based on genetic algorithms and particle swarm optimization, Expert Syst. Appl. **38**(9), 11217–11229 (2011)

78.16 R. Sepulveda, O. Castillo, P. Melin, A. Rodriguez-Diaz, O. Montiel: Experimental study of intelligent controllers under uncertainty using type-1 and type-2 fuzzy logic, Inf. Sci. **177**(10), 2023–2048 (2007)

78.17 H. Hagras: Hierarchical type-2 fuzzy logic control architecture for autonomous mobile robots, IEEE Trans. Fuzzy Syst. **12**, 524–539 (2004)

78.18 R. Martinez, A. Rodriguez, O. Castillo, L.T. Aguilar: Type-2 fuzzy logic controllers optimization using genetic algorithms and particle swarm optimization, Proc. IEEE Int. Conf. Granul. Comput. (2010) pp. 724–727

78.19 S.M.A. Mohammadi, A.A. Gharaveisi, M. Mashinchi: An evolutionary tuning technique for type-2 fuzzy logic controller in a non-linear system under uncertainty, Proc. 18th Iran. Conf. Electr. Eng. (2010) pp. 610–616

78.20 J.R. Castro, O. Castillo, L.G. Martinez: Interval type-2 fuzzy logic toolbox, Eng. Lett. **15**(1), 14 (2007)

78.21 O. Cordon, F. Gomide, F. Herrera, F. Hoffmann, L. Magdalena: Ten years of genetic fuzzy systems: Current framework and new trends, Fuzzy Sets Syst. **141**, 5–31 (2004)

78.22 O. Cordon, F. Herrera, P. Villar: Analysis and guidelines to obtain a good uniform fuzzy partition granularity for fuzzy rule-based systems using simulated annealing, Int. J. Approx. Reason. **25**, 187–215 (2000)

78.23 C. Wagner, H. Hagras: A genetic algorithm based architecture for evolving type-2 fuzzy logic controllers for real world autonomous mobile robots, Proc. IEEE Conf. Fuzzy Syst. (2007)

78.24 D. Wu, W.-W. Tan: Genetic learning and performance evaluation of interval type-2 fuzzy logic controllers, Eng. Appl. Artif. Intell. **19**(8), 829–841 (2006)

78.25 C.-F. Juang, C.-H. Hsu: Reinforcement interval type-2 fuzzy controller design by online rule generation and Q-value-aided ant colony optimization, IEEE Trans. Syst. Man Cybern. B **39**(6), 1528–1542 (2009)

Part G | 78

79. Pattern Recognition with Modular Neural Networks and Type-2 Fuzzy Logic

Patricia Melin

Interval type-2 fuzzy systems can be of great help in image analysis and pattern recognition applications. In particular, edge detection is a process usually applied to image sets before the training phase in recognition systems. This preprocessing step helps to extract the most important shapes in an image, ignoring the homogeneous regions and remarking the real objective to classify or recognize. Many traditional and fuzzy edge detectors can be used, but it is very difficult to demonstrate which one is better before the recognition results are obtained. In this chapter, we show experimental results where several edge detectors were used to preprocess the same image sets. Each resulting image set was used as training data for a modular neural network recognition system, and the recognition rates were compared. The goal of these experiments is to find the better edge detector that can be used to improve the training

data of a modular neural network for an image recognition system.

79.1 Related Work in the Area

In previous work, we have proposed extensions to the traditional edge detectors to improve their performance by using fuzzy systems [79.1–3]. The performed experiments have shown that the resulting images obtained with fuzzy edge detectors were visually better than the ones obtained with the traditional edge detection methods.

There is still work to be done on developing formal validation metrics for fuzzy edge detectors. In the literature, we can find comparison of edge detectors based on human observations [79.4–8], and some others that found the optimal values for parametric edge detectors [79.9].

Edge detectors can be used in recognition systems for different purposes, but in this work we are particularly interested in knowing, which is the best edge detector for a neural recognition system. In this chapter, we present some experiments which show that fuzzy edge detectors are a good method to improve the performance of neural recognition systems, and for this reason we propose that the recognition rate of the neural networks can be used as an edge detection performance index.

The rest of the chapter is organized as follows. Section 79.2 presents an overview of fuzzy edge detectors. Section 79.3 describes the experimental setup used to test the proposed fuzzy edge detectors in a modular neural recognition system. Section 79.4 presents the experimental results achieved with the proposed fuzzy edge detectors. Finally, Sect. 79.5 outlines the conclusions and future work.

79.2 Overview of Fuzzy Edge Detectors

In this section, an overview of the previously proposed fuzzy edge detectors is presented. First, the Sobel edge detector improved with fuzzy logic is presented. Second, the morphological gradient edge detector enhanced with fuzzy logic is also presented.

79.2.1 Sobel Edge Detector Improved with Fuzzy Logic

In the Sobel fuzzy edge detector we used the individual operators $Sobel_x$ and $Sobel_y$ as in the traditional method,

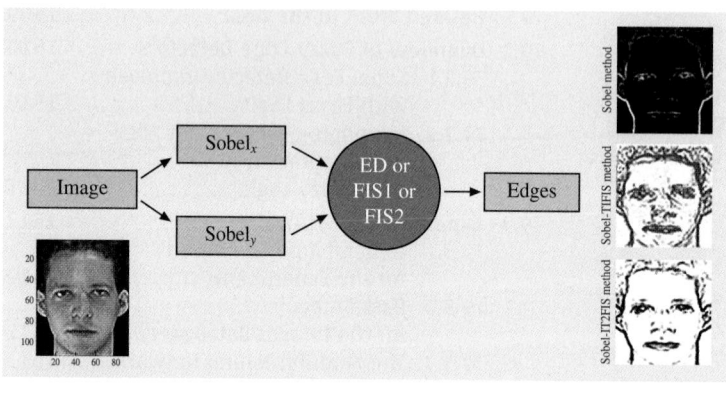

Fig. 79.1 Sobel edge detector enhanced with fuzzy logic

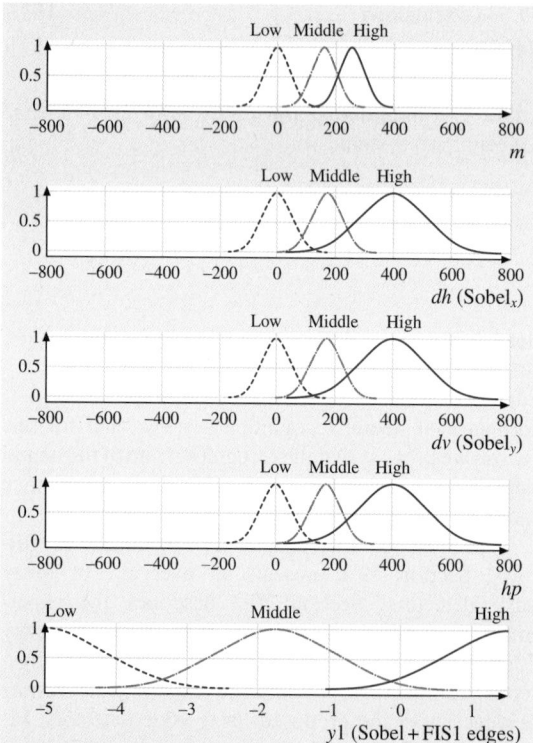

Fig. 79.2 Membership functions of the variables for the Sobel+FIS1 edge detector

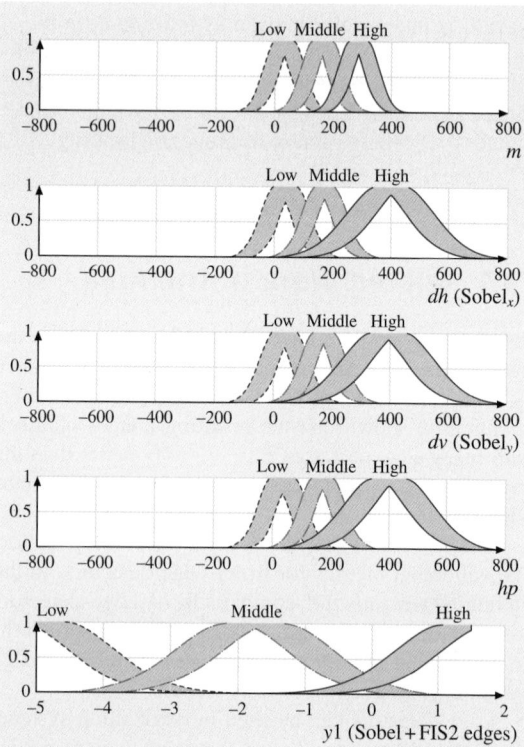

Fig. 79.3 Membership functions of the variables for the Sobel+FIS2 edge detector

and then we substitute the Euclidean distance of (79.1) by a fuzzy system, as shown in Fig. 79.1 [79.3].

$$\text{Sobel\_edges} = \sqrt{\text{Sobel}_x^2 + \text{Sobel}_y^2} \qquad (79.1)$$

The individual Sobel operators are the main inputs to the type-1 fuzzy inference system (FIS1) and type-2 fuzzy inference system (FIS2), and we have also considered adding two more inputs, which are filters that improve the final edge image. The fuzzy variables used in the Sobel+FIS1 and Sobel+ FIS2 edges detectors are shown in Fig. 79.2 and Fig. 79.3, respectively.

The use of the FIS2 [79.10, 11] provided images with better defined edges than the FIS1, which is a very important result in providing better inputs to the neural networks that will perform the recognition task.

The fuzzy rules for both the FIS1 and FIS2 are the same and are shown below:

1. If (dh is LOW) and (dv is LOW) then ($y1$ is HIGH)
2. If (dh is MIDDLE) and (dv is MIDDLE) then ($y1$ is LOW)
3. If (dh is HIGH) and (dv is HIGH) then ($y1$ is LOW)

4. If (dh is MIDDLE) and (hp is LOW) then ($y1$ is LOW)
5. If (dv is MIDDLE) and (hp is LOW) then ($y1$ is LOW)
6. If (m is LOW) and (dv is MIDDLE) then ($y1$ is HIGH)
7. If (m is LOW) and (dh is MIDDLE) then ($y1$ is HIGH)

The fuzzy rule base shown above infers the gray tone of each pixel for the edge image with the following reasoning: When the horizontal gradient d_h and vertical gradient d_v are LOW means that there is not enough difference between the gray tones in it's neighbors pixels, then the output pixel must belong of an homogeneous or not edges region, then the output pixel is HIGH or near WHITE. In the opposite case, when d_h and d_v are both HIGH this means that there is enough difference between the gray tones in its neighborhood, then the output pixel is an EDGE.

79.2.2 Morphological Gradient Edge Detector Improved with Fuzzy Logic

In the morphological gradient (MG), we calculated the four gradients as in the traditional method [79.12, 13], and substitute the sum of gradients in (79.2) with a fuzzy inference system, as shown in Fig. 79.4.

$$\text{MG\_edges} = D_1 + D_2 + D_3 + D_4 \qquad (79.2)$$

The linguistic variables used in the MG+FIS1 and MG+FIS2 edges detectors are shown in Fig. 79.5 and Fig. 79.6, respectively.

The rules for both the FIS1 and FIS2 are the same and are shown below:

1. If ($D1$ is HIGH) or ($D2$ is HIGH) or ($D3$ is HIGH) or ($D4$ is HIGH) then (E is BLACK)
2. If ($D1$ is MIDDLE) or ($D2$ is MIDDLE) or ($D3$ is MIDDLE) or ($D4$ is MIDDLE) then (E is GRAY)
3. If ($D1$ is LOW) and ($D2$ is LOW) and ($D3$ is LOW) and ($D4$ is LOW) then (E is WHITE)

After many experiments, we found that an edge exists when any gradient D_i is HIGH, which means that a difference of gray tones in any direction of the image must produce a pixel with a BLACK value or EDGE. The same behavior occurs when any gradient D_i is

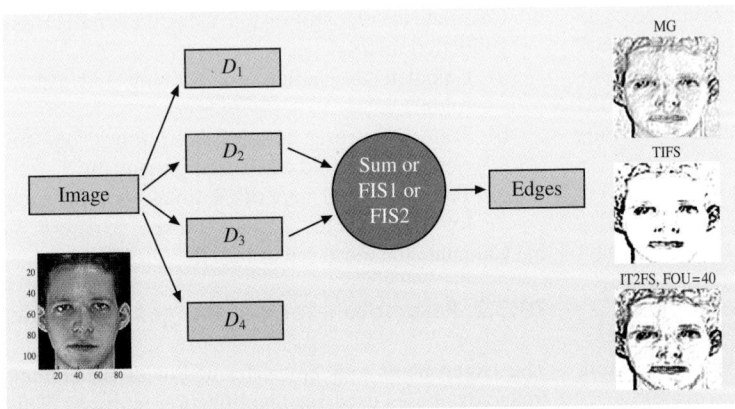

Fig. 79.4 Morphological gradient edge detector enhanced with fuzzy systems

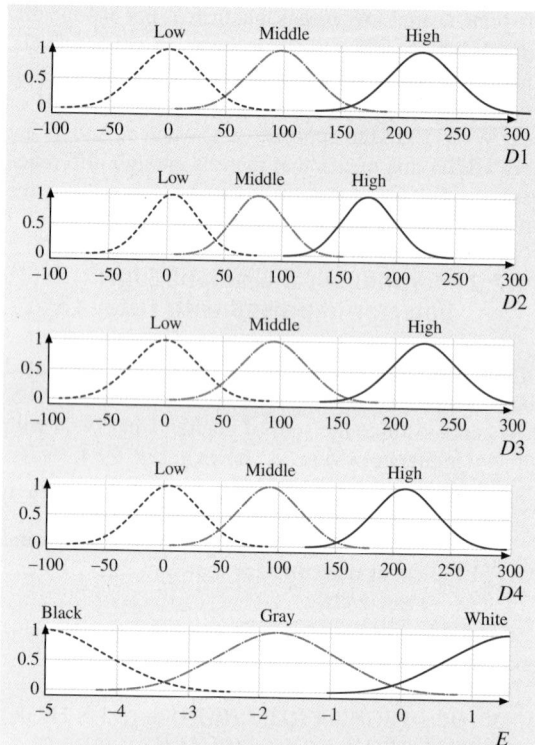

Fig. 79.5 Membership functions of the variables for the MG+FIS1 edge detector

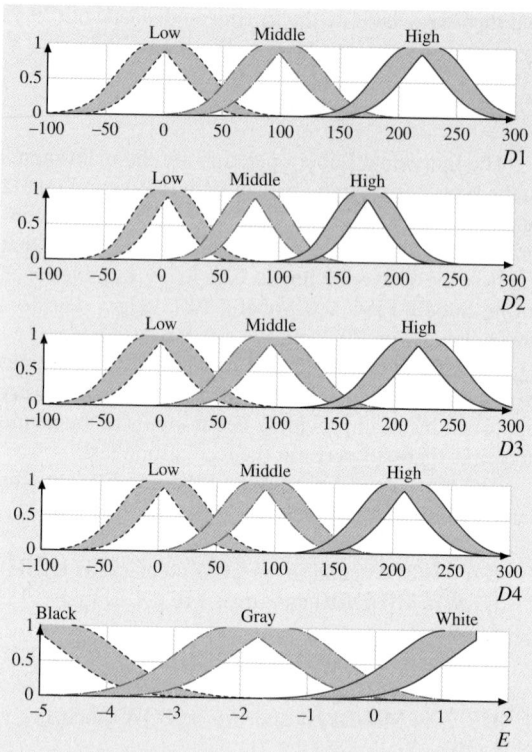

Fig. 79.6 Membership functions of the variables for the MG+FIS2 edge detector

MIDDLE, which means that even when the differences in the gray tones are not maximal, the pixel is an EDGE, then the only rule that found a non edge pixel is the

number 3, where only when all the gradients are LOW, the output pixel is WHITE, which means a pixel belonging to an homogeneous region.

79.3 Experimental Setup

The experiment consists on applying a neural recognition system using each of the previously presented edge detectors: Sobel, Sobel+FIS1, Sobel+FIS2, morphological gradient (MG), morphological gradient+FIS1 and morphological gradient+FIS2 and then comparing the results.

79.3.1 General Algorithm used for the Experiments

1. Define the database folder.
2. Define the edge detector.
3. Detect the edges of each image as a vector and store it as a column in a matrix.

4. Calculate the recognition rate using the k-fold cross-validation method.
   a) Calculate the indices for training and test k-folds.
   b) Train the neural network $k-1$ times, one for each training fold calculated previously.
   c) Test the neural network k times, one for each fold test set calculated previously.
5. Calculate the mean rate for all the k-folds.

79.3.2 Parameters for the Images Databases

The experiments can be performed with benchmark image databases used for identification purposes. This

Table 79.1 Particular information for the tested benchmark face databases

Database	Person number (p)	Samples number (s)	Fold size (m)	Training set size (i)	Test set size (t)
ORL	40	10	80	320	80
Cropped Yale	38	10	76	304	76
FERET	74	4	74	222	74

is the case of face recognition applications, then we used three of the most popular benchmark sets of images, the ORL face database [79.14], the Cropped Yale face database [79.15, 16], and the FERET face database [79.17].

For the three databases, we defined the variable p as the person number and s as number of samples for each person. The tests were made with k-fold cross-validation method, with $k = 5$ for the three databases. We can generalize the calculation of fold size m or number of samples in each fold, dividing the total number of samples for each person s by the fold number, and then multiplying the result by the person number p (3), then the train data set size i (4) can be calculated as the number of samples in $k-1$ folds m, and test data set size t (5) are the number of samples in only one fold.

$$m = (s/k)^{*}p \qquad (79.3)$$
$$i = m(k-1) \qquad (79.4)$$
$$t = m. \qquad (79.5)$$

The total number of samples used for each person were of 10 for the ORL and YALE databases; then if the

size m of each 5-fold is 2, the number of samples for training for each person is 8 and for testing is 2. For the experiments with the FERET face database, we use only the samples of 74 persons who have 4 frontal sample images. The particular information for each database is shown in Tab. 79.1.

79.3.3 The Modular Neural Network

In previous experiments with neural networks for image recognition, we have found a general structure with acceptable performance, even if it is not optimized. We used the same structure for multinet modular neural networks, in order to establish a standard for comparison for all the experiments [79.3, 18–23]. The general structure for the monolithic neural network is indicated below:

- Two hidden layers with 200 neurons.
- Learning algorithm: Gradient descent with momentum and adaptive learning rate backpropagation.
- Error goal 0.0001.

79.4 Experimental Results

In this section, we show the numerical results of the experiments. Table 79.2 contains the results for the ORL face database, Table 79.3 contains the results for the Cropped Yale database, and Table 79.4 contains the results for the FERET face database.

For a better appreciation of the results, we made plots for the values presented in Tables 79.2–79.4. Even

if this work does not pretend to make a comparison based on the training times as performance index for the edge detectors, it is interesting to note that the necessary time to reach the error goal is established for each experiment.

As we can see in Fig. 79.7, the lowest training times are for the morphological gradient+FIS2 edge

Table 79.2 Recognition rates for the ORL database of faces

Training set preprocessing method	Mean time (s)	Mean rate (%)	Standard deviation	Max rate (%)
MG+FIS1	1.2694	89.25	4.47	95.00
MG+FIS2	1.2694	90.25	5.48	97.50
Sobel+FIS1	1.2694	87.25	3.69	91.25
Sobel+FIS2	1.2694	90.75	4.29	95.00

Table 79.3 Recognition rates for the cropped Yale database of faces

Training set preprocessing method	Mean time (s)	Mean rate (%)	Standard deviation	Max rate (%)
MG+FIS1	1.76	68.42	29.11	100
MG+FIS2	1.07	88.16	21.09	100
Sobel+FIS1	1.17	79.47	26.33	100
Sobel+FIS2	1.1321	90	22.36	100

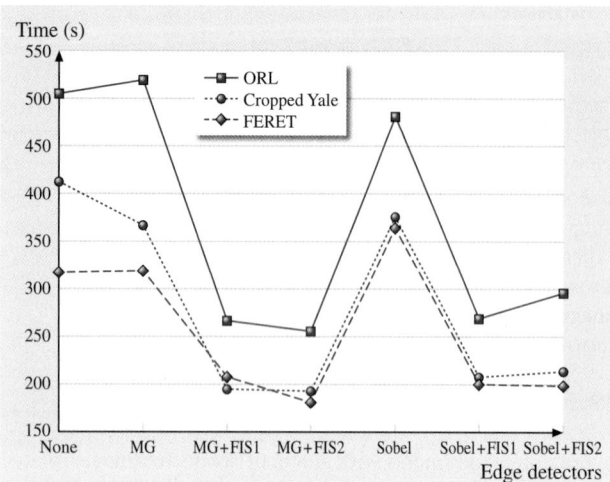

Fig. 79.7 Training time for the compared edge detectors tested with the ORL, Cropped Yale and FERET face databases

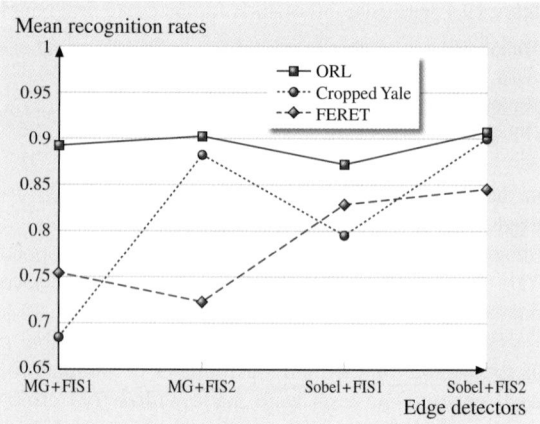

Fig. 79.8 Mean recognition rates for the compared edge detectors with ORL, Cropped Yale and FERET face databases

Table 79.4 Recognition rates for the FERET database of faces

Training set preprocessing method	Mean time (s)	Mean rate (%)	Standard deviation	Max rate (%)
MG+FIS1	1.17	75.34	5.45	79.73
MG+FIS2	1.17	72.30	6.85	82.43
Sobel+FIS1	1.17	82.77	00.68	83.78
Sobel+FIS2	1.17	84.46	03.22	87.84

detector and Sobel+FIS2 edge detector. That is because both edge detectors were improved with interval type-2 fuzzy systems and produce images with more homogeneous areas; which means a high frequency of pixels near the WHITE linguistic values.

However, the main advantages of the interval type-2 edges detectors are the recognition rates plotted in Fig. 79.8, where we can notice that the best mean performance of the neural network was achieved when it was trained with the data sets obtained with the MG+FIS2 and Sobel+FIS2 edge detectors.

Figure 79.9 shows that the recognition rates are also better for the edge detectors improved with interval type-2 fuzzy systems. The maximum recognition rates could not be the better parameter to compare the

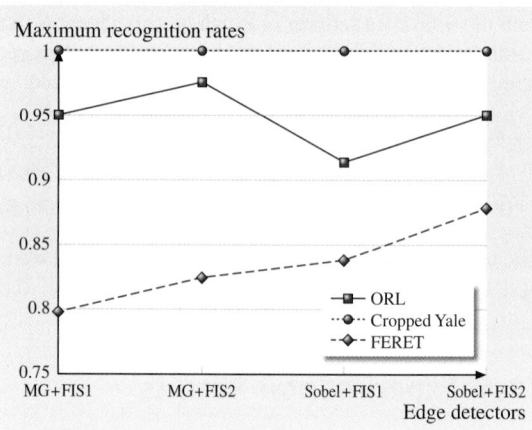

Fig. 79.9 Maximum recognition rates for the compared edge detectors with ORL, Cropped Yale and FERET face database

performance of the neural networks depending on the training set; but it is interesting to note that the maximum recognition rate of 97.5% was achieved when the neural network was trained with the ORL data set preprocessed with the MG+FIS2. This is important because in a real-world system, we can use this as the best configuration for images recognition, expecting to obtain good results.

79.5 Conclusions

This chapter is one of the first efforts to develop a comparison method for edge detectors as a function of their performance in different types of recognition systems. In this chapter, we show that Sobel and Morphologi- cal Gradient edge detectors improved with type-2 fuzzy logic have better performance than the type-1 fuzzy edge detector and traditional methods in an image recognition system based on neural networks.

References

79.1   O. Mendoza, P. Melin, G. Licea: A New Method for Edge Detection in Image Processing Using Interval Type-2 Fuzzy Logic, IEEE Int. Conf. Granular Comput. (GRC) (2007)

79.2   O. Mendoza, P. Melin, G. Licea: Fuzzy Inference Systems Type-1 and Type-2 for Digital Images Edges Detection, Eng. Lett., Int. Ass. Eng. 15(1), 45–52 (2007)

79.3   O. Mendoza, P. Melin, G. Licea: Interval type-2 fuzzy logic for edges detection in digital images, Int. J. Intell. Syst. 24(11), 1115–1134 (2009)

79.4   H. Bustince, E. Berrenechea, M. Pagola, J. Fernandez: Interval-valued fuzzy sets constructed from matrices: Application to edge detection, Fuzzy Sets Syst. 160(13), 1819–1840 (2009)

79.5   K. Revathy, S. Lekshmi, S.R. Prabhakaran Nayar: Fractal-based fuzzy technique for detection of active regions from solar, J. Solar Phys. 228, 43–53 (2005)

79.6   K. Suzuki, I. Horiba, N. Sugie, M. Nanki: Contour extraction of left ventricular cavity from digital subtraction angiograms using a neural edge detector, Syst. Comput. 34(2), 55–69 (2003)

79.7   L. Hua, H.D. Cheng, M. Zhang: A high performance edge detector based on fuzzy inference rules, Inf. Sci. 177(21), 4768–4784 (2007)

79.8   M. Heath, S. Sarkar, T. Sanocki, K.W. Bowyer: A robust visual method for assessing the relative performance of edge-detection algorithms, IEEE Trans. Pattern Anal. Mach. Intell. 19(12), 1338–1359 (1997)

79.9   Y. Yitzhaky, E. Peli: A method for objective edge detection evaluation and detector parameter selection, IEEE Trans. Pattern Anal. Mach. Intell. 25(8), 1027–1033 (2003)

79.10   J. Mendel: *Uncertain Rule-Based Fuzzy Logic Systems: Introduction and New Directions* (Prentice-Hall, Upper Saddle River 2000)

79.11   J.R. Castro, O. Castillo, P. Melin, A. Rodriguez-Diaz: Building fuzzy inference systems with a new interval type-2 fuzzy logic tool-box, Lect. Notes Comput. Sci. 4750, 104–114 (2008)

79.12   A.N. Evans, X.U. Liu: Morphological gradient approach for color edges detection, IEEE Trans. Image Process. 15(6), 1454–1463 (2006)

79.13   F. Russo, G. Ramponi: Edge extraction by FIRE operators Fuzzy Systems, Proc. 1st IEEE Conf. Evolu- tionary Computation (ICEC), Orlando, Florida (1994) pp. 249–253

79.14   AT & T Laboratories Cambridge, The ORL database of faces, http://www.cl.cam.ac.uk/research/dtg/ attarchive/facedatabase.html

79.15   A.S. Georghiades, P.N. Belhumeur, D.J. Kriegman: From few to many: Illumination cone models for face recognition under variable lighting and pose, IEEE Trans. Pattern Anal. Mach. Intell. 23(6), 643–660 (2001)

79.16   K.C. Lee, J. Ho, D. Kriegman: Acquiring linear subspaces for face recognition under variable lighting, IEEE Trans. Pattern Anal. Mach. Intell. 27(5), 684–698 (2005)

79.17   P.J. Phillips, H. Moon, S.A. Rizvi, P.J. Rauss: The FERET evaluation methodology for face-recognition algorithms, IEEE Trans. Pattern Anal. Mach. Intell. 22(10), 1090–1104 (2000)

79.18   O. Mendoza, P. Melin: The Fuzzy Sugeno Integral As A Decision Operator in The Recognition of Images with Modular Neural Networks. In: *Hybrid Intelligent Systems*, Studies in Fuzziness and Soft Computing, (Springer, Berlin, Heidelberg 2007) pp. 299–310

79.19   O. Mendoza, P. Melin, G. Licea: A Hybrid Approach for Image Recognition Combining Type-2 Fuzzy Logic, Modular Neural Networks and the Sugeno Integral, Inf. Sci. 179(13), 2078–2101 (2007)

79.20   O. Mendoza, P. Melin, G. Licea: Interval Type-2 Fuzzy Logic for Module Relevance Estimation in Sugeno Integration of Modular Neural Networks. In: *Soft Computing for Hybrid Intelligent Systems*, Studies in Computational Intelligence, Vol. 154, (Springer, Berlin, Heidelberg 2008) pp. 115–127

79.21   O. Mendoza, P. Melin, G. Licea: A hybrid approach for image recognition combining type-2 fuzzy logic, modular neural net-works and the Sugeno integral, Inf. Sci. 179(3), 2078–2101 (2009)

79.22   O. Mendoza, P. Melin, O. Castillo: Interval type-2 fuzzy logic and modular neural networks for face recognition applications, Appl. Soft Comp. J. 9(4), 1377–1387 (2009)

79.23   O. Mendoza, P. Melin, O. Castillo, G. Licea: Type-2 fuzzy logic for improving training data and response integration in modular neural networks for image recognition, Lect. Notes Comput. Sci. 4329, 604–612 (2007)

80. Fuzzy Controllers for Autonomous Mobile Robots

Patricia Melin, Oscar Castillo

This chapter addresses the tracking problem for the dynamic model of a unicycle mobile robot. A novel optimization method inspired from the chemical reactions is applied to solve this motion problem by integrating a kinematic and a torque controller based on fuzzy logic theory. Computer simulations are presented confirming that this optimization paradigm is able to outperform other optimization techniques applied to this particular robot application.

80.1 Fuzzy Control of Mobile Robots

Optimization is an activity carried out in almost every aspect of our life, from planning the best route of our way back home from work to more sophisticated approximations on the stock market, or the parameter optimization for a wave soldering process used in the manufacture of a printed circuit board assembly, optimization theory has gained importance over the last decades. From science to applied engineering (to name a few), there is always something to optimize and, of course, more than one way to do it.

In a generic definition, we may say that optimization aims to find the *best* available solution among a set of potential solutions in a defined search space. For almost every problem there exists a solution, not necessarily the best one, but we can always find an approximation to the *ideal solution*, and while in some cases or processes it is still common to use our own experience to qualify a process, a part of the research community has dedicated a considerable amount of time and effort to help find robust optimization methods for optima found in a vast range of applications.

That it is difficult to solve different problems by applying the same methodology, and even the most robust optimization approaches may be outperformed by other optimization techniques, depending on the problem to be solved.

When the complexity and the dimension of the search space make a problem unsolvable by a deterministic algorithm, probabilistic algorithms deal with this problem by going through a diverse set of possible solutions or candidate solutions. Many metaheuristic algorithms can be considered probabilistic because they apply probability tools to solve a problem; metaheuristic algorithms seek good solutions by mimicking natural processes or paradigms. Most of these novel optimization paradigms that were inspired by nature were conceived by mere observation of an existing process and their main characteristics were embodied as computational algorithms.

The importance of the optimization theory and its application has grown in the past few decades, from the well known genetic algorithm paradigm to particle swarm optimization (PSO), ant colony optimization

(ACO), harmonic search, deoxyribonucleic acid (DNA) computing, among others, abd they were all were introduced with the expectation that they would improve the results obtained with existing strategies.

There is no doubt that there could be some optimization strategies presented at some point that were left behind due their complexity and poor performance. Novel optimization paradigms should be able to perform well in comparison with another optimization techniques and must be *easily adaptable* to different kinds of problems.

Optimization based on chemical processes is a growing field that has been satisfactorily applied to several problems. In [80.1] a DNA-based algorithm was introduced to solve the small hitting set problem. A catalytic search algorithm was explored in [80.2], where some physical laws such as mass and energy conservation were taken into account. In [80.3], the potential roles of energy in algorithmic chemistries were illustrated. An energy framework was introduced, which keeps the molecules within reasonable length bounds, allowing the algorithm to behave thermodynamically and kinetically, similarly to real chemistry. A chemical reaction optimization was applied to a grid scheduling problem in [80.4], where molecules interact with each other aiming to reach the minimum state of free potential and kinetic energies. The main difference between these metaheuristics is the parameter representation, which can be explicit or implicit.

In this paper, we introduce an optimization method inspired by chemical reactions and its application for the optimization of the tracking controller of the dynamic model of the unicycle mobile robot.

The importance of applying this chemical optimization algorithm is that different methods have been applied to solve motion control problems. *Kanayama* et al. [80.5] propose a stable tracking control method for a nonholonomic vehicle using a Lyapunov function. *Lee* et al. [80.6] solved tracking control using backstepping and in [80.7] saturation constraints were used. Furthermore, most reported designs rely on intelligent control approaches such as fuzzy logic control [80.8–13] and neural networks [80.14, 15].

However, the majority of the publications mentioned above concentrated on kinematic models of mobile robots, which are controlled by the velocity input, while less attention has been paid to the control problems of nonholonomic dynamic systems, where forces and torques are the true inputs. *Bloch* and *Drakunov* [80.16] and *Chwa* [80.17] used sliding mode control for the tracking control problem. *Fierro* and *Lewis* [80.18] proposed a dynamical extension that makes the integration of kinematics and torque controller possible for a nonholonomic mobile robot. *Fukao* et al. [80.19] introduced an adaptive tracking controller for the dynamic model of mobile robots with unknown parameters using backstepping methodology, which has been recognized as a tool for solving several control problems [80.20, 21].

Motivated by this, herein a Mamdani fuzzy logic controller is introduced in order to drive the kinematic model to a desired trajectory in a finite time; considering the torque as the real input, a chemical reaction optimization paradigm is applied and simulations are shown.

Further publications [80.22–24] applied bio-inspired optimization techniques to find the parameters of the membership functions for the fuzzy tracking controller that solves the problem for the dynamic model of a unicycle mobile robot, using a fuzzy logic controller that provides the required torques to reach the desired velocity and trajectory inputs.

In this paper, the main contribution is the representation of the fuzzy controller in the chemical paradigm to search for the optimal parameters. Simulation results show that the proposed approach outperforms other nature-inspired computing paradigms, such as genetic algorithms, particle swarm, and ant colony optimization.

The rest of this chapter is organized as follows. Section 80.2 illustrates the proposed methodology. Section 80.3 describes the problem formulation and control objective. Section 80.4 describes the proposed fuzzy logic controller of the robot. Section 80.5 shows some experimental results of the tracking controller. In Sect. 80.6 some conclusions and future work are presented.

80.2 The Chemical Optimization Paradigm

The proposed chemical reaction algorithm is a metaheuristic strategy that performs a stochastic search for optimal solutions within a defined search space. In this optimization strategy, every solution is represented as an element (or compound), and the fitness or performance of the element is evaluated in accordance with

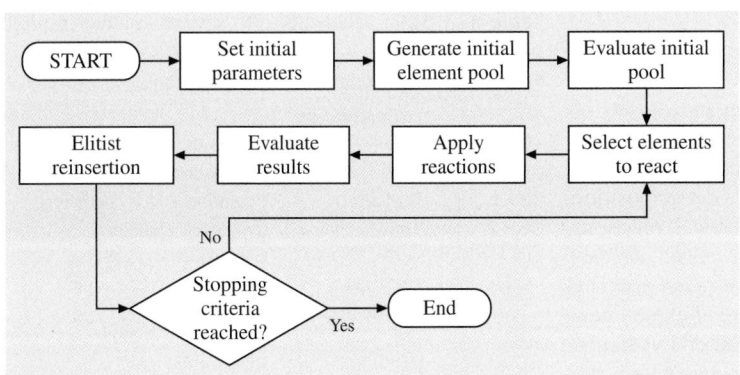

Fig. 80.1 General flowchart of the chemical reaction algorithm

the objective function. The general flowchart of the algorithm is shown in Fig. 80.1.

The main difference with other optimization techniques [80.1–4] is that no external parameters are taken into account to evaluate the results, while other algorithms introduce additional parameters (kinetic/potential energies, mass conservation, thermodynamic characteristics, etc.). This is a very straightforward methodology that takes the characteristics of the chemical reactions (synthesis, decomposition, substitution, and double-substitution) to find the optimal solution.

This approach is a static population-based metaheuristic that applies an abstraction of the chemical reactions as intensifiers (substitution, double substitution reactions) and diversifying (synthesis, decomposition reactions) mechanisms. The elitist reinsertion strategy allows the permanence of the best elements, and thus the average fitness of the entire element pool increases with every iteration. The algorithm may trigger only one reaction or all of them, depending on the nature of the problem to solve. For example, we may use only the decomposition reaction subroutine to find the minimum value of a mathematical function.

The pseudocode for the chemical reaction algorithm is as follows:

Algorithm 80.1 Chemical_Reaction_Algorithm
  Input: *problem_definition*, *objective_function*, *dimensions*,
  1: Assign values to variables: *pool_size*, *trials*, *upper_boundary*, *lower_boundary*, *synthesis_rate*, *decomposition_rate*, *singlesubstitution_rate*, *doublesubstitution_rate*.
  2: Generate randomly *Initial_Pool* in interval [*lower_boundary, upper_boundary*]
  3: Evaluate *Initial_Pool*

  4: Identify *best_solution*
  5: **while** (stopping criteria not met) **do**
  6:    Perform *Synthesis_Procedure*; Get *Synthesis_vector*
  7:    Perform *Decomposition_Procedure*; Get *Decomposition_vector*
  8:    Perform *SingleSubstitution_Procedure*; Get *SingleSubstitution_vector*
  9:    Perform *DoubleSubstitution_Procedure*; Get *DoubleSubstitution_vector*
 10:    Evaluate *Synthesis_vector*, *Decomposition_vector*, *SingleSubstitution_vector*, *DoubleSubstitution_vector*
 11:    Apply *elitist_reinsertion*; Get *improved_pool*
 12:    Update *best_solution*
 13: **end while**
      Output: *best_solution*

All nature-inspired paradigms have their own way to encode candidate solutions. When these parameters are defined, a set of processes or procedures are applied to lead the population to an optimal result. The main components of this chemical reaction algorithm are described below.

80.2.1 Elements/Compounds

These are the basic components of the algorithm. Each element or compound represents a solution within the search space. The initial definition of elements and/or compounds depends on the problem itself and can be represented as binary numbers, integer, floating, etc. They interact with each other implicitly; that is, the definition of the interaction is independent of the real molecular structure. In this approach the potential and kinetic energies and other molecular characteristics are not taken into account.

80.2.2 Chemical Reactions

A chemical reaction is a process in which at least one substance changes its composition and its sets of properties. In this approach, the chemical reactions behave as intensifiers (substitution, double substitution reactions) and diversifying (synthesis, decomposition reactions) mechanisms. The four chemical reactions considered in this approach are the *synthesis, decomposition, single and double-substitution reactions*. The objective of these operators is to explore or exploit new possible solutions within a slightly larger hypercube than the original elements/compounds, but within the previously specified range.

The *synthesis* and *decomposition* reactions are used to diversify the resulting solutions; these procedures were shown to be highly effective and to rapidly lead the results to a desired value. They can be described as follows.

80.2.3 Synthesis Reactions

This is a reaction of two reactants to produce one product. By combining two (or more) elements, this procedure allows us to explore higher-valued solutions within the search space. The result can be described as a compound $(B + C \rightarrow BC)$. The pseudocode for the synthesis reaction procedure is as follows:

Algorithm 80.2 Synthesis_Procedure
    Input: *selected_elements, synthesis_rate*
1: $n = $ size (*selected_elements*)
2: $i = $ floor ($n/2$)
3: **for** $j = 1$ to $i - 1$
4:     *Synthesis = selected_elements$_j$*
        + selected_elements$_{j+1}$
5:     $j = j + 2$
6: **end for**
    Output: *Synthesis_vector*

80.2.4 Decomposition Reactions

In this reaction, typically, only one reactant is given, which allows a compound to be decomposed into smaller instances $(BC \rightarrow B + C)$. The pseudocode for the decomposition reaction procedure is as follows:

Algorithm 80.3 Decomposition_Procedure
    Input: *selected_elements, decomposition_rate*
1: $n = $ size (*selected_elements*)

Table 80.1 Main elements of several nature-inspired paradigms

Paradigm	Parameter representation	Basic operations
GA	Genes	Crossover, mutation
ACO	Ants	Pheromone
PSO	Particles	Cognitive, social coefficients
GP	Trees	Crossover, mutation (In some cases)
CRM	Elements, Compounds	Reactions (combination, decomposition, Substitution, double-substitution)

2: Get *randval* randomly in interval [0, 1]
3: **for** $i = 1$ to n
4:     $Deco_1 = $ *selected_elements$_i$* $\times$ *randval*
5:     $Deco_2 = $ *selected_elements$_i$* $\times$ (1 − *randval*)
6:     $i = i + 1$
7: **end for**
    Output: *Decomposition_vector* ($Deco_1$, $Deco_2$)

The *single* and *double-substitution* reactions allow the algorithm to search for optima around a previously found good solution and they are described below.

80.2.5 Single-Substitution Reactions

When a free element reacts with a compound of different elements, the free element will replace one of the elements in the compound if the free element is more reactive than the element it replaces. A new compound and a new free element are produced. In the algorithm, a compound and an element are selected and a decomposition reaction is applied to the compound; two elements are generated from this operation. Then, one of the new generated elements is combined with the non-decomposed selected element $(C+ AB \rightarrow AC +B)$. The pseudocode for the single-substitution reaction procedure is as follows:

Algorithm 80.4 SingleSubstitution_Procedure
    Input: *selected_elements, singlesubstitution_rate*
1: $n = $ size (*selected_elements*)
2: $i = $ floor ($n/2$)
3: $a = $ *selected_elements$_1$, selected_elements$_2$, ..., selected_elements$_i$*
4: $b = $ *selected_elements$_{i+1}$, selected_elements$_{i+2}$, ..., selected_elements$_{i\times2}$*
5: Apply *Decomposition_Procedure* to a; Get $Deco_1$, $Deco_2$

6: Apply *Synthesis_Procedure* ($b+ Deco_1$); Get *Synthesis_vector* Output: *SingleSubstitution_vector* (*Synthesis_vector*, $Deco_2$)

80.2.6 Double-Substitution Reactions

Double-substitution or double-replacement reactions, also called double-decomposition reactions or metathesis reactions, involve two ionic compounds, most often in aqueous solution. In this type of reaction, the cations simply swap anions; in the algorithm, a similar process to that in the previous reaction happens. The difference is that in this reaction both of the selected compounds are decomposed, and the resulting elements are combined with each other ($AB + CD \rightarrow CB + AD$). The pseudocode for the double-substitution reaction procedure is as follows:

Algorithm 80.5 DoubleSubstitution_Procedure
  Input: *selected_elements*, *doublesubstitution_rate*
1: $n =$ size (*selected_elements*)
2: $i =$ floor ($n/2$)
3: $a =$ *selected_elements*$_1$, *selected_elements*$_2$, ..., *selected_elements*$_i$

4: $b =$ *selected_elements*$_{i+1}$, *selected_elements*$_{i+2}$, ..., *selected_elements*$_{i \times 2}$
5: Apply *Decomposition_Procedure* to a and b; Get ($Deco_1$, $Deco_2$), ($Deco_1'$, $Deco_2'$)
6: Apply *Synthesis_Procedure* ($Deco_1 + Deco_1'$), ($Deco_2 + Deco_2'$); Get *Synthesis_vector*$_1$, *Synthesis_vector*$_1'$
  Output: *SingleSubstitution_vector* (*Synthesis_vector*$_1$, *Synthesis_vector*$_1'$)

In this chemical reaction algorithm we may trigger only one reaction or all of them, depending on the nature of the problem to be solved, e.g., we can apply only the decomposition reaction subroutine to find the minimum value of a mathematical function.

Throughout the execution of the algorithm, whenever a new set of elements/compounds is created, an elitist reinsertion criterion is applied, allowing the permanence of the best elements, and thus the average fitness of the entire element pool increases through iterations.

In order to have a better picture of the general schema for this proposed chemical reaction algorithm, a comparison with other nature-inspired paradigms is shown in Table 80.1.

80.3 The Mobile Robot

Mobile robots are non-nonholonomic systems due to the constraints imposed on their kinematics. The equations describing the constraints cannot be integrated symbolically to obtain explicit relationships between robot positions in local and global coordinates' frames. Hence, control problems that involve them have attracted attention in the control community in recent years [80.25].

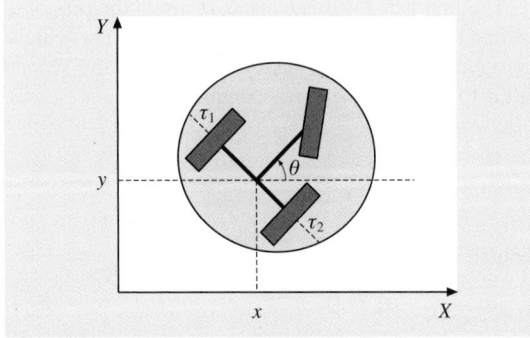

Fig. 80.2 Diagram of a wheeled mobile robot

The model considered is that of a unicycle mobile robot (see Fig. 80.2) that has two driving wheels fixed to the axis and one passive orientable wheel placed in front of the axis and normal to it [80.26].

The two fixed wheels are controlled independently by the motors, and the passive wheel prevents the robot from overturning when moving on a plane.

It is assumed that the motion of the passive wheel can be ignored from the dynamics of the mobile robot, which is represented by the following set of equations [80.18]

$$\dot{q} = \begin{vmatrix} \cos\theta & 0 \\ \sin\theta & 0 \\ 0 & 1 \end{vmatrix} \begin{vmatrix} v \\ w \end{vmatrix} \mathbf{M}(q)\dot{v} + V(q,\dot{q})v + \mathbf{G}(q)$$

$$= \boldsymbol{\tau} \,,$$

$$(80.1)$$

where $q = [x, y, \theta]^{\mathsf{T}}$ is the vector of generalized coordinates that describes the robot's position, (x, y) are the Cartesian coordinates, which denote the mobile center

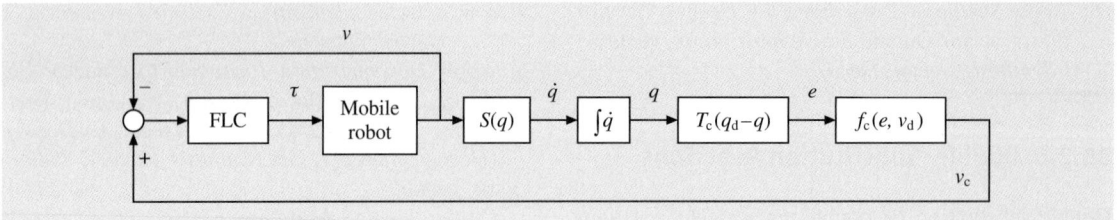

Fig. 80.3 Tracking control structure

of mass, and θ is the angle between the heading direction and the x-axis (which is taken in counterclockwise form); $v = [v, w]^T$ is the vector of velocities, v and w are linear and angular velocities respectively $\tau \in R^r$ is the input vector, $\mathbf{M}(q) \in R^{nxn}$ is a symmetric and positive-definite inertia matrix, $\mathbf{V}(q, \dot{q}) \in R^{nxn}$ is the centripetal and Coriolis matrix, and $\mathbf{G}(q) \in R^n$ is the gravitational vector. Equation (80.1) represents the kinematics or steering system of a mobile robot.

Notice the no-slip condition imposed a non-nonholonomic constraint described by (80.2), which means that the mobile robot can only move in the direction normal to the axis of the driving wheels.

$$\dot{y} \cos \theta - \dot{x} \sin \theta = 0 . \tag{80.2}$$

The control objective will be established as follows: given a desired trajectory $q_d(t)$ and the orientation of the mobile robot, we must design a controller that applies an adequate torque τ such that the measured positions $q(t)$ achieve the desired reference $q_d(t)$ represented as (80.3)

$$\lim_{t \to \infty} \| q_d(t) - q(t) \| = 0 . \tag{80.3}$$

To reach the control objective, the method is based on the procedure of [80.18], and we derive $\tau(t)$ of a specific $v_c(t)$ that controls the steering system (80.1) using a fuzzy logic controller (FLC). The general structure of a tracking control system is presented in Fig. 80.3.

The control is based on the procedure proposed by *Kanayama* et al. [80.5] and *Nelson* and *Cox* [80.27]

to solve the tracking problem for the kinematic model $v_c(t)$. Suppose that the desired trajectory q_d satisfies (80.4)

$$\dot{q}_d = \begin{vmatrix} \cos \theta_d & 0 \\ \sin \theta_d & 0 \\ 0 & 1 \end{vmatrix} \begin{vmatrix} v_d \\ w_d \end{vmatrix} . \tag{80.4}$$

Using the robot local frame (the moving coordinate system x-y in Fig. 80.1), the error coordinates can be defined as (80.5)

$$e = T_e(q_d - q), \begin{vmatrix} e_x \\ e_y \\ e_\theta \end{vmatrix}$$

$$= \begin{vmatrix} \cos \theta & \sin \theta & 0 \\ -\sin \theta & \cos \theta & 0 \\ 0 & 0 & 1 \end{vmatrix} \begin{vmatrix} x_d - x \\ y_d - y \\ \theta_d - \theta \end{vmatrix} . \tag{80.5}$$

Moreover, the auxiliary velocity control input that achieves tracking for (80.1) is given by (80.6)

$$v_c = f_c(e, v_d), \begin{vmatrix} v_c \\ w_c \end{vmatrix}$$

$$= \begin{vmatrix} v_d + \cos e_\theta + k_1 e_x \\ w_d + v_d k_2 e_y + v_d k_3 \sin e_\theta \end{vmatrix} , \tag{80.6}$$

where k_1, k_2 and k_3 are positive gain constants.

The first part for this work is to apply the proposed method to obtain the values of k_i ($i = 1, 2, 3$) to achieve the optimal behavior of the controller, and the second part is to optimize the fuzzy controller.

80.4 Fuzzy Logic Controller

The purpose of the fuzzy logic controller (FLC) is to find a control input τ such that the current velocity vector v is able to reach the velocity vector v_c, and this is

denoted as

$$\lim_{t \to \infty} \| v_c - v \| = 0 . \tag{80.7}$$

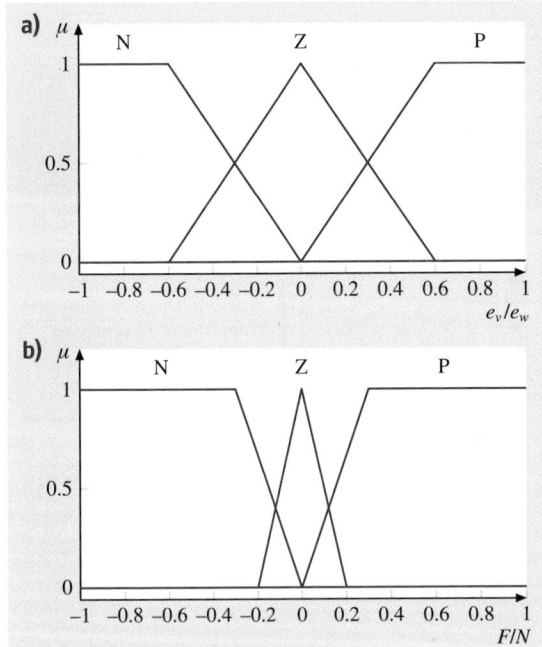

Fig. 80.4a,b Membership functions of (**a**) input e_v and e_w, and (**b**) output variables F and N

Table 80.2 Fuzzy rule set

e_v/e_w	N	Z	P
N	N/N	N/Z	N/P
Z	Z/N	Z/Z	Z/P
P	P/N	P/Z	P/P

the position and angular errors (denoted as e_v and e_w). The initial membership functions (MF) are defined by one triangular and two trapezoidal functions for each variable involved. Figure 80.4 depicts the MFs in which N, Z, P represent the fuzzy sets (negative, zero, and positive, respectively) associated to each input and output variable.

The rule set of the FLC contains nine rules, which govern the input–output relationship of the FLC, and this adopts the Mamdani-style inference engine. We use the center of gravity method to realize the defuzzification procedure. In Table 80.2 we present the rule set whose format is established as follows:

Rule i: If e_v is G1 and e_w is G2

then F is G3 and N is G4 ,

The input variables of the FLC correspond to the velocity errors obtained in (80.5) using the derivatives of

where G1 ... G4 are the fuzzy sets associated to each variable $i = 1...9$. In this case, P denotes *positive*, N denotes *negative*, and Z denotes *zero*.

80.5 Experimental Results

Several tests of the chemical optimization paradigm were made to test the performance of the tracking controller. First, we need to find the values of k_i ($i = 1, 2, 3$) shown in (80.6), which will guarantee convergence of the error e to zero.

To evaluate the constants obtained by the algorithm, the mobile robot tracking system, which consists in (80.5) and (80.6), was modeled using Simulink. Figure 80.5 shows the closed loop for the tracking controller.

The conditions to evaluate each result, which correspond to the final position error, are given by (80.8):

$$\text{EP} = \sum_{i=1}^{n} \frac{e_x(i) + e_y(i) + e_\theta(i)}{n} .\tag{80.8}$$

For the first set of experiments only the decomposition reaction mechanism was triggered and the decomposition factor was varied; this factor is the quantity of resulting elements after applying a decomposition reaction to a determined *compound*. The only restriction here is that x be the selected compound and x'_i ($i = 1 2, ..., n$) the resulting elements. The sum of all values found in the decomposition must be equal to the value of the original compound. This is shown in (80.9)

$$\sum_{i=1}^{n} x'_i = x .\tag{80.9}$$

Each experiment was executed 35 times and the test parameters for each set of experiments can be observed in Table 80.3.

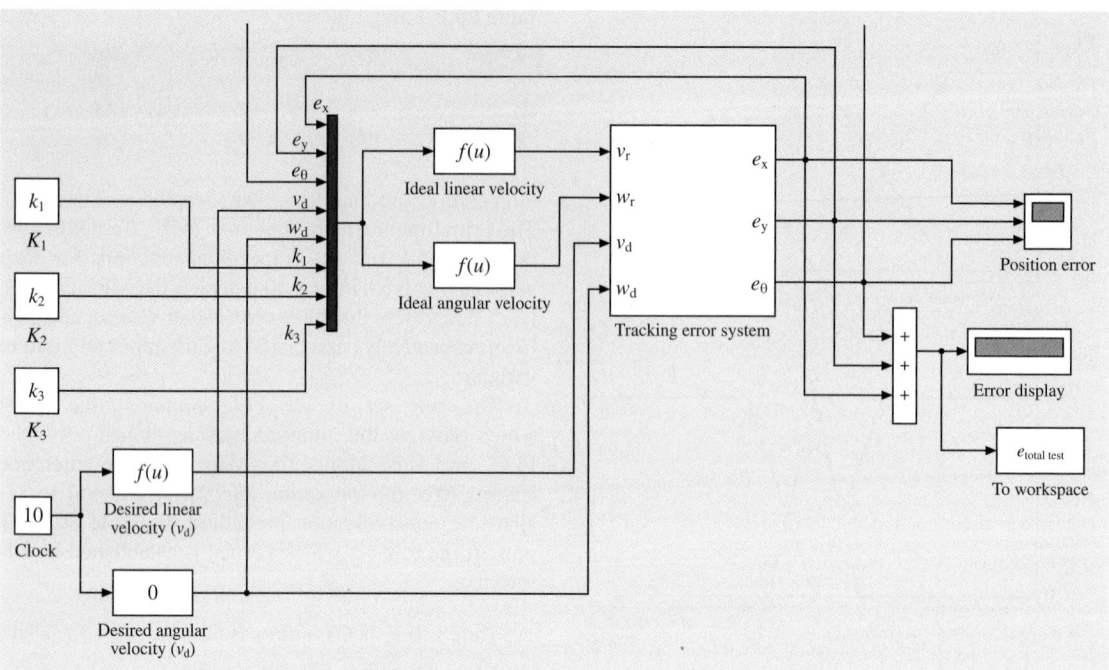

Fig. 80.5 Closed loop for the tracking controller system

The decomposition rate (Dec. rate) represents the percentage of the pool that are candidates for the decomposition and the decomposition factor (Dec. factor) is the number of elements that are to be decomposed into.

The selection strategy applied was stochastic universal sampling, which uses a single random value to

Table 80.3 Parameters of the chemical reaction optimization

No.	Elements	Iterations	Dec. factor	Dec. rate
1	2	10	2	0.3
2	5	10	3	0.3
3	2	10	2	0.4
4	2	10	3	0.4
5	5	10	2	0.4
6	5	10	3	0.4
7	5	10	2	0.5
8	10	10	2	0.5

Table 80.4 Experimental results of the proposed method for optimizing the values of the gains k_1, k_2, and k_3

No.	Best error	Mean	k_1	k_2	k_3
1	0.0086	1.1568	519	46	8
2	4.79×10^{-04}	0.1291	205	31	31
3	0.0025	0.5809	36	328	88
4	0.0012	0.5589	2	206	0
5	0.0035	0.0480	185	29	5
6	8.13×10^{-005}	0.0299	270	53	15
7	0.0066	0.1440	29	15	0
8	0.0019	0.1625	51	3	0

Position error in x

0.3
0.2
0.1
0
0   0.5   1   1.5   2   2.5   3   3.5

Position error in y

0.3
0.2
0.1
0
0   0.5   1   1.5   2   2.5   3   3.5

Position error in θ

0.5
0
−0.5
0   0.5   1   1.5   2   2.5   3   3.5

Fig. 80.6 Final position errors in x, y, and θ for experiment number 6

Table 80.5 Comparison of the best results

Parameters	Genetic algorithm	Chemical optimization algorithm
Individuals	5	2
Iterations	15	10
Crossover rate	0.8	N/A
Mutation rate	0.1	N/A
Synthesis rate	N/A	0.2
Decomposition rate	N/A	0.8
Substitution rate	N/A	0.6
Double substitution rate	N/A	0.6
k_1, k_2, k_3	43, 493, 195	36, 328, 88
Final error	0.006734	0.0025

Table 80.6 Parameters of the simulations for Type-1 FLC

Parameters	Value
Elements	10
Trials	15
Selection method	Stochastic universal sampling
k_1	117
k_2	226
k_3	137
Error	0.077178

Table 80.7 Parameters of the simulations for Type-2 FLC

Parameters	Value
Elements	10
Trials	10
Selection method	Stochastic universal sampling
k_1	117
k_2	226
k_3	137
Error	2.7736

sample all of the solutions by choosing them at evenly spaced intervals. In the example, for a pool containing five initial compounds, the vector length of decomposed elements when the decomposition factor is 3 and the decomposition rate is 0.4 will be six elements.

By applying this criterion, the initial pool of elements increased with every iteration. This is why the initial element pool was set to ten elements as the maximum. Table 80.4 shows the results after applying the chemical optimization paradigm.

As can be observed in Table 80.4, experiment number 6 seems to have the best result because it reached the smaller final error among all experiments. Figure 80.6 shows the final position errors in x, y, and θ for experiment number 6.

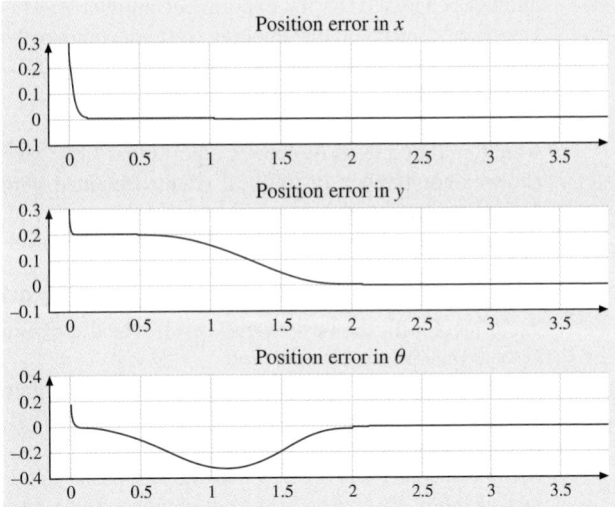

Fig. 80.7 Final position errors in x, y, and θ for experiment number 3

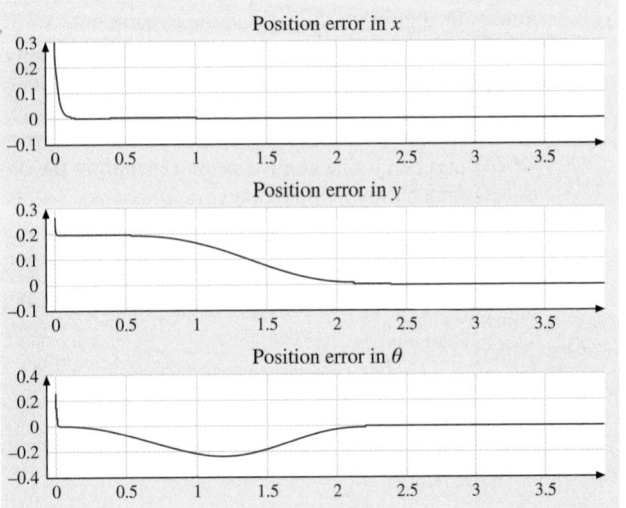

Fig. 80.8 Position errors in x, y, and θ of best result by applying GAs

By analyzing the graphical results of several sets of exercises, we noticed that the control obtained for some of them was *smoother* despite the average error value. This was the case for experiment number 3, in which the final error value was significantly higher than that obtained in experiment number 6. Figure 80.7 shows the final position errors in x, y, and θ for experiment number 3.

Comparing both graphics, we can observe that the average error obtained for θ is 0.0338 for experiment number 6 and 0.0315 for experiment number 3. This smoother control of the tracking system could make a big difference in the complete dynamic system of the mobile robot.

In previous work [80.28], the gain constant values were found by means of genetic algorithms. Table 80.5 shows a comparison of the best results obtained with both algorithms, and we can observe that the result with the chemical optimization outperforms the GA in finding the best gain values.

Figure 80.8 shows the result in Simulink for the experiment with the best overall result when applying GAs as the optimization method.

Once we have found optimal values for the gain constants, the next step is to find the optimal values for the input/output membership functions of the fuzzy controller. Our goal is that the lineal and angular velocities reach zero in the simulations. Table 80.6 shows the parameters of the simulations for Type-1 FLC.

Figure 80.9 shows the behavior of the chemical optimization algorithm throughout the experiment.

Figure 80.10 shows the resulted input and output membership functions found by the proposed optimization algorithm.

Figure 80.11 shows the trajectory obtained when simulating the mobile control system including the obtained input and output membership functions.

Figure 80.12 shows the best trajectory reached by the mobile robot when optimizing the input and output membership functions using genetic algorithms.

A Type-2 FLC was developed using the parameters of the membership functions found for Type-1 FLC. The parameters searched with the chemical reaction algorithm were for the footprint of uncertainty (FOU).

Table 80.7 shows the parameters used in the simulations and Fig. 80.13 shows the behavior of the chemical optimization algorithm throughout the experiment.

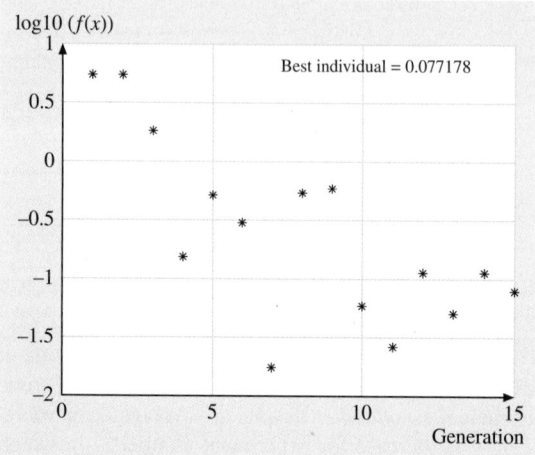

Fig. 80.9 Best simulation of experiments with the chemical optimization method

Fig. 80.10a-d Resulting input membership functions: (**a**) linear and (**b**) angular velocities and output (**c**) right and (**d**) left torque ▶

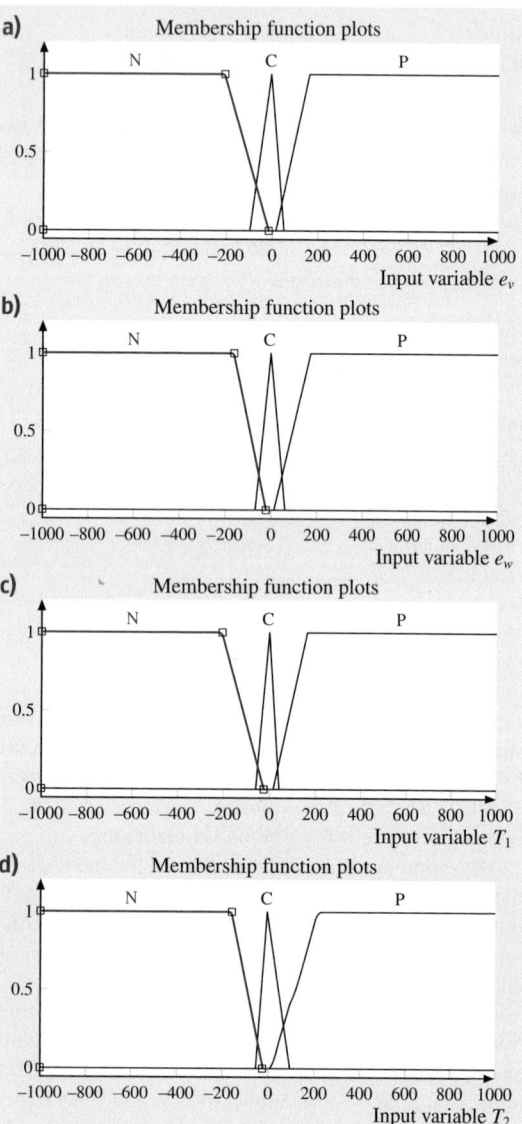

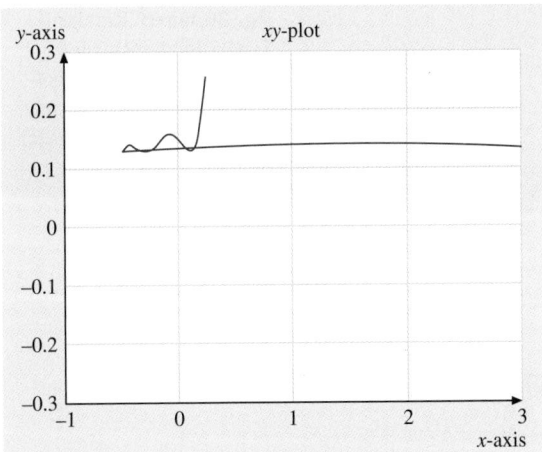

Fig. 80.11 Trajectory obtained when applying the chemical reaction algorithm

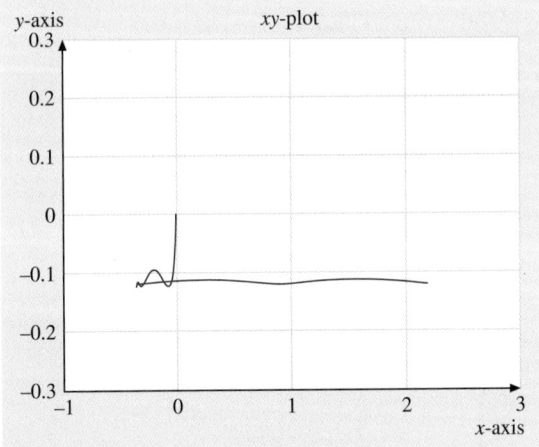

Fig. 80.12 Trajectory obtained using GAs

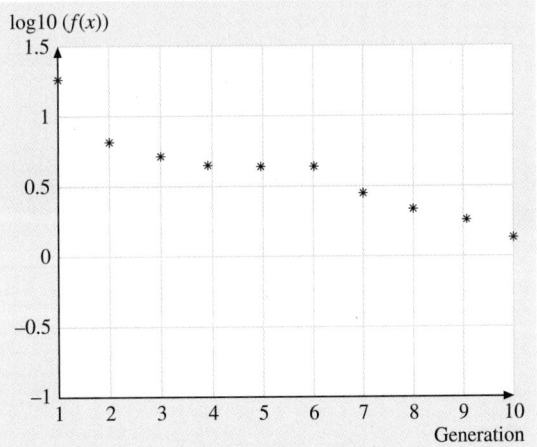

Fig. 80.13 Behavior of the algorithm when optimizing the Type-2 FLC

found were not adequate to make the FLC follow the desired trajectory.

In order to test the robustness of the Type-1 and Type-2 FLC, we added an external signal given by (80.10).

$$F_{\text{ext}}(t) = \varepsilon \times \sin \omega \times t . \tag{80.10}$$

This represents an external force applied in a period of 10 s to the trajectory obtained that will make the mobile robot move out of its path. The idea of adding this disturbance is to measure the errors obtained with the FLC and to test the behavior of the mobile robot under perturbed torques. Table 80.8 shows the parameters for the simulations and the errors obtained during the run of the simulation.

Figure 80.17 show the trajectories obtained for the Type-1 FLC optimized with GAs.

Figure 80.18 shows the trajectories obtained for the Type-1 FLC optimized with the chemical reaction algorithm.

Figure 80.19 shows the trajectories obtained for the Type-2 FLC optimized with the CRA method.

In Table 80.8 and Figs. 80.17 to 80.19 we can observe that the Type-2 FLC was able to maintain a more controlled trajectory in despite of the *large* error found by the algorithm ($e = 2.7736$). For larger epsilon (ε) values, it was difficult for the Type-1 FLCs to keep in the path, and in a determined time the controller was not able to return to the reference trajectory.

Figure 80.14 shows the resulting Type-2 input and output membership functions found by the proposed optimization algorithm and Fig. 80.15 shows the obtained trajectory reached by the mobile robot.

As observed in Table 80.7, the final error obtained is not smaller that the final error found for the Type-1 FLC. Despite this, the trajectory obtained, which is shown in Fig. 80.15, is acceptable, taking into account that the reference trajectory is a straight line. In Fig. 80.16 we can observe an *unacceptable* trajectory, which was found in the early attempts of optimization for the Type-1 FLC applying this chemical reaction algorithm. Here, we can observe that the parameters

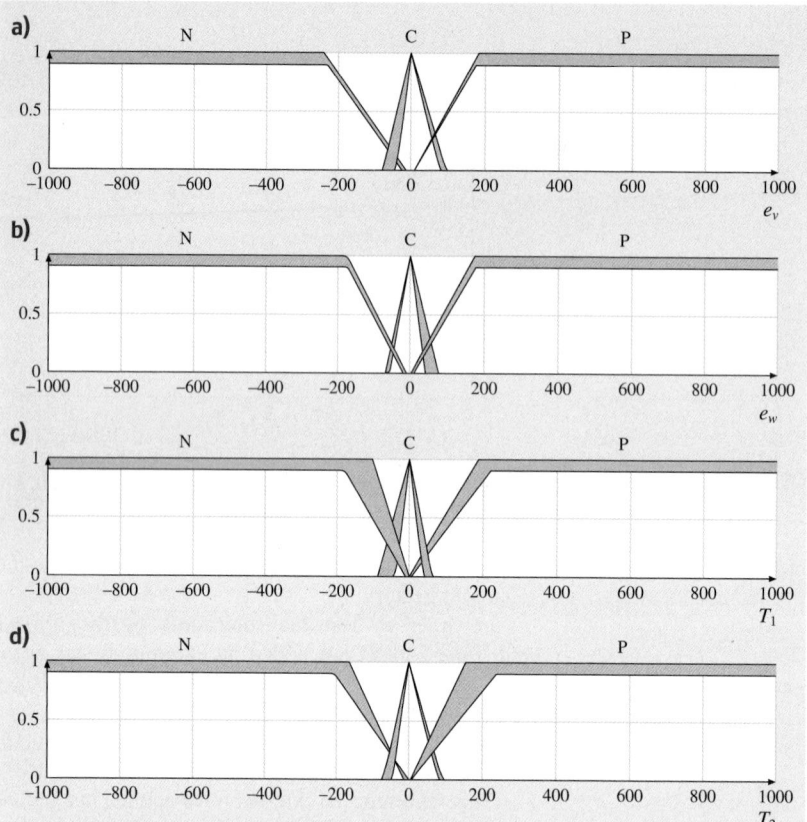

Fig. 80.14a–d Resulting Type-2 input membership functions, from top to bottom: (**a**) linear and (**b**) angular velocities and output (**c**) right and (**d**) left torque

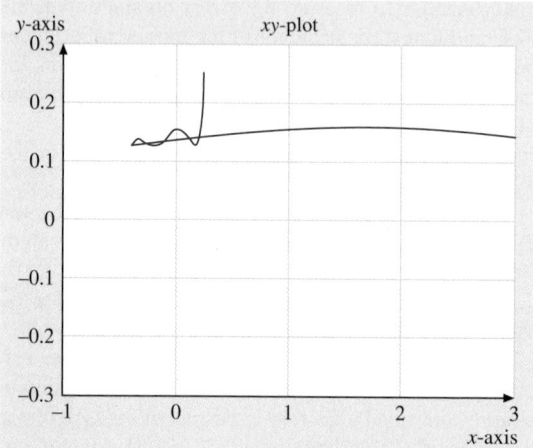

Fig. 80.15 Trajectory obtained for the mobile robot when applying the chemical reaction algorithm to the Type-2 FLC

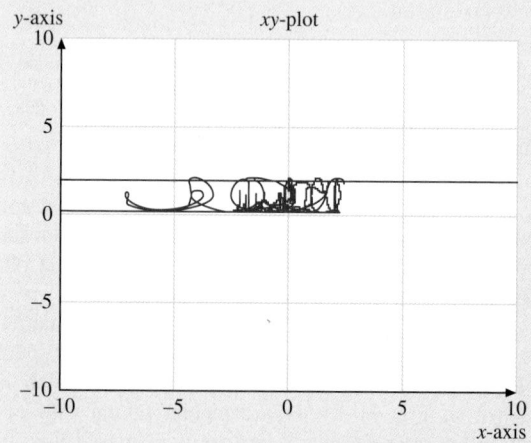

Fig. 80.16 Unacceptable trajectory resulting in early optimization trials

Table 80.8 Simulation parameters and errors obtained under disturbed torques

ε	Velocity errors	Type-1 (GA)	Type-1 (CRA)	Type-2 (CRA)
0.05	Final error	4.0997	0.9815	29.5115
	Average error	4.1209	1.5823	26.6408
5	Final error	4.1059	0.9729	29.52
	Average error	3.1695	1.8679	26.1646
10	Final error	4.1045	0.9745	29.51
	Average error	3.0985	1.7438	24.9467
30	Final error	4.0912	0.9783	29.51
	Average error	2.2632	1.9481	24.6032
32	Final error	3273	0.9748	29.52
	Average error	$3.4667 \times 10^{+003}$	2.8180	24.6465
34	Final error	$1.5705 \times 10^{+004}$	566.8	29.51
	Average error	$1.1180 \times 10^{+004}$	215.8198	24.9211
40	Final error	$2.534 \times 10^{+004}$	$3.5417 \times 10^{+04}$	29.51
	Average error	186.0611	$5.7492 \times 10^{+003}$	23.8938
41	Final error	8839	3168	685.1
	Average error	$2.0268 \times 10^{+004}$	$0.0503 \times 10^{+003}$	16.5257

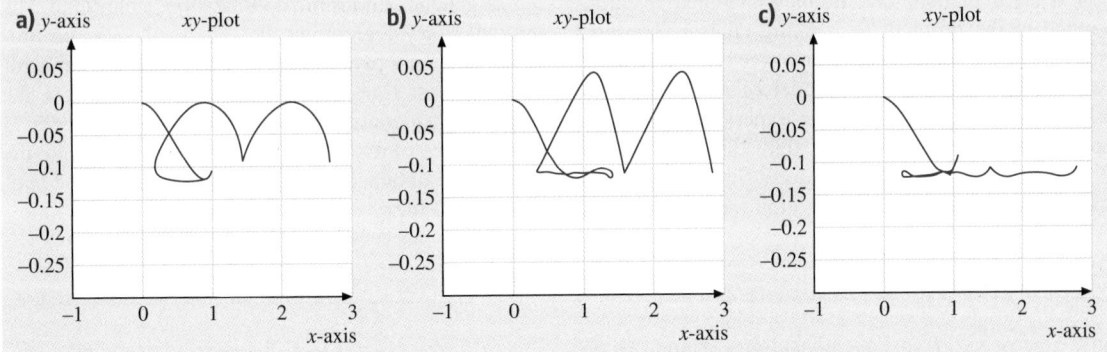

Fig. 80.17a-c From *left to right*: trajectory obtained with the Type-1 FLC optimized with GAs. (**a**) $\varepsilon = 30$, (**b**) $\varepsilon = 32$, (**c**) $\varepsilon = 34$

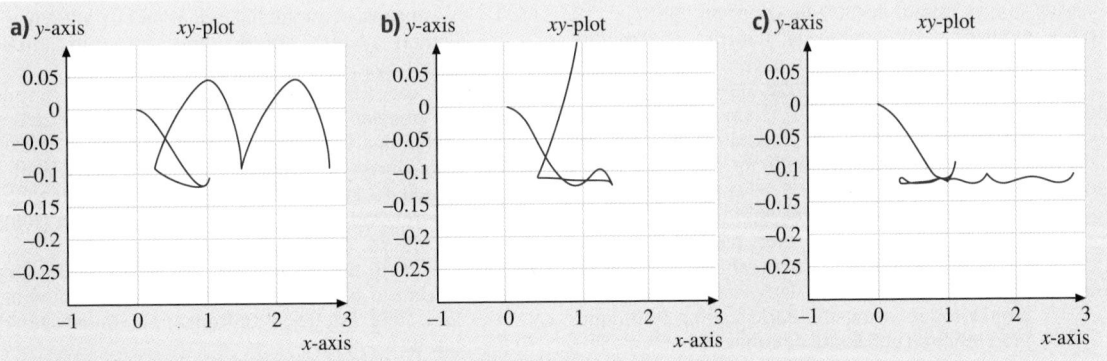

Fig. 80.18a-c From *left to right*: trajectory obtained with the Type-1 FLC optimized with CRA. (**a**) $\varepsilon = 30$, (**b**) $\varepsilon = 32$, (**c**) $\varepsilon = 34$

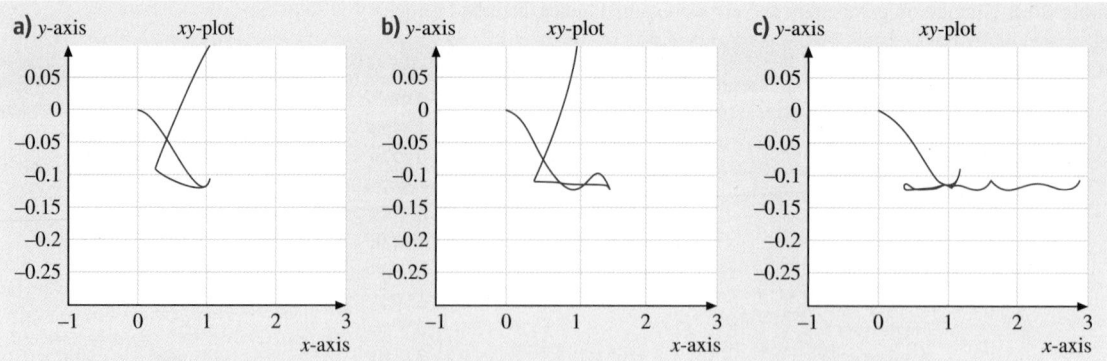

Fig. 80.19a–c From *left to right*: trajectory obtained with the Type-2 FLC optimized with CRA. (**a**) $\varepsilon = 30$, (**b**) $\varepsilon = 32$, (**c**) $\varepsilon = 34$

80.6 Conclusions

In this paper, we presented simulation results from an optimization method that mimics chemical reactions applied to the problem of tracking control. The goal was to find the gain constants involved in the tracking controller for the dynamic model of a unicycle mobile robot. In the figures of the experiments we were able to note the behavior of the algorithm and the solutions found through all the iterations. Simulation results show that the proposed optimization method is able to outperform the results previously obtained by applying a genetic algorithm optimization technique. The optimal fuzzy logic controller obtained with the proposed chemical paradigm is able to reach smaller error values in less time than genetic algorithms. Also, the Type-2 fuzzy controller was able to perform better in the presence of disturbance in this problem despite the *large* error obtained ($e = 2.7736$). The design of optimal Type-2 fuzzy controllers is performed at the time.

References

80.1   N.-Y. Shi, C.-P. Chu: A molecular solution to the hitting-set problem in DNA-based supercomputing, Inf. Sci. **180**, 1010–1019 (2010)

80.2   L. Yamamoto: Evaluation of a catalytic search algorithm, Proc. 4th Int. Workshop Nat. Inspired Coop. Strateg. Optim., NICSO 2010 (2010) pp. 75–87

80.3   T. Meyer, L. Yamamoto, W. Banzhaf, C. Tschudin: Elongation control in an algorithmic chemistry, Lect. Notes Comput. Sci. **5777**, 273–280 (2010)

80.4   J. Xu, A.Y.S. Lam, V.O.K. Li: Chemical reaction optimization for the grid scheduling problem, IEE Commun. Soc., ICC 2010 (2010) pp. 1–5

80.5   Y. Kanayama, Y. Kimura, F. Miyazaki, T. Noguchi: A stable tracking control method for a nonholonomic mobile robot, Proc. IEEE/RSJ Int. Workshop Intell. Robot. Syst., Osaka (1991) pp. 1236–1241

80.6   T.-C. Lee, C.H. Lee, C.-C. Teng: Tracking control of mobile robots using the backsteeping technique, Proc. 5th Int. Conf. Contr. Automat. Robot. Vis., Singapore (1998) pp. 1715–1719

80.7   T.-C. Lee, K. Tai: Tracking control of unicycle-modeled mobile robots using a saturation feedback controller, IEEE Trans. Control Syst. Techn. **9**(2), 305–318 (2001)

80.8   S. Bentalba, A. El Hajjaji, A. Rachid: Fuzzy control of a mobile robot: A new approach, IEEE Int. Conf. Control Appl., Hartford (1997) pp. 69–72

80.9   S. Ishikawa: A method of indoor mobile robot navigation by fuzzy control, Proc. Int. Conf. Intell. Robot. Syst., Osaka (1991) pp. 1013–1018

80.10   T.H. Lee, F.H.F. Leung, P.K.S. Tam: Position control for wheeled mobile robot using a fuzzy controller, 25th Annu. Conf. IEEE, San Jose (1999) pp. 525–528

80.11   S. Pawlowski, P. Dutkiewicz, K. Kozlowski, W. Wroblewski: Fuzzy logic implementation in mobile robot control, 2nd Workshop Robot Motion Control (2001) pp. 65–70

80.12   C.-C. Tsai, H.-H. Lin, C.-C. Lin: Trajectory tracking control of a laser-guided wheeled mobile robot, Proc. IEEE Int. Conf. Control Appl., Taipei (2004) pp. 1055–1059

80.13   S.V. Ulyanov, S. Watanabe, V.S. Ulyanov, K. Yamafuji, L.V. Litvintseva, G.G. Rizzotto: Soft computing for the intelligent robust control of a robotic uni-

80.14 R. Fierro, F.L. Lewis: Control of a nonholonomic mobile robot using neural networks, IEEE Trans. Neural Netw. **9**(4), 589–600 (1998)

80.15 K.T. Song, L.H. Sheen: Heuristic fuzzy-neural Network and its application to reactive navigation of a mobile robot, Fuzzy Sets Syst. **110**(3), 331–340 (2000)

80.16 A.M. Bloch, S. Drakunov: Tracking in nonholonomic dynamic system via sliding modes, Proc. IEEE Conf. Decis. Control, Brighton (1991) pp. 1127–1132

80.17 D. Chwa: Sliding-mode tracking control of nonholonomic wheeled mobile robots in polar coordinates, IEEE Trans. Control Syst. Tech. **12**(4), 633–644 (2004)

80.18 R. Fierro, F.L. Lewis: Control of a nonholonomic mobile robot: Backstepping kinematics into dynamics, Proc. 34th Conf. Decis. Control, New Orleans (1995) pp. 3805–3810

80.19 T. Fukao, H. Nakagawa, N. Adachi: Adaptive tracking control of a non-holonomic mobile robot, IEEE Trans. Robot. Autom. **16**(5), 609–615 (2000)

80.20 A.R. Sahab, M.R. Moddabernia: Backstepping method for a single-link flexible-joint manipulator using genetic algorithm, IJICIC **7**(7B), 4161–4170 (2011)

80.21 J. Yu, Y. Ma, B. Chen, H. Yu, S. Pan: Adaptive neural position tracking control for induction motors via backstepping, IJICIC **7**(7B), 4503–4516 (2011)

80.22 L. Astudillo, O. Castillo, L. Aguilar: Intelligent control for a perturbed autonomous wheeled mobile robot: A type-2 fuzzy logic approach, Nonlinear Stud. **14**(1), 37–48 (2007)

80.23 R. Martinez, O. Castillo, L. Aguilar: Optimization of type-2 fuzzy logic controllers for a perturbed autonomous wheeled mobile robot using genetic algorithms, Inf. Sci. **179**(13), 2158–2174 (2009)

80.24 O. Castillo, R. Martinez-Marroquin, P. Melin, J. Soria: Comparative study of bio-inspired algorithms applied to the optimization of type-1 and type-2 fuzzy controllers for an autonomous mobile robot, Stud. Comput. Intell. **256**, 247–262 (2009)

80.25 I. Kolmanovsky, N.H. McClamroch: Developments in nonholonomic nontrol problems, IEEE Control Syst. Mag. **15**, 20–36 (1995)

80.26 G. Campion, G. Bastin, B. D'Andrea-Novel: Structural properties and classification of kinematic and dynamic models of wheeled mobile robots, IEEE Trans. Robot. Autom. **12**(1), 47–62 (1996)

80.27 W. Nelson, I. Cox: Local path control for an autonomous vehicle, Proc. IEEE Conf. Robotics Autom. (1988) pp. 1504–1510

80.28 S. Oh, H. Jang, W. Pedrycz: A comparative experimental study of type-1/type-2 fuzzy cascade controller based on genetic algorithms and particle swarm optimization, Expert Syst. Appl. **38**(9), 11217–11229 (2011)

81. Bio-Inspired Optimization Methods

Fevrier Valdez

Although graphic processing units (GPUs) have been traditionally used only for computer graphics, a recent technique called general-purpose computing on graphics processing units allows GPUs to perform numerical computations usually handled by the CPU (central processing unit). The advantage of using GPUs for general purpose computation is the performance speedup that can be achieved due to the parallel architecture of these devices. This chapter describes the use of bio-inspired optimization methods as particle swarm optimization and genetic algorithms on GPUs to demonstrate the performance that can be achieved using this technology, primarily with regard to using CPUs.

81.1 Bio-Inspired Methods

In this chapter we describe the optimization of a set of mathematical functions using bio-inspired algorithms. We use genetic algorithms (GAs) and particle swarm optimization (PSO), simulated annealing (SA), and pattern search (PS) to optimize the functions. The main idea is to compare these metaheuristic methods using the CPU and GPUs. Nowadays several approaches have been taken to optimize mathematical functions, see, for example, [81.1–6]. Our approach, however, differs from these approaches because we make a comparison between the advantage of executing the methods on CPUs and GPUs with the aim of achieving the results quickly.

The main contribution of this work is the proposed approach for the implementation of bio-inspired optimization techniques on GPUs for optimization applications. The approach is illustrated with mathematical function optimization, but could be applicable to other problems.

The introduction to the proposed method, is followed by a description of bio-inspired methods in Sect. 81.2. In Sect. 81.3, a brief history of GPUs is presented, in Sect. 81.4 the experimental results are shown, and in Sect. 81.5 the conclusions are presented.

81.2 Bio-Inspired Optimization Methods

To compare the performance on a CPU or a GPU, it is necessary evaluate the methods with optimization problems. Some basic concepts of bio-inspired optimization are needed to understand the differences in the corresponding algorithms. Therefore, in this section we offer a brief description about the bio-inspired optimization methods used in this work. The methods used are described in the following sections.

81.2.1 Genetic Algorithms

Holland, from the University of Michigan initiated his work on genetic algorithms at the beginning of the 1960s. His first achievement was the publication of *Adaptation in Natural and Artificial Systems* [81.7] in 1975.

He had two goals in mind: to improve the understanding of the natural adaptation process and to design artificial systems having properties similar to natural systems [81.8].

The basic idea is as follows: the genetic pool of a given population potentially contains the solution, or a better solution, to a given adaptive problem. This solution is not *active* because the genetic combination on which it relies is split between several subjects. Only the association of different genomes can lead to the solution.

Holland's method is especially effective because it not only considers the role of mutation, but it also uses genetic recombination (crossover) [81.9]. The essence of the GA in both theoretical and practical domains has been well demonstrated [81.1, 10]. The concept of applying a GA to solve engineering problems is feasible and sound. However, despite the distinct advantages of a GA for solving complicated, constrained, and multiobjective functions where other techniques may have failed and the full power of the GA application is yet to be exploited [81.11, 12].

81.2.2 Particle Swarm Optimization

Particle swarm optimization (PSO) is a population-based stochastic optimization technique that was developed by *Eberhart* and *Kennedy* in 1995, inspired by the social behavior of bird flocking or fish schooling [81.3].

PSO shares many similarities with evolutionary computation techniques such as GAs [81.13]. The system is initialized with a population of random solutions and searches for optima by updating generations. However, unlike the GA, the PSO has no evolution operators such as crossover and mutation. In PSO, the potential solutions, called particles, fly through the problem space by following the current optimum particles [81.14].

Each particle keeps track of its coordinates in the problem space, which are associated with the best solution (fitness) it has achieved so far (the fitness value is also stored). This value is called *pbest*. Another *best* value that is tracked by the particle swarm optimizer is the best value obtained so far by any particle in the

neighbors of the particle. This location is called *lbest*. When a particle takes all the population as its topological neighbors, the best value is a global best and is called *gbest*.

The particle swarm optimization concept consists of, at each time step, changing the velocity of (accelerating) each particle toward its *pbest* and *lbest* locations (the local version of PSO). Acceleration is weighted by a random term, with separate random numbers being generated for acceleration toward *pbest* and *lbest* locations.

In the past several years, PSO has been successfully applied in many research and application areas. It is demonstrated that PSO obtains better results in a faster and cheaper way compared with other methods [81.15].

81.2.3 Simulated Annealing

SA is a generic probabilistic metaheuristic for the global optimization problem of applied mathematics, namely locating a good approximation to the global optimum of a given function in a large search space. It is often used when the search space is discrete (e.g., all tours that visit a given set of cities). For certain problems, simulated annealing may be more effective than exhaustive enumeration provided that the goal is merely to find an acceptably good solution in a fixed amount of time, rather than the best possible solution.

The name and inspiration come from annealing in metallurgy, a technique involving heating and controlled cooling of a material to increase the size of its crystals and reduce their defects. The heat causes the atoms to become unstuck from their initial positions (a local minimum of the internal energy) and wander randomly through states of higher energy; the slow cooling gives them more chances of finding configurations with lower internal energy than the initial one. By analogy with this physical process, each step of the SA algorithm replaces the current solution by a random *nearby* solution, chosen with a probability that depends both on the difference between the corresponding function values and also on a global parameter T (called the *temperature*), which is gradually decreased during the process. The dependency is such that the current solution changes almost randomly when T is large, but increasingly *downhill* as T goes to zero [81.16].

81.2.4 Pattern Search

Pattern search is a family of numerical optimization methods that do not require the gradient of the problem to be optimized, and PS can hence be used on functions that are not continuous or differentiable. Such optimization methods are also known as direct-search, derivative-free, or black-box methods.

The name pattern search was coined by *Hooke* and *Jeeves* [81.17]. An early and simple PS variant is attributed to Fermi and Metropolis when they worked at the Los Alamos National Laboratory as described by *Davidon* [81.18], who summarized the algorithm as follows:

They varied one theoretical parameter at a time by steps of the same magnitude, and when no such increase or decrease in any one parameter further improved the fit to the experimental data, they halved the step size and repeated the process until the steps were deemed sufficiently small.

81.3 A Brief History of GPUs

We have already looked at how central processors evolved in both clock speeds and core count. In the meantime, the state of graphics processing underwent a dramatic revolution. In late 1980s and early 1900s, the growth in popularity of graphically driven operating systems such Microsoft Windows helped create a market for a new type of processor. In the early 1990s, users began purchasing 2-D display accelerators for their personal computers. These display accelerators offered hardware-assisted bitmap operations to assist in the display and usability of graphical operating systems [81.19]. From a parallel-computing standpoint, NVIDIA's release of the GeForce 3 series in 2001 represents arguably the most important breakthrough in GPU technology. The GeForce 3 series was the computing industry's first chip to implement Microsoft's then new DirectX 8.0 standard. This standard required that

complaint hardware contain both programmable vertex and programmable pixel shading stages. For the first time, developers had some control over the exact computations that would be performed on their GPUs [81.19].

81.3.1 CUDA

In November 2006, NVIDIA unveiled the industry's first DirectX 10 GPU, the GeForce 8800 GTX. The GeForce 8800 GTX was also the first GPU to be built with NVIDIA's CUDA architecture. This architecture included several new components designed strictly for GPU computing and aimed to alleviate many of the limitations that prevented previous graphics processors from being legitimately useful for general-purpose computation [81.19].

81.4 Experimental Results

This section presents the experimental results obtained with the optimization methods analyzed in this research. The main contribution of this paper is to demonstrate the advantages of using GPUs to calculate complex processes.

To validate the proposed method we used a set of five benchmark mathematical functions; all functions were evaluated with different numbers of dimensions. In this case, the experimental results were obtained with 32 dimensions.

Table 81.1 shows the definitions of the mathematical functions used in this paper. The global minimum for the test functions is 0.

Tables 81.2 and 81.3 show the experimental results for the benchmark mathematical functions used in this research using the CPU and the GPU to process the GA. The table shows the experimental results of the evalua-

tions for each function with 32 dimensions; the best and worst values obtained with an average of 50 times can

Table 81.1 Mathematical functions

Function	Definition
De Jong's	$f_1 = \sum_{n=i}^{N} x_n^2$
Rotated hyper-ellipsoid	$f(x) = \sum_{i=1}^{n} \left(\sum_{j=1}^{i} x_j \right)^2$
Rosenbrock's valley	$f(x) = \sum_{i=1}^{n-1} 100 \left(x_{i+1} - x_i^2 \right)^2 + (1 - x_i)^2$
Rastrigin's	$f(x) = 10n + \sum_{i=1}^{n} \left(x_i^2 - 10\cos(2\pi x_i) \right)$
Griewank's	$f(x) = \sum_{i=1}^{n} \frac{x_i^2}{4000} - \cos\left(\frac{x_i}{\sqrt{i}} \right) + 1$

Part G | 81.4

Table 81.2 Experimental results with 32 dimensions with GA on a CPU

Function	Average	Best	Worst	Time (s)
De Jong's	0.00094	1.14×10^{-6}	0.0056	1.883603
Rotated hyper-ellipsoid	0.05371	0.00228	0.53997	2.015548
Rosenbrock's valley	3.14677173	3.246497	3.86201	3.001564
Rastrigin's	82.35724	46.0085042	129.548	1.452212
Griewank's	0.41019699	0.14192331	0.917367	2.548792

Table 81.3 Experimental results with 32 dimensions with GA on a GPU

Function	Average	Best	Worst	Time (s)
De Jong's	0.000084	1.14×10^{-8}	0.00040	0.360003
Rotated hyper-ellipsoid	0.005371	0.00228	0.53997	0.004590
Rosenbrock's valley	2.325468	1.97548	3.86201	0.005594
Rastrigin's	70.35724	41.54879	130.598	0.502254
Griewank's	0.31019699	0.04192331	0.917367	0.920154

Table 81.4 Experimental results with 32 dimensions with PSO on a CPU

Function	Average	Best	Worst	Time (s)
De Jong's	5.42×10^{-11}	3.40×10^{-12}	9.86×10^{-11}	2.5442154
Rotated hyper-ellipsoid	5.42×10^{-11}	1.93×10^{-12}	9.83×10^{-11}	1.2456487
Rosenbrock's Valley	3.2178138	3.1063	3.39178762	1.3659478
Rastrigin's	34.169712	16.14508	56.714207	3.569871
Griewank's	0.0114768	9.17×10^{-6}	0.09483	5.2654587

Table 81.5 Experimental results with 32 dimensions with PSO on the GPU

Function	Average	Best	Worst	Time (s)
De Jong's	2.20×10^{-11}	2.40×10^{-12}	9.86×10^{-11}	0.05040454
Rotated hyper-ellipsoid	4.20×10^{-11}	2.30×10^{-12}	9.83×10^{-11}	0.02045687
Rosenbrock's Valley	3.1071308	2.16020	3.39178762	0.03659470
Rastrigin's	34.199999	15.14508	53.802564	0.05678710
Griewank's	0.0201564	9.17×10^{-6}	0.094831	0.02654580

Table 81.6 Experimental results with 32 dimensions with SA on a CPU

Function	Average	Best	Worst	Time (s)
De Jong's	0.1210	0.0400	1.8926	3.0124
Rotated hyper-ellipsoid	0.9800	0.0990	7.0104	3.0215
Rosenbrock's Valley	1.2300	0.4402	10.790	2.9999
Rastrigin's	25.8890	20.101	33.415	3.2145
Griewank's	0.9801	0.2045	5.5678	4.0555

be seen after execution of the method. The processing time in seconds is also shown.

Tables 81.4 and 81.5 show the experimental results for the benchmark mathematical functions used in this research using the CPU and the GPU to process the PSO method. Table 81.4 shows the experimental results of the evaluations for each function with 32 dimensions when processing is performed on a CPU; the best and worst values obtained with the average of 50 times after execution of the method can be ob-

served. The processing time in seconds is also shown. Table 81.5 shows similar information, but for the PSO executed on the GPU. It is very easy to appreciate the differences in the results shown in both tables, which show that performance on the GPU is clearly superior.

Tables 81.6 and 81.7 show the experimental results for the benchmark mathematical functions used in this research using the CPU and the GPU to process the SA. The table shows the experimental results

Table 81.7 Experimental results with 32 dimensions with SA on a GPU

Function	Average	Best	Worst	Time (s)
De Jong's	0.10100	0.06012	1.2699	1.000124
Rotated hyper-ellipsoid	0.81200	0.0891	7.1003	1.001015
Rosenbrock's Valley	1.31200	0.40002	10.1290	1.018787
Rastrigin's	25.3256	21.100	32.2315	1.010145
Griewank's	0.99010	0.3050	6.50678	1.000325

Table 81.8 Experimental results with 32 dimensions with PS on the CPU

Function	Average	Best	Worst	Time (s)
De Jong's	0. 3528	0.2232	2.0779	4.2521
Rotated hyper-ellipsoid	16.2505	3.1667	25.782	6.2154
Rosenbrock's Valley	4.0568	3.0342	5.7765	5.2565
Rastrigin's	31.4203	25.7660	33.9866	3.25654
Griewank's	0.6897	0.0981	3.5061	2.1548

Table 81.9 Experimental results with 32 dimensions with PS on GPU

Function	Average	Best	Worst	Time (s)
De Jong's	0. 5208	0.1232	2.1579	1.1021
Rotated hyper-ellipsoid	16.5005	3.6197	25.182	2.1154
Rosenbrock's Valley	4.0588	3.00215	4.2565	2.5105
Rastrigin's	31.5203	25.4530	33.9866	1.6054
Griewank's	0.14970	0.00980	3.5061	1.4858

of the evaluations for each function with 32 dimensions; the best and worst values obtained with the average of 50 times after execution of the method can be seen. The processing time in seconds is also shown.

Tables 81.8 and 81.9 show the experimental results for the benchmark mathematical functions used in this research using the CPU and GPU to process the PS. The table shows the experimental results of the evaluations for each function with 32 dimensions; the best and worst values obtained with the average of 50 times after execution of the method can be seen. The processing time in seconds is also shown.

Figure 81.1 shows the comparison results between the processing time on the GPU and the CPU. The difference in time of each best time obtained in the experiments discussed in the paper is shown. The blue line represents the processing time in the CPU and the brown line represents the processing time in the GPU. Is

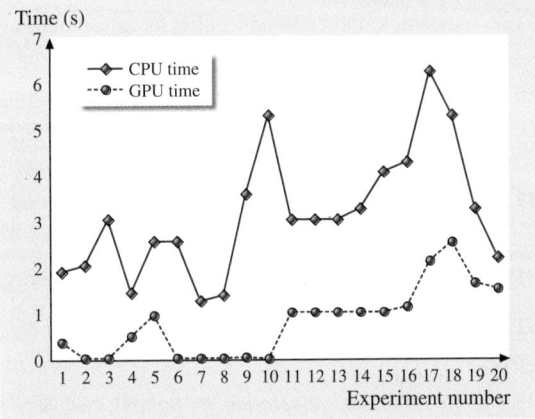

Fig. 81.1 Comparison of results between GPU and CPU

clear how the best time achieved is when the algorithms were executed on the GPU.

81.5 Conclusions

The analysis of the experimental results of the bio-inspired methods considered in this paper, the FPSO+FGA (FPSO: fuzzy particle swarm optimization; FGA: fuzzy generic algorithm), lead us to the conclusion that for the optimization of these benchmark mathematical functions execution on the GPU is a good alternative, because it is easier and very fast to optimize and achieve good results than to try it with PSO, GA, SA, and genetic pattern search (GPS) on the CPU [81.14], especially when the number of dimensions is increased. This is because processing on GPUs is faster than processing on CPUs. Also, the experimental results obtained with the use of GPUs in this research were compared with another similar approach [81.20, 21] and achieved good results quickly.

References

81.1   K.F. Man, K.S. Tang, S. Kwong: *Genetic Algorithms: Concepts and Designs* (Springer, Berlin, Heidelberg 1999)

81.2   R.C. Eberhart, J. Kennedy: A new optimizer using particle swarm theory, Proc. 6th Int. Symp. Micromach. Hum. Sci., Nagoya (1995) pp. 39–43

81.3   J. Kennedy, R.C. Eberhart: Particle swarm optimization, Proc. IEEE Int. Conf. Neural Netw., Piscataway (1995) pp. 1942–1948

81.4   O. Montiel, O. Castillo, P. Melin, A. Rodriguez, R. Sepulveda: Human evolutionary model: A new approach to optimization, Inf. Sci. **177**(10), 2075–2098 (2007)

81.5   D. Kim, K. Hirota: Vector control for loss minimization of induction motor using GA–PSO, Appl. Soft Comput. **8**, 1692–1702 (2008)

81.6   H. Liu, A. Abraham, A.E. Hassanien: Scheduling jobs on computational grids using a fuzzy particle swarm optimization algorithm, Future Gener. Comput. Syst. **26**(8), 1336–1343 (2010)

81.7   D.B. Fogel: An introduction to simulated evolutionary optimization, IEEE Trans. Neural Netw. **5**(1), 3–14 (1994)

81.8   D. Goldberg: *Genetic Algorithms* (Addison Wesley, Boston 1988)

81.9   C. Emmeche: *Garden in the Machine. The Emerging Science of Artificial Life* (Princeton Univ. Press, Princeton 1994) p. 114

81.10   J.H. Holland: *Adaptation in Natural and Artificial System* (Univ. of Michigan Press, Ann Arbor 1975)

81.11   T. Back, D.B. Fogel, Z. Michalewicz (Eds.): *Handbook of Evolutionary Computation* (Oxford Univ. Press, Oxford 1997)

81.12   O. Castillo, F. Valdez, P. Melin: Hierarchical Genetic Algorithms for topology optimization in fuzzy control systems, Int. J. Gen. Syst. **36**(5), 575–591 (2007)

81.13   O. Castillo, P. Melin: Hybrid intelligent systems for time series prediction using neural networks, fuzzy logic, and fractal theory, IEEE Trans. Neural Netw. **13**(6), 1395–1408 (2002)

81.14   F. Valdez, P. Melin: Parallel evolutionary computing using a cluster for mathematical function optimization, Fuzzy Information Processing Society (NAFIPS '07), San Diego (2007) pp. 598–602

81.15   P.J. Angeline: Using selection to improve particle swarm optimization, Proc. 1998 IEEE World Congr. Comput. Intell., Anchorage (1998) pp. 84–89

81.16   S. Kirkpatrick, C.J. Gelatt, M. Vecchi: Optimization by simulated annealing, Science **220**(4598), 671–680 (1983)

81.17   R. Hooke, T.A. Jeeves: *Direct search* solution of numerical and statistical problems, J. Assoc. Comput. Mach. (ACM) **8**(2), 212–229 (1961)

81.18   W.C. Davidon: Variable metric method for minimization, SIAM J. Optim. **1**(1), 1–17 (1991)

81.19   J. Sanders, E. Kandrot: *CUDA by Example: An Introduction to General-Purpose GPU Programming* (Addison Wesley, Boston 2011)

81.20   M.O. Ali, S.P. Koh, K.H. Chong, A.S. Hamoodi: Design a PID controller of BLDC motor by using hybrid genetic-immune, Mod. Appl. Sci. **5**(1), 75–85 (2011)

81.21   F. Valdez, P. Melin, O. Castillo: An improved evolutionary method with fuzzy logic for combining particle swarm optimization and genetic algorithms, Appl. Soft Comput. **11**(2), 2625–2632 (2011)

Acknowledgements

A.4 Aggregation Functions on [0,1]
by Radko Mesiar, Anna Kolesárová, Magda Komorníková

This work was supported by grants APVV–0073–10, VEGA 1/0143/11, and VEGA 1/0419/13.

A.5 Monotone Measures–Based Integrals
by Erich Klement, Radko Mesiar

This work was supported by grants APVV-0073-10, VEGA 1/0171/12, and GAČR P–402/11/0378.

A.6 The Origin of Fuzzy Extensions
by Humberto Bustince, Edurne Barrenechea, Javier Fernández, Miguel Pagola, Javier Montero

The work has been supported by projects TIN2013-40765-P and TIN2012-32482 of the Spanish Ministry of Science.

A.7 F–Transform
by Irina Perfilieva

This work relates to Department of the Navy Grant N62909-12-1-7039 issued by Office of Naval Research Global. The United States Government has a royalty-free license throughout the world in all copyrightable material contained herein. Additional support was also given by the European Regional Development Fund in the IT4Innovations Centre of Excellence project (CZ.1.05/1.1.00/02.0070).

A.8 Fuzzy Linear Programming and Duality
by Jaroslav Ramík, Milan Vlach

This work was supported by the European Regional Development Fund in the IT4Innovations Centre of Excellence project (CZ.1.05/1.1.00/02.0070).

A.9 Basic Solutions of Fuzzy Coalitional Games
by Tomáš Kroupa, Milan Vlach

The work of Tomáš Kroupa was supported by the grant P402/12/1309 of Czech Science Foundation. The work of Milan Vlach was supported by the Czech Science Foundation project No. P402/12/G097 DYME Dynamic Models in Economics.

B.10 Basics of Fuzzy Sets
by János Fodor, Imre Rudas

The authors have been supported in part by the Hungarian Scientific Research Fund OTKA under contract K-106392.

B.12 Fuzzy Implications: Past, Present, and Future
by Michał Baczynski, Balasubramaniam Jayaram, Sebastia Massanet, Joan Torrens

S. Massanet and J. Torrens acknowledge the support by the Spanish Grants MTM2009-10320 and TIN2013-42795-P, both with FEDER support. B. Jayaram would like to acknowledge the partial support given by the Department of Science and Technology, INDIA under the project SERB/F/2862/2011-12.

B.16 An Algebraic Model of Reasoning to Support Zadeh's CWW
by Enric Trillas

This chapter is partially supported by the Foundation for the Advancement of Soft Computing (Mieres, Asturias, Spain), and by MICINN/Government of Spain, under project TIN 2011-29827-C02-01. The author is in debt with Professors Claudio Moraga (Mieres), and Settimo Termini (Palermo), as well as with Dr. María G. Navarro (Madrid), for their comments and advises on the contents of this paper.

B.20 Application of Fuzzy Techniques to Autonomous Robots
by Ismael Rodríguez Fdez, Manuel Mucientes, Alberto Bugarín Diz

This work was supported by the Spanish Ministry of Economy and Competitiveness under grants TIN2011-22935 and TIN2011-29827-C02-02, and the Galician Ministry of Education under grant EM2014/012. I. Rodriguez-Fdez is supported by the Spanish Ministry of Education under the FPU national plan (AP2010-0627). This work was also partially supported by the European Regional Development Fund (ERDF/FEDER) under projects CN2012/151 and GRC2014/030 of the Galician Ministry of Education.

C.21 Foundations of Rough Sets
by Andrzej Skowron, Andrzej Jankowski, Roman Swiniarski

This work was supported by the Polish National Science Center, grants DEC-2011/01/B/ST6/03867, DEC-2011/01/D /ST6/06981, DEC-2012/05/B/ST6/03215, and DEC-2013/09/B/ST6/01568, as well as by the Polish National Center for Research and Development (NCBiR) under the grant SYNAT No. SP/I/1/77065/10 in the frame of the strategic scientific research and experimental development program: *Interdisciplinary System for Interactive Scientific and Scientific-Technical Information* and the grant No. O ROB/0010/03/001 in the frame of the Defence and Security Programmes and Projects: *Modern Engineering Tools for Decision Support for Commanders of the State Fire Service of Poland during Fire and Rescue Operations in Buildings.*

D.32 Kernel Methods
by Marco Signoretto, Johan Suykens

- EU: The research leading to these results has received funding from the European Research Council under the European Union's Seventh Framework Programme (FP7/2007-2013) / ERC AdG A-DATADRIVE-B (290923).
- Research Council KUL: GOA/10/09 MaNet, CoE PFV/10/002 (OPTEC), BIL12/11T; PhD/Postdoc grants
- Flemish Government:
  - FWO: projects: G.0377.12 (Structured systems), G.088114N (Tensor based data similarity); PhD/Postdoc grants
  - IWT: projects: SBO POM (100031); PhD/Postdoc grants
  - iMinds Medical Information Technologies SBO 2014
- Belgian Federal Science Policy Office: IUAP P7/19 (DYSCO, Dynamical systems, control and optimization, 2012-2017)

D.36 Cognitive Architectures and Agents
by Sebastien Hélie, Ron Sun

This research was supported by the ONR grant N00014-08-1-0068 to the second author. Requests for reprints should be addressed to Sébastien Hélie, Department of Psychological Sciences, Purdue University, West Lafayette, IN 47907-2081, e-mail: shelie@purdue.edu; or to Ron Sun, Department of Cognitive Science, Rensselaer Polytechnic Institute, email: rsun@rpi.edu.

D.40 Evolving Connectionist Systems: From Neuro-Fuzzy-, to Spiking- and Neuro-Genetic
by Nikola Kasabov

The work presented in this chapter is supported by the Knowledge Engineering and Discovery Research Institute (KEDRI, http://www.kedri.info) of the Auckland University of technology, New Zealand.

E.43 Genetic Programming
by James McDermott, Una-May O'Reilly

JMcD was funded for this research by the Irish Research Council for Science, Engineering, and Technology, co-funded by Marie Curie. U-MO'R acknowledges the support of the Li Ka Shing Foundation, General Electric, and the US Department of Energy. Thanks to Erik Hemberg for reading early drafts.

E.45 Estimation of Distribution Algorithms
by Martin Pelikan, Mark Hauschild, Fernando Lobo

This project was sponsored by the National Science Foundation under grants ECS-0547013 and IIS-1115352, by the University of Missouri in St. Louis through the High Performance Computing Collaboratory sponsored by Information Technology Services, and by the University of Missouri Bioinformatics Consortium (UMBC). Any opinions, findings, and conclusions or recommendations expressed in this material are those of the author(s) and do not necessarily reflect the views of the National Science Foundation.

E.46 Parallel Evolutionary Algorithms
by Dirk Sudholt

Part of this work was done while the author was a member of CERCIA, University of Birmingham, supported by EPSRC grant EP/D052785/1. The research leading to these results has received funding from the European Union Seventh Framework Programme (FP7/2007–2013) under grant agreement No. 618091 (SAGE).

E.49 Multi-Objective Evolutionary Algorithms
by Kalyanmoy Deb

This chapter is an updated version of a recent article by the author: K. Deb: Recent developments in evolutionary multi-objective optimization. In: *Trends in Multiple Criteria Decision Analysis Trends in Multiple Criteria Decision Analysis*, International Series in Operations Research & Management Science, Vol. 142, ed. by S.

Greco, M. Ehrgott and J. R. Figueira (Springer, New York 2010), pp. 339–368.

E.50 Parallel Multiobjective Evolutionary Algorithms
by Francisco Luna, Enrique Alba

This work has been partially funded by the Spanish Ministry of Science and Innovation and FEDER under contracts TIN2008-06491-C04-01 (the MSTAR project) and TIN2011-28194 (the roadME project), and by the Andalusian Government under contract P07-TIC-03044 (the DIRICOM project).

E.51 Many-Objective Problems: Challenges and Methods
by Antonio López Jaimes, Carlos Coello Coello

The second author acknowledges support from CONA-CyT project no. 103570.

E.52 Memetic and Hybrid Evolutionary Algorithms
by Jhon Amaya, Carlos Cotta Porras, Antonio Fernández Leiva

This work was partially supported by the Spanish MICINN under project ANYSELF (TIN2011-28627-C04-01) and by Junta de Andalucia under project TIC-6083.

E.54 Stochastic Local Search Algorithms: An Overview
by Holger Hoos, Thomas Stützle

This work has been supported by the Meta-X, an ARC project funded from the Scientific Research Directorate of the French Community of Belgium. Holger H. Hoos acknowledges support provided by a Discovery Grant from the Natural Sciences and Engineering Research Council of Canada (NSERC); Thomas Sttzle acknowledges support of the Belgian FNRS, of which he is a research associate.

E.56 How to Create Generalizable Results
by Thomas Bartz-Beielstein

This work was kindly supported by the Federal Ministry of Education and Research (BMBF) under the grants MCIOP (FKZ 17N0311) and CIMO (FKZ 17002X11).

In addition, the paper and the corresponding R code are based on *Chiarandini* and *Goegebeur*'s publication *Mixed models for the analysis of optimization algorithms* [81.3].

E.59 Modeling and Optimization of Machining Problems
by Dirk Biermann, Petra Kersting, Tobias Wagner, Andreas Zabel

This work was supported by the Deutsche Forschungsgemeinschaft (DFG) by founding the Collaborative Research Center Computational Intelligence (SFB 531).

E.64 Metaheuristic Algorithms and Tree Decomposition
by Thomas Hammerl, Nysret Musliu, Werner Schafhauser

The work was supported by the Austrian Science Fund (FWF): P20704-N18 and P24814-N23. Moreover, the research herein is partially conducted within the competence network Softnet Austria II (www.soft-net.at, COMET K-Projekt) and funded by the Austrian Federal Ministry of Economy, Family and Youth (bmwfj), the province of Styria, the Steirische Wirtschaftsfrderungsgesellschaft mbH. (SFG), and the city of Vienna in terms of the center for innovation and technology (ZIT).

F.66 Swarm Intelligence in Optimization and Robotics
by Christian Blum, Roderich Groß

This work was supported by grant TIN2012-37930-C02-02 of the Spanish Government. In addition, C. Blum acknowledges support from IKERBASQUE, the Basque Foundation for Science. R. Gro acknowledges support from the Engineering and Physical Sciences Research Council (EPSRC, grant no. EP/K033948/1).

F.68 Ant Colony Optimization for the Minimum-Weight Rooted Arborescence Problem
by Christian Blum, Sergi Mateo Bellido

This work was supported by grant TIN2007-66523 (FORMALISM) of the Spanish government.

F.71 Fundamental Collective Behaviors in Swarm Robotics
by Vito Trianni, Alexandre Campo

Vito Trianni acknowledges support from the H^2Swarm project, founded within the EUROCORES Programme *EuroBioSAS* of the European Science Foundation. The project is partially supported by funds from the Italian CNR and the Belgian F.R.S.-FNRS. Alexandre Campo acknowledges support from the CoCoRo project, funded by the Information and Communication Technologies programme (call FP7-ICT-2009-6) of the European Commission under grant number 270382.

About the Authors

Enrique Alba

Universidad de Malaga
E.T.S.I. Informática
Málaga, Spain
eat@lcc.uma.es

Chapter E.50

Enrique Alba is a Full Professor at the University of of Málaga (Spain) where received his degree in Engineering in 1992 and his PhD in Computer Science in 1999. He is an invited professor at INRIA, the University of Luxembourg, and the University of Ostrava. He has published 80 articles in ISI-ranked journals, 40 papers in LNCS, over 250 refereed conference papers, 11 books, and 39 book chapters.

Jose M. Alonso

European Centre for Soft Computing
Cognitive Computing
Mieres, Spain
jose.alonso@softcomputing.es

Chapter B.14

Jose Alonso received his MSc (2003) and PhD (2007) degrees in Telecommunication Engineering from the Technical University of Madrid, Spain. He has published more than 60 papers in international journals and as book chapters. His main research lines are soft computing, fuzzy logic, computing with perceptions, fuzzy modeling, interpretable fuzzy systems, knowledge extraction and representation, and development of free software tools.

Jhon Edgar Amaya

Universidad Nacional Experimental del Táchira
Dep. Electronic Engineering
San Cristóbal, Venezuela
jedgar@unet.edu.ve

Chapter E.52

Jhon Edgar Amaya received his degree in Engineering in 1997 from UNET and an MSc degree in Computation in 2003 from ULA. He obtained his PhD degree from the University of Malaga in 2011. He is an Associate Professor in the Department of Electronic Engineering at UNET. His research interests include evolutionary computation and soft computing applications in microelectronic devices.

Plamen P. Angelov

Lancaster University
School of Computing and Communications
Bailrigg, Lancaster, UK
p.angelov@lancaster.ac.uk

Chapter G.75

Plamen P. Angelov leads the Data Science Group at the School of Computing and Communications, Lancaster University, UK. He has authored or co-authored over 200 peer-reviewed publications in leading journals, five patents, two research monographs, and several edited books. His interests are computational intelligence and autonomous system modeling, identification, and machine learning.

Dirk V. Arnold

Dalhousie University
Faculty of Computer Science
Halifax, Nova Scotia, Canada
dirk@cs.dal.ca

Chapter E.44

Dirk Arnold is a Professor in the Faculty of Computer Science at Dalhousie University. His research interests include evolutionary computation, image processing, and computer graphics and animation.

Anne Auger

University Paris–Sud Orsay
CR Inria
Orsay Cedex, France
anne.auger@inria.fr

Chapter E.44

Anne Auger is a permanent researcher at the French National Institute for Research in Computer Science and Control (INRIA). She received her diploma (2001) and PhD (2004) in Mathematics from the Paris VI University. Prior to joining INRIA, she worked for 2 years (2004–2006) at ETH in Zurich. Her main research interest is stochastic continuous optimization, including theoretical aspects and algorithm designs.

Authors

Davide Bacciu

Università di Pisa
Dip. Informatica
Pisa, Italy
bacciu@di.unipi.it

Chapter D.31

Davide Bacciu received a Laurea degree in Computer Science from the University of Pisa in 2003 and a PhD in Computer Science and Engineering from IMT Lucca in 2008. Currently, he is with the CI and Machine Learning Group, the University of Pisa. His research interests include machine learning, graphical models, neural networks, learning in structured domains, and machine vision.

Michał Baczynski

University of Silesia
Inst. Mathematics
Katowice, Poland
michal.baczynski@us.edu.pl

Chapter B.12

Michał Baczynski received his MSc and PhD degrees in Mathematics from the University of Silesia, Poland in 1995 and 2000, respectively. He received the *Habilitation* degree from the Polish Academy of Sciences in 2010. His current research interests include fuzzy aggregation operations, chiefly fuzzy implications, approximate reasoning, fuzzy systems, and functional equations.

Edurne Barrenechea

Universidad Pública de Navarra
Dep. Automática y Computación
Pamplona (Navarra), Spain
edurne.barrenechea@unavarra.es

Chapter A.6

Edurne Barrenechea received the PhD degree in Computer Science from the Public University of Navarra in 2005 where she is an Associate Professor in the Department of Automatics and Computation. Her publications comprise more than 30 papers and about 20 book chapters. Her research interests include fuzzy techniques for image processing, and medical and industrial applications of soft computing techniques.

Thomas Bartz-Beielstein

Chapter E.56   For biographical profile, please see the section "About the Part Editors".

Lubica Benuskova

University of Otago
Dep. Computer Science
Dunedin, New Zealand
lubica@cs.otago.ac.nz

Chapter D.27

Lubica Benuskova is an Associate Professor at the Department of Computer Science at the University of Otago, Dunedin, New Zealand. She is also a Professor at Comenius University, Bratislava, Slovakia. Her research activities are mainly in the area of computational neuroscience. Currently she serves as a member of the Neural Networks Technical Committee of the of the IEEE/CIS.

Dirk Biermann

TU Dortmund University
Dep. Mechanical Engineering
Dortmund, Germany
biermann@isf.de

Chapter E.59

Dirk Biermann studied Mechanical Engineering and obtained his doctoral degree from the University of Dortmund (now TU Dortmund University). He has been Head of the Institute of Machining Technology (ISF) since 2007. He is an associate member of the International Academy of Production Engineering (CIRP) and of the German Academic Society for Production Engineering (WGP).

Sašo Blažič

University of Ljubljana
Faculty of Electrical Engineering
Ljubljana, Slovenia
saso.blazic@fe.uni-lj.si

Chapter G.75

Sašo Blažič received his BSc, MSc and PhD degrees in 1996, 1999, and 2002, respectively, from the Faculty of Electrical Engineering, University of Ljubljana, where he currently holds the position of Full Professor. His research interests include adaptive, fuzzy, and predictive control. Recently, he has focused on autonomous mobile systems, mobile robotics, and satellite systems.

Christian Blum

Chapters F.66, F.68   For biographical profile, please see the section "About the Part Editors".

Andrea Bobbio

Università del Piemonte Orientale
DiSit – Computer Science Section
Alessandria, Italy
andrea.bobbio@unipmn.it

Chapter F.69

Andrea Bobbio is Professor of Computer Science at the Università del Piemonte Orientale, Alessandria, Italy. He graduated from Politecnico di Torino in 1969, and in 1971 he joined the Istituto Elettrotecnico Nazionale Galileo Ferraris di Torino His activity focuses on the modeling and analysis of the performance and reliability of stochastic systems.

Josh Bongard

University of Vermont
Dep. Computer Science
Burlington, USA
josh.bongard@uvm.edu

Chapter D.37

Josh Bongard is an Associate Professor at the University of Vermont. He is the Director of the Morphology, Evolution, and Cognition Laboratory there, as well as the Vice Chair of the UVM Complex Systems Center. He is the recipient of an NSF PECASE award and has served as a Microsoft Research New Faculty Fellow.

Piero P. Bonissone

Piero P. Bonissone Analytics, LLC
San Diego, USA
bonissone@gmail.com

Chapter D.41

Piero Bonissone is a Coolidge Fellow and a Fellow of AAAI, IEEE, and IFSA. He has published over 150 articles and has received 65 US patents. Recently, he was bestowed the 2012 Fuzzy Systems Pioneer Award from the IEEE Computational Intelligence Society.

Dario Bruneo

Universita' di Messina
Dip. Ingegneria Civile, Informatica
Messina, Italy
dbruneo@unime.it

Chapter F.69

Dario Bruneo is Assistant Professor in the Department of Engineering at the University of Messina. His research activity focuses on the study of distributed systems in particular on the management of advanced service provisioning, system modeling, and performance evaluation. His current research topics include sensor networks, Internet of Things, monitoring and performance and reliability of complex systems.

Alberto Bugarín Diz

University of Santiago de Compostela
Research Centre for Information
Technologies
Santiago de Compostela, Spain
alberto.bugarin.diz@usc.es

Chapter B.20

Alberto Bugarín is a full Professor of Artificial Intelligence at the Research Centre for Information Technologies of the University of Santiago de Compostela (CiTIUS). His research interests focus on modeling intelligent systems with uncertainty (fuzzy rule-based systems) and their applications in adaptive business intelligence (planning/scheduling), automatic building of linguistic descriptions of data, and mobile robotics.

Humberto Bustince

Universidad Pública de Navarra
Dep. Automática y Computación
Pamplona (Navarra), Spain
bustince@unavarra.es

Chapter A.6

Humberto Bustince received his BSc degree in Physics from Salamanca University, Spain, and his PhD degree in Mathematics from the Public University of Navarra. He is a full Professor in the Department of Automatics and Computation at the Public University of Navarra. He has authored more than 120 journal papers and more than 100 contributions to international conferences. He has also co-authored four books.

Martin V. Butz

University of Tübingen
Computer Science, Cognitive Modeling
Tübingen, Germany
martin.butz@uni-tuebingen.de

Chapter E.47

Martin V. Butz is a full Professor of Cognitive Modeling at the University of Tübingen, Department of Computer Science and Department of Psychology. His research group focuses on how the brain develops anticipatory representations of the body, the surrounding space for controlling predictive interactions with the environment maximally effectively.

Authors

Alexandre Campo   Chapter F.71

Université Libre de Bruxelles
Unit of Social Ecology
Brussels, Belgium
alexandre.campo@ulb.ac.be

Alexandre Campo is a Postdoctoral Fellow at the Unit of Social Ecology at the Université Libre de Bruxelles. He received his PhD there in Applied Sciences in 2011. His research interests include the study and design of complex systems applied to swarm robotics. He has participated in several projects funded by the European Commission.

Angelo Cangelosi   Chapter D.37

Plymouth University
Centre for Robotics and Neural Systems
Plymouth, UK
A.Cangelosi@plymouth.ac.uk

Angelo Cangelosi is Professor of Artificial Intelligence and Cognition and directs the Centre for Robotics and Neural Systems at Plymouth University, UK. His main expertise is on language and cognitive modeling in humanoid robots. He has coordinated UK and FP7 projects (RobotDoC ITN, ITALK, BABEL), and is the 2012 Chair of the IEEE Autonomous Mental Development Technical Committee.

Robert Carrese   Chapter F.67

LEAP Australia Pty. Ltd.
Clayton North, Australia
robert.carrese@leapaust.com.au

Robert Carrese graduated as an Aerospace Engineer and then received his PhD degree in Aerodynamic Design and Optimization from RMIT University. Since 2012, he has been a Senior Engineer at LEAP Australia Pty. Ltd. His research interests include the development of evolutionary design processes for key aerodynamic deliverables and mitigation of adverse aeroelastic and aeroacoustic effects.

Ciro Castiello   Chapter B.14

University of Bari
Dep. Informatics
Bari, Italy
ciro.castiello@uniba.it

Ciro Castiello graduated in Informatics in 2001 and received his PhD in Informatics in 2005. He is currently a researcher in the Informatics Department at the University of Bari, Italy. His research interests include soft computing techniques, inductive learning mechanisms, and interpretability of fuzzy systems. He has published more than 50 peer-reviewed papers.

Oscar Castillo   Chapters G.78, G.80   For biographical profile, please see the section "About the Part Editors".

Davide Cerotti   Chapter F.69

Politecnico di Milano
Dip. Elettronica, Informazione e
Bioingegneria
Milano, Italy
davide.cerotti@polimi.it

Davide Cerotti obtained his degree in Computer Science from the University of Piemonte Orientale, Italy and his PhD in Computer Science in 2010 from the University of Turin, Italy. Currently he is a postdoctoral scholar at the Politecnico di Milano, Italy. His main research topics are Markovian agents, queueing networks, and performance evaluation of large-scale distributed systems.

Badong Chen   Chapter D.30

Xi'an Jiaotong University
Inst. Artificial Intelligence and Robotics
Xi'an, China
chenbd@mail.xjtu.edu.cn

Badong Chen received his PhD degree in Computer Science and Technology from Tsinghua University, China, in 2008. He was a Postdoctoral Associate at the University of Florida from 2010 to 2012. He is currently a Professor at Xi'an Jiaotong University, China. His research interests are in signal processing and machine learning, and their applications in cognition and neuroscience.

Ke Chen   Chapter D.28

The University of Manchester
School of Computer Science
Manchester, UK
chen@cs.manchester.ac.uk

Ke Chen has been an academic staff member at The University of Manchester since 2003. He has worked at Birmingham University, Peking University, Ohio State University, Kyushu Institute of Technology, and Tsinghua University. His main research interests lie in neural computation with an emphasis on deep and modular neural networks, machine learning, machine perception and their applications in intelligent systems.

Davide Ciucci   Chapter C.25

University of Milano-Bicocca
Dep. Informatics, Systems and
Communications
Milano, Italy
ciucci@disco.unimib.it

Davide Ciucci received a PhD in 2004 in Computer Science from the University of Milan and the Habilitation (HdR) from the University of Toulouse III in 2013. Since 2005, he has held a permanent research position at the University of Milano-Bicocca. His research activity concerns uncertainty management, with particular reference to rough sets, many-valued logics, and ontologies.

Carlos A. Coello Coello   Chapter E.51

For biographical profile, please see the section "About the Part Editors".

Chris Cornelis   Chapter C.26

Ghent University
Dep. Applied Mathematics and Computer
Science
Ghent, Belgium
chriscornelis@ugr.es

Chris Cornelis has an MSc and a PhD degree in Computer Science from Ghent University (Belgium). Currently, he is a Postdoctoral Fellow at the University of Granada, supported by the Ramón y Cajal program, as well as a Guest Professor at Ghent University. His current research interests include fuzzy sets, rough sets, and machine learning.

Nikolaus Correll   Chapter F.74

University of Colorado at Boulder
Dep. Computer Science
Boulder, USA
ncorrell@colorado.edu

Nikolaus Correll has been an Assistant Professor of Computer Science at the University of Colorado at Boulder since 2009. He obtained a PhD from EPFL in 2007 and did postdoctoral studies at MIT CSAIL. His research interests are modeling and design of large-scale distributed swarming systems and smart materials.

Carlos Cotta Porras   Chapter E.52

Universidad de Málaga
Dep. Lenguajes y Ciencias de la
Computación
Málaga, Spain
ccottap@lcc.uma.es

Carlos Cotta received his MSc and PhD degrees in Computer Science from the University of Málaga in 1994 and 1998, respectively. He has held a tenured professorship at this university since 2001. His main research areas involve metaheuristic optimization – in particular hybrid and memetic approaches – with a focus on both algorithmic and applied aspects (particularly combinatorial optimization) and complex systems.

Damien Coyle   Chapter D.39

University of Ulster
Intelligent Systems Research Centre
Derry, Northern Ireland, UK
dh.coyle@ulster.ac.uk

Damien Coyle is a Senior Lecturer at the School of Computing and Intelligent Systems, University of Ulster. His research interests include brain–computer interfaces, computational intelligence and neuroscience, neuroimaging, and biosignal processing. He is a recipient of the IEEE Computational Intelligence Society's Outstanding Doctoral Dissertation Award and the International Neural Network Society's Young Investigator of the Year Award.

Guy De Tré

Chapter B.19

Ghent University
Dep. Telecommunications and
Information Processing
Ghent, Belgium
guy.detre@ugent.be

Guy De Tré received his MSc in Computer Science in 1994 and his PhD in Engineering in 2000 from Ghent University (Belgium). He is Associate Professor in Fuzzy Information Processing in the Department of Telecommunications and Information Processing at Ghent University. His research is centred on soft computing techniques for database modeling, flexible querying, and decision support.

Kalyanmoy Deb

Chapter E.49

Michigan State University
Dep. Electrical and Computer Engineering
East Lansing, USA
kdeb@egr.msu.edu

Kalyanmoy Deb is a Koenig Endowed Chair Professor at Michigan State University. His research interests are in evolutionary optimization and their application in optimization, modeling, and machine learning. He has published 375 research papers with h-index 85. He is on the Editorial Board of 18 major international journals.

Clarisse Dhaenens

Chapter E.61

University of Lille
CRIStAL laboratory
Villeneuve d'Ascq Cedex, France
clarisse.dhaenens@univ-lille1.fr

Clarisse Dhaenens is Professor at the University of Lille. She obtained her PhD in 1998 from the Polytechnicum University of Grenoble (INPG). She became Associate Professor in 1999 at the University of Lille and a full Professor in 2006. Her work deals with operations research, combinatorial optimization with applications in knowledge discovery for bioinformatics and healthcare.

Luca Di Gaspero

Chapter E.62

Università degli Studi di Udine
Dip. Ingegneria Elettrica, Gestionale e
Meccanica
Udine, Italy
luca.digaspero@uniud.it

Luca Di Gaspero graduated and received a PhD in Computer Science from the University of Udine, where he is currently Senior Lecturer of Information Technology. In 2011 he was Visiting Professor at Vienna University of Technology, where he also received his Habilitation in 2014. His research focus is boosting metaheuristic techniques by hybridizing them with other optimization methods, mainly from the AI field.

Didier Dubois

Chapter A.3

Université Paul Sabatier
IRIT – Equipe ADRIA
Toulouse Cedex 9, France
dubois@irit.fr

Didier Dubois is a Research Advisor at IRIT, the Computer Science Department of Paul Sabatier University, Toulouse, France. He has co-authored 2 books, a monograph, and over 15 edited volumes on uncertain reasoning and fuzzy sets. His interests range from knowledge representation and reasoning to decision sciences and representation and processing of imprecise information.

Antonio J. Fernández Leiva

Chapter E.52

Universidad de Málaga
Dep. Lenguajes y Ciencias de la
Computación
Málaga, Spain
afdez@lcc.uma.es

Antonio J. Fernández-Leiva received his PhD degree in Computer Science from the University of Málaga in 2002 and later became an Associate Professor. In the past he has worked in private companies as a computer engineer. His main areas of research involve both the application of metaheuristics techniques to combinatorial optimization and the employment of computational intelligence in games.

Javier Fernández

Chapter A.6

Universidad Pública de Navarra
Dep. Automática y Computación
Pamplona (Navarra), Spain
fcojavier.fernandez@unavarra.es

Javier Fernández received his PhD in Mathematics from the University of the Basque Country in 2003. He was a postdoc researcher at the CNRS and currently he is a member of the GIARA research group at the Public University of Navarra. He has authored more than 30 papers in JCR journals. His main research interests are fuzzy sets and extensions, aggregation functions, image processing, and harmonic analysis.

Martin H. Fischer

University of Potsdam
Psychology Dep.
Potsdam OT Golm, Germany
martinf@uni-potsdam.de

Chapter D.37

Martin Fischer obtained a PhD from the University of Massachusetts in 1997 through graduate studies on motor control and visual attention. He then worked at LMU Munich for 3 years before moving to University of Dundee in Scotland. In 2011 he became Professor of Cognitive Sciences at the University of Potsdam in Germany, where he investigates embodied numerical cognition.

János C. Fodor

Óbuda University
Dep. Applied Mathematics
Budapest, Hungary
fodor@uni-obuda.hu

Chapter B.10

János Fodor is a Full Professor and the Rector of Óbuda University, Budapest, Hungary. He received his Master's degree in 1981 and his PhD in 1991, both in Mathematics. He is a Doctor of the Hungarian Academy of Sciences and is pursuing research in mathematical foundations of fuzzy logic, computational intelligence, and preference modeling. He has published 2 monographs and over 250 scientific papers.

Jairo Alonso Giraldo

Universidad de los Andes
Dep. Electrical and Electronics
Engineering
Bogotá, Colombia
ja.giraldo908@uniandes.edu.co

Chapter F.70

Jairo Giraldo is a PhD student in the Department of Electrical and Electronics Engineering at Universidad de los Andes, Colombia. His research interests include distributed control algorithms for power grids, multi-agent-based control, networked systems, and cybersecurity and privacy. He is a member of the IEEE Control Systems Society and the Power and Energy Society.

Siegfried Gottwald

Leipzig University
Inst. Philosophy
Leipzig, Germany
gottwald@uni-leipzig.de

Chapter A.2

Siegfried Gottwald was Full Professor for Nonclassical and Mathematical Logic at Leipzig University until 2008. He obtained a PhD in Mathematics in 1969. His research areas include many-valued and fuzzy logic, fuzzy sets, fuzzy relations and fuzzy control, and history of mathematics and logic. He is IFSA Fellow and he received the *Forschungspreis Technische Kommunikation* of the SEL-Alcatel foundation (1992).

Salvatore Greco

University of Catania
Dep. Economics and Business
Catania, Italy
salgreco@unict.it

Chapters C.22, C.24

Salvatore Greco is Professor at the University of Catania, Italy, and part-time Professor at the University of Portsmouth, UK. He is an active researcher in the area of rough set theory, multiple criteria decision aiding (MCDA), and non-additive integrals. He received the Multiple Criteria Decision Making Gold Medal in 2013.

Marco Gribaudo

Politecnico di Milano
Dip. Elettronica, Informazione e
Bioingegneria
Milano, Italy
marco.gribaudo@polimi.it

Chapter F.69

Marco Gribaudo is a Senior Researcher at the Politecnico di Milano, Italy. His current research interests are multi-formalism modeling, queueing networks, mean-field analysis and spatial models.

Roderich Groß   Chapter F.66   For biographical profile, please see the section "About the Part Editors".

Jerzy W. Grzymala-Busse   Chapter C.23

University of Kansas
Dep. Electrical Engineering and Computer
Science
Lawrence, USA
jerzygb@ku.edu

Jerzy W. Grzymala-Busse has been a Professor of Electrical Engineering and Computer Science at the University of Kansas since August 1993. His research interests include data mining, machine learning, knowledge discovery, expert systems, reasoning under uncertainty, and rough set theory. He is the author, co-author, or editor of 14 books, and has published over 300 articles.

Hani Hagras   Chapter B.18

University of Essex
The Computational Intelligence Centre
Colchester, UK
hani@essex.ac.uk

Hani Hagras is a Professor of Computational Intelligence, Director of the Computational Intelligence Centre, Head of the Fuzzy Systems Research Group and Head of the Intelligent Environments Research Group at the University of Essex, UK. He is a Fellow of IEEE and of IET. He has authored more than 250 papers. He was awarded the Oustanding Paper award of the IEEE Computational Intelligence Society, in 2004 and 2013.

Heiko Hamann   Chapter F.74

Universtity of Paderborn
Dep. Computer Science
Paderborn, Germany
heiko.hamann@uni-paderborn.de

Heiko Hamann is Assistant Professor at the Computer Science Department of the University of Paderborn. He spent his postdoctoral time at the University of Graz, Austria. He received his PhD from the University of Karlsruhe, Germany. The main focus of his research efforts is on swarm models, swarm robotics, and synthesis of robot controllers.

Thomas Hammerl   Chapter E.64

Vienna, Austria
thomas.hammerl@gmail.com

Thomas Hammerl is working as a self-employed software developer in Vienna, Austria. He received his Master's degree from Vienna University of Technology in 2009. While working on his thesis he did research on metaheuristics, constraint satisfaction, and optimization.

Julie Hamon   Chapter E.61

Ingenomix
Dep. Research and Development
Boisseuil, France
julie.hamon@ingenomix.fr

Julie Hamon graduated as a computer science and statistics engineer from the Ecole Polytechnique Universitaire de Lille (Polytech'Lille), and then received her PhD in Combinatorial Optimization and Statistics from Lille 1 University. She is currently a biostatistician at Ingenomix, a company specialized in genomic selection in bovine. Her main research interests include statistical models applied to genomic data.

Nikolaus Hansen   Chapter E.44

Universitè Paris-Sud
Machine Learning and Optimization
Group (TAO)
Orsay Cedex, France
hansen@lri.fr

Nikolaus Hansen is a Senior Researcher at Inria, France. After studies in medicine and mathematics he received his PhD in Civil Engineering from the Technical University Berlin in 1998 and his Habilitation in Computer Science from the University Paris-Sud in 2010. Before joining Inria he worked in applied artificial intelligence, evolutionary computation, and genomics in Berlin, and in computational science at the ETH Zurich.

Mark W. Hauschild   Chapter E.45

University of Missouri–St. Louis
Dep. Mathematics and Computer
Science
St. Louis, USA
markhauschild@gmail.com

Mark Hauschild received his PhD in Applied Mathematics from the University of Missouri in 2014. He is a Visiting Professor at the University of Missouri, where he teaches courses on AI. His research interests include estimation of distribution algorithms, genetic programming, and machine learning. He is currently a member of the Missouri Estimation of Distribution Algorithms Laboratory.

Sebastien Hélie   Chapter D.36

Purdue University
Dep. Psychological Sciences
West Lafayette, USA
shelie@purdue.edu

Sebastien Helie is Assistant Professor in the Department of Psychological Sciences at Purdue University. He received a PhD in Cognitive Science from the Université du Québec à Montreal. He uses computational cognitive neuroscience and neuroimaging methods to study categorization, automaticity, rule learning, sequence learning, and skill acquisition.

Jano I. van Hemert   Chapter E.65

Optos
Dunfermline, UK
jano@vanhemert.co.uk

Dr Jano van Hemert (MSc 1998, PhD 2002; Leiden University, The Netherlands) is a Senior Manager and the Academic Liaison at Optos plc. His main area of research is in computer science and its applications. From 2005 until 2011 he was a visiting researcher at the Human Genetics Unit in Edinburgh of the UK's Medical Research Council.

Holger H. Hoos      Chapter E.54

University of British Columbia
Dep. Computer Science
Vancouver, Canada
hoos@cs.ubc.ca

Holger H. Hoos is a Professor of Computer Science and a Faculty Associate at the Peter Wall Institute for Advanced Studies at the University of British Columbia (Canada). He is a Fellow of the Association for the Advancement of Artificial Intelligence (AAAI) and past President of the Canadian Artificial Intelligence Association (CAIAC). His main research interests span empirical algorithmics, artificial intelligence, bioinformatics, and computer music. His research has been published in numerous books and journals.

Tania Iglesias      Chapter B.11

University of Oviedo
Dep. Statistics and O.R.
Oviedo, Spain
iglesiasctania@uniovi.es

Tania Iglesias received her BSc degree in Mathematics from the University of Oviedo, Spain, in 2006 and her MSc degree from the University of Granada, Spain, in 2013. She is now a technician in the Statistical Consulting Unit of the University of Oviedo. Decision making in fuzzy sets is her main topic of research.

Giacomo Indiveri   Chapter D.38

University of Zurich and ETH Zurich
Inst. Neuroinformatics
Zurich, Switzerland
giacomo@ini.uzh.ch

Giacomo Indiveri is a Professor at the Faculty of Science of the University of Zurich, Switzerland. He obtained an MSc degree in Electrical Engineering and a PhD degree in Computer Science from the University of Genoa, Italy. His research interests lie in the study of real and artificial neural processing systems, and in the hardware implementation of neuromorphic cognitive systems.

Masahiro Inuiguchi   Chapter C.26

Osaka University
Dep. Systems Innovation, Graduate School
of Engineering Science
Toyonaka, Osaka, Japan
inuiguti@sys.es.osaka-u.ac.jp

Masahiro Inuiguchi received ME and DE degrees from Osaka Prefecture University in 1987 and 1991. He worked as a Research Associate at Osaka Prefecture University (1987–1992), Associate Professor at Hiroshima University (1992–1997) and Osaka University (1997–2003). At present, he is a Full Professor at Osaka University. His research interests include possibility theory, fuzzy mathematical programming, rough sets, and their applications to decision making.

Hisao Ishibuchi

Osaka Prefecture University
Dep. Computer Science and Intelligent
Systems, Graduate School of Engineering
Osaka, Japan
hisaoi@cs.osakafu-u.ac.jp

Chapter G.77

Hisao Ishibuchi has been a Professor at Osaka Prefecture University since 1999. His research interests include fuzzy rule-based systems, multiobjective optimization, and evolutionary games. He was the IEEE CIS Vice-President for Technical Activities for 2010–2013. Currently he is an IEEE CIS AdCom member (2014–2016), an IEEE CIS Distinguished Lecturer (2015–2017), and the Editor-in-Chief of the IEEE CI Magazine.

Emiliano Iuliano

CIRA, Italian Aerospace Research Center
Fluid Dynamics Lab.
Capua (CE), Italy
e.iuliano@cira.it

Chapter E.60

Emiliano Iuliano received the Laurea (MSc) degree and the Doctorate in Aerospace Engineering from the University of Naples in 2004 and 2012, respectively. He is currently a Senior Researcher at CIRA, the Italian Aerospace Research Center. His research interests include CFD analysis, aircraft aerodynamic design, surrogate-based optimization methods, and aircraft in-flight icing.

Julie Jacques

Alicante LAB
Seclin, France
julie.jacques@alicante.fr

Chapter E.61

Julie Jacques received her PhD degree in Computer Science from the University of Lille, France, in 2013. She has been working in the industry on applied research projects since 2008. Her research interests are operational research and data mining, with applications to health.

Andrzej Jankowski

Knowledge Technology Foundation
Warsaw, Poland
andrzej.adgam@gmail.com

Chapter C.21

Andrzej Jankowski, received his PhD from the University of Warsaw and is one of the founders of the Polish–Japanese Institute of Information Technology. He currently works for R&D and education AI projects at Warsaw University of Technology and the University of Warsaw. He has unique experience in managing complex IT projects in Central Europe and USA.

Balasubramaniam Jayaram

Indian Institute of Technology Hyderabad
Dep. Mathematics
Hyderabad, India
jbala@iith.ac.in

Chapter B.12

Balasubramaniam Jayaram received his MSc and PhD degrees in Mathematics from the Sri Sathya Sai Institute of Higher Learning, India, in 1999 and 2004. He has authored and co-authored more than 50 publications and a research monograph. His current research interests include fuzzy aggregation operations, approximate reasoning, clustering in high dimensions, and kernel learning methods.

Laetitia Jourdan

University of Lille 1
INRIA/UFR IEEA/laboratory CRIStAL/CNRS
Lille, France
laetitia.jourdan@univ-lille1.fr

Chapter E.61

Laetitia Jourdan is a full Professor of Computer Sciences at the University of Lille 1/LIFL. She holds a PhD in Combinatorial Optimization from the University of Lille 1 (France). Her areas of research are modeling datamining tasks as combinatorial optimization problems, solving methods based on metaheuristics, and incorporating learning in metaheuristics and multiobjective optimization.

Nikola Kasabov

Auckland University of Technology
KEDRI – Knowledge Engineering and
Discovery Research Inst.
Auckland, New Zealand
nkasabov@aut.ac.nz

Chapter D.40

Nikola K. Kasabov obtained his Master's and PhD degrees from the Technical University of Sofia, Bulgaria. He is the Director and the Founder of the Knowledge Engineering and Discovery Research Institute (KEDRI) and Professor of Knowledge Engineering at the School of Computing and Mathematical Sciences at Auckland University of Technology, New Zealand. He has published over 450 papers, books, and patents on informatics, computational intelligence, neural networks, bioinformatics, neuroinformatics.

Petra Kersting

TU Dortmund University
Dep. Mechanical Engineering
Dortmund, Germany
pkersting@isf.de

Chapter E.59

Petra Kersting studied Computer Science and finished her dissertation in Mechanical Engineering at TU Dortmund University. She is Head of the Division *Simulation and Optimization* and holds the junior professorship *Modeling methods for machining processes*. She is Research Affiliate of the International Academy of Production Engineering (CIRP) and Dorothea-Erxleben Visiting Professor 2014/2015 at the OvGU-Magdeburg.

Erich P. Klement

Johannes Kepler University
Dep. Knowledge-Based Mathematical
Systems
Linz, Austria
ep.klement@jku.at

Chapter A.5

Erich Peter Klement received his PhD degree from the University of Innsbruck. He is Professor of Mathematics at the Johannes Kepler University in Linz, Austria and the Head of the Fuzzy Logic Laboratory in Linz. He has (co-)authored 3 monographs and 80 journal papers and is a Member of the Editorial Boards of 16 journals.

Anna Kolesárová

Slovak University of Technology in
Bratislava
Faculty of Chemical and Food
Technology
Bratislava, Slovakia
anna.kolesarova@stuba.sk

Chapter A.4

Anna Kolesárová received her MSc degree in Mathematics and Physics from Comenius University in Bratislava and her PhD degree from the Slovak Academy of Sciences. In 2008 she became a Full Professor at the Slovak University of Technology in Bratislava. Her current research interests include aggregation functions, with special stress on copulas, measures and integrals, decision making, and fuzzy mathematics.

Magda Komorníková

Slovak University of Technology
Dep. Mathematics
Bratislava, Slovakia
magda@math.sk

Chapter A.4

Magda Komorníková graduated at Comenius University, Faculty of Mathematics and Physics in 1973 and received her PhD from the same faculty in 1979. Since 1990 she has been a member of the Department of Mathematics at the Slovak University of Technology in Bratislava. She has been a full professor since 2002. Her fields of interest are measure theory, uncertainty modeling, copulas, time series analysis, and aggregation and related operators.

Mark Kotanchek

Evolved Analytics LLC
Midland, USA
mark@evolved-analytics.com

Chapter E.57

Mark Kotanchek's diverse academic background (Engineering Science BSc, Acoustics MEng, Aerospace Engineering PhD, IEEE Senior Member) is consistent with the diversity of his professional experience. He founded Evolved Analytics in 2005.

Robert Kozma

University of Memphis
Dep. Mathematical Sciences
Memphis, USA
rkozma@memphis.edu

Chapter D.33

Robert Kozma has MSc degrees from Moscow Power Engineering University and Eötvös University (Budapest, Hungary), respectively. He received his PhD from Delft University of Technology (The Netherlands). He is a First Tennessee University Professor and receipient of the INNS Gabor Award. His research includes robust decision support for large-scale networks, autonmous robot control, sensor networks, brain networks, and brain–computer interfaces.

Tomáš Kroupa

Institute of Information Theory and
Automation
Dep. Decision-Making Theory
Prague, Czech Republic
kroupa@utia.cas.cz

Chapter A.9

Tomáš Kroupa obtained his PhD in Mathematical Engineering from the Czech Technical University in 2005. He is affiliated with the Institute of Information Theory and Automation, the Academy of Sciences of the Czech Republic. He has had a Senior Researcher position in the institute since 2014. His area of research is game theory and many-valued logics.

Rudolf Kruse

University of Magdeburg
Faculty of Computer Science
Magdeburg, Germany
kruse@iws.cs.uni-magdeburg.de

Chapter B.17

Rudolf Kruse obtained his PhD in Mathematics in 1980 from the University of Braunschweig. Since 1996 he has been a full professor at the University of Magdeburg where he is leading the computational intelligence research group. He is a Fellow of the International Fuzzy Systems Association and a Fellow of the Institute of Electrical and Electronics Engineers.

Tufan Kumbasar

Istanbul Technical University
Control Engineering Dep.
Maslak, Istanbul, Turkey
kumbasart@itu.edu.tr

Chapter B.18

Tufan Kumbasar is currently working as a Postdoctoral Research Fellow in the Control Engineering Department of Istanbul Technical University. His research interests lie predominately in the area of control engineering, particularly with respect to optimization, intelligent control, process control and mechatronic control methods, and their applications. He is particularly interested in the area of type-2 fuzzy logic systems, especially in their controller applications.

James T. Kwok

Hong Kong University of Science and
Technology
Dep. Computer Science and Engineering
Hong Kong, Hong Kong
jamesk@cse.ust.edu.hk

Chapter D.29

James Kwok received his PhD degree from the Hong Kong University of Science and Technology in 1996. He then joined the Department of Computer Science, Hong Kong Baptist University as an Assistant Professor. He is now a Professor in the Department of Computer Science and Engineering.

Rhyd Lewis

Cardiff University
School of Mathematics
Cardiff, UK
lewisR9@cf.ac.uk

Chapter E.63

Rhyd Lewis is a lecturer in operational research in the School of Mathematics at Cardiff University. He holds a PhD in Computer Science and Operational Research from Edinburgh Napier University and a degree in Computing from the University of Wales, Swansea. Dr Lewis is the Co-Founder and an Associate Editor of the International Journal of Metaheuristics.

Xiaodong Li

RMIT University
School of Computer Science and
Information Technology
Melbourne, Australia
xiaodong.li@rmit.edu.au

Chapter F.67

Xiaodong Li received his PhD degree in Artificial Intelligence from the University of Otago. He is currently an Associate Professor at the School of Computer Science and Information Technology, RMIT University. His research interests include evolutionary computation, machine learning, multiobjective optimization, and swarm intelligence.

Paulo J.G. Lisboa   Chapter D.31

Liverpool John Moores University
Dep. Mathematics & Statistics
Liverpool, UK
p.j.lisboa@ljmu.ac.uk

Paulo Lisboa has a PhD in Theoretical Physics from the University of Liverpool. He is a Professor in Industrial Mathematics at Liverpool John Moores University. He chairs the Medical Analysis Task Force in the Data Mining Technical Committee of the IEEE Computational Intelligence Society. His research focus is computational data analysis for medical decision support, in particular with interpretable machine learning models.

Weifeng Liu   Chapter D.30

Jump Trading
Chicago, USA
weifeng@ieee.org

Weifeng Liu received his PhD degree in Electrical and Computer Engineering from the University of Florida in 2008. He is currently a senior researcher with Jump Trading, Chicago, IL. His research interests include machine learning, adaptive signal processing and their applications to e-commerce, business, and finance.

Fernando G. Lobo   Chapter E.45

Universidade do Algarve
Dep. Engenharia Electrónica e
Informática
Faro, Portugal
fernando.lobo@gmail.com

Fernando Lobo is Associate Professor at the University of Algarve, Portugal. He received his PhD degree in 2000 from Universidade Nova de Lisboa, Portugal, and during that period was a regular visitor at the Illinois Genetic Algorithms Laboratory, UIUC. His major research interests are evolutionary computation, and computers and accessibility.

Antonio López Jaimes   Chapter E.51

CINVESTAV–IPN
Dep. Computación
México, Mexico
tonio.jaimes@gmail.com

Antonio López Jaimes received his BSc degree in Computer Science in 2002 from the Autonomous Metropolitan University, Mexico, and his MSc and PhD degrees in Computer Science from the National Polytechnic Institute, Mexico in 2005 and 2011. His research interests include evolutionary multiobjective optimization, parallel evolutionary algorithms, and interactive optimization methods. He is currently a Professor at the Autonomous Metropolitan University.

Francisco Luna   Chapter E.50

Centro Universitario de Mérida
Mérida, Spain
fluna@unex.es

Francisco Luna received his degree in Engineering and PhD in Computer Science in 2002 and 2008, from the University of Málaga, Spain. Since 2013, he has been Assistant Professor in the Universidad de Extremadura at the Campus Universitario de Mérida. His current research interests include the design and implementation of parallel and multi-objective metaheuristics, and their application to solving complex problems in the domain of telecommunications and combinatorial optimization.

Luis Magdalena   Chapter B.13   For biographical profile, please see the section "About the Part Editors".

Sebastia Massanet   Chapter B.12

University of the Balearic Islands
Dep. Mathematics and Computer Science
Palma de Mallorca, Spain
s.massanet@uib.es

Sebastia Massanet received his BS, MSc, and PhD degrees in Mathematics from the University of the Balearic Islands (UIB) in 2008, 2009, and 2012 where he is currently an Assistant Professor in the Department of Mathematics and Computer Science. His current research interests are fuzzy sets theory and related fields such as fuzzy connectives, fuzzy implications, functional equations, and fuzzy mathematical morphology.

Benedetto Matarazzo

University of Catania
Dep. Economics and Business
Catania, Italy
matarazz@unict.it

Chapter C.22

Benedetto Matarazzo is Chairman of the degree course in Corporate Finance at Catania University. He received the Gold Medal of International Society of Multiple Criteria Decision Making in 2009. His main research interests are in the fields of multiple criteria decision analysis, the rough set approach to decision analysis, and preference modeling.

Sergi Mateo Bellido

Polytechnic University of Catalonia
Dep. Computer Architecture
Barcelona, Spain
sergim@ac.upc.edu

Chapter F.68

Sergi Mateo Bellido obtained his degree in Computer Science from the Polytechnic University of Catalonia in 2011. Since then, he has been working in the field of high performance computing at Barcelona Supercomputing Center and recently at the Polytechnic University of Catalonia. He is interested in algorithms, parallel programming models and domain specific languages.

James McDermott

University College Dublin
Lochlann Quinn School of Business
Dublin 4, Ireland
jmmcd@jmmcd.net

Chapter E.43

James McDermott has research interests in evolutionary design and genetic programming. He was an IRC/Marie Curie Post-Doctoral Fellow at Massachusetts Institute of Technology and is now a lecturer and the Program Director in Business Analytics at University College, Dublin. He was Co-Chair of EvoMUSART 2013 and 2014 and Publication Chair of EuroGP 2015.

Patricia Melin

Chapters G.79, G.80

For biographical profile, please see the section "About the Part Editors".

Corrado Mencar

University of Bari
Dep. Informatics
Bari, Italy
corrado.mencar@uniba.it

Chapter B.14

Corrado Mencar is Assistant Professor in the Department of Informatics, at the University of Bari, Italy. His research interests are computational intelligence, with a special emphasis on fuzzy logic, granular computing, neuro-fuzzy systems, computational web intelligence, and intelligent data analysis. He has published more than 60 peer-reviewed papers.

Radko Mesiar

Chapters A.4, A.5

For biographical profile, please see the section "About the Part Editors".

Ralf Mikut

Karlsruhe Institute of Technology (KIT)
Inst. Applied Computer Science
Eggenstein-Leopoldshafen, Germany
ralf.mikut@kit.edu

Chapter B.17

Ralf Mikut graduated in automatic control at TU Dresden and received his PhD degree from the University of Karlsruhe in 1999. He is Associate Professor at the Karlsruhe Institute of Technology. His research interests include fuzzy systems, computational intelligence, and data mining in biological and technical applications.

Ali A. Minai

University of Cincinnati
School of Electronic & Computing Systems
Cincinnati, USA
ali.minai@uc.edu

Chapter D.35

Ali A. Minai is Professor in the School of Electronic and Computing Systems at the University of Cincinnati. He received his PhD in Electrical Engineering from the University of Virginia. His areas of research include complex systems, neural networks, cognitive models, computational neuroscience, and computational models of social systems. He is a senior member of IEEE and the International Neural Network Society, and a member of the Society for Neuroscience.

Authors

Sadaaki Miyamoto   Chapter B.15

University of Tsukuba
Risk Engineering
Tsukuba, Japan
miyamoto@risk.tsukuba.ac.jp

Sadaaki Miyamoto is a Professor of the University of Tsukuba, Japan. His current research interests include methodology for uncertainty modeling. In particular, he has been working on data clustering algorithms and the theory of generalized bags. He has published 3 books and over 300 research papers. In 2007, he became a Fellow of the International Fuzzy Systems Association.

Christian Moewes    Chapter B.17

University of Magdeburg
Faculty of Computer Science
Magdeburg, Germany
cmoewes@ovgu.de

Christian Moewes received his diploma in Computer Science from the University of Magdeburg, Germany in 2007. He has co-authored two textbooks, co-edited two proceedings, published three journal articles, and five peer-reviewed book chapters. Currently he focuses on the statistical analysis of dynamical brain networks by means of model-based methods.

Javier Montero    Chapter A.6

Complutense University, Madrid
Dep. Statistics and Operational Research
Madrid, Spain
monty@mat.ucm.es

Javier Montero is Professor at the Department of Statistics and Operational Research, Complutense University of Madrid. He holds a PhD in Mathematics. His research interests are aggregation operators, preference representation, multicriteria decision aid, group decision making, system reliability theory, image processing, and classification problems. He has been the President of the European Association for Fuzzy Logic and Technology.

Ignacio Montes   Chapter B.11

University of Oviedo
Dep. Statistics and O.R.
Oviedo, Spain
imontes@uniovi.es

Ignacio Montes received his BSc degree in Mathematics from the University of Oviedo, Spain, in 2009 and his MSc degree in 2010 also from the University of Oviedo. He is now a member of the Department of Statistics and Operational Research at the same university. Preference modeling with imprecise elements is his main topic of research.

Susana Montes   Chapter B.11

University of Oviedo
Dep. Statistics and O.R.
Oviedo, Spain
montes@uniovi.es

Susana Montes received her MSc degree in Mathematics from the University of Valladolid, Spain, in 1993, and her PhD degree from the University of Oviedo, Spain, in 1998. She is a Professor of Statistics and Operational Research at the University of Oviedo. She has published in international journals and international conference proceedings.

Oscar H. Montiel Ross    Chapter G.76

Mesa de Otay, Tijuana, Mexico
oross@citedi.mx

Oscar Humberto Montiel Ross received his MSc in Digital Systems in 1996 from the Instituto Politécnico Nacional-CITEDI, an MSc from Tijuana Institute of Technology in 2000, and his PhD in 2006 from the Universidad Autónoma of Baja California in Tijuana, México, both in Computer Science. He has published about 74 contributions in journals, book chapters, proceedings, and 4 books. His research interests include optimization, intelligent systems, and robotics.

Manuel Mucientes    Chapter B.20

University of Santiago de Compostela
Research Centre for Information Technologies
Santiago de Compostela, Spain
manuel.mucientes@usc.es

Manuel Mucientes is a Ramón y Cajal Research Fellow with the Research Centre for Information Technologies (CiTIUS) of the University of Santiago de Compostela. His current research interests are evolutionary algorithms, genetic fuzzy systems, motion planning and control in robotics, visual SLAM, web services, and process mining.

Nysret Musliu

Chapter E.64

Vienna University of Technology
Inst. Information Systems
Vienna, Austria
musliu@dbai.tuwien.ac.at

Nysret Musliu is Privat Dozent and Senior Researcher at Vienna University of Technology. He received his PhD in Computer Science from Vienna University of Technology in 2001 and his MSc degree from the University of Prishtina in 1996. His research interests include problem solving and search, metaheuristics, machine learning and optimization, constraint satisfaction, scheduling, and timetabling.

Yusuke Nojima

Chapter G.77

Osaka Prefecture University
Dep. Computer Science and Intelligent
Systems, Graduate School of Engineering
Osaka, Japan
nojima@cs.osakafu-u.ac.jp

Yusuke Nojima received his BS and MSc degrees from Osaka Institute of Technology, Japan, in 1999 and 2001, respectively, and his PhD from Kobe University, Hyogo, Japan, in 2004. Since 2004, he has been with Osaka Prefecture University, Japan, where he is currently an Associate Professor. His research interests include genetic fuzzy systems, evolutionary multiobjective optimization, and parallel distributed data mining.

Stefano Nolfi

Chapter D.37

Consiglio Nazionale delle Ricerche
(CNR-ISTC)
Inst. Cognitive Sciences and Technologies
Roma, Italy
stefano.nolfi@istc.cnr.it

Stefano Nolfi is Director of Research of the Institute of Cognitive Sciences and Technologies (CNR). He is one of the founders of evolutionary robotics. His research activities focus on the evolution and development of behavioral and cognitive skills in embodied agents. He has authored/co-authored more than 130 peer-reviewed scientific publications, including a monograph.

Una-May O'Reilly

Chapter E.43

Massachusetts Institute of Technology
Computer Science and Artificial
Intelligence Lab.
Cambridge, USA
unamay@csail.mit.edu

Una-May O'Reilly leads the AnyScale Learning For All (ALFA) group at Massachusetts Institute of Technology Computer Science and Artificial Intelligence Laboratory. She received the EvoStar Award for Outstanding Achievements in Evolutionary Computation in Europe in 2013 and serves as Vice-Chair of ACM SigEVO.

Miguel Pagola

Chapter A.6

Universidad Pública de Navarra
Dep. Automática y Computación
Pamplona (Navarra), Spain
miguel.pagola@unavarra.es

Miguel Pagola received his MSc and PhD degrees in Industrial Engineering from the Public University of Navarra, in 2000 and 2008. He is Associate Lecturer with the Department of Automatics and Computation at UPNa. His research interests include fuzzy techniques for image processing, fuzzy set theory, machine learning, and data mining. He is a member of the European Society for Fuzzy Logic and Technology.

Lynne Parker

Chapter F.72

University of Tennessee
Dep. Electrical Engineering and Computer
Science
Knoxville, USA
leparker@utk.edu

Lynne Parker is a Professor in the Department of Electrical Engineering and Computer Science at the University of Tennessee, Knoxville. She received her PhD degree from the Massachusetts Institute of Technology. She previously worked for several years as a full-time researcher at Oak Ridge National Laboratory. Her research focuses on distributed intelligent robotics, human–robot interaction, sensor networks, and machine learning.

Kevin M. Passino

The Ohio State University
Dep. Electrical and Computer Engineering
Columbus, USA
passino@ece.osu.edu

Chapter F.70

Kevin M. Passino is Professor of Electrical and Computer Engineering and Director of the Humanitarian Engineering Center at Ohio State University. He has been Vice-President of Technical Activities of IEEE Control Systems Society, an elected member of IEEE Control Systems Society Board of Governors, Program Chair of the 2001 IEEE Conference on Decision and Control, and is a Distinguished Lecturer for the IEEE Society on Social Implications of Technology.

Martin Pelikan

Sunnyvale, USA
martin@martinpelikan.net

Chapter E.45

Martin Pelikan received his PhD in Computer Science from the University of Illinois at Urbana-Champaign in 2002. He is now a software engineer at Google. Previously, he was an Associate Professor at the University of Missouri in St. Louis. His research in evolutionary computation focused mainly on estimation of distribution algorithms (EDAs), efficiency enhancement techniques, and scalability of EDAs.

Irina Perfilieva

University of Ostrava
Inst. Research and Applications of Fuzzy Modeling
Ostrava, Czech Republic
Irina.Perfilieva@osu.cz

Chapter A.7

Irina Perfilieva received her PhD from Moscow State University and is currently a Professor of Applied Mathematics at the University of Ostrava, Czech Republic. Her research interests are fuzzy transforms with applications to image processing and computer vision. She has published more than 300 journal and conference papers and is co-author of 5 books.

Henry Prade

Université Paul Sabatier
IRIT – Equipe ADRIA
Toulouse Cedex 9, France
prade@irit.fr

Chapter A.3

Henri Prade is a Research Director at CNRS at Paul Sabatier University. He is co-author of two monographs on fuzzy sets and possibility theory. His current research interests are uncertainty and preference modeling, non-classical logics, approximate, plausible and analogical reasoning with applications to artificial intelligence, and information systems. He is an ECCAI fellow, an IFSA fellow, a 2001 highly-cited ISI laureate, and received an IEEE pioneer award in 2002.

Mike Preuss

WWU Münster
Inst. Wirtschaftsinformatik
Münster, Germany
mike.preuss@tu-dortmund.de

Chapter E.58

Mike Preuss is Research Associate at ERCIS, University of Münster, Germany, and the Chair of Algorithm Engineering at TU Dortmund, Germany, where he received his PhD in 2013. His research interests focus on the field of evolutionary algorithms for real-valued problems, namely on multimodal and multiobjective optimization, and on computational intelligence methods for computer games.

José C. Principe

University of Florida
Dep. Electrical and Computer Engineering
Gainesville, USA
principe@cnel.ufl.edu

Chapter D.30

Jose C. Principe is currently a Distinguished Professor of Electrical and Biomedical Engineering at the University of Florida, Gainesville, USA. He is Founder and Director of the University of Florida Computational Neuro-Engineering Laboratory (CNEL). He is an IEEE Fellow and an AIMBE Fellow. He is involved in biomedical signal processing, in particular, the electroencephalogram (EEG) and the modeling and applications of adaptive systems.

Domenico Quagliarella

CIRA, Italian Aerospace Research Center
Fluid Dynamics Lab.
Capua (CE), Italy
d.quagliarella@cira.it

Chapter E.60

Domenico Quagliarella received his MSc degree and his PhD in Aerospace Engineering from the University of Naples, Italy. He is currently Senior Researcher at CIRA. His research interests are the application of hybrid multi-objective optimization methods to aerodynamic and multidisciplinary design. He is the author of about 65 publications.

Nicanor Quijano

Universidad de los Andes
Dep. Electrical and Electronics
Engineering
Bogotá, Colombia
nquijano@uniandes.edu.co

Chapter F.70

Nicanor Quijano received his PhD degree in Electrical and Computer Engineering from The Ohio State University in 2006. Since 2007 he has been with the Electrical Engineering Department, Universidad de los Andes, Colombia, where he is the Director of the Research Group on Control Systems. His research interests include hierarchical and distributed optimization methods using bio-inspired and game-theoretical techniques for dynamic resource allocation problems.

Jaroslav Ramík

Silesian University in Opava
Dep. Informatics and Mathematics
Karviná, Czech Republic
ramik@opf.slu.cz

Chapter A.8

Prof. Jaroslav Ramík, PhD, is a Professor of Mathematics, Statistics, and Operations Research at the Silesian University Opava, in Karvina, Czech Republic. His interests include optimization methods in economics and decision making. He is the author of 6 books and more than 50 papers listed in WoS. He is also active in the Czech Society for Operations Research and has served as its former president.

Ismael Rodríguez Fdez

University of Santiago de Compostela
Research Centre for Information
Technologies
Santiago de Compostela, Spain
ismael.rodriguez@usc.es

Chapter B.20

Ismael Rodríguez received his MSc in Computer Science in 2011 from the University of Santiago de Compostela. He is presently a PhD student at the Research Centre for Information Technologies of the University of Santiago de Compostela (CiTIUS). His current research interests are regression problems, evolutionary algorithms, and genetic fuzzy systems.

Franz Rothlauf

Johannes Gutenberg University Mainz
Gutenberg School of Management and
Economics
Mainz, Germany
rothlauf@uni-mainz.de

Chapter E.53

Franz Rothlauf received a Diploma in Electrical Engineering from the University of Erlangen, Germany, a PhD in Information Systems from the University of Bayreuth, Germany, and a Habilitation from the University of Mannheim, Germany. He is a Professor for Information Systems at the University of Mainz. His research activities include planning and optimization, evolutionary computation, e-business, and software engineering.

Jonathan E. Rowe

University of Birmingham
School of Computer Science
Birmingham, UK
J.E.Rowe@cs.bham.ac.uk

Chapter E.42

Jonathan Rowe received a degree in Mathematics and PhD in Computer Science from the University of Exeter. He has worked in the field of natural computation for 20 years. He joined the University of Birmingham in 2000 and is now the Head of the School of Computer Science. He is Associate Editor for Theoretical Computer Science and Natural Computing journals.

Imre J. Rudas

Óbuda University
Dep. Applied Mathematics
Budapest, Hungary
rudas@uni-obuda.hu

Chapter B.10

Imre J. Rudas received his Master's Degree in Mathematics in Budapest, and his Doctor of Science degree from the Hungarian Academy of Sciences. He is a full University Professor and Head of Óbuda University Research and Innovation Center. His present areas of research activities are robotics and computational intelligence. He has published 6 books and more than 690 scientific papers.

Günter Rudolph

Technische Universität Dortmund
Fak. Informatik
Dortmund, Germany
guenter.rudolph@cs.tu-dortmund.de

Chapter E.58

Günter Rudolph (PhD in Computer Science, 1996) has been a Professor of Computational Intelligence at the Department of Computer Science at TU Dortmund University since 2005. His research interests include the development and theoretical analysis of bio-inspired methods applied to difficult optimization problems encountered in engineering sciences, logistics, and economics.

Gabriele Sadowski

Technische Universität Dortmund
Bio- und Chemieingenieurwesen
Dortmund, Germany
gabriele.sadowski@bci.tu-dortmund.de

Chapter E.58

Gabriele Sadowski (PhD in Physical Chemistry, 1991) has been a Professor of Thermodynamics in the Department of Biochemical and Chemical Engineering at TU Dortmund since 2001. Her research interests include experimental investigation and thermodynamic modeling of phase behavior in systems containing complex molecules, like polymers, electrolytes, biomolecules, or pharmaceuticals.

Marco Scarpa

Universita' di Messina
Dip. Ingegneria Civile, Informatica
Messina, Italy
mscarpag@unime.it

Chapter F.69

Marco Scarpa received his Bachelor degree in Computer Engineering from the University of Catania, Italy and his PhD degree in Computer Science in 2000 from the University of Turin, Italy. He is currently Associate Professor of Computer Engineering at the University of Messina, Italy. His interests include performance and reliability modeling of distributed and real time systems and algorithms for their solution.

Werner Schafhauser

XIMES
Vienna, Austria
schafhauser@ximes.com

Chapter E.64

Werner Schafhauser is a Senior Consultant and a software developer at XIMES in Vienna where, amongst other things, he develops and applies optimization algorithms to real scheduling problems. His research interests include metaheuristic optimization, constraint satisfaction problems, structural decomposition methods, and scheduling. He has a PhD in Computer Science from Vienna University of Technology.

Roberto Sepúlveda Cruz

Mesa de Otay, Tijuana, Mexico
rsepulve@citedi.mx

Chapter G.76

Roberto Sepúlveda Cruz received his MSc from the Tijuana Institute of Technology, México, and his PhD from the Universidad Autónoma of Baja California, Tijuana, México, both in Computer Science in 1999 and 2006, respectively. His research interests include type-2 fuzzy systems, intelligent systems, and robotics. He is a member of the International Association of Engineers (IANG).

Jennie Si

Arizona State University
School of Electrical, Computer and Energy Engineering
Tempe, USA
si@asu.edu

Chapter D.34

Jennie Si received her BS and MSc degrees from Tsinghua University, Beijing, China, and her PhD from the University of Notre Dame. She has been on the faculty in the Department of Electrical Engineering at Arizona State University since 1991. Her research focuses on dynamic optimization using learning and neural network approximation approaches, namely approximate dynamic programming. She has served on the Executive Boards of several professional organizations.

Authors

Marco Signoretto   Chapter D.32

Katholieke Universiteit Leuven
Leuven, Belgium
marco.signoretto@esat.kuleuven.be

Marco Signoretto holds a PhD in Mathematical Engineering from Katholieke Universiteit Leuven, Belgium; a Laurea Magistralis in Electronic Engineering from the University of Padova, Italy; and an MSc in Methods for Management of Complex Systems from the University of Pavia, Italy. His research interests include mathematical modeling of structured data. His current work deals with methods based on (convex) optimization, structure-inducing penalties, and spectral regularization.

Andrzej Skowron   Chapter C.21

University of Warsaw
Faculty of Mathematics, Computer Science
and Mechanics
Warsaw, Poland
skowron@mimuw.edu.pl

Andrzej Skowron is a Full Professor at the Institute of Mathematics at the University of Warsaw. He received his PhD and DSc (Habilitation) from the University of Warsaw, and the title of Professor in 1991. He is an ECCAI Fellow. His area of expertise includes reasoning with imperfect data and knowledge, soft computing methods, rough sets, granular computing, data mining, adaptive and autonomous systems, perception-based computing, and interactive computational systems.

Igor Škrjanc   Chapter G.75

University of Ljubljana
Faculty of Electrical Engineering
Ljubljana, Slovenia
igor.skrjanc@fe.uni-lj.si

Igor Škrjanc is currently a Professor of Automatic Control with the Faculty of Electrical Engineering, the University of Ljubljana. His main research interests include intelligent, predictive control systems and autonomous mobile systems. In 2009 he received the Humboldt Research Fellowship for Experienced Researchers.

Roman Słowiński   Chapters C.22, C.24   For biographical profile, please see the section "About the Part Editors".

Guido Smits   Chapter E.57

Dow Benelux BV
Core R&D
NM Hoek, The Netherlands
gfsmits@dow.com

Guido F. Smits is a Data Scientist at Dow Chemical Company. His main area of interest and expertise is in innovative applications of computational intelligence to new product design and optimization. He has authored more than 70 papers and currently holds 15 patents. He has a PhD from the University of Leiden, NL.

Ronen Sosnik   Chapter D.39

Holon Institute of Technology (H.I.T.)
Electrical, Electronics and Communication
Engineering
Holon, Israel
ronens@hit.ac.il

Ronen Sosnik received his MSc degree in Neuroscience and his Research Doctorate in Neuroscience from Weizmann Institute of Science in 2000 and 2005. His research interests include computational motor control, motor learning, and neural substrates mediating the acquisition, representation, and generation of motion primitives. Currently, he is devising innovative experimental paradigms and mathematical methods for the construction of novel BCI systems.

Alessandro Sperduti   Chapters D.27, D.31

University of Padova
Dep. Pure and Applied Mathematics
Padova, Italy
sperduti@math.unipd.it

Alessandro Sperduti has a PhD in Computer Science from the University of Pisa. He has been Professor in Computer Science at the University of Padova since 2002 and Chair of the Data Mining and Neural Networks Technical Committees of IEEE CIS. His research interests include machine learning, neural networks, learning in structured domains, and data and process mining.

Kasper Støy

IT University of Copenhagen
Copenhagen S, Denmark
ksty@itu.dk

Chapter F.73

Kasper Stoy holds an Associate Professorship at The Maersk Mc-Kinney Moller Institute, University of Southern Denmark (USD). His research interests include design of modular robot systems and distributed control. He has authored a monograph. He holds an MSc in Computer Science and Physics, University of Aarhus, Denmark (1999) and a PhD in Computer System Engineering, USD (2003).

Harrison Stratton

Arizona State University & Barrow
Neurological Institute
Phoenix, USA
Harrison.Stratton@asu.edu

Chapter D.34

Harrison Stratton obtained his BSc in Physics from Virginia Polytechnic Institute and State University in 2008 and is currently completing his PhD at Arizona State University and Barrow Neurological Institute. His work focuses on the role of the endogenous cannabinoid system in regulating changes of emotion, memory, and learning.

Thomas Stützle

Université libre de Bruxelles (ULB)
IIRIDIA, CP 194/6
Brussels, Belgium
stuetzle@ulb.ac.be

Chapter E.54

Thomas Stützle is Senior Research Associate of the Belgian FRS-FNRS at the University of Brussels. He received his MSc from Université Karlsruhe (TH) and his PhD and Habilitation from the Technische Universität Darmstadt, Germany. His interests lie in stochastic local search, swarm intelligence, methodologies for engineering stochastic local search algorithms, multi-objective optimization, and automatic configuration of algorithms.

Dirk Sudholt

University of Sheffield
Dep. Computer Science
Sheffield, UK
d.sudholt@sheffield.ac.uk

Chapter E.46

Dirk Sudholt obtained Diploma and PhD degrees in Computer Science in 2004 and 2008, respectively, from the Technische Universität Dortmund, Germany. After holding postdoc positions in Berkeley, California, and Birmingham, UK, he joined the University of Sheffield, UK, as Lecturer in 2012. His research focuses on the computational complexity of randomized search heuristics.

Ron Sun

Rensselaer Polytechnic Institute
Cognitive Science Dep.
Troy, USA
rsun@rpi.edu

Chapter D.36

Ron Sun is Professor of Cognitive Sciences and Computer Science at Rensselaer Polytechnic Institute. He received his PhD from Brandeis University in 1992. His research interests center around the study of cognition. He has published many papers in these areas, as well as ten books.

Johan A. K. Suykens

Katholieke Universiteit Leuven
Leuven, Belgium
johan.suykens@esat.kuleuven.be

Chapter D.32

Johan A.K. Suykens is Professor at Katholieke Universiteit Leuven, Belgium, where he obtained a degree in Electro-Mechanical Engineering and a PhD in Applied Sciences. He is a senior IEEE member, has co/authored and edited several books, and received many prestigious awards.

Roman W. Swiniarski (deceased)

Chapter C.21

Authors

Authors

El-Ghazali Talbi

University of Lille
Computer Science CRISTAL
Villeneuve d'Ascq, France
el-ghazali.talbi@univ-lille1.fr

Chapter E.55

El-Ghazali Talbi received his Master's and PhD in Computer Science from the Institut National Polytechnique de Grenoble, France. He is a Full Professor at the University of Lille and the Head of the Optimization Team of the Computer Science Laboratory. His research interests are in the field of multi-objective optimization, parallel algorithms, metaheuristics, combinatorial optimization, cluster and grid computing.

Lothar Thiele

Swiss Federal Institute of Technology Zurich
Computer Engineering and Networks Lab.
Zurich, Switzerland
thiele@ethz.ch

Chapter E.48

Lothar Thiele received his Diploma and Dr.-Ing. degrees in Electrical Engineering from the Technical University of Munich where he also received his Habilitation. He joined ETH Zurich, Switzerland, as a Full Professor of Computer Engineering in 1994. His research interests include models, methods and software tools for the design of embedded systems, embedded software and bioinspired optimization techniques.

Peter Tino

University of Birmingham
School of Computer Science
Birmingham, UK
P.Tino@cs.bham.ac.uk

Chapter D.27

Peter Tino has a PhD in Computer Science (Slovak Academy of Sciences) and is a Reader in Complex and Adaptive Systems at the University of Birmingham, UK. He is a Vice-Chair of the Neural Networks Technical Committee of IEEE CIS. His main research interests include dynamical systems, machine learning, probabilistic modeling of structured data, evolutionary computation, and fractal analysis.

Joan Torrens

University of the Balearic Islands
Dep. Mathematics and Computer Science
Palma de Mallorca, Spain
jts224@uib.es

Chapter B.12

Joan Torrens received his BSc degree in Mathematics from Universitat Autònoma de Barcelona (1981) and his PhD degree in Computer Science from Universitat de les Illes Balears, Spain (1990). He is a Full Professor in the Department of Mathematics and Computer Science at Universitat de les Illes Balears. His research interests include fuzzy sets theory. He has published over 100 papers in journals.

Vito Trianni

Consiglio Nazionale delle Ricerche
Ist. Scienze e Tecnologie della Cognizione
Roma, Italy
vito.trianni@istc.cnr.it

Chapter F.71

Vito Trianni is a tenured researcher at ISTC-CNR, the Institute of Cognitive Sciences and Technologies of the Italian National Research Council. He received his PhD in Applied Sciences from the Université Libre de Bruxelles in 2006. He has thorough expertise, both theoretical and experimental, in the study and design of self-organizing behavior, especially applied to swarm robotics.

Enric Trillas

European Centre for Soft Computing
Fundamentals of Soft Computing
Mieres, Spain
enric.trillas@softcomputing.es

Chapter B.16

Enric Trillas, an Emeritus Researcher at ECSC, received a PhD in Mathematics from the University of Barcelona. From 1964–1984 he did research in ordered semigroups, probabilistic and generalized metric spaces, and in 1975 he began working in fuzzy logic. He has published over 400 papers, and published or edited several books. His current interests lie in fuzzy logic and in the mathematical analysis of natural language and commonsense reasoning.

Fevrier Valdez

Tijuana Institute of Technology
Tijuana, Mexico
fevrier@tectijuana.mx

Chapter G.81

Fevrier Valdez is a Professor in the Computer Science Department of Tijuana Institute of Technology. His research interests are bio-inspired optimization methods, parallel computing, fuzzy logic and neural networks. He has published several papers in journals, conference proceedings, and as book chapters.

Nele Verbiest

Ghent University
Dep. Applied Mathematics, Computer
Science and Statistics
Ghent, Belgium
nele.verbiest@ugent.be

Chapter C.26

Nele Verbiest holds a Master's degree in Mathematical Computer Science and a PhD in Computer Science, both from Ghent University. Her research interests include classification, evolutionary algorithms, instance selection, feature selection, and fuzzy rough set theory.

Thomas Villmann

University of Applied Sciences Mittweida
Dep. Mathematics, Natural and Computer
Sciences
Mittweida, Germany
thomas.villmann@hs-mittweida.de

Chapter D.31

Thomas Villmann received his PhD in Computer Science in 1996 and his Venia legendi in 2005 from Leipzig University. He has been a Full Professor of Computational Intelligence at the University of Applied Sciences Mittweida, Germany since 2009. His research focus includes the theory of prototype-based clustering and classification, non-standard metrics, information theory, and their application in pattern recognition for use in medicine, bioinformatics, remote sensing, and hyperspectral analysis.

Milan Vlach

Charles University
Theoretical Computer Science and
Mathematical Logic
Prague, Czech Republic
Milan.Vlach@mff.cuni.cz

Chapters A.8, A.9

Milan Vlach studied Mathematics at Charles University, Prague (1958–1960) and Moscow State University (1960–1963). Since graduating from Moscow State University (1960), he has been affiliated with Charles University. At present he is also affiliated with the Institute of Information Theory and Automation, Czech Academy of Sciences. His current area of interest includes game theory, fair division, and optimization theory.

Ekaterina Vladislavleva

Evolved Analytics Europe BVBA
Turnhout, Belgium
katya@evolved-analytics.com

Chapter E.57

Ekaterina Vladislavleva is CEO and Co-Founder of Evolved Analytics Europe, a Belgium-based advanced predictive analytics company – creator of DataStories.com. She received her PhD from CentER at Tilburg School of Economics (The Netherlands), her Doctorate in Engineering from Eindhoven University of Technology (The Netherlands) and MSc from Lomonosov Moscow State University (Russia).

Tobias Wagner

TU Dortmund University
Dep. Mechanical Engineering
Dortmund, Germany
wagner@isf.de

Chapter E.59

Tobias Wagner studied computer science and finished his Doctoral degree in Mechanical Engineering at the University of Dortmund (now TU Dortmund University). He has been a Postdoctoral Researcher (Akademischer Rat) at the Institute of Machining Technology (ISF) since 2013. His research interests include the empirical modeling and optimization of machining processes and process chains.

Jun Wang

The Chinese University of Hong Kong
Dep. Mechanical & Automation
Engineering
Hongkong, Hong Kong
jwang@mae.cuhk.edu.hk

Chapter D.33

Jun Wang is a Professor at the Chinese University of Hong Kong. He has been a National Thousand-Talent Chair Professor at Dalian University of Technology since 2011. He received his BSc degree in Electrical Engineering and his MSc degree in Systems Engineering from Dalian University of Technology, China and his PhD in Systems Engineering from Case Western Reserve University, Cleveland, Ohio, USA.

Simon Wessing

Technische Universität Dortmund
Fak. Informatik
Dortmund, Germany
simon.wessing@tu-dortmund.de

Chapter E.58

Simon Wessing has been a Research Associate in the Computational Intelligence Group, Technische Universität Dortmund since 2009. His research interest focuses on multimodal and global optimization. He develops new optimization algorithms for these problems based on evolutionary algorithms and other derivative-free methods.

Authors

Wei–Zhi Wu

Zhejiang Ocean University
School of Mathematics, Physics and
Information Science
Zhoushan, Zhejiang, China
wuwz@zjou.edu.cn

Chapter C.26

Wei-Zhi Wu received his BSc degree in Mathematics from Zhejiang Normal University, his MSc degree from East China Normal University, and his PhD degree from Xi'an Jiaotong University (2002). He is currently a Professor of Mathematics in the School of Mathematics, Physics, and Information Science, Zhejiang Ocean University. His current research interests include approximate reasoning, rough sets, random sets, formal concept analysis, and granular computing.

Lei Xu

The Chinese University of Hong Kong
Dep. Computer Science and Engineering
Hong Kong, Hong Kong
lxu@cse.cuhk.edu.hk

Chapter D.29

Lei Xu is a Professor at the Chinese University of Hong Kong. He is an IEEE Fellow, an IAPR Fellow, an Academician of the European Academy of Sciences. He has published well-cited papers on neural networks, machine learning, and pattern recognition.

JingTao Yao

University of Regina
Dep. Computer Science
Regina, Saskatchewan, Canada
jtyao@cs.uregina.ca

Chapter C.25

JingTao Yao is a Professor of Computer Science at the University of Regina. He has published over 100 papers on granular computing, rough sets, data mining, and Web-based support systems. He is the Elected Chair of the Steering Committee of the International Rough Set Society.

Yiyu Yao

Chapter C.24

For biographical profile, please see the section "About the Part Editors".

Andreas Zabel

TU Dortmund University
Dep. Mechanical Engineering
Dortmund, Germany
zabel@isf.de

Chapter E.59

Andreas Zabel studied Computer Science and finished his Doctoral degree in Mechanical Engineering at the University of Dortmund (now TU Dortmund University). He is Chief Engineer of the Institute of Machining Technology (ISF). His research includes the simulation of machining processes, modeling and analysis of tool wear, as well as augmented and virtual reality for process planning.

Sławomir Zadrożny

Polish Academy of Sciences
Systems Research Inst.
Warsaw, Poland
Slawomir.Zadrozny@ibspan.waw.pl

Chapter B.19

Sławomir Zadrożny is an Associate Professor (PhD 1994, DSc 2006) and a Deputy Director for Research at the Systems Research Institute, Polish Academy of Sciences. His current interests include applications of fuzzy logic in database management systems, information retrieval, decision support, and data analysis. He has authored and co-authored about 200 journal and conference papers. He is also a teacher at the Warsaw School of Information Technology.

Zhigang Zeng

Huazhong University of Science and
Technology
Dep. Control Science and Engineering
Wuhan, China
zgzeng@hust.edu.cn

Chapter D.33

Zhigang Zeng is Professor in the Department of Control Science and Engineering, Huazhong University of Science and Technology, Wuhan, Hubei, China. He received his BSc degree in Mathematics from Hubei Normal University, Huangshi, China, his MSc degree in Ecological Mathematics from Hubei University, Wuhan, China, in 1993 and 1996, and his PhD degree in Systems Analysis and Integration from Huazhong University of Science and Technology, Wuhan, China, in 2003.

Yan Zhang

University of Regina
Dep. Computer Science
Regina, Saskatchewan, Canada
zhang83y@cs.uregina.ca

Chapter C.25

Yan Zhang is currently a PhD student in the Department of Computer Science, the University of Regina, Canada. Her main research involves rough sets, granular computing, data analysis, and data mining. She has authored or co-authored more than 15 technical papers in international journals and conference proceedings. She is coauthor of two book chapters.

Zhi-Hua Zhou

Nanjing University
National Key Lab. for Novel Software
Technology
Nanjing, China
zhouzh@nju.edu.cn

Chapter D.29

Zhi-Hua Zhou received BSc, MSc, and PhD degrees in Computer Science from Nanjing University, China, in 1996, 1998 and 2000. He joined the Department of Computer Science and Technology of Nanjing University in 2001 and at present he is a Professor and Deputy Director of the National Key Lab for Novel Software Technology. His research interests include artificial intelligence, machine learning, data mining, pattern recognition and multimedia information retrieval.

Authors

Detailed Contents

Part A Foundations

30 Theoretical Methods in Machine Learning

31 Probabilistic Modeling in Machine Learning

32 Kernel Methods

Marco Signoretto, Johan A. K. Suykens

38 Neuromorphic Engineering

39 Neuroengineering

45 Estimation of Distribution Algorithms

Detailed Cont.

Detailed Cont.

Detailed Cont.

Index

Recently Published Springer Handbooks

Springer Handbook of Computational Intelligence (2015)
ed. by Kacprzyk, Pedrycz, 1633 p., 978-3-662-43505-2

Springer Handbook of Marine Biotechnology (2015)
ed. by Kim, 1512 p., 978-3-642-53970-1

Springer Handbook of Acoustics (2nd) (2015)
ed. by Rossing, 1286 p., 978-1-4939-0754-0

Springer Handbook of Spacetime (2014)
ed. by Ashtekar, Petkov, 887 p., 978-3-642-41991-1

Springer Handbook of Bio-/Neuro-Informatics (2014)
ed. by Kasabov, 1230 p., 978-3-642-30573-3

Springer Handbook of Nanomaterials (2013)
ed. by Vajtai, 1222 p., 978-3-642-20594-1

Springer Handbook of Lasers and Optics (2nd) (2012)
ed. by Träger, 1694 p., 978-3-642-19408-5

Springer Handbook of Geographic Information (2012)
ed. by Kresse, Danko, 1120 p., 978-3-540-72678-4

Springer Handbook of Medical Technology (2011)
ed. by Kramme, Hoffmann, Pozos, 1500 p., 978-3-540-74657-7

Springer Handbook of Metrology and Testing (2nd) (2011)
ed. by Czichos, Saito, Smith, 1229 p., 978-3-642-16640-2

Springer Handbook of Crystal Growth (2010)
ed. by Dhanaraj, Byrappa, Prasad, Dudley, 1816 p., 978-3-540-74182-4

Springer Handbook of Nanotechnology (3rd) (2010)
ed. by Bhushan, 1961 p., 978-3-642-02524-2